MOLECULAR DETECTION OF HUMAN BACTERIAL PATHOGENS

MOLECULAR DETECTION OF HUMAN BACTERIAL PATHOGENS

EDITED BY
DONGYOU LIU

CRC Press
Taylor & Francis Group
Boca Raton London New York

CRC Press is an imprint of the
Taylor & Francis Group, an **informa** business

CRC Press
Taylor & Francis Group
6000 Broken Sound Parkway NW, Suite 300
Boca Raton, FL 33487-2742

Printed in the United States of America on acid-free paper
10 9 8 7 6 5 4 3 2 1

International Standard Book Number: 978-1-4398-1238-9 (Hardback)

Visit the Taylor & Francis Web site at
http://www.taylorandfrancis.com

and the CRC Press Web site at
http://www.crcpress.com

This volume is dedicated to a group of bacteriologists whose insight, knowledge, and expertise have made the all-inclusive coverage of major human bacterial pathogens a reality.

Contents

Clostridia

SECTION IV Proteobacteria

Alphaproteobacteria

Betaproteobacteria

Gammaproteobacteria

Epsilonproteobacteria

SECTION V Spirochaetes

SECTION VI Pan-Bacterial Detection

Preface

Bacteria are small, unicellular organisms that are invisible to the naked eye but are nonetheless present ubiquitously and abundantly in all environments. Although a majority of bacteria are free-living and symbiotic, some are capable of leading a parasitic life, inducing a range of disease syndromes in human and animal hosts during the process. Among the most devastating bacterial pathogens, *Yersinia pestis*, the causative agent of bubonic plague, was responsible for three major human pandemics in history, killing 200 million people prior to the advent of antibiotics. The current, most common fatal bacterial diseases are tuberculosis (caused by *Mycobacterium tuberculosis*), killing about 2 million people a year alone, and cholera (caused by *Vibrio cholerae*). Other globally important bacterial diseases include pneumonia (caused by *Streptococcus* and *Pseudomonas*), tetanus, typhoid fever, diphtheria, syphilis, and leprosy.

Traditionally, bacteria have been identified and diagnosed with the help of various phenotypic procedures, such as Gram stain, morphological, biochemical, and serological examination. Since the phenotypic techniques are often slow and lack desired specificity and reproducibility, nucleic acid amplification technologies such as polymerase chain reaction (PCR) have played an increasingly prominent role in the laboratory diagnosis of bacterial infections. Given their ability to specifically detect a single copy of bacterial nucleic acid template in a matter of hours, PCR-based assays offer unsurpassed sensitivity, specificity, accuracy, precision, and result availability for bacterial identification. The recent advances in instrumentation automation and probe chemistries have facilitated the development of real-time PCR that provides a convenient platform for high throughput detection and quantitation of bacterial pathogens in clinical specimens.

Considering that numerous original molecular protocols and subsequent modifications have been described and scattered in various journals and monographs, it has become difficult if not impossible for someone who has not been directly involved in the development of original or modified protocols to know which are most appropriate to adopt for accurate identification of bacterial pathogens of interest. The purpose of this volume is to address this issue, with international scientists in respective bacterial pathogen research and diagnosis providing expert summaries on current diagnostic approaches for major human bacterial pathogens. Each chapter consists of a brief review of the classification, epidemiology, clinical features, and diagnosis of an important pathogenic bacterial genus; an outline of clinical sample collection and preparation procedures; a selection of representative stepwise molecular protocols; and a discussion on further research requirements relating to improved diagnosis. This book represents a reliable and convenient reference on molecular detection and identification of major human bacterial pathogens; an indispensable tool for upcoming and experienced medical, veterinary, and industrial laboratory scientists engaged in bacterial characterization; and an essential textbook for undergraduate and graduate students majoring in bacteriology.

A comprehensive and inclusive book such as this is undoubtedly beyond an individual's capacity. I am fortunate and honored to have a large panel of bacteriologists as chapter contributors, whose detailed knowledge and technical insights on human bacterial pathogen detection have greatly enriched this book. In addition, the professionalism and dedication of executive editor Barbara Norwitz and senior project coordinator Jill Jurgensen at CRC Press have enhanced its presentation. Finally, without the understanding and support of my family, Liling Ma, Brenda, and Cathy, the compilation of this all-encompassing volume would have not been possible.

Editor

Dongyou Liu, PhD, undertook his veterinary science education at Hunan Agricultural University, China. Upon graduation, he received an Overseas Postgraduate Scholarship from the Chinese Ministry of Education to pursue further training at the University of Melbourne, Australia, where he worked toward improved immunological diagnosis of human hydatid disease. During the past two decades, he has crisscrossed between research and clinical laboratories in Australia and the United States of America, with focuses on molecular characterization and virulence determination of microbial pathogens such as ovine footrot bacterium (*Dichelobacter nodosus*), dermatophyte fungi (*Trichophyton, Microsporum,* and *Epidermophyton*), and listeriae (*Listeria* species). He is the senior author of more than 50 original research and review articles in various international journals and the editor of the recently released *Handbook of* Listeria monocytogenes, *Handbook of Nucleic Acid Purification*, *Molecular Detection of Foodborne Pathogens*, and *Molecular Detection of Human Viral Pathogens*, as well as the forthcoming *Molecular Detection of Human Fungal Pathogens* and *Molecular Detection of Human Parasitic Pathogens*, all of which are published by CRC Press.

Contributors

Martina Adamek
Karlsruhe Institute of Technology
Institute of Functional Interfaces
Karlsruhe, Germany

Catherine C. Adley
Microbiology Laboratory
Department of Chemical and
Environmental Sciences
University of Limerick
Limerick, Ireland

Sascha Al Dahouk
Department of Internal Medicine III
RWTH Aachen University
Aachen, Germany

Rodrigo Alonso
Faculty of Pharmacy
Department of Immunology,
Microbiology and Parasitology
University of the Basque Country
Vitoria-Gasteiz, Spain

Maria Silvana Alves
Faculdade de Farmácia e Bioquímica
Universidade Federal de Juiz de Fora
Minas Gerais, Brazil

Paula Lopes Alves
Instituto de Biologia Experimental e
Tecnológica
Oeiras, Portugal

Emilio Aranda
Ciencia y Tecnología de los Alimentos
Escuela de Ingenierías Agrarias
Badajoz, Spain

Marco Arosio
USC Microbiology and Virology
A.O. Ospedali Riuniti
Bergamo, Italy

Aurora Fernández Astorga
Faculty of Pharmacy
Department of Immunology,
Microbiology and Parasitology
University of the Basque Country
Vitoria-Gasteiz, Spain

Frank W. Austin
College of Veterinary Medicine
Mississippi State University
Starkville, Mississippi

Mario J. Avila-Campos
Department of Microbiology
Institute of Biomedical Science
University of São Paulo
São Paulo, Brazil

R. Aznar
Department of Microbiology and
Ecology
University of Valencia
Valencia, Spain

Diprabhanu Bakshi
Syngene International
Bangalore, India

Stephan Bathe
German National Academic
Foundation
Bonn, Germany

Harsh Vardhan Batra
Defence R & D Establishment
Gwalior, India

Asim Bej
Department of Biology
University of Alabama at Birmingham
Birmingham, Alabama

Scott C. Bell
Department of Thoracic Medicine
The Prince Charles Hospital
Brisbane, Australia

María J. Benito
Ciencia y Tecnología de los Alimentos
Escuela de Ingenierías Agrarias
Badajoz, Spain

Patrick Boiron
Research group on Bacterial
Opportunistic Pathogens and
Environment
Université de Lyon
Lyon, France

Dianna J. Bopp
Wadsworth Center
New York State Department of Health
Albany, New York

Alejandra Bosch
Facultad de Ciencias Exactas
Centro de Investigación y Desarrollo
en Fermentaciones Industriales
Universidad Nacional de La Plata
La Plata, Argentina

Javier Enrique Botero
School of Dentistry
Grupo de Investigación Básica
y Clínica en Periodoncia y
Oseointegración
Universidad de Antioquia
Medellín, Colombia

Germán Bou
Servicio Microbiología
Complejo Hospitalario Universitario
La Coruña
La Coruña, Spain

June Brown
Actinomycete Reference Laboratory
Bacterial Zoonoses Branch
Division of Foodborne, Bacterial and
Mycotic Diseases
National Center for Zoonotic, Vector-
Borne, and Enteric Diseases
Centers for Disease Control and
Prevention
Atlanta, Georgia

Renato Corrêa Viana Casarin
Department of Oral Diagnostics
Division of Microbiology and
Immunology
Piracicaba Dental School
University of Campinas
Campinas, Brazil

Nicholas Cassai
Department of Pathology and
Laboratory Medicine
New York Harbor VA Medical Center
Brooklyn, New York

Marco Cassone
Sbarro Health Research Organization
College of Science and Technology
Temple University
Philadelphia, Pennsylvania

Bianca Castiglioni
Department of Veterinary Pathology, Hygiene and Public Health
University of Milan
Milan, Italy

Rita Castro
Instituto de Higiene e Medicina Tropical
Universidade Nova de Lisboa
Lisbon, Portugal

Tsung Chain Chang
Department of Medical Laboratory Science and Biotechnology
College of Medicine
National Cheng Kung University
Tainan, Taiwan

Rama Chaudhry
Department of Microbiology
All India Institute of Medical Sciences
New Delhi, India

Jianshun Chen
Institute of Preventive Veterinary Medicine
Zhejiang University
Zhejiang, China

Ameur Cherif
Faculté des Sciences de Tunis
Laboratoire Microorganismes et Biomolécules Actives
Campus Universitaire
Tunis, Tunisie

Bishwanath Kumar Chourasia
Department of Microbiology
All India Institute of Medical Sciences
New Delhi, India

Henrik Christensen
Faculty of Life Sciences
Department of Veterinary Pathobiology
Center for Applied Bioinformatics
Copenhagen University
Frederiksberg, Denmark

Deirdre L. Church
Departments of Pathology, Laboratory Medicine, and Medicine
University of Calgary
Alberta Health Services
and
Calgary Laboratory Services
Alberta, Canada

Lorenzo Ciceroni
Center for Research and Evaluation of Immunobiologicals
Istituto Superiore di Sanità
Rome, Italy

Alessandra Ciervo
Department of Infectious, Parasitic and Immune-Mediated Diseases
Istituto Superiore di Sanità
Rome, Italy

Lena Ciric
Microbial Diseases
UCL Eastman Dental Institute
London, United Kingdom

Ana Paula Vieira Colombo
Department of Medical Microbiology Institute of Microbiology
Federal University of Rio de Janeiro
Rio de Janeiro, Brazil

Andréa Vieira Colombo
Institute of Biomedical Sciences
State University of São Paulo
Sao Paulo, Brazil

Adolfo Contreras
School of Dentistry
Grupo de Medicina Periodontal
Universidad del Valle
Cali, Colombia

Piet Cools
Department of Clinical Biology, Immunology and Microbiology
Ghent University Hospital
Ghent, Belgium

Juan J. Córdoba
Facultad de Veterinaria
Higiene de los Alimentos
Universidad de Extremadura
Cáceres, Spain

María G. Córdoba
Ciencia y Tecnología de los Alimentos
Escuela de Ingenierías Agrarias
Badajoz, Spain

Yvon Cormier
Research Unit
Laval Hospital Centre of Pneumology
University of Laval
Quebec, Canada

Paola Cremonesi
Department of Veterinary Pathology, Hygiene and Public Health
University of Milan
Milan, Italy

Maria Teresa Barreto Crespo
Instituto de Biologia Experimental e Tecnológica
Oeiras, Portugal

Daniele Daffonchio
Dipartimento di Scienze e Tecnologie Alimentari e Microbiologiche
Università degli Studi di Milano
Milan, Italy

Sharvari Dalal
Department of Pathology
SUNY Downstate Medical Center
Brooklyn, New York

Maria Dalamaga
Laboratory of Clinical Biochemistry
Attikon General University Hospital
Medical School, National & Kapodistrian University of Athens
Athens, Greece

George N. Dalekos
Department of Medicine, Medical School
University of Thessaly
Larissa, Greece

Anupam Das
Department of Microbiology
All India Institute of Medical Sciences
New Delhi, India

Gregory A. Dasch
Rickettsial Zoonoses Branch
Division of Vector-Borne Diseases
Centers for Disease Control and Prevention
Atlanta, Georgia

Jonathan J. Dennis
Department of Biological Sciences
University of Alberta
Alberta, Canada

Rubens Clayton da Silva Dias
Faculdade de Medicina
Universidade Federal do Rio de Janeiro
Rio de Janeiro, Brazil

Xiaoming Dong
Beijing Wantai Biological Pharmacy Enterprise
Beijing, China

Caroline Duchaine
Research Unit
Laval Hospital Centre of Pneumology
University of Laval
Quebec, Canada

Francis Dziva
Division of Microbiology
Institute for Animal Health
Berkshire, United Kingdom

P. Elizaquível
Department of Microbiology and Ecology
University of Valencia
Valencia, Spain

Marina E. Eremeeva
Rickettsial Zoonoses Branch
Division of Vector-Borne Diseases
Centers for Disease Control and Prevention
Atlanta, Georgia

Emenike Ribs K. Eribe
Norwegian Geotechnical Institute
Oslo, Norway

Klaus Eschrich
Institute for Biochemistry
School of Medicine
University of Leipzig
Leipzig, Germany

Weihuan Fang
Institute of Preventive Veterinary Medicine
Zhejiang University
Zhejiang, China

Seamus Fanning
Centre for Food Safety
School of Agriculture, Food Science and Veterinary Medicine
Veterinary Sciences Centre
University College Dublin
Dublin, Ireland

María Fiuza
Faculty of Biology
Section of Microbiology
Department of Molecular Biology
University of León
León, Spain

Maria Fredriksson-Ahomaa
Ludwig-Maximilian University
Munich, Germany

Elerson Gaetti-Jardim Jr.
Department of Oral Pathology
São Paulo State University
São Paulo, Brazil

Ana Cristina Gales
Departamento de Medicina
Universidade Federal de São Paulo
São Paulo, Brazil

María-Luisa García-Lópcz
Veterinary Faculty
Department of Food Hygiene and Food Microbiology
University of León
León, Spain

Maria del Mar Garcia-Suarez
Area de Microbiologia
Departamento de Biologia Funcional Instituto
Universitario de Biotecnologia de Asturias
Asturias, Spain

José A Gil
Faculty of Biology
Section of Microbiology
Department of Molecular Biology
University of León Campus de Vegazana
León, Spain

Cecilia Girbau
Faculty of Pharmacy
Department of Immunology, Microbiology and Parasitology
University of the Basque Country
Vitoria-Gasteiz, Spain

John I. Glass
J. Craig Venter Institute
Rockville, Maryland

Maryna Golovchenko
Faculty of Science
University of South Bohemia
Ceské Budějovice, Czech Republic

Reginaldo Bruno Gonçalves
Department of Oral Diagnostics
Division of Microbiology and Immunology
Piracicaba Dental School
University of Campinas
Campinas, Brazil

Keith Grimwood
Queensland Paediatric Infectious Diseases Laboratory
Royal Children's Hospital Queensland Children's Medical Research Institute
Queensland, Australia

Libor Grubhoffer
Biology Centre
Institute of Parasitology
Academy of Sciences of the Czech Republic
Ceské Budějovice, Czech Republic

Miguel Gueimonde
Department of Microbiology and Biochemistry of Dairy Products
Instituto de Productos Lácteos de Asturias
Consejo Superior de Investigaciones Científicas
Asturias, Spain

Saija Hallanvuo
Finnish Food Safety Authority Evira
National Public Health Institute
Helsinki, Finland

Rudy A. Hartskeerl
WHO/FAO/OIE
and
National Collaborating Centre for Reference and Research on Leptospirosis
Department KIT Biomedical Research
Royal Tropical Institute
Amsterdam, the Netherlands

John P. Hays
Department of Medical Microbiology and Infectious Diseases
Rotterdam, the Netherlands

Alexander M. Hirschl
Division of Clinical Microbiology
Department of Laboratory Medicine
Medical University Vienna
Vienna, Austria

Jeffrey Hoorfar
National Food Institute
Technical University of Denmark
Søborg, Denmark

Kurt Houf
Faculty of Veterinary Medicine
Department of Veterinary Public Health and Food Safety
Ghent University
Ghent, Belgium

Satoru Inagaki
Department of Microbiology
Oral Health Science Center
Tokyo Dental College
Chiba, Japan

Kazuyuki Ishihara
Department of Microbiology
Oral Health Science Center
Tokyo Dental College
Chiba, Japan

Nahed Ismail
Department of Pathology
Meharry Medical College
Nashville, Tennessee

Carol Iversen
Centre for Food Safety
School of Agriculture
Food Science and Veterinary Medicine
Veterinary Sciences Centre
University College Dublin
Dublin, Ireland

Jari Jalava
Department of Infectious Disease Surveillance and Control
National Institute for Health and Welfare
Turku, Finland

Judith Jansen
Medical Faculty
Institute of Immunology
RWTH Aachen University
Aachen, Germany

Adriana Jaramillo
School of Dentistry
Grupo de Medicina Periodontal
Universidad del Valle
Cali, Colombia

Mathilde H. Josefsen
National Food Institute
Technical University of Denmark
Søborg, Denmark

Estelle Jumas-Bilak
Universitaire de Montpellier
Hôpital Arnaud de Villeneuve
Laboratoire de Bactériologie
Montpellier, France

Bernhard Kaltenboeck
Molecular Diagnostics Laboratory
Department of Pathobiology
College of Veterinary Medicine
Auburn University
Auburn, Alabama

Mireille M. Kattar
Department of Laboratory Medicine and Pathology
University of Alberta Hospital
Alberta, Canada

Timothy J. Kidd
Queensland Paediatric Infectious Diseases Laboratory
Royal Children's Hospital
Queensland Children's Medical Research Institute
Queensland, Australia

Rickard Knutsson
Swedish National Veterinary Institute
Uppsala, Sweden

Eija Könönen
University of Turku
Institute for Dentistry
Turku, Finland

Hannu Korkeala
Faculty of Veterinary Medicine
Department of Food and Environmental Hygiene
University of Helsinki
Helsinki, Finland

Kiersten J. Kugeler
Centers for Disease Control and Prevention
National Center for Zoonotic Vector-Borne and Enteric Diseases
Division of Vector-Borne Infectious Diseases, Bacterial Diseases Branch
Fort Collins, Colorado

Susanne Thisted Lambertz
Research and Development Department
National Food Administration
Uppsala, Sweden

K.A. Lampel
Food and Drug Administration
Division of Microbiology
College Park, Maryland

Brent A. Lasker
Actinomycete Reference Laboratory
Bacterial Zoonoses Branch
Division of Foodborne, Bacterial and Mycotic Diseases
National Center for Zoonotic, Vector-Borne, and Enteric Diseases
Centers for Disease Control and Prevention
Atlanta, Georgia

Susanna K.P. Lau
Department of Microbiology
The University of Hong Kong
Hong Kong

Kevin B. Laupland
Departments of Medicine, Critical Care Medicine, Community Health Sciences, and Pathology and Laboratory Medicine, and Centre for Antimicrobial Resistance
University of Calgary, Alberta Health Services, and Calgary Laboratory Services
Alberta, Canada

Frédéric Laurent
Faculté de Médecine Laennec
Lyon, France

Marc Lehmann
SIRS-Lab GmbH
Jena, Germany

Angelika Lehner
Vetsuisse Faculty
Institute for Food Safety and Hygiene
University of Zurich
Zurich, Switzerland

Manuel L. Lemos
Department of Microbiology and Parasitology
Institute of Aquaculture
University of Santiago de Compostela
Santiago de Compostela, Galicia, Spain

Michal Letek
Faculty of Biology
Section of Microbiology
Department of Molecular Biology
University of León Campus de Vegazana
León, Spain

Dongyou Liu
BioSecurity Quality Assurance Program
Royal College of Pathologists of Australasia
New South Wales, Australia

Charlotta Löfström
National Food Institute
Technical University of Denmark
Søborg, Denmark

José Maria Luengo
Department of Molecular Biology
University of León
León, Spain

Karlene H. Lynch
Department of Biological Sciences
University of Alberta
Alberta, Canada

Athanasios Makristathis
Division of Clinical Microbiology
Department of Laboratory Medicine
Medical University Vienna
Vienna, Austria

Fabiola Mancini
Department of Infectious, Parasitic and Immune-Mediated Diseases
Istituto Superiore di Sanità
Rome, Italy

Chaysavanh Manichanh
Digestive System Research Unit
University Hospital Vall d'Hebron
Bioinformatics and Genomics Program
Center for Genomic Regulation
Barcelona, Spain

Angel Manteca
Area de Microbiologia
Departamento de Biologia Funcional
Instituto Universitario de Biotecnologia de Asturias
Universidad de Oviedo
Asturias, Spain

Hélène Marchandin
Faculté de Pharmacie
Université Montpellier 1
Laboratoire de Bactériologie-Virologie-Contrôle Microbiologique
Montpellier, France

Abelardo Margolles
Department of Microbiology and Biochemistry of Dairy Products
Instituto de Productos Lácteos de Asturias
Consejo Superior de Investigaciones Científicas
Asturias, Spain

Luis M. Mateos
Faculty of Biology
Section of Microbiology
Department of Molecular Biology
University of León Campus de Vegazana
León, Spain

Johanna Maukonen
VTT Technical Research Centre of Finland
Espoo, Finland

Jere W. McBride
Department of Pathology
University of Texas Medical Branch at Galveston
Galveston, Texas

Andrew McDowell
Centre of Infection & Immunity
School of Medicine, Dentistry and Biomedical Sciences
Queen's University
Belfast, Northern Ireland

Francisco J. Mendez
Area de Microbiologia
Departamento de Biologia Funcional
Instituto Universitario de Biotecnologia de Asturias
Universidad de Oviedo
Asturias, Spain

Kåre Mølbak
Statens Serum Institut
Copenhagen, Denmark

Jens Kjølseth Møller
Department of Clinical Microbiology
Institute of Regional Health Services Research
University of Southern Denmark
Odense, Denmark

Beatriz Meurer Moreira
Instituto de Microbiologia
Universidade Federal do Rio de Janeiro
Rio de Janeiro, Brazil

Paolo Moroni
Quality Milk Production Services
Cornell University
Ithaca, New York

Benjamin D. Moser
Actinomycete Reference Laboratory Bacterial Zoonoses Branch
Division of Foodborne, Bacterial and Mycotic Diseases
National Center for Zoonotic, Vector-Borne, and Enteric Diseases
Centers for Disease Control and Prevention
Atlanta, Georgia

Dipalok Mukherjee
Biological Sciences
Mississippi State University
Starkville, Mississippi

Soma Mukherjee
Palli Shikha Bhavan
Sriniketan, Birbhum
West Bengal, India

Kimberlee A. Musser
Wadsworth Center
New York State Department of Health
Albany, New York

Beth Mutai
Global Emerging Infections Surveillance System
U.S. Army Medical Research Unit
Kenya Medical Research Institute
Narombi, Kenya

Germán Naharro
Department of Animal Health
University of León
León, Spain

Futoshi Nakazawa
Department of Oral Microbiology
School of Dentistry, Health Sciences
University of Hokkaido
Hokkaido, Japan

Elizabeth J. Nazarian
Wadsworth Center
New York State Department of Health
Albany, New York

Heinrich Neubauer
Friedrich-Loeffler-Institut
Institute of Bacterial Infections and Zoonoses
Jena, Germany

Juan P. Olano
Department of Pathology
University of Texas Medical Branch at Galveston
Galveston, Texas

James H. Oliver, Jr.
Georgia Southern University
The James H. Oliver, Jr. Institute of Arthropodology and Parasitology
Statesboro, Georgia

Ingar Olsen
Faculty of Dentistry
Institute of Oral Biology
University of Oslo
Oslo, Norway

Katharina E.P. Olsen
Statens Serum Institut
Copenhagen, Denmark

Efrén Ordoñez
Faculty of Biology
Section of Microbiology
Department of Molecular Biology
University of León
León, Spain

Andrés Otero
Veterinary Faculty
Department of Food Hygiene and Food Microbiology
University of León
León, Spain

Carlos R. Osorio
Department of Microbiology and Parasitology
Institute of Aquaculture
University of Santiago de Compostela
Compostela, Spain

Jie Ouyang
Department of Pathology
SUNY Downstate Medical Center
Brooklyn, New York

Vanya Paralanov
J. Craig Venter Institute
Rockville, Maryland

Sheila Patrick
Centre of Infection & Immunity
School of Medicine, Dentistry and Biomedical Sciences
Queen's University
Belfast, Northern Ireland

Biswajit Paul
Escorts Heart Institute and Research Centre
New Delhi, India

Debarati Paul
College of Veterinary Medicine
Mississippi State University
Starkville, Mississippi

Debosmita Paul
Jamia Millia Islamia
New Delhi, India

Ana Isabel Pelaez
Area de Microbiologia
Departamento de Biologia Funcional
Instituto Universitario de Biotecnologia de Asturias
Universidad de Oviedo
Asturias, Spain

Claire Pelletier
Laboratoire Dépatemental d'Analyses de Saône-et-Loire
Mâcon, France

J. Tony Pembroke
Molecular and Structural Biochemistry Laboratory
Department of Chemical and Environmental Sciences
University of Limerick
Limerick, Ireland

Filomena Martins Pereira
Instituto de Higiene e Medicina Tropical Universidade Nova de Lisboa
Lisbon, Portugal

Jeannine M. Petersen
Centers for Disease Control and Prevention
National Center for Zoonotic, Vector-Borne and Enteric Diseases
Division of Vector-Borne Infectious Diseases, Bacterial Diseases Branch
Fort Collins, Colorado

Efthimia Petinaki
Department of Microbiology
University of Thessaly Medical School
Larissa, Greece

Wolfgang Pfister
Institute for Medical Microbiology
University Hospital of Jena
Jena, Germany

Matthew R. Pincus
Department of Pathology
SUNY Downstate Medical Center
Brooklyn, New York

Giuliano Pisoni
Department of Veterinary Pathology, Hygiene and Public Health
University of Milan
Milan, Italy

Brunella Posteraro
Institute of Microbiology
Università Cattolica del Sacro Cuore
Rome, Italy

Patrizia Posteraro
Laboratory of Clinical Pathology and Microbiology
Ospedale San Carlo
Rome, Italy

Noura Raddadi
Dipartimento di Scienze e Tecnologie Alimentari e Microbiologiche
Università degli Studi di Milano
Milan, Italy

Peter Rådström
Department of Applied Microbiology
Lund University
Lund, Sweden

Annibale Raglio
USC Microbiology and Virology
A.O. Ospedali Riuniti
Bergamo, Italy

Derren Ready
Eastman Dental Hospital
UCLH NHS Foundation Trust
London, United Kingdom

D. Madhusudan Reddy
Palamuru University
Andhra Pradesh, India

Clara G. de los Reyes-Gavilán
Department of Microbiology and Biochemistry of Dairy Products
Instituto de Productos Lácteos de Asturias
Consejo Superior de Investigaciones Científicas
Asturias, Spain

Lothar Rink
Institute of Immunology
Medical Faculty
RWTH Aachen University
Aachen, Germany

Ignasi Roca
Department of Clinical Microbiology
School of Medicine
University of Barcelona
Barcelona, Spain

Camilla Rodrigues
P D Hinduja National Hospital & Medical Research Centre
Mumbai, India

Mar Rodríguez
Facultad de Veterinaria
Higiene de los Alimentos
Universidad de Extremadura
Cáceres, Spain

Veronica Rodriguez-Nava
Research group on "Bacterial Opportunistic Pathogens and Environment"
Université de Lyon
Lyon, France

Antoni Różalski
Department of Immunobiology of Bacteria
Institute of Microbiology and Immunology
University of Łódź
Łódź, Poland

Patricia Ruas-Madiedo
Department of Microbiology and Biochemistry of Dairy Products
Instituto de Productos Lácteos de Asturias
Consejo Superior de Investigaciones Científicas
Asturias, Spain

Pedro Rubio
Department of Animal Health
University of León
León, Spain

Nataliia Rudenko
Biology Centre
Institute of Parasitology
Academy of Sciences of the Czech Republic
České Budějovice, Czech Republic

Stefan Rupf
Clinic of Operative Dentistry
Periodontology and Preventive Dentistry
Saarland University Hospital
Homburg/Saar, Germany

Michael P. Ryan
Microbiology Laboratory
Department of Chemical and Environmental Sciences
University of Limerick
Limerick, Ireland

Maria Saarela
VTT Technical Research Centre of Finland
Espoo, Finland

Konrad Sachse
Friedrich-Loeffler-Institut (Federal Research Institute for Animal Health)
Institute of Molecular Pathogenesis
Jena, Germany

Atsushi Saito
Oral Health Science Center
Tokyo Dental College
Chiba, Japan

Daniel Saito
Department of Oral Diagnostics
Division of Microbiology and Immunology
Piracicaba Dental School
University of Campinas
Campinas, Brazil

Borja Sánchez
Department of Microbiology and Biochemistry of Dairy Products
Instituto de Productos Lácteos de Asturias
Consejo Superior de Investigaciones Científicas
Asturias, Spain

Ester Sánchez
Microbial Ecophysiology and Nutrition Research Group
Institute of Agrochemistry and Food Technology
Spanish Council for Scientific Research
Valencia, Spain

G. Sánchez
Department of Biotecnology
Institute of Agrochemistry and Food Technology
Valencia, Spain

Maurizio Sanguinetti
Institute of Microbiology
Università Cattolica del Sacro Cuore
Rome, Italy

Guido Lopes dos Santos Santiago
Department of Clinical Biology, Immunology and Microbiology
Ghent University Hospital
Ghent, Belgium

Jesús A. Santos
Veterinary Faculty
Department of Food Hygiene and Food Microbiology
University of León
León, Spain

Yolanda Sanz
Microbial Ecophysiology and Nutrition Research Group
Institute of Agrochemistry and Food Technology
Spanish Council for Scientific Research
Valencia, Spain

Roland P.H. Schmitz
SIRS-Lab GmbH
Jena, Germany

K.R. Schneider
Food Science and Human Nutrition Department
University of Florida
Gainesville, Florida

Bo Segerman
Swedish National Veterinary Institute
Uppsala, Sweden

Teresa Semedo-Lemsaddek
Faculdade de Ciências
Universidade de Lisboa
Center for Biodiversity Functional and Integrative Genomics
Edifício ICAT, Campus da FCUL
Lisbon, Portugal

Diego Omar Serra
Facultad de Ciencias Exactas
Centro de Investigación y Desarrollo en Fermentaciones Industriales
Universidad Nacional de La Plata
La Plata, Argentina

Nidhi Sharma
Department of Microbiology
All India Institute of Medical Sciences
New Delhi, India

Shubhada Shenai
P D Hinduja National Hospital & Medical Research Centre
Mumbai, India

Han Shiong Siah
TPP, Lambells Lagoon
Northern Territory, Australia

Shoo Peng Siah
Human Genetic Signatures
New South Wales, Australia

Maria R.L. Simionato
Department of Microbiology
Institute of Biomedical Science
University of São Paulo
São Paulo, Brazil

Mikael Skurnik
University of Helsinki
and
Helsinki University Central Hospital Laboratory Diagnostics
Helsinki, Finland

Renata Martins do Souto
Department of Medical Microbiology Institute of Microbiology
Federal University of Rio de Janeiro
Rio de Janeiro, Brazil

Paweł Stączek
Department of Immunobiology of Bacteria
Institute of Microbiology and Immunology
University of Łódź
Łódź, Poland

Paola Stefanelli
Department of Infectious, Parasitic and Immune-Mediated Diseases
Istituto Superiore di Sanità
Rome, Italy

Roger Stephan
Vetsuisse Faculty
Institute for Food Safety and Hygiene
University of Zurich
Zurich, Switzerland

L.K. Strawn
Department of Food Science
Cornell University
Ithaca, New York

Yi-Wei Tang
Departments of Pathology and Medicine
Vanderbilt University School of Medicine
Nashville, Tennessee

Marleen Temmerman
Department of Obstetrics and Gynaecology
Ghent University
Ghent, Belgium

Rogério Tenreiro
Faculdade de Ciências
Universidade de Lisboa
Center for Biodiversity Functional and Integrative Genomics
Edifício ICAT, Campus da FCUL
Lisbon, Portugal

Corinne Teyssier
Faculté de Pharmacie, BP 14491
Université Montpellier 1
Montpellier, France

María Mar Tomás
Servicio Microbiología
Complejo Hospitalario Universitario La Coruña
La Coruña, Spain

Herbert Tomaso
Friedrich-Loeffler-Institut
Institute of Bacterial Infections and Zoonoses
Jena, Germany

Martha E. Trujillo
Departmento de Microbiología y Genética
Edificio Departamental
Universidad de Salamanca
Salamanca, Spain

Sheng Kai Tung
Department of Medical Laboratory Science and Biotechnology
College of Medicine
National Cheng Kung University
Taiwan

Mario Vaneechoutte
Department of Clinical Biology, Immunology and Microbiology
Ghent University Hospital
Ghent, Belgium

A.C.M. Veloo
Department of Medical Microbiology
University Medical Center Groningen
University of Groningen
Groningen, the Netherlands

Suzanne J.C. Verhaegh
Department of Medical Microbiology and Infectious Diseases, Erasmus MC
Rotterdam, the Netherlands

Rita Verhelst
Department of Clinical Biology, Immunology and Microbiology
Ghent University Hospital
Ghent, Belgium

Hans Verstraelen
Department of Obstetrics and Gynaecology
Ghent University Hospital
Ghent, Belgium

Jordi Vila
Department of Clinical Microbiology
Hospital Clinic
University of Barcelona School of Medicine
Barcelona, Spain

Almudena F. Villadangos
Section of Microbiology
Department of Molecular Biology
Faculty of Biology
University of León
León, Spain

Georgia Vrioni
Department of Microbiology
National & Kapodistrian University of Athens Medical School
Athens, Greece

Ken B. Waites
Department of Pathology
University of Alabama at Birmingham
Birmingham, Alabama

Chengming Wang
Molecular Diagnostics Laboratory
Department of Pathobiology
College of Veterinary Medicine
Auburn University
Auburn, Alabama

B.R. Warren
Research, Quality and Innovation
ConAgra Foods, Inc.
Omaha, Nebraska

David M. Whiley
Queensland Paediatric Infectious Diseases Laboratory
Royal Children's Hospital
Queensland Children's Medical Research Institute
Queensland, Australia

Patrick C.Y. Woo
Department of Microbiology
The University of Hong Kong
Hong Kong

Li Xiao
Department of Pathology
University of Alabama at Birmingham
Birmingham, Alabama

Osvaldo Miguel Yantorno
Facultad de Ciencias Exactas
Centro de Investigación y Desarrollo en Fermentaciones Industriales
Universidad Nacional de La Plata
La Plata, Argentina

Ting Zhang
College of Veterinary Medicine
Mississippi State University
Starkville, Mississippi

1 Introductory Remarks

Dongyou Liu

CONTENTS

1.1 PREAMBLE

Bacteria (singular, bacterium) are small unicellular organisms that are classified taxonomically in the domain Bacteria (or Eubacteria), the kingdom Prokaryotae (or Prokaryota or Monera). The only other domain in the kingdom Prokaryotae covers Archaea (or Archaeobacteria for "ancient bacteria"). With sizes ranging from 10^{-7} to 10^{-4} mm, prokaryotes are bigger than viruses (10^{-8}–10^{-6} mm), but smaller than eukaryotes (10^{-5}–10^{3} mm). While both bacteria and archaea may have evolved independently from an ancient common ancestor, eukaryotes may have arisen from ancient bacteria entering into endosymbiotic associations with the ancestors of eukaryotic cells (possibly related to the archaea) to form either mitochondria or hydrogenosomes. A subsequent independent engulfment by some mitochondria-containing eukaryotes of cyanobacterial-like organisms may have led to the formation of chloroplasts in algae and plants.

In contrast to the organisms in the eukaryotic kingdoms Protista, Fungi, Plantae, and Animalia, those in the kingdom Prokaryotae lack nuclear membrane (with their DNA usually in a loop or coil), contain few independent membrane-bounded cytoplasmic organelles (e.g., vacuole, endoplasmic reticulum, Golgi apparatus, and mitochondria) apart from chromosome and ribosome, have no unique structures in their plasma membrane and cell wall, and do not undergo endocystosis and exocytosis. In other words, whereas eukaryotic chromosome resides within a membrane-delineated nucleus, bacterial chromosome is located inside the bacterial cytoplasm. This entails that all cellular events (e.g., translational and transcriptional processes, and interaction of chromosome with other cytoplasmic structures) in prokaryotes occur in the same compartment. Furthermore, while eukaryotic chromosome is packed with histones to form linear chromatin, bacterial chromosome assumes a highly compact supercoiled structure in circular form (and rarely in linear form).

Although archaea are similar to bacteria in most aspects of cell structure and metabolism, they differ from bacteria in that being extremophiles, they can live in extreme environments where no other life forms exist. This may be due to the unique structure in archaeal lipids in which the stereochemistry of the glycerol is the reverse of that found in bacteria and eukaryotes, possibly the result of an adaptation on the part of archaea to hyperthermophily. In addition, the archaeal cell wall does not contain muramic acid, which is commonly present in bacteria. The archaeal RNA polymerase core is composed of ten subunits in comparison with four subunits in bacteria. Besides possessing distinct tRNA and rRNA genes, archaea uses eukaryotic-like initiation and elongation factors in protein translation, and their transcription involves TATA-binding proteins and TFIIB as in eukaryotes.

Bacteria are ubiquitously distributed in virtually every habitat on earth, and are abundantly present in soil, fresh water, plants, and animals. With an estimated number of 5 nonillion (5×10^{30}), bacteria form much of the world's biomass. Bacteria play an essential role in chemical cycles, environmental maintenance, food production, and human wellbeing. However, some bacteria are pathogenic and capable of causing infectious diseases in humans, animals, and plants. Cholera, syphilis, anthrax, leprosy, bubonic plague, and tuberculosis are some of the examples of the deadly human diseases that are attributable to bacteria. Correct identification and detection of bacterial pathogens is not only fundamental to the study of these microorganisms but also critical to the control and prevention of the diseases they cause.

1.2 BACTERIAL CHARACTERISTICS

1.2.1 Classification

Bacteria are classified on the basis of their differences in morphology (e.g., rod, cocci, spirilla, and filament), cell wall structure (e.g., gram-negative and gram-positive), growth characteristics (e.g., aerobic and anaerobic), biochemical properties (e.g., fatty acids), and genetic features (e.g., 16S and 23S rRNA). Currently, the domain Bacteria (Eubacteria) is divided into 26 phyla (Acidobacteria, Actinobacteria, Aquificae, Bacteroidetes, Chlamydiae, Chlorobi, Chloroflexi, Chrysiogenetes, Cyanobacteria, Deferribacteres, Deinococcus-Thermus, Dictyoglomi, Fibrobacteres, Firmicutes, Fusobacteria, Gemmatimonadetes, Lentisphaerae, Nitrospira, Planctomycetes, Proteobacteria, Spirochaetes, Tenericutes, Thermodesulfobacteria, Thermomicrobia, Thermotogae, and Verrcomicrobia), whereas the domain Archaea (Archaeobacteria) is separated into two phyla (Crenarchaeota and Euryarchaeota). Among the 26 phyla in the domain Bacteria, Proteobacteria, and Firmicutes contain the largest numbers of genera and species followed by Cyanobacteria, Bacteroidetes, Spirochaetes, and Flavobacteria. Bacteria from other phyla are comparatively rare, from which fewer genera and species have been described. Most of human pathogenic bacteria are found in the phyla Actinobacteria, Bacteroidetes, Chlamydiae, Firmicutes, Fusobacteria, Proteobacteria, Spirochaetes, and Tenericutes (Table 1.1).

1.2.2 Morphology

Bacteria usually measure from 0.2 to 2.0 μm in width and 2–8 μm in length, and are 10 times smaller than eukaryotic cells. On one extreme, a few bacterial species (e.g., *Thiomargarita namibiensis*, and *Epulopiscium fishelsoni*) measure up to half a mm long and are visible to the naked eye. On the other extreme, the smallest bacteria in the genus *Mycoplasma* are only 0.3 μm in size, which are as small as the largest viruses.

Bacteria typically assume four distinctive forms: rod-like bacilli, spherical cocci, spiral bacteria (also called spirilla), and filamentous bacteria. Occasionally, a small number of bacterial species may appear tetrahedral or cuboidal in shape. While many bacterial species exist as

TABLE 1.1
Classification and Characteristics of Major Human Bacterial Pathogens

Phylum	Class	Brief Description	Notable Human Pathogens
Actinobacteria	Actinobacteria	Gram-positive bacteria with high G + C ratio; classification may be assisted through analysis of ferric uptake regulator (fur) and glutamine synthetase	*Actinomyces, Corynebacterium, Mycobacterium, Nocardia*
Firmicutes	Bacilli	Gram-positive cocci or rods with low G + C ratio; presence of cell wall	*Bacillus, Enterococcus, Listeria, Staphylococcus, Streptococcus*
	Clostridia	Gram-positive cocci or rods with low G + C ratio; presence of cell wall; some species (e.g., *Veillonella*) are gram-negative	*Clostridium, Eubacterium, Peptostreptococcus*
Tenericutes	Mollicutes	Small bacteria (0.2–0.3 μm in size) with low G + C ratio; absence of cell wall (outer membrane)	*Mycoplasma, Ureaplasma*
Bacteroidetes	Bacteroidia	Gram-negative, anaerobic bacteria; opportunistic pathogens	*Bacteroides, Porphyromonas, Prevotella*
	Flavobacteria	Gram-negative, anaerobic bacteria; opportunistic pathogens	*Elizabethkingia, Flavobacterium*
Chlamydiae	Chlamydiae	Gram-negative bacteria; obligate intracellular pathogens	*Chlamydia, Chlamydophila*
Fusobacteria	Fusobacteria	Gram-negative, filamentous, anaerobic bacteria	*Fusobacterium, Leptotrichia*
Proteobacteria	Alphaproteobacteria	Gram-negative, phototrophic bacteria, with symbiotic properties	*Bartonella, Brucella*
	Betaproteobacteria	Gram-negative, aerobic or facultative bacteria some of which are chemolithotrophs, or phototrophs	*Bordetella, Burkholderia, Neisseria*
	Gammaproteobacteria	Gram-negative, facultatively or obligately anaerobic bacteria, some of which are highly pathogenic	*Aeromonas, Klebsiella, Pseudomonas, Salmonella, Vibrio, Yersinia*
	Epsilonproteobacteria	Gram-negative, curved to spirilloid bacteria, inhabiting digestive tract	*Arcobacter, Campylobacter, Helicobacter*
Spirochaetes	Spirochaetes	Gram-negative, long, helically coiled (spiral-shaped), chemoheterotrophic, anaerobic bacteria, with length-wise flagella	*Borrelia, Leptospira, Treponema*

single cells, others present characteristic patterns such as diploids (pairs), chains, and clusters ("bunch of grapes"). In addition, some bacteria may be elongated to form filaments, which are often surrounded by a sheath containing many individual cells. The elaborated, branched filaments formed by *Nocardia* may even resemble fungal mycelia in appearance. Frequently, bacteria use quorum sensing to detect surrounding cells, migrate toward each other, and attach to solid surfaces to form dense aggregations called biofilms (bacterial mats, or fruiting bodies), which may measure a few micrometers in thickness to up to half a meter in depth, and which comprise multiple species of bacteria, archaea, and protists (numbering approximately 100,000 cells). The formation of biofilms protects bacteria from host defense mechanisms and antibiotic therapy, contributing to chronic bacterial infections and infections relating to implanted medical devices.

Structurally, a bacterial cell is surrounded by a rigid layer (cell wall) that is located externally to the lipid membrane. The cell wall provides structural support and protection, and acts as a filtering mechanism. In addition to prokaryotae, fungi and plantae also possess a cell wall, but animalia and most protista do not. While the bacterial cell wall is made up of peptidoglycan (also called murein, which in turn is composed of polysaccharide chain cross-linked by peptides containing D-amino acids), the archaeal cell wall consists of surface layer proteins (also known as S-layer), pseudopeptidoglycan (pseudomurein), and polysaccharides. By contrast, the fungal cell wall includes chitin, the algal cell wall has glycoprotein and polysaccharides, and the plant cell wall often incorporates cellulose and proteins such as extensins.

Based on the ability of bacterial cell wall to retain Gram stain (consisting of crystal violet as primary stain and Gram's iodine and basic fuchsin as subsequent stain), bacteria are divided into gram-positive and gram-negative categories. The gram-positive bacterial cell wall is composed of several layers of peptidoglycan (which is responsible for retaining the crystal violet dyes during the Gram staining procedure, leading to its purple color) surrounded by a second lipid membrane containing lipopolysaccharides and lipoproteins. Located outside of cytoplasmic membrane, peptidoglycan is a large polymer (formed by poly-*N*-acetylglucosamine and *N*-acetylemuramic acid) that contributes to the structural integrity of the bacterial cell wall in addition to countering the osmotic pressure of the cytoplasm. Peptidoglycan is predominant in the cell walls of high and low percentage G + C gram-positive organisms (e.g., actinobacteria and firmicutes). Also imbedded in the gram-positive cell wall are teichoic acids, some of which are lipid linked to form lipoteichoic acids. On the other hand, the gram-negative cell wall has a thin peptidoglycan layer adjacent to the cytoplasmic membrane that contributes to its inability to retain the crystal violet stain upon decolonization with ethanol during the Gram staining procedure (leading to its red or pink color after restaining with basic fuchsin). Apart from the thin peptidoglycan layer, the gram-negative cell wall also has an outer membrane that is formed by phospholipids and lipopolysaccharides.

Within the gram-positive bacterial category, there is another distinct group of bacteria (i.e., acid-fast bacteria such as *Mycobacterium* and *Nocardia*) that can resist decolorization with an acid-alcohol mixture during the acid-fast (or Ziehl–Neelsen) staining procedure and retain the initial dye carbol fuchsin and appear red. The acid-fast cell wall of *Mycobacterium* includes a large amount of glycolipids, especially mycolic acids that make up approximately 60% of the acid-fast cell wall in addition to a thin, inner-layer peptidoglycan. The presence of the mycolic acids and other glycolipids impede the entry of chemicals, causing the organisms to grow slowly and be more resistant to chemical agents and lysosomal components of phagocytes than most bacteria.

Whereas a vast majority of bacteria possess the gram-negative cell wall, the firmicutes and actinobacteria (previously known as the low percentage G + C and high percentage G + C gram-positive bacteria, respectively) have the gram-positive structure, and the tenericutes (e.g., the genus *Mycoplasma*) are devoid of a cell wall in spite of their similarity in G + C ratio to the firmicutes. The differences in the cell wall often determine the susceptibility and resistance of bacteria to antibiotics and other therapeutic reagents. Given that *Mycoplasma* species lack a cell wall, they are unaffected by such commonly used antibiotics such as penicillin and streptomycin that target cell wall synthesis. With their small size (0.3 μm), *Mycoplasma* species are often identified as a source of contaminating infection in the cell culture (where penicillin and streptomycin are incorporated in the culture media), causing retarded growth of cultured cell lines. The cell wall of bacteria forms part of pathogen-associated molecular patterns (or PAMPs), which are recognized by pattern-recognition receptors (or PRRs) in mammalian hosts to initiate and promote innate and adaptive immune defenses against invading bacteria.

Several recognizable extracellular structures are present in bacteria. These include flagella, pili, and fimbriae, which protrude from bacterial cell wall and are involved in bacterial twitching movement as well as interaction with one another and other organisms. Bacterial flagellum (measuring 20 nm in diameter and up to 20 μm in length) is a long, whip-like, and helical projection made up of repeating flagellin protein. The numbers and arrangements of flagella vary among bacterial genera and species. Monotrichous bacteria have a single flagellum, amphitrichous bacteria contain a single flagellum on each of cell poles, lophotrichous bacteria include multiple flagella that are located at one cell pole, and peritrichous bacteria have multiple flagella that are situated at several locations. Flagella in bacteria are powered by a flow of H+ ions (sometimes Na+ ions), and those in archaea are powered by adenosine 5'-triphoshate (ATP). Despite having a similar appearance, eukaryotic flagella (called cilia or undulipodia) differ from prokaryotic flagella in both structure and evolutionary origin. A eukaryotic flagellum is a bundle of nine fused pairs of microtubule doublets surrounding two single microtubules. Eukaryotic flagella are often arranged en masse at the surface of a stationary cell anchored within an organ, lashing back and forth and serving to move fluids along mucous membranes such as trachea. In addition, some eukaryotic cells (e.g., rod photoreceptor cells of eye, olfactory receptor cells of nose, and kinocilium in cochlea of ear) have immotile flagella that function as sensation and signal transduction devises.

Pilus and fimbria are proteinaceous, hair- or thread-like appendages in bacteria (particularly of gram-negative category) that are much shorter and thinner than flagellum. Bacteria have up to ten pili (typically 6–7 nm in diameter) whose main function is to connect the bacterium to another of the same or a different species to enable transfer of plasmids between the bacteria (i.e., conjugation). A fimbria (measuring 2–10 nm in diameter and up to several μm in length) is shorter than pilus. A bacterium possesses as many as 1000 fimbriae, which are deployed to attach to surface of another bacterium (to form a biofilm) or host cell (to facilitate invasion). Many pilin proteins are characteristic among bacterial species and subgroups, which have been exploited as targets for serological typing of bacteria (serotypes or serovars).

Many bacteria produce capsules or slime layers around their cells, which can protect cells from engulfment by eukaryotic cells (e.g., macrophages), act as antigens for cell recognition, and aid attachment to surfaces and the formation of biofilms. In addition, some gram-positive bacteria (e.g., *Bacillus, Clostridium*, and *Anaerobacter*) can form highly resistant, dormant structures called endospores, which contain a central core of cytoplasm with DNA and ribosomes surrounded by a cortex layer and protected by an impermeable and rigid coat. Endospores can survive extreme physical and chemical stresses (e.g., UV lights, γ-radiation, detergents and disinfectants, heat, pressure, and desiccation), and may remain viable for millions of years. Endospore-forming bacteria (e.g., *Bacillus anthrax* and *Clostridium tetanus*) are also capable of causing disease.

Underneath the lipid membrane is the cytoplasm, which is composed of nutrients (or nutrient storage granules such as glycogen, polyphosphate, sulfur, or polyhydroxyalkanoates), proteins, and other essential components. There is a notable absence of membrane-bound organelles (with the exception of chromosome and ribosome) in the bacterial cytoplasm, although certain subcellular compartments (prokaryotic cytoskeleton), such as carboxysome-containing polyhedral protein shells, have been detected. These polyhedral organelles compartmentalize bacterial metabolism, similar to the function performed by the membrane-bound organelles in eukaryotes. The bacterial chromosome consists of a single circular DNA molecule that is situated together with associated proteins and RNA in an irregularly shaped body called the nucleoid. The bacterial ribosomes are responsible for production of proteins.

1.2.3 Biology

Bacteria utilize many metabolic pathways (e.g., glycolysis, electron transport chains, chemiosmosis, cellular respiration, and photosynthesis) and thus virtually all carbon or energy supplies for their maintenance and growth. They are easily grown using either solid or liquid media (e.g., Luria Bertani broth). Solid growth media (e.g., agar plates) are useful for isolation of pure cultures of a bacterial strain, and liquid growth media are employed to generate bulk quantities of bacterial cells. In addition, selective media (containing specific nutrients and antibiotics) assist the isolation and identification of specific bacterial organisms.

As single-celled organisms, prokaryotes reproduce by asexual binary fission, which begins with DNA replication within the cell until the entire prokaryotic DNA is duplicated. The two chromosomes then separate as the cell grows and the cell membrane invaginates, splitting the cell into two daughter cells. This reproductive process is highly efficient and leads to exponential growth of bacteria. In fact, under optimal growth conditions, *Escherichia coli* cells can double every 20 min. Because bacteria are able to multiply rapidly with minimal nutritional requirements, they are abundant in virtually every habitat on earth. In soil, bacteria live by degrading organic compounds and assist in soil formation. In aquatic environments such as ponds, streams, lakes, rivers, seas, and oceans, bacteria such as cyanobacteria (sometimes called blue-green algae because of their color) utilize their chlorophylls to capture energy from the sunlight. In the depths of the sea, bacteria obtain energy from oxidizing or reducing naturally occurring sulfur compounds. In humans and animals, bacteria are found in large numbers on the skin, the respiratory and digestive tracts, and other parts of the body, constituting a normal microbiota in an essentially symbiotic relationship with mutual benefits. Although the vast majority of bacteria are harmless and sometimes even beneficial to their hosts, a few have the capacity to take advantage of temporary weakness in the host (e.g., injury and/or impaired immune function) to cause diseases of varying severity.

In a high-nutrient environment, the growth cycle of bacteria usually undergoes three phases. The first phase (the lag phase) is a period of slow growth with the bacterial cells adapting to the high-nutrient environment and preparing for fast growth. In the lag phase, the cell replicates its DNA and makes all the other molecules (e.g., ribosomes, membrane transport proteins) needed for the new cell. The second phase (the logarithmic phase or "log" phase, also known as the exponential phase) occurs when DNA replication stops, and is characterized by rapid cell division and exponential growth. The rate at which cells grow during this phase is known as the growth rate, and the time it takes the cells to double is known as the generation time. During the log phase, nutrients are metabolized at maximum speed until one of the nutrients is depleted, which poses a negative impact on growth. The final phase (the stationary phase) results from the depletion of nutrients. During the stationary phase, the cells decrease their metabolic activity and consume nonessential cellular proteins. As a transition from rapid growth to a stress response state, there is heightened expression of genes involved in DNA repair, antioxidant metabolism, and nutrient transport. Although the entire cycle of bacterial growth takes about an hour, a rapidly growing bacterial cell carries out multiple rounds of replication simultaneously, which helps to shorten the doubling time for most bacteria to about 20 min.

1.2.4 Genetics

Bacteria have a single circular chromosome that ranges in size from only 160,000 bp (base pairs) (e.g., *Candidatus Carsonella ruddii*) to 12,200,000 bp (e.g., *Sorangium cellulosum*). However, *Borrelia burgdorferi*, the causal agent for Lyme disease, contains a single linear chromosome. In addition, bacteria may possess small extrachromosomal DNA called plasmid, which ranges from 1 to 400 kb in size and comprises genes or gene cassettes for antibiotic resistance or virulence factors. As plasmids have at least an origin of replication (or *ori*)—a starting point for DNA replication—they are capable of autonomous replication independent of the chromosomal DNA. A plasmid that integrates into the chromosomal DNA is called episome, which permits its duplication with every cell division of the host. Some viruses (bacteriophages or phages) may also exist in bacteria, with some merely infecting and lysing their host bacteria, while others inserting into the bacterial chromosome. Phages are usually made up of a nucleic acid core (e.g., ssRNA, dsRNA, ssDNA, or dsDNA measuring 5–500 bp in length) with an outer protein hull. A phage containing particular genes may contribute to its host's phenotype, as illustrated by the evolution of *Escherichia coli* O157:H7 and *Clostridium botulinum*, which are converted from harmless ancestral bacteria into lethal pathogens through the integration of phages harboring toxin genes.

Being the key component of the ribosome, ribosomal RNA molecules (rRNA) consists of two complex folded subunits of differing sizes (small and large), whose main functions are to provide a mechanism for decoding messenger RNA (mRNA) into amino acids (at center of small ribosomal subunit) and to interact with transfer RNA (tRNA) during translation by providing petidyltransferase activity (large subunit). Whereas the two rRNA subunits in eukaryotes have sedimentation coefficiency values of 40S (Svedberg units) and 60S, those in bacteria measure 30S and 50S, respectively. In virtually all organisms, the small rRNA subunit (40S in eukaryotes and 30S in bacteria) contains a single RNA species (i.e., 18S rRNA in eukaryotes and 16S rRNA in bacteria); the large rRNA subunit (60S) in eukaryotes comprises three RNA species (5S, 5.8S, and 25/28S rRNA), while that (50S) in bacteria contains two RNA species (5S and 23S rRNA).

Although bacteria do not undergo meiosis or mitosis and do not require cellular fusion to initiate reproduction (as bacteria are not diploid), many bacteria do involve a cell-to-cell transfer of genomic DNA by various mechanisms. These mechanisms may range from the uptake of exogenous DNA from their environment (a process called transformation) and the integration of a bacteriophage introduces foreign DNA into the chromosome (a process called transduction), to the acquisition of DNA through direct cell contact (a process called conjugation). The incorporation of genes and DNA from other bacteria or the environment into the recipient cell's DNA is also called horizontal gene transfer. While DNA transfer occurs less frequently per individual bacterium than that among eukaryotes involving obligate sexual reproduction, the much shorter generation times and high numbers associated with bacteria can make the DNA transfer a significant contributor to the evolution of bacterial populations. Gene transfer is vital to the development of antibiotic resistance in bacteria as it allows the rapid transfer of resistance genes between different pathogens.

Regardless of genome size, most organisms show a mutation rate on the order of one mutation per genome per generation. Given their very short generation times (<1 h in culture media and a few hours in the wild) and small genomes (which are a 1000 times smaller than most eukaryotes), prokaryotes generally display 1000 times more mutations per gene, per unit time, and per individual than eukaryotes. Furthermore, with greater population sizes resulting in the absolute amount of mutational variation entering the population, prokaryotes have enormous capacity to adapt to and invade new niches, which are the key factors contributing to the evolutionary success of prokaryotes.

Genomic diversity in bacteria comes in two forms: (i) genetic heterogeneity wherein different strains have different alleles of the same gene and (ii) genomic plasticity where different strains have different genes. Recent studies indicate that each strain (serovar) within a bacterial species receives a unique distribution of genes from a population-based supragenome that is many times larger than the genome of any given strain.[1] Through the autocompetence and autotransformation mechanisms, bacterial strains (or serovars) within the species may evolve and generate diversity in vivo to enable them to persist in the face of myriad host defense mechanisms and environmental stresses. In other words, the strain (serovar)-specific genes (e.g., contingency genes) may provide for an increased number of genetic characters that facilitate the population as a whole to adapt rapidly to environmental factors, such as those experienced in the host during chronic infectious processes. There is evidence that under arduous external conditions, many bacteria form biofilms that often exchange DNA at rates several orders of magnitude greater than planktonic bacteria and that are responsible for many chronic bacterial infections in human patients. For example, biofilm-associated growth of *Pseudomonas aeruginosa* has been implicated in several chronic suppurative otitis media. It appears that the possession of a distributed genome is a common host interaction strategy.

1.2.5 Ecological and Medical Importance

Bacteria play a number of beneficial roles in the ecological balance of our planet. Bacteria are involved in the recycling of carbon dioxide (CO_2) to oxygen via photosynthesis;

in the decomposition of dead plant and animal matter, improving soil fertility; and in the fixation of nitrogen into the nitrogen compound ammonia for plant growth. Bacteria can help clean up oil spills, pesticides, and other toxic materials. Some bacteria are able to remove (leach) the copper from the ores (copper sulfides), while others are useful for food production such as yogurt, cheese, cider, and vinegar. Some bacteria (e.g., *Bacillus thuringiensis* or BT, a soil dwelling gram-positive bacterium) can be used as pesticides (trade names Dipel and Thuricide) in the biological control of Lepidopteran pest. In addition, as fast growers with relatively low demands for nutrients, bacteria represent ideal hosts for mass production of certain plastics, enzymes used in laundry detergents, and antibiotics such as streptomycin and tetracycline, as well as pharmaceuticals and fine chemicals upon genetic modification. Furthermore, some bacteria (e.g., *Listeria monocytogenes*, *Mycobacterium bovis* Baille Calmette-Guerin or BCG, *Salmonella, Shigella*, and *Escherichia coli*) are useful carriers for delivering vaccine molecules against microbial diseases and cancers.[2]

Although many bacteria are harmless or beneficial, a few can exert detrimental effects on food and plant production, as well as human and animal health. Some bacteria can cause food spoilage (e.g., *Lactobacillus*) and foodborne diseases (e.g., *Shigella, Campylobacter, Salmonella*, and *Listeria*), and others can harm agriculture because of the major diseases of plants and farm animals they cause. For instance, *Brachyspira hyodysenteriae* causes a type of severe diarrhea in pigs that can have disastrous consequences for pig farmers. Some bacteria are involved in metal corrosion (wearing away) through the formation of rust, especially on metals containing iron.

There are approximately ten times as many bacterial cells as human cells in the human body, with large numbers of bacteria on the skin and in the digestive tract. The communities of bacteria and other organisms that inhabit the body are sometimes referred to as the normal microflora or microbiota. Some bacteria in human body produce essential nutrients (e.g., vitamin K) that the body cannot make itself. Nonetheless, some bacteria are highly pathogenic and deadly. *Yersinia pestis* was responsible for the most fatal and devastating bacterial disease in history—the bubonic plague—which killed an estimated 200 million people prior to the advent of antibiotics. Currently, the most common fatal bacterial diseases are respiratory infections, with tuberculosis (caused by *Mycobacterium tuberculosis*) alone killing about 2 million people a year. One of the world's deadliest bacterial diseases today is cholera, which is caused by foodborne *Vibrio cholerae*. Other globally important bacterial diseases include pneumonia caused by *Streptococcus* and *Pseudomonas*, tetanus, typhoid fever, diphtheria, syphilis, and leprosy. Another common bacterial disease is tooth decay, which results from the acids bacteria produce from sugar via fermentation, which dissolves the enamel of the teeth and create cavities (holes) in the teeth.

1.3 BACTERIAL IDENTIFICATION

1.3.1 Current Diagnostic Approaches

Identification and characterization of bacteria and their roles in disease processes represent a critical step in the effective management of bacterial infections.

Traditionally, bacteria have been identified and diagnosed through examination of their phenotypic characteristics such as Gram staining, cell and colony morphology, and biochemical and serological properties. As a key component in the phenotypic identification scheme, in vitro culture is often time consuming, and its performance is affected by the changing features of bacterial metabolisms. Given the limited variations in bacterial cell and colony morphology, it is impossible to differentiate most bacterial species on morphological criteria alone. Biochemical and serological techniques are helpful in bacterial identification. However, overreliance on these techniques not only adds to the cost of identification, but also creates delay in the result availability. Furthermore, the intrinsic variability of phenotypic procedures remains a potential source of misdiagnosis.

In recent decades, DNA- or RNA-based genotypic (molecular) procedures have been increasingly utilized for microbial identification.[3] Besides comparison at the genome level (e.g., G + C content determination and DNA–DNA hybridization), detection of nucleotide differences among shared and uniquely present gene regions provides a practical means for bacterial characterization. Due to its crucial roles in cellular function and maintenance, the ribosomal RNA (rRNA) gene is highly conserved (i.e., the least variable) and abundant (with each living cell containing 10^4–10^5 copies of the 5S, 16S, and 23S rRNA molecules). The rRNA gene (and its genomic coding sequence rDNA) therefore offers a valuable target for confirmation of an organism's taxonomic status and species identity, and for estimation of the rates of species divergence.[4–8] In addition, a range of housekeeping-, species-, group-, and virulence-specific genes provide alternative targets for improved determination of bacterial organisms.[9,10]

The early molecular methods are largely nonamplified, as exemplified by G + C content determination, DNA–DNA hybridization, fluorescence in situ hybridization (FISH), ribotyping, and so forth. Because these methods often require large amounts of starting materials and are cumbersome to perform, they are now rarely used in routine clinical setting, and their applications are limited to the characterization and description of novel bacterial species/subspecies. The more recent molecular methods often involve nucleic acid amplification. These include polymerase chain reaction (PCR), ligase chain reaction (LCR), nucleic acid sequence-based amplification (NASBA), transcription-mediated amplification (TMA), strand displacement amplification (SDA), rolling circle amplification (RCA), cycling probe technology (CPT), branched DNA (bDNA), and loop mediated isothermal amplification (LAMP), and so forth. Due to their efficiency, simplicity, and robustness, PCR and its derivatives have been widely adopted in research and clinical laboratories for specific, sensitive and rapid identification and diagnosis of bacterial pathogens.[11]

In a standard (conventional) PCR, a pair of oligonucleotides (of about 20 bases in length) is typically used as primers to anneal and amplify a gene region of interest with the help of DNA polymerase. The resulting amplified products (or amplicons) are separated by agarose gel electrophoresis and visualized with a DNA-binding dye (e.g., ethidium bromide). Further refinements of standard PCR led to the development of nested PCR (in which two consecutive PCR are performed one after another), multiplex PCR (in which multiple pairs of primers are included for simultaneous amplification of several genes of interests from the same or different organisms), arbitrarily primed PCR (in which a single oligonucleotide of about 10 bases in length is used to amplify random regions in a genome), and reverse-transcriptase PCR (RT–PCR) (in which RNA instead of DNA is targeted).[12,13]

Increasingly, the formats of PCR product detection have moved away from gel electrophoresis to enzymatic signal detection (e.g., ELISA and flow cytometry), real-time detection (e.g., using intercalating fluorescent dye [SYBR® green], hydrolysis dual-labeled probes [TaqMan®], hybridization probes [LightCycler], molecular beacons, peptide nucleic acid [PNA] probes, TaqMan minor groove binding [MGB™] probes, locked nucleic acid [LNA®] primers and probes, and scorpions™), line probe assay (LiPA), microarray, sequencing analysis (e.g., pyrosequencing), mass spectrometry (MS), and so forth.[14] In particular, real-time PCR demonstrates superior sensitivity, rapidity, and broad dynamic range, eliminates postamplification handling steps (thus minimizing carryover contamination), and is amenable to automation for high throughput detection.[15] Furthermore, real-time PCR provides the option for melting curve analysis, permitting discrimination of the amplified product from nonspecific product or primer-dimmers.[14] For these reasons, real-time PCR is becoming a method of choice for microbial identification in clinical laboratories worldwide.

1.3.2 Performance Parameters

The performance of a diagnostic assay is often evaluated by using several key parameters, including detection limit, sensitivity, specificity, accuracy, intraassay precision, interassay precision, and linearity (as in the case of a quantitative assay). Detection limit (or limit of detection) is the lowest concentration or quantity of bacteria that can be detected by a given assay. Sensitivity is the percentage of samples containing bacteria of interest that are identified by the assay as positive for the bacteria. Specificity is the percentage of samples without bacteria of interest that are identified by the assay as negative for the bacteria. Accuracy (or trueness) is the degree of conformity of an assay's measurements to the actual (true) value. It is often estimated by analyses of reference materials or comparisons of results with those obtained by a reference method. The closer an assay's measurements to the accepted value, the more accurate the assay is. Precision is the degree of mutual agreement among a series of assay's individual measurements, values, or results. Usually characterized in terms of the standard deviation of the measurements, precision can be stratified into (i) repeatability (the variation arising using the same instrument and operator in a single rune—i.e., intraassay precision—or repeating during a short time period) and (ii) reproducibility (the variation arising using the same measurement process among different instruments and operators from one run to another—i.e., interassay precision—or over longer time periods). Linearity refers to the tendency of measurements by a quantitative assay to form a straight line when plotted on a graph. Data from linearity experiments may be subjected to linear regression analysis with an ideal regression coefficient of 1. In case of a nonlinear curve, other objective, statistically valid methods may be utilized.[16]

1.3.3 Result Interpretation

A positive result by a molecular assay for a given pathogen normally confirms the etiologic relationship if the clinical syndrome is compatible with the pathogen identified. Considering the sensitive nature of the amplified methods such as PCR, it is important to rule out the possibility of a false positive result. Occasionally, false positive results may originate from the low diagnostic specificity of the assay, in which primers bind to irrelevant sequences and occasionally a homologous sequence that is shared among related or unknown bacteria. More often, false positive results in the molecular testing come from contamination, which may arise during manual handling of the samples in the testing laboratory either at the pre or postextraction (while setting up the PCR) stages. This risk is heightened when a high copy number polynucleotide (or plasmid) is used as a quantification standard and distributed around the laboratory, contaminating reaction source. In addition, contamination may be attributable to samples referred from other laboratories that do not utilize manipulation techniques that are mandatory for the molecular testing. These may include the use of unplugged pipette tips, infrequent changing of gloves and using pipette for long periods without decontamination. Another cause of contamination is by amplification products from previous tests.[17–19] Contamination may also occur by leakage from tubes or microtiter plates with lids not tightly closed or by breakage of glass capillaries leading to spillage of the amplification mixture. Besides the adoption of stringent laboratory practice, the risk of contamination with PCR products may be reduced by replacing nucleotide dTTP with dUTP in PCR, and implementing a digestion step with Uracil-DNA-glycosylase (UNG) to remove previous PCR products containing dUTP prior to each amplification reaction. Furthermore, inclusion of multiple negative controls, such as no-template controls (NTC) and no-amplification controls (NAC) may help identify the likely source of contamination and prevent false positive results. Moreover, microbial DNA may come with PCR reagents.

Similarly, a negative result by a molecular assay for a given pathogen normally indicates the absence of the pathogen.

However, it is equally important to rule out the possibility of false negative results. One possible cause is due to the low sensitivity of the assay employed. Alternatively, an insufficient amount of bacteria may be present in the sample (due to sample degradation or prior antibiotic treatment). Another cause may be the impurity of the processed sample. Enzymes used in PCR and RT–PCR (e.g., DNA polymerase, reverse transcriptase) are impeded by components in blood and feces (e.g., heme, hemoglobulin, lactoferrin, immunoglobulin G, leukocyte DNA, polysaccharides, and urea), in foods (e.g., phenolics, glycogen, calcium ions, fat, and other organic substances), in environmental specimens (e.g., phenolics, humic acids, and heavy metals), and in added anticoagulants (e.g., EDTA and heparin) as well as nucleic acid purification reagents (e.g., detergents, lysozyme, NaOH, alcohol, EDTA, EGTA, phenol, and high salt concentrations).[20–24] Any impurities and contaminations present in the samples after nucleic acid isolation may contribute to false negative results. A useful way to determine the effectiveness of a nucleic acid purification procedure for removing inhibitory substances is to spike samples with well-defined DNA or RNA prior to and after sample preparation (as process and amplification internal controls). In light of the high sensitivity of PCR, the occurrence of false negative results is probably a truly underestimated problem.

Thus, before definitive diagnoses and treatment decisions are made, all molecular and immunological diagnostic results need be reviewed and critically interpreted in combination with the patient's clinical presentation.

1.3.4 Standardization, Quality Control, and Assurance

Molecular tests such as PCR offer improved sensitivity, specificity, accuracy, precision, and result availability for microbial identification and diagnosis. There is a clear trend toward the adoption and application of these methods in routine diagnostics. However, in view of the possibility of false positive and false negative results occurring in these highly sensitive tests, it is vital that molecular diagnostic methods are properly standardized and validated and appropriate quality control measures are put in place.

Standardization and Validation. Standardization of molecular tests addresses the need for standardized reagents and common units, contamination control mechanisms, inhibition control mechanisms, clinically relevant dynamic ranges and internal controls, and so forth. Validation helps to verify the sensitivity, specificity, accuracy, repeatability (intraassay precision), reproducibility (interassay precision), detection limit, and linearity (if quantitative) of molecular tests.[25]

Before validating a method, it is important to have all instruments calibrated and maintained throughout the testing process. The validation process may involve a series of steps, including (i) testing of dilution series of positive samples (or plasmid construct) to determine the limits of detection of the assay and their linearity over concentrations to be measured in quantitative test (using minimal number of reference calibrators such as previously tested patient samples or pooled sera); (ii) evaluating the sensitivity and specificity of the assay, along with the extent of cross-reactivity with other genomic material; (iii) establishing the day-to-day variation of the assay's performance; (iv) assuring the quality of assembled assays using quality control procedures that monitor the performance of reagent batches; and (v) aligning the in-house primer and probe sequences with a genome sequence databank to avoid extended specificity testing.[25–28]

Quality Control. Quality control strategies for nucleic acid–based tests include (i) designation of a "clean" area for reaction setup (e.g., room under negative air pressure, positive-displacement pipettes, aerosol-block pipette tips, UV-equipped PCR cabinet); (ii) use of personal protective equipment (PPE) (e.g., disposable gloves and lab coats to prevent introduction of contaminating DNA or nucleases); (iii) use of uracil-*N*-glycosylase (UNG) in real-time PCR (to eliminate crossover amplicon contamination); (iv) use of a "hot-start" method (to minimize false priming events by withholding a crucial reaction component until appropriate temperature is reached); (v) use of external positive and negative controls (to monitor reaction performance and contamination) and homologous or heterologous internal controls (to monitor presence of inhibitors).[29]

A variety of test controls may be considered for diagnostic PCR. These include (i) internal amplification control (IAC) (negative sample spiked with sufficient pathogen and processed throughout the entire protocol); (ii) processing positive control (PPC) (negative sample spiked with sufficient closely related, but nontarget, strain processed throughout the entire protocol); (iii) reagent control (blank) (containing all reagents, but no nucleic acid apart from the primers); (iv) Premises control (tube containing the master mixture left open in the PCR setup room to detect possible contaminating DNA in the environment [carried out at regular intervals as part of the quality assurance program]); (v) standard (3–4 samples containing a tenfold dilution series of known number of target DNA copies in a range).[29–32]

Quality Assurance. One way to assess preparedness of the diagnostic laboratories is through the conduct of an external quality assurance (EQA) program that provides characterized specimens containing pathogens of interest. The design of a quality assurance program has the following components: (i) internal quality control (IQC) materials are distributed every month and comprising three pools of clinical samples of known pathogen status (typically one negative, one positive containing 1 log10 over the lower limit of detection of the assay, and one low positive containing up to 1 $\log_{10}$ of the lower limit of detection of the assay). These are incorporated in test runs on a weekly basis. The purpose of IQC is to provide samples of known status for repeated testing in parallel with clinical samples to ensure reproducibility of the test system in an individual laboratory. (ii) EQA distributions of panels of five unknown samples distributed quarterly. Results are returned to the QA laboratory for assessment. EQA compares the performance of different testing sites using specimens of known but undisclosed content. (iii) Aliquots of all samples sent from the reference laboratory are posted back to Site A for repeat testing to check for integrity of the pools and for transport problems. (iv) A final element of the pilot program involves Sites B, C, and D sending an aliquot of every 50th sample to Site A to check for reproducibility. (v) A detailed record of distributions is kept to provide an audit trail.[33,34]

1.4 CONCLUSION

Molecular assays have the potential to increase the speed and accuracy of bacterial identification and diagnosis in research and clinical laboratories. When selecting and implementing a molecular test, a number of factors may be considered. These include (i) cost assessment; (ii) high throughput and automation capability (e.g., 96 or 384 well real-time PCR); (iii) detection limit; (iii) multiple target detection (e.g., multiplex PCR and microarray); (iv) broad range detection (e.g., 16S or 23S rRNA gene sequencing to identify poorly characterized, fastidious or noncultivable bacteria); (v) antimicrobial resistance detection; (vi) ease of use (commercial assays); (vii) sample preparation protocol (commercial kits). To overcome the limitations of molecular tests such as potential false positive and false negative results that impact significantly on patient management, laboratory space must be dedicated for instruments and sample preparation, contamination must be minimized, technicians must have proper training, and quality control procedures must be incorporated into routine laboratory workflow.[14]

There is a continuing trend toward miniature devices for microbial testing. One such promising device is biosensor. A biosensor incorporates a biological material (e.g., tissue, microorganisms, organelles, cell receptors, enzymes, antibodies, nucleic acids, natural products, etc.), a biologically derived material (e.g., recombinant antibodies, engineered proteins, aptamers, etc.) or a biomimic (e.g., synthetic catalysts, combinatorial ligands and imprinted polymers) in a physicochemical transducer or transducing microsystem (e.g., optical, electrochemical, thermometric, piezoelectric, magnetic, or micromechanical), leading to enhanced detection and identification of microbial pathogens in clinical, food, and environmental specimens.[35,36]

REFERENCES

1. Hogg, J.S. et al., Characterization and modeling of the *Haemophilus influenzae* core and supragenomes based on the complete genomic sequences of Rd and 12 clinical nontypeable strains, *Genome Biol.*, 8, R103, 2007.
2. Liu, D., *Listeria*-based anti-infective vaccine strategies, *Rec. Patents Anti-infect. Drug Disc.*, 1, 281, 2006.
3. Vaneechoutte, M., and Van Eldere, J., The possibilities and limitations of nucleic acid amplification technology in diagnostic microbiology, *J. Med. Microbiol.*, 46, 188, 1997.
4. Jonasson, J., Olofsson, M., and Monstein, H.J., Classification, identification and subtyping of bacteria based on pyrosequencing and signature matching of 16S rDNA fragments, *Apmis*, 110, 263, 2002.
5. Clarridge, J.E.III, Impact of 16S rRNA gene sequence analysis for identification of bacteria on clinical microbiology and infectious diseases, *Clin. Microbiol. Rev.*, 17, 840, 2004.
6. Barlaan, E.A. et al., Electronic microarray analysis of 16S rDNA amplicons for bacterial detection, *J. Biotechnol.*, 115, 11, 2005.
7. Cherkaoui, A. et al., Development and validation of a modified broad-range 16S rDNA PCR for diagnostic purposes in clinical microbiology, *J. Microbiol. Methods*, 79, 227, 2009.
8. Sontakke, S. et al., Use of broad range16S rDNA PCR in clinical microbiology, *J. Microbiol. Methods,* 76, 217, 2009.
9. Liu, D. et al., Characterization of virulent and avirulent *Listeria monocytogenes* strains by PCR amplification of putative transcriptional regulator and internalin genes, *J. Med. Microbiol.*, 52, 1066, 2003.
10. Liu, D. et al., A multiplex PCR for species- and virulence-specific determination of *Listeria monocytogenes*, *J. Microbiol. Methods*, 71, 133, 2007.
11. Lantz, P.G. et al., Biotechnical use of polymerase chain reaction for microbiological analysis of biological samples, *Biotechnol. Annu. Rev.,* 5, 87, 2000.
12. Keer, J.T., and Birch, L., Molecular methods for the assessment of bacterial viability, *J. Microbiol. Methods,* 53, 175, 2003.
13. Liu, D., Identification, subtyping and virulence determination of *Listeria monocytogenes,* in important foodborne pathogen, *J. Med. Microbiol.*, 55, 645, 2006.
14. Mothershed, E.A., and Whitney, A.M., Nucleic acid-based methods for the detection of bacterial pathogens: Present and future considerations for the clinical laboratory, *Clin. Chim. Acta*, 363, 206, 2006.
15. Smith, K., Diggle, M.A., and Clarke, S.C., Automation of a fluorescence-based multiplex PCR for the laboratory confirmation of common bacterial pathogens, *J. Med. Microbiol.*, 53, 115, 2004.
16. Hoorfar, J. et al., Practical considerations in design of internal amplification control for diagnostic PCR assays, *J. Clin. Microbiol.*, 42, 1863, 2004.
17. Millar, B.C., Xu, J., and Moore, J.E., Risk assessment models and contamination management: implications for broad-range ribosomal DNA PCR as a diagnostic tool in medical bacteriology, *J. Clin. Microbiol.*, 40, 1575, 2002.
18. Aslanzadeh, J., Preventing PCR amplification carryover contamination in a clinical laboratory, *Ann. Clin. Lab. Sci.*, 34, 389, 2004.
19. Borst, A., Box, A.T., and Fluit, A.C., False-positive results and contamination in nucleic acid amplification assays: Suggestions for a prevent and destroy strategy, *Eur. J. Clin. Microbiol. Infect. Dis.*, 23, 289, 2004.
20. Abu Al-Soud, W., and Rådström, P., Capacity of nine thermostable DNA polymerases to mediate DNA amplification in the presence of PCR-inhibiting samples. *Appl. Environ. Microbiol.*, 64, 3748, 1998.
21. Rådström, P. et al., Pre-PCR processing: strategies to generate PCR-compatible samples, *Mol. Biotechnol.,* 26, 133, 2004.
22. Wolffs, P. et al., Impact of DNA polymerases and their buffer systems on quantitative real-time PCR, *J. Clin. Microbiol.*, 42, 408, 2004.
23. Liu, D., Preparation of *Listeria monocytogenes* specimens for molecular detection and identification, *Int. J. Food Microbiol.*, 122, 229, 2008.
24. Handschur, M. et al., Preanalytic removal of human DNA eliminates false signals in general 16S rDNA PCR monitoring of bacterial pathogens in blood, *Comp. Immunol. Microbiol. Infect. Dis.*, 32, 207, 2009.

25. Malorny, B. et al., Standardization of diagnostic PCR for the detection of foodborne pathogens, *Int. J. Food Microbiol.*, 83, 39, 2003.
26. Anon, Microbiology of food and feeding stuffs—protocol for the validation of alternative methods. ISO 16140, first edition, ISO, Geneva, Switzerland, 2003.
27. Josefsen, M.H. et al., Validation of a PCR-based method for detection of foodborne thermotolerant campylobacters in a multi-center collaborative trial. *Appl. Environ. Microbiol.*, 70, 4379, 2004.
28. Hoorfar, J., Wolffs, P., and Radstrom, P., Diagnostic PCR: Validation and sample preparation are two sides of the same coin, *Apmis*, 112, 808, 2004.
29. Hoorfar, J. et al., Making internal amplification control mandatory for diagnostic PCR, *J. Clin. Microbiol.*, 41, 5835, 2003.
30. Rodriguez-Lazaro, D. et al., Construction strategy for an internal amplification control for real-time diagnostic assays using nucleic acid sequence-based amplification: Development and clinical application, *J. Clin. Microbiol.*, 42, 5832, 2004.
31. Trapmann, S. et al., Development of a novel approach for production of dried genomic DNA standards for quantitative PCR testing of food-borne pathogens, *Accred. Qual. Assur.*, 9, 695, 2004.
32. Burggraf, S., and Olgemoller, B., Simple technique for internal control of real-time amplification assays, *Clin. Chem.*, 50, 819, 2004.
33. Raggi, C.C. et al., External quality assurance program for PCR amplification of genomic DNA: An Italian experience, *Clin. Chem.*, 49, 782, 2003.
34. Cubie, H.A. et al., The development of a quality assurance programme for HPV testing within the UK NHS cervical screening LBC/HPV studies, *J. Clin. Virol.*, 33, 287, 2005.
35. Uslu, F., et al., Label-free fully electronic nucleic acid detection system based on a field-effect transistor device, *Biosens. Bioelectron.*, 19, 1723, 2004.
36. Lazcka, O., Del Campo, F.J., and Munoz, F.X., Pathogen detection: A perspective of traditional methods and biosensors, *Biosens. Bioelectron.*, 22, 1205, 2007.

Section I

Actinobacteria

2 *Actinomadura*

Martha E. Trujillo

CONTENTS

2.1 INTRODUCTION

The term actinomycete is an informal designation for filamentous gram-positive bacteria with high G + C content in their genome and that belong to the order *Actinomycetales* in the class *Actinobacteria*.[1,2] Most aerobic actinomycetes are soil saprophytes and are rarely encountered in clinical practice, but some are serious pathogens of humans and animals causing a number of diseases that include actinomycetoma, actinomycosis, and nocardiosis. The clinical manifestations and severity of the disease and the prognosis in an infected host are extremely variable and may be determined by factors such as the route of infection and the presence or absence of a properly functioning immune system.[3] The diagnosis of actinomycete infections has been hindered by a combination of clinical and microbiologic difficulties, including their often nonspecific clinical presentation, a requirement for invasive diagnostic biopsy procedures, difficulty in isolation, and incorrect identification.

Strategies to improve outcome for infected patients include a heightened awareness of clinicians and clinical microbiology personnel, which may enable the earliest possible diagnosis; standardization of antimicrobial susceptibility testing methods; and the evaluation of newer effective drug therapies for these patients. In addition, new developments in the classification of these microorganisms should serve as a framework for the identification of clinically significant species.

Actinomycetoma is a localized chronic, destructive, and progressive infection of the skin and subcutaneous tissues caused by aerobic actinomycetes.[3–5] The main species involved include *Actinomadura madurae*, *Actinomadura latina*, *Actinomadura pelletieri*, *Nocardia brasiliensis*, *Nocardia otitidiscaviarum*, *Nocardia transvalensis*, *Streptomyces somaliensis*, and *Streptomyces sudanensis*.[6,7] The disease is endemic in certain tropical and subtropical regions, where it has a devastating effect on patients as it frequently leads to deformities, disabilities, and eventually amputation of the affected organs.[7] Actinomycetoma can become dangerous to health, or even life, when treatment is inadequate or delayed.

This chapter will focus on the genus *Actinomadura* and its significance as a pathogen of actinomycetomas and nonmycetomic infections.

2.1.1 Taxonomy

The genus *Actinomadura* contains aerobic, gram-positive, nonacid-fast, nonmotile, and chemo-organotrophic actinomycetes that produce well-developed, nonfragmenting vegetative mycelia and aerial hyphae that differentiate into surface-ornamented spore chains of various lengths (10–50 spores).

Members of the genus *Actinomadura* are characterized chemotaxonomically by the presence of *meso*-diaminopimelic acid and madurose in their cell wall with peptidoglycan structures of the acetyl type, major proportions of hexahydrogenated menaquinones with nine isoprene units, complex fatty acid profiles, including hexadecanoic, 14-methylpentadecanoic, and 10-methyloctadecanoic acids as predominant components, and diphosphatidylglycerol and phosphatidylinositol as major phospholipids.[8]

The systematics of the genus *Actinomadura* has been significantly improved by the application of modern taxonomic methods.[8–14] On the basis of 16S and 23S rRNA gene sequence analyses, the genus is phylogenetically related to members of the family *Thermomonosporaceae*,[1,14] which also includes the genera *Actinocorallia*, *Spirillospora*, and *Thermomonospora*. The family *Thermomonosporaceae* belongs to the suborder *Streptosporangineae* in the order *Actinomycetales*.[2]

A. madurae was first described in 1894 by Vincent,[15] based on several strains isolated from an Algerian case of Madura foot, as "*Streptothrix madurae*." The organism was subsequently classified in the genus *Nocardia*,[16] then in the genus *Streptomyces*.[17] The taxonomic status of this microorganism remained controversial until Becker et al.[18] found that whole-organism hydrolysates of representative strains contained *meso*-diaminopimelic acid (*meso*-DAP) and a characteristic sugar identified as madurose.[19] In 1983, compelling evidence that the genus *Actinomadura* was heterogeneous was provided by Fischer et al.[10] when these authors assigned representative strains to two aggregate groups defined on the basis of chemical and nucleic acid pairing data—*A. madurae* group and *Actinomadura pusilla* group. The division of the genus *Actinomadura* into two aggregate groups was formally recognized by Kroppenstedt et al.,[8] who proposed that the genus *Actinomadura* be retained for *A. madurae* and related species, and that the *Actinomadura pusilla* group be reclassified in the genus *Microtetraspora*.

Taxonomic work carried out with a group of clinically significant *Actinomadura* strains received either as *A. madurae* or *A. pelletieri* revealed that a third species, *A. latina* should be officially recognized.[12] The genus currently comprises more than 40 validly described species (http://www.bacterio.net),[20] but only three—*A. latina*, *A. madurae*, and *A. pelletieri*—are considered pathogens. However, it is becoming increasingly evident that additional pathogenic actinomadurae need to be formally described as new species.[6,12,21]

2.1.2 Cell Morphology, Physiology, and Isolation

Actinomadura strains characteristically form nonfragmenting, extensively branched, substrate mycelia and aerial hyphae, which carry from 1 to 50 arthrospores. Spores are borne in curled, hooked, spiral, or straight chains and the spore surface may be folded, irregular, rugose, smooth, shiny, or warty.

Some strains, notably those from clinical sources, mostly lack aerial mycelium, with colonies exhibiting a cartilaginous or leathery appearance (Figure 2.1). However, members of most species form a spore-bearing, powdery aerial mycelium on media such as inorganic salts-starch (ISP 4), oatmeal (ISP 3), and yeast extract-malt extract agars (ISP 2) after cultivation for 10–14 days.[22] At maturity, aerial mycelia may be blue, cream, gray, green, pink, yellow, or white, thereby differing little from streptomycetes. Superficial similarity to streptomycetes is also reinforced by the morphology of the sporophores. Actinomadurae and streptomycetes can be distinguished by direct microscopic comparison of cultures. Most *Actinomadura* strains are conspicuous for the size of their spores. The diameter of spores noticeably exceeds the diameter of the hyphae, whereas in streptomycetes spores and hyphae are of similar diameter.

In general, actinomadurae are slow-growing microorganisms. These bacteria generally grow well on modified Bennett's agar,[23] glucose-yeast extract agar,[24] and on formulations used for the cultivation of *Streptomyces*.[22] *A. latina*, *A. madurae*, and *A. pelletieri* show moderate-to-good growth on Czapek-dox casamino acids.[25,26]

Actinomadura strains are strictly aerobic with an oxidative metabolism. Most species grow well between 25°C and 40°C but species such as *A. formosensis*, *A. rubrobrunea*, and *A. viridilutea* are thermophilic and grow between 45°C and 65°C. Actinomadurae can metabolize a wide range of sugars and amino acids as carbon sources for growth. Most strains hydrolyse aesculin, arbutin, casein, elastin, gelatine, testosterone, and Tweens 60 and 80.[6,27] Pectinase activity has been reported for *A. mexicana*, *A. napariensis*, and *A. verrucosispora*.[6,28] Pathogenic strains of *A. madurae* produce collagenolytic enzymes.[29] Most actinomadurae grow in the presence of NaCl 4% (w/v), but *A. atramentaria* can grow in NaCl concentrations of 7% (w/v).

Diverse culture media have been employed for the isolation of *Actinomadura* strains, especially from soil samples. The

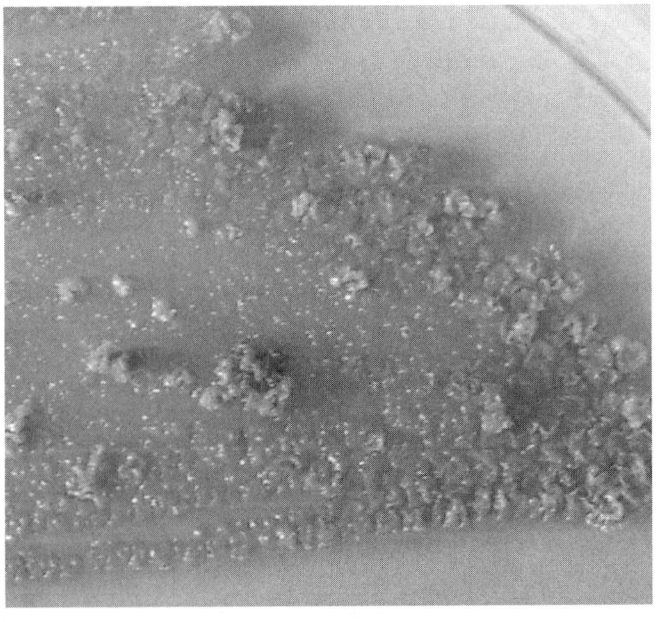

FIGURE 2.1 *Actinomadura* sp. grown on ISP 2 agar for 10 days. The strain does not produce aerial mycelium or diffusible pigments.

media that have proved most suitable include oatmeal agar (ISP 3), yeast extract-malt extract agar (ISP 2), starch-mineral salts agar (ISP 4), Bennett-sucrose agar, and glycerol-asparagine agar (ISP 5). Strains may be isolated from agar plates by dilution techniques after incubation for up to 6 weeks.

Trujillo and Goodfellow[6] successfully isolated new *Actinomadura* species from environmental samples collected in Hong Kong, Kenya, Mexico, South Africa, and Venezuela using the selective isolation procedure recommended by Athalye et al.[24]

Pathogen *Actinomadura* species can be isolated from clinical samples, including pus and biopsy material, using brain heart infusion agar,[30] Sabouraud dextrose agar,[31,32] and Columbia agar supplemented with 5% sheep blood.[33] All samples should be incubated aerobically between 27°C and 37°C for up to 3 weeks and examined both macroscopically and microscopically for growth every 2 days.

Actinomadurae can be recognized by their filamentous appearance, leathery colonies, and by the production of red prodiginine pigments. *A. madurae* and *A. pelletieri* produce prodigiosin-like pigments[34,35] that are similar to those of *Serratia marcescens*. Members of these species isolated from patients produce prodiginines characterized by a tripyrrole skeleton and identified as cyclononylprodiginine, nonylprodiginine, and undecylprodiginine.

A. latina, *A. madurae*, and *A. pelletieri* strains are mainly isolated from clinical material, though there is some evidence that members of *A. madurae* are widespread in organically rich soils. *A. madurae* strains isolated from environmental samples tend to lack the red endopigment of clinical isolates and sporulate more readily.[34–36]

2.1.3 Differentiation of Pathogen *Actinomadura* Species

Published descriptions of *Actinomadura* species are often incomplete since different investigators emphasize some phenotypic features and omit others, thereby making identification difficult. Nevertheless, most species can be separated using a combination of morphological and physiological properties, though in most cases only the type strain has been studied. However, even when several strains have been studied (e.g., *A. madurae*, *A. pelletieri*), the results tend to be variable or inconsistent when those from the literature are compared.

Enzymatic substrates based on the fluorophores 4-methylumbelliferone (4-MU) and 7-amino-methylcoumarin (7-AMC) were carried out by Trujillo and Goodfellow[6,12,36] to differentiate *Actinomadura* species. Encouraging results were obtained for differentiating between pathogenic *Actinomadura* species (Table 2.1).[6,36] Enzymatic tests were highlighted by Wink et al.[37] to differentiate *A. kijaniata* from *A. namibiensis*, as they share a DNA homology higher than 70%.

2.1.4 Pathogenicity and Clinical Features

Actinomycetoma. Actinomycetoma is a localized chronic, destructive infection of the skin and subcutaneous tissues caused by aerobic actinomycetes, *Actinomadura* being one of the main causal agents. The disease is characterized by progressive swelling of the infected area, distortion of the normal anatomy, and multiple draining sinuses and fistulae.[3–5,38–40] Purulent discharge containing the causative agent in the form of grains is characteristic of advanced stages of the disease. The grains vary in size and consistency and may be white, yellow, brown, red, or black depending on the causative agents. Almost 80% of the infections are through the lower extremitics of the body; this probably explains the etiology of "Madura foot" described by Gill of Madura in South India in 1842. In areas where actinomycetoma is endemic, it is a common habit to walk barefooted, exposing the skin to thorns and splinters on the soil; as a result, natural infection appears to be more frequent. Other sites of the human body may also be affected such as the back, head, knee, arm, and neck.[41,42] In Mexico, the back is the second most common location.[43]

The clinical picture of actinomycete mycetoma is almost uniform irrespective of the causal agent. In contrast to eumycetoma (caused by eumycetes), infections caused by actinomycetes are more aggressive and destructive and involve both muscle and bone at an early stage.[5,41] In general, actinomycetoma is painless, even when the disease is at an advanced stage.[41] However, respiratory, neurologic, or other symptoms may be present when the disease affects the chest, head, and neck, or spine. The invasion of the periosteum and adjacent bones may lead to osteomyelitis.[42]

The infection remains localized and constitutional disturbances are rare, but when they do occur they are generally due to septicemia or to immune depression. Actinomycetoma can produce many disabilities, distortion, and deformity. It can be fatal especially if it affects the skull.[41,44]

Nonmycetomic Infections. Nonmycetomic infections produced by *A. madurae* have been reported, including one involving an immunocompromised patient.[45] These observations suggested that *A. madurae* could play a role in pneumonitis or bronchitis. Although no detailed epidemiological studies have been carried out, Bar et al.[46] reported a growing number of patients with pneumonia caused by *A. madurae*. Most of the patients had been infected as a consequence of impaired immunity. These studies suggested that *A. madurae* should also be considered as a causative agent for nonmycetomic infections such as pneumonia. Since that time, no epidemiological studies on *Actinomadura* strains from nonmycetomic regions or clinical specimens such as sputum, bronchoalveolar lavage, and blood samples have been reported. However, in Japan between 1996 and 2004, 21 actinomycete strains were isolated from sputum and broncho alveolar lavage.[21] Molecular identification using 16S rRNA gene sequencing indicated that all strains belonged to the genus *Actinomadura* but were unrelated to *A. madurae* or *A. pelletieri*. Ninety-five percent of the clinical isolates showed a close phylogenetic relationship with the species *Actinomadura cremea* and *Actinomadura nitritigenes*, while the remaining strains could not be related to any other species due to their moderate 16S rRNA gene sequence similarity.

TABLE 2.1
Differential Characteristics That Separate Pathogenic *Actinomadura* Species

Species	*A. latina*	*A. madurae*	*A. pelletieri*
Enzyme tests			
Cleavage of 7-amino-4-methylcoumarin (7AMC-) substrates			
Glutaryl-L-phenylalanine-7AMC	−	+	+
L-pyroglutamide-7AMC	+	+	−
Cleavage of 4-methylumberlliferone (4MU-) substrates			
4MU-*N*-acetyl-D-galactosoamide	−	d	+
4MU-α-L-arabinopyranoside	+	+	−
4MU-β-D-cellopyranoside	+	+	−
4MU-β–D-galactoside	−	+	−
4MU-β-D-glucoside	+	+	−
4MU-β-D-glucuronide	+	+	−
4MU-β-D-lactoside	−	+	−
4MU-sulfate	−	+	−
4MU-xyloside	+	+	−
Growth on sole carbon sources (1%, w/v)			
Adonitol	−	+	−
Fructose	+	+	−
Galactose	+	+	−
Glycerol	+	+	−
Mannitol	+	+	−
Melezitose	+	−	−
Degradation tests			
Arbutin	+	+	−
Esculin	+	+	−
Starch	+	−	+

Source: Based on data from Trujillo, M.E., and Goodfellow, M., *Antonie van Leeuwenhoek*, 84, 39, 2003 and Trujillo, M.E., and Goodfellow, M., *Zbl. Bakt.*, 285, 212, 1997.

Epidemiology. It is not possible to state with any certainty the prevalence of *Actinomadura* infections (mainly actinomycetoma) in any part of the world, including the endemic areas, since there have not been any international surveillance efforts to ascertain the incidence of the disease. At most, information is limited to a few studies that have been made by individual researchers.[21,40,41,47]

Actinomycetomas have a worldwide distribution but occur mainly in the tropical and subtropical regions in the area between the latitudes of 15°S and 30°N. This area includes countries such India, Mali, Mexico, Senegal, Somalia, Sudan, and much of Central and South America.[5] Sudan appears to have the highest number of actinomycetomas in the world followed by Mexico with an average of 70 cases per year.[40] In these countries, the actinomycetoma infections are a major health problem, commonly affecting farmers and herdsmen. In general, the disease is four times more frequent in males than females and mainly affects adults between the ages of 16 and 40 years. Since these are the most active members of the society, the disease has a socioeconomic effect on dependent family members in underdeveloped countries.[48] Nevertheless, cases involving children and elderly men do occur.[49–51] Actinomycetoma infections have also been reported in the United States and Europe although in most cases, the patient contracted the disease in a different country.[38,52]

It is interesting to note that actinomycetoma becomes more active and aggressive in pregnant women. Hormonal changes and decreased immune response during pregnancy may be the explanation for this observation.[41]

There is evidence that climatic conditions may play a major role in the distribution of actinomycetomas. In the case of Mexico and Sudan, these countries have a rainy season from June to October; a dry, cool season from October to March; and hot and dry weather without rainfall from March to June. In Venezuela, most cases of actinomycetoma have been reported from semiarid areas colonized by trees, bushes, and cacti with thorns. The microorganisms are usually present in the soil. Traumatic inoculation of the subcutaneous tissues caused by sharp objects such as thorns or splinters is thought to be the route of entry. However, this theory has been recently disputed, as many patients have no history of trauma at the infection site.[41]

2.1.5 Diagnosis

The clinical diagnosis of actinomycetoma becomes more apparent in advanced disease with the development of characteristic sinuses and discharging grains; thus, the disease is often at an advanced stage when diagnosed.

2.1.5.1 Conventional Techniques

Microscopic Observation and Isolation. The diagnosis of the actinomycetoma is currently based on the isolation and identification of the causal microorganisms. In practice, the isolation of the causal agent is not always possible or can take a long time.

Direct microscopic examination of the pus from the lesions in the actinomycetomas with 10% KOH or saline reveals the presence of granules. The size, form, and color, together with the presence or absence of clubs or pseudoclubs, gives a clue to the identity of the etiologic agent. In the case of *A. madurae*, the granules can be seen without the aid of a microscope; in other species, the granules are smaller.[42]

Isolation of the microorganisms can be achieved by culture of the pus, granules, or tissue samples using various culture media such as Sabouraud, mycobiotic, or blood agar media.[33,42] The culture technique is often cumbersome and time consuming, and possible sample contamination may give a false positive result. This technique also requires experience to identify the causative microorganisms.

Fine Needle Aspiration Cytology. Actinomycetoma can be accurately diagnosed by fine needle aspiration cytology.[53–55] Mycetoma lesions have a distinct appearance in a cytology smear and are characterized by the presence of polymorphous inflammatory cells such as neutrophils, histiocytes, lymphocytes, plasma cells, macrophages, and foreign body giant cells. This allows differentiation from artifacts and inflammatory lesions caused by other bacteria and fungi. Diagnosis of actinomycetoma using fine needle aspiration cytology has proved to be as accurate as histopathological observations when the grains are present.[54]

This technique is fast, inexpensive, and easy to apply, and it is well tolerated by patients. It can be used in routine diagnosis and as an effective means of collection of material for culture and immunological studies. Due to the simplicity of the technique, it can be used in epidemiological surveys of mycetoma and for detection of early cases when radiological and serological techniques may not be useful.[41]

Radiology. The radiological features of mycetoma are complex. As the disease spreads to the bone, the earliest changes include periosteal erosion and adjacent sclerosis. There may be a soft tissue mass with obliteration of fascial planes. Bone cortex may be compressed from outside by the mass, producing scalloping, which is followed by variable amount of periosteal reaction. Sunray appearance and Codman's triangle may be present, producing a picture similar to that of osteogenic sarcoma. There may be multiple cavities within normal-density bone; these cavities are small and numerous and their edges not well defined. The cavity size is directly related to the size of the grains. Osteosporosis is common in advanced mycetoma due to compression of the bone and insufficient blood irrigation caused by pressure on the blood vessels.

Radiographs are helpful to determine the extent of the infection, although they are not diagnostic. Computed tomography or magnetic resonance imaging can also be helpful to determine the full extent of bone involvement and to delineate soft tissue involvement.[39]

Immunodiagnosis. Enzyme-linked immunosorbent assays have been used to screen for antibodies against *S. somaliensis* and *A. madurae* in two regions of Sudan.[56] These results indicated that a relatively large proportion of clinically asymptomatic individuals in one region were infected with *S. somaliensis* and, to a lesser degree, *A. madurae*. To date, however, there are no specific and reliable serologic or immunologic tests useful in the diagnosis of actinomycetoma. This is in part because of the lack of specific antigens that do not cross-react with antibodies of infections caused by other related microorganisms.[42] Attempts have been made to develop a delayed-type hypersensitivity test in mycetoma using microbial agents, but in most cases the assays have not been sensitive enough, or represented, as in the case of aerobic actinomycetic infection, cross-reactions with tuberculosis, and leprosy.[42]

2.1.5.2 Molecular Techniques

PCR-RFLP Analysis. Use of PCR coupled with restriction endonuclease and/or probe hybridization analyses of PCR products has been the focus of recent interest for the separation of mycobacteria from nocardiae as well as for the recognition of species within the genera *Mycobacterium* and *Nocardia*.[57] The method has proven to be sensitive, less time consuming, and less labor intensive than traditional biochemical methods for the identification of the clinically significant species.

This technique can be directly applied to clinical isolates provided that sufficient biomass is available for genomic DNA extraction. A 439-bp fragment corresponding to the 65-kDa heat shock protein gene is then amplified by PCR using primers TB11 and TB12,[57] and the PCR product is then digested using a combination of five commercially available restriction endonucleases (*Bst*EII, *Hae*III, *Msp*I, *Hinf*I, and *Bsa*HI). After digestion, the restriction fragments are electrophoresed on 3% Metaphor agarose containing ethidium bromide at 95 V for 1.5–2 h to obtain the RFLP band patterns. This PCR-RFLP methodology distinguished clinical isolates of aerobic actinomycetes, including *A. madurae*, with 96.8% accuracy. Thus, identification of clinical isolates can be accomplished within 24–48 h of receipt of pure cultures and therefore this system can be readily and economically implemented for routine clinical use.[57]

16S rRNA Gene Sequencing. Identification of *Actinomadura* actinomycetes from nonmycetomic clinical samples has also been carried out successfully using 16S rRNA gene sequencing. In this case, Hafany et al.[21] isolated *Actinomadura* strains from the sputa or the bronchoalveolar lavage fluid of patients with pulmonary infections.

For DNA sequencing, bacterial strains can be cultured in brain heart infusion (BHI) broth for 4 days at 32°C. Preparation of genomic DNA samples for PCR amplification can be achieved by applying almost any protocol described in the literature for bacterial DNA isolation. In some cases, crude cell extracts are sufficient to obtain DNA for PCR amplification. Thus, a cell suspension can be heated in boiling water for 10 min and a sample of the supernatant is used directly for PCR.[58]

16S rRNA genes can be amplified by PCR and directly sequenced using universal primers widely available in the literature.[14,58] The obtained sequences can then be compared to those deposited in public databases (e.g., EMBL, NCBI, Genbank, etc.) for identification at the genus level of the microorganism.

A useful tool is the EzTaxon.org server,[59] which contains a manually curated 16S rRNA database of type strains of prokaryotes, providing a workframe where nontype strains and environmental clones, which in many cases contain erroneous and mislabeled sequences, have been excluded. Therefore, sequence comparison and identification will only be performed with validated sequences.

2.1.6 Treatment of *Actinomadura* Infections

Medical Treatment. The most beneficial treatment begins at an early stage and consists of surgical debridement in conjuction with prolonged antibiotic medications. Combination drug therapy appears to achieve the best results because of a synergistic effect and reduces the likelihood of resistance.[39] Treatment periods are characteristically long, especially with bone and visceral involvement, and depend on clinical response. Cure may be defined by a lack of clinical activity response.

Several drugs have been used in the treatment of actinomycetoma caused by *Actinomadura* species, including streptomycin, dapsone, amikacin, and trimethoprim-sulfamelthoxazole.[60] Boiron et al.[61] used the disk diffusion method to assess the susceptibility of the pathogen species *A. madurae* and *A. pelletieri* against 29 different antibiotics. Favorable in vitro susceptibilities were obtained particularly with amikacin, imipenem, aminocycline, tobramycin, and vancomycin. The addition of aminoglycosides to cotrimoxazole gives better efficacy, and is associated with shorter treatment duration.

An extensive study on antibiotic susceptibility of the genus *Actinomadura* was carried out by Trujillo and Goodfellow.[6] Many strains were resistant to ampicillin and rifampicin. On the other hand, with the exception of *A. oligospora* and *A. yumaensis*, most strains studied were susceptible to tetracycline.

Vera-Cabrera et al.[62] examined 24 *A. madurae* strains isolated from patients with actinomycetoma for their sensitivity to new quinolones, including garenoxacin, gatifloxacin, moxifloxacin, and two oxazolidinones: linezolid and the compound DA-7867, and found that all but one of the strains showed high susceptibility to all of these agents; they were most active to DA-7867, showing a minimum inhibitory concentration of 0.06 μg/mL. However, further work needs to be done to determine whether these results can be extrapolated to antibiotic activity in subcutaneous abscesses and granulomas that are formed in cases of mycetoma.

Surgical Treatment. Surgical excision of early and well-defined lesions of mycetomas carries the best prognosis. Drug treatment is given before surgery to decrease lesion size, and continued afterwards to ensure complete clearance of infection. Nonetheless, relapse rates of 20%–90% have been reported. Where there is bone involvement, amputation of the affected limb often becomes unavoidable but must be an absolutely last resort. Unfortunately, in resource-poor regions where there is no available effective drug treatment, and where bone involvement may be complicated by secondary bacterial infection, it is sometimes the only option available to both doctors and patients.[41,42]

2.2 METHODS

2.2.1 Sample Preparation and Observation

Noncontaminated clinical samples including bone tissue, bronchial washings, cerebrospinal fluid, sinus discharge, urine, biopsy, and autopsy material can be collected for examination of the presence and isolation of *Actinomadura*. Clinical material, especially of sputa and sinus discharge, should be examined as soon as possible to prevent overgrowth by contaminants. If storage is unavoidable, samples should be kept at 4°C without the addition of preservative.

Fluid material (e.g., cerebrospinal, urine, blood, etc.) can be examined in wet mounts under the microscope without staining. Samples should be centrifuged (× 4000 rpm) and examined between slide and coverslip for the presence of branching filaments using greatly reduced light or phase contrast. Material from actinomycetomas usually contains granules that facilitate an immediate presumptive identification of the causative agent and provide a good source for isolation of the microorganism.

2.2.1.1 Microscopic Examination

1. Prepare a Gram stain smear from fresh material, from sediment after centrifugation, or from granules crushed between two slides. Gram-positive branched filaments (ca. 1 μm in diameter) are seen at high magnification (×1000).
2. Stain a second smear and examine for acid fastness. Actinomadurae are nonacid-fast.

2.2.1.2 Isolation and Culture

Clinical samples should be inoculated in media such as brain heart infusion (Oxoid CM261), Sabouraud dextrose,[31] and yeast extract agars[24] (see also section on isolation). Media should be transparent to facilitate direct microscopic examination of colonies. The grains should be cleaned several times in sterile saline and crushed with a sterile glass rod to obtain material for inoculation. Several plates should be streaked and incubated at 30°C–37°C. Colonies may grow after 7–10 days, but longer incubation periods are often necessary. Actinomadurae can be recognized by their filamentous appearance, leathery colonies, and by the production of red prodigiosin pigments. The colonies are usually folded, with an irregular surface and present a hard texture. Colony color may vary depending on the species, *A. madurae* and *A. latina* colonies are usually yellow-white or pink, while *A. pelletieri* colonies are red or purplish.

2.2.1.3 Isolation of DNA

There are many DNA extraction commercial kits available to isolate DNA for PCR amplification, however, DNA crude extract preparations also work well. We present two DNA extraction protocols that have been successfully applied to obtain DNA templates from *Actinomadura* strains.

Method 1. Strains of *Actinomadura* can be grown on Mueller-Hinton agar or in Mueller-Hinton broth supplemented with 0.4% Tween 80% and 10% oleic acid albumin-dextrose solution. Alternatively, the strains can be grown on any of the media mentioned in Section 2.1.2. The cells are harvested and suspended in 10 mM Tris-1 mM EDTA (pH 8.0) to a density equivalent to that of a MacFarland no. 4 standard, inactivated by submersion in a water bath at 80°C for 30 min, and disrupted for 5 min with zirconium beads (0.1–0.15 mm in diameter) in a Mini-BeadBeater. The resulting ground cell supernatants are subsequently clarified by centrifugation (7500 × *g* for 5 min). The supernatant is transferred to a clean tube and stored at 4°C.[63]

Method 2. Actinomycete strains are grown in 10 mL of tryptone yeast extract broth (ISP 1)[22] with agitation at 120 rpm for 24–48 h. Five mL of cultured cells are harvested by centrifugation (7500 × *g* for 5 min), washed once with 500 μL of TE buffer (10 mM Tris-HCl 1 mM EDTA pH 7.5), and resuspended in 500 μL TE buffer. The samples are heated in boiling water for 10 min, allowed to cool for 5 min and centrifuged (7500 × *g* for 5 min). The supernatant (300 μL) is transferred to a clean microtube and store at 4°C. If melanin or pigments are produced in ISP 1 broth, the microorganisms can be grown in Middlebrook 7H9 broth, as pigments may interfere with the PCR reaction.

2.2.2 Detection Procedures

(i) PCR Detection The application of PCR-based protocols has become and indispensable tool to aid in the identification of bacteria. Specifically, 16S rRNA gene sequencing provides and accurate and rapid way to identify clinically significant aerobic actinomycetes including *Actinomadura*.

Procedure A 50 μL reaction mixture can be prepared as follows: 5 μL PCR reaction buffer (10×, PerkinElmer); 3 μL $MgCl_2$ (25 mM); 2 μL bovine seroalbumin (0.1% p/v, Sigma); 0.5 μL primer SF1 (20 μM) (5′-AGAGTTTGATCMTG GCTCAG-3′); 0.5 μL primer SR5 (20 μM) (5′-AAGGAGGT GWTCCARCC-3′); 1.25 μL dNTP mix (10 mM each); 0.8 μL Gold *Taq* polymerase (5 U/μL, PerkinElmer); DNA template (4 μL), sterile MilliQ water (33 μL). The PCR amplification is performed with an initial denaturation step at 95°C for 9 min, followed by 35 cycles with denaturation at 94°C for 1 min, annealing at 55°C for 1 min, and extension at 72°C for 2 min, with a final extension of 72°C for 7 min.

PCR products are then electrophoresed on 1% agarose gels containing ethidium bromide (10 mg/mL) to ensure that the correct fragment has been amplified (~1500 bp). PCR fragments are subsequently excised from the agarose gel and purified using QIAquick PCR purification kit (Qiagen).

Sequence reactions are performed on an ABI377 sequencer (Applied Biosystems) using a BigDye terminator v3.0 cycle sequencing kit as supplied by the manufacturer. The following primers are used to obtain the complete 16S rRNA gene sequence (corresponding positions in the *E. coli* 16S rRNA gene sequence are given in parenthesis): (342); SR2 5′-GWATTACCGCGGCKGCTG-3′ (519); SR3, 5′-CCGTCAATTCMTTTRAGTTT-3′ (907); SR4, 5′-GGGTTGCGCTCGTTG-3′ (1100) and SR5, 5′-AAGGAGGTGWTCCARCC-3′ (1525). The assembled sequence is introduced in the public databases (e.g., GenBank/EMBL) for identification. Given the fact that GenBank/EMBL databases also contain a huge amount of environmental and uncultured bacterial sequences, it is sometimes difficult to obtain a clear identification of the microorganism if the user is not very experienced using these databases. To overcome this difficulty, we highly recommend to use the Eztaxon server (http://www.eztaxon.org), which offers a fully updated database of type strains with validly published prokaryotic names.[59]

Note 1. In many cases, identification at the genus level can be obtained by sequencing a hypervariable region within domain 1 of the gene using primer SR2. The ~450 bp sequenced DNA fragment is in most cases sufficient for correct identification of strains at the genus level.

Note 2. Couble et al.[64] have developed a 16S PCR-based assay for the rapid detection of *Nocardia* spp. directly from human clinical specimens such as skin biopsy samples, pus from abscesses, sputa, or bronchalveolar liquid. Although this method has not been used for the detection of *Actinomadura*, a similar approach could be applied if genus specific primers for *Actinomadura* are designed. Below we describe the published method for *Nocardia*.

Sample preparation for direct detection is prepared as follows: Resuspend 100 μL pus samples or 100 mg tissue biopsy specimens in 250 μL of sterile water (molecular grade). This mixture is incubated for 3 h at 55°C with proteinase K at 20 mg/mL (Sigma Aldrich) and inactivated for 15 min at 95°C. Nocardial DNA is then extracted with an MTB respiratory specimen preparation kit (Roche), according to the manufacturer's instructions. PCR amplification of a 16S rRNA gene fragment using *Actinomadura* specific primers could then be performed. PCR conditions given above can be used to amplify these fragments taking into consideration the *Tm* of the primers designed to optimize the annealing temperature.

(ii) Genotyping **Protocol 1**. PCR-restriction fragment length polymorphism (PCR-RFLP) of the 65-kDa heat shock protein gene.

PCR Amplification. Method 1 is recommended for DNA extraction. A 439-bp segment of the 65-kDa heat shock protein (HSP) is amplified from ground cell supernatants by PCR with 1.0 U of *Taq* DNA polymerase (Boehringer) in optimized buffer E (1.5 mM $MgCl_2$ [pH 9.0]; Invitrogen) containing 42 μM (each) dNTP mixture, 9% dimethyl sulfoxide, and 1 μM (each) primers TB11 (5′-ACCAACGATGGTGTGTCCAT-3′) and TB12 (5′-CTTGTCGAACCGCATACCCT-3′). PCR conditions are 45 cycles at 94°C, 55°C, and 72°C for 1 min each and then for a 10 min final extension at 72°C. Positive and negative controls should be included for comparison.

RFLP Analysis. Enzymatic digestion using commercially available restriction endonucleases *MspI* and *HinfI* is performed at incubation temperatures and using the buffers at the 10× concentration as recommended by the manufacturers. Restriction fragments are subsequently electrophoresed on 3% MetaPhor agarose containing ethidium bromide (0.625 μg/mL) in a Mini-Sub Cell electrophoresis system at 95V for 1.5–2.0 h. The PCR product digested with *MspI* DNA yields a two-band profile (265/180 bp). The PCR product digested with HinfI does not present any recognition sites. This combination should be sufficient to identify Actinomadura strains. However, it is very important to compare with appropriate controls.

Note. A timed protocol of PCR-RFLP has been presented by Wilson et al.[65] where by a clinical isolate can be identified within 2–3 working days.

Protocol 2. Amplified rRNA restriction analysis (ARDRA)

PCR Amplification. Method 2 is recommended for DNA extraction. PCR is carried out in 50 μL volumes containing 2 mM $MgCl_2$, 2 U *Taq* polymerase (Roche), 150 μM (each) dNTPs, 0.5 μM of each primer, and 2 μL template DNA. Primer F1 is 5′-AGAGTTTGATCITGGCTCAG-3′, I = inosine and primer R5 is 5′-ACGGITACCTTGTTACGACTT-3′. The primers amplify the nearly full-length 16S rRNA gene. PCR conditions: initial 96°C for 2 min; 30 cycles of 96°C for 45 s, 56°C for 30 s and 72°C for 2 min; and a final 72°C for 5 min. The PCR products are then electrophoresed on 1% agarose gels, containing ethidium bromide (10 g/mL), to ensure that a fragment of the correct size is amplified. PCR products are purified using the QIAquick PCR purification kit (Qiagen).

Resctriction Endonuclease Digestions and Analysis. The purified 16S rRNA gene should be digested using three different endonucleases, *Sau*3AI, *Asn*I, and *Pst*I. Digestion reactions are incubated at 37°C for 3–4 h. Samples are then electrophoresed on 1.5% agarose gels containing ethidium bromide. DNA digested with *Sau3*AI yields a large fragment of DNA between 980 and 1300 bp. The 16S rRNA gene should not be cut by *Asn*I. Digestion of the 16S rRNA gene with *Pst*I yields a band of 400–465 bp. The combination of the three results should identify the strain to the genus *Actinomadura*.[58]

2.3 CONCLUSIONS AND FUTURE PERSPECTIVES

Considerable challenges remain in the management of actinomycete infections, especially actinomycetomas. Unlike well-known tropical diseases such as leishmaniasis, malaria, and leprosy, actinomycetomas have received comparatively little international recognition. As a result, there has been little impetus for a new drug development, although they represent a group of tropical infections for which existing treatment options are characterized by high levels of inefficacy.

Correct species identification is important for initial diagnosis, treatment evaluation, and epidemiological studies. Modern molecular techniques such as PCR technology are valuable tools for identifying causative species where culture methods have failed. The availability of good drug-susceptibility testing is important given the high rates of resistance to commonly used first-line agents. Actinomycetomas are managed by a combination of antibiotic treatment, mainly sulphonamides, aminoglycosides, or tetracyclines. Carbapenems and oxazolidinones are promising treatment options. Nonetheless, in countries with significant disease burden, such as Mexico, Sudan, and India, physicians have used novel drug combinations in efforts to overcome drug resistance and improve efficacy. However, studies have not been randomized, species identification has not always been possible, clinical endpoints are not accurately stated, and follow-up periods are not defined.[43] Furthermore, the vast majority of studies have small patient numbers. Larger, multicenter trials are needed in order to better evaluate drug efficacy and safety. There are also relatively few published studies of current treatment regimes, and therefore they too need to be optimized by randomized blinded trials. New drugs will need to be made affordable to local populations, as this is a disease that primarily afflicts those of low socioeconomic status. Efforts continue to improve molecular diagnostic techniques that are not only sensitive and specific but also might readily be deployed in resource-poor settings. The availability of good drug-susceptibility testing and development of prognostic markers will also improve disease management.

Insufficient information is available on the pathogenicity mechanisms of *Actinomadura*, mainly due to lack of studies. Much work is also needed in this area. Finally, *Actinomadura* infections, especially actinomycetoma, can be treated given the presence of properly trained physicians, medical students, nurses, and paramedical staff in countries where these infections are endemic. It is also essential to provide health education to farm workers so that they can be protected. It is, however, important that rapid and efficient techniques are developed to identify the causal agents and to provide correct therapeutic treatment.

REFERENCES

1. Stackebrandt, E., Rainey, F.A., and Ward-Rainey, N.L., Proposal for a new hierarchic classification system, *Actinobacteria* classis nov. *Int. J. Syst. Bacteriol.*, 47, 379, 1997.
2. Zhi, X.-Y., Li, W.-J., and Stackebrandt, E., An update of the structure and 16S rRNA gene sequence-based definition of higher ranks of the class *Actinobacteria*, with the proposal of two new suborders and four new families and emended descriptions of the existing higher taxa, *Int. J. Syst. Evol. Microbiol.*, 59, 589, 2009.
3. McNeil, M.M., and Brown, J.M., The medically important aerobic actinomycetes: Epidemiology and microbiology, *Clin. Microbiol. Rev.*, 7, 357, 1994.
4. Develoux, M., Dieng, M.T., and Ndiaye, B., Mycetoma of the neck and scalp in Dakar, *J. Mycol. Med.*, 9, 179, 1999.
5. Fahal, A.H., and Hasan, M.A., Mycetoma, *Br. J. Surg.*, 79, 1138, 1992.
6. Trujillo, M.E., and Goodfellow, M., Numerical phonetic classification of clinically significant aerobic sporoactinomycetes and related organisms, *Antonie van Leeuwenhoek*, 84, 39, 2003.
7. Quintana, E.T. et al., *Streptomyces sudanensis* sp. nov., a new pathogen isolated from patients with actinomycetoma, *Antonie van Leeuwenhoek*, 93, 305, 2008.
8. Kroppenstedt, R.M., Stackebrandt, E., and Goodfellow, M., Taxonomic revision of the actinomycete genera *Actinomadura* and *Microtetraspora*, *Syst. Appl. Microbiol.*, 13, 148, 1990.
9. Athalye, M. et al., Numerical classification of *Actinomadura* and *Nocardiopsis*, *Int. J. Syst. Bacteriol.*, 35, 86, 1985.
10. Fischer, A., Kroppenstedt, R.M., and Stackebrandt, E., Molecular-genetic and chemotaxonomic studies on *Actinomadura* and *Nocardiopsis*, *J. Gen. Microbiol.*, 129, 3433, 1983.
11. Poschner, J. et al., DNA–DNA reassociation and chemotaxonomic studies on *Actinomadura*, *Microbispora*, *Microtetraspora*, *Micropolyspora* and *Nocardiopsis*, *Syst. Appl. Microbiol.*, 6, 264, 1985.
12. Trujillo, M.E., and Goodfellow, M., Polyphasic taxonomic study of clinically significant actinomadurae including the description of *Actinomadura latina* sp. nov, *Zbl. Bakt.*, 285, 212, 1997.
13. Zhang, Z., Wang, Y., and Ruan, J., Reclassification of *Thermomonospora* and *Microtetraspora*. *Int. J. Syst. Bacteriol.*, 48, 411, 1998.
14. Zhang, Z.S. et al., Clarification of the relationship between the members of the family *Thermomonosporaceae* on the basis of 16S rDNA, 16S–23S rRNA internal transcribed spacer and 23S rDNA sequences and chemotaxonomic analyses, *Int. J. Syst. Evol. Microbiol.*, 51, 373, 2001.
15. Vincent, H., Étude sur le parasite du pied le madura, *Ann. Inst. Pasteur,* 8, 129, 1894.
16. Pinoy, E., Isolement et culture d'une nouvelle oospora pathogene, in *Mycetome a grains rouges de la paroi thoracique*, Thiroux and Pelletier (Editors), *Bull. Soc. Path. Exot.*, 5, 585, 1912.
17. Waksman, S.A., and Henrici, A.T., Family III. Streptomycetaceae Waksman and Henrici. In Breed, Murray and Hitchens (Editors), *Bergey's Manual of Determinative Bacteriology*, 6th Ed., pp. 929–980, The Williams and Wilkins Co., Baltimore, 1948.
18. Becker, B., Lechevalier, M.P., and Lechevalier, H.A., Chemical composition of cell wall preparations from strains of various form genera of aerobic actinomycetes, *Appl. Microbiol.*, 13, 236, 1965.
19. Lechevalier, M.P., and Gerber, N.N., The identity of madurose with 3-O-methyl-galactose, *Carbohydr. Res.*, 13, 451, 1970.
20. Euzéby, J.P., List of bacterial names with standing in nomenclature: A folder available on the Internet, *Int. J. Syst. Bacteriol.*, 47, 590, 1997, (List of Prokaryotic names with standing in nomenclature. http://www.bacterio.net).
21. Hafany, A. et al., Majority of *Actinomadura* clinical isolates from sputa or bronchoalveolar lavage fluid in Japan belongs to the cluster of *Actinomadura cremea* and *Actinomadura nitritigenes*, and the description of *Actinomadura chibensis* sp. nov, *Mycopathologia*, 162, 281, 2006.
22. Shirling, E.B., and Gottlieb, D., Methods for characterization of *Streptomyces* species, *Int. J. Syst. Bacteriol.*, 16, 313, 1966.
23. Jones, K.L., Fresh isolates of actinomycetes in which the presence of sporogenous aerial mycelia is a fluctuating characteristic, *J. Bacteriol.*, 57, 141, 1949.
24. Athalye, M., Lacey, J., and Goodfellow, M., Selective isolation and enumeration of actinomycetes using rifampicin, *J. Appl. Bacteriol.*, 51, 289, 1981.
25. Cross, T., and Attwell, R.W., Recovery of viable thermoactinomycete endospores from deep mud cores. In Barker, Gould and Wolf (Editors), *Spore Research 1973*, p. 11, Academic Press, London, 1974.
26. Trujillo, M.E., Taxonomic revision of the genus *Actinomadura* and related taxa using rapid methods, Ph.D. thesis, University of Newcastle, Newcastle upon Tyne, England, UK, 1994.
27. Kroppenstedt, R.M., and Goodfellow, M., The family *Thermomonosporaceae*. In Starr, Stolp, Trüper, Balows, and Schlegel (Editors), *The Prokaryotes: A Handbook of Habitats, Isolation and Identification of Bacteria*, p. 1085, Springer-Verlag, Berlin, 1992.
28. Cook, A.E., le Roes, M., and Meyers, P.R., *Actinomadura napierensis* sp. nov., isolated from soil in South Africa, *Int. J. Syst. Evol. Microbiol.*, 55, 703, 2005.
29. Rippon, J.W., Extracellular collagenase produced by *Streptomyces madurae*, *Biochim. Biophys. Acta*, 159, 147, 1968.
30. Schaal, K.P., Zur mikrobiologisher Diagnostik der Nocardiose, *Zentralbl. Bakteriol. Parasitenkd. Infektionskr. Hyg. I. Abt. Orig.*, 220, 242, 1972.
31. Gordon, M.A., Aerobic pathogenic *Actinomycetaceae*. In Lennette, Spauling, and Truant (Editors), *Manual of Clinical Microbiology*, 4th Ed., p. 175, American Society for Microbiology, Washington D.C., 1974.
32. Maiti, P.K. et al., Mycetoma caused by a new red grain mycetoma agent in two members of a family, *J. Postgrad. Med.*, 49, 322, 2003.
33. De Palma, L. et al., A rare european case of madura foot due to actinomycetes, *Joint Bone Spine,* 73, 321, 2006.
34. Gerber, N.N., Prodigiosin-like pigments from *Actinomadura (Nocardia) pelletieri*, *J. Antibiot.*, 24, 636, 1971.
35. Gerber, N.N., Minor prodiginine pigments from *Actinomadura madurae* and *Actinomadura pelletieri*, *J. Heterocycl. Chem.*, 10, 925, 1973.
36. Goodfellow, M., Trujillo, M.E., and Alderson, G., Approaches towards the identification of sporoactinomycetes that cause mycetoma, *Biotechnologia*, 7–8, 271, 1995.
37. Wink, J. et al., *Actinomadura namibiensis* sp. nov, *Int. J. Syst. Bacteriol.*, 53, 721, 2003.

38. McIroy, J.A., Prestes, C.D., and Su, W.P., Mycetoma: Infection with tumefaction, draining sinuses, and "grains," *Cutis*, 49, 107, 1992.
39. Foltz, K.D., and Fallat, L.M., Madura foot: Atypical finding and case presentation, *J. Foot Ankle Surg*., 43, 327, 2004.
40. Mahmoudabadi, A.Z., and Zarrin, M., Mycetomas in Iran: A review article, *Mycopathologia*, 165, 135, 2008.
41. Fahal, A.H., Mycetoma: A thorn in the flesh, *Trans. R. Soc. Trop. Med. Hyg*., 98, 3, 2004.
42. Welsh, O., Vera-Cabrera, L., and Salinas-Carmona, M.C., Mycetoma, *Clin. Dermatol*., 25, 195, 2007.
43. Ameen, M., and Arenas, R., Development in the management of mycetomas, *Clin. Exp. Dermatol*., 34, 1, 2008.
44. Gumaa, S.A., Mahgoub, E.S., and El Sid, M.A., Mycetoma of the head and neck, *Am. J. Trop. Med. Hyg*., 35, 596, 1986.
45. McNeil, M.M. et al., Nonmycetomic *Actinomadura madurae* infection in a patient with AIDS, *J. Clin. Microbiol*., 30, 1008, 1992.
46. Bar, W. et al., Unusual gram positive rods, causing pneumonia, *Pneumologie*, 57, 259, 2003.
47. Serrano, J.A. et al., The actinomycetoma in Venezuela: A ten year study (1976–1986), *Rev. Inst. Med. Trop. Sao Paulo*, 30, 297, 1988.
48. Fahal, A.H., and Suliman, S.H., The clinical presentation of mycetoma, *Sudan Med. J*., 32, 46, 1994.
49. El Moghraby, I.M., Mycetoma in Gezira, *Sudan Med. J*., 9, 77, 1971.
50. Serrano, J.A. et al., Histological and ultrastructural studies on human actinomycetomas. In Szabó, Biró, and Goodfellow, M. (Editors), *Biological, Biochemical and Biomedical Aspects of Actinomycetes*, p. 647, Adakémiai Kiadó, Budapest, 1986.
51. Fahal, A.H., and Sabaa, A.H., Mycetoma in children in Sudan, *Trans. R. Soc. Trop. Med. Hyg*., 2009, Aug 27. (Epub ahead of print).
52. de Hoog, G.S. et al., Diagnostic problems with imported cases of mycetoma in The Netherlands, *Mycoses*, 36, 81, 1993.
53. Bapat, K.C., and Pandit, A.A., Actinomycotic mycetoma. Report of a case with diagnosis by fine needle aspiration, *Acta Cytol*., 35, 770, 1991.
54. El Hag, I.A., Fahal, A.H., and Khalil, E.A.G., Fine needle aspiration of cytology of mycetoma, *Acta Cytol*., 40, 461, 1996.
55. Fernandes, H. et al., Cytodiagnosis of actinomycetoma, *Diagn. Cytopathol*., 37, 506, 2009.
56. Taha, A., A serologic survey of antibiotics to *Streptomyces somaliensis* and *Actinomadura madurae* in the Sudan using enzyme linked immunosorbent assay (ELISA), *Trans. R. Soc. Trop. Med. Hyg*., 77, 49, 1983.
57. Steingrube, V.A. et al., Rapid identification of clinically significant species and taxa of aerobic actinomycetes, including *Actinomadura, Gordona, Nocardia, Rhodococcus, Streptomyces*, and *Tsukamurella* isolates, by DNA amplification and restriction endonuclease analysis, *J. Clin. Microbiol*., 35, 817, 1997.
58. Cook, A.E., and Meyers, P.R., Rapid identification of filamentous actinomycetes to the genus level using genus specific 16S rRNA gene restriction fragment patterns, *Int. J. Syst. Evol. Microbiol*., 53, 1907, 2003.
59. Chun, J. et al., EzTaxon: A web-based tool for the identification of prokaryotes based on 16S ribosomal RNA gene sequences, *Int. J. Syst. Evol. Microbiol*., 57, 2259, 2007.
60. Welsh, O. et al., Amikacin alone and in combination with trimethoprim-sulfamethoxazole in the treatment of actinomycetomic mycetoma, *J. Am. Acad. Dermatol*., 17, 443, 1987.
61. Boiron, P. et al., In vitro antibiotic susceptibility testing of agents of actinomycetoma. *Med. Microbiol. Lett*., 1, 38, 1992.
62. Vera-Cabrera, L. et al., Therapeutic effect of a novel oxazolidinone, DA-7867, in BALB/c mice infected with *Nocardia brasiliensis*, *PLoS Negl. Trop. Dis*., 2, 1, e289, 2004.
63. Steinbruge, V.A. et al., DNA amplification and restriction endonuclease análisis for differentiation of 12 Species and Taxa of *Nocardia*, including recognition of four new taxa within the *Nocardia asteroides* complex, *J. Clin. Microbiol*., 33, 3096, 1995.
64. Couble, A., et al., Direct detection of *Nocardia* spp. in clinical samples by a rapid molecular method, *J. Clin. Microbiol*., 43, 1921, 1998.
65. Wilson, R.W. et al., Clinical application of PCR-Restriction Enzyme Pattern Analysis for rapid identification of aerobic actinomycete isolates, *J. Clin. Microbiol*., 36, 148, 1998.

3 Actinomyces

Debarati Paul, D. Madhusudan Reddy, Dipalok Mukherjee, Biswajit Paul, and Debosmita Paul

CONTENTS

3.1 INTRODUCTION

3.1.1 Classification, Biology, and Epidemiology

The genus *Actinomyces* encompasses a large number of gram-positive, nonspore-forming, anaerobic, or facultatively anaerobic bacterial species belonging to the family *Actinomycetaceae*, order Actinomycetales, phylum Actinobacteria, domain Bacteria. While individual *Actinomyces* bacteria are rods of <1 μm in diameter, *Actinomyces* colonies on nutrient medium are irregular in form with radiant edges, consisting of fungus-like branched networks of hyphae.

Classification and speciation of *Actinomyces* are based on analysis of DNA and rRNA (e.g., DNA fingerprinting and ribotyping), cell wall structure, metabolic end products, biochemical reactions, and serology. Most species represent a distinct serologic group, often with subtypes. Slack et al. were the first to apply fluorescent antibody techniques to establish and revise four serological groups within *Actinomyces*.[1] By comparing long 16S rRNA sequences generated by reverse transcriptase, it was shown that all species formed a phylogenetically coherent cluster in which *Actinomyces bovis*, *Actinomyces viscosus*, *Actinomyces naeslundii*, *Actinomyces odontolyticus*, and *Actinomyces israelii* constituted genetically well-defined species.[2] To date, 36 or so *Actinomyces* species have been identified, of which 19 species are associated with humans and animals, with *A. israelii* being the most common cause of actinomycosis—a chronic suppurative and granulomatous disease of the cervico-facial, thoracic, or abdominal areas.

Actinomyces are often found in healthy humans in the oral cavity, lacunas of the tonsils, or on mucosa of the gastrointestinal tract. *Actinomyces*, in particular *A. israelii*, may cause infection when they are introduced into the soft tissues by trauma, surgery, or another infection. Once in the tissues, it may form an abscess that develops into a hard red to reddish purple lump. When the abscess breaks through the skin, it forms pus-discharging lesions. Some *Actinomyces* spp. are opportunist pathogens associated with other mixed anaerobic infections and with infections in severely compromised patients.[3]

The first published clinical description of the human form of the disease appeared in 1857.[4] The thoracic form was described 25 years later, but it was not until 1891 that *A. israelii*, the main species responsible for the human disease, was isolated.[5] Actinomycosis is endemic; sporadic cases involving humans and agricultural animals occur worldwide. According to previous surveys, the incidence of actinomycosis was about 1 per

300,000 persons in the Cleveland area in the 1970s, and about 1 per 100,000 persons in Germany and the Netherlands. A male-to-female infection ratio of 3:1 is encountered in most series. Unconfirmed explanations for this include poorer oral hygiene and augmented oral trauma in males. Cases of *Actinomyces* infestation in humans from sick humans or animals have not been described.

3.1.2 Clinical Features and Pathogenesis

Actinomycosis (also known as Rivalta disease, big jaw, clams, lumpy jaw, or wooden tongue) is an infection commonly of the face and neck that produces abscesses (collections of pus) and open-draining sinuses (tracts in the skin) and is chiefly caused by *Actinomyces israelii*. The word "*actinomycosis*" was derived from the Greek terms "*aktino,*" which refers to the radiating appearance of a sulphur granule, and "*mykos*," which labels the condition a mycotic disease.[5] Infection can develop in individuals of all ages; however, it is rare in the pediatric population and in patients older than 60 years. Most cases are encountered in individuals in the mid decades of life.

Actinomycosis is a chronic and progressive infection that does not show spontaneous remission.[6] Duration of an incubation interval is not known. It can fluctuate over a wide range and reach till several years (from time of a becoming infected development of demonstrative forms of an actinomycosis).

As *A. israelii* is a common colonizer of the vagina, colon, and mouth, its infection in host is facilitated by a breach of the mucosal barrier during various procedures (e.g., dental), aspiration, or pathologies such as diverticulitis. On a place of introduction of *Actinomyces* an infectious granuloma is formed that sprouts in surrounding tissues. In granulations there are abscesses that break forming fistulas. The skin lesion has secondary character. In the formation of pyeses, staphylococcal infection contamination mainly plays the secondary role. Being a chronic infectious disease, actinomycosis provokes a chronic productive and colliquative inflammatory reaction known as actinomycotic granuloma, with macroscopically visual yellowish granules produced by bacterial proliferation.

The basic clinical forms of actinomycosis include (i) actinomycosis of head, tongue, and neck; (ii) a thoracic/pulmonary actinomycosis; (iii) abdominal actinomycosis; (iv) actinomycosis of genitourinary organs; (v) skin actinomycosis; (vi) mycetoma; (vii) actinomycosis of the central excitatory system.[7–9] Classically, actinomycosis occurs in cervico-facial (55%), abdominopelvic (20%), thoracic (15%), and mixed organs (10%), including skin, brain, pericardium, and extremities.[10,11]

Cervico-facial actinomycosis (also known as lumpy jaw) is the most common and involves both the soft and hard tissue of the head and neck region. The various infection sites may be the scalp, the paranasal sinuses, the palate, the parotid gland, the temporal bone, the tear glands, the cheeks, the lower jaw, the tongue, the larynx, and the lower pharynx.[12,13]

Thoracic actinomycosis occurs in thoracic lumen, or lungs. Symptoms include chest pain, fever, and weight loss. Primary actinomycosis of the lung and chest wall has been less frequently reported.[11,14] Actinomyces is believed to invade the thoracic organ through the oropharynx, causing bronchopulmonary infection as well as pleural and chest wall involvement. The disease is usually proceeded by various forms of pulmonary infiltration or dense hilar lymphadenopathy.[4,15] Accurate diagnosis is seldom made at the time of hospital admission because the disease is rather uncommon and is often unsuspected. Pathologically, thoracic actinomycosis is a parenchymal infiltrate with involvement of the chest wall often with a rib eroded.[11] Due to lack of suspicion of the disease and unfamiliarity with the clinical manifestation, culture for the organisms is frequently not done at all or is incorrectly done without consideration for the microaerophilic needs of the bacterium.[11] Even among experienced physicians, sometimes despite pointers to the disease, delayed diagnosis, or misdiagnosis as tuberculosis, lung abscess, or lung cancer is common,[5] as it forms a mass that extends to the chest wall.

Abdominal actinomycosis leads to a number of disorders that are associated with lesions in the ileocecal region. Examples frequently encountered in developed countries include colon cancer, Crohn disease, and, less commonly, infection due to *Yersinia enterocolitica* and *Yersinia pseudotuberculosis*. Many other infectious, neoplastic, and drug-related causes of ileocecal lesions have been described. This condition is mostly preceded by surgery such as laparotomy for acute appendicitis, perforated ulcer, or gallbladder inflammation. Infection usually begins in the gastrointestinal tract and spreads to the abdominal wall. Spiking fever and chills, intestinal colic, vomiting, and weight loss, a palpable (can be felt) mass, and an external sinus are evident in this type of actinomycosis. This type of actinomycosis may be mistaken for Crohn disease, malignancy, tuberculosis, Amebiasis (an infection of the intestine or liver), or chronic appendicitis.

Genital actinomycosis and pelvic actinomycosis are rare. *Actinomyces* species exist in the normal flora of the gastrointestinal system and female genitourinary tracts. Abdominal actinomycosis accounts for 20% of all cases, and its diagnosis is difficult.[16] Pelvic actinomycosis affects the women's pelvic area and may cause lower abdominal pain, fever, and bleeding between menstrual periods. This form of the infection has been associated with the use of IUDs (intrauterine devices) that do not contain copper.

Actinomycosis of bones and joints seems rare. The clinical presentation of spinal actinomycosis may be nonspecific and back pain is the most consistent early symptom. A case of actinomycosis involving bones and joints has been recently documented.[17] The patient presented fever, pain in the neck and upper back, progressive weakness and numbness in all four limbs with difficulty ambulating, constipation, and uroschesis. Correct diagnosis is difficult because the clinical and radiological findings of actinomycosis closely resemble metastatic tumors and other infectious processes. Timely surgical debridement and decompression contributed to the prompt improvement of the patient's conditions, and histopathological demonstration of the inflammatory granulation tissue and gram-positive sulfur-containing filamentous bacteria led to the correct diagnosis of actinomycosis.[17]

Skin actinomycosis/primary cutaneous actinomycosis is very uncommon. Subcutaneous nodules usually occur on the exposed skin, extending slowly and breaking down to form sinuses. The infection elicits a granulomatous inflammatory response in the deep dermis and subcutaneous tissue as well the underlying bone, resulting in a clinical condition called mycetoma that is characterized by the formation of grains containing aggregates of the causative organisms that may be discharged onto the skin surface through multiple sinuses. Mycetoma was first described in the mid 1800s and initially named Madura foot, after the region of Madura in India where the disease was first identified. Mycetoma caused by microaerophilic actinomycetes is termed actinomycetoma, and mycetoma caused by true fungi is called eumycetoma.[18–20]

Actinomycosis treatment is long term, generally with up to 1 month of intravenous penicillin G followed by weeks to months of penicillin taken orally. In addition, surgical excision and drainage of abscesses may be necessary. Many patients are cured within <6 months of antibiotic therapy. If short-term antibiotic treatment is attempted, the clinical and radiological response should be closely monitored. Cervico-facial actinomycosis is especially responsive to brief courses of antibiotic treatment.[21] Table 3.1 shows some antibiotics that might be helpful in actinomycosis treatment. Conventional antibiotic therapy for actinomycosis consists of high-dose penicillin for a period of 6–12 months.[21,28,29] Good results with shorter courses of antibiotic treatment have been reported, but most were preceded by extensive surgery.[28,30]

3.1.3 Diagnosis

Actinomyces is usually part of the normal flora; its presence on sputum samples, bronchial washings, or cervicovaginal secretions may not be enough to make the diagnosis. *A. israelii* causes actinomycosis in which chronic granulomas become suppurative. Tissue pus contains sulfur granules, a tangled mass of branching bacteria, which provides a diagnostic feature for actinomycosis. Diagnosis of actinomycosis usually relies on the clinical picture and the presence of sulfur granules either observed macroscopically or microscopically. Since acitinomycosis presents varied manifestations that may complicate clinical diagnosis, it is helpful to apply laboratory means for accurate detection and identification of the organism. The conventional techniques for *Actinomyces* identification include Gram and other stains, microscopy, microbiological culture, and serology. In recent years, the application of molecular techniques has further improved the classification and identification of members of this genus.

3.1.3.1 Conventional Techniques

Actinomyces are identified as gram-positive rods that are non-acid fast in diphtheroidal arrangement. Hematoxylin-eosin staining of sulfur granules reveals basophilic masses with a radiating border of eosinophilic terminal clubs. Other useful stains used for detection are Grocott–Gomori methenamine-silver nitrate stain, periodic acid-Schiff, P-aminosalicylic acid, Goodpasture stain, and Brown–Brenn stain.[18–20]

For culture isolation, samples (e.g., tissue, pus, or sulfur granules) are transported on anaerobic transport media and inoculated on brain and heart infusion blood agar under anaerobic condition or enriched with carbon dioxide. Culture grows in 5–7 days and may take 2–4 weeks. *A. israelii* is a naerobic with clinical strains varying from obligate anaerobes to microaerophilic.

Identification of *Actinomyces* serologically was difficult before the introduction of fluorescent antibody (FA) and gel-diffusion procedures. These techniques have overcome the inherent difficulties in applying serological methods to filamentous organisms. The FA technique was first applied to *Actinomyces* by Slack and Moore, with the eventual establishment and revision of four serological groups by Slack and Gerencser.[1] *A. israeli* was designated as serological group D. The species specificity of the FA technique was confirmed by

TABLE 3.1
Commonly Used Antibiotics and Their Effectiveness in Treatment of Actinomycosis

Antibiotic	MIC (mg/l)		Effectiveness	Reference
	MIC_{50}	MIC_{90}		
Penicillin G	0.002	0.008	Efficacious	21, 22
Tetracycline (doxycycline)	0.047	0.25		21, 22
Erythromycin	0.016	0.016		21, 23
Clindamycin	0.032	0.064		21, 23
Imipenem	NR	NR	Probable efficacy	21, 24, 25
Ceftriaxone	0.008	0.094		21, 26, 27
Ciprofloxacin	>32	>32	Ineffective in vitro	21, 22
Metronidazole	NR	NR		21, 22
Aminoglycosides (amikacin)	NR	NR		21, 23

Lambert et al.,[31] who also reported two strains of *A. israeli* belonging to a second serotype. The gel-diffusion technique, first used in studying *Actinomyces* by King and Meyer,[32] has enabled identification of *A. israeli*[31] and elucidation of its generic relationships[33,34]; however, the technique is ordinarily less species specific than the FA test.[31] Landfried was able to place *Actinomyces* cultures into serological groups corresponding to those determined by FA, but observed more cross-reactions between groups with gel-diffusion than were found with the FA test. Pirtle et al.[35] reported isolation of a water-soluble, heat-stable antigen from culture supernatant fluid of *A. israelii*, which reacted specifically with *A. israeli* antiserum in gel-diffusion tests.

For diagnostic purpose, the intracutaneous test with actinolysathos has some value. However, it is necessary to accept only positive and sharply positive assays as weakly positive intracutaneous tests often lead to incorrect detection.

Other parameters for *Actinomyces* identification are based on resistance or sensitivity of growth in glycerol broth to lysozyme, urease activity, decomposition of the substrates casein, tyrosine, xanthine, and hypoxanthine. *A. israelii* may be identified definitively by detection with gas liquid chromatography (GLC) of acetic and lactic acid as end products of carbohydrate metabolism.

3.1.3.2 Molecular Techniques

Traditional ways of identification of anaerobic gram-positive nonsporulating bacilli by isolation of the organism and studying it phenotypically by elucidation of its morphologic and biochemical characteristics and metabolic end products are associated with a need for special equipment and expertise, as well as strains that are "unidentified" because of ambiguous biochemical profiles.[36] Identification by genotypic methods is particularly valuable for organisms such as *Actinomyces* spp., as they are poorly identified using conventional biochemical or serological methods. Some of the techniques popularly used are 16S ribosomal RNA gene sequencing, PCR–RFLP (polymerase chain reaction–restriction fragment length polymorphism)/ARDRA (amplified ribosomal DNA restriction analysis), DNA probes, and DNA–DNA hybridization.

Sequence information was used to develop an oligonucleotide probe for the *A. israelii* serotype 1 strains, which did not react with the serotype 2 strain or with rRNA from strains of eight *Actinomyces* species.[2] Species-specific oligonucleotide probes have been developed recently, however, due to high degree of intraspecific genetic diversity occurring within *Actinomyces* their utility is quite limited.[2,37] On the other hand PCR–RFLP/ARDRA has proved to be a highly discriminatory, practical, and cost-effective method for detection. Smaller case studies effectively used *Mnl*I,[38–40] whereas studies with more than 500 clinical isolates used separate digestions using *Hae*III and *Hpa*I.[39,41] These enzymes were not discriminatory when used alone, but when combined they were effective to differentiate all *Actinomyces*, *Arcanobacterium*, and *Actinobaculum* spp. to subspecies levels. These results corresponded to the conventional techniques as well. Only basic molecular expertise and lab facilities are required for conducting these tests.

3.2 METHODS

3.2.1 Sample Preparation

Frozen cultures can be thawed and suspended in distilled water to give a cell concentration of 1–3 mg of dry weight per mL. The cell suspensions should be diluted with distilled water, and 50 μL each of the serial dilutions can be subjected directly to PCR amplification. Alternatively, 40 μL of the suspensions can be mixed with 10 μL of a proteinase K solution (1 mg/mL) and 50 μL of 40 mM Tris buffer (pH 8.0) containing 1% Tween 20, 0.2% Nonidet P-40, and 0.2% mM EDTA. This buffer is a modification of PCR buffer used for sample preparation from blood cells and other fluids. The mixture should be incubated at 60°C for 20 min and then at 95°C for 10 min. Then, the protease treated sample may be centrifuged, and 5 μL of the resultant supernatant of it used for the PCR.

3.2.2 Detection Procedures

3.2.2.1 Single Round PCR

Cogulu et al.[42] utilized a pair of primers (Forward: 5′-AGA GTT TGA TCC TGG CTC AG-3′ and Reverse: 5′-CCA AAA CAC CAC AAA AGT GA-3′) in a single round PCR for amplification of an *A. israelii*-specific 230 bp product from endodontic samplings.

Procedure

1. Two sequential paper points are used to soak up the fluid in the canal and then transferred to sterile 2-mL Eppendorf tubes containing VMGA III transport medium. All samples are processed within 2 h. After thoroughly shaking the endodontic sample in a mixer for 60 s (Vortex, Scientific Industries Inc., Springfield, MA), all are frozen immediately at −20°C and stored until assayed by PCR.
2. Aliquots of each sample (1.0 mL) are centrifuged at 13,000 × *g* for 10 min. DNA is extracted with Genomic DNA Purification Kit (Fermentas). The DNA concentrations in clinical samples are determined by spectrophotometer measurement of the absorbance at 260 nm.
3. The PCR mixture (50 μL) is composed of 1 μM of each specific primer, 5 μL of 10× PCR buffer, 2 mM of $MgCl_2$, 1.25 U of DNA polymerase, and 0.2-mM concentrations of dNTP.
4. DNA amplification is performed in a thermal cycler GeneAmp PCR system (Applied Biosystems) with 30 cycles of 94°C for 1 min, 60°C for 1 min, and 72°C for 2.5 min.
5. PCR amplicons are analyzed by electrophoresis in a 1.5% agarose gel at 4 V/cm in Tris-borate EDTA buffer. The gels are stained for 15 min with 0.5 μg/mL

ethidium bromide and visualized under ultraviolet light. The identity of the *A. israelii*-specific 230 bp product is inferred by visual comparison with a molecular weight ladder (100 bp DNA ladder).

3.2.2.2 Nested PCR

Hansen et al.[43] described a seminested PCR targeting the 16S ribosomal RNA gene for specific detection of *A. israelii* in paraffin-embedded tissues. The first-round PCR utilizes oligonucleotide primers ACM-F-1 (5′-AAGTCGAACGGGTCTGCCTTG-3′, nt 5–25) and ACM-R-1 (5′-TCAAAGCCTTGGCAGGCCATC-3′, nt 221–241); the nested PCR uses primers ACM-F-2 (5′-TAACCTGCC CCTCACTTCTGG-3′, nt 72–92) and ACM-R-1.

First- and second-round PCR amplification is performed in the GeneAmp PCR System 9700 (Applied Biosystems) with an initial denaturation at 94°C for 5 min, followed by 30 cycles of 94°C for 30 s, 52°C for 30 s, and 72°C for 60 s. Two microliters of the first-round PCR product is used as template for the second-round PCR.

Because of a broad homology of bacterial ribosomal genes, the specificity of amplification products is confirmed by direct sequencing. The results suggest a nearly complete match to the 16S ribosomal RNA gene from *A. israelii* but no significant homology to other species.

Thus, this nested PCR protocol provides a highly sensitive method for amplification of *A. israelii* from nondecalcified, paraffin-embedded material. It is especially helpful in clinical cases with infected osteoradionecrosis (IORN), which do not reveal *Actinomyces* by means of light microscopy because the tissue specimens are small.

3.2.2.3 PCR–RFLP

The oligodeoxynucleotide primers are those designed to anneal to conserved regions of the eubacterial 16S rRNA[44]: forward, 5′-AGAGTTTGATCCTGGCTCAG-3′ (positions 8–27 of *E. coli* 16S rRNA); reverse, 5′-GGTTACCTTGTTACGACT-3′ (positions 1510–1492).

PCR is performed in a 100 μL volume by using a Perkin-Elmer Cetus thermocycler (GeneAmp PCR System 9600) and the GeneAmp PCR reagent kit with Taq polymerase (Perkin-Elmer Cetus Instruments) as specified by the manufacturer.

Cycling profiles consist of 30 cycles of 94°C for 1 min, 50°C for 1 min, and 72°C for 1.5 min, followed by a postrun of 72°C for 2 min.

The size of the products ~1600 bp is confirmed on 1.5% agarose gel using molecular size ladders (NEB).

A restriction enzyme cocktail is used to directly speciate organisms for RFLP. The restriction enzyme cocktail (5 U enzyme) is added to approximately 1 μg of DNA (10 μL PCR product) in PCR buffer that is adjusted to contain 100 mM NaCl, 1 mM dithiothreitol, and a final concentration of 7 mM $MgCl_2$. Restriction enzyme digests are performed at 37°C for 18 h.

The reaction is halted by freezing, and restriction fragments are analyzed on 10% TBE/polyacrylamide gels or 3.5% agarose gel in TAE buffer run at 5V/cm. A molecular weight ladder is included in each run. The gel is stained with ethidium bromide and visualized under UV illumination.

Dideoxy sequencing may be employed to further verify the identity of 16S rRNA PCR products. It is performed by using ABI kit with a cycle sequencing system as specified by the manufacturer. The modified cycle conditions are 25 cycles of 95°C for 30 s, 55°C for 52 s, and 72°C for 90 s. Automated electrophoresis and analysis of DNA sequence reactions with fluorescently labeled dideoxynucleotides are performed with a ABI Prism DNA sequencer.

3.2.2.4 Checkerboard DNA–DNA Hybridization

Checkerboard DNA–DNA hybridization as described by Scoransky et al.[45,46] is useful for the detection of *Actinomycetes* The protocol below is a rapid DNA–DNA hybridization method based on melting profiles in microplates (Figure 3.1).[47] Whole genomic DNA probes are labeled with digoxigenin coupled to dUTP, and anti-digoxigenin antibody is utilized for detection of the hybridization reaction.

Procedure

1. Genomic DNA of the reference and target strains are enzymatically digested with *Sau*3A, ligated to the double stranded linkers P (for the target DNA) and S (for the reference DNA). These linkers provide specific sites for PCR mediated amplification with target and reference primers P1 and S1, respectively. The oligonucleotide pairs P1 and P2 as well as S1 and S2 share complementary stretches and are designed for the formation of double stranded linkers carrying 5′-*Sau*3A-compatible overhangs.
2. For labeling of the PCR product, digoxygenin modified dUTP (Roche) is used in a ratio 1:19 of labeled to unlabeled dTTP.
3. Reference DNA, ligated with linker S, is amplified by PCR by applying the same conditions, except that primer S is used instead of primer P and that no labeled nucleotides are incorporated during PCR.
4. Aliquots (5 μL) of PCR products are electrophoretically separated on a 1% agarose gel and visualized by SYBR® Green (Cambrex Bio Science Rockland Inc.) staining.
5. PCR products are purified by using a QIAquick PCR purification kit (QIAGEN) and are eluted from the purification column in 50 μL sterile water.
6. The DNA of the reference strains is immobilized on MaxiSorp 96-well microplates (NalgenNunc). A total amount of 1 μg DNA in 50 μL PBS/$MgCl_2$ (8.0 mM Na_2HPO_4, 1.5 mM KH_2PO_4, 137 mM NaCl, 2.7 mM KCl pH 7.2; supplemented with $MgCl_2$ to a final concentration of 100 mM) is immobilized per well. Before immobilization, the DNA is denatured at 99°C for 10 min in PBS/$MgCl_2$ buffer, immediately transferred to the microplate, and incubated at 37°C for 60 min. The solution is discarded and the microplate dried for 2 h at 60°C. The coated plates

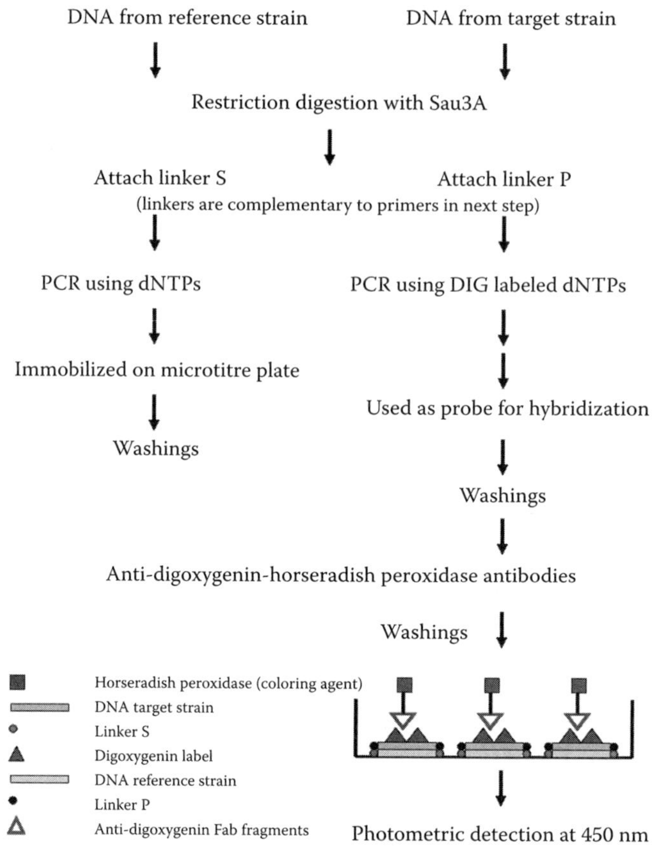

FIGURE 3.1 Schematic representation of the DNA–DNA hybridization procedure in microplates. DNA of interest is restricted with *Sau*3A, and ligated with the appropriated linker (P or S). During amplification of the restricted and ligated DNAs, the DNA of the target strain is labeled with Dig dUTP. Anti-digoxygenin-HRPO antibodies are used for detection of hybridized fragments. Photometric detection of color in the wells indicates the degree of hybridization between the target and reference strain.

can be stored at 4°C for 2–3 weeks sealed with adhesive film (NalgenNunc).

7. Prior to use, the plates are washed twice with 100 μL PBS to remove excess, not immobilized DNA. The appropriate hybridization temperature (THyb) is determined applying the following formula[10]: THyb = 81.5 + 16.6 × log M + 0.41 × (%G+C) – 0.7 × % formamide THyb = melting temperature [°C], M = concentration of monovalent cations).
8. Following prehybridization for 30 min at THyb with 50 μL hybridization solution [5× SSC, 1% blocking reagent (Roche), 0.1% n-laurylsarkosine, 0.02% SDS, 32% formamide], 20 ng denatured target DNA is diluted in 50 μL hybridization solution, and transferred to the coated microplates. The hybridization is performed for 2 h at THyb. The microplate cavities are washed twice for 6 min at THyb with 100 μL of the washing solutions.
9. Thermal denaturation curves are generated by applying washing solutions of increasing stringency. For the detection of DNA–DNA hybrids Commercial anti-digoxygenin-horseradish peroxidase Fab fragments (anti-DIG-POD) (Roche) stock solution is used after dilution 1:1000 by adding dilution buffer [1% blocking reagent (Roche) in PBS]. 100 μL of the antibody solution are added to each well. After incubating for 1 h at room temperature, the plates are washed three times with PBS (8.0 mM Na_2HPO_4, 1.5 mM KH_2PO_4, 137 mM NaCl, 2.7 mM KCl pH 7.2).
10. For colorimetric detection of the hybrids 100 μL of substrate solution (BM-Blue, Roche) are added to each well at room temperature. Color development is stopped after 20 min with 100 μL 1M sulfuric acid. The absorption is measured at 450 nm.

3.2.2.5 Real-Time PCR for *Actinomyces naeslundii* and *Actinomyces viscosus*

Suzuki et al.[48] described the use of TaqMan real-time PCR for detection of *A. naeslundii* and *A. viscosus* in dental plaque samples with primers derived from the 16S rRNA gene (Table 3.2).

Procedure

1. Real-time PCR mixture (20-μL) is composed of 1 μL of lysed cells, 1× qPCR Master Mix (Eurogentec), forward and reverse primers at 200 nM each, and the TaqMan probe at 250 nM. Separate PCR mixtures are prepared for *A. naeslundii* and *A. viscosus*.

TABLE 3.2
Primers and Probe for Real-Time PCR Detection of *A. naeslundii* and *A. viscosus*

Primer or Probe	Sequence (5′–3′)	Specificity
F1	TCG AAA CTC AGC AAG TAG CCG	*A. naeslundii*
R1	AGA GGA GGG CCA CAA AAG AAA	
Probe 1	FAM- GGG TAC TCT AGT CCA AAC TGG CGG ATA GCG -TAMRA	
F2	ATG TGG GTC TGA CCT GCT GC	*A. viscosus*
R2	CAA AGT CGA TCA CGC TCC G	
Probe 2	FAM- ACG GAG GTC GGG AAC GGT GGA AG -TAMRA	

2. Amplification is conducted at 50°C for 2 min and 95°C for 10 min, followed by 60 cycles of 95°C for 15 s, and 58°C for 1 min with the ABI PRISM 7700 Sequence Detection System (Applied Biosystems).

3.3 CONCLUSION

Actinomycosis is a rare disease but a serious one in case of late detection. It affects various body parts including neck/face, lungs, alimentary canal, or the pelvic/gonadial organs. Diagnosis is based on identification of characteristic sulfur granules in whole pus, histological staining, and preferably also by a positive culture. However, molecular methods of detection are more reliable and rapid. Treatment is usually with different antibiotics and is commonly associated with surgeries depending on the extent of spread of the disease. Beginning treatment promptly helps quicken the recovery and complete recovery is possible. Development of direct detection methods may aid patient management and further elucidate clinical associations.

ACKNOWLEDGMENTS

We thank Dr. Mark Lawrence and Dr. Shane Burgess for their support in writing the manuscript. We are also grateful to Dr. Rakesh K. Jain and Dr. Walter J. Diehl for their encouragement.

REFERENCES

1. Gerencser, M.A., and Slack, J.M., Serological identification of *Actinomyces* using fluorescent antibody techniques, *J. Dent. Res.*, 55, A184, 1976.
2. Stackebrandt, E., and Charfreitag, O., Partial 16S rRNA primary structure of five *Actinomyces* species: Phylogenetic implications and development of an *Actinomyces israelii*-specific oligonucleotide probe, *J. Gen. Microbiol.*, 136, 37, 1990.
3. Leek, J.H., Actinomycosis of the tympanomastoid, *Laryngoscope*, 84, 290–301, 1974.
4. Bennhoff, D.F., Actinomycosis: Diagnostic and therapeutic considerations and a review of 32 cases, *Laryngoscope*, 94, 1198, 1984.
5. Mabeza, G.F., and Macfarlane, J., Pulmonary actinomycosis, *Eur. Respir. J.*, 21, 545, 2003.
6. Sharkawy, A.A., Cervicofacial actinomycosis and mandibular osteomyelitis, *Infect. Dis. Clin. North Am.*, 21, 543, 2007.
7. Dhaliwal, U. et al., Clinical and cytopathologic correlation in chronic inflammations of the orbit and ocular adnexa: A review of 55 cases, *Orbit*, 23, 219, 2004.
8. Pittman, F.E. et al., Clinical, histochemical, and electron microscopic study of colonic histiocytosis, *Gut*, 7, 458, 1966.
9. Richtsmeier, W.J., and Johns, M.E., Actinomycosis of the head and neck, *CRC Crit. Rev. Clin. Lab. Sci.*, 11, 175, 1979.
10. Brown, J.R., Human actinomycosis. A study of 181 subjects, *Hum. Pathol.*, 4, 319, 1973.
11. Hsieh, M.J. et al., Thoracic actinomycosis, *Chest*, 104, 366, 1993.
12. Miller, M., and Haddad, A.J. Cervicofacial actinomycosis. *Oral Surg. Oral Med. Oral Pathol. Oral Radiol. Endod.*, 85, 496, 1998.
13. Belmont, M.J., Behar, P.M., and Wax, M.K. Atypical presentations of actinomycosis. *Head Neck*, 21, 264, 1999.
14. James, R., Heneghan, M.A., and Lipkansky, V., Thoracic actinomycosis, *Chest*, 87, 536, 1985.
15. Frank, P., and Strickland, B., Pulmonary actinomycosis, *Br. J. Radiol.*, 47, 373, 1974.
16. Karagulle, E. et al., Abdominal actinomycosis mimicking acute appendicitis, *Can. J. Surg.*, 51, E109, 2008.
17. Xu, G.P. et al., Cervical actinomycosis with spinal cord compression, case report and literature review, *Chemotherapy*, 54, 63, 2008.
18. Ramam, M. et al., A two-step schedule for the treatment of actinomycotic mycetomas. *Acta Derm. Venereol.*, 80, 378, 2000.
19. Tendolkar, U.M. et al., Microbiological study of mycetoma patients from Bombay with special reference to actinomyces immunofluorescence, *Indian J. Pathol. Microbiol.*, 36, 245, 1993.
20. Warren, N.G., Actinomycosis, nocardiosis, and actinomycetoma, *Dermatol. Clin.*, 14, 85, 1996.
21. Sudhakar, S.S., and Ross, J.J., Short-term treatment of actinomycosis: Two cases and a review, *Clin. Infect. Dis.*, 38, 444, 2004.
22. Wade, W.G., In-vitro activity of ciprofloxacin and other agents against oral bacteria, *J. Antimicrob. Chemother.*, 24, 683, 1989.
23. Baik, J.J. et al., Pulmonary actinomycosis in Korea, *Respirology*, 4, 31, 1999.
24. Edelmann, M. et al., Treatment of abdominothoracic actinomycosis with imipenem, *Eur J. Clin. Microbiol.*, 6, 194, 1987.
25. Yew, W.W. et al., Use of imipenem in the treatment of thoracic actinomycosis, *Clin. Infect. Dis.*, 19, 983, 1994.
26. Skoutelis, A., Petrochilos, J., and Bassaris, H., Successful treatment of thoracic actinomycosis with ceftriaxone, *Clin. Infect. Dis.*, 19, 161, 1994.

27. Rolfe, R.D., and Finegold, S.M., Comparative in vitro activity of ceftriaxone against anaerobic bacteria, *Antimicrob. Agents Chemother.*, 22, 338, 1982.
28. Trutnovsky, G., Tamussino, K., and Reich, O., Short-term antibiotic treatment of pelvic actinomycosis, *Int. J. Gynaecol. Obstet.*, 101, 203, 2008.
29. Fiorino, A.S., Intrauterine contraceptive device-associated actinomycotic abscess and Actinomyces detection on cervical smear, *Obstet. Gynecol.*, 87, 142, 1996.
30. Atad, J. et al., Pelvic actinomycosis. Is long-term antibiotic therapy necessary? *J. Reprod. Med.*, 44, 939, 1999.
31. Lambert, F.W., Jr., Brown, J.M., and Georg, L.K. Identification of Actinomyces israelii and *Actinomyces naeslundii* by fluorescent-antibody and agar-gel diffusion techniques. *J. Bacteriol.*, 94, 1287, 1967.
32. King, S., and Meyer, E., Gel diffusion technique in antigen-antibody reactions of *Actinomyces* species and "anaerobic diphtheroids," *J. Bacteriol.*, 85, 186,1963.
33. Slack, J.M., Landfried, S., and Gerencser, M.A., Morphological, biochemical, and serological studies on 64 strains of *Actinomyces israelii*, *J. Bacteriol.*, 97, 873, 1969.
34. Snyder, M.L. et al., Studies on oral filamentous bacteria. II. Serological relationships within the genera *Actinomyces*, *Nocardia*, *Bacterionema* and *Leptotrichia*, *J. Infect. Dis.*, 117, 341, 1967.
35. Pirtle, E.C., Rebers, P.A., and Weigel, W.W., Nitrogen-containing and carbohydrate-containing antigen from *Actinomyces bovis*, *J. Bacteriol.*, 89, 880, 1965.
36. Woo, P.C. et al., Diagnosis of pelvic actinomycosis by 16S ribosomal RNA gene sequencing and its clinical significance, *Diagn. Microbiol. Infect. Dis.*, 43, 113, 2002.
37. Jauh-Hsun, C., Vinh, T., Davies, J.K., and Figdor, D., Molecular approaches to the differentiation of *Actinomyces* species, *Oral Microbiol. Immunol.*, 14, 250, 1999.
38. Ruby, J.D. et al., Genetic characterization of the oral *Actinomyces*, *Arch. Oral Biol.*, 47, 457, 2002.
39. Sato, T. et al., Differentiation of oral *Actinomyces* species by 16S ribosomal DNA polymerase chain reaction-restriction fragment length polymorphism, *Arch. Oral Biol.*, 43, 247, 1998.
40. Tang, G., Samaranayake, L.P., and Yip, H.K. Genotypic diversity of oral *Actinomyces naeslundii* genospecies 1 and 2 in caries-active preschool children. *Oral Microbiol. Immunol.*, 19, 371, 2004.
41. Narayanan, S. et al., Biochemical and biological characterizations and ribotyping of *Actinomyces pyogenes* and *Actinomyces pyogenes*-like organisms from liver abscesses in cattle, *Vet. Microbiol.*, 61, 289, 1998.
42. Cogulu, D. et al., PCR-based identification of selected pathogens associated with endodontic infections in deciduous and permanent teeth, *Oral Surg. Oral Med. Oral Pathol. Oral Radiol. Endodontol.*, 106, 443, 2008.
43. Hansen, T. et al., *Actinomyces* in infected osteoradionecrosis—underestimated? *Hum. Pathol.*, 37, 61, 2006.
44. Weisburg, W.G. et al., 16S ribosomal DNA amplification for phylogenetic study, *J. Bacteriol.*, 173, 697, 1991.
45. Colombo, A.P. et al., Clinical and microbiological features of refractory periodontitis subjects, *J. Clin. Periodontol.*, 25, 169, 1998.
46. Socransky, S.S., Smith, C., and Haffajee, A.D., Subgingival microbial profiles in refractory periodontal disease, *J. Clin. Periodontol.*, 29, 260, 2002.
47. Mehlen, A. et al., Development of a fast DNA–DNA hybridization method based on melting profiles in microplates, *Syst. Appl. Microbiol.*, 27, 689, 2004.
48. Suzuki, N., Yoshida, A., and Nakano, Y. Quantitative analysis of multi-species oral biofilms by TaqMan real-time PCR, *Clin. Med. Res.*, 3, 176, 2005.

4 Atopobium

Piet Cools, Hans Verstraelen, Mario Vaneechoutte, and Rita Verhelst

CONTENTS

4.1 INTRODUCTION

4.1.1 CHARACTERISTICS OF *ATOPOBIUM* SPP.

Classification. The genus *Atopobium* was introduced by Collins and Wallbanks[1] to reclassify species formerly designated *Lactobacillus minutus* (Hauduroy 1937), *Lactobacillus rimae*, and *Streptococcus parvulus* (Weinberg 1937). In 1999, *Eubacterium fossor*[2] and a newly described species, *Atopobium vaginae*,[3] were placed in the genus.[4]

Although the genus *Atopobium*, belonging to the family *Coriobacteraceae*,[5] has a low G + C content, it represents a deep branch within the high G + C gram-positive bacteria, the Actinobacteria.[6] *Coriobacterium glomerans* is the closest phylogenetic relative.[3] Today, the genus houses five species: *A. vaginae* (type strain CCUG 38953^T), *A. rimae* (ATCC 49626^T), *A. parvulum* (ATCC 33793^T), *A. fossor* (NCTC 11919^T), and *A. minutum* (ATCC 33267^T), the latter being the type species.[1]

The first complete genome sequence of the genus *Atopobium* was published by Copeland in 2009. The *A. parvulum* type strain (IPP 1246T) has a 1,543,805 bp long single replicon genome with 1369 protein-coding and 49 RNA genes.

Morphology. Cells consist of short elliptical rods (*A. rimae*, *A. minutum*, *A. fossor*),[7] with central swellings (*A. rimae*)[7] or small cocci (*A. parvulum* and *A. vaginae*)[3,7] that may appear to be elliptical. Cells occur singly, in pairs, in clumps (*A. parvulum*),[7] and in short chains, and are gram-positive, nonmotile, and nonsporeforming.[1] Colonies on brain heart blood agar are pinpoint to 2 mm in diameter after 5 days of anaerobic incubation at 37°C. They appear translucent to transparent and are white, light yellow, or grey.[1]

Biology and Physiology. *Atopobium* species are obligately anaerobic.[1] Although originally described as facultative anaerobic,[3] Geissdörfer et al. and Ferris et al. found their *A. vaginae* isolates—including the type strain—to be strictly anaerobic.[8,9] *Atopobium* species are catalase negative and produce major amounts of lactic acid from glucose, together with acetic acid and formic acids. For *A. minutum*, *A. parvulum*, and *A. rimae*, trace amounts of succinic acid may be detected.[1,10] G + C content ranges from 39% (*A. parvulum*), over 44% (*A. vaginae*, *A. fossor*, and *A. minutum*), to 45%.

Growth Requirements. *Atopobium* spp. are fastidious and all need to be grown under anaerobic conditions (N_2 80%, H_2 10%, CO_2 10%) at 37°C.

A. parvulum strain IPP 1246^T, DSM 20469, can be grown in DSMZ medium 104 (modified PYG [http://www.dsmz.de/microorganisms/media_list.php]) or Fastidious Anaerobe Agar (FAA, LabM) supplemented with 5% horse blood.

Its growth is substantially stimulated by 0.02% (vol/vol) Tween 80 and by 10% (vol/vol) rabbit serum added to culture media.[7]

A. vaginae can be cultured on Columbia agar with 5% horse blood[3] or 5% sheep blood; tryptone soy agar supplemented with 5% sheep blood[11]; Brucella broth supplemented with vitamin K (1 μg/mL), hemin (5 μg/mL), and 5% laked horse blood[9]; Brucella agar with 5% sheep blood[9] or

Chocolate agar (Nikolaitchouk, personal communication, 2010) for 2–5 days.

A. fossor can be grown on DSMZ medium 110 (chopped meat medium with carbohydrates) or ATCC medium 1490 (modified chopped meat medium).

A. minutum can be grown on ATCC medium 593 (chopped meat medium).

A. rimae can be cultured on brain heart blood agar,[7] on fastidious anaerobe agar (FAA, LabM) supplemented with 5% horse blood,[12] in DSMZ medium 104, or on tryptic soy agar supplemented with 5% sheep blood.[13] Growth is stimulated by 0.02% Tween 80 and by 10% rabbit serum added to culture media.[7]

4.1.2 *A. vaginae* and Bacterial Vaginosis

Only recently, *A. vaginae* has been implicated as an important marker of bacterial vaginosis.[9,13–15] Bacterial vaginosis is the single most common cause of vaginitis among women of childbearing age.[16] Besides a nuisance problem, bacterial vaginosis is associated with adverse obstetric and gynecologic outcomes including premature labor, preterm birth, preterm prelabor rupture of membranes, miscarriage, intrauterine infection, postcesarean endometritis, chorioamnionitis, upper genital tract infections, and pelvic inflammatory disease.[17–24] Bacterial vaginosis also renders women more vulnerable to the acquisition of sexually transmitted diseases like gonorrhea, chlamydiosis,[25] genital herpes,[26] and HIV-1,[27] and it propagates viral replication and shedding of HIV-1[28] and HSV-2.[29]

Clinical Features. About half of all women with a diagnosis of bacterial vaginosis experience no symptoms at all. When present, characteristic symptoms are a profuse, homogeneous, off-white, and often malodorous vaginal discharge, which can cause considerable distress and discomfort. As bacterial vaginosis is rarely associated with inflammation, no inflammatory signs will be found and the discharge represents the only symptom, often accompanied with odor.

Association of *A. vaginae* with Bacterial Vaginosis. Although controversial,[30] bacterial vaginosis is typically regarded as a polymicrobial disease[31] with an unknown etiology[32] whereby the normal *Lactobacillus* dominated microflora is replaced by an overgrowth of mainly anaerobic bacteria.[33,34] The bacteria most frequently isolated from women with bacterial vaginosis include *Gardnerella vaginalis*, *A. vaginae*, *Mobiluncus* spp., *Bacteroides* spp., *Prevotella* spp., and mycoplasmas,[35,36] most of which can also be detected in women with normal microflora and by consequence are not specific for this syndrome.[35]

G. vaginalis is classically regarded as most associated with bacterial vaginosis, but is detected in some studies in 50%–60% of women without bacterial vaginosis.[37] Recent independent studies reveal a stronger association of *A. vaginae* with bacterial vaginosis than *G. vaginalis*.[9,13,14,32,38,39] This stronger association might be explained by the fact that *G. vaginalis* is more frequently detected in women with a normal vaginal microflora than *A. vaginae*, which is rarely (if ever) a component of a normal vaginal microflora.[9,13,32,39] *A. vaginae* not only appears to be more than twofold as specific for bacterial vaginosis, but its copresence with *G. vaginalis* was associated with significantly higher rates of recurrent bacterial vaginosis than those seen in women with only *G. vaginalis*.[32]

A. vaginae rarely occurs in the absence of *G. vaginalis*,[32] suggesting a possible synergism between the two organisms. This suggestion was strengthened by observations of high rates of co-occurrence of *A. vaginae* and *G. vaginalis*[13,39,40] and by the fact that *A. vaginae* was found in 50% of *G. vaginalis* biofilms in women with bacterial vaginosis.[15] In women with preterm labor, high vaginal concentrations of *A. vaginae* and *G. vaginalis* are significantly associated with the risk of preterm delivery.[24]

Since many bacterial vaginosis-associated bacteria have been found within the amnion and fetal membranes of women delivering prematurely,[41,42] suggesting that microbial invasion of the amniotic cavity results primarily from the ascent of bacteria from a disturbed vaginal microflora, it is worth noting that since *A. vaginae* has also been detected in upper reproductive tract disease,[3,8,43] it may play a role in adverse pregnancy outcomes. In postmenopausal women with bacterial vaginosis, *A. vaginae* was found to be the predominant species.[14]

***A. vaginae* and Bacterial Vaginosis Treatment.** Several studies pointed at *A. vaginae* as a metronidazole-resistant bacterium.[8,9,32] Though we found that metronidazole susceptibility varied to a significant extent across various *A. vaginae* strains,[11] the finding is intriguing, considering that metronidazole actually represents the standard therapy in case of bacterial vaginosis. Bradshaw et al. subsequently showed in a large prospective cohort study that rates of recurrence of bacterial vaginosis were strikingly higher among women harboring *A. vaginae* along with *G. vaginalis*, as compared to those carrying only *G. vaginalis*. Ferris et al. added another intriguing finding obtained from a treatment study with topical metronidazole gel. This study indicated that pretreatment *A. vaginae* concentrations were highest in patients who had completely or partially failed treatment, while concentrations in pretreatment *A. vaginae* were lowest in patients who were cured.[44]

In the pretreatment group, *A. vaginae* concentrations were highest in patients who completely or partially failed treatment, and lowest in patients who were cured.

Pathogenesis. Bacterial biofilms have been associated with several recalcitrant infections like cystic fibrosis[45] and chronic rhinosinusitis[46] and are estimated to account for over 80% of microbial infections in the body.[47] In addition, in bacterial vaginosis, the existence of a dense, adherent bacterial biofilm on the vaginal mucosa has been recently shown.[48] Moreover, this biofilm, composed for more than 90% of *G. vaginalis* and *A. vaginae*, is an obligate finding in bacterial vaginosis and is absent in women with normal vaginal microflora.

As a virulence mechanism, a biofilm prevents antimicrobial agents to reach bacteria—often present as latent cells within the biofilm—and enhances bacterial attachment to epithelial cells, allowing higher bacterial concentrations than in luminal fluid.[49] The biofilm formed in bacterial vaginosis and the synergism between *G. vaginalis* and *A. vaginae* may play a significant role in pathogenesis[50] and high recurrence rates of bacterial vaginosis.[15]

In response to *A. vaginae*, vaginal epithelial cells secreted IL-6 and IL-8 in an in vitro model. Furthermore, transcription of an antimicrobial β-defensin peptide occurred after triggering the Toll-like receptor 2.[51] This required live, adhering bacteria capable of protein synthesis. These data contribute indirect evidence for a putative role of *A. vaginae* in the pathogenesis of bacterial vaginosis.

4.1.3 *A. vaginae* in Pelvic Inflammatory Disease

Pelvic inflammatory disease (PID) is a condition that lacks a precise definition and may include infection of any or all of the following anatomic locations: the endometrium (endometritis), the Fallopian tubes (salpingitis), the ovaries (oophoritis), the uterine wall (myometritis), the uterine serosa and broad ligaments (parametritis), and the pelvic peritoneum.[52] In addition, a tuboovarian abscess may form. Some authors prefer the term salpingitis because, at a minimum, the oviduct (fallopian tube) is involved, and most of the long-term sequelae of PID result from destruction of the tubal architecture. Vaginal and endocervical infections can ascend and cause PID. Besides an indirect role for *A. vaginae* in the pathogenesis of PID by its putative role in pathogenesis of bacterial vaginosis, of which PID is one of the sequelae, *A. vaginae* has been recovered from patients with salpingitis,[43] as well as from a patient with a tuboovarian abscess.[8]

Although the etiology of salpingitis is often attributed to sexually transmitted organisms such as *Chlamydia trachomatis*[53] and *Neisseria gonorrhoeae*,[54] which colonize the cervix and then ascend to the upper genital tract and cause tissue damage, the majority of cases have no known etiology.[55] Recently, *A. vaginae* was detected in 2 of 11 Fallopian tube samples of women with salpingitis with bacterial etiology.[43] *A. vaginae* may partly be responsible for the association between bacterial vaginosis and salpingitis.[55]

In 2000, a 39-year-old woman had a transvaginal oocyte recovery as part of an in vitro fertilization program. In 2003, a woman with a 3-day history of severe pain in the lower abdomen was admitted to a clinic.[8] Laparotomy revealed a tuboovarian abscess. The infection spread from the abscess, causing inflammation of the peritoneum and appendix. *A. vaginae* was recovered from the abscess as the sole organism, defining its clinical significance. Other isolates of *A. vaginae*, listed in the culture collection at the University of Göteberg in Sweden, originated from blood and amniotic membrane infections.[14] A case of *Atopobium vaginae* causing uterine endometritis has been recently.[55a]

4.1.4 Cases of Sepsis by *Atopobium* Species

In 2007, a case of *A. rimae* bacteremia was reported in a 47-year-old man from Seoul, Korea with liver cirrhosis.[56] The patient was admitted to the hospital due to fever and chill. Abdominal and pelvic computed tomography showed a small abscess near the left adrenal gland.

In 2007, a 77-year-old woman with a history of right thoracotomy for pneumothorax was hospitalized for inhalation pneumonia caused by paralysis of the right vocal cord.[57] Septic shock and a fever of 38°C developed, which was complicated by acute respiratory failure and stroke. *A. rimae* was identified as causative agent of the sepsis. After transferring to the intensive care unit and 7 days treatment, the patient's condition improved and remained apyretic.

In 2008, a 38-year-old male was admitted to the medical intensive care unit with sepsis.[58] In addition to fever, elevated heart rate, and hypotension, physical examination revealed an extensive, necrotic decubitus ulcer of the left hip and poor oral hygiene. The *Atopobium* species associated with the sepsis likely originated from the ulcer. The *Atopobium* species reportedly showed maximum identity (99%) to *Atopobium* sp. oral clone C019, which has 98% sequence similarity to the type strain of *A. rimae*[59] and was provisionally named *Atopobium detroiti* by the authors.

4.1.5 *Atopobium* Species in Oral Health and Disease

Halitosis. Halitosis, or oral malodor, is a very common condition that may affect up to 30% of the population.[60,61] In most cases, halitosis arises from malodorous substances—mainly volatile sulfur compounds—produced in the oral cavity as a consequence of the metabolism of bacteria colonizing the oral surfaces, periodontal pockets, and especially the dorsal tongue surface. Species typically associated with halitosis are *Treponema denticola*, *Porphyromonas gingivalis*, *Porphyromonas endodontalis*, and *Eubacterium* species,[62] isolated from periodontal pockets. Whilst malodor is often associated with the presence of periodontitis, in many cases there is no such link, and evidence points to the importance of bacteria in tongue coatings that results in halitosis.[60] In characterizing the tongue dorsum microflora of subjects with and without halitosis, *A. parvulum* was one of the species most associated with halitosis in one study,[61] but with health in another.[63] Further research on the role of *A. parvulum* in halitosis and its ability to produce volatile sulfur compounds is needed to point out a potential contribution to the clinical presentation of halitosis. One study found *A. vaginae* in halitosis associated tongue dorsum microflora, but with low prevalence.[64]

Caries. Dental caries is a process that may take place on any tooth surface where dental plaque is allowed to develop over a period of time.[65] Formation of dental plaque—a biofilm—is a natural physiological process. There are three microbial hypotheses for the etiology of dental caries. The specific plaque hypothesis[66] states that only a few specific species, such as *Streptococcus mutans*, are actively involved in the disease, the nonspecific plaque hypothesis explains caries as the outcome of the overall activity of the total plaque microflora, composed of many bacterial species,[67] whereas the ecological plaque hypothesis suggests that caries is a result of a shift in the balance of the resident microflora driven by changes in local environmental conditions, like acid production.[68] Aas et al.[69] showed that 10% of subjects with dental caries did not have detectable levels of *S. mutans* and that other species, among which included *Atopobium* spp., dominated the microflora in caries progression.[69] Furthermore, in caries patients with *S. mutans*, other species, among which included *Atopobium* spp., were present at significantly higher levels than *S. mutans*. Keeping in mind their ability to produce high amounts of acid,[1] *Atopobium* species are likely to play important roles in caries progression and should be added to the list of putative etiological agents of caries.[69]

Gingivitis. The microflora within the plaque biofilm on surfaces of teeth at the gum (gingiva) margin can cause gingivitis, which can be defined as an inflammation of the gingiva without destruction of the supporting tissues. The microflora associated with dental plaque-induced gingivitis is different than that associated with health. In a model of experimental gingivitis, *A. rimae* was found among the species to be positively correlated with the initiation of gingivitis, whereas *A. minutum* was found among the species that seemed to be present as a result of gingivitis.[70–72] A clone with 98% sequence similarity to the type strain of *A. rimae* was isolated from a patient with acute necrotizing ulcerative gingivitis, and probably represents a new species.[59]

Periodontitis. Periodontitis is a bacterial driven inflammatory disease[73] and can be caused by an apical extension of gingivitis. On the other hand, if pulp necrosis, which can result from caries, occurs, oral bacteria can colonize the root canal system with the formation of a biofilm.[73a] When the colonization progresses apically, bacteria come into direct contact with the periradicular tissues, causing apical peridontitis.[74] These inflammatory reactions are short-lived and self-limiting if the colonization is transient in nature. In contrast, if bacterial contamination is persistent, immune reactions can cause destruction of surrounding tissues—periodontal ligament, cementum, and alveolar bone—leading to periradicular lesions, which can lead to tooth loss.

Although also found in healthy subjects, *A. minutum* and *A. rimae* were isolated from gingival plaque samples of patients with severe or juvenile periodontitis,[75–77] and *A. rimae* was found in periradicular lesions.[78]

Furthermore, *A. parvulum* and *A. rimae* were shown to be present in infected root canals.[79,80]

Rather than a single organism slipping past the host defense, it is the bacterial community within the biofilm that causes periodontal tissue breakdown, classifying this disease as polymicrobial. Individual contributions of bacteria in pathogenesis of periodontitis and periodontal lesions still need to be elucidated,[78] taking into consideration that *A. rimae* and *A. parvulum* are candidate pathogens.[77,80,81]

4.2 METHODS

4.2.1 Phenotypic Identification

Commercial microbial identification systems form the backbone of microbial identification in clinical microbiology laboratories. They exist of a series of biochemical tests that are based on detecting bacterial enzymes, products, or metabolic activity. In principle, the combination of the enzymatic or biochemical results of the different tests enables to identify the organisms to the species level.

Taken into account the fastidious nature of *Atopobium* species, and the fact that these tests frequently fail to identify many anaerobes, especially nonspore-forming gram-positive rods,[82] it is not surprising that in all cases of sepsis or abscesses involving *Atopobium* spp. the isolates could not be identified or were misidentified with these commercial biochemical identification systems, because *Atopobium* spp. are not included in the databases of these identification systems.

In three cases of sepsis, *A. rimae* or *Atopobium* sp. oral clone C019 could either not be identified or were misidentified using the RapID ANA II (Remel, Lexena, KS, US), API ANA strip, or Vitek ANI card (ANI code 1600201000) identification systems (bioMérieux, Marcy-l'Etoile, France).[56–58] *A. vaginae* isolates from a tuboovarian abscess or from patients with bacterial vaginosis were misidentified using the API ID32A (bioMérieux) or a RapID ANA II identification system as *Gemella morbillorum* (code 2002 0335 05).[8,9]

4.2.2 Genotypic Identification

This overview of molecular detection and identification of *Atopobium* spp. focuses on nucleic acid–based techniques, with emphasis on the 16S rRNA gene. The 16S rRNA gene contains regions that are highly conserved across bacterial species—reflecting absence of evolutionary divergence—and variable regions—reflecting evolutionary divergence among species. It has proven to be a very useful phylogenetic marker employed to identify bacteria.[83,84]

First, we discuss the techniques that are based on the hybridization of *Atopobium* spp. specific DNA probes—most commonly oligonucleotide probes—targeting the variable 16S rRNA alleles. Secondly, we review *Atopobium* spp.-specific conventional and quantitative real-time PCR (qPCR) assays, targeting the 16S rRNA gene region and the *cpn*60 gene. Finally, we comment on culture-dependent approaches using broad-range primers.

4.2.2.1 Sample Preparation

Atopobium spp. are gram-positive bacteria, for which most extraction methods have a relatively poor recovery of nucleic acids.[84] In order to detect or accurately quantify *Atopobium* spp. in biological samples, DNA extraction requires enzymatic pretreatment to specifically lyse the gram-positive cell wall. Mutanolysin, an enzyme that degrades the thick gram-positive peptidoglycan layer, has been proven to improve DNA recovery from gram positives.[85] In addition, in the pretreatment step, proteinase K can be used to break down contaminating proteins and nucleases, which are liberated during lysis of several oral bacteria.

4.2.2.2 Detection Procedures

4.2.2.2.1 *Fluorescence* In Situ *Hybridization*

Fluorescence in situ hybridization (FISH) enables direct visualization and detection of bacterial species. A typical FISH protocol includes four steps. First, prior to the hybridization, bacteria are fixed and permeabilized; this facilitates accessibility of the fluorescent probes to the target sequence and protects the RNA from degradation by endogenous RNAses.[86] Samples can be settled on membrane filters and covered with the fixing agent[87], or they can be mixed with the fixing agent, and stay in solution or be transferred and dried to glass slides.[88] After hybridization, a washing step removes unbound probe, and detection of the labeled bacteria is carried out by microscopy or flow cytometry.[89]

Probes can be labeled directly by synthesizing them with one or more modified dNTPs carrying a fluorochrome chemical side group.[86] Although this is the fastest and cheapest method—the fluorescent signal can be bound to the target in a single hybridization step[90]—it misses the advantage of simultaneous detection of several bacterial species provided by indirect labeling of separate probes with different fluorophores. When employing two or three dyes together, a color combination can be obtained that allows the detection of multiple targets by measuring the relative contribution of each color to the total signal.[91]

FISH has been used to study the composition of the microflora in human feces, including the detection of bacteria of the *Atopobium* cluster (i.e., belonging to the genera *Atopobium*, *Collinsella*, *Coriobacterium*, *Cryptobacterium*, and *Eggerthella*).[92–96] FISH has also been used for detection of this *Atopobium* cluster in vaginal fluids[33] or in vaginal biopsies.[15,48]

Sample Preparation and Detection. Whole stools are collected and kept at 4°C in sterile bags for no longer than 12 h.[92] The stools are kneaded mechanically at 4°C to distribute the sample evenly and 0.5 g portions are suspended in 4.5 mL PBS[92] or brain heart infusion broth.[93] Suspension is homogenized by vortexing with glass beads[92] or by using a magnetic stirrer.[93] After centrifugation, the supernatant is fixed by diluting 1:3 with 4% (wt/vol) paraformaldehyde in PBS (PFA),[95] and by keeping it at 4°C for 16 h. Samples can be stored in 50% (vol/vol) ethanol-PBS at –20°C[92] or –70°C.[93]

Vaginal biopsy specimens are fixed in nonaqueous Carnoy solution for 2 h, processed and embedded into paraffin blocks by standard techniques.[15,48] 4-µm thick microtome biopsy sections are placed on slides.[15,48] Vaginal fluid smears on glass slides are fixed in 95% ethanol.[33]

The probe with sequence 5′-GGTCGGTCTCTCA ACCC-3′, targeting the 16S rRNA gene of the *Atopobium* cluster,[15,48,92,93,95,96] is labeled with carbocyanine Cy3, carbocyanine Cy5, or fluorescein isothiocyanate (MWG Biotech, Ebersberg, Germany).[15,33,48] Hybridization on glass slides is carried out at 50°C without addition of formamide[92] or at 45°C with 10%–20% formamide concentrations, or in suspension at 35°C.[93] To enhance visualization after hybridization, bacteria can be counterstained with 4, 6-diamidino-2-phenylindole (DAPI).[15,34,48]

Detection of the fluorescent signal on the glass slides is performed with a fluorescence microscope (Nikon e600, Nikon) with a 100× oil immersion objective,[15,33,48] whereas a flow cytometer (Facs Calibur, Becton Dickinson) is used for the solution.[93,95]

4.2.2.2.2 *DNA Checkerboard Array*

The checkerboard DNA–DNA hybridization technique allows simultaneous analysis of several microbial species against up to 28 mixed microflora samples. It involves the extraction and labeling of total genomic DNA from a culturable microorganism for use as a probe in a hybridization assay with total genomic DNA extracted from clinical samples.[97] The sample DNA is fixed in thin lanes on a membrane, and after turning the membrane 90°, up to 40 whole-genome probes are loaded into the wells and allowed to hybridize with their target sequence. In two lanes, a mixture of whole genomic DNA standards, representing a known amount of cells of each species, are fixed. The checkerboard, with the columns representing samples and the rows representing species of interest, allows a comparison of spot darkness with that of the standards, making the technique semiquantitative.

This technique is being widely used for the detection of species associated with oral disease or health[98–104] and for detection of several species in vaginal samples.[105]

For the preparation of the probes, *A. vaginae*, *A. parvulum*, and *A. rimae* strains are grown on Trypticase soy agar supplemented with 5% defibrinated sheep blood or Columbia agar supplemented with 5% horse blood under anaerobic conditions at 35°C–37°C.[105,106] Mixtures are made with a concentration of 10^5 and 10^6 bacteria/mL in TE buffer. From these mixtures and the clinical samples, DNA is isolated using the method of Smith et al.[107] The whole genome probes are labeled with digoxigenin using "DIG-High Prime" (Roche, Germany).[105] The DNA samples are placed into the slots of a Minislot-30 apparatus (Immunetics, Cambridge, MA), concentrated onto a nylon membrane (Boehringer Mannheim) by vacuum, and fixed by cross-linking using UV light (Stratalinker 1800, Stratagene, La Jolla, CA), followed by baking at 120°C for 20 min.[105] The membrane with fixed DNA is placed in a Miniblotter-45 apparatus (Immunetics) with the lanes of DNA at 90° to the channels of the device. Probes are loaded into the wells, and hybridization takes place at 42°C overnight. After a series of wash steps, bound probes are detected by anti-digoxigenin antibody conjugated with an alkaline phosphatase and a chemifluorescent substrate (Boehringer Mannheim). Signal intensities of the samples are detected with a Lumi-Imager™ workstation (Boehringer Mannheim) or Storm FluorImager (Molecular Dynamics, Sunnyvale, CA), and semiquantitated by comparing to the standards.

4.2.2.2.3 *Reverse-Capture Checkerboard Hybridization Technology*

In a reverse-capture checkerboard hybridization assay, in contrast to the DNA–DNA checkerboard, the probe, rather than the sample, is first fixed to the membrane, hence the name reverse-capture checkerboard.[108] Probes are made with a polythymidine tail, which preferentially cross-links to a membrane, leaving the probe available for hybridization. Up to 30 probes that target regions of the 16S rRNA gene can be deposited on a nylon membrane in separate horizontal lanes.

The method is widely used in oral microbiology. A probe for *Atopobium* genomospecies C1 was included in an assay studying microbial communities associated with caries,[69] whereas a probe targeting *A. rimae* was used in studying periodontitis[81,109] and abscesses.[110]

DNA from caries samples is isolated using a bead beater apparatus (Biospec Products) and purified using a standard Geneclean protocol (Bio 101, Inc.).[69] For periodontal samples, a protocol using proteinase K is followed.[109] The 16S rRNA genes are amplified by PCR using universal primers, with the forward primer digoxigenin labeled at the 5′ end.[69,109] To achieve a better performance of the PCR, particularly for samples with low numbers of bacteria, a two-step hemi-nested approach can be used.[81,110] Here, using universal primers, a nearly full-length 16S rRNA gene fragment is amplified, which is used as template to run two sets of partial 16S rRNA gene amplification, with the forward primers 5′ digoxigenin labeled, resulting in two different fragments, which together encompass the nearly full-length 16S rRNA gene.

The checkerboard assay is performed with the Minislot-30 and Miniblotter-45 system (Immunetics, Cambridge, MA).[81,110] Probes and standards representing 10^5 and 10^6 (and 10^8) cells are brought into the wells and cross-linked to a nylon membrane (Amersham-Pharmacia Biotech) by UV irradiation using a Stratalinker 1800 (Stratagene).[69,81,110] The probe targeting *A. rimae* has the following sequence: 5′-$(T)_{20}$ACGGAAGACGTATTCTCC-3′.[81,110]

The membrane is prehybridized at 55°C for 1 h.[69] Labeled PCR products in 55°C preheated hybridization solution are denatured at 95°C for 5 min and loaded onto the membrane. Hybridization is carried out for 2 h at 55°C[110] or 54°C,[81] and is followed by a washing and blocking step. The membrane is incubated with anti-digoxigenin antibody conjugated with alkaline phosphatase (Roche Molecular Biochemicals) and ultrasensitive chemiluminescent substrate CDP Star is subsequently added (Roche Molecular Biochemicals). Standard chemifluorescence (Storm Technology Inc., Mountain View, CA)[69] or chemifluorescence (X-ray film)[81,110] is used for detection.

4.2.2.2.4 DNA Microarray

DNA microarrays, first described in 1995, can be regarded as a miniaturization of the reverse-capture checkerboard approach.[111] They consist of a high-density matrix of DNA probes, which are printed or synthesized on a glass or silicon slide.[112] Targets typically incorporate a fluorescent label, and when applied to the array, those targets that hybridize to complementary probes are detected.

Broad-range PCR amplicons of clinical DNA samples can be identified after hybridization to a microarray composed of hundreds to thousands of species-specific probes.[113]

Recently, a DNA microarray named the human oral microbe identification microarray (HOMIM) (http://mim.forsyth.org/) has been developed to detect approximately 300 of the most prevalent oral species, including unculturable ones.[114] The HOMIM, containing probes targeting *A. parvulum*, *A. rimae*, *Atopobium* sp. C019, and *Atopobium* sp. clone C3MLM018, is used to find bacterial associations in health and disease, to monitor effects of therapy on the oral ecology, and to perform microbial perturbation studies.[114–117]

For smaller samples (minimal pellet), a Ready-Lyse™ Lysozyme Solution protocol (Epicentre) is recommended for the HOMIM (http://mim.forsyth.org/homim.html). Alternatively, the QIAamp DNA Mini Kit (Qiagen) is used,[115,116] or samples are placed in 50 μL TE buffer, 0.5% Tween 20 and 1 μL Proteinase K, at 55°C for 2 h, followed by 5 min at 95°C.[117] Purification can be done using a MasterPure DNA Purification Kit (Epicentre). For larger samples (visible pellet), an Ultra Clean Microbial DNA Isolation Kit (MO BIO Laboratories, Inc.), using a bead-beating step, is recommended.

The 16S rRNA genes are amplified in two different PCR, respectively, using forward primer 5′-CCAGAGTTTGATYMTGGC-3′ and reverse primer 5′-GAAGGAGGTGWTCCARCCGCA-3′; and forward primer 5′-GACTAGAGTTTGATYMTGGC-3′ and reverse primer 5′-GYTACCTTGTTACGACTT-3′.[115,116]

Alternatively, a 1:1 ratio of the above-mentioned forward primers, together with either reverse primer 5′-GYTACCTTGTTACGACTT-3′ or 5′-GAAGGAGGTGW TCCADCC-3′, can be used.[117]

The two PCR products are combined, purified (QIAquick PCR Purification Kit, Qiagen), labeled via the incorporation of Cy3-deoxycytidine triphosphate (GE Healthcare Biosciences) during another amplification reaction, and purified again.[115–117]

PCR product is mixed with hybridization buffer, heated for 5 min at 100°C, and brought on the microarray. Hybridization is performed at 55°C for 16 h in a microarray incubator (Hybex Microarray Incubation system, SciGene, Sunnyvale, CA).[115,117] After washing, the microarray is scant with a high-resolution laser scanner (Axon 4000B, MDS Analytical Technologies, Sunnyvale, CA).[115–117] Data from individual HOMIM signals are translated to a "bar code" format, normalized, and scored 0–5.

4.2.2.2.5 Oligonucleotide-Coupled Fluorescent Microspheres

A variant of the microarray technology—where hybridization takes place on a two-dimensional surface—is a technology based on optically encoded microparticles. Pioneered by Luminex, these polystyrene microbeads are imbued with two fluorescent dyes and the ratio of these creates 100 unique spectral signatures. Each bead is coated with a different oligonucleotide probe. After PCR amplification of the targets, in which amplicons are labeled, only those beads that have a bound target will be detected by a flow cytometer. A 9-plex Luminex assay was designed for characterizing the vaginal microflora and for exploring molecular diagnostics of bacterial vaginosis.[118]

DNA is extracted using Instagene Matrix (BioRad Laboratories, Hercules, CA). The oligonucleotide capture probes are modified by addition of a 5-amino C12 linker at the 5′ end to facilitate covalent coupling to the carboxylated fluorescent microspheres (Luminex Corporation, Austin, TX).[118] The *A. vaginae* probe has the following sequence: 5′-CACTCTTCATTTCTTCTACCGCAGCATT AACA-3′. The vaginal swab DNA extracts are used as templates for PCR amplification of the 5′ end of the *cpn*60 UT using two universal primers. These primers are modified by the addition of biotin and four phosphorothioated bases at the 5′ end of the primer. The PCR product is subsequently treated with 20 U of T7 exonuclease at room temperature for 40 min, and the single-stranded PCR product is denatured at 95°C for 10 min. Hybridization of the oligonucleotide-coupled beads to the single-stranded PCR products proceeds at 60°C for 10 min, followed by the addition of streptavidin-R-phycoerythrin (Invitrogen). The mixture is analyzed on a Luminex instrument (Luminex, Austin, TX) that reports the mean fluorescence intensity (MFI) of the phycoerythrin reporter dye for each of the coupled bead types.

4.2.2.2.6 PCR Using Species-Specific Primers

Conventional PCR. In conventional PCR, detection and semiquantification of the amplified sequence are performed at the end of the reaction after the last PCR cycle and involves post-PCR analysis such as size fractionation of amplicons by gel electrophoresis and image analysis.

The technique is widely used for detection of *A. vaginae* or *Atopobium* spp. in studies characterizing the normal and disturbed (neo)vaginal microflora,[13,14,34,37,119,120] in studying the etiology, pathogenesis, and treatment of bacterial vaginosis,[121–124] as well as in exploring PCR as a molecular diagnostic tool for bacterial vaginosis.[37]

Quantitative PCR. Quantitative PCR (qPCR), also known as real-time PCR, is in essence the detection and quantification of a fluorescent signal directly proportional to the amount of amplicons generated by PCR. Detection is achieved by coupling a thermocycler with a fluorimeter, which measures the fluorescent signal after each amplification cycle.[125] This ability to monitor the amplification reaction during its exponential phase enables users to determine the initial amount of target with great precision by comparison with the signal strength obtained with a standard dilution curve, containing well-determined amounts of target DNA. The more nucleic acid templates present at the start of the reaction, the fewer amplification cycles are required to make sufficient product that can be detected. The cycle in which a significant increase in fluorescence above the threshold is measured—referred to as the "cycle of quantification" (C_q) value—can therefore be used to calculate the quantity of DNA in the sample.

Fluorescent markers can be incorporated in the amplicons by means of a intercalating fluorescent dye (SYBR® Green), dual-labeled hydrolysis probes (TaqMan®), hybridization probes (Light-Cycler), molecular beacons or by scorpion probes.

SybrGreen I binds in the minor groove of the DNA double helix, and fluoresces at a wavelength of 530 nm, while the fluorescence of the dye in solution is negligible.[126] The main drawback of these markers is their lack of specificity, because they will bind to any double-stranded DNA sequence, nonspecific amplicons and primer dimers included. The melting temperature T_m of the amplified products can be checked using DNA melting curves at the end of the PCR to guarantee the specificity of the signal.

Hydrolysis probes carry a fluorophore at the 5′ end and a quencher at the 3′ end.[127] The quencher can absorb the fluorescence emitted by the fluorophore and dissipate it in the form of heat. During extension, the exonuclease activity of the DNA polymerase hydrolyses the probe and releases the fluorophore at the 5′ end. The latter, now free from its quencher, will emit a characteristic fluorescence signal that can be measured at the end of each extension period.

Molecular beacons are probes with hairpin structure that, like hydrolysis probes, carry a fluorophore at the 5′ end and a quencher at the 3′ end.[128] In the nonhybridized state, the probe is folded in a hairpin structure owing to the partial internal complementarity of the probe.[129] In this configuration, the fluorophore and quencher are close to one another and no fluorescence is detected. During the annealing stage, the parts of the probe that form the loop anneal onto the target sequence, moving the fluorophore and its quencher apart, and the fluorescence can be detected.[130]

Species-specific quantitative PCR (qPCR) amplifies the same diagnostic amplicons as in conventional PCR, but in contrast to conventional PCR, amplification is followed during each replication cycle. qPCR has been widely used for the detection of *A. vaginae* or *Atopobium* spp. in studies characterizing the normal and disturbed vaginal microflora,[39,131,132] and in studying the etiology, pathogenesis, sequelae, treatment, and diagnosis of bacterial vaginosis.[24,32,130,131,133–135] qPCR is also used for detection of the *Atopobium* cluster in studies characterizing the microflora in human feces.[94,136]

Sample Preparation and Genomic DNA Extraction. Fecal samples (20 mg) are washed three times by suspending them in 1 mL PBS and centrifuging each preparation at 14,000 × *g* to remove possible PCR inhibitors.[94] Vaginal swabs samples are typically agitated or vortexed in 1 mL PBS[14,118] or 2 mL saline[37,121] to dislodge and wash the cells, followed by pelleting and resuspending the cells. Alternatively, swabs are vortexed in 400 μL lysis buffer.[13,39,120,122]

For DNA extractions, commercial kits or (semi) automated DNA extraction platforms are most often used, these include the DNeasy Tissue Kit (Qiagen),[132] AquaPure genomic DNA kit (Bio-Rad),[123] Wizard DNA purification kit (Promega),[137,138] QIAamp DNA Mini Kit (Qiagen)[38,135] with minor modifications involving mutanolysin pretreatment,[13,39] UltraClean Soil DNA Isolation kit,[37,121,133,134] QIAamp stool kit,[33] Instagene matrix (Bio-Rad Laboratories, Hercules, CA),[14,118] Automated

MagNA Pure LC with the DNA Isolation Kit I protocol (Roche Diagnostics),[32,131] MasterPure DNA purification kit (Epicentre),[124] the automated X-tractor Gene system (Corbett Robotics, Brisbane, Australia),[130] or the NucliSens EasyMAG platform (bioMérieux).[120] A DNA extraction method for fecal samples is described by Matsuki et al.[94]

Primers and Controls. The primers and probes applicable for amplification and detection of *Atopobium* spp. are listed in Table 4.1. Composition of the PCR mixtures and cycling conditions can be found in the accompanying references. The β-globine gene,[32,37,121,131,133] a human albumin gene,[24,135] or the human 18S rRNA gene[134] can be used as an internal control to assure that amplifiable DNA was successfully extracted and to monitor for PCR inhibitors. Alternatively, amplification of exogenous DNA from a jellyfish gene can be used to test for PCR inhibition, defined as a > two-cycle increase in the C_q for the extracted vaginal DNA sample compared to the no-template controls.[133,134]

Detection. After separation on 2% agarose gels, conventional PCR products are typically visualized by staining 5 μL of amplified product with ethidium bromide and placing the gel under UV light.[13,14,37,119–122,124] Specificity is confirmed by detection of the appropriate sized PCR product,[124] which can be sequenced as an additional control.[14,33,37]

For detection in qPCR assays, the LightCycler (Roche),[132] Stratagene MX 3000P,[24,135] RotorGene 3000,[130] and ABI platforms (Applied Biosystems)[39,94,118,133,134,136] have been used. In assays using SybrGreen I, analysis of the melting curve can be carried out after amplification to distinguish the targeted PCR product from aspecific nontargeted PCR products.

For quantification, a standard curve is made. DNA, extracted from *A. vaginae* cultured on TSA + 5% sheep blood (Becton Dickinson) and diluted in a 1/10 series, can be used as standards.[39,132] Alternatively, a serial diluted suspension of cloned 16S rRNA[24,133,135] or cpn60[118] plasmids can be used.

4.2.2.2.7 Sequence-Based Identification

In cases were *A. rimae, A. vaginae*, or *Atopobium* sp. oral clone C019 isolates could not be identified or were misidentified using conventional techniques, the 16S rRNA gene was (partially) sequenced to reach identification.[8,13,56–58]

TABLE 4.1
Primers and Probes Used for Amplification and Detection of *Atopobium* spp.

Primer and Probe Sequences (5′–3′)	Target Sequence	Target Species	Reference
GGTGAAGCAGTGGAAACACT ATTCGCTTCTGCTCGCGCA	16S rRNA gene	*A. vaginae*	39
CCCTATCCGCTCCTGATACC CCAAATATCTGCGCATTTCA VIC-GCAGGCTTGAGTCTGGTAGGGGA-TAMRA	16S rRNA gene	*A. vaginae*	24,135
GGTCGGTCTCTCAACCCGG TCATGGCCCAGAAGACCGCC FAM-CAGATTTAACTCCTGACCTAACAGACC-TAMRA	16S rRNA gene	*A. vaginae*	32,131
GCGAATATGGGAAAGCTCCG GAGCGGATAGGGGTTGAGC	16S rRNA gene	*A. vaginae*	120,122
TAGGTCAGGAGTTAAATCTG TCATGGCCCAGAAGACCGCC	16S rRNA gene	*A. vaginae*	123,124
CTTTGCAACCAACAATGACACC CAGCGCATACGGTAAGCGTAC	*Cpn60*	*A. vaginae*	118
CAGCGTTCCTGTTACTCC GCCACTTGAAGGTCTTGC cgcgatcATTCCACGGTTGCACATAACATGCgatcgcg[a]	16S rRNA gene	*A. vaginae*	130
GGTCGGTCTCTCAACCC CTCCTGACCTAACAGACC	16S rRNA gene	*A. vaginae*	14
GCAGGGACGAGGCCGCAA GTGTTTCCACTGCTTCACCTAA	16S rRNA gene	*Atopobium* spp.	37,121
TAGGCGGTYTGTTAGGTCAGGA CCTACCAGACTCAAGCCTGC *FAM-CTCAACCCCTATCCGCTCCTGAT-TAMRA*	16S rRNA gene	*Atopobium* spp.[b]	133,134
GGGTTGAGAGACCGACC CGGRGCTTCTTCTGCAGG	16S rRNA gene	*Atopobium* cluster[c]	94,132

[a] Uppercase letters indicate homology to the target sequence.
[b] *Atopobium* spp.: *Atopobium* sp. AY738657, AY738658.
[c] *Atopobium* cluster: Members of the genera *Atopobium*, *Collinsella*, *Eggerthella*, and *Coriobacterium*.

TABLE 4.2
Primers Used for Amplification of the 16S rRNA Gene

Primer	Sequence 5′–3′	Amplicon Size	Reference
fD1	AGAGTTTGATCCTGGCTCAG	1454 bp	57
rP2	ACGGCTACCTTGTTACGACTT		
8FPL	AGTTTGATCCTGGCTCAG	798 bp	56
806R	GGACTACCAGGGTATCTAAT		
10f	AGTTTGATCCTGGCTCAG	524 bp	13
534r	ATTACCGCGGCTGCTGG		
515FPL	TGCCAGCAGCCGCGGTAA	785 bp	56
13B	AGGCCCGGGAACGTATTCAC		

Anaerobic blood cultures or swabs can be (sub)cultured on a rich blood agar (e.g., Columbia agar with 5% horse blood[3]) and incubated anaerobically for 2–5 days at 37°C (see growth requirements). For DNA extraction, commercial kits such as the QIAamp tissue kit (QIAGEN)[57] or GenElute Bacterial Genomic DNA kit[56] can be used, preferentially after pretreatment with mutanolysin. Alternatively, DNA is extracted from a colony by simple alkaline extraction.[139] For sequencing purposes, 16S rRNA PCR amplification can be performed using 16S rRNA universal primers, as listed in Table 4.2.

To obtain pure amplicons, excess primers and nucleotides are removed using a commercial kit (QIAquick PCR Purification Kit, Qiagen).[13] Thereafter, the 16S rRNA gene amplicons are sequenced, either partially[58] or fully.[8,57,58] Commercial packages such as the MicroSeq® Full Gene or 500 16S rRNA Bacterial Identification Kit (PE Applied Biosystems) can be used.[58] For identification, the sequence is compared to a public database (e.g., Genbank) or a curated database (e.g., SmartGene IDNS Bacteria) by using analysis software (BLAST version 2.2.9[57] or SmartGene [www.idns-smartgene.com]).

4.3 FUTURE PERSPECTIVES

Highly sensitive high throughput analysis through deep pyrosequencing will further enlighten our knowledge of bacterial diversity and ecological dynamics associated with oral, reproductive tract, and intestinal tract infection(s). Knowledge obtained by this metagenomic approach might then be translated into rapid diagnostic assays for profiling a selected set of microbial communities in a lab-based setting. In case of bacterial vaginosis, distinct vaginal microflora patterns may be found in both intermediate and bacterial vaginosis microflora, thereby possibly differentiating intermediate microflora regressing to normal or developing to bacterial vaginosis and differentiating bacterial vaginosis microflora with different pathologic sequelae.

Furthermore, molecular diagnosis of bacterial vaginosis, with a central role for *A. vaginae*, is now pending.[37,118,133,135] Species-specific quantitative PCR techniques are being patented and will become commercially available as diagnostic assays. Also, ongoing sequence analysis of the whole genome of *A. vaginae* within the *A. vaginae* Genome Project (http://med.stanford.edu/sgtc/research/atopobium_vaginae.html) will probably reveal as yet undescribed virulence factors and will also improve our understanding of the pathogenic role of *A. vaginae* in reproductive tract infections.

To explore the role of *A. parvulum* and *A. rimae* in oral health and disease, microarrays like the TNO oral microarray (O-chip, TNO, Zeist, The Netherlands) or the HOMIM (http://mim.forsyth.org/homim.html) offer the possibility to detect about 300 oral species. In cases of bacteremia, even if *Atopobium* spp. are included in future commercial biochemical tests, sequencing the 16S rRNA gene or the cpn60 gene will remain the golden standard to identify these species, as they are generally slow growing and biochemically nonreactive.[12]

REFERENCES

1. Collins, M.D., and Wallbanks, S., Comparative sequence analyses of the 16S rRNA genes of *Lactobacillus minutus, Lactobacillus rimae* and *Streptococcus parvulus*: Proposal for the creation of a new genus *Atopobium*, *FEMS Microbiol. Lett.*, 74, 235, 1992.
2. Bailey, G.D., and Love, D.N., *Eubacterium-Fossor* sp. nov, an agar-corroding organism for normal pharynx and oral and respiratory-tract lesions of horses, *Int. J. Syst. Bacteriol.*, 36, 383, 1986.
3. Rodriguez Jovita, M. et al., Characterization of a novel *Atopobium* isolate from the human vagina: Description of *Atopobium vaginae* sp. nov., *Int. J. Syst. Bacteriol.*, 49, 1573, 1999.
4. Kageyama, A., Benno, Y., and Nakase, T., Phylogenic and phenotypic evidence for the transfer of *Eubacterium fossor* to the genus *Atopobium* as *Atopobium fossor* comb. nov., *Microbiol. Immunol.* 43, 389, 1999.
5. Rainey, F.A., Weiss, N., and Stackebrandt, E., *Coriobacterium* and *Atopobium* are phylogenetic neighbors within the *Actinomycetes* line of descent, *Syst. Appl. Microbiol.*, 17, 202, 1994.
6. Stackebrandt, E., Rainey, F.A., and WardRainey, N.L., Proposal for a new hierarchic classification system, *Actinobacteria* classis nov., *Int. J. Syst. Bacteriol.*, 47, 479, 1997.
7. Olsen, I.E. et al., *Lactobacillus uli* sp. nov. and *Lactobacillus rimae* sp. nov. from the human gingival crevice and emended descriptions of *Lactobacillus minutus* and *Streptococcus parvulus*, *Int. J. Syst. Bacteriol.*, 41, 261, 1991.

8. Geissdörfer, W. et al., Tuboovarian abscess caused by *Atopobium vaginae* following transvaginal oocyte recovery, *J. Clin. Microbiol.*, 41, 2788, 2003.
9. Ferris, M.J. et al., Association of *Atopobium vaginae*, a recently described metronidazole resistant anaerobe, with bacterial vaginosis, *BMC Infect. Dis.*, 4, 5, 2004.
10. Cato, E.P. et al., Transfer of *Peptostreptococcus parvulus* (Weinberg, Nativelle, and Prevot 1937) Smith 1957 to the gGenus *Streptococcus* as *Streptococcus parvulus* (Weinberg, Nativelle, and Prevot 1937) comb. nov., nom. rev., emend., *Int. J. Syst. Bacteriol.*, 33, 82, 1983.
11. De Backer, E. et al., Antibiotic susceptibility of *Atopobium vaginae*, *BMC Infect. Dis.*, 6, 51, 2006.
12. Downes, J. et al., Characterisation of *Eubacterium*-like strains isolated from oral infections, *J. Med. Microbiol.*, 50, 947, 2001.
13. Verhelst, R. et al., Cloning of 16S rRNA genes amplified from normal and disturbed vaginal microflora suggests a strong association between *Atopobium vaginae*, *Gardnerella vaginalis* and bacterial vaginosis, *BMC Microbiol.*, 4, 16, 2004.
14. Burton, J.P. et al., Detection of *Atopobium vaginae* in post-menopausal women by cultivation-independent methods warrants further investigation, *J. Clin. Microbiol.*, 42, 1829, 2004.
15. Swidsinski, A. et al., An adherent *Gardnerella vaginalis* biofilm persists on the vaginal epithelium after standard therapy with oral metronidazole, *Am. J. Obstet. Gynecol.*, 198, 97e1, 2008.
16. Sobel, J. D. et al., Bacterial vaginosis, *Annu. Rev. Med.*, 51, 349, 2000.
17. Hillier, S.L. et al., The role of bacterial vaginosis and vaginal bacteria in amniotic fluid infection in women in preterm labor with intact fetal membranes, *Clin. Infect. Dis.*, 20, S276, 1995.
18. Hillier, S.L. et al., Association between bacterial vaginosis and preterm delivery of a low-birth-weight infant. The Vaginal Infections and Prematurity Study Group, *N. Engl. J. Med.*, 333, 1737, 1995.
19. Hillier, S.L. et al., Role of bacterial vaginosis-associated microorganisms in endometritis, *Am. J. Obstet. Gynecol.*, 175, 435, 1996.
20. Hillier, S.L. et al., A case-control study of chorioamnionic infection and histologic chorioamnionitis in prematurity, *N. Engl. J. Med.*, 319, 972, 1988.
21. Leitich, H. et al., Bacterial vaginosis as a risk factor for preterm delivery: A meta-analysis, *Am. J. Obstet. Gynecol.*, 189, 139, 2003.
22. Sewankambo, N. et al., HIV-1 infection associated with abnormal vaginal flora morphology and bacterial vaginosis, *Lancet*, 350, 546, 1997.
23. Krohn, M.A. et al., The severity of fetal membrane infection and pregnancy complications, *Ann. Epidemiol.*, 3, 78, 1993.
24. Menard, J.P. et al., High vaginal concentrations of *Atopobium vaginae* and *Gardnerella vaginalis* in women undergoing preterm labor, *Obstet. Gynecol.*, 115, 134, 2010.
25. Wiesenfeld, H.C. et al., Bacterial vaginosis is a strong predictor of *Neisseria gonorrhoeae* and *Chlamydia trachomatis* infection, *Clin. Infect. Dis.*, 36, 663, 2003.
26. Cherpes, T.L. et al., Association between acquisition of herpes simplex virus type 2 in women and bacterial vaginosis, *Clin. Infect. Dis.*, 37, 319, 2003.
27. Atashili, J. et al., Bacterial vaginosis and HIV acquisition: A meta-analysis of published studies, *AIDS*, 22, 1493, 2008.
28. Sha, B.E. et al., Female genital-tract HIV load correlates inversely with *Lactobacillus* species but positively with bacterial vaginosis and *Mycoplasma hominis*, *J. Infect. Dis.*, 191, 25, 2005.
29. Cherpes, T.L. et al., Genital tract shedding of herpes simplex virus type 2 in women: Effects of hormonal contraception, bacterial vaginosis, and vaginal group B *Streptococcus* colonization, *Clin. Infect. Dis.*, 40, 1422, 2005.
30. Josey, W.E., and Schwebke, J.R., The polymicrobial hypothesis of bacterial vaginosis causation: A reassessment, *Int. J. STD AIDS*, 19, 152, 2008.
31. Hay, P. Recurrent bacterial vaginosis, *Curr. Opin. Infect. Dis.*, 22, 82, 2009.
32. Bradshaw, C.S. et al., The association of *Atopobium vaginae* and *Gardnerella vaginalis* with bacterial vaginosis and recurrence after oral metronidazole therapy, *J. Infect. Dis.*, 194, 828, 2006.
33. Fredricks, D.N., Fiedler, T.L., and Marrazzo, J.M., Molecular identification of bacteria associated with bacterial vaginosis, *N. Engl. J. Med.*, 353, 1899, 2005.
34. Fredricks, D.N., and Marrazzo, J.M., Molecular methodology in determining vaginal flora in health and disease: Its time has come, *Curr. Infect. Dis. Rep.*, 7, 463, 2005.
35. Hillier, S.L. et al., The normal vaginal flora, H2O2-producing lactobacilli, and bacterial vaginosis in pregnant women, *Clin. Infect. Dis.*, 16, S273, 1993.
36. Rosenstein, I.J. et al., Effect of intravaginal clindamycin cream on pregnancy outcome and on abnormal vaginal microbial flora of pregnant women, *Infect. Dis. Obstet. Gynecol.*, 8, 158, 2000.
37. Fredricks, D.N., Targeted PCR for detection of vaginal bacteria associated with bacterial vaginosis, *J. Clin. Microbiol.*, 45, 3270, 2007.
38. Thies, F.L., Konig, W., and Konig, B., Rapid characterization of the normal and disturbed vaginal microbiota by application of 16S rRNA gene terminal RFLP fingerprinting, *J. Med. Microbiol.*, 56, 755, 2007.
39. De Backer, E. et al., Quantitative determination by real-time PCR of four vaginal *Lactobacillus* species, *Gardnerella vaginalis* and *Atopobium vaginae* indicates an inverse relationship between *L. gasseri* and *L. iners*, *BMC Microbiol.*, 7, 115, 2007.
40. Verstraelen, H.M. et al., Culture-independent analysis of vaginal microflora: The unrecognized association of *Atopobium vaginae* with bacterial vaginosis, *Am. J. Obstet. Gynecol.*, 191, 1130, 2004.
41. Gravett, M.G. et al., Preterm labor associated with subclinical amniotic fluid infection and with bacterial vaginosis, *Obstet. Gynecol.*, 67, 229, 1986.
42. Watts, D.H. et al., The association of occult amniotic fluid infection with gestational age and neonatal outcome among women in preterm labor, *Obstet. Gynecol.*, 79, 351, 1992.
43. Hebb, J.K. et al., Detection of novel organisms associated with salpingitis, by use of 16S rDNA polymerase chain reaction, *J. Infect. Dis.*, 190, 2109, 2004.
44. Ferris, M.J. et al., Cultivation-independent analysis of changes in bacterial vaginosis flora following metronidazole treatment, *J. Clin. Microbiol.*, 45, 1016, 2007.
45. Moreau-Marquis, S., Stanton, B.A., and O'Toole, G.A., *Pseudomonas aeruginosa* biofilm formation in the cystic fibrosis airway, *Pulm. Pharmacol., Ther.*, 21, 595, 2008.
46. Prince, A.A. et al., Prevalence of biofilm-forming bacteria in chronic rhinosinusitis, *Am. J. Rhinol.*, 22, 239, 2008.

47. Davies, D. et al., Understanding biofilm resistance to antibacterial agents, *Nat. Rev. Drug Discov.*, 2, 114, 2003.
48. Swidsinski, A. et al., Adherent biofilms in bacterial vaginosis, *Obstet. Gynecol.*, 106, 1013, 2005.
49. Costerton, J.W., Stewart, P.S., and Greenberg, E.P., Bacterial biofilms: A common cause of persistent infections, *Science*, 284, 1318, 1999.
50. Hale, L.P., Swidsinski, A., and Mendling, W., Bacteria associated with bacterial vaginosis, *N. Engl. J. Med.*, 354, 202, 2006.
51. Libby, E.K. et al., *Atopobium vaginae* triggers an innate immune response in an in vitro model of bacterial vaginosis, *Microbes Infect.*, 10, 439, 2008.
52. Barrett, S., and Taylor, C., A review on pelvic inflammatory disease, *Int. J. STD AIDS* 16, 715; quiz 721, 2005.
53. Bebear, C., and de Barbeyrac, B., Genital *Chlamydia trachomatis* infections, *Clin. Microbiol. Infect.*, 15, 4, 2009.
54. Reyes, P.A. et al., Apoptosis related genes expressed in cultured Fallopian tube epithelial cells infected in vitro with *Neisseria gonorrhoeae, Biol. Res.*, 40, 319, 2007.
55. Soper, D.E. et al., Pelvic inflammatory disease, *Infect. Dis. Clin. North Am.*, 8, 821, 1994.
55a. Yamagishi, Y. et al., A case of uterine endometritis caused by *Atopobium vaginae, J. Infect. Chemother.*, 2010 Aug 14 [Epub ahead of print].
56. Chung, H.Y. et al., A case of bacteremia by *Atopobium rimae* in a patient with liver cirrhosis, *Korean J. Lab. Med.*, 27, 351, 2007.
57. Angelakis, E.M. et al., Human case of *Atopobium rimae* bacteremia, *Emerg. Infect. Dis.*, 15, 354, 2009.
58. Salimnia, H. et al., Sepsis associated with a new *Atopobium* species, provisionally named *Atopobium detroiti*: Case report and review of the current status of the species *Atopobium*, *Scand. J. Infect. Dis.*, 40, 679, 2008.
59. Dewhirst, F.E. et al., Characterization of novel human oral isolates and cloned 16S rDNA sequences that fall in the family *Coriobacteriaceae*: Description of *Olsenella* gen. nov., reclassification of *Lactobacillus uli* as *Olsenella uli* comb. nov. and description of *Olsenella profusa* sp. nov., *Int. J. Syst. Evol. Microbiol.*, 51, 1797, 2001.
60. Hughes, F.J., and McNab, R., Oral malodour—a review, *Arch. Oral Biol.*, 53, S1, 2008.
61. Kazor, C.E. et al., Diversity of bacterial populations on the tongue dorsa of patients with halitosis and healthy patients, *J. Clin. Microbiol.*, 41, 558, 2003.
62. Persson, S.J. et al., The formation of hydrogen sulfide and methyl mercaptan by oral bacteria, *Oral Microbiol. Immunol.*, 5, 195, 1990.
63. Riggio, M.P. et al., Molecular identification of bacteria on the tongue dorsum of subjects with and without halitosis, *Oral Dis.*, 14, 251, 2008.
64. Haraszthy, V.I. et al., Identification of oral bacterial species associated with halitosis, *J. Am. Dent. Assoc.*, 138, 1113, 2007.
65. Wolff, M.S., and Larson, C., The cariogenic dental biofilm: Good, bad or just something to control? *Braz. Oral Res.*, 23, S31, 2009.
66. Loesche, W.J. et al., The specific plaque hypothesis and the antimicrobial treatment of periodontal disease, *Dent. Update*, 19, 70, 1992.
67. Theilade, E. et al., The non-specific theory in microbial etiology of inflammatory periodontal diseases, *J. Clin. Periodontol.*, 13, 905, 1986.
68. Marsh, P.D. et al., Microbial ecology of dental plaque and its significance in health and disease, *Adv. Dent. Res.*, 8, 263, 1994.
69. Aas, J.A. et al., Bacteria of dental caries in primary and permanent teeth in children and young adults, *J. Clin. Microbiol.*, 46, 1407, 2008.
70. Moore, W.E. et al., Bacteriology of experimental gingivitis in young adult humans, *Infect. Immun.*, 38, 651, 1982.
71. Moore, W.E. et al., Bacteriology of experimental gingivitis in children, *Infect. Immun.*, 46, 1, 1984.
72. Moore, L.V. et al., Bacteriology of human gingivitis, *J. Dent. Res.*, 66, 989, 1987.
73. Socransky, S.S., and Haffajee, A.D., Microbial mechanisms in the pathogenesis of destructive periodontal diseases: A critical assessment, *J. Periodont. Res.*, 26, 195, 1991.
73a. Svensäter, G. et al., Biofilms in endodontic infections, *Endod. Topics*, 9, 27, 2004.
74. Ørstavik, D., and Pitt Ford, T.R., *Essential Endodontology: Prevention and Treatment of Apical Periodontitis*, 2nd edn, Blackwell Munksgaard, 2008.
75. Moore, W.E. et al., Bacteriology of severe periodontitis in young adult humans, *Infect. Immun.*, 38, 1137, 1982.
76. Moore, W.E. et al., Comparative bacteriology of juvenile periodontitis, *Infect. Immun.*, 48, 507, 1985.
77. Kumar, P.S. et al., Identification of candidate periodontal pathogens and beneficial species by quantitative 16S clonal analysis, *J. Clin. Microbiol.*, 43, 3944, 2005.
78. Subramanian, K., and Mickel, A.K., Molecular analysis of persistent periradicular lesions and root ends reveals a diverse microbial profile, *J. Endod.*, 35, 950, 2009.
79. Munson, M.A. et al., Molecular and cultural analysis of the microflora associated with endodontic infections, *J. Dent. Res.*, 81, 761, 2002.
80. Siqueira, J.F. et al., Novel bacterial phylotypes in endodontic infections, *J. Dent. Res.*, 84, 565, 2005.
81. Siqueira, J.F. et al., Bacteria in the apical root canal of teeth with primary apical periodontitis, *Oral Surg. Oral Med. Oral Pathol. Oral Radiol. Endod.*, 107, 721, 2009.
82. Kato, N., and Kato, H., Molecular detection and identification of anaerobic bacteria, *J. Infect. Chemother.*, 3, 5, 1997.
83. Olsen, G.J., and Woese, C.R., Ribosomal RNA: A key to phylogeny, *FASEB J.*, 7, 113, 1993.
84. de Boer, R. et al., Improved detection of microbial DNA after bead-beating before DNA isolation, *J. Microbiol. Methods*, 80, 209, 2010.
85. Fliss, I. et al., A rapid and efficient method of lysis of *Listeria* and other gram-positive bacteria using mutanolysin, *Biotechniques*, 11, 456, 1991.
86. Moter, A., and Gobel, U.B., Fluorescence in situ hybridization (FISH) for direct visualization of microorganisms, *J. Microbiol. Methods*, 41, 85, 2000.
87. Glockner, F.O., Fuchs, B.M., and Amann, R., Bacterioplankton compositions of lakes and oceans: A first comparison based on fluorescence in situ hybridization, *Appl. Environ. Microbiol.*, 65, 3721, 1999.
88. Amann, R.I., Krumholz, L., and Stahl, D.A., Fluorescent-oligonucleotide probing of whole cells for determinative, phylogenetic, and environmental studies in microbiology, *J. Bacteriol.*, 172, 762, 1990.
89. Amann, R., Fuchs, B.M., and Behrens, S., The identification of microorganisms by fluorescence in situ hybridisation, *Curr. Opin. Biotechnol.*, 12, 231, 2001.
90. Jain, K.K. et al., Current status of fluorescent in-situ hybridisation, *Med. Device Technol.*, 15, 14, 2004.
91. Levsky, J.M., and Singer, R.H., Fluorescence in situ hybridization: Past, present and future, *J. Cell Sci.*, 116, 2833, 2003.

92. Harmsen, H.J. et al., Development of 16S rRNA-based probes for the *Coriobacterium* group and the *Atopobium* cluster and their application for enumeration of *Coriobacteriaceae* in human feces from volunteers of different age groups, *Appl. Environ. Microbiol.*, 66, 4523, 2000.
93. Rigottier-Gois, L. et al., Fluorescent hybridisation combined with flow cytometry and hybridisation of total RNA to analyse the composition of microbial communities in human faeces using 16S rRNA probes, *FEMS Microbiol. Ecol.*, 43, 237, 2003.
94. Matsuki, T. et al., Use of 16S rRNA gene-targeted group-specific primers for real-time PCR analysis of predominant bacteria in human feces, *Appl. Environ. Microbiol.*, 70, 7220, 2004.
95. Mah, K.W. et al., Effect of a milk formula containing probiotics on the fecal microbiota of asian infants at risk of atopic diseases, *Pediatr. Res.*, 62, 674, 2007.
96. Kleessen, B. et al., Microbial and immunological responses relative to high-altitude exposure in mountaineers, *Med. Sci. Sports Exerc.*, 37, 1313, 2005.
97. Socransky, S.S. et al., "Checkerboard" DNA–DNA hybridization, *Biotechniques*, 17, 788, 1994.
98. Nascimento, C. et al., DNA checkerboard method for bacterial pathogen identification in oral diseases, *Int. J. Morphol.*, 24, 619, 2006.
99. Ximenez-Fyvie, L.A., Haffajee, A.D., and Socransky, S.S., Microbial composition of supra- and subgingival plaque in subjects with adult periodontitis, *J. Clin. Periodontol.*, 27, 722, 2000.
100. Siqueira, J.F. et al., Endodontic infections: Concepts, paradigms, and perspectives, *Oral Surg. Oral Med. Oral Pathol. Oral Radiol. Endod.*, 94, 281, 2002.
101. Socransky, S.S. et al., Microbial complexes in subgingival plaque, *J. Clin. Periodontol.*, 25, 134, 1998.
102. Sakamoto, M., Umeda, M., and Benno, Y., Molecular analysis of human oral microbiota, *J. Periodont. Res.*, 40, 277, 2005.
103. Colombo, A.P. et al., Clinical and microbiological features of refractory periodontitis subjects, *J. Clin. Periodontol.*, 25, 169, 1998.
104. Haffajee, A.D. et al., Subgingival microbiota in healthy, well-maintained elder and periodontitis subjects, *J. Clin. Periodontol.*, 25, 346, 1998.
105. Nikolaitchouk, N. et al., The lower genital tract microbiota in relation to cytokine-, SLPI- and endotoxin levels: Application of checkerboard DNA–DNA hybridization (CDH), *APMIS*, 116, 263, 2008.
106. Brito, L.C. et al., Use of multiple-displacement amplification and checkerboard DNA–DNA hybridization to examine the microbiota of endodontic infections, *J. Clin. Microbiol.*, 45, 3039, 2007.
107. Smith, G.L., Socransky, S.S., and Smith, C.M., Rapid method for the purification of DNA from subgingival microorganisms, *Oral Microbiol. Immunol.*, 4, 47, 1989.
108. Paster, B.J., Bartoszyk, I.M., and Dewhirst, F., Identification of oral streptococci using PCR-based, reverse-capture, checkerboard hybridization, *Methods Cell Sci.*, 20, 223, 1998.
109. Paster, B.J. et al., Bacterial diversity in necrotizing ulcerative periodontitis in HIV-positive subjects, *Ann. Periodontol.*, 7, 8, 2002.
110. Siqueira, J.F., Jr., and Rocas, I.N., The microbiota of acute apical abscesses, *J. Dent. Res.*, 88, 61, 2009.
111. Schena, M. et al., Parallel human genome analysis: Microarray-based expression monitoring of 1000 genes, *Proc. Natl. Acad. Sci. USA*, 93, 10614, 1996.
112. Mothershed, E.A., and Whitney, A.M., Nucleic acid-based methods for the detection of bacterial pathogens: Present and future considerations for the clinical laboratory, *Clin. Chim. Acta* 363, 206, 2006.
113. Palmer, C. et al., Rapid quantitative profiling of complex microbial populations, *Nucleic Acids Res.*, 34, 2006.
114. Olsen, I. et al., Cultivated and not-yet-cultivated bacteria in oral biofilms, *Microb. Ecol. Health Dis.*, 65, 2009.
115. Preza, D. et al., Microarray analysis of the microflora of root caries in elderly, *Eur. J. Clin. Microbiol. Infect. Dis.*, 28, 509, 2009.
116. Preza, D. et al., Diversity and site-specificity of the oral microflora in the elderly, *Eur. J. Clin. Microbiol. Infect. Dis.*, 28, 1033, 2009.
117. Colombo, A.P. et al., Comparisons of subgingival microbial profiles of refractory periodontitis, severe periodontitis, and periodontal health using the human oral microbe identification microarray, *J. Periodontol.*, 80, 1421, 2009.
118. Dumonceaux, T.J. et al., Multiplex detection of bacteria associated with normal microbiota and with bacterial vaginosis in vaginal swabs by use of oligonucleotide-coupled fluorescent microspheres, *J. Clin. Microbiol.*, 47, 4067, 2009.
119. Burton, J.P. et al., A preliminary survey of *Atopobium vaginae* in women attending the Dunedin gynaecology out-patients clinic: Is the contribution of the hard-to-culture microbiota overlooked in gynaecological disorders? *Aust. N. Z. J. Obstet. Gynaecol.*, 45, 450, 2005.
120. Weyers, S. et al., Microflora of the penile skin-lined neovagina of transsexual women, *BMC Microbiol.*, 9, 102, 2009.
121. Marrazzo, J.M. et al., Relationship of specific vaginal bacteria and bacterial vaginosis treatment failure in women who have sex with women, *Ann. Intern. Med.*, 149, 20, 2008.
122. Verstraelen, H. et al., Gene polymorphisms of Toll-like and related recognition receptors in relation to the vaginal carriage of *Gardnerella vaginalis* and *Atopobium vaginae*, *J. Reprod. Immunol.*, 79, 163, 2009.
123. Ferris, M.J., Masztal, A., and Martin, D.H., Use of species-directed 16S rRNA gene PCR primers for detection of *Atopobium vaginae* in patients with bacterial vaginosis, *J. Clin. Microbiol.*, 42, 5892, 2004.
124. Haggerty, C.L. et al., Clinical characteristics of bacterial vaginosis among women testing positive for fastidious bacteria, *Sex. Transm. Infect.*, 85, 242, 2009.
125. Higuchi, R. et al., Kinetic PCR analysis: Real-time monitoring of DNA amplification reactions, *Biotechnology (N Y)*, 11, 1026, 1993.
126. Arya, M. et al., Basic principles of real-time quantitative PCR, *Expert. Rev. Mol. Diagn.*, 5, 209, 2005.
127. Koch, W.H. et al., Technology platforms for pharmacogenomic diagnostic assays, *Nat. Rev. Drug. Discov.*, 3, 749, 2004.
128. Kim, Y., Sohn, D., and Tan, W., Molecular beacons in biomedical detection and clinical diagnosis, *Int. J. Clin. Exp. Pathol.*, 1, 105, 2008.
129. Abravaya, K. et al., Molecular beacons as diagnostic tools: Technology and applications, *Clin. Chem. Lab. Med.*, 41, 468, 2003.
130. Trama, J.P. et al., Rapid detection of *Atopobium vaginae* and association with organisms implicated in bacterial vaginosis, *Mol. Cell. Probes*, 22, 96, 2008.
131. Tabrizi, S.N. et al., Prevalence of *Gardnerella vaginalis* and *Atopobium vaginae* in virginal women, *Sex. Transm. Dis.*, 33, 663, 2006.

132. Biagi, E. et al., Quantitative variations in the vaginal bacterial population associated with asymptomatic infections: A real-time polymerase chain reaction study, *Eur. J. Clin. Microbiol. Infect. Dis.,* 28, 281, 2009.
133. Fredricks, D.N. et al., Changes in vaginal bacterial concentrations with intravaginal metronidazole therapy for bacterial vaginosis as assessed by quantitative PCR, *J. Clin. Microbiol.,* 47, 721, 2009.
134. Mitchell, C.M et al., Comparison of oral and vaginal metronidazole for treatment of bacterial vaginosis in pregnancy: Impact on fastidious bacteria, *BMC Infect. Dis.,* 9, 89, 2009.
135. Menard, J.P. et al., Molecular quantification of *Gardnerella vaginalis* and *Atopobium vaginae* loads to predict bacterial vaginosis, *Clin. Infect. Dis.,* 47, 33, 2008.
136. Matsuda, K. et al., Establishment of an analytical system for the human fecal microbiota, based on reverse transcription-quantitative PCR targeting of multicopy rRNA molecules, *Appl. Environ. Microbiol.,* 75, 1961, 2009.
137. Yamamoto, T. et al., Bacterial populations in the vaginas of healthy adolescent women, *J. Pediatr. Adolesc. Gynecol.,* 22, 11, 2009.
138. Zhou, X. et al., Characterization of vaginal microbial communities in adult healthy women using cultivation-independent methods, *Microbiol. Sgm.,* 150, 2565, 2004.
139. Vaneechoutte, M. et al., Isolation of *Moraxella canis* from an ulcerated metastatic lymph node, *J. Clin. Microbiol.,* 38, 3870, 2000.

5 *Bifidobacterium*

Abelardo Margolles, Patricia Ruas-Madiedo, Clara G. de los Reyes-Gavilán, Borja Sánchez, and Miguel Gueimonde

CONTENTS

5.1 INTRODUCTION

5.1.1 Classification and Taxonomy

The genus *Bifidobacterium* includes high G + C gram-positive nonspore-forming, nonmotile, and nonfilamentous polymorphic rod-shaped bacteria that can display slight bends or a large variety of branchings, the slightly bifurcated club-shaped or spatulated extremities being the most commonly found. They can be organized singly or in chains, in star-like aggregates, in "V," or in palisade arrangements when grown in vitro. They are strictly anaerobic, although some species, such as *Bifidobacterium animalis* and *Bifidobacterium psychraerophilum*, can tolerate moderately high oxygen concentrations, and they have a fermentative metabolism.[1,2] Bifidobacteria were first described at the beginning of the twentieth century and were included among the family *Lactobacillaceae*.[3] In 1924, *Lactobacillus bifidum* was reclassified as the new genus *Bifidobacterium* by Orla-Jensen.[1]

Currently, over 30 species are included in the genus *Bifidobacterium*. They form a close phylogenetic group and show high similarity (>93%) in the 16S rRNA sequences.[4] The *Bifidobacterium* genus is included in the phylum *Actinobacteria*, class *Actinobacteria*, subclass *Actinobacteridae*, order *Bifidobacteriales*, and family *Bifidobacteriaceae*. Phylogenetic studies, based on sequence comparison of 16S rRNA sequences and housekeeping genes, have clustered some *Bifidobacterium* species in six phylogenetic clusters named *Bifidobacterium boum*, *Bifidobacterium asteroides*, *Bifidobacterium adolescentis*, *Bifidobacterium longum*, *Bifidobacterium pullorum*, and *Bifidobacterium pseudolongum*.[5] Other species have displayed a large phylogenetic distance and have not been assigned to any specific cluster (Figure 5.1). The species included in the genus *Bifidobacterium* are *B. adolescentis, B. angulatum, B. animalis, B. asteroides, B. bifidum, B. bombi, B. boum, B. breve, B. catenulatum, B. choerinum, B. coryneforme, B. cuniculi, B. dentium, B. gallicum, B. gallinarum, B. indicum, B. longum, B. magnum, B. merycicum, B. minimum, B. mongoliense, B. pseudocatenulatum, B. pseudolongum, B. psychraerophilum, B. pullorum, B. ruminantium, B. saeculare, B. scardovii, B. subtile, B. thermacidophilum, B. thermophilum,* and *B. tsurumiense*.[6–9] Furthermore, *B. animalis* comprises of two subspecies (*animalis* and *lactis*), as well as *B. pseudolongum* (subsp. *globosum* and *pseudolongum*), and *B. thermacidophilum* (subsp. *thermoacidophilum* and *porcinum*), whereas the species *B. longum* is subdivided in the subspecies *longum, infantis,* and *suis* (Figure 5.1).

Bifidobacterium species can be isolated from a limited number of habitats—that is, human and animal gastrointestinal tract (GIT), food, insect intestine, and sewage.[10,11] Among the species most commonly found in human intestine and feces are *B. catenulatum*, *B. pseudocatenulatum*, *B. adolescentis*, *B. longum*, *B. pseudolongum*, *B. breve*, *B. angulatum*, *B. bifidum*, and *B. dentium*, whereas *B. animalis* is the species more often found in functional foods.[12] In this respect, it is of key importance to remark that a clear

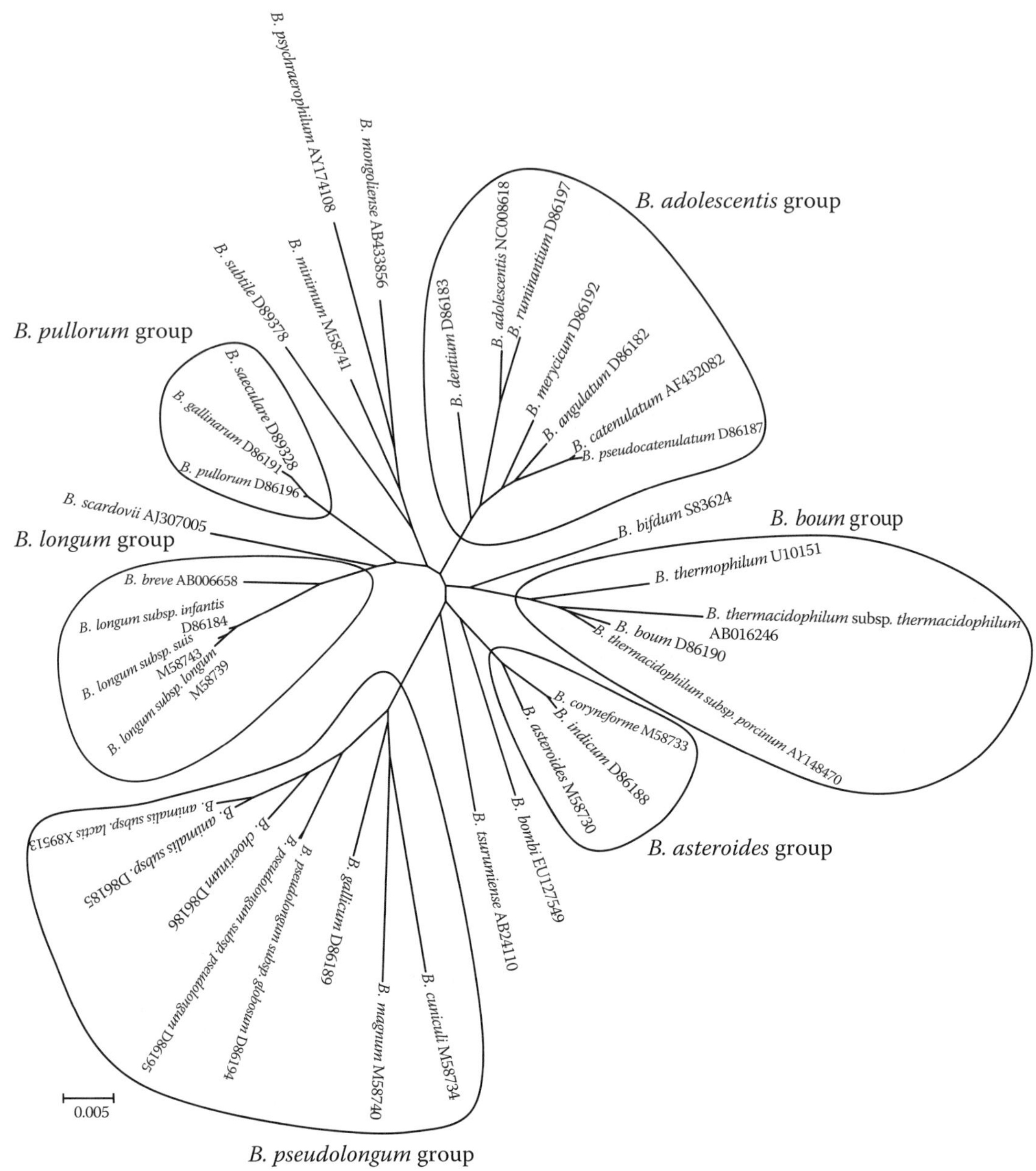

FIGURE 5.1 Evolutionary relationships of bifidobacteria.

ecological specialization seems to exist for different bifidobacterial taxa, and species distribution depends on host age and localization in the host (i.e., intestinal mucosa versus luminal/fecal environment).[13,14] Furthermore, novel metagenomic studies have analyzed the bifidobacterial microbiota in the human gut and revealed that its biodiversity was much broader than had been previously expected,[14,15] showing the existence of uncharacterized phylotypes that have passed unnoticed until recently, and highlighting the importance of novel culture-independent approaches to complement the more traditional culture-dependent techniques.

5.1.2 Biology and Physiology

As stated above, the natural habitat of the genus *Bifidobacterium* is the GIT of mammals, birds, and insects. The human GIT hosts a very complex microbiota that plays an important role on the health and wellbeing of the host.[16] This microbial community reaches one of the highest cell densities recorded for any ecosystem, although its taxonomical diversity is one of the lowest. A small number of taxa dominate this ecosystem in adult humans, mainly representatives of *Bacteroidetes* and *Firmicutes* divisions.[17] The phylum *Actinobacteria*, including the *Bifidobacteriaceae* family, is also present.

The GIT colonization of newborn babies starts at birth and increases in number and diversity in few days, bifidobacteria being one of the first and most abundant colonizers. Several factors affect this process, such as the type of delivery, the initial way of feeding (breast or infant formula), or the geographical region, among others. The number of bifidobacteria declines with the age of the host, remaining more or less stable during the adult life and tending to decrease in the elderly.[18] In addition, intersubject species variability in bifidobacterial populations has been noticed, but low diversity is found between different GIT regions of the same subject, denoting scarce intrasubject variability.[14]

At the beginning of the twentieth century, Tissier[3] described for the first time bifidobacterial strains (originally named *Bacillus bifidus*) isolated from breastfed infant feces. During recent years, an increasing number of articles have been published reporting on the bifidobacterial diversity of the GIT ecosystem. Commonly, these studies are based on fecal sampling, leading to the isolation of new strains. However, in the fecal material, transitory bacteria derived from the diet are usually present, and therefore not representing the mucosa-adhered microbiota.[19] For this reason, different GIT sections, such as colonic mucosa obtained from biopsies of healthy individuals, are also used to explore the bifidobacterial diversity of the human microbiota.[15] In addition, *Bifidobacterium* strains can also be isolated from human saliva[20] and breast milk,[21] as well as from sewage, in which bifidobacteria are used as indicators of fecal contamination.[22]

Regarding the physiology of this genus, the optimal growth temperature of most human isolated strains is about 36°C–38°C, being higher for that of animal origin (41°C–43°C). Viability of bifidobacteria at acidic pH values is very variable but, in general, it can be considered that resistance of bifidobacteria to acidic pH is weak, with the exception of *B. animalis*.[23] This species satisfactorily tolerates acidic pH and, in addition, the presence of oxygen, for these reasons being the most used *Bifidobacterium* spp. in fermented probiotic products. It has been suggested that the acid tolerance in *Bifidobacterium* is linked to the activity of the membrane-bound F_0F_1-ATPase, an enzyme responsible for the maintenance of the intracellular pH homeostasis in anaerobic bacteria through the ATP-dependent extrusion of protons.[24] The high F_0F_1-ATPase activity induction observed for this species appears to be responsible for its elevated acid resistance, this activity not being induced in nonresistant strains.[25]

Tolerance to bile salts is another crucial property for the persistence of bifidobacteria in the human GIT. Usually, deconjugated bile salts are more toxic than their corresponding taurine or glycine conjugates. The enzyme responsible for bile salts deconjugation is the bile salt hydrolase (BSH), but to date there is no solid evidence of a relationship between this enzyme and the levels of resistance to bile in bifidobacteria.[26]

Bifidobacteria have a specific pathway for the metabolism of sugars, the fructose-6-phosphate phosphoketolase (F6PPK) pathway or "bifido shunt" (Figure 5.2; Table 5.1), described for the first time in the work of Scardovi and Trovatelli.[27] The key enzyme of this pathway is a xylulose-5-phosphate/fructose-6-phosphate phosphoketolase (Xfp), which catalyzes the phosphorolytic cleavage of fructose-6-phosphate and/or xylulose-5-phosphate. To date, two types of Xfp activities have been described in prokaryotes and eukaryotes—fructose-6-phosphate phosphoketolase (F6PPK, EC 4.1.2.22), which catalyzes the conversion of fructose-6-phosphate to erythrose-4-phosphate and acetyl-phosphate; and xylulose-5-phosphate phosphoketolase (EC 4.1.2.9), which cleaves xylulose-5-phosphate, rendering acetyl-phosphate and glyceraldehyde-3-phosphate.[28] In bifidobacteria, in addition to the F6PPK, a dual xylulose-5-phosphate/fructose-6-phosphate phosphoketolase able to act over both fructose-6-phosphate and xylulose-5-phosphate has been described.[29] Bifidobacteria also present a transaldolase (Tal, EC 2.2.1.2) and a transketolase (Tkt, EC 2.2.1.1) able to convert the fructose-6-phosphate in acetyl-phosphate and glyceraldehyde-3-phosphate. An acetate kinase (AckA, EC 2.7.2.1) allows the conversion of acetyl-phosphate into acetate with the concomitant production of ATP. The glyceraldehyde-3-phosphate, through the last reactions of the Embden–Meyerhof–Parnas (EMP) pathway, is metabolized to pyruvate and the activity of lactate dehydrogenase NAD-dependent (Ldh2, EC 1.1.1.27) converts this intermediate metabolite in lactate. In addition, pyruvate in combination with Co-A is converted into formate and acetyl-CoA by the activity of the formate C-acetyltransferase (Pfl, EC 2.3.1.54). Finally, acetyl-CoA is metabolized to ethanol, by means of the bifunctional acetaldehyde-CoA/alcohol dehydrogenase (Adh2, EC 1.1.1.1), and/or to acetate by the activity of the phosphate acetyltransferase (Pta, EC 2.3.1.8). The theoretical ratio of the bifido shunt gives 1 mol of lactic acid and 1.5 moles of acetic acid per mol of glucose consumed, producing 2.5 moles of ATP without CO_2 formation, although the production of formic acid and ethanol can modify this fermentation balance. The energetic balance of the bifido shunt (2.5 moles ATP) is higher than that obtained by lactic acid bacteria after the hexose metabolism through the EMP (or homofermentative pathway, 2 moles ATP) or through the pentose-phosphoketolase (or heterofermentative pathway, 1 mol ATP).

The availability of several genome sequences of *Bifidobacterium* species provides genetic evidence to indicate that it is a prototrophic genus for some growth factors—that is, it is able to synthesize some vitamins (B complex, folic acid, thiamine, and nicotinate), aminoacids, and nucleotides. Genomic information also indicates that several species harbor genes required for metabolizing many monosaccharides or disaccharides through the fructose-6-phosphate phosphoketolase pathway. Analysis of the first bifidobacterial genome available, *B. longum* NCC2705,[30] indicated that this strain possesses all the enzyme-coding genes necessary to metabolize several plant- and milk-derived oligosaccharides, containing fructose, galactose, *N*-acetyl-glucosamine, *N*-acetyl-galactosamine, arabinose, xylose, ribose, sucrose, lactose, cellobiose, and melibiose through this shunt. In fact, nearly 10% of the genes contained in the genome of this strain encode for proteins that are predicted to be involved in the carbohydrate metabolism and transport. This supposes,

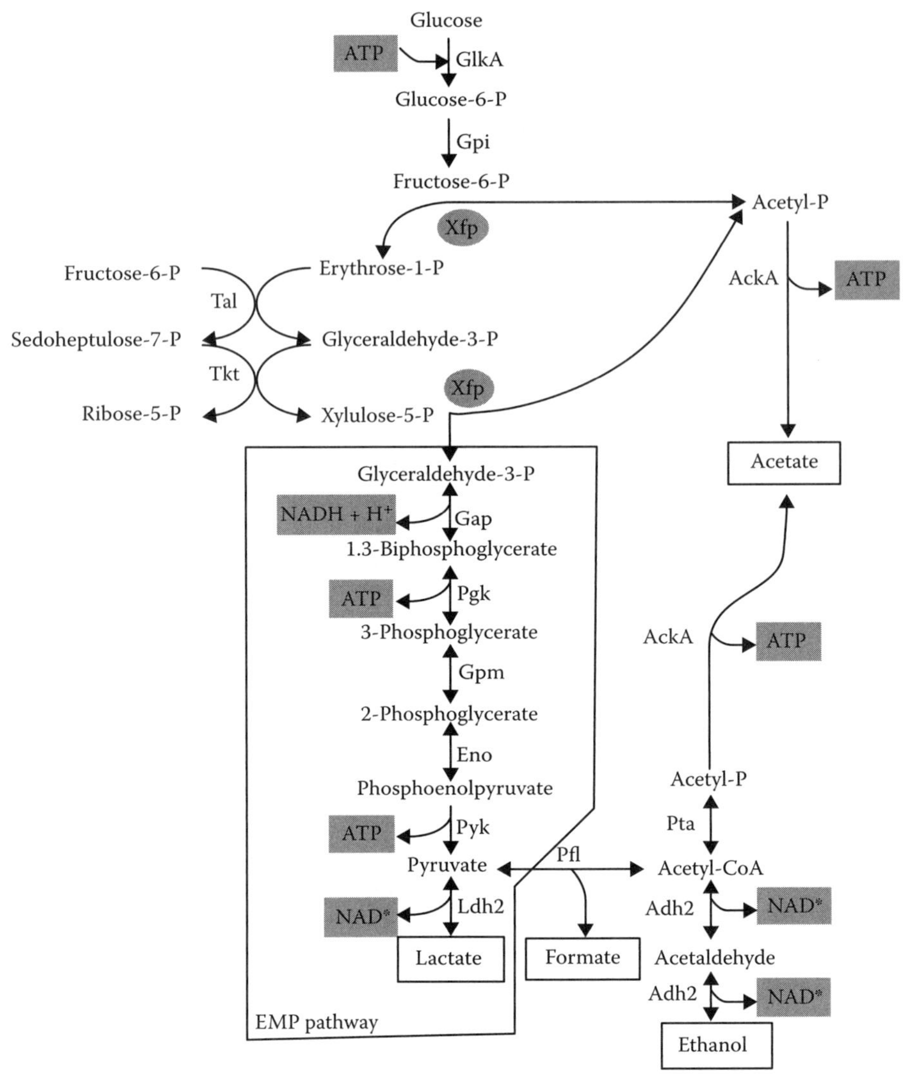

FIGURE 5.2 Fructose-6-phosphate phosphoketolase pathway "Bifid shunt."

theoretically, a selective advantage to colonize the colon, an environment poor in mono- and disaccharides because these are consumed by the host and microbiota in the upper GIT. Based on amino acid sequence identity, the presence of several glycolytic activities, including xylanases, arabinosidades, α-galactosidases, neopullanase, isomaltase, maltase, inulinase, β-galactosidases, β-glucosidases, hexosaminidases, and α-manosidases, as well as eight high-affinity oligosaccharide transporters, has been predicted.[30]

Similar gene arrangements have been seen after the genome annotation and comparison of the bifidobacterial species *B. adolescentis* ATCC15703, *B. animalis* subsp. *lactis* AD011, and *B. longum* subsp. *infantis* ATCC15697.[31,32] However, comparative genome analysis revealed some differences, such as the presence of a *fos* gene cluster in *B. animalis* subsp. *lactis* AD011, very similar to the one described in *B. breve* UCC2003.[33] This cluster is likely to be involved in fructooligosaccharide (FOS) catabolism, which is present in a panoply of food ingredients and supplements. FOS are composed of 4–60 fructose monomers linked by β(2–1) bonds, with a terminal glucose also connected by β(2–1) linkage. This β(2–1) bond is not recognized as a substrate by intestinal enzymes, and thus FOS are able to reach the colon, where they stimulate the growth of beneficial bacteria, including bifidobacteria.[34] This capacity of beneficially modulating the composition of the intestinal microbiota allows FOS to be considered as prebiotics.[35] The presence of *fos* cluster reflects the metabolic adaptation of bifidobacteria to the human GIT conditions.

Interestingly, comparative genomics revealed the presence of a specific genes cluster involved in the transport and metabolism of human milk oligosaccharides (HMO) in the strain *B. longum* subsp. *infantis* ATCC15697.[31] This strain lacked certain enzymatic activities related to the metabolism of plant-derived oligosaccharides, such as arabinose and xylose, which are abundant in adult diets. This fact, together with the evidence of the presence of similar HMO-related gene clusters in infant gut metagenomes, suggests a specific adaptation of *B. longum* subsp. *infantis* ATCC15697 to the infant GIT.[36]

Thus, bifidobacteria are able to ferment a wide variety of carbohydrates. In addition, they are capable of

TABLE 5.1
Accession Numbers of the Major Proteins Acting in Bifid Shunt in Some Available *Bifidobacterium* Genomes

	B. adolescentis ATCC 15703	*B. animalis* subsp. *lactis* AD011	*B. bifidum* NCIMB 41171	*B. longum* subsp. *longum* NCC2705	*B. longum* subsp. *infantis* ATCC 15697
GlkA	YP_909820	YP_002470082	ZP_03647167	NP_696841	YP_002322055
Gpi	YP_909094	YP_002469157	ZP_03645763	NP_695484	YP_002321916
Tal	YP_909693	YP_002470219	ZP_03646448	NP_695898	YP_002322563
Tkt	YP_909692	YP_002470218	ZP_03646447	NP_695899	YP_002322564
Xfp	YP_909550	YP_002470341	ZP_03646196	NP_696135	YP_002323176
Gap	YP_909942	YP_002469609	ZP_03645995	NP_696527	YP_002322378
Pgk	YP_909698	YP_002470227	ZP_03646458	NP_695890	YP_002322555
Gpm	YP_909241	YP_002469301	ZP_03647102	NP_696806	YP_002323591
Eno	YP_909508	YP_002469894	ZP_03646131	NP_696193	YP_002323289
Pyk	YP_909541	YP_002470352	ZP_03646166	NP_696160	YP_002323198
Ldh2	YP_909980	YP_002469660	ZP_03646455	NP_696472	YP_002322558
Pfl	YP_909855	YP_002469833	ZP_03646658	NP_696127	YP_002323169
Pta	YP_909551	YP_002470342	ZP_03646187	NP_696142	YP_002323183
AckA	YP_909552	YP_002470343	ZP_03646186	NP_696143	YP_002323184
Adh2	YP_909182	YP_002469236	ZP_03647027	NP_696730	YP_002323678

Additional information can be obtained through their corresponding accession number at http://www.ncbi.nlm.nih.gov/protein. GlkA, glucokinase; Gpi, glucose-6-phosphate isomerase; Tal, transaldolase; Tkt, transketolase; Xfp, xylulose-5-phosphate/fructose-6-phosphate phosphoketolase; Gap, glyceraldehyde-3-phosphate dehydrogenase C; Pgk, phosphoglycerate kinase; Gpm, phosphoglycerate mutase; Eno, enolase; Pyk, pyruvate kinase; Ldh2, lactate dehydrogenase; Pfl, formate acetyltransferase; Pta, Phosphate acetyltransferase; AckA, acetate kinase; Adh2, bifunctional acetaldehyde-CoA/alcohol dehydrogenase.

processing indigestible complex carbohydrates from our diet and from host origin.[6,30,37–39] Hence, the availability of genome sequences of intestinal *Bifidobacterium* spp., together with the use of physiological experiments, has provided considerable insight into the adaptation of this genus to the human gut environment, especially regarding the utilization of dietary carbohydrates, resistance to bile and acid conditions, and interaction with the host.

5.1.3 Clinical Features and Pathogenesis

It has been shown that *Bifidobacterium* is present at high levels in the gut microbiota mainly in infants but also in adults.[40] Reduced levels of these microorganisms in the intestine of subjects suffering from different diseases, including colon cancer, irritable bowel syndrome, inflammatory bowel disease, atopic disease, or celiac disease, have been observed.[41–45] For these reasons several health-promoting properties have been attributed to the microorganisms of this genus and increasing bifidobacterial levels in the GIT is often a treatment target of dietary intervention strategies.

So far, there have been no reports of sepsis related to bifidobacteria use in otherwise healthy individuals. Therefore, the current evidence strongly suggests that bifidobacteria are extremely safe to use in the general population. In fact, the most commonly used *Bifidobacterium* species are considered to be generally safe (GRAS), and in Europe such microorganisms have been granted a Qualified Presumption of Safety (QPS) status.[46] Despite the widespread use of the genus *Bifidobacterium*, in particular the species *B. animalis* subsp. *lactis*, in commercial probiotic products, these probiotics have never been related to sepsis associated with probiotic use, demonstrating the low pathogenic potential of bifidobacteria. This extensive use of bifidobacteria as probiotic agents, together with the numerous clinical trials conducted with these microorganisms, constitutes good proof of safety. As of July 2009, 43 undergoing or completed human clinical trials in which bifidobacteria were administered can be found registered at ClinicalTrials.gov. During last year alone, more than 25 human intervention studies, in which around 2000 volunteers—including healthy adults and elderly, but also full-term and preterm infants, HIV infected children, allergic subjects, pregnant women or IBS patients among others—received a test product containing bifidobacteria, have been published without reporting any detrimental effects.

In vitro assessments offer means to investigate the safety of bifidobacteria based on the intrinsic properties of the strains. Several different in vitro approaches have been used in the safety assessment of bifidobacteria. In vitro tests assessing the resistance to antibiotics and the presence of mobile antibiotic resistance genes are common and constitute perhaps the major safety concern regarding bifidobacteria. Several studies have attempted to identify the presence of relevant virulence determinants in bifidobacteria. However, to date such determinants have not been identified. It should be noted, however, that classical risk assessments, which are used for pathogens, might not always be directly applicable to microorganisms such as bifidobacteria, which become members of the normal healthy intestinal microbiota soon after birth and are also components of normal human diet. In pathogens, pathogenicity is normally a consequence of several properties of the strain. The possible presence of such a property in a strain of low infective potential and low clinical significance does not imply that the strain is pathogenic or poses a risk to health.[47] An example of this is the ability to adhere to human mucosa, which is a virulence factor in the case of true pathogens, but is also an essential feature of many commensal microbes with very low pathogenic potential, such as bifidobacteria. To date, no clear virulence factors similar to those associated with pathogenic microorganisms have been identified for bifidobacteria.[48] This is likely to result from the minimal infectivity of the bifidobacteria.

Animal models have been also used in the safety evaluation of *Bifidobacterium*.[49,50] Acute oral toxicity tests have been carried out without observing toxicity.[49,51] A number of studies have examined the effects of bifidobacteria in experimentally induced colitis in murine models. Despite disruption of the intestinal barrier in colitic animals, side effects due to increased bacterial translocation seem not to be apparent. In contrast, improvement of clinical condition by reducing the inflammatory colitis of these animals has been often reported. Immunocompromised animal models of both adult and young animals have also been tested to evaluate safety without observing detrimental effects. However, many of the animal models applied in the safety assessment of probiotics were originally developed to study pathogenic microorganisms in which virulence traits are present, and therefore they may not be optimal for studying translocation of non-virulent microorganisms such as bifidobacteria.

With regard to antibiotic resistance, *Bifidobacterium* species are resistant to aminoglycosides, metronidazole, and gram-negative spectrum antibiotics. They are also intrinsically resistant to mupirocin. In contrast, bifidobacteria are very susceptible to macrolides/lincosamides, vancomycin, rifampicin, spectinomycin, chloramphenicol, and β-lactams. The susceptibility to tetracyclines and cephalosporins varies among strains.[52] One of the current major concerns for the safe use of *Bifidobacterium* strains is the presence of tetracycline resistance genes, especially *tet*(W), although other genes, such as *tet*(M), *tet*(O), *tet*(L), *tet*(W/32/O), and *tet*(O/W), have been detected, albeit much less frequently.[53] In bifidobacteria, *tet* genes seem to be integrated in the chromosome, and they have not been associated with transposons or plasmids, but they are very often flanked by putative transposase genes.[53–55] Transposases are enzymes that catalyze the movement of DNA segments among different locations by recognizing insertion sequences in the DNA, and they are thought to be involved in the mobilization of *tet*(W) genes in bifidobacteria.[54]

The long history of safe consumption and the lack of toxicity and virulence determinants support the safety of *Bifidobacterium*. On the other hand, the low incidence rate of infections due to these microorganisms seems to support this hypothesis. However, it is important to underline that there is no "zero risk" in microbiology and safety assessment is always essential. In fact, although very rare, some bifidobacteria, probably of intestinal origin, have been associated with a few cases of bacteremia, usually in patients with underlying diseases.[56] Reduced immune function is the main common denominator in these patients and such patients are prone to bacteremia in general and not bifidobacterial infection in particular.

Of the about 2000 culture-positive cases of cervicofacial actinomycoses assessed by Pulverer et al., none was due to bifidobacteria.[57] The few cases of infection or sepsis reported have been mainly related to *Bifidobacterium eriksonii* (also known as *Actinomyces eriksonii*) which has been considered a synonym of *B. dentium*.[58] Often these cases are related with subjects with subjacent problems and correspond to polymicrobial infections.[59] A fatal case of pulmonary infection due to *B. eriksonii* in a 52-year-old alcoholic patient with periodontal disease was reported.[60] The presence of bifidobacteria was also observed in a case of infected cavitating lung tumor[61] and a case of pericarditis.[62] *B. adolescentis* was isolated in a case of urinary tract infection in a child.[63] The presence of *B. dentium* (named as *A. eriksonii*), often together with other microorganisms, in some rare cases of abscesses has also been observed.[64,65] *B. breve* was reported to be the causative agent of a neonatal meningitis case, being the second case of bifidobacterial meningitis reported to date.[66] Ha et al.[67] published a rare case of *B. longum* septicemia. The patient had undergone acupuncture for a lumbar hernia and small needles had been inserted in the lumbar area and left there. Other than the herniated intervertebral disk the subject was healthy, thus it seems that needle contamination or intestinal punctures by a needle were the likely causes of the infection.

Despite these few and isolated cases, the pathogenicity of bifidobacteria appears to be very low. Bourne et al.[68] carried out over 90,000 blood cultures from subjects with bacteremia symptoms, recovering over 9000 bacterial isolates. Only 10 out of those isolates were bifidobacteria, and in five cases it was possible to identify the strain as *B. eriksonii*. All these isolates were obtained from nine patients (in two of the cases, other microorganisms were also found) with underlying conditions (pregnancy complications, gastrointestinal disorders, systemic lupus).

To date, the major concern regards caries and mainly *B. dentium* and *Bifidobacterium*-related genera such as *Parascardovia* and *Scardovia*, as these microorganisms may be found forming part of the complex microbial community present in caries.[69,70]

Clinical trials using bifidobacteria, along with the widespread and safe use of this microbial genus worldwide, constitute the most compelling evidence of the safety of probiotics. Although very often the main target outcome in these intervention trials focuses on the health benefits of the strain, the safety of the administration and the potential adverse events, if any, are often reported as a secondary outcome. Clinical safety trials enable the in vivo evaluation of the effects in humans in a controlled manner, with a special focus on the attributes relevant to the safety of the administered strain and the factors contributing to possible adverse events. However, these clinical investigations are most commonly carried out with healthy volunteers. Considering the infection cases observed, it is clear that safety of bifidobacteria may be of particular interest to specific groups, such as neonates, who have compromised immune systems, or diseased patients. In neonates and low-birth weight infants, successful clinical interventions have been carried out,[71] and clinical trials suggest that these microorganisms are safe to use in follow-on formulas and growing-up milks.[72] Clinical evaluation in elderly populations is also of special interest, since elderly subjects commonly have health-related problems including infections and gastrointestinal problems. The safety and the lack of adverse events following the consumption of *Bifidobacterium* strains by elderly subjects have also been demonstrated. However, the possible risks may be elevated in certain high-risk populations, and the demonstration of the safety of a certain strain in the general population does not necessarily imply that the administration of the same strain is equally safe in high-risk populations.

In fact, despite the overwhelming evidence supporting bifidobacterial safety, there may be potentially adverse events, as demonstrated by a recent study carried out in severely ill patients with acute pancreatitis. Besselink et al.[73] investigated the effects of a probiotic mix containing *Lactobacillus*, *Lactococcus*, and *Bifidobacterium* strains in a randomized placebo-controlled trial that included 296 adults with a first episode of high-risk acute pancreatitis. The probiotic mix was administered by nasojejunal feeding tube, and had been specifically designed, on the basis of previous studies, to inhibit the growth of pathogens important in pancreatic necrosis.[74] The authors found a 2.53-fold increase in mortality risk in probiotic-treated participants, and nine cases of bowel ischemia (eight fatal) in the probiotic group, with no bowel ischemia in the placebo group. It is known that small bowel ischemia, increased intestinal permeability, and increased bacterial translocation are all associated with acute pancreatitis. In these cases, the direct application of probiotic bacteria to an already damaged small intestinal mucosa may have precipitated a local inflammatory response leading to increased risk of small bowel ischemia. However, it is not known why exactly mortality was higher in the treatment group, and although the mortality was associated with randomization, this does not necessarily imply that the probiotic itself was the causative factor.[75] In these cases, the gastrointestinal system plays a central role in the multisystem organ failure, leading to a breakdown of the intestinal barrier that increases translocation of bacteria and their components into the systemic circulation. In spite of that, translocation is difficult to induce in healthy humans, and if this occurs, detrimental effects have not been demonstrated for bifidobacteria.

Due to the general consideration as safe of bifidobacteria, most cases of the presence of these microorganisms in infection sites have been considered as contaminants and, much less probably, as pathogens.[76] In any case, despite the excellent overall safety record of bifidobacteria, they should be used with caution in certain specific patient groups with severe underlying diseases and reduced immune function. It is important to note that such patients in general have a high risk for bacteremia and, very often, intestinal bacteria are the origin of the infection.

5.1.4 Diagnosis

Other than the infection cases reported, bifidobacteria may have a profound effect on human health. Low levels of intestinal bifidobacteria population and the misbalance of intestinal microbiota are thought to play an important role in the etiology and pathogenesis of different inflammatory, metabolic, and autoimmune diseases. Thus, alterations in intestinal microbiota composition are found in patients with allergies or those suffering from different health disorders.[42,43,77,78] Whether these alterations on the microbiota composition and levels are the cause or consequence of illness is currently far from known. In any case, therapy with bifidobacteria and other probiotics have been successfully used for the alleviation of symptoms of these chronic disorders.[79,80]

The gut, particularly the colon, is largely the organ containing the highest population and diversity of *Bifidobacterium* species. Owing to this fact, methods for detection and enumeration of these microorganisms have mainly been developed focusing on intestinal bifidobacteria, taking into account the difficulty of isolation and characterization inherent in the high number and diversity of other microorganisms sharing the same environment. These methods have also been subsequently adapted and applied to the study of bifidobacteria in other locations. In this respect, it is worth mentioning the recent studies on *Bifidobacterium* species present in breast milk and the possible transmission of bifidobacterial populations, or their DNA, from mothers to children through placenta and breastfeeding.[21,81,82]

As deduced from that indicated above, although bifidobacteria have not been directly considered as the causative agent of any pathology, the development of quantitative and qualitative methods of detection and identification of these microorganisms has, and will, greatly contribute to gain insight into the relationship between levels and species composition of bifidobacteria populations and the health status of the host.

The correct identification and typification of *Bifidobacterium* strains is the starting point to establish its microbiological properties and safety. Thus, it is necessary to use adequate tools to provide proper strain identification, to track *Bifidobacterium* strains during food production, during their GIT passage and establishment, or to identify potential clinical isolates.

5.1.4.1 Phenotypic Techniques

The traditional identification of bifidobacteria has been based on phenotypic features. Cell morphology, production of metabolites, enzymatic activities and sugar fermentation profiles are the most commonly analyzed phenotypic characteristics. The association of a typical branched shape with the presence of fructose-6-phosphate phosphoketolase (F6PPK) activity is the first indication that a strain belongs to the genus *Bifidobacterium*.[83] However, not all strains have a V or clamped shape, and the levels of F6PPK activity vary enormously among species and strains.[84] Furthermore, the identification at species/strain level have some additional problems, and the classical phenotyping, such as sugar fermentation profiles, transaldolase serotyping, cell wall composition, and the study of the F6PPK isoforms, is clearly not discriminative enough to reach species, subspecies, and biotype-level identification with accuracy. In fact, most cases of bifidobacteria misidentifications are due to inappropriate phenotypic methods.[85]

Several different quantitative and qualitative methodologies have been used for *Bifidobacterium* microbiota assessment.[86,87] Traditional methods for the study of qualitative and quantitative composition of intestinal bifidobacteria have been carried out by the cultivation of feces. In some cases, it has also been considered that there are different mucosa-associated intestinal bacteria that can differ from feces in biopsies or samples taken from healthy individuals[88] or patients.[43,89] Due to the anaerobic nature of bifidobacteria, the samples must be handled in anaerobic conditions and processed immediately or shortly after collection. Bifidobacteria are nutritionally demanding microorganisms and therefore are cultured on suitable complex growth media (such as MRS, Rogosa, Beerens, and trypticase phytone yeast extract). Both nonselective and selective differential media have been used for growth and counting, with and without the addition of selective agents.[90–92] The choice of one medium or another is dependent on the other bacteria present in the sample and on the method used for subsequent identification.

Besides the characteristic "bifid" cell morphology and the F6PPK activity of the genus *Bifidobacterium*, another phenotypic method for identification at the species level is based on the determination of the cellular fatty acid composition[93] with the help of the chromatographic MIDI system directly in feces[42] or in previously isolated cultures.[92] In spite of the accuracy of the method, it is important to mention that the composition of fatty acids changes notably in bifidobacteria stressed cells,[94] which can lead to misidentifications if other alternative methods were not used in those cases.

The difficulty of cultivation of anaerobic intestinal microorganisms meant that until a few years ago, their study was restricted to the cultivable species and among them, only to the viable population. This led to the overestimation of some species and the underestimation of others. In addition, not all *Bifidobacterium* species are recovered from selective media with the same efficiency,[90,91] which could draw a distorted picture of the bifidobacteria population inhabiting complex human ecosystems.

5.1.4.2 Molecular Techniques

In the last few years, new molecular tools have been developed and applied, as alternative or complementary methods to traditional cultures, for the identification and typing of bifidobacteria. Most of these techniques use 16S rRNA and its encoding genes as target molecules, on which specific PCR primers and hybridization probes are designed to detect particular species or groups of microorganisms, including the genus *Bifidobacterium*[95–98] and some of its species.[40,96,97] Moreover, the use of 16S rRNA instead of the DNA (by obtaining cDNA from mRNA through reverse-transcriptase PCR reactions) can provide data on the activity and viability of microorganisms in their ecosystem rather than on cellular levels as occurs when DNA is the target molecule.

Among the most extensively used molecular tools for qualitative assessment of bifidobacteria is the sequence analysis of amplified 16S RNA genes, by PCR amplification from biological samples, followed by cloning and sequencing of the amplified DNA. Other widely used techniques are TGGE and DGGE, using primers for PCR amplification with one of these primers being tagged to a GC clamp to avoid complete dissociation of DNA strands. Amplified fragments are subsequently separated by denaturing gel electrophoresis through a gradient of temperature (TGGE) or denaturant agent (DGGE).[99] Furthermore, PCR amplification products can be separated by high-performance liquid chromatography (HPLC).[100] Analysis of terminal-restriction fragment length polymorphism (T-RFLP) by digestion of fluorescent labeled fragments with suitable restriction enzymes[101] or the use of Oligonucleotide arrays[102] may be also utilized.

Sequencing of ribosomal rRNA genes (rDNA) is nowadays the standard methodology for bifidobacteria identification. The bacterial rRNA genes are organized in *rrn* operons containing the genes for the three ribonucleic acids (5S RNA, 16S RNA, and 23S RNA), with bifidobacteria harboring from two to six operons, depending on the species.[10,30] The 16S and 23S rRNA genes, as well as the 16S-23S spacer region (Intergenic Transcribed Sequence, ITS), showed nucleotide fingerprints with different discriminatory levels. Sequencing of the 16S rRNA can be used for identification purposes or to establish phylogenetic relationships among species, subspecies, and biotypes,[5,97,103] whereas the 16S-23S rRNA ITS sequence is much more variable than the 16S rRNA structural gene, even within closely related taxa, which makes it a valuable target for species identification by using species-specific primers,[104] but also its higher discriminatory capacity may allow the differentiation of strains.[105]

Recently, multilocus sequence typing (MLST) methods have been applied for the typification of *Bifidobacterium* species and strains. MLST methods made use of a DNA sequencing procedure to characterize different housekeeping gene loci in the bacterial genome.[106] Its application in bifidobacteria has been shown to be highly discriminatory, the target genes studied for detailed identification and classification purposes down to subspecies level being numerous.[5,107–110]

Application of a variety of PCR protocols has enabled the differentiation between strains of the same species. Random amplification of polymorphic DNA (RAPD) techniques, amplified ribosomal RNA restriction analysis (ARDRA), ERIC (Enterobacterial Repetitive Intergenic Consensus sequence)-PCR, and REP (Repetitive Extrogenic Palindromic)-PCR are the most common PCR-based methods for bifidobacterial typing.[4,14,111–113] Other PCR-based typing approaches include T/DGGE (temperature/denaturant gradient gel electrophoresis), amplified fragment length polymorphism (AFLP), PCR coupled to enzyme-linked immunosorbent assay (ELISA), triplicate arbitrarily primer (TAP)-PCR, restriction fragment length polymorphism (RFLP)-PCR, and multiplex PCR.[108,111,114–116]

Some other methods for bifidobacteria identification and typing used DNA profiling, such as plasmid analysis, RFLP, or PFGE (pulsed field gel electrophoresis) of total DNA.[4,111,117] PFGE is often considered the best technique for strain-specific typification and has shown a high discriminatory power. Among the nucleic acids hybridization methods, ribotyping, microplate blot, and microarray analysis have also been used for bifidobacterial typing.[102,111,118] Finally, it is worthwhile to point out that other methods based on metabolite production, have also been applied to the identification and typing of *Bifidobacterium* such as chromatographic analysis of organic acids or the analysis of the intrinsic fluorescence of aromatic amino acids.[119,120]

Qualitative techniques provide a picture of the most abundant species and the diversity in microbial ecosystems but fail to provide data on the relative abundance of microbial groups as well as on the less abundant species. Among the quantitative techniques most extensively used for *Bifidobacterium* is fluorescent in situ hybridization (FISH), with specific probes coupled to flow cytometry or fluorescence microscopy detection. Although this technique has widely been used to study the evolution of intestinal populations,[45,95,121] it is extremely laborious and time consuming, and is influenced by cell permeability or the ribosome content of cells.

Quantitative real-time PCR is a faster tool to study complex microbial communities. Different PCR assays have been developed for fecal bifidobacteria and its species, or focused on the metabolically active microorganisms targeting 16S rDNA or rRNA, respectively, by using probes labeled with different fluorescent dyes. "Omics" (metagenomics, metaproteomics, and metabolomics) are very recent, high throughput, and promising techniques that can also be applied to the study of bifidobacteria in complex ecosystems. The sensitivity level of genetic methods for identification and typing of bacterial isolates is obviously limited by the previous sensitivity of the culture media and conditions used for isolating the microorganisms.

5.2 METHODS

5.2.1 Sample Preparation

Most of the studies evaluating bifidobacterial population composition and levels have used fecal samples as starting material. However, the same methods with slight modifications can be used to determine the presence or levels of bifidobacteria in other matrixes such as human tissues or fluids. Both culture-dependent and culture-independent methods are available to assess *Bifidobacterium* populations. Depending on the method used for the detection of the microorganism, different sample preparation procedures are needed.

When a culture-dependent method is going to be used, isolation of the microorganisms is required prior to identification by phenotypic or genotypic techniques. Given the complex microbial community present in fecal samples, isolation of bifidobacteria is a challenging issue. Typically, the fecal sample is diluted in an isotonic buffer (e.g., PBS) and homogenized using a blender. Decimal dilutions are then prepared and 0.1 mL of the appropriate dilution is plated in an adequate selective or differential culture media.[91,92] After the incubation of the plates at 37°C for 48 h under anaerobic conditions in an anaerobic chamber or a jar with an anaerobiosis system (e.g., Anaerocult A, Merck), putative bifidobacterial colonies may be picked up and their identity further confirmed by phenotypic or molecular techniques as stated in Section 5.1.4. Very often, molecular methods of identification focus on DNA analyses that require a step of extraction of DNA from the studied strain. Bifidobacteria are not easily lysed, and for this reason lysis buffers are often supplemented with lysozyme and mutanolysin to improve the lysis efficiency, thereby increasing the yield of the DNA extraction. Nowadays, in addition to the traditional methods of DNA extraction using phenol/chloroform, a number of DNA extraction kits are commercially available from the main molecular biology suppliers and are very often used for bacterial DNA extraction.

Culture-independent methods do not require a previous bacterial isolation step. In this case, the starting material may be a complex sample (i.e., fecal sample). The total DNA in the sample is extracted previously to the specific detection of bifidobacterial DNA in the complex DNA pool extract. Different DNA extraction methods for such complex samples have been reported in the literature[19,99] and commercial kits are also available (e.g., QIAamp DNA Stool Kit, Qiagen) to this end.

5.2.2 Detection Procedures

Among the easiest and most currently used molecular tools to detect the presence of bifidobacteria in complex samples is the use of PCR with specific primers. Several specific primers have been designed for the genus *Bifidobacterium*[97,98] and for different bifidobacterial species.[40,96,97,122] These primers have been extensively used for the culture-independent detection of bifidobacteria, mainly in fecal samples but also in other matrixes, including food products.

Procedure

1. Prepare the PCR mix (most commonly 25 or 50 μL) containing PCR buffer, DNA polymerase, dNTPs, and the appropriate primers (genus or species-specific depending on the aim of the study). Add DNA extract (most often 2–5 μL) or an equal amount of water (negative control).
2. Run the PCR program using the optimal conditions for the DNA polymerase used and the adequate annealing temperature (depending in the primers used).

3. After completion of PCR, add DNA loading buffer and visualize amplification products by standard agarose gel electrophoresis and ethidium bromide staining.

Note: Some of these primers have also been used for determination of bifidobacterial levels by means of quantitative real-time PCR. Two approaches may be used to this end; a combination of the primers with a DNA dying agent such as SYBR Green (SYBR Green assay),[15,21,44] or the combined use of the primers with a labeled internal oligonucleotide probe (5′nuclease assay).[40,98,122]

5.3 CONCLUSION AND FUTURE PERSPECTIVES

Bifidobacteria are normal inhabitants of the human intestinal tract, constituting an important part of the microbiota of the colon, where they reach very high numbers. These microorganisms have shown a high adaptation to their environment and are fastidious and difficult to cultivate in the laboratory. However, the development of techniques to grow anaerobic microorganisms, together with the recent development of culture-independent methods, has allowed an improvement of our knowledge on *Bifidobacterium* physiology and ecology.

Although some rare cases of infection have been reported, usually in subjects with underlying diseases, the major concern about *Bifidobacterium* safety regards the potential role of *B. dentium* in dental caries. In fact, these microorganisms have shown a very low pathogenic potential, as has been demonstrated by in vitro animal and human studies. Moreover, several studies have indicated that reduced intestinal bifidobacterial levels may be related to different diseases and several health-promoting properties have been attributed to these microorganisms. For these reasons, increasing bifidobacterial levels has often been taken as a treatment target. Different nutritional intervention strategies have been developed to this end, mainly using probiotic bacteria, including bifidobacteria, and prebiotic compounds. In this regard, it is important to underline that there is no "zero risk" and that the addition of any microorganisms to products for human or animal consumption requires a rigorous assessment of its safety.

In the near future, the recent developments on molecular methods and the availability of bifidobacterial genomes will improve our understanding of the biology of the genus *Bifidobacterium*, the interaction with the host, and the role in health and disease.

REFERENCES

1. Scardovi, V., Genus *Bifidobacterium.* Orla-Jensen 1924, 472[AL], In Sneath, P.H.A., Mair, N.S., Sharpe, M.E., and Holt, J.G. (Editors), *Bergey's Manual of Systematic Bacteriology,* Vol. 2, pp. 1418–1434, Williams and Wilkins, Baltimore, USA, 1986.
2. Simpson, P.J. et al., Intrinsic tolerance of *Bifidobacterium* species to heat and oxygen and survival following spray drying and storage, *J. Appl. Microbiol.*, 99, 493, 2005.
3. Tissier, M.H., Réchérches sur la flore intestinale normale et pathologique du nourisson (thesis), University of Paris, France, 1900.
4. Satokari, R.M. et al., Molecular approaches for the detection and identification of bifidobacteria and lactobacilli in the human gastrointestinal tract, *Syst. Appl. Microbiol.*, 26, 572, 2003.
5. Ventura, M. et al., Analysis of bifidobacterial evolution using a multilocus approach, *Int. J. Syst. Evol. Microbiol.*, 56, 2783, 2006.
6. Ventura, M. et al., Genomics of Actinobacteria: tracing the evolutionary history of an ancient phylum, *Microbiol. Mol. Biol. Rev.*, 71, 495, 2007.
7. Okamoto, M. et al., *Bifidobacterium tsurumiense* sp. nov., from hamster dental plaque, *Int. J. Syst. Evol. Microbiol.*, 58, 144, 2008.
8. Killer, J. et al., *Bifidobacterium bombi* sp. nov., from the bumblebee digestive tract, *Int. J. Syst. Evol. Microbiol.*, 59, 2020, 2009.
9. Watanabe, K. et al., *Bifidobacterium mongoliense* sp. nov., from airag, a traditional fermented mare's milk product from Mongolia, *Int. J. Syst. Evol. Microbiol.*, 59, 1535, 2009.
10. Ventura, M. et al., Insights into the taxonomy, genetics and physiology of bifidobacteria, *Antonie Van Leeuwenhoek*, 86, 205, 2004.
11. Felis, G.E., and Dellaglio, F., Taxonomy of lactobacilli and bifidobacteria, *Curr. Issues Intest. Microbiol.*, 8, 44, 2007.
12. Masco, L. et al., Culture-dependent and culture-independent qualitative analysis of probiotic products claimed to contain bifidobacteria, *Int. J. Food Microbiol.*, 102, 221, 2005.
13. Reuter, G., The *Lactobacillus* and *Bifidobacterium* microflora of the human intestine: composition and succession, *Curr. Issues Intest. Microbiol.*, 2, 43, 2001.
14. Turroni, F. et al., Exploring the diversity of the bifidobacterial population in the human intestinal tract, *Appl. Environ. Microbiol.*, 75, 1534, 2009.
15. Turroni, F. et al., Microbiomic analysis of the bifidobacterial population in the human distal gut, *ISME J.*, 3, 745, 2009.
16. Guarner, F., and Malagelada, J-R., Gut flora in health and disease, *Lancet*, 381, 512, 2003.
17. Bäckhed, F. et al., Host-bacterial mutualism in the human intestine, *Science,* 307, 1915, 2005.
18. Isolauri, E., Salminen, S., and Ouwehand, A.C., Probiotics, *Best Pract. Res. Clin. Gastroenterol.*, 18, 299, 2004.
19. Eckburg, P.B. et al., Diversity of the human intestinal microbial flora, *Science*, 308, 1635, 2005.
20. Beighton, D. et al., Isolation and identification of *Bifidobacteriaceae* from human saliva, *Appl. Environ. Microbiol.*, 74, 6457, 2008.
21. Martin, R. et al., Isolation of bifidobacteria from breast milk and assessment of the bifidobacterial population by PCR-denaturing gradient gel electrophoresis and quantitative real-time PCR, *Appl. Environ. Microbiol.*, 75, 965, 2009.
22. Lamendella, R. et al., Bifidobacteria in feces and environmental waters, *Appl. Environ. Microbiol.*, 74, 575, 2008.
23. Masco, L. et al., In vitro assessment of the gastrointestinal transit tolerance of taxonomic reference strains from human origin and probiotic product isolates of *Bifidobacterium*, *J. Dairy Sci.*, 90, 3572, 2007.
24. Cotter, P.D., and Hill, C., Surviving the acid test: Responses of gram-positive bacteria to low pH, *Microbiol. Mol. Biol. Rev.*, 67, 429, 2003.
25. Matsumoto, M., Ohishi, H., and Benno, Y., H+-ATPase activity in *Bifidobacterium* with special reference to acid tolerance, *Int. J. Food Microbiol.*, 93, 109, 2004.

26. Begley, M., Gahan, C.G., and Hill, C., The interaction between bacteria and bile, *FEMS Microbiol. Rev.*, 29, 625, 2005.
27. Scardovi, V., and Trovatelli, L.D., The fructose-6-phosphate shunt as peculiar pattern of hexose degradation in the genus *Bifidobacterium*, *Ann. Microbiol. Enzimol.*, 15, 19, 1965.
28. Sánchez, B. et al., Bacterial and eukaryotic phosphoketolases: Phylogeny, distribution and evolution, *J. Mol. Microbiol. Biotechnol.*, 18, 37, 2010.
29. Meile, L. et al., Characterization of the D-xylulose 5-phosphate/D-fructose 6-phosphate phosphoketolase gene (xfp) from *Bifidobacterium lactis*, *J. Bacteriol.*, 183, 2929, 2001.
30. Schell, M.A. et al., The genome sequence of *Bifidobacterium longum* reflects its adaptation to the human gastrointestinal tract, *Proc. Natl. Acad. Sci. USA*, 99, 14422, 2002.
31. Sela, D.A. et al., The genome sequence of *Bifidobacterium longum* subsp. *infantis* reveals adaptations for milk utilization within the infant microbiome, *Proc. Natl. Acad. Sci. USA*, 105, 18964, 2008.
32. Kim, J.F. et al., Genome sequence of the probiotic bacterium *Bifidobacterium animalis* subsp. *lactis* AD011, *J. Bacteriol.*, 191, 678, 2009.
33. Ryan, S.M., Fitzgerald, G.F., and van Sinderen, D., Transcriptional regulation and characterization of a novel beta-fructofuranosidase-encoding gene from *Bifidobacterium breve* UCC2003, *Appl. Environ. Microbiol.*, 71, 3475, 2005.
34. McKellar, R.C., and Modler, H.W., Metabolism of fructo-oligosaccharides by *Bifidobacterium* spp., *Appl. Microbiol. Biotechnol.*, 31, 537, 1989.
35. Gibson, G.R., and Roberfroid, M.B., Dietary modulation of the human colonic microbiota: Introducing the concept of prebiotics, *J. Nutr.*, 125, 1401, 1995.
36. Kurokawa, K. et al., Comparative metagenomics revealed commonly enriched gene sets in human gut microbiome, *DNA Res.*, 14, 169, 2007.
37. Klijn, A., Mercenier, A., and Arigoni, F., Lessons from the genomes of bifidobacteria, *FEMS Microbiol. Rev.*, 29, 491, 2005.
38. Barrangou, R. et al., Comparison of the complete genome sequences of *Bifidobacterium animalis* subsp. *lactis* DSM10140 and B1–04, *J. Bacteriol.*, 191, 4144, 2009.
39. Ventura, M. et al., Genome-scale analices of health promoting bacteria: Probiogenomics, *Nat. Rev. Microbiol.*, 7, 61, 2009.
40. Gueimonde, M. et al., Quantitative assessment of faecal bifidobacterial populations by using novel real-time PCR assays, *J. Appl. Microbiol.*, 102, 1116, 2007.
41. Favier, C. et al., Fecal beta-D-galactosidase production and Bifidobacteria are decreased in Crohn's disease, *Dig. Dis. Sci.*, 42, 817, 1997.
42. Kalliomäki, M. et al., Distinct patterns of neonatal gut microflora in infants in whom atopy was and was not developing, *J. Allergy Clin. Immunol.*, 107, 129, 2001.
43. Gueimonde, M. et al., Qualitative and quantitative analyses of the bifidobacterial microbiota in the colonic mucosa of patients with colorectal cancer, diverticulitis and inflammatory bowel disease, *World. J. Gastroenterol.*, 13, 3985, 2007.
44. Collado, M.C. et al., Imbalances in faecal and duodenal *Bifidobacterium* species composition in active and non-active coeliac disease, *BMC Microbiol.*, 8, 232, 2008.
45. Kerckhoffs, A.P. et al., Lower Bifidobacteria counts in both duodenal mucosa-associated and fecal microbiota in irritable bowel syndrome patients, *World J. Gastroenterol.*, 15, 2887, 2009.
46. EFSA, European Food Safety Authority Scientific Committee (EFSA) public consultation on the Qualified Presumption of Safety (QPS) approach for the safety assessment of micro-organisms deliberately added to food and feed. Annex 3: Assessment of gram positive non-sporulating bacteria with respect to a qualified presumption of safety. http://www.efsa.europa.eu/en/science/sc_commitee/sc_consultations/sc_consultation_qps.html. 2007.
47. Borriello, S.P. et al., Safety of probiotics that contain lactobacilli or bifidobacteria, *Clin. Infect. Dis.*, 36, 775, 2003.
48. Ouwehand, A.C., Saxelin, M., and Salminen, S., Assessment of potential risk factors and related properties of clinical, faecal and dairy *Bifidobacterium* isolates, *Biosci. Microflora*, 23, 37, 2004.
49. Zhou, J.S. et al., Acute oral toxicity and bacterial translocation studies on potentially probiotic strains of lactic acid bacteria, *Food Chem. Toxicol.*, 38, 153, 2000.
50. Zhou, J.S. et al., Safety assessment of potential probiotic lactic acid bacterial strains *Lactobacillus rhamnosus* HN001, *Lb. acidophilus* HN017, and *Bifidobacterium lactis* HN019 in BALB/c mice, *Int. J. Food Microbiol.*, 56, 87, 2000.
51. Kabeir, B.M. et al., Safety evaluation of *Bifidobacterium pseudocatenulatum* G4 as assessed in BALB/c mice, *Lett. Appl. Microbiol.*, 46, 32, 2008.
52. Zhou, J.S. et al., Antibiotic susceptibility profiles of new probiotic *Lactobacillus* and *Bifidobacterium* strains, *Int. J. Food Microbiol.*, 98, 211, 2005.
53. Florez, A.B. et al., Molecular analysis of tet(W) gene-mediated tetracycline resistance in dominant intestinal *Bifidobacterium* species from healthy humans, *Appl. Environ. Microbiol.*, 72, 7377, 2006.
54. Kazimierczak, K.A., Flint, H.J., and Scott, K.P., Comparative analysis of sequences flanking tet(W) resistance genes in multiple species of gut bacteria, *Antimicrob. Agents Chemother.*, 50, 2632, 2006.
55. van Hoek, A.H. et al., Resistance determinant erm(X) is borne by transposon Tn5432 in *Bifidobacterium thermophilum* and *Bifidobacterium animalis* subsp. *lactis*, *Int. J. Antimicrob. Agents*, 31, 544, 2008.
56. Gasser, F., Safety of lactic acid bacteria and their occurence in human clinical infection, *Bull. Ins. Pasteur*, 92, 45, 1994.
57. Pulverer, G., Schütt-Gerowitt, H., and Schall, K.P., Human cervicofacial actinomycoses: Microbiological data from 1997 cases, *Clin. Infect. Dis.*, 37, 490, 2003.
58. Biavati, B., Scardovi, V., and Moore, W.E.C., Electrophoretic patterms of proteins in the genus *Bifidobacterium* and proposal of four new species, *Int. J. Syst. Bacteriol.*, 32, 358, 1982.
59. Brook, I., Isolation of non sporing anaerobic rods from infections in children, *J. Med. Microbiol.*, 45, 21, 1996.
60. Green, S.L., Case-report: fatal anaerobic pulmonary infection due to *Bifidobacterium eriksonii*, *Postgrad. Med.*, 63, 187, 1978.
61. Liao, W-Y. et al., Bacteriology of infected cavitating lung tumor, *Am. J. Respir. Crit. Care Med.*, 161, 1750, 2000.
62. Brook, I., Pericarditis caused by anaerobic bacteria, *Int. J. Antimicrob. Agents*, 33, 297, 2009.
63. Brook, I., Urinary tract infection caused by anaerobic bacteria in children, *Urology*, 16, 596, 1980.
64. Cryan, B. et al., Epidural abscess due to streptococcus milleri and *Bifidobacterium* species, *J. Infect.*, 23, 214, 1991.

65. Civen, R., Vaisanen, M.L., and Finegold, S.M., Peritonsilar abscess, retropharyngeal adscess, mediastinitis, and nonclostridial anaerobic myonecrosis: A case report, *Clin. Infect. Dis.*, 16, S299, 1993.
66. Nakazawa, T. et al., Neonatal meningitis caused by *Bifidobacterium breve*, *Brain Dev.*, 18, 160, 1996.
67. Ha, G.Y. et al., Case of sepsis caused by *Bifidobacterium longum*, *J. Clin. Microbiol.*, 37, 1227, 1999.
68. Bourne, K.A. et al., Bacteremia due to *Bifidobacterium*, *Eubacterium* or *Lactobacillus*; twenty-one cases and review of the literature, *Yale J. Biol. Med.*, 51, 505, 1978.
69. Chhour, K-L. et al., Molecular analysis of microbial diversity in advanced caries, *J. Clin. Microbiol.*, 43, 843, 2005.
70. Mantzourani, M., Fenlon, M., and Beighton, D., Association between bifidobacteriaceae and the clinical severity of root caries lesions, *Oral Microbiol. Immunol.*, 24, 32, 2009.
71. Hoyos, A.B., Reduced incidence of necrotizing enterocolitis associated with enteral administration of *Lactobacillus acidophilus* and *Bifidobacterium infantis* to neonates in an intensive care unit, *Int. J. Infect. Dis.*, 3, 197, 1999.
72. Haschke, F. et al., Clinical trials prove the safety and efficacy of the probiotic strain *Bifidobacterium* Bb12 in follow-up formula and growing-up milks, *Monatsschr. Kinderheilkd.*, 1146, S26, 1998.
73. Besselink, M.G.H. et al., Probiotic prophylaxis in predicted severe acute pancreatitis: A randomized, double-blind, placebo-controlled trial, *Lancet,* 371, 651, 2008.
74. van Minnen, L.P. et al., Modification of intestinal flora with multispecies probiotics reduces bacterial translocation and improves clinical course in a rat model of acute pancreatitis, *Surgery*, 141, 470, 2007.
75. Reid, G. et al., Probiotic prophylaxis in predicted severe acute pancreatitis, *Lancet*, 372, 112, 2008.
76. Liong, M.T., Safety of probiotics: translocation and infection, *Nutr. Rev.*, 66, 192, 2008.
77. Sanz, Y. et al., Differences in faecal bacterial communities in coeliac and healthy children as detected by PCR and denaturing gradient gel electrophoresis, *FEMS Immunol. Med. Microbiol.*, 51, 562, 2007.
78. Swidsinski, A. et al., Active Crohn's disease and ulcerative colitis can be specifically diagnosed and monitored based on the biostructure of the fecal flora. *Inflamm. Bowel Dis.*, 14, 147, 2008.
79. Picard, C. et al., Review article: Bifidobacteria ans probiotic agents-physiological effects and clinical benefits, *Aliment. Pharmacol. Ther.*, 22, 495, 2005.
80. Zuccotti, G.V. et al., Probiotics in clinical practice: An overview, *J. Int. Med. Res.*, 36, 1A, 2008.
81. Gueimonde, M. et al., Breast milk: A source of bifidobacteria for infant gut development and maturation? *Neonatology*, 92, 64, 2007.
82. Satokari, R. et al., *Bifidobacterium* and *Lactobacillus* DNA in the human placenta, *Lett. Appl. Microbiol.*, 48, 8, 2009.
83. Ballongue, J., Bifidobacteria and probiotic action, In Salminen, S., von Wright, A., and Ouwehand, A. (Editors), *Lactic Acid Bacteria: Microbiological and Functional Aspects*, 3rd ed., pp. 67–124S, Marcel Dekker, New York, 2004.
84. Orban, J.I., and Patterson, J.A., Modification of the phosphoketolase assay for rapid identification of bifidobacteria, *J. Microbiol. Methods*, 40, 221, 2000.
85. Huys, G. et al., Accuracy of species identity of commercial bacterial cultures intended for probiotic or nutritional use, *Res. Microbiol.*, 157, 803, 2006.
86. Gueimonde, M., and de los Reyes-Gavilán, C.G., Detection and enumeration of gastrointestinal microorganisms, In Lee, Y.K., and Salminen, S. (Editors), *Handbook of Probiotics and Prebiotics*, 2nd ed., pp. 25–43, John Wiley and Sons, Inc., New Jersey, USA, 2009.
87. Margolles, A., Mayo, B., and Ruas-Madiedo, P., Screening, identification, and characterization of *Lactobacillus* and *Bifidobacterium* strain. In Lee, Y.K., and Salminen, S. (Editors), *Handbook of Probiotics and Prebiotics*, 2nd ed., pp. 4–24, John Wiley and Sons, Inc., New Jersey, USA, 2009.
88. Delgado, S., Suárez, A., and Mayo, B., Bifidobacterial diversity determined by culturing and by 16S rDNA sequence analysis in feces and mucosa from ten healthy Spanish adults, *Dig. Dis. Sci.*, 51, 1878, 2006.
89. Conte, M.P. et al., Gut-associated bacterial microbiota in paediatric patients with inflammatory bowel disease, *Gut*, 55, 1760, 2006.
90. Silvi, S., Rumney, C.J., and Rowland, I.R., An assessment of three selective media for bifidobacteria in faeces, *J. Appl. Bacteriol.*, 81, 561, 1996.
91. Apajalahti, J.H.A. et al., Selective plating underestimates abundance and shows differential recovery of bifidobacterial species from human feces, *Appl. Environ. Microbiol.*, 69, 5731, 2003.
92. Woodmansey, E. et al., Comparison of compositions and metabolic activities of fecal microbiotas in young adults and in antibiotic-treated and non-antibiotic treated elderly subjects, *Appl. Environ. Microbiol.*, 70, 6113, 2004.
93. Eerola, E., and Lehtonen, O.P., Optimal data processing procedure for automatic bacterial identification by gas-liquid chromatography of cellular fatty acids. *J. Clin. Microbiol.*, 26, 1745, 1998.
94. Ruiz, L. et al., Cell envelope changes in *Bifidobacterium animalis* ssp. *lactis* as a response to bile, *FEMS Microbiol. Lett.*, 274, 316, 2007.
95. Langendijk, P.S. et al., Quantitative fluorescence in-situ hybridization of *Bifidobacterium* spp. with genus-specific 16S ribosomal-RNA-targeted probes and its application in fecal samples, *Appl. Environ. Microbiol.*, 61, 3069, 1995.
96. Kaufmann, P. et al., Identification and quantification of *Bifidobacterium* species isolated from food with genus-specific 16S rRNA-targeted probes by colony hybridization and PCR, *Appl. Environ. Microbiol.*, 63, 1268, 1997.
97. Matsuki, T., Watanabe, K., and Tanaka, R., Genus- and species-specific PCR primers for the detection and identification of bifidobacteria, *Curr. Issues Intest. Microbiol.*, 4, 61, 2003.
98. Gueimonde, M. et al., New real-time quantitative PCR procedure for quantification of bifidobacteria in human fecal samples, *Appl. Environ. Microbiol.*, 70, 4165, 2004.
99. Favier, C.F., de Vos, W.M., and Akkermans, A.D.L., Development of bacterial and bifidobacterial communities in feces of newborn babies, *Anaerobe*, 9, 219, 2003.
100. Goldenberg, O. et al., Molecular monitoring of the intestinal flora by denaturing high performance liquid chromatography, *J. Microbiol. Meth.*, 68, 94, 2007.
101. Sakata, S. et al., Culture-independent analysis of fecal microbiota in infants, with special reference to *Bifidobacterium* species, *FEMS Microbiol. Lett.*, 243, 417, 2005.
102. Boesten, R.J., Schuren, F.H., and de Vos, W.M., A *Bifidobacterium* mixed-species microarray for high resolution discrimination between intestinal bifidobacteria, *J. Microbiol. Methods*, 76, 269, 2009.

103. Germond, J.E., Mamin, O., and Mollet, B., Species specific identification of nine human *Bifidobacterium* spp. in feces, *Syst. Appl. Microbiol.*, 25, 536, 2002.
104. Gürtler, V., and Stanisish, V.A., New approaches to typing and identification of bacteria using the 16S-23S rDNA spacer region, *Microbiol.*, 142, 3, 1996.
105. Leblond-Bourget, N. et al., 16S rRNA and 16S to 23S internal transcribed spacer sequence analyses reveal inter- and intraspecific *Bifidobacterium* phylogeny, *Int. J. Syst. Bacteriol.*, 46, 102, 1996.
106. Maiden, M.C. et al., Multilocus sequence typing: A portable approach to the identification of clones within populations of pathogenic microorganisms, *Proc. Natl. Acad. Sci. USA*, 95, 3140, 1998.
107. Requena, T. et al., Identification, detection, and enumeration of human *Bifidobacterium* species by PCR targeting the transaldolase gene, *Appl. Environ. Microbiol.*, 68, 2420, 2002.
108. Masco, L. et al., Polyphasic taxonomic analysis of *Bifidobacterium animalis* and *Bifidobacterium lactis* reveals relatedness at the subspecies level: Reclassification of *Bifidobacterium animalis* as *Bifidobacterium animalis* subsp. *animalis* subsp. nov. and *Bifidobacterium lactis* as *Bifidobacterium animalis* subsp. *lactis* subsp. nov., *Int. J. Syst. Evol. Microbiol.*, 54, 1137, 2004.
109. Yin, X. et al., The gene encoding xylulose-5-phosphate/fructose-6-phosphate phosphoketolase (xfp) is conserved among *Bifidobacterium* species within a more variable region of the genome and both are useful for strain identification, *FEMS Microbiol. Lett.*, 246, 251, 2005.
110. Ventura, M. et al., Molecular characterization of hsp20, encoding a small heat shock protein of *bifidobacterium breve* UCC2003, *Appl. Environ. Microbiol.*, 73, 4695, 2007.
111. Matto, J. et al., Genetic heterogeneity and functional properties of intestinal bifidobacteria, *J. Appl. Microbiol.*, 97, 459, 2004.
112. Ventura, M. et al., Molecular microbial analysis of *Bifidobacterium* isolates from different environments by the species-specific amplified ribosomal DNA restriction analysis (ARDRA), *FEMS Microbiol. Ecol.*, 36, 113, 2001.
113. Krizova, J., Spanova, A., and Rittich, B., Evaluation of amplified ribosomal DNA restriction analysis (ARDRA) and species-specific PCR for identification of *Bifidobacterium* species, *Syst. Appl. Microbiol.*, 29, 36, 2006.
114. Cusick, S.M., and O'Sullivan, D.J., Use of a single, triplicate arbitrarily primed-PCR procedure for molecular fingerprinting of lactic acid bacteria, *Appl. Environ. Microbiol.*, 66, 2227, 2000.
115. Favier, C.F. et al., Molecular monitoring of succession of bacterial communities in human neonates, *Appl. Environ. Microbiol.*, 68, 219, 2002.
116. Delcenserie, V. et al., Discrimination between *Bifidobacterium* species from human and animal origin by PCR-restriction fragment length polymorphism, *J. Food Prot.*, 67, 1284, 2004.
117. Alvarez-Martín, P., Flórez, A.B., and Mayo, B., Screening for plasmids among human bifidobacteria species: Sequencing and analysis of pBC1 from Bifidoba*cterium catenulatum* L48, *Plasmid*, 57, 165, 2007.
118. Wang, R.F. et al., DNA microarray analysis of predominant human intestinal bacteria in fecal samples, *Mol. Cell. Probes*, 18, 223, 2004.
119. Lee, K.Y., So, J.S., and Heo, T.R., Thin layer chromatographic determination of organic acids for rapid identification of bifidobacteria at genus level, *J. Microbiol. Meth.*, 45, 1, 2001.
120. Ammor, M.S. et al., Reagentless identification of human bifidobacteria by intrinsic fluorescence, *J. Microbiol. Meth.*, 69, 100, 2007.
121. Saulnier, D.M.A., Gibson, G.R., and Kolida, S., In vitro effects of selected synbiotics on the human faecal microbiota composition, *FEMS Microbiol. Ecol.*, 66, 516, 2008.
122. Haarman, M., and Knol, J., Quantitative real-time PCR assays to identify and quantify fecal *Bifidobacterium* species in infants receiving a prebiotic infant formula, *Appl. Environ. Microbiol.*, 71, 2318, 2005.

6 Corynebacterium

Luis M. Mateos, Michal Letek, Almudena F. Villadangos, María Fiuza, Efrén Ordoñez, and José A. Gil

CONTENTS

6.1 INTRODUCTION

6.1.1 Classification, Morphology, and Biology

The coryneform bacteria are a broad group of gram-positive, irregularly shaped rods (sometime oval- or club-shaped), aerobic or facultatively anaerobic, asporogenous, and non-partially-acid-fast microorganisms classically used in the production of primary metabolites (amino acids, nucleotides, vitamins, etc.).[1] The genus *Corynebacterium* was created in 1896 in order to assign a classification to *Corynebacterium diphtheriae* (the diphtheroid bacilli), as well as several animal pathogens and parasitic microorganisms.[2] During most of the twentieth century, many microorganisms were included in this group based on their staining, morphological, and metabolic properties; in fact, in *Bergey's Manual of Determinative Bacteriology*, the coryneform group comprised a broad and diverse number of bacteria of medical (saprophytic animal and human pathogens), agricultural (plant pathogens), and industrial (saprophytic nonpathogenic amino-acid-producing strains) interest. In the 1990s, phylogenetic (based on 16S and 23S rDNA sequence analyses) and chemotaxonomic (cell wall composition and lipid profiles) analyses resulted in the separation of some members of the coryneform group from those corresponding to plant pathogens.[3] Nonetheless, the number of species described for the genus *Corynebacterium* expanded, from 17 in the first edition of *Bergey's Manual of Systematic Bacteriology*[4] to more than 70 in the second edition (http://www.bergeys.org/). This large expansion of a bacterial genus is a rare occurrence in microbiology and reflects the enormous interest in these bacteria, mainly in medicine but also by industry. While very few data are available concerning the numbers of corynebacterial species present in nature, given what is known thus far about the genus *Corynebacterium*, it can be reasonably assumed that corynebacteria are abundantly disseminated within a broad range of different natural habitats. In turn, much species diversity remains to be identified and characterized.

The most prevalent human pathogen among the corynebacteria is *C. diphtheriae*, which continues to be a major cause of morbidity and mortality in undeveloped countries and increasingly in developing countries, where nontoxigenic members cause additional pathologies.[5] *Corynebacterium* species other than those belonging to the *C. diphtheriae* group can be opportunistic or nosocomial pathogens of humans.[6] Importantly, infections by these species are becoming increasingly common, mainly due to the remarkable increase in the number of susceptible individuals, especially patients with autoimmune disorders or undergoing immunosuppressive therapy as well as organ-transplanted patients.

Species belonging to the genus *Corynebacterium* constitute a reasonably monophyletic group from the suborder *Corynebacterineae*, in the order *Actinomycetales*. They have a GC content of 51%–65% and an optimal temperature for growth of 28°C–37°C. Two broad groups of species from the genus can be distinguished based on their habitats and the consequences of the host-bacteria interaction:

(i) Saprophytic members (nonpathogenic) are widely disseminated in different environments, such as soil, sea, dairy products, and animals. Of these, the main species used in industry is *C. glutamicum*. Additional species include *C. ammoniagenes, C. casei, C. callunae, C. efficiens, C. flavescens,* and *C. thermoaminogenes*.[7]

(ii) Among the *Corynebacterium* species of medical importance, the best described is *C. diphtheriae*, but during the last few decades dozens of others have been recognized as emergent human and animal pathogens. Their optimal temperature for growth is 35°C–37°C and most are auxotrophic, with requirements for vitamins, amino acids, and/or nitrogen bases. Frequently, two broad subgroups of *Corynebacterium* members are distinguished based on an essential requirement for a particular nutrient, better growth in the presence of lipids (lipophilic group), or equivalent growth in the presence or absence of lipids (nonlipophilic group). Among the lipophilic group, species such as *C. accolens, C. afermentans* (*lipophilum*), *C. appendicis, C. genitalium, C. hansenii, C. jeikeium, C. kroppenstedtii, C. lipophiloflavum, C. macginleyi, C. pseudogenitalium, C. resistens, C. sputi, C. tuberculostearicum, C. urealyticum*, and *C. ureicelerivorans* are the most representative. Within the nonlipophilic group, a broader number of human pathogens (ascertained or potential) have been described, including *C. diphtheriae, C. amycolatum, C. argentoratense, C. atypicum, C. auris, C. aurimucosum, C. confusum, C. coyleae, C. durum, C. freneyi, C. genitalum, C. glucuronolyticum, C. imitans, C. minutissimum, C. mucifaciens, C. pilbarense, C. pseudodiphtheriticum, C. pseudotuberculosis, C. resistens, C. riegelii, C. simulans, C. singulare, C. striatum, C. sundvallense, C. thomssenii, C. timonense, C. tuscanie, C. ulcerans*, and *C. xerosis*. In addition, the nonlipophilic group can be subdivided in fermentative and nonfermentative groups depending, respectively, on the ability or inability (to metabolize sugars under anoxic conditions). Table 6.1 summarizes most of the human-pathogenic *Corynebacterium* species isolated from clinical samples and characterized thus far, together with the most relevant information concerning these species.

6.1.2 Phenotypic and Phylogenetic Characterization

6.1.2.1 Phenotypic Analysis

In-depth characterization of new isolates as members of *Corynebacterium* and their description as species of this genus rely on phenotypic analyses and thus on a combination of (i) morphological, (ii) physiological, and (iii) chemotaxonomic criteria. Preliminary identification has been traditionally achieved by Gram staining, which reveals the distinctive morphology of *Corynebacterium* together with additional determination of the bacterium's physiological properties, such as growth in specific culture media, colony size and shape, pigmentation, cell motility, temperature, and pH. Further analyses include biochemical and enzymatic reactions, such as those determined in automated systems for rapid species identification. The three most frequently used systems are: (i) API Coryne (version 2.0; bioMérieux, France), (ii) BIOLOG (Biolog, Inc., Hayward, CA), and (iii) RapID CB Plus (Remel, Inc., Norcross, GA). Most of the biochemically based phenotypic analyses are carried out with the test included in the API Coryne system. A brief description of the test's reactions and corresponding protocol is provided in Section 6.2.2.2, and a description of the protocol for the BIOLOG system in Section 6.2.2.1. The performances of these systems have been evaluated in different studies with variable results because in most cases additional biochemical assays specific for corynebacteria, such as DNase activity and tyrosine hydrolysis, are necessary for definitive identification of the isolates in question.[47–49]

Most of the chemotaxonomic analyses exploit the presence in corynebacteria of the isomeric form of diaminopimelic acid[50] or the bacterial content of whole-cell sugars[51] or lipids. Alternatively, they are based on the bacteria's fatty acids profile or the presence of corynomycolic acids and other β-hydroxy-α-branched carboxylic acids.[52]

6.1.2.2 Protein Analysis

Multilocus Enzyme Electrophoresis (MEE). This technique allows the detection of amino acid substitutions in cellular enzymes that affect their conformation and overall charge. Mobility variants (electromorphs) of the same enzyme can be visualized by their activity in a gel (i.e., as bands of altered migration rates), and each electromorph

TABLE 6.1
***Corynebacterium* Species Described as Human Pathogens (or Potential Pathogens)**

Corynebacterium Species	Descriptions	Lipo	Fermentation	HMP Sequence/Size (Md)	GC%	Chemotype	Locus/Disease	Phenotype	Genotype	Reference
C. accolens	<10	L	++	Gi02699/2.4 Mb	59	Type IV CM+	Oral tract Endocard	Catalase+ Urease– Nit-Red+	ARDRA DNA Hyb	8
C. afermentans (subsp *lipophilum)s*	<10	L	NF	ND	68	Type IV CM+	Abcess/Valves infect.	Urease–	ARDRA DNA Hyb	9
C. amycolatum	>50	NL	NF*/+	Gi03352/2.5 Md	58	CM– Tubst–	Sepsis/Endocard	Catalase+	16S rDNA	10
C. argentoratense	<10	NL	+	ND	60	Type IV CM+	Throat	Catalase+ Urease–	DNA Hyb 16S rDNA	11
C. appendicis	<10	L	NF	Gi03635	ND	Type IV CM+ Tubst+	Oral infection	Catalase+	16S rDNA	12
C. atypicum	<10	NL	+++	Gi03105	ND	CM– Tubst–	Clinical sample	Catalase+ Halotolerant	16S rDNA	13
C.auris	>50	NL	NF	ND	70	Type IV CM+	Ear infection	Catalase+ Nit-Red–	16S rDNA	14
C. aurimucosum	<10	NL	++	Gc00991/2.7 Md	60	Type IV CM+ Tubst+	Bronchitis	Catalase+	DNA Hyb 16S rDNA	15
C. confusum	<10	NL	+	Gi03300	ND	Type IV CM+ Tubst+	Foot/Blood	Catalase+ Nit-Red+	16S rDNA	16
C. coyleae	<10	NL	NF*/+	Gi05148	62/64	Type IV CM+	Oral/Blood	Catalase+	DNA Hyb 16S rDNA	17
C. diphtheriae	>200	NL	++	Gc00173/2.8 Mb	53	Type IV CM+	Respiratory infection	Catalase+	16S rDNA DT-PCR-amp	2
C. durum	<10	NL	++	Gi03719	55	CM+	Respiratory infection	Catalase+ Nit-Red+	DNA Hyb 16S rDNA	18
C. freneyi	<10	NL	++	ND	ND	Type IV CM+	Ulcers/Pus	Catalase+	DNA Hyb ITS-PCR 16S rDNA	19
C. genitalium	<10	L	NF*/+	Gi02702/2.3 Mb	62	Type IV CM+	Urinary tract/Urethritis	Urease– Nit-Red–	DNA Hyb	20
C. glucuronolyticum	<10	NL	++	Gi03343/2.8 Md	59	CM+ Tubst–	Genital tract	Catalase+	DNA Hyb 16S rDNA	21
C. hansenii	<10	L	+++	Gi03314	ND	CM+	Liposarcom	Catalase+	DNA Hyb	22
C. imitans	<10	NL	++	Gi03315	ND	CM+	Human throat	Catalase+	Ribotyping PFGE 16S rDNA	23

continued

TABLE 6.1 (Continued)
***Corynebacterium* Species Described as Human Pathogens (or Potential Pathogens)**

Corynebacterium species	Descriptions	Lipo	Fermentation	HMP Sequence/Size (Md)	GC%	Chemotype	Locus/Disease	Phenotype	Genotype	Reference
C. jeikeium	>100	L	+	Gc00271/2.5 Mb	61	CM+	Nosoco-meningitis	Catalase+ Urease– Nit-Red–	DNA Hyb 16S rRNA	24
C. kroppenstedtii	<10	L	++	Gc00770/2.4Mb	57	CM– Tubst+	Sputum	Catalase+	16S rRNA	25
C. lipophiloflavum	<10	L	++	Gi02705/2.3Mb	65	Type IV CM+	Vaginosis	Catalase+ Urease–	16S rRNA	26
C. macginleyi	<10	L	+	Gi03316	58	Type IV CM+	Keratitis corneal/Oral tract	Catalase+ Urease– Nit-Red+	16S rRNA	27
C. massiliense	<10	NL	NF	ND	ND	Type IV CM+ Tubst+	Knee arthroplasty	Catalase+ Oxidase–	16S rRNA *rpoB*-ampl	28
C. minutissimum	>50	NL	NF*/+	Gi04644	58/60	Type IV CM+ Tubst+	Clinical sample	Catalase+	DNA Hyb 16S rRNA	15,29
C. mucifaciens	<10	NL	NF	Gi03722	64	CM+ Tubst+	Oral infection	Catalase+ Urease– Nit-Red–	16S rRNA	30
C. pilbarense	<10	NL	+	ND	ND	Type IV CM+ Tubst–	Clinical ankle aspirate	Catalase+ Oxidase–	DNA Hyb 16S rRNA	31
C. pseudodiphtheriticum	>50	NL	NF	Gi04695	55	Type IV	Pulmonary/Oral infection	Catalase+ Urease+	16S rRNA	2
C. pseudogenitalium	<10	L	+	Gi02707/2.6Mb	59	Type IV CM+	Urinary tract	Urease– Nit-Red–	16S rRNA	32
C. pseudotuberculosis	>200	NL	++	Gi04675/2.3Mb	52	Type IV CM+	Lynphaden/Granulom	Catalase+ Oxidase– Urease+	16S rRNA DT-PCR-amp	33
C. resistens	<10	L	+	Gi03107	55	Tubst+	Clinical samples	Catalase+	DNAHyb 16S rRNA	34
C. riegelii	<10	NL	+	Gi03318	ND	Type IV CM+	Urinary infection	Catalase+ Urease+	DNAHyb 16S rRNA	35
C. simulans	<10	NL	++	Gi04645	ND	Type IV+ CM+	Abcess and boil	Catalase+	16S rRNA ARDRA	36
C. singulare	<10	NL	++	Gi03319	ND	Type IV CM+	Semen specimen	Catalase+	DNAHyb 16S rRNA	37
C. striatum	>100	NL	+	Gi02710/2.7Mb	59		Urogenital tract	Catalase+ Nit-Red+	16S rRNA ARDRA	38

C. sputi	<10	L	+	ND	ND	Type IV CM+ Tubst+	Sputum	Catalase+ Oxidase– Urease+	16S rRNA	39
C. sundsvallense	<10	NL	+++	Gi04426	64	Type IV CM+	Clinical samples	Catalase+ Urease+	PAGE 16S rRNA	40
C. thomssenii	<10	NL	+++	ND	ND	Type IV CM+	Pleural fluids	Catalase+ Urease+	16S rRNA	41
C. timonense	<10	NL	++	ND	ND	Type IV CM+ Tubst–	Endocard	Catalase+ Oxidase–	16S rRNA *rpoB*-ampl	28
C. tuberculostearicum	<10	L	F	Gi03351/2.4Mb	60	CM+ Tubst+	Clinical samples	Catalase+ Urease– Oxidase–	PAGE 16S rRNA	42
C. tuscanie	<10	NL	++	ND	ND	CM+	Blood Endocard	Catalase+	16S rRNA	43
C. ulcerans	>100	NL	++	ND	ND	Type IV CM+	Clinical samples	Catalase+ Urease+	16S rRNA	44
C. urealyticum	>100	L	NF*/+	Gi01607/2.3Mb	64	Type IV CM+ Tubst+	Cystitis Pyelitis	Catalase+ Oxidase– Urease+	16S rRNA DNA Hyb	45
C. ureicelerivorans	<10	L	+	Gi03321	ND	Type IV CM+ Tubst+	Septicemia Blood	Catalase+ Urease+ Nit-Red–	16S rRNA *rpoB*-ampl	46
C. xerosis	>100	NL	+++	Gi04647	ND	Type IV CM+	Endocard	Catalase+ Urease–	16S rRNA ARDRA	2

Notes: Descriptions, number of References in PubMed; Lipo, L: lipophilic; NL: Nonlipophilic; fermentation, +++/++/+: relative amount of fermented substrates; NF, nonfermenters or variable*; HMP sequence/size, code from Human Microbione Project: http://www.hmpdacc.org/; chemotype, type IV: cell wall with meso-diaminopimelic acid, arabinose, and galactose; CM+/–, Corinomycolic acids presence or absence; Tubst+/–, tuberculostearic acids presence or absence; locus-disease, most frequent; phenotype, most characteristic; genotype, molecular technique used for description; reference, number from the references; ND, not determined.

potentially represents a distinct allele of that enzyme. The electromorph profile can be used to define the electrophoretic pattern of a corynebacterial strain, assuming that the number of bands analyzed is sufficient to allow intraspecies discrimination. The data are represented as a dendrogram and the distance between the strains determined. This method was used, for example, to distinguish different *C. diphtheriae* isolates.[53]

Protein Fingerprinting. SDS-PAGE protein patterns, from whole-cell soluble proteins or a purified particular fraction, can be generated for the strains under study and compared with those of reference strains. Strain-specific differentiation was previously used to characterize certain *Corynebacerium* strains in clinical isolates. Nowadays, however, it provides supplementary information to that gained through polyphasic approaches in order to fine-tune species identification.

6.1.2.3 Phylogenetic Analysis

Over the last two decades, a much broader range of taxonomic studies of bacteria has gradually replaced classical techniques. Molecular analyses of the informational units of cells (proteins, DNA and RNA) have been developed and used in the identification of *Corynebacterium* members. Modern molecular techniques are based on DNA analysis (DNA–DNA hybridization after gel electrophoresis, the presence of plasmids or integrated elements of uncertain origin, etc.), 16S-23S rRNA gene sequencing (full or partial), PCR-mediated amplification, and protein studies. Different genotyping techniques may be applied to *Corynebacterium* as tools for either species identification or the differentiation of strains within a species. DNA-based typing methods are highly discriminatory and can be used to distinguish closely related strains with similar phenotypic features. The most frequently used approach is based on PCR-mediated amplification of DNA units common to all bacteria but with internal specific changes. This polymorphism allows the separation of isolates belonging to closely related genera or of members of the same genus but from different species. Identification methods based on 16S rRNA and DNA-dependent RNA polymerase β-subunit (*rpoB*) genes are mostly employed together with "broad-range" primers.[54] However, the analysis of either 16S rRNA or *rpoB* often lacks specificity due to the absence of polymorphisms. Therefore, many clinical laboratories have sought to design specific primers and the respective protocols to specifically amplify a defined target within one species and thus, consequently, excluding other species.[55] In short, limited homology between target genes and broad-range primers on the one hand and insufficient polymorphism among closely related bacteria on the other can preclude the distinction of *Corynebacterium* members. Perhaps the most complete genotypic analyses of *Corynebacterium* species to date have been those of Khamis et al.[56,57] in which the effectiveness of sequencing the complete 16S rRNA was compared with the use of potentially more discriminatory targets, such as internal fragments of *rpoB*.

(i) DNA-Based Analysis **Single-Strand Conformation Polymorphism (SSCP).** In this classical form of mutation analysis, amplified double-stranded DNA is denatured to generate single-stranded DNA, which is then electrophoresed in a non-denaturing polyacrylamide gel. The migration rate of single-stranded DNA depends not only on its length but also on its secondary structural conformation, resulting in a specific SSCP pattern for each mutant gene. SSCP has been used to analyze important genes of *Corynebacterium*, such as *tox* and *dtxR*.[58]

Restriction Enzyme Analysis (REA) and Pulse-Field Gel Electrophoresis (PFGE). The digestion of chromosomal DNA (total DNA) with restriction endonucleases generates a series of fragments that are then usually separated by conventional agarose gel electrophoresis. By selecting the appropriate restriction enzyme or a set of enzymes species- or strain-significant, REA patterns are generated. Nonetheless, the resulting banding pattern (fragment sizes between 0.5 and 20 kb) is sufficiently complex such that additional analytic methods are usually required. A modification of REA is pulse-field gel electrophoresis (PFGE), in which a periodic change of the orientation of the electric field enables the separation of high-molecular-weight fragments. PFGE can be combined with the use of rare-cutting restriction endonucleases, which generate a low number of fragments, thus yielding a banding pattern that is more easily interpreted. The typical PFGE DNA fingerprint consists of 5–20 large (10–800 kb), well-resolved fragments and has high discriminatory and reproducible power.

Hybridization Analysis. Southern blot assays are used for species identification and are based on the use of randomly cloned heterologous or homologous DNA fragments as probes, which are then hybridized with total DNAs from the new isolates. Most of the newly identified pathogenic *Corynebacterium* species were initially characterized using other techniques, but hybridization analyses in which the isolates' total DNA is compared with that of closely related species remains the definitive test to confirm a new species.

Ribotyping and Restriction Fragment Length Polymorphism (RFLP). Ribotyping is based on the hybridization of digested chromosomal DNA with probes derived from specific rRNA genes. In the genomes of many bacteria, rRNA operons (*rrn*) are present in several copies and the resulting multiband patterns are frequently strain specific. Although the entire *rrn* operon can serve as a probe, nowadays a mixture of a few oligonucleotides from conserved regions of 16S–23S rRNA genes is used for this purpose, making the method a truly universal approach to bacterial strain typing. Two genetic patterns are considered identical when they are composed of the same number of fragments and, for homologous fragments, size differences are below a 5% threshold error value. Fingerprint patterns are more stable, more easily reproducible, and more readily interpreted than those obtained with REA. Ribotyping has been used with some success to study the diversity of *Corynebacterium* strains, although with low discriminatory power. The choice of a suitable restriction enzyme is important and different ribopatterns may be obtained by selecting different restriction enzymes. RFLP techniques are equivalent to ribotyping, the difference being that they make use of any DNA sequence (homologous or heterologous) for hybridization purposes.

16S-23S rRNA Gene-sequencing Analyses. All available *Corynebacterium* 16S rRNA, 23S rRNA, and 16S-23S rRNA spacer region sequences can be retrieved from two databases, EMBL and GenBank, and aligned using the Clustal program. Thus, potentially strain- and species-specific primers can be employed for specific purposes. The amplified rRNA fragment is further sequenced according to conventional techniques and the results are compared with the above-mentioned databases. As this is the molecular technique most widely used in *Corynebacterium* species identification and determination, a detailed protocol is described in the Methods section of this chapter.

Multilocus Sequence Typing (MLST). This method is based on the sequential amplification of several housekeeping genes (e.g., *adk*, *dnaA*, *fumC*, *gltA*, *gyrB*, *icd*, *purA*) obtained from the whole-genome sequences of four *Corynebacterium* species (*C. glutamicum*, *C. diphtheriae*, *C. efficiens*, and *C. jeikeium*). The design of a primer pair for each of the target genes is based on seven regions conserved among the four *Corynebacterium* species, as they are amplified under the same conditions. The nucleotide sequences obtained from these seven loci are concatenated, excluding the primer sequences, and then further aligned and compared using the Clustal X program to infer a phylogenetic tree.[59]

(ii) PCR-Based Analysis Here, the methods are mainly based on the use of either oligonucleotide sequences from widespread genes and conserved regions of DNA (mostly rRNA genes, but also *rpoB* genes and others) or specific oligonucleotides that amplify a certain gene or DNA fragment from a particular species that allows its direct identification. An example of the latter is amplification of rRNA intergenic spacer regions or the corynebacterial *tox* gene.

Randomly Amplified Polymorphic DNA (RAPD). PCR is carried out using short primers of random sequences under low-stringency annealing conditions, resulting in the amplification of randomly sized DNA fragments. This method requires carefully controlled conditions in order to obtain good reproducibility.

Denaturing Gradient Gel Electrophoresis (DGGE). Specific primers are used to amplify target sequences, frequently from rRNA genes. The PCR-amplified fragments are then separated in a sequence-specific manner by DGGE, yielding a profile of fragments that identify the microbial species under study. In most cases, PCR products of the same length from different species are well separated by DGGE.

Amplified Ribosomal DNA Restriction Analysis (ARDRA). This is a sequential method based on PCR amplification of the rRNA gene (frequently 16S rRNA) followed by restriction analysis of the amplified fragment with different restriction enzymes. ARDRA is a reliable and highly reproducible tool to discriminate very closely related species or isolates of the same species.

Multiplex PCR. With this method, more than one locus is simultaneously amplified in the same reaction. The method is suitable for in vivo studies and can be used for strain identification without the need for prior DNA extraction. Target sequences are frequently based on 16S-23S rRNA genes or their intergenic spacer region.

6.1.3 Pathogenesis and Diagnosis

6.1.3.1 *Corynebacterium Diphtheriae*

Corynebacterium diphtheriae is the causative microorganism of diphtheria, although, less often, *C. ulcerans* is involved. *C. diphtheriae* species comprise four biotypes (var.): *gravis*, *mitis*, *intermedius*, and *belfanti*; all of which, with the exception of *belfanti*, produce the lethal diphtheria exotoxin.[60] Diphtheria is usually classified according to its site of manifestation, which most often is the respiratory tract or the skin. *C. diphtheriae* releases the toxin after colonizing the upper respiratory tract and the skin. Once the toxin is absorbed by the circulatory system, it is distributed to distant organs (myocardium, peripheral nervous system, etc.), where it causes the typical symptoms of the disease. In respiratory diphtheria, the bacteria form a membrane in the posterior structures of the oral cavity (pharynx/larynx). The continued growth of the microorganism on the surface of this membrane generates the characteristic "pseudomembrane," which reduces airflow from the trachea and thus causes suffocation and possibly death. The genome sequence of *C. diphtheriae* biotype *gravis* ribotype "Sankt Petersburg" was published in 2003.[61] The bacterial toxin is encoded by a mobile temperate bacteriophage (corynephage) and inhibits protein synthesis in the host cell by ADP-ribosylation of elongation factor 2. However, *C. diphtheriae* also has a remarkable capacity to cause invasive bloodstream infections even in the absence of toxin production. Systemic complications of *C. diphtheriae* bacteremia are not unusual and include endocarditis, joint infections, and peripheral embolic disease. The evolutionary origin of nontoxigenic *C. diphtheriae* is unknown; however, it is very likely that the success of the diphtheria vaccine in preventing toxin-mediated disease resulted in the selection of new pathogenic *C. diphtheriae* variants expressing other virulence factors (e.g., siderophores or other iron-uptake proteins), or equipped with other disease-causing molecular mechanisms that allow the bacteria to overcome host conditions. The increased number of cases of endocarditis caused by these invasive clones testifies to their clinical importance.[62]

While used in the past, typing systems such as serotyping, phage typing, and bacteriocin typing have largely been replaced by molecular methods—most importantly, Southern hybridization with *tox* and IS probes, as well as cell wall protein analysis by SDS-PAGE, and gel separation of digested chromosomal DNA.[63] For *C. diphtheriae*, traditional molecular typing methods, including ribotyping, are straightforward and effective in distinguishing isolates of this species.[64]

An interesting molecular approach in the identification of *C. diphtheriae* strains and subspecies is the above-described multilocus enzyme electrophoresis (MEE). MEE was used to analyze the genetic diversity within a combined panel of preepidemic and epidemic *C. diphtheriae* strains isolated in 1985–1994[53] and for further studies; using this technique a clonal group of closely related strains was distinguished, whereas PFGE, RAPD, amplified restriction fragment polymorphism (AFLP), and ribotyping failed to do so. PCR-based methods are faster and simpler, although they sometimes have low discriminatory power and reproducibility. AFLP has been widely used for *C. diphtheriae* typing and was suggested as a replacement for ribotyping as the "gold standard" of typing methods, because it requires only a small amount of DNA, analyzes the entire genome, and previous sequence information on the target DNA is not necessary. In a recent study, four molecular typing methods (ribotyping, PFGE, RAPD, and AFLP) were used in the characterization of 755 isolates of *C. diphtheriae* in order to examine the global spread of the pathogen.[65] In that study, ribotyping was considered the gold standard for typing due to its discriminatory and reproducible power, followed by RAPD, PFGE, and AFLP. However, RAPD and AFLP are rapid methods that can be used as screening techniques prior to ribotyping, thus omitting the need for complex ribotyping analyses of strains that are unrelated.

Polymorphisms in the *tox* gene and its regulatory element, *dtxR*, in *C. diphtheriae* strains isolated in Russia and the Ukraine before and during an epidemic were studied by SSCP analysis,[58] while PCR–RFLP analyses of the 16S-23S rRNA internal transcribed spacer differentiated *C. diphtheriae* from *C. ulcerans*.[66]

Publication of the *C. diphtheriae* genomic sequence enabled a more thorough, precise, and comprehensive search of candidate polymorphic loci for the development of new typing methods of this pathogen. One such method is based on the clustered regularly interspaced short palindromic repeat (CRISPR) technique. CRISPR consists of short sequences (21–37 nucleotides long) that are repeated almost exactly and referred to as "units," which are separated by similarly sized nonrepetitive sequences called "spacers."[67] CRISPR sequences are present in many of the known bacteria and were advantageously used in hybridization macroarray-based methods to study the internal polymorphism of 156 strains of the epidemic-causing Sankt Petersburg clone in 1997–2002.[68]

6.1.3.2 *Corynebacterium* Species of the Lipophilic Group

The majority of the "lipid-requiring," or "lipophilic" strains of *Corynebacterium* are readily cultured in the laboratory as they are able to grow on most routinely used culture media, which contain at least minimal amounts of lipids and support the basal growth of lipophilic bacteria. Thus, *Corynebacterium* grows reasonably well on media enriched with a significant amount of lipids (e.g., 0.1%–1% Tween 80, or 2 drops of sterile serum). Lipophilic corynebacteria are usually fastidious, and slower growing than nonlipophilic strains. Most are nonhemolytic on blood Columbia agar, contain short corynemycolic acids (C22 to C36), arabinogalactans, and meso-diaminopimelic acid in their cell walls, and produce dehydrogenated menaquinones. In general, they lack the genes required for synthesizing the fatty acids present in the cell membrane and are resistant to a broad range of antibiotics.

***Corynebacterium urealyticum*.** This strictly aerobic species was previously included in *Corynebacterium* group D2 of the Centers for Disease Control (CDC, Atlanta, GA) classification due to its characteristics of very high urealytic activity, nonfermentative, nitrate-negative metabolism, and lipophilicity. However, DNA hybridization analysis invalidated this taxonomic assignment and resulted in the bacterium's formal designation as *C. urealyticum*. The description of this new species was also supported by phenotypic tests, such as API profiles, and by chemotaxonomic (presence of fatty acids, isoprenoids, quinones, and phospholipids) and molecular (rRNA gene profiling) analyses.[45]

The natural habitat of *C. urealyticum* is the human skin, but it also can be isolated from blood and wounds and readily colonizes the urinary tract. In fact, highly antibiotic-resistant strains of *C. urealyticum* have been associated with cases of encrusted cystitis and other urinary tract infections in hospital patients.[69]

C. urealyticum resembles *Corynebacterium jeikeium* in its colony and cellular morphologies. In addition, the two species have a common habitat and both are remarkably antibiotic resistant. However, both can be clearly distinguished to each other by amplification of the 16S-23S rRNA gene spacer regions.[70] Biochemical analyses of *C. urealyticum* indicate a closest relation to *Corynebacterium pseudodiphtheriticum* (due to its urealytic activity), although the latter is positive for nitrate reduction and nonlipophilic.

The genome sequence of *C. urealyticum* was recently obtained by pyrosequencing[71] and contains about 2020 coding sequences. Of these, 80% are orthologous to the corresponding genes of the *Corynebacterium jeikeium* genome,[72] indicating a close phylogenetic relationship between these two lipid-requiring species. A DNA region with 70 CRISPRs and eight CRISPR-associated genes was identified in the *C. urealyticum* genome. The DNA repeats are 28 bp in length, separated by variable 33-bp spacer sequences, and similar to the clustered repeats of CRISPRs in the *C. jeikeium* genome. The presence of *C. urealyticum* in patients with cystitis has been detected by PCR amplification of a 212-bp band characteristic for this specie.[73]

***Corynebacterium jeikeium*.** The bacterium is considered part of the normal flora of the human skin, where it predominantly colonizes the axillary, inguinal, and perineal areas. Initially, bacteria from "group JK corynebacteria" (CDC, Georgia At) were associated with cases of bacterial endocarditis that occurred following cardiac surgery and isolated as nondiphtherial corynebacteria from blood cultures. These bacteria were later named *C. jeikeium* and frequently confirmed as the causative agent of nosocomial infections, especially in immunocompromised patients.

Many of the reported *C. jeikeium* isolates are substantially multiresistant against clinically relevant antibiotics, limiting the therapeutic options in infected patients. Whole-genome sequencing of *C. jeikeium* showed that its multidrug resistance is encoded by chromosome-associated genes most likely acquired by bacterial horizontal transfer.[72] More than 50% of the *C. jeikeium* genome comprises genes orthologous to those of nonpathogenic corynebacteria, and a closest relation to the *C. urealyticum* genome has been reported.[71,72] The molecular techniques used to identify *C. jeikeium* are based on species-specific probes for its 16S rRNA.[74] Plasmid content and RFLP techniques rRNA-associated have also been used in strain identification, although polymorphisms due to minor changes in RFLP profiles were noted as well.[75]

Corynebacterium ureicelerivorans. This lipophilic coryneform bacterium generates organic acids from glucose and splits urea very rapidly. *C. ureicelerivorans*, most likely produces struvite stones, made up of ammonium, magnesium, and phosphates, as has been demonstrated in *C. urealyticum* and other urea-splitting organisms.

C. ureicelerivorans has a close genetic relationship with *C. mucifaciens* and *C. jeikeium*. New isolates of *C. ureicelerivorans* isolated from blood cultures obtained from patients with bacteremia were identified based on phenotypic (API and Biolog systems) and molecular (16S rDNA and *rpoB* sequencing) characteristics.[76] Colonies of these bacteria growing on blood Columbia agar are very small, whitish, dry, and circular, with intact edges. A misidentification of *C. ureicelerivorans* with *C. bovis* is possible when only the API system is used, as the identification code is the same for the two species. An equivalent uncertainty likewise results when the Biolog system is used for identification. Instead, molecular typing of *C. ureicelerivorans* is recommended and is based on *rpoB* sequence analyses.[76] According to sequencing database, the percent similarity of *C. ureicelerivorans* with *C. mucifaciens* is very high (91%–95%).

Corynebacterium macginleyi. This diphtheroid (named in honor of KJ McGinley, who made important contributions to the study of lipophilic diphtheroids) is oxidase-negative, facultatively anaerobic, and able to produce organic acids from sugars.[27] It has been implicated in conjunctivitis and endophthalmitis, but also in endocarditis and septicemia. One of the most interesting reports concerning *C. macginleyi* is the etiologic agent of keratitis following keratoplasty, described by Suzuki et al.[77] In their study, two *C. macginleyi* isolates were identified by 16S rRNA sequencing (99.2% similarity to the type strain) and further *rpoB* gene analyses. In the latter, the similarity of the isolates was 97.3% with *C. macginleyi* and 92% with the closely related *C. accolens*. Recently, new molecular techniques have been applied in the molecular identification of eye-associated pathogens. For example, multilocus sequence typing (MLST), with seven housekeeping genes (*adk, dnaA, fumC, gltA, gyrB, icd*, and *purA*) from sequenced species was used to design consensus primers for further DNA amplification from pathogenic strains. The MLST results were subsequently validated by RAPD analyses, thus confirming the technique's very good discriminatory power for *C. macginleyi* and *C. accolens*.[78]

Corynebacterium accolens. This species, formerly known as diphtheroid genomospecies II (CDC), is an inhabitant of the upper respiratory tract and is closely related to *C. macginleyi*. *C. accolens* was originally characterized by its satellite growth around a *Staphylococcus aureus* streak on blood agar.[8] When grown on Columbia blood agar with lipids, the bacterium forms small, gray, nonhemolytic, and transparent colonies. Many of the reported *Corynebacterium*-related cases of endocarditis have been due to lipophilic organisms such as *C. jeikeium*, C. *afermentans* (subsp. *lipophilum*), and *C. accolens*. Like *C. macginleyi*, which *C. accolens* most closely resembles, it is negative for maltose fermentation and urease production but positive for glucose fermentation and nitrate reductase production. However, *C. accolens* can be differentiated from *C. macginleyi* since the former is negative for alkaline phosphatase. In addition, the two species can be distinguished by ARDRA analyses based on *CfoI* digestion of amplified 16S rRNA gene fragments.[79]

6.1.3.3 *Corynebacterium* Species of the Nonlipophilic Group

Many of the members of this group were easily identified by ARDRA analyses based on their 16S rRNA genes.[80]

***C. xerosis, C. striatum*, and *C. amycolatum* Subgroup.** *C. xerosis* was one of the earliest *Corynebacterium* species identified as a human pathogen. It exhibits fermentative metabolism and α-glucosidase, alkaline phosphatase, and pyrazinamidase activities. Although this species was previously described as a definite human pathogen, most of the strains tentatively identified in the routine clinical laboratory as *C. xerosis* have since been reclassified as *C. striatum* or *C. amycolatum*. Nowadays, true *C. xerosis* clinical isolates are extremely rare,[6] although we have evidences that isolates obtained from clinical human samples are indeed *C. xerosis* (LM Mateos, unpublished results). All three species commonly occur in the normal human flora of the skin and the oropharynx, making it difficult to distinguish infection from contamination or colonization.

Phenotypically, *C. xerosis* strains produce yellowish colonies, express mycolic acids, have leucine arylamidase activity, produce lactic acid as a byproduct of glucose fermentation, and are susceptible to the vibriocidal agent O/129. *C. amycolatum* strains, in contrast, produce whitish-grayish colonies, lack mycolic acids, have either weak or no leucine arylamidase activity, produce large amounts of propionic acid from glucose fermentation, and are mostly resistant to O/129; a culture of *C. amycolatum* grown on Columbia agar with 5% blood can be seen in Figure 6.1. Accordingly, differentiation between *C. amycolatum* and *C. xerosis* is achieved by chemotaxonomic procedures, such as mycolic acid analysis and the detection of propionic acid production from glucose (*C. amycolatum*), whereas routine identification is difficult because the colonies formed by the two species are very similar in appearance. Macroscopically, *C. xerosis*

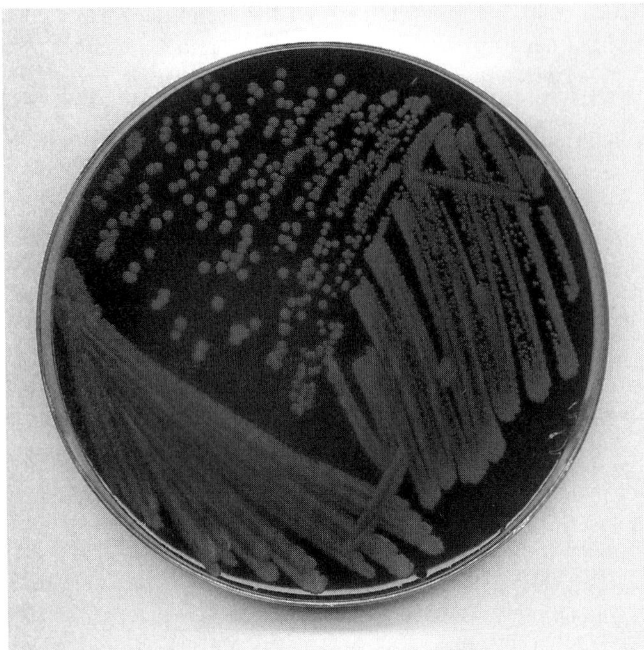

FIGURE 6.1 Culture of *C. amycolatum* grown on Columbia agar with 5% blood.

and *C. striatum* colonies can be distinguished because the colonies of *C. xerosis* are yellowish, dry, and granular, while those of *C. striatum* are whitish, moist, and smooth. As mentioned earlier, the two species can be differentiated by ARDRA[80] and other PCR-derived analyses.[55]

C. minutissimum. Colonies of this species are macroscopically convex and circular, about 1–1.5 mm in diameter (after 24 h in Columbia agar), with a shiny and moist surface and intact edges. *C. minutissimum* is the causative agent of erythrasma, an asymptomatic skin infection characterized by scaly, reddish, macular patches in intertriginous areas.[81] When tested with the API Coryne strip, *C. minutissimum* strains may be misidentified as *C. jeikeium* because the same code number is ascribed to both species; however, the latter requires lipids and is aerobic. *C. minutissimum* can also be misidentified as *C. amycolatum*, but certain phenotypic reactions (i.e., DNAase activity), the presence of tuberculostearic acid, and the products obtained from glucose fermentation can help to distinguish between the two. In addition, *C. minutissimum* is able to grow on medium containing 6.5% NaCl. Molecular analyses have been broadly used, targeting the PCR-amplified 16S rRNA gene. The polymorphism of the *divIVA* gene (after PCR amplification) has been successfully used to distinguish different nonlipophilic fermentative coryneform species.[55]

Corynebacterium freneyi. This strain, which is closely related to *C. amycolatum* and *C. xerosis*, was first described by Renaud et al.[19] Phenotypically, *C. freneyi* can be differentiated from *C. xerosis* by its distinct wrinkled colonies, whereas nearly all other routinely applied phenotypic tests do not allow unanimous definitive separation of the two species. Instead, molecular techniques—specifically, RFLP analysis of the 16S-23S rRNA intergenic spacer using the enzyme *CfoI*,[82] 16S rRNA gene sequencing, and DNA hybridization (% similarity)—must be applied. Polyphasic analyses employing the *rpoB* and 16S rRNA genes (among others) have been successfully used as well.[57]

6.2 METHODS

6.2.1 Sample Preparation

6.2.1.1 Culture Media for *Corynebacterium*

There are no selective or enrichment procedures that are specifically suitable for *Corynebacterium*; rather, a wide variety of media can be used for their culture as well as for macroscopic and microscopic studies. Most members of *Corynebacterium* are able to grow on modified peptone-yeast-glucose media such as *Corynebacterium* medium (DSMZ Medium 790), containing (per liter): 10.00 g peptone; 1.00 g yeast extract; 2.00 g $MgSO_4 \cdot 7H_2O$; and 2.00 g $(NH_4)_2SO_4$ at pH 7.0. Good complex media include brain heart infusion (BHI, Oxoid), tryptone-soy agar (TSA; Difco), and Columbia agar (Difco) supplemented with 5% horse or sheep blood, with incubation at 35°C–37°C in an air plus 5% CO_2-enriched atmosphere (when required). The composition of Columbia agar medium with 5% blood is described below.

Chemically defined media are available only for nonpathogenic corynebacteria, with the addition of certain growth factors to stimulate growth (e.g., dihydroxyphenolic compounds such as catechol, and vitamins biotin and thiamine)[83]; see below for the composition of BMCG medium.

Columbia Agar + 5% Sheep Blood (1 L). 5.0 g peptic digest of animal tissue; 5.0 g pancreatic digest of casein; 10.0 g yeast-enriched peptone; 3.0 g pancreatic digest of heart muscle; 1.0 g cornstarch; 5.0 g sodium chloride; 14.0 g agar; 50.0 mL sheep blood; and final pH of 7.3 ± 0.2 at 25°C.

Basal Medium for *Corynebacterium* Growth (Synthetic BMCG) (1 L). 7.0 g $(NH_4)_2SO_4$; 6.0 g Na_2HPO_4; 3.0 g KH_2P0_4; 0.5 g NaCl; 1.0 g NH_4Cl; 400 mg $MgSO_4{\cdot}7H_2O$; 20 mg $FeSO_4{\cdot}7H_2O$; 2 mg $MnSO_4{\cdot}H_20$; 0.2 mg $Na_2B_4O_7{\cdot}10H_2O$; 80 mg $(NH_4)_6Mo_7O_{24}{\cdot}4H_2O$; 20 mg $ZnSO_4{\cdot}7H_2O$; 0.5 mg $CuSO_4{\cdot}5H_2O$; 15 mg $MnCl_2{\cdot}4H_2O$; 0.2 mg $FeC1_3{\cdot}6H_2O$; 30 mg biotin; 2 mg thiamine; 3 mg nicotinic acid; 10 μM catechol; and 10 g of a carbon source (frequently glucose). All solutions are adjusted to neutral pH.

6.2.1.2 DNA Isolation

This protocol is based on the method described by Vaneechoutte et al.[80] Strains are grown on Columbia agar supplemented with 5% (v/v) sheep blood at 37°C (with or without CO_2 enrichment). A loopful of the corresponding colony (about 10–20 μL) is suspended in 300 μL of distilled water, boiled for 10 min, agitated thoroughly, and then centrifuged briefly in a benchtop microcentrifuge. The supernatant of the sample preparation is placed in a new Eppendorf tube and is ready to be used for further DNA analyses.

6.2.1.3 RNA Isolation

This protocol is a combination of the RNA extraction methods designed for Trizol (Tri Reagent, Sigma) and silica columns ("Qiagen RNeasy kit," Qiagen). As complete DNA digestion is crucial to obtain highly pure RNA, the samples are DNase treated twice, first using Turbo DNase (Ambion) and then again with the DNase provided in the column of the RNeasy kit (Qiagen), following the manufacturers' instructions. In brief, 50 mL of complex liquid medium (TSA or Columbia) is inoculated to reach an optical density at 600 nm (OD_{600}) of 0.05, and the bacteria are harvested at midexponential phase (OD_{600} = 0.25–0.32). One ml of the culture is transferred to a centrifuge tube and the cells are pelleted at 4000 × *g* for 5 min at 4°C. The supernatant is discarded, the pellet is mixed with 1 mL of Tri Reagent (Sigma), and the suspension transferred to a 2-mL screw-capped tube containing 500 µL DEPC-treated 0.1-mm zirconia-silica beads. The beads are beat for 2 × 30 s pulses at a speed setting of 6.5 and the tubes allowed to cool at room temperature (RT) for 5 min. After overnight storage at –80°C, the contents of the tubes are thawed on ice and centrifuged for 1 min at 16,100 × *g* at RT. The supernatant is transferred to a new tube, followed by the addition of 200 µL of chloroform, vigorous shaking for 15 s, and separation of the layers by allowing the tubes to stand for 15 min at RT. After centrifugation at 12,000 × *g* at 4°C for 15 min, the clear upper aqueous phase (≈ 700 µL) is transferred to a new tube and 500 µL of isopropanol is added. After adequate mixing, the tubes are allowed to stand at RT for 10 min to allow the RNA to precipitate. The precipitated RNA is collected by centrifugation at 12,000 × *g* at 4°C for 10 min. The supernatant is discarded, the pellet washed with 1.5 mL of 75% ethanol, and the sample centrifuged at 7500 × *g* at 4°C for 5 min. The supernatant is again discarded and the pellet dried for 10–15 min, resuspended in 90 µL of RNase-free water, and incubated at 65°C for 5 min. Following the addition of 10 µL of Ambion Turbo 10× buffer and 2 µL of Ambion Turbo DNase, the RNA is incubated at 37°C for 30 min and then thoroughly mixed with 350 µL of Qiagen RLT buffer. After the addition of 250 µL of 100% ethanol and thorough mixing, the sample is applied to the RNeasy column. The column is centrifuged for 15 s at 8000 × *g*, 350 µL of buffer RW1 is then added to the center of the column, and the column centrifuged again for 15 s at 8000 × *g*. In a separate tube, 10 µL of Qiagen DNase I is gently mixed with 70 µL of RDD buffer. The DNase mix is added to the center of the column, which is then incubated at RT for 15 min. Next, 350 µL of RW1 is loaded onto the center of the column, followed by centrifugation for 15 s at 8000 × *g*, the addition of 500 µL of buffer RPE to the center of the column, and centrifugation as before. The previous step is repeated, discarding the flow-through, and the sample centrifuged for a further 2 min at 8000 × *g*. The column is then transferred to a new Eppendorf tube, 50 µL of RNase-free water is added, and the column incubated at RT for 5 min followed by centrifugation for 1 min at 8000 × *g*. The previous step is repeated, and the RNA quantified at $OD_{260/280}$ to check for RNA integrity.

6.2.2 Detection Procedures

6.2.2.1 Biolog (GP2) System

The Biolog microbial identification system[84] is used to analyze gram-positive bacteria (GP2). The test reactions are carried out in a 96-well microtiter plate containing 95 dehydrated substrates. Organisms are prepared by subculturing them in the appropriate growth medium (supplemented with 5% sheep blood or 1% Tween 80 when required). After incubation, the isolates are subcultured on the surface of equivalent solid media. All plates are incubated for 24 h at 35°C under oxic conditions. A bacterial suspension is obtained by removing the bacterial colonies from the surface of the medium and mixing them with 5 mL of 0.85% saline solution. The bacterial suspension is then standardized to 35%–42% transmittance and a 150-µL aliquot of this suspension dispensed into the wells of a Biolog GP2 panel. After the panels have been incubated for 24 h at 35°C, they are read with the Biolog microplate reader (Biolog Microstation 3.5 software and the corresponding database). The Biolog software compares the results obtained with the test strain to the database and provides identification based on distance calculations.

6.2.2.2 API Coryne System

Version 2.0 of the API Coryne system[85] includes a gallery of 20 microcupules containing dehydrated substrates. The system provides the following biochemical tests: nitrate reduction, pyrazinamidase, pyrrolidonyl arylamidase, alkaline phosphatase, *β*-glucuronidase, *β*-galactosidase, α-glucosidase, *N*-acetyl-*β*-glucosaminidase, esculin, urease, gelatin hydrolysis, a fermentation (negative) control, glucose, ribose, xylose, mannitol, lactose, sucrose, and glycogen. All isolates should be tested using the procedures recommended by the manufacturer, except for isolates exhibiting lipid dependency. A suspension consisting of colony material from each isolate in 0.3 mL of sterile distilled water is swabbed onto the surface of a TSA agar plate supplemented with 5% sheep blood. For organisms exhibiting lipid dependency, colony material for each isolate is inoculated into 1 mL of TSB (tryptone soy broth, Difco) supplemented with 1% Tween 80, incubated for 4 h at 35°C under oxic conditions, and then spread over the entire surface of a TSA agar plate supplemented with 5% sheep blood. The plates are again incubated for 24 h at 35°C under oxic conditions and the resulting colonies are removed from the surface of the plate and suspended in 3.0 mL of sterile distilled water to a turbidity equivalent to a McFarland standard of 6. The first 11 microcupules of the test-strip gallery are inoculated with this suspension. For the nine microcupules of the fermentation test, a 0.5-mL aliquot of the culture is added to 3.0 mL of GP medium (provided in the system), mixed with a vortex, and the mixture used as inoculant. The urease, negative control, and fermentation-test microcupules are overlaid with sterile oil and all test strips are incubated oxically at 35°C for 24 h. Subsequently, the appropriate reagents are added to each well and the color reactions are read with the provided color chart after a 10-min incubation at RT. The previously determined catalase test results are used as the final test. Reactions are recorded on the provided test-result sheets and a seven-digit octal code is generated. This numerical profile is interpreted with the profile index provided by the manufacturer for a final identification.

6.2.2.3 Identifications Based on 16S rRNAs Genes

Sequencing of the 16S ribosomal gene (16S rRNA) is used for confirmatory molecular identification of a test organism. One primer pair is used routinely, with the upper and lower primers consisting of a consensus sequence with respect to specific parts of the amplified gene.[54] The size of the anticipated band depends on the primers used for PCR gene-amplification analyses. The PCR procedure used to amplify the rDNA sequence is frequently performed as follows: (i) A 1-μL aliquot of a frozen DNA stock (see the method describing total DNA extraction from corynebacteria) is added to 49 μL of a PCR mixture containing 22.7 μL of distilled water, 15.2 μL of 3.3× PCR buffer (provided by the commercial distributor), 4 μL of 10 mM dNTP mix, 1.6 μL of a 25 mM magnesium acetate solution, 2.5 μL of a 10 μM solution of the 5′ (upper) primer, 2.5 μL of a 10 μM solution of the 3′ (lower) primer, and 0.5 μL of the appropriate polymerase (frequently Taq or Pfu); (ii) the PCR amplification is performed in a thermal cycler under the following conditions: an initial 94°C for 1.5 min; 35 cycles of 94°C for 20 s; 55°C for 45 s; 72°C for 4 min; and a final 72°C for 7 min; however, the conditions should be optimized in each case to achieve the best results. The PCR-product mixture for each isolate is then purified using a commercial purification kit. The purified products are sequenced using the Sanger technique, with 3 μL of 1 μM primer, and then PCR-amplified using the previously described program. The completed sequences for each isolate are exported in a readable format and compared to the 16S rRNA gene sequence database using the National Center for Biotechnology Information taxonomy browser. An isolate is considered successfully identified to the species level if similarity with a reference sequence is ≥97%.

6.3 FUTURE PERSPECTIVES AND CONCLUSIONS

As corynebacteria constitute part of the normal skin flora, in most cases it is difficult to distinguish between infection, colonization, and contamination by these organisms. Despite this uncertainty, it is important to consider *Corynebacterium* species as causative organisms of abscesses, especially when these bacteria are isolated as a pure contaminant in samples obtained from different parts of the body or from different isolations of the same patient. The presence of gram-positive bacilli associated with polymorphonuclear cells in the isolated clinical specimen is a strong evidence for a causal role of corynebacteria-mediated disease.[6] Pure cultures support a possible disease association, as does the predominance of coryneform bacteria in material from normally sterile sites of the body.

Over the last two decades, a much broader range of taxonomic studies of bacteria has gradually replaced classical techniques based on morphological, physiological, and biochemical characterization. Nowadays, polyphasic taxonomy takes into account all available phenotypic and genotypic data, integrating them in a consensus-oriented classification within the framework of a general phylogeny derived from 16S rRNA gene sequence analysis. In some instances, consensus classification is obtained through the analysis of several parameters (usually phylogenetic ones) that are complemented by phenotypic or additional analyses. Molecular methods provide reliable and highly discriminatory solutions to bacterial characterizations, in contrast to traditional or classical typing techniques. Moreover, new approaches to characterization are being developed. For example, the analysis of large bacterial biomolecules such as proteins, nucleic acids, and sugars by MALDI/MALDI-TOF (mass spectrometry) allows direct identification of the suspected bacterium. Moreover, the technique can also be applied to crude cellular fractions or cellular suspensions, and thus to chemotaxonomic classification.

The use of molecular techniques such as PCR amplification in the direct detection of clinical pathogens has several advantages over phenotypic identification. Since a single small colony provides adequate starting material, final identification can be achieved within 8–12 h of culture; furthermore, identification by these approaches is less dependent on culture conditions. The specific detection of microorganisms from clinical samples often requires the use of species-specific primers and/or probes (mainly when the analyzed bacterium belongs to a previously sequenced organism). However, the use of universal primers (i.e., primers complementary to well-conserved regions of the bacterial genome) can be applied for specific fragment amplification in most bacterial species, in which case the identification relies on the relative polymorphism of the amplified fragment.

Accurate and timely identification of coryneform bacteria causing human infection is warranted for several reasons. Although most of these bacteria are generally of low virulence, the infections they cause may be severe or fatal. Also, the precise identification of coryneform bacterial strains is important because their antimicrobial susceptibilities are variable and an effective antimicrobial therapy optimizes disease outcome. The resistance of corynebacteria to multiple antimicrobial agents has increased over the past several decades and is even characteristic of species such as *C. jeikeium*, *C. urealyticum*, and *C. amycolatum*. For *C. diphtheriae*, the application of molecular typing/subtyping methods and continuous monitoring of bacterial spread (epidemic, endemic, or imported) can indicate the most effective preventive measures. Molecular-based analyses may also lead to the identification of novel organisms and to the discovery of new associations between established species and human disease.

More recently, *in silico* mining of complete-genome sequences of different coryneform bacteria has resulted in the identification of repeated sequences such as the CRISPR regions of *C. diphtheriae*[68] and multiple minisatellite DNA loci. These and techniques still under development (including those based on newly identified spacers and other such variants) will likely provide the next generation of typing procedures for coryneform bacteria.

ACKNOWLEDGMENTS

Research presented in this chapter was funded partially by grants from Junta Castilla y León (LE028A10-2) (2004/120) and Ministerio de Ciencia y Tecnología (BIO2005–02723 and BIO2008–00519) of Spain.

REFERENCES

1. Martín, J.F. et al., Cloning systems in amino acid-producing corynebacteria, *Biotechnology*, 5, 137, 1987.
2. Lehmann, K.B., and Neumann, R.O. In Lehmann, J.F. (Editor), *Atlas und Grundriss der bakteriologie und Lehrbuch der specillen bakteriologischen diagnostik*, 1st Ed., p. 1, München, 1896.
3. Goodfellow, M., Suprageneric classification of actinomycetes. In Williams, S.T., Sharpe, M.E., and Holt, J.G. (Editors), *Bergey's Manual of Systematic Bacteriology*, p. 2333, Williams and Wilkins, Baltimore, 1989.
4. Collins, M.D., and Cummins, C.S., Genus *Corynebacterium*. In Sneath, P.H.A., Mair, N.S., Sharpe, M.E., and Holt, J.G. (Editors), *Bergey's Manual of Systematic Bacteriology*, Vol. 2, p. 1266, Williams and Wilkins, Baltimore, 1986.
5. Puliti, M. et al., Experimental model of infection with non-toxigenic strains of *Corynebacterium diphtheriae* and development of septic arthritis, *J. Med. Microbiol.*, 55, 229, 2006.
6. Funke, G. et al., Clinical microbiology of coryneform bacteria, *Clin. Microbiol. Rev.*, 10, 125, 1997.
7. Liebl, W., *Corynebacterium* nonmedical. In Dworking, M., Falkow, S., Rosemberg, E., Scheifer, K., and Stackebrandt, E. (Editors), *The prokaryotes*, 3rd Ed., p. 796, Springer, Germany, 2006.
8. Neubauer, M. et al., *Corynebacterium accolens* sp. nov., a gram-positive rod exhibiting satellitism, from clinical material, *Syst. Appl. Microbiol.*, 14, 46, 1991.
9. Riegel, P. et al., Taxonomic study of *Corynebacterium* Group ANF-1 strains: Proposal of *Corynebacterium afermentans* sp. nov. containing the subspecies *C. afermentans* subsp. *afermentans* subsp. nov. and *C. afermentans* subsp. *lipophilum* subsp. nov., *Int. J. Syst. Bacteriol.*, 43, 287, 1993.
10. Collins, M.D., Burton, R.A., and Jones, D., *Corynebacterium amycolatum* sp. nov., a new mycolic acid-less *Corynebacterium* species from human skin, *FEMS Microbiol. Lett.*, 49, 349, 1988.
11. Riegel, P. et al., *Corynebacterium argentoratense* sp. nov. from the human throat, *Int. J. Syst. Bacteriol.*, 45, 533, 1995.
12. Yassin, A.F., Steiner, U., and Ludwig, W. *Corynebacterium appendicis* sp. nov., *Int. J. Syst. Evol. Microbiol.*, 52, 1165, 2002.
13. Hall, V. et al., *Corynebacterium atypicum* sp. nov. from a human clinical source, does not contain corynomycolic acids, *Int. J. Syst. Evol. Microbiol.*, 53, 1065, 2003.
14. Funke, G., Lawson, P.A., and Collins, M.D., Heterogeneity within human-derived centers for disease control and prevention (CDC) coryneform group ANF-1-like bacteria and description of *Corynebacterium auris* sp. nov., *Int. J. Syst. Bacteriol.*, 45, 735, 1995.
15. Yassin, A.F., Steiner, U., and Ludwig, W., *Corynebacterium aurimucosum* sp. nov. and emended description of *Corynebacterium minutissimum* Collins and Jones (1983), *Int. J. Syst. Evol. Microbiol.*, 52, 1001, 2002.
16. Funke, G. et al., *Corynebacterium confusum* sp. nov., isolated from human clinical specimens, *Int. J. Syst. Bacteriol.*, 48, 1291, 1998.
17. Funke, G., Ramos, C.P., and Collins, M.D., *Corynebacterium coyleae* sp. nov., isolated from human clinical specimens, *Int. J. Syst. Bacteriol.*, 47, 92, 1997.
18. Riegel, P. et al., *Corynebacterium durum* sp. nov., from human clinical specimens, *Int. J. Syst. Bacteriol.*, 47, 1107, 1997.
19. Renaud, F.N. et al., *Corynebacterium freneyi* sp. nov., alpha-glucosidase-positive strains related to *Corynebacterium xerosis*, *Int. J. Syst. Evol. Microbiol.*, 51, 1723, 2001.
20. Furness, G., and Evangelista, A.T., Infection of a nonspecific urethritis patient and his consort with a pathogenic species of nonspecific urethritis Corynebacteria, *Corynebacterium genitalium*, nov. sp., *Invest. Urol.*, 14, 202, 1976.
21. Funke, G. et al., *Corynebacterium glucuronolyticum* sp. nov. isolated from male patients with genitourinary infections, *Med. Microbiol. Lett.*, 4, 204, 1995.
22. Renaud, F.N. et al., *Corynebacterium hansenii* sp. nov., an alpha-glucosidase-negative bacterium related to *Corynebacterium xerosis*, *Int. J. Syst. Evol. Microbiol.*, 57, 1113, 2007.
23. Funke, G. et al., *Corynebacterium imitans* sp. nov. isolated from patients with suspected diphtheria, *J Clin Microbiol.*, 35, 1978, 1997.
24. Jackman, P.J.H. et al., Classification of corynebacteria associated with endocarditis (group JK) as *Corynebacterium jeikeium* sp. nov., *System. Appl. Microbiol.*, 9, 83, 1987.
25. Collins, M.D. et al., *Corynebacterium kroppenstedtii* sp. nov., a novel corynebacterium that does not contain mycolic acids, *Int. J. Syst. Bacteriol.*, 48, 1449, 1988.
26. Funke, G. et al., *Corynebacterium lipophiloflavum* sp. nov. isolated from a patient with bacterial vaginosis, *FEMS Microbiol. Lett.*, 150, 219, 1997.
27. Riegel, P. et al., Genomic diversity and phylogenetic relationships among lipid-requiring diphtheroids from humans and characterization of *Corynebacterium macginleyi* sp. nov., *Int. J. Syst. Bacteriol.*, 45, 128, 1995.
28. Merhej, V. et al., *Corynebacterium timonense* sp. nov. and *Corynebacterium massiliense* sp. nov., isolated from human blood and human articular hip fluid, *Int. J. Syst. Evol. Microbiol.*, 59, 1953, 2009.
29. Collins, M.D., and Jones, D., *Corynebacterium minutissimum* sp. nov., nom. rev., *Int. J. Syst. Bacteriol.*, 33, 870, 1983.
30. Funke, G., Lawson, P.A., and Collins, M.D., *Corynebacterium mucifaciens* sp. nov., an unusual species from human clinical material, *Int. J. Syst. Bacteriol.*, 47, 952, 1997.
31. Aravena-Roman, M. et al., *Corynebacterium pilbarense* sp. nov., a non-lipophilic species isolated from a Bactec anaerobic vial inoculated with an ankle aspirate, *Int. J. Syst. Evol. Microbiol.*, 60, 1484, 2010.
32. Furness, G., Sambury, S., and Evangelista, A.T., *Corynebacterium pseudogenitalium* sp. nov. commensals of the human male and female urogenital tracts, *Invest. Urol.*, 16, 292, 1979.
33. Peel, M.M. et al., Human lymphadenitis due to *Corynebacterium pseudotuberculosis*: Report of ten cases from Australia and review, *Clin. Infect. Dis.*, 24, 185, 1997.
34. Otsuka, Y. et al., *Corynebacterium resistens* sp. nov., a new multidrug-resistant coryneform bacterium isolated from human infections, *J. Clin. Microbiol.*, 43, 3713, 2005.

35. Funke, G., Lawson, P.A., and Collins, M.D., *Corynebacterium riegelii* sp. nov., an unusual species isolated from female patients with urinary tract infections, *J. Clin. Microbiol.,* 36, 624, 1998.
36. Wattiau, P., Janssens, M., and Wauters, G., *Corynebacterium simulans* sp. nov., a non-lipophilic, fermentative *Corynebacterium*, *Int. J. Syst. Evol. Microbiol.*, 50, 347, 2000.
37. Riegel, P. et al., *Corynebacterium singulare* sp. nov., a new species for urease-positive strains related to *Corynebacterium minutissimum*, *Int. J. Syst. Bacteriol.*, 47, 1092, 1997.
38. Coyle, M.B., Leonard, R.B., and Nowowiejski, D.J., Pursuit of the *Corynebacterium striatum* type strain, *Int. J. Syst. Bacteriol.*, 43, 848, 1993.
39. Yassin, A.F., and Siering, C., *Corynebacterium sputi* sp. nov., isolated from the sputum of a patient with pneumonia, *Int. J. Syst. Evol. Microbiol.*, 58, 2876, 2008.
40. Collins, M.D. et al., *Corynebacterium sundsvallense* sp. nov., from human clinical specimens, *Int. J. Syst. Bacteriol.,* 49, 361, 1999.
41. Zimmermann, O. et al., *Corynebacterium thomssenii* sp. nov., a *Corynebacterium* with N-acetyl-beta-glucosaminidase activity from human clinical specimens, *Int. J. Syst. Bacteriol.*, 48:489, 1998.
42. Feurer, C. et al., Taxonomic characterization of nine strains isolated from clinical and environmental specimens, and proposal of *Corynebacterium tuberculostearicum* sp. nov., *Int. J. Syst. Evol. Microbiol.,* 54, 1055, 2004.
43. Riegel, P. et al., Isolation of *Corynebacterium tuscaniae* sp. nov. from blood cultures of a patient with endocarditis, *J. Clin. Microbiol.*, 44, 307, 2006.
44. Riegel, P. et al., Taxonomy of *Corynebacterium diphtheriae* and related taxa, with recognition of *Corynebacterium ulcerans* sp. nov. nom. rev., *FEMS Microbiol. Lett.,* 126, 271, 1995.
45. Pitcher, D. et al., Classification of coryneform bacteria associated with human urinary tract infection (group D2) as *Corynebacterium urealyticum* sp. nov., *Int. J. Syst. Bacteriol.*, 42, 178, 1992.
46. Yassin, A.F., *Corynebacterium ureicelerivorans* sp. nov., a lipophilic bacterium isolated from blood culture, *Int. J. Syst. Evol. Microbiol.*, 57, 1200, 2007.
47. Funke, G. et al., Multicenter evaluation of the updated and extended API (RAPID) Coryne database 2.0, *J. Clin. Microbiol.*, 35, 3122, 1997.
48. Funke, G., Peters, K., and Aravena-Roman, M., Evaluation of the RapID CB plus system for identification of coryneform bacteria and *Listeria* spp., *J. Clin. Microbiol.,* 36, 2439, 1998.
49. Adderson, E.E., Boudreaux, J.W., and Hayden R.T., Infections caused by coryneform bacteria in pediatric oncology patients, *Pediatr. Infect. Dis. J.*, 27, 136, 2008.
50. Becker, B. et al., Rapid differentiation between *Nocardia* and *Streptomyces* by paper chromatography of whole cell hydrolysates, *Appl. Microbiol.,* 12, 421, 1964.
51. Lechevalier, M. P., Identification of aerobic actinomycetes of clinical importance, *J. Lab. Clin. Med.*, 71, 934, 1968.
52. Minnikin, D.E. et al., Thin-layer chromatography of methanolysates of mycolic acid-containing bacteria, *J. Chromatogr.,* 188, 221, 1980.
53. Popovic, T. et al., Molecular epidemiology of diphtheria in Russia, 1985–1994, *J. Infect. Dis.,* 174, 1064, 1996.
54. Relman, D.A., Universal bacterial 16S rDNA amplification and sequencing. In Persing, D.H., Smith, T.F., Tenover, F.C., and White, T.J. (Editors), *Diagnostic Molecular Microbiology; Principles and Applications*, p. 489, Mayo Foundation/American Society for Microbiology, Rochester, MN/Washington, DC, 1993.
55. Letek, M. et al., Identification of the emerging skin pathogen *Corynebacterium amycolatum* using PCR-amplification of the essential *divIVA* gene as a target, *FEMS Microbiol. Lett.*, 265, 256, 2006.
56. Khamis, A., Raoult, D., and La Scola, B., *rpoB* gene sequencing for identification of *Corynebacterium* species, *J. Clin. Microbiol.,* 42, 3925, 2004.
57. Khamis, A., Raoult, D., and La Scola, B., Comparison between *rpoB* and 16S rRNA gene sequencing for molecular identification of 168 clinical isolates of *Corynebacterium*, *J. Clin. Microbiol.*, 43, 1934, 2005.
58. Nakao, H. et al., Heterogeneity of diphtheria toxin gene, *tox,* and its regulatory element, *dtxR*, in *Corynebacterium diphtheriae* strains causing epidemic diphtheria in Russia and Ukraine, *J. Clin. Microbiol.,* 34, 1711, 1996.
59. Maiden, M.C., Multilocus sequence typing: A portable approach to the identification of clones within populations of pathogenic microorganisms, *Proc. Natl. Acad. Sci. USA,* 95, 3140, 1998.
60. Efstratiou, A., and George, R.C., Laboratory guidelines for the diagnosis of infections caused by *Corynebacterium diphtheriae* and *C. ulcerans*. World Health Organization, *Commun. Dis. Public Health*, 2, 250, 1999.
61. Cerdeño-Tárraga, A.M. et al., The complete genome sequence and analysis of *Corynebacterium diphtheriae* CTC13129, *Nucleic Acids Res.*, 31, 6516, 2003.
62. Moreira L.O. et al., Effects of iron limitation on adherence and cell surface carbohydrates of *Corynebacterium diphtheriae* strains, *Appl. Environ. Microbiol.,* 69, 5907, 2003.
63. Coyle, M.B. et al., The molecular epidemiology of three biotypes of *Corynebacterium diphtheriae* in the Seattle outbreak, 1972–1982, *J. Infect. Dis.*, 159, 670, 1989.
64. De Zoysa, A. et al., Molecular epidemiology of *Corynebacterium diphtheriae* from northwestern Russia and surrounding countries studied by using ribotyping and pulsed-field gel electrophoresis, *J. Clin. Microbiol.*, 33, 1080, 1995.
65. De Zoysa, A. et al., Comparison of four molecular typing methods for characterization of *Corynebacterium diphtheriae* and determination of transcontinental spread of *C. diphtheriae* based on BstEII rRNA gene profiles, *J. Clin. Microbiol.*, 46, 3626, 2008.
66. Mokrousov, I.V. et al., The genetic typing of *Corynebacterium diphtheriae* strains by the polymerase chain reaction with universal primers, *Zh. Mikrobiol. Epidemiol. Immunobiol.*, 5, 73, 1996.
67. Jansen, R. et al., Identification of genes that are associated with DNA repeats in prokaryotes, *Mol. Microbiol.*, 43, 1565, 2002.
68. Mokrousov, I., *Corynebacterium diphtheriae* spoligotyping based on combined use of two CRISPR loci, *Biotechnol. J.*, 2, 901, 2007.
69. Soriano, F. et al., In vitro and in vivo study of stone formation by *Corynebacterium* group D2 (*Corynebacterium urealyticum*), *J. Clin. Microbiol.,* 23, 691, 1986.
70. Aubel, D., Renaud, F.N.R., and Freney, J., Genomic diversity of several *Corynebacterium* species identified by amplification of the 16S-23S rRNA gene spacer regions, *Int. J. Syst. Bacteriol.*, 47, 767, 1997.
71. Tauch, A. et al., The lifestyle of *Corynebacterium urealyticum* derived from its complete genome sequence established by pyrosequencing, *J. Biotechnol.,* 136, 11, 2008.

72. Tauch, A. et al., Complete genome sequence and analysis of the multiresistant nosocomial pathogen *Corynebacterium jeikeium* K411, a lipid-requiring bacterium of the human skin flora, *J. Bacteriol.,* 187, 4671, 2005.
73. Simoons-Smit, A.M. et al., Chronic cystitis caused by *Corynebacterium urealyticum* detected by polymerase chain reaction, *Eur. J. Clin. Microbiol. Infect. Dis.,* 19, 949, 2000.
74. Hou, X.G. et al., Genetic identification of members of the genus *Corynebacterium* at genus and species levels with 16S rDNA-targeted probes, *Microbiol. Immunol.*, 41, 453, 1997.
75. Pitcher, D. et al., An investigation of nosocomial infection with *Corynebacterium jeikeium* in surgical patients using a ribosomal RNA gene probe, *Eur. J. Clin. Microbiol. Infect. Dis.,* 9, 643, 1990.
76. Fernández-Natal, M.I. et al., Isolation of *Corynebacterium ureicelerivorans* from normally sterile sites in humans, *Eur. J. Clin. Microbiol. Infect. Dis.*, 28, 677, 2009.
77. Suzuki, T. et al., Suture-related keratitis caused by *Corynebacterium macginleyi*, *J. Clin. Microbiol.*, 45, 3833, 2007.
78. Eguchi, H. et al., High-level fluoroquinolone resistance in ophthalmic clinical isolates belonging to the species *Corynebacterium macginleyi*, *J. Clin. Microbiol.*, 46, 527, 2008.
79. Claeys, G. et al., Endocarditis of native aortic and mitral valves due to *Corynebacterium accolens*: Report of a case and application of phenotypic and genotypic techniques for identification, *J. Clin. Microbiol.,* 34, 1290, 1996.
80. Vaneechoutte, M. et al., Evaluation of the applicability of amplified rDNA-restriction analysis (ARDRA) to identification of species of the genus *Corynebacterium*, *Res. Microbiol.,* 146, 633, 1995.
81. Darras-Vercambre, S. et al., Photodynamic action of red light for treatment of erythrasma: Preliminary results, *Photodermatol. Photo- immunol. Photomed.*, 22, 153, 2006.
82. Funke, G., and Frodl, R., Comprehensive study of *Corynebacterium freneyi* strains and extended and emended description of *Corynebacterium freneyi* Renaud, Aubel, Riegel, Meugnier, and Bollet 2001, *J. Clin. Microbiol.*, 46, 638, 2008.
83. Liebl, W. et al., High efficiency electroporation of intact *Corynebacterium glutamicum* cells, *FEMS Microbiol. Lett.*, 53, 299, 1989.
84. Lindenmann, K., von Graevenitz, A., and Funke, G., Evaluation of the Biolog system for the identification of asporogenous, aerobic gram-positive rods, *Med. Microbiol. Lett.,* 4, 287, 1995.
85. Funke, G., and Bernard, K.A., Coryneform gram-positive rods. In Baron, P.R., Jorgensen, E.J., Landry, J.H., and Pfaller, M.A. (Editors), *Murray Manual of Clinical Microbiology*, 9th Ed., p. 485, American Society for Microbiology, Washington, DC, 2007.

7 *Cryptobacterium*

Futoshi Nakazawa

CONTENTS

7.1 INTRODUCTION

7.1.1 Classification and Morphology

The genus *Cryptobacterium* was proposed by Nakazawa et al. in 1999 for strains of asaccharolytic, anaerobic, short gram-positive rods.[1] Initially, these strains were isolated from human oral cavities as *Eubacterium*-like organisms. The genus *Eubacterium* has been defined mainly on the basis of negative metabolic characteristics.[2] Because of its broad definition, a large number of novel gram-positive rods isolated from oral cavities have been classified previously as members of the genus *Eubacterium*. Therefore, it is inevitable that considerable heterogeneity exists among the species within the genus. In fact, it is described that the genus *Eubacterium* contains many gram-positive rods that are phylogenetically unrelated, and some of which qualify as members of the genus *Eubacterium* but could not be assigned to any of the named species.[2–8] In addition, several novel genera have been proposed for some of asaccharolytic, anaerobic, short gram-positive rods (e.g., *Cryptobacterium*, *Slackia*, *Eggerthella*, and *Mogibacterium*), and some species of *Eubacterium* have been transferred to the novel genera according to their phylogenetic characteristics.[1,9,10]

Cryptobacterium curturn is the sole species in the genus *Cryptobacterium* on the basis of morphological and biochemical characteristics, SDS-PAGE of whole-cell protein and Western immunoblotting analysis, G + C content, DNA–DNA hybridization, and 16S rRNA gene sequence data as member of the family *Coriobacteriaceae*.[1] To date, the family *Coriobacteriaceae* is defined only in terms of the 16S rRNA gene signature nucleotide, just like the subclass *Coriobacteridae* and the order *Coriobacteriales*.[11] Indeed, it would seem appropriate to examine the value of chemotaxonomy in improving the definition of the order and family. At the time of writing, the family *Coriobacteriaceae* proposed by Stackebrandt et al. comprises the following 13 genera: *Adlercreutzia*, *Asaccharobacter*, *Atopobium*, *Collinsella*, *Coriobacterium*, *Cryptobacterium*, *Denitrobacterium*, *Eggerthella*, *Enterorhabdus*, *Gordonibacter*, *Olsenella*, *Paraeggerthella*, and *Slackia*.[1,10,12–23]

Cryptobacterium cells are very short (0.4 × 0.8–1.0 μm), obligate anaerobic, nonmotile, nonspore forming rods that occur singly or in masses. The growing cells are gram-positive, but occasionally cells in older cultures are gram-negative. Transmission electron micrographs of thin sections of cell of *C. curtum* 12–3^T (=ATCC 700863) showed a single-layer cell wall approximately 10 nm in thickness (Figure 7.1), which is typical cell wall of gram-positive bacteria. Pili- or flagella-like structures were not found. The scanning electron micrograph of *C. curtum* 12–3^T, occurring in chains, is also available.[24]

On BHI-blood agar plates, *Cryptobacterium* cells form minute, circular, convex, and translucent colonies of less than 1 mm in diameter (about 0.3–0.5 mm) without hemolysis, even after prolonged incubation in an anaerobic glove box. Growth in broth media is very poor with or without carbohydrates. Also, it is described that *C. curtum* strain DSM 15641 (=strain 12–3^T) can also be grown anaerobically in DSMZ medium 78 (Chopped Meat Medium).[24,25]

7.1.2 Biology, Habitat, and Lifestyle

Strains of the genus *Cryptobacterium* are strictly anaerobic and asaccharolytic rods. They do not ferment adonitol, amygdalin, arabinose, cellobiose, erythritol, esculin, fructose,

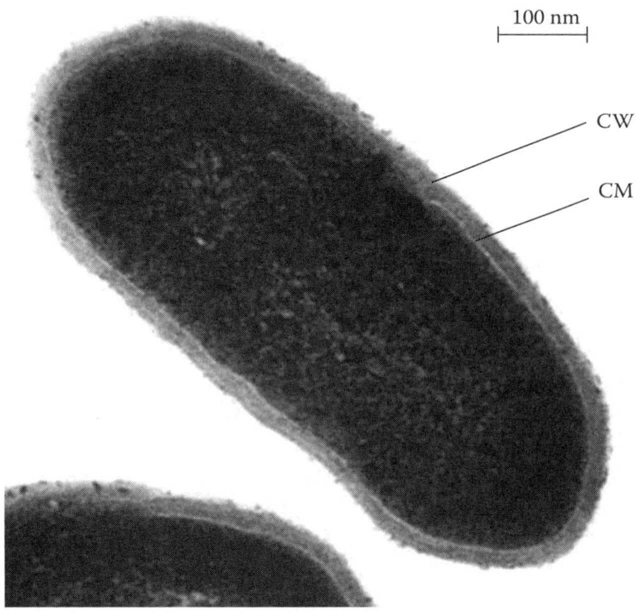

FIGURE 7.1 Transmission electron microphotograph of a thin section of cell of *C. curtum* 12–3^{T} (=ATCC 700863^{T}) showing the cytoplasmic membrane (CM) and typical gram-positive cell wall (CW).

galactose, glucose, glycogen, inositol, lactose, maltose, mannitol, mannose, melezitose, melibiose, rhamnose, ribose, salicin, sorbitol, starch, sucrose, trehalose, or xylose, and are inert in most conventional biochemical tests. They are positive for arginine hydrolysis. Nitrate reduction is negative and they do not hydrolyze esculin, starch, or gelatin. In addition, they do not produce indole, urease, or catalase. They also do not produce any metabolic end product in peptone-yeast extract-glucose broth.

C. curtum strains 12–3^{T} and KV43-B were isolated from periodontal pocket of a human adult with periodontal disease and necrotic dental pulp, respectively.[8,26] It has been suggested that *C. curtum* is one of the causative species identified by nonculture techniques in the microbial flora of an acute dental abscess with other anaerobic gram-positive rods such as *Eubacterium sulci*, *Mogibacterium timidum*, and *Mogibacterium vescum*.[27]

C. curtum and the closely related species such as *Slackia exigua* do not utilize carbohydrates, and are very difficult to cultivate. Optimal doubling time is 12 h.[28] Uematsu et al. described that amino acids are important metabolic substrates for growth of these asaccharolytic strains, especially arginine and lysine.[29,30] And they also reported that *C. curtum* metabolize arginine as source of energy via the arginine deiminase pathway and produce substantial amounts of citrulline and ornithine. Arginine may be provided by the enzymatic degradation of peptides, especially in the inflamed lesions where trypsin-like enzymes produce peptides containing arginine at the C-terminal. In this sense, it would appear that these infectious lesions, such as periodontal pocket, necrotic dental pulp, and dental abscess, have suitable environments for the *Cryptobacterium* together with other anaerobic and asaccharolytic strains. There are no chemotaxonomic data available to *C. curtum* strain 12–3^{T}.

7.1.3 Genetic and Genomic Features

As reported previously, the G + C contents of oral isolates that are asaccharolytic, anaerobic, and short gram-positive rods range from 38 to 62 mol%, as determined by HPLC.[1,3,5,6] Similarly, the G + C content of *C. curtum* 12–3^{T} and KV43-B are 50 and 51 mol%, respectively, which are at the same level as *Mogibacterium pumilum* (type species of the genus *Mogibacterium*) or *Eubacterium limosum* (type species of the genus *Eubacterium*).

The DNA–DNA relatedness data are very useful for identification of asaccharolytic, anaerobic, and gram-positive rods from human clinical specimens.[31] In fact, the DNA–DNA hybridization studies with the membrane filter method indicated that the DNA of *C. curtum* strain 12–3^{T} exhibited a high level (97% homology) of reassociation with the DNA of strain KV43-B. Moreover, the DNA from strains 12–3^{T} and KV43-B exhibited very low levels (1%–11% homology) of relatedness to type species of the other genera that are asaccharolytic, strictly anaerobic, and short gram-positive rods.[1,31]

The 16S rRNA gene sequence of *C. curtum* strain 12–3^{T} (1429 bases in length) was registered as accession number AB019260 in GenBank.[1]

According to the evolutionary tree constructed with the 16S rRNA gene sequence of *C. curtum* strain 12–3^{T} and other sequences of the related gram-positive bacteria selected from the GenBank database by the sequence similarity search using the BLAST and FASTA algorithms, the strain 12–3^{T} and KV43-B can be distinguished from the asaccharolytic *Eubacterium* species, including the type species of *Eubacterium*, *E. limosum*. These phylogenetic data completely agreed with previous results obtained by RFLP analysis of 16S rRNA.[8] Furthermore, although strain 12–3^{T} had a close phylogenetic relationship with *Coriobacterium glomerans* and *Denitrobacterium detoxzjicans,* the sequence data showed distinct lines exhibiting specific phylogenetic association with *C. glomerans* or *D. detoxijicans*.[1] These phylogenetic studies also indicated that strain 12–3^{T} and KV43-B belonged to a large group of the *Actinobacteria*, and form a cluster with the genus *Atopobium*, *Coriobacterium*, and *Denitrobacterium* (Figure 7.2). Other studies of the 16S rRNA gene sequences data also demonstrated that strain 12–3^{T} formed one cluster with the genus *Eggerthella*, *Collinsella*, *Slackia*, *Adlercreutzia*, *Asaccharobacter*, *Denitrobacterium*, *Slackia*, and *Coriobacterium,* all of which are members of the family *Coriobacteriaceae*.[24,31]

The strain 12–3^{T} (= DSM 15641 = ATCC 700683 = CCUG 43107) is the type strain of *C. curtum*, which is the sole species within the genus *Cryptobacterium*. It is interesting that *C. curtum* 12–3^{T} was initially isolated from human periodontal pocket as an asaccharolytic *Eubacterium*-like strain that is a member in the phylum *Firmicutes*. On the basis of a polyphasic taxonomic study—that notably uses 16S rRNA

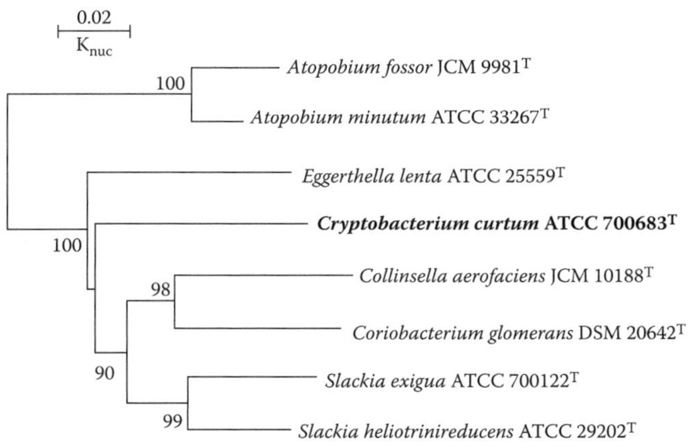

FIGURE 7.2 Evolutionary tree based on 16S rRNA gene sequence comparison showing the phylogenetic position of *C. curtum* 12–3^T (=ATCC 700863^T). The dendrogram was created by using the neighbor-joining method. The numbers on the tree indicate bootstrap values (%) calculated from 1000 replications.

gene sequence and DNA–DNA hybridization data—*C. curtum* 12–3^T showed very distant and isolated position within the family *Coriobacteriaceae* in the phylogenetic tree constructed with the 16S rRNA gene sequences. It was finally replaced as a species in the genus *Cryptobacterium,* family *Coriobacteriaceae,* phylum *Actinobacteria*.[1,24]

The whole genome sequence of *C. curtum* 12–3^T was reported in 2009 as the first complete genome sequence of the family *Coriobacteriaceae*.[24] According to the report, the genome size is 1,617,804 bp in length and comprises one main circular chromosome with a 50.9% DNA G + C content. Of the 1422 predicted genes, 1364 are protein-coding genes, 58 are RNA genes, and 1117 of these genes encode proteins with known function.[24]

7.1.4 Clinical Features and Pathogenesis

C. curtum was described as opportunistic pathogen with a typical occurrence in the oral cavity, and that is involved in such oral infections as periodontitis, inflammations, and abscesses.[24] *C. curtum* strain 12–3^T and VK43-B were initially isolated as the closest related rods to the genus *Eubacterium* taxonomically because of its strict anaerobic, asaccharolytic, and gram-positive rod.[1] In the same way, the numerous bacterial strains of asaccharolytic, anaerobic, and gram-positive rods isolated from human oral cavities have been classified as members of *Eubacterium* because of the broad definition of the genus. Furthermore, many of these strains have been documented as aetiologic agents of chronic periodontitis on the basis of their frequent isolation from diseased periodontal sites.[32–41]

In addition, sera of patients with periodontitis show higher antibody titers against some of these species than those of healthy people, suggesting that these bacteria cause immunological reactions in periodontal lesions.[42–45] Moreover, they have been recognized as pathogens in oral and dental infections such as advanced caries, pulpal infections, and dentoalveolar abscesses, and also in nonoral, clinically significant infections such as head and neck infections, abdominal infection bacteremia associated with malignancy, and genital tract infection associated with intrauterine devices.[7,46–55]

The published reports suggest that *Cryptobacterium* species may play some roles in oral pathological processes as important pathogens, like other bacterial species of asaccharolytic, anaerobic, and gram-positive rods in human hosts. In animal hosts, no data are currently available on pathogenesis of infections caused by *Cryptobacterium* species. Specific virulence factors remain unknown. The availability of the complete genome sequences of *C. curtum* strain 12–3^T should give new insights about the virulence of this microorganism.[24]

7.1.5 Diagnosis

Conventional Techniques. Strictly anaerobic conditions are required for the growth of the *Cryptobacterium* species. In our laboratory, *Cryptobacterium* strains are cultured in an anaerobic glove box (model AZ-Hard, Hirasawa, Japan) containing 80% N_2, 10% H_2, and 10% CO_2. Moreover, to ensure strictly anaerobic conditions in the box, the reduction of methyl viologen (at −446 mV in oxidation reduction potential) is carefully checked when culture procedures are carried out. Even under these anaerobic conditions, the growth of *Cryptobacterium* strains in broth media is very poor with or without carbohydrates because they are asaccharolytic and difficult to culture. Although no enrichment method and no specific selective medium have been reported for growth of *Cryptobacterium* strains, arginine and citrulline support the growth of the microorganisms. *Cryptobacterium* strains utilize arginine via the arginine deiminase pathway and produce substantial amounts of citrulline and ornithine.[28]

The type strain 12–3^T of *C. curtum* was isolated on brain heart infusion-yeast extract blood (sheep) agar (BHI-blood agar) plates at 37°C for 3 days in an anaerobic glove box.[8] Many other agar media or broths can be used for cultivation and isolation of *Cryptobacterium,* as Kavrommatis et al. demonstrated the growth of *C. curtum* on DSMZ medium 78.[24,25] Also, Wade et al. maintained the *Slackia* species, which are closely related to *Cryptobacterium,* on Fastidious Anaerobe Agar supplemented with 5% horse blood under anaerobic conditions.[10]

In the routine analysis of human oral samples, *Cryptobacterium* isolates are recognized on BHI-blood agar at 37°C for 5–7 days. *C. curtum* strain 12–3^T and KV43-B yield nonpigmented, nonhemolytic tiny colonies on this medium. After 96 h of incubation at 37°C in the anaerobic glove box, the colonies are 0.3–0.5 mm in diameter, and circular, convex, and whitish with a smooth surface.[1] Members of the genus *Cryptobacterium* are nonmotile and nonsporulating. Other biochemical tests are mainly negative for these microorganisms, except arginine hydrolysis, which is positive. *Cryptobacterium* strains show neither urease nor catalase activity. Nitrate reduction, gelatin and starch liquefaction, and esculin hydrolysis are negative. Indole is generally not produced. Although no an end product from peptone-yeast extract-glucose broth, *C. curtum* produces ornithine from BHI broth media with the addition of arginine or citrulline. Using rapid ID 32 A system, *Cryptobacterium* strains display the following positive enzymatic reactions: arginine arylamidase, leucyl glycine arylamidase, leucine arylamidase, alanine arylamidase, glycine arylamidase, and histidine arylamidase activities.

Several of the previously cited characteristics are variable among strains of asaccharolytic species making phenotypic identification difficult. Phenotypic identification of anaerobic and asaccharolytic gram-positive rods from human oral samples is mainly based on negative fermentation characteristics, which allow the identification of the genus *Eubacterium* mainly encountered in human clinical samples. In fact, strain 12–3^T and KV43-B were initially wrongly placed in the genus *Eubacterium*, and then relocated in the genus *Cryptobacterium* of the family *Coriobacteriaceae*. Strictly anaerobic gram-positive rods, which are asaccharolytic, have no end products, and are negative in most of the conventional biochemical tests, are often submitted to molecular identification.

Molecular Techniques. Molecular methods for identification of *Cryptobacterium* species include DNA–DNA hybridization and 16S rRNA gene sequencing, which are performed on DNA extracts obtained by Marmur protocol with slight modification.[3,56] The two strains of *C. curtum* each showed 97 mol% in homology of whole DNA relatedness. But they showed less than 11 mol% in the homologies against other strains in the family *Coriobacteriaceae*. Furthermore, these strains formed a cluster together with members of the family *Coriobacteriaceae* on the basis of their 16S rRNA gene sequences. The 16S rRNA gene sequence comparisons showed that *C. curtum* has a close phylogenetic relationship with *Denitrobacterium detoxzjicans* and *Coriobacterium glomerans*. But these sequence similarities were 88.5% with *D. detoxzjicans* and 86.6% with *C. glomerans*. The phylogenetic analysis indicated that *C. curtum* was clearly distinct from the other genera in the family *Coriobacteriaceae*.

Molecular techniques for taxonomic characterization of *Cryptobacterium* species particularly genomic structure analysis, are not detailed in this chapter but may be found in the following reference.[10,24]

7.2 METHODS

7.2.1 Sample Preparation

Cellular morphology is determined by examining cells grown on BHI-blood agar plates. Biochemical characteristics are tested by the procedures described previously by Holdeman et al.[25] No end products from peptone-yeast extract-glucose broth have been confirmed by gas chromatography.[57] Whole-cell protein profiles are examined by sodium dodecyl sulfate-poly acrylamide gel electrophoresis as described previously.[58] After these proteins are transferred to a nitrocellulose membrane, an immune rabbit antiserum is used as the first antibody and followed by incubation with goat antirabbit immunoglobulin G conjugated with peroxidase as the second antibody in the Western immunoblotting analysis.[59] A Rapid ID 32A kit (bioMérieux) is used for enzymic profile determination as recommended by the manufacturer.

DNA is isolated and purified by following a modification of the Marmur protocol.[3,56] Commercialized kits, such as the InstaGene DNA purification Matrix (BIO-RAD Laboratories), are also available for genomic DNA extraction. The G + C content of DNA is determined by high-performance liquid chromatography, as described previously.[60] DNA–DNA hybridization is performed by the membrane filter method according to the procedures described by Meyer et al., with multiprime DNA labeling kit (Amersham Pharmacia Biotech) and^{32}P-dCTP.[61]

7.2.2 Detection Procedures

Cryptobacterium could be identified by sequencing of the 16S rRNA gene and phylogenetic analysis. Also the data of complete genome sequence of *C. curtum* type strain may be useful for identification of *Cryptobacterium*.

Procedure

1. The 16S rRNA gene is amplified by PCR with a nucleotide primer set [27 forward primer; 5′-AGA GTT TGA TCM TGG CTC AG-3′, located at positions 8–27 (*Escherichia coli* numbering), and 1492 reverse primer; 3′-TTC AGC ATT GTT CCA TYG GCAT-5′, located at positions 1492–1513 (*E. coli* numbering)]. The PCR reaction is performed with Premix *Taq* (Takara-bio, Japan) according to the manufacturer's instructions in a thermal controller (e.g., model PTC-100; MJ Research, MA) using the following parameters: 35 cycles 94°C for 1 min, 45°C for 2 min, and 72°C for 2 min.
2. Amplification products are separated by 2.0% agarose gel electrophoresis and visualized by staining with ethidium bromide. According to these conditions, the amplification products of 1.5 kb are obtained from the agarose gel for 16S rRNA gene.
3. For sequence of 16S rRNA from the PCR, a Thermo Sequenase Fluorescent Labeled Primer Cycle Sequencing Kit (Amersham Pharmacia Biotech) is used with the 11 universal primers labeled with Cy-5, following the manufacturer's protocol. The sequences of the 16S rRNA are analyzed with a DNA sequencer (e.g., ALFexpress, Amersham Pharmacia Biotech, Buckinghamshire).
4. The sequences are aligned with each other and connected by using SEQMAN II of the LASERGENE program (DNASTAR). Programs MEGALIGN of LASERGENE, CLUSTAL W, and NJPLOT are used to compare sequences and to construct an evolutionary tree by the neighbor-joining method. Also, confidence intervals are assessed by NJPLOT and CLUSTAL W with bootstrap analysis.

Note. Matsuki et al. designed 16S rRNA gene-targeted group-specific primers for the member of the *Atopobium* cluster that includes *C. curtum*. These primers may be useful for rough identification of this microorganisms.[62]

7.3 CONCLUSION AND FUTURE PERSPECTIVES

Cryptobacterium species is isolated from human oral cavities and may act as opportunistic pathogens in infectious lesions. Only *C. curtum*, a type species of the genus *Cryptobacterium*, is known at the present time. The two strains of *C. curtum*, strain 12–3^T isolated from periodontal pocket and strain KV43-B isolated from necrotic dental pulp, can be differentiated from strains of other genera in the family *Coriobacteriaceae* by phenotypic characteristics including arginine hydrolysis, nitrate reduction, API ZYM code in Rapid ID32A system, protein profile of whole bacterial cell, and Western immunoblotting pattern.

16S rRNA gene sequencing is known as a useful tool for identification of the genus and the species levels, and is frequently used to identify such anaerobic gram-positive rods from human oral cavities. In fact, according to the results of 16S rRNA gene sequence, *Cryptobacterium* is established as genus in the family *Coriobacteriaceae*, which contains anaerobic, asaccharolytic, and gram-positive rods. Restriction fragment-length polymorphism analysis of the amplified 16S rRNA using the endonucleases *Alu*I and *Sma*I is also a convenient method of distinguishing *Cryptobacterium* from other genera.

Furthermore, DNA–DNA hybridization studies allow the genus *Cryptobacterium* to be distinguished from closely related taxa such as the genera *Collinsella*, *Eggerthella*, and *Slackia*. The genus *Cryptobacterium*, with *C. curtum* as the type species, is identified on the basis of a polyphasic taxonomic study, notably by using 16S rRNA gene sequence and DNA–DNA hybridization data.

Little is known about the role of *Cryptobacterium* species in human oral cavities. With the genome sequence of type species of the genus *Cryptobacterium* having been recently completed, further details regarding virulence factors or metabolism for the bacterium will be uncovered, contributing to increased knowledge of the roles of *Cryptobacterium* in human oral diseases.

REFERENCES

1. Nakazawa, F. et al., *Cryptobacterium curtum* gen. nov., sp. nov., a new genus of gram-positive anaerobic rod isolated from human oral cavities, *Int. J. Syst. Bacteriol.*, 49, 1193, 1999.
2. Moore, W.E.C., and Moore, L.V.H., Genus *Eubacterium*. In Sneath, P.H.A. et al. (Editors), *Bergey's Manual of Systematic Bacteriology*, Williams and Wilkins, Baltimore/London, 1986.
3. Nakazawa, F., and Hoshino, E., Genetic relationships among *Eubacterium* species, *Int. J. Syst. Bacteriol.*, 44, 787, 1994.
4. Cheeseman, S. et al., Phylogeny of oral asaccharolytic *Eubacterium* species determined by 16S ribosomal DNA sequence comparison and proposal of *Eubacterium infrmun* sp. nov. and *Eubacterium tardum* sp. nov, *Int. J. Syst. Bacteriol.*, 46, 957, 1996.
5. Poco, S.E. et al., *Eubacterium exiguum* sp. nov., isolated from human oral lesions, *Int. J. Syst. Bacteriol.*, 46, 1120, 1996.
6. Poco, S.E. et al., *Eubacterium minutum* sp. nov., isolated from human periodontal pockets, *Int. J. Syst. Bacteriol.*, 46, 31, 1996.
7. Wade, W.G. et al., An unclassified *Eubacterium* taxon in acute dentoalveolar abscess, *J. Med. Microbiol.*, 40, 115, 1994.
8. Sato, T. et al., Restriction fragment-length polymorphism analysis of 16S rDNA from oral asaccharolytic *Eubacterium* species amplified by polymerase chain reaction, *Oral. Microbiol. Immunol.*, 13, 23, 1993.
9. Nakazawa, F. et al., Description of *Mogibacterium pumilum* gen. nov., sp. nov. and *Mogibacterium vescum* gen. nov., sp. nov., and reclassification of *Eubacterium timidum* (Holdeman *et al.* 1980) as *Mogibacterium timidum* gen. nov., comb. nov, *Int. J. Syst. Evol. Microbiol.*, 50, 679, 2000.
10. Wade, W.G. et al., The family *Coriobacteriaceae*: Reclassification of *Eubacterium exiguum* (Poco *et al.* 1996) and *Peptostreptococcus heliotrinreducens* (Lanigan 1976) as *Slackia exigua* gen. nov., comb. nov. and *Slackia heliotrinreducens* gen. nov., comb. nov., and *Eubacterium lentum* (Prevot 1938) as *Eggerthella lenta* gen. nov., comb. nov., *Int. J. Syst. Bacteriol.*, 49, 595, 1999.
11. Stackebrandt, E. et al., Phylogenetic analysis of the genus *Desulfotomaculum*: Evidence for the misclassification of *Desulfotomaculum guttoideum* and description of *Desulfotomaculum orientis* as *Desulfosporosinus orientis* gen. nov., comb. nov., *Int. J. Syst. Bacteriol.*, 47, 1134, 1997.
12. Stackebrandt, E. et al., Proposal for a new hierarchic classification system, *Actinobacteria* classis nov., *Int. J. Syst. Bacteriol.*, 47, 479, 1997.
13. Naruo, T. et al., *Adlercreutzia equolifaciens* gen. nov., sp. nov., an equol-producing bacterium isolated from human faeces, and emended description of the genus *Eggerthella*, *Int. J. Syst. Evol. Microbiol.*, 58, 1221, 2008.
14. Minamida, K. et al., *Asaccharobacter celatus* gen. nov., sp. nov., isolated from rat caecum, *Int. J. Syst. Evol. Microbiol.*, 58, 1238, 2008.
15. Collins, M.D., and Wallbanks, S., Comparative sequence analyses of the 16S rRNA genes of *Lactobacillus minutus*, *Lactobacillus rimae* and *Streptococcus parvulus*: Proposal for the creation of a new genus *Atopobium*, *FEMS Microbiol. Lett.*, 74, 235, 1992.
16. Kageyama, A., and Benno, Y., Emendation of genus *Collinsella* and proposal of *Collinsella stercoris* sp. nov. and *Collinsella intestinalis* sp. nov., *Int. J. Syst. Evol. Microbiol.*, 50, 1767, 2000.
17. Haas, F., and König, H., *Coriobacterium glomerans* gen. nov., sp. nov. from the intestinal tract of the red soldier bug, *Int. J. Syst. Bacteriol.*, 38, 382, 1988.
18. Anderson, R.C. et al., *Denitrobacterium detoxificans* gen. nov., sp. nov., a ruminal bacterium that respires on nitrocompounds, *Int. J. Syst. Evol. Microbiol.*, 50, 633, 2000.
19. Kageyama, A. et al., Phylogenetic evidence for the transfer of *Eubacterium lentum* to the genus *Eggerthella* as *Eggerthella lenta* gen. nov., comb. nov., *Int. J. Syst. Bacteriol.*, 49, 1725, 1999.
20. Clavel, T. et al., *Enterorhabdus caecimuris* sp. nov., a novel member of the family *Coriobacteriaceae* isolated from a mouse model of spontaneous colitis, *Int. J. Syst. Evol. Microbiol.*, in Press, 2010.
21. Clavel, T. et al., Isolation of bacteria from the ileal mucosa of TNFdelta ARE mice and description of *Enterorhabdus mucosicola* gen. nov., sp. nov., *Int. J. Syst. Evol. Microbiol.*, 59, 1805, 2009.

22. Wurdemann, D. et al., *Gordonibacter pamelaeae* gen. nov., sp. nov., a new member of the Coriobacteriaceae isolated from a patient with Crohn's disease, and reclassification of *Eggerthella hongkongensis* Lau et al. 2006 as *Paraeggerthella hongkongensis* gen. nov., comb. nov., *Int. J. Syst. Evol. Microbiol.*, 59, 1405, 2009.
23. Dewhirst, F.E. et al., Characterization of novel human oral isolates and cloned 16S rDNA sequences that fall in the family *Coriobacteriaceae*: description of *Olsenella* gen. nov., reclassification of *Lactobacillus uli* as *Olsenella uli* comb. nov. and description of *Olsenella profusa* sp. nov., *Int. J. Syst. Evol. Microbiol.*, 51, 1797, 2001.
24. Mavrommatis, C. et al., Complete genome sequence of *Cryptobacterium curtum* type strain ($12–3^T$), *Standards Genom. Sci.*, 1, 93, 2009.
25. Holdeman, L.V., Cato, E.P., and Moore, W.E.C., *Anaerobe Laboratory Manual*, 4th Ed., Blacksburg, VA, Virginia Polytechnic Institute and State University, 1977.
26. Sundqvist, G., Bacteriological studies of necrotic dental pulps, *Umei University Dissertation*, 7, 1, 1976.
27. Robertson, D., and Smith, A.J., The microbiology of the acute dental abscess, *J. Med. Microbiol.*, 58, 155, 2009.
28. Uematsu, H. et al., Degradation of arginine by *Slackia exigua* ATCC 700122 and *Cryptobacterium curtum* ATCC 700683, *Oral Microbiol. Immunol.*, 21, 381, 2006.
29. Uematsu, H., and Hoshino, E., Degradation of arginine and other amino acids by *Eubacterium nodatum* ATCC 33099, *Microbiol. Ecol. Health Dis.*, 9, 305, 1996.
30. Uematsu, H. et al., Degradation of arginine and other amino acids by butyrate-producing asaccharolytic anaerobic gram-positive rods in periodontal pockets, *Arch. Oral. Biol.*, 48, 423, 2003.
31. Nakazawa, F., and Hoshino, E., DNA–DNA relatedness and phylogenetic positions of *Slackia exigua, Slackia heliotrinireducens, Eggerthella lenta*, and other related bacteria, *Oral Microbiol. Immunol.*, 19, 1, 2004.
32. Hill, G.B. et al., Characteristics and sites of infection of *Eubacterium nodatum*, *Eubacterium timidum*, *Eubacterium brachy*, and other asaccharolytic eubacteria, *J. Clin. Microbiol.*, 25, 1540, 1987.
33. Holdeman, L.V. et al., Description of *Eubacterium timidum* sp. nov., *Eubacterium brachy* sp. nov. and *Eubacterium nodatum* sp. nov., isolated from human periodontitis, *Int. J. Syst. Bacteriol.*, 30, 163, 1980.
34. Martin, S.A. et al., A comparison of the reactivity of *Eubacterium* species with localized and serum immunoglobulin from rapidly progressive and adult periodontitis, *J. Periodontol.*, 59, 32, 1986.
35. Moore, W.E.C. et al., Bacteriology of severe periodontitis in young adults, *Infect. Immun.*, 38, 1137, 1982.
36. Moore, W.E.C. et al., Bacteriology of moderate (chronic) periodontitis in mature adult humans, *Infect. Immun.*, 42, 510, 1983.
37. Moore, W.E.C. et al., Comparative bacteriology of juvenile periodontitis, *Infect. Immun.*, 48, 507, 1985.
38. Slayne, M.A., and Wade, W.G., The humoral response to asaccharolytic *Eubacterium* species in periodontitis, *Microbiol. Ecol. Health. Dis.*, **7**, 283, 1994.
39. Uematsu, H., and Hoshino, E., Predominant obligate anaerobes in human periodontal pockets, *J. Periodontal. Res.*, 27, 15, 1992.
40. Uematsu, H. et al., *Eubacterium saphenum* sp. nov., isolated from the human periodontal pockets, *Int. J. Syst. Bacteriol.*, 43, 302, 1993.
41. Wade, W.G. et al., The effects of antimicrobial acrylic strips on the subgingival microfora in chronic periodontitis, *J. Clin. Periodontol.*, 19, 127, 1992.
42. Gunsolley, J.C. et al., Serum antibodies to periodontal bacteria, *J. Periodontol.*, 61, 412, 1990.
43. Tew, J.G. et al., Serum antibody reactive with predominant organisms in the subgingival flora of young adults with generalized severe periodontitis, *Infect. Immun.*, 48, 303, 1985.
44. Tolo, K., and Jorkjend, L., Serum antibodies and loss of periodontal bone in patients with rheumatoid arthritis, *J. Clin. Periodontol.*, 17, 288, 1990.
45. Tolo, K., and Schenck, K., Activity of serum immunoglobulin G, A and M to six anaerobic oral bacteria in diagnosis of periodontitis, *J. Periodontal. Res.*, 20, 113, 1985.
46. Ando, N., and Hoshino, E., Predominant obligate anaerobes invading the deep layers of root canal dentine, *Int. Endod. J.*, 23, 20, 1990.
47. Edwardsson, S., Bacteriological studies on deep areas of carious dentine, *Odontol. Revy. Suppl.*, 32, 1, 1974.
48. Hoshino, E. et al., Bacterial invasion of non-exposed pulp, *Int. Endod. J.*, 25, 2, 1992.
49. Hoshino, E., Predominant obligate anaerobes in human carious dentin, *J. Dent. Res.*, 64, 1195, 1985.
50. Sato, T. et al., Predominant obligate anaerobes in necrotic pulps of human deciduous teeth, *Microbiol. Ecol. Health Dis.*, 6, 269, 1993.
51. Brook, I., and Frazier, E.H., Significant recovery of non-sporulating anaerobic rods from clinical specimens, *Clin. Infect. Dis.*, 16, 476, 1993.
52. Fanstein, V. et al., Bacteremia caused by non-sporulating anaerobes in cancer patients, *Medicine (Baltimore)*, 68, 151, 1989.
53. Brook, I., Anaerobic bacterial bacteremia: 12-year experience in two military hospitals, *J. Infect. Dis.*, 160, 1071, 1983.
54. Backer, E.D. et al., Antibiotic susceptibility of *Atopobium vaginae. BMC Infect. Dis.*, 6, 51, 2006.
55. Goldstein, E.J.C. et al., In vitro activities of daptomycin, vancomycin, quinupristin-dalfopristin, linezolid, and five other antimicrobials against 307 gram-positive anaerobic and 31 *Corynebacterium* clinical isolates. *Antimicrob. Agents Chemother.*, 47, 337, 2003.
56. Marmur, J., A procedure for the isolation of DNA from microorganisms, *J. Mol Biol.*, 3, 208, 1961.
57. Hoshino, E., and Sato, M., Production and degradation of formate by *Veillonella dispar* ATCC 17745, *J. Dent. Res.*, 65, 903, 1986.
58. Laemmli, U.K., Cleavage of structural proteins during the assembly of the head of bacteriophage T4, *Nature*, 227, 68, 1970.
59. Nakazawa, F., and Hoshino, E., Immunological specificity of oral *Eubacterium* species, *J. Gen. Microbiol.*, 39, 2635, 1993.
60. Katayama-Fujimura, Y. et al., Estimation of DNA base composition by high performance liquid chromatography of its nuclease PI hydrolysate, *Agric. Biol. Chem.*, 48, 3169, 1984.
61. Meyer, S.A., and Schleifer, K.H., Deoxyribonucleic acid reassociation in the classification of coagulase positive *Staphylococci*, *Arch. Microbiol.*, 7, 183, 1978.
62. Matsuki, T. et al., Use of 16S rRNA gene-targeted group-specific primers for real-time PCR analysis of predominant bacteria in human feces, *Appl. Environ. Microbiol.*, 70, 7220, 2004.

8 *Gardnerella*

Rita Verhelst, Hans Verstraelen, Piet Cools, Guido Lopes dos Santos Santiago, Marleen Temmerman, and Mario Vaneechoutte

CONTENTS

8.1 INTRODUCTION

The genus *Gardnerella*, consisting of a single facultatively anaerobic Gram-variable rod species, *Gardnerella vaginalis*, is classified under the family Bifidobacteriaceae, suborder Actinomycineae, phylum Actinobacteria. Although the bacterium possesses a thin gram-positive cell wall, it may appear either gram-positive or gram-negative under the microscope. The clinical significance of *G. vaginalis*, as well as its taxonomic position, has long been a matter of dispute largely due to variable Gram staining and lack of appropriate detection and identification methods (reviewed in Catlin[1]). Thus far, more than 1000 papers were addressed to this small (0.4 by 1.0–1.5 μm), pleomorphic, rod-shaped, fastidious bacterium organism.

Almost a century ago, Curtis et al.[2] reported that Gram-stained purulent vaginal discharges contained mainly gram-negative bacilli that he and his colleagues were unable to cultivate. It was not until 1953 that Leopold[3] succeeded in isolating such bacteria from men with prostatitis and women with cervicitis. The organism was named *Haemophilus vaginalis* by Gardner and Dukes in 1955[4] because it was a gram-negative, rod-shaped bacterium that could be isolated on blood agar and it was believed to be responsible for a characteristic vaginal discharge.

8.1.1 Cell Wall Structure and Chemical Composition

Based on electron microscopy data, Reyn et al.[5] concluded that, although the organism does not retain crystal violet during conventional Gram-staining procedures, the thin walls and septa of osmium-fixed *G. vaginalis* resemble those of gram-positive bacteria. Harper and Davis[6] and O'Donnell et al.[7] confirmed the gram-positive nature of the cell wall, based on the presence of the amino acids alanine, glutamic acid, glycine and lysine, and on the absence of the dibasic amino acid diaminopimelic acid (DAP) in the peptidoglycan. Also, based on electron microscopy studies, Sadhu et al.[8] concluded that the cell wall unequivocally has a gram-positive type of organization and that the unusual thinness of the cell walls (8–12 nm) accounts for the gram-negative staining tendency of *G. vaginalis*. Further evidence that *G. vaginalis* most properly groups with the gram-positive bacteria came from the study of Gelber et al.,[9] who found that this species expresses vaginolysin, a hemolysin belonging to the cholesterol-dependent cytolysins that have only been described in gram-positive organisms so far.

However, evidence for a gram-negative cell wall composition was gathered as well. Criswell et al.[10,11] argued for a gram-negative cell wall on the basis of the presence of a wide range of amino acids and because the cell wall has a trilaminar structure, consisting of two electron-dense layers separated by an electron-translucent layer and the presence of only low amounts of peptidoglycan (i.e., only 20% of the total cell wall weight). However, these studies have been faulted because the amino acid profiles for their control organisms were at variance with the currently accepted profiles.[6,7] Also, Greenwood and Pickett[12] interpreted the lamellar cell wall structure as more closely resembling that of gram-negative organisms and identified a lipopolysaccharide (LPS)-like cell wall component. However, the absence of DAP and hydroxy fatty acids, as well as the almost complete absence of endotoxin by the *Limulus* amebocyte lysate assay, indicate that *G. vaginalis* does not contain classical LPS.[6–8,13,14]

In addition, studies on antibiotic susceptibility led to contradicting conclusions on the nature of the cell wall. Although the antibiotic susceptibility pattern of *G. vaginalis* resembles that of gram-positive bacteria according to Reyn et al.[5] and Piot et al.,[15] Kharsany et al.[16] claimed that *G. vaginalis* is neither typically gram-positive nor typically gram-negative since antimicrobial agents regarded as specifically active against gram-positive or gram-negative organisms showed relatively poor activity. It should be mentioned that another important gram-positive member of disturbed vaginal microflora (i.e., *Lactobacillus iners*) typically stains gram-negative and that, also because the similarities of their colony morphology, *L. iners* has been misidentified on several occasions as *G. vaginalis* (unpublished data).

At present, on the basis of taxonomic data (see below), it is clear that *G. vaginalis* should be considered a gram-positive bacterium.

8.1.2 Taxonomy

Haemophilus vaginalis was renamed to *Corynebacterium vaginale* by Zinneman and Turner,[17] based on the absence of requirements for X and/or V factors (hemin and NAD, respectively), which are needed for the growth of accepted *Haemophilus* species, the tendency to retain the crystal violet dye in the Gram reaction, and a corynebacterium-like morphology. Reyn et al.[5] compared *H. vaginalis* with four possibly related species that had shown a similar degree of Gram-variability and suggested that *H. vaginalis* should be included in *Corynebacterium* or *Butyribacterium*, rather than in *Lactobacillus* and that this organism may be closely related to the genus *Propionibacterium*. However, based on volatile acid production and fermentation end product analysis, Moss and Dunkelberg[18] showed that this species did belong to neither the genera *Corynebacterium*, *Butyribacterium* nor *Lactobacillus*.

Three large taxonomic studies published in 1980[12,15] and 1996[19] analyzed data obtained by a variety of biochemical methods, DNA–DNA hybridization, 16S rRNA sequence analysis, and electron microscopy. Greenwood and Pickett[12] found that "*Corynebacterium vaginale*" formed a new taxon and proposed the name *Gardnerella vaginalis*, defining the organism as a facultative anaerobic, rod-shaped, catalase- and oxidase-negative, gram-negative to Gram-variable bacterium with laminated cell walls that produces acetic acid as the major end product of fermentation of carbohydrates. In the same year, Piot et al.[15] supported the creation of a new genus, but found little ground for *G. vaginalis* to be related to established gram-negative taxa. Pointing at differences in Gram-stain appearance depending on the growth condition and on the age of the culture, they found that *G. vaginalis* had an antibiotic susceptible pattern as found in gram-positive bacteria. Further phylogenetic analysis by Van Esbroeck et al.,[19] based on 16S rRNA gene sequence analysis, placed *G. vaginalis*, still the sole member of the genus *Gardnerella*, in the gram-positive family *Bifidobacteriales*. *G. vaginalis* is indeed closely related to the genus *Bifidobacterium*, with the highest level of similarity to the type species *Bifidobacterium bifidum* (i.e., 93.1%) and exhibits fructose-6-phosphate phosphoketolase activity, a feature that was assumed to be specific for the genus *Bifidobacterium*.[20] However, the difference in GC-content between *G. vaginalis* (42 mol%) and bifidobacteria (55–67 mol%),[21] is too large to consider *G. vaginalis* as a genuine *Bifidobacterium* species.

8.1.3 Biotypes

On the basis of the measurement of the activities of three enzymes (i.e., β-galactosidase, lipase, and hippurate hydrolase), *G. vaginalis* isolates were assigned to eight different biotypes by Piot et al.[22] Another scheme obtained by adding fermentation of arabinose, galactose, and xylose, revealed respectively 17 and 33 different biotypes in two studies.[23,24] Based on the latter scheme, different *G. vaginalis* biotypes were found in women with and without bacterial vaginosis.[25] Although, based on the scheme of Piot et al., the association of specific biotypes with bacterial vaginosis remains controversial until today. When the distributions of different *G. vaginalis* biotypes from women with and without bacterial vaginosis in three countries were examined, biotypes 1, 2, and 5 were found to be the most predominant groups present at all locations.[22] However, in a longitudinal study, *G. vaginalis* strains of biotypes 1, 5, and 6 were found to be most common in all clinical situations.[25] Of the 261 isolates examined in that study, the lipase-positive group, consisting of biotypes 1, 2, 3, and 4, was suggested to be more prevalent among women with bacterial vaginosis, a finding confirmed by Numanović et al.[26] Aroutcheva et al.[27] found diversity in the bacterial vaginosis group and a predominance of biotype 5 in women with normal vaginal microflora.

Recently, Moncla et al.[28] pointed at methodological differences in the previous studies and argued that 4-methylumbelliferyl-oleate and other 4-methylumbelliferyl-derivatives should not be used for the detection of bacterial lipase activity, and urged to reconsider possible associations between bacterial vaginosis and certain biotypes.

8.1.4 Pathogenesis: Virulence Factors

Several factors have been shown to contribute to the ability of bacteria to colonize and infect the host epithelium.

Pili. A key feature to vaginal colonization is the ability of *G. vaginalis* to adhere to the vaginal epithelium. Patterson et al.[29] recently compared bacterial adherence to vaginal epithelial cells, biofilm-producing capacities, and cytotoxic activity of bacterial vaginosis-associated organisms, suggesting that *G. vaginalis* is more virulent than many of the bacterial species frequently isolated from bacterial vaginosis.

G. vaginalis is undoubtedly equipped with multiple virulence properties and consequently, the idea that it is merely an initial colonizer that paves the way for additional species is being revisited.[9,29–31]

Bacterial adherence is clinically typically reflected by the presence of *clue cells* on microscopic examination of a vaginal smear. It was recently convincingly confirmed by FISH that *clue cells* are vaginal squamous epithelial cells coated with *G. vaginalis* and other anaerobic bacteria causing bacterial vaginosis.[31] *Clue cells* were first described by Gardner and Dukes in 1955 and were named as such because these cells provide an important "clue" to the diagnosis of bacterial vaginosis.[4]

At the ultrastructural level, *G. vaginalis* adherence is thought to relate to the presence of pili extending from the cell wall and to the presence of an exopolysaccharide (see below).

Johnson and Davies[32] documented by electron microscopy that at least some *G. vaginalis* strains carry narrow pili with a thickness between 3.0 and 7.5 nm. Boustouller et al. subsequently showed by electron microscopy that the pili were present among *G. vaginalis* strains freshly isolated from clinical specimens, whereas the expression of these pili tended to disappear upon repeated culture.[33] Accordingly, Fredricsson et al.[34] found that in vitro, *G. vaginalis* showed insufficient adherence to vaginal epithelial cells to produce so-called *clue cells*.

Exopolysaccharide. *G. vaginalis* has been shown to produce an electron-dense, web-like, fibrillar exopolysaccharide,[35] which was found to confer adherence of *G. vaginalis* cells to vaginal epithelial cells.[36]

Vaginolysin. *G. vaginalis* produces a cytolysin, a 60 kDa protein that is a member of the cholesterol-dependent cytolysin family of toxins, and that recognizes the complement regulatory molecule CD59 on the surface of human cells.[9,37] This species-specific lysis of human cells has limited mechanistic studies of bacterial vaginosis and its adverse consequences by the absence of a suitable animal model.

Vaginolysin is a pore-forming exotoxin that lyses endothelial cells, neutrophils, and erythrocytes (and therefore previously was called a hemolysin).[38–41] The lysis of neutrophils might explain the relative absence of neutrophils in bacterial vaginosis. Furthermore, vaginolysin triggers the immune system, induces interleukin-8 production by human epithelial cells, and a specific secretory immunoglobulin A targeting the *G. vaginalis* vaginolysin has been shown to be upregulated in the setting of bacterial vaginosis.[42,43]

Sialidase and Prolidase Activity. *G. vaginalis* produces sialidase and prolidase, two hydrolytic enzymes that may have a role in degrading several key mucosal protective factors, such as mucins, as well as contributing to exfoliation and detachment of vaginal epithelial cells.[44]

Phospholipase A_2 Activity. *G. vaginalis* has phospholipase A_2 activity, and hence may be directly implicated in the initiation of preterm labor through the conversion of arachidonic acid to prostaglandins xxx.[45]

Iron Acquisition. As iron is an essential growth factor, but also plays an important role in the virulence potential of many bacterial pathogens, a number of bacteria have developed high-affinity iron acquisition mechanisms. Basically, three types of mechanisms have been described (i.e., the secretion of high-affinity iron chelators or siderophores, the expression of cell-surface receptors that directly bind to iron-containing compounds, and hemolysis).[46] In *G. vaginalis*, there is evidence that all three mechanisms are present, including the production of siderophores,[46] the binding of iron carriers like catalase,[47] lactoferrin,[48] transferrin,[48] heme,[49] and hemoglobin,[50] and the production of the aforementioned vaginolysin leading to hemolysis with release of iron-rich compounds from red blood cells.

Biofilm Formation. New insights in the pathogenic role of *G. vaginalis* came from fluorescent in situ hybridization (FISH) studies on vaginal biopsies. Biopsies collected from women with and without bacterial vaginosis were hybridized with a variety of bacterial rRNA-targeted probes.[31] Typically, subjects with bacterial vaginosis had an adherent biofilm that was primarily composed with three bacterial groups, of which *G. vaginalis* was most prominent. *G. vaginalis* constituted 60%–90% of the biofilm mass, *Atopobium* accounted for 1%–40% of the biofilm mass, and lactobacilli were present between 1% and 5% in only 20% of the biopsy samples. While this adherent bacterial biofilm was present in 90% of bacterial vaginosis patients, only 10% of women with a normal vaginal microflora had a similar biofilm.

Patterson et al.[51] have highlighted the physiological differences of *G. vaginalis* bacteria grown under different conditions. In their study, planktonic *G. vaginalis* were more sensitive to hydrogen peroxide (5-fold) and lactic acid (4–8 fold) than *G. vaginalis* biofilm bacteria. Furthermore, the recognition that *G. vaginalis* forms vaginal biofilm provides a plausible explanation for the persistency and recurrence of bacterial vaginosis in many women, since bacteria in biofilms are more resistant to antibiotics compared to planktonic organisms.[52–55]

Saunders et al.[56] investigated the potential of lactobacilli to attempt clearance of the *G. vaginalis* biofilm. *G. vaginalis* biofilms grown in vitro could be displaced by the probiotic *Lactobacillus reuteri* RC-14 and to a limited extent by *Lactobacillus iners*, a nonhydrogen peroxide-producing *Lactobacillus* commonly found in the vaginal econiche, especially in disturbed vaginal microflora.[57,58]

Future studies on the structure and composition of biofilms will further improve our understanding of the pathogenic role of *G. vaginalis* in bacterial vaginosis.

8.1.5 Clinical Features

The most common infection caused by *G. vaginalis* is bacterial vaginosis in women. Bacterial vaginosis is a disruption of vaginal microflora characterized by a decrease in *Lactobacillus* species and increase in the quantity of anaerobic bacteria. *G. vaginalis* is present in up to 95% of cases of bacterial vaginosis[1] and nearly all women with bacterial vaginosis have high concentrations of *G. vaginalis*.[59–62] About half of all women microscopically confirmed bacterial vaginosis will present symptoms—the remainder being completely asymptomatic. When symptoms are present, women will present most typically with profuse, off-white, thin, and homogeneous discharge, quite different from the thick, white, curdy discharge observed with vaginal candidiasis. Vaginal discharge with bacterial vaginosis is typically accompanied by an unpleasant, fishy odor. This characteristic vaginal discharge is actually the only symptom—if present at all—and other vulvovaginal complaints such as soreness, pain, or itch, as often present with vulvovaginal candidiasis, are usually not associated with bacterial vaginosis. Similarly, on inspection of the vulvar area and the vaginal walls, no additional signs will be found, as bacterial vaginosis is not associated with (appreciable) vaginal inflammation, and hence no erythema or swelling occur.

In men, genital infection associated with *G. vaginalis* presents as balanoposthitis, which may be accompanied by diffuse erythema and pruritus of the glans, coronal sulcus, and meatus urethrae.[63,64] Urinary tract infection with *G. vaginalis* in men has also been described as urethritis or urethral syndrome.[65–68]

During pregnancy, *G. vaginalis* can be involved in upper genital tract infection, including chorioamnionitis and amniotic fluid infection,[69,70] infections associated with adverse pregnancy outcome such as late fetal loss, stillbirth, and preterm birth.

Procedures during delivery may also be complicated by *G. vaginalis* infection, such as wound infection following episiotomy[71] and following caesarean section,[72] and bacteremia occurring during caesarean section.[73]

Vertical transmission to the neonate can occur, and neonatal infections with *G. vaginalis* have been anecdotic reported, including *G. vaginalis*-infected scalp hematoma associated with electronic fetal monitoring,[74,75] neonatal conjunctivitis,[76] and disseminated neonatal infection presenting as meningitis[77] and neonatal sepsis.[78,79]

Sepsis and abscesses in multiple organ sites with *G. vaginalis* have also been described in immunocompetent men and women.[80–82]

Several case reports have also pointed at joint and musculoskeletal infections with *G. vaginalis*, including vertebral osteomyelitis and discitis,[83,84] and hip arthritis.[85]

G. vaginalis has also been associated with several autoimmune disorders, including reactive arthritis[86,87] and Reiter's syndrome.[88]

8.1.6 Epidemiology

The epidemiology of *G. vaginalis* in practice partly translates to the epidemiology of bacterial vaginosis. Bacterial vaginosis involves the overgrowth of *G. vaginalis* as a virtually absolute finding.

Among children and prepubertal girls bacterial vaginosis is rather rare, except in case of sexual abuse, even though vaginal carriage of *G. vaginalis* identified through PCR is common in young girls.[89] Swidsinski et al.[90] investigated *G. vaginalis* carriage in 50 premenarchal girls by FISH and found that dispersed growth of *Gardnerella* occurred in 10%, whereas none of the 50 girls showed cohesive *Gardnerella*. Higher rates of *G. vaginalis* have been found in sexually abused girls (7%–19%) as compared to nonabused girls (1%–4.2%) in most[91–93]—though not all—studies.[94]

Once beyond the menarche however, bacterial vaginosis is observed even among adolescent girls and sexually inexperienced girls, and young women in general, at an appreciable frequency, though at significantly lower rates as compared to sexually active adolescents and young women.[95,96] Shafer et al. reported that up to one-third of nonsexually active adolescents harbored *G. vaginalis*, compared to 60% of sexually active adolescents. Another culture-based study of 120 asymptomatic 14- to 17-year-old females found a twofold higher prevalence of *G. vaginalis* (34%) in sexually active females compared to women who reported no penetrative sexual activity (17%), albeit a nonsignificant difference. The largest sample up to date involved U.S. women entering the military, with a mean age of 19.1 years, and found bacterial vaginosis in women reporting to never have had vaginal intercourse to occur at a rate of 18%, whereas their sexually experienced counterparts had bacterial vaginosis at a rate of 28%.[97]

In adult, sexually experienced women, *G. vaginalis* is a common constituent of the normal vaginal microflora, albeit present in low numbers when compared to the predominant *Lactobacillus* species. In a European population, *G. vaginalis* was detected among pregnant women with normal vaginal microflora through species-specific PCR at a rate of 28.9%.[98] In a U.S. study involving women attending an STD clinic, *G. vaginalis* was detected with normal vaginal microflora through species-specific PCR at a rate of 58.7%.[99]

Women with lesbian orientation, or generally women-who-have-sex-with-women (WSW), present rather consistently with very high rates of bacterial vaginosis.[100–102]

Weyers et al.[103] characterized the microflora of the penile skin-lined neovagina of transsexual women and found that, although it is difficult to establish to what extent a bacterial vaginosis-like condition is present among these transsexual women, *G. vaginalis* was present in 30% of the women.

Carriage of *G. vaginalis* is also common in men. The two largest cohort studies on male carriage of *G. vaginalis* conducted until present—both involving male attendees of a sexually transmitted disease clinic—documented male urethral carriage of *G. vaginalis* at a rate of 11.4% (49/430) in the United Kingdom[104] and of 4.5% (10/309) in Sweden.[105] Male carriage of *G. vaginalis* may even be higher than estimated from the aforementioned studies as urethral sampling may not be the optimal approach to document it. Kinghorn et al.[63] found a significantly higher rate of *G. vaginalis* isolation from preputial than from urethral swabs. Swidsinski et al.[106] recently made a similar observation, reporting that desquamated epithelial cells loaded with *G. vaginalis* could only reliably be recovered in a urine specimen if the praeputium was not pulled back during voiding. These authors documented among 100 men admitted to a department of internal medicine that dispersed *Gardnerella* was found in 4% of males, while cohesive *Gardnerella* was present in 7%. Remarkably, using PCR, Schwebke et al.[107] found no difference in the prevalence rates of *G. vaginalis* between male sexual partners of women with and without bacterial vaginosis. However, this unexpected finding may be the result of the fact that both groups of men were recruited from the same high-risk cohort of patients attending an STD clinic.

In addition, several culture-based studies have also documented the presence of *G. vaginalis* in semen samples[108–113]—possibly pointing at a seminal or prostatic reservoir—whereby in one study *G. vaginalis* was recovered from semen in as much as 38% of 58 men attending an infertility clinic.[109] Evidence to sustain the theory that this microorganism can affect sperm characteristics, thereby causing infertility, was not found.[114] However, although *G. vaginalis* does not affect semen quality and does not cause an inflammatory reaction in infertile men, seminal colonization by this microorganism should be considered clinically relevant, since it may be transmitted to the female genital tract through coitus.[109,112,113,115]

Overall, *G. vaginalis* is closely linked to sexual behavior and activity, though it remains unclear through which mechanisms sexual activity enhances *G. vaginalis* carriage and colonisation.[89,116,117] Mechanistic studies of bacterial vaginosis and its adverse consequences have been limited by the absence of suitable animal models since *G. vaginalis* has been isolated from the reproductive tract of mares only.[118]

Interestingly, Piot et al.[22] found that biotypes isolated from women with bacterial vaginosis and from the urethras of their sexual partners were identical in 90% of the couples. Briselden et al.[25] found that women who acquire bacterial vaginosis are more likely to have a shift in biotype than women who had normal microflora at followup, suggesting that the *G. vaginalis* isolates recovered from women who develop bacterial vaginosis represent newly acquired strains rather than overgrowth of previously colonizing biotypes. Furthermore, Simoes et al.[30] found that *G. vaginalis* biotype 1 strains might be responsible for a higher rate of sexual transmission of HIV, since 50% of these strains were able to induce HIV expression in U1 cells.

8.2 METHODS

8.2.1 Phenotypic Identification

Several media possessing differential and/or selective features can be used to isolate *G. vaginalis* from clinical specimens (reviewed in Catlin[1]).[119,120] At present, human blood-bilayer-Tween (HBT) agar, developed by Totten et al.,[121] is the most satisfactory differential selective medium. It consists of a bottom layer of Columbia colistin-nalidixic acid agar, as described by Golberg and Washington,[119] supplemented with Proteose Peptone no. 3, amphotericin B, and Tween 80, and a top layer of the same composition to which 5% human blood is added. Often, minimal criteria such as the characteristics of colonies surrounded by zones of clear hemolysis with diffuse edges on HBT medium, the typical appearance of Gram-stained cells and a negative catalase test are used for the presumptive identification of *G. vaginalis* from vaginal fluid.[122] However, it should be noticed that HBT agar fails to inhibit a variety of gram-positive bacteria, such as lactobacilli, coryneforms, and streptococci.[122,123] We found that 3 out of 10 hemolytic strains isolated on HBT agar were identified through sequencing of the 16S rRNA gene as *L. iners* (unpublished data). Since *L. iners* strains are oxidase and catalase-negative, positive for β-galactosidase and hippurate hydrolase, and variable in their lipase activity, these strains may be misidentified as *G. vaginalis* biotypes 1 or 6. Moreover, as for *G. vaginalis*, short gram-positive *L. iners* rods can have a gram-negative stain appearance.[61] Misidentification of *L. iners* as *G. vaginalis* might have led to erroneous conclusions on the role of these species in bacterial vaginosis. Ison et al.[124] found that the addition of gentamicin (4 mg/L) instead of colistin to HBT medium increases the inhibition of non-*Gardnerella* species.

Rapid conventional methods for detection and identification of *G. vaginalis* were reviewed by Catlin.[1] It should be noted that the API-ZYM system for enzyme detection,[125] as well as the API-20A and Minitek systems,[126] were considered unreliable for the identification of *G. vaginalis* in the diagnostic laboratory. Also, several papers report the failure to detect *G. vaginalis* bacteremia using blood culture media containing the anticoagulant sodium polyanethole sulphate (SPS).[69,127–130] Reimer et al. have suggested that the incidence of bacteraemia caused by *G. vaginalis* may be higher than the number of reported cases since SPS, routinely added to most blood culture media, inhibits growth of *G. vaginalis* unless gelatin is also present.[131]

To tackle the limitations of conventional methods, molecular methods are increasingly used to identify microbes in clinical samples. However, one should realize that molecular methods also experience technical problems. For example, PCR amplification can fail due to the presence of PCR inhibitors. Ironically, initial attempts at amplification of the bacterial 16S rRNA gene from inoculated blood culture media failed due to the presence of SPS, the component that caused growth inhibition of *G. vaginalis* in SPS containing culture media.[132] Appropriate DNA extraction methods to remove PCR inhibitors are presented in the section "sample preparation." In addition, care should be taken not to miss certain species when "universal" primers are used, as discussed below (see section on sequencing).

8.2.2 Genotypic Identification

Material for identification of *G. vaginalis* can be obtained by self-swabbing[133–136] or after placement of a nonlubrificated speculum.[57] The swabs need to be rotated against the vaginal wall at the midportion of the vault and carefully removed to prevent contamination with the vulva and introitus microflora.

If culture media cannot be directly inoculated, the swabs should be placed in a transport medium. Swabs on solid transport medium are not preferred when performing direct molecular identification methods since the transport medium might cause PCR inhibition (unpublished observations). Flocked swabs are preferred for the purpose of direct molecular identification. The use of an ESwab (Copan, Brescia, It) with liquid Amies transport medium has the advantage that the same swab can be used for the purpose of Gram stain, culture, and PCR.[137] Alternatively, a flocked swab can be stored dry in an empty tube when frozen at –20°C within 5 h after collection.[61,138] Using dry swabs avoid problems of incompatibility of commercial collection kits and transport media for multiple testing of specimens with different nucleic acid amplification tests.[139]

8.2.2.1 Sample Preparation

Crude DNA Extracts of Cultured Isolates. Crude DNA extracts from cultured strains can be obtained by simple alkaline lysis as follows—resuspend one bacterial colony in 20 μL of lysis buffer (0.25% sodium dodecyl sulfate, 0.05 N NaOH) and heat at 95°C for 15 min. Spin down drops that are present on the lid and dilute the cell lysate by adding 180 μL of distilled water. Freeze at –20°C and use the supernatant for PCR analysis.

DNA Extracts from Clinical Samples. *G. vaginalis* are gram-positive bacteria, for which most conventional extraction methods have a relatively poor recovery of nucleic acids.[140] In order to detect or accurately quantify *G. vaginalis* in biological samples, DNA extraction requires chemical lysis, use of lytic enzymes, or physical disruption of the gram-positive cell wall. Several studies have reported promising results for bead beating to optimize isolation of microbial DNA.[140,141] Also, the muramidases lysozyme and mutanolysin, enzymes that degrade the gram-positive peptidoglycan layer, have been shown to improve DNA recovery from gram-positives.[57,142] In addition, proteinase K can be used to break down contaminating proteins and nucleases that are liberated during lysis of several bacteria. After this pretreatment step, DNA can be extracted using commercially available kits. However, a common limitation to PCR-based methods is failed amplification due to the presence of inhibitory substances in the sample. PCR inhibitors include heme compounds found in blood,[143] aqueous and vitreous humors of the eye,[144] heparin,[145] EDTA, urine,[146] and polyamines.[147] To deal with this problem, PCR inhibitors must be diluted, inactivated, or removed from the sample using an appropriate extraction method for each specific sample type. For example, Fredricks and Relman found that an organic extraction procedure with benzyl alcohol and guanidine hydrochloride buffer successfully removes SPS, yielding DNA that can be amplified by PCR without further processing or dilution.[132]

Here, we provide three examples of sample preparation methods for the detection and identification of *G. vaginalis* from clinical samples.

(i) Vaginal dry swabs and manual DNA extraction method[57,61] For DNA extraction from the dry vaginal swabs, the QIAamp DNA mini kit (Qiagen, Hilden, Germany) is used, according to the manufacturer's recommendations, with minor modifications. The dry swab specimen is swirled for 15 s in 400 μL of lysis buffer (20 mM Tris-HCl, pH 8.0; 2 mM EDTA; 1.2% Triton). Fifty units of mutanolysin (25 U/μL) (Sigma, Bornem, Belgium) are added and the samples are incubated for 30 min at 37°C. After the addition of 20 μL Proteinase K (20 mg/mL) and 200 μL AL buffer (Qiagen), samples are incubated for 30 min at 56°C. Next, 200 μL of ethanol is added and DNA is purified by adding the lysate to the Qiagen columns, as described by the manufacturer. Finally, the total bacterial DNA is eluted with 100 μL of AE buffer (Qiagen). DNA extracts are stored at –20°C.

(ii) Vaginal ESwab and semiautomated DNA extraction method Pretreatment of samples is carried out by adding 200 μL of buffer (20 mM Tris-HCl, pH 8.3, 0.5% SDS) and 2 μL of mutanolysin (25U/μL) to 200 μL of ESwab medium, followed by incubation for 15 min at 37°C. After adding 10 μL of proteinase K solution (25 mg/mL), the mixture is further incubated for 15 min at 55°C. The pretreated samples are stored at –80°C until further processing on the NucliSENS® easyMAG™ semiautomated DNA extraction automate (bioMérieux). Total bacterial DNA is eluted in 110 μL extraction buffer. DNA extracts are stored at –20°C.

(iii) Blood culture media and benzyl alcohol-guanidine hydrochloride organic extraction method[132] A total of 0.1 mL of the inoculated medium is added to 0.1 mL of lysis buffer and is briefly mixed with a vortex mixer. Lysis buffer consists of 5.0 M guanidine hydrochloride—100 mM Tris (pH 8.0) in sterile water. A total of 0.4 mL of water is added, followed by the addition of 0.8 mL of 99% benzyl alcohol (Sigma), and the sample is mixed again by vortexing. The sample is centrifuged at 7000 × *g* for 5 min. A total of 0.4 mL of the aqueous supernatant is removed and is placed in a new microcentrifuge tube. A total of 40 μL of 3.0 M sodium acetate is added, followed by the addition of 440 μL of isopropanol, and the sample is centrifuged at 16,000 × *g* for 15 min at 4°C. The precipitated DNA is washed with 70% ethanol and the pellet is air-dried. The DNA is resuspended in 0.1 mL of 10 mM Tris–0.1 mM EDTA buffer at pH 8.5.

8.2.2.2 Detection Procedures

8.2.2.2.1 Identification of Cultured Strains

Upon isolation of colonies, it is necessary to confirm the genus identity and also to further characterize to the species level. This characterization requires a battery of classical morphological and biochemical tests. The confidence level of the species identification will increase with the more tests that are carried out. Herein lies the greatest disadvantage of classical tools for identification of organisms, as even the most sophisticated array of tests can often lead to uncertainties in the classification of isolates.

DNA-based techniques have as an advantage that the genomic material of an organism is investigated in a direct way, independent of culture conditions, making the process of identification more reliable.

(i) Sequencing of 16S rRNA or cpn60 genes DNA sequencing is the process of determining the order of the bases A, T, C, and G in a DNA strand. In essence, the DNA is used as a template to generate a set of fragments that differ in length from each other by a single base. The fragments are then separated by size and the bases at the end are identified, recreating the original sequence of the DNA. Adequate genes for the purpose of bacterial species identification are the 16S ribosomal RNA gene (rRNA) or chaperonin-60 (*cpn*60; also known as *gro*EL or *hsp*60) genes. For identification, the sequence is compared to a public database (e.g., Genbank) or a curated database (e.g., SmartGene IDNS Bacteria) by using analysis software (BLAST[148] or SmartGene, www.idns-smartgene.com).

Sample Preparation and Applications for Detection of *G. vaginalis*. Although several universal primer pairs for amplification of the 16S ribosomal RNA gene have been published, not all of these primers are truly universal (reviewed by Srinivasan and Fredricks[149] and Kalra[150]). For example, the universal primer 10f (5′-AGTTTGATCCTGGCTCAG-3′) allows amplification of *G. vaginalis* DNA from cultured bacteria, although it has three mismatches, but it fails to amplify this species from a complex mixture of bacteria due to competition with targets fully complementary to the primer.[57]

In our hands, the species-specific primer set GV10f (5′-GGTTCGATTCTGGCTCAG-3′)—534r (5′-ATTACCG CGGCTGCTGG-3′) provides the most efficient amplification of *G. vaginalis*.[57,151] Amplification is performed in a 25 μL PCR mixture containing 0.1 μM of each primer, 12.5 μL of Promega master mix (Promega), 2.5 μL of DNA extract and distilled water. Thermal cycling consists of an initial denaturation of 5 min at 94°C, followed by 3 cycles of 1 min at 94°C, 2 min at 50°C, and 1 min at 72°C, followed by 35 cycles of 20 s at 94°C, 1 min at 50°C, and 1 min at 72°C, with a final extension of 10 min at 72°C and cooling to 10°C. PCR products are purified using the QIAquick PCR Purification Kit (Qiagen). Sequencing is done using the ABI Big Dye cycle sequencing reaction kit with AmpliTaq FS DNA polymerase (Applied Biosystems) with primer 534r. Sequencing reaction products are analyzed on an ABI 310 genetic analyzer (Applied Biosystems). Inspection of the electropherograms and comparison of the sequences to the 16S rRNA gene sequences in GenBank is done using SmartGene Software (http://www.smartgene.com/).

(ii) tRNA-intergenic polymorphism length analysis (tDNA–PCR) tDNA–PCR makes use of a single pair of universal primers—that is, primers complementary to conserved regions in the tRNA-genes of most bacterial groups—that are directed outwardly to amplify the spacers between the tRNA-genes. The tRNA-intergenic spacers are multiple per genome and have different lengths, such that amplification of the spacers from a bacterial genome results in a DNA-fingerprint consisting of multiple DNA-fragments immediately after amplification and electrophoresis.[152] The obtained patterns are highly similar for the different strains within the same species, but differ between species. As such, amplification of tRNA-intergenic spacers, combined with high resolution automated electrophoresis, as is possible on (e.g., the ABI310 capillary electrophoresis system),[153] makes it possible to construct a library with tDNA–PCR fingerprints of reference strains. Identification of unknown cultured strains can then be obtained by comparing the tDNA-fingerprint of the unknown with those present in the library.[154,155] Special software has been developed to do so with a minimum of hands on time.[156] We have developed such a library, covering most human bacterial species and several veterinary and environmental species. For most of the cultivable vaginal species, we have determined the corresponding tDNA–PCR patterns.[98,157]

Verhelst et al. and El Aila et al. applied tDNA–PCR identification for species cultured from the vaginal swabs of pregnant women for the purpose of microflora characterization.[98,158] Vaginal swabs on Amies transport medium or swirled in liquid Amies transport medium (ESwab) were streaked onto rich agar plates and incubated anaerobically at 37°C upon arrival at the microbiology laboratory. After 4 days of incubation, all the isolates with different colony morphology were selected for identification. DNA was extracted by simple alkaline lysis (see crude DNA extraction). tDNA–PCR and capillary electrophoresis were carried out. This method was found to be a reproducible, cheap, and fast identification method for most vaginal species, including *G. vaginalis*.

8.2.2.2.2 Culture-Independent Detection and Identification

(i) Conventional species-specific assays In conventional PCR, detection and semiquantification of the amplified sequence are performed at the end of the reaction after the last PCR cycle and involve postPCR analysis, such as size fractionation of amplicons by gel electrophoresis and image analysis.

Several assays for the detection of *G. vaginalis*, targeting respectively the 16S-23S rRNA spacer region[58,60,159] or the 16S rRNA gene,[160] were designed. Using species-specific PCR for *G. vaginalis*, targeting the 16S-23S rRNA spacer region, it could be shown that 40% of women, irrespective of their clinical status, and 10 out of 11 patients suffering from bacterial vaginosis, were carrying *G. vaginalis*.[60] Through conventional PCR analysis of 515 vaginal samples, Verhelst et al.[57,98] found that the presence of *A. vaginae* is a diagnostically more valuable marker for bacterial vaginosis than the presence of *G. vaginalis* since respectively 14.7% of normal vaginal microflora sample and 86.4% of bacterial vaginosis samples showed an amplicon while the percentage of positive samples for *G. vaginalis* specific PCR in both categories were respectively 28.9% and 86.4%. For this purpose, *G. vaginalis* species-specific primers F-GV1 (5′-TTACTGGTGTATCACTGTAAGG-3′) and R-GV-3 (5′-CCGTCACAGGCTGAACAGT-3′) as designed by Zariffard et al.[161] were used. Briefly, a 20 μL PCR mixture contained respectively 0.05 and 0.4 μM primers, 10 μL of Promega master mix (Promega), 2 μL of Qiagen DNA extract of the samples, and distilled water. Thermal cycling consisted of an initial denaturation of 10 min at 94°C, followed by 50 cycles of 5 s at 94°C, 45 s at 55°C, and 45 s at 72°C, and a final extension of 10 min at 72°C. Five μL of the amplified product was visualized on a 2% agarose gel.

(ii) Real-time assays Real-time polymerase chain reaction, also called quantitative polymerase chain reaction (qPCR), is a PCR procedure that enables both detection and quantification of a specific DNA target in a single tube, thereby limiting the risk of contamination and providing fast and reliable detection compared to classical end-point analysis PCR techniques. In contrast to conventional PCR, amplification is tracked during each replication cycle. Two common methods for detection of products in qPCR are (i) SYBRGreen, a fluorescent dye that intercalates with any double-stranded DNA, and (ii) sequence-specific DNA probes that are labeled with a fluorescent reporter that permits detection only after hybridization of the probe with its complementary DNA target. Using sequence-specific primers, the absolute number of chromosomes of a particular DNA target can be determined by extrapolating quantitative information obtained from a standard curve.

Primer sets targeting the 16S rRNA gene or the 16S-23S spacer region of *G. vaginalis* were applied to assess the number of chromosomes of these microorganisms in DNA extracts obtained from vaginal swab specimens.[61,159,161–165] De Backer et al.[61] confirmed by qPCR, using the primers designed by Zariffard et al.,[161] that the presence of *A. vaginae* is a diagnostically more valuable marker for bacterial vaginosis than the presence of *G. vaginalis*. The qPCR Core Kit for SYBR Green I (Eurogentec) was applied and analysis was performed on the ABI 7000 qPCR system (Applied Biosystems). PCR mixtures contained 2.5 μL of DNA extract, 2.5 μL of 10× Reaction Buffer, 3.5 mM $MgCl_2$, 0.2 mM dNTP mixture, 0.625 U HotGoldStar *Taq* polymerase, 0.75 μL SYBR® Green I, diluted tenfold in DMSO and 0.2 μM primers, and were adjusted with HPLC grade water to 25 μL. Thermal cycling consisted of an initial denaturation of 10 min at 95°C, followed by 50 cycles of 45 s at 94°C, 45 s at 55°C, and 45 s at 72°C. Using the same reaction conditions, El Aila et al. (in preparation) assessed the presence of *G. vaginalis* in the vaginal and rectal microflora of 71 pregnant women and found that *G. vaginalis* was present in 74.6% of the women, either vaginally only (9.8%), rectal only (15.4%), or in both sites (49.2%), and concluded that these results confirm that the rectum serves as a reservoir for *G. vaginalis* and other bacterial vaginosis associated species. Interestingly, these authors found also a quantitative correspondence between rectum and vagina for this and other species.

Other qPCR assays for quantitative assessment of *G. vaginalis* concentrations were published by Zariffard et al.,[161] Trama et al.,[162] Biagi et al.,[163] Fredricks et al.,[164] and Zozaya-Hinchliffe et al.[166] Furthermore, the utility of qPCR as a diagnostic tool for bacterial vaginosis has been explored by Sha et al.,[159] and Menard et al.[165] Using 203 samples, from both of women with and without bacterial vaginosis, Sha et al. reported that increasing levels of *G. vaginalis* and *M. hominis* and decreasing levels of lactobacilli were significantly associated with bacterial vaginosis, with a sensitivity and specificity of 83% and 78% when compared with Nugent score. Menard et al.[165] assessed the utility of *G. vaginalis* and *A. vaginae* loads as a diagnostic tool in 231 women and confirmed the results of De Backer et al.[61] that the presence of *A. vaginae* at $\geq 10^8$ copies/mL and *G. vaginalis* at $\geq 10^9$ copies/mL had a sensitivity and specificity of 95% and 99% when compared to Nugent scoring.

(iii) Commercial point-of-care test for the diagnosis of G. vaginalis The Affirm™ VP III Microbial Identification Test (BD Diagnostic Systems, NJ, USA) is a DNA hybridisation assay intended for detection and identification of *Candida* species, *G. vaginalis* and *Trichomonas vaginalis* in vaginal fluid specimens from patients with symptoms of bacterial vaginosis. The test uses two distinct single-stranded nucleic acid probes for each organism, a capture probe and a color development probe, that are complementary to unique genetic sequences of the target organisms. The capture probes are immobilized on beads embedded in a Probe Analysis Card (PAC), which contains a separate bead for each target organism. The color development probes are contained in a multiwell Reagent Cassette (RC). During sample preparation, the sample is treated with the Lysis Solution and heated. This process ruptures the walls of the organism, releasing the nucleic acid. A second solution, the Buffer Solution, is added. This solution stabilizes the nucleic acid and establishes the stringency conditions necessary for specific hybridization. Automated processing of the sample is performed using the BD MicroProbeTM Processor. The final step is reading the results of color development on each of the target organism beads and controls. The procedure requires between 30 and 45 min to complete and can be done in an office setting. This DNA hybridization test is positive only for concentrations of *G. vaginalis* in excess of 2×10^5 bacterial cells per ml of vaginal fluid[167] and therefore should be positive most often in women with bacterial vaginosis and rarely in women with normal vaginal microflora.

Promising results have been reported with this approach.[167] These authors concluded that the probe could be used as a supplement to the Amsel and Nugent criteria for the diagnosis of bacterial vaginosis.

(iv) Fluorescence in situ *hybridization (FISH)* FISH can be used to visualize bacteria. Fluorescent probes are hybridized to mRNA or DNA present in cells and are visualized using epifluorescence microscopy. Data can be collected in both quantitative and qualitative modes. For example, when

fluorescent probes are combined with flow cytometry, one can rapidly count and collect cells. With confocal scanning laser microscopy, one can visualize the spatial arrangement of cells in tissues or body fluids.

FISH can also be used to study the bacterial community structure and the spatial organization of bacteria on the epithelial surfaces of vaginal biopsy specimens.[31] Biopsy specimens are taken from the middle side wall of the vagina with biopsy forceps, fixed in nonaqueous Carnoy solution, and then processed and embedded into paraffin blocks. Four micrometer sections are placed on SuperFrost slides for FISH analysis. Oligonucleotide probes are synthesized with a carbocyanine dye (Cy3), fluorescein isothiocyanate, or Cy5 fluorescent dye at the 5′. The probe used for visualization of *G. vaginalis* (GardV) has the following sequence: 5′-CCA CCG TTA CAC CGC GAA-3. Hybridisation is performed using a washing temperature of 50°C and bacteria are detected by FISH and 4′,6-diamidino-2-phenylindole counter stain. This paper confirmed the strong link between *A. vaginae*, *G. vaginalis*, and bacterial vaginosis in a substantial manner, not only by showing the simultaneous presence of these two species in more than half of the cases of bacterial vaginosis, but also by showing that there was a spatial association in biofilms. Moreover, the realization that bacterial biofilms can be formed at the vaginal epithelium explains the presence of the well-known "clue cells," designated as such because they are a hallmark for bacterial vaginosis. Finally, the insight that vaginosis is associated with bacterial biofilms may also nicely explain the recurrent character of vaginosis in many women. A more recent FISH study explained the clinical observation that, despite the temporary improvement of clinical symptoms of bacterial vaginosis by antibiotic treatment, the symptoms recur.[106] It could be shown by FISH analysis of bacterial mRNA and DNA extracted from vaginal biopsies taken during treatment that the bacteria in the biofilm become dormant during treatment, but are not eradicated. After treatment, bacterial metabolic activity is regained and symptoms return.

(v) Checkerboard DNA hybridization A method allowing for simultaneous detection of multiple organisms is the checkerboard method.[168] In this case, DNA of the complex microflora is extracted, spotted in separate lanes on a membrane, and hybridized with multiple labeled whole genomic DNA probes.

A checkerboard for the simultaneous detection of *G. vaginalis*, *P. bivia*, *B. ureolyticus*, and *M. curtisii* was evaluated and it was concluded that hybridization was significantly associated with Gram-stain results since more species were detected by DNA hybridization as the Nugent score progressed from a normal to a disturbed microflora.[169]

Persson et al. assessed, by a semiquantitative DNA–DNA checkerboard hybridization assay of 74 bacterial species including *G. vaginalis*, whether the bacterial counts of oral and bacterial vaginosis species were higher in the vaginal microflora of women with gingivitis compared to healthy women. The load of *G. vaginalis* in the vaginal microflora of women with gingivitis was not elevated.[170]

(vi) Oligonucleotide-coupled fluorescent microsphere Hybridization can also take place on optically encoded microparticles.[171] Pioneered by Luminex, these polystyrene microbeads are imbued with two fluorescent dyes and the ratio of these creates 100 unique spectral signatures. Each bead is coated with a different oligonucleotide probe. After PCR amplification of the targets, wherein amplicons are labeled, only those beads that have bound target will be detected by a flow cytometer.

The main advantage of this approach is its ability to detect a large number of molecular targets simultaneously after a single PCR, making high-throughput analysis of multiple samples from large study groups or longitudinal studies technically feasible.

A nine-plex Luminex assay was designed for characterizing the vaginal microflora and for exploring molecular diagnostics of bacterial vaginosis.[172] DNA was extracted using Instagene Matrix (BioRad Laboratories). An amplicon for Luminex analysis was generated using two universal cpn60 primers sets. The *G. vaginalis* probe had the following sequence: 5′-CCAGCAACAACATTCTTTAAGCCTTCGT G-3′. The vaginal swab DNA extracts were used as templates for PCR amplification of the 5′ end of the *cpn*60 UT using two universal primers.

Using the presence of one or both of *A. vaginae* and *G. vaginalis* as the defining criterion for bacterial vaginosis, these authors found that the method showed a high sensitivity and specificity for diagnosing bacterial vaginosis compared to traditional microscopy.

8.3 CONCLUSIONS AND FUTURE PERSPECTIVES

While recent studies of the biofilm-forming potential and cytotoxic activity of *G. vaginalis*[32] have renewed interest in the virulence potential of this organism, ongoing sequence analysis of the whole genome within the Human Microbiome Project and the *G. vaginalis* Genome Project (http://med.stanford.edu/sgtc/research/gardnerella_vaginalis.html) will further improve our understanding of the pathogenic role of *G. vaginalis* in bacterial vaginosis.[173] This sensitive high-throughput analysis through deep pyrosequencing will further enlighten our knowledge of bacterial diversity and ecological dynamics associated with reproductive tract infection. Knowledge obtained by this metagenomic approach might then be translated into rapid diagnostic assays for profiling a selected set of microbial communities in a lab-based setting. To further explore the role of *G. vaginalis* in bacterial vaginosis, microarrays like the TNO vaginal microarray (V-chip, TNO, Zeist, Netherlands) offer the possibility to detect over 400 vaginal species.

In conclusion, molecular diagnosis of bacterial vaginosis, with a central role for *G. vaginalis* and *A. vaginae* (see Chapter 4) is now pending.[61,62,161,165,172] The study by Menard et al. that assessed the utility of quantitation of *G. vaginalis* and *A. vaginae* as a diagnostic tool for bacterial vaginosis clearly seems to provide a sound basis.[149] However, as has been recently reiterated by Kalra et al., distinct bacterial vaginosis profiles consisting of differing species might predispose to different health outcomes.[149] In cases of bacteremia, molecular detection methods should become the golden standard to detect and identify *G. vaginalis*, as blood culture media often contain inhibitors that prevent isolation of this species. Importantly, when changing traditional methods for molecular detection methods, special attention should be addressed to proper nucleic acid extraction methods.

REFERENCES

1. Catlin, B.W., Gardnerella vaginalis: Characteristics, clinical considerations, and controversies, *Clin. Microbiol. Rev.*, 5, 213, 1992.
2. Curtis, A.H., On the etiology and bacteriology of leucorrhea, *Surg. Gynecol. Obstet.*, 18, 299, 1914.
3. Leopold, S., Heretofore undescribed organism isolated from the genitourinary system, *US Armed Forces Med. J.*, 4, 263, 1953.
4. Gardner, H.L., and Dukes, C.D., *Haemophilus vaginalis* vaginitis: A newly defined specific infection previously classified non-specific vaginitis, *Am. J. Obstet. Gynecol.*, 69, 962, 1955.
5. Reyn, A., Birch-Andersen, A., and Lapage, S.P., An electron microscope study of thin sections of *Haemophilus vaginalis* (Gardner and Dukes) and some possibly related species, *Can. J. Microbiol.*, 12, 1125, 1966.
6. Harper, J.J., and Davis, G.H.G., Cell wall analysis of *Gardnerella vaginalis* (*Haemophilus vaginalis*), *Int. J. Syst. Bacteriol.*, 32, 48, 1982.
7. O'Donnell, A.G. et al., Fatty acid, polar lipid and wall amino acid composition of *Gardnerella vaginalis*, *Arch. Microbiol.*, 138, 68, 1984.
8. Sadhu, K. et al., *Gardnerella vaginalis* has a gram-positive cell-wall ultrastructure and lacks classical cell-wall lipopolysaccharide, *J. Med. Microbiol.*, 29, 229, 1989.
9. Gelber, S.E. et al., Functional and phylogenetic characterization of Vaginolysin, the human-specific cytolysin from *Gardnerella vaginalis*, *J. Bacteriol.*, 190, 3896, 2008.
10. Criswell, B.S. et al., *Haemophilus vaginalis* 594, a gram-negative organism? *Can. J. Microbiol.*, 17, 865, 1971.
11. Criswell, B.S. et al., Fine structure of *Haemophilus vaginalis*, *J. Bacteriol.*, 109, 930, 1972.
12. Greenwood, J.R., and Pickett, M.J., Transfer of *Haemophilus vaginalis* Gardner and Dukes to a new genus, *Gardnerella vaginalis* (Gardner and Dukes) comb. nov., *Int. J. Syst. Bacteriol.*, 30, 170, 1980.
13. Csango, P.A., Hagen, N., and Jagars, G., Characterization of *Gardnerella vaginalis* by gas chromatography, *Scand. J. Infect. Dis.*, 40, S19, 1983.
14. Jantzen, E., Berdal, B.P., and Omland, T., Cellular fatty acid composition of *Haemophilus* species, *Pasteurella multocida, Actinobacillus actinomycetemcomitans* and *Haemophilus vaginalis* (*Corynebacterium vaginale*), *Acta Pathol. Microbiol. Scand. B*, 88, 89, 1980.
15. Piot, P. et al., A taxonomic study of *Gardnerella vaginalis* (*Haemophilus vaginalis*) Gardner and Dukes 1955, *J. Gen. Microbiol.*, 119, 373, 1980.
16. Kharsany, A.B., Hoosen, A.A., and Van den Ende, J., Antimicrobial susceptibilities of *Gardnerella vaginalis*, *Antimicrob. Agents Chemother.*, 37, 2733, 1993.
17. Zinneman, K., and Turner, G.C., The taxonomic position of "*Haemophilus vaginalis*" (*Corynebacterium vaginale*), *J. Pathol. Bacteriol.*, 85, 213, 1963.
18. Moss, C.W., and Dunkelberg, W.E.J., Volatile and cellular fatty acids of *Haemophilus vaginalis*, *J. Bacteriol.*, 100, 544, 1969.
19. van Esbroeck, M. et al., Polyphasic approach to the classification and identification of *Gardnerella vaginalis* and unidentified *Gardnerella vaginalis*-like coryneforms present in bacterial vaginosis, *Int. J. Syst. Bacteriol.*, 46, 675, 1996.
20. Scardovi, V. Genus *Bifidobacterium* Orla-Jensen 1924, 472[AL]. In Sneath, P.H.A., Mair, N.S., Sharpe, M.E., and Holt, J.G. (Editors), *Bergey's Manual of Systematic Bacteriology*, vol 2, pp. 1418–1434, The Williams & Wilkins Co, Baltimore, 1986.
21. Bahaka, D. et al., Phenotypic and genomic analyses of human strains belonging or related to *Bifidobacterium longum, Bifidobacterium infantis*, and *Bifidobacterium breve*, *Int. J. Syst. Bacteriol.*, 43, 565, 1993.
22. Piot, P. et al., Biotypes of *Gardnerella vaginalis*, *J. Clin. Microbiol.*, 20, 677, 1984.
23. Benito, R. et al., A modified scheme for biotyping *Gardnerella vaginalis*, *J. Med. Microbiol.*, 21, 357, 1986.
24. Pedraza-Aviles, A.G. et al., *Gardnerella vaginalis* biotypes: modification of a proposed system, *Rev. Latinoam. Microbiol.*, 37, 19, 1995.
25. Briselden, A.M., and Hillier, S.L., Longitudinal study of the biotypes of *Gardnerella vaginalis*, *J. Clin. Microbiol.*, 28, 2761, 1990.
26. Numanovic, F. et al., Importance of isolation and biotypization of *Gardnerella vaginalis* in diagnosis of bacterial vaginosis, *Bosn. J. Basic Med. Sci.*, 8, 270, 2008.
27. Aroutcheva, A.A. et al., *Gardnerella vaginalis* isolated from patients with bacterial vaginosis and from patients with healthy vaginal ecosystems, *Clin. Infect. Dis.*, 33, 1022, 2001.
28. Moncla, B.J., and Pryke, K.M., Oleate lipase activity in *Gardnerella vaginalis* and reconsideration of existing biotype schemes, *BMC Microbiol.*, 9, 78, 2009.
29. Patterson, J.L. et al., Analysis of adherence, biofilm formation, and cytotoxicity suggest a greater virulence potential of *Gardnerella vaginalis* relative to other bacterial vaginosis-associated anaerobes, *Microbiology,* 156(pt 2), 392–399, 2010.
30. Simoes, J.A. et al., Human immunodeficiency virus type 1 stimulatory activity by *Gardnerella vaginalis*: relationship to biotypes and other pathogenic characteristics, *J. Infect. Dis.*, 184, 22, 2001.
31. Swidsinski, A. et al., Adherent biofilms in bacterial vaginosis, *Obstet. Gynecol.*, 106, 1013, 2005.
32. Johnson, A.P., and Davies, H.A., Demonstration by electron microscopy of pili on *Gardnerella vaginalis*, *Br. J. Vener. Dis.*, 60, 396, 1984.
33. Boustouller, Y.L., Johnson, A.P., and Taylor-Robinson, D., Pili on *Gardnerella vaginalis* studied by electronmicroscopy, *J. Med. Microbiol.*, 23, 327, 1987.
34. Fredricsson, B. et al., *Gardnerella*-associated vaginitis and anaerobic bacteria, *Gynecol. Obstet. Invest.*, 17, 236, 1984.
35. van der Meijden, W.I. et al., Descriptive light and electron microscopy of normal and clue-cell-positive discharge, *Gynecol. Obstet. Invest.*, 25, 47, 1988.
36. Scott, T.G., Curran, B., and Smyth, C.J., Electron microscopy of adhesive interactions between *Gardnerella vaginalis* and vaginal epithelial cells, McCoy cells and human red blood cells, *J. Gen. Microbiol.*, 135, 475, 1989.
37. Randis, T.M. et al., Antibody-based detection and inhibition of vaginolysin, the *Gardnerella vaginalis* cytolysin, *PLoS One*, 4, e5207, 2009.

38. Rottini, G. et al., Identification and partial characterization of a cytolytic toxin produced by *Gardnerella vaginalis*, *Infect. Immun.*, 58, 3751, 1990.
39. Moran, O. et al., Voltage-dependent cationic channels formed by a cytolytic toxin produced by *Gardnerella vaginalis*, *FEBS Lett.*, 283, 317, 1991.
40. Cauci, S. et al., Ionic factors regulating the interaction of *Gardnerella vaginalis* hemolysin with red blood cells, *Biochim. Biophys. Acta,* 1153, 53, 1993.
41. Cauci, S. et al., Pore-forming and haemolytic properties of the *Gardnerella vaginalis* cytolysin, *Mol. Microbiol.,* 9, 1143, 1993.
42. Cauci, S. et al., Specific immune response against *Gardnerella vaginalis* hemolysin in patients with bacterial vaginosis, *Am. J. Obstet. Gynecol.,* 175, 1601, 1996.
43. Cauci, S. et al., Correlation of local interleukin-1beta levels with specific IgA response against *Gardnerella vaginalis* cytolysin in women with bacterial vaginosis, *Am. J. Reprod. Immunol.,* 47, 257, 2002.
44. Cauci, S. et al., Among pregnant women with bacterial vaginosis, the hydrolytic enzymes sialidase and prolidase are positively associated with interleukin-1beta, *Am. J. Obstet. Gynecol.,* 198, 132.e1, 2008.
45. Bejar, R. et al., Premature labor. II. Bacterial sources of phospholipase, *Obstet. Gynecol.,* 57, 479, 1981.
46. Jarosik, G.P. et al., Acquisition of iron by *Gardnerella vaginalis*, *Infect. Immun.,* 66, 5041, 1998.
47. Jarosik, G.P., Binding of catalase by *Gardnerella vaginalis*, *FEMS Microbiol. Lett.,* 190, 191, 2000.
48. Jarosik, G.P., and Land, C.B., Identification of a human lactoferrin-binding protein in *Gardnerella vaginalis*, *Infect. Immun.,* 68, 3443, 2000.
49. Jarosik, G.P., and Land, C.B., Binding of heme by *Gardnerella vaginalis*, *J. Basic Microbiol.,* 41, 37, 2001.
50. Jarosik, G.P., Identification of a *Gardnerella vaginalis* hemoglobin-binding protein, *Curr. Microbiol.,* 42, 49, 2001.
51. Patterson, J.L. et al., Effect of biofilm phenotype on resistance of *Gardnerella vaginalis* to hydrogen peroxide and lactic acid, *Am. J. Obstet. Gynecol.,* 197, 170.e1, 2007.
52. Costerton, J.W., Stewart, P.S., and Greenberg, E.P., Bacterial biofilms: A common cause of persistent infections, *Science,* 284, 1318, 1999.
53. Xu, K.D., McFeters, G.A., and Stewart, P.S., Biofilm resistance to antimicrobial agents, *Microbiology,* 146, 547, 2000.
54. Mah, T.F., and O'Toole, G.A., Mechanisms of biofilm resistance to antimicrobial agents, *Trends Microbiol.,* 9, 34, 2001.
55. Fux, C.A. et al., Survival strategies of infectious biofilms, *Trends Microbiol.,* 13, 34, 2005.
56. Saunders, S. et al., Effect of Lactobacillus challenge on *Gardnerella vaginalis* biofilms, *Colloids Surf. B Biointerfaces,* 55, 138, 2007.
57. Verhelst, R. et al., Cloning of 16S rRNA genes amplified from normal and disturbed vaginal microflora suggests a strong association between *Atopobium vaginae, Gardnerella vaginalis* and bacterial vaginosis, *BMC Microbiol.,* 4, 16, 2004.
58. Hillier, S.L., Diagnostic microbiology of bacterial vaginosis, *Am. J. Obstet. Gynecol.,* 169, 455, 1993.
59. Obata-Yasuoka, M. et al., A multiplex polymerase chain reaction-based diagnostic method for bacterial vaginosis, *Obstet. Gynecol.,* 100, 759, 2002.
60. van Belkum, A. et al., Development of a species-specific polymerase chain reaction assay for *Gardnerella vaginalis*, *Mol. Cell. Probes,* 9, 167, 1995.
61. De Backer, E. et al., Quantitative determination by real-time PCR of four vaginal Lactobacillus species, *Gardnerella vaginalis* and *Atopobium vaginae* indicates an inverse relationship between *L. gasseri* and *L. iners*. *BMC Microbiol.,* 7, 115, 2007.
62. Fredricks, D.N. et al., Targeted PCR for detection of vaginal bacteria associated with bacterial vaginosis, *J. Clin. Microbiol.,* 45, 3270, 2007.
63. Kinghorn, G.R. et al., Balanoposthitis associated with *Gardnerella vaginalis* infection in men, *Br. J. Vener. Dis.,* 58, 127, 1982.
64. Burdge, D.R., Bowie, W.R., and Chow, A.W., *Gardnerella vaginalis*-associated balanoposthitis, *Sex. Transm. Dis.,* 13, 159, 1986.
65. Sturm, A.W., *Gardnerella vaginalis* in infections of the urinary tract, *J. Infect.,* 18, 45, 1989.
66. Lefevre, J.C. et al., Clinical and microbiologic features of urethritis in men in Toulouse, France, *Sex. Transm. Dis.,* 18, 76, 1991.
67. Smith, S.M., Ogbara, T., and Eng, R.H., Involvement of *Gardnerella vaginalis* in urinary tract infections in men, *J. Clin. Microbiol.,* 30, 1575, 1992.
68. Keane, F.E. et al., An association between non-gonococcal urethritis and bacterial vaginosis and the implications for patients and their sexual partners, *Genitourin Med.,* 73, 373, 1997.
69. Lee, W. et al., *Gardnerella vaginalis* chorioamnionitis: A report of two cases and a review of the pathogenic role of *G. vaginalis* in obstetrics, *Diagn. Microbiol. Infect. Dis.,* 8, 107, 1987.
70. Moran, D.J., and Payne, A., Subclinical intra-amniotic infection with *Gardnerella vaginalis* associated with preterm delivery. Case report, *Br. J. Obstet. Gynaecol.,* 96, 489, 1989.
71. Chowdhury, M.N., and Desilva, S.K., Episiotomy wound infection due to *Gardnerella vaginalis*, *Eur. J. Clin. Microbiol.,* 5, 164, 1986.
72. Sturm, A.W., de Leeuw, J.H., and de Pree, N.T., Postoperative wound infection with *Gardnerella vaginalis*, *J. Infect.,* 7, 264, 1983.
73. Boggess, K.A. et al., Bacteremia shortly after placental separation during cesarean delivery, *Obstet. Gynecol.,* 87, 779, 1996.
74. Nightingale, L.M. et al., Cephalhematoma complicated by osteomyelitis presumed due to *Gardnerella vaginalis*, *JAMA,* 256, 1936, 1986.
75. Eggink, B.H., Richardson, C.J., and Rowen, J.L., *Gardnerella vaginalis*-infected scalp hematoma associated with electronic fetal monitoring, *Pediatr. Infect. Dis. J.,* 23, 276, 2004.
76. Chowdhury, M.N., and Kambal, A.M., A case of conjunctivitis in a neonate due to *Gardnerella vaginalis*, *Trop. Geogr. Med.,* 37, 365, 1985.
77. Berardi-Grassias, L. et al., Neonatal meningitis due to *Gardnerella vaginalis*, *Eur. J. Clin. Microbiol. Infect. Dis.,* 7, 406, 1988.
78. Furman, L.M., Neonatal *Gardnerella vaginalis* infection, *Pediatr. Infect. Dis. J.,* 7, 890, 1988.
79. Amaya, R.A., Al-Dossary, F., and Demmler, G.J., *Gardnerella vaginalis* bacteremia in a premature neonate, *J. Perinatol.,* 22, 585, 2002.
80. Bastida Vila, M.T. et al., *Garderella vaginalis* bacteremia in an adult male, *Eur. J. Clin. Microbiol. Infect. Dis.,* 16, 400, 1997.
81. Calvert, L.D., Collins, M., and Bateman, J.R., Multiple abscesses caused by *Gardnerella vaginalis* in an immunocompetent man, *J. Infect.,* 51, E27, 2005.

82. Lagace-Wiens, P.R. et al., *Gardnerella vaginalis* bacteremia in a previously healthy man: Case report and characterization of the isolate, *J. Clin. Microbiol.*, 46, 804, 2008.
83. Hodge, T.W.J., Levy, C.S., and Smith, M.A., Disk space infection due to *Gardnerella vaginalis*, *Clin. Infect. Dis.*, 21, 443, 1995.
84. Graham, S. et al., Vertebral osteomyelitis and discitis due to *Gardnerella vaginalis*, *J. Med. Microbiol.*, 58, 1382, 2009.
85. Sivadon-Tardy, V. et al., *Gardnerella vaginalis* acute hip arthritis in a renal transplant recipient, *J. Clin. Microbiol.*, 47, 264, 2009.
86. Francois, S., Guyadier-Souguieres, G., and Marcelli, C., Reactive arthritis due to *Gardnerella vaginalis*. A case-report, *Rev. Rhum. Engl. Ed.*, 64, 138, 1997.
87. Schapira, D., Braun-Moscovici, Y., and Nahir, A.M., Reactive arthritis induced by *Gardnerella vaginalis*, *Clin. Exp. Rheumatol.*, 20, 732, 2002.
88. Toussirot, E., Plesiat, P., and Wendling, D., Reiter's syndrome induced by *Gardnerella vaginalis, Scand. J. Rheumatol.*, 27, 316, 1998.
89. Verstraelen, H. et al., The epidemiology of bacterial vaginosis in relation to sexual behaviour, *BMC Microbiol.*, 10, 81, 2010.
90. Swidsinski A, et al. Gardnerella biofilm involves females and males and is transmitted sexually. *Gynecol Obstet Invest.*, 70, 256, 2010.
91. Bartley, D.L., Morgan, L., and Rimsza, M.E., *Gardnerella vaginalis* in prepubertal girls, *Am. J. Dis. Child.*, 141, 1014, 1987.
92. Gardner, J.J., Comparison of the vaginal flora in sexually abused and nonabused girls, *J. Pediatr.*, 120, 872, 1992.
93. Steele, A.M., and de San Lazaro, C., Transhymenal cultures for sexually transmissible organisms, *Arch. Dis. Child.*, 71, 423, 1994.
94. Ingram, D.L. et al., *Gardnerella vaginalis* infection and sexual contact in female children, *Child Abuse Negl.*, 16, 847, 1992.
95. Shafer, M.A. et al., Microbiology of the lower genital tract in postmenarchal adolescent girls: Differences by sexual activity, contraception, and presence of nonspecific vaginitis, *J. Pediatr.*, 107, 974, 1985.
96. Bump, R.C., Sachs, L.A., and Buesching, W.Jr., Sexually transmissible infectious agents in sexually active and virginal asymptomatic adolescent girls, *Pediatrics,* 77, 488, 1986.
97. Yen, S. et al., Bacterial vaginosis in sexually experienced and non-sexually experienced young women entering the military, *Obstet. Gynecol.*, 102, 927, 2003.
98. Verhelst, R. et al., Comparison between Gram stain and culture for the characterization of vaginal microflora: Definition of a distinct grade that resembles grade I microflora and revised categorization of grade I microflora, *BMC Microbiol.*, 5, 61, 2005.
99. Fredricks, D.N., Fiedler, T.L., and Marrazzo, J.M., Molecular identification of bacteria associated with bacterial vaginosis, *N. Engl. J. Med.*, 353, 1899, 2005.
100. McCaffrey, M. et al., Bacterial vaginosis in lesbians: evidence for lack of sexual transmission, *Int. J. STD AIDS,* 10, 305, 1999.
101. Marrazzo, J.M. et al., haracterization of vaginal flora and bacterial vaginosis in women who have sex with women, *J. Infect. Dis.*, 185, 1307, 2002.
102. Evans, A.L. et al., Prevalence of bacterial vaginosis in lesbians and heterosexual women in a community setting, *Sex. Transm. Infect.*, 83, 470, 2007.
103. Weyers, S. et al., Microflora of the penile skin-lined neovagina of transsexual women, *BMC Microbiol.*, 9, 102, 2009.
104. Dawson, S.G. et al., Male carriage of *Gardnerella vaginalis*, *Br. J. Vener. Dis.*, 58, 243, 1982.
105. Holst, E., Goffeng, A.R., and Andersch, B., Bacterial vaginosis and vaginal microorganisms in idiopathic premature labor and association with pregnancy outcome, *J. Clin. Microbiol.*, 32, 176, 1994.
106. Swidsinski, A. et al., An adherent *Gardnerella vaginalis* biofilm persists on the vaginal epithelium after standard therapy with oral metronidazole, *Am. J. Obstet. Gynecol.*, 198, 97.e1, 2008.
107. Schwebke, J.R., Rivers, C., and Lee, J., Prevalence of *Gardnerella vaginalis* in male sexual partners of women with and without bacterial vaginosis, *Sex. Transm. Dis.*, 36, 92, 2009.
108. Chattopadhyay, B., and Teli, J.C., Isolation of *Gardnerella vaginalis* from routine genito-urinary tract specimens, *J. Infect.*, 8, 157, 1984.
109. Ison, C.A., and Easmon, C.S., Carriage of *Gardnerella vaginalis* and anaerobes in semen, *Genitourin Med.*, 61, 120, 1985.
110. Elsner, P., and Hartmann, A.A., *Gardnerella vaginalis* in the male upper genital tract: A possible source of reinfection of the female partner, *Sex. Transm. Dis.*, 14, 122, 1987.
111. Lam, M.H., Birch, D.F., and Fairley, K.F., Prevalence of *Gardnerella vaginalis* in the urinary tract, *J. Clin. Microbiol.*, 26, 1130, 1988.
112. Hillier, S.L. et al., Relationship of bacteriologic characteristics to semen indices in men attending an infertility clinic, *Obstet. Gynecol.*, 75, 800, 1990.
113. Kjaergaard, N. et al., Microbiology of semen specimens from males attending a fertility clinic, *APMIS,* 105, 566, 1997.
114. Andrade-Rocha, F.T., Colonization of *Gardnerella vaginalis* in semen of infertile men: Prevalence, influence on sperm characteristics, relationship with leukocyte concentration and clinical significance, *Gynecol. Obstet. Invest.*, 68, 134, 2009.
115. Virecoulon, F. et al., Bacterial flora of the low male genital tract in patients consulting for infertility, *Andrologia,* 37, 160, 2005.
116. Schwebke, J.R., New concepts in the etiology of bacterial vaginosis, *Curr. Infect. Dis. Rep.*, 11, 143, 2009.
117. Verstraelen, H., Bacterial vaginosis: a sexually enhanced disease, *Int. J. STD AIDS,* 19, 575, 2008.
118. Salmon, S.A. et al., Isolation of *Gardnerella vaginalis* from the reproductive tract of four mares, *J. Vet. Diagn. Invest.*, 2, 167, 1990.
119. Golberg, R.L., and Washington, J.A.I.I., Comparison of isolation of *Haemophilus vaginalis* (*Corynebacterium vaginale*) from peptone-starch-dextrose agar and Columbia colistin-nalidoxic acid agar, *J. Clin. Microbiol.*, 4, 245, 1976.
120. Piot, P. et al., Identification of *Gardnerella* (*Haemophilus*) *vaginalis*, *J. Clin. Microbiol.*, 15, 19, 1982.
121. Totten, P.A. et al., Selective differential human blood bilayer media for isolation of *Gardnerella* (*Haemophilus*) *vaginalis*, *J. Clin. Microbiol.*, 15, 141, 1982.
122. Piot, P., and Van Dyck, E., Isolation and identification of *Gardnerella vaginalis*, *Scand. J. Infect. Dis.*, 40, S15, 1983.
123. Holst, E., Reservoir of four organisms associated with bacterial vaginosis suggests lack of sexual transmission, *J. Clin. Microbiol.*, 28, 2035, 1990.
124. Ison, C.A. et al., Comparison of culture and microscopy in the diagnosis of *Gardnerella vaginalis* infection, *J. Clin. Pathol.*, 35, 550, 1982.

125. Taylor, E., and Phillips, I., The identification of *Gardnerella vaginalis*, *J. Med. Microbiol.*, 16, 83, 1983.
126. Bailey, R.K., Voss, J.L., and Smith, R.F., Factors affecting isolation and identification of *Haemophilus vaginalis* (*Corynebacterium vaginale*), *J. Clin. Microbiol.*, 9, 65, 1979.
127. Gibbs, R.S. et al., Microbiologic and serologic studies of *Gardnerella vaginalis* in intra-amniotic infection, *Obstet. Gynecol.*, 70, 187, 1987.
128. La Scolea, L.J.J., Dryja, D.M., and Dillon, W.P., Recovery of *Gardnerella vaginalis* from blood by the quantitative direct plating method, *J. Clin. Microbiol.*, 20, 568, 1984.
129. Legrand, J. C. et al., *Gardnerella vaginalis* bacteremia from pulmonary abscess in a male alcohol abuser, *J. Clin. Microbiol.*, 27, 1132, 1989.
130. Reimer, L.G., and Reller, L.B., *Gardnerella vaginalis* bacteremia: A review of thirty cases, *Obstet. Gynecol.*, 64, 170, 1984.
131. Reimer, L.G., and Reller, L.B., Effect of sodium polyanetholesulfonate and gelatin on the recovery of *Gardnerella vaginalis* from blood culture media, *J. Clin. Microbiol.*, 21, 686, 1985.
132. Fredricks, D.N., and Relman, D.A., Improved amplification of microbial DNA from blood cultures by removal of the PCR inhibitor sodium polyanetholesulfonate, *J. Clin. Microbiol.*, 36, 2810, 1998.
133. Boskey, E.R. et al., Acceptability of a self-sampling technique to collect vaginal smears for gram stain diagnosis of bacterial vaginosis, *Womens Health Issues*, 14, 14, 2004.
134. Jones, H.E. et al., Home-based versus clinic-based self-sampling and testing for sexually transmitted infections in Gugulethu, South Africa: randomised controlled trial, *Sex. Transm. Infect.*, 83, 552, 2007.
135. Baay, M.F. et al., Feasibility of collecting self-sampled vaginal swabs by mail: Quantity and quality of genomic DNA, *Eur. J. Clin. Microbiol. Infect. Dis.*, 28, 1285, 2009.
136. Forney, L.J. et al., Comparison of self-collected and physician-collected vaginal swabs for microbiome analysis, *J. Clin. Microbiol.*, 2010.
137. El Aila, N.A. et al., Genotyping of *Streptococcus agalactiae* (group B streptococci) isolated from vaginal and rectal swabs of women at 35–37 weeks of pregnancy, *BMC Infect. Dis.*, 9, 153, 2009.
138. Van Dyck, E. et al., Evaluation of the Roche *Neisseria gonorrhoeae* 16S rRNA PCR for confirmation of AMPLICOR PCR-positive samples and comparison of its diagnostic performance according to storage conditions and preparation of endocervical specimens, *J. Clin. Microbiol.*, 39, 2280, 2001.
139. Van Dyck, E. et al., Detection of *Chlamydia trachomatis* and *Neisseria gonorrhoeae* by enzyme immunoassay, culture, and three nucleic acid amplification tests, *J. Clin. Microbiol.*, 39, 1751, 2001.
140. de Boer, R. et al., Improved detection of microbial DNA after bead-beating before DNA isolation, *J. Microbiol. Methods*, 80, 209, 2010.
141. Fredricks, D.N., Smith, C., and Meier, A., Comparison of six DNA extraction methods for recovery of fungal DNA as assessed by quantitative PCR, *J. Clin. Microbiol.*, 43, 5122, 2005.
142. Fliss, I. et al., A rapid and efficient method of lysis of *Listeria* and other gram-positive bacteria using mutanolysin, *Biotechniques*, 11, 453, 456, 1991.
143. Akane, A. et al., Identification of the heme compound copurified with deoxyribonucleic acid (DNA) from bloodstains, a major inhibitor of polymerase chain reaction (PCR) amplification, *J. Forensic Sci.*, 39, 362, 1994.
144. Wiedbrauk, D.L., Werner, J.C., and Drevon, A.M., Inhibition of PCR by aqueous and vitreous fluids, *J. Clin. Microbiol.*, 33, 2643, 1995.
145. Holodniy, M. et al., Inhibition of human immunodeficiency virus gene amplification by heparin, *J. Clin. Microbiol.*, 29, 676, 1991.
146. Khan, G. et al., Inhibitory effects of urine on the polymerase chain reaction for cytomegalovirus DNA, *J. Clin. Pathol.*, 44, 360, 1991.
147. Ahokas, H., and Erkkila, M.J., Interference of PCR amplification by the polyamines, spermine and spermidine, *PCR Methods Appl.*, 3, 65, 1993.
148. Altschul, S.F. et al., Basic local alignment search tool, *J. Mol. Biol.*, 215, 403, 1990.
149. Srinivasan, S., and Fredricks, D.N., The human vaginal bacterial biota and bacterial vaginosis, *Interdiscip. Perspect. Infect. Dis.*, 2008, 750479, 2008.
150. Kalra, A. et al., Bacterial vaginosis: Culture- and PCR-based characterizations of a complex polymicrobial disease's pathobiology, *Curr. Infect. Dis. Rep.*, 9, 485, 2007.
151. Verstraelen, H. et al., Longitudinal analysis of the vaginal microflora in pregnancy suggests that *L. crispatus* promotes the stability of the normal vaginal microflora and that *L. gasseri* and/or *L. iners* are more conducive to the occurrence of abnormal vaginal microflora, *BMC Microbiol.*, 9, 116, 2009.
152. Welsh, J., and McClelland, M., Genomic fingerprints produced by PCR with consensus tRNA gene primers, *Nucleic Acids Res.*, 19, 861, 1991.
153. Vaneechoutte, M. et al., Comparison of PCR-based DNA fingerprinting techniques for the identification of *Listeria* species and their use for atypical *Listeria* isolates, *Int. J. Syst. Bacteriol.*, 48, 127, 1998.
154. Baele, M. et al., Application and evaluation of the interlaboratory reproducibility of tRNA intergenic length polymorphism analysis (tDNA PCR) for identification of *Streptococcus* species, *J. Clin. Microbiol.*, 39, 1436, 2001.
155. Baele, M. et al., tDNA-PCR followed by automated fluorescent capillary electrophoresis for identification of bacterial species, *CSH Protoc.*, 4, prot5196, 2009.
156. Baele, M. et al., Application of tRNA intergenic spacer PCR for identification of *Enterococcus* species, *J. Clin. Microbiol.*, 38, 4201, 2000.
157. Baele, M. et al., Identification of *Lactobacillus* species using tDNA-PCR, *J. Microbiol. Methods*, 50, 263, 2002.
158. El Aila, N.A. et al., Identification and genotyping of bacteria from paired vaginal and rectal samples from pregnant women indicates similarity between vaginal and rectal microflora, *BMC Infect. Dis.*, 9, 167, 2009.
159. Sha, B.E. et al., Utility of Amsel criteria, Nugent score, and quantitative PCR for *Gardnerella vaginalis, Mycoplasma hominis*, and *Lactobacillus* spp. for diagnosis of bacterial vaginosis in human immunodeficiency virus-infected women, *J. Clin. Microbiol.*, 43, 4607, 2005.
160. Nath, K., Sarosy, J.W., and Stylianou, S.P., Suitability of a unique 16S rRNA gene PCR product as an indicator of *Gardnerella vaginalis*, *Biotechniques*, 28, 222, 2000.
161. Zariffard, M.R. et al., Detection of bacterial vaginosis-related organisms by real-time PCR for *Lactobacilli, Gardnerella vaginalis* and *Mycoplasma hominis*, *FEMS Immunol. Med. Microbiol.*, 34, 277, 2002.
162. Trama, J.P. et al., Rapid detection of *Atopobium vaginae* and association with organisms implicated in bacterial vaginosis, *Mol. Cell. Probes*, 22, 96, 2008.

163. Biagi, E. et al., Quantitative variations in the vaginal bacterial population associated with asymptomatic infections: A real-time polymerase chain reaction study, *Eur. J. Clin. Microbiol. Infect. Dis.*, 28, 281, 2009.
164. Fredricks, D.N. et al., Changes in vaginal bacterial concentrations with intravaginal metronidazole therapy for bacterial vaginosis as assessed by quantitative PCR, *J. Clin. Microbiol.*, 47, 721, 2009.
165. Menard, J.P. et al., Molecular quantification of *Gardnerella vaginalis* and *Atopobium vaginae* loads to predict bacterial vaginosis, *Clin. Infect. Dis.*, 47, 33, 2008.
166. Zozaya-Hinchliffe, M. et al., Quantitative PCR assessments of bacterial species in women with and without bacterial vaginosis, *J. Clin. Microbiol.*, 48, 1812, 2010.
167. Briselden, A.M., and Hillier, S.L., Evaluation of affirm VP Microbial Identification Test for *Gardnerella vaginalis* and *Trichomonas vaginalis*, *J. Clin. Microbiol.*, 32, 148, 1994.
168. Socransky, S.S. et al., Use of checkerboard DNA–DNA hybridization to study complex microbial ecosystems, *Oral Microbiol. Immunol.*, 19, 352, 2004.
169. Boggess, K.A. et al., Use of DNA hybridization to detect vaginal pathogens associated with bacterial vaginosis among asymptomatic pregnant women, *Am. J. Obstet. Gynecol.*, 193, 752, 2005.
170. Persson, R. et al., The vaginal microflora in relation to gingivitis, *BMC Infect. Dis.*, 9, 6, 2009.
171. Dunbar, S.A., Applications of Luminex xMAP technology for rapid, high-throughput multiplexed nucleic acid detection, *Clin. Chim. Acta.*, 363, 71, 2006.
172. Dumonceaux, T.J. et al., Multiplex detection of bacteria associated with normal microbiota and with bacterial vaginosis in vaginal swabs using oligonucleotide-coupled fluorescent microspheres, *J. Clin. Microbiol.*, 47, 4067, 2009.
173. Turnbaugh, P.J. et al., The human microbiome project, *Nature*, 449, 804, 2007.

9 Gordonia

Brent A. Lasker, Benjamin D. Moser, and June Brown

CONTENTS

9.1 INTRODUCTION

The genus *Gordonia* was originally described in 1971 by Tsukamura following the isolation of organisms from sputa of patients with pulmonary disease and from soil.[1] The name of the genus was chosen to honor Ruth E. Gordon, an American microbiologist. There are currently 28 validly described species in this genus, and *Gordonia bronchialis* is the type species. According to the phylogenetic position of the genus based on the analysis of 16S rRNA sequences in 1997 by Stackebrandt et al., *Gordonia* is the type genus of the family *Gordoniaceae* within the suborder *Corynebacterineae* and of the order *Actinomycetales*.[2] In this same publication, these authors corrected the original spelling to *Gordonia*.[2] Most species are not yet associated with human infection because 15 of the 17 species identified in the last 10 years have been isolated from environmental sources. The clinically significant species are the primary focus of this chapter but the environmental species are included in comparative biochemical and susceptibility tables. An increased recognition of *Gordonia* species may result in delineation of species-specific differences, geographical distribution, disease associations, and greater awareness of antimicrobial susceptibility profiles. The biology of this genus has recently been reviewed with an emphasis on its potential usefulness in environmental and industrial biotechnology.[3]

9.1.1 CLASSIFICATION

In 1971, as originally described and named, the *Gordona* genus was composed of *Gordona bronchialis*, *G. rubra* (*Gordonia rubripertincta*), and *G. terrae*.[1] Tsukamura separated these species from rapidly growing mycobacteria by their slight acid fastness and their inability to produce arylsulfatase. In 1977, Goodfellow and Alderson transferred these members of *Gordona*, *Mycobacterium rhodochrous*, and the "rhodochrous" complex to the genus *Rhodococcus* based on extensive numerical computerized analysis of 92 characteristic traits.[4] The three species, originally assigned to the genus *Gordona*, were studied by Collins et al.[5] The study revealed that these species contained mycolic acids with a range from 48 to 66 carbon atoms and major amounts of dehydrogenated menaquinones with nine isoprene units [MK-9 (H_2)], therefore leaving the remaining species of the genus *Rhodococcus* more homogeneous, with their mycolic acids ranging from 34 to 52 carbon atoms and with a major menaquinone of only eight isoprene units [MK-8 (H_2)]. The mycolic acid and menaquinone analyses were in agreement with the data derived from the 16S rRNA gene sequencing of Stackebrandt et al. and led to the revival of the genus *Gordona* Tsukamura 1971.[6] *R. bronchialis*, *R. rubra*, and *R. terrae* were reclassified to the genus *Gordona* and, additionally, *Rhodococcus sputi* was reclassified as *Gordona sputi*.[6,7] The transferring of *Nocardia amarae* Lechevalier and Lechevalier 1974 and *Rhodococcus aichiensis* Tsukamura 1982 to the genus *Gordona* by Klatte et al. in 1994, based on 16S rRNA gene sequencing analysis and chemotaxonomic methods, further supported the integrity of the genus *Gordona*.[8–10] Two additional species added to the genus *Gordona* were *Gordona hydrophobica*[11] and *G. hirsuta*.[12] The correct epithet was introduced in 1997 as the genus *Gordonia*.[2] New species increased steadily from 1997 to 2009: *Gordonia rhizophera*,[13] *Gordonia alkanivorans*,[14] *Gordonia desulfuricans*,[15] *Gordonia polyisoprenivorans*,[16] *Gordonia amicalis*,[17] *Gordonia namibiensis*,[18] *Gordonia sihwensis*,[19] *Gordonia sinesedis*,[20] *Gordonia westphalia*,[21] *Gordonia paraffinivorans*,[22] *Gordonia otitidis*,[23] *Gordonia araii*,[24] *Gordonia defluvii*,[25] *Gordonia effusa*,[24] *Gordonia soli*,[26] *Gordonia shandongensis*,[27] *Gordonia malaquae*,[28] *Gordonia lacunae*,[29] *Gordonia cholesterolivorans*,[30] and *Gordonia kroppenstedtii*.[31] The taxonomic and clinical laboratory identification of the genus *Gordonia* today is dependent on phenotypic, chemotaxonomic, and molecular characterization.

9.1.2 Morphology

The microscopic and colonial morphology of the individual species of *Gordonia* is very similar to that of *Rhodococcus*. Colonial morphologies observed either microscopically or macroscopically are species dependent and range from slimy to smooth or glossy and may vary from smooth to rough after 2–3 days incubation on a variety of media (e.g., blood agar media, Middlebrook 7H11 agar, brain heart infusion and heart infusion agar). The color of the colonies, often not apparent until at least 7 days of incubation at optimal temperature, also varies through a wide range, from white, cream, beige, or pink, to the more classic orange-salmon color of most *Gordonia* species. The organisms are gram-positive with coccoid to bacillary forms that may be weakly acid fast (4–10 organisms per oil immersion field) with the modified Kinyoun (1% sulfuric acid) stain. Briefly, all members of the genus are aerobic, nonsporeforming, gram-positive nocardioforms with an oxidative carbohydrate metabolism.

9.1.3 Chemotaxonomic Characteristics

In addition to phenotypic features, there are a number of salient properties that are important in determining genus assignment. The chemotaxonomic characteristics that separate the genus *Gordonia* from related genera have been reviewed by Arenkötter et al.[3] and Takeuchi and Hatano.[13] Important differential traits include whole cell wall composition chemotype IV Lechevalier and Lechevalier[32] (contains arabinose and galactose as major sugars and *meso*-diaminopimelic acid as the only diamino acid); the presence of MK-9 (H_2) as the predominant menaquinone; saturated, unsaturated, and tuberculostearic acid in whole-cell fatty acid composition; mycolic acids with a range of 48–66 carbon atoms; *N*-glycolated muramic acid; presence of phosphatidylethanolamine; and G + C content of DNA ranging from 63 to 69 mol%.

9.1.4 Epidemiology

Members of the genus *Gordonia* were originally regarded as opportunistic pathogens and were isolated from sputum of humans with respiratory diseases (e. g., *G. bronchialis*, *G. aichiensis*, and *G. sputi*). Recently, more species have been isolated with the ability to degrade xenobiotics, environmental pollutants or slowly biodegradable natural polymers (e.g., benzothiophene and dibenzothiophene by *G. desulfuricans* and *G. amicalis*, alkanes by *G. alkanivorans*, malodorous animal emissions such as aldehydes and ketones by *G. hydrophobica* and *G. hirsuta*, polyisoprenoids and synthetic *cis*-1,4-polyisoprene by *G. polyisoprenivorans* and *G. westfalica*, nitrile compounds by *G. namibiensis*, hydrocarbons such as liquid paraffin by *G. paraffinivorans*,[22] and phenol compounds by *G. kroppenstedtii*).[31] Given the difficulties of accurately identifying *Gordonia* species by phenotypic tests alone, we may not be certain of the identification made by earlier investigators of *Gordonia* clinical isolates, especially those made without molecular studies. However, we present and review case reports that claim an association of certain gordoniae with clinical disease.

Gordonia araii. The type strain for this species, *G. araii* IFM 10211^T,[24] was first isolated from a sputum specimen in a 48-year-old male Japanese patient with bacterial pneumonia. The specimen underwent extensive polyphasic taxonomic analysis and was identified as a novel species based on 16S rRNA gene sequence analysis and DNA–DNA hybridization.[24] Pathogenic *G. araii* also appears in the literature as the disease causative agent in a 27-year-old following a sport-related injury of the right anterior cruciate ligament (ACL) and right medial meniscus tear.[33] After surgery, the patient complained of fever and knee pain, which was followed 10 weeks later by erythema and edema of the right knee. At 21 weeks after surgery, MRI revealed an effusion and edema in both the tibia and femur of the graft attachment site. All foreign materials related to the surgical procedure were removed and cultured; growth of a gram-positive bacillus was observed on brain heart infusion agar with 5% sheep blood after 3 days. This isolate was resistant to azithromycin (>256.0 ug/mL) and was treated with intravenous trimethoprim/sulfamethoxazole (175–875 mg twice daily). This was the first case of confirmed *G. araii* infection associated with a medical device in an immunocompetent individual.[33]

Gordonia aichiensis. *Gordonia aichiensis* (type strain E9028^T) was first isolated from human sputum and initially identified as *Rhodococcus aichiense*.[10] The transfer from *Rhodococcus* to the genus *Gordona* was described by Klatte et al.[8] Clinical presentation of *G. aichiensis* is rare in the literature. Two recorded cases are of a 74-year-old female and a 72-year-old male with *G. aichiensis* cultured from sputum and pus, respectively.[34]

Gordonia bronchialis. *Gordona (Gordonia) bronchialis* was first proposed as a new species by Tsukamura in 1971 after a number of organisms, isolated from patients with pulmonary disease, grew with intermediate characteristics between the genera *Mycobacterium* and *Nocardia*.[1] *Gordonia bronchialis* was identified as a human disease-causing agent in sternal wound infections following open heart surgery.[35] In these seven cases, Richet et al. determined that a noninfected circulating nurse and her dogs were the reservoirs for the organism. The nurse was inadvertently transmitting the infectious organism by touching a colonized area of her body, such as her scalp, and then accidentally transferred the organism to the water bath used in a standard clotting test during open heart surgery procedures and contaminated instruments and materials used on the patient.[35] In 2004, *G. bronchialis* was cultured from the blood of a 58-year-old woman with recently diagnosed diabetes mellitus presenting with breathlessness, fever, and chills related to a sequestrated lung.[36] Known to be pathogenic in immunocompromised individuals,[37] *G. bronchialis* has also been reported as the infectious agent in an immunocompetent individual with a recurrent breast abscess[38] and in a 45-day-old premature neonate with apnea and bradycardia.[39] Aoyama et al. characterized 31 *Gordonia* species clinical isolates from 1998 to 2008.[34] Of the isolated strains, 13 were identified as *G. bronchialis* and were isolated from sites such as pleural fluid, sputum, granuloma, blood, skin, and pus. This represents the largest series of *G. bronchialis* clinical infections.

Gordonia effusa. The type strain for *G. effusa* IFM 10200[T24] was first isolated from a sputum specimen from a 74-year-old male Japanese patient with kidney dysfunction. Described concurrently with *G. araii*, the initial isolated specimen underwent extensive polyphasic taxonomic analysis and was identified as a novel species based on 16S rRNA gene sequence analysis and DNA–DNA hybridization.[24] *Gordonia effusa* infection is rare and presents only this one time throughout the literature.

Gordonia otitidis. *Gordonia otitidis* is an uncommon clinical infection. The type strain of *G. otitidis,* IFM 10032[T], was first isolated from ear discharge in a 28-year-old Japanese female with external otitis in 2000.[23] Also, *G. otitidis* was isolated from pleural fluid in a 60-year-old Japanese male with bronchitis, assumed to be the same case identified by Aoyama et al. in their investigation of *G. otitidis* found in pleural fluid during their study of 31 *Gordonia* clinical isolates.[24] As recent as 2007, Blaschke et al. presented a case of infection in an 11-year-old who presented febrile with a history of central venous catheter use.[39]

Gordonia polyisoprenivorans. *Gordonia polyisoprenivorans* (type strain Kd2[T]) was first isolated from fouling water inside of an automobile tire.[16] Several other rubber-degrading *Gordonia* species have been previously identified in automobile tires and from the soil of a rubber tree plantation in Vietnam.[40,41] Although originally isolated from the environment, *G. polyisoprenivorans* appears in the literature as the disease-causing agent in clinical infection. Kempf et al. reported the first case of septicemia caused by *G. polyisoprenivorans* following a bone marrow transplant in a 26-year-old female.[42] The patient was suffering from chronic myelogenous leukemia and received an allogenic bone marrow transplantation in December 2001. Before transplantation, a Hickman catheter was inserted into the patient and used for a myeloablative condition regimen. In March 2002, after presenting with a fever, blood samples were cultured from her Hickman catheter revealing *G. polyisoprenivorans.*[42] Verma et al. describes an immunocompromised 78-year-old male who underwent surgery for an incarcerated left inguinal hernia.[43] The patient had a right internal jugular Hickman catheter inserted for blood transfusions during his procedure. Later, this patient underwent evaluation of recurrent anemia and a third-degree heart block. Echocardiography revealed vegetations on the aortic and mitral valve for which the patient underwent aortic valve replacement and debridement of the mitral valve, respectively. Histopathological examination showed acute endocarditis and blood culture grew a beaded gram-positive bacillus; later identified at Centers for Disease Control and Prevention (CDC) as *G. polyisoprenivorans*. Most recently, Gupta et al. presented the first case of pneumonia with associated bacteremia caused by *G. polyisoprenvirans* following long-term use of a central catheter.[44]

Gordonia rubripertincta. *Gordonia rubripertincta* (type strain ATCC 14352) was first described in clinical infection by Hart et al. reporting a case of a lung infection resembling tuberculosis caused by *Rhodococcus rubropertinctus.*[45] The patient showed no immunosuppression, which is unusual for disease caused by the "rhodochrous" complex.[45] The organism was identified as the primary pathogen in the abortion of 9-month-old fetus in a 13-year-old quarter horse. Samples of the lung following necropsy yielded the growth of a gram-positive, club-shaped branching organism.[46] *Rhodococcus rubropertinctus* was transferred into the genus *Gordona* in 1989 and renamed *Gordonia* in 1997.[2]

Gordonia sputi. *Gordonia sputi* is one of the most common disease-causing agents among the gordoniae. First described by Tsukamura in 1978 as *Rhodococcus sputi*, this organism has caused a series of human and animal infections. Tsukamura et al. describe a case of mesenteric lymphadenitis in a 6-month-old swine caused by *R. sputi.*[47] Riegel et al. reported on the first case of bacteremia due to a strain of *Gordona sputi* in an immunodeficient patient.[48] Riegel et al. describe a 34-year-old man with ocular melanoma and hepatic metastases, who was undergoing treatment prior to infection. Following 4 months of treatment with *cis*-platinum and intravenous interleukin-2, the patient developed a 40°C fever; blood culture grew a gram-positive coryneform bacterium. One week later, after another fever, three blood cultures yielded the same organism identified as *G. sputi.*[48] Kuwabara et al. described a case of mediastinitis caused by *G. sputi* infection in a 54-year-old man subsequent to coronary artery bypass grafting[49]; followed by a Lesens et al. report of the first case of endocarditis.[50] Lesens et al. described a 31-year-old woman with genetic blood disorders, hepatitis C infection, and hemochromatosis complicated by diabetes, adrenal insufficiency, and peripheral neuropathy presenting with mitral valve vegetations following the implantation of a central venous port. The site of the port showed no signs of infection; however, blood cultures from her central catheter and a peripheral source grew gram-positive coryneform bacterium (*G. sputi*). More recently, cases of *G. sputi* bacteremia were reported in Brust et al.,[37] as well as in 14 of the 31 clinical strains investigated by Aoyama et al.[34] *G. sputi* again appears in a series of infections caused by *Gordonia* species in a Taiwan medical center.[51] Of the 66 clinical isolates previously identified as *Rhodococcus* species, 15 were identified as *Gordonia* species; eight of them were *G. sputi*.[51]

Gordonia terrae. *Gordonia terrae* is a ubiquitous disease agent, causing brain abscesses,[52] central nervous symptoms,[53] and bacteremia[54–56] in humans. Drancourt et al. presented the case of an immunocompromised boy undergoing antineoplastic chemotherapy for a cerebral rhabdoid tumor that was complicated by the formation of a brain abscess caused by *G. terrae* infection.[52] A 40-year-old immunocompetent woman had *G. terrae* isolated from central spinal fluid (CSF) in her evaluation of hemiparesis and brain abscesses that proved difficult to treat.[53] A rare cause of mycetoma occurred in a 15-year-old boy in Sierra Leone following an injury to his left thumb from a thorn,[57] and an even rarer example of *G. terrae* infection occurred in 2009 following the nipple piercing of a 29-year-old female.[58] The granulomatous mastitis caused by the nipple piercing is the first reported *Gordonia* species-associated mastitis in the literature.[58] Aoyama et al. described *G. terrae* infection in 1 of the 31 investigated samples in their characterization study,[34] Lai et al. described *G. terrae* as the organism responsible for 7 of the 15 clinical *Gordonia* isolates in a Taiwan medical center,[51] and Blaschke et al. presented four episodes of *G. terrae* bacteremia in three children, all with central venous catheters.[39] The recurrence of *G. terrae* bacteremia in one child suggests that therapy may need to be prolonged compared to other organisms.

9.1.5 Diagnosis

9.1.5.1 Phenotypic, Biochemical, and Susceptibility Studies

With an increasing number of *Gordonia* species and related genera, the use of only a limited number of phenotypic traits to obtain exact species or genus level identification has become impossible. However, the production of aerial hyphae on solid media and the production of branched filaments by members of the genus *Nocardia* are useful in differentiating them from the gordoniae (except for *G. defluvii*, which has both aerial hyphae and branched filaments). Distinguishing *Gordonia* species from related genera *Corynebacterium*, *Mycobacterium*, *Rhodococcus*, and *Tsukamurella* is more problematic. *Gordonia* can be separated from *Mycobacterium* and *Tsukamurella* as *Gordonia* are not 100% acid fast with modified Kinyoun. In addition, *Gordonia* strains do not produce arylsulfatase, as do *Mycobacterium* strains. *Gordonia* species can be distinguished from *Corynebacterium* species by a combination of chemotaxonomic (*Gordonia* species have mycolic acids with 48–66 carbon atoms while *Corynebacterium* species contain 22–36 carbon atoms) and morphological features (*Gordonia* species usually have a rod-to-coccoid cycle while *Corynebacterium* species have club-shaped, palisade, or angular arrangements). The microscopic and colonial morphology of the individual species of *Gordonia* is very similar to that of *Rhodococcus* emphasizing the difficulty in differentiating gordoniae from the rhodococci without chemotaxonomic (menaquinone and HPLC analysis) and molecular methods.

Traditionally, the gordoniae were identified to the species level using a battery of biochemical tests including hydrolysis of adenine, casein, hypoxanthine, tyrosine, and xanthine.[59] The two major types of carbohydrate utilization tests used to distinguish among the species of *Gordonia* included those of Gordon (assimilation and assimilation with acid production) and those of Goodfellow (assimilation only).[60] However, because of the generally nonreactive nature of most species to these tests and because of the increasing number of described species, species assignment using biochemical methods alone does not provide reliable identification of all the currently recognized species. (See Table 9.1 for useful tests to establish a tentative identification of the gordoniae including the major clinically relevant species.)

The Clinical Laboratory Standards Institute (CLSI, formerly NCCLS) has published an approved standard for susceptibility testing of mycobacteria, nocardiae, and other aerobic actinomycetes.[61] The recommended method for gordoniae is broth microdilution.[61] Antimicrobial agents recommended for testing include amikacin, amoxicillin-clavulanate, ceftriaxone, ciprofloxacin, clarithromycin, imipenem, linezolid, minocycline, and trimethoprim/sulfamethoxazole. Interpretive breakpoints of these agents are provided and a recommended quality control procedure is specified.[61] The antimicrobial agent vancomycin was added to our panel; it is not included in the recommendations of the approved standard M24-A.[61] Since no interpretive criteria for the breakpoints for vancomycin have been recommended for *Gordonia* species, those for bacteria that grow aerobically were used.[62] Our susceptibility studies of 23 available type strains are shown in Table 9.2. All species tested were susceptible to amikacin, imipenem, linezolid, and minocycline; all species were susceptible to vancomycin except *G. effusa*, which was intermediate; all species were susceptible to clarithromycin except *G. bronchialis*; all species were susceptible to ciprofloxacin except *G. otitidis* and *G. sputi*; most species were susceptible to amoxicillin/clavulanate except *G. sinesedis*, *G. soli*, and *G. terrae*; and most species were susceptible to ceftriaxone except *G. nambiensis*, *G. sihwensis*, and *G. soli*. More than half (61%) of the type strains of *Gordonia* species were resistant to trimethoprim/sulfamethoxazole. Until recently, there were few studies on the antimicrobial susceptibility data of clinical isolates of *Gordonia*. Comparisons of our study with these studies are difficult because either no type strains were included or the species identification was not given. In 2007, Blaschke et al.[39] reported their susceptibility studies (n = 6) as well as data that were previously published for a total of 24 isolates. The authors deducted from their analysis that initial therapy for *Gordonia* could include a carbapenem or a fluoroquinolone in combination with an aminoglycoside. These authors also noted that 35% of the isolates were resistant to trimethoprim/sulfamethoxazole (often used to treat *Nocardia* infection) and 11% of the isolates were resistant to vancomycin (often used for treatment of *Rhodococcus* infection).[39] Our results

TABLE 9.1
Phenotypic Characteristics of the Type Strains of the Genus *Gordonia*

Characteristic	1	2	3	4	5	6	7[a]	8[a]	9	10	11	12	13[a]	14[a]	15[a]	16	17	18	19	20	21	22	23	24	25	26	27	28
Acid-Fast[b]	−	−	−	+	+	−	+	+	−	−	−	+	NT	+	−	−	+	−	−	−	+	−	−	+	−	+	−	+
Color	S	O	Y	O	W	O	B	NP	B	B	Y	Y	W	P/O	C	O	S	O	Y	S	O	Y	Y	Y	O	O	O	S
Assimilation w/acid production																												
Adonitol	−	−	−	−	−	−	NT	NT	−	−	−	−	NT	W	−	−	−	−	+	−	+	−	−	−	−	−	−	−
L-Arabinose	−	−	−	−	−	−	NT	−	−	−	−	−	NT	+	−	−	−	−	+	−	−	−	−	−	−	−	−	+
Cellobiose	−	−	−	−	−	−	NT	−	−	−	−	−	NT	+	−	−	−	−	−	+	−	−	−	−	−	−	−	+
Dulcitol	−	−	−	−	−	−	NT	NT	−	−	−	−	NT	NT	NT	−	−	−	−	−	−	−	−	−	−	−	−	+
i-Erythritol	−	−	−	−	−	−	NT	NT	−	−	−	−	NT	NT	−	−	−	−	+	−	−	−	−	−	−	−	−	−
D-Fructose	+	+	+	+	+	+	NT	−	+	+	+	+	NT	+	NT	+	−	−	+	+	+	−	+	+	+	+	+	+
D-Galactose	−	−	−	−	−	−	−	−	−	−	−	−	−	W	−	−	−	−	+	−	+	+	−	−	−	−	−	+
D-Glucose	+	W+	+	+	+	+	NT	NT	+	+	+	+	NT	+	+	+	−	+	+	+	+	+	+	+	+	+	+	+
Glycerol	+	+	−	+	+	+	NT	−	+	+	+	−	NT	NT	NT	+	+	+	+	+	+	−	+	+	+	+	−	+
i-*myo*-Inositol	−	−	+	−	−	+	NT	NT	−	−	+	−	+	W	−	−	−	−	+	+	+	−	−	+	−	−	−	−
Lactose	−	−	−	−	−	−	NT	NT	−	−	−	−	NT	+	−	−	−	−	−	−	−	−	−	−	−	−	−	+
Maltose	+	+	+	−	−	−	NT	−	−	−	−	+	NT	NT	−	−	−	−	−	+	−	+	−	+	+	+	+	+
D-Mannitol	−	−	+	+	−	−	NT	NT	−	+	−	−	NT	+	−	+	+	−	+	+	+	−	−	−	+	+	−	+
Mannose	+	+	+	−	+	+	NT	−	+	+	+	−	NT	+	NT	+	+	+	+	+	+	−	+	+	+	+	−	+
Melibiose	−	−	−	−	−	−	NT	NT	−	−	−	−	NT	W	NT	−	−	−	−	−	−	−	−	−	−	−	−	+
Raffinose	−	−	−	−	−	−	NT	NT	−	−	−	−	NT	W	−	−	−	−	−	−	−	−	−	−	−	−	−	+
L-Rhamnose	−	−	+	−	−	−	NT	−	−	−	−	−	−	+	−	−	−	−	+	+	+	−	−	−	−	−	−	+
Salicin	−	−	+	−	+	−	NT	−	−	−	−	−	NT	W	NT	−	−	−	−	−	−	−	−	−	−	−	−	+
D-Sorbitol	−	−	−	+	−	−	NT	NT	−	−	−	−	NT	NT	−	+	+	−	+	−	+	−	−	−	+	+	+	−
Sucrose	+	+	+	+	−	+	NT	NT	+	+	+	+	+	+	+	+	+	+	+	+	+	−	+	+	+	+	+	+
Trehalose	+	+	+	+	+	+	NT	NT	+	+	+	+	NT	+	+	+	+	+	+	+	+	+	+	+	+	+	+	+
D-Xylose	−	−	−	−	−	−	NT	NT	−	−	−	−	NT	+	+	−	+	−	+	−	+	−	−	−	−	−	−	+
Growth at																												
25°C	+	+	+	+	+	+	NT	NT	+	+	+	+	NT	NT	NT	+	+	+	+	+	+	+	+	+	+	+	+	+
35°C	+	+	+	+	+	+	NT	NT	+	+	+	+	NT	NT	NT	+	+	+	+	+	+	+	+	+	+	+	+	+
45°C	−	−	−	−	−	−	NT	NT	−	−	−	−	NT	NT	NT	−	−	+	−	−	−	−	+	+	−	−	−	−

continued

TABLE 9.1 (Continued)
Phenotypic Characteristics of the Type Strains of the Genus *Gordonia*

	1	2	3	4	5	6	7[a]	8[a]	9	10	11	12	13[a]	14[a]	15[a]	16	17	18	19	20	21	22	23	24	25	26	27	28
Hydrolysis of																												
Casein	−	−	−	−	−	−	NT	NT	−	−	−	−	−	+	−	−	−	−	−	−	−	−	−	−	−	−	−	−
Esculin	+	+	+	−	+	−	NT	−	−	+	−	−	+	−	−	−	−	−	−	−	−	−	−	−	−	+	−	+
Hypoxanthine	−	−	−	−	−	−	NT	−	−	−	−	−	NT	−	−	−	−	−	−	−	−	−	−	+	−	−	−	+
Tyrosine	−	−	−	−	−	−	NT	−	−	−	−	−	NT	−	−	−	−	−	−	−	−	−	−	+	−	−	−	−
Urea	+	+	+	−	−	+	NT	NT	+	+	−	+	+	W	+	+	+	+	+	+	+	−	−	+	+	+	+	+
Utilization of																												
Acetamide	+	+	+	−	−	+	NT	−	+	−	−	+	−	NT	−	−	+	+	+	−	+	+	+	−	−	−	−	+
Citrate	−	+	−	−	−	+	−	NT	+	+	−	+	−	+	+	+	+	+	+	+	+	−	+	+	+	−	+	+
Growth in lysozyme	+	−	−	−	−	−	NT	NT	−	+	−	+	NT	NT	NT	−	+	−	+	−	−	+	−	+	−	−	−	−
Arylsulfatase production	−	−	−	−	+	−	NT	NT	−	−	−	−	NT	NT	NT	−	+	−	−	+	−	−	−	−	−	−	−	−
Nitrate reduction	+	+	+	−	+	−	NT	+	−	+	+	+	−	+	NT	+	+	+	−	+	−	+	+	+	+	+	+	+

Notes: 1, *G. aichiensis*; 2, *G. alkanivorans*; 3, *G. amarae*; 4, *G. amicalis*; 5, *G. araii*; 6, *G. bronchialis*; 7, *G. cholesterolivorans*[30]; 8, *G. defluvii*[25]; 9, *G. desulfuricans*; 10, *G. effusa*; 11, *G. hirsuta*; 12, *G. hydrophobica*; 13, *G. kroppenstedti*[31]; 14, *G. lacunae*[29]; 15, *G. malaquae*[28]; 16, *G. namibiensis*; 17, *G. otitidis*; 18, *G. paraffinivorans*; 19, *G. polyisoprenivorans*; 20, *G. rhizosphera*; 21, *G. rubripertincta*; 22, *G. shandongensis*; 23, *G. sihwensis*; 24, *G. sinesedis*; 25, *G. soli*; 26, *G. sputi*; 27, *G. terrae*; 28, *G. westfalica*.

All species showed a rod to cocci microscopic morphology, except for *G. cholesterolivorans* and *G. defluvii*; these two species additionally showed hyphae.

Abbreviations and symbols: −, negative; +, positive; NT, not tested; S, salmon; O, orange; Y, yellow; W, white; B, beige; NP, no pigment; P, pink; C, cream; W+, weak positive.

[a] Assimilation/utilization data from previously published original manuscripts; not acid production.

[b] Acid-fast organisms usually from 4 to 10 per oil immersion field.

TABLE 9.2
Antimicrobial Susceptibility Results of Type Strains of the Genus *Gordonia*[a,b]

Species	Type Strain No.	Amikacin	Amox/Clav	Ampicillin	Ceftriaxone	Clarithromycin	Ciprofloxacin	Imipenem	Linezolid	Minocycline	TMP/SMX	Vancomycin[c]
		Antimicrobial Agent										
G. aichiensis	ATCC 33611	S	S	S	S	S	S	S	S	S	S	S
G. alkanivorans	DSM 44369	S	S	S	S	S	S	S	S	S	S	S
G. amarae	ATCC 27808	S	S	S	S	S	S	S	S	S	S	S
G. amicalis	DSM 44461	S	S	S	S	S	S	S	S	S	S	S
G. araii	IFM 10211	S	S	I	S	S	S	S	S	S	R	S
G. bronchialis	ATCC 25592	S	S	S	S	R	S	S	S	S	R	S
G. desulfuricans	DSM 44462	S	S	S	S	S	S	S	S	S	S	S
G. effusa	IFM 10200	S	S	S	S	S	S	S	S	S	R	I
G. hirsuta	ATCC 700255	S	S	S	S	S	S	S	S	S	S	S
G. hydrophobica	ATCC 700089	S	S	S	S	S	S	S	S	S	S	S
G. namibiensis	DSM 44568	S	S	R	R	S	S	S	S	S	S	S
G. otitidis	DSM 44809	S	S	S	S	S	R	S	S	S	R	S
G. paraffinivorans	DSM 44604	S	S	S	S	S	S	S	S	S	R	S
G. polyisoprenivorans	ATCC BAA-14	S	S	S	S	S	S	S	S	S	R	S
G. rhizosphera	DSM 44383	S	S	S	S	S	S	S	S	S	S	S
G. rubripertincta	ATCC 14352	S	S	S	S	S	S	S	S	S	R	S
G. shandongensis	JCM 13907	S	S	S	S	S	S	S	S	S	R	S
G. sihwensis	DSM 44576	S	S	S	R	S	S	S	S	S	R	S
G. sinesedis	DSM 44455	S	R	S	S	S	S	S	S	S	R	S
G. soli	DSM 44995	S	R	R	R	S	S	S	S	S	R	S
G. sputi	ATCC 29627	S	S	S	S	S	R	S	S	S	R	S
G. terrae	ATCC 25594	S	R	R	S	S	S	S	S	S	R	S
G. westfalica	DSM 44215	S	S	S	S	S	S	S	S	S	R	S

[a] *Abbreviations*: amox/clav, amoxicillin/clavulanic acid; TMP/SMX, trimethoprim/sulfamethoxazole; S, susceptible; I, intermediate; R, resistant.
[b] Based on interpretations of MICs, using CLSI breakpoints.[61]
[c] Interpretive breakpoints for vancomycin are derived from M7-A7, CLSI.[62]

were comparable to the studies of Aoyama et al.[34] except that in their report all 31 strains of clinical *Gordonia* had MICs of >2/38 for trimethoprim/sulfamethoxazole that were incorrectly interpreted as susceptible. This incorrect interpretation has been corrected (Mikami, Y., personal communication, 2009). Although Aoyama et al.[34] determined susceptibility studies for 31 molecularly identified strains (14 *G. sputi*, 13 *G. bronchialis*, two strains of *G. aichiensis,* and one strain each of *G. otitidis* and *G. terrae*); these authors did not use any type strains. In another series of nine clinical isolates (five isolates of *G. terrae* and four isolates of *G. sputi*) by Lai et al. in 2009,[51] results were comparable to our study of 23 type strains except Lai et al. found all nine isolates were susceptible to trimethoprim/sulfamethoxazole.[51] These discordant results may reflect different testing methods or difficulties in the interpretation of breakpoints for trimethoprim/sulfamethoxazole.

9.1.5.2 Molecular Techniques for Identification and Phylogenetic Analysis

Need for Molecular Identification Techniques. In the past two decades, the need by clinical and reference laboratories for a rapid and accurate method for identifying *Gordonia* clinical isolates has increased due to the rising number of reports of human infections in part as well as an increasing awareness of these infections by clinicians. Identification of *Gordonia,* as well as for other clinically or environmentally relevant bacteria, using conventional biochemical and physiologic methods is difficult due to the potentially long incubation times needed for identification, the availability of only a few discriminatory tests, and the high degree of expertise required to perform the tests.[63] *Gordonia* clinical isolates have often been misidentified or regarded as nocardioform actinomycete contamination when cultures grew *Gordonia*. Accurate identification of patient specimens is important for understanding its epidemiology, identifying its clinical importance, defining its spectrum of disease, assessing its pathogenicity, and establishing risk factors for infection. Molecular methods for identification, especially DNA sequencing, have become increasingly an important adjunct to conventional phenotypic identification methods. A polyphasic approach provides accurate and reliable identification for existing and newly characterized species. DNA sequencing generally provides a more rapid and definitive identification than can be obtained by biochemical analysis alone. Molecular characterization of bacteria may also provide clinically relevant information on the phylogenetic relationship between pathogenic *Gordonia* species and other bacteria. Molecular identification has been aided by technological advances in DNA sequencing. These advances include the availability of automated sequencers, commercially available reagents, improved software for analysis, and more comprehensive and reliable gene databases.

DNA RFLP Analysis. Restriction fragment length polymorphism (RFLP) analysis provided an early method for molecular identification of *Gordonia* clinical isolates. Briefly, a 439 bp DNA fragment of the 65 kDa heat shock protein gene (*gro*E) was amplified by PCR. The PCR product was digested with a battery of restriction endonucleases and the restriction fragments were resolved by MetaPhor agarose gel electrophoresis. Following electrophoresis, gels were stained and identification was made by comparing the restriction fragment profiles obtained from the clinical isolate to the profiles of other type or reference strains.[64] PCR–RFLP was able to distinguish between the profiles obtained from clinical isolates of actinomycetes, including *G. bronchialis* and *G. sputi*. However, interpretation was difficult due to multiple patterns obtained from isolates of *G. bronchialis*.

RFLP analysis of the 16S rRNA gene-containing fragments or ribotype analysis has also been used for the identification of *Gordonia* clinical isolates. Ribotyping starts by digestion of purified genomic DNA with a restriction endonuclease. The DNA fragments are separated by agarose gel electrophoresis and the DNA fragments are transferred by the method of Southern to a solid support such as a nitrocellulose or nylon filter. Southern blots are hybridized with a labeled 16S rRNA probe. Identification is made following comparison of the RFLP hybridization profiles of 16S rRNA gene-containing fragments obtained from the unknown isolate to the hybridization profiles obtained from reference or type strains.[65,66] Ribotyping was used by Buchman et al.[65] as an adjunct method to identify an isolate obtained from a case patient with central venous catheter sepsis since traditional biochemical identification was ambiguous between *G. rubripertincta* and *G. terrae*. The profile of 16S rRNA gene-containing fragments was similar to the rRNA gene fragment profile obtained from the *G. terrae* type strain and the profile clearly differed from the 20 other type strains. However, identification of the isolate from a second patient using ribotyping was not reliable due to heterogeneity of the profiles for *G. sputi* and *G. bronchialis* isolates. Ribotype analysis has been used as an adjunct method for identification of *G. terrae* in a brain abscess[52] and in a mycetoma patient in Sierra Leone.[57] Ribotyping has been successful as an adjunct method for identification and typing of a cluster of patient isolates identified as *G. bronchialis* obtained from sternal wound infections after coronary bypass surgery.[35] In addition, ribotyping confirmed the results obtained by DNA–DNA hybridization, and was instrumental in the reevaluation of the taxonomic position of *Rhodococcus chubuensis* to *G. sputi*.[67]

Gordonia environmental isolates have been identified by use of PCR-denaturing gradient gel electrophoresis (DGGE).[68–70] Briefly, *Gordonia*-specific primers, G268F and G1096R, were used to amplify 16S rRNA genes by PCR. rRNA-containing bands were either excised from agarose gels for direct DNA sequencing or the excised sample was further resolved by DGGE analysis. Following DGGE, bands were excised and subjected to DNA sequence analysis. Identification is made by comparing 16S rRNA gene sequence obtained from the unknown organism to DNA sequences within an appropriate sequence database with the high degree of sequence similarity. PCR–DGGE has detected and identified *Gordonia* in oil-contaminated soils and environmental samples but has never been applied to the identification of clinical samples. DGGE analysis is more sensitive than conventional methods for recovery of rRNA gene sequences; however, the extra steps required for DGGE increase the risk of cross contamination and may be too technically demanding for most clinical laboratories. PCR–RFLP and ribotyping have been previously useful for identification, but have many weaknesses. First, there are problems associated with ensuring complete digestion of genomic DNA and accurate size determination of restriction fragments. For example, each sample should be run with internal standards to correct for lane-to-lane or between run variability that would make intra- and interlaboratory comparisons especially difficult. A second obstacle is the lack of a standard protocol and criteria for interpretation. Interpretation of profiles is critical since variant profiles may be obtained by a new species or by intraspecies heterogeneity. Third, due to the increasing number of restriction endonucleases that would be required to resolve all validated species and recognize newly recovered species, the method rapidly becomes more complex. As may be expected, failure of the method to reliably identify *Gordonia* clinical isolates has previously been documented.[48,53]

DNA Sequencing. Of all the gene targets found useful for identification and or phylogenetic analysis of bacteria, 16S rRNA gene sequences are used most widely. Their widespread utilization is due to several factors. First, the 16S rRNA gene is universally present in all bacteria and the biological function of the gene is well known. Second, 16S rRNA genes evolve at a slower rate relative to that observed for house keeping genes. The slower rate is due to the constraints of the three-dimensional structure required for a functional 16S rRNA molecule. Therefore, the gene includes both conserved and variable sequence motifs. Conserved sequence motifs may provide consensus family- or genus-specific sequences whereas variable regions may provide for species-specific identification. The 16S rRNA gene has been found to be appropriate for phylogenetic analysis due to these slowly evolving conserved sequences. Lastly, the gene is relatively easy to sequence and its length—approximately 1.4 kb—has been determined to be large enough for phylogenetic analysis. Both clinical and environmental isolates of *Gordonia* spp. have been identified using 16S rRNA gene sequences. Isolates of *G. aichiensis* and *G. amarae* were among the first pathogenic strains to be identified and classified using analysis of 16S rRNA gene sequences.[8,71]

To expedite the rapid detection and identification of *Gordonia* species in complex environmental samples, Patel et al.[72] developed primer pair, G268F and G1096R, to amplify by PCR an 828 bp 16S rRNA gene fragment.[68] The primer pair was able to amplify the appropriate sized fragments from 19 *Gordonia* type strains and *Gordonia*-specific signature nucleotides were observed. The primer pair may be *Gordonia*-specific since it failed to amplify the 16S gene for other species of actinomycetes. Analysis of 16S rRNA gene sequences for 47 strains of *Gordonia*, including pathogenic strains of *G. bronchialis*, *G. terrae*, *G. polyisoprenivorans*, and *G. sputi*, formed a distinct clade that was separate from phylogenetically related actinomycete species. However, not all species of *Gordonia* were distinguishable using 16S rRNA gene sequences. Reliable identification was not always possible since the fragment length was short, at 828 bp. Longer 16S rRNA gene sequences have been shown to improve the accuracy and reliability of molecular identification for other species of bacteria.[73]

Identification of clinical isolates by 16S rRNA gene analysis is an increasingly important adjunct method to tradition biochemical diagnostics and polyphasic characterization. However, there are several difficulties that may impede accurate identification. For instance, public and private sequence databases have been known to harbor sequences containing errors, ambiguous bases, or misidentified strains due to lack of quality control or a protocol to remove errors. The collection of reference and type species may be either incomplete or requires updating. A second problem is the lack of a universal standard for interpretation, especially for identification of species with nearly identical 16S sequences. Simply, how similar should sequences be for reliable and accurate identification? Many species have recently evolved and have nearly indistinguishable sequences. These recently evolved species are difficult to identify due to lack of sequence divergence in a short time period or gene sequences made homogeneous due to recombination. For example, the sequence of a 450 bp 16S rRNA gene fragment from a patient with bacteremia and endocarditis was 100% identical to the 16S rRNA gene sequence of *G. sputi* but the relatedness of the isolate to *G. sputi* by DNA–DNA hybridization analysis was only 55%, far below the threshold of 70% recommended for the definition of same species.[56] Likewise, the nearly full-length 16S rRNA gene sequence obtained from a patient with bacteremia differed by the 16S rRNA gene sequence of *G. sputi* and *G. aichiensis* by 1 and 3 bp, respectively, whereas the isolate was determined to be 60% or 21% related by DNA–DNA hybridization analysis to *G. sputi* or *G. aichiensis*, respectively.[48] These investigations demonstrate the practical difficulties of reliable identification of *Gordonia* spp. based solely on analysis of 16S rRNA gene sequences. Accurate identification may be mired by the presence of multiple copies of 16S rRNA genes resulting in ambiguous bases or sequence heterogeneity since the majority rule consensus sequence ignores real allelic differences. Other issues involve the inability to identify samples obtained from mixed cultures, the high cost of supplies, technical expertise, and equipment, all of which may be a significant factor delaying procurement of routine DNA sequencing by many clinical laboratories.

To circumvent the difficulties associated with analysis of 16S gene sequences, an alternative gene target, DNA gyrase subunit B (*gyrB*), was sequenced and directly compared to analysis using 16S rRNA gene sequences for identification and phylogenetic analysis of *Gordonia* isolates.[74] Briefly, genomic DNA was purified and a ~1.2 kb *gyrB* gene fragment was amplified by PCR using primers GYRB1 and GYRB2 for 25 different strains representing 17 different type strains of *Gordonia*. GyrB amino acid and *gyrB* nucleotide sequences were aligned by Clustal X software and analyzed using MEGA 2.1 software. The range of sequence similarity for 12 *Gordonia* type strains *gyrB* nucleotide sequences was 79.3%–97.2%, corresponding to 270–41 bp differences. In comparison, a sequence difference of 0.3%–3.8% was observed by 16S rRNA gene sequencing for the same 12 type strains. The branching orders of the 16S rRNA and *gyrB* gene trees were similar, but *gyrB* gene sequences showed a higher degree of separation between species. Enhanced discrimination may be due to the higher degree of nucleotide substitution rate observed for the *gyrB* gene than observed for 16S rRNA gene sequences. Interestingly, the correlation coefficients of GyrB amino acid sequences with DNA–DNA homology and 16S rRNA gene sequences with DNA–DNA homology were both identical ($R^2 = 0.865$). Nucleotide substitutions were scattered over the entire length of the *gyrB* gene. In contrast, nucleotide substitutions in the 16S rRNA gene were concentrated to a few variable regions, thereby allowing *gyrB* to provide better discrimination between species than obtained by 16S rRNA gene sequences. In summary, Shen et al.[74] reported *gyrB* gene sequences were a good alternative and appropriate marker for identification of *Gordonia* species.

In a similar investigation, Kang et al.[75] compared the phylogenetic relationships obtained between 23 *Gordonia* type strains using three different gene sequences: *gyrB*, *sec*A1, and 16S rRNA. Kang et al. amplified an approximate 1230 bp *gyrB* gene fragment using PCR primer pairs, F1 and R1296, and fragments were then sequenced. Similarity of the GyrB amino acid sequence and *gyrB* gene sequence ranged from 75.7% to 99.8%, and 77.5% to 97.6%, respectively. The *gyrB* gene tree branching order obtained by Kang et al.[75] was consistent with the investigation by Shen et al.[68] A 496 bp gene fragment, representing approximately 160 amino acids of the *sec*A1 gene, was amplified by PCR and sequenced. Interspecies sequence similarity for the *sec*A1 gene sequence and the SecA1 amino acid sequence was 81.9%–98.0%, and 83.0%–100%, respectively. Additionally, a *secA1* signature sequence for *Gordonia* species, IKKK, located between amino acid residues 120–130, was identified. In comparison, the interspecies sequence similarity for 16S rRNA gene sequences ranged from 94% to 99.8%; this is higher than obtained for *gyrB* and *secA1* gene sequences. Most of the nucleotide variability in the 16S rRNA gene resides between nucleotide positions (based on *Escherichia coli* strain K-12) 136–229 and 996–1028.[3] The branching order for the three gene trees was highly similar but not identical, suggesting a variable rate of evolutionary change for each gene. Both *gyrB* and *secA1* gene sequences provided a higher degree of discrimination than that obtained by 16S rRNA gene sequences, whereas the highest degree of discrimination and identification was observed with *gyrB* gene sequences. Phylogenetic analysis using the combination of these three gene sequences allowed for the unambiguous identification and resolution of all 23 *Gordonia* type strains and resulted in a higher degree of resolution obtainable than by using a sequence from a single gene. Kang et al.[75] concluded that both *gyrB* and *secA1* gene sequences are appropriate molecular markers for identification and phylogenetic analysis of *Gordonia* species isolates.

At present, only a few clinical or reference laboratories utilize *gyrB* and *secA1* gene sequences for identification of *Gordonia* clinical isolates. One difficulty is that not all species of *Gordonia* are capable of being amplified using a single standard amplification regimen with primers F1 and R1296.[75] Other difficulties include establishment of a similarity cutoff criteria for reliable and accurate identification since intraspecies sequence variability has been observed. For example, the 16S rRNA gene similarity ranged from 99.6% to 99.9% for six isolates of *G. terrae*. However, analysis of *gyrB* gene sequences demonstrates two subclades. One isolate has 98.0% sequence similarity to the type strain whereas the sequence similarity for the remaining four isolates was 87.0%, suggesting that *G. terrae* may be heterogeneous, composed of a physiological similar but genetically different set of isolates. DNA–DNA hybridization or analysis of other molecular markers may be required to determine whether *G. terrae* is a heterogeneous species.

Identification of Clinical Isolates Using DNA Sequencing. There are several reports of *Gordonia* clinical isolates misidentified by routine biochemical tests such as API Coryne or by other phenotypic tests that rely primarily on the analysis of 16S rRNA gene sequences for accurate identification.[33,36,38,39,43,54,76] For instance, an isolate obtained from a patient with a sequestrated lung[34] and from a patient with a recurrent breast abscess[38] were incorrectly identified as a *Rhodococcus* species by the API Coryne, but were identified correctly as *G. bronchialis* by analysis of 16S rRNA gene sequences. Similarly, a blood culture isolate from a patient with endocarditis was initially identified as a *Corynebacterium*. This isolate was later identified as *G. polyisoprenivorans* since 1444 of 1445 bp of the 16S rRNA gene sequences were identical to the *G. polyisoprenivorans* type strain.[43] In a large retrospective study in Japan, 14 of 34 clinical isolates of *Gordonia* identified by biochemical profiles were then confirmed by analysis of 16S rRNA gene sequences as *G. sputi*[34]; however, these isolates may be heterogeneous since 16S rRNA gene analysis detected up to four genetic groups. *G. sputi*, identified in sputum samples using morphologic and biochemical tests (before the availability of routine 16S gene sequencing), were considered a source of pulmonary infections.[45,49,63] However, it may be interesting to reevaluate these studies with 16S rRNA gene sequences to provide a more robust and accurate identification. A clinical isolate of *G. araii*, associated with an orthopedic device in an immunocompetent patient, used both 16S rRNA and *gyrB* sequences for identification, since biochemical analysis was unreliable for accurate identification.[33] Analysis of 16S rRNA gene sequences may also be used for the initial identification and characterization of novel species of *Gordonia*. For instance, two recently described pathogenic species of *Gordonia*, *G. araii*, and *G. effusa*, isolated from sputum of patients with bacterial pneumonia or kidney dysfunction, respectively, were identified based on differences in 16S rRNA gene similarity to other known *Gordonia* species.[24]

Similarly, analysis of 16S rRNA gene was used for definitive identification of a strain obtained from an ear discharge and was later identified as a new species, *G. otitidis*.[23] Despite the known limitations of solely using 16S rRNA gene analysis for species identification, routine use of these gene sequences presently provides a useful and practical tool to identify rare or unknown clinical isolates.

Microarrays and Whole Genome Sequencing. Microarrays or microchips generally consist of a complex mixture of short, specific oligonucleotide sequences immobilized on a solid support. The immobilized oligonucleotides are hybridized with a mixture of labeled nucleic acid probes allowing simultaneous detection and quantitative measurement of a complex mixture of specific gene sequences. DNA microarrays have the potential to be used for identification of pathogenic *Gordonia* or other bacteria. Microarrays may also be fabricated to determine antimicrobial resistance or detect patterns of transcription during infection. As a proof of concept for rapid species identification using microarray technology, a microarray was fabricated with 45 random, unsequenced, genomic DNA sequences of unknown function. All were 200–1500 bp in length and were obtained from *G. amarae* genomic DNA.[77] *G. amarae* microarrays were hybridized with labeled genomic DNA, purified from thirteen different species of bacteria or DNA purified from *G. amarae* itself. The microarray was observed to accurately identify *G. amarae* since positive signals were obtained only from *G. amarae* DNA and showed no cross hybridization by other species of bacteria. Positive signals were also observed using *G. amarae* DNA mixed with heterologous DNA probe. Sequencing of the entire *Gordonia* species genomes or particular gene sequences may provide species-specific sequences useful for the fabrication of *Gordonia*-specific microarrays.

9.2 METHODS

9.2.1 Sample Preparation

Presently, molecular identification of *Gordonia* isolates requires culturing of clinical samples on appropriate growth medium since DNA used for molecular analysis is obtained from pure cultures. Clinical samples are commonly cultured aerobically on rich medium including trypticase soy agar, brain heart infusion agar, and chocolate or blood agars. Species of *Gordonia* have been isolated from a vast array of clinical samples including blood,[39,42,43,48,55,56] sputum,[34,45,63] biopsied tissue,[33,38,46,57,58] CSF,[39,53,54] and a lymph node.[47]

Nucleic acids for molecular analysis are purified from colonies grown on the appropriate solid agar or obtained from cells grown in broth. Cells may be lysed and DNA purified using a combination of enzymatic digestion and deproteinization by repeated phenol-chloroform-isoamyl alcohol extractions; lysis using sodium dodecyl sulfate (SDS) in the presence of proteinase K[35]; or lysis with guanidine thiocyanate.[23,24] DNA has been obtained by cell rupture using repeated cycles of freezing in liquid nitrogen and boiling[53] or with a beadbeater using zirconium beads.[64] Recently, high purity DNA has been obtained using commercially available DNA isolation kits such as the soil DNA Extraction Kit,[78] the UltraClean Soil DNA Isolation Kit,[69,78] the UltraClean Microbial Genomic DNA Isolation Kit,[74] the ISOPLANT I Kit,[75] or the DNAeasy Tissue Kit.[36]

9.2.2 Detection Procedures

(i) Phylogenetic analysis of 16S rRNA gene sequences Phylogenetic relationships among *Gordonia* species isolates can be determined following the amplification, alignment, and analysis of near full-length 16S rRNA gene sequences.

Procedure

1. Amplification of near full-length 16S rRNA gene is performed in a 50 μL reaction mixture consisting of approximately 10 ng of purified genomic DNA, 200 μM of each dNTP, 5% (vol/vol) dimethyl sulfoxide, 2.5 U of Expand High-Fidelity DNA Polymerase (Roche Diagnostics Corp), and 100 nM each of primers fD1 (5′-CCGAATTCGTCG ACAACAGAGTTTGATCCTGGCTCAG-3′) and rP2 (5′-CCCGGGATCCAAGCTTACGGCTACC TTGTTACGACTT-3′) in the buffer supplied by the manufacturer.
2. Amplification conditions consist of an initial denaturation step at 94°C for 5 min followed by 35 cycles of amplification at 94°C for 15 s, 50°C for 15 s, and 72°C for 1.5 min, with a final extension at 72°C for 5 min.
3. PCR products are confirmed by visualization on 1.2% agarose gels.
4. Primers and nucleotides are removed with ExoSap-It (USB Corporation) using the reagents and protocol supplied by the manufacturer.
5. Purified amplicons are then sequenced in the forward and reverse directions by cycle sequencing with the ABI Prism BigDye terminator cycle sequencing ready reaction kit (Perkin Elmer/Applied Biosystems). Sequencing reactions are then loaded and resolved by capillary electrophoresis using an ABI 3730 DNA analyzer (Applied Biosystems).
6. Chromatograms are edited, assembled, and trimmed using Seqmerge software (Wisconsin Package version 10.3; Accelrys, Inc., San Diego, CA).
7. High-quality consensus sequences can be submitted to GenBank for comparative analysis using BLASTn software available at the BLAST Web site http://www.ncbi.nlm.nih.gov/BLAST. Since there is no universally recognized criterion for identification to the genus or species levels using 16S rRNA gene sequences, isolates were identified using the identity of the type strain with the highest degree of sequence similarity.
8. Phylogenetic analysis of 16S rRNA gene sequences is performed following alignment of nucleotide

sequences using CLUSTAL W software. The aligned sequences are then used to construct a neighbor-joining tree using MEGA software.

9. Tree topology is evaluated by bootstrap analysis based on 1000 resamplings using MEGA software.

Additional 16S rRNA primers for phylogenetic analysis of *Godonia* spp. are listed in Table 9.3.

(ii) Phylogenetic analysis using gyrB gene sequences Phylogenetic relationships among *Gordonia* species isolates can be determined following the alignment and analysis of *gyrB* gene sequences.

Procedure

1. Amplification of the DNA gyrase β-subunit (*gyrB*) are performed using 20 ρmol of primers GYRB1 (5′-ATGACNCARYTNCAYGCNGGN-3′) and GYRB2 (5′-SAYGATCTTGTKRTASCGMAAYT T-3′) in a volume of 20 μL reaction mixture using the protocol and amplification conditions described by Shen et al.[74] Alternatively, amplification was performed with 20 ρmol each of primers GYRB1 and GYRB2, 0.2 mM of each of the four dNTPs, 5% (vol/vol) dimethyl sulfoxide, and 0.2 U of iProof High-Fidelity DNA Polymerase in iProof GC buffer (BioRad).
2. Amplification conditions consisted of an initial denaturation of template DNA at 98°C for 1 min, followed by 30 cycles, consisting of 98°C for 15 s, 59°C for 30 s, and 72°C for 1 min, followed by a final extension at 72°C for 5 min.
3. PCR products are confirmed by visualization on 1.2% agarose gels.
4. Primers and nucleotides are removed with ExoSap-It (USB Corporation) using the reagents and protocol supplied by the manufacturer.
5. Purified amplicons are then sequenced in the forward and reverse directions by cycle sequencing with the ABI Prism BigDye terminator cycle sequencing ready reaction kit (Perkin Elmer/Applied Biosystems) using sequencing primers GYRB1, GYRBF1, GYRBF2, GYRB2, and GYRBR1, described previously by Shen[74] and a new reverse sequencing primer, R2 (5′-GAGATG GCGTTCCTSAACAAG-3′). Sequencing reactions are then loaded and resolved by capillary electrophoresis using an ABI 3730 DNA analyzer (Applied Biosystems).
6. Chromatograms are edited, assembled, and trimmed using Sequencher version 4.8 software (Gene Codes Corp, Ann Arbor, MI).
7. High-quality consensus sequences can be submitted to GenBank for comparative analysis using BLASTn

TABLE 9.3
16S rRNA Gene Primers for Amplification and Sequencing *Gordonia* spp.

Primer	Sequence (5′–3′)	Application	Orientation[a]	Reference[b]
fD1	CCGAATTCGTCG ACAACAGAGTTTGATCCTGGCTCAG	PCR	F	Weisburg et al., 1991
rP2	CCCGGGATCCAA GCTTACGGCTACCTTGTTACGACTT	PCR	R	Weisburg et al., 1991
fD1-5p	CCGAATTCGTCGACAACAG	Sequencing	F	Morey et al., 2006
F27	AGAGTTTGATCMTGGCTCAG	Sequencing	F	Heuer et al., 1997
F785	GGATTAGATACCCTGGTA	Sequencing	F	Weisburg et al., 1991
R357	CTGCTGCCTCCCGTA	Sequencing	R	Weisburg et al., 1991
R802	TACCAGGGTATCTAATCC	Sequencing	R	Weisburg et al., 1991
F357	TACGGGAGGCAGCAG	Sequencing	F	Weisburg et al., 1991
R530	GTATTACCGCGGCTGCTG	Sequencing	R	Weisburg et al., 1991
R519	GWATTACCGCGGCKGCTG	Sequencing	R	Weisburg et al., 1991
F530	CAGCAGCCGCGGTAATAC	Sequencing	F	Weisburg et al., 1991
BSF917	GAATTGACGGGGRCCC	Sequencing	F	Wuyts et al., 2002
BSR926	CCGTCAATTYYTTTRAGTTT	Sequencing	R	Wuyts et al., 2002
BSF1099	GYAACGAGCGCAACCC	Sequencing	F	Wuyts et al., 2002
BSR1114	GGTTGCGCTCGTTRC	Sequencing	R	Wuyts et al., 2002
BSF1391	TGTACACACCGCCCGTC	Sequencing	F	Wuyts et al., 2002
BSR1407	GACGGGCGGTGTGTRC	Sequencing	R	Wuyts et al., 2002
rP2-5p	CCCGGGATCCAAGCTTAC	Sequencing	R	Wuyts et al., 2002

[a] F, forward; R, reverse.
[b] Additional sources:

Weisburg, W.G. et al., 16S ribosomal DNA amplification for phylogenetic study. *J. Bacteriol.*, 173, 697, 1991.

Morey, R.E. et al., Species-specific identification of *Leptospiraceae* of 16S rRNA gene sequencing. *J. Clin. Microbiol.*, 44, 3510, 2006.

Heuer, H. et al., Analysis of actinomycete communities by specific amplification of genes encoding 16S rRNA and gel-electrophoretic separation in denaturing gradients. *Appl. Environ. Microbiol.*, 63, 3233, 1997.

Wuyts, J. et al., The European database on small subunit ribosomal RNA. *Nucleic Acids Res.*, 30, 183, 2002.

software available at the BLAST Web site http://www.ncbi.nlm.nih.gov/BLAST. Since there is no universally recognized criterion for identification to the genus or species levels using *gyrB* gene sequences, isolates are identified using the identity of the type strain with the highest degree of sequence similarity.

8. Phylogenetic analysis of *gyrB* gene sequences is performed following alignment of nucleotide sequences using CLUSTAL W software and then used to construct a neighbor-joining tree using MEGA software.
9. Tree topology is evaluated by bootstrap analysis based on 1000 resamplings using MEGA software.

9.3 CONCLUSIONS AND FUTURE PERSPECTIVES

A method for identification of *Gordonia* in tissues would help to eliminate unnecessary culturing and shorten the length of time for identification. One potential method for identification in tissues is by fluorescence in situ hybridization (FISH). Briefly, species- and or genus-specific fluorescently labeled oligonucleotide probes are designed to specifically hybridize to 16S rRNA in cells due to their relative abundance. Cells are hybridized following a permeabilizing step using enzymatic or acid treatment of fixed cells. The specific fluorescence of target cells, visualized with a fluorescent microscope, readily distinguishes target cells from nontarget cells and debris. At present, FISH has been used to identify only *G. amarae* in environmental samples, including activated sludge at wastewater treatment plants.[79,80] FISH may be a useful method for rapid identification, estimating cell viability, or direct detection in paraffin-embedded tissue samples. De Los Reyes et al.[79] magnified the fluorescent signal for detection of *Gordonia* in environmental samples by simultaneous fluorescent antibody staining and FISH. Presently, FISH has not been used for identification of clinical specimens or tissues. While FISH may provide rapid identification, laboratories would require a complete array of species-specific probes and would require additional probes for new species.

As discussed above, microarrays have the potential for rapid and specific identification and detection of *Gordonia* species isolates in either environmental or clinical samples. Species-specific and genus-specific oligonucleotide sequences may be acquired from whole genome sequencing projects or genome sequencing surveys. Microarrays can also be fabricated for investigations of gene expression during infection and disease prognosis, host interactions, identification of antibiotic resistance markers and resistant alleles, response to therapy, and obtaining transcription profiles during infection. Microarrays specific for *G. polyisoprenivorans* may be useful for determining the molecular mechanism for biofilm formation and propagation in catheter materials.

There is an urgent need for a rapid, reliable, and accurate gene target for the molecular identification of pathogenic bacteria, including *Gordonia* species. Several potential gene targets are being investigated.[81,82] Criteria for selecting a new molecular target are that the target gene must be easy to amplify using a standard protocol, the target gene must be present in the bacteria of interest, the observed sequence variation must be high enough for species identification but not too conserved that species may artificially cluster together, and the target must be easy to interpret for unambiguous identification. Alternatively, a strategy using multilocus sequencing may provide better identification and robust phylogenetic signals than obtainable by analysis of a single target gene. The phylogenetic reconstruction and identification scheme using the combination of 16S rRNA, *secA1,* and *gyrB* gene sequences by Kang et al.[75] demonstrates the promise of this approach. Multilocus sequencing analysis is more labor intensive and costly than analysis of a single gene sequence. The method does, however, have the advantage of increased reliability and accuracy for species identification and in determining the phylogenetic relatedness of known clinical isolates and in identifying new species. However, at present, the optimal gene targets or the number of genes is not known. Pyrosequencing may provide an alternative method to conventional gene sequencing and has been shown to be useful for identification of species phylogenetically related to *Gordonia* including *Mycobacterium.*[83] Pyrosequencing is based on real-time detection of DNA synthesis using a mixture of four enzymes. During DNA synthesis, when a deoxynucleotide has been incorporated into a DNA strand by Klenow Polymerase, pyrophosphate is liberated. ATP sulfurylase converts the pyrophosphate to ATP, which is then used in a light-producing reaction by luciferase.[84] Pyrosequencing is faster and simpler than conventional DNA sequencing since the method does not rely on gels or expensive dyes for resolution of bases. The short read length of approximately 100 bp does pose a serious disadvantage, but may be compensated by sequencing of hypervariable regions of a gene.

Purification of DNA is considered to be the most labor-intensive and tedious step in molecular identification, and is therefore the most likely step to be improved by automation. A simple and efficient method of purification of genomic DNA from clinical specimens or cultures would greatly facilitate down stream methods for molecular diagnostics and DNA–DNA hybridization analysis. Likewise, rapid advances in DNA sequencing technology, automated systems, and lower costs may allow in-house molecular diagnostics for most clinical laboratories, including the ability to assemble and annotate entire genomes. Sequencing of entire *Gordonia* genomes may provide further information on unique gene targets for identification or phylogenetic analysis, for potential mechanisms of antibiotic resistance and for the discovery of unique metabolic pathways.

The frequent misidentification of *Gordonia* species suggests the potential underestimation and thus under reporting of these emerging pathogens. Accurate and rapid identification by molecular methods is important for a reliable description of pathogenesis, epidemiologic investigations, application of appropriate antimicrobial therapy, and patient management. Together, these results strongly suggest using a polyphasic approach of combining morphological, biochemical, and molecular analysis for strain identification.

REFERENCES

1. Tsukamura, M., Proposal of a new genus, *Gordona*, for slightly acid-fast organisms occurring in sputa of patients with pulmonary disease and in soil, *J. Gen. Microbiol.*, 68, 15, 1971.
2. Stackebrandt, E., Rainey, F.A., and Ward-Rainey, N.L., Proposal for a new hierarchic classification system, *Actinobacteria* classis nov., *Int. J. Syst. Bacteriol.*, 47, 479, 1997.
3. Arenskötter, M., Bröker, D., and Steinbüchel, A., Biology of the metabolically diverse genus *Gordonia*, *Appl. Envir. Microbiol.*, 70, 3195, 2004.
4. Goodfellow, M., and Alderson, G., The actinomycete—genus: A home for the "rhodochrous" complex, *J. Gen. Microbiol.*, 100, 99, 1977.
5. Collins, M.D. et al., Menaquinones composition of mycolic acid-containing actinomycetes and some sporoactinomycetes, *J. Appl. Bacteriol.*, 58, 77, 1985.
6. Stackebrandt, E., Smida, J., and Collins, M.D., Evidence of phylogenetic heterogeneity within the genus *Rhodococcus*: revival of the genus *Gordona* (Tsukamura), *J. Gen. Appl. Microbiol.*, 34, 341, 1988.
7. Tsukamura, M., Numerical classification of *Rhodococcus* (formerly *Gordona*) organisms recently isolated from sputa of patients: Description of *Rhodococcus sputi* Tsukamura sp. nov., *Int. J. Syst. Bacteriol.*, 28, 169, 1978.
8. Klatte, S., Rainey, F.A., and Kroppenstedt, R.M., Transfer of *Rhodococcus aichiensis* Tsukamura 1982 and *Nocardia amarae* Lechevalier and Lechevalier 1974 to the genus *Gordona* as *Gordona aichiensis* comb. nov. and *Gordona amarae* comb. nov., *Int. J. Syst. Bacteriol.*, 44, 769, 1994.
9. Lechevalier, M.P., and Lechevalier, H.A., *Nocardia amarae* sp. nov, an actinomycete common in foaming activated sludge, *Int. J. Syst. Bacteriol.*, 24, 278, 1974.
10. Tsukamura, M., Numerical analysis of the taxonomy of nocardiae and rhodococci. Division of *Nocardia asteroides* sensu stricto into two species and descriptions of *Nocardia paratuberculosis* sp. nov. Tsukamura (formerly the Kyoto-I group of Tsukamura), *Nocardia nova* sp. nov. Tsukamura, *Rhodococcus aichiensis* sp. nov., Tsukamura, *Rhodococcus chubuensis* sp. nov., Tsukamura, and *Rhodococcus obuensis* sp. nov., Tsukamura, *Microbiol. Immunol.*, 26, 1101, 1982.
11. Bendinger, B. et al., *Gordona hydrophobica* sp. nov., isolated from biofilters for waste gas treatment, *Int. J. Syst. Bacteriol.*, 45, 544, 1995.
12. Klatte, S. et al., *Gordona hirsuta* sp. nov., *Int. J. Syst. Bacteriol.*, 46, 876, 1996.
13. Takeuchi, M., and Hatano, K., *Gordonia rhizosphera* sp. nov. isolated from the mangrove rhizosphere, *Int. J. Syst. Bacteriol.*, 48, 907, 1998.
14. Kummer, C., Schumann, P., and Stackebrandt, E., *Gordonia alkanivorans* sp. nov., isolated from tar-contaminated soil, *Int. J. Syst. Bacteriol.*, 49, 1513, 1999.
15. Kim, S.B. et al., *Gordonia desulfuricans* sp. nov., a benzothiophene-desulphurizing actinomycete, *Int. J. Syst. Bacteriol.*, 49, 1845, 1999.
16. Linos A. et al., *Gordonia polyisoprenivorans* sp. nov., a rubber-degrading actinomycete isolated from an automobile tyre, *Int. J. Syst. Bacteriol.*, 49, 1785, 1999.
17. Kim, S.B. et al., *Gordonia amicalis* sp. nov., a novel dibenzothiophene-desulphurizing actinomycete, *Int. J. Syst. Evol. Microbiol.*, 50, 2031, 2000.
18. Brandão, P.F.B. et al., *Gordonia namibiensis* sp. nov., a novel nitrile metabolising actinomycete recovered from an African sand, *Syst. Appl. Microbiol.*, 24, 510, 2001.
19. Kim, K.K. et al., *Gordonia sihwensis* sp. nov., a novel nitrate-reducing bacterium isolated from a wastewater-treatment bioreactor, *Int. J. Syst. Evol. Microbiol.*, 53, 1427, 2003.
20. Maldonado, L.A. et al., *Gordonia sinesedis* sp. nov., a novel soil isolate, *Antonie van Leeuwenhoek*, 83, 75, 2003.
21. Linos, A. et al., *Gordonia westfalica* sp. nov., a novel rubber-degrading actinomycete, *Int. J. Syst. Evol. Microbiol.*, 52, 1133, 2002.
22. Xue, Y. et al., *Gordonia paraffinivorans* sp. nov., a hydrocarbon-degrading actinomycete isolated from an oil-producing well, *Int. J. Syst. Evol. Microbiol.*, 53, 1643, 2003.
23. Lida, S. et al., *Gordonia otitidis* sp. nov., isolated from a patient with external otitis, *Int. J. Syst. Evol. Microbiol.*, 55, 1871, 2005.
24. Kageyama, A. et al., *Gordonia araii* sp. nov. and *Gordonia effusa* sp. nov., isolated from patients in Japan, *Int. J. Syst. Evol. Microbiol.*, 56, 1817, 2006.
25. Soddell, J.A. et al., *Gordonia defluvii* sp. nov., an actinomycete isolated from activated sludge foam, *Int. J. Syst. Evol. Microbiol.*, 56, 2265, 2006.
26. Shen, F.T. et al., *Gordonia soli* sp. nov., a novel actinomycete isolated from soil, *Int. J. Syst. Evol. Microbiol.*, 56, 2597, 2006.
27. Luo, H. et al., *Gordonia shandongensis* sp. nov., isolated from soil in China, *Int. J. Syst. Evol. Microbiol.*, 57, 605, 2007.
28. Yassin, A.F. et al., *Gordonia malaquae* sp. nov., isolated from sludge of a wastewater treatment plant, *Int. J. Syst. Evol. Microbiol.*, 57, 1065, 2007.
29. Le Roes, M., Goodwin, C.M., and Meyers, P.R., *Gordonia lacunae* sp. nov., isolated from an estuary, *Syst. Appl. Microbiol.*, 31, 17, 2008.
30. Drzyzga, O. et al., *Gordonia cholesterolivorans* sp. nov., a cholesterol-degrading actinomycete isolated from sewage sludge, *Int. J. Syst. Evol. Microbiol.*, 59, 1011, 2009.
31. Kim, K.K. et al., *Gordonia kroppenstedtii* sp. nov., a phenol-degrading actinomycete isolated from a polluted stream, *Int. J. Syst. Evol. Microbiol.*, 59, 1992, 2009.
32. Lechevalier, M.P., and Lechevalier, H.A., Chemical composition as criterion in the classification of aerobic actinomycetes, *Int. J. Syst. Bacteriol.*, 20, 435, 1970.
33. Jannat-Khah, D.P. et al., *Gordonia araii* infection associated with an orthopedic device and review of the literature on medical device-associated *Gordonia* infections, *J. Clin. Microbiol.*, 47, 499, 2009.
34. Aoyama, K. et al., Characterization of clinical isolates of *Gordonia* species in Japanese clinical samples during 1998–2008, *Mycopathologia*, 168, 175, 2009.
35. Richet, H.M. et al., A cluster of *Rhodococcus* (*Gordona*) *bronchialis* sternal-wound infections after coronary-artery bypass surgery, *N. Engl. J. Med.*, 324, 104, 1991.
36. Sng, L.H. et al., Bacteremia caused by *Gordonia bronchialis* in a patient with sequestrated lung, *J. Clin. Microbiol.*, 42, 2870, 2004.
37. Brust, J.C. et al., Five cases of bacteraemia due to *Gordonia* species, *J. Med. Microbiol.*, 58, 1376, 2009.
38. Werno, A.M. et al., Recurrent breast abscess caused by *Gordonia bronchialis* in an immunocompetent patient, *J. Clin. Microbiol.*, 43, 3009, 2005.
39. Blaschke, A.J. et al., *Gordonia* species: emerging pathogens in pediatric patients that are identified by 16S ribosomal RNA gene sequencing, *Clin. Infect. Dis.*, 45, 483, 2007.
40. Linos, A., and Steinbuchel, A., Investigations of the microbial breakdown of natural and synthetic rubbers, *DECHEMA Monogr.*, 133, 279, 1996.

41. Linos, A., and Steinbuchel, A., Microbial degradation of natural and synthetic rubbers by novel bacteria belonging to the genus *Gordonia*, *Kautsch. Gummi. Kurstst.*, 51, 496, 1998.
42. Kempf, V.A. et al., *Gordonia polyisoprenivorans* septicemia in a bone marrow transplant patient, *Eur. J. Clin. Microbiol. Infect. Dis.*, 23, 226, 2004.
43. Verma, P. et al., Native valve endocarditis due to *Gordonia polyisoprenivorans*: Case report and review of literature of bloodstream infections caused by *Gordonia* species, *J. Clin. Microbiol.*, 44, 1905, 2006.
44. Gupta, M., et al., A rubber-degrading organism growing from a human body, *Int. J. Infect. Dis.*, 12, e332, 2009.
45. Hart, D.H.L. et al., Lung infection caused by *Rhodococcus*, *Aust. NZ J. Med.*, 18, 790, 1988.
46. Edwards, J.F., and Simpson, R.B., Nocardioform actinomycete (*Rhodococcus rubropertinctus*)-induced abortion in a mare, *Vet. Pathol.*, 25, 529, 1988.
47. Tsukamura, M. et al., Mesenteric lymphadenitis of swine caused by *Rhodococcus sputi*, *J. Clin. Microbiol.*, 26, 155, 1988.
48. Riegel, P. et al., Bacteremia due to *Gordona sputi* in an immunocompromised patient, *J. Clin. Microbiol.*, 34, 2045, 1996.
49. Kuwabara, M. et al., Mediastinitis due to *Gordona sputi* after CABG, *J. Cardiovasc. Surg.*, 40, 675, 1999.
50. Lesens, O. et al., Bacteremia and endocarditis caused by a *Gordonia* species in a patient with a central venous catheter, *Emerg. Infect. Dis.*, 6, 382, 2000.
51. Lai, C.C. et al., Infections caused by *Gordonia* species at a medical center in Taiwan, 1997 to 2008, *Clin. Microbiol. Infect.*, "Accepted Article" doi: 10.1111/j.1469-0691.200903333085.x
52. Drancourt, M. et al., Brain abscess in an immunocompromised child due to *Gordona terrae*: Case report and review of *Gordona terrae* infections, *Clin. Infect. Dis.*, 19, 258, 1994.
53. Drancourt, M. et al., *Gordona terrae* central nervous system infection in an immunocompetent patient, *J. Clin. Microbiol.*, 35, 379, 1997.
54. Gil-Sande, E. et al., Etiological misidentification by routine biochemical tests of bacteremia caused by *Gordona terrae* infection in the course of an episode of acute cholecystitis, *J. Clin. Microbiol.*, 44, 2645, 2006.
55. Grisold, A.J. et al., Isolation of *Gordonia terrae* from a patient with catheter-related bacteraemia, *J. Med. Microbiol.*, 56, 1687, 2007.
56. Pham, A.S. et al., Catheter-related bacteremia caused by the nocardioform actinomycete *Gordonia terrae*, *Clin. Infect. Dis.*, 36, 524, 2003.
57. Bakker, X.R., Spauwen, P.H.M., and Dolmans, W.M.V. Mycetoma of the hand caused by *Gordona terrae*: A case report, *J. Hand Surg.*, 29B, 188, 2009.
58. Zardawi, I.M. et al., *Gordonia terrae*-induced suppurative granulomatous following nipple piercing, *Pathology*, 36, 275, 2009.
59. Berd, D., Laboratory identification of clinically important aerobic actinomycetes, *Appl. Microbiol.*, 25, 665, 1973.
60. Conville, P.S., and Witebsky, F.G., *Nocardia*, *Rhodococcus*, *Gordonia*, *Actinomadura*, *Streptomyces*, and other aerobic actinomycetes. In Murray, P. R., Baron, E. J., Jorgensen, J. H., Landry, M.I., and Pfaller, M.A. (Editors), *Manual of Clinical Microbiology*, 9th Ed., p. 515, American Society for Microbiology, Washington, D.C., 2007.
61. Clinical and Laboratory Standards Institute. Susceptibility testing of mycobacteria, nocardiae, and other aerobic actinomycetes. Approved standard M24-A. National Committee for Clinical Laboratory Standards, Wayne, PA, 2003.
62. Clinical and Laboratory Standards Institute. Methods for dilution antimicrobial susceptibility tests for bacteria that grow aerobically; approved standard. M7-A7. Clinical and Laboratory Standards Institute, Wayne, PA, 2006.
63. Osoagbaka, O.U., Evidence for the pathogenic role of *Rhodococcus* species in pulmonary disease, *J. Appl. Bacteriol.*, 66, 497, 1989.
64. Steingrube, V.A. et al., Rapid identification of clinically significant species and taxa of aerobic actinomycetes, including *Actinomadura*, *Gordona*, *Nocardia*, *Rhodococcus*, *Streptomyces*, and *Tsukamurella* isolates, by DNA amplification and restriction endonuclease analysis, *J. Clin. Microbiol.*, 35, 817, 1997.
65. Buchman, A.L. et al., Central venous catheter sepsis caused by unusual *Gordona* (*Rhodococcus*) species: identification with a digoxigenin-labeled rDNA probe, *Clin. Infect. Dis.*, 15, 694, 1992.
66. Lasker, B.A. et al., Identification and epidemiological typing of clinical and environmental isolates of the genus *Rhodococcus* with use of a digoxigenin-labeled rDNA gene probe, *Clin. Infect. Dis.*, 15, 223, 1992.
67. Riegel, P. et al., *Rhodococcus chubuensis* Tsukamura 1982 is a later subjective synonym of *Gordona sputi* [Tsukakmura 1978] Stackebrandt 1989 comb.nov., *Int. J. Syst. Bacteriol.*, 44, 764, 1994.
68. Shen, F-T., and Young, C-C., Rapid detection and identification of the metabolically diverse genus *Gordonia* by 16S rRNA-gene-targeted genus-specific primers, *FEMS Microbiol. Lett.*, 250, 221, 2005.
69. Shen, F-T. et al., Detection of filamentous genus *Gordonia* in foam sample using genus-specific primers combined with PCR—denaturing gradient gel electrophoresis analysis, *Can. J. Microbiol.*, 53, 768, 2007.
70. Shen, F-T. et al., Molecular detection and phylogenetic characterization of *Gordonia* species in heavily oil-contaminated soils, *Res. Microbiol.*, 159, 522, 2008.
71. Ruimy, R. et al., A phylogeny of the genus *Nocardia* deduced from the analysis of small-subunit ribosomal DNA sequences, including transfer of *Nocardia amarae* to the genus *Gordona* and *Gordona amarae* comb. nov., *FEMS Microbiol. Lett.*, 123, 261, 1994.
72. Patel, J.B. et al., Sequence-based identification of aerobic actinomycetes, *J. Clin. Microbiol.*, 42, 2530, 2004.
73. Cloud, J.L. et al., Evaluation of partial 16S ribosomal DNA sequencing for identification of *Nocardia* species by using the MicroSeq 500 system with an expanded database, *J. Clin. Microbiol.*, 42, 578, 2004.
74. Shen, F-T. et al., Phylogenetic analysis of members of the metabolically diverse genus *Gordonia* based on proteins encoding the *gyrB* gene, *Res. Microbiol.*, 157, 367, 2006.
75. Kang, Y. et al., Phylogenetic studies of *Gordonia* species based on *gyrB* and *sec*A1 gene analysis, *Mycopathologia*, 167, 95, 2009.
76. Blanc, V. et al., *Gordonia terrae*: a difficult-to-diagnose emerging pathogen? *J. Clin. Microbiol.*, 45, 1076, 2007.
77. Kim, B.C., Park, J.H., and Gu, M.B., Multiple and simultaneous detection of specific bacteria in enriched bacterial communities using a DNA microarray chip with randomly generated genomic DNA probes, *Anal. Biochem.*, 77, 2311, 2005.
78. Kim, B.C., Park, J.H., and Gu, M.B., Multiple and simultaneous detection of specific bacteria in enriched bacterial communities using a DNA microarray chip with randomly generated genomic DNA probes, *Anal. Biochem.*, 77, 2311, 2005.

79. De los Reyes, F.L. III. et al., Characterization of filamentous foaming in activated sludge systems using oligonucleotide hybridization probes and antibody probes, *Water Sci. Tech.*, 37, 485, 1998.
80. Carr, E.L. et al., Improved permeabilization protocols for fluorescence in situ hybridization (FISH) of mycolic-acid-containing bacteria found in foams, *J. Microbiol. Methods*, 61, 47, 2005.
81. Santos, S.R., and Ochman, H., Identification and phylogenetic sorting of bacterial lineages with universally conserved genes and proteins, *Environ. Microbiol.*, 6, 754, 2004.
82. Zeigler, D.R., Gene sequences useful for predicting relatedness of whole genomes in bacteria, *Int. J. Syst. Evol. Microbiol.*, 53, 1893, 2003.
83. Heller, L.C., Jones, M., and Widen, R.H., Comparison of DNA pyrosequencing with alternative methods for identification of mycobacteria, *J. Clin. Microbiol.*, 46, 2092, 2008.
84. Ahmadian, A., Ehn, M., and Hober, S., Pyrosequencing: history, biochemistry and future, *Clin. Chim. Acta.*, 363, 83, 2006.

10 *Micrococcus* and *Kocuria*

Dongyou Liu

CONTENTS

10.1 INTRODUCTION

10.1.1 Classification and Biology

The genera *Micrococcus* and *Kocuria* belong to the family Micrococcaceae, order Actinomycetales, class Actinobacteria, phylum Actinobacteria. Other recognized genera in the family Micrococcaceae include *Acaricomes*, *Arthrobacter*, *Citricoccus*, *Nesterenkonia*, *Renibacterium*, *Rothia*, *Sinomonas*, *Stomatococcus*, and *Zhihengliuella*.

The genus *Micrococcus*, first described in 1872, once contained a diverse group of bacteria that was later dissected into the genera *Kocuria*, *Micrococcus*, *Nesterenkonia*, *Kytococcus*, and *Dermacoccus*.[1,2] At the present, the genus consists of six species: *Micrococcus antarcticus*, *Micrococcus endophyticus*, *Micrococcus flavus*, *Micrococcus luteus*, *Micrococcus lylae*, and *Micrococcus yunnanensis*.[1–7] The species *Micrococcus luteus* is subdivided into three biovars on the basis of their distinct chemotaxonomic and biochemical traits: biovar I (represented by strain DSM 20030T); biovar II (represented by strains D7, 3, 6, 7, 13C2, 38, 83, and 118); and biovar III (represented by strain Ballarat = DSM 14235 = CCM 4960).[3]

The genus *Kocuria* was created in 1995 to cover a number of gram-positive, catalase positive, coagulase negative, strictly aerobic cocci that were classified previously within the genus *Micrococcus*.[2] *Kocuria* spp. are clearly separable from *Micrococcus* and other related gram-positive cocci (e.g., *Arthrobacter* and *Rothia*) on the basis of distinct 16S rRNA gene sequences.[2] As it currently stands, the genus *Kocuria* (named after Miroslav Kocur) is composed of 16 species: *Kocuria aegyptia*, *Kocuria atrinae*, *Kocuria carniphila*, *Kocuria flava*, *Kocuria gwangalliensis*, *Kocuria halotolerans*, *Kocuria himachalensis*, *Kocuria koreensis*, *Kocuria kristinae*, *Kocuria marina*, *Kocuria palustris*, *Kocuria polaris*, *Kocuria rhizophila*, *Kocuria rosea*, *Kocuria salsicia*, *Kocuria turfanensis*, and *Kocuria varians*.[8–18]

Members of the genera *Micrococcus* and *Kocuria* are distributed in a wide variety of natural environments such as soil, the rhizosphere, freshwater, marine sediments, and fermented foods. In fact, *Micrococcus* and *Kocuria* spp. are often used as starter cultures in the manufacture of fermented fish and dry sausages to promote the desired red color development and stability (through reduction of nitrate to nitrite), to prevent rancidity (through peroxide decomposition), to maintain hygienic safety (through nitrite permanence), and to contribute to the flavor of fermented meat products (through formation of esters and other aromatic compounds from amino acids). In addition, *Micrococcus* and *Kocuria* spp. are also found in the skin, mucosae, and oropharynx of humans as commensal microflora, and several of them have been associated with human diseases.

10.1.2 Morphology and Biochemical Properties

Members of the genus *Micrococcus* are gram positive, nonmotile, nonendospore forming, aerobic, and chemoorganotrophic cocci. *Micrococcus* cells are catalase and oxidase positive, mesophilic, and nonhalophilic. The peptidoglycan of *Micrococcus* spp. contains amino acid L-lysine, which is characteristic of the genus. The predominant menaquinones in *Micrococcus* cells are either MK-8 and MK-8(H2), although minor amounts of MK-8(H), MK-7 or MK-7(H2), and MK-9(H2) may be present. *Micrococcus* cells lack mycolic acids and teichoic acids, but may contain teichuronic acids as an amino sugar in the cell wall polysaccharide. The G + C content of *Micrococcus* DNA is 69–76 mol%. *Micrococcus* spp. are primarily found in mammalian skin. The type species of the genus is *Micrococcus luteus*.[1–3]

Members of the genus *Kocuria* are gram-positive, nonencapsulated, nonendospore forming, chemoorganotrophic, and strictly aerobic cocci. *Kocuria* spp. are catalase positive, nonhalophilic, and mesophilic. They usually grow on simple media, and are characterized by the presence of menaquinones MK-7(H_2) and MK-8(H_2), lysine-based A3α-type peptidoglycan, PI-type phospholipids, and saturated branched fatty acids in their cell envelopes.[2] The G + C content of *Kocuria* DNA ranges from 66 to 75 mol%. The type species is *Kocuria rosea*.[10]

Among the pathogenic *Kocuria* spp., *Kocuria rosea* (*rosea*, rose colored; formerly *Micrococcus rosea*) cells are spherical (with a diameter of 1–1.5 μm) and occur in pairs, tetrads, and clusters. Colonies are circular, slightly convex, smooth (occasionally rough), and pink or red, and colony morphology and color tend to become more distinct with age. *K. rosea* grows well at temperature between 25°C and 37°C and in the presence of NaCl concentrations of up to 7.5% NaC1. Its growth is stimulated by cysteine and thiamine or cysteine, thiamine, and pantothenic acid. In nutrient broth, *K. rosea* demonstrates moderate turbidity and a deposit. Most *K. rosea* do not grow on inorganic nitrogen agar and on Simmons citrate agar. *K. rosea* is oxidase test negative (rarely weakly positive) and benzidine test positive. The bacterium can reduce nitrate, but not nitrite. In addition, it is negative for arginine dihydrolase, phenylalanine deaminase, urease, lecithinase, phosphatase, p-galactosidase, and coagulase. Furthermore, *K. rosea* does not hydrolyse esculin and gelatin, and its hydrolysis of Tween 80 and starch is variable. Many *K. rosea* strains can utilize rhamnose, xylose, and glucitol, but form no acid from mannose, galactose, lactose, and glycerol. The G + C content of *K. rosea* DNA is 66–75 mol%. *K. rosea* is saprophytic, nonhemolytic, and is often isolated from soil and water, and also occasionally from human infections.[2]

Kocuria varians (*varians*, varying; formerly *Micrococcus varians*) cells are spherical (with a diameter of 0.9–1.5 μm) and occur in tetrads and irregular clusters of tetrads or, rarely, in packets or as single cells. The optimal growth temperature of *K. varians* ranges from 22°C to 37°C. Colonies (up to 4 mm in diameter) are circular, slightly convex, smooth and glistening, and yellow in color. Some *K. varians* strains may form rough, wrinkled, matt, dry colonies. Colony morphology and color often become more distinct with age. The bacterium usually grows on Simmons citrate agar but not on inorganic nitrogen agar. In nutrient broth, *K. varians* shows slight turbidity and a deposit. It can grow on nutrient agar containing up to 7.5%–10% NaCl, and its growth is stimulated by methionine or cysteine and thiamine. *K. varians* is oxidase test negative; benzidine test positive, in addition to being usually positive for urease and nitrate and nitrite reductases. It is negative for arginine dihydrolase, coagulase, p-galactosidase, phosphatase, lecithinase, phenylalanine deaminase, and DNase. *K. varians* does not hydrolyse esculin and Tween 80, and may show varied hydrolysis of starch and gelatin. In addition, it is indole and hydrogen sulfide negative. While the bacterium produces acid from glucose, xylose, and fructose, it does not do so from mannose, maltose, arabinose, raffinose, galactose, rhamnose, ribitol, lactose, mannitol, glycerol, glucitol, inositol, and salicin. The G + C content of *K. varians* DNA is 66–72 mol%. The bacterium is saprophytic, nonhemolytic, and isolated from mammalian skin, soil, and water.[2]

Kocuria kristinae (*kristinae*, of Kristin; formerly *Micrococcus kristinae*) cells are 0.7–1.1 μm in diameter and occur in tetrads or clusters. The optimal growth temperature of *K. kristinae* ranges from 25°C to 37°C. Colonies are circular, entire or crenate, convex, smooth or rough, up to 2 mm in diameter, and pale cream to pale orange in color, which becomes more intense with time. *K. kristinae* is lightly facultatively anaerobic, and grows well on nutrient agar containing up to 10% NaCl, but very weakly in the presence of 15% NaC1. Its growth is stimulated by leucine, lysine, valine, tyrosine, and niacin. *K. kristinae* is unable to grow on Simmons citrate agar, and rarely grows on inorganic nitrogen agar. Being oxidase and benzidine test positive, the bacterium is negative for arginine dihydrolase, lecithinase, and phosphatase activity, and variable for urease P-galactosidase activity. Most *K. kristinae* strains do not reduce nitrate. It is positive for hydrolysis of esculin but negative for hydrolysis of gelatin, Tween 80, and starch. It can produce acid aerobically from glucose, fructose, mannose, sucrose, and glycerol, and usually from maltose and glucitol, although it cannot utilize rhamnose, xylose, ribose, arabinose, raffinose, melibiose, mannitol, ribitol, and galactitol; usually does not utilize galactose and lactose. The G + C content of *K. kristinae* DNA is 67 mol%. It is nonhemolytic, and its primary habitat is human skin. *K. kristinae* has been associated with catheter-related bacteremia.[2]

Kocuria rhizophila (rhiza, a root; philos, loving; rhizophila root-loving) cells are gram-positive, nonacid-fast, nonmotile, nonendospore forming, aerobic, chemoorganotrophic, and spherical with a diameter of 1.0–1.5 μm, and occur in pairs, tetrads, and packets. Colonies are 1.5–2.5 mm in diameter, and are opaque, smooth with irregular edges with yellow pigmentation. The bacterium grows well between pH 5.7–7.7, in the presence of 10% NaCl, and on inorganic nitrogen and Simmons' citrate, but does not grow at temperature above 40°C. It is positive for catalase, benzidine gelatinase, phosphatase, and Tween 80 hydrolysis, but negative for oxidase, indol, urease, arginine dihydrolase and phenylalanine deaminase, and esculin and starch hydrolysis. The bacterium does not reduce nitrate to nitrite. It produces acid production from D-glucose, D-fructose, D-mannose, and saccharose, and also utilizes adonitol, L-arabinose, D-fructose, L-fucose, D-glucose, turanose, xylitol, methylpyruvate, glucuronamide, dextrin, glycogen, Twee 40, Tween 80, and *N*-acetyl-D-glucosamine. The major menaquinones of *K. rhizophila* are MK-7(H2) and MK-8(H2). The genome of *K. rhizophila* DC2201 (NBRC 103217) has been shown to harbor a single circular chromosome of 2,697,540 bp in length with an average G + C content of 71.16%. Of the 2357 predicted protein-coding genes, 87.7% encode proteins that are orthologous to actinobacterial proteins.[19] *K. rhizophila* was first isolated from the rhizosphere of Typha angustifolia from a floating mat on the Soroksar tributary of the river Danube, Hungary[8] and a human case of *K. rhizophila* infection was documented recently.

Kocuria marina (*marina*, of the sea) cells are gram-positive, aerobic, nonmotile, halotolerant cocci that grow at 4°C–43°C. The bacterium is positive for catalase, β-galactosidase, and urease, but negative for arginine dihydrolase, lysine and ornithine decarboxylase, oxidase, and alkaline phosphatase. It reduces nitrate, but does not produce hydrogen sulfide. In addition, it produces acid from L-fucose, although it is negative for indole and acetoin (Voges–Proskauer reaction). It can hydrolyse casein, gelatin, and Tween 40, but not agar, alginate, cellulose, DNA, starch, Tween 20, and Tween 80. Furthermore, it utilizes glucose, lactose, mannose, and sucrose as sole carbon sources, but not adonitol, L-arabinose, *meso*-inositol, mannitol, sorbitol, citric acid, or malonic acid. The G + C content of *K. marina* DNA is 60 mol%.[11] The bacterium was first isolated from marine environment, but a case of human infection with this bacterium was recently reported.[11]

10.1.3 Clinical Features and Pathogenesis

Micrococcus and *Kocuria* spp. are widespread in nature and are frequently found as normal skin flora in humans and other mammals. *Micrococcus* strains such as *M. luteus* have been associated with catheter-related bacteremia in patients undergoing hemodialysis or leukaemia treatment, pneumonia, endocarditis, intracranial abscesses, continuous ambulatory dialysis peritonitis, septic arthritis, and meningitis.[20–27] Since *Kocuria* spp. were contained in the genus *Micrococcus* prior to 1995, some of the early reports documenting the roles of *Micrococcus* strains in human infections may be attributable to *Kocuria* spp.

In recent years, a number of *Kocuria*-related infections have been documented. Basaglia et al.[28] described the first case of a catheter-related recurrent bacteremia due to *Kocuria kristinae* in a 51-year-old woman with a poorly differentiated ovarian carcinoma. The patient had a permanent central venous catheter (CVC) implanted for chemotherapy and developed neutropenia and sepsis a few months afterwards. *Kocuria kristinae* isolates grew from three blood samples cultured aerobically onto sheep blood agar plates at 37°C for 18 h. They were nonhemolytic, catalase positive, strictly aerobic, nonmotile, and unable to reduce nitrate to nitrites or to hydrolyze gelatin, arginine, and esculin. The bacterial isolates produced acid from trehalose, glucose, fructose, mannose, glycerol, and saccharose but not from mannitol, raffinose, arabinose, lactose, or ribose. This identification was confirmed with the commercially available ID32 Staph ATB system (Biomerieux), as well as sequencing analysis of the partial 16S rRNA gene (rDNA; 642 bp). The patient recovered after removal of CVC and antibiotic treatment.

Altuntas et al.[29] reported the first case of a catheter-related bacteremia caused by *Kocuria rosea* in a 39-year-old man undergoing peripheral blood stem cell transplantation due to relapsed Hodgkin disease. The patient presented chills, high fever (temperature 38.5°C), tachycardia (heart rate 108 beats/min), and increased respiratory rate 3 days after implantation of a Hickman-type central venous catheter. Serial blood cultures taken from the catheter and a peripheral vein revealed the presence of *Kocuria rosea* as confirmed with the commercial tests (ID32 Staph ATB system, and mini-API, Biomerieux). The patient recovered after treatment with vancomycin and by catheter removal.

Ma et al.[30] described the first case of *K. kristinae*-related acute cholecystitis in a patient who underwent laparoscopic cholecystectomy. The patient developed postoperative fever and recovered after levofloxacin treatment. Becker et al.[31] documented the first case of a *Kocuria rhizophila* infection in a boy with methylmalonic aciduria due to a noncobalamin-responsive deficiency of methylmalonyl coenzyme A mutase. The patient developed sepsis two years after a subcutaneous implantable vascular-access port (Port-A-Cath; Vital-Port) was placed in his left internal jugular vein. The patient showed the signs of acute pancreatitis and fever. A *Kocuria rhizophila* strain was isolated from blood samples drawn through a port system and from peripheral veins during septic episodes. The bacterium resembles *Micrococcus* species as assessed by the Bactec 9240 system (Becton Dickinson). It is gram-positive, catalase positive and oxidase negative, and appears in pairs, tetrads, and packets. Colonies are smooth and circular with a yellow tinge and appeared dull and creamy on Mueller-Hinton blood agar under strictly aerobic conditions. 16S rRNA gene sequence analysis of the bacterium confirmed its *K. rhizophila* identity.

Lee et al.[32] reported the first two cases of *Kocuria marina* peritonitis in patients undergoing continuous ambulatory peritoneal dialysis. Furthermore, Lai et al.[33] isolated *Kocuria* bacteria in blood specimens from five immunocompromised patients. Of these patients, three had catheter-related bacteremia and one had infective endocarditis caused by *K. kristinae*, and one had a *K. marina* isolate.

10.1.4 Diagnosis

Micrococcus and *Kocuria* spp. are cultured on sheep blood agar, MacConkey agar, and chocolate agar at 35°C for 48 h. Gram-positive cocci from pale cream colonies after two days incubation appear in tetrads. The organisms are nonhemolytic, catalase positive, coagulase negative, and nonmotile, and can be characterized by using bioMérieux ID32 Staph ATB system, BD Phoenix PMC/ID-13 system (Becton Dickinson), and Vitek 2 (bioMérieux) as well as by their antibiotic profiles.[34]

Further confirmation is often achieved by using molecular techniques. Sequencing analysis of the partial or full 16S rRNA gene has proven valuable. Other molecular techniques for *Micrococcus* and *Kocuria* spp. include ribotyping (e.g., automated ribotyping with *Eco*RI and *Pvu*II as restriction enzymes based on the RiboPrinter® Microbial

Characterization System (DuPont-Qualicon, Wilmington, DE), denatured gradient gel electrophoresis (DGGE) analysis of PCR amplified 16S rDNA V3 region, the 16S-23S rRNA intergenic spacer region (ISR)-PCR, random amplified polymorphic DNA (RAPD), pulsed-field gel electrophoresis (PFGE), and *rpoB* gene sequencing.[35]

10.2 METHODS

10.2.1 Sample Preparation

BacT/Alert blood-culture bottles are inoculated with 8–10 mL blood and incubated for 7 days. Gram-positive isolates from primary cultures are grown on 5% blood agar and MacConkey agar for 48 h at 37°C under aerobic conditions.[36]

The blood bottles are kept at –20°C until the DNA extraction is performed. Bacterial DNA is extracted from the bottles by the alkali wash/lysis method. Extracted DNA is stored at –20°C prior to PCR.[36]

10.2.2 Detection Procedures

(i) Sequencing analysis of complete 16S rRNA gene Paster et al.[37] described the application of universal primers 27F (5′-AGAGTTTGATCCTGGCTCAG-3′, corresponding to positions 8–27 in *Escherichia coli* 16S rRNA gene) and 1492R (5′-TACCTTGTTACGACTT-3′, corresponding to positions 1492–1507) for amplification of nearly the entire length of the 16S rRNA gene from bacteria including *Kocuria* and *Micrococcus* spp.

Procedure

1. PCR mixture (100 μL) is made up of 50 mM KCl, 10 mM Tris pH 8.3, 1.5 mM $MgCl_2$, 2 mM each of dNTP, 1 μM (each) primers 37F and 1492R, 2.5 U of *Taq* DNA polymerase, and 1 pg to 100 ng of template DNA.
2. The mixture is subjected to an initial denaturation at 94°C for 5 min, 25 cycles of 94°C for 60 s, 50°C for 30 s, and 72°C for 2 min, and a final extension step at 72°C for 5 min.
3. The resulting PCR products are then sequenced by the use of ABI technology with the same universal primers. The nucleotide sequence is analyzed with BLAST program.

(ii) Sequencing analysis of partial 16S rRNA gene Karahan et al.[36] utilized primers fD1 (5′ AGAGTT TGATCCTGGCTCAG 3′) and 800R (5′ GAGTACC AGGGTATCTAATCC 3′) to amplify an 800 bp region of the gene encoding the 16S rRNA of eubacteria.

Procedure

1. PCR mixture (50 μL) is made up of 5 μL extracted bacterial DNA, 25 pmol of each primer, 200 μmol each dNTP, 1.5 mM $MgCl_2$ and 1.5 U *Taq* DNA polymerase (Fermentas) and appropriate 1 × PCR buffer.
2. PCR amplification is conducted with an initial denaturation at 94°C for 5 min, 35 cycles of 94°C for 1 min, 60/58/55°C for 1 min (2/2/31 cycles), and 72°C for 1 min, and a final extension at 72°C for 4 min.
3. A portion (10 μL) of the PCR products are subjected to 2% (w/v) agarose gel electrophoresis, stained with ethidium bromide, and visualized under UV.
4. The remaining PCR products (40 μL) are purified by the EZNA Cycle-Pure Kit (Omega Bio-tek), and subjected to automated sequencing with a CEQ2000XL sequencer (Beckman Coulter) by using the fD1 and 610R (5′-TACCGCGGCTGCTGGCAC-3′) primers.
5. The resulting nucleotide sequences are matched to existing GenBank sequences by nucleotide-nucleotide BLAST analysis.

Alternatively, the 16S and 23S rRNA intergenic region may be amplified and sequenced. Another useful target is the housekeeping gene *rpoB*, which harbors sufficient nucleotide differences for discrimination among members of *Micrococcus* and *Kocuria* spp.[35]

10.3 CONCLUSION

Members of the genera *Micrococcus* and *Kocuria* are gram-positive, catalase positive, coagulase negative, strictly aerobic cocci that are found ubiquitously in the environments and are present in the skin, mucosae, and oropharynx of humans and animals as commensal microflora. Several *Micrococcus* and *Kocuria* species are opportunistic human pathogens that have been associated with catheter-related bacteremia, pneumonia, intracranial abscesses, septic arthritis, meningitis, peritonitis, and endocarditis in human patients, particularly immunocompromised individuals. As *Micrococcus* and *Kocuria* species demonstrate phenotypic variability, they are difficult to identify with standard culture and biochemical tests,[38] and the true incidence of *Micrococcus* and *Kocuria* infections may have been underestimated. Recent development and application of PCR and DNA sequencing techniques have facilitated rapid, precise identification and determination of these bacteria. Consequently, an increasing number of clinical cases due to *Micrococcus* and *Kocuria* species have been reported. This highlights the value of molecular tests in the clinical diagnosis of hard-to-identify organisms, such as *Micrococcus* and *Kocuria* spp.

REFERENCES

1. Kloos, W.E., Tornabene, T.G., and Schleifer, K.H., Isolation and characterization of micrococci from human skin, including two new species: *Micrococcus lylae* and *Micrococcus kristinae*, *Int. J. Syst. Bacteriol.*, 24, 79, 1974.
2. Stackebrandt, E. et al., Taxonomic dissection of the genus *Micrococcus*: *Kocuria* gen. nov., *Nesterenkonia* gen. nov., *Kytococcus* gen. nov., *Dermacoccus* gen. nov., and *Micrococcus* Cohn 1872 gen. emend., *Int. J. Syst. Bacteriol.*, 45, 682, 1995.

3. Wieser, M. et al., Emended descriptions of the genus *Micrococcus*, *Micrococcus luteus* (Cohn 1872) and *Micrococcus lylae* (Kloos et al. 1974), *Int. J. Syst. Evol. Microbiol.*, 52, 629, 2002.
4. Liu, H. et al., Characterization of *Micrococcus antarcticus* sp. nov., a psychrophilic bacterium from Antarctica, *Int. J. Syst. Evol. Microbiol.*, 50, 715, 2000.
5. Liu, X.Y. et al., *Micrococcus flavus* sp. nov., isolated from activated sludge in a bioreactor, *Int. J. Syst. Evol. Microbiol.*, 57, 66, 2007.
6. Chen, H.H. et al., *Micrococcus endophyticus* sp. nov., isolated from surface-sterilized *Aquilaria sinensis* roots, *Int. J. Syst. Evol. Microbiol.*, 59, 1070, 2009.
7. Zhao, G.Z. et al., *Micrococcus yunnanensis* sp. nov., a novel actinobacterium isolated from surface-sterilized *Polyspora axillaris* roots, *Int. J. Syst. Evol. Microbiol.*, 59, 2383, 2009.
8. Kovacs, G. et al., *Kocuria palustris* sp. nov. and *Kocuria rhizophila* sp. nov., isolated from the rhizoplane of the narrow-leaved cattail (*Typha angustifolia*), *Int. J. Syst. Bacteriol.*, 49, 167, 1999.
9. Tang, J.S., and Gillevet, P.M., Reclassification of ATCC 9341 from *Micrococcus luteus* to *Kocuria rhizophila*, *Int. J. Syst. Evol. Microbiol.*, 53, 995, 2003.
10. Reddy, G.S. et al., *Kocuria polaris* sp. nov., an orange-pigmented psychrophilic bacterium isolated from an Antarctic cyanobacterial mat sample, *Int. J. Syst. Evol. Microbiol.*, 53, 183, 2003.
11. Kim, S.B. et al., *Kocuria marina* sp. nov., a novel actinobacterium isolated from marine sediment, *Int. J. Syst. Evol. Microbiol.*, 54, 1617, 2004.
12. Tvrzova, L. et al., Reclassification of strain CCM 132, previously classified as *Kocuria varians*, as *Kocuria carniphila* sp. nov., *Int. J. Syst. Evol. Microbiol.*, 55, 139, 2005.
13. Li, W.J. et al., *Kocuria aegyptia* sp. nov., a novel actinobacterium isolated from a saline, alkaline desert soil in Egypt, *Int. J. Syst. Evol. Microbiol.*, 56, 733, 2006.
14. Mayilraj, S. et al., *Kocuria himachalensis* sp. nov., an actinobacterium isolated from the Indian Himalayas, *Int. J. Syst. Evol. Microbiol.*, 56, 1971, 2006.
15. Zhou, G. et al., *Kocuria flava* sp. nov. and *Kocuria turfanensis* sp. nov., airborne actinobacteria isolated from Xinjiang, China, *Int. J. Syst. Evol. Microbiol.*, 58, 1304, 2008.
16. Park, E.J. et al., *Kocuria atrinae* sp. nov., isolated from traditional Korean fermented seafood, *Int. J. Syst. Evol. Microbiol.*, 60, 914, 2010.
17. Seo, Y.B. et al., *Kocuria gwangalliensis* sp. nov., an actinobacterium isolated from seawater, *Int. J. Syst. Evol. Microbiol.*, 59, 2769, 2009.
18. Yun, J.H. et al., *Kocuria salsicia* sp. nov., isolated from salt-fermented seafood, *Int. J. Syst. Evol. Microbiol.*, 2010 March 12.
19. Takarada, H. et al., Complete genome sequence of the soil Actinomycete *Kocuria rhizophila*, *J. Bacteriol.*, 190, 4139, 2008.
20. Souhami, L. et al., *Micrococcus luteus* pneumonia: A case report and review of the literature, *Med. Pediatr. Oncol.*, 7, 309, 1979.
21. Fosse, T. et al., Meningitis due to *Micrococcus luteus*, *Infection*, 13, 280, 1985.
22. Wharton, M. et al., Septic arthritis due to *Micrococcus luteus*, *J. Rheumatol.*, 13, 659, 1986.
23. Selladurai, B. et al., Intracranial suppuration caused by *Micrococcus luteus*, *Br. J. Neurosurg.*, 7, 205, 1993.
24. von Eiff, C. et al., *Micrococcus luteus* as a cause of recurrent bacteremia, *Pediatr. Infect. Dis. J.*, 15, 711, 1996.
25. Peces, R. et al., Relapsing bacteraemia due to *Micrococcus luteus* in a haemodialysis patient with a Perm-Cath cathether, *Nephrol. Dial. Transplant.*, 12, 2428, 1997.
26. Shanks, D. et al., Fatal *Micrococcus* sp. infection in a child with leukaemia—a cautionary case, *Med. Pediatr. Oncol.*, 37, 553, 2001.
27. Ramos, E.R. et al., The crucial role of catheters in micrococcal bloodstream infections in cancer patients, *Infect. Control Hosp. Epidemiol.*, 30, 83, 2009.
28. Basaglia, G. et al., Catheter-related bacteremia due to *Kocuria kristinae* in a patient with ovarian cancer, *J. Clin. Microbiol.*, 40, 311, 2002.
29. Altuntas, F. et al., Catheter-related bacteremia due to *Kocuria rosea* in a patient undergoing peripheral blood stem cell transplantation, *BMC Infect. Dis.*, 4, 62, 2004.
30. Ma, E.S. et al., *Kocuria kristinae* infection associated with acute cholecystitis, *BMC Infect. Dis.*, 5, 60, 2005.
31. Becker, K. et al., *Kocuria rhizophila* adds to the emerging spectrum of micrococcal species involved in human infections, *J. Clin. Microbiol.*, 46. 3537, 2008.
32. Lee, J.Y. et al., Two cases of peritonitis caused by *Kocuria marina* in patients undergoing continuous ambulatory peritoneal dialysis, *J. Clin. Microbiol.*, 47, 3376, 2009.
33. Lai, C.C. et al., Catheter-related bacteremia and infective endocarditis caused by *Kocuria* species, *Clin. Microbiol. Infect.*, 2010 [Epub ahead of print].
34. Szczerba, I., Susceptibility to antibiotics of bacteria from genera *Micrococcus*, *Kocuria*, *Nesternkonia*, *Kytococcus* and *Dermacoccus*, *Med. Dosw. Mikrobiol.*, 55, 75, 2003.
35. Hellmark, B., Söderquist, B., and Unemo, M., Simultaneous species identification and detection of rifampicin resistance in staphylococci by sequencing of the rpoB gene, *Eur. J. Clin. Microbiol. Infect. Dis.*, 28, 183, 2009.
36. Karahan, Z.C. et al., PCR evaluation of false-positive signals from two automated blood-culture systems, *J. Med. Microbiol.*, 55, 53, 2006.
37. Paster, B.J. et al., Prevalent bacterial species and novel phylotypes in advanced noma lesions, *J. Clin. Microbiol.*, 40, 2187, 2002.
38. Ben-Ami, R. et al., Erroneous reporting of coagulase-negative staphylococci as *Kocuria* spp. by the Vitek 2 system, *J. Clin. Microbiol.*, 43, 1448, 2005.

11 Mycobacterium

Shubhada Shenai and Camilla Rodrigues

CONTENTS

11.1 INTRODUCTION

The genus mycobacterium comprises more than 125 species, some of which are pathogenic or potentially pathogenic to human and animals and some of which are saprobes. A variety of diseases, including pulmonary disease, skin and soft tissue infections, lymphadenitis, and disseminated disease, are caused by mycobacteria. The most important pathogen of this genus is *M. tuberculosis* (MTB) complex, the causative agent of tuberculosis (TB).[1] Current global estimates indicate that about one-third of the world's population is infected with MTB; 8.8 million individuals develop the disease annually. In addition, there are increasing reports of nontuberculous mycobacteria (NTM) infections not only in the immunocompromised but also in immunocompetent hosts.[2,3] NTM may cause both asymptomatic infection and symptomatic disease in humans. They infect many sites within the body but primarily cause pulmonary disease, cervical lymphadenopathy, and localized skin and soft tissue lesions.

11.1.1 Classification, Morphology, and Epidemiology

Mycobacteria are believed to be among the oldest bacteria on earth, and are ubiquitous in the environment (Table 11.1). Mycobacteria are slender rods that sometimes show branching filamentous forms resembling fungal mycelium, hence the name "mycobacteria," meaning fungus like. (The Latin prefix "*myco*" means both *fungus* and *wax*; its use here relates to the "waxy" compounds in the cell wall.)

All the members of this genus are aerobic nonspore forming, noncapsulated, nonmotile, single cell bacteria and cell shape varies from that of coccobacilli to elongated rods, approximately 0.2–0.8 μm wide by 1–10 μm long. They are classified as gram-positive bacteria but most of them do not stain well with Gram-stain.[4] The unique character of mycobacteria is that they are acid fast. Acid-fastness is essentially due to the uniquely thick cell wall, which is composed of an interlacing layer of lipids, peptidoglycans, and arabinomannans.[4,5]

Many mycobacterium species adapt readily to growth on simple substrates, using ammonia or amino acids as nitrogen sources and glycerol as a carbon source in the presence of mineral salts. Optimum growth temperatures vary widely according to the species and range from 23°C to over 50°C. Some species can be difficult to culture and are fastidious in their growth requirements, sometimes taking over two years to cultivate in culture.[6] Furthermore, some species also have extremely long reproductive cycles—*M. leprae*, may take more than 20 days for one division cycle.[1]

In 1954, Timpe and Runyon classified all mycobacteria into four groups on the basis of growth rate, pigmentation, colonial morphology, and biochemical test characterization. Organisms that took less than 7 days to grow isolated colonies on solid media were called rapidly growing mycobacteria (RGM) and those that took more than 7 days were called as slowly growing mycobacteria (SGM).

MTB is an obligate pathogen of humans and is rarely identified in other mammals. It is transmitted from person to person, and it has no significant environmental reservoirs. A number of clinically important NTM are found in soil, water,

TABLE 11.1
Scientific Classification

Kingdom	Bacteria
Phylum	Actinobacteria
Order	Actinomycetales
Suborder	Corynebacterineae
Family	Mycobacteriaceae
Genus	*Mycobacterium*
Clinically relevant species	*M. tuberculosis* complex, *M. leprae*, *M. avium*, *M. intracellulare*, *M. kansasii*, *M. marinum*, *M. ulcerans*, *M. simiae*, *M. gordonae*, *M. genavense*, *M. szulgai*, *M. scrofulaceum*, *M. haemophilum*, *M. fortuitum* complex, *M. chelonae*, *M. abscessus*, etc.

domestic and wild animals, milk and other food materials. They may contaminate clinical specimens from the environment or transiently colonize body surfaces and sections. Most infections with NTM appear to be acquired by aspiration or inoculation of the organisms from the natural reservoir. There is no firm evidence of human-to-human or animal-to-human transmission of NTM disease.[7] The geographic distribution of some species does not appear to be uniform. Depending on the locality, infections caused by other mycobacteria have been reported to account for 0.5%–30% of all mycobacterial infections.[8–10]

11.1.2 Clinical Features and Pathogenesis

There are five main types of disease caused by mycobacteria: TB, pulmonary infections resembling TB, localized skin and soft tissue infections, localized lymph node involvement, and disseminated disease.

TB. The clinical manifestation and eventual outcome of TB depends on the virulence of the pathogen and the nature of the host's immune response. When a person inhales air that contains droplets, most of the larger droplets are lodged in the upper respiratory tract (i.e., nose and throat), where infection is unlikely to develop. However, the droplet nuclei may reach the small air sacs of the lung (the alveoli) where infection begins. When organisms reach the lungs, there are four potential outcomes. (i) The initial host response can be completely effective and kill all the organisms; in this case the patient has no chance of developing TB. (ii) The organism can begin to multiply and grow immediately after the infection, causing clinical disease known as primary tuberculosis. (iii) The bacilli may remain dormant and do not cause disease. This is referred to as latent infection indicated only by a positive tuberculin skin test. (iv) There could be reactivation of the latent organisms causing post-primary infection.[11]

Some bacilli may be killed immediately by the process of phagocytosis, which involves binding of the bacterium to the host cell, internalization, and finally growth inhibition or killing.[11] If the bacilli are able to survive the initial defenses, they multiply within the alveolar macrophages and a small number of bacilli enter the bloodstream and disseminate. Intracellular multiplication of the bacillus is interrupted only with the development of specific cellular immunity, which sets in about 6–8 weeks after infection. Once the immune response called cell-mediated hypersensitivity is established, the T cells usually respond to the tuberculin skin test (PPD test) and produce a characteristic red welt. Some of the T cell signals produce inflammatory reactions; other signals recruit and activate specialized cells to kill bacilli and wall off infected macrophages in tiny, hard capsules known as tubercles.

A tubercle consists of a few small lymphocytes and a compact collection of activated macrophages, which sometimes differentiate into epitheloid cells or multinucleated giant cells. The massive activation of macrophages that occurs within tubercles often results in the concentrated release of lytic enzymes.[4] These enzymes destroy nearby healthy cells, resulting in a circular region of necrotic tissue, which eventually form a lesion with a caseous consistency. As these caseous lesions heal, they calcify and are visible on X-rays, where they are called Ghon complexes.

For majority of individuals with the normal immune function, the initial infection with *M. tuberculosis* appears to be arrested once cell-mediated immunity develops, although small number of viable bacilli within the granuloma may persist. If the cell-mediated immunity is unable to control the infection, the individual progresses to active disease called progressive primary tuberculosis.[12] Individuals infected with MTB have a 10% lifetime risk of developing the disease. The remaining 90% will stay infected, but be free of disease for the rest of their lives.[13]

Pulmonary Infections Resembling TB. Chronic pulmonary disease is the most common localized clinical manifestation of NTM. MTB complex is the most frequent pathogen causing lung disease, followed by *M. avium* complex (MAC), *M. kansasii*, *M. abscessus*, *M. fortuitum*, *M. szulgai*, *M. simiae*, *M. xenopy*, *M. malmoense*, *M. celatum*, *M. asiaticum*, and *M. shimodii*. Respiratory infections due to NTM are often associated with various conditions such as chronic obstructive pulmonary disease, cystic fibrosis of lung, bronchiectasis, emphysema of lung, previously treated pulmonary TB, and lung cancer. Other predisposing conditions include diabetes mellitus, leukemia, systemic lupus erythematosus, chronic alcoholism, AIDS, and transplant patients.[14,15] The interpretation of NTM in the sputum of HIV-positive patients presents a particular problem, as these patients are frequently infected with NTM without evidence of pulmonary disease.[15,16]

Localized Skin and Soft Tissue Infections. Skin and soft tissue infections in immunocompetent patients are the result of direct exposure to contaminated water sources through inoculation after accidental trauma, surgery, and injection. Watery skin lesions may be caused due to the inoculation of opportunistic mycobacteria into superficial abrasions. Such infections are usually caused by *M. marinum* and occur in users of swimming pools and keepers of tropical fish, hence the names "swimming pool granuloma" and "fish fancier's finger."[17–19] Other species such as *M. kansasii*, *M. abscessus*, and *M. chelonae* cause similar lesions. Infections due to *M. ulcerans* lead to necrosis of subdermal tissue and secondary skin ulceration.

Post injection mycobacterial abscesses are usually due to the rapidly growing species *M. fortuitum*, *M. abscessus*, and *M. chelonae*. These may occur sporadically in diabetics,[20] for example, or in "epidemics" when several individuals are injected with a contaminated batch of a vaccine or drug.[21] Similar infections, caused by rapid growers, results in penetrating injuries surgery. Serious outbreaks of sternal infections have followed open heart surgery.[22] A number of cases of corneal infection by rapid growing species have also occurred, presumably as a result of direct implantation.[23] *M. haemophilium* is a rare cause of nodular or ulcerative skin lesions and all reported infections have occurred in immunosuppressed individuals, particularly recipients of renal transplants.[17,24]

Localized Lymph Node Involvement. Lymphadenopathy due to NTM is usually cervical, unilateral, and self-limiting. Most cases occur in children under the age of 5 years.[25] The most common species found in lymphadenitis in children are *M. avium*, *M. intracellulare*, *M. scrofulaceum*, *M. malmoense*, *M. kansasii*, *M. haemophilum*, and *M interjectum*.[25] In adults, the infection may occur as part of a disseminated infection.

Disseminated Disease. This form of nonpulmonary disease is similar to that caused by MTB complex. Thus, single or multiple lesions may occur in bone, urinary tract, central nervous system, lymph nodes and skin. In other cases, the infection resembles cryptic disseminated TB with multi-organ involvement. In such cases there is little or no histological evidence of an immune response: indeed, the bone marrow may be teeming with acid-fast bacilli while appearing normal when stained with the usual hematological stains.[25] Such disseminated disease usually occurs in association with congenital or acquired immunodeficiency affecting cell-mediated immunity, including therapeutic immunosuppresion following transplant surgery[26] and AIDS.[16] Most cases of disseminated disease are due to MAC or *M. chelonae*.

11.1.3 Diagnosis

Presumptive diagnosis of mycobacterial infection is generally made on the basis of patient's history, clinical presentation of disease, radiological findings, and presence of acid-fast bacilli (AFB) in the patient's specimen. The most commonly used methods for diagnosis of mycobacterial diseases are listed in Table 11.2. Clinical symptoms and radiological findings, though useful for preliminary diagnosis, may not always provide absolute prediction of the disease. Serological tests for TB are often inconclusive. Microscopy is rapid, easy to perform, and specific; however, it lacks sensitivity, requiring >10,000 bacilli/mL of sputum to become positive. Further, microscopy cannot distinguish between viable and nonviable cells, between different species of mycobacteria and drug-susceptible or drug-resistant *M. tuberculosis* complex.[27]

Accurate, rapid microbiological diagnosis of TB and other mycobacterial infections begin with proper specimen collection and the rapid transport of the specimen to the laboratory. Clearly labeled specimen must be transported to the laboratory quickly because some tests, such as AFB smears and nucleic acid amplification (NAA) procedures, can reliably be reported within 24–48 h of receipt. If processing is delayed, the specimens should be refrigerated properly. As NTMs are ubiquitous in nature and may sometimes be isolated as laboratory contaminants, the isolation of these organisms from specimens should meet certain criteria and guidelines to confirm their etiological significance.[14]

11.1.3.1 Phenotypic Techniques

Culture. Isolation of mycobacteria by culture is sensitive and regarded as the gold standard. It has been estimated to detect 10–100 viable mycobacteria per ml of sample.[28] Another advantage of culture is that it allows specific species identification and testing for recognition of drug susceptibility patterns. With the emergence of drug-resistant tuberculosis, accurate, rapid, and cost-effective tests to detect MDRTB and XDRTB presents a challenge to TB control activities.

Most laboratories in disease endemic region rely on conventional Lowenstein & Jensen (L.J) media for culture followed by use of different biochemical tests for identification of mycobacteria. These conventional culture and identification

TABLE 11.2
Most Commonly Used Methods for Diagnosis of Mycobacteria

Method	Advantages	Disadvantages
Clinical signs and symptoms	Rapid	Not specific, not conclusive, not always present
Radiology (X-ray, CT, MRI)	Readily available	Nonspecific and inconclusive
Serology	Rapid, cost effective	Not sensitive and nonspecific
Microscopy	Rapid, cost effective	Limited sensitivity and specificity
Conventional culture (egg/agar based)	Specific, cost effective	Time consuming (4–10 weeks)
Automated liquid culture	Specific	Slow (2–3 weeks), expensive
Commercial hybridization probe assay	Rapid	Expensive, demanding technical expertise and not available for all species
PCR (in-house/kit based)	Rapid	Expensive, demanding technical expertise and quality control at each step
Sequencing	Rapid, gold standard	Expensive

tests are complex, laborious, time consuming, and many times fail to provide precise identification. Newer phenotypic techniques aim for a more rapid detection of growth by using metabolic activities of growing bacteria. In contrast, automated liquid culture-based techniques can detect growth dependent changes such as CO_2 production [BACTEC 460 (Becton Dickinson, Sparks, MD), MB/BacT (bioMerieux)] or O_2 consumption [Mycobacterium Growth Indicator Tube (MGIT (Becton Dickinson, Sparks, MD)) 960, Versa TREK (Trek Diagnostics)] and help in rapid diagnosis of disease. Liquid culture systems, when combined with DNA probes or the *Capilia TB* assay (TAUNS, Numazu, Japan) for rapid species identification, are capable of producing positive results in 2 weeks for the vast majority of sputum smear-positive specimens, and within 3 weeks for smear-negative specimens.[11]

Mycobacteriophages. The technology uses bacteriophages to infect live MTB and detect the bacilli using either the phage-amplification (FASTPlaque TB™) method[29–31] or the detection of light (Luciferase reporter phage technique).[32] Amplification-based assays are commercially available and have specificity of 98.7%–99% and sensitivity of 70%–75% compared to conventional L.J method.[29–31] In general, phage assays have a turn around time of 2–3 days and require a laboratory infrastructure similar to that required for performing mycobacterial cultures.

High Performance Liquid Chromatography (HPLC). HPLC is based on the observation that each mycobacterium species synthesizes a unique set of mycolic acids, β-hydroxy-α-fatty acids, that are components of the cell wall.[33,34] HPLC can produce a pattern that reliably identifies and distinguishes more than 50 Mycobacterium species. It can be performed in a few hours from pure cultures. The initial equipment cost for HPLC limits its availability.

Due to various limitations of phenotypic methods and HPLC in recent years, molecular techniques have become increasingly popular, as they are rapid, reliable, highly sensitive, and can be used directly for large number of clinical specimens.

11.1.3.2 Molecular Biological Techniques

Molecular methods have several advantages over culture-based techniques and include a shorter turn around time, no requirement for culture of an organism, the possibility of direct application in clinical specimens, less biohazard risk, and feasibility for automation. Nucleic acid amplification tests (NAA), also called direct amplification tests, are available either as commercial kits or also widely used with inhouse assays.

COBAS Amplicor. The Amplicor MTB test (Roche Molecular Systems, Basel, Switzerland) relies on the standard polymerase chain reaction (PCR). A 584-bp fragment of the 16S ribosomal RNA gene, comprising a species-specific region flanked by genus-specific sequences, is amplified using biotinylated primers. Then the use of magnetic beads coated with specific probes allows the removal of any other DNA by washing. The detection of the specific amplification product is performed by adding an avidin-enzyme conjugate and a chromogenic subject. The system uses uracil-*N*-glycosylase in the master mix for contamination control. The enzyme uracil-*N*-glycosylase, added to the samples before amplification, destroys any amplicons resulting from previous amplifications without damaging the uracil free target DNA. From the literature review, specificity of this USFDA-approved test is close to 100% and sensitivity ranges from 90% to 100% in smear-positive samples and from 50% to 95.9% in smear-negative ones.[35,36]

BD Probe Tec ET. The BD Probe Tec ET system (Becton Dickinson, Sparks, MD) uses DNA polymerase and isothermal strand displacement amplification to produce multiple copies of IS6110, an insertion element to MTB complex. The turn around time is 3.5–4 h. The literature reports a rate of sensitivity ranging from 98.5% to 100% for smear-positive samples and variable (0.33%–100%) for smear-negative ones.[37,38]

Inhouse PCR assays are used predominantly in developing countries, where commercial NAA are expensive, and they vary widely in their target gene sequences, protocols, and laboratory methods. The accuracy of inhouse PCR test has been more heterogeneous than commercial kits.[39] Inhouse tests are laboratory PCR is highly sensitive (96%) and specific (97%–100%). However, its relationship with the clinical disease needs to be ascertained.[40,41]

Molecular identification of mycobacteria targeting different genes. like IS6110, 16S rRNA, 65 kiloDalton (kDa) heat-shock protein gene (*hsp* 65), 32-kDa protein genes, the gene encoding 126-kDa fusion protein, RD region, and *rpoβ* gene have been reported.[42–46] The most commonly used mycobacterial DNA sequences for PCR are IS6110 (IS986), 16S rRNA, and the gene encoding 65 kDa. An insertion sequence (IS) designated IS6110 is specific for *M. tuberculosis* complex and generally occurs in 1–20 copies per cell, making it an ideal target for amplification in inhouse PCR protocols. PCR detection of IS6110 in pulmonary and extrapulmonary TB, when compared to culture, has a sensitivity, specificity, and positive predictability of 83.5%, 99%, and 94.2% respectively.[46]

PCR Restriction Enzyme Analysis (PCR RE). The PCR RE method is based on the amplification of a fragment of the target gene by PCR followed by the digestion of the amplified product with different restriction enzymes (RE).[47–51] The products of the digestion reaction are then separated and visualized on agarose gel electrophoresis. In several recent studies PCR RE has performed well, as compared to the conventional methods.[47–51] This simple and cost-effective method has been applied to several genes, such as 16S ribosomal RNA (rRNA), *rpoβ*, hsp65, gyrB, and dnaJ, for the rapid differentiation of closely related clinical isolates such as *M. avium subspecies avium* and *Mycobacterium avium subspecies paratuberculosis*, and subspecies of *M. kansasii*.[47–51] In 2000, Lee et al. amplified 360-bp fragment within *rpoβ* gene and digested with restriction enzymes *Msp I*, *Hae III*, *Sau 3 AI*, and *Hinc II* for differentiation of most of the mycobacteria to the species and subspecies level.[49] Besides differentiating among,

M. abscessus-M. chelonae-M. fortuitum, *M. avium-M. paratuberculosis-M. intracellulare*, *M. marinum-M. ulcerans*, and *M. celatum* types 1 and 2, *rpoβ* PCR–RE allowed the differentiation of six subspecies of *M. kansasii*.

PCR-Based Sequencing. PCR-based sequencing has become the gold standard for identification of mycobacterial species. The method consists of PCR amplification of mycobacterial DNA with genus-specific primers and sequencing of the amplicons. The organism is identified by comparison of the nucleotide sequence with reference sequences. Usually, only one sequencing reaction is needed for a definite identification. This method allows for direct detection of mycobacterial species that cannot be grown on conventional laboratory culture media, and several previously unrecognized species have been identified.[37,52,53]

The target most commonly used is the gene coding for the 16S rRNA. This gene is present in all bacterial species and contains both conserved and variable regions, making it an ideal target for taxonomic purposes. Sequencing of two hyper variable regions of the 16S rRNA gene allows for identification of the majority of mycobacterial species. However, the members of the MTB complex cannot be distinguished. For example, *M. kansasii* has a sequence identical to that of nonpathogenic species *M. gastri*.[37] Several other target genes viz. The 32-kDa protein,[45] the 65-kDa heat-shock protein,[54] and the 16S-23S rRNA internal transcribed spacer[43] have been well characterized for species identification by sequencing. In 1999, Kim et al. reported that comparative sequence analysis of the RNA polymerase gene encompassing the region related to RIF resistance in MTB, is useful for the differentiation of mycobacterial species.[52]

Loop-Mediated Isothermal Amplification (LAMP). LAMP is a novel, rapid, and simplified NA platform developed by Eiken Chemical Co., Ltd. (Tokyo, Japan).[55] Its major advantages are speed, simplicity, and lack of requirement for a thermal cycler, thus facilitating high throughput. The technology uses four different primers specifically designed to recognize six distinct regions on the target gene, and the reaction process proceeds at a constant temperature (isothermal) using strand-displacement reaction. Amplification and detection of the gene products can be completed in a single step, by incubating the mixture of samples, primers, DNA polymerase with strand-displacement activity, and substrates at a constant temperature. Currently, there is limited evidence on the accuracy of LAMP for the detection of TB. Studies are underway to define the test performance and compare its performance against conventional test.

Real-Time PCR. Real-time PCR has been successfully applied directly to clinical specimens for rapid detection of mycobacteria and drug-resistance testing. Different probes have been used, like the TaqMan probe, fluorescence resonance energy transfer (FRET) probes, molecular beacons, and bioprobes. The main advantages of real-time PCR techniques are the speed of the test and a lower risk of contamination. Major disadvantages are the requirement for expensive equipment and reagents, as well as the need for skilled technical personnel. Results could be obtained in an average of 1.5–2 h after DNA extraction. Real-time PCR could eventually be implemented in reference laboratories with the required capacity to properly set up the technique and in settings where it can contribute to the management of TB patients.

DNA Probe Hybridization. The DNA probe technology for identification of fastidious organisms is still one of the most successful molecular diagnostic procedures worldwide. Accuprobe (Gen-Probe, San Diego, CA), is a commercial method based on species-specific DNA probes that hybridize to rRNA for the identification of limited number of important mycobacteria, including MTB complex, *M. avium*, *M. intracellulare*, MAC, *M. kansasii*, and *M. gordonae*. The probes have been extensively evaluated in a clinical setting and have shown good sensitivity and specificity with results in 2 h. Overall, DNA probes show >99% sensitivity and specificity for MTB complex isolates and approximately 95% sensitivity and >99% specificity for MAC isolates.[56–61] Although rare, some instances of cross-reaction between MTB complex probe and *M. terrae* and *M. celatum* isolates and *M. avium* complex probe with recently described species *M. palustre*, *M. parascrofulaceum*, and *M. saskatchewanense* have been reported.[62,63]

Line Probe Assays. The line probe assay uses the reverse line blot hybridization technology with differently specific DNA probes immobilized in parallel lines on a paper strip or a nylon membrane. The target DNA is PCR amplified using biotinylated primers and finally incubated with a strip or membrane coated with species-specific probes. Finally hybrids are detected through a colorimetric or enhanced chemiluminiscence (ECL) reaction.[36,37,64] Three commercial kits for species identification are currently available in the market for detection of mycobacteria and drug-resistance testing: INNO-LiPA, MYCOBACTERIA (Innogenetics NV, Gent, Belgium)[65,66]; Geno Type Mycobacterium (Hain, Germany)[36,37]; and Geno Type MTBC (Hain, Germany).

Many studies conducted on application of LiPA assay on MTB isolates for detection of RIF resistance showed greater than 95% sensitivity and 100% specificity. Very few studies that applied LiPA directly to clinical specimens had sensitivity ranged from 80%–100%.[65,66] The GenoType MDBDR, on the other hand, detects resistance to INH and RIF in culture samples based on the most common mutations in the *katG* and *rpoβ* genes, respectively.

DNA Arrays. More recently, DNA microarrays or high-density oligonucleotide arrays (DNA chip) have been applied for species identification of mycobacteria. DNA chip generally comprises a glass surface on which multiple DNA probes with known identities are fixed for molecular hybridization with specific PCR amplified products, which allows the examination of parallel gene expression or genotyping.[67] This method allows the simultaneous analysis of thousands of genes in a short assay time and as such is useful for phylogenetic analysis and species identification. For the identification of bacteria, this method may involve the labeling of in vitro RNA transcribed from a target gene from bacteria in specimens, subsequent hybridization of the labeled in vitro transcribed RNA to species-specific oligonucleotide probes on a microarray, and detection of the label, usually by fluorescence. The Affymetrix Genechip, which uses large sets of oligonuleotides that are synthesized rather than spotted onto a glass substrate, has been successfully applied using 16S rRNA genes as a target for the identification of mycobacterium species isolates.[68,69]

The recently described CombiChip Mycobacteria Drug-Resistance detection DNA chip is an oligonucleotide microchip coupled with PCR for the detection of resistance to INH and RIF. It was compared with sequencing and drug susceptibility testing in 69 INH and/or RIF resistant and 27 drug-susceptible MTB isolates.[70] It allowed identification of 84% of INH resistant isolates based on *katG* codon 315 and *inhA*-15 mutations, and 100% of RIF-resistant isolates based on seven codons on *rpoβ* gene viz. 511, 513, 516, 522, 526, 531, and 533. The overall specificity was 100% and 95.3% for detecting INH and RIF resistance, respectively. Due to the high cost involved, for the time being the use of microarrays for detecting drug resistance in TB is still beyond the reach of most clinical mycobacteriology laboratories.

GeneXpert. Xpert(R) MTB/RIF is a test that can simultaneously identify MTB and resistance to rifampicin (RIF). It was developed through a partnership between Cepheid and the Foundation for Innovative New Diagnostics (FIND), the University of Medicine and Dentistry of New Jersey. The system works on the principle of real-time nested PCR and the detection of amplicons by using molecular beacons. It can deliver an accurate diagnosis of the disease in less than 2 h. This system is basically designed for clinicians so that proper treatment can be initiated rapidly.

11.2 METHODS

11.2.1 Sample Preparation

DNA Extraction. The ability of molecular assays to detect mycobacteria in clinical specimens is dependent on both the target sequence selected and the efficiency of the DNA extraction procedure. The most commonly used methods for DNA extraction are Tris-EDTA (TE) boil extraction, lysis extraction (Infectio Diagnostics, Inc. Quebec, Canada), QIAGEN QIAmp DNA mini kit (QIAGEN, Inc., Valencia, CA), sodium dodecyl sulfate (SDS)-Triton X extraction, and SDS-Triton X plus sonication.

11.2.2 Detection Procedures

(i) Standard PCR

Principle. In PCR, DNA is amplified using two primers and one enzyme (Figure 11.1). The repeated thermal cycling of PCR begins with denaturation of the double stranded DNA amplicon followed by primer annealing to target DNA and extension of PCR products by Taq polymerase in presence of excess of deoxynucleoside triphosphates (DNTPs). Each cycle theoretically doubles the number of target nucleic acid molecules; therefore, after 20 cycles, million-fold amplification is achieved, and after 30 cycles, a billion-fold amplification is observed.[71]

Procedure. A basic PCR set up requires several components and reagents viz. DNA template, two primers complementary to the 3′ends of each of the sense and antisense strand of the DNA target, Taq polymerase, deoxynucleoside triphosphates (DNTPs), Buffer solution, divalent cations, magnesium or manganese ions, and monovalent cation potassium ions. The PCR is commonly carried out in a reaction volume of 10–200 μL in small reaction tubes (0.2–0.5 mL volumes) in a thermal cycler. The PCR usually consists of a

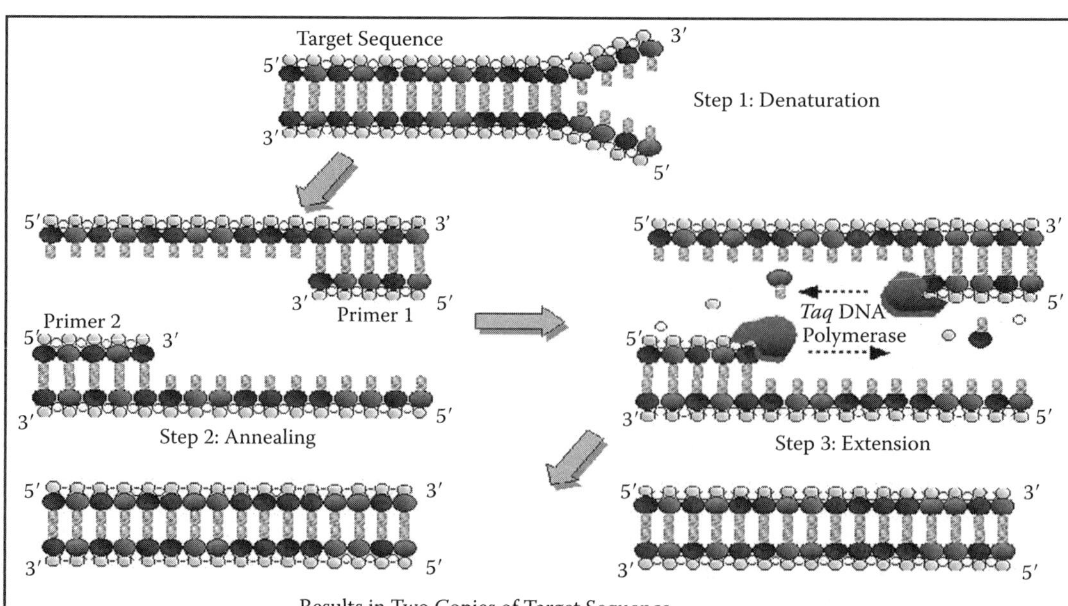

FIGURE 11.1 PCR—Amplification of DNA.

series of 20–40 cycles. Detection of amplified PCR products can be done by gel electrophoresis. The size of PCR product is determined by comparison with a DNA ladder (molecular weight marker), which contains DNA fragments of known size, run on the gel alongside the PCR products.

Note: Incorporation of a negative control, positive control, and internal control is essential with each batch of PCR.

(ii) Real-Time PCR

Principle. Real-time PCR methods employ fluorescent dyes or fluorophore-containing DNA probes to measure the amount of amplified product. TaqMan probe consists of two types of fluorophores, which are the fluorescent parts of reporter proteins. The probe is attached or unattached to the template DNA, and before the polymerase acts, the quencher fluorophore (usually a long-wavelength colored dye, such as red) reduces the fluorescence from the reporter fluorophore (usually a short-wavelength colored dye, such as green). It does this by the use of fluorescence resonance energy transfer, which is the inhibition of one dye caused by another without emission of a proton. The reporter dye is found on the 5′ end of the probe and the quencher at the 3′end. There are two other types of real-time PCR methods, the molecular beacon method and the SYBR® Green method. The molecular beacon method utilizes a reporter probe that is wrapped around into a hairpin. It also has a quencher dye that must be in close contact to the reporter to work. An important difference of the molecular beacon method in comparison to the TaqMan® method is that the probe remains intact throughout the PCR product, and is rebound to the target at every cycle.

Procedure. Real-time PCR is an exceptionally powerful research tool. With the correct kit reagent and experimental design, it enables rapid generation of high quality meaningful data. A suitable pair of primers and probes can be designed based on empirical design guidelines and using primer design softwares. General quantitative real-time PCR protocol is summarized in Figure 11.2.

Note: Multiple housekeeping genes should be validated for normalization in each experiment and incorporation of an external standard (alien RNA) is also needed to identify possible PCR inhibitors between treatments and runs.

(iii) Genotyping

Principle. Genotyping refers to the process of determining the genotype of a MTB complex genome by the use of biological assays. Current methods include PCR DNA sequencing, Allele Specific Oligonucleotide (ASO) probes and hybridization to DNA microarrays or beads. Mycobacterial genotyping technique is used to compare DNA from different sources or to identify a genome by comparison with a known standard.

Procedure. The genotyping laboratories can use three genotyping methods: spoligotyping, Mycobacterium Interspersed Repetitive Units (MIRU) analysis, and *IS6110*-based Restriction Fragment Length Polymorphism (RFLP) analysis, also known as fingerprinting. Spoligotyping and MIRU analysis are based on PCR. Together, these two methods are referred to as the PCR genotyping tests.

RFLP. RFLP analysis was performed by the internationally standardized procedure.[72] In brief, MTB genomic DNA is extracted and digested with PvuII. The resulting DNA fingerprints are separated by gel electrophoresis, transferred to nylon membrane by the Southern blotting and probed with a horseradish peroxidase-labeled 245 bp sequence of IS6110 DNA probe. Hybridization of probe to a piece of minisatellite

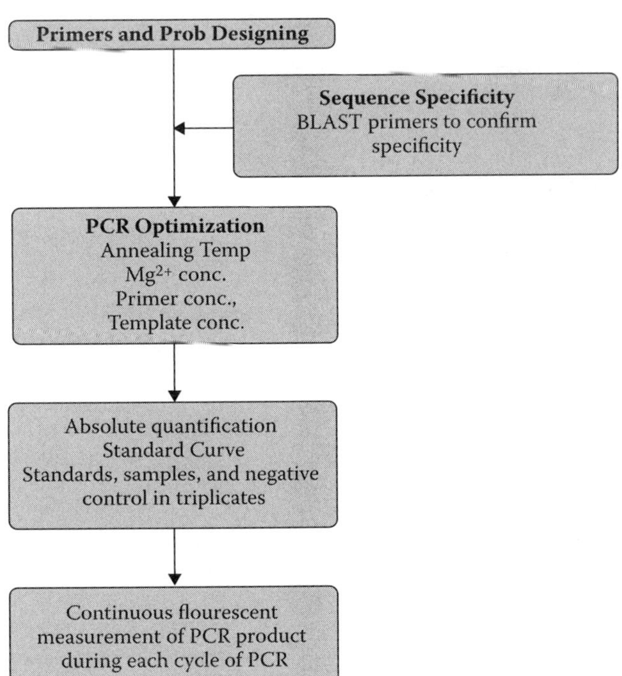

FIGURE 11.2 General Protocol for Standardization of Quantitative Real-Time PCR.

DNA results in a dark band on an autoradiograph of the nitrocellulose paper. The resulting patterns are unique to each individual.

Spoligotyping. Spoligotyping is generally performed with a commercially available kit (Isogen Bioscience). Briefly, the procedure involves amplification of the direct repeat (DR) region of *M. tuberculosis* genome by PCR using specific biotinylated primers.[73] The amplified products are then hybridized onto a nylon membrane blotted with 43 spacer sequences. The spacer sequences are specific to the DR region and the strains differed in presence or absence of these spacer. The results of hybridization are detected by ECL detection system.[72,74]

MIRU-VNTR. Mycobacterial interspersed repetitive unit-variable number tandem repeat (MIRU) typing is a MTB complex specific name of a multiple locus variable number of tandem repeats (VNTR) analysis. This method relies on PCR amplification using primers specific for the flanking regions of the VNTRs, and on determination of the sizes of the amplicons, which reflect the numbers of the amplified VNTR copies. Their use provides typing data in a portable numerical format, adequate for studying the global molecular epidemiology of infectious agents.[75,76]

Note: Genotyping of MTB isolates is useful in TB control for confirming the dynamics of TB outbreaks, laboratory cross-contamination, and identified MTB strains, transmission sites, and accurate epidemiologic links.

11.3 CONCLUSION AND FUTURE PERSPECTIVES

It is important to realize that there is no "stand alone" assay for the identification of mycobacteria especially NTM. Many new species may not be recognized in all assays. This plethora of species poses an additional challenge for the clinical mycobacteriology laboratory to provide timely services. Furthermore, some of the species (e.g., *M. haemophilum*, *M. genavense*) require special growth conditions, necessitating an exquisite collaboration between the clinician requesting the test and the laboratory professional performing the test.

To overcome these shortcomings of conventional methods, in recent years molecular techniques have become increasingly popular, as they are rapid, reliable, highly sensitive, and specific and can be used for large number of samples. Accuprobe (Gen-Probe, San Diego, Calif.) and PCR-based assays are more widely used in developed countries for species identification of mycobacteria than phenotypic tests and have significantly improved turnaround time and commonly used. However, results of these tests should be correlated with the clinical presentation and the clinician should notify the laboratory when results are not consistent. The DNA microarray method holds promise for the future because it is easy to perform, can be readily automated, and allows for identification of a large number of mycobacterial species in one reaction.

However, due to the requirement of additional equipment and trained personnel, these methods are expensive and have not yet gained easy access to the routine procedures performed in the clinical mycobacteriology laboratory, especially in low-income TB endemic countries. In these countries with limited resources, the diagnosis of mycobacterial infections is established empirically on clinico-radiological basis or only by sputum AFB smear examination. Due to lack of facilities most laboratories in disease endemic countries do not identify the mycobacteria by culture and infections caused by NTM are underdiagnosed.

Today, with techniques and equipment becoming increasingly complex and expensive, it is difficult to upgrade local laboratories to perform these assays. As specimen delivery and communication of results can be rapidly and easily achieved, utilization of reference laboratories for rarely performed sophisticated molecular tests would be a more practical approach.

REFERENCES

1. Soini, H., and Musser, J.M., Molecular diagnosis of mycobacteria, *Clin. Chem.*, 47, 809, 2001.
2. Rosenzweig, D.Y., Nontuberculous mycobacterial disease in the immunocompetent adult, *Semin. Respir. Infect.*, 11, 252, 1996.
3. Saubolle, M.A. et al., *Mycobacterium haemophilium:* Microbiology and expanding clinical and geographical spectra of disease in humans, *Clin. Microbiol. Rev.*, 9, 435, 1996.
4. Ananthanarayan, R., and Pannikar, C.K.J., *Mycobacterium tuberculosis*. In Ananthanarayan, R., and Pannikar, C.K.J. (Editors), *Textbook of Microbiology*, pp. 324–337, Orient Longman, Madras, 1996.
5. Jenkins, P.A., The microbiology of tuberculosis. In Davis, P.D.O. (Editor), *Clinical Tuberculosis*, p. 33, Chapman & Hall, London, 1994.
6. Dodge, O.G., Mycobacterial skin ulcers in Uganda: Histopathological aspects, *J. Pathol. Bacteriol.*, 88, 169, 1964.
7. Wallace, R.J., Jr., O'Brein, R., Glassroth, J., Raleigh, J., and Dutta, A., Diagnosis and treatment of disease caused by non tuberculous mycobacteria, *Am. Rev. Respir. Dis.*, 142, 940, 1990.
8. O'Brian, R.J., Geiter, L.J., and Snider, D.E., The epidemiology of non tuberculous mycobacteria disease in the United States: Results from national survey, *Am. Rev. Respir. Dis.*, 135, 1007, 1987.
9. Matoss, E.D. et al., Nontuberculous mycobacteria at a multiresistant tuberculosis reference centre in Bahia: Clinico-epidemiological aspects, *Braz. J. Infect. Dis.*, 8, 296, 2004.
10. Choudhari, S. et al., Clinical significance of nontuberculous mycobacteria isolates in a Canadian tertiary care center, *Clin. Infect. Dis.*, 21, 128, 1995.
11. Schluger, N.W., and Rom, W.N., The host immune response to tuberculosis, *Am. J. Resp. Crit. Care Med.*, 157, 679, 1998.
12. Danneenberg, A.M.J., and Tomashifski, J.F., Jr., Pathogenisis of pulmonary tuberculosis. In Fishmann, A.P. (Editor), *Pulmonary Diseases and Disorders*, 3rd ed., p. 2447, McGraw Hill, New York, 1998.
13. Bloom, B.R., and Murray, C.J.L., Tuberculosis: Commentary on a reemergent killer, *Science*, 257, 1055, 1992.

14. Katoch, V.N., Infections due to non tuberculous mycobacteria (NTM), *Indian J. Med. Res.*, 120, 290, 2004.
15. Graybill, J.R., et al., Disseminated mycobacteriosis due to *Mycobacterium abscessus* in two recipients of renal homograft, *Am. Rev. Respir. Dis.*, 109, 4, 1974.
16. Iseman, M.D. et al., Disease due to *Mycobacterium avium-intracellulare*, *Chest*, 87, 139S, 1985.
17. ATS guidelines 2007, *David E. Griffith, Timothy Aksamit, Barbara A. Brown-Elliott.* An Official ATS/IDSA statement: Diagnosis, treatment, and prevention of nontuberculous mycobacterial diseases, *Am. J. Respir. Crit. Care Med.*, 175, 367, 2007.
18. Collins, C.H. et al., *Mycobacterium marinum* infectious in man, *J. Hyg. (Lond).*, 94, 135, 1985.
19. Grange, J.M., Infections caused by opportunist mycobacteria: A review, *J. Royal Soc. Med.*, 79, 226, 1986.
20. Jackson, P.G. et al., Injection abscesses due to *Mycobacterium chelonae* occurring in a diabetic patient, *Tubercle*, 62, 277, 1981.
21. Borghans, J.G., and Standford, J.L., *Mycobacterium chelonae* in abscesses after injection of diphtheria-pertussis-tetanus-polio vaccine, *Am. Rev. Respir. Dis.*, 107, 1, 1973.
22. Kuritaky, J.N. et al., Sternal wound infections and endocarditis due to organisms of the *Mycobacterium fortuitum* complex, *Ann. Intern. Med.*, 98, 938, 1983.
23. Gangadharam, P.R.J., Lanier, J.D., and Jones, D.E., Keratitis due to *Mycobacterium kansasii*, *Tubercle.*, 59, 55, 1978.
24. Moulsdale, M.T. et al., Infection by *Mycobacterium haemophilum*, a metabolically fastidious acid-fast bacillus, *Tubercle.*, 64, 29, 1983.
25. Cohen, R.J. et al., Occult infections with *M. intracellulare* in bone-marrow biopsy specimens from patients with AIDS, *N. Engl. J. Med.*, 308, 1475, 1983.
26. Graybill, J.R. et al., Disseminated mycobacteriosis due to *Mycobacterium abscessus* in two recipients of renal homograft, *Am. Rev. Respir. Dis.*, 109, 4, 1974.
27. Gordin, F., and Slutlin, G., The validity of acid-fast smears in the diagnosis of tuberculosis, *Arch. Pathol. Lab. Med.*, 114, 1025, 1990.
28. Kent, and Kubica, G.P., Public health mycobacteriology a guide for level III lab. Atlanta, GA: U.S. Department of health and human services, Public health services. Center for disease control, 64, 1985.
29. Mole, R.J., and WO'C Maskell, T., Review phage as a diagnostic—the use of phage in TB diagnosis, *J. Chem. Technol. Biotechnol.*, 76, 683, 2001.
30. Shenai, S., Rodrigues, C., and Mehta, A.P., Evaluation of a new phage amplification technology for the rapid diagnosis of tuberculosis, *Int. J. Med. Microbiol.*, 20, 194, 2002.
31. Kalantri, S. et al., Bacteriophage-based tests for the detection of *Mycobacterium tuberculosis* in clinical specimens: A systematic review and meta-analysis, *BMC Infect. Dis.*, 5, 59, 2005.
32. Jacob, W.R. et al., Rapid assessment of drug susceptibilities of *Mycobacterium tuberculosis* by means of luciferase reporter phages, *Science*, 260, 819, 1993.
33. Butler, W.R., and Kilburn, J.O., Identification of major slowly growing pathogenic *Mycobacteria* and *Mycobacteria gordonae* by high performance liquid chromatography of their mycolic acids, *J. Clin. Microbiol.*, 26, 50, 1988.
34. Butler, W.R., and Guthertz, L.S., Mycolic acid analysis by high-performance liquid chromatography for identification of *Mycobacterim* species, *Clin. Microbiol. Rev.*, 14, 704, 2001.
35. Eing, B.R., Becker, A., Sohns, A., and Ringelmann, R. Comparison of Roche Cobas Amplicor *Mycobacterium tuberculosis* assay with in-house PCR and culture for detection of *M. tuberculosis, J. Clin. Microbiol.*, 36, 2023, 1998.
36. Pai, M., Kalantri, S., and Dheda, K., New tools and emerging technologies for the diagnosis of tuberculosis: Part II. Active tuberculosis and drug resistance, *Expert Rev. Mol. Diagn.*, 6, 423, 2006.
37. Tortoli, E., and Palomino, J.C., New diagnostic methods. In *Tuberculosis*, Chapter 14, p. 441, 2007.
38. Bergmann, J.S., and Woods, G.L., Clinical evaluation of the BD-ProbeTec Strand Displacement Amplification Assay for rapid diagnosis of tuberculosis, *J. Clin. Microbiol.*, 36, 2766, 1998.
39. Piersimoni, C., and Scarparo, C., Relevance of commercial amplification methods for direct detection of *Mycobacterium tuberculosis* complex in clinical samples, *J. Clin. Microbiol.*, 41, 5355, 2003.
40. Chengtao, H., and Weishilbourn, R.M., The polymerase chain reaction (PCR) is the most widely used technique for the study of DNA. Applications of PCR have been extended significantly by the development of long PCR, *Curr. Issues Mol. Biol.*, 1, 77, 1999.
41. Johansen, I.S., Rapid diagnosis of mycobacterial diseases and their implication on clinical management, *Dan. Med. Bull.*, 53, 28, 2006.
42. Flores, L.L. et al., In-house nucleic acid amplification tests for detection of *Mycobacterium tuberculosis* in sputum specimens meta analysis and meta-regression, *BMC Microbiol.*, 5, 55, 2005.
43. Roth, A. et al., Differentiation of phylogenetically related slowly growing mycobacteria based on 16S-23S rRNA gene internal transcribed spacer sequences, *J. Clin. Microbiol.*, 36, 139, 1998.
44. Verma, A., Rattan, A., and Tyagi, J.S., Development of 23S rRNA based PCR for the detection of mycobacteria, *Indian J. Biochem. Biophys.*, 31, 288, 1994.
45. Soini, H., Bottger, E.C., and Viljanen, M.K., Identification of mycobacteria by PCR-based sequence determination of the 32-kilodalton protein gene, *J. Clin. Microbiol.*, 32, 2944, 1994.
46. Cave, M.D. et al., Stability of DNA fingerprint pattern produced with *IS6110* in strains of *Mycobacterium tuberculosis*, *J. Clin. Microbiol.*, 32, 262, 1994.
47. Telenti, A. et al., Rapid identification of mycobacteria to the species level by polymerase chain reaction and restriction enzyme analysis, *J. Clin. Microbiol.*, 31, 175, 1993.
48. Vaneechoutte, M. et al., Identification of *Mycobacterium* species by using amplified ribosomal DNA restriction analysis, *J. Clin. Microbiol.*, 31, 2061, 1993.
49. Lee, H. et al., Species identification of mycobacteria by PCR–restriction fragment length polymorphism of the *rpo*β gene, *J. Clin. Microbiol.*, 38, 2966, 2000.
50. Roth, A. et al., Novel diagnostic algorithm for identification of mycobacteria using genus-specific amplification of the 16S-23S rRNA gene spacer and restriction endonucleases, *J. Clin. Microbiol.*, 38, 1094, 2000.
51. Magalhaes, V.D. et al., Reliability of hsp65-RFLP análisis for identification of species in cultured strains and clinical specimens, *J. Microbiol. Methods*, 49, 295, 2002.
52. Kim, B.J. et al., Identification of mycobacterial species by comparative sequence analysis of the RNA polymerase gene (rpoB), *J. Clin. Microbiol.*, 37, 1714, 1999.

53. Kapur, V. et al., Rapid Mycobacterium species assignment and unambiguous identification of mutations associated with antimicrobial resistance in *Mycobacterium tuberculosis* by automated DNA sequencing, *Arch. Pathol. Lab. Med.*, 119, 131, 1995.
54. McNabb, A. et al., Assessment of partial sequencing of the 65-kiodalton heat shock protein gene (hsp 65) for routine identification of mycobacterium species isolated from clinical sources, *J. Clin. Microbiol.*, 42, 3000, 2004.
55. Zhu R.Y., et al. Use of visual loop-mediated isothermal amplification of rimM sequence for rapid detection of *Mycobacterium tuberculosis and Mycobacterium bovis, J Microbiol. Methods.*, 78(3), 339, 2009.
56. Bull, T.J., and Shanson, D.C. Evaluation of a commercial chemiluminescent Gen Probe system "AccuProbe" for the rapid identification of mycobacteria including "MAIC X", isolated from blood and other sites, from patients with AIDS. *J. Hosp. Infect.*, 21, 143, 1992.
57. Tortoli, E. et al. Evaluation of a commercial DNA probe assay for the identification of *Mycobacterium kansasii, Eur. J. Clin. Microbiol. Infect. Dis.*, 13, 264, 1994.
58. Tortoli, E., Simonetti, M.T., and Lavinia, F., Evaluation of reformulated chemiluminescent DNA probe (Accuprobe) for culture identification of *Mycobacterium kansasii, J. Clin. Microbiol.*, 34, 2838, 1996.
59. Saito, H. et al., Identification and partial characterization of *Mycobacterium avium* and *Mycobacterium intracellulare* by using DNA probes, *J. Clin. Microbiol.*, 27, 994, 1989.
60. Gonzales, R., and Hanna, B.A., Evaluation of Gen-Probe DNA hybrodozatopn system for the identification of *Mycobacterium tuberculosis* and *Mycobacterium avium-intracellulare, Diagn. Microbiol. Infect. Dis.*, 8, 69, 1987.
61. Shenai, S., Rodrigues, C., and Mehta, A., Rapid diagnosis of tuberculosis using transcription mediated amplification, *Indian J. Med. Microbiol.*, October, 19, 184, 2001.
62. Torkko, P. et al., *Mycobacterium palustre* sp nov., a potentially pathogenic slow-growing mycobacterium isolated from veterinary and clinical specimens, and Finnish stream water, *Int. J. Syst. Evol. Microbiol.*, 52, 1519, 2002.
63. Turenne, C.Y. et al., Mycobacterim, parascrofulaceum sp, nov., a novel slowly growing, scotochromogenic clinical isolates related to Mycobacterium interjectum and Accuprobe positive for Mycobacterium avium complex, *Int. J. Syst. Evol. Microbiol.*, 54, 659, 2004.
64. Shenai, S., Rodrigues, C., and Mehta, A., Rapid speciation of 15 clinically relevant mycobacteria with simultaneous detection of resistance to rifampicin, isoniazid and streptomycin in *M. tuberculosis* complex, *Int. J. Infect. Dis.*, 13, 46, 2009.
65. Tortoli, E., Mariottini, A., and Mazzarelli, G., Evaluation of INNO-LiPA MYCOBACTERIA V2: Improved reverse hybridization multiple DNA probe assay for mycobactrial identification, *J. Clin. Microbiol.*, 41, 4418, 2003.
66. Martin, A., and Porteals, F., Drug resistance and drug resistance detection. In *Tuberculosis*, Chapter 19, p. 635, 2007.
67. Fukushima, M. et al., Development of DNA microarray method to detect and identify pathogenic bacteria, *Jpn. J. Electrophoresis*, 45, 235, 2004.
68. Troesch, A. et al., *Mycobacterium* species identification and rifampicin resistance testing with high-density DNA probe arrays, *J. Clin. Microbiol.*, 37, 1079, 1999.
69. Fukushima, M. et al., Detection and identification of Mycobacterium species isolates by DNA microarray, *J. Clin. Microbiol.*, 41, 2605, 2003.
70. Caoli, J.C. et al., Evaluation of the TB-Biochip oligonucleotide microarray system for rapid detection of rifampin resistance in *Mycobacterium tuberculosis*, *J. Clin. Microbiol.*, 44, 2378, 2006.
71. Erlich, H.A., *PCR Technology: Principles and Applications for DNA Amplifications*, Stockton Press, New York, 1989.
72. van Embden, J.D. et al., Strain identification of *Mycobacterium tuberculosis* by DNA fingerprinting: Recommendations for a standardized methodology, *J. Clin. Microbiol.*, 31, 406, 1993.
73. Kamerbeek, J. et al., Simultaneous detection and strain differentiation of *Mycobacterium tuberculosis* for diagnosis and epidemiology, *J. Clin. Microbiol.*, 35, 907, 1997.
74. Groenen, P.M.A. et al., Nature of DNA polymorphism in the direct repeat cluster of *Mycobacterium tuberculosis*: Application for strain differentiation by a novel typing method, *Mol. Microbiol.*, 10, 1057, 1995.
75. Supply, P. et al., Automated high-throughput genotyping for study of global epidemiology of *Mycobacterium tuberculosis* based on mycobacterial interspersed repetitive units, *J. Clin. Microbiol.*, 39, 3563, 2001. doi: 10.1128/JCM.39.10.3563–3571.2001.
76. Klevytska, A. et al., Identification and characterization of variable-number tandem repeats in the *Yersinia pestis* genome, *J. Clin. Microbiol.*, 39, 3179, 2001.

12 *Nocardia*

Veronica Rodriguez-Nava, Frédéric Laurent, and Patrick Boiron

CONTENTS

12.1 INTRODUCTION

Bacteria of the *Nocardia* genus are ubiquitous in the environment. They have been isolated in soil, plants, and compost, as well as in humans. The infections they cause generally originate from the inhalation of the bacteria or by their contamination of wounds. Their proliferation within the body gives rise to a localized or disseminated granulomatous and suppurative infection, referred to as nocardiosis.[1]

Nocardia infections have long been considered to be rare, especially since the diagnosis of nocardiosis, whether direct, indirect, or immunologic, has long been out of the reach of a large number of laboratories. In recent years, the situation has dramatically changed. The number of patients suffering from nocardiosis worldwide is constantly increasing. Although a majority of patients have risk factors (generally an immunosuppressive situation), it must be noted that 30% of cases were observed in subjects who were apparently immunocompetent. Due to the nature and seriousness of the infection, the accurate and rapid detection and identification of such bacteria, as well as an assessment of their sensitivity levels to antibiotics, are essential. Diagnostic methods to replace conventional ones were proposed to microbiology laboratories in order to help them more quickly detect and identify *Nocardia* species whose number has tripled within the past few years and which display different epidemiological and pathogenic characteristics.

12.1.1 CLASSIFICATION

Bacteria with positive Gram staining form one of the ten major suprageneric groups[2] which is further divided into two principle groups depending on DNA base composition. The actinomycete group consists of bacteria whose DNA is rich in guanine (G) and cytosine (V), characterized by a G + C% > 55. The use of 16S rRNA sequencing techniques has allowed for the identification of phylogenetic relations between prokaryotes groups.[3]

The standard chemical identification process of actinomycetes relies on Lechevalier's chemotaxonomic system,[4] which allows one to distinguish between 10 chemotypes. This system is based on the presence of parietal amino acids [2,4-diaminobutyric acid, *meso*-2,6-diaminopimelic acid (*meso*-DAP), glycine, lysine, ornithine, aspartic acid] and parietal lipids (fatty acids, phospholipids, menaquinones, mycolic acids), as well as cellular sugars (arabinose, galactose, madurose, xylose) whose single or combined predominant presence provides a possible indication of a particular genus.

Within the past two decades, improvements in methods for the identification of bacteria, such as chemical analysis techniques (major cellular and membranous component) and molecular biology techniques (G + C%, DNA–DNA homologies, DNA–RNA, 16S rRNA sequencing), have lead to fundamental changes in the classification and identification of such bacteria.

The *Nocardia* genus shares two types of features with other actinomycete genera. The first is morphological: its fragile transient primary (aerial) mycelium is split into bacillary and coccoid elements, which makes *Nocardia* bacteria resemble other nocardioform actinomycete genera, according to Lechevalier definition (*Rhodococcus*, *Faenia*, *Pseudonocardia*, *Saccharomonospora*, *Saccharopoly-spora*, *Actinopolyspora*, *Amycolata*, *Amycolatopsis*, *Oerskovia*, *Promicromonospora*, *Nocardioides*, *Intrasporangium*). Aside from *Rhodococcus*, these genera resemble the *Nocardia* genus neither from a phylogenetic nor biochemical point of view.[4]

In terms of phylogenetic and biochemical features, the *Nocardia* genus resembles the *Rhodococcus*, *Gordonia*, *Tsukamurella*, *Dietzia*, and *Skermania* genera forming the *Nocardiaceae* family and the *Mycobacterium* and *Corynebacterium* genera within a suprageneric group of nocardioform actinomycetes, according to the definition of Goodfellow.[5,6]

12.1.2 Epidemiology

Habitat. Nocardia bacteria live as saprophytes in the soil and can be found in decomposing vegetable matter, in fresh or salt water, and in atmospheric dust. They are vastly present in soil, where as a saprophyte they contribute to the decomposition of vegetal, organic matter. These bacteria can be easily isolated in fresh and salt water, compost, plant matter, or atmospheric dust.

Several authors have demonstrated the abundance of actinobacteria in caves,[7–9] which leads to the conclusion that such a habitat is particularly prosperous for the development of this group of bacteria. Diverse new species of the *Nocardia* genus described in the recent years, among which *N. jejuensis*, *N. speluncae*, and *N. altamirensis* were isolated from subterraneous environments. The identification of these species indicates that such environments constitute an ecological niche with a high potential for the study of biodiversity and the description of new species.

Nocardia bacteria are not part of the standard microbial flora of humans. Although the presence of *Nocardia* on the skin and inside the respiratory tract has been reported, their isolation usually indicates a patent or latent infection, particularly among immunosuppressed patients, and must therefore undergo clinical analysis.[10,11]

Contamination Modes. The contamination by *Nocardia* species generally occurs through the inhalation of spores, originating from telluric bacteria or occasionally from a skin trauma. A metastatic presence in the central nervous system is in most cases the consequence of a hematogenous dissemination from an initial pulmonary or cutaneous contamination.[10] Entry through the digestive tract is much more rare and leads to isolated abdominal forms.

Interhuman transmission between immunosuppressed patients has been reported. Several species of the *Nocardia* genus were found in domestic and hospital air and dust. This observation may be an indication of a risk of nosocomial infections although only a few cases have been reported.[12]

Influence of Age and Gender. Nocardia infections mostly occur in adults aged 60–80 years and with no significant difference between the genders.

Prevalence and Incidence of the Disease. Five hundred to one thousand cases of nocardiosis were diagnosed annually in the in the United States during the 1970s,[13] but the actual incidence was probably higher due to a low "suspicion index" (inaccurate diagnosis techniques, confusion with other infections, and intercurrent pathologies sometimes masking diagnosis).

A more recent epidemiological study of nocardiosis conducted by Saubolle et al. (2003)[14] in the state of Arizona reported 455 cases diagnosed between 1998 and 2002, the most frequent etiologic agent being the *N. asteroides*, present in 307 cases.

A survey conducted in France between 1987 and 1990 reported 150–250 new cases annually.[15] Nearly 1000 strains were sent to the French Nocardiosis Observatory between 2000 and 2007. More than 75% (880 cases) belong to the actinobacteria family. Among these, 607 cases of nocardiosis originating from all regions of France (mainland and overseas) were identified. The most frequent species found in 75% of the clinical isolates are *N. farcinica* (26%), *N. nova* (20%), *N. abscessus* (18%), and *N. cyriacigeorgica* (12%).[16]

The apparent increase in incidences of nocardiosis can be explained by the aging of the general population and by the use of aggressive immunosuppressive therapeutics. A better knowledge of nocardiosis and improvements in diagnostic methods are also possible explanations for the increases of diagnosed cases.

Predisposition Factors. Patients presenting primary or secondary immunosuppression are particularly predisposed to *Nocardia* infections. Long-term corticotherapy appears to be the main factor of predisposition as it is found in nearly a third of all patients. Transplantations and hematology also constitute major risk factors. Neoplasia, and particularly those affecting the tracheobronchial tree (oto-rhino-laryngological and pulmonary cancers), bronchial dilation, and HIV also constitute predisposing factors. No predisposing factor was found in one-third of all cases.[10]

A number of debilitating factors were described: tuberculosis, silicosis, disseminated acute erythematous lupus, periarteritis nodosa, polyarthralgia rheumatica, pemphigoid, asthma, emphysema, bronchial and obstructive chronic bronchopneumopathy, pulmonary alveolar proteinosis, sarcoidosis, diabetes, malnutrition, intestinal lipodystrophy, colite ulcerosa, chronic alcoholism, toxicomania, and pregnancy.[17]

12.1.3 Clinical Features and Pathogenesis

12.1.3.1 Localized Pulmonary Nocardiosis

Clinical Presentation. The lung is the most commonly affected organ, touched in 60%–80% of the cases.[18] Pulmonary nocardiosis generally takes the form of chronic pneumopathy but can become acute following a rapid degradation. A pulmonary lesion is a localized pneumopathy surrounded by an inflammatory response. Occasionally, an abscess can form and progressively spread. The clinical expression of nocardiosis may also include forms such as bronchitis or pharyngitis.

The clinical pulmonary signs are polymorphous. Some evoke a bronchial syndrome, with dyspnea, thoracic pain of pleuropulmonary origin, and possible respiratory distress. Other symptoms can be observed, including fever, shivering, asthenia, anemia, anorexia, general state alteration, and night sweats.

Hemoptysis is rare and expectorating pulmonary purulence or mucupurulence (often of the color green) can appear. Nocardiosis can provoke a mediastinitis with a large venous obstruction and a superior caval syndrome.

Biology. Biological signs are neither obvious nor specific. A polynuclear neutrophils hyperleukocytosis (whose importance is unrelated to the seriousness of the infection) can be found occasionally. However, a moderate anemia associated to other inflammatory signs is generally observed.

Radiology. Radiological signs of pulmonary nocardiosis, highly polymorphous and unspecific, do not allow for a diagnosis.

Nocardiosis can lead to a wide variety of radiological images with one or both lungs being affected, most commonly with apical localization: infiltrates, segmented, lobar or diffuse condensation, necrotizing pneumopathy, small or large abscesses, pleural effusion, cavities of all sizes, single or multiple nodular images with military aspect, fistulous tracts (in case of the spread of the infection to the thoracic wall), and, more rarely, hilary lesions, adenopathies, or calcifications.

Differential Diagnosis. The clinical differential diagnosis has to be conducted for all typical or atypical pneumopathies, whether of bacterial, mycobacterial, viral, mycotic, or toxic origin, whether particular or medicinal. The differential diagnosis must exclude the pulmonary expression of a systematic illness, as well as primary or secondary neoplastic disorders. These diagnostic difficulties may in some cases lead to a pulmonary biopsy that can help distinguish between various hypotheses.

Evolution. The *Nocardia* infection has a tendency to spread from its primary pulmonary site to other tissues through the lymphatic or circulatory system. The central nervous system constitutes the most frequent metastatic localization. The skin and subcutaneous tissues are less frequently touched, along with the pleural cavity and the thoracic wall, the eye, the liver, the lymphatic ganglions and other sites: intestines, bones, heart, pericardium, endocardium, kidneys, endocrine glands, and so forth.

The pulmonary infection is sometimes asymptomatic and nocardiosis is revealed through pulmonary radiography or through the emergence of one or several infectious, metastatic sources.

12.1.3.2 Extrapulmonary Localized Nocardiosis

(i) Neuromeningeal Forms They are most frequently secondary with a pulmonary source.[19]

Clinical Presentation. The infection of the central nervous system generally takes the form of one or several abscesses that can be situated anywhere in the brain (cerebellum, pons, rachidian bulb, Gasser's ganglion) with a tendency to form satellite extensions. A meningitic infection, whether acute or chronic, generally originates from a rupture of a cerebral abscess in the ventricular system. Pure meningitis without the presence of any cerebral abscesses is more rarely observed.

The clinical situation results from the intracerebral spread of the infection, with variable signs and focal symptoms depending of the affected region of the brain: cephalea (often chronic, preceding the emergence of other signs by several weeks or months, and being the most frequent symptom), nausea and vomiting (resulting from an intracranial hypertension), neck stiffness, photophobia (more specific to an meningitic infection), sensitive or motor (most commonly hemiparesis), possibly associated with aphasia or dysarthria, sensory disorder: visual (dilopia, hemianopsia) or auditory, affected cranial nerves, convulsive crisis (signs of cortical irritation), vertigo, instability, ataxia (in case of cerebellar affection), alteration of the superior functions (ranging from confusion and amnesia to coma), behavioral disorder: irritability and frontal syndrome.

It must be emphasized that in a number of cases of nocardiosis in cerebral sites, no patent neurological symptoms are found and only a systematic search through imaging will allow one to identify this type of localization. When such a systematic search was conducted, a cerebral localization was identified in more than a third of all patients.

Biology. A normal cerebrospinal fluid (CSF) is common in *Nocardia* infections. However, in cases of meningeous infection, the examination of the CSF may be indicative of an increase in white blood cells (mostly polynuclear neutrophils), a hypoglycorrhachia, and a hyperproteinorachia.

Imaging. Cerebral tomodensitometry allows for the diagnosis of the intracerebral spread of the infection as to localization, size, and number. In cerebral nocardiosis, it is essential to carry out imaging after the intravascular injection of a contrast-enhancing agent. In 80% of all cases, cerebral abscesses are seen in the form of concentric circles with a hypodense central zone and peripheral halo. Magnetic resonance imaging (MRI) allow for the identification of the liquid nature of the lesion.

Differential Diagnosis. The images obtained (abscess displaying a smooth capsule of uniform thickness and hypodense content) may be compatible with other lesions, mainly a glioma (in which case the central zone appears to be denser), a primitive or metastatic tumor (in which case the edge is more regular), a cerebrovascular accident, or multiple sclerosis.

Therefore, MRI imaging may turn out to be insufficient for differentiating an abscess from another liquid tumor. In this case, Indium-111 leucocyte scintigraphy may be very helpful.

When one or several cerebral abscesses are associated with an extra cerebral source of infection (in particular pulmonary), several other etiologic agents may be suspected: *Pneumococcus*, *Streptococcus*, *Bacteroides*, *Mycobacterium tuberculosis*, *Cryptococcus*, *Coccidioides*, *Histoplasma*, *Blastomyces*, *Aspergillus*. The sterility of the CRF is an argument in favor of nocardiosis.

(ii) Cutaneous and Subcutaneous Infections Cutaneous nocardiosis can be observed in two different epidemiologic and clinical situations.

Primitive, cutaneous nocardiosis occurs following a traumatism in health patients. It generally lies on exposed parts of the skin. The lesions are acute or subacute and may resemble pyodermatitis, cellulite, subcutaneous abscess, or lymphangitis. There is no systemic dissemination of the infection and a spontaneous healing of the lesions is possible.

Secondary cutaneous nocardiosis is observed primarily in immunosuppressed patients and manifests itself though single or multiple subcutaneous abscesses that can fistulate to the skin, or through cellulites. It originates (in 10%–15% of cases) from hematogenous spread of a source, and is often pulmonary. It may also originate from a direct extension of a pulmonary source through the thoracic wall.

(iii) Other Sites of Infection Almost all organs can be touched secondarily, each giving rise to particular clinical signs: the eye (keratitis, dacryocystitis, chorioretinitis, endophthalmitis), liver (abscess), pancreas (abscess, even pancreatitis), heart (pericarditis, endocarditis), bones (osteitis, osteomyelitis, arthritis), kidneys, suprarenal glands, spleen, intestine, peritoneum (particularly among dialyzed patients), lymphatic ganglions, thyroid, ear canal, amygdales, pharynx, buccal cavity, and trachea (particularly through contamination following transtracheal aspiration).

12.1.4 Diagnosis

A diagnosis of nocardiosis must be evoked (along with common bacterial and viral infections, tuberculosis, and fungal opportunistic infection) in cases of febrile, suppurative disease or in the presence of a pathological pulmonary radiographic image for patients whose immune system was altered by certain therapeutic agents or by an underlying pathology.[20,21] The diagnosis of nocardiosis relies on bacteriologic identification in patient samples occasionally obtained using intrusive methods. The histology may also be informative.

12.1.4.1 Phenotypic Techniques

The identification of *Nocardia* and the diagnosis of nocardiosis represent great difficulties for nonspecialized microbiological laboratories.[22,23] About 30 years ago, conventional identification techniques were developed: strict aerobic respiratory type; macroscopic and microscopic morphology [presence of ramified, gram-positive filaments that split into bacillary and coccoid elements and produce (rarely) short chain conidia]; and chemotaxonomic, biochemical, and physiological characteristics.[24]

Microscopy. Thin smears prepared from the pure pathologic mixture or from centrifugation sediment as well as the juxtaposition of biopsies can be performed for a microscopic examination. The presence of ramified filaments can be observed, although in a majority of cases it is difficult to detect the often rare bacteria through direct observation.

Nocardia display diverse morphological characteristics. The classical appearance is the ramified filament form, but *Nocardia* filaments are fragile and often break into shorter fragments during the sampling or smearing process. The "zigzagging" aspect of the ramified filaments can be observed in a low-brightness phase-contrast direct examination.

The Gram staining is positive, but often weakly and irregularly so (dotted, stained, or striped appearance). Upon Ziehl–Neelsen staining (discoloration by 1% or 2% sulfuric acid during 1 min), the acido-alcohol resistance may vary. Such a partially acid-alcohol fast character allows one to differentiate between *Nocardia* and bacteria of the *Actinomyces* and *Streptomyces* genera. Gomori staining (methenamine silver) may be used to look for *Nocardia* during histopathological examination, but also stains fungus and *Pneumocystis jiroveci*.

These sole morphological features in an aerobic microorganism have long been considered sufficient to establish the identification of *Nocardia*. To date, these criteria are not considered sufficient to exclude other bacteria, and more reliable criteria (i.e., analysis of the cellular and parietal constituents) are now required. A negative microscopic examination does not rule out a nocardiosis diagnosis and may be followed by a positive culture if the incubation is sufficiently long.[22]

Culture. Culturing is easy on substrates free of antibiotics (5% sheep blood agar, enriched blood agar, BCP medium, BACTEC® or BACTalertEC® systems) as well as culture substrates for fungus or mycobacteria (Sabouraud, Löwenstein-Jensen, Bennett). The use of a selective paraffin medium or a normal or selective BCYE (buffered charcoal yeast extract) medium may eliminate the conventional fast-growing bacteria, which occasionally mask the presence of *Nocardia*.

Cultures are incubated at a temperature ranging from 32°C to 37°C in aerobiosis or possibly in the presence of 5%–10% of CO_2. The growth of colonies generally takes between 4 and 5 days but may take up to 2 weeks.

Routinely, the colonies are flat or slightly dome shaped, often with a corrugated surface, cerebriform, and covered within a few days with a short duvet of aerial mycelium that masks the color of the colony and confers it a chalky white appearance. The colonies are encrusted in the agar gel (solid medium). Certain colonies appear to be pigmented, their color ranging from yellowish (or even whitish) beige to orange or red. Their consistency is generally firm although crumbly. Upon microscopic examination *Nocardia* appear to be ramified and split into bacillary or even coccoid elements within 24–48 h.

Biochemical Characterization. Several biochemical properties of *Nocardia* are useful for their characterization. These include the substrate decomposition profiles (tyrosine, xanthine, uric acid, adenine, hypoxanthine, casein, urea, and testosterone), the growth capacity on carbon-rich substrates (citrate, acetamide, rhamnose, mannitol, sorbitol, etc.), enzymatic activity (arylsulfatase, catalase, acetamidase, galactosidase, urease, etc.), and sensitivity characteristics or resistance to antibiotics (ampicillin, carbenicilline, amoxicillin + clavulanic acid, cefamandole, kanamicine, etc.). The analysis of chemotaxonomical markers such as fatty acids, mycolic acid, or menaquinones is based on the system of Lechevalier et al.[25,26] and Goodfellow et al.,[6] and essentially allows the identification of a new type of *Nocardia*. Being laborious and expensive to carry out, this is limited to specialized laboratories.

Phenotypical techniques based on the macroscopic and microscopic features of colonies as well as biochemical characteristics allow one to identify the *Nocardia* species most responsible for infection but fail to bring new species to light. In addition, given the great variety of phenotypical tests applied to *Nocardia* and described in the literature, no precise choice has been generally agreed upon. Today, they only complement molecular diagnostic techniques for *Nocardia*, or are limited to a premature diagnosis, according to the minimal criteria for orientation.[24]

12.1.4.2 Molecular Techniques

The emergence of molecular identification methods has enabled the development of several techniques for definitive identification of members of the *Nocardia* genus.[27–30]

DNA Hybridization. DNA hybridization is one of the molecular bacteriological methods useful for the identification of microorganisms. Brownell et al.[31] developed a system of probes for the identification of *N. asteroides* from a genomic bank obtained with a cloned GUH-2 strain in the Lambda EMBL3 vector. This work has not benefited from real development, and no other technique of this type has since been proposed.

Ribotyping. The ribotyping technique was evaluated by Laurent et al.[32] for the *Nocardia* genus. The comparative study of restriction profiles of 21 strains belonging to complex *N. asteroides* confirmed both the heterogeneity of this species and the high level of discrimination and reproducibility of the technique for the identification of isolates belonging to this complex. Nevertheless and despite its promising character, ribotyping has never seen diagnostic application due to its long and fastidious implementation.

In 2007, Kalpoe et al.[33] developed a typing technique based on molecular epidemiology (Amplified Fragment Length Polymorphism, AFLP) in order to study a possible epidemic, observed in the renal transplant service of the Leiden Medical Center in the Netherlands, caused by the *N. farcinica* species. They compared this technique with others, such as RAPD (Randomly Amplified Polymorphic DNA) and pulsed field electrophoresis. Results showed that the AFLP analysis seems to be an effective alternative to traditional and poorly reproducible methods, such as RAPD for epidemiological typing of *N. farcinica* strains.

PCR–Restriction Fragment Length Polymorphism (PCR–RFLP). The technique for the identification of actinomycetes of medical interest used by Steingrube et al.[29,34] based on the number of restriction sites, their positions and the sizes of fragments generated from amplified product of a short segment (440 base pairs) of the gene coding for the heat shock protein (hsp65) and digested by the enzymes (*Bst* EII, *Msp* I, *Hinf* I, and *Bsa* HI) leading to the establishment of a decision tree for identification was reevaluated and applied to the new taxonomical context of *Nocardia* (43 species tested). The results clearly showed its inability to distinguish the ensemble of *Nocardia* species known to date. This technique can nevertheless distinguish without ambiguity species of clinical interest if it is complemented by appropriate phenotypical tests. This method is, therefore, useful by default.[35]

The work of Laurent et al.[27] has enabled the selection of primers within the 3′ region of the ribosomal 16S DNA of *Nocardia* that allow the specific amplification of a fragment of 596 base pairs for strains belonging to the *Nocardia* genus. The specificity of the amplified fragment of 596 base pairs can be verified thanks to the position of restriction sites of two enzymes specific to the *Nocardia* genus: all *Nocardia* strains tested showed a restriction site specific to the *Mnl* I enzyme and no restriction site for the enzyme, *Sac*I. This technique is directly applicable to clinical samples (bronchoalveolar lavage, deep puss, biopsies, CSF or puncture of cerebral abscess) without the need for cell cultures.[28]

With the same aim as Steingrube et al., Conville et al.,[30] have proposed a PCR–RFLP technique based on the part of the gene coding the 16S rRNA for the identification of the species of the *Nocardia* genus. Their goal was to use this technique to optimize the speed and the accuracy of the identification, as well as to remove the possible ambiguities in the identification obtained by the RFLP of the hsp65 gene. The limited success of this technique can probably be attributed to the taxonomic inflation observed in recent years as well as the polymorphism of the gene coding for 16S rRNA within the *Nocardia* genus. All of this work constituted a major advance in the identification of species within the *Nocardia* genus. However, these techniques remain inadequate given the new taxonomic context of the *Nocardia* genus and novel diagnosis means must be proposed.[36]

DNA Sequencing. In recent years, the development of novel molecular biology techniques has allowed for the large scale sequencing of DNA. The DNA sequences obtained have been used as taxonomic signatures and allow one to ascribe a microorganism to a particular species of taxonomic group. The current taxonomical complexity of the *Nocardia* genus, which includes over 70 species to date of which at least half are found in clinical settings, requires the use of bioinformatic analysis of ribosomal 16S DNA sequences.[37] The combined used of sequences of multiple genes (*hsp65*, *rpo*B, *sod*) should enable one to better discriminate between species in the long term. In fact, the multilocus approach should enable the development of a phylogenetic metatree that should further improve the resolution of the identification within different genera of this family.

16S rRNA Sequencing. To date, the analysis by partial (500 bp) or complete (1500 bp) sequencing of 16S rRNA remains the reference molecular method for the identification of a number of microorganisms. Identification by sequencing is becoming an essential diagnosis tool. Although its cost is limiting for certain clinical biology laboratories, it is a useful alternative for the identification of slow-growing bacteria for which specialized identification techniques are required. However, the precise identification of a microorganism by sequencing will require the availability of sequences in databases open to the general public. Indeed, a monitored database of quality is essential for the successful identification of microorganisms.

The use of such a marker for the identification of *Nocardia* was studied by several authors.[30,38–42] The analysis of the complete gene showed that the study of the first 500 base pairs on the 5′ end was insufficient to identify the species of most actinomycetes of medical interest. According to Stackebrandt et al.[43] it is now commonly accepted that when the homology between the 16S RNA sequences of two strains is less than 97%, these belong to different species. Several groups have studied its application with the aim to identify bacterial strains within the same species.

The work of Mellman et al.[40] on a fragment of 500 bp of the 5′ end of the 16S rRNA gene of 30 strains of *Nocardia* have showed an interspecies polymorphism sufficient for the identification of all tested species, except for *N. soli* and *N. cummidelens*. Phenotypic tests were proposed to doubtlessly differentiate these species. The sequences obtained for all these type strains were used to build a genomic bank dedicated to *Nocardia*. The identification by sequencing of 91 clinical strains originally identified though conventional tests were compared to reference sequences of type strains in the database. The results suggested that a similarity index greater than 99.12% is required to classify a strain as a given species with a statistical probability of error of 1%. Indeed, identification by sequencing of a 500 bp fragment of 16S rRNA of clinical strains yielded excellent results in a majority of isolates. These results were compared with other genomic databases in order to test the degree of accuracy and quality of the identification. With the same goal, Cloud et al.[39] studied the identification of *Nocardia* species of medical interest by sequencing that same 500 bp fragment of 16S rRNA, having chosen a similarity criterion of 99% with a 3% statistical error probability.

Nevertheless, in our work on 5′-end sequencing of the 16S gene of all type strains of the *Nocardia* genus (43 species described at the time this work was conducted, versus 30 species used in the work of Mellman), we observed the limitations of the use of the gene coding the 16S rRNA as a genotypic marker for the identification of all species. For instance, *N. cummidelens* and *N. soli* on the one hand, and *N. shimofusensis* and *N. higoensis* on the other, display identical sequences. Similarly, the sequences of *N. abscessus* and *N. asiatica* differ by only one nucleotide.

Clarridge et al.[44] clearly indicate that it is impossible to set a definitive similarity or dissimilarity index for the identification of species by 16S rRNA sequencing, partly because different values are obtained from different databases and using different methods. These facts highlight the need for controlled, monitored, and corrected reference databases.

Sequencing of the hsp65 Gene. The *hsp65* gene seems to be an interesting target according to studies published on *Mycobacterium*.[45] Rodriguez-Nava et al.[42] described a novel molecular technique to differentiate and identify 43 reference species of the *Nocardia* genus by the amplification and sequencing of a 440 bp fragment of the *hsp65* gene. The sequencing of that gene in the *Nocardia* genus was evaluated through comparison with the sequencing of 16S rRNA. They computed an evolution distances matrix in order to evaluate the interspecies polymorphism of each genetic target and observed an interspecies variability. Results showed sequences similarities on the order of 88%–100% (93.7% on average) for hsp65 on one hand and a 90.5%–100% sequences similarity (94.8% on average) for the gene coding for 16S rRNA. These results demonstrate the utility of hsp65 sequencing: it allows one to differentiate between species having identical 16S rRNA molecules.

The study of other genotypic markers among which *hsp65*, *rpo*B, *rec*A, *sod*, and *gyr*B (used for the identification of mycobacteria) may be considered to solve certain differentiation issues of species within the *Nocardia* genus.

Databases and Biocomputing Tools. The importance of generalized databases is on the rise (EMBL,[46] GenBank,[47] and DDBJ[48]). The majority of scientific journals in the molecular biology field systematically require the access number given by the genomic databases when publishing a sequence. Although sequencing methods have been much improved today, the methods used in the 1990s did not always offer quality sequences. Moreover, the lack of quality controls during the submission of sequences to databases inevitably leads to the introduction of erroneous sequences. Specialized databases were therefore created in order to remedy the introduction of errors in general databases.

Specialized databases are dedicated to a given biological field. Indeed, they consist of sequences and data that are specific to a given topic. Several specialized genomic databases have been developed in recent years that offer efficient bioinformatic systems for the identification of microorganisms. In the field of actinomycetes, RIDOM, MicroSeq, and, more recently, BIBI[49] databases were developed for *Mycobacteria* and *Nocardia* genera.

MicroSeq (Microbial Identification System). MicroSeq is a genomic database developed by Perkin Elmer Applied Biosystems. It was created for the identification of filamentous fungi, yeast, and eubacteria. Two kits are commercially available for the identification of bacteria—the MicroSeq500[R] (which generates a fragment of 500 bp at the 5′ end) and, the most often used, the 16S rRNA (which generates a 1500 bp fragment). MicroSeq contains all reactants and protocols necessary to the sequencing and analysis of results.

The database associated with this system includes about 1400 sequences issued from different collections of bacterial strains, notably the most typical strains, including 82 nucleotidic sequences of the 16S rRNA gene of 82 species belonging to the *Mycobacterium* genus[50] and 28 sequences corresponding to 28 species of *Nocardia*.[39]

The MicroSeq system of identification functions as follows: an unknown isolate is identified by partial sequencing of the 16S rRNA gene using the MicroSeq kit, and its species is confirmed if the similarity in the sequences is ≥99% as compared to that of a type strain of the MicroSeq database.

Regarding *Nocardia*, Cloud et al.[39] tested the MicroSeq system. In order to evaluate the interest of this database, the authors tested 94 strains of clinical origins belonging to the *Nocardia* genus. The results obtained via sequencing were compared to other molecular studies including restriction enzyme analysis of the gene coding for 16S rRNA or the gene *hsp65* and biochemical methods and antibiotic sensitivity. The MicroSeq system was revealed to be capable of correctly identifying only six species that are pathogenic for humans (*N. brasiliensis*, *N. cyriacigeorgica*, *N. farcinica*, *N. nova*, *N. otitidiscaviarum*, and *N. veterana*). However, neither type IV *N. abscessus* and *N. asteroides* nor the other species in the genus were clearly identified.

One disadvantage of MicroSeq is that it is a closed system. Only bacteria of medical interest have been integrated, which makes it impossible to identify environmental species for which a new implication in human pathology has been observed. In addition, access to the database requires payment.

RIDOM (Ribosomal Differentiation of Medical Microorganisms). RIDOM was developed in 1999 by Harmsen et al.[51] The aim was to provide bioinformatic services for the identification of microorganisms of medical interest via sequencing. The RIDOM site is directly accessible on the web at http://www.ridom.de.

The operation of the intuitive HTML interface of RIDOM is based on two programs: the FASTA program allows one to compare often bulky sequences using a heuristic method; the CLUSTAL W program is used for aligning homologous sequences.

The database was constructed using the 16S rRNA sequences of several strain collections of medical interest belonging to several genera including *Mycobacterium* [in all, 199 sequences of 16S rRNA and 84 sequences of internal transcribed spacers (ITS)[52]] and *Nocardia* (in all, 30 sequences of 16S rRNA from type strains[40]).

RIDOM's diagnostic procedure is based on amplifying and sequencing a part of the gene coding for the small subunit of 16S rRNA corresponding to the 5′ region of an unknown specimen. The diagnostic is linked to the analysis of similarities in the sequences sought in the database by the user. The percentage of similarities of sequences is calculated and must be[3] 99.12% with respect to a type strain in order for a species name to be obtained.

In case identification at the species level cannot be provided, the RIDOM system proposes the necessary information to the user that enables him or her to apply a polyphasic approach (combining phenotypical, chemotaxonomic, or molecular approaches). This information is available through hypertext links to relevant online web pages.

A study[40] sought to evaluate the quality of results provided by two specialized databases (RIDOM and MicroSeq 500) and by the generalized database, GenBank. The results for the three databases were compared with phenotypical identification. This study was carried out via the sequencing of 91 clinical strains using the two systems, RIDOM and MicroSeq separately. A similarity value of 99.12% was set as the identification criterion by these authors. The results obtained using RIDOM showed that among the 91 isolates tested, 65 sequences (71.4%) were identical. On the other hand, 11 isolates (12.1%) showed sequence similarities lower than 99.12%. Meanwhile, the MicroSeq system showed perfect agreement for only 24 isolates (26.4%). Finally, the sequencing results for the ensemble of isolates were compared to those of Genbank. These results suggested that sequences were 100% similar for 54 isolates, although 38 of these isolates are not exclusive to one species.

The ensemble of the results obtained through the comparison of these three systems indicates the superiority of RIDOM. Nevertheless, the main disadvantages of RIDOM concerning the Nocardia genus must be noted. First, it is not exhaustive with respect to diagnostic needs (only 30 species are indexed). Second, it is currently impossible to access data concerning Nocardia on the Web site.

BIBI (BioInformatic Bacterial Identification). BIBI database aims at facilitating the identification of bacteria, especially in the clinical domain.[49] The database uses several bioinformatic tools to automate the analysis of DNA sequences. Two types of databases stem from BIBI:

- General databases: the major database, "Bacteria," gathers all the sequences from the GenBank and includes all genes and bacteria species.
- Specialized and controlled databases: there are five of these databases, corresponding to 16S rRNA, *hsp65*, *rpo*B, *sod*, and *gyr*B genes.

BIBI uses several similar search programs, such as BLAST and CLUSTAL W, applied to the various subsets of sequences from Genbank. These sequences are filtered and stored in a new database designed for bacterial identification.

The analysis of an unknown sequence is conducted in four steps: search of known sequences similar to the unknown one, extraction of the selected sequences, alignment of the sequence, and display of the results.[49]

The first step consists of submitting the unknown sequence to the server. The search for similar sequences is conducted through BLAST using a database chosen by the user. The results given by BLAST are then filtered by BIBI and entered in a FASTA file. The unknown sequence and those similar to it are aligned using CLUSTAL W, which performs a multiple alignment, creating three separate files that contain the alignment of sequences, the phylogenetic tree in the NEWICK format, and the distance matrix.

The final result is presented in the form of a table of phylogenetic distances between the unknown sequence and all the other similar sequences selected by BLAST. The closest sequences correspond to the identified species. All files produced during this process are accessible online or can be downloaded by FTP "File Transfer Protocol."

BIBI (http://pbil.univ-lyon1.fr/bibi) is a free, simple, and user-friendly bioinformatic tool for the identification of bacteria that quickly gives the user a result while allowing him/her to monitor the satisfactory progress of every step of the process. As a controlled database, it contains perfect and nonambiguous sequences of normalized size and uses a standardized annotation system.

As for the *Nocardia* and *Mycobacterium* genera, this specialized database includes all partial sequences of 16S rRNA, *hsp65*, *rpo*B, and *sod* of the type strains of both genera.

The *Mycobacterium* genus contains a total of 388 sequences of 97 species and subspecies that correspond to the four studied genes, 16S rRNA, *hsp*65, *rpo*B, and *sod*.[45] The *Nocardia* genus contains over 200 nucleotidic sequences from over 60 species corresponding to the same four genes.

12.2 METHODS

12.2.1 Sample Preparation

Samples can have various origins: bronchoalveolar lavage, aspirations, expectorations, abscess puss, blood (hemocultures are rarely positive, even in the case of disseminated nocardiosis), CSF, pleural liquid, bone tissue, lymphatic ganglions, peritoneal fluid effusion, hepatic biopsy, cutaneous biopsy, pulmonary biopsy, and so forth, and can be collected by puncture, biopsy, during the course of a surgical procedure or even postmortem.

In the case of a localized pulmonary infection, bacteria may occasionally be isolated from spital or from material collected through a bronchoscopy, transtracheal aspiration, transcutaneous pulmonary biopsy, or even thoracotomy. However, the presence of *Nocardia* may only be the consequence of a simple colonization. It is important to note that the enzymatic digestion processes and alkaline decontamination of the expectoration products used for the isolation of mycobacteria alter the viability of *Nocardia* and greatly reduce the chances of obtaining a culture.

Once pulmonary nocardiosis has been diagnosed, a checkup must always be performed to determining whether the infection has spread and should at least include a cerebral scan, even in the absence of any neurological symptoms.

Given cerebral lesions, guided stereotaxic procedures allow for the draining of abscesses and lead to a diagnosis, even for cerebral sites not easily accessible by conventional surgery.

Following the diagnosis of an extracerebral nocardiosis, the lesions observed on a scan are presumed to be of the same origin. One must therefore avoid puncturing such lesions for the sole purpose of diagnosis, given the risk of meningitic contamination, except for cases involving AIDS patients for which multiple germ abscesses are common.

12.2.2 Detection Procedures

12.2.2.1 Genus-Specific Identification

The work of Laurent et al.[27] has enabled the selection of primers within the 3′ region of the ribosomal 16S RNA of *Nocardia* that allow the specific amplification of a fragment of 596 base pairs for strains belonging to the *Nocardia* genus. The specificity of the amplified fragment of 596 base pairs can be verified thanks to the position of restriction sites of two enzymes specific to the *Nocardia* genus: all *Nocardia* strains tested showed a restriction site specific to the *Mnl* I enzyme and no restriction site for the enzyme, *Sac* I.

This technique is directly applicable to clinical samples (bronchoalveolar lavage, deep puss, biopsies, CSF or puncture of cerebral abscess) without the need for cell cultures.[28]

12.2.2.2 Species-Specific Identification

Several molecular approaches have been developed for which the reliability has been called into question recently.[29,30] The current taxonomical complexity of the *Nocardia* genus, which includes over 70 species to date of which at least half are found in clinical settings, requires the use of bioinformatic analysis of ribosomal 16S RNA sequences.[37] The combined used of sequences of multiple genes (*hsp65*, *rpo*B, *sod*) should enable one to better discriminate between species in the long term. In fact, the multilocus approach should enable the development of a phylogenetic metatree that should further improve the resolution of the identification within different genera of this family.

12.3 CONCLUSIONS AND FUTURE WORK

Nocardiosis is an infection which principally affects immunosuppressed patients although immunocompetent subjects may be touched as well. Incidences of *Nocardia* infection worldwide have been on the rise in recent years, mostly due to the aging of the population and to the usage of aggressive immunosuppressive therapeutics. In France, such infections are often associated with the *N. farcinica* (26%) and *N. nova* (20%) species, which are particularly resistant to a number of antibiotics.

Phenotypical techniques used some years ago for the identification of *Nocardia* were soon replaced by molecular biological methods. These methods have lead to the complete modification of the taxonomy and one can now differentiate more quickly and more accurately between a much larger number of species of the *Nocardia* genus via specific genetic markers (16S rRNA, *rpo*B, *sod*, *hsp65*).

Several difficulties are encountered in the isolation and culture of *Nocardia* such as their slow growth which can be masked by the presence of other fast-growing microorganisms. To date, this often-tedious procedure can be replaced by a standard real-time PCR reaction carried out directly from a clinical sample without need of culture.

Besides, in order to reduce the costs of the analysis and to bring down the time needed for diagnosis to several hours, it would be conceivable the direct detection of *Nocardia* from a clinical sample or a more complex sample as a soil one by real-time PCR: this could turn out to be of great interest for the specific and quantitative detection of *Nocardia*. However the culture must not be neglected because it is necessary for the antibiotic susceptibility testing report.

Moreover, other techniques (including mass spectrometry), dedicated to the identification of microorganisms through comparison of their protein fingerprints with the aid of a database, may turn out to have interesting implications for the diagnosis of *Nocardia*.

REFERENCES

1. Boiron, P., and Laurent, F., *Nocardia* et nocardiose. In Bricaire, L. and Bricaire, F. (Editors), *Maladies Infectieuses*, pp. 1–8, EMC Elsevier SAS, Paris, 2007.
2. Woese, C.R. et al., The phylogeny of purple bacteria: The alpha subdivision, *Syst. Appl. Microbiol.*, 5, 315, 1984.
3. Stackebrandt, E., Rainey, F.A., and Ward-Rainey, N.L., Proposal for a new hierarchic classification system, *Actinobacteria* classis nov., *Int. J. Syst. Bacteriol.*, 47, 479, 1997.
4. Lechevalier, H.A., Nocardioforms actinomycetes. In Williams, S.T., Sharpe, M.E., and Holt, J.G. (Editors), *Bergey's Manual of Systematic Bacteriology*, pp. 1348–2404, Williams and Wilkins Co, Baltimore, 1989.
5. Chun, J., and Goodfellow, M., A phylogenetic analysis of the genus *Nocardia* with 16S rRNA gene sequence, *Int. J. Syt. Bacteriol.*, 45, 240, 1995.
6. Goodfellow, M., and Lechevalier, M.P., Genus *Nocardia* Trevisan 1889, 9AL. In Williams, S.T., Sharpe, M.E., and Holt, J.G. (Editors), *Bergey's Manual of Systematic Bacteriology*, vol. 4, pp. 2350–2361, Williams and Wilkins, Baltimore, 1989.
7. Groth, I., and Saiz-Jimenez, C., Actinomycetes in hypogean environments, *Geomicrobiol. J.*, 16, 1, 1999.
8. Schabereiter-Gurtner, C. et al., Phylogenetic 16S rRNA analysis reveals the presence of complex and partly unknown bacterial communities in Tito Bustillo cave, Spain, and on its Palaeolithic paintings, *Environ. Microbiol.*, 4, 392, 2002.
9. Jurado, V. et al., *Agromyces salentinus* sp. nov. and *A. neoliticus* sp. nov., *Int. J. Syst. Evol. Microbiol.*, 55, 153, 2005.
10. Beaman, B.L., and Beaman, L., *Nocardia* species: Host-parasite relationships, *Clin. Microbiol. Rev.*, 8, 213, 1995.
11. McNeil, M.M., and Brown, J.M., The medically important aerobic actinomycetes: Epidemiology and microbiology, *Clin. Microbiol. Rev.*, 7, 357, 1994.
12. Exmelin, L. et al., Molecular study of nosocomial nocardiosis outbreak involving heart transplant recipients, *J. Clin. Microbiol.*, 34, 1014, 1996.
13. Beaman, B.L. et al., Nocardial infections in the United States, 1972–1974, *J. Infect. Dis.*, 134, 286, 1976.
14. Saubolle, M.A., and Sussland, D., Nocardiosis: Review of clinical and laboratory experience, *J. Clin. Microbiol.*, 41, 4497, 2003.
15. Boiron, P. et al., Review of nocardial infections in France 1987 to 1990, *Eur. J. Clin. Microbiol. Infect. Dis.*, 11, 709, 1992.
16. Rodriguez-Nava, V. et al., Molecular identification of clinical isolates of *Nocardia* and related Actinomycetes in the French Observatory for Nocardiosis (OFN)—New epidemiology data of nocardiosis in France between 2000 and 2007. ESCCMID, Abstract number: P1718, Barcelona, Spain, 2008.
17. Rippon, J.W., Nocardiosis. In Rippon, J.W. (Editor), *Medical Mycology. The Pathogenic Fungi and the Pathogenic Actinomycetes*, 3rd ed., pp. 53–68, WB Saunders Company, Philadelphia, 1988.
18. Filice, G.A., Nocardiosis. In Sarosi, G.A., and Davies, S.F. (Editors), *Fungal Diseases of the Lung*, 2nd ed., pp. 191–204, Raven Press, New York, 1993.
19. Loeffler, J.M. et al., Nocardial brain abscess: observation of treatment strategies and outcome in Switzerland from 1992 to 1999, *Infection*, 29, 337, 2001.
20. Laurent, F., Freney, J., and Boiron, P., *Nocardia* et actinomycètes aérobies apparentés. In Freney, J. et al. (Editors), *Actualités en bactériologie clinique théorique et pratique*, pp. 1–21, Editions ESKA, Editions Alexandre Lacassagne, 2003.
21. Beaman, B., The pathogenesis of *Nocardia*. In Fiscetti, V.A., Novick, R.P., Ferretti, J.J., Pottnoy, D.A., and Rood, J.I. (Editors), *Gram Positive Pathogens*, pp. 594–606, ASM Press, Washington DC, 2000.
22. Boiron, P., Provost, F., and Dupont, B., Technical protocols. In Boiron, P., Provost, F. and Dupont, B. (Editors), *Methodes de laboratoire pour le diagnostic de la nocardiose*, pp. 107–126, Institut Pasteur, Paris, 1993.
23. Goodfellow, M., The family *Nocardiaceae*. In Balows, A. et al. (Editors), *The Prokaryotes. A Handbook on the Biology of Bacteria: Ecophysiology, Isolation, Identification, Applications*, 2nd ed., Vol II, pp. 1188–1213, Springer-Verlag, Berlin, 1992.
24. Wauters, G. et al., Distribution of *Nocardia* species in clinical samples and their routine rapid identification in the laboratory, *J. Clin. Microbiol.*, 43, 2624, 2005.
25. Lechevalier, H., and Lechevalier, M.P., Classification of aerobic actinomycetes based on their morphology and their chemical composition, *Ann. Inst. Pasteur (Paris)*, 108, 662, 1965.
26. Lechevalier, M.P., and Lechevalier, H.A., Chemical composition as a criterion in the classification of aerobic actinomycetes, *Int. J. Syst. Bacteriol.*, 20, 435, 1970.
27. Laurent, F., Provost, F., and Boiron, P., Rapid identification of clinically relevant *Nocardia* species to genus level by 16S rRNA gene PCR, *J. Clin. Microbiol.*, 37, 99, 1999.
28. Couble, A. et al., Direct detection of *Nocardia* spp. in clinical samples by a rapid molecular method, *J. Clin. Microbiol.*, 43, 1921, 2005.
29. Steingrube, V.A. et al., Rapid identification of clinically significant species and taxa of aerobic actinomycetes, including *Actinomadura*, *Gordona*, *Nocardia*, *Rhodococcus*, *Streptomyces*, and *Tsukamurella* isolates, by DNA amplification and restriction endonuclease analysis, *J. Clin. Microbiol.*, 35, 817, 1997.
30. Conville, P.S. et al., Identification of *Nocardia* species by restriction endonuclease analysis of an amplified protein of the 16S rRNA gene, *J. Clin. Microbiol.*, 38, 158, 2000.

31. Brownell, G.H., and Belcher, K.E., DNA probes for the identification of *Nocardia asteroides*, *J. Clin. Microbiol.*, 28, 2082, 1990.
32. Laurent, F. et al., Ribotyping: A tool for taxonomy and identification of the *Nocardia asteroides* complex species, *J. Clin. Microbiol.*, 34, 1079, 1996.
33. Kalpoe, J.S. et al., Molecular typing of a suspected cluster of *Nocardia farcinica* infections by use of randomly amplified polymorphic DNA, pulsed-field gel electrophoresis, and amplified fragment length polymorphism analyses, *J. Clin. Microbiol.*, 45, 4048, 2007.
34. Steingrube, V.A. et al., DNA amplification restriction endonuclease analysis for differentiation of 12 species and taxa of *Nocardia*, including recognition of four new taxa within the *Nocardia asteroides* complex, *J. Clin. Microbiol.*, 33, 3096, 1995.
35. Rodriguez-Nava, V., Devulder, G., and Laurent, F., Taxonomie et identification des *Nocardia*: évolution et perspectives, *Bull. Soc. Fr. Microbiol.*, 18, 177, 2003.
36. Rodriguez-Nava, V. et al., *Nocardia ignorata*: A new agent of human nocardiosis isolated from respiratory specimens in Europe and soil sample from Kuwait, *J. Clin. Microbiol.*, 43, 6167, 2005.
37. Patel, J.B. et al., Sequence-based identification of aerobic actinomycetes, *J. Clin. Microbiol.*, 42, 2530, 2004.
38. Cloud, J.L. et al., Identification of *Mycobacterium* spp. by using a commercial 16S ribosomal DNA sequencing kit and additional sequencing libraries, *J. Clin. Microbiol.*, 40, 400, 2002.
39. Cloud, J.L. et al., Evaluation of partial 16S ribosomal DNA sequencing for identification of *Nocardia* species by using the MicroSeq 500 system with an expanded database, *J. Clin. Microbiol.*, 42, 578, 2004.
40. Mellmann, A. et al., Evaluation of RIDOM, MicroSeq, and Genbank services in the molecular identification of *Nocardia* species, *Int. J. Med. Microbiol.*, 293, 359, 2003.
41. Patel, J.B. et al., Sequence-based identification of aerobic actinomycetes, *J. Clin. Microbiol.*, 42, 2530, 2004.
42. Rodríguez-Nava, V. et al., Use of PCR-restriction enzyme pattern analysis and sequencing database for *hsp65* gene based identification of *Nocardia* species, *J. Clin. Microbiol.*, 44, 536, 2006.
43. Stackebrandt, E. et al., Report of the ad hoc committee for the re-evaluation of the species definition in bacteriology, *Int. J. Syst. Evol. Microbiol.*, 52, 1043, 2002.
44. Clarridge, J.E., Impact of 16S rRNA gene sequence analysis for identification of bacteria on clinical microbiology and infectious diseases, *Clin. Microbiol. Rev.*, 17, 840, 2004.
45. Devulder, G., de Montclos, M., and Flandrois, J.P., A multigene approach to phylogenetic analysis using the genus *Mycobacterium* as a model, *Int. J. Syst. Evol. Microbiol.*, 55, 293, 2005.
46. Baker, W. et al., The EMBL nucleotide sequence database, *Nucleic Acids Res.*, 28, 19, 2000.
47. Benson, D.A. et al., GenBank, *Nucleic Acids Res.*, 28, 15, 2000.
48. Tateno, Y. et al., DNA Data Bank of Japan (DDBJ) in collaboration with mass sequencing teams, *Nucleic Acids Res.*, 28, 24, 2000.
49. Devulder, G. et al., BIBI, a bioinformatics bacterial identification tool, *J. Clin. Microbiol.*, 41, 1785, 2003.
50. Hall, L. et al., Evaluation of the MicroSeq system for identification of mycobacteria by 16S ribosomal DNA sequencing and its integration into a routine clinical mycobacteriology laboratory, *J. Clin. Microbiol.*, 41, 1447, 2003.
51. Harmsen, D. et al., Intuitive hypertext-based molecular identification of micro-organisms, *Lancet*, 353, 291, 1999.
52. Harmsen, D. et al., RIDOM: Comprehensive and public sequence database for identification of *Mycobacterium* species, *BMC Infect. Dis.*, 3, 26, 2003.

13 *Propionibacterium*

Andrew McDowell and Sheila Patrick

CONTENTS

13.1 INTRODUCTION

13.1.1 Classification, Morphology, and Biology

The genus *Propionibacterium* is currently composed of the following 11 species: *P. freudenreichii* (*P. freudenreichii* subsp. *freudenreichii* and *P. freudenreichii* subsp. *shermanii*), *P. jensenii*, *P. thoenii*, *P. acidipropionici*, *P. cyclohexanicum*, *P. microaerophilum*, *P. acnes*, *P. avidum*, *P. granulosum*, *P. australiense*, and *P. propionicum*. The species *Propionibacterium lymphophilum* and *Propionibacterium innocuum*, previously described as part of the *Propionibacterium* genus, have now been reclassified within the family *Propionibacteriaceae* as *Propionimicrobium lymphophilum* and *Propioniferax innocua*, respectively.[1,2] Bacteria within the *Propionibacterium* genus are gram-positive, anaerobic-to-aerotolerant-to-microaerophilic, nonspore-forming pleomorphic rods. They are 0.2–1.5 μm in diameter × 1–5 μm in length and have a high G + C content [57–68 mol% (Tm)]. They can appear diphtheroid or club shaped, with one end rounded and the other tapered or pointed, but may also be coccoid, bifid, branched, or filamentous with lengths of up to 20 μm.[3] Propionibacteria are chemoorganotrophs with complex nutritional requirements and produce large amounts of propionic and acetic acids, and lesser amounts of isovaleric, formic, succinic or lactic acids, and carbon dioxide as the end products of their fermentation reactions.[3] Although predominantly anaerobic, most propionibacteria are catalase positive. Tetrahydrogenated menaquinones with nine isoprene units [MK-9(H4)] are the major respiratory quinones. Their optimal growth temperature is between 30°C and 37°C, and on solid media colonies will appear smooth, rough, or convex, and may be white, grey, pink, red, yellow, or orange in color.[3]

Within the genus, species have been traditionally differentiated into two groups, the "classical" or "dairy" and the "cutaneous" propionibacteria. While molecular analyses have revealed that the phylogenetic relationships amongst propionibacteria do not exactly mirror their classification based on habitat (Figure 13.1), it is still a useful and convenient way to group the species. The "classical" or "dairy" propionibacteria includes the species *P. freudenreichii*, *P. jensenii*, *P. thoenii*, and *P. acidipropionici*, which are isolated mainly from cheese and other dairy products and are also used as starter cultures by the dairy industry and

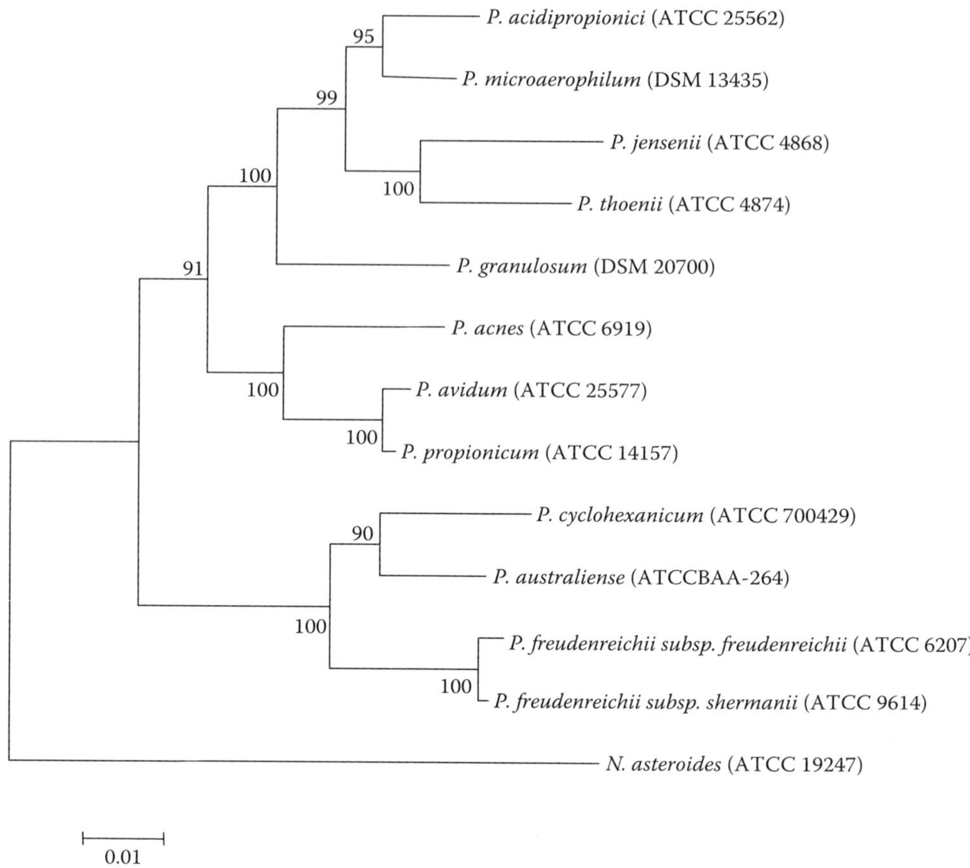

FIGURE 13.1 Phylogenetic tree of 16S rRNA sequences for the genus *Propionibacterium*. The neighbor-joining tree was constructed using the Jukes-Cantor-based algorithm. The sequence input order was randomized, and bootstrapping resampling statistics were performed using 100 data sets. The 16S rRNA sequence from *Nocardia asteroides* was used as a distant outgroup to root the tree as it also belongs to the order *Actinomycetales*. Bootstrap values are shown at each node of the tree.

as protective biopreservatives and probiotics.[4] These organisms have also been isolated from silage, fermenting olives, and soil.[5–8] The "cutaneous" propionibacteria, which include the species *P. acnes*, *P. granulosum*, and *P. avidum*, are isolated from human skin, although they also colonize the mouth, upper respiratory tract, large intestine, and genitourinary tract.[3,9] There have also been reports in the literature describing their isolation from animal species such as cows, pigs, and dogs.[10–12] Levels of *P. acnes* colonization on the skin vary from person to person and from the region of the body sampled.[13] *P. acnes* is especially prevalent in sebaceous gland-rich areas on the skin, with the neck, forehead, and shoulder showing some of the highest concentrations compared to other sites, such as the abdomen, hip, knee, and chest where levels are lower.[14] *Propionibacterium granulosum* is present on the sebum-rich, oily areas of the skin, and is especially common in the ala nasi (wing of the nose), although the total numbers present are much less than *P. acnes*.[15] The distribution of *P. avidum* appears more restricted and is found principally in moist areas, including vestibule of the nose, the axilla, and the perineum.[15] The "cutaneous" group has also been known as the "anaerobic coryneforms," "anaerobic diphtheroids," "acnes group strains," or "cutaneous propionbacters." The more recently described *Propionibacterium* species, *P. microaerophilum*, *P. cyclohexanicum*, and *P. australiense*, do not conveniently fit into either the "classical" or the "cutaneous" groups based on habitat.[16–18] On the basis of 16S rRNA sequence, however, *P. microaerophilum*, recovered from a decantation reservoir of olive mill wastewater, clusters with the "dairy" propionibacteria, while *P. cyclohexanicum* and *P. australiense*, recovered from spoiled orange juice and granulomatous bovine lesions, respectively, cluster with one another and are more closely related to *P. freudenreichii* than any of the other propionibacteria.[16–18] The species *P. propionium*, previously known as *Arachnia propionica*, clusters with the "cutaneous" species based on 16S rRNA sequence and is present in the oral cavity of humans including the mucosa, tonsils, and dental plaque, and may also be present in the oral cavity of other animals.[19] It has also been isolated from the gastrointestinal and female genital tract, but it is not believed to be part of the indigenous microbiotia in these areas.[19]

Morphologically, the "classical" propionibacteria tend to be shorter and thicker than the "cutaneous" group, which are more often irregular and curved, although all the strains may be variable in morphology, especially in early exponential phase cultures.[3] Strains of *P. acnes* can give rise to longer and more slender irregular rods in young cultures, while in stationary phase cultures all strains tend to be more coccoid.[3] The cell dimensions of *P. acnes* range from 0.5 to 1.5 μm in width and approximately 1–5 μm in length.[20] *Propionibacterium propionicum* and strains of *P. acnes* phylotype III can form filaments.[21,22] With *P. propionicum*, these filaments are branched and are especially evident in young liquid media cultures, but commonly break up into short rods in older cultures. Morphologically and phenotypically, *P. propionicum* resembles *Actinomyces israelii*. Some strains of propionibacteria (e.g., *P. thoenii*; *P. avidum*) appear capsulated, and it seems that a considerable number of strains of all species may produce extracellular slime not organized in the form of clear-cut individual capsules.[23,24] The extracellular material is carbohydrate in nature.[23,25] The cell lipids of both "classical" and "cutaneous" strains contain large amounts of C_{15} branched-chain fatty acids.[26]

Both groups of propionibacteria tend to have similar nutritional needs, including a requirement for pantothenate. Many also need biotin and some require thiamine and *p*-aminobenzoic acid. For others, thiamine and *p*-aminobenzoic acid are not essential but do stimulate growth, as does nicotinamide and oleate.[27–29] In contrast, *P. microaerophilum* does not require any of these growth factors and is therefore prototrophic. Amino acid requirements are complex for the "cutaneous" group.[28] Most strains of *P. acnes* and *P. granulosum* require a full complement of 18 amino acids for good growth, which can be enhanced further by the addition of lactate, pyruvate, ketoglutarate, as well as guanine and adenine (20/1 ratio).[28] Some strains may also require additional factors, such as heme and vitamin K. In contrast, after several transfers *P. avidum* can grow in a simple medium containing salts, glucose, and vitamins in the absence of amino acids.[28]

Propionibacterium acnes has been grouped into three distinct clades, known as phylotypes I, II, and III, based on the analysis of the *recA* housekeeping gene and a more variable hemolysin/cytotoxin gene.[22,30] While the term phylotype has a somewhat arbitrary definition, we have used it in a straightforward context to describe the previous *P. acnes* types resolved phylogenetically. The phylotype I grouping can also be further subdivided into two very closely related clusters known as IA (including type strain ATCC6909) and IB. While these different lineages can be identified upon sequence analysis of *recA*, they share at least 99.8% identity in their 16S rRNA sequences (1484 bp section).[22,30]

Earlier studies by Johnson and Cummins demonstrated that strains now known to represent *P. acnes* phylotypes I and II share DNA–DNA hybridization values of 88%–99%.[31] Combined with more recent 16S rRNA sequence data, this demonstrates that these lineages represent distinct sequence clusters within *P. acnes* and not novel species, at least based on current definitions of a bacterial species.[32,33] The phylotypes of *P. acnes* can also be distinguished by serological agglutination tests and monoclonal antibody-labeling methods, which reflect differences in the composition of their cell walls.[22,30,31] Previous work has shown that phylotype I strains contain glucose, mannose, and galactose in their cell wall, while phylotype II strains contain only glucose and mannose.[31] Data on the composition of type III cell walls is currently not available. While types I and II both contain LL-diaminopimelic acid (LL-DAP) in their cell wall peptidoglycan, some strains of phylotype II may occasionally contain *meso*-DAP. Other phenotypic differences between these groupings include variation in their susceptibility to bacteriophage infection and distinct fermentation profiles with sugar and sugar alcohols.[34,35] The "cutaneous" species *P. avidum* can also be split into two distinct serovars.[31,36] By DNA–DNA hybridization, sequences within each group are at least 90% similar, but between the two groups the average similarity is 80%.[37] The cell wall composition of each group mirrors that found in *P. acnes* phylotypes I and II, with *P. avidum* type I containing glucose, mannose, galactose, and LL-DAP, while phylotype II contains glucose, mannose, and mostly *meso*-DAP, but sometimes LL-DAP. Strong cross reaction between *P. acnes* phylotype II strains and sera to *P. avidum* type II has been observed.[31,36] To date, phylogenetic differences between the *P. avidum* serovars have not been investigated. In contrast to both *P. acnes* and *P. avidum* species, *P. granulosum* appears to belong to a single serological type.[36] Two serovars of *P. propionicum* have been identified using fluorescent antibody methods with the type strain ATCC14157, representing serovar 1.[38,39] No cross reaction between the serovars has been observed by cell wall agglutination tests, but some reaction between serovar 1 and *P. avidum* has been observed by immunofluorescence.[38] The two serovars of *P. propionicum* share very little DNA homology (1%) and also form distinct subclusters by numerical phonetic analyses suggesting they may actually represent distinct species.[40]

13.1.2 Propionibacteria and Disease

Within the *Propionibacterium* genus, the "classical" or "dairy" propionibacteria are nonpathogenic, but strains of the "cutaneous" species and *P. propionicum* are associated with opportunistic human infection and disease.[4] *Propionibacterium acnes* represents the major opportunistic human pathogen within the "cutaneous" group and although *P. granulosum* has been found in sarcoid tissue and associated with acne; endocarditis, endophthalmitis, and septicemia in an immunocompromised patient; and *P. avidum* with septicemia and severe infection after invasive procedures; they are still considered a relatively rare cause of human disease based on literature reports (although more cases of *P. granulosum* infections have been described compared to *P. avidum*).[41–47] *Propionibacterium australiense* has been shown to cause granulomatous lesions in cattle from different geographical sites in northern and central Queensland, Australia.[48] These lesions are present in large numbers throughout the animal, including the external surface of the rumen and the reticulum, tongue, and peritoneal surface of the gastrointestinal tract. The lesions consist of a characteristic fibrous outer capsule that surrounds thick yellow viscous pus with caseating granulomas. To date, there are no reports of this organism causing human infections. Since both *P. acnes* and *P. propionicum* represent the major human pathogens within the *Propionibacterium* genus, their association with infection and disease will now be described in more detail.

Propionibacterium acnes is a widespread opportunist most noted for its potential involvement in the inflammatory condition acne. It is also a cause of biofilm-associated infections of indwelling medical devices, such as prosthetic hip, knee, and shoulder joints, spinal implants, prosthetic heart valves, intravascular catheters, and central nervous system shunts.[49–58] Infection of orthopaedic implants with *P. acnes* can be insidious with delayed presentation of symptoms; indeed, many of the traditional signs of underlying infection are absent.[52,55] It can, however, also cause acute infection associated with severe tissue damage and sinus tract formation.[59] The bacterium is also responsible for endophthalmitis posteye surgery and has been linked to synovitis acne pustulosis hyperostosis osteitis (SAPHO) syndrome, a chronic condition that describes an association of characteristic bone, joint, and skin lesions, and sarcoidosis, a multisystem granulomatous inflammatory disease that is characterized by noncaseating granulomas.[60–62] The bacterium may also have a role to play in primary biliary cirrhosis and the development of prostate cancer via chronic low-grade infection of the prostate gland.[63,64] In addition to its role in human infection, *P. acnes* has recently been identified as the cause of necrosuppurative placentitis and abortion in an adult Holstein cow.[10] An emerging number of studies have reported the association between the phylotypes of *P. acnes* and different infections. Studies by Sampedro et al. found that the total distribution of phylotypes I (collectively IA & IB) or II between prosthetic hip, knee, shoulder, and spine implants, removed because of aseptic failure or underlying infection, was similar, although all isolates recovered from knee, shoulder, and spine prostheses were predominately phylotype I (IA & IB), with equal numbers of types I and II being isolated from prosthetic hips.[54] They did show, however, that within the phylotype I grouping, isolates of type IB were much more commonly associated with the implant infection cohort compared with aseptic failure, where equal number of types IA and IB were found. The isolation of *P. acnes* in association with implants that fail aseptically may reflect colonization of the device prior to the development of infection.[51] Studies of *P. acnes* isolates recovered from the forehead skin of healthy volunteers versus those isolated from infected tissue adjacent to orthopaedic implants found the distribution of *P. acnes* phylotypes to be similar, although type IA was the predominant phylotype isolated.[65] Studies by Shannon et al. found that phylotypes IB and II were more frequently isolated from the radical prostatectomy specimens of patients with prostate cancer, although no significant difference in the prevalence of these *P. acnes* types in the urinary tract of patients with prostate pathology compared with normal control men was observed.[66] Types IB and II were, however, more common in adult men compared with adolescents suggesting an age-related increase.[66]

Propionibacterium propionicum causes human infections that may be indistinguishable from *Actinomyces* species based on clinical symptoms and pathogenesis. The bacterium is a recognized cause of actinomycosis, a disease that is characterized by a chronic and subacute granulomatous infection that commonly affects the cervicofacial area (face and neck), but can also affect the thorax, abdomen, or pelvis.[67] Aggressive long-term, high-dose, antibiotic therapy is required and if left untreated the infection may spread to the brain or other organs to become life threatening. It is also a common cause of lacrimal canaliculitis, a disease where the tear duct becomes blocked with the bacterium and concomitant oral or nasal microbiota.[68] This condition presents mainly in females over 40 years old and can respond poorly to topical antimicrobial treatment. The organism is also associated with primary and persistent ondodontic infection as well as intrauterine contraceptive devices and has been described as the cause of a finger infection following a human bite, and as a cause of a renal abcess.[67,69,70]

13.1.3 Pathogenic Features of Propionibacteria

Previous studies indicate that strains of *P. granulosum* have the ability to produce active deoxyribonuclease, lecithinase, hyaluronidase, and a lipase that is considerably more active than that of *P. acnes*.[71–73] The organism does not, however, produce gelatinase and it is also considered to be nonhaemolytic.[3] In contrast, *P. avidum* strains do produce hemolytic and gelatinase activities as well as deoxyribonuclease and proteinase. Most are, however, negative for lecithinase, hyaluronidase, and chondroitin sulfatase.[71,72] *Propionibacterium acnes* also produces extracellular enzymes involved in tissue degradation, including ribonuclease, neuraminidase, hyaluronidase, acid phosphatase, lecithinase, lipase, and proteinase.[71,74–80] More detailed insights into the pathogenic potential of *P. acnes* have been determined by analysis of the whole genome sequence of the phylotype IB strain KPA171202 and the draft genome sequence of the phylotype IA strain ATCC6919.[81,82] This has revealed a gene family consisting of five homologues of the *Streptococcus agalactiae* cohemolysin or CAMP (Christie Atkins Munch Peterson) factor. The gene family is present in all *P. acnes* phylogenetic groupings described to date and the cohemolytic reaction, a synergistic lysis of erythrocytes by the interaction of the CAMP factor with *Staphylococcus aureus* sphingomyelinase C, has been observed amongst isolates representing types IA, IB, and II.[82] The CAMP reaction in phylotype III strains has not been examined. Differential production of the different CAMP factor proteins amongst *P. acnes* phylotypes has been found, with strains of phylotype II and a commonly isolated phylotype IB strain (ST-10 by MLST; McDowell & Patrick, unpublished) producing large quantities of CAMP factor 1 compared to phylotype IA organisms.[82] In addition, CAMP factor proteins have been shown to bind the Fc fragment of immunoglobulins of the IgG and IgM class and also to act as pore-forming toxins.[83] They are also lethal when administered to rabbits and mice.[84] A number of putative hemolysin genes have also been identified within the genome sequences, including homologues of *tlyC* and *tlyA*.[30,81] Variable hemolysis of phylotype I strains on horse blood has been observed, but phylotype II and phylotype III strains are nonhemolytic under these conditions.[22] Genes coding for surface proteins with cell adherent properties have also been described. These proteins contain the C-terminal–LPXTG—cell wall sorting signal that is acted on by the transpeptidase sortase. Three genes have also been described that contain contiguous stretches of 12–16 guanine or cytosine residues within the promoter or 5′ end regions. These homopolymeric tracts may cause slipped strand mispairing during replication, leading to the expression of proteins with altered molecular masses, thus leading to phenotypic variation and possibly immune evasion or adaptation to environmental changes.[85] Phase variation due to slipped strand mispairing has been observed in several notable pathogens (e.g., *Neisseria meningitis*).[86] The strong proinflammatory nature of *P. acnes* has been known for some time. Indeed, the immunostimulatory activity and adjuvant properties of *P. acnes* isolates previously referred to as "*Corynebacterium parvum*" are exploited in studies of the inflammatory response.[87] Furthermore, *P. acnes* has also been used as a nonspecific immune stimulant in the equine industry (EqStim®).[88] It can activate complement by both the classical and alternate pathways leading to inflammatory processes through antibody-dependent and independent mechanisms.[89] Homologues of GroEL, DnaK, and several other heat shock proteins, which are major targets for the immune system, are also present in the genome.[81,90] Three gene clusters putatively involved in extracellular polysaccharide/lipoglycan biosynthesis have also been identified in the genome sequence and may also be important in modulating immunogenicity. The production of porphyrins by *P. acnes* may lead to inflammation as well as generate toxic reduced oxygen species that cause cell and tissue damage. It has been suggested that porphyrin production by the bacterium may also be important in the skin condition progressive macular hypomelanosis.[91] Studies in vitro have shown that *P. acnes* can stimulate the production of antimicrobial peptide and proinflammatory mediators from human keratinocytes and sebocytes and differences have also been noted in the capacity of strains representing phylotypes IA and IB to modulate the viability and differentiation of these cell types.[92] However, further studies with a larger number of strains will be required to demonstrate that these differences are truly phylotype-specific. The production of biofilm is a well-established virulence determinant used by pathogenic bacteria, and the formation of biofilm by *P. acnes* has been demonstrated by various studies, although the underlying molecular details of biofilm regulation in *P. acnes* have not yet been determined.[93–94] It has been shown in vitro that isolates associated with orthopaedic implant infections produce greater levels of biofilm when compared to isolates from the skin of healthy volunteers.[65] Furthermore, *P. acnes* within a biofilm is not only more resistant to antibiotics, but the production of lipase by the bacterium is upregulated by this mode of growth.[94] *P. acnes* is also resistant to killing and degradation by human phagocytic cells and can thus survive intracellularly. It may, therefore, persist and cause a chronic low-grade infection. Despite its pathogenic potential, *P. propionicum* appears unremarkable in its production of specific virulence factors.[3] Strains cause β-hemolysis on human blood agar and are variable in the production of lipase. No hyaluronidase, deoxyribonuclease, or proteolytic activity is produced, although some strains may liquefy gelatin.

A study of *P. acnes* isolates causing surgical, foreign body, and septicemia infections in Europe has found that 3% are resistant to tetracycline, 15% to clindamycin, and 17% to erythromycin, while none are resistant to linezolid, benzylpenicillin, or vancomycin.[95] The use of topical formulations of erythromycin and clindamycin to treat acne has resulted in cross-resistant propionibacteria strains in Europe and Asia.[96–98] Strains of *P. acnes* are resistant to metronidazole, sulfonamides, and aminoglycosides.[99–101] The current antimicrobial susceptibility of *P. granulosum* isolates has not been as widely investigated as *P. acnes*, but studies with two isolates responsible for septicemia and endocarditis in compromised patients revealed susceptibility to a range of compounds.[42,44] The septicemia isolate was sensitive to penicillin G, ampicillin, erythromycin, cefotaxime, clindamycin, rifampin, and imipenem, but resistance was found to metronidazole, sulphonamide, fosfomycin, and tobramycin.[44] Gentamicin, streptomycin, amikacin, pefloxacin, and chloramphenicol had poor activity. This data appeared to substantiate previous studies.[102] The endocarditis isolate was susceptible to penicillin, clindamycin, augmentin, piperacillin, chloramphenicol, cefotaxime, and cefazolin, but resistant to metronidazole.[42] Erythromycin resistant strains of *P. granulosum* have been isolated from the skin of acne patients treated with oral erythromycin and topical clindamycin.[103] Little information is available on the current susceptibility of *P. avidum* isolates. Susceptibility testing of a *P. avidum* isolate responsible for a breast abscess following reduction surgery revealed sensitivity to penicillin G, ampicillin, erythromycin, tetracycline, and vancomycin, but resistance to metronidazole.[47] Much earlier studies by Hoeffler et al. found that isolates were generally sensitive to a range of drugs including benzylpenicillin, ampicillin, cephalothin, rifampin, clindamycin, erythromycin, and minocycline.[102] As with other "cutaneous" propionibacteria, erythromycin-resistant strains have been isolated from the skin of acne patients treated with oral erythromycin and topical clindamycin.[103] *Propionibacterium propionicum* appears sensitive to a wide range of antibiotics such as penicillins, cephalosporins, chloramphenicol, tetracyclines, erythromycin, clindamycin, and imipenem, but fluoroquinolones, aztreonam, aminoglycosides, and metronidazole are ineffective.[67]

13.1.4 Diagnosis of Human Propionibacterial Infections

13.1.4.1 Conventional Techniques

Isolation of *Propionibacterium* Species from Clinical Samples. Good growth of all propionibacteria can be achieved using a medium with the following composition; trypticase (1% w/v), yeast extract (0.5% w/v), glucose (1% w/v), $CaCl_2$ (0.002% w/v), $MnSO_4$ (0.002% w/v), $NaCl_2$ (0.002% w/v), K-phosphate buffer (0.05 M), Tween-80 (0.05% w/v), sodium dithionite (0.05% w/v), $NaHCO_3$ (0.1% w/v; added as sterile solution along with inoculum), and pH 7.0.[104] Complex media, such as brain heart infusion broth or agar, is often used for the growth of "cutaneous" strains which will also grow very well on anaerobic blood agar. A medium containing sodium oleinate for the growth of propionibacteria and inhibition of *Actinomyces* strains from human skin swabs has also been described.[105] The addition of bromocresol purple to the medium permits identification of *P. acnes*, which appears as yellow or slightly yellow opaque colonies with yellow zones of lysis due to the fermentation of glycerol, while *Staphylococcus epidermidis* appears as white translucent colonies. Propionibacteria can also be recovered using TYEG agar (2% w/v tryptone, 1% w/v yeast extract, 0.5% w/v glucose) containing furazolidone (6 μg/ml) to inhibit staphylococcal growth.[96] Although some strains of propionibacteria may be aerotolerant, a culture methodology similar to that used for the primary isolation for strict anaerobes should still be adopted, such as that detailed in the Wadsworth KTL Anaerobic Bacteriology Manual.[106] It is important that growth media and diluents are pre-reduced and pre-equilibrated under anaerobic conditions. Furthermore, the incorporation of L-cysteine hydrochloride (0.05% w/v) into liquid media and diluents provides a suitable reducing agent. An anaerobic gas atmosphere within an anaerobic jar or cabinet, from which oxygen has been scavenged by palladium catalysts, should be used for culture. A lack of adherence to strict anaerobiosis when handling and processing clinical specimens will result in reduced detection of pathogenic *Propionibacterium* species.[49,57] Furthermore, as *P. acnes* causes biofilm-associated infection of prosthetic implants, it is imperative that removed devices undergo sonication to detach any adherent bacteria.[49,50,53] Failure to do so will result in poor diagnostic sensitivity. It is worth highlighting that sonication is superior to alternate "scraping" methods for the detachment of bacteria growing on titanium and steel surfaces in vitro.[107] The "cutaneous" group strains grow at 36°C–37°C, in contrast to the "classical" propionibacteria, which grow best at 30°C–32°C. In general, while maximum growth in complex media of pure cultures is attained in 48 h, primary isolation from clinical samples may require incubation for 7–14 days.

Biochemical Analyses. The "cutaneous" propionibacteria and *P. propionicum* can be differentiated by using a number of simple biochemical tests (Table 13.1). These include sucrose and maltose fermentation, gelatin liquefaction, esculin and casein hydrolysis, and catalase, indole, and nitrate production. A simple thin-layer chromatography (TLC) method has also been described for the differentiation of *P. acnes* from *P. propionicum* based on orcinol glycolipid profiles.[108] A number of studies have evaluated commercial phenotypic identification kits for their ability to differentiate "cutaneous" propionibacteria, although the number of species and strains examined were limited. The RapID™ ANA II System (Remel, Thermo Fisher Scientific) correctly identified 21/23 *P. acnes* strains (91%), while the Rapid ID 32A system (bioMérieux) was able to correctly identify all 10 *P. acnes* strains (100%) under investigation.[109,110] Similarly, the BBLCrystal™ Anaerobe (ANR) identification system (Benton Dickson) correctly identified all *P. acnes* (n = 10), *P. avidum* (n = 9), and *P. granulosum* (n = 7) isolates analysed.[111] In contrast, problems have been encountered with the API (RAPID) Coryne system and database 2.0, which covered 49 taxa.[112] Of six *P. acnes* strains tested, only one was correctly identified with additional tests, while four were not identified, and one was misidentified. With four *P. avidum* strains, two were correctly identified, and two could only be identified with additional investigations. Within *P. acnes*, phylotype I strains can be identified from those of type II and III based on the fermentation of sorbitol and also clear zones of β-hemolytic activity on anaerobic blood agar plates. While these simple tests

TABLE 13.1
Biochemical Tests to Differentiate *Propionibacterium* Species Associated with Human Infection[a]

Characteristic	*P. acnes*	*P. granulosum*	*P. avidum*	*P. propionicum*
Sucrose fermentation	–	+	+	+
Maltose fermentation	–	d+	+	+
Gelatin hydrolysis	+	d–	–	–
Esculin hydrolysis	–	–	+	–
Casein hydrolysis	+	–	+	+
Catalase production	d+	+	+	–
Indole production[b]	d+	–	–	–
Nitrate reduction[b]	d+	–	–	+

[a] *Symbols*: +, positive reaction in 90% or more of isolates; –, negative reaction in 90% or more of isolates; d+, positive reaction in 40%–90% of isolates; d –, positive reaction in 10%–40% of isolates.

[b] Indole production and nitrate reduction are the main characteristics used to differentiate *P. acnes* and while the majority of isolates are positive for these reactions, it is important to be aware that isolates negative for these tests do exist. For example, Sampedro et al. found that 15% of *P. acnes* isolates were negative for indole production.[54]

are 100% specific for type I organisms, they are not 100% sensitive. Types I and II can also be differentiated from each other based on cell wall sugar composition.[31] As highlighted in Section 13.1.1, the serotypes of *P. acnes*, *P. avidum*, and *P. propionicum* can be differentiated using cell wall agglutination and fluorescent antibody-labeling methods.

13.1.4.2 Molecular Detection

Molecular-based detection methods for bacterial infections have a number of notable advantages over traditional culture-based methods, including increased sensitivity and the opportunity to detect viable but nonculturable (VBNC) bacteria that arise due to stress or unfavorable growth conditions. Although dormant, such organisms may retain their pathogenicity and have the capacity to "resuscitate" and initiate infection.[113] In addition, components of nonculturable bacteria may drive an inflammatory response. Bacteria can enter the VBNC state when growing as a biofilm, which may impact on the cultivation of medical device-related infections. The administration of prophylactic antibiotics prior to surgery may also interfere with the growth of pathogenic species in tissue specimens giving rise to false-negative culture reports. Antibiotic challenge may also induce a dormant state akin to VBNC cells.[114] Another significant advantage of a molecular approach is the shorter time between receipt of a clinical sample and confirmation of infection, leading to earlier antimicrobial intervention and patient management/care. This is especially important when dealing with organisms, such as *P. acnes*, which can require extended anaerobic growth conditions, or infections due to *Mycobacterium tuberculosis*, which can be slow and cumbersome to detect using conventional culture methods.[115] Identification of bacteria based on phenotypic properties can also be unstable at times and dependent upon changes in environmental conditions such as growth substrate, temperature, and pH levels.[116]

Molecular-based detection of propionibacteria in clinical samples via conventional or real-time PCR amplification of the small subunit ribosomal RNA (16S rRNA) gene has been the predominant approach described in the literature (Table 13.2). Studies have utilized either species-specific 16S rRNA primers to directly target the bacteria or have used universal or broad-range primers that amplify all bacterial sequences present in the sample, followed by nested PCR with "internal" species-specific 16S rRNA primers or, more commonly, separation of the mixed amplicons by cloning and identification by comparison of the nucleotide sequences to those present in online databases, such as GenBank and the Ribosomal Database project.[119,125,129–131] While broad-range 16S rRNA PCR can facilitate the identification of multiple species causing polymicrobial infections, the method can be challenging since bacterial or DNA contamination can occur at any stage in the protocol, potentially leading to false-positive results. Sources of contamination range from surgical or laboratory personnel, nonsterile sites surrounding the clinical specimen, the specimen container, DNA extraction and PCR setup reagents and vials, pipettes, PCR cabinets, and aerosols.[132] Contamination may also occur due to "carryover" amplicons from previous PCR experiments, particularly when using a nested-PCR methodology. To avoid such problems, a range of preventative measures are required including the use of molecular grade reagents, protective clothing, sterile gloves, rooms specifically designated for DNA extraction and PCR setup, and containing equipment and reagents that must stay on site, filtered tips, HEPA-filtered safety cabinets with UV light for DNA degradation, chemical decontamination of work areas, incorporation of dUTP into PCR reactions and enzymatic treatment of new samples with Uracil-DNA-glycosylase, and the addition of appropriate control samples.

Scrupulous adherence to aseptic processing protocols and the inclusion of multiple negative control checks is particularly important for the detection of "cutaneous" propionibacteria in samples, even when adopting a species-specific rather than broad-range PCR methodology. While not conclusive, such an approach will provide some reassurance that the organisms detected are not contaminants. In addition, if *Propionibacterium* species are repeatedly isolated from an infection, this should provide very strong evidence that the bacterium is not a contaminant. For definitive evidence of propionibacterial infection, however, a polyphasic approach should be adopted where possible, including parallel culture experiments on agar plates (to assess bacterial counts) as well as other nonculture-based detection methods, such as Gram staining, immunofluorescence microscopy (IFM), and/or in situ hybridization (ISH). Furthermore, the potential presence of PCR inhibitors in clinical samples also means that internal amplification controls targeting human genes, such as the *β*-globulin gene, should also be included to rule out false-negative results.

In addition to PCR, ISH-based methodologies have also been used for the molecular identification of *P. acnes* in prostate and sarcoid tissue and to screen isolates from human blood cultures (Table 13.2). This method, which normally uses a single oligonucleotide probe (15–30 nucleotides) that is covalently linked to an enzymatic (e.g., horseradish peroxidase) or fluorescent (e.g., fluorescein isothiocyanate) reporter molecule (FISH), targets rRNA sequences within the bacterial ribosome.[133] As with IFM, it provides the opportunity to directly observe, in situ, microbial cells in a sample, and when used with confocal laser scanning microscopy can provide information on the morphology and spatial arrangement of infecting organisms.[134] Inherent problems can arise, however, including autofluorescence from tissue samples, photobleaching, insufficient penetration of the probe into the bacterial cell, and weak signals due to low ribosome number. The latter can be a particular problem with slow-growing or dormant organisms. Inaccessibility of the target sequence for hybridization can also occur due to the formation of hairpin loops, and poor signal intensities may further result from self-annealing and hairpin formation of the probe itself.[135]

In addition to detecting propionibacteria in clinical samples, molecular methods have also been used to characterize isolated organisms (Table 13.2). Differentiation of *P. acnes* isolates from other "cutaneous" propionibacteria has been achieved by PCR targeting of the 16S rRNA and lipase gene

TABLE 13.2
Species-Specific Oligonucleotide Primers and Probes for the Detection of *Propionibacterium* Species

Species Targeted	Gene/Method[a]	Primers/Probes	Sequences (5′–3′)	Anneal/Hybrid Temperature (°C)	Amplicon Size (bp)	Sample(s) Analyzed	Reference
P. acnes	16S rRNA;	PArA-1	AAGCGTGAGTGACGGTAATGGGTA	55–62	677	Prosthetic shoulder sonicate; isolates	53,54
	RT–PCR	PArA-2	CCACCATAACGTGCTGGCAACAGT				
P. acnes	16S rRNA;	PAC16S598	GCCCCAAGATTACACTTCCG	46	–	Isolates	117
	FISH						
P. acnes	16S rRNA;	Pa-F	AACCGCTTTCGCCTGTGA	62	182	Bronchoalveolar	118
	RT–PCR	Pa-R	ACGCTCAGGGTTAAGCCC	60	102	Lavage cells	
P. granulosum	16S rRNA;	Pg-F	ACATGGATCCGGGAGCTTC				
	RT–PCR	Pg-R	ACCCAACATCTCACGACACG				
P. acnes	16S rRNA;	PA-F′	GGGTTGTAAACCGCTTTCGCCT	54	587	Prostate; Sarcoid lymph/lung tissue;	41,119
	PCR	PA-R	GGCACACCCATCTCTGAGCAC	58	426	Vitreous fluid	120,121
P. granulosum	16S rRNA;	PG-1	GGAAACCTGATGCTTAACGT	50	587	Sarcoid lymph/lung tissue;	122
	PCR	PG-2	GGCACCAACCATCTCTGGAA			Vitreous fluid	41,120
P. acnes	16S rRNA;	$PA\text{-}F'^{b}_{M}$	GGGTTGTAA(A/T)CCGCTTTCGCCTG			Liver tissue	121
	PCR	$PA\text{-}R_M$	GGGACACCCATCTCTGAGCAC				63
P. acnes	23S rRNA; FISH	PAC23S[c]	GAGTGTGTGAACCGATCATGTAGTAGGCAA	75°C followed by 37°C	–	Prostate tissue	123
P. acnes	16S rRNA;	PAS9	CCCTGCTTTTGTGGGGTG	57	946/573[d]	Isolates	124
	SNt-PCR	PAS10	CGCCTGTGACGAAGCGTG	57	515/277[e]	Isolates	124
P. acnes/	Lipase;	PAS11	CGACCCCAAAAGAGGGAC				
P. avidum	SNt-PCR	Lip1	TCACTGATGAAGATCAACGCAC				
		Lip2	ACTTCTCCTGCAAGGTCAAG				
		Lip4	TGCGAATGTCCGACGAAGTCGA				
P. propionicum	16S rRNA;	PP-1	GACGGTAGCAGTAGAAGAAGCAC	55	567[f]	Isolates;	69,125
	Nt-PCR	PP-2	CTGTAAACCGACCAAAAAGG			Root canal samples	
P. acnes	16S rRNA;	PAC-F	GCGTGAGTGACGGTAATGGGTA	60	131	Sarcoid lymph tissue	126
	RT–PCR	PAC-R	TTCCGACGCGATCAACCA	42	–	Sarcoid lymph tissue	
	16S rRNA; ISH	PAC16S	ATGGATCGGGAGTGCTCAGAGATGGG				
P. acnes	16S rRNA;	Pa1	AAGGCCCTGCTTTTGTGG	62	387/160[g]	Vitreous fluid; Aqueous humour	127,128
	SNt-PCR	rPa2	TCCATCCGCAACCGCCGAA				
		rPa3	ACTCACGCTTCGTCACAG				

[a] RT–PCR = Real-time PCR; FISH = Fluorescent in situ hybridization; SNt-PCR = Semi-nested PCR; Nt-PCR = Nested-PCR; ISH = In situ hybridization.
[b] Primers $PA\text{-}F'_M$ and $PA\text{-}R_M$ are modification of primers PA-F' AND PA-R.[120]
[c] FISH probe PAC23S used with four probes generated from *P. acnes* strain KPA171202 by PCR and cloning (see Section 2.2.3).
[d] Amplicon sizes of 946/573 bp correspond to first round (PAS9/PAS11) and second round (PAS9/PAS10) products, respectively.
[e] Amplicon sizes of 515/277 bp correspond to first round (Lip1/Lip4) and second round (Lip1/Lip2) products, respectively. The SNt-PCR detects both *P. acnes* and *P. avidum*.
[f] Second round amplicon size after Nt-PCR with PP-1/PP-2. First round PCR utilizes universal 16S rRNA primers to give a ~1500 bp product (see Section 2.2.2).
[g] Amplicon sizes of 387/160 bp correspond to first round (Pa1/rPa3) and second round (Pa1/rPa2) products, respectively. The SNt-PCR will also give an amplicon with *Actinomyces israelii*, but the product can be differentiated from the *P. acnes* product by digestion with *Xho*I.

sequences.[124] As mentioned earlier, nucleotide sequence analysis of the *P. acnes recA* and *tly* gene loci facilitates unambiguous determination of phylotype status (IA, IB, II, III).[22,30] Specific PCR primers have now also been described for differentiation of *P. acnes* phylotypes, with type IA-specific primers targeting a solute-binding protein (SBP) gene, and type IB primers amplifying part of a putative Y40U fragment which brackets a point of sequence divergence between type IB and types IA and II. Type II can be differentiated from IA and IB by primers, which amplify within putative RHS-family proteins, and bracket a ~3450-bp deletion in type II.[66]

Molecular methods have also been applied to propionibacteria in an attempt to differentiate the species and type individual strains. These methods have included ribotyping, Random Amplification of Polymorphic DNA (RAPD), which is a low stringency PCR that uses one or sometimes several primers of short length to generate a unique genetic fingerprint, and Amplified Fragment Length Polymorphism Analysis (AFLP), a technique based on the selective amplification of DNA fragments produced from a total restriction enzyme digest of genomic DNA.[136–138] Macrorestriction digestion of genomic DNA followed by pulse field gel electrophoresis has also been used to type propionibacteria isolates.[64,95,139]

13.2 METHODS

13.2.1 Sample Preparation

13.2.1.1 Preparation of Sonicate from Retrieved Medical Devices

To significantly improve the detection of *P. acnes* infections associated with indwelling medical devices, such as prosthetic joints, catheters, and heart valves, it is imperative that samples are processed under strict anaerobic conditions and adherent bacteria removed by sonication. The procedure described is based on a modification of Tunney et al.[140]

Collection of Devices

1. When immediately removed from the patient, each component of an implant (e.g., femoral and acetabular components of a prosthetic hip) should be placed aseptically into a sterile bag and inserted into an anaerobic jar (e.g., Oxoid Anaerobic Jar HP0011, Oxoid Ltd.) containing a catalyst (e.g., Oxoid Low Temperature Catalyst BR0042, Oxoid Ltd.) and a Gaspak (e.g., Oxoid Gas Generating Kit BR0038, Oxoid Ltd.).
2. The jar should be transported immediately to the laboratory and placed inside an anaerobic cabinet (e.g., Don Whitley Mk III, Don Whitley Scientific Ltd.) containing an atmosphere of 80% N_2, 10% H_2, and 10% CO_2 (v/v/v), where it can be opened.

Sample Processing

1. Pre-reduce quarter-strength Ringer's solution (25%, v/v) by the addition of L-cysteine (0.05%, w/v) and remove oxygen by boiling at 100°C for 1 h or microwaving at full power for 5 min. Upon cooling, store the solution in the anaerobic cabinet prior to use.
2. Add pre-reduced Ringer's solution (100 mL) to the sterile bag containing the removed device and seal the bag with a sterile plastic clip. Gently agitate the bag by hand for 1–2 min to remove any extraneous blood and tissue and aspirate and discard the resulting solution. Add fresh sterile pre-reduced Ringer's solution (100 mL) and reseal the bag.
3. Remove the bag from the cabinet and dislodge any adherent bacteria on the implant into the Ringer's solution by mild ultrasonication (5 min in a 150 W ultrasonic bath operating at a frequency of 50 kHz). Transfer the Ringer's solution from the sealed bag into 50 mL sterile centrifuge tubes inside the cabinet. Centrifuge the tubes for 20 min at 10,000 × *g* and 4°C and transfer back to the cabinet. Collectively resuspend the resulting pellets in a total volume of 5 mL Ringer's solution (twenty fold concentration).
4. Total viable counts are performed by spread plating (within the anaerobic cabinet) 0.5 mL volumes of the resulting concentrated sonicate sample onto 3 × blood agar plates [BA; CM0271; Oxoid Ltd.; 5% (v/v) horse blood] and 3 × anaerobic blood agar plates [ABA; CM0972; Oxoid Ltd.; 5% (v/v) horse blood], which have been stored in the anaerobic cabinet prior to use. Incubate all BA and ABA plates at 37°C aerobically and anaerobically, respectively, for up to 14 days. The use of BA plates will allow the simultaneous identification of pathogenic aerobes in addition to propionibacteria and other anaerobes.
5. For molecular detection, extract DNA from 1 mL of sonicate using the methods described below. Store any remaining sample at –80°C for further analyses.

13.2.1.2 DNA Extraction from Clinical and Bacterial Samples

Different methods have been used for the extraction of bacterial DNA from clinical samples associated with propionibacteria infections, including prostate, lymph node, liver and prosthetic joint-associated tissue, vitreous and aqueous humors, root canal and sonicate samples, and bronchoalveolar lavage cells (Table 13.3).[41,49,51,53,63,118–120,125–127,129,140–142] DNA has been extracted from freshly collected samples and has also been isolated from paraffin-embedded tissue samples. Methods have included the preparation of "crude" DNA from root canal samples by simple boiling to purification of genomic DNA from tissue; the latter often based on the use of commercially available kits. These methods can also be applied to the isolation of DNA from pure cultures of propionibacteria.

(i) Purification of Bacterial Genomic DNA The following method is based on that previously described by Tunney et al.

TABLE 13.3
Summary of Different Methods Used to Extract DNA from Clinical Samples Associated with *Propionibacterium* Infections

Reference[a]	Extraction Method[b]	Sample Types	Disease
53	Qiagen QIAamp DNA mini Kit	Sonicate from failed prosthetic shoulder implant	Prosthetic joint infection
118	Roche MagNA Pure Compact Nucleic acid isolation kit	Human vitreous fluid; aqueous humor; cell suspensions	Bacterial endophthalmitis; corneal ulcers
141	Qiagen QIAamp DNA mini Kit; Phenol:Chloroform extraction	Bronchoalveolar lavage cells	Sarcoidosis
119	Qiagen QIAamp DNA mini Kit; MasterPure DNA purification kit	Core samples from thawed prostate tissue and seminal vesicle samples	Adenocarcinoma of the prostate gland
	Qiagen QIAamp DNA mini Kit; Puregene DNA purification kit; Phenol:Chloroform extraction	Section of human prostate tissue from frozen tissue block	
63	DEXPAT™ kit and analysis of crude extract[a]	Laser capture microdissection of paraffin-embedded human liver tissue	Primary biliary cirrhosis
129	Qiagen QIAamp DNA Blood Mini Kit	Section of paraffin-embedded human prostate tissue	Benign prostatic hyperplasia
142	Phenol:Chloroform extraction	Human vitreous fluid	Uveitis associated with sarcoidosis
69,125	Boiling and analysis of a crude supernatant	Human root canal samples	Periodontitis
126	Qiagen QIAamp Tissue kit	Section of paraffin-embedded human lymph node tissue	Sarcoidosis
41	Phenol:Chloroform extraction	Section of paraffin-embedded human lung and lymph node tissue	Sarcoidosis
50	Phenol:Chloroform extraction	Sonicate from failed prosthetic hip implant	Prosthetic joint infection
120	DEXPAT™ kit and Phenol:Chloroform extraction[a]	Section of paraffin-embedded human lymph node tissue	Sarcoidosis
127	Phenol:Chloroform extraction	Human vitreous fluid; aqueous humor	Bacterial endophthalmitis

[a] References are given in chronological order starting with most recent.
[b] DEXPATTM = DNA Extraction from Paraffin-embedded Tissue.

for the purification of bacterial DNA from sonicate samples but could be applied to other sample types or pure cultures.[140]

1. Centrifuge 1 mL of sample at 10,000 × *g* for 15 min at room temperature and discard the supernatant.
2. Add lysozyme (1–2 mg/mL) in 10 mM Tris pH 8.0 and 50 mM glucose to aid extraction of the bacterial DNA.
3. Resuspend the pellet by vortex mixing in 200 µL cell lysis buffer [10 mM Tris pH 8.0, containing 5 mM EDTA, 0.5% (w/v) sodium dodecyl sulfate, and 100 µg/mL proteinase K] and incubate for 3 h at 55°C. Adjust the temperature to 37°C and leave overnight. The next morning, raise the temperature again to 55°C for 1 h.
4. Add an equal volume of saturated phenol/chloroform to the sample and mix by inversion before centrifugation at 10,000 × *g* for 15 min. This step can be repeated.
5. Remove the aqueous supernatant to a fresh tube making sure the interface is left behind.
6. Add ethanol (100%; 2.5 vols), 3 M sodium acetate, pH 5.2 (0.1 vol), and 2 µL SeeDNA (GE Healthcare) and mix by inversion followed by centrifugation at 10,000 × *g* for 15 min at room temperature.
7. Discard the supernatant and wash the pelleted DNA with 1 mL of 70% (v/v) ethanol. Centrifuge the sample at 10,000 × *g* for 15 min at room temperature and remove the resulting supernatant. Dry the pellet in a vacuum dryer.
8. Dissolve the extracted DNA in TE buffer (10 mM Tris pH 8.0, 1 mM EDTA) and determine the concentration by UV absorbance at 260/280. The sample can be stored at –20°C.

Note: In previous studies with pure cultures, *P. acnes* has been grown in the presence of penicillin G to help weaken its thick gram-positive cell wall prior to DNA extraction.[137] The use of additional enzymes such as lysostaphin or achromopeptidase in addition to lysozyme may improve cell wall lysis of propionibacteria.[123]

(ii) Preparation of Bacterial Genomic DNA by Boiling If DNA from propionibacteria isolates is intended only for PCR, then preparation by boiling is much faster, unless a high-quality DNA template is required for PCR.[66]

1. Wash a small loop full of bacteria by suspension in 1 mL of phosphate buffered saline, pH 7.2, followed by centrifugation at 10,000 × *g* for 5 min to pellet the bacteria.
2. Resuspend the pellet in 100–200 µL of ultrapure water, boil for 10 min, and then chill on ice.

3. Centrifuge at 15,000 × *g* for 2 min to remove cell debris and store the supernatant at –20°C pending analysis.

13.2.1.3 Preparation of Samples for Fluorescent In Situ Hybridization

Preparations of samples for FISH are as described by Poppert et al. for pure bacterial cultures and by Alexeyev et al. for sections of tissue.[117,123]

1. For analysis of pure cultures, bacteria can be grown in broth or agar plates. Wash the bacteria and resuspend to the required concentration in 0.9% (w/v) saline. Apply approximately 10 μL of the bacterial suspension to an 8 well-glass slide, air dry, and fix for 10 min in pure methanol. Note: as a guide, an OD (600 nm) of 0.3 corresponds to approximately 10^8 CFU/ml for *P. acnes*. Remove the slides, air dry, and analyze immediately or store at –20°C for several weeks.
2. To help permeabilize the peptidoglycan layer prior to hybridization, treat fixed cultures with 6 μL of a solution of lysozyme (2 mg/mL) and lysostaphin (0.1 mg/mL) in 10 mM Tris-HCl pH 8.0, for 10 min at 36°C. Stop the reaction by incubation in methanol for 3 min.
3. For tissue sections embedded in paraffin, dewaxing in xylene is mandatory. Dewax specimens in 100% xylene for 20 min and then rehydrate in an ethanol series (100%, 70%, 30%, and 0%) in 10 mM Tris, pH 7.5. Treat sections with lysozyme (1 mg/mL; Sigma-Aldrich) and achromopeptidase (30 U/mL; Sigma-Aldrich) in 10 mM Tris, pH 7.5 at 37°C for 25 min, wash for 5 min in 10 mM Tris, pH 7.5; and dehydrate in an ethanol series (30%, 70%, and 100%) in 10 mM Tris, pH 7.5.

13.2.2 Detection Procedures

13.2.2.1 PCR Detection of *Propionibacterium* Species

Principle. Based on the primers and protocols described for the detection of *P. acnes* and/or *P. granulosum* in human prostate tissue, sarcoid tissue and vitreous fluid from sarcoidic eyes with uveitis, and *P. propionicum* in human root canal samples.[69,119,120,121,122]

Procedure

1. The PCR mixture consists of 1× PCR buffer (20 mM Tris-HCl, pH 8.4, 50 mM KCl), 200 μM dNTPs, 300 nM each primer for *P. acnes* (PA-F'/PA-R) and *P. granulosum* (PG-1/PG-2) (Table 13.2) and 1 μM each primer for *P. propionicum* (PP-1/PP-2; Table 13.2), 2 mM $MgCl_2$, 1.25 U Platinum Taq Hotstart DNA polymerase (Invitrogen Life Technologies), 1 μg of extracted genomic DNA and ultrapure water to a final volume of 25 μL. A negative control sample (water in place of DNA) should be included in each PCR run.
2. Run the following programs on a thermal cycler: for *P. acnes* and *P. granulosum*, initial denaturation at 95°C for 3 min; 40 cycles of 95°C for 60 s, primer pair annealing temperature (Table 13.2) for 60 s, 72°C for 60 s; and a final extension step at 72°C for 10 min. For *P. propionicum*, initial denaturation at 95°C for 2 min, 36 cycles of 95°C for 30 s, 55°C for 60 s, 72°C for 60 s; and a final extension step at 72°C for 2 min.
3. After PCR amplification, analyze 5–10 μL of each amplicon on a 1.5% (w/v) agarose gel in 1× Tris-acetate-EDTA buffer containing 0.5 μg/mL ethidium bromide or, alternatively, 1:10,000 dilution of GelRed (Cambridge Bioscience). A DNA molecular size marker should be run alongside and the resolved products visualized and photographed under UV light.

Note: PCR method can be applied to clinical samples and used to identify isolates.

13.2.2.2 Nested and Semi-Nested PCR Detection of *Propionibacterium propionicum* and *Propionibacterium acnes*

Principle. Based on the primers and protocols described for nested-PCR detection of *Propionibacterium propionicum* in human root canal samples and semi-nested PCR detection of *Propionibacterium acnes* in human vitreous samples.[125,128]

Procedure

1. For nested-PCR detection of *P. propionicum*, the first round consists of the amplification of nearly full length 16S rRNA using the universal forward primer 5′-AGAGTTTGATCCTGGCTCAG-3′ and reverse primer 5′-ACGGCTACCTTGTTACGACTT-3′. The PCR mixture consists of 1× PCR buffer (20 mM Tris-HCl pH 8.4, 50 mM KCl), 200 μM dNTPs, 200 nM each primer, 2 mM $MgCl_2$, 1.25 U *Tth* DNA polymerase (Biotools), 2 μL DNA preparation, and ultrapure water to a final volume of 25 μL. A negative control sample (water in place of DNA) should be included in each PCR run.
2. Run the following program on a thermal cycler: initial denaturation at 97°C for 60 s; 26 cycles of 97°C for 45 s, 55°C for 45 s, 72°C for 60 s; and a final extension step at 72°C for 4 min.
3. After first round amplification, 1 μL of the resulting product is used as the template for the nested species-specific reaction. PCR set up conditions are as described in (1) but with 100 nM of the *P. propionicum* primers PP-1/PP-2 (Table 13.2). Amplification conditions are initial denaturation at 95°C for 2 min; 26 cycles of 95°C for 30 s, 55°C for 60 s, 72°C for 60 s; and a final extension step at 72°C for 2 min.
4. For seminested PCR detection of *P. acnes*, the first round PCR mixture consists of 1× PCR buffer (10 mM Tris-HCl, pH 8.3, 50 mM KCl, 0.01% (w/v) gelatin, 1.5 mM $MgCl_2$), 200 μM dNTPs, 1 μM each of primers Pa1/rPa3 (Table 13.3), 0.5 U *Taq* DNA polymerase

(Stratagene), 10 ng DNA, and ultrapure water to a final volume of 25 μL. A negative control sample (water in place of DNA) should be included in each PCR run.

5. Run the following program on a thermal cycler: initial denaturation at 94°C for 5 min; 34 cycles of 94°C for 30 s, 62°C for 60 s, 72°C for 2 min; and a final extension step at 72°C for 8 min.
6. After first round amplification, 1 μL of the resulting product is used as the template for the seminested reaction. PCR set up and cycling conditions are as described in (4) and (5) but using the primer combination Pa1/rPa2 (Table 13.2) and only 24 amplification cycles.
7. After PCR amplification, analyze 5–10 μL of each amplicon on a 1.5% (w/v) agarose gel in 1 × Tris-acetate-EDTA buffer containing 0.5 μg/mL ethidium bromide or, alternatively, 1:10,000 dilution of GelRed (Cambridge Bioscience). A DNA molecular size marker should be run alongside and the resolved products visualized and photographed under UV light.

Note: When setting up the PCR master mix for broad-range 16S rRNA PCR, all reagents except the primers and template DNA can be UV irradiated for 15 min to destroy any artificial DNA.[143] This may, however, reduce the detection limit of the PCR. Particular care must also be taken to ensure that samples are not contaminated by "carry over" products from first round amplification. The seminested PCR described for *P. acnes* will also generate an amplicon from *Actinomyces israelii*, but the product can be differentiated by digestion with the restriction enzyme *Xho*I.

13.2.2.3 Fluorescent In Situ Hybridization for the Detection of *Propionibacterium acnes*

Principle. Based on the hybridization protocol described for the detection of *P. acnes* in prostate tissue.[123]

Procedure

1. Incubate approximately 100–120 ng each of the *P. acnes*-specific Cy3.5-tagged probe PAC23S (Table 13.2), plus four additional *P. acnes*-specific probes generated by PCR amplification and cloning (see notes), all in 10 μL of 40 mM sodium phosphate buffer, pH 7.0, containing 50% (v/v) formamide, 2× Saline-Sodium Citrate buffer (SSC) (1× SSC is 0.15 M NaCl plus 0.015 M sodium citrate), 10% (w/v) dextran sulfate, 0.1% (w/v) sodium dodecyl sulfate, and 1× Denhardt's solution, on the slide at 75°C for 5 min, followed by overnight incubation at 37°C. Denhardt's solution comprises high molecular weight polymers that are capable of saturating nonspecific binding sites and artificially increasing the concentration of probe available for hybridization.
2. After hybridization, wash the slides with 2% SSC containing 0.3% Nonidet P-40 (Biochem) and stain with 4,6-diamidino-2-phenylindole dihydrochloride (DAPI; Sigma).
3. Wash off the DAPI stain from the slide with phosphate buffered saline and allow to air dry. Mount the slides in Citi-fluor antifade solution (Citi-fluor, England) and seal coverslips with nail varnish.
4. Examine slides under a fluorescence microscope and capture different fields of view electronically.

Note: In addition, Alexeyev et al. also utilized four large *P. acnes* probes (size range, 5–6 kb) that overlapped a 14,313 bp segment of the KPA171202 genome.[123] These four fragments were synthesized by using Elongase enzyme mix (Invitrogen) and the following primer sets: Primers 1L (CCAAAGACAATTCTGGGAAGATTACAG, nucleotides 320829–320855) and 1R (ATACAACTTC GGCCTTTACACTACAGC, nucleotides 326085–326059); Primers 2L (AAAATGTGAAGTGTGAGGTGATTCTGC, nucleotides 325533–325559) and 2R (ATTAAGGTG GAATGGTTCTACTGGTTG, nucleotides 331266–331240); Primers 3L (ATAATGAGGATGGTGCAGACCAGTAG, nucleotides 330317–330342) and 3R (GATGGAGGA AAT CACATATAGCGAGAC, nucleotides 335700–335674); Primers 4L (GAGTGGATCTTCTTCGTCAACGTC, nucleotides 335119–335142) and 4R (AGAACTCGTCCTTCTTCA ACTTCCAAC; nucleotides 341090–341064). Products were cloned by using a Topo XL PCR cloning kit (Invitrogen) and the plasmids isolated by use of the QIAGEN midi kit. The libraries were then directly labeled with diethylaminocoumarin-dUTP (DEC) (Perkin-Elmer) by using a nick translation kit (Vysis Inc.).

13.2.2.4 Nucleotide Sequence Analysis for Differentiation of *Propionibacterium acnes* Phylotypes

Principle. Based on sequence analysis of the *recA* housekeeping gene and the putative hemolysin/cytotoxin gene *tly*.[22,30]

Procedure

1. The PCR mixture consists of 1× PCR buffer (20 mM Tris-HCl pH 8.4, 50 mM KCl), 200 μM dNTPs, 200 nM each of primers PAR-1/PAR-2 (Table 13.4), 1.5 mM $MgCl_2$, 1.25 U Platinum Taq Hotstart DNA polymerase (Invitrogen), 2 μL of DNA (prepared by boiling method), and ultrapure water to a final volume of 25 μL. A negative control sample (water in place of DNA) should be included in each PCR run.
2. Run the following program on a thermal cycler: one cycle at 95°C for 3 min; 35 cycles of 95°C for 60 s, 55°C for 30 s, 72°C for 90 s; and a final extension step at 72°C for 10 min.
3. After PCR amplification, analyze 5–10 μL of each amplicon on a 1.5% (w/v) agarose gel in 1× Tris-acetate-EDTA buffer containing 0.5 μg/mL ethidium bromide or, alternatively, 1:10,000 dilution of GelRed (Cambridge Bioscience). A DNA molecular size marker should be run alongside and the resolved products visualized and photographed under UV light.

4. Prior to sequencing, purify PCR products (5–10 μL) using a QIA-quick PCR purification kit (Qiagen). This kit provides spin columns, buffers, and collection tubes for silica membrane-based purification of PCR products >100 bp in size. After purification, estimate the concentration of the PCR product by running a small amount (1 μL) on a 1.5% (w/v) agarose gel (as described earlier) alongside a DNA ladder with known mass amounts per fragment.
5. To sequence, use a BigDye Terminator v 3.1 Cycle Sequencing Kit, which provides all the necessary reagents in a premixed, ready reaction format (Applied Biosystems). As described by the manufacturer, the sequencing mixture should consist of ×1 BigDye sequencing buffer, ×1 Ready Reaction Premix, 3.2 pmol of the relevant sequencing primer, 10–20 ng of the purified PCR product and ultrapure water to a final volume of 20 μL, all in a 0.2 mL microcentrifuge tube. Keep all samples on ice during preparation. Place samples on a thermocycler and carry out sequencing under the following conditions: one cycle at 96°C for 1 min; 25 cycles of 96°C for 10 s, 50°C for 5 s, 60°C for 4 min.
6. Purify sequenced samples using a Montage SEQ_{96} Sequencing Reaction Cleanup Kit (Millipore) according to the manufacturer's instructions and analyze on an ABI PRISM® analyzer, such as the ABI PRISM® 3100 genetic analyzer. After analysis, open the electropherogram files using a sequence analysis software program such as Chromas or EditView. Phylotype status can be determined by comparison of resulting sequences with those in online databases.

Note: The primers PAR-1 and PAR-2 target sequences which flank the 5′ and 3′ ends of the *recA* open reading frame, respectively, while PAT-1 and PAT-2 similarly target sequences that flank the 5′ and 3′ ends of the *tly* open reading frame. Complete gene sequences can be obtained by using both the forward and reverse primers for each locus. Type-specific polymorphisms in the *recA* and *tlyA* genes will facilitate phylotyping of each isolate; type I isolates (IA and IB) differ from type II in 10 specific bases within the *recA* locus, while the *tly* gene shows greater variation with 18 specific differences between type I (IA and IB) and type II.[30] Type IB strains contain an additional polymorphism at nucleotide position 897 in *recA* and two polymorphisms at positions 375 and 766 in *tly* (which match those in type IIs) compared to type IA. Some strains of IB may also contain an additional polymorphism in *recA* (McDowell & Patrick, unpublished). Strains within the type III clade contain a combination of four type I-specific polymorphisms (common to IA and IB), six type II-specific polymorphisms, and five polymorphisms unique to the group.[22]

13.2.2.5 PCR Differentiation of *Propionibacterium acnes* Types IA, IB, and II

Principle. Based on the method of Shannon et al.[66] Variations in DNA sequence between types IA, IB, and II were detected by RAPD using a range of random primers. Amplified bands specific to each phylotype fingerprint were then sequenced and used as the basis for the design of type-specific primers.

Procedure

1. The PCR mixture consists of 1× PCR buffer containing 1.5 mM $MgCl_2$ (Qiagen), 200 μM dNTPs, 500 nM each primer (Table 13.4), 0.75 U Hotstar Taq DNA polymerase (Qiagen), 2 μL of purified genomic DNA (Qiagen DNA Extraction Kit), and ultrapure water to a final volume of 20 μL. A negative control sample (water in place of DNA) should be included in each PCR run.
2. Run the following program on a thermal cycler: 15 min at 95°C to activate the HotStar Taq; 35 cycles of 30 s at 94°C, 30 s at annealing temperature (65°C for PR245/247; 55°C for the other primer sets), 72°C for 1 min; and a final extension step at 72°C for 10 min.
3. After PCR amplification, analyze 10–20 μL on a 1.5% (w/v) agarose gel in 1× Tris-acetate-EDTA buffer containing 0.5 mg/mL ethidium bromide or, alternatively, 1:10,000 dilution of GelRed (Cambridge

TABLE 13.4
PCR Primers for Phylotyping of *Propionibacterium acnes* Isolates

Reference	Specificity	Primer	Sequences (5′–3′)	Amplicon Size (bp)	Target Loci	Note
22,30	All phylotypes	PAR-1	AGCTCGGTGGGGTTCTCTCATC	1201	*recA*	Identify phylotype by nucleotide sequencing of amplicons. Primers target sequences flanking the open reading frames of both genes.
		PAR-2	GCTTCCTCATACCACTGGTCATC			
	All phylotypes	PAT-1	CAGGACGTGATGGCAATGCGA	909	*tly*	
		PAT-2	TCGTTCACAAGACCACAGTAGC			
66	Type IA	PR213	GACGATTACTGGGGCAAGAA	584	*sbp*	With the primer pair PR245/247, type II isolates generate a much smaller PCR product compared with IA and IB
	Type IB	PR216	GGGGTGAACTTGACGTTGAT	725	*Y40U*	
	Type IA, IB, II	PR256	GTCATCGGCTTCTGGATCAT	4027 (IA, IB)	*RHS*	
		PR257	CCTGGTTTCCACGATGTCTT			
		PR245	ATAAACCCATCGGCGGCTCGAT	573 (II)		
		PR247	AGCAAGGGAGCATTCAGTGG			

Bioscience). A DNA molecular size marker should be run alongside and the resolved products visualized and photographed under UV light.

Note: For optimal performance of these type-specific PCRs, and especially for successful amplification of the 4027-bp fragment from types IA and IB with PR245/247, it is important that purified genomic DNA is used as a good quality template. The primers PR213/PR216 amplify a portion of the solute binding-protein gene (*sbp*) of a peptide uptake operon that is within an 8.7-kb region present between nucleotides 1579146 and 1579405 in type IA isolates only (GenBank accession number DQ208967). The primer combination of PR256/PR257 amplify part of a putative Y40U fragment (nucleotides 948634–949358), which brackets a region of sequence divergence between types IA and IB, with the PR256 binding site being present only in type IB and either absent or significantly different in sequence in type II. Finally, the primer pair PR245/PR247 amplify putative RHS-family proteins (nucleotide 246705–250731) and bracket a ~3450-bp deletion in type II, giving rise to a smaller amplicon compared to types IA and IB.

13.3 CONCLUSIONS AND FUTURE PRESPECTIVES

Within the genus *Propionibacterium*, only those organisms that are members of the "cutaneous" group are associated with human disease. Although *P. granulosum* and *P. avidum* have been found to cause opportunistic human infections, in practical terms it is *P. acnes* and to a lesser extent *P. propionicum* that are associated with the overwhelming majority of propionibacteria infections in humans. The relationship between the "cutaneous" propionibacteria and humans is, under normal circumstances, a positive one. They are an important component of the normal microbiota on human skin where they help maintain a healthy ecosystem and prevent colonization with more pathogenic microbes.[144] Due to their potent immunomodulatory properties, their use as adjuvants may also be beneficial in the treatment of various cancers.[145] In terms of disease, *P. acnes* is probably the most well-known species within the *Propionibacterium* genus due to its association with acne. There is, however, a growing body of evidence to suggest that its involvement in other infections and diseases has been significantly underestimated, especially in relation to infections of indwelling medical devices where it will persist as an adherent biofilm.[49,53] This has occurred, in part, due to the traditional and overly simplistic view that *P. acnes* is a relatively harmless member of the human microbiota and therefore its presence in clinical samples simply reflects contamination. In addition, failure of prosthetic orthopaedic implants due to infection is often not suspected with *P. acnes* (or looked for postoperatively) due to subtle signs and symptoms and, consequently, patients are often misdiagnosed preoperatively as presenting with aseptic loosening. To compound this, in some laboratories anaerobic processing and culture of specimens may either not occur or be inappropriate resulting in poor diagnostic sensitivity for *P. acnes* and other propionibacterial infections. To improve detection, samples should be processed under strict anaerobic conditions, retrieved indwelling devices sonicated to ensure adherent organisms are removed, and cultures incubated for at least 7 days, and ideally up to 14 days, to prevent a false-negative result. While culture-based detection remains the mainstay of the diagnostic bacteriologist, molecular-based analysis of specimens is a valuable adjunct and, due to much shorter processing times, can lead to faster diagnosis as well as the capacity to identify VBNC organisms. This may be especially important in relation to prosthetic hip joint revision arthroplasty, where rapid and reliable identification of propionibacteria infection on the removed device may significantly impact on patient treatment and management. If available, the application of real-time PCR for the detection of propionibacteria in samples can offer notable improvements on conventional PCR methods due to reduced assay times, postamplification analysis, and increased specificity and sensitivity. Background amplification products can also be identified based on quantification and melting curve analysis. Problems can be encountered when attempting to optimally extract DNA from propionibacteria due to the thickness of their gram-positive cell envelope and low sensitivity to lysozyme and proteinase K. This could impact on detection limits for PCR or FISH diagnosis when using tissue samples and therefore attention should be paid to the use of appropriate enzymatic treatments that may be more appropriate for lysis, such as achromopeptidase.[123]

The molecular identification of novel phylotypes within *P. acnes* has stimulated investigations of whether these distinct groups vary in their ability to cause disease, and are, therefore, clinically important. Their discovery has also raised important issues regarding how we now compare and contrast data generated in previous *P. acnes*-related studies where the phylotype status of isolates was not known. Continued and improved molecular epidemiological studies will hopefully help to address these issues. The molecular detection of phylotypes associated with greater pathogenicity could prove especially valuable information for the diagnostic bacteriologist as they attempt to confirm whether a *P. acnes* isolate is clinically relevant. The availability of the *P. acnes* complete genome sequence will undoubtedly facilitate the development of improved molecular detection and typing platforms for the bacterium, and it is hoped that the complete genome sequences of other "cutaneous" species and *P. propionicum* will eventually be determined. This will facilitate direct genomic comparisons, give important and much needed insights into the pathogenic potential of these other organisms, and provide a foundation for improved diagnostics. The development of DNA-microarrays based on such *Propionibacterium* genome sequences would also facilitate transcriptomic profiling and the potential identification of propionibacterial genes upregulated during disease. These loci could then provide novel diagnostic platforms for the detection of infection versus contamination by reverse-transcriptase PCR.

ACKNOWLEDGMENTS

The authors acknowledge the Northern Ireland Health and Social Care Research & Development Office (NIHSC R&D Office), Arthritis Research UK, the British Orthopaedic Society Wishbone Trust and The Prostate Cancer Charity UK, for funding to investigate the role of *Propionibacterium acnes* in orthopaedic-related infections and prostate cancer.

REFERENCES

1. Stackebrandt, E. et al., *Propionimicrobium* gen. nov., a new genus to accommodate *Propionibacterium lymphophilum* (Torrey 1916) Johnson and Cummins 1972, 1057AL as *Propionimicrobium lymphophilum* comb. nov., *Int. J. Syst. Evol. Microbiol.*, 52, 1925, 2002.
2. Yokota, A. et al., Transfer of *Propionibacterium innocuum* Pitcher and Collins 1991 to *Propioniferax* gen. nov. as *Propioniferax innocua* comb. nov., *Int. J. Syst. Bacteriol.*, 44, 579, 1994.
3. Patrick, S., and McDowell, A., *Propionibacterium* Orla-Jensen 1909, 337[AL]. In Goodfellow, M., Kämpfer, P., Busse, H.J., Trujillo, M., Suzuki, K., Ludwig, W., and Whitman, W.B. (Editors), Volume 5, *The Actinobacteria, Bergey's Manual of Systematic Bacteriology*, 2nd ed., Springer, New York (in press).
4. Meile, L., Le Blay, G., and Thierry, A., Safety assessment of dairy microorganisms: *Propionibacterium* and *Bifidobacterium*, *Int. J. Food Microbiol.*, 126, 316, 2008.
5. Hayashi, S., and Furusaka, C., Enrichment of *Propionibacterium* in paddy soils by addition of various organics substances, *Ant. v. Leeuwenhoek*, 46, 313, 1980.
6. Plastourgos, S., and Vaughn, R.H., Species of *Propionibacterium* associated with zapatera spoilage of olives, *Appl. Microbiol.*, 5, 267, 1957.
7. Cancho, F.G., Navarro, L.R., and de la Borbolla y Alcala, R., La formacion de acido propionico durante la conservacion de las aceitunas verdes de mesa. III. Microorganismos responsables, *Grasas Aceites*, 31, 245, 1980.
8. Prévot, A.-R., and Fredette, V., *Manual for the Classification and Determination of the Anaerobic Bacteria*, 345 pp, Lea & Febiger, Philadelphia, PA, 1966.
9. Bollgren, I. et al., Periurethral anaerobic microflora of healthy girls, *J. Clin. Microbiol.*, 4, 419, 1979.
10. Lyons, J.A. et al., Isolation of *Propionibacterium acnes* from a case of placentitis and abortion in a cow, *J. Vet. Diagn. Invest.*, 21, 274, 2009.
11. Oomi, H. et al., *Propionibacterium* from liver and lung of pigs and guinea pigs, 1: Physiological identification, *Anaerobe*, 8, 223, 2002.
12. Harvey, R.G., Noble, W.C., and Lloyd, D.H., Distribution of propionibacteria on dogs: A preliminary report of the findings on 11 dogs, *J. Small Anim. Pract.*, 32, 80, 2008.
13. McLorinan, G.C. et al., *Propionibacterium acnes* wound contamination at the time of spinal surgery, *Clin. Orthop. Relat. Res.*, 437, 67, 2005.
14. Patel, A. et al., *Propionibacterium acnes* colonization of the human shoulder, *J. Shoulder Elb. Surg.*, 2009.
15. McGinley, K.J., Webster, G.F., and Leyden, J.J., Regional variations of cutaneous propionibacteria, *Appl. Environ. Microbiol.*, 35, 62, 1978.
16. Koussemon, M. et al., *Propionibacterium microaerophilum* sp. nov., a microaerophilic bacterium isolated from olive mill wastewater, *Int. J. Syst. Evol. Microbiol.*, 51, 1373, 2001.
17. Kusano, K. et al., *Propionibacterium cyclohexanicum* sp. nov., a new acid-tolerant omega-cyclohexyl fatty acid-containing *Propionibacterium* isolated from spoiled orange juice, *Int. J. Syst. Bacteriol.*, 47, 825, 1997.
18. Bernard, K.A. et al., *Propionibacterium australiense* sp. nov. derived from granulomatous bovine lesions, *Anaerobe*, 8, 41, 2002.
19. Bowden, G.H., Actinomyces and arachnia. In Duerden, B.I., and Drasar, B.S. (Editors), *Anaerobes in Human Disease*, p. 132, Edward Arnold, London, 1991.
20. Willis, A.T., *Anaerobic Bacteriology: Clinical and Laboratory Practice*, 3rd ed., Cambridge University Press, London, Butterworths, 1977.
21. Buchanan, B.B., and Pine, L., Characterization of a propionic acid producing actinomycete, *Actinomyces propionicus*, sp. nov., *J. Gen. Microbiol.*, 28, 305, 1962.
22. McDowell, A. et al., A new phylogenetic group of *Propionibacterium acnes*, *J. Med. Microbiol.*, 57, 218, 2008.
23. Skogen, L.O., Reinbold, G.W., and Vedamuthu, E.R., Capsulation of *Propionibacterium*, *J. Milk Food Technol.*, 37, 314, 1974.
24. Stackebrandt, E., Cummins, C.S., and Johnson, J.L., Family *Propionibacteriaceae*: The genus *Propionibacterium*. In Dworkin, M., Falkow, S., Rosenberg, E., Schleifer, K., and Stackebrandt, E. (Editors), *The Prokaryotes*, p. 400, Springer Verlag, New York, 2006.
25. van Schalkwyk, C., Joubert, H., and Britz, T.J., Extracellular polymer production and potential for aggregate formation by classical propionibacteria, *World J. Microbiol. Biotechnol.*, 19, 285, 2003.
26. Moss, C.W. et al., Cultural characteristics and fatty acid composition of propionibacteria, *J. Bacteriol.*, 97, 561, 1969.
27. Delwiche, E.A., Vitamin requirements of the genus *Propionibacterium*, *J. Bacteriol.*, 58, 395, 1949.
28. Ferguson, D.A., and Cummins, C.S., Nutritional requirements of anaerobic coryneforms, *J. Bacteriol.*, 135, 858, 1978.
29. Holland, K.T., Greenman, J., and Cunliffe, W.J., Growth of cutaneous propionibacteria on synthetic medium; growth yields and exoenzyme production, *J. Appl. Bacteriol.*, 47, 383, 1979.
30. McDowell, A. et al., *Propionibacterium acnes* types I and II represent phylogenetically distinct groups, *J. Clin. Microbiol.*, 43, 326, 2005.
31. Johnson, J.L., and Cummins, C.S., Cell wall composition and deoxyribonucleic acid similarities among the anaerobic coryneforms, classical propionibacteria, and strains of *Arachnia propionica*, *J. Bacteriol.*, 109, 1047, 1972.
32. Vandamme, P. et al., Polyphasic taxonomy, a consensus approach to bacterial systematics, *Microbiol. Rev.*, 60, 407, 1996.
33. Cohan, F.M., What are bacterial species? *Annu. Rev. Microbiol.*, 56, 457, 2002.
34. Webster, G.F., and Cummins, C.S., Use of bacteriophage typing to distinguish *Propionibacterium acne* types I and II, *J. Clin. Microbiol.*, 7, 84, 1978.
35. Kishishita, M. et al., Biotyping of *Propionibacterium acnes* isolated from normal human facial skin, *Appl. Environ. Microbiol.*, 38, 585, 1979.
36. Cummins, C.S., Identification of *Propionibacterium acnes* and related organisms by precipitin tests with trichloracetic acid extracts, *J. Clin. Microbiol.*, 2, 104, 1975.
37. Goodsell, M.E. et al., Two types of *Propionibacterium avidum* with different isomers of diaminopimelic acid, *Curr. Microbiol.*, 22, 225, 1991.

38. Holmberg, K., and Forsum, U., Identification of *Actinomyces*, *Arachnia*, *Bacterionema*, *Rothia*, and *Propionibacterium* species by defined immunofluorescence, *Appl. Microbiol.*, 25, 834, 1973.
39. Gerencser, M.A., and Slack, J.M., Isolation and characterization of *Actinomyces propionicus*, *J. Bacteriol.*, 94, 109, 1967.
40. Schofield, G.M., and Schaal, K.P., A numerical taxonomic study of members of the *Actinomycetaceae* and related taxa, *J. Gen. Microbiol.*, 127, 237, 1981.
41. Gazouli, M. et al., Assessment of mycobacterial, propionibacterial, and human herpesvirus 8 DNA in tissues of Greek patients with sarcoidosis, *J. Clin. Microbiol.*, 40, 3060, 2002.
42. Chaudhry, R. et al., *Propionibacterium granulosum*: A rare cause of endocarditis, *J. Infect.*, 3, 284, 2000.
43. Walker, J. et al., Postoperative *Propionibacterium granulosum* endophthalmitis, *Arch. Ophthalmol.*, 108, 1073, 1990.
44. Branger, C., Bruneau, B., and Goullet, P., Septicemia caused by *Propionibacterium granulosum* in a compromised patient, *J. Clin. Microbiol.*, 12, 2405, 1987.
45. Million, M. et al., Septic arthritis of the hip with *Propionibacterium avidum* bacteremia after intraarticular treatment for hip osteoarthritis, *Joint Bone Spine,* 75, 356, 2008.
46. Vohra, A. et al., Splenic abscess caused by *Propionibacterium avidum* as a complication of cardiac catheterization, *Clin. Infect. Dis.*, 26, 770, 1998.
47. Panagea, S. et al., Breast abscess caused by *Propionibacterium avidum* following breast reduction surgery: Case report and review of the literature, *J. Infect.*, 51, 253, 2005.
48. Forbes-Faulkner, J.C. et al., Novel *Propionibacterium* infection in cattle, *Aust. Vet. J.*, 78, 175, 2000.
49. Tunney, M.M. et al., Improved detection of infection in hip replacements. A currently underestimated problem. *J. Bone Joint Surg. Br.*, 80, 568, 1998.
50. Tunney, M.M. et al., Detection of prosthetic hip infection at revision arthroplasty by immunofluorescence microscopy and PCR amplification of the bacterial 16S rRNA gene, *J. Clin. Microbiol.*, 37, 3281, 1999.
51. Zeller, V. et al., *Propionibacterium acnes*: An agent of prosthetic joint infection and colonization, *J. Infect.*, 2, 119, 2007.
52. Söderquist, B., Holmberg, A., and Unemo, M., *Propionibacterium acnes* as an etiological agent of arthroplastic and osteosynthetic infections—two cases with specific clinical presentation including formation of draining fistulae, *Anaerobe*, 16, 304, 2010.
53. Piper, K.E. et al., Microbiologic diagnosis of prosthetic shoulder infection by use of implant sonication, *J. Clin. Microbiol.*, 46, 1878, 2009.
54. Sampedro, M.F. et al., Species of *Propionibacterium* and *Propionibacterium acnes* phylotypes associated with orthopedic implants, *Diagn. Microbiol. Infect. Dis.*, 2, 138, 2009.
55. Dodson, C.C. et al., *Propionibacterium acnes* infection after shoulder arthroplasty: A diagnostic challenge, *J. Shoulder Elbow Surg.*, 19, 303, 2010.
56. Clayton, J.J. et al., Endocarditis caused by *Propionibacterium* species: A report of three cases and a review of clinical features and diagnostic difficulties, *J. Med. Microbiol.*, 55, 981, 2006.
57. Martin-Rabadan, P. et al., *Propionibacterium acnes* is a common colonizer of intravascular catheters, *J. Infect.*, 56, 257, 2008.
58. Brook, I., and Frazier, E.H., Infections caused by *Propionibacterium* species, *Rev. Infect. Dis.*, 13, 819, 1991.
59. Zappe, B. et al., *Propionibacterium* spp. in prosthetic joint infections: A diagnostic challenge, *Arch. Orthop. Trauma. Surg.*, 10, 1039, 2008.
60. Aldave, A.J. et al., Treatment strategies for postoperative *Propionibacterium acnes* endophthalmitis, *Ophthalmology*, 106, 2395, 1999.
61. Kotilainen, P. et al., *Propionibacterium acnes* isolated from sternal osteitis in a patient with SAPHO syndrome, *J. Rheumatol.*, 23, 1302, 1996.
62. Eishi, Y. et al., Quantitative analysis of mycobacterial and propionibacterial DNA in lymph nodes of Japanese and European patients with sarcoidosis, *J. Clin. Microbiol.*, 40, 198, 2002.
63. Harada, K. et al., Molecular identification of bacterial 16S ribosomal RNA gene in liver tissue of primary biliary cirrhosis: Is *Propionibacterium acnes* involved in granuloma formation? *Hepatology,* 33, 530, 2001.
64. Cohen, R.J. et al., *Propionibacterium acnes* associated with inflammation in radical prostatectomy specimens: A possible link to cancer evolution? *J. Urol.*, 173, 1969, 2005.
65. Holmberg, A. et al., Biofilm formation by *Propionibacterium acnes* is a characteristic of invasive isolates, *Clin. Microbiol. Infect.*, 8, 787, 2009.
66. Shannon, B.A., Cohen, R.J., and Garrett K.L., Polymerase chain reaction-based identification of *Propionibacterium acnes* types isolated from the male urinary tract: Evaluation of adolescents, normal adults and men with prostatic pathology, *BJU Int.*, 2, 388, 2006.
67. Hall, V., Anaerobic *Actinomyces* and related organisms. In Gillespie, S.H. and Hawkey, P.M. (Editors). *Principles and practice of Clinical Bacteriology*, 2nd ed., p. 575, John Wiley & Sons. Ltd, Chichester, UK, 2006.
68. Brazier, J.S., and Hall, V., *Propionibacterium propionicum* and infections of the lacrimal apparatus, *Clin. Infect. Dis.*, 17, 892, 1993.
69. Siqueira, J.F., Jr., and Rôcas, I.N., Polymerase chain reaction-based analysis of microorganisms associated with failed endodontic treatment, *Oral Surg. Oral. Med. Oral. Pathol. Oral. Radiol. Endod.*, 1, 85, 2004.
70. Brock, D.W. et al., Actinomycosis caused by *Arachnia propionica*: Report of 11 cases, *Am. J. Clin. Pathol.*, 59, 66, 1973.
71. Hoeffler, U., Production of hyaluronidase E.C.3.2.1.36 by propionibacteria from different origins, *Zentralbl. Bakteriol. Parasit.. Infekt. Hyg. I. Abt. Orig. A.*, 245, 1, 1980.
72. Greenman, J., Holland, K.T., and Cunliffe, W.J., Effects of pH on biomass, maximum specific growth rate and extracellular enzyme production by three species of cutaneous propionibacteria grown in continuous culture, *J. Gen. Microbiol.*, 129, 1301, 1983.
73. Greenman, J., Holland, K.T., and Cunliffe, W.J., Effects of glucose concentration n biomass, maximum specific growth rate and extracellular enzyme production by three species of cutaneous *Propionibacteria* grown in continuous culture, *J. Gen. Microbiol.*, 127, 376, 1981.
74. Smith, R.F., Role of extracellular ribonuclease in growth of *Corynebacterium acnes*, *Can. J. Microbiol.*, 15, 749, 1969.
75. von Nicolai, H., Hoffler, U., and Zilliken, F., Isolation, purification, and properties of neuraminidase from *Propionibacterium acnes*, *Zentralbl. Bakteriol. A.*, 247, 84, 1980.
76. Ingham, E. et al., Purification and partial characterization of an acid phosphatase (EC 3.1.3.2) produced by *Propionibacterium acnes*, *J. Gen. Microbiol.*, 118, 59, 1980.

77. Werner, H., Lipase and lecithinase activities of aerobic and anaerobic *Corynebacterium* and *Propionibacterium* species, *Zentralbl. Bakteriol. Orig*., 204, 127, 1967.
78. Smith, R.F., and Willett, N.P., Lipolytic activity of human cutaneous bacteria, *J. Gen. Microbiol*., 52, 441, 1968.
79. Ingham, E. et al., Partial purification and characterization of lipase (EC 3.1.1.3) from *Propionibacterium acnes*, *J. Gen. Microbiol*., 124, 393, 1981.
80. Ingram, E. et al., Studies of the extracellular proteolytic activity produced by *Propionibacterium acnes*, *J. Appl. Bacteriol*., 54, 263, 1983.
81. Brüggemann, H. et al., The complete genome sequence of *Propionibacterium acnes*, a commensal of human skin, *Science*, 305, 671, 2004.
82. Valanne, S. et al., CAMP factor homologues in *Propionibacterium acnes*: A new protein family differentially expressed by types I and II, *Microbiology*, 151, 1369, 2005.
83. Lang, S., and Palmer, M., Characterization of *Streptococcus agalactiae* CAMP factor as a pore-forming toxin, *J. Biol. Chem*., 278, 38167, 2003.
84. Brüggemann, H., Insights in the pathogenic potential of *Propionibacterium acnes* from its complete genome, *Semin. Cutan. Med. Surg*., 24, 67, 2005.
85. Lodes, M.J. et al., Variable expression of immunoreactive surface proteins of *Propionibacterium acnes*, *Microbiology*, 152, 3667, 2006.
86. Hammerschmidt, S. et al., Capsule phase variation in *Neisseria meningitidis* serogroup B by slipped-strand mispairing in the polysialyltransferase gene (siaD): Correlation with bacterial invasion and the outbreak of meningococcal disease, *Mol. Microbiol*., 20, 1211, 1996.
87. Fink, H. et al., Systemic inflammatory response syndrome increases immobility-induced neuromuscular weakness, *Crit. Care Med*., 36, 910, 2008.
88. Flaminio, M.J., Rush, B.R., and Shuman, W., Immunologic function in horses after non-specific immunostimulant administration, *Vet. Immunol. Immunopathol*., 63, 303, 1998.
89. Shannon, B.A., Garrett, K.L., and Cohen, R.J., Links between *Propionibacterium acnes* and prostate cancer, *Future Oncol*., 2, 225, 2006.
90. Graham, G.M. et al., Proinflammatory cytokine production by human keratinocytes stimulated with *Propionibacterium acnes* and *P. acnes* GroEL, *Br. J. Dermatol*., 150, 421, 2004.
91. Westerhof, W. et al., *Propionibacterium acnes* and the pathogenesis of progressive macular hypomelanosis, *Arch. Dermatol*., 140, 210, 2004.
92. Nagy, I. et al., *Propionibacterium acnes* and lipopolysaccharide induce the expression of antimicrobial peptides and proinflammatory cytokines/chemokines in human sebocytes, *Microbes Infect*., 8, 2195, 2006.
93. Tunney, M.M. et al., Biofilm formation by bacteria isolated from retrieved failed prosthetic hip implants in an in vitro model of hip arthroplasty antibiotic prophylaxis, *J. Orthop. Res*., 25, 2, 2007.
94. Coenye, T., Peeters, E., and Nelis, H.J., Biofilm formation by *Propionibacterium acnes* is associated with increased resistance to antimicrobial agents and increased production of putative virulence factors, *Res. Microbiol*., 158, 386, 2007.
95. Oprica, C., and Nord, C.E., European surveillance study on the antibiotic susceptibility of *Propionibacterium acnes*, *Clin. Microbiol. Infect*., 11, 204, 2005.
96. Ross, J.I. et al., Antibiotic-resistant acne: Lessons from Europe, *Br. J. Dermatol*., 148, 467, 2003.
97. Tan, H.H. et al., Antibiotic sensitivity of *Propionibacterium acnes* isolates from patients with acne vulgaris in a tertiary dermatological referral centre in Singapore, *Ann. Acad. Med. Singapore*, 30, 22, 2001.
98. Ishida, N. et al., Antimicrobial susceptibilities of *Propionibacterium acnes* isolated from patients with acne vulgaris, *Microbiol. Immunol*., 52, 621, 2008.
99. Chow, A.W., Patten, V., and Guze, L.B., Susceptibility of anaerobic bacteria to metronidazole: Relative resistance of non-spore-forming gram-positive baccilli, *J. Infect. Dis*., 131, 182, 1975.
100. Pochi, P.E., and Strauss, J.S., Antibiotic sensitivity of *Corynebacterium acnes* (*Propionibacterium acnes*). *J. Invest. Dermatol*., 36, 423, 1961.
101. Wang, W.L. et al., Susceptibility of *Propionibacterium acnes* to seventeen antibiotics, *Antimicrob. Agents Chemother*., 11, 171, 1977.
102. Hoeffler, U., Ko, H.L., and Pulverer, G., Antimicrobiol susceptibility of *Propionibacterium acnes* and related microbial species, *Antimicrob. Agents Chemother*., 10, 387, 1976.
103. Eady, E.A. et al., Erythromycin resistant propionibacteria in antibiotic treated acne patients: Association with therapeutic failure, *Br. J. Dermatol*., 121, 51, 1989.
104. Malik, A.C., Reinbold, G.W., and Vedamuthu, E.R., An evaluation of the taxonomy of *Propionibacterium*, *Can. J. Microbiol*., 14, 1185, 1968.
105. Kishishita, M. et al., New medium for isolating propionibacteria and its application to assay of normal flora of human facial skin, *Appl. Environ. Microbiol*., 40, 1100, 1980.
106. Jousimies-Somer, H. et al., *Wadsworth-KTL Anaerobic Bacteriology Manual*, 6th ed., Star Publishing, Belmont, 2002.
107. Bjerkan, G., Witsø, E., and Bergh K., Sonication is superior to scraping for retrieval of bacteria in biofilm on titanium and steel surfaces in vitro, *Acta Orthop*., 80, 245, 2009.
108. Mordarska, H., and Paściak, M., A simple method for differentiation of *Propionibacterium acnes* and *Propionibacterium propionicum*, *FEMS. Microbiol. Lett*., 123, 325, 1994.
109. Celig, D.M., and Schreckenberger, P.C., Clinical evaluation of the RapID-ANA II panel for identification of anaerobic bacteria, *J. Clin. Microbiol*., 29, 457, 1991.
110. Looney, W.J., Gallusser, A.J., and Modde H.K., Evaluation of the ATB 32 A system for identification of anaerobic bacteria isolated from clinical specimens, *J. Clin. Microbiol*., 28, 1519, 1990.
111. Cavallaro, J.J., Wiggs, L.S., and Miller, J.M., Evaluation of the BBL Crystal Anaerobe identification system, *J. Clin. Microbiol*., 35, 3186, 1997.
112. Funke, G. et al., Multicenter evaluation of the updated and extended API (RAPID) Coryne database 2.0, *J. Clin. Microbiol*., 35, 3122, 1997.
113. Oliver, J.D., The viable but nonculturable state in bacteria, *J. Microbiol*., 43, 93, 2005.
114. Brown, M.R., and Williams P., The influence of environment on envelope properties affecting survival of bacteria in infections, *Annu. Rev. Microbiol*., 39, 527, 1985.
115. Palomino, J.C. et al., Rapid culture-based methods for drug-resistance detection in *Mycobacterium tuberculosis*, *J. Microbiol. Methods*, 75, 161, 2008.
116. Janda, J.M., and Abbott, S.L., Bacterial identification for publication: When is enough? *J. Clin. Microbiol*., 40, 1887, 2002.

117. Poppert, S., Riecker, M., and Essig A., Rapid identification of *Propionibacterium acnes* from blood cultures by fluorescence in situ hybridization, *Diagn. Microbiol. Infect. Dis.*, 66, 214, 2010.
118. Goldschmidt, P. et al., New test for the diagnosis of bacterial endophtalmitis, *Br. J. Ophthalmol.*, 93, 1089, 2009.
119. Sfanos, K.S. et al., A molecular analysis of prokaryotic and viral DNA sequences in prostate tissue from patients with prostate cancer indicates the presence of multiple and diverse microorganisms, *Prostate*, 68, 306, 2008.
120. Ishige I. et al., Quantitative PCR of mycobacterial and propionibacterial DNA in lymph nodes of Japanese patients with sarcoidosis, *Lancet*, 354, 120, 1999.
121. Yasuhara, T., et al., The presence of *Propionibacterium* spp. in the vitreous fluid of uveitis patients with sarcoidosis, *Acta. Ophthalmol. Scand.*, 83, 364, 2005.
122. Sfanos, K.S., and Isaacs, W.B., An evaluation of PCR primer sets used for detection of *Propionibacterium acnes* in prostate tissue samples, *Prostate*, 68, 1492, 2008.
123. Alexeyev, O.A. et al., Direct visualization of *Propionibacterium acnes* in prostate tissue by multicolor fluorescent in situ hybridization assay, *J. Clin. Microbiol.*, 45, 3721, 2007.
124. Nakamura, M. et al., Identification of *Propionibacterium acnes* by polymerase chain reaction for amplification of 16S ribosomal RNA and lipase genes, *Anaerobe,* 9, 5, 2003.
125. Siqueira, J.F., Jr., and Rôças, I.N., Polymerase chain reaction detection of *Propionibacterium propionicus* and *Actinomyces radicidentis* in primary and persistent endodontic infections, *Oral Surg. Oral Med. Oral Pathol. Oral Radiol. Endod.*, 96, 215, 2003.
126. Yamada, T. et al., In situ localization of *Propionibacterium* acnes DNA in lymph nodes from sarcoidosis patients by signal amplification with catalysed reporter deposition, *J. Pathol.*, 198, 541, 2002.
127. Therese, K.L., Anand, A.R., and Madhavan, H.N., Polymerase chain reaction in the diagnosis of bacterial endophthalmitis, *Br. J. Ophthalmol.*, 82, 1078, 1998.
128. Hykin, P.G. et al., The diagnosis of delayed post-operative endophthalmitis by polymerase chain reaction of bacterial DNA in vitreous samples, *J. Med. Microbiol.*, 40, 408, 1994.
129. Alexeyev, O. et al., Association between the presence of bacterial 16S RNA in prostate specimens taken during transurethral resection of prostate and subsequent risk of prostate cancer (Sweden), *Cancer Cause Control*, 17, 1127, 2006.
130. Clarridge, III, J.E., Impact of 16S rRNA gene sequence analysis for identification of bacteria on clinical microbiology and infectious diseases, *Clin. Microbiol. Rev.*, 17, 840, 2004.
131. Kolbert, C.P., and Persing, D.H., Ribosomal DNA sequencing as a tool for identification of bacterial pathogens, *Curr. Opin. Microbiol.*, 2, 299, 1999.
132. McDowell, A., and Patrick, S., Evaluation of nonculture methods for the detection of prosthetic hip biofilms, *Clin. Orthop. Relat. Res.*, 437, 74, 2005.
133. Moter, A., and Göbel, U.B., Fluorescence in situ hybridization (FISH) for direct visualization of microorganisms, *J. Microbiol. Methods*, 41, 85, 2000.
134. Wagner, M., Horn, M., and Dairns, H., Fluorescence in situ hybridization for the identification and characterization of prokaryotes, *Curr. Opin. Microbiol.*, 6, 302, 2003.
135. Fuchs, B.M. et al., Unlabeled helper oligonucleotides increase the in situ accessibility to 16S rRNA of fluorescently labeled oligonucleotide probes, *Appl. Environ. Microbiol.*, 66, 3606, 2000.
136. de Carvalho, A.F., Gautier, M., and Grimont F., Identification of dairy *Propionibacterium* species by rRNA gene restriction patterns, *Res. Microbiol.*, 145, 667, 1994.
137. Perry A.L. et al., Analysis of clinical isolates of *Propionibacterium acnes* by optimised RAPD, *FEMS. Microbiol. Lett.*, 228, 51, 2003.
138. Mohammadi, T. et al., Amplified-fragment length polymorphism analysis of *Propionibacterium* isolates implicated in contamination of blood products, *Br. J. Haematol.*, 131, 403, 2005.
139. Gautier, M., de Carvalho, A., and Rouault, A., DNA fingerprinting propionibacteria strains by pulsed field gel electrophoresis, *Curr. Microbiol.*, 32, 17, 1996.
140. Tunney, M.M. et al., Detection of prosthetic joint biofilm infection using immunological and molecular techniques, *Methods Enzymol.*, 310, 566, 1999.
141. Ichikawa, H. et al., Quantitative analysis of propionibacterial DNA in bronchoalveloar lavage cells from patients with sarcoidosis, *Sarcoidosis Vasc. Diffuse Lung Dis.*, 25, 15, 2008.
142. Yasuhara, T. et al., The presence of *Propionibacterium* spp. in the vitreous fluid of uveitis patients with sarcoidosis, *Acta Ophthalmol. Scand.*, 83, 364, 2005.
143. Millar, B.C., Xu, J., and Moore J.E., Risk assessment models and contamination management: Implications for broad-range ribosomal DNA PCR as a diagnostic tool in medical bacteriology, *J. Clin. Microbiol.*, 40, 1575, 2002.
144. Bojar, R.A., and Holland, K.T., Acne and *Propionibacterium acnes*, *Clin. Dermatol.*, 22, 375, 2004.
145. Isenberg, J. et al., Preoperative immunostimulation by *Propionibacterium* granulosum KP-45 in colorectal cancer, *Anticancer Res.*, 14, 1399, 1994.

14 *Rhodococcus*

Debarati Paul, Dipaloke Mukherjee, Soma Mukherjee, and Debosmita Paul

CONTENTS

14.1 INTRODUCTION

The genus name "*Rhodococcus*" was first used by Zopf in 1891 to describe two species of red-pigmented bacteria.[1–3] The taxon *Rhodococcus* is of interest for a variety of reasons, including its several metabolic abilities, as they can degrade a wide range of environmental pollutants and transform or synthesize a number of compounds with possible useful applications. Rhodococci are virtually ubiquitous, which is evident by their presence in a large variety of sources including soils, rocks, boreholes, groundwater, marine sediments, animal dung, the guts of insects, and in healthy and diseased animals and plants.[4–6] The commercial potential of *Rhodococcus* species is increasingly being recognized. Their outstanding ability to synthesize several products such as surfactants, flocculants, amides, and polymers, as well as their capacity to degrade or transform a wide range of chemicals make rhodococci actually or potentially useful in environmental and industrial biotechnology. This increasing interest is reflected in patenting. Finnerty lists 10 patents relating to rhodococci up to 1990,[1] Warhurst and Fewson report that a further 20 patent families were submitted in the following two years, while from 1993 to November 1996, a further 80 patents were submitted to the World Patent Index.[6]

Interestingly, two strains of *Rhodococcus triatome* have been identified as symbionts in bloodsucking bugs of the genus *Triatoma*.[7] There has also been an increased interest in infections caused by rhodococci in humans, animals, and plants.[4,8–10] Rhodococci are nonpathogenic with the exception of two strains, namely *R. fascians*[11] and *R. equi*.[12] *Rhodococcus fascians* is the only described phytopathogen in this genus that causes leafy gall disease and is a biotrophic organism with broad host range. Through the secretion of potent morphogens, the bacteria stimulate the formation of new shoots from differentiated tissues.[13–18] However, since it is out of the scope of this chapter, we would limit our discussion on this pathogen.

The other pathogenic Rhodococcus is *R. equi*, which causes infections in both horses and humans. It is a dangerous pathogen of young foals aged up to 3–5 months.[19]

14.1.1 Classification and Epidemiology

The current systematics of this genus in the NCBI database is as follows: Actinobacteria, Actinobacteridae, Actinomycetales, Corynebacterineae, Nocardiaceae, and Rhodococcus. However, this is unclear, partly because many members were originally included before the application of a polyphasic taxonomic approach, central to which is the acquisition of 16S rRNA sequence data. Bacteria of the order Actinomycetales were originally classified as fungi because of their hyphal and mycelial morphology, and include the mycolic acid-containing Corynebacterineae, which presently contain the genera *Gordonia*, *Rhodococcus*, and *Nocardia*.[2] *Rhodococcus* was revived and redefined in 1977 to accommodate the "*rhodochrous*" complex, which comprised a number of strains that resembled but did not belong to the established genera *Nocardia*, *Corynebacterium*, and *Mycobacterium*. Rhodococci are described as aerobic, gram-positive, nonmotile (with exception of *R. agilis*), mycolate-containing, nocardioform actinomycetes. The term "nocardioform" is morphologically descriptive and refers to mycelial growth with fragmentation into rod-shaped or coccoid elements. There are currently 12 established *Rhodococcus* species, namely *R. coprophilus*, *R. equi*, *R. fascians*, *R. erythropolis*, *R. globerulus*, *R. marinonascens*, *R. opacus*, *R. percolatus*, *R. rhodnii*, *R. rhodochrous*, *R. ruber*, and *R. zopfii*.[20] Some strains bearing other generic and specific names, such as *Arthrobacter picolinophilus*, *Corynebacterium hoagii*, *Nocardia restricta*, and *Nocardia calcarea* have been shown recently to belong to established *Rhodococcus* species. *Nocardia corynebacteroides* also appears to be a *Rhodococcus* but does not belong to any of the currently valid species.[9,21,22] One of the most important characteristic features that mark the differences between Nocardia species from that of Rhodococcus is the presence of an MK-8(H_4) type cyclic component in menaquinone. All the species under the genus Nocardia exhibits this feature with the exception of the strain *N. corynebacteroides*.[9] Other invalidated species names, such as *R. longus*, *R. minimus*, and *R. luganensis*,[2] occur in the literature, while there are many strains that are not identified to species level.[2]

In order to provide a framework for a taxonomic approach based on multiple genetic loci, a survey of the known genome characteristics of members of the genus Rhodococcus was undertaken. This included (i) genetics of cell envelope biosynthesis; (ii) virulence genes; (iii) gene clusters involved in metabolic degradation and industrially relevant pathways; (iv) genetic analysis tools; (v) rapid identification of bacteria including rhodococci with specific gene RFLPs; and (vi) genomic organization of *rrn* operons. Genes encoding virulence factors have been characterized for *Rhodococcus equi* and *Rhodococcus fascians*. Based on peptide signature comparisons deduced from gene sequences for cytochrome P-450, mono- and dioxygenases, alkane degradation, nitrile metabolism, proteasomes, and desulfurization, phylogenetic relationships can be deduced for *Rhodococcus erythropolis*, *Rhodococcus globerulus*, *Rhodococcus ruber*, and a number of undesignated *Rhodococcus* spp. that may distinguish the genus Rhodococcus into two further genera.[3]

R. equi is a gram-positive, pleomorphic, intracellular bacillus that is an obligatory aerobe, sometimes weakly acid-alcohol resistant, and with distinctive biochemical features. *R. equi* is ubiquitous in soil, as well as in the feces of herbivores, birds, dogs, and so forth. It is a common pathogen in domestic animals that occasionally affects humans. The pathogenicity of *R. equi* is expressed mostly in hosts who are immunosuppressed.[23] Infection in other animal species is rare and usually the result of immunosuppression.

Rhodococcus equi is an important cause of AIDS-associated pneumonia in HIV-infected humans.[24] This pathogen was originally isolated by Magnusson in 1923 from granulomatous lung infections in young horses.[25] Infection in humans may result from environmental exposure[26,27] because the organism is abundantly present in soil.[28] The stool of horses and other animals is the source of soil contamination. While early cases occurred mostly in persons with a history of contact with horses, only 20%–30% of recent cases can be traced to such contact.[29] *R. equi* infection in a human was first reported in 1967 in a 29-year-old man with plasma cell hepatitis who was receiving immunosuppressant medications. Since then, *R. equi* has become an important opportunistic pathogen in immunocompromised patients, especially those with AIDS.

14.1.2 Clinical Features

R. equi is a rare opportunistic pathogen found in severely immunocompromised patients and also in human immunodeficiency virus (HIV)-infected persons, and is associated with significant mortality. In earlier cases, patients receiving immunosuppressant therapy were more likely to be successfully treated with antimicrobial agents as compared to the cases in AIDS patients.[30] Most often, patients have a slowly progressive granulomatous pneumonia, with lobar infiltrates, frequently developing to cavitating lesions visible on chest X-ray (Figure 14.1a and 14.1b). Other sites of infection include abscesses of the central nervous system, pelvis, subcutaneous tissue, and lymphadenitis.[29–31] Delays in accurate diagnosis of *R. equi* are still very common[27,29] despite increased awareness of this organism as an opportunistic pathogen in humans. Factors for delayed diagnosis include the insidious onset of disease, clinical resemblance of the infection to mycobacterial, fungal, and actinomycotic infections, and the relatively nondescript bacteriologic profile of *R. equi*. Table 14.1 lists some of the previously reported clinical manifestations of *R. equi* infection.

The sites of infection and the specimens from which *R. equi* can be isolated include, in a roughly declining order of frequency[29,31–37]:

(i) Pulmonary (usually the primary site of infection)—sputum, bronchial washings, lung tissue or abscess contents, pleural fluid (empyema)
(ii) Blood (predominantly dissemination from lung infection)
(iii) Central nervous system (predominantly hematogenous spread)—brain abscesses, cerebrospinal fluid
(iv) Subcutaneous and other soft tissue or organ abscesses (predominantly hematogenous spread)
(v) Wound drainage, infected joints, vitreous fluid (usually due to traumatic inoculation)
(vi) Implanted indwelling devices such as peritoneal dialysis and intravenous catheters
(vii) Other sites such as pharynx, middle ear, lymph nodes, bone, and stool.

Pulmonary infections are the most common form of human disease caused by *R. equi*. Cases of lung infection caused by inhalation and cutaneous lesions caused by wound contamination have been documented; the latter are almost the only *R. equi* infections reported in healthy persons, frequently children.[35] In one large review of 72 cases, pneumonia occurred in 76% of patients, and the lung was the sole site of infection in 82% of them.[35] Pneumonia was accompanied by extrapulmonary infection in 18% of cases, while

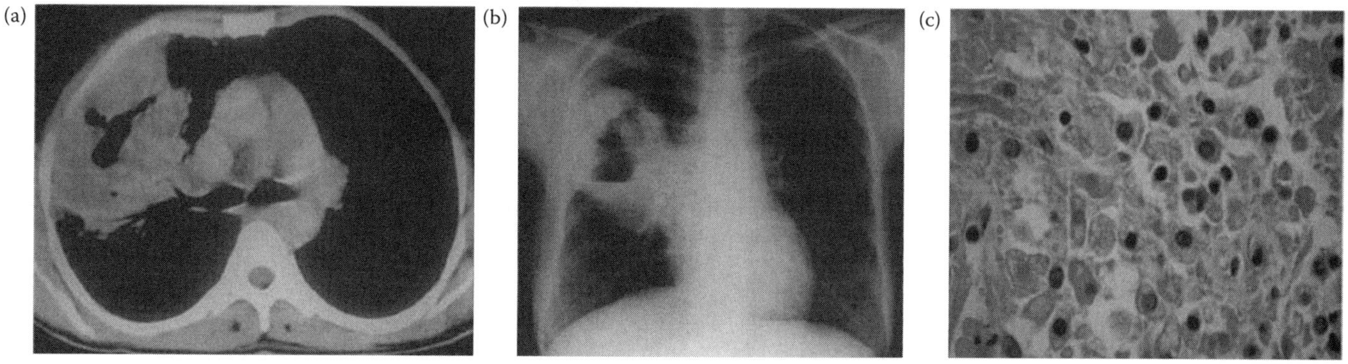

FIGURE 14.1 (a) Thorax CT-scan and (b) Rx-ray of *R. equi* bacteremia with lung abscess. (c) Lung histological section with sheets of macrophages with foamy or abundant granular eosinophilic cytoplasm that contain Michaelis-Gutmann bodies, characterized by visible PAS-positive and diastase-resistant basophilic structures with targetoid-like concentric laminations (HE-400×).

infection at extrapulmonary sites occurred without evidence of pulmonary involvement in 24%. Morphology, partial acid fastness, and a distinctive histopathologic profile in bronchial specimens (Figure 14.1c) contribute to accurate diagnosis. Numerous polymorphonuclear leukocytes with intracellular pleomorphic gram-positive bacteria, microabscesses, pseudotumors, and malakoplakia are noted on tissue.[29,35] Malakoplakia is a relatively rare granulomatous inflammation not typically associated with histology of lung infection and can be of help in forming a differential diagnosis.[35,38,39] There have been few reports of pulmonary infection caused by *Rhodococcus equi* in immunocompetent hosts. A couple of cases of infection by *R. equi* in patients without impaired immunity have been reported in the literature, and four of these cases have involved pulmonary disease that showed very different radiological and developmental forms:[23,35,40,41] multiple pulmonary nodules and a chronic course; multiple bilateral lobar affectation and acute respiratory failure requiring mechanical ventilation; and atypical community pneumonia. All these patients had cavitation; only one of them required open lung biopsy for diagnosis.

TABLE 14.1
Clinical Manifestations of *R. equi* Infection as Shown by Various Reports

Organ	Clinical Manifestations
Lung	Pulmonary nodules, lung abscess, pneumonia, pneumothorax
Skin	Spontaneous keratitis and endophthalmitis
Kidney	Renal abscess, hepatic abscess
Bladder	Urinary tract infection, bacterimia, isolated and catheter related
Bone	Septic arthritis, osteomyelitis, infected bone marrow
Cervix	Adenitis along with mesenteric adenitis and peritonitis
Thyroid	Thyroid abscess
Spleen	Spleen abscess
Prostrate glands	Prostrate abscess
Colon	Pedunculated and sessile colonic polyps
Liver	Hepatic abscess
Heart	Ventricular shunt infection, sternal-wound infection following coronary bypass grafting

Case studies of *R. equi* infections in human beings indicate diseases related to other organs of the body, too. One study described a kidney transplant recipient who had undergone renal transplantation at the age of 43 years because of end-stage renal disease secondary to IgA nephropathy.[42] A chest X-ray revealed that the patient had an enlarged left hilus and left-sided pleural effusion. On the patient's admission to hospital, bronchoscopy findings were macroscopically normal and no mycobacteria were detected in his sputum samples, bronchoalveolar lavage fluid, or pleural fluid. One week after his admission, cultures from these samples and from blood samples returned positive for *Rhodococcus equi*. The case presented here illustrates the importance of extensive diagnostic procedures in kidney transplant recipients. The clinical presentation of a disease can be atypical in immunocompromised patients, and the spectrum of potential pathologies (e.g., carcinomas) and possible disease-causing microbial agents (e.g., viruses, fungi, and bacteria such as *Mycobacterium tuberculosis*) in such patients is large. Continued investigations are, therefore, warranted until a definite diagnosis can be made.[42]

A case study wherein *R. equi* was the cause of endophthalmitis in an immunocompetent individual showed that the causative agent was a rare *R. equi* Prescott serotype 4, having been isolated by J. F. Prescott from a few pigs and by the Japanese from horse manure.[43] A case of chronic scalp wound infection due to *R. equi* in an immunocompetent individual following a heavily contaminated traumatic injury was successfully treated by combining antibiotics and surgery.[44] This patient was probably exposed to the organism due to the heavy contamination of his wound with soil and his poor personal hygiene. This case helps to elicit the pathogenic potential of *R. equi* even in the nonimmunosuppressed, as well as to appreciate the variable microbial aetiologies in chronic soft tissue infections and the value of identifying the causative agent. Cases of brain abscesses caused by *R. equi* have been reported in HIV negative individuals.[45,46] In one case, a middle-aged person with a history of chronic hepatitis C and alcohol abuse was admitted to the hospital with a 5-day history of hemoptysis and shortness of breath. A magnetic resonance imaging scan of the brain was obtained that showed four ring enhancing lesions in the cerebral cortex. This case emphasizes the role of *R. equi* in the combined lung-brain abscess syndrome. The study emphasized that all individuals who present with cavitary lung lesions and brain abscesses should be screened for *R. equi* and treated accordingly.

Deaths exceed 50% among AIDS patients with documented *R. equi* pneumonia and are almost always preceded by multiple relapses, which are common even when successful treatment is achieved.

14.1.3 Pathogenesis and Virulence

R. equi infection in animals and in humans show major differences.[47] The ability of *R. equi* to persist in and eventually to destroy macrophages is the basis of its pathogenicity. The pathological change elicited by *R. equi* is similar to that of some other parasites within the group *Corynebacterium-Mycobacterium-Nocardia* because they have lipid-rich cell wall components.[26,48] In nearly all infected foals, the virulence of *R. equi* is mediated by a surface antigen, VapA, (Virulence-Associated Protein), which is associated with an 85 to 90-kbp plasmid.[49,50] Foals infected with VapA-positive isolates develop severe bronchopneumonia, whereas plasmid-cured derivatives are harmless. In pigs, nearly all isolates are of intermediate virulence and express a 20-kDa antigen associated with five large plasmids.[51] On the other hand, only 20%–25% of isolates recovered from humans express VapA.[28,52–54] Expression of the 20-kDa antigen is highly variable and may differ between *R. equi* strains from different geographical areas. The pathogenesis of infection with intermediately virulent and avirulent strains appears to involve mechanisms different from those associated with VapA-positive isolates.[28]

The majority of laboratory studies have investigated the virulence and immunity associated with *R. equi* infection using VapA-positive isolates. Therefore, conclusions about the pathogenesis of *R. equi* in humans cannot be drawn from these studies. In animals, both humoral and cell-mediated immunity play a role in the control of *R. equi* infection. Transfer of plasma from immunized horses to infected foals increases opsonization and clearance of *R. equi*.[52] Although normal and CD8_ T cell-deficient mice can clear *R. equi* infection,[55] mice affected with severe combined immunodeficiency (SCID) remain persistently infected but do not develop granulomas.[56,57] Nude mice that are given a CD4_ Th1 cell line transferred from immune mice are able to express IFN-g and clear *R. equi* from their lungs. On the other hand, mice that receive CD4_ Th2 cells but do not receive Th1 cells express IL-4 but not IFN-g, fail to clear the infection, and develop large granulomas with eosinophils in the lung.[58,59] Thus, the Th1 response, acting through IFN-g, appears sufficient to effect pulmonary clearance of *R. equi* while the Th2 response is nonprotective.

This VapA is susceptible to trypsin digestion and accessible to biotin labeling, which suggests that it is present on the outer surface of the cell wall. Radiolabeled VapA was detected as a lipid modified protein anchored to the cell wall.[60,61] In gram-positive bacteria, lipoproteins are generally involved in transport of solutes into the cytosol, adherence to surfaces, or have enzymatic activity.[62] The signal sequence of lipoproteins contains a "lipobox," which consists of the amino acid sequence (S/T/G/L/A/V)L(A/S)(G/A)C. The cysteine residue is the first amino acid of the mature protein that is lipid modified when a diacylglycerol moiety is transferred to its sulfhydryl group from a glycerophospholipid molecule. Analysis of the genomic sequence of *Mycobacterium tuberculosis* revealed the presence of 65 putative lipoproteins containing a lipobox, suggesting that mycolic acid-containing actinomycetes employ a conventional lipid modification system as found in other bacteria.[62] However, the VapA protein contains neither cysteine nor a lipobox, suggesting that *R. equi* may employ an additional mechanism to lipid-modifying proteins. VapA is encoded by the *vapA* gene that was found to be located on an indigenous *R. equi* plasmid of approximately 80–85 kb.[61,63] Isolates from the submaxillary lymph nodes of infected pigs also contain a large plasmid. However, these plasmids do not encode VapA, but a related protein of 20 kDa (VapB). Interestingly, pig isolates are less virulent in mice than foal isolates harboring the *vapA* gene, suggesting host specificity of strains harboring different plasmids.[51] These studies clearly establish that the virulence plasmid is essential for virulence in foals and pigs, and therefore encodes one or more virulence factors. However, *R. equi* isolates from cattle and goats lack a virulence plasmid and human isolates contain either VapA or VapB encoding plasmids or lack a virulence plasmid altogether.[53,64,65] This suggests that the pathogenesis of *R. equi* infection in these hosts is different than in foals or pigs.

Nucleotide sequencing of the virulence plasmid (Figure 14.2a) of two foal isolates revealed the presence of 69 ORF.[66] Based on comparisons with genes previously identified in other organisms three functional regions of the virulence plasmid were recognized. Two of these harbor genes that are similar to those encoding proteins involved in conjugation and in plasmid replication, stability, and segregation. ORF 34, 38, 39, and 40 of the virulence plasmid are similar to the open reading frames found on an indigenous plasmid of *Micrococcus* sp 28 (Accession number AY034092), and are organized in a similar fashion suggesting a common origin.[24] The identification of genes that resemble genes required for conjugation suggest that the virulence plasmid can be mobilized from virulent to avirulent strains. Another segment of ~27.5 kb bears the pathogenicity island (Figure 14.2b) characterized by a significantly lower G + C content than the rest of the plasmid, flanked by genes similar to transposon resolvases, and harbors the *vapA* gene. Analysis of the pathogenicity island revealed six *vapA* homologues (*vapC, D, E, F, G, H*). With exception of *vapF*, which is not functional, all *vap* genes encode proteins with a clear signal sequence, indicating that these are extracellular proteins. Byrne et al.

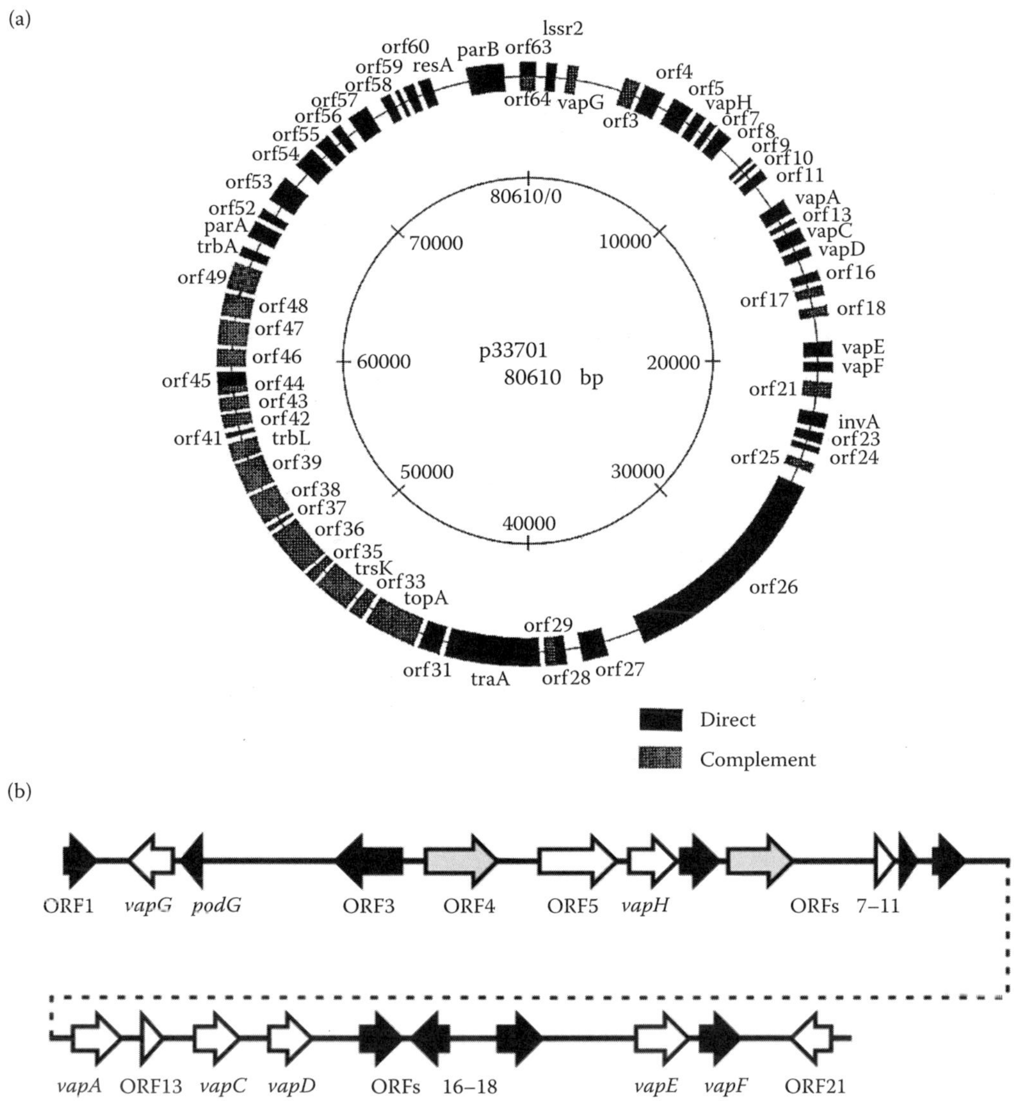

FIGURE 14.2 (a) Map of the *R. equi* virulence plasmid p33701. The circle shows orientation of the ORFs, with the direction of reading denoted by shading. (b) Schematic view of the 27.5 kb pathogenicity island of the virulence plasmid of *R. equi*. The arrows indicate the position and direction of transcription of the genes located within the pathogenicity island. White arrows indicate that the genes encode proteins with a signal sequence and may therefore be secreted. Grey arrows indicate the two transcriptional regulators encoded by the pathogenicity island.

showed that VapC, D, and E are secreted, but in contrast to VapA, they are not anchored to the cell wall.[67] In addition to the six *vap* genes, four more genes encode proteins with a clear signal sequence; so actually there are 10 total extracellular proteins encoded by the pathogenicity island. As extracellular proteins can interact with the host environment, these 10 proteins could be important virulence factors.

ORF8 encodes a response regulator, which usually interacts with a sensor kinase protein to regulate gene expression. The later autophosphorylates conserved histidine residue in response to an environmental signal; the phosphate is transferred to a conserved aspartate residue in the response regulator resulting in transcriptional activation. Although the virulence plasmid encodes a response regulator, it does not encode a sensor kinase, indicating that the ORF8 gene product interacts with a chromosomally encoded sensor kinase protein. The PhoP/PhoQ system is an example of a two-component system in *Salmonella typhimurium* that is induced within the phagosome of the macrophage.[68] This regulatory system has also been identified in *Yersinia pestis* and was shown to be important for survival under conditions of macrophage-induced stress.[69] ORF4 encodes a protein homologous to the LysR type transcriptional regulators (LTTR). LTTR are, after the two-component response regulators, the second largest class of bacterial regulatory proteins.[70] A number of LTTR have been identified that are associated with virulence gene regulation. For example, SpvR encoded by the 90 kb virulence plasmid of *Salmonella typhimurium*, regulates the *spv* operon in response to conditions such as late growth phase.[71] Interestingly, the *spv* operon has been shown to be induced in bacteria within macrophages.[72] Three other genes, ORF 3, 5, and 21, encode proteins with clear homologues in other bacteria. The protein encoded by ORF 21, which has a *M. tuberculosis* homologue (Rv1885c), is a secreted chorismate mutase, an enzyme of the shikimate pathway for biosynthesis of aromatic amino acids. The extracellular location of this protein is enigmatic, as this pathway is usually located inside the cytoplasm. The protein encoded by ORF 5 is predicted to be an integral membrane protein, displaying a strong similarity to sugar permeases (Pfam 00083). ORF 3 encodes an O-methyl-transferase protein that lacks a signal sequence and is therefore likely to be located in the cytoplasm. Although it has not yet been shown whether these genes are required for virulence, their presence in the pathogenicity island might suggest that virulent *R. equi* either produces a metabolite that is secreted by the cell or that *R. equi* takes up a modified host metabolite. The former could serve to alter the macrophage response; the latter could be involved in signaling.

Studies show that the regulation of expression of genes within the pathogenicity island is complex, and depends on at least five environmental signals: temperature, pH, oxidative stress, magnesium, and iron. It seems very likely that the two transcriptional regulators encoded by ORF4 and ORF8 play an important role in transducing these signals to the transcription apparatus, although this has not yet been established. To date, two transcriptional repressors, Fur and IdeR, controlling iron-dependent gene expression have been identified.[73] An *ideR* homologue of *R. equi* was recently cloned and overexpressed in *E. coli*. It was shown that the *R. equi* IdeR protein is functional; it binds to a consensus IdeR recognition sequence and regulates gene expression in an Fe^{2+} dependent manner.[74] Interestingly the promoter region of the *vapA* gene contains an IdeR consensus binding site,[75] suggesting that the iron-dependent expression of this gene may be controlled by IdeR.

14.1.4 Cell Biology

It is understood that nonspecific degranulation of lysosomes in *R. equi*-infected macrophages was suggested to be the major survival and pathogenic mechanism of infection in horses.[76] Such degranulation in vivo would contribute to the tissue destruction and neutrophil influx that is characteristic of *R. equi* lung lesions.[76] The basis of pathogenicity of *R. equi* is its ability to multiply in and eventually to destroy alveolar macrophages. Since opsonization of *R. equi* with specific antibody is followed by increased phagosome-lysosome fusion and by significantly enhanced *R. equi* killing by equine macrophages,[76,77] macrophage entry through non-Fc receptors may be important in determining the fate of the bacteria. Once bacteria are internalized, the phagosome undergoes a series of fusion and fission events with vesicles from the endocytic pathway, a complex maturation process that leads to the formation of phagolysosomes. During maturation, the phagosome membrane increasingly changes to resemble that of late endosomes and lysosomes. For bacteria that do not interfere with the process, phagocytosis to phagolysosome formation takes about 5 min.[78] Phagocytosed bacteria are degraded in the late phagolysosome by the same mechanisms employed by lysosomes, involving a wide range of acid-resistant hydrolases within the acid environment of the phagolysosome. Intramacrophage *R. equi* appear to be located exclusively within membrane-enclosed vacuoles and persistence correlates with the absence of phagosome-lysosome fusion.[76] It has been shown that plasmid-encoded products contribute to the ability of *R. equi* to survive and replicate in macrophages. Preliminary observations suggested that maturation of the phagosome is more efficiently diverted in strains possessing the virulence plasmid compared to plasmid-negative isogenic strains, and that cytotoxicity of *R. equi* for J774E murine macrophages is strongly upregulated by the possession of the virulence plasmid.[24]

14.1.5 Diagnosis

14.1.5.1 Phenotypic Identification

Isolation and Colonial Morphology. *R. equi* grows readily when incubated aerobically at 37°C on nonselective media routinely used in clinical microbiology laboratories. At 24 h of incubation, colonies are 1–2 mm in diameter and are not distinctive. By 48 h of incubation on nonselective media such as Trypticase soy blood agar, they developed their characteristic appearance[26] exhibited by irregularly round, smooth, semi-transparent, glistening, coalescing, mucoid, teardrop colonies with entire edges. The colonies vary in size from 2 to 4 mm, although coalesced colonies may appear larger. The organism does not grow on most types of MacConkey agar. Rare isolates of *R. equi* may grow poorly at 37°C.[79] Colony variation is present in fresh cultures. The classical viscous-mucoid coalescing colony is usually the predominant type present, but less mucoid forms may also be seen. Cultures may have a slightly earthy smell. Pigment production is rarely marked in cultures that are less than 4 days old[80] and it may surprise those expecting all rhodococci (red-pigmented cocci) to show obviously pink or red colonial pigmentation immediately on isolation. After 4–7 days on nonselective medium such as Trypticase soy blood agar, colonies may develop a delicate shade of salmon pink, although they may be nonpigmented or appear slightly yellow.[26,80] Perhaps the best description of the typical colony pigment on blood agar is as light fawn colored. Cultures kept on slopes for prolonged periods without subculture commonly become rough, dry, and orange-red but revert to classical colonies on subculture.[81] The macroscopic colony morphology of the type culture ATCC 6939[81] is pigmented than typical *R. equi*.

Isolation of *R. equi* from the infection site is so far the best method of diagnosing infection. Sputum is an often but not always useful specimen in humans with pneumonic disease.[82] Diagnosis is more reliably made by bronchial brushings, percutaneous thoracic aspiration, or, more drastically, open lung biopsy during lobectomy.[83] On occasion, bronchial brushings in human patients have failed to yield *R. equi*.[83] This failure supports the findings of Martens et al. that culture of bronchial secretions obtained by transtracheal aspiration from experimentally infected foals does not consistently yield *R. equi*.[84] They attributed this inconsistency to either periodic shedding into airways or the inability of organisms inside phagocytes to be isolated by standard culture techniques. Although most culture-positive foals with *R. equi* pneumonia yield pure cultures of organisms from lung aspirates, up to one-third may also yield a variety of opportunist pathogens.[26] Thus, *R. equi* may be present in mixed culture and may be ignored in part because it does not show its distinctive colony size and mucoid appearance until after 48 h of incubation. Blood cultures from febrile human patients with pneumonic disease have yielded *R. equi* (sometimes repeatedly) in about one-third of cases reported.[85]

Microscopic Studies and Staining Properties. *R. equi* is a gram-positive pleomorphic coccobacillus that varies from distinctly coccoid to bacillary depending on growth conditions. It is usually coccoid, both on solid media and in purulent material from patients, but in liquid media, particularly in young cultures, it forms long rods or short filaments that may show rudimentary branching. The occasional reports that *R. equi* is acid fast in the Ziel–Neelsen stain appear to depend on staining technique, the older age of the cultures, and the growth medium.[26,86] Most report a failure to demonstrate acid fastness of *R. equi*. Metachromatic granules have occasionally been described,[26] but like acid-fast staining, this is an inconsistent feature of the organism.

Biochemical Characterization. Characteristics that can be used in routine clinical microbiology laboratories to identify *R. equi* are shown in Table 14.2. The organism is generally biochemically unreactive. *R. equi* is an obligate aerobe with simple growth requirements. It can utilize carbon from a wide variety of sole carbon sources, including acetic, pyruvic, butyric, and propionic acids, and nitrogen from ammonium sulfate or potassium nitrate.[26] The organism is nonmotile and nonflagellated and may produce small numbers of pili.[86] Some strains may produce a thin pellicle that is readily disrupted. It fails to oxidize or ferment most carbohydrates or alcohols and is nonproteolytic. No acid metabolic products from glucose were detected by gas-liquid chromatography.[87] It is catalase positive, almost invariably urease positive, and oxidase negative. *R. equi* produces equi factors that interact with the phospholipase D of *Corynebacterium pseudotuberculosis*, the beta-toxin of *S. aureus*, and an uncharacterized partial hemolysin of *L. monocytogenes* to give an area of complete hemolysis with sheep or cattle erythrocytes. While this characteristic is not unique for *R. equi*, it is distinctive for this organism and should always be used in identification since no equi factor-negative isolates of *R. equi* have been described. Antimicrobial drug susceptibility is consistent and might be a useful adjunct in identification. *R. equi* produces lipase and phosphatase, but not DNase, elastase, lecithinase, or protease. The API ZYM (Analytab Products Inc., Plainview, NY) gave distinctive enzyme profiles of value in identifying *R. equi*. All isolates tested were positive for leucine arylamidase, acid phosphatase, and phosphamidase; >90% were positive for valine amylamidase, esterase lipase, and alpha-glucosidase.[26]

TABLE 14.2
Biochemical Characteristics for Identifying *R. equi* in Comparison to *Corynebacterium* and Other Closely Related Pathogens

Characteristic	Reaction
Beta-hemolysis	–
Esculin hydrolysis	–
Nitrate reduction	+
Urease	+
Casein digestion	–
Glucose	–
Maltose	–
Sucrose	–
Catalase	+
Cytochrome coxidase	–
Carbohydrate fermentation	–
Alcohol fermentation	–
Gelatin hydrolysis	–
Indole	–
H_2S	Variable
Hippurate hydrolysis	–
Nitrate reduction	+
equi factors (CAMP test)	+
Lipase	+
Phosphatase	+

Capsular Serotypes. *R. equi* possesses a distinct, antigenically variable, lamellar polysaccharide capsule, which has provided the basis of a capsular serotyping system.[88] At least 27 different capsular serotypes have been identified, of which Prescott capsular serotype 1 (or its equivalent in other typing schemes) is the most common worldwide.[26,88] No relationship between serotype and virulence is apparent.

A number of workers have investigated alternative, immunological approaches to diagnosis in foals but none of those can be recommended as replacing cultural approaches. In immunosuppressed human patients, including those with AIDS, immunological tests cannot be depended on for diagnosis.[26] Pathologic findings in patients with *R. equi* infection are relatively consistent, despite the broad range of tissues involved. Specimens are typically necrotic and contain a dense, histiocytic infiltrate with an eosinophilic, granular cytoplasm, and intrahistiocytic coccobacilli. Multiple microabscesses that contain abundant neutrophils may be present. In some cases, concentrically layered basophilic inclusions, called "Michaelis–Gutman bodies," are prominent. The inclusions are thought to result from impaired macrophage lysosomal function that leads to failure of lysophagosomal fusion and ineffective killing of ingested organisms.[89] This characteristic pathologic finding, together with Michaelis-Gutman bodies, is termed "malakoplakia," and has been described in other bacterial infections.[29,47] In rare cases, a dense proliferation of spindle cells can mimic either Kaposi's sarcoma or lymphosarcoma.[29,47]

A synergistic hemolysis-inhibition assay for antibody to equi factors distinguished foals with naturally occurring *R.*

equi pneumonia from healthy foals, although it failed to identify foals in the early stages of experimentally induced disease.[90] An agar gel immunoprecipitation test for antibody to equi factors was also shown to identify foals with *R. equi* pneumonia, although a small proportion of apparently healthy foals as also positive in the test.[26] Using more sensitive synergistic-hemolysis inhibition and immunoprecipitation assays than described previously, Skalka demonstrated the widespread nature of subclinical infection or exposure to *R. equi* in foals.[26,91] In general, ELISA has been used to distinguish infected from healthy foals.[92] In one study, a Tween 20 extract from the type strain, ATCC 6939, gave the broadest cross reactivity of the serotypes tested.[26,92] The mean values of anti-*R. equi* IgG antibody in foals known or suspected to be ill with *R. equi* pneumonia were strikingly higher than in normal foals in the limited number of animals tested, but there were no differences in mean values of anti-*R. equi* IgM or IgA antibodies.[92] This ELISA was more sensitive than agar gel diffusion or indirect hemagglutination tests.[92] In two unrelated studies, using different antigens, no correlation was found between ELISA antibody levels and clinical disease.[26] The antigens used in ELISA are important in determining their value for serological diagnosis.[92] Supernatants of autoclaved cells or of homogenized cells were relatively poor antigens compared with Tween 20-extracted or sonicated materials.[92] Two unique features of *R. equi* help distinguish it from other organisms. First, synergistic hemolysis occurs when *R. equi* cultured on sheep blood agar are cross-streaked with other bacteria, including *Staphylococcus aureus*, *Listeria monocytogenes*, and *Corynebacterium pseudotuberculosis*.[26] Second, antagonism between imipenem and other β-lactam antibiotics has been documented in vitro. This antagonism is widespread among *R. equi* isolates and is novel among gram-positive bacteria.[47,93] Serologic assays have not been clinically validated and are not commercially available.

14.1.5.2 Molecular Identification

Polymerase chain reaction (PCR) detection of *R. equi* can offer rapid detection of this pathogenic bacterium, however, only a few rapid molecular methods have been developed for *R. equi* identification. Clinical isolates from foals can be identified by detection of the VapA antigen with monoclonal antibodies, or alternatively, by PCR detection of its gene, *vapA*, which is present on an 85 kb virulence plasmid.[66,94,95] One such study has shown that PCR of tracheal wash using primers that recognized the vapA virulence plasmid of *R. equi* has a diagnostic sensitivity of 100% and specificity of 90.6%. In contrast, sensitivity and specificity are only 57.1% and 93.8%, respectively, for standard microbiologic culture of tracheal wash and only 62.5% and 75.9%, respectively, for serology.[96] However, the virulence plasmid is not present in all strains of human and environmental origin (in general, in nonequine isolates),[53] thus limiting the general usefulness of the *vapA*-based identification of *R. equi* by PCR. In another study, species-specific primers for the PCR for the detection of *Rhodococcus equi* were developed. These primers were based on unique DNA fragments produced from *R. equi* reference strains and field isolates. Following random amplification of polymorphic DNA from *R. equi* and *R. rhodochrous* with a set of 40 arbitrary 10 bp primers, a pair of species-specific primers was designed to detect a unique 700 bp fragment of *R. equi* chromosomal DNA. This PCR product was limited to *R. equi* and was not detectable in other *Rhodococcus* species or in a panel of additional gram-positive and gram-negative bacteria.[97] Other PCR-based molecular methods have been used to amplify the 16S rRNA,[98–100] but their validities have been assessed with only a small number of strains. Although 16S rRNA sequencing is accepted as a general means for species differentiation, some heterogeneity can exist between different isolates of the same species.[100] A recent study of the 16S rRNA sequences of several representative strains showed that *R. equi* is a very heterogeneous taxon, with variations in 16S rRNAs of up to 4%.[101] On the other hand, closely related species may have identical or almost identical 16S rRNA sequences.[102] Therefore, it is useful to have other species-specific targets, such as *choE*, to undertake an assay for the reliable identification of a bacterial species. Ladron et al. reported a specific PCR assay for the rapid and reliable identification of *R. equi* based on the amplification of a fragment of the *choE* gene encoding cholesterol oxidase. The *choE*-based PCR was assessed by using a panel of strains comprising 132 isolates from different sources and of different geographical origins, all initially identified biochemically as *R. equi*, and 30 isolates of representative non-*R. equi* actinomycete species, including cholesterol oxidase producers. The expected 959 bp amplicon was observed only with *R. equi* isolates, as confirmed by sequencing of a variable region of the 16S rRNA gene from a random sample of 20 PCR-positive isolates. All *R. equi* isolates gave a positive *choE*-based PCR result, which correlated with a high degree of conservation of the *choE* gene.[103]

Finally, a PCR-RFLP method targeting a 65-kDa heat shock protein gene was shown to discriminate *R. equi* strains.[104] This test was primarily devised for the identification of mycobacteria. However, this assay is too laborious and lengthy because, due to the conserved nature of the 65-kDa heat shock protein gene, the amplification product is of the same size for all actinomycete species and *R. equi* can be discriminated only by restriction analysis of the amplicon. Advanced techniques, such as multiplex PCR was used for molecular characterization of this bacterium for epidemiological studies.[105,106] Therefore, these studies suggest that the molecular detection of this pathogen is an area of current research interest.

14.1.6 Treatment and Control

Treatment. A proposed approach for the treating patients with *Rhodococcus equi* infections based on in vitro susceptibility data and published case reports[47] has been outlined in Figure 14.3. These infections can be successfully treated using lipophilic antimicrobial drugs that can penetrate the macrophages or neutrophils in which the organisms survive.[26,107] A number of in vitro studies are being carried out to study the

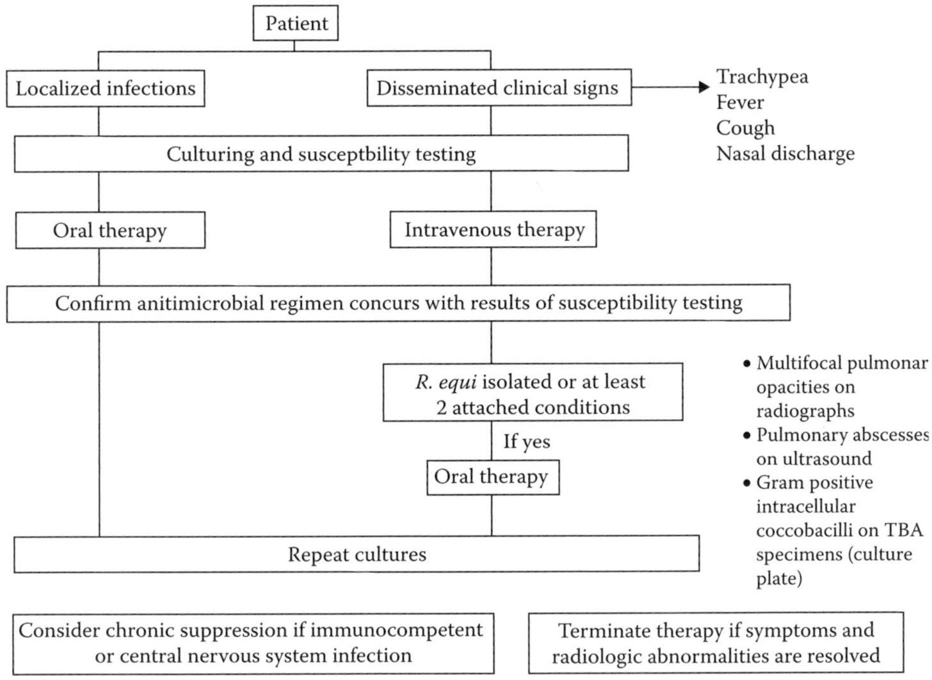

FIGURE 14.3 Proposed approach for the treatment of *R. equi* infection. Intravenous antibiotics are typically required for initial treatment, except in the case of localized infection in an immunocompetent host. Antimicrobial agents are rated as "preferred" or "alternative" on the basis of their in vitro activity, the results of animal studies, clinical experience, and published case reports.

antimicrobial susceptibility of clinical isolates since they cause life-threatening infections in severely immunocompromised persons.[43,108,109] While variations exist, most strains were susceptible to inhibition by glycopeptide antibiotics (including vancomycin and teicoplanin) and rifampin. Macrolide antibiotics, such as erythromycin and clarithromycin, were also inhibitory to many strains. Resistance to β-lactam antibiotics (with the exception of carbapenems, specifically imipenem) was generally reported, and is not related to the production of a β-lactamase. The standard treatment in pneumonic foals is oral administration of a combination of erythromycin estolate and rifampin.[110] These drugs have also been used alone or in combination, usually with good clinical response, in many human infections.[26,111,112] Unfortunately, however, there is no standard treatment protocol for pulmonary and/or systemic *R. equi* infections as relapse in spite of treatment is common in a majority of cases,[108] followed by high mortality rate, especially among AIDS patients (Table 14.1). A proposed regimen involves parenteral glycopeptide plus imipenem for at least 3 weeks, followed by an oral combination of rifampin, plus either macrolides or tetracycline.[43] Examples of efficacious protocols for parenteral treatment are available (Table 14.1). Susceptibility testing should be performed for all *R. equi* isolates, to avoid the use of antibiotics to which the isolate has in vitro resistance. *R. equi* is usually susceptible in vitro to a number of substances, including erythromycin, rifampin, flouroquinolones, aminoglycosides, glycopeptides, and imipenem. Susceptibility to cotrimoxazole, tetracycline, chloramphenicol, clindamycin, and cephalosporins is variable. Isolates are typically resistant to penicillins, but use of penicillins is not recommended because of rapid acquisition of resistance.[12,35,113] For immunocompromised patients and patients with serious infections, intravenous therapy with two-drug or three-drug regimens that include vancomycin, imipenem, aminoglycosides, ciprofloxacin, rifampin, and/or erythromycin are appropriate. Regimens that include 12 drugs have been successful for some patients, but no data are available to indicate that they are superior. Most patients will require intravenous antibiotics for a minimum of 2 weeks, at which point clinical improvement should be evident. Oral antibiotics can then be substituted and continued until all culture results are negative and the patient's symptoms and radiologic abnormalities have resolved.[26,114] The duration of therapy depends on the site(s) and extent of infection, underlying immunocompetence of the host, and the clinical response to therapy. A minimum of 6 months of antibiotic therapy is typically required for immunocompromised patients.

Besides the use of antimicrobial drugs, the approach used in treatment of human infection involves drainage of suppurative lesions, surgical resection of granulomatous tissue, and control of predisposing factors such as decreasing the dose of concurrent immunosuppressive drugs or control of underlying malignancies. Careful and repeated culture and susceptibility testing during treatment is required to discover acquired resistance, in a manner similar to the treatment of mycobacterial infection.

Prevention and Control. Primary prophylaxis against *Rhodococcus equi* is not routinely recommended, because sufficient data is not available to support its efficiency and also the infection is rare. Macrolide prophylaxis against

Mycobacterium avium complex infection may offer some protection against *R. equi* infection for patients with AIDS. Immunocompromised patients with significant exposure to domesticated animals should be cautioned regarding the possible risk of *R. equi* infection. Some investigators have advocated isolation of hospitalized patients with *R. equi* pneumonia to prevent nosocomial spread.[115,116]

There is no effective vaccine currently available for the prevention of *R. equi* pneumonia. DNA vaccines are known to offer specific advantages over conventional vaccines. Some recombinant DNA vaccine candidates, namely *pcDNA3-Re1*, *pcDNA3-Re3*, and *pcDNA3-Re5*, were designed by combining a heat shock protein GroEL2 to a virulence-associated protein A (VapA) from *R. equi* to protect C3H/He mice against the *R. equi* infection. VapA was shown to be strongly recognized by sera from pneumonic foals. All vaccines elicited at least a doubling of the IgG2a/IgG1 ratio in comparison to the controls, indicating a bias to the Th1 response, which is postulated to be crucial for bacterial clearance and protective immunity against intracellular pathogens, including *R. equi*.[117] Improved methods to effectively control this pathogen and treat infections caused by it are areas of current research and holds potential challenges.

14.2 METHODS

14.2.1 Sample Preparation

DNA can be extracted from tracheal aspirate samples of infected foals as follows:[118]

A volume of 300 μL of sodium iodide solution containing 6 M NaI, Tris hydrochloride (pH 8), 13 mM EDTA, 0.5% sodium *N*-lauroylsarcosine, and 10 μg of glycogen is added to tubes, each containing 100 μL of tracheal aspirate sample. The tubes are incubated at 60°C for 15 min. A volume of 400 μL of 100% isopropanol is added to each tube, mixed thoroughly, and left at room temperature for 15 min. The tubes are centrifuged at a speed of 10000 × *g* for 15 min and the supernatant is completely discarded. A volume of 1 mL of Washing Solution B is used to suspend the pellet in each tube, followed by vigorous vortexing. Each tube is briefly centrifuged at 10000 × *g* for 5 min. DNA is resuspended in TE buffer after the completely discarding supernatant from each tube. This protocol is actually the DNA extraction protocol followed in the manufacturer's instructions of the DNA Extractor kit produced by Wako Pure Chemical Industries, Japan.

14.2.2 Detection Procedures

(i) PCR PCR is a very efficient tool to detect rapidly the presence of *R. equi*. Sellon et al.[96] demonstrated that detection of *R. equi* by performing PCR on tracheal wash samples of infected foals using primers that recognized the vapA virulence plasmid of *R. equi* is 100% more sensitive and 90.6% more specific, as compared to the 57.1% sensitivity and 93.8% specificity found respectively in the standard microbiologic culture procedures of tracheal wash. For serology, the sensitivity and specificity were found to be only 62.5% and 75.9%, respectively.

The PCR protocol described here was originally reported by Ladron et al.[103]:

1. The oligonucleotide primers are synthesized flanking the segment to be amplified.
2. PCR mixture is prepared by mixing 10 μL of template, 40 pmol of each oligonucleotide primer, 4 U of *Taq* DNA polymerase (Bioline, London, United Kingdom), 0.2 mM each deoxynucleoside triphosphate, 1.5 mM $MgCl_2$, and 10 μL of 10× PCR amplification buffer [160 mM $(NH_4)_2SO_4$, 670 mM Tris-HCl (pH 8.8), 0.1% Tween 20]. Finally, double-distilled water is added to the PCR mixture to make a final volume of 100 μL.
3. Run the following cycling program in a PCR machine: one cycle of 95°C for 5 min; 30 cycles of 95°C for 1 min, 55°C for 1 min, and 72°C for 1 min; and one cycle of 72°C for 10 min.
4. The amplified DNA is visualized generally on a 0.7% agarose gel by UV fluorescence (approximately 5 μL sample), following staining with ethidium bromide.[118]

(ii) Restriction Fragment Length Polymorphism (RFLP) Analysis The PCR products may be purified by use of the Gel Extraction Purification Kit (Qiagen) and subsequently digested with the following restriction enzymes for generating restriction patterns: *Ava*I, *Bam*HI, *Bgl*I, *Hin*PI, *Pvu*I, *Xho*I, and *Xmn*I (New England Biolabs). The digestion products are resolved by electrophoresis on 1.2% agarose gels in Tris borate buffer and visualized by ethidium bromide staining.

(iii) Multiplex PCR Assay for the vap Gene Family The primers for PCR amplification for *vap* genes were designed with Vector NTI 8.0 software (InforMax Inc., 2002) based on sequences published through GenBank.[119] PCR amplification is performed as described previously[118] with some modifications.

1. For each reaction 4 μL (50 ng) of DNA preparation is used in a 25 μL reaction mixture containing 10 mM Tris, 50 mM KCl, 2.5 mM $MgCl_2$, 200 mM (each) deoxynucleoside triphosphates, 1 μM of each primer (30 pmol), and 1 U of Taq DNA polymerase.
2. The PCR mixture is subjected to 30 cycles of 94°C of 90 s, 55°C for 1 min, and 72°C for 2 min, with a final extension at 72°C for 10 min.
3. The amplification products (15 μL) is separated by 1.5% agarose gel electrophoresis for 60 min at 80 V. DNA fragments are visualized by UV fluorescence after staining with ethidium bromide.

TABLE 14.3
Shuttle Vectors for Use in *R. equi* and Other Species

Name	Plasmids Used in Constructing Vector		Use	Reference
pRE-1	pACYC177 (*E. coli*)	pOTS (*ori* from *R. equi*)	Regulation of virulence gene expression in *R. equi*	121
pRE-7	pRE-1	pBS (vapA subcloned in MCS)		
pREV5	pAL5000, pMF1 (*ori* from *M. fortuitum*)	Several *E. coli* vectors and pHP45X-apra	These vectors were compatible in *R. equi*. Used for analysis of gene functions in trans	122
pREV2	pMF1 (*ori* from *M. fortuitum*)	pSUM36 and pEP2 (*E. coli*)		
pMS2	pMS1 (*Corynebacterium* strain)	pBR322 (*E. coli*)	Study gene function in a wide range of bacterium: *R. equi*, *Corynebacterium*, *E. coli*, *B. subtilis*, *S. aureus*	Katsumata, R. et al., U.S. Pat. No. 4,500,640
pMS20	pMS2	pE194 (*S. aureus* and *B. subtilis*)		
pMS21	pMS2	pTV543 (*B. subtilis*)		
pJW1484	pIAM1484 (*R. erythropolis*)	pBSk (-) *E. coli*	Relatively small in size, can replicate in a variety of species of *Rhodococcus* and has high replication stability in a host	United States Patent Application 20060263868
pLG1484	pJW1484 was modified for higher replication stability			
pDA37	Nocardioform plasmid	*E. coli*-positive selection vector constructed by M. Zabeau et al.	Cloning DNA without prior alkaline phosphatase treatment	123

(iv) Sequencing Control strains (without *vap* genes) may be sequenced in an automatic auto sequencer (MegaBace 500). Each *vap* gene should be separately submitted for PCR and the products obtained may be purified with polyethyleneglycol (PEG 8000). The sequences can be analyzed by submitting them to a consensus analysis based on chromatogram reliability obtained using the Staden Package Gap 4 program.[120] Thereafter, samples may be compared to *vap* gene sequences stored at GenBank using the Chromas Proprogram, version 1.34.

14.3 CONCLUSION AND FUTURE PERSPECTIVES

Much has been learned in the last decades about *R. equi* infections in foals that can be applied to the diagnosis and treatment of infection in humans. *R. equi* is a rare but recognized pathogen in humans and has emerged as an important cause of morbidity and mortality among immunocompromised patients. Increasing awareness of *R. equi* infection improves the likelihood of its accurate and timely diagnosis. Although considerable progress has been made in understanding important aspects of virulence of *R. equi*, there is an almost complete lack of genetic tools for this bacterium that will be required to make further progress. To address this problem *R. equi–E. coli* shuttle vectors have been developed to facilitate transformation and random as also site-specific mutagenesis procedures (Table 14.3). The shuttle vectors that replicate in *R. equi* are based on different replicons so that they can be maintained in the cell at the same time. Random and transposon-based site directed mutagenesis techniques have also been applied to study virulence of this pathogen.

There is no effective vaccine currently available for the prevention of *R. equi* pneumonia. Chimeric DNA vaccines are being formulated and tested for their efficiency in mice. Vaccines targeting the *vap*A and heat shock proteins[117] in this organism have been successful in preventing pneumonia in mice, although further studies are required to develop effective vaccines in foals and humans.

Molecular detection of the pathogen in clinical samples and knowledge on molecular basis of pathogenesis and epidemiology are also primary areas of future research work. In recent times, due to the upsurge of various molecular techniques, study of such diseases has become potentially easier and quicker. Therefore, foolproof vaccines and other means of cure can be anticipated to be available soon.

ACKNOWLEDGMENTS

We thank Dr. Debdas Mukherjee and Dr. Biswajit Paul for useful suggestions. Debarati Paul is grateful to Dr. Mark Lawrence and Dr. Shane Burgess for their support.

REFERENCES

1. Finnerty, W.R., The biology and genetics of the genus *Rhodococcus*, *Annu. Rev. Microbiol.*, 46, 193, 1992.
2. Bell, K.S. et al., The genus Rhodococcus, *J. Appl. Microbiol.*, 85, 195, 1998.
3. Gurtler, V., Mayall, B.C., and Seviour, R. Can whole genome analysis refine the taxonomy of the genus *Rhodococcus*? *FEMS Microbiol. Rev.*, 28, 377, 2004.
4. Larkin, M.J. et al., Applied aspects of *Rhodococcus* genetics, *Antonie Van Leeuwenhoek*, 74, 133–153, 1998.
5. Warhurst, A.M., et al., Metabolism of styrene by *Rhodococcus rhodochrous* NCIMB 13259, *Appl. Environ. Microbiol.*, 60, 1137–1145, 1994.
6. Warhurst, A.M., and Fewson, C.A., Biotransformations catalyzed by the genus *Rhodococcus*, *Crit. Rev. Biotechnol.*, 14, 29–73, 1994.
7. Yassin, A.F., *Rhodococcus triatomae* sp. nov., isolated from a blood-sucking bug, *Int. J. Syst. Evol. Microbiol.*, 55, 1575–1579, 2005.
8. Dabbs, E.R., Cloning of genes that have environmental and clinical importance from rhodococci and related bacteria, *Antonie Van Leeuwenhoek*, 74, 155–167, 1998.

9. Rainey, F.A. et al., Dietzia, a new genus including *Dietzia maris* comb. nov., formerly *Rhodococcus maris*, *Int. J. Syst. Bacteriol.*, 45, 32–36, 1995.
10. Goodfellow, M., Alderson, G., and Chun, J., Rhodococcal systematics: Problems and developments, *Antonie Van Leeuwenhoek*, 74, 3–20, 1998.
11. Goethals, K. et al., Leafy gall formation by *Rhodococcus fascians*, *Annu. Rev. Phytopathol.*, 39, 27–52, 2001.
12. Kedlaya, I., Ing, M.B., and Wong, S.S., *Rhodococcus equi* infections in immunocompetent hosts: Case report and review, *Clin. Infect. Dis.*, 32, E39–E46, 2001.
13. Cornelis, K. et al., Virulence genes of the phytopathogen *Rhodococcus fascians* show specific spatial and temporal expression patterns during plant infection, *Mol. Plant Microbe. Interact.*, 15, 398–403, 2002.
14. Crespi, M. et al., Fasciation induction by the phytopathogen *Rhodococcus fascians* depends upon a linear plasmid encoding a cytokinin synthase gene, *Embo J.*, 11, 795–804, 1992.
15. de, O.M.C.L. et al., De novo cortical cell division triggered by the phytopathogen *Rhodococcus fascians* in tobacco, *Mol. Plant Microbe. Interact.*, 14, 189–195, 2001.
16. Depuydt, S. et al., An integrated genomics approach to define niche establishment by *Rhodococcus fascians*, *Plant Physiol.*, 149, 1366–1386, 2008.
17. Temmerman, W. et al., The fas locus of the phytopathogen *Rhodococcus fascians* affects mitosis of tobacco BY-2 cells, *FEBS Lett.*, 492, 127–132, 2001.
18. Vandeputte, O. et al., Biosynthesis of auxin by the gram-positive phytopathogen *Rhodococcus fascians* is controlled by compounds specific to infected plant tissues, *Appl. Environ. Microbiol.*, 71, 1169–1177, 2005.
19. Prescott, J.F., Epidemiology of *Rhodococcus equi* infection in horses, *Vet. Microbiol.*, 14, 211–214, 1987.
20. Boone, D.R. et al., *Bergey's Manual of Systematic Bacteriology,* Springer, New York, 2001.
21. Ruimy, R. et al., *Nocardia pseudobrasiliensis* sp. nov., a new species of *Nocardia* which groups bacterial strains previously identified as *Nocardia brasiliensis* and associated with invasive diseases, *Int. J. Syst. Bacteriol.*, 46, 259–264, 1996.
22. Koch, C., Schumann, P., and Stackebrandt, E., Reclassification of *Micrococcus agilis* (Ali-Cohen 1889) to the genus *Arthrobacter* as *Arthrobacter agilis* comb. nov. and emendation of the genus *Arthrobacter*, *Int. J. Syst. Bacteriol.*, 45, 837–839, 1995.
23. Linares, M.J., Lopez-Encuentra, A., and Perea, S., Chronic pneumonia caused by *Rhodococcus equi* in a patient without impaired immunity, *Eur. Respir. J.*, 10, 248–250, 1997.
24. Meijer, W.G., and Prescott, J.F., *Rhodococcus equi*, *Vet. Res.*, 35, 383–396, 2004.
25. Vyslouzil, L., Seidl, K., Svarcova, J., and Landsmannova, V., Findings of *Corynebacterium equi* Magnusson 1923 in connection with foal mortality in the Eastern Bohemia Region, *Vet. Med. (Praha)*, 29, 563–568, 1984.
26. Prescott, J.F., *Rhodococcus equi*: An animal and human pathogen, *Clin. Microbiol. Rev.*, 4, 20–34, 1991.
27. McNeil, M.M., and Brown, J.M., The medically important aerobic actinomycetes: Epidemiology and microbiology, *Clin. Microbiol. Rev.*, 7, 357–417, 1994.
28. Takai, S. et al., Virulence of *Rhodococcus equi* isolates from patients with and without AIDS, *J. Clin. Microbiol.*, 32, 457–460, 1994.
29. Scott, M.A. et al., *Rhodococcus equi*—an increasingly recognized opportunistic pathogen. Report of 12 cases and review of 65 cases in the literature, *Am. J. Clin. Pathol.*, 103, 649–655, 1995.
30. Doig, C., Gill, M.J., and Church, D.L., *Rhodococcus equi*—an easily missed opportunistic pathogen, *Scand. J. Infect. Dis.*, 23, 1–6, 1991.
31. Harvey, R.L., and Sunstrum, J.C., *Rhodococcus equi* infection in patients with and without human immunodeficiency virus infection, *Rev. Infect. Dis.*, 13, 139–145, 1991.
32. Sirera, G. et al., Relapsing systemic infection due to *Rhodococcus equi* in a drug abuser seropositive for human immunodeficiency virus, *Rev. Infect. Dis.*, 13, 509–510, 1991.
33. Vestbo, J. et al., Severe *Rhodococcus equi* pneumonia: Case report and literature review, *Eur. J. Clin. Microbiol. Infect. Dis.*, 10, 762–768, 1991.
34. Hsueh, P.R. et al., Report of invasive *Rhodococcus equi* infections in Taiwan, with an emphasis on the emergence of multidrug-resistant strains, *Clin. Infect. Dis.*, 27, 370–375, 1998.
35. Verville, T.D. et al., *Rhodococcus equi* infections of humans. 12 cases and a review of the literature, *Medicine (Baltimore)*, 73, 119–132, 1994.
36. Torres-Tortosa, M. et al., Prognosis and clinical evaluation of infection caused by *Rhodococcus equi* in HIV-infected patients: A multicenter study of 67 cases, *Chest*, 123, 1970–1976, 2003.
37. Hoque, S. et al., *Rhodococcus equi* in CAPD-associated peritonitis treated with azithromycin, *Nephrol. Dial. Transplant.*, 11, 2340–2341, 1996.
38. Sutor, G.C. et al., Poststenotic cavitating pneumonia due to *Rhodococcus equi* in HIV infection, *Aids*, 10, 339–340, 1996.
39. Goupil, F. et al., Pulmonary malacoplakia caused by Rhodococcus equi in AIDS: A case report. *Rev. Pneumol. Clin.*, 55, 171–174, 1999.
40. Egawa, T. et al., Human pulmonary infection with *Corynebacterium equi*, *Eur. Respir. J.*, 3, 240–242, 1990.
41. Moyer, D.V., and Bayer, A.S., Progressive pulmonary infiltrates and positive blood cultures for weakly acid-fast, gram-positive rods in a 76-year-old woman, *Chest*, 104, 259–261, 1993.
42. Speck, D. et al., A pulmonary mass caused by *Rhodococcus equi* infection in a renal transplant recipient, *Nat. Clin. Pract. Nephrol.*, 4, 398–403, 2008.
43. Ebersole, L.L., and Paturzo, J.L., Endophthalmitis caused by *Rhodococcus equi* Prescott serotype 4, *J. Clin. Microbiol.*, 26, 1221–1222, 1988.
44. Nasser, A.A., and Bizri, A.R., Chronic scalp wound infection due to *Rhodococcus equi* in an immunocompetent patient, *J. Infect.*, 42, 67–68, 2001.
45. Corne, P., Rajeebally, I., and Jonquet, O., *Rhodococcus equi* brain abscess in an immunocompetent patient, *Scand. J. Infect. Dis.*, 34, 300–302, 2002.
46. Kamboj, M., Kalra, A., and Kak, V., *Rhodococcus equi* brain abscess in a patient without HIV, *J. Clin. Pathol.*, 58, 423–425, 2005.
47. Weinstock, D.M., and Brown, A.E., *Rhodococcus equi*: An emerging pathogen, *Clin. Infect. Dis.*, 34, 1379–1385, 2002.
48. Bernard, K. et al., Cellular fatty acid composition and phenotypic and cultural characterization of CDC fermentative coryneform groups 3 and 5, *J. Clin. Microbiol.*, 32, 1217–1222 1994.

49. Takai, S. et al., Association between a large plasmid and 15- to 17-kilodalton antigens in virulent *Rhodococcus equi*, *Infect. Immun.*, 59, 4056–4060, 1991.
50. Giguere, S. et al., Role of the 85-kilobase plasmid and plasmid-encoded virulence-associated protein A in intracellular survival and virulence of *Rhodococcus equi*, *Infect. Immun.*, 67, 3548–3557, 1999.
51. Takai, S. et al., Identification of intermediately virulent *Rhodococcus equi* isolates from pigs, *J. Clin. Microbiol.*, 34, 1034–1037, 1996.
52. Nordmann, P., Ronco, E., and Nauciel, C., Role of T-lymphocyte subsets in *Rhodococcus equi* infection, *Infect. Immun.*, 60, 2748–2752, 1992.
53. Takai, S. et al., Identification of virulence-associated antigens and plasmids in *Rhodococcus equi* from patients with AIDS, *J. Infect. Dis.*, 172, 1306–1311, 1995.
54. Caterino-De-Araujo, A. et al., Search for an antibody profile of *Rhodococcus equi* infection in AIDS patients despite the diversity of isolates and patient immune dysfunction, *Microbes Infect.*, 1, 663–670, 1999.
55. Kanaly, S.T., Hines, S.A., and Palmer, G.H., Cytokine modulation alters pulmonary clearance of *Rhodococcus equi* and development of granulomatous pneumonia, *Infect. Immun.*, 63, 3037–3041, 1995.
56. Antinori, S. et al., Disseminated *Rhodococcus equi* infection initially presenting as foot mycetoma in an HIV-positive patient, *Aids*, 6, 740–742, 1992.
57. Akan, H. et al., *Rhodococcus equi* and *Nocardia brasiliensis* infection of the brain and liver in a patient with acute nonlymphoblastic leukemia, *Eur. J. Clin. Microbiol. Infect. Dis.*, 17, 737–739, 1998.
58. Kanaly, S.T., Hines, S.A., and Palmer, G.H., Failure of pulmonary clearance of *Rhodococcus equi* infection in CD4+ T-lymphocyte-deficient transgenic mice, *Infect. Immun.*, 61, 4929–4932, 1993.
59. Nordmann, P., Ronco, E., and Guenounou, M., Involvement of interferon-gamma and tumor necrosis factor-alpha in host defense against *Rhodococcus equi*, *J. Infect. Dis.*, 167, 1456–1459, 1993.
60. Tan, C. et al., Molecular characterization of a lipid-modified virulence-associated protein of *Rhodococcus equi* and its potential in protective immunity, *Can. J. Vet. Res.*, 59, 51–59, 1995.
61. Takai, S. et al., Virulence-associated 15- to 17-kilodalton antigens in *Rhodococcus equi*: Temperature-dependent expression and location of the antigens, *Infect. Immun.*, 60, 2995–2997, 1992.
62. Sutcliffe, I.C., and Russell, R.R., Lipoproteins of gram-positive bacteria, *J. Bacteriol.*, 177, 1123–1128, 1995.
63. Tkachuk-Saad, O., and Prescott, J., *Rhodococcus equi* plasmids: Isolation and partial characterization, *J. Clin. Microbiol.*, 29, 2696–2700, 1991.
64. Flynn, O. et al., Virulence-associated protein characterisation of *Rhodococcus equi* isolated from bovine lymph nodes, *Vet. Microbiol.*, 78, 221–228, 2001.
65. Takai, S., Epidemiology of *Rhodococcus equi* infections: A review, *Vet. Microbiol.*, 56, 167–176, 1997.
66. Takai, S. et al., DNA sequence and comparison of virulence plasmids from *Rhodococcus equi* ATCC 33701 and 103, *Infect. Immun.*, 68, 6840–6847, 2000.
67. Byrne, B.A. et al., Virulence plasmid of *Rhodococcus equi* contains inducible gene family encoding secreted proteins, *Infect. Immun.*, 69, 650–656, 2001.
68. Alpuche Aranda, C.M. et al., *Salmonella typhimurium* activates virulence gene transcription within acidified macrophage phagosomes, *Proc. Natl. Acad. Sci. USA*, 89, 10079–10083, 1992.
69. Oyston, P.C. et al., The response regulator PhoP is important for survival under conditions of macrophage-induced stress and virulence in *Yersinia pestis*, *Infect. Immun.*, 68, 3419–3425, 2000.
70. Schell, M.A., Molecular biology of the LysR family of transcriptional regulators, *Annu. Rev. Microbiol.*, 47, 597–626, 1993.
71. Caldwell, A.L., and Gulig, P.A., The *Salmonella typhimurium* virulence plasmid encodes a positive regulator of a plasmid-encoded virulence gene, *J. Bacteriol.*, 173, 7176–7185, 1991.
72. Rhen, M., Riikonen, P., and Taira, S., Transcriptional regulation of *Salmonella enterica* virulence plasmid genes in cultured macrophages, *Mol. Microbiol.*, 10, 45–56, 1993.
73. Hantke, K., Iron and metal regulation in bacteria, *Curr. Opin. Microbiol.*, 4, 172–177, 2001.
74. Boland, C.A., and Meijer, W.G., The iron dependent regulatory protein IdeR (DtxR) of *Rhodococcus equi*, *FEMS Microbiol. Lett.*, 191, 1–5, 2000.
75. Ren, J., and Prescott, J.F., Analysis of virulence plasmid gene expression of intra-macrophage and in vitro grown *Rhodococcus equi* ATCC 33701, *Vet. Microbiol.*, 94, 167–182, 2003.
76. Hietala, S.K., and Ardans, A.A., Interaction of Rhodococcus equi with phagocytic cells from *R. equi*-exposed and non-exposed foals, *Vet. Microbiol.*, 14, 307–320, 1987.
77. Zink, M.C. et al., Electron microscopic investigation of intracellular events after ingestion of *Rhodococcus equi* by foal alveolar macrophages, *Vet. Microbiol.*, 14, 295–305, 1987.
78. Mayorga, L.S., Bertini, F., and Stahl, P.D., Fusion of newly formed phagosomes with endosomes in intact cells and in a cell-free system, *J. Biol. Chem.*, 266, 6511–6517, 1991.
79. Muller, F. et al., Characterization of Rhodococcus equi-like bacterium isolated from a wound infection in a noncompromised host, *J. Clin. Microbiol.*, 26, 618–620, 1988.
80. Elissalde, G.S., Renshaw, H.W., and Walberg, J.A., Corynebacterium equi: An interhost review with emphasis on the foal, *Comp. Immunol. Microbiol. Infect. Dis.*, 3, 433–445, 1980.
81. Williams, G.D., Flanigan, W.J., and Campbell, G.S., Surgical management of localized thoracic infections in immunosuppressed patients, *Ann. Thorac. Surg.*, 12, 471–482, 1971.
82. Golub, B., Falk, G., and Spink, W.W., Lung abscess due to *Corynebacterium equi*. Report of first human infection, *Ann. Intern. Med.*, 66, 1174–1177, 1967.
83. Savdie, E., Pigott, P., and Jennis, F., Lung abscess due to *Corynebacterium equi* in a renal transplant recipient, *Med. J. Aust.*, 1, 817–819, 1977.
84. Martens, R.J., Fiske, R.A., and Renshaw, H.W., Experimental subacute foal pneumonia induced by aerosol administration of *Corynebacterium equi*, *Equine Vet. J.*, 14, 111–116, 1982.
85. Berg, R. et al., Corynebacterium equi infection complicating neoplastic disease, *Am. J. Clin. Pathol.*, 68, 73–77, 1977.
86. Yanagawa, R., and Honda, E., Presence of pili in species of human and animal parasites and pathogens of the genuscorynebacterium, *Infect. Immun.*, 13, 1293–1295, 1976.
87. Reddy, C.A., and Kao, M., Value of acid metabolic products in identification of certain corynebacteria, *J. Clin. Microbiol.*, 7, 428–433, 1978.

88. Nakazawa, M.K. et al., Serogrouping of *Rhodococcus equi*, *Microbiol. Immunol.*, 27, 837–846, 1983.
89. Drancourt, M. et al., *Rhodococcus equi* infection in patients with AIDS, *J. Infect.*, 24, 123–131, 1992.
90. Prescott, J.F., Coshan-Gauthier, R., and Barksdale, L., Antibody to equi factor(s) in the diagnosis of Corynebacterium equi pneumonia of foals, *Can. J. Comp. Med.*, 48, 370–373, 1984.
91. Skalka, B., and Svastova, A., Two techniques for detection of antibodies against *Corynebacterium (Rhodococcus) equi* in horse sera, *Vet. Microbiol.*, 10, 293–300, 1985.
92. Takai, S., Kawazu, S., and Tsubaki, S., Enzyme-linked immunosorbent assay for diagnosis of *Corynebacterium (Rhodococcus) equi* infection in foals, *Am. J. Vet. Res.*, 46, 2166–2170, 1985.
93. Nordmann, P., Nicolas, M.H., and Gutmann, L., Penicillin-binding proteins of *Rhodococcus equi*: Potential role in resistance to imipenem, *Antimicrob. Agents Chemother.*, 37, 1406–1409, 1993.
94. Takai, S. et al., Evaluation of a monoclonal antibody-based colony blot test for rapid identification of virulent *Rhodococcus equi*, *J. Vet. Med. Sci.*, 56, 681–684, 1994.
95. Takai, S. et al., Identification of virulent *Rhodococcus equi* by amplification of gene coding for 15- to 17-kilodalton antigens, *J. Clin. Microbiol.*, 33, 1624–1627, 1995.
96. Sellon, D.C. et al., Comparison of nucleic acid amplification, serology, and microbiologic culture for diagnosis of *Rhodococcus equi* pneumonia in foals, *J. Clin. Microbiol.*, 39, 1289–1293, 2001.
97. Arriaga, J.M. et al., Detection of *Rhodococcus equi* by polymerase chain reaction using species-specific nonproprietary primers, *J. Vet. Diagn. Invest.*, 14, 347–353, 2002.
98. Bell, K.S. et al., Identification of *Rhodococcus equi* using the polymerase chain reaction, *Lett. Appl. Microbiol.*, 23, 72–74, 1996.
99. Sellon, D.C. et al., Nucleic acid amplification for rapid detection of *Rhodococcus equi* in equine blood and tracheal wash fluids, *Am. J. Vet. Res.*, 58, 1232–1237, 1997.
100. Kirschner, P., and Bottger, E.C., Microheterogeneity within rRNA of *Mycobacterium gordonae*, *J. Clin. Microbiol.*, 30, 1049–1050, 1992.
101. McMinn, E.J. et al., Genomic and phenomic differentiation of *Rhodococcus equi* and related strains, *Antonie Van Leeuwenhoek*, 78, 331–340, 2000.
102. Patel, R. et al., Determination of 16S rRNA sequences of enterococci and application to species identification of nonmotile *Enterococcus gallinarum* isolates, *J. Clin. Microbiol.*, 36, 3399–3407, 1998.
103. Ladron, N. et al., Rapid identification of *Rhodococcus equi* by a PCR assay targeting the choE gene, *J. Clin. Microbiol.*, 41, 3241–3245, 2003.
104. Taylor, T.B. et al., Routine use of PCR-restriction fragment length polymorphism analysis for identification of mycobacteria growing in liquid media, *J. Clin. Microbiol.*, 35, 79–85, 1997.
105. Ocampo-Sosa, A.A. et al., Molecular epidemiology of *Rhodococcus equi* based on traA, vapA, and vapB virulence plasmid markers, *J. Infect. Dis.*, 196, 763–769, 2007.
106. Krewer, C.D.C. et al., Molecular characterization of *Rhodococcus equi* isolates of horse breeding farms from an endemic region in South of Brazil by multiplex PCR. Vol. 39 188–193 (scielo, 2008).
107. Hillidge, C.J., Review of *Corynebacterium (Rhodococcus) equi* lung abscesses in foals: Pathogenesis, diagnosis and treatment, *Vet Rec.*, 119, 261–264, 1986.
108. Bernheimer, A.W., Linder, R., and Avigad, L.S., Stepwise degradation of membrane sphingomyelin by corynebacterial phospholipases, *Infect. Immun.*, 29, 123–131, 1980.
109. Eales, L.J., and Parkin, J.M., Current concepts in the immunopathogenesis of AIDS and HIV infection, *Br. Med. Bull.*, 44, 38–55, 1988.
110. Hillidge, C.J., Use of erythromycin-rifampin combination in treatment of *Rhodococcus equi* pneumonia, *Vet. Microbiol.*, 14, 337–342, 1987.
111. Hillerdal, G. et al., Infection with *Rhodococcus equi* in a patient with sarcoidosis treated with corticosteroids, *Scand. J. Infect. Dis.*, 20, 673–677, 1988.
112. Samies, J.H. et al., Lung abscess due to *Corynebacterium equi*. Report of the first case in a patient with acquired immune deficiency syndrome, *Am. J. Med.*, 80, 685–688, 1986.
113. Ellenberger, M.A., Kaeberle, M.L., and Roth, J.A., Equine cell-mediated immune response to *Rhodococcus (Corynebacterium) equi*, *Am. J. Vet. Res.*, 45, 2424–2427, 1984.
114. Aihara, H., Watanabe, K., and Nakamura, R., Characterization of production of cholesterol oxidases in three *Rhodococcus* strains, *J. Appl. Bacteriol.*, 61, 269–274, 1986.
115. Arlotti, M. et al., *Rhodococcus equi* infection in HIV-positive subjects: A retrospective analysis of 24 cases, *Scand. J. Infect. Dis.*, 28, 463–467, 1996.
116. Scotton, P.G. et al., *Rhodococcus equi* nosocomial meningitis cured by levofloxacin and shunt removal, *Clin. Infect. Dis.*, 30, 223–224, 2000.
117. Phumoonna, T. et al., Chimeric vapA/groEL2 DNA vaccines enhance clearance of *Rhodococcus equi* in aerosol challenged C3H/He mice, *Vaccine*, 26, 2457–2465, 2008.
118. Takai, S. et al., Detection of virulent *Rhodococcus equi* in tracheal aspirate samples by polymerase chain reaction for rapid diagnosis of R. equi pneumonia in foals, *Vet. Microbiol.*, 61, 59–69, 1998.
119. Monego, F. et al., Molecular characterization of *Rhodococcus equi* from horse-breeding farms by means of multiplex PCR for the vap gene family, *Curr. Microbiol.*, 58, 399–403, 2009.
120. Staden, R., Beal, K.F., and Bonfield, J.K., The Staden package, 1998, *Methods Mol. Biol.*, 132, 115–130, 2000.
121. Zheng, H., Tkachuk-Saad, O., and Prescott, J.F., Development of a Rhodococcus equi-Escherichia coli plasmid shuttle vector, *Plasmid*, 38, 180–187, 1997.
122. Mangan, M.W., Byrne, G.A., and Meijer, W.G., Versatile *Rhodococcus equi–Escherichia coli* shuttle vectors, *Antonie van Leeuwenhoek*, 87, 161–167, 2005.
123. Dabbs, E.R., Gowan, B., and Andersen, S.J., Nocardioform arsenic resistance plasmids and construction of *Rhodococcus* cloning vectors, *Plasmid*, 23, 242–247, 1990.

15 *Saccharopolyspora*

Caroline Duchaine and Yvon Cormier

CONTENTS

15.1 INTRODUCTION

15.1.1 Classification

In agricultural environments, thermophilic actinomycetes grow in heated moist hay, grain, and straw, and are thus present in the bioaerosols of dairy barns during material handling. "Farmer's lung," one of the most common forms of hypersensitivity pneumonitis,[1,2] is usually caused by *Saccharopolyspora rectivirgula*, a thermophilic actinomycete.[2,3] *S. rectivirgula* has also been associated with a respiratory disease in horses known as heaves.[4,5] Aerobiological and environmental studies are frequently performed to evaluate the presence of *S. rectivirgula* in the work environment of farmers and thus also to assess exposure.[6–8] Identification of actinomycetes is difficult using traditional microbiological approaches since most of their differential characteristics are structural and identification requires fastidious techniques such as electron microscopy or slide culture. *S. rectivirgula* can be differentiated from the closest affiliated species by a combination of tests. Nowadays, molecular biology approaches, including sequencing of the 16S rRNA gene, allow for the identification of isolates.

In 1987, gram-positive bacteria with high G + C content were grouped to form a new class (Actinomycetes) composed of three groups: Bifidobacteria, Propionibacteria, and Actinomycetales. The latter included several families, such as Pseudonocardiaceae, which shelters the genus *Saccharopolyspora*. The actinomycetes, which were included in the Actinomycetales family, were grouped for their ability to form branched hyphae during one of their division stage. *S. rectivirgula* mycelium, when mature, is fragmented into rods and cocci by septation and hyphae disarticulation. Actinomycetes harboring this characteristic are called "nocardioforms," since species from the *Nocardia* genus exhibit the same characteristics.[9]

The genus *Micropolyspora* was created in 1961 to group actinomycetes producing short chains of spores in the aerial and substrate mycelium.[10] Pepys and collaborators established the relationship between a thermophilic actinomycete and Farmer's lung disease for the first time in 1963.[1] *Thermopolyspora polyspora* was the first name given to this Farmer's lung agent.[11] This name was subsequently rejected because the species description was obtained using contaminated strains.[12] *Micropolyspora faeni* was the proposed name for this thermophilic actinomycete (meaning: isolated from hay ("*foin*" in French) by Cross et al.[13]

During the same period, Russian scientists isolated thermophilic actinomycetes from soil samples.[14] These actinomycetes were placed in the *Thermopolyspora* genus following morphological analyses as well as cell wall chemical composition. The isolated strains named *T. rectivirgula* were later placed in the *Micropolyspora* genus,[15] and, following new analyses, were named *Micropolyspora rectivirgula*. *M. rectivirgula* had never been implicated in any lung affection.

In 1977, type species *Micropolyspora brevicatena* was reclassified because of the presence of mycolic acid in its cell wall.[16] The loss of the type species in the genus *Micropolyspora* was followed by the dissolution of the genus and the overall reclassification of the other species, including the overall comparison and classification of *Micropolyspora faeni* and *Micropolyspora rectivirgula*. The two species showing high resemblance and similarities in their morphological features,[17] phage susceptibility,[18] comparative taxonomic and serologic studies were performed in 1979 and the two species were considered as synonymous.[19] The International Code of Nomenclature of Bacteria indicates that the earliest name has priority when synonyms are discovered. However, even if the epithet *rectivirgula* had been published prior to the epithet *faeni*, the synonym species description was proposing the epithet *faeni* priority: this name was thus proposed as *nomen conservandum* according to International Code of Nomenclature of Bacteria of 1975. Finally, at this time, even the presence of more than 200 peer-reviewed publications on Farmer's lung disease and involving the species *M. faeni* did not weigh enough for the *nomen conservandum* to be accepted and the request was rejected. The earliest name was thus conserved and *Micropolyspora rectivirgula* was kept.

Since the type species of the *Micropolyspora* genus (*M. brevicatena*) had just been reclassified in the *Nocardia* genus (and named *Nocardia brevicatena*), the genus was no more valid and *M. rectivirgula* had to be placed in a new genus and its name had to be changed again. A new genus was proposed based on this species (*Faenia*) and *Faenia rectivirgula* was proposed as the type species for the new name.[20] The genus *Faenia* was described as an actinomycete producing branched aerial and substrate mycelium, 0.8–1.2 μm diameter, with spore chains (2–15) on both mycelia, gram-positive, and growing from 30°C to 60°C. The cell wall is of type IV: *meso*- diaminopimelic acid (*meso*-DAP), galactose, arabinose, and no nocardiomycolic acids. *F. rectivirgula* (type species) was described in 1983. In addition to the genus characteristics, it has smooth, ovoid, or spherical spores that are 0.7–1.5 μm diameter and form straight chains (never spirals). Older hyphae and spores may exhibit a negative response to Gram staining. Colonies are yellow-brown to pale orange and aerial mycelium is white. Growth is possible between 33°C and 60°C and optimal growth temperature is between 50°C and 55°C.

In 1984, Goodfellow established a classification of actinomycetes and grouped all the genera containing *meso*-DAP, arabinose and galactose, and lacking mycolic acids in a group named micropolyspora (unofficial name).[21] In 1988, Embley et al.[22] revised this unofficial group and created the *Pseudonocardiaceae* family. It now includes *Actinopolyspora*, *Amylocolatopsis*, *Faenia*, *Pseudonocardia*, *Saccharomonospora*, and *Saccharopolyspora*. This family exhibits all the characteristics proposed in 1984.

Actinomycetes phylogeny was established using the 16S rRNA gene analyses.[22,23] The same authors[24] established that *Faenia rectivirgula*, *Saccharopolyspora hirsute*, and *Pseudonocardia thermophila* formed a very tight phylogenetic group of actinomycetes, and possessed more affinity between each other than with the other members of *Pseudonocardiacea*e family. Chemotaxonomic analyses also demonstrated high similitude between *Saccharopolyspora* and *Faenia*.[25]

In 1989, Korn-Wendish et al. performed a comprehensive comparison of *Faenia rectivirgula* with species of the genus *Saccharopolyspora* (*S. taberi*, *S. erythrea*, and *S. hirsuta*), as well as with *Pseudonocardia thermophila*, *Saccharomonospora viridis*, and *Amylocalatopsis azurea*. This study included morphological, physiological, biochemical (DAP), ADN G + C content, phage sensitivity, enzyme and protein profiles, and restriction enzyme digestion polymorphism data demonstrated that *F. rectivirgula* showed enough similarities with members of the genus *Saccharopolyspora* to be transferred to it.[26] The *Saccharopolyspora rectivirgula* species description was finally published; its main features are that it is aerobic, gram-positive, nonacid-fast, with a type IV cell wall that includes *meso*-DAP, arabinose and galactose, and no mycolic acid. Substrate mycelium is yellow to orange on tryptic soy agar and is well developed, branched, and septed with a diameter from 0.5 to 0.8 μm. Aerial mycelium is white to pale pink and develops in the presence of 0.5% NaCl. Spores are present on both mycelium types and are in chains on short, unbranched, lateral or terminal sporophores. Spore surface is rough. Strains of *S. rectivirgula* are thermophilic and grow at temperatures between 37°C and 63°C, with optimum growth at 50°C. SR grows in the presence of 10% NaCl and is lysozyme sensitive. Esculin, arbutin, and urea are hydrolyzed. The G + C content is 70.4 mol%.

Recently, Yassin identified a novel *Saccharopolyspora* specie (i.e., *Saccharopolyspora rosea*) from a human patient with bronchial carcinoma.[27] The bacterium demonstrates morphological and chemotaxonomic characteristics consistent with those of members of the genus *Saccharopolyspora*, but is clearly distinguishable from the recognized members of the genus *Saccharopolyspora* by various biochemical tests. Comparative analysis of the 16S rRNA gene sequence revels that the type strain IMMIB L-1070(T) (=DSM 45226(T) = CCUG 56401(T)) displays 93.5%–96.9% sequence similarity with respect to members of the genus *Saccharopolyspora*, thus confirming its distinct phyletic position.[27]

15.1.2 Clinical Features and Pathogenesis

Although *S. rectivirgula* is not a pathogenic bacterium per se, it is involved in an immunologic disease called hypersensitivity pneumonitis (HP), or extrinsic allergic alveolitis. HP is a very complex affection with clinical signs and symptoms varying in intensity and induced by a wide range of inhaled organic antigen. Its classical clinical presentation includes three major forms: acute, subacute, and chronic. Inhaled antigens can be derived from various sources such as hay, straw, bedding, water reservoirs, and animal dander or proteins. The most common forms of HP are "bird fancier's disease" (induced by bird antigens) and farmer's lung disease (induced by *S. rectivirgula*). HP is recognized by different clinical features. In the acute form, patients develop cough, shortness of breath (dyspnea), fever, and chills 3–8 h after exposure to the causative antigen. The subacute form is more insidious, usually lacking the fever and chills, and evolves into progressive dyspnea over weeks to months. The chronic form is usually reserved for permanent sequela of long-term ongoing subacute disease or repeated bouts of acute HP. The pathophysiology of HP is still complex and incompletely understood. Basically, HP is a delayed cellular immune response characterized by the accumulation of activated lymphocytes in the lungs.[28] The pathology is similar to nonspecific interstitial pneumonitis (NSIP), with the presence of ill-formed granulomas. In more chronic forms lung fibrosis[29] and/or emphysemateous destruction can occur.[30] The major HP treatment is still antigen contact avoidance by environmental quality control. In severe acute forms, oral corticosteroids will accelerate the initial recovery but corticosteroids probably do not alter the long-term outcome.[31]

15.1.3 Diagnosis

Validated diagnostic criteria were recently published by Lacasse et al.[32] These criteria are based on clinical presentation, environmental exposure, and specific serum antibodies to the suspected antigen. Bronchoalveaolar lavage, which yields large numbers of lymphocytes[28] and typical interstitial patterns on high-resolution computed tomography,[33] is usually done to confirm the clinical diagnosis.

The isolation of *S. rectivirgula* from a patient with farmer's lung has never been successful since this actinomycete is thermophilic and does not invade human lung tissues or grow in the lungs. Bronchoalveolar lavage will show inflammatory cells but when cultured, *S. rectivirgula* is not recovered. However, a recent report shows the detection of b-glucans, a mould surface antigen, in the BAL of a patient with HP due to moulds exposure.[34] The causative agent can be searched for in the work environment of a patient suffering farmer's lung through air or bulk samples (hay, straw, bedding material, compost, etc.). The search for the SR is however usually not required; its presence and exposure are indirectly confirmed by the identification of specific IgG antibodies in the serum of the exposed subject.

To confirm exposure to *S. rectivirgula* infection, air or bulk samples are often obtained from the dairy barn or animal house where patient has been exposed. The samples are plated on Tryptic soy agar and incubated at 52°C for 5 days. Colonies of *S. rectivirgula* are recognized by their yellowish/brown color and white aerial mycelium. Biochemical identification of *S. rectivirgula* relies on the metabolic reactions, cell membrane lipids, and colony and cell morphology. Cellular fatty acids (CFA) are structural components present in the membrane or cell wall. When cells are cultured in standard conditions, lipid composition is reproducible at the genus, species, or subspecies levels. CFA are analyzed using gas chromatography and comparison databases are available (MIDI Labs).[35]

Soluble proteins profile analysis has been described.[36] Briefly, polyacrylamide gel electrophoresis in denaturing conditions of a protein suspension obtained after cell breaking using glass beads or ultrasonication is used for profile comparison purposes. This type of analysis may lead to numerical analysis[37] and the determination of dendrograms and similarity profiles. Biochemical tests were validated.[38] These authors proposed an identification algorithm using colony and cell morphologies, macromolecule degradation, utilization of sugars, acid production from the sugars, and antibiotic resistance. Those tests allow the differentiation between the agents most frequently responsible for HP cases *Saccharopolyspora*, *Thermoactinomyces*, and *Saccharomonospora*. Serological techniques such as electrosyneresis (ES), ouchterlony double diffusion (DD), ELISA, and Western blot (WB) have also been employed for identification of *Saccharopolyspora rectivirgula* associated with farmer's lung disease.[39]

Due to the limited value and the lack of reproducibility of some classical biochemical tests, Duchaine et al.[40] proposed a simple and rapid scheme for the identification of *S. rectivirgula*: optimal growth temperature (55°C); colony appearance based on morphology (filamentous) and color (beige to orange-brown); microscopic morphology (chains of spores on both aerial and substrate mycelium); growth on NaCl 10%; cell wall analysis (type IV); and the verification of antibody response with serum from a patient with farmer's lung. This last criterion is important to confirm the immunogenicity of the strains identified as *S. rectivirgula*. This scheme provides an accurate and efficient way of identifying *S. rectivirgula* strains and evaluating exposure to this bacterium.

When *Faenia rectivirgula* was transferred into the *Saccharopolyspora* genus, Korn-Wendisch et al.[26] published a complete description of *S. rectivirgula* in comparison with the other species regarding various markers (morphology, biochemistry, DNA, physiology, phage susceptibility patterns, enzyme and protein patterns, plasmids, and restriction fragments length polymorphisms). Given the difficulty and time consuming nature of differentiating members of the family Pseudonocardiaceae on the basis of micromorphology and biochemical properties, molecular techniques have been increasingly applied. Morón et al.[41] reported the use of genus-specific primers from 16S rRNA gene for PCR identification of *Saccharopolyspora*. Random amplified ribosomal DNA restriction analysis approach has been published[42] in the differentiation of closely related thermophilic actinomycetes involved in HP disease but has not been used in any other published report. Although this information is available, some authors used to base their identification of *S. rectivirgula* on colony morphology.[43] Amplified ribosomal DNA restriction analysis (ARDRA) is also valuable for identification of environmental actinomycete-like organisms, including *Saccharopolyspora rectivirgula* and *Saccharomonospora viridis*.[43]

15.2 METHODS

15.2.1 Sample Preparation

Air or bulk samples are obtained from dairy barn or animal house where the patient has been exposed to farmer's lung. Several air sampling approaches have been described.[7]

15.2.2 Detection Procedures

(i) Genus-Specific PCR Morón et al.[41] employed genus-specific primers for PCR identification of *Saccharopolyspora*. The primers are derived from the conserved region of the 16S rRNA gene—that is, forward primers SP1 (5′-GTGGAACCCATCCCCACACC-3′) and SP2 (5′-GTGGAAMCAGTCCCCACACC-3′)

both at positions 801–819, and reverse primer SP3 (5′-GGTGACGGTAGGTGTAGAAG-3′) at positions 450–469. The expected size of the PCR product using either SPl/SP3 or SP2/SP3 as amplification primers is 371 bp.

Procedure

1. PCR mixture (25 μL) is made up of 0.2 mM each of the four dNTPs, 0.1 μM each primer, 5 μL extracted DNA, and 0.5 U *Taq* polymerase (Appligene) with the appropriate reaction buffer. Controls without bacterial DNA are included for each PCR run.
2. Amplification is conducted in a Perkin Elmer Cetus DNA Thermal Cycler 480 with 40 cycles of 93°C for 30 s, 53°C for 30 s, and 72°C for 2 min, followed by 10 min at 72°C.
3. The resulting amplicons are analyzed by electrophoresis (0.25 V/cm) in 1.2% (w/v) agarose gels, and stained with ethidium bromide.

Note: The product of 371 bp is obtained with both pairs of primers, SPl/SP3 and SP2/SP3, from all the *Saccharopolyspora* strains tested, including *Saccharopolyspora erythraea*, *Saccharopolyspora gregorii*, *Saccharopolyspora hirsuta* subsp. *hirsuta*, *Saccharopolyspora hirsuta kobensis*, *Saccharopolyspora hordei*, *Saccharopolyspora rectivirgula*, *Saccharopolyspora spinosa*, and *Saccharopolyspora taberil.* Two strains of *Saccharopolyspora rectivirgula* examined contain sequences around positions 450–469 that do not perfectly match with SP3 (nucleotide positions 464 and 465), but still produce the 371 bp amplification fragment. The effect of this divergence is only observed when the annealing temperature is increased by at least 4°C. In this case, experiments done at annealing temperatures of 57°C or higher fail to produce amplification fragments.

(ii) 16S rRNA Gene Sequence Analysis Paster et al.[44] reported a pair of universal primers, 27F (5′-AGAGTTTG ATCCTGGCTCAG-3′, positions 8 to 27 in *Escherichia coli* 16S rRNA gene) and 1492R (5′-TACCTTGTTACGACTT-3′, positions 1492 to 1507), which cover nearly the entire length of the gene the 16S rRNA from bacteria. Sequencing analysis of the resulting amplicon facilitates the identification of bacteria including *Saccharopolyspora* spp. Alternative 16S rRNA gene primers described by Frank et al.[45] may be also utilized.

Procedure

1. PCR mixture (100 μL) is made up of 50 mM KCl, 10 mM Tris pH 8.3, 1.5 mM $MgCl_2$, 2 mM each of dNTP, 1 μM (each) primers 37F and 1492R, 2.5 U of *Taq* DNA polymerase, and 1 pg to 100 ng of template DNA.
2. The mixture is subjected to an initial denaturation at 94°C for 5 min, 25 cycles of 94°C for 60 s, 50°C for 30 s, and 72°C for 2 min, and a final extension step at 72°C for 5 min.
3. The resulting PCR products are then sequenced by use of the ABI technology with the same universal primers. The nucleotide sequence is analyzed with BLAST program.

15.3 CONCLUSION

Saccharopolyspora rectivirgula (*Micropolyspora faeni*) is one of the major agents responsible for farmer's lung disease, a form of hypersensitivity pneumonitis that results from repeated inhalation of antigens from mouldy hay or straw. Although *S. rectivirgula* is capable of inducing a complex immunologic reaction in patients with farmer's lung disease, and is detectable in the surrounding environment of patients, its precise role in the pathogenesis of farmer's lung disease is still unclear. Due to its variable phenotypic characteristics, identification of *Saccharopolyspora* strains on the basis of classical tests is difficult and often inconclusive. In particular, microscopic and macroscopic examinations require specialized skills and the assessment of growth properties; macromolecule degradation, citrate utilization, and acid production from carbohydrates are time consuming and expensive. Although fatty acid analysis is very useful for the identification of *Saccharopolyspora* strains, it is not readily available outside of well-resourced research or reference laboratories. In recent years, molecular techniques have been applied for rapid and specific detection of *Saccharopolyspora* bacteria. Future development of real-time PCR will not only allow convenient detection and quantitation of *Saccharopolyspora* bacteria, but also reveal new insights on the role of these bacteria in farmer's lung disease.

REFERENCES

1. Pepys, J. et al., Farmer's lung: Thermophilic actinomycetes as a source of "farmer's lung hay" antigen, *Lancet*, 1, 12, 1963.
2. Lacasse, Y., and Cormier, Y., Hypersensitivity pneumonitis, *Orphanet J. Rare Dis.*, 1, 25, 2006.
3. Lacasse, Y. et al., Classification of hypersensitivity pneumonitis: A hypothesis, *Int. Arch. Allergy Immunol.*, 149, 161, 2009.
4. Couetil, L.L. et al., Inflammatory airway disease of horses, *J. Vet. Intern. Med.*, 21, 356, 2007.
5. Derksen, F.J. et al., Airway reactivity in ponies with recurrent airway obstruction (heaves), *J. Appl. Physiol.*, 58, 598, 1985.
6. Cormier, Y. et al., Airborne microbial contents in two types of swine confinement buildings in Quebec, *Am. Ind. Hyg. Assoc. J.*, 51, 304, 1990.
7. Duchaine, C., Grimard, Y., and Cormier, Y., Influence of building maintenance, environmental factors, and seasons on airborne contaminants of swine confinement buildings, *Aihaj*, 61, 56, 2000.
8. Kang, Y.-J., and Frank, J.F., Biological aerosols: A review of airborne contamination and its measurement in dairy processing plants, *J. Food Prot.*, 52, 512, 1989.
9. Prauser, H., Nocardioides, a new genus of order Actinomycetales, *Int. J. Syst. Bacteriol.*, 26, 58, 1976.
10. Lechevalier, H.A., Solotorovsky, M., and McDurmont, C.L., A new genus of actinomycetales: *Micropolyspora* gen. nov., *J. Gen. Microbiol.*, 26, 11, 1961.

11. Corbaz, R., Gregory, P.H., and Lacey, M.E., Thermophilic and mesophilic *Actinomycetes* in mouldy hay, *J. Gen. Microbiol.*, 32, 449, 1963.
12. Henssen, A., Importance of thermophilic microorganisms for the decomposition of stable manure, *Arch. Mikrobiol.*, 27, 63, 1957.
13. Cross, T., Maciver, A.M., and Lacey, J., The thermophilic actinomycetes in mouldy hay: Micropolyspora faeni sp. nov., *J. Gen. Microbiol.*, 50, 351, 1968.
14. Krassilnikov, N.A., and Agre, N.S., On two new species of *Thermopolyspora*, *Hindustan Antibiotics Bull.*, 6, 97, 1964.
15. Lechevalier, H.A., and Lechevalier, M.P., Biology of actinomycetes, *Ann. Rev. Microbiol.*, 21, 71, 1967.
16. Collins, M.D. et al., Distribution of menaquinones in actinomycetes and corynebacteria, *J. Gen. Microbiol.*, 100, 221, 1977.
17. Dorokhova, L.A. et al., Morphology of 2 cultures belonging to the *Micropolyspora* genus, *Mikrobiologiia*, 39, 95, 1970.
18. Prauser, H., and Momirova, S., Phage sensitivity, cell wall composition and taxonomy of various thermophilic actinomycetes, *Zeitschrift Allgemeine Mikrobiol.*, 10, 219, 1970.
19. Jones, M.P., McCarthy, A.J., and Cross, T. Taxonomic and serologic studies on *Micropolyspora faeni* and *Micropolyspora* strains from soil bearing the specific epithet rectivirgula, *J. Gen. Microbiol.*, 115, 343, 1979.
20. Kurup, V.P., and Agre, N.S., Transfer of *Micropolyspora rectivirgula* (Krassilnikov and Agre 1964) Lechevalier, Lechevalier, and Becker 1966 to *Faenia* gen. nov., *Int. J. Syst. Bacteriol.*, 33, 663, 1983.
21. Goodfellow, M., and Cross, T. Classification. In Goodfellow, M., Mordarski, M., and Williams, S.T. (Editors), *The Biology of Actinomycetes*, pp. 7–164, Academic Press, London, 1984.
22. Embley, M.T., Smida, J., and Stackebrandt, E., The phylogeny of Mycolate-less wall chemotype IV Actinomycetes and description of *Pseudonocardiaceae* fam. nov., *Syst. Appl. Microbiol.*, 11, 44, 1988.
23. Fox, G.E., and Stackebrandt, E., The application of 16 S rRNA cataloguing and 5 S rRNA sequencing in bacterial systematics, *Methods Microbiol.*, 19, 405, 1987.
24. Embley, T. et al., Chemotaxonomy of wall type IV actinomycetes which lack mycolic acids, *J. Gen. Microbiol.*, 134, 953, 1988.
25. Embley, T.M., Smida, J., and Stackebrandt, E., Reverse transcriptase sequencing of 16S ribosomal RNA from *Faenia rectivirgula*, *Pseudonocardia thermophila* and *Saccharopolyspora hirsuta*, three wall type IV actinomycetes which lack mycolic acids, *J. Gen. Microbiol.*, 134, 961, 1988.
26. Korn-Wendish, F. et al., Transfer of *Faenia rectivirgula* Kurup and Agre 1983 to the genus *Saccharopolyspora* Lacey and Goodfellow 1975, elevation of *Saccharopolyspora hirsuta* subso. *taberi* Labeda 1987 to species level, and emended description of the genus *Saccharopolyspora*, *Int. J. Syst. Bacteriol.*, 39, 430, 1989.
27. Yassin, A.F., *Saccharopolyspora rosea* sp. nov., isolated from a patient with bronchial carcinoma, *Int. J. Syst. Evol. Microbiol.*, 59, 1148, 2009.
28. Cormier, Y. et al., Bronchoalveolar lavage in farmers' lung disease: diagnostic and physiological significance, *Br. J. Ind. Med.*, 43, 401, 1986.
29. Perez-Padilla, R. et al., Mortality in Mexican patients with chronic pigeon breeder's lung compared with those with usual interstitial pneumonia, *Am. Rev. Respir. Dis.*, 148, 49, 1993.
30. Lalancette, M. et al., Farmer's lung. Long-term outcome and lack of predictive value of bronchoalveolar lavage fibrosing factors, *Am. Rev. Respir. Dis.*,148, 216, 1993.
31. Monkare, S., Influence of corticosteroid treatment on the course of farmer's lung, *Eur. J. Respir. Dis.*, 64, 283, 1983.
32. Lacasse, Y. et al., Clinical diagnosis of hypersensitivity pneumonitis, *Am. J. Respir. Crit. Care Med.*, 168, 952, 2003.
33. Lynch, D.A. et al., Hypersensitivity pneumonitis: Sensitivity of high-resolution CT in a population-based study, *Ajr*, 159, 469, 1992.
34. Ashitani, J. et al., Elevated levels of beta-D-glucan in bronchoalveolar lavage fluid in patients with farmer's lung in Miyazaki, Japan, *Respiration; Int. Rre. Tthoracic Dis.*, 75, 182, 2008.
35. Bernard, K.A., Bellefeuille, M., and Ewan, E.P., Cellular fatty acid composition as an adjunct to the identification of asporogenous, aerobic gram-positive rods, *J. Clin. Microbiol.*, 29, 83, 1991.
36. Moore, W.E. et al., Polyacrylamide slab gel electrophoresis of soluble proteins for studies of bacterial floras, *Appl. Environ. Microbiol.*, 39, 900, 1980.
37. Kersters, K., and De Ley, J., Identification and grouping of bacteria by numerical analysis of their electrophoretic protein patterns, *J. Gen. Microbiol.*, 87, 333, 1975.
38. Kurup, V.P., and Fink, J.N., A scheme for the identification of thermophilic actinomycetes associated with hypersensitivity pneumonitis, *J. Clin. Microbiol.*, 2, 55, 1975.
39. Reboux, G. et al., Assessment of four serological techniques in the immunological diagnosis of farmers' lung disease, *J. Med. Microbiol.*, 56, 1317, 2007.
40. Duchaine, C. et al., *Saccharopolyspora rectivirgula* from Quebec dairy barns: Application of simplified criteria for the identification of an agent responsible for farmer's lung disease, *J. Med. Microbiol.*, 48, 173, 1999.
41. Morón, R., González, I., and Genilloud, O., New genus-specific primers for the PCR identification of members of the genera *Pseudonocardia* and *Saccharopolyspora*, *Int. J. Syst. Bacteriol.*, 49, 149, 1999.
42. Harvey, I. et al., Random amplified ribosomal DNA restriction analysis for rapid identification of thermophilic Actinomycete-like bacteria involved in hypersensitivity pneumonitis, *Sys. Appl. Microbiol.*, 24, 277, 2001.
43. Kurtboke, D.I., Murphy, N.E., and Sivasithamparam, K., Use of bacteriophage for the selective isolation of thermophilic actinomycetes from composted eucalyptus bark, *Can. J. Microbiol.*, 39, 46, 1993.
44. Paster, B.J. et al., Prevalent bacterial species and novel phylotypes in advanced noma lesions, *J. Clin. Microbiol.*, 40, 2187, 2002.
45. Frank, J.A. et al., Critical evaluation of two primers commonly used for amplification of bacterial 16S rRNA genes, *Appl. Environ. Microbiol.*, 74, 2461, 2008.

16 Streptomyces

Angel Manteca, Ana Isabel Pelaez, Maria del Mar Garcia-Suarez, and Francisco J. Mendez

CONTENTS

16.1 INTRODUCTION

16.1.1 Classification, Morphology, Biology, and Epidemiology

Actinomycetales are gram-positive, mycelium-forming, soil bacteria that play an important role in mineralization processes in nature. *Streptomyces* is an extremely important bacterium in biotechnology because approximately two-thirds of industrial antibiotics are synthesized by members of this genus.[1] Streptomycetes present a complex development cycle that includes programmed cell-death processes, as well as differentiation phenomena that lead to sporulation[2–4] (Figure 16.1). After spore germination, a fully compartmentalized mycelium starts to grow and early on suffers an orderly death process in which live and dead segments alternate in the same hyphae.[2–4] Subsequently, the viable segments start to enlarge as a multinucleated mycelium that grows in successive waves that determine the characteristic complex growth curves of this microorganism. Two stages of secondary multinucleated mycelium have been defined: early secondary mycelium that does not yet have the hydrophobic layers characteristic of aerial hyphae and late secondary mycelium that starts to express the genes encoding for the chaplin and rodlin proteins, which are mainly responsible for the formation of the hydrophobic layer characteristic of aerial structures.[5] The traditionally denominated substrate-vegetative mycelium corresponds to the early secondary multinucleated mycelium, which still lacking the hydrophobic layers characteristic of the aerial mycelium and would match therefore to the late secondary mycelium which has acquired those hydrophobic covers.[6] A vegetative role has been postulated to these compartmentalized hyphae by contrast to the secondary multinucleated mycelium that should be considered the reproductive structure, since it is destined to sporulate.[7] The secondary mycelium has been described also as the antibiotic producing.[7]

Streptomyces displays an unequivocally morphology characterized by the formation of branched, thin hyphae of 0.5–2 μm in diameter, and spore chains. Isolated colonies of this bacterium are also easy to identify as they exhibited grayish-white appearance, and a medium size around 3 mm in diameter (Figure 16.2a).

Streptomycetes are infrequently encountered in clinical practice but are important potential causes of serious human and animal infections.[10,11] Their natural habitats are mainly natural soils, and they are universally distributed in environment. The ubiquitous nature and the low pathogenicity of *Streptomyces* organisms make most clinical isolates contaminants or colonizers. Despite that, at least in our experience, *Streptomyces* strains are often not recognized in hospitals, and several *Streptomyces* infections and colonizations are not being identified. In the Oviedo Central Hospital (Spain), we have described three such clinical cases during last 5 years (2004–2009)[8,9,12]; in the Section of Clinical Microbiology, University of Texas, six cases have been described during 9 years (1997–2005)[11]; and probably many more cases are either not being identified or they are considered as contaminants.

16.1.2 Clinical Features and Pathogenesis

As stated earlier, Streptomycetes are rarely documented as cause of human disease. The most common manifestation

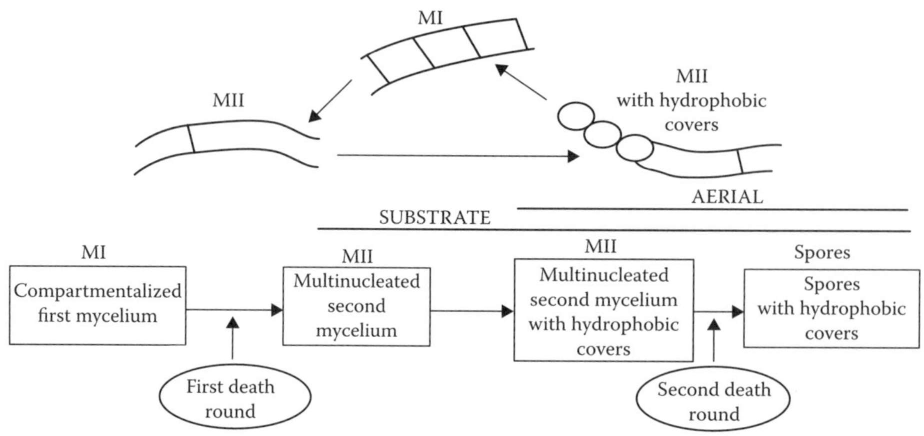

FIGURE 16.1 Cell-cycle features of the *Streptomyces* development. Mycelial structures (MI, first compartmentalized mycelium; MII, second multinucleated mycelium), cell-death processes, spores, and traditional denomination ("substrate" and aerial) are indicated.

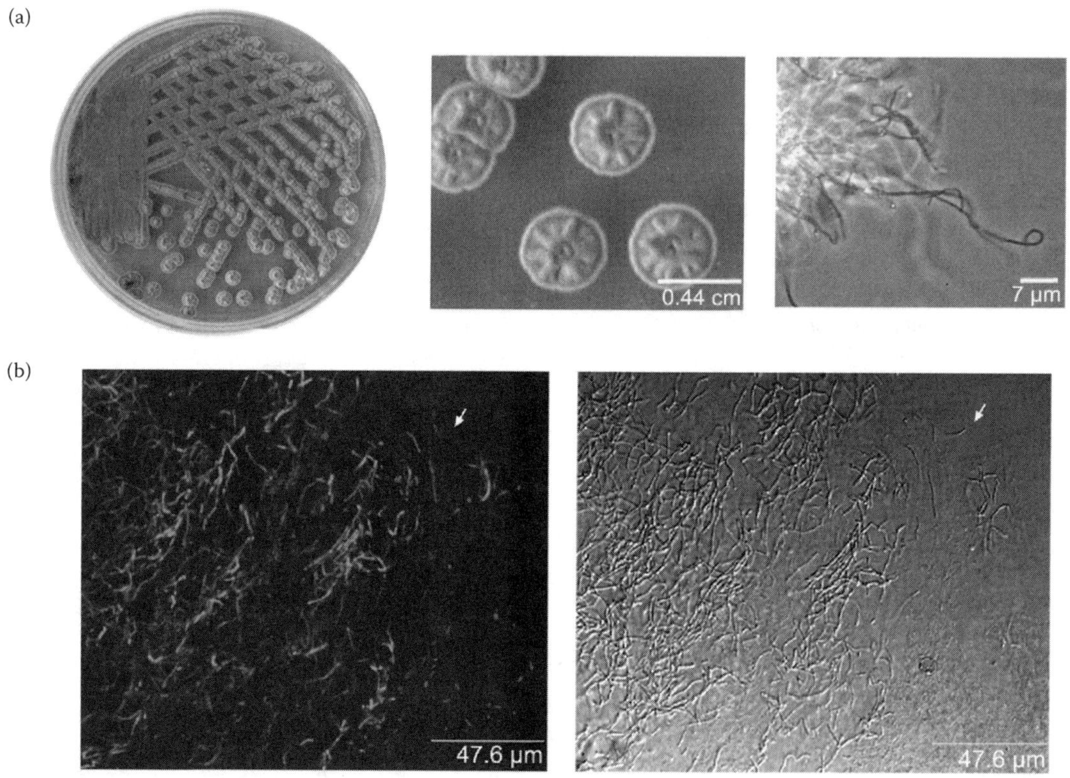

FIGURE 16.2 *Streptomyces* morphology. (a) Colonies of *Streptomyces cinereoruber* isolated from a patient with severe chronic obstructive pulmonary disease.[8] (b) Laser confocal observations of silicone prothesis infected with *Streptomyces* spp.[9] Samples are stained with the viability assay LIVE/DEAD Bac-Light Bacterial Viability kit (Molecular Probes) (left), or in differential interference contrast mode (right). Arrows illustrate the border of the silicone biopsy.

of Actinobacteria infections are actinomycetomas which are generally easy to identify by their clinical presentation: chronic subcutaneous inflammatory tumor with discharging sinuses. The commonest affected sites are the foot, knee, thigh, hand, and arm. Lung-associated actinomycetoma are extremely rare and had been attributed to members of the *Nocardia* genus,[13] and some *Streptomyces* species, such as *Streptomyces pneumoniae* or *Streptomyces cinereoruber*.[8,14] Recently, a silicone mammary implant actinomycetoma has been also reported to *Streptomyces*.[9] Most of these cases had in common some degree of immunosuppression. In the last 20 years, at least 10 species belonging to the *Streptomyces* genus have been isolated from mycetomic and nonmycetomic infections.[14,15] *S. somaliensis* is a well-documented agent in actinomycetomas.[8,16] *S. albus* has been also cited as a human pathogen and the recent literature includes cases of extrinsic alveolitis and skin hypersensitivity type II in farm workers.[12] Pulmonary infections due to *Streptomyces* in non-immunocompromised hosts have been also reported[9,14,17,18] (Table 16.1).

TABLE 16.1
Examples of Well-Documented *Streptomyces* Infections

Isolate	Type of Infection	Immunosuppression and/or Antineoplastic Treatments	Reference
S. albus	Actinomycetoma forearm	Yes	12
S. cinereoruber	Lung coinfection	No	8
Streptomyces spp.	Silicone mammary implant infection	Yes	9
S. maritimus	Lung nodule	Yes	11
S. albus	Lung nodule	Yes	11
S. nobilis	CVC-related boodstream infection	Yes	11
S. humidus	CVC-related boodstream infection	Yes	11
S. indigoferus	CVC-related boodstream infection	Yes	11
Streptomyces spp.	Hypersensitivity pneumonitis	Yes	11
S. penumonia	Lung infection	No	14
S. coelicolor	Brain abscess	NR	19

Abbreviations: CVC, central venous catheter; NR, non reported.

Streptomycetes are most commonly limited to causing actinomycotic mycetoma in areas of the body that come into contact with soil, or those traumatized gain access to subcutaneous tissues through abrasions in the skin. Initial formation of painless nodules is followed by fluctuance and progression to formation of draining sinuses. The sinuses begin to drain, with the exudates often containing hard, compacted, filamentous microcolonies or granules. Additional sinus tracts and nodules may form, and the involved areas eventually become deformed by fibrous tissue. The process remains characteristically pyogenic rather than granulomatous and may reach deeper connective tissue, muscle, and bone. The disease rarely spreads hematogenously via the lymphatics.[11]

The treatment of *Streptomyces* infections remains a problem; there are no approved breakpoint antibiotics for *Streptomyces* species, and the correlation between the data obtained in vitro and the clinical evolution remains unclear.[10] Many drugs are being used with variable results. The drugs used most frequently include trimethoprim-sulfamethoxazole (TS), dapsone, amikacin alone or combined with TS, and amoxicillin-clavulanic acid.[15] All treatments have in common the need for long courses of therapy to obtain satisfactory results (Table 16.2). Our experience with three clinical cases is that prolonged treatment with TS (several weeks) is very effective (Table 16.2).[8,9,12]

16.1.3 Diagnosis

Conventional Techniques. The first approach to diagnosticate a *Streptomyces* infection/colonization is a direct examination of a biopsy or drainage material collected. Usually, granules (averaging 1 mm in diameter) are observed. A microscopic analysis can show gram-positive fine filaments.[8,9,12] One of the main differences with fungal hyphae is that *Streptomyces* hyphae are much finer, having the diameter of a bacterium (0.5–2 μm) (Figure 16.2b). However, *Streptomyces* hyphae are sometimes very fragmented, and mycelia appearance is not evident.[12]

Streptomycetes are bacteria that are easily cultivable almost in all microbial culture media. However, and in order to facilitate the full differentiation and sporulation that facilitate colonies differentiation, poor media are preferred. Incubation temperatures of 28°C–30°C are recommended, and long incubation periods are necessary (around 1 week). In this sense, a good, easy-to-prepare media is GAE.[8] Alternative culture media for *Streptomyces* cultures are detailed in Kieser et al. (Reference 20; see Methods.)

Biochemical markers of taxonomic relevance in *Streptomyces* are, for example, utilization of cysteine, valine, phenylalanine, histidine, hydroxyproline, sucrose, mannitol, raffinose, dextran, or xylitol.[1] Other activities with taxonomy relevance are lecithinase activity, lipolysis, pectin hydrolysis, nitrate reduction, elastin degradation, and more.[1]

Molecular Techniques. Identification of *Streptomyces* genus by the conventional techniques described in the preceding material is relatively easy. However, morphological, physiological, and biochemical approaches are not sufficient to

TABLE 16.2
Antibiotic Susceptibility Data for the Cases Indicated in Table 16.1. The Number of Sensible Strains (11 in Total) and the Percentage for Each Antibiotic Are Indicated

Antibiotic	Susceptibility
Imipenem	8 (73%)
Trimethroprim-sulfamethoxazole	6 (54%)
Erythromycin	5 (45%)
Clarithromycin	4 (36%)
Amikacin	5 (45%)
Amoxicilin-clavulanate	4 (36%)
Ceftriaxone	3 (27%)
Ampicilin	3 (27%)
Vancomycin	3 (27%)
Minocycline	2 (18%)
Tetracycline	2 (18%)
Vancomycin	2 (18%)

characterize strains and species in this genus, which is characterized by a broad biochemical and genetic diversity that traditionally has attributed to the description of more species than really have been validated by molecular techniques.[20] For all these reasons, molecular techniques are beginning to be especially important for definitive identification and taxonomic characterization of *Streptomyces* strains. Also, molecular techniques are much faster, which can be also a critical advantage in the identification of this slow-growing bacterium.

A first molecular approach to *Streptomyces* identification can be the analysis of amplification restriction of the 65-kDa fragment coding for the HSP.[21] This kind of analysis can differentiate *Streptomyces* of other Actinomycetes and also works to identify some *Streptomyces* species, such as *S. albus*, *S. griseus*, and *S. somaliensis*.[21]

The fastest and most reliable methods in bacterial taxonomy and identification, and for identification of *Streptomyces* in particular, are those based in the analysis of 16S ribosomal DNA sequence: sequencing of 16S ribosomal RNA (rRNA)[22] and analysis of the α-region of the 16S rRNA in the case of Streptomycetes (positions 158 to 276).[23] The rRNA genes are essential for the survival of all organisms and are highly conserved in the bacterial and other kingdoms. Consequently, characterization of the 16s rRNA gene is now well established as a standard method for the identification of species, genera, and families of bacteria.[24] This is reflected in the large body of information that has been accumulated on the phylogenetic analysis of microorganisms using this gene.[25] In consequence, the full sequence of 16S rRNA can be considered as the master molecule for bacterial taxonomy.[25] A rapid identification of 16S ribosomal DNA can be performed using the primers pA (5′-AGAGTTTGATCCTGGCTCAG-3′) and pH (5′-AAGGAGGTGATCCAGCCGCA-3′), as described previously.[26] The α-region of the 16S rRNA has been described as one of the most variable in *Streptomyces*, and it is very useful for species identification.[23] For this reason, when partial sequences of 16S rRNA are used for identification of this bacterium, they should include at least the α-region. Usually, correct 16S rRNA sequence analysis needs the elaboration of phylogenetic trees, including type 16S rRNA sequences from databases. For this end, phylogenetic trees based on maximum likelihood methods are the most recommended.

For bacterial identification, 16S rRNA sequencing is definitive. It is the easiest, fastest, and most reliable technique for *Streptomyces* genus identification. However, when identification at the species level is necessary, additional experiments are usually needed. For example, the HSP RFLP indicated above, morphological and biochemical markers; RFLP from total DNA; DNA: DNA hybridizations; analysis of membrane fatty acids; enzyme patterns; and more.[27]

16.2 METHODS

16.2.1 Sample Preparation

***Streptomyces* Cultivation and Isolation.** The microorganisms should be isolated from the biological sample by growing them in solid media in which they have an entire life cycle with abundant sporulation, facilitating isolation and identification. We recommended poor media as GAE (glucose, asparagine, yeast extract)[28] or SFM (Soya flour–mannitol).[20] After agar addition, solid cultures are prepared in Petri dishes (8.5-cm diameter) as lawns on solid medium (30 mL/plate). Plates were incubated at 30°C at least during 7 days. Strains should be conserved as lyophilized spores.[20]

Microscopy. Isolated colonies or original sample (biopsy, sputum, exudates, etc.) can be observed directly at phase contrast microscope, or stained with Gram or other colorants. An alternative staining very useful in *Streptomyces* is the viability assay (LIVE/DEAD Bac-Light Bacterial Viability Kit, Molecular Probes, L-13152)[2] (Figure 16.2b); a viability assay, based on staining cells in a population with propidium iodide and SYTO 9.[2] Propidium iodide is specific for nucleic acids of damaged (leaky) cells, whereas SYTO 9 stains those of viable cells. The stain mixture is added directly to the samples on the slide. The coverslip is placed on top after staining for at least 10 min in the dark, and samples are observed under a fluorescence or confocal microscope at wavelengths of 488 and 568 nm for excitation, and 530 nm (green) or 630 nm (red) for emission. This fluorescence staining combined with confocal microscopy allows direct observation of *Streptomyces* hyphae in the original infected tissue, even if this tissue is very thick or dirty. An example of this kind of observation has been published recently for a *Streptomyces* infection in a silicone mammary implant.[12]

Physiological/Biochemical Phenotypes and Susceptibility Tests. Physiological and biochemical phenotypes can be determined using standard bacterial identification techniques.[1,29] Antimicrobial susceptibility testing can be performed simultaneously by Kirby Bauer disc diffusion[30] or E-tests (AB Biodisk, Sweden). In the case of E-tests, a protocol for *Streptomyces* has been described by Martin et al.[12]

16.2.2 Detection Procedures

Streptomyces DNA can be obtained from isolated colonies using standard methods.[20] To obtain a rapid strain identification, PCR can be performed directly from a dilution of the isolated colonies, or even from a dilution of the original sample, biopsy, or wherever. The best primers for amplification of 16S rRNA are pA and pH.[26] The 50 μL reaction mixture should contain 0.2 mM dNTPs mixture, 30 pmol/μL of each primer (pA and pH); 1.5 mM $MgCl_2$; 1x PCR buffer; 0.2 μL (1 U) *Eco* Taq DNA polymerase (Ecogen). The PCR program is as follows: denaturation at 95°C for 5 min; 30 cycles of 95°C for 1 min, 57°C for 1 min, and 72°C for 90 s; and final extension at 72°C for 10 min. The amplification products (1500 bp) is analyzed by horizontal gel electrophoresis in 0.8% agarose gel in Tris-Borate-EDTA (TBE), followed by ethidium bromide staining and visualization under UV light. PCR products from the extracted bands are purified by GFX™ PCR DNA and Gel Band Purification Kit (Amersham Biosciences). The amplicon is sequenced in an automatic sequencing as for example ABI Prism™ 3100 Genetic Analyzer (Perking Elmer).

Sequence identification is done using the Gapped BLAST database search program facility of the National Center for Biotechnology Information (NCBI).[31] For a more in-depth comparison between 16S rRNA sequences, phylogenetic trees are very useful. The most adequate phylograms are those based in nonparametric analysis [maximum likelihood methods (ML)]. An easy and robust free online platform for ML phylogenetic analyses can be found at http://phylemon.bioinfo.cipf.es/.[32]

Preliminary information can be obtained from isolates before sequencing. The 16S rRNA from different *Streptomyces* isolates can be compared by fragment length polymorphism (RFLP).[33] Here, 16S rRNA are digested with restriction enzymes as *Hae*III and *Alu*I. An aliquot of each of the PCR-amplified 16S rRNA genes is digested at 37°C for 4 h, and the resulting restriction fragments separated on a 3% agarose electrophoresis gel. Identical patterns of bands are assigned to the same microorganism, whereas different patterns confirm those that are different *Streptomyces* species.

16.3 CONCLUSION AND FUTURE PERSPECTIVES

The infections caused by *Streptomyces* species are rare and in general not very well documented. In this chapter, we described some of the most significant cases of *Streptomyces* infections as well as the methods for identification of these bacteria, with special emphasis on molecular techniques, which are especially important for the correct identification of this unusual pathogenic bacterium. Morphological, physiological, and biochemical approaches are not enough to characterize strains and species in this genus.[20] The 16S rRNA sequencing and analysis is a culture-independent technique that can be performed directly from the primary material (i.e., biopsies, sputa, exudates). It is one of the most reliable and faster techniques for the correct identification of *Streptomyces* infections. Of all the most recent advances in molecular biology, 16S rRNA analysis is undoubtedly the technique of choice for identification of bacteria in general, and *Streptomyces* in particular, and in the near future, it will be increasingly implanted for bacterial clinical diagnosis. It should be noted that for identification at the species level, additional experiments are usually necessary, such as the HSP RFLP indicated previously; morphological and biochemical markers; RFLP from total DNA; DNA:DNA hybridizations; analysis of membrane fatty acids; or enzymatic patterns.

Other innovative techniques that are being implanted in basic research will be progressively extended to clinical microbiological studies, including pathogenic *Streptomyces* strains. For instance, confocal microscopy, or Fourier transform infrared spectroscopy (FTIR) microscopy and even mass spectrometry techniques are becoming very important in the study of the biology of *Streptomyces*.[2,3,34–36] These techniques will be very useful for the analysis of the clinical *Streptomyces* infections from direct biopsies, or other kind of primary samples. FTIR can be also applied to species identification.[34]

ACKNOWLEDGMENTS

This work was supported by grant MEC-06-BI02006-15336-C04-02 and MEC-07-BIO2007-66313 from the DGI, Subdireccion General de Proyectos de Investigacion, Ministerio de Ciencia e Innovación, Spain. The authors would like to acknowledge Professor Jesus Sanchez (Area Microbiologia, Universidad de Oviedo) for the critical reading of the manuscript and help with phylogenetic analysis. A.M. was supported by a postdoctoral grant from Ministerio de Ciencia e Innovación, Spain; A.I.P. by grant CENIT-07-CLEAM, Ministerio de Industria, Turismo y Comercio (CDTI), Spain; and M.M.G.S. was supported by grant MEC-06-BIO2006–1533-C04–02.

REFERENCES

1. Locci, R. et al., Streptomycetes and related genera. In S.T. Williams, M.E. Sharpe and J.G. Holt (Editors), *Bergey's Manual of Systematic Bacteriology*, Vol. 4, pp. 2451–2508, Williams & Wilkins Co., 1984.
2. Manteca, A., Fernandez, M., and Sanchez. J., A death round affecting a young compartmentalized mycelium precedes aerial mycelium dismantling in confluent surface cultures of *Streptomyces antibioticus*, *Microbiology*, 151, 3689, 2005.
3. Manteca, A., Fernandez, M., and Sanchez, J., Mycelium development in *Streptomyces antibioticus* ATCC11891 occurs in an orderly pattern which determines multiphase growth curves, *BMC Microbiol.*, 5, 51, 2005.
4. Manteca, A. et al., A proteomic analysis of *Streptomyces coelicolor* programmed cell death, *Proteomics*, 6, 6008, 2006.
5. Manteca, A. et al., Aerial hyphae in surface cultures of *Streptomyces lividans* and *Streptomyces coelicolor* originate from viable segments surviving an early programmed cell death event, *FEMS Microbiol. Lett.*, 274, 118, 2007.
6. Flärdh, K., and Buttner, M. J., *Streptomyces* morphogenetics: Dissecting differentiation in a filamentous bacterium, *Nat. Rev. Microbiol.*, 7, 36, 2009.
7. Manteca, A., and Sanchez, J., *Streptomyces* development in colonies and soils, *Appl. Environ. Microbiol.*, 75, 2920, 2009.
8. Manteca, A. et al., A rare case of lung coinfection by *Streptomyces cinereoruber* and *Haemophilus influenzae* in a patient with severe chronic obstructive pulmonary disease: Characterization at species level using molecular techniques, *Diagn. Microbiol. Infect. Dis.*, 60, 307, 2008.
9. Manteca, A. et al., A rare case of silicone mammary implant infection by *Streptomyces* spp. in a patient with breast reconstruction after mastectomy: Taxonomic characterization using molecular techniques, *Diagn. Microbiol. Infect. Dis.*, 63, 390, 2009.
10. McNeil, M.M., and Brown, J.M., The medically important aerobic actinomycetes: Epidemiology and microbiology, *Clin. Microbiol, Rev.*, 7, 357, 1994.
11. Kapadia, M., Rolston, K.V., and Han, X.Y. Invasive *Streptomyces* infections: Six cases and literature review, *Am. J. Clin. Pathol.*, 127, 619, 2007.
12. Martin, M.C. et al., *Streptomyces albus* isolated from a human actinomycetoma and characterized by molecular techniques, *J. Clin. Microbiol.*, 42,5957, 2004.

13. Fuentes, A. et al., Report of five cases treated with imipenem or imipenem plus amikacin, *Gac. Med. Mex.*, 142, 247, 2006.
14. Dunne, E.F., Burman, W.J., and Wilson, M.L., *Streptomyces pneumonia* in a patient with human immunodeficiency virus infection: Case report and review of the literature on invasive streptomyces infections, *Clin. Infect. Dis.*, 27, 93, 1998.
15. Trujillo, M.E., and Goodfellow, M., Numerical phenetic classification of clinically significant aerobic sporoactinomycetes and related organisms, *Antonie Leeuwenhoek*, 84, 39, 2003.
16. Dieng, M.T. et al., Actinomycetomas in Senegal: Study of 90 cases, *Bull Soc. Pathol. Exot.*, 98, 18, 2005.
17. Gugnani, A.C., Unaogu, I.C., and Emeruwa, C.N., Pulmonary infection due to *Streptomyces griseus*, *J. Commun. Dis.*, 25, 38, 1993.
18. Nichols, W.G. et al., Fatal pulmonary infection associated with a novel organism, "para-streptomyces abscessus," *J. Clin. Microbiol.*, 43, 5376, 2005.
19. Clarke, P.R. et al., Brain abcess due to *Streptomyces griseus*, *J. Neurol. Neurosurg. Psychiatry*, 27, 553, 1964.
20. Kieser, T. et al., Practical streptomyces genetics, *Practical streptomyces genetics*, pp. 2–5. The John Innes Foundation, Norwich, England, 2000.
21. Steingrube, V.A. et al., Rapid identification of clinically significant species and taxa of aerobic actinomycetes, including *Actinomadura*, *Gordona*, *Nocardia*, *Rhodococcus*, *Streptomyces*, and *Tsukamurella* isolates, by DNA amplification and restriction endonuclease analysis, *J. Clin. Microbiol.*, 35, 817, 1997.
22. Edwards, U. et al., Isolation and direct complete nucleotide determination of entire genes. Characterization of a gene coding for 16S ribosomal DNA, *Nucleic Acids. Res.*, 17, 7843, 1989.
23. Kataoka, M. et al., Application of the variable region in 16S rDNA to create an index for rapid species identification in the genus *Streptomyces*, *FEMS Microbiol. Lett.*, 151, 249, 1997.
24. Amann, R.I., Ludwig, W., and Schleifer K.H., Phylogenetic identification and in situ detection of individual microbial cells without cultivation, *Microbiol. Rev.*, 59, 143, 1995.
25. Cole, J. R. et al., The Ribosomal Database Project: Improved alignments and new tools for rRNA analysis, *Nucleic Acids Res.*, 37, 141, 2009.
26. Bruce, K.D. et al., Amplification of DNA from native populations of soil bacteria by using the polymerase chain reaction, *App. Environ. Microbiol.*, 58, 3413, 1992.
27. Vandamme, P. et al., Polyphasic taxonomy, a consensus approach to bacterial systematic, *Microbiol. Rev.*, 60, 407, 1996.
28. Mendez, C. et al., Role of substrate mycelium in colony development in *Streptomyces*, *Can. J. Microbiol.*, 31, 446, 1985.
29. Murray, P. et al., *Manual of Clinical Microbiology*, 9th ed., American Society for Microbiology, Washington, DC, 2007.
30. Biemer, J.J., Antimicrobial susceptibility testing by the Kirby-Bauer disc diffusion method, *Ann. Clin. Lab. Sci.*, 3, 135, 1973.
31. Zheng, Z. et al., A greedy algorithm for aligning DNA sequences, *J. Comput. Biol.*, 7, 203, 2000.
32. Tarraga, J. et al., A suite of web tools for molecular evolution, phylogenetics and phylogenomics, *Nucleic Acids. Res.*, 35, 38, 2007.
33. Broda, D.M., Musgrave, D.R. and Bell, R.G., Use of restriction fragment length polymorphism analysis to differentiate strains of psychrophilic and psychrotropic clostridia associated with blown pack' spoilage of vacuum-packed meats, *J. Appl. Microbiol.*, 88, 107, 2000.
34. Zhao, H. et al., Differentiation of Micromonospora isolates from a coastal sediment in Wales on the basis of Fourier transform infrared spectroscopy, 16S rRNA sequence analysis, and the amplified fragment length polymorphism technique, *Appl. Environ. Microbiol.*, 70, 6619, 2004.
35. Huleihel, M., Pavlov, V., and Erukhimovitch, V., The use of FTIR microscopy for the evaluation of anti-bacterial agents activity, *J. Photochem. Photobiol. B.*, 96, 17, 2009.
36. Freiwald, A., and Sauer, S. Phylogenetic classification and identification of bacteria by mass spectrometry, *Nat. Protoc.*, 4, 732, 2009.

17 *Tropheryma*

Shoo Peng Siah, Han Shiong Siah, and Dongyou Liu

CONTENTS

17.1 INTRODUCTION

Whipple's disease is a rare illness that occurs in less than one in 1,000,000 people and affects multiple organs and organ systems, predominantly in middle-aged men.[1,2] It was first described postmortem by Dr. George H. Whipple in 1907 in a 36-year-old male, who died following a 5-year illness with fever, chronic cough, lymphadenopathy, and progressive weight loss.[3] Up until 1947, Whipple's disease was only diagnosed at autopsy, and in 1949, the bacterium was first detected as a period acid-Schiff positive (red staining) inclusion in foamy macrophages from an intestinal biopsy.[4] Definitive confirmation of the pathogen was first obtained in 1961 by electron microscopic examination of the rod-shaped organism. Following the analysis of a short segment of the 16S ribosomal RNA (16S rRNA) gene by broad-range polymerase chain reaction (PCR) and sequencing, the identity of the bacterium was confirmed as a member of the actinomycetes, and the name *Tropheryma whippeli* was proposed in 1992.[5] "*Tropheryma*" is derived from the Greek words for food (*trophi*) and barrier (*eryma*). The current name *Tropheryma whipplei* was adopted in 2001[6] after its successful culture and propagation in a human embryonic lung fibroblast cell line.[7] Another group also reported the cultivation of the bacterium from cerebrospinal fluid.[8] The completion of the whole genome sequences of two *T. whipplei* strains in 2003[9,10] rendered it possible to devise a cell-free culture system supplemented with amino acids for in vitro growth of the bacterium.[11]

17.1.1 Classification, Morphology, and Genome Structures

Tropheryma whipplei was the only member of the genus *Tropheryma*, which is classified in the family Cellulomonadaceae, suborder Micrococcineae, order Actinomycetales, subclass Actinobacteridae, class Actinobacteria, phylum Actinobacteria.

Sequence analysis of its 23S rRNA, 5S rRNA, DNA-dependent RNA polymerase (*rpoB*), 65-kDa heat-shock protein (*groEL2*), elongation factor TU (*tuf*), ATPase beta subunit (*atpD*), and RNase P RNA (*rnpb*) genes placed *Tropheryma whipplei* phylogenetically among the high G + C content gram-positive bacterial class Actinobacteria.[8,12,13]

T. whipplei is a gram-positive bacterium although it is poorly stained due to its atypical envelope and thick cell wall.[5] As such, it is commonly found to be gram-negative or gram-indeterminant. The bacterium has a unique ultrastructure unlike any other bacteria. Its membrane may be of bacterial or host origin as it lacks lipopolysaccharide, a feature that is shared by the macrophage plasma membrane.[6,14] The bacterium measures 0.2–0.25 μm by 1.0–2.5 μm and occurs as periodic acid-Schiff positive inclusions in macrophages.[15,16]

Forming rope-like aggregates, the bacterium can be found intact extracellularly but are in various stages of degradation intracellularly.[6] The bacterium multiplies extracellularly by perpendicular cell division but rarely within the macrophages.[17] Electron microscopy showed that the extracellular bacteria in cell cultures adhere to each other via extracellular matrix and tend to occur as aggregates. The bacterium is neither covered by flagella nor motile. Within infected cells, particularly in macrophages, it may be found in various stages of degradation within the vacuoles.[6]

The *Tropheryma whipplei* genome is the smallest within the Actinobacteria and DNA sequencing of the *T. whipplei* Twist and TW08/27 strains indicates that the circular genome is approximately 0.93 Mb in size and has a G + C content of approximately 46%, which is the lowest amongst the high G + C content gram-positive bacteria.[9,10,18] The genome contains 808 predicted open reading frames. The pathway involved in synthesis of nine amino acids (histidine, tryptophan, leucine, arginine, proline, lysine, methionine, cysteine, and asparagine) are absent and synthesis of another seven (glutamate, glutamine, aspartate, threonine, valine, isoleucine, and phenylalanine) is partially deficient. The genes that encode components of the tricarboxylic acid cycle and homologues for the thioredoxin pathways (glutaredoxin, thioredoxin, and thioredoxin reductase), which are involved in various essential redox reaction in the cell) are also absent.[9,10,19] A recent comparative genomic analysis of 16 different *T. whipplei* strains, including the Twist isolate, revealed that the coding sequences in this bacterium are highly conserved (ranging between 97.76% and 99.88%).[20]

17.1.2 Epidemiology

Whipple's disease has been reported to have affected 619 patients (in addition to 79 unpublished cases) between the period 1907 to 1986.[4] In 2008, the number of confirmed cases rose to 664.[2] Whipple's is a predominately a male disease (accounting for up to 86% of the reported cases).[2,4,17] Age is another major factor in understanding the disease, and most cases only appear in patients between the ages of 40 and 60.[2,4,17] Over the last few decades, the mean age of diagnosis of Whipple's disease has been increasing, from 48.7 years (1965–1975), 50.7 years (1976–1985) to 57 years (1986–1995).[2]

Whipple's disease is also likely to be race specific, as a majority of the patients are of European descent, chiefly in North America and Europe.[2,4,17] Whipple's disease is usually detected more frequently in farmers and carpenters, and the frequency of the disease is estimated to be less than one case per million of population per year.[2,4] There has been no confirmed human-to-human transmission of disease.[2] Furthermore, a limited number of PCR studies on a range of animals' intestines suggested that none of the animals has *T. whipplei* present.[17]

17.1.3 Clinical Features

Whipple's disease is known as a multisystem disorder, with a majority of organs and organ systems being affected, although its effect on the gastrointestinal (GI) system is particularly notable.[4] The manifestation of the "classic" Whipple's disease includes arthralgia, weight loss, diarrhea, abdominal pain, and low-grade fever or chills. It may be years before all these clinical features of Whipple's disease fully manifest in a patient. Symptoms that manifest involve the GI tract, the central nervous system (CNS), the joints, and the cardiovascular systems.

Gastrointestinal Manifestations. GI symptoms of Whipple's disease are usually described as weight loss, malabsorption, diarrhea, and abdominal pain, which may in turn progress to abdominal lymphadenopathy and a severe wasting syndrome. Weight loss is the most common manifestation (80%–90%) of Whipple's disease, with most patients losing a net weight of between 5 and 15 kg per year. The most severe cases may lead to cachexia. Gastrointestinal bleeding is also common and may be caused by hematochezia. Furthermore, in some patients, the presenting intestinal lesion may appear similar to the lesions of untreated celiac disease. Stools samples from patients tend to be voluminous and fatty due to malabsorption.[4,17,21]

Central Nervous System Manifestations. Neurological manifestation is observed in 10%–40% of patients reported to have Whipple's disease. Most often, neurological manifestations occur as a relapse after antibiotic treatment of intestinal and other manifestations of Whipple's disease. CNS manifestation appearing with the intestinal disease is less frequent. However, approximately 50%–70% of patients are reported to have dementia, altered levels of consciousness, memory impairment, and confusion. Whipple's disease affects ocular muscle movement; it is particularly present as progressive supranuclear ophthalmophegia with oculomasticatory myorhythemia or oculofacioskeletal myorhythemia. Although less well known, Whipple's disease has been shown to produce slow rhythmical movement of facial and other muscles. Presence of abnormalities is obvious on neuroimaging scans, although there is no typical pattern or distribution to the CNS manifestations. Other neurological symptoms that may appear are epilepsy, focal cerebral lesion, ataxia seizures, headache, and meningitic features.[4,17,22–26]

Joint Manifestation. Articular symptoms account for the largest group (79%–90%), present in patients outside the GI manifestations. Whipple's disease is harder to diagnosis in earlier stages because joint manifestations may precede GI or other systemic symptoms for several years. Arthragias and arthritis are the most common joint manifestations in patients, most of them occurring in knees, elbows, fingers, ankles, and shoulder joints.[4,17,27]

Cardiovascular System. Cardiovascular manifestations, including endo-, myo-, and pericarditis, appear in one-third of patients with Whipple's disease. It was thought that endocarditis with major valve destruction was common (58%); however, with the use of PCR, a recent case study found that in 19 cases, *T. whipplei* infection was a factor in culture-negative endocarditis in Whipple's disease.[28,29]

Other Manifestations. Another symptom of Whipple's disease is the appearance of grayish-brown skin hyperpigmentation in about half of all cases. Ocular manifestations including uveitis, vitritis, retinitis, retrobulbar neuritis, and papilledema may also present; however, it is rare (5%). Pulmonary manifestations occur in 30%–40% of patients with Whipple's disease, including pluritic chest pain, chronic cough, and dyspnea.[4,17]

17.1.4 Diagnosis

17.1.4.1 Phenotypic Techniques

Histopathological Examination. When Whipple's disease—even if the primary presentation does not involve the GI tract—is suspected, the first approach of investigation is to undertake an upper GI endoscopy in which multiple biopsies from the lower duodenum and upper ileum are taken and examined by microscopy. Involvement of the intestine is seen in the majority of Whipple's disease, so illnesses presenting with symptoms are rarely negative. However, approximately 10%–15% of cases may not show gastrointestinal symptoms and biopsies are negative, although PCR investigations of other organs may be positive. As such, biopsies of other organs may be useful.[2]

Because neurological signs occur in between 10% and 40% of patients with Whipple's disease, periodic acid-Schiff (PAS) staining of cerebrospinal fluid or brain biopsies for PAS-positive inclusion may be informative in patients with CNS symptoms. Other biopsy samples that may be investigated include synovial fluid and/or tissues; soft tissue of patients with arthralgia or spondyodiscitis; cardiac valves of patients with culture negative endocarditis; bone marrow; or biopsies of the skin, liver, muscle, and eye.[2]

Microscopy and Electron Microscopy. Whipple's disease was first described by George Whipple in 1907 and was first stained as sickled-form particules in macrophages using the periodic acid-Schiff stain in 1949. Periodic acid selectively oxidizes the glucose residues and creates aldehyde that reacts with the Schiff reagent, resulting in a purple magenta color reaction. The stain is used in histology to stain carbohydrate-rich macromolecules, such as glycogen, glycoprotein, and proteoglycans, commonly found in connective tissues and basal laminae. Detection of the bacterium with a stain predominantly used in histology and pathology was consistent with the observation of polysaccharide rich inner layer of its trilaminar cell wall by electron microscopy. The samples are typically digested with diastase to eliminate staining of glycogen in the biopsy samples, which may otherwise lead to a false-positive result.[17]

Electron microscopy examination indicated that the trilaminar layers consist of the aforementioned polysaccharide inner layer, an electron-translucent middle layer, and an electron-dense outer membrane that is somewhat similar to the membrane of a gram-negative bacterium that is absent of polysaccharides.[17]

Histopathologically, the bacterium can be detected in the foamy macrophages in the lamina propria of duodenal endoscopy biopsy or in the abdominal lymph nodes. PAS staining of the inclusion is not diagnostic of *T.whipplei* infection because other organisms, including yeast cells and fungal hyphae in tissues, may also be stained (http://www.hardydiagnostics.com/catalog2/hugo/Stains.htm). As such, other confirmatory tests are required.

Culture of the Bacterium. *T. whipplei* was successfully cultivated and maintained in human embryonic fibroblast cell line (HEL) following shell vial centrifugation technique, which was used with the intent of enhancing adhesion of the bacteria to the HEL cells[7] after many previous failed attempts.[15] The first strain, Twist-Marseille^{T} (=CNCM I-2202^{T}), from an aortic valve of 42-year-old male diagnosed with endocarditis, was confirmed by PAS staining, amplification, and sequencing of the 16S rRNA gene and electron microscopy.[6,7,30,31] Another strain (TW08/27) was isolated by a second group from cerebrospinal fluid of a women after presentation with severe weight loss, who grew the pathogen in murine fibroblast cell line, MRC-5, and primary human foreskin fibroblast cells.[9,32,33]

Up until 2003, the bacterium had only been successfully cultured in eukaryotic cells systems but the introduction of a cell-free axenic medium supplemented with amino acids enabled the faster isolation of *T. whipplei*.[11] Specifically, the supernatant of *T. whipplei*–infected fibroblasts was mixed with DMEM/F12 medium and supplemented with fetal calf serum, L-glutamine, and human nonessential amino acids.[11] Using the initial isolation methods of co-culturing the pathogen with eukaryotic cell lines, the doubling time of the bacterium ranged from between 4 and 17 days.[7,9] In the cell-free axenic medium, the bacterium doubling time decreased to approximately 28 h.[11] However, due to the difficulties associated with isolation and propagation of the bacterium, only two research laboratories in the world are involved in culturing the bacterium to date.

Serological Assays. Antibodies to *T. whipplei* may be detected by western blot. The assay tests total immunoglobulins (IgT), including IgG, IgM, and IgA heavy and light chains. The production of rabbit polyclonal antibodies has also facilitated its detection in tissues by immunohistochemical staining.[30,34–37]

17.1.4.2 Genotypic Techniques

A number of *T. whipplei* gene regions have been targeted for investigation in Whipple's disease. These include the 16S rRNA operons, 16S-23S rRNA spacer region (or internal transcribed spacer), 5S rRNA, heat-shock protein 65 gene (*hsp65 or groEL2*).[8,12,16,33,36,38–64]

The 16S rRNA gene (16S rDNA) serves as a good tool for evolutionary analysis of bacteria strains because the gene accumulates mutations over time so that the phylogenetic distance of one bacterium from another can be calculated based on the mutational rates. However, some regions of the 16S rRNA gene are conserved among all bacteria. It was using these conserved regions that enabled confirmation of the etiologic agent for Whipple's disease directly from DNA extracted from tissue.[5,65]

In a preliminary study, the 16S-23S rRNA intergenic spacer was reported to be homogenic,[49] suggesting that it may be useful for diagnosing *T. whipplei*. However, a revised study of the same intergenic spacer region showed that this region is more hypermutable than its flanking structural genes (16S rRNA and 23S rRNA) and therefore is useful for species identification and genotyping purpose.[49]

The 16S-23S rRNA intergenic spacer serves as a useful target for genotyping *T. whipplei* strains, and seven different spacer types have been identified so far; most spacers differ by only a few nucleotides along the 293–294 bp sequence.[18] *T. whipplei* possesses several stretches of variable-number tandem repeat (VNTR) sequences; which can be also exploited for typing studies.

After completion of the entire *Tropheryma* genome sequencing, other genes have been identified for molecular characterization of the bacterium. These include genes that encode the DNA-dependent RNA polymerase (*rpoB*), elongation factor TU (*tuf*), ATPase beta subunit (*atpD*), RNase P RNA (*rnpb*) and WiSP protein, and heat-shock protein 6. The *rpoB* gene, which is highly conserved in many bacteria, encodes the β subunit of the bacterial RNA polymerase, which catalyses the synthesis of RNA. The *tuf* gene, which contains a conserved P-loop motif of IGHVDHGKTT, encodes the elongation factor Tu that delivers aminoacyl-tRNA to A-site of the ribosomes during protein biosynthesis. The *atpD* gene is involved in F1F0-type ATP synthesis and is involved in regulating intracellular pH or synthesizing ATP in many enteric bacteria. The WiSP family of proteins is involved in surface protein interactions. The heat-shock protein 6 is implicated in the folding of proteins in prokaryote and eukaryotic cells and is required for normal cell growth. Being stress-induced, it acts to stabilize or protect disassembled polypeptides under heat-shock conditions.

Using some of these genes as targets, Maiwald et al.[8] identified the closest relatives phylogenetically for *T.whipplei*, which differ depending on the genes analyzed.[18] For instance, the closest relatives for *T. whipplei* are *Nocardioides jenseni* (for *rpnB*), *Propionibacterium acnes* (for *groEL2*), *Micrococcus luteus* (for *tuf*), and *S. coelicolor* (for *atpD* and *rpoB*).[18]

Real-time quantitative PCR (qPCR) detection of *T. whipplei* is based on the amplification of the target sequence using specific primers and the recognition and hybridization of a probe specific for a target sequence in the DNA target, followed by the displacement of the probe as the target sequence is amplified.[66] Fenollar et al.[67] used quantitative PCR to investigate the frequency of *T. whipplei* infection in the feces of 150 healthy Senegalese children.

17.2 METHODS

17.2.1 Sample Preparation

DNA from bronchoalveolar lavage (BAL) fluid sample may be extracted in a MagnaPure LC workstation (Roche Diagnostics) by using the MagnaPure LC DNA Isolation Kit II (Roche Diagnostics). Briefly, BAL sample (500 μL) is centrifuged in a 1.5 mL tube for 15 min at 14,000 × *g*, and 400 μL of the supernatant is discarded. The pellet is resuspended in the remaining 100 μL, mixed with 100 μL of lysis buffer and 50 μL of proteinase K, incubated overnight at 56°C, disrupted for 1 min with glass beads in a MagnaLyser (Roche Diagnostics), and then processed on the MagnaPure LC workstation according to the manufacturer's recommendations.

Alternatively, saliva (300 μL) is first liquefied with 300 μL of 3% NALC (*N*-acetyl-l-cysteine; Sigma), centrifuged for 10 min at 14,000 × *g*, and the supernatant discarded. Cerebrospinal fluid specimens, gastric aspirates, and rinse fluid of knee are centrifuged for 10 min at 14,000 × *g*, and supernatants discarded. The pellets are suspended in 200 μL of digestion buffer containing 200 μg of proteinase K per ml and incubated at 55°C for 1.5 h with agitation. DNA is extracted with QIAamp DNA binding columns (Qiagen) according to the manufacturer's protocol except that in the final step only 100 μL (instead of 200 μL) of elution buffer AE is used. 5 μL of the eluate is used for PCR.[68]

Biopsy specimens are suspended directly in 200 μL of digestion buffer containing 200 μg of proteinase K per ml and incubated at 55°C for 1.5 h with agitation. DNA is extracted with QIAamp DNA binding columns as above. The paraffin-embedded tissue (two to five 5- to 10-μm-thick sections per specimen) is suspended in 200 μL of digestion buffer (50 mM Tris-HCl pH 8.5, 1 mM EDTA, 0.5% sodium dodecyl sulfate) without proteinase K. Mixtures are incubated at 95°C for 10 min with moderate agitation in a thermomixer (Eppendorf) followed by centrifugation at 14,000 × *g* for 30 min at room temperature. The resulting paraffin layers are removed with sterile pipette tips. Proteinase K (Roche Diagnostics) is then added to a final concentration of 200 μg per mL for further processing as for biopsy specimens.[68]

One loop of *T. whipplei* bacterial biomass is suspended in 200 μL of 4% CHELEX 100 (Bio-Rad). After incubation at 95°C for 15 min and centrifugation at 14,000 × *g* for 10 min, 1 μL of the supernatant is used for PCR.[68]

17.2.2 Detection Procedures

(i) Nested PCR Morgenegg et al.[68] designed a nested PCR assay for identification of *Tropheryma whipplei* with primers from the heat-shock protein 65 gene (*hsp65*). In the first PCR, primers whipp-frwl and whipp-rev generate a fragment of 500 bp; and in the second PCR, primers whipp-frw2 and whipp-rev generate a fragment of 357 bp from *T. whipplei* isolate (Table 17.1).

Procedure

1. The first-round PCR mixture (50 μL) comprises 200 μM each of dATP, dCTP, dGTP, and dTTP, 25 pmol each of primers (whipp-frwl and whipp-rev), 5 μL of 10× PCR buffer, 2.5 U of Ampli*Taq*Gold polymerase (Perkin-Elmer), and 1 μL DNA template.
2. The first-round PCR mixture is subjected to an initial activation-denaturation step of 12 min at 95°C; 40 cycles of 1 min at 95°C, 57°C, and 72°C, respectively; and a final extension step of 10 min at 72°C, on a Gene Amp PCR System 9600 (Perkin-Elmer).
3. The resulting PCR product (10 μL) is separated by 1.5% agarose gel electrophoresis, stained with ethidium bromide, and visualized under UV light (a specific fragment of 500 bp is expected).

TABLE 17.1
Primers for Specific Amplification of *T. whipplei hsp65* Gene

Primer	Sequence (5′→3′)	Position[a]	Amplicon Size (bp)
	First-Round PCR		
whipp-frw1	TGACGGGACCACAACATCTG	728–747	500
whipp-rev	ACATCTTCAGCAATGATAAGAAGTT	1231–1207	
	Second-Round PCR		
whipp-frw2	CGCGAAAGAGGTTGAGACTG	872–881	357
whipp-rev	As above	1231–1207	

[a] Corresponding to the *hsp65* gene sequence of *E. coli* (GenBank accession no. X07850).

4. The second-round PCR mixture (50 μL) is made up of 200 μM each of dATP, dCTP, dGTP, and dTTP, 25 pmol each of primers (whipp-frw2 and whipp-rev), 5 μL of 10× PCR buffer, 2.5 U of Ampli*Taq*Gold polymerase (Perkin-Elmer) and 1 μL of the first-round PCR product as template. Amplification is carried out with an initial 95°C for 12 min; 30 cycles of 1 min at 95°C, 60°C, and 72°C, respectively; and a final 72°C for 10 min. The amplified product is examined as above (a specific fragment of 357 bp is expected).
5. Optionally, the second-round PCR product (357 bp) may be verified by restriction enzyme digestion (RFLP analysis). Briefly, 10 μL PCR product is added with 2 μL of 10× restriction buffer, 1 μL of endonuclease *Xho*I (10 U in buffer H; Roche Diagnostics), and 7 μL of distilled water. The mixture is incubated at 37°C for 3 h. The digest (20 μL) and 10 μL of the undigested products are electrophoresed on a 3% agarose gel (2% NuSieveGTP agarose, FMC BioProducts, and 1% UltraPure agarose, Life Technologies), stained with ethidium bromide, and photographed under UV light. Three fragments of 202, 92, and 63 bp are expected from the digested product compared to the single fragment of 357 bp from the undigested product.

Note: The *T. whipplei*–specific seminested PCR assay, based on the partial *hsp65* gene sequence, demonstrates high sensitivity and specificity. Seventeen patients shown to be positive by 16S rDNA PCR and/or internal transcribed spacer (ITS) PCR were also positive by the nested *hsp65* PCR. To prevent possible crossover contamination from previous amplicons, dUTP instead of dTTP may be used in the first-round PCR along with 2% Tween 20 and uracil *N*-glycosylase (0.25U per 50 μL; Inotech, Dottikon, Switzerland).

(ii) Quantitative PCR Fenollar et al.[69] outlined a testing scheme for identification of *Tropheryma whipplei* from clinical samples (e.g., 1 g stool, 200 μL saliva, 1 biopsy, or 200 μL body fluid) processed by QIAamp DNA MiniKit (Qiagen). The scheme involved use of two quantitative PCR (qPCR) with primers from the repetitive sequences of the bacterium.

From October 2003 to March 2004, the *T. whipplei*–specific qPCR assay (targeting a 164 bp sequence) incorporated primers 53.3F (5′-AGAGAGATGGGGTGCAGGAC-3′) and 53.3R (5′-AGCCTTTGCCAGACAGACAC-3′) in a FastStart DNA Master SYBR Green kit (Roche). If the result of this first assay was positive, it was systematically confirmed by a second qPCR using primers targeting a different gene region: 342F (5′-AGATGATGGATCTGCTTTCTTATCTG-3′) and 492R (5′-AACCCTGTCCTGCACCCC-3′).

Since April 2004, the *T. whipplei*–specific qPCR (targeting a 155 bp sequence) incorporated the primers TW27F (5′-TGTTTTGTACTGCTTGTAACAGGATCT-3′) and TW182R (5′-TCCTGCTCTATCCCTCCTATC ATC-3′), and a TaqMan probe (27F–182R, 6-FAM-AGAGATACATTTGTGTTAGTTGTTACA-TAMRA). If the result of this first assay was positive, it was systematically confirmed by a second PCR using primers that targets a different 150 bp region: TW13F (5′-TGAGTGATGGTAGTC TGAGAGATATGT-3′) and TW163R (5′-TCCATAACAAA GACAACAACCAATC-3′) together with a TaqMan probe (13F–163R, 6-FAM-AGAAGAAGATGTTACGGGTTG-TAMRA).

(iii) Quantitative Real-Time PCR Bousbia et al.[70] described a quantitative real-time PCR using the QuantiTect Probe PCR Kit on LightCycler instrument (Roche Diagnostics) with cycling conditions specified by the manufacturer. The specimens are first tested with the primers Twhi3F: 5′-TTGTGTATTTGGTATTAGATGAAACAG-3′ and Twhi3R: 5′-CCCTACAATATGAAACAGCCTTTG-3′ and the specific TaqMan probe Twhi3: 6-FAM-GGGATAGAGCAGGAGGTGTCTGTCTGG-TAMRA.

When the specimen is positive in the above assay, the result is confirmed by a second quantitative PCR with the primers Twhi2F: 5′-TGAGGATGTATCTGTGTATGGGACA-3′ and Twhi2R: 5′-TCCTGTTACAAGCAGTACAAAA CAAA-3′ and the Twhi2 probe: 6-FAM-GAGAGATGGGG TGCAGGACAGGG-TAMRA.

(iv) 16S rRNA Sequencing Bousbia et al.[70] utilized a broad-range primer-based PCR targeting the 16S rRNA gene followed by sequencing to identify *Tropheryma whippeli* associated with pneumonia. The eubacterial broad-range 16S rRNA primers for amplification consist of 536F: 5′-CAGCAGCCGCGGTAATAC-3′ and rp2: 5′-ACGGCTACCTTGTTACGACTT-3′.

Procedure

1. The PCR mixture (50-μL) is composed of 5 μL extracted DNA, 1× PCR buffer (5 μL), 2 μL of 25 mM $MgCl_2$, 200 μM of each dNTP, 0.2 mM of each 536F and rp2 primer, and 1 U of HotStar Taq DNA Polymerase (Qiagen).
2. Amplification is performed with an initial denaturation and activation step at 95°C for 15 min; 35 cycles of 95°C for 1 min, 62°C for 30 s, and 72°C for 90 s; and an extension step at 72°C for 10 min, in an ABI themocycler (Applied Biosystems).
3. The PCR products are purified by using the NucleoFast 96 PCR Kit (Macherey-Nagel), and 4 μL of purified PCR product is sequenced in 20 μL final volume containing sequencing buffer, 3.2 pmol of forward (536F) or reverse (rp2) primer, 3 μL of Big Dye Terminator V1.1 mix (Applied Biosystems), and 8 μL of deionized water with cycling conditions specified by the manufacturer.
4. Sequencing reactions are then purified using Sephadex Gel-Filtration (Sigma-Aldrich), and the purified products are sequenced on an ABI PRISM 3130xl genetic analyzer (Applied Biosystems). Sequences obtained are analyzed with Autoassembler software (Perkin-Elmer) and compared with those available in the GenBank database using the BLAST program (www.ncbi.nlm.nih.gov/BLAST).

Note: When the BAL fluid is polymicrobial, 3 μL of purified PCR product (from step 3) is cloned into PCR II TA cloning vector (Invitrogen) according to the manufacturer, and 56–64 white colonies are screened for each specimen. The universal M13 primers (M13F: 5′-GTAAAACGACGGCCAG-3′ and M13R: 5′-CAGGAAACAGCTATGAC-3′) are then employed to sequence the cloned inserts from both directions.

(v) Genotyping Li et al.[71] developed a genotyping system for *Tropheryma whipplei* on the basis of four highly variable genomic sequences (HVGSs). The HVGS 1 marker separates *T. whipplei* into eight genotypes (1–8); the HVGS 2 marker also separates *T. whipplei* into eight genotypes (1–8); the HVGS 3 marker into six genotypes (1–6); and the HVGS 4 marker into four genotypes (1–4) (Table 17.2).

Procedure

1. PCR mixture (25 μL) is made up of 5 μL of 3 ng/μL DNA, 50 pM each primer, 200 μM (each) dNTPs, 1.5 U Hotstart Taq DNA polymerase (Qiagen), 2.5 μL 10× PCR buffer; and 1 μL 25 mM $MgCl_2$. Sterile water is used as a negative control in each PCR run.
2. The mixture is subjected to an initial 95°C for 15 min; 40 cycles of 94°C for 30 s, 55°C for 30 s, and 72°C for 1 min; and a final 72°C for 5 min, in a PTC-200 automated thermal cycler (MJ Research).
3. PCR products are purified by using the MultiScreen PCR filter plate (Millipore). Amplicons are sequenced in both directions using BigDye 1.1 chemistry (Applied Biosystems) on an ABI 3130*xl* automated sequencer (Applied Biosystems). All sequences are checked twice in both directions to ensure the reliability of the typing method.
4. Nucleotide sequences are edited using the Autoassembler package (Perkin-Elmer). Multiple alignments of sequences are carried out using the CLUSTALW software. For phylogenetic analysis, sequences of the four HVGSs are concatenated. Phylogenetic relationships among *T. whipplei* genotypes are inferred using the unweighted pair group

TABLE 17.2
Primers for Amplification and Sequencing of *Tropheryma whippeli* Highly Variable Genomic Sequences (HVGSs)

HVGS	Content of HVGS	Genotype	Size (bp)	Forward and Reverse Primers (5′–3′)
HVGS 1	TWT133 and intergenic spacer (TWT133-TWT134)	8	227	GCTGCGCGAAGTAATTTG AGATACATGCGGAGATACT
HVGS 2	proS/prolyl-tRNA synthetase (TW183)	8	318	GCCTTGACTATGACATAATCAA TCGGACTAAAAGTGCGACAC
HVGS 3	secA-hp (TWT131) intergenic spacer	6	150	TTTGTCATAGGCATTTCTGTAG AGACCTCACTGTTATACGGAT
HVGS 4	TW183 and intergenic spacer (TWT183-TWT184)	4	162	CGGATCTTCACGAAATGTCC ATAACAAGAAGCTGGATATGC

Source: Adapted from Li, W. et al., *Microbiology*, 154, 521, 2008.

method with arithmetic mean (UPGMA), neighbor-joining and maximum-parsimony methods within the MEGA 3.1 software.

Note: For each HVGS, a sequence type is defined as a sequence exhibiting unique mutation(s). HVGS genotypes are defined as unique combinations of the four HVGS types.

17.3 CONCLUSION

Tropheryma whipplei is a gram-positive, nonmotile bacterium that is responsible for Whipple's disease—a systemic chronic infection displaying several clinical traits, including histological lesions in the gastrointestinal tract, endocarditis with negative blood cultures, and isolated neurological infection.[72,73] The disease predominantly affects middle-aged men. Although Whipple's disease is rare, it can be fatal without antibiotic treatment. Because the culture is notably time consuming and has only been successfully achieved in two laboratories, the current diagnostic scheme for Whipple's disease relies on diastase-resistant periodic acid-Schiff staining together with PCR, involving organ/tissue biopsy (small intestine, cardiac valve, and cerebrospinal fluid) as well as saliva and stool specimens.[74] When classic Whipple's disease with gastrointestinal symptoms is suspected, qPCR screening of saliva and stool specimens provides a valuable noninvasive way to confirm the infection, especially if bacterial load in stool is $>10^4$ CFU/g. For localized Whipple's disease (often accounting for nearly one-quarter of cases), use of broad-spectrum and specific qPCR assay is essential for diagnosis.[69]

REFERENCES

1. Swartz, M.N., Whipple's disease: past, present, and future, *N. Engl. J. Med.*, 342, 648, 2000.
2. Schneider, T. et al., Whipple's disease: New aspect of pathogenesis and treatment, *Lancet Infect. Dis.*, 8, 179, 2008.
3. Whipple, G.H., A hitherto undescribed disease characterized anatomically by deposits of fat and fatty acids in the intestinal and mesenteric lymphatic tissues, *Bull. Johns Hopkins Hosp.*, 18, 382, 1907.
4. Maiwald, M., Update on Whipple's disease and *Tropheryma whipplei*, *Clin. Microbiol. Newsl.*, 28, 177, 2006.
5. Relman, D.A. et al., Identification of the uncultured bacillus of Whipple's disease, *N. Engl. J. Med.*, 327, 293, 1992.
6. La Scola, B. et al., Description of *Tropheryma whipplei* gen nov, sp nov, the Whipple's disease bacillus, *Int. J. Syst. Evol. Microbiol.*, 51, 1471, 2001.
7. Raoult, D. et al., Cultivation of the bacillus of Whipple's disease, *N. Engl. J. Med.*, 342, 620, 2000.
8. Maiwald, M., Lepp, P.W., and Relman, D.A., Analysis of conserved non-rRNA genes of *Tropheryma whipplei*, *Syst. Appl. Microbiol.*, 26, 3, 2003.
9. Bentley, S.D. et al., Sequencing and analysis of the genome of the Whipple's disease bacterium *Tropheryma whipplei*, *Lancet*, 361, 637, 2003.
10. Raoult, D. et al., *Tropheryma whipplei* twist: A human pathogenic actinobacteria with a reduced genome, *Genome Res.*, 13, 1800, 2003.
11. Renesto, P. et al., Genome-based design of a cell-free culture medium for *Tropheryma whipplei*, *Lancet*, 362, 447, 2003.
12. Drancourt, M., Carlioz, A., and Raoult, D., rpoB sequence analysis of cultured *Tropheryma whippelii*, *J. Clin. Microbiol.*, 39, 2425, 2001.
13. Marth, T., and Raoult, D., Whipple's disease, *Lancet*, 361, 239, 2003.
14. Silva, M.T., Macedo, P.M., and Moura Nunes, J.F., Ultrastructure of bacilli and the bacillary origin of the macrophagic inclusions in Whipple's disease, *J. Gen. Microbiol.*, 131, 1001, 1985.
15. Dobbins, W.O. 3rd, Whipple's disease: An historical perspective, *Q. J. M.*, 56, 523, 1985.
16. Maiwald, M. et al., *Tropheryma whippelii* DNA is rare in the intestinal mucosa of patients without other evidence of Whipple disease, *Ann. Intern. Med.*, 134, 115, 2001.
17. Dutly, F., and Altwegg, M., Whipple's disease and "*Tropheryma whippelii*," *Clin. Microbiol. Rev.*, 14, 561, 2001.
18. Maiwald, M. et al., Organization, structure, and variability of the rRNA operon of the Whipple's disease bacterium (*Tropheryma whippelii*), *J. Bacteriol.*, 182, 3292, 2000.
19. Crapoulet, N. et al., *Tropheryma whipplei* genome at the beginning of the post-genomic era, *Curr. Genom.*, 6, 195, 2005.
20. La, M.V. et al., Comparative genomic analysis of Tropheryma whipplei strains reveals that diversity among clinical isolates is mainly related to the WiSP proteins, *BMC Genomics*, 8, 349, 2007.
21. Amsler, L. et al., Prevalence of *Tropheryma whipplei* DNA in patients with various gastrointestinal diseases and in healthy controls, *Infection*, 31, 81, 2003.
22. Rickman, L.S. et al., Uveitis caused by *Tropheryma whippelii* (Whipple's bacillus), *N. Engl. J. Med.*, 332, 363, 1995.
23. Louis, E.D. et al., Diagnostic guidelines in central nervous system Whipple's disease, *Ann. Neurol.*, 40, 561, 1996.
24. Helliwell, T.R. et al., Dermatomyositis and Whipple's disease, *Neuromuscul. Disord.*, 10, 46, 2000.
25. Messori, A. et al., An unusual spinal presentation of Whipple disease, *Am. J. Neuroradiol.*, 22, 1004, 2001.
26. Matthews, B.R. et al., Cerebellar ataxia and central nervous system Whipple disease, *Arch. Neurol.*, 62, 618, 2005.
27. Ayoub, W.T. et al., Bone destruction and ankylosis in Whipple's disease, *J. Rheumatol.*, 9, 930, 1982.
28. Richardson, D.C. et al., *Tropheryma whippelii* as a cause of afebrile culture-negative endocarditis: The evolving spectrum of Whipple's disease, *J. Infect.*, 47, 170, 2003.
29. Marin, M., *Tropheryma whipplei* infective endocarditis as the only manifestation of Whipple's disease, *J. Clin. Microbiol.*, 45, 2078, 2007.
30. Raoult, D. et al., Culture and immunological detection of *Tropheryma whippelii* from the duodenum of a patient with Whipple disease, *JAMA*, 285, 1039, 2001.
31. Raoult, D., Fenollar, F., and Birg, M.L., Culture of *T whipplei* from the stool of a patient with Whipple's disease, *N. Engl. J. Med.*, 355, 1503, 2006.
32. Drancourt, M. et al., Culture of *Tropheryma whippelii* from the vitreous fluid of a patient presenting with unilateral uveitis, *Ann. Intern. Med.*, 139, 1046, 2003.
33. Fenollar, F., and Raoult, D., Molecular techniques in Whipple's disease, *Expert Rev. Mol. Diagn.*, 1, 299, 2001.
34. Baisden, L. et al., Diagnosis of Whipple disease by immunohistochemical analysis: A sensitive and specific method for the detection of *Tropheryma whipplei* (the Whipple bacillus) in paraffin-embedded tissue, *Am. J. Clin. Pathol.*, 118, 742, 2002.

35. Lepidi, H. et al., Whipple's disease: Immunospecific and quantitative immunohistochemical study of intestinal biopsy specimens, *Hum. Pathol.*, 34, 589, 2003.
36. Lepidi, H. et al., Cardiac valves in patients with Whipple endocarditis: Microbiological, molecular, quantitative histologic, and immunohistochemical studies of 5 patients, *J. Infect. Dis.*, 190, 935, 2004.
37. Kowalczewska, M. et al., Identification of candidate antigen in Whipple's disease using a serological proteomic approach, *Proteomics*, 6, 3294, 2006.
38. Lowsky, R. et al., Diagnosis of Whipple's disease by molecular analysis of peripheral blood, *N. Engl. J. Med.*, 331, 1343, 1994.
39. Maiwald, M. et al., Detection of *Tropheryma whippelii* DNA in a patient with AIDS, *J. Clin. Microbiol.*, 33, 1354, 1995.
40. Cohen, L. et al., Polymerase chain reaction of cerebrospinal fluid to diagnose Whipple's disease, *Lancet*, 347, 329, 1996.
41. Maiwald, M. et al., Reassessment of the phylogenetic position of the bacterium associated with Whipple's disease and determination of the 16S-23S ribosomal intergenic spacer sequence, *Int. J. Syst. Bacteriol.*, 46, 1078, 1996.
42. von Herbay, A., Ditton, H.J., and Maiwald, M. Diagnostic application of a polymerase chain reaction assay for the Whipple's disease bacterium to intestinal biopsies, *Gastroenterology*, 110, 1735, 1996.
43. Dauga, C., Miras, I., and Grimont, P.A. Strategy for detection and identification of bacteria based on 16S rRNA genes in suspected cases of Whipple's disease, *J. Med. Microbiol.*, 46, 340, 1997.
44. Ramzan, N.N. et al., Diagnosis and monitoring of Whipple disease by polymerase chain reaction, *Ann. Intern. Med.*, 126, 520, 1997.
45. Celard, M. et al., Polymerase chain reaction analysis for diagnosis of *Tropheryma whippelii* infective endocarditis in two patients with no previous evidence of Whipple's disease, *Clin. Infect. Dis.*, 29, 1348, 1999.
46. Delanty, N. et al., Synovial fluid polymerase chain reaction as an aid to the diagnosis of central nervous system Whipple's disease, *Ann. Neurol.*, 45, 137, 1999.
47. Ehrbar, H.U. et al., PCR-positive tests for *Tropheryma whippelii* in patients without Whipple's disease, *Lancet*, 353, 2214, 1999.
48. Hinrikson, H.P. et al., Detection of three different types of "*Tropheryma whippelii*" directly from clinical specimens by sequencing, single-strand conformation polymorphism (SSCP) analysis and type-specific PCR of their 16S-23S ribosomal intergenic spacer region, *Int. J. Syst. Bacteriol.*, 49, 1701, 1999.
49. Hinrikson, H.P., Dutly, F., and Altwegg, M. Homogeneity of 16S-23S ribosomal intergenic spacer regions of *Tropheryma whipplei* in Swiss patients with Whipple's disease, *J. Clin. Microbiol.*, 37, 152, 1999.
50. Hinrikson, H.P., Dutly, F., and Altwegg, M., Analysis of the actinobacterial insertion in domain III of the 23S rRNA gene of uncultured variants of the bacterium associated with Whipple's disease using broad-range and "*Tropheryma whippelii*"-specific PCR, *Int. J. Syst. Evol. Microbiol.*, 50, 1007, 2000.
51. Hinrikson, H.P., Dutly, F., and Altwegg, M. Evaluation of a specific nested PCR targeting domain III of the 23S rRNA gene of "*Tropheryma whippelii*" and proposal of a classification system for its molecular variants, *J. Clin. Microbiol.*, 38, 595, 2000.
52. Lange, U., and Teichmann, J., Diagnosis of Whipple's disease by molecular analysis of synovial fluid, *Arthritis Rheum.*, 42, 1777, 1999.
53. Misbah, S.A. et al., Anonymous survey of blood donors by polymerase chain reaction for *Tropheryma whippelii*, *Q. J. M.*, 92, 61, 1999.
54. Pron, B. et al., Diagnosis and follow-up of Whipple's disease by amplification of the 16S rRNA gene of *Tropheryma whippelii*, *Eur. J. Clin. Microbiol. Infect. Dis.*, 18, 62, 1999.
55. Street, S., Donoghue, H.D., and Neild, G.H., *Tropheryma whippelii* DNA in saliva of healthy people, *Lancet*, 354, 1178, 1999.
56. Fenollar, F. et al., Quantitative detection of *Tropheryma whipplei* DNA by real-time PCR, *J. Clin. Microbiol.*, 40, 1119, 2002.
57. Fenollar, F. et al., Culture of *Tropheryma whipplei* from human samples: A 3-year experience (1999 to 2002), *J. Clin. Microbiol.*, 41, 3816, 2003.
58. Fenollar, F. et al., Use of genome selected repeated sequences increases the sensitivity of PCR detection of *Tropheryma whipplei*, *J. Clin. Microbiol.*, 42, 401, 2004.
59. Fredricks, D.N., and Relman, D.A., Localization of *Tropheryma whippelii* rRNA in tissues from patients with Whipple's disease, *J. Infect. Dis.*, 183, 1229, 2001.
60. Geißdörfer, W. et al., Detection of a new 16S-23S rRNA spacer sequence variant (type 7) of *Tropheryma whippelii* in a patient with prosthetic aortic valve endocarditis, *Eur. J. Clin. Microbiol. Infect. Dis.*, 20, 762, 2001.
61. Bruhlmann, P., Michel, B.A., and Altwegg, M., Diagnosis and therapy monitoring of Whipple's arthritis by polymerase chain reaction, *Rheumatology*, 39, 1427, 2000.
62. Maibach, R.C., Dutly, F., and Altwegg, M., Detection of *Tropheryma whipplei* DNA in feces by PCR using a target capture method, *J. Clin. Microbiol.*, 40, 2466, 2002.
63. Puéchal, X. et al., Polymerase chain reaction testing for *Tropheryma whippelii* in unexplained isolated cases of arthritis, *Arthritis Rheum.*, 44, 1130, 2002.
64. Rolain, J.M., Fenollar, F., and Raoult, D., False positive PCR detection of *Tropheryma whipplei* in the saliva of healthy people, *BMC Microbiol.*, 7, 48, 2007.
65. Wilson, K.H. et al., Phylogeny of Whipple's disease associated bacterium, *Lancet*, 338, 474, 1991.
66. Sloan, L.M., Rosenblatt, J.E., and Cockerill, III, F.R., Detection of *Tropheryma whipplei* DNA in clinical specimens by LightCycler real-time PCR, *J. Clin. Microbiol.*, 43, 3516, 2005.
67. Fenollar, F. et al., *Tropheryma whipplei* in Fecal Samples from Children, Senegal, *Emerg. Infect. Dis.*, 15, 922, 2009.
68. Morgenegg, S., Dutly, F., and Altwegg, M., Cloning and sequencing of a part of the heat shock protein 65 gene (*hsp65*) of "*Tropheryma whippelii*" and its use for detection of "*T. whippelii*" in clinical specimens by PCR, *J. Clin. Microbiol.*, 38, 2248, 2000.
69. Fenollar, F. et al., Value of Tropheryma whipplei quantitative polymerase chain reaction assay for the diagnosis of Whipple disease: usefulness of saliva and stool specimens for first-line screening, *Clin. Infect. Dis.*, 47, 659, 2008.
70. Bousbia, S. et al., *Tropheryma whipplei* in patients with pneumonia, *Emerg. Infect. Dis.*, 16, 258, 2010.
71. Li, W. et al., Genotyping reveals a wide heterogeneity of *Tropheryma whipplei*, *Microbiology*, 154, 521, 2008.
72. Fenollar, F., Puechal, X., and Raoult, D., Whipple's disease, *N. Engl. J. Med.*, 356, 55, 2007.
73. Fleming, J.L., Wiesner, R.H., and Shorter, R.G., Whipple's disease: Clinical, biochemical, and histopathologic features and assessment of treatment in 29 patients, *Mayo Clin. Proc.*, 63, 539, 1988.
74. Kowalczewska, M., and Raoult, D., Advances in Tropheryma whipplei research: The rush to find biomarkers for Whipple's disease, *Future Microbiol.*, 2, 631, 2007.

18 *Tsukamurella*

Mireille M. Kattar

CONTENTS

18.1 INTRODUCTION

The genus *Tsukamurella* was created in 1988 in honor of Michio Tsukamura (a Japanese microbiologist) to accommodate mesophilic gram-positive weakly acid fast aerobic actinomycetes with highly unsaturated mycolic acids in their cell walls, encompassing chains 62–78 carbon atoms in length. Their predominant menaquinone is of the MK-9 type. It currently includes 10 validly named species, among which five have been associated with human infections, whereas the other five have been solely isolated from environmental sources.

18.1.1 Classification, Morphology, and Epidemiology

Classification and Epidemiology. The genus *Tsukamurella* belongs to the family Tsukamuralleceae in the suborder Corynebacterineae. The suborder Corynebacterineae belongs to the order Actinomycetales within the class Actinobacteria.[1] Tsukamurellae are soil organisms and have been found in treated and untreated drinking water.[2] The genus *Tsukamurella* was first established in 1988 to accommodate a group of chemically unique organisms characterized by a series of very long chain (62–78 carbon atoms), highly unsaturated (2–7 double bonds) mycolic acids, in addition to possessing *meso*-diaminopimelic acid, arabinose, and galactose, which are common to the genus *Corynebacterium*.[3] The latter features are characteristic of the cell-wall chemotype IV.[4] The type species *Tsukamurella paurometabola*, described by Steinhaus in 1941 as *Corynebacterium paurometabolum*, was originally isolated from the mycetomes and ovaries of bed bugs (*Cimex lectularius*).[5] In a comprehensive numerical phenetic study by Jones et al.,[6] the type strain *C. paurometabolum* ATCC 8368 was found to be distinct from *Corynebacterium* species. The first reported human isolate of *Tsukamurella*, named *Gordona aurantiaca*, was identified in the sputum of a patient with tuberculosis, but its causal role in the patient's disease was doubtful.[7] In 1985, *G. aurantiaca* was renamed *Rhodococcus aurantiacus* by Tsukamura and assigned a new type strain, ATCC 25938.[8] Chemical and numerical taxonomic studies of *C. paurometabolum* and *G. aurantiaca* revealed that they had similar cellular composition.[9,10] Both possess highly unsaturated mycolic acids containing 62–78 carbon atoms. They also contain fully unsaturated menaquinones, which include MK-8, MK-9, and MK-10, with predominant 9 isoprene units (MK-9). This is in contrast to other mycolic acid-containing taxa such as *Mycobacterium*, *Nocardia*, *Rhodococcus*, and *Corynebacterium* species, which possess dihydrogenated menaquinones.[11] Menaquinones are present in the cell membranes of most bacteria, where they function in electron transport, oxidative phosphorylation, and perhaps active transport. Based on these results and the more than 99% similarity in their 16S rRNA gene sequences, Collins et al. proposed in 1988 that *C. paurometabolum* and *G. aurantiaca* be reduced to a single species and assigned to a new genus, *Tsukamurella*.[3] The type species is *Tsukamurella paurometabolum*, and strain ATCC 8368 was designated as the type strain. Yassin later correctly renamed *T. paurometabolum* as *T. paurometabola*.[12] Four other species associated with human infections were subsequently described. Two strains of *T. inchonensis* were initially isolated from multiple blood cultures of a patient who had ingested hydrochloric acid and from the lung tissue of a patient who had a lung tumor, respectively.[12] *T. pulmonis* was first isolated from the sputum of a 92-year-old woman diagnosed three years previously with pulmonary tuberculosis.[13] Strains of *T. tyrosinosolvens* were initially recovered from elderly patients; two patients had cardiac pacemaker implants, and the organism grew from multiple blood cultures; in another two patients with chronic lung infections, the organisms were obtained from sputum cultures.[14] *T. strandjordii* was isolated from multiple blood cultures of a 5-year-old child with acute myelogenous leukemia and an indwelling catheter.[15] Five other species of *Tsukamurella* were proposed from 2003 to 2009, all of which were discovered in environmental samples. *T. sunchonensis*[16] was isolated from activated sludge foam in Korea, and *T. spumae*[17] and *T. pseudospumae*[18] in the United Kingdom. In this respect, *Tsukamurella* constitute one of several genera of aerobic actinomycetes that have been associated with sludge foams. The activated sludge process is the most commonly used method for the treatment of wastewater. Activated sludge foams may represent a public health threat because of the potential spread of pathogens through aerosols.[19] *T. spongiae* was isolated from a deep-water marine sponge off the coast of Curaçao in the Netherlands Antilles.[20] The most recently described species is *T. carboxydivorans*, which was isolated from a soil sample collected from a roadside in Seoul, Korea.[21] The Guanine + Cytosine content of *Tsukamurella* species ranges from 67% to 77% mol, with the highest G + C content observed in *T. carboxydivorans.* In contrast to the other species, *T. carboxydivorans* contains the longest chain mycolic acids with 81–95 carbon atoms, whereas *T. spongiae* contains the shortest with 58–75 carbon atoms.

Morphology. *Tsukamurella* species are gram-positive, long, straight to slightly curved rod-shaped and partially acid alcohol fast measuring 0.5–0.8 μm by 1–5 μm.[12–14] Acid fastness requires use of a modified acid-fast stain with a weaker decolorizer (1% sulfuric acid H_2SO_4). Clinical isolates have occasionally been reported to be acid-fast negative.[22] On scanning electron microscopy, bacterial cells fragment into three to four parts, which separate and grow as independent cells. Growth is obligately aerobic. Visible colony growth from clinical specimens and dilute inocula occurs within 2 days. Growth in blood culture bottles from clinical specimens occurs within 1–2 days. Colonies show rudimentary to no branching. The cells are nonmotile and do not form spores, capsules, or aerial hyphae. All *Tsukamurella* species usually grow at temperatures ranging from 25°C to 35°C. *T. paurometabola* and *T. pseudospumae* also grow at 10°C, and *T. inchonensis* grows at 42°C. The optimal growth temperature for *T. carboxydivorans* is 30°C. After 24–48 h of incubation, colony size is 2–5 mm, and colonies are easily emulsified. Depending on the species, colony color is either cream white or pigmented. Colonies of *T. paurometabola* type strain are convex dry creamy white with entire edges. Colonies of *T. pulmonis, T. strandjordii, T. spongiae,* and *T. carboxydivorans* are also cream-colored rough with fringed edges and raised, wrinkled centers. Colonies of *T. inchonensis* and *T. tyrosinosolvens* are also rough and dry, with fringed edges, but are yellow to orange with no diffusible pigment. Colonies of *T. sunchonensis*, *T. spumae*, and *T. pseudospumae* have irregular margins and are orange to red.

18.1.2 Clinical Features and Pathogenesis

The pathogenic role of Tsukamurellae was initially a subject of debate.[23,24] Tsukamura and Mizuno first reported that since 1971, they had isolated *Gordona aurantiaca* from sputa of 14 patients with tuberculosis or "tuberculosis-like" illness.[7,25] They stated, however, that this organism had not caused disease in humans. In 1982, Tsukamura and Kawakami reported the first case of a lung infection caused by *Gordona aurantiaca* that mimicked tuberculosis.[25] Since then, 24 case reports and six short case series of human infection have been published, and these cases occurred nearly equally in male and female patients.[22,26–51] Most cases occurred in adult patients, but there were three cases in pediatric patients.[37,52] These infections occurred predominantly in immunocompromised patients and, occasionally, in immunocompetent hosts. *Tsukamurella* species appear to have a low pathogenic potential and have been mostly linked to opportunistic infections in patients with foreign medical devices, which account for over 80% of cases reported. The majority of cases consisted of bloodstream infections secondary to central venous catheters placed several months earlier for treatment of cancer.[30,34,36,37,40,45,48] The organisms were frequently cultured from the catheter tips and peripheral blood, as well as from blood cultures drawn through the central catheters. Infections sometimes accompanied a febrile neutropenic episode. None of the patients had tunnel or catheter exit-site infection. In virtually all these cases, positive blood cultures persisted despite treatment with various antibiotics, thereby necessitating the removal of the catheter, which resulted in the patient's prompt recovery and subsequently negative blood cultures. No firm conclusions can be drawn concerning the efficacy of antibiotics because therapy varied widely in terms of class and duration (reviewed in Ref. 48). Five cases occurred in patients with chronic renal failure.[22,27,37,38,42] Two of these patients had indwelling catheters for hemodialysis. Another two patients had peritonitis secondary to peritoneal dialysis catheters. The organism was grown from blood or peritoneal fluid in all patients. It also grew from the catheter tip in one patient. In all five patients, initial antibiotic therapy with one or multiple drugs failed to control the infection, which only resolved upon removal of the catheter. Three patients had cardiac pacemaker implants.[14,41] One patient had a chronic wound infection following prosthetic knee replacement, which resolved after 2 months of treatment with clarithromycin and ciprofloxacin, to which the organism was susceptible.[35]

The second most common syndrome associated with Tsukamurellae is pulmonary infection mimicking tuberculosis. Ten such cases have been reported, 70% of which were in patients with no evident predisposing factor.[13,14,25,29,39,43,44,46,47]

The organism was often grown repeatedly from sputum and, less commonly, from lung tissue. Two patients had lung cancer, and one patient was HIV positive.[12,39,43] Treatment and follow-up information was available for four patients. Three patients responded to a combination of a quinolone and a rifamycin (rifampin or rifabutin) for 3–6 months, and one patient achieved cure following 4 months of treatment with trimethoprim and sulfamethoxazole. In one report, a HIV-positive female patient was colonized in the sputum with *Tsukamurella* species but did not develop any respiratory symptoms after a follow-up period of 4 months.[53]

Three cases of skin and soft tissue infection due to Tsukamurellae have been reported in immunocompetent adult patients. One case occurred in a 30-year-old female patient following surgical treatment for constrictive tenosynovitis.[28] She developed a wound infection with multiple abscesses and arm gangrene that failed to respond to a combination of a quinolone, rifampin, and minocycline and eventually required amputation of her arm. Another patient had a chronic superficial infection that responded to topical hexamidine and fusidic acid.[32]

There have been two reported cases of central nervous system infection due to Tsukamurellae. One case of fatal meningitis occurred in a 43-year-old male patient with hairy cell leukemia.[26] A 36-year-old female patient developed skull osteomyelitis and a brain abscess following purulent chronic otitis media.[49] The infection responded to a 6-week combination therapy with amikacin and imipenem, followed by oral ciprofloxacin, clarithromycin, and thrimetroprim-sulfamethoxazole.

Lastly, *Tsukamurellae* species have been associated with ocular infections in immunocompetent adults in China. Three patients had mild conjunctivitis that responded to a 10-day course of topical chloramphenicol or polymyxin B and neomycin.[51] Two patients developed keratitis, which responded to topical gentamicin or levofloxacin for 10 days to 3 weeks. Both patients had microtrauma to the corneal surface, which predisposed them to developing a deeper eye infection.[50]

In 1992, Auberbach et al. reported an outbreak of pseudoinfection with *T. paurometabola* involving 10 patients that occurred at a single hospital in South Carolina over a 30-month period.[23] These isolates were grown from eight tissue specimens, a bronchial washing, and swab from a fistula. The isolates from the outbreak were similar by ribotyping. The specimens from the outbreak were more likely to have been processed in the mycobacteriology laboratory and to have been handled by one technician.

Prior to 1995, infections with Tsukamurellae were all attributed to *T. paurometabola*, previously known as *Gordona aurantiaca* or *Rhodococcus aurantiacus*. However, the organism descriptions in most of these reports, including the pseudooutbreak, did not fit that of *T. paurometabola* type strain but rather strain ATCC 25938, which has been demonstrated to be a misnamed *T. inchonensis* strain.[15]

Following the original descriptions of *T. inchonensis*, *T. pulmonis*, and *T. tyrosinosolvens* from 1995 to 1997, and *T. strandjordii* in 2001, all reported human infections where species identification were performed, were attributed to either *T. pulmonis* or *T. tyrosinosolvens*. Species identification of the isolates was achieved by phenotypic methods or 16S rRNA gene sequencing, or a combination of the two methods (see below).

In terms of antimicrobial susceptibility patterns, and prior to the publication of the Clinical and Laboratory Standards Institute (CLSI) in vitro susceptibility testing guidelines for aerobic actinomycetes in 2003, testing was not standardized. Prior to 2003, published reports included such methods as disk diffusion, broth microdilution, broth macrodilution, E-test, and agar dilution. The reference method recommended by CLSI is broth microdilution.[54] The primary antibiotic panel includes amikacin, tobramycin, amoxicillin-clavulanic acid, ceftriaxone, ciprofloxacin, clarithromycin, imipenem, linezolid, minocycline, and trimetroprim-sulfamethoxazole. The secondary antibiotic panel includes cefepime, cefotaxime, doxycycline, and gentamicin.

Tsukamurellae are resistant in vitro to agents used for the treatment of tuberculosis, which include streptomycin, isoniazid, ethambutol, cycloserine, protionamide, and capreomycin.[12–14] They are also resistant to a wide range of antibiotics of different classes. The mechanisms of antimicrobial resistance in Tsukamurellae are unknown. When grown on some hydrophobic carbon sources, they may produce glycolipids that are secreted but are also found attached to the bacterial cell wall. These glycolipids are believed to have antimicrobial properties.[55] They are usually susceptible to quinolones, amikacin, clarithromycin, trimethoprim-sulfamethoxazole, minocycline, and rifampin. They are reportedly resistant to penicillin, ampicillin, ampicillin/sulbactam, amoxicillin/clavulanic acid, carbenicillin, piperacillin/tazobactam, oxacillin, first-generation cephalosporins, cefoxitin, moxalactam, and vancomycin. They may be either susceptible or resistant to erythromycin, clindamycin, doxycycline, gentamicin, tobramycin, third-generation cephalosporins, imipenem, and chloramphenicol.

Histotological examination of tissues infected with Tsukamurellae usually shows granulomatous inflammation without central necrosis, reminiscent of the tissue response encountered in some cases of tuberculosis.[28,32] This has been attributed in a mouse model to the secretion of interferon-γ by recruited CD4+ T lymphocytes.[56]

18.1.3 Diagnosis

18.1.3.1 Conventional Techniques

The morphological features on Gram staining, positive modified acid-fast staining, strictly aerobic growth, and colonial morphology represent the first indication that an organism belongs to the aerobic actinomycetes. Using phenotypic methods, members of the genus *Tsukamurella* may be confused with, and have been variously misidentified, as rapidly growing *Mycobacterium* species, *Nocardia*, *Rhodococcus*, or *Corynebacterium* species.[15,22,30,32,36,57] Although *Tsukamurella* species are considered to be weakly positive on acid-fast staining, some original descriptions state that some of the members may appear strongly acid fast.[12] In addition, they grow on media used in the clinical laboratory for the isolation of *Mycobacterium* species, including Lowenstein-Jensen and Middlebrook 7H11 and similar to *Mycobacterium* species, are resistant to lysozyme. Furthermore, they frequently react in the same conventional phenotypic tests used for the identification of *Mycobacterium* species.[37] *Tsukamurella* species also grow on sheep blood agar, chocolate, and MacConkey without crystal violet. Commercially available panels have limited value for identification of aerobic actinomycetes, which are generally not included in the database of these systems. For example, *Tsukamurella* species are invariably misidentified as *Rhodococcus* species by the API Coryne.[15,48] Once a clinical isolate has been recognized as a *Tsukamurella*, standard phenotypic methods have limited usefulness for species-level identification, and the interpretation of these tests may be problematic for a number of reasons: only one or few strains have been included in the original species descriptions and the few subsequent articles; studies describing phenotypic properties of novel species have used predominantly home-brew methods and, occasionally, commercial media of different basic composition; recent characterizations of novel species have only included a limited set of phenotypic markers; there is conflicting information about certain differential physiological and biochemical characteristics of the various species; different media, whether commercial or home-brew, may yield inconsistent results, with either positive or negative reactions for the same phenotypic marker[15]; some biochemical reactions are variable among strains within the same species and even among different isolates of the same strain.[28]

Tsukamurella species are able to utilize a number of substrates including sugars as a sole carbon source and amino acids as a sole carbon and nitrogen source. *Tsukamurella paurometabola* is the most biochemically inert, whereas *T. tyrosinosolvens*, *T. inchonensis*, and *T. carboxydivorans* are the most active. In contrast to other Tsukamurellae, *T. carboxydivorans* possesses a carbon monoxide dehydrogenase enzyme (CO-DH), which can be upregulated, and which enables it to grow under a CO-rich atmosphere of up to 300,000 p.p.m. *T. tyrosinosolvens* hydrolyzes tyrosine, a property shared by *T. carboxydivorans*. Other differential phenotypic properties are summarized in Table 18.1.

Mycolic acids are very long-chain branched α-alkyl-β-hydroxy fatty acids and the most characteristic component in the cell-wall lipids of aerobic actinomycetes that include Tsukamurellae. *Tsukamurella* species possess mycolic acids generally ranging in length from 62 to 78 carbon atoms with 1–7 double bonds.[11] *T. carboxydivorans* contains longer mycolic acids with 81–95 carbons. Whole cell acid and alkaline methanolysates produce two adjacent mycolic acid spots, designated α and α', on thin-layer chromatography, which are unique to the genus Tsukamurellae.[11–14] Reverse-phase high-performance liquid chromatographic analysis (HPLC) of mycolic acid esters yields a single cluster of poorly resolved peaks with retention times of 6.25–7.50 min.[15] This profile is distinctive for members of the genus Tsukamurellae and allows their separation from other genera among the aerobic actinomycetes.[11,15] However, this profile is fairly similar among *Tsukamurella* species; thus, HPLC analysis is not suitable for species-level identification.[15] Quantitative analyses of pyrolysis products of mycolic acid methyl esters by gas chromatography and mass spectrometry using a standardized procedure on the Microbial Identification System (MIS Inc., Del.) yield straight-chain saturated and monounsaturated fatty acids ranging from 12 to 20 carbons in length and frequently tuberculostearic acid.[15,20,21] The most abundant fatty acids are $C_{16:0}$ and $C_{18:1}\omega9c$, which account for 65%–75% of fatty acid content in all species. Using the MIS library and pattern-recognition software, the short-chain fatty acid profiles of *Tsukamurella* isolates overlapped with that of other members of the aerobic actinomycetes, such as *Nocardia*, *Rhodococcus*, or *Gordonia* species with high similarity indices[15,34]; thus, MIS appeared not to be useful for genus-level identification. However, using a heuristic library approach developed with the MIS software from profiles of a large number of reference and clinical strains of aerobic actinomycetes, McNabb et al. were able to correctly assign eight clinical isolates to the genus *Tsukamurella* and to determine that they were closely related to *T. inchonensis* strain ATCC 25983 but not to *T. paurometabola* strain ATCC 8368.[58]

18.1.3.2 Molecular Techniques

Novel genomic species of *Tsukamurella* have been characterized using DNA–DNA hybridization studies and 16S rRNA gene sequencing. The minimum level of DNA–DNA relatedness between strains required to delineate a novel species is 70%.[59] Original species descriptions that included more than one strain invariably showed levels of DNA–DNA reassociation ≥70% among strains of the same species and positive hybridization, albeit at levels below this cutoff, among strains of different species. DNA–DNA hybridization methods, although a standard for taxonomists, are typically not performed in the clinical laboratory. The phylogenetic relationships among well-characterized reference and type strains of Tsukamurellae based on 16S rRNA sequences are shown in Figure 18.1. The 16S rRNA sequences of *Tsukamurella* species are closely related, with similarity levels exceeding 99% among most species. The lowest similarity index of 98.8% is observed between *T. spongiae* type strain and *T. strandjordii*, *T. spumae*, and *T. pseudospumae*. *Tsukamurella sunchonensis* and *T. pseudospumae* share an identical 16S rRNA sequence and very likely represent the same species. The most closely related species are *T. tyrisonosolvens* and *T. carboxydivorans*, with a 16S similarity index of 99.8%, followed by *T. spumae* and *T. pseudospumae* at 99.7%. Although the 16S sequences of isolates of the same species are usually identical, limited intraspecific polymorphisms have been observed. For example, in the original description by Yassin et al., one isolate of *T. tyrosinosolvens* showed only a 99.7% sequence similarity to the three other strains, including the type strain, which had identical sequences. Similarly,

TABLE 18.1
Selected Differential Phenotypic Properties of *Tsukamurella* Species

	T. paurometabola	*T. pulmonis*	*T. inchonensis*	*T. tyrosinosolvens*	*T. strandjordii*	*T. sunchonensis*	*T. spumae*	*T. pseudospumae*	*T. spongiae*	*T. carboxydivorans*
Growth at										
10°C	(+)	(–)	(–)	(–)	(–)	(–)	(–)	(+)	(–)	(–)
25–28°C	(+)	(+)	(+)	(+)	(+)	(+)	(+)	(+)	(+)	(+)
35–37°C	(+)	(+)	(+)	(+)	(+)	(+)	(+)	(+)	(+)	(+)
42–45°C	(–)	(–)	(+)	(–)	(–)	(–)	(–)	(–)	(–)	(–)
Hydrolysis										
Esculin	(+)	(+)	(+)	(+)	(–)	(+)	(–)	(–)	NA	(–)
Urea	(+)	(+)	(+)	(–)	(+)	(–)	(–)	(–)	NA	(+)
Casein	(–)	(–)	(–)	(–)	(–)	(–)	NA	NA	NA	(+)
Hypoxanthine	(–)	(+)	(+)	(+)	(–)	(+)	(+)	(+)	NA	(+)
Xanthine	(–)	(–)	(–)	(–)	(–)	(–)	NA	NA	NA	(+)
Tyrosine	(–)	(–)	(–)	(+)	(–)	(+)	(+)	(+)	NA	(+)
Sugar utilization as sole carbon source										
D-Arabinose	(–)	(+)	(+)	(+)	(–)	(–)	(–)	(+)	(–)	(+)
L-Arabinose	(–)	(+)	(–)	(–)	(–)	NA	(+)	(+)	(+)	(+)
D-Cellobiose	(–)	(+)	(–)	(+)	(–)	NA	(–)	(–)	(+)	(+)
Dulcitol	(–)	(+)	(–)	(+)	(–)	NA	(+)	(–)	(–)	(+)
meso-Erythritol	(–)	(+)	(–)	(+)	(–)	(–)	(+)	(–)	(–)	(+)
D-Fructose	(+)	(+)	(–)	(+)	(–)	(+)	(+)	(+)	(+)	(+)
D-Maltose	(–)	(+)	(+)	(+)	(–)	(+)	(+)	(+)	(–)	(+)
D-Mannitol	(+)	(–)	(+)	(+)	(+)	NA	(+)	(–)	(+)	(+)
D-Melezitose	(–)	(–)	(+)	(+)	(–)	(+)	(+)	(+)	(+)	(+)
D-Ribose	(+)	(+)	(+)	(+)	(+)	NA	(+)	(+)	(+)	(+)
D-Sorbitol	(–)	(–)	(+)	(+)	(+)	(–)	(+)	(–)	(+)	(+)
D-Xylose	(+)	(+)	(+)	(+)	(–)	(–)	(+)	(–)	(+)	(+)

Notes: (+): Positive; (–): Negative; NA: Not available.

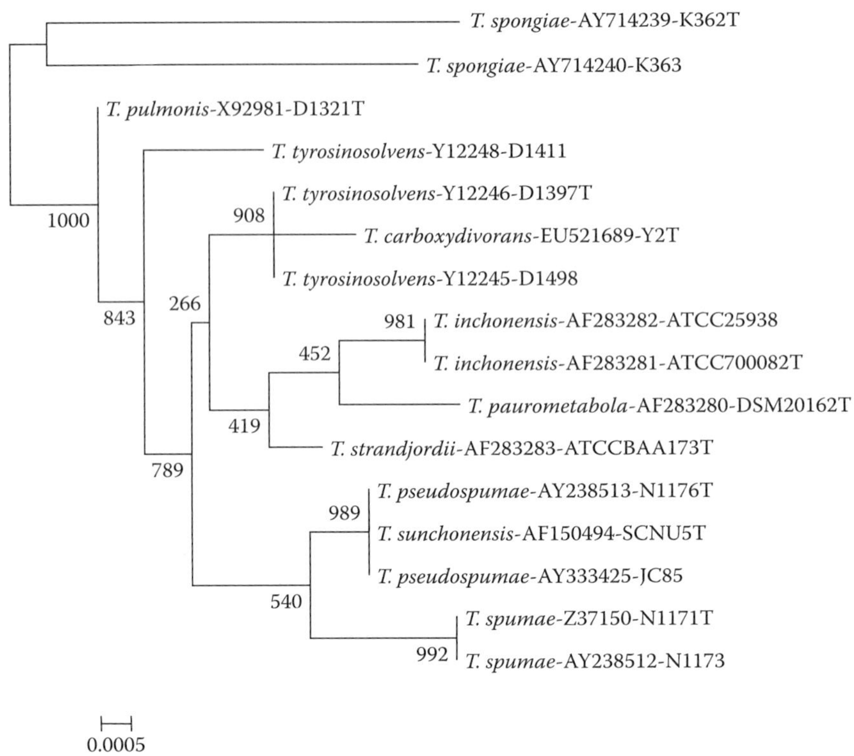

FIGURE 18.1 Neighbor-joining phylogenetic tree based on 16S rRNA sequences of reference and type strains of *Tsukamurella* species. Numbers at the nodes indicate bootstrap values. The scale at the bottom indicates the genetic distance. Species names are followed by GenBank accession numbers and strain names. "T" indicates the type strains.

the published sequences of the two strains of *T. spongiae* have only 98.9% sequence similarity (Figure 18.1). In one study, partial sequencing of the first 500 bp of the 16S rRNA gene and comparison with sequences in a commercial database and in GenBank allowed genus-level identification of Tsukamurellae but species identification could only be achieved for 62% of isolates.[60] However, alignment of full-length 16S rRNA sequences of well-characterized strains of *Tsukamurella* disclosed species-specific signature sequences that could potentially be used for definitive species-level identification of unknown isolates (Table 18.2). Ribotyping has revealed that *Tsukamurellae* species may harbor up to four ribosomal operons.[17]

The sequences of other housekeeping genes such *hsp65*, *rpoB*, *gyrB*, *groEL*, and 16S-23S internal transcribed spacer were demonstrated to be phylogenetically informative and more discriminatory for species definition of other aerobic actinomycete genera, such as *Mycobacterium* and *Nocardia*. However, the number of published sequences for these genes is insufficient to draw similar conclusions regarding *Tsukamurella* species. The published *groEL* sequences of *T. paurometabola* (AF352578) and *T. tyrosinosolvens* (U90204) show a difference of 5.4%. Restriction fragment length polymorphism (RFLP) analysis of *hsp65* has shown that this method is reliable in distinguishing *Tsukamurella paurometabola* from other aerobic actinomycetes.[61] The published *gyrB* sequences of *T. paurometabola* and *T. inchonensis* display a difference of 7.6%. These results indicate that building a library of more polymorphic sequences of housekeeping genes could potentially improve the confidence level for identifying *Tsukamurella* species.

18.2 METHODS

16S rRNA gene sequencing for identification of *Tsukamurella* species from bacterial cultures is detailed below.

18.2.1 Sample Preparation

Bacteria grown on any media are suitable for the test. Depending on the size of the colonies, 1–20 colonies from a pure culture are scraped off the agar with a disposable or flamed wire loop without disrupting the agar surface. Agar media contain substances that may inhibit the PCR reaction. Colonies of any age can be used, although freshly subcultured colonies are preferred. Bacteria can also be collected from a broth culture (e.g., blood culture bottle, trypticase soy broth, or brain heart infusion broth). About 1.5 ml of broth is collected with a sterile pipette in a 2 mL tube. Heat the tube at 95°C for 15 min in a dry heat block to inactivate the bacteria. Spin at 13,000 rpm for 5 min. The supernatant is discarded, leaving the bacterial pellet. The pellet is washed twice in 500 μL of 1× sterile Phosphate Buffered Saline (PBS) before proceeding to DNA extraction. The washing step is optional for bacteria collected from the surface of solid agar media. The pellet size should ideally not exceed 3 mm^3. If the samples are not to be processed the same day, cell pellets can be stored frozen at –70°C.

TABLE 18.2
16S rRNA Gene Sequence Signatures That Differentiate *Tsukamurella* Species

Position	*T. paurometabola*	*T. strandjordii*	*T. pulmonis*	*T. inchonensis*	*T. tyrosinosolvens*	*T. carboxydivorans*	*T. spumae*	*T. pseudospumae*	*T. sunchonensis*	*T. spongiae*
86	T	T	C	T	T	T	T	T	T	C
88	T	T	T	T	T	T	C	C	C	T
89	T	C	C	C	C	C	Del	Del	Del	C
90	T	G	G	G	G	G	Del	Del	Del	G
145	T	C	C	C	C	C	C	C	C	C
190	C	C	C	C	C	C	T	T	T	C
191	T	T	C	T	T/C	T	C	C	C	C
203	G	G	G	G	G	G	T	T	T	G
215	C	C	C	C	T/C	T	C	C	C	C
381	C	A	A	C	A/G	A	A	A	A	A
546	A	A	G	G	G	G	G	G	G	G
560	A	A	T	T	T	T	T	T	T	T
595	A	A	A	G	A	A	A	A	A	A
737	C	C	C	C	C	C	C	C	C	T
739	C	C	C	C	C	C	C	C	C	T
743	G	G	G	G	G	G	G	G	G	C
923	A	A	A	A	A	G	A	A	A	A
1011	C	G	G	C	G	G	C	G	G	G
1018	G	T	T	G	T	T	G	T	T	T
1019	C	C	C	T	C	C	T	C	C	C
1136	A	A	A	T	A	A	T	A	A	A
1260	G	G	G	G	G	G	G	A	A	G
1267	T	T	C	T	T	T	T	T	T	C
1365	G	G	G	G	G	A	G	G	G	G

Notes: Positions are based on *E. coli* numbering (Ref. 65: Brosius et al. 1978)
Del: Deletion

DNA is extracted with the Qiagen QIAamp DNA Mini Kit as follows:

Add 180 μL of Buffer ATL (Lysis Buffer from the Qiagen QIAamp DNA Mini Kit) and 20 μL Proteinase K to the bacterial pellet and vortex vigorously.
Incubate overnight at 56°C in a dry heat block.
Proceed with the extraction using the tissue protocol of QIAamp DNA Mini Kit as described by the manufacturer.

18.2.2 Detection Procedures

(i) PCR Amplification The PCR mix consists of:

HotStart Taq Polymerase	0.5 μL
10× PCR buffer	10 μL
P1*: 50 pmoles (Stock 25 μM)	2 μL
P10*: 50 pmoles (Stock 25 μM)	2 μL
dNTP mixture (25 mM each)	0.8 μL
$MgCl_2$ (25 mM stock)	12 μL
Molecular biology grade ddH_2O	62.7 μL
Bacterial DNA solution	10 μL
Total reaction volume	100 μL

The PCR and sequencing primers are:

1. P1* = 8FPL (F: 8–27): 5′, AG AGT TTG ATC CTG GCT CAG 3′,
2. P2 = 516F (F: 516–533): 5′, T GCC AGC AGC CGC GGT AA 3′,
3. P3 = 516R (R: 516–533): 5′, TT ACC GCG GCT GCT GGC A 3′,
4. P4 = 806F (F: 787–806): 5′, ATT AGA TAC CCT GGT AGT CC 3′,
5. P5 = 806R (R: 787–806): 5′, GG ACT ACC AGG GTA TCT AAT 3′,
6. P6 = 1175F (F: 1175–1194): 5′, GAG GAA GGT GGG GAT GAC GT
7. P7 = 1175R (R: 1175–1194): 5′, AC GTC ATC CCC ACC TTC CTC 3′,
8. P10* = DG74 (R: 1522–1540): 5′, A GGA GGT GAT CCA RCC GCA 3′.

(*) indicates PCR primers. Numbers in parentheses correspond to *E. coli* numbering positions.[62] Primer sequences are derived from references.[63–65] F = Forward; R = Reverse. For primer P10, R represents A or G.

Each run includes a negative control, 10 μL of water added to the PCR mix instead of DNA, and a positive control, 10 μL of DNA prepared from a laboratory strain of *E. coli*.

The thermocycling parameters are as follows: initial denaturation at 95°C for 10 min; followed by 30 cycles of 95°C for 30 s; 68°C for 30 s and 72°C for 45 s; and a final 72°C for 10 min.

If the PCR product is to be used the same day, it can be stored at 4°C or on ice; otherwise, storage at −20°C is recommended.

(ii) DNA Gel Electrophoresis Prepare 1% agarose gel by mixing the agarose powder with 1× Tris/Borate EDTA (TBE). Dissolve by microwaving on high until the solution is clear. Add 5 μL of Ethidium Bromide (10 mg/mL) per 100 mL of agarose solution. Add 10 μL of PCR product to each well, mixed, with 10 μL of 2× loading dye. Add 1× TBE to the electrophoresis chamber to cover the gel surface and run for 45 min at 80 V with a DNA ladder of the appropriate size. If a product of expected size is visualized (approximately 1500 bp), proceed to PCR product purification and sequencing.

(iii) PCR Product Purification Purify the PCR product with Qiagen MinElute PCR Purification Kit or equivalent PCR purification kit. Follow instructions in package insert.

(iv) Standard Cycle Sequencing Eight sequencing reactions are prepared for each sample with primers P1 to P7 and P10 (primer sequences are detailed above).

The reaction final volume is 10 μL and consists of 4 μL of purified PCR product solution, 4 μL of Big Dye Ready Reaction mix (Applied Biosystems) and 2 μL of primer diluted to 1 μM (final concentration = 0.2 μM).

The sequencing parameters are as follows: 1 cycle at 96°C for 1 min, followed by 25 cycles of 96°C for 10 s, 50°C for 5 s, and 60°C for 4 min.

The sequencing reactions are purified using the Qiagen DyeEx Spin Kit or equivalent. Reactions are analyzed by capillary electrophoresis on an ABI 3130 Genetic Analyzer (Applied Biosystems) or equivalent.

(v) Sequence Analysis The sequencing chromatogram contigs are assembled and edited using sequencing analysis software, for example, ChromasPro (Technelysium Pty Ltd., Australia) in order to resolve ambiguous positions.

The sequence of an unknown isolate is saved in FASTA format and searched against GenBank using BLAST (http://www.ncbi.nlm.nih.gov/BLAST). If the unknown sequence matches that of a *Tsukamurella* species, place it in a single file together with 16S rRNA sequences of reference and type strains of *Tsukamurella* species. The GenBank accession numbers of these sequences are shown in Figure 18.1. Align these sequences using the CLUSTAL algorithm with CLUSTAL_X,[66] BioEdit,[67] (http://www.mbio.ncsu.edu/BioEdit/BioEdit.html), or MEGA software.[68] These are freely available on the world-wide Web. Perform phylogenetic analysis with the neighbor-joining algorithm for the purpose of species identification. Phylogenetic trees can be constructed and viewed with either MEGA (www.megasoftware.net) or PHYLIP (http://evolution.genetics.washington.edu/phylip.html). Phylogenetic trees generated in CLUSTAL_X using the neighbor-joining algorithm can be viewed using either MEGA or the Treeview software (http://taxonomy.zoology.gla.ac.uk/rod/treeview.html). Alternatively, align the

Tsukamurella sequences with the *E. Coli* 16S rRNA sequence J01695.[62] The polymorphic 16S rDNA sequence signatures as outlined in Table 18.2 may then be used to derive the species identification.

18.3 CONCLUSIONS AND FUTURE PERSPECTIVES

Tsukamurella species are soil organisms that have emerged over the past decade as a rare but significant cause of opportunistic infections associated with indwelling foreign medical devices in immunocompromised patients. Infections in immunocompetent patients have been sporadically reported and present mostly as a tuberculosis-like illness. Despite the fact that Tsukamurellae are multidrug-resistant organisms, the mechanisms of antimicrobial resistance have not been studied. It is uncertain whether they harbor plasmids. Conventional phenotypic methods allow genus-level presumptive identification at best. The small number of reported *Tsukamurella* strains that have been studied phenotypically and their physiological variability make it difficult to fit a given strain within a described species. There are no standardized molecular typing methods for *Tsukamurella* species. In a study investigating a pseudooutbreak that spanned a 30-month period at a hospital in South Carolina, the authors used ribotyping, which demonstrated that the 12 outbreak isolates had a similar profile, which was distinct from other sporadic nonoutbreak-related isolates.[23] Common methods for molecular typing of other bacterial species such as pulsed field gel electrophoresis (PFGE), multilocus sequence typing (MLST), and multilocus variable number tandem repeat analysis (MLVA) need to be developed for Tsukamurellae. Such progress is hampered by the paucity of sequence information in public databases. As of August 27, 2009, there were 105 nucleotide sequences of named *Tsukamurella* species, 90% of which are 16S rRNA sequences. Alignment of the 16S sequences of well-characterized strains enabled us to identify polymorphic signature sequences that could potentially serve as probe targets for the design of real-time PCR assays for rapid species-level identification (Table 18.2). The genome of *Tsukamurella paurometabola* type strain has been recently sequenced, but its annotation has not yet been completed. Out of 8862 published Tsukamurellae protein sequences in GenBank, 8848 belong to the *Tsukamurella paurometabola* type strain. Sequencing of a *T. tyrosinosolvens* isolate genome is currently in progress. As more sequence information becomes available, new methods based on comparative genomics can potentially be developed for more rapid and definitive species identification and molecular typing of Tsukamurellae in a clinical setting.

REFERENCES

1. Zhi, X.Y., Li, W.J., and Stackebrandt, E., An update of the structure and 16S rRNA gene sequence-based definition of higher ranks of the class Actinobacteria, with the proposal of two new suborders and four new families and emended descriptions of the existing higher taxa, *Int. J. Syst. Evol. Microbiol.*, 59, 589, 2009.
2. Pavlov, D. et al., Potentially pathogenic features of heterotrophic plate count bacteria isolated from treated and untreated drinking water, *Int. J. Food Microbiol.*, 92, 275, 2004.
3. Collins, M.D. et al., *Tsukamurella* gen. nov., harboring *Corynebacterium paurometabolum* and *Rhodococcus aurantiacus*, *Int. J. Syst. Bacteriol.*, 38, 385, 1988.
4. Lechevalier, M.P., and Lechevalier, H.A., Chemical composition as a criterion in the classification of aerobic actinomycetes, *Int. J. Syst. Bacteriol.*, 20, 435, 1970.
5. Steinhaus, E.A., A study of the bacteria associated with thirty species of insects, *J. Bacteriol.*, 42, 757, 1941.
6. Jones, D., A numerical taxonomic study of coryneform and related bacteria, *J. Gen. Microbiol.*, 87, 52, 1975.
7. Tsukamura, M., and Mizuno, S., A new species *Gordona aurantiaca* occurring in sputa of patients with pulmonary disease, *Kekkaku*, 46, 93, 1971.
8. Tsukamura, M., and Yano, I., *Rhodococcus sputi* sp. nov., nom. rev., and *Rhodococcus aurantiacus* sp. nov., nom. rev., *Int. J. Syst. Bacteriol.*, 35, 364, 1985.
9. Collins, M.D., and Jones, D., Lipid composition of *Corynebacterium paurometabolum*, *FEMS Microbiol. Lett.*, 13, 13, 1982.
10. Goodfellow, M. et al., Chemical and numerical taxonomy of strains received as *Gordona aurantiaca*, *J. Gen. Microbiol.*, 109, 57, 1978.
11. Koerner, R.J., Goodfellow, M., and Jones, A.L., The genus *Dietzia*: A new home for some known and emerging opportunist pathogens, *FEMS Immunol. Med. Microbiol.*, 55, 296, 2009.
12. Yassin, A.F. et al., *Tsukamurella inchonensis* sp. nov., *Int. J. Syst. Bacteriol.*, 45, 522, 1995.
13. Yassin, A.F. et al., *Tsukamurella pulmonis* sp. nov., *Int. J. Syst. Bacteriol.*, 46, 429, 1996.
14. Yassin, A.F. et al., *Tsukamurella tyrosinosolvens* sp. *nov.*, *Int. J. Syst Bacteriol.*, 47, 607, 1997.
15. Kattar, M.M. et al., *Tsukamurella strandjordae* sp. nov., a proposed new species causing sepsis, *J. Clin. Microbiol.*, 39, 1467, 2001.
16. Seoung, C.N. et al., *Tsukamurella sunchonensis*, sp. nov., a bacterium associated with foam in activated sludge, *J. Microbiol.*, 41, 83, 2003.
17. Nam, S.W. et al., *Tsukamurella spumae* sp. nov., a novel actinomycete associated with foaming in activated sludge plants, *Syst. Appl. Microbiol.*, 26, 367, 2003.
18. Nam, S.W. et al., *Tsukamurella pseudospumae* sp. nov., a novel actinomycete isolated from activated sludge foam, *Int. J. Syst. Evol. Microbiol.*, 54, 1209, 2004.
19. Goodfellow, M. et al., Activated sludge foaming: The true extent of actinomycete diversity, *Water Sci. Technol.*, 37, 511, 1998.
20. Olson, J.B. et al., *Tsukamurella spongiae* sp. nov., a novel actinomycete isolated from a deep-water marine sponge, *Int. J. Syst. Evol. Microbiol.*, 57, 1478, 2007.
21. Park, S.W., et al., *Tsukamurella carboxydivorans* sp. nov., a carbon monoxide-oxidizing actinomycete, *Int. J. Syst. Evol. Microbiol.*, 59, 1541, 2009.
22. Jones, R.S. et al., Persistent bacteremia due to *Tsukamurella paurometabolum* in a patient undergoing hemodialysis: Case report and review, *Clin. Infect. Dis.*, 18, 830, 1994.
23. Auerbach, S.B. et al., Outbreak of pseudoinfection with *Tsukamurella paurometabolum* traced to laboratory contamination: Efficacy of joint epidemiological and laboratory investigation, *Clin. Infect. Dis.*, 14, 1015, 1992.
24. Haft, R.F. et al., Pseudoinfection due to *Tsukamurella paurometabolum*, *Clin. Infect. Dis.*, 15, 883, 1992.

25. Tsukamura, M., and Kawakami, K., Lung infection caused by *Gordona aurantiaca (Rhodococcus aurantiacus)*, *J. Clin. Microbiol.*, 16, 604, 1982.
26. Prinz, G. et al., Meningitis caused by *Gordona aurantiaca (Rhodococcus aurantiacus)*, *J. Clin. Microbiol.*, 22, 472, 1985.
27. Casella, P., Tommasi, A., and Tortorano, A.M., Peritonie da *Gordona aurantiaca* (*Rhodococcus aurantiacus*) in dialisi peritoneale ambulatoria continua, *Microbiologia Medica*, 2, 47, 1987.
28. Tsukamura, M. et al., Severe progressive subcutaneous abscesses and necrotizing tenosynovitis caused by *Rhodococcus aurantiacus*, *J. Clin. Microbiol.*, 26, 201, 1988.
29. Osoagbaka, O.U., Evidence for the pathogenic role of Rhodococcus species in pulmonary diseases, *J. Appl. Bacteriol.*, 66, 497, 1989.
30. Shapiro, C.L. et al., *Tsukamurella paurometabolum*: A novel pathogen causing catheter-related bacteremia in patients with cancer. *Clin. Infect. Dis.*, 14, 200, 1992.
31. Lai, K.K., A cancer patient with central venous catheter-related sepsis caused by *Tsukamurella paurometabolum* (*Gordona aurantiaca*), *Clin. Infect. Dis.*, 17, 285, 1993.
32. Granel, F. et al., Cutaneous infection caused by *Tsukamurella paurometabolum*, *Clin. Infect. Dis.*, 23, 839, 1996.
33. Rey, D. et al., *Tsukamurella* infections. Review of the literature apropos of a case. *Pathol Biol (Paris)*, 45, 60, 1997.
34. Chong, Y. et al., *Tsukamurella inchonensis* bacteremia in a patient who ingested hydrochloric acid, *Clin. Infect. Dis.*, 24, 1267, 1997.
35. Larkin, J.A. et al., Infection of a knee prosthesis with *Tsukamurella* species, *South Med. J.*, 92, 831, 1999.
36. Maertens, J. et al., Catheter-related bacteremia due to *Tsukamurella pulmonis*, *Clin. Microbiol. Infect.*, 4, 51, 1998.
37. Schwartz, M.A. et al., Central venous catheter-related bacteremia due to *Tsukamurella* species in the immunocompromised host: A case series and review of the literature, *Clin. Infect. Dis.*, 35, e72, 2002.
38. Sheridan, E.A. et al., *Tsukamurella tyrosinosolvens* intravascular catheter infection identified using 16S ribosomal DNA sequencing, *Clin. Infect. Dis.*, 36, e69, 2003.
39. Alcaide, M.L., Espinoza, L., and Abbo, L., Cavitary pneumonia secondary to *Tsukamurella* in an AIDS patient. First case and a review of the literature, *J. Infect.*, 49, 17, 2004.
40. Elshibly, S. et al., Central line-related bacteraemia due to *Tsukamurella tyrosinosolvens* in a haematology patient, *Ulster Med. J.*, 74, 43, 2005.
41. Almehmi, A. et al., Implantable cardioverter-defibrillator infection caused by *Tsukamurella*, *W.V. Med. J.*, 100, 185, 2004.
42. Shaer, A.J., and Gadegbeku, C.A., *Tsukamurella* peritonitis associated with continuous ambulatory peritoneal dialysis, *Clin. Nephrol.*, 56, 241, 2001.
43. Perez, V.A., Swigris, J., and Ruoss, S.J., Coexistence of primary adenocarcinoma of the lung and *Tsukamurella* infection: A case report and review of the literature, *J. Med. Case Reports*, 2, 207, 2008.
44. Maalouf, R. et al., First case report of community-acquired pneumonia due to *Tsukamurella pulmonis*, *Ann. Intern. Med.*, 150, 147, 2009.
45. Shim, H.E. et al., A case of catheter-related bacteremia of *Tsukamurella pulmonis*, *Korean J. Lab. Med.*, 29, 41, 2009.
46. Moore, J.E., Xu, J., and Millar, B.C., Analysis of 16S-23S intergenic spacer regions of the rRNA operons in *Tsukamurella pulmonis*, *Br. J. Biomed. Sci.*, 63, 25, 2006.
47. Matsumoto, T. et al., *Tsukamurella tyrosinosolvens* cultured from sputum of a patient who received total gastrectomy for gastric cancer, *Kekkaku*, 81, 487, 2006.
48. Bouza, E. et al., *Tsukamurella*: A cause of catheter-related bloodstream infections, *Eur. J. Clin. Microbiol. Infect. Dis.*, 28, 203, 2009.
49. Sheng, W.H. et al., Brain abscess caused by *Tsukamurella tyrosinosolvens* in an immunocompetent patient, *J. Clin. Microbiol.*, 47, 1602, 2009.
50. Woo, P.C. et al., First report of *Tsukamurella* keratitis: Association between *T. tyrosinosolvens* and *T. pulmonis* and ophthalmologic infections, *J. Clin. Microbiol.*, 47, 1953, 2009.
51. Woo, P.C. et al., *Tsukamurella* conjunctivitis: A novel clinical syndrome, *J. Clin. Microbiol.*, 41, 3368, 2003.
52. Clausen, C., and Wallis, C.K., Bacteremia caused by *Tsukamurella* species, *Clin. Microbiol. Newsl.*, 16, 6, 1994.
53. Rey, D. et al., Tsukamurella and HIV infection, *AIDS*, 9, 1379, 1995.
54. NCCLS: M24-A. Susceptibility testing of mycobacteria, nocardiae, and other aerobic actinomycetes; Approved standard. Wayne, PA: NCCLS, 2003.
55. Vollbrecht, E. et al., Production and structure elucidation of di- and oligosaccharide lipids (biosurfactants) from *Tsukamurella* sp. nov., *Appl. Microbiol. Biotechnol.*, 50, 530, 1998.
56. Asano, M. et al., Reciprocal action of interferon-gamma and interleukin-4 promotes granulomatous inflammation induced by *Rhodococcus aurantiacus* in mice, *Immunology*, 88, 394, 1996.
57. Stanley, T. et al., The potential misidentification of *Tsukamurella pulmonis* as an atypical *Mycobacterium* species: A cautionary tale, *J. Med. Microbiol.*, 55, 475, 2007.
58. McNabb, A. et al., Fatty acid characterization of rapidly growing pathogenic aerobic actinomycetes as a means of identification, *J. Clin. Microbiol.*, 35, 1361, 1997.
59. Wayne, L.G., Brenner, D.J., and Colwell, R.R., International committee on systematic bacteriology report of the *ad hoc* committee on reconciliation of approaches to bacterial systematics, *Int. J. Syst. Bacteriol.*, 37, 463, 1987.
60. Patel, J.B. et al., Sequence-based identification of aerobic actinomycetes, *J. Clin. Microbiol.*, 42, 2530, 2004.
61. Steingrube, V.A. et al., Rapid identification of clinically significant species and taxa of aerobic actinomycetes, including *Actinomadura, Gordona, Nocardia, Rhodococcus, Streptomyces,* and *Tsukamurella* isolates, by DNA amplification and restriction endonuclease analysis, *J. Clin. Microbiol.*, 35, 817, 1997.
62. Brosius, J. et al., Complete nucleotide sequence of a 16S ribosomal RNA gene from *Escherichia coli*, *Proc Natl Acad Sci USA*, 75, 4801, 1978.
63. Fredericks, D.N., and Relman, D.A., Sequence-based identification of microbial pathogens: A reconsideration of Koch's postulates, *Clin. Microbiol. Rev.*, 9, 18, 1996.
64. Fredricks, D.N., and Relman, D.A., Improved amplification of microbial DNA from blood cultures by removal of the PCR inhibitor sodium polyanetholesulfonate, *J. Clin. Microbiol.*, 36, 2810, 1998.
65. Greisen, K. et al., PCR primers and probes for the 16S rRNA gene of most species of pathogenic bacteria, including bacteria found in cerebrospinal fluid, *J. Clin. Microbiol.*, 32, 335, 1994.
66. Thompson, J.D. et al., The CLUSTAL_X windows interface: Flexible strategies for multiple sequence alignment aided by quality analysis tools, *Nucleic Acids Res.*, 25, 4876, 1997.
67. Hall, T.A., BioEdit: A user-friendly biological sequence alignment editor and analysis program for Windows 95/98/NT, *Nucleic Acids Symposium Series*, 41, 95, 1999.
68. Tamura, K. et al., MEGA4: Molecular Evolutionary Genetics Analysis (MEGA) software version 4.0. *Mol. Biol. Evol.*, 24, 1596, 2007.

Section II

Firmicutes and Tenericutes

Bacilli

19 *Abiotrophia*

Marco Arosio and Annibale Raglio

CONTENTS

19.1 INTRODUCTION

The genus *Abiotrophia*, nowadays, is well known, but its definition has undergone extensive change. Only the application of molecular biology has made it possible to verify the correct classification of genus and species.

The word "*Abiotrophia*" means "life nutrition deficiency" and refers to the species' requirements for supplemented media for growth.[1]

The genus *Abiotrophia* was proposed to include microorganisms initially known as nutritionally variant streptococci (NVS), also called "satelliting streptococci," which are fastidious bacteria that need additional growth factors to survive in complex media: L-cysteine or pyridoxal hydrochloride (vitamin B6) has been shown to favor growth.[2]

Frenkel and Hirsch described first *Abiotrophia* as ungroupable viridans streptococci that grew as satellite colonies around other bacteria.[3]

Because of sugar fermentation and cell wall composition, these bacteria appeared to be mutant subspecies of *Streptococcus mitis*, but later, important differences in the enzymatic properties and patterns of PBPs between NVS and *S. mitis* appeared.[4]

Many synonyms were given to these fastidious microorganisms, such as satelliting *Streptococcus*, thiol-requiring *Streptococcus*, vitamin B6–dependant *Streptococcus*, piridoxal-dependant *Streptococcus*, symbiotic *Streptococcus*, and nutritionally variant streptococci, and only later were they classified as a new genus, *Abiotrophia*.[5]

Following the study by Collins and Lawson, *Abiotrophia defectiva* has remained the only *Abiotrophia* species, and a new genus was proposed as *Granulicatella* for the species found in human beings: *Granulicatella adiacens*, *Granulicatella para-adiacens* and *Granulicatella elegans*, and a species *Granulicatella balaenopterae* originating from a whale.[6]

A. defectiva is present in a normal human oral cavity, urogenital tract, and intestinal flora.[7]

Like other viridans group streptococci *A. defectiva* causes bacteremia and sepsis and is also an etiological agent of endocarditis and other infectious diseases.[8–13]

The diagnosis is difficult but not impossible since *A. defectiva* identification starts from the suspect to its isolation. We will describe the traditional techniques for *A. defectiva* isolation and identification and underline the application of molecular biology to confirm the identification of a cultivated strain or whether it is present in sterile body fluid.

19.1.1 Classification, Morphology, and Biology

Classification. The so-called NVS were classified in different genera (*Abiotrophia* and *Granulicatella*) and were differentiated in five species by molecular biology.

At the beginning, Bouvet et al. considered these microorganisms to be members of two novel streptococcal species, and so they called them *Streptococcus defectivus* and *Streptococcus adjacens* and showed the presence of two different species within the NVS by chromosomal DNA–DNA hybridisation.[14]

Bouvet's hybridization studies revealed that *S. defectivus* and *S. adjacens* shared less than 10% DNA homology with each other and with other streptococcal species.[6]

The phylogenetic differentiation of these microorganisms from streptococci group, together with nutritional considerations, satellitism, and pyrrolidonyl arylamidase productions, led to the definition of the new genus *Abiotrophia*.

It was only after the application of molecular biology and the analysis of 16S rRNA sequences that Kawamura et al. proposed the creation, within genus *Abiotrophia*, of two new species *A. defectiva* and *Abiotrophia adiacens*.[15]

The Kawamura study using 16S rRNA sequence analysis allowed the exclusion of these NVS organisms from the genus *Streptococcus* based on phylogenetic features (e.g., 16S rRNA sequence is different from streptococcal species by more than 11%) and also whether they shared a higher 16S rRNA relatedness with other catalase-negative bacteria (e.g., *Aerococcus*, *Carnobacterium*, *Enterococcus*) than with streptococci.[6]

Later, two further species of genus *Abiotrophia* were characterized: *Abiotrophia elegans* from a patient with endocarditis and *Abiotrophia balaenopterae* from a whale (*Balaenoptera acutorostrata*).[16,17]

These new species presented a closer phylogenetic affinity with *Abiotrophia adiacens* (16S rRNA sequence differed about 3%) than with *A. defectiva* (sequence differed about 7%); therefore, two distinct lines were defined within the genus *Abiotrophia*, namely, *A. defecti*va and a group including *Abiotrophia adiacens*, *Abiotrophia elegans*, and *Abiotrophia balaenopterae* (not human).[6]

Kanamoto et al. observed the heterogeneity among *Abiotrophia* strains and proposed a fourth human pathogen, *Abiotrophia para-adiacens*.[18]

In the same year, on the basis of 16S rRNA studies, Collins and Lawson noted that the genus *Abiotrophia* could be restricted to species *A. defective*, which is phylogenetically closer to several non-*Abiotrophia* genera (*Globicatella*, *Facklamia*, and *Eremococcus*), while the new genus *Granulicatella* was proposed for the other four species *Granulicatella adiacens*, *Granulicatella para-adiacens*, *Granulicatella elegans*, and *Granulicatella balaenopterae*.[6]

Morphology. *A. defectiva* observation by electron microscope presents a normal streptococcal ultrastructure if the cells are cultured in appropriate conditions. However, if the conditions are not optimal, the cell wall will show abnormal features considered by some authors as L-forms organisms.[1]

Colonies of *A. defecti*va are small, measuring 0.2–0.5 mm in diameter, and are either nonhemolytic or α-hemolytic on supplemented blood agar culture. On Gram stain, cell may be pleomorphic and present variable staining features.

The aspect of colonies as well as the cellular morphology to Gram stain could be very variable because of nutritional state of the microorganisms.[4,19]

Biology. *A. defective*, like all the NVS, is defined by its requirement of pyridoxal or thiol group supplementation for growth. It also presents the particular feature of forming satellite colonies, usually nonhemolytic or alpha-hemolytic, and a variety of gram-positive and gram-negative bacteria (staphylococci, streptococci, and family Enterobacteriaceae strains), as well as yeasts can support their growth.[1]

A. defectiva is able to grow in most blood culture media (with the notable exception of unsupplemented tryptic soy broth) because the pyridoxal present in human blood is sufficient to permit growth.[4]

For subculture, however, solid media must be supplemented with 0.001% pyridoxal or 0.01% L-cysteine. Alternatively, the subculture plate can be cross-streaked with *Staphylococcus aureus* to provide these factors and support the growth of satellite colonies.

Some studies suggest that blood cultures suspected of harboring NVS should be subcultured within 48 h because viability of the organism may decline with continued incubation.[19]

The original efforts to identify the factors required for growth of NVS focused on sulfhydryl-containing compounds. A lot of authors studied which supplement should have been the best. In particular, Frenkel and Hirsch reported that cysteine, thioglycolate, reduced glutathione, and thiomalic acid were able to help the growth of NVS.[3]

Other studies demonstrated that L-cysteine was a useful supplement for NVS growth. Cayeux et al. used brain heart infusion as a basal medium and obtained a good growth by either L-cysteine or glutathione in association with either thioglycolate or dithiothreitol.[20]

A number of authors showed that vitamin B6 analogs were able to help the growth of NVS. George found pyridoxine (vitamin B6) supplementation allowed for growth of NVS on horse blood agar.[21]

Carey and Robert found that pyridoxal and pyridoxamine (active forms of vitamin B6) were effective, but pyridoxine was inhibitory for an NVS strain they studied.[1]

These apparent discrepancies were resolved because the studies of Shiller and Roberts suggested that pyridoxine is not assimilated by NVS and can inhibit their growth in media supplemented with other forms of vitamin B6.[1,22]

Regarding biochemical and phenotypical classification *Abiotophia* shares the following characteristics: optochin resistance, vancomycin susceptibility, leucine aminopeptidase (LAP), L-pyrrolidonyl-beta-naphthylamide (PYR) and arylamidase production, negative reaction for gas production in De Man, Rogosa, and Share (MRS) broth, grows in 6.5% NaCl or at 10°C and 45°C, and no reaction on bile-esculin medium (Table 19.1).

The genera *Abiotrophia* could be differentiated from *Granulicatella* by their patterns of carbohydrate fermentation and production of B-galactosidase. However, not all isolates of NVS can be easily classified at species level by biochemical and phenotypical tests. Only the application of molecular biology allows accurate species typing.

19.1.2 Clinical Features

A. defectiva is a part of a normal flora of the oral cavity and the urogenital and intestinal tracts.

Like other NVS normally resident in the oral cavity, *Abiotrophia* has been identified as agent of endocarditis involving both native and prosthetic valves.

In addition to their involvement with endocarditis, *Abiotrophia* has been implicated in a variety of other infections. In particular, before the final typing of *Abiotrophia defective*, all the clinical cases were reported as caused by NVS.[10–12,23–36]

TABLE 19.1
Identification of *Abiotrophia* and *Granulicatella* Species[a]

Species	Arg[b]	Hip	Pul	Suc	Tag	Tre	α-Gal	β-Glu	β-Gal	Origin
A. defectiva	–	–	+	+	v	+	+	–	+	human
G. adiacens	–	–	–	+	+	–	–	v	–	human
G. paraadiacens	–	–	–	+	–	–	–	–	–	human
G. elegans	+	v	–	+	–	–	–	–	–	human

Source: Adapted from: http://www.cdc.gov/ncidod/biotech/strep/strep-doc/section2.htm#table7

[a] All strains are positive for PYR and LAP and sensitive to vancomycin; negative reactions for gas production in MRS broth, growth in 6.5% NaCl or at 10 and 45°C. No reaction on bile-esculin medium. All strains require pyridoxal for growth and satillite around staphylococcus culture on blood agar plates.

[b] *Abbreviations*: Arg, deamination of arginine; Hip, hyrolysis of hippurate; Pul, Suc, Tag, Tre, acid production from pullulan, sucrose, tagatose, & trehalose broth respectively; α-gal, production of α-galactosidase; β-glu, production of β-glucouronidase; β-gal, production of -β-galactosidase + = 85% or more of the strains positive, – = 15% or less than the strains positive, v = variable reactions (16 to 84% positive).

Reports documented the isolation of NVS from the blood of patients with postpartum or postabortal sepsis, cirrhosis, and pancreatic abscess.[37]

Isolation has also been made from patients with otitis media, otitis externa, wound infection, and vaginal discharge. NVS have been implicated in conjunctivitis and infectious crystalline keratopathy.[3,21]

In the recent literature, it is possible to find that *A. defectiva* has also been isolated in cases of ophthalmic infections (keratitis and postoperative endophthalmitis); a brain abscess following neurosurgery; peritonitis; osteomyelitis (multiple discitis, sacroiliitis); arthritis; sepsis in immunocompromised patients and iatrogenic meningitis following myelography; and spinal anesthesia.[9,13,38–47]

Endocarditis. Because *Abiotrophia* spp. are fastidious organisms, it is likely that most cases are misdiagnosed as culture-negative endocarditis; therefore, their role in endocarditis may be underestimated.

Christensen et al. (2001) reported that *G. adiacens* and *A. defectiva* accounted for 3% of bacteremia due to α-hemolytic streptococcus-related species other than *S. pneumoniae*.[48,49]

Roberts underlined that NVS are the cause of 5%–6% endocarditis, and these strains could justify a culture-negative result.[50]

Stein and Nelson showed that the mortality rate of NVS endocarditis was higher than that reported for endocarditis caused by other streptococci or enterococci. The NVS endocarditis was also more difficult to treat, and the authors proposed that this could be due to the slow growth of NVS.[51]

Probably, the prolonged L-phase that *Abiotrophia* seems to be able to live that has been reported by some authors, can explain this difficulties encountered in treatment as a mechanism of noninherited resistance.

Other Site Infections. Due to the *Abiotrophia* growth characteristics, its role in different infection sites may be underestimated, as for endocarditis. Only recently, since the development of enriched agar and broth media and improvements in molecular biology, it has been possible to detect some clinical cases in a more precise manner.

In a recent report, corneal scrapings of four patients with severe infectious keratitis were analyzed by culture and PCR of 16S ribosomal RNA (rRNA), followed by direct sequencing of the resulting amplicon. The medical history of the four patients included laser-assisted in situ keratomileusis, penetrating keratoplasty, and trauma. Only PCR was able to identify a possible pathogen in all four cases, while bacterial cultures were either negative or did not correspond to the clinical picture. The authors concluded that analysis of corneal scraping by 16S rRNA PCR should be considered as a supplement to standard microbiological procedures. However, the results of this relatively new method have to be interpreted carefully.[57]

Interestingly, *A. defectiva* and *Gnanulicatella* were identified by molecular methods from the same maxillary sinus. Recent results from a multicenter study corroborate that *Abiotrophia* and *Granulicatella* species are associated with chronic maxillary sinusitis and may contribute to antibiotic treatment failure.[53]

Other studies confirmed that molecular techniques can be useful to pinpoint *Abiotrophia* as a cause of brain abscess and meningitis.[54,55]

19.1.3 Diagnosis

The diagnosis is not easy because *A. defectiva* is a fastidious microorganism, and it is likely that most cases are misdiagnosed as culture negative. As discussed above, the diagnosis can be done by conventional methods and/or by molecular biology.

A. defectiva should also be suspected when Gram stains of biological specimens reveal microbial cells but are culture negative.

The following critical points are worth emphasizing: (i) microbiologists should remember that *A. defectiva* cells may exhibit variable morphology and staining characteristics; (ii) supplementation of laboratory media is desirable and often necessary to encourage the growth of these organisms; (iii) once a suspected *A. defectiva* is cultured, its identity should be confirmed by establishing its requirement for pyridoxal or cysteine; (iv) final identification is possible using biochemical tests but molecular technique should be preferred because it is more accurate; and (v) molecular technique is also useful to identify the presence of *A. defectiva* in sterile body fluids.

FIGURE 19.1 Colonies of *A. defectiva* on chocolate agar with 5% horse blood after 18–24 hours.

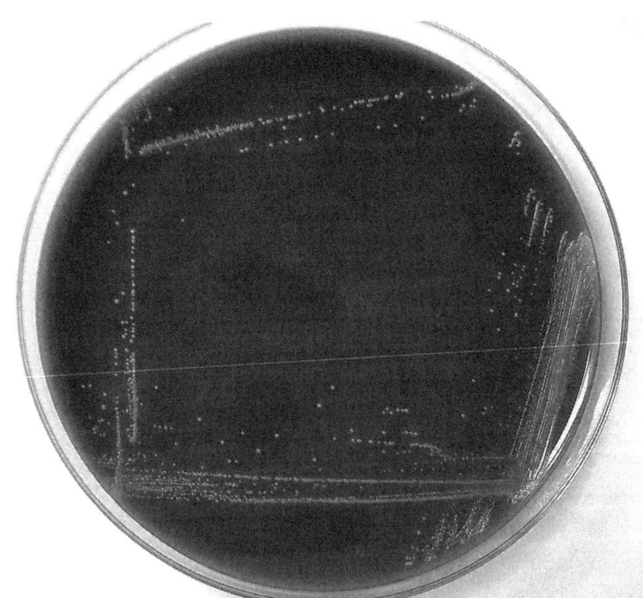

FIGURE 19.2 Colonies of *A. defectiva* on chocolate agar with 5% horse blood after 36–48 hours.

Conventional Techniques. *A. defectiva* is not able to grow on the majority of agar or broth media normally used in routine microbiology laboratory practice. In fact, *A. defectiva* can be isolated only by media that are supplemented with vitamin B6; pyridoxal hydrochloride (10–100 mg/L); or L-cysteine (100 mg/L) under an aerobic or anaerobic atmosphere.

In particular, it does not grow on sheep blood agar or chocolate agar with trypticase soy or Mueller-Hinton or Todd-Hewitt, but it can be isolated by brain heart infusion, thioglycolate broth, Schaedler, Columbia and Brucella, or chocolate agar with 5% horse blood.[56]

The biochemical tests useful for the identification of *A. defectiva* are reported in Table 19.1. Tiny alpha-hemolytic or nonhemolytic colonies may appear after 18 h on the adequate solid media (Figures 19.1–19.3).

Alternatively a *Staphylococcus aureus* strain is proposed by the Centers for Disease Control and Prevention (CDC) as helper to induce satellite growth (Figure 19.4, Table 19.2).

Following improvement of the culture media by including cysteine, *A. defectiva* can be detected in routine blood cultures in 2 or 3 days.

The microscopic examination reveals both morphological pleiomorphism and variable Gram staining (Figures 19.5 and 19.6).

Conventional current methods of identification rely on enzymatic and physiological properties, using commercially available identification systems such as Rapid ID 32 STREPT (bioMerieux) or automatic systems. In literature, these methods are not so accurate and often do not discriminate the species level. It is underlined that better results can be obtained when a dense inoculum id used.[57]

Molecular Techniques. The development of molecular techniques was very helpful for the diagnosis of *Abiotrophia* infections.[2]

In particular, it is possible to identify *Abiotrophia* from its colonies or by looking for its DNA in clinical samples with negative cultures.

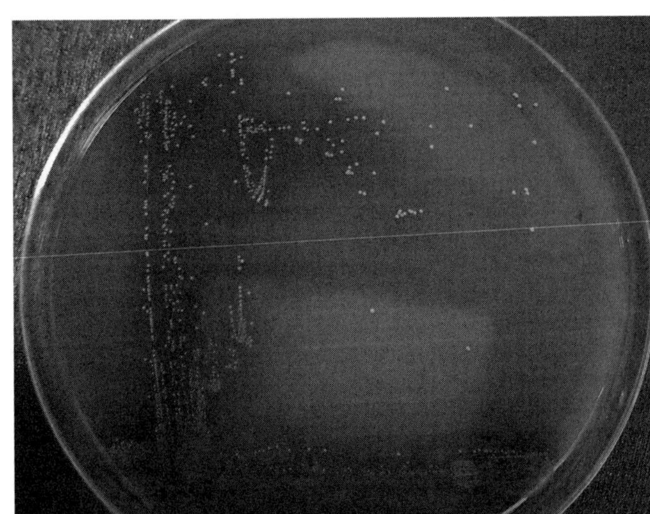

FIGURE 19.3 Colonies of *A. defectiva* on columbia agar with 5% horse blood after 66–72 hours.

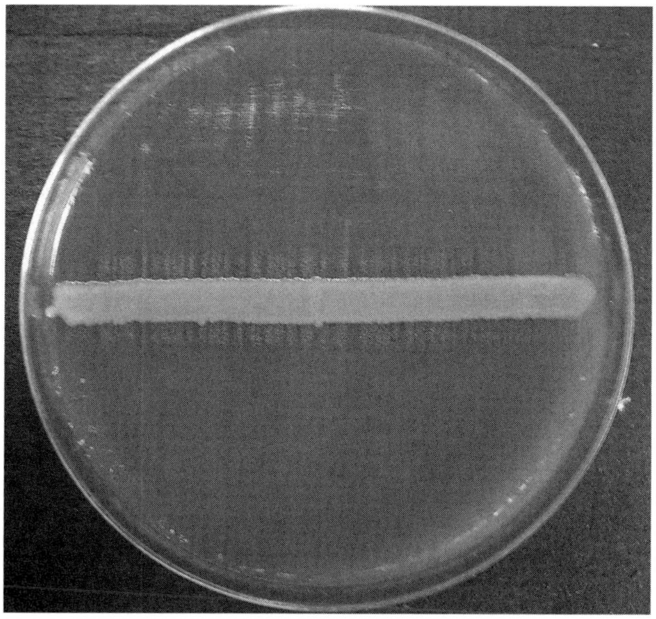

FIGURE 19.4 Satellite growth on blood agar after 18–24 hours.

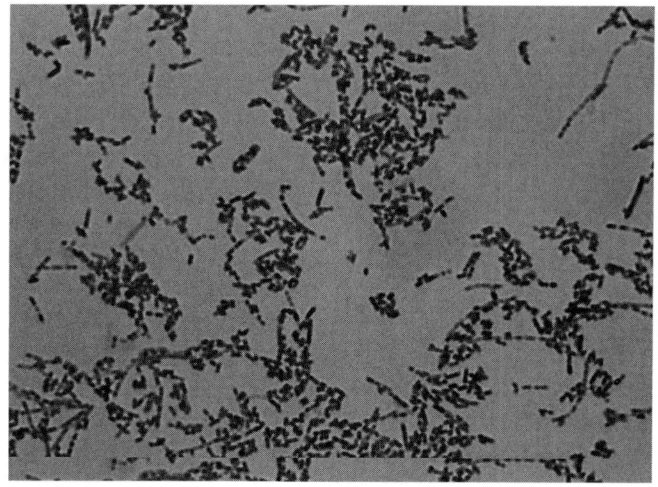

FIGURE 19.5 Gram stain from colonies of *A. defectiva* on chocolate agar with 5% horse blood after 18–24 hours.

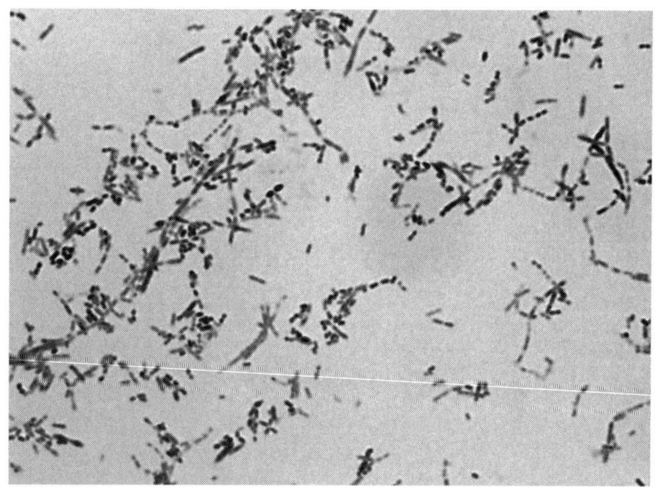

FIGURE 19.6 Gram stain from colonies of *A. defectiva* on chocolate agar with 5% horse blood after 66–72 hours.

TABLE 19.2
Satellite Test (SAT)

I. **Principle**

Abiotrophia and *Granulicatella* are usually fastidious and will grow only on supplemented media or enriched chocolate agar. The purpose of the SAT test is to determine if the unknown streptococcal strain is a nutritionally variant streptococci (NVS). These strains require pyridoxal for growth. Growth of *Staphylococcus aureus* on a TSA sheep blood agar plate releases these factors allowing NVS to grow. An alternative to performing the SAT test is to test for the requirement of vitamin B6 (pyridoxal).

II. **Inoculum**

The cultures are typically received on chocolate agar slants.

III. **Reagents and Materials**

1. *Staphylococcus aureus* ATCC-25923
2. TSA-sheep blood agar plate

IV. **Procedure**

1. Streak the submitted culture over the entire surface of a TSA-sheep blood agar plate with an inoculating loop.
2. A single streak of *Staphylococcus aureus* ATCC-25923 is then made across the middle of the agar plate.
3. Incubate the plate in a candle extinction jar or CO_2 incubator for 24 h or more.

V. **Reading and Interpretation**

If the strain is a NVS growth will appear only adjacent to the staphylococcus streak, about 2 to 5 mm wide. Streptococci that are not NVS will grow over the entire surface of the agar plate.

Source: Adapted from: http://www.cdc.gov/ncidod/biotech/strep/strep-doc/section1.htm

In the literature of 1990s, it is possible to confirm the identity of *Abiotrophia* by 16S rRNA gene PCR and restriction fragment length polymorphism analysis. This method was able to discriminate only *A. defectiva* from *Granulicatella* (formerly *Abiotrophia*) *adjacens*.[58]

Nowadays, the most commonly used technique is nucleotide-sequencing analysis of 16S rRNA and 740-bp *rpoB* gene fragment. In particular, 16S rRNA analysis is performed by two different approaches: a commercial method, MicroSeq 500–1500 (Applied Biosystems) and homemade tests using different primers[59–61] (Table 19.3).

The Microseq 500 16S rRNA–based bacterial identification system is useful for identification of most clinically important bacterial strains with ambiguous biochemical profiles.

The PCR amplification of a 740-bp *rpoB* gene fragment followed by sequencing is a suitable molecular approach for the identification of *Streptococcus*, *Enterococcus*, *Gemella*, *Abiotrophia*, and *Granulicatella* isolates at the species level.[62]

TABLE 19.3
Primers and Probes for PCR Detection of *A. defectiva*

Primers for PCR (5′–3′)	Primers for Sequencing (5′–3′)	Target Gene	Reference
LPW200-F: GAGTTGCGAACGGGTGAG **LPW205-R**: CTTGTTACGACTTCACCC	**LPW200-F**: GAGTTGCGAACGGGTGAG **LPW205-R**: CTTGTTACGACTTCACCC **LPW99-F**: TTATTGGGCGTAAAGCGA **LPW273-R**: TTGCGGGACTTAACCCAAC	16S rRNA	PC-Y Woo et al.[60]
31-F: GCCTTAGGACCTGGTGGTTT **830-R**: GTTGTAACCTTCCAWGTCAT	**2333-F**: AARYTIGGMCCTGAAGAAAT **3073-R**: TGIARTTTRTCATCAACCATGTG	*rpoB*	M. Drancourt et al.[62]
F: AGAGTTTGATCCTGGCTCAG **R**: GGCTACCTTGTTACGACTT	Analysis by RFLP	16S rRNA	Ohara-Nemoto et al.[58]
F: TATTACCGCGGCTGCTGGCA **R**: TCAGATTGAACGCTGGCGGC	**F**: TATTACCGCGGCTGCTGGCA **R**: TCAGATTGAACGCTGGCGGC	16S rRNA	Mikkelsen et al.[44]
Bact-338-F: ACTCCTACGGGAGGCAGC **Bact-1525-R**: AAGGAGGTGATCCAGCC	**M13-F**: GTAAAACGACGGCCAGT **M13-R**: GGAAACAGCTATGACCATG	16S rRNA	Paju et al.[53]
13B-F: GTGAATACGTTCCCGGGCCT **6-R**: GGGTTTYCCCCRTTCRGAAAT	**13B-F**: GTGAATACGTTCCCGGGCCT **6-R**: GGGTTTYCCCCRTTCRGAAAT	16S rRNA	Tung et al.[56]

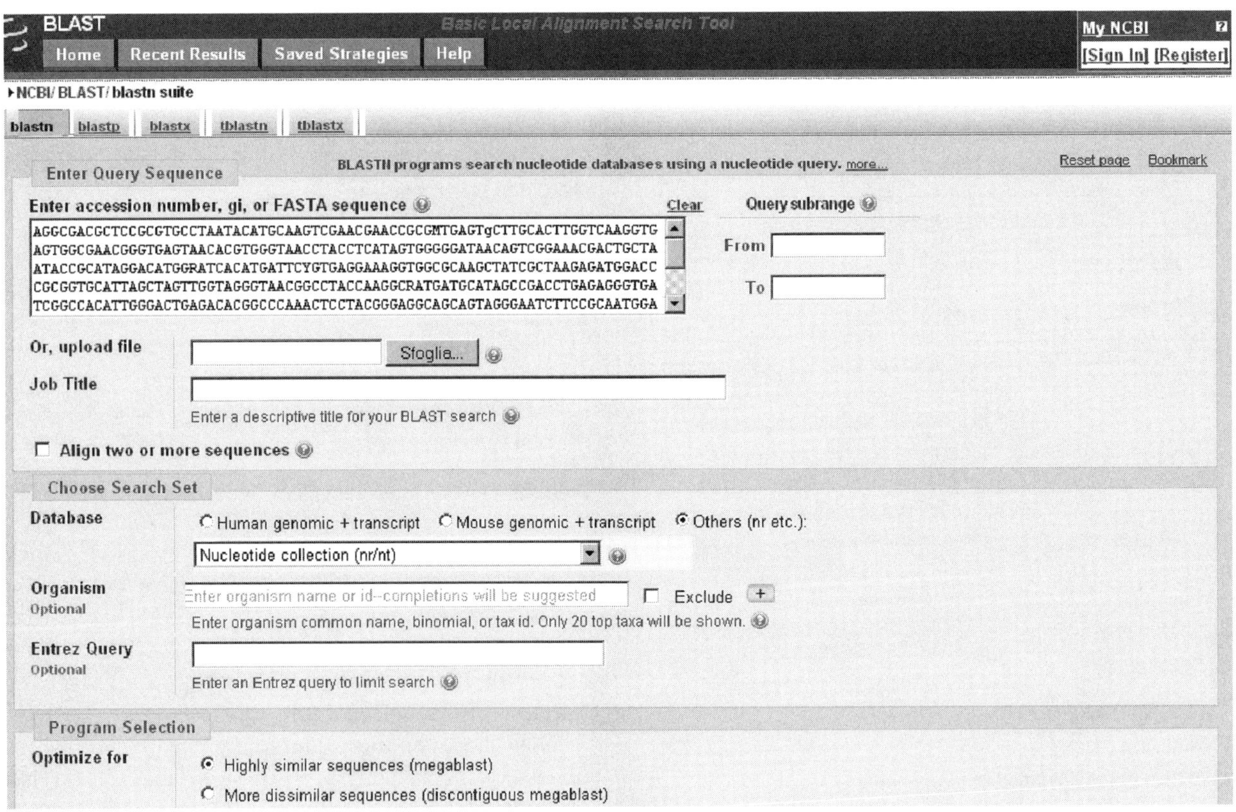

FIGURE 19.7 Public bank web page where is possible to enter the sequence.

A critical step in these techniques is the DNA extraction, particularly from sterile body fluids (blood, cerebrospinal fluid [CSF], and others). The sequence analysis can be performed by a public bank or using commercial software programs: the former has more data but is not always well controlled, while the latter present a low number of gene sequences of particular bacteria in the database (Figures 19.7 and 19.8).

19.1.4 Antibiotic Susceptibility

Because of their growth requirements, uncommon isolation from human clinical specimens, and lack of standardized testing methodology and interpretation, there are limited antimicrobial susceptibility data for the fastidious microorganisms available.[63]

It often happens that the clinician asks for a susceptibility test, and the clinical microbiologist has trouble producing it.

```
>gb|AY879306.1|  Abiotrophia defectiva strain 99383068 16S ribosomal RNA gene,
partial sequence
Length=1517

 Score =  902 bits (488),  Expect = 0.0
 Identities = 499/506 (98%), Gaps = 1/506 (0%)
 Strand=Plus/Plus

Query  13   GCGTGCCTAATACATGCAAGTCGAACGAACCGCG-MTGAGTGCTTGCACTTGGTCAAGGT  71
            ||||||||||||||||||||||||||||||||||  ||||||||||||||||||||||||
Sbjct  34   GCGTGCCTAATACATGCAAGTCGAACGAACCGCGACTGAGTGCTTGCACTTGGTCAAGGT  93

Query  72   GAGTGGCGAACGGGTGAGTAACACGTGGGTAACCTACCTCATAGTGGGGGATAACAGTCG  131
            ||||||||||||||||||||||||||||||||||||||||||||||||||||||||||||
Sbjct  94   GAGTGGCGAACGGGTGAGTAACACGTGGGTAACCTACCTCATAGTGGGGGATAACAGTCG  153

Query  132  GAAACGACTGCTAATACCGCATAGGACATGGRATCACATGATTCYGTGAGGAAAGGTGGC  191
            ||||||||||||||||||||||||||||||| |||||||||||| |||||||||||||||
Sbjct  154  GAAACGACTGCTAATACCGCATAGGACATGGAATCACATGATTCCGTGAGGAAAGGTGGC  213

Query  192  GCAAGCTATCGCTAAGAGATGGACCCGCGGTGCATTAGCTAGTTGGTAGGGTAACGGCCT  251
            ||||||||||||||||||||||||||||||||||||||||||||||||||||||||||||
Sbjct  214  GCAAGCTATCGCTAAGAGATGGACCCGCGGTGCATTAGCTAGTTGGTAGGGTAACGGCCT  273

Query  252  ACCAAGGCRATGATGCATAGCCGACCTGAGAGGGTGATCGGCCACATTGGGACTGAGACA  311
            |||||||| |||||||||||||||||||||||||||||||||||||||||||||||||||
Sbjct  274  ACCAAGGCAATGATGCATAGCCGACCTGAGAGGGTGATCGGCCACATTGGGACTGAGACA  333

Query  312  CGGCCCAAACTCCTACGGGAGGCAGCAGTAGGGAATCTTCCGCAATGGACGCAAGTCTGA  371
            ||||||||||||||||||||||||||||||||||||||||||||||||||||||||||||
Sbjct  334  CGGCCCAAACTCCTACGGGAGGCAGCAGTAGGGAATCTTCCGCAATGGACGCAAGTCTGA  393

Query  372  CGGAGCAACGCCGCGTGAGTGAAGAAGGTCTTCGGATCGTAAAGCTCTGTTGTTAGAGAA  431
            ||||||||||||||||||||||||||||||||||||||||||||||||||||||||||||
Sbjct  394  CGGAGCAACGCCGCGTGAGTGAAGAAGGTCTTCGGATCGTAAAGCTCTGTTGTTAGAGAA  453

Query  432  GAACAGCGCATAGAGTAACTGTTATGCGTGTGACGGTATCTAACCAGAAAGCCACGGCTA  491
            ||||||||||||||||||||| ||||||||||||||||||||||||||||||||||||||
Sbjct  454  GAACAGCGCATAGAGTAACTGATATGCGTGTGACGGTATCTAACCAGAAAGCCACGGCTA  513

Query  492  ACTACGTGCCAGCAGCCGCGTTAATA  517
            |||||||||||||||||||| |||||
Sbjct  514  ACTACGTGCCAGCAGCCGCGGTAATA  539
```

FIGURE 19.8 Sequence of *A. defectiva*.

Fortunately, in 2006 the Clinical and Laboratory Standards Institute (CLSI) published new guidelines on antimicrobial susceptibility testing of infrequently encountered or fastidious bacteria in the document M45-A "Methods for Antimicrobial Dilution and Disk Susceptibility Testing of Infrequently Isolated or Fastidious Bacteria. Approved Guidelines."[64]

The testing conditions reported by CLSI document are as follows:

Medium: cation-adjusted Mueller-Hinton broth with lysed horse blood (19.5%–5% v/v) and 0.001% (i.e., 1 mcg/mL) pyridoxal hydrochloride.
Inoculum: direct colony suspension, equivalent to a 0.5 Mc Farland standard.
Incubation: 35°C, ambient air, 20–24 h.

The minimal quality-control recommendation is to utilize *Streptococcus pneumoniae* ATCC 49619.

The agents to consider for primary testing are penicillin, vancomycin, and cefotaxime or ceftriaxone.

It is also possible to utilize the Etest method (AB BIODISK) for the chemo-antibiotic susceptibility test, although in the United States it is for investigational use only.

The testing conditions reported by Etest application sheet EAS 008–2007 are as follows:

Agar medium: chocolate Mueller-Hinton + 0.001% pyridoxal HCl + 0.01% cysteine or Isosensitest + 5% human blood + 0.001% pyridoxal Hcl + 0.01% cysteine
Inoculum: suspension in broth to 1 McFarland
Incubation: 35°C/5% CO_2/20–24 h.

The quality-control recommendation is to utilize *Streptococcus pneumoniae* ATCC 49619.

A susceptibility test is usually not necessary for isolates from respiratory sources or wounds, while *A. defectiva* isolated from normally sterile sources may be justified in immunodeficient patients.

A. defectiva may present decreased susceptibility to penicillin, resulting in greater difficulty in the treatment of patients with endocarditis. In an isolate from an immunosuppressed patient was reported fluoroquinolone resistance.[64]

For isolates of *A. defective*, a synergistic effect has been demonstrated between beta-lactam agents and aminoglycosides.[65]

Tuohy et al. observed the susceptibilities of all strains to clindamycin, rifampicin, levofloxacin, ofloxacin, and quinupristin-dalfopristin, and the percentage of isolates of *A. defectiva* examined that were susceptible to other agents tested was as follows: penicillin, 8%; amoxicillin, 92%; ceftriaxone, 83%; and meropenem, 100%.[66]

Zheng et al. referred high rates of beta lactam and macrolide resistance in a pediatric population of *A. defectiva* isolates.[7]

Because of the use of macrolides as alternatives in prophylaxis for infective endocarditis in patients allergic to penicillin, the observation of increasing macrolide resistance may lead to failure in an important proportion of these patients undergoing dental or other invasive procedures.

19.2 METHODS

The conventional methods described in Section 19.1.3 are quite similar to those usually applied in a traditional microbiology laboratory.

Also, the molecular biology methods applied for the diagnosis of *Abiotrophia* are the same ones usually proposed for the identification of a strain from a positive culture or for the research of the DNA of a strain in a broth culture or in a usually sterile body fluid. Here, we will describe the principal steps of the common molecular biology procedures.[55]

19.2.1 Sample Preparation

Adequate Samples. (i) Colonies from a culture plate; (ii) positive broth or blood culture; (iii) sterile body fluid such as: CSF, samples from abscess, peritoneal, pericardial, and joint fluid; and (iv) biopsy such as corneal scrapings or native or prosthetic valves.

DNA Extraction. Usually it is sufficient to select one colony from a culture plate or 1 mL of culture broth or 500 μL–1 mL of sterile body fluid. It is important to underline that in some cases the sterile body fluid extract should be diluted assuming the presence of inhibitors and in some others it is better to concentrate due to the risk of low DNA load.

The extraction from colonies is easy and some authors propose a simple lysis procedure with suspension in distilled water heated at 100°C and centrifuged at 20.000 × *g* for 10 min. DNA preparation from body fluid, biopsy, or blood may be more time consuming.

There are different types of extractions, some manual and some by automatic systems. There are a large number of companies in the market producing DNA-extraction kits and systems that perform well. The most commonly used manual methods are PrepMan Ultra Sample Preparation Reagent Protocol (Applied Biosystems) and QIAamp Mini Kit (Quiagen), and for both the procedure is very simple. An increasing number of automatic systems are available in the market due to the interest of the companies in this new diagnostic field.

19.2.2 Detection Procedures

The amplification can be performed using both homemade and commercial methods. There are a lot homemade protocols. Among the commercial techniques, MicroSeq 500/1500 16S rRNA Bacterial Identification PCR kit (Applied Biosystem) is widely used.

DNA Extraction. QIAamp Mini Kit (QIAGEN) can be used for DNA extraction for all samples on duplicate. The total DNA must be eluted in 50 μL of AE buffer by modifying the procedure described by manufacturer (QIAGEN) and stored at −80°C until measurement in the PCR. Because some extracted DNAs do not amplify, it can be useful to dilute 1:3 supposing the presence of inhibitors.

PCR Amplification and Sequencing. Using MicroSeq500 16S rRNA Bacterial Identification PCR kit (Applied Biosystems), 15 μL must be examined by amplification 500 bp to the 5′end of 16S rRNA gene. Amplification of the bacterial 16S rRNA is carried out in a GeneAmp 9600 thermal cycler (Applied Biosystems). Amplicon detection, although not mandatory, could be verified by 2% agarose gel electrophoresis previous purification with YM-100 Microconcentrators (Celera Diagnostics).

The sequencing reactions could be performed with MicroSeq500 16S rRNA Bacterial Identification Sequencing Kit. Cycle reactions are purified by the use of Centri-Sep Spin Columns (Applied Biosystems), and the cycle-sequencing products are analyzed on 3130 Genetic Analyzer (Applied Biosystems) according to the manufacturer's instructions.

Data Analysis. Raw data are analyzed with the Sequencing Analysis software v. 5.2 (Applied Biosystems). Sequences from both strands were aligned by using NCBI BLAST 2 Sequence (http://www.ncbi.nlm.nih.gov/blast/bl2seq/wblast2.cgi). The resulting consensus sequence, obtained from the double-stranded portion of the alignment, is aligned with sequences stored in GenBank (BLAST; http://www.ncbi.nlm.nih.gov/BLAST).

Criteria for Identification. For designation to species or genus level, we propose using the identification criteria formulated by Drancourt et al.: the species level is defined as a 16S rRNA sequence similarity of ≥99% with the prototype strain sequence in GenBank; the genus level is defined as a similarity of ≥97% with the prototype strain sequence in GenBank; a score lower than 97% should not be considered acceptable.

After amplification, there some methods available to identify the organism from which DNA is derived. Some authors proposed restriction fragment length polymorphism (RFLP) analysis, but the sequence analysis followed by comparison of the nucleotide sequence with a database is easy and widely used. There are public databases such as GeneBank or the EMBL, where the limit is the risk of sequence not well controlled.

19.3 CONCLUSIONS AND FUTURE PERSPECTIVES

The microbiological history of *A. defectiva* is fascinating and at the same time very instructive. From the general definition of NVS in 1960s, we are able today to reach the correct diagnosis.

Since their original description in 1961, the NVS have evolved from laboratory curiosities into streptococcal species in 1990s, and in 2000, into well-defined genera (e.g., *Abiotrophia*, *Granulicatella and Globicatella*, *Facklamia* and *Eremococcus*) and species (*A. defectiva*, *Granulicatella adiacens*, *Granulicatella para-adiacens*, *Granulicatella elegans*, and *Granulicatella balaenopterae*).

There are numerous reports of *Abiotrophia* as cause of endocarditis, and in recent years there has been an increasing number of cases related to different sites of infection: brain abscess, meningitidis, peritonitis, osteomyelitis, arthritis, otitis, and keratitis. It is now well recognized that attempts should be made to look for *Abiotrophia* in Gram stain–positive and culture-negative samples.

The traditional microbiology of *Abiotrophia* is well defined, and the correct solid or liquid media are available. The satellite test is now proposed and standardized by CDC. The 2006 CLSI document M45-A published the guidelines for the antimicrobial susceptibility testing and the agents to consider. The diffusion of these indications will improve the knowledge about the real susceptibility of *Abiotrophia* and the clinical activity of penicillin and macrolides.

The application of DNA amplification methods is useful for the correct classification of genera and species, and the sequence analysis of 16S rRNA allows identification of *Abiotrophia* as a cause of culture-negative endocarditis or other sites' infections. Molecular biology opens new fields of study: this is true above all for the so-called fastidious or infrequently isolated bacteria such as *Abiotrophia*.

At the moment, there are no studies on resistance genes by the application of molecular biology techniques in strains of *Abiotrophia*, but it is probable that we will be able to confirm their presence in the near future. Further studies will also provide insights into the mechanisms by which they produce disease or are able to live in the L-phase.

REFERENCES

1. Ruoff, K.L., Nutritionally variant streptococci, *Clin. Microbiol. Rev.*, 4, 184, 1991.
2. Boutoille, D. et al., *Abiotrophia* related endocarditis: Contribution of molecular biology, *Presse Med.*, 28, 2149, 1999.
3. Frenkel, A., and Hirsch, W., Spontaneous development of L forms of streptococci requiring secretion of other bacteria or sulphydryl compounds for normal growth, *Nature*, 191, 728, 1961.
4. Johson, C.C., and Tunkel, A.R., Viridans streptococci and groups C and G streptococci. In Mandell, G.L., Bennett, J.E., Dolin, R. (Editors), *Principles and practice of Infectious Diseases*, 4 ed., p. 1851, Churchill Livingstone, New York, 1995.
5. Brouqui, P., and Raoult, D., Endocarditis due to rare and fastidious bacteria, *Clin. Microbiol. Rev.*, 14, 177, 2001.
6. Collins, M.D., and Lawson, P.A., The genus *Abiotrophia* (Kawamura et al.) is not monophyletic: Proposal of *Granulicatella* gen. nov., *Granulicatella adiacens* comb. nov., *Granulicatella elegans* comb. nov. and *Granulicatella balaenopterae* comb. nov., *Int. J. Syst. Evol. Microbiol.*, 50, 365, 2000.
7. Zheng, X., Antimicrobial susceptibilities of invasive pediatric *Abiotrophia* and *Granulicatella* isolates, *J. Clin. Microbiol.*, 42, 4323, 2004.
8. Layland, J. et al., *Abiotrophia* endocarditis: The great masquerader, *Aust. N.Z. J. Psychiatry*, 42, 991, 2008.
9. O'Connor, K.M., Williams, P., and Pergam, S.A., An unusual case of knee pain: Pseudogout and *Abiotrophia defectiva* infection, *South. Med. J.*, 101, 961, 2008.
10. Kiernan, T.J., *Abiotrophia defectiva* endocarditis and associated hemophagocytic syndrome—a first case report and review of the literature, *Int. J. Infect. Dis.*, 12, 478, 2008.
11. Yerebakan, C., Aortic valve endocarditis due to *Abiotrophia defectiva*: A rare etiology, *Wien Med. Wochenschr.*, 158, 152, 2008.
12. Assche, A.F., and Stephens, D.P., Infective endocarditis with *Abiotrophia defectiva*: The first Australian experience, *Crit. Care Resusc.*, 10, 54, 2008.
13. Taylor, C.E., and Fang, M.A., Septic arthritis caused by *Abiotrophia defectiva*, *Arthritis Rheum.*, 55, 976, 2006.
14. Bouvet, A. et al., *Streptococcus defectivus* sp. nov. and *Streptococcus adjacens* sp. nov. Nutritionally variant streptococci from human clinical specimens, *Int. J. Syst. Bacteriol.*, 39, 290, 1989.
15. Kawamura, Y. et al., Transfer of *Streptococcus adjacens* and *Streptococcus defectivus* to *Abiotrophia* gen. nov. as *Abiotrophia adiacens* comb. nov. and *Abiotrophia defectiva* comb. nov., respectively, *Int. J. Syst. Bacteriol.*, 45, 798, 1995.
16. Lawson, P.A. et al., *Abiotrophia balaenopterae* sp. nov., isolated from the minke whale (*Balaenoptera acutorostrata*), *Int. J. Syst. Bacteriol.*, 49, 503, 1999.
17. Roggenkamp, A. et al., *Abiotrophia elegans* sp. nov., a possible pathogen in patients with culture-negative endocarditis, *J. Clin. Microbiol.*, 36, 100, 1998.
18. Kanamoto, T., Sato, S., and Inoue, M., Genetic heterogeneities and phenotypic characteristics of strains of the genus *Abiotrophia* and proposal of *Abiotrophia para-adiacens* sp. nov., *J. Clin. Microbiol.*, 8, 492, 2000.
19. Gross, K.C. et al., Evaluation of blood culture media for isolation of pyridoxal-dependent *Streptococcus mitis*, *J. Clin. Microbiol.*, 14, 266, 1981.
20. Cayeux, P. et al., Bacterial persistance in streptococcal endocarditis due thiol-requiring mutants, *J. Infect. Dis.*, 124, 247, 1971.
21. George, R.H., The isolation of symbiotic streptococci, *J. Med. Microbiol.*, 7, 77, 1974.
22. Schiller, N.L., and Roberts, R.B., Vitamin B6 requirements of nutritionally variant *Streptococcus mitior* (*mitis*), *J. Clin. Microbiol.*, 15, 740, 1982.
23. Lopardo, H., Mastroianni, A., and Casimir, L., Bacteremia due to *Abiotrophia defectiva* in a febrile neutropenic pediatric patient, *Rev. Argent. Microbiol.*, 39, 93, 2007.
24. Takayama, R. et al., A case of isolated tricuspid valve infective endocarditis caused by *Abiotrophia defectiva*, *Int. J. Cardiol.*, 118, 3, 2007.
25. Senn, L., Entenza, J.M., and Prod'hom, G., Adherence of *Abiotrophia defectiva* and *Granulicatella* species to fibronectin: is there a link with endovascular infections? *FEMS Immunol. Med. Microbiol.*, 48, 215, 2006.
26. Yemisen, M. et al., Abiotrophia defectiva: A rare cause of infective endocarditis, *Scand. J. Infect. Dis.*, 38, 939, 2006.
27. Senn, L. et al., Bloodstream and endovascular infections due to *Abiotrophia defectiva* and *Granulicatella* species, *BMC Infect. Dis.*, 6, 9, 2006.

28. Sharaf, M.A., and Shaikh, N., *Abiotrophia* endocarditis: Case report and review of the literature, *Can. J. Cardiol.*, 21, 1309, 2005.
29. Lainscak, M. et al., Infective endocarditis due to *Abiotrophia defectiva*: A report of two cases, *J. Heart Valve Dis.*, 14, 33, 2005.
30. Hashimoto, T. et al., A woman with infectious endocarditis caused by *Abiotrophia defectiva*, *Intern. Med.*, 43, 1000, 2004.
31. Raff, G.W. et al., Aortitis in a child with *Abiotrophia defectiva* endocarditis, *Pediatr. Infect. Dis. J.*, 23, 574, 2004.
32. Chang, H.H. et al., Endocarditis caused by *Abiotrophia defectiva* in children, *Pediatr. Infect. Dis. J.*, 21, 697, 2002.
33. Ray, M. et al., Infective endocarditis in a child due to *Abiotrophia defectivus*, *Indian Pediatr.*, 39, 388, 2002.
34. Wijetunga, M., Bello, E., and Schatz, I., *Abiotrophia* endocarditis: A case report and review of literature, *Hawaii Med. J.*, 61, 10, 2002.
35. Leonard, M.K., Pox, C.P., and Stephens, D.S., *Abiotrophia* species bacteremia and a mycotic aneurysm in an intravenous drug abuser, *N. Engl. J. Med.*, 344, 233, 2001.
36. Okada, Y. et al., Endocardiac infectivity and binding to extracellular matrix proteins of oral *Abiotrophia* species, *FEMS Immunol. Med. Microbiol.*, 27, 257, 2000.
37. McCarthy, L.R., and Bottone, E.J., Bacteremia and endocarditis caused by satelliting streptococci, *Am. J. Clin. Pathol.*, 61, 585, 1974.
38. Arslan, U. et al., First case of peritonitis due to *Abiotrophia defectiva*, *Perit. Dial. Int.*, 26, 725, 2006.
39. Esteban, J. et al., Postoperative endophthalmitis due to *Abiotrophia defectiva*, *Enferm. Infecc. Microbiol. Clin.*, 23, 455, 2005.
40. Wilhelm, N. et al., First case of multiple discitis and sacroiliitis due to *Abiotrophia defectiva*, *Eur. J. Clin. Microbiol. Infect. Dis.*, 24, 76, 2005.
41. Zenone, T., and Durand, D.V., Brain abscesses caused by *Abiotrophia defectiva*: Complication of immunosuppressive therapy in a patient with connective-tissue disease, *Scand. J. Infect. Dis.*, 36, 497, 2004.
42. Cerceo, E. et al., Central nervous system infections due to *Abiotrophia* and *Granulicatella* species: an emerging challenge? *Diagn. Microbiol. Infect. Dis.*, 48, 161, 2004.
43. Ince, A. et al., Total knee arthroplasty infection due to *Abiotrophia defectiva*, *J. Med. Microbiol.*, 51, 899, 2002.
44. Mikkelsen, L., Theilade, E., and Poulsen, K., *Abiotrophia* species in early dental plaque, *Oral. Microbiol. Immunol.*, 15, 263, 2000.
45. Michelow, I.C. et al., *Abiotrophia* spp. brain abscess in a child with Down's syndrome, *Pediatr. Infect. Dis. J.*, 19, 760, 2000.
46. Namdari, H. et al., *Abiotrophia* species as a cause of endophthalmitis following cataract extraction, *J. Clin. Microbiol.*, 37, 1564, 1999.
47. Schlegel, L. et al., Iatrogenic meningitis due to *Abiotrophia defectiva* after myelography, *Clin. Infect. Dis.*, 28, 155, 1999.
48. Christensen, J.J., and Facklam, R.R., *Granulicatella* and *Abiotrophia* species from human clinical specimens, *J. Clin. Microbiol.*, 39, 3520, 2001.
49. Christensen, J.J., Gruhn, N., and Facklam, R.R., Endocarditis caused by *Abiotrophia* species, *Scand. J. Infect. Dis.*, 31, 210, 1999.
50. Roberts, R.B. et al., Viridans streptococcal endocarditis: The role of various species, including pyridoxal-dependent streptococci, *Rev. Infect. Dis.*, 1, 955, 1979.
51. Stein, D.S., and Nelson K.E., Endocarditis due to nutritionally deficient streptococci: Therapeutic dilemma, *Rev. infect. Dis.*, 9, 908, 1987.
52. Thiemo, R. et al., 16S rDNA PCR analysis of infectious keratitis: A case series, *Acta Ophthalmol. Scand.*, 82, 463, 2004.
53. Paju, S. et al., Molecular analysis of bacterial flora associated with chronically inflamed maxillary sinuses, *J. Med. Microbiol.*, 52, 591, 2003.
54. Biermann, C. et al., Isolation of Abiotrophia adiacens from a brain abscess which developed in a patient after neurosurgery, *J. Clin. Microbiol.*, 37, 769, 1999.
55. Arosio, M. et al., Evaluation of the MicroSeq 500 16 S rDNA-based gene sequencing for the diagnosis of culture-negative bacterial meningitis, *New Microbiol.*, 31, 343, 2008.
56. Koneman, E. et al., *Color Atlas and Textbook of Diagnostic Microbiology*, 4th ed., p. 389, J.B. Lippincott Company, Philadelphia, 1992.
57. Ruoff, K.L., Aerococcus, Abiotrophia, and other aerobic catalase-negative, gram-positive cocci. In Murray P.R. et al. (Editors), *Manual of Clinical Microbiology*, 9th ed., vol. 1, p. 443, ASM press, Washington DC, 2007.
58. Ohara-Nemoto, Y. et al., Identification of *Abiotrophia adiacens* and *Abiotrophia defectiva* by 16S rRNA gene PCR and restriction fragment length polymorphism analysis, *J. Clin. Microbiol.*, 35, 2458, 1997.
59. Tung, S.K. et al., Identification of species of *Abiotrophia*, *Enterococcus*, *Granulicatella* and *Streptococcus* by sequence analysis of the ribosomal 16S-23S intergenic spacer region, *J. Med. Microbiol.*, 56, 504, 2007.
60. Woo, P.C. et al., *Granulicatella adiacens* and *Abiotrophia defectiva* bacteraemia characterized by 16S rRNA gene sequencing, *J. Med. Microbiol.*, 52, 137, 2003.
61. Roggenkamp, A. et al., PCR for detection and identification of *Abiotrophia* spp., *J. Clin. Microbiol.*, 36, 2844, 1998.
62. Drancourt, M. et al., rpoB gene sequence-based identification of aerobic Gram-positive cocci of the genera *Streptococcus*, *Enterococcus*, *Gemella*, *Abiotrophia*, and *Granulicatella*, *J. Clin. Microbiol.*, 42, 497, 2004.
63. Jorgensen, J.H., and Hindler, J.F., New consensus guidelines from the clinical and laboratory standards institute for antimicrobial susceptibility testing of infrequently isolated or fastidious bacteria, *Clin. Infect. Dis.*, 44, 280, 2007.
64. CLSI, Methods for antimicrobial dilution and disk susceptibility testing of infrequently isolated or fastidious bacteria; approved guideline. M45-A, 2006.
65. Murray, C.K. et al., *Abiotrophia* bacteremia in a patient with neutropenic fever and antimicrobial susceptibility testing of *Abiotrophia* isolates, *Clin. Infect. Dis.*, 32, E140, 2001.
66. Tuohy, M.J., Procop, G.W., and Washington, J.A., Antimicrobial susceptibility of *Abiotrophia adiacens* and *Abiotrophia defectiva*, *Diagn. Microbiol. Infect. Dis.*, 38, 189, 2000.

20 *Aerococcus*

Debarati Paul, D. Madhusudan Reddy, and Dipalok Mukherjee

CONTENTS

20.1 INTRODUCTION

20.1.1 Classification, Morphology, and Biology

Aerococcus is a genus of aerobic gram-positive cocci that resemble enterococci but do not form chains. They are frequently isolated as airborne saprophytes in hospitals and as a pathogen of lobsters, cause greening in blood agar, and grow in the presence of 40% bile. In humans, they are found in endocarditis and in urinary tract infections and also some other health-related problems such as spondylodiscitis, lymphadenitis, and so forth.[1–3]

The genus *Aerococcus* was initially described by Williams.[4] *Aerococcus viridans* has been isolated from a broad range of habitats such as air, vegetation, dust, soil, and meat-curing brines, and has also been associated, although rarely, with human infections. Despite its presence in such diverse sources, the taxonomic kinship of strains has been consistently demonstrated.[5] Later on, *Aerococcus urinae* was proposed as a new species based on analysis of 16S rRNA sequences.[6] *A. urinae* was isolated from urine of elderly persons suffering from urinary tract infections and from blood of patients with endocarditis and urosepticemia. *Aerococcus* spp. is a rarely reported pathogen, possibly due to difficulties in the identification of the organism. *A. urinae* is a gram-positive coccus that grows in pairs and clusters, produces alpha-hemolysis on blood agar, and is negative for catalase and pyrrolidonyl aminopeptidase. *A. viridans* organisms are gram-positive cocci with a strong tendency to form tetrads. *A. urinae* and *A. viridans* slightly differ in their reactions in the LAP and pyronylaminopeptidase (PYR) disk tests. *A. urinae* has the characteristics of being LAP-positive and PYR-negative, whereas *A. viridans* is LAP-negative and PYR-positive.[7,8] More recently, three additional species of *Aerococcus* have been described, based on 16S rRNA sequences and polyphasic taxonomy.[9] *Aerococcus urinaehominis* has been recovered from a urine culture,[5] *Aerococcus sanguicola* has been isolated from urine and blood,[10] and *A. christensenii* has been recovered from a vaginal specimen.[11]

In general, the genus *Aerococcus* consists of bacterial strains that are gram-positive, nonspore-forming cocci that occurred as single cells, in pairs, tetrads, or in small groups. The cocci are facultatively anaerobic and catalase- and oxidase-negative. When grown on Columbia horse blood agar for 24 h, small (<1 mm diameter), entire, and grey colonies appear. Cells are nonmotile, and *a*-hemolysis is observed on blood agar.[12] *A. urinae* is found in a constant but low percentage of urine specimens (0 ± 8%), but nearly all patients present clinical signs of acute infection. Although *A. viridans* has been shown to be responsible for a number of clinical conditions, a large proportion of strains isolated from clinical samples fail to exhibit any clinical manifestations.[5,13] *A. urinaehominis*, also originated from a human urine sample, but no further clinical details were available on the organism, and its pathogenic potential remains unknown. *A. christensenii* is another strain that was not reported to cause disease and would not be discussed further.

The absence of some of the important distinguishing characteristic features of this genus from the databases of most commercial identification systems may allow *A. urinae* to be misidentified as a staphylococci, streptococci, enterococci, pediococci, lactococci, or leuconostocs. Bosley et al. conducted experiments to differentiate among these genera.[14]

The cellular fatty acid profile of aerococci is distinguishable from other genera; two relatively novel fatty acids found in the aerococci are identified as C16:1 omega 9c and C16:1 omega 9t. DNA–DNA reassociation studies using 11 strains of aerococci with the type strain *A. viridans* ATCC 11563 show that DNA from eight of these strains are more than 75% related to the type strain and had 1%–4% divergence in related sequences. The remaining three strains are 60%–70% related to the type strain, have 7%–11.5% divergence, and may represent a second species, *Aerococcus* genospecies 2, which may now be known as *A. urinae*. Beta-glucuronidase, alpha-galactosidase, and beta-galactosidase are useful in characterizing the aerococci.

20.1.2 Clinical Features and Pathogenesis

Patients infected with *Aerococcus urinae* tend to show one or more symptoms of urinary tract infection: dysuria (50%), pollakisuria (60%), abdominal pain (32.5%), flank tenderness (10%), fever, and general discomfort. Most patients have local or systemic conditions predisposing to urinary tract infections. Men have significantly more local problems, such as prostrate or urethral hyperplasia. Women have more systemic problems, such as dementia or cerebrovascular accidents. Men are in better physical condition than women.[15]

A retrospective chart review is performed by Sierra-Hoffman et al. on 54 patients with positive urine cultures during a 1-year period to assess the clinical significance of *A. urinae*.[9] The patients are classified as either having a UTI (urinary tract infection) or being "colonized," depending on clinical and laboratory data. Patients are considered to have a UTI if they have >5 urine white blood cells and $\geq 10^5$ CFU/mL on urine culture. They also are required to have at least two clinical symptoms (or only one symptom if catheterized). Clinical symptoms included are dysuria, frequency, nocturia, fever, myalgias, chills, altered mental status, and suprapubic pain. Patients with *A. urinae* recovered, but not meeting these criteria are classified as being "colonized." There is an overall female predominance (91% in the UTI group and 95% in the colonized group). However, there are no differences in the age, sex, and underlying conditions in patients whose isolates are judged to represent a UTI as opposed to being colonized, with the exception that more patients with a urinary catheter are noted in the UTI group ($P < .01$). Both groups have a significant but equal incidence of underlying urologic conditions.

- *A. sanguinicola* is similar to *A. urinae* in clinical implication, as both are first described as causes of urinary tract infections and are later found to be associated with bacteremia, sepsis, and endocarditis. *A. viridans*, the first *Aerococcus* species described, has been isolated from patients with meningitis, endocarditis, arthritis,[16] and bacteremia in neutropenia and from HIV patients.
- *A. viridans* is rarely associated with human infections; however, it could be a potential causative agent of bacteremia, especially in immunocompromised patients. Jafar et al. reported the first case of *A. viridans* bacteremia in a patient with HIV infection.[17] The presumed route of infection in this patient is esophageal ulcers.
- Infections due to this organism presumably seem to occur in previously damaged tissues or may be nosocomial in association with a prolonged hospitalization, antibiotic treatment, invasive procedures, presence of foreign bodies, or a neutropenic state.[18,19] Kern and Vanek[20] suggested that both granulocytopenia and oral mucositis may be the major risk factors for aerococcal bacteremia. Swanson et al.[21] also described that *A. viridans* might be a significant pathogen in patients with functional asplenia, in that the organism is encapsulated with an acidic polysaccharide, and the strains with heavier encapsulation are more virulent.[22] Recently, Nasoodi et al.[23] reported spondylodiscitis due to *A. viridans* infection. The patient had low-grade back pain. The authors hypothesized that this man was immunocompromised due to his liver cirrhosis and became bacteremic following inoculation from a fish hook injury or following invasion from his gastrointestinal system after eating infected imported crustaceans. Infection is common in lobsters, causing Gaffkaemia,[19] and other crustaceans, such as crab and some prawn species.[23]

Apart from urinary tract infections, *A. urinae* also may be associated with severe infections, such as endocarditis[24] and septicemia. Astudillo et al. reported the first case of *A. urinae* spondylodiscitis,[2] followed by Tekin et al.[24] In this case, severe back pain was the major condition that the patient paid attention to and sought for help for. Surprisingly, despite the fact that urinary tract is the main route of entry of the organism, urine culture was found to be sterile. Even though *A. urinae* endocarditis is clinically indistinguishable from endocarditis caused by other bacterial agents, infective endocarditis due to *A. urinae* has a poor prognosis. Late diagnosis is a possibility because of the subacute character of the disease and the unspecific symptoms of the elderly with comorbidities.[25] Santos et al. reported a case of lymphadenitis caused by *A. urinae*.[3] As for other urinary tract pathogens, the gastrointestinal tract could be a natural reservoir for *A. urinae*, though this has not been studied. Urinary tract infection is almost always the focus for an evolving bacteremia.

An 80-year-old male with preexisting coronary artery disease and dementia and was admitted to the hospital for altered mental status. The patient developed slowly worsening urinary symptoms, including nocturia and frequency, over the 3 weeks prior to admission. Urine cultures grew $>10^5$ CFU of *A. urinae*, the only isolate, per mL. His urinary symptoms resolved and his mental status improved by hospital day 3.[26] In another case, a 58-year-old person suffered from several days of dysuria, increased urinary frequency, and nocturia. Urinalysis revealed no bacteria, trace amounts of leukocyte esterase, and negative results for nitrites, protein, ketones, bilirubin, and blood. A second urinalysis demonstrated a moderate level of leukocyte esterase, and negative urine chemistries. The urine culture on the second urine sample later grew >10^5 CFU of *A. urinae*, the only isolate, per mL.[26] The case studies indicate that *A. urinae* is a rarely reported pathogen, possibly due to difficulties in the identification of the organism.[26] Several conventional and molecular methods are therefore used for detection of the pathogen.

In a case of urinary tract infections caused by *A. urinae*, the patient was treated empirically with intravenous ciprofloxacin.[26] Urine cultures grew >10^5 CFU of *A. urinae*, the only isolate, per mL. His urinary symptoms resolved and his mental status improved by hospital day 3. The patient was discharged after oral ciprofloxacin was prescribed. Results of repeated urinalysis after discharge were normal.

A. viridans strains are naturally susceptible to penicillin, macrolides and related drugs, tetracyclines, and chloramphenicol, and they are intrinsically resistant at a low level to aminoglycosides.[1] In a case of endocarditis caused by *A. viridans* in a 10-year-old child, he received ampicillin (200 mg/kg/day) and gentamicin (7.5 mg/kg/day) besides decongestive measures. Since the patient's condition deteriorated during hospital stay, amikacin (15 mg/kg/day) and norfloxacin (15 mg/kg/day) were started on sixth day based on antibiogram. Improvement was noticed from 11th day of hospital stay with a gradual return of fever to baseline. However, at this point the patient developed severe chest pain and a pericardial rub. Echocardiography confirmed the presence of a pericardial effusion. The pericardial fluid was tapped. It was found to be sterile on culture, and steroids were started. The effusion gradually disappeared over the next 5 days, and the cardiac failure was also controlled. Amikacin and norfloxacin were continued for a total duration of 3 weeks.[27]

These case studies indicate that the treatment is often case sensitive and also depends on the species and its antibiotic susceptibility. *A. viridans* was found to be resistant to at least one antibiotic amongst eight wild type strains: erythromycin (six strains), tetracycline and minocycline (five strains), chloramphenicol (one strain), and high levels of streptomycin (one strain).[1] Augustine et al.[27] reported a case of endocarditis caused by *A. viridans* with multidrug resistance, that is, resistance to penicillin, ampicillin, cefotaxime, gentamicin, and intermediate resistant to ciprofloxacin, but they did not discuss minimum inhibitory concentrations (MICs). According to the antimicrobial susceptibility data of 30 aerococcal isolates obtained from Centers for Disease Control and Prevention,[14] the MICs for nine strains were 0.5 g or more of penicillin per mL, with MICs for five strains being more than 1 g/mL; therefore, approximately 46% of aerococci tested were either relatively resistant or resistant to penicillin. Moreover, Christensen et al.[28] recently documented that penicillin resistance should be the peculiar characteristic of *A. viridans* capable of differentiating it from *Aerococcus*-like organisms.[19] Certainly, further investigations are needed to establish the optimal treatment for this pathogen.

In a case of endocarditis caused by *A. urinae*, treatment modalities were available for 10/11 patients.[25] Among nine of these patients treated with a combination of a beta-lactam and an aminoglycoside, seven died. The remaining patient was treated with a beta-lactam alone and survived with neurologic sequelae. In all lethal cases, death occurred during the first 4 weeks (mean, 12 days). It is suggested that all patients are treated intravenously for a minimum of 6 weeks, since all surviving patients, including the presented patient, were treated for this time period either with a beta-lactam alone or with a beta-lactam and an aminoglycoside.

20.1.3 Diagnosis

The concept of polyphasic taxonomy integrates different kinds of data and information on microorganisms, that is, phenotypic, genotypic, and phylogenetic, and this method has gained increasing acceptance since its introduction.[29] This method includes transmission electron microscopy, fatty acid methyl ester analysis, multilocus enzyme electrophoresis, antimicrobial susceptibility tests, and comparison of 16S rRNA and/or other housekeeping gene(s) sequence data, and ribotyping. After the discovery of *A. viridans*, all the other strains were discovered on the basis of polyphasic taxonomy.

Colony Morphology, Staining, and Biochemical Tests. The presumptive identification of aerobic gram-positive alpha-hemolytic cocci and the decision on whether to more fully identify the isolates are frequently based on Gram stain, colony appearance, and catalase reaction. Because *A. urinae* is catalase-negative, it could be mistaken for alpha-hemolytic streptococci or enterococci, which are more common urine isolates. At 24 h, the colony morphology resembles that of an alpha-hemolytic streptococcus or lactobacillus; at 48 h, it is similar to that of an *Enterococcus*. The Gram stain should be differential because *A. urinae* forms pairs, tetrads, and clusters. However, since *A. urinae* shows smaller cocci and fewer tetrads than *A. viridans* does, it could be confused with pediococci or densely packed streptococci or enterococci.

The most important routine tests are detection of leucine arylamidase, β-glucuronidase, PYR, hydrolysis of hippurate, and antibiotic susceptibility patterns.[7,11,18,26,30,31] Rapid PYR testing is useful for distinguishing between *A. viridans* or enterococci (both PYR positive) and *A. urinae* (PYR negative). A Gram stain should be carefully examined for the characteristic arrangement in clusters and tetrads to rule out lactobacillus and other streptococcus. Pediococci are PYR negative and have a Gram stain morphology similar to that of *A. urinae*; however, they differ in their resistance to vancomycin and in their positive bile esculin test result.[18,26]

Transmission Electron Microscopy. The transmission electron microscope (TEM) operates on the same basic principles as the light microscope but uses electrons as "light source." Their much lower wavelength makes it possible to get a resolution a thousand times better than with a light microscope. Cells of both *A. urinae* and *A. viridans* had a mean diameter of 0.90–1.10 μm, and strains of both species showed a surface layer.[32] In contrast to *A. viridans*, about 30% of all *A. urinae* cells showed signs of asymmetrical cell division.[32]

Sample preparation for TEM requires cultures in log phase. The cells can be fixed for 2 h in 3% glutaraldehyde in 0.1 M cacodylate buffer, pH 7.2, and then pelleted (8800 × *g* for 2 min). The precipitates should be enrobed in melted Noble agar (Difco) and small cubes with visible clusters of cells can be transferred to 3% glutaraldehyde and fixed overnight at 4°C. The specimens may be post fixed for 90 min in 1% (w/v) OsO_4 in cacodylate buffer and stained with 2% (w/v) uranyl acetate in barbiturate buffer. The procedures for the dehydration, embedding in Epon (glycide ether) and the further preparation of thin sections can be performed as described by Blom et al.[33] The stained and loaded samples may be viewed using TEM.

Antimicrobial Susceptibility Tests. For this, the antimicrobial contained in a reservoir is allowed to diffuse out into the medium (e.g., blood agar) and interact in a plate freshly seeded with the test organisms. A variety of antimicrobial containing reservoirs are used, but the antimicrobial impregnated absorbent paper disc is by far the commonest type used (disc diffusion method). The results of in vitro antibiotic susceptibility testing guide clinicians in the appropriate selection of initial drugs used for individual patients in specific situations. The selection of an antibiotic panel for susceptibility testing is based on the commonly observed susceptibility patterns and is revised periodically. According to the antimicrobial susceptibility data of 30 aerococcal isolates obtained from Centers for Disease Control and Prevention,[14] the MICs for nine strains were 0.5 g or more of penicillin per mL, with MICs for five strains being more than 1 g/mL; therefore, approximately 46% of aerococci tested were either relatively resistant or resistant to penicillin. Moreover, Christensen et al.[19,28] documented that penicillin resistance should be the peculiar characteristic of *A. viridans*, capable of differentiating it from *Aerococcus*-like organisms. Table 20.1 shows the antibiotic susceptibilities of *A. urinae*, *A. viridans*, and *A. sanguinicola*.

TABLE 20.1
Antibiotic Susceptibilities of *A. urinae* and *A. viridans*. Data Has Been Adapted from Observations Reported Earlier[a]

Antibiotic	*A. urinae*	*A. viridans*	*A. sanguinicola*
Penicillin G	S	R	S
Trimethoprim-sulfamethoxazole	R	S	NA
Vancomycin	S	S	S
Gentamicin	R	S/R	NA
Tetracycline	S	S	NA
Ciprofloxacin	S	R	NA
Ampicillin	S	R	S
Erythromycin	R	R	S
Cefuroxime	S	R	S
Chloramphenicol	S	R	NA

Abbreviations: S, susceptible; R, resistant; NA, not available.
Source: Uh, Y. et al., *J. Korean Med. Sci.*, 17, 113, 2002; Zhang, Q. et al., *J. Clin. Microbiol.*, 38, 1703, 2000; Skov, R.L. et al., *Diagn. Microbiol. Infect. Dis.*, 21, 219, 1995; Ibler, K. et al., *Scand. J. Infect. Dis.*, 40, 761, 2008; Colakoglu, S. et al., *Infection*, 36, 288, 2008.

Fatty Acid Methyl Ester Analysis. Microorganisms can be classified based on gas chromatographic analysis of extracted microbial fatty acid methyl esters (FAMEs). Microbial fatty acid profiles are unique from one species to another, and this has allowed for the creation of very large microbial libraries. Location of the double-bond position of monounsaturated fatty acids of *A. viridans* was accomplished by combined gas chromatography (GC)–mass spectrometry analysis of dimethyl disulfide (DMDS) derivatives. The monoenoic fatty acids from whole bacterial cells were converted to methyl esters and then to DMDS adducts and analyzed by capillary GC–mass spectrometry.[37,38]

Multilocus Enzyme Electrophoresis (MLEE). Isolates are characterized by the relative electrophoretic mobilities of a large number of water-soluble cellular enzymes. Because the net electrostatic charge and, hence, the rate of migration of a protein during electrophoresis are determined by its amino acid sequence, mobility variants (electromorphs or allozymes) of an enzyme can be directly equated with alleles at the corresponding structural gene locus.[39] Christensen et al.[32] used various enzymes for generating MLEE patterns and finally constructed UPGMA and Minimal Evolution trees based on the results. They found that the tree concurs with the ribotyping data and distinguished *A. urinae* strains as a group of strains distinct from the type strain of *A. viridans*.

Sample preparation methods for MLEE have been adapted from those described by Selander et al.[39] For this technique, whether cells are grown in broth or on agar plates is unimportant, since growth conditions do not affect the electrophoretic mobilities of enzymes. However, some enzymes may vary in activity (but not in mobility) on gels, depending on the type of medium in which the cells were grown. The activity level of certain enzymes may also vary with the composition of the culture medium, but, again, electrophoretic mobilities are not affected. Cells are normally lysed by sonication; however, any method of lysis that does not denature proteins may be used. After lysis and centrifugation at 30,000 × *g* for 20 min, aliquots of the several milliliters of lysate (supernatant) are transferred to three or four culture tubes and stored at −70°C until used for electrophoresis. At that temperature, lysates can be stored for several years without significant loss of activity of most enzymes. The stability of the enzymes varies markedly among species of bacteria, and so this step may be avoided. Apparatus for horizontal starch-gel electrophoresis is used. Starch gels are preferred over polyacrylamide gels. To prepare a gel, a suspension of starch and gel buffer is heated just beyond the boiling point.[39] The suspension is immediately poured into a lucite gel mold. After the gel has cooled at room temperature for 2 h, it is wrapped in plastic film to prevent desiccation. Gels are used within 24 h of preparation. In loading a gel, pieces of Whatman no. 3 filter paper (9 mm by 6 mm) are individually dipped into samples of lysate, blotted on filter paper to remove excess liquid, and then inserted at 3-mm intervals in a continuous slit cut in the gel. Up to 20 lysates can be electrophoresed on a single gel. Pieces of filter paper dipped in amaranth dye are inserted at one or both ends of the slit to mark the migration front of the buffer line. During electrophoresis, a constant voltage is maintained, and the gel is cooled by a pan of ice supported above the gel mold on a thin plate of glass. Gels may be run in cold rooms or refrigerators at 4°C. Following electrophoresis, three or four horizontal slices (1–2 mm thick) are cut from the gel with a thin wire and incubated individually at 37°C in various selective enzyme staining solutions.[39] The activity level and degree of separation of electromorphs of an enzyme may depend on the type and pH of the buffer system used and, to some extent, on the concentration of starch in the gel. Thus, an enzyme may appear to be monomorphic or polymorphic for only two or three electromorphs in one buffer system but exhibit 5–10 electromorphs in another buffer system. Occasionally, the relative mobilities of certain electromorphs are reversed in different buffer systems.

Comparisons of the mobilities of enzymes from different isolates are made visually against one another on the same gel slice. It is not sufficient to compare relative mobilities of an enzyme in different strains by measuring distances of migration from the origin, even when the enzymes have been electrophoresed on the same gel. Each isolate is characterized by its combination of electromorphs over the number of enzymes assayed, and distinctive profiles of electromorphs, corresponding to unique multilocus genotypes, are designated electrophoretic types (ETs), which are equivalent to allele profiles. For analysis, a distance matrix in MEGA format may be generated using the computer program ETMEGA of T.S. Whittam (www.foodsafe.msu.edu/whittam/#programs) as mentioned by Chritensen et al.[32] ETMEGA calculates the genetic distance between pairs of "electrophoretic types," which were defined by distinct patterns of electrophoretic mobilities of the 11 housekeeping enzymes. Distance is measured as the proportion of mismatched loci between pairs of ETs. Null alleles are not used in the calculation of pair wise distances. Phylogenetic and molecular evolutionary analyses using this distance matrix may be conducted using MEGA version 4.[32,40–42]

Ribotyping. Ribotyping is a method that can identify and classify bacteria based upon differences in rRNA. It generates a highly reproducible and precise fingerprint that can be used to classify bacteria from the genus through and beyond the species level. DNA is extracted from a colony of bacteria and then restricted into discrete-sized fragments. In experiments performed by Christensen et al.[32] ribotyping of the 26 *A. urinae* strains and the type strain of *A. viridans* with *Eco*R1 and *Hae*III resulted in large DNA fragments of more than 40 kb, which could not be resolved using the RiboPrinter system. Ribotyping these strains with *Xba*I resulted in seven distinct ribotype patterns, each comprising between 1 and 12 strains. They showed that strains of *A. urinae* were clearly distinct from *A. viridans*^{T}.

20.2 METHODS

20.2.1 Sample Preparation

For extraction of *Aerococcus* DNA, a Wizard SV genomic DNA purification system (Promega, Madison, Wis.) may be used with some modifications. The bacterial pellet is resuspended in 400 µl enzymatic lysis solution (47 mM EDTA, 25 mg/ml lysozyme [Sigma, St. Louis, Mo.], 20 µg/ml lysostaphin [Sigma]) and incubated at 37°C for 2 h. Then, proteinase K (19.2 mg/ml; Roche) is added to a final concentration of 0.4 mg/ml and the mixture is incubated at 55°C for 1 h. Nuclei Lysis solution (Promega) and RNase solution (Promega) are added, and after mixing, the reaction solution is incubated at 80°C for 10 min. Further purification steps are done according to the instructions of the manufacturer. The final elution volume is 200 µl. For each PCR, 2 µl of this DNA extract is applied.

20.2.2 Detection Procedures

(i) Aerococcus Viridans-Specific PCR Martín et al. utilized *A. viridans*-specific primers AC2 (5′-GTGC TTGCACTTCTGACGTTAGC-3′) and AC4 (5′-TGAGC CGTGGGCTTTCACAT-3′) to amplify a 540 bp fragment from 16S rRNA gene.[43] Furthermore, Grant et al. developed a rapid protocol for identification of *A. viridans* through PCR amplification of a 16S rRNA gene fragment, followed by detection with an oligonucleotide probe.[44]

(ii) PCR and Sequencing Sequencing analysis of 16S rRNA gene allows identification of all *Aerococcus* species.[45] In this protocol, a large fragment of the 16S rRNA gene (corresponding to positions 30 ± 1521 of the *Escherichia coli* 16S rRNA gene) is amplified by PCR using conserved primers close to the 3′ and 5′ ends of the gene. The PCR product is directly sequenced using a *Taq* Dye-Deoxy Terminator Cycle Sequencing kit (Applied Biosystems) and an automatic DNA sequencer.

(iii) PCR Ribotyoing PCR ribotyping was developed by Kostman et al.[46] and Gurtler et al.[47] in the 1990s as a response in part to the need in the clinical microbiology laboratory setting for expeditious epidemiological discrimination among pathogenic microorganisms without the use of probes, thus making the analysis more widely applicable.[48] Primers complementary to the 3′ end of the 16S rRNA gene and the 5′ end of the 23S rRNA gene were used such that the amplified fragment revealed length heterogeneity of the internal spacer regions (ISR) located between the 16S and 23S genes of the six rRNA gene operons.[49] However, PCR ribotyping proved not to be universally applicable.[50] In some species, ISR length is variable within and between isolates, in others ISR lengths are limited to one or two sizes, usually dependent on the number of tRNAs present.[48] In an attempt to further subtype isolates for which PCR ribotyping was nondiscriminatory, Ryley et al.[51] and Shreve et al.[52] performed PCR amplification of ISRs, as described above, followed by restriction endonuclease

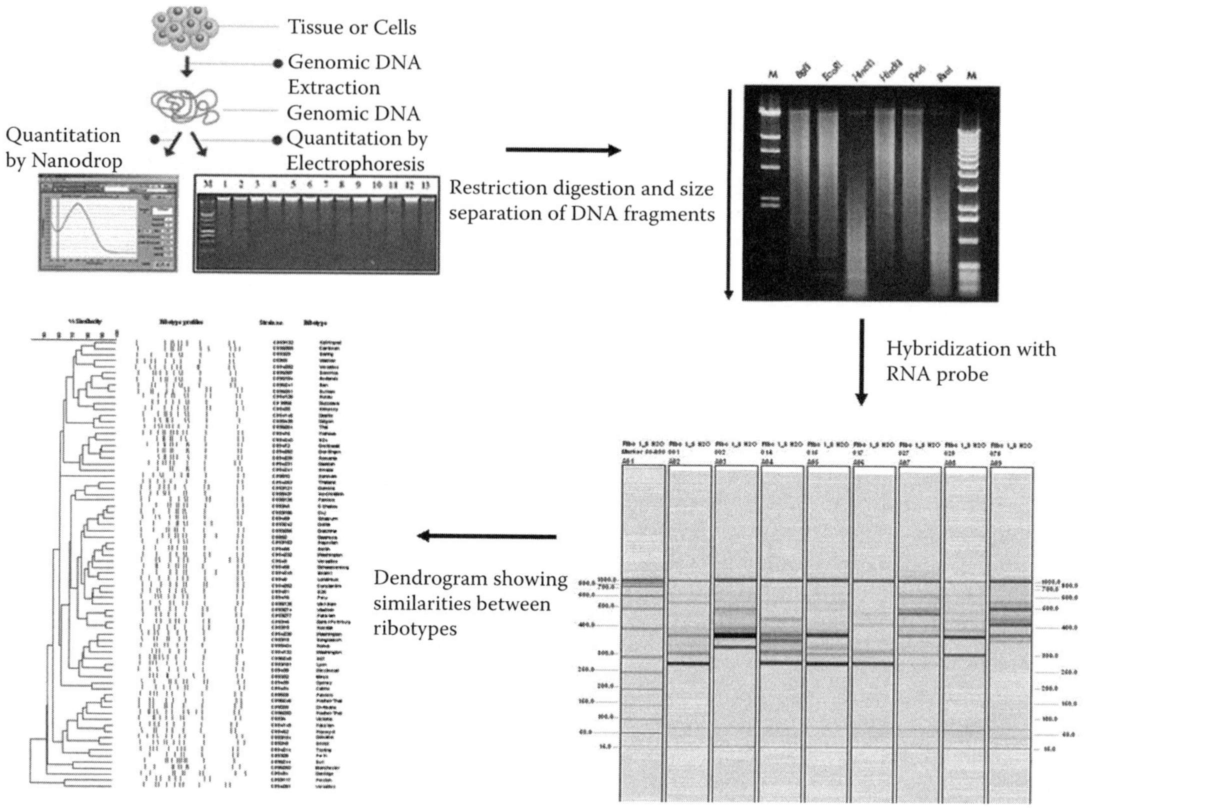

FIGURE 20.1 Demonstration of ribotyping. DNA is isolated from a cultured isolate, restricted and size separated in an agarose gel. The gel is then hybridized with labeled rRNA probe from a type strain of the species or genus, which binds to fragments containing copies of the rRNA operon. Following probe detection, fragments with bound probe are visualized, forming a characteristic RFLP. The advantages of ribotyping over other RFLP methods is that both highly variable and conserved sequences are highlighted by the probe, which further makes the method suitable to type strains over a broad phylogenetic range, and the same process can be used for analysis of strains within any species.

digestion. Christensen et al.[32] found the following enzymes suitable for *Aerococcus*: *Xba*I, *Eco*R1, and *Hae*III. An alternate variation of PCR ribotyping, amplified rRNA gene restriction analysis (ARDRA) is based on PCR amplification of the 16S rRNA gene followed by restriction digestion. Jayarao et al.[53] developed this technique to determine subspecies of *Streptococcus uberis*, as a means of avoiding methods involving DNA hybridization or sequencing. Later, as described by Dupont Qualicon, the RiboPrinter automated ribotyping system came as a technological breakthrough with respect to convenience, reproducibility, and speed. The speed in this case appears to be two- to threefold greater than that for manual ribotyping, with far less labor-intensive procedures. For polyphasic characterization of *A. urinae*, automated ribotyping method (Figure 20.1) was used successfully with the Riboprinter (Qualicon Inc., Wilmington, DE).[32]

20.3 CONCLUSION

The genus *Aerococcus* shows distinct interspecies differences, but constitutes a well-circumscribed genus as revealed by 16S rRNA sequence and other data. Phenotypic and molecular analyses demonstrate that *A. urinae* is to some extent a heterogeneous species belonging to the genus *Aerococcus*. At least two distinct lineages of *A. urinae* exist, which can be identified by phenotypic markers. Future recognition of these biovars may reveal potential differences in ecology and clinical significance. In daily routine practice, species identification of *Aerococcus* causes problems. Routine diagnostic procedures should be able to identify this bacteria, and clinicians should be aware of the pathogenic potential of this organism.

ACKNOWLEDGMENTS

We thank Dr. Mark Lawrence and Dr. Shane Burgess for their support in writing the manuscript. We are also grateful to Dr. Amit Ghosh, and Dr. Walter J. Diehl for encouragement.

REFERENCES

1. Buu-Hoi, A., Le Bouguenec, C., and Horaud, T., Genetic basis of antibiotic resistance in *Aerococcus viridans*, *Antimicrob. Agents Chemother.*, 33, 529, 1989.
2. Astudillo, L. et al., Spondylodiscitis due to *Aerococcus urinae*: A first report, *Scand. J. Infect. Dis.*, 35, 890, 2003.
3. Santos, R. et al., Lymphadenitis caused by *Aerococcus urinae* infection, *Scand. J. Infect. Dis.*, 35, 353,2003.
4. Williams, R.E., Hirch, A., and Cowan, S.T., *Aerococcus*, a new bacterial genus, *J. Gen. Microbiol.*, 8, 475, 1953.

5. Lawson, P.A., Falsen, E., Ohlen, M., and Collins, M.D., *Aerococcus urinaehominis* sp. nov., isolated from human urine, *Int. J. Syst. Evol. Microbiol.*, 51, 683, 2001.
6. Aguirre, M., and Collins, M.D., Phylogenetic analysis of some Aerococcus-like organisms from urinary tract infections: Description of *Aerococcus urinae* sp. nov., *J. Gen. Microbiol.*, 138, 401, 1992.
7. Christensen, J.J. et al., Bacteremia/septicemia due to *Aerococcus*-like organisms: Report of seventeen cases. Danish ALO Study Group, *Clin. Infect. Dis.*, 21, 943, 1995.
8. Christensen, J.J., Korner, B., and Kjaergaard, H., *Aerococcus*-like organism—an unnoticed urinary tract pathogen, *Apmis*, 97, 539, 1989.
9. Sierra-Hoffman, M., Watkins, K., Jinadatha, C., Fader, R., and Carpenter, J.L., Clinical significance of *Aerococcus urinae*: A retrospective review, *Diagn. Microbiol. Infect. Dis.*, 53, 289, 2005.
10. Lawson, P.A., Falsen, E., Truberg-Jensen, K., and Collins, M.D., *Aerococcus sanguicola* sp. nov., isolated from a human clinical source, 51, 475, 2001.
11. Collins, M.D., Jovita, M.R., Hutson, R.A., Ohlen, M., and Falsen, E., *Aerococcus christensenii* sp. nov., from the human vagina, *Int. J. Syst. Bacteriol.*, 49 Pt 3, 1125, 1999.
12. Collins, M.D., Jovita, M.R., Hutson, R.A., Ohlen, M., and Falsen, E., Note: *Aerococcus christensenii* sp. nov., from the human vagina, *Int. J. Syst. Bacteriol.*, 49, 1125, 1999.
13. Christensen, J.J. et al., *Aerococcus urinae*: Intraspecies genetic and phenotypic relatedness, *Int. J. Syst. Bacteriol.*, 47, 28, 1997.
14. Bosley, G.S. et al., Phenotypic characterization, cellular fatty acid composition, and DNA relatedness of aerococci and comparison to related genera, *J. Clin. Microbiol.*, 28, 416, 1990.
15. Schuur, P.M.H. et al., Urinary tract infections with *Aerococcus urinae* in the south of The Netherlands, *J. Urol.*, 161, 1414, 1999.
16. Martin, V. et al., Characterization of *Aerococcus viridans* isolates from swine clinical specimens, *J. Clin. Microbiol.*, 45, 3053, 2007.
17. Razeq, J.H., Thomas, G.M., and Alexander, D., The first reported case of *Aerococcus* bacteremia in a patient with HIV infection, *Emerg. Infect. Dis.*, 5, 838, 1999.
18. Facklam, R., and Elliott, J.A., Identification, classification, and clinical relevance of catalase-negative, gram-positive cocci, excluding the streptococci and enterococci, *Clin. Microbiol. Rev.*, 8, 479, 1995.
19. Uh, Y., Son, J.S., Jang, I.H., Yoon, K.J., and Hong, S.K., Penicillin-resistant *Aerococcus viridans* bacteremia associated with granulocytopenia, *J. Korean Med. Sci.*, 17, 113, 2002.
20. Kern, W., and Vanek, E., *Aerococcus* bacteremia associated with granulocytopenia, *Eur. J. Clin. Microbiol.*, 6, 670, 1987.
21. Swanson, H., Cutts, E., and Lepow, M., Penicillin-resistant *Aerococcus viridans* bacteremia in a child receiving prophylaxis for sickle-cell disease, *Clin. Infect. Dis.*, 22, 387, 1996.
22. Hermansson, K. et al., Structural studies of the capsular polysaccharide from *Aerococcus viridans* var. homari, *Carbohydr. Res.*, 208, 145, 1990.
23. Nasoodi, A., Ali, A.G., Gray, W.J., and Hedderwick, S.A., Spondylodiscitis due to *Aerococcus viridans*, *J. Med. Microbiol.*, 57, 532, 2008.
24. Tekin, A., Tekin, G., Turunc, T., Demiroglu, Z., and Kizilkilic, O., Infective endocarditis and spondylodiscitis in a patient due to *Aerococcus urinae*: first report, *Int. J. Cardiol.*, 115, 402, 2007.
25. Ebnother, C., Altwegg, M., Gottschalk, J., Seebach, J.D., and Kronenberg, A., *Aerococcus urinae* endocarditis: Case report and review of the literature, *Infection*, 30, 310, 2002.
26. Zhang, Q., Kwoh, C., Attorri, S., and Clarridge, J.E., *Aerococcus urinae* in urinary tract infections, *J. Clin. Microbiol.*, 38, 1703, 2000.
27. Augustine, T., Thirunavukkarasu, Bhat, B.V., and Bhatia, B.D., *Aerococcus viridans* endocarditis. Case report, *Indian Pediatr.*, 31, 599, 1994.
28. Christensen, J.J., Korner, B., Casals, J.B., and Pringler, N., *Aerococcus*-like organisms: Use of antibiograms for diagnostic and taxonomic purposes, *J. Antimicrob. Chemother.*, 38, 253, 1996.
29. Vandamme, P. et al., Polyphasic taxonomy, a consensus approach to bacterial systematics, *Microbiol. Rev.*, 60, 407, 1996.
30. Christensen, J.J., Vibits, H., Ursing, J., and Korner, B., *Aerococcus*-like organism, a newly recognized potential urinary tract pathogen, *J. Clin. Microbiol.*, 29, 1049, 1991.
31. Kristensen, B., and Nielsen, G., Endocarditis caused by *Aerococcus urinae*, a newly recognized pathogen, *Eur. J. Clin. Microbiol. Infect. Dis.*, 14, 49, 1995.
32. Christensen, J.J. et al., *Aerococcus urinae*: Polyphasic characterization of the species, *Apmis*, 113, 517, 2005.
33. Blom, J., Mansa, B., and Wilk, A., A study of Russell bodies in human monoclonal plasma cells by means of immunofluorescence and electron microscopy, *Acta Pathol. Microbiol. Scand.*, *A* 84, 335, 1976.
34. Skov, R.L., Klarlund, M., and Thorsen, S., Fatal endocarditis due to *Aerococcus urinae*, *Diagn. Microbiol. Infect. Dis.*, 21, 219, 1995.
35. Ibler, K. et al., Six cases of *Aerococcus sanguinicola* infection: Clinical relevance and bacterial identification, *Scand. J. Infect. Dis.*, 40, 761, 2008.
36. Colakoglu, S. et al., Three cases of serious infection caused by *Aerococcus urinae*: A patient with spontaneous bacterial peritonitis and two patients with bacteremia, *Infection*, 36, 288, 2008.
37. Moss, C.W., and Daneshvar, M.I., Identification of some uncommon monounsaturated fatty acids of bacteria, *J. Clin. Microbiol.*, 30, 2511, 1992.
38. Moss, C.W., Wallance, P.L., and Bosley, G.S., Identification of monounsaturated fatty acids of *Aerococcus viridans* with dimethyl disulfide derivatives and combined gas chromatography-mass spectrometry, *J. Clin. Microbiol.*, 27, 2130, 1989.
39. Selander, R.K. et al., Methods of multilocus enzyme electrophoresis for bacterial population genetics and systematics, *Appl. Environ. Microbiol.*, 51, 873, 1986.
40. Kumar, S., Nei, M., Dudley, J., and Tamura, K., MEGA: A biologist-centric software for evolutionary analysis of DNA and protein sequences, *Brief Bioinform.*, 9, 299, 2008.
41. Kumar, S., Tamura, K., and Nei, M., MEGA3: Integrated software for Molecular Evolutionary Genetics Analysis and sequence alignment, *Brief Bioinform.*, 5, 150, 2004.
42. Tamura, K., Dudley, J., Nei, M., and Kumar, S., MEGA4: Molecular Evolutionary Genetics Analysis (MEGA) software version 4.0, *Mol. Biol. Evol.*, 24, 1596, 2007.
43. Martín, V. et al., Characterization of *Aerococcus viridans* isolates from swine clinical specimens, *J. Clin. Microbiol.*, 45, 3053, 2007.
44. Grant, K.A. et al., Rapid identification of *Aerococcus viridans* using the polymerase chain reaction and an oligonucleotide probe, *FEMS Microbiol. Lett.*, 95, 63, 2006.
45. Lawson, P.A. et al., *Aerococcus urinaehominis* sp. nov., isolated from human urine, *Int. J. Syst. Evol. Microbiol.*, 51, 683, 2001.

46. Kostman, J.R. et al., A universal approach to bacterial molecular epidemiology by polymerase chain reaction ribotyping, *J. Infect. Dis.*, 171, 204, 1995.
47. Gurtler, V., and Stanisich, V.A., New approaches to typing and identification of bacteria using the 16S-23S rDNA spacer region, *Microbiology*, 142, 3, 1996.
48. Bouchet, V., Huot, H., and Goldstein, R., Molecular genetic basis of ribotyping, *Clin. Microbiol. Rev.*, 21, 262, 2008.
49. Gurtler, V., and Barrie, H.D., Typing of *Staphylococcus aureus* strains by PCR-amplification of variable-length 16S-23S rDNA spacer regions: characterization of spacer sequences, *Microbiology*, 141, 1255, 1995.
50. Cartwright, C.P., Polymerase chain reaction ribotyping: A "universal" approach? *J. Infect. Dis.*, 172, 1638, 1995.
51. Ryley, H.C., Millar-Jones, L., Paull, A., and Weeks, J., Characterisation of *Burkholderia cepacia* from cystic fibrosis patients living in Wales by PCR ribotyping, *J. Med. Microbiol.*, 43, 436, 1995.
52. Shreve, M.R., Johnson, S.J., Milla, C.E., Wielinski, C.L., and Regelmann, W.E., PCR ribotyping and endonuclease subtyping in the epidemiology of *Burkholderia cepacia* infection, *Am. J. Respir. Crit. Care. Med.*, 155, 984, 1997.
53. Jayarao, B.M., Dore, J.J., Jr.. and Oliver, S.P., Restriction fragment length polymorphism analysis of 16S ribosomal DNA of *Streptococcus* and *Enterococcus* species of bovine origin, *J. Clin. Microbiol.*, 30, 2235, 1992.

21 *Bacillus*

Noura Raddadi, Ameur Cherif, and Daniele Daffonchio

CONTENTS

21.1 INTRODUCTION

21.1.1 Classification, Morphology, and Biology of *Bacillus* spp.

Bacteria in the genus *Bacillus* are gram-positive, spore formers, and rod-shaped. They are very diverse in terms of physiology, ecological niche, genes sequences, and regulation. Their impact on human activity varies from probiotic effect[1,2] to severe pathogenecity. The oldest, most infectious, and potentially lethal human disease is the anthrax that is caused by *Bacillus anthracis*.[3] However, many other species within this genus have emerged as new human pathogens associated with foodborne diseases that can cause severe and even fatal infections. These include *B. cereus*, *B. weihenstephanenesis*, *B. pumilus*, *B. mojavensis*, *B. licheniformis*, *B. subtilis*, and *B. circulans*. All *Bacillus* species, due to endospore formation, can survive heat treatment and disinfection procedures and hence pose serious health concerns as contaminants of food or in hospital settings as a cause of nosocomial infections,[4–10] as well as severe illnesses, especially in immunocompromised patients or immunocompetent individuals with risk factors such as intravenous drug use, hemodialysis, and leukemia. The production of toxins and enzymes by these bacteria is considered the causative agent of several human diseases.

In this chapter, we survey the different human pathologies caused by *Bacillus* spp., mainly those of the *B. cereus* group. Before giving details on the clinical features, toxicology, and the methods of detection of these bacteria and their corresponding pathogenecity factors, we will briefly introduce the complex ecology and the identification challenges that characterize several *Bacillus* species. The *B. cereus* group, which shows a "bivalent face" in terms of its important impact on human activity will be used as a paradigm.[11,12]

B. anthracis is the most serious human pathogen in this group and the major subject of this chapter. This bacterium is the etiological agent of anthrax, a potentially fatal disease for humans and animals, and a bioterrorism agent.

B. cereus is the most known foodborne pathogen that has been associated with food-poisoning illnesses[13,14] and other kinds of clinical infections that are reported in this chapter.

B. thuringiensis is an insect pathogen widely used as biopesticide, and differently from *B. cereus*, has a very useful impact on human activities being widely used in agriculture for insect/pest biocontrol.

B. weihenstephanensis, a psychrotolerant species capable of growing at temperatures as low as 4°C–6°C is implicated in food spoilage.[15]

In addition to *B. anthracis*, *B. cereus*, *B. thuringiensis*, and *B. weihenstaephanenesis*, the *B. cereus* group encompasses two other species, *B. mycoides* and *B. pseudomycoides*, which are typically isolated from soil and plant rhizosphere.

The six species are known to be strictly related phylogenetically, as has been shown by DNA–DNA hybridization studies[16–18] and the sequencing of the ribosomal RNA genes.[19–21] However, a marked variability is always observed when large collections of strains are examined by DNA fingerprinting methods that target the whole genome[22–26] and/or discrete genes.[27–31] Hence, the phylogenetic and taxonomic relationships among these species are still open to debate. The genetic variability observed within and between these species has raised several interesting questions and poses a challenge to microbiologists: (i) What is the evolutionary pathway of these species and how are they differentiated during evolution? (ii) What is the genetic baseline driving the different ecotypes and the virulence?

21.1.2 Genetic Baseline of *B. cereus* Group Ecotypes

The ecology of these bacteria is still far from clear, despite several recent analyses that have shown that a major environmental niche for these bacteria is the invertebrate intestinal tract.[32–34] In a general model, these species commonly live associated to the intestine of invertebrates, assuming a symbiotic life style. But, occasionally, they escape such an ecotype, becoming invasive for other animal hosts that are in specific insects, arthropods, and nematodes for *B. thuringiensis*, and in mammals and humans for *B. anthracis* and *B. cereus*.[35] Apart from this general scheme, the ecology of these species outside the invertebrate host is not completely clear, and the ability of several environments with which these bacteria have been commonly associated, to continuously support their life cycle has been questioned. It has been shown that with respect to other soil-inhabiting bacteria, such as those of the *B. subtilis* group that have genomes harboring many genetic determinants for the metabolism of plant-derived sugars, *B. cereus* group bacteria are rich in genes for protein metabolism. This suggests that they evolved in relation to nutrient-rich environments, such as animal guts or animal tissues and fluids, rather than the plant environment.[36,37] Specialization on different "animal" niches has lead to different evolution of the species in the *B. cereus* group.

For instance, *B. anthracis* has been proposed to diverge from the other members of the group by specializing as a lethal mammal pathogen. It has been supposed that *B. anthracis* has evolved along two possible pathways[38]: In the first pathway, one considers *B. anthracis* to be a relatively ancient organism with a low growth rate, determined by the previously mentioned ecological constraints, and evolving separately from a common ancestor with *B. cereus* and other relatives. In the second pathway, it has been supposed to be derived relatively recently from *B. cereus* through the acquisition and rearrangement of plasmids resulting in the actual pXO plasmid pattern responsible for lethal virulence.

21.1.3 Species Discrimination in *B. cereus* Group

The similarity among such closely related bacteria as the species of the *B. cereus* group posed serious challenges for species discrimination. This is an important point to address, considering the differences in the virulence potential of these microorganisms. Typically, *B. anthracis* can be differentiated from most of *B. cereus* and *B. thuringiensis* strains by specific tests, such as penicillin G and gamma phage sensibility, lack of motility, and beta-hemolysis on blood agar. However, a problem still exists for isolates borderline between species, such as, for instance, the pathogenic *B. cereus* G9241, for which these tests fail to recognize its pathogenic potential.[39] This strain has been confirmed to be a *B. cereus*, harboring a virulence plasmid very similar to plasmid pXO1 of *B. anthracis* and a second plasmid encoding for a capsule synthesis. However, such a second plasmid and the capsule-coding genes were completely different from the pXO2 plasmid and the typical capsule of *B. anthracis*.[39] Thus, considering the virulence potential of strains that are genetically the near neighbors of *B. anthracis*, approaches that might rapidly identify these strains are of great interest. From a safety perspective, these strains could represent alternative hosts for *B. anthracis* toxin genes.[39] Several *B. cereus* strains resulted, strictly related to *B. anthracis*. For example, Keim et al.[38] and Radnedge et al.[40] individuated by amplified fragment-length polymorphism (AFLP) *B. cereus* and *B. thuringiensis* strains closely related to *B. anthracis*. Radnedge et al.[41] tried to identify, by suppression subtractive hybridization, genomic regions of *B. anthracis* absent in these closely related strains. Apart from AFLP, several other methods based on whole-genome fingerprinting have been used for typing *B. anthracis* and for the identification of species borderline strains, like, among others, rep-PCR[12] and multilocus sequence typing (MLST)[42] or, recently, comparative genomics[43] and microarray analysis.[44] Alternative approaches have been based on length or sequence polymorphisms in variable-number tandem repeats in multiple loci (multilocus VNTR analysis [MLVA])[45] or on signature single nucleotide polymorphisms (SNPs) in the genome.[46] MLVA has been used as a gold standard for subtyping *B. anthracis* isolated worldwide (see among others[35,45,47,48]). Together with MLVA, SNP analysis is greatly contributing in typing and tracing *B. anthracis* isolates, especially by the whole-genome SNP analysis.[49] Besides those identified in the whole genome, SNPs in housekeeping genes have been nicely exploited to identify strains related to certain species or genetic types. For example, Prüss et al.[50] showed that certain nucleotides in the 16S rRNA gene and their relative prevalence among the different ribosomal operons in the genome correlate with the psychrotolerance of *B. cereus* strains, and hence, are signature of the species *B. weihenstephanensis*. Gierczyński et al.[51] developed a simple and cost-effective restriction-site-insertion PCR (RSI-PCR)–based assay that precisely detects the *B. anthracis*-specific nonsense mutation in *plcR* gene by restriction digestion with the endonuclease *Ssp*I.

21.1.4 Clinical Features and Toxicology of Human Pathogenic *Bacillus* spp.

21.1.4.1 *Bacillus anthracis*

B. anthracis, the causative agent of anthrax disease, has been well studied for over 150 years and continues to be a threat as a biological weapon.[3] This bacterium, forms spores (1–2 μm in size) able to withstand several kinds of stresses that are usually lethal to vegetative cells, such as ultraviolet and ionizing radiation, heat, various chemicals,[52] and oxidative stress.[53] Anthrax is primarily a disease affecting herbivores, but it can be transmitted to all mammals, including humans. Typically, animals become infected with *B. anthracis* by ingesting spores while grazing on contaminated soil or feed and also through the skin via biting insects.[54] Transmission of the spores to humans occurs through three routes of entry: (i) cutaneously, by direct contact with infected animals or by handling infected animal products[55]; (ii) respiratory, by inhalation of sufficient quantity of spores; and (iii) oropharyngeal and/or gastrointestinal (GI), by the ingestion of undercooked, contaminated meat.[56] All three forms of anthrax can be fatal if not treated in time, but the cutaneous form is often self-limiting.[57] A typical anthrax skin lesion corresponds to an ulcer covered by a characteristic black eschar.[58] The pathogenecity of *B. anthracis* depends on two major factors: a poly-γ-D-glutamic acid capsule and the three-component secreted anthrax toxin codified by plasmids pXO2 and pXO1, respectively. In addition to protecting the bacteria from the phagocytosis, the capsule allows bacterial proliferation and subsequent toxemia to develop in the bloodstream because of its poor immunogenic properties. The anthrax toxin is composed of three proteins known as protective antigen (PA), lethal factor (LF), and edema factor (EF). Once in the body, the PA binds to receptors that are present on most cells, including macrophages, and after proteolytic activation, it combines to LF and EF, allowing their delivery inside the cell.[59] Transcription of the genes *pag*, *lef*, and *cya* encoding the three-toxin proteins PA, LF, and EF, respectively, is regulated by a gene known as the anthrax toxin activator (*atxA*); whereas the capsule proteins, codified by the *capBCADE* operon, are regulated by AcpA and AcpB.[60,61]

Once in the host and after their phagocytosis by local macrophages, spores germinate and multiply in their way to the regional lymph nodes.[41] Along with other mechanisms, two spore-bound superoxide dismutases (SOD15 and SODA1), enzymes that are capable of detoxifying oxygen radicals,[62] were recently reported to play a crucial role in protecting the spores against the oxidative burst of macrophages in mouse model.[53] Following spore germination and bacterial proliferation, the capacity of the lymph nodes is exceeded; hence, the vegetative bacteria enter the bloodstream, which leads to bacteremia and the distinctive pathology of edema, hemorrhage, and tissue necrosis.[59] Dissemination of *B. anthracis* by hematogenous or lymphatogenous spread from a cutaneous, mediastinal, or GI focus can lead to infection at other sites, such as the brain, allowing entry into the central nervous system (CNS) and development of anthrax meningitis.[63] Another form of anthrax meningitis in which no primary focus is found has been reported and is known as "primary anthrax meningitis."[64]

With regard to the epidemiology, anthrax infection has been reduced dramatically in developed countries as the result of the widespread use of vaccines in high-risk people and animals, as well as the improvements in industrial hygiene. Currently, only very few cases of human anthrax occur naturally in these countries. However, the scenario is different in developing countries such as Africa and Asia, where the disease is endemic; this is due in part to difficulties of implementing vaccination programs. Of the three forms of the disease, cutaneous anthrax is the most frequent, followed by the inhalational and GI or oropharyngeal form.[55,57,65,66] Only few case reports have been published regarding anthrax meningitis epidemiology, and most of them report anthrax meningitis in settings of one of the three forms of the disease.[57,64,67,68]

21.1.4.2 *Bacillus cereus* and Other Human Pathogenic *Bacillus* spp.

Due to their ubiquitous presence in the environment, *Bacillus* spp., other than *B. anthracis*, isolated from clinical specimens have been often considered as a contamination; hence, most laboratories do not identify these isolates further. It has been reported that *Bacillus* spp. are isolated from 0.1% to 0.9% of blood cultures, but only 5%–10% of these are clinically significant infections.[69–71] However, in recent years there has been a growing number of reports dealing with their implication in severe and even lethal infections. *Bacillus* spp. other than *B. anthracis* were shown to cause nosocomial bacteremia, pneumonia, brain and liver abscesses, infection of CNS, periodontitis, peritonitis, endophthalmitis, soft-tissue infection, and meningitis.

Nosocomial Bacteremia. *B. cereus* is the most frequently isolated of *Bacillus* spp. involved in the development of nosocomial infections.[7–10,72,73] Van Der Zwet[72] described an outbreak of invasive *B. cereus* infections in a neonatal intensive care unit caused by contaminated balloons used for manual ventilation. These balloons had been cleaned with detergent, which was insufficient to kill spores of *B. cereus*. Kalpoe et al.[9] reported an outbreak of vancomycin-resistant *B. cereus* respiratory tract colonization in pediatric intensive care unit (PICU) patients due to contamination of reusable ventilator air-flow sensors. *B. cereus* spore contamination of alcohol used to disinfect the air-flow sensors was postulated to be the source of the dissemination. Dohmae et al.[8] reported on five bloodstream infections in five patients related to catheter use. *B. cereus* contamination of reused towels, which would have occurred during the wash process, was considered to be the origin of this outbreak. Ohsaki et al.[7] reported on the occurrence of three consecutive cases of *B. cereus* bacteremia in patients with hematologic malignancies after hospital renovation. The authors considered that filters used in the heating, ventilation, and air-conditioning systems, as well as towels and gowns, were the probable sources of the *B. cereus* outbreak. Kuroki et al.,[10] demonstrated that *B. cereus* and *B. thuringiensis* species previously implicated in nosocomial bacteremia by catheter infection were able to form biofilm after 24 h of incubation. To date, there is one report describing the occurrence of fatal *B. thuringiensis* bacteremia, in a neutropenic patient who suffered from severe pulmonary disease,[74] indicating that this entomopathogenic bacterium may behave as an opportunistic pathogen in immunocompromised individuals. *B. cereus* together with *B. licheniformis* and *B. pumilus* were shown to be implicated in 3.4% of 350 bacteremic episodes that occurred during a 5-year period in patients having hemotological malignancies.[75] To date, only one case of *B. licheniformis* nosocomial infection, affecting a very-low-birth-weight neonate in a neonatal intensive care unit, has been reported.[76] In immunocompetent individuals, *B. licheniformis* bacteremia has been reported in association with central venous catheters.[77] *B. pumilus* has also been implicated in persistent bacteremia in a patient with cholangitis and in nosocomial bacteremia via catheter infection.[78,79] In addition to antibiotic treatment, the central venous catheter may need to be removed for complete cure.

Taken together, these reports indicate that, although cases of *Bacillus* spp. other than *B. anthracis* and *B. cereus* bacteremia are rare in healthy individuals, it is important to consider the potential pathogenicity of *Bacillus* spp. in immunocompromised hosts, including premature infants and neonates, as well as in patients with various risk factors.

Endophthalmitis. Infectious endophthalmitis, the inflammation of intraocular tissues or fluids due to intraocular infection, is a major public health concern. In addition to fungi and gram-negative bacteria[80]; different *Bacillus* species, such as *B. megaterium*,[81] *B. subtilis*, *B. mycoides*, *B. pumilus*, *B. flexus*, *and B. thuringiensis*, have been reported as causative agents of infectious endophthalmitis.[82] However, the most important eye pathogen in this genus is *B. cereus* because of its ability to blind rapidly. Endophthalmitis associated to this bacterium often results in significant vision loss or loss of globe architecture in 1–2 days.[5,83] The eye pathogenesis of *B. cereus* is due to rapid growth, migration, and toxin production by this microorganism, which lead to the failure of the host immune defenses.[82,84] In addition to toxin production by the offending bacterium, the pathogenesis of *B. cereus* was also attributed to the intraocular inflammatory response and, in particular, to tumor necrosis factor-alpha (TNFα) in experimental endophthalmitis.[85] Recently, *B. cereus* was also shown to induce the permeability of the blood-ocular barrier in experimental endophthalmitis.[86]

Foodborne Illnesses. *Bacillus cereus* is well known as a common cause of emetic and diarrheal food poisoning. The symptoms of these foodborne illnesses are usually mild and the disease is often self-limiting. However, this opportunistic human pathogen can be associated with severe intoxications and infections that can be fatal, even in otherwise healthy individuals. Indeed, three fatal cases due to fulminant liver failure following intoxication with cereulide-producing *B. cereus* strains and liver abscess have been reported to date.[13,87,88] Other *Bacillus* spp. have also been implicated in foodborne illnesses, including *B. weihenstephanensis*, *B. subtilis*, *B. licheniformis*, and *B. pumilus*; while other species, such as *B. mojavensis*, *B. firmus*, *B. megaterium*, *B. simplex*, and *B. fusiformis* were shown to produce toxins with a cereulide-like mode of action.[14]

Pneumonia. *Bacillus cereus* has also been associated with fulminant pneumonia in five otherwise healthy metalworkers: a single nonfatal case in 1994 and four fatal cases: two in 1997 and two in 2003.[89–91] The *B. cereus* strains responsible for these cases of fulminant pneumonia were shown to produce a capsule and to have the genetic determinants that codify for the anthrax toxins.[92] A sixth case of fatal pneumonia caused in a leukemic patient by carbapenem-resistant *B. cereus* has also been reported recently.[93] The fact that B. *cereus* has been implicated in such unusual severe fatal pneumonia in immunocompetent individuals highlights the need for increased awareness of the potential for this opportunistic human pathogen to cause serious systemic infections, even in patients who appear to be otherwise healthy. Thus, a better understanding of the environmental and occupational risk factors that may increase the susceptibility of individuals to infection by this bacterium should be considered.

Periodontitis. The genus *Bacillus* has received relatively little attention in oral microbiology, and the *Bacillus* spp. isolated from human oral samples have been often regarded as transient microflora or as external contaminants. However, there have been several reports on the implication of *Bacillus* species, such as *B. cereus/B. thuringiensis*,[94] *B. pumilus*,[95] biofilm-forming *B. subtilis*,[96] *B. circulans*, *B. laterosporus*, *B. lentus*, and *B. sphaericus*[97] in oral disease, mainly gingivitis and periodontitis. On the other hand, cryptic (having no definable origin) pyogenic infections of the central nervous system (CNS) secondary to dental affections have been reported in which *B. subtilis* and *B. circulans* are implicated.[98] These cases indicate that, although dental and paradental infections seem to be a rare cause of intracerebral or intraspinal infections in otherwise healthy individuals, it is recommended that a dental focus should always be considered in the evaluation and treatment of these kinds of cryptic infections of the CNS.

Meningoencephalitis and Meningitis. In addition to *B. anthracis*, to the best of our knowledge, only *B. cereus* has been reported to be implicated in cases of meningoencephalitis in preterm infants[99] and neonates[100] or in cases of meningitis in neonates[101,102] and immunocompromised individuals.[103,104]

Soft-tissue Infection. To date, there have been two case reports that have implicated *B. cereus* in necrotizing fasciitis and myositis that mimick clostridial myonecrosis in immunocompromised patients.[105,106] This is a severe infection of skin and soft tissue by *B. cereus* that has symptoms similar to clostridial gas gangrene: in both cases the infection is caused by a gram-positive rod, and gas production and muscles inflammation (myositis) could be observed. However, to select the appropriate therapy, it is very important to differentiate between the two kinds of infections. Indeed, the treatment of clostridial gas gangrene is usually based on penicillin, an antibiotic to which *B. cereus* is resistant because it produces β-lactamase. *B. cereus* has to be suspected as a pathogen when the skin and soft-tissue infection occurs with sepsis in immunocompromised patients; whereas, *Clostridium* have to be considered when the infection occurs after trauma in otherwise healthy individuals. Another criterion for the differentiation between the two conditions is the site of infection: for clostridial gas gangrene, the primary site of inflammation is the muscle parenchyma, in contrast to *B. cereus* infection, where inflammation often develops from the subcutaneous tissues to muscles.[106]

Peritonitis. *B. cereus*,[107,108] *B. circulans*,[109] and B. *licheniformis*[110] have also been implicated in peritonitis in patients undergoing peritoneal dialysis. There is also one case report on *B. cereus* peritonitis after cesarean section.[111]

21.1.5 Diagnosis Techniques

The diagnosis of a human illness is based on the combination of different clues including the symptoms of the illness, the history of the patient, the possible risk factors, and the detection and identification of the suspected agent and/or its pathogenicity factors. Conventional detection procedures of *Bacillus* human pathogens are often based in the first step on the isolation and identification of the bacterium on selective cultural media from clinical specimens based on biochemical, phenotypic, and physiological characteristics. Otherwise, detection of the whole microorganism or its pathogenecity factors could be performed by immunological assays. Nucleic acid–based approaches are, however, most-often applied due to their sensitivity and rapidity.

21.1.5.1 Conventional Diagnosis Techniques

Fast and reliable identification of the pathogenic microorganism implicated in the illness is key to a timely and adequate treatment of a patient. Typically, *B. anthracis* can be identified from clinical specimens and differentiated from other *Bacillus* spp. by classical microbiological tests that are recommended by the World Health Organization (WHO) and by the Centers for Disease Control and Prevention (CDC). Like the other species of the *B. cereus* group, *B. anthracis* is positive to Gram staining and produces spores; however, it can be distinguished from the nonanthrax group (*B. cereus* and *B. thuringiensis*) by its capacity to produce a polypeptide capsule in vivo and in vitro when grown under suitable conditions, its sensitivity to gamma phage and penicillin G, and by the lack of motility and β-hemolytic activity on sheep or horse blood agar medium.[57]

In addition to blood agar medium, *B. anthracis* can be isolated on the selective polymyxin-lysozyme-ethylenediaminetetraacetic acid-thallium acetate (PLET) agar[112] or on anthracis chromogenic agar (ACA).[113] On PLET medium, *B. anthracis* colonies appear small, white, domed, and circular. Anthracis chromogenic agar contains the chromogenic substrate 5-bromo-4-chloro-3-indoxyl-choline phosphate that yields teal colonies upon hydrolysis by phosphatidylcholine-specific phospholipase C (PC-PLC), also known as lecithinase C. Upon incubation at 35°C–37°C, *B. cereus* and *B. thuringiensis* colonies are dark teal-blue within 24 h; whereas, *B. anthracis* colonies need at least 48 h to develop this color because of the slower rate of synthesis of this enzyme as a result of the truncated *plcR* gene and PlcR regulator. However, the blood-agar medium is easier and more cost effective, and hence is recommended for the isolation of *B. anthracis* from samples not overloaded with background bacilli, such as clinical specimens. The selective media PLET and ACA are suitable for the isolation of *B. anthracis* by well trained and expert personal, from samples with high numbers of bacteria but taking into account that there are limitations.[114]

However, this classical identification procedure is time-consuming and needs at least 24–48 h to be accomplished. In addition, there are several other challenges in the detection of *B. anthracis*: (i) not all the *B. anthracis* isolates fit all the above described characteristics, and atypical *B. anthracis* isolates lacking pXO2 (and hence capsule) have been identified; (ii) pathogenic capsule-producing *B. cereus* strains harboring plasmids similar to those of *B. anthracis* have been identified in cases of human fatal pneumonia and from the carcasses of great apes in Africa[115]; and (iii) blood cultures do provide *B. anthracis* identification, whereas cultures from skin or lesions are not reliable as they yield positive results in only 60%–65% of cases, especially after antibiotic treatment.[58]

The limitations of culture-based diagnosis of pathogenic *Bacillus* spp. causing human illness make necessary the application of quicker and more specific approaches, especially for the detection of *B. anthracis*. Several antibody-based tests targeting spores, vegetative cells, and anthrax toxin proteins have been developed for the specific detection of *B. anthracis*.[59,116–118] Recently, several other approaches targeting the *B. anthracis* capsular antigen detection by latex agglutination[119] or the entire organism mediating phage bioluminescence[120] have been reported. However, these antibody-based assays bear the limitation of sensitivity since low concentration of antigens could not be detected in the first steps of bacterial infection.

21.1.5.2 Molecular Diagnosis Techniques

DNA isolation from clinical specimens is a critical step in molecular biology–based diagnosis techniques because of the presence of high levels of inhibitors and host DNA in relation to the amount of DNA from the pathogenic bacterium, especially in the first step of the infection. Also, safety is an important consideration for the personnel handling specimens that are suspected to contain infectious *B. anthracis* spores, which are known to be difficult to inactivate. Several extraction kits are commercially available for the rapid isolation of DNA to be used in nucleic acid–based detection assays. In particular, the UltraClean Microbial DNA Isolation Kit (Mo Bio Laboratories, Inc.) was shown to offer the best method for depleting viable *B. anthracis* spores from samples.[121] Dauphin et al.[122] reported that gamma irradiation of DNA preparations allowed inactivation of residual *B. anthracis* spores; their complete removal could be reached by centrifugal filtration with 0.1-micron filter unit.[123]

21.2 METHODS

21.2.1 Sample Preparation

Bacterial Isolation. *Bacillus* spp. are often isolated from clinical specimens on 5%–7% horse or sheep blood agar medium. After overnight incubation at 35°C, nonanthrax bacteria of the *B. cereus* group are gram-positive, rod-shaped, spore formers, motile, and beta-hemolytic; whereas *B. anthracis* is nonmotile and nonhemolytic. *B. anthracis* is also sensitive to gamma phage and penicillin G. Media like PLET or ACA could also be used for the selective isolation of *B. anthracis*.

DNA Isolation. DNA template for PCR reactions can be prepared from the purified isolates by a simple boiling method as following. Cells are collected by centrifugation (5000 × *g* for 2 min) from 1 mL of a 4 h-culture grown in nutrient broth at 35°C. The cell pellet is washed once with 500 μL of sterile MilliQ water or TE buffer (pH = 8), centrifuged, resuspended in 100 μL of sterile MilliQ water or TE buffer, and boiled for 10–15 min. The boiled samples are centrifuged at 10,000 × *g* for 5 min to precipitate cell debris, and the supernatants, which contain nucleic acids, are collected and stored at –20°C until use. There are also several commercial kits available for total DNA extraction from overnight-grown pure bacterial cultures as well as from spores.[121] The phenol-chloroform extraction protocol is also useful to obtain pure high-molecular-weight genomic DNA.[124]

Total DNA from clinical specimens can be isolated using commercial DNA extraction kits, such as the ChargeSwitch gDNA Mini Bacteria Kit (Invitrogen); DNeasy Blood and Tissue kit (Qiagen); NucliSens Isolation Kit (Biomerieux); Puregene Genomic DNA Purification Kit (Qiagen); QIAamp DNA Blood Mini Kit (Qiagen); MagNA Pure LC (MPLC) microbiology kit MGrade (Roche Diagnostics GmbH); MagSi-DNA isolation kit for blood (Magna-Medics Diagnostics B.V.), and the UltraClean Microbial DNA Isolation Kit (Mo Bio Laboratories), following the manufacturers' instructions. However, this step of sample processing may be hampered by the low concentration of the pathogen and the dominance of human DNA, especially in samples such as blood, cerebrospinal fluid, or in other primary sterile sites. Horz et al.[125] developed two new kits, MolYsis® (Molzym GmbH & Co. KG, Bremen, Germany) and Pureprove® (SIRS-Lab GmbH, Jena, Germany) which is also known as Looxster® Universal kit, for the enrichment of bacterial DNA from blood samples that were also shown to be efficient using oral samples from plaque periodontitis and caries. Hansen et al.[126] found that pretreatment of blood samples with MolYsis Basic prior to DNA isolation with MolYsis Complete Kit improves the detection of the DNA of pathogenic bacteria from blood samples allowing a detection limit of 50 CFU/mL in multiplex real-time PCR.

21.2.2 Detection Procedures

Culture-dependent techniques for direct isolation and identification of Bacillus spp. based on traditional methods, looking to metabolism and physiology, is time consuming and laborious. Also, it is not possible to rely on bacterial culture for the diagnosis of cutaneous human anthrax, for which it has been reported that the bacterial culture–based diagnosis yields positive results only in 60%–65% of cases.[58] Furthermore, the presence of atypical nonencapsulated *B. anthracis* strains and pathogenic capsule-producing *B. cereus* isolates represents a further challenge for a specific and accurate identification based on this phenotypic character. Thus, a combination of classical microbiological assays and nucleic acid–based methods is often necessary to confirm the identity of the infectious bacterial agent. Actually, there is a wide variety of molecular-based techniques for the detection of *B. anthracis*, including PCR, real-time PCR, and multiplex PCR. All these techniques are based on the specific amplification of the genes *pag*, *lef*, and *cap*. However, all these genetic determinants reside on plasmids; hence *B. anthracis* isolates that lack one of the plasmids can not be easily identified based on these techniques. For this reason, chromosomal-based *B. anthracis*–specific markers, such as BA813, rpoB, gyrA, gyrB, saspB, plcR, and BA5345 have been identified.[51,127–129]

Recently, Antwerpen et al.[129] reported the development of a TaqMan real-time–based assay that targets the locus BA_5345, and that specifically detected *B. anthracis*, as was shown by testing 328 *Bacillus* strains. Within this sequence, they amplified a 96-bp large fragment by using the primers dhp61_183- 113F (5′-CGTAAGGACAATAAAGCCGTTG T-3′), dhp61_183–208R (5′-CGATACAGACATTTATTGGG AACTACAC-3′) and TaqMan -probe dhp61_183–143T (5′-6-FAM-TGCAATCGATGAGCTAATGAACAATGACCCT–TAMRA-3′). The TaqMan® PCR assay is performed in a 20 μL PCR-reaction mixture containing 1× ABITaqMan® mix (Applied Biosystems, Darmstadt, Germany); 0.9 μM primer dhp61_183–113F; 0.3 μM of primer dhp61_183–208R and 0.25 μM of labeled probe dhp61_183–143T; and 2 μL of template DNA. Amplification is carried out on the MX3000 Real-Time PCR System (Stratagene) with thermal conditions as follows: an initial denaturation step of 10 min at 95°C is followed by 45 cycles consisting of 15 s at 95°C and 1 min at 55°C. This assay is reported to have a detection limit of 12.7 copies/reaction.

Another molecular assay that can be applied for the rapid and sensitive/quantitative detection of *B. anthracis* has been reported recently.[130] This molecular assay is based on the combination of molecular beacons and real-time PCR technologies. Molecular beacons are highly specific nucleic acid probes that, when bound to their target DNA sequences, act as switches by emitting fluorescence as a result of the alteration of their conformation.[130,131] The molecular beacon–based real-time PCR assay was performed by designing five molecular beacons targeting *B. anthracis capA*, *capB*, *capC*, *lef*, and *pag* alleles that were used in five uniplex assays for the detection of these genes in a broad range of samples.[130] The assay is conducted in 25 μL reaction mixture containing 5.0 μL DNA, 4.5 μL distilled water, 12.5 μL Platinum Quantitative PCR SuperMix-UDG, 1.0 μL of one of the five molecular beacons and 1.0 μL (20 pmol/μL) of each of the forward and reverse primers. The assay is run on a 7900HT Real-Time PCR System (Applied Biosystems) with the following thermal cycling program: denaturation (94°C for 10 min), followed by 40 cycles of amplification (94°C denaturation for 15 s; 50°C annealing and data collection for 30 s; and 72°C polymerization for 30 s). During each cycle of amplification, the fluorescence emission in each sample is automatically recorded at 490 nm.

Another innovative powerful gene-amplification technique that is becoming very popular and emerging as a simple and rapid diagnostic tool for the detection and identification of microbial diseases is the loop-mediated isothermal amplification (LAMP) that was described for the first time by Notomi et al.[132] LAMP is based on the principle of strand-displacing DNA synthesis by the Bst DNA polymerase, with six distinct primers that recognize eight distinct sequences of a target gene. DNA amplification performed under isothermal conditions (60°C–65°C) using a heating block and/or water bath, is usually completed in less than 1 h. The gene-amplification products can be analyzed by agarose gel electrophoresis or by real-time monitoring of the turbidity of the reaction mixture in a turbidimeter. In this later case, it is also possible to quantify the gene copy number based on a standard curve generated from different concentrations of the gene copy number plotted against time of positivity. Moreover, gene amplification can be monitored by the naked eye either as turbidity or in the form of a color change when a fluorescent dsDNA intercalating dye is used. LAMP assays have been used for rapid detection of several pathogenic viruses, bacteria, and blood protozoa.[133,134]

Recently, a specific LAMP assay targeting the chromosomal *sap* gene that codifies for the S-layer constituent protein and *cap* and *pag* plasmid genes was designed by Kurosaki et al.[135] as a method for *B. anthracis* detection. Loopamp DNA Amplification Kit (Eiken Chemical, Tokyo, Japan) is used for performing LAMP reaction following the manufacturer's instructions. Briefly, a 25 μL final volume reaction mixture containing 40 pmol of each forward inner primer (FIP) and a reverse inner primer (BIP); 5 pmol of each of the outer primers F3 and B3; 20 pmol of each of the two loop primers (LF and LB); 12.5 μL of 2× reaction mix; 1.0 μL of Bst DNA polymerase; and 5.0 μL of DNA sample are incubated at 63°C for 60 min. The LAMP products are analyzed on 1% agarose gel and ehidium bromide staining. Otherwise, for real-time monitoring of amplification, the reaction mixtures are incubated and observed by spectrophotometric analysis using a real-time turbidimeter (LA-200; Teramecs, Kyoto, Japan). The time of positivity observed through real-time LAMP assay is determined as the time at which the turbidity reached the threshold value fixed at 0.1, which is double the average turbidity value of negative controls in several replicates. The authors reported that this assay is 10- to 100-fold more sensitive than conventional PCR-based methods, allowing the detection of 10 fg of bacterial DNA per reaction and 3.6 CFU of bacterial spores in blood samples from experimentally infected mice.

21.3 CONCLUSIONS AND FUTURE PERSPECTIVES

Spores produced by *Bacillus* spp., other than *B. anthracis*, are ubiquitous in the environment owing to their resistance to harsh environmental conditions as well as to disinfection procedures. Their presence in food samples or clinical specimens has often been considered as a contamination. However, *Bacillus* spp. can cause serious and even lethal infections or intoxications not only in immunocompromised individuals or patients with other high-risk factors but also in otherwise healthy individuals. Conventional diagnostic procedures are based mainly on phenotypic and biochemical characteristics. For example, in the *B. cereus* group, the differentiation of the well-known opportunistic human pathogen *B. cereus* and the anthrax agent *B. anthracis* is based in part on the presence of a poly-gamma-D-glutamic acid capsule. However, both encapsulated *B. cereus* that has been implicated in fatal pneumonia and *B. thuringiensis* strains producing a capsule similar to the *B. anthracis* one[136] have been discovered recently. This finding underlines the importance of using multiple tests for a reliable identification of isolates associated with human disease.

Different *Bacillus* spp. outbreaks in hospital environments where the origin of the disseminated isolate was the reusable equipment, such as towels, gowns, and hand ventilator balloons, have been reported. Also, *Bacillus* spp. have been isolated from alcoholic cotton and detergents. These reported cases highlight the importance of using disposal or wrapping steamed towels and wrapping alcoholic cotton as preventive measures to avoid nosocomial bacteremia caused by these spore formers. They also underscore the importance of using professionally performed sterilization procedures for all reusable equipment and that should include eradication of spores. For this purpose, autoclaving is imperative and especially for reusable equipment in intensive care units and for material used by immunocompromised patients. The 70%–90% alcohol that is extensively used in hospitals is suitable as germicide but does not affect spores. On the other hand, regular cleaning of the filters of the air-conditioning system and the inlets and outlets of the air-ventilation system, as well as using the recommended concentration of sodium hypochlorite in the laundry process would be important measures to avoid nosocomial bacteremia by spore formers.

Several molecular detection methods are actually available for *B. anthracis* and its pathogenecity factors, including the secreted anthrax toxins and the capsule, as well as for *B. cereus* and its toxins. These include conventional PCR, multiplex PCR and RT–PCR. However, although highly specific and sensitive, these molecular methods have some drawbacks. The conventional PCR bears the risk of false-positive results due to sample contamination. The multiplex PCR could be a good alternative for detecting different pathogens or different toxins of the same pathogenic bacterial species simultaneously. However, in some cases, it is possible to get false-negative results because of the presence of different primer pairs in the same reaction mix or different annealing temperatures. In addition, this method is qualitative and does not allow quantification of the pathogenic bacterium or of its toxins. In order to quantify the presence of a specific pathogenic bacterium or its toxins, TaqMan or SYBR green-based RT–PCRs could be applied. However, although highly sensitive, this technique could be biased by the presence of PCR inhibitors in clinical specimens. Also, the presence of dominant concentrations of human DNA in comparison to the DNA from the pathogenic bacterium in samples such as blood and cerebrospinal fluid, in particular, during the first step of the infection, can deeply affect the sensitivity of the PCR. Another practical drawback of RT–PCR is the necessity of expensive and sophisticated real-time platforms and reagents that are not available in most laboratories especially in developing countries. During this last decade, and since its description in 2000, LAMP technique has been largely used for the diagnosis of human pathogenic bacteria as well as viruses and protozoa. Only one recent report described the application of this assay for the diagnosis of *B. anthracis* in experimentally infected mice. The advantages of this technique consist of rapid amplification at one set temperature without the need of a thermal cycler, simple operation, and easy detection by a turbidimeter or naked eye. Also, combining LAMP assay with boiling extraction, it is possible to quickly detect very low copies of the target gene and hence to perform an efficient on-site diagnosis in less than 1 h without the need for culture growth or precise DNA extraction. Hence, LAMP can be considered an innovative technique that has potential applications for both clinical diagnosis and surveillance of infectious diseases in developing countries, without requiring sophisticated equipment or skilled personnel. It is, however, important to take into account that one of the major drawbacks of this assay is that designing the primers presents a difficulty because at least six distinct primers that recognize eight distinct sequences of a target gene are required for the performance of the assay.

In conclusion, the most serious and important human pathogen within the genus *Bacillus* is *B. anthracis*. However, other species should also be considered in the diagnosis of septicemic infections, pneumonia, meningitis, endophthalmitis, and cutaneous infections. Among others, *Bacillus cereus* is being acknowledged as an emerging pathogen, especially in immunocompromised individuals. Also, although rare, capsule-producing pathogenic *B. cereus* strains with fatal anthrax-like pathology exist and should be considered, especially in patients with high-risk occupational factors, such as metal workers, and in particular, in zones where anthrax was endemic. As far as the detection of *Bacillus* spp. in clinical specimens, a combination of culture-based and molecular methods, including PCR amplification of the virulence factor encoding genes is necessary for the confirmation of the diagnosis and hence the choice of the appropriate therapy. To avoid misdiagnosis and unnecessary treatment, performing more than one set of cultures from clinical specimens is recommended before assuming that an isolate is a pathogen or dismissing a pathogen as a contaminant, which can cause complications and even mortality due to delay in the treatment.[71]

REFERENCES

1. Kim, K.M. et al., Characterization of *Bacillus polyfermenticus* KJS-2 as a Probiotic, *J. Microbiol. Biotechnol.*, 19, 1013, 2009.
2. Sánchez, B. et al., Identification of surface proteins involved in the adhesion of a probiotic *Bacillus cereus* strain to mucin, *Microbiology*, 155, 1708, 2009.
3. Schwartz, M., Dr. Jekyll, and Mr. Hyde, A short history of anthrax, *Mol. Aspects. Med.*, 30, 347, 2009.
4. Shimoyama, R. et al., Contamination of *Bacillus cereus* in an apheresis-derived platelet concentrate, *J. Clin. Apheresis*, 12, 103, 1997.
5. Pinna, A. et al., *Bacillus cereus* keratitis associated with contact lens wear, *Ophthalmology*, 108, 1830, 2001.
6. Störmer, M. et al., Spore-forming organisms in platelet concentrates: A challenge in transfusion bacterial safety, *Transfus. Med.*, 18, 371, 2008.
7. Ohsaki, Y. et al., Undetected *Bacillus* pseudo-outbreak after renovation work in a teaching hospital, *J. Infect.*, 54, 617, 2007.
8. Dohmae, S. et al., *Bacillus cereus* nosocomial infection from reused towels in Japan, *J. Hosp. Infect.*, 69, 361, 2008.
9. Kalpoe, J.S. et al., Dissemination of *Bacillus cereus* in a paediatric intensive care unit traced to insufficient disinfection of reusable ventilator air-flow sensors, *J. Hosp. Infect.*, 68, 341, 2008.
10. Kuroki, R. et al., Nosocomial bacteremia caused by biofilm-forming *Bacillus cereus* and *Bacillus thuringiensis*, *Intern. Med.*, 48, 791, 2009.
11. Daffonchio, D., Cherif, A., and Borin, S., Homoduplex and heteroduplex polymorphisms of the amplified ribosomal 16S-23S internal transcribed spacers describe genetic relationships in the "*Bacillus cereus* group," *Appl. Environ. Microbiol.*, 66, 5460, 2000.
12. Cherif, A. et al., Genetic relationship in the "*Bacillus cereus* group" by rep-PCR fingerprinting and sequencing of a *Bacillus anthracis*-specific rep-PCR fragment, *J. Appl. Microbiol.*, 94, 1108, 2003.
13. Dierick, K. et al., Fatal family outbreak of *Bacillus cereus*-associated food poisoning, *J. Clin. Microbiol.*, 43, 4277, 2005.
14. Raddadi, N. et al., *Bacillus*. In Liu, D. (Editor), *Molecular Detection of Food-borne Pathogens*, 1st ed., chapter 10, Taylor & Francis Group, New York, 2009.
15. Baron, F. et al., Isolation and characterization of a psychrotolerant toxin producer, *Bacillus weihenstephanensis*, in liquid egg products, *J. Food Prot.*, 70, 2782, 2007.
16. Nakamura, L.K., DNA relatedness among *Bacillus thuringiensis* serovars, *Int. J. Syst. Bacteriol.*, 44, 125, 1994.
17. Nakamura, L.K., and Jackson, M.A., Clarification of the taxonomy of *Bacillus mycoides*, *Int. J. Syst. Bacteriol.*, 45, 46, 1995.
18. Nakamura, L.K., *Bacillus pseudomycoides* sp. nov., *Int. J. Syst. Bacteriol.*, 48, 1031, 1998.
19. Ash, C. et al., Phylogenetic heterogeneity of the genus *Bacillus* revealed by comparative analysis of small-subunit-ribosomal RNA sequences, *Lett. Appl. Microbiol.*, 13, 202, 1991.
20. Ash, C. et al., Comparative analysis of *Bacillus anthracis*, *Bacillus cereus*, and related species on the basis of reverse transcriptase sequencing of 16S rRNA, *Int. J. Syst. Bacteriol.*, 41, 343, 1991.

21. Ash, C., and Collins, M.D., Comparative analysis of 23S ribosomal RNA gene sequences of *Bacillus anthracis* and emetic *Bacillus cereus* determined by PCR-direct sequencing, *FEMS Microbiol. Lett.*, 94, 75, 1992.
22. Brousseau, R. et al., Arbitrary primer polymerase chain reaction, a powerful method to identify *Bacillus thuringiensis* serovars and strains, *Appl. Environ. Microbiol.*, 59, 114, 1993.
23. Carlson, C.R., Grønstad, A., and Kolstø, A.-B., Physical maps of the genomes of three *Bacillus cereus* strains, *J. Bacteriol.*, 174, 3750, 1992.
24. Carlson, C.R., and Kolstø, A.-B., A complete physical map of a *Bacillus thuringiensis* chromosome, *J. Bacteriol.*, 175, 1053, 1993.
25. Carlson, C.R., Caugant, D.A., and Kolstø, A.-B., Genotypic diversity among *Bacillus cereus* and *Bacillus thuringiensis* strains, *Appl. Environ. Microbiol.*, 60, 1719, 1994.
26. Carlson, C.R., Johansen, T., and Kolstø, A.-B., The chromosome map of *Bacillus thuringiensis* subsp. *canadensis* HD224 is highly similar to that of the *Bacillus cereus* type strain ATCC 14579, *FEMS Microbiol. Lett.*, 141, 163, 1996.
27. Daffonchio, D. et al., Restriction Site Insertion-PCR (RSI-PCR) for rapid discrimination and typing of closely related microbial strains, *FEMS Microbiol. Lett.*, 180, 77, 1999.
28. Daffonchio, D. et al., A randomly amplified polymorphic DNA marker specific for the *Bacillus cereus* group is diagnostic for *Bacillus anthracis*, *Appl. Environ. Microbiol.*, 65, 1298, 1999.
29. Kim, Y.R., Czajka, J., and Batt, C.A., Development of a fluorogenic probe-based PCR assay for detection of *Bacillus cereus* in nonfat dry milk, *Appl. Environ. Microbiol.*, 66, 1453, 2000.
30. Schraft, H., and Griffith, M.W., Specific oligonucleotide primers for detection of lecitinase-positive *Bacillus* spp. by PCR, *Appl. Environ. Microbiol.*, 61, 98, 1995.
31. Schraft, H. et al., Epidemiological typing of *Bacillus* spp. isolated from food, *Appl. Environ. Microbiol.*, 62, 4229, 1996.
32. Margulis, L. et al., The *Arthromitus* stage of *Bacillus cereus*: Intestinal symbionts of animals, *Proc. Natl. Acad. Sci. USA*, 95, 1236, 1998.
33. Jensen, G.B. et al., The hidden lifestyles of *Bacillus cereus* and relatives, *Environ. Microbiol.*, 5, 631, 2003.
34. Swiecicka, I., and Mahillon J., Diversityof commensal Bacillus cereus sensu latoisolated from the commonsowbug (Porcellio scaber, Isopoda), *FEMS Microbiol. Ecol.*, 56, 132, 2006.
35. Ciammaruconi, A. et al., Fieldable genotyping of *Bacillus anthracis* and *Yersinia pestis* based on 25-loci multi locus VNTR analysis, *BMC Microbiol.*, 8, 21, 2008.
36. Ivanova, N. et al., Genome sequence of *Bacillus cereus* and comparative analysis with *Bacillus anthracis*, *Nature*, 423, 87, 2003.
37. Gohar, M. et al., A comparative study of *Bacillus cereus*, *Bacillus thuringiensis* and *Bacillus anthracis* extracellular proteomes, *Proteomics*, 5, 3696, 2005.
38. Keim, P. et al., Molecular evolution and diversity in *Bacillus anthracis* as detected by amplified fragment length polymorphism markers, *J. Bacteriol.*, 179, 818, 1997.
39. Radnedge, L. et al., Genome differences that distinguish *Bacillus anthracis* from *Bacillus cereus* and *Bacillus thuringiensis*, *Appl. Environ. Microbiol.*, 69, 2755, 2003.
40. Read, T.D. et al., The genome sequence of *Bacillus anthracis* Ames and comparison to closely related bacteria, *Nature*, 423, 81, 2003.
41. Sanz, P. et al., Detection of *Bacillus anthracis* spore germination in vivo by bioluminescence imaging, *Infect. Immun.*, 76, 1036, 2008.
42. Helgason, E. et al., Multilocus sequence typing scheme for bacteria of the *Bacillus cereus* group, *Appl. Environ. Microbiol.*, 70, 191, 2004.
43. Zwick, M.E. et al., Microarray-based resequencing of multiple *Bacillus anthracis* isolates, *Genome Biol.*, 6, R10, 2005.
44. Keim, P. et al., Multiple-locus variable-number tandem repeat analysis reveals genetic relationships within *Bacillus anthracis*, *J. Bacteriol.*, 182, 2928, 2000.
45. Read, T.D. et al., Comparative genome sequencing for discovery of novel polymorphisms in *Bacillus anthracis*, *Science*, 296, 2028, 2002.
46. Gierczyński, R. et al., Intriguing diversity of *Bacillus anthracis* in eastern Poland—the molecular echoes of the past outbreaks, *FEMS Microbiol. Lett.*, 239, 235, 2004.
47. Merabishvili, M. et al., Diversity of *Bacillus anthracis* strains in Georgia and of vaccine strains from the former Soviet Union, *Appl. Environ. Microbiol.*, 72, 5631, 2006.
48. Prüss, B.M. et al., Correlation of 16S ribosomal DNA signature sequences with temperature-dependent growth rates of mesophilic and psychrotolerant strains of the *Bacillus cereus* group, *J. Bacteriol.*, 181, 2624, 1999.
49. Van Ert, M.N. et al., Global genetic population structure of *Bacillus anthracis*, *PLoS ONE*, 2, e461, 2007.
50. Prüss, B.M. et al., Correlation of 16S ribosomal DNA signature sequences with temperature-dependent growth rates of mesophilic and psychrotolerant strains of the *Bacillus cereus* group, *J. Bacteriol.*, 181, 2624, 1999.
51. Gierczyński, R. et al., Specific *Bacillus anthracis* identification by a plcR–targeted restriction site insertion-PCR (RSI-PCR) assay, *FEMS Microbiol. Lett.*, 272, 55, 2007.
52. Nicholson, W.L. et al., Resistance of *Bacillus* endospores to extreme terrestrial and extraterrestrial environments, *Microbiol. Mol. Biol. Rev.*, 64, 548, 2000.
53. Cybulski, R.J. et al., Four superoxide dismutases contribute to *Bacillus anthracis* virulence and provide spores with redundant protection from oxidative stress, *Infect. Immun.*, 77, 274, 2009.
54. Hugh-Jones, M., and Blackburn, J. The ecology of *Bacillus anthracis*, *Mol. Aspects. Med.*, 30, 356, 2009.
55. Karahocagil, M.K. et al., Cutaneous anthrax in Eastern Turkey: A review of 85 cases, *Clin. Exp. Dermatol.*, 33, 406, 2008.
56. Ozdemir, H. et al., Anthrax of the gastrointestinal tract and oropharynx: CT findings, *Emerg. Radiol.*, DOI 10.1007/s10140–009-0821-y, 2009.
57. World Health Organization, *Anthrax in Humans and Animals*, 4th ed., WHO, Geneva. Available from http://www.who.int/csr/resources/publications/ anthrax_webs.pdf, 2008.
58. Dixon, T. et al., Anthrax, *N. Engl. J. Med.*, 341, 815, 1999.
59. Edwards, K. et al., *Bacillus anthracis*: Toxicology, epidemiology and current rapid-detection methods, *Anal. Bioanal. Chem.*, 384, 73, 2006.
60. Drysdale, M., Bourgogne, A., and Koehler, T.M. Transcriptional analysis of the *Bacillus anthracis* capsule regulators, *J. Bacteriol.*, 187, 5108, 2005.
61. Fouet, A., and Mock, M., Regulatory networks for virulence and persistence of *Bacillus anthracis*, *Curr. Opin. Microbiol.*, 9, 160, 2006.
62. Fang, F.C., Antimicrobial reactive oxygen and nitrogen species: Concepts and controversies, *Nat. Rev. Microbiol.*, 2, 820, 2004.

63. van Sorge, N.M. et al., Anthrax toxins inhibit neutrophil signaling pathways in brain endothelium and contribute to the pathogenesis of meningitis, *PLoS One*, 3, doi:10.1371/journal.pone.0002964, 2008.
64. Sejvar, J.J., Tenover, F.C., and Stephens, D.S. Management of anthrax meningitis, *Lancet. Infect. Dis.*, 5, 287, 2005.
65. Simonson, T.S. et al., *Bacillus anthracis* in China and its relationship to worldwide lineages, *BMC Microbiol.*, 9, 71, 2009.
66. Doganay, M., and Metan, G., Human Anthrax in Turkey from 1990 to 2007, *Vector-born. Zoonot.*, 9, DOI: 10.1089/vbz.2008.0032, 2009.
67. Lanska, D., Anthrax meningoencephalitis, *Neurology*, 59, 327, 2002.
68. Narayan, S.K. et al., Anthrax meningoencephalitis—declining trends in an uncommon but catastrophic CNS infection in rural Tamil Nadu, South India, *J. Neurol. Sci.*, 281, 41, 2009.
69. Weber, D.J. et al., Clinical significance of *Bacillus* species isolated from blood cultures, *South. Med. J.*, 82, 705, 1989.
70. Lee, C.C. et al., Clinical significance of potential contaminants in blood cultures among patients in a medical center, *J. Microbiol. Immunol. Infect.*, 40, 438, 2007.
71. Ashkenazi-Hoffnung, L. et al., Seasonality of *Bacillus* species isolated from blood cultures and its potential implications, *Am. J. Infect. Control*, 37, 495, 2009.
72. Van Der Zwet, W.C. et al., Outbreak of *Bacillus cereus* infections in a neonatal intensive care unit traced to balloons used in manual ventilation, *J. Clin. Microbiol.*, 38, 4131, 2000.
73. Hernaiz, C. et al., Nosocomial bacteremia and catheter infection by *Bacillus cereus* in an immunocompetent patient, *Clin. Microbiol. Infect.*, 9, 973, 2003.
74. Ghelardi, E. et al., *Bacillus thuringiensis* pulmonary infection: Critical role for bacterial membrane-damaging toxins and host neutrophils, *Microbes Infect.*, 9, 591, 2007.
75. Ozkocaman, V. et al., *Bacillus* spp. among hospitalized patients with haematological malignancies: Clinical features, epidemics and outcomes, *J. Hosp. Infect.*, 64, 169, 2006.
76. Lépine, A. et al., *Bacillus licheniformis* septicemia in a very-low-birth-weight neonate: A case report, *Infection*, 37, 156, 2009.
77. Blue, S.R., Singh, V.R., and Sanbolle, M.A., *Bacillus licheniformis* bacteraemia: Five cases associated with indwelling central venous catheters, *Clin. Infect. Dis.*, 20, 629, 1995.
78. Haymore, B.R. et al., A case of persistent *Bacillus pumilis* bacteremia associated with cholangitis, *J. Infect.*, 52, 154, 2006.
79. Bentur, H.N., Dalzell, A.M., and Riordan, F.A.I. Central venous catheter infection with *Bacillus pumilus* in an immunocompetent child: A case report, *Ann. Clin. Microbiol. Antimicrob.*, 6, 12, 2007.
80. Ramakrishnan, R. et al., Microbiological profile of culture proven cases of exogenous and endogenous endophthalmitis: A 10-year retrospective study, *Eye*, 23, 945, 2009.
81. Ramos-Esteban, J.C. et al., *Bacillus megaterium* delayed onset lamellar keratitis after LASIK, *J. Refract. Surg.*, 22, 309, 2006.
82. Callegan, M.C. et al., Virulence factor profiles and antimicrobial susceptibilities of ocular bacillus isolates, *Curr. Eye. Res.*, 31, 693, 2006.
83. Kopel, A.C., Carvounis, P.E., and Holz E.R. *Bacillus cereus* endophthalmitis following intravitreous bevacizumab injection, *Ophthalmic. Surg. Lasers. Imaging.*, 39, 153, 2008.
84. Callegan, M.C. et al., *Bacillus* endophthalmitis: Roles of bacterial toxins and motility during infection, *Invest. Ophthalmol. Vis. Sci.*, 46, 3233, 2005.
85. Ramadan, R.T. et al., A role for tumor necrosis factor-alpha (TNFα) in experimental *Bacillus cereus* endophthalmitis pathogenesis, *Invest. Ophthalmol. Vis. Sci.*, 49, 4482, 2008.
86. Moyer, A.L. et al., *Bacillus cereus*-induced permeability of the blood-ocular barrier during experimental endophthalmitis, *Invest. Ophthalmol. Vis. Sci.*, 50, 3783, 2009.
87. Mahler, H. et al., Fulminant liver failure in association with the emetic toxin of *Bacillus cereus*, *N. Engl. J. Med.*, 336, 1142, 1997.
88. Latsios, G. et al., Liver abscess due to *Bacillus cereus*: A case report, *Clin. Microbiol. Infect.*, 9, 1234, 2003.
89. Hoffmaster, A.R. et al., Identification of anthrax toxin genes in a *Bacillus cereus* associated with an illness resembling inhalation anthrax, *Proc. Natl. Acad. Sci. USA*, 101, 8449, 2004.
90. Miller, J.M. et al., Fulminating bacteremia and pneumonia due to *B. cereus*, *J. Clin. Microbiol.*, 35, 504, 1997.
91. Avashia, S.B. et al., Fatal pneumonia among metalworkers due to inhalation exposure to *Bacillus cereus* containing *Bacillus anthracis* toxin genes, *Clin. Infect. Dis.*, 44, 414, 2007.
92. Hoffmaster, A.R. et al., Characterization of *Bacillus cereus* isolates associated with fatal pneumonias: strains are closely related to *Bacillus anthracis* and harbor *B. anthracis* virulence genes, *J. Clin. Microbiol.*, 44, 3352, 2006.
93. Katsuya, H. et al., A patient with acute myeloid leukemia who developed fatal pneumonia caused by carbapenem-resistant *Bacillus cereus*, *J. Infect. Chemother.*, 15, 39, 2009.
94. Helgason, E. et al., Genetic structure of population of *Bacillus cereus* and *B. thuringiensis* isolates associated with periodontitis and other human infections, *J. Clin. Microbiol.*, 38, 1615, 2000.
95. Johnson, B.T. et al., Extracellular proteolytic activities expressed by *Bacillus pumilus* isolated from endodontic and periodontal lesions, *J. Med. Microbiol.*, 57, 643, 2008.
96. Yamane, K. et al., Identification and characterization of clinically isolated biofilm-forming gram-positive rods from teeth associated with persistent apical periodontitis, *J. Endod.*, 35, 347, 2009.
97. Sunde, P.T. et al., Microbiota of periapical lesions refractory to endodontic therapy, *J. Endod.*, 28, 304, 2002.
98. Ewald C., Kuhn, S., and Kalff, R. Pyogenic infections of the central nervous system secondary to dental affection—a report of six cases, *Neurosurg. Rev.*, 29, 163, 2006.
99. Lequin, M.H. et al., *Bacillus cereus* meningoencephalitis in preterm infants: Neuroimaging characteristics, *Am. J. Neuroradiol.*, 26, 2137, 2005.
100. Manickam, N., Knorr, A., and Muldrew, K.L. Neonatal meningoencephalitis caused by *Bacillus cereus*, *Pediatr. Infect. Dis. J.*, 27, 843, 2008.
101. Evreux, F. et al., A case of fatal neonatal *Bacillus cereus* meningitis, *Arch. Pediatr.*, 14, 365, 2007.
102. Lebessi, E. et al., *Bacillus cereus* meningitis in a term neonate, *J. Matern. Fetal. Neonatal. Med.*, 22, 458, 2009.
103. de Almeida, S.M. et al., Fatal *Bacillus cereus* meningitis without inflammatory reaction in cerebral spinal fluid after bone marrow transplantation, *Transplantation*, 76, 1533, 2003.
104. Haase, R. et al., Successful treatment of *Bacillus cereus* meningitis following allogenic stem Cell transplantation, *Pediatr. Transplant.*, 9, 338, 2005.
105. Meredith, F.T. et al., *Bacillus cereus* necrotizing cellulitis mimicking clostridial myonecrosis: Case report and review of the literature, *Scand. J. Infect. Dis.*, 29, 528, 1997.
106. Sada, A., Necrotizing fasciitis and myonecrosis "synergistic necrotizing cellulitis" caused by *Bacillus cereus*, *J. Dermatol.*, 36, 423, 2009.

107. Pinedo, S., Bos, A.J., and Siegert, C.E. Relapsing *Bacillus cereus* peritonitis in two patients on peritoneal dialysis, *Perit. Dial. Int.*, 22, 424, 2002.
108. Monteverde, M.L. et al., Relapsing *Bacillus cereus* peritonitis in a pediatric patient on chronic peritoneal dialysis, *Perit. Dial. Int.*, 26, 715, 2006.
109. Berry, N. et al., *Bacillus circulans* peritonitis in a patient treated with CAPD, *Perit. Dial. Int.*, 24, 488, 2004.
110. Park, D.J. et al., Relapsing *Bacillus licheniformis* peritonitis in a continuous ambulatory peritoneal dialysis patient, *Nephrology*, 11, 21, 2006.
111. Shimoni, Z. et al., *Bacillus cereus* peritonitis after Cesarean section, *Surg. Infect.*, 9, 105, 2008.
112. Knisely, R.F., Selective medium for *Bacillus anthracis*, *J. Bacteriol.*, 92, 784, 1966.
113. Juergensmeyer, M.A. et al., A selective chromogenic agar that distinguishes *Bacillus anthracis* from *Bacillus cereus* and *Bacillus thuringiensis*, *J. Food. Protect.*, 69, 2002, 2006.
114. Marston, C.K. et al., Evaluation of two selective media for the isolation of *Bacillus anthracis*, *Lett. Appl. Microbiol.*, 47, 25, 2008.
115. Klee, S.R. et al., Characterization of *Bacillus anthracis*-like bacteria isolated from wild great apes from Cote d'Ivoire and Cameroon, *J. Bacteriol.*, 188, 5333, 2006.
116. Hao, R. et al., Rapid detection of *Bacillus anthracis* using monoclonal antibody functionalized QCM sensor, *Biosens. Bioelectron.* 24, 1330, 2009.
117. Tang, S. et al., Detection of anthrax toxin by an ultrasensitive immunoassay using europium nanoparticles, *Clin. Vaccine. Immunol.*, 16, 408, 2009.
118. Duriez, E. et al., Femtomolar detection of the anthrax edema factor in human and animal plasma, *Anal. Chem.*, 81, 5935, 2009.
119. AuCoin, D.P. et al., Rapid detection of the poly-γ-D-glutamic acid capsular antigen of *Bacillus anthracis* by latex agglutination, *Diagn. Microbiol. Infect. Dis.*, 64, 229, 2009.
120. Schofield, D.A., and Westwater, C., Phage-mediated bioluminescent detection of *Bacillus anthracis*, *J. Appl. Microbiol.*, 107, 1468, 2009.
121. Dauphin, L.A., Moser, B.D., and Bowen, M.D. Evaluation of five commercial nucleic acid extraction kits for their ability to inactivate *Bacillus anthracis* spores and comparison of DNA yields from spores and spiked environmental samples, *J. Microbiol. Methods*, 76, 30, 2009.
122. Dauphin, L.A. et al., Gamma irradiation can be used to inactivate *Bacillus anthracis* spores without compromising the sensitivity of diagnostic assays, *Appl. Environ. Microbiol.*, 74, 4427, 2008.
123. Dauphin, L.A., and Bowen, M.D., A simple method for the rapid removal of *Bacillus anthracis* spores from DNA preparations, *J. Microbiol. Methods*, 76, 212, 2009.
124. Sambrook, J., Fritsch, E.F., and Maniatis, T., *Molecular Cloning: A Laboratory Manual*, 2nd ed., Cold Spring Harbor Laboratory Press, Cold Spring Harbor, NY, 1989.
125. Horz, H.P. et al., Selective isolation of bacterial DNA from human clinical specimens, *J. Microbiol. Methods*, 72, 98, 2008.
126. Hansen, W.L., Bruggeman, C.A., and Wolffs, P.F. Evaluation of new preanalysis sample treatment tools and DNA isolation protocols to improve bacterial pathogen detection in whole blood, *J. Clin. Microbiol.*, 47, 2629, 2009.
127. Ramisse, V. et al., The Ba813 chromosomal DNA sequence effectively traces the whole *Bacillus anthracis* community, *J. Appl. Microbiol.*, 87, 224, 1999.
128. Qi, Y. et al., Utilization of the *rpoB* gene as a specific chromosomal marker for real-time PCR detection of *Bacillus anthracis*, *Appl. Environ. Microbiol.*, 67, 3720, 2001.
129. Antwerpen, M.H. et al., Real-time PCR system targeting a chromosomal marker specific for *Bacillus anthracis*, *Mol. Cell. Probes*, 22, 313, 2008.
130. Hadjinicolaou, A.V., and Victoria, L., Use of molecular beacons and multi-allelic real-time PCR for detection of and discrimination between virulent *Bacillus anthracis* and other *Bacillus isolates*, *J. Microbiol. Methods*, 78, 45, 2009.
131. Li, Y. et al., Molecular beacons: An optimal multifunctional biological probe, *Biochem. Biophys. Res. Comm.*, 373, 457, 2008.
132. Notomi, T. et al., Loop-mediated isothermal amplification of DNA, *Nucleic Acids Res.*, 28, E63, 2000.
133. Mori, Y. et al., Detection of loop-mediated isothermal amplification reaction by turbidity derived from magnesium pyrophosphate formation, *Biochem. Biophys. Res. Comm.*, 289, 150, 2001.
134. Parida, M. et al., Loop mediated isothermal amplification (LAMP): A new generation of innovative gene amplification technique; perspectives in clinical diagnosis of infectious diseases, *Rev. Med. Virol.*, 18, 407, 2008.
135. Kurosaki, Y. et al., A simple and sensitive method for detection of *Bacillus anthracis* by loop-mediated isothermal amplification, *J. Appl. Microbiol.*, 107, 1947, 2009.
136. Cachat, E. et al., A *Bacillus thuringiensis* strain producing a polyglutamate capsule resembling that of *Bacillus anthracis*, *FEMS Microbiol. Lett.*, 285, 220, 2008.

22 Enterococcus

Teresa Semedo-Lemsaddek, Paula Lopes Alves, Rogério Tenreiro, and Maria Teresa Barreto Crespo

CONTENTS

22.1 INTRODUCTION

22.1.1 Classification and Morphology

Enterococci are mainly viewed as commensal bacteria inhabiting the gastrointestinal tract of humans and animals, but in fact, these microorganisms have been recognized as potentially pathogenic for humans for more than a century. In Thiercelin's[1] original description, these bacteria had been isolated from patients with enteritis, appendicitis, and meningitis.[2]

The classification and nomenclature of the genus *Enterococcus* has raised interest and constant discussion over the years, ever since Thiercelin first proposed the name "*enterocoques*" in 1902. In 1970, Kalina[3] proposed that *Streptococcus faecalis* and *S. faecium* should be renamed as *Enterococcus faecalis* and *E. faecium*, respectively. These proposals were not officially recognized for many years, and the genus *Enterococcus* did not appear in the Approved List of Bacterial Names until the 1980s.

In the 1980s, Schleifer and Kilpper-Bälz[4] resurrected the genus *Enterococcus* after performing DNA–DNA and DNA–rRNA hybridizations. Other studies carried out in the same decade, including 16S rRNA sequencing,[5] also demonstrated that enterococci form a separate genus. Analysis of these molecular data divided the streptococci *sensu lato* into three genera: (i) *Streptococcus sensu stricto*, comprising the majority of the known species; (ii) *Enterococcus*, for the enteric group; and (iii) *Lactococcus*, for the lactic streptococci.

Since then, the genus *Enterococcus* has been accepted as valid, and today it comprises 33 species. In 2007, Kohler[6] performed an extensive review of the publications that lead to each of the species currently accepted and included in TOBA (Taxonomic Outline of the Bacteria and Archaea),[7] based on a comprehensive taxonomy using 16S rRNA gene phylogeny.

Members of the genus *Enterococcus* are gram-positive ovoid cells that appear singly, in pairs, or in short chains; are catalase- and oxidase-negative; facultative anaerobes; and homofermentative, with lactic acid as the end product of glucose fermentation. The majority of the enterococci grow from 10°C to 45°C, with 6.5% NaCl and at pH 9.6, hydrolyze esculin in the presence of 40% bile salts and the G + C content ranges from 37% to 45%.[8,9] However, not all enterococci exhibit these characteristics; several authors have analyzed and reported differences in physiological properties among different species and strains belonging to the same species,[10–15] making a reliable identification at species level difficult.

22.1.2 Clinical Features, Epidemiology, and Pathogenesis

Although enterococci have traditionally been regarded as low-grade pathogens, in the last decades they have emerged as an increasingly important cause of nosocomial infections, mainly in patients with underlying disease. All kinds of severe immunosupression (e.g., oncology, hematology, nephrology, or transplantation units), long hospital stays,[16] or the presence of urinary or vascular catheters constitute notable risk factors.[17]

The infections most frequently caused by enterococci are urinary tract infections (UTIs),[18] followed by intraabdominal/intrapelvic abscesses or postsurgery wound and bloodstream infections (BSIs).[19] Enterococci are also infrequent causes of central nervous system (CNS) infections, neonatal infections, respiratory tract infections, osteomyelitis, or cellulitis.[20]

The importance of enterococci in the hospital setting is increasing every year. Enterococci are the third most common cause of nosocomial blood infections in the United States,[21] and in Europe, the fourth (http://www.earss.rivm.nl/). Regarding urinary tract infections, they rank second in both the United States and Europe.[22,23]

Although more than thirty species have been identified,[7] only two are responsible for the majority of human infections. The species *E. faecalis* and *E. faecium* account for 80% and 20% of the enterococcal infections, respectively. The proportion of *E. faecium* infections is increasing, and other species are being more frequently isolated from clinical samples, namely, *E. gallinarum*, *E. casseliflavus*, and *E. raffinosus*.[22,24,25]

The inherent tolerance of enterococci to unfavorable conditions allows them to survive for long periods in hospital settings, even on dry surfaces, and medical staff and instruments can be responsible for the transmission between hospitalized patients.[17,26–28] These characteristics contribute to an increased risk of enterococcal nosocomial infections and emphasize the importance of bacterial control in the health care environment. Enterococcal infections are an important health concern, and as nosocomial pathogens, they have a direct and significant economic impact by prolonging hospital stays and requiring additional therapeutic treatments.[29]

Another important factor contributing to the increasing number of enterococcal infections is the uncontrolled use of antimicrobial agents, which enhances the selective pressure for bacteria with natural and/or acquired antibiotic resistances, such as enterococci. Administration of antimicrobial agents that have little effect on enterococci results in the elimination of many endogenous microbiota and may open new niches for the overgrowth of pathogenic strains.

Enterococci have intrinsic resistance to several antimicrobial agents (e.g., cephalosporins, lincosamides, many β-lactams and low levels of aminoglycosides); frequently acquire plasmids and transposons carrying antimicrobial resistance genes (e.g., chloramphenicol, erythromycin, tetracyclin, and glycopeptides such as vancomycin) that can subsequently be transferred to other strains sharing the same ecological niche.[2,16,26,28]

Vancomycin resistance gains increased importance because this antimicrobial is considered a last resource treatment for many human infections. Vancomycin-resistant enterococci (VRE) were first detected in Europe (United Kingdom, and France) in 1988, and in the following year in the United States.[30] In the last decades, the incidence of VRE rose, mainly due to *E. faecium* isolates, and vancomycin-resistant strains have already been reported in countries throughout the world, such as Australia, Belgium, Canada, Denmark, Germany, Italy, Malaysia, Netherlands, Spain, Sweden, and Portugal.[31]

While the increase of VRE in the United States is attributed to the uncontrolled usage of antimicrobials in hospitals,[31] in Europe the reservoir of resistant strains seems to be linked with the use of avoparcin as a feed supplement for various livestock, and contamination via the food chain has been suggested.[32–35] The usage of pulsed field gel electrophoresis has also demonstrated that acute outbreaks are typically related to a single VRE clone or a limited number of clones.[36–38] Asymptomatic patients with VRE colonizing the gastrointestinal tract are usually the reservoir, and the dissemination between patients is mainly attributed to the indirect contact of the hands of health care workers.

Surprisingly, little is known about the factors that contribute to the ability of enterococci to cause infection. In *E. faecalis*, several putative virulence factors have been described over the last years, involved either in the attachment to host cells or to extracellular matrix proteins, or implicated in cell and tissue damage.

For gastrointestinal commensals such as enterococci, adhesins that promote binding to host cell receptors are expected to play an important role in the establishment and maintenance of colonization.[39] The enterococcal surface protein (Esp) is a high molecular weight extracellular surface protein of unknown function whose frequency is increased among infection-derived isolates.[40] This protein contributes to colonization and persistence of enterococci in the urinary tract[41] and also to biofilm formation on abiotic surfaces.[42] Aggregation substance (AS) is a pheromone-inducible surface protein that mediates the binding of donor cells to plasmid free recipients.[43,44] This protein is essential for high-efficiency conjugation of sex pheromone plasmids, contributing to the dissemination of antimicrobial resistance and virulence determinants. Different functions have been attributed to AS in addition to bacterial cell aggregation, including adherence to host tissues[45,46] and internalization,[47] adhesion to fibrin and increased cell-surface hydrophobicity,[48] resistance to killing by polymorphonuclear leukocytes and macrophages,[46,49] and increased vegetation size in experimental endocarditis[50] and endophthalmitis models.[51] Another adhesion-associated protein was identified while analyzing serum from patients with *E. faecalis* endocarditis (named *Efa*A for *E. faecalis* antigen A).[52,53]

Regarding cell and tissue damage, the first virulence factor to be studied was cytolysin, a posttranslationally modified protein toxin that causes a β-hemolytic reaction on certain blood erythrocytes and also possesses bactericidal activity against a broad range of gram-positive bacteria.[54–56] Gelatinase is an extracellular zinc-endopeptidase capable of hydrolysing gelatin, collagen, casein, and other small biologically active peptides, and its potential contribution to virulence has been suggested in several studies.[57–60]

As regards *E. faecium*, only three putative virulence genes have been identified; they encode an enterococcal surface protein (esp_{efm}), a putative hyaluronidase (hyl)[61,62] and an endocarditis-related adhesin *Efa*Afm.[52,53]

The putative virulence factors described for the genus *Enterococcus* can either be encoded by mobile genetic elements, such as plasmids or transposons, or by the bacterial chromosome. These genomic region clustering genes that contribute, directly or indirectly, to the pathogenic potency of the harboring microorganism are termed pathogenicity islands or PAIs.[63] The generalized sequencing of genomes revealed that PAIs are much more widespread than previously thought.[63] Regarding the genus *Enterococcus*, two PAIs have been described: one in a vancomycin-resistant strain of *E. faecalis* that caused a nosocomial outbreak,[64] and the other in a group of nosocomial strains of *E. faecium*.[65] More recently, the worldwide dissemination of the *E. faecalis* PAI among isolates from diverse locations and origins was demonstrated.[66]

22.1.3 Diagnosis

22.1.3.1 Phenotypic Techniques

Detection and reliable identification, especially regarding VRE, is of major concern since VRE leaves very few options for the treatment of enterococcal infections.[16,28] At the present time in Europe, VRE infections are increasing; meanwhile, in the United States, colonization of hospitalized patients is endemic in many tertiary care hospitals and increasingly causes infections.[16]

Conventional culture methods remain the most reliable and accurate techniques for the detection of bacterial pathogens such as enterococci. Traditionally, enterococcal identification relied on various cultivation procedures associated with the analysis of phenotypic profiles including Gram staining, colony morphology, motility, pigmentation, growth requirements, and enzymatic and/or metabolic activities. However, this culture-based identification is time consuming and frequently limited by the variability of atypical strains, the lack of sufficient data on recently described species, and the low reproducibility of some tests. Since *Enterococcus* is neither a phylogenetically nor phenotypically coherent and homogeneous genus, manual or automated systems based on phenotypic characteristics do not always allow the correct strain identification; therefore these problems assume great importance.[9]

Another disadvantage of the standard culture methods is their lack of ability to detect a substantial proportion of the bacterial population that is nonculturable. These viable but nonculturable (VBNC) microorganisms demonstrate metabolic activity,[67,68] maintain their pathogenicity[69,70] and their antibiotic resistance traits,[71] can express[72] and exchange genes,[73] and are able to reestablish division upon restoration of favorable conditions.[74,75]

It has been previously demonstrated that *E. faecalis* can enter the VBNC state,[72] thus representing a risk in hospital settings and emphasizing the importance of molecular detection in substitution of standard culture methods. Consequently, the optimization of molecular methods for microbial analysis is becoming more important every day, with special relevance for rapid, easy-to-implement, and reliable procedures that may be applied in routine laboratories.

22.1.3.2 Molecular Techniques

Nucleic acid–based identification and detection methods have been developed for nearly all bacterial pathogens.[76] Rapid and sensitive DNA-based assays that are applicable for the direct detection of enterococci in human clinical samples may improve the quickness and the accuracy of the detection of these bacteria.

In spite of all the advantages associated with molecular-based methodologies, the detection of microorganisms in clinical samples can come across several problems, such as PCR inhibition by sample-associated components, cross-contamination in sensitive PCR assays, and detection of living as well as dead cells (this specific problem can be solved by using mRNA as amplification target instead of DNA in RT–PCR).

Although PCR amplification of species-specific targets enables precise identification of known microbial species with short reporting times, it is also important to highlight that positive amplification only demonstrates the presence of the appropriate target DNA sequences in the sample and does not imply that the target organism is viable. This problem can be overcome by the evaluation of cell viability (mRNA detection by RT–PCR) or visualization of living cells in mixed bacterial populations (in situ PCR/hybridization).

Several molecular diagnostic methods based on the detection of bacterial nucleic acids have already been described for enterococci. The majority of these methodologies relies on PCR due to the low amounts of DNA and/or cells needed, as well as the detection of a specific DNA sequence against a large background of prokaryotic and eukaryotic cells and organic material present in mixed samples.[77] Under ideal conditions, with 100% efficiency and unlimited resources, a single copy of target gene can be amplified to approximately 3.4×10^{10} copies after 35 cycles of PCR. However, it becomes important to emphasize that the ability to design oligonucleotide primers is limited by our knowledge of a microorganism's genome, as well as by the representativeness of publicly available sequence databases in terms of all variants of that microbial group.

Distinct molecular approaches have already been described for enterococci including broad-range PCR,[78] multiplex PCR,[79–81] species-specific PCR,[82–87] real-time PCR,[76,88,89] direct detection of bacteria using fluorescent in situ hybridization (FISH) probes,[90–92] and amplification of 16S and 23S ribosomal gene sequences for hybridization to a microarray.[93,94]

Conventional PCR assays detect only the presence or absence, rather than the quantity, of a target microorganism and cannot distinguish between nonviable and viable cells. To deal with these limitations several modifications of standard PCR have recently been developed.

Real-time quantitative PCR (qPCR), such as the TaqMan system, relies on the release and detection of a signal after cleavage of a fluorescent-labeled probe by the 5′–3′-exonuclease activity of Taq DNA polymerase. A fluorogenic oligonucleotide probe, which specifically binds within the two PCR primer regions, is degraded in each PCR cycle, thereby monitoring the amplification of the target gene by an increasing fluorescence signals in real time. The signal is related to the amount of amplicon present during each cycle and will increase as the amount of specific amplicon increases.[76,88,89,95,96] The specificity of the probe introduces a high level of confidence concerning the detection of target microorganisms in mixed samples.

Another approach with known advantages is hybridization of whole bacterial cells with fluorescent oligonucleotides (FISH) targeted to broad-range sequences (16 or 23S rRNA) or specific genes.[91–93] This method is rapid and easy-to-perform and allows the in situ observation of the abundance, spatial distribution, and morphology of bacterial cells.[92] FISH assays can significantly shorten time to retrieve results and to initiate control measures in the food industry.

Microarray technology is also a powerful tool that can be used for simultaneous detection of large sets of different genes of DNA (or RNA) targets on a glass slide.[76] By processing multiple samples and analyzing up to tens of thousands of genes in a single assay, this method is a promising alternative for monitoring environmental and food bacteria, but it requires expensive microarray and laser-scanner apparatus. When applying microarrays for microorganism detection, PCR amplification of target genes is often involved in order to enhance detection sensitivity.[76,97–99] Microarray methodology has already been applied for the detection of enterococci in environments such as wastewater[76] and artificially contaminated milk samples.[94]

A summary of available probes for the most relevant enterococcal human pathogens that can be used in qPCR, RT–qPCR, hybridization assays, FISH, and microarrays is presented in Table 22.1.

As stated before, a variety of molecular methods have been developed for enterococcal detection and identification at the genus or species level. The detected DNA targets include genes encoding for rRNA,[76,84,88,89,103,104] the alpha subunit of RNA polymerase and the phenylalanyl-tRNA synthase,[106] the elongation factor EF-Tu,[83] the D-alanine:D-alanine ligase,[80] a penicillin-binding protein,[77] the muramidase,[87] the alpha subunit of the bacterial ATP synthase,[107] the manganese-dependent superoxide dismutase,[82,108] and the heat-shock proteins (groESL),[76,86,109,110] as well as genus or species-specific antimicrobial determinants.[80,81,89]

A short description of each method is presented below and additional information about broad-range PCR primers (designed for eubacteria and lactic acid bacteria and to be used in positive control reactions) and enterococcal PCR primers (designed for selective amplification at genus and species level) can be found in Table 22.2. Regarding specific primers for identification at species level, Table 22.2 only includes the most relevant human pathogens (*E. faecalis*, *E. faecium*, *E. casseliflavus/E. flavescens*, and *E. gallinarum*) and for information about other enterococcal species, see Semedo-Lemsaddek et al.[140]

Ribosomal RNA. rRNA is a universal component of bacterial ribosomes and 5S, 16S, and 23S rRNAs are present at the small (30S) or large (50S) subunits that comprise the complete active ribosome (70S). Since rRNAs are found in high copy numbers (10^3–10^4 molecules) per actively growing cell and their coding genes are usually present in more than two copies, targeting the rRNAs or their coding genes has the potential to increase the detection sensitivity compared to assays based on the detection of single-copy genes.[121]

Small subunit ribosomal RNA (16S rRNA) gene sequences are widely used as a target for the development of PCR primers and/or specific hybridization probes due to their phylogenetic informational content and wide availability in databases. By comparison, although the large subunit rRNA (23S rRNA) gene contains more sequence variations than the 16S rRNA gene, and thus may provide more useful targets for microorganism identification at species or intraspecies level, it has been less frequently used, mainly because of the limited availability of sequence information. Since both 16S rRNA and 23S rRNA are part of the ribosome, and, due to evolutionary constraints, there is a minimal variation within these regions. So, examining the 16S-23S rRNA intergenic noncoding region (ITS) may be necessary to distinguish very similar microorganisms.

Because of the universality of rRNA genes, the selection of specific primers or probes has to be criteriously performed. The use of such gene targets will be more difficult when analyzing products with complex mixtures of bacteria, making specificity analysis fundamental for validation of the method.

Nowadays, several databases compiling full and partial rRNA sequences are available online and updated regularly, allowing fast and easy comparison of data from unknown organisms. These databases include the Ribosomal Database Project II Release 10 (update 14, released on August 31, 2009) that consists of 1074,075 aligned and annotated 16S rRNA sequences, along with seven online analysis tools (http://rdp.cme.msu.edu/).

Amplification of the rRNA genes (e.g., PCR or RT–PCR) and the use of probes directed to these regions (e.g., FISH or microarrays) have been widely applied for the detection and/or identification of *Enterococcus* as a genus or for particular species. Since *E. faecalis* is the most commonly encountered species, the majority of the primers and probes are specific for its identification.

RNA Polymerase and Phenylalanyl-tRNA Synthase. In 2005, Naser[106] and coworkers reported that all enterococcal species analyzed were clearly differentiated on the basis of their RNA polymerase (*rpoA*) and phenylalanyl-tRNA synthase (*pheS*) gene sequences. Strains of the same enterococcal species had at least 99% *rpoA* and 97% *pheS* sequence similarity, whereas different enterococcal species had a maximum of 97% *rpoA* and 86% *pheS* sequence similarity, pointing to the usefulness of this gene as a target for species-level

TABLE 22.1
Probes for Detection and Identification of Enterococci

Organism	Target	Sequence (5′–3′)	Method	Reference
Eubacteria	16S rRNA	EUB338- GCTGCCTCCCGTAGGAGT	FISH	100
	16S rRNA	GTACAAGGCCCGGGAACGTATTCACC	Microarrays	101
Gram-positive bacteria	16S rRNA	GACATAAGGGGCATGATGATTTGACGT	Microarrays	101
Enterococcus spp.	16S rRNA	Encl 31- CCCCTTCTGATGGGCAGG	Microarrays	96
	16S rRNA[a]	Encl 45- GGGATAACACTTGCAAAC	Microarrays	96
	16S rRNA	Encl 259- GAAGTCGCGAGGCTAAGC	Microarrays	96
	16S rRNA	ENC221- CACCGCGGGTCCATCCATCA	FISH	92
	16S rRNA	ENF191- GAAAGCGCCTTTCACTCTTATGC	FISH	92
	23S rRNA	TGGTTCTCTCCGAAATAGCTTTAGGGCTA	TaqMan PCR (qPCR)	88
	23S rRNA	GPL813TQb- TGGTTCTCTCCGAAATAGCTTTAGGGCTA	TaqMan PCR (qPCR)	88
	23S rRNA	Enc01aV- AGGTTAAGTGAATAAGGG	Microarrays	96
	23S rRNA	ENC176- CAGTTCTCTGCGTCTACCTC	FISH	92
vanA enterococci	*vanA* gene	vana3- CAACTAACGCGGCACTGTTTCCCAAT	TaqMan PCR (qPCR)	89
vanC enterococci	16S rRNA	EGAC183- CAACTTTCTTCCATGCGGAAAAT	FISH	92
E. casseliflavus	23S rRNA	Eca58- AGCTTGTCCGTACAGGTA	Microarrays	94
E. faecalis	16S rRNA	EFA1-TGATTTGAAAGGCGCTTTCGGGTGTCGCTGATGGATGGAC	Microarrays	84
	16S rRNA	EFA2- GAAGAACAAGGATGAGAGTAACTGTTCATCCCTTGACGG	Microarrays	84
	16S rRNA	EFA3- GAAGTACAACGAGTTGCGAAGTCGCGAGGCTAAGCTAAT	Microarrays	84
	16S rRNA	CAATTGGAAAGAGGAGTGGCGGACG	TaqMan PCR (qPCR)	102
	16S rRNA	TGCCGCATGGCATAAGAGTGAAAGGCGCTTTCGGGTGTC	Microarrays	103
	16S rRNA	CAAGGACGTTAGTAACTGAACGTCCCCTGACGGTATCTAA	Microarrays	103
	16S rRNA	GAAGTACAACGAGTCGTTAGACCGCGAGGTCATGCAAATC	Microarrays	103
	16S rRNA	CCTCGCGGTCTAGCAGCTCGTTGTGCTT	Microarrays	101
	16S rRNA	Enfl84- UGCACUCAAUUGGAAAGAGG	FISH	91
	16S rRNA	efs129 CCCTCTGATGGGTAGGTT	Microarrays	96
	ITS region	Efs1- TATTTATTGATTAACCTTCT(T)	Hybridization	104
	ITS region	Efs2- AAGAAGTGATCAAGACCCA(T)	Hybridization	104
	23S rRNA	Efs18i- CGAAATGCTAACAACACC	Microarrays	modified by 93
	23S rRNA	Efa54- CAAAAACAACTGGTACAG	Microarrays	96
E. faecium	16S rRNA	TGATTTGAAAGGCGCTTTCGGGTGTCGCTGATGGATGGAC	Microarrays	103
	16S rRNA	GAAGAACAAGGATGAGAGTAACTGTTCATCCCTTGACGG	Microarrays	103
	16S rRNA	GAAGTACAACGAGTTGCGAAGTCGCGAGGCTAAGCTAAT	Microarrays	103
	ITS region	Efm1- TTTTATGAGACGATCGAT(T)	Hybridization	104
	ITS region	Efm2- TCTTGATCTAACTTCTAT(T)	Hybridization	104
	23S rRNA	Efm09i- GGATGTTACGATTGTGTG	Microarrays	modified by 93
	23S rRNA	Efm09- CACACAATCGTAACATCC	Microarrays	96
	23S rRNA	Efi58- TGACTCCTCTCCAGACTT	Microarrays	96
	23S rRNA	ATCATACGATCAGCCGCAGTGAATA	RT–qPCR	76
	23S rRNA	ENU140- TTCACACAATCGTAACATCCTA	FISH	92
	23S rRNA	ENU1470- GACTCCTTCAGACTTACTGCTTGG	FISH	92
E. flavescens	23S rRNA	Efl58i- TTCTACCTATACGGACAA	Microarrays	96
E. gallinarum	ITS region	Egal- GAGTGGACAAGTTAAAGA(T)	Hybridization	104
	23S rRNA	Ega09- CACAACTGTGTAACATCC	Microarrays	96
	23S rRNA	EGA141- ATTCACAACTGTGTAACATCCTAT	FISH	92
E. flavescens/ E. gallinarum	23S rRNA	Eacdfg57- AGACATATCCATCAGTCT	Microarrays	96
E. casseliflavus/ E. flavescens	ITS region	Ecasc- GAGTTGAAATGTTAAAAGAG(T)	Hybridization	104
	23S rRNA	Ecafl09i- GGATGTTACGTCTGCGTG	Microarrays	96

continued

TABLE 22.1 (Continued)
Probes for Detection and Identification of Enterococci

Organism	Target	Sequence (5′–3′)	Method	Reference
E. casseliflavus/ E. gallinarum/ E. flavescens	16S rRNA	CGF- CCGTATAACACTATTTTCCGC	Hybridization	105
E. faecium	23S rRNA	Edfm57- CTGCTTGGACAGACATTT	Microarrays	96
E. faecium	16S rRNA	FMDUR- CATTCAGTTGGGCACTCTAGCAAGA	Hybridization	105
E. faecium	16S rRNA	Enfm93- CCGGAAAAAGAGGAGUGGC	FISH	91
E. gallinarum/ E. casseliflavus/ E. flavescens	16S rRNA	Ecg191- GCGCCTTTCAACTTTCTT	Microarrays	96
E. faecium	16S rRNA	Enc93- GCCACTCCTCTTTTTCCG	Microarrays	96
E. casseliflavus/ E. flavescens	16S rRNA	Ecf459- GGGATGAACATTTTACTC	Microarrays	96

[a] Not *E. faecalis*.

TABLE 22.2
Broad-Range Control and Enterococcal Genus- and Species-Specific PCR Primers

Primers (5′–3′)	Product size (bp)	Observations	Reference
	Eubacteria		
16S rRNA			
For- GGATTAGATACCCTGGTAGTCC	320		111 in 112
Rev- TCGTTGCGGGACTTAACCCAAC			
Sm785 For- GGATTAGATACCCTGGTAGTC	~1500	Sm785 For + 422Rev	113
For- GCTGGATCACCTCCTTTC	676/578		114
Uni331 For- TCCTACGGGAGGCAGCAGT	466		115
Uni797 Rev- GGACTACCAGGGTATCTAATCCTGTT			
For- GAGAGTTTGATYCTGGCTCAG	1554		84, 103
Rev- AAGGAGGTGATCCARCCGCA			
1020 For- TTAAACTCAAAGGAATTGACGG	170		79
1190 Rev- CTCACGRCACGAGCTGACGAC			
Ef16 For- AGAGTTTGATCCTGGCTCA	1522		86
Ef16 Rev- GGTTACCTTGTTACGACTTC			
U1 For- CCAGCAGCCGCGGTAATACG	996		116
U2 Rev- ATCGG(C/T)TACCTTGTTACGACTTC			
23S rRNA			
422 Rev- GGAGTATTTAGCCTT	~1500	Sm785 For + 422Rev	113
L189 Rev- GGTACTTABATGTTTCAGTTC	~1000	ENFE + L189	117
ECST784 For- AGAAATTCCAAACGAACTTG	92		118
ENC854 Rev- CAGTGCTCTACCTCCATCATT			
For- AGGAKGTTGGCTTAGAAGCAG	900		76
Rev- CGCTACCTTAGGACCGTTATAGTTAC			
rrn			
13B For- GTGAATACGTTCCCGGGCCT	~600	16S rRNA + 23S rRNA + ITS	104
6 Rev-GGGTTYCCCCRTTCRGAAAT			
Chaperonin 60			
H279A For- GAIIIIGCIGGIGA(TC)GGIACIACIAC	600		109
H280A Rev- (TC)(TG)I(TC)(TG)ITCICC(AG)AAICCIGGIGC(TC)TT			

continued

TABLE 22.2 (Continued)
Broad-Range Control and Enterococcal Genus- and Species-Specific PCR Primers

Primers (5′–3′)	Product Size (bp)	Observations	Reference
590 For- GGNGACGGNACNACNACNGCAACNGT	589		110
590 Rev- TCNCCRAANCCNGGYGCNTTNACNGC			
Elongation factor EF-Tu (*tuf*)			
U1 For- AAYATGATIACIGGIGCIGCICARATGGA	803		83
U2 Rev- AYRTTITCICCIGGCATIACCAT			
	Lactic acid bacteria		
α subunit of ATP synthase (atpA)			
atpA-20 For- TAYRTYGGKGAYGGDATYGC	1102		107
atpA-27 Rev- CCRCGRTTHARYTTHGCYTG			
Phenylalanyl-tRNA synthase (pheS)			
pheS-21 For- CAYCCNGCHCGYGAYATGC	494		106
pheS-22 Rev- CCWARVCCRAARGCAAARCC			
pheS-21 For- CAYCCNGCHCGYGAYATGC	431		106
pheS-23 Rev- GGRTGRACCATVCCNGCHCC			
RNA polymerase a subunit (rpoA)			
rpoA-21 For- ATGATYGARTTTGAAAAACC	815		106
rpoA-23 Rev- ACHGTRTTRATDCCDGCRCG			
	Genus *Enterococcus*		
16S rRNA			
E1 For- TCAACCGGGGAGGGT	733		119
E2 Rev- ATTACTAGCGATTCCGG			
16S rRNA			
Ec-ssu1 For- GGATAACACTTGGAAACAGG	115		120
Ec-ssu1 Rev- TCCTTGTTCTTCTCTAACAA			
16S rRNA			
For- ATCAGAGGGGGATAACACTT	337		121
Rev- ACTCTCATCCTTGTTCTTCTC			
23S rRNA			
ECST784 For - AGAAATTCCAAACGAACTTG	92		88
ENC854 Rev- CAGTGCTCTACCTCCATCATT			
ITS region			
For- CAAGGCATCCACCGT	320 + 420		122,123 in 124
Rev- GAAGTCGTAACAAGG			
Chaperonin 60 GroES			
EntGroES For- TTAAAACCATTAGGCGATCG	1609		110
EntGroES Rev- CCCATNCCCATNGANGGRTCCAT			
Elongation factor EF-Tu *(tuf)*			
Ent1 For- TACTGACAAACCATTCATGATG	112		83
Ent2 Rev- AACTTCGTCACCAACGCGAAC			
Manganese-dependent superoxide dismutase *(sodA)*			
D1 For- CCITAYICITAYGAYGCIYTIGARCC	480		82
D2 Rev- ARRTARTAIGCRTGYTCCCAIACRTC			
	E. faecalis		
16S rRNA			
Efs130 For- AACCTACCCATCAGAGGG	360		125
Efs490 Rev- GACGTTCAGTTACTAACG			126
For- CGCTTCTTTCCTCCCGAGT	143		102
Rev- GCCATGCGGCATAAACTG			

continued

TABLE 22.2 (Continued)
Broad-Range Control and Enterococcal Genus- and Species-Specific PCR Primers

Primers (5′–3′)	Product Size (bp)	Observations	Reference
For- TACTGACAAACCATTCATGATG	112		127
Rev- AACTTCGTCACCAACGCGAAC			
72 For- CCGAGTGCTTGCACTCAATTGG	138		79
210 Rev- CTCTTATGCCATGCGGCATAAAC			
ENFE For- GTCGCTAGACCGCGAGGTCATGA	~1000	ENFE + L189	128
Chaperonin 60 GroESL			
EfGroES For- GGAATTGTTCTTGCATCCGT	185		110
EfGroES Rev- ACAATTAAGTATTCTACGCC			
For- TGTGGCAACAGGGATCAAGA	64		76
Rev- TTCAGCGATTTGACGGATTG			
Efes For- GTGTTAAAACCATTAGGCGAT	650		86
Efes Rev- AAGCCTTCACGAACAATGG			
D-Alanine/D-alanine ligase *(ddl)*			
For- ATCAAGTACAGTTAGTCTTTATTAG	941		80
Rev- ACGATTCAAAGCTAACTGAATCAGT			
DD13 For- CACCTGAAGAAACAGGC	475		129
DD13 Rev- ATGGCTACTTCAATTTCACG			
Iron sulfur–binding protein			
Efis For- ATGCCGACATTGAAAGAAAAAATT	803		86
Efis Rev- TCAATCTTTGGTTCCATCTCT			
Manganese-dependent superoxide dismutase *(sodA)*			
FL1 For- ACTTATGTGACTAACTTAACC	360		108
FL2 Rev- TAATGGTGAATCTTGGTTTGG			
Penicillin-binding protein *(pbp5)*			
For- CATGCGCAATTAATCGG	444		77
Rev- CATAGCCTGTCGCAAAAC			
Ef0027- unknown function			
For- GCCACTATTTCTCGGACAGC	518		130
Rev- GTCGTCCCTTTGGCAAATAA			
vane			
EE1 For- TGTGGTATCGGAGCTGCAG	430		129
EE2 Rev- ATAGTTTAGCTGGTAAC			
For- TGTGGTATCGGAGCTGCAG	513		131
Rev- GTCGATTCTCGCTAATCC			
vanG			
For- CGGTTGTGCCGTACTTGGC	810		132
Rev- GGGTAAAGCCATAGTCTGGGGC			
EG1 For- CGGCATCCGCTGTTTTTGA	941		129
EG2 Rev- GAACGATAGACCAATGCCTT			
E. faecalis* and *E. faecium			
vanA			
vana3 For- CTGTGAGGTCGGTTGTGCG	64		89
vana3 Rev- TTTGGTCCACCTCGCCA			
vanB			
For- GTGACAAACCGGAGGCGAGGA	433		133
Rev- CCGCCATCCTCCTGCAAAAAA			
For- CATCGCCGTCCCCGAATTTCAAA	298		81
Rev- GATGCGGAAGATACCGTGGCT			

continued

TABLE 22.2 (Continued)
Broad-Range Control and Enterococcal Genus- and Species-Specific PCR Primers

Primers (5′–3′)	Product Size (bp)	Observations	Reference
EB3 For- ACGGAATGGGAAGCCGA	647		129
EB4 Rev- TGCACCCGATTTCGTTC			
For- CATCGCCGTCCCCGAATTTCAAA	297		134
Rev- GATGCGGAAGATACCGTGGCT			
vanD			
ED1 For- TGTGGGATGCGATATTCAA	500		129
ED2 Rev- TGCAGCCAAGTATCCGGTAA			
For- GARGATGGITSCATMCARGGW	630		135
Rev- MGTRAAICCIGGCAKRGTRTT			
For-TAAGGCGCTTGCATATACCG	461		136
Rev- TGCAGCCAAGTATCCGGTAA			
vanD-U1 TATTGGAATCACAAAATCCGG	628		137
vanD-U2 CGGCTGTGCTTCCTGATG			
	E. faecium		
23S rRNA			
Ef Rev- CACACAATCGTAACATCCTA	676/578		95
D-Alanine/D-alanine ligase *(ddl)*			
For- GCAAGGCTTCTTAGAGA	550		80
Rev- CATCGTGTAAGCTAACTTC			
FAC1 For- GAGTAAATCACTGAACGA	1091		129
FAC2 Rev- CGCTGATGGTATCGATTCAT			
Manganese-dependent superoxide dismutase *(sodA)*			
FM1 For- GAAAAAACAATAGAAGAATTAT	215		108
FM2 Rev- TGCTTTTTTGAATTCTTCTTTA			
Unknown region			
For- TTGAGGCAGACCAGATTGACG	658		138
Rev- TATGACAGCGACTCCGATTCC			
	Other enterococcal species		
Manganese-dependent superoxide dismutase (sodA)			
CA1 For- TCCTGAATTAGGTGAAAAAAC	288	*E. casseliflavus*	108
CA2 Rev- GCTAGTTTACCGTCTTTAACG			
FV1 For- GAATTAGGTGAAAAAAAAGTT	284	*E. flavescens*	108
FV2 Rev- GCTAGTTTACCGTCTTTAACG			
GA1 For- TTACTTGCTGATTTTGATTCG	173	*E. gallinarum*	108
GA2 Rev- TGAATTCTTCTTTGAAATCAG			
vanA			
For- CATGAATAGAATAAAAGTTGCAATA	1030	Several enterococcal species	133
Rev- CCCCTTTAACGCTAATACGATCAA			
For- TCTGCAATAGAGATAGCCGC	377	Several enterococcal species	81
Rev- GGAGTAGCTATCCCAGCATT			
For- GCTATTCAG CTGTACTC	783	Several enterococcal species	134
Rev- CAGCGGCCATCATACGG			
vanC1			
For- GGTATCAAGGAAACCTC	822	*E. gallinarum* and *E. casseliflavus/E. flavescens*	80
Rev- CTTCCGCCATCATAGCT			
For- GACCCGCTGAAATATGAAG	438	*E. gallinarum* and *E. casseliflavus/E. flavescens*	81
Rev- CGGCTTGATAAAGATCGGG			

continued

TABLE 22.2 (Continued)
Broad-Range Control and Enterococcal Genus- and Species-Specific PCR Primers

Primers (5′–3′)	Product Size (bp)	Observations	Reference
For- GGTATCAAGGAAACCTC Rev- CTTCCGCCATCATAGCT	822	*E. gallinarum* and *E. casseliflavus/E. flavescens*	134
vanC1/C2			
EC5 For- ATGGATTGGTAYTKGTAT EC8 Rev- TAGCGGGAGTGMCYMGTAA	815/827	*E. gallinarum* and *E. casseliflavus/E. flavescens*	129
vanC2			
For- CTCCTACGATTCTCTTG Rev- CGAGCAAGACCTTTAAG	430	*E. gallinarum* and *E. casseliflavus/E. flavescens*	81
vanC2/C3			
For-CGGGGAAGATGGCAGTAT Rev- CGCAGGGACGGTGATTTT	484	*E. gallinarum* and *E. casseliflavus/E. flavescens*	139
For- CTCCTACGATTCTCTTG Rev- CGAGCAAGACCTTTAAG	439	*E. gallinarum* and *E. casseliflavus/E. flavescens*	134

molecular identification. However, no species-specific primers have yet been developed or applied to this genus.

Elongation Factor EF-Tu. The *tuf* gene, encoding elongation factor EF-Tu, is involved in peptide chain formation and is an essential constituent of the bacterial genome[83,141] making this region a good target for diagnostic purposes. Ke and coworker[83] reported in 1999 a PCR-based assay that targets the *tuf* gene and detects most enterococcal species with excellent sensitivity and acceptable specificity.

D-Alanine:D-Alanine Ligase. A PCR assay developed by Dutka-Malen and coworkers[80] enabling the identification at the species level (species *E. faecium* and *E. faecalis*) and targeting the D-alanine:D-alanine ligase (*ddl*) gene was applied to characterize clinically relevant enterococci. Primers for species-specific PCR of *E. durans* and *E. hirae* targeting the *ddl* gene were also developed and validated by Knijff and coworkers.[142]

Penicillin-Binding Proteins. Penicillin-binding proteins (PBPs) are the enzymes involved in terminal stages of peptidoglycan synthesis, and PBP1 and PBP5 are the prevalent PBPs. Thus, these proteins are potential targets for bacterial cell detection. The gene coding for PBP5 (*pbp5*) has previously been used for the development of an *E. faecalis* species-specific probe[143] and was subsequently used in competitive PCR for quantification of nonculturable *E. faecalis* cells.[77]

Muramidase. In 2006, Arias and coworkers[87] searched the GenBank for sequences to be applied for the identification of *E. hirae*. During this process, an *E. durans* muramidase gene (*mur*) with 82% homology to *E. hirae* was detected; primers were developed for both genes and used for multiplex PCR. This method appears to provide a rapid and accurate identification of these two very similar enterococcal species.

ATP Synthase. *atpA* codes for the alpha subunit of the bacterial ATP synthase, which functions in ATP synthesis coupled to proton transport.[144] The aim of the study developed by Naser and coworkers[107] was to analyze the usefulness of *atpA* gene sequences for the reliable identification of *Enterococcus* species. All species were differentiated, with a maximum of 92% similarity. The intraspecies *atpA* sequence similarities for all species varied from 98.6% to 100%, except for *E. faecium* strains that showed a lower *atpA* sequence similarity of 96.3%. This study clearly showed that *atpA* provides an alternative tool for the phylogenetic study and identification of enterococci, but no genus- or species-specific primers have been described to demonstrate this potential.

Manganese-Dependent Superoxide Dismutase. In 2000, Poyart and coworkers[82] sequenced the manganese-dependent superoxide dismutase gene (*sodA*) of 19 species of the genus *Enterococcus*. Since variations in their sequences appeared to be greater between species and lower within species, this gene was used to develop species-specific primers[108] that can be applied to identify 23 validly published enterococcal species.

Heat-Shock Proteins. The *groESL* genes (also known as chaperonin 60 genes, *cpn10/60* or *hsp10/60*), which encode 10-kDa (GroES) and 60-kDa (GroEL) heat-shock proteins, are ubiquitous and evolutionarily highly conserved among bacteria.[145] Goh and coworkers[109] demonstrated that reverse checkerboard hybridization methodology, based on an approximately 600-bp region of the chaperonin 60 gene, can be an efficient method for the identification of *Enterococcus* species. Teng and coworkers[110] determined the *groESL* gene's full-length sequence of *E. faecalis* and used this information to develop primers specific for this species.

Antimicrobial Resistance. Over the years, several authors used multiplex PCR to both detect the presence of vancomycin resistance determinants and identify enterococci to the species level.[80,112,129,139,146] Glycopeptide resistance has been detected in several species of enterococci, with seven glycopeptide resistance genotypes already described. Five of these genotypes (*vanA*, *vanB*, *vanD*, *vanE*, and *vanG*) are acquired mechanisms and the other two (*vanC1* and *vanC2/C3*) are intrinsic properties. *vanA* and *vanB* are the most common resistance genotypes.

The *vanA* genes encode proteins that confer high-level resistance to vancomycin and teicoplanin[147] and have been found in several enterococcal species.[148] The *vanB* genes confer resistance to various concentrations of vancomycin but not teicoplanin[149]; *vanB*-type glycopeptide resistance has been described for *E. faecalis* and *E. faecium*. The *vanD* phenotype is characterized by resistance to moderate levels of vancomycin and to low levels of teicoplanin;[136,150] *vanD* determinants are constitutively expressed in *E. faecalis* but inducible in *E. faecium*.[136] The *vanE* gene is induced by low levels of vancomycin[131] and has only been detected in one strain of *E. faecalis*. The *vanC1* and *vanC2/C3* genes are specific to the motile VRE species *E. gallinarum* and *E. casseliflavus/E. flavescens*, respectively.[80] Recently, a new vancomycin resistance locus, *vanG*, has been detected in four *E. faecalis* clinical isolates with a moderated level of resistance to vancomycin and full susceptibility to teicoplanin.[132]

22.2 METHODS

22.2.1 Sample Preparation

It is often difficult to establish whether an *Enterococcus* strain is contributing to the infection or whether it is colonizing a human or an animal. Bacteremia as major cause of nosocomial infection represents an important source of morbidity and mortality. Gram-positive microorganisms, especially *Staphylococcus* and *E. faecalis*, account for the majority of episodes of bacteremia in critically ill patients in intensive care units (ICU). The rapid and reliable detection and subsequent identification of microorganisms, namely, *Enterococcus*, from blood is critical so that appropriate antimicrobial therapy can be provided. The same principle has to apply to the rapid and reliable detection of microbial samples from stool samples, oral samples, urine, pus, or exudates.

The appropriate collection and transport of specimens are critical to ensure that proper diagnostic information is given by the microbiology laboratory.[151,152] Also the initial processing of clinical specimens, which is a multifaceted endeavor as one has to consider specimen type and its anatomical origin, is a critical step. Description of the pretreatment and treatment of the specimens received in laboratory for aerobic microbiology in which *Enterococcus* is included have been gathered by York in 2007,[153] but generally they start by microscopical observation, Gram stain and isolation from the clinical samples in liquid and solid media, followed by identification procedures. Selective agents that are commonly used in selective media to isolate enterococci include: sodium azide, thallous acetate, kanamycin, and gentamicin. Other limiting ingredients or factors are crystal violet, Tween 80, carbonate, and bile salts. Special growth conditions may be also used, like low pH (e.g., 6.0 or 6.2) or elevated incubation temperature (42°C or 45°C). Nutritional requirements vary among the members of the genus, but it usually an advantage to use peptones, beef or yeast extract. Considering the mentioned factors it is easy to understand why there are so many varieties of selective media and discrepant results in the literature. ABA (esculin bile azide agar), KAA (kanamycin aesculin azide agar), CATC (citrate azide tween carbonate agar), ME (m-*Enterococcus* agar), TITG (thallous acetate tetrazolium glucose agar), and KA (crystal violet azide agar) are commonly used to cultivate enterocci.[154–156] Laboratories vary on the media they use depending on the type of sample; for instance, some use KAA agar to isolate enterococci from feces and cysteine-lactose-electrolyte-deficient (CLED) agar can be used in bacteriological studies of urine.[157]

The detection and identification of vancomycin-resistant enterococci (VRE) led to the development of special media and procedures to detect and identify them. Some authors demonstrated that VRE grew on Mueller-Hinton agar with 6–12 μg/mL of vancomycin, although 6 μg/mL of vancomycin seems to be the optimum concentration.[12,158]

Chromogenic media are now also widely used namely for cell counts and identification of enterococci from urine or stool samples, either if they are vancomycin-resistant or not. Media like BBL CHROMagar (BD Diagnostics), Chromogenic VRE (AES Laboratoire), chromID VRE (bioMérieux), VRE agar (Oxoid) or even bile esculin agar supplemented with 6 μg/mL of vancomycin (BEV), and bile esculin azide vancomycin (BEAV) agar have been used.[159–162]

After isolation of potential *Enterococcus* strains from clinical samples, using any of the media already mentioned, strains can be grown in trypticase-soy-5% sheep blood agar, brain heart infusion-5% sheep blood agar, or any blood agar base containing 5% animal blood, as well as brain heart infusion (BHI), MRS agar or broth (DeMan, Rogosa, Sharpe) or M17agar or broth to be further characterized or identified by PCR. All enterococci grow at 35°C–37°C and do not require an atmosphere containing increased levels of carbon dioxide.

22.2.2 Detection Procedures

22.2.2.1 Semiautomatic and Automatic Identification

The blood culture is the gold standard for diagnosing bacteria, but it may take more than two days for results to be available. Semiautomatic or automatic methods to test blood culture or other body fluids and to identify enterococci directly from samples, like BACTEC Peds PLUS/F and BACTEC 9240 that use spectrophotometry to monitor the CO_2 being produced during growth, are now commonly used, and the results have been compared to other systems of identification.[163,164] Other systems used for the identification of clinically isolated enterococci are kits like API 20 Strep and API Rapid ID 32 STREP (bioMérieux) and BBL Crystal gram-positive and BBL Crystal Rapid gram-positive (BD Diagnostics), that were already compared in terms of specificity.[165] There are also systems that identify and at the same time test the antibiotic susceptibility of the strains like the Vitek 2 and Vitek Compact Systems (bioMerieux). Also very used are the identification systems Bact/ALERT 3D (bioMérieux), BACTEC 9000 (BD Diagnostics), Versa Trek (Trek), or Phoenix 100 (Becton Dickson).[165–168]

22.2.2.2 Multiplex PCR for the Detection of *Enterococcus* Species

Although most of the routine identification of pathogens in clinical samples start with bacterial isolation, the advent of molecular approaches opens the road for direct application of such methods without culturing. In this context, the experimental procedure hereby described may be applied using DNA extracted both from isolated strains and clinical samples. A list of kits that can be used to extract DNA, directly from biological samples or from isolated colonies, has already been compiled.[140]

After DNA extraction, it is proposed that the detection/identification of *Enterococcus* can follow the steps shown in the flow chart of Figure 22.1. The procedure is a genus- and species-specific multiplex PCR, adapted from the 23-species detection methodology of Jackson and coworkers,[108] enabling the detection of *E. faecalis*, *E. faecium*, *E. casseliflavus*, and *E. gallinarum*.

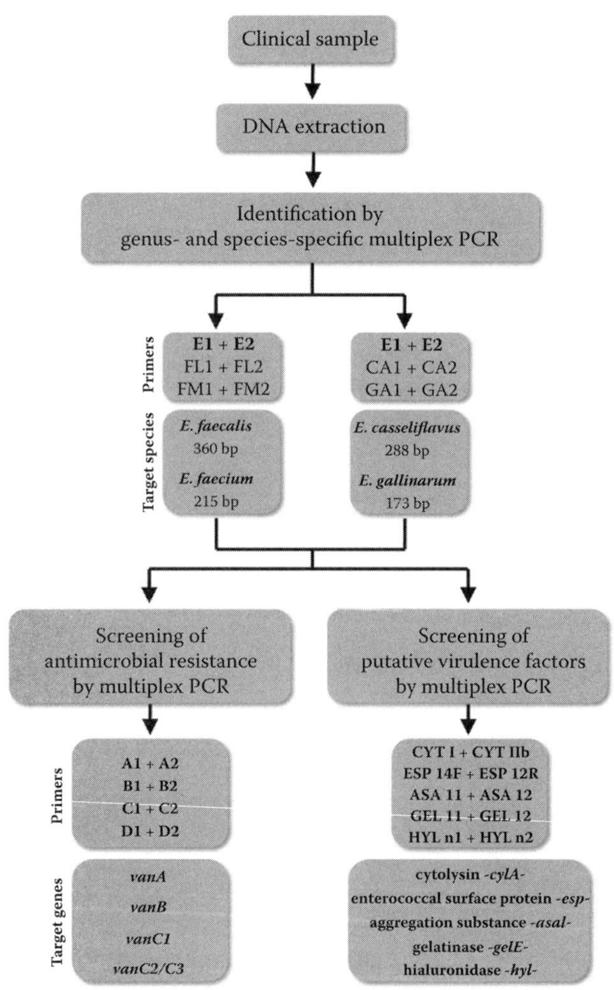

FIGURE 22.1 Flowchart of the multiplex-PCR-based procedures for the detection and/or identification of enterococci at genus and species level, as well as for the assessment of antimicrobial and virulence determinants.

Because this approach combines the genus-specific primers (E1/E2) described by Deasy and coworkers[119] with primers for the superoxide dismutase (*sod*) gene[108] specific for *E. faecalis* (FL1/FL2), *E. faecium* (FM1/FM2), *E. casseliflavus* (CA1/CA2), and *E. gallinarum* (GA1/GA), it provides a rapid and reliable method for the detection and/or identification of the most important human pathogenic enterococcal species.

Amplification procedure consists of the preparation of a PCR master mix (60 μL) containing 3 mM $MgCl_2$, 0.2 mM dNTP mix, 3.5 U of high fidelity DNA polymerase, and 3.75 μL (16 μM) of genus-specific primer set (E1/E2). The master mix should be equally divided in three separated vials in order to use the primer combinations depicted in Figure 22.1 (for sequence see Table 22.2) and have a positive control of the reaction. To vial number 1 add 2.5 μL (16 μM) of primers FL1/FL2 and 1.25 μL (16 μM) of primers FM1/FM2; to vial number 2 add 2.5 μL (16 μM) of primers GA1/GA2 and 1.25 μL (16 μM) of primers CA1/CA2; to vial number 3 (positive control) add DNA from a nonpathogenic strain of *Enterococcus*. PCR reactions are performed in a final volume of 22.5 μL after the addition to each vial of 2.5 μL of test DNA, extracted either from clinical sample or isolated strain.

The PCR amplification program includes an initial denaturation at 95°C for 4 min; 30 cycles of 95°C for 30 s, 55°C for 1 min, and 72°C for 1 min; a final extension step at 72°C for 7 min. After PCR amplification an aliquot of 10 μL of product is electrophoresed on a 2% 1× Tris-acetate-EDTA agarose gel containing 2 μg/mL ethidium bromide. A DNA molecular size marker (preferably with a 10 bp resolution) should be used as standard. Following electrophoresis, the agarose gels are visualized and photographed under UV light.

Since the presence of enterococcal isolates carrying virulence and antimicrobial resistance determinants turns the pathogenicity potential of these bacteria higher, we suggest an integrative approach leading to species identification and assessment for the presence of vancomycin resistance and virulence determinants (Figure 22.1). Although this approach is not essential for treating enterococcal infections, it may contribute to the implementation of more adequate antimicrobial treatment and/or to the implementation of measures aimed at the elimination of more virulent enterococci from hospital settings. Screening for the presence of antimicrobial resistance and virulence determinants can also be performed by multiplex PCR, using the methodologies described by Dutka-Malen et al.[80] and Vankerckhoven et al.,[169] respectively.

Using the molecular size marker included in the agarose gel, the size of the amplicons can be estimated and compared with the expected ones (Table 22.2 and Figure 22.1). If the obtained fragments show the expected molecular size for the genus and species primer set, the presence of the target species is detected in the sample (or the isolate is identified at species level if working with pure cultures isolated from clinical samples). When only the genus-specific amplicon is visualized, the presence of another enterococcal species is inferred. If no amplicons are observed, with the exception of the positive control reaction tube (which demonstrates the absence of PCR inhibitors), this means that either no enterococci are present in the analyzed sample at detectable levels or the existing strains show sequence variations in *sod* gene that prevent primer annealing (the false-negative problem can occur despite the applied methodology and has to be taken in account).

22.3 CONCLUSIONS AND FUTURE PERSPECTIVES

Enterococcus are generally not as virulent as other gram-positive bacteria but can occur as a polymicrobial infection in debilitated hosts.[170] Resistance of enteroccci to multiple antimicrobial agents, including vancomycin, can occur in health care centers. Considering that the most problematic enterocci are VREs, the infection control has to pass through careful analysis on VRE spread. Since colonized patients are the primary reservoir of VRE, a patient-to-patient transmission occurs in the first place. As hospital staff and environmental contamination, through medical devices or surroundings of colonized patients, also plays a critical role in spread, all these aspects have to be taken into consideration.[170] The Hospital Infection Control Practices Advisory Committee (HIPAC) has published in 1995 "Recommendations for Preventing the Spread of Vancomycin Resistance" (Centers for Disease Control and Prevention. Recommendations for Preventing the Spread of Vancomycin Resistance).[171] These measures include the appropriate use of vancomycin, the establishment of education programs, the clear understanding of the role of microbiology laboratory, the prevention and control of nosocomial transmission of VRE, and the isolation precautions to prevent patient-to-patient transmission.

More recently, research on antimicrobial stewardship, meaning the optimal selection dose and duration of an antimicrobial that results in best clinical outcome for the treatment and prevention of infection, with minimum toxicity to the patient and minimal impact on subsequent resistance and development, has produced more and more data.[172–174] Health care facilities should then adopt antimicrobial stewardship programs (ASPs) and infection control programs (ICPs) to monitor antimicrobial use while simultaneously optimizing treatment, outcome, and cost. The IDSA/SHEA guidelines for developing an institutional program to enhance antimicrobial stewardship can be used as a starting point to control the threat.[175,176]

All these evaluations have to take on board, not only healthy people, who seem to be unaffected by enterococcal strains, but also the elderly population, infants, and immunocompromised people. The microbiological risk assessment of emerging pathogens has to be performed, and global data indicate that the epidemiology of human diseases is changing because of changes in lifestyles and improved surveillance and recovery/detection methods.[177,178] Pathogen control, and especially VRE, depends on the laboratory's ability to detect, identify, and recognize those agents; therefore, in the case of *Enterococcus*, a detailed combination of a more specific genus and species method is proposed here based on the work of Jackson and coworkers.[108] PCR methods in general, and real-time PCR methods in particular, are increasingly used for the detection and quantification of pathogens in clinical samples.

Nonetheless, the new species that are constantly appearing in the literature, calling for new and/or reassessed methods of identification and quantification as well as for the need to always be alert to new antibiotic resistances, are a constant challenge and constitute new routes of needed research in the area of work with *Enterococcus*.

ACKNOWLEDGMENTS

The authors acknowledge the partial funding from different projects on *Enterococcus* or food quality and safety that have financed their research on the subject, namely, Fundação para a Ciência e Tecnologia through Program PRAXIS XXI, project 2/2.1/BIO/1121/95 and Program SAPIENS, project POCTI/AGR/36165/99 and EU, through Specific Supported Action project BIOSAFE SSPS-CT-2006–022725. T. Semedo-Lemsaddek thanks the financial support of the FCT Pos-Doctoral fellowship SFRH/BPD/20892/2004.

REFERENCES

1. Thiercelin, E., and Jouhaud, L., Reproduction de l'entérocoque; taches centrales; granulation périphériques et microblastes, *C. R. Soc. Biol.*, 55, 686, 1903.
2. Murray, B.E., The life and times of the *Enterococcus*, *Clin. Microbiol. Rev.*, 46, 65, 1990.
3. Kalina, A.P., The taxonomy and nomenclature of enterococci, *Int. J. Syst. Bacteriol.*, 20, 185, 1970.
4. Schleifer, K.H., and Kilpperbalz, R., Transfer of *Streptococcus faecalis* and *Streptococcus faecium* to the Genus *Enterococcus* nom. rev. as *Enterococcus faecalis* comb. nov. and *Enterococcus faecium* comb. nov., *Int. J. Syst. Bacteriol.*, 34, 31, 1984.
5. Ludwig, W. et al., The phylogenetic position of *Streptococcus* and *Enterococcus*, *J. Gen. Microbiol.*, 131, 543, 1985.
6. Kohler, W., The present state of species within the genera *Streptococcus* and *Enterococcus*, *Int. J. Med. Microbiol.*, 297, 133, 2007.
7. Garrity, G.M. et al., The Bacteria: Phylum Firmicutes: Class "*Bacilli*." In *Taxonomic Outline of the Bacteria and Archea (TOBA)*, Release 7.7, Part 9, 2007.
8. Devriese, L.A. et al., Identification of aesculin-hydrolyzing streptococci, lactococci, aerococci and enterococci from subclinical intramammary infections in dairy cows, *Vet. Microbiol.*, 70, 87, 1999.
9. Devriese, L.A., Baele, M., and Butaye, P., The genus *Enterococcus*, In Dworkin, M. et al. (Editors), *The Prokaryotes*, pp. 163–174, Springer, Heidelberg, 2002.
10. Holt, J.G. et al., *Bergey's Manual of Determinative Bacteriology*, 9th ed., William & Wilkins, Baltimore, MD, 1994.
11. Devriese, L.A. et al., Identification of *Enterococcus* species isolated from foods of animal origin, *Int. J. Food Microbiol.*, 26, 187, 1995.

12. Facklam, R.R., Carvalho, M.G.S., and Teixeira L.M., History, taxonomy, biochemical characteristics, and antibiotic susceptibility testing of enterococci. In Gilmore, M.S. et al. (Editors), *The Enterococci: Pathogenesis, Molecular Biology and Antibiotic Resistance*, pp. 1–46, ASM Press, Washington, D.C., 2002.
13. Doming, K.J., Mayer, H.K., and Kneifel, W., Methods used for the isolation, enumeration, characterization and identification of *Enterococcus* spp. 2. Pheno- and genotypic criteria, *Int. J. Food Microbiol.*, 88, 165, 2003.
14. Foulquié Moreno, M.R. et al., The role and application of enterococci in food and health, *Int. J. Food Microbiol.*, 106, 1, 2006.
15. Ogier, J.C., and Serror, P., Safety assessment of dairy microorganisms: The *Enterococcus* genus, *Int. J. Food Microbiol.*, 126, 291, 2008.
16. Kayser, F.H., Safety aspects of enterococci from the medical point of view, *Int. J. Food Microbiol.*, 88, 255, 2003.
17. Lund, B. et al., Frequent transmission of enterococcal strains between mechanically ventilated patients treated at an intensive care unit, *J. Clin. Microbiol.*, 40, 2084, 2002.
18. Low, D.E. et al., Clinical prevalence, antimicrobial susceptibility, and geographic resistance patterns of enterococci: Results from the SENTRY Antimicrobial Surveillance Program, 1997–1999, *Clin. Infect. Dis.*, 32, S133, 2001.
19. Schaberg, D.R., Culver, D.H., and Gaynes, R.P., Major trends in the microbial etiology of nosocomial infection, *Am. J. Med.*, 91, S72, 1991.
20. Murray, B.E., Diversity among multidrug-resistant enterococci, *Emerg. Infect. Dis.*, 4, 37, 1998.
21. Wisplinghoff, H. et al., Nosocomial bloodstream infections in US hospitals: Analysis of 179 cases from a perspective nation wide surveillance study, *Clin. Infect. Dis.*, 39, 309, 2004.
22. Huycke, M.M., Sahm, D.F., and Gilmore, M.S., Multiple drug resistant enterococci: The nature of the problem and an agenda for the future, *Emerg. Infect. Dis.*, 4, 239, 1998.
23. Bouza, E. et al., A European perspective on nosocomial urinary tract infections II. Report on incidence, clinical characteristics and outcome (ESGNI-004 study). European Study Group on Nosocomial Infection, *Clin. Microbiol. Infect.*, 7, 532, 2001.
24. Reynolds, R. et al., Antimicrobial susceptibility of the pathogens of bacteraemia in the UK and Ireland 2001–2002: The BSAC Bacteraemia Resistance Surveillance Programme, *J. Antimicrob. Chemother.*, 53, 1018, 2004.
25. Coque, T.M. et al., Population structure of *Enterococcus faecium* causing bacteremia in a Spanish University Hospital: Setting the scene for a future increase in vancomycin resistance? *Antimicrob. Agents Chemother.*, 49, 2693, 2005.
26. Giraffa, G., Enterococci from foods, *FEMS Microbiol. Rev.*, 26, 163, 2002.
27. Klein, G., Taxonomy, ecology and antibiotic resistance of enterococci from food and the gastro-intestinal tract, *Int. J. Food Microbiol.*, 88, 123, 2003.
28. Mundy, L.M., Sahm, D.F., and Gilmore, M., Relationships between enterococcal virulence and antimicrobial resistance, *Clin. Microbiol. Rev.*, 13, 513, 2000.
29. Wenzel, R.P., and Edmond, M.B., The impact of hospital-acquired bloodstream infections, *Emerg. Infect. Dis.*, 7, 174, 2001.
30. Uttley, A.H.C. et al., Vancomycin-resistant enterococci, *Lancet*, 1, 57, 1988.
31. Menichetti, F., Current and emerging serious Gram-positive infections, *Clin. Microbiol. Infect.*, 11, 22, 2005.
32. Klare, I. et al., *Enterococcus faecium* strains with vanA mediated high-level glycopeptides resistance isolated from animal foodstuffs and fecal samples of humans in the community, *Microb. Drug Resist.*, 1, 265, 1995.
33. Aarestrup, F.M., Occurrence of glycopeptide resistance among *Enterococcus faecium* isolates from conventional and ecological poultry farms, *Microb. Drug Resist.*, 1, 255, 1995.
34. Bager, F. et al., Avoparcin used as a growth promoter is associated with the occurrence of vancomycin resistant *Enterococcus faecium* on Danish poultry and pigs farms, *Prev. Vet. Med.*, 31, 95, 1997.
35. Wegener, H.C. et al., Use of antimicrobial growth promoters in food animals and *Enterococcus faecium* resistance to therapeutic antimicrobial drugs in Europe, *Emerg. Infect. Dis.*, 5, 329, 1999.
36. Vergis, E.N. et al., Determinants of vancomycin resistance and mortality rates in enterococcal bacteremia: A prospective multicenter study, *Ann. Intern. Med.*, 135, 484, 2001.
37. Byers, K.E. et al., A hospital epidemic of vancomycin-resistant *Enterococcus*: Risk factors and control, *Infect. Control Hosp. Epidemiol.*, 22, 140, 2001.
38. Falk, P.S. et al., Outbreak of vancomycin-resistant enterococci in a burn unit, *Infect. Control Hosp. Epidemiol.*, 21, 575, 2000.
39. Jett, B.D., Huycke, M.M., and Gilmore, M.S., Virulence of enterococci, *Clin. Microbiol. Rev.*, 7, 462, 1994.
40. Shankar, V. et al., Infection derived-*Enterococcus faecalis* strains are enriched in esp, a gene encoding a novel surface protein, *Infect. Immun.*, 1, 193, 1999.
41. Shankar, N. et al., Role of *Enterococcus faecalis* surface protein Esp in the pathogenesis of ascending urinary tract infection, *Infect. Immun.*, 69, 4366, 2001.
42. Toledo-Arana, A. et al., The enterococcal surface protein, Esp, is involved in *Enterococcus faecalis* biofilm formation, *Appl. Environ. Microbiol.*, 67, 4538, 2001.
43. Dunny, G.M., Leonard, B.A.B., and Hedberg, P.J., Pheromone inducible conjugation in *Enterococcus faecalis*. Interbacterial and host parasite chemical communication, *J. Bacteriol.*, 177, 871, 1995.
44. Wirth, R., The Sex pheromone system of *Enterococcus faecalis:* More than just a plasmid collection mechanism, *Eur. J. Biochem.*, 222, 235, 1994.
45. Sartingen, S. et al., Aggregation substance increases adherence and internalization, but not translocation, of *Enterococcus faecalis* through different intestinal epithelial cells in vitro, *Infect. Immun.*, 68, 6044, 2000.
46. Suβmuth, S.D. et al., Aggregation substance promotes adherence, phagocytosis, and intracellular survival of *Enterococcus faecalis* within human macrophages and suppresses respiratory burst, *Infect. Immun.*, 68, 4900, 2000.
47. Wells, C.L. et al., Inducible expression of *Enterococcus faecalis* aggregation substance surface protein facilitates bacterial internalization by cultured enterocytes, *Inf. Immun.*, 68, 7190, 2000.
48. Hirt, H., Erlandsen, S.L., and Dunny, G.M., Heterologous inducible expression of *Enterococcus faecalis* pCF10 aggregation substance Asc10 in *Lactococcus lactis* and *Streptococcus gordonii* contributes to cell hydrophobicity and adhesion to fibrin, *J. Bacteriol.*, 182, 2299, 2000.
49. Rakita, R.M. et al., *Enterococcus faecalis* bearing aggregation substance is resistant to killing by human neutrophils despite phagocytosis and neutrophil activation, *Infect. Immun.*, 67, 6067, 1999.
50. Schlievert, P.M. et al., Aggregation and binding substances enhance pathogenicity in rabbit models of *Enterococcus faecalis* endocarditis, *Infect. Immun.*, 66, 218, 1998.

51. Jett, B.D., Atkuri, R.V., and Gilmore, M.S., *Enterococcus faecalis* localization in experimental endophthalmitis: Role of plasmid-encoded aggregation substance, *Infect. Immun.*, 66, 843, 1998.
52. Lowe, A.M., Lambert, P.A., and Smith, A.W., Cloning of an *Enterococcus faecalis* endocarditis antigen homology with adhesins from some oral streptococci, *Infect. Immun.*, 63, 703, 1995.
53. Singh, K.V. et al., In vivo testing of an *Enterococcus faecalis* efaA mutant and use of efaA homologs for species identification, *FEMS Immunol. Med. Microbiol.*, 21, 323, 1998.
54. Haas, W., and Gilmore, M.S., Molecular nature of a novel bacterial toxin: The cytolysin of *Enterococcus faecalis*, *Med. Microbiol. Immun.*, 187, 183, 1999.
55. Ike, Y., Hashimoto, H., and Clewell, D.B. Hemolysin of *Streptococcus faecalis* subspecies *zymogenes* contributes to virulence in mice, *Infect. Immun.*, 45, 528, 1994.
56. Johnson, A.P., The Pathogenicity of enterococci, *J. Antimicrob. Chemoth.*, 33, 1083, 1994.
57. Coque, T.M. et al., Incidence of hemolysin, gelatinase, and aggregation substance among enterococci isolated from patients with endocarditis and other infections and from feces of hospitalized and community based persons, *J. Infect. Dis.*, 171, 1223, 1995.
58. Nakayama, J., Kariyama, R., and Kumon, H., Description of a 23.9-kilobase chromosomal deletion containing a region encoding fsr genes which mainly determines the gelatinase negative phenotype of clinical isolates of *Enterococcus faecalis* in urine, *Appl. Environ. Microbiol.*, 68, 3152, 2002.
59. Pillai, S.K. et al., Prevalence of the fsr locus in *Enterococcus faecalis* infection, *J. Clin. Microbiol.*, 40, 2651, 2002.
60. Qin, X. et al., Effects of *Enterococcus faecalis* fsr genes on production of gelatinase and a serine protease and virulence, *Infect. Immun.*, 68, 2579, 2000.
61. Willems, R.J.L. et al., Variant esp gene as a marker of a distinct genetic lineage of vancomycin-resistant *Enterococcus faecium* spreading in hospitals, *Lancet*, 357, 853, 2001.
62. Rice, L.B. et al., A potential virulence gene, hyl (Efm), predominates in *Enterococcus faecium* of clinical origin, *J. Infect. Dis.*, 2003, 187, 508, 2003.
63. Hacker, J., and Kaper, J.B., Pathogenicity islands and the evolution of microbes, *Annu. Rev. Microbiol.*, 54, 641, 2000.
64. Shankar, N., Baghdayan, A.S., and Gilmore, M.S., Modulation of virulence within a pathogenicity island in vancomycin-resistant *Enterococcus faecalis*, *Nature*, 417, 746, 2002.
65. Leavis, H. et al., A novel putative enterococcal pathogenicity island linked to the esp virulence gene of *Enterococcus faecium* and associated with epidemicity, *J. Bacteriol*, 186,672, 2004.
66. Nallapareddy, S.R. et al., Molecular characterization of a widespread, pathogenic, and antibiotic resistance-receptive *Enterococcus faecalis* lineage and dissemination of its putative pathogenicity island, *J. Bacteriol.*, 187, 5709, 2005.
67. Rahman, I. et al., Methionine uptake and cytopathogenicity of viable but nonculturable *Shigella dysenteriae* type 1, *Appl. Environ. Microbiol.*, 60, 3573, 1994.
68. Lleo, M.D., Tafi, M.C., and Canepari, P., Nonculturable *Enterococcus faecalis* cells are metabolically active and capable of resuming active growth, *Syst. Appl. Microbiol.*, 21, 333, 1998.
69. Colwell, R.R., and Huq, A., Vibrios in the environment: Viable but nonculturable *Vibrio cholerae*. In Kaye, T. (Editor), *Vibrio and Cholera: Molecular Global Perspectives*, pp. 117–133, ASM Press, Washington D.C., 1994.
70. Pruzzo, C. et al., In vitro adhesion to human cells by viable but nonculturable *Enterococcus faecalis*, *Curr. Microbiol.*, 45, 105, 2002.
71. Lleo, M.D. et al., Vancomycin resistance is maintained in enterococci in the viable but nonculturable state and after division is resumed, *Antimicrob. Agents Chemother.*, 47, 1154, 2003.
72. Lleo, M.D. et al., mRNA detection by reverse transcription-PCR for monitoring viability over time in an *Enterococcus faecalis* viable but nonculturable population maintained in a laboratory microcosm, *Appl. Environ. Microbiol.*, 66, 4564, 2000.
73. Arana, I. et al., Detection and enumeration of viable but nonculturable transconjugants of *Escherichia coli* during the survival of recipient cells in river water, *J. Appl. Microbiol.*, 83, 340, 1997.
74. Oliver, J.D., Formation of viable but nonculturable cells. In Kjelleberg, S. (Editor), *Starvation in Bacteria*, pp. 239–272, Plenum Press, New York, 1993.
75. Lleo, M.M. et al., Resuscitation rate in different enterococcal species in the viable but non-culturable state, *J. Appl. Microbiol.*, 91, 1095, 2001.
76. Lee, D.Y., Shannon, K., and Beaudette, L.A., Detection of bacterial pathogens in municipal wastewater using an oligonucleotide microarray and real-time quantitative PCR, *J. Microbiol. Methods*, 65, 453, 2006.
77. Lleo, M.D. et al., Competitive polymerase chain reaction for quantification of nonculturable *Enterococcus faecalis* cells in lake water, *FEMS Microbiol. Ecol.*, 30,345, 1999.
78. Monstein, H.J. et al., Division of the genus *Enterococcus* into species groups using PCR-based molecular typing methods, *Microbiology*, 144, 1171, 1998.
79. Sedgley, C.M. et al., Quantitative real-time PCR detection of oral *Enterococcus faecalis* in humans, *Arch. Oral Biol.*, 50, 575, 2005.
80. Dutka-Malen, S., Evers, S., and Courvalin, P., Detection of glycopeptideresistance genotypes and identification to the species level of clinically relevant enterococci by PCR, *J. Clin. Microbiol.*, 33, 1434, 1995.
81. Klare, I., and Witte, W., Glycopeptide resistant Enterococci—origin, spread, transferable resistance and significance, *Wien. Klin. Wochenschr.*, 109, 293, 1997.
82. Poyart, C., Quesnes, G., and Trieu-Cuot, P., Sequencing the gene encoding manganese dependent superoxide dismutase for rapid species identification of enterococci, *J. Clin. Microbiol.*, 38, 415, 2000.
83. Ke, D.B. et al., Development of a PCR assay for rapid detection of enterococci, *J. Clin. Microbiol.*, 37, 3497, 1999.
84. Wang, R.F. et al., Development of a membrane-array method for the detection of human intestinal bacteria in fecal samples, *Mol. Cell. Probes*, 16, 341, 2002.
85. Sedgley, C., Buck, G., and Appelbe, O., Prevalence of *Enterococcus faecalis* at multiple oral sites in endodontic patients using culture and PCR, *J. Endodont.*, 32, 104, 2006.
86. Mahmoudpour, A. et al., Isolation and identification of *Enterococcus faecalis* from necrotic root canals using multiplex PCR, *J. Oral Sci.*, 49, 221, 2007.
87. Arias, C.A. et al., Rapid identification of *Enterococcus hirae* and *Enterococcus durans* by PCR and detection of a homologue of the *E. hirae* mur 2 gene in *E. durans*, *J. Clin. Microbiol.*, 44, 1567, 2006.
88. Frahm, E., and Obst, U., Application of the fluorogenic probe technique (TaqMan PCR) to the detection of *Enterococcus* spp. and *Escherichia coli* in water samples, *J. Microbiol. Methods*, 52, 123, 2003.

89. Volkmann, H. et al., Detection of clinically relevant antibiotic-resistance genes in municipal wastewater using real-time PCR (TaqMan), *J. Microbiol. Methods*, 56, 277, 2004.
90. Kempf, V.A J., Trebesius, K., and Autenrieth, I.B. Fluorescent in situ hybridization allows rapid identification of microorganisms in blood cultures, *J. Clin. Microbiol.*, 38, 830, 2000.
91. Waar, K. et al., Fluorescent in situ hybridization with specific DNA probes offers adequate detection of *Enterococcus faecalis* and *Enterococcus faecium* in clinical samples, *J. Med. Microbiol.*, 54, 937, 2005.
92. Wellinghausen, N. et al., Rapid identification of clinically relevant *Enterococcus* species by fluorescence in situ hybridization, *J. Clin. Microbiol.*, 45, 3424, 2007.
93. Betzl, D., Ludwig, W., and Schleifer, K.H., Identification of lactococci and enterococci by colony hybridization with 23S ribosomal RNA targeted oligonucleotide probes, *Appl. Environ. Microbiol.*, 56, 2927, 1990.
94. Lehner, A. et al., Oligonucleotide microarray for identification of *Enterococcus* species, *FEMS Microbiol. Lett.*, 246, 133, 2005.
95. Frahm, E. et al., Application of 23S rDNA targeted oligonucleotide probes specific for enterococci to water hygiene control, *Syst. Appl. Microbiol.*, 21, 450, 1998.
96. Behr, T. et al., A nested array of rRNA targeted probes for the detection and identification of enterococci by reverse hybridization, *Syst. Appl. Microbiol.*, 23, 563, 2000.
97. Chizhikov, V. et al., Microarray analysis of microbial virulence factors, *Appl. Environ. Microbiol.*, 67, 3258, 2001.
98. Sergeev, N. et al., Multipathogen oligonucleotide microarray for environmental and biodefense applications, *Biosens. Bioelectron.*, 20, 684, 2004.
99. Warsen, A.E. et al., Simultaneous discrimination between 15 fish pathogens by using 16S ribosomal DNA PCR and DNA microarrays, *Appl. Environ. Microbiol.*, 70, 4216, 2004.
100. Amann, R.I., Krumholz, L., and Stahl, D.A. Fluorescent oligonucleotide probing of whole cells for determinative, phylogenetic, and environmental studies in microbiology, *J. Bacteriol.*, 172, 762, 1990.
101. Jin, L.Q. et al., Detection and identification of intestinal pathogenic bacteria by hybridization to oligonucleotide microarrays, *World J. Gastroent.*, 11, 7615, 2005.
102. Santo Domingo, J.W., Siefring, S.C., and Haugland, R.A., Real time PCR method to detect *Enterococcus faecalis* in water, *Biotechnol. Lett.*, 25, 261, 2003.
103. Wang, R.F. et al., DNA microarray analysis of predominant human intestinal bacteria in fecal samples, *Mol. Cell. Probes*, 18, 223, 2004.
104. Tung, S.K. et al., Array based identification of species of the genera Abiotrophia, Enterococcus, Granulicatella, and Streptococcus, *J. Clin. Microbiol.*, 44, 4414, 2006.
105. Manero, A., and Blanch, A.R., Identification of *Enterococcus* spp. based on specific hybridisation with 16S rDNA probes, *J. Microbiol. Methods*, 50, 115, 2002.
106. Naser, S.M. et al., Application of multilocus sequence analysis (MLSA) for rapid identification of *Enterococcus* species based on rpoA and pheS genes, *Microbiology*, 151, 2141, 2005.
107. Naser, S. et al., Phylogeny and identification of enterococci by atpA gene sequence analysis, *J. Clin. Microbiol.*, 43, 2224, 2005.
108. Jackson, C.R., Fedorka-Cray, P.J., and Barrett, J.B., Use of a genus- and species-specific multiplex PCR for identification of enterococci, *J. Clin. Microbiol.*, 42, 3558, 2004.
109. Goh, S.H. et al., Identification of *Staphylococcus* species and subspecies by the chaperonin 60 gene identification method and reverse checkerboard hybridization, *J. Clin. Microbiol.*, 35, 3116, 1997.
110. Teng, L.J. et al., Determination of *Enterococcus faecalis* groESL full length sequence and application for species identification, *J. Clin. Microbiol.*, 39, 3326, 2001.
111. Van de Klundert, J.A.M., and Vliegenthart, J.S., PCR detection of genes coding for aminoglycoside-modifying enzymes. In Persing, D.H. et al. (Editors), *Diagnostic Molecular Microbiology*, pp. 547–552, ASM Press, Washington, D.C., 1993.
112. Kariyama, R. et al., Simple and reliable multiplex PCR assay for surveillance isolates of vancomycin-resistant enterococci, *J. Clin. Microbiol.*, 38, 3092, 2000.
113. McClellan, D.L., Griffen, A.L., and Leys, E.J. Age and prevalence of *Porphyromonas gingivalis* in children, *J. Clin. Microbiol.*, 34, 2017, 1996.
114. Berthier, F., and Ehrlich, S.D., Rapid species identification within two groups of closely related lactobacilli using PCR primers that target the 16S/23S rRNA spacer region, *FEMS Microbiol. Lett.*, 161, 97, 1998.
115. Nadkarni, M.A. et al., Determination of bacterial load by real-time PCR using a broad-range (universal) probe and primers set, *Microbiology*, 148, 257, 2002.
116. Lu, J.J. et al., Use of PCR with universal primers and restriction endonuclease digestions for detection and identification of common bacterial pathogens in cerebrospinal fluid, *J. Clin. Microbiol.*, 38, 2076, 2000.
117. Kumar, P.S. et al., New bacterial species associated with chronic periodontitis, *J. Dent. Res.*, 82, 338, 2003.
118. Volkmann, H. et al., Evaluation of inhibition and cross-reaction effects on real-time PCR applied to the total DNA of wastewater samples for the quantification of bacterial antibiotic resistance genes and taxon specific targets, *Mol. Cell. Probes*, 21, 125, 2007.
119. Deasy, B.M. et al., A rapid PCR based method to distinguish between *Lactococcus* and *Enterococcus*, *Syst. Appl. Microbiol.*, 23, 510, 2000.
120. Matsuda, K. et al., Sensitive quantitative detection of commensal bacteria by rRNA-Targeted reverse Transcription-PCR, *Appl. Environ. Microbiol.*, 73, 6695, 2007.
121. Matsuda, K. et al., Establishment of an analytical system for the human fecal microbiota, based on reverse transcription-quantitative PCR targeting of multicopy rRNA molecules, *Appl. Environ. Microbiol.*, 75, 1961, 2009.
122. Barry, T., The 16S723S ribossomal spacer region for DNA probes to identify eubacteria, *PCR Methods Appl.*, 1, 51, 1991.
123. Jensen, M.A., Webster, J.A., and Straus, N., Rapid identification of bacteria on the basis of polymerase chain reaction-amplified ribosomal DNA spacer polymorphisms, *Appl. Environ. Microbiol.*, 59, 945, 1993.
124. Molander, A. et al., A protocol for polymerase chain reaction detection of *Enterococcus faecalis* and *Enterococcus faecium* from the root canal, *Int. Endod. J.*, 35, 1, 2002.
125. Meier, H. et al., Detection of enterococci with rRNA targeted DNA probes and their use for hygienic drinking water control, *Water Sci. Technol.*, 35, 437, 1997.
126. Bartosch, S. et al., Characterization of bacterial communities in feces from healthy elderly volunteers and hospitalized elderly patients by using real-time PCR and effects of antibiotic treatment on the fecal microbiota, *Appl. Environ. Microbiol.*, 70, 3575, 2004.

127. Siqueira, J.F., and Rocas, I.N., Polymerase chain reaction based analysis of microrganisms associated with failed endodontic treatment, *Oral Surg. Oral Med. Oral Pathol. Oral Radiol. Endodontol.*, 97, 85, 2004.
128. Gomes, B.P.F.A. et al., *Enterococcus faecalis* in dental root canals detected by culture and by polymerase chain reaction analysis, *Oral Surg. Oral Med. Oral Pathol. Oral Radiol. Endodontol.*, 102, 247, 2006.
129. Depardieu, F., Perichon, B., and Courvalin, P., Detection of the van alphabet and identification of enterococci and staphylococci at the species level by multiplex PCR, *J. Clin. Microbiol.*, 42, 5857, 2004.
130. Liu, D. et al., PCR amplification of a species-specific putative transcriptional regulator gene reveals the identity of *Enterococcus faecalis*, *Res. Microbiol.*, 156, 944, 2005.
131. Fines, M. et al., VanE, a new type of acquired glycopeptide resistance in *Enterococcus faecalis* BM4405, *Antimicrob. Agents Chemother.*, 43, 2161, 1999.
132. McKessar, S.J. et al., Genetic characterization of vanG, a novel vancomycin resistance locus of *Enterococcus faecalis*, *Antimicrob. Agents Chemother.*, 44, 3224, 2000.
133. Clark, N.C. et al., Characterization of glycopeptide-resistant enterococci from United States hospitals, *Antimicrob. Agents Chemother.*, 37, 2311, 1993.
134. Miele, A., Bandera, M., and Goldstein, B.P., Use of primers selective for vancomycin-resistance genes to determine van genotype in enterococci and to study gene organization in vanA isolates, *Antimicrob. Agents Chemother.*, 39, 1772, 1995.
135. Gold, H.S. et al., A Gene conferring resistance to vancomycin but not teicoplanin in isolates of *Enterococcus-faecalis* and *Enterococcus-faecium* demonstrates homology with VanB, VanA, and VanC genes of enterococci, *Antimicrob. Agents Chemoter.*, 37, 1904, 1993.
136. Perichon, B., Reynolds, P., and Courvalin, P., VanD type glycopeptide-resistant *Enterococcus faecium* BM4339, *Antimicrob. Agents Chemother.*, 141, 2016, 1997.
137. Boyd, D.A. et al., *Enterococcus faecium* N03–0072 carries a new VanD-type vancomycin resistance determinant: Characterization of the VanD5 operon, *J. Antimicrob. Chemother.*, 54, 680, 2004.
138. Cheng, S.Q. et al., A PCR assay for identification of *Enterococcus faecium*, *J. Clin. Microbiol.*, 35, 1248, 1997
139. Satake, S. et al., Detection of vancomycin-resistant enterococci in fecal samples by PCR, *J. Clin. Microbiol.*, 35, 2325, 1997.
140. Semedo-Lemsaddek, T. et al., *Enterococcus*. In Liu, D. (Editor), *Molecular Detection of Foodborne Pathogens*, p. 157, CRC Press, Taylor & Fancis Group, New York, 2009.
141. Grunberg-Manago, M., Regulation of the exoression of aminoacyl tRNA synthetases and translation factors. In Neidhardt, F. et al. (Editors), *Escherichia coli and Salmonella: Cellular and Molecular Biology*, 2nd ed., pp. 1386–1391, ASM Press, Washington, D.C., 1996.
142. Knijff, E. et al., Rapid identification of *Enterococcus durans* and *Enterococcus hirae* by PCR with primers targeted to the ddl genes, *J. Microbiol. Methods*, 47, 35, 2001.
143. Robbi, C. et al., The gene encoding for penicillin-binding protein 5 of *Enterococcus faecalis* is useful for development of a species-specific DNA probe, *Microb. Drug Resist.*, 2, 215, 1996.
144. Muller, V., and Gruber, G., ATP synthases: structure, function and evolution of unique energy converters, *Cell. Mol. Life Sci.*, 60, 474, 2003.
145. Ellis, R.J., *Champeronins*: introductory perspective. In Ellis, R.J. (Editor), *The Chaperonins*, pp. 2–25, Academic Press, Inc., San Diego, CA, 1996.
146. Li, S., Zhang, Z., and Mi, Z.H., Vancomycin-resistant enterococci in a Chinese hospital, *Curr. Microbiol.*, 55, 125, 2007.
147. Arthur, M., Reynolds, P., and Courvalin, P., Glycopeptide resistance in enterococci, *Trends Microbiol.*, 4, 401, 1996.
148. Mendéz-Alvarez, S., Pérez-Hernández, X., and Claverie-Martin, F., Glycopeptide resistance in enterococci, *Int. Microbiol.*, 3, 71, 2000.
149. Evers, S., Sahm, D.F., and Courvalin, P., The vanB gene of vancomycin-resistant *Enterococcus faecalis* V583 is structurally related to genes encoding D-Ala D-Ala ligases and glycopeptide-resistance proteins VanA and VanC, *Gene*, 124, 143, 1993.
150. Ostrowsky, B.E. et al., A cluster of VanD vancomycin-resistant *Enterococcus faecium*: Molecular characterization and clinical epidemiology, *J. Infect. Dis.*, 180, 1177, 1999.
151. Linscott, A.J., Collection, transport, and manipulation of clinical specimens and initial laboratory concerns. In Garcia, L.S., and Isenberg, H.D. (Editors in Chief), *Clinical Microbiology Procedures Handbook*, Volume 1, 2nd ed., pp. 1–36, ASM, Washington D.C., 2007.
152. Thompson, R.B. Jr., Specimen collection, transport, and processing: Bacteriology. In Murray, P.R. et al. (Editors), *Manual of Clinical Microbiology*, Volume 1, 9th ed., pp. 291–333, ASM, Washington, D.C., 2007.
153. York, M.K., Aerobic bacteriology. In Garcia, L.S., and Isenberg, H.D. (Editors in Chief), *Clinical Microbiology Procedures Handbook*, Volume 1, 2nd ed., p. 37, ASM, Washington D.C., 2007.
154. Reuter, G., Culture media for enterococci and group D-streptococci. In Corry J.E.L. et al. (Editors), *Handbook of Culture Media for Food Microbiology*, Volume 34, pp. 51–62, Elsevier Sciences B.V., Amsterdam, The Netherlands, 1995.
155. Reuter, G., and Klein, G., Culture media for enterococci and group D-streptococci. In Corry J.E.L. et al. (Editors), *Handbook of Culture Media for Food Microbiology*, Volume 37, pp. 111–126, Elsevier Sciences B.V., Amsterdam, The Netherlands, 2003.
156. Atlas, R.M., *Handbook of Microbial Media for the Examination of Food*, CRC Press, USA, 1995.
157. Solway, S. et al., Isolation of streptogramin resistant *Enterococcus faecium* from human and non human sources in a rural community, *J. Antimicrob. Chemoth.*, 52, 707, 2003.
158. Swenson, J.M. et al., Development of a standardized screening method for detection of vancomycin-resistant enterococci, *J. Clin. Microbiol.*, 32, 1700, 1994.
159. Asir, K. et al., Evaluation of chromogenic media for the isolation of vancomycin-resistant enterococci from stool samples, *Lett. Appl. Microbiol.*, 48, 230, 2009.
160. Kuch, A. et al., New selective and differential chromogenic agar medium, chromID VRE, for screening vancomycin-resistant *Enterococcus* species, *J. Microbiol. Methods.*, 77, 124, 2009.
161. Ledeboer, N.A., Tibbetts, R.J., and Dunne, W.A. A new chromogenic agar medium, chromID VRE, to screen for vancomycin-resistant *Enterococcus faecium* and *Enterococcus faecalis*, *Diagn. Microbiol. Infect. Dis.*, 59, 477, 2007.
162. Cuzon, G. et al., A novel chromogenic medium for vancomycin resistant *Enterococcus faecium* and *Enterococcus faecalis* detection, *Int. J. Antimicrob. Agents*, 29, S22, 2007.
163. Hughes, J.G. et al., Culture with BACTEC Peds Plus/F bottle compared with conventional methods for detection of bacteria in synovial fluid, *J. Clin. Microbiol.*, 39, 4468, 2001.

164. Qian, Q. et al., Direct identification of bacteria from positive blood cultures by amplification and sequencing of the 16S rRNA gene: Evaluation of BACTEC 9240 instrument true positive and false positive results, *J. Clin. Microbiol.*, 39, 3578, 2001.
165. Hamilton-Miller, J.M.T., and Shah, S., Identification of clinically isolated vancomycin-resistant enterococci: Comparison of API and BBL Crystal systems, *J. Med. Microbiol.*, 48, 695, 1999.
166. Peters, R.P.H. et al., Rapid identification of pathogens in blood cultures with a modified fluorescence in situ hybridization assay, *J. Clin. Microbiol.*, 44, 4186, 2006.
167. Miyoung, K. et al., Comparison of the MicroScan, VITEK 2, and Crystal GP with 16S rRNA sequencing and MicroSeq 500 v2.0 analysis for coagulase-negative staphylococci, *BMC Microbiol.*, 8, 233, 2008.
168. Pendle, S. et al., Difficulties in detection and identification of *Enterococcus faecium* with low-level inducible resistance to vancomycin, during a hospital outbreak, *Clin. Microbiol. Infec.*, 14, 853, 2008.
169. Vankerckhoven, V. et al., Development of a multiplex PCR for the detection of asa1, gelE, cylA, esp, and hyl genes in enterococci and survey for virulence determinants among European hospital isolates of *Enterococcus faecium*, *J. Clin. Microbiol.*, 42, 4473, 2004.
170. Chen, A.Y., and Zervos, M.J., *Enterococcus*: Antimicrobial resistance in enterococci epidemology, treatment, and control. In Mayers, D.L. (Editor), *Antimicrobial Drug Resistance*, pp. 715–734, Humana Press, 2009.
171. Centers for Disease Control and Prevention. *Recommendations for Preventing the Spread of Vancomycin Resistance Recommendations of the Hospital Infection Control Practices Advisory Committee* (HICPAC). *MMWR*, 44, 1, 1995.
172. Owens, R.C. Jr., and Ambrose, P.G., Antimicrobial stewardship and the role of pharmacokinetics-pharmacodynamics in the modern antibiotic era, *Diagn. Microbiol. Infect. Dis.*, 57, 77S, 2007.
173. Owens, R.C. Jr., Antimicrobial stewardship: concepts and strategies in the 21st century, *Diagn. Microbiol. Infect. Dis.*, 61, 110, 2008.
174. Patel, D., Lawson, W., and Guglielmo, B.J., Antimicrobial stewardship programs: Interventions and associated outcomes, *Expert Rev. Anti. Infect. Ther.*, 6, 209, 2008.
175. Dellit, T. et al., Infectious Diseases Society of America and the Society for Healthcare Epidemiology of America guidelines for developing an institutional program to enhance antimicrobial stewardship, *Clin. Infect. Dis.*, 44, 159, 2007.
176. Drew, R.H., Antimicrobial stewardship programs: How to start and steer a successful program, *J. Manag.Care Pharm.*, 15, S18, 2009.
177. Brown, M., and McClure, P., Microbiological risk assessment for emerging pathogens. In Motarjemi, Y., and Adams, M. (Editors), *Emerging Foodborne Pathogens*, pp. 130–152, Woodhead Publishing Limited and CRC Press LLC, England, 2006.
178. Skovgaard, N., New trends in emerging pathogens, *Int. J. Food Microbiol.*, 120, 217, 2007.

23 *Granulicatella*

Sheng Kai Tung and Tsung Chain Chang

CONTENTS

23.1 INTRODUCTION

23.1.1 Classification, Morphology, and Biology

Bacteria in the genus *Granulicatella* are gram-positive, catalase-negative, facultative anaerobic cocci. Species in the genus can cause opportunistic infections in immunocompromised patients and are infrequently isolated from clinical laboratories.[1] The genera *Abiotrophia* and *Granulicatella* have been proposed to accommodate microorganisms previously known as nutritionally variant streptococci or satelliting streptococci.[1–3] By chromosomal DNA–DNA hybridizations, Bouvet et al.[4] demonstrated that nutritionally variant streptococci were really members of two novel streptococcal species, *Streptoccoccus defectivus* and *Streptoccoccus adjacens*. Based on the 16S rRNA gene sequence, Kawamura et al.[3] transferred *Streptoccoccus defectivus* and *Streptoccoccus adjacens*, respectfully, to two new species, *Abiotrophia defectiva* and *Abiotrophia adiacens*. A third species, *Abiotrophia elegans*, a possible pathogen in patients with endocarditis, was described in 1998 by Roggenkamp et al.[5] In 2000, Collins and Lawson[2] proposed a new genus, *Granulicatella*, with *Granulicatella adiacens* and *Granulicatella elegans* encompassing strains formerly recognized as *Abiotrophia adiacens* and *Abiotrophia elegans*, respectfully. *Abiotrophia defectiva* remains as the sole species in *Abiotrophia*.[2] A third species in *Granulicatella* is *Granulicatella balaenopterae*; however, there was no report of human infection caused by this microorganism.

Bacteria in the genus *Granulicatella* are nonmotile cocci that occur as single cells, in pairs, or in short chains. Cell morphology depends on culture conditions. Elongated or swollen cells may be observed, especially when grown under less optimal nutritional conditions.[6] The G + C content of *Granulicatella* DNA is approximately 36–37.5 mol%. Members of *Granulicatella* are normal flora of the oral cavity or upper respiratory tract[1] and are non-betahemolytic on blood-based solid media. *Granulicatella* exhibit satellitism around the colonies of other bacteria, such as *Staphylococcus epidermidis* and other streptococci.[7] *Granulicatella* species usually have low virulence and are normally pathogenic in immunocompromised hosts.[1] Predisposing factors that lead to *Granulicatella* infections include previously damaged tissues (e.g., heart valves), invasive procedures, prolonged hospitalization, and the presence of foreign bodies. *Granulicatella* can be isolated from a variety of clinical samples, including blood, urine, cerebrospinal fluid, wound tissues, and others specimens.[1] Infections caused by *Granulicatella* might be due to some microorganisms entering the bloodstream and disseminating to other tissues.[8,9]

23.1.2 Clinical Features

Granulicatella species have been reported as etiologic agents of bacteremia, infectious endocarditis, brain abscess, postpartum or postabortal sepsis, pancreatic abscess, wound infection, vertebral osteomyelitis, conjunctivitis, cirrhosis, endophthalmitis, infectious crystalline keratopathy, septic arthritis, otitis media, central nervous system infections, breast implant–associated infection, and peritoneal dialysis–related peritonitis.[5,7,10–17] Using a molecular technique, the proportion of *Granulicatella elegans*, *Granulicatella adiacens*, and *Abiotrophia defectiva* was found to be 1:11:1 in the human mouth.[18] Exopolysaccharides that can adhere to human tissues are produced by *Granulicatella*.[7]

All members of *Granulicatella* are susceptible to vancomycin. However, there are still no standardized methods and interpretation criteria for antibiotic susceptibility testing results in the documents of CLSI (Clinical and Laboratory Standards Institute). Most investigators have employed streptococcal (other than *Streptococcus pneumoniae*) interpretive criteria for determining the results of susceptibility.[1] Since *Granulicatella* species are fastidious, Mueller-Hinton medium supplemented with blood and pyridoxal hydrochloride (0.001%) is used for susceptibility testing, and a CO_2-enriched atmosphere[1] is required for incubation. Tuohy et al.[19] examined a collection of 21 *Granulicatella adiacens* isolates for their susceptibilities to a variety of antibiotics. They found that all strains were susceptible to clindamycin, rifampin, levofloxacin, ofloxacin, and quinupristin-dalfopristin. The percentage of isolates that were susceptible to other agents was as follows: penicillin, 55%; amoxicillin, 81%; ceftriazone, 63%; and meropenem, 96%. However, Zheng et al.[20] observed high rates of beta-lactam and macrolide resistance in a collection of pediatric *Granulicatella* isolates, and hence they proposed that antimicrobial susceptibility testing should be systematically done to achieve appropriate antimicrobial therapy. In severely ill patients or those with a suboptimal response to initial therapy with beta-lactam antibiotics, treatment with vancomycin should be considered.[21] The combination therapy with vancomycin and an aminoglycoside should be initiated empirically in patients with central nervous system infections.[13]

23.1.3 Diagnosis

23.1.3.1 Conventional Techniques

Phenotypic characteristics that can be used as a guide to identify *Granulicatella* and *Abiotrophia* species are listed in Table 23.1. Members of *Granulicatella* are likely to be isolated on rich and nonselective media. They can grow on chocolate agar, in thioglycolate broth, and on brucella agar with 5% horse blood, but not on Typticase soy agar with 5% sheep blood.[1] Cell growth requires thiol compounds such as L-cysteine or the active form of vitamin B_6, pyridoxal phosphate, in media. Growth does not occur at 10°C or 45°C. *Granulicatella* in clinical samples may be overlooked when media without nutritional supplements are used. Gram stain morphology, although prone to be subjective in interpretation, is a major decision point in the identification protocols. Gram stain morphology of *Granulicatella* spp. resembles that of streptococci, with cells being cocci or coccobacilli in pairs and chains. Cells grown in broth, such as thioglycollate, should be used for determination of cellular morphology. Besides the characteristics listed in Table 23.1, additional tests are recommended to make a final identification, especially for isolates recovered from important clinical specimens.[1] Additional tests useful for *Granulicatella* spp. differentiation can be found in several studies.[4,6,22] Commercial identification kits, such as API 20 Strep, rapid ID32 STREP, and the VITEK 2 system (all from BioMérieux, Marcy l'Etoile, France), could be used for identification of *Granulicatella* spp. However, *Granulicatella elegans* cannot be identified by both API 20 Strep and rapid ID32 STREP kits since the species is not included in the databases of the two kits. *Granulicatella adiacens* and *Granulicatella elegans* are phenotypically similar; several tests (hydrolysis of arginine and hippurate, and production of β-glucuronidase) can be used to differentiate the two species (Table 23.1). Most clinical *Granulicatella adiacens* isolates could be identified by the API 20 STREP system with > 80%–90% confidence, whereas both the Vitek System (GPI) and ID32 STREP identified only a small portion of *Granulicatella adiacens* isolates.[23]

It should be noted that the growth of *Granulicatella elegans* is only supported by L-cysteine hydrochloride and not by pyridoxal phosphate when the organism is grown in Todd-Hewitt or casein-soy peptone broth.[5] The media of blood culture systems supplemented only with pyridoxal phosphate

TABLE 23.1
Phenotypic Characteristics of *Abiotrophia defectiva* and *Granulicatella* Species[a]

Characteristics	*A. defectiva*	*G. adiacens*	*G. balaenopterae*	*G. elegans*
Production of acid from				
Lactose	+	–	–	–
Maltose	+	+	+	–
Pullulan	+	–	+	–
Sucrose	+	+	–	+
Tagatose	–	+	–	–
Trehalose	+	–	+	–
Hydrolysis of				
Arginine	–	–	+	+
Hippurate	–	–	–	+
Production of				
α-Galactosidase	+	–	–	–
β-Galactosidase	+	–	–	–
β-Glucuronidase	–	+	–	–
N-Acetyl-β-glucosaminidase	–	–	+	–
Pyrrolidonyl arylamidase	+	+	ND	+
Leucine aminopeptidase	+	+	ND	+
Satelliting behavior	+	+	ND	+

Symbols: –, negative reactions ≤15%; +, positive reactions ≥85%; ND, not determined.

[a] Data compiled from References 1, 2, 5, 6, 44, and 45.

may fail to support the growth of *Granulicatella elegans*, and thus, these systems might not be able to detect this microorganism as a possible pathogen.

23.1.3.2 Molecular Techniques

Molecular methods developed for identification of members of *Granulicatella* are limited. Sequence analysis of the 16S rRNA gene has been widely used for bacterial identification.[24,25] The technique is becoming popular for identifying biochemically unidentified bacteria or for providing reference identifications for unusual strains.[24] Sequencing of the 16S rRNA gene can be successfully used to identify *Granulicatella* species. Multiple sequences of the 16S rRNA genes for *Granulicatella adiacens* and *Granulicatella elegans* are available in the nucleotide database of National Center for Biotechnology Information, USA. The 16S rRNA gene sequences of *Granulicatella adiacens* and *Abiotrophia defectiva* formed a distinct cluster; this cluster is not closely related to other viridans streptococcal species and *Abiotrophia*.[3,5] Woo et al.[23] found that 16S rRNA gene sequencing is the technique of choice for identifying *Granulicatella adiacens* and proposed that early surgical intervention should be considered when endocarditis caused by this microorganism is diagnosed.

The ribosomal 16S-23S intergenic spacer (ITS) region has been suggested as a good candidate for bacterial identification and strain typing.[26–29] Sequences of the ITS region in bacteria have low intraspecies variation and high interspecies divergence.[26,27,29] The feasibility of using ITS sequence to identify *Granulicatella adiacens* and *Granulicatella elegans* was established by Tung et al.[30] It should be noted that in the genome of *Granulicatella* spp. possesses multiple ITS fragments having different lengths and sequences, therefore the PCR products amplified from the ITS region of an isolate cannot be directly sequenced. Before sequencing, the amplicons should be eluted from agarose gel or cloned into a plasmid and reamplified. Once the ITS sequence is determined, species identification can be done by searching databases using the BLAST (http://www.ncbi.nlm.nih.gov/blast/) sequence analysis tool. The *rpoB* gene, encoding the highly conserved subunit of the bacterial RNA polymerase, has been demonstrated to be a suitable target for identification of a wide range of bacteria.[31] The gene was shown to be more discriminative than the 16S rRNA gene for identifying enteric bacteria.[31] Partial *rpoB* gene sequence (740 bp) had been applied to identify streptococci and *Granulicatella adiacens*.[32]

In recent years, DNA array technology has been successfully used to identify microorganisms that are difficult to be determined to the species level; these bacteria include viridans streptococci,[33] nonfermenting gram-negative bacteria,[34] *Legionella*,[35] *Mycobacterium*[36,37] and foodborne bacterial pathogens.[38] The method consists of PCR amplification of a specific gene region, followed by hybridization of the PCR product labeled with a fluorescent dye (or with a molecular technique that could be recognized by antibody) to a panel of oligonucleotide probes immobilized on a solid support such as glass slide or nylon membrane. The test sensitivity and specificity of the arrays are largely determined by the sequences of the oligonucleotide probes. The whole procedure of array hybridization takes between 8 and 24 h, starting from isolated colonies or clinical specimens. The hybridization patterns could be read by a fluorescence scanner if a fluorescent label is used or by the naked eye if colorimetric reaction is used. By using the oligonucleotide array platform, five oligonucleotide probes designed from the ITS regions were successfully used to identify *Granulicatella adiacens*, *Granulicatella elegans*, and *Granulicatella balaenopterae*.[39] The array method used a protocol encompassing DNA extraction, PCR amplification of the ITS regions, and hybridization of the PCR products to species-specific probes on the array.

23.2 METHODS

23.2.1 Sample Preparation

The boiling method described by Millar et al. could be used to extract DNA from the bacterial colonies.[40] Briefly, one to several colonies from pure cultures are suspended in 50 μL of sterilized water and heated at 100°C for 15 min in a heating block. After centrifugation in a microcentrifuge (6000 × *g* for 3 min), the supernatant contains bacterial DNA. Alternatively, a variety of DNA extraction kits may be also used.

23.2.2 Detection Procedures

23.2.2.1 16S rRNA Gene Sequencing

Principle. Relman[25] utilized primer pair 8FPL (5′-GTTT GATCCTGGCTCAG-3′) and 1492RPL (5′-GGTTACCT TGTTACGACTT-3′) for PCR amplification of the nearly complete length of the 16S rRNA gene. Sequencing analysis of the resulting fragment reveals the identity of the bacteria under investigation.

Procedure

1. Prepare PCR mixture (50 μL) consisting of 10 mM Tris-HCl pH 8.3, 50 mM KCl, 1.5 mM $MgCl_2$, 1 mM dNTPs (0.25 mM each), 0.1 μM (each) primer, 1 U of *Taq* DNA polymerase, and 5 μL (1–5 ng) of template DNA.
2. Conduct PCR amplification with 1 cycle of 94°C for 5 min; 5 cycles of 94°C for 1 min, 52°C for 1 min, and 72°C for 1.5 min; 25 cycles of 94°C for 1 min, 55°C for 1 min, and 72°C for 1.5 min; and a final 72°C for 5 min.
3. Purify and sequence PCR products in both directions by using the above two primers and an additional primer, 1055r (5′-CACGAGCTGACGACAG CCAT-3′).[25,27]
4. Compare the determined sequences to known sequences of the 16S rRNA genes in the databases of the National Center for Biotechnology Information using the BLASTN algorithm.

Note: Species identification is determined from the best-scoring reference sequence of the BLAST output that has a sequence similarity of ≥99% with the query sequence. All positions showing differences from the best-scoring reference sequence could be visually inspected in the electropherogram, and the sequence is corrected if needed, that is, when obvious sequencing software errors occur, or when undetermined nucleotides in the sequence can be determined by visual inspection of the electropherogram. Afterwards, a second search is done with the BLAST algorithm. Ideally, 1300–1500 bp are sequenced with <1% position ambiguities, with a minimum of 500–525 bp being sequenced.[24] The criteria for species identification is at least 99% sequence similarity; ideally, >99.5% sequence similarity, with sequence match to type strain or reference strain of species that has undergone DNA-relatedness studies.

23.2.2.2 Sequencing of the Ribosomal 16S-23S Intergenic Spacer (ITS) Region

Principle. Gürtler and Stanisich[28] designed the bacteria-specific universal primer pair 13BF (5′-GTGAATACGTTC CCGGGCCT-3′) and 6R (5′-GGGTTYCCCCRTTCR GAAAT-3′) (where Y is C or T and R is A or G) for amplification of a DNA fragment that encompassed ITS, a portion of the 3′ end of the 16S rRNA gene, and a portion of the 5′ end of the 23S rRNA gene.[27,30] The 5′ end of primer 13BF is located at position 1371 of the 16S rRNA gene, and the 5′ end of primer 6R is located at position 108 downstream of the 5′ end of the 23S rRNA gene (*Escherichia coli* numbering). The identity of the bacteria concerned is determined following nucleotide sequencing analysis of the ITS fragment.

Procedure

1. Prepare PCR mixture (50 μL) consisting of 10 mM Tris-HCl (pH 8.3), 50 mM KCl, 1.5 mM $MgCl_2$, 0.8 mM deoxyribonucleoside triphosphates (0.2 mM each), 0.1 μM primer (each), 1 U of *Taq* DNA polymerase, and 5 μL (1–5 ng) of template DNA.
2. Perform PCR amplification with 1 cycle of 94°C for 5 min; 5 cycles 94°C for 1 min; 55°C for 1 min, and 72°C for 1 min; 30 cycles of 94°C for 1 min, 60°C for 1 min, and 72°C for 1 min; and a final 72°C for 3 min.
3. Purify and sequence PCR products in both directions by using the above two primers.

Note: Species of *Granulicatella* produce multiple ITS amplicons in PCR with different length and sequence.[27] Only the shortest ITS fragment of a strain, which is usually the dominant band, is eluted from the agarose gel and sequenced. In this way, the ITS of *Granulicatella elegans* can be sequenced. However, the amplicons of *Granulicatella adiacens* eluted from gels cannot be directly sequenced due to the presence of different amplicons with similar size in the eluted DNA. For this reason, the eluted DNA from gel can be cloned with a Topo TA cloning kit (Invitrogen, Carlsbad, California, USA). The ITS regions of positive clones are reamplified and sequenced. The determined sequences are compared to known sequences in the databases of the National Center for Biotechnology Information using the BLASTN algorithm. Species identity is determined from the best-scoring reference sequence of the BLAST output and whether the best-scoring reference sequence in the databases has a sequence identity of ≥98% with the query sequence.[27]

23.2.2.3 Partial *rpoB* Gene Sequencing

Principle. Drancourt et al.[32] employed primers Strepto F (5′-AARYTIGGMCCTGAAGAAAT-3′) and Strepto R (5′-T GIARTTTRTCATCAACCATGTG-3′) (where M is A or C, R is A or G, and W is A or T) to amplify a portion (740 bp) of the *rpoB* gene from *Streptococcus*, *Enterococcus*, *Gemella*, *Abiotrophia*, and *Granulicatella*.

Procedure

1. Prepare PCR mixture (50 μL) consisting of 2.5 U of *Taq* polymerase, 1× *Taq* buffer, 1.8 mM $MgCl_2$ (Gibco-BRL/Life Technologies, Cergy Pontoise, France), 0.2 mM concentrations of dATP, dTTP, dGTP, and dCTP, and 0.2 μM concentrations of each primer, and 5 μL (1–5 ng) of template DNA.
2. Subject the PCR mixture to one cycle of 95°C for 2 min; 35 cycles of 94°C for 30 s, 52°C for 30 s, and 72°C for 60 s; and a final 72°C for 5 min.
3. Purify the amplified product by using a QIAquick spin PCR purification kit (Qiagen, Hilden, Germany) and determine the nucleotide sequence of the amplicon.

Note: This primer pair Strepto F and Strepto R is shown to be almost specific for streptococcal species, since no amplification products were obtained from 58 other bacterial isolates belonging to nonstreptococcal species, including other gram-positive cocci.[32]

23.2.2.4 Array-Based Method

Principle. Tung et al.[39] reported an array-based method for identification of *Abiotrophia* (1 species), *Enterococcus* (18 species), *Granulicatella* (3 species), and *Streptococcus* (31 species and 6 subspecies). The method consists of PCR amplification of the ribosomal DNA intergenic spacer (ITS) regions, and hybridization of the digoxigenin-labeled PCR products to a panel of 88 oligonucleotide probes (16- to 30-mers) immobilized on a nylon membrane. The probes are designed either from the ITS regions or from the 3′ ends of the 16S rRNA genes, and belong to three categories: species specific, group specific, and supplemental probes.

Procedure

1. Amplify the ITS region for hybridization with primer pair 8FPL (5′-GTTTGATCCTGGCTCAG-3′) and 1492RPL (5′-GGTTACCTTGTTACGACTT-3′) as shown in Section 23.2.2.2, except that the reverse primer 6R (5′dig-GGGTTYCCCCRTTCRGAAAT3′) is labeled with a digoxigenin molecule at its 5′ end.[39] Labeling of both primers with digoxigenin is encouraged because this can increase the hybridization signal.

2. Fabricate oligonucleotide array:
   (a) Customer synthesize *Granulicatella adiacens*–specific probe (5′-TATCACAACAAATAACC AATTAA-3′), *Granulicatella elegans* probes (5′-GAGGTTAACTCTCAACTCGACCT TTGAAAA-3′, 5′-TTTAACAAGAAGCAACG CGACCATA-3′) and *Granulicatella balaenopterae*-specific probes (5′-ATAACGGA ACCTACCAAGTTCACTTC-3′, 5′-TGAGAGA TTAATTCTCTCTAGACTTTGATC-3′, 5′-GCAAACGCGAATCATATTGAGAC TTAA-3′)[30];
   (b) Add multiple thymine bases (5–15) to the 3′ ends of the above probes to increase the binding of these probes to the nylon membrane that is a solid support for immobilizing probes[41];
   (c) Design a positive control probe (5′-GTCGT AACAAGGTAGCCGTA-3′) from a conserved region at the 3′ end of the 16S rRNA gene (GenBank AB023575) to verify all the procedures including PCR, hybridization, antibody reaction, and colorimetric reaction;
   (d) Dilute the oligonucleotide probes 1:1 (final concentration, 10 μM) with a tracking dye solution (30% [vol/vol] glycerol, 40% [vol/vol] dimethyl sulfoxide, 1 mM EDTA [disodium salt], 0.15% [wt/vol] bromophenol blue, 10 mM Tris-HCl [pH 7.5]).
   (e) Draw the probe solutions into different wells of a round-bottom microtiter plate and spotted onto a positively charged nylon membrane (Roche, Mannheim, Germany) with an automatic arrayer (SR-A300; Ezspot, Taipei, Taiwan) by using a solid pin (400–1000 μm in diameter); alternatively, spot the probe solutions on the membrane using the solid pin by hand; use dye solution without probe as a negative control;
   (f) Dry and expose the membrane to short-wave UV light (Stratalinker 1800, California, USA) for 30 s; Remove unbound oligonucleotides by two washes (2 min each) at room temperature in 0.5 × SSC (1 × SSC is 0.15 M NaCl plus 0.015 M sodium citrate)-0.1% sodium dodecyl sulfate (SDS);
   (g) Air dry and store the arrays at room temperature under desiccation for further use.
3. Hybridize the labeled PCR products to probes:
   (a) Prehybridize each array for 2 h with 1 mL of hybridization solution (5 × SSC, 1% [wt/vol] blocking reagent (Roche), 0.1% *N*-laurylsarcosine, 0.02% SDS) in an individual well of a 24-well cell culture plate.
   (b) Heat the digoxigenin-labeled PCR product amplified from an isolate in a boiling water bath for 5 min and immediately cool on an ice bath.
   (c) Hybridize at 45°C for 90 min in an incubator with a shaking speed of 60 rpm.
   (d) Wash the array three times (5 min each) in 1 mL of washing buffer (2× SSC, 0.1% SDS) and one wash (1 min) in 1 mL of a second washing buffer (0.5× SSC, 0.1% SDS).
   (e) Block for 1 h with 1 mL of blocking solution supplied in the DIG Nucleic Acid Detection kit (Roche) (dissolve 1% [wt/vol] blocking reagent in maleic acid buffer [0.1 M maleic acid, 0.15 M NaCl, pH 7.5]).
   (f) Remove the blocking solution, add 0.5 mL of alkaline phosphatase-conjugated sheep anti-digoxigenin antibodies (diluted 1:2500 in blocking solution) to each well, and incubate for 1 h.
   (g) Wash the array three times (15 min each) in 1 mL of washing solution (0.3% [vol/vol] Tween 20 in maleic acid buffer), wash once in 1 mL of detection buffer (0.1 M Tris-HCl, 0.15 M NaCl, pH 9.5) for 5 min.
   (h) Add 0.5 mL of alkaline phosphatase substrate (a stock solution of nitroblue tetrazolium chloride and 5-bromo-4-chloro-3-indolylphosphate diluted 1:50 in detection buffer) (Roche) to each well, and incubate the plate at 37°C in dark (without shaking).

Note: Color development is clearly visible between 30 min and 1 h after the start of the reaction; then the hybridized spot can be easily recognized by the naked eye. A strain is identified as one of the *Granlicatella* species when the probe (or all probes) specified for that species is hybridized.

23.3 CONCLUSION AND FUTURE PERSPECTIVES

Members of the genus *Granulicatella* bacteria are gram-positive, catalase-negative, and facultatively anaerobic cocci. Currently, the genus contains three species, *Granulicatella adiacens*, *Granulicatella elegans*, and *Granulicatella balaenopterae*, with the first two being human pathogens. *Granulicatella* spp. normally display low virulence to hosts and are infrequently isolated as opportunistic pathogens.[1] These bacteria may enter the bloodstream and disseminate to other tissues and thus cause infections.[8,9] A wide spectrum of infections caused by *Granulicatella* has been reported.[7,10,11,13–17,42,43]

They resemble other better known streptococcal isolates and consequently may be misidentified as these bacteria. *Granulicatella* spp. are part of the normal flora of human pharynx, urogenital, and intestinal tracts.[1,7] They form tiny colonies on blood agar supplemented with pyridoxal phosphate or L-cysteine and exhibit satellitism around the colonies of *Staphylococcus epidermidis* and other streptococci.[7] *Granulicatella* may be overlooked because of their poor growth on unsupplemented media and may not be accurately identified by conventional diagnostic procedures or commercial kits. Molecular techniques are good alternatives for identification of doubtful organisms.

Molecular methods used to identify *Granulicatella* species include 16S rRNA gene sequencing, ribosomal 16S-23S intergenic spacer (ITS) sequencing, partial sequencing of the *rpoB* gene, and oligonucleotide probe hybridization. Sequencing of the 16S rRNA gene seems to be a prior choice to identify *Granulicatella* spp. since their respective 16S rRNA gene sequences are available in GenBank. Sequencing of the ribosomal 16S-23S intergenic spacer (ITS) is another alternative. However, *Granulicatella* spp. possess multiple ITS fragments with different length and sequence; therefore, the amplified ITS should be eluted from agarose gel or cloned before sequencing.[30] For this reason, ITS sequencing seems to be impractical for identification of *Granulicatella*. This drawback can be overcome by the hybridization technique, with probes being designed from the ITS region. Five oligonucleotide probes were successfully used to identify the three *Granulicatella* species.[39] The array format originally developed by Tung et al.[39] can be easily transformed to the dot hybridization format without sacrifice of the probe efficacy.

Finally, partial sequencing (740 bp) of *rpoB*, encoding the subunit of the bacterial RNA polymerase, could be used to identify *Granulicatella adiacens*.[32] Unfortunately, there is no *rpoB* sequence of *Granulicatella elegans* deposited in public databases. Accurate identification of members of *Granulicatella* may be of great help in elucidating the epidemiology of these bacteria.

REFERENCES

1. Ruoff, K.L., *Aerococcus*, *Abiotrophia*, and other aerobic catalase-negative, gram-positive cocci. In Murray, P. R. et al. (Editors), *Manual of Clinical Microbiology*, 9th ed., pp. 443–454, American Society for Microbiology, Washington, DC, 2007.
2. Collins, M.D., and Lawson, P.A., The genus *Abiotrophia* (Kawamura et al.) is not monophyletic: Proposal of *Granulicatella* gen. nov., *Granulicatella adiacens* comb. nov., *Granulicatella elegans* comb. nov. and *Granulicatella balaenopterae* comb. nov., *Int. J. Syst. Evol. Microbiol.*, 50, 365, 2000.
3. Kawamura, Y. et al., Transfer of *Streptococcus adjacens* and *Streptococcus defectivus* to *Abiotrophia* gen. nov. as *Abiotrophia adiacens* comb. nov. and *Abiotrophia defectiva* comb. nov., respectively, *Int. J. Syst. Bacteriol.*, 45, 798, 1995.
4. Bouvet, A., Grimont, F., and Grimont, P.A.D., *Streptococcus defectivus* sp. nov. and *Streptococcus adjacens* sp. nov., nutritionally variant streptococci from human clinical specimens, *Int. J. Syst. Bacteriol.*, 39, 290, 1989.
5. Roggenkamp, A. et al., *Abiotrophia elegans* sp. nov., a possible pathogen in patients with culture-negative endocarditis, *J. Clin. Microbiol.*, 36, 100, 1998.
6. Christensen, J.J., and Facklam, R.R., *Granulicatella* and *Abiotrophia* species from human clinical specimens, *J. Clin. Microbiol.*, 39, 3520, 2001.
7. Ruoff, K.L., Nutritionally variant streptococci, *Clin. Microbiol. Rev.*, 4, 184, 1991.
8. Coykendall, A.L., Classification and identification of the viridans streptococci, *Clin. Microbiol. Rev.*, 2, 315, 1989.
9. Strom, B.L. et al., Dental and cardiac risk factors for infective endocarditis. A population-based, case-control study, *Ann. Intern. Med.*, 129, 761, 1998.
10. Altay, M. et al., A novel agent of peritoneal dialysis-related peritonitis: *Granulicatella adiacens*, *Perit. Dial. Int.*, 28, 96, 2008.
11. Biermann, C. et al., Isolation of *Abiotrophia adiacens* from a brain abscess which developed in a patient after neurosurgery, *J. Clin. Microbiol.*, 37, 769, 1999.
12. Bouvet, A., Human endocarditis due to nutritionally variant streptococci: *Streptococcus adjacens* and *Streptococcus defectivus*, *Eur. Heart. J.*, 16 Suppl B, 24, 1995.
13. Cerceo, E. et al., Central nervous system infections due to *Abiotrophia* and *Granulicatella* species: An emerging challenge? *Diagn. Microbiol. Infect. Dis.*, 48, 161, 2004.
14. Heath, C.H. et al., Vertebral osteomyelitis and discitis associated with *Abiotrophia adiacens* (nutritionally variant streptococcus) infection, *Aust. N. Z. J. Med.*, 28, 663, 1998.
15. Hepburn, M.J. et al., Septic arthritis caused by *Granulicatella adiacens*: Diagnosis by inoculation of synovial fluid into blood culture bottles, *Rheumatol. Int.*, 23, 255, 2003.
16. Namdari, H. et al., *Abiotrophia* species as a cause of endophthalmitis following cataract extraction, *J. Clin. Microbiol.*, 37, 1564, 1999.
17. Ormerod, L.D. et al., Infectious crystalline keratopathy. Role of nutritionally variant streptococci and other bacterial factors, *Ophthalmology*, 98, 159, 1991.
18. Sato, S. et al., *Abiotrophia elegans* strains comprise 8% of the nutritionally variant streptococci isolated from the human mouth, *J. Clin. Microbiol.*, 37, 2553, 1999.
19. Tuohy, M.J. et al., Antimicrobial susceptibility of *Abiotrophia adiacens* and *Abiotrophia defectiva*, *Diagn. Microbiol. Infect. Dis.*, 38, 189, 2000.
20. Zheng, X. et al., Antimicrobial susceptibilities of invasive pediatric *Abiotrophia* and *Granulicatella* isolates, *J. Clin. Microbiol.*, 42, 4323, 2004.
21. Senn, L. et al., Bloodstream and endovascular infections due to *Abiotrophia defectiva* and *Granulicatella* species, *BMC Infect. Dis.*, 6, 9, 2006.
22. Beighton, D. et al., Analysis of enzymatic activities for differentiation of two species of nutritionally variant streptococci, *Streptococcus defectivus* and *Streptococcus adjacens*, *J. Clin. Microbiol.*, 33, 1584, 1995.
23. Woo, P.C. et al., *Granulicatella adiacens* and *Abiotrophia defectiva* bacteraemia characterized by 16S rRNA gene sequencing, *J. Med. Microbiol.*, 52, 137, 2003.
24. Janda, J.M., and Abbott, S.L., 16S rRNA gene sequencing for bacterial identification in the diagnostic laboratory: Pluses, perils, and pitfalls, *J. Clin. Microbiol.*, 45, 2761, 2007.
25. Relman, D.A., Universal bacterial 16S rDNA amplification and sequencing. In Persing, D.H. et al. (Editors), *Diagnostic Molecular Microbiology*, pp. 489–495, American Society for Microbiology, Washington, DC, 1993.
26. Chang, H.C. et al., Species-level identification of isolates of the *Acinetobacter calcoaceticus–Acinetobacter baumannii* complex by sequence analysis of the 16S–23S rRNA gene spacer region, *J. Clin. Microbiol.*, 43, 1632, 2005.
27. Chen, C.C. et al., Identification of clinically relevant viridans group streptococci by sequence analysis of the 16S-23S ribosomal DNA spacer region, *J. Clin. Microbiol.*, 42, 2651, 2004.
28. Gürtler, V., and Stanisich, V.A., New approaches to typing and identification of bacteria using the 16S–23S rDNA spacer region, *Microbiology*, 142, 3, 1996.
29. Hassan, A.A. et al., Inter- and intraspecies variations of the 16S–23S rDNA intergenic spacer region of various streptococcal species, *Syst. Appl. Microbiol.*, 26, 97, 2003.

30. Tung, S.K. et al., Identification of species of *Abiotrophia*, *Enterococcus*, *Granulicatella* and *Streptococcus* by sequence analysis of the ribosomal 16S-23S intergenic spacer region, *J. Med. Microbiol.*, 56, 504, 2007.
31. Mollet, C., Drancourt, M., and Raoult, D., *rpoB* sequence analysis as a novel basis for bacterial identification, *Mol. Microbiol.*, 26, 1005, 1997.
32. Drancourt, M. et al., *rpoB* gene sequence-based identification of aerobic gram-positive cocci of the genera *Streptococcus*, *Enterococcus*, *Gemella*, *Abiotrophia*, and *Granulicatella*, *J. Clin. Microbiol.*, 42, 497, 2004.
33. Chen, C.C. et al., Identification of clinically relevant viridans streptococci by an oligonucleotide array, *J. Clin. Microbiol.*, 43, 1515, 2005.
34. Su, S.C. et al., Identification of nonfermenting Gram-negative bacteria of clinical importance by an oligonucleotide array, *J. Med. Microbiol.*, 58, 596, 2009.
35. Su, H.-P. et al., Identification of *Legionella* species by an oligonucleotide array, *J. Clin. Microbiol.*, 47, 1386, 2009.
36. Fukushima, M. et al., Detection and identification of *Mycobacterium* species isolates by DNA microarray, *J. Clin. Microbiol.*, 41, 2605, 2003.
37. Park, H. et al., Detection and genotyping of *Mycobacterium* species from clinical isolates and specimens by oligonucleotide array, *J. Clin. Microbiol.*, 43, 1782, 2005.
38. Lin, M.C. et al., Identification of six foodborne pathogens and *Psedomonas aeruginosa* grown on selective media by an oligonucleotide array, *J. Food Prot.*, 68, 2278, 2005.
39. Tung, S.K. et al., Array-based identification of species of the genera *Abiotrophia*, *Enterococcus*, *Granulicatella*, and *Streptococcus*, *J. Clin. Microbiol.*, 44, 4414, 2006.
40. Millar, B.C. et al., A simple and sensitive method to extract bacterial, yeast and fungal DNA from blood culture material, *J. Microbiol. Methods*, 42, 139, 2000.
41. Brown, T.J., and Anthony, R.M., The addition of low numbers of 3′ thymine bases can be used to improve the hybridization signal of oligonucleotides for use within arrays on nylon supports, *J. Microbiol. Methods*, 42, 203, 2000.
42. del Pozo, J.L. et al., *Granulicatella adiacens* breast implant-associated infection, *Diagn. Microbiol. Infect. Dis.*, 61, 58, 2008.
43. Gephart, J.F., and Washington, J.A. 2nd, Antimicrobial susceptibilities of nutritionally variant streptococci, *J. Infect. Dis.*, 146, 536, 1982.
44. Kanamoto, T. et al., Genetic heterogeneities and phenotypic characteristics of strains of the genus *Abiotrophia* and proposal of *Abiotrophia* para-adiacens sp. nov., *J. Clin. Microbiol.*, 38, 492, 2000.
45. Ohara-Nemoto, Y. et al., Infective endocarditis caused by *Granulicatella elegans* originating in the oral cavity, *J. Clin. Microbiol.*, 43, 1405, 2005.

24 *Lactobacillus*

Ester Sánchez and Yolanda Sanz

CONTENTS

24.1 INTRODUCTION

The genus *Lactobacillus* belongs to the lactic acid bacteria (LAB) group, which comprises a broadly defined gram-positive bacterial species of nonspore-forming rods or cocobacilli, and which is characterized by the formation of lactic acid as the sole or main end product of carbohydrate metabolism. They are catalase-negative, strictly fermentative, aerotolerant or anaerobic, aciduric or acidophilic, and have complex nutritional requirements and a low GC (guanine + cytosine) content (<50 mol%).[1]

This genus is one of the most important taxa in food microbiology and human health. Lactobacilli are found in a variety of habitats, where rich carbohydrate-containing substances are available. These include the mucosal membranes of humans and animals (oral cavity, intestine, and vagina), plants and material of plant origin, and man-made habitats such as sewage and fermented or spoiled food. *Lactobacillus* species play a crucial role in the production and preservation of fermented foods, including vegetables, meats, and dairy products, where they are used as starters, adjunct starters, or protective cultures.[2] *Lactobacillus* species are also found in the human the gastrointestinal tract soon after birth. Some strains are considered probiotics, which confer health benefits to the consumer, and are commercialized in the form of functional foods or food supplements.

Because of the variety of habitats colonized by *Lactobacillus* species and their impact on food safety, technology, and health, the availability of reliable tools for their detection and identification is of utmost importance. Traditionally, the identification of *Lactobacillus* strains was based on physiological and metabolic phenotypic traits, such as their carbohydrate fermentation ability. These culture-dependent methods are time consuming and often lead to erroneous results because of the great complexity and phenotypic variability of this genus. Modern developments in molecular biology and increased knowledge of the genome sequences of many bacterial groups have turned molecular methods into the most commonly applied, as they give more rapid and reliable identification results. This chapter includes an overview of the ecology and taxonomy of the *Lactobacillus* genus and the molecular techniques developed and applied for their detection at species and strain level in complex ecosystems of relevance to human activities.

24.1.1 Taxonomy

The genus *Lactobacillus* was first described by Beijerinck in 1901.[3] It belongs to the phylum Firmicutes, class *Bacilli*, order

Lactobacillales, and family *Lactobacillaceae*. The closest relatives are grouped within the same family and include the genera *Paralactobacillus* and *Pediococcus*. The phylogenetically closest family appears to be *Leuconostocaceae*, which includes the genera *Leuconostoc*, *Oenococcus*, and *Weissella*.[4]

Members of the genus *Lactobacillus* are gram-positive, nonspore-forming bacteria, and rods or cocobacilli in shape. They are catalase-negative, fermentative, aerotolerant or anaerobic, chemoorganotrophic, aciduric or acidophilic, and have complex nutritional requirements (carbohydrates, amino acids, peptides, fatty acid esters, salts, nucleic acid derivatives and vitamins). *Lactobacillus* species show a GC content ranging from 32 to 54 mol%.

The genus *Lactobacillus* was initially subdivided into different groups, based on phenotypic characteristics. The first classification was based on their optimal growth temperatures and hexose fermentation pathways.[5] Based on 16S rRNA studies, the genus *Lactobacillus* and related genera were subdivided into three groups: *Lactobacillus delbrueckii* group, *Lactobacillus casei-Pediococcus* group, and *Leuconostoc* group, which also contained some lactobacilli.[6] The historical subdivisions of the genus *Lactobacillus* based on the type of fermentation were reviewed by Hammes and Vogel (1995).[1] They divided *Lactobacillus* species into: (i) obligate homofermentative lactobacilli that ferment hexose sugars almost exclusively (>85%) to lactic acid by the Embden–Meyerhof–Parnas (EMP) pathway, but not pentoses and gluconate because they lack phosphoketolase activity; (ii) facultative heterofermentative lactobacilli that degrade hexoses to lactic acid by the EMP pathway and also ferment pentoses (and often gluconate) because they possess both aldolase and phosphoketolase activities; and (iii) obligately heterofermentative lactobacilli that degrade hexoses by the phosphogluconate pathway producing lactate, ethanol or acetic acid, and carbon dioxide in equimolar amounts; moreover, they ferment pentoses by the same pathway. However, the division of lactobacilli according to the type of fermentation is not in accordance with their phylogenetic clustering as revealed by analysis of their 16S rRNA.

The application of polyphasic approaches, which consist of integrating different types of data (phenotypic, genotypic, and phylogenetic) of microorganisms, has been shown to be very useful for taxonomy.[7] The use of polyphasic taxonomy has led to taxonomic changes within the *Lactobacillus* genus. For example, it has led to the reclassification of species, such as *L. sobrius* strain DSM 16698T as *L. amylovorus*;[8] and genera, such as *Pediococcus dextrinicus* as *L. dextrinicus*.[9] It has also led to the identification of many new species (e.g., species isolated from sunki, a traditional Japanese pickle: *L. kisonensis*, *L. otakiensis*, *L. rapi*, and *L. sunkii*).[10] At the time of writing (August 2009), in the Approved List of Bacterial Names[11] the genus *Lactobacillus* includes 133 validly described species and 18 subspecies, which reflects its complexity. The description of a large number of new species and the following phylogenetic reexamination of the genus has caused some confusion recently, and earlier lists of identified lactobacilli will continue being modified.[12]

24.1.2 Ecology

Lactobacilli are found in environments where carbohydrates are available, such as food (dairy products, fermented meat, sourdough, vegetables, fruits, and beverages); sewage and plant material; and the respiratory, gastrointestinal, and genital tracts of humans and animals.

Lactobacilli are important in the production of foods that involve lactic acid fermentation, notably dairy products (e.g., yogurt and cheese), fermented vegetables (e.g., olives, pickles and sauerkraut), fermented meats (sausages), and sourdough bread. The use of lactobacilli as starter cultures in the food industry has a long history.[13] Starter cultures play critical roles in the fermentation processes, development of flavor and texture, and preservation of different fermented foods. For instance, the expression of 15 genes and/or operons was induced in *L. sakei* during raw-sausage fermentation, reflecting the functional adaptation and roles of this microorganism to the fermentation sausage environment.[14]

Some *Lactobacillus* species can also contribute to food spoilage, which are cause of concern in the food industry. The overgrowth of nonstarter species can cause acidification, smear, and out flavors in different products, such as cooked ham and cheeses.[15,16]

Lactobacillus species are normal residents of the gastrointestinal tract of humans and animals, and have received considerable attention for their possible use as probiotics.[17,18] Several recent studies have highlighted the benefits and limitations of the use of specific *Lactobacillus* strains in different health-related areas, including maintenance of Crohn's disease remission,[19,20] stimulation of the immune system,[21,22] prevention of allergic diseases,[23] and reduction of the recurrence of *Clostridium difficile* infections.[18,24] For the time being, dairy products are the main vehicle used to supply humans with probiotic strains and, especially, fermented milks, which allow their growth and survival better than other food matrixes.

In general, *Lactobacillus* are avirulent or show low virulence potential; however, members of this genus have been associated with infections in immunosuppressed subjects and patients with health complications, such as bacteremia, endocarditis, and vascular graft infection.[25–27] Lactobacilli have also been isolated from carious dentin lesions and are associated with the development of dental caries due to their aciduric characteristics.[28]

24.1.3 Phenotypic Characterization

The phenotypic characterization of lactobacilli usually includes evaluation of morphology, mode of glucose fermentation, growth at different temperatures, configuration of the produced lactic acid, type of fermented carbohydrates, production of methyl esters from fatty acids, and the pattern of whole-cell or cell-wall proteins. However, the use of these traditional methods for the identification of lactobacilli is time-consuming, requires cultivation, and may lead to erroneous results. In general, phenotypic tests show poor reproducibility and ambiguous results. The expression of phenotypic features depends on the environmental conditions contributing to the variability of the results; for example, growth at different temperatures changes the lipid composition of the membrane of *L. acidophilus* CRL 640.[29] Moreover, lactobacilli share many biochemical and phenotypical characteristics within species of the same genus and other related genera. Altogether, these drawbacks adversely affect the reliability of phenotype-based methods as unique tools for identification of lactobacilli at genus and at species level.

Commercial kits based on biochemical and enzymatic analysis has been used for long time as simple and fast methods for the identification of *Lactobacillus* at species level. Nevertheless, the results of these analyses, for instance using the API 50 CH strips, did not have good reproducibility, probably due to their susceptibility to different environmental and physiological conditions. These data suggest that the use of the current API 50 CH database and other similar methods for identification of *Lactobacillus* species could lead to misidentification, when applied as the only identification tool.[30]

Anyway, phenotypic characterization is important for a preliminary classification as well as to get basic knowledge of the properties of *Lactobacillus* strains. Salt and pH tolerance, growth at certain temperatures, configuration of the produced lactic acid, antibiotic resistance, and bacteriocin production are important characteristics for selecting *Lactobacillus* strains as possible starter or protective cultures and as probiotics.

24.1.4 Genotypic Characterization

Many different genotyping techniques can be applied for species identification, strain differentiation, and quantification. The major advantages of these DNA-based typing methods lie in their discriminatory power, universal applicability, and culture independency.[31] Species and closely related strains with similar phenotypic features can be distinguished by DNA-based molecular fingerprinting techniques, including restriction fragment length polymorphism (RFLP), pulse field gel electrophoresis (PFGE), randomly amplified polymorphic DNA (RAPD), and ribotyping: identification and quantification of species and groups can be achieved by molecular techniques based on the use of specific primers and probes, including conventional polymerase chain reaction (PCR), real-time PCR hybridization techniques, and sequencing of rRNA and other gene markers. Moreover, detection and evolution of species in complex ecosystems can be achieved by denaturing gradient gel electrophoresis (DGGE) and temperature gradient gel electrophoresis (TGGE). These methodologies and examples of their application to the genus *Lactobacillus* are described below.

24.1.4.1 Molecular Fingerprinting

Many common typing methods have been adapted to the genus *Lactobacillus*. These techniques are applied to differentiate between species and strains, which may belong to the same species, previously isolated on a culture medium. The techniques used for the fingerprinting of *Lactobacillus* species are summarized in Figure 24.1.

Restriction Fragment Length Polymorphism. In RFLP analysis, the DNA is digested by restriction enzymes and the resulting restriction fragments are separated according to their lengths by gel electrophoresis. The banding patterns that result are referred as DNA fingerprinting. Because of the high specificity of restriction enzymes and the stability of chromosomal DNA, a reproducible pattern of fragments is obtained after the complete digestion of the chromosomal DNA by a particular enzyme. One general criticism about this method is the complexity of banding patterns. However, this technique can be applied on a selected genetic marker by its amplification with specific primers (PCR–RFLP), reducing its complexity. Then, the generated DNA fingerprinting and the discrimination power between different bacterial species or subspecies, will be dependent on the restriction enzymes used and the length of the amplified fragment.

Amplified ribosomal DNA restriction analysis (ARDRA) is essentially PCR–RFLP analysis of the amplified 16S rRNA gene. ARDRA patterns are highly reproducible and comparable between laboratories, but the discriminatory power of this technique is generally low because of the conserved nature of this gene. Other highly conserved genes, such as *rpoB*, *hsp60*, or the internal spacer region between the 16S and 23S rRNA genes (ISR), and their restriction by different enzymes have been used for *Lactobacillus* identification by PCR–RFPL. Chenoll et al. (2006)[32] identified *L. tucceti* as a novel specie based on their singular ISR-*DdeI* and ISR-*HaeIII* profiles that allowed its differentiation from 68 lactic acid bacteria reference strains and showed that the discriminatory power of the ISR digestion was better than that of the 16S rRNA gene.[33] Furthermore, RFLP analysis of the *rpoB* sequence amplified by PCR has successfully been used to differentiate 12 lactobacilli at special level after two or three digestions (*Aci*I, *Hinf*I, and *Mse*I).[32]

Pulse Field Gel Electrophoresis. PFGE employs an alternating field of electrophoresis to allow the separation of large DNA fragments obtained from restriction digests of genomic DNA with rare-cutting enzymes (i.e., *Sfi*I, *Not*I, *Apa*I, *Sgr*A, and *Sma*I). PFGE is performed as a standard gel electrophoresis, but the voltage is periodically switched among three directions; one that runs through the central axis of the gel, and two that run at an angle of 120 degrees either side.[34] The technique can be more time consuming than other fingerprinting strategies (4–5 days). However, the profile generated by PFGE has a very good discriminatory power. Indeed, PFGE technique has been successfully used for *Lactobacillus* species and subspecies differentiation.[34–36] For example, the diversity and dynamics of *Lactobacillus* populations were monitored throughout the manufacturing of Camembert cheese. *L. paracasei* was the species most frequently isolated and these isolates (39) were classified by PFGE into 21 different profiles among the same species.[37]

The typing efficacy of this technique can be increased by the use of several single restriction enzymes, followed by numerical analysis of the combined patterns. Sánchez et al (2004).[35]

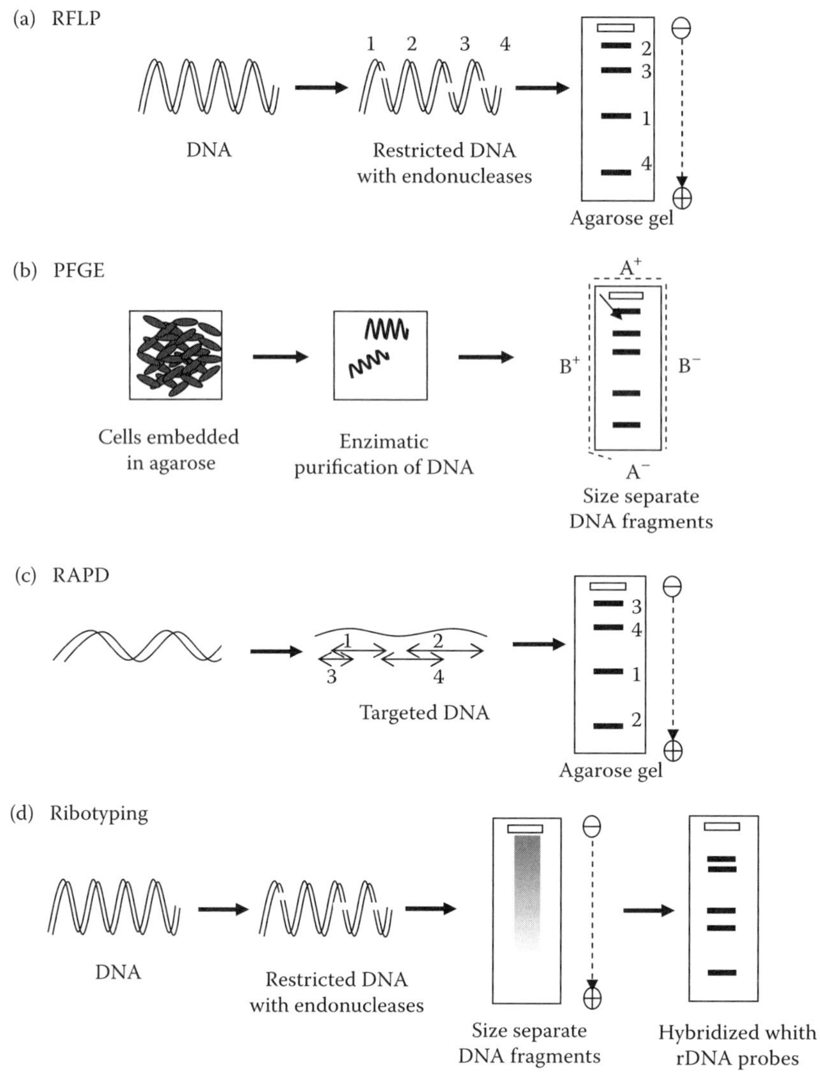

FIGURE 24.1 Fingerprinting techniques for *Lactobacillus* species and strains (a) Restriction fragment length polymorphism (RFLP). The DNA fragment is restricted into pieces (for example 1–4) by a restriction enzyme. The restricted DNA is then loaded into a well in an agarose gel and the fragments are separated by electrophoresis, based on fragment size. (b) Pulse field gel electrophoresis (PFGE). Whole cells are embedded in agarose and DNA is enzymatically purified and restricted by rare-cutting enzymes in situ. The agarose plug is then inserted into a well in an agarose gel and the large restricted fragments are separated by an electric current which pulses from different angles. (c) Random amplification of polymorphic DNA (RAPD). PCR is run with arbitrary sequence primers that amplify segments of DNA of unknown identity. The amplified DNA is then loaded into an agarose gel and the fragments are separated by electrophoresis, based on fragment size. (d) Ribotyping. DNA is isolated from a cultured isolate, restricted and size separated in an agarose gel. The gel is then hybridized with a labeled rRNA/rDNA probe, which binds to fragments containing copies of the specific rRNA. Following probe detection, fragments with bound probe are visualized.

used five rare-cutting enzymes (*Sfi*I, *Not*I, *Apa*I, *Sgr*A, and *Sma*I) for PFGE analysis of 149 lactobacilli isolated from fermented vegetables. The use of *Sfi*I and *Sma*I led to the most informative digestion patterns, and their combined numerical analysis improved the discriminatory power of the technique.[35] Some comparative studies of different typing techniques have indicated that isolates that are indistinguishable by PFGE are unlikely to demonstrate substantial differences by other typing techniques.[38,39] Nevertheless, PFGE patterns can be used for bacterial identification only when a set of known strains is included in the same analysis and when a suitable and objective method of analysis of the patterns is applied.[40]

Random Amplification of Polymorphic DNA Fingerprinting. The RAPD technique is a PCR-based discrimination method in which short arbitrary primers (8–12 nucleotides) anneal to multiple random target sequences of the genomic DNA, resulting in patterns of diagnostic value. Low-stringency annealing conditions are used in PCR reactions, which results in the amplification of randomly sized DNA fragments. RAPD–PCR is a molecular-typing method that is very suitable for identification and differentiation of *Lactobacillus* strains. It is less time consuming than PFGE, but the construction of a database is nearly impossible due to the high variability of the results. If an unidentified strain has to be investigated a set of known strains has to be included in the same RAPD procedure. As the reproducibility of RAPD patterns is occasionally poor, this method needs to be performed under carefully controlled conditions.

RAPD technique has been reported as a very useful tool for *Lactobacillus* differentiation at strain level,[15,16,41,42] but other complementary tools are required for species identification, such as sequencing of molecular marker genes, DGGE, or TGGE. In this context, for example, Dimitonova et al. (2008), compared the RAPD patterns of 20 vaginal lactobacilli isolates, generated with the M13V primer. Two main clusters were generated and divided into subgroups. Identical RAPD patterns were shown in different isolates, suggesting that they were multiple isolates from a single species and identical strains. ARDRA-*Hae*III analysis and 16S rRNA gene sequencing allowed the identification at species level of seven strains, three as *L. fermentum*, two as *L. gasseri*, one as *L. brevis*, and one as *L. salivarius*.[43]

Ribotyping. A ribotype is essentially a RFLP profile consisting of the restriction fragments from a particular genome, which contain rRNA genes. The bacterial chromosomal DNA is totally restricted into multiple fragments, using a specific restriction enzyme. The restricted fragments are then separated by agarose gel electrophoresis and hybridized with 16S and 23S rRNA probes. The hybridization can be carried out directly in the gel using gel hybridization techniques or on a nylon or nitrocellulose membrane following Southern transfer DNA from the gel to the membrane.[44] Following probe detection, restriction bands containing copies of the rRNA genes are visualized and the pattern of the band sizes represents a characteristic fingerprint. Ribotyping fingerprint patterns are more stable and more easily interpretable than those obtained by only restriction enzyme analysis; furthermore, they are highly reproducible, and a single rRNA probe can be used for typing all bacteria because of the similarity of ribosomal genes.[45] Ribotyping technique also allows the differentiation of *Lactobacillus* isolates at strain level. In this context, for example, the probiotic strain *L. rhamnosus* 35 was characterized at molecular level and compared with seven reference strains from the *L. casei* group by ribotyping.[46] This molecular technique is also a useful tool for the identification of new species, for instance, *Lactobacillus similis*.[47] Automated ribotyping has also been used for the characterization of *Lactobacillus*. For instance, 91 type and reference stains were classified in different ribotypes. Strains of the *L. casei* group were divided into five ribotyping geno types, and strains of the *L. acidophilus* group were divided into two major clusters, one containing *L. acidophilus*, *L. amylovorus*, *L. crispatus* and *L. gallinarum*, and another containing *L. gasseri* and *L. johnsonii*.[48]

Several studies have compared the discriminatory power of different molecular-typing methods.[35,49,50] In general, the most commonly applied techniques can be placed in the following order with respect to their discriminatory power: PFGE > ribotyping > RAPD > ARDRA. The choice of the most appropriate typing method also depends on other features such as cost, high throughput capacity, and reproducibility of the fingerprints.

24.1.4.2 PCR and Hybridization Techniques Based on Specific Primers and Probes

Specific oligonucleotide probes and primers targeting rRNA or rDNA have been designed for detection and quantification of different species of the *Lactobacillus* genus using diverse PCR and hybridization techniques. Lactobacilli are phylogenetically heterogeneous, which causes difficulties in the design of a very specific genus probe, but some group-specific probes and primers have been described.[51] Frequently, *Lactobacillus* group-specific primers cover, in addition to *Lactobacillus*, *Leuconostoc*, *Pediococcus*, and *Weissella* genera.[52,53] Species-specific probes and primers have also been designed for the identification of *Lactobacillus* species, but the probe panel is still incomplete, and the detection of some species by these techniques is not possible yet.[51]

Conventional and Real-Time PCR. Amplification of DNA by PCR is one of the easiest and most rapid ways of detecting and identifying specific bacterial sequences with the use of adequate primers. A summary of primers and conditions designed for the detection of members of the *Lactobacillus* genus is shown in Table 24.1. The conventional PCR is sufficiently sensitive to detect *Lactobacillus* at genus[57,58] and at specie level.[52–56] More recently, multiple PCR approaches have been designed to detect simultaneously different species in a single PCR reaction. For example, it has been possible to simultaneously distinguish *L. plantarum*, *L. pentosus* and *L. paraplantarum* species by using a multiplex PCR protocol. The size of the amplicons was 318 bp for *L. plantarum*, 218 bp for *L. pentosus*, and 107 bp for *L. paraplantarum*.[59] Conventional PCR can be used for semi-quantitative analyses by making dilutions of known amounts of DNA almost identical to the target.[60]

Real-time PCR is a DNA-based technique that monitors the amplification of the target DNA in real time by fluorescence detection. The amount of DNA corresponding to gene copy numbers or bacterial cells in a sample is determined by measuring the cycle number at which the increase in fluorescence (and therefore cDNA) is exponential. Real-time PCR can be used to quantify bacteria from various samples including milk, feces, food, and water. Contemporary, real-time PCR allows the monitoring of the complete amplification and a more accurate quantification. Even though, the determination of bacterial numbers is difficult due to factors affecting the amplification reaction and variation of the copy number and the ribosomal content of cells in different bacteria.[61] In spite of these limitations, sets of species-specific 16S rRNA primers and real-time PCR have been used successfully for monitoring and quantifying, for example, the genus *Lactobacillus* in feces of patients with celiac disease and healthy controls.[62,63] Scheirlinck et al. (2009)[64] also studied the predominant sourdough LAB species in the production of sourdough bakeries. Real-time PCR assays were developed using species-specific primers targeting the *pheS* gene of *L. plantarum* and *L. sanfranciscensis* to detect, quantify, and monitor them in sourdough and bakery environments.[64]

TABLE 24.1
16S rRNA Targeted PCR Primers Designed for the Detection of *Lactobacillus* Species

Specificity	Sequence (5′–3′)	Target Site	Reference
L. acidophilus	GATCGCATGATCAGCTTATA	205–224	52
	AGTCTCTCAACTCGGCTATG	328–309	
	AGCTGAACCAACAGATTCAC	70–89	53
	ACTACCAGGGTATCTAATCC		
	GCGGACGGGTGAGTAACACG		53
	GCTTACGCCATCTTTCAGCCAA		
L. casei	GCGGAAATAGTAGTGTGACGATC		54
	GCGGAAATAGGTGCATAGGCG		
	CTCAAAGCCGTGACGGTC		54
	ACGTGGTGCTAATAATCCTAGTG		
	GTAATGACGTTAGGAAAGCG		53
	ACTACCAGGGTATCTAATCC		
	GTGCTTGCACTGAGATTCGACTTA	61–84	52
	TGCGGTTCTTGGATCTATGCG	185–165	
	TCTTGATTTAATTTTG	68	55
L. casei/paracasei	GCACCGAGATTCAACATGG	86–104	56
L. casei/paracasei/ rhamnosus/zeae	GCGGACGGGTGAGTAACACG	93–112	52
	GCTTACGCCATCTTTCAGCCAA	213–192	
	GGTTCTTGGATYTATGCGGTATTAG	196–169	56
L. crispatus	ACATGAATCGCATGATTCAAG		52
	AACTCGGCTACGCATCATTG		
	AGCGAGCGGAACTAACAGATTTAC	65–89	56
L. delbrueckii	GCGGACGGGTGAGTAACACG	185–205	52
	GCTTACGCCATCTTTCAGCCAA	305–286	
	GGRTGATTTGTTGGACGCTAG	92–102	56
	GCCGCCTTTCAAACTTGAATC	219–199	56
L. fermentum	GCACCTGATTGATTTTGGTCG	86–94	56
	GTTGTTCGCATGAACAACGCTTAA	160	55
L. gallinarum	CGGTAATGACGCTGGGGAC	78–97	56
L. gasseri	CAGTTACTACCTCTATCTTTCTTCACTAC	470–442	56
L. johnsonii	GAGCTTGCCTAGATGATTTTA		53
	ACTACCAGGGTATCTAATCC		
	CACTAGACGCATGTCTAGAG	170–189	52
	AGTCTCTCAACTCGGCTAT	296–277	
L. paracasei	GTGCTTGCACCGAGATTCAACATG	60–83	52
	TGCGGTTCTTGGATCTATGCG	185–165	
	CCGAGATTCAACATGG	69	55
L. plantarum	ATCATGATTTACATTTGAGTG	96	55
L. reuteri	TGAATTGACGATGGATCACCAGTG	91	55
L. rhamnosus	TGCTTGCATCTTGATTTAATTTTG	81–104	56
	GTGCTTGCATCTTGATTTAATTTT	88–111	52
	TGCGGTTCTTGGATCTATGCG	213–194	
L. salivarius	CGAAACTTTCTTACACCGAATGC	67–91	56
	ATTCACTCGTAAGAAGT	95	55

Hybridization Techniques. Hybridization techniques with specific probes are also useful for the detection and quantification of lactobacilli. Of these techniques, the most commonly used are dot blot hybridization and fluorescence in situ hybridization (FISH).

Dot blot hybridization is used for the identification of specific nucleic acid sequences or genes with varying degree of homology and for the estimation of relative amounts of nucleic acid with known homology. DNA or RNA is extracted from a sample or cultured bacteria, fixed to a nylon or nitrocellulose membrane and hybridized with a labeled oligonucleotide or DNA fragment complementary to the nucleic acids of interest. For example, dot blot hybridization analysis has been used for screening the presence of *Lactobacillus reuteri* and the gene encoding for the enzyme responsible for reuterin production (*gldC*).[65]

In FISH, the detection of rRNA sequences within morphologically intact cells is achieved using fluorescently labeled oligonucleotide probes. Special permeabilization procedures may be required to aid passage of the probe into the cells of some bacteria. The multiple rRNA molecules present in bacterial cells capture the probe, the signal is intensified, and the cells can be detected and numerated by using an epifluorescence microscope or flow cytometry.[66,67] For example, using fluorescent *L. plantarum*-specific targeted oligonucleotide probes, the presence of *L. plantarum* L2 on the jejunum, ileum and colon of rats was demonstrated, after its oral administration.[68] The use of FISH coupled with flow cytometry analysis has also demonstrated that *Lactobacillus* population is significantly reduced in the intestinal microbiota of patients with celiac disease compared with healthy children.[69]

24.1.4.3 Sequencing of rRNA and Other Molecular Marker Genes

Relatedness among organisms is estimated through the comparison of molecular sequences, mainly 16S rRNA encoding genes.[70] It is based on several major assumptions: rRNA genes are highly conserved because of the fundamental role of ribosomes in the protein biosynthesis that was developed in the early stages of the evolution of microorganisms. Horizontal gene transfer phenomena among organisms have not involved those genes, and the amount of similarity of these sequences between different individuals is representative of the variation of their genomes. The 16S rRNA gene is a well-conserved universal marker, but there are some shortcomings associated with its use. First, the 16S rRNA genes are so well conserved that it has limited resolving power. Second, although the 16S rRNA gene is a universal marker, different bacterial species have different copy numbers of the gene and this leads to an over- and underrepresentation of some bacterial species. In general, if the 16S rRNA gene sequence identity of two microorganisms is lower than 97%, they are poorly related at the genomic level and, therefore, they are considered to belong to different species. If the 16S rRNA gene sequence identity of two microorganisms is higher than 97%, even if they are identical sequences, they have to be considered closely related, and only total DNA–DNA hybridization data and/or analysis of other more discriminative gene sequences are decisive for the identification at the species level. Phylogenetic and alignment analyses are largely preferred over DNA hybridization tests because the latter are time consuming, complex, and expensive. In contrast, identification through sequence analysis is based on specific DNA amplification, sequencing reaction, and comparison with public databases, which are very fast, more reproducible and much less expensive.[12]

In general, 16S rRNA gene analysis has been used for the identification of different *Lactobacillus* species,[71,72] but the high similarities of the 16S rRNA gene in this bacterial group usually require the confirmation by the analysis of other markers. Thus, the sequencing of several other genes, such as the *recA* (encoding recombinase A)[42,61,73]; *tuf* gene (encoding elongation factor Tu, involved in protein biosynthesis)[74,75]; hsp60 (groEL, encoding a 60-kDa heat shock protein)[76]; *mal* (encoding malolactic enzyme)[77]; and *rpoB* (encoding RNA polymerase beta subunit)[32] has been used for the identification of lactobacilli.

24.1.4.4 Temperature and Denaturing Gradient Gel Electrophoresis (TGGE and DGGE)

For DGGE and TGGE analysis, RNA or DNA is extracted from an environmental sample and is amplified with specific primers. In DGGE as well as in TGGE, DNA fragments of the same length, but with different sequences, can be separated by electrophoresis through a polyacrylamide gel containing a linear gradient of DNA denaturants formed by urea and formamide (DGGE) or a linear temperature gradient (TGGE). Separation is based on the melting of rRNA genes at specific denaturing points based on their sequence; therefore, each individual sequence will begin to melt at a characteristic denaturing point. The melting changes the conformation of the DNA molecule, slowing its migration though the gel.

By using DGGE or TGGE, 50% of the sequence variants can be detected in DNA fragments up to 500 bp. This percentage can be increased to nearly 100% by the attachment of a GC-rich sequence to one side of the DNA fragment. A GC clamp is generally included at the five-end of one of the PCR primers, co-amplified, and thus introduced into the amplified DNA fragments in order to maintain the double-stranded integrity of the amplified products, which improves the detection.

The gene sequences from bacterial species in a mixed culture are first amplified using conserved bacterial primers (universal or group-specific primers) that target a hypervariable region, producing amplicons of the same length but with differing sequences that are specific at specie level. DGGE and TGGE allow the separation of these amplicons, producing a molecular fingerprint of the bacterial species present in a sample.

The design of primers based on sequence variability in 16S rRNA, 23S rRNA, and other molecular marker genes together with DGGE and TGGE techniques have improved our understanding of the lactobacilli communities present in a variety of complex ecosystems, including the intestinal tract[52,78,79] and fermented food.[32,42] DGGE and TGGE techniques are useful tools for monitoring both dynamic changes in mixed populations over time and the diversity of *Lactobacillus* communities because they are rapid and economical. However, they also have several limitations.[80] First, only relatively small fragments can be separated by DGGE and TGGE, limiting the amount of sequence information for phylogenetic inferences. Second, the use of different regions of the 16S rRNA and different DGGE and TGGE conditions might result in different resolutions of separation. Finally, DGGE and TGGE only display the DNA fragments obtained from the predominant species present in the community. Several studies revealed that lactobacilli make up less than 1% of the gastrointestinal community, and in this case, the use of a nested approach improves the detection of bands in the denaturing gradient gels. The nested-PCR approach consisting in a first amplification with universal bacterial primers and a second amplification with specific primers for *Lactobacillus* and closely related species.[81]

24.2 METHODS OF THE MOST COMMONLY USED TECHNIQUES FOR *LACTOBACILLUS* DETECTION

24.2.1 DNA Sample Preparation

Lactobacillus DNA can be extracted with conventional DNA extraction procedures using lysozyme, proteinase K, and sodium dodecyl sulfate, with slight modifications depending on the type of starting material.[82] Commercialized kits may also be used for genomic DNA extraction from fresh or frozen human stool or other sample types, using a rapid and simplified procedure. DNA extracts from reference bacterial strains can also be obtained from a high, concentrated cell suspension in 50 μL of sterile water by boiling for 10 min and freezing for 5 min at –20°C. To verify whether a particular DNA extraction method is effective for a specific matrix, it should be tested on samples containing a well-known number of lactobacilli cells.

24.2.2 Detection Procedures

24.2.2.1 Identification of *Lactobacillus* Species by PCR and DNA Sequencing

Principle. *Lactobacillus* spp. can be identified by PCR amplification and sequencing analysis of 16S rRNA using the universal primers 27f and 1492R. The primer sequences are: 27f, 5′-AGAGTTTGATCCTGGCTCAG-3′[83]; and 1495R, 5′-CTACGGCTACCTTGTTACGA-3′.[84]

Procedure. The reaction mixture (50-μL) is performed as follows: 5 μL of extracted DNA, 5 μL 10× buffer, 0.2 mM of dNTP, 2.5 mM $MgCl_2$, 10 pmol of each primer, and 1 U of *Taq* DNA polymerase. Amplification is done using a PCR System programmed as follows: 94°C for 5 min; 30 cycles of 94°C for 1 min, 58°C for 1 min and 72°C for 2 min; and finally, 72°C for 10 min. PCR products (1.5 kb) are electrophoresed on a 1% agarose gel in 1× Tris-Borate-EDTA, and visualize by UV transillumination after ethidium bromide staining. Sequence the PCR products by using the forward primer and the reverse primer.

For bacterial identification, partial 16S rRNA gene sequences are compared with sequences deposited in the GenBank database, using the program standard nucleotide-nucleotide Basic Local Alignment Search Tool (BLAST). A threshold of 97% similarity in 16S rRNA gene sequence has been proposed for the bacterial-species delineation based on a correlation study between DNA–DNA hybridization results and similarity values for 16S rRNA gene sequences.

24.2.2.2 Differentiation of *Lactobacillus* Species by Multiplex or Simplex PCR

Principle. The species *Lactobacillus plantarum*, *Lactobacillus pentosus*, and *Lactobacillus paraplantarum* can be differentiated by doing a unique PCR reaction with *recA* derived primers. The multiplex PCR assay can be performed with the primers paraF (5′-GTCACAGGCATTACGAAAAC-3′), pentF (5′-CAGTGG CGCGGTTGATATC-3′), planF (5′-CC GTTTATGCGGAACACCTA-3′), and pREV (5′-TCGGG ATTACCAAACATCAC-3′).[59] Lactobacillus species can also be detected by simplex PCR using primers specifics for each specie in each reaction, as those summarized in Table 24.1.

Procedure. The PCR mixture (20 μL) is composed of 5 μL of DNA, 2 μL 10× buffer, 1.5 mM $MgCl_2$, 0.12 mM of the primer planF and 0.25 mM of the primers paraF, pentF, and pREV, 0.3 mM of dNTPs, and 1 U of Taq DNA polymerase. PCRs are performed using a PCR System programmed as follows: 94°C for 3 min, 30 cycles of 94°C for 30 s, 56°C for 10 s, and 72°C for 30 s; and final extension at 72°C for 5 min. The PCR products are then separated electrophoretically on an agarose gel and stained with ethidium bromide. *L. plantarum*, *L. pentosus*, and *L. paraplantarum* strains generate a 318-, 218-, and 107-bp fragments, respectively. Similarly, the amplification products of a simple PCR reaction can be detected on an agarose gel and identified according to its expected size by ethidium bromide staining.

24.2.2.3 Detection and Identification of *Lactobacillus* Species by DGGE

Principle. Fecal *Lactobacillus* communities can be monitored by PCR amplification and DGGE analysis of using 16S rRNA group specific primers.

Procedure. Genomic DNA is amplified using the primers Lacl (5′-AGCAGTAGGGAATCTTCCA-3′); and Lac2 (5′-ATTYCACCGCTACACATG-3′). A GC clamp (5′-C GCCCGGGGCGCGCCCCGGGCGGCCCGGGGGCA CCGGGGG-3′) was attached to the reverse primer (Lac2) to obtain PCR fragments suitable for DGGE analysis.[52] Amplification can be carried out using a GeneAmp 2400 Thermocycler (Perkin-Elmer). The reaction mixture (50 μL) contained 25-pmol amounts of each primer, 0.2 mM of each dNTP, reaction buffer, 2 mM $MgCl_2$, 25 μg of bovine serum albumin, 2.5 U of *Taq* polymerase (Amersham Pharmacia Biotech), and 1 μL of DNA solution. The amplification program is 94°C for 2 min; 35 cycles of 94°C for 30 s, 61°C for 1 min, and 68°C for 1 min; and finally, 68°C for 7 min.

DGGE analysis of PCR amplicons is carried out on the Dcode Universal Mutation Detection System (Bio-Rad, Richmond, CA). The denaturing gradient of urea and formamide used for PCR products separation is 25%–45%. A 100% denaturant corresponds to 7M urea and 40% (v/v) formamide.

DGGE ladders for the identification of *Lactobacillus*-like populations are prepared by mixing equal amounts of amplicons obtained from selected reference *Lactobacillus* species using primers Lacl and Lac2-GC.

Selected unknown DGGE bands can be excised from the acrylamide gels and reamplified with primers Lacl and Lac2. After purification, sequencing is caring by DNA sequencing using an ABI PRISM-3130XL Genetic Analyzer (Applied

Biosystems, California, USA). Alternatively isolated bands can be cloned into a vector, checked by PCR, and sequenced afterwards. Sequences identities are determined in GenBank database.

24.2.2.4 Quantification of *Lactobacillus* by Real-Time PCR (SYBR Green)

Principle. Genomic DNA has been used to detect and quantify *Lactobacillus* spp. with high sensibility by using group-specific primers and real-time PCR (SYBR Green). The primer-ssequences used are: 5′-AGCAGTAGGGAATCTTCCA-3′, and 5′-CACCGCTACACATGGAG-3′.[85]

Procedure. Quantitative PCR can be performed with an iCycler iQ apparatus (Bio-Rad, USA) associated with the iCycler optical system interface software (version 2.3; Bio-Rad) or other equivalent equipments. All PCRs are performed in a volume of 25 μL, using 96-well optical grade PCR plates and an optical sealing tape (Bio-Rad).

Reaction mixtures for the optimized SYBR Green assays consist of 1:75,000 dilution of SYBR Green (Molecular Probes); 10 mM Tris-HCl (pH 8.8); 150 mM KCl; 0.1% Triton X-100; 2 mM $MgCl_2$; 100 μM each dNTP; 0.5 μM of each primer and 0.02 U Dynazyme II μL^{-1} (Finnzymes, Finland); and either 5 μL template or water (no-template control). The thermal cycling conditions are an initial DNA denaturation step at 95°C for 5 min, followed by 35 cycles of denaturation at 95°C for 15 s; primer annealing at 58°C for 20 s; extension at 72°C for 30 s, and an additional incubation step at 80°C–85°C for 30 s to measure the SYBR Green fluorescence. Finally, melt curve analysis was performed by slowly cooling the PCRs from 95°C to 60°C (0.3°C per cycle) with simultaneous measurement of the SYBR Green signal intensity.

The bacterial concentration from each sample is calculated by comparing the Ct (cycle threshold) values obtained from standard curves. Standard curves can be created by using serial tenfold dilution of pure culture DNA corresponding to 10^2–10^9 cells.

24.3 CONCLUSIONS

This decade has seen the emergence of numerous molecular approaches for the analysis of diverse aspects of *Lactobacillus* populations in complex ecosystems, including the gastrointestinal tract and different foods, and for the identification of new species. Quantitative and qualitative detection can be optimally achieved by the combined use of different methods that complement each other. Thus, numbers can be quantified by using labeled group-specific probes and fluorescent in situ hybridization (FISH), coupled with either microscopy or flow cytometry detection, or by real-time PCR. The species diversity can be subsequently determined by PCR in combination with DGGE, TGGE or DNA sequencing. Strain differentiation could be possible by using PFGE, ribotyping, and RAPD. The application of these techniques is contributing to a comprehensive analysis of the ecological distribution, functional roles, and adaptation of *Lactobacillus* strains to different environments. In the near future, it can be anticipated that the increasing use of massive sequencing approaches to the analysis of whole bacterial genomes, metagenomics, and transcripomics will lead to the identification of novel species and roles of *Lactobacillus* strains in human health and food safety and technology.

ACKNOWLEDGMENTS

This work was supported by grants AGL2007–66126-C03–01/ALI and Consolider Fun-*C*-Food CSD2007–00063 from the Spanish Ministry of Science and Education (MEC). The scholarship to E. Sánchez from Institute Danone is fully acknowledged.

REFERENCES

1. Hammes, W.P., and Vogel, R.F., The genus *Lactobacillus*. In Wood, B.J.B., and Holzapfel, W.H. (Editors), *The Lactic Acid Bacteria*, vol. 2, p. 19, Blackie Academic and Professional, London, United Kingdom, 1995.
2. Bernardeau, M., Vernoux, J.P., Henri-Dubernet, S., and Gueguen, M., Safety assessment of dairy microorganisms: The *Lactobacillus* genus, *Int. J. Food Microbiol.*, 126, 278, 2008.
3. Beijerinck, M.W., Sur les ferments lactiques de l'industrie, *Arch. Néerland. Sci. Exact. Nat.*, 6, 212, 1901.
4. Hammes, W.P., and Hertel, C., The Genera *Lactobacillus* and *Carnobacterium*. In Dworkin, M. (Editor), *The Prokaryotes: An Evolving Electronic Resource for the Microbiological Community*, vol. 3.15, Springer-Verlag, New York, 2003.
5. Orla-Jensen, S., The lactic acid bacteria, *Mem. Acad. Roy. Sci. Denmark Sec. Sci.*, 5, 81, 1919.
6. Collins, M.D. et al., The phylogenetic analysis of the genus *Lactobacillus* and related acid lactic bacteria as determined by reverse transcriptase sequencing of 16Sr RNA, *FEMS Microbiol. Lett.*, 77, 5, 1991.
7. Vandamme, P. et al., Polyphasic taxonomy, a consensus approach to bacterial systematics, *Microbiol. Rev.*, 60, 407, 1996.
8. Jakava Viljanen, M. et al., *Lactobacillus sobrius* Konstantinov et al. 2006 is a later synonym of *Lactobacillus amylovorus* Nakamura 1981, *Int. J. Syst. Evol. Microbiol.*, 58, 910, 2008.
9. Haakensen, M. et al., Reclassification of *Pediococcus dextrinicus* (Coster and White 1964) back 1978 (Approved Lists 1980) as *Lactobacillus dextrinicus* comb. nov., and emended description of the genus *Lactobacillus*, *Int. J. Syst. Evol. Microbiol.*, 59, 615, 2009.
10. Watanabe, K. et al., *Lactobacillus kisonensis* sp. nov., *Lactobacillus otakiensis* sp. nov., *Lactobacillus rapi* sp. nov. and *Lactobacillus sunkii* sp. nov., heterofermentative species isolated from sunki, a traditional Japanese pickle, *Int. J. Syst. Evol. Microbiol.*, 59, 754, 2009.
11. List of prokaryotic names with standing in nomenclature, www.bacterio.cict.fr/
12. Felis, G.E., and Dellaglio, F., Taxonomy of Lactobacilli and Bifidobacteria, *Curr. Issues Intest. Microbiol.*, 8, 44, 2007.
13. Jay, J.M., *Modern Food Microbiology*, 5th ed. Chapman and Hall, New York, NY, 1996.
14. Hufner, E. et al., Identification of *Lactobacillus sakei* genes induced during meat fermentation and their role in survival and growth, *Appl. Environ. Microbiol.*, 73, 2522, 2007.

15. Sánchez, I. et al., Genetic diversity, dynamics, and activity of *Lactobacillus* community involved in traditional processing of artisanal Manchego cheese, *Int. J. Food. Microbiol.*, 107, 265, 2006.
16. Antonsson, M., Molin, G., and Ardo, Y., *Lactobacillus* strains isolated from Danbo cheese as adjunct cultures in a cheese model system, *Int. J. Food Microbiol.*, 85, 159, 2003.
17. Lebeer, S., Vanderleyden, J., and De Keersmaecker, S.C., Genes and molecules of lactobacilli supporting probiotic action, *Microbiol. Mol. Biol. Rev.*, 72, 728, 2008.
18. Guarino, A., Lo, V.A., and Canani, R.B., Probiotics as prevention and treatment for diarrhea, *Curr. Opin. Gastroenterol.*, 25, 18, 2009.
19. Scott, A.C., Young, R.F., and Fedorak, P.M., Comparison of GC-MS and FTIR methods for quantifying naphthenic acids in water samples, *Chemosphere*, 73, 1258, 2008.
20. Geier, M.S., Butler, R.N., and Howarth, G.S., Inflammatory bowel disease: Current insights into pathogenesis and new therapeutic options; probiotics, prebiotics and synbiotics, *Int. J. Food Microbiol.*, 115, 1, 2007.
21. del Giudice, M.M., and Brunese, F.P., Probiotics, prebiotics, and allergy in children: What's new in the last year? *J. Clin. Gastroenterol.*, 42, 205, 2008.
22. Reid, G., Devillard, E., Probiotics for mother and child, *J. Clin. Gastroenterol.*, 38, 94, 2004.
23. Isolauri, E., and Salminen, S., Probiotics: Use in allergic disorders: A Nutrition, Allergy, Mucosal Immunology, and Intestinal Microbiota (NAMI) Research Group Report, *J. Clin. Gastroenterol.*, 42, 91, 2008.
24. Sazawal, S. et al., Efficacy of probiotics in prevention of acute diarrhoea: A meta-analysis of masked, randomised, placebo-controlled trials, *Lancet Infect. Dis.*, 6, 374, 2006.
25. Svec, P. et al., Identification of lactic acid bacteria isolated from human blood cultures, *FEMS Immunol. Med. Microbiol.*, 49, 192, 2007.
26. Chazan, B. et al., Bacteremia and pyelonephritis caused by *Lactobacillus jensenii* in a patient with urolithiasis, *Isr. Med. Assoc. J.*, 10, 164, 2008.
27. Tommasi, C. et al., Diagnostic difficulties of *Lactobacillus casei* bacteraemia in immunocompetent patients: A case report, *J. Med. Case Rep.*, 2, 315, 2008.
28. Teanpaisan, R., and Dahlen, G., Use of polymerase chain reaction techniques and sodium dodecyl sulfate-polyacrylamide gel electrophoresis for differentiation of oral *Lactobacillus* species, *Oral Microbiol. Immunol.*, 21, 79, 2006.
29. Murga, M.L. et al., Influence of growth temperature on cryotolerance and lipid composition of *Lactobacillus acidophilus*, *J. Appl. Microbiol.*, 88, 342, 2000.
30. Boyd, M.A., Antonio, M.A., and Hillier S.L., Comparison of API 50 CH strips to whole-chromosomal DNA probes for identification of *Lactobacillus* species, *J. Clin. Microbiol.*, 43, 5309, 2005.
31. Farber, J.M., An introduction to the hows and whys of molecular typing, *J. Food Prot.*, 59, 1091, 1996.
32. Claisse, O., Renouf, V., and Lonvaud-Funel, A., Differentiation of wine lactic acid bacteria species based on RFLP analysis of a partial sequence of rpoB gene, *J. Microbiol. Methods*, 69, 387, 2007.
33. Chenoll, E., Macián, M.C., and Aznar R., *Lactobacillus rennini* sp. nov., isolated from rennin and associated with cheese spoilage, *Int. J. Syst. Evol. Microbiol.*, 56, 449, 2006.
34. Klein, G. et al., Taxonomy and physiology of probiotic lactic acid bacteria, *Int. J. Food Microbiol.*, 41, 103, 1998.
35. Sánchez, I., Sesena, S., and Palop, LL., Polyphasic study of the genetic diversity of lactobacilli associated with "Almagro" eggplants spontaneous fermentation, based on combined numerical analysis of randomly amplified polymorphic DNA and pulsed-field gel electrophoresis patterns, *J. Appl. Microbiol.*, 97, 446, 2004.
36. Roy, D. et al., Molecular identification of potentially probiotic lactobacilli, *Curr. Microbiol.*, 40, 40, 2000.
37. McLeod, A. et al., Diversity of *Lactobacillus sakei* strains investigated by phenotypic and genotypic methods, *Syst. Appl. Microbiol.*, 31, 393, 2008.
38. Gordillo, M.E., Singh, K.V., and Murray, B.E., Comparison of ribotyping and pulsed-field gel electrophoresis for subspecies differentiation of strains of *Enterococcus faecalis*, *J. Clin. Microbiol.*, 31, 1570, 1993.
39. Olsen, J.E. et al., Clonal lines of *Salmonella enterica* serotype Enteritidis documented by IS200-, ribo-, pulsed-field gel electrophoresis and RFLP typing, *J. Med. Microbiol.*, 40, 15, 1994.
40. Nigatu, A., Ahrne, S., and Molin, G., Randomly amplified polymorphic DNA (RAPD) profiles for the distinction of *Lactobacillus* species, *Antonie Van Leeuwenhoek*, 79, 1, 2001.
41. Catzeddu, P. et al., Molecular characterization of lactic acid bacteria from sourdough breads produced in Sardinia (Italy) and multivariate statistical analyses of results, *Syst. Appl. Microbiol.*, 29, 138, 2006.
42. Siragusa, S. et al., Taxonomic structure and monitoring of the dominant population of lactic acid bacteria during wheat flour sourdough type I propagation using *Lactobacillus sanfranciscensis* starters, *Appl. Environ. Microbiol.*, 75, 1099, 2009.
43. Dimitonova, S.P. et al., Phenotypic and molecular identification of lactobacilli isolated from vaginal secretions, *J. Microbiol. Immunol. Infect.*, 41, 469, 2008.
44. Wiedmann, M., Molecular subtyping methods for *Listeria monocytogenes*, *J. AOAC Int.*, 85, 524, 2002.
45. Mohania, D. et al., Molecular approaches for identification and characterization of lactic acid bacteria, *J. Dig. Dis.*, 9, 190, 2008.
46. Coudeyras, S. et al., Taxonomic and strain-specific identification of the probiotic strain *Lactobacillus rhamnosus* 35 within the *Lactobacillus casei* group, *Appl. Environ. Microbiol.*, 74, 2679, 2008.
47. Kitahara, M., Sakamoto, M., and Benno, Y., *Lactobacillus similis* sp. nov., isolated from fermented cane molasses, *Int. J. Syst. Evol. Microbiol.*, 60, 187, 2009.
48. Ryu, C.S. et al., Characterization of the *Lactobacillus casei* group and the *Lactobacillus acidophilus* group by automated ribotyping, *Microbiol. Immunol.*, 45, 271, 2001.
49. Ruiz, P. et al., Intraspecific genetic diversity of lactic acid bacteria from malolactic fermentation of Cencibel wines as derived from combined analysis of RAPD-PCR and PFGE patterns, *Food Microbiol.*, 25, 942, 2008.
50. Tynkkynen, S. et al., Comparison of ribotyping, randomly amplified polymorphic DNA analysis, and pulsed-field gel electrophoresis in typing of *Lactobacillus rhamnosus* and *L. casei* strains, *Appl. Environ. Microbiol.*, 65, 3908, 1999.
51. Satokari, R.M. et al., Molecular approaches for the detection and identification of bifidobacteria and lactobacilli in the human gastrointestinal tract, *Syst. Appl. Microbiol.*, 26, 572, 2003.
52. Furet, J.P., Quénée, P., and Tailliez, P., Molecular quantification of lactic acid bacteria in fermented milk products using real-time quantitative PCR, *Int. J. Food Microbiol.*, 97, 197, 2004.

53. Walter, J. et al., Detection and identification of gastrointestinal *Lactobacillus* species by using denaturing gradient gel electrophoresis and species-specific PCR primers, *Appl. Environ. Microbiol.*, 66, 297, 2000.
54. Fujimoto, J. et al., Identification and quantification of *Lactobacillus casei* strain Shirota in human feces with strain-specific primers derived from randomly amplified polymorphic DNA, *Int. J. Food Microbiol.*, 126, 210, 2008.
55. Chagnaud, P. et al., A. Rapid PCR-based procedure to identify lactic acid bacteria: Application to six common *Lactobacillus* species, *J. Microbiol. Methods*, 44, 139, 2001.
56. Byun, R. et al., Quantitative analysis of diverse *Lactobacillus* species present in advanced dental caries, *J. Clin. Microbiol.*, 42, 3128, 2004.
57. Walter, J. et al., Detection of *Lactobacillus*, *Pediococcus*, *Leuconostoc*, and *Weissella* species in human feces by using group-specific PCR primers and denaturing gradient gel electrophoresis, *Appl. Environ. Microbiol.*, 67, 2578, 2001.
58. Heilig, H.G. et al., Molecular diversity of *Lactobacillus* spp. and other lactic acid bacteria in the human intestine as determined by specific amplification of 16S ribosomal DNA, *Appl. Environ. Microbiol.*, 68, 114, 2002.
59. Torriani, S., Felis, G.E., and Dellaglio, F., Differentiation of *Lactobacillus plantarum*, *L. pentosus*, and *L. paraplantarum* by *recA* gene sequence analysis and multiplex PCR assay with *recA* gene-derived primers, *Appl. Environ. Microbiol.*, 67, 3450, 2001.
60. Morrison, C., and Gannon, F., The impact of the PCR plateau phase on quantitative PCR, *Biochim. Biophys. Acta*, 1219, 493, 1994.
61. von, W.F., Gobel, U.B., and Stackebrandt, E., Determination of microbial diversity in environmental samples: pitfalls of PCR-based rRNA analysis, *FEMS Microbiol. Rev.*, 21, 213, 1997.
62. Collado, M.C. et al., Specific duodenal and faecal bacterial groups associated with paediatric celiac disease, *J. Clin. Pathol.*, 62, 264, 2009.
63. De Palma, G. et al., Effects of a gluten-free diet on gut microbiota and immune function in healthy adult human subjects, *Br. J. Nutr.*, 102, 1154, 2009.
64. Scheirlinck, I. et al., Molecular source tracking of predominant lactic acid bacteria in traditional Belgian sourdoughs and their production environments, *J. Appl. Microbiol.*, 106, 1081, 2009.
65. Cadieux, P. et al., Evaluation of reuterin production in urogenital probiotic *Lactobacillus reuteri* RC-14, *Appl. Environ. Microbiol.*, 74, 4645, 2008.
66. Amann, R.I. et al., Combination of 16S rRNA-targeted oligonucleotide probes with flow cytometry for analyzing mixed microbial populations, *Appl. Environ. Microbiol.*, 56, 1919, 1990.
67. Zoetendal, E.G., Akkermans, A.D., and de Vos, W.M., Temperature gradient gel electrophoresis analysis of 16S rRNA from human fecal samples reveals stable and host-specific communities of active bacteria, *Appl. Environ. Microbiol.*, 64, 3854, 1998.
68. Wang, B. et al., Isolation of adhesive strains and evaluation of the colonization and immune response by *Lactobacillus plantarum* L2 in the rat gastrointestinal tract, *Int. J. Food Microbiol.*, 132, 59, 2009.
69. Nadal, I. et al., Imbalance in the composition of the duodenal microbiota of children with celiac disease, *J. Med. Microbiol.*, 56, 1669, 2007.
70. Woese, C.R., Bacterial evolution, *Microbiol. Rev.*, 51, 221, 1987.
71. Duan, Y. et al., Identification and characterization of Lactic Acid Bacteria isolated from Tibetan Qula cheese, *J. Gen. Appl. Microbiol.*, 54, 51, 2008.
72. Ennahar, S., Cai, Y., and Fujita, Y., Phylogenetic diversity of lactic acid bacteria associated with paddy rice silage as determined by 16S ribosomal DNA analysis, *Appl. Environ. Microbiol.*, 69, 444, 2003.
73. Felis, G.E. et al., Comparative sequence analysis of a *recA* gene fragment brings new evidence for a change in the taxonomy of the *Lactobacillus casei* group, *Int. J. Syst. Evol. Microbiol.*, 51, 2113, 2001.
74. Ventura, M. et al., Analysis, characterization, and loci of the *tuf* genes in lactobacillus and bifidobacterium species and their direct application for species identification, *Appl. Environ. Microbiol.*, 69, 6908, 2003.
75. Chavagnat, F. et al., Comparison of partial *tuf* gene sequences for the identification of lactobacilli, *FEMS Microbiol. Lett.*, 217, 177, 2002.
76. Blaiotta, G. et al., *Lactobacillus* strain diversity based on partial *hsp60* gene sequences and design of PCR-restriction fragment length polymorphism assays for species identification and differentiation, *Appl. Environ. Microbiol.*,74, 208, 2008.
77. Groisillier, A., and Lonvaud-Funel, A., Comparison of partial malolactic enzyme gene sequences for phylogenetic analysis of some lactic acid bacteria species and relationships with the malic enzyme, *Int. J. Syst. Bacteriol.*, 49, 1417, 1999.
78. Scanlan, P.D. et al., Culture-independent analyses of temporal variation of the dominant fecal microbiota and targeted bacterial subgroups in Crohn's disease, *J. Clin. Microbiol.*, 44, 3980, 2006.
79. Sanz, Y. et al., Differences in faecal bacterial communities in celiac and healthy children as detected by PCR and denaturing gradient gel electrophoresis, *FEMS Immunol. Med. Microbiol.*, 51, 562, 2007.
80. Muyzer, G., and Smalla, K., Application of denaturing gradient gel electrophoresis (DGGE) and temperature gradient gel electrophoresis (TGGE) in microbial ecology, *Antonie Van Leeuwenhoek*, 73, 127, 1998.
81. Nielsen, D.S. et al., Case study of the distribution of mucosa-associated *Bifidobacterium* species, *Lactobacillus* species, and other lactic acid bacteria in the human colon, *Appl. Environ. Microbiol.*, 69, 7545, 2003.
82. Yu, Z., and Morrison, M., Improved extraction of PCR-quality community DNA from digesta and fecal samples, *Biotechniques*, 36, 808, 2004.
83. Mora, D. et al., Genotypic characterization of thermophilic bacilli: A study on new soil isolates and several reference strains, *Res. Microbiol.*, 149, 711, 1998.
84. Jensen, M.A., Webster, J.A., and Straus N., Rapid identification of bacteria on the basis of polymerase chain reaction-amplified ribosomal DNA spacer polymorphisms, *Appl. Environ. Microbiol.*, 59, 945, 1993.
85. Malinen, E. et al., Analysis of the fecal microbiota of irritable bowel syndrome patients and healthy controls with real-time PCR, *Am. J. Gastroenterol.*, 100, 373, 2005.

25 *Leuconostoc*

María Mar Tomás and Germán Bou

CONTENTS

25.1 INTRODUCTION

Members of the genus *Leuconostoc* (*leucus*, clear, light; *nostoc*, algal generic name; *leuconostoc*, colorless nostoc) are facultatively anaerobic, catalase-negative, gram-positive cocci organized in pairs and chains.[1] They are heterofermentative lactic acid bacteria found naturally in milk, grass, herbage, grapes, many vegetables, meat, and sugar.[2,3] Members of this group are used in dairy fermentations to produce aromatic compounds.[4]

25.1.1 Taxonomy of Genus *Leuconostoc*

Species of *Leuconostoc* are the only catalase negative cocci that are vancomycin resistant, PYR and LAP negative, and that produce CO_2 from glucose. They could appear morphologically similar to:

1. Genus *Weissella*, which includes organisms previously classified as *Leuconostocs* and the species formerly named *Lactobacillus confusus*.
2. Alpha-hemolytic or nonhemolytic *Enterococcus* or *Lactococcus* spp. It happens when these bacteria grown on the surface of 5% sheep blood agar plates. In the past it was known that *Leuconostoc* spp. was associated with human diseases[5] because these bacteria were generally identified as variants of established human pathogens, such as viridans streptococci or enterococci, or more frequently, were reported as unidentified gram-positive cocci. As many as 31% of *Leuconostoc* spp. react with the group D antiserum and may be reported as unidentified *Enterococcus* spp. The nonserogroup D bacteria have characteristics that sometimes lead to their identification as *Streptococcus uberis*, *S. bovis*, *S. sanguis*, *S. equinus*, *S. anginosus* group, or *Aerococcus*-like or *Lactobacillus*-like bacteria.
3. The heterofermentative vancomycin-resistant *Lactobacillus* spp. However, species of *Leuconostoc* spp tend to be more coccoid-like than the *Lactobacillus* spp. Cautious interpretation of the Gram stain coloring from growth of these bacteria in thioglycolate broth is important for differentiation of leuconostocs and gas-forming lactobacilli.

With the recognition of *Leuconostoc* spp. as human pathogens, such mistakes in identification are now odd because of the unique results produced by leuconostocs in the tests listed in Table 25.1.[6]

25.1.2 Isolation and Identification of *Leuconostoc* spp.

Isolation Procedure. Generally, there are no special requirements for isolation of the group of bacteria discussed here; general recommendations for the culture of blood, body fluids, and other specimens should be followed. These organisms are likely to be isolated on rich, nonselective media (e.g., blood and chocolate agars and thioglycolate broth) since they are nutritionally fastidious. If selective isolation of members of the vancomycin-resistant genera *Leuconostoc* and *Pediococcus* is desired, Thayer–Martin medium may be used to inhibit normal flora or other contaminating microorganisms.[34]

TABLE 25.1
Basic Phenotypic Characteristics of Catalase-Negative, Gram-Positive Cocci

Organisms (reference[s])	PYR	LAP	NaCl	BE	ESC	MOT	HIP	SAT	ARG	BGUR	MORPH	VAN
Abiotrophia defectiva[7–9]	+	+	–	–		–	–	+	–	–	Chains	S
Aerococcus christensenii[10,11]	–	+	+	–	–	–	+	–	–	–	Clusters	S
Aerococcus sanguinola[11,12]	+	+	+	+	+	ND	+	–	–	+	Clusters	ND
Aerococcus urinae[11,13,14]	–	+	+	–	V	–	+	–	–	+	Clusters	S
Aerococcus urinaehominis[11,15]	–	–	+	–	+	ND	+	–	–	+	Clusters	ND
Aerococcus viridans[a,5,11]	+	–	+	V	+[b]	–	V	–	–	ND	Clusters	S
Dolosicoccus[16]	+	–	–	ND	ND	–	ND	–	ND	ND	Chains	S
Dolosigranulum[11,17,18]	+	+	+	–	+	–	–	–	–	–	Clusters	S
Enterococcus[c,d,5]	+	+	+	+	+	V	ND	–	ND	ND	Chains	S/R
Facklamia languida[e,18,19]	+	+	+	ND	–	–	–	–	–	–	Clusters	S
Facklamia spp[e,18,20–22]	+	+	+	ND	–	–	+	–	V	ND	Chains	S
Gemella haemolysans[f,5]	+	+	–	–	–	–	ND	–	ND	ND	Clusters	S
Gemella spp[f,5,23,24]	+	+	–	–	ND	–	ND	–	–	ND	Chains	S
Globicatella[5,25–27]	V	–	+	+	+	–	+	–	–	ND	Chains	S
Granulicatella adiacens[g,7–9]	+	+	–	–	ND	–	–	+	–	+	Chains	S
Granulicatella elegans[g,7–9]	+	+	–	–	ND	–	+	+	+	–	Chains	S
Helcococcus kunzii[a,5,28]	+	–	+	–	+	–	–	–	–	–	Clusters	S
Ignavigranum[18,26,29]	+	+	+	–	–	–	–	V	–	ND	Chains	S
Lactococcus[d,h,5]	V	+	V	+	+	–	ND	–	ND	ND	Chains	S
Leuconostoc[5]	–	–	V	V	ND	–	ND	–	ND	ND	Chains	R
Pediococcus[5]	–	+	V	+	ND	–	ND	–	ND	ND	Clusters	R
Streptococcus[h,i,5]	V	+	V	V	V	–	ND	–	ND	ND	Chains	S
Tetragenococcus[5]	–	+	+	+	ND	–	ND	ND	ND	ND	Clusters	S
Vagococcus[c,5]	+	+	V	+	+	+	ND	–	ND	ND	Chains	S
Weissella[30,31]	V	ND	ND	+	+	–	ND	ND	+	ND	Short rods coccobacilli in pairs and chains	R

Source: From Murray, P.R., Baron, E.J., Jorgensen, J.H., Landry, M.L., and Pfaller, M.A. (Editors), *Manual of Clinical Microbiology*, Vol. 1, 31: 443–454, ASM PRESS, Washington, DC, 2007. With permission from American Society of Microbiology.

Abbreviations and symbols: PYR, production of pyrrolidonyl arylamidase; LAP, production of leucine aminopeptidase; NaCl, growth in 6.5%NaCl; BE, hydrolysis of esculin in the presence of 40% bile; ESC, hydrolysis of esculin; MOT, motility; HIP, hydrolysis of hippurate; SAT, satelliting behavior; ARG, hydrolysis of arginine; BGUR, production of β-glucuronidase; MORPH, cellular arrangement; VAN, susceptibility to vancomycin; +, ≥90% of strains positive; –, ≤10% of strains positive; V, variable (60%–90% of strains positive); ND, no data; chains, cells arranged primarily in pairs and chains; clusters, cells arranged primarily in clusters, tetrads, or irregular groups; S, susceptible; R, resistant.

[a] Although *H. hunzii* shares some phenotypic traits with *A. viridans*, *H. kunzii* is facultative and usually nonhemolytic, in contrast to *A. viridans*, which favors an anaerobic growth atmosphere and is alpha-hemolytic. Two additional species of *Helcococcus* (*H. sueciensis* and "*H. pyogenes*") have been proposed[32,33] both based on the isolation of single strains. These new species display negative reactions in the PYR test, in contrast to *H. kunzii*.

[b] LaClaire and Facklam[18] note that strains of *Dolosigranulum* are esculin hydrolysis positive when tested on conventional media, but the original description of this organism notes a lack of esculin hydrolysis activity.

[c] Most enterococcal strains are capable of growth at 45°C, which differentiates them from vagococci have been reported as testing positive with a commercially available nucleic acid probe for members of the genus *Enterococcus*.

[d] Phenotypically similar strains of enterococci and lactococci can be differentiated with a commercially available nucleic acid probe for members of the genus *Enterococcus*.

[e] *F. languida* cells are characteristically arranged in clusters, and cells of other *Facklamia* species are arranged in pairs and chains.

[f] *G. haemolysans* cells are arranged in pairs and clusters, and the cells of other *Gemella* species are arranged in pairs and chains.

[g] Formerly classified as a member of the genus *Abiotrophia*.

[h] Most lactococcal strains are capable of growth at 10°C, which differentiates them from streptococci, which may be phenotypically similar.

[i] *Streptococcus pyogenes* and some strains of *S. pneumoniae* are PYR positive.

Attempts have been made to develop media for the isolation and enumeration of these organisms, and both selective and differential media have been described; however, no medium has proven satisfactory.

Comprehensive reviews of differential and selective media for *Leuconostoc* have been published by Garvie,[35] Teuber and Geis,[36] and Cogan.[37]

Most media are based on the ability of leuconostocs to utilize citrate, which is recognized by the presence of halos around colonies growing on media containing insoluble calcium citrate.[38] Such differentiation is not accurate since not all leuconostocs utilize citrate; furthermore, other bacteria associated with green plants, such as *Lactobacillus* species[39] and *Lactococcus lactis* subsp. Lactis biovar. diacetylactis, utilize citrate.[40] As *Leuconostoc* species are resistant to vancomycin and *Lactococcus* species and some lactobacilli are sensitive to vancomycin, we considered inclusion of this antibiotic in a *Leuconostoc* selective medium (LUSM medium) for the isolation of *Leuconostoc* spp. The proposed medium contains 1.0% glucose, 1.0% Bacto Peptone (Difco), 0.5% yeast extract (BBL), 0.5% meat extract (Difco), 0.25% gelatin (Difco), 0.5% calcium lactate, 0.05% sorbic acid, 75 ppm of sodium azide (Sigma), 0.25% sodium acetate, 0.1% (vol/vol) Tween 80, 15% tomato juice, 30 μg of vancomycin (Sigma) per mL, 0.20 μg of tetracycline (Serva) per mL, 0.5 mg of cysteine hydrochloride per mL, and 1.5% agar (Difco). The LUSM medium has been successfully used for the isolation and enumeration of *Leuconostoc* spp. in dairy products and vegetables.[2]

Identification of Species. The physiological reactions that are useful for identifying the five *Leuconostoc* spp. that have been isolated from humans infections are listed in Table 25.1.[5] All of the isolates were PYR, Voges-Proskauer, and LAP negative, produced gas from glucose in MRS broth, and were resistant to vancomycin.[5] Identification of *Leuconostoc* species that fails to grow in the laboratory at 35°C to 37°C is not included in Table 25.2 (infections in humans by these bacteria are unlikely).

Leuconostoc mesenteroides and *L. pseudomesenteroides* species are sometimes confused with each other. Prolonged incubation of NaCl broth may be necessary for correct differentiation between the 6.5% NaCl-tolerant *L. mesenteroides* and the 6.5% NaCl-intolerant *L. pseudomesenteroides*. Any increase in the number of bacteria in the NaCl broth signifies that the bacterium is *L. mesenteroides*.

Moreover, *L. paramesenteroides* is considered a species of *Leuconostoc* with characteristics similar to those of *L. mesenteroides*. Dextran production by *L. mesenteroides* is used to differentiate these species. However, current phylogenetic analysis has indicated that *L. paramesenteroides* is more closely related to the lactobacilli than to members of the genus *Leuconostoc*,[41,42] and there has been an informal proposal to reclassify this and related species into a new genus, *Weisella*.[42]

Morphologically, there are some characteristic differentiates between *Lactobacillus* spp. and *Leuconostoc* spp. One of them is that *Leuconostoc* spp. appear as chains of cocci.

Leuconostoc citreum is also clearly distinguishable from the other clinically significant species of *Leuconostoc* because it is the only one that does not ferment either raffinose or melibiose. As its name implies, this bacterium also produces a characteristic yellow pigment that can be observed in bacteria grown in MRS broth.[5]

The only clinical isolate of *Leuconostoc* spp. that does not hydrolyze esculin is *Leuconostoc lactis* (Table 25.2). As with *L. paramesenteroides*, this species does not produce dextran from glucose. These characteristics clearly differentiate this species from other *Leuconostoc* species.

Finally, the other features to distinguish among these five species of *Leuconostoc* are (Table 25.2):

1. Alpha hemolysis on Trypticase soy agar with 5% sheep blood (BBL, Cockeysville, Md.).
2. Susceptibility to bacitracin.
3. Serogroup D antigen.

Methods of identifying *Leuconostoc cremoris*, *L. carnosum*, *L. oenos*, *L. fallax*, *L. argentinum*, and *L. gelidum* have been studied in different works.[43,44]

A new species, *Leuconostoc fallax* was investigated by reverse transcriptase sequencing of 16S rRNA in 1991.[43]

TABLE 25.2
Positive Reactions of Centers for Disease Control and Prevention Clinical Isolates of *Leuconostoc* spp.

	% Positive in Given Test							
Species	**ESC**	**LM**	**RAF**	**MEL**	**NaCl**	**ARA**	**TRE**	**SUCA**
L. mesenteroides	100	11	94	100	89	97	97	74
L. pseudomesenteroides	100	54	77	100	0	62	92	67
L. paramesenteroides	100	50	100	10	100	100	100	0
L. citreum	100	4	8	0	63	92	100	83
L. lactis	4	37	96	100	48	52	48	15

Source: From Facklam, R., and Elliott, J.A., *Clin. Microbiol. Rev.*, 8, 479, 1995. With permission from American Society of Microbiology.
Abbreviations: ESC, esculin hydrolysis; LM, acid and clot production in litmus milk; NaCl, growth in broth containing 6.5%NaCl; SUCA, dextran production on sucrose agar; Acid formation from; RAF, raffinose; MEL, melibiose; ARA, arabinose; TRE, trehalose.

25.1.3 Virulence and Clinical Significance

The Center for Disease Control and Prevention (CDC)[5] carried out a study of the virulence of *Leuconostoc* species. For the majority of the isolates studied, there was a lack of information about underlying disease in patients, and the possibility that *Leuconostoc* spp. are primarily opportunistic pathogens of debilitated individuals could not be confirmed. However, considering the wide distribution of these species in the environment and the relatively few infections that they cause, it is assumed that they are poorly virulent for healthy humans.

It has been demonstrated that dextran-producing *Leuconostoc* strains are able to inhibit the formation of *Streptococcus mutans* biofilm in vitro.[45] Moreover, *Leuconostoc mesenteroides* and *Lactococcus lactis* have been used as probiotic strains for *Aeromonas salmonicida* infection in brown trout.[46]

Handweger and colleagues[47] noted host defense impairment, invasive procedures breaching the integument, gastrointestinal symptoms, and prior antibiotic treatment as common features among adult patients infected with *Leuconostoc spp.* The clinical significance of infection with these bacteria in immunocompetent patients is not clear. For example, in one study the bacteria were eliminated by several patients without the use of antimicrobial agents.[48] However, a clinic case involving pulmonary abscess caused by *Leuconostoc* species in an immunocompetent patient has been reported recently.[49] Infections in immunocompromised patients can be severe, and identification of the bacterium as a *Leuconostoc* spp. is essential to ensure that an appropriate antimicrobial agent is used to treat the patient.[50]

Invasive procedures recognized as risk factors associated with this infection include total parenteral nutrition through a central venous catheter, and gastrostomy for enteral feeding.[51–53] Gastrointestinal symptoms identified as risk factors include short gut syndrome. Several studies have described a link between short bowel syndrome and *Leuconostoc* bacteremia in pediatric patients.[51–53] Finally, prior administration of antibiotics, particularly vancomycin, has been implicated in infections by *Leuconostoc*. The bacteria are often associated with a polymicrobial infection and are isolated after the patient has been treated with vancomycin to eliminate the suspected pathogen.[47,54–56] However, a review of *Leuconostoc* bacteremia in pediatric patients[52] concluded that in the cases of pediatric patients with short bowel syndrome, prior administration of vancomycin was not a required risk factor for development of this infection.

Moreover, predisposition to *Leuconostoc* bacteremia has been observed among neonates, suggesting that infants may become colonized during delivery, by Leuconostoc spp. inhabiting the maternal genital tract.

Leuconostoc spp. have occasionally been isolated from humans in cases of bacteremia or septicemia,[5,48,50,57] meningitis,[58,59] and peritonitis. Some reports have implicated leuconostocs as agents of infection in abdominal abscess,[60] osteomyelitis,[61] ventriculitis,[62] postsurgical endophthalmitis,[63] hepatic dysfunction,[53] and pneumonia.[49] An outbreak of nosocomial urinary tract infection due to *Leuconostoc mesenteroides* was reported at a tertiary care centre in northern India.[64]

These species have not previously been considered as agents that cause severe hospital outbreaks threatening the lives of large numbers of persons. However, three previous reports have described hospital transmission of *Leuconostoc* spp. Two of these affected a small number of patients, and no epidemiological studies were conducted to clarify the genetic relationship among the bacterial strains involved or the source of the nosocomial infection. The third outbreak occurred in Spain, in 2008,[50] and was the largest nosocomial outbreak caused by *Leuconostoc* spp. worldwide. In this latest example, parenteral nutrition was suggested as the most likely source of infection.

Treatment with high-dose penicillin or clindamycin[47,54,55,58] or with tobramycin[55] has been reported to be successful. Swenson et al.[65] found that none of the isolates were resistant to imipenem, minocycline, chloramphenicol, gentamicin, or daptomycin. Daptomycin has also been used to treat infections caused by *Leuconostoc* spp.[66]

25.1.4 Antibiotic Susceptibilities

The lack of standardized methods and interpretation criteria for antibiotic susceptibility testing led Swenson et al.[65] to carry out a study of susceptibility in *Leuconostoc mesenteroides*, *L. paramesenteroides*, *L. citreum*, *L. pseudomesenteroides*, *L. lactis*, and *L. dextranicum* with 79 samples (Table 25.3). However, information on the in vitro antimicrobial susceptibility of *L. garvieae* is scarce.

Glucopeptides. Members of the vancomycin-resistant genera *Leuconostoc* and *Pediococcus* are also resistant to teicoplanin. Moreover, most strains are susceptible to daptomycin.[65,67]

β-lactam Agents. MICs of ampicillin were high, but MICs of penicillin were in the range of 0.25–2.0 μg/mL in 96% of the isolates tested.[65] Among cephalosporins, cephalothin was overall the most active cephalosporin tested, but cefuroxime, ceftizoxime, and cefotaxime were more active against strains of *L. citreum*. Except for strains of *L. mesenteroides* and *L. pseudomesenteroides*, the other *Leuconostoc* spp. are usually susceptible to impenem. To date, no β-lactamase has been detected in any of the strains.

Macrolides. All strains of *Leuconostoc* were susceptible to erythromycin, roxithromycin, and josamycin. Clindamycin MICs were higher (0.12–2 μg/mL) in some strains but these samples did not display greater resistance to erythromycin.

Aminoglycosides. Gentamicin was the most active of the four aminoglycosides tested and displayed the greatest activity against leuconostocs.

25.1.5 Molecular Identification of *Leuconostoc* spp.

PCR-Based Methods. In 2000, Lee et al.[68] developed a multiplex PCR-based method for rapid and reliable identification of *Leuconostoc* species, by using species-specific primers

TABLE 25.3
Range of MICs by Species for Selected Antimicrobial Agents

Organism *(n)*	Range of MICs (μg/mL)								
	Penicillin	Cephalothin	Cefuroxime	Ceftriaxone	Imipenem	Erythromycin	Rifampin	Gentamicin	Ciprofloxacin
Leuconostoc mesenteroides (19)	0.5–1	2–16	8–32	8–64	2–8	≤0.015–0.06	0.5–8	≤0.25–0.5	1–4
L. citreum (12)	0.12–0.5	0.5–2	≤0.25–1	≤0.25–4	1–2	0.03–0.06	1–2	≤0.25	0.5–2
L. pseudomesenteroides (4)	0.12–0.5	0.5–4	0.5–16	1–16	1–8	0.06	1–2	≤0.25–0.5	2–4
L. lactis (3)	0.5	2–4	2–8	4–16	2–4	0.06	1–8	≤0.25–0.5	2
L. dextranicum (1)	0.03	0.12	0.5	1	0.12	≤0.015	0.06	≤0.25	1
L. paramesenteroides (1)	0.5	32	8	4	≤0.06	0.06	64	≤0.25	2
Leuconostoc species (7)	0.25–2	1–32	0.5–16	2–128	0.12–4	0.06	1–64	≤0.25–0.5	1–4

Source: Swenson, J.M., et al., *Antimicrob. Agents Chemother.*, 34, 543, 1990. With permission from American Society of Microbiology.

targeted to the genes encoding 16S rRNA. This assay allowed the detection and differentiation of *Leuconostoc* species in mixed populations from natural sources, as well as from pure cultures, within 3 hours.

Arbitrarily primed (AP)-PCR can be used to generate typical DNA band profiles. A single triplicate AP-PCR (TAP-PCR) procedure with an 18-mer oligonucleotide, which enabled fingerprinting of major genera of lactic acid bacteria at the strain level, was developed in 2000.[69] The authors used this assay to discriminate *Leuconostoc cremoris* from other lactic acid bacteria, such as *Lactobacillus caseii*, *Bifidobacterium breve*, *Lactobacillus plantarum*, *Pediococcus pentosaceus*, and *P. cerevisiae*. Moschetti et al.[70] also used randomly amplified polymorphic DNA analysis with a 10-mer oligonucleotide to characterize *Leuconostoc* sp. strains. All except two strains of *Leuconsotoc mesenteroides* subsp. *mesenteroides*, as well as two strains formerly identified as *L. gelidum*, and one strain of *Leuconostoc* showed a common band at about 1.1 Kb. This significant band was absent in all other strains of the genus *Leuconostoc*. In conclusion, in this study the set of primers from RAPD-PCR allowed identification of strains belonging to *L. mesenteroides* subsp. *mesenteroides*, and could be used for biotechnological purposes to monitor this microorganism in mixed populations of microflora present in fermented feed and food products.

Similar conclusions were obtained in another study in which a reliable, reproducible, and rapid molecular RAPD-PCR fingerprinting method was developed to identify the three subspecies of *Leuconostoc mesenteroides* (*mesenteroides*, *cremoris*, and *dextranicum*).

rRNA-Based Techniques. Ribosomal RNA-based techniques including analysis of profiles generated by ISR amplification, ISR restriction, and ARDRA were evaluated to identify *Leuconostoc*. In a study by Chenoll et al.,[71] 13 *Leuconostoc* isolates were included and challenged for characterization by rRNA-based techniques. These include *L. argentinum*, *L. carnosum*, *L. citreum*, *L. fallax*, *L. ficulneum*, *L. fructosum*, *L. gasicomitatum*, *L. gelidum*, *L. lactis*, *L. mesenteroides* supsp. *cremoris*, *L. mesenteroides* sups. *dextranicum*, *L. mesenteroides* subsp. *mesenteroides*, and *L. pseudomesenteroides*.

The 16S-23S ISR amplified from genomic DNA in all *Leuconostoc* strains, except *L. fallax*, shared a single ISR amplified fragment of about 700 bp, which showed an ISR amplification pattern with four bands of sizes 1450, 830, 730, and 650 bp. ISR restriction analysis with *Dde*I of Leuconostoc revealed four to six bands. Six species with 93% similarity showed single profiles, whereas the remainder were clustered in two groups. One group included *L. gelidum* and *L. gasicomitatum*, and the other group clustered *L. lactis* and *L. citreum* together with the three subspecies of *L. mesenteroides*. The *Hae* III restriction enzyme failed to digest the amplified ISR of *L. fallax*, whereas ISR-Hae III band patterns of the remaining Leuconostoc species yielded between one and five bands. Seven profiles with 90% similarity were included in a single cluster and the other grouped into two clusters. One cluster included *L. gelidum* and *L. pseudomesenteroides*. The three subspecies of *L. mesenteroides* shared a profile with *L. gasicomitatum*.

On the other hand, the amplified 16S ribosomal DNA restriction analysis (ARDRA) with *Hae* III of the above-mentioned strains of Leuconostoc generated seven profiles of two to four bands. Some of these band patterns were shared by a number of strains: (i) *L. ficulneum* and *L. fructosum*; (ii) *L. carnosum*, *L. gelidum*, and *L. lactis*; (iii) *L. pseudomesenteroides* and its three subspecies. *Leuconostoc argentinum*, *L. citreum*, *L. fallax*, and *L. gasicomitatum* also yielded specific band patterns. When ARDRA was performed with *Dde* I, four unique profiles were recognized (*L. argentinum*, *L. carnosum*, *L. fallax*, and *L. fructosum*), along with a profile that was shared by *L. ficulneum*, *L. gelidum*, and *L. mesenteroides* subsp. mesenteroides, and another profile that was the same for the rest of strains. When the method was used with both restriction enzymes (*Hae*III and *Dde*I) all species were discriminated, except *L. pseudomesenteroides*, which was obtained by the BioNumerics Composite Data Set tool.

Epidemiological Studies. *Leuconostoc* spp. are rarely isolated from hospital environments because of their epidemiology and virulence features. Very few reports of hospital-associated cases associated with *Leuconostoc* spp. have been described in the relevant literature.[50,72,73]

In the first example, *L. pseudomesenteroides* was described as a cause of nosocomial urinary tract infections in five patients.[73] Analysis of chromosomal DNA by pulsed-field-gel electrophoresis (PFGE) after treatment with *Sma*I indicated a clonal relationship among the isolates, and the report emphasized the possibility of nosocomial transmission of this unusual opportunistic, vancomycin-resistant pathogen.

In another example, 42 patients became infected between July 2003 and October 2004 by strains of *L. mesenteroides* subsp. *mesenteroides* (genotype 1) in different departments of the Juan Canalejo Hospital in northwest Spain. During 2006, six inpatients in different departments of the same hospital became infected (genotypes 2–4) by the same *L. mesenteroides* subsp. *mesenteroides*. After an epidemiological study and molecular analysis by PFGE with *Apa*I as restriction enzyme, it was concluded that parenteral nutrition was the most likely source of infection, and it was suggested that products manufactured in the centralized hospital pharmacy may have been the origin of the unexpected nosocomial outbreak.[50]

25.2 METHODS

25.2.1 Sample Preparation

Sample Collection and Culture. Freshly collected fecal samples are diluted tenfold in phosphate buffer (0.05 M pH 7.0) and stored at −82°C in 1-ml aliquots for later DNA extraction. For bacteriological culture, fresh fecal samples are homogenized and diluted tenfold in prereduced diluent (containing 8.5 g of NaCl, 1.0 g of peptone, and 0.1 g of cysteine per liter, pH 7.0) in an anaerobic chamber. Viable cell counts of lactobacilli are determined by plating on Rogosa SL agar (Difco) after incubation for 2 days. Colonies are picked randomly from an agar plate containing 30–300 colonies. These subcultured isolates represent the predominant strains comprising the population selected on Rogosa agar.[74]

DNA Extraction. Frozen fecal sample (1 mL) is thawed on ice and centrifuged. The pellet is resuspended in 100 μL of lysis buffer (6.7% sucrose, 50 mM Tris-HCl pH 8.0, 10 mM EDTA, 20 mg of lysozyme per mL, 1000 U of mutanolysin per ml, 100 μg of RNaseA per ml). After incubation for 1 h at 37°C, 6 μL of sodium dodecyl sulfate (20%) and 5 μL of proteinase K solution (15 mg/mL) are added, and the mixture was further incubated for 15 min at 60°C until the cells lysed. After cooling on ice, 400 μL of Tris-HCl (pH 8.0) is added and the mixture is extracted once with phenol-chloroform-isoamyl alcohol (25:24:1) and twice with chloroform. After ethanol precipitation the DNA is dissolved in 100 μL of Tris HCl pH 8.0).[74]

To extract DNA from cultured isolates, 2 mL of inoculated culture medium is incubated overnight at 30°C, and cells are concentrated by centrifugation at 4500 × *g* for 5 min. The pellet is resuspended in 180 μL lysis buffer (20 mM Tris-HCl, pH 8.0, 2 mM EDTA, 1.2% Triton X-100, 20 mg/mL lysozyme, 100 U/mL mutanolysin) and incubated for 30 min at 37°C. Afterwards, the DNeasy tissue kit (QIAGEN) is applied.[73]

25.2.2 Detection Procedures

We present below a real-time PCR protocol for detection and quantitation of *Leuconostoc* spp. that was developed by Friedrich and Lenke (2006).[73] Other PCR procedures may be also utilized.[68,74–80]

The quantitative real-time PCR of Friedrich and Lenke[73] are based on 16S rRNA gene primers and probe.

Primer or Probe	Sequence (5′→3′)	Target
Bac1108R	CGCTCGTTGCGGGAC	Bacterial 16S rRNA gene
Leuc986F	CAGGTCTTGACATCCTTTGAAG	*Leuconostoc* spp. 16S rRNA gene
PLeuc1026F	Cy5-CTCTTCGGAGACAAAGTGA CAGGT-BHQ2	*Leuconostoc* spp. 16S rRNA gene

Procedure

1. Oligonucleotides are diluted in ultraPure distilled water (Invitrogen) to stock concentrations of 100 μM.
2. PCR mixture (20 μL) is composed of 1× PCR buffer (QIAGEN), 5 mM $MgCl_2$, 0.9 μM of each of the primers (Leuc986F and Bac1108R), 0.2 μM of the dual-labeled probe PLeuc1026F, 0.8 mM of each dNTPs, 2.5 U of HotStarTaq DNA polymerase (QIAGEN), and 4 μL of template.
3. Amplification and detection are carried out on a Rotor-Gene model 3000 quantitative PCR cycler (Corbett Life Science), applying a first denaturation at 95°C for 15 min, followed by 50 cycles at 94°C for 20 s and 58°C for 45 s.
4. Fluorescence data are acquired at the end of each elongation step at 58°C in the channel Cy5.

25.3 CONCLUSION

Although *Leuconostoc* spp. are present in dairy products and are relatively nonpathogenic to human hosts, infections with these bacteria do occur occasionally, especially in immunocompromised individuals, pediatric patients, and the elderly who have a catheter or ventilator implanted. The clinical symptoms range from bacteremia/sepsis, brain and liver abscesses, peritonitis, endophthalmitis, endocarditis, osteomylitis, to nosocomial urinary tract infections. Recent development and application of molecular diagnostic procedures have done much to give *Leuconostoc* spp. long-overdue recognition as important opportunistic pathogens. It is envisaged that with their high sensitivity, specificity, and rapid result availability, molecular tests will play an even more prominent role in the diagnosis and management of clinical diseases due to *Leuconostoc* spp. in future.

REFERENCES

1. Garvie, E.I., Genus *Leuconostoc*. In Sneath, P.H.A., Mair, N.S., and Sharp, M.E. (Editors), *Bergey's Manual of Systematic Bacteriology*, Vol. 2, pp. 1075–1079, The Williams & Wilkins Co., Baltimore, 1986.
2. Benkerroum, N. et al., Development and use of a selective medium for isolation of *Leuconostoc* spp. from vegetables and dairy products, *Appl. Environ. Microbiol.*, 59, 607, 1993.
3. Elliott, J.A., and Facklam, R.R., Identification of *Leuconostoc* spp. by analysis of soluble whole-cell protein patterns, *J. Clin. Microbiol.*, 31, 1030, 1993.
4. Teuber, M., The family Streptoccaceae (nonmedical aspects). In Starr, M.P. (Editor), *The Prokaryotes*, Vol. 2, pp. 1614–1630, Springer-Verlag, New York, 1981.
5. Facklam, R., and Elliott, J.A., Identification, classification, and clinical relevance of catalase-negative, gram-positive cocci, excluding the streptococci and enterococci, *Clin. Microbiol. Rev.*, 8, 479, 1995.
6. Ruoff, K.L., Aerococcus, Abiotrophia, and other aerobic catalase-negative, gram-positive cocci. In Murray, P.R., Baron, E.J., Jorgensen, J.H., Landry, M.L., and Pfaller, M.A. (Editors), *Manual of Clinical Microbiology*, Vol. 1, 31: 443–454, ASM PRESS, Washington, DC, 2007.
7. Christensen, J.J., and Facklam, R.R., *Granulicatella* and *Abiotrophia* species from human clinical specimens, *J. Clin. Microbiol.*, 39, 3520, 2001.
8. Collins, M.D., and Lawson, P.A., The genus *Abiotrophia* (Kawamura et al.) is not monophyletic: Proposal of *Granulicatella* gen. nov., *Granulicatella adiacens* comb. nov., *Granulicatella elegans* comb. nov. and *Granulicatella balaenopterae* comb. nov., *Int. J. Syst. Evol. Microbiol.*, 50, 365, 2000.
9. Kawamura, Y. et al., Transfer of *Streptococcus adjacens* and *Streptococcus defectivus* to *Abiotrophia* gen. nov. as *Abiotrophia adiacens* comb. nov. and *Abiotrophia defectiva* comb. nov., respectively, *Int. J. Syst. Bacteriol.*, 45, 798, 1995.

10. Collins, M.D. et al., *Aerococcus christensenii* sp. nov., from the human vagina, *Int. J. Syst. Bacteriol.*, 49, 1125, 1999.
11. Facklam, R. et al., Phenotypic description and antimicrobial susceptibilities of *Aerococcus sanguinicola* isolates from human clinical samples, *J. Clin. Microbiol.*, 41, 2587, 2003.
12. Lawson, P.A. et al., *Aerococcus sanguicola* sp. nov., isolated from a human clinical source, *Int. J. Syst. Evol. Microbiol.*, 51, 475, 2001.
13. Christensen, J.J. et al., *Aerococcus*-like organism, a newly recognized potential urinary tract pathogen, *J. Clin. Microbiol.*, 29, 1049, 1991.
14. Christensen, J.J. et al., *Aerococcus urinae*: intraspecies genetic and phenotypic relatedness, *Int. J. Syst. Bacteriol.*, 47, 28, 1997.
15. Lawson, P.A. et al., *Aerococcus urinaehominis* sp. nov., isolated from human urine, *Int. J. Syst. Evol. Microbiol.*, 51, 683, 2001.
16. Collins, M.D. et al., *Dolosicoccus paucivorans* gen. nov., sp. nov., isolated from human blood, *Int. J. Syst. Bacteriol.*, 49, 1439, 1999.
17. Aguirre, M., and Collins, M.D., Phylogenetic analysis of *Alloiococcus otitis* gen. nov., sp. nov., an organism from human middle ear fluid, *Int. J. Syst. Bacteriol.*, 42, 79–83, 1992.
18. LaClaire, L.L., and Facklam, R.R., Comparison of three commercial rapid identification systems for the unusual gram-positive cocci *Dolosigranulum pigrum*, *Ignavigranum ruoffiae*, and *Facklamia* species, *J. Clin. Microbiol.*, 38, 2037, 2000.
19. Lawson, P.A. et al., *Facklamia languida* sp. nov., isolated from human clinical specimens, *J. Clin. Microbiol.*, 37, 1161, 1999.
20. Collins, M.D. et al., Phenotypic and phylogenetic characterization of some *Globicatella*-like organisms from human sources: Description of *Facklamia hominis* gen. nov., sp. nov., *Int. J. Syst. Bacteriol.*, 47, 880, 1997.
21. Collins, M.D. et al., *Facklamia sourekii* sp. nov., isolated from human sources, *Int. J. Syst. Bacteriol.*, 49, 635, 1999.
22. Collins, M.D. et al., *Facklamia ignava* sp. nov., isolated from human clinical specimens, *J. Clin. Microbiol.*, 36, 2146, 1998.
23. Collins, M.D. et al., *Gemella bergeriae* sp. nov., isolated from human clinical specimens, *J. Clin. Microbiol.*, 36, 1290, 1998.
24. Collins, M.D. et al., Description of *Gemella sanguinis* sp. nov., isolated from human clinical specimens, *J. Clin. Microbiol.*, 36, 3090, 1998.
25. Collins, M.D. et al., *Globicatella sanguis* gen.nov., sp.nov., a new gram-positive catalase-negative bacterium from human sources, *J. Appl. Bacteriol.*, 73, 433, 1992.
26. Facklam, R., What happened to the streptococci: Overview of taxonomic and nomenclature changes, *Clin. Microbiol. Rev.*, 15, 613, 2002.
27. Shewmaker, P.L. et al., DNA relatedness, phenotypic characteristics, and antimicrobial susceptibilities of *Globicatella sanguinis* strains, *J. Clin. Microbiol.*, 39, 4052, 2001.
28. Collins, M.D. et al., Phylogenetic analysis of some *Aerococcus*-like organisms from clinical sources: Description of *Helcococcus kunzii* gen. nov., sp. nov., *Int. J. Syst. Bacteriol.*, 43, 425, 1993.
29. Collins, M.D. et al., *Ignavigranum ruoffiae* sp. nov., isolated from human clinical specimens, *Int. J. Syst. Bacteriol.*, 49, 97, 1999.
30. Flaherty, J.D. et al., Fatal case of endocarditis due to *Weissella confusa*, *J. Clin. Microbiol.*, 41, 2237, 2003.
31. Olano, A. et al., *Weissella confusa* (basonym: *Lactobacillus confusus*) bacteremia: A case report, *J. Clin. Microbiol.*, 39, 1604, 2001.
32. Collins, M.D. et al., *Helcococcus sueciensis* sp. nov., isolated from a human wound, *Int. J. Syst. Evol. Microbiol.*, 54, 1557, 2004.
33. Panackal, A.A. et al., Prosthetic joint infection due to "*Helcococcus pyogenes*" [corrected], *J. Clin. Microbiol.*, 42, 2872, 2004.
34. Ruoff, K.L. et al., Vancomycin-resistant gram-positive bacteria isolated from human sources, *J. Clin. Microbiol.*, 26, 2064, 1988.
35. Garvie, E.I., *Leuconostoc oenos* sp.nov., *J. Gen. Microbiol.*, 48, 431, 1967.
36. Geis, A., Singh, J., and Teuber, M., Potential of lactic streptococci to produce bacteriocin, *Appl. Environ. Microbiol.*, 45, 205, 1983.
37. Cogan, T.M., O'Dowd, M., and Mellerick, D., Effects of pH and sugar on acetoin production from citrate by *Leuconostoc lactis*, *Appl. Environ. Microbiol.*, 41, 1–8, 1981.
38. Harvey, R.J., and Collins, E.B., Role of citritase in acetoin formation by *Streptococcus diacetilactis* and *Leuconostoc citrovorum*, *J. Bacteriol.*, 82, 954, 1961.
39. Edwards, C.G. et al., *Lactobacillus kunkeei* sp. nov.: A spoilage organism associated with grape juice fermentations, *J. Appl. Microbiol.*, 84, 698, 1998.
40. Sesma, F. et al., Cloning of the citrate permease gene of *Lactococcus lactis* subsp. *lactis* biovar diacetylactis and expression in *Escherichia coli*, *Appl. Environ. Microbiol.*, 56, 2099, 1990.
41. Martinez-Murcia, A.J., and Collins, M.D., A phylogenetic analysis of the genus *Leuconostoc* based on reverse transcriptase sequencing of 16 S rRNA, *FEMS Microbiol. Lett.*, 58, 73, 1990.
42. Martinez-Murcia, A.J., Harland, N.M., and Collins, M.D., Phylogenetic analysis of some leuconostocs and related organisms as determined from large-subunit rRNA gene sequences: Assessment of congruence of small- and large-subunit rRNA derived trees, *J. Appl. Bacteriol.*, 74, 532, 1993.
43. Martinez-Murcia, A.J., and Collins, M.D., A phylogenetic analysis of an atypical leuconostoc: Description of *Leuconostoc fallax* sp. nov., *FEMS Microbiol. Lett.*, 66, 55, 1991.
44. Schleifer, K.H., Gram-positive cocci. In P.H.A. Sneath, NSM, M.E. Sharp, and J.G. Holt (Editors), *Bergey's Manual of Systematic Bacteriology*, Vol. 2, pp. 999–1002, The Williams & Williams Co., Baltimore, 1986.
45. Kang, M.S. et al., Effect of *Leuconostoc* spp. on the formation of *Streptococcus mutans* biofilm, *J. Microbiol.*, 45, 291, 2007.
46. Balcázar, J.L. et al., Effect of *Lactococcus lactis* CLFP 100 and *Leuconostoc mesenteroides* CLFP 196 on *Aeromonas salmonicida* Infection in brown trout (Salmo trutta), *J. Mol. Microbiol. Biotechnol.*, 17, 153, 2009.
47. Handwerger, S. et al., Infection due to *Leuconostoc* species: Six cases and review, *Rev. Infect. Dis.*, 12, 602, 1990.
48. Ling, M.L., *Leuconostoc* bacteraemia, *Singapore Med. J.*, 33, 241, 1992.
49. Camarasa, A., Chiner, E., and Sancho-Chust, J.N., Pulmonary abscess due to *Leuconostoc* species in an immunocompetent patient, *Arch. Bronconeumol.*, 45, 471, 2009.
50. Bou, G. et al., Nosocomial outbreaks caused by *Leuconostoc mesenteroides* subsp. *mesenteroides*, *Emerg. Infect. Dis.*, 14, 968, 2008.

51. Dhodapkar, K.M., and Henry, N.K., *Leuconostoc* bacteremia in an infant with short-gut syndrome: Case report and literature review, *Mayo Clin. Proc.*, 71, 1171, 1996.
52. Florescu, D. et al., *Leuconostoc* bacteremia in pediatric patients with short bowel syndrome: Case series and review, *Pediatr. Infect. Dis. J.*, 27, 1013, 2008.
53. Jofre, M.L. et al., *Leuconostoc* infections in patients with short gut syndrome, parenteral nutrition and continuous enteral feeding, *Rev. Chil. Infectol.*, 23, 340, 2006.
54. Bernaldo de Quiros, J.C. et al., *Leuconostoc* species as a cause of bacteremia: Two case reports and a literature review, *Eur. J. Clin. Microbiol. Infect. Dis.*, 10, 505, 1991.
55. Isenberg, H.D. et al., Clinical laboratory challenges in the recognition of *Leuconostoc* spp., *J. Clin. Microbiol.*, 26, 479, 1988.
56. Rubin, L.G. et al., Infection with vancomycin-resistant "streptococci" due to *Leuconostoc* species, *J. Infect. Dis.*, 157, 216, 1988.
57. Peters, V.B. et al., *Leuconostoc* species bacteremia in a child with acquired immunodeficiency syndrome, *Clin. Pediatr. (Phila)*, 31, 699, 1992.
58. Coovadia, Y.M., Solwa, Z., and van den Ende, J., Meningitis caused by vancomycin-resistant *Leuconostoc* sp., *J. Clin. Microbiol.*, 25, 1784, 1987.
59. Friedland, I.R., Snipelisky, M., and Khoosal, M., Meningitis in a neonate caused by *Leuconostoc* sp., *J. Clin. Microbiol.*, 28, 2125, 1990.
60. Montejo, M. et al., Abdominal abscess due to *Leuconostoc* species in a liver transplant recipient, *J. Infect.*, 41, 197, 2000.
61. Zaoui, A. et al., *Leuconostoc* osteomyelitis, *Joint Bone Spine*, 72, 79, 2005.
62. Deye, G. et al., A case of *Leuconostoc* ventriculitis with resistance to carbapenem antibiotics, *Clin. Infect. Dis.*, 37, 869, 2003.
63. Kumudhan, D., and Mars, S., *Leuconostoc mesenteroids* as a cause of post-operative endophthalmitis—a case report, *Eye*, 18, 1023, 2004.
64. Taneja, N. et al., Nosocomial urinary tract infection due to *Leuconostoc mesenteroides* at a tertiary care centre in north India, *Indian J. Med. Res.*, 122, 178, 2005.
65. Swenson, J.M., Facklam, R.R., and Thornsberry, C., Antimicrobial susceptibility of vancomycin-resistant *Leuconostoc*, *Pediococcus*, and *Lactobacillus* species, *Antimicrob. Agents Chemother.*, 34, 543, 1990.
66. Golan, Y. et al., Daptomycin for line-related *Leuconostoc* bacteraemia, *J. Antimicrob. Chemother.*, 47, 364, 2001.
67. de la Maza, L., Ruoff, K.L., and Ferraro, M.J., In vitro activities of daptomycin and other antimicrobial agents against vancomycin-resistant gram-positive bacteria, *Antimicrob. Agents Chemother.*, 33, 1383, 1989.
68. Lee, H.J., Park, S.Y., and Kim, J., Multiplex PCR-based detection and identification of *Leuconostoc* species, *FEMS Microbiol. Lett.*, 193, 243, 2000.
69. Cusick, S.M., and O´Sullivan, D., Use of single, triplicate arbitrarily primed-PCR procedure for molecular fingerprinting of lactic acid bacteria, *Appl. Environ. Microbiol.*, 66, 2227, 2000.
70. Moschetti, G. et al., Specific detection of *Leuconostoc mesenteroides* subs. *mesenteroides* with DNA primers identified by randomly amplified polymorphic DNA analysis, *Appl. Environm. Microbiol.*, 66, 422, 2000.
71. Chenoll, E., Macian, M.C., and Aznar, R., Identification of *Carnobacterium*, *Lactobacillus*, *Leuconostoc* and *Pediococcus* by rDNA-based tecniques, *Syst. Appl. Microbiol.*, 26, 546, 2003.
72. Scano, F. et al., *Leuconostoc* species: A case-cluster hospital infection, *Scand. J. Infect. Dis.*, 31, 371, 1999.
73. Cappelli, E.A. et al., *Leuconostoc pseudomesenteroides* as a cause of nosocomial urinary tract infections, *J. Clin. Microbiol.*, 37, 4124, 1999.
74. Friedrich, U., and Lenke, J., Improved enumeration of lactic acid bacteria in mesophilic dairy starter cultures by using multiplex quantitative real-time PCR and flow cytometry-fluorescence in situ hybridization, *Appl. Environ. Microbiol.*, 72, 4163, 2006.
75. Walter, J. et al., Detection of *Lactobacillus*, *Pediococcus*, *Leuconostoc*, and *Weissella* species in human feces by using group-specific PCR primers and denaturing gradient gel electrophoresis, *Appl. Environ. Microbiol.*, 67, 2578, 2001.
76. Heilig, H.G. et al., Molecular diversity of *Lactobacillus* spp. and other lactic acid bacteria in the human intestine as determined by specific amplification of 16S ribosomal DNA, *Appl. Environ. Microbiol.*, 68, 114, 2002.
77. Jang, J. et al., A rapid method for identification of typical *Leuconostoc* species by 16S rDNA PCR-RFLP analysis, *J. Microbiol. Methods*, 55, 295, 2003.
78. Walczak, P., Konopacka, M., and Otlewska, A., Genetic diversity among *Lactococcus* sp. and *Leuconostoc* sp. strains using PCR-RFLP of insertion sequences ISS1-type, IS904, and IS982, *Pol. J. Microbiol.*, 54, 183, 2005.
79. Elizaquível, P., Chenoll, E., and Aznar, R., A TaqMan-based real-time PCR assay for the specific detection and quantification of *Leuconostoc mesenteroides* in meat products, *FEMS Microbiol. Lett.*, 278, 62, 2008.
80. Schillinger, U. et al., A genus-specific PCR method for differentiation between *Leuconostoc* and *Weissella* and its application in identification of heterofermentative lactic acid bacteria from coffee fermentation, *FEMS Microbiol. Lett.*, 286, 222, 2008.

26 *Listeria*

Dongyou Liu and Ting Zhang

CONTENTS

26.1 INTRODUCTION

26.1.1 Classification, Morphology, and Genome Organization

Classification. The genus *Listeria* is classified taxonomically in the family Listeriaceae, order Bacillales, class Bacilli, phylum Furmicutes. The only other genus in the family Listeriaceae is *Brochothrix* (consisting of two species, *Brochothrix campestris* and *Brochothrix thermosphacta*). Until recently, six closely related, gram-positive bacterial species have been recognized in the genus *Listeria*—that is, *L. monocytogenes*, *L. ivanovii*, *L. seeligeri*, *L. innocua*, *L. welshimeri*, and *L. grayi*, with *L. ivanovii* being further separated into two subspecies, *L. ivanovii* subsp. *londoniensis* and *L. ivanovii* subsp. *ivanovii*.[1,2] Whereas *L. monocytogenes* is a facultative intracellular pathogen for humans and animals, *L. ivanovii* (formerly *L. monocytogenes* serotype 5) mainly infects ungulated animals (e.g., sheep and cattle), and the other four species are essentially saprophytes that have adapted for free living in soil and decaying vegetation.[1,3] In late 2009, two additional *Listeria* species were described. Having originated from soil and water samples at the Finger Lakes National Forest in New York, *L. marthii* sp. nov. demonstrates a close phylogenetic relatedness to *L. monocytogenes* and *L. innocua*, and a more distant relatedness to *L. welshimeri, L. seeligeri, L. ivanovii*, and *L. grayi*. Given its absence of a homologue of the *L. monocytogenes* virulence gene island, *L. marthii* is likely nonpathogenic.[4] *L. rocourtiae* sp. nov. was identified from precut lettuce in Austria. This species is distinguishable from other *Listeria* species by using phenotypic tests, and its type strain is avirulent as assessed by in vitro cell culture and in vivo mouse model.[5]

On the basis of serological interactions between *Listeria* somatic (O)/flagellar (H) antigens and their corresponding antisera, *Listeria* is differentiated into at least 15 serovars, with *L. monocytogenes* consisting of serovars 1/2a, 1/2b, 1/2c, 3a, 3b, 3c, 4a, 4ab, 4b, 4c, 4d, 4e, and 7; *L. ivanovii* of serovar 5; *L. innocua* of serovars 4ab, 6a, and 6b; *L. welshimeri* of serovars 6a and 6b; *L. seeligeri* of serovars 1/2b, 4c, 4d, and 6b; and *L. grayi* of serovar Grayi (Table 26.1).[6] Upon phylogeneitc analyses, *L. monocytogenes* is grouped into three evolutionary lineages, with lineage I comprising serovars 1/2b, 3b, 4b, 4d, and 4e; lineage II covering serovars 1/2a, 1/2c, 3a, and 3c; and lineage III containing serovars 4a and 4c.[7,8] It was recently shown that *L. monocytogenes* lineage III can be subdivided into three distinct genetic subgroups, IIIA, IIIB, and IIIC, with subgroup IIIA consisting of typical rhamnose-positive avirulent serovar 4a and virulent serovar 4c strains; subgroup IIIC of atypical rhamnose-negative virulent serovar 4c strains; and subgroup IIIB of

TABLE 26.1
Compositions of Somatic (O)/Flagellar (H) Antigens in *Listeria* Serovars

Serovar	O-antigen	H-antigen
1/2a	I, II	A, B
1/2b	I, II	A, B, C
1/2c	I, II	B, D
3a	II, IV	A, B
3b	II, IV	A, B, C
3c	II, IV	B, D
4a	(V), VII, IX	A, B, C
4ab	V, VI, VII, IX	A, B, C
4b	V, VI	A, B, C
4c	V, VII	A, B, C
4d	(V), VI, VIII	A, B, C
4e	V, VI (VIII), (IX)	A, B, C
7	XII, XIII	A, B, C
5	(V), VI, (VIII), X	A, B, C
6a	V, (VI), (VII), (IX), XV	A, B, C
6b	(V), (VI), (VII), IX, X, XI	A, B, C

atypical rhamnose-negative virulent nonserovar 4a and nonserovar 4c strains, some of which may be related to serovar 7 (Table 26.2).[9–14] Subsequent study revealed that subgroup IIIB (including serovar 7) forms a separate lineage (IV) within *L. monocytogenes*.[15]

In accordance with its virulence potential in mammalian hosts, *L. monocytogenes* is also separated into pathogenic (virulent) and nonpathogenic (avirulent) strains. It is most noteworthy that *L. monocytogenes* serovars 4b, 1/2a, 1/2b, and 1/2c account for over 98% isolations from clinical cases of human listeriosis, with serovar 4b alone being responsible for 49% of *Listeria*-related foodborne diseases during a survey of French patients between 2001 and 2003 (Table 26.3).[16] In experimental mouse models, *L. monocytogenes* serovars 4b, 1/2a, 1/2b, and 1/2c tend to display higher infectivity than other serovars through intragastric inoculation. However, all *L. monocytogenes* serovars except 4a have the capability to cause mouse mortality via intraperitoneal route (Table 26.4).[9,10,12,17,18]

Morphology. *Listeria* is a gram-positive, nonspore-forming, motile, facultatively anaerobic, rod-shaped bacterium of 0.4–0.5 μm × 1–1.5 μm. In direct smears, *Listeria* may appear coccoid, resembling streptococci, and longer cells may look like corynebacteria. This bacterium exhibits its characteristic tumbling motility when examined by light microscopy. The organism is actively motile by means of peritrichous flagella at room temperature (20°C–25°C), but it does not synthesize flagella at body temperatures (37°C).

Being nutritionally undemanding, *Listeria* spp. grow well on a number of nonselective microbiological media such as tryptone soy broth (TSB) and brain heart infusion (BHI) broth, although the availability of biotin, riboflavin, thiamine, thioctic acid, cysteine, glutamine, isoleucine, leucine, and valine, as well as carbohydrates (e.g., glucose), aids their optimal growth. Colonies of 0.5–1.5 mm in diameter usually emerge after 24–48 h incubation on agar media (e.g., blood agar or BHI agar), and appear as round, translucent, low convex with a smooth surface and a crystalline center. After 3–7 days, older and larger colonies of 3–5 mm in diameter may give a more opaque appearance. Upon extended incubation, some rough colonies with a sunken center may develop occasionally; and some *Listeria* cells from older cultures may lose their ability

TABLE 26.2
Characteristics of *L. monocytogenes* Lineages I–IV[a]

			PCR Reactivity									
Lineage	Serovar	Rhamnose Activity	*inlA*	*lmo0733*	*lmo2672*	*inlJ*	*inlC*[b]	*lmo1134*	ORF2819	ORF2110	*lmo0737*	*lmo1118*
I	1/2b	+	+	+	+	+	+	+	+	–	–	–
	3b	+	+	+	+	+	+	+	+	–	–	–
	4b	+	+	+	+	+	+	+	+	+	–	–
	4d	+	+	+	+	+	+	+	+	+	–	–
	4e	+	+	+	+	+	+	+	+	+	–	–
II	1/2a	+	+	+	+	+	+	+	–	–	+	–
	1/2c	+	+	+	+	+	+	+	–	–	+	+
	3a	+	+	+	+	+	+	+	–	–	+	–
	3c	+	+	+	+	+	+	+	–	–	+	+
IIIA	4a	+	+	+	–	–	–	–	–	–	–	–
	4c	+	+	+	–	+	±	–	–	–	–	–
IIIC	4c	–	+	+	+	+	+	–	–	–	–	–
IV (IIIB)	7 and unusual 4a, 4b, 4c	–	+	–	+	–	+	–	+	–	–	–

[a] Summarized from Liu et al.[9–12]; Doumith et al.[13]; Roberts et al.[14]
[b] *inlC* is also found in some *L. ivanovii* strains.[12]

TABLE 26.3
***L. monocytogenes* Serovars Causing Human Listeriosis**[a]

Serotype	No. of Isolates (%)	Tendency to Cause
4b	294/603 (49%)	CNS infections > M/N diseases > Bacteremia
1/2a	163/603 (27%)	Bacteremia > M/N diseases > CNS infections
1/2b	120/603 (20%)	M/N diseases > Bacteremia > CNS infections
1/2c	22/603 (4%)	Bacteremia > CNS infections > M/N diseases
3a/3b	4/603 (<1%)	Bacteremia

[a] Based on the analysis of 603 *L. monocytogenes* isolates from 603 French patients during 2001–2003.[16]
M/N diseases, maternal-neonatal diseases; CNS infections, central nervous system infections.

to retain the Gram stain and fail to display characteristic violet color. The optimal growth temperatures for *Listeria* are between 30°C and 37°C; *Listeria* are motile with peritrichous flagella at 20°C–28°C, but become nonmotile at 37°C.

Genomic Organization. The genus *Listeria* possesses genomes of relatively low G + C contents of 36%–39%, which implies its relatedness to other members of the phylum Firmicutes such as the genera *Bacillus*, *Brochothrix*, *Clostridium*, *Enterococcus*, *Lactobacillus*, *Staphylococcus*, and *Streptococcus*. The recent publication of whole genome sequences of several *Listeria* spp. have helped uncover valuable new details on the differential genetic compositions among *Listeria* bacteria.[19–22] These include two *L. monocytogenes* serovar 1/2a (EGD-e and F6854) and two serovar 4b (F2365 and H7858) strains, one *L. innocua* (CLIP 11262), one *L. seeligeri* (SLCC3954), and one *L. welshimeri* strain (SLCC 5334) (Table 26.5). Interestingly, the nonpathogenic *L. seeligeri* type strain (of serovar 1/2b) (SLCC3954) harbors the smallest genome of the *Listeria* species sequenced to date.[22] The sequencing analyses of 15 other *L. monocytogenes* strains covering serovars 1/2a (7), 1/2b (4), 1/2c (1), 4b (1), and unknown (2) are in progress.

Comparison of the six *Listeria* genomes at the nucleotide and predicted protein levels shows that in addition to many shared genetic components, a total of 51, 97, 69, and 61 strain-specific genes are identifiable, respectively, from *L. monocytogenes* serovar 1/2a strains EGD-e and F6854, and from serovar 4b strains F2365 and H7858 (Table 26.5). Further analysis reveals that 83 of these genes are limited to the serovar 1/2a strains (of lineage II), and 51 genes are limited to the serovar 4b strains (of lineage I). In addition, 149 and 311 species-specific genes are recognized in *L. innocua* CLIP 11262 and *L. welshimeri* SLCC 5334, respectively (Table 26.5).[19–21]

The fact that some *L. monocytogenes* strain-specific genes tend to have atypical base composition implies their acquisition through horizontal gene transfer. In addition, some *L. monocytogenes* strain-specific genes encode putative surface-associated proteins such as internalins, which may contribute to the increased pathogenicity of the strains concerned.[20] Surprisingly, 37 (44%) of the 83 serovar 1/2a-specific and 33 (65%) of the 51 serovar 4b-specific genes encode hypothetical proteins for which no biochemical information is available. The serovar 1/2a-specific genes include three clusters that encode pathways for the transport and metabolism of carbohydrates, an operon that encodes the biosynthetic pathway for the antigenic rhamnose substituents that decorate the cell wall-associated teichoic acid polymer in serovar 1/2a strains, five glycosyl transferases, and an adenine-specific DNA methyltransferase.[20]

While the genomes of serovar 4b strains (F2365 and H7858) do not contain intact insertion sequence (IS)

TABLE 26.4
***L. monocytogenes* Serovars and Relative Virulence**

			PCR[a]			
Strain	Source	Serovar	inlJ	inlC	LD_{50}[b]	Relative Virulence (%)[c]
HCC8	Catfish brain	1	+	+	$<7.0 \times 10^8$	70
EGD	Guinea pig	1/2a	+	+	$<1.1 \times 10^7$	100
ATCC 19112	Human	2	+	+	1.6×10^9	30
ATCC 19114	Ruminant brain	4a	–	–	1.9×10^{10}	0
HCC25	Catfish kidney	4a	–	–	3.5×10^{10}	0
ATCC 19115	Human	4b	+	+	6.0×10^8	70
ATCC 19116	Chicken	4c	+	–	2.6×10^8	100
874	Cow brain	4c	+	+	$<8.0 \times 10^7$	100
ATCC 19117	Sheep	4d	+	+	8.8×10^8	40
ATCC 19118	Chicken	4e	+	+	7.8×10^9	50
R2-142	Food	7	–	+	$<5 \times 10^7$	100

[a] PCR was performed using primers targeting inlJ and inlC genes.[9,10,12]
[b] LD_{50} (medium lethal dose) values was determined in A/J mouse strain via intraperitoneal route.[9,17]
[c] Relative virulence (%) was calculated by dividing the number of dead mice with the total number of mice tested using EGD as reference.[17]

TABLE 26.5
Comparison of *Listeria* Genomes[a]

Feature	*L. monocytogenes* EGD-e	*L. monocytogenes* F6854	*L. monocytogenes* F2365	*L. monocytogenes* H7858	*L. innocua* CLIP 11262	*L. welshimeri* SLCC 5334
Serovar	1/2a	1/2a	4b	4b	6a	6b
Genome size (bp)	2,944,528	2,950,285	2,905,187	2,972,254	3,011,208	2,814,130
G + C content (%)	39	37.8	38	38	37.4	36.4
Protein-coding genes (ORFs)	2853	2973	2821	3024	2973	2780
Strain-specific genes	61	97	51	69	149	311

[a] Adapted from Glaser et al.[19]; Nelson et al.[20]; and Hain et al.[21]

elements, they possess four copies of transposase ORFA of the IS3 family. Similarly, the genomes of serovar 1/2a strains (F6854 and EGD-e) comprise three copies of the same transposase ORFA in a corresponding location. The extra copy of the transposase ORFA in the serotype 4b strains F2365 and H7858 may have resulted from a duplication event (along with the associated regions). Furthermore, an intact IS element (ISLmo1) is present in the serotype 1/2a strains F6854 (two copies) and EGD-e (three copies), respectively. Because all *L. monocytogenes* genomes sequenced to date harbor copies of transposase ORFA, it is probable that transposase ORFA gene originated from the genome of an ancestral *Listeria* before the strains diverged, in comparison to ISLmo1, which may represent a recent acquisition in serovar 1/2a strains.[20]

Besides possessing many unique genes, *Listeria* species/serovars also show differences among their shared genes. Southern blot analysis with a virulence-specific internalin *inlJ* probe revealed that genomic DNA from the more pathogenic *L. monocytogenes* serovars (e.g., 1/2a, 1/2b, 1/2c, and 4b) tends to form a 5.0 kb *Hin*dIII band, and that from the less pathogenic serovars (e.g., 3a and 4c) forms a 1.5 or 2.0 kb *Hin*dIII fragment. On the other hand, genomic DNA from the nonpathogenic *L. monocytogenes* serovars (i.e., 4a) forms no band.[10,11] Although being present in all *L. monocytogenes* serovars, transcriptional regulator gene *lmo0733* is recognized as a *Hin*dIII band of 5.0 or 6.0 kb in the more pathogenic serovars (e.g., 1/2a, 1/2c, and 4b), and as a *Hin*dIII band of 1.0 or 1.5 kb in the less pathogenic or nonpathogenic serovars (e.g., 4a and 4c).[10,11]

Recent studies indicate that there are some unusual *Listeria* strains possessing genetic elements that normally belong to other *Listeria* species. Examination of several hemolytic *L. innocua* strains (e.g., PRL/NW 15B95 and J1–023) disclosed the existence of *Listeria* pathogenicity island 1 (LIPI-1, or *Prf*A-regulated virulence gene cluster) in these atypical *L. innocua* strains, contributing to their positive hemolytic reactions.[23,24] The *Prf*A-regulated virulence gene cluster was previously only found in *L. monocytogenes*, *L. ivanovii*, and *L. seeligeri*. In addition, these atypical *L. innocua* strains also harbor *L. monocytogenes* internalin gene *inlA* and partial internalin gene *inlB* sequences. The *L. innocua* species status of these strains was only resolved after DNA–DNA hybridization, microarray, and PCR experiments targeting a host of *L. innocua*-specific genes.[23,24] Similarly, *L. monocytogenes* serovars 4a and 4c strains harboring *L. innocua*-specific genes *lin0464*, *lin0372*, and *lin1073* have also been described.[25–27] These findings suggest that *L. innocua* may have descended from *L. monocytogenes* through successive gene deletion, and that atypical hemolytic strains PRL/NW 15B95 and J1–023 may represent some intermediates that have yet to completely eliminate or lose the remaining *L. monocytogenes* genes.[24,27] An ancestral *L. monocytogenes* strain (probably of serovar 1/2c) might have evolved to become serovar 4b through gene deletion, which latter gave rise to serovar 4a possibly via serovar 4c, and finally to *L. innocua* via additional gene deletion events.[13,28] The fact that *L. monocytogenes* serovar 4c and 4a strains harboring *L. innocua*-specific genes *lin0464*, *lin0372*, and *lin1073* indicates the possibility that during their transition to *L. innocua,* some *L. monocytogenes* strains may pick up certain *L. innocua* genes while shedding their own.[27] Apart from horizontal gene transfer, deletion of key virulence-associated genes in the genome of a common *Listeria* ancestor may lead to subsequent changes in the corresponding phenotypes, including pathogenicity. Obviously, additional studies are required to determine what other *L. innocua* genes that have been gained by these unusual *L. monocytogenes* 4a and 4c strains, and what *L. monocytogenes* genes have been lost by these strains. The availability of these data will lead to a better understanding of *Listeria* evolution and yield new details on the phylogenetic relationship among individual *Listeria* species.

26.1.2 Biology and Epidemiology

Listeria spp. are capable of adaptation and survival in a variety of arduous environments. Besides the ability to grow from 0°C to 45°C and sustain various heat treatments, *Listeria* spp. also grow between pH 6 and pH 9, and recover after treatment with 100 mM Tris buffer at pH 3 and pH 12 as well as gastric fluids.[29,30] In addition, *Listeria* spp. tolerate a range of osmolarity and pressure, and are able to grow in the presence of 10% NaCl and withstand from a 20 h treatment with 7 M NaCl.[29,30]

Given their ability to endure external stresses including extreme pH, temperature, and salt ranges, *Listeria* species are abundantly present in all the environments (e.g., soil, water, effluents, and foods). Although *Listeria* spp. are rarely isolated from unpolluted seawater, groundwater, and spring water, where the availability of nutrients is limited—with an average presence of about 1%—they are commonly found in river and bay water, sewage, and industrial and farming effluents (with an average presence of 20% and higher). The incidence of *Listeria* spp. in waters appears to increase with animal and human activity, as polluted water and sewage/sludge contain various nutrients essential for optimal growth of *Listeria*.[26]

Listeria spp. are also routinely isolated in soil and plants (with an average presence of 20%) where they exist as saprophytes and scavengers. *L. monocytogenes* (especially serovars 1/2a, 1/2b, and 1/2c, which happen to be commonly associated with human foodborne listeriosis) has the ability to adhere to the surfaces of food processing benches, machineries, and floors, in which it develops into biofilm matrix with increased resistance to detergents and sterilizers that are commonly used to clean the surfaces of food processing facilities.

Frequently, *Listeria* spp. are isolated from fish, shellfish, and other seafoods caught in waters with farms and human settlements.[31] A relatively high prevalence (often in the range of 20%–65%) of *L. monocytogenes* and *Listeria* spp. has been documented in meat products (e.g., chicken, pork, and beef), which provide nutritional media for *Listeria* growth and maintenance. On the other hand, the presence of *Listeria* spp. and *L. monocytogenes* in raw bulk tank milk, dairy milk, and vegetables is generally low (often <5%). Considering the tendency toward increased consumption of processed and ready to eat (RTE) food products in recent decades, it is no surprise that human listeriosis is reported with high regularity.

Between 15% and 20% of wildlife and domestic animals are found to be infected with *Listeria* spp. (including pathogenic *L. monocytogenes* and *L. ivanovii*) via ingestion of *Listeria*-contaminated feed (e.g., silage and grasses) and water. At least 37 mammalian species and 17 bird species have been shown to harbor *Listeria* bacteria.[32]

The prevalence of *Listeria* spp. in asymptomatic healthy adults ranges from 1% to 5%. The incidence of disease due to *L. monocytogenes* is at about 3.4%–7.7% cases of listeriosis per million as reported in the United States and France. However, the incidence of *L. monocytogenes* infections is notably higher among pregnant women (accounting for about 17%–24% of clinical cases of listeriosis) and the elderly (50% of the cases involving people aged >70 years), as well as individuals with underlying diseases (e.g., cancer, organ transplantation, AIDS, chronic hepatic disorder, and diabetes) who are taking medications (e.g., systemic steroids or cyclosporine) or receiving radiation treatment or chemotherapy.

Epidemic listeriosis is characterized by the appearance of clusters of suspected or confirmed cases within a specified time frame and in a relatively large number. Sporadic listeriosis tends to occur in an unspecified time frame involving a smaller number of patients, where dietary risk factors (consumption of uncooked or nonreheated hot dogs or undercooked chicken, soft cheeses or delicatessen foods, unpasteurized milk or eating pate) may be responsible.

26.1.3 Clinical Features and Pathogenesis

Clinical Features. *L. monocytogenes* (originally named *Bacterium monocytogenes*) was first identified by E.G.D. Murray in 1926 in the United Kingdom as a cause of infection with monocytosis in laboratory rodents. This bacterium has been long regarded largely as an animal pathogen affecting ruminants, pigs, dogs, cats, and so forth, with the infected animals showing encephalitis signs (e.g., uncoordinated posture and circular movement, the so-called circling disease).

L. monocytogenes emerged as a significant foodborne pathogen in the late 1970s and early 1980s, when several large outbreaks of food-related human listeriosis cases with significant morbidity and mortality occurred. Listeriosis in humans comes in two forms, noninvasive gastrointestinal listeriosis and invasive listeriosis. The noninvasive listeriosis tends to occur in immunocompetent individuals, in whom a typical febrile gastroenteritis takes place with fever, watery diarrhea, nausea, headache, and joint and muscle pain being the most common symptoms. Invasive listeriosis is often seen in immunocompromised adults, such as the elderly, patients receiving immunosuppressive agents, pregnant women, and infants, with clinical manifestations ranging from sepsis, meningitis, rhombencephalitis, perinatal infections, abortions, and occasional death.[1] The mortality rates of invasive listeriosis may approach 30%, which far exceeds those caused by other common foodborne pathogens such as *Salmonella enteritidis* (with a mortality of 0.38%) and *Campylobacter* species (0.02%–0.1%).[33,34]

Thus, the severity and prognosis of human listeriosis are very much dependent on the immune status of individuals. In immunocompetent individuals, the invading *L. monocytogenes* bacteria are valiantly fought off and eliminated by host's cell-mediated immune network, and the disease symptoms often disappear with a short period (a matter of days). However, in immunocompromised individuals such as pregnant women, neonates, the elderly, and individuals under chemotherapy that suppresses immune functions, *L. monocytogenes* has a propensity to cause especially severe problems, given *L. monocytogenes*' ability to cross the host's intestinal, blood-brain, and fetoplacental barriers, affecting the uterus at pregnancy, the central nervous system, or the bloodstream.

In early-onset fetomaternal/neonatal listeriosis, *L. monocytogenes* enters to the bloodstream of pregnant women (who may be asymptomatic), colonizes the placenta, and invades the fetus, leading to abortion within the last third of the gestation period, with a stillborn fetus or the birth of a baby showing a severe, most often fatal septicemic syndrome known as "granulomatosis infantiseptica" (with physical retardation as a possible outcome). In the late-onset perinatal listeriosis resulting either from low-level transplacental infection or horizontal transmission at neonatology wards, newborns between weeks one and eight postpartum often show a febrile syndrome associated with meningitis, although gastroenteritis (nausea, vomiting, and diarrhea) and pneumonia can also be observed. The mortality of late-onset listeriosis is generally lower than that in early-onset listeriosis.

Pathogenesis. In a typical mode of infection, *L. monocytogenes* enters the host via contaminated foods. In the stomach, *L. monocytogenes* has to overcome a range of nonspecific host defense mechanisms such as oxygen tension, acidity (pH 2.0), the sterilizing effects of bile salts, antimicrobial peptides, proteolytic enzymes (largely through the actions of several stress-response genes, *opuCA*, *lmo1421*, and *bsh*), and related proteins. Afterwards, *L. monocytogenes* crosses the intestinal epithelium (via enterocytes and/or the Peyer's patches) with the assistance of a family of surface proteins called internalins. The most notable internalins are InlA and InlB. Whereas InlA (an 88 kDa protein encoded by *inlA*) interacts with E-cadherin to mediate *L. monocytogenes* entry into epithelial cells, InlB (a 65 kDa protein encoded by InlB) recognizes C1q-R (or Met) to facilitate *L. monocytogenes* entry into a much broader range of host-cell types including hepatocytes, fibroblasts, and epithelioid cells. Besides InlA and InlB, other internalins (e.g., InlC and InlJ) also appear to be involved in the later (i.e., postintestinal) stages of *L. monocytogenes* infection.[35,36] In fact, by binding to a carboxy-terminal SH3 domain in Tuba, which normally engages the human actin regulatory protein N-WASP, InlC promotes protrusion formation by inhibiting Tuba and N-WASP activity, probably by impairing binding of N-WASP to the Tuba SH3 domain. Thus, InlC enhances bacterial dissemination by relieving cortical tension, thereby enhancing the ability of motile bacteria to deform the plasma membrane into protrusions.[36] *L. monocytogenes* also passively enters the host's monocytes, macrophages, or polymorphonuclear leukocytes via phagocytosis. *Listeria* induces macrophage phagocytic uptake by displaying D-galactose in its teichoic acids that bind to macrophage's polysaccharide receptors. This internalization process protects *L. monocytogenes* from host immune surveillance functions.

Following its uptake by host cells, *L. monocytogenes* is primarily located in single-membraned vacuoles (phagosomes). Two virulence-associated molecules are responsible for lysis of the primary single-membraned vacuoles and subsequent escape by *L. monocytogenes*, listeriolysin O (LLO) and phosphatidylinositol-phospholipase C (PI-PLC or PlcA). LLO (a 529 amino acids, 58 kDa protein encoded by *hly*) is a pore-forming, thiol-activated toxin that is essential for *L. monocytogenes* virulence. PI-PLC (a 33 kDa protein encoded by *plcA*), acting in synergy with phosphatidylcholine-phospholipase C (PC-PLC, or PlcB, a 29 kDa protein encoded by *plcB*), aids LLO in lysing the primary vacuoles.[7]

After lysis of the primary single-membraned vacuoles, *L. monocytogenes* is released to the cytosol, where it starts multiplying intracellularly shortly afterwards (with generation time of 40–60 min, which is comparable to those observed in rich broth culture). The host-cell cytoplasm hence allows listerial growth with high efficiency. The intracellular mobility and cell-to-cell spread of *L. monocytogenes* require another surface protein, *Act*A (a 67 kDa protein encoded by *actA*), which is cotranscribed with PC-PLC (*Plc*B) and mediates the formation of polarized actin tails that propel the bacteria toward the cytoplasmic membrane. At the membrane, bacteria become enveloped in filopodium-like structures that are recognized and engulfed by adjacent cells, resulting in the formation of secondary double-membraned vacuoles. A successful lysis of the secondary double-membraned vacuoles signals the beginning of a new infection cycle, which is dependent on PC-PLC (PlcB) upon activation by Mpl (a 60 kDa metalloprotease encoded by *mpl*).[1]

The virulence-associated proteins PlcA (PI-PLC), LLO, Mpl, ActA, and PlcB (PC-PLC) are encoded by their underlying genes located in a 9.6 kb virulence gene cluster,[37] which is primarily regulated by a pleiotropic virulence regulator, PrfA (a 27 kDa protein encoded by *prfA*). The *prfA* gene is situated immediately downstream of, and sometimes cotranscribed with, the *plcA* gene. *prfA* possesses two separate transcription binding sites in its promoter region, with prfAp1 being the usual prfA box and prfAp2 strongly resembling the consensus sequence from the −35 and −10 regions of the alternative sigma factor σB-dependent promoters. Thus, *prfA* gene is also partially regulated by σB via prfAp2, in addition to its self-regulation. Positioned elsewhere in the genome, the *inlA* and *inlB* genes possess a transcription binding site similar to that recognized by PrfA, they may also be partially regulated by PrfA.[1]

In addition to the aforementioned virulence-associated genes and proteins, many other proteins and genes are involved in *L. monocytogenes* adaptation to host environments (e.g., GAD, BSH, BilZ, BtlB, OpuC, and OppA), adhesion to host cells (e.g., Ami, DltA, and FbpA), internalization (e.g., SrtA, SrtB, Auto, Vip, and LpeA), intracellular survival and replication (e.g., Lsp, SipX, SipZ, Hpt, LplA, Fri, HupC, MnSOD, RelA, and Lgt), cell-to-cell spread (SecA2), evasion of host immune attack (PgdA, p60, and MprF), and regulation of virulence gene expression (SigmaB, MogR, CtsR, Per, Fur, LisRK, AgrA, VirR, DegU, StP, and Htq).[1,35,38–40]

26.1.4 Diagnosis

26.1.4.1 Phenotypic Techniques

In Vitro Culture. As *Listeria* spp. are indistinguishable from other gram-positive bacteria microscopically, and are rarely present in sufficient numbers in clinical samples for direct detection by serological and molecular procedures, in vitro culture enrichment and isolation are often undertaken.

For isolation of *Listeria* spp. from nonsterile samples such as feces, a range of selective broths and agars are available. PALCAM- and OXFORD-selective agars are recommended by ISO 11290–1 for primary and secondary plating media, respectively, for isolation of all *Listeria* spp. Chromogenic medium ALOA is also useful for *Listeria* isolation. This agar medium incorporates 5-bromo-4-chloro-3-indolyl-β-D-glucopyranosid as an enzyme substrate for β-D-glucosidase, which is secreted by all *Listeria* spp., and L-α-phosphatidylinositol as a second substrate for phospholipase C, which is a virulence factor produced by pathogenic *L. monocytogenes* and *L. ivanovii,* leading to the formation of typical blue-turquoise colonies, with an outer white zone of precipitation. Application of selective chromogenic *L. monocytogenes* plating media facilitates the presumptive detection of *L. monocytogenes* within 24 or 48 h. Subsequent confirmation of *Listeria* identity is then achieved with biochemical, immunological, and molecular tests.

Biochemical Tests. *Listeria* species are catalase positive, indole and oxidase negative, and are able to hydrolyse aesculin, but not urea.[41] However, *Listeria* species show significant variations in their ability to hemolyse horse or sheep red blood cells, and in their ability to produce acid from L-rhamnose, D-xylose, and α-methyl-D-mannoside. In particular, *L. monocytogenes* (producing listeriolysin or LLO), *L. ivanovii* (producing ivanolysin or ILO), and *L. seeligeri* (producing seeligeriolysin or SLO) are known to be hemolytic, while *L. innocua, L. welshimeri,* and *L. grayi* are generally not. This difference in hemolytic activity is utilized as one of the criteria for discrimination among *Listeria* species. Furthermore, *L. ivanovii* is differentiated biochemically from *L. monocytogenes* and other *Listeria* species by its production of a wide, clear, or double zone of hemolysis on sheep or horse blood agar, a positive Christie-Atkins-Munch-Petersen (CAMP) reaction with *Rhodococcus equi* but not with hemolytic *Staphylococcus aureus*, and fermentation of D-xylose but not L-rhamnose. *L. ivanovii* is also distinguished from *L. monocytogenes* by its strong lecithinase reaction with or without charcoal in the medium, in comparison to *L. monocytogenes*, which requires charcoal for its lecithinase reaction.[42] Similarly, *L. innocua* is identified from *L. monocytogenes* on the basis of its negative CAMP reaction and its failure to cause β-hemolysis or to show PI-PLC activity on chromogenic media. *L. welshimeri* is differentiated from other *Listeria* species by its negative β-hemolysis and CAMP reactions, and by its acid production from D-xylose and α-methyl-D-mannoside. Commercial kits (e.g., API-*Listeria*, *Listeria* ID, Microbact 12L, and MICRO-ID *Listeria*) incorporating a range of biochemical tests are often employed these days for characterization of *Listeria* spp.

Immunological Methods. Immunological (serological) methods (e.g., agglutination and enzyme-based immunoassay or EIA) employ polyclonal and monoclonal antibodies to *Listeria* antigens provide a straightforward approach for species- and serovar-specific identification of *L. monocytogenes* and other *Listeria* spp. A number of commercial serological assays are available for listerial detection, including Assurance *Listeria* EIA, Diatheva *Listeria* ELISA, Dynabeads anti-*Listeria*, EIA Foss *Listeria*, ListerTest *Listeria* Rapid Test, *Listeria*-Tek, TECRA Listeria Visual Immuno Assay, Transia Plate *Listeria*, Vidas LIS, Vidas LMO, and VIP for *Listeria*, as well as Denken serotyping kit.

Serotyping offers a useful means to track *L. monocytogenes* isolates involved in listeriosis outbreaks.[43] However, considering that both *L. monocytogenes* and *L. seeligeri* containing serovars 1/2a, 1/2b, 3b, 4a, 4b, 4c, and 6b, the inability of serotyping methods to correlate serovars directly with species identities limits their potential for clinical application. Furthermore, due to antigen sharing among various *L. monocytogenes* serovars, with 1/2a and 3c both possessing H antigens A and B; 4a–d, 1/2b, and 3b all having H antigens A, B, and C; 1/2c and 3a both harboring H antigens B and D; and multiple, common O antigens being present in different serotypes (Table 26.1), it can be a challenge to conclusively determine the serovar of some *L. monocytogenes* strains, especially among serovars 4a, 4b, and 4c, which demonstrate contrasting pathogenicity.[11] The high cost of acquiring subtype-specific antisera also makes the use of serotyping methods unattractive.

Besides their use for identification purpose, immunological reactions can be exploited for specific adsorption and separation of *Listeria* bacteria. For example, immunomagnetic separation (IMS) based on antibody-coated magnetic beads is often utilized as a pretreatment and/or preconcentration step to capture and extract the targeted pathogen from the clinical, food, and environmental specimens. The captured *Listeria* bacteria are then detected by other methods such as optical, magnetic force microscopy and magneto resistance (bead array counter).

Other In Vitro Techniques. Phage typing exploits naturally occurring bacteriophages to lyse closely related *Listeria* independently of the bacterial species and serovar identities. Through examination of bacteriophage induced, host-specific lysis of *Listeria* on agar plates, *Listeria* strains can be separated into distinct phage groups and phagovars, which are useful for tracking the origin and course of listeriosis outbreaks. Multilocus enzyme electrophoresis (MLEE) is a protein-based, isoenzyme typing method that correlates specific protein band patterns with genotypes. Esterase typing targets a class of heat-stable enzymes (esterases) that hydrolyse carboxylic acid esters to measure the esterase activity from cell extracts of individual *L. monocytogenes* strains on starch gels following electrophoresis.

Recently, a classification system based on Fourier transform infrared (FTIR) spectroscopy combined with artificial neural network analysis was used to differentiate 12 serovars of *L. monocytogenes*. Out of 166 *L. monocytogenes* strains tested, the O antigens (serogroup) of 164 (98.8%) and the H antigens of 152 (91.6%) strains were identified correctly. Importantly, 40 out of 41 potentially epidemic serovar 4b strains were distinguished unambiguously. FTIR analysis thus has potential for the rapid, simultaneous identification of both *Listeria* species and serovar by simply measuring a whole-cell infrared spectrum.[44]

In another development, matrix-assisted laser desorption ionization-time of flight mass spectrometry (MALDI-TOF MS) showed promise for rapid identification of *Listeria* species, typing, and clonal lineages. The method involves a short inactivation/extraction and drying of cell material followed by MALDI-TOF MS analysis of the chemistry of major proteins, yielding profile spectra consisting of a series of peaks, a characteristic "fingerprint" mainly derived from ribosomal proteins. As specimen preparation from plate or liquid culture takes only a few minutes and spectrum is available within 1 minute, MALDI-TOF MS fingerprinting may have potential for improving *Listeria* identification and subtyping.[45]

In Vivo Assays. Because *L. monocytogenes* serovars demonstrate varied infectivity, pathogenicity, and virulence in mammalian hosts, in vivo assays based on animal models such as rodents offer a valuable way to confirm their virulence potential as well as their identity, considering that *L. innocua*, *L. seeligeri*, *L. welshimeri*, and *L. grayi* are nonpathogenic. While *L. monocytogenes* epidemic strains appear to be more invasive than environmental strains via intragastric route,[46] all *L. monocytogenes* serovars but 4a are capable of causing mouse mortality via intraperitoneal inoculation (Table 26.4).[9,10,12,17] In addition, in vivo mouse models represent an indispensable tool for the studies of *L. monocytogenes* pathogenesis and virulence mechanisms that are not easily obtained with in vitro techniques.

26.1.4.2 Genotypic Techniques

(i) Nonamplified Methods The early approaches for genotypic identification and characterization of *Listeria* spp. are based largely on nonamplified methods. For example, the relative proportion of guanine and cytosine (or G + C content, usually expressed as a percentage) among *Listeria* spp. (ranging from 36% to 39%) and other bacteria (e.g., *Esherichia coli* at about 50%) is measured with a spectrophotometry after thermal denaturation (melting) of DNA double helix. The reassociation value of duplex DNA formed between a reference strain and test strain is determined by DNA–DNA hybridization to confirm species identity. A stretch of labeled nucleotides may be used as probe in dot blot, Southern blot, or fluorescence in situ hybridization (FISH) procedure for species- and type-specific determination. One of the most widely applied nonamplified methods for typing is pulse-field gel electrophoresis (PFGE), which differentiates *L. monocytogenes* strains on the basis of characteristic band patterns (consisting of 8 and 25 large DNA bands of 40–600 kb in size) upon enzymatic digestion of genomic DNA and electrophoresis.[47] Combining PFGE with a ribosomal DNA probe (e.g., *E. coli rrnB* rRNA operon), ribotyping simplifies the typing of *L. monocytogenes* strains.[2]

(ii) Amplified Methods Following the introduction of polymerase chain reaction (PCR) in the later 1980s, and other amplification procedures (e.g., NASBA and LAMP) in subsequent years, rapid, sensitive and specific identification and typing of *Listeria* have become a reality. While agarose gel electrophoresis has been routinely used to detect the amplified nucleic acid products, refinements in instrument automation and probe technologies have enabled real-time monitoring of amplification process and instant result availability.

Species Identification. Nonspecific, shared, and specific gene regions may be targeted for genotypic identification of *Listeria* spp. Examples of nonspecific gene regions include random primer sites and repetitive elements (e.g., repetitive extragenic palindromes, or REP; enterobacterial repetitive intergenic consensus sequences, or ERIC; variable-number tandem repeats, or VNTR).[48–50] The shared gene regions useful for *Listeria* identification comprise ribosomal RNA genes (i.e., small-subunit 16S and large subunit 23S rRNA genes as well as their intergenic region), housekeeping genes (*groESL*, *rpoB*, *recA*, *gyrB*, and *prs*), genes encoding for invasion-associated protein (*iap*), flagellin A (*flaA*), and fibronectin-binding protein (*fbp*). The species-specific genes include *hly*, *plcA*, *plcB*, *actA*, *mpl*, *inlA*, *inlB*, delayed-type hypersensitivity gene, aminopeptide gene and putative transcriptional regulator gene *lmo0733* among many others.[2,9,12,51,52] The development of PCR assays with specificity for other *Listeria* spp. has contributed to the improved discrimination of *L. monocytogenes* from nonmonocytogenes species.[26,53–65]

The amplification formats have also evolved from single round to nested, from singleplex to multiplex, and the detection formats from gel electrophoresis to real time. Due to its convenience and reduced possibility of carryover contamination, real-time PCR has become increasingly adopted in clinical laboratories for identification of *L. monocytogenes*.[52,66–82] Real-time PCR that targets a number of genes have been reported, including 16S rRNA gene,[83,84] the intergenic region spacer (IGS) between the 16S and 23S rRNA genes,[79] *prfA* gene,[87] *iap* gene,[88] and *ssrA* gene.[89]

Typing. Although random primer sites, REP, ERIC, and VNTR can be targeted for typing *L. monocytogenes*, rRNA and other specific genes tend to give more accurate genotype details.[90,91] Amplified fragment length polymorphism (AFLP) utilizes adaptors (which are ligated first to restriction enzyme-digested DNA) to generate distinct amplification products that are detectable after electrophoretic separation. PCR-restriction fragment length polymorphism (RFLP) undertakes PCR amplification of one or more *L. monocytogenes* housekeeping or virulence-associated genes (e.g., *hly*, *act*A, and *inl*A) followed by digestion with selected restriction enzymes (e.g., *Hha*I, *Sac*I, or *Hin*fI) and separation by agarose gel electrophoresis.[50] Multilocus sequence typing (MLST) can also be used for genetic subtyping of *L. monocytogenes*. Parisi et al.[92] compared AFLP with MLST for typing 103 *L. monocytogenes* isolates from food and environmental sources, and showed that MLST was advantageous over AFLP in terms of the unambiguity of sequence data despite AFLP's cost effectiveness and easy performance.

Use of group-specific primers in multiplex PCR formats, *L. monocytogenes* serovar groups have been readily distinguished.[93–96] Recently, Kérouanton et al.[97] designed two multiplex PCR assays to cluster *L. monocytogenes* strains into five molecular serogroups: IIa, IIb, IIc, IVa, IVb in agreement with the most commonly encountered serotypes. The first multiplex PCR recognizes *L. monocytogenes* serotypes 1/2a, 1/2c, 1/2b, and 4b (showing 90.6%, 97.8%, and 100% concordance between conventional and molecular methods), together with the *prf*A gene primers for *L. monocytogenes* species confirmation. The second multiplex PCR, incorporating the *fla*A gene primers (specific for 1/2a and 3a strains) and *prs* gene primers (specific for *Listeria* genus), resolves a small number of IIa and IIc molecular serogroup strains (consisting of serotypes atypical 1/2a, 3a, and 1/2c strains) that give equivocal results in the first multiplex PCR, leading to a total agreement between molecular and conventional serotyping methods. Similarly, Nightingale et al.[98] combined a previously reported multiplex PCR with sigB allelic typing to classify the four major serotypes (i.e., 1/2a, 1/2b, 1/2c, and 4b) into unique genetic subgroups. This combined approach also facilitated differentiation of lineage I serotype 4b isolates from the genetically distinct serotype 4b isolates classified into lineage III.

Additionally, by using primers from *inlA* for species-specific recognition, and those from *inlJ* (or *lmo2821*) and *inlC* for virulence determination in a multiplex PCR, it is possible to differentiate *L. monocytogenes* naturally avirulent serovar 4a strains from other serovars with potential to cause mouse mortality via intraperitoneal route.[12,61] Use of primers targeting *L. monocytogenes* serovar 4b strains allows specific detection of most relevant clinical clones. Recently, Miya et al.[49] employed multilocus variable-number tandem repeat analysis (MLVA) to characterize 60 serotype 4b isolates from various sources. Compared to automated *Eco*RI ribotyping, single enzyme pulsed-field gel electrophoresis (PFGE) with *Apa*I, and a multilocus sequence typing (MLST) scheme targeting six virulence and virulence-associated genes, MLVA shows a higher discriminatory power and is easy to interpret, although MLVA, PFGE, and MLST all consistently grouped the same isolates into three major clusters, corresponding to the three major *L. monocytogenes* epidemic clones (i.e., ECI, ECIa, and ECII).

26.2 METHODS

26.2.1 Sample Preparation

Bacterial Isolation. As Gram-staining and microscopic examination of CSF or meconium permit only a presumptive diagnosis, it is important to isolate the pathogen as a prerequisite for confirmation. A positive culture from a normally sterile site is indicative of listeriosis. *L. monocytogenes* is routinely cultured from clinical specimens such as blood, cerebrospinal fluid (CSF), amniotic fluid, placenta, meconium, lochia, gastric washings, or ear swabs from newborns by directly plating the material onto blood agar plates and incubating overnight at 35°C in an ambient atmosphere. Stool specimens (other than meconium) needs be selectively enriched before being plated on selective agar media. For any pregnant woman presenting with fever, especially if the fever is accompanied by flu-like or gastrointestinal symptoms, blood cultures for *L. monocytogenes* should be considered.

To selectively enrich *L. monocytogenes* bacteria, typically a 25 g stool is mixed in 225 mL of culture medium (e.g., half Fraser broth, which contains half a volume of acriflavine and nalidixic acid), and incubated at 30°C for 24 h. Afterwards, a 0.1 mL of the primary enriched culture is added to 10 mL of full strength Fraser broth (which contains one volume of acriflavine and nalidixic acid) and incubated at 37°C for 24 h. A loopful of the secondary enrichment culture is then streaked onto BCM® *Listeria monocytogenes* plating medium at 37°C for 24 h for production of turguoise colonies.

DNA Extraction. *Listeria* DNA can be prepared by boiling if it is intended for PCR only. Namely, suspend one to two colonies in 100 μL of 50 mM NaOH, vortex, heat at 100°C for 10 min, add 10 μL of 1 M Tris pH 8.0, centrifuge at 13,000 × *g* for 1 min, and use 2–5 μL in 25 μL PCR mixture. Alternatively, *Listeria* DNA can be extracted by using commercial DNA purification kits or phenol-chloroform.[99–101]

26.2.2 Detection Procedures

26.2.2.1 Multiplex PCR Identification of *Listeria* Species

Principle. Based on the six *Listeria* species-specific genes described previously,[59–63] Huang et al.[25] designed two multiplex PCR, with one focusing on *L. welshimeri*, *L. monocytogenes*, and *L. seeligeri*; and another on *L. ivanovii*, *L. grayi*, and *L. innocua* (Table 26.6).

TABLE 26.6
Identities and Sequences of *Listeria* Primers for Species-Specific Identification

Target Gene	Specificity	Primer Sequences (5′→3′)[a]	PCR Product (bp)
		Pool 1	
lwe7-571 (encoding phosphotransferase system enzyme IIBC)	*L. welshimeri*	TCCCACCATTGGTGCTACTCA TTGGCGTACCAAAGAAATACG	608
Lmo0733 (encoding transcriptional regulator)	All *L. monocytogenes* lineages but lineage subgroup IIIB	CGCAAGAAGAAATTGCCATC TCCGCGTTAGAAAAATTCCA	453
lse24-315 (encoding internalin)	*L. seeligeri*	CGCGACGGCTAAAGTACTAA ATTGCTCGCTTTGAAGTCGT	375
		Pool 2	
liv22-228 (encoding N-acetylmuramidase)	*L. ivanovii*	$(N)_{15}$CGAATTCCTTATTCACTTGAGC $(N)_{15}$GGTGCTGCGAACTTAACTCA	493
lgr20-246 (encoding oxidoreductase)	*L. grayi*	CTGCACGATCAAGGTCAATC CGTATTGCGCACCAGTGATA	420
iap (encoding invasion-associated protein)	*L. innocua*	TTGCTACTGAAGAAAAAGCA TCTGTTTTTGCTTCTGTAGC	250

[a] N, random nucleotide.

Procedure

1. Prepare the PCR mixture (25 μL) containing 0.5 U *Taq* Gold DNA polymerase (Applied Biosystems), 1× PCR buffer (10 mM Tirs pH 8.3, 50 mM KCl, 2.5 mM $MgCl_2$), 200 μM dNTPs, 25 pmol each primer (*L. welshimeri, L. monocytogenes*, and *L. seeligeri* in pool 1; *L. ivanovii, L. grayi*, and *L. innocua* in pool 2), 2 μL DNA (prepared by boiling method), and distilled water (to 25 μL). Include a tube containing PCR mixture without DNA template as a negative control in each run.
2. Run the following cycling program in a GeneAmp PCR System 9700 (Perkin Elmer): 1 cycle of 95°C for 9 min; 30 cycles of: 94°C for 30 s, 60°C for 30 s, and 72°C for 1 min; and 1 cycle of 72°C for 7 min.
3. After completion of PCR cycles, add 3 μL of 6×DNA loading buffer into each tube; load 15 μL of PCR product on 2.0% agarose gel containing 0.5 μg/mL ethidium bromide. Include a DNA molecular marker in a spare lane. Run at 5 V per cm gel for about 30 min. Visualize under UV light box and photograph.

Note: In pool 1, *L. welshimeri* forms a band of 608 bp, *L. monocytogenes* a band of 453 bp, and *L. seeligeri* a band of 375 bp; in pool 2, *L. ivanovii* shows a band of 493 bp, *L. grayi* a band of 420 bp, and *L. innocua* a band of 250 bp. Nevertheless, it has been demonstrated recently that the *lmo0733* gene encoding transcriptional regulator appears to be absent in *L. monocytogenes* lineage subgroup IIIB strains, which belong to atypical rhamnose-negative virulent non4a and non4c strains, some of which may be related to serovar 7 and have been designated as novel evolutionary lineage IV in a more recent study.[15] Therefore, if it is desirable to detect all *L. monocytogenes* serovars and strains, primers targeting *inl*A gene may be employed.

26.2.2.2 Multiplex PCR Differentiation of *L. monocytogenes* Serogroups

Principle. Doumith et al.[94] developed a multiplex PCR that incorporates *L. monocytogenes lmo0737* gene primers for recognition of serovars 1/2a, 1/2c, 3a, and 3c; *lmo1118* gene primers for detection of serovars 1/2c and 3c; ORF2819 primers for serovars 1/2b, 3b, 4b, 4d, and 4e; ORF2110 primers for serovars 4b, 4d, and 4e; and *prs* primers as an internal amplification control covering all *L. monocytogenes* serovars (Table 26.7).

Procedure

1. Prepare PCR mixture (100 μL) containing 2 U of *Taq* DNA polymerase (Roche, Boehringer), 0.2 mM deoxynucleoside triphosphates, 1× PCR buffer (50 mM Tris-HCl-10 mM KCl-50 mM $(NH_4)_2SO_4$-2 mM $MgCl_2$, pH 8.3), and the five primer sets (1 μM each for *lmo0737*, ORF2819, and ORF2110); 1.5 μM each for *lmo1118*; and 0.2 μM each for *prs* (Table 26.7).
2. Perform PCR amplification in a thermocycler for 1 cycle of 94°C for 3 min; 35 cycles of 94°C for 25 s, 53°C for 1 min 10 s, and 72°C for 1 min and 10 s; and 1 cycle of 72°C for 7 min.
3. After completion, mix 5 μL of PCR mixture with 3 μL of loading buffer and separate on a 2% agarose gel in a TBE buffer (90 mM Trizma base, 90 mM boric acid, 2 mM EDTA, pH 8.3).
4. Stain the gel with ethidium bromide, and visualize with UV transilluminator.

Note: All *Listeria* species (including serovars 4a and 4c) form a 370 bp fragment with the *prs* gene primers; serovars 1/2a and 3a show a 691 bp fragment with the *lmo0737* primers only; serovars 1/2c and 3c display a 691 bp fragment with the *lmo0737* primers and a 906 bp fragment with the *lmo1118* primers; serovars 1/2b, 3b, and 7 show a 471 bp fragment with the ORF2819 primers; serovars 4b, 4d, and 4e form a 471 bp fragment with the ORF2819 primers and a 597 bp fragment with the ORF2110 primers.

26.2.2.3 Multiplex PCR Determination of *L. monocytogenes* Virulence

Principle. Liu et al.[12] reported a multiplex PCR incorporating *inlA* gene primers for species-specific confirmation

TABLE 26.7
Identities and Sequences of *L. monocytogenes* Primers for Serogrouping

Target Gene	Specificity	Primer Sequences (5′→3′)	PCR Product (bp)
lmo0737 (encoding protein of unknown function)	*L. monocytogenes* serovars 1/2a, 1/2c, 3a, and 3c	AGGGCTTCAAGGACTTACCC ACGATTTCTGCTTGCCATTC	691
lmo1118 (encoding protein of unknown function)	*L. monocytogenes* serovars 1/2c and 3c	AGGGGTCTTAAATCCTGGAA CGGCTTGTTCGGCATACTTA	906
ORF2819 (encoding transcriptional regulator)	*L. monocytogenes* serovars 1/2b, 3b, 4b, 4d, and 4e	AGCAAAATGCCAAAACTCGT CATCACTAAAGCCTCCCATTG	471
ORF2110 (encoding secreted protein)	*L. monocytogenes* serovars 4b, 4d, and 4e	AGTGGACAATTGATTGGTGAA CATCCATCCCTTACTTTGGAC	597
prs (encoding phosphoribosyl pyrophosphate synthetase)	All *Listeria* species	GCTGAAGAGATTGCGAAAGAAG CAAAGAAACCTTGGATTTGCGG	370

(with an 800 bp product), and *inlC* and *inlJ* gene primers for virulence determination of *L. monocytogenes* (with 517 bp and 238 bp bands [Table 26.8]). The inclusion of both the *inlC* and *inlJ* gene primers in the multiplex PCR has an added advantage in that it provides a double verification of *L. monocytogenes* virulence for most cases. The usefulness of this multiplex PCR for virulence determination is supported by the observations that *L. monocytogenes* virulent strains with the potential to cause mouse mortalities via intraperitoneal inoculation of mice are positive with the *inlC* and/or *inlJ* gene primers, while naturally avirulent strains unable to produce mouse mortality are negative with these primers.

Procedure

1. Prepare the PCR mixture (25 μL) containing 0.8 U *Taq* DNA polymerase (Fisher Scientific), 1× PCR buffer, 200 μM dNTPs, 40 pmol each of *inlA* primers, 30 pmol each of *inlC* primers, 20 pmol each of *inlJ* primers, 10 ng (1 μL) of DNA, and distilled water (to 25μL). Include a tube containing PCR mixture without DNA template as a negative control in each run.
2. Run the following cycling program in a GeneAmp PCR System 9700 (Perkin Elmer) or other cycler: 1 cycle of 94°C for 2 min; 30 cycles of: 94°C for 20 s, 55°C for 20 s, and 72°C for 50 s; and 1 cycle of 72°C for 2 min.
3. After completion of PCR cycles, add 3 μL of 6× DNA loading buffer into each tube; separate the PCR products on 1.5% agarose gel containing 0.5 μg/mL ethidium bromide. Include a DNA molecular marker in a spare lane.
4. Visualize under UV light box and photograph using a ChemiImager 5500 documentation system (BSI, Stafford, TX).

Note: The species identity of *L. monocytogenes* strains is confirmed through the formation of an 800 bp band with the *inlA* gene primers, and the virulence potential is determined by the production of 517 bp and/or 238 bp bands with the *inlC* and *inlJ* gene primers in the multiplex PCR. Even though *L. ivanovii* strains also harbor a gene with homology to *L. monocytogenes inlC* and cross-react with *L. monocytogenes inlC* gene primers they can readily excluded by their negative reaction with *L. monocytogenes inlA* gene primers included in the multiplex PCR.

Although the presence of *inlC* and/or *inlJ* genes in a given *L. monocytogenes* strain implies its potential virulence and its ability to cause mouse mortality via the intraperitoneal route, it does not indicate the certainty of the strains harboring these genes to produce disease in humans via the conventional oral ingestion. Considering that only three serotypes (1/2a, 1/2b, and 4b) are isolated from human listeriosis cases, many serotypes (e.g., 1/2c, 3a, 3b, 3c, 4c, 4d, and 4e) that are recognized as having virulence potential by the multiplex PCR may encounter difficulty crossing the host's intestinal or other barriers during infection. Obviously, for a *L. monocytogenes* strain to be fully infective to humans via oral ingestion, it requires involvement of many other known and uncharacterized virulence genes and proteins. Nonetheless, an implication from the multiplex PCR results is that while *L. monocytogenes* serotypes other than 1/2a, 1/2b, and 4b may not cause disease via oral route in immunocompetent individuals, they (except serovar 4a) may be potentially infectious via open wound or blood transfusion, especially in immunosuppressed hosts.

26.2.2.4 Multiplex PCR Determination of *L. monocytogenes* Serotypes 1/2a and 4b and Epidemic Clones I, II, and III

Principle. Chen and Knabel[96] described a multiplex PCR for simultaneous detection of *L. monocytogenes* serotypes 1/2a and 4b, and epidemic clones I, II, and III. Of the four known major epidemic clones of *L. monocytogenes* (ECI, ECII, ECIII, and ECIV), three (ECI, ECII, and ECIV) belong to serotype 4b and one (ECIII) belongs to serotype 1/2a. A total of seven pairs of primers are included in multiplex PCR for detection of *Listeria* genus, *L. monocytogenes* species, and serotypes 1/2a and 4b, and epidemic clones I, II, and III of *L. monocytogenes* (Table 26.9). The ECIV isolates react with Serotype 4b-specific (ORF2110) primers, but not with epidemic clones ECI-specific (LMOf2365_2798) and ECII-specific (LMOh7858_0487.8 to *inlA*) primers in this PCR.

TABLE 26.8
Identities and Sequences of *L. monocytogenes* Internalin Gene Primers for Species- and Virulence Determination

Target Gene	Specificity	Primer Sequences (5′→3′)	Nucleotide Positions	PCR Product (bp)
inlA	All *L. moocytogenes* serovars	ACGAGTAACGGGACAAATGC	94612–94631	800
		CCCGACAGTGGTGCTAGATT	95411–95392	
inlC	All *L. moocytogenes* serovars but 4a and some 4c	AATTCCCACAGGACACAACC	107306–107325	517
		CGGGAATGCAATTTTTCACTA	107822–107802	
inlJ	All *L. moocytogenes* serovars but 4a and some IIIB	TGTAACCCCGCTTACACAGTT	188989–189009	238
		AGCGGCTTGGCAGTCTAATA	189226–189207	

TABLE 26.9
Identity and Sequences of Primers for Multiplex PCR Identification of *L. monocytogenes* Serotypes 1/2a and 4b and Epidemic Clones I, II, and III

Target Gene	Specificity	Primer Sequence (5′–3′)	Amplicon Size (bp)	Final Concentration (nM)
LMOf2365_2798	ECI	(F) AATAGAAATAAGCGGAAGTGT	303	50
		(R) TTATTTCCTGTCGGCTTAG		50
LMOh7858_0487.8 to *inlA*	ECII	(F) ATTATGCCAAGTGGTTACGGA	889	90
		(R) ATCTGTTTGCGAGACCGTGTC		90
LMOF6854_2463.4	ECIII	(F) TTGCTAATTCTGATGCGTTGG	497	30
		(R) GCGCTAGGGAATAGTAAAGG		30
ORF2110	Serotype 4b	(F) AGTGGACAATTGATTGGTGAA	597	100
		(R) CATCCATCCCTTACTTTGGAC		100
lmo0737	Serotype 1/2a	(F) GAGTAATTATGGCGCAACATC	724	30
		(R) CCAATCGCGTGAATATCGG		30
lmo2234	*L. monocytogenes*	(F) TGTCCAGTTCCATTTTTAACT	420	30
		(R) TTGTTGTTCTGCTGTACGA		30
Iap	*Listeria* spp.	(F) ATGAATATGAAAAAAGCAAC	1450–1600	35
		(R) TTATACGCGACCGAAGCCAAC		35

Procedures

1. *L. monocytogenes* isolates are grown overnight on trypticase soy yeast extract agar (Difco Laboratories, Becton Dickinson) and genomic DNA is extracted using an UltraClean Microbial DNA extraction kit (Mo Bio Laboratories) and stored at –20°C before use.
2. PCR mixture (10 μL) is prepared using QIAGEN multiplex PCR kit (QIAGEN) containing 1 μL of DNA 1 and 30–100 nM of primers (Table 26.9).
3. Amplification is carried out in a Mastercycler thermocycler (Eppendorf Scientific), with an initial activation step for 15 min at 95°C; 15 cycles of 1 min at 94°C, 1 min with a touchdown from 55°C to 51°C (3 cycles per temperature), and 1 min at 72°C; 15 cycles of 1 min at 94°C, 1 min at 50°C, and 1 min at 72°C, and 1 final cycle for 8 min at 72°C.
4. The 10-μL reaction mixture is mixed with 3 μL of loading buffer and separated on a 1.5% agarose gel in a 0.5× Tris-boratc-EDTA buffer with a 1-h run time. The PCR product is visualized by ethidium bromide staining.

26.2.2.5 Real-Time PCR Detection of *L. monocytogenes*

Murry et al.[102] described a duplex real-time 5′ nuclease PCR assays for the detection of the *hly* gene of *L. monocytogenes*[85] and a green fluorescent protein (*gfp*) gene of *E. coli* BLR-GFP strain included as internal control (IC) for DNA extraction and amplification (Table 26.10).

Procedure

1. Real-time PCR mixture (25 μL) is made up of 12.5 μL qPCR mastermix (Eurogentec), 5 μL of template DNA, forward and reverse primers (200–300 nM) and probes (50–200 nM) for *hly* and *gpf* genes. All reactions are performed in duplicate for each sample, with positive and nontemplate controls used in every experiment.
2. PCR amplification is carried out in 96-well plates and sealed with optical adhesive covers (Applied Biosystems) on either an ABI Prism™7000 or a 7700 Sequence Detector System machine (Applied Biosystems) with the following cycling parameters: 50°C for 2 min, 95°C for 10 min and then 45 cycles of 95°C for 20 s, and 60°C for 1 min.
3. PCR products are detected by monitoring the increase in fluorescence of the reporter dye at each PCR cycle.

Note: Applied Biosystems software plots the normalized reporter signal, ΔRn, (reporter signal minus background) against the number of amplification cycles and also determines the threshold cycle (Ct) value (i.e., the PCR cycle number at which fluorescence increases above a defined threshold level). Ct values of N40 are taken as no PCR amplification detected and Ct values of b40 are taken as positive for PCR amplification for *L. monocytogenes*. For the *gfp* component of the assay, a mean Ct value of b32.1 is interpreted as no PCR inhibition, a mean Ct value of N40, as total inhibition of PCR, and a mean Ct value of between ≥32.2 and ≤40, as a loss of sensitivity equivalent to or greater than one log reduction.

26.3 CONCLUSION AND FUTURE PERSPECTIVES

Being widely distributed in the environment and highly tolerant of external stresses such as wide temperature, pH, and salt ranges, which often underscore many food processing routines, *L. monocytogenes* has emerged as an important foodborne pathogen in recent decades. Infants, pregnant

TABLE 26.10
Identities and Sequences of Primers and Probes for Real-Time PCR Detection of *L. monocytogenes* and Internal Control

Organism	Target	Primer/Probe	Sequences (5′→3′)
L. monocytogenes	*Hly* gene	hly forward primer (300 nM)	TGC AAG TCC TAA GAC GCC A CAC TGC
		hly reverse primer (300 nM)	ATC TCC GTG GTA TAC TAA
		hly probe (200 nM)[a]	CGA TTT CAT CCG CGT GTT TCT TTT CG
E. coli BLR-GFP	*Gfp* gene	gpfF (200 nM)	CCT GTC CTT TTA CCA GAC AAC CA
		gpfR (200 nM)	GGT CTC TCT TTT CGT TGG GAT CT
		gpf probe (50 nM)[b]	TAC CTG TCC ACA CAA TCT GCC CTT TCG

[a] Probe labeled with FAM™ = carboxyfluorescein (reporter dye) and TAMRA = 6-carboxy-N,N,N′,N′-tetramethylrhodamine (quencher).
[b] Probe labeled with VIC™ (reporter dye) and TAMRA = 6-carboxy-N,N,N′,N′-tetramethylrhodamine (quencher).

women, the elderly, and immunocompromised persons are especially vulnerable to *Listeria* infection due to their impaired immune function. Without accurate diagnosis and prompt antibiotic therapy, an easily treated malaise can turn into a deadly disease in a short time, with the mortality rate approaching 30%.

Although conventional, phenotype-based tests have played essential roles in the laboratory identification and diagnosis of *L. monocytogenes,* they suffer from the shortcomings of being slow and variable at times. The development of in vitro nucleic acid amplification techniques such as PCR has made it possible to achieve sensitive, rapid, and specific determination of *L. monocytogenes*. While gel-based detection format is still widely utilized due to its relative low expense, real-time monitoring of amplification and detection has come increasingly popular, which reduces the potential carryover contamination. It is envisaged that with the instrumentation and probes becoming more affordable in future, real-time PCR will play an even more dominant role in clinical laboratories for diagnosis of listeriosis and other infectious diseases.

Several recent technological developments have shown promise for further improving the identification and typing of *L. monocytogenes*. For example, besides offering a precise way for identification and typing, microarrays present new opportunity to the analysis of *L. monocytogenes* gene activation in host-pathogen relationship. This permits the investigation of the in/activation (down/up-regulation) of *L. monocytogenes* virulence-associated genes during distinct phases of infection in mammalian hosts and under other strenuous external conditions. Biosensors incorporating a biological material (e.g., tissue, microorganisms, organelles, cell receptors, enzymes, antibodies, nucleic acids, natural products, etc.), a biologically derived material (e.g., recombinant antibodies, engineered proteins, aptamers, etc.) or a biomimic (e.g., synthetic catalysts, combinatorial ligands, and imprinted polymers) in a physicochemical transducer or transducing microsystem (e.g., optical, electrochemical, thermometric, piezoelectric, magnetic, or micromechanical) are another powerful analytical device for enhancing the detection and identification of microbial pathogens.[103] Future research along these lines will lead to innovations that further improve the sensitivity, rapidity and specificity of molecular diagnostic assays for *L. monocytogenes.*

REFERENCES

1. Vazquez-Boland, J.A. et al., *Listeria* pathogenesis and molecular virulence determinants, *Clin. Microbiol. Rev.*, 14, 584, 2001.
2. Liu, D., Identification, subtyping and virulence determination of *Listeria monocytogenes,* in important foodborne pathogen, *J. Med. Microbiol.*, 55, 645, 2006.
3. Low, J.C., and Donachie, W., A review of *Listeria monocytogenes* and listeriosis, *Vet. J.*, 153, 9, 1997.
4. Graves, L.M. et al., *Listeria marthii* sp. nov., isolated from the natural environment, Finger Lakes National Forest. *Int. J. Syst. Evol. Microbiol.*, 60, 1280, 2010.
5. Leclercq, A., *Listeria rocourtiae* sp. nov., *Int. J. Syst. Evol. Microbiol.*, 60, 2210, 2010.
6. Seeliger, H.P.R., and Jones, D., *Listeria*. In Sneath, P.H.A. et al. (Editors), *Bergey's Manual of Systematic Bacteriology*, vol. 2, pp. 1235–1245, Williams and Wilkins, Baltimore, 1986.
7. Rasmussen, O.F. et al., *Listeria monocytogenes* exists in at least three evolutionary lines: Evidence from flagellin, invasive associated protein and listeriolysin O genes, *Microbiology*, 141, 2053, 1995.
8. Wiedmann, M. et al., Ribotypes and virulence gene polymorphisms suggest three distinct *Listeria monocytogenes* lineages with differences in pathogenic potential, *Infect. Immun.*, 65, 2707, 1997.
9. Liu, D. et al., Characterization of virulent and avirulent *Listeria monocytogenes* strains by PCR amplification of putative transcriptional regulator and internalin genes, *J. Med. Microbiol.*, 52, 1066, 2003.
10. Liu, D. et al., *Listeria monocytogenes* subgroups IIIA, IIIB and IIIC delineate genetically distinct populations with varied pathogenic potential, *J. Clin. Microbiol.*, 44, 2229, 2006.
11. Liu, D. et al., *Listeria monocytogenes* serotype 4b strains belonging to lineages I and III possess distinct molecular features, *J. Clin. Microbiol.*, 44, 214, 2006.
12. Liu, D. et al., A multiplex PCR for species- and virulence-specific determination of *Listeria monocytogenes*, *J. Microbiol. Methods*, 71, 133, 2007.
13. Doumith, M. et al., New aspects regarding evolution and virulence of *Listeria monocytogenes* revealed by comparative genomics and DNA arrays, *Infect. Immun.*, 72, 1072, 2004.

14. Roberts, A. et al., Genetic and phenotypic characterization of *Listeria monocytogenes* lineage III, *Microbiology*, 152, 685, 2006.
15. Ward, T.J. et al., Multilocus genotyping assays for single nucleotide polymorphism-based subtyping of *Listeria monocytogenes* isolates, *Appl. Environ. Microbiol.*, 74, 7629, 2008.
16. Goulet, V., Surveillance of human listeriosis in France, 2001–2003, *Euro Surveill.*, 11, 79, 2006.
17. Liu, D., *Listeria monocytogenes*: Comparative interpretation of mouse virulence assay, *FEMS Microbiol. Lett.*, 233, 159, 2004.
18. Liu, D. et al., Toward an improved laboratory definition of *Listeria monocytogenes* virulence, *Int. J. Food Microbiol.*, 118, 101, 2007.
19. Glaser, P. et al., Comparative genomics of *Listeria* species, *Science*, 294, 849, 2001.
20. Nelson, K.E. et al., Whole genome comparisons of serotype 4b and 1/2a strains of the food-borne pathogen *Listeria monocytogenes* reveal new insights into the core genome components of this species, *Nucleic Acids Res.*, 32, 2386, 2004.
21. Hain, T. et al., Whole-genome sequence of *Listeria welshimeri* reveals common steps in genome reduction with *Listeria innocua* as compared to *Listeria monocytogenes*, *J. Bacteriol.*, 188, 7405, 2007.
22. Steinweg, C. et al., The complete genome sequence of *L. seeligeri*, a non-pathogenic member of the genus *Listeria*, *J. Bacteriol.*, 190, 1473, 2010.
23. Johnson, J. et al., Natural atypical *Listeria innocua* strains with *Listeria monocytogenes* pathogenicity island 1 genes, *Appl. Environ. Microbiol.*, *70*, 4256, 2004.
24. Volokhov, D.V. et al., The presence of the internalin gene in natural atypically hemolytic *Listeria innocua* strains suggests descent from *L. monocytogenes*, *Appl. Environ. Microbiol.*, *73*, 1928, 2007.
25. Huang, B. et al., Comparison of multiplex PCR with conventional biochemical methods for the identification of *Listeria* species isolates from food and clinical samples in Queensland, Australia, *J. Food Prot.*, 70, 1874, 2007.
26. Liu, D., Lawrence, M.L., and Hitchins, A.D., Molecular characterization of *Listeria monocytogenes* strains harboring *Listeria innocua* putative transcriptional regulator gene lin0464, *J. Rapid Methods Automat. Microbiol.*, 16, 412, 2008.
27. Chen, J. et al., *Listeria monocytogenes* serovar 4a is a possible evolutionary intermediate between *L. monocytogenes* serovars 1/2a and 4b and *L. innocua*, *J. Microbiol. Biotechnol.*, 19, 238, 2009.
28. Schmid, M.W. et al., Evolutionary history of the genus *Listeria* and its virulence genes, *Syst. Appl. Microbiol.*, 28, 1, 2005.
29. Sleator, R.D., Gahan, C.G.M., and Hill, C., A postgenomic appraisal of osmotolerance in *Listeria monocytogenes*, *Appl. Environ. Microbiol.*, 69, 1, 2003.
30. Liu, D. et al., Comparative assessment of acid, alkali and salt tolerance in *Listeria monocytogenes* virulent and avirulent strains, *FEMS Microbiol. Lett.*, 243, 373, 2005.
31. Chen, J. et al., Molecular characteristics and virulence potential of *Listeria monocytogenes* isolates from Chinese food systems, *Food Microbiol.*, 26, 103, 2009.
32. Liu, D., Epidemiology. In Liu, D. (Editor), *Handbook of Listeria monocytogenes*, Chapter 2, pp. 27–59, Taylor and Francis CRC Press, Boca Raton, FL, 2008.
33. Altekruse, S.F., Cohen, M.L., and Swerdlow, D.L., Emerging foodborne diseases, *Emerg. Infect. Dis.*, 3, 285, 1997.
34. Mead, P.S. et al., Food-related illness and death in the United States, *Emerg. Infect. Dis.*, 5, 607, 1999.
35. Sabet, C. et al., LPXTG protein InlJ, a newly identified internalin involved in *Listeria monocytogenes* virulence, *Infect. Immun.*, 73, 6912, 2005.
36. Rajabian, T. et al., The bacterial virulence factor InlC perturbs apical cell junctions and promotes cell-to-cell spread of *Listeria*, *Nat. Cell Biol.*, 11, 1212, 2009.
37. Gouin, E., Mengaud, J., and Cossart, P., The virulence gene cluster of *Listeria monocytogenes* is also present in *Listeria ivanovii*, an animal pathogen, and *Listeria seeligeri*, a non-pathogenic species, *Infect. Immun.*, 62, 3550, 1994.
38. Liu, D., Listeria-based anti-infective vaccine strategies, *Recent Patents Anti-infect. Drug Disc.*, 1, 281, 2006.
39. Liu, D. et al., Characteristics of cell-mediated, anti-listerial immunity induced by a naturally avirulent *Listeria monocytogenes* serotype 4a strain HCC23, *Arch. Microbiol.*, 188, 251, 2007.
40. Camejo, A. et al., In vivo transcriptional profiling of *Listeria monocytogenes* and mutagenesis identify new virulence factors involved in infection, *PLoS Pathog.*, 5, e1000449, 2009.
41. Rocourt, J., and Catimel, B., Biochemical characterization of species in the genus *Listeria*, *Zbl. Bakteriol. Mikrobiol. Hyg. [A]*, 260, 221, 1985.
42. Ermolaeva, S. et al., A simple method for the differentiation of *Listeria monocytogenes* on induction of lecithinase activity by charcoal, *Int. J. Food Microbiol.*, 82, 87, 2003.
43. Palumbo, J.D. et al., Serotyping of *Listeria monocytogenes* by enzyme-linked immunosorbent assay and identification of mixed-serotype cultures by colony immunoblotting, *J. Clin. Microbiol.*, 41, 564, 2003.
44. Rebuffo-Scheer, C.A., Schmitt, J., and Scherer, S., Differentiation of *Listeria monocytogenes* serovars by using artificial neural network analysis of Fourier-transformed infrared spectra, *Appl. Environ. Microbiol.*, 73, 1036, 2007.
45. Barbuddhe, S.B. et al., Rapid identification and typing of *Listeria* species by matrix-assisted laser desorption ionization–time of flight mass spectrometry, *Appl. Environ. Microbiol.*, 74, 5402, 2008.
46. Kim, S.H. et al., Oral inoculation of A/J mice for detection of invasiveness differences between *Listeria monocytogenes* epidemic and environmental strains, *Infect. Immun.*, 72, 4318, 2004.
47. Halpin, J.L. et al., Re-evaluation, optimization, and multilaboratory validation of the PulseNet-standardized pulsed-field gel electrophoresis protocol for *Listeria monocytogenes*, *Foodborne Pathog. Dis.*, 7, 293, 2010.
48. Jersek, B. et al., Typing of *Listeria monocytogenes* strains by repetitive element sequence-based PCR, *J. Clin. Microbiol.*, 37, 103, 1999.
49. Miya, S. et al., Development of a multilocus variable-number of tandem repeat typing method for *Listeria monocytogenes* serotype 4b strains, *Int. J. Food Microbiol.*, 124, 239, 2008.
50. Vaneechoutte, M. et al., Comparison of PCR-based DNA fingerprinting techniques for the identification of *Listeria* species and their use for atypical *Listeria* isolates, *Int. J. Syst. Bacteriol.*, 48, 127, 1998.
51. Paillard, D. et al., Rapid identification of *Listeria* species by using restriction fragment length polymorphism of PCR-amplified 23S rRNA gene fragments, *Appl. Environ. Microbiol.*, 69, 6386, 2003.
52. Navas, J. et al., Evaluation of effects of primary and secondary enrichment for the detection of *Listeria monocytogenes* by real-time PCR in retail ground chicken meat, *Foodborne Pathog. Dis.*, 3, 347, 2006.

53. Bubert, A. et al., Detection and differentiation of *Listeria* spp. by a single reaction based multiplex PCR, *Appl. Environ. Microbiol.*, 65, 4688, 1999.
54. Bubert, A., Kohler, S., and Goebel, W., The homologous and heterologous regions within the *iap* gene allow genus- and species-specific identification of *Listeria* spp. by polymerase chain reaction, *Appl. Environ. Microbiol.*, 58, 2625, 1992.
55. Czajka, J. et al., Differentiation of *Listeria monocytogenes* and *Listeria innocua* by 16S rRNA genes and intraspecies discrimination of *Listeria monocytogenes* strains by random amplified polymorphic DNA polymorphisms, *Appl. Environ. Microbiol.*, 59, 304, 1993.
56. Cocolin, L. et al., Direct identification in food samples of *Listeria* spp. and *Listeria monocytogenes* by molecular methods, *Appl. Environ. Microbiol.*, 68, 6273, 2002.
57. Gilot, P., and Content, J., Specific identification of *Listeria welshimeri* and *Listeria monocytogenes* by PCR assays targeting a gene encoding a fibronectin-binding protein, *J. Clin. Microbiol.*, 40, 698, 2002.
58. Garrec, N. et al., Heteroduplex mobility assay for the identification of *Listeria* spp and *Listeria monocytogenes* strains: Application to characterization of strains from sludge and food samples, *FEMS Immunol. Med. Microbiol.*, 38, 57, 2003.
59. Liu, D. et al., Identification of *Listeria innocua* by PCR targeting a putative transcriptional regulator gene, *FEMS Microbiol. Lett.*, 223, 205, 2003.
60. Liu, D. et al., Use of PCR primers derived from a putative transcriptional regulator gene for species-specific identification of *Listeria monocytogenes*, *Int. J. Food Microbiol.*, 91, 297, 2004.
61. Liu, D. et al., Species-specific PCR determination of *Listeria seeligeri*, *Res. Microbiol.*, 155, 741, 2004.
62. Liu, D. et al., PCR detection of a putative N-acetylmuramidase gene from *Listeria ivanovii* facilitates its rapid identification, *Vet. Microbiol.*, 101, 83, 2004.
63. Liu, D. et al., Identification of a gene encoding a putative phosphotransferase system enzyme IIBC in *Listeria welshimeri* and its application for diagnostic PCR, *Lett. Appl. Microbiol.*, 38, 151, 2004.
64. Liu, D. et al., Isolation and PCR amplification of a species-specific, oxidoreductase-coding gene region in *Listeria grayi*, *Can. J. Microbiol.*, 51, 95, 2005.
65. Liu, D., Lawrence, M.L., and Ainsworth, A.J., A novel PCR assay for *Listeria welshimeri* targeting species-specific transcriptional regulator gene lwe1801, *J. Rapid Methods Automat. Microbiol.*, 16, 154, 2008.
66. Cox, T. et al., Rapid detection of *Listeria monocytogenes* in dairy samples utilizing a PCR-based fluorogenic 50 nuclease assay, *J. Ind. Microbiol. Biotechnol.*, 21, 167, 1998.
67. Schneider, A. et al., Real-time detection of *Listeria monocytogenes* with the LightCycler system, *Biochemica*, 4, 21, 2002.
68. Koo, K., and Jaykus, L.A., Detection of *Listeria monocytogenes* from a model food by fluorescence resonance energy transfer-based PCR with an asymmetric fluorogenic probe set, *Appl. Environ. Microbiol.*, 69, 1082, 2003.
69. Jothikumar, N., Wang, X., and Griffiths, N.W., Real-time multiplex SYBR green I-based PCR assay for simultaneous detection of *Salmonella* serovars and *Listeria monocytogenes*, *J. Food Prot.*, 66, 2141, 2003.
70. Nguyen, L. et al., Detection of *Escherichia coli* O157:H7 and *Listeria monocytogenes* in beef products by real-time polymerase chain reaction, *Foodborne Pathog. Dis.*, 1, 231, 2004.
71. Rodríguez-Lázaro, D. et al., Rapid quantitative detection of *Listeria monocytogenes* in meat products by real-time PCR, *Appl. Environ. Microbiol.*, 70, 6299, 2004.
72. Rodríguez-Lázaro, D. et al., Rapid quantitative detection of *Listeria monocytogenes* in salmon products: evaluation of pre-real-time PCR strategies, *J. Food Prot.*, 68, 1467, 2005.
73. Rodríguez-Lázaro, D., and Hernandez, M., Isolation of *Listeria monocytogenes* DNA from meat products for quantitative detection by real-time PCR, *J. Rapid Methods Aut. Microbiol.*, 14, 395, 2006.
74. Wang, X., Jothikumar, N., and Griffiths, M., Enrichment and DNA extraction protocols for the simultaneous detection of *Salmonella* and *Listeria monocytogenes* in raw sausage meat with multiplex real-time PCR, *J. Food Prot.*, 67, 189, 2004.
75. Rudi, K. et al., Detection of viable and dead *Listeria monocytogenes* on gouda-like cheeses by real-time PCR, *Lett. Appl. Microbiol.*, 40, 301, 2005.
76. Berrada, H. et al., Quantification of *Listeria monocytogenes* in salads by real-time quantitative PCR, *Int. J. Food Microbiol.*, 107, 202, 2006.
77. Pan, Y., and F. Breidt, Jr., Enumeration of viable *Listeria monocytogenes* cells by real-time PCR with propidium monoazide and ethidium monoazide in the presence of dead cells, *Appl. Environ. Microbiol.*, 73, 8028–8031, 2007.
78. Long, F. et al., Development of a quantitative polymerase chain reaction method using a live bacterium as internal control for the detection of *Listeria monocytogenes*, *Diagn. Microbiol. Infect, Dis.*, 62, 374, 2008.
79. Rantsiou, K. et al., Detection, quantification and vitality of *Listeria monocytogenes* in food as determined by quantitative PCR, *Int. J. Food Microbiol.*, 121, 99, 2008.
80. Delibato, E. et al., PCR experion automated electrophoresis system to detect *Listeria monocytogenes* in foods, *J. Sep. Sci.*, 32, 3817, 2009.
81. D'Urso, O.F. et al., A filtration-based real-time PCR method for the quantitative detection of viable *Salmonella enterica* and *Listeria monocytogenes* in food samples, *Food Microbiol.*, 26, 311, 2009.
82. Vanegas, M.C. et al., Detection of *Listeria monocytogenes* in raw whole milk for human consumption in Colombia by real-time PCR, *Food Control*, 20, 430, 2009.
83. De Martinis, E.C., Duvall, R.E., and Hitchins, A.D., Real-time PCR detection of 16S rRNA genes speeds most-probable-number enumeration of foodborne *Listeria monocytogenes*, *J. Food Prot.*, 70, 1650, 2007.
84. Aparecida de Oliveira, M. et al., Quantification of *Listeria monocytogenes* in minimally processed leafy vegetables using a combined method based on enrichment and 16S rRNA real-time PCR, *Food Microbiol.*, 27, 19, 2006.
85. Nogva, H.K. et al., Application of 5′-nuclease PCR for quantitative detection of *Listeria monocytogenes* in pure cultures, water, skim milk, and unpasteurized whole milk. *Appl. Environ. Microbiol.*, 66, 4266, 2000.
86. Singh, J., Batish, V.K., and Grover, S., A molecular beacon-based duplex real-time polymerase chain reaction assay for simultaneous detection of *Escherichia coli* O157:H7 and *Listeria monocytogenes* in milk and milk products, *Foodborne Pathog. Dis.*, 6, 1195, 2009.
87. Rossmanith, P. et al., Detection of *Listeria monocytogenes* in food using a combined enrichment/real-time PCR method targeting the *prfA* gene, *Res. Microbiol.*, 157, 763, 2006.
88. Hein, I. et al., Detection and quantification of the iap gene of *Listeria monocytogenes* and *Listeria innocua* by a new real-time quantitative PCR assay, *Res. Microbiol.*, 152, 37, 2001.
89. O'Grady, J. et al., Rapid real-time PCR detection of *Listeria monocytogenes* in enriched food samples based on the ssrA gene, a novel diagnostic target, *Food Microbiol.*, 25, 75, 2008.

90. den Bakker, H.C., Fortes, E.D., and Wiedmann, M., Multilocus sequence typing of outbreak-associated *Listeria monocytogenes* isolates to identify epidemic clones, *Foodborne Pathog. Dis.*, 7, 257, 2010.
91. Tamburro, M. et al., Typing of *Listeria monocytogenes* strains isolated in Italy by inlA gene characterization and evaluation of a new cost-effective approach to antisera selection for serotyping, *J. Appl. Microbiol.*, 108, 1602, 2010.
92. Parisi, A. et al., Amplified fragment length polymorphism and multi-locus sequence typing for high-resolution genotyping of *Listeria monocytogenes* from foods and the environment, *Food Microbiol.*, 27, 101, 2010.
93. Jinneman, K.C., and Hill, W.E., *Listeria monocytogenes* lineage group classification by MAMA-PCR of the listeriolysin gene, *Curr. Microbiol.*, 43, 129, 2001.
94. Doumith, M. et al., Differentiation of the major *Listeria monocytogenes* serovars by multiplex PCR, *J. Clin. Microbiol.*, 42, 3819, 2004.
95. Doumith, M. et al., Multicenter validation of a multiplex PCR assay for differentiating the major *Listeria monocytogenes* serovars 1/2a, 1/2b, 1/2c and 4b: toward an international standard, *J. Food Prot.*, 68:2648, 2005.
96. Chen, Y., and Knabel, S.J., Multiplex PCR for simultaneous detection of bacteria of the genus *Listeria*, *Listeria monocytogenes*, and major serotypes and epidemic clones of *L. monocytogenes*, *Appl. Environ. Microbiol.*, 73, 6299, 2007.
97. Kérouanton, A. et al., Evaluation of a multiplex PCR assay as an alternative method for *Listeria monocytogenes* serotyping, *J. Microbiol. Methods*, 80, 134, 2010.
98. Nightingale, K. et al., Combined sigB allelic typing and multiplex PCR provide improved discriminatory power and reliability for *Listeria monocytogenes* molecular serotyping, *J. Microbiol. Methods*, 68, 52, 2007.
99. Liu, D., Preparation of *Listeria monocytogenes* specimens for molecular detection and identification, *Int. J. Food Microbiol.*, 122, 229, 2008.
100. Liu, D., Isolation of bacterial DNA from cultures. In Liu, D. (Editor), *Handbook of Nucleic acid Purification*, Chapter 5, pp. 85–105, Taylor and Francis CRC Press, Boca Raton, FL, 2009.
101. Liu, D., and Busse, H.-J., Listeria. In Liu, D. (Editor), *Molecular Detection of Foodborne Pathogens*, pp. 207–220, Taylor and Francis CRC Press, Boca Raton, FL, 2009.
102. Murphy, N.M. et al., Construction and evaluation of a microbiological positive process internal control for PCR-based examination of food samples for *Listeria monocytogenes* and *Salmonella enterica*, *Int. J. Food Microbiol.*, 120, 110, 2007.
103. Lazcka, O., Del Campo, F.J. and Munoz, F.X., Pathogen detection: A perspective of traditional methods and biosensors. *Biosens. Bioelectron.*, 22, 1205, 2007.

27 *Paenibacillus*

Matthew R. Pincus, Nicholas Cassai, Jie Ouyang, and Sharvari Dalal

CONTENTS

27.1 INTRODUCTION

27.1.1 Paenibacilli vs Bacilli

The genus *Bacillus* covers a large group of gram-positive to gram-variable bacilli that characteristically form endospores and generally undergo aerobic metabolism. Because *B. anthracis*, a cause of serious human infection especially from spores and a potential agent in biological terrorist attacks, is a member of this genus, there has been considerable interest in the bacilli. Other members of the bacilli such as *B. cereus* and *B. sublilis* are also known to cause human disease that often results in sepsis. It came to be recognized that the bacilli contained a highly heterogeneous group of bacteria, a substantial number of which were gram-variable bacilli, often presenting as gram-negative, and which were facultatively anaerobic. This distinct group was referred to as "Group 3 bacilli" and included *B. polymyxa* and closely related species.

To determine if the subgroup of bacilli with these unusual characteristics constituted a distinct genus, the sequences of the DNA's encoding the 16S rRNA's of several type strains of the members of this group were amplified with general primers using classical PCR, and the amplified products sequenced.[1] The results showed that the Group 3 members contained strongly homologous sequences giving between 92% and 99.8% sequence identity, whereas there was much less identity to sequences from other members of the genus *Bacillus*.[1] Using distance matrix methods, members of the Group 3 bacilli were found to exist in a distinct cluster that included such organisms as *B. polymyxa*, *B. validus*, *B. alvei*, and *B. larvae*. The six basic clusters that resulted are shown in Figure 2 of Reference 1. Interestingly, organisms such as *Listeria* and *Enterococcus* clustered in Group 5 have some sequence identity to the Group 3 organisms, but contain multiple, distinctly different sequences.[1]

For Group 3 bacilli, a 22-base oligonucleotide called BG3 was found in the 16S rRNA that was highly conserved in the Group 3 bacilli but differed markedly from the homologous segments from the other non-Group 3 bacilli as shown in Table 27.1.[1] Using a probe that consisted of this sequence, the 16S rRNA from the members of the Group 3 bacilli were all found to hybridize with this probe, unlike that of other *Bacillus* group.[1] These findings established that the Group 3 bacilli constitute a distinct genus, which was given the name "Paeni" (from the Latin word "paene," meaning "almost") bacillus. Members of the *Paenibacillus* genus are nonpigmented rods that form ellipsoidal endospores, stain gram variably, and are motile using peritrichous flagellae.

It should be noted that just because two species are closely related by their 16S rRNA sequences it does not imply that they exhibit growth and biochemical characteristics that will also be similar or identical. An example of this is furnished in Table 27.2 for *P. thiaminolyticus* and *P. popilliae*.[2] Column 5 shows that the correlation in rRNA sequences can be as high as 0.97 between two strains, the cutoff for species identity. Yet, these two species differ vastly in growth characteristics (i.e., *P. popilliae* is fastidious requiring nutritional supplementation). The growth of *P. thiaminolyticus* is not fastidious and does not require supplementation. Their biochemical profiles also differ substantially. Conversely, biochemical reactions for organisms that are now classified as Paenibacilli may be more similar between a *Bacillus* species and a *Paenibacillus* species than between two Paenibacilli.[1]

TABLE 27.1
Unique Nucleotide Sequence in Group 3 Bacilli (Paenibacilli)[a]

Group	Nucleotide Sequence																				
Paenibacilli	T	C	G	A	T	A	C	C	C	T	T	G	G	T	G	C	C	G	A	A	G	T
1	T	C	C	A/G[b]	C	X[c]	C	C	T/C	T	T	A	G	T	G	C	T	G	A/C	A	G	T/C
2	T	C	C	G	C	X	C	C	C	T	T	A	G	T	G	C	T	G	C	A	G	C
4	T	C	A	A	T	A	C	C	C	T	T/C	A	G	T	G	C	C	G	C	A	G	C
5	C	A	C/T	A	C	X	C	C	T	T	T	A	G	T	G	C	T	G	C	A	G	X

[a] Adapted from Table 3 of Ref. 1.
[b] A/G means that either A or G can occur at this position.
[c] X means a nucleotide deletion.

TABLE 27.2
Similarity between 16S rRNA Genes of *Paenibacillus* Test and Type Strains[a]

Test Strains		Type Strains					
		P. thiaminolyticus[b]		*P. popilliae*[c]		*P. alvei*[d]	
Strain Number	**Number of Clones**	**AJ320490[e]**	**AB073197**	**AB073198**	**AF071859**	**AJ320491**	**AB073200**
P. thiaminolyticus[f]	12	0.953–0.979	**0.963–0.993**	0.947–**0.971**	0.929–0.949	<0.770	<0.770
P. popilliae[g]	12	0.926–0.942	<0.931–0.957	**0.981–1.000**	0.962–0.981	<0.770	<0.770
P. alvei[h]	10	<0.791	0.795–0.813	0.791–0.809	0.779–0.798	0.930–0.945	**0.940–0.955**

[a] Adapted from Table 4 of ref. 2. Highest similarities are indicated in bold.
[b] DSM7262 (IFO15656).
[c] ATCC14706.
[d] DSM29 (IAM1258).
[e] GenBank accession number.
[f] ATCC11377.
[g] ATCC14706.
[h] ATCC14040.

27.1.2 Human Infections Caused by the Paenibacilli

Human infection by Paenibacilli is rare. As may be noted from Tables 27.3 and 27.4, presently only 13 of more than 100 known Paenibacilli have been found to cause human disease. However, over the past decade, there have been increasing numbers of human cases reported. More than half of human infections by Paenibacilli have been caused by *P. licheniformis*.[3–13] The types of infections caused by this organism are summarized in Table 27.3. Interestingly, almost 75% of the infections were blood-borne (bacteremia/septicemia). Many of these were associated with indwelling catheters in patients with cancers (see entries 1–6) who were all immunosuppressed. One of the causes of bacteremia was self-injection of organic drain fluid containing this organism. Table 27.4 summarizes the cases of human infection caused by the other *Paenibacillus* species. As can be seen in this table, there were 17 cases of human infection, most of which have been reported since the year 2000.[2,5,14–26] This may be due to increased accuracy of diagnosis, both from API strips and from more sensitive and accurate cloning methods (see below). As it happens, most of the Paenibacilli are sensitive to a rather wide range of antimicrobial agents, allowing for effective treatment even though the exact identity of the organism may have been in question. As may be seen in Tables 27.3 and 27.4, there appears to be no known age- or sex-related predisposition in the acquisition of human disease caused by these organisms.

There are several striking features in Table 27.4. First, the causative organism in more than one-third of these cases has been *P. alvei*. This organism was known to cause infections in bees in hives, and, until relatively recently, was not known to cause human disease. The second feature is that, as with *P. licheniformis*, almost half of the cases involve sepsis (i.e., blood-borne disease). Interestingly, most of the cases caused by *P. alvei*[14–19] were not blood borne. As can be seen in both Tables 27.3 and 27.4, the Paenibacilli listed cause serious diseases including bacteremia/sepsis, severe endocarditis—caused both by *P. licheniformis*[4] and *P. popilliae* and that induced heart block[24]—and brain abscess caused by *P. macerans*.[21] Organisms that were recently discovered (sp. nov.) using cloning and phylogenetic analysis (i.e., *P. sanguinis*, *P. massiliensis*, and *P. timonensis*)[25] were found to cause

TABLE 27.3
Human Infections Caused by *P. licheniformis*

No.	Age	Sex	Condition	Source	Reference
1	17	F	Stage III Hodgkin's lymphoma	Blood; Broviac catheter	3
2	9	F	Acute lymphoblastic leukemia	Blood; Hickman catheter	3
3	4	M	Solid tissue tumor	Blood; Broviac catheter	3
4	54	M	Solid tissue tumor; inflammatory bowel disease	Blood; Hickman catheter	3
5	2	M	Leukemia	Blood; Broviac catheter	3
6	62	M	Leukemia	Blood; Hickman catheter	3
7	73	M	Acute prosthetic aortic valve dehiscence	Aortic valve abscess	4
8	18	F	Munchausen's Syndrome	Blood	5
9	NA[a]	F	Pregnancy; acute fibrinolysis; catheter	Blood, bacteremia	6
10	57	M	Volvulus, perforation of upper small bowel	Blood: Bacteremia, peritonitis	7
11	49 cases of elderly individuals who received food from Meals-on-Wheels	M and F	Diarrheal illness	Food poisoning	8
12	46	F	Eye injury	Corneal ulcer	9
13	30	M	Ventriculitis	Meningioma	10
14	63	M	Arteriography	Blood: septicemia	11
15	41	M	Penetrating orbital injury	Cerebral abscess	12
16	36	F	Self-injection of organic drain cleaner	Blood; septicemia	13

[a] NA, not available.

severe bacteremia. We recently reported *P. thiaminolyticus*, an organism found in soil that was not described previously as causing human disease, to have been the agent that caused sepsis in an octogenarian.[2]

Thus, it appears that, overall, while human infections by Paenibacilli are rare, they can cause serious human disease. Members of this genus may be suspected as a possible cause of sepsis when gram-positive or gram-variable bacilli are identified in blood cultures.

27.1.3 Antibiotic Resistance

Sets of genes that confer resistance of organisms to the aminoglycoside antibiotics vancomycin and teichoplanin occur in Group D enterococci, particularly in *S. faecalis* and *S. faecium*.[27] Transposons containing these genes are readily transferable horizontally to other species including various bacilli (e.g., *B. circulans*, *S. bovis*, and, disturbingly, *Staphylococcus aureus*).[28] Surprisingly, recent studies have found the occurrence of vancomycin and teichoplanin resistance genes in a number of Paenibacilli.[28,29] This raises the important questions as to whether the occurrence of these resistance genes in Paenibacilli will increase the virulence or Paenibacilli infections in humans and whether these organisms can serve as a "reservoir" for this transferable resistance.

Bacterial Cell Wall Synthesis. Before discussing these possibilities, it would be wise to review the mechanisms involved in aminoglycoside resistance in bacteria. These antibiotics are directed against cell wall synthesis in bacteria. Figure 27.1 summarizes the known major steps that are involved.[30] The cell wall consists of long chains of repeating units of *N*-Acetylglucosamine (GlcNAc) and *N*-acetylmuramic acid (MurNAc) in beta-1,4-linkage. These chains contain pendant oligopeptide sequences that become cross-linked as described below. The carbohydrate-peptide structure is referred to as a peptidoglycan. Attached to the 3-O-lactyl side chain of the MurNAc carbohydrate units is a pentapeptide consisting of pentapeptide L-Ala-D-Gln-L-Lys-D-Ala-D-Ala; the Lys residue is connected by an amide bond on its epsilon-amino group to pentaglycyl, as shown in the figure. The free alpha-amino group of the pentaglycyl sequence of one chain forms an amide bond with the penultimate D-alanine residue of another chain, liberating the terminal D-Ala residue in a transpeptidation reaction, resulting in the formation of an interchain cross-link. Multiple cross-links like this give strength to the cell wall.

Sites for Inhibition of Cell Wall Synthesis.[30] The step in cell wall synthesis at which penicillin and its derivatives and the aminoglycosides act is the transpeptidation reaction as indicated in Figure 27.1. Penicillin and its derivatives block the enzyme, peptidyl transferase, which catalyzes this reaction as shown in the figure, thereby blocking the cross-linking of the chains, which weakens the cell wall and makes it permeable to intracellular contents and susceptible to osmotic lysis.[27] In contrast the aminoglycosides bind not to the enzyme but actually to the D-Ala-D-Ala dipeptide sequence itself, thereby blocking the transpeptidation reaction as also shown in this figure.[27] Resistance genes confer on bacteria the ability to substitute D-lactic acid or D-serine for the terminal D-Ala residue in the synthesis of the peptidoglycan.[27,28] The presence of the lactic acid residue significantly lowers

TABLE 27.4
Summary of Human Infections Caused by *Paenibacilli* Other than *P. licheniformis*[a]

Patient						
Age	Sex	Condition	Species	Source	Method of Identification	Reference
n/a[b]	n/a[b]	Meningitis	*P. alvei*	Cerebrospinal fluid	Biochemical tests	17
20 days	F	Meningitis	*P. alvei*	Cerebrospinal fluid	Biochemical tests	19
26 years	F	Sickle cell disease and hip prosthesis infection	*P. alvei*	Blood (hip prothesis)	Biochemical tests	18
22 years	M	Traumatic injury and endophthalmitis	*P. alvei*	Foreign body from eye	Biochemical tests	14
62 years	M	Pneumonitis and pleuritis	*P. alvei*	Pleural fluid	Biochemical tests	15
62 years	M	Lower extremity cellulitis	*P. alvei*	Culture from site	Biochemical tests	16
93 years	F	Cerebral infarction	*P. polymyxa*	Blood	Biochemical tests	20
18 years	F	Munchausen's Syndrome	*P. polymyxa Bacillus licheniformis Bacillus pumilus*	Blood from self-injected compound	Biochemical tests	5
52 years	M	Periorbital trauma	*P. macerans*	Brain abscess	Biochemical tests	21
9 years	M	Neutropenic fever	*P. hongkongensis*	Pseudobacteremia	16S rRNA	23
57 years	M	Endocarditis	*P. popilliae*	Blood (heart valve)	Biochemical tests	24
49 years	M	Carcinoma of oropharynx	*P. sanguinis*	Blood	Biochemical tests; 16S rRNA	25
13 years	M	Acute lymphoblastic Leukemia	*P. massiliensis*	Blood	Biochemical tests; 16S rRNA	25
75 years	F	Nephropathy, Hemodialysis	*P.timonensis*	Blood	Biochemical tests; 16S rRNA	25
35 years	F	Drug abuse, night sweats, pulmonary nodules	*P. provencensis*	Cerebrospinal fluid	Biochemical tests; 16S rRNA	26
54 years	M	Cerebellar syndrome; Whipple's Disease	*P. urinalis*	Urine	Biochemical tests; 16S rRNA	26
80 years	M	Colon cancer, renal failure, hemodialysis	*P. thiaminolyticus*	Blood (possible Permacath)	Biochemical tests; 16S rRNA	2

[a] Adapted from Table 5 of ref. 2. One case of *P. licheniformis* is reported in this table as one of three causative organisms in a patient.
[b] Neonate age and sex information not available.

the affinity of the aminoglycosides for the D-Ala-lactyl unit, called a depsipeptide.[27,28]

Structure of Vancomycin-Resistance Genes.[28] The operons for the genes coding for vancomycin resistance have been partially elucidated using polymerase chain reaction (PCR) techniques as described below for several of the Paenibacilli. There are at least two prototypical gene sequences for vancomycin resistance, called vanA and vanB, as found in *E. faecium* and *E. faecalis*, respectively, shown in Figure 3 of Reference 28. Common to all sequences going in the 5′-to-3′ direction are two 5′-terminal control open reading frames, called vanR and vanS. All sequences further contain a downstream sequence of vanH-van{A or B or F}-vanX. vanH encodes a dehydrogenase, vanA, B, or F encode a ligase, and vanX encodes a D,D-dipeptidase. In the *E. faecium* operon, vanY, encoding a D,D-carboxypeptidase, follows the vanH-vanA-vanX sequence while it precedes the vanH-vanB-vanX sequence in the *E. faecalis* operon (Fig. 3 of Ref. 28). vanA genes have been found to be associated with the transposon TN1547 and vanB genes with transposons TN1549 and TN5382.[28]

Several other *Paenibacillus* species (i.e., *P. thiaminolyticus* and *P. apiarius*) have been discovered to contain vancomycin-resistance genes that are strongly homologous to those occurring in *E. faecium* and *E. faecalis*.[28] The gene sequences in the vancomycin-resistance operons of these two organisms were mapped using a combination of restriction fragment length polymorphism (RFLP) PCR and successive thermal asymmetric interlaced (TAIL) PCR.[28] The latter was used to obtain sequences of the 5′- and 3′-flanking regions of the vanH and vanX regions. *P. thiaminolyticus* has a vanA-like operon structure and has therefore been termed $vanA_{PT}$. Interestingly, the operon structure of *P. apiarius*, which has not been found as yet to cause human infection, is a hybrid of vanA and vanB since vanW, another member of the vancomycin-resistance genes, precedes vanH (as occurs in vanB), while it otherwise follows the structure of vanA.[28] Because of its closer similarity to vanA, it is referred to as $vanA_{PA}$.[28]

Significantly, aminoglycoside resistance genes have been discovered in one of the soil-borne Paenibacilli, *P. popilliae*. This species, easily recognizable by the paraspore crystals it contains (as shown in the electron microscopic picture in

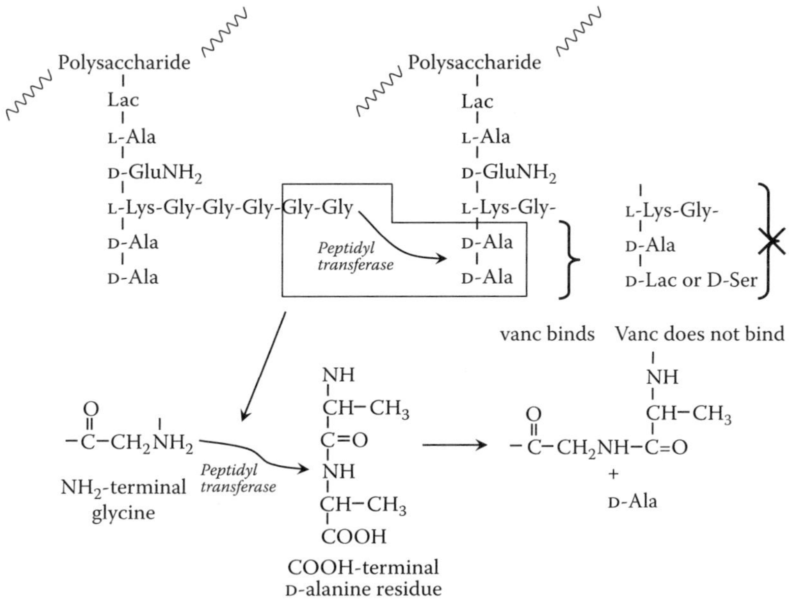

FIGURE 27.1 Basis for vancomycin resistance. Cell wall biosynthesis showing how interchain cross links form. Polysaccharide consists of chains of block copolymers of (*N*-Acetylglucosamine-*N*-Acetyl muramic acid) or $(GlnNAc\text{-}MurNAc)_N$ in beta-1,4-linkage. The MurNAc units contain a D-lactyl moiety on the 3-O position. Esterified to this is a pentapeptide as shown, the lysine (Lys) residue of which is linked via its epsilon-amino group to a pentaglycyl chain. The free alpha amino group of the pentaglycyl unit displaces the terminal D-Ala residue of the pentapeptide of an adjacent chain to form a cross-link in a trans peptidation reaction which is blocked by penicillin and its derivatives that "mimic" the terminal D-Ala-D-Ala residues. This allows penicillin and its derivatives to inhibit the transpeptidase competitively. Vancomycin and other aminoglycosides also inhibit transpeptidation but do so by binding directly to the D-Ala-D-Ala units. This binding blocks the transpeptidation reaction. In bacteria that have developed vancomycin resistance, the terminal D-Ala is replaced with another amino acid like D-serine (Ser) or by D-lactic acid. Vancomycin and other aminoglycosides can no longer bind to the terminal peptide units as represented by the "X" in the figure. (Adapted from Davis, B.D. et al., *Microbiology*, 2nd ed., p. 114, Harper and Row, Hagerstown, MD, 1973. With permission.)

Figure 27.2), has been found to cause fatal milky disease in larvae of the Japanese beetle and has therefore been used as a biopesticide. The operon sequence is shown as the third entry of Figure 3 of Reference 28. The ligase-encoding gene in this sequence, named vanF, has relatively low homology to those in vanA and vanB.[28] The overall structure of the *P. popilliae* resistance gene operon is similar to that of vanB, with vanZ substituting for vanW. Thus, *P. popilliae*, which, besides occurring throughout the world in soil, is being used industrially as a pesticide, harbors a vancomycin-resistance gene complex. Since this organism has been described as a human pathogen in one patient, causing endocarditis (Table 27.4, ref. 24), this finding may have important medical implications.

Transposase in Paenibacilli. An important aspect of the above findings was the sequence of the vanR domain in *P. thiaminolyticus*. This sequence revealed homology (39% identity and 58% similarity) to the putative transposase of *Lactobacillus lactis*, suggesting that the vancomycin-resistance genes of *P. thiaminolyticus* may be transposable.[28] In the one case of human infection by this organism that we reported recently,[2] the organism was susceptible to all antibiotics—except for clindamycin—tested in the antibiogram, including vancomycin.[2] Thus, there may be heterogeneity for presence of the vancomycin-resistance genes in these bacteria.

This type of heterogeneity has been found for vanF in *P. popilliae*. Most strains of this species in North America have been found to carry vanF and are resistant to vancomycin. On the other hand, strains from Central and South America have been found to be sensitive to this antimicrobial.[28]

Can Paenibacilli Transfer Aminoglycoside Resistance? While these findings suggest that the Paenibacilli can transfer vancomycin-resistance genes to other bacteria, there are findings that suggest that this may not be the case at least in some instances. Currently, evidence has accumulated that the aminoglycoside resistance genes have been present on the chromosomes and not as plasmids of the Paenibacilli for a long period of time. Attempts to transfer the $vanA_{PT}$ genes in in vitro conjugation experiments have been unsuccessful, and no plasmids have been recovered from any of the Paenibacilli that are so large as to accommodate the entire aminoglycoside resistance minigenome.[28] Thus, the aminoglycoside resistance genes may be integral parts of the chromosomes of Paenibacilli, making their transfers unlikely.[28] However, conclusive proof of chromosomal location of the aminoglycoside resistance genes awaits genomic sequencing. At present, it is not clear whether Paenibacilli are active in transferring vancomycin resistance to other microbes.

Origins of Vancomycin Resistance. Questions have arisen as to the origin of the vancomycin-resistance genes. One hypothesis is that they arose in bacteria that produced aminoglycoside antibiotics. Evidence against this view is furnished by the finding that the resistance genes in these

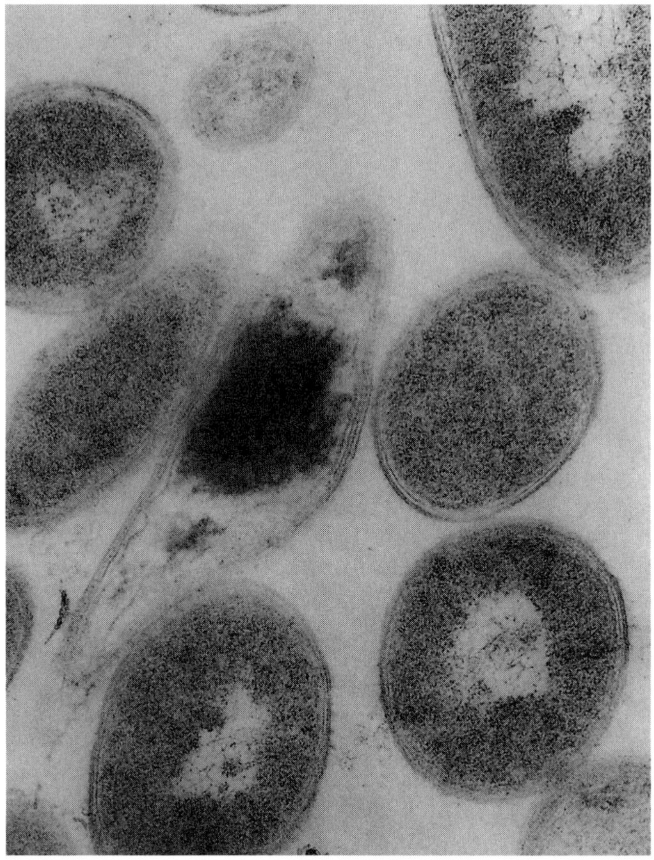

FIGURE 27.2 Electron micrograph of *P. popilliae*, showing characteristic endospores and a paraspore crystal near the endospore in the organism in the lower middle part of the figure.

organisms have relatively low homology to those found in organisms affecting humans, although mutations may have occurred that explain these differences. Another hypothesis is that they arose in bacteria in soil to ensure their survival in the presence of these antibiotics from the organisms producing them.[28] This hypothesis is hard to prove since it is not known if aminoglycoside production occurs extensively in soil or if lethal concentrations can be accumulated in soil. Yet, another theory proposes that the van operons evolved from initial functions that were not related to aminoglycoside resistance. For example, the D,D-dicarboxypeptidase enzyme encoded by this complex is thought to contribute to germination and sporulation.[28] Since vanA_{PT} and vanA_{PA} have strongly similar sequences, it has been suggested that these sequences arose from a common ancestor. Because the presence of these genes appears to be chromosomal in these Paenibacilli, but is associated with mobile genetic elements in the enterococci, the vanA elements may have first occurred in the Paenibacilli.[28]

27.1.4 Paenibacilli That Cause Human Infection

Criteria for Infection. Of the over 100 species belonging to this genus, 13 thus far have been found to cause human disease, many of which have been described over the past 10 years. It is important to distinguish between these bacteria as true infectious agents as opposed to being merely contaminants since the Paenibacilli are ubiquitously distributed in soil, in food and on devices, tubes,[31] and so forth, and until relatively recently were generally not thought to cause human disease. One criterion for deciding whether the presence of Paenibacilli are infectious agents in human specimens such as blood cultures is to have repeated positive results, making the chances of the organism(s) being merely contaminants unlikely. In addition, if the patients' conditions improve through the use of antibiotics to which the Paenibacilli are sensitive, this is a further indication that the organisms are truly infectious. Finally, the occurrence of one or more of these organisms in surgical specimens that show evidence of an infectious process is considered evidence of their being infectious agents.[4] We now describe the species that have been found to cause human disease by one or more of the above criteria, giving characteristics that are unique to each that can facilitate detection of their presence in human specimens.

P. licheniformis.[3–13] This organism is the most common cause of human disease among the Paenibacilli (see Table 27.3). Like *P. polymyxa*, it occurs mainly in soil and is also present in food products. Multiple cases of infection by this organism have been reported over the years, the majority of them on the basis of biochemical identification but not on 16S rRNA sequences. Three cases of *P. licheniformis* and one of *P. Pumilus* were identified in immunosuppressed patients undergoing chemotherapy.[5] One case of bacterial endocarditis, much like the case of *P. popilliae* described below, was reported in an aortic valve requiring a porcine valve replacement.[4] In addition, five cases of bacteremia with *P. licheniformis* as the causative agent were described in immunosuppressed patients with indwelling catheters.[3] In the five cases (age range 4–62 years), two patients had leukemia or lymphoma and three had solid tissue tumors; of these, one patient was neutropenic, but all were considered to be immunosuppressed.[3] Between the years 1976 and 1992, at least seven more cases of infection by *P. licheniformis* were described. As shown in Table 27.3, three of these cases involved septicemia or bacteremia, the three cases involved pregnancy, volvulus with perforation, and were a possible result of arteriography. This organism was also found to cause the outbreak of diarrhea in a group of 49 recipients of food from the Meals-on-Wheels program (entry 4, Table 27.3). The other cases involved a corneal ulcer from an eye injury, cerebral ventriculitis associated with a meningioma, and a cerebral abscess resulting from a penetrating orbital injury. Interestingly, originally thought not to be infectious in humans, this species is present in some organic drain cleaners.[13] A case has been reported of an individual who attempted suicide by consuming organic drain cleaner containing *P. licheniformis* who then became septic from this organism, again confirming that this organism can be infectious in humans.[13]

Like *P. polymyxa* and *P. pumilus*, both discussed below, *P. licheniformis* is a gram-positive rod with central, nondistending spores (they are paracentral in *P. pumilus* and subterminal in *P. polymyxa*).[3–13] All three organisms are motile and have peritrichous flagella. They are all capable of anaerobic growth. *P. licheniformis* produces acid from a wide variety of sugars and metabolites, including glucose, fructose, mannose, mannitol, arabinose, salicin, trehalose, and glycerol.[3–13] Both *P. polymyxa* and *P. pumilus* likewise produce acid from these compounds, except that *P. polymyxa* does not produce acid from glycerol.[3–13] It is capable of growth at 50°C but not at 60°C and can grow in 7% NaCl solution and is hemolytic on blood agars.[3–13] It is further catalase positive, indole negative, and Voges-Proskauer positive. It reduces nitrate, as does *P. polymyxa*, but not *P. pumilus*.[5]

P. alvei. Next to *P. licheniformis*, these organisms are the most common cause of *Paenibacillus* infections in humans. Of the at least six reported cases of infections due to *P. alvei*, as shown in Table 27.4, only one of these was blood borne, unlike many of the other Paenibacilli causing human infection. Of the reported cases, three involved occurrence of *P. alvei* in body fluids, two in cerebrospinal fluid (CSF), each in a case of meningitis and one in pleural fluid in a case of pneumonitis. One case involved endophthalmitis from trauma, as described below for *P. macerans*, and another was cultured directly from a case of cellulitis. The one case of bacteremia resulted from a hip prosthesis infection.

P. alvei is unique among the Paenibacilli for occurring in the larvae of bees as well as in the honey produced by bees.[32] It shares this property with *P. apiarius*, the cause of American foulbrood disease of bees, which also harbors vancomycin-resistance genes as described above.[28] *P. alvei* are gram positive rods containing oval, nondistending spores that are motile with peritrichous flagellae. A distinguishing characteristic of this species is the swarming of cultures on blood agar. While there is considerable heterogeneity among different strains of *P. alvei*, all of these isolates in one study were found to be catalase, oxidase, and Voges-Proskauer positive and nitrate reduction negative, and all produced acid from glycerol, esculin, and maltose. However, the isolates produced variable results for 16 of the 49 biochemical tests on the API 50 CH test system; negative reactions were recorded in another 30 assays.[32] The genetic and biochemical heterogeneity in *P. alvei* isolates has been postulated to result from their adaptation to differing conditions in various areas of the world in which they originated.[32]

P. alvei is also unique in producing a lytic toxin called alveolysin.[33] This is an amphiphilic protein of Mr 63 kD that induces lysis of eukaryotic but not human cells. The lytic activity is sulfhydryl-dependent since oxidation of the sulfhydryls to disulfides destroys its activity; activation can be reversibly restored by treatment with thiols. In its activity, it resembles streptolysin O and a number of other bacterial lysins.[33] Antibodies have been raised against this protein and can be used in its identification and, if necessary, in the identification of the organism. In addition, *P. alvei* produces tryptophanase that can also be useful in identification.[34]

P. popilliae is a gram-variable aerobic and facultatively anaerobic motile rod with peritrichous flagellae, having terminal or subterminal ellipsoidal spores contained in swollen sporangia. Characteristic of this species is the presence of paraspore crystals, an example of which is shown in Figure 27.2, an electron microscopic (EM) image of this organism. This feature helps in the identification of the species.[35] These organisms do not produce acid from any sugars except glucose and trehalose and are negative for oxidase, catalase, indole release, and nitrate reduction, and are also negative for the Voges-Proskauer reaction.[35] *P. popilliae* is a gram-negative, spore-forming rod, 1.3–5.2 μm × 0.5–0.8 μm. It is a fastidious organism that grows only on rich media containing yeast extract, casein hydrolysate or an equivalent amino acid source, and sugars.[35] Several amino acids are known to be required for growth, as are the vitamins thiamine and barbituric acid. Trehalose, the sugar found in insect hemolymph, is a favored carbon source.[35]

The spores of *P. popilliae* infect larvae (grubs) of Japanese beetles and cause the so-called milky disease that eventually kills the larvae, preventing their development into adult beetles.[28,29,35] It has been used as a pesticide active ingredient for more than 50 years. The spores also infect larvae of some closely related beetles, but do not infect other nontarget organisms such as other insects, birds, mammals, earthworms, and plants.[35] As discussed in Section 27.1.3, *P. popilliae* have been found to harbor the vanF vancomycin resistance gene; this seems to be present in strains in South America but is much less common in North American strains.[28,29] The fact that this gene complex exists is disquieting in view of the widespread use of this organism as a pesticide.

There is one reported case of *P. popilliae*-induced bacterial endocarditis in a patient, occurring on the bicuspid aortic valve and complicated with prolonged (>30 days) complete heart block.[24] The organism was susceptible to a wide range of antibiotics and was treated successfully. The fact that this organism can cause human disease and can harbor vancomycin resistance genes may have implications for human disease in the future.

P. polymyxa. *P. polymyxa* is a gram-negative, plant-growth-promoting rhizobacterium that is used as a soil inoculant in agriculture and horticulture.[5,31,36] It is capable of fixing nitrogen. It is also ubiquitously present in food products.[5,31,36] A unique feature of this species is its ability to kill infectious agents that block plant growth, in part by producing the antimicrobial, polymyxin E, which is effective against a range of gram-negative bacteria.[36] Quite recently, a newly described strain of this species, called OSY-DF, has been found to produce polymyxin E and a new lantibiotic (i.e., a peptide that contains naturally occurring L-amino acids and dehydrated and nonamino acids in its primary sequence) that kills gram-positive bacteria.[36] The 16S rRNA of this strain was found to have >99% identity in sequence to that of *P. polymyxa*.[36]

P. polymyxa are gram-positive rods with subterminal, nondistending spores that are motile with peritrichous flagellae.[5] They are negative for oxidase, urease, and indole production but are positive for catalase, Voges-Proskauer reaction and nitrate reduction. They also ferment glucose, hydrolyze esculin and starch, and liquefy gelatin.[5]

It would seem that this organism would be unlikely to cause any kind of human disease. However, two cases have been reported. The first is that of an 18-year-old female who was admitted to the hospital for acute retroorbital headache, nonproductive cough, and poorly localized abdominal pain.[5] One day postadmission, she became febrile, tachycardic, and hypotensive, and was found to have an elevated white blood cell count and serum C-reactive protein level. Numerous blood cultures revealed an unusual finding of three paenibacilli, *polymyxa*, *licheniformis*, and *pumilis*. The patient was known to have a psychiatric history of Munchausen syndrome and to have injected herself with soil.[5] Since all three of these organisms occur in soil, self-injection with soil was a possible cause of the sepsis. The patient responded to vancomycin, ciprofloxacin, and metronidazole. It seems that the most effective of these three antibiotics was vancomycin, since withdrawal of vancomycin with maintenance of ciprofloxin and metronidazole resulted in recurrence of blood-borne *P. pumilis*. Restoration of vancomycin resulted in the patient's becoming afebrile and her blood cultures becoming negative.[5]

The second reported case is that of a 93-year-old female who was admitted to the hospital for a stroke but who then became febrile. Blood cultures were positive for *P. polymyxa*. The patient was known to perform weeding around her house. Thus, soil, introduced possibly through opened skin, was postulated to be the cause of the sepsis.[20]

P. pumilus. This organism occurs in soil and was identified as a causative agent of sepsis in the 18-year-old patient with a history of Munchausen syndrome described above for *P. polymyxa*.[5] Interestingly, of the three cultured from the patient's blood cultures, it was this organism that recurred when vancomycin was omitted from the multiple antibiotic regimen. Thus, this antibiotic is effective in treating *P. pumilus*.[5] As far as it is known, this organism does not harbor vancomycin-resistance genes. These organisms share many of the features of *P. polymyxa* and *P. licheniformis*. They are gram-positive rods that contain paracentral, nondistending spores that are motile with peritrichous flagella and are hemolytic on blood agar. Like *P. polymyxa* and *P. licheniformis*, they produce acid from arabinose, glycerol, salicin, trehalose, and mannitol, and are catalase positive, indole negative, and Voges-Proskauer negative. Unlike these latter two related organisms, they do not usually reduce nitrates.[5]

P. thiaminolyticus. This organism was recently found to cause sepsis in an 80-year-old male on hemodialysis for end stage renal disease; the first reported case of human infection by this species.[2] The patient has a history of congestive heart failure, hypertension, and Parkinson disease, and is status-post subtotal colectomy for colon cancer. His subclavian vein permacath for hemodialysis has become irritated, and he has become febrile with an elevated white cell count. Two successive blood cultures were positive for gram-variable bacilli that were subsequently identified as *P. thiaminolyticus*. The patient responded to single doses of vancomycin and gentamicin after each dialysis treatment and to iv zosyn (piperacillin and tazobactam), amikacin, and then oral p.p. levofloxacin. There was evidence of osteomyelitis and subclavian vein thrombosis, both of which have remitted after antibiotic treatment.

As we noted above in the section on antibiotic resistance, *P. thiaminolyticus* harbors the $vanA_{PT}$ vancomycin-resistance genes.[28] However, the recovered bacteria in this case were sensitive to vancomycin and, except for clindamycin, to all of the other antimicrobials in the antibiogram, including ampicillin, cefazolin, cefotaxime, cefuroxime, levofloxacin, erythromycin, gentamicin, imipenem, linezolid, penicillin, rifampin, synercid, tetracyclin, and trimeth/sulfa.[2]

As noted above, *P. thiaminolyticus* has been recovered from the soil from the larvae of dead honeybees and human feces.[37,38] It is a gram-variable ellipsoidal bacterium that is motile with peritrichous flagella and, like virtually all of the Paenibacilli, has terminal or subterminal ellipsoidal spores contained in the swollen sporangia. It can grow anaerobically and at 44°C.[37,38] It is Voges-Proskauer test negative.[37]

A highly distinguishing characteristic of *P. thiaminolyticus* is its production of the enzyme thiaminase, which catalyzes the displacement by nucleophiles of the thiazolium ring of thiamine, resulting in inactivation of this cofactor.[39] Cattle infected with *P. thiaminolyticus* that produces this enzyme in their central nervous systems develop cerebrocortical necrosis, and sheep infected with this organism develop polioencephalomalacia.[39] The etiological agent for both of these conditions is bacterially produced thiaminase. If this organism were to cause disease in humans as a result of thiaminase activity, symptoms similar to those of Wiernecke's encephalopathy, as observed in some alcoholics, might be expected.

In addition, *P. thiaminolyticus* also uniquely produces an enzyme, termed YZGD, that has dual activities as a pyridoxal phosphatase (PLPase) and as a so-called Nudix, or nucleoside diphosphate hydrolase.[40] Both the PLPase and the Nudix hydrolase require a divalent cation for activity. However, Nudix hydrolase activity requires Mn ion while, in addition to Mn, PLPase activity can utilize Co^{2+}, Mg^{2+}, and Zn^{2+}. Interestingly, YZGD's phosphatase activity has a pH optimum at an acidic pH (5), while the pH optimum for Nudix activity is at the alkaline pH of 8.5. It is not known how this dual enzyme functions in the bacterium or if it is a causative factor of disease as thiaminase is. Nonetheless, both enzymes can be helpful in identifying this organism. The dendrogram based on the comparison of the sequences of the rDNA's of these organisms shows that *P. thiaminolyticus* is closely related to both *P. alvei* and *P. popilliae*.

P. macerans. Unlike most of the other Paenibacilli, this species is not commonly found to occur in soil. Rather, it occurs in plant materials and has been found in cattle abortions. It was the cause of a fatal frontal lobe brain abscess that resulted from trauma to the right orbit of a 52-year-old man by a hedge branch.[21] These organisms were susceptible to a wide range of antibiotics, including susceptibility to amoxicillin, ticarcillin, piperacillin, cefalotin, cefoxitin, cefotaxime, gentamicin, tobramycin, netilmicin, amikacin, minocycline, pefloxacin, and trimethoprim-sulfamethoxazole, and were resistant to colistin. This organism was also recovered from a wound infection following removal of a melanoma.[39] As with *P. polymyxa*, *P. licheniformis* and *P. pumilus*, *P. macerans* was identified by the use of morphology, growth, and biochemical reaction characteristics, not with 16S rRNA sequencing and sequence comparisons. These facultatively anaerobic organisms are gram-variable rods with subterminal spores in swollen sporangia that are nonhemolytic on blood agar. They are motile with polytrichous flagellae. They are catalase positive, indole negative and Voges-Proskauer negative. Like the three related species *P. polymyxa*, *P. licheniformis*, and *P. pumilus*, they utilize glucose, as well as xylose, arabinose, mannitol, and galactose, but not inositol and sorbitol. These organisms grow at 50°C but not at 60°C and do not grow in 7% NaCl.[21]

New Species Associated with Human Infection. Over the past 5 years (2004–2009), five new species of Paenibacilli have been identified that are associated with human infection. These are P. *massiliensis*, *P. sanguinis*, *P. timonensis*, *P. provencensis*, and *P. urinalis*.[25,26] Several criteria were followed in determining that these organisms isolated from human tissue constituted new species as opposed to different type strains within a species. These were that the 16S rRNA sequences were <97% identical to the known sequences of the other Paenibacilli and hybridization of 16S rRNA from the new organism to that of a given species closest in 16S rRNA sequence to that of the new species was <50%.[25,26]

P. sanguinis was isolated from blood cultures from an afebrile 49-year-old man being given chemoradiation for an epidermoid carcinoma.

P. massiliensis was isolated from the blood culture of a 13-year-old boy who was being treated with chemotherapy and an allograft for acute lymphoblastic leukemia. He developed graft-versus-host disease with cutaneous and hepatic manifestations and became febrile and neutropenic.[25]

P. timonensis was isolated from the blood culture of a 75-year-old female with a history of chronic interstitial nephropathy who was afebrile but had a productive cough.[25]

P. urinalis was isolated from the urine of a 36-year-old female with a past history of IV drug abuse who presented in an emaciated state with a chronic cough, sweats, inflammatory syndrome, and dysphagia and was found to have pulmonary nodules on chest X-ray. No growth of mycobacteria was obtained on cultures. It was felt that *P. urinalis* that grew in the urine sample might have been a contaminant.

P. provencensis was cultured from the cerebrospinal fluid of a 58-year-old male with Whipple disease and cerebellar syndrome and diarrhea. DNA from *Tropheryma whipplei* was found in the patient's CSF, from which *P. provencensis* was cultured. This organism was also thought to be a contaminant, although it is not clear how it would have been introduced into a CSF sample.[26]

One common characteristic to all of these patients is that each was immunocompromised, a state that seems to be true of many patients with infections by other species of *Paenibacillus*. Also common to all five of these organisms is that they are rods (bacilli), the first three being gram positive and the last two being gram negative.[25,26] They are all motile via peritrichous flagella and have terminal or subterminal ellipsoidal spores contained in swollen sporangia. All, except for *P. provencensis*, exhibit the ability to grow anaerobically and to grow at 44°C. They are all nonhemolytic, are Voges-Proskauer test negative, and do not liquefy gelatin. None of these organisms produce acid from D-turanose, L-fucose, and D-arabitol.[25,26]

P. hongkongensis. This organism was discovered as the cause of sepsis in a 9-year Asian boy with a history of medulloblastoma that was excised but recurred with the bone metastasis for which the patient received five courses of chemoradiation.[23] After the sixth course, the patient became febrile with a low white blood cell count; blood cultures revealed gram-negative rods that grew on horse blood agar as nonhemolytic, grey colonies of 1 mm in diameter after 24 h of incubation at 37°C in ambient air.[23] No enhancement of growth was seen in 5% CO_2. These bacteria grew at 50°C as pinpoint colonies after 72 h of incubation, but did not grow at higher temperatures (e.g., 65°C) or on MacConkey agar.[23] Empirical treatment with ceftazidime and amikacin resulted in negative blood cultures and remission of the fever at the end of 1 week, and the patient then underwent a seventh course of chemoradiation.[23]

Sequencing of the 16S rRNA revealed 92% identity to the sequence of *P. macerans*. Since this was less than the

97% sequence identity cut-off, this organism was classified as a new species, *P. hongkongensis*, strain HKU3.[23] It was also related by sequence to a number of other Paenibacilli (i.e., *P. borealis*, *P. ehimensis*, and *P. amylolyticus*). Of these, only *P. macerans* has been found—thus far—to cause human infection (see above).[21,23] Unlike most of the other Paenibacilli, it was found to be nonmotile. It is oxidase, catalase positive, and reduces nitrates but does not induce indole production. It produces β galactosidase, hydrolyses esculin, and utilizes sodium acetate.[23]

27.2 METHODS

27.2.1 Sample Preparation

Paenibacillus bacteria are grown on blood agar at 37°C in ambient air for 24–48 h. A DNA is prepared from *Paenibacillus* colonies using a rapid protocol. Briefly, add 80 µL of 50 mM NaOH to 20 µL of bacterial cells suspended in distilled water, incubate the mixture at 60°C for 45 min, add 6 µL of 1 M Tris/HCl (pH 7.0) to the mix, dilute the mix 100×, and use 5 µL of the diluted extract in PCR mix (total volume 50 µL).[23]

27.2.2 Detection Procedures

As specific primers for most of *Paenibacillus* species are not yet available, the current method for identification of *Paenibacilli* relies on the use of 16S RNA gene primers (e.g., (LPW55: 5′-AGTTTGATCCTGGCTCAG-3′ and LPW324: 5′-TTGTTACGACTTCACCCCA-3′) for amplification followed by sequencing analysis of the resulting product.[23]

Procedure

1. Prepare PCR mixture (50 µL) containing 5 µL of bacterial DNA template, PCR buffer (10 mM Tris/HCl pH 8.3, 50 mM KCl, 2mM $MgCl_2$, and 0.01% gelatin), 200 µM of each dNTP, and 1.0 U Taq polymerase (Boehringer Mannheim), 0.5 µM of each primer (LPW55 and LPW324).
2. Amplify in a thermal cycler (Perkin Elmer Cetus) with 1 cycle of 94°C for 5 min; 40 cycles of 94°C for 1 min, 55°C for 1 min, and 72°C for 2 min; a final 72°C for 10 min.
3. Electrophorese a 10 µL aliquot of each amplified product in 1.0% agarose gel with a molecular size marker (e.g., λ DNA *Ava*II digest) in parallel; stain the gel with ethidium bromide (0.5 µg/mL) for 15 min, rinse, and photograph under ultraviolet light illumination.
4. Purify the PCR product from gel using the QIAquick PCR purification kit; sequence both strands of the PCR product twice with an ABI377 automated sequencer (Perkin-Elmer), using the PCR primers (LPW55 and LPW324) and additional sequencing primers (LPW278: 5′-CCCTTATGACCTGGGCTAC-3′ and LPW279: 5′-CTGGCAACAGAGCTTTACG-3′).
5. Compare the sequence of the PCR product with known 16S rRNA gene sequences in GenBank by multiple sequence alignment using the CLUSTAL W program.

Note: The sequence of the PCR product showing more than 97% nucleotide identity, with the 16S rRNA gene of the known *Paenibacillus* species confirms its identity. On the other hand, if the sequence of the PCR product exhibits less than 97% nucleotide identity with the 16S rRNA gene of all previously described *Paenibacillus* species, it may suggest a novel species.[23]

27.3 CONCLUSIONS

Although *Paenibacillus* infection in humans is rare and is caused by a relatively small percentage of the organisms in this genus, there has been both an increase in the number of reported infections by these bacteria and an increasing number of the members of this genus. Many of these infections are blood borne, causing either bacteremia or septicemia. Since these organisms occur in soil and on surfaces, opportunistic infections can be acquired especially by immunocompromised individuals. In addition, introduction of the organisms to the gastrointestinal tract or to the blood stream, as occurs in cases of self-immolation or from skin-piercing trauma, results in severe infection, proving that organisms hitherto considered to be benign to humans are highly infectious and can cause severe disease. Surprisingly, a number of Paenibacilli harbor aminoglycoside resistance genes that contain elements that encode transposing enzymes, admitting of the possibility of genetic transfer of the resistance genes to other organisms and of their own resistance to antimicrobial therapy. Thus, the discovery of gram-variable, facultatively anaerobic, endospore-forming organisms in human specimens may represent serious infection. The expanding number of new members of this genus that are associated with human disease requires accurate identification of causative organisms (i.e., biochemical reaction profiles and sequence of their 16S rRNA). While many of the causative organisms from this genus have a susceptibility to a wide range of antibiotics thus far, some, such as *P. pumilus*, have a more restricted susceptibility. This finding, in addition to the antibiotic resistance genes in a number of the Paenibacilli, may herald treatment problems in the future.

REFERENCES

1. Ash, C., Priest, F.G., and Collins, M.D., Molecular identification of rRNA group 3 bacilli (Ash, Farrow, Wallbanks and Collins) using a PCR probe test, *Antonie van Leeuwenhoek*, 64, 253, 1993.
2. Ouyang, J. et al., *Paenibacillus thiaminolyticus:* A new cause of human infection, inducing bacteremia in a patient on hemodialysis, *Ann. Clin. Lab. Sci.*, 38, 393, 2008.
3. Blue, S.R., Singh, V.R., and Saubolle, M.A., *Bacillus licheniformis* bacteremia: Five cases associated with indwelling central venous catheters, *Clin. Infect. Dis.*, 20, 629, 1995.

4. Santini, F. et al., *Bacillus licheniformis* prosthetic aortic valve endocarditis, *J. Clin. Microbiol.*, 33, 3070, 1995.
5. Galanos, J. et al., Bacteremia due to three *Bacillus* species in a case of Munchausen's syndrome, *J. Clin. Microbiol.*, 41, 2247, 2003.
6. Peloux, Y., Charrel-Taranger, C., and Gouin, F., Nouvelle affection opportuniste a Bacillus: un cas de bacteriemie a Bacillus licheniformis, *Pathol. Biol.*, 24, 97, 1976.
7. Sugar, A.M., and McCloskey, R.V., Bacillus licheniformis sepsis, *JAMA*, 238, 1180, 1977.
8. Jephcott, A.E. et al., An unusual outbreak of food poisoning associated with meals-on-wheels, *Lancet*, ii, 129, 1977.
9. Tabbara, K.F., and Tarrabay, N., *Bacillus licheniformis* corneal ulcer, *Am. J. Ophthalmol.*, 87, 717, 1979.
10. Young, R.F. et al., Postoperative neurosurgical infections due to *Bacillus* species, *Surg. Neurol.*, 18, 271, 1982.
11. Hardy, A.J. et al., Etat septique après arteriographie. Du role d'un germe repute non pathogene: *Bacillus licheniformis*, *J. Med.* (Strasbourg), 17, 18, 1986.
12. Jones, B., Hanson, M.F., and Logan, N.A., Isolation of *Bacillus licheniformis* from a brain abscess following a penetrating orbital injury, *J. Infect.*, 24, 103, 1992.
13. Hannah, W.N., and Ender, P.T., Persistent *Bacillus licheniformis* bacteremia associated with an intentional injection of organic drain cleaner, *Chron. Infect. Dis.*, 29, 659, 1999.
14. Antonello, A., and Weinstein, G.W., Successful treatment of *Bacillus alvei* endophthalmitis, *Am. J. Ophthalmol.*, 108, 454, 1989.
15. Coudron, P.E., Payne J.M., and Markowitz, S.M., Pneumonia and empyema infection associated with a *Bacillus* species that resembles *B. alvei*, *J. Clin. Microbiol.*, 29, 1777, 1991.
16. Shin, J.H. et al., A case of *Paenibacillus alvei* cellulitis in immunocompetent patient, *Korean J. Lab. Med.*, 25, 53, 2005.
17. Park, S.J., Cong, W., and Lee, S.Y., Bacterial and fungal species from cerebrospinal fluid in the past five years, *Korean J. Pathol.*, 10, 137, 1976.
18. Reboli, A.C., Bryan, C.S., and Farrar, W.E., Bacteremia and infection of a hip prosthesis caused by *Bacillus alvei*, *J. Clin. Microbiol.*, 27, 1395, 1989.
19. Wiedermann, B.L., Non-anthrax *Bacillus* infections in children, *Pediatr. Infect. Dis. J.*, 6, 218, 1987.
20. Nasu, Y. et al., A case of *Paenibacillus polymyxa* bacteremia in a patient with cerebral infarction, *Kansenshogaku Zasshi*, 77, 844, 2003.
21. Bert, F., Ouahes, O., and Lambert-Zechovsky, N., Brain abscess due to *Bacillus macerans* following a penetrating periorbital injury, *J. Clin. Microbiol.*, 33, 1950, 1995.
22. Barrero, F. et al., Catheter-associated infection by *Bacillus macerans* in a patient with acute leukemia, *Enferm. Infect. Microbiol. Clin.*, 14, 628, 1996.
23. Teng, J.L. et al., Pseudobacteraemia in a patient with neutropenic fever caused by a novel paenibacillus species: *Paenibacillus hongkongensis* sp. nov., *Mol. Pathol.*, 56, 29, 2003.
24. Wu, Y.J. et al., *Bacillus popilliae* endocarditis with prolonged complete heart block, *Am. J. Med. Sci.*, 317, 263, 1999.
25. Roux, V., and Raoult, D., *Paenibacillus massiliensis* sp. nov., *Paenibacillus sanguinis* sp. nov. and *Paenibacillus timonensis* sp. nov., isolated from blood cultures, *Int. J. Syst. Evol. Microbiol.*, 54, 1049, 2004.
26. Roux, V., Fenner, L., and Raoult, D., *Paenibacillus provencensis* sp. nov., isolated from human cerebrospinal fluid, and *Paenibacillus urinalis* sp. nov., isolated from human urine, *Int. J. Syst. Evol. Microbiol.*, 58, 682, 2008.
27. Winn, W., Jr. et al. (Editors), *Koneman's Color Atlas and Textbook of Diagnostic Microbiology*, 6th ed., Chapter 17, p. 945, Lippincott, Williams and Wilkins, New York, 2006.
28. Guardabassi, L. et al., Glycopeptide resistance vanA operons in *Paenibacillus* strains isolated from soil, *Antimicrob. Agents Chemother.*, 49, 4227, 2005.
29. Patel, R. et al., The biopesticide *Paenibacillus popilliae* has a vancomycin resistance gene cluster homologous to the enterococcal vanA vancomycin resistance gene cluster, *Antimicrob. Agents Chemother.*, 44, 705, 2000.
30. Davis, B.D. et al., *Microbiology*, 2nd ed., p. 114, Harper and Row, Hagerstown, MD, 1973.
31. Gardener, B.B.M., Ecology of *Bacillus* and *Paenibacillus* spp. in agricultural systems, *Phytopathology*, 94, 1252, 2004.
32. Djordjevic, S.R. et al., Genetic and biochemical diversity among isolates of *Paneibacillus alvei* cultured from Australian honeybee (*Apis mellifera*) colonies, *Appl. Environ. Microbiol.*, 66, 1098, 2000.
33. Geoffroy, C., and Alouf, J.E., Selective purification by thiol-disulfide interchange chromatography of alveolysin, a sulhydryl-actiavted toxin of *Bacillus alvei*, *J. Biol. Chem.*, 258, 9968, 1983.
34. Hoch, S.O., and DeMoss, R.D., Catalytic studies on tryptophanase from *Bacillus alvei*, *J. Bacteriol.*, 114, 341, 1973.
35. Pettersson, B., Rippere, K.E., and Yousten, A.A., Transfer of *Bacillus lentimorbus* and *Bacillus popilliae* to the genus *Paenibacillus* with emended descriptions of *Paenibacillus lentirnorbus* comb. nov. and *Paenibacillus popilliae* comb. nov., *Int. J. System. Bacteriol.*, 49, 531, 1999.
36. He, Z. et al., Isolation and identification of a *Paenibacillus* strain that co-produces a novel lantibiotic and polymyxin, *Appl. Environ. Microbiol.*, 73, 168, 2007.
37. Shida, O. et al., Transfer of *Bacillus alginolyticus*, *Bacillus chondroitinus*, *Bacillus curdlanolyticus*, *Bacillus glucanolyticus*, *Bacillus kobensis*, and *Bacillus thiaminolyticus* to the genus *Paenibacillus* and emended description of the genus *Paenibaci*, *Int. J. System. Bacteriol.*, 47, 289, 1997.
38. Kuno, Y., *Bacillus Thiaminolyticus*, a new thiamin-decomposing bacterium, *Proc. Jpn. Acad.*, 27, 362, 1951.
39. Ihde, D.C., and Armstrong, D., Clinical spectrum of infection due to *Bacillus* species, *Am. J. Med.*, 55, 839, 1973.
40. Tirell, M. et al., YZGD from *Paenibacillus thiaminolyticus*, a pyridoxal phosphatase of the HAD (haloacid dehalogenase) superfamily and a versatile member of the Nudix (nucleoside diphosphate x) hydrolase superfamily YZGD from *Paenibacillus thiaminolyticus*, a pyridoxal phosphatase of the HAD (haloacid dehalogenase) superfamily and a versatile member of the Nudix (nucleoside diphosphate x) hydrolase superfamily, *Biochem. J.*, 394, 665, 2006.

28 *Staphylococcus*

Paolo Moroni, Giuliano Pisoni, Paola Cremonesi, and Bianca Castiglioni

CONTENTS

28.1 INTRODUCTION

28.1.1 Classification, Morphology, and Biology

Staphylococcus (from the Greek "σταφυλή"/"staphylē," "bunch of grapes," and "κόκκος"/"kókkos," "granule") is a genus of gram-positive bacteria belonging to the family Staphylococcaceae, order Bacillales, class Bacilli, phylum Firmicutes. Of the six genera in the family Staphylococcaceae (i.e., *Gemella*, *Jeotgalicoccus*, *Macrococcus*, *Nosocomiicoccus*, *Salinicoccus*, and *Staphylococcus*), the genus *Staphylococcus* is noted for its complexity (consisting of >40 species) and for its disease causing potential (with *Staphylococcus aureus* being a most common pathogen of man and animals), even though most *Staphylococcus* species are harmless and normally reside on the skin and mucous membranes of humans and other organisms or represent a small component of soil microbial flora.

Under the microscope, *Staphylococcus* bacteria appear round (cocci), approximately 1 μm in diameter, and form in grape-like clusters that are indicative of the ability to divide in more than one plane. *Staphylococcus* spp. are catalase positive (meaning that it can produce the enzyme "catalase") and able to convert hydrogen peroxide (H_2O_2) to water and oxygen, which makes the catalase test useful to distinguish staphylococci from enterococci and streptococci.

The main classification of *Staphylococcus* spp. is based on the presence or absence of coagulase production. Coagulase is an enzyme that causes plasma clot formation.

The coagulase-positive staphylococci (CPS) are represented by *Staphylococcus aureus*, *S. intermedius*, *S. pseudointermedius*, *S. schleiferi* subsp. *coagulans*, *S. hyicus* subsp. *hyicus*, and *S. lutrae*.[1–4] *S. aureus* is the best known among CPS and has been frequently implicated in the etiology of a group of infections and intoxications in humans. However, while the majority of *S. aureus* are coagulase positive, some may be atypical in that they do not produce coagulase.

A great number of the *Staphylococcus* species distinct from *S. aureus* belong to the group known as coagulase-negative staphylococci (CNS), so named for their inability to clot plasma due to the lack of production of the secreted enzyme coagulase.[5]

CNS, representing the majority of the staphylococcal species often occurring as skin commensals, have been considered for long time to be saprophytic or, rarely, opportunistic pathogens with low virulence. Currently, several species of CNS are considered to be potential pathogens, mainly causing nosocomial infections and often being involved in infections related to implanted medical devices such as intravenous catheters, prosthetic heart valves, pace maker electrodes, urinary tract catheters, orthopedic implants, and a range of other polymer and metal implants. CNS are now a leading cause of bacteremia in the United States, Canada, South America, and Europe[6,7] and many of these CNS are resistant to multiple classes of antimicrobial agents.[8] Despite this, and in contrast to the case of *S. aureus*, infections caused by CNS typically manifest as less severe and subacute diseases that are infrequently associated with mortality. *Staphylococcus epidermidis*, *S. haemolyticus*, and *S. saprophyticus* are the most frequently caused diseases in humans. Other significant opportunistic pathogens are *Staphylococcus hominis*, *S. schleiferi* subsp. *schleiferi*, *S. warneri*, *S. caprae*, *S. capitis*, *S. simulans*, *S. cohnii*, *S. xylosus*, *S. saccharolyticus*, *S. auricularis*, *S. pasteuri*, *S. lentus*, *S. chromogenes*, and *S. lugdunensis*.[9–14]

Although staphylococci can be found in the environment (dust and soil), they are isolated primarily from mammals, in which they may be either transient or permanent residents. Colonization with staphylococci occurs very early after birth and most babies are colonized with multiple strains within the first week of life, particularly in the nose, throat, and umbilicus.[15]

S. aureus is generally a commensal bacteria present on the human epithelia of the skin and mucosae. It has its primary ecological site in the vestibulum nasi, a region in the forefront of the nose. Extranasal sites that typically harbor the organism include the skin, perineum, and pharynx.[16–19] Other carriage sites, including the gastrointestinal tract,[20] vagina,[21] and axillae,[22] harbor *S. aureus*, but less frequently.

S. aureus carriage can be considered as either persistent or intermittent. When two consecutive nasal cultures taken approximately a week apart are positive, with over 10^3 colony-forming units (CFUs), a person is classified as a persistent carrier.[23] When only one of two cultures is positive or when both cultures are positive with low CFUs, the person is considered an intermittent carrier. There is clinical relevance in this distinction. When the *S. aureus* infection rates were determined for persistently and intermittently carrying patients on chronic ambulatory peritoneal dialysis, it appeared that only the persistently colonized patients were at significant risk of becoming autoinfected, thus requiring more antibiotics.[24] About 20% of the population are persistent carriers of *S. aureus*.[25]

On the other side, CNS normally inhabit the skin and mucous membranes in concentrations up to 10^4–10^6 CFU/cm,[2,26] and certain staphylococcal species often show characteristic ecology. For example, *S. capitis* is found almost exclusively on the adult head, and *S. cohnii* on the feet, while *S. auricularis* is isolated only from the external auditory meatus.[27] *Staphylococcus epidermidis* is the predominant human species (comprises 60%–90% of all staphylococci recovered from human sources) and is considered a typical commensal of the skin; it is isolated from a wide variety of anatomic sites, including mucous membranes, moist regions like the groin or axilla, and dry, exposed skin surfaces where bacterial counts usually are much lower (10–10^3 CFU/cm^2). Furthermore, occurrence of *S. epidermidis* was significantly detected in saliva from healthy adults.[28] The second most common species is *S. hominis*, which is widely distributed on the dry, glabrous skin of the arms, legs, and trunk. In a large study of blood culture isolates, 12% of all CNS were identified as *S. hominis*.[29] The indigenous flora of the perineal and inguinal region is very similar to other moist areas of the body, with *S. epidermidis* predominating; however, *S. saprophyticus* is found frequently, even as part of the normal vaginal flora, and represent the source for inoculation of the urinary tract, which leads to urinary tract infections, especially in young women.

Staphylococcus lugdunensis is an integral part of the normal skin flora, but it is more common in the lower than the upper part of the body; 32% and 15% of investigated individuals were positive in the inguinal and perineal areas, respectively, and 28% were positive in the nail bed of the first toe and/or between the first and second toe.[30]

Staphylococcus caprae is usually associated with animal skin or mammary gland and milk—especially in goats[31]—but shows an increasing trend as a human pathogen infecting implanted foreign bodies.[32]

28.1.2 Clinical Features and Pathogenesis of *Staphylococcus aureus*

Staphylococcus aureus is responsible for a variety of suppurative (pus-forming) infections and intoxication in humans. It causes superficial skin lesions such as pimples, impetigo, boils, styes and furuncules, cellulitis folliculitis, carbuncles, staphylococcal scalded skin syndrome (SSSS), and abscesses; more serious infections such as pneumonia, mastitis, phlebitis, meningitis, urinary tract infections (UTIs), and septicemia; and deep-seated infections such as osteomyelitis and endocarditis. *S. aureus* is a major cause of nosocomial (hospital-acquired) infection of surgical wounds and infections associated with indwelling medical devices. *S. aureus* causes also food poisoning (SFP) by producing enterotoxins in food that has been contaminated with the pathogen, as well as toxic shock syndrome (TSS) by release of superantigens into the blood stream.

S. aureus produces many potential virulence factors: (i) surface proteins that promote colonization of host tissues; (ii) proteins that promote bacterial spread in tissues (leukocidin, kinases, hyaluronidase); (iii) biochemical properties that enhance their survival in phagocytes (carotenoids, catalase production); (iv) surface factors that inhibit phagocytic engulfment (capsule, protein A); (v) immunological disguises (protein A, coagulase, clumping factor, biofilm production); (vi) membrane-damaging toxins that lyse eucaryotic cell membranes (hemolysins, leukotoxin, leukocidin); (vii) exotoxins that damage host tissues or otherwise provoke symptoms of disease (staphylococcal enterotoxins SEA to SEQ, toxic shock syndrome toxin-1 TSST-1, exfoliatin toxins ET); and (viii) mechanisms of resistance to antimicrobial agents.

Some virulence factors are expressed by genes that are located on mobile genetic elements called pathogenicity islands (e.g., TSST and some enterotoxins)[33] or lysogenic bacteriophages (e.g., Panton–Valentine leukocidin [PVL])[33,34] and factors associated with suppressing innate immunity such as the chemotaxis inhibitory protein and staphylokinase,[35] which are integrated in the bacterial chromosome.

Given the pathogenesis of the diseases caused by *S. aureus* is multifactorial, it is consequently difficult to identify exactly the role of each virulence factor. However, there are correlations between strains isolated from particular diseases and the expression of particular virulence determinants, which suggest their role in such diseases.

28.1.2.1 Tissue Colonization

In order to start infection, the pathogen must gain access to the host and attach to host cells or tissues. *S. aureus* expresses several surface proteins, termed MSCRAMMs (microbial surface components recognizing adhesive matrix molecules), which promote adherence and attachment to host cell proteins (fibronectin, fibrinogen, collagen, vitronectin, laminin, and other ligands) that constitute the extracellular matrix of epithelial and endothelial surfaces.[36]

In addition, most strains express a fibrin/fibrinogen-binding protein (clumping factor) that promotes attachment of bacteria to blood clots and traumatized tissue, typical of traumatic and chirurgical wounds. In addition, the importance of a collagen-binding adhesin in the pathogenesis of septic arthritis and osteomyelitis has been demonstrated.[37] Interaction with collagen may also be important in promoting bacterial attachment to damaged tissue where the underlying layers have been exposed.

Infections associated with indwelling medical devices ranging from simple intravenous catheters to prosthetic joints and replacement heart valves can be caused by *S. aureus* because immediately after biomaterial is implanted in the human body it becomes coated with a complex mixture of host proteins and platelets.

28.1.2.2 Invasion

The invasion of host tissues by staphylococci involves the production of a wide array of extracellular proteins, some of which may occur also as cell-associated proteins. These proteins described below have different roles in the invasive process.

Alpha-Toxin (Alpha-Hemolysin). Alpha-hemolysin is the best studied of the *S. aureus* cytotoxins. A great percentage of strains produce this toxin, and it is toxic to a wide range of mammalian cells. It is particularly active against rabbit erythrocytes, and it is also dermonecrotic and neurotoxic. Staphylococcal alpha-toxin is involved in the induction of host cell death[38–40] and phagolysosomal escape.[41] Furthermore, alpha-toxin has been described to be a key effector for intracellular pathogen survival within professional phagocytes.[42] It functions as a pore former that increases the permeability of cells or vesicles[43,44] and exhibits enhanced hemolytic activity at low pH.[45] The toxin is secreted by *S. aureus* in a monomeric 27 kDa form. On target membranes, the subunits oligomerize to form heptameric rings with a central pore that perforate the membrane for ions and macromolecules.[46] In humans, platelets, erythrocytes, and monocytes are particularly sensitive to alpha toxin.

Beta-Toxin (Beta-Hemolysin; Sphingomyelinase C). This toxin is a sphingomyelinase that damages membranes rich in this lipid. The majority of human isolates of *S. aureus* do not express β-toxin while it is produced in large quantity by a number of *S. aureus* strains of animal origin. The role of beta-toxin in disease is not clearly understood. The high level of expression in animal strains indicates that beta-toxin producers have some selective advantage from toxin secretion.

Delta-Toxin (Delta-Hemolysin; Delta-Lysin). This is a very small peptide toxin produced by most strains (97%) of *S. aureus*. It is also hypothesized that 50%–70% of coagulase-negative staphylococci produce this toxin. For example, this toxin is also produced by *S. epidermidis*. Although delta-toxin causes a wide range of cytotoxic effects, its importance in disease etiology remains unclear. Delta-toxin is able of lysing erythrocytes and other mammalian cells, as well as subcellular structures such as membrane-bound organelles, spheroplasts, and protoplasts.[47]

Gamma-Toxin and Panton–Valenine–Leukocidin (Two-Component Toxins). Two types of bicomponent toxins are made by *S. aureus*, gamma-toxin and PVL. Each of these toxins is made as two nonassociated secreted proteins, referred to as S and F subunits (for slow- and fast-eluting proteins in an ion-exchange column),[48] which act together to damage membranes. These toxins form a heterooligomeric transmembrane pore composed of four F and four S subunits (called lukF and lukS for PVL), thereby forming an octameric pore in the affected membrane. Gamma-toxin is made by virtually every strain of *S. aureus*, while PVL is made by 2%–3% of strains. The toxins affect neutrophils and macrophages, and gamma-toxin is additionally able to lyse many varieties of mammalian erythrocytes. Nearly 90% of the strains isolated from severe dermonecrotic lesions express PVL, which suggests that it is an important factor in necrotizing skin infections.

Staphylocoagulase (Coagulase). Coagulase is an extracellular protein that binds to prothrombin in the host to form a complex called staphylothrombin. The protease activity characteristic of thrombin is activated in the complex, resulting in the conversion of fibrinogen to fibrin.[49,50] A small amount of coagulase is tightly bound on the bacterial cell surface, where it can react with prothrombin leading to fibrin clotting.[51] Coagulase is regarded as a traditional marker for discriminating *S. aureus* from other less-pathogenetic staphylococci, CNS. However, there is no evidence that it represents a virulence factor, although it is reasonable to speculate that the bacteria could protect themselves from phagocytic and immune defense by causing localized clotting.

Clumping Factor. Clumping factor A (clfA) belongs to a class of cell wall-localized proteins that are covalently anchored to the peptidoglycan,[52–54] and it is the major staphylococcal fibrinogen (Fg)-binding protein responsible for the observed clumping of *S. aureus* in blood plasma.[55,56] Essentially all *S. aureus* clinical strains carry the *clfA* gene.[57] The interaction between bacteria and fibrinogen in solution results in the instantaneous clumping of bacterial cells. The affinity is very high, and clumping occurs in low concentrations of fibrinogen. It has been shown that clfA also promotes adherence of *S. aureus* to immobilized fibrinogen, to blood clots, to conditioned biomaterial *ex vivo,* and to damaged heart valves in vivo.[58–60] It has been described a novel fibrinogen-binding protein of *S. aureus* called clumping factor B (clfB), which has significant similarity to the previously described clfA.[61]

Staphylokinase. Many strains of *S. aureus* express a plasminogen activator called staphylokinase.[62] Staphylokinase is a bacteriophage-encoded protein expressed by lysogenic strains of *S. aureus.* Present understanding of the role of staphylokinase during bacterial infection is based on its interaction with the host proteins, alpha-defensins and plasminogen. Alpha-defensins are bactericidal peptides originating from human neutrophils. Binding of staphylokinase to alpha-defensins abolishes their bactericidal properties, which makes staphylokinase a vital tool for staphylococcal resistance to host innate immunity. A complex formed between staphylokinase and plasminogen activates plasmin-like proteolytic activity causes dissolution of fibrin clots and facilitates bacterial penetration into the surrounding tissues.

Other Extracellular Enzymes. *S. aureus* can produce many other extracellular enzymes such as proteases, lipases, deoxyribonuclease (DNAse), hyaluronidase, collagenase,[63] and a fatty acid modifying enzyme (FAME).[64] These enzymes can convert organic compounds and macromolecules of host tissues into nutrients required for bacterial growth,[65] and are regarded as important pathogenic factors that contribute to the invasiveness and infection of *S. aureus.*

28.1.2.3 Avoidance of Host Defense

S. aureus expresses a number of factors that have the potential to interfere with host defense mechanisms (innate and adaptive immunity). This includes both structural and soluble elements of the bacterium.

Capsular Polysaccharide. *S. aureus* produces various surface polysaccharides and most strains express capsular polysaccharides (CPs) in vivo or under defined culture conditions.[66] Although at least 11 serotypes based on CPs have been proposed, only four CPs (CP1, CP2, CP5, and CP8) have been characterized.[67] Phagocytosis and killing by neutrophil granulocytes play key roles in defense against *S. aureus* infections. Most clinical isolates from both human and animal infections produce either CP5 or CP8. Although CP does impede phagocytosis, it has been demonstrated that lack of CP expression results in an increased capacity for persistence in the host. This capacity was attributed to more effective interaction between unmasked staphylococcal surface ligands and cell receptors, which lead to internalization into epithelial cells.[68]

Protein A. Protein A is a surface protein of *S. aureus* and is usually linked covalently to the peptidoglycan,[69] but mutants and clinical strains have been isolated that secrete all of their protein A into the growth medium.[70] It is primarily known for its ability to bind to the Fc region of IgGs and inhibit opsonophagocytosis.[71,72] The consequence of this interaction is the coating of the bacterial surface with IgG molecules in the incorrect orientation to be recognized by the neutrophil Fc receptor.[52]

Superantigens: Enterotoxins and Toxic Shock Syndrome Toxin. Superantigens are a class of highly potent immunostimulatory molecules produced by *S. aureus.*[73] These toxins possess the unique ability to interact simultaneously with MHC class II molecules and T cell receptors, forming a trimolecular complex that induces profound T cell proliferation. In this way, up to 30% of the T cell population can be stimulated.[74] The resultant massive cytokine release causes epithelial damage and leads to capillary leak and hypotension. The staphylococcal superantigens are designated staphylococcal enterotoxins (SEs) A, B, C (and antigenic variants), D, E, the recently discovered enterotoxins G to R and U, and toxic shock syndrome toxin-1 (TSST-1).

The SEs are a group of extracellular proteins with molecular weights in the range 27,000–30,000 Da, and are of similar composition and biological activity. To date, 19 types of staphylococcal enterotoxins (SEs) have been described.[75] On the basis of their antigenicity and mode of action in the host, SEs have been divided into two groups. The members of group 1, the classical emetic toxins designated SEA, SEB, SEC1, SECbov, SED, and SEE, are the cause of about 95% of staphylococcal food poisoning (SFP) in humans. These toxins are resistant to gastrointestinal proteases and after their intestinal absorption the patient can develop symptoms of food poisoning depending on the SE amount ingested.[76] Group 2 includes toxins possibly involved in the remaining 5% of SFP outbreaks (i.e., other recently identified SEs, designated SEG, SEH, SEI, SEJ, SEK, SEL, SEM, SEN, SEO, SEP, SEQ, SER, and SEU), the emetic activities of which are not currently fully understood.

TSST-1 was the first marker toxin identified for toxic shock syndrome (TSS).[77] TSS is defined as an acute and potentially fatal illness characterized by high fever, a diffuse erythematous rash, desquamation of the skin 1–2 weeks after onset, hypotension and involvement of three or more organ systems. TSS is usually classified into two categories: menstrual TSS, originally described in association with tampon use (this toxin is currently accepted as the cause of 100% of menstruation-associated TSS cases), and nonmenstrual TSS, which occurs in a variety of clinical settings and has a tendency to recur. TSS can occur as a sequel to any staphylococcal infection if TSST-1 is released systemically and the host lacks appropriate neutralizing antibodies. TSS-like systemic symptoms that are indistinguishable from typical TSS might be caused by toxins other than TSST-1, especially SEB.[65]

Exfoliatin Toxins. The exfoliative toxins (ET) of *S. aureus* are responsible for the skin lesions of staphylococcal scalded skin syndrome (SSSS).[78] The exfoliatin toxin causes separation within the epidermis, between the living layers and the superficial dead layers. The separation is through the stratum granulosum of the epidermis. The exfoliative toxins are responsible for a spectrum of disease ranging from localized blisters to extensive exfoliation. It is not clear how this process benefits the organism, but it has been suggested that damaging the protective epidermis allows the organism to propagate and invade deeper tissues.[79] There are two antigenically distinct forms of the toxin, ETA and ETB. Epidemiological studies have shown that ETA is the predominant serotype in Europe and the United States, while ETB is more prevalent in Japan.[80] *S. aureus* also produces exfoliative toxin C (ETC), a heat-labile toxin that was isolated from a horse with skin infection.[81] Recently, another serotype, termed ETD, was identified whilst screening the genome of clinical *S. aureus* isolates from diseased patients with probes for the *eta* and *etb* genes.[82] The toxins have esterase and protease activity and apparently target a protein that is involved in maintaining the integrity of the epidermis. Target for the toxins has recently been identified as desmoglein-1, a desmosomal glycoprotein that plays an important role in maintaining cell-to-cell adhesion in the superficial epidermis.[79]

28.1.2.4 Antibiotic Resistance

The evolution of increasingly antimicrobial-resistant *S. aureus* strains depend on a multitude of factors, including the widespread and sometimes inappropriate use of antimicrobials, the extensive use of these agents as growth enhancers in animal feed, and, with the increase in regional and international travel, the relative ease with which antimicrobial-resistant bacteria cross geographic barriers. *S. aureus* isolates from intensive care units across the country and from blood culture isolates worldwide are increasingly resistant to a greater number of antimicrobial agents.[8]

Penicillin has been the drug of choice for treatment of infections caused by this organism; however, resistant *S. aureus* were reported as early as 1944.[83] By the late 1960s, more than 80% of both community and hospital acquired staphylococcal isolates were resistant to penicillin. This pattern of resistance, first emerging in hospitals and then spreading to the community, is now a well-established pattern that recurs with each new wave of antimicrobial resistance.[84] Staphylococcal resistance to penicillin is mediated by *blaZ*, the gene that encodes β-lactamase. This predominantly extracellular enzyme, synthesized when staphylococci are exposed to β-lactam antibiotics, hydrolyzes the β-lactam ring, rendering the β-lactam inactive.[85]

Methicillin was the first of the semisynthetic penicillinase-resistant penicillins, but its introduction was rapidly followed by the appearance of methicillin-resistant *S. aureus* (MRSA) in 1961.[86] MRSA have since become a major nosocomial pathogen worldwide. The reservoir of MRSA is infected and colonized patients, and the major mode of transmission from patient to patient is the contaminated hands of health care workers.[87] Methicillin resistance requires the presence of the chromosomally localized *mecA* gene.[88] *mecA* is responsible for synthesis of penicillin-binding protein 2a (PBP2a; also called PBP2′).[89] PBPs are membrane-bound enzymes that catalyze the transpeptidation reaction that is necessary for cross-linkage of peptidoglycan chains. PBP2a substitutes for the other PBPs and, because of its low affinity for all β-lactam antibiotics, enables staphylococci to survive exposure to high concentrations of these agents. Thus, resistance to methicillin confers resistance to all β-lactam agents, including cephalosporins.[90]

An additional concern is the emergence of vancomycin-intermediate *S. aureus* (VISA) and, more recently, vancomycin-resistant *S. aureus* (VRSA).[91]

28.1.3 Clinical Features and Pathogenesis of Coagulase-Negative Staphylococci

Among CNS, *S. epidermidis* is the predominant pathogen in intravascular catheter-related infections, nosocomial bacteremia, endocarditis, urinary tract and surgical wounds infections, central nervous system shunt infections, ophthalmologic infections, peritoneal dialysis-related infections and infections of prosthetic joints.[5] *S. haemolyticus* has been implicated in native valve endocarditis (NVE), septicemia, urinary tract infections, peritonitis, and wound, bone, and joint infections.[5] *S. saprophyticus* is associated with urinary tract infections in young females. *S. lugdunensis* has been implicated in arthritis, catheter infections, bacteremia, urinary tract infections, prosthetic joint infections, and endocarditis.[5] Other CNS species have been implicated in a variety of infections. Other infrequent human *S. caprae* infections include bone and joint infection, urinary tract infection, sepsis/bacteremia, recurrent sepsis, endocarditis, and acute otitis externa.[92–94]

28.1.3.1 Virulence Factors

The role of biofilm formation as a determinant of CNS infection has been the subject of ongoing study since the observation that some isolates of *S. epidermidis* produced mucoid growth in vitro that adhered to the walls of culture tubes.[95] This "slime" or biofilm is the most important virulence factor of *S. epidermidis*. The biofilm formation enables colonization and persistence on prosthetic material, resistance to the effects of antibiotics, and evasion of the immune system.[96] Biofilm production has been reported for *S. hominis* and *S. haemolyticus*,[97] *S. warneri*,[98] *S. saprophyticus*, and *S. simulans*.[99]

Biofilm formation is thought to occur in three distinct steps. First, the cells bind to the surface of the device in a reversible manner as a result of nonspecific forces such as polarity and hydrophobicity. Next, more specific adherence occurs as a result of the production of surface proteins (adhesins, autolysin/adhesins *Atl*E and Aae, the fibrinogen-binding protein Fbe/Sdrg, the fibronectin-binding protein Embp, and the lipase *Geh*D) that specifically bind to human host proteins that coat intravascular catheters or other devices. Once bound, the cells produce a thick biofilm consisting mainly of polysaccharide intercellular adhesion (PIA).[100] Cell aggregation and biofilm accumulation are mediated by the products of the chromosomal intercellular adhesion (*ica*) operon.[101] This operon is composed of the *icaR* regulatory gene and *icaADBC* biosynthetic genes.

Other potential virulence factors of *S. epidermidis* include the following extracellular enzymes and toxins: metalloprotease with elastase activity, cysteine protease, serine protease, lipase, FAME, and the δ-toxin. CNS also produce lantibiotics.[96] Lantibiotics are bacteriocins that are produced by *S. epidermidis* and other gram-positive bacteria—for example, *Bacillus subtilis* and lactobacilli—but not by *S. aureus*. Lantibiotics, such as the well-characterized epidermin, Pep5, epilancin K7, and epicidin 280, are antibiotic peptides active against gram-positive bacteria.[102] Lantibiotic production may play a substantial part in bacterial interference on skin and mucous membranes by excluding competing organisms that are sensitive to their bactericidal activities. *Staphylococcus lugdunensis* and *S. schleiferi* are more pathogenic than other CNS species. Like *S. aureus*, they may express a clumping factor and/or produce a thermostable DNAse.[103] A variable percentage of these strains produces extracellular slime or glycocalix, esterase, FAME, protease, lipase and a hemolysin. *S. lugdunensis* does not produce α-, β-, or γ-hemolysins, but a heat-stable, δ-like hemolysin that shows synergistic hemolytic activity with *S. aureus* when grown in proximity to each other on sheep's blood agar.[104] The main virulence factor of *S. saprophyticus* in UTI is the ability to attach to uroepithelial cells by means of surface-associated proteins—the autolysin/adhesin Aas, Ssp, and Sdrl. Once colonization has been established, several invasion factors are produced: urease, elastase, lipase, and FAME.[103]

28.1.4 Diagnosis

28.1.4.1 Conventional Techniques

Depending on the kind of infection, an appropriate specimen should be obtained accordingly and sent to the laboratory for identification. Specimens may include exudates, pus from abscesses, mastitic milk, skin scraping, urine, blood, and affected tissue samples.

Preparation and examination of Gram-stained smears of blood, pus, or exudates are first performed to guide the way, which should show gram-positive cocci in the typical cluster formation.

The isolation of staphylococci from clinical specimens is usually not difficult, since staphylococci grow easily on commonly used media (e.g., blood agar) within a broad range of growth conditions.

The routine medium for inoculation of specimens are sheep or bovine blood (5%) agar plates incubated aerobically at 37°C for 24–48 h. Colonies usually appear in 24 h. After 48 h incubation, colonies can reach 3–4 mm in diameter. The colonies are round, smooth, and on blood agar tend to appear substantial and opaque compared to the smaller, translucent colonies of beta-hemolytic streptococci. *S. aureus* strains usually have a golden-yellow pigment and also some of the CNS produce pigment, especially strains of *S. chromogenes* whose colonies are orange-yellow, while the colonies of other CNS are not pigmented. Blood agar plates prepared with either ovine or bovine erythrocytes are preferable as the red cells from both these animal species are susceptible to the α- and β-hemolysins that are produced commonly by staphylococcal isolates. *S. aureus* and *S. intermedius* are usually hemolytic and often produce both the α- and β-hemolysins, and thus exhibit a double halo of hemolysis. Hemolytic activity in CNS is variable and often slow in appearing.

Species determination of staphylococcal isolates is essential to distinguish *S. aureus* from coagulase-negative staphylococci (CNS).

Species identification of CNS isolated from clinical infections is difficult to perform and expensive. According to Huebner et al.[9] CNS species identification is in general not necessary for good patient management. However, recent literature underlines that accurate species identification is necessary to provide a better understanding of the role of the different pathogenic CNS species.[105,106]

On the other hand, different tests can be used to identify *S. aureus*, including growth on selective media, production of protein A, cell-bound clumping factor, extracellular coagulase, and heat-stable nuclease.

Selective Media. Mannitol salt agar and Baird-Parker medium are specifically selective media used for *S. aureus* identification. Mannitol salt agar, which is a selective medium with 7%–9% NaCl, allows *S. aureus* to grow producing yellow-colored colonies as a result of mannitol fermentation and subsequent drop in the medium's pH.

Baird-Parker agar is a type of agar used for the selective isolation of gram-positive *Staphylococcus* species. This medium contains lithium chloride and tellurite to inhibit the growth of accompanying microbial flora, whereas pyruvate and glycine selectively stimulate the growth of staphylococci.

Staphylococcus colonies grow with two distinct features on this opaque (because of its egg-yolk content) medium—characteristic zones and rings formed as a result of lipolysis and proteolysis and the reduction of tellurite to tellurium produces a black coloration.

Tube Coagulase Test. Free (extracellular) coagulase clots plasma in the absence of calcium. 0.5 mL of rabbit plasma is placed in a small test tube. One to two drops of an overnight broth culture of the *Staphylococcus* or a suspension made from a colony on an agar plate in sterile water are added. Tubes are examined after incubation for 4 and 24 h.[107,108] Tests negative after 4 h should be reexamined at 24 h because a small proportion of strains require longer incubation for clot formation. Some other species of staphylococci, including *S. schleiferi*, *S. intermedius* and *S. pseudointermedius*, may also give positive results in tube coagulase tests but are not commonly isolated from human infections. In addition, rare strains of *S. aureus* are negative in coagulase tests.

Slide Coagulase Test. Clumping factor (bound coagulase) differs from free coagulase in that it is cell-bound and requires only fibrinogen.[107] A loopful of the staphylococcal culture is suspended in a drop of water on a microscope slide. A drop of rabbit plasma is added and mixed with the bacterial suspension. The slide is gently rocked, and within 1 or 2 min, clumping will indicate a positive reaction. The slide agglutination test for clumping factor is very rapid, but up to 15% of *S. aureus* strains are negative,[5] so negative isolates should be confirmed with a tube agglutination test. Some less common species of staphylococci, including *S. schleiferi* and *S. lugdunensis*, may give positive results in the slide coagulase test.

Latex Agglutination Tests. A simple slide agglutination test for the simultaneous detection of protein A and clumping factor uses latex particles coated with human plasma.[109] Several commercial kits based on this principle are available for the rapid identification of *S. aureus*. These tests are not sensitive with some MRSA that produce little or no clumping factor and protein A.[110] Later formulations of latex tests include protein A and/or clumping factor but also detect various surface antigens, which improved the sensitivity of the tests but at some expense to specificity due to cross-reaction with CNS.[111–113] In addition, any test including clumping factor may give false-positive results with *S. lugdunensis* and *S. schleiferi*.[108,114]

DNAse Tests. DNAse agar media are used to determine the production of deoxyribonuclease by bacteria, particularly *S. aureus*.[115] After incubation, the inoculated plates are flooded with 1 N HCl. Bacteria possessing the enzyme degrade the DNA into nucleotide fractions that are not affected by the acid, and a clear zone in the media occurs. If no DNAse is produced, the plate will appear opaque, as the HCl combines with nucleate salts to yield free nucleic acid, which is cloudy. DNAse medium can be modified by adding the toluidine blue into the agar.[116] When DNA is hydrolyzed, the dye becomes complexed with oligonucleotides and mononucleotides, and the absorption spectrum of the dye shifts, resulting in a metachromatic (pink) color in the agar.

DNAse plates can be used to screen *S. aureus* isolates, but, as various amounts of DNAse are produced by CNS, positives should be confirmed with an additional test.

Commercial Biochemical Tests. There are many commercial kits available that include identification of *S. aureus* and differentiation of CNS. Methods include commercial test systems such as the API 20 Staph system (bioMérieux), API ID 32 Staph (bioMérieux), Staph-Zym (Rosco), the Vitek system (bioMérieux), and other combinations of biochemical tests, which may not be available in commercial formats. While performance of these tests may be good, they are slower and technically more time consuming or more expensive than tests such as coagulase and latex agglutination.

28.1.4.2 Molecular Techniques

The detection of nucleic acid targets, with its high sensitivity and specificity, provides an alternative technique for the accurate identification and classification of *Staphylococcus* species. In general, genotypic methods have higher discriminatory power, reproducibility and typeability than phenotypic methods. Since its discovery in the 1980s, PCR has become the cornerstone of molecular biology, and currently available molecular diagnostic tests also rely heavily on this technology. Benefits afforded by PCR over conventional culture include detection limits below those possible with culture, high-throughput screening (96 specimens/run), and, importantly, shorter times to detection. Many efforts particularly focus on the timely identification of pathogenic staphylococci.

Genotypic methods used for identification at the species level and for strain typing (i.e., differentiation of isolates at the subspecies level) of *Staphylococcus* species include amplified fragment length polymorphism (AFLP) analysis,[117,118] ribotyping,[119] and DNA sequencing.[105,106]

Across bacterial genera, the most common target for DNA-sequencing is the 16S rRNA gene.[120] The sequence variability of the 16S rRNA genes of staphylococci enables identification at species level. The 16S rRNA gene represents a ubiquitous, highly conserved gene in which discrete regions have accumulated point mutations during the evolution of individual species. Comparison of the 16S rRNA sequence of an unknown organism to a sequence repository, such as the GenBank database maintained by the National Center for Biotechnology Information (http://www.ncbi.nlm.nih.gov/GenBank/index.html) or the Ribosomal Database Project (http://rdp.cme.msu.edu/html/), can usually give an accurate identification. Thus, PCR amplification and sequencing of this gene have become a tool for molecular identification of pathogenic bacteria in the clinical microbiology laboratory.[121] Real-time PCR assays utilizing fluorescent resonance energy transfer probes that bind to regions of the staphylococcal 16S rRNA gene following amplification with broad-range primers have also been developed.[122] Such methods may enable identification of organisms in a shorter time than is needed to amplify and sequence the 16S rRNA gene.

Many *Staphylococcus* species are closely related, and 16S rRNA sequence-based typing may not have sufficient discriminatory power to differentiate all of them.[123,124] Therefore, species identification systems based on the 16S-23S rRNA intergenic spacer region[125] and on the housekeeping genes *cpn60* (chaperonin or heat-shock protein 60),[126] *dnaJ* (heat-shock protein 40),[124] *rpoB* (beta subunit of RNA polymerase),[127,128] *sodA* (superoxide dismutase A),[105,106,129] *tuf* (elongation factor Tu),[105] *femA*,[130] and *gap*[131,132] have been developed and implemented for *Staphylococcus* species identification.

For analysis of sequence data, the interpretation criteria are gene specific. For the highly conserved 16S rRNA gene, 98% or 99% has been used as cut-off value.[123,133,134] Other housekeeping genes, such as *sodA* and *tuf*, are less conserved, which allows them to be used in sequence-based strain typing methods.[135] For these genes, homology values of 97% or more are considered acceptable.[105,106] Within-species heterogeneity of housekeeping genes differs between bacterial species.[136] Thus, both within species-variability and between-species variability of sequence data should be considered to decide on appropriate cut-off values for homology.[120] Some authors specifically include criteria for difference in sequence identity from the next closest species (e.g., 5% or more) in guidelines for interpretation of DNA sequence data of *Staphylococcus* species.[106]

Members of the genus *Staphylococcus* are major human pathogens, causing a wide variety of hospital- and community-acquired infections worldwide. *S. aureus*, in particular, is responsible for diseases that range from mild skin infections to life-threatening illnesses, such as deep postsurgical infections, septicemia, and toxic shock syndrome.[8] Therefore, it is important to characterize and distinguish *S. aureus* strains and CNS. Furthermore, both *S. aureus* and CNS are characterized by the ability to develop and accumulate antibiotic resistance determinants, resulting in the formation of multidrug-resistant strains. Such strains have demonstrated resistance to penicillins, clindamycin, tetracyclins, macrolides, linezolid, methicillin and, more recently, mupirocin and vancomycin. This resistance limits therapeutic intervention and substantially increases patient morbidity and mortality.

Simple and rapid identification and discrimination of *S. aureus* and CNS, the detection of methicillin resistance (MR) in *S. aureus* (MRSA) and mupirocin resistance (MuR) in CNS (MRCNS) are essential for the choice of effective antimicrobial chemotherapy and for limiting the unnecessary use of certain classes of antibiotics.

Since MR (encoded by *mecA* gene) and high-level MuR (H-MuR, encoded by *mupA* gene) are mediated by the expression of PBP2a and by the expression of altered isoleucyl-tRNA synthetase, respectively, PCR detection of *mecA* and *mupA* is considered the "gold standard" for the detection MR and HMuR.[137–139] Recently, Perez-Roth et al.[140] demonstrated the feasibility of a multiplex PCR for the simultaneous identification of *S. aureus* (by means of *femB*, a gene found in all *S. aureus* strains) and detection of MR and MuR (using *mecA* and *mupA*, respectively), while Zangh et al.[141] developed a novel quadriplex PCR assay to identify most staphylococci, to discriminate *S. aureus* from CNS and other bacteria, and to simultaneously detect MR resistance and MuR. The quadriplex PCR assay targeted the 16S rRNA, *nuc*, *mupA*, and *mecA* genes.

Al-talib et al.[142] recently developed a pentaplex PCR assay for the rapid detection of MRSA. The assay simultaneously detected five genes, namely 16S rRNA of the *Staphylococcus* genus, *femA* of *S. aureus*, *mecA*, *lukS* that encodes production of Panton–Valentine leukocidin (PVL), a necrotizing cytotoxin, and one internal control. The sensitivity and specificity of the pentaplex PCR assay was evaluated by comparing it with the conventional method. The sensitivity of the pentaplex PCR at the DNA level was found to be 10 ng DNA. The pentaplex PCR assay developed was rapid and gave results within 4 h, which is essential for the identification of *Staphylococcus* spp., virulence and their resistance to methicillin.

For epidemiological studies, numerous techniques are available to differentiate *S. aureus*, specifically MRSA isolates. Initial molecular techniques compared restriction endonuclease patterns of chromosomal or plasmid DNA.[143,144] The second generation of genotyping methods included Southern blot hybridization using gene specific probes, ribotyping,[119] pulsed-field gel electrophoresis (PFGE), and polymerase chain reaction (PCR)-based approaches.[143,145,146] Multiple-locus variable-number tandem-repeat analysis (MLVA) is a method for bacterial typing based on the detection of short sequence repeats that vary in copy number, such as variable numbers of tandem repeats (VNTR) at different regions of the microbial genome. These VNTRs are often highly polymorphic, with variation in both the number of repeat units and by sequence heterogeneity among individual units.[147,148]

Multilocus sequence typing (MLST) is the reference method used to identify clones within populations of *S. aureus*, which is characterized by a dozen major clonal complexes and comprises several hundred clonal lineages or sequence types (STs).[149] Genetic variation is analyzed directly by nucleotide sequencing the internal fragments of seven housekeeping genes. The sequences of each gene are defined as distinct alleles, and the ST of each isolate is defined by the alleles at each of the seven housekeeping loci. A major advantage of MLST is the ability to compare sequence data between laboratories through the MLST Web site (http://www.mlst.net/). However, for epidemiological typing, MLST has only moderate discriminatory power, and it remains too labor-intensive and costly for use as a primary typing tool.

Additional molecular tests should be applied to subtype the mobile methicillin-resistance element, called the staphylococcal chromosomal cassette mec (SCCmec), which differentiates MRSA clones of common ancestry but distinct epidemiological origin.[150]

Pulsed-field gel electrophoresis (PFGE) has been accepted as the reference method for molecular strain typing of MRSA. PFGE is known to be highly discriminatory and therefore frequently used for outbreak analysis.[151] However, this strategy is labor intensive, time consuming, and its technical instability has an adverse effect on reproducibility.[152] Therefore, results are not comparable between laboratories. Other typing methods such as manual sequence-based PCR (rep-PCR)[152,153] and randomly amplified polymorphic DNA (RAPD) analysis generate similar problems.[154] Rep-PCR uses primers that target noncoding repetitive sequences interspersed in bacterial and fungal genomes.[155–157] When separated by electrophoresis, the amplified DNA fragments constitute a genomic fingerprint that can be employed for subspecies discrimination and strain delineation of bacteria and fungi.[158] The development of a commercially available, automated rep-PCR assay system, the DiversiLab System (bioMérieux, Boxtel, Netherlands), offers advances in standardization and reproducibility over manual fingerprint generating systems.[159]

The determination of sequence polymorphism in the variable X region of the *spa* gene encoding the staphylococcal surface protein A (*spa* sequence typing) has become one of the most popular MRSA typing system. Its popularity is attributable to its many practical advantages, including high throughput and full portability of data, thanks to its absolute reproducibility, which allows internet-based type assignment and comparison with worldwide databases (http://www.spaserver.ridom.de/ and http://www.seqnet.org/).[160,161] It has therefore become one of the primary typing methods for MRSA surveillance schemes. Nevertheless, this technique also has limitations, which necessitate the use of additional tests for reliable typing in some circumstances. Combinations of methods including multiple-locus sequence typing (MLST), *spa* typing, and SCCmec typing have been used to characterize MRSA for epidemiological purposes but none has substituted PFGE as a reference method for bacterial typing.[162]

Finally, single nucleotide polymorphism (SNP) detection methods offer an efficient alternative to the above-mentioned typing methods. These methods have been developed for rapid identification of major *S. aureus* clonal complexes and STs, as well as for detailed discrimination of regionally emerging MRSA subclones.[163] Additional mutation discovery studies in different *S. aureus* lineages should allow the development of a SNP dataset that could be used for high-throughput SNP typing systems. In the meantime, PCR- or microarray-based detection of clone-specific SNPs, genes, or genomic islands can be useful for rapid identification of epidemiologically important MRSA clones.

28.2 METHODS

28.2.1 Sample Preparation

Sample Collection and Processing. Screening for *Staphylococcus* spp. infection in patients is generally performed by exudates, pus from abscesses, mastitic milk, skin scraping, superficial sites swabbing including nares, axillae/groin, throat, perineum, rectum, sputum of ventilated patients, wounds, and/or catheter insertion sites, urine, and blood.

The specimen for analysis should be collected in sterile conditions and placed in a buffered mediam for transportation to the laboratory by the most rapid mode available. The use of liquid rather than gel transport medium may increase the recovery of bacteria and the sensitivity of detection.

The bacterium is isolated on selective medium and further identified by biochemical and/or molecular methods by DNA extraction from cultures.

DNA Extraction. Working protocols for rapid DNA extraction directly from whole blood, buffy coat, or cultured cells are described in the literature.[163–166]

The following protocol, based on the ability of silica-resin to bind DNA in the presence of a high concentration of guanidine thiocyanate chaotropic agent that guarantees an excellent disruption of bacterial cells from the main gram-positive pathogens (such as *S. aureus*), can be used for DNA extraction both from whole blood or cultured cells.

1. For cultured cells samples, transfer about 10^9 cells from culture into a 1.5 mL vial, centrifuge at 10,000 × *g* for 1 min and remove the supernatant. Wash the pellet once with 500 μL of sterilized saline solution (NaCl 0.9% w/v); centrifuge at 10,000 × *g* for 1 min and discard the supernatant. Add to the pellet 50 μL of sterilized saline solution. Resuspend the pellet vortexing (20 s). For whole blood samples, transfer 300 μL of whole blood into a 1.5 mL vial.
2. Add 200 μL of lysis buffer (3 M guanidine thiocyanate, 20 mM EDTA, 10 mM Tris-HCl pH 6.8, 40 mg/mL Triton X-100, 10 mg/mL DL-dithiothreitol) and 100 μL of binding solution (40 mg/mL silica from Sigma Aldrich, directly suspended in the lysis buffer) to the sample.
3. Mix and incubate for 3 min at room temperature. Centrifuge for 30 s at 450 × *g* (or 1 min at 200 × *g* for blood sample) and discard the supernatant.
4. Add 100 μL of lysis buffer and resuspend by vortexing (20 s). Centrifuge for 30 sec at 450 × *g* (or 1 min at 200 × *g* for blood sample) and discard the supernatant. Repeat this step once.
5. Add 100 μL of washing solution (25% w/v ethanol, 25% w/v isopropanol, 100 mM NaCl, 10 mM Tris-HCl pH 8) and resuspend by vortexing (20 s). Centrifuge for 30 s at 450 × *g* (or 1 min at 200 × *g* for blood sample) and discard the supernatant. Repeat this step once.
6. Add 100 μL of absolute ethanol solution and resuspend by vortexing (20 s). Centrifuge for 30 s at 450 × *g* (or 1 min at 200 × *g* for blood sample) and discard the supernatant.
7. Dry the pellet in an Eppendorf heat block at 56°C for 10 min (or at room temperature for at least 15–20 min).
8. Add 70 μL of elution buffer (10 mM Tris-HCl pH 8.0, 1 mM EDTA), resuspend the pellet, vortex for 20 s and incubate for 10 min at 65°C.
9. Centrifuge for 5 min at 500 × *g* (or 3 min at 500 × *g* for cheese sample) and transfer the supernatant, containing the DNA, into a clean tube.
10. To increase the DNA yield, a second elution step (with 5 min heating) may be performed.
11. The solution contains pure DNA useful for molecular biology techniques. It can be used immediately or stored at –20°C for up to 6 months.

28.2.2 Detection Procedures

In the present section, a multiplex-PCR-based protocol with four primer sets is described to identify methicillin-resistant *Staphylococcus aureus*. The primer sets include the *mecA* gene, a structural determinant that encodes PBP2a, considered as a useful molecular marker of putative methicillin-resistance in *S. aureus* and CNS[140,167]; *fem*A and *fem*B, implicated in cell wall metabolism and essential for the expression of methicillin-resistance in *S. aureus*[168] and IS431, a transposable element of methicillin-resistance Staphylococci, related to antimicrobial resistance determinants in bacterial cells through its ability to enhance gene transfer or rearrangement affecting gene expression via the recombination process.[169]

All primers used in the study, ranging from 20 to 21-mers, were designed using the Primer3 program (http://www-genome.wi.mit.edu/cgibin/primer/primer3_www.cgi). The primer sequences and the PCR product lengths are shown in Table 28.1.

TABLE 28.1
Multiplex PCR Primers for the Identification of Methicillin-Resistant *Staphylococcus aureus*

Primer	Sequence (5′–3′)	Amplicon Size (bp)	Target Gene
IS431-F376	TCAAGAATATGCCCCGATTT	468	IS431
IS431-R843	AGAGACCTGCGGTTCTTTTT		IS431
MECA-F141	GGCTATCGTGTCACAATCGTT	302	*mecA*
MECA-R442	TGGAACTTGTTGAGCAGAGG		*mecA*
FEMA-F346	GAGGTCATTGCAGCTTGCTT	213	*femA*
FEMA-R558	CTCGCCATCATGATTCAAGT		*femA*
FEMB-F100	CGTGAGAATGATGGCTTTGA	388	*femB*
FEMB-R487	TTAATACGCCCATCCATCGT		*femB*

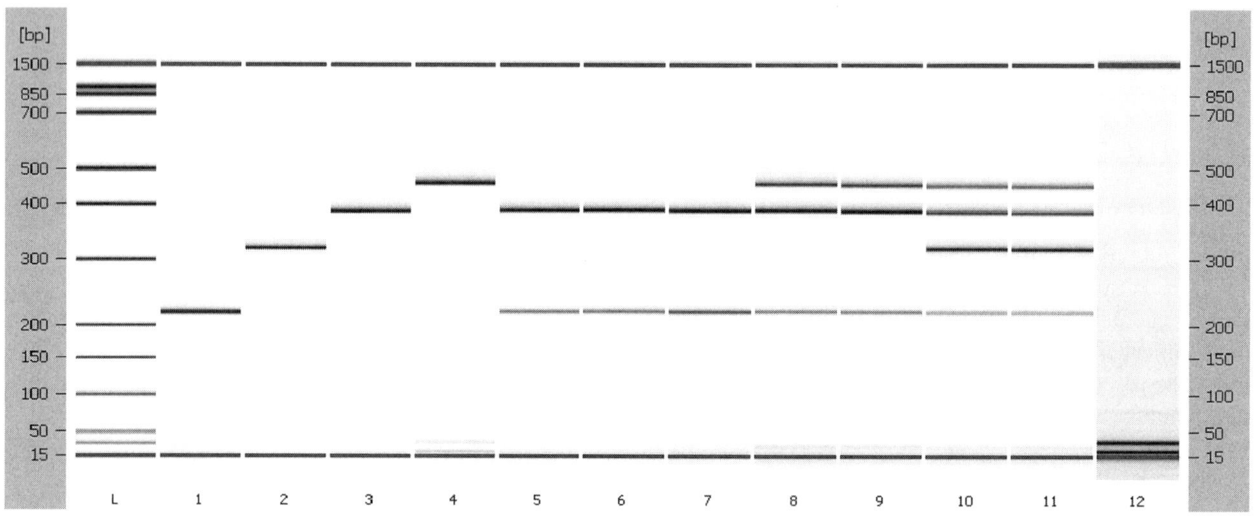

FIGURE 28.1 Single and multiplex PCR analysis for the detection of methicillin-resistant *Staphylococcus aureus* strains isolated from different patients by using the Agilent 2100 Bioanalyser software.

From lane 1 to lane 4 the gel like image shows the single PCR results obtained from the amplification of *fem*A (213 bp), *mec*A (302 bp), *fem*B (388 bp), IS431 (488 bp), respectively.

From lane 5 to lane 10 all the samples analyzed by multiplex PCR are positive for *fem*A (213 bp) and *fem*B (388 bp), three samples are positive also for IS431 (488 bp), and only one sample is positive for *mec*A (302 bp) gene.

Lane 11 reference strain MRSA ATCC 43300. Lane 12: negative control. L: DNA 500 ladder.

1. PCR mixture (50 μL) is composed of DNA template, 2 U of AccuPrime™ *Taq* DNA polymerase (Invitrogen,), 5 μL of 10X AccuPrime™ PCR buffer II containing 2 mM of each dNTPs (Invitrogen), 20 μM of the primer pairs FEMA-F346 and FEMA-R558, FEMB-F100 and FEMB-R487, MECA-F141 and MECA-R442, 40 μM of the primer pairs IS431-F376 and IS431-R843 and double-distilled water.
2. Amplification is carried out in a thermal cycler GeneAmp PCR System 2700 (Applied Biosystems) with an initial denaturation at 94°C for 5 min; 30 cycles of 94°C, 56°C, and 68°C for 1 min each; and a final extension at 72°C for 7 min.
3. The amplified PCR products are visualized simultaneously by standard gel electrophoresis in a 3% w/v agarose gel (GellyPhor, Euroclone), stained with ethidium bromide (0.05 μg/μL; Sigma Aldrich), and photographed under ultra violet light using the BioProfile system (Mitsubishi). Molecular size markers (100-bp and 1-kb DNA ladder; Finnzymes) are included in each agarose gel (Figure 28.1).

28.3 CONCLUSIONS AND FUTURE PERSPECTIVES

Staphylococci are considered one of the most important and leading cause of community- and hospital-acquired infections. Thus, the results of diagnostic assays can be a key factor for medical decision making, including antimicrobial choice for patient therapy, and epidemiological studies.

Today, identification approaches by the use of molecular tests not only perform comparably to conventional tests but also have the advantage of providing results more rapidly by eliminating the time needed to recover and grow colonies on solid agar media. However, one of the limitations for the use of molecular methods as a routine detection practice is the high cost associated with its use. Other limitations are the workload of specimen processing and the need for skilled personnel to perform the methodology, thereby making it difficult to maintain the service 24 h/7 days a week. Thus, additional technological enhancements are required to provide molecular techniques with more flexibility in processing samples and, in general, more ease of use.

The public health burden caused by MRSA infections is today widely recognized, and is a cause of public alarm. Effective MRSA risk management in the health care system as well as in the community should rely on accurate detection of reservoirs and sources of transmission, as well as on close monitoring of the impact of interventions on disease incidence and bacterial dissemination. MRSA carrier screening and disease surveillance are key information tools for integrated MRSA control and individual risk assessment. Consequently, rapid detection and identification of MRSA are absolute goals. Until recently, microbiological methods dedicated to MRSA identification were based on the utilization of selective growth media, which are time consuming and preclude same-day diagnosis. For more than one decade, nucleic acid–based identification assays have demonstrated their usefulness and robustness for the detection of hardly cultivable, noncultivable and even killed microorganisms, as well as for the identification of specific pathogens against the background of a complex microflora. Molecular assays based on target nucleic acid amplification, and especially real-time PCR, have proven rapid, affordable and successful in terms of sensitivity and specificity. However, the current view is still that molecular methods should be coupled to conventional methods, because bacterial cultures are still required for further antimicrobial susceptibility testing or epidemiological typing. As the molecular epidemiology of MRSA may substantially vary from country to country, and possibly also between different regions, this underlines the importance of having an epidemiologic molecular tool ideally as close as possible between labs.

Current challenges for MRSA screening are based on the choice of the most appropriate assay, both in terms of feasibility (costs, technical expertise), sensitivity and specificity, and assay performance. One has to be especially careful when embarking on detection strategies that are based on mobile and highly variable genetic regions, such as the SCCmec insertion site. Indeed, iterative changes in the detection protocol to adapt for emerging variants might not only affect the performance of the assay but also open unpredictable and systematic failure in the infection surveillance and prevention.

Future developments of molecular methods will be directed at ensuring that, whichever assay methodology is used, there is a high degree of specificity for MRSA detection.

For example, such methods can be applied for the direct detection of MRSA from nonsterile specimens such as nasal samples without the previous isolation, capture, or enrichment of MRSA because these samples often contain both CNS and *S. aureus*, either of which can carry *mecA*.

Eventually, it may be likely that in clinical microbiology laboratories the chip arrays, which are currently used mainly in a research setting,[170] will be developed for the rapid mass screening of bacterial isolates for both species identification and antimicrobial resistance genes detection. However, they are expensive and require dedicated platforms and software that restrict their use to a few specialized research teams. Recently, microarrays targeting a limited number of well-known virulence and resistance genes have been described that allow high-throughput characterization of clinical isolates of *S. aureus*.[171,172]

Recent advances for use in research laboratories include pyrosequencing,[173] denaturing high performance liquid chromatography (dHPLC)[174] and nucleic acid analysis by mass spectrometry.[175] These high-throughput technologies will allow rapid detection of known single nucleotide polymorphisms (SNPs), including those associated with antibiotic resistance phenotypes, and can also be used for novel SNP discovery. Further technological advances, which use microfluidics and nanolitre volumes, promise amplification and detection of resistance genes (or other targets) in minutes.[176]

REFERENCES

1. Igimi, S., Takahashi, E., and Mitsuoka, T., *Staphylococcus schleiferi* subsp. *coagulans* subsp. *nov*., isolated from the external auditory meatus of dogs with external ear otitis, *Int. J. Syst. Bacteriol*., 40, 409, 1990.
2. Foster, G., et al., *Staphylococcus lutrae* sp. *nov*., a new coagulase-positive species isolated from otters, *Int. J. Syst. Bacteriol*., 47, 724, 1997.
3. Komori, Y. et al., Characterization of coagulase from *Staphylococcus intermedius*, *J. Nat. Toxins*., 10, 111, 2001.
4. Devriese, L.A. et al., *Staphylococcus pseudointermedius* sp. nov., a coagulase-positive species from animals, *Int. J. Syst. Evol. Microbiol*., 55, 1569, 2005.
5. Bannerman, T.L., and Peacock, S.J., *Staphylococcus*, *Micrococcus*, and other catalase-positive cocci. In Murray, P.R., Baron, E.J., Jorgensen, J.H., Landry, M.L., and Pfaller, M.A. (Editors), *Manual of Clinical Microbiology*, 9th ed., vol. 1, pp. 390–411, ASM Press, Washington, DC. 2007.
6. Fluit, A.C., Verhoef, J., and. Schmitz, F.J., Frequency of isolation and antimicrobial resistance of gram-negative and gram-positive bacteria from patients in intensive care units of 25 European university hospitals participating in the European arm of the SENTRY Antimicrobial Surveillance Program 1997–1998, *Eur. J. Clin. Microbiol. Infect. Dis*., 20, 617, 2001.
7. Pfaller, M.A. et al., Survey of blood stream infections attributable to gram-positive cocci: frequency of occurrence and antimicrobial susceptibility of isolates collected in 1997 in the United States, Canada, and Latin America from the SENTRY Antimicrobial Surveillance Program, *Diagn. Microbiol. Infect. Dis*., 33, 283, 1999.
8. Diekema, D.J. et al., Survey of infections due to *Staphylococcus* species: Frequency of occurrence and antimicrobial susceptibility of isolates collected in the United States, Canada, Latin America, Europe, and the Western Pacific region for the SENTRY Antimicrobial Surveillance Program, 1997–1999, *Clin. Infect. Dis*., 32, 114, 2001.

9. Huebner, J., and Goldmann, D.A., Coagulase-negative staphylococci: Role as pathogens, *Annu. Rev. Med.*, 50, 223, 1999.
10. Kawamura, Y. et al., Distribution of *Staphylococcus* species among human clinical specimens and emended description of *Staphylococcus caprae*, *J. Clin. Microbiol.*, 36, 2038, 1998.
11. Kumar, D. et al., Case of *Staphylococcus schleiferi* subspecies *coagulans* endocarditis and metastatic infection in an immune compromised host, *Transpl. Infect. Dis.*, 9, 336, 2007.
12. Koksal, F., Yasar, H., and Samasti, M., Antibiotic resistance patterns of coagulase-negative *Staphylococcus* strains isolated from blood cultures of septicemic patients in Turkey, *Microbiol Res.*, 164, 404, 2009.
13. Piette, A., and Verschraegen, G., Role of coagulase-negative staphylococci in human disease, *Vet. Microbiol.*, 134, 45, 2009.
14. Rogers, K.L., Fey, P.D., and Rupp, M.E., Coagulase-negative staphylococcal infections, *Infect. Dis. Clin. North Am.*, 23, 73, 2009.
15. Goldmann, D.A., Leclair, J., and Macone, A., Bacterial colonization of neonates admitted to an intensive care environment, *J. Pediatr.*, 93, 288, 1978.
16. Williams, R.E., Healthy carriage of *Staphylococcus aureus*: Its prevalence and importance, *Bacteriol. Rev*, 27, 56, 1963.
17. Armstrong-Esther, C.A., and Smith, J.E., Carriage patterns of *Staphylococcus aureus* in a healthy non-hospital population of adults and children, *Ann. Hum. Biol.*, 3, 221, 1976.
18. Wertheim, H.F. et al., Effect of mupirocin treatment on nasal, pharyngeal, and perineal carriage of *Staphylococcus aureus* in healthy adults, *Antimicrob. Agents Chemother.*, 49, 1465, 2005.
19. Ridley, M., Perineal carriage of *Staphylococcus aureus*, *Br. Med. J.*, 34, 270, 1959.
20. Rimland, D., and Roberson, B., Gastrointestinal carriage of methicillin-resistant *Staphylococcus aureus*, *J. Clin. Microbiol.*, 24, 137, 1986.
21. Guinan, M.E. et al., Vaginal colonization with *Staphylococcus aureus* in healthy women: A review of four studies, *Ann. Intern. Med.*, 96, 944, 1982.
22. Dancer, S.J., and Noble, W.C., Nasal, axillary, and perineal carriage of *Staphylococcus aureus* among women: Identification of strains producing epidermolytic toxin, *J. Clin. Pathol.*, 44, 681, 1991.
23. Nouwen, J.L. et al., Predicting the *Staphylococcus aureus* nasal carriage state: Derivation and validation of a culture rule, *Clin. Infect. Dis.*, 39, 806, 2004.
24. Nouwen, J.L. et al., Persistent (not intermittent) nasal carriage of *Staphylococcus aureus* is the determinant of chronic peritoneal dialysis-related infections, *Kidney Int.*, 67, 1084, 2005.
25. Kluytmans, J., van Belkum, A., and Verbrugh, H., Nasal carriage of *Staphylococcus aureus*: Epidemiology, underlying mechanisms, and associated risks, *Clin. Microbiol. Rev.*, 10, 505, 1997.
26. Noble, W.C., and Pitcher, D.G., Microbial ecology of the human skin, *Adv. Microbiol. Ecol.*, 2, 245, 1978.
27. Kloos, W.E., Taxonomy and systematics of staphylococci indigenous to humans. In Crossley, K.B., and Archer, G.L. (Editors), *The Staphylococci in Human Disease*, vol. 1, pp. 113–137, Churchill Livingston, New York, 1997.
28. Ikeda, Y. et al., PCR-based identification of *Staphylococcus epidermidis* targeting gseA encoding the glutamic acid-specific protease, *Can. J. Microbiol.*, 50, 493, 2004.
29. Jarlov, J.O. et al., Coagulase-negative staphylococci in Danish blood cultures: Species distribution and antibiotic susceptibility, *J. Hosp. Infect.*, 32, 217, 1996.
30. Bieber, L., and Kahlmeter, G., *Staphylococcus lugdunensis* in several niches of the normal skin flora, *Clin Microbiol Infect.*, 16, 385, 2010.
31. Moroni, P. et al., Characterization of *Staphylococcus aureus* isolated from chronically infected dairy goats, *J. Dairy Sci.*, 88, 3500, 2005.
32. Blanc, V. et al., Infection after total hip replacement by *Staphylococcus caprae*. Case report and review of the literature, *Pathol. Biol.*, 47, 409, 1999.
33. Novick, R.P., Mobile genetic elements and bacterial toxinoses: The superantigen-encoding pathogenicity islands of *Staphylococcus aureus*, *Plasmid*, 49, 93, 2003.
34. Narita, S. et al., Phage conversion of Panton–Valentine leukocidin in *Staphylococcus aureus*: Molecular analysis of a PVL-converting phage, phiSLT, *Gene*, 268, 195, 2001.
35. de Haas, C.J. et al., Chemotaxis inhibitory protein of *Staphylococcus aureus*, a bacterial antiinflammatory agent, *J. Exp. Med.*, 199, 687, 2004.
36. Patti, J.M. et al., MSCRAMM-mediated adherence of microorganisms to host tissues, *Annu. Rev. Microbiol.*, 48, 585, 1994.
37. Patti, J.M. et al., The *Staphylococcus aureus* collagen adhesin is a virulence determinant in experimental septic arthritis, *Infect. Immun.*, 62, 152, 1994.
38. Essmann, F. et al., *Staphylococcus aureus* alpha-toxin-induced cell death: Predominant necrosis despite apoptotic caspase activation, *Cell Death Differ.*, 10, 1260, 2003.
39. Haslinger-Löffler, B. et al., Multiple virulence factors are required for *Staphylococcus aureus*-induced apoptosis in endothelial cells, *Cell. Microbiol.*, 7, 1087, 2005.
40. Haslinger-Löffler, B. et al., *Staphylococcus aureus* induces caspase-independent cell death in human peritoneal mesothelial cells, *Kidney Int.*, 70, 1089, 2006.
41. Jarry, T. M., Memmi, G., and Cheung, A.L., The expression of alpha-haemolysin is required for *Staphylococcus aureus* phagosomal escape after internalization in CFT-1 cells, *Cell. Microbiol.*, 10, 1801, 2008.
42. Kubica, M. et al., A potential new pathway for *Staphylococcus aureus* dissemination: The silent survival of *S. aureus* phagocytosed by human monocyte-derived macrophages, *PLoS One*, 3, 1409, 2008.
43. Forti, S., and Menestrina, G., Staphylococcal alpha-toxin increases the permeability of lipid vesicles by cholesterol- and pH-dependent assembly of oligomeric channels, *Eur. J. Biochem.*, 181, 767, 1989.
44. Walev, I. et al., Staphylococcal alpha-toxin kills human keratinocytes by permeabilizing the plasma membrane for monovalent ions, *Infect. Immun.*, 61, 4972, 1993.
45. Bhakdi, S., Muhly, M., and Fussle, R., Correlation between toxin binding and hemolytic activity in membrane damage by staphylococcal alpha-toxin, *Infect. Immun.*, 46, 318, 1984.
46. Bhakdi, S., and Tranum-Jensen, J. Alpha-toxin of *Staphylococcus aureus*, *Microbiol. Rev.*, 55, 733, 1991.
47. Freer, J.H., and Birbeck, T.H., Possible conformation of delta-lysin, a membrane-damaging peptide of *Staphylococcus aureus*, *J. Theor. Biol.*, 94, 535, 1982.
48. Woodin, A.M., Purification of the two components of leukocidin from *Staphylococcus aureus*, *Biochem. J.*, 75, 158, 1960.
49. Hemker, H.C., Bas, B.M., and Muller, A.D., Activation of a pro-enzyme by a stoichiometric reaction with another protein. The reaction between prothrombin and staphylocoagulase, *Biochim. Biophys. Acta*, 379, 180, 1975.

50. Kawabata, S. et al., Isolation and characterization of staphylocoagulase chymotryptic fragment. Localization of the procoagulant- and prothrombin-binding domain of this protein, *J. Biol. Chem.*, 261, 1427, 1986.
51. McDevitt, D., Vaudaux, P., and Foster, T.J., Genetic evidence that bound coagulase of *Staphylococcus aureus* is not clumping factor, *Infect. Immun.*, 60, 1514, 1992.
52. Foster, T.J., Immune evasion by staphylococci, *Nat. Rev. Microbiol.*, 3, 948, 2005.
53. Marraffini, L.A., Dedent, A.C., and Schneewind, O., Sortases and the art of anchoring proteins to the envelopes of gram-positive bacteria, *Microbiol. Mol. Biol. Rev.*, 70, 192, 2006.
54. Mazmanian, S.K., Ton-That, H., and Schneewind, O., Sortase-catalysed anchoring of surface proteins to the cell wall of *Staphylococcus aureus*, *Mol. Microbiol.*, 40, 1049, 2001.
55. O'Brien, L. et al., Multiple mechanisms for the activation of human platelet aggregation by *Staphylococcus aureus*: Roles for the clumping factors ClfA and ClfB, the serine-aspartate repeat protein SdrE and protein A, *Mol. Microbiol.*, 44, 1033, 2002.
56. McDevitt, D. et al., Characterization of the interaction between the *Staphylococcus aureus* clumping factor (ClfA) and fibrinogen, *Eur. J. Biochem.*, 247, 416, 1997.
57. Peacock, S.J. et al., Virulent combinations of adhesin and toxin genes in natural populations of *Staphylococcus aureus*, *Infect. Immun.*, 70, 4987, 2002.
58. McDevitt, D. et al., Molecular characterization of the clumping factor (fibrinogen receptor) of *Staphylococcus aureus*, *Mol. Microbiol.*, 11, 237, 1994.
59. Moreillon, P., Role of *Staphylococcus aureus* coagulase and clumping factor in pathogenesis of experimental endocarditis, *Infect. Immun.*, 63, 4738, 1995.
60. Vaudaux, et al., Use of adhesion-defective mutants of *Staphylococcus aureus* to define the role of specific plasma proteins in promoting bacterial adhesion to canine arteriovenous shunts, *Infect. Immun.*, 63, 585, 1995.
61. Ní Eidhin, D. et al., Clumping factor B (ClfB), a new surface-located fibrinogen-binding adhesin of *Staphylococcus aureus*, *Mol. Microbiol.*, 30, 245, 1998.
62. Bokarewa, M.I., Jin, T., and Tarkowski, A., *Staphylococcus aureus*: Staphylokinase, *Int. J. Biochem. Cell Biol.*, 38, 504, 2006.
63. Bronner, S., Monteil, H., and Prévost, G., Regulation of virulence determinants in *Staphylococcus aureus*: Complexity and applications, *FEMS Microbiol. Rev.*, 28, 183, 2004.
64. Mortensen, J.E., Shryock, T.R., and Kapral, F.A., Modification of bactericidal fatty acids by an enzyme of *Staphylococcus aureus*, *J. Med. Microbiol.*, 36, 293, 1992.
65. Dinges, M.M., Orwin, P.M., and Schlievert, P.M., Exotoxins of *Staphylococcus aureus*, *Clin. Microbiol. Rev.*, 13, 16, 2000.
66. O'Riordan, K., and Lee, J.C., *Staphylococcus aureus* capsular polysaccharides, *Clin. Microbiol. Rev.*, 17, 218, 2004.
67. Lee, C.Y., Capsule production. In Honeyman, A.L., Friedman, H., and Endinelli, M. (Editors), *Staphylococcus aureus Infection and Disease*, pp. 35–45, Kluwer Academic/Plenum Publishers, New York, 2001.
68. Buzzola, F.R. et al., Differential abilities of capsulated and noncapsulated *Staphylococcus aureus* isolates from diverse agr groups to invade mammary epithelial cells, *Infect. Immun.*, 75, 886, 2007.
69. Sjoquist, J. et al., Localization of protein A in the bacteria, *Eur. J. Biochem.*, 30, 190, 1972.
70. Movitz, J., Masuda, S., and Sjoquist, J., Physico- and immunochemical properties of staphylococcal protein A extracellularly produced by a set of mutants from *Staphylococcus aureus* Cowan I., *Microbiol. Immunol.*, 23, 51, 1979.
71. Forsgren, A., and Sjoquist, J., "Protein A" from *S. aureus*. I. Pseudo-immune reaction with human gamma-globulin, *J. Immunol.*, 97, 822, 1966.
72. Uhlen, M. et al., Expression of the gene encoding protein A in *Staphylococcus aureus* and coagulase-negative staphylococci, *J. Bacteriol.*, 15, 9713, 1984.
73. Papageorgiou, A.C., and Acharya, K.R., Microbial superantigens: From structure to function, *Trends Microbiol.*, 8, 369, 2000.
74. Issa, N.C., and Thompson, R.L., Staphylococcal toxic shock syndrome, *Postgrad. Med.*, 110, 55, 2001.
75. Thomas, D. et al., Diversity in *Staphylococcus aureus* enterotoxins, *Chem. Immunol. Allergy*, 93, 24, 2007.
76. Bergdoll, M.S., Enterotoxins. In Easmon, C.S.F., and Adlam, C. (Editors), *Staphylococci and Staphylococcal Infections*, Chapter 2, Academic Press, London, 1983.
77. Bergdoll, M.S., and Schlievert, P.M., Toxic-shock syndrome toxin, *Lancet*, 8404, 691, 1984.
78. Ladhani, S., and Evans, R.W., Staphylococcal scalded skin syndrome, *Arch. Dis. Child.*, 78, 85, 1998.
79. Amagai, M. et al., Toxin in bullous impetigo and staphylococcal scalded-skin syndrome targets desmoglein-1, *Nat. Med.*, 6, 1275, 2000.
80. Ladhani, S. et al., Clinical, microbial and biochemical aspects of the exfoliative toxins causing staphylococcal scalded skin syndrome, *Clin. Microbiol. Rev.*, 12, 224, 1999.
81. Sato, H. et al., A new type of staphylococcal exfoliative toxin isolated from a *Staphylococcus aureus* strain isolated from a horse with phlegmon, *Infect. Immun.*, 62, 3780, 1994.
82. Yamaguchi, T. et al., Identification of the *Staphylococus aureus* etd pathogenicity island which encodes a novel exfoliative toxin, ETD, and EDIN-B, *Infect. Immun.*, 70, 5835, 2002.
83. Kirby, W.M.M., Extraction of a highly potent penicillin inactivator from penicillin resistant staphylococci, *Science*, 99, 452, 1944.
84. Chambers, H.F., The changing epidemiology of *Staphylococcus aureus*? *Emerg. Infect. Dis.*, 7, 178, 2001.
85. Dyke, K., and Gregory, P.D., Resistance to β-lactam antibiotics. Resistance mediated by β-lactamase. In Crossley, K.B., and Archer, G.L. (Editors), *The Staphylococci in Human Disease*, pp. 139–157, Churchill Livingstone, New York, NY, 1997.
86. Jevons, M.P., Celbenin-resistant staphylococci, *Br. Med. J.*, 1, 124, 1961.
87. Thompson, R.L., Cabezudo, I., and Wenzel, R.P., Epidemiology of nosocomial infections caused by methicillin-resistant *Staphylococcus aureus*, *Ann. Intern/ Med.*, 97, 309, 1982.
88. Chambers, H.F., Methicillin resistance in staphylococci: Molecular and biochemical basis and clinical implications, *Clin. Microbiol. Rev.*, 10, 781, 1997.
89. Ubukata, K. et al., Expression and inducibility in *Staphylococcus aureus* of the mecA gene, which encodes a methicillin-resistant *S. aureus*-specific penicillin-binding protein, *J. Bacteriol.*, 171, 2882, 1989.
90. Lowy, F.D., Antimicrobial resistance: The example of *Staphylococcus aureus*, *J. Clin. Invest.*, 111, 1265, 2003.
91. Centers for Disease Control and Prevention (CDC). *Staphylococcus aureus* resistant to vancomycin, *JAMA*, 288, 825, 2002.

92. Shuttleworth, R. et al., Human isolates of *Staphylococcus caprae*: Association with bone and joint infection, *J. Clin. Microbiol.*, 35, 2537, 1997.
93. Vandenesch, F. et al., Identification and ribotypes of *Staphylococcus caprae* isolates isolated as human pathogens and from goat milk, *J. Clin. Microbiol.*, 33, 888, 1995.
94. Ross, T.L. et al., Methicillin-resistant *Staphylococcus caprae* in a neonatal intensive care unit, *J. Clin. Microbiol.*, 43, 363, 2005.
95. Bayston, R., and Penny, S.R., Excessive production of mucoid substance in *Staphylococcus* SIIA: A possible factor in colonisation of Holter shunts, *Dev. Med. Child. Neurol.*, 27, S25, 1972.
96. Vuong, C., and Otto, M., *Staphylococcus epidermidis* infections, *Microbes Infect.*, 4, 481, 2002.
97. Jones, J.W. et al., A study of coagulase-negative staphylococci with reference to slime production, adherence, antibiotic resistance patterns and clinical significance, *J. Hosp. Infect.*, 22, 217, 1992.
98. Hall, R.T. et al., Characteristics of coagulase-negative staphylococci from infants with bacteremia, *Pediatr. Infect. Dis. J.*, 6, 377, 1987.
99. Deighton, M.A. et al., Species identification, antibiotic sensitivity and slime production of coagulase-negative staphylococci isolated from clinical specimens, *Epidemiol. Infect.*, 101, 99, 1988.
100. O'Gara, J.P., and Humphreys, H., *Staphylococcus epidermidis* biofilms: Importance and implications, *J. Med. Microbiol.*, 50, 582, 2001.
101. Heilmann, C.O. et al., Molecular basis of intercellular adhesion in the biofilm-forming *Staphylococcus epidermidis*, *Mol. Microbiol.*, 20, 1083, 1996.
102. Jack, R.W., and Jung. G., Lantibiotics and microcins: Polypeptides with unusual chemical diversity, *Curr. Opin. Chem. Biol.*, 4, 310, 2000.
103. von Eiff, C. et al., Infections associated with medical devices, pathogenesis, management and prophylaxis, *Drugs*, 65, 179, 2005.
104. Vandenesch, F. et al., Delta-like haemolysin produced by *Staphylococcus lugdunensis*, *FEMS Microbiol. Lett.*, 62, 65, 1991.
105. Heikens, E. et al., Comparison of genotypic and phenotypic methods for species-level identification of clinical isolates of coagulase-negative staphylococci, *J. Clin. Microbiol.*, 43, 2286, 2005.
106. Sivadon, V. et al., Use of genotypic identification by sodA sequencing in a prospective study to examine the distribution of coagulase-negative *Staphylococcus* species among strains recovered during septic orthopedic surgery and evaluate their significance, *J. Clin. Microbiol.*, 43, 2952, 2005.
107. Cookson, B., *Staphylococcus aureus*. In Emmerson, M., Kibbler, C., and Hawkey, P. (Editors), *Principles of Clinical Bacteriology*, pp. 109, John Wiley, Oxford, 1997.
108. Wichelhaus, T.A. et al., Evaluation of modern agglutination tests for identification of methicillin-susceptible and methicillin-resistant *Staphylococcus aureus*, *Eur. J. Clin. Microbiol. Infect. Dis.*, 18, 756, 1999.
109. Essers, L., and Radebold, K., Rapid and reliable identification of *Staphylococcus aureus* by a latex agglutination test, *J. Clin. Microbiol.*, 12, 641, 1980.
110. Kuusela, P., Hilden, P., and Savolainen, K. et al., Rapid detection of methicillin-resistant *Staphylococcus aureus* strains not identified by slide agglutination tests, *J. Clin. Microbiol.*, 32, 143, 1994.
111. Blake, J.E., and Metcalfe, M.A., A shared noncapsular antigen is responsible for false-positive reactions by *Staphylococcus epidermidis* in commercial agglutination tests for *Staphylococcus aureus*, *J. Clin. Microbiol.*, 39, 544, 2001.
112. van Griethuysen, A. et al., International multicenter evaluation of latex agglutination tests for identification of *Staphylococcus aureus*, *J. Clin. Microbiol.*, 39, 86, 2001.
113. Gupta, H. et al., Comparison of six rapid agglutination tests for identification of *Staphylococcus aureus*, including methicillin-resistant strains, *Diagn. Microbiol. Infect. Dis.*, 31, 333, 1988.
114. Peacock, S.J. et al., *Staphylococcus schleiferi* subsp. *schleiferi* expresses a fibronectin-binding protein, *J. Clin. Microbiol.*, 67, 4272, 1999.
115. Weckman, B.G., and Catlin, B.W., Deoxyribonuclease activity of micrococci from clinical sources, *J. Bacteriol.*, 73, 747, 1957.
116. Schreier, J.B., Modification of deoxyribonuclease test medium for rapid identification of *Serratia marcescens*, *Am. J. Clin. Pathol.*, 51, 711, 1969.
117. Taponen, S. et al., Clinical characteristics and persistence of bovine mastitis caused by different species of coagulase-negative staphylococci identified with API or AFLP, *Vet. Microbiol.*, 115, 199, 2006.
118. Taponen, S. et al., Bovine intramammary infections caused by coagulase-negative staphylococci may persist throughout lactation according to amplified fragment length polymorphism-based analysis, *J. Dairy Sci.*, 90, 3301, 2007.
119. Carretto, E. et al., Identification of coagulase-negative staphylococci other than *Staphylococcus epidermidis* by automated ribotyping, *Clin. Microbiol. Infect.*, 11, 177, 2005.
120. Lan, R., and Reeves, P.R., When does a clone deserve a name? A perspective on bacterial species based on population genetics, *Trends Microbiol.*, 9, 419, 2001.
121. Clarridge, J.E., Impact of 16S rRNA gene sequence analysis for identification of bacteria on clinical microbiology and infectious diseases, *Clin. Microbiol. Rev.*, 17, 840, 2004.
122. Skow, A. et al., Species-level identification of staphylococcal isolates by real-time PCR and melt curve analysis, *J. Clin. Microbiol.*, 43, 2876, 2005.
123. CLSI, Interpretive Criteria for Microorganism Identification by DNA Target Sequencing; Proposed Guideline. Clinical and Laboratory Standards Institute, Wayne, PA (CLSI document MM18-P, ISBN 1–56238-646–8), 2007.
124. Shah, M.M. et al., dnaJ gene sequence-based assay for species identification and phylogenetic grouping in the genus *Staphylococcus*, *Int. J. Syst. Evol. Microbiol.*, 57, 25, 2007.
125. Maes, N. et al., Rapid and accurate identification of Staphylococcus species by tRNA intergenic spacer length polymorphism analysis, *J. Clin. Microbiol.*, 35, 2477, 1997.
126. Kwok, A.Y. et al., Species identification and phylogenetic relationships based on partial HSP60 gene sequences within the genus *Staphylococcus*, *Int. J. Syst. Bacteriol.*, 49, 1181, 1999.
127. Drancourt, M., and Raoult, D., rpoB gene sequence-based identification of *Staphylococcus* species, *J. Clin. Microbiol.*, 40, 1333, 2002.
128. Mellmann, A. et al., Sequencing and staphylococci identification, *Emerg. Infect. Dis.*, 12, 333, 2006.
129. Poyart, C. et al., Rapid and accurate species-level identification of coagulase-negative staphylococci by using the sodA gene as a target, *J. Clin. Microbiol.*, 39, 4296, 2001.

130. Vannuffel, P. et al., Molecular characterization of femA from *Staphylococcus hominis* and *Staphylococcus saprophyticus*, and femA-based discrimination of staphylococcal species, *Res. Microbiol.*, 150, 129, 1999.
131. Yugueros, J. et al., Glyceraldehyde-3-phosphate dehydrogenase-encoding gene as a useful taxonomic tool for *Staphylococcus* spp. *J. Clin. Microbiol.*, 38, 4351, 2000.
132. Yugueros, J. et al., Identification of *Staphylococcus* spp. by PCR-restriction fragment length polymorphism of gap gene, *J. Clin. Microbiol.*, 39, 3693, 2001.
133. Nelson, K.E. et al., Phylogenetic analysis of the microbial populations in the wild herbivore gastrointestinal tract: Insights into an unexplored niche, *Environ. Microbiol.*, 5, 1212, 2003.
134. Gill, J.J. et al., Characterization of bacterial populations recovered from the teat canals of lactating dairy and beef cattle by 16S rRNA gene sequence analysis, *FEMS Microbiol. Ecol.*, 56, 471, 2006.
135. Zadoks, R.N., Schukken, Y.H., and Wiedmann, M., Multilocus sequence typing of *Streptococcus uberis* provides sensitive and epidemiologically relevant subtype information and reveals positive selection in the virulence gene pauA, *J. Clin. Microbiol.*, 43, 2407, 2005.
136. Loch, I.M., Glenn, K., and Zadoks, R.N., Macrolide and lincosamide resistance genes of environmental streptococci from bovine milk, *Vet. Microbiol.*, 111, 133, 2005.
137. Anthony, R.M. et al., Use of the polymerase chain reaction for rapid detection of high-level mupirocin resistance in staphylococci, *Eur. J. Clin. Microbiol. Infect. Dis.*, 18, 30, 1999.
138. Fujimura, S., Watanabe, A., and Beighton, D., Characterization of the mupA gene in strains of methicillin-resistant *Staphylococcus aureus* with a low level of resistance to mupirocin, *Antimicrob. Agents Chemother.*, 45, 641, 2001.
139. NCCLS, MIC interpretive standards for *Staphylococcus* spp. An NCCLS global informational supplement, vol. 22, pp. 96–125. NCCLS, Wayne, PA, 2002.
140. Perez-Roth, E. et al., Multiplex PCR for simultaneous identification of *Staphylococcus aureus* and detection of methicillin and mupirocin resistance, *J. Clin. Microbiol.*, 39, 4037, 2001.
141. Zangh, K. et al., New Quadriplex PCR assay for detection of methicillin and mupirocin resistance and simultaneous discrimination of *Staphylococcus aureus* from Coagulase-Negative Staphylococci, *J. Clin. Microbiol.*, 42, 4947, 2004.
142. Al-Talib, H. et al., A pentaplex PCR assay for the rapid detection of methicillin-resistant *Staphylococcus aureus* and Panton–Valentine Leucocidin, *BMC Microbiol.*, 28, 9, 2009.
143. Stefani, S., and Agodi, A., Molecular epidemiology of antibiotic resistance, *Int. J. Antimicrob. Agents*, 13, 143, 2000.
144. Hookey, J.V., Richardson, J.F., and Cookson, B.D., Molecular typing of *Staphylococcus aureus* based on PCR restriction fragment length polymorphism and DNA sequence analysis of the coagulase gene, *J. Clin. Microbiol.*, 36, 1083, 1998.
145. Trindade, P.A. et al., Molecular techniques for MRSA typing: Current issues and perspectives, *Braz. J. Infect. Dis.*, 7, 32, 2003.
146. Mepherso, M.J., and Moller, S.G., *PCR*, 1st ed., BIOS Scientific, Trowbridge, UK, 2000.
147. Oliveria, A.M., and Ramos, M.C., PCR-based ribotyping of *Staphylococcus aureus*, *Braz. J. Med. Biol. Res.*, 35, 175, 2002.
148. Kostyal, D.A., and Camp, C.J., Pulsed-field gel electrophoresis (PFGE), a molecular tool for monitoring nosocomial infections, *Gut. J.*, 70, 15, 2001.
149. Enright, M.C. et al., Multilocus sequence typing for characterization of methicillin-resistant and methicillin-susceptible clones of *Staphylococcus aureus*, *J. Clin. Microbiol.*, 38, 1008, 2000.
150. Robinson, D.A., and Enright, M.C., Multilocus sequence typing and the evolution of methicillin-resistant *Staphylococcus aureus*, *Clin. Microbiol. Infect.*, 10, 92, 2004.
151. Tenover, F.C., Interpreting chromosomal DNA restriction patterns produced by pulsed-field gel electrophoresis: criteria for bacterial strain typing, *J. Clin. Microbiol.*, 33, 2233, 1995.
152. van Belkum, A., Assessment of resolution and intercenter reproducibility of results of genotyping *Staphylococcus aureus* by pulsed-field gel electrophoresis of SmaI macrorestriction fragments: A multicenter study, *J. Clin. Microbiol.*, 36, 1653, 1998.
153. Deplano, A. et al., Multicenter evaluation of epidemiological typing of methicillin-resistant *Staphylococcus aureus* strains by repetitive-element PCR analysis, The European Study Group on Epidemiological Markers of the ESCMID, *J. Clin. Microbiol.*, 38, 3527, 2000.
154. Bart-Delabesse, E. et al., Comparison of restriction fragment length polymorphism, microsatellite length polymorphism, and random amplification of polymorphic DNA analyses for fingerprinting Aspergillus fumigatus isolates, *J. Clin. Microbiol.*, 39, 2683, 2001.
155. Koeuth, T., Versalovic, J., and Lupski, J.R., Differential subsequence conservation of interspersed repetitive *Streptococcus pneumoniae* BOX elements in diverse bacteria, *Genome Res.*, 5, 408, 1995.
156. Stern, M.J. et al., Repetitive extragenic palindromic sequences: A major component of the bacterial genome, *Cell*, 37, 1015, 1984.
157. Versalovic, J., Koeuth, T., and Lupski, J.R., Distribution of repetitive DNA sequences in eubacteria and application to fingerprinting of bacterial genomes, *Nucleic Acids Res.*, 19, 6823, 1991.
158. Versalovic, J., and Lupski, J.R., Molecular detection and genotyping of pathogens: More accurate and rapid answers, *Trends Microbiol.*, 10, S15, 2002.
159. Healy, M. et al., Microbial DNA typing by automated repetitive-sequence-based PCR, *J. Clin. Microbiol.*, 43, 199, 2005.
160. Harmsen, D. et al., Typing of methicillin-resistant *Staphylococcus aureus* in a university hospital setting using a novel software for spa-repeat determination and database management, *J. Clin. Microbiol.*, 41, 5442, 2003.
161. Friedrich, A.W. et al., SeqNet.org: A European laboratory network for sequence-based typing of microbial pathogens. *Euro Surveill.*, 11, E060112.4, 2006.
162. Tenover, F.C. et al., Multiple-locus variable-number tandem repeat assay analysis of methicillin-resistant *Staphylococcus aureus* strains, *J. Clin. Microbiol.*, 45, 2215, 2007.
163. Nubel, U. et al., Frequent emergence and limited geographic dispersal of methicillin-resistant *Staphylococcus aureus*, *Proc Natl Acad Sci USA*, 105, 14130, 2008.
164. Malferrari, G. et al., Automated system for nucleic acid extraction, purification and analysis, *Nucleic Acids Purif.*, 6, 52, 2004.
165. Chang, J.J., Zhang, S.H., and Li, L., Comparing and evaluating six methods of extracting human genomic DNA from whole blood, *Fa Yi Xue Za Zhi*, 25, 109, 2009.

166. Hansen, W.L., Bruggeman, C.A., and Wolffs, P.F., Evaluation of new pre-analytical sample treatment tools and DNA isolation protocols to improve bacterial pathogen detection from whole blood, *J. Clin. Microbiol.*, 47, 2629, 2009.
167. Swenson, J.M., and Tenover, F.C., Results of disk diffusion testing with cefoxitin correlate with presence of *mec*A in *Staphylococcus* spp., *J. Clin. Microbiol.*, 43, 3818, 2005.
168. Vannuffel, P. et al., Specific detection of methicillin-resistant *Staphylococcus* species by multiplex PCR, *J. Clin. Microbiol.*, 33, 2864, 1995.
169. Kobayashi, N., Alam, M., and Urasawa, S., Analysis on distribution of insertion sequence IS431 in clinical isolates of staphylococci, *Diagn. Microbiol. Infect. Dis.*, 39, 61, 2001.
170. El Garch, F. et al., StaphVar-DNA microarray analysis of accessory genome elements of community-acquired methicillin-resistant *Staphylococcus aureus*, *J. Antimicrob. Chemother.*, 63, 877, 2009.
171. Saunders, N.A. et al., A virulence-associated gene microarray: A tool for investigation of the evolution and pathogenic potential of *Staphylococcus aureus*, *Microbiology*, 150, 3763, 2004.
172. Monecke, S., Slickers, P., and Ehricht, R., Assignment of *Staphylococcus aureus* isolates to clonal complexes based on microarray analysis and pattern recognition, *FEMS Immunol. Med. Microbiol.*, 53, 237, 2008.
173. Zhu, W. et al., Use of pyrosequencing to identify point mutations in domain V of 23S rRNA genes of linezolid-resistant *Staphylococcus aureus* and *Staphylococcus epidermidis*, *Eur. J. Clin. Microbiol. Infect. Dis.*, 26, 161, 2007.
174. Jury, F. et al., Rapid cost-effective subtyping of methicillin-resistant *Staphylococcus aureus* by denaturing HPLC, *J. Med. Microbiol.*, 55, 1053, 2006.
175. Wolk, D.M. et al., Pathogen profiling: Rapid molecular characterization of *Staphylococcus aureus* by PCR/ESI-MS and correlation with phenotype, *J. Clin. Microbiol.*, 47, 3129, 2009.
176. Boedicker, J.Q. et al., Detecting bacteria and determining their susceptibility to antibiotics by stochastic confinement in nanoliter droplets using plug-based microfluidics, *Lab. Chip.* 8, 1265, 2008.

29 *Streptococcus*

Judith Jansen and Lothar Rink

CONTENTS

29.1 INTRODUCTION

Streptococcus is a very heterogeneous genus consisting of several different groups of bacteria causing a diversity of diseases. Frequently encountered streptococcal infections include pneumonia, otitis media, sinusitis, and pharyngitis, with less common but more severe diseases being streptococcal toxic shock syndrome (STSS) and necrotizing fasciitis, which are potentially lethal.[1,2]

Since conventional laboratory techniques for diagnosis of streptococcal infections suffer from a delay of at least one day in results availability, progress has been made in the development and application of molecular methods for improved identification and differentiation of streptococcal subgroups.[3]

Infections caused by streptococci are usually treated with antibiotics. Emergence of streptococcal resistance to different types of antibiotics has made the selection of the right antimicrobial therapy increasingly difficult.[4]

In view of these therapeutic difficulties, vaccines have been developed for certain groups of streptococci to prevent rather than treat streptococcal infections. For other groups, potential vaccines are still under investigation, with the varied nature of streptococcal antigens being the main obstacle to be overcome.[1,2,5]

29.1.1 Classification, Morphology, and Biology/Epidemiology

Streptococci (from the Greek word *streptos*, meaning twisted, and *kokkos*, meaning berry) belong to the genus *Streptococcus*, which covers more than 50 species of gram-positive, spherical, and nonmotile bacteria that frequently appear in chains or pairs.[6] Streptococci are facultative anaerobes, with the ability to grow in the presence and in the absence of oxygen. They are mostly cultured on media containing sterile sheep blood under enhanced CO_2 conditions. Colonies may appear domed and "mucoid," which indicates that the bacteria are encapsulated. Otherwise, the colonies show a collapsed and "matte" surface, implicating a missing capsule.[7]

Given the disease-causing potential of streptococci in both humans and animals, a large number of laboratory methods have been developed that permit discrimination and classification of streptococci pathogenic for animals on the one hand and for humans on the other.

The earliest method for differentiating streptococci was probably devised by Schottmüller in 1903, who analyzed the occurrence of β-hemolysis along with the growth of some strains of streptococci on blood agar.[8] Hemolysis is the destruction of red blood cells that can be observed on blood agar plates. The type of hemolysis (α-, β-, γ-hemolysis) is still relevant for the classification of streptococci today. α-hemolysis, as displayed by *Streptococcus pneumoniae*, is characterized by the presence of a greenish halo around bacterial colonies due to the production of hydrogen peroxide by the streptococci, which causes oxidation of hemoglobin in erythrocytes.[9–11] When the red blood cells are completely lysed, a clear halo can be seen around the colonies as the blood agar becomes transparent. This is called β-hemolysis. The most important streptococci causing β-hemolysis are *Streptococcus pyogenes* and *Streptococcus agalactiae*.[6] γ-Hemolysis describes the lack of any type of hemolysis. Therefore, the appropriate streptococci are called nonhemolysing (NH) streptococci. The distinction of α- and γ-hemolysis is not that important, since the composition of the medium and the incubation atmosphere determine if α- or γ-hemolysis occurs. For example, the lack of oxygen impairs the formation of hydrogen peroxide and consequently there will be no observable α-hemolysis.[10]

Other tests that had been developed mainly for differentiating streptococci of human and animal origins were based on biochemical characteristics such as the final hydrogen-ion concentration in streptococci-inoculated glucose broth and the capacity of streptococci to hydrolyze sodium hippurate.[12] In 1933, Lancefield introduced a precipitation assay that allowed for the classification of hemolytic streptococci into five groups (A–E) based on group-specific carbohydrate antigens in the cell wall known as the "C substance."[6,13,14] To date, the "Lancefield groups" contain the groups A–H and K–V.[6]

The only clinically relevant member of the group A streptococci (GAS) is *Streptococcus pyogenes*, a β-hemolytic streptococcus being highly pathogenic for humans.[6,10] Although *S. pyogenes* is carried asymptomatically in 10%–70% of humans depending on season, geographical location, and age, it is capable of inducing a variety of diseases ranging from pharyngitis, scarlet fever, and skin diseases to life-threatening diseases such as necrotizing fasciitis, toxic shock, and sepsis.[15,16] Toxins produced by *S. pyogenes*, which belong to the family of superantigens (SAgs), are responsible for scarlet fever, necrotizing fasciitis, and streptococcal toxic shock syndrome (STSS) (see Section 29.1.2.1).[1,17]

Bacterial pharyngitis is most commonly caused by *S. pyogenes*, and accounts for 15%–30% of cases of acute pharyngitis in children and 5%–10% in adults.[18] Moreover, *S. pyogenes* infections may result in several delayed sequelae, including acute rheumatic fever, acute poststreptococcal glomerulonephritis, and reactive arthritis.[1,19]

Further classification of *S. pyogenes* is based on (i) the T antigen, a trypsin-resistant surface protein; (ii) the ability to produce a serum opacity factor; and (iii) the M protein, which allows the differentiation of serotypes or M types. Approximately 80 M-types have been defined serologically.[14,15,20,21] Since the distinct N-termini of the M proteins are M type-specific, the so-called 5′ emm sequences are exploited for determination of new M serotypes. With this method, a total of 124 M/emm serotypes could be identified.[21]

Lancefield's group B streptococci (GBS) comprise *Streptococcus agalactiae*, a β-hemolytic organism that causes neonatal sepsis as well as maternal chorioamnionitis, puerperal endometritis, pneumonia, and meningitis.[5,10,16,22] Before confirmation as a human pathogen, *S. agalactiae* was mostly found as the causative pathogen of bovine mastitis.[23] In contrast to GAS, the rings of hemolysis on blood agar are rather thin.[5] A subclassification based on the immunogenic type of the polysaccharide capsule defines *S. agalactiae* serotypes. At present, nine serotypes have been described, with serotype III causing the majority of infections in neonates.[16,24] Colonization of the vagina/rectum with GBS is observed in approximately 20%–30% of all pregnant women. About 50% of newborns with mothers carrying GBS will be exposed to GBS and colonized during vaginal delivery, but invasive disease will emerge in only 1%–2% of the infants.[22,23]

Other β-hemolytic streptococci belong to the Lancefield group C, which comprises *Streptococcus dysgalactiae* and *Streptococcus equi*, and to the Lancefield group G, which includes *Streptococcus canis*. These streptococci cause infections in animals, but have also been isolated from humans suffering from pharyngitis and one case of streptococcal meningitis caused by *S. equi* subsp. *zooepidemicus* following contact to an infected horse.[3,10,25] Thus, these streptococci can be considered as zoonostic pathogens with the capacity to transfer from animals to humans.

Group D streptococci are nonhemolytic streptococci that comprise, among others, *Streptococcus bovis* and *Streptococcus equinus*, which are mainly responsible for infections in cattle and horses.[10] However, a case of *S. equinus* endocarditis in a patient with histiocytosis X has been described.[26]

Streptococcus suis is an α-hemolytic bacterium that usually belongs to the group R streptococci. It leads to infections in pigs and humans, with infectious diseases like meningitis, arthritis, endocarditis, endophthalmitis, and sepsis being reported in persons who have had close contact with pigs. Hearing loss is a possible sequela.[10,27,28]

One rather inhomogeneous group is called "viridans streptococci," referring to the greenish halo that can be observed on blood agar as a result of α-hemolysis, which is characteristic of most of the streptococci belonging to this group. Synonymously, the expression "oral streptococci" is used because these bacteria mainly inhabit the oral cavities and upper respiratory tracts of humans as commensales without causing any disease.[29] This group consists of five subgroups: *Streptococcus mutans* group, *Streptococcus salivarius* group, *Streptococcus anginosus* group, *Streptococcus sanguinis* group, and *Streptococcus mitis* group.[10]

Streptococcus pneumoniae, or "pneumococcus," is present asymptomatically in the nasopharynx of up to 50% of children and up to 5% of adults and is still the major cause of community-acquired pneumonia.[2,10,16] This α-hemolytic pathogen usually consists of a polysaccharide capsule, which allows the definition of immunologically distinct serotypes/capsular types, with 91 serotypes having been identified so far. *S. pneumoniae* cannot be classified according to the group antigens introduced by Lancefield.[2,6,11,30] Because of its arrangement in pairs, it is also referred to as "diplococcus."

S. pneumoniae causes acute otitis media, sinusitis, pneumonia, bacteremia, and meningitis. Pneumococci that are not encapsulated are nontypable and rarely seem to cause invasive disease.[2,30]

The incidence of pneumococcus-derived disease has two peaks: one occurs in the first years of life and the other in elderly people.[30] The highest mortality due to invasive pneumococcal diseases can be observed in patients aged ≥65 years.[2] Although all serotypes can cause various diseases typical of pneumococcosis, certain serotypes seem to be associated with particular disease presentations, age groups, and populations.[30] The rising resistance of pneumococci to antibiotics has been an important issue lately, thus complicating treatment of pneumococcal infections.[4] An overview of streptococci is given in Table 29.1.

29.1.2 Clinical Features and Pathogenesis

29.1.2.1 Group A Streptococci (*Streptococcus pyogenes*)

The pathogenic potential of *S. pyogenes* is due to several virulence factors regulated by genetic mechanisms that respond to environmental changes, thus adapting the virulence factors to the environmental challenges.[7,13,19]

The hyaluronate capsule (the degree of encapsulation is not consistent) and the M-protein both contribute to resistance to phagocytosis, the latter by binding to complement factors.[7,14,19,20] In addition, the capsule can induce the opening of intercellular junctions, which facilitates invasion of host tissues, and the M protein plays a role in the process of intracellular entry. Antibodies directed against the variable part of the M protein confer type-specific immunity and thus do not protect against infection resulting from another serotype.[16,19]

Other virulence factors include molecules that mediate adhesion (e.g., Lipoteichoic acid, fibronectin-binding proteins such as protein F1, and the M protein) as well as factors that interface with the complement system. These factors include a C5a peptidase that cleaves the chemotactic complement fragment C5a, an endoglycosidase that interacts with bound immunoglobulin to prevent initiation of the classical pathway, and the streptococcal inhibitor of complement (SIC), which prevents complement-induced lysis and possibly inhibits antimicrobial peptides, lysozyme, and the secretory proteinase inhibitor.[16,19]

Two hemolysins have been characterized. Streptolysin O (SLO), which is named after its oxygen lability, is lytic to cells with cholesterol present in the plasma membrane and it is antigenic. Streptolysin S (SLS) is a very potent cytotoxin associated with ulceration of tissue. SLS is not susceptible to oxygen, but it is thermolabile. A naturally occurring antibody against SLS could not be detected.[16,19] Other secreted proteins that facilitate the spread of the bacterium are DNases A, B, C, and D; hyaluronidase, which is involved in the degradation of connective tissue; streptococcal pyrogenic exotoxin B (SpeB), a cysteine protease; and streptokinase, which induces fibrinolysis by activating plasminogen. Antibodies directed against SLO, DNase B, hyaluronidase, and streptokinase can be used to serologically diagnose a GAS infection.[16,19]

The streptococcal pyrogenic exotoxins (Spes) are SAgs and consist of a total number of 11 types, which have different numbers of allelic variants: the bacteriophage-encoded SpeA 1–5 and SpeC—which are associated with scarlet fever—SpeG, SpeH, SpeI, SpeJ, SpeK, SpeL, SpeM, streptococcal superantigen A (SSA), and streptococcal mitogenic exotoxin Z (SMEZ) 1–47, with SMEZ 2 being the most potent superantigen described so far.[16,17,19, 20,53,54] SpeB was among the first characterized SAgs, but its observed mitogenic activity, which is typical of SAgs, was due to contamination with another unknown SAg. This SAg was named "SPEX" at the time of its detection, but it could be shown that it belongs to the SMEZ allele group.[55,56] Certain SAg profiles are more strongly correlated to invasiveness of GAS than the emm type (unpublished data). SAgs are potent immunostimulators capable of binding to the major histocompatibility complex (MHC) class II molecule and the T cell receptor (TCR) Vβ domain. This leads to activation of a large set of T cells resulting in increased secretion of cytokines such as tumor necrosis factor α, interleukin (IL)-1β and IL-2, and interferon-γ.[19,57] In turn, complement activation, coagulation/fibrinolysis, shock, and, finally, multiorgan failure are induced. These are the characteristics of the streptococcal toxic shock syndrome (STSS), which usually begins with severe headache, fever, myalgias, and arthralgias. It has a fatality rate of about 30%.[17,19] Clinical signs of severity include hypotension, renal impairment, coagulopathy, liver involvement, adult respiratory distress syndrome (ARDS), generalized erythematous macular rash that may desquamate, and soft tissue necrosis.[31] Absence of antitoxin antibodies is a strong susceptibility factor in STSS, thus one possible therapeutic option is the application of intravenous immunoglobulin (IVIG) to neutralize streptococcal superantigens.[17,19]

Necrotizing fasciitis, which is also associated with the presence of SAgs, is a painful infection that mainly destroys fascia and fat while skin lesions are of minor importance. The lack of severe skin destruction leads to underestimation of the infection and increased mortality. It is commonly associated with high fever. Rapid and thorough surgical debridement is necessary in order to control the infection.[1,31]

Pneumonia can also be a result of infection with *S. pyogenes* and, together with necrotizing fasciitis, it is the major cause of STSS.[31] Other diseases caused by *S. pyogenes* are pharyngitis and scarlet fever.[16] Scarlet fever is mostly associated with streptococcal throat infection, but it can also be a result of infections at other sites. The clinical symptoms of scarlet fever like strawberry tongue, rash, and desquamation of the skin are due to the presence of SpeA and/or SpeC.[1,19]

GAS that cause impetigo by invading the skin differ in M-protein serotype from those responsible for throat infections. Impetigo is restricted to the epidermis and it is highly contagious. When the epidermis is penetrated through skin lesions, it may result in erysipelas or cellulitis. Erysipelas is characterized by

TABLE 29.1
Overview of Streptococci

Organism	Group	Number of Serotypes	Hemolysis	Diseases in Humans	Virulence Factors	Annual Incidence	Resistance	Diagnostic Methods
S. pyogenes	A (Ref. 10)	124 (Ref. 21)	β (Ref. 6)	Pharyngitis; scarlet fever; pneumonia; impetigo; erysipelas; cellulitis STSS; necrotizing fasciitis; sepsis; sequelae: rheumatic fever, glomerulonephritis, reactive arthritis (Refs. 1,16,31)	Hyaluronate capsule; M protein; M-like proteins: for example, MRP and Enn; antiopsonic Mac; lipoteichoic acid; fibronectin-binding proteins: for example, PrtF1 = SfbI, PrtF2, SfbII, FBP54 (PavA-like, see *S. pneumoniae*), PFBP; C5a peptidase; GRAB; SPN; peptidoglycan; endoglycosidase; SIC; hemolysins: SLO and SLS; DNase A-D (streptodornase); streptokinase; hyaluronidase; SpeB (cysteine protease); pyrogenic exotoxins: SpeA 1–5, SpeC, SpeG, SpeH, SpeI, SpeJ, SpeK, SpeL, SpeM, SSA, SMEZ 1–47 (Refs. 16,19,32)	Invasive infections (USA): 3.6/100,000 (Ref. 19); pharyngitis in children (more developed countries): ca 0.15/child; pharyngitis in adults: ca 0.04/adult (Ref. 33)	Macrolides (14.0%) (Ref. 34), fluoroquinolones (8.9%) (Ref. 35), clindamycin (3.2%), minocycline (11.6%) (Ref. 36)	Microscopy, culture, latex agglutination, EIA, optical immunoassay, chemiluminescent DNA-probe hybridization, FISH, emm-typing, 16S rRNA sequence typing, PCR (Refs. 3,37–40)
S. agalactiae	B (Ref. 10)	9 (Ref. 16)	β (Ref. 6)	Urinary tract infection; maternal chorioamnionitis; puerperal endometritis; septicemia; in neonates: pneumonia, meningitis, sepsis (Refs. 5,16,22)	Polysaccharide capsule; C5a peptidase; Lmb; fibronectin-binding proteins: FbsA, PavA-like protein (see *S. pneumoniae*); further LPXTG-anchored[a] surface proteins; hemolysin (CylE); α-proteins; β-proteins (Ref. 16)	Early-onset disease (USA): 0.5/1,000 live births (Ref. 5); late-onset disease (USA): 0.4/1,000 live births (Ref. 23); adults (USA): 7.9/100,000 (Ref. 24)	Erythromycin (25%), clindamycin (15%) (Ref. 22)	Microscopy, culture, latex fixation/agglutination, FISH, 16S rRNA sequence typing, PCR (Refs. 5,22,37,40,41)
S. pneumoniae	Non-typeable (Ref. 6)	91 (Ref. 2)	α (Ref. 11)	Acute otitis media, sinusitis, pneumonia, bacteremia, meningitis (Refs. 2,30)	Polysaccharide capsule; hemolysin (pneumolysin); peumococcal choline-binding surface proteins: autolysin Lyt A, PspA, PspC = CbpA, CbpD, CbpE, CbpG, LytB, LytC;	Invasive pneumococcal disease (Germany): 4.6/100,000 < 65 years; 16.2/100,000 > 65 years (Ref. 2)	Penicillin (24.6%), macrolides (28.0%), tetracycline (25.2%), fluoroquinolones (0%–15.2%,	Microscopy, culture, Quellung reaction, latex agglutination, coagulation, counter immuno-electrophoresis, ELISA, flow cytometry,

					LPXTG-anchored[a] surface proteins: hyaluronidase, neuraminidase; further surface proteins: PsaA, PavA; enolase; superoxide dismutase; NADH oxidase; hydrogen peroxide; components of iron uptake systems (Refs. 16,42)		according to the country) (Refs. 4,43)	immunochromatographic tests, DNA-probe hybridization, 16S rRNA sequence typing, PCR (Refs. 11,37,40,44,45)
S. dysgalactiae subsp. *equisimilis/dysgalactiae*	C (Ref. 10)	N/A	β (subsp. *equisimilis*); α (subsp. *dysgalactiae*) (Ref. 6)	subsp. *equisimilis*: pharyngitis, rarely deep soft tissue infections; subsp. *dysgalactiae*: — (Refs. 3,10,39)	Several proteins binding to components of plasma or host tissues, for example, protein G, FnB A, FnB B, M-like proteins, hyaluronidase, fibrinolysin (Refs. 46,47)	N/A	subsp. *equisimilis*: erythromycin (29.2%), tetracycline[b] (58.0%), clindamycin (5.7%), fluoroquinolones (1.0%) (Ref. 48)	Microscopy, culture, biochemical methods, agglutination assays, FISH, 16S rRNA sequence typing, emm-typing (subsp. *equisimilis* only) (Refs. 39,48)
S. equi subsp. *equi/zooepidemicus*	C (Ref. 10)	N/A	β (Ref. 10)	subsp. *equi*: —; subsp. *zooepidemicus*: pharyngitis, nephritis, bacteremia, meningitis (Refs. 3,10,25)	Nonantigenic, antiphagocytic hyaluronic acid capsule; hyaluronidase; streptokinase; hemolytic SLS; IgG-Fc-receptor proteins; pyrogenic, complement-activating peptidoglycan; adhesins; leukocidal toxin • subsp. *equi* only: exotoxins: SePE-I, -H, -K, -L; antiphagocytic M protein SeM, SzPSe protein (variant of the SzP protein described below) • subsp. *zooepidemicus* only: opsonogenic M-like SzP protein (Ref. 47)	N/A	subsp. *zooepidemicus*: amikacin (98.5%, in horses) (Ref. 49)	Microscopy, culture, biochemical characterization, agglutination assays, FISH, 16S rRNA sequence typing (Refs. 25,37,40,47)
S. suis	R, S, T (Ref. 10)	35 (Ref. 10)	α (Ref. 10)	Meningitis, arthritis, endocarditis, endophthalmitis, sepsis, sequela: permanent hearing loss (Ref. 28)	Capsular polysaccharide, muramidase-released protein, extracellular protein factor, hemolysin (suilysin), adhesins (Ref. 50)	3.0/100,000 among abattoir workers and pig breeders; 1.2/100,000 among butchers (Netherlands) (Ref. 28)	Ofloxacin, tetracycline, clindamycin, erythromycin (Ref. 28)	Microscopy, culture, biochemical characterization, FISH, 16S rRNA sequence typing (Refs. 28,40)

continued

TABLE 29.1 (Continued)
Overview of Streptococci

Organism	Group	Number of Serotypes	Hemolysis	Diseases in Humans	Virulence Factors	Annual Incidence	Resistance	Diagnostic Methods
Viridans streptococci: *S. mutans* group, *S. salivarius* group, *S. anginosus* group, *S. sanguinis* group, *S. mitis* group	Variable (Ref. 29)	*S. mutans*: 8 (a–h) (Ref. 29)	Mainly α (Ref. 29)	Carious lesions, abscesses, endocarditis (Ref. 29)	*S. mutans*: acid-production/-tolerance; hemolysin; proteases; adhesins: SpaP = antigen I/II, glucosyltransferases and glucan-binding proteins A and C, WapA, SloC, PavA-like protein (see *S. pneumoniae*) (Ref. 16)	Endocarditis (USA): 1.7–3.5/100,000 (Ref. 51)	Erythromycin (60.7%), tetracycline (21.4%), penicillin (39.3%), clindamycin (14.3%), ceftriaxone (10.7%) (Ref. 52)	Microscopy, culture, biochemical characterization, FISH, 16S rRNA sequence typing (Refs. 29,37,40)

Cbp, choline-binding protein; EIA, enzyme immunoassay; ELISA, enzyme-linked immunosorbent assay; FBP54, fibronectin-binding protein 54; FbsA, fibrinogen-binding protein from S. agalactiae; FISH, fluorescence in situ *hybridization; FnB, fibronectin-binding protein; GRAB, streptococcal surface protein G-related* α_2*-macroglobulin-binding protein; Ig, immunoglobulin; Lmb, laminin-binding protein; Mac, Mac1-like protein; MRP, M-related protein; NAD(H), nicotine adenine dinucleotide oxidized (reduced) form; PavA, pneumococcal adhesion and virulence A; PCR, polymerase chain reaction; PFBP, pyogenes fibronectin-binding protein; PrtF1/2, protein F1/2; Psa, pneumococcal surface antigen; Psp, pneumococcal surface protein; RADT, rapid antigen detection test; SePE, S. equi pyrogenic exotoxin; SfbI/II, streptococcal fibronectin binding protein I/II; SIC, streptococcal inhibitor of complement; SLO, streptolysin O; SloC, S. mutans LraI (lipoprotein receptor antigen I) operon C; SLS, streptolysin S; SMEZ, streptococcal mitogenic exotoxin Z; SpaP, streptococcal protein antigen P; Spe, streptococcal pyrogenic exotoxin; SPN, S. pyogenes NAD-glycohydrolase; WapA, wall associated protein A.*

[a] Proteins containing an LPXTG amino acid motif near the carboxy-terminal end are transported and covalently bound to the outside of the cell surface, which is mediated by a sortase-enzyme.[16]

[b] Only intermediate strains.

raised and erythematous superficial layers of skin, while subcutaneous tissues are affected by cellulitis.[1]

S. pyogenes may lead to delayed sequelae, such as acute poststreptococcal glomerulonephritis, acute rheumatic fever, and reactive arthritis. Acute poststreptococcal glomerulonephritis can occur as a response to a pharyngeal or skin infection with *S. pyogenes* (or rarely with group C or G streptococci), but only specific M-serotypes seem to be nephritogenic.[1,19] Clinical symptoms emerging after a latent period of 1–4 weeks include fever, edema, hypertension, hematuria, urinary sediment abnormalities, and a decrease in serum complement. Antibodies against DNase B or hyaluronidase are elevated and antiSLO antibodies are usually raised when glomerulonephritis occurs after throat infection. The development of chronic glomerulonephritis or hypertension is a rare scenario in children, but may be more frequent in adults. The causal mechanisms leading to glomerulonephritis comprise deposition of immune complexes, direct complement activation, presence of antibodies cross-reactive with streptococcal and glomerular antigens, and modification of glomerular tissues by streptococcal products.[1]

The development of rheumatic fever is dependent on prior upper respiratory tract infection with GAS.[19] Five clinical manifestations, which are subsumed under the so-called Jones criteria, are characteristic: migratory, polyarticular arthritis; carditis with affection of the heart valves; Syndeham's chorea, a neurological disorder causing involuntary movements; erythema marginatum; and/or subcutaneous nodules. These symptoms emerge after a latency period of 1–5 weeks, and are probably mediated by an autoimmunological mechanism based on molecular mimicry, inducing antibodies (e.g., antistreptococcal antimyosin antibodies) as well as T cells. Antibodies directed against DNase B and SLO are elevated, the latter being considered an indicator of antecedent streptococcal infection. Treatment involves high-dose aspirin.[1,19]

Reactive arthritis is a nonpurulent arthritis that persists for several weeks or months following a group A streptococcal infection. Aspirin or salicylates are not effective for treatment.[1] Finally, obsessive-compulsive disorders and Tourette syndrome seem to be associated with GAS infections, PANDAS (pediatric autoimmune neuropsychiatric disorders associated with streptococcal infections) being one subgroup of tics and obsessive compulsions.[1]

Group A streptococci are still exquisitely sensitive to penicillin, which is first choice concerning antibiotic treatment and prophylaxis of rheumatic fever, whereas increasing resistance to macrolides has been reported.[1,31,34] The development of a vaccine has been worked on for years and several possible antigens (e.g., parts of the M protein, the C5a peptidase, and an attenuated SpeA–SpeB fusion protein) have been evaluated.[1,58,59]

29.1.2.2 Group B Streptococci (*Streptococcus agalactiae*)

S. agalactiae colonizes the recta and vaginas of pregnant women and is transferred to the newborn during vaginal delivery.[22] Survival in the host is promoted by several virulence factors such as (i) the polysaccharide capsule, which prevents deposition of complement thus protecting from phagocytosis; (ii) a C5a peptidase; (iii) adhesins such as laminin-binding and fibronectin-binding proteins; and (iv) a hemolysin (CylE) causing β-hemolysis, which mediates invasion and tissue damage. Further virulence factors are the α- and β-proteins. The β-protein binds to the Fc-fragment of IgA and to complement factor H, thus compromising the immune system, whereas the αC-protein seems to mediate GBS internalization by binding to integrins and glycosaminoglycan, which are present on epithelial cells, and by inducing transcytosis.[16,60,61]

GBS colonization of the lower gastrointestinal and genital tract is usually asymptomatic, but can lead to urinary tract infection, endometritis, chorioamnionitis, and septicemia. Neonatal infection caused by GBS can lead to septicemia, pneumonia, and/or meningitis.[22]

In neonates, early- and late-onset diseases are distinguished. Early-onset disease occurs in the first week after birth predominantly resulting in pneumonia and/or sepsis. Late-onset disease is defined as occurring between days 7 and 90 after birth and is not as closely related to maternal vaginal colonization as early-onset disease. Hence, the main mechanism of GBS-transfer seems to be contact between mother or personal staff and child. Meningitis and/or septicemia are common clinical manifestations and the mortality rate is lower compared to early-onset disease.[22]

The recommended strategy to prevent the transfer of GBS from mother to child during vaginal delivery (and thus early-onset disease) is the administration of intravenous antibiotics during parturition. Penicillin and ampicillin are the antimicrobial agents of choice for GBS strains that show resistance to erythromycin and clindamycin.[5,22,24] A GBS vaccine consisting of the GBS polysaccharide conjugated with tetanus toxoid was shown to induce an "excellent immune response."[5] A desirable vaccine will not only reduce bacterial colonization, but also elicit humoral and mucosal immunity.[5]

29.1.2.3 Viridans Streptococci

Infections caused by "viridans streptococci" or "oral streptococci," which physiologically inhabit the oral cavity, include bacterial endocarditis, especially in patients with prosthetic valves, abscesses, and various infections in immunocompromised patients—also called opportunistic infections (for virulence factors, refer to Table 29.1). *S. mutans* and *S. sanguis* can frequently be isolated from carious lesions.[29] Clinical trials have been conducted to determine safety and effects of vaccines against dental carious based on *S. mutans* antigens such as antigen I/II, glucosyltransferases, and/or glucan-binding proteins[62] (see Table 29.1).

29.1.2.4 *Streptococcus pneumoniae*

A major virulence factor of *S. pneumoniae* is the polysaccharide capsule, which protects the bacterium from phagocytosis. A pore-forming hemolysin (pneumolysin) stimulates the production of inflammatory cytokines and suppresses the response of the immune system by inhibiting phagocyte and lymphocyte function and by interfering with the complement system. Surface proteins such as pneumococcal surface protein (Psp) A and Psp C, both being members of the choline-binding protein family, also have the capability of influencing the complement system. PspC is known to mediate adhesion of pneumococci to the nasopharynx and translocation across epithelial cells and it might play a role in recruitment of immune cells as well. The autolysin Lyt A, a surface protein belonging to the choline-binding protein family, probably exerts its virulent effects by releasing components from the bacterial cell wall or other virulence factors. Hyaluronidase and neuraminidase contribute to virulence by their enzymatic activity, resulting in, for example, destruction of connective tissue and colonization, respectively. Pneumococcal surface antigen A (PsaA) and the fibronectin-binding protein PavA (pneumococcal adhesion and virulence A) are adhesins that mediate the attachment of *S. pneumoniae* to host cells, thereby supporting invasion of the host (for further virulence factors, see Table 29.1).[16,47]

Asymptomatic carriage of certain pneumococci rarely leads to invasive disease. The infecting strain, potentially causing acute otitis media, sinusitis, pneumonia, bacteremia, or meningitis, has usually been acquired recently.[2,30,44]

Spreading of the bacteria to other organs like the lung can occur by aspiration or via bacteremia. Adhesive types are responsible for localized infections whereas less adhesive types predominantly induce invasive disease.[44]

Amoxicillin and amoxicillin/clavunate (β-lactamase-inhibitor) form the first line therapy for pneumonia.[63,64] In recent years, the development and spread of multidrug-resistant clones have emerged as a significant problem. About 40% of *S. pneumoniae* are found to be highly drug resistant, with the percentage of resistant clones being highly variable among different countries.[65] Furthermore, the increasing use of fluoroquinolones and a subsequent increase in resistance rates are worrisome.[4] Antibiotic resistance seems to be acquired by pneumococci through gene transfer. Penicillin-resistant *S. pneumoniae* contain mosaic genes encoding penicillin-binding proteins (PBPs) crucial for penicillin resistance, which may have been transferred from *S. oralis* and *S. mitis* (oral streptococci belonging to the *S. mitis* group).[6,66,67]

29.1.3 Diagnosis

29.1.3.1 Conventional Techniques

Several body fluids and variable methods can be used to diagnose a streptococcal infection. Specimens allowing the detection of streptococci include sputum or other respiratory secretions, blood, urine, cerebrospinal fluid (CSF), and swabs from throat, vagina, and rectum depending on the type of infection.[5,11,44,68]

A simple method for confirmation of a streptococcal infection is microscopy. Streptococci appear as gram-positive, spherical bacteria that arrange in chains or pairs, the latter being typical of pneumococci. The so-called Quellung reaction is the standard typing method for pneumococci, which enhances a microscopic detection of the capsule through formation of a halo after reaction of the pneumococci with streptococcal antisera.[11]

The diagnostic gold standard still is the culture on 5% sheep blood agar, which may be supplemented with gentamicin in order to suppress growth of nonstreptococcal bacteria. Incubation is carried out at 35°C for 24–48 h. The atmosphere may be aerobic, aerobic with elevated CO_2, or anaerobic.[11,69] Colonies will present "mucoid" or "matte" according to the degree of encapsulation, and the kind of hemolysis, which is specific for the different groups of streptococci, can be examined.[7,10] Further typing of the grown streptococcal colonies can be accomplished by biochemical tests, for example, Rapid ID32 Strep API system (bioMérieux, Marcy-l'Etoile, France), or latex agglutination tests allowing detection of the streptococcal cell wall carbohydrate antigens as described below.[37,70,71] Identification of *S. pneumoniae* is usually based on its susceptibility to optochin, which is an antibacterial agent, and on its bile solubility. These two reactions especially help to distinguish *S. pneumoniae* from viridans streptococci.[11]

The major disadvantage of culture is the delay in results availability.[3] Therefore, rapid antigen detection tests (RADTs) have been developed that provide results within 15 min and that are easy to handle.[3,72] However, the shortcomings of RADTs include the possible lack of sensitivity, the sometimes challenging interpretation of the result, and the high cost.[73]

Latex agglutination is a commonly used RADT.[74] The antigen is first extracted from the organisms with either nitrous acid or an enzyme. Then a solution of latex particles labeled with group-specific antibodies is added and agglutination can be observed if the group-specific antibodies bind the group antigen forming part of the bacterial cell wall. Most of those commercially available kits are capable of detecting group A, B, C, D, F, and G antigens. Examples are Prolex (Pro-Lab Diagnostica), a latex agglutination kit containing modified nitrous acid reagents; Streptex (Murex Diagnostics Ltd.), an enzyme extraction kit; and PathoDx (Diagnostic Products Corporation), another nitrous acid kit not capable of detecting the group D antigen.[74]

Besides latex agglutination, other RADTs (e.g., enzyme immunoassay and optical immunoassay) have also been utilized for detection of GAS. The QuickVue In-Line One-Step Strep A test (Quidel Corporation) is an enzyme-linked immunoassay kit for examination of throat samples with regard to the presence of GAS.[75] Due to the low sensitivity of RADTs, it is recommended that for detection of *S. pyogenes*, negative results obtained with RADTs should be confirmed by subsequent culture.[3]

Recently, a rapid immunochromatographic test (ICT) has been developed to detect the C polysaccharide cell wall antigen that is common to all strains of *S. pneumoniae*, the NOW® *S. pneumoniae* urinary antigen test (Binax Inc.). Although it is designed for urine, this test can be used with other body fluids like CSF or pleural fluid, and it has been proven valuable as an alternative to culture allowing diagnosis of pneumococcal bacteremia from positive blood cultures.[11,76] The results of the test are obtained within 15 min.[45,76]

As a more advanced method, a rapid multiplex assay for serotyping pneumococci using a flow cytometer consisting of 24 assays specific for 36 serotypes has been designed and evaluated. This method allows the handling of a large number of bacteria.[77] Further diagnostic tests are based on the detection of pneumolysin, another protein characteristic for pneumococci, but its superiority to the C polysaccharide still has to be demonstrated.[11]

29.1.3.2 Molecular Techniques

For accurate diagnosis and differentiation of streptococci, molecular techniques are available. The analysis of 16S rRNA sequences facilitates phylogenetic analysis and species identification.[37] Multilocus enzyme electrophoresis (MLEE) analyzes the electrophoretic mobility of streptococcal housekeeping enzymes on starch gels and equates the different charge variants of each enzyme with alleles at the corresponding genetic locus. This allows definition of electrophoretic types (ET) differing by certain alleles.[20,78] As a modification of MLEE, multilocus sequence typing (MLST) is based on nucleotide sequencing of internal fragments of seven housekeeping genes instead of electrophoretic mobility. This method defines sequence types (ST) that describe the allelic profile of streptococcal isolates and is valuable for global epidemiology.[78]

Polymerase chain reaction (PCR) is a nucleic acid amplification procedure that has proven powerful for detection and identification of GAS, GBS, and *S. pneumoniae*.[3,5,11,44,45] Although it requires special equipment and takes 1 h instead of 15 min (as in the case of RADTs),[5,73] the PCR-based tests demand minimal amounts of starting materials, are unaffected by prior antibiotic therapy, and are independent of viability of the pathogen.[11]

PCR approaches for determination of the emm type/serotype of *S. pyogenes* by targeting the genes encoding the M protein (emm sequences) allow rapid and practical characterization of the pathogen. This method is not dependent on the expression of the M protein in contrast to serological M typing, and it makes possible the detection of new emm types that are not yet registered (see Section 29.2.2.1).[21,38,79] The recent development of a multiplex PCR enables the simultaneous detection of 11 SAgs of *S. pyogenes* (see Section 29.2.2.2).[53]

For diagnosis of pneumococcal infection, the pneumolysin gene (*ply*) is a common PCR target. Since *ply* is also present in some nonpneumococcal viridans streptococci, this test is not highly specific. Hence, a positive result does not exclusively indicate the presence of *S. pneumoniae* in the analyzed specimen, but it may also be due to viridans streptococci present in the probe, which physiologically belong to the human mouth flora. Thus, inclusion of other targets (e.g., the autolysin-encoding *lytA* gene) is desirable to improve specificity.[11,16,45]

Molecular typing of pneumococci, which is based on sequencing and serotype(s)/group(s)-specific PCR targeting the *cps* (capsular polysaccharide synthesis) gene cluster, offers an alternative to conventional serotyping. When this method was extended to 90 serotypes in 2005, it could identify serotypes of approximately 73% of isolates and serogroups of another 22%.[80] Pai et al. have developed a sequential multiplex PCR for determination of capsular pneumococcal serotypes. This approach uses short sequences of the *cps* locus that are specific to serotypes or serogroups, including 28 primer pairs distributed among seven multiplex PCRs.[81]

Recent application of a real-time PCR based on the detection of the pneumococcal capsular *wzg* gene allowed a more rapid diagnosis compared to conventional PCR.[82] Considering the rising antibiotic resistance of pneumococci, PCR approaches for determination of susceptibility to certain antimicrobial agents have also been developed. A real-time PCR detecting *mef*(A), *erm*(A), and *erm*(B), which are responsible for macrolide resistance in *S. pneumoniae* and *S. pyogenes*, makes rapid susceptibility testing possible.[34] Harris et al. combined detection of *S. pneumoniae* with determination of penicillin susceptibility by developing a duplex real-time PCR method targeting the *lytA* gene and the gene of the penicillin-binding protein 2b (*pbp2b*), whose absence or modification is responsible for penicillin resistance (see Section 29.2.2.3).[83]

Quantitative PCR permits discrimination between colonization and infection with the higher bacterial burden present during infection.[11,45] Moreover, fluorescence in situ hybridization (FISH) targeted to ribosomal RNA is a molecular technique appropriate for detection of streptococci.[40,41]

Two molecular tests have been developed for detection of *S. pyogenes*: a chemiluminescent single-stranded DNA-probe detecting GAS-specific rRNA sequences (Group A Streptococcus Direct Test, Gen-Probe, Inc.) and a one-rapid-cycle real-time PCR method (LightCycler Strept-A assay, Roche Applied Science).[3,71,73] The Group A Streptococcus Direct Test is a reliable and easy to perform technique designed to screen throat swabs from patients possibly infected with *S. pyogenes*. The method is a DNA-probe assay based on nucleic acid hybridization. Specific rRNA sequences unique to *S. pyogenes* are bound by a chemiluminescent single-stranded DNA probe, which results in a stable double-stranded hybrid that can be measured in a luminometer in relative light units (RLU). The RLU are directly proportional to the amount of hybridization and thus to the amount of *S. pyogenes* present in the probe.[71] The LightCycler Strept-A assay is a real-time PCR that amplifies a 198 bp fragment of the *ptsI* (phosphotransferase) gene of GAS, which is detected by fluorescent resonance energy transfer (FRET). Differentiation of GAS and group G and C streptococci, which are potentially associated with pharyngitis, is possible. A great advantage of this test is that only 3 min of personnel time are required for performance. It is considered as adequate test for detection of GAS from throat swabs.[73]

Because most molecular techniques take 1.5–2 h to generate the results, they are not suitable for point-of-care applications. Nonetheless, these tests provide time saving in comparison with culture, allowing initiation of antimicrobial therapy on the day the throat swab is taken, consequently reducing the risk for sequelae such as rheumatic fever.[3,73] Moreover, use of molecular tests overcomes the problem of false-negative cultures due to prior antibiotic therapy since their results are not influenced by antimicrobial therapy.[11] For an overview of diagnostic methods, refer to Table 29.1.

29.2 METHODS

29.2.1 Sample Preparation

In order to analyze streptococci with molecular methods, usually the preparation of genomic DNA is necessary. Prior to DNA isolation, streptococci can be grown in Todd Hewitt broth and incubated overnight at 37°C. From this medium, bacteria can be quantified if necessary, and DNA can be directly extracted by using commercially available DNA isolation kits such as the DNeasy kit (Qiagen).[53]

Alternatively, DNA can be prepared by a rapid boiling method: (i) pick half a loop of a fresh overnight culture and resuspend in lysis solution (0.25% w/v SDS, 0.05N NaOH); (ii) boil for 5 min in water bath; (iii) add 900 μL of distilled water and centrifuge at 11,000 × *g* for 5 min.

29.2.2 Detection Procedures

29.2.2.1 PCR-Based Emm-Typing of *S. Pyogenes*

Principle. The genetic sequence encoding the M protein—the so-called emm sequence—can be used to genetically determine the serotypes of *S. pyogenes*. By targeting the variable and specific 5'emm sequence, 104 emm types (designated "emm + number"—e.g., emm1 or emm23) have been identified.[21,79] The variety of emm types emphasizes the difficulty in designing specific primer pairs for every single emm type. Therefore, a primer pair based on highly conserved sequences was designed ("all M" primer), which has the potential of amplifying a great proportion of the described emm genes. Primers used by Podbielski et al. are the following[38]:

"all M" forward: 5′ ATA AGG AGC ATA AAA ATG GCT 3′
"all M" reverse: 5′ AGC TTA GTT TTC TTC TTT GCG 3′

With these primers, gene products of 0.9–1.85 kb in size depending on the M type are obtained. The *N*-terminal 250–300 nucleotides covering the variable part of the emm-sequence with specificity for each serotype are also amplified. Sequencing of this gene region allows determination of the serotype after comparison with the registered variable emm-sequences entered into the M type-specific sequence database available at the internet page of the Centers for Disease Control and Prevention (CDCP).[38,79]

Procedure

1. Prepare a PCR mixture (100 μL) containing 5 μL of template DNA (as extracted in Section 29.2.1), 10 μL of tenfold PCR buffer (500 mM KCl, 100 mM Tris-HCl pH 8.3, 15 mM $MgCl_2$, and 0.01% w/v Gelatin), 5 μL of dNTPs (10 mM each), 5–20 pmol each of all forward and reverse primers, 1 μL (5 U) Taq polymerase, and distilled water (to 100 μL).
2. Run the cycling program as follows: 1 cycle of 94°C for 2 min; 30 cycles of 94°C for 1 min, 50°C for 1 min, and 72°C for 1.5 min; keep the tubes at 4°C until they are removed from the cycler.
3. Purify the amplified PCR products with a standard PCR purification kit (e.g., Quiaquick, Quiagen) and sequence the products using standard DNA sequencing techniques.
4. Compare the resulting emm-sequences with the emm-database at the Centers for Disease Control (http://www.cdc.gov/ncidod/biotech/strep/strep-blast.htm) and assign an emm-type.

An alternative protocol for emm-type determination is available at http://www.cdc.gov/ncidod/biotech/strep/protocol_emm-type.htm.

29.2.2.2 Multiplex PCR for Detection of Group A Streptococcal Superantigens

Principle. SAgs are proteins secreted by *S. pyogenes* that can lead to activation of a great number of T cells and a subsequent cytokine storm after having cross-linked MHC class II and the TCR,[17,57] with the life-threatening STSS (described in Section 29.1.2.1) as a possible consequence. Recent development of two multiplex PCR facilitates detection of all 11 Sags identified to date. The first multiplex PCR covers SpeA 1–3 + 5, speC, speG, speJ, speK, and speL, and the second multiplex PCR detects speA1–4, speH, speI, speM, SSA, and SMEZ.[53] The multiplex PCR utilizes *S. pyogenes* DNA purified by the DNeasy kit (see Section 29.2.1) and a multiplex PCR kit (Qiagen). With the aim of excluding false-negative PCR results, 16S rRNA primers are included as an internal positive control. The primer sequences are shown in Table 29.2.[53]

Procedure

1. Prepare a PCR mixture (50 μL) containing 25 μL master mix (DNA polymerase, dNTP mix, PCR buffer), 5 μL Q-solution (to generate bands of similar intensity), 100 ng template DNA, 200 nM of each SAg gene primer pair, and 20 nM 16S rRNA primers (Table 29.2).
2. Amplify the DNA in a thermocycler with the following protocol: an initial denaturation at 95°C for 15 min, 35 cycles of amplification (30 s at 94°C, 90 s at 57°C, 90 s at 72°C), and a final extension at 72°C 10 min.
3. Analyze the samples on a 2.5% agarose gel.

Note: The PCR can also be carried out as simple standard PCR using the same conditions with only one primer pair, or as real-time PCR in a real-time cycler with little changes to the PCR mixture: the reaction mix contains additional 1.3 μL SYBR-Green (10×) (Cambrex, Verviers, Belgium) and 0.5 μL ROX (100×) (TIB MOLBIOL), and the template DNA is added as the last component. For real-time PCR, the number of cycles is increased from 35 to 40 cycles. Using real-time

TABLE 29.2
Primer Mixes for Detection of Group A Streptococcal Superantigens

Primer Mix	Gene	Primer Sequences (5′–3′)	Product Size (bp)
Mix 1	16S rRNA	f: GTGAGTAACGCGTAGGTAACCTACCTCATAG	784
		r: CCCAGGCGGAGTGCTTAATG	
	*Spe*C	f: GGTAAATTTTTCAACGACACACACATTAAA	258
		r: TGTTGAGATTCTCCCGAAATAAATAGAT	
	*Spe*G	f: GCTATGGAAGTCAATTAGCTTATGCAGAT	393
		r: TTATGCGAACAGCCTCAGAGG	
	*Spe*J	f: CAATTAAATTACGCATACGAAATCATACCAGTA	207
		r: ACGAGTAAATATGTACGGAAGACCAAAAATA	
	*Spe*K	f: TATCGCTTGCTCTATACACTACTGAGAGT	233
		r: CCAAACTGTAGTATTTTCATCCGTATTAAA	
	*Spe*L	f: GGACGCAAGTTATTATGGATGCTCA	460
		r: TTAAATAAGTCAGCACCTTCCTCTTTCTC	
	*Spe*A 1–3+5	f: GGTATTTGCTCAACAAGACCCCGAT	643
		r: TGTGTTTGAGTCAAGCGTTTCATTATCT	
Mix 2	16S rRNA	f: GGTGAGTAACGCGTAGGTAACCTACC	1042
		r: GCCCAACTTAATGATGGCAACTAACA	
	*Spe*H	f: TCTATCTGCACAAGAGGTTTGTGAATGTCCA	338
		r: GCATGCTATTAAAGTCTCCATTGCCAAAA	
	*Spe*I	f: AAGGAAAAATAAATGAAGGTCCGCCAT	217
		r: TCGCTTAAAGTAATACCTCCATATGAATTCTTT	
	*Spe*M	f: GCTTTAAGGAGGAGGAGGTTGATATTTATGCTCTA	411
		r: CAAAGTGACTTACTTTACTCATATCAATCGTTTC	
	SMEZ	f: CAATAATTTCTCGTCCTGTGTTTGGAT	309
		r: GATAAGGCGTCATTCCACCATAG	
	SSA	f: AATTATTATCGATTAGTGTTTTTGCAAGTA	570
		r: AGCCTGTCTCGTACGGAGAATTATTGAACTC	
	*Spe*A 1–4	f: CAAGAAGTATTTGCTCAACAAGACCCCA	512
		r: TTAGATGGTCCATTAGTATATAGTTGCTTGTTATC	

f, forward primer; r, reverse primer.
Source: From Lintges, M. et al., *Int. J. Med. Microbiol.*, 297, 471, 2007.

PCR, it can be rapidly determined whether or not the strain under investigation possesses a SAg gene and the risk of post PCR contamination is excluded due to the fact that the tubes do not have to be opened after amplification.[53]

29.2.2.3 Detection of *S. pneumoniae* and Determination of Penicillin Susceptibility by Duplex Real-Time PCR

Principle. Since penicillin resistance among pneumococci keeps rising, a method allowing rapid determination of penicillin susceptibility is desirable. Altered PBPs, especially PBP2b, with reduced affinity to penicillin seem to be responsible for penicillin resistance. Therefore, the *pbp2b* gene is used in combination with the *lytA* gene, which is a common PCR target for identification of *S. pneumoniae*, with the aim of rapidly detecting *S. pneumoniae* and at the same time determining its susceptibility to penicillin by real-time PCR.[83] DNA can be extracted by a boiling method that slightly differs from the one described in Section 29.2.1, or by mechanical disruption.[83,84] The *pbp2b* gene fragment amplified by this PCR comprises 89 bp, the *lytA* gene fragment is 101 bp in size. As positive control, DNA isolated from a penicillin susceptible *S. pneumoniae* isolate is included and 5 μL of water serves as a negative control. Sequences of primers and probes are listed in Table 29.3.[83]

Procedure

1. Prepare a PCR mixture containing 1× Quantitect multiplex master mix (Qiagen); 0.1 μM of each of the primers LytA-F, LytA-R, Pbp-2b-F, and Pbp-2b-RMOD (Table 29.3); 0.1 μM each of the probes LytA-probe and Pbp-2b-probe (Table 29.3); 5 μL of DNA or positive/negative control reagents; and PCR-grade water up to a final volume of 40 μL.
2. Amplify the DNA in a real-time PCR system according to the following protocol: 50°C for 2 min, followed by 95°C for 10 min, and 45 cycles of 95°C for 15 s and 60°C for 1 min.
3. Analyze the results using appropriate software.

Note: The detection limit for both genes is 0.5 colony-forming units (CFU) per reaction. The *lytA* PCR can be either positive or negative showing the presence or absence of *S. pneumoniae* in the probe. The *pbp2b* real-time PCR as part of

TABLE 29.3
Primers and Probes for Detection of *S. pneumoniae*

Primer/Probe[a]	Sequence[b]	Reference
LytA-F	5′-ACG CAA TCT AGC AGA TGA AGC-3′	85
LytA-R	5′-TGT TTG GTT GGT TAT TCG TGC-3′	85
LytA-probe	FAM 5′-TTT GCC GAA AAC GCT TGA TAC AGG G-3′ TAMRA	85
Pbp-2b-F	5′-ATT CTT GGT ATA CTC AGG CT-3′	86
Pbp-2b-RMOD[c]	5′-TGT TTG GAC CAT ATA GGT ATT-3′	83
Pbp-2b-probe	VIC 5′-CAC AGC GGT CCA AGC TCT-3′ TAMRA	86

[a] F, forward primer; R, reverse primer.

[b] FAM: 6-carboxyfluorescein; TAMRA: 6-carboxytetramethylrhodamine.

[c] Pbp-2b-RMOD is a modified version of the primer published by Kearns et al.[86]

the duplex assay should be interpreted as follows: a penicillin susceptible result is a *pbp2b* PCR positive with a C_T value maximal three cycles greater ($\Delta C_T \leq 3$) than that of the simultaneously run *lytA* PCR; a negative *pbp2b* PCR result is interpreted as reduced penicillin susceptibility. Every PCR result not matching one of these two criteria is equivocal and does not allow interpretation of penicillin susceptibility. Susceptibility is defined as minimal inhibitory concentration (MIC) $\leq$ 0.06 mg/L, and reduced penicillin susceptibility is defined as MIC > 0.06 mg/L. Although the *lytA* gene is also present in oral streptococci, leading to false positive results, this may not be a drawback from a clinical perspective, since most of the detected oral streptococci cause serious diseases.[83]

29.3 CONCLUSION AND FUTURE PERSPECTIVES

Streptococci are common pathogens that cause a variety of human diseases, ranging from pharyngitis and impetigo (GAS) to meningitis and pneumonia in newborns (GBS), pneumonia (*S. pneumoniae*), and necrotizing fasciitis and STSS (*S. pyogenes*).[1,22,44] The fact that *S. pyogenes* may be the cause of nonsuppurative sequelae like poststreptococcal glomerulonephritis and rheumatic fever emphasizes the necessity of an early antibiotic therapy.[1]

Since the diagnostic gold standard for identification of different streptococci is culture, providing the result the next day at the earliest, RADTs have been developed to overcome this obstacle.[3,11,22] Although RADTs are usually more expensive than culture, they are advantageous in helping implement early treatment, which in the case of GAS infection reduces the risk of the spread of the infection, allows early return to school or work, and shortens the symptomatic period.[3] The problem arising from RADTs is their low sensitivity, leading to false-negative results. Therefore, at least for GAS, it is recommended that a negative RADT should be confirmed by culture.[3] While the molecular methods offer the prospects of rapid, sensitive, and specific detection and identification of streptococci, they are not yet suitable for point-of-care applications because of the requirement of laboratory equipment and personnel time. In addition, in most cases, the results delivered by the molecular tests are too detailed and not essential for clinical evaluation and treatment.

Difficulties that have to be overcome concerning the treatment of streptococcal infections are the rising resistance rates to antibiotics, especially concerning *S. pneumoniae*. The resistance rates to macrolides, penicillin, and fluoroquinolones vary among different countries and data strongly suggest that resistant clones—even multiresistant clones—emerge with increasing use of antibiotics.[4] If *S. pyogenes* is resistant to penicillin, therapy of GAS infections will become very difficult, resulting in rising mortality due to GAS associated diseases. Thus, although early antibiotic treatment is favorable, the use of antimicrobial agents should be reduced in order to prevent development of antimicrobial resistance. This is only possible when highly sensitive and specific diagnostic tools are available. Furthermore, international cooperation is desirable.

It will be possible in the future, as it is already the case for *S. pneumoniae*, to prevent streptococcal diseases by vaccination.[1,2,5,30] As a consequence, the problems concerning diagnosis and treatment might not be as relevant as they are today, but different problems may arise, considering that the strains included in the vaccine may be replaced by other strains as already observed for pneumococci today.[2] Therefore, diseases caused by streptococci will remain a challenge for physicians and laboratories concerning diagnosis, treatment, and prevention.

REFERENCES

1. Cunningham, M.W., Pathogenesis of group A streptococcal infections, *Clin. Microbiol. Rev.*, 13, 470, 2000.
2. Pletz, M.W. et al., Pneumococcal vaccines: mechanism of action, impact on epidemiology and adaption of the species, *Int. J. Antimicrob. Agents*, 32, 199, 2008.
3. Gerber, M.A., and Shulman, S.T., Rapid diagnosis of pharyngitis caused by group A streptococci, *Clin. Microbiol. Rev.*, 17, 571, 2004.
4. Reinert, R.R. et al., Antimicrobial susceptibility of *Streptococcus pneumoniae* in eight European countries from 2001 to 2003, *Antimicrob. Agents Chemother.*, 49, 2903, 2005.
5. Larsen, J.W., and Sever, J.L., Group B *Streptococcus* and pregnancy: A review, *Am. J. Obstet. Gynecol.*, 198, 440, 2008.
6. Köhler, W., The present state of species within the genera *Streptococcus* and *Enterococcus*, *Int. J. Med. Microbiol.*, 297, 133, 2007.
7. Stollerman, G.H., and Dale, J.B., The importance of the group A *Streptococcus* capsule in the pathogenesis of human infections: A historical perspective, *Clin. Infect. Dis.*, 46, 1038, 2008.
8. Schottmüller, H., Die Artunterscheidung der für den Menschen pathogenen Streptokokken durch Blutagar, *Muench. Med. Wochenschr.*, 50, 849, 1903.

9. Barnard, J.P., and Stinson, M.W., The alpha-hemolysin of *Streptococcus gordonii* is hydrogen peroxide, *Infect. Immun.*, 64, 3853, 1996.
10. Facklam, R., What happened to the streptococci: Overview of taxonomic and nomenclature changes, *Clin. Microbiol. Rev.*, 15, 613, 2002.
11. Werno, A.M., and Murdoch, D.R., Medical microbiology: Laboratory diagnosis of invasive pneumococcal disease, *Clin. Infect. Dis.*, 46, 926, 2008.
12. Edwards, P.R., The biochemical characters of human and animal strains of hemolytic streptococci, *J. Bacteriol.*, 23, 259, 1932.
13. Lancefield, R.C., A serological differentiation of human and other groups of streptococci, *J. Exp. Med.*, 59, 441, 1933.
14. Lancefield, R.C., Current knowledge of type-specific M antigens of group A streptococci, *J. Immunol.*, 89, 307, 1962.
15. Kehoe, M.A. et al., Horizontal gene transfer among group A streptococci: Implications for pathogenesis and epidemiology, *Trends Microbiol.*, 4, 436, 1996.
16. Mitchell, T.J., The pathogenesis of streptococcal infections: From tooth decay to meningitis, *Nat. Rev. Microbiol.*, 1, 219, 2003.
17. Fraser, J.D., and Proft, T., The bacterial superantigen and superantigen-like proteins, *Immunol. Rev.*, 225, 226, 2008.
18. Bisno, A.L., Acute pharyngitis, *N. Engl. J. Med.*, 344, 205, 2001.
19. Bisno, A.L., Brito, M.O., and Collins, C.M., Molecular basis of group A streptococcal virulence, *Lancet Infect. Dis.*, 3, 191, 2003.
20. Reid, S.D. et al., Group A *Streptococcus*: Allelic variation, population genetics, and host-pathogen interactions, *J. Clin. Invest.*, 107, 393, 2001.
21. Facklam, R.F. et al., Extension of the Lancefield classification for group A streptococci by addition of 22 new M protein gene sequence types from clinical isolates: Emm103 to emm124, *Clin. Infect. Dis.*, 34, 28, 2002.
22. Pettersson, K., Perinatal infection with Group B streptococci, *Semin. Fetal Neonatal Med.*, 12, 193, 2007.
23. Mullaney, D.M., Group B streptococcal infections in newborns, *J. Obstet. Gynecol. Neonatal Nurs.*, 30, 649, 2001.
24. Phares, C.R., Active Bacterial Core surveillance/Emerging Infections Program Network. Epidemiology of invasive group B streptococcal disease in the United States, 1999–2005, *JAMA*, 299, 2056, 2008.
25. Downar, J., Streptococcal meningitis resulting from contact with an infected horse, *J. Clin. Microbiol.*, 39, 2358, 2001.
26. Sechi, L.A., De Carli, S., and Ciani, R., *Streptococcus equinus* endocarditis in a patient with pulmonary histiocytosis X, *Am. J. Med.*, 108, 522, 2000.
27. Lütticken, R., Meningitis caused by *Streptococcus suis*: Case report and review of the literature, *Infection*, 14, 181, 1986.
28. Huang, Y.T. et al., *Streptococcus suis* infection, *J. Microbiol. Immunol. Infect.*, 38, 306, 2005.
29. Whiley, R.A., and Beighton, D., Current classification of the oral streptococci, *Oral Microbiol. Immunol.*, 13, 195, 1998.
30. Hausdorff, W.P., Feikin, D.R., and Klugman, K.P., Epidemiological differences among pneumococcal serotypes, *Lancet Infect. Dis.*, 5, 83, 2005.
31. Baxter, F., and McChesney, J., Severe group A streptococcal infection and streptococcal toxic shock syndrome, *Can. J. Anaesth.*, 47, 1129, 2000.
32. Madden, J.C., Ruiz, N., and Caparon, M., Cytolysin-mediated translocation (CMT): A functional equivalent of type III secretion in gram-positive bacteria, *Cell*, 104, 143, 2001.
33. Carapetis, J.R. et al., The global burden of group A streptococcal diseases, *Lancet Infect. Dis.*, 5, 685, 2005.
34. Reinert, R.R. et al., Molecular characterisation of macrolide resistance mechanisms of *Streptococcus pneumoniae* and *Streptococcus pyogenes* isolated in Germany, 2002–2003, *Int. J. Antimicrob. Agents*, 24, 43, 2004.
35. Malhotra-Kumar, S. et al., Emergence of high-level fluoroquinolone resistance in emm6 *Streptococcus pyogenes* and in vitro resistance selection with ciprofloxacin, levofloxacin and moxifloxacin, *J. Antimicrob. Chemother.*, 63, 886, 2009.
36. Funahashi, K. et al., T serotypes and antimicrobial susceptibilities of group A *Streptococcus* isolates from pediatric pharyngotonsillitis, *Jpn. J. Infect. Dis.*, 61, 454, 2008.
37. Brandt, C.M. et al., Characterization of blood culture isolates of *Streptococcus dysgalactiae* subsp. *equisimilis* possessing Lancefield's group A antigen, *J. Clin. Microbiol.*, 37, 4194, 1999.
38. Podbielski, A., Melzer, B., and Lütticken, R., Application of the polymerase chain reaction to study the M protein(-like) gene family in beta-hemolytic streptococci, *Med. Microbiol. Immunol.*, 180, 213, 1991.
39. Sing, A., Trebesius, K., and Heesemann, J., Diagnosis of *Streptococcus dysgalactiae* subspecies *equisimilis* (Group C streptococci) associated with deep soft tissue infections using fluorescent in situ hybridization, *Eur. J. Clin. Microbiol. Infect. Dis.*, 20, 146, 2001.
40. Gescher, D.M. et al., Fluorescence in situ hybridisation (FISH) accelerates identification of Gram-positive cocci in positive blood cultures, *Int. J. Antimicrob. Agents*, 32, S51, 2008.
41. Artz, L.A., Kempf, V.A., and Autenrieth, I.B., Rapid screening for *Streptococcus agalactiae* in vaginal specimens of pregnant women by fluorescent in situ hybridization, *J. Clin. Microbiol.*, 41, 2170, 2003.
42. Gosink, K.K. et al., Role of novel choline binding proteins in virulence of *Streptococcus pneumoniae*, *Infect. Immun.*, 68, 5690, 2000.
43. Goldsmith, C.E. et al., Increased incidence of ciprofloxacin resistance in penicillin-resistant pneumococci in Northern Ireland, *J. Antimicrob. Chemother.*, 41, 420, 1998.
44. Obaro, S., and Adegbola, R., The pneumococcus: carriage, disease and conjugate vaccines, *J. Med. Microbiol.*, 51, 98, 2002.
45. Klugman, K.P., Madhi, S.A., and Albrich, W.C., Novel approaches to the identification of *Streptococcus pneumoniae* as the cause of community-acquired pneumonia, *Clin. Infect. Dis.*, 47, S202, 2008.
46. Lindgren, P.E. et al., Two different genes coding for fibronectin-binding proteins from *Streptococcus dysgalactiae*. The complete nucleotide sequences and characterization of the binding domains, *Eur. J. Biochem.*, 214, 819, 1993.
47. Timoney, J.F., The pathogenic equine streptococci, *Vet. Res.*, 35, 397, 2004.
48. Broyles, L.N. et al., Population-based study of invasive disease due to beta-hemolytic streptococci of groups other than A and B, *Clin. Infect. Dis.*, 48, 706, 2009.
49. Luque, I., Molecular typing and anti-microbial susceptibility of clinical isolates of *Streptococcus equi* ssp. *zooepidemicus* from equine bacterial endometritis, *J. Vet. Med. B. Infect. Dis. Vet. Public Health*, 53, 451, 2006.
50. Gottschalk, M., and Segura, M., The pathogenesis of the meningitis caused by *Streptococcus suis*: The unresolved questions, *Vet. Microbiol.*, 76, 259, 2000.
51. Tleyjeh, I.M. et al., Temporal trends in infective endocarditis: A population-based study in Olmsted County, Minnesota, *JAMA*, 293, 3022, 2005.

52. Yap, R.L., Mermel, L.A., and Maglio, J., Antimicrobial resistance of community-acquired bloodstream isolates of viridans group streptococci, *Infection*, 34, 339, 2006.
53. Lintges, M. et al., A new closed-tube multiplex real-time PCR to detect eleven superantigens of *Streptococcus pyogenes* identifies a strain without superantigen activity, *Int. J. Med. Microbiol.*, 297, 471, 2007.
54. Maripuu, L., Eriksson, A., and Norgren, M., Superantigen gene profile diversity among clinical group A streptococcal isolates, *FEMS Immunol. Med. Microbiol.*, 54, 236, 2008.
55. Braun, M.A. et al., Stimulation of human T cells by streptococcal "superantigen" erythrogenic toxins (scarlet fever toxins), *J. Immunol.*, 150, 2457, 1993.
56. Gerlach, D. et al., Purification and biochemical characterization of a basic superantigen (SPEX/SMEZ3) from *Streptococcus pyogenes*, *FEMS Microbiol. Lett.*, 188, 153, 2000.
57. Rink, L. et al., Microbial superantigens stimulate T cells by the superantigen bridge and independently by a cytokine pathway, *J. Interferon Cytokine Res.*, 17, 489, 1997.
58. McNeil, S.A. et al., Safety and immunogenicity of 26-valent group A *Streptococcus* vaccine in healthy adult volunteers, *Clin. Infect. Dis.*, 41, 1114, 2005.
59. Ulrich, R.G., Vaccine based on a ubiquitous cysteinyl protease and streptococcal pyrogenic exotoxin A protects against *Streptococcus pyogenes* sepsis and toxic shock, *J. Immune Based Ther. Vaccines*, 6, 8, 2008.
60. Baron, M.J. et al., Identification of a glycosaminoglycan binding region of the alpha C protein that mediates entry of group B Streptococci into host cells, *J. Biol. Chem.*, 282, 10526, 2007.
61. Bolduc, G.R., and Madoff, L.C., The group B streptococcal alpha C protein binds alpha1beta1-integrin through a novel KTD motif that promotes internalization of GBS within human epithelial cells, *Microbiology*, 153, 4039, 2007.
62. Taubman, M.A., and Nash, D.A., The scientific and public-health imperative for a vaccine against dental caries, *Nat. Rev. Immunol.*, 6, 555, 2006.
63. McCracken, G.H., Diagnosis and management of pneumonia in children, *Pediatr. Infect. Dis. J.*, 19, 924, 2000.
64. Örtqvist, A., Hedlund, J., and Kalin, M., *Streptococcus pneumoniae*: Epidemiology, risk factors, and clinical features, *Semin. Respir. Crit. Care. Med.*, 26, 563, 2005.
65. Van Bambeke, F. et al., Multidrug-resistant *Streptococcus pneumoniae* infections: Current and future therapeutic options, *Drugs*, 67, 2355, 2007.
66. Hakenbeck, R. et al., Acquisition of five high-Mr penicillin-binding protein variants during transfer of high-level beta-lactam resistance from *Streptococcus mitis* to *Streptococcus pneumoniae*, *J. Bacteriol.*, 180, 1831, 1998.
67. Hakenbeck, R. et al., Mosaic genes and mosaic chromosomes: Intra- and interspecies genomic variation of *Streptococcus pneumoniae*, *Infect. Immun.*, 69, 2477, 2001.
68. Farrington, M., and Rubenstein, D., Antigen detection in pneumococcal pneumonia, *J. Infect.*, 23, 109, 1991.
69. Kellogg, J.A., Suitability of throat culture procedures for detection of group A streptococci and as reference standards for evaluation of streptococcal antigen detection kits, *J. Clin. Microbiol.*, 28, 165, 1990.
70. Kaufhold, A., and Ferrieri, P., The microbiologic aspects, including diagnosis, of beta-hemolytic streptococcal and enterococcal infections, *Infect. Dis. Clin. North Am.*, 7, 235, 1993.
71. Pokorski, S.J. et al., Comparison of Gen-Probe Group A Streptococcus Direct Test with culture for diagnosing streptococcal pharyngitis, *J. Clin. Microbiol.*, 32, 1440, 1994.
72. Leung, A.K. et al., Rapid antigen detection testing in diagnosing group A beta-hemolytic streptococcal pharyngitis, *Expert Rev. Mol. Diagn.*, 6, 761, 2006.
73. Uhl, J.R. et al., Comparison of LightCycler PCR, rapid antigen immunoassay, and culture for detection of group A streptococci from throat swabs, *J. Clin. Microbiol.*, 41, 242, 2003.
74. Petts, D.N., Evaluation of a modified nitrous acid extraction latex agglutination kit for grouping beta-hemolytic streptococci and enterococci, *J. Clin. Microbiol.*, 33, 1016, 1995.
75. Nerbrand, C., Jasir, A., and Schalén, C., Are current rapid detection tests for Group A Streptococci sensitive enough? Evaluation of 2 commercial kits, *Scand. J. Infect. Dis.*, 34, 797, 2002.
76. Petti, C.A., Woods, C.W., and Reller, L.B., *Streptococcus pneumoniae* antigen test using positive blood culture bottles as an alternative method to diagnose pneumococcal bacteremia, *J. Clin. Microbiol.*, 43, 2510, 2005.
77. Yu, J. et al., Rapid multiplex assay for serotyping pneumococci with monoclonal and polyclonal antibodies, *J. Clin. Microbiol.*, 43, 156, 2005.
78. Enright, M.C., and Spratt, B.G., Multilocus sequence typing, *Trends Microbiol.*, 7, 482, 1999.
79. Beall, B., Facklam, R., and Thompson, T., Sequencing emm-specific PCR products for routine and accurate typing of group A streptococci, *J. Clin. Microbiol.*, 34, 953, 1996.
80. Kong, F. et al., A molecular-capsular-type prediction system for 90 *Streptococcus pneumoniae* serotypes using partial cpsA-cpsB sequencing and wzy- or wzx-specific PCR, *J. Med. Microbiol.*, 54, 351, 2005.
81. Pai, R., Gertz, R.E., and Beall, B., Sequential multiplex PCR approach for determining capsular serotypes of *Streptococcus pneumoniae* isolates, *J. Clin. Microbiol.*, 44, 124, 2006.
82. Tarragó, D. et al., Identification of pneumococcal serotypes from culture-negative clinical specimens by novel real-time PCR, *Clin. Microbiol. Infect.*, 14, 828, 2008.
83. Harris, K.A. et al., Duplex real-time PCR assay for detection of *Streptococcus pneumoniae* in clinical samples and determination of penicillin susceptibility, *J. Clin. Microbiol.*, 46, 2751, 2008.
84. Harris, K.A., and Hartley, J.C., Development of broad-range 16S rDNA PCR for use in the routine diagnostic clinical microbiology service, *J. Med. Microbiol.*, 52, 685, 2003.
85. McAvin, J.C. et al., Sensitive and specific method for rapid identification of *Streptococcus pneumoniae* using real-time fluorescence PCR, *J. Clin. Microbiol.*, 39, 3446, 2001.
86. Kearns, A.M. et al., Rapid real-time PCR for determination of penicillin susceptibility in pneumococcal meningitis, including culture-negative cases, *J. Clin. Microbiol.*, 40, 682, 2002.

Section II

Firmicutes and Tenericutes

Clostridia

30 Acidaminococcus

Hélène Marchandin and Estelle Jumas-Bilak

CONTENTS

30.1 INTRODUCTION

30.1.1 Classification and Morphology

Classification. The genus *Acidaminococcus* was proposed in 1969 by Rogosa to cover a group of anaerobic, gram-negative cocci.[1] Initial description of the genus *Acidaminococcus* was based on diplococci from the pig alimentary tract previously reported by Fuller in 1966 and included in a unique species, *Acidaminococcus fermentans*, type species of the genus.[1,2] Type strain of *A. fermentans* is strain VR4^T = ATCC 25085^T = CCUG 9996^T = CIP 106432^T = DSM 20731^T. An emended description of *A. fermentans* was published by Cook et al. in 1994.[3] In 2007, a second species was described in the genus, *Acidaminococcus intestini*, to accommodate gram-negative cocci isolated from human clinical samples that displayed distinctive features of *A. fermentans*.[4] Type strain of *A. intestini* is strain ADV 255.99^T = AIP 283.01^T = CIP 108586^T = CCUG 50930^T. Thus, to date, the genus comprises two validated species.

Initially classified in the family *Neisseriaceae*, the genus encountered several taxonomic reappraisals; it was transferred by Rogosa to the family *Veillonellaceae*, also called *Sporomusa*, subbranch of the *Clostridium* subphylum of the gram-positive bacteria.[5,6] Thus, despite its gram-negative type of cell wall, *A. fermentans* was classified in the phylum *Firmicutes* (low G + C gram-positive bacteria), in the class *Clostridia*, and in the order *Clostridiales*.[7] Recently, a large phylogenetic analysis based on 16S rRNA gene sequences of members of the phylum *Firmicutes* and other phyla indicated that the family *Veillonellaceae* formed a robust lineage clearly separated from those of the classes *Bacilli*, *Clostridia*, *Thermolithobacteria*, and *Erysipelotrichia* recognized in the phylum *Firmicutes*.[8] This study highlighted that the family *Veillonellaceae* was a class-level taxon in the phylum *Firmicutes*, for which the name *Negativicutes* classis nov. was proposed, based on the gram-negative type of cell wall of its members, with type order *Selenomonadales* ord. nov.[8] In this order, a novel family, *Acidaminococcaceae* fam. nov., is proposed and description of the family *Veillonellaceae* is amended. The family *Acidaminococcaceae* included the validated genera *Acidaminococcus*, *Phascolarctobacterium*, *Succinispira*, and *Succiniclasticum* (Figure 30.1). The four other known genera of anaerobic gram-negative cocci, *Anaeroglobus*, *Megasphaera*, *Negativicoccus*, and *Veillonella*, also belong to the phylum *Firmicutes* but to the emended family *Veillonellaceae* together with both genera *Allisonella* and *Dialister* (Figure 30.1).[8,9]

Morphology. Members of the genus *Acidaminococcus* are gram-negative, nonspore-forming, and nonmotile cocci (Figures 30.2a and 30.2b). *A. fermentans* was described as oval or kidney-shaped diplococci.[1] *A. intestini* cells are also coccoid but usually smaller than those of *A. fermentans*, with cells size of 0.5–0.6 μm versus 0.6–1 μm (Table 30.1).[1,4] They occurred as single cells or as diplococci.[4] Rogosa first described the presence of a multilayered cell wall membrane characteristic of gram-negative organisms after electron microscopy (EM) of thin sections of *A. fermentans*.[1] Similar surface layers typical of an outer membrane were also observed about 40 years later for *A. intestini* on ultrathin sections (Figure 30.2c), as described for several other

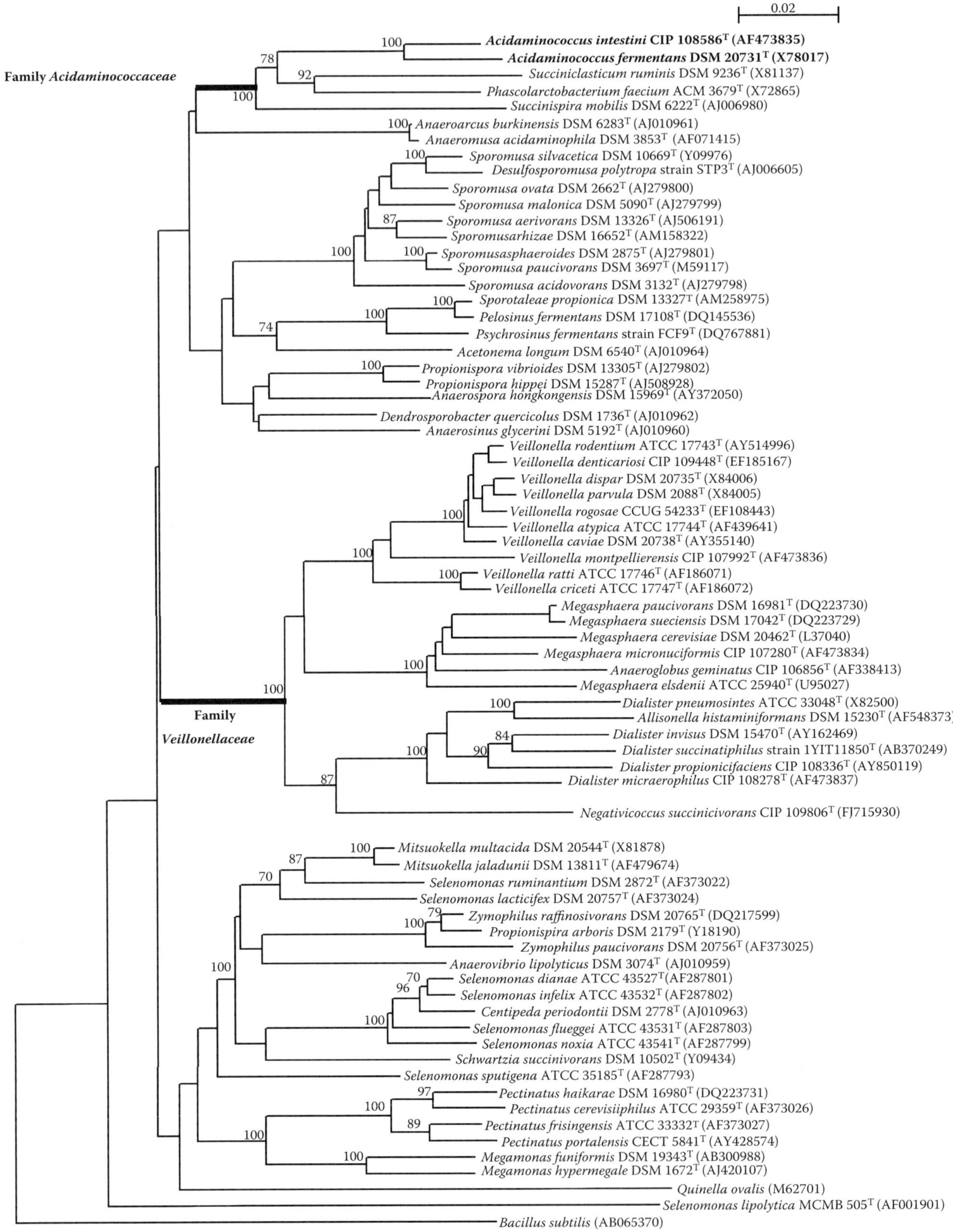

FIGURE 30.1 Phylogenetic placement of the genus *Acidaminococcus* among the class *Negativicutes* (formerly family *Veillonellaceae*). The tree based on partial 16S rRNA gene sequences (1211 nt), was obtained using the neighbor-joining method with F84 substitution model for distance calculation, as previously described.[8] Roots of the family *Acidaminococcaceae* and of the emended family *Veillonellaceae* are indicated by bold lines. GenBank accession numbers are given in parentheses. *Bacillus subtilis* was the outgroup organism. Numbers at nodes are bootstrap values based on 1000 resamplings and are indicated when >70 only. Bar, 0.02 substitutions per site.

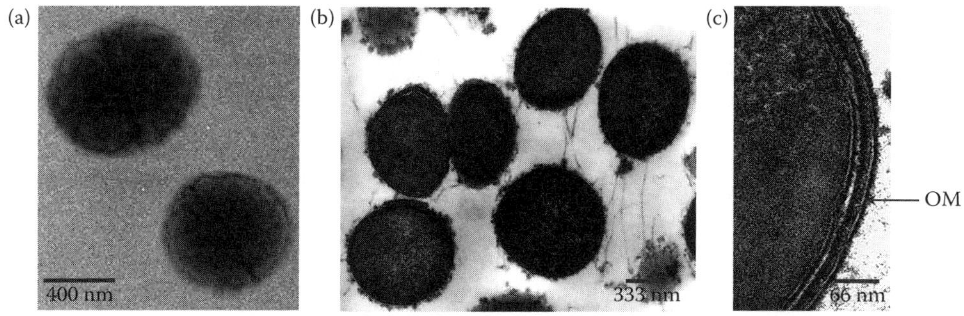

FIGURE 30.2 Electron microscopy of *Acidaminococcus intestini* strain ADV 255.99^T. (a) General morphology after negative staining; (b) general morphology of ultrathin sections; (c) ultrastructure after EM of ultrathin sections showing the gram-negative type cell wall with an outer membrane (OM).

TABLE 30.1
Phenotypic Characteristics That Differentiate *Acidaminococcus* spp. from Other Genera of Anaerobic Gram-Negative Cocci

	Acidaminococcus					
Characteristic	***A. fermentans***	***A. intestini***	***Negativicoccus***	***Megasphaera***	***Veillonella***	***Anaeroglobus***
Cell size (μm)	0.6–1.0	0.5–0.6	0.4	0.4–2.6	0.3–0.5	0.5–1.1
Catalase[a]	–	–	–	–	+/–	–
Growth in microaerophilic conditions	–	–	+	–	–	–
Nitrate reduction	–	–	–	–	+	–
Indole production	–	–/+	–	–	–	–
Ability to ferment carbohydrates	–[b]	–	–	+	–	+
Fructose	–	–	–	–/+	–/+	–
Galactose	–	–	ND	–	–	+
Glucose	–[b]	–	–	–/+	–	–
Maltose	–	–	–	–/+	–	
Mannose	–	–	–		–	+
Fermentation of lactate	–	–	–	+	+	–
Decarboxylation of succinate	–	ND	+	–	+	–
Amino acids are the main source of energy	+	+	–	–	–	–
Gas production	+	+	–	+/–	+/–	–
Metabolic end products	A, B	A, B, P, (L) (trace amounts 2-OH-B, 2-OH-V)	A,P (L), trace amounts 2-OH-V	(A), (P), (iB), B, iV, (V), C, (iC), (Pha)	A, P	A, P, iB, B, iV

Data are from References 4, 8, and 11–14.

[a] Some strains produce an atypical catalase also called pseudocatalase lacking porphyrin.

[b] About 40% of *A. fermentans* strains ferment glucose but the reaction is weak.

A, acetic acid; B, butyric acid; P, propionic acid; L, lactic acid; 2-OH-B, 2-hydroxybutyric acid; 2-OH-V, 2-hydroxyvaleric acid; iB, isobutyric acid; iV, isovaleric acid; V, valeric acid; C, caproic acid; iC, isocaproic acid; Pha, 2-phenylacetic acid; parentheses indicated a variable production among strains or species; ND, not determined.

Variable characteristics among species of the genus are indicated by +/– for genus with a majority of species presenting the characteristic or by –/+ for genus with a majority of species lacking the characteristic.

Variable characteristics among strains of a species are indicated by –/+ for a species with a majority of strains lacking the characteristic.

members of the class *Negativicutes*.[4,10] In both species, the outer layer has a tendency to strip away from the cell as in many other gram-negative organisms.[1] Endotoxinic activity was detected by a positive Schwarzman reaction in rabbits for cell extracts of *A. fermentans* whereas the reaction was negative for a gram-positive comparator.[1] Altogether, these observations explained the negative reaction to Gram stain for members of the genus *Acidaminococcus*, in spite of their belonging to the low G + C gram-positive/*Firmicutes* phylum.

30.1.2 Biology, Habitat, and Lifestyle

Biology. The strictly anaerobic members of the genus *Acidaminococcus* possess a fermentative type of metabolism and complex nutritional requirements.[1,11] Etymology of the genus name is "the amino acid coccus," revealing the ability to use amino acids as the sole source of energy for growth. These amino acids, particularly glutamic acid, are the main sources for energy. Other energy sources are not used; particularly, lactate is not fermented, succinate is not decarboxylated, and polyols are not attacked.[1,4,11] Regarding carbohydrates, *Acidaminococcus* spp. are mainly unable to ferment carbohydrates, but about 40% of *A. fermentans* strains are able to weakly catabolize glucose.[1,11] These are useful characteristics for differentiating *Acidaminococcus* from the other genera of anaerobic, gram-negative cocci (Table 30.1).[15] *A. fermentans* ferments glutamate via the 2-hydroxyglutarate pathway to ammonium, carbon dioxide, acetate, butyrate, and molecular hydrogen.[16] Among the unusual enzymes involved in this pathway, the key enzyme is 2-hydroxyglutaryl-coenzyme A dehydratase.[11,17] Cook et al. then demonstrated that *A. fermentans* was also able to metabolize citrate and to convert *trans*-aconitate to acetate.[5] Optimum temperature and optimum pH for growth are 30°C–37°C and 7.0, respectively. Main metabolic end products produced by both species are acetate and butyrate, with *A. intestini* also producing propionate.[1,4,11] Putrescine, cadaverine, and spermidine are the major polyamines in the cell wall peptidoglycan.[18] Salt tolerance of *A. fermentans* is estimated to 4% by Chatterjee and Chakraborti.[19]

Habitat and Lifestyle. *Acidaminococcus* spp. are part of the normal intestinal microbiota and were demonstrated either by cultivation or by cultivation-independent molecular approaches in gut of man and animals.[3,20,21] By cultivation, *A. fermentans* was found in 25% of normal human fecal samples and was usually present in relatively large amounts.[20] Cultivation-independent studies of human fecal samples found several uncultured bacterium clones representative of both *Acidaminococcus* species on the basis of 16S rRNA gene sequence identity level >99% with the type strain of the species. For example, several uncultured *A. fermentans* clones as well as several uncultured *A. intestini* clones were reported by Mai et al., including clone 001B-a1 (GenBank accession number DQ904735) and clone 014C-G10 (GenBank accession number DQ905693), respectively.[22] Several clones of *A. intestini* were also recovered during two studies on gut microbiome in obese patients and from human intestinal biopsies in patients with Crohn disease and ulcerative colitis; for respective examples, clones RL186_aah59c08 (DQ059641), TS57_a02g03 (FJ370581), and CD33 (DQ441284).[23–25]

In animals, *A. fermentans* was cultured from rumen dairy cow, pig intestine, and rumen wall in lambs.[26] A recent cultivation-independent study evaluating the bacterial diversity in the feces of cattle demonstrated that *Acidaminococcus* spp. were present in all the 20 cow samples analyzed and represents 0.15%–1.16% of the total bacterial population.[27] In cattle, diet influenced the bacterial diversity observed along the digestive tract. *A. fermentans* only occurred in the colon of animals receiving high-fiber diet while it was isolated at all digestive sites (rumen, ileum contents, ileum wall, colon contents, colon wall, and feces) from grain-fed animals.[28] *A. fermentans* is not normally a predominant ruminal bacterium, its concentration in the rumen being estimated to 10^2 colony forming units/mL.[3] However, its ability to oxidize *trans*-aconitate to nontoxic acetate is associated to the decrease of the accumulation of tricarballylate, a toxic end product of ruminal fermentation. This phenomenon resulting from metabolic competition between ruminal bacteria may well contribute to grass tetany prevention in ruminants.[29]

16S rRNA gene sequences showing identity levels ranging from 94% to 97% with *A. fermentans* or *A. intestini* are also deposited in the databases, suggesting that other uncharacterized taxa related to the genus *Acidaminococcus* may be discovered in the future either in the mammal gut or in the environment. For example, uncultured clones retrieved during metagenomic studies of gut microbiota of different mammalian species may represent novel *Acidaminococcus* species like uncultured bacterium clone orang1_h10_1 detected in the feces of orangutan (GenBank accession number EU777540, about 97% of 16S rRNA gene sequence identity with *A. fermentans*) and uncultured bacterium clone CAL_b10_1 detected in the feces of Goeldi's marmoset (GenBank accession number EU773666, about 96% of 16S rRNA gene sequence identity with *A. intestini*).[30] In the environment, sequences displaying about 94.3% of 16S rRNA gene sequence identity with *A. fermentans* were reported from microbial assemblages associated with high-temperature petroleum reservoirs (uncultured bacterium 018G7 and 016B1, GenBank accession numbers AF220334 and AF220326, respectively), suggesting that the habitat of the members of the genus *Acidaminococcus* may be larger than the mammalian gut.[31]

30.1.3 Genetic and Genomic Features

The 16S and 23S rRNA gene sequences are available for the two characterized *Acidaminococcus* species. These two species shared about 95.8% of their 16S rRNA gene sequences and less than 88.9% of their 23S rRNA gene sequences.[4]

Several genomic features were studied in the genus *Acidaminococcus*, including DNA G + C content determination and a large-scale chromosome structure study based on pulsed-field gel electrophoresis.[4,9,11] DNA G + C content as well as *rrn* skeleton were shown to be distinctive between species.[4] Indeed, DNA G + C content of *A. fermentans* was 54.7–57.4 mol% while a lower value of 49.3 mol% was reported for the type strain of *A. intestini*.[4,11]

As previously reported for several members of the class *Negativicutes*, the *rrn* skeleton was also useful for distinguishing between species since *A. fermentans* and *A. intestini* possess 6 and 3 *rrn* operons, respectively.[4,8,10,32] The two species displayed a similar genomic size of about 2.9 Mb ± 140 kb and harbored a unique chromosome with circular topology.[4]

The annotated genome sequence of the strain *Acidaminococcus* sp. D21 isolated from the human gut microbiota is now available as a part of the Human Microbiome Project (http://www.broadinstitute.org). One of the goals of this project is to sequence 1000 microbial genomes of bacteria associated with the human body. The strain D21 probably belonged to *A. intestini* on the basis of its 16S rRNA gene sequence. The genome contains 2005 genes on 2.24 Mb. The G + C content is 50.2 mol%. The total sequencing of the type strain of *A. fermentans* is currently in process.

30.1.4 Clinical Features and Pathogenesis

Among the anaerobic gram-negative cocci isolated in human beings, *Veillonella* are the most encountered while *Acidaminococcus* spp. remain rare.[9,33] A study conducted over a 3-year period (1999–2001) in the Teaching Hospital of Montpellier, France allowed us to consecutively collect 120 isolates of gram-negative anaerobic cocci from various human clinical isolates.[34] Among them, 116 were affiliated to the genus *Veillonella* by phenotypic and/or molecular-based methods for identification. Among the four remaining strains, one *Anaeroglobus geminatus*, two *Megasphaera* spp., and one *A. intestini* were identified using 16S rRNA gene sequencing. These data confirmed the relatively low rate of *Acidaminococcus* spp. recovery from human clinical samples during the routine microbiological analysis.

A. fermentans strains reported in the literature were isolated in mixed aerobic-anaerobic cultures during various infectious processes. *A. fermentans* was reported from pyogenic infections in children.[35] It was recognized as one of the most common nonsporing anaerobe isolated from infected postoperative wounds in gynecologic and obstetrical patients.[36] A case of bacteremia associating *A. fermentans*, *Bacteroides fragilis*, and *Escherichia coli* was reported in a patient with carcinoid syndrome by Peraino et al.[37] Two strains were isolated from aspirated specimens of abdominal pus during human intraabdominal infections.[38] *A. fermentans* was also isolated from pus of maxillary sinus during chronic sinusitis or maxillary cyst, from a closed abdominal abscess and from a perianal abscess.[20,39,40] The species *A. intestini* was described on the basis of 11 clinical isolates recovered from various sites in diverse clinical situations; strains were mainly isolated from intraabdominal and abscess fluid samples but also during mandible necrosis and infection of parietal hematoma.[4]

A task of our laboratory is to identify most of the colonies isolated from polymicrobial human specimens by conventional methods complemented by 16S rRNA gene-based identification. Since 1999, we have identified 24 clinical isolates corresponding to *Acidaminococcus* members. To our knowledge, we have assembled the largest described collection of *Acidaminococcus* clinical isolates. These strains and their isolation characteristics are presented in Table 30.2 (Jean-Pierre, H., Jumas-Bilak, E., and Marchandin, H., unpublished data, 2009). In our experience, *A. intestini* was recovered fivefold more frequently than *A. fermentans*. Indeed, among the 24 *Acidaminococcus* strains isolated over a 9-year period, we identified 4 *A. fermentans*, 19 *A. intestini*, and a strain that may represent a novel species in the genus since its 16S rRNA gene sequence identity level was 93.2% with *A. fermentans*.

Published data as well as data from our collection of *Acidaminococcus* spp. suggest that these organisms may play a role in human pathological processes as opportunistic pathogens. In animals, *A. fermentans* was essentially recovered from ovine foot root in mixed cultures.[41]

In animal or human hosts, no data are currently available on pathogenesis of infections caused by *Acidaminococcus* spp. Potential virulence factors remains also unknown except for lipopolysaccharide with endotoxinic activity.[1,9] However, pathogenicity of *A. fermentans* was estimated to be lower than that of other nonclostridial anaerobes in ischemic mouse thigh model, both mortality and thigh swelling being less pronounced for *A. fermentans* than for the other nonclostridial anaerobes tested.[42] The availability of the complete genome sequences of *Acidaminococcus* D21 and *A. fermentans* type strain should give new insights about the virulence of these bacteria.

30.1.5 Antimicrobial Susceptibility Pattern

Studies investigating the antimicrobial susceptibility pattern of *A. fermentans* are rare and included a very limited number of isolates. Seven strains recovered from ovine foot rot were submitted to antimicrobial susceptibility assay against 28 antimicrobial agents by agar dilution method.[43] A full susceptibility was demonstrated to β-lactam agents tested. Resistance to trimethoprim was noted for all the seven strains. Proportion of strains displaying susceptibility to other agents varied according to drug, from 29% for nalidixic acid to 86% for chloramphenicol, sulfametoxypyridiazine, and fosfomycin.[43] It is noteworthy that only 57% of the strains of this study were susceptible to metronidazole.

Regarding isolates of human origin, a β-lactamase-producing strain of *A. fermentans* was described from a perianal human abscess.[40] Multiresistance was conferred by the β-lactamase named ACI-1, the first one described in an anaerobic, gram-negative cocci. Resistance was noted against several antimicrobial agents including amoxicillin, ticarcillin, cefotaxime, and ceftazidime. ACI-1 was susceptible to β-lactamase inhibitor and thus susceptibility to

TABLE 30.2
Characteristics of the 24 *Acidaminococcus* spp. Strains Isolated from Human Clinical Samples in the Montpellier University Hospital over a 9-Year Period[a]

Species/Strain[b]	Date of Isolation (Year/Month)	Age (Years)/ Sex of Patient	Unit Category	Origin	Associated Bacteria[c]	
					Aerobic	Anaerobic
***Acidaminococcus fermentans* (*n* = 4)**						
ADV2297.03	2003/01	81/F	Surgery General	Drain	2	*Bacteroides uniformis*
ADV6092.03	2003/03	0/F (newborn)	Medicine General	Gastric fluid	*Streptococcus agalactiae*	*Fusobacterium* sp.
ADV1338.05	2005/12	0/M (newborn)	Medicine General	Gastric fluid	*Escherichia coli*	*Bacteroides fragilis* group
23/01/08-B-3310	2008/01	76/M	Surgery General	Jejunal stoma	3	2
***Acidaminococcus intestini* (*n* = 19)**						
ADV255.99[T]	1999/06	45/M	Surgery General	Peritoneal fluid	*Escherichia coli*	*Bacteroides fragilis*
ADV5206.02	2002/09	88/M	Medicine General	Rectum	2	>3
ADV2290.04	2004/04	75/M	Surgery General	Abdominal fluid	>3	2
ADV5199.04	2004/06	56/M	Medicine General	Mandible necrosis	3	>3
ADV1190.04	2004/12	52/M	Medicine ICU	Sacral pressure ulcer	3	>3
04/02/05-B-5309	2005/02	33/F	Surgery General	Abscess	CNS	>3
23/02/05-B-3397	2005/02	72/M	Surgery General	Gangrene	3	2
01/03/05-B-1397	2005/03	69/M	Surgery ICU	Drain	>3	*Bacteroides distasonis*
29/03/05-B-2093	2005/03	55/M	Surgery General	Abscess (buttock)	2	>3
10/08/05-B-3202	2005/08	42/M	Surgery General	Drain (T-tube)	2	0
14/10/05-B-5175	2005/10	30/M	Medicine General	Semen	*Streptococcus* sp.	*Prevotella* sp.
06/10/05-B-4321	2005/10	39/M	Surgery General	Mandibular wound	3	>3
10/10/05-B-1347	2005/10	26/F	Medicine General	Ileon biopsy	2	>3
01/12/05-B-4276	2005/12	25/M	Surgery General	Drain	>3	3
02/03/07-B-5168	2007/03	42/M	Surgery General	Drain	>3	>3
29/08/07-B-3093	2007/08	56/F	Medicine General	Anus	3	0
28/08/07-B-2316	2007/08	35/M	Medicine General	Pressure ulcer	3	>3
06/12/07-B-4271	2007/12	65/M	Surgery General	Colonic stoma	>3	2
10/10/08-B-5118	2008/10	76/M	Surgery General	Abscess under stoma	3	*Bacteroides thetaiotaomicron*
Unknown *Acidaminococcus* sp. (*n* = 1)						
24/07/05-B-7007	2007/05	86/M	Medicine ICU	Peritoneal fluid	*Streptococcus* sp.	*Bacteroides vulgatus*

[a] Data are from Reference 4 and Jean-Pierre, H., Jumas-Bilak, E., and Marchandin H., unpublished data, 2009.
[b] Strains labeled "ADV" were previously published.[4]
[c] Number of species indicated in case of polymorphic aerobic and/or anaerobic cultures; species name indicated in case of pure culture. Main associated species were *Bacteroides* spp., mainly species of the *B. fragilis* group; *Enterobacteriaceae*, mainly *E. coli, and Enterococcus* spp.
Species identification: >99% of 16S rRNA gene sequence identity level with type strain of the species.
F, female; M, male; ICU, intensive care unit; CNS, coagulase-negative *Staphylococcus*.

amoxicillin + clavulanate as well as to ticarcillin + clavulanate was observed.[40] Described in 2000, ACI-1 remains the only β-lactamase hitherto described among strict anaerobic gram-negative cocci. The G + C content of *aci1* gene (42.1 mol%) differed from that of *A. fermentans* total DNA (54.7–57.4 mol%) suggesting that this gene was probably acquired by horizontal gene transfer. Highest sequence similarity levels were observed with class A β-lactamases from *Haemophilus influenzae* and from some gram-positive organisms, particularly belonging to the genus *Bacillus*. Further investigations on flanking regions of the *aci1* gene revealed the existence of transposable genetic elements in the *Acidaminococcus* strain (transposase and putative resolvase), suggesting that this gene may be part of a transposable element.[44]

In a large study on ertapenem activity on anaerobes, Goldstein et al. reported that one out of the two *A. fermentans* included in their work displayed resistance to ertapenem with minimum inhibitory concentration ≥16 μg/mL.[38]

30.1.6 Diagnosis

Conventional Techniques. Strictly anaerobic conditions are required for growth of the *Acidaminococcus* species. For example, *Acidaminococcus* spp. grew well in anaerobic conditions obtained in an anaerobic jar with AnaeroGen System® (Oxoid, Basingstoke, UK). No enrichment method and no specific selective medium exist for growth of *Acidaminococcus* spp.[1,4,9,11] Many agar media or broths can be used for cultivation and isolation of *Acidaminococcus* spp. Rogosa described the growth of *A. fermentans* on Reinforced Clostridial Medium (RCM; Oxoid, Basingstoke, UK).[1] Sugihara et al. isolated *A. fermentans* on several media including kanamycin-vancomycin blood agar, neomycin blood agar, *Veillonella* agar with neomycin, *Fusobacterium* agar, blood agar, China-blue agar, and rifampin-vancomycin blood agar.[20] However, in the routine analysis of human clinical samples, *Acidaminococcus* spp. isolates grew well on Columbia sheep blood agar incubated at 37°C during 2–5 days. Both species yielded nonpigmented and nonhemolytic colonies on this medium, and colony morphology is as follows after 48 h of incubation at 37°C: for *A. fermentans*, colonies are 1–2 mm in diameter, round, entire, slightly raided, and whitish gray or nearly transparent; for *A. intestini*, colonies are about 0.5 mm in diameter, circular, convex, and whitish with a smooth surface.[1,4]

Members of the genus *Acidaminococcus* are gas producers. Other biochemical tests are mainly negative for these species. *Acidaminococcus* spp. do not possess oxidase nor catalase. Urease activity, nitrate reduction, gelatin liquefaction, milk modification, and aesculin hydrolysis are negative. H_2S is not produced. Indole is generally not produced.[1,4,9,11] Presumptive identification tests showed that all the strains displayed susceptibility to discs that contained 4 µg metronidazole or 10 µg colistin (Rosco, Taastrup, Denmark) and resistance to discs that contained 5 µg vancomycin. Susceptibility to discs that contained 1 mg kanamycin or 1 mg bile is noted for strains of *A. fermentans* and for the majority of *A. intestini* isolates. However, *A. intestini* strains with resistance to kanamycin or bile discs were described.[4] Using rapid ID 32 A system, *Acidaminococcus* species displayed the following positive enzymatic reactions: arginine arylamidase, leucine arylamidase, glycine arylamidase, and histidine arylamidase activities. Two enzymatic reactions may differentiate the two species of the genus, pyroglutamic acid arylamidase, and leucyl glycine arylamidase activities.[4] The metabolic end products produced by both species are acetic and butyric acids. In addition, *A. intestini* strains produced propionic acid and variable amounts of lactic acid, 2-hydroxybutyric, and 2-hydroxyvaleric acids.[1,4]

Several of the previously cited characteristics were found to be variable among strains of a same species making phenotypic identification difficult. *In fine*, phenotypic identification of anaerobic gram-negative cocci is mainly based on nitrate reduction detection, which allowed the identification of the genus *Veillonella* mainly encountered in human clinical samples. Anaerobic gram-negative cocci, which did not reduce nitrate, are often submitted to molecular identification.[33]

Molecular Techniques. Molecular methods for identification included 16S and 23S rRNA genes sequencing that may be performed on DNA extracts rapidly obtained by a boiling–freezing method.[4] The two species of the genus *Acidaminococcus* each formed a very tight group on the basis of their rRNA gene sequences. Indeed, intraspecific sequences identity level was >99.5% for both species and both genes. Interspecific 16S and 23S rRNA gene sequence identity levels were <95.8% and <88.9%, respectively.[4] As previously reported for other species of medical importance, the amplified variable region of the 23S rRNA gene shows more variation between species than 16S rRNA gene.[45] Thus, the two species of the genus *Acidaminococcus* are clearly separated from each other on the basis of their rRNA genes sequences. Therefore, ribosomal genes are markers of choice for species identification in this genus.

Molecular techniques for taxonomic characterization of *Acidaminococcus* spp., particularly genomic structure analysis, are not detailed in this chapter but may be found in the following reference.[46]

30.2 METHODS

30.2.1 Sample Preparation

Biochemical reactions are performed according to the procedures described by Holdeman et al.[47] For gas formation detection, cultures are observed for areas of disruption in loosely covered trypticase/glucose/yeast extract (TGY) deep agar and for gas bubbles in TGY broth. Special potency discs are used as described by Jousimies-Somer et al.[48] A Rapid ID 32 A kit (bioMérieux, Marne-la-Coquette, France) is used for enzymic profile determination as recommended by the manufacturer. Metabolic end products are assayed by quantitative gas chromatography as described previously.[49]

DNA can be extracted with conventional DNA extraction procedures using lysozyme, proteinase K, and sodium dodecyl sulfate, as previously described.[50] Commercialized kits may also be used for genomic DNA extraction like AquaPure genomic DNA isolation kit (Bio-Rad, Laboratories, Marne La Coquette, France). However, crude DNA extracts can also be obtained from a high concentrated bacterial suspension in 50 µL of sterile water by boiling for 10 min and freezing for 5 min at –20°C.

30.2.2 Detection Procedures

Principle. *Acidaminococcus* spp. can be identified by PCR amplification and sequencing analysis of 16S and/or 23S rRNA genes.

Primer	Sequence (5′–3′)[a]	Position[b]	Specificity
27f	GTGCTGCAGAGAGTTT GATCCTGGCTCAG	8–36	16S rRNA gene
1492r	CACGGATCCTACGGGT ACCTTGTTACGACTT	1478–1508	
6f	GCGATTTCYGAAY GGGGRAACCC	108–130	The 5′ end of 23S rRNA gene
10r	TTCGCCTTTCCCTCA CGGTACT	436–457	

[a] Where Y is C or T and R is A or G.[45]
[b] Based on *E. coli* numbering.

Procedure

1. Prepare PCR mixture (50 μL) containing 200 nM of each primer, 200 μM each dNTP, 2.5 mM $MgCl_2$, 1 U of *Taq* polymerase (Promega) in the appropriate reaction buffer, 50 pmol of primers 27f and 1492r (for 16S rRNA gene), or 6f and 10r (for 23S rRNA gene), and 5 μL of extracted DNA.
2. To amplify the 16S rRNA gene, carry out 30 cycles of 94°C for 1 min and 65°C for 1 min and 72°C for 2 min[50]; to amplify the 23S rRNA gene, perform 35 cycles of 94°C for 1 min, 60°C for 1 min, and 72°C for 2.5 min.
3. Electrophorese the products on a 1.5% agarose gel in 0.5× Tris-Borate-EDTA, and visualize with ethidium bromide stain (in these conditions, amplification products of 1.4 kb and 350 pb are obtained for 16S and 23S rRNA gene, respectively).
4. Sequence the PCR products by using the forward primer.

Note: For bacterial identification, partial 16S rRNA gene sequences ranging from 500 to 700 bp or partial 23S rRNA gene sequences of about 350 pb containing <1% undetermined positions were compared with sequences deposited in the GenBank database using the program standard nucleotide-nucleotide Basic Local Alignment Search Tool (BLAST) (http://www.ncbi.nlm.nih.gov/BLAST/).[51] Sequences were also matched against our own database constituted by non-deposited 16S rRNA gene sequences compiled during our routine activity of identification of clinical isolates.

A threshold of 97% similarity in 16S rRNA gene sequence has been proposed for the bacterial species delineation based on correlation study between DNA–DNA hybridization results and similarity values for 16S rRNA gene sequences.[52] More recently, this level has been reevaluated to 98.7%, always on the basis of correlation between DNA–DNA hybridization values and 16S rRNA gene sequences identity levels.[53] However, several authors recommended a >99% similarity value for 16S rRNA gene-based microbial identification in routine exercise.[54] For the two currently recognized species in the genus *Acidaminococcus*, the 16S rRNA gene sequences differed by more than 4%. Therefore, identification to the species level could be performed whatever the threshold value used. However, near the cut-off value of 97%, one should pay attention to the potential isolation of a nondescribed species. No cut-off value has been yet defined for identification based on 23S rRNA gene sequencing but in the genus *Acidaminococcus*, we showed that a value of 99% might also be retained. However, this value could be artificially high due to the limited size of the analyzed region (350 bp) and should be reevaluated for a larger sequence of the 23S rRNA gene.

30.3 CONCLUSION AND FUTURE PERSPECTIVES

Acidaminococcus spp. belong to human and animal gut microbiotae and may also act as opportunistic pathogens in mixed aerobic-anaerobic cultures. The two species of the genus, *A. fermentans* and *A. intestini*, can be differentiated by phenotypic characteristics including pyroglutamic acid arylamidase and leucyl glycine arylamidase activities detected using the rapid ID 32 A system and metabolic end products assayed by gas chromatography. 16S and 23S rRNA gene sequencing constitute useful tools for identification to the genus and to the species levels and 16S rRNA gene sequencing is frequently used to identify such anaerobic gram-negative cocci which do not reduce nitrate. Little is known about the role of these bacteria in mammal gut. The butyrate-producing *Acidaminococcus* organisms may contribute to healthy status. Indeed, butyrate is a short-chain fatty acid, which is considered as an important energy source for colonocytes on the basis of in vitro experiments.[55] Moreover, it possesses anti-inflammatory properties; particularly the ability to decrease lipopolysaccharide-induced cytokine response.[56] However, *Acidaminococcus* spp. may contribute to the spread of resistance genes encoding resistance to medically important antimicrobial agents in one of the most complex microbial ecosystems known.[40] Virulence of *Acidaminococcus* spp. appears to be low but remains to be explored.

The genome of each of the two species of the genus is currently being totally sequenced. Total sequences analysis will bring a lot of information regarding virulence factors or metabolism for these bacteria and will contribute to increase our knowledge on their role in human microbiotae.

ACKNOWLEDGMENTS

The authors are very grateful to Dr Hélène Jean-Pierre from the Anaerobe Laboratory of the Montpellier Teaching Hospital for her contribution to the data reported in this chapter and to Bernard Gay for electron microscopy.

REFERENCES

1. Rogosa, M., *Acidaminococcus* gen. n., *Acidaminococcus fermentans* sp. n., anaerobic Gram-negative diplococci using amino acids as the sole energy source for growth, *J. Bacteriol.*, 98, 756, 1969.

2. Fuller, R., Some morphological and physiological characteristics of gram-negative anaerobic bacteria isolated from the alimentary tract of the pig, *J. Appl. Bacteriol.*, 29, 375, 1966.
3. Cook, G.M. et al., Emendation of the description of *Acidaminococcus fermentans*, a *trans*-aconitate and citrate-oxidizing bacterium, *Int. J. Syst. Bacteriol.*, 44, 576, 1994.
4. Jumas-Bilak, E. et al., *Acidaminococcus intestini* sp. nov., isolated from human clinical samples, *Int. J. Syst. Evol. Microbiol.*, 57, 2314, 2007.
5. Rogosa, M., Transfer of *Veillonella* Prévot and *Acidaminococcus* Rogosa from *Neisseriaceae* to *Veillonellaceae* fam. nov. and the inclusion of *Megasphaera* Rogosa in *Veillonellaceae*, *Int. J. Syst. Bacteriol.*, 21, 231, 1971.
6. Both, B. et al., Phylogenetic and chemotaxonomic characterization of *Acidaminococcus fermentans*, *FEMS Microbiol. Lett.*, 76, 7, 1992.
7. Ludwig, W., Schleifer, K.-H., and Whitman, W.B., Revised road map to the phylum *Firmicutes*. In De Vos, P. et al. (Editors), *Bergey's Manual of Systematic Bacteriology*, 2nd ed., vol. 3 (The *Firmicutes*), pp. 1–13, Springer-Verlag, New York, 2009.
8. Marchandin, H. et al., *Negativicoccus succinicivorans* gen. nov., sp. nov., isolated from human clinical samples, emended description of the family *Veillonellaceae* and description of *Negativicutes* classis nov., *Selenomonadales* ord. nov., and *Acidaminococcaceae* fam. nov. in the bacterial phylum *Firmicutes*, *Int. J. Syst. Evol. Microbiol.*, 60, 127, 2010.
9. Rogosa, M., Family I. *Veillonellaceae*. In Krieg, N.R., and Holt, J.G. (Editors), *Bergey's Manual of Systematic Bacteriology*, vol. 1, p. 680, Williams and Wilkins, Baltimore, 1984.
10. Marchandin, H. et al., Phylogenetic analysis of some *Sporomusa* sub-branch members isolated from human clinical specimens: description of *Megasphaera micronuciformis* sp. nov., *Int. J. Syst. Evol. Microbiol.*, 53, 547, 2003.
11. Russel, J.B., Genus III. *Acidaminococcus*. In De Vos, P. et al. (Editors), *Bergey's Manual of Systematic Bacteriology*, 2nd ed., vol. 3 (The *Firmicutes*), p. 1067, Springer-Verlag, New York, 2009.
12. Carlier, J.-P. et al., *Anaeroglobus geminatus* gen. nov., sp. nov., a novel member of the family *Veillonellaceae*, *Int. J. Syst. Evol. Microbiol.*, 52, 983, 2002.
13. Carlier, J.-P., Genus I. *Veillonella*. In De Vos, P. et al. (Editors), *Bergey's Manual of Systematic Bacteriology*, 2nd ed., vol. 3 (The *Firmicutes*), pp. 1059–1064, Springer-Verlag, New York, 2009.
14. Marchandin, H., Juvonen, R., and Haikara A., Genus XIII. *Megasphaera*. In De Vos, P. et al. (Editors), *Bergey's Manual of Systematic Bacteriology*, 2nd ed., vol. 3 (The *Firmicutes*), pp. 1082–1089, Springer-Verlag, New York, 2009.
15. Holt, J.G. et al., *Bergey's Manual of Determinative Bacteriology*, 9th ed., p. 347, Williams & Wilkins, Baltimore, 1994.
16. Buckel, W., and Barker, H.A., Two pathways of glutamate fermentation by anaerobic bacteria, *J. Bacteriol.*, 117, 1248, 1974.
17. Hans, M., Buckel, W., and Bill, E., Spectroscopic evidence for an all-ferrous [4Fe-4S]0 cluster in the superreduced activator of 2-hydroxyglutaryl-CoA dehydratase from *Acidaminococcus fermentans*, *J. Biol. Inorg. Chem.*, 13, 563, 2008.
18. Hamana, K. et al., Covalently linked polyamines in the cell wall peptidoglycan of *Selenomonas*, *Anaeromusa*, *Dendrosporobacter*, *Acidaminococcus* and *Anaerovibrio* belonging to the *Sporomusa* subbranch, *J. Gen. Appl. Microbiol.*, 48, 177, 2002.
19. Chatterjee, B.D., and Chakraborti, C.K., Rapid diagnosis of anaerobic gram-positive cocci by salt tolerance, *J. Indian Med. Assoc.*, 92, 255, 1994.
20. Sugihara, P.T. et al., Isolation of *Acidaminococcus fermentans* and *Megasphaera elsdenii* from normal human feces, *Appl. Microbiol.*, 27, 274, 1974.
21. Moore, W.E.C., and Holdeman, L.V., Human fecal flora: the normal flora of 20 Japanese-Hawaiians, *Appl. Microbiol.*, 27, 961, 1974.
22. Mai, V. et al., Effect of bowel preparation and colonoscopy on post-procedure intestinal microbiota composition, *Gut*, 55, 1822, 2006.
23. Ley, R.E. et al., Microbial ecology: Human gut microbes associated with obesity, *Nature*, 444, 1022, 2006.
24. Turnbaugh, P.J. et al., A core gut microbiome in obese and lean twins, *Nature*, 457, 480, 2009.
25. Gophna, U. et al., Differences between tissue-associated intestinal microfloras of patients with Crohn's disease and ulcerative colitis, *J. Clin. Microbiol.*, 44, 4136, 2006.
26. Rieu, F., Fonty, G., and Gouet, P., Colony counts and characterization of bacteria adherent to the rumen wall and desquamated epithelial cells in conventional young lambs, *Can. J. Microbiol.*, 35, 698, 1989.
27. Dowd, E. et al., Evaluation of the bacterial diversity in the feces of cattle using 16S rDNA bacterial tag-encoded FLX amplicon pyrosequencing (bTEFAP), *BMC Microbiol.*, 8, 125, 2008.
28. Krause, D.O. et al., Diet influences the ecology of lactic acid bacteria and *Escherichia coli* along the digestive tract of cattle: Neural networks and 16S rDNA, *Microbiology*, 149, 57, 2003.
29. Cook, G.M., Wells, J.E., and Russell, J.B., Ability of *Acidaminococcus fermentans* to oxidize *trans*-aconitate and decrease the accumulation of tricarballylate, a toxic end product of ruminal fermentation, *Appl. Environ. Microbiol.*, 60, 2533, 1994.
30. Ley, R.E. et al., Evolution of mammals and their gut microbes, *Science*, 320, 1647, 2008.
31. Orphan, V.J. et al., Culture-dependent and culture-independent characterization of microbial assemblages associated with high-temperature petroleum reservoirs, *Appl. Environ. Microbiol.*, 66, 700, 2000.
32. Jumas-Bilak, E. et al., *Dialister micraerophilus* sp. nov. and *Dialister propionicifaciens* sp. nov., isolated from human clinical samples, *Int. J. Syst. Evol. Microbiol.*, 55, 2471, 2005.
33. Marchandin, H., Les cocci anaérobies à Gram négatif. In Freney, J., Renaud, F., Leclercq, R., and Riegel, P. (Editors), *Précis de bactériologie clinique*, 2nd ed., Chapter 113, Eska, Paris, 2007.
34. Marchandin, H., Organisation génomique, phylogénie et taxonomie polyphasique des bactéries du genre *Veillonella* et des genres apparentés du sous-groupe *Sporomusa*. PhD thesis, University of Montpellier 1, 2001.
35. Brook, I., Anaerobic infections in childhood, *Rev. Infect. Dis.*, 1, S187, 1984.
36. Chatterjee, B.D., and Chakraborti, C.K., Non-sporing anaerobes in certain surgical group of patients, *J. Indian Med. Assoc.*, 93, 333, 1995.
37. Peraino, V.A., Cross, S.A., and Goldstein, E.J.C., Incidence and significance of anaerobic bacteremia in a community hospital, *Clin. Infect. Dis.*, 16, S288, 1993.
38. Goldstein, E.J. et al., Comparative in vitro activities of ertapenem (MK-0826) against 1,001 anaerobes isolated from human intra-abdominal infections, *Antimicrob. Agents Chemother.*, 44, 2389, 2000.

39. Nakashima, M. et al., Cefotiam passage into the maxillary sinus tissues, *Jpn. J. Antibiot.*, 36, 2730, 1983.
40. Galán, J.C., et al., ACI-1 from *Acidaminococcus fermentans*: Characterization of the first β-lactamase in anaerobic cocci, *Antimicrob. Agents Chemother.*, 44, 3144, 2000.
41. Piriz Duran, S. et al., Isolation and identification of anaerobic bacteria from ovine foot rot in Spain, *Res. Vet. Sci.*, 49, 245, 1990.
42. Chatterjee, B.D., and Chakraborti, C.K., Ischaemic mouse thigh model for evaluation of pathogenicity of non-clostridial anaerobes, *Indian J. Med. Res.*, 89, 36, 1989.
43. Piriz, S. et al., Susceptibilities of anaerobic bacteria isolated from animals with ovine foot rot to 28 antimicrobial agents, *Antimicrob. Agents Chemother.*, 36, 198, 1992.
44. Galán, J.C. et al., Genetic analysis of the ACI-1 beta-lactamase flanking region of *Acidaminococcus fermentans* RYC-MR95: A new putative transposable element, p. 42, presented at *Intersci. Conf. Antimicrob. Agents Chemother.*, San Francisco, September 26–29, 1999.
45. Anthony, R.M., Brown, T.J., and French, G.L., Rapid diagnosis of bacteremia by universal amplification of 23S ribosomal DNA followed by hybridization to an oligonucleotide array, *J. Clin. Microbiol.*, 38, 781, 2000.
46. Marchandin, H. et al., Intra-chromosomal heterogeneity between the four 16S rRNA gene copies in the genus *Veillonella*: Implications for phylogeny and taxonomy, *Microbiology*, 149, 1493, 2003.
47. Holdeman, L.V., Cato, E.P., and Moore, W.E.C., *Anaerobe Laboratory Manual*, 4th ed., Blacksburg, VA, Virginia Polytechnic Institute and State University, 1977.
48. Jousimies-Somer, H.R. et al., *Wadworth-KTL Anaerobic Bacteriology Manual*, 6th ed., Belmont, CA, Star Publishing Company, 2002.
49. Carlier, J.-P., Gas chromatography of fermentation products: Its application in diagnosis of anaerobic bacteria, *Bull. Inst. Pasteur*, 83, 57, 1985.
50. Marchandin, H., and Jumas-Bilak, E., 16S rRNA gene sequencing: interest and limits for identification and characterization of novel taxa within the family *Acidaminococcaceae*. In McNamara, P.A. (Editor), *Trends in RNA Research*, p. 225, Nova Science, New York, 2006.
51. Altschul, S.F. et al., Gapped BLAST and PSI-BLAST: A new generation of protein database search programs, *Nucleic Acids Res.*, 25, 3389, 1997.
52. Stackebrandt, E., and Goebel, B.M., Taxonomic note: a place for DNA–DNA reassociation and 16S rRNA sequence analysis in the present species definition in bacteriology, *Int. J. Syst. Bacteriol.*, 44, 846, 1994.
53. Stackebrandt, E., and Ebers, J., Taxonomic parameters revisited: tarnished gold standards, *Microbiol. Today*, 33, 152, 2006.
54. Janda, J.M., and Abbott, S.L., 16S rRNA gene sequencing for bacterial identification in the diagnostic laboratory: Pluses, perils, and pitfalls, *J. Clin. Microbiol.*, 45, 2761, 2007.
55. Roediger, W.E., Utilization of nutrients by isolated epithelial cells of the rat colon, *Gastroenterology*, 83, 424, 1982.
56. Segain, J.P. et al., Butyrate inhibits inflammatory responses through NFkappaB inhibition: Implications for Crohn's disease, *Gut*, 47, 397, 2000.

31 *Anaerococcus, Parvimonas,* and *Peptoniphilus*

Dongyou Liu and Frank W. Austin

CONTENTS

31.1 INTRODUCTION

31.1.1 Classification

The term "gram-positive anaerobic cocci" (GPAC) commonly refers to a heterogeneous group of gram-positive cocci that are classified within the genera *Peptostreptococcus*, *Peptococcus*, *Ruminococcus*, *Coprococcus*, and *Sarcina* under the families Peptostreptococcaceae, Peptococcaceae, Ruminococcaceae, Lachnospiraceae, and Clostridiaceae; order Clostridiales; class Clostridia; phylum Firmicutes, and that are unable to grow in the presence of oxygen.[1] Alternative terms for GPAC include "anaerobic *Streptococcus*," "anaerobic coccus," and "*Peptococcus* and *Peptostreptococcus*." Peptococci are considered as the anaerobic equivalent of staphylococci, which appear in clusters; peptostreptococci are considered as the anaerobic equivalent of streptococci, which appear in chains. Peptococci and most peptostreptococci are capable of utilizing the products of protein decomposition as their sole energy source. Most peptostreptococci are heterofermentative, whereas streptococci are homofermentative, producing lactic acid from carbohydrates. Although both ruminococci and sarcinae ferment carbohydrates, ruminococci are further defined by their digestion of cellulose and their fermentation of cellobiose and sarcinae by the conspicuous arrangement of cells in packets. Coprococci, like ruminococci, use fermentable carbohydrates for growth, but they form different metabolic end products, notably, butyric acid.[1,2]

Forming part of the normal flora of all mucocutaneous surfaces, GPAC are often isolated from infections such as deep-organ abscesses, obstetric and gynecological sepsis, and intraoral infections. Previously, using phenotypic techniques, most GPAC isolates have been assigned to the genus *Peptostreptococcus*, family Peptostreptococcaceae, which once consisted of a large number of species (e.g., *P. anaerobius*, *P. asaccharolyticus*, *P. barnesae*, *P. harei*, *P. heliotrinreducens*, *P. hydrogenalis*, *P. indolicus*, *P. ivorii*, *P. lacrimalis*, *P. lactolyticus*, *P. magnus*, *P. micros*, *P. octavius*, *P. prevotii*, *P. productus*, *P. tetradius*, and *P. vaginalis* as well as the "trisimilis" group and the βGAL group).[1,3–6]

Following recent 16S rRNA gene sequencing analyses, it is clear that the various genera of the GPAC belong to distinct *Clostridium* rRNA groups.[6,7] For example, *Peptostreptococcus anaerobius*, the type species of the genus *Peptostreptococcus*, constitutes a member of *Clostridium* rRNA group XI, whereas other peptostreptococci are located in *Clostridium* rRNA group XIII and have been now assigned to separate genera (*Anaerococcus*, *Finegoldia*, *Gallicola*, *Parvimonas*, and *Peptoniphilus*).[8,9] Notably, *Peptostreptococcus magnus* and *Peptostreptococcus micros* were reclassified as *Finegoldia magnus* and *Parvimonas micra*, respectively[8–10]; *Peptostreptococcus productus* was transferred to the genus *Ruminococcus*.[11] Furthermore, three new genera were created to accommodate former *Peptostreptococcus* spp.: *Anaerococcus* (*an*, "without"; *aer*, "air"; *coccus*, "berry";

Anaerococcus, anaerobic coccus), *Peptoniphilus* (*peptonum*, "peptone"; *philos* "liking"; *Peptoniphilus* refers to the use of peptone as a major energy source), and *Gallicola* (*gallus*, "rooster or chicken"; *cola*, "inhabitant").[9,12,13] The genus *Anaerococcus* covers the saccharolytic, butyrate-producing species: *A. hydrogenalis*, *A. lactolyticus*, *A. octavius*, *A. prevotii*, *A. tetradius*, and *A. vaginalis* as well as *A. murdochii*. The genus *Peptoniphilus* comprises the nonsaccharolytic, butyrate-producing species: *Pn. asaccharolyticus*, *Pn. harei*, *Pn. indolicus*, *Pn. ivorii*, and *Pn. lacrimalis* as well as *Pn. gorbachii* and *Pn. olsenii*. The genus *Gallicola* contains a single species, *G. barnesae*.[9,12,13] On the other hand, *Sarcina* forms a member of *Clostridium* rRNA group I, *Coprococcus* is related to *Clostridium* rRNA group XVI, and *Ruminococcus* is situated in two different rRNA lines.[14]

Currently, the only species remaining in the genus *Peptostreptococcus* are *Peptostreptococcus anaerobius*[15] and *Peptostreptococcus stomatis*, a newly described species from the human oral cavity.[16] As *P. anaerobius* and *P. stomatis* are discussed in Chapter 77, *Finegoldia magnus* covered in Chapter 47, and *Gallicola barnesa* is found so far in chicken feces, this chapter will focus on the human-related GPAC that were known previously as peptostreptococci but now fall under the genera *Anaerococcus* (*A. hydrogenalis*, *A. lactolyticus*, *A. murdochii*, *A. octavius*, *A. prevotii*, *A. tetradius*, and *A. vaginalis*), *Parvimonas* (*Pa. micra*) and *Peptoniphilus* (*Pn. asaccharolyticus*, *Pn. gorbachii*, *Pn. harei*, *Pn. indolicus*, *Pn. ivorii*, *Pn. Lacrimalis*, and *Pn. olsenii*).

31.1.2 Morphology and Biochemical Properties

(i) The Genus Anaerococcus The genus *Anaerococcus* comprises multiple species of strictly anaerobic, gram-positive, nonmotile cocci that measure 0.5–1.8 μm in diameter and occur in pairs, tetrads, irregular masses, or chains. *Anaerococcus* spp. form grey-white convex colonies of 1–3 mm in diameter after 5-day incubation on enriched blood agar (e.g., brucella blood agar). Members of the *Anaerococcus* genus have the capacity to metabolize peptones and amino acids and generate butyric acid, lactic acid, propionic, and succinic acids as the major metabolic end products from peptone–yeast extract–glucose (PYG) medium. Most *Anaerococcus* species ferment glucose, fructose, sucrose, and lactose but do not produce indole. The G – C content of *Anaerococcus* DNA is 25 ± 33 mol%(*T*m), and the type species is *Anaerococcus prevotii*.[1,9]

A. hydrogenalis cells are of 0.5–1.8 μm in diameter and are usually arranged in clusters and occasionally in tetrads and short chains. On enriched blood agar, *A. hydrogenalis* forms grey-white convex colonies of 2–3 mm in diameter after 5 days, with some *A. hydrogenalis* strains producing an unpleasant odor, which may be also generated by *A. vaginalis* strains. *A. hydrogenalis* strains are strongly saccharolytic, produce indole (in contrast to other *Anaerococcus* spp.), but are weakly proteolytic; these distinct features facilitate their identification through the proteolytic enzyme profile (PEP) analysis. Some strains also produce urease, alkaline phosphatase, and α-glucosidase but are negative for arginine dihydrolase and coagulase. Besides forming butyrate as the terminal volatile fatty acid (VFA), *A. hydrogenalis* generates acid from glucose, lactose, raffinose, and mannose as well as hydrogen from peptone–yeast extract. While the "*trisimilis*" group previously classified under the *Peptostreptococcus* also comprises saccharolytic indole- and butyrate-producing strains, it can be differentiated from *A. hydrogenalis* by its slower colony growth and formation of the proteolytic enzyme pyroglutamyl arylamidase. In addition, the "*trisimilis*" group produces acid from trehalose and hydrolyzes hippurate as assessed with the Rapid ID 32 STREP kit (API-bioMérieux). *A. hydrogenalis* corresponds to Hare group III on the basis of fermentation reactions, gas production from five carbohydrates and five organic acids, as well as serological reactions, and is associated phylogenetically with other butyrate-producing, saccharolytic species such as *A. prevotii* and *A. tetradius* in *Clostridium* cluster XIII, as revealed by 16S rRNA sequence analysis.[2,9]

A. lactolyticus cells range from 0.5 to 1.5 μm in diameter and appear in clusters and short chains. *A. lactolyticus* grows slowly on enriched blood agar and forms translucent colonies 1 mm in diameter after incubation for 5 days. *A. lactolyticus* strains are strongly saccharolytic and proteolytic, produce urease, but are negative for indole, alkaline phosphatase, arginine dihydrolase, and coagulase. *A. lactolyticus* also forms butyrate as the terminal VFA but does not ferment raffinose and ribose. It is closely related to *A. vaginalis*, *A. prevotii*, and *A. tetradius* on the basis of 16S rRNA sequence data and cell-wall peptidoglycan type.[2,9]

A. murdochii cells are ≥0.7 μm in diameter and form gray, flat or low convex, entire, circular, often matt colonies of 1–2 mm in diameter with whiter centers after 5-day incubation on brucella blood agar. *A. murdochii* strains are obligately anaerobic and negative for indole, catalase, coagulase, urease, and nitrate. *A. murdochii* produces butyric acid and acetic acid in peptone–yeast extract broth and peptone–yeast extract–glucose broth. In the Rapid ID 32A test (API-bioMérieux), *A. murdochii* is positive for arginine dihydrolase, β-galactosidase, alkaline phosphatase, arginine arylamidase, glycine arylamidase (strong or weak), histidine arylamidase (strong or weak), leucine arylamidase, and pyroglutamic acid arylamidase, although it displays variable activity for β-glucosidase, leucyl glycine arylamidase, alanine arylamidase, and serine arylamidase. It ferments mannose but not raffinose.[13]

A. octavius cells measure from 0.7 to 0.9 μm in diameter and occur in clusters. Colonies of *A. octavius* are 1–2 mm in diameter after 5 days of incubation on enriched blood agar. These are yellowish-white, glistening, circular, and raised with an entire margin. *A. octavius* is moderately saccharolytic but weakly proteolytic, producing a very distinctive VFA profile: large amounts of butyric acid, small amounts of isovaleric acid, and a large terminal peak of *n*-caproic acid. *A. octavius* ferments glucose, ribose, and mannose but is negative for indole, urease, alkaline phosphatase, arginine dihydrolase and coagulase. It corresponds to Hare group VIII and is closely related to *A. lactolyticus*, *A. prevotii*, and *A. tetradius* on the basis of 16S rRNA sequencing studies.[2,9]

A. prevotii cells vary from 0.6 to 1.5 μm in diameter and often appear in clusters or tetrads. Colonies of *A. prevotii* are typically 2 mm in diameter, matte grey, and low convex after 5-day incubation on enriched blood agar. *A. prevotii* produces urease, is weakly saccharolytic and proteolytic, and forms butyrate as the terminal VFA. *A. prevotii* is distinguishable from *A. tetradius* by its production of α-galactosidase and of acid from raffinose, ribose, and mannose. Being negative for indole, alkaline phosphatase, and coagulase, *A. prevotii* is differentiated from indole-negative strains of *Pn. asaccharolyticus* by the latter's distinctive colony morphology and inability to ferment sugars. In addition, *A. prevotii* is separated from strains of *A. lactolyticus*, *A. vaginalis*, and other saccharolytic, butyrate-producing GPAC through proteolytic enzyme profile (PEP) analysis. *A. prevotii* is closely related to *A. tetradius*, *A. hydrogenalis*, *A. lactolyticus*, and *A. vaginalis* in *Clostridium* cluster XIII on the basis of 16S rRNA sequence data and peptidoglycan structure.[2,9]

A. tetradius cells are of 0.5–1.8 μm in diameter and arranged in clusters and tetrads. The colonies of *A. tetradius* are matte grey, low convex, and measure up to 2 mm in diameter after 5-day incubation. Being strongly saccharolytic but weakly proteolytic, *A. tetradius* is negative for indole, alkaline phosphatase, arginine dihydrolase, and coagulase. *A. tetradius* ferments glucose and mannose, and generates butyric acid from PYG as major metabolic end product. Production of urease and the saccharolytic enzymes α-glucosidase and β-glucuronidase is characteristic of *A. tetradius* and *A. prevotii* in comparison with other GPAC species. Use of PEPs facilitates further discrimination of *A. prevotii* and *A. tetradius* from other saccharolytic groups. *A. tetradius* is separated from *A. prevotii* by its ability to ferment a wider range of sugars and to form alkaline phosphatase.[3]

A. vaginalis cells are of 0.5–1.5 μm in diameter and are usually arranged in clusters or tetrads. *A. vaginalis* forms grey-white, low-convex colonies of 2–3 mm in diameter after 5-day incubation. Similar to *A. hydrogenalis*, some *A. vaginalis* strains may generate an unpleasant odor. *A. vaginalis* demonstrates moderate saccharolytic activity, uses peptones and oligopeptides as the major energy source, and forms butyrate is VFA. *A. vaginalis* is readily identifiable by its distinctive PEP with the ATB 32A kit, with most *A. vaginalis* strains being positive for arginine dihydrolase, arginine, leucine, and histidine arylamidase; negative for urease and coagulase; and variable for indole, mannose, catalase, and alkaline phosphatase. Comparisons of DNA–DNA hybridization and 16S rRNA sequence data suggest that *A. vaginalis* is closely related to *A. hydrogenalis*, *A. prevotii*, and *A. tetradius*; and the cell-wall peptidoglycan structure of *A. vaginalis* is similar to that of *A. hydrogenalis*.[2,9]

(ii) The Genus Parvimonas The genus *Parvimonas* is currently composed of a single gram-positive, coccus-shaped species, *Parvimonus micros*. The cells of *Pa. micros* are 0.3–0.7 μm in diameter and usually arranged in pairs and chains, as well as in clusters when cultured on solid media. *Pa. micros* forms highly characteristic white (or grey), glistening, and domed colonies 1 mm in diameter after 5-day incubation, and its colonies are often surrounded by a distinctive yellow-brown halo of discolored agar up to 2 mm wide. Two morphotypes (i.e., rough [Rg] and smooth [Sm] morphotypes) are identified in *Pa. micros*; both morphotypes demonstrate identical PEPs, and the Rg morphotype readily changes to a Sm-like variant (Rg^{Sm}) in broth culture. Analyses using PAGE, PyMS, and 16S rRNA sequencing indicate that the two morphotypes constitute distinct clusters. *Pa. micros* is strongly proteolytic, relatively aerotolerant, and asaccharolytic. It forms alkaline phosphatase, and occasionally, acetic acid. *Pa. micros* periodontal strains are consistently positive for phenylalanine and glutamylglutamyl arylamidase and variable in relation to formation of proline and leucylglycine arylamidase; most *Pa. micros* vaginal strains form leucylglycyl arylamidase but not proline, phenylalanine, or glutamylglutamyl arylamidase. *Pa. micros* is identified by a combination of colony morphology (presence of haloes), PEP, and cell size (0.3–0.7 μm). The ability of *Pa. micros* to use the reduced form of glutathione to form hydrogen sulfide has been exploited for its identification via a selective and differential medium named PMM.[1,8,10]

(iii) The Genus Peptoniphilus The genus *Peptoniphilus* consists of multiple non spore–forming, gram-positive, obligately anaerobic coccal species. *Peptoniphilus* cells are 0.4–1.5 μm in diameter, and appear in pairs, short chains, tetrads, or small clusters. Colonies of *Peptoniphilus* spp. are gray or whitish to lemon-yellow, flat or low convex, circular, opaque or glistening, ranging from 1 to 3 mm in diameter after 5-day incubation on enriched blood agar or brucella blood agar. *Peptoniphilus* spp. produce butyric acid from PYG but do not ferment carbohydrates. The G – C content of *Peptoniphilus* DNA is 30 ± 34 mol% and the type species is *Peptoniphilus asaccharolyticus*.[1,9]

Pn. asaccharolyticus demonstrates distinctive cellular and colony morphology. Its cells vary from 0.5 to 0.9 μm in diameter and appear more uniform than most other GPAC species. Arranged in clusters, *Pn. asaccharolyticus* cells retain Gram's stain poorly, resembling *Neisseriae* strains. *Pn. asaccharolyticus* forms glistening, low convex, usually whitish to lemon-yellow colonies of 2–3 mm in diameter after 5 days on enriched blood agar, often accompanied by a characteristic musty odor. *Pn. asaccharolyticus* strains are asaccharolytic, do not produce acid from carbohydrates, and are negative for urease, alkaline phosphatase, arginine dihydrolase and coagulase. *Pn. asaccharolyticus* strains form butyric acid as their terminal VFA, and most strains (80%–90%) are indole positive. *Pn. asaccharolyticus* shares a close genetic relationship to *Pn. indolicus* in *Clostridium* cluster XIII as analyzed by 16S rRNA sequencing.[2,8]

Pn. gorbachii cells are of ≥0.7 μm in diameter and form gray, flat or low convex, circular, entire, opaque colonies of 1–3 mm in diameter after 5-day incubation on brucella blood agar. Being obligately anaerobic and asaccharolytic, *Pn. gorbachii* is unable to produce acid from glucose. In the Rapid ID 32A test (API-bioMérieux), *Pn. gorbachii* isolates show positive reactions for arginine arylamidase, phenylalanine arylamidase, leucine arylamidase, tyrosine arylamidase, glycine arylamidase, glutamyl glutamic acid arylamidase, histidine arylamidase, and serine arylamidase. *Pn. gorbachii* strains are indole-variable, and catalase-, coagulase-, urease-, and nitrate-negative. In peptone–yeast extract broth and peptone–yeast extract–glucose broth, *Pn. gorbachii* isolates produce acetic acid, butyric acid, and propionic acid.[13]

Pn. harei cells are of 0.5–1.5 μm in diameter, and may be circular, oval, or elliptical in shape. *Pn. harei* strains form flat, translucent, nonhemolytic colonies of 1 mm in diameter after 5-day incubation on enriched blood agar. *Pn. harei* is moderately proteolytic, and indole- and catalase-variable. Also it is unable to ferment carbohydrates, and is negative for urease, arginine dihydrolase, alkaline phosphatase, and coagulase. *Pn. harei* produces butyric acid as the terminal VFA. *Pn. harei* strains can be differentiated from other GPAC species (than *Pn. asaccharolyticus*) by PEP analysis; *Pn. harei* and *Pn. asaccharolyticus* are distinguished by their clearly different cell and colony morphology. *P. harei* belongs to *Clostridium* cluster XIII and is closely related to *Pn. lacrimalis* and *Pn. asaccharolyticus* based on DNA–DNA hybridization and 16S rRNA sequence data.[6]

Pn. indolicus shows similar cell and colony morphology as well as biochemical activity to that of *Pn. asaccharolyticus*. Both species are moderately proteolytic activity but unable to ferment sugars. Most *Pn. indolicus* strains are positive for coagulase, alkaline phosphatase, and indole, and negative for urease and arginine dihydrolase. *Pn. indolicus* is distinguished from *Pn. asaccharolyticus* by its production of peptocoagulase, reduction of nitrate to nitrite, and formation of propionate from lactate. *Pn. indolicus* produces butyrate as the terminal VFA. *Pn. indolicus* appears closely related to but distinct from *Pn. asaccharolyticus* on the basis of nucleic acid studies.

Pn. ivorii cells are of 0.4–1.5 μm in diameter and arranged in clusters. *Pn. ivorii* strains form yellow-white, low-convex colonies of 1–2 mm in diameter after 5-day incubation. *Pn. ivorii* does not ferment carbohydrates, and the only proteolytic enzyme *Pn. ivorii* produces is proline arylamidase. *Pn. ivorii* strains produce several VFAs, including butyric and isovaleric acids, and are negative for urease, arginine dihydrolase, alkaline phosphatase, and coagulase. Because few species of GPAC form proline arylamidase (except for strains of *Pn. octavius*), *Pn. ivorii* is distinguishable by PEP analysis. *Pn. ivorii* is differentiated from *Pn. octavius* by the latter's ability to ferment several carbohydrates. *Pn. ivorii* is related to *Pn. anaerobius* phylogenetically and is assigned to *Clostridium* cluster XIII as a distinct lineage.

Pn. lacrimalis cells are 0.5–0.7 μm in diameter and appear in short chains or clusters. It forms pink-white colonies 1–2 mm in diameter after 5-day incubation. *Pn. lacrimalis* strains are asaccharolytic, strongly proteolytic, and produce butyrate as the terminal VFA. In addition to its inability to ferment carbohydrates, *Pn. lacrimalis* is negative for indole, urease, arginine dihydrolase, alkaline phosphatase, and coagulase. In the ATB 32A test (API-bioMérieux), *Pn. lacrimalis* shows distinctive PEP and is readily identified.[4]

Pn. olsenii cells are of ≥ 0.7 μm in diameter, and form gray, convex, circular, entire, opaque colonies 2–3 mm in diameter with a whiter central peak after 5-day incubation. Being obligately anaerobic, *Pn. olsenii* is indole-variable, and catalase, coagulase, urease, and nitrate negative. Additionally, *Pn. olsenii* is asaccharolytic, and unable to produce acid from glucose. In peptone–yeast extract broth and PYG broth, *Pn. olsenii* produces acetic acid, butyric acid, and propionic acid. In the Rapid ID 32A test (API-bioMérieux), *Pn. olsenii* is positive for alkaline phosphatase, arginine arylamidase, leucyl glycine arylamidase, phenylalanine arylamidase, leucine arylamidase, tyrosine arylamidase, alanine arylamidase, glycine arylamidase, histidine arylamidase, and serine arylamidase.[13]

Since *A. hydrogenalis*, *Pn. asaccharolyticus*, *Pn. Indolicus*, and *Pn. harei* are all indole positive, their differentiation is based mainly on cellular and colony morphology as well as smell, in addition to biochemical properties. *A. hydrogenalis* is strongly saccharolytic while *Pn. asaccharolyticus* is asaccharolytic. *Pn. indolicus* is distinctive in its production of coagulase. *Pn. harei* shows characteristic cell and colony morphology. For some indole-negative *Pn. asaccharolyticus* strains, the characteristic cell and colony morphology together with examination of the whole-cell composition by PyMS are used to verify their identity.[1,6]

31.1.3 Clinical Features and Pathogenesis

Being part of the normal flora on all mucocutaneous surfaces, GPAC are found in the mouth (e.g., the plaque and the gingival sulcus), upper respiratory and gastrointestinal tracts, female genitourinary system (and vaginal discharges), and skin (and superficial wounds). These bacteria have been associated with a variety of human infections, including oral and respiratory tract infections, intraabdominal infections, infections of the genitourinary tract, gynecological sepsis, obstetric infections, superficial and soft-tissue infections, central nervous system infections, and musculoskeletal infections. Indeed, GPAC account for approximately one-third of all infections involving anaerobic bacteria. Most GPAC-related infections are polymicrobial.[1,17–19]

A. hydrogenalis has been isolated in human vaginal specimens, nose, skin, purulent sternotomy wound, infected pacemaker site, cellulitic leg ulcer, ingrowing toenails, and inflamed foot ulcers. *A. lactolyticus* is found in human vaginal discharges. *A. murdochii* has been identified from foot/

thumb abscess, plantar ulcer, chronic sternal wound infection, and severe soft-tissue infection of the neck. *A. octavius* is detected in nasal specimens, the skin, vagina, and infected hip prosthesis. *A. prevotii* is isolated from the fecal and vaginal samples. *A. tetradius* has been detected in endometrial biopsy specimen. *A. vaginalis*, considered as one of the more pathogenic species of GPAC, is isolated from the vaginal flora, ovarian abscess, postsurgical wound infections (of the leg, inguinal region, and trunk), pleural empyema, infected skin graft on the leg, and superficial abscess of the upper arm.

Pa. micros forms part of the normal flora of the gingival crevice and the gastrointestinal tract. Increasingly, *Pa. micros* is recognized as an important oral pathogen, and it has been isolated from many anatomical sites outside the mouth including brain, lung, jaw, head, neck and bite abscesses, spinal fluid, blood, skin, abscesses of the diaphragm and intraabdominal, pleural empyemas, anorectal abscesses, paravertebral abscess, native valve endocarditis, intrauterine contraceptive devices, and chronic sinusitis. *Pa. micros* appears well adapted to the serum-rich, oxygen-depleted subgingival environment, where the crevicular fluid contains relatively high concentrations of glycoproteins and peptides but low concentrations of carbohydrates, providing a major source of nutrients for asaccharolytic flora. *Pa. micros* is capable of forming hydrogen sulfide (a cytotoxic agent) from glutathione (a tripeptide involved in the intracellular defense against reactive oxygen metabolites). As a strongly proteolytic organism, *Pa. micros* is typically isolated from intraoral sepsis and extraoral abscesses and associated with microaerophilic streptococci and fastidious anaerobes.[8]

Pn. asaccharolyticus is part of the normal flora of the genitourinary and gastrointestinal tracts as well as the skin. It represents one of the most common and important species of GPAC in human clinical material, involving obstetric and gynecological infections, skin abscess, peritoneal abscess, submandibular sebacous cyst, and ischiorectal abscess. *Pn. gorbachii* and *Pn. olsenii* are isolated from human infections of mild to moderate severity (e.g., dry gangrene with superficial infection; cellulitis, abscess, and osteomyelitis). *Pn. harei* is found in pus from sacral sore, antral washouts, and the peritoneal cavity, abscesses of the face, breast, thigh, and submandibular tissues. Being an important veterinary pathogen, *Pn. indolicus* is infrequently isolated from human pathological specimens. The only documented *Pn. indolicus* isolations were from the skin lesion of a sheepherder with a superficial abscess. *Pn. ivorii* is isolated from leg ulcer, preputial sac in a patient with balanitis, and intrauterine contraceptive device. It is not as clinically significant as *Pn. harei*. *Pn. lacrimalis* is cultured from human lachrymal gland abscess.[8]

The pathogenicity of most GPAC spp. is often expressed through synergistic interactions with other facultative organisms. Anaerobic bacteria are known to elaborate a variety of virulence factors, such as superoxide dismutase, immunoglobulin proteases, coagulation-spreading factors (coagulase and hyaluronidase), adherence factors, and extracellular enzymes (DNase, RNase, and hemolysins). GPAC have the ability to form capsules, and encapsulated organisms are able to seed more successfully to distant organs.[1]

31.1.4 Diagnosis

(i) Phenotypic Techniques A number of phenotypic features of GPAC are useful for their identification. These include microscopic appearance, in vitro colonial morphology, ability to form gas from carbohydrates and organic acids, carbohydrate fermentation reactions, proteolytic enzyme profile (PEP), and volatile fatty acids (VFA) as detected by gas-liquid chromatography (GLC).[20–24]

Culture. GPAC are cultivated on enriched blood agar such as brucella blood agar supplemented with 5% sheep blood, hemin, and vitamin K_1 and incubated at 37°C under N_2 (86%), H_2 (7%), and CO_2 (7%) gas phase. Primary plates are examined after incubation for 48 h and 5 days. Distinctive colony morphology and smell after 5-day incubation are helpful for distinguishing *Pn. harei* from *Pn. asaccharolyticus*. The resulting isolates may be tested for metronidazole susceptibility in which GPAC show zones of inhibition of 15 mm or greater after incubation for 48 h; whereas microaerophilic strains such as *Streptococcus* spp., *Staphylococcus saccharolyticus,* and *Gemella morbillorum* show no zones. Further identification requires subculture onto enriched blood agar, and incubation in an aerobic atmosphere containing 6% CO_2 to detect capnophilic organisms (after 5 days). A broth culture is inoculated if GLC is available for the detection of VFAs. PEP kits can be employed for identification. The ATB 32A kit (API-bioMérieux) is also useful.

Gas-Liquid Chromatography. GLC detects fatty acid by-products of metabolism and separates GPAC into three groups: (i) an acetate group containing notably *Finegodia magnus* and *Pa. micros*, which produce acetic acid or no VFAs; (ii) a butyrate group, which produce butyric acid as their major terminal VFA; and (iii) a caproate group, which produce large quantities of longer-chain VFAs. GLC may be also useful for identifying *Peptococcus niger* and *A. octavius*, which produce *n*-caproic acid, and *Pn. ivorii*, which produces a terminal peak of isovaleric acid.[2,8] Being time consuming and requiring expensive equipment, GLC is beyond the capabilities of most clinical diagnostic laboratories.

Biochemical Tests. Most GPAC spp. are asaccharolytic and strongly proteolytic, utilizing the products of protein decomposition as a major energy source. Many GPAC strains are positive for peptidase reactions (which correlate well with VFA production) and negative for carbohydrate fermentation reactions. Several commercial preformed enzyme kits incorporating a range of saccharolytic and proteolytic enzymes, carbohydrate fermentation reactions, indole production and nitrate reduction, and alkaline phosphatase and arginine dihydrolase activity are available. These include RapID ANA (Innovative Diagnostic Systems, Atlanta, GA), Vitek ANI card (Vitek Systems, Hazelwood, MO), and ATB 32A (Rapid ID 32 A) (API-bioMérieux). These kits are useful for differentiation of other related bacteria from GPAC. Proteolytic enzyme profiles (PEPs) distinguish clearly and reproducibly among GPAC species, and commercial PEP kits offer speedy, simple, and discriminatory means for GPAC identification.[1] Recently, a biochemical scheme involving minimum essential biochemical tests has been proposed for accurate identification of clinically important GPAC.[24]

(ii) Molecular Techniques Molecular techniques provide a rapid, sensitive, and specific alternative to the established phenotypic methods for GPAC identification. Analyses of signature nucleotides in 16S rRNA genes allow the unambiguous, definitive determination of GPAC.[6,12,13,25] Amplification and sequencing of the 16S-23S intergenic spacer region (ISR) are also valuable for discrimination among different species of the former peptostreptococci.[26] Use of DNA probes targeting the 16S rRNA gene has facilitated the precise detection and differentiation of *Pa. micros* and other GPAC spp.[27,28] Application of PCR assays enables the accurate detection of *F. magna* and *Pa. micra* in oral clinical specimens,[29,30] and a multiplex PCR assay using genus- and species-specific primers offers a valuable tool for the rapid identification of GPAC spp.[13]

For instance, Hill et al.[26] employed universal primers complementary to the 16S rRNA gene (primer A; 5′ TTGTACACACCGCCCGTCA 3′) and the 23S rRNA gene (primer B; 5′ GGTACCTTAGATGTTTCAGTTC 3′) of *Escherichia coli* in PCR, followed by sequencing to identify and distinguish 17 of the currently recognized species or groups in the former *Peptostreptococcus* genus, in addition to two closely related strains, *Peptococcus niger* and *Ruminococcus productus*. The results suggest that members of *Peptoniphilus lacrimalis*, *Pn. ivorii*, *Anaerococcus octavius*, *Peptostreptococcus anaerobius*, and *Micromonas micros* produced distinct amplification patterns, enabling their rapid differentiation. While *A. vaginalis* demonstrates considerable intraspecies variations, *Pn. ivorii* is relatively conserved in the 16S-23S ISR.

31.2 METHODS

31.2.1 Sample Preparation

Culture. Clinical GPAC isolates are cultured on brucella blood agar supplemented with 5% sheep blood, hemin, and vitamin K_1 and incubated at 37°C under N_2 (86%), H_2 (7%), and CO_2 (7%) gas phase. The resulting isolates are phenotypically identified by using the Rapid ID 32A (API-bioMérieux). Glucose fermentation tests are performed using prereduced, anaerobically sterilized peptone–yeast extract–glucose broth tubes (Anaerobe Systems, Morgan Hill, CA). The strains are grown in peptone–yeast extract broth and peptone–yeast extract–glucose broth (Anaerobe Systems, Morgan Hill, CA) for metabolic end product (short-chain volatile and nonvolatile fatty acids) analysis by GLC.

DNA Extraction. One or two colonies of GPAC isolates are suspended in 50 μL of Tris/HCl/EDTA/saline pH 8.0, heated for 10 min at 95°C, and centrifuged at 18,600 × *g* for 2 min. The supernatant is used as DNA template for PCR.[12] Alternatively, crude DNA extracts are prepared from each isolate by inoculating two loopfuls of bacterial cells from broth cultures into 100 μL of sterile water, boiling for 10 min, and removing cell debris by centrifugation. The supernatant is retained for subsequent PCR analysis.[30] In addition, genomic DNA may be extracted from cells in the midlogarithmic-growth phase by using a QIAamp DNA Mini Kit (Qiagen).

31.2.2 Detection Procedures

31.2.2.1 Multiplex PCR for Genus- and Species-Specific Identification

Song et al.[12] developed a two-step multiplex PCR assay for identification of 14 GPAC species belonging to *Anaerococcus hydrogenalis*, *Anaerococcus lactolyticus*, *Anaerococcus octavius*, *Anaerococcus prevotii*, *Anaerococcus tetradius*, *Anaerococcus vaginalis*, *Finegoldia magna*, *Parvimonas micros*, *Peptoniphilus asaccharolyticus*, *Peptoniphilus harei*, *Peptoniphilus indolicus*, *Peptoniphilus ivorii*, *Peptoniphilus lacrimalis*, and *Peptostreptococcus anaerobius*.

In the first step, genus-specific primers are designed from the 16S rRNA gene. These include ANAC (for the genus *Anaerococcus*), PEPN (for the genus *Peptoniphilus*), PEPA (for the genus *Peptostreptococcus*), MICR (for the genus *Micromonas* or *Parvimonas*), and FIGD (for the genus *Finegoldia*) (Table 31.1). Use of primers ANAC, PEPN, FIGD, MICR, and PEPA together with reverse primer 1392B in multiplex PCR-G enables amplification of specific fragments from the genera *Anaerococcus*, *Peptoniphilus*, *Peptostreptococcus*, *Parvimonas*, and *Finegoldia*.

In the second step, species-specific primers targeting six species of the genus *Anaerococcus* and five species of the genus *Peptoniphilus* are also selected from the 16S rRNA sequence (Table 31.1). Use of primers OCTA, HYDR, and LACT, together with reverse primer 1392B in multiplex PCR-Ia, leads to the identification of *A. octavius*, *A. hydrogenalis*, and *A. lactolyticus*; use of primers TET-f, PRE-f, and VAG-f, and reverse primers 341B and 1392B in multiplex PCR-Ib distinguishes *A. tetradius*, *A. prevotii*, and *A. vaginalis*; use of primers HAR-f and ASA-f and reverse primers 341B and HAR-r in multiplex PCR-IIa differentiates *Pn. harei* and *Pn. asaccharolyticus*; and use of primers INDO, IVOR, and LACR and reverse primer 1392B in multiplex PCR-IIb separates *Pn. indolicus*, *Pn. Ivorii*, and *Pn. lacrimalis*. Primers 8UA/341B are derived from a conserved region of the 16S rRNA gene that is present in all eubacteria, and they are used to verify the purity and quality of DNA prepared from clinical specimens (Table 31.1).

Procedure

1. PCR mixture (50 μL) is made up of 1.25 U *Taq* DNA polymerase (Promega), 50 mM KCl, 10 mM Tris/HCl pH 9.0, 0.1% Triton, 2.5 mM $MgCl_2$, 0.2 mM dNTPs, and 5 μL of bacterial lysate as the template DNA. Primer concentration is 0.25 μM (each),

TABLE 31.1
Genus- and Species-Specific Primers for GPAC Bacteria

Primer	Sequence (5′→3′)	Target	Amplicon Size (bp)	Multiplex PCR
ANAC	GCGTGATTTAGAAGGC	*Anaerococcus* spp.	980	G
PEPN	TGCGCAAGCATGAAA	*Peptoniphilus* spp.	510	G
PEPA	GTAGTTAGCCTCCGAAA	*P. anaerobius*	780	G
MICR	TCGGGACAACTATACAG	*M. micros* (*Pa. micros*)	380	G
FIGD	GCATAAAATCGTAGAAACAC	*F. magna*	1200	G
HYDR	ATTTCCTTACGACTTCGGTC	*A. hydrogenalis*	400	Ia
LACT	CAGAGCAGCTAAACAGCGATGTCA	*A. lactolyticus*	150	Ia
OCTA	AACTGGAGACCTTGAGTAATGG	*A. octavius*	760	Ia
PRE-f	GGTCTAGAGATAGACTCTTA	*A. prevotii*	400	Ib
TET-f	TGACATAAACTCTTCGCAT	*A. tetradius*	150	Ib
VAG-f	CTGTTAGGCTTGAGAGATGA	*A. vaginalis*	760	Ib
ASA-f	ATGAAAATCAAACAGAACC	*Pn. asaccharolyticus*	300	IIa
HAR-f	TGAGAAGAGTAGAGGTAAGT	*Pn. Harei*	600	IIa
HAR-r	CGGTCCTCAGTATGTCA	*Pn. Harei*	600	IIa
INDO	TGAAATCGTAGTGACACATGT	*Pn. indolicus*	1350	IIb
IVOR	GGAAACTGTCGAGCTTGAGTG	*Pn. Ivorii*	760	IIb
LACR	GGCTTGACATATAAGAGACGAACT	*Pn. lacrimalis*	420	IIb
341B	CCGTCAATTCMTTTRAGTTT	Eubacteria		Ib, IIa
8UA	AGAGTTTGATCCTGGCTCAG	Eubacteria		
1392B	ACGGGCGGTGTGTAC	Eubacteria		G, Ia, Ib, IIb

except for MICR and PEPA (0.125 μM each) and FIGD (0.5 μM) in multiplex PCR-G.

2. PCR amplification is conducted for 35 cycles. Each cycle consists of denaturation for 20 s at 95°C; annealing for 1 min at 50°C for multiplex PCR-G, 62°C for multiplex PCR-Ia and multiplex PCR-Ib, and 53°C for multiplex PCR-IIa and multiplex PCR-IIb; extension for 30 s at 72°C for all. A final extension for 5 min at 72°C is added to the end of cycles.
3. PCR products are electrophoresed on a 2% agarose gel, stained with ethidium bromide and visualized under UV light.

Note: In the multiplex PCR-G, the six reference strains of *A. hydrogenalis*, *A. lactolyticus*, *A. octavius*, *A. prevotii*, *A. tetradius*, and *A. vaginalis* all produced a unique DNA fragment of about 980 bp in size; the five reference strains of *Pn. asaccharolyticus*, *Pn. harei*, *Pn. indolicus*, *Pn. Ivorii*, and *Pn. lacrimalis* formed a specific DNA fragment of about 510 bp in size; and the species of the genera *Micromonas* (*M. micros*), *Peptostreptococcus* (*P. anaerobius*), and *Finegoldia* (*F. magna*) generated specific DNA fragments of sizes 380, 780, and 1200 bp, respectively.

Three species of the genus *Anaerococcus* (*A. lactolyticus*, *A. hydrogenalis*, and *A. octavius*) yielded unique DNA fragments (of 150, 400, and 760 bp in size, respectively) in the multiplex PCR-Ia; the other three species of the genus *Anaerococcus* formed specific DNA fragments (*A. tetradius*, 150 bp; *A. prevotii*, 400 bp; *A. vaginalis*, 760 bp) in the multiplex PCR-Ib; two species of the genus *Peptoniphilus* displayed characteristic DNA fragments (*Pn. asaccharolyticus*, 300 bp; *Pn. harei*, 600 bp) in the multiplex PCR-IIa; and the other three species of the genus *Peptoniphilus* showed signature DNA fragments (*Pn. lacrimalis*, 420 bp; *Pn. ivorii*, 750 bp; *Pn. indolicus*, 1350 bp) in multiplex PCR-IIb.

It was noteworthy that an amplicon was generated from *Pn. asaccharolyticus* multiplex in the PCR-IIb, which is similar in size to that from *Pn. indolicus*. Nonetheless, the identification achieved by the multiplex PCR assays is in 100% agreement with 16S rRNA sequencing identification. Therefore, this multiplex PCR scheme provides a simple, rapid, and reliable method for the identification of GPAC species.

31.2.2.2 PCR–RFLP

Riggio et al.[30] described PCR-restriction fragment length polymorphism (RFLP) for identification of GPAC, including *A. prevotii*, *F. magnus*, *P. anaerobius*, *Pa. micros*, *Pn. asaccharolyticus*, *Pn. indolicus*, and *R. productus*. The primers targeting conserved regions of the 16S rRNA gene consist of 5′-AGAGTTTGATCMTGGCTCAG-3′ (27f; *Escherichia coli* positions 8 to 27) and 5′-ACGGGCGGTGTGTRC-3′ (1392r; *E. coli* positions 1405 to 1391), where M = C + A and R = A + G. Use of these primers results in the amplification of an expected product of about 1500 bp.

Procedure

1. PCR mixture (50 μL) is composed of 10 μL of bacterial DNA crude extract and 40 μL of reaction mixture containing 1× PCR buffer (10 mM Tris-HCl pH 9.0, 50 mM KCl, 1.5 mM $MgCl_2$, 0.1% Triton

X-100), 1.0 U of *Taq* DNA polymerase (Promega), 0.2 mM each dNTPs, and each primer at a concentration of 0.2 μM.
2. Amplification is carried out with an initial denaturation step of 94°C for 5 min, 40 cycles of 94°C for 1 min, 60°C for 1 min, and 72°C for 1 min, followed by a final step at 72°C for 10 min.
3. Ten μL of each PCR product is electrophoresed on a 2% agarose gel, and stained with ethidium bromide (0.5 μg/mL) and visualized under UV light. A fragment of about 1500 bp is observable.
4. Ten μL of each PCR product is digested separately in a total volume of 20 μL with 5 U of each of the restriction enzymes *Cfo*I, *Hin*fI, *Rsa*I, and *Mnl*I at 37°C for 3 h. Restriction fragments are visualized by agarose gel electrophoresis as above.

Note: Distinct electrophoretic patterns are obtained with restriction enzymes *Cfo*I, *Hin*fI, and *Rsa*I for *A. prevotii*, *F. magnus*, *P. anaerobius*, *Pa. micros*, and *R. productus*. However, *Pn. asaccharolyticus* and *Pn. indolicus* required another restriction enzyme (*Mnl*I). *Rsa*I and *Hin*fI appeared to be the most discriminatory, generating six distinct RFLP groups from seven type strain species (Table 31.2).

31.2.2.3 Real-Time PCR for *Parvimonas micros*

(i) Protocol of Martin et al.[31] Martin et al.[31] reported a real-time PCR for *Parvimonas micros*, with primers and probe from the 16S rRNA gene (rDNA) sequence (Table 31.3). The probe is labeled with fluorescent dyes 6-carboxyfluorescein (FAM) at the 5′ end, and 6-carboxytetramethylrhodamine (TAMRA) at the 3′end.

Procedure

1. Bacteria are cultured to the late exponential phase, harvested by centrifugation (14,000 × *g* at 18°C–20°C for 2 min), washed, and resuspended in 10 mM phosphate buffer (pH 6.7) containing 1 mg of lysozyme/mL, 1 mg of mutanolysin/mL, and 5 mM $ZnCl_2$. After incubation at 60°C for 30 min, DNA is extracted and purified with a QIAamp DNA Mini Kit (QIAGEN). The DNA concentration (A_{260}) and purity (A_{260}/A_{280}) are measured.
2. To extract DNA from the anaerobic bacteria present in homogenized carious dentine, frozen suspensions are thawed on ice and 80-μL samples are combined with 100 μL of ATL buffer (QIAGEN) and 400 μg of proteinase K (QIAGEN). The samples are vortexed for 10 s prior to incubation at 56°C for 40 min, with vortexing every 10 min to lyse the cells. Following the addition of 200 μg of RNase (Sigma), the samples are incubated for a further 10 min at 37°C before the DNA is finally purified with a QIAamp DNA Mini Kit.
3. Real-time PCR mix (25-μL) is prepared with the TaqMan Universal PCR Master Mix containing 100 nM each forward primer, reverse primer, and probe and between 10 and 100 pg of template DNA/μL.
4. The real-time PCR conditions are set at 50°C for 2 min and 95°C for 10 min, followed by 40 cycles at 95°C for 15 s and 60°C for 1 min in an ABI PRISM 7700 Sequence Detection System (Applied Biosystems). Data analysis used Sequence Detection software (version 1.6.3), supplied by Applied Biosystems.
5. The sensitivity of real-time PCR in detecting DNA is determined in duplicate with DNA extracted from *Parvimonas micros* by using the appropriate homologous DNA as a standard (ranges, 2.0 fg to 2.0 ng, 82.9 fg to 8.29 ng, 3.6 fg to 3.6 ng, 24.1 fg to 2.41 ng, and 77.2 fg to 7.72 ng, respectively).

(ii) Protocol of Boutaga et al.[32] Boutaga et al.[32] reported a real-time PCR with the 16S rRNA gene primers and probe for detection of *Parvimonas micros* in subgingival plaque samples from adult patients with periodontitis (Table 31.4). The probe incorporates fluorescent dyes 6-carboxyfluorescein (FAM) at the 5′ end and 6-carboxytetramethylrhodamine (TAMRA) at the 3′end. The C_t parameter is defined as the fractional cycle number at which the fluorescence of the reporter dye generated by cleavage of the probe crosses

TABLE 31.2
PCR–RFLP Analyses of 16S rRNA Genes of GPAC Species

	Fragment Size (bp)			
Species	***Cfo*I**	***Hin*fI**	***Rsa*I**	***Mnl*I**
Peptostreptococcus anaerobius	700, 430, 340	1500	460, 360, 350, 150, 120	ND
Peptoniphilus asaccharolyticus	870, 430, 220	750, 260, 220, 120	530, 490, 420, 120	420, 350, 270, 180, 130
Peptoniphilus indolicus	870, 430, 220	750, 260, 220, 120	530, 490, 420, 120	350, 270, 190, 170, 130
Anaerococcus prevotii	700, 430, 380	470, 380, 250, 220, 150, 125	370, 330, 180, 150, 120	ND
Finegoldia magnus	1100, 430	1100, 380	460, 360, 180, 150, 120	ND
Parvimonus micros	1100, 430	520, 390, 240, 200, 180	560, 360, 180, 150, 120	ND
Ruminococcus productus	940, 430, 200	950, 380, 100	530, 490, 420, 120	ND

ND, not determined.

TABLE 31.3
Sequences of Oligonucleotide Primers and Probe for *Parvimonas micros*

Primer or Probe	Sequence (5′→3′)	*Tm* (°C)
Forward	AGTGGGATAGCCGTTGGAAA	58.1
Reverse	GACGCGAGCCCTTCTTACAC	58.5
Probe	ACCGCATGAGACCACAGAATCGCA	68.6

TABLE 31.4
Primers and Fluorogenic Probe for the Specific Detection of *Parvimonas micros*

Primer	Sequence (5′→3′)
Forward	GCC GTA AAC GAT GAG TGC TAG G
Reverse	CCA GGC GGA ATG CTT AGT GT
Probe	FAM-TGG GAG TCA AAT CTC GGT GCC G-TAMRA

an arbitrarily defined threshold within the logarithmic phase. Hence, this parameter can be used to compare different amplification reactions.

Procedure

1. PCR mixture (25 μL) is made up of 12.5 μL of 2× TaqMan universal PCR master mixture (PCR buffer, deoxynucleoside triphosphates, AmpliTaq Gold, an internal reference signal [6-carboxy-X-rhodamine], uracil *N*-glycosylase, $MgCl_2$; Applied Biosystems), 300 nM forward primer, 900 nM reverse primer, 100 nM probe, 5 μL of purified DNA from plaque samples. The negative control is 5 μL of sterile H_2O.
2. The mixture is subjected to an initial amplification cycle of 50°C for 2 min and 95°C for 10 min, followed by 45 cycles at 95°C for 15 s and 60°C for 1 min. The data are analyzed with ABI 7000 Sequence Detection System software.
3. The analytical sensitivity of each of the 5 real-time PCR sets is determined in triplicate with DNA isolated from *Parvimonas micros* strain. Serial tenfold dilutions of homologous DNA are used as standard curves.

31.2.2.4 PCR for *Peptoniphilus* spp. and *Peptoniphilus lacrimalis*

Fredricks et al.[33] designed primers from 16S rRNA gene for specific detection of *Peptoniphilus* spp. and *Peptoniphilus lacrimalis* in vaginal specimens (Table 31.5).

Procedure

1. Vaginal swabs are placed in 15-mL conical vials with 2 mL of saline and vortex mixed for 1 min to dislodge cells. The sample solution is centrifuged at 14,000 rpm for 10 min, and the pellet is resuspended in 100 μL supernatant. DNA is extracted from the pellet using the Ultra Clean Soil DNA Kit (MoBio, Carlsbad, CA). DNA is eluted from silica columns in a volume of 150 μL buffer. Sham digests using a swab without human contact are performed with each round of DNA extraction (every 10–25 samples) to control for contamination that might arise from kit reagents or collection swabs.
2. PCR mixture (50-μL) is made up of 1× PCR buffer II, 2 mM $MgCl_2$, 0.8 mM dNTPs, 1 U AmpliTaq Gold DNA polymerase (Applied Biosystems), 0.2 μM each of forward and reverse primers Pepton-1003F and Pepton-1184R (or P.lacri-999F and Pepton-1184R), and 1 μL of template DNA.
3. The mixture is subjected to a premelt at 95°C for 10 min; 40 cycles of 95°C for 30 s, 55°C for 30 s, and 72°C for 30 s; and a final extension at 72°C for 7 min.
4. The PCR products are electrophoresed in 2% agarose gels, stained with ethidium bromide, and visualized under UV light.

Note: A *β*-globin PCR protocol may be used to check the quality of extracted DNA and to monitor for PCR inhibitors, with the following primers: GH2O, 5′-GAA GAGCCAAGGACAGGTAC-3′, and PCO4, 5′-CAACT TCATCCACGTTCACC-3′. The PCR conditions are the same as described above.

31.2.2.5 Sequencing Analysis of 16S rRNA Gene

Song et al.[13] employed universal 16S rRNA gene primers 8UA (positions 8 to 28, *Escherichia coli* numbering) and 1485B (positions 1485 to 1507)[25] for PCR amplification and sequencing analysis of GPAC isolates of human origin, leading to the identification of three novel species: *Peptoniphilus gorbachii* sp. nov., *Peptoniphilus olsenii* sp. nov., and *Anaerococcus murdochii* sp. nov.

Procedure

1. PCR is performed for 35 cycles of 30 s at 95°C, 30 s at 45°C, and 1.5 min at 72°C, with a final extension at 72°C for 5 min.

TABLE 31.5
PCR Primers for *Peptoniphilus* spp. and *Peptoniphilus lacrimalis*

Target	Primer	Sequence (5′→3′)
Peptoniphilus spp.	Pepton-1003F	GACCGGTATAGAGATATACCCT
	Pepton-1184R	CACCTTCCTCCGATTTATCATC
P. lacrimalis	P.lacri-999F	AAGAGACGAACTTAGAGATAAGTTTT
	Pepton-1184R	As above

2. A portion of the PCR product is checked on 1% agarose gel. The remaining amplified product is purified by using the QIAamp PCR purification kit (QIAGEN), and both strands of the PCR products are directly sequenced with an ABI 3100 Avant Genetic System (Applied Biosystems).
3. The closest known relatives of the new isolates are determined by performing database searches using BLAST software.
4. Almost-full-length sequences of the 16S rRNA gene sequences (>1400 nucleotides) of the unidentified bacteria and of closely related bacteria are aligned using CLUSTAL-W software. A phylogenetic tree is reconstructed using PAUP* 4.0 DNA analysis software (Sinauer Associates, Inc., Sunderland, MA). The stability of the groupings is estimated by bootstrap analysis (1000 replications) using the same program.

Note: When the 16S rRNA gene sequence divergence is approximately 3% or more between the test isolate and reference strains in GenBank, it is likely that the organism in question is phylogenetically distinct. A combination of 16S rRNA gene PCR and DNA sequencing is also useful for identification of *Peptococcus niger*, a related gram-positive anaerobic coccus that is found occasionally in human polymicrobial infections.[34]

31.3 CONCLUSION

Gram-positive anaerobic cocci (GPAC) constitute part of commensal organisms that are present on various human mucocutaneous surfaces and that have been frequently isolated from clinical specimens. Due to their heterogeneous nature, classification and identification of GPAC using phenotypical techniques have proven difficult and time consuming in spite of the fact that a streamlined biochemical testing scheme has been developed.[24] This may have contributed in part to the inadequate appreciation of the roles of GPAC in polymicrobial infections and diseases in the human population until recently.

It is only in the past decade or two that the taxonomical status of GPAC has become clear after comparative analysis of the 16S rRNA gene sequences. The once all-encompassing genus *Peptostreptococcus* has been shown to contain several phylogenetically diverse groups of organisms, for which independent genera (i.e., *Anaerococcus*, *Finegodia*, *Gallicola*, *Parvimonas*, and *Peptoniphilus*) have been created. This has facilitated the development of PCR assays for genus- and species-specific detection and determination. Application of these genotypic techniques enables rapid and accurate characterization of GPAC from clinical samples and provides a valuable tool for evaluating the effectiveness of antibiotic therapy.

REFERENCES

1. Murdoch, D.A., Gram-positive anaerobic cocci, *Clin. Microbiol. Rev.*, 11, 81, 1998.
2. Conrads, G. et al., Taxonomic update and clinical significance of species within the genus *Peptostreptococcus*, *Clin. Infect. Dis.*, 25, S94, 1997.
3. Ezaki, T. et al., Transfer of *Peptococcus indolicus*, *Peptococcus asaccharolyticus*, *Peptococcus prevotii*, and *Peptococcus magnus* to the genus *Peptostreptococccus* and proposal of *Peptostreptococcus tetradius* sp. nov., *Int. J. Syst. Bacteriol.*, 33, 683, 1983.
4. Li, N. et al., Three new species of the genus *Peptostreptococcus* isolated from humans: *Peptostreptococcus vaginalis* sp. nov., *Peptostreptococcus lacrimalis* sp. nov., and *Peptostreptococcus lactolyticus* sp. nov., *Int. J. Syst. Bacteriol.*, 42, 602, 1992.
5. Murdoch, D.A., and Kelly, M., *Pepstostreptococcus prevotii* or *Peptostreptococcus vaginalis*? Identification and clinical importance of a new species of *Peptostreptococcus*, *Anaerobe*, 3, 23, 1997.
6. Murdoch, D.A. et al., Description of three new species of the genus *Peptostreptococcus* from human clinical specimens: *Peptostreptococcus harei* sp. nov., *Peptostreptococcus ivorii* sp. nov. and *Peptostreptococcus octavius* sp. nov., *Int. J. Syst. Bacteriol.*, 47, 781, 1997.
7. Li, N., Hashimoto, Y., and Ezaki, T., Determination of 16S rRNA ribosomal RNA sequences of all members of the genus *Peptostreptococcus* and their phylogenetic position, *FEMS Microbiol. Lett.*, 116, 1, 1994.
8. Murdoch, D.A., and Shah, H.N., Reclassification of *Peptostreptococcus magnus* (Prevot 1933) Holdeman and Moore 1972 as *Finegoldia magna* comb. nov. and *Peptostreptococcus micros* (Prevot 1933) Smith 1957 as *Micromonas micros* comb. nov., *Anaerobe*, 5, 555, 1999.
9. Ezaki, T. et al., Proposal of the genera *Anaerococcus* gen. nov., *Peptoniphilus* gen. nov. and *Gallicola* gen. nov. for members of the genus *Peptostreptococcus*, *Int. J. Syst. Evol. Microbiol.*, 51, 1521, 2001.
10. Tindall, B.J., and Euzéby, J.P., Proposal of *Parvimonos* gen. nov. and *Quatrionicoccus* gen. nov. as replacements for the illegitimate, prokaryotic, generic names *Micromonas*

Murdoch and Shah 2000 and *Quadricoccus* Maszenan *et al.* 2002, respectively, *Int. J. Syst. Evol. Microbiol.*, 56, 2711, 2006.
11. Ezaki, T. et al., 16S ribosomal DNA sequences of anaerobic cocci and proposal of *Ruminococcus hansenii* comb. nov. and *Ruminococcus productus* comb. nov., *Int. J. Syst. Bacteriol.*, 44, 130, 1994.
12. Song, Y. et al., Rapid identification of gram-positive anaerobic coccal species originally classified in the genus *Peptostreptococcus* by multiplex PCR assays using genus- and species-specific primers, *Microbiology*, 149, 1719, 2003.
13. Song, Y., Liu, C., and Finegold, S.M., *Peptoniphilus gorbachii* sp. nov., *Peptoniphilus olsenii* sp. nov., and *Anaerococcus murdochii* sp. nov. isolated from clinical specimens of human origin, *J. Clin. Microbiol.*, 45, 1746, 2007.
14. Falsen, E. et al., *Fastidiosipila sanguinis* gen. nov., sp. nov., a new gram-positive, coccus-shaped organism from human blood, *Int. J. Syst. Evol. Microbiol.*, 55, 853, 2005.
15. Murdoch, D.A. et al., Proposal to restrict the genus *Peptostreptococcus* (Kluyver & van Niel 1936) to *Peptostreptococcus anaerobius*, *Anaerobe*, 6, 257, 2002.
16. Downes, J., and Wade, W.G., *Peptostreptococcus stomatis* sp. nov., isolated from the human oral cavity, *Int. J. Syst. Evol. Microbiol.*, 56, 751, 2006.
17. Finegold, S.M., Anaerobic infections in humans: An overview, *Anaerobe*, 1, 3, 1995.
18. Wren, M.W.D., Anaerobic cocci of clinical importance, *Br. J. Biomed. Sci.*, 53, 294, 1996.
19. Simmon, K.E. et al., Genotypic diversity of anaerobic isolates from bloodstream infections, *J. Clin. Microbiol.*, 46, 1596, 2008.
20. Murray, P.R., Weber, C.J., and Niles, A.C., Comparative evaluation of three identification schemes for anaerobes, *J. Clin. Microbiol.*, 22, 52, 1985.
21. Murdoch, D.A., and Mitchelmore, I.J., The laboratory identification of gram-positive anaerobic cocci, *J. Med. Microbiol.*, 34, 295, 1991.
22. Wilson, M.J. et al., Evaluation of a phenotypic scheme for identification of the "butyrate-producing" *Peptostreptococcus* species, *J. Med. Microbiol.*, 49, 747, 2000.
23. Jousimies-Somer, H.R. et al., *Anaerobic Bacteriology Manual*, 6th ed., Star Publishing Co., Belmont, CA, 2002.
24. Song, Y., Liu, C., and Finegold, S.M., Development of a flow chart for identification of gram-positive anaerobic cocci in the clinical laboratory, *J. Clin. Microbiol.*, 45, 512, 2007.
25. Song, Y. et al., 16S ribosomal DNA sequence-based analysis of clinically significant gram-positive anaerobic cocci, *J. Clin. Microbiol.*, 41, 1363, 2003.
26. Hill, K.E. et al., Heterogeneity within the gram-positive anaerobic cocci demonstrated by analysis of 16S-23S intergenic ribosomal RNA polymorphisms, *J. Med. Microbiol.*, 51, 949, 2002.
27. Yasui, S., Development and clinical application of DNA probe specific for *Peptostreptococcus micros*, *Bull. Tokyo Med. Dent. Univ.*, 36, 49, 1989.
28. Wildeboer-Veloo, A.C. et al., Development of 16S rRNA-based probes for the identification of gram-positive anaerobic cocci isolated from human clinical specimens, *Clin. Microbiol. Infect.*, 13, 985, 2007.
29. Riggio, M.P., Lennon, A., and Smith, A., Detection of *Peptostreptococcus micros* DNA in clinical samples by PCR, *J. Med. Microbiol.*, 50, 249, 2001.
30. Riggio, M.P., and Lennon, A., Identification of oral *Peptostreptococcus* isolates by PCR-restriction fragment length polymorphism analysis of 16S rRNA genes, *J. Clin. Microbiol.*, 41, 4475, 2003.
31. Dowd, S.E. et al., Survey of bacterial diversity in chronic wounds using pyrosequencing, DGGE, and full ribosome shotgun sequencing, *BMC Microbiol.*, 8, 43, 2008.
32. Martin, F.E. et al., Quantitative microbiological study of human carious dentine by culture and real-time PCR: Association of anaerobes with histopathological changes in chronic pulpitis, *J. Clin. Microbiol.*, 40, 1698, 2002.
33. Boutaga, K. et al., Periodontal pathogens: A quantitative comparison of anaerobic culture and real-time PCR, *FEMS Immunol. Med. Microbiol.*, 45, 191, 2005.
34. Fredricks, D.N. et al., Targeted PCR for detection of vaginal bacteria associated with bacterial vaginosis, *J. Clin. Microbiol.*, 45, 3270, 2007.

32 Catabacter

Susanna K.P. Lau and Patrick C.Y. Woo

CONTENTS

32.1 INTRODUCTION

32.1.1 TAXONOMY

The genus *Catabacter* consists of a recently described, gram-positive, anaerobic, bacterial species *Catabacter hongkongensis*. Being only distantly related to other anaerobic gram-positive bacilli, which traditionally included species of the genera *Actinomyces*, *Bifidobacterium*, *Clostridium*, *Eggerthella*, *Eubacterium*, *Lactobacillus*, and *Propionibacterium*, *C. hongkongensis* forms a distinct lineage peripherally associated with clusters I, III, and XIVb of the clostridia on phylogenetic tree upon 16S ribosomal RNA gene analysis (Figure 32.1). Based on its phylogenetic position and distinct phenotypic characteristics, *C. hongkongensis* is classified as the sole species under the genus *Catabacter*, family Catabacteriaceae, order Clostridiales, class Clostridia, phylum Firmicutes.[1]

32.1.2 MORPHOLOGY AND BIOLOGY

Catabacter hongkongensis is a motile, catalase-positive, strictly anaerobic, nonsporulating, gram-positive coccobacillus. The name *Catabacter* represents "catalase-positive bacterium." Bacterial cells appeared as short, straight rods tapered at both ends with a tuft of flagellae inserted on one side on scanning electron microscopy (Figure 32.2). The bacterium grows in standard anaerobic blood culture bottles and on sheep blood agar as nonhemolytic colonies at 37°C in anaerobic environment.[1] It utilizes arabinose, glucose, mannose, and xylose.[1] Interstrain variations have been observed with glycerol and rhamnose fermentation and leucine arylamidase.[1] *C. hongkongensis* is susceptible to bile and kanamycin but resistant to colistin.[1] The current commercially available bacterial identification systems did not include *C. hongkongensis* in their databases and most reported the result "unidentified" on previous isolates. The observed code profile by the ATB Expression system (ID32A) has been 0012000010, and that by the API system has been (20A) 40415042 or 40414052.[1]

The G + C content of the type strain has been estimated to be 40.2, using the thermal denaturation method as described previously.[1–3] Briefly, the temperature of the genomic DNA in SSC (0.15 M NaCl with 0.015 M sodium citrate) buffer (25 μg/mL) was increased slowly (0.5°C/min) from 25°C and the absorbance of the solution at 260 nm was monitored continuously against a blank containing SSC buffer only. The T_m of the DNA is defined as the temperature at 50% hyperchromicity. The G + C content of the genomic DNA was calculated by the formula: $(G + C)\% = 2.44T_m - 169$.

32.1.3 EPIDEMIOLOGY AND CLINICAL MANIFESTATIONS

Catabacter hongkongensis was first discovered in patients with bacteremic sepsis.[1] It has been isolated from the blood of patients in Hong Kong and Canada with sepsis associated with intestinal obstruction, acute appendicitis, and biliary sepsis.[1] In the first report on four cases of *C. hongkongensis* bacteremia, three patients had significant underlying diseases including end-stage renal failure, plasmacytoma, and metastatic carcinoma of lung. One case occurred in a previously healthy young individual with perforated acute appendicitis. Fever and leukocytosis were common.[1] The source of *C. hongkongensis* bacteremia was believed to be the gastrointestinal tract. Therefore, *C. hongkongensis* may be part of the human gut flora, similar to other nonsporulating, anaerobic gram-positive bacilli such as *Bifidobacterium*, *Eggerthella*, *Eubacterium*, and *Lactobacillus*. Besides the four reported

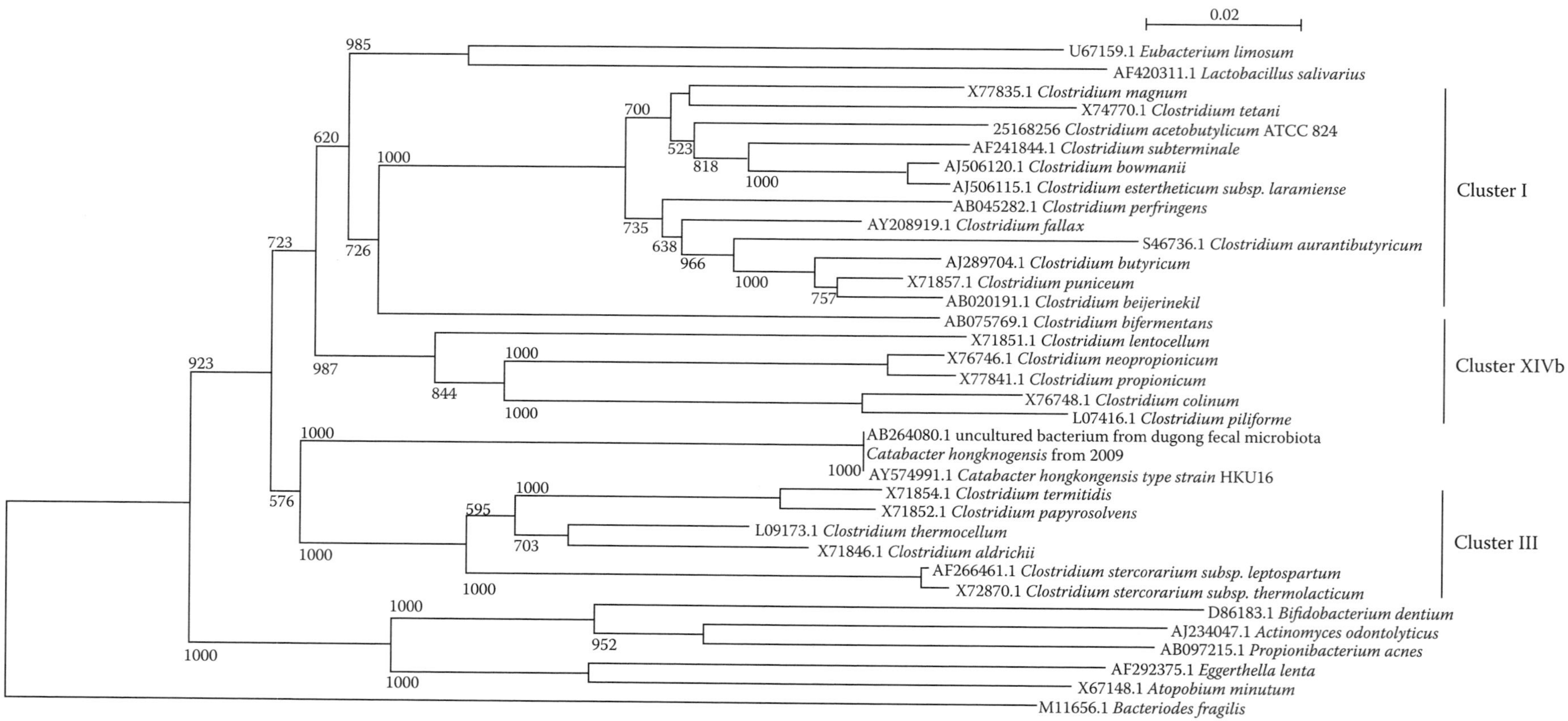

FIGURE 32.1 Phylogenetic tree showing the relationships of *Catabacter hongkongensis* to related anaerobic gram-positive bacteria. The type strain HKU16 was isolated from the blood culture of a patient with abdominal sepsis. Another clinical isolate was recently from the blood of a patient in Hong Kong. The sequence AB264080.1 was detected in the fecal microflora of a dugong in Japan. The tree was constructed by using the neighbor-joining method using Kimura's two-parameter correction and *Bacteroides fragilis* as the root. A total of 1238 nucleotide positions were included in analysis. Bootstrap values were calculated from 1000 trees. The scale bar indicates the estimated number of substitutions per 50 bases. Names and accession numbers are given as cited in the GenBank database.

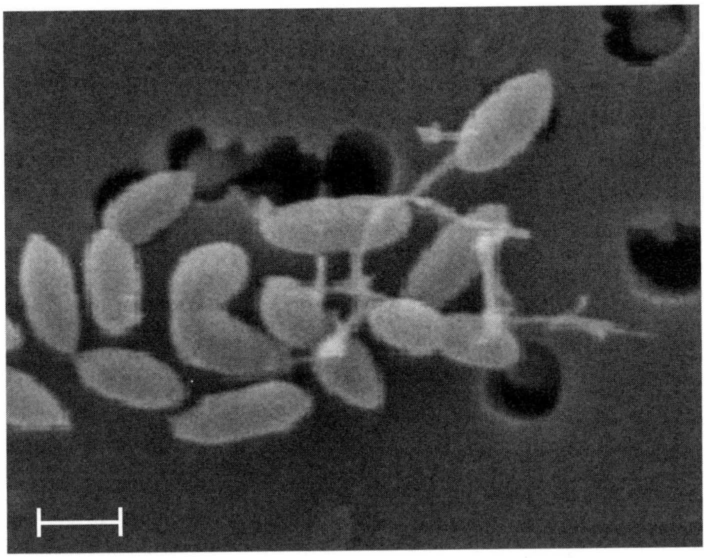

FIGURE 32.2 Scanning electron micrograph of *Catabacter hongkongensis*. The bacterium is short and straight rod with tapered ends, which multiplied by longitudinal division. Cells vary in length from 0.71 to 1.12 μm and diameter from 0.36 to 0.48 μm (mean = 0.96 μm × 0.42 μm, $n = 20$). Bar, 0.5 μm.

cases, more cases of *C. hongkongensis* bacteremia have subsequently been identified in Hong Kong (unpublished data) (Figure 32.1). Although most patients with *C. hongkongensis* infection responded to systemic antibiotics, mortality has occurred in two patients with underlying advanced malignancy.[1]

Apart from causing bacteremia, a *Ruminococcus*-like organism from a human clinical source in the United Kingdom has been found to possess 16S rRNA sequence with >99% identity to *C. hongkongensis* (GenBank accession no. AJ318864). However, the true identity and disease association of this isolate has not been reported. It is likely that this isolate is also a strain of *C. hongkongensis* that has been misidentified as a *Ruminococcus* strain. The rare report of *C. hongkongensis* may be a reflection of the difficulty in accurately identifying anaerobic gram-positive bacilli in clinical laboratories. Now that 16S rRNA gene analysis is more readily available, more strains of *C. hongkongensis* will be uncovered, which will help better define its epidemiology and clinical spectrum of disease.

Since the first report of *C. hongkongensis* infections, 16S rRNA sequence related to *C. hongkongensis* has subsequently been detected in various environmental samples. In a study on urban aerosols collected in the United States, 16S rRNA sequences belonging to a diverse bacterial population, including genus *Catabacter*, were detected, although the degree of sequence similarity was not mentioned.[4] Similar findings were also observed in the mangrove sediment in China.[5] In another study from Japan, among microbial communities from rice-paddy-field soil used for microbial fuel cells, 16S rRNA sequences of diverse bacteria, including one with 93% identity to that of *C. hongkongensis*.[6] However, since in none of these reports were bacteria isolated for further characterization, and since the identity of these sequences with that of *C. hongkongensis* was not very high or unknown, it remains to be determined if *C. hongkongensis* or some closely related bacteria were present in these environments. Interestingly, in a recent study on the fecal microflora of a dugong (*Dugong dugong*), an aquatic herbivorous mammal in captivity at Toba Aquarium, Japan, 16S rRNA sequence clones belonging to diverse bacterial phyla, including one that possessed 100% homology with *C. hongkongensis*, were identified (Figure 32.1).[7] This suggests that *C. hongkongensis* may be one of the gut commensals in this animal, which supports the speculation that the gastrointestinal tract is a source of infection in humans. In addition, *C. hongkongensis* 16S rRNA has been detected in lake water and anaerobic wastewater treatment reactors.[8,9]

32.1.4 Pathogenicity and Therapy

Given its ability to invade into bloodstream causing systemic sepsis, *C. hongkongensis* may possess virulence factors that are yet to be defined. One possible factor may be related to its ability to produce catalase.[1] Despite being strictly anaerobic, *C. hongkongensis* is able to survive in the presence of oxygen for 3–4 days.[1] Such aerotolerance may be related to the production of catalase, which may protect *C. hongkongensis* from oxidative stress.[1] In *B. fragilis*, such aerotolerance has been described as a potential virulence factor, as clinical isolates were more aerotolerant than fecal isolates.[10]

C. hongkongensis is most susceptible to metronidazole with minimal inhibitory concentration (MIC) of <0.016 μg/mL.[1] It demonstrates variable susceptibility to penicillin with MIC as high as 4 μg/mL.[1] It is also susceptible to vancomycin but resistant to cefotaxime.[1] In previous cases, two patients have responded to intravenous cefuroxime/metronidazole combinations, and one to oral ciprofloxacin.[1] However, given the uncertain activity of quinolones, metronidazole should be considered the drug of choice for *C. hongkongensis* infections.

32.2 METHODS

Considering the difficulty in identification by phenotypic characteristics, real-time PCR and 16S rDNA gene sequencing can be utilized for accurate and definitive identification of *C. hongkongensis*. When a catalase-positive, motile, nonsporulating, anaerobic, gram-positive bacillus is encountered in clinical microbiology laboratories, the possibility of *C. hongkongensis* should be considered and the isolate is subjected to real-time PCR and 16S rRNA gene PCR and sequencing.[1,11]

32.2.1 Sample Preparation

To extract bacterial DNA, 80 μL of NaOH (0.05 M) is added to 20 μL of bacterial cells suspended in distilled water, and the mixture is incubated at 60°C for 45 min, followed by addition of 6 μL of Tris-HCl (pH 7.0), achieving a final pH of 8.0.[1] The resultant mixture is diluted 100× and 5 μL of the diluted extract is used for PCR.

32.2.2 Detection Procedures

(i) Real-time PCR Codony et al.[11] designed a pair of *Catabacter*-specific PCR primers from the 16S rRNA sequence (GenBank accession No. AY574991) for detection of *Catabacter hongkongensis* in polluted water samples. Use of forward primers (GTC GAA CGA AGT TGC TCT TT) and reverse primer (CAT GCG GTT TCG TGG TC) together with Quantifast SYBR Green Mix permits real-time amplification and detection of a specific 153 bp fragment from *C. hongkongensis*.

Water (100 mL) is concentrated by membrane filtration with a nylon membrane (0.45 μm porous diameter; Millipore). Cells are resuspended in 5 mL of sterile saline solution by vigorous vortexing with 15 glass beads (5 mm diameter) for 60 s and sonicated for 3 min (6 L, 150 W; Selecta). The resulting suspension (4 mL) is concentrated to 200 μL by centrifugation (10,000 × *g*, 5 min) and DNA is extracted with the DNeasy Tissue Kit (Qiagen).

Real-time PCR mixture (20 μL) is composed of 10 μL of Quantifast SYBR Green Mix (Qiagen), 0.5 μmol/L each of reverse and forward primers, 0.2 U of uracil-DNA-glycosylase (New England BioLabs), and 10 μL of sample.

Amplification and detection is conducted in the LightCycler 1.5 PCR system (Roche Applied Science), with one 2-min step at 50°C to allow uracil-DNA-glycosylase to break down the possible contaminating amplicons, one 5-min step at 95°C for *Taq* polymerase activation, 45 cycles of 95°C for 10 s, and 60°C for 30 s for PCR amplification, and a final melting temperature ramp from 65°C to 95°C at 0.1°C per s.

The PCR products from positive samples may be confirmed by DNA sequencing after purification with the Cycle Pure Kit (Omega Bio-Tek).

(ii) 16 rRNA Gene Sequence Analysis To examine the 16S rRNA gene, bacterial DNA extracts can be amplified with primers (LPW57 5′-AGTTTGATCCTGGCTCAG-3′ and LPW205 5′-CTTGTTACGACTTCACCC-3′).[1] The PCR mixture (50 μL) should contain bacterial DNA, PCR buffer (10 mM Tris-HCl pH 8.3, 50 mM KCl, 2 mM $MgCl_2$ and 0.01% gelatin), 200 μM of each dNTPs, and 1.0 U *Taq* polymerase (Boehringer, Mannheim, Germany). The mixtures should be amplified in 40 cycles of 94°C for 1 min, 55°C for 1 min, and 72°C for 2 min, and a final extension at 72°C for 10 min. Both strands of the PCR products should be sequenced twice using the PCR primers (LPW57 and LPW205) and additional sequencing primers LPW284 5′-GTTTACAACCCGAAGGCC-3′ and LPW306 5′-TGAGATGTTGGGTTAAGT-3′.

32.3 CONCLUSION

Catabacter hongkongensis is a recently described motile, catalase-positive, strictly anaerobic, nonsporulating, gram-positive bacillus which has been isolated from blood cultures of patients with bacteremic sepsis. It has been isolated from the blood of patients in Hong Kong and Canada with sepsis associated with intestinal obstruction, acute appendicitis, and biliary sepsis. *C. hongkongensis* is susceptible to metronidazole and vancomycin, but demonstrates variable susceptibility to penicillin and resistance to cefotaxime. Although most patients with *C. hongkongensis* infection responded to systemic antibiotics, mortality has occurred in two patients with underlying advanced malignancy. The source of *C. hongkongensis* bacteremia was believed to be the gastrointestinal tract. The rare report of *C. hongkongensis* infections may reflect the difficulty in accurately identifying anaerobic gram-positive bacilli in clinical laboratories. Because currently available commercial identification systems fail to identify the bacterium, PCR and 16S rRNA gene sequencing remain the most accurate methods for definitive identification of *C. hongkongensis*. Based on its phylogenetic position and distinct phenotypic characteristics, it is classified as the sole species under the genus *Catabacter*, family Catabacteriaceae. When a catalase-positive, motile, nonsporulating, anaerobic, gram-positive bacillus is encountered in clinical microbiology laboratories, the possibility of *C. hongkongensis* should be considered and metronidazole should be considered the drug of choice.

REFERENCES

1. Lau, S.K. et al., Catabacter hongkongensis gen. nov., sp. nov., isolated from blood cultures of patients from Hong Kong and Canada, *J. Clin. Microbiol.*, 45, 395, 2007.
2. Woo, P.C.Y. et al., *Streptococcus sinensis sp. nov.*, a novel *Streptococcus* species isolated from a patient with infective endocarditis, *J. Clin. Microbiol.*, 40, 805, 2002.
3. Yuen, K.Y. et al., *Laribacter hongkongensis* gen. nov., sp. nov., a novel Gram-negative bacterium isolated from a cirrhotic patient with bacteremia and empyema, *J. Clin. Microbiol.*, 39, 4227, 2001.

4. Brodie, E.L. et al., Urban aerosols harbor diverse and dynamic bacterial populations, *Proc. Natl. Acad. Sci. USA*, 104, 299, 2007.
5. Liang, Jun-Bin et al., Recovery of novel bacterial diversity from mangrove sediment, *Mar. Biol.*, 150, 739, 2007.
6. Ishii, S., Hotta, Y., and Watanabe, K., Methanogenesis versus electrogenesis: morphological and phylogenetic comparisons of microbial communities, *Biosci. Biotechnol. Biochem.*, 72, 286, 2008.
7. Tsukinowa, E. et al., Fecal microbiota of a dugong (Dugong dugong) in captivity at Toba Aquarium, *J. Gen. Appl. Microbiol.*, 54, 25, 2008.
8. El Saied, H.E., Molecular genetic monitoring of bacterial communities in Manzala lake, Egypt, based on 16S rRNA gene analysis, *Egypt. J. Aquatic Res.*, 33, 179, 2007.
9. Fernández, N. et al., Analysis of microbial community during biofilm development in an anaerobic wastewater treatment reactor, *Microb. Ecol.*, 56, 121, 2008.
10. Tally, F.P. et al., Oxygen tolerance of fresh clinical anaerobic bacteria, *J. Clin. Microbiol.*, 1, 161, 1975.
11. Codony, F. et al., Detection of *Catabacter hongkongensis* in polluted European water samples, *J. Zhejiang Univ. Sci. B*, 10, 867, 2009.

33 *Clostridium*

Juan J. Córdoba, Emilio Aranda, María G. Córdoba, María J. Benito, Dongyou Liu, and Mar Rodríguez

CONTENTS

33.1 INTRODUCTION

33.1.1 Classification, Morphology, and Biology/Epidemiology of *Clostridium*

The genus *Clostridium* comprises a highly heterogeneous group of gram-positive, endospore-producing, rod-shaped anaerobes, showing diverse physiology, and genetics.[1] For instance, the optimal growth temperature for clostridia varies from 34°C to 37°C for *C. beijerinckii* to 75°C–78°C for *C. thermohydrosulfuricum*. The guanosine and cytosine content (GC composition) of *Clostridium* DNA also varies from 24% in *C. pasterianum* to 55% in *C. barkeri*.[1] This genus currently contains 146 validly species, which are grouped into 19 clusters.[2,3] Cluster I consists of the majority of pathogenic species that have been regarded as the true representatives of the genus (i.e., *Clostridium* sensu stricto). It has been shown recently that the species of *Clostridium* cluster I are not only phylogenetically distinct from the remaining cluster of clostridia but also share many unique molecular characteristics.[4]

Clostridium spp. are distributed ubiquitously in the environment (e.g., soil, sewage, and marine sediments) and exist in the gastrointestinal tract of humans and domestic and feral animals.[5] Several species show remarkable morphological differences, such as terminal spore formation in *C. cadaversis*,[6] subterminal spore formation in *C. sporogenes*, and the central spore formation in *C. bifermentan*. As obligate anaerobes, some clostridia ferment by pathways that generate organic solvents, such as butyric acid, acetic acid, butanol, ethanol, and acetone, as well as large amounts of gas (CO_2 and H_2) during fermentation of sugars. Such species are named "solventogenic bacteria," and so far five of them are known, namely, *C. acetobutylicum*, *C. beijerinckii*, *C. saccharoperbutylacetonicum*, C. *saccharobutilicum*, and *C. ljungdahlii*. In addition, clostridia generate a variety of malodorous compounds during the fermentation of amino acids and fatty acids. The capability of clostridia to degrade large biological molecules (e.g., proteins, lipids, collagen, and cellulose) in the environment into fermentable components is undoubtedly assisted by their production of a wide range of extracellular enzymes. Thus, besides being useful for industrial production of alcohols and commercial solvents, clostridia play an important role in biodegradation and the carbon cycle. Furthermore, a few species (e.g., *C. butyricum* and *C. pasteurianum*) are capable of fixing nitrogen. Under stressful conditions, such as a lack of nutrients or unfavorable temperatures, clostridia tend to produce spores that facilitate their persistence in the environment.

Although most *Clostridium* species are saprophytes, 34 are considered pathogenic to man and animals.[3] More than half of the pathogenic species are members of cluster I, including those considered the major pathogenic agents, that is, *C. botulinum*, *C. barati*, *C. haemolyticum*, *C. novyi*, *C. perfringens*, *C. tetani*, *C. septicum*, and *C. chauvoei*. The only exception among these major pathogenic species is *C. difficile*, which is located in cluster XI. These pathogenic species often secrete an array of invasins, exotoxins, and other extracellular enzymes (e.g., proteases, lipases, collagenase, and hyaluronidase). Various toxins are responsible for such human diseases as tetanus, botulism, and gas gangrene.

In this chapter, we focus on the main human pathogenic *Clostridium* species (i.e., *C. botulinum* and other botulinum neurotoxin-producer species such as some strains of *C. barati* and *C. butyricum*; *C. perfringens*; *C. tetani*, and *C. difficile*). Molecular methods for their rapid detection and identification are presented.

Botulinum neurotoxin (BoNT) is produced by some species of the genus *Clostridium*, particularly *C. botulinum*, but some strains of *C. barati* and *C. bytiricum*. BoNTs can also be separated into one of seven distinct serotypes (A–G) on the basis of the immunological characteristics of its botulinum toxin. Moreover, four immunologically or genetically distinct BoNT/A subtypes, five BoNT/B subtypes, and six BoNT/E subtypes have been identified.[7,8] Some strains of *C. butyricum* and *C. barati* have been shown to produce type E and F neurotoxins, respectively.

C. botulinum is a diverse species. The taxonomic denominator for *C. botulinum* is the production of BoNT. *C. botulinum* is subdivided into four distinct groups based on physiological and genetic criteria, with Group I (proteolytic *C. botulinum*) strains producing one or two toxins of types A, B, or F; Group II (nonproteolytic *C. botulinum*) strains producing toxins of types B, E, or F; Group III strains producing toxins of types C or D; and Group IV strains producing toxin of type G.[9] For the Group IV organisms, the species name *C. argentinense* has been adopted. Generally, groups I and II cause botulism in humans, while group III is involved with animal botulism, but exceptions occur. Group IV has not generally associated with illness.[10]

Phenotypically, groups I and II *C. botulinum* strains differ from each other significantly. Group I organisms seem to be more terrestrial in origin and are present in temperature climates, whereas group II strains, particularly those producing E BoNT, are frequently found in aquatic environments in the Northern hemisphere.[10]

The strains of Group I *C. botulinum* are proteolytic and capable of utilizing amino acids as an energy source. The minimum growth temperature of Group I *C. botulinum* is 13°C–16°C and the optimum 37°C–42°C.[11] Under otherwise favorable conditions, growth is typically inhibited by a water activity of 0.94, corresponding to approximately 10% NaCl (w:v) in brine. Growth does not occur under a pH of 4.5. Group II *C. botulinum* is nonproteolytic and saccharolytic. The minimum growth temperature is 3°C.[9] The inhibitory water activity is 0.97, corresponding to 5% NaCl in brine. The spores of Group I *C. botulinum* possess a very high heat resistance. The spores of Group II *C. botulinum* are less heat resistant than those of Group I. As Group I and II strains differ in their epidemiologies, discrimination between the two groups should be done whenever an outbreak strain is isolated.

The genome of *C. botulinum* Group I (proteolytic) strain Hall A (ATCC 3502) consists of a chromosome of 3,886,916 bp and a plasmid of 16,344 bp, with 3650 and 19 predicted genes, respectively.[12] Being adapted to a saprophytic lifestyle in both soil and aquatic environments, *C. botulinum* relies on the toxins and extracellular enzymes it produces to kill prey species, to gain access to nutrient sources, and to degrade rotting or decayed tissues.

C. perfringens (formerly called *C. welchii*) is considered to be one of the common microorganisms that cause human and animal diseases.[13] Depending on the production of the four major toxins alpha, beta, epsilon and iota, they are divided into the five toxigenic types A, B, C, D, and E[9,14] (Table 33.1). Other toxins, like $\beta 2$ or enterotoxin, are found in several toxigenic types and may contribute to disease in various animals, but their finding is not necessarily linked with disease.[15,16] The α-toxin is present in all types of *C. perfringens* and lies within chromosome of the bacterial DNA. The β, $\beta 2$, ε- and ι-toxin genes of reference strains were described to be plasmid borne, whereas the enterotoxin gene may be present on either chromosome or plasmid.[14]

C. tetani is the culprit for tetanus (lockjaw) in both humans and animals. The natural habitat of this microorganism is soil, dust, and the intestinal tracts of various animals. *C. tetani* can be distinguished into 11 groups on the basis of its reaction with flagellar antigens.[9] The spores of *C. tetani* are extremely stable, and although boiling for 15 min kills most, some will survive unless autoclaved at 120°C, 1.5 bar, for 15 min.[17] When *C. tetani* spores enter the skin via a puncture wound (e.g., stepping on a rusty nail), they germinate and form *C. tetani* vegetative cells. *C. tetani* produce a tetanus toxin which comprises up to 5%–10% of the bacterial cell weight and is heat labile (destroyed at 56°C in 5 min) and O_2 sensitive. The genetic factor controlling the production of tetanus toxin is associated with an extrachromosomal plasmid.[18,19]

TABLE 33. 1
***C. perfringens* Toxinotypes and Their Major Toxins**

Toxinotype	Alpha	Beta	Epsilon	Iota
A	+			
B	+	+	+	
C	+	+		
D	+		+	
E	+			+

C. difficile belongs to the normal microbiota of the mammalian gastrointestinal tract. *C. difficile* proliferation and infection in the human colon often occur after use of broad-spectrum antibiotics.[9] Through release of two toxins (A and B), *C. difficile* destroys the intestinal lining and causes antibiotic-associated diarrhea (AAD), colitis, and pseudomembranous colitis. *C. difficile* toxin A is an enterotoxin that induces fluid accumulation in the bowel, and toxin B is a cytopathic toxin that is extremely lethal. Both toxins are highly unstable and tend to degrade at room temperature.

33.1.2 Clinical Features and Pathogenesis

C. botulinum produces botulinum neurotoxin (BoNT) during growth, which is a single polypeptide chain of around 150 kDa. After nicking by bacterial protease (or gastric proteases), it turns into a light chain of 50 kDa and a heavy chain of 100 kDa. Being specific for peripheral nerve endings at the point where a motor neuron stimulates a muscle, BoNT binds to the neuron and prevents the release of acetylcholine across the synaptic cleft. The heavy chain mediates binding to presynaptic receptors via its carboxy terminus and forms a channel through the membrane of the neuron via its amino terminus, facilitating entrance of the light chain. Inside a neuron, the toxin specifically cleaves SNARE (soluble NSF-attachment protein receptors; NSF stands for *N*-ethylmaleimide-sensitive fusion protein) complex proteins, synaptobrevin (VAMP), synaptosomal protein (SNAP-25), and syntaxin, which are involved in neurotransmitter release from synaptic vesicles.[9] This abolishes the ability of affected cells to release a neurotransmitter, thus paralyzing the motor system.

In structural and functional terms, BoNT is similar to tetanus toxin, as both are zinc-dependent endopeptidases that cleave a set of proteins in the synaptic vesicles of neurons involved in the excretion of neurotransmitters. However, while BoNT displays a preference for stimulatory motor neurons at a neuromuscular junction in the peripheral nervous system, causing weakness or flaccid paralysis, tetanus toxin (tetanospasmin) acts on inhibitory motor neurons of the central nervous system, causing rigidity and spastic paralysis.

Clinically, botulism often shows one of the four following manifestations: (i) foodborne botulism, which is caused by the ingestion of foods containing BoNT; (ii) infant botulism, which occurs in infants under one year of age after ingestion of *C. botulinum* spores that then colonize the intestinal tract and produce BoNT; (iii) wound botulism, which is caused by infection of *C. botulinum* in a wound, with BoNT being produced and spread to the body via the bloodstream; and (iv) intestinal colonization of *C. botulinum* in adults, which resembles infant botulism in its etiology.

The paralysis typically starts in ocular and facial nerves and then proceeds to the throat, chest, and extremities, potentially leading to respiratory difficulty and death by asphyxia.

C. perfringens type A causes gas gangrene (myconecrosis), diarrhea, and foodborne illness in humans[20]; whereas type B and type D strains are responsible for fatal enterotoxemia in domestic animals and occasionally in humans,[21] and type C strains induce severe necrotic enteritis in humans. The pathogenicity of type E strains of *C. perfringens* is unclear and has rarely been isolated in humans.[21] On the molecular level, the pathogenic effects of enterotoxins of *C. perfringens* are based on its pore-forming capability in eukaryotic cells. Enterotoxin of *C. perfringens* is a single polypeptide of 35 kDa (containing 319 amino acids), which binds to protein receptors belonging to the claudin family (proteins involved in tight junction structure and function) to form a large complex of 155 kDa. The continuing presence of the toxin in intoxicated cells leads to the formation of another large complex of 200 kDa together with a tight junction protein called occludin. These enterotoxin-containing complexes cause changes in cellular membrane permeability, resulting in death of the cells via either the apoptotic or oncotic pathways due to calcium influx through *C. perfringens* enterotoxin pores.[22]

C. tetani causes tetanus, which is one of the most dramatic and globally prevalent diseases of humans and vertebrate animals. The organism secretes tetanus toxin, which is the second most poisonous substance known, with a human lethal dose of ≈1 ng/kg. The tetanus toxin or tetanospasmin released by *C. tetani* cells migrates along neural paths and blocks the release of neurotransmitters (glycine and gamma-amino butyric acid) from the spinal cord and brainstem that regulate muscle contraction. It binds to nerve cells, penetrates the cytosol, and blocks neurotransmitter release causing spastic paralysis.[23] This leads to continuous muscle contraction, primarily in the neck and jaw muscles (lockjaw), and respiratory failure, with mortality rates ranging from 40% to 78%. There are between 800,000 and 1 million deaths due to tetanus each year, of which about 400,000 are due to neonatal tetanus.[17] Eighty per cent of these deaths occur in Africa and Southeast Asia, and it remains endemic in 90 countries world wide. In South Africa approximately 300 cases occur each year, from 12 to 15 cases are reported each year in Britain, and between 50 and 70 in the United States.[24]

C. difficile is the most common cause of nosocomial, or institution-acquired, diarrhea.[25] The spectrum of disease caused by *C. difficile* is broad, ranging from acute watery diarrhea with abdominal pain, low-grade fever, and leukocytosis, to the major complications of dehydration, hypotension, toxic megacolon, septicemia perforation, and death.[25] Typically, *C. difficile*-associated diarrhea occurs in hospitalized patients following antibiotic treatment; it is debilitating and prolongs hospitalization. However, cases of infection have also been reported in patients outside of the usual affected groups, such as younger people and people not in a hospital or institutional environment.[26] This has been a great cause of concern in the medical community, as the new strains appear to cause a more severe disease. The pathogenicity depends on the production of enterotoxin A and cytotoxin B.[27]

33.1.3 Diagnosis of *Clostridium*

Although some species, such as *C. botulinum* and *C. tetani*, produce well defined and very distinctive symptoms, other species, such as *C. difficile* and *C. perfringens*, tend to cause similar clinical signs and their differentiation relies on the use of laboratory procedures. Various physiological and morphological methods have been traditionally used in the taxonomic determination of *Clostridium*. However, traditional identification methods based on phenotypic properties are laborious and lack discriminatory power, and misidentification may occur occasionally. Progress in molecular biology in the last decade has opened up the possibility of characterizing *Clostridium* at the genomic level.

33.1.3.1 *C. botulinum*

Rapid and accurate detection of *C. botulinum* and BoNTs is vital for implementing early supportive care, such as early administration of antitoxin, leading to a lower fatality rate and a shorter disease in human patients.

Conventional Techniques. The complexity in the *C. botulinum* diagnostics arises from the lack of validation of sensitive and specific in vitro methods. The standard detection and isolation of *C. botulinum* is based on the presumptive identification of the toxigenic strains through the mouse bioassay.[10,28] In the mouse bioassay, mice are injected with culture supernatant of microorganism or with samples from patients exhibiting symptoms of botulism. This technique is highly sensitive and specific but requires specialized facilities and personnel. In addition, it is time consuming, costly, and raises ethical concerns due to its reliance on experimental animals.

Culture of *C. botulinum* requires strict anaerobic conditions. All culture media must be deoxygenated by heating in a boiling water bath or by the continuous flow of an anaerobic gas mixture, and the use of anaerobic jars and anaerobic workstations is a necessity for successful diagnostics.[10] The detection of BoNTs production in vitro is, therefore, a preferred approach for the identification of *C. botulinum*, including its distinction from genetically close relatives such as *C. sporogenes* and *C. novyi*.[29]

Several sample prepreparation steps can be utilized to enhance the isolation of *C. botulinum* from clinical samples (e.g., serum and feces). These include: (i) pretreatment with ethanol to eliminate vegetative bacteria and recover spores; (ii) heating to eliminate nonspore-forming bacteria (i.e., 80°C for 10 min for Group I spores or 60°C for 10–20 min for Group II spores); and (iii) inclusion of lysozyme (5 μg/mL) or other heat-resistant lytic enzymes in the culture medium to facilitate germination of heat-stressed spores.[10]

Routine liquid media for cultivation of *C. botulinum* include chopped-meat–glucose-starch medium; cooked-meat medium; broths containing various combinations of tryptone, peptone, glucose, yeast extract, and trypsin (e.g., tryptone-peptone-glucose–yeast extract medium); reinforced clostridial medium; and fastidious anaerobe broth.[30–32] All of these media are nonselective and thus allow the growth of a range of other bacteria.

Blood agar and egg yolk agar (EYA)[33] are the most common unselective plating media, with EYA enabling the lipase reaction typical of *C. botulinum*. However, many other clostridia produce lipase and may therefore confuse the identification.[34] EYA medium alone does not contain any inhibitory compounds, but when supplemented with cycloserine, sulfamethoxazole, and trimethoprim, EYA-based selective media (e.g., *C. botulinum* isolation medium) have been reported to select Group I *C. botulinum*.[35,36]

A dilemma arising from the different physiologies of Group I and II strains is the selection of the correct incubation temperature. Group I strains grow optimally at 35°C–37°C, whereas Group II strains favor lower temperatures of 25°C–30°C.[37] However, use of replicate samples at 26°C–30°C and at 35°C–37°C is preferable to ensure optimal growth for both groups. The optimal incubation time varies with the sample material and the detection method selected to identify the presence of *C. botulinum* in the culture tube.

Because the prevalence of *C. botulinum* in naturally contaminated samples is generally low (10–1000 spores/kg) and proper selective media are not available, quantification of *C. botulinum* in samples containing other bacteria by use of plate-count procedures is difficult, if not impossible.[10] Therefore, alternative detection methods such as enzyme-linked immunosorbent assay (ELISA)[38] or a mass spectrometric-based endopeptidase method[39] have been developed. Recently, surface plasmon resonance (SPR) sensors have also been applied to the detection of three BoNTs serotypes: A, B, and F.[40]

Molecular Techniques. Polymerase chain reaction (PCR) represents the most popular molecular technique for microbial detection as the sensitivity of PCR can be as low as one copy of the target sequence by using two oligonucleotide primers.

PCR identification of *C. botulinum* often targets the BoNT gene *bot* although other types of toxin genes are also of diagnostic value.[10] Use of multiplex PCR assays enables the simultaneous detection of two or more types of BoNT genes.[41–43] However, PCR detection of the neurotoxin genes does not provide details on the physiological group and epidemiology of *C. botulinum* isolates. For differentiation between Group I (proteolytic) and Group II (nonproteolytic) strains of *C. botulinum*, several molecular typing methods are useful.[44–46] Furthermore, Janda and Whitehouse[47] have developed a multilocus PCR followed by electrospray ionization mass spectrometry to identify *C. botulinum*.

Another strategy is to utilize universal primers that recognize all BoNT genes but the nucleotide diversity among the BoNT gene may present a potential problem.[48] Nonetheless, BoNTs are generated as part of a progenitor toxin complex and a conserved component among serotypes is the nontoxic nonhemagglutinin (NTNH). East and Collins[49] demonstrated that the gene encoding NTNH reveals a high level of similarity, is present in all strains that produce BoNTs, and absent from strains that are nontoxic. Aranda et al.[50] developed a PCR protocol to detect all BoNTs, producing strains using a single set of degenerate primers. This protocol yields a single PCR product in agarose gel, providing a more specific detection than hybridation with a BoNT probe.

Several real-time PCR protocols[51] along with biosensor technology[52] and DNA microarray[53] have also been reported for determination of *Clostridium*. Biosensor technology combines the accuracy and sensitivity of standard approaches with improvement in rapidity of detection. Real-time-PCR, biosensors, and DNA microarray offer the possibility of continuous and real-time monitoring of the environment for the presence of infectious agents, but nowadays they have not been applied as the main techniques for *Clostridium* detection.

One of the problems when using PCR with clinical samples or pathogenic isolates is the possibility of detecting naked DNA derived from dead and degrading cells, which results in false-positives. An approach to ensure that only live cells are detected is reverse-transcription PCR (RT–PCR), where gene expression, rather than the gene itself, is detected. RT–PCR protocols for *C. botulinum* have been described.[54–56]

For genetic characterization of *C. botulinum* isolates, a variety of PCR procedures (PCR–RFLP, RAPD, Q–PCR, Rep–PCR, etc.) are used.[10] Other nonamplified molecular procedures for the characterization of *C. botulinum* consist of DNA sequencing, pulsed-field gel electrophoresis (PFGE), and ribotyping.[10] These methods are only used to type *C. botulinum* isolates, but do not allow detection of this microorganism directly from clinical or food samples.

33.1.3.2 *C. perfringens*

The most accepted criterion in establishing a definitive diagnosis of *C. perfringens* is the detection of its toxins or genes encoding these toxins.[13,57]

Conventional Techniques. On blood agar plates, *C. perfringens* colonies are flat, rough, and translucent, with regular or irregular margins, an inner zone of hemolysis, and an outer zone of less complete clearing. Occasionally, strains without inner zone hemolysis are seen. On Nagler agar containing 5%–10% egg yolk, *C. perfringens* generates a characteristic white precipitate as a result of its alpha toxin interacting with the lipids in egg yolk.[9]

Several solid media have been developed for cultivation and isolation of *C. perfringens* from different sample materials. These include neomycin blood agar, sulfite polymyxin sulfadiazine (SPS) agar, tryptone sulfite neomycin (TSN) agar, Shahidi Ferguson *perfringens* (SFP) agar, oleandomycin polymyxin sulfadiazine *perfringens* (OPSP) agar, and tryptose sulfite cycloserine (TSC) agar with or without egg yolk.[58–61] Isolation of typical colonies from media is followed by confirmation of *C. perfringens* colonies with different tests to reveal phenotypic characteristics. Observation of nitrate reduction, gelatin liquefaction, lactose fermentation, or lack of motility in suspected *C. perfringens* isolates is usually sufficient to distinguish *C. perfringens* from other organisms.[34] Identification of *C. perfringens* may be also carried out using biochemical test kits such as API.[61] Several international method standards are available for confirmation of *C. perfringens* in different samples, including methods published by the Association of Official Analytical Chemists (AOAC), the International Standardization Organization (ISO), and the Nordic Committee on Food Analysis (NCFA).

Given that not all *C. perfringens* isolates present in food or feces produce toxins that are related to diseases or cause food poisoning, the most accepted criterion in establishing a definitive conventional diagnosis is the detection of toxins.[62] Among the techniques for detection of the major toxins of *C. perfringens* are the mouse neutralization test (MNT),[62] ELISAs,[63–65] counterimmnunoelectrophoresis,[66] and latex agglutination test.[67,68]

Molecular Techniques. The PCR offers a reliable diagnostic tool for typing *C. perfringens* and major toxin genes. Many PCR protocols focus on individual *C. perfringens* toxin genes,[21] in particular the *cpe* gene.[69] Multiplex PCR protocols targeting several genes have also been described.[70–73] Baums et al.[74] developed a reliable species-specific multiplex PCR for the detection of the *cpa*, *cpb*, *cpb2*, *cpe*, *etx*, and *iap* genes of *C. perfringens* isolates without DNA purification. Furthermore, Joshy et al.[75] reported a multiplex PCR to simultaneously detect the enterotoxin gene of *C. perfringens* and BoNT genes of *C. botulinum*. In addition to PCR, other approaches such as DNA microarray have also been reported for detecting toxin genes of *C. perfringes*.[13]

Furthermore, transcriptional gene *cpf0765* appears to be uniquely present in *C. perfringens*, offering another valuable target for *C. perfringens*-specific identification. Using primers Cpf0765F (5′-CAAAGGGTTTGGGAATTCTT-3′ corresponding to nt 907739–907759) and Cpf0765R (5′-GAA ATGCAAATATTGTGGAAAAA-3′ corresponding to nt 908206–908228) in PCR, a specific band of 490 bp is detectable from *C. perfringens* only (regardless of its toxinotypes), upon assessment of a collection of bacterial strains consisting of *C. perfringens* (n = 29) of toxinotypes A through E, *C. difficile* (n = 2), *C. chauvoei* (n = 1), *C. septicum* (n = 1), *C. sordelli* (n = 1), and other bacterial strains (n = 44, representing 38 distinct species) (D. Liu, F.W. Austin, C.C. Wu, and L.A. Hanson, unpublished data). Because transcriptional regulator genes play critical roles in the regulation of bacterial responses to changing external conditions, further characterization of the transcriptional regulator gene *cpf0765* and its related protein will yield new details on the molecular mechanisms of *C. perfringens* survival and adaptation in its niche environment.

The PCR–DGGE of the ribosomal DNA is a variation of PCR used for the study of microbial ecology. Denaturing gradient gel electrophoresis (DGGE) is based on the separation of PCR amplicons of the same size but different sequences. This method has been used for the differentiation of *C. perfringens* from other composition of ileal bacterial microbiota of broiler chicken.[76]

The real-time PCR allows the quantification of the copy numbers of plasmid-borne toxin genes.[76] Gurjar et al.[77] designed a dual-labeled fluorescence hybridization probe

(TaqMan®)-based real-time multiplex PCR assay for detection of toxin genes α (*cpa*), β (*cpb*), ι (*iap*), ε (*etx*), $\beta 2$ (*cpb2*), and enterotoxin (*cpe*) of *C. perfringens* directly from cattle feces (Table 33.2). Similarly, three real-time fluorogenic (TaqMan®) multiplex PCR has been used for the correct detection of six toxic genes of *C. perfringens*.[14,78]

Other molecular typing tools for *C. perfringens* include DNA sequencing, pulsed-field gel electrophoresis (PFGE), plasmid profiling, and ribotyping.[79–82] These methods are only used to type *C. perfringens* isolates but do not allow detection of this microorganism directly from clinical or food samples.

33.1.3.3 *C. tetani*

The diagnosis of tetanus relies mainly on clinical observation, although laboratory confirmation is helpful with the detection of toxin in serum or material from infected wounds, and isolation of *C. tetani* in the infected wounds.[83] There have been few molecular techniques for *C. tetani*, probably because, traditionally, the diagnosis of tetanus is made clinically, and tetanus is easily preventable by active immunization with the tetanus toxoid.

Conventional Techniques. Similar to other *Clostridium* spp., it is a challenge to identify *C. tetani* based on anaerobic in vitro culture followed by other phenotypic procedures. The standard detection of *C. tetani* is dependent on the production of toxin and the conventional toxin neutralization test.[84]

Thus, failure to culture *C. tetani* does not rule out the disease because the toxin is contained in a plasmid and the presence of the bacterium does indicate infection, as not all strains possess the plasmid.[24,85] The normal media are nonselective and include chopped-meat–glucose-starch medium and cooked-meat medium at 37°C.[20] On blood agar plates, *C. tetani* colonies produce a zone of β-hemolysis.

The isolation of the tetanus toxin-producing *C. tetani* from the infected wound is useful, but frequently unsuccessful. The same is true for the demonstration of toxin in the serum of the patient.[86]

Laboratory confirmation includes the detection of the toxin in body fluids, usually serum or material from an infected wound, which traditionally relies on the use of mouse bioassay. These tests, although highly sensitive, are slow, labor intensive, and require the use of live experimental animals. As in cases of wound botulism, the laboratory diagnosis of tetanus is problematic. Appropriate samples are not always collected, that is, samples are too small for toxin tests, the serum is not drawn before the administration of antitoxin, and wound samples are not collected before antibiotic therapy; the levels of toxin in serum can be below the

TABLE 33.2
Primers and Probes for Determination of *C. botulinum, C. perfringens, C. tetani,* and *C. difficile* Toxinotypes

Species	Target Genes	Primers and Probes (5′–3′)	Procedure	Reference
C. botulinum	BoNT A, B, E, F	BoNT1·1:CC(C/A)TAT(A/G)TAGG(A/T)C(T/C) TGCTTTAAATATAGG(A/T)A(A/T)T	PCR	50
		BoNT2:TTAGT(T/A)ATAGTTACAAAAATCCA(T/C)(T/C)T(A/G) TTTATATATA		
		ProbeConBoNT:ATAATTC(A/G)GG(A/C)TGG(A/C)TGGAAA(G/A)TATC		
C. perfringens	*Cpa, cpb, cpb2, etx, ia, cpe s*	cpa F: TGCACTATTTTGGAGATATAGATAC	Multiplex real-time PCR	77
		cpa R: CTGCTGTGTTTATTTTATACTGTTC		
		cpa Probe: HEX-TCCTGCTAATGTTACTGCCGTTGA-TAMRA		
		cpb F: ATTTCATTAGTTATAGTTAGTTCAC		
		cpb R: TTATAGTAGTAGTTTTGCCTATATC		
		cpb Probe: HEX-AACGGATGCCTATTATCACCAACT-TAMRA		
		cpb2: TAACACCATCATTTAGAACTCAAG		
		cpb2 R: CTATCAGAATATGTTTGTGGATAAAC		
		cpb2 Probe: HEX-TGCTTGTGCTAGTTCATCATCCCA-TAMRA		
		etx F:TTAACTAATGATACTCAACAAGAAC		
		etx R: GTTTCATTAAAAGGAACAGTAAAC		
		etx Probe: FAM-TGCTTGTATCGAAGTTCCCACAGT-TAMRA		
		ia F: CAAGATGGATTTAAGGATGTTTC		
		ia R: TTTTGGTAATTTCAAATGTATAAGTAG		
		ia Probe: FAM-TTTCATCGCCATTACCTGGTTCAT-TAMRA		
		cpe F: AACTATAGGAGAACAAAATACAATAG		
		cpe R: TGCATAAACCTTATAATATACATATTC		
		cpe Probe: HEX-TCTGTATCTACAACTGCTGGTCCA-TAMRA		
C. tetani	TeNT	TQ Tet1: CCTAGTTTCAAAACTTATTGGCTTATGTAA	Real-time PCR	87
		TQ Tet2: CATAATTCTCCTCCTAAATCTGTTAATGATG		
		Tet Probe: FAM-TATACCACCAACAAATATAAGAG		
C. difficile	*TcdB*	TcdB-1F: GAAGGTGGTTCAGGTCATAC	Real-time PCR	116
		TcdB-1R: CATTTTCTAAGCTTCTTAAACCTG		

level of detection, and the appropriate wound is not always sampled.[87,88]

Molecular Techniques. Laboratory diagnosis of tetanus is enhanced with the development of molecular techniques for the *C. tetani* neurotoxin gene, which is closely related to BoNT in terms of structure and function.[20] Therefore, it is necessary to design primers that do not react with BoNT genes. Several conventional PCR[19,89,90] and real-time PCR protocols have been utilized for detection of the tetanus toxin gene.[19,87] Plourde-Owobi et al.[19] demonstrated the feasibility of using PFGE for the analysis of *C. tetani* strains. DNA microarrays containing species-specific oligonucleotide probes from regions of the 16S rRNA gene have also been applied successfully.[91]

33.1.3.4 *C. difficile*

Most diagnostic methods detect *C. difficile* toxin (CDT) A, with or without CDT B. The limitations of the available testing methods may exact a negative impact on treatment of *C. difficile*. Early initiation of therapy may be critical because some tests have long turnaround times and others have poor sensitivity.

Conventional Techniques. The reference method for the detection of *C. difficile* is cytotoxicity of stool in cell culture that may be neutralized by antisera.[92–94] This test takes from 2–3 days to complete. Some researchers recommend the additional culture of stool for *C. difficile* with confirmation of toxin production; this may improve sensitivity but is time consuming.[95,96]

George et al.[95] in 1979 developed a selective and differential agar that contains cycloserine, cefoxitin, fructose, and egg yolk (CCFA), with and without taurocholate, to facilitate the isolation of *C. difficile* because other selective media for clostridia are inhibitory to the growth of this microorganism. Other culture media are fastidious anaerobic agar with sheep blood (FAASB), *C. difficile* selective agar (CDSA), and *C. difficile* brucella agar (CDBA).[97] CDMN (*C. difficile*, moxalactam, norfloxacin) agar is another media used for the detection of *C. difficile* in food samples.[98,99]

The direct detection of the antigens of toxins A or B, or both, by commercially produced enzyme immunoassays (EIAs) is rapid and simple and is the method most widely applied in clinical laboratories.[100] Several studies reported a ≥90% sensitivity for EIAs.[94] The following assays, which detect CDT A and B, have been used: Meridian Premier (ELISA), TechLab Tox A/B II (ELISA), TechLab Tox A/B Quik Chek (rapid antigen capture), Remel Xpect (rapid chromatographic immunoassay), bioMérieux VIDAS (EIA), and Meridian Immunocard (rapid EIA).[100,101] However, these techniques show variations in terms of sensitivity and specificity.[102,103]

Molecular Techniques. The emergence of the hypervirulent strain has increased interest in *C. difficile* typing and stimulated the application of newer genotype-based methods, such as PCR-ribotyping,[104] amplified fragment length polymorphism (AFLP),[105] multilocus sequence typing (MLST),[106] multilocus variable-number tandem-repeat analysis (MLVA),[107] and surface layer protein A gene sequence typing (slpAST).[108] All techniques are capable of detecting outbreak strains, but Killgore et al.[109] showed that only REA and MLVA gave sufficient discrimination among the strains from different outbreaks. Overall, these assays are more time consuming than PCR assay targeting the *C. difficile* toxin genes *tcdA* and *tcdB*.[110]

PCR assays have been widely used for the detection of *C. difficile* (irrespective of toxin type) and toxigenic *C. difficile*, as they provide an easier method for *C. difficile* identification and verification of the ability to produce toxin.[86] An obstacle encountered with PCR technology is its inability to directly detect *C. difficile* in clinical specimens or feces because of the presence of amplification inhibitors.[111] For this reason, some reports have proposed protocols for on-chip extraction of nucleic acids from clinical samples.[25,112]

Conventional PCR methods targeting the 16S rRNA, toxin A (*tcd*A), toxin B (*tcd*B), and toxin C (*tcdC*) gene have been described.[86,99,100,113,114] Further improvement relates to the development of multiplex PCR assays that enable the simultaneous detection of four of toxins genes.[115]

Rapid real-time PCR methods used directly on stool specimens have been described for the diagnosis of *C. difficile* infection.[100,110,116,117] These assays target the genes associated with toxin B, toxin A and B, or toxin C genes and display high concordance with the results from either cytotoxicity of tissue cell lines or toxigenic cultures.[27,116,117] A real-time PCR assay (BD GeneOhm Cdiff) has become commercially available, and it is a more rapid and sensitive (less than 3 h) method than the cytotoxicity assays for the detection of toxigenic *C. difficile* in stool samples.[118,119]

A variety of nonamplified and amplified genetic typing techniques has been also applied for typing international epidemic strains of *C. difficile*: restriction endonuclease analysis, PFGE, PCR-ribotyping, multilocus sequence typing, multilocus variable-number tandem-repeat analysis, amplified fragment length polymorphism, and surface layer protein A gene sequence typing.[109,120–122]

Typing methods began to evolve from phenotype-based methods toward genotype-based methods, such as restriction endonuclease analysis (REA) of the total bacterial genome,[123] pulsed-field gel electrophoresis (PFGE),[105,124] and arbitrarily primed PCR (AP-PCR).[125]

33.2 METHODS

33.2.1 Sample Preparation

DNA Extraction. For detection of *C. botulinum*, *C. perfringens,* and *C. tetani*, clinical or food samples (2–10 g) are homogenized with 90 mL of TYPCG broth containing trypticase (30 g/L), yeast extract (12 g/L), peptone (10 g/L), cysteine (0.5 g/L), glucose (5 g/L), sodium chloride (5 g/L), starch (1 g/L), and sodium acetate (3 g/L) for 2 min with a Stomacher. For the detection of *C. difficile,* 2–10 g of samples are homogenized in 90 mL of CCF broth[126] containing peptone (40 g/L), sodium phosphate (5 g/L), potassium phosphate (1 g/L), sodium chloride (2 g/L), magnesium sulfate (0.1 g/L), fructose (6 g/L), cycloserine (500 mg/L), and cefoxitin (15 mg/L).

The homogenized suspensions are incubated in anaerobic conditions at 37°C for 18 h for *C. botulinum*, *C. perfringens*, and *C. tetani*, and at 35°C for 18 h for *C. difficile* before DNA recovery (Figure 33.1).

Cells are harvested from 1.5 mL of TYPCG broth by centrifugation at 9840 × *g* for 10 min. The pelleted cells are suspended in 500 μL of TES (50 mM Tris–Cl, 5 mM EDTA, 50 mM NaCl) buffer containing 2 mg/mL lysozyme. To remove proteins and RNA, 25 μL proteinase K (10 mg/mL) and 25 μL RNAase (10 mg/mL) are added, and the mixture is heated at 65°C for 30 min. Then, 50 μL 20% sodium dodecyl sulfate (SDS) are added. DNA is further purified by using phenol–chloroform–isoamyl alcohol (25:24:1) and precipitated with ethanol. The final pellet is dissolved in 50 mL TE buffer. For PCR, 10 μL of this DNA is used (Figure 33.1).

Alternatively, clostridial DNA is collected from presumptive colonies isolated in cultures media for PCR using a boiling method. Namely, one to two colonies are suspended in 100 μL of water, heated at 95°C for 5 min, and centrifuged at 13,000 × *g* for 5 min, after which 1–5 μL is added to the PCR mixture.

33.2.2 Detection Procedures

33.2.2.1 PCR for *C. botulinum*

Principle. A conventional PCR method was developed by Aranda et al.[50] for the detection of BoNT genes of *C. botulinum*, and *C. butyricum* and *C barati* producing BoNT E and F. This method employs two degenerate primers from the conserved region of BoNT genes and detects down to 1 CFU/g of sample (Table 33.2).

Procedure

1. Prepare the PCR mixture in 50 μL volume containing 1 μL of a deoxyribonucleotide mixture (10 mM each), 1 μL of each primer (20 mM), 10 μL of the extracted DNA (50–300 ng) and 1 U of Taq DNA polymerase.
2. Run the following cycling program in a PCR machine: one cycle of 94°C for 3 min; 29 cycles of

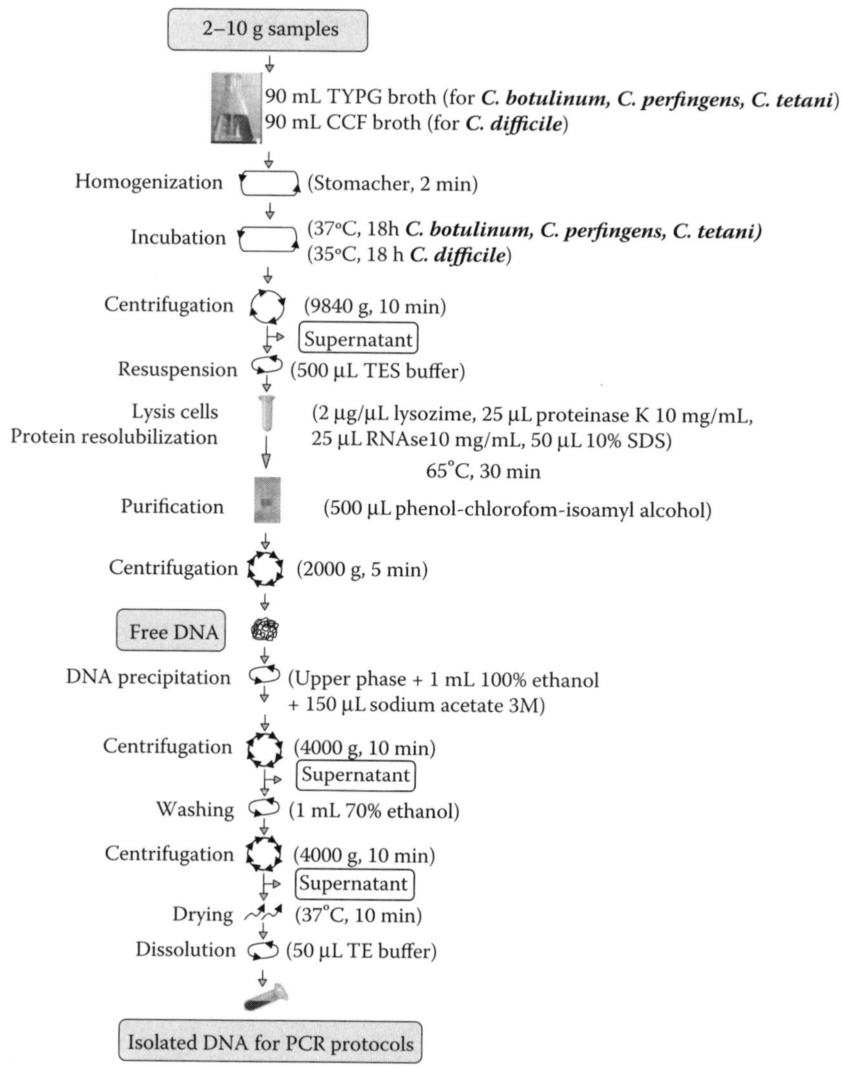

FIGURE 33.1 Protocol for DNA isolation of *C. botulinum*, *C. perfringens*, *C. tetani*, and *C. difficile* for PCR applications.

92°C for 1 min, 47°C for 1 min, and 58°C for 5 min. A final extension is performed at 58°C for 5 min.
3. Separate the PCR products on a 1% agarose gel containing 0.5 μg/mL ethidium bromide and visualized under a UV light box and photograph. A product of 1.1 kb is expected for all types of *C. botulinum*.
4. Alternatively the PCR products are detected by a specific DNA probe, Con.BoNT (Table 33.2). Namely, DNA amplicons from BoNT genes are denatured by boiling for 5 min in 1× SSC (0.15 mM NaCl and 0.015 mM trisodium citrate). The denatured DNA fragments are transferred to a Nylon membrane Hybond H+ (Amershan) with a slot blot apparatus and fixed in 0.4 mM NaOH. The blot is hybridized with probe Con.BoNT labeled with fluorescein-dUTP for 2 h and washed in 1× SSC – 0.1% SDS (w/v) at 42°C for 30 min. The hybridizations are detected using an enhanced chemiluminescence kit (Amersham), exposing the blot to X-ray film for 5 min, though other hybridization detection system could be used.

33.2.2.2 Multiplex Real-Time PCR for *C. perfringens*

Principle. A multiduplex real-time PCR was reported previously for the detection of the toxin genes encoding for the toxins of *C. perfringens*: alpha, beta, beta2, epsilon, iota, and enterotoxin, with primers from the underlying toxin genes *cpa*, *cpb*, *cpb2*, *etx*, *iap*, and *cpe*[77] (Table 33.2). Three duplex real-time PCR reactions are performed for the following toxin genes: reaction 1: *cpa* + *cpb2*; reaction 2: *cpb* + *etx*; reaction 3: *iap* + *cpe*.

Procedure

1. Prepare three duplex PCR reaction mix in 24 μL containing each one 5.5 μL H_2O, 12.5 μL of 2× Brilliant Multiplex QPCR Master Mix (Stratagene), 1 μL of forward and reverse primers (10 nM) of respective toxin genes, 1 μL (100 nM) of Taqman hybridization probe for each toxin gene, and 1 μL of target DNA template was added to each reaction.
2. Run the following cycling conditions in a real-time PCR machine: initial 95°C for 10 min and 40 cycles of 95°C for 30 s and 55°C for 1 min.
3. PCR products are detected by monitoring the increase in fluorescence due to release of reporter dyes from the probes due to the thermodependent 5′exonuclease activity of Taq polymerase at 515 and 556 nm wavelength during the annealing step. Positive reactions are assigned a cycle threshold (Ct) value equal to the cycle number where the fluorescence observed in the experimental tube was first detectable above background. A negative result was assigned when no amplification occurred or the Ct value higher than 36 cycles.

33.2.2.3 Real-Time PCR for *C. tetani*

Principle. In routine practice, for detection of *C. tetani* it is necessary to use molecular methods because the toxins are contained in a plasmid and the presence of the bacterium does not indicate infection. The real-time PCR is the most sensitive and specific method for detection of this microorganism. For development of the real-time PCR assay for the tetanus toxin gene (tetX), primers and probe of Akbulut et al.[87] are used (Table 33.2).

Procedure

1. Prepare the PCR mixture containing 25 μL of 1× qPCR Master Mix, 0.1 μM concentrations of fluorescent probe, 0.3 μM concentrations each of the appropriate reverse and forward oligonucleotide primers, 5 μL of DNA template, and distilled water to 50 μL.
2. Run the following cycling program in a real-time PCR machine: one cycle of 2 min at 55°C and 10 min at 95°C, followed by 40 cycles of 95°C for 15 s and finally by 60°C for 1 min.
3. PCR products are detected like in Section 33.2.2.2. Samples are designated as positive when Ct value of minor 35 is obtained.

33.2.2.4 Real-Time PCR for *C. difficile*

Principle. A clinically validated real-time PCR test was described by Peterson et al.[116] for the sensitive and specific detection of toxin A-negative, toxin B-positive, and toxin A and B–positive *C. difficile*. Primers used target a region of the toxin B gene (*tcd*B) that is highly conserved among multiple strains of toxigenic *C. difficile*[116] (Table 33.2).

Procedure

1. Prepare the PCR mixture containing in each 20 μL reaction 2 μL of sample, 2 μL of 10× Fast Start DNA Master SYBR Green mix, 5 mM $MgCl_2$ (final concentration), 2% DMSO and 0.4–0.5 μM of each primer.
2. Run the following cycling program in a real-time PCR machine: initial denaturing at 95°C for 10 min, followed by amplification for 40 cycles of 3 s at 95°C, 5 s at 62°C, 20 s at 72°C, and 5 s at 76°C with fluorescence acquisition.
3. A Melt Curve is performed after amplification consisting of 60 s at 95°C, 60 s at 65°C, and a stepwise temperature increase of 0.2°C per s until reaching 90°C, with fluorescence acquisition at each temperature transition.
4. For analysis amplification Ct and confirmation of melt curve will determine a positive result.

33.3 CONCLUSION

Species of genus *Clostridium* are widely present in the environment (e.g., soil, sewage, and marine sediments) and in the gastrointestinal tract of humans and domestic and feral animals. Although most *Clostridium* species are saprophytes, 34 species have been considered pathogenic to man and animals. These pathogenic species often secrete an array of invasins, exotoxins, and other extracellular enzymes and various toxins that are responsible for such human diseases as tetanus, botulism, gas gangrene, foodborne illnesses, and diarrhea. Although some species, such as *C. botulinum* and *C. tetani* symptoms, are well defined and very distinctive, other species, such as *C. difficile* and *C. perfringens*, would exhibit similar symptoms, and molecular differentiation is needed. Different physiological and morphological methods have been traditionally used in taxonomic differentiation of *Clostridium*. However, several studies have shown that traditional identification methods based on phenotypic properties are laborious and lack discriminatory power, and misidentification occurs frequently. Progress in the molecular biology in the last decade has opened up the possibility of characterizing *Clostridium* at the genomic level. The application of molecular procedures has greatly improved laboratory diagnostics of these important bacterial pathogens. We review the main conventional methods, as well as the molecular techniques for detecting the main human pathogenic *Clostridium* species (i.e., *C. botulinum* and other botulinum neurotoxin-producer species such as some strains of *C. barati* and *C. butyricum*; *C. perfringens*; *C. tetani*, and *C. difficile*). Sample processing and molecular methods for their rapid detection and identification are presented, such as a PCR for *C. botulinum* encoding BoNT A, B, E, and F, a multiduplex real-time PCR for the detection of the toxin genes encoding for the toxins of *C. perfringens*, a real-time PCR assay for the tetanus toxin gene, and a real-time PCR test for detection of toxigenic *C. difficile*. There is no doubt that continuing refinement in the sample-processing procedures and automation in the target amplification and detection will make molecular identification of *C. botulinum*, *C. perfringens*, *C. tetani*, and *C. difficile* an indispensable tool in clinical testing laboratories, and further strengthen our capacity in the tracking and prevention of illnesses resulting from these pathogenic clostridia.

ACKNOWLEDGMENTS

This work was funded by the Spanish Comisión Interministerial de Ciencia y Tecnología (project Carnisenusa CSD2007–00016, Consolider Ingenio 2010).

REFERENCES

1. Berezima, O.V. et al., Extracellular glycosyl hydrolase activity of the *Clostridium* strains producing acetone, butanol, and ethanol, *Appl. Biochem. Microbiol.*, 44, 42, 2008.
2. Collins, M.D. et al., The phylogeny of the genus *Clostridium*: Proposal of five new genera and eleven new species combination, *Int. J. Syst. Bacteriol.*, 44, 812, 1994.
3. Stackebrandt, E. et al., Phylogenetic basis for a taxonomic dissection of the genus *Clostridium*, *FEMS Inmunol. Medical Microbiol.*, 24, 253, 1999.
4. Gupta, R.S., and Gao, B., Phylogenomic analyses of clostridia and identification of novel protein signatures that are specific to the genus *Clostridium sensu stricto* (Cluster I), *Int. J. Syst. Evol. Microbiol.*, 59, 285, 2009.
5. Lauro, F.M., Bertoloni, G., and Obraztsova, A., Pressure effects on *Clostridium* strains isolated from a cold deep-sea environment, *Extremophiles*, 8, 169, 2004.
6. Madigan, M.T. et al., *Biology of Microorganisms*, 8th ed., Prentice Hall, Englewood Cliffs, NJ, 1997.
7. Hill, K.K. et al., Genetic diversity among botulinum neurotoxin-producing clostridial strains, *J. Bacteriol.*, 189, 818, 2007.
8. Chen, Y. et al., Sequencing the botulinum neurotoxin gene and related genes in *Clostridium botulinum* type E strains reveals orfx3 and a novel type E neurotoxin subtype, *J. Bacteriol.*, 189, 8643, 2007.
9. Heikinheimo, A. et al., *Clostridium.* In Liu, D. (Editor), *Molecular Detection of Foodborne Pathogens*, pp. 145–155, Taylor and Francis, Boca Raton, FL, 2009.
10. Lindström, M., and Korkeala, H., Laboratory diagnostics of botulism, *Clin. Microbiol. Rev.*, 19, 298, 2006.
11. Hinderink, K., Lindström, M., and Korkeala, H., Group I *Clostridium botulinum* strains show significant variation in growth at low and high temperature, *J. Food Prot.*, 72, 375, 2008.
12. Sebaihia, M. et al., Genome sequence of a proteolytic (Group I) *Clostridium botulinum* strain Hall A and comparative analysis of the clostridial genomes, *Genome Res.*, 17, 1082, 2007.
13. Al-Khaldi, S.F. et al., Identification and characterization of *Clostridium perfringens* using single target DNA microarray chip, *Int. J. Food Microbiol.*, 91, 289, 2004.
14. Albini, S. et al., Real-time multiplex PCR assays for reliable detection of *Clostridium perfringens* toxin genes in animal isolates, *Vet. Microbiol.*, 127, 179, 2008.
15. Songer, J.G., and Uzal, F.A., Clostridial enteric infections in pigs, *J. Vet. Diagn. Invest.*, 17, 528, 2005.
16. Smedley, J.G. et al., The enteric toxins of *Clostridium perfringens*, *Rev. Physiol. Biochem. Pharmacol.*, 152, 183, 2005.
17. Farrar, J.J. et al., Neurological aspects of tropical disease: Tetanus, *J. Neurol. Neurosur. Ps.*, 69, 292, 2000.
18. Hara, T., Matsuda, M., and Yoneda, M., Isolation and some properties of nontoxigenic derivatives of a strain of *Clostridium tetani*, *J. Res. Inst. Microb. Dis.*, 20, 105, 1977.
19. Plourde-Owobi, L. et al., Molecular characterization of *Clostridium tetani* strains by pulsed-field gel electrophoresis and colony PCR, *Appl. Environ. Microbiol.*, 71, 5604, 2005.
20. Hatheway, C.L., Toxigenic clostridia, *Clin. Microbiol. Rev.*, 3, 66, 1990.
21. Wu, J. et al., Detection and toxin typing of *Clostridium perfringens* in formalin-fixed, paraffin-embedded tissue samples by PCR, *J. Clin. Microbiol.*, 47, 807, 2009.
22. Smedley, J.G. et al., Identification of a prepare large-complex stage in the mechanism of action of *Clostridium perfringens* enterotoxin, *Infect. Immun.*, 75, 2381, 2007.
23 Herr, A.E. et al., On-chip native gel electrophoresis-based immunoassays for tetanus antibody and toxin, *Anal. Chem.*, 77, 585, 2005.
24. Cook, T.M., Protheroe, R.T., and Handel, J.M., Tetanus: A review of the literature, *Br. J. Anaesth.*, 87, 477, 2001.
25. Gillers, S. et al., Microscale sample preparation for PCR of *C. difficile* infected stool, *J. Microbiol. Methods*, 78 203, 2009.
26. Warny, M. et al., Toxin production by an emerging strain of *Clostridium difficile* associated with outbreaks of severe disease in North America and Europe, *Lancet*, 366, 1079, 2005.
27. Wolff, D., Brüning, T., and Gerritzen, A., Rapid detection of the *Clostridium difficile* ribotype 027 tcdC gene frame shift mutation at position 117 by real-time PCR and melt curve analysis, *Eur. J. Clin. Microbiol. Infect. Dis.*, 28, 959, 2009.

28. Hatheway, C.L., and Johnson E.A., *Clostridium*: The spore bearing anaerobes. In Balows, A., and Duerden, B.I. (Editors), *Microbiology and Microbial Infections*, 9th ed., pp. 731–782, Topley and Wilson's, Arnold, London, 1998.
29. Collins, M.D., and East, A.K., Phylogeny and taxonomy of the food-borne pathogen *Clostridium botulinum* and its neurotoxins, *J. Appl. Microbiol.*, 84, 5, 1998.
30. Lilly, T. et al., An improved medium for detection of *Clostridium botulinum* type E, *J. Milk Food Technol.*, 34, 492, 1971.
31. Quagliaro, D.A., An improved cooked meat medium for the detection of *Clostridium botulinum*, *J. AOAC*, 60, 563, 1977.
32. Saeed, E.M.A., Studies on isolation and identification of *Clostridium botulinum* investigating field samples specially from equine grass sickness cases. Doctoral thesis, University of Göttingen, Göttingen, Germany, 2005.
33. Hauschild, A.H.W., and Hilsheimer, R., Enumeration of *Clostridium botulinum* spores in meats by a pour-plate procedure, *Can. J. Microbiol.*, 23, 829, 1977.
34. Cato, E., George, W.L., and Finegold, S., Genus *Clostridium*. In Sneath, P.H.A. et al. (Editors), *Bergey's Manual of Systematic Bacteriology*, pp. 1141–1200, Williams and Wilkins, Baltimore, MD, 1986.
35. Mills, D.C., Midura, T.F., and Arnon, S.S., Improved selective medium for the isolation of lipase-positive *Clostridium botulinum* from feces of human infants, *J. Clin. Microbiol.*, 2, 947, 1985.
36. Silas, J.C. et al., Selective and differential medium for detecting *Clostridium botulinum*, *Appl. Environ. Microbiol.*, 50, 1110, 1985.
37. Smith, L.D.S., and Sugiyama, H., *Botulism. The Organism, Its Toxins, the Disease*, Charles C. Thomas, Springfield, IL, 1988.
38. Ferreira, J.L. et al., Detection of botulinal neurotoxins A, B, E, and F by amplified enzyme-linked immunosorbent assay: Collaborative study, *J. AOAC Int.*, 86, 314, 2003.
39. Kalb, S.R. et al., The use of Endopep–MS for the detection of botulinum toxins A, B, E, and F in serum and stool samples, *Anal. Biochem.*, 351, 84, 2006.
40. Ladd, J. et al., Detection of botulinum neurotoxins in buffer and honey using a surface plasmon resonance (SPR) sensor, *Sens. Actuators B*, 130, 129, 2008.
41. Lindström, M.R. et al., Multiplex PCR assay for detection and identification of *Clostridium botulinum* types A, B, E, and F in food and fecal material, *Appl. Environ. Microbiol.*, 67, 5694, 2001.
42. Kasai, Y. et al., Quantitative duplex PCR of *Clostridium botulinum* types A and B neurotoxin genes, *J. Food Hyg. Soc. Jpn.*, 48, 19, 2007.
43. Sakuma, T. et al., Rapid and simple detection of *Clostridium botulinum* types A and B by loop-mediated isothermal amplification, *J. Appl. Microbiol.*, 106, 1252, 2009.
44. Hyytiä, E. et al., Characterisation of *Clostridium botulinum* groups I and II by randomly amplified polymorphic DNA analysis and repetitive element sequence-based PCR, *Int. J. Food Microbiol.*, 48, 179, 1999.
45. Paul, C. et al., A unique restriction site in the flaA gene allows rapid differentiation of group I and group II *Clostridium botulinum* strains by PCR-Restriction fragments length polymorphism analysis, *J. Food Prot.*, 70, 2133, 2007.
46. Dahlsten, E. et al., PCR assay for differentiating between Group I (proteolytic) and Group II (nonproteolytic) strains of *Clostridium botulinum*, *Int. J. Food Microbiol.*, 124, 108, 2008
47. Janda, J.M., and Whitehouse, C.A., Usefulness of multilocus polymerase chain reaction followed by electrospray ionization mass spectrometry to identify a diverse panel of bacterial isolates, *Diagn. Microbiol. Infec. Dis.*, 63, 403, 2009.
48. Hauser, D. et al., Botulinal neurotoxin C1 complex genes, clostridial neurotoxin homology, and genetic transfer in *Clostridium botulinum*, *Toxicon*, 33, 515, 1995.
49. East, A.K., and Collins, M.D., Conserved structure of genes encoding components of botulinum neurotoxin complex M and the sequence of the gene coding for the nontoxic component in nonproteolytic *Clostridium botulinum* type F, *Curr. Microbiol.*, 29, 69, 1994.
50. Aranda, E. et al., Detection of *Clostridium botulinum* types A, B, E, and F in foods by PCR and DNA probe, *Lett. Appl. Microbiol.*, 25, 186, 1997.
51. Raphael, B.H., and Andreadis, J.D., Real-time PCR detection of the nontoxic nonhemagglutinin gene as a rapid screening method for bacterial isolates harboring the botulinum neurotoxin (A–G) gene complex, *J. Microbiol. Methods*, 71, 343, 2007.
52. Dover, J.E. et al., Recent advances in peptide probe-based biosensors for detection of infectious agents, *J. Microbiol. Meth.*, 78, 10, 2009.
53. Yoo, M. et al., High-throughput identification of clinically important bacterial pathogens using DNA microarray, *Mol. Cell. Probe*, 23,171, 2009.
54. Lövenklev, M. et al., Relative neurotoxin gene expression in *Clostridium botulinum* type B, determined using quantitative reverse transcription-PCR, *Appl. Environ. Microbiol.*, 70, 2919, 2004.
55. McGrath, S., Dooley, J.S.G., and Haylock, R.W., Quantification of *Clostridium botulinum* toxin gene expression by competitive reverse transcription-PCR, *Appl. Environ. Microbiol.*, 66, 1423, 2000.
56. Chen, Y. et al., Quantitative real-time reverse transcription-PCR analysis reveals stable and prolonged neurotoxin cluster gene activity in a *Clostridium botulinum* type E strain at refrigeration temperature, *Appl. Environ. Microbiol.*, 74, 6132, 2008.
57. Uzal, F.A., Diagnosis of *Clostridium perfringens* intestinal infections in sheep and goats, *Anaerobe*, 10, 135, 2004.
58. Harmon, S.M., Kautter, D.A., and Peeler, J.T., Improved medium for enumeration of *Clostridium perfringens*, *Appl. Microbiol.*, 22, 688, 1971.
59. Hauschild, A.H., and Hilsheimer, R., Enumeration of foodborne *Clostridium perfringens* in egg yolk-free tryptose-sulfite-cycloserine agar, *Appl. Microbiol.*, 27, 521, 1974.
60. Hauschild, A.H., Hilsheimer, R., and Griffith, D.W., Enumeration of fecal *Clostridium perfringens* spores in egg yolk-free tryptose-sulfite-cycloserine agar, *Appl. Microbiol.*, 27, 527, 1974.
61. Labbé, R., *Clostridium perfringens*. In Lund, B., Baird-Parker, T. and Gould, G. (Editors), *The Microbiological Safety and Quality of Food*, pp. 1110–1135, Aspen Publishers, Gaithersburg, MD, 2000.
62. Sterne, M., and Batty, I., *Pathogenic Clostridia*, Butterworths, London, 1975.
63. Piyankarage, R.H. et al., Sandwich enzyme-linked immunosorbent assay by using monoclonal antibody for detection of *Clostridium perfringens* enterotoxin, *J. Vet. Med. Sci.*, 61, 45, 1999.
64. El Idrissi, A.H., and Ward, G.E., Development of double sandwich ELISA for *Clostridium perfringens* beta and epsilon toxins, *Vet. Microbiol.*, 31, 89, 1992.

65. Nagahama, M. et al., Enzyme-linked immunosorbent assay for rapid detection of toxins from *Clostridium perfringens*, *FEMS Microbiol. Lett.*, 84, 41, 1991.
66. Petit, L., Gibert, M., and Popoff, M.R., Detection of enterotoxin of *Clostridium perfingens.* In Batt, C.A., Patel, P., and Robinson, R.K. (Editors), *Encyclopedia of Food Microbiology*, pp. 438–444, Academic Press, London, 1999.
67. Marks, S.L. et al., Evaluation of methods to diagnose *Clostridium perfringens* associated diarrhea in dogs, *J. Am. Vet. Med. Assoc.*, 214, 357, 1999.
68. Berry, P.R. et al., Evaluation of ELISA, RPLA, and Vero cell assays for detecting *Clostridium perfringens* enterotoxin in faecal specimens, *J. Clin. Pathol.*, 41, 458, 1988.
69. Nakamura, M. et al., PCR identification of the plasmid-borne enterotoxin gene (cpe) in *Clostridium perfringens* strains isolated from food poisoning outbreaks, *Int. J. Food Microbiol.*, 294, 261, 2004.
70. Meer, R., and Songer, G., Multiplex polymerase chain reaction assay for genotyping *Clostridium perfringens*, *Am. J. Vet. Res.*, 58, 702, 1997.
71. Heikinheimo, A., and Korkeala, H., Multiplex PCR assay for toxinotyping *Clostridium perfringens* isolates obtained from Finnish broiler chickens, *Lett. Appl. Microbiol.*, 40, 407, 2005.
72. Uzal, F.A., and Songer J.G., Diagnosis of *Clostridium perfringens* intestinal infections in sheep and goats, *J. Vet. Diag. Invest.*, 20, 253, 2008.
73. Goncuoglu, E.M. et al., Molecular typing of *Clostridium perfringens* isolated from turkey meta by multiplex PCR, *Lett. Appl. Microbiol.*, 47, 31, 2009.
74. Baums, C.G. et al., Diagnostic multiplex PCR for toxin genotyping of *Clostridium perfringens* isolates, *Vet. Microbiol.*, 100, 11, 2004.
75. Joshy, L., Chaurdhry, R., and Chandel, D.S., Multiplex PCR for the detection of *Clostridium botulinum* and *C. perfringens* toxin genes, *Indian J. Med. Res.*, 128, 206, 2008.
76. Feng, Y. et al., Identification of changes in the composition of ileal bacterial microbiota of broiler chickens infected with *Clostridium perfringens*, *Vet. Microbiol.*, 140, 2009.
77. Gurjar, A.A. et al., Real-time multiplex PCR assay for rapid detection and toxintyping of *Clostridium perfringens* toxin producing strains in feces of dairy cattle, *Mol. Cell. Probe.*, 22, 90, 2008.
78. Abildgaard, L. et al., Sequence variation in the a-toxin encoding plc gene of *Clostridium perfringens* strains isolated from diseased and healthy chickens, *Vet. Microbiol.*, 136, 293, 2009.
79. Ridell, J. et al., Prevalence of the enterotoxin gene and clonality of *Clostridium perfringens* strains associated with food-poisoning outbreaks, *J. Food Prot.*, 61, 240, 1998.
80. Maslanka, S.E. et al., Molecular subtyping of *Clostridium perfringens* by pulsed-field gel electrophoresis to facilitate food-borne-disease outbreak investigations, *J. Clin. Microbiol.*, 37, 2209, 1999.
81. Eisgruber, H., Wiedmann, M., and Stolle, A., Use of plasmid profiling as a typing method for epidemiologically related *Clostridium perfringens* isolates from food poisoning cases and outbreaks, *Lett. Appl. Microbiol.*, 20, 290, 1995.
82. Schalch, B. et al., Ribotyping for strain characterization of *Clostridium perfringens* isolates from food poisoning cases and outbreaks, *Appl. Environ. Microbiol.*, 63, 3992, 1997.
83. Turillazzi, E. et al., An immunohistochemical study on a tetanus fatal case using toxin fragment C (TTC). Should it be a useful diagnostic tool? *Neuropathology*, 29, 68, 2009.
84. Davis, B.D., *Microbiology: Including Immunology and Molecular Genetics*, Harper & Row Hagerstown, MD, 1980.
85. Greene, C.E., *Tetanus*. In Greene, C.E. (Editor), *Infectious Diseases of the Dog and Cat*, W.B. Saunders Co, Philadelphia, 1998.
86. Kato, N., and Kato, H., Molecular detection and identification of anaerobic bacteria. *J. Infect. Chemother.*, 3, 5, 1997.
87. Akbulut, D., Grant, K.A., and McLauchlin, J., Improvement in laboratory diagnosis of wound botulism and tetanus among injecting illicit-drug; users by use of real time PCR assay for neurotoxin gene fragments, *J. Clin. Microbiol.*, 43, 4342, 2005.
88. Goonetilleke, A., and Harris, J.B., Clostridial neurotoxins, *J. Neurol. Neurosur. Ps.*, 75, 35, 2004.
89. Nagao, K. et al., Detection of the tetanus toxin gene by polymerase chain reaction: A case study, *Jpn. J. Infect. Dis.*, 60, 149, 2007.
90. Campbell, J.I. et al., Microbiologic characterization and antimicrobial susceptibility of *Clostridium tetani* isolated from wounds of patients with clinically diagnosed tetanus, *Am. J. Trop. Med. Hyg.*, 80, 827, 2009.
91. Franke-Whittle, I.H., Klammer, S.H., and Insam, H., Design and application of an oligonucleotide microarray for the investigation of compost microbial communities, *J. Microbiol. Methods*, 62, 37, 2005.
92. Bartlett, J.G., Clinical practice. Antibiotic-associated diarrhea, *N. Engl. J. Med.*, 346, 334, 2002.
93. Poutanen, S.M., and Simor, A.E., *Clostridium difficile*-associated diarrhea in adults, *CMAJ*, 171, 51, 2004.
94. Musher, D.M. et al., Detection of *Clostridium difficile* toxin: Comparison of enzyme immunoassay results with results obtained by cytotoxicity assay, *J. Clin. Microbiol.*, 45, 2737, 2007.
95. George, W.L. et al., Selective and differential medium for isolation of *Clostridium difficile*, *J. Clin. Microbiol.*, 9, 214,1979.
96. Bartlett, J.G., and Gerding, D.N., Clinical recognition and diagnosis of *Clostridium difficile* infection, *Clin. Infect. Dis.*, 46, S12, 2008.
97. Nerandzic, M.M., and Donskey, C.J., Effective and reduced-cost modified selective medium for isolation of *Clostridium difficile*, *J. Clin. Microbiol.*, 47, 397, 2009.
98. Rodriguez-Palacios, A. et al. *Clostridium difficile* in retail ground meat, Canada. *Emerg Infect. Dis.*, 13, 485, 2007.
99. Bakri, M.M. et al., *Clostridium difficile* in ready-to-eat salads, Scotland, *Emerg. Infect. Dis.*, 15, 817, 2009.
100. Sloan, L.M. et al., Comparison of real-time PCR for detection of the tcdC gene with four toxin immunoassays and culture in diagnosis of *Clostridium difficile* infection, *J. Clin. Microbiol.*, 46, 1996, 2008.
101. Planche, T. et al., Diagnosis of *Clostridium difficile* infection by toxin detection kits: A systematic review, *Lancet Infect. Dis.*, 8, 777, 2008.
102. Eltringham, I.J., Diagnosis of *Clostridium difficile* infection by toxin detection kits, *Lancet Infect. Dis.*, 9, 141, 2008.
103. Ould-Hocine, Z.F. et al., Comparison of three enzyme immunoassays for detection of *Clostridium difficile* toxins A and B, *Pathol. Biol.*, 56, 424, 2008.
104. Sadeghifard, N. et al., The mosaic nature of Iitergenic 16S-23S rRNA spacer regions suggests rRNA operon copy number variation in *Clostridium difficile* strains, *Appl. Environ. Microbiol.*, 72, 7311, 2006.
105. Klaassen, C.H.W., van Haren, H.A., and Horrevorts, A.M., Molecular fingerprinting of *Clostridium difficile* isolates: Pulsed-field gel electrophoresis versus amplified fragment length polymorphism, *J. Clin. Microbiol.*, 40, 101, 2002.

106. Lemee, L. et al., Multilocus sequence typing analysis of human and animal *Clostridium difficile* isolates of various toxigenic types, *J. Clin. Microbiol.*, 42, 2609, 2004.
107. Van den Berg, R.J. et al., Typing and subtyping of *Clostridium difficile* isolates by using multiple-locus variable-number tandem-repeat analysis, *J. Clin. Microbiol.*, 45, 1024, 2007.
108. Kato, H., Yokoyama, T., and Arakawa, Y., Typing by sequencing the slpA gene of *Clostridium difficile* strains causing multiple outbreaks in Japan, *J. Med. Microbiol.*, 54, 167, 2005.
109. Killgore, G. et al., Comparison of seven techniques for typing international epidemic strains of *Clostridium difficile*: Restriction endonuclease analysis, pulsed-field gel electrophoresis, PCR-ribotyping, multilocus sequence typing, multilocus variable-number tandem-repeat analysis, amplified fragment length polymorphism, and surface layer protein A gene sequence typing, *J. Clin. Microbiol.*, 46, 431, 2008.
110. Belanger, S.D. et al., Rapid detection of *Clostridium difficile* in feces by real-time PCR, *J. Clin. Microbiol.*, 41, 730, 2003.
111. Clarke, W.C., and Dufour, D.R., *Contemporary Practice in Clinical Chemistry*, AACC Press, Washington, DC, 2006.
112. Bjerketorp, J. et al., Rapid lab-on-a-chip profiling of human gut bacteria, *J. Microbiol. Methods*, 72, 82, 2008.
113. Lemee, L. et al., Multiplex PCR targeting tpi (triose phosphate isomerase), tcdA (toxin A), and tcdB (toxin B) genes for toxigenic culture of *Clostridium difficile*, *J. Clin. Microbiol.*, 42, 5710, 2004.
114. Pituch, H. et al., Detection of binary-toxin genes (cdtA and cdtB) among *Clostridium difficile* strains isolated from patients with *C. difficile*-associated diarrhoea (CDAD) in Poland, *J. Med. Microbiol.*, 54, 143, 2005.
115. Persson, S., Torpdahi, M., and Olsen, K.E.P., New multiplex PCR method for the detection of *Clostridium difficile* toxin A (tcdA) and toxin B (tcdB) and the binary toxin (cdtA/cdtB) genes applied to a Danish strain collection, *Clin. Microbiol. Infect.*, 14, 1057, 2008.
116. Peterson, L.R. et al., Detection of toxigenic *Clostridium difficile* in stool samples by real-time polymerase chain reaction for the diagnosis of *C. difficile*-associated diarrhea, *Clin. Infect. Dis.*, 45, 1152, 2007.
117. van den Berg, R.J. et al., Evaluation of real-time PCR and conventional diagnostic methods for the detection of *Clostridium difficile*-associated diarrhea in a prospective multicentre study, *J. Med. Microbiol.*, 56, 36, 2007.
118. Stamper, P.D. et al., Comparison of a commercial real-time PCR assay for tcdB detection to a cell culture cytotoxicity assay and toxigenic culture for direct detection of toxin-producing *Clostridium difficile* in clinical samples, *J. Clin. Microbiol.*, 47, 373, 2009.
119. Barbut, F. et al., Rapid detection of toxigenic strains of *Clostridium difficile* in diarrheal stools by Real-Time PCR, *J. Clin. Microbiol.*, 47,1276, 2009.
120. Joost, I. et al., Characterisation of *Clostridium difficile* isolates by slpA and tcdC gene sequencing, *Int. J. Antimicrob. Agents*, 33, S13, 2009.
121. Balassiano, I.T. et al., Characterization of *Clostridium difficile* strains isolated from immunosuppressed inpatients in a hospital in Rio de Janeiro, Brazil, *Anaerobe*, 15, 61, 2009.
122. Gerding, D.N., *Clostridium difficile* 30 years on: What has, or has not, changed and why? *Int. J. Antimicrob. Agents*, 33, S1, 2009.
123. Clabots, C.R. et al., Development of a rapid and efficient restriction endonuclease analysis typing system for *Clostridium difficile* and correlation with other typing systems, *J. Clin. Microbiol.*, 31, 1870, 1993.
124. Fawley, W.N., and Wilcox, M.H., Pulsed-field gel electrophoresis can yield DNA fingerprints of degradation-susceptible *Clostridium difficile* strains, *J. Clin. Microbiol.*, 40, 3546, 2002.
125. Killgore, G.E., and Kato, H., Use of arbitrary primer PCR to type *Clostridium difficile* and comparison of results with those by immunoblot typing, *J. Clin. Microbiol.*, 32, 1591, 1994.
126. Peterson, L.R., Kelly, P.J., and Nordbrock, H.A., The role of culture and toxin detection in laboratory testing for diagnosis of *Clostridium difficile*–associated diarrhea, *Eur. J. Clin. Microbiol. Infect. Dis.*, 15, 330, 1996.

34 *Dialister*

Ana Paula Vieira Colombo, Renata Martins do Souto, and Andréa Vieira Colombo

CONTENTS

34.1 INTRODUCTION

Dialister pneumosintes belongs to a group of anaerobic, gram-negative, small rods that are unreactive to commonly used biochemical tests. Over the years, because of difficulties in precisely identifying this microorganism by phenotypical means, the role of this species in several human infections has been underestimated. More recently, the development and improvement of molecular techniques of microbial identification have shown *D. pneumosintes* to be an important human bacterial pathogen. This chapter describes characteristics of this pathogen and related infections, as well as molecular techniques for its identification in clinical samples.

34.1.1 Classification, Morphology, and Epidemiology

Dialister (*ex* Bergey et al.),[1] Moore and Moore,[2] gen. nov., nom. Rev; type species: *Dialister pneumosintes*,[2,3] emend. Downes et al.[4] emend. Jumas-Bilak et al.[5]

Dialister pneumosintes (*pneuma*, breathed air; *sintes*, a spoiler[6,7]) was first described by Olitsky and Gates as *Bacterium pneumosintes* and was first isolated from nasopharyngeal secretions of patients with influenza during the epidemic of 1918 through 1921.[3] The species was later placed in the genus *Dialister*,[1] and subsequently transferred to the genus *Bacteroides*.[8] As a result of the taxonomic revision of the genus *Bacteroides* in the 1980s, Shah and Collins[9] proposed removing the microorganism from this genus. In 1994, the species was reclassified as *D. pneumosintes*.[2] This species was formerly included in the family Bacteroidaceae by virtue of being a gram-negative, anaerobic bacillus, but a phylogenetic analysis of 16S rRNA gene sequences by Willems and Collins[10] showed that the species belonged to the *Sporomusa* subbranch of the gram-positive bacterial phylum, and was closely related to the genus *Veillonella*.

D. pneumosintes is a small, obligately anaerobic, nonmotile, nonsporing, nonfermentative, gram-negative rod.[11–13] Cells of the type strain are 0.2–0.4 by 0.3–0.6 µm, arranged singly, in pairs, or in very short chains. Gram stains usually give the impression of very small cocci.[12,13] Cells from glucose broth may be 0.5–1 µm long after several transfers. Broth cultures usually are not turbid, but numerous very small cells can be seen in stains from the cultures. Because the cells are so small, stains have to be studied carefully to distinguish cells from background stain.[11–13] Deep agar colonies are punctiform, granular, and white with no evidence of gas production. Surface colonies on blood agar are punctiform, circular, entire, convex, clear, transparent, shiny, smooth, and show several white spots at the center.[12,13] Although the colony morphology of *D. pneumosintes* resembles that of *Eubacterium* species, the negative Gram stain reaction excludes the *Eubacterium* species. Trace amounts of acetate (mean, 0.07 meq/100 mL; range, 0–0.37 meq/100 mL), propionate (mean, 0.04 meq/100 mL; range, 0–0.06 meq/100 mL), lactate (mean, 0.06 meq/100 mL; range, 0–0.23 meq/100 mL), and succinate (mean, 0.01 meq/100 mL; range, 0–0.03 meq/100 mL) may be produced in peptone yeast glucose (PYG) broth cultures.[2] The G + C content of the DNA is 35 mol%.[4] The major cellular constituents of *D. pneumosintes* include fatty acids $C_{18:1}$ *cis*-9 fatty acid, $C_{16:0}$ fatty acid, $C_{18:0}$ fatty acid, and $C_{16:1}$ *cis*-9 fatty acid.[2]

Although *D. pneumosintes* is considered a member of the human normal microbiota of the upper respiratory tract, vaginal flora, and oral cavity,[2,12,14,15] this species has been implicated in several infections, including periodontal diseases,[13,15–21] root canal infections,[22,23] dentoalveolar[24] and brain abscesses,[25] infections of the respiratory tract,[12,26] bite wounds,[27] and bacteremia.[12,25,28] This microorganism has also been recovered from the amniotic fluid and placentas of pregnant women with premature rupture of the amniotic sac,[14] mastoid infections, decubitus ulcers, and peritoneal and vaginal infections.[28] The largest described collection of *Dialister* clinical isolates, assembled by Morio and coworkers in 2007,[29] has allowed evaluation of the relative clinical significance and sources of the *Dialister* species, as well as their antimicrobial susceptibilities. Demographic information showed that the majority of *Dialister* isolates were identified from male patients (66.2%) and adults (mean age of recovery, 47 years; range, 21 months to 92 years). The isolates were recovered from various clinical samples, including blood cultures, bone, and supra- and infradiaphragmatic pus. The majority of the strains were isolated from cutaneous and soft-tissue infections (52.7%). *D. pneumosintes* was the only species recovered from blood cultures in the study (2.7%), and was the predominant species in intraabdominal (11 out of 13 isolates) and respiratory (7 out of 8 isolates) samples. Except for one strain from blood, all isolates of *Dialister* spp. were recovered from mixed flora, mainly associated with anaerobic gram-positive cocci and *Streptococcus* spp.

34.1.2 Clinical Features and Pathogenesis

Clinical data regarding *D. pneumosintes* is scanty probably due to the lack of identification of this bacterium in routine. This species may be considered an opportunistic pathogen associated with a wide range of infections in the human body.[12]

Lepargneur et al.[28] reported a case of a young woman admitted postpartum with high fever and lumbar pain. The patient had a previous episode of vaginosis and presented suppurative thrombosis of the ovarian vein. *D. pneumosintes* was isolated as a single agent in blood culture. In two cases of brain abscesses, this species was isolated by anaerobic culture of blood and pus samples.[25] *D. pneumosintes* has also been isolated from purulent infections associated with a mixed microbiota, particularly in the oral cavity. Endodontic and periodontal diseases are the most frequent oral infections related to *D. pneumosintes*. Periodontal diseases are infections characterized by chronic inflammatory lesions and destruction of the supporting periodontal tissues that may lead to eventual tooth loss.[30] Clinical features include swelling, redness, and bleeding of the marginal gingiva (gingivitis), usually associated with large amounts of visible dental plaque and calculus (tartar). Gingivitis may progress to periodontitis which is characterized by pathological deepening of the gingival sulcus (periodontal pocket) due to destruction of the periodontal ligament and alveolar bone, presence of subgingival plaque and calculus, bleeding, suppuration and mobility of the tooth (Figure 34.1).[31]

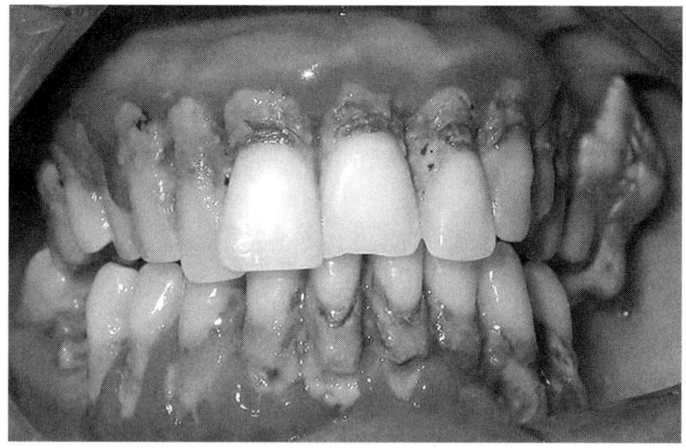

FIGURE 34.1 Clinical features of chronic periodontitis. Large amounts of dental plaque and calculus, gingival inflammation and severe loss of periodontal attachment are shown.

Severe forms of periodontitis (localized and generalized aggressive periodontitis, and necrotizing periodontitis) have a more rapid progression, and are usually associated with impairment of the immune system.[31] Although these are chronic inflammatory diseases, they may become acute inflammatory processes (periodontal abscesses) characterized by pain, edema, suppuration and bleeding of the gingiva, and tooth mobility.[31] Endodontic infections are also inflammatory diseases of microbial etiology primarily caused by infection of the root canal system.[32] Under normal conditions, the dental pulp is sterile and isolated from oral microorganisms by overlying enamel and dentin. When these protective layers are breached (e.g., due to caries, trauma, restorative procedures, etc.), the pulp becomes exposed to the oral microbiota, which will lead to intraradicular inflammation and necrosis. Bacteria within a necrotic pulp will propagate and migrate toward the periapical tissues, resulting in inflammation and development of apical periodontitis (extraradicular infection).[33] Symptoms may vary from transient pain under stimuli to continuous and spontaneous severe pain, sensitivity to biting or percussion, abscess and swelling. Radiographic features range from slight thickening of periodontal ligament to extensive apical radiolucent areas.[33]

Periodontal and endodontic infections are polymicrobial diseases normally associated with microbial complexes of the supra- and subgingival biofilms.[32,34] Most of these pathogens are gram-negative anaerobic rods and spirochetes.[32,34] In particular, *D. pneumosintes* has been found in 46% of cases of persistent apical periodontitis,[32] 62% of chronic periodontitis lesions,[21] and 42% of periodontal sites that continued to show attachment loss after treatment.[35] This microorganism is pathogenic for rabbits when injected intratracheally[36]; however, the exact mechanisms of virulence of this species in the pathogenesis of periodontal and endodontic diseases are unknown. In these diseases, oral pathogens may cause tissue destruction directly by invading the tissues and producing an array of enzymes that will degrade bone and connective tissues. However, the major destruction observed in these pathologies is due to the potent host inflammatory and immune response triggered by these bacterial pathogens.[30,33]

Among several bacterial virulence factors, lipopolysaccharide (LPS), a major constituent of the cell wall of gram-negative bacteria, plays an important role in the pathogenesis of periodontal and endodontic diseases. LPS activates cells to release pro-inflammatory cytokines, prostaglandins and several matrix metalloproteinases that contribute to the destruction of periodontal/pulp connective tissues and resorption of alveolar bone.[30,33] Although highly conserved, LPS contains important structural differences among different bacterial species that can significantly alter host responses.[37] For instance, LPS from the periodontal pathogen *Porphyromonas gingivalis* induces the secretion of a variety of inflammatory mediators, but inhibits endothelial cell expression of E-selectin and interleukin-8 triggered by other bacteria.[37] This may allow *P. gingivalis* to evade innate host defense mechanisms, resulting in colonization and chronic disease. No information about the activities of *D. pneumosintes* LPS on the pathogenesis of these infections is available. However, it is possible that this species may be protected from the immune system when associated with other bacterial pathogens and/or viruses in mixed periodontal[18,38,39] and endodontic infections.[22] In fact, studies have suggested that *D. pneumosintes* contained in a mixed flora may behave as a deadly pathogen.[25]

34.1.3 Diagnosis

34.1.3.1 Conventional Techniques

For determination of Gram reaction, a buffered Gram's stain procedure of cells from young culture is recommended.[12] A complex nonselective medium containing 4.3% brucella agar (BBL Microbiology Systems), 0.3% agar, 0.2% yeast extract, 5% defibrinated sheep blood, 0.2% hemolyzed sheep erythrocytes, 0.0005% hemin, and 0.00005% menadione is recommended for isolation of *D. pneumosintes* in culture.[12] Growth of this species is most rapid at 37°C in an anaerobic chamber containing 85% N_2, 10% H_2, and 5% CO_2.[13] *D. pneumosintes* is asaccharolitic and nonreactive in conventional biochemical tests. This species is nitrate-negative; it doesn't hydrolyze esculin and urea, and it doesn't produce indole or catalase.[11] Moreover, this species does not modify milk or gelatin. There is no growth in 20% bile,[5] and no DNase or phosphatase is detected.[12] Enzyme profiles can be generated with rapid tests like Rapid ID 32A (API® bioMerieux). *D. pneumosintes* shows arginine arylamidase, leucine arylamidase, glycine arylamidase, and histidine arylamidase activities (profile 0000012401).[5] *D. pneumosintes* is susceptible to kanamycin and resistant to vancomycin and colistin.[5] Morio et al.[29] observed strains that showed decreased susceptibility to several agents like metronidazole, erythromycin, pristinamycin, rifampin, piperacillin, and levofloxacin.

34.1.3.2 Molecular Techniques

Polymerase Chain Reaction (PCR). PCR is a technique to amplify a single or few copies of a gene, a part of a gene, or a noncoding sequence of DNA across several orders of magnitude, generating millions or more copies of a particular DNA sequence.[40] PCR has been used routinely for the rapid diagnosis of many infectious diseases, particularly those caused by uncultivable or slow-growing microorganisms. Doan et al.[13] developed a specific and sensitive 16S rRNA-based PCR method for identification of *D. pneumosintes* in subgingival plaque samples from chronic periodontitis lesions. Primers were designed based on the 16S rRNA sequence of *D. pneumosintes*.[41] Several other studies employed this methodology for detection of this species in saliva, dental plaque, infected pulpal and periradicular samples.[16,18,21–23,38,39] Our group, for instance, observed a significantly higher frequency of *D. pneumosintes* in subgingival plaque of individuals with periodontal diseases compared to periodontally healthy subjects.[21] Surprisingly, this species was detected in only 3% of the saliva samples from the same subjects, suggesting that this pathogen colonizes primarily the anaerobic and protein-rich subgingival environment. Conventional PCR methods, however, do not allow for accurate quantification of the target DNA (or infectious agent), and thus miss important diagnosis and treatment aspects. Recent advances in the synthesis of fluorogenic probes and development of instrumentation for monitoring of fluorescence have facilitated the development of quantitative PCR (real-time PCR) assays for specific, automated detection and quantification of amplified bacterial products.[42] The first system developed, SYBR® Green, measures the increase in fluorescence caused by the binding of SYBR® Green dye to double-stranded DNA. The TaqMan® system measures the accumulation of PCR products with a fluorogenic probe.[42] Real-time PCR assays for quantification of periodontal pathogens in plaque samples including *D. pneumosintes* have also been developed.[19,20,43] Using this technique, Gomes et al.[20] reported significantly higher numbers of *D. pneumosintes* in subgingival plaque of smokers than nonsmokers with periodontitis.

Cloning and DNA Sequencing. Bacteria contain conserved nucleic acid sequences that can be used as targets in assays that are based on DNA amplification. Detection of these conserved sequences in a clinical sample suggests the presence of the bacteria. The small subunit (16S) of the ribosomal RNA genes has proven to be the most valuable and widely used target for bacterial detection and identification.[44] The 16S rRNA genes contain numerous regions of highly conserved sequences which can be target with PCR primers. Amplification of the more variable sequence between the two priming sites can be sequenced and used for bacterial identification, particularly in mixed microbial flora.[17,44] Cloning and sequencing have been widely used for identification of *D. pneumosintes* directly in clinical specimens or isolates.[13,17,25,26,28,45–47] Basically, genomic DNA from bacterial isolates or clinical specimens is extracted, purified, and PCR amplified with universal primers for the 16S rRNA. The multiple 16S rRNA sequence types are cloned by inserting these amplicons into bacterial vectors (e.g., plasmids) that will be transformed into *Escherichia coli* to construct a 16S rRNA genomic library. Vectors containing the cloned sequences are extracted from *E. coli*, purified, and then sequenced. For identification, sequences are analyzed and compared to the 16S rRNA sequences in several databases such as the Ribosomal Database Project[48] and GenBank.[49] 16S rRNA gene and *dna*K sequencing have also been employed for differentiating *D. pneumosintes* from other species of the genus *Dialister* including *Dialister invisus*, *Dialister micraerophilus*, and *Dialister propionicifaciens*.[4,5]

Checkerboard DNA–DNA Hybridization. The checkerboard DNA–DNA hybridization technique was first described in 1994 for the detection of periodontal pathogens directly in samples of subgingival dental plaque.[50] This method is based on hybridization of whole genomic or oligonucleotide DNA probes with DNA obtained from bacterial isolates or clinical samples. The checkerboard format allows for the simultaneous detection of various species in a large number of samples (Figure 34.2), being mostly useful for determining the composition or profile of complex microbial ecosystems.[51] In this methodology, individual DNA samples are placed into channels of a slot-blot apparatus over a nylon membrane. Controls corresponding to 10^5 and 10^6 bacterial cells containing DNA of all the test species are placed on the same membrane. A total of 30 samples can be deposited on the membrane. The membrane with fixed DNA is placed in a blotter with the lanes of DNA at 90° to the 45 channels of the blotter. A 30 × 45 checkerboard pattern is produced. Each channel is used as a hybridization chamber for separate DNA probes. Up to 45 DNA probes can be employed (Figure 34.2).

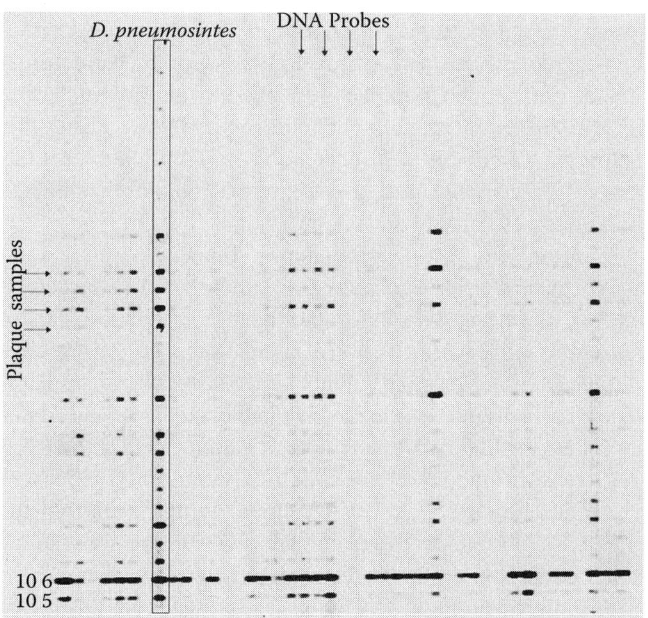

FIGURE 34.2 Example of the checkerboard DNA–DNA hybridization method used to detect bacterial species in 28 subgingival plaque samples from a single individual. The horizontal lanes are the plaque samples from each tooth and the two last lanes are standards containing either 10^6 or 10^5 cells of test species. Vertical lanes represent the whole genomic DNA probes. Lane 8 corresponds to the *D. pneumosintes* DNA probe. A signal at the intersection of the vertical and horizontal lanes indicates the presence of a species. The intensity of the signal is related to the number of organisms of that species in the sample.

After hybridization, positive signals on the membrane are detected by scanning of fluorescence or chemiluminescence and autoradiography. A modification of the checkerboard using a PCR-based, reverse-capture checkerboard hybridization methodology has also been described.[52] Conversely to the first checkerboard technique described, in this method up to 30 reverse-capture probes that target regions of 16S rRNA genes are deposited on a nylon membrane in separate horizontal lanes of the slot-blot apparatus. 16S rRNA genes of DNA from up to 45 samples or pure cultures are PCR amplified and labeled. The labeled samples are placed perpendicularly to the lanes of probes fixed on the membrane. Hybridization and signal detection are performed as previously described.[50,51] The checkerboard is a rapid, sensitive, and relatively inexpensive technique. It overcomes many of the limitations of cultural methods, including the loss of viability of microorganisms and the growth of fastidious or uncultivable species. The entire sample may be employed without dilution or amplification (with appropriate regard to total sample size), overcoming problems in quantification imposed by either serial dilution or amplification procedures. DNA probes for specific species may be removed by stripping and the samples reprobed by another set of new probes, increasing the array of species that can be evaluated. Finally, it provides quantitative data which may be important in the diagnosis of certain infections where species levels and proportions may be crucial for determining the prognosis and treatment.[51] The checkerboard DNA–DNA hybridization methodology has been employed in several studies evaluating the microbial profile of different periodontal and endodontic infections.[35,53,54] More recently, our group introduced a whole genomic probe for *D. pneumosintes*. In an ongoing study of microbial epidemiology of periodontal diseases, this species has been detected in subgingival plaque samples of approximately 40% and 20% of chronic and aggressive periodontitis subjects, respectively (data not published). It should be pointed out, though, that detection of *D. pneumosintes* in clinical samples by this technique is indicated when looking for this pathogen in association with other potential pathogenic species in mixed infections.[51]

Human Oral Microbe Identification Microarray (HOMIM). Based on 16S rRNA sequencing of a large collection of oral samples and isolates, Dr. Bruce Paster and Dr. Floyd Dewhirst from The Forsyth Institute, Boston, Massachusetts, have identified over 600 human oral species or phylotypes in the oral cavity.[17] Using this information, in 2008 they developed the Human Oral Microbe Identification Microarray (HOMIM),[55] which allows for the simultaneous detection of about 300 of the most prevalent oral bacterial species, including "uncultivable" microorganisms. HOMIM is a reverse-capture technique in which 16S rRNA-based, reverse-capture oligonucleotide probes are printed on aldehyde-coated glass slides in duplicate. Positive and negative controls, as well as universal probes for the bacterial 16S rRNA genes, are also included on the slide. DNA from clinical samples or bacterial isolates is extracted and the 16S rRNA genes PCR amplified using universal forward and reverse primers. Purified PCR products are labeled via the incorporation of a fluorescent dye in a second nested PCR. The labeled 16S amplicons are hybridized overnight with probes on the slides. Up to five samples can be hybridized simultaneously against probes to 300 different species per slide. After washing, the microarray slides are scanned and crude data is extracted using image analysis software (Figure 34.3). Data are normalized by comparing individual signal intensities to the average of signals from universal probes. The detection of a particular species in a sample is determined by the presence of a fluorescent spot for that unique probe. Recent studies have employed this technique for analyzing the microbiota associated with root caries[56] and periodontal diseases.[35] Of interest, Colombo et al.[35] reported a significantly higher prevalence of *D. pneumosintes* in subjects with refractory periodontitis (i.e., subjects who did not respond to periodontal treatment) compared to successfully treated or periodontally healthy individuals by HOMIM. Moreover, periodontal sites showing disease progression presented a greater frequency of this pathogen than healthy sites or sites that showed periodontal attachment gain suggesting that this species may be a pathogen associated with treatment failure and recurrence of disease.[35] The HOMIM method should also be used for identification of bacterial species in mixed infections.

34.2 METHODS

34.2.1 Sample Preparation

Efficient methods of sample collection and DNA extraction that produces pure and high-quality DNA are crucial for the success of several molecular methods of microbial identification. Genomic DNA of *D. pneumosintes* can be extracted from pure culture or directly from clinical samples by a great variety of conventional methods as well as commercial kits (Wizard® SV Genomic DNA Purification Kit, Promega, Madison, WI; QIAamp DNA Mini Kit, QIAGEN, Valencia, CA; DNA UltraClean™, MoBio laboratories, Inc, Carlsbad, CA).[13,16,18–23,26,35,56]

Sample Collection. Saliva can be obtained by rinsing out the mouth with 10 mL of 0.9% sterile saline for 1 min, or by spitting directly into a sterile plastic recipient.[21,57] Supragingival plaque can be removed by sterile periodontal curettes. Subgingival plaque can be collected by inserting a sterile periodontal curette[21] or paper points into the subgingival site.[16] Samples from root canal are collected with sterile paper points using strict asepsis.[22,23] Samples from oral mucosa and tongue are acquired by wiping a sterile swab on the mucosal surfaces. Samples are placed into microcentrifuge tubes containing TE (pH 7.6), homogenized by vortex mixing for 1 min and stored at –20°C until use.[21,57,58] For cultural identification, samples are placed into tubes with VMGA III (viability preserving medium no. III, University of Gothenburg, Sweden)[59] or trypticase-soy broth with 5% dimethyl sulfoxide (TSB-DMSO; Difco, Detroit, MI, US).[53] A 100 μL aliquot is serially diluted in broth consisting of 0.5% NaCl, 0.25% Thiotone E-peptone (BBL), 0.25% tryptose (Difco), and plated on brucella blood agar medium.[12]

Genomic DNA Extraction

Protocol 1[57]

1. Samples are centrifuged at 300 × *g* for 10 min. The pellet is washed twice in 0.9% saline (0.9 g NaCl in 100 mL dH_2O), resuspended in 100 μL of 50 mM NaOH, and boiled for 10 min.
2. Samples are then neutralized with 14 μL of 1 M Tris (pH 7.5) and centrifuged at 14,000 × *g* for 3 min. Supernatants are collected and stored at –20°C.

Protocol 2[58]

1. Warm samples to 37°C for 10 min (for samples in transport media), and mix well.
2. Take 0.2 mL of the suspension and wash three times with distilled water by centrifugation at 10,000 × *g* for 2 min.
3. Resuspend the bacterial pellet in 0.1 mL of distilled water, boil for 10 min and place it on ice. Centrifuge at 12,000 × *g* for 5 min to remove cell debris, and transfer the supernatant containing DNA into a sterile microcentrifuge tube. Store DNA at –20°C.

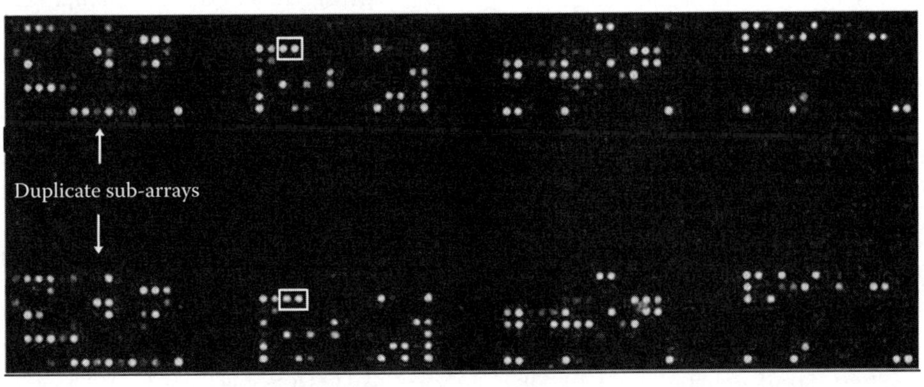

FIGURE 34.3 Image of a scanned glass slide of HOMIM showing four 8 × 15 duplicate subarrays Probes are organized phylogenetically on each subarray. This image shows one plaque sample hybridized against 400 probes to 300 taxa. The two duplicate spots inside the boxes correspond to positive signals to *D. pneumosintes* probes X78 and AA90.

Protocol 3[35]

1. Place sample in a microcentrifuge tube containing 44 μL of TE (pH 7.6), add 5 μL of 5% Tween 20, and then 1 μL of Proteinase K stock solution (10 mg/mL).
2. Incubate at 55°C for 2 h, and at 95°C for 10 min. Store DNA at –20°C.

34.2.2 Detection Procedures

(i) PCR Identification[13]

1. Prepared PCR mixture (50 μL) containing 5 μL (100–200 ng) of DNA template, 5 μL of 10× PCR buffer (200 mM Tris-HCl pH 8.0; 500 mM KCl), 1 μL of 10 mM dNTP mixture (0.2 mM final concentration of each dNTP), 1.5 μL of 50 mM $MgCl_2$, 0.5 μL (1 μM) of each primer (forward: 5′-TTC TAA GCA TCG CAT GGT GC-3′, reverse: 5′-GAT TTC GCT TCT CTT TGT TG-3′), 0.25 μL (1.25 U) of *Taq* DNA polymerase, and 36.25 μL of ultrapure distilled water.
2. Amplify with an initial step at 95°C for 2 min; 36 cycles of 94°C for 30 s, 55°C for 1 min and 72°C for 2 min; and a final step at 72°C for 10 min.
3. Mix 10 μL of each PCR product with 5 μL of loading buffer and place into the wells of a 1.5% agarose gel in 1X TAE running buffer. Run electrophoresis at 98V for ≅ 1 h. Stain the gel with ethidium bromide (0.5 μg/mL).

Note: A product of 1505 bp visualized in the gel under UV indicates the presence of *D. pneumosintes* DNA (Figure 34.4). A positive control with DNA from reference strain ATCC 33048 of *D. pneumosintes* should be included in all PCR amplifications.

(ii) Real-Time PCR Detection[19]

1. Prepared PCR mixture (25 μL) containing 2.5 μL of DNA template (100–200 ng), 2.5 μL of 10× TaqMan Universal PCR Master Mix with the reference dye carboxy-X-rhodamine (ROX) (qPCR Core Kit, Eurogentec, Cologne, Germany), 1.5 μL of 50 mM $MgCl_2$, 1 μL of dNTP mixture (Eurogentec), 0.25 μL (12.5 pmol) of each primer (forward: 5′-GAG GGG TTT GCG ACT GAT TA-3′, reverse: 5′-CCG TCA GAC TTT CGT CCA TT-3′), 3.75 pmol of TaqMan probe (5′-FAM[6-carboxyfluorescein reporter dye]-CAC CAA GCC GAC GAT CAG TAG CCG-3′-TAMRA[6-carboxytetramethylrhodaminequencher dye]), and 13.25 μL of ultrapure distilled water.
2. Amplify with an initial 95°C for 10 min; 40 cycles of 95°C for 15 s and 60°C for 1 min.

(iii) Sequencing Analysis of 16S rRNA Gene[17]

1. Prepare PCR mixture (50 μL) containing 1 μL of DNA template (100–200 ng), 5 μL of 10× Platinum Taq® PCR buffer (Invitrogen, Carlsbad, CA, US), 1.5 μL of 50 mM $MgCl_2$, 4 μL of 10 mM dNTP mixture, 1 μL (20 μM) of each primer (9 forward: 5′-GAG TTT GAT YMT GGC TCA G-3′, 1541 reverse: 5′-AAG GAG GTG WTC CAR CC-3′), 0.5 μL (2.5 U) Platinum® Taq DNA Polymerase (Invitrogen), 36 μL of ultrapure distilled water.
2. Amplify with an initial 94°C for 4 min, 30 cycles of 94°C for 45 s, 60°C for 45 s, 72°C for 90 s, plus an additional s per cycle; and a final 72°C for 15 min.
3. Mix 20 μL of the product with 5 μL of loading buffer, run on 1% agarose gel stained with ethidium bromide, and visualized by UV light. A band of approximately 1500 bp is expected.
4. Cut band out of gel and place into a sterile microcentrifuge tube. Purify the amplified 16S rRNA product by using spin columns (QIAquick gel extraction kit, Qiagen).
5. Clone the purified PCR products into plasmids and transformed into competent *E. coli* TOP10 using a TOPO cloning kit (Invitrogen).
6. Amplify the clone with the correct insert size (1500 bp) using universal M13 primers, with a mix containing 1 μL of each clone (cell suspension in TE), 5 μL of 10× Platinum Taq® PCR buffer (Invitrogen), 1.5 μL of 50 mM $MgCl_2$, 4 μL of 10 mM dNTP, 1 μL (20 μM) of each primer [M13 forward (-40): 5′-GTT TTC CCA GTC ACG AC-3′, M13 reverse: 5′-CAG GAA ACA GCT ATG ACC-3′)], 0.25 μL (1.25 U) Platinum® Taq DNA Polymerase, 36.25 μL of ultrapure distilled water; and cycling conditions of 94°C

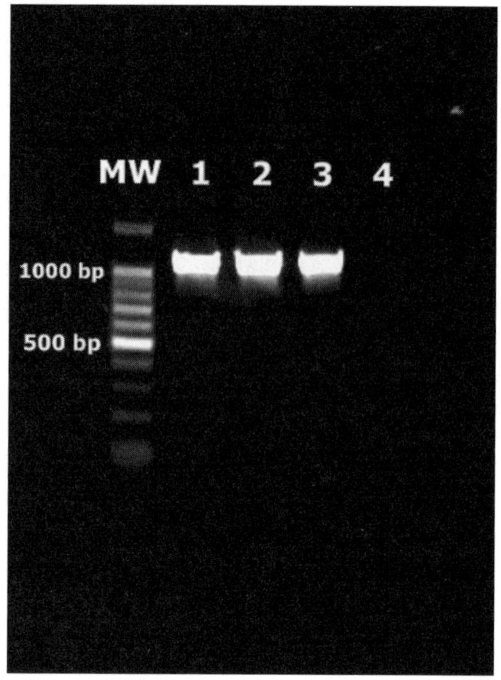

FIGURE 34.4 PCR amplification of *D. penumosintes* genomic DNA shows a 1105 bp product. Lane MW, 1Kb ladder marker; lane 1, reference strain ATCC 33048; lane 2, saliva sample, and lane 3, subgingival plaque sample of a patient with periodontitis; lane 4, subgingival plaque of a patient with periodontal health.

for 4 min, 30 cycles of 94°C for 45 s, 60°C for 45 s, 72°C for 90 s, plus an additional s per cycle; and a final 72°C for 15 min. Check the PCR products by running 5 μL of PCR product in 2 μL of loading buffer on a 1% agarose gel.

7. Prior to sequencing the amplicons, remove residual primers with exonuclease I (Exo)/shrimp alkaline phosphatase (SAP) treatment (GE Healthcare Bio-Sciences). Add 2 μL containing 2 U Exo and 0.2 U SAP to 5 μL of PCR product, and incubate at 37°C for 15 min, 80°C for 15 min, and store at 4°C.
8. Add the PCR products to the cycle sequence reaction (the BigDye® Terminator kit protocol, PE Applied Biosystems), containing 7 μL Exo/SAP PCR product, 0.5 μL Big Dye Mix, 4 μL sequencing buffer, 8 μL ultrapure distilled water, and 0.5 μL of (3.2 μM) primer 533R (5′-TTA CCG CGG CTG CTG-3′).
9. Amplify with 25 cycles of 96°C for 10 s, 60°C for 5 s, and 60°C for 4 min.
10. Purify each clone with Sephadex™ G-50 (GE Healthcare). Resuspend in Hi-Di™ Formamide (Applied Biosystems, Foster City, CA, US) and put directly onto the sequencer.

(iv) Checkerboard DNA–DNA Hybridization[50,51]

Applying Samples on Membranes

1. Cut 10 15 × 15 cm pieces of Whatman paper filters (GE Healthcare), place them on the acrylic lower part of the Miniblotter® (Immunetics, Cambridge, MA, US), and lay a nylon membrane (15 × 15 cm) (GE Healthcare) over it. Set up the upper part of the MiniSlot™ (Immunetics) apparatus over the membrane. Place samples in individual microcentrifuge tubes containing 150 μL of TE (pH 7.6). Add 150 μL of 0.5 M NaOH (freshly made) and boil samples in a hot water bath for 5–10 min. Add 800 μL of freshly made 5M NH_4 acetate.

Note: Two pools of DNA standards corresponding to 10^5 and 10^6 bacterial cells of the test species (equivalent to 1 and 10 ng of DNA, respectively) should also be prepared in TE and boiled together with the samples.

2. Using disposable pipettes, mix and place the two standards (usually the two last slots of the MiniSlot™), and then the 28 samples into each slot of the MiniSlot™.
3. Take the membrane and fix the DNA samples by UV cross-linking for 30 s and baking in oven at 120°C for 20 min. The membrane can be stored in a plastic bag until use.

Hybridization with DNA Probes

4. Wet the nylon membrane with fixed DNA samples in 2× SSC buffer, then place it in a bag with 50 mL of prehybridization buffer and incubate at 42°C for at least 1 h.
5. Construct DNA probes using the DIG DNA Labeling Kit (Roche Applied Science) according to manufacturers' instructions. Dilute each DNA probe in a final volume of 150 μL of hybridization buffer (20 ng/mL of probe). Boil probes for 10 min and place on ice.
6. Take the membrane and place it under the upper part of the Miniblotter® (Immunetics). Put a plastic film and a pad cushion under the membrane and screw the lower part of the apparatus tightly. Rotate the membrane 90° so that the lanes with the fixed DNA are perpendicular to the channels of the blotter.
7. Take 136–150 μL of each probe and fill each channel of the blotter. Wrap the whole apparatus with plastic film and put into a plastic bag with 50 mL of distilled water to avoid dry out of the membrane. Leave it in the oven at 42°C for at least 16 h.

Note: The makeup of key buffers used in this step:

20× SSC: 350.64 g NaCl and 176.5 g Na_3 Citrate in 2 L dH_2O. Adjust pH to 7.0.
2× SSC: Dilute 50 mL of 20× SSC in 450 mL dH_2O and autoclave for 30 min.
Formamide: Add AG 501-X8 40.0 g (Analytical Grade Mixed Bed Resin) to 800 mL 98% formamide. Stir for 30 min and filter using a Whatman filter. Keep at −20°C.
Denhardt's Solution: Mix 5 g Ficoll, 5g polyvinylpyrrolidone and 5 g bovine serum albumin Fraction V in 500 mL dH_2O. Keep in aliquots at −20°C.
Yeast RNA: Mix 1 g of yeast RNA in 100 mL dH_2O, boil until dissolved and keep in aliquots at −20°C.
Prehybridization Stock Solution: Mix 88 g NaCl, 44 g NaCitrate and 6.9 g NaH_2PO_4 in 500 mL dH_2O. Adjust pH to 6.5 and autoclave for 20 min.
Prehybridization Solution (p/membrane)**:** Mix 25 mL formamide, 5 mL Denhardt's solution, 3 mL yeast RNA, 12.5 mL prehybridization stock solution, 5 mL 10% casein.
Hybridization Stock Solution: Mix 30.7 g NaCl, 16.1 g NaCitrate, and 2 g NaH_2PO_4 in 300 mL dH_2O. Adjust pH to 6.5. Autoclave for 20 min.
Hybridization Solution (p/membrane)**:** Mix 10 mL formamide, 0.4 mL Denhardt's solution, 0.4 mL yeast RNA, 8.2 mL hybridization stock solution, 2 mL 10% Casein, 2 g dextran sulfate.

Washing and Detection

8. Using a manifold device (Immunetics), aspirate the probes by vacuum suction.
9. Wash the membrane 2× for 20 min with PO_4 buffer in a circulating water bath at 68°C.
10. Wash the membrane in maleic acid buffer for 5 min, and then with 500 mL of blocking buffer for 1 h on rotator.
11. Incubate the membrane in 50 mL of blocking buffer containing anti-digoxigenin-AP (Roche Applied

Science) at 1:20,000 (for scanner imaging detection) or at 1:10,000 (for X-ray film) on rotator for 30 min at room temperature.

12. Wash the membrane 2× with maleic acid buffer for 20 min, and detection buffer for 5 min.
13. For detection on X-ray film, dilute 1 mL of Lumi-Phos 530 (Lumigen, Inc., Southfield, MI, US) in 4 mL of detection buffer. For detection using the Storm, dilute 1 mL of ECF substrate (GE Healthcare) in 4 mL of detection buffer (made up of 100 mL diethanolamine solution and 100 mL $MgCl_2$ solution).
14. Put the membrane in a small plastic box and pour the diluted Lumi-Phos or ECF solution on the membrane. Make sure the solution covers the membrane evenly. Place membrane in a plastic reaction folder and seal it.
15. For detection with an imaging system scanner, the membrane can be scanned next day. For X-ray detection, the membrane can be placed between aluminum foils at 37°C for 30–60 min. In a dark room, put the membrane in an autoradiograph film holder cassette and place the X-ray film over it. Leave it for 30 min, and then proceed to film developing.

Note: The makeup of the key buffers used in this step:

PO_4 Buffer: Mix 30 g SDS, 1.1 g EDTA, 8.4 g Na_2HPO_4 in 3 L dH_2O. Autoclave for 30 min.

Maleic Acid Buffer: Mix 34.8 g maleic acid, 522 g NaCl, 24 g NaOH in 3 L dH_2O until completely dissolved. Adjust pH to 8.0, add 9 mL of Tween 20 and autoclave for 30 min.

Casein: Mix 70 g of casein in 630 mL maleic acid buffer and boil until dissolved. Autoclave for 20 min and store at 4°C.

Blocking Buffer: Mix 50 mL casein in 450 mL maleic acid buffer. Take 50 mL for making the blocking with anti-DIG at 1:10,000 (5 μL) or 1:20,000 (2.5 μL).

Diethanolamine Solution: Mix 16.8 g diethanolamine 16.8 g in 800 mL dH_2O. Adjust pH to 9.5 and autoclave for 20 min.

$MgCl_2$: Mix 0.325 g $MgCl_2$ in 800 ml dH_2O and autoclave for 20 min.

Detection Buffer: Mix 100 mL diethanolamine solution and 100 mL $MgCl_2$ solution.

(v) Human Oral Microbe Identification Microarray (HOMIM)[35]

Amplification of 16S rRNA Genes by PCR

1. Amplify the 16S rRNA genes of each sample in two separate PCR mixtures using different primer sets: 4F: 5′-CCAGAGTTTGATYMTGGC-3′ and 6F: 5′-GACTAGAGTTTGATYMTGGC–3′; 1492R: 5′-GYTACCTTGTTACGACTT-3′; 1541R: 5′-GAAG GAGGTGWTCCADCC-3′.
2. The first PCR mixture (25 μL) consists of 1 μL of sample DNA, 2.5 μL of 10× PCR buffer, 0.5 μL of dNTP (10 mM), 1 μL (2 mM) $MgSO_4$, 15 pmol of primers 4F/6F mix (1:1 ratio), and 15 pmol of primer 1541R, 1.25U Platinum® Taq DNA Polymerase High Fidelity (Invitrogen), and ultrapure distilled water.
3. The second PCR mixture (25 μL) consists of 1 μL of sample DNA, 2.5 μL of 10× PCR buffer, 0.5 μL of dNTP (10 mM), 1 μL (2 mM) $MgSO_4$, 15 pmol of primers 4F/6F mix (1:1 ratio), and 5 pmol of primer 1492R, 1.25U Platinum® Taq DNA Polymerase High Fidelity (Invitrogen), and ultrapure distilled water.
4. Perform both PCR with an initial 94°C for 2 min; 32 cycles of 94°C for 30 s, 55°C for 30 s, 68°C for 1.5 min with an additional 1 s for each cycle; and a final 68°C for 10 min.
5. Examine the PCR products on 1.5% agarose gel electrophoresis in 1× TAE buffer with ethidium bromide staining and visualize under short-wavelength UV light. A fragment of about 1500 bp is expected.
6. After amplification, combine the two products of each sample giving a 50 μL PCR product mix that is then purified using a QIAquick® PCR Purification Kit (Qiagen).

Labeling of Samples

7. Label the purified PCR products via the incorporation of Cy3-dCTP (GE Healthcare) fluorescent dye during another PCR with primer set 9F: 5′-GAG TTT GAT YMT GGC TCA G-3′ and 1492R.
8. The PCR mixture (25 μL) is composed of 7 μL of purified PCR product, 2.5 μL of 10× PCR buffer, 2.5 μL of 10 mM dNTP mixture without dCTP, 2.5 μL of dCTP mixture (1 μL of 1mM Cy3-dCTP + 4 μL of 2 mM dCTP), 1 μL (2 mM) $MgSO_4$, 15 pmol of the 9F primer and 5 pmol of the 1492R primer, 0.25 μL 10% Triton X, 1.25U *Taq* polymerase, and ultrapure distilled water.
9. The cycling conditions comprise an initial 94°C for 2 min; 40 cycles of 94°C for 30 s, 55°C for 30 s, 68°C for 1.5 min with an additional 1 s for each cycle; and a final 68°C for 10 min.
10. Visualize PCR products on 1.5% agarose gel electrophoresis and then purify using QIAquick® PCR Purification Kit (Qiagen).

Capture Probes and Slide Printing

11. Synthesize 16S rRNA-based oligonucleotide reverse-capture probes with a 5′-(C6)-amine modified base, eight spacer thymidines and 18–20-nucleotides of target sequence.
12. Print probes on 25 × 76 mm aldehyde-coated glass slides as four 8 × 15 duplicate subarrays (Figure 34.3). Five copies of the array are printed per slide, allowing the hybridization of up to five samples per slide.
13. Positive controls, that is, "universal" probes that hybridize with all or most bacterial species, providing both array orientation and labeling efficiency, as

well as negative controls to determine array background levels should be also printed on the slices. For monitoring signal linearity, the inclusion of a universal 16S rRNA probe in a series of concentrations is suggested. Moreover, probes should be designed to have the same melting temperature (51°C–53°C), and G + C content.

Microarray Hybridization and Reading

14. Block slides with sodium borohydride in order to reduce nonreactive aldehyde groups and fluorescent background. Set up the slides placing one glass cover slips on each array (five cover slips per slide).
15. Mix 14 μL of each labeled and purified sample with 6 μL of hybridization buffer and boil for 5 min. Load slowly 10 μL of each sample solution at one corner and under the cover slip, allowing capillary action to pull the solution over the array.
16. Place slides on incubation racks in a humidified hybridization chamber and put it in a microarray incubator (Hybex® Microarray Incubation System, SciGene Corporation, Sunnyvale, CA, US) at 55°C for 16 h.
17. Remove slides from the chamber and sink into a water bath at 55°C containing a washing solution (buffer 1) to remove the cover slips.
18. Wash slides using as automated microarray robotic instrument (Little Dipper™ Microarray Processor, SciGene Corporation) according to the following washing program: washing in buffer 1 at 55°C for 2 min, washing in buffer 2 for 2 min, and buffer 3 for 2 min at room temperature.
19. Once the slides are finished spinning in the centrifuge, they are removed and placed in a light-proof container until scanning. Scan the microarray slides using a scanner Axon GenePix® 4000B (MDS Analytical Technologies, Sunnyvale, CA, US). Images are captured by imaging software (MDS Analytical Technologies), and files analyzed using a *Data Analysis Procedure of the HOMIM online tool.*[60]

Note: The makeup of key buffers used in this procedure:

Sodium Borohydride: Mix 1 g $NaBH_4$, 300 mL 1× PBS (8 g NaCl, 0.2 g KCl, 1.4 g Na_2HPO_4, 0.24 g KH_2PO_4 in 1 L of dH_2O; pH 7.4) and 100 mL 99% ethanol.

HOMIM Hybridization Buffer: 2× SSC, 0.2 μg/μL yeast tRNA and 0.1% SDS.

10% Lauryl Sulfate (SDS): Mix 10 g SDS in 100 mL dH_2O. Store at room temperature.

Washing Buffers: 1 (2× SSC and 10% SDS); 2 (2× SSC) and 3 (0.1× SSC).

34.3 CONCLUSION

A large number studies employing modern molecular technologies for microbial identification have indicated the potential role of *D. pneumosintes* in the etiology and pathogenesis of various bacterial infections in humans, particularly mixed infections of the oral cavity. However, the implication of this species in opportunistic infections remains unclear. Moreover, very little information about virulence factors of this organism is available. Thus, future investigations should focus, not only on the rapid and precise identification of this species in different infections, but also on unraveling the mechanisms of pathogenicity promoted by this pathogen in the human body. Consequently, diagnostic as well as preventive and therapeutic strategies can be developed for the controlling of infections caused by *D. pneumosintes*.

REFERENCES

1. Bergey, D.H. et al., *Bergey's Manual of Determinative Bacteriology*, p. 271, The Williams & Wilkins Co., Baltimore, 1923.
2. Moore, L.V.H., and Moore, W.E.C. *Oribaculum catoniae* gen. nov., sp. nov.; *Catonella mo&i* gen. nov., sp. nov.; *Halklla seregens* gen. nov., sp. nov.; *Johnsonella ignava* gen. nov., sp. nov.; and *Dialisterpneumosintes* comb. nov., nom. rev., anaerobic gram-negative bacilli from the human gingival crevice, *Int. J. Syst. Bacteriol.*, 44, 187, 1994.
3. Olitsky, P., and Gates, F.L. Experimental studies of the nasopharyngeal secretions from influenza patients, *J. Exp. Med.*, 33, 713, 1921.
4. Downes, J., Munson, M., and Wade, W.G. *Dialister invisus* sp. nov., isolated from the human oral cavity, *Int. J. Syst. Evol. Microbiol.*, 53, 1937, 2003.
5. Jumas-Bilak, E. et al., *Dialister micraerophilus* sp. nov. and *Dialister propionicifaciens* sp. nov., isolated from human clinical samples, *Int. J. Syst. Evol. Microbiol.*, 55, 2471, 2005.
6. Hitchens, A.P., *Bergey's Manual of Determinative Bacteriology*, 7th ed., p. 440, The Williams & Wilkins Co., Baltimore, 1957.
7. Cato, E.P., Holdeman, L.V., and Moore, W.E.C. Proposal of neotype strains for seven non-saccharolitic Bacteroides species, *Int. J. Syst. Bacteriol.*, 29, 427, 1979.
8. Holdeman, L.V., and Moore, W.E.C., Bacteroides. In Cato, E.P. et al. (Editors), *Outline of Clinical Methods in Anaerobic Bacteriology*, 2nd rev, p. 33, Virginia Polytechnic Institute and State University, Blacksburg, VA, 1970.
9. Shah, H.N., and Collins, M.D. Proposal to restrict the genus *Bacteroides* (Castellani and Chalmers) to *Bacteroides fragilis* and closely related species, *Int. J. Syst. Bacteriol.*, 39, 85, 1989.
10. Willems, A., and Collins, M.D. Phylogenetic placement of *Dialister pneumosintes* (formerly *Bacteroides pneumosintes*) within the *Sporomusa* subbranch of the *Clostridium* subphylum of the Gram-positive bacteria, *Int. J. Syst. Bacteriol.*, 45, 403, 1995.
11. Holt, J.G. et al., *Bergey's Manual of Determinative Bacteriology*, 9th ed., p. 317, The Williams & Wilkins Co., Baltimore, 1994.
12. Holdeman, L.V., Kelly, W.E., and Moore, W.E.C. *Bergey's Manual of Systematic Bacteriology*, vol. 1, p. 628, The Williams & Wilkins Co., Baltimore, 1984.
13. Doan, N. et al., Molecular detection of *Dialister pneumosintes* in subgingival plaque of humans, *J. Clin. Microbiol.*, 38, 3043, 2000.
14. Evaldon, G. et al., Microbiological findings in pregnant women with rupture of the membranes, *Med. Microbiol. Immunol.*, 168, 283, 1980.

15. Moore, W.E.C., and Moore, L.V., The bacteria of periodontal diseases, *Periodontol. 2000*, 5, 66, 1994.
16. Contreras, A. et al., Importance of *Dialister pneumosintes* in human periodontitis, *Oral Microbiol. Immunol.*, 15, 269, 2000.
17. Paster, B.J. et al., Bacterial diversity in human subgingival plaque, *J. Bacteriol.*, 183, 3770, 2001.
18. Ghayoumi, N., Chen, C., and Slots, J. *Dialister pneumosintes*, a new putative periodontal pathogen, *J. Periodont. Res.*, 37, 75, 2002.
19. Nonnenmacher, C. et al., Quantitative detection of periodontopathogens by real-time PCR, *J. Microbiol. Methods*, 59, 117, 2004.
20. Gomes, S.C. et al., Periodontal status in smokers and never-smokers: Clinical findings and real-time polymerase chain reaction quantification of putative periodontal pathogens, *J. Periodontol.*, 77, 1483, 2006.
21. Ferraro, C.T.L. et al., Detection of *Dialister pneumosintes* in the subgingival biofilm of subjects with periodontal disease, *Anaerobe*, 13, 244, 2007.
22. Siqueira, J.F., and Rôças, I.N., Positive and negative associations involving *Dialister pneumosintes* in primary endodontic infections, *J. Endod.*, 20, 438, 2003.
23. Rôças, I.N., and Siqueira, J.F., Characterization of Dialister species in infected root canals, *J. Endod.*, 32, 1057, 2006.
24. Wade, W.G. et al., Molecular detection of novel anaerobic species in dentoalveolar abscesses, *Clin. Infect. Dis.*, 25, 235, 1997.
25. Rousse, J.M. et al., *Dialister pneumosintes* associated with human brain abscesses, *J. Clin. Microbiol.*, 40, 3871, 2002.
26. Bahrani-Mougeot, F.K. et al., Molecular analysis of oral and respiratory bacterial species associated with ventilator-associated pneumonia, *J. Clin. Microbiol.*, 45, 1588, 2007.
27. Goldstein, E.J.D., Citron, D.M., and Finegold, S.M. Role of anaerobic bacteria in bite-wound infections, *Rev. Infect. Dis.*, 6, 177, 1984.
28. Lepargneur, J.P., Dubreuil, L., and Levy, J., Isolation of *Dialister pneumosintes* isolated from a bacteremia of vaginal origin, *Anaerobe*, 12, 274, 2006.
29. Morio, F. et al., Antimicrobial susceptibilities and clinical sources of *Dialister* species, *Antimicrob. Agents Chemoter.*, 51, 4498, 2007.
30. Page, R.C. et al., Advances in the pathogenesis of periodontitis: Summary of developments, clinical implications and future directions, *Periodontol. 2000*, 14, 216, 1997.
31. Armitage, G.C., Development of a classification system for periodontal diseases and conditions, *Ann. Periodontol.*, 4, 1, 1999.
32. Siqueira, J.F., and Rôças, I.N., Endodontic microbiology. In Torabinejad, M., and Walton, R. (Editors), *Endodontics: Principles and Practice*, p. 38, Saunders/Elsevier, St. Louis, 2009.
33. Torabinejad, M., and Shababang, S., Pulp and periapical pathosis. In Torabinejad, M., and Walton, R. (Editors), *Endodontics: Principles and Practice*, p. 49, Saunders/Elsevier, St. Louis, 2009.
34. Socransky, S.S. et al., Microbial complexes in subgingival plaques, *J. Clin. Periodontol.*, 25, 134, 1998.
35. Colombo, A.P.V. et al., Subgingival microbiota of refractory periodontitis determined by microarray (HOMIM), *J. Periodontol.*, 80, 1424, 2009.
36. Prévot, A.R., Turpin, A., and Kaiser, P., *Les bactéries anaérobies*, Dunod, Paris, 1967.
37. Wang, P.-L., and Ohura, K., *Porphyromonas gingivalis* lipopolysaccharide signaling in gingival fibroblasts-CD14 and toll-like receptors, *Crit. Rev. Oral Biol. Med.*, 13, 132, 2002.
38. Kamma, J.J., Contreras, A., and Slots, J., Herpes viruses and periodontopathic bacteria in early-onset periodontitis, *J. Clin. Periodontol.*, 28, 879, 2001.
39. Slots, J., Sugar, C., and Kamma, J.J., Cytomegalovirus periodontal presence is associated with subgingival *Dialister pneumosintes* and alveolar bone loss, *Oral Microbiol. Immunol.*, 17, 369, 2002.
40. Mullis, K.B., and Faloona, F.A., Specific synthesis of DNA in vitro via a polymerase-catalyzed chain reaction, *Methods Enzymol.*, 155, 335, 1987.
41. Larsen, N. et al., The Ribosomal Database Project, *Nucleic Acids Res.*, 21, 3021, 1993.
42. Higuchi, R. et al., Kinetic PCR: Real-time monitoring of DNA amplification reactions, *BioTechniques*, 11, 1026, 1993.
43. Sakamoto, Y.T. et al., Rapid detection and quantification of five periodontopathic bacteria by real-time PCR, *Microbiol. Immunol.*, 45, 39, 2001.
44. Fredricks, D.N., and Relman, D.A., Sequence-based identification of microbial pathogens: A reconsideration of Koch's postulates, *Clin. Microbiol. Rev.*, 9, 18, 1996.
45. Kroes, I., Lepp, P.W., and Relman, D.A., Bacterial diversity within the human subgingival crevice, *Proc. Natl. Acad. Sci. USA*, 96, 14547, 1999.
46. Munson, M.A. et al., Molecular and cultural analysis of the microflora associated with endodontic infections, *J. Dent. Res.*, 81, 761, 2002.
47. Kumar, P.S. et al., Identification of candidate periodontal pathogens and beneficial species by quantitative 16S clonal analysis, *J. Clin. Microbiol.*, 43, 3944, 2005.
48. Ribosomal Database Project. Available at: http://wdcm.nig.ac.jp/RDP/html. Accessed July 31, 2009.
49. GenBank. Available at: http://www.ncbi.nlm.nih.gov/. Accessed July 31, 2009.
50. Socransky, S.S. et al., Checkerboard DNA–DNA hybridization, *BioTechniques*, 17, 788, 1994.
51. Socransky, S.S. et al., Use of checkerboard DNA–DNA hybridization to study complex microbial ecosystems, *Oral Microbiol. Immunol.*, 19, 352, 2004.
52. Paster, B.J., Bartoszyk, I.M., and Dewhirst, F.E., Identification of oral streptococci using PCR-based, reverse-capture, checkerboard hybridization, *Meth. Cell Sci.* 20, 223, 1998.
53. Siqueira, J.F. et al., Comparison of 16S rDNA-based PCR and checkerboard DNA–DNA hybridization for detection of selected endodontic pathogens, *J. Med. Microbiol.*, 51, 1090, 2002.
54. Colombo, A.P. et al., Subgingival microbiota of Brazilian Subjects with untreated chronic periodontitis, *J. Periodontol.*, 73, 360, 2002.
55. Human Oral Microbe Identification Microarray (HOMIM). Available at: http://mim.forsyth.org/homim.html. Accessed July 31, 2009.
56. Preza et al., Microarray analysis of the microflora of root caries in elderly, *Eur. J. Clin. Microbiol. Infect. Dis.*, 46, 2015, 2008.
57. Laine, M.J. et al., The mouthwash: a noninvasive sampling method to study cytokine gene polymorphisms, *J. Periodontol.*, 71, 1315, 2000.
58. Ashimoto, A. et al. Polymerase chain reaction detection of 8 putative periodontal pathogens in subgingival plaque of gingivitis and advanced periodontitis lesions, *Oral Microbiol. Immunol.*, 11, 266, 1996.
59. Dahlén, G. et al., A comparison between two transport media for saliva and subgingival samples, *Oral Microbiol. Immunol.*, 8, 375, 1993.
60. Human Oral Microbe Identification Microarray (HOMIM). Available at: http://bioinformatics.forsyth.org/homim/. Accessed July 31, 2009.

35 Eubacterium

Johanna Maukonen and Maria Saarela

CONTENTS

35.1 INTRODUCTION

Eubacterium species are found in the mammalian intestinal tract and oral cavity, in addition to the environment (Table 35.1). The genus *Eubacterium* was originally established to describe a group of fecal bacteria that were considered to be beneficial.[65] However, some species might be considered to be opportunistic,[66,67] even though no primary-disease-causing species of *Eubacterium* have been isolated and no specific virulence factors have been identified. Since this book covers human bacterial pathogens, only those *Eubacterium* species that have been found in humans are discussed in detail.

35.1.1 Taxonomy and Phylogeny

Taxonomy. Cells of *Eubacterium* species are strictly anaerobic, gram-positive, nonsporing, irregular rods—the shapes vary from cocci to long rods—often with swollen or tapered ends, and sometimes even curved. The DNA G + C content varies between 30 and 57 mol%. The cells of *Eubacterium* species are usually arranged one by one, in pairs, or in chains. The motility is variable. Young cultures usually stain gram-positive.[68] However, some *Eubacterium* species (e.g., *Eubacterium plautii*) stain gram-negative with occasional weak gram-positive areas,[69] or they may decolorize rather easily (e.g., *Eubacterium cylindroides*).[70] This variable staining has caused confusion; for example, *Eubacterium sulci* was originally described as *Fusobacterium sulci*.[71,72] Therefore, these species can be phenotypically classified correctly after confirming the absence of lipopolysaccharides and after the analysis of the outer membrane proteins.[73]

The genus *Eubacterium* includes strains that do not produce (i) propionic acid as major acid (contrary to *Propionibacterium*); (ii) lactic acid alone (contrary to *Lactobacillus*); (iii) more acetic acid than lactic acid with and without formic acid (contrary to *Bifidobacterium*); (iv) succinic acid (in the presence of carbon dioxide) and lactic acid with small amounts of acetic or formic acid (contrary to Actinomyces).[74] Because of the above-mentioned rather loose genus definition, the genus *Eubacterium* is very heterogeneous. Spore formation is the principal characteristic by which strains have been (and still are) placed within *Clostridium* species rather than *Eubacterium* species. Therefore, when phenotypic identification methods have been used, clostridial strains that have lost their capacity to form spores or that require very special conditions for the induction of spore formation may have incorrectly been classified as *Eubacterium* species.[70] Furthermore, the intermingling of the species of these two genera throughout the phylum Firmicutes suggests either that sporulation has emerged frequently throughout evolution or that it is a feature that has been lost from some species over time.[75] Besides being phylogenetically closely related, many phenotypic features are shared between the genera *Eubacterium* and *Clostridium*. Most of the saccharolytic *Eubacterium* species produce butyric acid and hydrogen gas, as do many saccharolytic *Clostridium* species.[70] In addition to ethanol, butanol is formed by *Eubacterium saburreum* and *Eubacterium tenue*, as they are also by some solvent-producing clostridia.[76] Furthermore, *Eubacterium* species have the ability to degrade amino acids, since isobutyric and isovaleric acids have been shown to be produced by *Eubacterium acidaminophilum*, *Eubacterium brachy*, *Eubacterium callandri*, *Eubacterium combesii*, *Eubacterium limosum*, and *Eubacterium tenue*,[74,77,78] and phenylacetic acid by *Eubacterium timidum*.[66]

Phylogeny. The genus *Eubacterium* is phylogenetically diverse. Because of its loose taxonomic definition, the *Eubacterium* species are widely scattered along the phylogenetic tree,[79] having species in families Erysipelotrichaceae,

TABLE 35.1
Habitats of the *Eubacterium* Species. The Species Transformed from Genus *Eubacterium* Are Denoted with "*"

Species	Human Intestine	Human Oral Cavity	Human Infections	Infection Site/s (Habitat of the Nonhuman Species)
Eubacterium acidaminophilum	–	–	–	(Environment)
Eubacterium aggregans	–	–	–	(Environment)
Eubacterium angustum	–	–	–	(Environment)
Eubacterium barkeri	•[a,b]	–	–	(Environment)
Eubacterium biforme	•	–	–	
Eubacterium brachy	–	•	•	Periodontitis,[1–3] deep caries,[4] vertebral abscess,[5] liver abscess,[6] and infections of head, neck and lung[6–8]
Eubacterium budayi	•[c]	–	–	(rumen)
Eubacterium callanderi	–	–	•	Bacteremia[9] (Environment)
Eubacterium cellulosolvens	•[c]	–	–	(Rumen)
Eubacterium combesii	•[a]	–	•	Penile ulcerations (a single case),[10] bacteremia after dental extraction[11]
Eubacterium contortum	•	–	•	Anaerobic empyema[12] and bacteremia after dental extraction[11]
Eubacterium coprostanoligenes	–	–	–	(Environmental)
Eubacterium cylindroides	•	–	–	Abdominal wounds, gynecological, and head infections[13]
Eubacterium desmolans	•	–	–	
Eubacterium dolichum	•	–	–	
Eubacterium eligens	•	–	–	
Eubacterium fissicatena	–	–	–	(Alimentary tract of goat)
Eubacterium hadrum	•	–	–	
Eubacterium hallii	•	–	–	
Eubacterium infirmum	–	•	•	Endodontic infections[14,15] and periodontitis[16,17]
Eubacterium limosum[d]	•	–	•	Bacteremia,[13,18] abscess, chronic mastoiditis, burns, bites,[19] perapical lesions,[20,21] and penile ulcerations (a single case)[10]
Eubacterium minutum	–	•	•	Periodontal pockets[22] and endodontal infections[8]
Eubacterium moniliforme	•	–	•	Abscess (a single case),[19] wounds, head infections, bacteremia,[13] and penile ulcerations[10]
Eubacterium multiforme	•[e]	–	•	Apical periodontitis, decubitus ulcers (only a single case or isolate)[19,23] (Environment)
Eubacterium nitritogenes	•	–	–	
Eubacterium nodatum	–	•	•	Periodontitis,[1,17,24–29] endodontic infections,[8,30] intrauterine device-associated infections; endometrium or IUD,[6,31] various infections of head, neck and lungs, liver and pelvic abscess; thumb wound, necrotizing fascitis, and amniotic fluid[6]
Eubacterium oxidoreducens	–	–	–	(Rumen)
Eubacterium plautii	–	–	•	Fulminant infection following dog bite[32] and infection in a kidney transplant recipient (involving pleura effusion)[33]
Eubacterium plexicaudatum	–	–	–	(Cecum of rodents)
Eubacterium pyruvativorans	–	–	–	(Rumen)
Eubacterium ramulus	•	–	–	
Eubacterium rectale	•	–	•	Apical periodontitis (only a single isolate)[23] and penile ulceration (a single case)[10]
Eubacterium ruminantium	–	–	–	(Rumen)
Eubacterium saburreum	•[a]	•	•	Endodontic infection,[34,35] periodontitis,[36] penile ulcers,[10] and bacteremia after dental extraction (a single case)[37]
Eubacterium saphenum	–	•	•	Periodontal[3,38,39] and endodontal infections[8,40]
Eubacterium siraeum	•	–	–	
Eubacterium sulci	–	•	•	Dorsum of the tongue,[41] endodontal infections,[8] and periodontitis[3,42]
Eubacterium tarantellae	–	–	–	(Fish pathogen)
(*Eubacterium tardum* →) *Eubacterium minutum*	–	•	•	Type strain isolated from periodontal pocket[16]
Eubacterium tenue	•	•	•	Bacteraemia,[11,43] pulmonary aspiration pneumonia, paronychia, gastrostomy site, decubitus ulcers, conjunctivitis,[19] and culdocentensis (a single case)[44]

continued

TABLE 35.1 (Continued)
Habitats of the *Eubacterium* Species. The Species Transformed from Genus *Eubacterium* Are Denoted with "*"

Species	Human Intestine	Human Oral Cavity	Human Infections	Infection Site/s (Habitat of the Nonhuman Species)
Eubacterium tortuosum	•	–	•	Bacteremia,[45] head infections,[13] and peritoneal fluid (a single case)[46] (Animal pathogen; liver granulomas)[47]
Eubacterium uniforme	–	–	–	(sheep rumen)
Eubacterium ventriosum	•	–	•	Bacteremia[45]
Eubacterium xylanophilum	–	–	–	(sheep rumen, pig GI tract)
Eubacterium yurii	–	•	•	Subgingival dental plaque[48]
Eubacterium yurii subsp. *margaretiae*	–	•	•	Subgingival dental plaque[48]
Eubacterium yurii subsp. *schtitka*	–	•	•	Subgingival dental plaque[49]
Eubacterium yurii subsp. *yurii*	–	•	•	Subgingival dental plaque[48]
*Actinobaculum suis**	–	–	–	(Infections in sows and pigs)
*Atopobium fossor**	–	–	–	(Respiratory pathogen of horses)
*Collinsella aerofaciens**	•	•	•	Tonsillitis,[50] endodontic infections,[51] HIV-related periodontitis,[52] bacteremia after dental extraction,[11,37] abdominal wound (a single case),[13] peritoneal fluid (a single case),[46] penile ulcers (a single case),[10] and culdocentensis (a single case)[44]
*Dorea formicigenerans**	•	–	–	
*Eggerthella lenta**	•	–	•	Bacteremia,[11,13,37,43,53,54] abscess, bite, peritonitis, conjunctivitis,[19] endodontic infections,[20,55] wounds, gynecological and head infections,[13] and penile ulcers[10]
*Mogibacterium timidum**	–	•	•	Periodontitis,[1,38] dental abscess,[56] and endodontic infections[40]
*Pseudoramibacter alactolyticus**	–	•	•	Apical periodontitis,[57] endodontic infections,[55,58–61] and bacteremia[37,53]
*Slackia exigua**	–	•	•	Periodontal disease[28] and endodontic infections[40]

[a] Detected only in reference 62.
[b] Type strain isolated from river mud.
[c] Detected only in reference 63.
[d] Type species.
[e] Detected only in reference 64.

Ruminococcaceae, Clostridiaceae, Peptostreptococcaceae, Eubacteriaceae, Lachospiraceae,[79] and Incertae Sedis XIII[80] within phylum Firmicutes. At present, nearly full-length 16S rRNA sequences have been determined for 38 *Eubacterium* spp. type strains (Figure 35.1). *Eubacterium limosum*, the type species of genus *Eubacterium*, forms a family of Eubacteriaceae with *Eubacterium callanderi*, *Eubacterium aggregans*, and *Eubacterium barkeri* along with, for example, *Acetobacterium* spp.,[79] and it has been proposed that these *Eubacterium* species should form a genus *Eubacterium* sensu stricto.[81] Other major clusters of *Eubacterium* spp. include species within the family Erysipelotrichaceae (*Eubacterium biforme*, *Eubacterium cylindroides*, *Eubacterium dolichum*, and *Eubacterium tortuosum*; clostridial cluster XVI[82]),[79,83] species within the family Lachnospiraceae (*Eubacterium cellulosolvens*, *Eubacterium rectale*, *Eubacterium eligens*, *Eubacterium ventriosum*, *Eubacterium xylanophilum*, *Eubacterium ruminantium*, and *Eubacterium hallii*; clostridial cluster XIVa[82])[79]; species within the family Ruminococceae (*Eubacterium siraeum*, *Eubacterium plautii*, and *Eubacterium desmolans*; clostridial cluster IV[82]),[79] species within the family Clostridiaceae (*Eubacterium multiforme*, *Eubacterium nitritogenes*, *Eubacterium budayi*, and *Eubacterium combesii*,[79] in addition to *Eubacterium tarantellae* and *Eubacterium moniliforme*[83]; clostridial cluster I[82]), and the oral asaccharolytic *Eubacterium* species within the family Incertae Sedis XIII (*Eubacterium minutum*, *Eubacterium nodatum*, *Eubacterium pyruvativorans*, *Eubacterium infirmum*, *Eubacterium sulci*, *Eubacterium brachy*, and *Eubacterium saphenum*[80]). The species belonging to the family Clostridiaceae fall within the genus *Clostridium* sensu stricto, which highlights again the loose definition of *Eubacterium* species. Furthermore, *Eubacterium yurii* has been proposed to belong to clostridial cluster XIVb,[83] but in the recent all-species living tree that has been constructed from all type strains of all bacterial species with validly published names,[79] *Eubacterium yurii* subsp. *schtitka* and *Eubacterium yurii* subsp. *margaretiae* fall in the middle of the family Peptostreptococceae (clostridial cluster XI[82]).[79] The rest of the *Eubacterium* species are either scattered among the Firmicutes, having *Clostridium* species as close relatives, or did not have a validly 16S rRNA sequenced type strain to be included in the all-species living tree in August 2008.[79]

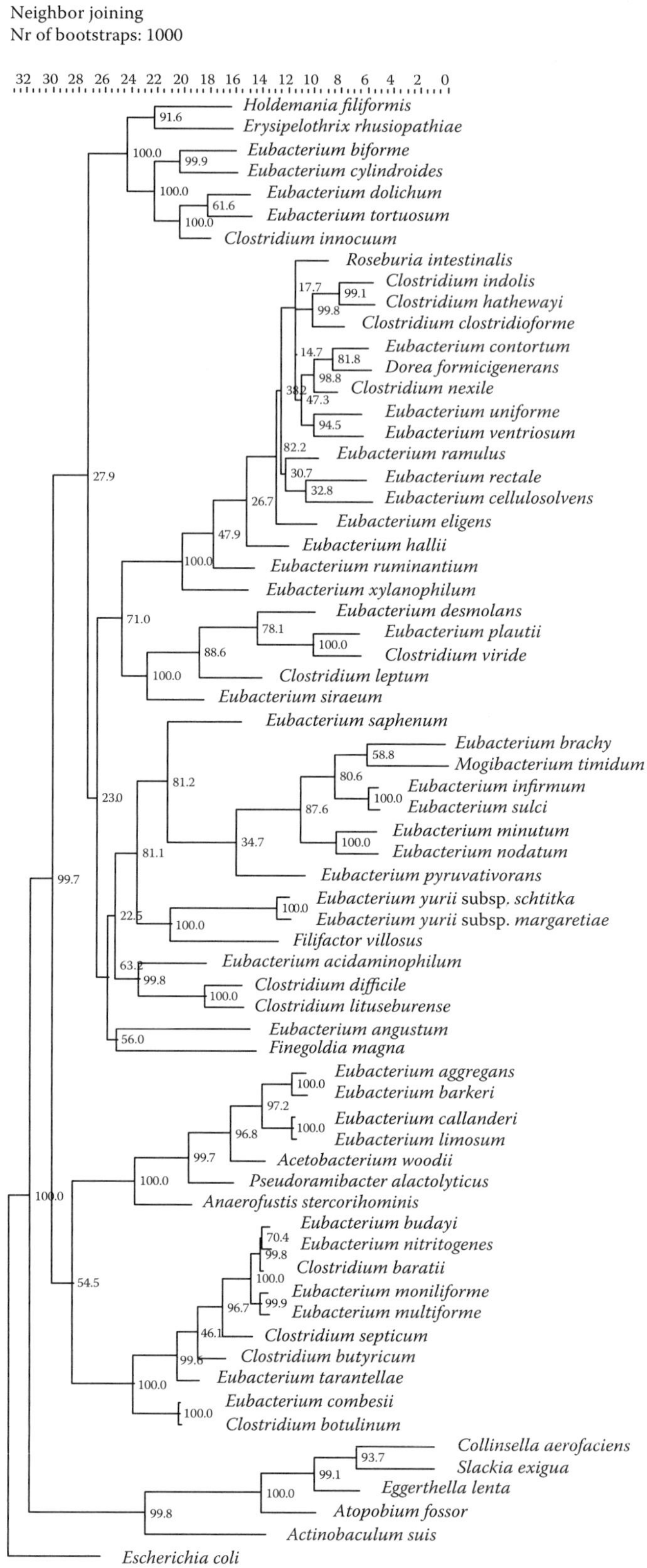

FIGURE 35.1 Phylogenetic tree of the *Eubacterium* species, showing the relationship between the *Eubacterium* species within for example, the families Erysipelotrichaceae, Ruminococcaceae, Clostridiaceae, Peptostreptococcaceae, Eubacteriaceae, Lachospiraceae, and Incertae Sedis XIII within phylum Firmicutes. The phylogenetic tree was constructed from evolutionary distances of nearly full-length 16S rRNA sequences of type strains by the neighbor-joining method using the Kodon software (Applied Maths, Belgium).

There are several species that have been transferred (Table 35.1)—based on their phylogeny—from the genus *Eubacterium* to other genera within the phylum Actinobacteria (a phylum of gram-positive bacteria with mainly high G + C content), such as *Eggerthella lenta*,[84] *Collinsella aerofaciens*,[85] *Pseudoramibacter alactolyticus*,[81] *Slackia exigua*,[84] *Dorea formigenerans*,[86] *Atopobium fossor*,[87] and *Mogibacterium timidum*.[88] *Eubacterium suis* has been reclassified twice; at first, it was reclassified as *Actinomyces suis*,[89] and thereafter as *Actinobaculum suis*.[90] In addition, *Eubacterium tardum*[91] and *Eubacterium minutum*[22] were described within the same year, and since phenotypic and genetic analyses of type strains were identical, *Eubacterium minutum* had precedence by prior publication.[92] Furthermore, since majority of the current *Eubacterium* species do not belong in the genus *Eubacterium* sensu stricto,[81] they most likely will be moved to other existing or novel genera in the future.[75]

35.1.2 Clinical Features and Pathogenesis

Eubacterium species are present in the mammalian intestinal tract and oral cavity as well as the environment (Table 35.1). The genus *Eubacterium* was originally established to describe a group of fecal bacteria that were considered to be beneficial.[65] Many *Eubacterium* species produce butyrate,[42,93] which is the preferred energy source for the epithelial cells of the colon, playing an important role in the energy metabolism and normal development of these cells,[94] in addition to its protective role in colonic diseases.[95–97] In early cultivation based quantitative studies of human intestinal microbiota, *Eubacterium* species were found from 94% of subjects, with a median count of 5×10^{10} CFU/g being the second most numerous bacterial group.[62] However, 58% of the isolates, accounting for 3×10^9 CFU/g could not be speciated with phenotypic methods.[62] Furthermore, *C. aerofaciens* and *Eg. lenta* were detected from over 40% of the subjects and *E. contortum*, *E. cylindroides*, and *E. rectale* from over 20% of subjects. Since *C. aerofaciens* and *Eg. lenta* were classified as *Eubacterium* species at that time, the average number of *Eubacterium* species present in human intestinal microbiota is lower than the numbers reported by Finegold et al.[62] Furthermore, in a molecular based study of Schwiertz et al.[98] in which fluorescent in situ hybridization (FISH) was used for quantification of 10 different *Eubacterium* species, the counts were at least 10–100 times lower than with the cultivation based detection.[62] *E. barkeri*, *E. dolichum*, *E. limosum*, and *E. moniliforme* were not detected at all with the detection limit of 10^7 CFU/mL.[98] The numbers of *E. ramulus* ($\sim 7 \times 10^8$ CFU/g) have been similar with both conventional and molecular techniques.[61,99] *E. ramulus*, in particular, has raised interest since it is capable to degrade flavonoids, which are regarded to be beneficial to health.[99,100] In addition, an environmental *E. coprostanoligenes* species is capable of reducing cholesterol to coprostanol, but so far it has only been used in animal experiments.[101–103]

E. limosum, *E. combesii*, *E. contortum*, *E. moniliforme*, *E. multiforme*, *E. rectale*, *E. ventriosum*, and *E. tortuosum*, which are part of the indigenous intestinal microbiota, have been detected in human infections (Table 35.1). *E. limosum* is the only intestinal (present) *Eubacterium* species which has been detected from several infection sites, namely, abscess, chronic mastoiditis, burns, bites,[19] penile ulcerations,[10] and bacteremia[18]; whereas the other intestinal *Eubacterium* species mentioned above have only been detected in sporadical cases[11,12,19,23,45] *Eubacterium tenue*, on the other hand, has been detected from both gastro-intestinal and oral microbiota besides several infection sites, namely, bacteremia, pulmonary aspiration pneumonia, paronychia, gastrostomy site, conjunctivitis, and decubitus ulcers.[11,19,104]

Most of the *Eubacterium* species isolated from human oral cavity (Table 35.1) are quite asacchorolytic and difficult to cultivate with standard media. A pathogenic role may be indicated by the high incidence (35%–42% of samples cultured) and concentration (3%–57% of cultivable microbiota) of *E. brachy*, *E. nodatum*, and *E. timidum*.[1,66] Oral *Eubacterium* species *E. brachy*, *E. infirmum*, *E. minutum*, *E. nodatum*, *E. saburreum*, *E. saphenum*, *E. sulci*, *E. yurii* subsp. *yurii*, *E. yurii* subsp. *schtitka*, and *E. yurii* subsp. *margaretiae* have all been found from various oral infections. Besides oral infections, *E. brachy* has been detected in, for example, vertebral abscess,[5] infections of head, neck, and lung[6–8] and liver abscess[66] and *E. nodatum* in intrauterine device–associated infections,[31,66] and from various infections of head, neck and lungs, liver abscess, pelvic abscess, thumb wound, necrotizing fascitis, and amniotic fluid.[66]

When fecal or oral *Eubacterium* species have been studied, typically only one of the two sites has been sampled from the volunteers/patients. Therefore there is sparse knowledge on the potential simultaneous presence of the same *Eubacterium* species in both sites, although it seems that several *Eubacterium* species are repeatedly isolated only from fecal or oral samples. In a report by Maukonen et al.[105] both fecal and salivary samples from the same subjects were studied with denaturing gradient gel electrophoresis (DGGE) to define the temporal stability, similarity, and diversity of the clostridial cluster XIVa, which includes several *Eubacterium* species. It was found that the profiles of fecal and salivary samples were completely different, and since Maukonen et al.[106] showed that each band position in DGGE-gels contained only one phylotype, it may be assumed that the phylotypes present in the feces and saliva of a given person are mostly different, as has been shown in studies where oral samples or fecal samples alone have been analyzed from different people.

There are also environmental *Eubacterium* species, which have been detected in infections in humans. *E. callanderi* has been identified from a single case of bacteremia in a patient with a bladder carcinoma[9] and *E. plautii* from a fulminant infection following a dog bite[32] and from an infection in a kidney transplant recipient involving pleural effusion.[33]

35.1.3 Diagnosis

Eubacterium species are typically detected only from polymicrobial infections. In addition, since this genus is part of the human indigenous microbiota and only an opportunistic pathogen, there are no clear symptoms which would indicate that the infection in question involves *Eubacterium* species. In addition, no specific virulence factors have been characterized from *Eubacterium* species. *Eubacterium* species are most likely to be recovered from oral infections,[66,67] abscess, body fluids, or wound samples in mixed cultures, along with gram-negative rods or facultative organisms.[107] *Eubacterium* species have been detected, for example, in appendicitis, kidney abscess, bladder abscess, Bartholin gland abscess, scrotal or testicular abscess, periapical abscess, breast abscess, miscellaneous skin soft-tissue and bone infections, orofacial infections, head and neck infections, thorax infections, pelvis infections, periodontitis, blood, and peritoneal fluid.[19,66,67,107] Since *Eubacterium* species are also part of the endogenous microbiota of the oral cavity, genital tract, and gut, contamination by normal microbiota during specimen collection is often a problem. Therefore, communication with the person submitting the specimen for culture is desirable and sometimes essential.[107]

The most common mistake when wound, abscess, or body-fluid samples from urogenital and oral sites are handled is the failure to maintain stringent anaerobic condition that is critical for isolation of fastidious and slow-growing bacteria such as *Eubacterium* species.[107] Special attention should be paid to the uninterrupted anaerobic isolation and incubation to ensure an opportunity to detect these bacteria, since most of the *Eubacterium* species are highly sensitive to oxygen.[70] Only *E. desmolans* can tolerate a 4-h exposure to air.[108] The most important parameter is, therefore, the time elapse between specimen collection and anaerobic handling.[107] In addition, specimens should be collected in a way that avoids contamination by the normal microbiota and extra exposure to air.[67] Furthermore, extended (over 48 h) anaerobic incubation is required to accurately define anaerobic etiological agents.[107] There are also some special requirements for some species, for example, *E. yurii* is especially sensitive to detergent residues on glassware, and successful cultivation requires intensive rinsing of all glassware used.[70]

If phenotypic methods are used for the identification of possible *Eubacterium* isolates, it should be remembered that certain *Eubacterium* species cannot be properly differentiated, as, for example, *E. budayi* and *E. nitrogenes* have very close phenotypes.[70] In addition, if the strains of those species are also able to hydrolyze gelatin, they may be misclassified as *Clostridium perfringens*, which is also possible for *E. moniliforme* and *E. ventriosum*. Further, variations in acid formation from maltose or sucrose and the amount of H_2 produced may lead to the identification of *Lachnospira multipara* instead *of Collinsella aerofaciens* or of *Eubacterium eligens* and *Eubacterium tarantellae* according to the differentiation scheme.[74] Nonsaccharolytic species should be positively characterized, instead of exclusion identification, by their individual abilities, as has been attempted for *E. brachy*, *E. nodatum*, and *E. timidum*.[66] In addition, many *Eubacterium* species are nonreactive in the commonly used biochemical and physiological tests, and therefore particular caution should be exercised when commercially available rapid identification kits are used.[75] In a study by Downes et al.,[109] 105 isolates from oral infections which had been phenotypically identified as *Eubacterium* species were phenotypically and genotypically rechecked. Out of the 105 isolates, only 91 isolates belonged to previously described species and only 11 isolates (10%) belonged to the genus *Eubacterium* as it is described at present. Moreover, only 43 isolates (41%) belonged to the genus *Eubacterium* as it was described in 1990. In an separate study by Petti et al.,[110] where phenotypic identification indicated that the bacteremia was caused by *Eubacterium tenue*, the correct infective agent was *Cardiobacterium valvarum* as verified with 16S rRNA sequencing. These examples highlight the difficulty of phenotypic identification of the genus *Eubacterium*.

Since the genus *Eubacterium* is phylogenetically highly diverse, it is not possible to design genus-specific probes/primers for a direct molecular detection and quantification. Therefore, the *Eubacterium* species of interest has to be detected individually or by using group-specific probes/primers that also target other members of the given clostridial cluster,[82] besides the intended *Eubacterium* species (Table 35.2). However, 16S rRNA sequencing of isolates is a reliable way to identify *Eubacterium* species. Since the genus is phylogenetically diverse, all the different species are separable by 16S rRNA sequencing. It has to be kept in mind that with conventional culture-based techniques only viable bacteria are detectable, whereas with DNA-based methods both dead and alive bacteria are detected. Furthermore, the validation of the molecular techniques to be used is of utmost importance. Even though there are "ready-to-use" protocols in the literature, each method should be validated in the individual laboratory to ensure correct and specific identification.

35.2 METHODS

35.2.1 Sample Preparation and Isolation

Since *Eubacterium* species are strict anaerobes, anaerobic techniques should be utilized throughout the sampling, transportation, isolation, and culture stages when culture-based methods are applied. Anaerobic isolation procedures are described in detail in, for example, *Wadsworth-KTL Anaerobic Bacteriology Manual*,[67] *The Prokaryotes*,[83] and in the *Manual of Clinical Microbiology*.[107]

A number of complex media are available for the isolation and cultivation of nonspore-forming gram-positive bacteria. Blood agars (5% sheep blood or rabbit blood) supplemented with hemin and vitamin K are most often recommended. Other media include CDC anaerobe blood agar, brain heart infusion agar, brucella agar, modified Schaedler agar, phenylethyl alcohol blood agar, and heart infusion

TABLE 35.2
***Eubacterium* Species- and Group-Specific Primers and Probes**

Fluorescent In Situ Hybridization	**Probe**	**Sequence**	**Reference**
E. barkeri and *E. aggregans*	Ebar1237	3′-TTA CCC AAC CCT GTT TCC-5′	98
E. biforme	Ebif462	3′-CCC TTA CTA CTC ACT CAC-5′	98
E. contortum	Econ1122	3′-GGC CTA AAC CGC TGG CTA CT-5′	98
E. cylindroides	Ecyl461	3′-TCC CTT ACT AGG CAC CCA-5′	98
E. cylindroides	Ecyl466	3′-TAC GAT ACA CCC ACT GCC-5′	98
E. desmolans	Edes635	3′-AAA GTT TTG ACR ACC AGA-5′	111
E. dolichum	Edol183	3′-TGT CTC CGA GAT GCC TCG-5′	98
E. hadrum	Ehad579	3′-GAC TTG CCA TAC CAC CTA CG-5′	98
E. hallii	Ehal578	5′-TTG CAC TGC CAC CTA CGC-3′	112
E. limosum and *E. callenderi*	E.lim1433	3′-TCG GAC ACT CTC TTG GCG-5′	98
E. moniliforme	E.mon84	3′-CCG CTA ATC CAT TTC CCG-5′	98
E. multiforme	Emul183	5′-GTT CCT TCA TGC GAA GGT-3′	113
E. ramulus	E.ram-0997	3′-GGG CCA CTG ACT TGT ACA-5′	99
E. rectale	Erec834	5′-CGA GAA GCA AUG CUU CCC-3′	113
E. rectale	Erec838	5′-CGG CAC CGA GAA GCA ATG CT-3′	113
E. saburreum	Esab1467	5′-AGT TAT CCT CCC TGC CTT-3′	113
E. tortuosum	Etor727	5′-AGA CCA GGC AAC CGC CTT-3′	113
E. tortuosum	Etor129	5′-CCA GTT ACA TGG GTA GGT-3′	113
E. ventriosum	Even66	3′-TCT GTC CAA GGT GCT TCG-5′	98
E. hallii subgroup[a]	Ehal1469	3′-CCA GTT ACC GGC TCC ACC-5′	114
Lachonospira multipara subgroup[b]	Lach571	3′-GCC ACC TAC ACT CCC TTT-5′	114
E. cylindroides subgroup[c]	Ecyl387	3′-CGC GGC ATT GCT CGT TCA-5′	114
R. intestinalis subgroup[d]	Rint623	5′-TTC CAA TGC AGT ACC GGG-3′	112
C. viride subgroup[e]	Cvir141	3′-ACT CTC RGC CCT TGT GGG-5′	111
E. cylindroides subgroup[f]	Ecyl387	3′-ACT TGC TCG TTA ACG GCG C-5′	111
Roseburia and *Eubacterium* group (subgroup of cluster XIVa)[d]	Rrec584	5′-TCA GAC TTG CCG (C/T)AC CGC-3′	115
E. rectale–*C. coccoides* clostridial cluster XIV[g]	Erec482	5′-GCT TCT TAG TCA RGT ACC G-3′	116
C. leptum–clostridial cluster IV[h]	Clept866	3′-GUG UUA UUC AWU AGG UGG-5′	111
C. leptum–clostridial cluster IV[h]	Clept1240	5′-GTT TTR TCA ACG GCA GTC-3′	117
PCR-Detection	**Primer**	**Sequence**	**Reference**
E. biforme[i]	BIFO-F	5′-ATA GGT AGC AGA CAA GCA TT-3′	118
	BIFO-R	5′-GAA CAG TTT CCA GAG CCA GTA C-3′	118
E. cylindroides[i]	CYLIN-F	5′-GAG AGA TCG CAT GAA CTT TC-3′	118
	CYLIN-R	5′-ACG GAT CAT TCC CTA TCC GT-3′	118
E. eligens[i]	ELI-F	5′-CCG CAT AAG CGC ACA ATG TT-3′	118
	ELI-R	5′-CCC TCC TGC ACT CCA GCC TT-3′	118
E. dolichum[i]	DOLI-F	5′-GGC TTC GAG GCA TCT CGG AGA C-3′	118
	DOLI-R	5′-ACT ATC ATG TCA TTC CCT AC-3′	118
E. limosum[i]	LIMO-F	5′-TGG ATC CTT CGG GTG ACA TT-3′	118
	LIMO-R	5′-CTC ATT GGG TAC CGT CAT TC-3′	118
E. rectale[i]	REC-F	5′-CAT TGC TTC TCG GTG CCG TC-3′	118
	REC-R	5′-ATT TGC TCG GCT TCA CAG CT-3′	118
E. saphenum	sap156F	5′-AAC CAC ATA AAA TCA TAG G-3′	40
	sap552F	5′-TCT ACT AAG CGC GGG GTG A-3′	41
	sap964R	5′-ATA CCC GAT TAA GGG TAC-3′	41
E. saphenum[j]	E.saphenum	5′-CCT CTG ACG TAC CCT TAA-3′	39
	L189	5′-GGT ACT TAG ATG TTT CAG TTC-3′	39
E. tortuosum[i]	TORT-F	5′-GGT AAC CTA CCC ATG TAA CT-3′	118
	TORT-R	5′-GGT AAG CTA CCG TCA CTT GT-3′	118
E. rectale–*C. coccoides* clostridial cluster XIVa[g]	Ccoc-F	5′-AAA TGA CGG TAC CTG ACT AA-3′	119
	Ccoc-R	3′-AAG CGT TCT TAC TTT GAG TTT C-5′	119

continued

TABLE 35.2 (Continued)
***Eubacterium* Species- and Group-Specific Primers and Probes**

PCR-Detection	Primer	Sequence	Reference
E. rectale–C. coccoides clostridial cluster XIVa[g] (with a TaqMan probe)	F_Ccoc07	5′-GAC GCC GCG TGA AGG A-3′	120
	R_Ccoc14	5′-AGC CCC AGC CTT TCA CAT C-3′	120
	P_Erec482	5′-dye-CGG TAC CTG ACT AAG AAG-3′	120
C. leptum–clostridial cluster IV[h]	Clept-F	5′-GCA CAA GCA GTG GAG T-3′	121
	Clept-R3	3′-AAC TGT TTT GCC TCC TTC-5′	121
C. leptum–clostridial cluster IV[h] (with a TaqMan probe)	F_Clept09	5′-CCT TCC GTG CCG SAG TTA-3′	120
	R_Clept08	5′-GAA TTA AAC CAC ATA CTC CAC TGC TT-3′	120
	P_Clept01	5′-dye-CAC AAT AAG TAA TCC ACC-3′	120
E. brachy subgroup[k]	27F	5′-AGAGTTTGATCCTGGCTCAG-3′	17
	557R	3′-TTC GCR TCC CAA ATT CCG-5′	17

[a] Includes, for example, *E. hallii* species.
[b] Includes, for example, *E. eligens* species.
[c] Includes, for example, *E. biforme*, *E. cylindroides*, *E. tortuosum*, and *E. dolichum* species.
[d] Includes, for example, *E. rectale* and *E. ramulus* species.
[e] Includes, for example, *E. plautii* species.
[f] Includes, for example, *E. cylindroides*, *E. biforme*, and *E. dolichum* species.
[g] Includes, for example, *E. cellulosolvens*, *E. rectale*, *E. eligens*, *E. ventriosum*, *E. xylanophilum*, *E. ruminantium*, and *E. hallii* species.
[h] Includes, for example, *E. siraeum*, *E. plautii*, and *E. desmolans* species.
[i] The species-specific PCR-reactions were performed as a nested protocol after universal PCR with primers 27F (5′-AGAGTTTGATCCTGGCTCAG-3′) and 1492R (5′-GGTTACCTTGTTACGACTT-3′).
[j] The species-specific PCR-reaction was performed as a nested protocol after universal PCR with primers 422 (5′-GGAGTATTTAGCCTT-3′) and 785 (5′-GGATTAGATACCCTGGTAGTC-3′).
[k] Includes, for example, *E. brachy*, *E. infirmum*, *E. nodatum*, and *E. tardum* species.

agar. Anaerobic blood culture media, enriched thioglycolate medium and chopped meat—glucose medium will also support the growth of these bacteria. However, media containing vancomycin should not be used.[107] In addition, the phosphate concentration should be kept below 5 mM, since some species are inhibited by higher concentrations.[78] In most cases, the above-mentioned media are adequate for the isolation of *Eubacterium* species. However, due to the partly unresolved nutritional requirements of several *Eubacterium* species, anaerobic complex media outlined in, for example, *Wadsworth-KTL Anaerobic Bacteriology Manual*[67] and *Bergey's Manual of Systematic Bacteriology*[74] may have to be supplemented with extra complex ingredients (vitamins, rumen fluid, volatile fatty acids, and/or special substrates such as cellobiose or maltose) to obtain reasonable growth for all *Eubacterium* species.[70] In addition, most media that are used for the isolation of *Eubacterium* species contain a salt solution, including $CaCl_2$, $MgCl_2$, potassium phosphate, $NaHCO_3$, and NaCl.

When only DNA-based molecular techniques are used for the detection and quantification of *Eubacterium* species, there is no need for anaerobic techniques. However, special attention should be paid on the aseptic working methods to prevent contamination of the samples. There are numerous published methods for both RNA and DNA extraction of different sample materials. In addition, there are commercial kits for different sample types. For DNA extraction of fecal samples, soil kits are usually suitable, with minor modifications,[106] in addition to the fecal kits. There are also several commercial kits for blood and tissue samples. Regardless, whether a commercial kit or a "manual" nucleic acid isolation procedure is used, it should be validated in the laboratory first. Especially with gram-positive bacteria, the cell disruption has to be rigorous enough to lyse the cell wall. If FISH is used, a detailed procedure is described in, for example, *Molecular Microbial Ecology Manual*.[122]

35.2.2 Detection Procedures

Direct microscopic examination of clinical samples after Gram staining is an excellent presumptive method for the detection of anaerobic gram-positive rods and should be used whenever possible.[107] Special attention should, however, be paid on, for example, the species *E. plautii*, which stains gram-negative,[69] and *E. cylindroides*, which decolorizes easily, leading to possible misidentification as *Faecalibacterium prausnitzii*.[70] *Wadsworth-KTL Anaerobic Bacteriology Manual* (Tables 6–16)[67] and *Manual of Clinical Microbiology* (Table 8)[107] contain details showing cellular characteristics of each clinically interesting *Eubacterium* species.

If phenotypic methods are used for the identification of possible *Eubacterium* isolates, the methods described by, for example, Holdeman et al.[123] can be used.[75] In addition, *Wadsworth-KTL Anaerobic Bacteriology Manual*

(Figures 6–24)[67] and *Manual of Clinical Microbiology* (Table 10)[107] contain flow charts for the phenotypic identification of *Eubacterium* species and in *The Prokaryotes*[70,75] there are tables showing different biochemical reactions of nearly all *Eubacterium* species. However, since *Eubacterium* genus is a genus of convenience based on negative identification and thus is very heterogeneous, phenotypic identification is not the first recommended identification method. Instead, 16S rRNA sequencing should be utilized for the reliable identification of the isolates.[105] The use of 16S rRNA sequencing enables also the identification of all the other genera; consequently, the correct identification will also be obtained in those cases in which the isolate does not belong to the genus *Eubacterium*.

Because *Eubacterium* species are opportunistic pathogens typically detected in multispecies infections, the use of PCR, quantitative real-time PCR (qPCR), or FISH for the direct identification and quantification of *Eubacterium* species from a sample is not necessarily cost-effective, especially since each *Eubacterium* species has to be detected separately. However, if a sample has only been prepared for molecular analysis, there are several published procedures for identification/quantification of *Eubacterium* species to a species or group level (Table 35.2). In addition, a checkerboard DNA–DNA hybridization method, including *E. nodatum*, has been developed for oral microbiota.[124]

Since 16S rRNA sequencing is the most useful molecular technique for the identification of *Eubacterium* spp., it is the only method that is presented in detail here.

(i) DNA Extraction Any commercial kit may be used for the extraction of DNA from the bacterial isolates. However, if fresh isolates are used, the fastest procedure is as follows: transfer one isolate into 20 μL of TE-buffer (100 μL of 0.5 M EDTA, 500 μL of 1M Tris; fill up to 50 mL with Milli-Q water, filter sterilize 0.22 μm). Lyze the bacterial cells in a 95°C water bath for 15 min. Centrifuge the lysate for 1 min at 13.000 rpm, and thereafter use the supernatant. If isolates have been stored at +4°C, heat lysis may not be sufficient enough. In that case, the following procedure, besides commercial kits, is recommended: suspend bacterial mass from one agar plate into sterile 500 μL of glass bead–water suspension (0.1 g of glass beads, Sigma G-1145/500 μL of water). Lyse the cells with, for example, a FastPrep machine (QBIOgene; 60 s, speed 6.5 m/s). Centrifuge the lysate and glass beads for 3 min at 13.000 rpm, and thereafter use the supernatant. Store the isolated DNA at −80°C.

(ii) 16S rRNA PCR Amplify 16S rRNA using primers 8F (5′-AGAGTTTGATYMTGGCTCAG-3′)[125] and BSR1541/20 (5′-AAGGAGGTGATCCAGCCGCA-3′)[126] in a PCR mixture containing 0.2 mM of each dNTP, 0.2 μM primer, and 3 U of Dynazyme II DNA polymerase (Finnzymes) in 10× Dynazyme buffer. For PCR amplification use the following conditions: 1 cycle of 94°C for 5 min; 35 cycles of 94°C for 45 s, 56°C for 45 s, and 72°C for 2 min; and final extension for 10 min at 72°C.[105] Check the amplification products using agarose gel electrophoresis. Purify the PCR products with the Qiaquick PCR purification kit (Qiagen). Check the purified amplification products using agarose gel electrophoresis.

(iii) Sequencing Reaction In most cases, a single sequencing reaction with a primer 8F, producing a 500–800 base pair long sequence, is adequate for identification purposes. However, if the whole 16S rRNA sequence is needed, sequencing reactions with the following primers are also performed: 8F, 341F (5′-CCTACGGGAGGCAGCAG-3′),[127] U772f (5′-AACAGGATTAGATACCCTGGTAGTCC-3′),[128] BSF1099/16 (5′-GYAACGAGCGCAACCC-3′),[126] BSR1541/20 (5′-AAGGAGGTGATCCAGCCGCA-3′),[126] BSR1114/16 (5′-GGGTTGCGCTCGTTRC-3′),[126] and 534R (5′-ATTACCGCGGCTGCTGG-3′).[127] PCR for sequencing is performed with the ABI PRISM BigDye terminator Cycle sequencing kit v.3.1 (Applied Biosystems) using one or all of the above-mentioned primers (one primer per reaction tube). Thereafter, ethanol/EDTA/sodium acetate precipitation is performed according to the manufacture's instructions (ABI PRISM BigDye terminator Cycle sequencing kit v.3.1). All the sequences are analyzed with ABI PRISM 3100 automated capillary DNA cycle sequencer (Applied Biosystems). The resulting sequences are checked, edited, and assembled with, for example, the Chromas program (Technelysium Pty Ltd.). The sequences are thereafter identified through the GenBank database (www.ncbi.nlm.nih.gov) using the BLAST (Basic Local Alignment Search Tool) algorithm[129] and/or using the Classifier tool of the Ribosomal Database Project (RDP) II (rdp.cme.msu.edu/index.jsp).[130]

35.3 CONCLUSION AND FUTURE PERSPECTIVES

Genus *Eubacterium*, as presently defined, is a genus of convenience based on negative identification, which leads to both phylogenetical and phenotypical diversity. It can be anticipated that numerous species reclassifications will occur when the phylum Firmicutes is phylogenetically reorganized. Furthermore, the *Eubacterium* species that presently fall within the genus *Clostridium* senso stricto are probably clostridia that have either lost their ability to produce spores or did not produce spores at the time of the isolation and identification, and therefore were placed within the genus *Eubacterium*. Therefore, the total number of species belonging to the genus *Eubacterium* will in all likelihood decrease even though new species will probably also be identified. Since species of genus *Eubacterium* are part of the human indigenous microbiota and only opportunistic pathogens, with no known specific virulence factors, there are no clear symptoms/markers that would indicate that the infection in question involves *Eubacterium* species. In addition, since *Eubacterium* species have typically been detected from polymicrobial infections, it is unlikely that in clinical settings one would expect *Eubacterium* species to act as the primary infective agent and would therefore start the microbial identification with *Eubacterium*-specific procedures. If *Eubacterium* species are suspected to be involved in a clinical case, it has to be remembered that since these bacteria are strict anaerobes, and also fairly slow growing, proper anaerobic techniques have to be used throughout the isolation procedure. Commercial and low-cost sequencing services enable clinical laboratories to use direct sequencing of the 16S rRNA gene from clinical isolates as a routine tool for the correct identification of suspected *Eubacterium* isolates instead of laborious, and in many cases inconclusive, phenotypic identification methods.

REFERENCES

1. Holdeman, J.V. et al., Descriptions of *Eubacterium timidum* sp. nov., *Eubacterium brachy* sp. nov., and *Eubacterium nodatum* sp. nov. isolated from human periodontitis, *Int. J. Syst. Bacteriol.*, 30, 163, 1980.
2. Martin, S.A. et al., A comparison of the reactivity of *Eubacterium* species with localized and serum immunoglobulins from rapidly progressive and adult periodontitis patients, *J. Periodontol.*, 59, 32, 1988.
3. Paster, B.J. et al., The breadth of bacterial diversity in the human periodontal pocket and other oral sites, *Periodontology 2000*, 42, 80, 2006.
4. Hahn, C., and Falkler Jr., W.A., Antibodies in normal and diseased pulps reactive with microorganisms isolated from deep caries, *J. Endodontol.*, 18, 28, 1992.
5. De Vries, J.J.C., Arents, N.L.A., and Manson, W.L., *Campylobacter* species isolated from extra-oro-intestinal abscesses: A report of four cases and literature review, *Eur. J. Clin. Microbiol. Infect. Diseases*, 27, 1119, 2008.
6. Hill, G.B., Ayers, O.M., and Kohan, A.P., Characteristics and sites of infection of *Eubacterium nodatum*, *Eubacterium timidum*, *Eubacterium brachy*, and other asaccharolytic eubacteria, *J. Clin. Microbiol.*, 25, 1540, 1987.
7. Rochford, J.C., Pleuropulmonary infection associated with *Eubacterium brachy*, a new species of *Eubacterium*, *J. Clin. Microbiol.*, 12, 722, 1980.
8. Munson, M.A. et al., Molecular and cultural analysis of the microflora associated with endodontic infections, *J. Dent. Res.*, 81, 761, 2002.
9. Thiolas, A. et al., *Eubacterium callanderi* bacteremia: Report of the first case, *J. Clin. Microbiol.*, 41, 2235, 2003.
10. Chapel, T. et al., The microbiological flora of penile ulcerations, *J. Infect. Dis.*, 137, 50, 1978.
11. Hall, G., Nord, C.E., and Heimdahl, A., Elimination of bacteraemia after dental extraction: Comparison of erythromycin and clindamycin for prophylaxis of infective endocarditis, *J. Antimicrob. Chemother.*, 37, 783, 1996.
12. Thadepalli, H., and Gangopadhyay, P.K., Rapid diagnosis of anaerobic empyema by direct gas-liquid chromatography of pleural fluid, *Chest*, 77, 507, 1980.
13. Rodloff, A.C. et al., Identification and susceptibility patterns of anaerobic bacteria isolated from clinical specimens during a one-year period, *Infection*, 12, 395, 1984.
14. Rôças, I.N., Hülsmann, M., and Siqueira Jr., J.F., Microorganisms in root canal-treated teeth from a German population, *J. Endodontol.*, 34, 926, 2008.
15. Fouad, A.F. et al., Molecular characterization of the presence of *Eubacterium* spp. and *Streptococcus* spp. in endodontic infections, *Oral Microbiol. Immunol.*, 18, 249, 2003.
16. Cheeseman, S.L. et al., Phylogeny of oral asaccharolytic *Eubacterium* species determined by 16S ribosomal DNA sequence comparison and proposal of *Eubacterium infirmum* sp. nov. and *Eubacterium tardum* sp. nov., *Int. J. Syst. Bacteriol.*, 46, 957, 1996.
17. Spratt, D.A., Weightman, A.J., and Wade, W.G., Diversity of oral asaccharolytic *Eubacterium* species in periodontitis—identification of novel phylotypes representing uncultivated taxa, *Oral Microbiol. Immunol.*, 14, 56, 1999.
18. Sungkanuparph, S., Chansirikarnjana, S., and Vorachit, M., *Eubacterium* bacteremia and colon cancer, *Scand. J. Infect. Dis.*, 34, 941, 2002.
19. Brook, I., Isolation of non-sporing anaerobic rods from infections in children, *J. Med. Microbiol.*, 45, 21, 1996.
20. Vigil, G.V. et al., Identification and antibiotic sensitivity of bacteria isolated from periapical lesions, *J. Endodontol.*, 23, 110, 1997.
21. Lin, S., Sela, G., and Sprecher, H., Periopathogenic bacteria in persistent periapical lesions: An in vivo prospective study, *J. Periodontol.*, 78, 905, 2007.
22. Poco Jr., S.E. et al., *Eubacterium minutum* sp. nov., isolated from human periodontal pockets, *Int. J. Syst. Bacteriol.*, 46, 31, 1996.
23. Molander, A., Reit, C., and Dahlén, G., The antimicrobial effect of calcium hydroxide in root canals pretreated with 5% iodine potassium iodide, *Dental Traumatol.*, 15, 205, 1999.
24. Faveri, M. et al., Microbiological profile of untreated subjects with localized aggressive periodontitis, *J. Clin. Periodontol.*, 36, 739, 2009.
25. Haffajee, A.D. et al., Comparison between polymerase chain reaction-based and checkerboard DNA hybridization techniques for microbial assessment of subgingival plaque samples, *J. Clin. Periodontol.*, 36, 642, 2009.
26. Haffajee, A.D. et al., Clinical and microbiological changes associated with the use of combined antimicrobial therapies to treat "refractory" periodontitis, *J. Clin. Periodontol.*, 31, 869, 2004.
27. Haffajee, A.D., Teles, R.P., and Socransky, S.S., Association of *Eubacterium nodatum* and *Treponema denticola* with human periodontitis lesions, *Oral Microbiol. Immunol.*, 21, 269, 2006.
28. Booth, V. et al., Gram-positive anaerobic bacilli in human periodontal disease, *J. Periodontol. Res.*, 39, 213, 2004.
29. Colombo, A.P.V. et al., Subgingival microbiota of Brazilian subjects with untreated chronic periodontitis, *J. Periodontol.*, 73, 360, 2002.
30. Vianna, M.E. et al., Microarrays complement culture methods for identification of bacteria in endodontic infections, *Oral Microbiol. Immunol.*, 20, 253, 2005.
31. Hill, G.B., *Eubacterium nodatum* mimics *Actinomyces* in intrauterine device-associated infections and other settings within the female genital tract, *Obstet. Gynecol.*, 79, 534, 1992.
32. Garre, M. et al., Fulminant *Eubacterium plautii* infection following dog bite in asplenic man, *Lancet*, 338, 384, 1991.
33. Orlando, G. et al., *Eubacterium plautii* infection in a kidney transplant recipient: A noteworthy case of pleural effusion and fever, *Clin. Transplant.*, 22, 520, 2008.
34. Sassone, L.M. et al., A microbiological profile of symptomatic teeth with primary endodontic infections, *J. Endodontol.*, 34, 541, 2008.
35. Sassone, L. et al., Evaluation of the microbiota of primary endodontic infections using checkerboard DNA–DNA hybridization, *Oral Microbiol. Immunol.*, 22, 390, 2007.

36. Brennan, R.M. et al., Bacterial species in subgingival plaque and oral bone loss in postmenopausal women, *J. Periodontol.*, 78, 1051, 2007.
37. Hall, G. et al., Prophylactic administration of penicillins for endocarditis does not reduce the incidence of postextraction bacteremia, *Clin. Infect. Dis.*, 17, 188, 1993.
38. Mayanagi, G. et al., Detection frequency of periodontitis-associated bacteria by polymerase chain reaction in subgingival and supragingival plaque of periodontitis and healthy subjects, *Oral Microbiol. Immunol.*, 19, 379, 2004.
39. Kumar, P.S. et al., New bacterial species associated with chronic periodontitis, *J. Dent. Res.*, 82, 338, 2003.
40. Hashimura, T., Sato, M., and Hoshino, E., Detection of *Slackia exigua*, *Mogibacterium timidum* and *Eubacterium saphenum* from pulpal and periradicular samples using the Polymerase Chain Reaction (PCR) method, *Int. Endod. J.*, 34, 463, 2001.
41. Kazor, C.E. et al., Diversity of bacterial populations on the tongue dorsa of patients with halitosis and healthy patients, *J. Clin. Microbiol.*, 41, 558, 2003.
42. Uematsu, H. et al., Degradation of arginine and other amino acids by butyrate-producing asaccharolytic anaerobic gram-positive rods in periodontal pockets, *Arch. Oral Biol.*, 48, 423, 2003.
43. Lau, S.K.P. et al., Anaerobic, non-sporulating, gram-positive bacilli bacteraemia characterized by 16S rRNA gene sequencing, *J. Med. Microbiol.*, 53, 1247, 2004.
44. Monif, G.R.G. et al., Cul de sac isolates from patients with endometritis salpingitis peritonitis and gonococcal endocervicitis, *Obstet. Gynecol.*, 126, 158, 1976.
45. DeHertogh, D., and York, M.K., *Eubacterium tortuosum* and *Eubacterium ventriosum* Bacteremia, *Clin. Microbiol. Newsl.*, 3, 60, 1981.
46. Lou, M.A., Thadepalli, H., and Mandal, A.K., Safety and efficacy of mezlocillin: A single-drug therapy for penetrating abdominal trauma, *J. Trauma*, 28, 1541, 1988.
47. Williams, S.M., Hafner, S., and Sundram, Y., Liver granulomas due to *Eubacterium tortuosum* in a seven-week-old bobwhite quail, *Avian Dis.*, 51, 797, 2007.
48. Margaret, B.S., and Krywolap, G.N., *Eubacterium yurii* subsp. *yurii* sp. nov. and *Eubacterium yurii* subsp. *margaretiae* subsp. nov.: Test tube brush bacteria from subgingival dental plaque, *Int. J. Syst. Bacteriol.*, 36, 145, 1986.
49. Margaret, B.S., and Krywolap, G.N., *Eubacterium yurii* subsp. *schtitka* subsp. nov.: Test tube brush bacteria from subgingival dental plaque, *Int. J. Syst. Bacteriol.*, 38, 207, 1988.
50. Rajasuo, A. et al., Bacteriologic findings in tonsillitis and pericoronitis, *Clin. Infect. Dis.*, 23, 51, 1996.
51. Gomes, B.P.F.A., Lilley, J.D., and Drucker, D.B., Associations of endodontic symptoms and signs with particular combinations of specific bacteria, *Int. Endod. J.*, 29, 69, 1996.
52. Rams, T.E. et al., Microbiological study of HIV-related periodontitis, *J. Periodontol.*, 62, 74, 1991.
53. Bourne, K.A. et al., Bacteremia due to *Bifidobacterium*, *Eubacterium* or *Lactobacillus*. Twenty-one cases and review of the literature, *Yale J. Biol. Med.*, 51, 505, 1978.
54. Chan, R.C., and Mercer, J., First Australian description of *Eggerthella lenta* bacteraemia identified by 16S rRNA gene sequencing, *Pathology*, 40, 409, 2008.
55. Sundqvist, G., Associations between microbial species in dental root canal infections, *Oral Microbiol. Immunol.*, 7, 257, 1992.
56. Robertson, D., and Smith, A.J., The microbiology of the acute dental abscess, *J. Med. Microbiol.*, 58, 155, 2009.
57. Siqueira Jr., J.F. et al., Bacteria in the apical root canal of teeth with primary apical periodontitis, *Oral Surg., Oral Med., Oral Pathol., Oral Radiol. Endodontol.*, 107, 721, 2009.
58. Jacinto, R.C. et al., Bacterial examination of endodontic infections by clonal analysis in concert with denaturing high-performance liquid chromatography, *Oral Microbiol. Immunol.*, 22, 403, 2007.
59. Siqueira Jr., J.F. et al., Cultivable bacteria in infected root canals as identified by 16S rRNA gene sequencing, *Oral Microbiol. Immunol.*, 22, 266, 2007.
60. Preza, D. et al., Bacterial profiles of root caries in elderly patients, *J. Clin. Microbiol.*, 46, 2015, 2008.
61. Preza, D. et al., Microarray analysis of the microflora of root caries in elderly, *Eur. J. Clin. Microbiol. Infect. Dis.*, 28, 509, 2009.
62. Finegold, S.M., Sutter, V.L., and Mathisen, G.E., Normal indigenous intestinal flora. In Hentges, D.J. (Editor), *Human Intestinal Microflora in Health and Disease*, pp. 3–31, Academic Press, New York, NY, 1983.
63. Finegold, S.M., Attebery, H.R., and Sutter, V.L., Effect of diet on human fecal flora: Comparison of Japanese and American diets, *Am. J. Clin. Nutr.*, 27, 1456, 1974.
64. Finegold, S.M. et al., Fecal bacteriology of colonic polyp patients and control patients, *Cancer Res.*, 35, 3407, 1975.
65. Prevot, A.R., Études de systématique bactérienne, *Ann.Inst. Pasteur (Paris)*, 60, 285, 1938.
66. Hill, G.B., Ayers, O.M., and Kohan, A.P., Characteristics and sites of infection of *Eubacterium nodatum*, *Eubacterium timidum*, *Eubacterium brachy*, and other asaccharolytic eubacteria, *J. Clin. Microbiol.*, 25, 1540, 1987.
67. Jousimies-Somer, H.R. et al., *Wadsworth - KTL Anaerobic Bacteriology Manual*, 6th ed., Star Publishing Company, Belmont, CA, 2002.
68. Holt, J.G. et al., Genus *Eubacterium*. In Hensyl W.R. (Editor), *Bergey's Manual of Determinative Bacteriology*, 9th ed., p. 577, Williams & Wilkins, Baltimore, MD, 1994.
69. Hofstad, T., and Aasjord, P., *Eubacterium plautii* (Seguin 1928) comb. nov., *Int. J. Syst. Bacteriol.*, 32, 346, 1982.
70. Andreesen, J.R., The genus *Eubacterium*. In Balows, A. et al. (Editors), *The Prokaryotes*, 2nd ed., pp. 1914–1924. Springer-Verlag, New York, NY, 1992.
71. Cato, E.P., Moore, L.V.H., and Moore, W.E.C., *Fusobacterium alocis* sp. nov. and *Fusobacterium sulci* sp. nov. from the human gingival sulcus, *Int. J. Syst. Bacteriol.*, 35, 475, 1985.
72. Jalava, J., and Eerola, E., Phylogenetic analysis of *Fusobacterium alocis* and *Fusobacterium sulci* based on 16S rRNA gene sequences: Proposal of *Filifactor alocis* (Cato, Moore and Moore) comb. nov. and *Eubacterium sulci* (Cato, Moore and Moore) comb. nov., *Int. J. Syst. Bacteriol.*, 49, 1375, 1999.
73. Baardsen, R. et al., Outer membrane protein pattern of *Eubacterium plautii*, *J. Gen. Microbiol.*, 134, 1561, 1988.
74. Moore, W.E.C., and Holdeman-Moore, L.V., Genus *Eubacterium*. In Sneath, P.H.A. et al. (Editors), *Bergeys Manual of Systematic Bacteriology*, 2nd ed., pp. 1353–1373, Williams & Wilkins, Baltimore, MD, 1986.
75. Wade, W.G., The genus *Eubacterium* and related genera. In Stackebrandt, E. et al. (Editors), *Prokaryotes*, 4th ed., pp. 823–835, Springer, New York, NY, 2006.
76. Bahl, H., and Gottschalk, G., Microbial production of butanol/acetone. In Rehm, H.J., and Reed, G. (Editors), *Biotechnology*, 6b, p. 1, VCH Verlagsgesellschaft, Weinheim, Germany, 1988.

77. Mountfort, D.O. et al., *Eubacterium callanderi* sp. nov. that demethoxylates O-methoxylated aromatic acids to volatile fatty acids, *Int. J. Syst. Bacteriol.*, 38, 254, 1988.
78. Zindel, U. et al., *Eubacterium acidaminophilum* sp. nov., a versatile amino acid-degrading anaerobe producing or utilizing H2 or formate. Description and enzymatic studies, *Arch. Microbiol.*, 150, 254, 1988.
79. Yarza, P. et al., The All-Species Living Tree project: A 16S rRNA-based phylogenetic tree of all sequenced type strains, *Syst. Appl. Microbiol.*, 31, 241, 2008.
80. Cole, J.R. et al., The Ribosomal Database Project: Improved alignments and new tools for rRNA analysis, *Nucleic Acids Res.*, 37, D141, 2009.
81. Willems, A., and Collins, M.D., Phylogenetic relationships of the genera *Acetobacterium* and *Eubacterium* sensu stricto and reclassification of *Eubacterium alactolyticum* as *Pseudoramibacter alactolyticus* gen. nov., comb. nov., *Int. J. Syst. Bacteriol.*, 46, 1083, 1996.
82. Collins, M.D. et al., The phylogeny of the genus *Clostridium*: Proposal of five new genera and eleven new species combinations, *Int. J. Syst. Bacteriol.*, 44, 812, 1994.
83. Wiegel, J., Tanner, R., and Rainey, F.A., An introduction to the family Clostridiaceae. In Stackebrandt, E. et al. (Editors), *Prokaryotes*, 4th ed., pp. 654–678, Springer, New York, NY, 2006.
84. Wade, W.G. et al., The family Coriobacteriaceae: Reclassification of *Eubacterium exiguum* (Poco et al, 1996) and *Peptostreptococcus heliotrinreducens* (Lanigan 1976) as *Slackia exigua* gen. nov., comb. nov. and *Slackia heliotrinireducens* gen. nov., comb. nov., and *Eubacterium lentum* (Prevot 1938) as *Eggerthella lenta* gen. nov., comb. nov., *Int. J. Syst. Bacteriol.*, 49, 595, 1999.
85. Kageyama, A., Benno, Y., and Nakase, T., Phylogenetic and phenotypic evidence for the transfer of *Eubacterium aerofaciens* to the genus *Collinsella* as *Collinsella aerofaciens* gen. nov., comb. nov., *Int. J. Syst. Bacteriol.*, 49, 557, 1999.
86. Taras, D. et al., Reclassification of *Eubacterium formicigenerans* Holdeman and Moore 1974 as *Dorea formicigenerans* gen. nov., comb. nov., and description of *Dorea longicatena* sp. nov., isolated from human faeces, *Int. J. Syst. Evol. Microbiol.*, 52, 423, 2002.
87. Kageyama, A., Benno, Y., and Nakase, T., Phylogenic and phenotypic evidence for the transfer of *Eubacterium fossor* to the genus *Atopobium* as *Atopobium fossor* comb. nov., *Microbiol. Immunol.*, 43, 389, 1999.
88. Nakazawa, F. et al., Description of *Mogibacterium pumilum* gen. nov., sp. nov. and *Mogibacterium vescum* gen. nov., sp. nov., and reclassification of *Eubacterium timidum* (Holdeman et al. 1980) as *Mogibacterium timidum* gen. nov., comb. nov., *Int. J. Syst. Evol. Microbiol.*, 50, 679, 2000.
89. Ludwig, W. et al., Phylogenetic evidence for the transfer of *Eubacterium suis* to the genus *Actinomyces* as *Actinomyces suis* comb. nov., *Int. J. Syst. Bacteriol.*, 42, 161, 1992.
90. Lawson, P.A. et al., Characterization of some *Actinomyces*-like isolates from human clinical specimens: Reclassification of *Actinomyces suis* (Soltys and Spratling) as *Actinobaculum suis* comb. nov. and description of *Actinobaculum schaalii* sp. nov., *Int. J. Syst. Bacteriol.*, 47, 899, 1997.
91. Cheeseman, S.L. et al., Phylogeny of oral asaccharolytic *Eubacterium* species determined by 16S ribosomal DNA sequence comparison and proposal of *Eubacterium infirmum* sp. nov. and *Eubacterium tardum* sp. nov., *Int. J. Syst. Bacteriol.*, 46, 957, 1996.
92. Wade, W.G. et al., *Eubacterium minutum* is an earlier synonym of *Eubacterium tardum* and has priority, *Int. J. Syst. Bacteriol.*, 49, 1939, 1999.
93. Barcenilla, A. et al., Phylogenetic relationships of butyrate-producing bacteria from the human gut, *Appl. Environ. Microbiol.*, 66, 1654, 2000.
94. Cummings, J.H., and Macfarlane, G.T., Colonic microflora: Nutrition and health, *Nutrition*, 13, 476, 1997.
95. Hague, A., and Paraskeva, C., The short-chain fatty acid butyrate induces apoptosis in colorectal tumour cell lines, *Eur. J. Cancer Prevent.*, 4, 359, 1995.
96. Abrahamse, S.L., Pool-Zobel, B.L., and Rechkemmer, G., Potential of short chain fatty acids to modulate the induction of DNA damage and changes in the intracellular calcium concentration by oxidative stress in isolated rat distal colon cells, *Carcinogenesis*, 20, 629, 1999.
97. Rosignoli, P. et al., Protective activity of butyrate on hydrogen peroxide-induced DNA damage in isolated human colonocytes and HT29 tumour cells, *Carcinogenesis*, 22, 1675, 2001.
98. Schwiertz, A., Le Blay, G., and Blaut, M., Quantification of different *Eubacterium* spp. in human fecal samples with species-specific 16S rRNA-targeted oligonucleotide probes, *Appl. Environ. Microbiol.*, 66, 375, 2000.
99. Simmering, R., Kleessen, B., and Blaut, M., Quantification of the flavonoid-degrading bacterium *Eubacterium ramulus* in human fecal samples with a species-specific oligonucleotide hybridization probe, *Appl. Environ. Microbiol.*, 65, 3705, 1999.
100. Schneider, H. et al., Anaerobic transformation of quercetin-3-glucoside by bacteria from the human intestinal tract, *Arch. Microbiol.*, 171, 81, 1999.
101. Freier, T.A. et al., Characterization of *Eubacterium coprostanoligenes* sp. nov., a cholesterol-reducing anaerobe, *Int. J. Syst. Bacteriol.*, 44, 137, 1994.
102. Li, L. et al., Effect of orally administered *Eubacterium coprostanoligenes* ATCC 51222 on plasma cholesterol concentration in laying hens, *Poult. Sci.*, 75, 743, 1996.
103. Li, L. et al., Effect of feeding of a cholesterol-reducing bacterium, *Eubacterium coprostanoligenes*, to germ-free mice, *Lab. Anim. Sci.*, 48, 253, 1998.
104. Lau, S.K.P. et al., Anaerobic, non-sporulating, gram-positive bacilli bacteraemia characterized by 16S rRNA gene sequencing, *J. Med. Microbiol.*, 53, 1247, 2004.
105. Maukonen, J. et al., Intra-individual diversity and similarity of salivary and faecal microbiota, *J. Med. Microbiol.*, 57, 1560, 2008.
106. Maukonen, J. et al., PCR DGGE and RT-PCR DGGE show diversity and short-term temporal stability in the *Clostridium coccoides—Eubacterium rectale* group in the human intestinal microbiota, *FEMS Microbiol. Ecol.*, 58, 517, 2006.
107. Hillier, S.L., and Moncla, B.J., *Peptostreptococcus*, *Propionibacterium*, *Eubacterium*, and other nonsporeforming anaerobic gram-positive bacteria. In Murray, P.R. et al. (Editors), *Manual of Clinical Microbiology*, 6th ed., pp. 587–602, American Society for Microbiology, Washington D.C., 2005.
108. Morris, G.N., Winter, J., and Cato, E.P., *Eubacterium desmolans* sp. nov., a steroid desmolase-producing species from cat fecal flora, *Int. J. Syst. Bacteriol.*, 36, 183, 1986.
109. Downes, J. et al., Characterisation of *Eubacterium*-like strains isolated from oral infections, *J. Med. Microbiol.*, 50, 947, 2001.

110. Petti, C.A., Polage, C.R., and Schreckenberger, P., The role of 16S rRNA gene sequencing in identification of microorganisms misidentified by conventional methods, *J. Clin. Microbiol.*, 43, 6123, 2005.
111. Lay, C. et al., Design and validation of 16S rRNA probes to enumerate members of the *Clostridium leptum* subgroup in human faecal microbiota, *Environ. Microbiol.*, 7, 933, 2005.
112. Hold, G.L. et al., Oligonucleotide probes that detect quantitatively significant groups of butyrate-producing bacteria in human feces, *Appl. Environ. Microbiol.*, 69, 4320, 2003.
113. Schwiertz, A. et al., Influence of resistant starch on the SCFA production and cell counts of butyrate-producing *Eubacterium* spp. in the human intestine, *J. Appl. Microbiol.*, 93, 157, 2002.
114. Harmsen, H.J.M. et al., Extensive set of 16S rRNA-based probes for detection of bacteria in human feces, *Appl. Environ. Microbiol.*, 68, 2982, 2002.
115. Walker, A.W. et al., pH and peptide supply can radically alter bacterial populations and short-chain fatty acid ratios within microbial communities from the human colon, *Appl. Environ. Microbiol.*, 71, 3692, 2005.
116. Franks, A.H. et al., Variations of bacterial populations in human feces measured by fluorescent in situ hybridization with group-specific 16S rRNA-targeted oligonucleotide probes, *Appl. Environ. Microbiol.*, 64, 3336, 1998.
117. Sghir, A. et al., Quantification of bacterial groups within human fecal flora by oligonucleotide probe hybridization, *Appl. Environ. Microbiol.*, 66, 2263, 2000.
118. Kageyama, A., and Benno, Y., Rapid detection of human fecal *Eubacterium* species and related genera by nested PCR method, *Microbiol. Immunol.*, 45, 315, 2001.
119. Matsuki, T. et al., Development of 16S rRNA-gene-targeted group-specific primers for the detection and identification of predominant bacteria in human feces, *Appl. Environ. Microbiol.*, 68, 5445, 2002.
120. Furet, J.-P. et al., Comparative assessment of human and farm animal faecal microbiota using real-time quantitative PCR, *FEMS Microbiol. Ecol.*, 68, 351, 2009.
121. Matsuki, T. et al., Use of 16S rRNA gene-targeted group-specific primers for real-time PCR analysis of predominant bacteria in human feces, *Appl. Environ. Microbiol.*, 70, 7220, 2004.
122. Amann, R.I., In situ identification of microorganisms by whole cell hybridization with rRNA-targeted nucleic acid probes. In *Molecular Microbial Ecology Manual*, p. 1, Kluwer Academic Publishers, The Netherlands, 1995.
123. Holdeman, L.V., Cato, E.P., and Moore, W.E.C. (Editors), *Anaerobe Laboratory Manual*, 4th ed., Anaerobe Laboratory, Virginia Polytechnic Institute and State University, Blacksburg, VA, 1977.
124. Socransky, S.S. et al., Checkerboard' DNA–DNA hybridization, *BioTechniques*, 17, 788, 1994.
126. Wilmotte, A., Van Der Auwera, G., and De Wachter, R., Structure of the 16 S ribosomal RNA of the thermophilic cyanobacterium Chlorogloeopsis HTF ("*Mastigocladus laminosus* HTF") strain PCC7518, and phylogenetic analysis, *FEBS Lett.*, 317, 96, 1993.
127. Muyzer, G., De Waal, E.C., and Uitterlinden, A.G., Profiling of complex microbial populations by denaturing gradient gel electrophoresis analysis of polymerase chain reaction-amplified genes coding for 16S rRNA, *Appl. Environ. Microbiol.*, 59, 695, 1993.
125. Ben-Dov, E. et al., Advantage of using inosine at the 3′ termini of 16S rRNA gene universal primers for the study of microbial diversity, *Appl. Environ. Microbiol.*, 72, 6902, 2006.
128. Nadkarni, M.A. et al., Determination of bacterial load by real-time PCR using a broad-range (universal) probe and primers set, *Microbiology*, 148, 257, 2002.
129. Altschul, S.F. et al., Basic local alignment search tool, *J. Mol. Biol.*, 215, 403, 1990.
130. Wang, Q. et al., Naïve Bayesian classifier for rapid assignment of rRNA sequences into the new bacterial taxonomy, *Appl. Environ. Microbiol.*, 73, 5261, 2007.

36 *Finegoldia*

A.C.M. Veloo

CONTENTS

36.1 INTRODUCTION

36.1.1 Classification, Morphology, and Epidemiology

Classification. The genus *Finegoldia* belongs to the class of clostridia, phylum Firmicutes. *Finegoldia magna*, the only species present in this genus, is the most commonly isolated gram-positive anaerobic coccus (GPAC) from human clinical specimens.[1] The taxonomy of this species has changed dramatically. In 1933, *F. magna* was first described by Prevot[2] and named *Diplococcus magna* based on the cellular appearance. Holdeman and Moore[3] added the species to the genus *Peptococcus*. In 1983, Ezaki et al.[4] transferred *Peptococcus magnus* to the genus *Peptostreptococcus*. This transfer was based on DNA base composition, DNA–DNA hybridization data, and cellular fatty acid profiles. Classification of GPAC using pyrolysis mass spectrometry (PMS) showed that *P. magnus* was distinct from other GPAC species.[5] In addition, 16S ribosomal sequence analyses indicated that *Peptostreptococcus magnus* is phylogenetically distant from the other GPAC. Therefore, the genus *Finegoldia* was proposed to cover the species *F. magna*.[6]

Morphology. Cells of different strains of *F. magna* can vary in size. Gram staining shows clumps of cells, mostly in tetrad formation, with a diameter of 0.8–1.6 μm, which are stained gram-positive. A typical clump consists of large cocci in the middle, surrounded by cells smaller in size. The largest cells have the strongest color, while the smaller cells can be decolorized.[7]

After 48 h of incubation on blood agar plates, colonies are approximately 1 mm in diameter. After five days of incubation >90% of the strains had colonies >2 mm in diameter. The colonies can vary in size and color on one plate. Some colonies are convex and whitish, while others are flat and translucent. This might give the impression that it is a mixed culture. However, colonies always have a perfect round shape, and are usually slightly convex.[7]

Epidemiology. *F. magna* is one of the most frequently recovered anaerobes from clinical infectious material. In such infections approximately 30% of the anaerobes are GPAC of which around 30% is *F. magna*.[8] Remarkably, *F. magna* can be isolated from clinical specimens in pure culture and can be surpassed by *Bacteroides fragilis* in this aspect only.

Neut et al.[9] analyzed the presence of GPAC in normal oral, fecal, and vaginal microbiota. *F. magna* was found in the vaginal and fecal microbiota (especially healthy adults), but not in the oral microbiota. Riggio et al.[10] determined the presence of *F. magna* in oral samples using a specific PCR. In total, 33 subgingival plaque samples of patients with adult periodontal disease and 60 pus aspirates from patients with acute dentoalveolar abscesses were analyzed. *F. magna* DNA was encountered in two subgingival plaque samples. These results show that *F. magna* is not a major pathogen in adult periodontal disease and dentoalveolar abscesses. Gao et al.[11] analyzed the superficial skin bacterial biota of human forearm of six subjects. A total of 91 genera were found, of which six were observed in all subjects. One of these was *Finegoldia* AB109769, suggesting that *F. magna* is part of the normal skin microbiota. Higaki et al.[12] analyzed the anaerobes isolated from infectious skin diseases. The most commonly isolated anaerobes were GPAC (66 of 106 strains, 62%), among these *F. magna* was the most frequently isolated GPAC (27 strains, 41%). *F. magna* shows pathogenic features similar to *Staphylococcus aureus*. In patients with infected breast abscesses, *F. magna* can be isolated alone or together with *S. aureus*.[12] Since both organisms display synergism of pathogenicity, it is more difficult to cure such infections.

Brook[13] evaluated the recovery of anaerobes, among them *F. magna*, from clinical specimens during a 12-year period. Of all GPAC isolated, 18.4% were identified as *F. magna*. The majority of these *F. magna* strains (65.7%) were isolated from abscesses, obstetrical and gynecological infections, and wounds. The highest frequency of recovery of *F. magna* was shown in bone and chest infections. In pediatric patients 680 GPAC were recovered from 598 clinical specimens, from 554 patients.[14] From all these strains, 10.9% were *F. magna*. The majority was isolated from abscesses. Bourgault et al.[15] evaluated the clinical significance of *F. magna* in 222 patients. Of these patients, 183 had an infection in which *F. magna* played an important role. In 17.5% of these cases, *F. magna* was isolated as a pure culture from infections of bone and joint (56.3%), soft tissue (37.5%), and vascular (6.3%). In mixed infections in which *F. magna* involved, the most frequently isolated facultative bacteria were group D *Streptococcus*, *Staphylococcus epidermis*, *Escherichia coli*, and *S. aureus*. The most frequently found anaerobic bacteria were: *Prevotella melaninogenica*, *B. fragilis*, and *Bacteroides* sp. These mixed infections were mainly infections of soft tissue (37.7%), bone and joint (21.2), and foot ulcers (19.2%). From this data it can be concluded that there is a strong association between *F. magna* found in pure culture and orthopedic procedures and postoperative wound infection.

36.1.2 Clinical Features and Pathogenesis

F. magna is capable of producing several virulence factors (Table 36.1). In 1984, Brook et al.[23] examined the pathogenicity of GPAC in mixed infections. Abscesses caused by two organisms, including one strain of GPAC, were larger compared with abscesses caused by one organism. From their experiments it was concluded that *F. magna* and *Peptostreptococcus anaerobius* were equally important or more important than the other bacteria in mixed infections. This supports the hypothesis that bacteria in mixed infections may have a synergistic nature. The collagenase production of *F. magna* is associated with the site of infection.[16] Collagen is abundantly present in the skin, tendons, and cartilage and is an organic component of bones, teeth, and the cornea. The breakdown of collagen will result in loss of tissue integrity and disease progression, thereby providing an environment suitable for growth of anaerobic bacteria. The production of collagenase may also be important for the growth of asaccharolytic bacteria, such as *F. magna*, since during collagen breakdown amino acids are released which may be necessary for growth and survival.[24] Ng et al.[25] determined the aminopeptidase activities of some GPAC strains. *F. magna* together with *P. micra* were found to degrade most substrates. There was a correlation between gelatin hydrolysis and the number of aminopeptidases produced. The authors state that gelatin hydrolysis reflects the pathogenetic potential of a strain. The growth of the bacteria can be correlated to the amount of aminopeptidases produced by protein degradation.

Myhre[17] described that 42% of the *F. magna* strains is able to bind human serum albumin (HSA). This ability was originally described for different streptococcal species. In group C and G streptococci this is mediated by protein G, and in group A streptococci by protein M.[26] De Chateau et al.[27] demonstrated that some strains of *F. magna* express protein peptostreptococcal albumin binding (PAB) on their cell surface. Sequence analyses revealed homology with the HSA-binding domain of protein G and to the framework regions of protein L (described later). This suggests an interspecies exchange of an HSA-binding protein module. In general, host binding cell wall proteins of gram-positive bacteria share a common structure, including a (from the distal NH_2 terminus)[28]:

- Signal sequence
- Variable NH_2-terminal region
- Varying number of repeated domains that independently bind different plasma proteins
- Proline-rich region supposedly intercalating the protein in the gram-positive cell wall
- COOH-terminal cell wall sorting signal, required to anchor the protein to the cell wall

TABLE 36.1
Virulence Factors of *F. magna*

Virulence Factor	Function	Reference
Collagenase	Breakdown of collagen	16
PAB	Binding to human serum albumin	17
Protein L	Immunoglobulin (Ig)-binding protein	18
	Release of *de novo*-synthesized mediators	19
SufA	Degradation of fibrinogen	20
	Degradation of antibacterial peptides LL-37 and MIG/CXCL9	20,21
	Release of FAF from bacterial cell wall	21
FAF	Mediation of bacterial aggregation through protein-protein interactions between FAF molecules on neighboring *F. magna* bacteria	22
	Binding with BM40, a noncollagenous glycoprotein, present on the skin	
	Blocking the activity of LL-37, an antibacterial peptide	
	Inactivation of the antibacterial peptide MIG/CXCL9	21

PAB contains a GA module (protein *G*-related albumin-binding module). This is a centrally located domain of 45 amino acid residues, which is responsible for the binding of HSA. This domain is subject to module shifting. The predecessor of the PAB protein is urPAB. This protein does not contain the shuffled GA module, but has a uGA domain in the NH_2-terminal region. This domain shows 38% similarity with the GA module and binds HSA to a lesser extent. PAB also contains an analogous uGA domain, which indicates a second binding site for HSA. The affinity for HSA differs between the GA modules. A reason for bacteria to acquire the GA module is that the older uGA domain has lost its function due to the difference in affinity for HSA. The binding affinity for HSA is not only found on the cell surface, but also in the culture supernatant. The growth of HSA-binding strains is stimulated by the addition of HSA to the growth medium.[29] This selective advantage increases the virulence of HSA-binding *F. magna* strains.

Felten et al.[30] studied the binding of 14 *F. magna* strains isolated from bone and joint infections to collagen, fibrinogen, and fibronectin after implantation of a foreign body. From these strains, 81% bound to collagen, 69% to fibrinogen, and 46% to fibronectin. When these results were compared to the binding abilities from *F. magna* strains from other infections, a correlation was found between fibrinogen binding and bone and joint infections (69% against 20%). Krepel et al.[31] tried to elucidate the role of *F. magna* in three different polymicrobial environments: intraabdominal infections, nonpuerperal breast abscesses, and diabetic foot ulcers. An association was made between phenotypic characteristics and the site of infection. In total, 336 clinical specimens were examined: 222 were intraabdominal, from which 11 *F. magna* strains were isolated; 58 nonpuerperal breast abscesses, from which 21 *F. magna* strains were isolated; and 56 diabetic foot ulcers, from which 18 *F. magna* strains were isolated. From the *F. magna* strains the hippurate hydrolase, collagenase, and gelatinase production was determined. Strains with the lowest enzymatic activity were isolated from intraabdominal infections. The most proteolytic strains were predominantly isolated from soft-tissue infections. These are the kind of infections that tend to be chronic and heal slowly. Edmiston et al.[32] showed that *F. magna* is the most common anaerobe isolated from nonpuerperal breast infections. *F. magna* strains isolated from nonpuerperal breast abscesses and diabetic foot infections were shown to have a higher collagenase production compared to *F. magna* strains isolated from intraabdominal infections.[16] Stephens et al.[33] determined the impact of the presence of GPAC present in deep tissues of chronic wounds. Clinical samples of 18 patients with chronic venous leg ulcers were cultured. Six of these patients had *F. magna*. None of these *F. magna* strains had any hydrolytic enzyme activity or affected the endothelial cell proliferation. All inhibited fibroblast proliferation and keratinocyte wound repopulation.

Björk[18] was the first to describe a novel bacterial cell-wall protein that is able to bind with Ig light chains (L chains); therefore, this protein was named protein L. L chains are shared between the different Ig classes. Protein L was found to have affinity with IgG, IgM, IgA, $F(ab')_2$, and Fab fragments, and with κ and λ L chains. The reaction with λ L chains is very weak compared with κL chains. Nilson et al.[34] described that protein L binds exclusively to the V_L domain of Ig and not to the C_L domain. This binding strongly depends on the three-dimensional structure of the V_L domain, indicating that several sites of V_L are involved. It requires the spatial proximity of the κI, κIII, and κIV light chain molecules. κL chains represent 65% of human immunoglobulins, and of the entire κ chain population, κI, κIII, and κIV proteins represent 60%, 28%, and 2%, respectively.[35] Protein L has five highly homologous domains that are involved in the binding of Ig. These domains interact with the framework regions of the V_L domain.[36] The strength of the binding of protein L with κ chains is less when compared with the binding of the complete Ig. The conformation resulting from the interaction between heavy and light chains in the Ig provides a more favorable binding site for protein L.[37] The binding site of Ig is close to the antigen-binding site, but the interaction between protein L and Ig was not obstructed by occupation of the antigen-binding site.[37,38] Åkerström et al.[37] showed that protein L has no disulfide bond or a subunit structure, and that protein L has two non-Ig-binding fragments that were found to be unique. This was confirmed by Graille et al.[39] A single protein L domain can react with the variable regions of κL chains of two Fab molecules, in a sandwich fashion. The contact residues in the variable region are remote from the hypervariable loops. It was suggested that the two binding sites on protein L have a different affinity for Ig. In vivo experiments by Smith et al.[40] showed that protein L prefers to target B cells. This is due to the interaction with Ig on the surface of these cells. This interaction strongly activates the B cells, which results in an upregulation of MHC-II and CD86. These surface molecules are important in initiating an antibody response. No specific binding of protein L with other splenocytes, like T cells and certain dendritic cell subsets, was observed. The activation of B cells also results in an increased expression of the target immunoglobulin. When mature B cells are exposed to protein L, a reduction of splenic marginal zone B cells and peritoneal B1 cells was observed.[41] These two B-cell subsets are involved in the first-line immune response against foreign invaders. They have a high antigen-presenting capacity and secrete preferentially potentially protective natural IgM. B1 cells are located in the cavities of the body and are important in contributing to the production of natural antibodies and T-cell independent immune responses. Marginal zone B cells are located at the periphery of the splenic periarteriolar lymphoid sheath at the border of white and red pulp, and they are the first to encounter blood-borne antigens.

The overall design of protein L is similar to those of protein A from staphylococci and protein G from streptococci, but the primary structure is different. When the amino acid compositions of these proteins are compared, protein L has a higher amount of glycine and a lower amount of lysine. No amino acid sequence homology was demonstrated between these three proteins, apart from the carboxyl-terminal transmembrane region. Some similarity was seen between the W-region of protein G and the amino acid sequence of one of the tryptic peptides of protein L. It has been proposed that this W-region anchors protein G to the cell wall. Protein L is much smaller in size than protein G, which facilitates tissue penetration.[38] One common feature between the three proteins is that they all possess multiple copies of Ig-binding domains. In each protein these domains are highly conserved.[39]

Since protein L is able to bind with all human Ig, it is also able to bind with the κ L chain of IgE. Since the binding of anti-IgE with the Fc portion of IgE stimulates the release of histamine from human basophils, it is possible that the binding of protein L with IgE also stimulates the release of histamine, which will trigger an inflammatory response. Histamine and tryptase are both involved in allergic reactions. This stimulation of basophils was described by Patella et al.[19] The release of histamine is dependent on the concentration of protein L. The interaction with IgE present on the surface of basophils mediates the release of protein L. The stimulation by protein L on the basophils is greater than the stimulation by anti-IgE. Patella et al.[19] also described the release of the preformed *de novo*-synthesized mediators leukotriene C4 (LTC4) from basophils, or PGD2 from human skin mast cells, both chemical mediators of human inflammatory cells. Genovese et al.[42] described the release of histamine from human heart mast cells. They found a significant correlation between histamine release and tryptase release, and the release of LTC_4. It is interesting to note that mast cells tend to accumulate at the site of a chronic infection.[43] The release of *de novo*-synthesized mediators may contribute to the pathogenesis of the infecting strain. It is hypothesized that this may cause mycocardial damage in patients with bacterial infections.[42] Protein L helps the bacteria to adhere to the wound surfaces. The covering of the bacterial cell wall by host proteins allows the bacteria to evade the immune response of the host.[38]

Protein L is expressed at the surface of ±10% of the *F. magna* strains.[38,44] Some protein L molecules are released into the growth medium, but most molecules are associated with the cell wall. The protein L from the growth medium shows a considerable heterogeneity in size. This indicates proteolytic degradation of the released protein.[37] Kastern et al.[44] found that only *F. magna* strains that express protein L possess the protein L encoding gene. In nonexpressing protein L strains, this gene is not present rather than being downregulated. The features described for protein L may explain why protein L-expressing *F. magna* strains are more often associated with clinical infections than are nonexpressing strains.

Kastern et al.[44] determined the presence of protein L in 30 *F. magna* strains, all derived from clinical specimens. Four of these strains expressed protein L, and all of these four strains were isolated from women with bacterial vaginosis. The negative strains were from healthy women (n = 19), men (n = 4), and women with bacterial vaginosis (n = 3). These results indicate that there is a correlation between protein L-expressing *F. magna* strains and bacterial vaginosis. This was confirmed by de Château et al.[29] They determined the presence of protein L and HSA-binding protein in 48 *F. magna* strains. Thirty of these strains were isolated from suppurative infections. One of them expressed protein L, and 16 strains (53%) were binding HSA. Eight of the 48 strains were isolated from patients with bacterial vaginosis. None of these strains showed HSA binding, and five were expressing protein L. The remaining ten strains of the original 48 were commensal strains, and none of them was expressing protein L or binding HSA. These results confirm the correlation between protein L-expressing *F. magna* strains and bacterial vaginosis. Furthermore, it also shows that *F. magna* strains isolated from localized suppurative infections preferentially express HSA-binding protein. It is striking to notice that no strains were found that expressed protein L and HSA-binding protein in combination. This was also noticed by Ng et al.[45] A total of 32 *F. magna* strains, from different origins and countries, were analyzed. Strains were found to be protein L-expressing, HSA-binding protein expressing, or expressing neither of them. No strains were found to express both proteins. Molecular typing of these strains showed that protein L and HSA-binding strains are associated with genotypic clusters.

Recently, two other proteins that enhance the virulence of *F. magna* have been described, that is, a subtilisin-like proteinase (SufA) and a *F. magna* adhesion factor (FAF). FAF is expressed by more than 90% of the *F. magna* strains. This protein is cell-wall bound and can be released in the growth medium. The soluble form of FAF causes large protein aggregates, and the cell-wall bound FAF induces bacterial aggregation through protein-protein interactions between the FAF proteins of the different *F. magna* cells.[22] FAF has the typical features of surface proteins of gram-positive bacteria; a C-terminal part with a cell wall spanning region, a membrane anchor, and an intracellular charged tail. *F. magna* strains which express FAF interact with BM40 and colonize the skin in the same way as SufA-expressing *F. magna* strains. Another feature of FAF is that it protects *F. magna* against LL-37, an antimicrobial peptide.[22] BM40 is able to stimulate wound healing and it increases the albumin transport across the endothelium. The increase of albumin will stimulate PAB-expressing *F. magna* strains in their growth.

SufA is the first described proteinase for *F. magna*.[20] It is associated with the cell wall, but is also released in the growth medium. Most *F. magna* strains possess homologs of SufA. SufA is able to degrade the antibacterial peptides LL-37 and MIG/CXCL9. Thus *F. magna* enhances its own growth and can spread to commensal areas, where it is not present under normal conditions. Since SufA is a subtilisin, it first has to undergo autocatalytic maturation before it can be active. For full enzymatic activity, dimer formation of SufA is required. The antibacterial peptide MIG/CXCL9 binds with the CXCR3 receptor, which activates the G-protein.[46] This receptor is expressed on eosinophils, NK cells, activated T cells, and endothelial cells. Remarkably, MIG/CXCL9 degraded by SufA is still able to bind to the CXCR3 receptor.[21] The fragments of MIG/CXCL9 are still able to kill *Streptococcus pyogenes*, but *F. magna* remains unaffected. This may be explained by the fact that the dimer formation is affected due to the fragmentation of MIG/CXCL9. This may result in a reduced antibacterial activity against *F. magna*. During infections caused by *S. pyogenes*, an ecological niche can be created for *F. magna*. SufA is also able to release the FAF protein from the cell wall of *F. magna*. Karlsson et al.[21] described that FAF is able to bind with MIG/CXCL9 with a high affinity. The release of FAF from the cell wall results in a decrease of the antibacterial activity of MIG/CXCL9, thereby promoting the growth of *F. magna* during inflammation. In human plasma, SufA degrades fibrinogen, a major clotting enzyme.[47] It increases the thrombin-induced coagulation time in a dose-dependent manner. Fibrinogen is cleaved by thrombin to create fibrin, which forms a temporary matrix in which cells can proliferate during wound repair. In its soluble form SufA forms dimers and/or multimers, which are proteolytically more active when compared with the monomers.[47] Fibrinogen consists of three pairs of nonidentical chains Aα, Bβ, and γ.[48] First, SufA removes the C-terminal portion of fibrinogen Aα chains. Second, the NH_2-terminal part of the Bβ chains is attacked. At higher concentrations of SufA, the Aα chains are further processed, thereby removing the central polymerization sites. SufA associated with the cell wall prevents the formation of fibrin networks by binding to keratinocytes.[47] When the skin is damaged or infected, *F. magna* SufA-expressing bacterial cells will be in contact with plasma proteins. The fibrinogen present in the plasma will be broken down by SufA present on the cell wall. The formation of a fibrin network is delayed. The fibrinopeptides (FPA and FPB) that are released during the cleavage of fibrinogen are chemotactic agents for neutrophils, macrophages, and fibroblasts. They also exert antibacterial properties against gram-negative and gram-positive bacteria.[49] It seems that besides the clotting, other fibrinogen-mediated processes are also disturbed by SufA. FAF- and SufA-expressing strains might impair wound healing, as has been described by Stephens et al.[33]

36.1.3 Genome Organization

Recently, the genome of *F. magna* ATCC29328 was assessed by Goto et al.[50] It consists of a circular chromosome (1.797.577 bp, average G + C content 32.3%) and a plasmid pPEP1 (189.163 bp, average G + C content 29.7%). Complete gene sets for the biosynthesis of glycine, serine, threonine, aspartate, and asparagine were present. There were no carbohydrate phosphotransferase system (PTS) genes present for glucose, maltose, mannose, glucitol, cellobiose, and lactose. PTS genes for mannitol, galactitol, and sucrose were incomplete. A lot of genes encoding aminoacid/oligopeptide transporters were found on the genome. This enables *F. magna* to take up amino acids from the environment for growth and survival. Genes for superoxide reductase, NADH oxidase, and putative NADH dehydrogenase were also present. They probably are important for the survival of *F. magna* in intermediate aerobic conditions, such as mucosa and the skin. Virulence factors for antiphagocytosis were encoded by genes present on the chromosome and the plasmid, one on each. In total, four genes encoding albumin-binding protein homologs were present, three on the chromosome, and one on the plasmid. In total, 10 genes encoding collagen-binding proteins were found, five on the chromosome, and five on the plasmid. In total, 20 genes encoding *N*-acetylmuramoyl-L-alanine amidase homologs (Cwp66) were encountered, most of them located on the chromosome. These proteins play a role in the adherence of bacteria to host cells. The presence of genes encoding sortase was assessed on the chromosome and plasmid. On the chromosome four genes encoding sortase, homologs were present, and on the plasmid, seven. The presence of seven sortase homologs on the plasmid is especially interesting. It is the highest number of sortase homologs present on a plasmid, for as far as genome sequences are determined for gram-positive bacteria. This feature might be unique for *F. magna*. Since plasmids are considered to be of foreign origin, the amount of sortase homologs present on a plasmid may play an important role in the pathogenesis of *F. magna*. Sortases are extracellular transpeptidases that catalyze the cell anchoring of cell-wall proteins. Sortases can be grouped into four or five different classes.[51,52] Each subgroup has its own preference for substrates, depending on the amino acids present in the cell-wall sorting signal pentapeptide motif. Sortase A is the most important one and catalyzes the highest number of substrates. The precursor of a cell-wall-bound protein is synthesized in the cytoplasm with an N-terminal signal peptide and a C-terminal sorting signal (Figure 36.1). The cell-wall sorting signal consists out of a pentapeptide motif (for sortase A: LPXGT motif) a hydrophobic region, and a tail of charged residues.[53] The N-terminal signal peptide directs the precursor to the membrane for translocation. It is assumed that, after cleavage by a signal peptidase, the hydrophobic region and the positively charged tail retain the precursor within the secretory pathway until the sortase has recognized the substrate.[28] The membrane-anchored sortase A cleaves between the threonine and glycine residues of the LPXTG-like motif.[54] An amide-bond is formed between the carboxyl group produced by the cleavage of the LPXGT motif and peptidoglycan of the cell wall. Since most virulence factors are displayed on the cell wall, sortases play an important role in the virulence of bacteria. The amount of cell-wall anchored protein will be enriched and more varied. This may result in an enhancement of interaction between host tissue and other bacteria in mixed infections. Therefore, sortases are a possible new target for the development of new therapeutic drugs against bacterial infections.[55]

36.1.4 Diagnosis

Phenotypical Techniques. *F. magna* strains are difficult to identify, since the strains do not show any saccharolytic activity and only produce acetate as volatile fatty acids (VFAs). Identification is therefore based on negative reactions.

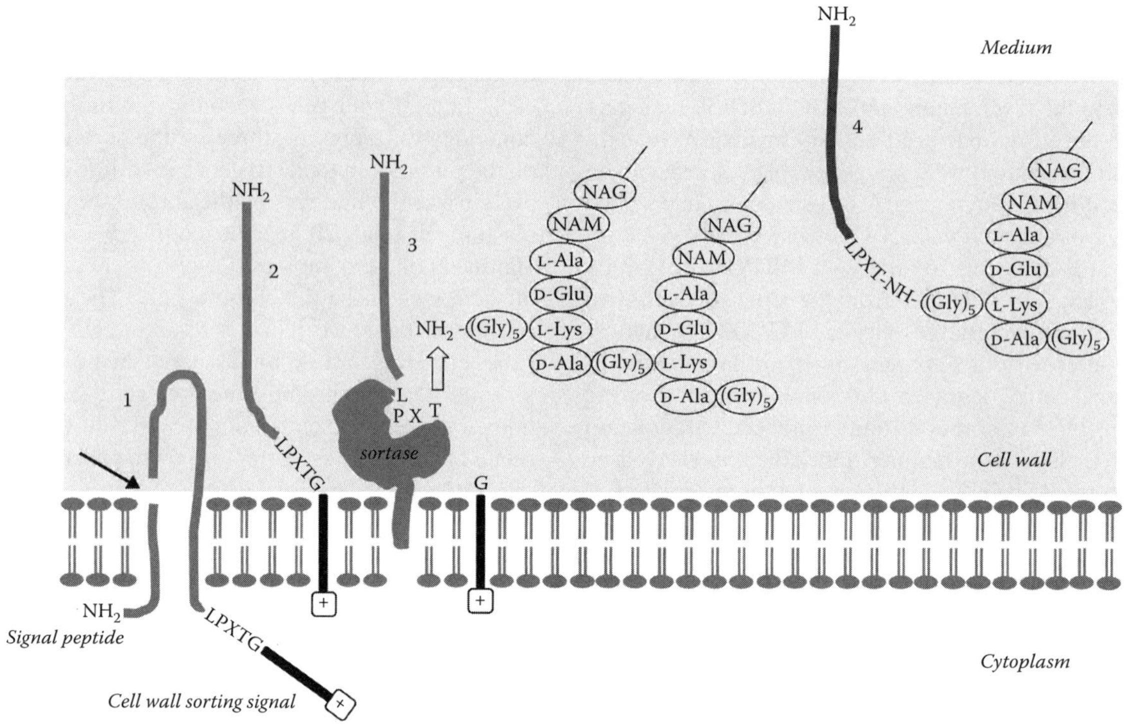

FIGURE 36.1 A schematic overview of the mechanism of covalent binding of surface proteins. (Adapted from Cossart, P., and Jonquières, R., *Proc. Natl. Acad. Sci. USA*, 97, 5013, 2000 [Copyright 2000 National Academy of Sciences, USA].)

1. The precursor protein is directed by the N-terminal region of the signal peptide to the cell membrane, after which it is cleaved by a signal peptidase (arrow).
2. The precursor retains within the secretory pathway, probably due to the positively charged tail and the hydrophobic region.
3. The pentapeptide motif LPXTG is cleaved by sortase between the threonine and glycine residues. An amide-bond is formed between the produced carboxyl and peptidoglycan in the cell wall.
4. The surface protein is anchored in the cell wall.

Microscopic appearance may be rather variable and therefore is not suitable for identification. Ezaki et al.[56] showed that GPAC can easily be identified by their amidase and oligopeptidase activities. Murdoch et al.[7] characterized the described species in further detail, among them *F. magna*. In total, nine reference strains of *F. magna* and 78 clinical isolates were analyzed. All strains showed similar proteolytic activity. Strongly positive reactions were obtained for arginine, leucine, and pyroglutamyl aminopeptidase. Negative reactions were obtained for proline, phenylalanine, and glutamylglutamyl aminopeptidases. Variations were observed for other proteolytic enzymes. No acid production from mannose, raffinose, glucose, and trehalose was found. A minor part of the strains (22%) was able to produce a small amount of acid from fructose. Most of the strains were negative for alkaline phosphatase, but 7% was found to be positive, including the type strain. Nineteen percent of the strains were able to produce catalase. All strains were resistant to sodium polyanethol sulfonate (SPS). If a metronidazole disk is used to exclude anaerobic staphylococci or capnophilic streptococci, it should be noted that *F. magna* is able to produce colonies in the inhibition zone after 48 hours of incubation. Recently, Song et al.[57] developed a flow chart that made the identification of GPAC easier. In this flow chart, *Parvimonas micra* and *F. magna* are differentiated from the other GPAC by their inability to produce β-glucuronidase and to ferment glucose and their ability to produce pyroglutamyl arylamidase. *P. micra* and *F. magna* are differentiated from each other by the production of proline arylamidase; *P. micra* is a producer and *F. magna* is not.[57]

Two current methods to phenotypically identify anaerobes are the Vitek system and Matrix-Assisted Laser Desorption Ionisation Time of Flight Mass Spectometry (MALDITOF-MS). Recently, BioMérieux (Marcy, France) has developed a new colorimetric identification card (ANC card), which can be used in the Vitek 2 system for the identification of anaerobes, including *F. magna*. Evaluation of this card showed that *F. magna* can be reliably identified using this method.[58,59] MALDITOF-MS is a promising new method for identifying bacteria; however, no evaluation studies have been published yet that describe the suitability of the method for the identification of GPAC, including *F. magna*.

Genotypical Techniques. Nowadays, nucleic acid–based techniques are available to improve the identification of bacteria. Song et al.[60] evaluated if 16S rRNA sequencing is suitable for the identification of GPAC. They established that the quality of the sequences in the public databases can be poor and may lead to misidentification. They sequenced the type strain of *F. magna* and 36 clinical isolates. When the sequences of the clinical isolates were compared with the sequenced type strain, the homology was >99%. However, when they were compared with the sequences available in the public databases, homology was <98%. When analyzing sequences of GPAC, it is best to compare them with the sequences published by Song et al.[60]

PCR methods to identify *F. magna* were developed by Song et al.[61] and by Riggio et al.[10] Primers specific for *F. magna* were designed and validated. Song et al.[61] validated the primer set on the type strain of *F. magna* and 60 sequenced clinical isolates. The primers were shown to be specific. Riggio et al.[10] validated the specificity of their primers with five *F. magna* strains. Furthermore, they applied their PCR method on subgingival plaque samples from patients with adult periodontal disease and on 60 pus aspirates from subjects with acute dentoalveolar abscesses. No nonspecific reactions were encountered.

A species-specific 16S rRNA-based probe was developed by Wildeboer-Veloo et al.[62] The fluorescently labeled probe was used to rapidly and reliably identify *F. magna*, using fluorescent in situ hybridization (FISH). The probe was designed and validated using sequences of reference strains of *F. magna* and 26 clinical isolates, and was shown to be specific. The probe was applied on 100 unknown clinical isolates of GPAC, of which 29 strains were identified as *F. magna*. Permeabilization of the cell wall of *F. magna* strains using proteinase K was necessary prior to hybridization. The probe has the potential to be suitable for direct application on clinical material.

36.2 METHODS

36.2.1 Sample Preparation

No special sample preparation is necessary for the isolation of *F. magna*. The usual sample preparation conditions for the isolation of anaerobic bacteria will be sufficient.[63] Clinical specimens should be collected, avoiding contamination with commensal skin or mucus surface microbiota. Transportation of specimens should take place in an appropriate transport medium for anaerobic bacteria. There is no selective medium available for the isolation of *F. magna*. *F. magna*, like most GPAC, will grow on a standard medium, like Brucella blood agar (BBA) and Schaedler blood agar. Mostly, clinical specimens suspected of containing anaerobic bacteria are cultured on a set of plates. A universal medium (e.g., BBA), phenyl alcohol blood agar (PEA), which prevents the swarming of bacteria and media selective for gram-negative anaerobes like bacteroides bile esculin agar (BBE) and kanamycin vancomycin laked blood agar (KVLB). The latter promotes also the pigmentation of bacteria. *F. magna* will grow on the universal medium and PEA.[63] Strains suspected to be *F. magna*, after being tested for aerotolerance, can be identified as described in Section 36.1.3.

36.2.2 Detection Procedures

F. magna can be recognized by its colony form and cell morphology, as described above. However, this is not sufficient for identification. Other biochemical features that can be used are its slow growth, odor, resistance to SPS, nitrate reduction, coagulase production, no indole production, and enzyme profile.[57,63,64]

Bassetti et al.[65] reported a case of endocarditis caused by *F. magna*. However, in this case the blood cultures were initially negative. They showed that the detection of *F. magna* in blood depends on the type of blood-culture systems used. Bact/ALERT FA and FN bottles were still negative after 4 weeks of incubation. SEPTI-CHEK BHI-S bottles and the ISOLATOR system already showed growth after two days of incubation. This may account for the fact that bacteremia caused by *F. magna* is rare.[66] Microbiologists should be aware of the fact that a bacteremia caused by *F. magna* can be missed when the blood-culture system Bact/ALERT is used.

We present below three molecular-based techniques for identification and confirmation of *F. magna*.

(i) Protocol of Song et al.[61]

Principle. A multiplex PCR assay was developed to rapidly identify GPAC. Genus- and species-specific primers were used. Since *F. magna* is the only species present in the genus *Finegoldia*, it can be identified directly to species level.

Procedure

1. Suspend one or two colonies of the bacterial strain in 50 μL Tris-HCl/EDTA/saline (pH 8.0), and incubate for 10 min at 95°C.
2. Centrifuge for 2 min at 18600 *g*, and decant the supernatant.
3. Resuspend the pellet.
4. Prepare the PCR mixture (50 μL) containing: 1.25 U *Taq* polymerase (Promega), 50 mM KCl, 10 mM Tris-HCl (pH 9.0), 0.1% Triton, 2.5 mM $MgCl_2$, 0.2 mM dNTPs, 5 μL of bacterial lysate, 0.25 μM of the universal primer 1392B and 0.5 μM of the genus/species-specific primer FIGD.
5. Perform the PCR for 35 cycles; denaturation at 95°C for 20 s, annealing at 50°C for 1 min, and extension at 72°C for 30 s. To the final extension add a cycle of 72°C for 5 min.
6. Analyze PCR products on a 2% agarose gel, followed by ethidium-bromide staining.
7. Visualize under UV illumination.

Note: *F. magna* strains will yield an amplicon size of 1200 bp.

(ii) Protocol of Riggio et al.[10]

Principle. This PCR assay was developed for the direct detection of *F. magna* in subgingival plaque sample.

Procedure

1. Take a subgingival plaque sample using a sterile curette and put it in 0.5 mL freshly prepared fastidious anaerobe broth (Bioconnections). Mix for 30 s.
2. Add 3 μL achromopeptidase (20 U/μL in 10 mM Tris-HCl, 1 mM EDTA, pH 7.0) to 100 μL of plaque sample. Incubate at 56°C for 30 min, boil for 5 min and store at –70°C.
3. Prepare the PCR mixture (50 μL) containing: 45 μL 1×PCR buffer (10 mM Tris-HCl, pH 9.0, 50 mM KCl, 1.5 mM $MgCl_2$, 0.1% Triton X-100), 1 U *Taq* DNA polymerase (Promega), 0.2 mM dNTPs, 0.2 μM of each primer, and 5 μL lysed plaque sample.
4. Perform the PCR, starting with a denaturation step of 94°C of 5 min followed by 35 cycles of denaturation of 1 min at 94°C, annealing of 1 min at 60°C, and extension of 1 min at 72°C. To the final extension add a cycle of 72°C for 10 min.
5. Analyze the PCR products on a 2% agarose gel containing ethidium bromide, by adding 2 μL gel loading dye [0.25% (w/v) bromophenol blue, 50% (v/v) glycerol, 100 mM EDTA, pH 8.0] to 10 μL PCR product.
6. Visualize under UV illumination.

Note: *F. magna* strains will yield an amplicon size of 553 bp. A positive and negative control were added to the assay, consisting of genomic *F. magna* DNA and water, respectively. An appropriate DNA ladder should be included during electrophoresis.

(iii) Protocol of Wildeboer-Veloo et al.[62]

Principle. *F. magna* strains are identified by FISH, using a fluorescently labeled species-specific 16S rRNA-based probe.

Procedure

1. Take some colonies of bacterial cells and suspend them in phosphate-buffered saline (8 g NaCl, 0.2 g KCl, 1.44 g Na_2HPO_4, 0.24 g KH_2PO_4/L, pH 7.4). Add an equal volume of 96% v/v ethanol. Mix and store at –20°C.
2. Spot the ethanol-fixed cells on three gelatin-coated glass slides and dry at room temperature.
3. Immerse the slides in 96% v/v ethanol for 10 min and allow them to dry.
4. Add 100 μL proteinase K (500 mg/L in 50 mM Tris-HCl, pH 7.6) to each slide and cover it with a coverslip. Incubate for 10 min at room temperature.
5. Stop the enzymatic reaction by incubating the slides in 96% v/v ethanol for 2 min. Allow the slides to dry.
6. Prepare three different hybridization mixtures; one for the positive control (probe EUB338), one for the specific probe (Fmag1250), and one for the negative control (probe non-EUB338). Each mixture contains 10 μL probe (100 ng/μL) and 110 μL hybridization buffer (0.9 M NaCl, 20 mM Tris-HCl, pH 7.2, SDS 0.1% w/v).
7. Add to each slide approximately 100 μL of a probe-hybridization mixture and cover the slide with a coverslip.
8. Incubate overnight at 50°C in a dark humid chamber/box.
9. Wash the slides in washing buffer (0.9 M NaCl, 20 mM Tris-HCl, pH 7.2) for 15 min at 50°C. Rinse the slides quickly in milli-Q and dry with compressed air.
10. Add approx. 10 μL vectashield® (Vector laboratories, Burlingame, USA) and a coverslip. Fix the coverslip with nail polish.
11. Examine the slides under an epifluorescence microscope and compare the signal of the specific probe with the positive and negative control.

Note: The proteinase K treatment is necessary to permeabilize the cell wall of *F. magna* to enable the probe to access the bacterial cell. However, this treatment also increases loss of bacterial cells from the slide. It is therefore recommended to put sufficient bacterial cells on the slide. This technique has the potential for direct application on clinical material.

36.3 CONCLUSION AND FUTURE PERSPECTIVES

F. magna is part of the commensal microbiota and clearly an important pathogen. *F. magna* is able to express different virulence factors: collagenase, PAB, protein L, SufA, and FAF. The sequencing of the genome of *F. magna* ATCC29328 revealed a total of four albumin-binding proteins, 10 collagen-binding proteins, 20 Cwp66 homologs, and 11 sortase homologs. Besides these proteins, some other putative virulence factors may have been revealed, for example, hemolysin and bacteriocin. The high number of sortase homologs enables *F. magna* to enrich and vary cell wall–anchored proteins, hereby increasing its virulence.

Each of these virulence factors increases the pathogenicity of *F. magna*. Until now, it is unclear how many virulence factors one *F. magna* strain can express at the same time. Especially the high number of sortase homologs enables *F. magna* to acquire several virulence factors in one strain. Ng et al.[45] studied 32 *F. magna* strains and found none of the strains expressed both protein L and HSA-binding protein. Molecular typing of these *F. magna* strains expressing protein L or protein PAB revealed that these strains form their own genotypic cluster. More studies should be performed whether *F. magna* strains expressing certain virulence factors, for example, collagenase, and so forth are genotypically different from each other. It will also be interesting to know if more expressed virulence factors are associated with genotypic clusters of *F. magna*.

Several authors[16,27,29,30,32,33] have shown that there is a relationship between certain virulence factors of *F. magna* and the site of infection. For example, *F. magna* strains isolated from nonpuerperal breast abscesses and diabetic foot infections have a higher collagenase production when compared with strains isolated from intraabdominal infections.[16] This site-specific infection and expression of virulence factors have also been shown for protein L and HSA-binding protein.[44] However, these conclusions are based on a small number of studies, and therefore on a relatively small number of strains. Whether there is a site-specific infection for collagenase-, SufA-, and FAF-expressing *F. magna* strains is still unknown. Further studies are necessary to elucidate all virulence factors of *F. magna* and their role in certain infections.

It is also shown that *F. magna* has a negative influence on healing of chronic wounds.[33] It is not clear whether these strains express only one virulence factor or several, or whether there is a coherence between the different virulence factors, as has been described by Karlsson et al.[21] between SufA and FAF. In order to eludicate the role of *F. magna* and the expression of virulence factors, microbiologists should pay more attention to the identification of *F. magna* from clinical specimens. In general, GPAC are susceptible to the antibiotics used to treat an anaerobic infection.[1] However, microbiologists should realize that *F. magna* has one of the highest resistance rates of the GPAC[67] (unpublished data), and that there seem to be geographical differences in resistance.[1,68] To get a coherent picture of *F. magna* and its pathogenicity, studies with larger groups of strains should be performed, thereby studying all the specific virulence factors of *F. magna* at the same time and taking the site of infection into account.

REFERENCES

1. Murdoch, D.A., Gram-positive anaerobic cocci, *Clin. Microbiol. Rev.*, 11, 81, 1998.
2. Prevot, A.R., Études des systématique bactérienne. I. Lois générales. II. Cocci anaérobies, *Ann. Sci. Nat.*, 15, 23, 1933.
3. Holdeman Moore, L.V., Johnson, J.L. and Moore, W.E.C., Genus *Peptococcus*. In Sneath, P.H.A. (Editors), *Bergey's Manual of Systematic Bacteriology* 2, p. 1082, Williams and Wilkins, Baltimore, 1986.
4. Ezaki, N. et al., Transfer of *Peptococcus indolicus, Peptococcus asaccharolyticus, Peptococcus prevotii* and *Peptococcus magnus* to the genus *Peptostreptococcus* and proposal of *Peptostreptococcus tetradius* sp. nov., *Int. J. Syst. Bacteriol.*, 33, 683, 1983.
5. Murdoch, D.A., and Magee, J.T., A numerical taxonomic study of the gram-positive anaerobic cocci, *J. Med. Microbiol.*, 43, 148, 1995.
6. Murdoch, D.A., and Shah, H.N., Reclassification of *Peptostreptococcus magnus* (Prevot 1933) Holdeman and Moore 1972 as *Finegoldia magna* comb. nov. and *Peptostreptococcus micros* (Prevot 1933) Smith 1957 as *Micromonas micros* comb. nov., *Anaerobe*, 5, 555, 1999.
7. Murdoch, D.A., and Mitchelmore, I.J., The laboratory identification of gram-positive anaerobic cocci, *J. Med. Microbiol.*, 34, 295, 1991.
8. Wren, M.W.D. et al., The anaerobic culture of clinical specimens; a 14-month study, *J. Med. Microbiol.*, 10, 49, 1977.
9. Neut, C. et al., Analysis of gram-positive anaerobic cocci in oral, fecal and vaginal flora, *Eur. J. Clin. Microbiol. Infect. Dis.*, 4, 435, 1985.
10. Riggio, M.P., and Lennon, A., Specific PCR detection of *Peptstreptococcus magnus*, *J. Med. Microbiol.*, 52, 309, 2003.
11. Gao, Z. et al., Molecular analysis of human forearm superficial skin bacterial biota, *Proc. Natl. Acad. Sci.*, 104, 2927, 2007.
12. Higaki, S. et al., Anaerobes isolated from infectious skin diseases, *Anaerobe*, 5, 583, 1999.
13. Brook, I., Recovery of anaerobic bacteria from clinical specimens in 12 years at two military hospitals, *J. Clin. Microbiol.*, 26, 1181, 1988.
14. Brook, I., Peptostreptococcal infection in children, *Scand. J. Infect. Dis.*, 26, 503, 1994.
15. Bourgault, A-M., Rosenblatt, J.E., and Fitzegerald, R.H., *Peptococcus magnus*: A significant human pathogen, *Ann. Intern. Med.*, 93, 244, 1980.
16. Krepel, C.J. et al., Anaerobic pathogenesis: Collagenase production by *Peptostreptococcus magnus* and its relationship to site of infection, *J. Infect. Dis.*, 163, 1148, 1991.
17. Myrhe, E.B., Surface receptors for human serum albumin in *Peptostreptococcus magnus* strains, *J. Med. Microbiol.*, 48, 189, 1984.
18. Björck, L., Protein L, a novel bacterial cell wall protein with affinity for Ig L chains, *J. Immunol.*, 140, 1194, 1988.
19. Patella, V. et al., Protein L, a bacterial Ig-binding protein that activates human basophils and mast cells, *J. Immunol.*, 145, 3054, 1990.
20. Karlsson, C. et al., SufA—a novel subtisilin-like serine proteinase of *Finegoldia magna*, *Microbiology.*, 153, 4208, 2007.
21. Karlsson, C. et al., SufA of the opportunistic pathogen *Finegoldia magna* modulates actions of the antibacterial chemokine MIG/CXCL9, promoting bacterial survival during epithelial inflammation, *J. Biol. Chem.*, 284, 29499, 2009.
22. Frick, I.-M. et al., Identification of a novel protein promoting the colonization and survival of *Finegoldia magna*, a bacterial commensal and opportunistic pathogen, *Mol. Microbiol.*, 70, 695, 2008.
23. Brook, I., and Walker, R.I., Pathogenicity of anaerobic gram-positive cocci, *Infect. Immun.*, 45, 320, 1984.
24. Harrington, D.J., Bacterial collagenases and collagen-degrading enzymes and their potential role in human disease, *Infect. Immun.*, 64, 1885, 1996.
25. Ng., J. et al., Aminopeptidase activities in *Peptostreptococcus* spp. are statistically correlated to gelatin hydrolysis, *Can. J. Microbiol.*, 44, 303, 1998.
26. Myrhe, E.B., and Kronvall, G., Demonstration of specific binding sites for human serum albumin in group C and G streptococci, *Infect. Immun.*, 27, 6, 1980.
27. de Château, M., and Björck, L., Protein PAB, a mosaic albumin-binding bacterial protein representing the first comtemporary example of module shuffling, *J. Biol. Chem.*, 269, 12147, 1994.
28. Schneewind, O., Model, P., and Fischetti, V.A., Sorting of protein A to the staphylococcal cell wall, *Cell*, 70, 267, 1992.
29. de Château, M., Holst, E., and Björck, L., Protein PAB, an albumin-binding bacterial surface protein promoting growth and virulence, *J. Biol. Chem.*, 271, 26609, 1996

30. Felten, A. et al., *Peptostreptococcus magnus* osteoarticular infections after orthopedic surgery. 14 cases and pathogenicity factors, *Pathol. Biol. (Paris)*, 46, 442, 1998.
31. Krepel, C.J., Enzymatically active *Peptostreptococcus magnus*: Association with site of infection, *J. Clin. Microbiol.*, 30, 2330, 1992.
32. Edmiston, C.E. et al., The nonpuerperal breast infection: Aerobic and anaerobic microbial recovery from acute and chronic disease, *J. Infect. Dis.*, 162, 695, 1990.
33. Stephens, P. et al., Anaerobic cocci populating the deep tissues of chronic wounds impair cellular wound healing responses in vitro, *Br. J. Dermatol.*, 148, 456, 2003.
34. Nilson, B.H.K. et al., Protein L from *Peptostreptococcus magnus* binds to the κ light chain variable domain, *J. Biol. Chem.*, 267, 2234, 1992.
35. Solomon, A., Bence-Jones proteins and light chains of immunoglobulins (first of two parts), *N. Engl. J. Med.*, 294, 17, 1976.
36. Wikström, M. et al., Mapping of the immunoglobulin light chain-binding site of protein L, *J. Mol. Biol.*, 250, 128, 1995.
37. Åkerström, B., and Björck, L., Protein L: An immunoglobulin light chain-binding bacterial protein, *J. Biol. Chem.*, 264, 19740, 1989.
38. Housden, N.G. et al., Immunoglobulin-binding domains: Protein L from *Peptostreptococcus magnus*, *Biochem. Soc. Trans.*, 31, 716, 2003.
39. Graille, M. et al., Complex between *Peptostreptococcus magnus* protein L and a human antibody reveals structural convergence in the interaction modes to Fab binding proteins, *Structure*, 9, 679, 2001.
40. Smith, D. et al., Whole-body autoradiography reveals that the *Peptostreptococcus magnus* immunoglobulin-binding domains of protein L preferentially target B lymphocytes in the spleen and lymp nodes in vivo, *Cell. Microbiol.*, 6, 609, 2004.
41. Viau, M. et al., Specific in vivo deletion of B-cell subpopulations expressing human immunoglobulins by the B-cell superantigen protein L, *Infect. Immun.*, 72, 3515, 2004.
42. Genovese, A. et al., Bacterial immunoglobulin superantigen proteins A and L activate human heart mast cells by interacting with immunoglobulin E, *Infect. Immun.*, 68, 5517, 2000.
43. Malaviya, R. et al., Mast cells process bacterial Ags through an phagocytic route for class I MHC presentation to T cells, *J. Immunol.*, 156, 1490, 1996.
44. Kastern, W. et al., Protein L, a bacterial immunoglobulin-binding protein and possible virulence determinant, *Infect. Immun.*, 58, 1217, 1990.
45. Ng, J. et al., Differentiation of protein L-containing and albumin-binding *Peptostreptococcus magnus* isolates by DNA amplification, ribotyping, and pulsed field gel electrophoresis, *Anaerobe*, 2, 95, 1996.
46. Loetscher, M. et al., Chemokine receptor specific for IP10 and Mig: Structure, function, and expression in activated T-lymphocytes, *J. Exp. Med.*, 184, 963, 1996.
47. Karlsson, C. et al., Suf A—a bacterial enzyme that cleaves fibrinogen and blocks fibrin network formation, *Microbiology*, 155, 238, 2009.
48. Mosesson, M.W., Fibrinogen and fibrin structure and functions, *J. Thromb. Haemost.*, 3, 1894, 2005.
49. Tang, Y.Q., Yeaman, M.R., and Selsted, M.E., Antimicrobial peptides from human platelets, *Infect. Immun.*, 70, 6524, 2002.
50. Goto, T. et al., Complete genome sequence of *Finegoldia magna*, an anaerobic opportunistic pathogen, *DNA Res.*, 15, 39, 2008.
51. Comfort, D., and Clubb, R.T., A comparative genome analysis identifies distinct sorting pathways in gram-positive bacteria, *Infect. Immun.*, 72, 2710, 2004.
52. Dramsi, S., Trieu-Cuot, P., and Bierne, H., Sorting sortases: A nomenclature proposal for the various sortases of gram-positive bacteria, *Res. Microbiol.*, 156, 289, 2005.
53. Schneewind, O., Mihaylova-Petkov, D., and Model, P., Cell wall sorting signals in surface proteins of gram-positive bacteria, *EMBO J.*, 12, 4803, 1993.
54. Mazmanian, S.K. et al., *Staphylococcus aureus* sortase, an enzyme that anchors surface proteins to the cell wall, *Science*, 285, 760, 1999.
55. Maresso, A.W., and Schneewind, O., Sortase as target of anti-infective therapy, *Pharmacol. Rev.*, 60, 128, 2008.
56. Ezaki, T., and Yabuuchi, E., Oligopeptidase activity of gram-positive anaerobic cocci used for rapid identification, *J. Gen. Appl. Microbiol.*, 31, 255, 1985.
57. Song, Y., Liu, C., and Finegold, S.M., Development of a flow chart for identification of gram-positive anaerobic cocci in the clinical laboratory, *J. Clin. Microbiol.*, 45, 512, 2007.
58. Rennie, R.P. et al., Multi-center evaluation of the VITEK 2 anaerobes & corynebacteria (ANC) identification card, *J. Clin. Microbiol.*, 46, 2646, 2008.
59. Mory, F. et al., Evaluation of the new Vitek 2 ANC card for identification of medically relevant anaerobic bacteria, *J. Clin. Microbiol.*, 47, 1923, 2009.
60. Song, Y. et al., 16S ribosomal DNA sequence-based analysis of clinically significant gram-positive anaerobic cocci, *J. Clin. Microbiol.*, 41, 1363, 2003.
61. Song, Y. et al., Rapid identification of gram-positive anaerobic coccal species originally classified to the genus *Peptostreptococcus* by multiplex PCR assays using genus- and species-specific primers, *Microbiology*, 149, 1719, 2003.
62. Wildeboer-Veloo, A.C.M. et al., Development of 16S rRNA-based probes for the identification of gram-positive anaerobic cocci isolated from clinical specimens, *Clin. Microbiol. Infect.*, 13, 985, 2007.
63. Jousimies-Somer, H.R. et al., *Anaerobic Bacteriology Manual*, 6th ed., Belmont, CA, Star Publishing Co., 2002.
64. Letournel-Glomoud, C., Houssaye, S., and Peyrard, N., Identification of gram-positive anaerobic cocci in a clinical laboratory using a simplified flowchart, *Anaerobe.*, 5, 153, 1999.
65. Bassetti, S. et al., Endocarditis caused by *Finegoldia magna* (formerly Peptostreptococcus magnus): diagnosis depends on the blood culture system used, *Diagn. Microbiol. Infect. Dis.*, 47, 359, 2003.
66. Topiel, M.S., and Simon, G.L., *Peptococcaceae* bacteremia, *Diagn. Microbiol. Infect. Dis.*, 4, 109, 1986.
67. Brazier, J.S. et al., Antibiotic susceptibilities of gram-positive anaerobic cocci: Results of a sentinel study in England and Wales, *J. Antimicrob. Chemother.*, 52, 224, 2003.
68. Brazier, J. et al., European surveillance study on antimicrobial susceptibility of gram-positive anaerobic cocci, *Int. J. Antimicrob. Agents*, 31, 316, 2008.

37 *Mogibacterium*

Reginaldo Bruno Gonçalves, Daniel Saito, and Renato Corrêa Viana Casarin

CONTENTS

37.1 INTRODUCTION

37.1.1 Classification

The genus *Eubacterium* is composed of obligately anaerobic, asaccharolytic, gram-positive, nonspore-forming, rod-shaped bacteria.[1] The genus is distinguished from other genera generally on the basis of the negative metabolic characteristics of its representative species.[2] In addition to the nonreactivity in conventional biochemical tests, *Eubacterium* species are difficult to culture, presenting slow growth even on artificial media designed for fastidious microorganisms and under strict anaerobic conditions.[3]

Due to a relatively high looseness in its taxonomical definition, the *Eubacterium* genus has become, over the years, a taxonomic repository of species that could not be placed into other genera.[4,5] Because of its broad definition, the genus has covered a large number of phenotypically diverse organisms.[5,6] It has been reported that the bacterial species belonging to the genus were not phylogenetically homogeneous, and several species were in fact closely related to *Clostridium*.[7,8] In addition, many oral asaccharolytic isolates that qualified as members of *Clostridium* but could not be assigned to any of the existing species were eventually assigned to *Eubacterium*.[9–12]

In 2000, the new genus *Mogibacterium* was proposed to include the novel species *M. pumilum* and *M. vescum*, as well as former *Eubacterium timidum* by taxonomic reassignment.[13] Members of the genus *Mogibacterium* are described as strictly anaerobic and asaccharolytic, gram-positive, rod-shaped bacteria. Cells are nonmotile and do not form spores. Their growth in broth media is very poor. Metabolic end product in PYG medium is phenyl-acetate- and catalase-negative. The DNA base composition is 45–46 G – C mol%. Some members of this genus can be distinguished from each other by 16S rRNA gene sequences and DNA–DNA homology values. The type species of the genus is *Mogibacterium pumilum*.

37.1.2 Morphology and Biology

Mogibacterium pumilum, whose nomenclature refers to the tiny colonies formed in culture plates, forms minute colonies <1 mm in diameter on BHI-blood agar plates, even after prolonged incubation in an anaerobic chamber.[13] Colonies are circular, convex, and translucent. Individual cells are about 0.2–0.3 μm by 1.0 μm, occurring singly, in short chains, or in clumps. Growth in PY broth media is very poor, with or without carbohydrates. No hemolysis occurs around colonies on BHI-blood agar plates. Cells are inert in most biochemical tests. Starch and aesculin are not hydrolyzed, and nitrate is not reduced. No liquefaction of gelatin occurs. Indole, urease, and catalase tests are also negative. Ammonia is not produced from arginine. All strains are nonfermentative and do not utilize adonitol, amygdalin, arabinose, cellobiose, erythritol, aesculin, fructose, galactose, glucose, glycogen, inositol, lactose, maltose, mannitol, mannose, melezitose, melibiose, rhamnose, ribose, salicin, sorbitol, starch, sucrose, trehalose, or xylose.[13] Nonetheless, they produce phenyl acetate (about 3 mM) as the sole metabolic end product in PY or PYG broth. The type strain is ATCC 700696, which was originally isolated from human periodontal pockets.[13]

Mogibacterium vescum, designated due to poor growth in broth media with or without carbohydrates, was originally isolated from human periodontal pockets.[13] This strain can be distinguished from *M. pumilum* by 16S rRNA gene sequence similarity and DNA–DNA relatedness values. The type strain is ATCC 700697. *Mogibacterium timidum*

(*Eubacterium timidum*[14]), nominated due to its slight or slow growth in clumps, is composed of obligately anaerobic, nonspore-forming, nonmotile, gram-positive rods that are regular to slightly diphtheroid in shape. Cells from actively growing cultures are gram-positive, but older cultures may stain as gram-negative. The type strain is ATCC 33093, which was isolated from a subgingival area with periodontitis.[14]

Today, the genus *Mogibacterium* also includes *M. diversum* and *M. neglectum*.[4] *Mogibacterium diversum*, nominated due to its novel lineage in the evolutionary tree, was first isolated from human tongue plaque. The type strain is ATCC 700923. *Mogibacterium neglectum*, nominated because of the poor growth and tiny colonies that caused it to be neglected scientifically for a long time, was originally isolated from human necrotic dental pulps. According to the nearly full nucleotide sequence of the 16S rRNA gene and the DNA–DNA relatedness, this species can be distinguished from *M. pumilum*, *M. vescum*, and *M. diversum*. The type strain is strain ATCC 700924.

37.1.3 Characterization and Identification

The slow-growing characteristics and fastidious nutritional requirements of *Mogibacterium* spp. have potentially overshadowed their true prevalence in oral infections. However, with the establishment of DNA-based detection techniques, cultivable and noncultivable *Mogibacterium* species and phylotypes can now be detected with relatively high frequency.[15–19]

A number of phenotypic studies have successfully contrasted *Mogibacterium timidum* from *Eubacterium* and other *Mogibacterium* species. In a comparative report, Wade et al.[1] used protein profile analysis, metabolic end-product analysis, and API analysis to characterize 31 strains of *Eubacterium*, including *M. timidum*. Although high coherence was found among all methodologies, only protein profile analysis was able to unambiguously differentiate all tested strains. According to Nakazawa and Hoshino,[6] Western immunoblotting of whole bacterial cell extracts with rabbit antisera can also be a good method for discriminating *M. timidum* from *Eubacterium* species due to their high heterogeneity in whole-cell protein composition.

Mogibacterium timidum does not produce volatile fatty acid metabolic products, that is, isobutyric, butyric, isovaleric, and isocaproic acids. However, it does produce phenylacetic acid as a major nonvolatile fatty acid end product when grown on PYG broth.[20,21] This phenotypic marker is sufficient to differentiate *M. timidum* from *E. nodatum* and *E. brachy*.[1,20] In addition, *M. timidum* may produce low quantities of lactic and succinic acids, which *E. brachy* does not produce.[20] *M. timidum* can also be easily differentiated from *E. lentum* because the former utilizes pyruvate as a substrate, whereas the latter uses nitrate.[21]

Differentiation of *M. timidum* from other *Mogibacterium* species by phenotypic approaches can be problematic because all produce high amounts of phenylacetic acid while presenting negative results in many other biochemical assays.[20] According to Downes et al.,[20] both *M. timidum* and *M. vescum* show proline arylamidase activity when cultured under aerobic conditions, as determined by use of the Rapid ID 32A identification system (bioMérieux). However, in anaerobic conditions, *M. timidum* can be differentiated from *M. vescum* because the former displays pyroglutamate arylamidase activity, whereas the latter does not. Differences in colony morphology could also be detected after incubation for 7 days on Fastidious Anaerobe Agar. *M. timidum* colonies were 0.3–0.5 mm in diameter, circular, entire, convex, and translucent. *M. vescum* colonies were 0.6–0.8 mm in diameter, circular, entire, and umbonate, with an opaque off-white center.

Characterization of *Mogibacterium* species can be also achieved by genotypic analysis. Determination of G + C molecular content may be used to differentiate particular subgroups of *Mogibacterium*. Accordingly, while *M. neglectum* displays a 42% value, *M. pumilum* and *M. vescum* display 46%, and *M. timidum* 50%. In addition, whole genome relatedness assays may also be used, although hybridization levels between certain strains may not be sufficiently discriminatory (Table 37.1). Alternatively, DNA sequencing analysis can be employed. However, it should be noted that 16S rRNA gene analysis may not provide sufficient phylogenetic discriminatory power to easily distinguish *M. diversum* and *M. neglectum* from *M. vescum*,[4] although some authors have successfully distinguished *M. timidum* from *M. vescum* by this approach.[20] A phylogenetic tree of 16S rRNA gene nucleotide sequences from *Mogibacterium* species and phylotypes from the Ribosomal Database Project II (http://rdp.cme.msu.edu/) is displayed in Figure 37.1.

According to Nakazawa et al.[4] whole-cell protein profile analysis harbors enough analytic resolution to differentiate all species of *Mogibacterium*. In effect, protein profiles generated by SDS page analysis may reveal typical gel bands with no common major bands, which could therefore be used for species fingerprinting. In addition, western immunoblotting with rabbit antisera can be used to identify *Mogibacterium* species due to the existence of species-specific antigens. For instance, particular strains of *M. diversum* and *M. neglectum* have been shown to possess antigens not present in any other species. Also, it has been suggested that 16S rRNA gene RFLP analysis with *Hpa*II restriction endonuclease can generate

TABLE 37.1
DNA–DNA Hybridization Range Levels among Different Strains of *Mogibacterium* Species

	M. diversum	*M. neglectum*
M. diversum	76%–106%	–
M. pumilum	37%–38%	29%–39%
M. vescum	23%–30%	26%–29%
M. neglectum	30%–38%	96%–100%
M. timidum	16%	17%–18%

Source: Nakazawa, F. et al., *Int. J. Syst. Evol. Microbiol.*, 52, 115, 2002.

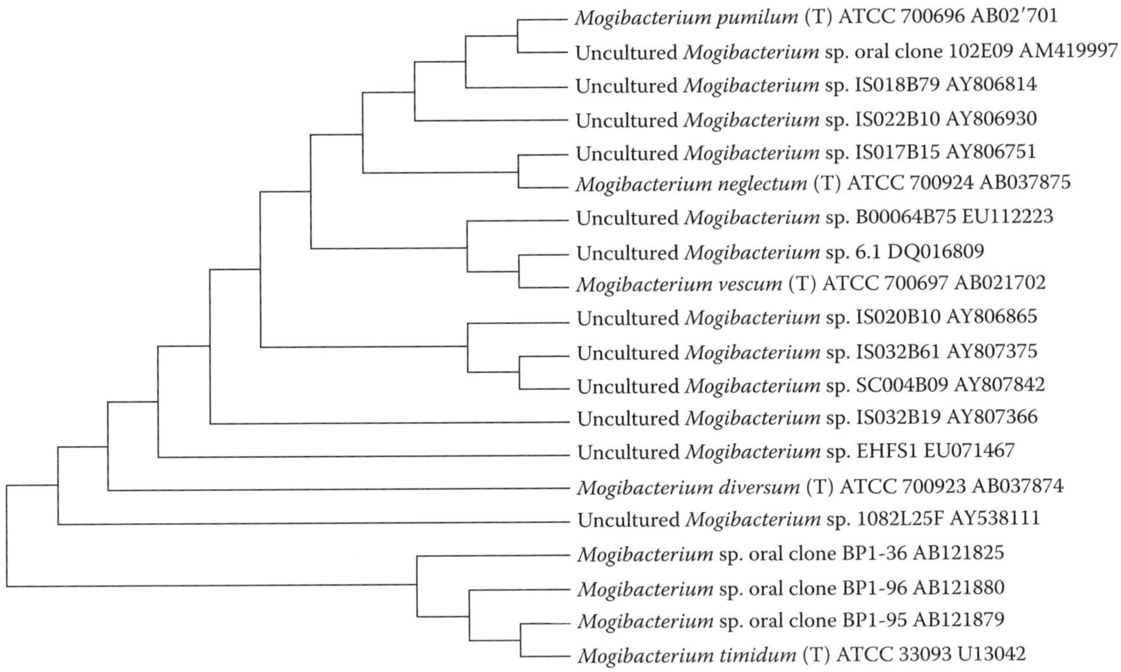

FIGURE 37.1 Phylogenetic relationships of 16S rRNA gene sequences from different *Mogibacterium* species and phylotypes downloaded from the Ribosomal Database Project II.[22] Sequences were aligned with ClustalW.[23] The tree was constructed with the MEGA 4 software,[24] according to the neighbor-joining algorithm and based on the Jukes and Cantor nucleotide substitution model.

TABLE 37.2
Broad Detection and Species-Specific Oligonucleotide Primers Used for PCR Amplification and Sequencing Analysis of *Mogibacterium timidum* and *Mogibacterium* spp.

Primer Pair	Oligonucleotide Sequence	Purpose	Specificity	Sample Source	Reference
27F 1492R	AGAGTTTGATC[A/C]TGGCTCAG TACGG[C/T]TACCTTGTTACGACTT	16S rRNA clonal analysis	*Mogibacterium* spp.	Endodontic infections	Rolph et al.[19]
		Phylogenetic analysis	*Mogibacterium* spp.	Type strains and periodontitis	Cheeseman et al.[5]
		16S rRNA clonal analysis	*Mogibacterium* spp.	Acute dental abscess	Sakamoto et al.[27]
fD1 rD1	AGAGTTTGATCCTGGCTCAG AAGGAGGTGATCCAGCC	16S rRNA clonal analysis	*Mogibacterium* spp.	Primary endodontic infections	Saito et al.[16]
27F 1525R	AGAGTTTGATC[A/C]TGGCTCAG AAGGAGGTG[A/C]TCCAGCC	16S rRNA clonal analysis	*M. timidum* *M. vescum*	Endodontic infections	Munson et al.[18]
		Characterization		Various oral infections	Downes et al.[20]
63F 1387R	CAGGCCTAACACATGCAAGTC GGGCGGWGTGTACAAGG	RFLP analysis	*M. timidum* and *Eubacterium* spp.	Type strains	Sato et al.[25]
	AAG CTT GGA AAT GAC GC CCT TGC GCT TAG GTA A	PCR detection	*M. timidum*		Mayanagi et al.[17] Hashimura et al.[26]

unambiguous profiles for *Mogibacterium* spp. and could constitute a valuable species-specific identification technique.[25]

A number of molecular techniques have been used in the detection of *Mogibacterium timidum* in oral clinical samples. By use of the polymerase chain reaction (PCR) with species-specific oligonucleotides, *M. timidum* has been detected in endodontic infections[26] and periodontitis.[17] Other techniques have been also applied, such as checkerboard DNA–DNA hybridization and 16S rRNA clone library analysis, with broad detection primers. Primer pair combinations that have been successfully used to amplify DNA from *Mogibacterium* species are shown in Table 37.2.

37.1.4 Prevalence and Clinical Implications

Mogibacterium timidum was initially detected in periodontal pockets by Holdeman et al.[14] Since this study, several others have included this anaerobic, gram-positive, rod-shaped bacteria as a target bacteria or have detected it in periodontal disease, endodontic infections, and other infection foci. Table 37.3 displays the prevalence of *M. timidum* in different infection foci.

Endodontic Infections. Considering endodontic infections, studies detected *M. timidum* in frequencies ranging from 3% to 70% of teeth; however, detection rates varied according to technique used and the pulpal condition.

Wade et al.[11] identified *M. timidum* in 3% of sites presenting acute endodontic lesions, and Downes et al.[20] detected it in 13% of endodontically infected teeth by protein and enzymatic profile tests. Sakamoto et al.[27] identified *M. timidum* in one of seven cases of symptomatic lesions, and in three of four cases of chronic endodontic lesions, through T-RFLP coupled with 16S rRNA gene cloning analysis. Hashimura et al.,[26] using PCR technique, observed that *M. timidum* presented superior detection rates in teeth presenting with necrotic endodontic lesions (71%) than cases of pulpitis or refractory (failed treatment) endodontic lesions (29% and 25%, respectively). Considering the high prevalence of *M. timidum* in endodontic lesions, especially when target-ended techniques were used in detection, more studies should be done to clarify its potential in the etiology in pulpal pathologies, as well as to determine whether this oral pathogen could be a reliable marker for endodontic infections.

TABLE 37.3
***M. timidum* Prevalence in Different Forms of Oral and Body Infections**

Author	Technique	Site	Prevalence
Endodontic Lesions			
Wade et al.[11]	Culture	Pulpal abscess	3%
Hashimura et al.[26]	PCR	Pulpitis	29%
		Necrosis	71%
		Retreatment	25%
Downes et al.[20]	Culture/protein and enzymatic profile	Odontogenic lesions	13%
Sakamoto et al.[27]	TRFLP/16s clonning	Symptomatic endodontic lesion	1 of 7
		Asymptomatic endodontic lesion	3 of 4
Periodontal Lesions			
Holdeman et al.[14]	–	Periodontitis	–
Moore et al.[28]	Lab assays	Periodontitis	70%
Moore et al.[29]	Culture/lab assays	Adult gingivitis	9%
		Child gingivitis	3%
Moore et al.[30]	Culture/lab assays	Juvenile peridontitis	58%
		Health individual	11%
Booth et al.[31]	Slot-Blot hybridization	Periodontitis	2,5%
		Health individual	5%
Mayanagi et al.[17]	PCR	Periodontitis supragingival	65%
		Periodontitis subgingival	70%
		Health supragingival	15%
		Health subgingival	10%
Chronic Periodontitis			
Colombo et al.[32]	Microarray	Refractory periodontitis Periodontal health	
Gonçalves et al.[33]	PCR	Chronic periodontitis	40%
		Chronic periodontitis + uncontroled diabetic	48%
		Aggressive periodontitis	35%
Other Sites			
Hill et al.[21]	Culture/enzymatic tests	Head and neck, lung and liver abscesses	–
Brook[34]	–	Genital tract	–
Brook and Frazier[35]	–	Abdominal and head and neck infectious	–
Downes et al.[20]	Culture/enzymatic tests	Peri-implantitis	15%
Tyrrel et al.[36]	Culture/enzymatic tests	Tongue dorsum/bad breath	25%

Periodontal Infections. Concerning periodontal infections, *M. timidum* has been studied in different disease forms and with different techniques. However, differently from endodontic lesions, *M. timidum* appears to be more frequently and constantly detected in subgingival biofilm associated with periodontal disease.

Holdeman et al.[14] was the first to provide evidence of the presence of *M. timidum* in periodontal pockets. After that, Moore et al.[28–30] tried to determine the potential associations between *M. timidum* and periodontal disease. In their first study, the detection rates in periodontal pockets of patients with chronic periodontitis (formerly, adult periodontitis) achieved 70% of the affected population. In addition, by applying the experimentally induced gingivitis model, the above-mentioned authors have determined that the prevalence of this microorganism was superior in adults compared with youngsters (9% and 3%, respectively).[28,29] Interestingly, *M. timidum* detection frequencies increased as the clinical parameters for gingivitis were shown to be more severe.[29] Considering their results, the authors have suggested that *M. timidum* may contribute to the increased susceptibility of adults to experimental gingivitis and to periodontitis. In a sequential study, Moore et al.[30] compared the presence of several microorganisms in the supragingival and subgingival biofilm in patients with different forms of periodontitis and in healthy individuals. In juvenile periodontitis (aggressive periodontitis), *M. timidum* was present in 58% of the studied subjects; in chronic severe periodontitis, 70% of the patients presented this specie in their subgingival biofilm, whereas Moore detected its presence in only in 11% of the healthy individuals. In spite of the potential relationships between *M. timidum* and periodontitis, no further studies were conducted to explore the possible pathogenic mechanisms of this oral species as well as from the other species in the *Mogibacterium* genus.

In 2004, Booth et al.[31] detected *M. timidum* in one shallow pocket of chronic periodontitis patients through a noncultivable technique (slot-blot hybridization). However, since the detection limit of the methodology was 10^4 bacteria, the results should be carefully considered, as noted by the authors. Mayanagi et al.[17] evaluated the presence of *M. timidum* in sub- and supragingival biofilm of periodontitis and healthy individuals using a PCR technique. Patients with periodontal disease frequently presented superior detection rates compared to healthy individuals. In supragingival biofilm, while only 15% of healthy individuals presented *M. timidum*, the species was detected in 65% of periodontitis patients. Concerning subgingival biofilm, 10% and 70% of healthy and periodontitis patients, respectively, were infected by this microorganism. Recently, Colombo et al.,[32] using a microarray methodology, revealed that periodontitis patients harboring higher amounts of *M. timidum* displayed a refractory type of periodontitis, resulting in lower gain of clinical attachment level. The authors suggested, as previously observed by Moore et al.,[28–30] that *M. timidum* is positively associated with periodontal disease, and its importance in pathogenesis should be investigated.

In order to fully screen *M. timidum* in periodontal disease, we have evaluated its presence in subgingival biofilm in three different periodontal conditions using a PCR technique.[33] *M. timidum* was detected in the deep pockets of 48% of chronic periodontitis patients, and in 35% of aggressive periodontitis patients. In addition, 48% of chronic periodontitis patients with a diagnostic of diabetes type II were infected by *M. timidum*. The results of this study highlight the considerably high prevalence of the species in subgingival biofilm of periodontitis patients and the need for additional studies addressing the potential pathogenic of all species in the *Mogibacterium* genus.

Altogether, considering the possible virulence factors and its high prevalence in periodontitis biofilm, *M. timidum* harbors a high potential to initiate and/or increase the inflammatory response in periodontal tissues. In this sense, *M. timidum* could be considered as a periodontal pathogen enrolled in destructive disease.

Other Oral Sites. In a halitosis population, *M. timidum* was present in 25% of tongue dorsum's patients.[36] Usually, bad breath is associated with sulfur volatile compounds (SCV)–producing bacteria, but the potential of *M. timidum* or other *Mogibacterium* species to produce these SVCs still is unknown.

Extraoral Infections. *Mogibacterium timidum* was also identified in other oral and extraoral infection sites. Hill et al.[21] have employed microbial culture and biochemical tests, in order to evaluate biodiversity in head and neck abscesses (Ludwig's angina) and also in lung and liver acute infections. They were able to identify *M. timidum* in many of those cases. Moreover, Brook and Brook and Frazier et al.[34,35] detected this pathogen in head and neck infections, as well as in abdominal abscess and in the genital tract.

37.2 METHODS

37.2.1 Sample Preparation and Collection

As discussed previously, and considering its fastidious characteristics, molecular detection is the most predictable technique for detecting and identifying *Mogibacterium* in human samples. Therefore, some caution should be used in sample collection and preparation.

To determine the presence of this genus in the oral microbiota in sites other than subgingival area and endodontic infections, a swab can be used. To identify its presence on subgingival or endodontic microbiota, a sterile paper point can be introduced into subgingival area or dental canal and maintained for 30–60 s. Alternatively, a periodontal scaler can also be useful to collect the periodontal biofilm. Both techniques were largely used in previous studies and allowed *Mogibacterium* detection in subgingival area by molecular techniques.[13,17,19,20,25–27,32,36]

After sample collection, in swab, paper points, or scaling instruments, the sample should be maintained in an adequate medium or buffer to posterior DNA extraction. Usually, Tris-EDTA (0.1 M, pH 8.0) is used to maintain the sample until DNA extraction. Regarding DNA isolation, several possibilities are presented in literature and many ready-to-use commercial kits are available. As alternative to DNA extractions kits, more traditional DNA extraction techniques can be used. We have used with success a sequential extraction technique that presented consistent results concerning the DNA quality.[37–39] A short description of the technique is presented below.

Initially, tubes containing swab, paper points, or subgingival material are vortexed thoroughly for 1 min. Immediately, in a flux chamber, the swab or the paper point is removed from tube, and then it is centrifuged again at 12,000 × *g* for 1 min. The supernatant is discarded and the pellet is resuspended in 700 μL of extraction buffer (NaCl 1.4 M, Tris-HCl 100 mM, EDTA 20 mM, PVP-40 1%, CTAB 2%, *β*-mercaptoethanol 0.2%) containing 100 μg/mL of proteinase K. The tube is vortexed for 30 s and incubated at 65°C for 30 min. Then, 650 μL of CIA 24:1 (chloroform:isoamylic alcohol) is added to the tube, homogenized and centrifuged at 12,000 × *g* for 7 min. The supernatant is collected and mixed with 200 μL extraction buffer (without proteinase K) and 650 μL of CIA. The tube is centrifuged once again at 12,000 × *g* for 7 min and supernatant is removed to another tube. It is mixed with 650 μL CIA, homogenized and centrifuged again. The supernatant is then added to a tube containing 700 μL of isopropanol, homogenized, and centrifuged at 12,000 × *g* for 10 min.

After removal of the supernatant, the pellet is washed with 300 μL of 70% alcohol, centrifuged at 12,000 × *g* for 4 min. This step is repeated two times. Afterwards, the pellet is dried at room temperature and resuspended with 30 μL of TE (0.01 M) containing 10 μg/mL of RNAse. The tube is incubated at 37°C for 30 min and frozen until use.

37.2.2 Detection Procedures

(i) PCR Detection While traditional PCR procedure is useful for detection of *Mogibacterium* genus,[26] nested PCR appears show high specificity and sensitivity.[17] Nested PCR is a modification of traditional PCR, which can reduces the contamination due to amplification of unexpected primer binding sites. Nested PCR involves two sets of primers, used in two successive runs of PCR, with the second set intended to amplify a secondary target within the first run product, resulting in a very specific PCR amplification. The advantage of nested PCR is that if a wrong PCR fragment is amplified in the first PCR cycle, the probability of wrong sequence amplification will be quit low in the second round of PCR.

Procedure. In the first amplification, the 16S rRNA genes can be amplified by PCR with universal primers[17] 27F (5′-AGAGTTTGATCCTGGCTCAG-3′) and 1492R (5′-TACGGGTACCTTGTTACGACTT-3′). PCR amplification parameters are an initial heat activation of 15 min at 95°C, 35 cycles of 1 min at 94°C for denaturation, 1 min at 60°C for annealing, 1.5 min at 72°C for extension, and 10 min at 72°C for final extension. This amplified DNA is subjected a second PCR with a new set of primers, for *Mogibacterium timidum* (5′-AAGCTTGGAAATGACGC-3′ and 5′-CCTTGCGCTTAGGTAA-3′) can be used. The second PCR is performed initially with 15 min at 95°C; 35 cycles of 1 min at 94°C, 1 min at 55°C, and 1.5 min at 72°C; and finally 10 min at 72°C.

(ii) Cloning Among the different techniques available to detect *Mogibacterium* genus in oral samples, molecular approaches independent of cultivation, such as cloning of 16S rRNA gene and terminal restriction fragment length polymorphism (TRFLP or sometimes T-RFLP), offer a rapid and sensitive alternative to traditional methods to identify *Mogibacterium*.

The 16S ribosomal gene (16S rDNA) has a unique combination of qualities that makes it applicable for bacterial identification to various environmental conditions.[42] In recent decades, various techniques geared to the analysis of microbial communities were developed based on this gene, including analysis of clonal libraries, denaturing gradient gel electrophoresis (DGGE), denaturing high performance liquid chromatography (DHPLC) and terminal restriction fragment length polymorphism (TRFLP).[16,40–45] These techniques, associated with the existence of extensive databases encompassing ribosomal DNA sequences, such as GenBank (www.nlm.nih.gov) and the Ribosomal Database Project II (http://rdp8.cme.msu.edu), allow the molecular identification of microorganism for phylogenetic or ecological studies.

The TRFLP technique is an automated and sensitive fingerprinting method that uses fluorescently labeled primers in PCR, followed by digestion with restriction enzymes and analysis of terminal fragments with a DNA sequencer.[41,42] The sequencer recognizes only the fluorescently labeled terminal fragments, and therefore, in principle, each fragment represents a unique operational taxonomic unit (OTU) in the sample.[45] Cloning of 16S rRNA gene represents a possibility of determining the fingerprint of a sample through an insertion of amplified DNA, using a vector, in a competent-cell, which will be cultured in a specific medium. Colonies formed in this medium represent a specific bacterial DNA within the cell, and a selection of each colony allows identification of the bacteria type (through DNA sequencing).[16]

Both techniques represent open-ended identification techniques, that is, the identification occurs without a target or a specific bacteria focus. They have the advantage of determining the biodiversity of different samples, providing information on the bacteria, fungal, or viral species present in a determined environment. However, at same time, if the aim is to detect a specific genus or specie, those techniques should be adapted to properly identify the bacterium in the sample.

Procedure. T-RFLP analysis initiate with a traditional PCR reaction, using a universal primer for amplification of 16S rRNA gene sequences. One primer should be labeled at the 5′ end. Usually, 6′-carboxyfluorescein (6-FAM) is the most useful label. In a previous study,[27] primers pair (1492R and 27F+6-FAM) was used and was able to identify *Mogibacterium* genus. In this study, a total volume of 50 μL containing 1–5 μL of DNA, 1.25 U of Taq polymerase, 5 μL of 10× Taq buffer, 4 μL of dNTP mixture (2.5 mm each), and 10 pmol of each primer. 16S rRNA genes were amplified, initiating with 3 min at 95°C, and 30 cycles consisting of 95°C for 30 s, 50°C for 30 s, and 72°C for 1.5 min, and a final extension of 10 min at 72°C. PCR products are purified and digested with several restriction enzymes. *Hpa*II restriction endonuclease appears to generate unambiguous profiles for *Mogibacterium* spp. For a digestion, 20 U of endonuclease in a total volume of 10 μL is incubated at 37°C for 3 h. The digest product of the restriction (1 μL) is mixed with 8 μL of Hi-Di Formamide and 1 μL of DNA fragment length standard. The length of T-RF is then determined on an ABI PRISM 3100 Genetic Analyzer (Applied Biosystems) in GeneScan mode (15 kV, 100 mA, and 60°C for 40 min for each sample). Fragment sizes are estimated and predicted T-RFLP patterns of the 16S rRNA genes of known bacterial species are obtained using the specified programs.[41,42,45]

For 16S rRNA cloning, extracted DNA is amplified following the same protocol previously described, using universal primers (27F and 1492r, both not labeled). DNA is purified using ready-to-use kit (QIAquick PCR purification kit; Qiagen) and amplicons are ligated to vectors. Nowadays, different kits, containing different vectors, are available and some differences in methodology can occur. 4 μL of purified DNA is mixed to 1 μL of pCR®2.1-TOPO® vector (Invitrogen). After ligation, 2 μL of prepared TOPO TA vector is added to one shot TOP10 competent cells and gently mixed. After a heat-shock (30 s at 42°C), 250 μL of SOC medium (2% tryptone, 0.5% yeast extract, 10 mM NaCl, 2.5 mM KCl, 10 mM $MgCl_2$, 10 mM $MgSO_4$, 20 mM glucose), is added and shaken at 37°C for 1 h. The mix is then spread on Luria Bertani plates (2% tryptone, 0.5% yeast extract, 10 mM NaCl) and incubated at 37°C overnight. White colonies are picked and cultured, overnight in LB medium. After centrifugation at 12,000 × *g* for 10 min, plasmids are extracted from the cells using commercial kits (PureLink™ Quick Plasmid Miniprep Kit, Invitrogen). Sequencing is conducted using DNA sequencers, and each one is checked for possible chimeric artifacts (Chimera Check program of the Ribosomal Database Project II [RDP-II]) and similarity in specific programs (GenBank, EMBL, and DDBJ databases).[16]

Note: Although TRFLP analysis can be used as a single technique to identify bacterial species, an association between it and 16S cloning techniques would be useful in order to validate the TRFLP results. It was recently used by Sakamoto et al.[27] and represents an advance in fidelity of bacterial diversity analysis.

37.3 CONCLUSION

The fastidious nutritional requirements and their slow-growing characteristics have eclipsed the accurate prevalence *Mogibacterium* spp. in oral infections. It was only with the establishment of molecular-based detection techniques that *Mogibacterium* species started to be detected with relatively high frequency in clinical samples. In addition, to a high prevalence in oral infections (mainly *Mogibacterium timidum*), the species from *Mogibacterium* genus can have several possible virulence factors with high potential to initiate and increase inflammatory response. In this sense, seems very important to carry out additional studies to understand the true role of *Mogibacterium* species in oral and extra oral infections.

REFERENCES

1. Wade, W.G., Slayne, M.A., and Aldred, M.J., Comparison of identification methods for oral asaccharolytic *Eubacterium* species, *J. Med. Microbiol.*, 33, 239, 1990.
2. Moore, W.E.C., and Moore, L.V.H., Genus *Eubacterium* Prevot 1938, 294[AL]. In Sneath, P.H.A., Mair, N.S., Sharpe, M.E., and Holt, J.G. (Editors), *Bergey's Manual of Systematic Bacteriology*, vol. 2, pp. 1353–1373, Williams & Wilkins, Baltimore, 1986.
3. Uematsu, H., and Hoshino, E., Predominant obligate anaerobes in human periodontal pockets, *J. Periodont. Res.*, 27, 15, 1992.
4. Nakazawa, F. et al., Taxonomic characterization of *Mogibacterium diversum* sp. nov. and *Mogibacterium neglectum* sp. nov., isolated from human oral cavities, *Int. J. Syst. Evol. Microbiol.*, 52, 115, 2002.
5. Cheeseman, S.L. et al., Phylogeny of oral asaccharolytic *Eubacterium* species determined by 16S ribosomal DNA sequence comparison and proposal of *Eubacterium infirmum* sp. nov. and *Eubacterium tardum* sp. nov., *Int. J. Syst. Bacteriol.*, 46, 957, 1996.
6. Nakazawa, F., and Hoshino, E., Immunological specificity of oral *Eubacterium* species, *J. Gen. Microbiol.*, 139, 2635, 1993.
7. Collins, M.D. et al., The phylogeny of the genus *Clostridium*: proposal of new genera and eleven new species combinations, *Int. J. Syst. Bacteriol.*, 44, 812, 1994.
8. Willems, A., Amat-Marco, M., and Collins, M.D., Phylogenetic analysis of *Butyrivibrio* strains reveals three distinct groups of species within the *Clostridium subphylum* of the gram-positive bacteria, *Int. J. Syst. Bacteriol.*, 46, 195, 1996.
9. Poco, S.E. et al., *Eubacterium exiguum* sp. nov., isolated from human oral lesions, *Int. J. Syst. Bacteriol.*, 46, 1120, 1996.
10. Poco, S.E. et al., *Eubacterium minutum* sp. nov., isolated from human periodontal pockets, *Int. J. Syst. Bacteriol.*, 46, 31, 1996b.
11. Wade, W.G. et al., An unclassified *Eubacterium* taxon in acute dento-alveolar abscess, *J. Med. Microbiol.*, 40, 115, 1994.
12. Sato, T. et al., Predominant obligate anaerobes in necrotic pulps of human deciduous teeth, *Microb. Ecol. Health Dis.*, 6, 269, 1993.
13. Nakazawa, F. et al., Description of *Mogibacterium pumilum* gen. nov., sp. nov. and *Mogibacterium vescum* gen. nov., sp. nov., and reclassification of *Eubacterium timidum* (Holdeman et al. 1980) as *Mogibacterium timidum* gen. nov., comb. nov., *Int. J. Syst. Evol. Microbiol.*, 50, 2679, 2000.
14. Holdeman, L.V. et al., Description of *Eubacterium timidum* sp. nov., *Eubacterium brachy* sp. nov. and *Eubacterium nodatum* sp. nov., isolated from human periodontitis, *Int. J. Syst. Bacteriol.*, 30, 163, 1980.

15. Robertson, D., and Smith, A.J., The microbiology of the acute dental abscess, *J. Med. Microbiol.*, 58, 155, 2009.
16. Saito, D. et al., Identification of bacteria in endodontic infections by sequence analysis of 16S rDNA clone libraries, *J. Med. Microbiol.*, 55, 101, 2006.
17. Mayanagi, G. et al., Detection frequency of periodontitis-associated bacteria by polymerase chain reaction in subgingival and supragingival plaque of periodontitis and healthy subjects, *Oral Microbiol. Immunol.*, 19, 379, 2004.
18. Munson, M.A. et al., Molecular and cultural analysis of the microflora associated with endodontic infections, *J. Dent. Res.*, 81,761, 2002. Erratum in: *J. Dent. Res.*, 82, 69, 2003.
19. Rolph, H.J. et al., Molecular identification of microorganisms from endodontic infections, *J. Clin. Microbiol.*, 39, 3282, 2001.
20. Downes, J. et al., Characterization of *Eubacterium*-like strains isolated from oral infections, *J. Med. Microbiol.*, 50, 947, 2001.
21. Hill, G.B., Ayers, O.M., and Kohan, A.P., Characteristics and sites of infection of *Eubacterium nodatum*, *Eubacterium timidum*, *Eubacterium brachy*, and other asaccharolytic eubacteria, *J. Clin. Microbiol.*, 25, 1540, 1987.
22. Cole, J.R. et al., The Ribosomal Database Project (RDP-II): Sequences and tools for high-throughput rRNA analysis, *Nucleic Acids Res.*, 33, 294, 2005.
23. Thompson, J.D., Gibson, T.J., and Higgins, D.G., Multiple sequence alignment using ClustalW and ClustalX, in *Curr. Protoc. Bioinformatics.*, 2002 Aug; Chapter 2. Unit 2.3.
24. Tamura, K. et al., MEGA4: Molecular Evolutionary Genetics Analysis (MEGA) software version 4.0., *Mol. Biol. Evol.*, 24, 1596, 2007.
25. Sato, T. et al., Restriction fragment-length polymorphism analysis of 16S rDNA from oral asaccharolytic *Eubacterium* species amplified by polymerase chain reaction, *Oral Microbiol. Immunol.*, 13, 23, 1998.
26. Hashimura, T., Sato, M., and Hoshino, E., Detection of *Slackia exigua*, *Mogibacterium timidum* and *Eubacterium saphenum* from pulpal and periradicular samples using the Polymerase Chain Reaction (PCR) method, *Int. Endod. J.* 34, 463, 2001.
27. Sakamoto, M. et al., Molecular analysis of bacteria in asymptomatic and symptomatic endodontic infections, *Oral Microbiol. Immunol.*, 21, 112, 2006.
29. Moore, W.E. et al., Bacteriology of experimental gingivitis in children, *Infect. Immun.*, 46, 1, 1984.
28. Moore, W.E. et al., Bacteriology of experimental gingivitis in young adult humans, *Infect. Immun.*, 38, 651, 1982.
30. Moore, W.E. et al., Comparative bacteriology of juvenile periodontitis, *Infect. Immun.*, 48, 507, 1985.
31. Booth, V. et al., gram-positive anaerobic bacilli in human periodontal disease, *J. Periodont. Res.*, 39, 213, 2004.
32. Colombo, A.P. et al., Comparisons of subgingival microbial profiles of refractory periodontitis, severe periodontitis, and periodontal health using the human oral microbe identification microarray, *J. Periodontol.*, 80, 1421, 2009.
33. Gonçalves, R.B., Saito, D., and Casarin, R.C.V., unpublished data, 2009.
34. Brook, I., Anaerobic bacterial bacteremia: 12-year experience in two military hospitals, *J. Infect. Dis.*, 160, 1071, 1989.
35. Brook, I., and Frazier, E.H., Anaerobic osteomyelitis and arthritis in a military hospital: A 10-year experience, *Am. J. Med.*, 94, 21, 1993.
36. Tyrrell, K.L. et al., Anaerobic bacteria cultured from the tongue dorsum of subjects with oral malodor, *Anaerobe*, 9, 243, 2003.
37. Del Peloso Ribeiro, E. et al., Periodontal debridement as a therapeutic approach for severe chronic periodontitis: A clinical, microbiological and immunological study, *J. Clin. Periodontol.*, 35, 789, 2008.
38. Casarin, R.C.V. et al., Levels of *A. actinomycetemcomitans*, *P. gingivalis*, inflammatory cytokines and species-specific immunoglobulin G in generalized aggressive and chronic periodontitis, *J. Periodont. Res.*, accepted, 2010.
39. Klein, M., and Gonçalves, R.B., Detection of *Tannerella forsythensis* (Bacteroides forsythus) and *Porphyromonas gingivalis* by polymerase chain reaction in subjects with different periodontal status, *J. Periodontol.*, 74, 798, 2003.
40. Goldenberg, O. et al., Molecular monitoring of the intestinal flora by denaturing high performance liquid chromatography, *J. Microbiol. Methods*, 68, 94, 2007.
41. Saito, D. et al., Assessment of intraradicular bacterial composition by terminal restriction fragment length polymorphism analysis, *Oral Microbiol. Immunol.*, 24, 369, 2009.
42. Marsh, T.L. et al., Terminal restriction fragment length polymorphism analysis program, a web-based research tool for microbial community analysis, *Appl. Environ. Microbiol.*, 66, 3616, 2000.
43. Li, Y. et al., Survey of oral microbial diversity using PCR-based denaturing gradient gel electrophoresis, *J. Dent. Res.*, 84, 559, 2005.
44. Liu, W.T. et al., Characterization of microbial diversity by determining terminal restriction fragment length polymorphisms of genes encoding 16S rRNA, *Appl. Environ. Microbiol.*, 63, 4516, 1997.
45. Marsh, T.L., Terminal restriction fragment length polymorphism (TRFLP): An emerging method for characterizing diversity among homologous populations of amplification products, *Curr. Opin. Microbiol.*, 2, 323, 1999.

38 Peptostreptococcus

Eija Könönen and Jari Jalava

CONTENTS

38.1 INTRODUCTION

38.1.1 CLASSIFICATION, MORPHOLOGY, BIOLOGY, AND EPIDEMIOLOGY

The genus *Peptostreptococcus*, currently including two species of gram-positive, obligately anaerobic cocci, *Peptostreptococcus anaerobius* and *Peptostreptococcus stomatis*, belongs to the family *Peptostreptococcaceae* in the phylum *Firmicutes*.

The genus was described in 1936 by Kluyver and van Niel,[1] *P. anaerobius* being chosen as the type species of the genus, and this was confirmed in the Approval Lists of Bacterial Names in 1980.[2] At that time, the genus was still composed of a widely diverse group of gram-positive coccal organisms, including four species: *P. anaerobius*, *Peptostreptococcus micros*, *Peptostreptococcus parvulus*, and *Peptostreptococcus productus*.[2] Since then, the latter two species have been transferred to other genera twice, currently locating in novel genera as *Atopobium parvulum* (formerly *Streptococcus parvulus*)[3] and *Blautia producta* (formerly *Ruminococcus productus*),[4] respectively. In contrast, the former *Peptococcus* species, except *Peptococcus niger*, were transferred to the genus *Peptostreptococcus*,[5] and a considerable number of new species have been recovered and described as members of *Peptostreptococcus*.[5–7]

With the introduction of molecular methods in microbiology, including the taxonomy of gram-positive anaerobic cocci, this wide and heterogeneous group has gone through remarkable taxonomic revisions based on differences in their 16S rRNA sequences.[8] New genera have been suggested and formed from species previously belonging to the genus *Peptostreptococcus*. In 1999, *Peptostreptococcus magnus* and *P. micros* were reclassified as *Finegoldia magna* and *Micromonas micros*, respectively, representing each of these new genera as a single species.[9] However, the name *Micromonas* proved to be illegitimate; thus, many researchers went back to using the name *P. micros* for this species until a new name, *Parvimonas micra*, was published in 2006.[10] In 2001, Ezaki et al.[11] made a radical revision among the genus *Peptostreptococcus* and reclassified nonsaccharolytic peptostreptococci that use peptone as their energy source and produce butyrate as a major metabolite of their glucose fermentation as *Peptoniphilus* species, and saccharolytic, butyrate-producing peptostreptococci as *Anaerococcus* species. Furthermore, the former *Peptostreptococcus barnesae* was reclassified as *Gallicola barnesae*.[11] After these taxonomic revisions, *P. anaerobius*, which had been the type species of the genus despite its distinct genetic position and biochemical features, remained as a single representative of the genus. Indeed, as suggested first by Huss et al.[12] and later confirmed by other research groups,[8,13,14] *P. anaerobius* proved to be more closely related to some *Clostridium* and *Eubacterium* species than to the other gram-positive anaerobic cocci. In 2006, Downes and Wade described a novel species, *P. stomatis*, to be included in the genus *Peptostreptococcus*.[15] *P. stomatis* is both phenotypically and phylogenetically very similar to *P. anaerobius*. Moreover, the use of 16S rRNA sequence analysis has led to detection of novel "*Peptostreptococcus*" candidates in clinical specimens.[16–19] In part, this may be due to the improper usage of the genus name. Indeed, some have suggested that "*Peptostreptococcus*" phylotypes hardly belong to the genus *Peptostreptococcus*, and that, instead, their phylogenetic position refers to other genera, such as *Parvimonas*.

Peptostreptococcus species are strict anaerobes, that is, during testing of their atmospheric requirements, they fail to grow, not only in aerobic, but also in microaerobic (5% oxygen) conditions. *Peptostreptococcus* species appear on blood agar as relatively wide colonies, up to 4 mm in diameter.[15] *Peptostreptococcus* cells stain as gram-positive and are coccal or coccobacillary and have relatively large (up to 1 μm) cells, often occurring in pairs and chains. However, pleomorphic forms are also common, in particular, when fresh cultures (2- to 3-days' old), as recommended, are not used for slide preparation.[20] Outside the peptidoglycan layer, *P. anaerobius* (sensu lato) cells have a crystalline surface protein layer, that is, S-layer[21]; its role in virulence and microbial ecology is not well known.

Unlike many gram-positive anaerobic cocci, *P. anaerobius* and *P. stomatis* are weakly saccharolytic.[15,20] A rather unique feature is their production of isocaproic acid as one of the major metabolic end products of their glucose fermentation. Other volatile fatty acids produced in considerable amounts are acetic, isovaleric, and isobutyric acids.[15,22] Cellular fatty acid composition, mainly including iso-branched-chain acids 10:0, 12:0, 14:0, and 16:0, is typical for *P. anaerobius* (sensu lato).[23] A useful diagnostic characteristic is their sensitivity to sodium polyanethol sulfonate (SPS),[15,20,24–26] a special-potency antimicrobial used in presumptive identification of gram-positive anaerobic cocci. The guanine-cytosine content of the two *Peptostreptococcus* species varies between 33 and 36 mol%.[15,20]

Gram-positive anaerobic cocci, including *Peptostreptococcus* species, are common members of the microbiota in humans, residing mucosal surfaces of the mouth, intestine, and the female genital tract as well as skin.[20,27] Previously, *P. anaerobius* was considered a common species in both the upper and lower parts of the body. For example, *P. anaerobius* (sensu lato) has been isolated as one of the most frequent recoveries from dorsal tongue surfaces of subjects with oral malodor.[28] However, in a study in which a PCR assay specific for *P. anaerobius* was developed, the assay failed to detect any *P. anaerobius* in oral specimens, including 60 subgingival plaque samples and 43 pus samples from dentoalveolar abscesses.[29] Due to the recent recovery of *P. stomatis*,[15] it is rather clear that these two phenotypically and phylogenetically similar *Peptostreptococcus* species have been mixed in studies run, at least, before 2006. Therefore, when no separation between *P. anaerobius* and *P. stomatis* has been made, the term "*P. anaerobius* sensu lato" is used in this chapter. On the basis of identifying all seven oral isolates as *P. stomatis* and all six nonoral isolates from samples collected below the waistline as *P. anaerobius*, Downes and Wade[15] suggested that these two closely related species may have distinct habitats. In DNA-based studies, *P. stomatis* (= *Peptostreptococcus* sp. oral clone CK035) has been detected in samples from the healthy oral cavity[16] as well as in those collected from various oral infections.[19,30–32]

Indeed, besides their role as commensals, gram-positive anaerobic cocci are considered clinically significant recoveries from a wide variety of mixed infections involving anaerobic bacteria.[20,22,27,33,34] *F. magna* seems to have the highest pathogenic potential, but the role of both *Peptostreptococcus* species as significant infectious organisms is also evident.[15,20,25,27,33–35]

38.1.2 Clinical Features and Pathogenesis

Although most infections where anaerobic bacteria are involved have their origin in the resident microbiota, only a limited number of species cause these infections in immunocompetent individuals. Gram-positive anaerobic cocci are among the most common anaerobic organisms isolated from polymicrobial infections.[20,27,33] In addition, *F. magna*, in particular, but also *P. anaerobius* sensu lato have been detected as monocultures.[20,34] As presented in Table 38.1, *P. anaerobius* sensu lato has been reported in a wide variety of human infections all over the body, including the head,[15,19,20,25,30–33,36–38] respiratory tract,[17,20,27,39–42] gastrointestinal tract,[20,22,25,34,43] genitourinary tract,[15,18,20,25,34,44–46] skin and soft tissue,[15,20,22,25,27,33,34,47–55] bone and joints,[20,22,33,34,56] and cardiovascular sites.[20,25,33,57–59] For instance, a comprehensive study by Brook,[33] including nearly 16,000 clinical specimens collected from two military hospitals in the United States between June 1973 and June 1985, reported the common presence of *P. anaerobius* sensu lato especially in obstetric infections, abscesses, and wounds. In an English tertiary hospital, all *P. anaerobius* sensu lato were from specimens collected below the diaphragm, mainly from the gastrointestinal and female genitourinary tracts.[34] According to the experience of the Wadsworth VA Medical Center, Los Angeles, between years 1991 and 1997, this organism has been isolated, in particular, from miscellaneous skin, soft tissue, and bone infections; peritoneal fluid; and appendicidal tissue or abscesses.[22] Abscesses with involvement of *P. anaerobius* sensu lato can locate in various anatomical sites; for example, nasal septum, dentoalveolar region, pharynx, pleural cavity, breast, abdominal region, genital tract, and perirectal area.[20,34,42,43,52–55] Connective tissue destruction in skin and soft-tissue infections and abscesses may be partly explained by proteolytic activities of this organism.[60] In an in vitro study, where the examined strains originated from vaginal specimens, it was shown that increased availability of amino acids by *Prevotella bivia* supports the growth of *P. anaerobius* sensu lato.[61] The observed synergistic interplay strengthens their involvement in polymicrobial infections of the female genital tract. Surprisingly, although *P. anaerobius* is considered a common species in infections

TABLE 38.1
Detection of *Peptostreptococcus* Organisms in Human Infections

Site of Detection and Disease Association	*P. anaerobius* sensu lato	*P. anaerobius* sensu stricto	*P. stomatis*	Novel Phylotype(s)	Reference(s)
Eye infections	x				33
Ear infections	x				20,33
Mouth					
Endodontic infection		x	x		15,31,36,37
Periodontal diseases			x	x	15,19,30,32
Pericoronitis	x		x		15,38
Abscess	x		x		2,15
Pharynx					
Peritonsillar abscess	x				20
Respiratory Tract					
Sinusitis, chronic	x				20,39
Pleural empyema	x				4,20
Aspiration pneumonia	x				27
Ventilator-associated pneumonia		x	x	x	17
Cystic fibrosis	x				41
Lung abscess	x				42
Abdomen/Intestine					
Appendicitis	x				22
Peritonitis	x				22
Abscess	x				20,22,34,43
Genitourinary Tract					
Bacterial vaginosis	x	x			2,15
Pelvic inflammatory disease	x	x	x	x	18,20,44
Intraamniotic infection				x	45
Septic abortion	x				46
Urinary tract infection		x			15
Abscess	x				20,34
Skin and/or Soft Tissue					
Necrotizing soft-tissue infections	x				20
Diabetic foot infections	x	x			20,27,47–50
Chronic wounds	x	x			15,22,33,50
Bite wound infections, animal	x				51
Abscess above the waist	x				20,22,52–54
Abscess below the waist	x	x			15,20,22,34,55
Bone and/or joint infections	x	x			20,22,33,34,56
Cardiovascular Sites					
Bacteremia/septicemia	x				25,33,57,58
Endocarditis	x				20,59

of the female genital tract,[15,18,20,44–46] it was not detected in vaginal swab specimens associated with bacterial vaginosis by using a semiquantitative multiplexed bead-based flow cytometry.[62] This method is rapid and easy to use; however, the quantity of organisms in the sample needs to be quite high to be detected. *P. anaerobius* sensu lato is considered an important anaerobic organism in mixed surgical infections.[49] Unfortunately, in many earlier culture-based studies, no attempts have been made to identify gram-positive anaerobic cocci even to the genus level, but, despite this, the term "*Peptostreptococcus* spp." has been used as a synonym for this group of organisms. For instance, gram-positive anaerobic cocci are among the most common recoveries from bacteremias in which anaerobes are involved[63]; however, it is not clear what the impact of true *Peptostreptococcus* species is. In this context, it is noteworthy that automated blood culture sets, where SPS is a common component, may be inhibitory to the growth of *P. anaerobius* sensu lato due to its sensitivity to SPS: therefore, additional methods for blood cultures have been recommended in patients whenever septicemia or endocarditis caused by gram-positive anaerobic cocci is suspected.[22,64]

There are still confusing data on the role of *P. anaerobius* sensu stricto in oral infections. In a PCR-based study using broad-range 16S rRNA gene primers,[36] this organism was reported as one of the most frequent findings in root canal microbiotas with chronic apical periodontitis. Also, when using a DNA–DNA hybridization checkerboard method where *P. anaerobius* was one of the target organisms, it was frequently detected in samples taken from necrotic pulp of endodontally infected teeth.[31] In another study on endodontic infections,[37] *Peptostreptococcus* sp. clone CK035 (currently *P. stomatis*) was found in refractory cases but not in de novo cases. Furthermore, when Kumar et al.[19] used a quantitative ribosomal 16S cloning and sequencing, *Peptostreptococcus* proved to belong to most numerous organisms in periodontal pockets; especially *Peptostreptococcus* sp. clone CK035 and an uncultivated *Peptostreptococcus* sp. clone (notably, not close to true *Peptostreptococcus* species, but rather to *Parvimonas*) were recovered from subgingival biofilms significantly associated with chronic periodontitis, but *P. anaerobius* was also detected. Interestingly, both *P. anaerobius* and *Peptostreptococcus* sp. oral clone CK035, that is, *P. stomatis*, have been detected in bronchoalveolar lavage samples from patients with ventilator-associated pneumonia where the most potential source is the oral cavity, via aspiration of infectious material.[17] According to a prospective case series on infected dog and cat bite wounds, conducted at 18 university-affiliated emergency departments in the United States,[51] *P. anaerobius* sensu lato was isolated from cutaneous wounds of 8% and 5% of the patients bitten by a dog or a cat, respectively. In animals, the isolation of this organism seems to be limited to canine and feline specimens.[65]

Since *P. stomatis* was just lately distinguished from *P. anaerobius*,[15] the interpretation of their individual roles in human infections is somewhat hampered. Until now, the only accurate data on the involvement of *P. stomatis* in infectious processes, not only in the oral cavity, but also at different body sites come from a study by Könönen et al.[25]; *P. anaerobius* sensu lato strains collected from oral samples and various clinical specimens from other sites of the body between 1985 and 2005 were reexamined. As seen in Table 38.2, most *P. anaerobius* sensu stricto strains originated from ulcer and skin specimens of the lower extremities and pus specimens of the genitourinary tract, and only occasionally from specimens of the head and gastrointestinal tract. In contrast, *P. stomatis* strains originated from oropharyngeal specimens, but also from gastrointestinal specimens, such as infected appendicidal tissue and feces.[25]

Although antimicrobials commonly used for treating anaerobic infections are generally effective against gram-positive anaerobic cocci, intermediate and resistant *Peptostreptococcus* strains have been increasingly reported. In a Finnish strain collection,[25] for instance, no resistant strains were observed during 1980s, while during 1990s and early 2000s, the prevalence of resistant strains (nearly all of them were *P. anaerobius* sensu stricto) was 4% and 31%, respectively. Penicillin, amoxicillin/clavulanic acid, ticarcillin/clavulanic acid, and ceftobiprole have been among the drugs, which have shown to be less effective against *P. anaerobius* sensu lato than other gram-positive anaerobic cocci.[35,66–70] No beta-lactamase production has been detected but, instead, penicillin-binding proteins have been suggested as the mechanism of resistance.[20] Of some concern may be the frequent presence of *nim* genes, which encode nitroimidazole resistance in gram-positive anaerobic cocci, including *Peptostreptococcus*.[71] In general, the most potential drugs against the genus *Peptostreptococcus* organisms include carbapenems, chloramphenicol, and metronidazole[25,35,66–70,72] as well as drugs with enhanced activity against gram-positive organisms, such as vancomycin, dalbavancin, daptomycin, and linezolid.[35,66–68,72] Also, clindamycin and some fluoroquinolones, for example, moxifloxacin, show relatively good susceptibility rates.[25,66,67,69,70] However, antimicrobial resistance of gram-positive anaerobic cocci varies depending on geographic location, as can be seen between European countries.[35] Furthermore, in a recent survey in Taiwan, worrying resistance rates were reported among gram-positive anaerobic cocci (13/32 of the tested strains were *P. anaerobius* sensu lato): 13% for moxifloxacin, 9% for clindamycin and metronidazole, and 6% for chloramphenicol.[73] Therefore, gathering information on local susceptibility patterns would help to choose optimal antimicrobial therapy whenever these organisms are involved. For example, very high resistance rates were reported in a national survey run in New Zealand: 45% of the *P. anaerobius* sensu lato isolates showed resistance to penicillin and 35% to the combination of amoxicillin/clavulanic acid.[70] It is notable that antimicrobial susceptibilities can also vary considerably among the species within the same genus. The only data currently available on antimicrobial susceptibilities of *P. anaerobius* (sensu stricto) and *P. stomatis* separately come from a study, where amoxicillin, amoxicillin/clavulanic acid, cefoxitin, ertapenem, azithromycin, clindamycin, metronidazole, and moxifloxacin were tested.[25] Except for one strain with a reduced susceptibility to clindamycin, the *P. stomatis* strains were susceptible to all these drugs, whereas 13% of *P. anaerobius* strains demonstrated intermediate/resistant MICs to one or more drugs, including amoxicillin, amoxicillin/clavulanic acid, cefoxitin, azithromycin, and/or moxifloxacin. Indeed, between the two species within the genus *Peptostreptococcus*, *P. anaerobius* had constantly higher MIC_{50} and MIC_{90} values than did *P. stomatis*.[25] This observation underlines the importance of precise identification of causative organisms to obtain optimal treatment outcome.

38.1.3 Diagnosis

To enable reliable detection of causative organism(s), appropriate specimen collection techniques are extremely important. Aseptically taken blood, tissue biopsies, aspirates (e.g., cerebrospinal fluid, joint fluids, and pus), wound curettages,

TABLE 38.2
***P. anaerobius* (sensu stricto) and *P. stomatis* Strains from 63 Subjects Divided according to Their Isolation Site**

Anatomical Site	No. and Isolation Site of *P. anaerobius* Strains (n = 30)		No. and Isolation Site of *P. stomatis* Strains (n = 33)	
Mouth	2	odontogenic abscess, lip	6	saliva (n = 3), periodontal pocket, pericoronitis, odontogenic abscess
Nasopharynx	1	sinus puncture	6	nasopharyngeal aspirate, tonsilla (n = 4), peritonsillar abscess
Gastrointestinal tract	3	stomach ulcer, appendix, abdominal pus	15	appendix (n = 12), feces (n = 3)
Female genital tract	7	Bartholin cyst, vagina, cervix, intrauterine device (n = 2), pelvic abscess (n = 2)	1	perineal abscess
Urinary tract	1	urethra	0	
Upper extremities	1	skin	0	
Lower extremities	10	cyst, bursitis, skin (n = 4), ulcer (n = 4)	2	skin, ulcer
Unspecified	4	abscess, surgical wound, skin/soft tissue	1	skin
Blood	1		2	

Source: Data adapted from Könönen, E. et al., *Antimicrob. Agents Chemother.*, 51, 2205, 2007.

bronchoalveolar lavage, root canal exudates, and subgingival plaque are proper specimens for detecting anaerobes, whereas mucosal/cutaneous swabs are not recommended.[22] Moreover, when diagnostics is based on culture, proper transportation and laboratory processing methods are warranted to support the survival of anaerobic organisms in the specimen and to isolate and identify them accurately. Concerning the survival of anaerobic organisms, such as *P. anaerobius*, the effectiveness of different transporters varies, in particular, when the transport media are held at room temperature.[74]

In general, the diagnostics of gram-positive anaerobic cocci are challenging, especially when relying solely on phenotypic methods. According to Song et al.,[75] only 65% of the 190 gram-positive anaerobic coccal isolates, collected from clinical specimens and identified by phenotypic methods, agreed with the identification made by 16S rRNA sequencing or multiplex PCR with genus- and species-specific primers. It is notable, however, that the two *Peptostreptococcus* species, *P. anaerobius* and *P. stomatis*, have some phenotypic key characteristics, which can be used to distinguish them from other gram-positive anaerobic cocci and are relatively easy to perform without advanced technology. These include the demonstration of their sensitivity to SPS and testing of α-glucosidase and proline arylamidase reactions.[15,25,26] Matrix-assisted laser desorption ionization time-of flight (MALDI-TOF) mass spectrometry has recently been suggested as a rapid and inexpensive method in comparison to biochemical identification for identification of bacterial isolates.[76] Molecular methods are increasingly used for identifying isolated strains and, more importantly, for detecting bacteria directly from clinical specimens. In this context, the 16S rRNA gene has proven to be invaluable because of its conserved and variable domains, which can be used for detecting differences at various levels of bacterial taxonomy, including the taxonomic changes and identification of gram-positive anaerobic cocci.[8,11,77] Although DNA-based methods are especially useful for identification of fastidious organisms, culture is still necessary for antibiotic-sensitivity testing.

38.1.3.1 Conventional Techniques

The conventional approach for detecting bacteria from clinical specimens includes culture and the identification of isolated organisms. Disadvantages are due to the methodology, that is, the need for expensive equipment and high expertise in the laboratory as well as rather laborious and time-consuming techniques.

A selection of solid nonselective media and selective media, which should be freshly prepared or prereduced and anaerobically sterilized (PRAS), is preferable for effective isolation of anaerobic species.[22] *Peptostreptococcus* species grow on the routine, nonselective, blood base media used for anaerobes, such as enriched brucella agar and fastidious anaerobe agar (FAA). In addition, phenylethyl alcohol (PEA) sheep blood agar may increase their detection frequency from purulent specimens and in the case of mixed infections.[22] After inoculation, the plates are immediately incubated at 36°C–37°C in an anaerobic chamber or in jars for 48 h and then examined. The total viable counts are determined and colony types isolated and subcultured for identification procedures. However, a further incubation time of 3–5 days and reexamination of the plates are recommended for slow-growing organisms, including gram-positive anaerobic cocci.

Sometimes the characteristic smell of *P. anaerobius* sensu lato, that is, a sweet, pungent odor due to the production of isocaproic acid, gives a hint for its presence and detection in the specimen.[20,22] *P. anaerobius* strains grow as large (2–4 mm in diameter), convex, off-white colonies, while *P. stomatis* strains form smaller, highly convex, and creamy to off-white colonies after 5 days of incubation on FAA.[15] The susceptibility pattern to special-potency antimicrobial disks (a zone size ≥10 mm is considered susceptible), containing vancomycin (5 μg), kanamycin (1000 μg), colistin (10 μg), and SPS (5%) is useful in presumptive identification of anaerobic taxa.[22] Gram-positive anaerobic cocci are usually susceptible to vancomycin and kanamycin and resistant to colistin and SPS. However, *P. anaerobius* sensu lato is typically susceptible to SPS and many strains exhibit resistance also to kanamycin.[22] Indeed, measuring the diameter of the growth-inhibition zone surrounding an SPS disk offers a practical method for identification of *P. anaerobius* sensu lato among gram-positive anaerobic cocci.[20,22] Figure 38.1 presents a flowchart where SPS-sensitive *P. anaerobius* and *P. stomatis*, besides occasional strains of *P. micra*, can be readily separated from SPS-resistant *Anaerococcus*, *Finegoldia*, and *Peptoniphilus* species. SPS-sensitive *P. micra* strains show an inhibition zone of <12 mm and, in addition, their glistening white, halo-surrounding colonies and small (0.6 μm) cells differ from those of the two *Peptostreptococcus* species.[26] The sizes of the inhibition zone, that is, ≥12–19 mm and >19 mm, may further differentiate *P. anaerobius* and *P. stomatis*, respectively, from each other.[15,25] It is notable, however, that frozen and thawed strains that are retested show smaller zones of inhibition than freshly examined strains.[24]

Rapid, easy-to-use test systems, including commercially available preformed enzyme kits, which can be manually or automatically handled, and individual diagnostic tablets, are widely used in clinical microbiology laboratories. For example, the use of the Rapid ID 32A (bioMeriéux), RapID ANA (Remel), BBL Crystal ANR ID (Becton Dickinson) test kits as well as the colorimetric ANC identification card (bioMeriéux), together with an automated Vitek 2 ANC system software, may offer a relatively reliable identification for human and veterinary peptostreptococci.[15,25,46,78–80] However, one problem with their use is related to the incomplete databases of the products due to the changing taxonomy of gram-positive anaerobic cocci. It is noteworthy that a heavy inoculum from 2- to 3-day-old cultures is needed for testing to obtain optimal reactions. Concerning *Peptostreptococcus* species, this is easily achieved, since they grow well on solid blood media. A positive α-glucosidase reaction, together with the sensitivity to SPS, refer to *P. anaerobius* sensu lato,[20,22] while a positive proline arylamidase reaction can be used to distinguish *P. anaerobius* from proline arylamidase-negative *P. stomatis* (Figure 38.1).[15,25,26] The corresponding Rapid ID 32A codes are 0400020000 and 0400000000.

In addition, the determination of isocaproic acid as a major end product of glucose metabolism by gas liquid chromatography (GLC) is useful in identification of *Peptostreptococcus* species[22,26]; however, GLC is seldom available in clinical microbiology laboratories.

38.1.3.2 Molecular Identification

The present classification of bacteria is based on their 16S rRNA sequences. This phylogenetic classification of bacteria offers a well-defined framework that can be used to develop molecular tools for identification of bacteria. The 16S rRNA molecules consist of highly conserved sequences interspersed with regions of more variable sequences. It is possible to rationally design primers and probes covering a desired phylogenetic group (a certain division, genus, or species). So, it is not surprising that molecular detection, identification, and also quantification of *Peptostreptococcus* species rely on 16S rRNA genes. Several PCR methods have been used to detect *P. anaerobius*, but only few probes specific for the other *Peptostreptococcus* species, *P. stomatis*, have been described. This might reflect the fact that *P. stomatis* was just recently differentiated from *P. anaerobius*.[15] As can be seen in Figure 38.2,[15] both species can be identified using 16S rRNA gene sequences.

38.2 MOLECULAR METHODS

38.2.1 Sample Preparation

Proper sampling and storing of the clinical specimens is as essential for DNA-based microbiological analyses as it is for culture-based analyses. The quantitative recovery of total DNA, representing the whole bacterial community, is a prerequisite for all subsequent DNA-based analyses, especially, if quantitative results are expected.

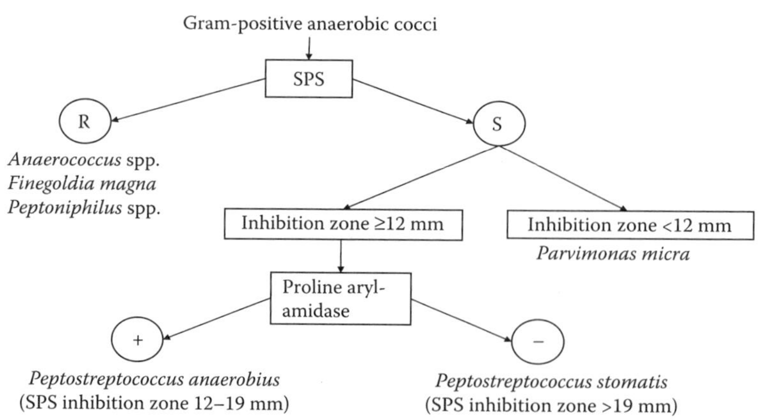

FIGURE 38.1 Flowchart of phenotypic reactions for detection of the two *Peptostreptococcus* species (SPS = sodium polyanethol sulfonate; R, resistant; S, susceptible; +, positive; –, negative).

```
                                          51                                              100
P. anaerobius NCTC 11460T         (25)  CACATGCAAGTCGAGCG----------------CGTCTGATTTGATGCTT
P. anaerobius ATCC 27337          (51)  CACATGCAAGTCGAGCG----------------CGTCTNATTTGATGCTT
P. stomatis strain W2278T         (25)  CACATGCAAGTCGAGCGAGGGTTTGCTCAGTATTGAGTATTCTAAGACTA
Peptostreptococcus sp. CK035      (46)  CACATGCAAGTCGAGCGAGGGTTTGCTCAGTATTGAGTATTCTAAGACTA
Uncultured bacterium FJ982988     (46)  CACATGCAAGTCGAGCGAGGGTTTGCTCAGTATTGAGTATTCTAAGACTA
Variable positions                                       ***************** ** ** * ****  *
Variable region                                          ---------------------------------
PAN-1                                                                    CGTCTWATTTNATGCTT
rcc-2#                                                           GCTCAGTATTGAGTATTCTAAGACT

                                          101                                             150
P. anaerobius NCTC 11460T         (63)  GCAT---------TNATGAAAGATGAGCGGCGGACGGGTGAGTAACGCGT
P. anaerobius ATCC 27337          (89)  GCNT---------TAATGAAAGATGAGCGGCGGACGGGTGAGTAACGCGT
P. stomatis strain W2278T         (75)  GAATGTTCAATTCTGAGCAAAACCAAGCGGCGGACGGGTGAGTAACGCGT
Peptostreptococcus sp. CK035      (96)  GAATGTTCAATTCTGAGCAAAACCAAGCGGCGGACGGGTGAGTAACGCGT
Uncultured bacterium FJ982988     (96)  GAATGTTCAATTCTGAGCAAAACCAAGCGGCGGACGGGTGAGTAACGCGT
Sequence differences                     ** ********* ****   *****
Variable region                         -------------
PAN-1 continues                         GCA

                                          151                                             200
P. anaerobius NCTC 11460T        (100)  GGGTAACCTGCCCTATACACATGGATAACATACTGAAAAGTTTACTAATA
P. anaerobius ATCC 27337         (126)  GGGTAACCTGCCCTATACACATGGATAACATACTGAAAAGTTTACTAATA
P. stomatis strain W2278T        (125)  GGGTAACCTGCCCTATACACATGGATAACATACTGAAAAGTTTACTAATA
Peptostreptococcus sp. CK035     (146)  GGGTAGCCTGCCCTATACACATGGATAACATACTGAAAAGTTTACTAATA
Uncultured bacterium FJ982988    (146)  GGGTAGCCTGCCCTATACACATGGATAACATACTGAAAAGTTTACTAATA
Sequence differences                         *
                                          201                                             250
P. anaerobius NCTC 11460T        (150)  CATGATAATATATATTTACGGCATCGTAGATATATCAAAGTGTTAGCGGT
P. anaerobius ATCC 27337         (176)  CATGATAATATATATTTACGGCATCGTAGATATATCAAAGTGTTAGCGGT
P. stomatis strain W2278T        (175)  CATGATAATATATATTTGCGGCATCGCAGATATATCAAAGTGTTAGCGGT
Peptostreptococcus sp. CK035     (196)  CATGATAATATATATTTGCGGCATCGCAGATATATCAAAGTGTTAGCGGT
Uncultured bacterium FJ982988    (196)  CATGATAATATATATTTGCGGCATCGCAGATATATCAAAGTGTTAGCGGT
Sequence differences                                     *        *

...
                                          401                                             450
P. anaerobius NCTC 11460T        (350)  AATGGGCGCAAGCCTGATGCAGCAACGCCGCGTGAACGATGAAGGTCTTC
P. anaerobius ATCC 27337         (376)  AATGGGCGCAAGCCTNATGCAGCAACGCCGCGTGAACGATGAAGGTCTTC
P. stomatis strain W2278T        (375)  AATGGGCGAAAGCCTGATGCAGCAACGCCGCGTGAACGATGAAGGTCTTC
Peptostreptococcus sp. CK035     (396)  AATGGGCGAAAGCCTGATGCAGCAACGCCGCGTGAACGATGAAGGTCTTC
Uncultured bacterium FJ982988    (396)  AATGGGCGAAAGCCTGATGCAGCAACGCCGCGTGAACGATGAAGGTCTTC
Sequence differences                            *

...
                                          551                                             600
P. anaerobius NCTC 11460T        (500)  GCGTTATCCGGATTTACTGGGCGTAAAGGGTGCGTAGGTGGTCTTTCAAG
P. anaerobius ATCC 27337         (526)  GCGTTATCCGGATTTACTGGGCGTAAAGGGTGCGTAGGTGGTCTTTCAAG
P. stomatis strain W2278T        (525)  GCGTTATCCGGATTTACTGGGCGTAAAGGGTGCGTAGGTGGTCCTTCAAG
Peptostreptococcus sp. CK035     (546)  GCGTTATCCGGATTTACTGGGCGTAAAGGGTGCGTAGGTGGTCCTTCAAG
Uncultured bacterium FJ982988    (546)  GCGTTATCCGGATTTACTGGGCGTAAAGGGTGCCTAGGTGGTCCTTCAAG
Sequence differences                                                       *

                                          601                                             650
P. anaerobius NCTC 11460T        (550)  TCGGTGGTTAAAGGCTACGGCTCAACCGTAGTTAGCCTCCGAAACTGGAA
P. anaerobius ATCC 27337         (576)  TCGGTGGTTAAAGGCTACGGCTCAACCGTAGTTAGCCTCCGAAACTGGAA
P. stomatis strain W2278T        (575)  TCGGTGGTTAAAGGCTACGGCTCAACCGTAGTAAGCCGCCGAAACTGGAG
Peptostreptococcus sp. CK035     (596)  TCGGTGGTTAAAGGCTACGGCTCAACCGTAGTAAGCCGCCGAAACTGGAG
Uncultured bacterium FJ982988    (596)  TCGGTGGTTAAAGGCTACGGCTCAACCGTAGTAAGCCGCCGAAACTGGAG
Sequence differences                                                    *    *           *
PEPA                                                               GTAGTTAGCCTCCGAAA
rcc-1#                                                            CGTAGTTAGCCTCCGAAAC

...
```

FIGURE 38.2 An alignment of the 16S rRNA genes of *Peptostreptococcus* species and phylotypes. The accession numbers for the sequences are as follows: *P. anaerobius* NCTC 11460^T, AY326462; *P. anaerobius* ATCC 27337, L04168; *P. stomatis* W2278^T, DQ160208; *Peptostreptococcus* sp. CK035, AF287763; Uncultured bacterium, FJ982988. Variable region was identified by Downes and Wade.[15] Only the regions with differences are shown, and the three dots indicate those parts of the alignment, which do not contain any differences between the sequences and had been removed. The number above and in the beginning of the sequences indicate the relative position of the nucleotides in the alignment. Only species- or genus-specific primers are shown here. Mismatches between oligonucleotide primes or probes and the target sequences are highlighted by underlining the corresponding nucleotides. (# Primers sequence are shown in the forward direction.)

```
                                     801                                            850
P. anaerobius NCTC 11460T     (750)  GGTAGTCCACGCTGTAAACGATGAGTACTAGGTGTCGGGGGTTACCCCCC
P. anaerobius ATCC 27337      (776)  GGTAGTCCACGCTGTAAACGATGAGTACTAGGTGTCGGGG-TTNCCCCCC
P. stomatis strain W2278T     (775)  GGTAGTCCACGCTGTAAACGATGAGTACTAGGTGTCGGGGGTTACCCCCC
Peptostreptococcus sp. CK035  (796)  GGTAGTCCACGCTGTAAACGATGAGTACTAGGTGTCGGGGGTTACCCCCC
Uncultured bacterium FJ982988 (796)  GGTAGTCCACGCTGTAAACGATGAGTACTAGGTGTCGGGGGTTACCCCCC
Sequence differences                                                         *
RT-PEPAf                                                                            GC

                                     851                                            900
P. anaerobius NCTC 11460      (800)  TCGGTGCCGCAGCTAACGCATTAAGTACTCCGCCTGGGGAGTACGCACGC
P. anaerobius ATCC 27337      (825)  TCGGTGCCGCAGCTAACGCATTAAGTACTCCGCCTGGGGAGTACGCACGC
P. stomatis strain W2278      (825)  TCGGTGCCGCAGCTAACGCATTAAGTACTCCGCCTGGGGAGTACGCACGC
Peptostreptococcus sp. CK035  (846)  TCGGTGCCGCAGCTAACGCATTAAGTACTCCGCCTGGGGAGTACGCACGC
Uncultured bacterium FJ982988 (846)  TCGGTGCCGCAGCTAACGCATTAAGTACTCCGCCTGGGGAGTACGCACGC
RT-PEPAf continues                   TCGGTGCCTTCACTAACG

                                     901                                            950
P. anaerobius NCTC 11460T     (850)  AAGTGTGAAACTCAAAGGAATTGACGGGGACCCGCACAAGTAGCGGAGCA
P. anaerobius ATCC 27337      (875)  AAGTGTGAAACTCAAAGGAATTGACGGGGACCCGCACAAGTAGCGGAGCA
P. stomatis strain W2278T     (875)  AAGTGTGAAACTCAAAGGAATTGACGGGGACCCGCACAAGTAGCGGAGCA
Peptostreptococcus sp. CK035  (896)  AAGTGTGAAACTCAAAGGGATTGACGGGGACCCGCACAAGTAGCGGAGCA
Uncultured bacterium FJ982988 (896)  AAGTGTGAAACTCAAAGGGATTGACGGGGACCCGCACAAGTAGCGGAGCA
Sequence differences                                   *
                                     951                                           1000
P. anaerobius NCTC 11460T     (900)  TGTGGTTTAATTCGAAGCAACGCGAAGAACCTTACCTAAGCTTGACATCC
P. anaerobius ATCC 27337      (925)  TGTGGTTTAATTCGAAGCAACGCGAAGAACCTTACCTAAGCTTGACATCC
P. stomatis strain W2278T     (925)  TGTGGTTTAATTCGAAGCAACGCGAAGAACCTTACCTAAGCTTGACATCC
Peptostreptococcus sp. CK035  (946)  TGTGGTTTAATTCGAAGCAACGCGAAGTACCTTACCTAAGCTTGACATCC
Uncultured bacterium FJ982988 (946)  TGTGGTTTAATTCGAAGCAACGCGAAGTACCTTACCTAAGCTTGACATCC
Sequence differences                                            *
                                     1001                                          1050
P. anaerobius NCTC 11460T     (950)  CTTAGACCGGTGTTTAATCACACCTTCCCTTCGGGGCTGAGGTGACAGGT
P. anaerobius ATCC 27337      (975)  CTTAGACCGGTGTTTAATCACACCTT-CCTTCGGGGCTGAGGTGACAGGT
P. stomatis strain W2278T     (975)  CTCGGACAGGTGTTTAATCACACCCTTCCTTCGGGACTGAGGTGACAGGT
Peptostreptococcus sp. CK035  (996)  CTCGGACAGGTGTTTAATCACACCCTTCCTTCGGGACTGAGGTGACAGGT
Uncultured bacterium FJ982988 (996)  CTCGGACAGGTGTTTAATCACACCCTTCCTTCGGGACTGAGGTGACAGGT
Sequence differences                   **   *                * *        *
PAN-2# and RT-PEPAr#                                   CACACCTTCCCTTCGGGGCT

...                                  1101                                          1150
P. anaerobius NCTC 11460T     (1050) CAACGAGCGCAACCCTTGTCTTTAGTTGCCAGCATTCAGTTGGGCACTCT
P. anaerobius ATCC 27337      (1074) CAACGAGCGCAACCCTNGTCTTTAGTTGCCAGCATTCAGTTGGGCACTCT
P. stomatis strain W2278T     (1075) CAACGAGCGCAACCCTTGTCTTTAGTTGCCAGCATTCAGTTGGGCACTCT
Peptostreptococcus sp. CK035  (1096) CAACGAGCGCAACCCTTGTCTTTAGTTGCCAGCATTTAGTTGGGCACTCT
Uncultured bacterium FJ982988 (1096) CAACGAGCGCAACCCTTGTCTTTAGTTGCCAGCATTTAGTTGGGCACTCT
Sequence differences                                                     *
                                     1151                                          1200
P. anaerobius NCTC 11460T     (1100) AGAGAGACTGCCAGGGATAACCTGGAGGAAGGTGGGGATGACGTCAAATC
P. anaerobius ATCC 27337      (1124) AGAGAGACTGCCAGGGATAACCTGGAGGAAGGTGGGGATGACGTNNAATC
P. stomatis strain W2278T     (1125) AGAGAGACTGCCAGGGATAACCTGGAGGAAGGTGGGGATGACGTCAAATC
Peptostreptococcus sp. CK035  (1146) AGAGAGACTGCCAGGGATAACCTGGAGGAAGGTGGGGATGACGTCAAATC
Uncultured bacterium FJ982988 (1146) AGAGAGACTGCCAGGGATAACCTGGAGGAAGGTGGGGATGACGTCAAATC
Sequence differences

...
                                     1251                                          1300
P. anaerobius NCTC 11460T     (1200) GAGGGTTGCCAAACCGTGAGGTGGAGCTAATCCCTTAAAGCCATTCTCAG
P. anaerobius ATCC 27337      (1224) GAGGGTTGCCAAACCGTGAGGTGGAGCTAATCCCTTAAAGCCATTCTCAG
P. stomatis strain W2278T     (1225) GAGGGTTGCCAAACCGTGAGGTGGAGCCAATCCCTTAAAGCCATTCTCAG
Peptostreptococcus sp. CK035  (1246) GAGGGTTGCCAAACCGTGAGGTGGAGCCAATCCCTTAAAGCCATTCTCAG
Uncultured bacterium FJ982988 (1246) GAGGGTTGCCAAACCGTGAGGTGGAGCCAATCCCTTAAAGCCATTCTCAG
Sequence differences                                            *

...
                                     1401                                          1450
P. anaerobius NCTC 11460T     (1350) GCCCGTCACACCATGGGAGTCGGAAACACCCGAAGCCGATTATCCAACCG
P. anaerobius ATCC 27337      (1374) GCCCGTCACACCATGGGAGTCGGAAACACCCGAAGCCGATTATCCAACCG
P. stomatis strain W2278T     (1375) GCCCGTCACACCACGGGAGTCGGAAACACCCGAAGCCGATTATCCAACCG
Peptostreptococcus sp. CK035  (1396) GCCCGTCACACCACGGGAGTCGGAAACACCCGAAGCCGATTATCCAACCG
Uncultured bacterium FJ982988 (1396) GCCCGTCACACCACGGGAGTCGGAAACACCCGAAGCCGATTATCCAACCG
Sequence differences                   *
```

FIGURE 38.2 (Continued)

The first step in sample preparation is the quantitative lysis of bacterial cells. Because the structure of the cell wall of different bacteria can vary a lot, mechanical rather than enzymatic methods are preferred, although also enzymatic methods have been used.[16] One example of the mechanical methods is so-called bead-beating method, which is known for its efficient lysis of bacteria. DNA isolation from clinical specimens is done using ceramic or silica particles. These particles are mixed together with the specimen in a small test tube. The reciprocal motion (e.g., 7000 rpm) of the lyser causes high-energy collisions between bacterial cells and the particles. These collisions disrupt bacterial cells.

The next step is the purification of the bacterial DNA. First, protein molecules are denatured and separated from total DNA. Clinical specimens can contain several substances that inhibit PCR that should be removed during subsequent purification steps. Several different kinds of DNA purification methods have been described. Older methods are based on organic solvents like phenol, which denature proteins. Denatured proteins are removed and DNA precipitated to remove any phenol residues. Finally, DNA is dissolved to water or buffer. Phenol and other toxic chemicals are nowadays more or less replaced with commercial DNA purification kits. The kits developed for DNA isolation utilize several different methods some of those especially developed for isolation and purification of bacterial DNA. Zoetendal et al.[81] have presented very detailed procedures, including commercial DNA purification kits that can be used for isolation of bacterial DNA from different kinds of specimens.

38.2.2 Genus- and Species-Specific Detection Procedures

Several PCR and hybridization primers and probes have been described for the detection of *Peptostreptococcus* species. Most of them, however, were designed before the 16S rRNA gene sequence for *P. stomatis* was available. The specificity of the published molecular tools has not been properly tested with the type strains of the two *Peptostreptococcus* species, that is, it is unclear what the real specificity of one particular primer or probe is.

38.2.2.1 Conventional PCR Methods

Several PCR primers that are specific for *P. anaerobius* have been described (Table 38.3). These primers have been used in conventional PCR to identify isolated *Peptostreptococcus* species[75] and to detect *P. anaerobius* directly in clinical specimens. Although these PCR primers were designed before *P. stomatis* was described,[15] they seem to be specific only for *P. anaerobius* (Figure 38.2). For example, the primer PAN-1 has several mismatches when compared to the 16S rRNA gene sequence of *P. stomatis* (Figure 38.2). The number and the position of the mismatches near the 3′-end of the primer indicate that it is unlikely that PAN-1 could be used for amplification of 16S rRNA genes of *P. stomatis*. This, of course, depends on the amplification conditions and cannot be deducted from the sequences only. However, a direct amplification of *P. stomatis* 16S rRNA genes from clinical specimens might be difficult. In fact, Riggio and Lennon[29] were not able to detect *P. anaerobius* 16S rRNA genes in either subgingival plaque specimens or pus aspirates from dentoalveolar abscesses when they used the primers PAN-1 and PAN-2. The primer PAN-2 has also mismatches in comparison to the *P. stomatis* 16S rRNA gene sequence (Figure 38.2). The authors concluded that *Peptostreptococcus* organisms do not play an important role in pathogenesis of oral infections.[29] This may be true for *P. anaerobius* sensu stricto, but, obviously, the conclusion is not correct because the primers used were not optimal for detection of *P. stomatis*. Later, it has also been shown that *Peptostreptococcus* isolates from dentoalveolar abscesses have identical 16S rRNA gene sequences when compared to those of the type strain of *P. stomatis*.[15]

Song et al.[75] have designed a PCR method for identification of *P. anaerobius*. This method utilizes one specific and one universal primer. The *P. anaerobius*-specific primer PEPA has a perfect match with the 16S rRNA gene sequence of *P. anaerobius* but has only two mismatches with the 16S rRNA gene sequence of *P. stomatis* (Figure 38.2). Because the other primer used in this PCR method is universal and can

TABLE 38.3
PCR Primers Used to Detect and Identify *Peptostreptococcus* Species

PCR	Primer	Specificity	Direction	Size (bp)	Label or Modification	Sequence (5′→3′)	Method	Reference
8f/519r	8f	Bacteria, universal	forward		digoxigenin	AGAGTTTGATYMTGGC	reverse-capture checkerboard	36
	519r	Bacteria, universal	reverse			GTGCCAGCAGCCGCGGTMA		36
515f/1492r	515f	Bacteria, universal	forward		digoxigenin	GTGCCAGCAGCCGCGGTMA	reverse-capture checkerboard	36
	1492r	Bacteria, universal	reverse			GYTACCTTGTTACGACTT		36
PEPA/1392B	PEPA	*P. anaerobius*	forward	780		GTAGTTAGCCTCCGAAA	conventional PCR	75
	1392B	Bacteria, universal	reverse			ACGGGCGGTGTGTAC		75
PAN-1/PAN-2	PAN-1	*P. anaerobius*	forward	943		CGTCTWATTTNATGCTTGCA	conventional PCR	29
	PAN-2[a]	*P. anaerobius*	reverse			AGCCCCGAAGGGAAGGTGTG		29
RT-PEPA	RT-PEPAf	*P. anaerobius*	forward	188	SYBR green 490	GCTCGGTGCCTTCACTAACG	real-time PCR	82
	RT-PEPAr[a]	*P. anaerobius*	reverse			AGCCCCGAAGGGAAGGTGTG		82

[a] PAN-2 and RT-PEPAr have identical sequences.

be used for amplification of almost all bacterial 16S rRNA genes, it is difficult to estimate the specificity of the PCR. The two mismatches may be enough for distinguishing the two *Peptostreptococcus* species, but further laboratory experiments are needed to confirm their specificity.

38.2.2.2 Real-Time PCR

Ahmed et al.[82] described a real-time PCR assay, which can be used for direct detection and quantification of *P. anaerobius* 16S rRNA genes in clinical specimens. By using this method, they successfully measured the amount of *P. anaerobius* in mucosal tissue of the terminal ileum and large intestine. However, probe match results in the Ribosomal Database Project (RDP)[83] pointed out that the real-time PCR primer RT-PEPAf, used in the study of Ahmed,[82] has a perfect match with only one 16S rRNA gene of *P. anaerobius* (Jalava, J., unpublished observation). There are several 16S rRNA sequences available from the type strain of *P. anaerobius*; however, it is important to understand that these sequences are not identical. One particular 16S rRNA gene (accession number D14150) of the type strain of *P. anaerobius* ATCC 27337^T differs from other 16S rRNA gene sequences and it contains quite a lot of ambiguous positions. The 16S rRNA gene sequences of the type strain, stored in culture collections as *P. anaerobius* NCTC 11460^T (accession number: AY326462) and as *P. anaerobius* ATCC 27337^T (accession number: L04168), are almost but not completely identical (Figure 38.2). The quality of the sequences is quite good, that is, there are few ambiguous positions, while the real-time PCR primer RT-PEPAf has five mismatches with these sequences (Figure 38.2). On the other hand, the primer PAN-1 has a perfect match with 44 different 16S rRNA gene sequences of the RDP database, 43 of these sequences being obtained from bacteria that belong to the genus *Peptostreptococcus* (Jalava, J., unpublished observation). Although most of these sequences were actually originating from uncultured bacteria, the primer PAN-1 seems to work better with *Peptostreptococcus* species than the primer RT-PEPAf. In other words, the amplification of *P. anaerobius* 16S rRNA genes with the primer RT-PEPAf may not be optimal.

38.2.2.3 Reverse-Capture Checkerboard Hybridization Assay

A reverse-capture checkerboard hybridization assay is a molecular method, which can be used for direct detection of bacteria in clinical specimens. It utilizes universal 16S rRNA gene primers and species-specific capture probes.[84] In this method, 16S rRNA genes are first amplified using universal primers 8f and 1492r (Table 38.3). After this, a second round of amplification is performed with two different sets of universal primers (pair 8f-519r and pair 515f-1492) that are labeled with digoxigenin (Table 38.3). These PCR products are used as hybridization probes in subsequent steps. The specificity of the method is obtained by the capture probes, which are immobilized via polythymidine tails on a nylon membrane. As many as 1350 hybridizations can be performed simultaneously using this method.[36]

Reverse-capture checkerboard hybridization has been used for studying the oral microbiota, and capture probes specific for *P. anaerobius* and *P. stomatis* have been described.[36] The *P. anaerobius*-specific capture probe, rcc-1, has a perfect match with *P. anaerobius* sequences and two mismatches when compared to 16S rRNA gene sequences of *P. stomatis* (Figure 38.2). Similarly, the *P. stomatis*-specific capture probe, rcc-2, has several mismatches compared to *P. anaerobius* sequences and a perfect match to target sequences (Figure 38.2). When Rôças and Siqueira[36] used this reverse-capture checkerboard hybridization method for analyzing root canal specimens from 43 subjects with chronic apical periodontitis, they were able to detect 16S rRNA genes of both *Peptostreptococcus* species. In fact, 54% of the specimens were positive for *P. anaerobius*, being more frequent than *P. stomatis* (less than 30% of the specimens). These results indicate that, in addition to *P. stomatis*, *P. anaerobius* 16S rRNA genes can be detected in specimens obtained from the oral microbiota.

38.3 CONCLUSIONS AND FUTURE PERSPECTIVES

Culture, direct DNA-based techniques, and cloning and sequencing of specific genes are among the methods that can be chosen for studying bacterial etiology of human and animal infections. The information on the genus *Peptostreptococcus* obtained by culture is limited to cultivable species, which can be identified by using biochemical tests or, more rapidly, by MALDI-TOF mass spectrometry. Since PCR, DNA–DNA checkerboard hybridization, or microarrays are designed for one or more selected target organism(s), the bacterial detection using these molecular methods is limited to already known species. When using a broad-range, open-ended molecular approach, it is possible to detect both cultivable species and known but uncultivable phylotypes, as well as unknown phylotypes with novel sequences. In the current literature, several *Peptostreptococcus* sp. clones have already been detected in a variety of clinical specimens. Conceivably, novel *Peptostreptococcus* will be described as soon as isolate(s) obtained by culture can be characterized and connected to these certain phylotypes detected by molecular methods.

It is often challenging to decide whether an organism isolated from a polymicrobial infectious lesion has a real causative role in the infection. Gram-positive anaerobic cocci, however, have proven to be significant pathogens. The two cultivable *Peptostreptococcus* species so far, *P. anaerobius* and *P. stomatis*, are involved in a variety of human infections, including bacteremia. In addition, several *Peptostreptococcus* phylotypes have been connected to clinically relevant specimens. Despite close phylogenetic and/or phenotypic similarities, their characteristics, such as preferred habitats, links to specific infections, virulence factors, as well as antimicrobial susceptibility patterns, can vary. Therefore, it is of outmost importance that clinicians and clinical microbiology laboratories are aware of these differences. If the species identification is not considered, this leads to incomplete diagnosis and may result in poor therapy outcomes.

In future, molecular biology methods will also be increasingly used in clinical microbiology laboratories. Indeed, molecular methods based on 16S rRNA genes have shown to be useful tools in detection and identification of *Peptostreptococcus* species. However, most of the primers and probes have been designed only for *P. anaerobius*. Although relatively few *P. stomatis*-specific primers and probes have been tested thus far, the 16S rRNA gene sequences available indicate that there is enough variation in the 16S rRNA molecules so that this species also can be identified. Certainly, more 16S rRNA gene sequences from properly characterized clinical isolates of *P. stomatis* are needed for building up a sequence data collection that would enable to analyze variation in the 16S rRNA gene sequences. Without these sequence data, it is not possible to deduce the specificity of the molecular tools.

In the public sequence databases, there are several 16S rRNA gene sequences annotated to be from *Peptostreptococcus* species. However, BLAST searches[85] done with these sequences show that they actually do not originate from *Peptostreptococcus* species. Because this kind of inaccurate sequence data can disturb and complicate molecular analyses, it is of the utmost importance that the information in current databases will be updated.

REFERENCES

1. Kluyver, A.J., and Van Niel, C.B., Prospects for a natural system of classification of bacteria, *Zentralblatt für Bacteriologie, Parasitenkunde, Infektionskrankheiten und Hygiene, Abteilung II*, 94, 369, 1936.
2. Skerman, V.B.D., McGowan, V., and Sneath, P.H.A., Approved lists of bacterial names, *Int. J. Syst. Bacteriol.*, 30, 225, 1980.
3. Collins, M.D., and Wallbanks, S., Comparative sequence analyses of the 16S rRNA genes of *Lactobacillus minutus*, *Lactobacillus rimae* and *Streptococcus parvulus*: proposal for the creation of a new genus, *Atopobium*, *FEMS Microbiol. Lett.*, 95, 235, 1992.
4. Liu, C. et al., Reclassification of *Clostridium coccoides*, *Ruminococcus hansenii*, *Ruminococcus hydrogenotrophicus*, *Ruminococcus luti*, *Ruminococcus productus* and *Ruminococcus schinkii* as *Blautia coccoides* gen. nov., comb. nov., *Blautia hansenii* comb. nov., *Blautia hydrogenotrophica* comb. nov., *Blautia luti* comb. nov., *Blautia producta* comb. nov., *Blautia schinkii* comb. nov. and description of *Blautia wexlerae* sp. nov., isolated from human feces, *Int. Syst. Evol. Microbiol.*, 58, 1896, 2008.
5. Ezaki, T. et al., Transfer of *Peptococcus indolicus*, *Peptococcus asaccharolyticus*, *Peptococcus prevotii*, and *Peptococcus magnus* to the genus *Peptostreptococcus* and proposal of *Peptostreptococcus tetradius* sp. nov., *Int. J. Syst. Bacteriol.*, 33, 683, 1983.
6. Ezaki, T. et al., *Peptostreptococcus hydrogenalis* sp. nov.from human fecal and vaginal flora, *Int. J. Syst. Bacteriol.*, 40, 305, 1990.
7. Li, N. et al., Three new species of the genus *Peptostreptococcus* isolated from humans: *Peptostreptococcus vaginalis* sp. nov., *Peptostreptococcus lacrimalis* sp. nov., and *Peptostreptococcus lactolyticus* sp. nov., *Int. J. Syst. Bacteriol.*, 42, 602, 1992.
8. Ezaki, T. et al., 16S ribosomal DNA sequences of anaerobic cocci and proposal of *Ruminococcus hansenii* comb.nov. and *Ruminococcus productus* comb.nov., *Int. J. Syst. Bacteriol.*, 44, 130, 1994.
9. Murdoch, D.A., and Shah, H.N., Reclassification of *Peptostreptococcus magnus* (Prevot 1933) Holdeman and Moore 1972 as *Finegoldia magna* comb. nov. and *Peptostreptococcus micros* (Prevot 1933) Smith 1957 as *Micromonas micros* comb. nov., *Anaerobe*, 5, 555, 1999.
10. Tindall, B.J., and Euzeby, J.P., Proposal of *Parvimonas* gen. nov. and *Quatrionicoccus* gen. nov. as replacements for the illegitimate, prokaryotic, generic names *Micromonas* Murdoch and Shah 2000 and *Quadricoccus* Maszenan et al. 2002, respectively, *Int. J. Syst. Evol. Microbiol.*, 56, 2711, 2006.
11. Ezaki, T. et al., Proposal of the genera *Anaerococcus* gen. nov., *Peptoniphilus* gen. nov. and *Gallicola* gen. nov. for members of the genus *Peptostreptococcus*, *Int. J. Syst. Evol. Microbiol.*, 51, 1521, 2001.
12. Huss, V.A.R., Festl, H., and Schleifer, K.H., Nucleic acid hybridization studies and deoxyribonucleic acid base compositions of anaerobic, gram-positive cocci, *Int. J. Syst. Bacteriol.*, 34, 95, 1984.
13. Paster, B.J. et al., Phylogeny of the ammonia-producing ruminal bacteria *Peptostreptococcus anaerobius*, *Clostridium sticklandii*, and *Clostridium aminophilum* sp. nov., *Int. J. Syst. Bacteriol.*, 43, 107, 1993.
14. Collins, M.D. et al., The phylogeny of the genus *Clostridium*: Proposal of five new genera and eleven new species combinations, *Int. J. Syst. Bacteriol.*, 44, 812, 1994.
15. Downes, J., and Wade, W.G., *Peptostreptococcus stomatis* sp. nov., isolated from the human oral cavity, *Int. J. Syst. Evol. Microbiol.*, 56, 751, 2006.
16. Aas, J.A. et al., Defining the normal bacterial flora of the oral cavity, *J. Clin. Microbiol.*, 43, 5721, 2005.
17. Bahrani-Mougeot, F.K. et al., Molecular analysis of oral and respiratory bacterial species associated with ventilator-associated pneumonia, *J. Clin. Microbiol.*, 45, 1588, 2007.
18. Hebb, J.K. et al., Detection of novel organisms associated with salpingitis, by use of 16S rDNA polymerase chain reaction, *J. Infect. Dis.*, 190, 2109, 2004.
19. Kumar, P. et al., Identification of candidate periodontal pathogens and beneficial species by quantitative 16S clonal analysis, *J. Clin. Microbiol.*, 43, 3944, 2005.
20. Murdoch, D.A., Gram-positive anaerobic cocci, *Clin. Microbiol. Rev.*, 11, 81, 1998.
21. Kotiranta, A. et al., Crystalline surface protein of *Peptostreptococcus anaerobius*, *Microbiology*, 140, 1065, 1995.
22. Jousimies-Somer, H.R. et al., *Wadsworth–KTL Anaerobic Bacteriology Manual*, 6th ed., Star Publishing, Belmont, CA, 2002.
23. Moss, C.W., Lambert, M.A., and Lombard, G.L., Cellular fatty acids of *Peptostreptococcus variabilis* and *Peptostreptococcus anaerobius*, *J. Clin. Microbiol.*, 5,665, 1977.
24. Wideman, P.A. et al., Evaluation of the sodium polyanethol sulfonate disk test for the identification of *Peptostreptococcus anaerobius*, *J. Clin. Microbiol.*, 4, 330, 1976.
25. Könönen, E. et al., Antimicrobial susceptibilities of *Peptostreptococcus anaerobius* and the newly described *Peptostreptococcus stomatis* isolated from various human sources, *Antimicrob. Agents Chemother.*, 51, 2205, 2007.
26. Song, Y., Liu, C., and Finegold, S.M., Development of a flow chart for identification of gram-positive anaerobic cocci in the clinical laboratory, *J. Clin. Microbiol.*, 45, 512, 2007.

27. Finegold, S.M., Anaerobic infections in humans: An overview, *Anaerobe*, 1, 3, 1995.
28. Tyrrell, K.L. et al., Anaerobic bacteria cultured from the tongue dorsum of subjects with oral malodour, *Anaerobe*, 9, 243, 2003.
29. Riggio, M.P., and Lennon, A., Development of a PCR assay specific for *Peptostreptococcus anaerobius*, *J. Med. Microbiol.*, 51, 1097, 2002.
30. Aas, J.A. et al., Subgingival plaque microbiota in HIV positive patients, *J. Clin. Periodontol.*, 34, 189, 2007.
31. Brito, L.C.N. et al., Use of multiple-displacement amplification and checkerboard DNA–DNA hybridization to examine the microbiota of endodontic infections, *J. Clin. Microbiol.*, 45, 3039, 2007.
32. Faveri, M. et al., Microbiological diversity of generalized aggressive periodontitis by 16S rRNA clonal analysis, *Oral Microbiol. Immunol.*, 23, 112, 2008.
33. Brook, I., Recovery of anaerobic bacteria from clinical specimens in 12 years at two military hospitals, *J. Clin. Microbiol.*, 26, 1181, 1988.
34. Murdoch, D.A., Mitchelmore, I.J., and Tabaqchali, S., The clinical importance of gram-positive anaerobic cocci isolated at St Bartholomew's Hospital, London, in 1987, *J. Med. Microbiol.*, 41, 36, 1994.
35. Brazier, J. et al., European surveillance study on antimicrobial susceptibility of gram-positive anaerobic cocci, *Int. J. Antimicrob. Agents*, 31, 316, 2008.
36. Rôças, I.N., and Siqueira, J.F., Jr., Root canal microbiota of teeth with chronic apical periodontitis, *J. Clin. Microbiol.*, 46, 3599, 2008.
37. Rolph, H.J. et al., Molecular identification of microorganisms from endodontic infections, *J. Clin. Microbiol.*, 39, 3282, 2001.
38. Rajasuo, A. et al. Bacteriologic findings in tonsillitis and pericoronitis, *Clin. Infect. Dis.*, 23, 51, 1996.
39. Finegold, S.M. et al., Bacteriologic findings associated with chronic bacterial maxillary sinusitis in adults, *Clin. Infect. Dis.*, 35, 428, 2002.
40. Boyanova, L. et al., Anaerobic microbiology in 198 cases of pleural empyema: A Bulgarian study, *Anaerobe*, 10, 261, 2004.
41. Worlitzsch, D. et al., Antibiotic-resistant obligate anaerobes during exacerbations of cystic fibrosis, *Clin. Microbiol. Infect.*, 15, 454, 2009.
42. Wang, J.-L. et al., Changing bacteriology of adult community-acquired lung abscess in Taiwan: *Klebsiella pneumoniae* versus anaerobes, *Clin. Infect. Dis.*, 40, 915, 2005.
43. Alvarez, J.A. et al., Anaerobic liver abscesses as initial presentation of silent colonic cancer, *HPB*, 6, 41, 2004.
44. Haggerty, C.L. et al., Bacterial vaginosis and anaerobic bacteria are associated with endometritis, *Clin. Infect. Dis.*, 39, 990, 2004.
45. Han, Y.W. et al., Uncultivated bacteria as etiologic agents of intra-amniotic inflammation leading to preterm birth, *J. Clin. Microbiol.*, 47, 38, 2009.
46. Ng, J. et al., Identification of five *Peptostreptococcus* species isolated predominantly from the female genital tract by using the Rapid ID32A system, *J. Clin. Microbiol.*, 32, 1302, 1994.
47. Citron, D.M. et al., Bacteriology of moderate-to-severe diabetic foot infections and in vitro activity of antimicrobial agents, *J. Clin. Microbiol.*, 45, 2819, 2007.
48. Dowd, S.E. et al., Polymicrobial nature of chronic diabetic foot ulcer biofilm infections determined using bacterial tag encoded FLX amplicon pyrosequencing (bTEFAP), *PLoS One*, 3, e3326, 2008.
49. Edmiston, C.E. et al., Anaerobic infections in the surgical patient: microbial etiology and therapy, *Clin. Infect. Dis.*, 35 (suppl 1), S112, 2002.
50. Wall, I.B. et al., Potential role of anaerobic cocci in impaired human wound healing, *Wound Rep. Reg.*, 10, 346, 2002.
51. Talan, D.A. et al., Bacteriologic analysis of infected dog and cat bites, *N. Engl. J. Med.*, 340, 85, 1999.
52. Brook, I., Recovery of anaerobic and aerobic bacteria from a post-traumatic septal abscess: A report of two cases, *Ann. Otol. Rhinol. Laryngol.*, 107, 959, 1998.
53. Brook, I., Recovery of anaerobic bacteria from 3 patients with infection at a pierced body site, *Clin. Infect. Dis.*, 33, e12, 2001.
54. Moazzez, A. et al., Breast abscess bacteriologic features in the era of community-acquired methicillin resistant *Staphylococcus aureus* epidemics, *Arch. Surg.*, 142, 881, 2007.
55. Brook, I., and Frazier, E.H., The aerobic and anaerobic bacteriology of perirectal abscesses, *J. Clin. Microbiol.*, 35, 2974, 1997.
56. Fenollar, F. et al., Analysis of 525 samples to determine the usefulness of PCR amplification and sequencing of the 16S rRNA gene for diagnosis of bone and joint infections, *J. Clin. Microbiol.*, 44, 1018, 2006.
57. Rajasuo, A. et al., Bacteremia following surgical dental extraction with an emphasis on anaerobic strains, *J. Dent. Res.*, 83, 170, 2004.
58. Konstantinou, E. et al., Difficult intubation provokes bacteremia, *Surg. Infect.*, 9, 521, 2008.
59. Cone, L.A., Battista, B.A., and Shaeffer, C.W. Jr., Endocarditis due to *Peptostreptococcus anaerobius*: Case report and literature review of peptostreptococcal endocarditis, *J. Heart Valve Dis.*, 12, 411, 2003.
60. McGregor, J.A. et al., Protease production by microorganisms associated with reproductive tract infection, *Am. J. Obstet. Gynecol.*, 154, 109, 1986.
61. Pybus, V., and Onderdonk, A.B., A commensal symbiosis between *Prevotella bivia* and *Peptostreptococcus anaerobius* involves amino acids: Potential significance to the pathogenesis of bacterial vaginosis, *FEMS Immunol. Med. Microbiol.*, 22, 317, 1998.
62. Dumonceaux, T.J. et al., Multiplex detection of bacteria associated with normal microbiota and with bacterial vaginosis in vaginal swabs using oligonucleotide-coupled fluorescent microspheres, *J. Clin. Microbiol.*, 47, 4067, 2009.
63. Lassmann, B. et al., Reemergence of anaerobic bacteremia, *Clin. Infect. Dis.*, 44, 895, 2007.
64. van der Vorm, E.R. et al., Apparent culture-negative prosthetic valve endocorditis caused by *Peptostreptococcus magnus*, *J. Clin. Microbiol.*, 38, 4640, 2000.
65. Hirsh, D.C., Biberstein, E.L., and Jang, S.S., Obligate anaerobes in clinical veterinary practice, *J. Clin. Microbiol.*, 10, 188, 1979.
66. Brazier, J.S. et al., Antibiotic susceptibilities of gram-positive anaerobic cocci: Results of a sentinel study in England and Wales, *J. Antimicrob. Chemother.*, 52, 224, 2003.
67. Ednie, L., Shapiro, S., and Appelbaum, P.C., Antianaerobe activity of ceftobiprole, a new broad-spectrum cephalosporin, *Diagn. Microbiol. Infect. Dis.*, 58, 133, 2007.
68. Goldstein, E.J.C. et al., In vitro activity of ceftobiprole against aerobic and anaerobic strains isolated from diabetic foot infections, *Antimicrob. Agents Chemother.*, 50, 3959, 2006.
69. Koeth, L.M. et al., Surveillance of susceptibility patterns in 1297 European and US anaerobic and capnophilic isolates to co-amoxiclav and five other antimicrobial agents, *J. Antimicrob. Chemother.*, 53, 1039, 2004.

70. Roberts, S.A. et al., Antimicrobial susceptibility of anaerobic bacteria in New Zealand, *J. Antimicrob. Chemother.*, 57, 992, 2006.
71. Theron, M.M., Janse van Rensburg, M.N., and Chalkley, L.J., Nitroimidazole resistance genes (*nimB*) in anaerobic gram-positive cocci (previously *Peptostreptococcus* spp.), *J. Antimicrob. Chemother.*, 54, 240, 2004.
72. Goldstein, E.J.C. et al., In vitro activities of dalbavancin and nine comparator agents against anaerobic gram-positive species and corynebacteria, *Antimicrob. Agents Chemother.*, 47, 1968, 2003.
73. Liu, C.-Y. et al., Increasing trends in antimicrobial resistance among clinically important anaerobes and *Bacteroides fragilis* isolates causing nosocomial infections: Emerging resistance to carbapenems, *Antimicrob. Agents Chemother.*, 52, 3161, 2008.
74. Stoner, K.A. et al., Quantitative survival of aerobic and anaerobic microorganisms in Port-A-Cul and Copan transport systems, *J. Clin. Microbiol.*, 46, 2739, 2008.
75. Song, Y. et al., Rapid identification of gram-positive anaerobic coccal species originally classified in the genus *Peptostreptococcus* by multiplex PCR assays using genus- and species-specific primers, *Microbiology*, 149, 1719, 2003.
76. Seng, P. et al., Ongoing revolution in bacteriology: Routine identification of bacteria by matrix-assisted laser desorption ionization time-of-flight mass spectrometry, *Clin. Infect. Dis.*, 49, 543, 2009.
77. Song, Y. et al., 16S ribosomal DNA sequence-based analysis of clinically significant gram-positive cocci., *J. Clin. Microbiol.*, 41, 1363, 2003.
78. Adney, W.S., and Jones, R.L. Evaluation of the RapID-ANA system for identification of anaerobic bacteria of veterinary origin, *J. Clin. Microbiol.*, 22, 980, 1985.
79. Cavallaro, J.J., Wiggs, L.S., and Miller J.M., Evaluation of the BBL Crystal Anaerobe Identification system, *J. Clin. Microbiol.*, 35, 3186, 1997.
80. Mory, F. et al., Evaluation of the new Vitek 2 ANC card for identification of medically relevant anaerobic bacteria, *J. Clin. Microbiol.*, 47, 1923, 2009.
81. Zoetendal, E.G. et al., Isolation of DNA from bacterial samples of the human gastrointestinal tract, *Nat. Protoc.*, 1, 870, 2006.
82. Ahmed, S. et al., Mucosa-associated bacterial diversity in relation to human terminal ileum and colonic biopsy samples, *Appl. Environ. Microbiol.*, 73, 7435, 2007.
83. Cole, J.R. et al., The Ribosomal Database Project: Improved alignments and new tools for rRNA analysis, *Nucleic Acids Res.*, 37, D141, 2009.
84. Socransky, S.S. et al., "Checkerboard" DNA–DNA hybridization, *Biotechniques*, 17, 788, 1994.
85. Altschul, S.F. et al., Basic local alignment search tool, *J. Mol. Biol.*, 215, 403, 1990.

39 Tannerella

Derren Ready and Lena Ciric

CONTENTS

39.1 INTRODUCTION

39.1.1 Classification, Habitat, and Phylotypes

Classification. *Tannerella forsythia* was originally isolated in the 1970s from dental plaque collected from subjects diagnosed with progressing advanced periodontitis at the Forsyth Institute and was initially described as "fusiform *Bacteroides*."[1] This oral bacterium was primarily placed into the *Bacteroides* genus and classified as *Bacteroides forsythus*,[2] and the species name referred to the Forsyth family, who founded the Forsyth Institute. A phylogenetic analysis based on the 16S rRNA gene led to a reclassification of this bacterium, and it could no longer remain in the *Bacterioides* genus. *B. forsythus* also could not be placed in the *Porphyromonas* or *Prevotella* genera, which resulted in reclassification of this bacterium to the genus *Tannerella*, within the phylum *Bacteroidetes*. The genus name was chosen in recognition of Anne Tanner who described this microorganism in 1986[2] and this bacterium was then reclassified to *T. forsythensis*, and to ensure grammatical consistency, it was finally named *T. forsythia*.[3]

Habitat. *T. forsythia* is predominantly found in the oral cavity and has been isolated from supragingival and subgingival plaque, periodontal pockets, endodontic lesions, and oral tissues. In addition to this human oral reservoir, it has also been isolated from subgingival plaque samples from monkeys (*Macaca fascicularis*) and from infected cat and dog bites of humans, suggesting that it is likely to be found in the oral cavity of a variety of animal hosts. The small-subunit rRNA gene sequence analysis indicated that the three animal bite–wound isolates had 99.93% homology with each other and 98.9% and 99.22% homology with a human (ATCC 43037) and monkey oral strain, respectively.[4]

Phylotypes. Three distinct *Tannerella* phylotypes have been described from human oral sites, *T. forsythia*, and two phylotypes from a clone library BU045 and BU063[5]; currently, there are no cultured viable isolates of the BU045 and BU063 clones. In a study of samples collected from subjects with periodontitis and from healthy control subjects, the presence of *T. forsythia* and/or the oral clone BU063 was investigated. This study used an oral sampling approach that included every tooth and a PCR-based detection method. Workers reported that the BU063 clone was associated with oral health, whereas the presence of *T. forsythia* was strongly associated with periodontitis. In addition, *T. forsythia* and BU063 were found together less often than would be expected by chance alone.[6]

39.1.2 Clinical Features and Pathogenesis

Role for *Tannerella forsythia* in Periodontal Diseases. *T. forsythia* belongs to the "red complex" of bacteria that are considered to be a major etiological contributor to periodontitis.[7] Periodontitis is among mankind's most widespread microbially driven diseases,[8] and the patients affected by this chronic inflammatory disease experience the loss of supporting connective tissue and alveolar bone around their teeth. Additional signs and symptoms may include gingival bleeding, pain, swelling, oral malodor, gingival recession, and tooth mobility and migration. Progression of periodontitis can lead to bone destruction along the affected root surface, which may present as horizontal bone loss (destruction of the entire thickness of the alveolar bone), or vertical bone loss (only a portion of the bone is resorbed). If left untreated, periodontitis may lead to acute periodontal abscess and subsequent tooth loss.

The complexity of the microbial population harbored within oral biofilms in periodontal pockets, combined with the diversity of the clinical diseases, has made the identification of specific microbial etiological agents especially challenging. However, in 1996, three bacteria, *Tannerella forsythia*, *Aggregatibacter* (formerly *Actinobacillus*) *actinomycetecomitans*, and *Porphyromonas gingivalis*, were officially designated as etiological agents of periodontitis.[9] The inclusion of *T. forsythia* in this list resulted from a variety of sources. In 1982, *T. forsythia* was isolated from subgingival and supragingival plaque of young adults with severe periodontitis. The numbers of *T. forsythia* isolated from subgingival sites were in higher proportions than those isolated from supragingival sites.[10,11] Other studies have also demonstrated an association with the presence of *T. forsythia* and periodontal diseases,[12–15] and data from molecular-based studies have reported the detection of *T. forsythia* in 64%–88% of patients with periodontitis.[6,16–19] *T. forsythia* is not only linked to active periodontitis but has also been reported to play an important role in the progression of periodontitis, as demonstrated in a study investigating the composition of the subgingival microbiota of periodontitis subjects.[20] The use of molecular methods to characterize the microbial community present in healthy and periodontally diseased sites is beginning to alter our understanding of the etiology of periodontitis, and we may need to consider that not all *T. forsythia* strains will be equally pathogenic, and that their role in disease may be modified dependent on the presence and the absence of cultivable and other noncultivable microorganisms. In a UK-based study investigating the prevalence of *T. forsythia*, *A. actinomycetemcomitans*, and *P. gingivalis* in subjects with generalized aggressive periodontitis,[16] the presence of a single periodontal pathogen was shown to correlate with a greater number of periodontal pockets when compared to subjects that harbored two or three periodontal pathogens (including *T. forsythia*). Further work is needed in this area, but these initial findings suggest that the presence or absence of periodontal pathogens may not be the only indicator of disease risk, and that different combinations of bacteria may also affect the clinical severity of these periodontal diseases.

Virulence Factors. *T. forsythia* has been reported to harbor several potential virulence factors, including a sialidase,[21] a trypsin-like protease[22]; alpha-D-glucosidase, and *N*-acetyl-beta-glucosaminidase-D-glucosidase,[23] a hemagglutinin[24]; apoptosis-inducing activity[25]; components of the bacterial S-layer[26]; methylglyoxal production[27]; and a cell surface–associated, as well as secreted, leucine-rich repeat protein BspA. The BspA protein has been shown to bind fibronectin and fibrinogen[28] and has been reported to play a role in bacterial adherence and inflammation by triggering the release of proinflammatory cytokines from monocytes, and chemokines from osteoblasts, resulting in inflammation and bone resorption. Sharma et al.[29] demonstrated alveolar bone loss in mice infected with a wild-type strain of *T. forsythia*, whereas a BspA mutant was impaired. Invasion of epithelial cells by periodontopathogens is also considered to be an important virulence mechanism for evasion of the host defense mechanisms and in addition provides a protected microbial reservoir. In a study by Inagaki et al.,[30] epithelial cell attachment and invasion by *T. forsythia* was shown to be dependent on a cell surface–associated protein BspA. Furthermore, the presence of *P. gingivalis* or its outer membrane vesicles enhanced the attachment and invasion of *T. forsythia* to epithelial cells.

39.1.3 Diagnosis

39.1.3.1 Recovery of *Tannerella forsythia* Using Culture-Based Methods

Culture methods have the advantage of allowing for antibacterial susceptibility testing to be carried out on isolated bacteria. The culture and isolation of *T. forsythia* from a clinical sample may not be a great expense due to low consumable costs; however, the methodology is labor intensive, time consuming, and requires an experienced oral microbiologist; an anaerobic atmosphere; and a suitable culture growth medium is necessary, in addition to the sample requiring appropriate collection and transportation. Furthermore, national and international antibiotic standards rarely include details for susceptibility testing for oral bacteria.

Culture Media and Colonial Morphology. *T. forsythia* is an obligate anaerobe that can be isolated and cultured from routine clinical samples.[31–33] On conventional blood agar plates, colonies are tiny or may be seen as satellite colonies near a feeder strain, such as *Fusobacterium nucleatum* or *Staphylococcus aureus*. The poor growth on conventional bacterial culture media led to Wyss[34] determining that *T. forsythia* requires exogenous *N*-acetylmuramic acid (NAM), an acetylated amino acid sugar and structural component of peptidoglycan, as a growth factor. Thirty-three human-derived and four monkey-derived strains of *T. forsythia* were characterized by Braham and Moncla.[21] All strains developed circular, slightly convex, pale specked-pink colonies on blood agar plates after 6–7 days of anaerobic incubation of primary subgingival plaque samples. It was also noted that on subculture all strains grew in a satellite pattern near a *Staphylococcus aureus* feeder strain. Routine culture from clinical samples can be achieved on fastidious anaerobic agar supplemented with 7% defibrinated horse blood and 10 mg/L NAM,[34] if incubated in an anaerobic atmosphere. Colonies may be visible after 48–72 h incubation and are initially circular, tiny, and opaque. However, from 5 to 10 days of incubation may be required before the colonies demonstrate the characteristic pale pink morphology.

Cellular Morphology. Figure 39.1 shows a transmission electron micrograph image of *T. forsythia*. *T. forsythia* is a gram-negative rod that may appear as fusiforms, frequently with spheroids and swelling forms. *T. forsythia* has an outer layer known as the S-layer, and an inner and outer membrane. The cell morphology has been shown to alter dependent on nutrient availability. Gmur et al.[35] investigated the prevalence of *T. forsythia* in 64 subgingival plaque clinical specimens sampled from gingival or periodontal pockets with probing depths ranging from 1 to 8 mm. Samples were examined using specific monoclonal antibodies in conjunction with an indirect immunofluorescence technique. *T. forsythia* was seen in samples from pockets as shallow as 2 mm, but the percentage of positive samples increased in relation to the depth of the periodontal pocket. The morphology of these cells varied from medium to long rods or short filaments, whereas on NAM, deficient media cells appear pleomorphic.

Species Confirmation. Frequently, biochemical tests require bacteria to have been initially grown on an agar plate to achieve sufficient growth. *T. forsythia* is able to metabolize a range of substrates; it can ferment lactose (MUG test positive), hydrolyses aesculin, is bile sensitive, and can hydrolyze the trypsin-like benzyol-DL-arginine-2-naphthylamide (BANA).[21,36] The trypsin-like activity is useful because the majority of oral gram-negative anaerobic species are unable to produce this enzyme. Other oral bacteria that can be found in periodontal pockets and that can produce trypsin-like activity, including *Capnocytophaga* spp., *Porphyromonas gingivalis*, and *Treponema denticola*. *T. forsythia* has been shown to give positive results for *N*-acetylglucosaminidase, alpha-glucosidase, beta-glucosidase, alpha-fucosidase, phosphatase, alanine aminopeptidase, and sialidase. *T. forsythia* is oxidase- and catalase-negative. In addition, *T. forsythia* strains have been shown to be negative for alpha-arabinosidase, alpha-galactosidase, arginine utilization, leucine amino peptidase, praline amino

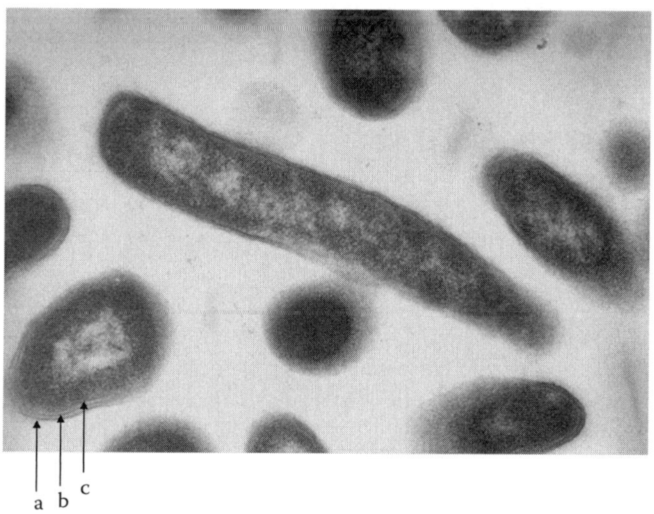

FIGURE 39.1 Transmission electron micrograph of *T. forsythia* cells. (a) Outer layer (S-layer); (b) outer membrane; and (c) inner membrane. (Image kindly provided by Dr N. Mordan at UCL Eastman Dental Institute, London, UK.)

peptidase, pyroglutamic acid arylamidase, and tyrosine aminopeptidase.[21,36] Biochemical analysis has shown that the monkey-derived strains are similar in profile to the human-derived strains. The main exception to this is indole production; human-derived strains failed to produce indole, whereas monkey-derived strains are indole positive. Interestingly the bite wound strains, unlike human and monkey strains, did not require NAM supplementation for growth.[37]

Species confirmation may be achieved by PCR-based methods, by amplifying specific regions within the bacterial small-subunit 16S rRNA gene. Primers have been designed to amplify the hypervariable regions of the 16S rRNA gene found in *T. forsythia* with the production of amplicons of known sizes. These methods have been shown to be simple to perform, rapid, reliable, and suitable for the confirmation of cultured *T. forsythia* or direct identification of this bacterium in clinical samples.[16,38,39]

39.1.3.2 Molecular Techniques

Of the variety of molecular biology tools that have been used for the detection and identification of *T. forsythia*, the majority fall into either probe-based or PCR-based methods. Molecular techniques have been used for the detection of *T. forsythia* in clinical samples for 30 years since Murray and French succeeded in detecting a number of periodontal pathogens using whole genomic probes.[40] In the following years, *T. forsythia* was detected in clinical samples using more stable recombinant DNA probes[41] and the checkerboard DNA–DNA hybridization technique was developed by Haffajee and Socransky in the early 1990s[42,43] which was capable of detecting 40 taxa simultaneously. As time has gone on, DNA microarrays targeting a number of specific *T. forsythia* genes have also been developed.[4,44] Simple-to-perform PCR methods have also been used in the detection of the pathogen providing a diagnosis in matter of hours.[17,45–47] Direct detection of *T. forsythia* by PCR using species-specific 16S rRNA gene primers was first carried out a few years after the probe methods were developed.[17] Since then many primer sets have been developed for the detection of *T. forsythia* in clinical samples.[16,48] The use of quantitative PCR (qPCR) for the detection and enumeration of *T. forsythia* in periodontal samples has also become popular in recent years.[47,49]

Today, more and more microbiological testing is being carried out using molecular techniques and a number of commercial kits have even been developed for the rapid detection of multiple pathogens in a matter of hours. Some examples include the microIDent®plus kit (Hain Diagnostics, Germany, http://www.hain-diagnostics.com/product-mid.html), ParoCheck™ (Greiner Bio-One GmbH, Germany) and the Pado Test 4.5® (IAI, Switzerland), all of which use DNA-based detection of periodontal pathogens, including *T. forsythia*. Another commercially available kit, BANA periodontal test (OraTec Corporation, USA), detects the enzymatic activity of *A. actinomycetemcomitans*, *T. forsythia*, and *P. gingivalis*.[50]

However, despite the availability of a number of nucleic acid–based techniques and kits, culture analysis of periodontal samples still remains the most frequently used method for the detection and enumeration of *T. forsythia*. One of the main pitfalls of using culture methods is that not all viable bacteria found in the periodontal pockets are cultivable and that bacterial counts do not reflect the in vivo conditions.[50] *T. forsythia* is a nutritionally fastidious organism which requires the inclusion of all its various growth-stimulation factors in a single medium and will not grow on media suitable for other members of the "red complex."[51] The species in the "red complex," *P. gingivalis*, *T. denticola*, and *T. forsythia*, are often isolated together as they stimulate each others growth in vivo.[2] Hence, in vitro, each requires its own specialized medium. The colony morphology of *T. forsythia* also varies with growth conditions; therefore, it is not always simple to identify.[2] Being an anaerobic organism, the growth times required for efficient identification of the organism are rather long. Once cultured the colonies undergo a number of functional assays to ascertain their identity.[21,52,53] Immunological tests such as bacterial concentration fluorescence immunoassays (BCFIAs), immunofluorescence microscopy, and the enzyme-linked immunosorbent assay (ELISA) have also been used to detect periodontal pathogens.[54–58] However, the use of these methods is time-consuming and has become less frequent over time due to the ease and speed of application of molecular methods.

Molecular methods, like all others, are not without fault, however. Nonspecific binding and the production of primer dimers and other artifacts during PCR can lead to a false-positive result.[59–62] There have been a number of comparative studies looking at molecular and culture methods. The great majority of molecular probe assays were found to be more sensitive than culture methods.[63–65] Comparisons of the sensitivity and specificity of PCR methods with culture methods have found some inconsistent results: although in most cases, culture-negative samples were found to be PCR-positive,[47,66–69] some studies found culture-positive samples to be PCR-negative.[70,71] It is likely that the culture-positive but PCR-negative samples were in fact detecting other closely related taxa. The sensitivity of enzyme and immunological assays used for the detection of *T. forsythia* has also been found to be lower than in DNA-based tools.[4,50] On the whole, the majority of studies have found genomic methods to be more sensitive and specific than other detection techniques, and they have the advantage of being able to detect difficult-to-culture or as yet unculturable taxa.

Comparing the probe and PCR-based methods has shown advantages and disadvantages for both. While PCR methods are quicker to execute, they can only detect a small amount of organisms at a time, whereas the more labor-intensive probe methods can now detect over 200 taxa at once. On the other hand, PCR-based methods are on the whole more sensitive than probe methods, and the detection limit of qPCR can be as low as one cell in a sample.

39.2 METHODS

39.2.1 Sample Preparation

Sample Collection and Storage. Suitable samples for the isolation of *T. forsythia* include dental plaque, tissue biopsies, aspirates, and swabs from bite wounds. Clinical specimens collected from sites within the oral cavity need to be taken carefully to avoid contamination with the indigenous oral microbiota present in supragingival plaque and saliva. The number of samples collected from each patient and the sites selected for sampling are important when aiming to provide a representative sample of the subgingival microbiota.

In the majority of previous studies, the presence of *T. forsythia* has been investigated in periodontal pockets. Prior to sampling, the patient is examined and sample sites are identified. These usually comprise the deepest pockets, over 5 mm in depth.[66,72] Other clinical measures are also taken, such as bleeding upon pocket probing, plaque index, and attachment level.[66,72] Once the pockets are identified, the supragingival plaque is removed prior to sampling the subgingival plaque to prevent cross contamination in the microbiological samples. Most studies have used one of two methods for the collection of microbiological samples from periodontal pockets: sterile paper points or Gracey's curettes. Of the two, paper points are more frequently used. While there has been some debate over the ideal method for the sampling of subgingival plaque,[73,74] a recent comparative study of the two methods found that, although the use of curettes yielded higher total bacterial counts, both methods were as effective with regard to the investigation of plaque composition when using molecular methods.[75]

The method for the use of paper points varies a little from study to study but does generally consist of the following consecutive steps. Single or multiple sterile paper points are inserted into the periodontal pocket and are either held stationary or agitated within the pocket for a short time, ranging from 5 to 15 s.[67,76–78] In the case of Gracey's curettes, the process involves the insertion of the tip of the curette to the base of the pocket. It is then pressed against the tooth surface, and the subgingival plaque is then pulled toward the orifice of the pocket.[72,79,80]

The next stage of the process is the same whichever method is used. The microbiological sample is placed into either an empty sterile tube[67] or one containing sterile saline solution,[76] TE buffer (10 mM Tris-HCl pH 8 and 1 mM EDTA pH 8)[74] or transport medium.[66] The transport media which have previously been used range from Reduced Transport Medium,[31,66,68] VGMA III Medium,[32,33,77] and BHI Transport Medium.[78,81] The collected samples are, ideally, taken to the laboratory for processing within 30 min of collection, processed immediately or stored frozen at either –20°C[66,72] or –70°C[76] for subsequent analysis.

A small number of studies have tried to detect periodontal pathogens, including *T. forsythia* in supragingival plaque and saliva as a predictive measure of periodontal disease.[19,82–84] A number of endodontic studies have also detected the oral pathogen. In these studies, No. 15 K-type file with the handle removed were used to collect a microbiological sample.[85,86] The samples were then processed with the periodontal sampling methods.

Nucleic Acid Extraction. As *T. forsythia* is a gram-negative organism, nucleic extraction is a relatively simple process, unlike with some gram-positive pathogens. This makes it an ideal candidate for molecular detection. Previous studies have used a range of methods in order to carry out DNA extraction from the microbiological samples.

The simplest method skips the DNA extraction step and uses the bacterial suspension in phosphate buffered saline directly in the molecular method of choice.[49,80] The most popular and slightly more complex method used consists of boiling the bacterial suspension at 94°C–121°C for 5–20 min.[46,67,72,76,87] A few microliters of the supernatant are then used for molecular detection.

A number of traditional DNA extraction methods have also been used consisting of a series of steps that carry out cell lysis, protein extraction, and final DNA precipitation and wash steps.[41,46,88,89] The cells are lyzed either mechanically using bead beating[49] or enzymatically using lysozyme, proteinase K, or pronase E.[41,46,88,89] Sodium dodecyl sulfate has also been used to aid cell lysis,[41,88] and some studies have added a freeze-thaw step to promote lysis.[88] For protein treatment, ammonium acetate can be used.[46] The mixture is then often extracted using phenol:chloroform:iso-amyl alcohol (12:12:1 or 25:24:1).[41,88,89] Finally, the nucleic acids are precipitated using isopopanol or 70% ethanol, and resuspended in TE buffer or water.[41,46,88,89] The resulting DNA can then be used in the molecular detection methods of choice straight away or stored.

A number of commercially available kits have also been frequently used for DNA isolation in molecular studies detecting *T. forsythia* in microbiological samples. The most frequently used are the following: MagNA Pure DNA Isolation Kit III (Roche, USA)[47,66,69]; High Pure PCR Template Preparation Kit (Boehringer),[68] Puregene DNA Isolation Kit,[38,49,90,91] QIAamp Kit (Qiagen),[39,46,85] Viogene Blood and Tissue Genomic DNA Miniprep System (Sunnyvale, USA),[83,84] and the Masterpure™ kit (Epicentre).[82]

39.2.2 Detection Procedures

39.2.2.1 Probe-Based

Probe-based systems were the first molecular methods to be applied to the investigation of periodontal pathogens. The initial techniques used whole genomic probes, which allowed for the identification of bacterial species from pure culture.[40] Occasionally, the high genomic similarity between some periodontal pathogens posed problems in the detection of species in polymicrobial samples,[9] and, consequently, the technique is still employed but less frequently.[81,92–94] As time has gone on, species-specific DNA fragments, oligonucleotides, and microarrays targeting a number of specific *T. forsythia* genes have been developed.[4,9,41,44,85,95] In this case, it is first necessary to amplify the genome by PCR using nonspecific primers before applying it to the probe.[42]

The specific details of the techniques vary from study to study, but all center on the same principle. First the probes are immobilized onto a matrix and labeled. This is followed by prehybridization, the hybridization itself, a wash step, and, finally, the detection. The matrix usually consists of a nylon membrane[9,41,95,96] or glass slides in the case of microarrays.[44] The probes can be labeled using ^{32}P,[41] Lumigen™ PPD,[9] digoxigenin,[85,95,96] and fluorophores like Cy3 and Cy5 have been used in microarrays,[44] Detection systems vary from autoradiography,[41] chemiluminescent detection,[9,95,96] or laser scanners.[44]

Genomic probes, like the ones used in the checkerboard technique,[42] use the whole genome of the organism. In the case of DNA probes, the gene used most frequently is the 16S rRNA gene, targeting the hypervariable regions,[41,44,85,95,96] One study has also used arbitrary primed PCR, which does not focus on any particular genes and produces good results in the detection of oral pathogens.[9]

Some commercial probe-based tests that include *T. forsythia* have been developed. The Human Oral Microbe Identification Microarray (HOMIM, Forsyth Institute, USA, http://mim.forsyth.org/homim.html) is now available. The system allows the detection of over 200 oral species based on 16S rRNA gene sequences using Cy3 labeled reverse-capture oligonucleotide probes printed onto a glass slide, and it has been used in a recent study comparing the taxa present during periodontitis and health.[97] ParoCheck™ (Greiner Bio-One, http://www.greinerbioone.com/en/germany/files/84226/070090_ParoCheck_LabInfo.PDF) is another microarray system that allows for the simultaneous detection of 10 or 20 periodontal pathogens. The Pado Test 4.5® (IAI, Switzerland, http://www.institut-iai.ch/Padotest.html) identifies four periodontal pathogens by direct hybridization of species-specific probes on the ribosomal RNA of the bacteria. The BD Affirm™ DP Microbial Identification Test (Becton Dickinson), a chair-side, semiautomated procedure that uses oligonucleotide DNA probes, has been used in some studies but is no longer available.[81]

39.2.2.2 PCR-Based

The use of end-point PCR and qPCR has now become a common method for the detection and quantification of periodontal pathogens due to its ease of application. These methods are relatively straightforward and give results in a matter of hours. Now qPCR is increasingly being used due to the quantitative basis of the results: rather than simply confirming the presence of a pathogen, it allows for the enumeration of the number of cells within the sample.

(i) End-Point PCR for Detection of T. forsythia Unlike the probe-based methods that consist of a number of steps carried out over a number of days, the PCR methods can be performed within a morning or afternoon. The DNA sample is added to a mix containing oligonucleotides, polymerase, buffers, and dNTPs and placed on a thermal cycler, where amplification occurs. The details of the procedure vary from study to study, but the only real variation is the choice of oligonucleotides. A large number of studies have been performed using PCR in the detection of *T. forsythia* but the majority use oligonucleotides, which were designed many years ago by the pioneers of this method, in the detection of periodontal pathogens. The oligonucleotide sets most frequently used are, as with the probe methods, based on the hypervariable regions of the 16S rRNA gene.[17,38,39,46,76,77,80,98,99] The most popular sets are those designed by Ashimoto et al. in 1996,[17] Slots et al. in 1995,[77] Conrads et al. in 1999,[46] and Tran and Rudney in 1999.[39] A number of studies have also

designed their own oligonucleotide sets.[66,87,90,100,101] More often than not, these are also based on the 16S rRNA gene, though other genes have also been used, such as the *bsp*A gene[69,100] and *waa*A gene.[102] Most PCR assays have involved the use of a single oligonucleotide set for the detection of *T. forsythia* per reaction, though some multiplex studies capable of detecting a handful of pathogens simultaneously have also been carried out.[16,39,46,103]

Procedure. The procedure developed by Ashimoto et al. in 1996[17] is one of the most commonly used end-point PCR methods in the detection of *T. forsythia* in clinical samples.[19,78,86,88] Amplification is carried out in a final volume of 50 μL, containing 5 μL 10× reaction buffer, 1.25 U of Taq DNA polymerase, and 0.2 mM of each deoxyribonucleotide. The final concentration of Mg^{2+} is 1.5 mM. Each reaction is supplemented with 1 μM of each oligonucleotide. The oligonucleotide sequences are as follows: F 5′-GCG TAT GTA ACC TGC CCG CA-3′ and R 5′-TGC TTC AGT GTC AGT TAT ACC T-3′. The resulting amplicon is 641 bp in length, covering base positions 120–760 of the *T. forsythia* 16S rRNA gene. The reaction conditions consist of an initial denaturation step of 95°C for 2 min, then 36 cycles of 95°C for 30 s, 60°C for 1 min, and 72°C for 1 min, followed by a final step of 72°C for 2 min.

(ii) Quantitative PCR for the Detection of T. forsythia
Nearly all qPCR studies have used Taqman probes to measure fluorescence, or amplification,[66,76,87,90,100,101,104] though SYBR Green or fluorescin assays have also been used as a measure.[82,88] Another variation of traditional PCR, loop-mediated isothermal amplification (or LAMP), has also been used in the detection of periodontal pathogens.[105,106] The method offers high sensitivity and enables simple visual judgment, using the naked-eye, of DNA amplification through a color change of the reaction mixture. The meridol Perio Diagnostics test (GABA International, Sweden, http://www.gaba.com/htm/568/en/meridol_Perio_Diagnostics.htm?Productgroup=MeridolPerioDiagnostics) and the microIDent®plus kit (Hain Diagnostics, Germany, http://www.hain-diagnostics.com/product-mid.html) are both commercial detection methods which use qPCR to detect and quantify a number of periodontal pathogens including *T. forsythia*.

Procedure. The protocol developed by Kuboniwa et al. 2004,[92] which describes the detection *T. forsythia* and five other periodontal pathogens by qPCR, is one of the most frequently cited. Briefly, oligonucleotides are based on the 16S rRNA gene sequences of the pathogen. The sequences are as follows: F 5′-AGC GAT GGT AGC AAT ACC TGT C-3′; R 5′-TTC GCC GGG TTA TCC CTC-3′; and P 5′-FAM-TGA GTA ACG CGT ATG TAA CCT GCC CGC-TAMRA-3′. The resulting amplicon is 88 bp in length. The final reaction volume is 20 μL, containing a final concentration of 500 nM of the forward and reverse primers, and 200 nM of the TaqMan probe. Mg^{2+} is added to a final concentration of 3 mM. The research group uses the Roche LightCycler ™ and the appropriate qPCR mix. These can be substituted according to the instrument being used; however, oligonucleotide concentrations will have to be optimized for best amplification efficiency on alternative instrument.

39.3 CONCLUSIONS AND FUTURE PERSPECTIVES

Periodontal infections are common and represent an economic burden for patient treatment. Currently, microbiological diagnostic tools are underused in these patients, and there is the potential for patients to be overtreated (unnecessary antibiotics administered) and undertreated (prolonging infection and inflammation). Molecular-based methods have already shown the potential to provide a high throughput diagnostic tool for periodontitis patients. The use of molecular methods may improve our understanding of these diseases and provide useful input into appropriate patient management.

We are now entering the era of next-generation deep-sequencing technology, where 400 Mb of sequence data can be collected in one run, 80 times the size of the average bacterial genome. As we learn more about bacterial genotypes, better primer and probe design will be possible, allowing for better sensitivity and specificity at higher throughput. The Roche Lighcycler is already capable of performing runs in less than 1 h, and new smaller microarray chips capable of detecting higher numbers of taxa as well as specific genes are in development. This will allow for the rapid detection and identification of more taxa allowing for real-time epidemiology and the monitoring of treatment success. These methods will also allow us to learn more about the community dynamics within systems. This is very important because most infectious diseases involving bacterial pathogens are polymicrobial.

The entire *Tannerella forsythia* genome has not yet been sequenced. When available, this information will allow us to compare strains and closely related taxa more thoroughly thorough phylogenetic analysis. This would allow for the documentation of the spread of strains and for associations between strains and disease to become more apparent. It is possible that rare or newly acquired antibiotic resistance and virulence genes, as well as those that have already been well characterized, will be uncovered, allowing for the selection and application of better treatment strategies.

The large amount of data collected could be entered into an interactive database, which could be used for the prediction of likelihood of successful treatment using nonsurgical therapy. Finally, the data gained about pathogens could be linked to human genetic polymorphisms in order to predict disease susceptibility.

REFERENCES

1. Tanner, A.C. et al., A study of the bacteria associated with advancing periodontitis in man, *J. Clin. Periodontol.*, 6, 278, 1979.
2. Tanner, A.C.R. et al., *Bacteroides forsythus* sp. nov, a slow-growing, fusiform *Bacteroides* sp. from the human oral cavity, *Int. J. Syst. Bacteriol.*, 36, 213, 1986.

3. Maiden, M.F., Cohee, P., and Tanner, A.C., Proposal to conserve the adjectival form of the specific epithet in the reclassification of *Bacteroides forsythus* Tanner et al. 1986 to the genus *Tannerella* Sakamoto et al. 2002 as *Tannerella forsythia* corrig., gen. nov., comb. nov. Request for an opinion, *Int. J. Syst. Evol. Microbiol.*, 53, 2111, 2003.
4. Tanner, A.C.R., and Izard, J., *Tannerella forsythia*, a periodontal pathogen entering the genomic era, *Periodontology 2000*, 42, 88, 2006.
5. Paster, B.J. et al., Bacterial diversity in human subgingival plaque, *J. Bacteriol.*, 183, 3770, 2001.
6. Leys, E.J. et al., Association of *Bacteroides forsythus* and a novel *Bacteroides* phylotype with periodontitis, *J. Clin. Microbiol.*, 40, 821, 2002.
7. Socransky, S.S. et al., Microbial complexes in subgingival plaque, *J. Clin. Periodontol.*, 25, 134, 1998.
8. Pihlstrom, B.L., Michalowicz, B.S., and Johnson, N.W., Periodontal diseases, *Lancet*, 366, 1809, 2005.
9. Guillot, E., and Mouton, C., A PCR-DNA probe assay specific for Bacteroides forsythus, *Mol. Cell. Probes*, 10, 413, 1996.
10. Moore, W.E.C. et al., Bacteriology of severe periodontitis in young-adult humans, *Infect. Immun.*, 38, 1137, 1982.
11. Tanner, A., and Bouldin, H., The microbiota of early periodontitis lesions in adults, *J. Clin. Periodontol.*, 16, 467, 1989.
12. Dzink, J. L. et al., Gram negative species associated with active destructive periodontal lesions, *J. Clin. Periodontol.*, 12, 648, 1985.
13. Dzink, J.L., Socransky, S.S., and Haffajee, A.D., The predominant cultivable microbiota of active and inactive lesions of destructive periodontal diseases, *J. Clin. Periodontol.*, 15, 316, 1988.
14. Tanner, A.C., Socransky, S.S., and Goodson, J.M., Microbiota of periodontal pockets losing crestal alveolar bone, *J. Periodont. Res.*, 19, 279, 1984.
15. Lai, C.H. et al., *Bacteroides forsythus* in adult gingivitis and periodontitis, *Oral Microbiol. Immunol.*, 2, 152, 1987.
16. Ready, D. et al., Disease severity associated with presence in subgingival plaque of *Porphyromonas gingivalis*, *Aggregatibacter actinomycetemcomitans*, and *Tannerella forsythia*, singly or in combination, as detected by nested multiplex PCR, *J. Clin. Microbiol.*, 46, 3380, 2008.
17. Ashimoto, A. et al., Polymerase chain reaction detection of 8 putative periodontal pathogens in subgingival plaque of gingivitis and advanced periodontitis lesions, *Oral Microbiol. Immunol.*, 11, 266, 1996.
18. van Winkelhoff, A.J. et al., *Porphyromonas gingivalis*, *Bacteroides forsythus* and other putative periodontal pathogens in subjects with and without periodontal destruction, *J. Clin. Periodontol.*, 29, 1023, 2002.
19. Okada, M., Hayashi, F., and Nagasaka, N., PCR detection of 5 putative periodontal pathogens in dental plaque samples from children 2 to 12 years of age, *J. Clin. Periodontol.*, 28, 576, 2001.
20. Faveri, M. et al., Microbiological profile of untreated subjects with localized aggressive periodontitis, *J. Clin. Periodontol.*, 36, 739, 2009.
21. Braham, P.H., and Moncla, B.J., Rapid presumptive identification and further characterization of *Bacteroides forsythus*, *J. Clin. Microbiol.*, 30, 649, 1992.
22. Saito, T. et al., Cloning, expression, and sequencing of a protease gene from Bacteroides forsythus ATCC 43037 in *Escherichia coli*, *Infect. Immun.*, 65, 4888, 1997.
23. Hughes, C.V. et al., Cloning and expression of alpha-D-glucosidase and N-acetyl-beta-glucosaminidase from the periodontal pathogen, *Tannerella forsythensis* (*Bacteroides forsythus*), *Oral Microbiol. Immunol.*, 18, 309, 2003.
24. Murakami, Y. et al., *Bacteroides forsythus* hemagglutinin is inhibited by N-acetylneuraminyllactose, *Oral Microbiol. Immunol.*, 17, 125, 2002.
25. Arakawa, S. et al., Novel apoptosis-inducing activity in *Bacteroides forsythus*: A comparative study with three serotypes of *Actinobacillus actinomycetemcomitans*, *Infect. Immun.*, 68, 4611, 2000.
26. Sabet, M. et al., The surface (S-) layer is a virulence factor of *Bacteroides forsythus*, *Microbiology*, 149, 3617, 2003.
27. Maiden, M. F., Pham, C., and Kashket, S., Glucose toxicity effect and accumulation of methylglyoxal by the periodontal anaerobe *Bacteroides forsythus*, *Anaerobe*, 10, 27, 2004.
28. Sharma, A. et al., Cloning, expression, and sequencing of a cell surface antigen containing a leucine-rich repeat motif from *Bacteroides forsythus* ATCC 43037, *Infect. Immun.*, 66, 5703, 1998.
29. Sharma, A. et al., *Tannerella forsythia*-induced alveolar bone loss in mice involves leucine-rich-repeat BspA protein, *J. Dent. Res.*, 84, 462, 2005.
30. Inagaki, S. et al., *Porphyromonas gingivalis* vesicles enhance attachment, and the leucine-rich repeat BspA protein is required for invasion of epithelial cells by "*Tannerella forsythia*," *Infect. Immun.*, 74, 5023, 2006.
31. Syed, S.A., and Loesche, W.J., Survival of human dental plaque flora in various transport media, *Appl. Microbiol.*, 24, 638, 1972.
32. Slots, J., Rapid identification of important periodontal microorganisms by cultivation, *Oral Microbiol. Immunol.*, 1, 48, 1986.
33. Moller, A.J., Microbiological examination of root canals and periapical tissues of human teeth. Methodological studies, *Odontol. Tidskr.*, 74, S1, 1966.
34. Wyss, C., Dependence of proliferation of *Bacteroides forsythus* on exogenous N-acetylmuramic acid, *Infect. Immun.*, 57, 1757, 1989.
35. Gmur, R., Strub, J.R., and Guggenheim, B., Prevalence of *Bacteroides forsythus* and *Bacteroides gingivalis* in subgingival plaque of prosthodontically treated patients on short recall, *J. Periodont. Res.*, 24, 113, 1989.
36. Loesche, W.J., The identification of bacteria associated with periodontal disease and dental caries by enzymatic methods, *Oral Microbiol. Immunol.*, 1, 65, 1986.
37. Hudspeth, M.K. et al., Characterization of *Bacteroides forsythus* strains from cat and dog bite wounds in humans and comparison with monkey and human oral strains, *J. Clin. Microbiol.*, 37, 2003, 1999.
38. Gafan, G.P. et al., Prevalence of periodontal pathogens in dental plaque of children, *J. Clin. Microbiol.*, 42, 4141, 2004.
39. Tran, S.D., and Rudney, J.D., Improved multiplex PCR using conserved and species-specific 16S rRNA gene primers for simultaneous detection of *Actinobacillus actinomycetemcomitans*, *Bacteroides forsythus*, and *Porphyromonas gingivalis*, *J. Clin. Microbiol.*, 37, 3504, 1999.
40. Murray, P.A., and French, C.K. DNA probe detection of periodontal pathogens. In Meyers, W.M. (Editor), *New Biotechnology in Oral Research*, pp. 33–53, Karger, Basel, 1989.
41. Dix, K. et al., Species-specific oligodeoxynucleotide probes for the identification of periodontal bacteria, *J. Clin. Microbiol.*, 28, 319, 1990.
42. Socransky, S.S. et al., Checkerboard DNA–DNA hybridization, *Biotechniques*, 17, 788, 1994.
43. Haffajee, A.D. et al., The use of DNA probes to examine the distribution of subgingival species in subjects with different levels of periodontal destruction, *J. Clin. Periodontol.*, 19, 84, 1992.

44. Huyghe, A. et al., Novel Microarray design strategy to study complex bacterial communities, *Appl. Environ. Microbiol.*, 74, 1876, 2008.
45. Umeda, M. et al., Risk indicators for harboring periodontal pathogens, *J. Periodontol.*, 69, 1111, 1998.
46. Conrads, G. et al., Simultaneous detection of *Bacteroides forsythus* and *Prevotella intermedia* by 16S rRNA gene-directed multiplex PCR, *J. Clin. Microbiol.*, 37, 1621, 1999.
47. Boutaga, K. et al., The additional value of real-time PCR in the quantitative detection of periodontal pathogens, *J. Clin. Periodontol.*, 33, 427, 2006.
48. Cortelli, J.R. et al., Etiological analysis of initial colonization of periodontal pathogens in oral cavity, *J. Clin. Microbiol.*, 46, 1322, 2008.
49. Suzuki, N. et al., Real-time TaqMan PCR for quantifying oral bacteria during biofilm formation, *J. Clin. Microbiol.*, 42, 3827, 2004.
50. Suchett-Kaye, G., Morrier, J.J., and Barsotti, O., Clinical usefulness of microbiological diagnostic tools in the management of periodontal disease, *Res. Microbiol.*, 152, 631, 2001.
51. Socransky, S.S. et al., Microbial complexes in subgingival plaque, *J. Clin. Periodontol.*, 25, 134, 1998.
52. Loesche, W.J., Kazor, C.E., and Taylor, G.W., The optimization of the BANA test as a screening instrument for gingivitis among subjects seeking dental treatment, *J. Clin. Periodontol.*, 24, 718, 1997.
53. Wikstrom, M. et al., Detection of *Porphyromonas gingivalis* in gingival exudate by a dipeptide enhanced trypsin-like activity, *J. Periodontol.*, 65, 47, 1994.
54. Sakellari, D., Menti, S., and Konstantinidis, A., Free soluble receptor activator of nuclear factor-kappa b ligand in gingival crevicular fluid correlates with distinct pathogens in periodontitis patients, *J. Clin. Periodontol.*, 35, 938, 2008.
55. Yang, H.W., Huang, Y.F., and Chou, M.Y., Occurrence of *Porphyromonas gingivalis* and *Tannerella forsythensis* in periodontally diseased and healthy subjects, *J. Periodontol.*, 75, 1077, 2004.
56. Genco, R.J. et al., Relationship of stress, distress, and inadequate coping behaviors to periodontal disease, *J. Periodontol.*, 70, 711, 1999.
57. Grossi, S.G. et al., Treatment of periodontal disease in diabetics reduces glycated hemoglobin, *J. Periodontol.*, 68, 713, 1997.
58. Ellwood, R. et al., Prevalence of suspected periodontal pathogens identified using ELISA in adolescents of differing ethnic origins, *J. Clin. Periodontol.*, 24, 141, 1997.
59. Choi, B.K. et al., Diversity of cultivable and uncultivable oral spirochetes from a patient with severe destructive periodontitis, *Infect. Immun.*, 62, 1889, 1994.
60. Ford, J.E., McHeyzerwilliams, M.G., and Lieber, M.R., Chimeric molecules created by gene amplification interfere with the analysis of somatic hypermutation of murine immunoglobulin genes, *Gene*, 142, 279, 1994.
61. Meyerhans, A., Vartanian, J.P., and Wainhobson, S., DNA recombination during PCR, *Nucleic Acids Res.*, 18, 1687, 1990.
62. Pallansch, L. et al., Use of an RNA folding algorithm to choose regions for amplification by the polymerase chain reaction, *Anal. Biochem.*, 185, 57, 1990.
63. Tsai, C.Y. et al., A rapid DNA probe test compared to culture methods for identification of subgingival plaque bacteria, *J. Clin. Periodontol.*, 30, 57, 2003.
64. Maiden, M.F.J. et al., "Checkerboard" DNA-probe analysis and anaerobic culture of initial periodontal lesions, *Clin. Infect. Dis.*, 25, S230, 1997.
65. Papapanou, P.N. et al., "Checkerboard" versus culture: a comparison between two methods for identification of subgingival microbiota, *Eur. J. Oral Sci.*, 105, 389, 1997.
66. Boutaga, K. et al., Periodontal pathogens: A quantitative comparison of anaerobic culture and real-time PCR, *Fems Immunol. Med. Microbiol.*,45, 191, 2005.
67. Lau, L. et al., Quantitative real-time polymerase chain reaction versus culture: A comparison between two methods for the detection and quantification of *Actinobacillus actinomycetemcomitans*, *Porphyromonas gingivalis* and *Tannerella forsythensis* in subgingival plaque samples, *J. Clin. Periodontol.*, 31, 1061, 2004.
68. Eick, S., and Pfister, W., Comparison of microbial cultivation and a commercial PCR-based method for detection of periodontopathogenic species in subgingival plaque samples, *J. Clin. Periodontol.*, 29, 638, 2002.
69. Boutaga, K. et al., Comparison of subgingival bacterial sampling with oral lavage for detection and quantification of periodontal pathogens by real-time polymerase chain reaction, *J. Periodontol.*, 78, 79, 2007.
70. Jervoe-Storm, P.M. et al., Comparison of culture and real-time PCR for detection and quantification of five putative periodontopathogenic bacteria in subgingival plaque samples, *J. Clin. Periodontol.*, 32, 778, 2005.
71. D'Ercole, S. et al., Comparison of culture methods and multiplex PCR for the detection of periodontopathogenic bacteria in biofilm associated with severe forms of periodontitis, *New Microbiol.*, 31, 383, 2008.
72. Darby, I.B. et al., Microbial comparison of smoker and non-smoker adult and early-onset periodontitis patients by polymerase chain reaction, *J. Clin. Periodontol.*, 27, 417, 2000.
73. Renvert, S. et al., Comparative study of subgingival microbiological sampling techniques, *J. Periodontol.*, 63, 797, 1992.
74. Kornman, K.S., Sampling od micro-organisms associated with periodontal disease, *Oral Microbiol. Immunol.*, 1, 21, 1986.
75. Jervoe-Storm, P.M. et al., Comparison of curet and paper point sampling of subgingival bacteria as analyzed by real-time polymerase chain reaction, *J. Periodontol.*, 78, 909, 2007.
76. Shelburne, C.E. et al., Quantitation of *Bacteroides forsythus* in subgingival plaque—comparison of immunoassay and quantitative polymerase chain reaction, *J. Microbiol. Methods*, 39, 97, 2000.
77. Slots, J. et al., Detection of putative periodontal pathogens in subgingival specimens by 16S ribosomal DNA amplification with the polymerase chain reaction, *Clin. Infect. Dis.*, 20, S304, 1995.
78. Takeuchi, Y. et al., Prevalence of periodontopathic bacteria in aggressive periodontitis patients in a Japanese population, *J. Periodontol.*, 74, 1460, 2003.
79. Socransky, S.S. et al., Use of checkerboard DNA–DNA hybridization to study complex microbial ecosystems, *Oral Microbiol. Immunol.*, 19, 352, 2004.
80. Meurman, J. H. et al., Identification of *Bacteroides forsythus* in subgingival dental plaque with the aid of a rapid PCR method, *J. Dental Res.*, 76, 1376, 1997.
81. Yano-Higuchi, K. et al., Prevalence of *Bacteroides forsythus*, *Porphyromonas gingivalis* and *Actinobacillus actinomycetemcomitans* in subgingival microflora of Japanese patients with adult and rapidly progressive periodontitis, *J. Clin. Periodontol.*, 27, 597, 2000.
82. Rudney, J.D., Chen, R., and Pan, Y., Endpoint quantitative PCR assays for *Bacteroides forsythus*, *Porphyromonas gingivalis*, and *Actinobacillus actinomycetemcomitans*, *J. Periodont. Res.*, 38, 465, 2003.

83. Kulekci, G. et al., Salivary detection of periodontopathic bacteria in periodontally healthy children, *Anaerobe*, 14, 49, 2008.
84. Leblebicioglu, B. et al., Salivary detection of periodontopathic bacteria and periodontal health status in dental students, *Anaerobe*, 15, 82, 2009.
85. Rocas, I.N., and Siqueira, J.F., Root canal microbiota of teeth with chronic apical periodontitis, *J. Clin. Microbiol.*, 46, 3599, 2008.
86. Siqueira, J.F. et al., Comparison of 16S rDNA-based PCR and checkerboard DNA–DNA hybridisation for detection of selected endodontic pathogens, *J. Med. Microbiol.*, 51, 1090, 2002.
87. Price, R. R. et al., Targeted profiling of oral bacteria in human saliva and in vitro biofilms with quantitative real-time PCR, *Biofouling*, 23, 203, 2007.
88. Sakamoto, M. et al., Rapid detection and quantification of five periodontopathic bacteria by real-time PCR, *Microbiol. Immunol.*, 45, 39, 2001.
89. Suzuki, N. et al., Identification of *Actinobacillus actinomycetemcomitans* serotypes by multiplex PCR, *J. Clin. Microbiol.*, 39, 2002, 2001.
90. Kuboniwa, M. et al., Quantitative detection of periodontal pathogens using real-time polymerase chain reaction with TaqMan probes, *Oral Microbiol. Immunol.*, 19, 168, 2004.
91. Nibali, L. et al., Gene polymorphisms and the prevalence of key periodontal pathogens, *J. Dental Res.*, 86, 416, 2007.
92. Persson, G. R. et al., *Tannerella forsythia* and *Pseudomonas aeruginosa* in subgingival bacterial samples from parous women, *J. Periodontol.*, 79, 508, 2008.
93. Tanner, A. et al., Microbiota of health, gingivitis, and initial periodontitis, *J. Clin. Periodontol.*, 25, 85, 1998.
94. Haffajee, A.D. et al., Microbial complexes in supragingival plaque, *Oral Microbiol. Immunol.*, 23, 196, 2008.
95. Choi, B.K. et al., Detection of major putative periodontopathogens in Korean advanced adult periodontitis patients using a nucleic acid-based approach, *J. Periodontol.*, 71, 1387, 2000.
96. Paster, B.J., Bartoszyk, I.M., and Dewhirst, F.E., Identification of oral streptococci using PCR-based, reverse-capture, checkerboard hybridization, *Methods Cell Sci.*, 20, 223, 1998.
97. Colombo, A.P.V. et al., Comparisons of subgingival microbial profiles of refractory periodontitis, severe periodontitis, and periodontal health using the human oral microbe identification microarray, *J. Periodontol.*, 80, 1421, 2009.
98. Gersdorf, H. et al., Identification of *Bacteroides forsythus* in subgingival plaque from patients with advanced periodontitis, *J. Clin. Microbiol.*, 31, 941, 1993.
99. Maeda, H. et al., Quantitative real-time PCR using TaqMan and SYBR Green for *Actinobacillus actinomycetemcomitans*, *Porphyromonas gingivalis*, *Prevotella intermedia*, tetQ gene and total bacteria, *Fems Immunol. Med. Microbiol.*, 39, 81, 2003.
100. Morillo, J.M. et al., Quantitative real-time polymerase chain reaction based on single copy gene sequence for detection of periodontal pathogens, *J. Clin. Periodontol.*, 31, 1054, 2004.
101. Suzuki, N. et al., Quantitative microbiological study of subgingival plaque by real-time PCR shows correlation between levels of Tannerella forsythensis and Fusobactetium spp., *J. Clin. Microbiol.*, 42, 2255, 2004.
102. Hyvarinen, K. et al., Detection and quantification of five major periodontal pathogens by single copy gene-based real-time PCR, *Innate Immun.*, 15, 195, 2009.
103. Morikawa, M. et al., Comparative analysis of putative periodontopathic bacteria by multiplex polymerase chain reaction, *J. Periodont. Res.*, 43, 268, 2008.
104. Sawamoto, Y. et al., Detection of periodontopathic bacteria and an oxidative stress marker in saliva from periodontitis patients, *Oral Microbiol. Immunol.*, 20, 216, 2005.
105. Yoshida, A. et al., Loop-mediated isothermal amplification method for rapid detection of the periodontopathic bacteria *Porphyromoas ginvivalis*, *Tannerella forsythia*, and *Treponema denticola*, *J. Clin. Microbiol.*, 43, 2418, 2005.
106. Miyagawa, J. et al., Rapid and simple detection of eight major periodontal pathogens by the loop-mediated isothermal amplification method, *Fems Immunol. Med. Microbiol.*, 53, 314, 2008.

40 *Veillonella*

Dongyou Liu

CONTENTS

40.1 INTRODUCTION

40.1.1 Classification, Morphology, and Biology

The genus *Veillonella* encompasses multiple species of gram-negative, anaerobic cocci belonging taxonomically to the family Veillonellaceae, order Clostridiales, class Clostridia, phylum Firmicutes. At the present, a total of 11 species are recognized within the genus, including, *Veillonella atypica*, *Veillonella caviae*, *Veillonella criceti*, *Veillonella denticariosi*, *Veillonella dispar*, *Veillonella magna*, *Veillonella montpellierensis*, *Veillonella parvula*, *Veillonella ratti*, *Veillonella rodentium*, and *Veillonella rogosae*.[1–5]

Members of the genus *Veillonella* (*Veillonella*, named after Adrien Veillon) lack flagella, spores, and capsule. Apart from *V. criceti*, *Veillonella* species do not ferment carbohydrates (to lactic acid) or amino acids, and they rely on the fermentation of lactate, pyruvate, malate, fumarate, and/or oxaloacetate as a source of carbon and energy.[6]

Veillonella spp. are present in the oral cavities, genitourinary, respiratory, and intestinal tracts of humans and animals as commensal organisms, and some of these species are opportunistic pathogens, with the ability to take advantage of host's weakened immune functions and injury (surgery) to cause infections.[7,8] Of the 11 known species, *V. atypica*, *V. denticariosi*, *V. dispar*, *V. parvula*, and *V. rogosae* have been isolated from human oral cavities only, and these oral *Veillonella* spp. are chiefly located in the tongue, dental plaque, and buccal mucosa. Together with other carbohydrate-fermenting strains of *Actinomyces*, *Streptococcus*, and *Lactobacillus* in the polymicrobial community of dental plaque, *Veillonella* spp. are known to produce volatile sulfur-compounds, which represent a cause of oral malodor. In addition, *V. montpellierensis* and a strain of the *V. ratti*– *V. criceti* group have also been reported from human patients. *V. parvula*, and other oral *Veillonella* species have been identified in intraradicular abscesses involving the apical root canal and the dentinal tubules.[9] Other *Veillonella* species are mostly found in animals.

V. parvula is a small, gram-negative, strictly anaerobic coccus of 0.3–0.5 μm in diameter that typically appears in pairs, single cells, groups of cells, or chains. It has the ability to ferment organic acids (e.g., pyruvate and lactate) into acetate, propionate, CO_2, and hydrogen as source of energy. Forming part of the commensal microbiota of the oral cavity, gastrointestinal tract, and vagina in humans, *V. parvula* has been detected in primary intraradicular infections (e.g., abscesses in the apical root canal and in the dentinal tubule). It is particularly common in periradicular lesions from teeth after failed endodontic treatment. The G + C content of *V. parvula* is 38–41 mol%.

V. rogosae (*rogosae*, named in honor of Morrison Rogosa) is gram-negative, nonmotile, nonsporulating, strictly anaerobic, and oxidase-negative coccus of 0.3–0.5 μm in diameter that occurs singly and in short chains. On *Veillonella* agar, colonies measure 2–4 mm in diameter and are of domed shape with an entire edge. *V. rogosae* is able to reduce nitrate, but unable to hydrolyze esculin or arginine. In addition, the bacterium does not generate acids from carbohydrates or exhibit extracellular glycosidic enzyme activities. *V. rogosae* is positive for pyroglutamic acid arylamidase activity and variable for alkaline phosphatase activity. It produces acetic and propionic acids as its major acid end products.[4]

V. denticariosi (*dens, dentis* tooth; *cariosus* rotten, decayed; *denticariosi*, of a decayed tooth) is a gram-negative, strictly anaerobic, nonmotile, nonspore-forming coccus of 0.4 μm in diameter (with a convoluted surface) and is typically arranged singly or in pairs, as well as in irregular masses in trypticase–glucose–yeast extract (TGY) broth. After 48 h at 37°C in an anaerobic atmosphere on *Veillonella* agar, colonies measure approximately 2 mm in diameter and appear circular, smooth, beige colored, opaque, and convex. *V. denticariosi* is catalase-negative and arginine dihydrolase positive. Also, it is able to reduce nitrates and ferment lactate, with acetate and propionate as the major metabolic end products. However, the bacterium does not produce gas.[3]

V. montpellierensis (*montpellierensis* pertaining to Montpellier in the south of France, where the type strain and two other strains were isolated) is gram-negative, strictly anaerobic, nonmotile, nonsporulating coccus (of spherical or slightly elongated shape) with a convoluted surface. *V. montpellierensis* cells measure 0.3–0.5 μm in diameter and occur singly, in pairs or in short chains. On Columbia blood agar, colonies are 1–3 mm in diameter and appear smooth, opaque, and grayish-white. *V. montpellierensis* is oxidase-negative and catalase-variable. It is able to reduce nitrate and produce gas, with acetic and propionic acids as the major metabolic end products.[2]

40.1.2 Clinical Features and Pathogenesis

Veillonella spp. are gram-negative anaerobic organisms that constitute part of the normal microflora in the oral cavity and the genitourinary, respiratory, and intestinal tracts of humans and animals.[10–13] Given their common occurrence in the tongue, dental plaque, and buccal mucosa, *Veillonella* spp. are frequently identified in the polymicrobial community involved in periodontitis.[14–19] Furthermore, a number of *Veillonella* spp. have been shown as rare causes of serious infections such as meningitis, osteomyelitis, prosthetic joint infection, pleuropulmonary infection, endocarditis, and bacteremia.[20–25] For example, *V. parvula* has been identified in cases of primary endodontic infections.[26] *V. parvula* discitis and vertebral osteomyelitis have been also documented. Risk factors for *Veillonella* infections include periodontal disease, immunodeficiency, intravenous drug use, premature birth, and possibly colonoscopy, as well as other biopsy procedures.

Veillonella spp. play a variety of roles in the polymicrobial disease. By utilizing lactate produced by carbohydrate-fermenting lactobacilli, *Streptococcus mutans*, and *Actinomyces* spp.,[6] *Veillonella* spp. reduce the concentration of lactate and increase the number of organisms. By converting lactate into the weaker acetic and propionic acids, *Veillonella* spp. modulate the acid flux and ameliorate the caries process, as the weaker acids are less capable of decalcifying enamel.

V. parvula pathogenicity is related to its lipopolysaccharide (LPS), which demonstrates an in vivo toxicity comparable to LPS from classic enteric organisms.[27] By activating the complement system, *V. parvula* LPS enhances the release of proinflammatory byproducts.[28] *V. parvula* also generates hydrogen sulfide from L-cysteine in the presence of human serum, which is toxic to host cells and contributes to its disease-inducing potential. In addition, menaquinones produced by *V. parvula* supply the specific nutritional requirement for *Porphyromonas* and *Prevotella* species. *V. parvula* also provides peptidase, which complements the activity of *Treponema denticola*, enhancing the growth of *T. denticola* in co-cultures. Thus, the presence of *V. parvula* is critical for the ecological balance of endodontic mixed consortium and for the pathogenesis of periodontal diseases.[29–31]

40.1.3 Diagnosis

Phenotypic Identification. *Veillonella* spp. are isolated on *Veillonella* agar, which is composed of Bacto tryptone (5 g/L, Difco), Bacto yeast extract (5 g/L, Difco), sodium thioglycolate (0.75 g/L, Sigma), Bacto basic fuchsin (0.002 g/L, Difco), 60% sodium lactate (21 mL/L Sigma), and Bacto agar (15 g/L Difco). The pH of the media is adjusted to 7.5 prior to autoclaving; and vancomycin (7.5 μg/mL, Sigma) is added after autoclaving.[32] The inoculated plates are incubated in an atmosphere of 90% (v/v) nitrogen, 5% (v/v) hydrogen, and 5% (v/v) carbon dioxide at 37°C for 4 days. Isolates are presumptively identified as *Veillonella* spp. on the basis of their ability to grow on the selective medium; their typical opaque or grayish-white colonies of 2–4 mm in diameter, regular, and slightly domed in shape with an entire edge; their gram-negative spherical cells of 0.3–0.5 μm under microscope; and their absence of hemolytic activity in blood agar.[2]

The resulting *Veillonella* isolates may be further examined under electron microscopy. Namely, *Veillonella* cells are fixed in Karnovsky's fixative for 1 h. The bacteria are then washed in phosphate buffer and dehydrated using graded ethanol and embedded in fresh Spurr's resin. Ultrathin sections are cut using a Reichert Ultracut E microtome and picked up on 400-mesh copper grids. The sections are stained with uranyl acetate and lead citrate and observed under a Philips CM10 electron microscope at high magnification to distinguish cellular structures. The ultrathin sections reveal structural components (outer membrane, thin peptidoglycan layer and cytoplasmic membrane) that are characteristic of a gram-negative cell wall and consistent with the genus *Veillonella*.[2]

Conventional biochemical analysis of *Veillonella* isolates is often performed with the Rapid ID 32A system (bioMérieux) using cells that are grown for 24 h on *Veillonella* agar at 37°C in an anaerobic atmosphere.[2] In addition, *Veillonella* spp. may be characterized by their interactions with polyclonal antibody and their co-aggregation abilities.[12]

Genotypic Identification. As phenotypic techniques do not allow reliable discrimination among members of the genus *Veillonella*, molecular procedures such DNA–DNA hybridization, 16S ribosomal RNA gene sequencing, and PCR have been developed for improved identification of these bacteria.[9,33–38] In contrast to its usefulness in the identification of other bacteria, 16S rRNA gene comparison has not proven as powerful for the differentiation of *Veillonella*, spp. This is due in part to the fact that *Veillonella* spp. demonstrate a high level of sequence conservation in their 16S rRNA gene. In addition, a relatively high level of intrachromosomal heterogeneity is also observed in some oral *Veillonella* isolates.[38] For instance, while *V. atypica* and *V. montpellierensis* are clearly distinguishable through 16S rRNA gene sequence comparison, it is often problematic when using this approach to try to differentiate among *V. dispar*, *V. parvula*, and *V. denticariosi*.[2,3]

For this reason, several housekeeping genes, including *dnaK*, *rpoB*, and *gyrB* have been targeted for more accurate identification of *Veillonella* spp.[39] The *dnaK* gene encoding a 70-kDa heat-shock protein shows a higher interspecies variability than the 16S rRNA gene.[38] Use of *dnaK* gene sequence comparison facilitates speciation of all known members of the *Veillonella* genus.[2] Interestingly, *rpoB* gene sequencing appears to be more discriminatory than *gyrB* gene sequencing for *Veillonella* identification.[3] In fact, by designing five individual forward primers and a shared reverse primer from a highly variable region (positions 2500–3100) of the *rpoB* gene, five oral *Veillonella* species (i.e., *Veillonella atypica*, *V. denticariosi*, *V. dispar*, *V. parvula*, and *V. rogosae*) can be readily identified via two parallel PCR.[40]

More recently, real-time PCR has been reported for rapid and convenient detection and quantitation of *Veillonella* bacteria. Using primers and probe derived from the 16S rRNA sequences, *Veillonella* and other oral bacteria can be quantitatively determined in human saliva and saliva-derived plaque through a Taqman PCR assay.[41] In addition, a real-time PCR based on the SYBR green dye has also been described for specific detection of the 16S rRNA gene in *Veillonella* spp.[31]

40.2 METHODS

40.2.1 Sample Preparation

Sample Collection. After removal of the superficial plaque above the infected dentin, plaque samples are collected from the sound occlusal surface of first molar tooth and from the buccal surface of the same tooth with the use of sterile dental excavators. The collected samples are transferred into 1 mL of Fastidious Anaerobe Broth (LabM Ltd.) and stored on ice. Alternatively, subgingival plaque samples are collected on sterile endodontic paper points (Caulk-Dentsply) following isolation and supragingival plaque removal. Plaque is collected and pooled from the mesial sulcus of tooth. Samples are placed in 1.5 mL microcentrifuge tubes and frozen until further analysis.

Microbiological Isolation. *Veillonella* spp. are isolated on *Veillonella* agar under anaerobical conditions for four days, and the number of presumptive *Veillonella* in each sample is by counting typical colonies. Presumptive *Veillonella* spp. are subcultured onto Fastidious Anaerobe Agar (LabM) supplemented with 5% (v/v) horse blood and grown anaerobically for 48 h for subsequent characterization.

DNA Extraction. Bacteria are recovered from the paper points by adding 750 μL of sampling buffer and vortexing for 1 min. The paper points are then removed, the sample is centrifuged and the supernatant discarded. The pellet is suspended in 1% sodium dodecyl sulfate (SDS) in TE, proteinase K is added, and the samples are incubated overnight. DNA is isolated on glass beads and eluted in TE.

In addition, DNA may be extracted from bacterial cells after treatment with sodium dodecyl sulfate, proteinase K, phenol/chloroform/isoamyl alcohol extraction, and ethanol precipitation. This is followed by RNAse treatment, phenol/chloroform/isoamyl alcohol extraction, and ethanol precipitation.

Alternatively, DNA may be prepared from cultured bacteria by simple alkaline lysis. One colony is suspended in 20 μL of lysis buffer (0.25% SDS, 0.05 N NaCl, and 95 mL sterile distilled water) and heated for 15 min at 95°C. After brief centrifugation, 180 μL of sterile water is added. The sample is centrifuged again for 5 min at 16,300 × *g*.

40.2.2 Detection Procedures

(i) Nested PCR for V. parvula Rôças and Siqueira[26] described a nested PCR for identification of *V. parvula*. The first PCR utilizes a pair of universal 16S rRNA gene primers: 27f (5′- AGA GTT TGA TCC TGG CTC AG -3′) and 1,492r (5′- ACG GCT ACC TTG TTA CGA CTT -3′), generating a 1466 bp fragment. The second PCR targets *V. parvula* species-specific signatures of the 16S rRNA gene with primers VPf (5′- TGA AAG GTG GCC TCT ATT TAT -3′, positions 181–201) and VPr (5′- CAA TCC TTC TAA CTG TTC GCA AG -3′, positions 450–472), producing a fragment of 292 bp.

Procedure

1. The first PCR mixture (50 μL) is made up of 20 nM each of primers 27f and 1,492r, 200 μM of deoxynucleotide triphosphates, 1 U of Taq polymerase and 5 μL of the DNA extracts from clinical samples. Positive control (genomic DNA extract of the reference strains) and negative control (sterile ultrapure water) are included in each run.
2. The cycling conditions consist of 22 cycles of 95°C for 1 min, 42°C for 2 min, and 72°C for 3 min followed by an extension at 72°C for 10 min.
3. The second PCR mixture (50 μL) contains 1 μM of each VPf and VPr primers, 5 μL of 10× PCR buffer, 2 mM $MgCl_2$, 1.25 U of *Tth* DNA polymerase (Biotools) and 0.2 mM of each deoxyribonucleoside triphosphate, and 1 μL of the first PCR product.
4. The second PCR mixture is subjected to an initial denaturation step at 95°C for 2 min, and a touchdown PCR procedure as follows: denaturing temperature of each cycle at 95°C for 30 s, annealing temperature initially set at 63°C and then lowered 0.5°C every other cycle until it reached 60°C. Additional 21 cycles are carried out at 60°C. Primer annealing is performed using this scheme for 30 s, and primer extension is carried out at 72°C for 1 min. The final extension step is at 72°C for 5 min.

5. The second round PCR products are separated by electrophoresis in a 1.5% agarose gel, stained with 0.5 μg/mL ethidium bromide and visualized under ultraviolet transillumination. A specific band of 292 bp is expected.

(ii) PCR for Veillonella spp. Igarashi et al.[40] designed five species-specific forward primers and a genus-specific reverse primer from the partial sequences of the *rpoB* genes for PCR differentiation of *Veillonella atypica*, *V. denticariosi*, *V. dispar*, *V. parvula*, and *V. rogosae*. The species-specific forward primers consist of DENF for *V. denticariosi*, PARF for *V. parvula*, ROGF for *V. rogosae*, ATYF for *V. atypica*, and DISF for *V. dispar* (Table 40.1). Together with the genus-specific reverse primer VR, PARF amplifies a 623 bp fragment from *V. parvula*; ATYF amplifies a 384 bp fragment from *V. atypica*; DISF amplifies a 321 bp fragment from *V. dispar*; ROGF amplifies a 604 bp fragment from *V. rogosae*, and DENF amplifies a 693 bp fragment from *V. denticariosi* (Table 40.1).

Due to the similarity in their product sizes, these primers are applied in two separate PCR to identify all five species of oral *Veillonella* spp. The first PCR contains four primers, PARF, ATYF, DISF, and VR, leading to the identification of *Veillonella atypical*, *V. dispar*, and *V. parvula*. The second PCR contains three primers, DENF, ROGF, and VR, leading to the discrimination of *V. denticariosi* and *V. rogosae*.

Procedure

1. PCR mixture (50 μL) is composed of 1 μL template, 2.5 μL of each primer (10 pmol/μL), 19 μL sterile water, and 25 μL master mix from a HotStarTaq Master mix kit (Qiagen).
2. The mixture is subjected to preheating at 94°C for 15 min followed by 20 cycles of 92°C for 1 min, annealing at 57°C for 1 min, and extension at 72°C for 1 min, with a final extension at 72°C for 5 min.
3. The PCR products are applied to a 1.5% agarose gel, stained with ethidium bromide, and visualized under UV light.

Note: These primers provide a rapid and reliable way to identify the recognized oral *Veillonella* species.

(iii) Real-Time PCR for Veillonella spp. Periasamy and Kolenbrander[31] developed a real-time PCR for quantitative detection of *Veillonella* spp. in biofilm, with primers from the 16S rRNA gene: forward primer 5′-CCGTGATGGGATGGAAACTGC-3′ and reverse primer 5′-CCTTCGCCACTGGTGTTCTTC-3′.

Procedure

1. DNA is extracted from biofilms by immersing biofilm-covered pegs in 40 μL of sterile ultrapure water plus 160 μL of 0.05 M sodium hydroxide. After incubation at 60°C for 60 min, 18.4 μL of 1 M Tris-HCl pH 7.0 is added to neutralize the pH. The extract is used as the template DNA for the q-PCR analyses. The bacterial genomic DNA used for standard curves is extracted from overnight cultures of *Veillonella* sp. strain PK1910, with a DNA extraction kit (Qiagen). Genomic DNA is stored at −20°C.
2. Quantification of *Veillonella* sp. in the biofilms is performed by q-PCR analysis using the SYBR green dye. The q-PCR mixture (25 μL) is made up of 3 μL of template, 3.5 μL of diethyl pyrocarbonate-treated ultrapure water, 12.5 μL of Power SYBR green PCR master mixture (Applied Biosystems), and 3 μL each of the forward and reverse primers (375 nM).
3. q-PCR is conducted in an MX3005P thermocycler (Stratagene) with the following thermal cycle recommended for the Power SYBR green PCR master mixture: 95°C for 10 min and then 40 cycles of 30 s at 95°C and 1 min at 56°C. Dissociation curves are generated by incubating reaction products at 95°C for 1 min and at 56°C for 30 s and then incrementally increasing the temperature to 95°C. Fluorescence data are collected at the end of the 56°C primer annealing step for 40 amplification cycles and throughout the dissociation curve analysis.
4. Analysis of the melting curves with both primer sets reveals a single sharp peak. DNA concentrations (ng/mL) are calculated based on standard curves obtained by using tenfold serial dilutions of bacterial DNA isolated with a DNA extraction kit (Qiagen) and quantified with the PicoGreen fluorescence assay (Invitrogen). To convert ng DNA to numbers of cells, the following weights and genome sizes were used: 3.05 fg/genome and 3.0 Mb for veillonellae.

TABLE 40.1
Identity and Sequence of *Veillonella*-Specific Primers

Specificity	Primer	Sequence (5′→3′)	PCR Product (bp)
V. denticariosi	DENF	GAAAGAAGCGCGCACCGACAG	693
V. parvula	PARF	GAAGCATTGGAAGCGAAAGTTTCG	623
V. rogosae	ROGF	ATTGCAGAAGATGTAACAGTAAGC	604
V. atypica	ATYF	TCTCTTGTTGAAGAATTAGAACGC	384
V. dispar	DISF	AACGCGTTGAAATTCGTCATGAAC	321
Veillonella spp.	VR	GTGTAACAAGGGAGTACGGACC	

(iv) Analysis of rpoB and dnaK Gene Sequences of Veillonella spp. Jumas-Bilak et al.[2] reported a pair of primers (Veill-rpoBF, 5′- GTAACAAAGGTGTCGTTTCTCG-3′ and Veill-rpoBR, 5′-CACCRTCAAATACAGGTGTAGC-3′) targeting a 608 nt region of the *rpoB* gene for identification of *Veillonella* spp. Following PCR amplification, the 608 bp fragment is sequenced by using the Big Dye Ready Reaction Termination Mix (ABI). Each sequence is aligned by CLUSTAL W in Bioedit (www.mbio.ncsu.edu/BioEdit/page2.html), and a phylogenetic tree is constructed by using DNAdist with the Jukes-Cantor model. This technique enables precise determination of all *Veillonella* spp.

In addition, Jumas-Bilak et al.[2] utilized primers Veill-dnaKF, 5′-TATGGAAGGYGGCGAACC-3′ and Veill-dnaKR, 5′-GCAGCRSTYAATGTTACATCC-3′ for PCR amplification and sequence analysis of part of *dnaK* gene and. This allows identification of a majority of *Veillonella* spp.

40.3 CONCLUSION

The genus *Veillonella* covers 11 gram-negative, anaerobic, coccal species that are commonly distributed in the oral cavities and in the genitourinary, respiratory, and intestinal tracts of humans and animals. Several *Veillonella* spp. (e.g., *V. atypica*, *V. denticariosi*, *V. dispar*, *V. parvula*, and *V. rogosae*) are colonizers of human oral cavities, and assume a vital function in the ecological balance of endodontic mixed consortium as well as the pathogenesis of periodontal diseases. Besides their well-defined role in periodontitis, *Veillonella* spp. may also cause meningitis, osteomyelitis, prosthetic joint infection, pleuropulmonary infection, endocarditis, and bacteremia occasionally. Identification of *Veillonella* spp. is possible with phenotypic techniques. However, molecular techniques targeting 16S rRNA and house keeping genes (e.g., *dnaK*, *rpoB*, and *gyrB*) provide a much more specific, rapid alternative for differentiation of *Veillonella* spp. The application of real-time PCR also permits convenient detection and quantitation of *Veillonella* spp. directly from clinical specimens, contributing to the improved epidemiological investigation and control of diseases due to *Veillonella* spp.

REFERENCES

1. Mays, T.D. et al., Taxonomy of the genus *Veillonella* Prevot, *Int. J. Syst. Bacteriol.*, 32, 28, 1982.
2. Jumas-Bilak, E. et al., *Veillonella montpellierensis* sp. nov., a novel anaerobic gram-negative coccus isolated from human clinical samples, *Int. J. Syst. Evol. Microbiol.*, 54, 1311, 2004.
3. Byun, R. et al., *Veillonella denticariosi* sp. nov., isolated from human carious dentine, *Int. J. Syst. Evol. Microbiol.*, 57, 2844, 2007.
4. Arif, N. et al., *Veillonella rogosae* sp. nov., an anaerobic gram-negative coccus isolated from dental plaque, *Int. J. Syst. Evol. Microbiol.*, 58, 581, 2008.
5. Kraatz, M., and Taras, D., *Veillonella magna* sp. nov., isolated from the jejunal mucosa of a healthy pig, and emended description of *Veillonella ratti*, *Int. J. Syst. Evol. Microbiol.*, 58, 2755, 2008.
6. Delwiche, E.A., Pestka, J.J., and Tortorello, M.L., The veillonellae: gram-negative cocci with a unique physiology, *Annu. Rev. Microbiol.*, 39, 175, 1985.
7. Aas, J.A., Bacteria of dental caries in primary and permanent teeth in children and young adults, *J. Clin. Microbiol.*, 46, 1407, 2008.
8. Arif, N. et al., Diversity of *Veillonella* spp. from sound and carious sites in children, *J. Dent. Res.*, 87, 278, 2008.
9. Marchandin, H. et al., Molecular identification of the first human isolate belonging to the *Veillonella ratti–Veillonella criceti* group based on 16S rRNA and *dnaK* gene sequencing, *Res. Microbiol.*, 156, 603, 2005.
10. Hughes, C.V. et al., Coaggregation properties of human oral *Veillonella* spp.: Relationship to colonization site and oral ecology, *Appl. Environ. Microbiol.*, 54, 1957, 1988.
11. Mager, D.L. et al., Distribution of selected bacterial species on intraoral surfaces, *J. Clin. Periodontol.*, 30, 644, 2003.
12. Palmer, R.J. Jr., Diaz, P.I., and Kolenbrander, P.E., Rapid succession within the *Veillonella* population of a developing human oral biofilm in situ, *J. Bacteriol.*, 188, 4117, 2006.
13. Aas, J.A. et al., Defining the normal bacterial flora of the oral cavity, *J. Clin. Microbiol.*, 43, 5721, 2005.
14. Wittgow, W.C. Jr., and Sabiston, C.B. Jr., Microorganisms from pulpal chambers of intact teeth with necrotic pulps, *J. Endod.*, 1, 168, 1975.
15. Sundqvist, G., Associations between microbial species in dental root canal infections, *Oral Microbiol. Immunol.*, 7, 257, 1992.
16. Marchandin, H. et al., Prosthetic joint infection due to *Veillonella dispar*, *Eur. J. Clin. Microbiol. Infect. Dis.*, 20, 340, 2001.
17. Peters, L.B. et al., Viable bacteria in root dentinal tubules of teeth with apical periodontitis, *J. Endod.*, 27, 76, 2001.
18. Faveri, M. et al., Microbiota of the dorsum of the tongue after plaque accumulation: An experimental study in humans, *J. Periodontol.*, 77, 1539, 2006.
19. Preza, D. et al., Bacterial profiles of root caries in elderly patients, *J. Clin. Microbiol.*, 46, 2015, 2008.
20. Brook, I., *Veillonella* infections in children, *J. Clin. Microbiol.*, 34, 1283–1285, 1996.
21. Brook, I., and Frazier, E.H., Infections caused by *Veillonella* species, *Infect. Dis. Clin. Pract.*, 1, 377, 1992.
22. Singh, N., and Yu, V.L., Osteomyelitis due to *Veillonella parvula*: Case report and review, *Clin. Infect. Dis.*, 14, 361, 1992.
23. Liu, J.W. et al., Two fatal cases of *Veillonella* bacteraemia, *Eur. J. Clin. Microbiol. Dis.*, 17, 62, 1998.
24. Isner-Horobeti, M. et al., *Veillonella* discitis. A case report, *Joint Bone Spine*, 73, 113, 2005.
25. Marriott, D., Stark, D., and Harkness, J., *Veillonella parvula* discitis and secondary bacteremia: A rare infection complicating endoscopy and colonoscopy? *J. Clin. Microbiol.*, 45, 672, 2007.
26. Rôças, I.N., and Siqueira, J.F. Jr., Culture-independent detection of *Eikenella corrodens* and *Veillonella parvula* in primary endodontic infections, *J. Endod.*, 32, 509, 2006.
27. Nygren, H., Dahlen, G., and Nilsson, L.A., Human complement activation by lipopolysaccharides from *Bacteroides oralis*, Fusobacterium nucleatum, and *Veillonella parvula*, *Infect. Immun.*, 26, 391, 1979.
28. Matera, G. et al., Biological effects of *Veillonella parvula* and *Bacteroides intermedius* lipopolysaccharides, *Microbiologica*, 14, 315, 1991.
29. Bhatti, M.A., and Frank, M.O., *Veillonella parvula* meningitis: Case report and review of *Veillonella* infections, *Clin. Infect. Dis.*, 31, 839, 2000.

30. Washio, J. et al., Hydrogen sulfide-producing bacteria in tongue biofilm and their relationship with oral malodour, *J. Med. Microbiol.*, 54, 889, 2005.
31. Periasamy, S., and Kolenbrander, P.E., *Aggregatibacter actinomycetemcomitans* builds mutualistic biofilm communities with *Fusobacterium nucleatum* and *Veillonella* species in saliva, *Infect. Immun.*, 77, 3542, 2009.
32. Rogosa, M. et al., Improved medium for selective isolation of *Veillonella*, *J. Bacteriol.*, 76, 455, 1958.
33. Smith, G., Socransky, S.S., and Smith, C.M., Rapid method for the purification of DNA from subgingival microorganisms, *Oral Microbiol. Immunol.*, 4, 47, 1989.
34. Sato, T. et al., PCR-restriction fragment length polymorphism analysis of genes coding for 16S rRNA in *Veillonella* spp., *Int. J. Syst. Bacteriol.*, 47, 1268, 1997.
35. Sato, T. et al., Differentiation of *Veillonella atypica*, *Veillonella dispar* and *Veillonella parvula* using restricted fragment-length polymorphism analysis of 16S rDNA amplified by polymerase chain reaction, *Oral Microbiol. Immunol.*, 12, 350, 1997.
36. Rolph, H.J. et al., Molecular identification of microorganisms from endodontic infections, *J. Clin. Microbiol.*, 39, 3282, 2001.
37. Siqueira, J.F. Jr. et al., Microbiological evaluation of acute periradicular abscesses by DNA–DNA hybridization, *Oral Surg. Oral Med. Oral Pathol. Oral Radiol. Endod.*, 92, 451, 2001.
38. Marchandin, H. et al., Intra-chromosomal heterogeneity between the four 16S rRNA gene copies in the genus *Veillonella*: Implications for phylogeny and taxonomy, *Microbiology*, 149, 1493, 2003.
39. Beighton, D. et al., The predominant cultivable *Veillonella* spp. of the tongue of healthy adults identified using rpoB sequencing, *Oral Microbiol. Immunol.*, 23, 344, 2008.
40. Igarashi, E. et al., Identification of oral species of the genus *Veillonella* by polymerase chain reaction, *Oral Microbiol. Immunol.*, 24, 310, 2009.
41. Price, R.R., Targeted profiling of oral bacteria in human saliva and in vitro biofilms with quantitative real-time PCR, *Biofouling*, 23, 203, 2007.

Section II

Firmicutes and Tenericutes

Mollicutes

41 *Mycoplasma*

Rama Chaudhry, Bishwanath Kumar Chourasia, and Anupam Das

CONTENTS

41.1 INTRODUCTION

Mycoplasmas (class Mollicutes) are the self replicating, smallest free-living bacteria in terms of both cellular dimensions and genome size that are capable of a cell-free existence.[1] They are unique among bacteria because they differ from the other common bacteria having their minute size, lack of cell wall, and ability to pass through filters with pore sizes that retain other bacteria and that were earlier thought to be viruses.[2,3] But their difference from viruses was realized by finding both RNA and DNA in them and their ability to grow in cell-free media (Table 41.1).

The smallest mycoplasmas are 0.1–0.3 μm in size. They are highly pleomorphic because they lack a rigid cell wall and instead are bounded by a triple-layered "unit membrane" that contains sterols. They require sterol as a key requirement for their growth, which is not a component of the bacterial cell or viruses. Sterols provide some structural support to the osmotically fragile mycoplasmas as they are lacking in the capacity to synthesize cell wall peptidoglycan or its precursors.[4] The lack of peptidoglycan makes them inherently resistant to beta-lactam antibiotics.[5]

41.1.1 Classification, Morphology, and Biology

Classification. Mycoplasmas belong to class Mollicutes, order Mycoplasmatales, containing five families. To date, about 200 established species have been described within the class Mollicutes, and this number is increasing day by day.[6,7]

Family *Mycoplasmatacae*: The members of this family are the ones that infect and colonize human beings. They cannot synthesize their own cholesterol and require it as their growth factor in culture medium. This family contains the most number of Mycoplasma species.[6] More than 100 of the species are divided into two genera. Members of the genus (i) *Mycoplasma* can utilize either glucose or arginine, but are unable to hydrolyze urea. Of the several species in this genus, 13 are implicated in human disease.[8,9] Members of the genus (ii) *Ureaplasma* can hydrolyze urea. These were earlier named as T-strains, as they form tiny colonies (ureaplasma colonies 15–30 μm as compared to mycoplasma colonies 200–500 μm in diameter).

Family *Acholeplasmataceae*: The members of this family can synthesize carotenol as a substitute for cholesterol, but can incorporate cholesterol if provided in the medium, though this is not an absolute requirement. These are mostly saprophytic organisms and some species had been isolated from birds.

Family *Spiroplasmataceae*: Plant mycoplasmas that require sterol for their growth.

TABLE 41.1
Characteristics of Mycoplasmas Compared to Bacteria, Chlamydiae, and Viruses

Character	Mycoplasma	Bacteria	Virus	Chlamydiae
Size	0.3 μm	1–2 μm	<0.5 μm	0.3 μm
Cell wall	Absent	Present	Absent	Present
Nucleic acid	DNA and RNA	DNA and RNA	DNA or RNA	DNA and RNA
Multiplication in cell-free medium	Yes	Yes	No	No
Susceptibility to antibiotics acting on cell	R	S	R	S

R, Resistant; S, Susceptible.

TABLE 41.2
Association of Mycoplasmas with Different Clinical Diseases

Site	Organism	Occurrence	Disease
Respiratory tract infections	*M. pneumoniae*	Common	URI, pneumonia, tracheobronchitis
	M. orale (Oropharynx)	Common	Unknown
	M. salivarium (Oropharynx)	Common	? Periodontal disease
	M. buccale	Uncommon	Unknown
	M. faucium	Uncommon	Unknown
	M. lipophilum	Rare	Unknown
Respiratory Genitourinary tract infections	*M. primatum*	Rare	Unknown
Genitourinary tract infections	*M. hominis*	Common	Pyelonephritis, Cervicitis, Vaginitis, Post partum fever
	M. genitalium	Uncommon	Urethritis
	M. penetrans	Rare	Unknown
	M. spermatophilum	Rare	Unknown
	M. fermentans	Uncommon	Unknown
	Ureaplasma	Common	Urethritis, upper genitourinary infections, prematurity
Blood stream associated	*M. fermentans* (Incognitus strain)	Uncommon	? Fulminant disseminated disease in patients with AIDS
	M. pirum	Rare	Septicemia in HIV positive patients

Family *Anaeroplasmataceae*: Strict anaerobes. No human association had been identified, isolated from intestine of cattle and birds. They require sterol for growth.

Family *Entomoplasmataceae:* Infects insects or plants.[8,9]

Mycoplasmas were earlier called pleuropneumonia-like organisms (PPLOs), because of their similarity to *Mycoplasma mycoides*, the causative organism of bovine pleuropneumonia. Later they were renamed as mycoplasma. The first mycoplasma isolated from humans was from Bartholin's gland abscess by Dienes and Edsall in 1937, which was probably *Mycoplasma hominis*. More than 200 *Mycoplasma* species have been identified in humans, animals, plants, and arthropods, and since then a few of them have been associated with human disease (*Mycoplasma pneumoniae*, *M. hominis*, *Ureaplasma urealyticum*, *Mycoplasma genitalium*, *Mycoplasma fermentans*, *Mycoplasma salivarium*, *Mycoplasma orale*, etc.).[10] Among these, *M. pneumoniae* is the most important pathogen, which is responsible for 10%–20% of the cases of community-acquired pneumonia and has been associated with acute exacerbations of asthma.[11] Some of the species, such as *M. orale* and *M. salivarium*, can occur as commensal in mucosa of human beings[8] (Table 41.2).

Morphology. Cell morphology varies with mycoplasmal species, environmental conditions, and stage of growth cycle. Pleomorphic shape varies from spherical to branching filaments.[12] On agar mycoplasmas produce colonies that have a "fried egg appearance" when viewed from above, which is due to its agar embedded center and a halo due to spreading of the colony around on the agar surface, while colonies of *M. pneumoniae* are spherical without the halo on initial isolation, coined as Mulberry appearance.[13] Mycoplasma colonies vary in size from 200 to 500 μm (Ureaplasma colonies 15–30 μm).

Mycoplasmas are characterized by their small genomes consisting of a single circular double-stranded DNA molecule in an unbounded nucleus having genome sizes ranging from 0.58 to 2.2 Mbp, with a low G + C content (23–40 mol%). *M. genitalium* have the smallest genome size of 577 kb. Owing to their small genome size, mycoplasmas have limited biosynthetic capability and have special requirements to be added in the media in which they grow (e.g., nucleic acid precursors, sterols, etc.). Though DNA homology studies have failed to demonstrate any significant relationship between mycoplasmas and any known bacteria, they are thought derive from gram-positive bacteria through reductive evolution.[14,15]

Mycoplasmas stain poorly with Gram stain and are regarded as gram negative. Special stains used for visualization of mycoplasma cells are Dienes' stain, Overnight Giemsa stain, and Acridine orange, a fluorochrome stain that stains nucleic acids. They can also be visualized by electron microscopy, dark ground microscopy, or phase contrast microscopy in fluid cultures.[9,16]

Some of the species of Mycoplasma have a tip-like projection, which is responsible for adherence to cells. Though mycoplasmas are devoid of flagella, they are usually nonmotile. The gliding motility that has been seen in some species is attributed to their tip-like projection. Mycoplasmas replicate by binary fission.

Biology. Members of the genus *Mycoplasma* are aerobic and facultative anaerobic, fastidious organisms with ability to grow in cell-free media, but with special ingredients. Mycoplasmas are grown in soft agar or in broth media with a high concentration of 10%–20% serum as a source of cholesterol and lipids, thallium acetate and Penicillin as selective agents, glucose, and phenol red as an indicator.[9,16,17]

The temperature range for growth is 22°C–41°C; the parasitic species grow optimally at around 35°C, while saprophytic species grow optimally at around 22°C. Anaeroplasma is strictly anaerobic and oxygen sensitive.[6,16,17] Though cell culture is not necessary, mammalian cell lines may be used for isolation of mycoplasmas, such as *M. pneumoniae* from clinical specimens, which may be earlier then when cell-free media is used. Though *M. pneumoniae* had been cultured in hamsters, cotton rats, and in chick embryos, animal inoculation is not used for diagnosis of mycoplasmas.[16]

Different species of mycoplasmas vary in their biochemical activity. Many of the species ferment carbohydrates, some hydrolyze arginine and some, such as ureaplasmas, hydrolyze urea. These metabolic activities of mycoplasma are utilized as an indicator of growth in liquid media containing suitable pH indicators.[9]

Mycoplasmas may be divided on the basis of biochemical activity into those that (i) obtain energy by fermentation of glucose (e.g., *M. pneumoniae*, *M. genitalium*); (ii) split arginine (e.g., *M. salivarium*, *M. orale*); (iii) hydrolyze urea (e.g., *Ureaplasma*); and (iv) few species have the capacity both to ferment glucose and split arginine (e.g., *M. fermentans*).

Both protein and glycolipid antigens have been identified in mycoplasmas. *M. pneumoniae* possesses both protein and glycolipid antigens that elicit antibody responses in infected individuals. Antibodies against these antigens are detected by various serological tests such as the complement fixation test and ELISA. P1 adhesin protein of *M. pneumoniae* has been shown to be immunogenic.[16–18]

Epidemiology and Clinical Features. Mycoplasmas are ubiquitous in the environment, and can be parasites, commensals, saprophytes, or pathogens for plants, humans, animals, and insects.[14,19] They appear to be host specific. Of the various species of mycoplasmas associated with human beings, most are considered to be a part of normal flora and have been isolated from human mucous membranes and tissues. *M. orale*, *M. salivarium*, and some other mycoplasma species are present as a normal flora of healthy adults.

Species of mycoplasma like *M. pneumoniae*, *M. hominis*, *M. genitalium*, *Ureaplasma urealyticum*, and so forth, are associated with human diseases and are present in various proportions in different groups.[8,19] Infections due to *M. pneumoniae* are endemic all over the world. Though most cases occur singly or as family outbreaks, it can occur as mini epidemics in a population with close contacts.[8]

Mycoplasmas produce a multitude of clinical disease manifestations, such as respiratory tract infections (e.g., pneumonia, tracheobronchitis, bronchiolitis, and pharyngitis) and genitourinary infections (e.g., urethritis, prostatitis, epididymitis, pyelonephritis, PID [Table 41.2], infertility, chorioamnionitis, and spontaneous abortion). Genital mycoplasmas such as ureaplasmas can also cause CLD (chronic lung disease) in preterm babies and *M. hominis* has been associated with extragenital infections in immunocompromised patients (e.g., wound infections). Apart from these various autoantibodies developed in mycoplasmal infections are responsible for various autoimmune phenomenons, like Raynaud disease.[17,20]

Genital mycoplasmas like *Ureaplasma* and *M. hominis* are frequent in healthy sexually active females, with isolation of *Ureaplasma* being up to 70% and *M. hominis* being up to 10% in this group. Higher prevalence rates had been noted in individuals presenting with sexually transmitted diseases.

Mycoplasmas are primarily extracellular parasites. Most of them exhibit flask-shaped or filamentous structures composed of a network of cytadherence proteins, known as adhesins. They attach to the surface of ciliated and nonciliated epithelial cells with the help of these adhesins (e.g., P1 adhesin protein of *M. pneumoniae* and P140 adhesin protein of *M. genitalium*). These interacts with the sialyated glycoprotein receptors at the base of the cilia on the epithelial cell surface The ciliated epithelial cells are destroyed by direct cytotoxicity of some substances liberated, such as hydrogen peroxide, or they may cause cytolysis via an inflammatory response mediated through chemotaxis of mononuclear cells or antigen antibody interaction. This loss of cilia and ciliated epithelial cells interferes with normal clearance of upper airways, and in turn the lower respiratory tract is infected with microbes leading to pneumonia.[16,17]

Immune Response. Mycoplasmal proteins can act as a superantigen, which does not require to be presented with MHC restriction and thus stimulates a large number of T cells, resulting in widespread immune activation (e.g., MAS protein of animal mycoplasma). *Mycoplasma arthritidis*, which causes acute polyarthritis in mice, has been described as a potent mitogen, which acts as a superantigen. This causes inflammatory cells to migrate to site of infection and release of various cytokines, contributing to disease manifestations. The superantigenic activity of MAS protein of *M. arthritidis* differs from other superantigens (e.g., TSST-1 and Exfoliative Toxin-A of *Staphylococcus aureus*) in its inability to induce TNF-α in peripheral mononuclear cells.[16,17]

M. pneumoniae infection leads to formation of several classes of antibodies, which either serve to neutralize the agent or act as autoantibodies. These autoantibodies can act against various tissues of human body (e.g., lung, brain, and smooth muscle), explaining various clinical manifestations due to *M. pneumoniae* infection, such as Raynaud phenomenon, chronic renal failure, Gullain Barre syndrome.[17]

In about 50% of patients with *M. pneumoniae* infection, IgM antibodies (i.e., cold agglutinins) develop. These antibodies agglutinate human erythrocytes at 4°C but not at 37°C, hence the name cold agglutinins. They appear 1–2 weeks after initial infection and are nonspecific. One theory of development of cold agglutinins is that they are the result of cross-reactive autoantibodies developed against I antigen of human erythrocytes during acute mycoplasmal infection. The other theory states that these antibodies develop directly as a result of antigenic alteration of human erythrocytes during acute mycoplasmal infection. Though these antibodies are nonspecific in nature, they can arise in patients with other atypical pneumonias, infectious mononucleosis, CMV, and lymphoma as well. They were earlier used for presumptive diagnosis of mycoplasmal infection.[17]

41.1.2 *Mycoplasma pneumoniae*

41.1.2.1 General Characteristics

Among the human mycoplasmas, *M. pneumoniae* is by far the best-known and most carefully studied one. The organism was earlier known as Eaton agent, as it was first isolated in tissue culture from the sputum of a patient with primary atypical pneumonia by Eaton et al, in 1944. Though first isolated in tissue culture, Channock et al. were successful in culturing the organism in a cell-free medium, and proposed the designation of *M. pneumoniae* in 1963.[21,22]

M. pneumoniae is one of the common causes of atypical pneumonia, a term used to describe a group of pneumonia with pronounced systemic manifestations, lesser respiratory symptoms, and interstitial pattern on chest radiography and failure to identify the causative organism by conventional techniques used (i.e., Gram stain and culture). The other causes of atypical pneumonia being *Chlamydia*, *Legionella*, *Pneumocystis carinii*, and viruses such as Influenza virus, adeno virus, respiratory syncitial virus, cytomegalo virus, and the newly described coronavirus variant that causes severe acute respiratory syndrome (SARS).[8,23]

M. pneumoniae unlike most other human mycoplasmas can grow aerobically and ferments glucose producing acid. Also it has the ability to reduce the dye tetrazolium, absorbs guinea pig and chick erythrocytes on its colonies and can lyze red blood cells, which serves in its identification as other mycoplasmas are lacking in these properties.[24]

M. pneumoniae divides by binary fission with a doubling time of more than six hours. Consequently, they grow very slow with a range of 5–20 days to grow in culture. Colonies of *M. pneumoniae* are different from those of other mycoplasmas and are lacking in outer halo and appear dense mulberry shaped in the beginning.[24]

41.1.2.2 Epidemiology and Pathogenesis

Respiratory disease caused by *M. pneumoniae* is worldwide in distribution. Though all age groups are susceptible, higher incidence has been seen in children from 5 to 20 years old. The high attack rate in children as compared to the adults may be explained by the fact that most adults have developed partial immunity due to previous exposures and second attacks are infrequent. Though *M. pneumoniae* infection is common in children, infections may occur endemically and, occasionally, epidemically in older persons.[16] The infection rate is more in populations with close contact approaching 50%–90%, but rate of pneumonitis due to the infection is lesser and variable with the range of 3%–30%. Studies have shown that incidence of pneumonia due to *M. pneumoniae* increases with age, as pneumonia due to *M. pneumoniae* is more commonly seen in elderly persons when compared to upper respiratory infections and bronchitis in younger patients, especially those less than 3 years of age. No sexual or racial predilection has been seen.[8]

Though higher incidence had been shown during the summer and fall, most surveys have shown little or no seasonal preponderance. The increase incidence in summer and fall may be due to comparative rise in cases of pneumonia due to other infectious agents during cold months.[8]

M. pneumoniae infection spreads by respiratory secretions. Transmission requires close association with the index case. *M. pneumoniae* can cause sporadic infections as well as institutional outbreaks. Since the organism tends to be associated with desquamated epithelial cells, relatively large droplets may be required for transmission by respiratory route, as evidenced by the close personal contact of outbreak settings (e.g., military barracks, schools).[16] The infection commonly spreads among family members within a household. Approximately 30% of family contacts may eventually become infected with *M. pneumoniae* from an index case. In view of the long incubation period due to slow generation time, a gap of 1–3 weeks between each case in a household is typical.[25]

In *M. pneumoniae* infections, adherence to sialoglycoprotein receptors in respiratory epithelium is the major virulence factor for the pathogenesis. This cytadherence of this organism to cells in respiratory epithelium is an essential and crucial step in tissue colonization and subsequent disease pathogenesis.[26,27] *M. pneumoniae* has a polar, tapered cell extension at one of the poles containing an electron dense core in the cytoplasm, called as tip organelle, consisting of a network of several interactive proteins and accessory proteins present on the attachment organelle, which attaches to sialoglycoprotein receptors of the tracheal epithelium.[16] It is assumed that hydrogen peroxide and free radical O_2^- ions secreted from *M. pneumoniae* tip at the time of adherence cause the cytolysis of the epithelial cells.[28,29]

41.1.2.3 Genomic Structure

Genomes of *M. pneumoniae* have been sequenced in 1996 and were shown to consist of 816,394 bp with 687 genes, which when compared was found to be one-fifth in size to the genome of *E. coli*.[30,31] *M. pneumoniae* has an exceptional position among the Mollicutes since its DNA has the highest G + C content (40 mol%), whereas the genomes of most of the other Mollicutes have a G + C content below 30 mol%. The *M. pneumoniae* adhesin genes, P1 and ORF6, and their repetitive sequences exhibit a G + C content as high as 56 mol%, while on the other extreme the origin of replication of this mycoplasma has a G + C content of only 26 mol%, compared to 40 mol% of the entire *M. pneumoniae* genome. Consequently, many of the mycoplasmal intergenic regions have a higher A + T content than the coding regions, reaching values as high as 80–90 mol%. The variable G + C content of coding regions within the mycoplasmal genome has phylogenetic relevance, indicating the highly conserved nature of the rRNA and tRNA genes and the possible exogenous origin of the adhesin genes.

M. pneumoniae uses UGA codon as tryptophan instead of a stop codon (because of presence of UGA codons the translation of mycoplasmal proteins terminates prematurely [truncation] when cloned genes are expressed in heterologous system such as *E. coli*, hence interfering in expression), requires the absence of the peptide chain release factor 2 (RF2) and the presence of the release factor 1 (RF1). The latter recognizes the stop codons UAG and UAA and RF2 recognizes the stop codons UGA and UAA. Since the UGA codon is frequently located within a gene, it is essential to exclude RF2 to prevent the premature termination of proteins.

41.1.2.4 Clinical Disease

Respiratory Tract Infection. Infections caused by *M. pneumoniae* are usually clinically apparent diseases. Respiratory infections may progress to tracheobronchitis and pneumonia. Pneumonia caused by *M. pneumoniae* is typically described as "walking pneumoniae" as the patient is not very sick usually.[16] It is characterized by more of systemic manifestations and lesser respiratory signs. *M. pneumoniae* infection of respiratory tract is characterized by fever, malaise, headache and cough, with nonproductivity being the clinical hallmark. Examination of the chest in a typical case is inconclusive as there are not many respiratory findings even cases with severe cough, though there is usually radiological evidence of pneumonia.[8] There is gradual onset of symptoms as compared to those that are caused by viruses. Though rare, *M. pneumoniae* infection is associated with acute respiratory distress syndrome (ARDS).

Extrapulmonary Manifestations. Various autoantibodies developed in *M. pneumoniae* infection, which are responsible for various autoimmune phenomenons, like Raynaud disease.

1. There may be various skin manifestations in *M. pneumoniae* infection ranging from macular, morbilliform, papulovesicular eruptions, urticaria, and erythema nodosum. These manifestations are thought to be due to immunological phenomenon, but contradicting this view there are reports of isolation of *M. pneumoniae* from these lesions.[32,33]
2. Raynaud phenomenon is a syndrome of transient reversible vasospasm of the digits on exposure to the cold, and occurs mostly in young women without any association with infection. It has been reported in patients with acute pneumonia due to *M. pneumoniae*. Although the pathophysiology of this correlation is unclear, it might be related to cold agglutinin formation in *M. pneumoniae* infection.[34–36]
3. Cardiac abnormalities are the most commonly reported extrapulmonary manifestations. The mechanism of heart damage is unclear but it manifests as arrhythmia, congestive failure, chest pain, and electrocardiographic abnormalities. Cardiac complications are responsible for prolonged illness and death.[8,37]
4. Neurological manifestations are usually reversible but with very high mortality. These manifest as aseptic meningitis, transverse myelitis, brainstem dysfunction, Gullain Barre syndrome, and peripheral neuropathy.[8,38] Patients with Gullain Barre syndrome after *M. pneumoniae* infection often have antibodies to galactocerebroside (GalC), and, in certain cases, anti-GM1 antibodies. Studies in individual aged <35 years with Gullain Barre syndrome have revealed a significant association with *M. pneumoniae* infection. In a study conducted by Gorthi S.P. et al., 2006, *M. pneumoniae* seropositivity was found in more than 50% of patients with GBS.[39]
5. *M. pneumoniae* infection is also associated with various musculoskeletal, renal, and hematological abnormalities. These are attributed to immune mechanisms. These commonly manifest as polyarthralgia, immune complex deposition of kidneys, and aplastic anaemia.[8]

41.1.2.5 Diagnosis

Culture. Specimens are inoculated immediately in PPLO broth and transported as soon as possible. On arrival at the laboratory, these are inoculated in PPLO agar and PPLO broth and incubated at 35°C–37°C in presence of 5% CO_2. The plates are examined once weekly for 1 month. The liquid media is observed once daily for color change (orange to yellow color change of phenol red–containing media due to fermentation of glucose) till 1 month.[40] If any color change is noticed, the PPLO broth is subcultured into a PPLO agar plate and observed for colonies. If no color change is seen for 1 week, the broth is blindly subcultured into a PPLO agar plate to look for colonies.

Growth Inhibition Test. It is performed by spreading mycoplasma positive culture all over the plate and drying them at 37°C. Filter paper discs containing mycoplasma antisera are placed on the seeded plates. Plates are incubated at 37°C and observed for zone of inhibition of colonies around discs.[24]

Tetrazolium Reduction Test. It is based on the ability of *M. pneumoniae* to reduce triphenyl tetrazolium chloride, which is a colorless to red colored formazan compound, while other mycoplasmas cannot. Standard mycoplasma agar is prepared by incorporating 2 mL solution of 1% 2,3,5-triphenyll tetrazolium chloride. A block of agar containing suspected *M. pneumoniae* colonies is placed colony-side down on tetrazolium plate and the plate is reincubated. In 3–6 days, the colony-containing block becomes pink or reddish if *M. pneumoniae* is present.

Hemadsorption of Sheep Erythrocytes. Culture plates with suspected colonies are flooded with 1% suspension of sheep erythrocytes in saline. After a contact period of 30 minutes, the erythrocyte suspension is aspirated and the colonies washed with saline and observed under microscope. Erythrocytes are seen to be adhered to positive colonies of *M. pneumoniae*.[24]

Serological diagnosis of *M. pneumoniae* infection had been the cornerstone, as culture is tedious and time consuming. It is recommended to test both IgG and IgM simultaneously in paired specimens collected 2–3 weeks apart for accurate diagnosis of recent or current diagnosis of *M. pneumoniae* infection. A fourfold or greater rise in antibody titer indicates recent or current infection various tests available for serology of *M. pneumoniae* are cold agglutinin assay, complement fixation test (CFT), IFA, particle agglutination assay, and ELISA.[24]

Cold Agglutinin Assay. About half of the patients infected with *M. pneumoniae* develop cold hemagglutinins to their own or group "O" erythrocytes and a smaller proportion develops agglutinins to Streptococcus MG. These heterogenetic reactions probably depend on fortuitous similarities between glycolipid haptens in the mycoplasma membrane and carbohydrate determinants of the streptococcal cell or in the "I" antigen of the erythrocyte.[41–43] However, cold agglutinin assay detects nonspecific antibodies. Testing is done by combining a patient's serum and type O erythrocytes and incubated it at 4°C for several minutes, noting the presence of hemaglutination. The highest dilution showing hemaglutination at 4°C is regarded as the agglutinin titer. A titer of 1:32 or more is highly suggestive of infection with *M. pneumoniae* infection. The clumping is reversed at 37°C. As cold hemagglutinins develop rapidly it may be of value to estimate them as a bedside test, as well as subsequently testing for complement fixing or other antibody to *M. pneumoniae*.[24,44]

Complement Fixation Test (CFT). Detects antibodies produced in response to *M. pneumoniae* infection. But false-positive reactions also occur, especially with *M. genitalium*, as their antigens are crossreacting. Recent *M. pneumoniae* infection is indicated by a titer of >32 or a fourfold rise in titer in paired serum samples.

Immunofluorescence Assay. IFA for *M. pneumoniae* detects specific antibody in serum with the help of antigen affixed to glass slides. The results are accurate, but interpretations are subjective and a fluorescence microscope is necessary.

Particle Agglutination Tests. These tests utilize latex and gelatin as a carrier particle of antigens, which, when incubated with test serum containing specific antibodies, are agglutinated, and which can be visualized with the naked eye. These tests can be performed to provide qualitative as well as quantitative results. Cut-off titer should be standardized according to the geographic location.[24]

ELISA. ELISA is one of the most widely used tests for diagnosis of *M. pneumoniae* infection. ELISA can detect very small amount of antibody, is more sensitive for detecting acute infection than culture, and can be comparable in sensitivity to polymerase chain reaction (PCR).[24]

Antigen Detection Techniques. Rapid assays available for antigen detection of *M. pneumoniae* in respiratory tract specimens are direct immunofluorescence, counterimmunoelectrophoresis, immunoblotting, and antigen capture enzyme immunoassay. These tests are of low sensitivity and have cross reactivity with other mycoplasmas present in respiratory tract.

DNA Probes. DNA hybridization techniques using 16sRNA genes were used earlier. These have same sensitivity as the antigen detection assays but are of lower sensitivity than PCR-based assays, so they are not commonly used.

PCR. PCR-based assays are rapid and more sensitive than culture of *M. pneumoniae*. PCR-based assays may detect DNA in clinical specimens before the serological tests become positive. They use ATPase gene, P1 adhesin gene, and conserved regions of 16sRNA as targets.

41.1.3 *Mycoplasma hominis*

M. hominis is commonly isolated from genitourinary tract of healthy as well as diseased persons, though their isolation from healthy, sexually active women is lesser (10%) when compared to *Ureaplasma* (66%). *M. hominis* is commonly associated with bacterial vaginosis. *M. hominis* is also associated with mild, self-limiting postpartum fevers, infection of kidneys, joints, surgical wounds, and other sites, especially in immunocompromised patients. It grows in PPLO broth containing serum and arginine with ease in 2–5 days. They cannot ferment glucose but metabolize arginine to ornithine with the production of ammonia and a consequent change in pH of media to alkaline side, which can be visualized with the help of a suitable pH indicator.

Detection of *M. hominis* in the vagina or male urethra in an asymptomatic person can be considered as a normal flora, but isolation from sites other than these is significant.

41.1.4 *Mycoplasma fermentans*

M. fermentans was first isolated by Lo et al. from the blood, organs, and Kaposi sarcoma lesions of patients with AIDS.

Earlier it was thought to be a cofactor in the development of AIDS, as suggested by some in vitro studies. Later on, it was shown to be equally associated with healthy patients in addition to those with AIDS. It has also been associated with rheumatoid arthritis. *M. fermentans* is rarely isolated from genitourinary tract.

41.1.5 Mycoplasma genitalium

M. genitalium has the smallest genome (580 kb) amongst all mycoplasmas and is closely related to *M. pneumoniae*. *M. genitalium* utilizes glucose with the production of acid but does not metabolize arginine. In contrast to other genital mycoplasmas (*M. hominis*, *Ureaplasma*), it grows very slowly and produces microscopically visible colonies. Its growth in broth is indicated by change in pH of the medium due to the acid production as a result of fermentation of glucose. As a consequent to their slow growth in conventional culture, PCR assay is a better alternative for their diagnosis.

M. genitalium is isolated from genitourinary tract, but infrequently when compared to *U. urealyticum* and *M. hominis*. *M. genitalium* is regarded as one of the common causes of nongonococcal urethritis. Studies have shown that *M. genitalium* is strongly associated with pelvic inflammatory disease, cervicitis, and endometritis.

Detection of *M. genitalium* in vagina or male urethra cannot be ignored as normal flora and implies potential disease. It should be treated with concern.

The mycoplasmas are also called as stealth pathogens or cell-wall-deficient forms. There is growing interest in understanding their role in emerging global infection.

41.2 METHODS

41.2.1 Sample Preparation

General laboratory feature are nonspecific for diagnosis of mycoplasmal infection. The patients may have leukocytosis, elevated ESR. Radiologically, the patient presents with diffuse reticular infiltrates of bronchopneumonia in perihilar region or lower lobes, usually with a unilateral distribution and with a hilar lymphadenopathy.[16]

Microbiological diagnosis of mycoplasmal infections can be done by cultural methods and by noncultural methods. Microscopy is of no use as they stain poorly as are devoid of cell wall.[9] The following types of specimens can be used: throat swabs, sputum, bronchoalveolar lavage, synovial fluid, blood, pericardial fluid, and cerebrospinal fluid.

In Vitro Isolation. The specimens for isolation of *M. pneumoniae* are throat washing, bronchial washings, or expectorated sputum; throat swabs and washings being the commonest samples as patients with *M. pneumoniae* infection produce minimal or no sputum. For isolation of genital mycoplasmas samples such as urine, vaginal/cervical swabs, urethral swabs, and swabs from placental membranes are required. The specimens should be inoculated into mycoplasma broth medium as soon as possible after collection as the organisms are very delicate and cannot withstand drying.[9]

The most common media used for isolation of mycoplasmas is PPLO broth, composed of nutrient broth supplemented with 20% horse serum to act as a source of cholesterol, Penicillin (1000 IU/mL) to inhibit the growth of bacteria, phenol red to act as an indicator to visualize change in color of the media in case of any mycoplasmal growth. The media also contains glucose or arginine or urea each at a final concentration of 0.1%.

Plates of solid PPLO agar or containers of PPLO broth are incubated in 35°C–36°C in a CO-2 incubator (5%). Growth of mycoplasmas in liquid media does not produce any turbidity due to their small size, but they produce color change on the basis of presence of glucose or arginine or urea in the medium, which can be metabolized by different mycoplasma species giving rise to change in pH and a consequent color change due to phenol red indicator.[45]

Growth of *M. pneumoniae* in liquid culture media is indicated by change in color of media from red to yellow within 1 week as a consequence to hydrolysis of glucose resulting in acidic shift. *M. hominis* metabolizes arginine, reducing ammonia, thus, the color of the media changes from yellow to pink. *U. urealyticum* breaks down urea-producing ammonia, hence color changes from yellow to pink.[45]

Growth of *Ureaplasma urealyticum* is visible within 1–2 days of *M. hominis* as evidenced by color change. *M. pneumoniae* takes longer time to grow and cultures have to be incubated for 4–6 weeks before being discarded as negative. Whenever any growth is detected in liquid media, these are subcultured in PPLO agar and incubated at 35°C–37°C in presence of 5% CO_2. The culture plates are examined there after under microscope for any growth of characteristic fried egg- (*M. hominis*, *Ureaplasma*) or mulberry-shaped (*M. pneumoniae*) colonies. The colonies can be confirmed with the help of Dienes' stain.

Phenotypic Identification. Colonies on PPLO agar can be identified with the help of Dienes' stain, which is a mixture of methylene blue 2.5 g, azure 1.25 g, maltose 10 g, sodium carbonate 0.25 g, and benzoic acid 0.20 g in 100 mL of distilled water. The final solution is diluted to 1:10. There are two methods for staining the colonies.

The stain may be dried on a cover slip, which is then applied on a block of agar bearing colonies. The portion of the plate containing suspected mycoplasma colonies might be flooded with Dienes' stain. Either of these preparations performed are observed under low-power microscopy. Colonies of mycoplasma retain the stain for at least 2 days while those of bacterial colonies lose the color within 30 min. Mycoplasmas stain an intense royal blue in the center and lighter in the periphery.

Diagnosis by noncultural methods is advantageous as these do not require a stringent procedure for maintaining viability and are also rapid. Various serological methods had been tried for detection of mycoplasmal infections. The most commonly used procedures for the detection of antibodies are the complement fixation test, particle agglutination, immunofluorescence, and ELISA. Though antibody detection had been applied for *M. pneumoniae* as well as genital mycoplasmas, antigen detection techniques had not been developed for genital mycoplasmas. Detection of *M. pneumoniae*-specific antigens directly in throat washing, swabs, and nasopharyngeal aspirates had been done with the use of an antigen capture ELISA.[24]

DNA Extraction by Phenol–Chloroform Method. A total of 1 mL of a late exponential phase culture of mycoplasma (PPLO broth) is centrifuged for 30 min at 10,000 × *g*. The pellet is suspended in 100 μL of STE buffer (10 mM NaCl, 20 mM Tris-HCl pH 8.0, 1 mM EDTA) and incubated overnight at 37°C with 1% SDS and 50 μg of proteinase K per mL. DNA is extracted three times with an equal volume of phenol saturated with TE (10 mM Tris-HCl pH 7.5, 1 mM EDTA), once with phenol-chloroform-isoamylic alcohol (25:24:1; vol/vol/vol), and once with chloroform-isoamyl alcohol (24:1; vol/vol). The aqueous phase is made to 0.3 M CH_3COONa, and the DNA is precipitated with two volume of absolute ethanol. The nucleic acid is suspended in TE buffer and is treated with DNase-free RNase A (50 μg/mL) at 37°C for 30 min. After another extraction with phenol and ethanolic precipitation, the DNA is suspended in TE buffer.[46]

DNA Extraction by Boiling Method. This is the simplest and a rapid method of DNA-preparation for PCR. 200 μL of PPLO broth in a micro centrifuge tube is centrifuge at 14,000 rpm for 30 min to pellet the cell. The cell pellet is washed twice with PBS (pH 7.2) by giving spin at 14,000 rpm for 1 min. After decanting the supernatant, pellet is resuspended in 50 Ml sterile milli-Q water. The tube is placed in a boiling water bath for 10 min and after boiling give short spin (30 s). The supernatant (contain DNA) is transferred in a fresh tube immediately.[47]

41.2.2 Detection Procedures

41.2.2.1 Standard PCR Detection

Rapid confirmation of the diagnosis of Mycoplasma is important for clinical and epidemiological reasons,[48] but culture of Mycoplasma is slow, technically demanding, and relatively insensitive.[49] Currently, diagnosis of Mycoplasma infection usually relies on serology, which also has major limitations.[49,50] Among the methods of in vitro amplification of nucleic acid, the simplest and most widely laboratory-tested method is PCR, and it has been accepted as a valuable method for diagnosis of mycoplasma infection (Figure 41.1).[49,51]

Genomic DNA (as a positive control) is best for PCR, but it is well known that mycoplasmas are fastidious organism and sometimes it becomes difficult to get genomic DNA. Cloning is a good option to clone the target gene, which is used for PCR amplification as positive control.

(i) PCR for M. pneumoniae P1 Gene All *M. pneumoniae* contain the P1 gene, which is responsible for receptor recognition, as the probable adhesion protein.[52] PCR amplification for the detection of P1 gene is suitable and a good option. The PCR reaction is performed with primer sequence Mpn P1-F 5′ CAA GCC AAA CAC GAG CTC CGG CC 3′ and Mpn P1-R 5′ CCA GTG TCA GCT GTT TGT CCT TCC CC 3′ in a reaction volume of 50 μL containing 1× PCR buffer

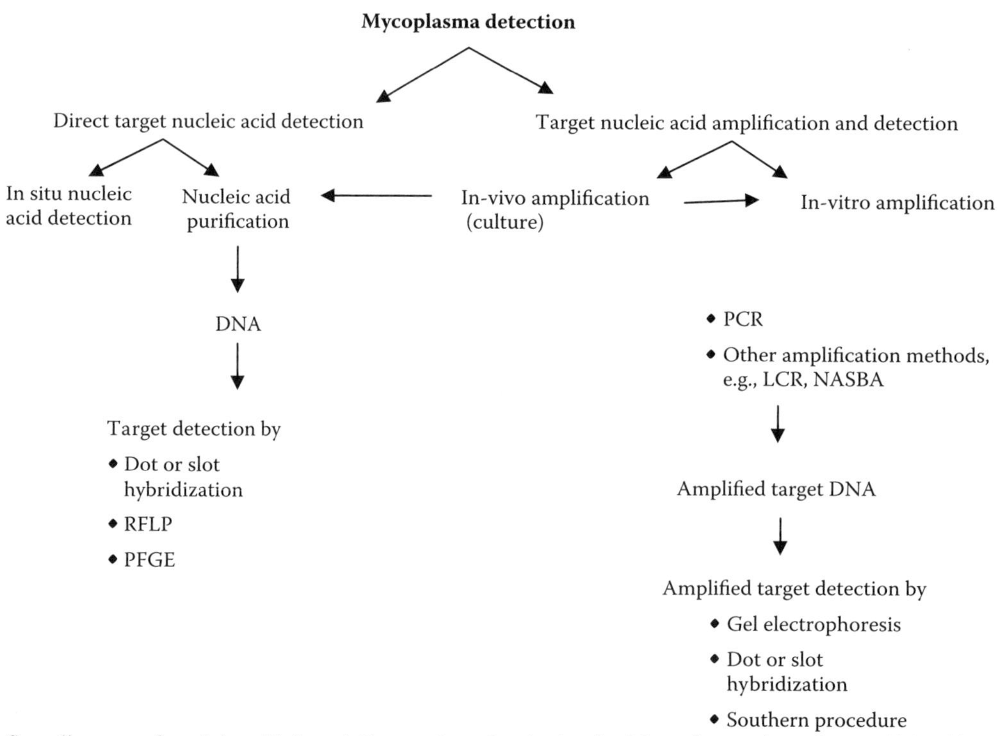

FIGURE 41.1 A flow diagram of nucleic acid–based diagnostic technologies for *Mycoplasma* detection and identification.

(100 mM Tris-HCl pH 9.0, 500 mM KCl and 1% gelatin), 200 μM dNTPs, 1.5 mM $MgCl_2$, 20 pmol of each primer, 1U Taq Polymerase, and 50–100 ng of DNA. Amplification is carried out by denaturation at 94°C for 1 min, annealing at 55°C for 1 min and extension at 72°C for 2 min for 35 cycles. A final extension step of 10 min at 72°C is done to ensure that the polymerization of every amplified fragment is completed. The size of the amplified product is 543 bp.[53]

(ii) PCR M. pneumoniae P30 Gene P30 gene offer valuable targets for molecular detection of *M. pneumoniae*.[54] Designed a pair of primers P30-PF 5′ATG AAG TTA CCA CCT CGA AGA AGC 3′; and P30-PR 5′ TTA GCG TTT TGG TGG AAA ACC GGG TTG 3′ that amplify a fragment of 825 bp.

Prepare PCR mixture (50 μL) containing 5 μL of 10× PCR buffer without $MgCl_2$, 200 μM each deoxynucleoside triphosphate, 3 μL of 25 mM $MgCl_2$, 10 pmole of each primers (P30-PF and P30-PR), 1 U of Taq DNA polymerase, and 6 μL (50 ng) of DNA template.

Conduct PCR amplification for one cycle of 94° for 10 min, 30 cycles of 94°C for 30 s, 54°C for 45 s, and 72°C for 1 min 20 s, and one cycle of 72°C for 10 min.

Portion (20 μL) of the PCR products is analyzed on a 1.5% agarose gel by electrophoresis in 0.5× Tris-borate-EDTA buffer. Electrophoresis is performed at 100 V for 1 h, then the gel is stained with a solution containing 0.5 μg ethidium bromide per mL for 30 min. DNA fragment is visualized by UV illumination at 312 nm and size of the PCR product is compared with marker.

(iii) PCR for M. hominis PCR reaction is performed with specific oligonucleotide primers. These are chosen in published sequences in the 16S rRNA gene of *M. hominis*.[55] The sequences of the oligonucleotides are 5′-CAA TGG CTA ATG CCG GAT ACG C-3′ (RNAH1; sense), 5′-GGT ACC GTC AGT CTG CAA T-3′ (RNAH2; antisense). Amplification is performed in 50 μL of reaction mixture containing 1× PCR buffer, 2.5 mM $MgCl_2$, 200 μM dNTPs, 1.25 units of Taq polymerase, 20 pmole of each primer and 50–100 ng of sample DNA. The thermal profile involved an initial denaturation step at 94°C for 3 min followed by 30 cycles of denaturation at 94°C for 1 min, primer annealing at 62°C for 1 min, and primer elongation at 72°C for 1 min. The cycling is followed by a final extension step at 72°C foe 10 min. Aliquots of amplified sample (10 μL) is analyzed by electrophoresis on a 1.5% agarose gel stained with ethidium bromide.

(iv) PCR for M. genitalium PCR amplification is carried out with the published primer sequences in the Pa gene of *M. genitalium*.[53] Primer sequences are Mge Pa primer 1: 5′-GGT GGC TCC TCC AAA CCA ACC ACC 3′ and Mge Pa primer 2: 5′-GCA ACA GTT GAT TGC GCT GCG GG 3′. Amplification is performed in 50 μL of reaction mixture containing 1× PCR buffer, 1.5 mM $MgCl_2$, 200 μM dNTPs, 2 U of Taq polymerase, 20 pmole of each primer, and 50 ng of sample DNA. The PCR cycle involved an initial denaturation step at 94°C for 3 min followed by 30 cycles of denaturation at 94°C for 1 min, primer annealing at 60°C for 1 min, and primer elongation at 72°C for 2 min. The cycling is followed by a final extension step at 72°C for 10 min. The size of the PCR product is 121 bp. Aliquots of amplified sample (10 μL) is analyzed by electrophoresis on a 1.5% agarose gel stained with ethidium bromide.

(v) PCR Detection of Mycoplasmas Contaminating Cell Culture Contamination of cell culture (primary and continuous eukaryotic cell lines) by of class Mollicutes (including *Mycoplasma* and *Acholeplasma* species) can lead to unreliable experimental results, represent major problems of economic and biological importance in basic research, diagnosis, and biotechnological production.[56–58] The common contaminants are two bovine mollicutes, *Mycoplasma arginini* and *Acholeplasma laidlawii*; two human mollicutes, *M. orale*, and *M. fermentans*; and a porcine mollicute, *Mycoplasma hyorhinis*.[59] Although contamination originates from laboratory personnel and commercial animal sera used in culture media, the main source of contamination of clean culture are mollicute-infected cultures.[60–62]

Numerous methods for detecting mollicute infection have been developed. Direct tests are based on microbiological culture, and indirect tests are based on measurement of specific markers or characteristics associated with mollicutes, including DNA fluorochrome staining, DNA probes, and enzyme-linked immunosorbent assay (ELISA), immunofluorescence, electron microscopy, autoradiography, and biochemical assays.[59,61,63,64] Although efforts have focused on the improvement of these techniques, detection of mollicutes in cell cultures remains a serious problem. Recently, the application of PCR-based methods of detection has attracted much attention because of their extreme sensitivity and specificity.

Mycoplasmic 16S rRNA sequences have been determined and provide a basis for a systematic phylogenetic analysis of mollicutes.[57] Typically, PCR targets are chosen within the gene coding for this evolutionarily conserved 16S rRNA (Tables 41.3 and 41.4). Computer alignment studies of mollicute 16S rRNA sequences have revealed the existence of regions with highly conserve sequences and regions with sequence variability at the genus and species levels, allowing the selection of genus and species-specific primers and, thus, the detection and identification of mollicutes.

DNA Extraction. One mL of cell culture suspension, bovine serum, or propagated mycoplasma culture is taken in a microcentrifuge tube and centrifuge for 10 min at 13,000 × *g*. The pellet is washed twice with 1 mL PBS by short spin and resuspended in 100 μL PBS. After incubation for 15 min at 95°C, the Wizard DNA Extraction kit (Promega) is used for DNA isolation.[65]

PCR Amplification. Amplification is carried out in a volume of 25 μL in 1× PCR buffer containing 1.5 mM $MgCl_2$, 200 μM of each dNTP, 0.33 μM of each oligonucleotide, 100 ng extracted DNA, and 1 U Taq DNA polymerase. Primers were complementary to regions of the 16S rRNA genes of mycoplasma species, which are most commonly

TABLE 41.3
Primers Used in Detection by PCR of *Mycoplasma*-Contaminated Cell Culture

Primer Sequence 5′→3′
1-P1 GTGCCAGCAGCCGCGGTAATAC
2-P2 TCTACGCATTTCACCGCTACAC
1-P3 CTTGTACACACCGCCCGTCACACCATG
2-P4 TACCTTGTTACGACTTCACCCCA
1-P3 CTTGTACACACCGCCCGTCACACCATG
2-IP3 ATCGCTAGTCCTACCTTGGG
1-P3 CTTGTACACACCGCCCGTCACACCATG
2-IP′3 GTCACCAGTCCTACCTTAGG
1MCGpF11 ACACCAAGGGAG(C/T)TGGTAAT
2-R2–1R CTCCTAGTGCCAAG(C/G)CAT(C/T)C
1-F1 ACACCATGGGAG(C/T)TGGTAAT
2-R1 CTTC(A/T)TCGACTT(C/T)CAGACCCAAGGCAT
1-Myco9 (C/T)GCCTG(A/G)GTAGTA(C/T)(A/G)(T/C)(T/A)CGC
2-Myco3 GCGGTGTGTACAA(G/A)(A/C)CCCGA
1-My 1 GCTGTGTGCCTAATACATGCAT
2-My 2 TGGTAGACAGTGAGACAATTGGAG
1-Molli 1 TACGGGAGGCAGCAGTA
2-Molli 2a TCAAGATAAAGTCATTTCCT
1-Molli 1 TACGGGAGGCAGCAGTA
2-Molli 2b TACCGTCAATTTTTAATTTTT
1-RnA5 AGAGTTTGATCCTGGCTCAGGA
2-RNA3 CGAGCTGACGACAACCATGCAC
1-GPO3 GGGAGCAAACAGGATTAGATACCCT
2-MGSO TGCACCATCTGTCACTCTGTTAACCTC

found in cell cultures (*A. laidlawii*, *M. arginini*, *M. bovis*, *M. hyorhinis*, *M. hominis*, *M. fermentans*, *M. orale*). The resulting PCR product is separated by electrophoresis in 1.3% agarose gel, stained with ethidium bromide, visualized under UV light and documented by photography. The length of the amplification product is 282 bp for the primer set GPO 3 and MGSO while length of the amplification product of rest of the primer set is between 502 and 520 bp depending on mycoplasma species.[65]

41.2.2.2 Real-Time PCR Detection

Molecular methods such as conventional PCR offer a better approach for rapid detection, but suffer from several drawbacks, such as being technically demanding and prone to carry over contamination during the manipulation of post-amplification products. These drawbacks limit the practical application of conventional PCR.

Real-time quantitative PCR method offers an attractive alternative to conventional PCR in the clinical diagnostic laboratory.[66–69] This technique is less time consuming, it has decreased risk of false-positive results, and is able to quantify the amplified product.

(i) Real-Time PCR for M. pneumoniae (P1 Gene) The primers and probe sequences are designed from the internal sequence of the 543 bp of P1 gene. The primers and probe sequences are: forward primer 5′ AAC CTC GCG CCT AAT ACT AAT ACG 3′ reverse primer 5′ TTG CGG CGT TGC TTT CAG 3′ Probe 5′ AAA GTC GAC CAA CCC C 3′. These primers amplify 73 bp segments.[70]

TABLE 41.4
Species-Specific Primers from 16S rRNA for Identification by PCR of *Mycoplasma*-Contaminated Cell Culture

Primer Sequence 5′→3′ Position[a] Specificity
ARG2[b] TCAACCAGGTGTTCTTTCCC 460–440 *M. arginini*
ACH3[b] AGCCGGACTGAGGGTCTAC 277–296 *A. laidlawii*
FER[b] AAGAAGCGTTTCTTCGCTGG 203–222 *M. fermentans*
HYO[b] GAAAGGAGCTTCACAGCTTC 198–217 *M. hyorhinis*
ORA[b] GGAGCGTTTCGTCCGCTAAG 199–218 *M. orale*
PIR[b] GTCCGTTTGGACCGCTATAG 203–222 *M. pirum*
SAL[b] GCTGCGTCAACAGTTCTCTG 849–830 *M. salivarium*
PNEU-GEN[b] CCTGCAAGGGTTCGTTATTT 204–223 *M. Pneumoniae/M. genitalium*
moli2b[c] TACCGTCAATTTTTAATTTT 451–471 *A. laidlawii*
P1[c] AAGGACCTGCAAGGGTTCGT NR *M. Pneumoniae*
P2[c] CTCTAGCCATTACCTGCTAA NR *M. Pneumoniae*
P1[d] TGAAAGGCGCGTTAAGGCGC NR *M. hominis*
P2[d] GTCTGCAATCATTTCCTATTGCAAA NR *M. hominis*
P1[e] GAAGCCTTTCTTCGCTGGAG NR *M. fermentans*
P2[e] ACAAAATCATTTCCTATTCTGTC NR *M. fermentans*
P1[f] AGCGTTTGCTTCACTTTGAA NR *M. pulmonis*
P2[f] GGGCATTTCCTCCCTAAGCT NR *M. pulmonis*
RNA2[b] TTCTATAGCTTTGCCAAG NR *M. pirum*
Phyo[b] TTCACAGCTTCACTTAAAA 207–225 *M. hyorhinis*
Pora[b] GCGTTTCGTCCGCTAAGA 202–219 *M. orale*
Pacho[b] AACACATTTAAAGATTTA 189–206 *A. laidlawii*
Psal[b] GGGCCTTTAAAGCTCCAC 200–217 *M. salivarium*
Parg[b] GCGAGGTTCTTTTGAACC 68–85 *M. arginini*
Pfer[b] TTTCTTCGCTGGAGGAGCG 206–224 *M. fermentans*

[a] IUB *E. coli* 16S rRNA portion. E
[b] Used with a genus-specific primer to perform PCR.
[c–f] Primer pairs used together to perform PCR.
NR, Not reported.

The reactions are performed in a final volume of 20 μL containing 20× assay mix (primers and probe), 2× TaqMan® Universal Master Mix (enzyme, buffer, and dNTPs), and template DNA diluted in RNase free water. Amplification and product detection are performed with the detection system (real-time PCR machine).

Amplification is carried out by initial activation of enzyme at 55°C for 2 min and denaturation at 95°C for 10 min followed by denaturation at 95°C for 15 s and annealing at 60°C for 1 min for 40 cycles.

Once the reaction for standard curve is finalized, similar reactions are performed for DNA extracted from samples.

Positive Control. In order to facilitate bacterial quantification, plasmids containing target sequence are constructed. The plasmid containing target sequence of 543 bp fragment of P1 protein gene cloned in pGEMTEasy vector is used as a positive control. Real-time PCR assay is performed to prepare the standard curve for *M. pneumoniae* positive control DNA template. The plasmid DNA is quantified spectrophotometrically. The copy number calculation is done using the following formula:

$$\text{Copy number (copies/}\mu\text{L)} = \frac{6.023 \times 1023 \text{ (copies/mol)} \times \text{conc. of std (g/}\mu\text{L)}}{\text{MW (g/mol)}}$$

where Conc. of std, obtained spectrophotometrically; MW, molecular weight of each pGEMTEasy plasmid with cloned gene.

Using this copy number concentration, a set of dilutions is prepared to make standard curve.

(ii) Real-Time PCR for M. pneumoniae (CARDS Gene) Recently identified ADP-ribosylating toxin gene encoding the CARDS (community-acquired respiratory distress syndrome) toxin is used for real-time PCR. The primers and probe sequences are designed from CARDS toxin gene. The primers and probe sequences are: forward primer 5′ TTTGGTAGCTGGTTACGGGAAT 3′ reverse primer 5′ GGTCGGCACGAATTTCATATAAG 3′ Probe 5′TGTACCAGAGCACCCCAGAAGGGCT 3′. These primers amplify 73 bp segments.[71]

The real-time PCR mixture is prepared in a total volume of 25 μL. Each PCR mixture contained the following per reaction: 12.5 μL of platinum quantitative PCR SuperMix-UDG, 1.5 μL of 50 mM MgCl2, 0.5 uM final concentrations of each primer, a 0.1 uM final concentration of the probe, 1.25 U of platinum *Taq* DNA polymerase (5 U/μL), 1 μL of 10 mM PCR nucleotide mix, 5 μL of extracted nucleic acid (100 pg to 1 fg) from each specimen, and nuclease-free water to achieve a 25 μL final volume. Real-time PCR for target gene is performed under the following conditions: initial activation of 95°C for 2 min, followed by 45 cycles of 95°C for 10 s, and 60°C for 30 s.

(iii) Real-Time PCR for M. hominis The primers and probes that are used for detection of *M. hominis* has been taken from the published nucleotide sequence of gap gene of *M. hominis* type strain PG21 (BMC MICROBIOLOGY). This gene belongs to the housekeeping genes and is therefore very conservative in all organisms. Primers and probes sequences are: forward primer 5′ GGA AGA TAT GTA ACA AAA GAA GGT GCT G 3′, reverse primer 5′ TTT ATC TTC TGG CGT AAT GAT ATC TTC G 3′. Probes I 5′ AGC AGG TGC TAA AAA GGT GTT TAT TAC TGC TCC-FL-3′, probe II 5′-LCred705-GCT AAA AGC GAA GGT GTT AAA ACA GTT GTT TAT TCA GTA-3′. One hybridization probe is labeled with fluorescein (FL) in the 3′ end, and the other with LightCycler Red705 (LCred705) in the 5′ end. When the probe is hybridized less than five nucleotides apart, fluorescence resonance energy transfer (FRET) is induced. The distance between the two probes is one nucleotide after annealing to template DNA allowing FRET light to be measured.

Real-time PCR is performed in glass capillary tubes. The reaction mixture is composed of 0.5 μM of each primer, 0.2 μM of each probe, 5 mM of $MgCl_2$ (PCR buffer), 2 μL of ready to use Fast Start DNA Master Hybridization probes (contains a hotstart *Taq* DNA polymerase and reaction mixture), 1.5 μL Uracil-DNA Glycocylase (heat labile), and DNA template. Water is added up to 20 μL. The PCR product is 144 bp in size.

Fluorescence emitted at 705 nm is measured at reach annealing step since the fluorescence signal is emitted when both probes are hybridized. The result is interpreted with LightCycler software Vers. 3.5 (Roche diagnostics). Quantification software performs all additional steps for generation of a standard curve.[72]

41.3 CONCLUSION AND FUTURE PERSPECTIVE

Mycoplasmas being devoid of cell walls are resistant to antibiotics that specifically inhibit bacterial cell wall synthesis, like penicillins and beta-lactam antibiotics. Mycoplasmas are sensitive to tetracyclins, macrolides, and the newer quinolones; which are recommended in the treatment of diseases caused by these organisms.

Antibiotic resistance in mycoplasmas may be due to two mechanisms; first they might be innately resistant to the antibiotics, which act on the cell wall (e.g., β lactam antibiotics), a property conferred to them due to their cell wall-deficient form. Second, they might become resistant to antibiotics to which they are usually considered sensitive. Tetracycline resistance had been observed in *M. hominis*, up to the extent of 30% in some areas. This resistance to tetracycline had been attributed to acquisition of streptococcal *tetM* gene. Resistance to tetracycline had also been noted in ureaplasmas to the extent of 10% in some areas.

Some strains of *M. hominis* and *ureaplasmas* are seen to be resistant to erythromycin. These resistant organisms are seen to be sensitive to clindamycin. Antibiotic susceptibility of genital mycoplasma is mandatory for effective management of these cases.

Live attenuated and killed vaccines against *M. pneumoniae* had been tried but had not given optimum results. Various antigens of *M. pneumoniae* including P1 antigen is under evaluation as a potent vaccine candidate. Antibiotic prophylaxis, especially Azithromycin, have been shown to reduce the secondary attack rate of *M. pneumoniae* infection.

As *M. pneumoniae* infection spreads by close contact; isolation of the index case can be applied for prevention of spread of infection to other persons. The future of diagnostic test for *M. pneumoniae* and genital mycoplasma is to have simultaneous detection of all respiratory pathogens/genital pathogens using microarray or chip technology.[73,74]

REFERENCES

1. Wilson, M.H. et al., Ultrastructure study of *Mycoplasma pneumoniae* in organ culture, *J. Bacteriol.*, 125, 332–339, 1976.
2. Eaton, M.D. et al., Studies on the etiology of primary atypical pneumonia. II. Properties of the virus isolated and propagated in chick embryos, *J. Exp. Med.*, 82, 317, 1945.
3. Elford, W.J. et al., Ultrafiltration methods and their application in bacteriological and pathological studies, *Br. J. Exp. Pathol.*, 10, 126, 1929.

4. Rodwell, A.W. et al., Nutrition growth and reproduction, In Barile, M. F. et al. (Editors), *The Mycoplasmas*, vol. 1, pp. 103–139, Academic Press, New York, 1979.
5. Johansson, K.E. et al., Taxonomy of mollicutes. In Razin, S. et al. (Editors), *Molecular Biology and Pathogenicity of Mycoplasma*, pp. 1–29, Kluwer Academic/Plenum Publishers, New York, 2002.
6. Razin, S. *Mycoplasma*. In Borriello, P.S. et al. (Editors), *Topley, et al. Microbiology and Microbial Infections: Bacteriology*, Vol. 2, 10th ed., pp. 1957–2005, American Society for Microbiology, Washington DC, 2005.
7. Neimark, H. et al., Proposal to transfer some members of the genera *Hemobartonella* and *eperythrozoon* to the genus *Mycoplasma* with description of "*Candidatus* Mycoplasma haemofelis," "*Candidatus* Mycoplasma haemomuris," "*Candidatus* Mycoplasma haemosuis," and "*Candidatus* Mycoplasma wenyonii," *Int. J. Syst. Evol. Microbiol.*, 51, 891–899, 2001.
8. Baum, S.G., *Mycoplasma pneumoniae* and atypical pneumoniae. In Mandel, G.L. et al. (Editors), *Mandell, Douglas and Bennett, Principles and Practice of Infectious Diseases*, vol. 2, 6th ed., Churchill Livingstone, Philadelphia, PA, 2005.
9. Waites, K.B. et al., *Mycoplasma* and *Ureaplasma*. In Murray, P.R. et al. (Editors), *Manual of Clinical Microbiology*, 9th ed., Vol. 1, pp. 1004–1020, American Society for Microbiology, Washington DC, 2007.
10. Atkinson, T.P. et al., Epidemiology, clinical manifestations, athogenesis & laboratory detection of *Mycoplasma pneumoniae* infections, *FEMS Microbiol Rev.*, 32, 956–973, 2008.
11. Gill, J.C. et al., Isolation of *Mycoplasma pneumoniae* from asthmatic patients, *Ann. Allerg.*, 70, 23–25, 1993.
12. Taylor-Robinson, D., Mycoplasmas atypical pneumonia; genital tract infection. In David Greenwood, et al. (Editors), *David, Richard, and John, Medical Microbiology A Guide to Microbial Infections: Pathogenesis, Immunity, Laboratory diagnosis and Control*, 5th ed., Churchill Livingston, Philadelphia, PA, 1997.
13. Murray, R. et al., *Mycoplasma* and *Ureaplasma*. In *Medical Microbiology*, 5th ed., Elsevier Mossy, Philadelphia, PA, 2005.
14. Razin, S. et al., Molecular biology and pathogenicity of Mycoplasmas, *Microbiol. Mol. Biol. Rev.*, 62, 1094–1156, 1998.
15. Maniloff, J., Phylogeny of mycoplasmas. In Maniloff, J. et al. (Editors), *Mycoplasmas Molecular Biology and Pathogenesis*, pp. 549–559, American Society for Microbiology, Washington, DC, 1992.
16. Waites, K.B. et al., *Mycoplasma pneumoniae* and its role as a human pathogen, *Clin. Microbiol. Rev.*, 17(4), 697–728, 2004.
17. Baum, S.G., Introduction to mycoplasma diseases. In Mandel, G.L. et al. (Editors), *Mandell, Douglas and Bennett, Principles and Practice of Infectious Diseases*, vol. 2, 6th ed., Churchill Livingstone, Philadelphia, PA, 2005.
18. Chaudhry, R. et al., Expression and immunological characterization of the carboxy-Terminal Region of the P1 Adhesin Protein of *Mycoplasma pneumoniae*, *J. Clin. Microbiol.*, 43, 321–325, 2005.
19. Blanchard, A., *Ureaplasma urealyticum* urease genes; use of a UGA tryptophan codon, *Mol. Microbiol.*, 4, 669–676, 1990.
20. Taylor-Robinson, D., Infections due to species of Mycoplasma and Ureaplasma: An update, *Clin. Infect. Dis.*, 23, 671–684, 1996.
21. Chanock, R.M. et al., *Mycoplasma pneumoniae*: Proposed nomenclature for atypical pneumonia organism (Eaton agent), *Science*, 140, 662, 1963.
22. Chanock, R.M. et al., Growth on artificial medium of an agent associated with atypical pneumonia and its identification as a PPLO, *Proc. Natl. Acad. Sci. USA.*, 48, 41–49, 1962.
23. Drosten, C. et al., Identification of a novel coronavirus in patients with severe acute respiratory syndrome, *N. Engl. J. Med.*, 348, 1967–1976, 2003.
24. Marmion, B.P. et al., *Mycoplasma pneumoniae* and other medically important members of the family mycoplasmataceae. In Collee, J.G. et al. (Editors), *Mackie & McCartney, Practical Medical Microbiology*, 14th ed., Churchill Livingstone, Philadelphia, PA, 1996.
25. Foy, H.M. et al., Epidemiology of *Mycoplasma pneumoniae* infection in families, *JAMA*, 197, 859, 1966.
26. Zhang, Q., Glycolipid receptors for attachment of *Mycoplasma hypopneumoniae* to porcine respiratory ciliated cells, *Infect. Immun.*, 62, 4367–4373, 1994.
27. Razin, S. et al., Mycoplasma adhesion, *J. Gen. Microbiol.*, 138, 407–422, 1992.
28. Almagor, M. et al., Protective effects of glutathione redox cycle and vitamin E on cultured fibroblasts infected by *Mycoplasma pneumoniae*, *Infect Immun.*, 52, 240–244, 1986.
29. Tryon, V.V. et al., Pathogenic determinants and mechanisms. In Maniloff, J. (Editor), *Mycoplasmas: Molecular Biology and Pathogenesis*, pp. 457–471, American Society for Microbiology, Washington, DC, 1992.
30. Himmelreich, R. et al., Complete sequence analysis of the genome of the bacterium *Mycoplasma pneumoniae*, *Nucleic Acids Res.*, 24, 4420–4449, 1996.
31. Rudd, K.E., Ecogene: A genome sequence database for *Escherichia coli* K-12, *Nucleic Acid Res.*, 28, 60–64, 2000.
32. Cherry, J.D. et al., *Mycoplasma pneumoniae* infections and Exanthems, *J. Pediatr.*, 87, 369, 1975.
33. Murray, H.W. et al., The protean manifestations of *Mycoplasma pneumoniae* in adults, *Am. J Med.*, 58, 229, 1975.
34. Furioli, J. et al., *Mycoplasma pneumoniae* infection: Manifestation in a 3 years old child by Raynaud's phenomenon, *Arch. Fr. Pediatr.*, 42, 313, 1985.
35. Pandey, A. et al., Acute lower respiratory tract infections in Indian children with special reference to *Mycoplasma pneumoniae*, *Trop. Pediatr.*, 46, 371–374, 2000.
36. Schubothe, H., The cold hemagglutinin disease, *Semin haematol.*, 3, 27, 1966.
37. Goyal, P. et al., Association of common chronic infections with coronary artery disease in patients without any conventional risk factors, *Indian J. Med. Res.*, 125, 129–136, 2007.
38. Steele, J.C. et al., *Mycoplasma pneumoniae* as a determitant of the Guillain–Barre syndrome, *Lancet*, 2, 719, 1969.
39. Gorthi, S.P. et al., Guillain Barre Syndrome association with Campylobacter jejuni and *Mycoplasma pneumoniae* infection in India, *Nat. Med. J. India.*, 19, 137–139, 2006.
40. Daxboeck, F. et al., Laboratory diagnosis of *Mycoplasma pneumoniae* infection, *Clin. Microbiol. Infect.*, 9, 263–273, 2003.
41. Costea, N. et al., Inhibition of cold agglutinins (anti one) by *Mucoplasma pneumoniae* antigens, *Proc. Soc. Exp. Biol.*, 134, 476–479, 1972.
42. Marmion, B.P. et al., Immunochemichal analysis of *Mycoplasma pneumoniae*. 1. Methods of extraction and reactions of fractions from *M. pneumoniae* and from *M. mycoides* with homologous antisera and antisera against streptococcus, *M. G. Aust. J. Exp. Biol. Med. Sci.*, 45, 163–187, 1967.
43. Plackett, P. et al., Immunochemical analysis of *Mycoplasma pneumoniae*. 3. Separation and identification o serologically active lipids, *Aust. J. Exp. Biol. Med. Sci.*, 47, 171–195, 1969.

44. Barile, M.F., Mycoplasma–tissue cell interactions. In Tully, J.G. et al. (Editors), *The Mycoplasmas II. Human and Animal Mycoplasmas*, Vol. 2, pp. 425–474, Academic Press, New York, NY, 1979.
45. Waites, K.B. et al., *Cumitech 34*, *Laboratory Diagnosis of Mycoplasmal Infections*, American Society for Microbiology, Washington, DC, 2001.
46. Bernet, C. et al., Detection of *Mycoplasma pneumonia* by using the polymerase chain reaction, *J. Clin. Microbiol.*, 27, 2492–2496, 1989.
47. Peerahe, S.N. et al., Comparison of culture with the polymerase chain reaction for detection of genital mycoplasma, *Eur. J. Gen. Med.*, 5(2), 107–111, 2008.
48. Wadowsky, R.M. et al., Inhibition of PCR-based assay for *Bordetella pertussis* by using calcium alginate fiber and aluminum shaft components of a nasopharyngeal swab, *J. Clin. Microbiol.*, 32, 1054–1057, 1994.
49. Dorigo-Zetsma, J.W. et al., Comparison of PCR, culture, and serological tests for diagnosis of *Mycoplasma pneumoniae* respiratory tract infection in children, *J. Clin. Microbiol.*, 37, 14–17, 1999.
50. Sillis, M., The limitations of IgM assay in the serological diagnosis of *Mycoplasma pneumoniae* infections, *J. Med. Microbiol.*, 33, 253–258, 1990.
51. Abele-Horn, M. et al., Molecular approaches to diagnosis of pulmonary diseases due to *Mycoplasma pneumoniae*, *J. Clin. Microbiol.*, 36, 548–551, 1998.
52. Feldner, J. et al., *Mycoplasma pneumoniae* adhesion localized to tip structure by monoclonal antibody, *Nature*, 298, 765–767, 1982.
53. Williamson, J. et al., Laboratory diagnosis of *Mycoplasma pneumoniae* infection. 4. Antigen capture and PCR-gene amplification for detection of the mycoplasma: Problems of clinical correlation, *Epidemiol. Infect.*, 109, 519–537, 1992.
54. Chaudhry, R. et al., Adhesion proteins of *Mycoplasma pneumoniae*, *Front Biosci.*, 12, 690–699, 2007.
55. Blanchard, A. et al., Evaluation of interspecies genetic variation within the 16S rRNA gene of *Mycoplasma hominis* and detection by polymerase chain reaction, *J. Clin. Microbiol.*, 31, 1358–1361, 1993.
56. Pawar, V. et al., Trends in the incidence and distribution of mycoplasma contamination detected in cell lines and their products, *Int. Organ. Mycoplasmol.Lett.*, 3, 77, 1994.
57. Weisburg, W.G. et al., A phylogenetic analysis of the mycoplasma: Basis for their classification, *J. Bacteriol.*, 171, 6455–6467, 1989.
58. White, B.A., PCR protocols. Current methods and applications. In *Methods in Molecular Biology*, vol. 15. Humana Press, Totowa, NJ, 1993.
59. Barile, M.F., Mycoplasmas in cell cultures. In Kahane, I. et al. (Editors), *Rapid Diagnosis of Mycoplasmas*, pp. 155–193, Plenum Press, New York, 1993.
60. Arai, S. et al., Comparative studies to detect mycoplasma contamination in bioindustrial materials, for validating standard method, *Int. Organ. Mycoplasmol Lett.*, 3, 48–49, 1994.
61. Hay, R.J. et al., Mycoplasma infection of cultured cells, *Nature*, 339, 487–488, 1989.
62. Somerson, N.L. et al., The mycoplasma flora of human and nonhuman primates. In Tully, J. G. (Editors), *The Mycoplasmas*, vol. 2, pp. 191–216, Academic Press, New York, 1979.
63. McGarrity, G.J. et al., Cell culture mycoplasmas. In Razin, S. (Editors), *The Mycoplasmas*, vol. 4, pp. 353–390, Academic Press, New York, 1985.
64. Uphoff, C.C. et al., Sensitivity and specificity of five different mycoplasma detection assays, *Leukemia*, 6, 335–341, 1992.
65. Dvorakova, H. et al., Detection of mycoplasma contamination in cell cultures and bovin sera, *Vet. Med.-Czech.*, 50(6), 262–268, 2005.
66. Costa, C. et al., Development of two real-time quantitative Taq-Man PCR assays to detect circulating *Aspergillus fumigatus* DNA in serum, *J. Microbiol. Methods*, 44, 263–269, 2001.
67. Edwards, M.C. et al., Multiplex PCR: Advantages, development, and applications, *PCR Methods Appl.*, 3, S65–S75, 1994.
68. Hayden, R.T. et al., Direct detection of legionella species from bronchoalveolar lavage and open lung biopsy specimens: Comparison of lightcycler PCR, in situ hybridization, direct fluorescence antigen detection, and culture, *J. Clin. Microbiol.*, 39, 2618–2626, 2001.
69. Kearns, A.M. et al., Development and evaluation of a real-time quantitative PCR for the detection of human cytomegalovirus, *J. Virol. Methods.*, 95, 121–131, 2001.
70. Chaudhry, R. et al., Diagnosis of *Mycoplasma pneumoniae* by Serology, PCR and Real-Time PCR. Presented in 109th ASM Annual Meeting May 17–21, 2009 in Philadelphia, Pennsylvania.
71. Jonas, M.W. et al., Evaluation of three real-time PCR assays for detection of *Mycoplasma pneumoniae* in an outbreak investigation, *J. Clin. Microbiol.*, 46, 3116–3118, 2008.
72. Baczynska, A. et al., Development of real-time PCR for detection of *Mycoplasma hominis*, *BMC Microbiol.*, 4, 35, 2004.
73. Mallard, K. et al., Development of real time PCR for the differential detection and quantification of *Ureaplasma urealyticum* and *Ureaplasma parvum*, *J. Microbiol. Methods*, 60, 13–19, 2005.
74. Saiki, R.K. et al., Primer-directed enzymatic amplification of DNA with a thermostable DNA polymerase, *Science*, 239, 487–491, 1988.

42 Ureaplasma

Ken B. Waites, Li Xiao, Vanya Paralanov, and John I. Glass

CONTENTS

42.1 INTRODUCTION

Since the first description of the organisms now known as ureaplasmas more than 50 years ago, there has been considerable progress in understanding their biological characteristics, and many clinical and basic investigations have documented their contributions as pathogens in eukaryotic hosts. Even though cultures are generally reliable when properly performed, procedures for growing the organisms from clinical specimens are somewhat complex, expensive, and still not widely available outside of large hospitals, reference laboratories, or research facilities. Culture alone cannot distinguish the two *Ureaplasma* species that infect humans from one another; nor can it identify and distinguish among individual serovars. Such discriminations can be valuable in studies of epidemiology and mechanisms of disease pathogenesis. Analysis of data obtained from the determination of the complete genome sequences of all 14 serotypes of human *Ureaplasma* spp. and development of molecular-based assays have further enhanced capabilities for detection of ureaplasmas in clinical specimens and performance of research studies aimed at understanding their roles as pathogens in many different conditions.

This chapter provides background information, descriptions, and methodologies for the performance of molecular-based assays for the detection and characterization of human ureaplasmas that can be applied in clinical and research settings. Molecular-based procedures that have been developed and published thus far vary both in their gene targets and in the basic assay methods. Therefore, anyone desiring to perform molecular detection assays for ureaplasmas must design and perform internal validation procedures to ensure accuracy and use rigorous quality control for all parts of the process. Knowledge of the basic biology of ureaplasmas and the conditions with which they have been associated is important for understanding the basis for how molecular assays for their detection and characterization have been developed and applied.

42.1.1 Classification, Morphology, and Biology

Mycoplasmas and ureaplasmas represent the smallest self-replicating organisms, both in cellular dimensions and genome size that are capable of a cell-free existence. They are included within the phylum *Tenericutes* and the class *Mollicutes*, which is comprised of four orders, five families, eight genera, and about 200 known species. Shepard provided the first description of ureaplasmas, initially known as "T-strain" mycoplasmas, when he cultivated them in vitro from the urethras of men with nongonococcal urethritis (NGU).[1] The genus *Ureaplasma*, established in 1974,[2] comprises those members of the family *Mycoplasmataceae* that hydrolyze urea and use it as a metabolic substrate for generation of ATP. This genus currently has seven recognized species that have been isolated from humans and various animals: *U. canigenitalium* (dogs), *U. cati* (cats), *U. diversum* (cattle), *U. felinum* (cats), *U. gallorale* (chickens), *U. parvum* (humans), and *U. urealyticum* (humans). Numerous other ureaplasmas of animal origin have been described, but they have not been given species designations. Human ureaplasmas were considered to belong to a single species, *U. urealyticum*, comprised of 14 serovars distributed between two genetically distinct biovars until 2002, when the two biovars were reclassified as two separate species, *U. parvum* and *U. urealyticum*, based on the sequences of 16S rRNA gene, the 16S-23S rRNA intergenic region, the urease gene, and DNA–DNA hybridization experiments.[3] *U. parvum* contains the serotypes 1, 3, 6, and 14 while the remaining serovars 2, 4, 5, 7–13 are assigned to *U. urealyticum*.

The small genome of 760–1140 kb and limited biosynthetic capacity of human *Ureaplasma* spp. are responsible for many of their biological characteristics and fastidious growth requirements. Lack of a rigid cell wall prevents them from staining by Gram stain, confers pleomorphism on their cells and makes them very susceptible to dehydration, thereby limiting them to a parasitic existence in association with eukaryotic cells of their host. Ureaplasmas are coccoid cells approximately 0.2–0.3 μm in diameter with a generation time of about 1 h. The organisms can be cultivated overnight in liquid media such as Shepard's 10 B broth, containing a mycoplasma broth base supplemented with 20% horse serum, 10% urea, DNA, yeast extract, and L-cysteine.[4] They will produce granular brown urease-positive colonies approximately 15–60 μm in diameter on A8 agar after 24–48 h of incubation at 37°C due to calcium chloride incorporated into the media.[4]

As many as 40%–80% of healthy adult women may harbor ureaplasmas in their cervix or vagina that are readily transmitted venereally as well as vertically; with a transmission rate to infants born to colonized mothers as high as 90%.[5] Their occurrence is somewhat less in the lower urogenital tract of healthy men (approximately 20%–29%).[6,7] *U. parvum* is considerably more common than *U. urealyticum* as a colonizer of the male and female urogenital tracts and in the neonatal respiratory tract.[5]

42.1.2 Clinical Features and Pathogenesis

Human ureaplasmas reside primarily on the mucosal surfaces of the urogenital tracts of adults or the respiratory tracts in infants and are capable of attaching to a variety of cell types such as urethral epithelial cells, spermatozoa, and erythrocytes.[5] The adhesins of ureaplasmas have not been characterized completely, but current evidence suggests the receptors are sialyl residues and/or sulfated compounds similar to what have been observed for other mycoplasmas.[5] A major family of surface proteins, the multiple banded antigens (MBA), contain serotype-specific and cross-reactive epitopes that are immunogenic during ureaplasmal infections and have been used as a basis for the development of reagents for diagnostic purposes.[6–9] Although there is no evidence ureaplasmas produce toxins, they do produce an IgA protease, a hemolysin, and release ammonia through urealytic activity, all of which are considered virulence factors.[5] An intact humoral immune response appears to be important in containing ureaplasmas and limiting their invasion and dissemination beyond mucosal surfaces as evidenced by their tendency to cause chronic respiratory infections and arthritis in persons with hypogammaglobulinemia and to cause invasive disease in preterm neonates.[5] The capacity of ureaplasmas to induce inflammation has been shown conclusively through animal inoculation studies in which the organisms upregulated amniotic fluid leukocytes, proinflammatory cytokines, prostaglandins, metalloproteinases, and uterine activity to induce chorioamnionitis, a systemic fetal inflammatory response, and contribute to preterm labor and fetal lung injury.[10]

A great deal of information concerning the basic biology of *U. parvum* was derived from studies of its genome, which was sequenced by Glass et al. and published in 2000.[11] More extensive comparative genomic characterization of human ureaplasmas has been made possible by the recent genome sequencing of all 14 serovars of *U. parvum* and *U. urealyticum*. This information has facilitated identification of additional targets for molecular-based assays to distinguish between the two *Ureaplasma* species as well as the 14 serovars.[6]

Despite the fact that ureaplasmas occur commonly in the lower urogenital tracts of healthy adults, there is evidence that these organisms play etiologic roles in some genital and extragenital diseases of adults and neonates. However, a complete understanding of the role of ureaplasmas in human diseases has been unattainable thus far for several reasons, which include the following:

1. Their high prevalence as colonizers in healthy persons
2. Poorly designed research studies that attempted to relate the presence of ureaplasmas in the lower urogenital tract to pathology in the upper tract or in offspring
3. Failure to consider other multifactorial aspects of some clinical conditions and potential confounders (e.g., bacterial vaginosis)
4. Unfamiliarity of microbiologists with the conditions necessary for their in vitro cultivation
5. Considering these organisms only as a last resort in conditions thought to be most likely due to other microorganisms
6. Lack of suitable diagnostic tests that can identify and characterize the two species and their respective serovars that could account for differential pathogenicity.

Improved understanding of the role of ureaplasmas as human pathogens that has been evident in recent years is due to several factors. These include the more widespread availability of commercial media formulations that has enabled more clinical microbiology laboratories to offer ureaplasma cultures; the development of sensitive and specific molecular-based diagnostic tests that do not require viable organisms and can differentiate between *U. parvum* and *U. urealyticum* in clinical specimens; and simulation of human ureaplasmal infections using rodent and primate models. Table 42.1 summarizes the various human diseases with which *Ureaplasma* spp. have been implicated. The number of published case reports implicating *Ureaplasma* spp. in a variety of systemic infections involving persons with and without impaired host

TABLE 42.1
Human Diseases Associated with or Caused by *Ureaplasma* Species

Disease	Association
Urethritis	+
Prostatitis	±
Epididymitis	±
Urinary calculi	+
Pyelonephritis	±
Bacterial vaginosis	±
Infertility	±
Chorioamnionitis	+
Spontaneous abortion	+
Prematurity/low birth weight	+
Intrauterine growth retardation	±
Postpartum/postabortion fever	+
Arthritis, osteomyelitis & other extragenital infections	+
Congenital or neonatal pneumonia	+
Chronic lung disease of prematurity	±
Neonatal bacteremia	+
Neonatal meningitis	+

± significant association but causal role not proven; + causal role proven.

defenses has increased in recent years as a result of a more widespread utilization of universal PCR primers for bacterial 16S rRNA when infection is suspected and no conventional microbes are detected by culture.

Since *Ureaplasma* spp. may occur singularly or simultaneously in various body sites as commensals or potentially as pathogens, the availability of molecular-based assays that can differentiate between them has stimulated interest in whether there is a differential pathogenicity between the two species, subgroups, or individual serovars within them. Prior to the availability of PCR assays, there were numerous studies in which ureaplasmas grown from culture were typed using polyclonal or monoclonal antibodies, immunofluorescence, immunoperoxidase, agar growth inhibition, or an indirect approach that used serology to ascertain differential pathogenicity at the serovar level.[5] Results were varied and inconsistent due to the inefficient, imprecise, and nonreproducible methods, occurrence of multiple cross-reactions, and the fact that many persons may have had more than one serovar in the presence or absence of disease.[5] Much more data have been reported in the past several years since PCR was introduced for detecting and characterizing ureaplasmas in clinical specimens. Since 1990, nearly 300 peer-reviewed publications have described basic and clinical investigations involving ureaplasmas using PCR and various other molecular-based techniques.

Kim et al.[12] found no difference in pregnancy outcome and magnitude of intraamniotic inflammatory response, chorioamnionitis, birthweight or gestational age at delivery, or neonatal morbidity in women whose amniotic contained ureaplasmas that was attributable to an individual species. In contrast, Deguchi et al.[13] noted a significantly higher prevalence of *U. urealyticum* in men with NGU as compared to *U. parvum* for which prevalence was not significantly different from men without NGU. According to Abele-Horn et al.,[14] *U. urealyticum* was predominant in patients with pelvic inflammatory disease, as well as in patients who had had a miscarriage, and it seemed to have more adverse effects on pregnancy outcome, as assessed by birth weight, gestational age, and preterm delivery, than *U. parvum*. These investigators[14] also observed that *U. urealyticum* was more likely than *U. parvum* to be associated with development of neonatal chronic lung disease, but others have not confirmed this association.[15,16] When PCR was used to compare the distribution of species and prevalence of tetracycline resistance mediated by *tetM*, no differences were seen by some investigators,[5,17] but others[14,18] have found more tetracycline resistance in *U. urealyticum* as opposed to *U. parvum*. Based on these limited observations, whether the two *Ureaplasma* species are likely to have differential pathogenicity or susceptibility to tetracycline is unresolved and will require carefully designed studies of larger numbers of specimens from defined patient populations that can be tested for the presence of each species and their respective serovars.

Among the ureaplasmas of animal origin, the bovine species *U. diversum* has been studied most extensively. Culture and PCR-based amplification of 16S rRNA have demonstrated its frequent occurrence in the vagina of cows and it is a recognized cause of bovine urogenital infections, including granular vulvitis, endometritis, salpingitis, spontaneous abortion, and infertility.[19,20] *U. diversum* may also be involved in bovine respiratory infections.[20] The importance of other ureaplasmas of animal origin as agents of disease in their respective hosts has not been investigated to any significant degree and only a few reports of their isolations have been documented. All further discussion of detection methods for ureaplasmas will focus on the human species *U. parvum* and *U. urealyticum*.

42.1.3 Diagnosis

Conventional Techniques. Culture is considered the reference standard for detection of *Ureaplasma* spp. in clinical specimens and remains the most widely used method for laboratory diagnosis. Some laboratories prefer to prepare their own liquid and solid media to reduce costs, but commercial sources are now available in many countries.[21] If specimens have to be shipped to an off-site reference laboratory for culture, particular care must be taken to ensure the organisms remain viable, which means the specimen must be inoculated into an appropriate transport medium and shipped frozen on dry ice. Detailed methods and quality control procedures for culture-based detection of ureaplasmas are available in reference texts and have not been reiterated here since the focus of this publication is on molecular-based technology.[22] Due to their relatively rapid growth rates, confirmed culture results for human ureaplasmas are usually available within 2–5 days, exclusive of specimen transport time and shipment if an off-site reference laboratory is used. Human ureaplasmas can be positively identified to genus level by their colonial morphology and urease production on A8 agar, but genotypic or serologic procedures are necessary for species designation. For routine diagnostic and patient management purposes, it is not necessary to determine which species are present in a clinical specimen. A single culture procedure utilizing 10 B broth and A8 agar can detect *Ureaplasma* spp. as well as *Mycoplasma hominis*. Culture has an additional advantage in that it provides an isolate that can be studied further, and on which antimicrobial testing can be performed if needed. Up to 50% of *Ureaplasma* isolates in the United States may contain *tetM* and exhibit tetracycline resistance.[5]

While serology has been an important tool in diagnosis of *Mycoplasma pneumoniae* infections for many years, it has not proven useful for ureaplasmal infections, largely because of ubiquity of these organisms in healthy persons makes interpretation of antibody titers difficult and the mere existence of antibodies alone cannot reliably distinguish among colonization, current, or prior infection.[4] Even though serological assays for detection of antibody responses to human ureaplasmas have been applied in research settings, such assays have never been standardized and are not generally available for diagnostic purposes.

Molecular Techniques. The advances in molecular biology over the past 25 years have greatly improved diagnostic methods for infectious diseases and provided the opportunity for significantly reduced turnaround time, though often with greater expense, and the requirement for having complex molecular biology laboratory equipment and personnel with sufficient education and training to employ these techniques correctly.

The most important advance in molecular technology is the PCR assay, which was first described in its modern form in 1988.[23] By the end of that decade the widespread use of the PCR assay was bringing about a revolution in clinical diagnostics. No area of diagnostic medicine was more enabled by PCR than clinical microbiology. Improved methods for detection of the *Ureaplasma* spp. that infect humans are part of that advance. Shortly after the publication of the *U. urealyticum* urease gene sequences in 1990,[24] Willoughby and colleagues reported the first use of PCR to detect ureaplasmas.[25]

According to the detection strategy, all PCR methods can be classified into conventional PCR or real-time PCR. Conventional PCR measures the end-stage PCR products, using gels or other methods, while real-time PCR detects and quantifies the products simultaneously with amplification. Real-time PCR is able to amplify and simultaneously detect and quantify the target DNA molecule as it accumulates during the reaction in *real time* after each amplification cycle. Real-time PCR is frequently combined with reverse transcription to quantify messenger RNA (mRNA). Multiplex PCR uses several pairs of primers annealing to different target sequences and allows the simultaneous analysis of multiple targets in a single sample, thereby enabling detection of multiple bacterial species in a single assay.

While conventional PCR methods for detection of *Ureaplasma* spp. can take between 2 and 3 days, about the same turnaround time as culture, real-time PCR can potentially provide results the same day a specimen is received. Since organism viability does not have to be maintained for PCR-based detection, specimen collection, handling, and transport for *Ureaplasma* detection by PCR is much simpler than for culture. The improved specificity of real-time PCR compared to conventional PCR is mainly due to the use of an additional oligonucleotide probe that binds to the target sequence. Usually, the probe is designed to correspond to a sequence located between the forward and reverse primers. Generally, the probe is labeled with a fluorophore and/or quencher. In the case of Taqman probe when the quencher is in close proximity to the fluorophore, the fluorescence is inhibited or emitted at a different wavelength. Utilizing a fluorescently labeled probe allows a signal to be detected by the instrument only if the probe has annealed to its complementary target sequence and the DNA polymerase has amplified the target region, therefore increasing the distance between the quencher and the fluorophore. The use of a labeled probe minimizes the probability of cross-reaction and detection of undesired amplicons. Such probes are also specifically useful when differentiating very closely related organisms is desired, as is the case of determining which of the 14 *Ureaplasma* serovars are present in a clinical specimen. Another feature of real-time PCR that contributes to the specificity of the assay is that the amplicon melting temperature is determined at the end of the assay, and can be used to verify whether the desired PCR product is being detected.

The third important factor, analytical sensitivity, can also be improved when using real-time PCR assays. Real-time PCR diagnostic assays are often designed to produce shorter amplicons with an optimal length of under 300 bp as compared to conventional PCR. The smaller the amplicon, the more efficient is the PCR assay and the lower the probability for the target templates to be sheared during handling. Based on careful primer selection and optimization of reaction conditions, an increased sensitivity of real-time PCR for detection of ureaplasmas may be observed compared to assays designed for conventional PCR.[6]

An added bonus of real-time PCR over conventional PCR is that real-time PCR can be used to quantitate the starting amount of DNA molecules and therefore the pathogen load of the patient can be estimated. This is especially useful when detecting organisms such as ureaplasmas, which can be part of the human flora but can be harmful when their load is higher than normal. In such cases, a qualitative positive result from conventional PCR may not be sufficient to decide whether a person needs medical treatment. Theoretically, each commensal organism can have its own established cycle threshold (C_t) or cross-point (C_p) values range in which it would be considered harmless based on accumulated clinical data. The lower the C_t or C_p value, the higher the load of the organism in the patient.

Various PCR methods, as well as their respective gene targets, have been reported to detect *Ureaplasma* spp.

The earlier gel-based conventional PCR assays targeted sequences of 16S rRNA gene and 16S rRNA -23S rRNA intergenic spacer regions,[13,26–30] the *urease* gene,[24,31] and the *mba* genes.[6,7,9,29,32,33] Published real-time PCR assays have targeted primarily the *urease* gene and its subunits[34–36] as well as the *mba* gene.[6,37] A summary of published PCR assays and gene targets used for *Ureaplasma* PCR assays for detection, speciation, and serotyping of the organisms is provided in Tables 42.2, 42.3, and 42.4. PCR primers for specific *Ureaplasma* gene targets may behave differently when used with the various thermocyclers, protocols, and reaction mixtures. Thus, each primer set needs to be customized for the instrument to be used. Primer development software for the respective instruments is useful to guide this process.

Published PCR assays to characterize the *Ureaplasma* spp. at the serovar level have mainly focused on primers based on the *mba* gene and its 5′-end upstream regions. In combination with direct sequencing or restriction enzyme analysis, these assays were capable of distinguishing the 4 serovars of *U. parvum* and dividing the 10 serovars of *U. urealyticum* into different subgroups.[7,32,33,37,41] Because of limited sequence variation in the *mba* genes, earlier PCR-based methods lacked the capacity for complete separation of all 14 serovars. Moreover, whole genome sequencing of all 14 serovars has shown *mba* to be part of a large gene family present in many variations in different serovars and the gene is phase variable.[43,44] Thus, while the *mba* gene is a valid target for serovar-specific PCR, attention must be paid to make certain that targets do not span sites at which chromosome rearrangements occur during phase variation. Keeping these limitations in mind, we evaluated the complete genome sequences for all 14 serovars and designed primers and probes targeting

TABLE 42.2
Primer and Probe Sequences for *Ureaplasma* Detection and Species Differentiation

Target	Primer/Probe	Sequence (5′→3′)	Probe Type	Reference
UU063	UP063#1F	TGCGGTGTTTGTGAACT		Xiao[6]
	UP063#1R	TGATCAAACTGATATCGCAATTATAGA		Xiao[6]
	UP063#1 Probe 1	TGGTTTAACGTGTTTTTGAAGTGCTACAAAAT- fluorescein	FRET	Xiao[6]
	UP063#1 Probe 2	LC Red 640-CCCATTTCAGCCATGGTGCCATCA-phosphate	FRET	Xiao[6]
UUR10_0680	UU127#1F	FGGATTTGTTAGATATCGTCAAGG		Xiao[6]
	UU127#1R	TCATCTTTTAAAGCTCCAACATTATTAGT		Xiao[6]
	UU127#1 Probe 1	AACACGAGTATGGATGAAATCAAAATCATCAA-fluorescein	FRET	Xiao[6]
	UU127#1 Probe 2	LC Red 705-AATAACGGTGGTTCAGCTATTTGAGTATG AGC-phosphate	FRET	Xiao[6]
16S rRNA	P1	AGAGTTTGATCCTGGCTCAGGA		Robertson[27]
	U3	TAGAAGTCGCTCTTTGTGG		Robertson[27]
	U8	GAAGATGTAGAAAGTCGCGTTTGC		Robertson[27]
	P6	GGTAGGGATACCTTGTTACGACT		Robertson[27]
	UPS1	ATGAGAAGATGTAGAAAGTCGCTC		Kong[29]
	UPA	TTAGCTACAACACCGACCCATTC		Kong[29]
	UPA1	CGTTAGCTACAACACCGACCCA		Kong[29]
	UPS	CATCATTAAATGTCGGCCCGAATGG		Kong[29]
	UUA	CTACAACACCGACTCGTTCGAG		Kong[29]
	UUS	CATTAAATGTCGGCTCGAACGAG		Kong[29]
	Ure-1S	CTAGATGCTTAACGTCTAGCTGTATCAA		Yoshida[62]
	Ure-1A	GCCGACATTTAATGATGATCGT		Yoshida[62]
	Ure-P1	FAM-AAGGCGCCAACTTGGACTATCACTGAC-TAMRA	Taqman	Yoshida[62]
16S-23S rRNA intergenic	MCGpFll	AAACTATGGGAGCTGGTAAT		Harasawa[38]
	R16-2	GTGGGGATGGATCACCTCCT		Harasawa[38]
	MCGpR2	CCATTCACCATAAACTCTT		Harasawa[38]
	R23-1R	CTCCTAGTGCCAAGGCATC/TC		Harasawa[38]
	F1	ACACCATGGGAG(C/T)TGGTAAT		Harasawa[38]
	R1	CTTC(A/T)TCGACTT(C/T)CAGACCCAAGGCAT		Harasawa[38]
	UPSA	TAGAATCCGACCATATGAATTTTTA		Kong[29]
	UUSA	AGAGTCCGACCATATGAACTTTTG		Kong[29]
Urease gene and adjacent regions	UUS1	CACAGATGTCCTTGATGTAC		Blanchard[24]
	UUA1	TACTTCACGAGCAGATTGCA		Blanchard[24]
	U1A	GATGGTAAGTTAGTTGCTGAC		Blanchard[24]
	U1B	ACGACGTCCATAAGCAACT		Blanchard[24]

continued

TABLE 42.2 (Continued)
Primer and Probe Sequences for *Ureaplasma* Detection and Species Differentiation

Target	Primer/Probe	Sequence (5→3′)	Probe Type	Reference
	U5	CAATCTGCTCGTGAAGTATTAC		Blanchard[39]
	U4	ACGACGTCCATAAGCAACT		Blanchard[39]
	U9 (probe)	GAGATAATGATTATATGTCAGGATCA		Blanchard[39]
	biovar 1 probe	AAATTGACTTGATGATCCTG	LH, biotin-labeled	Povlsen[31]
	biovar 2 probe	ATTTGTTTCAAATAAGTGG	LH, biotin-labeled	Povlsen[31]
	UU-1524R	TTCCTGTTGCCCCTCAGTCT		Yi[35]
	UU-1613F	AAGGTCAAGGTATGGAAGATCCAA		Yi[35]
	UU-parvo	FAM-TCCACAAGCTCCAGCAGCAATTTG-TAMRA	Taqman	Yi[35]
	UU-T960	VIC-ACCACAAGCACCTGCTACGATTTGTTC-TAMRA	Taqman	Yi[35]
	UPureF-M	GATCACATTTTCACTTGTTTGAAGTG		Mallard[36]
	UPureR-M	AACGTCGTCCATAAGCAACTTTG		Mallard[36]
	UPure1MGB	6-FAM-AGGAAATGAAGATAAAGAAC-MGB	Taqman MGB	Mallard[36]
	UUureF-M	GATCACATTTCCACTTATTTGAAACA		Mallard[36]
	UUureR-M	AAACGACGTCCATAAGCAACTTTA		Mallard[36]
	Uuure2MGB	6-FAM-AAACGAAGACAAAGAAC-MGB	Taqman MGB	Mallard[36]
	UPureF-C	CATTGATGTTGCACAAGGAGAAA		Cao[34]
	UPureR-C	TTAGCACCAACATAAGGAGCTAAATC		Cao[34]
	UPureFP	6-FAM-TTGACCACCCTTACGAG-MGB	Taqman MGB	Cao[34]
	UUureF-C	ATCGACGTTGCCCAAGGGGA		Cao[34]
	UUureR-C	TTAGCACCAACATAAGGAGCTAAATC		Cao[34]
	UUureFP	6-FAM-TTGTCCGCCTTTACGAG-MGB	Taqman MGB	Cao[34]
	UPS2	CAG GAT CAT CAA GTC AAT TTA G		Kong[29]
	UPA2	AAC ATA ATG TTC CCC TTT TTA TC		Kong[29]
	UUS2	CAG GAT CAT CAA ATC AAT TCA C		Kong[29]
	UUA2	CAT AAT GTT CCC CTT CGT CTA		Kong[29]
mba gene and upstream regions	UMS-125	GTATTTGCAATCTTTATATGTTTTCG		Teng[9]
	UMS51	CTGAGCTATGACATTAGGTGTTACC		Teng[9]
	UMS226	CAGCTGATGTAAGTGCAGCATTAAATTC		Teng[9]
	UMA427	ACCTGGTTGTGTAGTTTCAAAGTTCAC		Teng[9]
	UMS-170	GTATTTGCAATCTTTATATGT TTTCG		Teng[9,33]
	UMA263	TTTGTTGTTGCGTTTTCTG		Teng[9,33]
	UMA1213	CTAAAGTAATTATTTTCCAGTAGTTTC		Zheng[8]
	UMA1586	GATAATCATTCATCTTCTCTTAATTGTC		Zheng[8]
	UMA269'	CCAAATGACCTTTTGTAACTAGAT		Kong[32]
	UMA269	CTAAATGACCTTTTTCAAGTGTAC		Kong[32]
	UMS-54	CTTAGTGTTCATATTTTTTACTAG		Kong[32]
	UMSu1	TTACTGTAGAAATTATGTAAGATTGC		Ren[40]
	UMSu3	ACTGTAGAAATTATGTAAGATTACC		Ren[40]
	UMSu6	CTTAGTGTTCATATTTTTTACTAG		Ren[40]
	UMSu14	ATTACTGTAGAAATTATGTAAGATTAA		Ren[40]
	UMS-57	(T/C)AAATCTTAGTGTTCATATTTTTTAC		Kong[29]
	UMA222	GTAAGTGCAGCATTAAATTCAATG		Kong[29]
	UMS-61	TTTGCAAAACTATAAATAGA AC		Kong[29]
	UMS-83	TACTGTAGAAATTATGTAAGATTGC		Kong[29]
	UMA219	GTAATTGCAACATGGAATTCAGCTTCG		Kong[29]
	UMA219′	GTAATTGCAAC TGGAATTCAGTTTCA		Kong[29]
	UMS-112	GATTAAACAAAATCTTAATGTTGTTA		Kong[29]
	UMS-1129	GATTAAACAAAATCTTAATGTTGTTG		Kong[29]
	UMA194	CGTTTAATGCTTTTTTATCATTTTCAG		Kong[29]
	UMA194′	CGTTTAATGCTTTTTTATCATTTTCAT		Kong[29]
	UMSPS1	CCTCGTGAACCAAAACCTAATG		Kong[7]
	UMSPS2	GGATTAATCAAGACTTCAGGTTTG		Kong[7]
	UMS3S	TTACTGTAGAAATTATGTAAGATTACC		Kong[7]

continued

TABLE 42.2 (Continued)
Primer and Probe Sequences for *Ureaplasma* Detection and Species Differentiation

Target	Primer/Probe	Sequence (5′→3′)	Probe Type	Reference
	UMS14S	AATTACTGTAGAAATTATGTAAGATTAAT		Kong[7]
	UMA314A	GTTGTTCTTTACCTGGTTGTGTAG		Kong[7]
	UMA314A′	TG/TGTTGTTCTTTACCTGGTTGTGTA		Kong[7]
	UMA6A	CCTGGTTCTTGAGTTTTCGGAG		Kong[7]
	UMA6A′	TTTACCTGGTTCTTGAGTTTTCGG		Kong[7]
	UMA1A	TTTCTTTTGGTTCTTCAGTTTTTGAAG		Kong[7]
	UMA1A′	ATTTTCTTTTGGTTCTTCAGTTTTTGA		Kong[7]
	UMSUS	GTTTACGACATTGAAAATTTCGATG		Kong[7]
	UMAUA	GGG(G/T)(A/T)GTT(G/T)(A/C/T)ACCA(C/T)T(G/T)CCTGGTT		Kong[7]
	UMSUS1	AACTGCATCT(C/T)TAG(C/T ATTACCT		Kong[7]
	UMSUS2	CCTGATAATTT(G/T)AATTATCAAACAG		Kong[7]
	UMAUA1	GCCCAATTCATAGGCTATTAATTG		Kong[7]
	UMAUA2	AAAAAAATAGCCCAATTCATAGGC		Kong[7]
	UMA2A1	TTCCTGGTTTTGTTTCAAAACCTAT		Kong[7]
	UMA2A2	CACTTCCTGGTTTTGTTTCAAAAC		Kong[7]
	UMA2A	CCACTTCCTGGTTTTGTAGTTTC		Kong[7]
	UMA10A	CCACTTCCTGGTTGTGTAGTTGA		Kong[7]
	UMA4A1	TTGCCTGGTTGTGTTTCGAACTC		Kong[7]
	UMA4A2	CATTGCCTGGTTGTGTTTCGAAC		Kong[7]
	UMA9A1	CTGGAGTTGGTGTAGGCGCAT		Kong[7]
	UMA9A2	TTCTGGAGTTGGTGTAGGCGC		Kong[7]
	UMA7A1	GTAATTGCAACATGGAATTCAGTTTCA		Kong[29]
	UMA7A2	GGTTCTGGTGTATGAGTGCTTTT		Kong[7]
	UMA7A3	GTTGGTTCTGGTGTATGAGTGC		Kong[7]
	UP F	GTATTTGCAATCTTTATATGTTTTCG		Cao[37]
	UP R	TCCAGCTCCAACTAAGGTAAC		Cao[37]
	UP FP1	6-FAM-TGTAAGATTGCTAAATC-MGB	Taqman MGB	Cao[37]
	UP FP3	HEX-TGTAAGATTACCAAATC-MGB	Taqman MGB	Cao[37]
	UP FP6	6-FAM-AGTGTTCATATTTTTTACTAG-MGB	Taqman MGB	Cao[37]
	UP FP14	HEX-TCTTAGCTATGACATTAG-MGB	Taqman MGB	Cao[37]
	UU F	GTATTTGCAATCTTTATATGTTTTCG		Cao[37]
	UU R	CCCTGCTCCCACTAAAGTTAC		Cao[37]
Member of *mba* gene family but not *mba*	UP1F1-2	GATAATTTGAATTATCAAACAGAAAAAGTG		Xiao[6]
	UP1R1-2	TGTTCTTTACCTGGTTGTTGT		Xiao[6]
	UP1 probe 1	Fluorescein-SPC-GAACCAAAAGAAAATGGTGGAGAACAACC-Phosphate	SPC	Xiao[6]
Putative lipoprotein	UU2_F_1	CTTGTTGATAAGCAAAAGGATTAC		Xiao[6]
	UU2_R_1	GTTTTGGTTCTGGATTGTTTGAC		Xiao[6]
Conserved hypothetical protein	UP3F1-2	AATAATGCAAATTATGATGTTAAATTAACC		Xiao[6]
	UP3R1-2	TGTTTCTATGTTAACATTTAACAATTTAGC		Xiao[6]
Intergenic	Uu04_1F	CTAAAGCGTGCTTAATGTGT		Xiao[6]
	Uu04_1R	GTTAGTTTGGGAACCACCT		Xiao[6]
	Uu04_1 probe 1	TGTTATTATTATAGAAATTGTTGTAAGAATCAAAG-FL	FRET	Xiao[6]
	Uu04_1 probe 2	LC Red 610-ATAATTTCGTATAATGTCATGTTGTTCACCTCCTT-Phosphate	FRET	Xiao[6]
ATP-dependent RNA helicase	Uu05-3F	ATTATGAAAAATTAAACTCTCATTACTCG		Xiao[6]
	Uu05-3R	TTCTCATATTGAAAAGAAAATGAATGCTC		Xiao[6]
	Uu05-3 probe	Fluorescein-SPC-AGGAAGAATAAAACATTTTAATTTATATCCACGAA-Phosphate	SPC	Xiao[6]
mba	UP6F1	GAACTTTGAAACAGCTCCG		Xiao[6]
	UP6R1	CCTGGTTCTTTACCTGGTTCTTTA		Xiao[6]
mba	UU7_F_1	CAAACAAAGCATTAACAGCTTCAAAA		Xiao[6]
	UU7_R_1	TTATTAACTTCTTCCTTTGTTGTAAGTGTAGC		Xiao[6]

continued

TABLE 42.2 (Continued)
Primer and Probe Sequences for *Ureaplasma* Detection and Species Differentiation

Target	Primer/Probe	Sequence (5′→3′)	Probe Type	Reference
Intergenic	UU8_F_1	AGTTTTAATTATTTCGTTGTTTAAGTAGC		Xiao[6]
	UU8_R_1	AGTTTTTACGTGGTAAGTGGTT		Xiao[6]
FtsK/spoIIIE family protein	UU9_F_1	CGAAGCGGAAGTCGCAGGT		Xiao[6]
	UU9_R_1	TATTGCCACACCAGCCAGCA		Xiao[6]
Member of *mba* gene family but not *mba*	UU10_F_4	TGCATCAACATCGCTAAATTCA		Xiao[6]
	UU10_R_4	ATACGCCTACTCCGACT		Xiao[6]
	UU10_P_4	Fluorescein-SPC-TTGGTGTAGGTGTTGGTTGTGG-Phosphate	SPC	Xiao[6]
Intergenic	UU11_F_1	AACCTTATCAATTCTATTAATCATAGTTCA		Xiao[6]
	UU11_R_1	TGAAAACAAAAAACACGCTCC		Xiao[6]
	UU11_P_1	Fluorescein-SPC-AAACCAAACTAACACATTAAGCACGC-Phosphate	SPC	Xiao[6]
Conserved hypothetical protein	Uu12_1F	AATTGGTCAAACAACATATCGTG		Xiao[6]
	Uu12_1R	ATTTAACACGATTAACATCTTCACGTTTA		Xiao[6]
	Uu12_1 probe 1	TTTGCGAAGCATTACCGCCAAATAC-FL	FRET	Xiao[6]
	Uu12_1 probe 2	LC Red 705-CGCTTAGTTATGAACCAGCTGTTATTAATATA GCG-Phosphate	FRET	Xiao[6]
Conserved domain protein Intergenic	UU13_F_1	CCAAGACCAGCACCTACTGCC		Xiao[6]
Putative single-strand binding protein	UU13_R_1	AGAAAACGAGGAGGAATAAATAATGGATT		Xiao[6]
Intergenic	UP14F1	TCCACACTACGGTAAGTAGTTT		Xiao[6]
	UP14R1	CGCGCAGACCCTTGAATA		Xiao[6]

FRET, fluorescence resonance energy transfer; SPC, simple probe chemistry; MGB, minor groove binder.

TABLE 42.3
Primer/Probe Pairs Used to Detect and Differentiate *Ureaplasma* spp.

Specificity	Target Gene/Region	Primer Pair	Probe	PCR Type	Reference
Up and Uu	16S rRNA	P1, P6		Conventional	Robertson[27]
	16S-23S rRNA spacer regions	MCGpFl1, R23-1R		Conventional	Harasawa[38]
	16S-23S rRNA spacer regions	R16-2, MCGpR2		Conventional	Harasawa[38]
	Urease gene subunits and adjacent regions	U2B, U2C		Conventional	Ruifu[42]
	Urease gene subunits and adjacent regions	UUSP, UCA2		Conventional	Ruifu[42]
	Urease gene subunits and adjacent regions	U5, U4	U9	Conventional	Blanchard[39]
	mba gene and upstream regions	UMS-125, UMA226		Conventional	Teng[9,33]
	16S rRNA	Ure-1S, Ure-1A	Ure-P1	Real-time	Yoshida[38]
Up	16s rRNA gene and 16S-23S rRNA intergenic region	P6, U3		Conventional	Robertson[27]
	16s rRNA gene and 16S-23S rRNA intergenic region	UPS1, UPA		Conventional	Kong[29]
	16s rRNA gene and 16S-23S rRNA intergenic region	UPS1, UPA1		Conventional	Kong[29]
	16s rRNA gene and 16S-23S rRNA intergenic region	UPS, UPSA		Conventional	Kong[29]
	16s rRNA gene and 16S-23S rRNA intergenic region	UPS2, UPA2		Conventional	Kong[29]
	mba gene and upstream regions	UMS51, UMA427		Conventional	Teng[9,33]
	mba gene and upstream regions	UMS-57, UMA222		Conventional	Kong[29]
	Urease gene subunits and adjacent regions	UPure F-C, Upure R-C	Upure FP	Conventional	Cao[34]

continued

TABLE 42.3 (Continued)
Primer/Probe Pairs Used to Detect and Differentiate *Ureaplasma* spp.

Specificity	Target Gene/Region	Primer Pair	Probe	PCR Type	Reference
	Urease gene subunits and adjacent regions	UPure F-M, Upure R-M	UPure1MGB	Real-time	Mallard[36]
	Urease gene subunits and adjacent regions	UU-1524R, UU-1613F	UU-parvo	Real-time	Yi[35]
	Urease gene subunits and adjacent regions	U5, U4	biovar 1 probe	Conventional	Povlsen[31]
	UU063	UU063#1F, UU063#1R	UP063#1 Probe1, UP063#1 Probe2	Multiplex Real-time	Xiao[6]
Uu	16S rRNA	P6, U8		Conventional	Robertson[27]
	16S rRNA	U8, UUA		Conventional	Robertson[27]
	16S rRNA	UUS, UUSA		Conventional	Robertson[27]
	Urease gene subunits and adjacent regions	UUS1, UUA1		Conventional	Blanchard[24]
	Urease gene subunits and adjacent regions	UUS2, UUA2		Conventional	Blanchard[24]
	mba gene and upstream regions	UMS-170, UMA263		Conventional	Teng[9,33]
	mba gene and upstream regions	UMS-61, UMA263		Conventional	Teng[9,33]
	mba gene and upstream regions	UU F, UU R		Real-time	Cao[37]
	Urease gene subunits and adjacent regions	Uuure F-C, Uuure R-C	Uuure FP	Real-time	Cao[34]
	Urease gene subunits and adjacent regions	Uuure F-M, Uuure R-M	Uuure2MGB	Real-time	Mallard[36]
	Urease gene subunits and adjacent regions	UU-1524R, UU-1613F	UU-T960	Real-time	Yi[35]
	Urease gene subunits and adjacent regions	U5, U4	biovar 2 probe	Conventional	Povlsen[31]
	UUR10_0680	UU127#1F,UU127#1R	UU127#1 Probe 1, UU127#1 Probe 2	Multiplex Real-time	Xiao[6]
Serovar 1	*mba* gene and upstream regions	UMS-83, UMA269′		Conventional	Kong[29]
Serovar 1	*mba* gene and upstream regions	UMSu1, UMA269′		Conventional	Ren[40]
Serovar 1	*mba* gene and upstream regions	UMS-83, UMA1A		Conventional	Kong[7]
Serovar 1	*mba* gene and upstream regions	UMS-83, UMA1A′		Conventional	Kong[7]
Serovar 1	*mba* gene and upstream regions	UMS-83, UMA269′		Conventional	Kong[7]
Serovar 1	*mba* gene and upstream regions	UP F, UP R	UP FP1	Real-time	Cao[37]
Serovar 1	Member of *mba* gene family but not *mba*	UP1F1-2, UP1R1-2	UP1 probe 1	Real-time	Xiao[6]
Serovar 2	putative lipoprotein	UU2_F_1, UU2_R_1		Real-time	Xiao[6]
Serovar 3	*mba* gene and upstream regions	UMSu3, UMA269		Conventional	Ren[40]
Serovar 3	*mba* gene and upstream regions	UMS3S, UMA269		Conventional	Kong[7]
Serovar 3	*mba* gene and upstream regions	UMS3S, UMA314A		Conventional	Kong[7]
Serovar 3	*mba* gene and upstream regions	UMS3S, UMA314A′		Conventional	Kong[7]
Serovar 3	*mba* gene and upstream regions	UP F, UP R	UP FP3	Real-time	Cao[37]
Serovar 3	conserved hypothetical protein	UP3F1-2, UP3R1-2		Real-time	Xiao[6]
Serovar 3 and 14	*mba* gene and upstream regions	UMS-125, UMA269		Conventional	Kong[29]
Serovar 4	Intergenic	Uu04_1F, Uu04_1R	Uu04_1 probe 1, Uu04_1 probe 2	Real-time	Xiao[6]
Serovar 5	ATP-dependent RNA helicase	Uu05-3F, Uu05-3R	Uu05-3 probe	Real-time	Xiao[6]
Serovar 6	*mba* gene and upstream regions	UMS-54, UMA269′		Conventional	Kong[29]
Serovar 6	*mba* gene and upstream regions	UP F, UP R	UP FP6	Real-time	Cao[37]
Serovar 6	*mba* gene and upstream regions	UMSu6, UMA269′		Conventional	Ren[40]
Serovar 6	*mba* gene and upstream regions	UMS-54, UMA6A		Conventional	Kong[7]
Serovar 6	*mba* gene and upstream regions	UMS-54, UMA6A′		Conventional	Kong[7]
Serovar 6	*mba* gene and upstream regions	UMS-54, UMA269′		Conventional	Kong[7]
Serovar 6	*mba*	UP6F1, UP6R1		Real-time	Xiao[6]
Serovar 7	*mba*	UU7_F_1, UU7_R_1		Real-time	Xiao[6]
Serovar 8	Intergenic	UU8_F_1, UU8_R_1		Real-time	Xiao[6]
Serovar 9	FtsK/spoIIIE family protein gene	UU9_F_1, UU9_R_1		Real-time	Xiao[6]
Serovar 10	Member of *mba* gene family but not *mba*	UU10_F_4, UU10_R_4	UU10_P_4	Real-time	Xiao[6]

continued

TABLE 42.3 (Continued)
Primer/Probe Pairs Used to Detect and Differentiate *Ureaplasma* spp.

Specificity	Target Gene/Region	Primer Pair	Probe	PCR Type	Reference
Serovar 11	Intergenic	UU11_F_1, UU11_R_1	UU11_P_1	Real-time	Xiao[6]
Serovar 12	conserved hypothetical protein	Uu12_1F, Uu12_1R	Uu12_1 probe 1, Uu12 _1 probe 2	Real-time	Xiao[6]
Serovar 13	conserved domain protein, intergenic region and putative single-strand binding protein	UU13_F_1, UU13_R_1		Real-time	Xiao[6]
Serovar 14	*mba* gene and upstream regions	UMSu14, UMA269		Conventional	Ren[40]
Serovar 14	*mba* gene and upstream regions	UMS14S, UMA269		Conventional	Kong[7]
Serovar 14	*mba* gene and upstream regions	UMS14S, UMA314A		Conventional	Kong[7]
Serovar 14	*mba* gene and upstream regions	UMS14S, UMA314A′		Conventional	Kong[7]
Serovar 14	*mba* gene and upstream regions	UP F, UP R	UP FP14	Real-time	Cao[37]
Serovar 14	Intergenic	UP14F1, UP14R1		Real-time	Xiao[6]
UU Subtype 1 and 2	*mba* gene and upstream regions	UMS-61, UMA219		Conventional	Kong[29]
UU Subtype 3	*mba* gene and upstream regions	UMS-61, UMA219′		Conventional	Kong[29]
UU Subtype 1 and 3	*mba* gene and upstream regions	UMS-112, UMA194		Conventional	Kong[29]
UU Subtype 2	*mba* gene and upstream regions	UMS-112, UMA194′		Conventional	Kong[29]
UU A/B	*mba* gene and upstream regions	UMS-61, UMA2A1		Conventional	Kong[7]
UU A/B	*mba* gene and upstream regions	UMSUS, UMA2A1		Conventional	Kong[7]
UU A/B	*mba* gene and upstream regions	UMS-61, UMA2A2		Conventional	Kong[7]
UU A/B	*mba* gene and upstream regions	UMSUS, UMA2A2		Conventional	Kong[7]
UU A	*mba* gene and upstream regions	UMS-61, UMA2A		Conventional	Kong[7]
UU A	*mba* gene and upstream regions	UMSUS, UMA2A		Conventional	Kong[7]
UU B	*mba* gene and upstream regions	UMS-61, UMA10A		Conventional	Kong[7]
UU B	*mba* gene and upstream regions	UMSUS, UMA10A		Conventional	Kong[7]
UU C	*mba* gene and upstream regions	UMS-61, UMA4A1		Conventional	Kong[7]
UU C	*mba* gene and upstream regions	UMSUS, UMA4A1		Conventional	Kong[7]
UU C	*mba* gene and upstream regions	UMS-61, UMA4A2		Conventional	Kong[7]
UU C	*mba* gene and upstream regions	UMSUS, UMA4A2		Conventional	Kong[7]
UU D	*mba* gene and upstream regions	UMS-61, UMA9A1		Conventional	Kong[7]
UU D	*mba* gene and upstream regions	UMSUS, UMA9A1		Conventional	Kong[7]
UU D	*mba* gene and upstream regions	UMS-61, UMA9A2		Conventional	Kong[7]
UU D	*mba* gene and upstream regions	UMSUS, UMA9A2		Conventional	Kong[7]
UU E	*mba* gene and upstream regions	UMS-61, UMA7A1		Conventional	Kong[7]
UU E	*mba* gene and upstream regions	UMS-61, UMA7A2		Conventional	Kong[7]
UU E	*mba* gene and upstream regions	UMSUS, UMA7A2		Conventional	Kong[7]
UU E	*mba* gene and upstream regions	UMS-61, UMA7A3		Conventional	Kong[7]
UU E	*mba* gene and upstream regions	UMSUS, UMA7A3		Conventional	Kong[7]

new specific regions for *Ureaplasma* speciation and serotyping by real-time PCR. One multiplex species-specific PCR assay and 14 serovar-specific monoplex PCR were developed and shown to accurately distinguish between the two species and among all 14 serovars without cross-reactivities.[6] *UU063* gene (NP_077893), which encodes a conserved hypothetical protein that is identical in all 4 *U. parvum* serovars, was used as target for *U. parvum*; a 15,072 bp open reading frame (ORF), UUR10_0680, that is conserved (>99.97%) in all 10 *U. urealyticum* serovars (serovar 10, GenBank ID ACI60066.1), was selected as target for *U. urealyticum*. Utilization of serovar-specific PCR assays to study large numbers of clinical specimens and/or *Ureaplasma* isolates may eventually answer the question of differential pathogenicity.

At least one molecular-based assay that includes detection of *Ureaplasma* spp. in clinical specimens is commercially available in some countries, but not in the United States at present. Seegene, Inc. (Rockville, MD, http://www.seegene.com) markets their product STD6 ACE Detection that simultaneously detects *Trichomonas vaginalis*, *M. hominis*, *Mycoplasma genitalium*, *Chlamydia trachomatis*, *Neisseria gonorrhoeae*, and *U. urealyticum* in endocervical/urethral

TABLE 42.4
Comparison of Real-Time PCR Assays for Detection of *Ureaplasma* spp.

Real-time PCR	Mallard[36]		Yi[35]	Cao[34]		Cao[37]		Xiao[6]
Format	Not reported		Allelic Discrimination	Not reported		Not reported		Hybriprobe
Amplicon Size	*U. urealyticum* 100 bp	*U. parvum* 99 bp	90 bp	90 bp	110 bp	UU2,4,5,7-14: 261 bp UP1,3,14: 216 bp UP6: 217 bp		152 bp
Reaction volume	50 µL		20 µL	20 µL		20 µL		20 µL
Buffer				1× PCR Buffer		1× PCR Buffer	1× PCR Buffer	4 µL 5× Multiplex DNA Master HybProbe buffer
$MgCl_2$	2× TaqMan Universal PCR Master Mix		10 µL 2× TaqMan PCR Master Mix	3.5 mmol/L		2.5 mmol/L	3.0 mmol/L	3mM
dNTPs				0.2 mmol/L		0.2 mmol/L	0.2 mmol/L	in buffer
DNA Pol				0.5 U Hotstart Taq DNA Pol		0.5 U Hotstart Taq DNA Pol	0.5 U Hotstart Taq DNA Pol	in buffer
Primer F	500 nM UUureF	700 nM UPureF	1 µL UU1613F	0.3 µmol/L UUureF	0.35 µmol/L UPureF	0.45 µmol/L UPF		0.3 µM UP063#1F and 0.2uM UU127#1F
Primer R	500 nM UUureR	700 nM UPureR	1 µL UU1524R	0.3 µmol/L UUureR	0.35 µmol/L UPureR	0.45 µmol/L UPR		0.5 µM UP063#1R and 0.2 µM UU127#1R
Probe(s)	0.250 mM UUure2 TaqMan (5′-6-FAM—MGB-3′)	0.250 mM UPure1 TaqMan (5′-6-FAM—MGB-3′)	250 nM UU-parvo (5′-FAM—TAMRA-3′) 250 nMUU-T960 (5′-VIC—TAMRA-3′)	0.20 µmol/L Uuure TaqMan (5′-6-FAM—MGB-3′)	0.20 µmol/L Upure TaqMan (5′-6-FAM—MGB-3′)	0.30 µmol/L UPFP14 (5′-HEX—MGB-3′) 0.25 µmol/L UPFP1 (5′-6-FAM—MGB-3′)	0.35 µmol/L UPFP3 (5'-HEX—MGB-3') 0.30 µmol/L UPFP6 (5′-6-FAM—MGB-3′)	0.15 µM UP063#1 Probe1 (3'-FL) 0.15 µM UP063#1 Probe2 (5'-LC Red 640) 0.2 µM UU127#1 Probe1 (3′-FL) 0.2 µM UU127#1 Probe2 (5′-LC Red 705)

continued

TABLE 42.4 (Continued)
Comparison of Real-Time PCR Assays for Detection of *Ureaplasma* spp.

Real-Time PCR	Mallard[36]		Yi[35]		Cao[34]		Cao[37]		Xiao[6]	
Template	5 µL		6 µL		1 µL		2 µL		2 µL	
Cycling Instrument	7900HT Sequence Detection System ABI		7900HT Sequence Detection System ABI		Not reported		iCycler iQ multicolor real-time PCR detection system Bio-Rad		Roche LightCycler 2.0	
Initial	50°C	2 min	56°C	2 min	94°C	10 min	95°C	5 min	40°C	10 min
	95°C	10 min	95°C	10 min					95°C	10 min
# cycles	45×		40×		40×		40×		45×	
	95°C	15 s	95°C	15 s	94°C	15 s	94°C	15 s	95°C	15 s
	60°C	60 s	60°C	60 s	58°C	30 s	52°C	60 s	55°C	10 s
Final Extension									72°C	9 s
positive when C_t	Not reported		≤35		≤35		≤35		not used with Light Cycler	
Gene Target	Urease (*UreB*)		Urease (*UreC*)		Urease (*UreG*)		*mba* (some in gene, some outside)		*U. parvum:UU063*; *U. urealyticum: UU127*	
Discrimination Level	Species		Species		Species		*U. parvum* serovars		Species	

swabs. The novel feature of the Seegene technology is a dual priming oligonucleotide system that contains two separate priming regions linked by a polydeoxyinosine spacer. The polydeoxyinosine has no binding specificity or priming, but instead forms a bubble or tiny D-loop that results in more stable priming. The Seegene STD6 ACE kit works with any thermocycler, and the post PCR assay is designed for either manual or automated gel electrophoresis.[45] The current *Ureaplasma* assay is not species specific. It amplifies a 130 bp region of the *ureD* gene cassette.[46] Seegene plans to market a new version of their kit for detecting urogenital pathogens that differentiates *U. urealyticum* from *U. parvum*, targeting the *U. urealyticum ureD* and *U. parvum ureC* genes.

PCR can be used in combination with other nucleic acid–based detection (NABD) techniques including reverse line hybridization blotting combined with multiplex PCR (mPCR/RLB)[47] and microarrays.[48] These methods require a visualization step based on specific interaction between biotin and conjugated streptavidin.

The mPCR/RLB technique integrates multiplex PCR (mPCR) using biotin-labeled primer pairs or biotin-dNTPs to generate labeled PCR products that hybridize with highly specific, membrane-bound, amine-labeled oligonucleotide probes by reverse line blot hybridization.[49] To visualize the hybridized PCR products, the membrane is incubated with peroxidase-labeled streptavidin and chemiluminescent substrate—for example, electrochemiluminescence (ECL) detection liquid. Chemiluminescence results can be detected by a light-sensitive film or a lumino-imager.[49] The mPCR/RLB technique has been used to develop a multiplex assay that detects 14 urogenital pathogens, including both *Ureaplasma* spp.[47]

The microarray used for the detection of *Ureaplasma* spp. and some other microorganisms is based on immobilizing probes on glass surface that hybridize to their complimentary biotin labeled-DNA targets produced by previous PCR procedure. The microarray is then incubated with gold-conjugated streptavidin. The interaction of biotin and streptavidin leads to silver precipitation onto the streptavidin-bound nanogold particles. This is visualized as black spots on the microarray.[48]

There are other NABD techniques that do not use the PCR that have been used for detection of pathogenic organisms such as loop-mediated isothermal amplification (LAMP),[50] ligase chain reaction (LCR),[51] nucleic acid sequence-based amplification (NASBA),[52] strand displacement amplification (SDA)[53] and in situ hybridization (ISH).[54] In some cases, their chief attribute is that their technology is not subject to patents surrounding PCR. Among these, only ISH has been specifically utilized for detection of ureaplasmas in published investigations. ISH allows visualization of the localized gene expression within the context of tissue morphology. ISH was shown to be useful in detection of *Ureaplasma* in lung tissue samples collected from intranasally inoculated newborn mice. After fixation of the tissue samples, the slides were incubated with biotinylated DNA probes specific for an internal nucleotide sequence within the urease gene. The hybridization signal was amplified by binding of peroxidase-conjugated streptavidin to biotin and byotinyl tyramide incubations. A chromogenic dye producing a brown precipitate at the hybridization sites was used to visualize the hybridization signal.[54]

Selection of Culture versus Molecular-Based Tests for *Ureaplasma* Detection. While there is no question that molecular-based tests have been extremely important for research studies investigating *Ureaplasma* pathogenesis, epidemiology, and disease over the past two decades, culture still plays a significant role in routine diagnostic testing of clinical specimens obtained from patients in whom a clinically significant *Ureaplasma* infection is suspected. The discussion above makes a strong and clear-cut case for the many advantages of real-time PCR over conventional PCR for detection and speciating ureaplasmas. Thus, the decision is quite straightforward in selecting real-time PCR over conventional PCR if financial resources permit. It is somewhat more complex deciding whether PCR or culture-based detection should be employed on a routine basis. Some important comparative aspects of culture versus real-time PCR are summarized in Table 42.5. Even though real-time PCR is theoretically more sensitive than culture and can provide same-day results, it may not be a practical methodology for daily runs in a clinical laboratory that receives only an occasional specimen submitted for *Ureaplasma* detection or which does not have suitable physical facilities, molecular diagnostic equipment, or personnel with the necessary education and training to perform NABD techniques. Use of real-time PCR for research purposes when samples can be collected and assayed in batches is quite different from a clinical setting in which patient specimen volume and frequency of submission may not be predictable. Each PCR assay requires reagents for positive and negative and internal controls as well as the specimen

TABLE 42.5
Comparison of Culture versus Real-Time PCR for Detection of *Ureaplasma* spp.

Variable	Culture	Real-time PCR
Turnaround time	2–5 days	1 day
Organism viability required	Yes	No
Isolate available for further tests	Yes	No
Useful to assess efficacy of antimicrobial eradication	Yes	No[a]
Quantitative	Yes	Yes
Practical to test specimens one at a time	Yes	No
Analytical sensitivity (number of CFUs detected)	10–100	≤1
Differentiates *Ureaplasma* species	No	Yes
Detects *Mycoplasma hominis* simultaneously	Yes	Yes[b]

[a] PCR assay may amplify DNA from nonviable organisms that have been killed by antibiotics for unknown periods but it may be of some value to monitor organism load over time.

[b] *M. hominis* can be detected only by using a second set of multiplex PCR primers.

to be tested, so the cost per test decreases if more tests are performed simultaneously. In contrast, the cost of culture is essentially the same whether one specimen or multiple specimens are processed at the same time. The reagent cost of running a single real-time PCR assay for *Ureaplasma* detection and speciation is approximately fourfold greater than the costs for liquid and solid media for culture-based diagnosis, assuming the media is prepared on site. If purchased commercially, the cost of media is greater but is still considerably less than the cost of running a single PCR assay.

42.2 METHODS

42.2.1 Sample Preparation

Sample Collection. Any clinical specimens suitable for culture, including body fluids such as cerebrospinal fluid, peritoneal fluid, pleural fluid, semen, urine, and blood; nasal and endotracheal aspirates; urethral or cervical swabs; endometrial or lung tissues; bone; and the bacterial isolates themselves, are also suitable for diagnostic testing for ureaplasmas by molecular methods if they are collected, stored, and processed correctly. For the best results, specimens to be analyzed by a PCR assay should be collected in sterile tubes or vials with caps secured tightly to prevent leakage, frozen at −80°C within one hour of collection, transported on dry ice to the laboratory where the test will be performed, and remain frozen until DNA isolation. Blood (minimum 0.5 mL) should be collected in a vacutainer tube containing acid citrate dextrose. PCR can be performed directly on undiluted fluids. If an insufficient volume of original specimen necessitates dilution, culture transport media such as Shepard's 10 B broth[22] or PBS (phosphate buffered saline, pH 7.4, without calcium and magnesium) buffer can be used. Specimens should be inoculated into 0.3–0.7 mL of transport media or PBS buffer at time of collection or as soon as possible thereafter. For swab specimens, use only Dacron or polyester swabs. Calcium alginate and cotton swabs can be inhibitory. Swab specimens should be rinsed in either culture transport media or PBS buffer and the swab extracted. Excess fluid should be removed from the swab by pressing the swab against the inside of the tube or cryovial before removing and discarding the swab. We have determined that the 10 B broth used for culture does not have any deleterious effect on performance of the PCR assay we utilize, so it is suitable as a transport medium. However, it is possible that this culture broth could be inhibitory when using other primers and reaction conditions. Thus, it is mandatory to verify that culture broth is not inhibitory before using it for PCR transport. The advantage of using culture broth for PCR transport is that it can be used interchangeably for specimen transport whether culture, PCR, or both assays are to be performed on the same specimen. Use of transport medium is of greater importance for detection of ureaplasmas by culture than by PCR, since viability of the organisms must be maintained for variable periods of time, depending on proximity of the laboratory to the clinical facility where specimens are obtained.

DNA Extraction. The simple lysis and proteinase K treatment usually yields PCR-detectable DNA from *Ureaplasma* unless the specimen is inhibitory.[39] Suitable specimens for this procedure include body fluids (other than blood) and transport media containing material obtained from swabs. Potentially inhibitory specimens including blood, tissue samples, lower respiratory secretions, and subcultures of *Ureaplasma* isolates should be purified by the QIAamp® DNA Blood Mini Kit (Qiagen) or other commercial genomic DNA purification kits. Tissues should be minced in 0.5 mL 10 B broth or PBS buffer using sterile scissors prior to DNA extraction.

Automated or semiautomated DNA isolation methods such as Qiagen BioRobot EZ1 (Qiagen), NucliSENS® (bioMérieux), or MagNaPure LC (Roche Applied Science) can also be used to prepare DNA samples. Each laboratory must perform an evaluation of every assay component from sample collection (swab type), transport media, DNA extraction method, to final PCR amplification and detection procedures using the specific primers and probes, reaction conditions, and controls applicable to the assay to ensure the techniques are valid and there is no PCR inhibition at any step.

To avoid contamination, DNA preparation should be carried out in a biological safety cabinet and separated from the PCR set up area. Fresh gloves must be worn when changing from patient specimens to controls and vice versa. Dedicated filtered pipette tips for each processing area should be used in all PCR-related procedures. Clinical samples should be processed prior to processing positive controls.

DNA Extraction Procedure Based on Proteinase K Digestion

1. Chill a refrigerated bench top centrifuge to 4°C. Make certain two incubators have equilibrated to 60°C and 95°C, respectively. Label the tubes for sample processing.
2. In the biosafety cabinet, thaw proteinase K lysis buffer (0.5 mg/mL, stored in −80°C freezer) and samples to room temperature. Once all the reagents and samples are thawed, keep them on ice.
3. In the biosafety cabinet, vortex each sample and transfer the sample to the sample processing tube.
4. Centrifuge at 14,000 × *g* for 20 min at 4°C. Make sure the tubes are oriented in a manner such that the pellet is easily observed.
5. In the biosafety cabinet, discard the supernatant into the biohazard waste bag using a sterile filtered tip. Remove as much supernatant as possible.
6. Add 200 μL proteinase K lysis buffer to the tube and vortex briefly at high speed to dissolve the pellet.
7. Incubate at 60°C for 1 h then heat-inactivate the proteinase K by incubating samples at 95°C for 10 min.
8. Store the deproteinized samples at −80°C until use.

42.2.2 Detection Procedures

Publications describing real-time PCR for detection and characterization of ureaplasmas have utilized the ABI Prism

7900HT (Applied Biosystems)[36,37]; the iCycler iQ (Bio-Rad),[35,41] and the LightCycler 2.0 (Roche).[6] A customized protocol for the Roche LightCycler 2.0 for detection and speciation of *Ureaplasma* spp. in clinical specimens using primers and probes we developed is provided below.

Procedure

1. Set up PCR in a specifically designated biosafety cabinet. Clean cabinet first with 70% ethanol or 10% chlorine bleach. Prepare a work sheet. Calculate total reaction numbers and reagent volumes. Each sample is run in duplicate. Include a single reaction with external control to test for possible inhibitors. This protocol utilizes DNA extracted as described earlier. Turn on the computer, LightCycler 2.0, and LightCycler Carousel Centrifuge. Open LightCycler Software 4 and setup programs. Name and save a new file for each run. Run self-test. Prepare materials for PCR: (i) Pipettes: a set of 2, 20, 200 and 1000 μL pipettes, wipe before and after using with 70% ethanol or 10% chlorine bleach; (ii) Corresponding Pipette tips; (iii) LightCycler PCR capillaries (20 μL); (iv) LightCycler Centrifuge Adaptor (precooled at 4°C). Put the glass capillaries into the adaptor; (v) Precooled (4°C) racks (Benchtop freezer); and (vi) Small Biohazard waste bag.
2. Prepare Master Mix. Collect reagents from freezers. Let frozen reagents thaw on ice. Add all ingredients together in the listed sequence below; mix well by pipetting up and down, spin down for 5 s. Aliquot 16 μL of the Master Mix to all PCR capillaries.

PCR-grade H_2O	3.65 μL
25 mM $MgCl_2$	1.60 μL
Primer 1: UP063#1F (10 μM)	0.60 μL
Primer 2: UP063#1R (10 μM)	1.00 μL
Primer 3: UU127#1F (10 μM)	0.40 μL
Primer 4: UU127#1R (10 μM)	1.00 μL
Probe 1: UP063#1 probe 1 (4 μM)	0.75 μL
Probe 2: UP063#1 probe 2 (4 μM)	0.75 μL
Probe 3: UU127#1 probe 1 (4 μM)	1.00 μL
Probe 4: UU127#1 probe 2 (4 μM)	1.00 μL
LightCycler® Uracil-DNA Glycosylase (2U/μL)	0.25 μL
5× LightCycler® Multiplex DNA Master HybProbe	4.00 μL
Total	16.00 μL

3. Add 4 μL of PCR-grade H_2O to the negative control capillary and cap it.
4. Transfer clinical samples to the biosafety cabinet. Add 2 μL of PCR-grade H_2O to all the other capillaries except those for internal controls. Add 2 μL of clinical sample to the corresponding capillary. Discard tips in the biohazard waste bag. Cap all of the duplicate-reaction capillaries. Make sure not to cross contaminate the samples. Decontaminate pipette and change gloves frequently.
5. Remove negative and clinical samples from the cabinet. Transfer positive controls to the hood. Change gloves when switching form patient samples to positive controls, or vice versa. Add 2 μL of positive control (mixture of *U. parvum* serovar 3 and *U. urealyticum* serovar 10 genomic DNA) and external control (which is also positive control) to the corresponding capillaries, and cap them.
6. Insert the capillaries into the corresponding holes of PCR carousel. Centrifuge the carousel using the LightCycler Carousel Centrifuge 2.0.
7. Check to be sure all liquid sinks to the bottom of the capillary. Put the carousel into the LightCycler 2.0 and close the lid.
8. PCR program:

Program	Cycles	Analysis Mode	Target (°C)	Hold (hh:mm:ss)	Slope (°C/s)	Acquisition Mode
Preincubation	1	None	40	00:10:00	20	None
			95	00:10:00	20	None
Amplification	50	Quantification	95	00:00:15	20	None
			55	00:00:10	20	Single
			72	00:00:09	20	None
Melting	1	Melting Curves	95	00:00:00	20	None
			60	00:00:30	20	None
			95	00:00:00	0.1	Continuous
Cooling	1	None	40	00:00:30	20	None

9. Click "Start Run" to start the PCR program.
10. Generate color compensation file. The crosstalk between individual channels must be compensated using a color compensation object. For each set of parameters (new lot of oligonucleotides), a Color Compensation object has to be prepared and has to be generated in a separate run.

    Master Mix for Color Compensation object:

PCR-grade H_2O	6.15 μL
25 mM $MgCl_2$	1.60 μL
Primer1: UP063#1F (10 μM)	0.60 μL
Primer 2: UP063#1R (10 μM)	1.00 μL
Primer 3: UU127#1F (10 μM)	0.40 μL
Primer 4: UU127#1R (10 μM)	1.00 μL
LightCycler® Uracil-DNA Glycosylase (2U/μL)	0.25 μL
5× LightCycler® Multiplex DNA Master HybProbe	4.00 μL
Total	15.00 μL

Multiply the amount by the number of reactions (×5). Pipet 15 μL of master mix to the capillary. Add additional water and probes to the corresponding capillary.

Capillary	Water (μL)	Probe (μL)
Blank	5.00	
Probes	4.00	0.50 μL of UP063#1 Probe 1; 0.50 μL of UU127#1 Probe1
LC-Red 640 Probe		5.00
LC-Red 705 Probe		5.00

Cycle the samples with the programs above. When the experiment is finished, select "Color Compensation" from the "Analysis" menu, and click "Save CC Object."

11. After completion of PCR, use LCS4 software to analyze results. First, perform "Absolute Quantification" analysis. When probes annealed to the target DNA, fluorescence (640 or 705 nm) emitted by fluorescence resonance transfer (FRET) from donor probes to the acceptor probes (UP063#1 Probel to UP063#1 Probe2 or UU127#1 Probel to UU127#1 Probe2) are measured in channel 640/back530 (for *U. parvum*) and channel 705/back530 (for *U. urealyticum*). Apply valid Color Compensation file to the analysis. Check the cross-point (C_p) value of each PCR sample. Positive control should show a C_p value around 30. Negative control should not show a C_p value. Then, perform "Tm Calling" analysis. Settings are set as "high sensitive" and "2 peaks or less." Positive control and samples with internal control should show similar Tm values, which are 67.56 ± 0.51°C for *U. parvum* serovar 3 and 65.72 ± 0.30°C for *U. urealyticum* serovar 10 (Figure 42.1). Positive samples should also show similar Tms as positive controls. Negative control and samples should not have Tm values in this range. An inhibited PCR is determined by attenuated or no amplification of internal control in the presence of clinical sample.
12. Clean the pipettes and the biosafety cabinet with 70% ethanol or 10% chlorine bleach.
13. If necessary, load potentially inhibited samples and controls to a 2% agarose gel to verify the PCR results. The positive and internal control amplicon size should be 152 bp, and the negative control should not show the 152 bp amplicon.

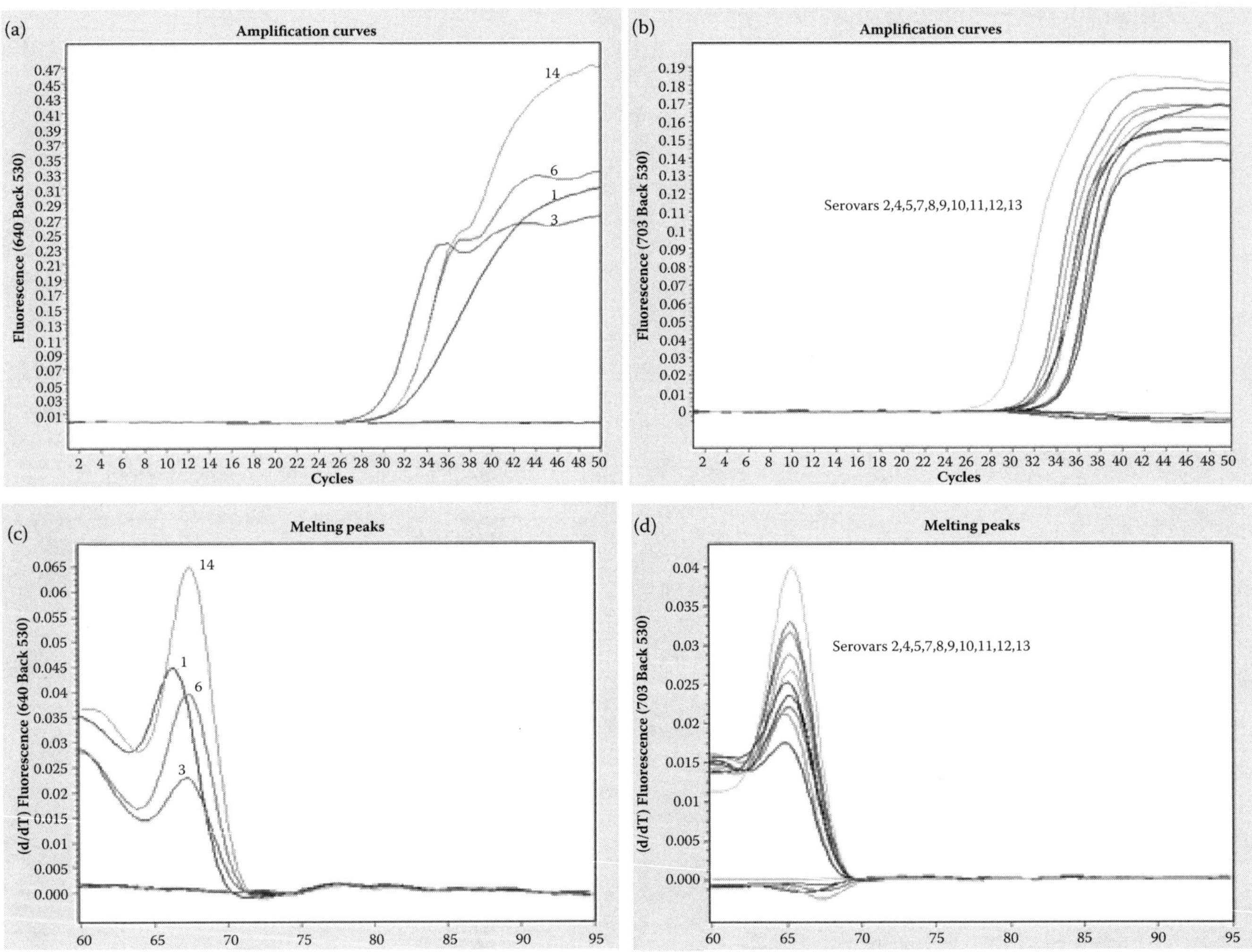

FIGURE 42.1 Multiplex real-time PCR detection of *U. parvum* and *U. urealyticum*. Amplification curves (a, b) and melting peaks for *Ureaplasma* species-specific multiplex real-time PCR assays tested against 14 ATCC serovar type strains demonstrating simultaneous distinction between *U. parvum* (a,c) and *U. urealyticum* (b,d) and detection of all serovars of each species. The melting temperature was about 67.5°C for *U. parvum* and 65.7°C for *U. urealyticum*. (From Xiao, L., et al., *J. Clin. Microbiol.*, 48, 2715–2723, 2010. With permission of American Society of Microbiology.)

Quality Control. Good quality control procedures should limit the risk of false-positive and false-negative results of PCR assays. The false-positive results from contamination are a major problem for conventional PCR but have now been minimized with real-time PCR. The carryover contamination can be eliminated by uracil-DNA glycosylase (UNG).[55] Personnel training and laboratory design also help minimize contamination.[56]

In addition to human errors, reasons for false-negative results include the presence of PCR inhibitors in the clinical specimen, suboptimal reagent preparation and reaction conditions, and inefficient extraction of the target DNA. Many substances present in clinical samples, such as hemoglobin, and certain compounds used for DNA extraction, such as ethanol or detergents, are potent amplification inhibitors.[57,58] These inhibitory factors and suboptimal PCR conditions can be detected by simply mixing a positive control DNA with the sample after purification.[59] However, this external control strategy cannot reveal inefficient extraction of target DNA. In this case, use of an internal control overcomes the limitations of the external control. The internal control can be a plasmid or oligonucleotide containing a sequence similar to or not related to the target, but can be differentiated from the assay PCR.[31,60] The internal control is added directly to the crude sample and coprocessed for purification and amplification with the sample. This type of internal control is thought to be the most accurate method to modulate the important steps of diagnostic PCR protocols.[60,61]

Determination of Assay Sensitivity and Specificity. The PCR analytical sensitivity test should be performed against serial dilutions of template DNA, either bacterial genomic DNA from a defined inoculum titer or a plasmid containing the target sequence and can be expressed in terms of amount of DNA detected or numbers of organisms (CFU). The analytical specificity should also be tested against other pathogens that appear in the same body locations or show sequence similarities to the targets. Human genomic DNA should always be included in the test because of its presence in clinical specimens and possible inhibitory effects. Due to different calculation units used in various published studies, the sensitivity of reported *Ureaplasma* PCR assays cannot be compared directly with one another. For *Ureaplasma* spp. detection and differentiation, early conventional PCR assays showed analytical sensitivities of about 1–10 colony forming units (CFU) for detecting any *Ureaplasma* spp.[39]; or 0.12–1.2 color changing units (CCU) per μL of the original sample for detecting *U. parvum*, and 12–120 CCU/μL of the original sample for detecting *U. urealyticum*.[31] More recently developed real-time PCR assays revealed detection limits of 5–10 copies/reaction mixture.[34,36,62] The detection limits established for our multiplex assay described above were 2.8×10^{-2} CFU/μL PCR mixture for *U. parvum* and 4.1×10^{-2} CFU/μL PCR mixture for *U. urealyticum*.[6] It is suggested that before a laboratory can begin routine PCR-based diagnostic work, it must be able to demonstrate that its molecular results compare favorably with conventional techniques to establish a clinical sensitivity for the assay.[56]

42.3 CONCLUSION AND FUTURE PERSPECTIVES

The first culture methods developed for human ureaplasmas in the 1950s were further refined over the next several years so that basic and clinical studies could be performed to investigate their pathogenic role in disease. Commercial marketing of agar and broth media based on the original formulations described by researchers facilitated the adoption of ureaplasma cultures for diagnostic purposes by larger hospital laboratories and reference laboratories, so now it is a somewhat commonly performed procedure in the United States. Widespread availability of diagnostic kits in several European and Asian countries has also contributed to more broad use of culture-based diagnosis for ureaplasmal infections. The development of PCR-based detection systems has been obligatory for very fastidious species such as *M. genitalium* and are now the tests of choice for *M. pneumoniae*. However, the need for PCR-based detection of *Ureaplasma* spp. and *M. hominis* has not been as critical because of their relative ease of in vitro cultivation and rather rapid growth. PCR-based studies have been necessary to provide data to support the separation of *U. parvum* and *U. urealyticum* as individual species and to enable complete characterization of all 14 serovars in a rapid and reproducible manner. Since the question of differential pathogenicity of the two human *Ureaplasma* species and their respective serotypes has not been answered conclusively, there is no requirement for identification of ureaplasmas to the species or serovar level for diagnostic purposes. Therefore, laboratories should consider the relative attributes of culture versus PCR for routine detection and make a decision based on their physical facilities, financial abilities, and staffing. We anticipate that molecular-based technology is likely to become much more widely available and simplified in the future, making it appeal to a broad range of clinical laboratories. If more diagnostic companies choose to include gene targets for *Ureaplasma* spp. in commercial PCR kits, especially if such products can be used to detect other pathogens in the same assay, their use is likely to increase. Availability of the complete genomic sequence data for all 14 serovars will facilitate development of such additional assays.

ACKNOWLEDGMENTS

The authors have received support from the National Institute of Allergy and Infectious Diseases, National Institutes of Health, Department of Health and Human Services under grant RO1A1072577 and contract N01-AI-30071. We thank Lynn Duffy for her critical review of the chapter and helpful suggestions concerning laboratory diagnosis of ureaplasmal infections.

REFERENCES

1. Shepard, M.C., The recovery of pleuropneumonia-like organisms from Negro men with and without nongonococcal urethritis, *Am. J. Syph. Gonor. Vener. Dis.*, 38, 113, 1954.

2. Shepard, M. et al., *Ureaplasma urealyticum* gen. nov., sp. nov.: Proposed nomenclature for the human T (T-strain) mycoplasmas, *Int. J. Syst. Bacteriol.*, 24, 160, 1974.
3. Robertson, J.A. et al., Proposal of *Ureaplasma parvum* sp. nov. and emended description of *Ureaplasma urealyticum* (Shepard et al. 1974), *Int. J. Syst. Evol. Microbiol.*, 52, 587, 2002.
4. Waites, K.B., and Taylor-Robinson, D., *Mycoplasma* and *Ureaplasma*. In Murray, P.R. et al. (Editors), *Manual of Clinical Microbiology*, 9th ed., pp. 1004–1020, ASM Press, Washington DC, 2007.
5. Waites, K.B., Katz, B., and Schelonka, R.L., Mycoplasmas and ureaplasmas as neonatal pathogens, *Clin. Microbiol. Rev.*, 18, 757, 2005.
6. Xiao, L. et al., Detection and characterization of human *Ureaplasma* species and serovars by real-time PCR, *J. Clin. Microbiol.*, 48, 2715, 2010.
7. Kong, F. et al., Molecular genotyping of human *Ureaplasma* species based on multiple-banded antigen (MBA) gene sequences, *Int. J. Syst. Evol. Microbiol.*, 50, 1921, 2000.
8. Zheng, X. et al., Small repeating units within the *Ureaplasma urealyticum* MB antigen gene encode serovar specificity and are associated with antigen size variation, *Infect. Immun.*, 63, 891, 1995.
9. Teng, L.J. et al., *Ureaplasma urealyticum* biovar specificity and diversity are encoded in multiple-banded antigen gene, *J. Clin. Microbiol.*, 32, 1464, 1994.
10. Novy, M.J. et al., *Ureaplasma parvum* or *Mycoplasma hominis* as sole pathogens cause chorioamnionitis, preterm delivery, and fetal pneumonia in rhesus macaques, *Reprod. Sci.*, 16, 56, 2009.
11. Glass, J.I. et al., The complete sequence of the mucosal pathogen *Ureaplasma urealyticum*, *Nature*, 407, 757, 2000.
12. Kim, M. et al., Biovar diversity of *Ureaplasma urealyticum* in amniotic fluid: Distribution, intrauterine inflammatory response and pregnancy outcomes, *J. Perinat. Med.*, 31, 146, 2003.
13. Deguchi, T. et al., Association of *Ureaplasma urealyticum* (biovar 2) with nongonococcal urethritis, *Sex. Transm. Dis.*, 31, 192, 2004.
14. Abele-Horn, M. et al., Association of *Ureaplasma urealyticum* biovars with clinical outcome for neonates, obstetric patients, and gynecological patients with pelvic inflammatory disease, *J. Clin. Microbiol.*, 35, 1199, 1997.
15. Heggie, A.D. et al., Identification and quantification of ureaplasmas colonizing the respiratory tract and assessment of their role in the development of chronic lung disease in preterm infants, *Pediatr. Infect. Dis. J.*, 20, 854, 2001.
16. Katz, B. et al., Characterization of ureaplasmas isolated from preterm infants with and without bronchopulmonary dysplasia, *J. Clin. Microbiol.*, 43, 4852, 2005.
17. Martinez, M.A. et al., Occurrence and antimicrobial susceptibility of *Ureaplasma parvum* (*Ureaplasma urealyticum* biovar 1) and *Ureaplasma urealyticum* (*Ureaplasma urealyticum* biovar 2) from patients with adverse pregnancy outcomes and normal pregnant women, *Scand. J. Infect. Dis.*, 33, 604, 2001.
18. Domingues, D. et al., Genital mycoplasmas in women attending a family planning clinic in Guine-Bissau and their susceptibility to antimicrobial agents, *Acta Trop.*, 86, 19, 2003.
19. Sanderson, M.W. et al., Prevalence and reproductive effects of *Ureaplasma diversum* in beef replacement heifers and the relationship to blood urea nitrogen level, *Theriogenology*, 54, 401, 2000.
20. Vasconcellos Cardoso, M. et al., Detection of *Ureaplasma diversum* in cattle using a newly developed PCR-based detection assay, *Vet. Microbiol.*, 72, 241, 2000.
21. Waites, K.B., Talkington, D.F., and Bebear, C.M., Mycoplasmas. In Truant, A.L. (Editor), *Manual of Commercial Methods in Clinical Microbiology*, pp. 201–224, American Society for Microbiology, Washington DC, 2002.
22. Waites, K.B. et al., *Mycoplasma* and *Ureaplasma*. In Isenberg, H. (Editor), *Clinical Microbiology Procedure Handbook*, pp. 3.15.1–3.15.17 ASM Press, Washington DC, 2004.
23. Saiki, R.K. et al., Primer-directed enzymatic amplification of DNA with a thermostable DNA polymerase, *Science*, 239, 487, 1988.
24. Blanchard, A., *Ureaplasma urealyticum* urease genes; use of a UGA tryptophan codon, *Mol. Microbiol.*, 4, 669, 1990.
25. Willoughby, J.J. et al., Isolation and detection of urease genes in *Ureaplasma urealyticum*, *Infect. Immun.*, 59, 2463, 1991.
26. Birch, D.E., Simplified hot start PCR, *Nature*, 381, 445, 1996.
27. Robertson, J.A. et al., Polymerase chain reaction using 16S rRNA gene sequences distinguishes the two biovars of *Ureaplasma urealyticum*, *J. Clin. Microbiol.*, 31, 824, 1993.
28. Yoshida, T. et al., Phylogeny-based rapid identification of mycoplasmas and ureaplasmas from urethritis patients, *J. Clin. Microbiol.*, 40, 105, 2002.
29. Kong, F. et al., Species identification and subtyping of *Ureaplasma parvum* and *Ureaplasma urealyticum* using PCR-based assays, *J. Clin. Microbiol.*, 38, 1175, 2000.
30. Harasawa, R., and Kanamoto, Y., Differentiation of two biovars of *Ureaplasma urealyticum* based on the 16S-23S rRNA intergenic spacer region, *J. Clin. Microbiol.*, 37, 4135, 1999.
31. Povlsen, K., Jensen, J.S., and Lind, I., Detection of *Ureaplasma urealyticum* by PCR and biovar determination by liquid hybridization, *J. Clin. Microbiol.*, 36, 3211, 1998.
32. Kong, F. et al., Comparative analysis and serovar-specific identification of multiple-banded antigen genes of *Ureaplasma urealyticum* biovar 1, *J. Clin. Microbiol.*, 37, 538, 1999.
33. Teng, L.J. et al., Rapid detection and biovar differentiation of *Ureaplasma urealyticum* in clinical specimens by PCR, *J. Formos. Med. Assoc.*, 94, 396, 1995.
34. Cao, X. et al., Real-time TaqMan polymerase chain reaction assays for quantitative detection and differentiation of *Ureaplasma urealyticum* and *Ureaplasma parvum*, *Diagn. Microbiol. Infect. Dis.*, 57, 373, 2007.
35. Yi, J., Yoon, B.H., and Kim, E.C., Detection and biovar discrimination of *Ureaplasma urealyticum* by real-time PCR, *Mol. Cell. Probes*, 19, 255, 2005.
36. Mallard, K., Schopfer, K., and Bodmer, T., Development of real-time PCR for the differential detection and quantification of *Ureaplasma urealyticum* and *Ureaplasma parvum*, *J. Microbiol. Methods*, 60, 13, 2005.
37. Cao, X. et al., Two multiplex real-time TaqMan polymerase chain reaction systems for simultaneous detecting and serotyping of *Ureaplasma parvum*, *Diagn. Microbiol. Infect. Dis.*, 59, 109, 2007.
38. Harasawa, R. et al., Detection and tentative identification of dominant mycoplasma species in cell cultures by restriction analysis of the 16S-23S rRNA intergenic spacer regions, *Res. Microbiol.*, 144, 489, 1993.
39. Blanchard, A. et al., Detection of *Ureaplasma urealyticum* by polymerase chain reaction in the urogenital tract of adults, in amniotic fluid, and in the respiratory tract of newborns, *Clin. Infect. Dis.*, 17 Suppl 1, S148, 1993.
40. Ren, Y., and Zhu, X., Investigation on biovars and genotypes of *Ureaplasma urealyticum* in the cervix in a Chinese gynecologic check-up population and sex workers, *Acta Derm. Venereol.*, 83, 175, 2003.

41. Knox, C.L., Giffard, P., and Timms, P., The phylogeny of *Ureaplasma urealyticum* based on the mba gene fragment, *Int. J. Syst. Bacteriol.*, 48 Pt 4, 1323, 1998.
42. Ruifu, Y. et al., Biovar diversity is reflected by variations of genes encoding urease of *Ureaplasma urealyticum*, *Microbiol. Immunol.*, 41, 625, 1997.
43. Zimmerman, C.U. et al., Alternate phase variation in expression of two major surface membrane proteins (MBA and UU376) of *Ureaplasma parvum* serovar 3, *FEMS Microbiol. Lett.*, 292, 187, 2009.
44. Bastien, P., Procop, G.W., and Reischl, U., Quantitative real-time PCR is not more sensitive than "conventional" PCR, *J. Clin. Microbiol.*, 46, 1897, 2008.
45. Chun, J.Y. et al., Dual priming oligonucleotide system for the multiplex detection of respiratory viruses and SNP genotyping of CYP2C19 gene, *Nucleic Acids Res.*, 35, e40, 2007.
45. Kim, T.H., Detection of cryptic microorganisms in patients with chronic prostatitis by multiplex polymerase chain reaction, *Korean J. Urol.*, 3, 304, 2007.
47. McKechnie, M.L. et al., Simultaneous identification of 14 genital microorganisms in urine by use of a multiplex PCR-based reverse line blot assay, *J. Clin. Microbiol.*, 47, 1871, 2009.
48. Cao, X. et al., Visual DNA microarrays for simultaneous detection of *Ureaplasma urealyticum* and *Chlamydia trachomatis* coupled with multiplex asymmetrical PCR, *Biosens. Bioelectron.*, 22, 393, 2006.
49. Kong, F., and Gilbert, G.L., Multiplex PCR-based reverse line blot hybridization assay (mPCR/RLB)—a practical epidemiological and diagnostic tool, *Nat. Protoc.*, 1, 2668, 2006.
50. Mori, Y., and Notomi, T., Loop-mediated isothermal amplification (LAMP): A rapid, accurate, and cost-effective diagnostic method for infectious diseases, *J. Infect. Chemother.*, 15, 62, 2009.
51. Wiedmann, M. et al., Ligase chain reaction (LCR)—overview and applications, *PCR Methods Appl.*, 3, S51, 1994.
52. Loens, K. et al., Development of real-time multiplex nucleic acid sequence-based amplification for detection of *Mycoplasma pneumoniae*, *Chlamydophila pneumoniae*, and *Legionella* spp. in respiratory specimens, *J. Clin. Microbiol.*, 46, 185, 2008.
53. Barken, K.B., Haagensen, J.A., and Tolker-Nielsen, T., Advances in nucleic acid-based diagnostics of bacterial infections, *Clin. Chim. Acta*, 384, 1, 2007.
54. Benstein, B.D. et al., *Ureaplasma* in lung. 1. Localization by in situ hybridization in a mouse model, *Exp. Mol. Pathol.*, 75, 165, 2003.
55. Burkardt, H.J., Standardization and quality control of PCR analyses, *Clin. Chem. Lab. Med.*, 38, 87, 2000.
56. Syed, F.F., Millar, B.C., and Prendergast, B.D., Molecular technology in context: A current review of diagnosis and management of infective endocarditis, *Prog. Cardiovasc. Dis.*, 50, 181, 2007.
57. Saiki, R.K. et al., Enzymatic amplification of beta-globin genomic sequences and restriction site analysis for diagnosis of sickle cell anemia, *Science*, 230, 1350, 1985.
58. Association for Molecular Pathology statement. Recommendations for in-house development and operation of molecular diagnostic tests, *Am. J. Clin. Pathol.*, 111, 449, 1999.
59. Kwok, S., and Higuchi, R., Avoiding false positives with PCR, *Nature*, 339, 237, 1989.
60. Burggraf, S., and Olgemoller, B., Simple technique for internal control of real-time amplification assays, *Clin. Chem.*, 50, 819, 2004.
61. Al-Soud, W.A., and Radstrom, P., Purification and characterization of PCR-inhibitory components in blood cells, *J. Clin. Microbiol.*, 39, 485, 2001.
62. Yoshida, T. et al., Quantitative detection of *Ureaplasma parvum* (biovar 1) and *Ureaplasma urealyticum* (biovar 2) in urine specimens from men with and without urethritis by real-time polymerase chain reaction, *Sex. Transm. Dis.*, 34, 416, 2007.

Section III

Bacteroidetes, Chlamydiae, and Fusobacteria

43 Bacteroides

Rama Chaudhry and Nidhi Sharma

CONTENTS

43.1 INTRODUCTION

43.1.1 Description of *Bacteroides*

The genus *Bacteroides* belongs taxonomically to the family Bacteroidaceae, order Bacteroidales, class Bacteroidia, phylum Bacteroidetes. Of the six genera (i.e., *Acetofilamentum*, *Acetomicrobium*, *Acetothermus*, *Anaerorhabdus*, *Bacteroides*, and *Capsularis*) within the family Bacteroidaceae, the genus *Bacteroides* (*bacter*, rod; *-oides*, from *eidos*, shape or figure; *Bacteroides*, rodlike) is not only very complex (with a large number of individual species) but also includes several important pathogens of humans and animals.

Bacteroides are strict anaerobes, and many species are obligate parasite of humans and animals. The genus Bacteroides is the most important constituents of normal human flora of gastrointestinal tract. In 1863, Pasteur reported a variety of anaerobic bacteria that could only be grown in gaseous environment having substantially reduced oxygen tension. With the improvement in anaerobic technology over the past 20–30 years, a high incidence of anaerobic organisms has been demonstrated in clinical specimens and better understanding of these organisms has evolved. In humans anaerobic bacteria are normally prevalent in the oral cavity, gastrointestinal tract and genital organ. The *Bacteroides* are gram-negative, nonspore-forming bacilli. *Bacteroides* are commonly found in the human intestine, where they have a symbiotic host-bacterial relationship with humans. They assist in breaking down food and in producing valuable nutrients and energy that the body needs. However, when *Bacteriodes* are introduced to parts of the body other than the gastrointestinal area, they can cause or exacerbate abscess. From these endogenous sources, a wide variety of human infection may occur, the common infections caused by these bacteria are periodontal disease; postaspiration, pleuro-pulmonary infection; genital tract infection in women, and intraabdominal abscesses. *Bacteroides fragilis* is the most common species and its putative presence has implications for antimicrobial selection.[1]

Anaerobic bacteria can cause infection when a normal barrier is damaged due to surgery, injury, or disease. Usually, the immune system kills invading bacteria, but sometimes the bacteria are able to grow and cause infection. Body sites that have tissue destruction (necrosis) or poor blood supply are low in oxygen and favor the growth of anaerobic bacteria such as *Bacteroides* and *Clostridia* spp.[2] This chapter highlights the classification, biology, epidemiology, and pathogenesis of *Bacteroides* spp. and their molecular detection methods.

43.1.2 Classification

Professor WEC Moore and his colleagues at Virginia Polytechnic Institute of Anaerobic Laboratory in the United States carried out the meticulous systemic and taxonomic studies of nonclostridial anaerobes. Professors Moore and Holdmen are the major contributors in Bergey's manual for the section dealing with classification of *Bacteroides*, *fusobacterium*, *propionibacterium*, and *Eubacterium*.[3] Its classification had undergone many changes since Veillon and Zuber named their isolates, *Bacillus fragilis* and *Bacillus fusiformis*. Castellani and Chalmers changed it to genus *Bacteroides* in the year 1994. In the eighth edition of Bergey's manual, all the member of *Bacteroides* were put in a single species; *B. fragilis* with five subspecies (*fragilis*, *vulgatus*, *distasonis*, *ovatus*, and *thetaiotamicron*). In the ninth edition of Bergey's manual, 42 species were added to genus Bacteroides, and later on, two more genera of clinical importance were identified *Prevotella* and *Porphyromonas*. Finally, the genus includes species that were formally described as the *B. fragilis* group, contain more than 15 new genera, and have been proposed for organisms classified as *Bacteroides*, most of which are monospecific. The important species of *Bacteroides* are as follows: *Bacteroides caccae*, *Bacteroides distasonis*, *Bacteroides eggerthii*, *B. fragilis*, *Bacteroides merdae*, *Bacteroides ovatus*, *Bacteroides stercoris*, *Bacteroides thetaiotaomicron*, *Bacteroides uniformis*, and *Bacteriodes vulgatus*.

Anaerobes make up the majority of bacterial flora found in the human colon; the predominant bacteria being *Bacteroides*. The colon contains over 400 species of organisms and has more than 10^{11} organisms per gram of wet weight. *Bacteroides* sp. constitutes nearly 10^{11} organisms per gram of feces (dry weight). These anaerobic bacteria enhance health of the human host by helping catabolize complex molecules such as fucosylated glycans.[5]

Bacteroides spp. are playing a fundamental role in this ecosystem by processing complex molecules into simpler compounds (used by the human host as well as the *Bacteroides*). *Bacteroides* spp. are known to break down a wide variety of indigestible dietary plant polysaccharides, such as amylose, amylopectin, and pullulan. *B. thetaiotaomicron* is a dominant member of normal intestinal flora of humans and animals. It has a capacity to use a variety of host derived glycan, including chondroitin sulfate, mucin, hyaluronate, and heparin; 61% of its glycosylhydrolases are present in periplasm or outer membrane or extracellular space. These enzymes are not only important for fulfilling its needs but also improving the metabolic milieu of intestinal ecosystem as a source of energy (calories as fermentation products of dietary polysaccharides). The ability to convert complex polysaccharides into usable compounds might allow *Bacteroides* to be more competitive than bacteria that must rely on other sources of energy. Furthermore, *Bacteroides* actually stimulates the gut lining to produce fucosylated glycans. *B. thetaiotaomicron* also stimulate angiogenesis (formation of blood vessels) in the newborn epithelium, enhancing host capacity of absorbing nutrients. Thus, *Bacteroides* bacteria have a complex and generally beneficial relationship with their host—so long as they are retained within the gut lumen.[6,7]

43.1.3 Biology

Bacteroides are gram-negative, nonspore-forming, anaerobic, rod-shaped bacteria. They have an outer membrane, a peptidoglycan layer, and a cytoplasmic membrane. The main byproducts of their anaerobic respiration are acetic acid, isovaleric acid, and succinic acid. They are involved in many important metabolic activities in the human colon, including the fermentation of carbohydrates, the utilization of nitrogenous substances, and the biotransformation of bile acids and other steroids. Most intestinal bacteria are saccharolytic, which means they obtain carbon and energy by hydrolysis of carbohydrate molecules.[8]

It is estimated that only about 2% of simple sugars make it past the upper gastrointestinal tract and to the *Bacteroides*. Thus, simple sugars are probably not *Bacteroides* main source of energy. However, polysaccharides from plant fibers, such as cellulose, xylan, arabinogalactan, and pectin, and vegetable starches such as amylose and amylopectin, are much more prevalent in the colon. These polysaccharides have also been shown to induce a variety of glucosidase activities of *Bacteroides* including a β-1,3-glucosidase activity responsible for laminarin degradation and a variety of α- and β-1,4 and -1,6 xylosidase and glucosidase activities. A large part of the *Bacteroides* 4779-member proteome includes proteins that hydrolyze these polysaccharides.[9]

B. thetaiotaomicron have been shown to bind to polysaccharides with their outer membrane receptor system before pulling the polysaccharides into the periplasm for monosaccharide degradation. This technique may help insure that the polysaccharides are not stolen by other intestinal organisms or lost in the intestine by diffusion. *Bacteroides* polysaccharide utilization genes are thought to be controlled by repressor/inducer mechanisms.[10]

Bacteroides vulgatus is a member of the normal distal human gut microbiota. The distal gut microbiota contain more bacterial cells than all of our body's other microbial communities combined. More than 90% of phylogenetic types belong to two divisions, the bacteroidetes and the firmicutes, with the remaining types distributed among eight other divisions. *B. vulgatus* is the only sequenced bacteroidetes that possesses a gene coding for a xylanase. It has the largest and most complete set of enzymes that target pectin.[8] Although *Bacteroides* are gram-negative, as are most bacteria in the human colon, and live in the same environment as *E. coli*, the two bacteria are actually less closely related than *Bacteroides* are to gram-positive bacteria.

Genomic and proteomic advancement have upgraded the knowledge and understanding of the unique adaptive nature of *Bacteroides* species. The genome of the circular chromosome of many *Bacteroides* species has been studied; research is being done on sequencing *Bacteroides* species in order to understand their pathogenic properties. Almost all species of *Bacteroides* have 40%–48% G – C composition. The genome of *B. fragilis* (NCTC9343) is 5,205,140 bp long and has 43.19% G – C content. Much of the genome is controlled by sigma factor, which responds to environmental factors. A large part of the proteins made by the *Bacteroides* genome goes to breaking down polysaccharides and metabolizing their sugars. There have been a total of three genome projects done on two different species of *Bacteroides*. The genomes of *B. thetaiotaomicron* VPI-5482, *B. fragilis* YCH46, and *B. fragilis* NCTC 9343 were sequenced.[9]

The completion of the sequencing projects for *B. thetaiotaomicron* in 2003 and *B. fragilis* in 2004–2005,[11,12] and subsequent proteomic analyses, have vastly added to the understanding of the manner in which these organisms adapt to, and thrive in, the human gut, such as (i) complex systems to sense the nutrient available and tailor nutrient-metabolizing systems accordingly, (ii) multiple pump systems to rid the bacteria of toxic substances, and (iii) the ability to control the environment by interacting with the host immune system so that it controls other (competing) pathogens.[13]

43.1.4 Epidemiology and Transmission

Bacteroides are the predominant species in the oral cavity, gastrointestinal tract and vaginal tract in bacterial vaginosis, otherwise lactobacilli are the predominant in vagina. Most infection is acquired endogenously, when the integrity of the colonized mucosa or lumen is breached by trauma or during surgery. *B. vulgatus* and *B. thetaiotaomicron* are more commonly isolated from feces. An enterotoxin producing *B. fragilis* (ETBF) has been associated with diarrhea. When *Bacteroides* escape the gut, they are responsible for many types of infections and abscesses that can occur all over the body, including the central nervous system, the head, the neck, the chest, the abdomen, the pelvis, the skin, and the soft tissues (Figure 43.1).

The widely accepted model for abdominal infections goes as follows: after disruptions of the intestinal wall, bacterial flora infiltrate the cavity, aerobes (the most active part in infection) like *E. coli* start the preliminary tissue destruction and decrease the oxidation-reduction potential of the oxygenated tissue (low oxidation-reduction potential favors anaerobe growth); anaerobic *Bacteroides* start to replicate, and then dominate the infection.[14]

Along with diarrhea and abscesses, *Bacteroides* have been known to be involved in cases of meningitis and shunt infections, especially in children. Any tissue not normally colonized with *Bacteroides* has a possibility of infection when introduced to mucus or other materials containing *Bacteroides*. Diagnosis and treatment are complicated due to the slow growth of *Bacteroides*, the increasing resistance to antibiotics, and the polymicrobial nature of infection with *Bacteriodes* (most infections deemed *Bacteroides*, can actually involve between 5–10 organisms). *B. fragilis*, which occurs with the most in clinical infections, along with *B. distasonis*, *B. ovalus*, *B. thetaiotaomicron*, and *B. vulgatus* are resistant to penicillins, mostly due to the production of beta-lactamase.

43.1.5 Clinical Features and Pathogenesis

Bacteroides are prevalent in the gastrointestinal area and are responsible for most intraabdominal infections, such as perirectal abscesses and decubitus ulcers. In addition, *Bacteroides* present a huge problem as a source of infection during gastrointestinal surgeries. Thus, proper measures and

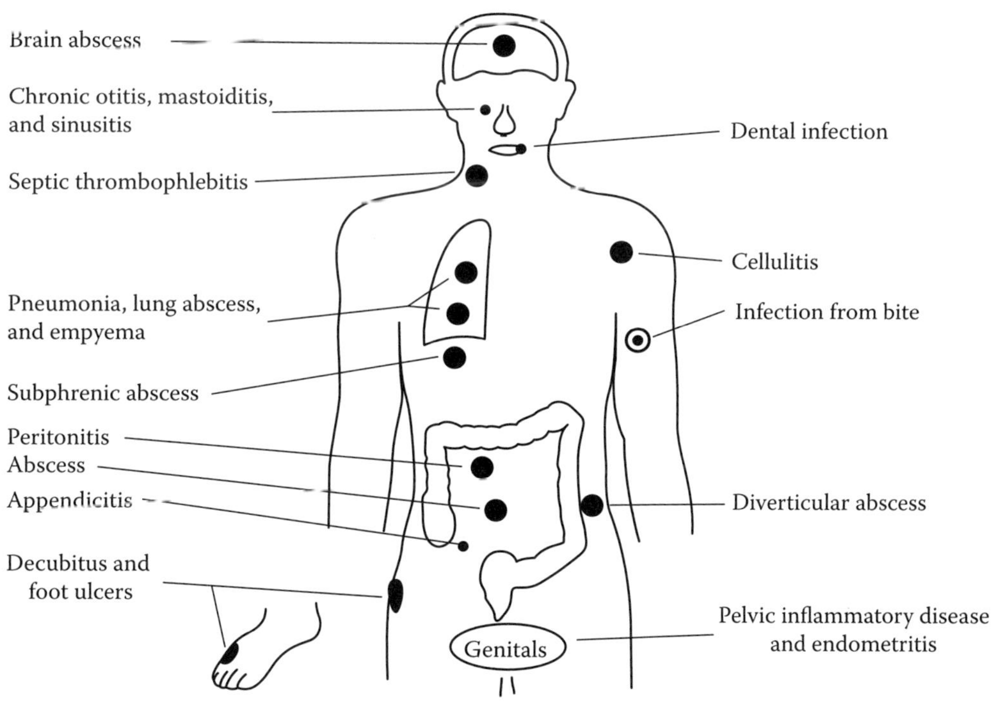

FIGURE 43.1 Common sites of *Bacteroides* and other anaerobic bacterial infections. (From www.ncbi.nlm.gov.)

efficient surgical draining must take place in order to lower the risk of infection.[15]

B. fragilis accounts for only 0.5% of the human colonic flora. It is the most commonly isolated anaerobic pathogen, due to potent virulence factors such as (i) adherence to tissues, (ii) protection from the host immune response (such as oxygen toxicity and phagocytosis), or (iii) destruction of tissues. The fimbriae and agglutinins of *B. fragilis* allow them to be established in the host tissue. The polysaccharide capsule, LPS, and a variety of enzymes protect it from the host immune response. The capsule is responsible for abscess formation, and histolytic enzymes found in *B. fragilis*, can mediate tissue destruction. The actual formation of the abscess is an example of a pathological host response to the invading bacterium: a fibrous membrane localizes invading bacteria and surrounds a mass of cellular debris, dead polymorphonuclear leukocytes, and a mixed population of bacteria. Abscesses left untreated can expand and can even cause intestinal obstruction, erosion of resident blood vessels, and ultimately fistula formation. Abscesses may also rupture and result in bacteremia and disseminated infection. *B. fragilis* is the only bacterium that has been shown to induce abscess formation as the sole infecting organism. Abscess formation has been clearly linked to the *B. fragilis* capsule in an animal model.[16] This has been attributed to the presence of a polysaccharide capsule or to the action of one or more of the extracellular- or membrane-associated enzymes that it forms: proteinases, including collagenase, fibrinolysin, hemolysin, neuraminidase, phosphatase, DNase, hyaluronidase, chondroitin sulfatase, and heparinase, which appears to play its role in periodontal disease and abscess formation.

These organisms carry gene encoding for virulent products involved in pathogenicity (e.g., enzymes), synthesizing capsular polysaccharides, which inhibit phagocytosis and abscess formation, fimbriae, and pili, promoting adherence to epithelial cells and mucus, neuraminidase, and enterotoxin.[17–19] Some *B. fragilis* strains produce a potent zinc-dependent metalloprotease (enterotoxin) with a variety of pathological effects on intestinal mucosal cells. This *ETBF* has been associated with diarrhea in animal and human.[20,21]

B. fragilis forms less than 10% of the *Bacteroides* in normal human feces, yet it is by far the most common species of bacteroides group to be isolated from intestinal infections, blood, and abscesses. It thus appears to be particularly pathogenic to human beings. *B. thetaiotaomicron* is a common commensal in normal human feces and the second most common species of the genus isolated from clinical infections.[22]

List of Bacteroides spp.

- *Bacteroides acidifaciens*
- *Bacteroides barnesiae*
- ***Bacteroides caccae***
- *Bacteroides capillosus*
- *Bacteroides cellulosilyticus*
- *Bacteroides coprocola*
- *Bacteroides coprophilus*
- *Bacteroides coprosuis*
- *Bacteroides denticanum*
- *Bacteroides dorei*
- ***Bacteroides distasonis***
- ***Bacteroides eggerthii***
- *Bacteroides finegoldii*
- ***Bacteroides fragilis***
- *Bacteroides galacturonicus*
- *Bacteroides gallinarum*
- *Bacteroides helcogenes*
- *Bacteroides heparinolyticus*
- *Bacteroides intestinalis*
- ***Bacteroides merdae***
- *Bacteroides nordii*
- ***Bacteroides ovatus***
- *Bacteroides pectinophilus*
- *Bacteroides plebeius*
- *Bacteroides pyogenes*
- *Bacteroides salanitronis*
- *Bacteroides salyersiae*
- ***Bacteroides stercoris***
- *Bacteroides tectus*
- ***Bacteroides thetaiotaomicron***
- ***Bacteroides uniformis***
- ***Bacteroides vulgatus***
- *Bacteroides xylanolyticus*
- *Bacteroides zoogleoformans*

43.1.6 Laboratory Identification

43.1.6.1 Phenotypic Techniques

Isolation and Identification. The genus *Bacteroides* comprises small, nonmotile, gram-negative bacilli and cocco bacilli. They are moderately pleomorphic. They grow well on brain heart infusion agar (BHIA) and *Bacteroides* bile esculin agar (BBE), when the plates are incubated anaerobically at 37°C for 24–48 h.

Pure colonies of *B. fragilis* on BBE and BHIA are circular, low convex (1–3 mm in diameter) with an entire edge. They are usually smooth, shiny, translucent, or semiopaque. They are nonhemolytic. A granular precipitate around the colonies is an important characteristic of *B. fragilis* sp. They are further identified by the usage of conventional biochemical tests or by using API 20A/API 32 A system (bioMérieux). Some of the species within the group are very similar biochemically and therefore difficult to differentiate (Wardsworth Anaerobic Manual).

Biochemical Tests. *B. fragilis* is catalase negative and grows well in 20% bile. It has resistance to kanamycin, vancomycin, and colistin by special potency identification antibiotic discs. It hydrolyses esculin and ferments three major sugars viz, glucose, sucrose, and lactose (Table 43.1). In fact, the above-mentioned methods are quiet helpful, but due to long processing time to complete the identification, these tests are no longer to be considered a choice. Hence the introduction of rapid tests based on molecular methods are the techniques of choice.[23,24]

TABLE 43.1
Biochemical Reactions to Identify the *B. fragilis* Group

Biochemical	*B. caccae*	*B distasonis*	*B. fragilis*	*B. merdae*	*B. ovatus*	*B. stercoris*	*B. thetaiotaomicron*	*B. uniformis*	*B. vulgatu*	*B. eggerthii*
Growth in 20% bile	+	+	+	+	+	+	+	w+	+/–	+
Indole	–	–	–	–	+	+	+	+	–	+
Catalase	–/+	+/–	+	–	+/–	–	+	–/+	–/+	–
Esculin-hydrolysis	+	+	+	+	+	+/–	+	+	–/+	+
Arabinose	+	–/+	–	–/+	+	–/+	+	+	+	+
Cellobiose	+	+	+/–	v	+	–/+	+/–	+	–	–/+
Rhamnose	+	v	–	+	+	+	+	–/+	+	+/–
Salicin	–/+	+	–	+	+	–/+	–/+	+/–	–	–
Sucrose	+	+	+	+	+	+	+	+	+	–
Trehalose	+	+	–	+	+	–	+	+	–	–
Xylose	+	+	+	+	+	+	+	+	+	+
Xylan	–	–	–	–	+	v	–	–	–/+	+
α-fructosidase	+	–	+	–	+/–	v	+	+	+	–
Glucose	+	+	+	+	+	+	+	+	+	+
Lactose	+	+	+	+	+	+	+	+	+	+
Mannitol	–	–	–	–	+	–	–	–	–	–

43.1.6.2 Molecular Techniques

Over the past several years, the development and application of molecular diagnostic techniques have initiated a revolution in the diagnosis and monitoring of infectious diseases. Molecular methods such as nucleic acid amplification with polymerase chain reaction and organic probe have been used profusely over the last decades for direct detection of fastidious organisms. PCR methods are usually more sensitive than culturing. PCR methods' gene amplification has also been used for the detection of pathogenic bacteria and to differentiate pathogenic strains from nonpathogenic strains.

Other nucleic acid techniques, such as plasmid profiling, various methods for generating restriction fragment length polymorphisms, and PCR, are making increasing inroads into clinical laboratories. PCR-based systems that detect the etiologic agents of disease directly from clinical samples without the need for culture have been useful in the rapid detection of uncultivable or fastidious microorganisms. In addition, sequence analysis of amplified microbial DNA allows for identification and better characterization of the pathogen. Subspecies variation, identified by various techniques, has been shown to be important in the prognosis of certain diseases.[25]

The identification of *B. fragilis* in clinical microbiology laboratory is sometimes difficult, as it is a fastidious organism and takes 3–5 days to grow in culture. Therefore, it becomes essential to apply a rapid and sensitive molecular diagnostic technique to detect neuraminidase gene specific for *B. fragilis* and enterotoxin gene to detect ETBF directly from fecal samples of patients with diarrhea. Neuraminidase enzyme is produced by all *B. fragilis* strains in vivo and in vitro at elevated levels. The neuraminidase gene *nan*H of *B. fragilis* has been partially sequenced. To date, this has been one of the methods of choice for the detection of *B. fragilis*. The direct detection of *B. fragilis* from clinical specimens by using the PCR method for amplifying a specific fragment of the neuraminidase gene and enterotoxin gene from *B. fragilis* were already described.[23]

The other species that are most frequently isolated from that *B. fragilis* group are *B. vulgatus*, *B. thetaiotaomicron*, *B. distasonis* and, less frequently, *B. eggerthii*. The *B. fragilis* group is commonly associated with a variety of human infections other than the diarrhea described above, such as intraabdominal abscesses, wound infections, and bacteremia. The *B. fragilis* group bacteremia contribute significantly to morbidity and mortality. In fact, *B. fragilis* accounts for 63% of *Bacteroides*-related clinical infections; *B. thetaiotaomicron* for 14%; *B. vulgatus* and *B. ovatus* for 7% each; *B. distasonis* for 6%; and *B. uniformis* for 2%.[26,27]

In the past decades, molecular techniques have been developed for specific detection of *B. fragilis* group, and some of these techniques have been shown to have adequate sensitivity and specificity. Kuwaara et al.[28] assessed PCR amplification and sequencing analysis of structural variation in 16S-23S rRNA internal transcribed spaces region (ITS) among *Bacterodes* for detection of some medically important strains of Bacteroides. Seven strains of *Bacteroides* (i.e., *B. distasonis*, *B. eggerthii*, *B. fragilis*, *B. ovatus*, *B. thetaiotaomicron*, *B. uniformis*, and *B. vulgatus*) produced one to three ITS amplification products with size ranging from 810 to 815 bp.[29] Other Bacteroides strain could be differentiated at species level on the basis of ITS amplification pattern and restriction fragment length polymorphism (RFLP) analysis using nucleotide-recognizing enzyme *Msp* I.

A multiplex PCR was recently developed by Liu et al.[30] to identify the 10 *B. fragilis* group: *B. caccae*, *B. distasonis*, *B. eggerthii*, *B. fragilis*, *B. merdae*, *B. ovatus*, *B. stercoris*, *B. thetaiotaomicron*, *B. uniformis*, and *B. vulgatus*. Based on the sequence analysis of the 16S rRNA gene, the 16S-23S rRNA ISR, and a variable region of the 23S rRNA gene, the 10 *B. fragilis* group species, a two-step multiplex PCR identification scheme was established to rapidly and accurately identify the *B. fragilis* group species. All these species were divided into three subgroups by multiplex PCR-G followed by three multiplex PCR assay with three species-specific primer mixtures for identification to the species level. The primers were designed from nucleotide sequences of the species 16S rRNA, the 16S-23S rRNA intergenic spacer region (ISR), and part of the 23S rRNA gene.

43.2 METHODS

43.2.1 Sample Preparation

43.2.1.1 Sample Collection and Processing

Collection and Transportation of Clinical Specimen. Specimens must be protected from deleterious effects of oxygen until they can be cultured in proper anaerobic transport medium; anaerobic bacteria can survive for up to several days. In the case of a fecal specimen, as it contains enzymes that degrade the bacterial component, it should be collected in wide-mouthed container and inoculated as soon as possible on respective media within 2 h of collection.

Specimen Processing. Specimens can be processed on the open bench-top with immediate incubation under anaerobic conditions. For *B. fragilis*, clinical samples streak on bacteroides bile esculin agar (BBE), brain heart infusion agar (BHIA) supplemented with hemin and Vit K (Diffco), and incubated anaerobically at 37°C for 24 h and 48–72 h. Anaerobiosis for the growth of *B. fragilis* and other anaerobes was generated by using ANOXOMAT System (MART Sint-Genesius-Rode, Belgium) at an anaerobic gaseous atmosphere (N_2_80%, H_2_10%, and CO_2_10%) in an anaerobic jar.

BBE agar is useful for the rapid isolation and presumptive identification of the *B. fragilis* group. The medium contains 100 g/mL of gentamicin to inhibit most of aerobic organisms and 20% bile to inhibit other anaerobes. As differential agents, esculin and iron have been incorporated into the agar to aid in detecting the esculin-positive *B. fragilis* group organisms. BBE agar is also a useful medium due to H_2S production. Other organisms that may grow on this medium are *Fusobacterium mortiferum*, *F. varium*, gentamicin-resistant *Entreobacteriaceae*, *enterococci*, *pseudomonads*, *staphylococci*, and yeast. However, the colony size of the facultative anaerobic organism is usually less than 1 mm in diameter. Selective media (i.e., BBE) may be examined after 24 h of incubation, as selective organisms grow rapidly.

Examination of Culture. After incubation, observe colonial characteristics and record the colony description in detail (i.e., color, hemolysis, florescence, pigment, and shape and size with peripheral striations for crenate/entire edge). On BBE agar, they hydrolyze esculin to blacken the agar. When all colony types are confirmed as pure anaerobes, do the Gram stain and further subculture on BHIA for special potency disc (kanamycin 1 mg, colistin 10 μg, vancomycin 5 μg, sodium polyanethosulfonate 1 mg, penicillin 2 U). A zone size of >10 mm is considered as sensitive or <10mm considered as resistant. The special potency antibiotic disc pattern (colistin-resistant and vancomycin-sensitive) will demonstrate the gram positivity of the isolates.

43.2.1.2 DNA Extraction

(i) Extraction of DNA from Bacterial Strains by Phenol–Chloroform Method Suspend one or two pure colonies from the culture plate in 100 μL of 1× TE buffer (10 mM Tris-HCl pH 7.5; 1 mM EDTA), and 200 μL of lysis buffer [200 mM NaCl, 20 mM EDTA, 20 mM Tris-HCl pH 8.0, 4% (w/v) SDS, 1 mg/ml proteinase K] in a microcentrifuge tube. Incubate it in a water bath at 60°C for 2 h. After cell lyses, extract DNA by phenol-chloroform-isoamyl alcohol (PCI) method (25:24:1).[31,32] The step-wise procedure is given in detail.

Take 100 μL of clinical sample (Stool/pus) in 1.5 mL microcentrifuge tube (MCT)

↓

Add equal volume of 1× TE buffer (100 μL)

↓

Mix with 200 μL of lysis buffer

↓

Keep at water bath (60°C) for 1 h

↓

Add phenol-chloroform-isoamyl alcohol (equal volume at room temp.) and mix gently by inverting

↓

MCT till the milky consistency then centrifuge at 6000 rpm for 10 min

↓

A complex is formed between aqueous and organic phase

↓

Transfer aqueous phase into another MCT

↓

Add equal vol. of PCI to the aqueous phase and mix properly till the milky consistency

↓

Transfer aqueous phase into another MCT

↓

Centrifuge at 6000 rpm for 10 min

↓

Add equal vol. of PCI to the aqueous phase and mix properly till the milky consistency (repeat above steps once more)

Transfer supernatant to fresh tube and add cold absolute alcohol (double the volume) and 4 M NaCl (1/10th of the volume), invert the tubes 5–10 times (a thread-like structure will appear)

↓

Keep it at –20°C for overnight

↓

Centrifuge at high speed for 15 min at room temperature

↓

Decant off supernatants and wash with 70% alcohol and spin for 15 min

↓

Decant supernatant and dry the pellet at 37°C for 1 h

↓

Dissolve in 30 μL 1× TE buffer and stored at 4°C or use immediately for PCR

(ii) Extraction of DNA from Bacterial Strains by Boiling Method Extraction by boiling is the simplest and a rapid method of DNA-preparation for PCR. Briefly, colonies isolated and identified as *B. fragilis* are picked up with sterile loop and suspended in a sterile phosphate buffer saline (PBS) pH 7.2 in 1.5 mL microcentrifuge tube and wash the cells with 1 × PBS by giving spin at 4000 × *g* for 1 min (2–3 times). Pellet in 50 μL sterile double distilled water and boil it for 3–5 min, then give a short spin (30 s) in a centrifuge tube (eppendorf) and collect the resuspended supernatant (40 μL) in another tube that contains DNA.

Culture the strain on BHI agar/Robertson Cooked Meat medium (RCM)

↓

Take few colonies and suspended in sterile PBS (1 mL) or 1 mL of cultured RCM in 1.5 mL tube and spin at 4000 × g for 1 min

↓

Discarded the supernatant and add PBS for washing the cells (repeat this step twice or thrice)

↓

Resuspend the pallet in 50 μL sterile distilled water and boil it at 100°C for 3 min

↓

Give short spin (30 s), collect the supernatant (containing DNA) in another 1.5 mL tube for use in PCR

(iii) Extraction of DNA from Clinical (Pus/Stool) Samples Homogenize one loop full sample with 200 μL sterile PBS or normal saline and vortex it for 1 min. Take 100 μL from the sample in 1.5 mL tube. Add 100 μL of TE buffer [10 mM Tris-HCl pH 7.5, 1 mM EDTA] and 200 μL of lysis buffer [200 mM NaCl, 20 mM EDTA, 20 mM Tris-HCl pH 8.0, 4% (w/v) SDS, 1 mg/mL proteinase K] in tube. Incubate it in water bath at 60°C for 2 min. After cell lyses, extract DNA by phenol-chloroform-isoamyl alcohol method and proceed for PCR as described.[21]

Take one loop full of stool* sample (fresh within 10 min of collection) in 1.5 mL tube, add 200 μL sterile PBS or NS and homogenize by adding glass beads (2–3 autoclaved) and vortex for 1 min

↓

Keep the tubes in racks for 1 min and after the fecal matter settles, take 100 μL of supernatant in 1.5 mL tube

↓

Add 100 μL of TE buffer and 200 μL of lysis buffer to tube containing clinical sample

↓

Incubate it in water bath at 60°C for 2 h

↓

After cell lyses, extract DNA by phenol–chloroform method (25:24:1) and then proceed for PCR

* If it is diarrhea sample than do not add saline, use 100 μL of sample for DNA extraction
(Extraction should include spiked stool sample)

43.2.2 Detection Procedures

43.2.2.1 PCR Assay for Neuraminidase Gene of *B. fragilis*

Neuraminidase enzyme is produced by all *B. fragilis* strains at elevated levels in comparison to other *Bacteroides* sp. considered as genus specific (where failure of cultivation of this bacteria from fecal material), PCR assay for detection of neuraminidase gene of *B. fragilis* is a suitable and alternative option. It is carried out by using primers, designed to amplify a 262 bp sequence, specific for neuraminidase gene as reported by Jatwani et al.[33] Forward primer GAI 11 (5′-GCC GGT CAG AATGGG AGT AGG AGA CC-3′) and reverse primer GAI 12 (5′-CCC GAC GAG CCG GAC CTT GCA ACA GA-3′) are used. The PCR reaction mixture consists of 50 pM of each primer, 100 μM of deoxynucleotide triphosphate (dNTP) each, 1 U Taq DNA polymerase, 1× PCR buffer containing 1.5 $MgCl_2$, and 5 μL of DNA template in 25 μL of reaction volume. PCR cycle consists of 35 cycles of initial denaturation at 94°C for 1 min, denaturation at 94°C for 1 min, annealing at 62°C for 1 min, extension at 72°C for 1 min, and final extension at 72°C for 5 min. The amplified PCR product (262 bp) is separated by agarose gel electrophoresis, stained with ethidium bromide and detected under UV light.

After the establishment of ETBF as a causative organism in diarrhea, it has now been well understood that mere detection of *B. fragilis* in the stool specimen is not sufficient. Therefore, currently PCR-based techniques are targeting the enterotoxin gene for confirmation of the presence of ETBF.

43.2.2.2 PCR Assay for Enterotoxin Gene of *B. fragilis*

DNA extraction from clinical specimens is similar to the process used for detection of neuraminidase gene. PCR assay for detection of enterotoxin gene of *B. fragilis* is carried out by using primers, designed to amplify a 294 bp sequence, specific for neuraminidase gene as reported by Pantosti et al.[34] Forward primer BF1 (5′-GAC GGT GTA TGTGAT TTG TCT GAG AGA-3′) and reverse primer BF2 (5′-ATC CCT AAG ATT TAT TATCCC AAG TA-3′) was used for PCR. The reaction mixture contained 100 μM of each dNTP, 50 pM of each primer, 1 U of DNA polymerase, and 10 μL of DNA template in final volume of 25 μL of 1× PCR buffer containing 1.5 mM $MgCl_2$. Samples are subjected to initial denaturation at 94°C for 3 min; 35 cycles of 94°C, 58°C, and 72°C for 1 min each; and a final extension at 72°C for 5 min. PCR product is electrophoresed in 1.6% agarose gel, stained with ethidium bromide and examined under Gel Doc System.

43.2.2.3 Two-Step Multiplex PCR Assays of the *B. fragilis* Group Species

Multiplex PCR utilizes multiple primer sets to amplify sequences of different length. Using a multiplex PCR, it is possible to amplify variety of target genes in single PCR, depending on the species and the presence of multiple genes, a two-step multiplex PCR identification scheme was established to rapidly and accurately identify the *B. fragilis* group species, based on the sequence analysis of the 16S rRNA gene, the 16S-23S rRNA ISR, and a variable region of the 23S rRNA gene of the 10 ATCC type strains representing the 10 *B. fragilis* group species.[30]

Based on the multialignment analysis data, a potential *B. fragilis*-specific primer pair, Bfr-F and Bfr-R, was selected from the 16S rRNA gene and two potential subgroup-specific downstream primers, BFR-G2 (for subgroup-II, which includes *B. eggerthii*, *B. stercoris*, and *B. uniformis*) and BFR-G3 (for subgroup-III, which includes *B. distasonis* and *B. merdae*), were selected from the 23S rRNA gene. In addition, 10 species-specific primer pairs were designed from the 16S-23S rRNA ISR region. These primers were designed with minimal differences in their annealing temperature within each primer set, and to yield amplification products that ranged between 200 and 700 bp and differed by at least 50 bp (Tables 43.2 and 43.3).

Ten species of the *B. fragilis* group were first grouped by multiplex PCR (designated multiplex PCR-G) and then further identified to the species level by three multiplex PCR assays (named multiplex PCR-I, multiplex PCR-II, and multiplex PCR-III).

One or two pure colonies form culture plate is resuspended in 1× TE buffer [10 mM Tris-HCl pH 8.0, 1 mM EDTA], incubated for 10 min at 95°C and centrifuged at 13,000 × *g* for 2 min. The lysate is used as DNA template.

PCR amplification was carried out in a total volume of 50 μL containing 1.25 U of Taq polymerase (Promega), 50 mM KCl, 10 mM Tris-HCl pH 9.0, 0.1% Triton, 2.5 mM $MgCl_2$, 0.5 mM (each) primer, 0.2 mM dNTPs, and 3 μL of bacterial lysate as the DNA template. PCR was carried out for 35 cycles. Each cycle consisted of 95°C for 20 s for denaturation, annealing for 1 min at 52°C for multiplex PCR-G, 62°C for multiplex PCR-I, 60°C for multiplex PCR-II, and 55°C for multiplex PCR-III; extension was performed at 72°C for 30 s. A cycle of 72°C for 5 min was added to the

TABLE 43.2
PCR Product to Detect *B. fragilis* Group

Method	Gene	Primer	Fragment Size	Species
				Sub gr. I
1-PCR-G	16S rRNA	Bfr-F & Bfr-R	230 bp	*B. caccae*
				B. frgilis
				B. ovatus
				B. thetaiotamicron
				B. vulgatus
				Sub gr. II
		Bfr-G2 & Bfr-G3	450 bp	*B. eggerthii*
				B. stercoris
				B. uniformis
				Sub gr. III
	23S rRNA	G23S-1 & G23S-2	400 bp	*B. merdae*
				B. distasoni
2-PCR-I	16S-23S rRNA ISR	1392A		**Sub gr. I**
		Bth-R	180 bp	*B. thetaiotamicron*
		Bcae-R	250 bp	*B. vulgatus*
		Bvul-R	420 bp	*B. fragilis*
		Bfra-R	500 bp	*B. caccae*
		Bova-R	610 bp	*B. ovatus*
3-PCR-II		23 R4		**Sub gr. II**
		Begg-F	250 bp	*B. eggerthii*
		Buni-F	350 bp	*B. uniformis*
		Bste-F	400 bp	*B. stercoris*
4-PCR-III				**Sub gr. III**
		Bdis-F & R	220 bp	*B. distasoni*
		Bmer-F&R	310 bp	*B. merdae*

TABLE 43.3
Primer Sequences

Bfr-F	CTGAACCAGCCAAGTAGCG
Bfr-R	CCGCAAACTTTCACAACTGACTTA
Bfr-G2	ATCAGGTTCGACTCTTGCT
Bfr-G3	CCGTCAGCTGGCAGGA
G23S-1	GTTGGCTTAGAAGCAGC
G23S-2	CATTTTGCCGAGTTCCTT
1392A	GTACACACCGCCCGT
Bth-R	ACCTATGAAATCGTTGTTACG
Bvul-R	GGCTTCTTACTTTCTCTCTTCCG
Bfra-R	GCTAATCCCCCAATCATAC
Bcac-R	TCGTTTCCCATTGCTGG
Bova-R	AATAATGCGTACTCGAACAC
23R4	GGGTTBCCCCATTCGG
Begg-F	GTCATATTAACGGTGGCG
Buni-F	TCCGTTTTCCACTTATAAGA
Bste-F	CTACGACATAGTCTTGGTGAG
Bdis-F	TGATCCCTTGTGCTGCT
Bdis-R	ATCCCCCTCATTCGGA
Bmer-F	GAGGTATGTAGCTCTCTGGTA
Bmer-R	TTTTTACCCCTTACGGAG

final extension. PCR products were analyzed by electrophoresis on a 6% polyacrylamide gel followed by ethidium bromide staining.[30]

43.2.2.4 Multiplex PCR Assay for Subtyping *B. fragilis*

ETBF produce a 20-kDa zinc metalloprotease toxin, BFT (*B. fragilis* toxin), which can be separated into three distinct subtypes. A multiplex PCR was reported recently for detection of three *B. fragilis* subtypes targeting the enterotoxin (*bft*) gene, with a common forward primer (GBF-201, 5′-GAA CCT AAA ACG GTA TAT GT-3′) and three reverse primers (BFET-TYPE1, 5′-ATT GAA CCA GGA CAT CCC T-3′, specific for *bft-1*; GBF-322, 5′-CGC TCG GGC AAC TAT-3′, specific for *bft-2*; and BFET-TYPE3, 5′-CGT GTG CCA TAA CCC CA-3′, specific for *bft-3*).[35]

B. fragilis isolates are grown on brucella blood agar under anaerobic conditions at 37°C for 48 h. A bacterial colony is taken from blood agar culture and resuspended in 200 μL of distilled water in a microcentrifuge tube. The sample is then boiled for 10 min and centrifuged for 1 min at 14,000 rpm, and 2 μL of supernatant is used for subsequent PCR.

PCR mixture (30-μL) is prepared with 1× buffer; 2 mM $MgCl_2$; 200 mM deoxynucleoside triphosphate; 20 pmol each of GBF-201, BFET-TYPE1, GBF-322, and BFET-TYPE3; 0.5 U of *Taq* DNA polymerase (Promega); and 2 μL of bacterial sample DNA.

The mixture is subjected to an initial denaturation at 94°C for 5 min; 34 cycles of 94°C for 3 min, 63°C for 1 min, and 72°C for 1 min; and a final extension at 72°C for 5 min.

The PCR products are separated by 6% polyacrylamide gel (45 min at 125 V) and visualized after ethidium bromide staining. ETBF subtypes are ascertained by the characteristic band patterns observed.[35]

43.3 CONCLUSION

Bacteroides spp. are obligately anaerobic, gram-negative, and saccharolytic; contain enzymes of the hexose monophosphate shunt-pentose phosphate pathway, have a DNA-base composition of about 40%–48% G – C content, and *Bacteroides* membranes contain a mixture of long-chain fatty acids and mainly straight chain-saturated, anteiso-methyl, and isomethyl branched acids. The *B. fragilis* group is one of the most important pathogens in polymicrobial infections. The distribution of the different members of the *B. fragilis* group in clinical infections varies. *B. fragilis* accounts for 63% of the entire group isolates, *B. thetaiotaomicron* for 14%, *B. vulgatus* and *B. ovatus* for 7% each, *B. distasonis* for 6%, and *B. uniformis* for 2%. All members of the group induce bacteremia that was associated with an average mortality of 27%. The *B. fragilis* group resists beta-lactam antibiotics by producing the enzyme beta-lactamase. These organisms carry genes encoding for virulent products involved in pathogenicity, abscess formation, fimbriae, and pili, promoting adherence to epithelial cells and mucus, neuraminidase, and enterotoxin. Some *B. fragilis* strains produce a potent zinc-dependent metalloprotease (enterotoxin) with a variety of pathological effects on intestinal mucosal cells. This *ETBF* has been associated with diarrhea.

Infectious diarrhea is mainly caused by both aerobic and anaerobic bacteria and viruses and parasites, but groups of organisms, especially *B. fragilis*, have been overlooked due to lack of awareness and suboptimal methods of diagnosis. Therefore, this chapter highlights the role of molecular methods especially PCR assay for detection of enterotoxigenic strains of *B. fragilis* and multiplex PCR for all 10 species of *B. fragilis* group. On the basis of culture technique, it is difficult to differentiate enterotoxigenic strain of *B. fragilis* from nonenterotoxigenic strains of *Bacteroides*. Molecular methods will be very useful not only in diagnosis but also in epidemiologic studies of *Bacteriodes* spp.

Nucleic acid amplification assays, and the methods to detect the product of amplification, have evolved rapidly in the past decade, culminating in the exiting application of real-time PCR and similar technologies. The number of molecular application for the detection and characterization of all types of microorganism increases each year, with many commercial venders seeking FDA approval for their application. Development of DNA microarray chips will be helpful for the detection of multiple pathogens simultaneously without any need of sophisticated equipments. As diarrhegenic agents are potential components of bioterrorism, with water as the transmission vehicle, chip-based technology will be valuable for their sensitive, specific, and rapid detection.

REFERENCES

1. Hardie, J.M., and Shah, H.N., Bacteroides and other non sporing anaerobes. In *Medical Microbiology*, 14th ed., ELBS Churchill Livingstone, Philadelphia, PA, 1992.
2. Sutter, V.L. et al., General consideration. In Jousimies-Somer, H. et al. (Editors), *Anaerobic Bacteriology Manual*, 4th ed., pp. 1–9, Star Publishing Company, Belmont, CA, 1986.

3. Holdeman, L.V.R., Kelly, R.W., and Moore, W.E.C., Genus I. Bacteroides Castellani and Chalmers 1919. 959AL. In Krieg, N.R., and Holt, J.G. (Editors), *Bergey's Manual of Systematic Bacteriology*, vol. 1, pp. 604–631, The Williams and Wilkins Co., Baltimore, 1984.
4. Shah, H.N., Gharbia, S.E., and Olsen, I., *Bacteroides, Prevotella*, and *Porhyromonas.* In Peter Borriello, S., Patrick, R., and Gudo Funke (Editors), *Bacteriology*, 10th edn, Vol 2, pp. 1913–1938, Hodder Arnold, London, 2005.
5. McCarthy, R.E., Pajeau, M., and Salyers, A.A., Role of starch as a substrate for *Bacteroides vulgatus* growing in the human colon, *Appl. Environ. Microbiol.*, 54, 1911, 1988.
6. Hooper, L.V., and Gordon, J.I., Commensal host-bacterial relationships in the gut, *Science*, 292, 115, 2001.
7. Xu, J. et al., Evolution of symbiotic bacteria in the distal human intestine, *PLoS Biol.*, 5, 1574, 2007.
8. Reeves, A.R., Wang G.R., and Salyers, A.A., Characterization of four outer membrane proteins that play a role in utilization of starch by *Bacteroides thetaiotaomicron*, *J. Bacteriol.*, 179, 643, 1997.
9. Xu, J. et al., A Genomic view of the Human–*Bacteroides thetaiotaomicron* symbiosis, *Science*, 299, 2074, 2003.
10. Reeves, A.R. et al., A *Bacteroides thetaiotaomicron* outer membrane protein that is essential for utilization of maltooligosaccharides and starch, *J. Bacteriol.*, 178, 823, 1996.
11. Kuwahara, T., A Genomic analysis of *Bacteroides fragilis* reveals extensive DNA inversions regulating cell surface adaptation, *Proc. Natl. Acad. Sci. USA*, 101, 14919, 2004.
12. Cerdeno-Tarraga, A.M. et al., Extensive DNA inversions in the *B. fragilis* genome control variable gene expression, *Science*, 307, 1463, 2005.
13. Wexler, H.M., *Bacteroides*: The good, the bad, the nitty-gritty, *Clin. Mirobiol. Rev.*, 20, 593, 2007.
14. Wang, Y. et al., Structural basis of the abscess-modulating polysaccharide A2 from *Bacterodes fragilis*, *Proc. Natl. Acad. Sci. USA*, 97, 13478, 2000.
15. Citron, D.M., Poxton, I., and Baron, E.J., *Bacteroides, Porphyromonas, Prevotella, Fusobacterium*, and other anaerobic gram-negative Rods. In Murray, P.R. et al. (Editors), *Manuals of Clinical Microbiology*, 9th ed., vol 1, ASM Press, Washington DC.
16. Tzianabos, A.O., and Kasper, D.L., *Anaerobic Infections: General concepts. Principles and practice of Infectious Diseases*, 6th ed., Vol. 2, Elsevier, Churchill Livingstone, Philadelphia, PA, 1993.
17. Comstock, L.E. et al., Analysis of a capsular polysaccharide biosynthesis locus of *Bacteroides fragilis*, *Infect. Immun.*, 67, 3525, 1999.
18. Tzianabos, A.O. et al., Structure function relationships for polysaccharide-include intra—abdominal abscess, *Infect. Immun.*, 62, 3590, 1994.
19. Coyne, M.J. et al., Polysaccharide biosynthesis locus required for virulence of *Bacteroides fragilis*, *Infect. Immun.*, 69, 4342, 2001.
20. Sack, R.B. et al., Enterotoxigenic *Bacteroides fragilis*; epidemiologic studies of its role as a human diarrhoeal pathogen, *J. Diarr. Dis. Res.*, 10, 4, 1992.
21. Sharma, N., and Chaudhry, R., Rapid detection of enterotoxigenic *Bacteroides fragilis* in diarrhoeal faecal samples, *Indian J. Med. Res.*, 124, 575, 2006.
22. Polk, B.F., and Kasper, D.L., *Bacteroides fragilis* subspecies in clinical isolates, *Ann. Intern. Med.*, 86, 569, 1977.
23. Chaudhry, R., Pandey, A., and Sharma, A., *Bacteroides*. In Liu, D. (Editors), *Molecular Detection of Foodborne Pathogens*, Chapter 22, pp. 307–316, CRC Press, FL, 2009.
24. Somer, J.H. et al., Advanced identification methods. In Jousimies-Somer, H. et al. (Editors), *Wadsworth-KTL Anaerobic Bacteriology Manual 2002*, Chapter 6, 6th edn, pp. 81–99 Star Publishing Company, Belmont, CA.
25. Nolte, F.S., and Caliendo, A.M., Molecular detection and identification of microorganisms. In Murray, P.R. et al. (Editors), *Manuals of Clinical Microbiology*, 9th ed., vol 1, ASM Press, Washington DC, 2007.
26. Redondo, M.C. et al., Attributable mortality of bacteremia associated with the *Bacteroides fragilis* group, *Clin. Infect. Dis.*, 20, 1492, 1995.
27. Brook, I., Pathogenicity of the *Bacteroides fragilis* group, *Ann. Clin. Lab. Sci.*, 19, 360, 1989.
28. Kuwahara, T. et al., Genetic variation in 16S-23S rDNA internal transcribed spacer regions and the possible use of this genetic variation for molecular diagnosis of *Bacteroides* species, *Micribiol. Immunol.*, 45, 191, 2001.
29. Linda, K., Host distributions of uncultivated fecal *Bacteroidales* bacteria reveal genetic markers for fecal source identification, *Appl. Envron. Microbiol.*, 71, 3184, 2005.
30. Liu, C. et al., Rapid detection of the species of the *Bacteroides fragilis* group by multiplex PCR assays using group and species-specific primers, *FEMS Microbiol. Lett.*, 222, 9, 2003.
31. Hans, R.J. et al., Nucleic acid technique in diagnostic microbiology: Section A. In Collee, J.G. et al. (Editors), *A Practical Medial Microbiology of Mackie and McCartney*, 16th ed., pp. 205–242, Churchill Livingstone, Philadelphia, PA, 1991.
32. Sambrook, J., Fritsch, E.F., and Maniatis, T., *Molecular Cloning: A Laboratory Manual*, 2nd ed., Cold Harbor Laboratory Press, Cold Spring Harbour, 1989.
33. Jotwani, R. et al., Detection of *Bacteroides fragilis* in clinical spacimen by polymerase chain reaction amplification of the gene neuraminidase gene, *Curr. Microbiol.*, 31, 235,1995.
34. Pantosti, A. et al., Detection of enterotoxigenic *Bacteroides fragilis* and its toxin in stool samples from adults and children in Italy, *Clin. Infect. Dis.*, 24, 12, 1997.
35. Avila-Campos, M.J. et al., Determination of *bft* gene subtypes in *Bacteroides fragilis* clinical isolates, *J. Clin. Microbiol.*, 45, 1336, 2007.

44 *Capnocytophaga*

Kazuyuki Ishihara, Satoru Inagaki, and Atsushi Saito

CONTENTS

44.1 INTRODUCTION

44.1.1 Classification, Morphology, and Biology

Capnocytophaga is a genus in the family Flavobacteriaceae[1] and members belong to the *Flavobacterium-Cytophaga* ribosomal RNA homology group.[2] *Capnocytophaga* species were first isolated from the human oral cavity by Prevot in 1956 as *Fusobacterium nucleatum* var. *ochraceus*. The term "Capnocytophaga" means "eater of carbon dioxide"[3] and comprises a group of capnophilic, facultatively anaerobic as well as gram-negative fusiform rods. These microorganisms are short-to-elongated flexible rods or filaments, 0.42–0.6 μm in diameter and 2.5–5.7 μm in length. The ends of the cells are usually rounded or tapered (Figure 44.1a). *Capnocytophaga* species can be pleomorphic and they do not form capsules and sheaths. The colonial morphology is usually flat, presenting concentrically spreading growth from a central point of inoculation (Figure 44.1b). The key characteristics of this genus are listed in Table 44.1.

Capnocytophaga species grow on Tryptic soy agar containing 5% sheep blood (TS blood agar), although blood is not required for growth. Occasionally on blood agar, this microorganism exhibits a gliding characteristic observed for swarming colonies of the family Proteus. This spreading growth can occur in the absence of blood and appears to be seen most often in media containing 3% agar.[1] Unlike Proteus, the gliding property of *Capnocytophaga* species is not dependent upon flagella, fimbriae, or pili, since none of these appendages have been observed by electron microscopy.[4] Primary isolation and initial in vitro growth require CO_2, although some strains are reported to grow aerobically without CO_2. The optimum temperature for their growth is 35°C–37°C.

Capnocytophaga species are capable of utilizing several carbohydrates as fermentable substrates and energy sources.[3] Polysaccharides such as dextran, glycogen, inulin, or starch may be fermented by the majority of these species. The major acid end products from glucose of these microorganisms are succinic acid and acetic acids, with trace amounts of propionic acid and isovaleric acid. Species isolated from humans are catalase- and oxidase-negative whereas those isolated from animals are both positive in these properties.[5] The mol% GC of the DNA from these strains is 38–44 mol%.[5–7]

Capnocytophaga species are found in association with animal and human hosts. Although the pathogenicity of these organisms is not fully defined, they are frequently isolated from pulmonary lesions and abscesses, as well as from healthy oral sites. *Capnocytophaga* species are classified into *Capnocytophaga canimorsus*, *Capnocytophaga cynodegmi*, *Capnocytophaga gingivalis*, *Capnocytophaga granulosa*, *Capnocytophaga haemolytica*, *Capnocytophaga leadbetteri*, *Capnocytophaga ochracea*, and *Capnocytophaga sputigena*. In addition, *Capnocytophaga* genospecies AHN8471 has also been described. The former two species have been isolated from animals and the latter seven were isolated from the human oral cavity.

(a)

(b)

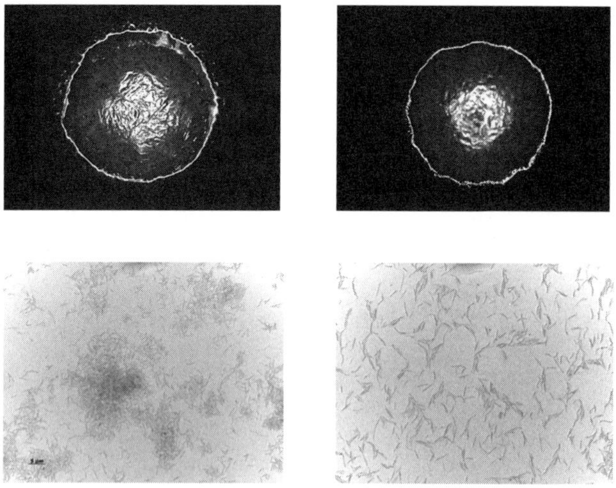

C. ochracea C. gingivalis

FIGURE 44.1 Typical colony morphology and Gram staining of *Capnocytphaga* species. (a) Appearance of *Capnocytophaga ochracea* and *gingivalis* on Typricase soy blood agar. (b) Gram stain of *Capnocytophaga ochracea* and *gingivalis.*

TABLE 44.1
Key Characteristics of *Capnocytophaga* spp.

Characteristics	Result
Yellow or orange pigmentation of colony	+
Surface translocation; gliding motility	+
Gram stain	–
Fusiform to rod-shaped cells	+
CO_2 required for initial growth	+
Catalase	–[a]
Oxidase	–[a]
Acetic acid	+
Succinic acid	+
Fermentation	
D-Glucose	+
D-Lactose	+
D-Maltose	+
D-Mannitol	–
D-Mannose	+
D-Sucrose	+[a]
D-Xylose	–
H_2S production	–
Indole production	–
Inhibition of growth by bile (10%, wt/vol)	+

[a] Variations are observed in some strains.

C. ochracea is the type species of the genus *Capnocytophaga* and is the most biologically active among human oral species (Table 44.2). The species name is derived from the word "ochra," which means "yellow." This species was formerly called *Bacteroides ochraceus.* As for the species *C. sputigena*, "sputigena" refers to the site of origin, meaning "produced by sputum." *C. sptigena* is intermediate between *C. ochracea* and *C. gingivalis. C. gingivalis* was also named to indicate its oral origin; gingivalis means "of the gum." In the case of *C. haemolytica*, "hemolytica" refers to *β*-hemolysis on blood agar plates. No other species with the exception of *C. cynodegmi* exhibits *β*-hemolysis. About half of the *C. cynodegmi* strains show beta-like hemolytic activity. *C. granulosa* was named in reference to the granular inclusions, which are observed at the ends and in the middle of Gram-stained cells that are stained by basic fucshin, *C. leadbetteri* was designated in honor of E.R. Leadbetter, an American microbiologist who (as first author) described and proposed the genus *Capnocytophaga.*[8] This species appears less metabolically active than other species of this genus.

In 1976, a case report of a "previously undescribed Gram-negative bacillus causing septicemia and meningitis" appeared.[9] The isolated microorganisms were designated as Centers for Disease Control (CDC) group dysgonic fermenter 2 (DF-2) and later became known as the "dog-bite" organism.[10] Microorganisms in this group were designated as *C. canimorsus* and *C. cynodegmi.* Both canimorsus and synodegmi mean "dog bite" in Latin. These microorganisms have been isolated as part of the normal flora in dogs, cats, and from patients after dog or cat bites.

Comparison of the characteristics of these species is listed in Table 44.2 based on the previous reports.[5,8] Several studies indicated difficulties in distinguishing the species of *Capnocytophaga* by phenotypic tests.[4,11,12] *C. ochrachea* and *C. sputigena* are so closely related that distinction between them by phenotypic comparison is difficult. As a recent study clearly confirmed, *C. ochracea* and *C. sputigena* are separate species[6] and analysis beyond mere routine phenotypic examination is required for definite identification. One of the key characteristics in initially discriminating between these species by phenotypic examination is their relative ability to ferment cellobiose and also *β*-glucosidase activities. *C. gingivalis* and *C. granulose* group can be distinguished from the *C. ochracea* and *C. sputigena* group by fermentation of amygdalin and hydrolysis of dextran. Although discrimination between *C. gingivalis* and *C. glanulosa* is difficult, fermentation of lactose and raffinose and the presence of trypsin-like and *β*-glucosidase activities are the best choices for separation. *C. leadbetteri* consists of weakly saccharolytic strains. Lack of fermentation is a key factor for distinguishing this species from the others. *Capnocytophaga* genospecies AHN8471 showed the same phenotypic characteristics as *C. ochracea* and *C. sputigena.* However, partial 16S rRNA gene sequence and multilocus enzyme electrophoresis indicated genetic differences between these species.[6] Oligo-peptidase activities have been detected in *Capnosytophaga* species. Most of the species show especially strong activity against lysyl-alanine- and lysyl-proline-4-methoxy-*β*-naphthylamide as well as tripeptide substrates such as alanyl-alanyl-phenylalanine-4-methoxy- and glycyl-prolyl-leucine-*β*-naphthylamide. Aminopeptidase activity was also reported for these species using alanine-*β*-naphthylamide, arginine-*β*-naphthylamide, leucine-*β*-naphthylamide, methionine-*β*-naphthylamide, and serine-*β*-naphthylamide.[13,14] *C. haemolitica* can be distinguished from the other species by their *β*-hemolytic

TABLE 44.2
Comparison of Phenotypic Characteristics of *Capnocytophaga* Species

Characteristics	*C. l*	*C. ge*	*C. o*	*C. s*	*C gi*	*C. gr*	*C. h*	*C. ca*	*C. cy*
Hemolysis	–	–	–	–	–	–	β	–	β-like
Catalase	–	–	–	–	–	–	–	+	+
Oxidase	–	–	–	–	–	–	–	+	+
Fermentation of									
Amygdalin	–	+	+	+	–	–	w	NT	NT
Cellobiose	–	+	+	+	+	–	–	–	+
Fructose	–	+	+	+	+	+	–	–	+
Galactose	w	+	+	+	+	+	+	+	+
Glucose	w	+	+	+	+	+	+	+	+
Lactose	w	+	+	+	+	+	+	+	+
Raffinose	–	+	+	+	+	w	+	–	+
Sucrose	–	+	+	+	+	+	+	–	+
Hydrolysis of								NT	NT
Aesculin	–	+	+	+	–	+	+	+	+
Dextran	–	+	+	+	–	–	–	NT	NT
Starch	v	+	+	+	+	+	+	+	+
API ZYM								NT	NT
C4-esterase	v	+	+	w	+	w	–	NT	NT
C8-esterase lipase	w	w	w	w	w	w	+	NT	NT
Cystine arylamidase	+	+	+	+	+	+	–	NT	NT
Trypsin	–	w	w	w	+	w	+	NT	NT
α-Chymotrypsin	W	w	w	w	w	w	w	NT	NT
Acid phosphatase	+	+	+	+	w	+	+	NT	NT
β-Galactosidase	+	w	w	w	+	+	w	NT	NT
β-Glucosidase	–	w	w	w	w	+	–	NT	NT
N-acetyl-β-Glucosaminidase	+	+	+	+	w	–	+	NT	NT
IgA1 cleavage	+	+	+	+	+	+	–	NT	NT

C. l: *C. leadbetteri*, *C. ge*: *Capnocytophaga genospecies*, *C. o*: *C. ochracea*, *C. s*: *C. sputigena*, *C. gi*: *C. gingivalis*, *C. gr*: *C. granulosa*, *C. h*: *C. haemolytica*, *C. ca*: *C. canimorsus*, *C. cy*: *C. cynodegmi*; w: weak positive, NT: not tested.

activity and the absence of aminopeptidase activity.[7] *C. canimorsus* and *C. synodegmi* are distinguished from other *Capnocytophaga* species in human oral cavity by their catalase and oxidase activities. *C. cynodegmi* can be distinguished from *C. canimorsus* by its ability to ferment sucrose, raffinose, melibiose, and inulin.[5]

44.1.2 Clinical Features and Pathogenesis

44.1.2.1 *Capnocytophaga* Species Isolated from Human Oral Cavity

In the human oral cavity, *C. gingivalis*, *C. ochracea*, *C. leadbetteri*, *C. sputigena*, *C. granulosa*, and *C. haemolytica* colonize in dental plaque. The detection rate of *C. gingivalis*, *C. ochrasea*, and *C. sputigena* in 5-year-old children is more than 30%.[15,16] This prevalence suggested that *Capnocytophaga* species are part of the commensal bacterial flora. Recent epidemiological data indicated that high numbers of *Capnocytophaga* species and anaerobes were present in diseased sites in diabetes mellitus (DM)-periodontitis subjects versus non-DM-periodontitis, as well as at diseased sites versus healthy sites. Analysis of individual species revealed that the outcome varied with the site and DM status.[17] Pathogenesis of *Capnocytophaga* species in the human oral cavity has been focused mainly on *C. gingivalis*, *C. ochracea*, and *C. sputigena*. Recently, more information has become available with regard to *C. leadbetteri*, *C. granulosa*, and *C. haemolytica*. *Capnocytophaga* species in the human oral cavity were previously associated with aggressive periodontitis (formerly juvenile periodontitis).[18] However, contradictory data have been reported,[19–23] suggesting that *Capnocytophaga* might not be the major pathogen in the disease. This microorganism has been implicated as a focal infection agent especially in immunocompromised patients. *C. ochracea* and *C. sputigena* have also been detected from vertebral osteomyelitis in patients with cancer.[24] Bacteremia by *C. ochracea* and *C. gingivalis* was reported in patients with leukemia, acute erythroblastic leukemia, and Hodgkin disease.[25–28] Cases of sepsis and purpura fulminans induced by *C. ochracea* were reported in an immunocompetent patient.[29] Bacterial endocarditis by *C. haemolitica* in a patient who had undergone two previous aortic valve replacements was also reported.[30] Other than bacteremia, pneumonia caused by *C. sputigena* in a 13-year-old teenager wearing orthodontic braces after influenza symptoms,[31] pneumonia and sepsis due to *C. gingivalis* after autologous transplantation,[32] and pyogenic arthritis caused by *C. gingivalis* in an immunocompromised 3-year-old male were reported.[33] Bacteremia caused by dental plaque biofilms in the gingival crevice or periodontal pocket has also been reported.[34] It is possible that spontaneous contamination with *Capnocytophaga* species is a cause of the disease in immunocompromised patients.

Immunosuppression was not clearly indicated in several cases of systemic illness caused by *Capnocytophaga* species. *C. ochracea* was detected in the case of chorioamnionitis with intact membranes[35] and an intraamniotic infection caused by *C. sputigena* was also reported.[36] In the latter case, *C. sputigena* was found on membranes, cord, amniotic fluid, and the placenta. It was also identified in maternal endocervix cultures. *Capnocytophaga* genospecies AHN8471 was isolated form a patient with bacteremia and possible endocarditis.[37] Brain abscesses caused by multidrug-resistant *C. ochracea* has also been reported.[38]

Several potential virulence factors of *Capnocytophaga* species have been suggested. *Capnocytophaga* species were reported to produce cell-associated immunoglobulin A1 (IgA1) protease activity. The activities measured may have been influenced by concomitant glycosidase activity degrading IgA1 glycans.[8] The protease was detected from all oral *Capnocytophaga* species except *C. haemolytica*.[6] The protease activities from *C. ochracea*, *C. gingivalis*, and *C. sputigena* were inhibited by metal chelaters, suggesting that this protease is a metalloprotease. Sonic extracts of *C. ochracea* have been reported to have immunosuppressive effects against mice spleen cells.[39] These activities may be involved in evasion from host defenses. LPS from *C. gingivalis* also exhibits hemagglutinating activity in contrast to LPS from *C. ochracea*,[40] suggesting its contribution to colonization of the host. Moreover, sonic extracts of *C. sputigena* were also reported to inhibit the proliferation of human fibroblasts.[41] Heterogeneity of virulence factors among isolates suggests that pathogenic potential may be associated with some of the species and possibly only with certain members of the species.[4,42] Further analysis required to clarify the activity of these factors in the isolated strains from the lesion. Given the difficulty in identification of the species by phenotypic tests, the identity of virulence factors in each species should be confirmed in combination with molecular identification of the species.

Most *Capnocytophaga* strains are susceptible to penicillin G, ampicillin, clindamycin, chloramphenicol, tetracyclin, metronidazole, and erythromycin, but are resistant to aminoglycosides and trimetoprim.[3] In the case of pneumonia induced by *C. gingivalis* following stem cell or marrow transplantation, fluoroquinolone-resistant strains,[32] and in regards to bacteremia induced by *C. sptigena*, CfxA3-beta-lactamase-producing strains,[28] were identified. In brain abscesses associated with *C. ochracea*, resistance to cephalosporins, including cefuroxime, ceftazidime, ceftriaxone, cefepime, and cefixime (MIC > 256 μg/mL) was observed. Since systemic illnesses caused by *Capnocytophaga* species can be accompanied by immunosuppression, we need to be aware of the susceptibility of the isolated *Capnosytophaga* species to antibiotics.

44.1.2.2 *Capnocytophaga* Species in Bite-Related Infections Caused by Cats and Dogs

C. canimorsus and *C. cynodegmi* have been isolated from septicemia after dog or cat bites and following contact with such animals.[10,43] Among the isolates submitted to the CDC for identification, *C. canimorsus* was much more prevalent and has caused more serious infections than *C. cynodegmi*.[44] Of the 150 strains of *C. canimorsus* examined by the CDC, 88% were from blood, 5% from spinal fluid, 2% from wounds, and 2% from dog mouths, whereas in contrast, the eight *C. cynodegmi* strains were isolated form either dog mouths or localized wound infections following dog bites.[45] Sepsis is caused by infection by *C. canimorsus*, *Pasteurella multocida*, Staphylococcus species including MRSA, and Streptococcus species after dog bite. Endocarditis, meningitis, and peritonitis can also complicate bite wound infections. *C. canimorsus* is a major cause of nonstaphylococcal or nonstreptococcal septicemic infection. This species is generally transmitted by bites, scratches, or mere contact with animals. Moreover, animals are responsible for the infection in 89.5% of the cases.[10] The majority of the infections are associated with dogs (91%) or cats (8%).

At first, the spectrum of *C. canimorsus* infections can range from cellulitis to meningitis and endocarditis.[9,46] After these early reports, gangrene, sepsis, meningitis, pneumonia, and endocarditis have been reported.[10,47] Most cases are associated with an underlying immune disorder, including splenectomy, chronic alcohol use, or cirrhosts.[48] However, no identifiable risk factor has been found in 40% of cases.[49] Generally, the prognosis for infection by *C. canimorsus* is poor and the mortality rate is approximately 30%. *C. canimorsus* was retrieved from the saliva of 61 dogs out of 106 sampled. As with human clinical isolates, all dog strains exhibited sialidase activity and 61% of the isolates from dogs blocked the killing of phagocytosed *Escherichia coli* by macrophages.[50] This microorganism can feed on polymorphonuclear cells by deglucosylating the glycans of host cells.[51] Macrophages infected with *C. canimorsus* Cc5 (a clinical isolate from a patient with fetal septicemia) do not release tumor necrosis factor, interleukin 1α (IL-1α), IL-6, IL-8, gamma interferon, macrophage inflammatory protein 1β, nor nitric oxide.[52] This is due to the inability of Toll-like receptors to respond to *C. canimorsus* Cc5.[52] In addition, this microorganism is resistant to complement killing as well to PMN-mediated phagocytosis and killing.[53] C3b accumulation occurred on the surface of the microorganism but proper localization of the membrane attack complex (MAC) did not occur, suggesting that the resistance was due to protection against accumulation of active MAC on the bacterial surface. A mutant sensitive to phagocytosis isolated from a library of transposon Tn4351-induced mutants of Cc5 was isolated and analyzed for accumulation of complement. In the mutant, accumulation of MAC was significantly increased. The transposon was inserted after the first 26 bp of a gene encoding a putative glycosyltransferase in the mutant. The inactivation of the glucosyltransferase induced alterations in the surface structure of the Cc5 mutant. The A polysaccharide structure, likely to be LPS, protects this microorganism from deposition of the complement MAC which induces bacteriolysis. These immune modulating activities may be associated with evasion from host defense and be involved in the systemic pathogenicity of the microorganisms.

C. cynodegmi has been isolated from humans with infected dog or cat bite wounds. However, no evidence for factors responsible for systemic infection has been reported at the time when the species was identified.[5] Recent reports indicate that this microorganism can cause cellulitis, bacteremia, and pneumonitis in patients with diabetes[54]; peritonitis in peritoneal dialysis patient,[43] sepsis, and meningitis after splenectomy.[55] The number of reports, however, are low compared with those for *C. canimorsus* and no virulence factors for the former organism have been identified. Further analysis is required to clarity the pathogenesis of this microorganism.

44.1.3 Diagnosis

44.1.3.1 Conventional Techniques

For isolation of *Capnocytophaga* species, Thayer–Martin selective medium or Trypticase soy-blood agar containing bacitracin (50 μg/mL) and polymixin B (100 μg/mL)[56] are used. *Capnocytophaga* species also can be isolated by standard laboratory procedures employing Trypticase soy-blood-agar-based media, such as TS blood agar or chocolate agar, although they do not grow on MacConkey agar.[3] When transportation of clinical samples is required, reduced transport fluid[57] can be used.

Agar plates, on which the sample is inoculated, are incubated 3–5 days at 35°C–37°C in an anaerobic chamber (5% CO_2, 10% H_2, 85% N_2) or anaerobic jars (under 5% CO_2, 95% air atmosphere). After incubation, distinct colony morphology is observed—concentrically spreading films from a central colony or point of inoculation (Figure 44.1b). Colonies of *Capnocytophaga* species are surface spreading (translocating), pigmented speckled, and they have been observed to form pits in the agar. To obtain pure isolates, the colonies organized on the agar surface can be easily removed and restreaked onto Thayer–Martin or TS blood agar. One of the key features for identification is cell morphology. This microorganism is distinguished from other fastidious gram-negative bacilli by its shape. The long, delicate, fusiform appearance of *Capnocytophaga* species is distinguishable from the coccobacillary appearance of *Moraxella*, *Kingella*, *Haemophilus*, *Actinobacillus*, and *Aggregatibacter*. *Capnocytophaga* species also exhibit the straight narrow, rod forms resembling *Eikenella* species.[3] As described earlier, identification of the species by conventional phenotypic characterization is shown in Figure 44.2. Inter- or intrastrain identification can be very difficult without molecular investigation because of the complex similarities and differences in characteristics among strains.[6]

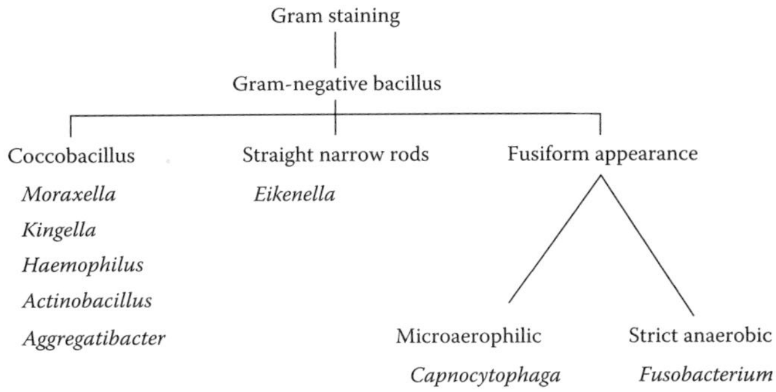

FIGURE 44.2 Determinative key to the *Capnocytophaga* species from oral gram negative bacteria.

44.1.3.2 Molecular Techniques

Investigation into possible associations of individual species within this genus with various pathological conditions has been hampered by the lack of a reliable scheme for species identification because these species share similar morphological and biological characteristics. For example, *C. genospecies* AHN8471 demonstrates nearly identical phenotypic features to *C. ochracea* and *C. sputigena*, although the genetic variations among these species are noticeable.[6] To analyze the habitat of *Capnocytophaga* species, detection techniques with high sensitivity and specificity for all strains need to be employed. Recent progress in molecular diagnostics offers new avenues for improved discrimination among various *Capnocytophaga* species. Indeed, PCR and DNA–DNA hybridization have been successfully applied for identification and quantitation of *Capnocytophaga* species in the oral specimens.

In the DNA–DNA hybridization checkerboard system,[58] DNA isolated from clinical samples is applied onto membranes, followed by addition of digoxigenin labeled whole genome DNA probe. *Capnocytophaga* species in the sample are then detected by alkaliphosphatase-labeled anti-digoxigenin antibody. Although only *C. gingivalis*, *C. ochracea*, and *C. sputigena* are currently detected by using this system, it may be possible to identify other species with minimal modifications. As whole genome probes are being used in the DNA–DNA hybridization, cross-reactions among closely related *C. haemolytica*, *C. granulose*, *C. gingivalis*, *C. ochracea*, and *C. sputigena* strains are observed. It was noted that the highest cross-reactivity among the species was under 31% apart from some strains of *C. ochracea* and *C. sputigena*.[7] This system detected all 3200 pure isolates from subgingival samples of 64 subjects and heterologous species cross-reaction was less than 5%.[58] Thus, the checkerboard DNA–DNA hybridization system provides a very useful means to detect *Capnocytophaga* species in the presence of other oral microbiota. However, the possibility of cross-reaction among *Capnocytophaga* species should always be kept in mind.

PCR is the most specific and sensitive method for detecting *Capnocytophaga* species. Primers for detection are designed based on the sequence of 16S ribosomal RNA (16S rRNA) sequence. Previously, detection of *C. gingivalis*, *C. granulosa*, *C. haemolytica*, *C. ochracea*, *C. sputigena*, *C. canimorsus*, and *C. cynodegmi* by PCR was described,[59,60] and species-specific primers were reported only for *C. gingivalis*, *C. ochracea*, and *C. sputigena*.[16,59] For other species, PCR-restriction fragment length polymorphism analysis or sequencing of the 16S rRNA sequences after PCR were performed for identification.[60,61] Genomic restriction analysis, conventional ribotyping, and 16S-23S intragenetic spacer gene sequencing are currently available for intragenetic differentiation of bacterial species. 16S rRNA sequencing is regarded as the "gold standard" for bacterial identification. However, these methods are time consuming and expensive. Compared to these methods, PCR–RFLP analysis is relatively cost effective and represents a convenient method for the detection of *Capnocytophaga* species.

44.2 METHODS

44.2.1 Sample Preparation

For identification of bacteria by PCR or DNA hybridization, a standard mini prep method[62] as modified by Smith et al.[63] is used. Briefly, samples are isolated from subgingival plaque, body fluids, or infected tissue by curette or paper points and precipitated by centrifugation. The cells are suspended in 567 µL TE buffer and proteinase K and SDS are added to a final concentration of 100 µg/mL and 0.5%, respectively, and incubated for 1 h at 37°C. After the addition of 150 µL of 5 M NaCl, 85 µL of CATB/NaCl solution (10% hexadecyltrimethyl ammonium bromide in 0.7 M NaCl) is added to the solution and incubated for 10 min at 65°C. DNA is extracted with chloroform/isoamyl alcohol (24:1, v/v) and centrifuged for 5 min in a microcentrifuge. The aqueous phase is then transferred to a fresh tube and DNA is extracted with phenol/chloroform/isoamyl alcohol (25:24:1, v/v/v). The aqueous phase is next transferred to a fresh tube and DNA is precipitated with isopropanol. The DNA obtained is dissolved with 300 µL TE containing 2 µL of 10 mg/mL RNase and heated at 50°C for 60 min. Each suspension is finally mixed with 30 µL of 3M sodium acetate buffer (pH 7.6) and reextracted with chloroform/isoamyl alcohol followed by DNA precipitation with ethanol.

To extract sample DNA quickly for PCR, a few bacterial colonies can be aseptically suspended in sterile water in an Eppendorf tube. The tube is then vigorously agitated and boiled for 2 min. Immediately after boiling, the tubes are placed on ice. A 1:10 dilution of this bacterial suspension in sterile water is subsequently prepared and kept on ice. It can be used later as the nucleic acid template for PCR amplification.[17]

44.2.2 Detection Procedures

44.2.2.1 PCR Identification

Among *Capnocytophaga* species, *C. gingivalis*, *C. ochracea*, and *C. sputigena* can be distinguished by PCR with universal forward and species-specific reverse primers.[59] Specifically, universal 16S rRNA forward primer (5′-AGA GTT TGA TCC TGG CTC AG-3′)[64] recognizing all *Capnocytophaga* species is applied in separate reactions with three species-specific reversed primers (T_A: annealing temperature) (i.e., *C. ochracea* reverse primer: 5′-GAT GCC GTC CCT ATA TAC TAT GGG G-3′ [T_A = 55°C]; *C. sputigena* reverse primer: 5′-GAT GCC GCT CCT ATA TAC CAT TAG G-3′ [T_A = 55°C]; *C. gingivalis* reverse primer: GGA CGC ATG CCC ATC TTT CAC CAC CGC [T_A = 70°C]).

PCR mixture (50 µL) is prepared with 1× PCR buffer, 2 U of Taq polymerase, 0.2 mM of each dNTP, 25 pM of universal forward primer and one of the species-specific primer, and 5 µL of extracted DNA. The amplification is conducted in a thermocycler with 30 cycles of 94°C for 1 min, an annealing temperature (see T_A above; different for each specific primer used) for 1 min and 72°C for 2.5 min. The sensitivity of this PCR is about 1.5 pg of the total genomic DNA (50–100 cells).

Alternatively, *Capnocytophaga* species can be detected and differentiated by PCR-16S rRNA PCR-restriction fragment length polymorphism (PCR–RFLP) analysis.[60] This method employs the following primers: 5′-AGAGTTTGATCMTGGCTCAG-3′, 5′-TACGGYTACCT TGTTACGACTT-3′, 5′-CTCCTACGG GAGGCAGCAG-3′ (in the sequence, M stands for C or A, and Y stands for C or T); and it allows separation of all *Capnocytophaga* species designated up to 2005.

DNA template is added to the PCR tube (50 µL final volume) containing 1× PCR buffer, 0.2 mM each dNTP, 25 pmol of each PCR primer, and 1 U of Taq polymerase. The bacterial DNA template and the PCR mix are incubated under the following conditions: 1 cycle of 95°C for 5 min; 29 cycles at 94°C for 1 min, 54°C for 1 min, and 72°C for 1 min; and a final extension at 72°C for 5 min. After PCR, 21.5 µL of the PCR amplicon is mixed with 1 U of restriction enzyme (either *Cfo*I, *Hae*III, or *Rsa*I) and 2.5 µL of the corresponding 10× restriction enzyme buffer. The mixture is then incubated overnight at 37°C. The resulting PCR–RFLP digests are analyzed by electrophoresis on 2% agarose gels. This method can be utilized for detection of this species from systemic disease such as meningitis. The RFLP on *rpoB* and *rrs* has been carried out to identify the species in blood from the patients and oral cavity of dog and cat.[65]

44.2.2.2 DNA–DNA Hybridization

To detect *Capnocytophaga* species in a sample with other oral microorganisms, checker board DNA–DNA hybridization is often applied.[58] *C. gingivalis*, *C. ochracea*, and *C. sputigena* have been successfully detected but the detection of other species has not yet been reported.

Briefly, genomic DNA of the target species are purified as above and 1–3 μg of genomic DNA is labeled with the digoxigenin random primer technique.[66] These digoxigenin-labeled genomic DNA samples serve as efficient probes for checkerboard DNA–DNA hybridization. To examine the bacterial composition including *Capnocytophaga* species in the sample, test samples or control samples (10^5 and 10^6 cells of target species) are placed in 0.15 mL of TE. Next, 0.15 mL of 0.5 M NaOH is added to each sample and boiled in a water bath for 5 min. The samples are neutralized using 0.8 mL of 5 M ammonium acetate. The released DNA is placed into the extended slots of a Minislot 30 (Immunetics, Cambridge, MA) and then concentrated onto a nylon membrane under vacuum followed by fixation to the membranes by cross-linking using ultraviolet light (e.g., Stratalinker 1800, Stratagene, La Jola, CA) with baking at 120°C for 20 min. The membrane with fixed DNA is next placed in a miniblotter 45 (Immunetics), with the "lanes" of DNA at 90°C to the channels of the device. Each channel is used as a hybridization chamber for separate DNA probes prepared above. After application of heat denatured DNA probe, the membrane is hybridized at 42°C for 1 h in 50% formamide, 5 × SSC, 1% casein, 5 × Denhardt's reagent, 25 mM sodium phosphate (pH 6.5) and 0.5 mg/mL yeast RNA. After hybridization, membranes are washed at low stringency to remove loosely bound probe and then at high stringency (68°C, 0.1 × SSC, 0.1% SDS, 20 min twice). Hybridized probe is detected with 1:25,000 dilution of anti-digoxigenin antibody conjugated with alkaline phosphatase using the modification described by Engler-Blum et al.[67] The washed membranes are lastly incubated with AttoPhos (Promega, Madison, WI) overnight at room temperature and signals are detected using a Storm Fuloroimager (Molecular Dynamics, Sunnyvale, CA).

The sensitivity of this assay is set at 10^4 cells of the target species by adjusting the concentration of each DNA probe. If a Fluoroimager is not available, after washing the membranes may be incubated in Lumiphos 530 (Lumigen, Southfield) at 37°C for 45 min, placed in a film cassette with Reflection NEF film (Dupont, Boston, MA) at 37°C for 1 h and then developed. Signals can be evaluated visually by comparison with standards for the target species. Compared to anaerobic culturing, whole genomic DNA probes are not as sensitive. In addition, they have the potential to cross-hybridize at low levels to related species. If this cross-reactivity is undesirable, cloned genes or short oligomers may be used instead as specific probes.

44.3 CONCLUDING REMARKS

Capnocytophaga species are commensal microorganisms in humans or animals. Although these microorganisms do not possess clearly identified virulence factor(s), they have been implicated in human diseases such as endocarditis, meningitis, and sepsis, mainly in immunocompromised hosts. In addition, *C. canimorsus*-related diseases can sometimes be life threatening. To investigate virulence factors in *Capnocytophaga* species, precise identification to the species level by molecular techniques is required. In fact, recent molecular analysis revealed four distinct species based upon genospecies identification. These species may be the key for understanding the complex virulence properties of these microorganisms. A combination of molecular and conventional methods as a part of comprehensive study of the microbiology of the organisms should greatly enhance the characterization of various virulence factors in *Capnocytophaga* species. The new era of molecular microbiology provides unprecedented opportunities to develop and implement improved diagnostic tools that simultaneously test both microbiological and host variables, thereby leading to a better understanding of infections due to *Capnocytophaga* species.

REFERENCES

1. Leadbetter, E.R. et al., *Capnocytophaga*: New genus of gram-negative gliding bacteria. I. General characteristics, taxonomic considerations and significance, *Arch. Microbiol.*, 122, 9, 1979.
2. Vandamme, P. et al., Polyphasic analysis of strains of the genus *Capnocytophaga* and Centers for Disease Control group DF-3, *Int. J. Syst. Bacteriol.*, 46, 782, 1996.
3. Holt, S.C., and Kinder, S.A., Genus II. *Capnocytophaga*. In Staley, J.T. et al. (Editors), *Bergy's Manual of Systematic Bacteriology*, vol. 3, pp. 2050, Williams and Wilkins, Baltimore, 1989.
4. Laughon, B.E. et al., API ZYM system for identification of *Bacteroides* spp., *Capnocytophaga* spp., and spirochetes of oral origin, *J. Clin. Microbiol.*, 15, 97, 1982.
5. Brenner, D.J. et al., *Capnocytophaga canimorsus* sp. nov. (formerly CDC group DF-2), a cause of septicemia following dog bite, and *C. cynodegmi* sp. nov., a cause of localized wound infection following dog bite, *J. Clin. Microbiol.*, 27, 231, 1989.
6. Frandsen, E.V. et al., Diversity of *Capnocytophaga* species in children and description of *Capnocytophaga leadbetteri* sp. nov. and *Capnocytophaga* genospecies AHN8471, *Int. J. Syst. Evol. Microbiol.*, 58, 324, 2008.
7. Yamamoto, T. et al., *Capnocytophaga haemolytica* sp. nov. and *Capnocytophaga granulosa* sp. nov., from human dental plaque, *Int. J. Syst. Bacteriol.*, 44, 324, 1994.
8. Frandsen, E.V. et al., Enzymatic and antigenic characterization of immunoglobulin A1 proteases from *Bacteroides* and *Capnocytophaga* spp, *Infect. Immun.*, 55, 631, 1987.
9. Bobo, R.A., and Newton, E.J., A previously undescribed gram-negative bacillus causing septicemia and meningitis, *Am. J. Clin. Pathol.*, 65, 564, 1976.
10. Lion, C. et al., *Capnocytophaga canimorsus* infections in human: Review of the literature and cases report, *Eur. J. Epidemiol.*, 12, 521, 1996.
11. Kristiansen, J.E. et al., Rapid identification of *Capnocytophaga* isolated from septicemic patients, *Eur. J. Clin. Microbiol.*, 3, 236, 1984.
12. Khwaja, K.J. et al., Protein profiles of *Capnocytophaga* species, *J. Appl. Bacteriol.*, 68, 385, 1990.
13. Suido, H. et al., Arylaminopeptidase activities of oral bacteria, *J. Dent. Res.*, 65, 1335, 1986.
14. Suido, H. et al., Identification of periodontopathic bacteria based upon their peptidase activities, *Adv. Dent. Res.*, 2, 304, 1988.

15. Kobayashi, N. et al., Colonization pattern of periodontal bacteria in Japanese children and their mothers, *J. Periodont. Res.*, 43, 156, 2008.
16. Hayashi, F. et al., PCR detection of *Capnocytophaga* species in dental plaque samples from children aged 2 to 12 years, *Microbiol. Immunol.*, 45, 17, 2001.
17. Ciantar, M. et al., *Capnocytophaga* spp. in periodontitis patients manifesting diabetes mellitus, *J. Periodontol.*, 76, 194, 2005.
18. Newman, M.G. et al., Studies of the microbiology of periodontosis, *J. Periodontol.*, 47, 373, 1976.
19. Dzink, J.L. et al., The predominant cultivable microbiota of active and inactive lesions of destructive periodontal diseases-disease, *J. Clin. Periodontol.*, 15, 316, 1988.
20. Moore, W.E., and Moore, L.V., The bacteria of periodontal diseases, *Periodontology*, 2000, 5, 66, 1994.
21. Colombo, A.P. et al., Clinical and microbiological features of refractory periodontitis subjects, *J. Clin. Periodontol.*, 25, 169, 1998.
22. Paster, B.J. et al., Bacterial diversity in human subgingival plaque, *J. Bacteriol.*, 183, 3770, 2001.
23. Kumar, P.S. et al., New bacterial species associated with chronic periodontitis, *J. Dent. Res.*, 82, 338, 2003.
24. Duong, M. et al., Vertebral osteomyelitis due to *Capnocytophaga* species in immunocompetent patients: Report of two cases and review, *Clin. Infect. Dis.*, 22, 1099, 1996.
25. Garcia-Cia, J.I. et al., Mixed bacteremia with *Capnocytophaga sputigena* and *Escherichia coli* following bone marrow transplantation: Case report and review, *Eur. J. Clin. Microbiol. Infect. Dis.*, 23, 139, 2004.
26. Mantadakis, E. et al., *Capnocytophaga gingivalis* bacteremia detected only on quantitative blood cultures in a child with leukemia, *Pediatr. Infect. Dis. J.*, 22, 202, 2003.
27. Bilgrami, S. et al., *Capnocytophaga* bacteremia in a patient with Hodgkin's disease following bone marrow transplantation: Case report and review, *Clin. Infect. Dis.*, 14, 1045, 1992.
28. Matsumoto, T. et al., First case of bacteremia due to chromosome-encoded CfxA3-beta-lactamase-producing *Capnocytophaga sputigena* in a pediatric patient with acute erythroblastic leukemia, *Eur. J. Med. Res.*, 13, 133, 2008.
29. Desai, S.S. et al., *Capnocytophaga ochracea* causing severe sepsis and purpura fulminans in an immunocompetent patient, *J. Infect.*, 54, e107, 2007.
30. Gutierrez-Martin, M.A. et al., Aortic valve endocarditis by *Capnocytophaga haemolytica*, *Ann. Thorac. Surg.*, 84, 1008, 2007.
31. Atmani, S. et al., *Capnocytophaga sputigena* pleuropneumonitis: A case report, *Arch. Pediatr.*, 15, 1535, 2008.
32. Geisler, W.M. et al., Pneumonia and sepsis due to fluoroquinolone-resistant *Capnocytophaga gingivalis* after autologous stem cell transplantation, *Bone Marrow Transplant.*, 28, 1171, 2001.
33. Rodgers, G.L. et al., Pyogenic arthritis caused by *Capnocytophaga gingivalis* in an immunocompetent three-year-old male, *J. Clin. Rheumatol.*, 7, 265, 2001.
34. Lockhart, P.B. et al., Bacteremia associated with toothbrushing and dental extraction, *Circulation*, 117, 3118, 2008.
35. Douvier, S. et al., Chorioamnionitis with intact membranes caused by *Capnocytophaga sputigena*, *Eur. J. Obstet. Gynecol. Reprod. Biol.*, 83, 109, 1999.
36. Alanen, A., and Laurikainen, E., Second-trimester abortion caused by *Capnocytophaga sputigena*: case report, *Am. J. Perinatol.*, 16, 181, 1999.
37. Mills, J.M. et al., A patient with bacteraemia and possible endocarditis caused by a recently-discovered genomospecies of *Capnocytophaga: Capnocytophaga genomospecies* AHN8471: A case report, *J. Med. Case Rep.*, 2, 369, 2008.
38. Wang, H.K. et al., Brain abscess associated with multidrug-resistant *Capnocytophaga ochracea* infection, *J. Clin. Microbiol.*, 45, 645, 2007.
39. Ochiai, K. et al., Purification of immunosuppressive factor from *Capnocytophaga ochracea*, *J. Med. Microbiol.*, 47, 1087, 1998.
40. Okuda, K., and Kato, T., Hemagglutinating activity of lipopolysaccharides from subgingival plaque bacteria, *Infect. Immun.*, 55, 3192, 1987.
41. Stevens, R.H. et al., Detection of a fibroblast proliferation inhibitory factor from *Capnocytophaga sputigena*, *Infect. Immun.*, 27, 271, 1980.
42. Irving, J.T. et al., Histological changes in experimental periodontal disease in rats monoinfected with gram-negative organisms, *J. Periodontal. Res.*, 13, 326, 1978.
43. Pers, C. et al., *Capnocytophaga cynodegmi* peritonitis in a peritoneal dialysis patient, *J. Clin. Microbiol.*, 45, 3844, 2007.
44. Gill, V.J., *Capnocytophaga*. In Mandell, G.L. et al. (Editors), *Mandell, Douglas, and Bennett's Principle and Practice of Infectious Diseases*, pp. 2441, Churchill Livingstone, New York, 2000.
45. Bremmelgaard, A. et al., Susceptibility testing of Danish isolates of *Capnocytophaga* and CDC group DF-2 bacteria, *APMIS*, 97, 43, 1989.
46. Butler, T. et al., Unidentified gram-negative rod infection. A new disease of man, *Ann. Intern. Med.*, 86, 1, 1977.
47. Wilde, B. et al., *Capnocytophaga* spp. DF-1 pneumonia in an immune-competent 56-year old man post-CABG, *J. Chemother.*, 20, 126, 2008.
48. Oehler, R.L. et al., Bite-related and septic syndromes caused by cats and dogs, *Lancet Infect. Dis.*, 9, 439, 2009.
49. Meyer, S. et al., *Capnocytophaga canimorsus* resists phagocytosis by macrophages and blocks the ability of macrophages to kill other bacteria, *Immunobiology*, 213, 805, 2008.
50. Mally, M. et al., Prevalence of *Capnocytophaga canimorsus* in dogs and occurrence of potential virulence factors, *Microbes Infect.*, 11, 509, 2009.
51. Mally, M. et al., *Capnocytophaga canimorsus*: A human pathogen feeding at the surface of epithelial cells and phagocytes, *PLoS Pathog.*, 4, e1000164, 2008.
52. Shin, H. et al., Escape from immune surveillance by *Capnocytophaga canimorsus*, *J. Infect. Dis.*, 195, 375, 2007.
53. Shin, H. et al., Resistance of *Capnocytophaga canimorsus* to killing by human complement and polymorphonuclear leukocytes, *Infect. Immun.*, 77, 2262, 2009.
54. Sarma, P.S., and Mohanty, S., *Capnocytophaga cynodegmi* cellulitis, bacteremia, and pneumonitis in a diabetic man, *J. Clin. Microbiol.*, 39, 2028, 2001.
55. Khawari, A.A. et al., Sepsis and meningitis due to *Capnocytophaga cynodegmi* after splenectomy, *Clin. Infect. Dis.*, 40, 1709, 2005.
56. Mashimo, P.A. et al., Selective recovery of oral *Capnocytophaga* spp. with sheep blood agar containing bacitracin and polymyxin B, *J. Clin. Microbiol.*, 17, 187, 1983.
57. Syed, S.A., and Loesche, W.J., Survival of human dental plaque flora in various transport media, *Appl. Microbiol.*, 24, 638, 1972.
58. Socransky, S.S. et al., Use of checkerboard DNA–DNA hybridization to study complex microbial ecosystems, *Oral Microbiol. Immunol.*, 19, 352, 2004.

59. Conrads, G. et al., PCR reaction and dot-blot hybridization to monitor the distribution of oral pathogens within plaque samples of periodontally healthy individuals, *J. Periodontol.*, 67, 994, 1996.
60. Ciantar, M. et al., Molecular identification of *Capnocytophaga* spp. via 16S rRNA PCR-restriction fragment length polymorphism analysis, *J. Clin. Microbiol.*, 43, 1894, 2005.
61. Gottwein, J. et al., Etiologic diagnosis of *Capnocytophaga canimorsus* meningitis by broad-range PCR, *Eur. J. Clin. Microbiol. Infect. Dis.*, 25, 132, 2006.
62. Willson, K., Preparation of genomic DNA from bacteria. In Ausubel, F.M. et al. (Editors), *Current Protocols in Molecular Biology*, pp. 2.4.1, Wiley Interscience, Philaderphia, 1990.
63. Smith, G.L. et al., Rapid method for the purification of DNA from subgingival microorganisms, *Oral Microbiol. Immunol.*, 4, 47, 1989.
64. Edwards, U. et al., Isolation and direct complete nu determination of entire genes. Characterization of a ge ing for 16S ribosomal RNA, *Nucleic Acids Res.*, 17, 1989.
65. van Dam, A.P. et al., Molecular characterization of *C cytophaga canimorsus* and other canine Capnocytophaga and assessment by PCR of their frequencies in dogs, *J. Microbiol.*, 47, 3218, 2009.
66. Feinberg, A.P., and Vogelstein, B., A technique for radiola ing DNA restriction endonuclease fragments to high spec activity, *Anal. Biochem.*, 132, 6, 1983.
67. Engler-Blum, G. et al., Reduction of background problems nonradioactive northern and Southern blot analyses enabl higher sensitivity than 32P-based hybridizations, *Ana Biochem.*, 210, 235, 1993.

45 *Chlamydia*

Jens Kjølseth Møller

CONTENTS

45.1 INTRODUCTION

Chlamydia is an obligate intracellular parasite of eukaryotic cells, and can only be propagated in cell culture in the laboratory. Initially, *Chlamydia* was considered a virus because of its inability to grow on artificial media and its capability of passing through filters generally retaining bacteria. *Chlamydia* has a unique growth cycle involving two distinct developmental forms: the infectious, extracellular nonreplicating elementary body (EB), and the noninfectious, intracellular replicating reticulate body (RB). An association between *Chlamydia* and disease was established in 1907, when the transmission of trachoma from a human to another primate was experimentally performed by inoculating the eyes of orangutans with human conjunctival scrapings.[1] A relationship between the "virus" causing trachoma and the agents causing the nongonococcal urethritis and cervicitis, and conjunctivitis of the newborn, was demonstrated. These syndromes are all characterized by epithelial cell inclusions that are morphologically indistinguishable from those of trachoma. Their similar infective agent (now *Chlamydia trachomatis*) was successfully cultured in embryonated hen eggs in 1957 by Tang et al.[2] In the 1960s, electron microscopy studies of these microorganisms and evidence for bacterial rRNA made it clear that the former inclusion conjunctivitis virus was in fact a bacterium. The genus *Chlamydia* was established in 1966.[3]

The main focus of this chapter is the molecular detection of *C. trachomatis*, but it partly covers other species of the family *Chlamydiaceae*.

45.1.1 Taxonomy, Surface Structures, and Genome Organization

Chlamydiaceae are coccoid bacteria measuring 200–300 nm. Molecular insight into *Chlamydiaceae* has been hampered because these bacteria grow only within eukaryotic host cells and a system for gene transfer is lacking in these organisms. Thus, proteins involved in transformation and with the competence to acquire exogenous DNA have not been identified. Furthermore, no genes have been found that encode DNA restriction endonuclease or modification systems, which may suggest that the intracellular and intravacuolar growth of *Chlamydiaceae* has precluded frequent genetic exchange with other organisms.[4] The unique biphasic development cycle of *Chlamydiaceae* with EB and RB has important consequences for the epidemiology, clinical course, and laboratory diagnosis of chlamydial infections.

Taxonomy. In 1971 it was proposed that organisms of the genus *Chlamydia*, family *Chlamydiaceae*, were classified in a separate order, *Chlamydiales ord. nov.*[5] The principal argument for this change was that *Chlamydia* multiply by a developmental cycle that is unique among bacteria. Two species (*Chlamydia trachomatis* and *Chlamydia psittaci*) were originally recognized and separated from each other by their capability to accumulate glycogen in the inclusions in infected cells. *Chlamydiales* was the only bacterial order that had just one family and one genus (*Chlamydiaceae* and *Chlamydia*, respectively) until a taxonomic revision of the family *Chlamydiaceae* was proposed by Everett et al. in 1999.[6] This revision, based on biochemical, ecological, and molecular data, divided the *Chlamydiaceae* into two genera, *Chlamydia* and *Chlamydophila*, comprising nine species (Table 45.1). Comparing complete genome sequences of species within the *Chlamydiaceae* reveals that subtle variations in gene content can cause marked differences in host range and disease process. Read et al.[7] found that 798 of a total of 1009 *C. caviae* genes (open reading frames, ORFs) had orthologs in three other examined *Chlamydiaceae* species; 787 belonged to perfect clusters of orthologous groups. These genes probably represent the core genome required for the developmental cycle and intracellular survival of the *Chlamydiaceae*. Based on these findings, strains of *Chlamydiaceae* from divergent hosts thus seem closely related biologically and ecologically, therefore it has recently been proposed that the *Chlamydiaceae* are reunited into a single genus.[8]

Surface Structures of *Chlamydia*. Like other obligate intracellular parasites, a number of essential steps are necessary before *Chlamydia* can infect the host. These include adhering to and entering host cells, establishing an intracellular inclusion-vacuole for replication, exiting the host cell for subsequent infection of neighboring cells, and evading host defense mechanisms. Adhesion to and invasion of the host cell, and the ability to defeat host defense mechanisms are likely to depend on the surface structure of the EB. Both EBs and RBs are surrounded by inner and outer membranes. The outer membrane of EB is composed of protein, phospholipid, and lipopolysaccharide (LPS). It has been proposed that *Chlamydia* lack peptidoglycan (PG) because muramic acid has not been biochemically detected or is detected in relatively small amounts.[9] However, genes encoding proteins for the entire pathway for PG synthesis and membrane assembly are present in the chlamydial genome.[10] By scrutinizing the expressed proteome of a *C. trachomatis* L2 strain, an enzyme essential for cell wall biosynthesis has been detected in the RBs, suggesting that PG is produced during growth within host cells.[11] Notably, it is of clinical importance that *Chlamydia* contains penicillin-binding proteins and is sensitive to beta-lactam antibiotics that inhibit PG synthesis.[12]

Surface-exposed proteins of microorganisms are major mediators in interactions between pathogens and host cells. A genomic approach for analysis of surface proteins in *Chlamydia* may provide a systematic analysis of proteins of importance for pathogenicity and provide opportunities for development of specific diagnostic assays for assessment of chlamydial infections.[13] Antigen detection of *Chlamydia* can be obtained by use of antibodies against major outer membrane components.[14]

The outer membrane of *C. trachomatis* contains LPS, which is a major surface component of the chlamydial cell. The structure of LPS in *C. trachomatis* is simple and consists only of lipid A and 3-deoxy-D-manno-octulosonic acid. The chlamydial LPS cannot be used to distinguish the individual chlamydial species, and share antigenic determinants with LPS of other gram-negative organisms.[14]

The major outer membrane protein (MOMP or OMP1) is the predominant protein at the surface of both EB and RB forms of *Chlamydia*.[15] MOMP has a molecular mass of ~40 kDa and makes up 60% of the total outer membrane protein. Polyclonal and monoclonal antibodies specific for OMP1 have been used for direct Immunofluorescence detection of *C. trachomatis* in urogenital specimens.[16,17]

TABLE 45.1
The Family *Chlamydiaceae*

Genus	Species	Host	Human Disease	Clade[a] and Serovars
Chlamydia	*C. trachomatis*	Man	Trachoma	Ocular trachoma: A, B, C
			Urogenital infection	
			Urogenital infection	STD: D, E, F, G, H, I, Ia, J, Ja, K
			Lymphogranuloma venereum	LGV: L1, L2, L3
	C. suis	Swine		
	C. muridarum	Mice (rodents)		
Chlamydophila	*Cp. pneumoniae*	Man	Pneumonia	
	Cp. pecorum	Cattle and swine		
	Cp. psittaci	Birds and mammals	Pneumonia	
	Cp. abortus	Sheep/goats		
	Cp. felis	Cats		
	Cp. caviae	Guinea pigs		

[a] Based on Stothard et al.[19]

Sequencing the genome of *C. trachomatis* serovar D revealed a family of 9 polymorphic membrane proteins (Pmp)[10] designated PmpA to PmpI. PmpD is an autotransporter protein that is posttranslationally processed and secreted outside *Chlamydia*, and coincides with the development of EBs.[18] The *Pmp*A to *Pmp*I genes have been used for typing, but the discriminating capacity is limited.[19] Based upon RFLP studies, however, the Pmp genes cluster *C. trachomatis* serovars into three clades[19] reflecting specific disease groups (Table 45.1).

The Genome of *Chlamydia*. The size of the *C. trachomatis* genome is ~1.04 Mbp compared to ~1.23 Mbp for *Cp. Pneumoniae*.[20] The complete sequence of *C. trachomatis* serotype L2 was described in 1998.[10] The *C. trachomatis* genome encodes about 875 proteins. Genomic sequencing indicates that approximately 70 proteins are found only in *C. trachomatis*. Conversely, 214 proteins are unique to *Cp. Pneumoniae*.[20] An extraordinary conservation of gene order has been shown in the genomes from the different *Chlamydia* species.[21] *C. trachomatis* oculotropic and genitotropic strains are highly related with more than 99.6% nucleotide identity.[10,22] Thus, only minor genome differences determine the various clinical manifestations of different groups of *C. trachomatis* strains[23] (Table 45.1). For example, differences in the genes at the locus that encode tryptophan synthase, strictly correlate with ocular and genital tract tropism for *C. trachomatis* serovars causing trachoma and sexually transmitted diseases, respectively.[24]

Chlamydial Plasmids. Clinical strains of *C. trachomatis* harbor a common plasmid,[25] which was sequenced in 1987.[26] Plasmids from *C. trachomatis* are all about 7500 nucleotides in size and highly conserved with less than 1% nucleotide sequence variation.[27] The *C. trachomatis* plasmids contain eight ORFs or predicted coding sequences (*CDSs*).[28] Plasmid-free variants of *C. trachomatis* have been reported,[29,30] which suggests that the common plasmid in *C. trachomatis* may not be crucial for the survival.[31] Yet, it is remarkable that plasmid-free strains seem to be rare, which may indicate that the plasmid do play a role for the infectivity of the pathogen.[32] On the other hand, the new variant *C. trachomatis* strain, with a 377 bp deletion in the *CDS1* gene of the plasmid, has recently established itself firmly in Sweden,[33,34] which indicates that not all plasmid genes are necessary for the infectivity of *Chlamydia*.

Neither human *Cp. pneumoniae* strains nor some *C. psittaci* analyzed so far share a common plasmid.[27] Among other *C. psittaci* isolates, four different plasmids were associated with avian-, feline-, equine-, and guinea pig-related strains. All four *C. psittaci* plasmid types differed from the plasmid found in *C. trachomatis*.[35]

The *C. trachomatis* plasmid is a favored target for DNA-based detection of the *C. trachomatis* pathogen because generally there are 2–10 copies of the plasmid present per chlamydial cell.[25,28,36] This natural target amplification factor enhances the possibility of detecting the bacterium. Seth-Smith et al.[28] looked at candidate plasmid regions for improved diagnostic targets and found that *CDS2* was the most conserved gene in the *C. trachomatis* plasmid. This speaks for the *CDS2* gene as a possible target for screening assays to avoid diagnostic failures.

45.1.2 Epidemiology, Clinical Features, and Pathogenesis

A marked rise in the positive rate of *C. trachomatis* in clinical specimens has recently been noted in several countries.[37] The uniform rise after a period with a decrease in the positive rate seems to reflect a true increase in the rate of urogenital *Chlamydia* infections and not only the increased use of the more sensitive Nucleic Acid Amplification Tests (NAATs). The number of *C. trachomatis* positive specimens (and probably also the prevalence of infection) is very dependent on age and gender. The majority of *Chlamydia* positive specimens are seen in women aged 18–25.[38,39] Young men in general seem to be a few years older before they acquire *Chlamydia* infections (later sexual debut), and are, furthermore, tested less often than females. Yet, more men seem to be tested when urine sampling is used compared to swabs.

Most countries perform only opportunistic screening for chlamydial infections in young people. Controlled intervention trials are needed to ascertain if a systematic screening can reduce the number of related reproductive complications (ectopic pregnancy and infertility). Two randomized trials have shown a reduced number of cases of pelvic inflammatory disease (PID) after one year of follow-up in women screened for *C. trachomatis*.[38,40] A major obstacle to conducting controlled intervention studies of reproductive complications is that the damage to reproductive organs caused by a *C. trachomatis* urogenital infection may first be noted several years later when the women want to reproduce.

Chlamydiaceae are associated with a large variety of diseases in man and animal (Table 45.1). Psittacosis is a zoonotic disease caused by infection with *Cp. psittaci*, its main reservoir being birds. *C. trachomatis* causes trachoma, the world's leading cause of preventable blindness, which still affects tens of millions of people—especially in developing countries.[41] Urogenital infections including urethritis, epididymitis, cervicitis, salpingitis, and PID form another major group of diseases caused by *C. trachomatis*.[42] Severe sequelae may follow a urogenital infection, causing ectopic pregnancy and infertility.[42] Lymphogranuloma venereum (LGV) is a severe genital infection caused by *C. trachomatis* serovar L1–3, which is still endemic in many developing countries. LGV lately has spread in men who have sex with men in Western countries.[43]

Chlamydia seems to be not acutely toxigenic, and the chlamydial infection appears in many cases asymptomatic and persistent. Chlamydial infection is a complex process including attachment to and growth within target cells, mechanisms of dissemination and persistence, directed remodeling of the intracellular environment, and the nature of the host immune response to infection.[44] The inflammatory processes of chlamydial disease are elicited by infected host cells and are probably necessary and sufficient to account for chronic and intense inflammation as well as the promotion of cellular proliferation, tissue remodeling and scarring, which is the ultimate cause of disease sequelae.[45]

45.1.3 Diagnosis

The etiological diagnosis of chlamydial infection took a first step in 1907 when Halberstaedter and von Prowazek identified inclusions in conjunctival scrapings from Trachoma patients by means of Giemsa staining. Presently, detection of *Chlamydiaceae* and in particular *C. trachomatis* can be accomplished by a variety of different principles and methods, including culture and antigen and nucleic acid detection.[14] *Chlamydia* diagnostics have evolved from simple light microscopy of cell inclusions to serology, cell culture, antigen detection, DNA probe test, and, finally, NAAT.

The decision about which assays the laboratory should use is complex, involving not only the laboratory but also the physicians who depend on the test results.[46] The ideal test has a reproducibly high sensitivity and specificity (no false-negative and false-positive results). A confirmatory test can compensate for a screening assay with a high sensitivity but lower specificity. Most decisions about use of assays have the underlying assumption that the assay chosen detects current infection only. It should be born in mind that the various diagnostic principles/methods of chlamydial detection are not equal in respect to their target. Only culture detects live bacteria, the other assays may also reflect the presence, either current or past, of a marker for the particular organism. Analytical sensitivity of conventional and molecular assays for *Chlamydia* ranges from 1 to >1000 organisms per sample.[14]

Since the late 1990s, NAATs have become the preferred methods for detection of *C. trachomatis* infection. NAATs provide enhanced sensitivity and noninvasive diagnostic/screening options. For example, the possibility of testing urine samples or self-collected vaginal swabs, which facilitates otherwise cumbersome screening. Today, NAATs are recommended for all types of urogenital samples except in cases with medico-legal implications such as rape or sexual abuse, where additionally culture of *C. trachomatis* is mandatory.[46]

The fundamental feature of NAATs is their ability to multiply and, in principle, detect down to a single copy of the target DNA or RNA. However, the detection of a nucleic acid is not to be taken as a sign of a current infection. A positive NAAT could reflect a variety of states, including (i) a current clinically active infection, (ii) the presence of residual DNA from a previous infection, (iii) the presence of residual DNA as a result of stochastic or systematic contamination in the laboratory, or (iv) a genuine false-positive result.[47] NAATs may therefore have a lower positive predictive value for detecting current infection than generally expected.

Test-of-cure may be indicated when symptoms persist in patients treated appropriately with antibiotics. NAATs detect nucleic acids from dead *C. trachomatis* and may lead to false-positive results in a cured patient within 4 weeks after completion of treatment. Culture or a nonamplification assay should therefore be used as test-of-cure within that period, bearing in mind that these assays may miss a persistent infection because they often are less sensitive than NAATs. Persistent symptoms may also be caused by untreated concomitantly sexually transmitted pathogens (e.g., *Neisseria gonorrhoeae* and *Mycoplasma genitalium*).

45.1.3.1 Conventional Techniques

Cell Culture. Historically, cell culture has been the reference standard for detection of *C. trachomatis*. A breakthrough in the detection of chlamydial infections by culture *was* made in 1965, when Gordon and Quan published their paper on the use of irradiated McCoy cells for isolation of the Trachoma Agent in cell culture.[48] This culture technique was simple in comparison to the existing isolation of microorganisms from embryonated hen eggs. The use of the irradiated McCoy cell system and a sensitive, further simplified technique using cycloheximide-treated McCoy cells[49] made it possible to screen for *Chlamydia* in the cervix and fallopian tubes of patients with acute salpingitis.[50] A disadvantage of cell culture is, however, the potentially suboptimal sensitivity in areas where fast transportation of samples and adequate culture techniques/facilities cannot be met. The main advantage of cell culture is that the specificity is 100% per definition. This "eliminates" the risk of false-positive results, which is of major concern when low-prevalence populations are tested or when medico-legal aspects are involved.

Today, nucleic acid based methods have, in general, replaced cell culture. Though, if optimal transport and culture techniques/facilities are used, cell culture may be as sensitive as NAATs.[51]

Direct Microscopy. Until the beginning of the 1980s, the routine diagnosis of chlamydial infection was based on the demonstration by direct microscopy of pathognomonic cell inclusions of cell cultures inoculated with clinical material. Chlamydial inclusions were initially demonstrated, either by simple staining with iodine (*C. trachomatis* only) or Giemsa stain. Direct microscopy of EBs in clinical specimens became possible when the monoclonal antibody technology facilitated the use of Immunofluorescence in the early 1980s. The advantage was that DFA did not depend on viable *C. trachomatis* particles and a rapid cold transport chain. A disadvantage was, however, the need of fluorescence microscopy equipment, and careful staff skilled in microscopy. The labor-intensive and eye-tiring methodology precluded a more comprehensive screening for *C. trachomatis* in clinical samples. This became possible by the use of monoclonal antibodies and the introduction of the enzyme immunoassay technique.

Serology. Serology is of limited value in the diagnosis of current chlamydial infection. One problem is that systemic and local antibodies may persist for years after the infection has resolved. Another problem was the low sensitivity and specificity of earlier serology assays based on whole chlamydial antigens. Chlamydial antibodies have, however, been useful for demonstrating associations between *C. trachomatis* and clinical conditions such as ectopic pregnancy and tubal factor infertility.[52]

Antibodies directed against epitopes on MOMP encoded by the *ompl* gene divide *C. trachomatis* strains into 15 serovars A–L[53] (Table 45.1). The distribution of serovars varies, but in general, serovar E seems to be dominating in most countries.[53] Among men who have sex with men, certain subtypes of serovars D, G, and J seem to predominate.[54] The serovars L1–L3 are associated with Lymphogranuloma venereum and have previously been mainly confined to developing countries. In recent years, a spread in Europe and North America of serovar L2 among men who have sex with men has been observed, albeit with a somewhat changed symptomatology.[55]

The commercially available Enzyme Immunoassays (EIAs) all use the LPS as antigen. Cross-reacting organisms such as strains of *Acinetobacter calcoaceticus*, *Escherichia coli*, *Gardnerella vaginalis*, *Neisseria gonorrhoeae*, and group B streptococci make the EIA assays less suitable for samples taken from anatomical sites with an extensive normal flora (e.g., the pharynx and rectum).[56,57] Diagnostic performance (specificity) can be improved by testing positive samples by another assay (confirmatory test).[58] The analytical sensitivity of EIA has been improved by signal amplification.[59]

Rapid enzyme immunoassay-based diagnostic tests for chlamydial antigen were developed for point-of-care testing in the doctor's surgery. A recent study on an internet-available *C. trachomatis* test for home use showed an assay yielding more false-positive (18/193) than true-positive (7/38) results.[60]

45.1.3.2 Molecular Techniques

Molecular detection assays can be separated into nonamplified DNA hybridization (probe) tests and amplification tests.[14] In probe tests, an oligonucleotide sequence (probe) hybridizes to a target within bacterial DNA or RNA present in the clinical specimens, and a signal is detected by Chemiluminescence. The NAAT assays utilize oligonucleotide sequences (primers) to hybridize to and flank the target in the genomic DNA or RNA and then multiply exponentially the target sequence by various molecular techniques.

Probe Tests. The PACE2 assay from Gen-Probe is a probe test targeting ribosomal RNA (rRNA) of *C. trachomatis*. The Gen-Probe PACE2 assay uses the intrinsic "amplification" of the target rRNA molecule, which is present in hundreds of copies per *Chlamydia* cell. The Digene Hybrid Capture II probe test uses RNA probes that recognize and hybridize with specific DNA sequences of *C. trachomatis*. The Digene test has been cleared for use on an automated instrument. The probe tests are being phased out in favor of NAATs.

NAATs. Genomic or plasmid DNA sequences or rRNA are the main targets for molecular detection of *C. trachomatis*. Plasmid-free *C. trachomatis* isolates have occasionally been reported,[30,61] but they seem to be a negligible problem. However, the new *C. trachomatis* variant with a 377 bp deletion in the target region for the Abbott and Roche assays[62] has led to modifications of the two assays using a dual target approach (Table 45.2). Ribosomal RNA is used as target in the Gen-Probe Aptima Combo 2 assay (23S rRNA), the Gen-Probe single analyte APTIMA *C. trachomatis* assay (16S rRNA). The advantage of using ribosomal RNA molecules is the presence of targets in high numbers in the chlamydial cell leading to an enhanced analytical sensitivity. Furthermore, since ribosomal RNA molecules are among the most conserved sequences, the likelihood of genetic changes leading to false-negative results should be minimal. In-house *C. trachomatis* PCRs based on genes encoding 16S rRNA have also been described.[63,64]

Real-time PCR allows for the quantification of bacterial load. Michel et al.[65] evaluated the *C. trachomatis* load in matched specimens from different anatomic sites. For men bacterial load was highest for urine (1200 EBs per 100 μL) and significantly lower for urethral swabs (821 EBs per 100 μL). The mean organism loads for infected women were 2,231, 773, 162, and 47 EBs per 100 μL for endocervical swabs, self-collected vaginal swabs, urethral swabs, and first void urine samples, respectively. NAAT assays offer a reliable method for the use of noninvasive samples such as urine (males) and self-collected vulvo-vaginal specimens, which makes larger screening programs possible.

Genotyping. Serotyping and genotyping based on single genes of *C. trachomatis* have only been of limited success in the studies of *C. trachomatis* epidemiology. Genotype

TABLE 45.2
Targets for Selected Commercial *C. trachomatis* NAAT Assays

Company and Assay	NAAT Principle	Sequence within Target Specified
Abbott RealTime CT new formulation	Real-time PCR (Dual plasmid targets)	A: ORF 1[a]; B: ORF 3
Becton Dickinson ProbeTec™ CT	Strand Displacement Amplification	ORF 4
Gen-Probe APTIMA COMBO 2®	Transcription Mediated Amplification	23S rRNA molecule
Gen-Probe APTIMA® CT	Transcription Mediated Amplification	16S rRNA molecule
COBAS® TaqMan® CT Test, v2.0	Real-time PCR (Dual targets)	A: ORF 1[a]; B: chromosomal *ompl* gene
Siemens VERSANT™ kPCR CT	Real-time PCR	ORF 2

[a] Within the deletion area of the Swedish plasmid mutant.
ORF, Open reading frame.

based on PCR and sequencing of *ompl* coincides with the MOMP serotype, and genotyping is now replacing serotyping. The subject was recently reviewed by Pedersen et al.[53] Typing of *C. trachomatis* may be wanted for several purposes. A major aspect is the analysis of transmission patterns in sexual networks, and to look for possible association between types and specific disease entities or syndromes. An interesting but unresolved issue is to discriminate between persistence and reinfection in persons with repeatedly positive *Chlamydia* tests. Genotyping may also be crucial in investigations of sexual assaults. However, developing highly discriminating genotyping methods is difficult because of the highly conserved genome of *C. trachomatis*. With the aim of analyzing evolutional changes, a Multilocus sequence typing (MLST) system for the entire *Chlamydiaecae* family based on seven housekeeping genes was developed and could discriminate between reference strains of *C. trachomatis*, but provided almost no resolution for *Cp. pneumoniae*.[66]

Recently, two alternative detection systems have been reported to provide high resolution typing of *C. trachomatis*, and of other *Chlamydiaceae* species. High resolution MLST typing of *C. trachomatis* strains has been achieved by sequencing five target regions[67]; analysis of 47 clinical isolates of representative serotypes resulted in 32 genetic variants among 12 *ompl* variants. An alternative multiple-locus variable-number tandem repeat analysis (MLVA) system for high resolution genotyping of *C. trachomatis* describes an analysis of variable numbers of tandem repeats (VNTR) in three loci combined with *ompl* sequencing that reached a significantly higher diversity index than *ompl* alone.[68] The combined *ompl* and VNTR genotyping system was readily performed on samples submitted in the Gen-Probe transport media for the Aptima Combo 2 assay for *C. trachomatis*.

45.2 METHODS

Three main issues need to be addressed before the selection of a molecular *Chlamydia* assay: target population; specimen types and handling (collection and transport); and the practical implementation of the assay.

Before implementation of an assay, the laboratory should consider handling of false-positive results, control of sample inhibition and contamination, quality control of the assay, and reporting (results and/or interpretations).

An adequate number of epithelial cells infected with *C. trachomatis* in the specimen collected is crucial for any assay. No test, however, is better than the weakest link in the assay procedure. For NAATs this comprises sample preparation (extraction and purification of nucleic acids), amplification, and detection.

45.2.1 Sample Preparation

Urine in particular but also other specimen types may contain inhibitors of the NAATs. Nucleic acid extraction methods, therefore, play an important role in the overall performance of an assay. The extraction can be generic (i.e., purifying all nucleic acids in the sample), or target specific (i.e., purifying nucleic acids containing the target of the assay only). Use of target capture thus increases the specificity of the particular assay but the generic methods provide purified material for other NAAT assays. Automated extraction methods generally apply magnetic particles to which the nucleic acids are bound by various principles. A magnet retains the particles in the sample tube during the washing procedures whereas potential inhibitors and other substances are removed.

Urogenital Samples. All commercial NAAT assays are FDA approved for urogenital swabs. Vaginal swab specimens are the preferred specimen type for chlamydial screening of females. Vaginal swab specimens are as clinical sensitive and specific as cervical swab specimens.[69] Furthermore, a vaginal swab, as with a urine sample may be self-collected at home or in the STD clinic or the GPs surgery. Cervical samples are still preferred by most doctors when full pelvic examinations are carried out, but vaginal swab specimens may constitute an appropriate sample type. Urine samples combined with a cervical swab specimen have been shown to significantly improve the yield when compared with urine samples only.[70] Liquid-based Pap screening for cervical cancer may also provide sample material for testing by FDA approved *C. trachomatis* assays.[71]

For males, a urine specimen is preferred for two reasons: (i) NAATs are at least as sensitive on urine as for a urethral swab[72] and (ii) urine sampling is noninvasive, which may motivate more men to be examined. However, urine specimens must contain an adequate number of cells with *C. trachomatis* and hence a "first-void" urine sample should be collected preferably at a time when the last urination was no less than 1 h earlier.

Extragenital Samples. The NAAT assays are not FDA approved for use on specimens from extragenital sites such as the eye, pharynx, and rectum. An in-house validation should therefore be performed if specimens from extragenital sites are tested. In the final report, the laboratory should state that the specimen is not approved for the assay. Furthermore, the result should be interpreted with caution, if an independent NAAT has not been used for confirmation. It is surprising that commercial companies have not validated their assays for pharyngeal and rectal swabs considering the potential market for these tests.

45.2.2 Detection Procedures

All diagnostic tests should be evaluated for their analytic and clinical performance before routine use. The analytic specificity of a *Chlamydia* NAAT is determined by testing a variety of bacteria, parasites, fungi, viruses, yeast, and so forth that may be isolated from the urogenital tract or other sites of sampling. The analytic sensitivity is determined by testing a serial dilution of all serovars of *C. trachomatis*. The clinical validity of a test is a measure of its ability to detect the associated disease or disorder. This may be the difficult part due to the lack of a clinical standard for *Chlamydia* infection, especially when no clinical signs or symptoms are present.

45.2.2.1 Commercial Assays

Manufacturers of commercial NAAT assays detecting *C. trachomatis* have different NAAT principles, targets, and hardware solutions (Table 45.2). Real-time PCR is used by Abbott, Roche and Siemens, Strand Displacement Amplification (SDA) by Becton Dickinson, and Transcription Mediated Amplification (TMA) by Gen-Probe.

Choice of assay platform depends on a series of practical and economical factors of which the performance can be scrutinized through External Quality Assessment (EQA) programs. These are supplied by, for example, the College of American Pathologists or by European organizations such as United Kingdom National External Quality Assessment Service (UK NEQAS), and Quality Control for Molecular Diagnostics (QCMD). In surveys or reports from these EQA programs, the rate of false-positive and false-negative results according to the product manufacturer can be compared.

45.2.2.2 Combo-Testing

Dual detection of the both *C. trachomatis* and *N. gonorrhoeae* in the same urogenital sample has been FDA approved for the major commercially available NAATs. However, the advantage of the combo-testing may be outweighed by its inappropriate use in testing populations with a low prevalence of for example, gonorrhea. Even with an acceptable specificity of the *N. gonorrhoeae* assay, the number of false-positive tests may be unacceptably high. The BD Viper™ and the TIGRIS® DTS™ System from Gen-Probe both offer a high throughput automated platform for combined testing capable of analyzing more than 350 samples for both *C. trachomatis* and *N. gonorrhoeae* per 8-h shift. Compared to manual platforms these robots significantly reduce manual labor and repetitive movements for the laboratory staff. These two platforms were evaluated and compared with the more recent Abbott RealTime PCR system *m*2000,[73] which has a lower throughput (186 specimens per 8-h shift). It was concluded that all three assays were suitable for the detection of the two microorganisms.[73]

45.2.2.3 Sensitivity and Specificity of Commercial NAATs

A systematic review assessing the sensitivity and specificity of commercial NAATs for detection of *C. trachomatis* in urine and cervical (women) and urethral (men) swab samples, respectively, was published by Cook et al. in 2005.[74] The sensitivities were between 83.3% and 96.7% with specificities in the range 93.8%–99.6%. Assays based on TMA and SDA found significantly more *C. trachomatis* than PCR assays in cervix specimens. In female urine, TMA found significantly more *C. trachomatis* than SDA. For male samples (urethra and urine), no significant difference in sensitivity was observed for PCR, SDA, and TMA. Results of TMA and PCR were nearly identical for urine and cervix specimens, whereas the sensitivity of SDA for *C. trachomatis* in cervix and urine specimens was significantly different.

Notably, with a decreasing prevalence of the *Chlamydia* disease, even specificity figures >99% may create increasing problems with false-positive tests and a decreasing positive predictive value (PPV) of the assay. With a disease prevalence of 1%, the false-positive rate (1-PPV) of an assay with 90% sensitivity and 99% specificity is then more than 50%.

45.2.2.4 In-House PCR Assays

In-house or "home-brewed" PCR tests for *Chlamydia* can be useful as a supplement to the commercial *C. trachomatis* assay or as the only option when a commercial assay is not available (e.g., in detection of other *Chlamydia* species including *Cp. Pneumoniae*).[75–78] Using a second target, in-house PCRs may serve as confirmatory testing of samples with equivocal test results in commercial assays. In-house assays may also be used for identification of plasmid-free strains[61] or the new variant of *C. trachomatis* that was discovered in Sweden.[79] There are also several important applications for in-house PCR assays where methods for subspeciation or genotyping are wanted[53] or when the plasmids and genomes of the various *Chlamydia* species are compared.[28] A specific real-time in-house PCR test for detection of *C. trachomatis* serovar L1–L3 causing Lymphogranuloma venereum has been described.[80]

Principle. *C. trachomatis* can be identified and genotyped by PCR amplification and sequencing analysis of the *ompl* gene[53] and three VNTRs[68] using the primers shown in the Table 45.3.

Procedure

1. To amplify the *ompl* gene, prepare PCR mixture (50 μL) comprising 5 μL Extensor Hi-fidelity PCR Mix 10× (Thermo Fisher), 0.8 μL dNTP (25 mmol of each), 1 μL (100 pmol) each of forward and reverse primers, 0.25 μL of Extensor polymerase, 36.95 μL H_2O, and 5 μL of extracted DNA. The primers used for generating an approximately 1.1-kb fragment of the *ompl* gene are PCTM3 (F) and NRI (R).
2. The thermo cycler program is as follows: 120 s at 94°C, followed by a touch-down protocol of 5 cycles of (denaturation at 94°C for 10 s, annealing at 65°C–1°C/cycle for 30 s, extension at 68°C for 60 s), 45 cycles of (10 s at 94°C, 30 s at 60°C, and 60 s at 68°C), and finally one cycle of 7 min at 68°C.
3. To amplify all three VNTR genes, prepare a multiplex PCR mixture (50 μL) consisting of 5 μL AmpliTaq Gold® buffer 10× (Applied Biosystems), 10 μL of MgCl 25mM, 0.8 μL dNTP (25 mmol of each), 1 μL (100 pmol) of each forward (CT1291 F, CT1335 F) and reverse (CT1291 F, CT1335 F) primers, and 0.25 μL each of the CT1299 F and CT1299 R primers, 24.2 μL H_2O, 0.5 μL AmpliTaq Gold® polymerase, and 5 μL of extracted DNA
4. The thermo cycler program is 10 min at 94°C, 40 cycles of (45 s at 94°C, 45 s at 56°C, and 45 s at 72°C), and finally one cycle of 10 min at 72°C.

TABLE 45.3
Primers for Identification and Sequence Analysis of *C. trachomatis*

Primer	Sequence (5′→3′)	Position[a]	Specificity
PCTM3	TCCTTGCAAGCTCTGCCTGTGGGGAATCCT	58–84	*Ompl* gene
NRI	CCGCAAGATTTTCTAGATTTC	1073–1053	
CT1291 F	GCCAAGAAAAACATGCTGGT	195, 536–195, 555	Hypothetical protein
CT1291 R	AGGATATTTCCCTCAGTTATTCG	195, 760–195, 738	CT172.1
CT1299 F	TTGTGTAAAGAGGGTCTATCTCCA	291, 758–291, 781	Noncoding region
CT1299 R	AAGTCCACGTTGTCATTGTACG	291, 945–291, 924	
CT1335 F	TCATAAAAGTTAAATGAAGAGGGACT	737, 225–737, 250	DNA topoisomerase
CT1335 R	TAATCTTGGCTGGGGATTCA	737, 377–737, 358	

[a] According to the *ompl* nucleotide sequence of *C. trachomatis* serovar strain D/UW-3/Cx.[11]

5. Electrophorese the products on a 2% (w/v) agarose gel, and visualize with ethidium bromide stain.
6. Purify PCR products before sequencing is performed. Excise the band of the expected size from the 2% agarose gel, and purify DNA using the QIAquick Gel Extraction Kit (QIAgen). An insertion mutation may have occurred in the VNTR CT1291 region since in some strains the band is 224, and in others approximately 510 bp.
7. Perform sequencing reactions with the BigDye® v 3.1 Cycle Sequencing Kit (Applied Biosystems) according to the manufacturer's instructions using 6 μL master mix (2 μL BigDye, 0.4 μL (10 pmol/μL) primer (F or R), 3.6 μL H_2O) and 4 μL template. The thermo cycler program is 120 s at 94°C, 25 cycles of 10 s at 94°C, 5 s at 50°C, and 4 min at 60°C, and hold at 8°C. Sequence the *ompl* once with the forward and once with the reverse primer, creating an overlap of approximately 250 bp in the mid region. Genotype the *ompl* gene sequences using the program standard nucleotide-nucleotide Basic Local Alignment Search Tool (BLAST) (http://www.ncbi.nlm.nih.gov/BLAST/) "BLAST two sequences" approach. Use accession numbers for the *ompl* reference types listed in paper by Pedersen et al.[68] Investigate visually the sequences of the VNTRs (only the repeated part of the sequences) for differences.[68]

45.2.2.5 Quality Control

Essentially, quality control is performed to secure that assays do not produce false-negative and false-positive results (retain highest possible sensitivity and specificity). To ensure good laboratory practice, the performance of NAATs should be scrutinized by a series of internal and external controls. Failure to follow the regulations and requirements for performance, monitoring, auditing, recording, analysis, and reporting of test results can be devastating to patients and laboratory.

An internal certification should be performed even if an approved commercial test is implemented. This can be done by use of preselected panels of test samples with known results or initially by running the new assay in parallel with the old routine system, bearing in mind that a new more sensitive assay may require confirmatory testing. Use of an in-house assay necessitates a more thorough validation. The analytical sensitivity should be calculated by determining the limit of detection through serial dilutions of known amounts of purified DNA and cultured bacteria spiked into the relevant specimen matrix. Spiked samples should be purified by the new sample extraction procedure before testing. The analytical specificity of the in-house assay should be checked with isolates covering a broad geographical and temporal spectrum, as well as different patient populations.

Internal controls are useful in the daily runs. Correct test results for the positive and negative controls monitor the general performance of the assay. An internal amplification control (IAC) (i.e., a small amount of a known target) may be added to each reaction in order to demonstrate lack of inhibition in the individual clinical sample.[81] The IAC should be added before the NA extraction procedure in order to also control for accidental loss of the specimen during purification. Lack of amplification of the IAC may also document partial reagent or instrument failure affecting only a subset of the specimens. Periodically, the laboratory can perform an internal quality assessment by selecting a panel of specimens with known results and relabel them before repeat testing as for ordinary clinical specimens.

Confirmatory testing of positive results has been recommended, although the procedure may be unnecessary.[82] There are three supplemental testing approaches: (i) repeating the original test on the original specimen, (ii) retesting the original specimen with a different NAAT, and (iii) performing a different test on a duplicate specimen. However, as pointed out by Hadgu and Sternberg,[83] if a test result cannot be confirmed by itself, it is not important whether it is confirmed or not by another test. On the other hand, failure to repeat positive results could reflect the variability inherent with low copy numbers of targets.

Use of external controls (blind samples) is an important quality control, and laboratories should participate in organized EQA schemes. It is essential that the quality control samples are tested according to the standard method used in the laboratory including all pre- and postexamination procedures. Exchanging specimens between laboratories using different NAAT methodologies may also be of help in validation of NAAT results.

Last but not least, laboratories should provide guidelines for proper sampling and educate health care personnel to ensure correct use of the assays and interpretation of results.

45.3 CONCLUSIONS AND FUTURE PERSPECTIVES

Detection methods based on nucleic acid amplification led to *Chlamydia* being detected in novel tissues and cells, for example, human joints, atherosclerotic plaques, brains, and amoebae. With improved molecular identification and typing methods for *Chlamydia*, new insight into host-parasite relations and spread of these pathogens can be envisaged. The epidemiology of *C. trachomatis* infection in young people worldwide indicates that we still have more to learn about preventing the spread of this pathogen and the sequelae of its infection. We may also have to reevaluate the meaning of (e.g., a positive NAAT assay in asymptomatic patients and consider whether this result reflects true infection with *C. trachomatis* and whether the patient needs to be treated with antibiotics). Induction of protective immunity by vaccination against *C. trachomatis* may hopefully constitute a future option in the prevention of sexually transmitted infections with *Chlamydia*.

Today, cell culture detection of *Chlamydia* is rarely used. This may, in combination with the lack of nonculture assays for drug susceptibility test prevent disclosure of inadequate antimicrobial treatment of asymptomatic infections. Drug resistant *C. trachomatis* strains have been described,[84] but they do not at present seem to be a widespread clinical problem. However, the rapid spread of a mutant variant of *C. trachomatis* in Sweden[62] speaks for surveillance also for drug resistant mutants.

Future developments are likely to include multiplex STD panels combined with detection systems based upon a single detection chip or a liquid array technology. By creating multiplex PCR detection systems that not only detect all relevant microorganisms but also determine their drug resistance pattern and genotype, we may have concomitant information for treatment and prevention of sexually transmitted infections.

REFERENCES

1. Collier, L.H., Recent advances in the virology of trachoma, inclusion conjunctivitis and allied diseases, *Br. Med. Bull.*, 15, 231, 1959.
2. Tang, F.F. et al., Studies on the etiology of trachoma with special reference to isolation of the virus in chick embryo, *Chin. Med. J.*, 75, 429, 1957.
3. Page, L.A., Revision of the family *Chlamydiaceae* Rake (*Rickettsiales*): Unification of the Psittacosis-Lymphogranuloma venereum-Trachoma group of organisms in the genus *Chlamydia* Jones, Rake and Stearns, 1945, *Int. J. Syst. Bacteriol.*, 16, 223, 1966.
4. Wagar, E.A., Safarians, S., and Pang, M., Analysis of genomic DNA from *Chlamydia trachomatis* for Dam and Dcm methylation, *FEMS Microbiol. Lett.*, 77, 161, 1992.
5. Storz, J., and Page, L.A., Taxonomy of the *Chlamydiae*: Reasons for classifying organisms of the genus *Chlamydia*, family *Chlamydiaceae*, in a separate order, *Chlamydiales* ord. nov., *Int. J. Syst. Bacteriol.*, 21, 1971.
6. Everett, K.D., Bush, R.M., and Andersen, A.A., Emended description of the order *Chlamydiales*, proposal of *Parachlamydiaceae* fam. nov. and *Simkaniaceae* fam. nov., each containing one monotypic genus, revised taxonomy of the family *Chlamydiaceae*, including a new genus and five new species, and standards for the identification of organisms, *Int. J. Syst. Bacteriol.*, 49, 415, 1999.
7. Read, T.D. et al., Genome sequence of *Chlamydophila caviae* (*Chlamydia psittaci* GPIC): Examining the role of niche-specific genes in the evolution of the *Chlamydiaceae*, *Nucleic Acids Res.*, 31, 2134, 2003.
8. Stephens, R.S. et al., Divergence without difference: Phylogenetics and taxonomy of *Chlamydia* resolved, *FEMS Immunol. Med. Microbiol.*, 55, 115, 2009.
9. Fox, A. et al., Muramic acid is not detectable in *Chlamydia psittaci* or *Chlamydia trachomatis* by gas chromatography-mass spectrometry, *Infect. Immun.*, 58, 835, 1990.
10. Stephens, R.S. et al., Genome sequence of an obligate intracellular pathogen of humans: *Chlamydia trachomatis*, *Science*, 282, 754, 1998.
11. Skipp, P. et al., Shotgun proteomic analysis of *Chlamydia trachomatis*, *Proteomics*, 5, 1558, 2005.
12. Barbour, A.G. et al., *Chlamydia trachomatis* has penicillin-binding proteins but not detectable muramic acid, *J. Bacteriol.*, 151, 420, 1982.
13. Montigiani, S. et al., Genomic approach for analysis of surface proteins in *Chlamydia pneumoniae*, *Infect. Immun.*, 70, 368, 2002.
14. Black, C.M., Current methods of laboratory diagnosis of *Chlamydia trachomatis* infections, *Clin. Microbiol. Rev.*, 10, 160, 1997.
15. Caldwell, H.D., Kromhout, J., and Schachter, J., Purification and partial characterization of the major outer membrane protein of *Chlamydia trachomatis*, *Infect. Immun.*, 31, 1161, 1981.
16. Caldwell, H.D., and Schachter, J., Antigenic analysis of the major outer membrane protein of *Chlamydia* spp., *Infect. Immun.*, 35, 1024, 1982.
17. Wang, S.P. et al., Immunotyping of *Chlamydia trachomatis* with monoclonal antibodies, *J. Infect. Dis.*, 152, 791, 1985.
18. Kiselev, A.O. et al., Expression, processing, and localization of PmpD of *Chlamydia trachomatis* serovar L2 during the chlamydial developmental cycle, *PLoS One*, 2, e568, 2007.
19. Stothard, D.R., Toth, G.A., and Batteiger, B.E., Polymorphic membrane protein H has evolved in parallel with the three disease-causing groups of *Chlamydia trachomatis*, *Infect. Immun.*, 71, 1200, 2003.
20. Kalman, S. et al., Comparative genomes of *Chlamydia pneumoniae* and *C. trachomatis*, *Nat. Genet.*, 21, 385, 1999.
21. Read, T.D. et al., Genome sequences of *Chlamydia trachomatis* MoPn and *Chlamydia pneumoniae* AR39, *Nucleic Acids Res.*, 28, 1397, 2000.
22. Carlson, J.H. et al., Comparative genomic analysis of *Chlamydia trachomatis* oculotropic and genitotropic strains, *Infect. Immun.*, 73, 6407, 2005.
23. Kari, L. et al., Pathogenic diversity among *Chlamydia trachomatis* ocular strains in nonhuman primates is affected by subtle genomic variations, *J. Infect. Dis.*, 197, 449, 2008.

24. Caldwell, H.D. et al., Polymorphisms in *Chlamydia trachomatis* tryptophan synthase genes differentiate between genital and ocular isolates, *J. Clin. Invest.*, 111, 1757, 2003.
25. Palmer, L., and Falkow, S., A common plasmid of *Chlamydia trachomatis*, *Plasmid*, 16, 52, 1986.
26. Sriprakash, K.S., and Macavoy, E.S., Characterization and sequence of a plasmid from the trachoma biovar of *Chlamydia trachomatis*, *Plasmid*, 18, 205, 1987.
27. Thomas, N.S. et al., Plasmid diversity in *Chlamydia*, *Microbiology*, 143, 1847, 1997.
28. Seth-Smith, H.M. et al., Co-evolution of genomes and plasmids within *Chlamydia trachomatis* and the emergence in Sweden of a new variant strain, *BMC Genomics*, 10, 239, 2009.
29. Stothard, D.R. et al., Identification of a *Chlamydia trachomatis* serovar E urogenital isolate which lacks the cryptic plasmid, *Infect. Immun.*, 66, 6010, 1998.
30. An, Q. et al., Infection with a plasmid-free variant *Chlamydia* related to *Chlamydia trachomatis* identified by using multiple assays for nucleic acid detection, *J. Clin. Microbiol.*, 30, 2814, 1992.
31. Peterson, E.M. et al., The 7.5-kb plasmid present in *Chlamydia trachomatis* is not essential for the growth of this microorganism, *Plasmid*, 23, 144, 1990.
32. Carlson, J.H. et al., Comparative genomic analysis of *Chlamydia trachomatis* oculotropic and genitotropic strains, *Infect. Immun.*, 73, 6407, 2005.
33. Ripa, T., and Nilsson, P., A variant of *Chlamydia trachomatis* with deletion in cryptic plasmid: Implications for use of PCR diagnostic tests, *Euro Surveill.*, 11, E061109.2, 2006.
34. Møller, J.K., Pedersen, L.N., and Persson, K., Comparison of Gen-probe transcription-mediated amplification, Abbott PCR, and Roche PCR assays for detection of wild-type and mutant plasmid strains of *Chlamydia trachomatis* in Sweden, *J. Clin. Microbiol.*, 46, 3892, 2008.
35. Lusher, M., Storey, C.C., and Richmond, S.J., Plasmid diversity within the genus *Chlamydia*, *J. Gen. Microbiol.*, 135, 1145, 1989.
36. Tam, J.E. et al., Location of the origin of replication for the 7.5-kb *Chlamydia trachomatis* plasmid, *Plasmid*, 27, 231, 1992.
37. Fenton, K.A., and Lowndes, C.M., Recent trends in the epidemiology of sexually transmitted infections in the European Union, *Sex. Transm. Infect.*, 80, 255, 2004.
38. Østergaard, L. et al., Home sampling versus conventional swab sampling for screening of *Chlamydia trachomatis* in women: A cluster-randomized 1-year follow-up study, *Clin. Infect. Dis.*, 31, 951, 2000.
39. Richardus, J.H., and Götz, H.M., Risk selection and targeted interventions in community-based control of chlamydia, *Curr. Opin. Infect. Dis.*, 20, 60, 2007.
40. Scholes, D. et al., Prevention of pelvic inflammatory disease by screening for cervical chlamydial infection, *N. Engl. J. Med.*, 334, 1362, 1996.
41. Kasi, P.M. et al., Blinding trachoma: A disease of poverty, *PLoS Med.*, 1, e44, 2004.
42. Cates, W. Jr., and *Wasserheit*, J.N., Genital chlamydial infections: Epidemiology and reproductive sequelae, *Am. J. Obstet. Gynecol.*, 164, 1771, 1991.
43. Klint, M. et al., Lymphogranuloma venereum prevalence in Sweden among men who have sex with men and characterization of *Chlamydia trachomatis* ompA genotypes, *J. Clin. Microbiol.*, 44, 4066, 2006.
44. Rockey, D.D., Lenart, J., and Stephens, R.S., Genome sequencing and our understanding of Chlamydiae, *Infect. Immun.*, 68, 5473, 2000.
45. Stephens, R.S., The cellular paradigm of chlamydial pathogenesis, *Trends Microbiol.*, 11, 44, 2003.
46. Centers for Disease Control and Prevention, Workowski, K.A., and Berman, S.M., Sexually transmitted diseases treatment guidelines, 2006, *MMWR Recomm Rep.*, 55, 1, 2006.
47. Hadgu, A., Issues in *Chlamydia trachomatis* testing by nucleic acid amplification test, *J. Infect. Dis.*, 193, 1335, 2006.
48. Gordon, F.B., and Quan, A.L., Isolation of the trachoma agent in cell culture, *Proc. Soc. Exp. Biol. Med.*, 118, 354, 1965.
49. Ripa, K.T., and Mårdh, P.A., Cultivation of *Chlamydia trachomatis* in cycloheximide-treated mccoy cells, *J. Clin. Microbiol.*, 6, 328, 1977.
50. Mårdh, P.A. et al., *Chlamydia trachomatis* infection in patients with acute salpingitis, *N. Engl. J. Med.*, 296, 1377, 1977.
51. Møller, J.K., Østergaard, L.J., and Hansen, J.T., Clinical evaluation of four non-related techniques for detection of *Chlamydia trachomatis* in endocervical specimens obtained from a low prevalence population, *Immunol. Infect. Dis.*, 4, 191, 1994.
52. Persson, K., The role of serology, antibiotic susceptibility testing and serovar determination in genital chlamydial infections, *Best Pract. Res. Clin. Obstet. Gynaecol.*, 16, 801, 2002.
53. Pedersen, L.N., Herrmann, B., and Møller, J.K., Typing *Chlamydia trachomatis*: From egg yolk to nanotechnology, *FEMS Immunol. Med. Microbiol.*, 55, 120, 2009.
54. Mossman, D. et al., Genotyping of urogenital *Chlamydia trachomatis* in Regional New South Wales, Australia, *Sex. Transm. Dis.*, 35, 614, 2008.
55. van de Laar, M.J., The emergence of LGV in Western Europe: What do we know, what can we do? *Euro Surveill.*, 11, 146, 2006.
56. Hammerschlag, M.R., Rettig, P.J., and Shields, M.E., False positive results with the use of chlamydial antigen detection tests in the evaluation of suspected sexual abuse in children, *Pediatr. Infect. Dis. J.*, 7, 11, 1988.
57. Taylor-Robinson, D., Thomas, B.J., and Osborn, M.F., Evaluation of enzyme immunoassay (Chlamydiazyme) for detecting *Chlamydia trachomatis* in genital tract specimens, *J. Clin. Pathol.*, 40, 194, 1987.
58. Østergaard, L., and Møller, J.K., Use of PCR and direct immunofluorescence microscopy for confirmation of results obtained by Syva MicroTrak *Chlamydia* enzyme immunoassay, *J. Clin. Microbiol.*, 33, 2620, 1995.
59. Okadome, A. et al., Reactivity of a dual amplified chlamydia immunoassay with different serovars of *Chlamydia trachomatis*, *Int. J. STD AIDS*, 10, 460, 1999.
60. Michel, C.E. et al., Pitfalls of internet-accessible diagnostic tests: Inadequate performance of a CE-marked *Chlamydia* test for home use, *Sex. Transm. Infect.*, 85, 187, 2009.
61. Magbanua, J.P. et al., *Chlamydia trachomatis* variant not detected by plasmid based nucleic acid amplification tests: Molecular characterisation and failure of single dose azithromycin, *Sex. Transm. Infect.*, 83, 339, 2007.
62. Ripa, T., and Nilsson, P.A., A *Chlamydia trachomatis* strain with a 377-bp deletion in the cryptic plasmid causing false-negative nucleic acid amplification tests, *Sex. Transm. Dis.*, 34, 255, 2007.
63. Mahony, J.B. et al., Evaluation of the NucliSens Basic Kit for detection of *Chlamydia trachomatis* and *Neisseria gonorrhoeae* in genital tract specimens using nucleic acid sequence-based amplification of 16S rRNA, *J. Clin. Microbiol.*, 39, 1429, 2001.

64. Madico, G. et al., Touchdown enzyme time release-PCR for detection and identification of *Chlamydia trachomatis*, *C. pneumoniae*, and *C. psittaci* using the 16S and 16S-23S spacer rRNA genes, *J. Clin. Microbiol.*, 38, 1085, 2000.
65. Michel, C.E. et al., *Chlamydia trachomatis* load at matched anatomic sites: Implications for screening strategies, *J. Clin. Microbiol.*, 45, 1395, 2007.
66. Pannekoek, Y. et al., Multi locus sequence typing of *Chlamydiales*: Clonal groupings within the obligate intracellular bacteria *Chlamydia trachomatis*, *BMC Microbiol.*, 8, 42, 2008.
67. Klint, M. et al., High-resolution genotyping of *Chlamydia trachomatis* strains by multilocus sequence analysis, *J. Clin. Microbiol.*, 45, 1410, 2007.
68. Pedersen, L.N., Pødenphant, L., and Møller, J.K., Highly discriminative genotyping of *Chlamydia trachomatis* using omp1 and a set of variable number tandem repeats, *Clin. Microbiol. Infect.*, 14, 644, 2008.
69. Schachter, J. et al., Vaginal swabs are the specimens of choice when screening for *Chlamydia trachomatis* and *Neisseria gonorrhoeae*: Results from a multicenter evaluation of the APTIMA assays for both infections, *Sex. Transm. Dis.*, 32, 725, 2005.
70. Airell, A. et al., *Chlamydia trachomatis* PCR (Cobas Amplicor) in women: Endocervical specimen transported in a specimen of urine versus endocervical and urethral specimens in 2-SP medium versus urine specimen only, *Int. J. STD AIDS*, 11, 651, 2000.
71. Chernesky, M. et al., Abilities of APTIMA, AMPLICOR, and ProbeTec assays to detect *Chlamydia trachomatis* and *Neisseria gonorrhoeae* in PreservCyt ThinPrep Liquid-based Pap samples, *J. Clin. Microbiol.*, 45, 2355, 2007.
72. Sugunendran, H. et al., Comparison of urine, first and second endourethral swabs for PCR based detection of genital *Chlamydia trachomatis* infection in male patients, *Sex. Transm. Infect.*, 77, 423, 2001.
73. Levett, P.N. et al., Evaluation of three automated nucleic acid amplification systems for detection of *Chlamydia trachomatis* and *Neisseria gonorrhoeae* in first-void urine specimens, *J. Clin. Microbiol.*, 46, 2109, 2008.
74. Cook, R.L. et al., Systematic review: noninvasive testing for *Chlamydia trachomatis* and *Neisseria gonorrhoeae*, *Ann. Intern. Med.*, 142, 914, 2005.
75. Mitchell, S.L. et al., Evaluation of two real-time PCR chemistries for the detection of *Chlamydophila pneumoniae* in clinical specimens, *Mol. Cell. Probes*, 23, 309, 2009.
76. Robertson, T. et al., Characterization of *Chlamydiaceae* species using PCR and high resolution melt curve analysis of the 16S rRNA gene, *J. Appl. Microbiol.*, 107, 2017, 2009.
77. Tang, Y.W. et al., Qualitative and quantitative detection of *Chlamydophila pneumoniae* DNA in cerebrospinal fluid from multiple sclerosis patients and controls, *PLoS One*, 4, e5200, 2009.
78. McKechnie, M.L. et al., Simultaneous identification of 14 genital microorganisms in urine by use of a multiplex PCR-based reverse line blot assay, *J. Clin. Microbiol.*, 47, 1871, 2009.
79. Catsburg, A. et al., TaqMan assay for Swedish *Chlamydia trachomatis* variant, *Emerg. Infect. Dis.*, 13, 1432, 2007.
80. Morre, S.A. et al., Lymphogranuloma venereum diagnostics: From culture to real-time quadriplex polymerase chain reaction, *Sex. Transm. Infect.*, 84, 252, 2008.
81. Hoorfar, J. et al., Practical considerations in design of internal amplification controls for diagnostic PCR assays, *J. Clin. Microbiol.*, 42, 1863, 2004.
82. Moncada, J., Donegan, E., and Schachter, J., Evaluation of CDC-recommended approaches for confirmatory testing of positive *Neisseria gonorrhoeae* nucleic acid amplification test results, *J. Clin. Microbiol.*, 46, 1614, 2008.
83. Hadgu, A., and Sternberg, M., Reproducibility and specificity concerns associated with nucleic acid amplification tests for detecting *Chlamydia trachomatis*, *Eur. J. Clin. Microbiol. Infect. Dis.*, 28, 9, 2009.
84. Somani, J. et al., Multiple drug-resistant *Chlamydia trachomatis* associated with clinical treatment failure, *J. Infect. Dis.*, 181, 1421, 2000.

46 *Chlamydophila*

Chengming Wang, Bernhard Kaltenboeck, and Konrad Sachse

CONTENTS

46.1 INTRODUCTION

46.1.1 Classification, Morphology, and Biology

The traditional classification of chlamydiae was mainly based on host susceptibility and the biological properties of the pathogens.[1,2] But in the past three decades, molecular phylogenetic investigations have revolutionized our view of the taxonomic relationships of microorganisms.[3–5] Studies of phylogenetic relationships between the major outer membrane protein (*ompA*) genes of chlamydiae laid the groundwork for the classification based on genetic relatedness.[6] Further phylogenetic analyses of other chlamydial genes, most notably the ribosomal RNA genes, demonstrated essentially identical evolutionary relationships among chlamydiae. A subsequent classification scheme, based on the sequence differences of the 16S and 23S ribosomal RNA genes, separated the family *Chlamydiaceae* into two genera, *Chlamydia* and *Chlamydophila*[7–10] (Figure 46.1). The reclassification of *Chlamydiaceae* into two genera did not meet with wide approval from scientific colleagues, public health workers, and funding agencies.[11,12] Therefore, reclassification of *Chlamydiaceae* into a single genus *Chlamydia* has been proposed, containing nine species named by the current species epithets.[12]

The *Chlamydiales* are obligate intracellular bacterial pathogens of higher cells that infect eukaryotic cells, from single-celled organisms such as amoebae to virtually any cell type of multicellular organisms. The chlamydial elementary body (EB) is the infectious stage of the chlamydial developmental cycle, and functions as a tough "spore-like" body whose purpose is to permit chlamydial survival in the nonsupportive environment outside the host cell. The EB is near the limit of light microscopic visibility being approximately 0.3 μm in diameter and round or, occasionally pear-shaped, containing electron-dense structures made up of tightly coiled genomic DNA covered by histone like proteins. The ultrastructure of EBs has been extensively studied,[13] and spikes in the membrane are thought to be the tubular apparatus of a chlamydial type III secretion system.

The chlamydial reticulate body (RB) represents the noninfectious intracellular developmental stage with a diameter of approximately 1 μm. The RB is metabolically active and the cytoplasm is rich in ribosomes, which are required for protein synthesis. RBs replicate exclusively in an intracellular vacuole termed an "inclusion." As the RBs begin to differentiate into EBs at the end of the chlamydial developmental cycle, sites of recondensation of nucleic acid appear in the inclusion's cytoplasm. In the maturing inclusion, chlamydial particles appear to be packed tightly within the inclusion membrane. Development of chlamydiae is highly dependent on nutrient supply and the metabolic status of host cells. Nutrient deficiencies such as low glucose, iron, or amino acid levels lead to delayed development and aberrant chlamydial organisms within the inclusion. These aberrant inclusions are thought to represent persistent chlamydial forms that are important for maintaining the infection in a host population, particularly in the absence of overt clinical signs.

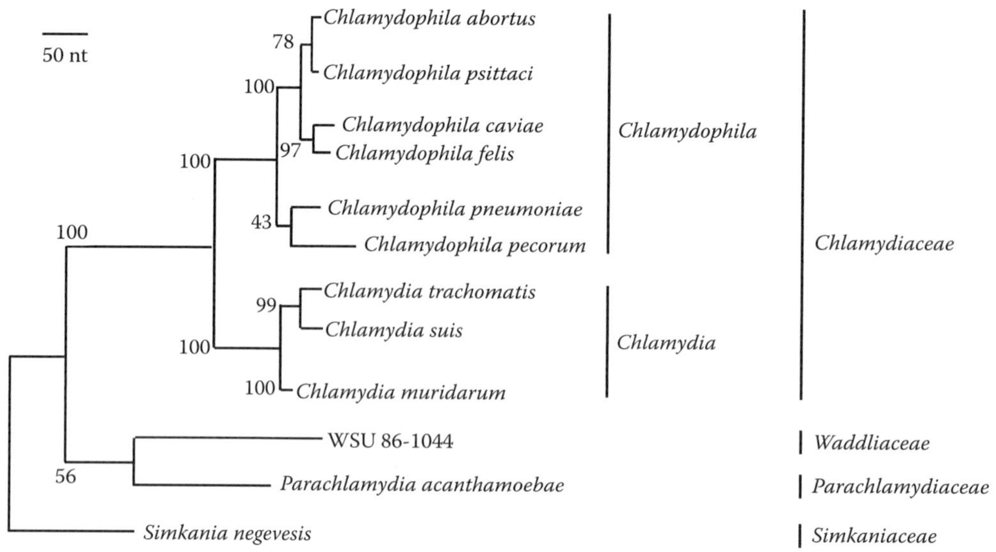

FIGURE 46.1 Phylogeny of the order *Chlamydiales* based on full-length 16S rRNA genes.[7,9] Branch lengths are measured in nucleotide substitutions, and numbers show branching percentages in bootstrap replicates. Only bacteria of the family *Chlamydiaceae* have fully established pathogenic potential for vertebrates. Strains of the remaining families *Waddliacae*, *Parachlamydiaceae*, and *Simkaniaceae* have been variably and inconsistently associated with diseases.

While sharing the unique biphasic developmental lifecycle, *Chlamydophila* spp. demonstrate assorted epidemiological characteristics. *C. pneumoniae*, first described in 1986,[14] is mainly a human pathogen but also infects koalas, horses, and frogs.[15–18] This pathogen is distributed worldwide and is arguably among the most common human pathogens. Acute, chronic, and asymptomatic infections with *C. pneumoniae* occur with high frequency in virtually all humans during their lifetime.[19,20] The 5- to 20-year age group shows the most rapid rise in prevalence.[19] *C. pneumoniae* contributes to 10% or more of the respiratory diseases in young adults, and to more serious respiratory disease in senior individuals. Kleemola et al.[21] reported epidemic outbreaks of *C. pneumoniae* infection among military recruits in Finland between 1957 and 1985. Each outbreak lasted approximately 6 months, and epidemics occurred during all seasons of the year. The infection rate varied from 60 to 80 per 1000 men. Transmission was slow, with the case-to-case interval averaging 30 days and ranging as high as 3 months.[21]

It is possible that all avian species are the natural hosts for *C. psittaci* as subclinical carriers.[22,23] Stress factors such as crowding and shipping enhance shedding of the organisms in birds, resulting in high prevalence of the infection. Fatal human *C. psittaci* infection was first described as "psittacosis" in Switzerland in the 1870s, and it attracted more public attention in 1929 and 1930 during a widespread outbreak.[24] Since most avian species are subclinical carriers of *C. psittaci*, persons with continual contact with birds are at great risk for psittacosis. This includes poultry farmers and processors, zoo workers, owners of pet shops and of racing pigeons, and laboratory workers.[25,26] More recently, an outbreak of psittacosis in a poorly managed poultry flock involving two different genotypes of *C. psittaci* spread to about 100 small flocks in the surrounding region and led to infection of 24 individuals, with one lethal case.[27] Laroucau et al.[28] reported five severe cases of psittacosis in individuals associated with duck farms in France in 2006, where human samples and duck isolates exhibited identical genetic characteristics in their PCR–RFLP restriction patterns.

C. abortus causes infectious abortion and mastitis in sheep, goat, and cattle, as well as pneumonia in pigeons, turkeys, and sparrows.[24] Abortion in sheep (ovine enzootic abortion, [OEA]) caused by *C. abortus* was first described by Grieg in Scotland in 1936,[29] and chlamydial abortion has subsequently been recognized as one of the most important causes of abortion in sheep. Infection with *C. abortus* has also been reported to be associated with abortion in women.[30,31] In most cases of *C. abortus*-induced human abortion, direct contact to infected sheep or goats occurred. The outcome of human infection in the first trimester of pregnancy is likely to be spontaneous abortion, whereas later infection causes stillbirths or preterm labor.[32] Therefore, pregnant women must avoid all contact with lambing ewes and newborn lambs, and should not handle contaminated clothing from those working with these animals. Immunocompromised individuals also must avoid contact with potential sources of infection at lambing time.[30]

46.1.2 Clinical Features and Pathogenesis

C. pneumoniae is primarily a human pathogen of the respiratory tract and causes acute or chronic bronchitis and pneumonia.[33] Infections are mainly asymptomatic and unrecognized, or mildly symptomatic illnesses. The most frequently recognized disease manifestations associated with *C. pneumoniae* infections are pneumonia and bronchitis. In adults, 10% of cases of pneumonia and approximately 5% of bronchitis and sinusitis cases have been attributed to the organism.[19,33] Severe systemic infections with *C. pneumoniae*, while uncommon, do occur. The onset of disease is gradual, and illness is more frequent in young adults and older debilitated individuals.[24] The incubation period of the symptoms may extend over several weeks. In these cases, pharyngitis and laryngitis precede lower respiratory disease. If lower respiratory tract symptoms occur, radiography usually demonstrates pneumonia. Numerous studies have demonstrated strong links between *C. pneumoniae* infection and obstructive pulmonary disease, metabolic syndrome, insulin resistance, Alzheimer's disease, and cardiovascular disease.[19,20,34–41] However, antibiotic prevention in large clinical trials, including a 1-year course of weekly azithromycin administration in a 4012-patient trial, failed to reduce secondary coronary events.[42,43] In an obese mouse model, Wang et al.[41] examined the longitudinal progression of insulin resistance and diabetes under influence of *C. pneumoniae* infection, genetic background, and dietary fat concentration. They concluded that murine *C. pneumoniae* infection enhances insulin resistance and diabetes in a genetically nutritionally restricted manner only in obese C57BL/6, but not A/J mice, via circulating inflammatory mediators such as TNF-α.[41]

C. psittaci primarily infects birds, but is readily transmissible to humans. *C. psittaci* infections in birds are often systemic with intermittent shedding of the organism, and may take a clinically inapparent, chronic, or acute course. Most organs, such as conjunctiva and the respiratory and gastrointestinal tracts, become infected. Stress will commonly trigger the onset of severe symptoms, and result in the rapid deterioration and death of the animals. Until recently, nine serovars of *C. psittaci* were distinguished, all of which were defined by epitopes on variable domains of the major outer membrane protein OmpA.[44,45] Seven serovars were thought to predominantly occur in a particular order or class of *Aves* and two in nonavian hosts, that is, type A in psittacine birds; B in pigeons; C in ducks and geese; D in turkeys; E in pigeons, ducks, and others; E/B in ducks; F in parakeets; WC in cattle; and M56 in rodents. However, comprehensive analysis of all available *ompA* gene sequences revealed the limitations of this classification[46] and the relative character of the assumed host specificity.[46,47] Consequently, adjustments to the typing scheme were suggested, which include the introduction of subgroups to the more heterogeneous genotypes A, E/B, and D, as well as six new genotypes representing so-far untypable strains.[46]

Most of the avian genotypes have also been identified sporadically in cases of zoonotic transmission. Human psittacosis is relatively rarely diagnosed in the United States and Europe, but cases with mild symptoms may often be overlooked when species-specific testing for *C. psittaci* is not conducted. Most cases are associated with infected birds, and parrots or parakeets are the most common source of infections. The infection is transmitted to humans by the respiratory route during contact with sick birds or from the contaminated environment, particularly by inhalation of aerosolized dried bird feces. Human case-to-case transmission is very rare. Nevertheless, in a remarkable episode of psittacosis in the Bayou region of Louisiana, 18 individuals became infected by secondary or tertiary transmission from the wife of a trapper, and eight of them died from interstitial pneumonia.[48] The incubation period for psittacosis ranges from 1 to 4 weeks. The initial symptoms may include malaise and pain in the limbs, and, more often, high fever, headache, and cough. The pulse rate is typically slow in relation to the temperature, and there is little or no evidence of chest consolidation. The spleen can be enlarged. Meningitis, meningoencephalitis, myocarditis, and endocarditis occur in severe cases, and the patients may become drowsy if the central system is affected.[24] Patients may recover without treatment and the infection usually resolves within 2 to 3 weeks. Death, due to respiratory or cardiovascular failure, has been rare since the advent of antibiotics. Before the antibiotic era, the death rate was as high as 50% in psittacosis patients older than 50 years.[24] Suppressing the avian sources of infection, such as import bans for psittacine birds during the 1929 psittacosis pandemic,[49] effectively controls psittacosis. Similarly, tetracycline treatment of imported birds during a 30-day quarantine period, as practiced in the United States, effectively reduces *C. psittaci* shedding and controls psittacosis.[24]

C. abortus strains are widespread in ruminants, efficiently colonize the placenta, and are primarily associated with cases of abortion and weak neonates. It has been reported that human *C. abortus* infections can result from contact with infected sheep,[50–52] goats,[31,53,54] or cattle.[55–57] The majority of these reports referred to abortions in pregnant sheep farmers, who were exposed to *C. abortus* during lambing. For example, in a Swiss farm, abortion affected 50% of the goats during the lambing season, and *C. abortus* was identified as the causing pathogen. During this time, a pregnant woman had contact with aborting goats, developed a severe generalized infection, and aborted herself. Her placenta contained *C. abortus* as shown by both immunohistochemistry and polymerase chain reaction (PCR).[54] Thus, pregnant women should stay away from sheep, especially during lambing. Inhalation of infected material from sheep can also result in respiratory disease in nonpregnant humans, but the products of the sheep dairy industry are usually not a hazard to human health.[24]

C. pecorum is associated with polyarthritis, enteritis, pneumonia, and urogenital infections in cattle, sheep, goats, koala, and swine.[58–60] *C. felis* was isolated from cat pneumonia in 1944 and is associated with pneumonia and conjunctivitis in cats that may occasionally be transmitted to owners, resulting in mild conjunctivitis.[24,61–63] *C. caviae* was originally isolated from conjunctival scrapings of guinea pigs,[64] and may also cause fertility disorders in these animals. A recent report has associated this species with rhinitis and conjunctivitis in horses.[65]

The disease outcome of chlamydial infection is determined by the interaction between chlamydial replication and the host response, and a series of pathogen- and host-associated factors.[41,66] Different efficiencies in infectivity and replication of the pathogen determine consistent, but not absolute, associations between chlamydial strain, host, and disease manifestation. These bacteria produce only minimally toxic or nontoxic products and regulate their growth based on the availability of host cytoplasmic nutrients. The initial events that elicit the host's innate immune response may determine the type of repair response and ensure the outcome of a chlamydial infection.[66] Chronic granulomatous lesions of mononuclear cell aggregates and fibrosis are the most common disease manifestations for chlamydial infections. Repeated infection and host genetic susceptibility are the main factors that exacerbate chlamydial infections.[41]

Cell-mediated immunity plays an important role in immune protection as well as disease development of chlamydial infection, with $CD4^+$ helper lymphocytes assuming the key role in a protective host response.[67–69] Th1-type effector cytokines, such as IFN-γ, restrict chlamydial replication, contribute to a protective delayed-type hypersensitivity response,[70,71] and exert the main proinflammatory effects via stimulation of the production of free radical molecules such as reactive oxygen (ROS) and reactive nitrogen oxide species (RNOS).[72] These central inflammatory mechanisms play a pivotal role in pathogen elimination, but also in cellular signaling and regulation of the immune response to chlamydial infection.[73,74] Ultimately, they are also the cause of disease by inducing "collateral" tissue damage in an excessive response to the typically minor threat of chlamydial infection. The macro- and micronutrient content of diets influences free radical production and outcomes of infectious diseases in rodent models.[75–80] Wang et al.[41] quantitatively dissected the immune and disease response to repeated *C. pneumoniae* lung infection by multivariate modeling at dichotomous levels of four effects: mouse strain (A/J or C57BL/6), dietary protein content (14% protein, 0.3% L-cysteine/0.9% L-arginine or 24% protein, 0.5% L-cysteine/2.0% L-arginine); dietary antioxidant content (90 IU α-tocopherol/kg vs. 450 IU α-tocopherol/kg and 0.1% g L-ascorbate), and time course (3 or 10 days postinfection). Contrast analyses indicated that chlamydial disease is characterized by a profound early T helper cell suppression. Thus, the mechanisms that regulate the host T-cell population are the critical determinants of chlamydial pathogenesis.[41]

Still, we know very little about the mechanisms of the disease production by *Chlamydophila* infections. One of the major restrictions is that we do not yet have the genetic methodology to pursue the function of individual chlamydial genes.[81] It is imperative to further explore the mechanisms of chlamydial diseases at the cellular and molecular levels of both pathogen and host.

46.1.3 Diagnosis

46.1.3.1 Conventional Techniques

Since chlamydiae are obligate intracellular bacteria, *tissue culture* techniques are required for isolation and propagation. A number of different cell lines can be used, such as McCoy, Buffalo Green Monkey Kidney (BGMK), HeLa 229, and Hep-2.[82–84] Embryonated hens' eggs are also occasionally used, particularly when high yields are required. Cell culture is usually considered 100% specific, since the presence of chlamydial inclusions is an unambiguous demonstration of infection, but sensitivity is considerably lower (60%–80%) since not all strains are easy to culture. Results are available within 48–72 h for well-growing isolates, but may be delayed for several days in the case of more difficult samples. Infection of chlamydiae in cell culture can be enhanced by centrifugation and/or by chemical treatment of cultured cells, before or during infection, using for example cycloheximide. As isolation is necessary to demonstrate the viability of a field strain and also facilitates detailed characterization by molecular and biochemical methods, cell culture is still widely regarded as the gold standard in chlamydial diagnosis.[85] However, diagnosis by cell culture requires very experienced laboratory personnel and strictly standardized protocols and, due to time and cost constraints, will remain a domain of specialized laboratories.

The characteristic intracytoplasmic inclusions of chlamydiae can be visualized in smears and tissue preparations from pathological specimens using *histochemical and immunohistochemical staining* under light microscopy. Depending on the manifestation of the infection, typical samples include smears prepared from biopsies, bronchoalveolar lavage, and placental membranes, as well as swabs taken from nose, ear, pharynx, vagina, or the moist coats of aborted fetuses. Prepared smears can be stained for detection of chlamydiae using one of several staining procedures, for example, modified Machiavello, modified Giménez, Giemsa, or modified Ziehl-Neelsen stain.[86,87] The latter is considered the most satisfactory method, where small coccoid EBs stain red/pink against a counterstained blue or green cellular background. However, while the procedure works well with heavily infected tissue, samples with low levels of infection can be easily overlooked.

Furthermore, chlamydiae can be demonstrated in histological preparations using a variety of staining procedures. A simple method involves the histochemical staining of thin tissue sections (≤4 μm) with Giemsa after fixation in fluids such as Bouin and Carnoy.[86] Dark-ground methylene blue staining, which has been shown to be a more reliable method for detecting *C. abortus* in fetal membranes than Giemsa, can also be used.[88] However, both these techniques are nonspecific and can cross-react with other bacterial species. Therefore, care must be taken with interpretation of results.

Immunohistochemical staining procedures that utilize monoclonal antibodies (mAbs) against chlamydial surface antigens, such as LPS or MOMP, are more sensitive in comparison to histochemical staining. A direct immunoperoxidase method for detecting *C. abortus* in formalin-fixed tissues has proved to be a very rapid and sensitive test.[89] An alternative approach involves the use of a fluorescein-conjugated anti-mouse antiserum in combination with the mAb. Enhanced labeling can be achieved using the more complex streptavidin-biotin method.[90] All in all, immunoperoxidase staining of formalin-fixed, paraffin-embedded or cryostat tissue sections is commonly used for diagnostic purposes, as well as for epidemiological and pathogenesis studies.[91–93]

Most of the commercial *enzyme immunoassays* (EIAs) were designed for *Chlamydia trachomatis*. However, since they are based on the family-specific LPS antigen, in principle, they should also be suitable for detection of *Chlamydophila* spp., albeit with some qualification on performance. These assays include direct fluorescent antibody tests (e.g., IMAGEN, Celltech; Chlamydia-Direct IF, BioMerieux; Vet-IF, Cell Labs), plate-based ELISA (Chlamydiazyme, Abbott; IDEIA, Dako; IDEIA PCE, Dako; Pathfinder, Kallestad; Chlamydia-EIA, Pharmacia), and solid-phase ELISA (Clearview Chlamydia MF, Unipath; Surecell, Kodak).

While a well-validated EIA can be a useful and easy-to-handle diagnostic tool, many comparative studies have revealed limitations in terms of sensitivity and specificity.[94–96] The performance of individual assays can depend on sample type and antigen load. Typical findings varied between 60% and 70% for both sensitivity and specificity as compared to culture and PCR.

Generally speaking, labeled mAbs against chlamydial LPS for immunofluorescent microscopy and ELISAs kits using chlamydial LPS antigen work well, whereas capture ELISAs tend to show considerable proportions of false positives.

The *detection of specific antibodies* can be conducted using the microimmunofluorescence (MIF) test.[97] The fact that it utilizes purified elementary bodies as antigen facilitates identification of species-specific antibodies, although practical experience is required to differentiate the signal from family-specific reactions. The test was found to be useful in serological surveys, but a recent study raised serious doubts on its suitability for diagnosing human *C. psittaci* infection.[98] The complement fixation test (CFT) was the first described LPS-based assay and, despite its rather complex setup and limited sensitivity, is still used in some laboratories. More recently developed antibody tests include the recombinant LPS-based rELISA (www.medac.de), which detects *Chlamydiaceae*-specific antibodies at higher sensitivity than CFT, and the recomLine Chlamydia IgG and IgA (www.mikrogen.de), a strip immunoassay allowing detection of antibodies to *C. pneumoniae*, *C. psittaci*, and *Chlamydia trachomatis*. The general significance of serology in diagnosis of acute infections is limited because chlamydial antibodies appear only after 7–10 days and early antibiotic therapy can suppress the regular humoral response.

46.1.3.2 Molecular Techniques

Conventional DNA Amplification Methods. The broad use of the PCR and other DNA amplification methods, which began in the 1990s, revolutionized the diagnosis of infectious diseases by allowing rapid identification from clinical specimens and DNA-based inter- and intraspecies differentiation. The majority of published conventional PCR methods are based on targets in the ribosomal RNA operon[7,8,99] or the *ompA* gene.[100,101]

An optimized nested PCR assay based on the elaborate primer system of Kaltenboeck et al.[100] was modified and described in detail by Sachse and Hotzel.[102] The first round of amplification generates a *Chlamydiaceae*-specific product, which serves as template in the second round, where species-specific primers are used. This assay was found to be very robust for routine use and proved the most sensitive among several protocols.

Another well-validated assay for detection of *Chlamydia trachomatis*, *Chlamydia pneumoniae*, and *Chlamydia psittaci* (old nomenclature), which can also be run in triplex mode, was published by Reference 103. The high sensitivity of this procedure was attained through a touchdown enzyme time release methodology featuring the use of hot-start DNA polymerase, a touchdown protocol for annealing temperatures to improve primer binding specificity, and an enzyme time release protocol to allow 60 cycles to be run for improved sensitivity.

Although a large number of PCR tests for chlamydiae have been published, it is not always clear from the paper whether the test has been properly validated. The use of poorly validated or unvalidated test protocols may lead to the generation of invalid data due to poor performance parameters of that test. This issue was addressed in a review by Apfalter et al.[104] The authors emphasized that preanalytical procedures, sample preparation and DNA extraction, assay design and setup, as well as interpretation and confirmation of results should be subject to validation in order to ensure high accuracy of data and high specificity and sensitivity. If considering the use of a published amplification assay, the prospective user should always check the data for adherence to the general rules on test validation.

Real-Time PCR. In addition to information on the presence or absence of a given pathogen, real-time PCR provides the possibility to quantitate the amount of the agent present in the sample. This can be achieved by cumulative measurement of the fluorescent signal generated by a specifically annealed dual-labeled fluorogenic probe upon exonuclease digestion.[105,106] As a major advantage, real-time PCR does not require post-PCR sample handling, which precludes potential PCR product carryover contamination and results in more rapid and high-throughput assays. The procedure has a large dynamic range of target molecule detection comprising at least five orders of magnitude.

Quantitative real-time PCR is based on measurement in the log-linear phase. Fluorescence signals are generated by specific binding of dye molecules, such as SYBR Green I, to double-stranded DNA[107] or by labeled oligonucleotide hybridization probes.[105] As fluorescence readings of PCR in the log-linear phase can be correlated to the initial number of target gene copies, quantitation of the number of microbial cells present in the sample can be accomplished. Practically, DNA concentration (as copy number) is calculated from the number of amplification cycles necessary to generate a fluorescence signal of a given threshold intensity.

A summary of published real-time PCR assays for chlamydiae is given in Table 46.1. Most of the family-specific real-time PCR tests are targeting the 23S rRNA gene. The methodologies developed by DeGraves et al.[109] and Ehricht et al.[108] have been validated and used in routine testing. Details are described in Section 46.2.2. Detection limits were

TABLE 46.1
Real-Time PCR Methods for *Chlamydiae*

Specificity	Target Gene	Reference
Chlamydiaceae	23S rRNA	Everett et al.[7,8]
Chlamydiaceae	23S rRNA	Ehricht et al.[108]
Chlamydiaceae, *Chlamydia psittaci*,[a] *C. pneumoniae*, *C. pecorum*, *Chlamydia trachomatis*	23S rRNA	DeGraves et al.[109]
C. psittaci	*omp*A	Geens et al.[110,b]
C. psittaci	*omp*A	Pantchev et al.[111,c]
C. psittaci	*inc*A	Menard et al.[112]
C. abortus	*omp*A	Pantchev et al.[111,c]
C. felis	*omp*A	Helps et al.[113]
C. felis	*omp*A	Pantchev et al.[114]
C. pecorum	*omp*A	Pantchev et al.[114]
C. caviae	*omp*A	Pantchev et al.[114]
Chlamydia suis	23S rRNA	Pantchev et al.[114]

[a] Based on four-species classification of the family *Chlamydiaceae*.
[b] Includes primer sets for genotype-specific amplification.
[c] Includes internal amplification control.

reported to be in the order of 1 to 10 target copies. However, Ehricht et al.[108] pointed out that the actual detection limit may be dependent on the integrity of target DNA.

DeGraves et al.[115] developed a highly sophisticated real-time PCR platform for sensitive detection of chlamydiae, which is distinguished by its optimized nucleic acid–extraction protocol ensuring high template yield, its step-down thermal cycling profile ensuring high product yield, and its design for the high-throughput regime of a diagnostic laboratory. This comprehensive and robust system is based on fluorescence resonance energy transfer (FRET) technology run on the LightCycler and allows detection of the pathogens both at genus and species level, according to the traditional four-species nomenclature.

Geens et al.[110] also used LightCycler technology to develop a panel of real-time PCR tests for *C. psittaci*. While the species-specific assay is based on SYBR Green detection and therefore is slightly less sensitive than fluorescent probe-based assays, the authors managed to design individual tests for the *C. psittaci* genotypes A, B, C, D, E, F, and E/B, respectively. Interestingly, some of the assays had to be designed as competitive reactions. As the closely related genotypes A, B, and E could not be distinguished by individual TaqMan probes, nonfluorescent competitor oligonucleotides had to be included alongside the probes to attain the necessary specificity.

Using discriminatory target segments in the *omp*A gene and TaqMan technology, Pantchev et al.[111,114] developed individual assays for *C. psittaci*, *C. abortus*, *C. felis*, *C. caviae*, *C. pecorum*, and *Chlamydia suis* (for details, see Section 46.2.2). In view of the close genetic relatedness between *C. psittaci* and *C. abortus* it was necessary to design a minor groove-binding (MGB™) probe. Furthermore, these assays include an internal amplification control,[116] a sensitive indicator of amplification efficiency and of the presence or absence of DNA polymerase inhibitors.

Based on *inc*A gene sequence analysis of five *C. psittaci* strains, Menard et al.[112] published a TaqMan real-time PCR protocol for detection of this species from clinical samples. They also used a MGB probe to exclude cross-reactions with *C. abortus*.

DNA Microarray Technology. The new possibilities provided by this methodological approach may be particularly beneficial for laboratory diagnosis of infectious diseases. This highly parallel approach allows DNA samples to be simultaneously examined by a large number of probes, which may be derived from a polymorphic gene segment and/or from different genomic regions. A specific microarray hybridization test can be regarded as an equivalent to resequencing the target site, so that DNA microarray-based tests can attain far higher discriminating power than PCR.

Sachse et al.[117] developed a microarray assay for the detection and differentiation of *Chlamydia* spp. and *Chlamydophila* spp. The test uses the commercially available ArrayTube™ (AT) system (Clondiag Chip Technologies, www.clondiag.com), a less expensive system for processing low- and high-density DNA arrays that involves spotted DNA chips of 3 × 3 mm size assembled onto the bottom of 1.5-mL plastic microreaction tubes. In contrast to other microarray equipment, hybridization and signal processing can be conducted in an easy and rapid fashion on standard laboratory equipment without additional devices, such as hybridization chambers. Hybridization signals are amplified by an enzyme-catalyzed precipitation reaction. A CCD camera integrated in a light transmission reader is used to monitor DNA duplex formation by kinetic measurement of the precipitation reaction at each spot.

Hybridization probes for all *Chlamydiaceae* spp. were designed on the basis of a multiple sequence alignment, from which a highly discriminatory segment in domain I of the 23S rRNA gene was identified. Several rounds of optimization and refinement led to the present version of the chip,[118] which is carrying 28 species-specific probes (for all nine chlamydial species); three genus-specific probes for *Chlamydia* and *Chlamydophila*, respectively; five probes identifying the closest relatives *Simkania negevensis* and *Waddlia chondrophila*, as well as four hybridization controls (consensus probes) and a staining control (biotinylated oligonucleotide).

The AT assay has been used for the direct detection of chlamydiae from clinical tissue as shown in a validation study.[118] The sensitivity was shown to be equivalent to that of real-time PCR,[108] thus rendering the test suitable for use in the diagnostic lab.

Genotyping by Microarray. The great potential of the AT microarray technology for rapid intraspecies characterization has been demonstrated by the recent development of a genotyping assay for *C. psittaci*.[46,83,117] This assay was shown to discriminate all established genotypes and to identify so-far untyped strains. Its high specificity, which allows detection of single-nucleotide polymorphisms, is due to the parallel approach consisting in the use of 35 hybridization probes derived from VD2 and VD4 of the *omp*A gene. This new test represents a promising diagnostic tool for tracing epidemiological chains, exploring the dissemination of genotypes and identifying nontypical representatives of *C. psittaci*.

46.2 METHODS

46.2.1 Sample Preparation

A large number of manufacturers offer DNA extraction kits for clinical samples from different matrices. To verify whether a particular kit is suitable for a given panel of specimens, it should be tested on a series of spiked samples containing defined numbers of *Chlamydia* cells. The first step with commercial kits is usually the lysis of bacterial and tissue cells using a special buffer ("lysis buffer"), the effectiveness of which is decisive for the kit's performance. An optional RNase digestion is intended to remove cellular RNA. Subsequently, the lysate is centrifuged through a minicolumn, where the released DNA is selectively bound to a solid phase (modified silica, hydroxyl apatite, or special filter membrane). After washing, the DNA can be eluted with elution buffer or water. These DNA preparations are usually of sufficient purity and virtually free of DNA polymerase inhibitors. Large diagnostic laboratories will probably prefer to use robotic systems for DNA extraction.

In the following paragraphs, we describe alternative DNA preparation protocols that proved satisfactory in our hands. They can be used if, for some reason, kits are not an option.

Swabs (e.g., Nasal, Vaginal, Conjunctival), Mucus, Bronchoalveolar Lavage, and Sputum. The swab is placed into a 2-mL Safe-Lock tube containing 500 μL of lysis buffer (100 mM Tris-base, pH 8.5, 0.05% [v/v] Tween 20). The closed tube is vortexed thoroughly for 1 min and centrifuged at 12,000 × *g* for 30 s. To squeeze out the remaining liquid, the swab is placed into a 1-mL pipette tip whose lower half was cut off and transferred into a fresh tube. After centrifugation at 12,000 × *g* for 1 min, the liquid is combined with that in the first tube from the previous centrifugation. (In the case of mucus, bronchoalveolar lavage, or sputum samples, the previous steps are omitted.) The fluid or mucus is centrifuged at 12,000 × *g* for 15 min, the supernatant is discarded and the pellet resuspended in 50 μL of lysis buffer containing 20 μL of proteinase K (10 mg/mL in water). After incubation at 60°C for 2 h, proteinase K is inactivated by heating at 97°C for 15 min. Finally, the mix is centrifuged at 12,000 × *g* for 5 min to remove debris, and 5 μL of the supernatant is used as template for PCR.

Tissue from Lung, Tonsils, Lymph Nodes, Spleen, Liver, and Other Organs. One hundred mg of homogenized tissue are boiled in 200 μL water in a plastic tube for 10 min. Subsequently, the tube is allowed to cool to room temperature. Optionally, proteinase digestion can be conducted to increase the final yield of DNA: 200 μL SDS (10 mg/mL) and 20 μL of proteinase K (10 mg/mL) are added and the mix is incubated at 55°C for 1 h. Following the addition of 200 μL phenol (saturated solution in TE buffer), the vessel is vortexed vigorously for 1 min and centrifuged at 12,000 × *g* for 5 min. The (upper) aqueous phase is transferred into a fresh tube, 200 μL of chloroform-isoamyl alcohol (24:1, v/v) are added, and the mix is vortexed at highest intensity for 1 min. Again, the aqueous phase is pipetted into a fresh tube. To precipitate the DNA, 120 μL of isopropanol are added, the reagents are thoroughly mixed and incubated at room temperature for 10 min. Following centrifugation at 12,000 × *g* for 10 min, the supernatant is discarded, the DNA pellet is allowed to air dry for 30 min and redissolved in 20 μL TE buffer. One μL is used as template for an amplification reaction.

Feces. One hundred mg of sample is mixed with 200 μL water by vortexing for 1 min. The following steps are identical to the protocol for tissue.

46.2.2 Detection Procedures

46.2.2.1 Real-Time PCR (TaqMan®) Detection of *Chlamydiaceae* and *Chlamydophila* spp.

Principle. The assays were designed as duplex amplification reactions, all of which include an internal amplification control (IAC). The latter was developed by Hoffmann et al.[116] as a universal heterologous amplification control system for real-time PCR. It consists of a plasmid carrying a 712-bp fragment of the EGFP gene, which serves as the IAC template (Intype IC-DNA, Labordiagnostik, Leipzig), as well as primers EGFP1-F/EGFP10-R and probe EGFP-HEX. The oligonucleotide sequences are given in Table 46.2. In addition, a primer-probe system targeting β-actin can be used as an efficiency control of DNA extraction (to be run in duplex mode with the chlamydia PCR or in triplex including the IAC).

Procedure. The assays are performed in 96-well optical microtiter plates on an Mx3000P Real-Time PCR System (Stratagene). The final 25-μL reaction mixture contains 12.5 μL of 2× TaqMan Gene Expression Mastermix with ROX (Applied Biosystems), each primer at a final concentration of 0.3 μM, each probe of 0.2 μM, 2500 copies of the IAC template and 1 μL of sample DNA. The following cycling parameters are used: initial cycle of 95°C for 10 min; 45 cycles of 95°C for 15 s; and 60°C for 1 min. For quantitative determinations, each plate run should include decimal serial dilutions of purified genomic DNA from the respective chlamydial species containing 2×10^{-1} to 2×10^{5} IFU per PCR reaction, which are used as copy standards. The cycle threshold value (Ct) is calculated automatically by the MxPro4.01 software (Stratagene).

Note: In a routine diagnostic setting, the *Chlamydiaceae*-specific real-time PCR assay has been successfully used as a screening test that is followed by the ArrayTube (AT) test (described in Section 46.2.2.3) to identify the chlamydial species involved,[47,83] that is, samples are first examined by real-time PCR, and the positives are subjected to the AT assay.

TABLE 46.2
Primers and Probes for Real-Time PCR Detection of *Chlamydiae*

Specificity	Primers, Probe and Sequences (5′→3′)	Amplicon Size (bp)	Target Gene	Reference
Chlamydiaceae	Ch23S-F: CTGAAACCAGTAGCTTATAAG CGGT	111	23S rRNA	Ehricht et al.[108]
	Ch23S-R: ACCTCGCCGTTTAACTTAACTCC			
	Ch23S-p: FAM-CTCATCATGCAAAAGGCACGCCG-TAMRA			
C. psittaci	CppsOMP1-F: CACTATGTGGGAAGGTGCTTCA	76	*omp*A	Pantchev et al.[111]
	CppsOMP1-R: CTGCGCGGATGCTAATGG			
	CppsOMP1-S: FAM-CGCTACTTGGTGTGAC-TAMRA (MGB® probe)			
C. abortus	CpaOMP1-F: GCAACTGACACTAAGTCGGCTACA	82	*omp*A	Pantchev et al.[111]
	CpaOMP1-R: ACAAGCATGTTCAATCGATAAGAGA			
	CpaOMP1-S: FAM-TAAATACCACGAATGGCAAGTTGGTTTAGCG-TAMRA			
C. pecorum	CppecOMP1-F: CCATGTGATCCTTGCGCTACT	76	*omp*A	Pantchev et al.[114]
	CppecOMP1-R: TGTCGAAAACATAATCTCCGTAAAAT			
	CppecOMP1-S: FAM-TGCGACGCGATTAGCTTACGCGTAG-TAMRA			
C. felis	CpfOMP1-F: TCGGATTGATTGGTCTTGCA	78	*omp*A	Pantchev et al.[114]
	CpfOMP1-R: GCTCTACAATGCCTTGAGAAATTTC			
	CpfOMP1-S: FAM-ACTGATTTCGCCAATCAGCGTCCAA-TAMRA			
C. caviae	CpcavOMP1-F: GAATAACATAGCCTACGGCAAACATA	84	*omp*A	Pantchev et al.[114]
	CpcavOMP1-R: CGATCCCAAATGTTTAATGCTAAGA			
	CpcavOMP1-S: FAM-CAAGATGCAGAATGGTCCACAAACGC-TAMRA			
Chlamydia suis	Csuis23S-F: CCTGCCGAACTGAAACATCTTA	118	23S rRNA	Pantchev et al.[114]
	Csuis23S-R: CCCTACAACCCCTCGCTTCT			
	Csuis23S-S: FAM-CGAGCGAAAGGGGAAGAGCCTAAACC-TAMRA			
IAC	EGFP1-F: GAC CAC TAC CAG CAG AAC AC	117	EGFP	Hoffmann et al.[116]
	EGFP10-R: CTT GTA CAG CTC GTC CAT GC			
	EGFP-HEX: HEX-AGC ACC CAG TCC GCC CTG AGC A-BHQ1			
Chlamydiaceae	CPEC23SUP: GGG GTT GTA GGG TCG ATA ACG TGA GAT C	168	23S rRNA	DeGraves et al.[109]
	CTR23SUP: GGG GTT GTA GGR TTG RGG AWA AAG GAT C			
	CHL23SDN: GAG AGT GGT CTC CCC AGA TTC ARA CTA			
	CHL23SLCR640: LCR640-CCT GAG TAG RRC TAG ACA CGT GAA AC-Phos			
	CHL23SFLU: GRA YGA HAC AGG GTG ATA GTC CCG TA-6FAM			

46.2.2.2 Real-Time FRET PCR (LightCycler®) Detection of *Chlamydiaceae* spp.

Principle. An integrated nucleic acid isolation and FRET PCR platform was developed to specifically detect, differentiate, and quantify all *Chlamydiaceae* spp. with high sensitivity.[109,115] Sampling into guanidinium-based buffer for maximum preservation of nucleic acids and optimized nucleic acid extraction has vastly improved detection of low numbers of chlamydial genomes.[109,115] Step-down thermal cycling and an excess of hot-start *Taq* polymerase vastly improved the robustness and sensitivity of the real-time PCR. The requirement for simultaneous side-by-side annealing of two FRET probes used for real-time detection of the amplification product ensures essentially 100% specificity. The amplification of *Chlamydiaceae* 23S rRNA allowed for the differentiation of chlamydial species and was more robust than amplification of the *ompA* gene at low target numbers.

Procedure. Total nucleic acid extraction is performed with the High-Pure PCR Template Preparation Kit® (Roche Molecular Biochemicals) by glass fiber matrix binding and low-volume elution in 40 μL total volume.[109] The PCR buffer is 4.5 mM $MgCl_2$, 50 mM KCl, 20 mM Tris-HCl, pH 8.4, supplemented with 0.05% each Tween 20 and Nonidet P-40, and 0.03% acetylated bovine serum albumin (Roche Molecular Biochemicals). Nucleotides are used at 0.2 mM (dATP, dCTP, dGTP) and 0.6 mM (dUTP). Primer CHL23SDN is used at 1 μM, primers CPEC23SUP and CTR23SUP at 0.5 μM each, the LightCycyler Red 640 probe CHL23SLCR640 at 0.2 μM, and the fluorescein probe CHL23SFLU at 0.1 μM. For each 20 μL total reaction volume, 1.5 U hot-start Platinum *Taq* DNA polymerase (Invitrogen) and 0.2 U heat-labile uracil-DNA glycosylase (Roche Molecular Biochemicals) are used.[109] Thermal cycling consists of a 2 min at 95°C followed by 18 high-stringency step-down thermal cycles, 40 low-stringency fluorescence acquisition cycles, and melting curve determination between 50°C and 80°C. The thermal protocol for the 23S rRNA qPCR is 6 × 12 s @ 64°C, 8 s @ 72°C, 0 s @ 95°C; 9 × 12 s @ 62°C, 8 s @ 72°C, 0 s @ 95°C; 3 × 12 s @ 60°C, 8 s @ 72°C, 0 s @ 95°C; 40 × 8 s @ 54°C and fluorescence acquisition, 8 s @ 72°C, 0 s @ 95°C.[109,115]

Notes: In a routine diagnostic setting, the real-time FRET PCR (LightCycler®) platform[109,115] has been successfully applied to quantify and differentiate the *Chlamydiaceae* spp. in clinical samples with high specificity and sensitivity.

46.2.2.3 Detection of *Chlamydophila* spp. Using the ArrayTube® Test

Principle. Prior to the AT test, sample DNA is prepared by standard extraction protocols and amplified by a consensus PCR using a biotinylated primer (U23F-19: 5′-ATT GAM AGG CGA WGA AGG A-3′ and 23R-22: 5′-biotin-GCY TAC TAA GAT GTT TCA GTT C-3′). After hybridization and staining, unique species-specific hybridization patterns for all nine species of the family *Chlamydiaceae* are obtained and immediately processed by the Iconoclust software (Clondiag).

Procedure. In biotinylation PCR, the reaction mix is composed of 1 μL DNA template, 0.2 μL (1 U) *Taq* DNA Polymerase (Fermentas), 2 μL 10× *Taq* Buffer with KCl, 1 μL $MgCl_2$ (25 mM), 2 μL dNTP mix (1 mM each), 1 μL of each primer (0.5 μM, U23F-19 and 23R-22), and 12.8 μL deionized water. The sample DNA is amplified with 40 cycles of 94°C for 30 s, 55°C for 30 s, and 72°C for 30 s.

The hybridization protocol starts with conditioning of the AT vessel by washing twice with 500 μL of Hybridization buffer 1 (Clondiag) at 30°C for 5 min. All incubations are conducted upon slight shaking (550 rpm) on a heatable horizontal tube shaker (Thermomixer comfort, Eppendorf). For denaturation, 1 μL of the biotinylated PCR product is diluted with 99 μL hybridization buffer in a separate tube, heated at 95°C for 5 min and put on ice. After transfer into the AT, hybridization is allowed to proceed at 58°C for 60 min. Subsequently, the supernatant is discarded, and the tube is washed with 500 μL 2 × SSC/0.01% Triton X-100 (40°C, 5 min), 500 μL 2 × SSC (30°C, 5 min) and 500 μL 0.2 × SSC (20°C, 5 min). Vacant binding sites of the microarray are blocked by incubation with a 2% solution of Blocking Reagent (Roche) in hybridization buffer at 30°C for 15 min. Subsequently, the AT is incubated with 100 μL of a 1:10,000 dilution of streptavidin-conjugated horseradish peroxidase (AT Staining Kit, Clondiag) followed by three-step washing as above. Finally, 100 μL of the peroxidase substrate (AT Staining Kit) is added. Immediately afterwards, measurement of the hybridization signal is started (at 25°C, using the ATR-01 array tube reader), and the final image is recorded after 10 min of continuous precipitation. Processing of hybridization signals is done using the Iconoclust software (Clondiag).

Note: In a routine diagnostic setting, the AT test has been successfully combined with the *Chlamydiaceae*-specific real-time PCR protocol described in Section 46.2.2.1,[83] that is, samples are first examined by real-time PCR, and the positives are AT tested to identify the species involved.

46.3 CONCLUSION AND FUTURE PERSPECTIVES

Chlamydophila spp. are intracellular pathogens of humans and animals that have evolved in the face of unique and complex long-term relationships with their hosts.[2,24,119–121] Many pioneering chlamydia researchers have contributed to extraordinary advances in the last 20 years, such as robust detection by PCR, ligase chain reaction, as well as serology, improved animal models, and whole-genome sequencing of multiple chlamydial strains and species.[81,109,122–125] This chapter mainly dealt with three human disease–related *Chlamydophila* species: *C. pneumoniae*, *C. psittaci*, and *C. abortus*. While *C. pneumoniae* is the dominant endemic human *Chlamydophila* species, *C. psittaci* and *C. abortus* also infect humans, breaking out from their respective endemically infected host populations. When diagnosed, infections by *C. psittaci* and *C. abortus* in humans are typically severe and sometimes lead to fatal diseases. However, tests for these infections are not routinely performed, and, therefore, the true prevalence and clinical picture are poorly understood. The availability of complete genome sequences of all chlamydial species has opened the possibility of sensitive monospecific ELISA tests for each species. Such tests will be required for a complete assessment of the prevalence of these infections in humans.

In animals, recent diagnostic advances have dramatically increased detection frequencies of chlamydial infections.[126–130] Thus, the paradigm has shifted from rare but severe *Chlamydophila* spp. infections to widespread, low-level endemic infections. While evidence concerning the effect on animal health of these ubiquitous clinically inapparent infections is just emerging, it remains incomplete. Most likely, these infections are associated with several production diseases that limit profits in agriculture, such as fertility disorders, increased somatic cell counts in milk (subclinical mastitis), and reduced weight gains and fitness of young animals.[119,131,132]

Similarly, in humans, the paradigm of *C. pneumoniae* infection has shifted from acute respiratory disease to that of a virtually ubiquitous, clinically inapparent infection that nevertheless is strongly epidemiologically linked to chronic inflammatory diseases, such as coronary atherosclerosis and metabolic syndrome.[122] The current prevailing notion for the involvement of *C. pneumoniae* in atherosclerosis stipulates that *C. pneumoniae*-infected macrophages disseminate to early atherosclerotic lesions and exacerbate pathological processes.[122] To explore these mechanisms, Wang et al.[41] quantitatively examined, in an obese mouse model, the influence of *C. pneumoniae* infection on progression of insulin resistance, a central feature of metabolic syndrome that also underlies atherosclerosis. They demonstrated conclusively that low-copy *C. pneumoniae* organisms dispersed to secondary tissues are irrelevant to the progression of insulin resistance. Rather, acute infection of the lung, the primary target of *C. pneumoniae*, causes an increase in circulating cytokines that drive the long-term exacerbation of insulin resistance and accelerate onset of type 2 diabetes, metabolic syndrome, and presumably, atherosclerosis.[41] The concept of circulating inflammatory mediators released from the main infection site is consistent with the mechanisms enhancing insulin resistance proposed for other prevalent pathogens that typically reside only at predilection infection sites, such as *H. pylori*,[133] *P. gingivalis*,[134] and Hepatitis C virus.[135]

Given the widespread health effects of chlamydial infections[122] and the failed antibiotic prophylaxis of *C. pneumoniae*-induced exacerbation of insulin resistance and cardiovascular disease,[42,43] vaccination may be an attractive and realistic approach to control most of the *Chlamydophila*-associated diseases in human beings. Such vaccination approaches may have to balance the potential for exacerbated tissue damage at the primary infection site against the benefits gained by limiting chlamydial infections and thereby minimizing whole-host negative effects. In summary, infections with *Chlamydophila* spp. are associated with a wide range of typically asymptomatic disease conditions in both humans and animals.[24,122] One of the key advances in the understanding of chlamydial infections has been the development of molecular detection methods for these pathogens with high sensitivity and specificity.[109,126,136] Research leading to effective strategies for prevention and therapy of chlamydial infections, such as development of effective vaccines, will have an enormous impact on the health of humans and animals.

While tremendous progress in the understanding of basic biology and molecular make-up of chlamydiae has been made over the last two decades, this progress has not translated into clinically significant improvements in prophylaxis and therapy. The main progress in understanding both human and animal chlamydial diseases, with actionable consequences in the clinical setting, has come from improvements in diagnostics. The ideal diagnostic tool should allow a rapid diagnosis at low cost, and with high specificity and sensitivity. The main challenge for PCR-based detection methods is the efficient extraction of nucleic acids from low-copy specimens and their efficient transfer into the amplification mix, as well as sensitive detection and quantification of low target copy amplicons in a single-tube assay. Well-engineered platform technologies and advanced analytical methods will further increase the potential of molecular diagnostic techniques and make them reliable and universally applied tools for cost-effective rapid detection of Chlamydiae including simultaneous differentiation and typing of these bacteria.

REFERENCES

1. Moulder, J.W. et al., Genus I. *Chlamydia* Jones, Rake and Stearns 1945, 55[AL]. In Krieg, N.R., and Holt, J.G. (Editors), *Bergy's Manual of Systemic Bacteriology*, vol. 1, pp. 729–739, Williams and Wilkins, Baltimore, 1984.
2. Stephens, R.S., Introduction. In Stephens, R.S. (Editor), *Chlamydia, Intracellular Biology, Pathogenesis, and Immunity*, American Society for Microbiology, Washington, DC, 1999.
3. Amann, R.N. et al., Obligate intracellular bacterial parasites of acanthamoebae related to *Chlamydia* spp., *Appl. Environ. Microbiol.*, 63, 115, 1997.
4. Pace, N.R., A molecular view of microbial diversity and the biosphere, *Science*, 276, 734, 1997.
5. Hugenholtz, P., Goebel, B.M., and Pace, N.R., Impact of culture-independent studies on the emerging phylogenetic view of bacterial diversity, *J. Bacteriol.*, 180, 4765, 1998.
6. Kaltenboeck, B., Kousoulas, K.G., and Storz, J., Structures of and allelic diversity and relationships among the major outer membrane protein (*ompA*) genes of the four chlamydial species, *J. Bacteriol.*, 175, 487, 1993.
7. Everett, K.D.E., Bush, R.M., and Andersen, A.A., Emended description of the order *Chlamydiales*, proposal of *Parachlamydiaceae* fam. nov. and *Simkaniaceae* fam. nov., each containing one monotypic genus, revised taxonomy of the family *Chlamydiaceae*, including a new genus and five new species, and standards for the identification of organisms, *Int. J. Syst. Bacteriol.*, 49, 415, 1999.
8. Everett, K.D., and Andersen, A.A., Identification of nine species of the *Chlamydiaceae* using PCR-RFLP, *Int. J. Syst. Bacteriol.*, 49, 803, 1999.
9. Bush, R.M., and Everett, K.D.E., Molecular evolution of the *Chlamydiaceae*, *Int. J. Syst. Evol. Microbiol.*, 51, 203, 2001.
10. Horn, M. et al., *Neochlamydia hartmannellae* gen. nov., sp. nov. (*Parachlamydiaceae*), an endoparasite of the amoeba *Hartmannella vermiformis*, *Microbiology*, 146, 1231, 2000.
11. Schachter, J. et al., Radical changes to chlamydial taxonomy are not necessary just yet, *Int. J. Syst. Evol. Microbiol.*, 51, 249, 2001.
12. Stephens, R.S. et al., Divergence without difference: Phylogenetics and taxonomy of *Chlamydia* resolved, *FEMS Immunol. Med. Microbiol.*, 55, 115, 2009.
13. Matsumoto, A., Electron microscopic observations of surface projections on *Chlamydia psittaci* reticulate bodies, *J. Bacteriol.*, 151, 1040, 1982.
14. Grayston, J.T. et al., A new *Chlamydia psittaci* strain, TWAR, isolated in acute respiratory tract infections, *N. Engl. J. Med.*, 315, 161, 1986.
15. Glassick, T., Giffard, P., and Timms P, Outer membrane protein 2 gene sequences indicate that *Chlamydia pecorum* and *Chlamydia pneumoniae* cause infections in koalas, *Syst. Appl. Microbiol.*, 19, 457, 1996.
16. Storey, C., Evidence for *Chlamydia pneumoniae* of nonhuman origin, *J. Gen. Microbiol.*, 139, 2621, 1993.
17. Berger, L. et al., *Chlamydia pneumoniae* in a free-ranging giant barred frog (*Mixophyes iteratus*) from Australia, *J. Clin. Microbiol.*, 37, 2378, 1999.
18. Hotzel, H. et al., Genetic characterization of a *Chlamydophila pneumoniae* isolate from an African frog and comparison to currently accepted biovars, *Syst. Appl. Microbiol.*, 24, 63, 2001.
19. Kuo, C.C. et al., *Chlamydia pneumoniae* (TWAR), *Clin. Microbiol. Rev.*, 8, 451, 1995.
20. Boman, J., et al., High prevalence of *Chlamydia pneumoniae* DNA in peripheral blood mononuclear cells in patients with cardiovascular disease and in middle-aged blood donors, *J. Infect. Dis.*, 178, 274, 1998.
21. Kleemola, M. et al., Epidemics of pneumonia caused by TWAR, a new *Chlamydia* organism, in military trainees in Finland, *J. Infect. Dis.*, 157, 230, 1988.
22. Mayer, K.F., The host spectrum of psittacosis-lymphogranuloma venereum (PL) agents, *Am. J. Ophthalmol.*, 63, 1225, 1967.
23. Kaleta, E.F., and Taday, E.M., Avian host range of *Chlamydophila* spp. based on isolation, antigen detection and serology, *Avian Pathol.*, 32, 435, 2003.
24. Schachter, J., Infection and disease epidemiology. In Stephens, R.S. (Editor), *Chlamydia, Intracellular Biology, Pathogenesis, and Immunity*, p. 139, American Society for Microbiology, Washington, DC, 1999.
25. Mayer, K.F., and Eddie, B., Human carrier of the psittacosis virus, *J. Infec. Dis.*, 88, 109, 1951.

26. Palmer, S.R., Psittacosis in man-recent developments in UK: A review, *J. R. Soc. Med.*, 75, 262, 1982.
27. Gaede, W. et al., *Chlamydophila psittaci* infections in humans during an outbreak of psittacosis from poultry in Germany, *Zoonoses Public Health*, 55, 184, 2008.
28. Laroucau, K., Chlamydial infections in duck farms associated with human cases of psittacosis in France, *Vet. Microbiol.*, 135, 82, 2009.
29. Greig, J.R., Enzootic abortion in ewes: A preliminary note, *Vet. Rec.*, 42, 1225, 1936.
30. Entrican, G., Buxton, D., and Longbottom, D., Chlamydial infection in sheep: Immune control versus fetal pathology, *J. R. Soc. Med.*, 94, 273, 2001.
31. Walder, G. et al., An unusual cause of sepsis during pregnancy: Recognizing infection with *Chlamydophila abortus*, *Obstet. Gynecol.*, 106, 1215, 2005.
32. Hyde, S.R., and Benirschke K., Gestational psittacosis: Case report and literature review, *Mod. Pathol.*, 10, 602, 1997.
33. Grayston, J.T., *Chlamydia pneumoniae*, strain TWAR pneumoniae, *Annu. Rev. Med.*, 43, 317, 1992.
34. Gupta, S. et al., Elevated *Chlamydia pneumoniae* antibodies, cardiovascular events, and azithromycin in male survivors of myocardial infarction, *Circulation*, 96, 404, 1997.
35. Campbell, L.A. et al., Preclinical models for *Chlamydia pneumoniae* and cardiovascular disease: Hypercholesterolemic mice, *Clin. Microbiol. Infect.*, 4, S23, 1998.
36. Ekesbo, R. et al., Combined seropositivity for *H. pylori* and *C. pneumoniae* is associated with age, obesity and social factors, *J. Cardiovasc. Risk*, 7, 191, 2000.
37. Smieja, M. et al., Association of circulating *Chlamydia pneumoniae* DNA with cardiovascular disease: A systematic review, *BMC Infect. Dis.*, 2, 21, 2001.
38. Dart, A.M., Martin, J.L., and Kay, S., Association between past infection with *Chlamydia pneumoniae* and body mass index, low-density lipoprotein particle size and fasting insulin, *Int. J. Obes. Relat. Metab. Disord.*, 26, 464, 2002.
39. Nabipour, I. et al., The association of metabolic syndrome and *Chlamydia pneumoniae*, *Helicobacter pylori*, cytomegalovirus, and herpes simplex virus type 1: The Persian Gulf Healthy Heart Study, *Cardiovasc. Diabetol.*, 5, 25, 2006.
40. Fernández-Real, J.M. et al., Burden of infection and insulin resistance in healthy middle-aged men, *Diabetes Care*, 29, 1058, 2006.
41. Wang, C., Gao, D., and Kaltenboeck, B., Acute *Chlamydia pneumoniae* reinfection accelerates the development of insulin resistance and diabetes in obese C57BL/6 mice, *J. Infect. Dis.*, 200, 279, 2009.
42. Cannon, C.P. et al., Antibiotic treatment of *Chlamydia pneumoniae* after acute coronary syndrome, *N. Engl. J. Med.*, 352, 1646, 2005.
43. Grayston, J.T. et al., Azithromycin for the secondary prevention of coronary events, *N. Engl. J. Med.*, 352, 1637, 2005.
44. Vanrompay, D., et al., Serotyping of European isolates of *Chlamydia psittaci* from poultry and other birds, *J. Clin. Microbiol.*, 31, 134, 1993.
45. Andersen, A.A., Two new serovars of *Chlamydia psittaci* from North American birds, *J. Vet. Diag. Invest.*, 9, 1997.
46. Sachse, K. et al., Genotyping of *Chlamydophila psittaci* using a new DNA microarray assay based on sequence analysis of *omp*A genes, *BMC Microbiol.*, 8, 63, 2008.
47. Sachse, K et al., DNA microarray-based genotyping of *Chlamydophila psittaci* strains from cell culture and clinical samples, *Vet. Microbiol.*, 135, 22, 2009.
48. Olson, B.J., and Treuting, W.L., An epidemic of a severe pneumoniae in the Bayou Region of Louisiana. IV. A preliminary note on etiology, *Public Health Rep.*, 131, 376, 1944.
49. Schachter, J., and Dawson, C.R., *Human Chlamydial Infections*, PSG Publishing Co., Littleton, 1978.
50. Giroud, P., and Jadin, J., Premiers resultats concernant le virus des Bashi isole dans la provence du Kivu au Congo Belge, *Bull. Soc. Pathol. Exot.*, 47, 578, 1954.
51. Mare, C.J., Mammalian chlamydiosis. In Beran, G.W. (Editor), *Handbook of Zoonoses*, p. 403, CRC Press, Boca Raton, 1994.
52. Jorgensen, D.M., Gestational psittacosis in a Montana sheep rancher, *Emerg. Infect. Dis.*, 3, 191, 1997.
53. Villemonteix, P. et al., Pregnancy complicated by severe *Chlamydia psittaci* infection acquired from a goat flock: Case report, *Eur. J. Obstet. Gynecol. Reprod. Biol.*, 37, 91, 1990.
54. Pospischil, A. et al., Abortion in humans caused by *Chlamydophila abortus* (*Chlamydia psittaci* serovar 1), *Schweiz Arch. Tierheilkd*, 144, 463, 2002.
55. Enright, J.B., and Sadler, W.W., Presence in human sera of complement-fixing antibodies to the virus of sporadic bovine encephalomyelitis, *Proc. Soc. Exp. Biol. Med.*, 85, 466, 1954.
56. Barnes, M.G., and Brainerd, H., Pneumonitis with alveolar-capillary block in a cattle rancher exposed to epizootic bovine abortion, *N. Engl. J. Med.*, 271, 981, 1964.
57. Page, L.A., and Smith, P.C., Placentitis and abortion in cattle inoculated with chlamydiae isolated from aborted human placental tissue, *Proc. Soc. Exp. Biol. Med.*, 146, 269, 1974.
58. Kaltenboeck, B., Kousoulas, K.G., and Storz, J., Detection and strain differentiation of *Chlamydia psittaci* mediated by a two-step polymerase chain reaction, *J. Clin. Microbiol.*, 29, 1969, 1991.
59. Fukushi, H., and Hirai, K., Proposal of *Chlamydia pecorum* sp. nov. for *Chlamydia* strains derived from ruminants, *Int. J. Syst. Bacteriol.*, 42, 306, 1992.
60. Jackson, M. et al., Epizootiology of *Chlamydia* infections in two free-range koala populations, *Vet. Microbiol.*, 65, 255, 1999.
61. Baker, J.A., A virus causing pneumonia in cats and producing elementary bodies, *J. Exp. Med.*, 79, 159, 1944.
62. Cello, R.M., Ocular infections in animals with PLT (*Bedsonia*) group agents, *Am. J. Opthalmol.*, 63, 1270, 1967.
63. Shewen, P.E., Povey, R.C., and Wilson, M.R., Feline chlamydial infection, *Can. Vet. J.*, 19, 289, 1978.
64. Murray, E.S., Guinea pig inclusion conjunctivitis virus. I. Isolation and identification as a member of psittachosis-granuloma-trachoma group, *J. Infect. Dis.*, 114, 1, 1964.
65. Gaede, W. et al., Detection of *Chlamydophila caviae* and *Streptococcus equi* subsp. *zooepidemicus* in horses with signs of rhinitis and conjunctivitis, *Vet. Microbiol.*, 142, 440, 2010.
66. Ramsey, K.H., Alternative mechanisms of pathogenesis. In Bavoil, P.M., and Wyrick, B. (Editors), *Chlamydia Genomics and Pathogenesis*, p. 435, Horizon Bioscience Press, Norfolk, 2006.
67. Morrison, R.P., Feilzer, K., and Tumas, D.B., Gene knockout mice establish a primary protective role for major histocompatibility complex class II-restricted responses in *Chlamydia trachomatis* genital tract infection, *Infect. Immun.*, 63, 4661, 1995.
68. Rank, R.G., The role of the CD4 T cell in the host response to *Chlamydia*. In Bavoil, P.M., and Wyrick, B. (Editors), *Chlamydia Genomics and Pathogenesis*, p. 365, Horizon Bioscience Press, Norfolk, 2006.
69. Natividad, et al., Genetic variation at the TNF locus and the risk of severe sequelae of ocular *Chlamydia trachomatis* infection in Gambians, *Genes Immun.*, 8, 288, 2007.

70. Perry, L.L., Feilzer, K., and Caldwell, H.D., Immunity to *Chlamydia trachomatis* is mediated by T helper 1 cells through IFN-gamma-dependent and -independent pathways, *J. Immunol.*, 158, 3344, 1997.
71. Rottenberg, M.E. et al., Regulation and role of IFN-gamma in the innate resistance to infection with *Chlamydia pneumoniae*, *J. Immunol.*, 164, 4812, 2000.
72. Swindle, E.J., and Metcalfe, D.D., The role of reactive oxygen species and nitric oxide in mast cell-dependent inflammatory processes, *Immunol. Rev.*, 217, 186, 2007.
73. Huang, J. et al., The quantity of nitric oxide released by macrophages regulates *Chlamydia*-induced disease, *Proc. Natl. Acad. Sci. USA*, 99, 3914, 2002.
74. Ramsey, K.H. et al., Role for inducible nitric oxide synthase in protection from chronic *Chlamydia trachomatis* urogenital disease in mice and its regulation by oxygen free radicals, *Infect. Immun.*, 69, 7374, 2001.
75. Li, P. et al., Amino acids and immune function, *Br. J. Nutr.*, 98, 237, 2007.
76. Cunningham-Rundles, S., McNeeley, D.F., and Moon, A., Mechanisms of nutrient modulation of the immune response, *J. Allergy Clin. Immunol.*, 115, 1119, 2005.
77. Field, C.J., Johnson, I.R., and Schley, P.D., Nutrients and their role in host resistance to infection, *J. Leukoc. Biol.*, 71, 16, 2002.
78. Wintergerst, E.S., Maggini, S., and Hornig, D.H., Contribution of selected vitamins and trace elements to immune function, *Ann. Nutr. Metab.*, 51, 301, 2007.
79. Chan, J. et al., Effects of protein calorie malnutrition on tuberculosis in mice, *Proc. Natl. Acad. Sci. USA*, 93, 14857, 1996.
80. McMurray, D.N., and Bartow, R.A., Immunosuppression and alteration of resistance to pulmonary tuberculosis in guinea pigs by protein undernutrition, *J. Nutr.*, 122, 738, 1992.
81. Ward, M.E., Mechanisms of *Chlamydia*-induced disease. In Stephens, R.S. (Editor), *Chlamydia, Intracellular Biology, Pathogenesis, and Immunity*, p. 171, American Society for Microbiology, Washington, DC, 1999.
82. Thewessen, E.A. et al., Comparison of HeLa 229 and McCoy cell cultures for detection of *Chlamydia trachomatis* in clinical specimens, *J. Clin. Microbiol.*, 27, 1399, 1989.
83. Johnston, S.L., and Siegel, C., Comparison of Buffalo green monkey kidney cells and McCoy cells for the isolation of *Chlamydia trachomatis* in shell vial centrifugation culture, *Diagn. Microbiol. Infect. Dis.*, 15, 355, 1992.
84. Roblin, P.M. et al., Use of HEp-2 cells for improved isolation and passage of *Chlamydia pneumoniae*, *J. Clin. Microbiol.*, 30, 1968, 1992.
85. Thejls, H. et al., Expanded gold standard in the diagnosis of *Chlamydia trachomatis* in a low prevalence population: Diagnostic efficacy of tissue culture, direct immunofluorescence, enzyme immunoassay, PCR and serology, *Genitourin. Med.*, 70, 300, 1994.
86. Stamp, J.T. et al., Enzootic abortion in ewes: Transmission of the disease, *Vet. Rec.*, 62, 251, 1950.
87. Aitken, I.D., and Longbottom, D., Enzootic abortion of ewes (ovine chlamydiosis). In OIE Biological Standards Commission (Editor), *Manual of Diagnostic Tests and Vaccines for Terrestrial Animals (Mammals, Birds and Bees)*, p. 635, Office International des Epizooties, Paris, 2004.
88. Dagnall, G.J., and Wilsmore, A.J., A simple staining method for the identification of chlamydial elementary bodies in the fetal membranes of sheep affected by ovine enzootic abortion, *Vet. Microbiol.*, 21, 233, 1990.
89. Finlayson, J. et al., Direct immunoperoxidase method for demonstrating *Chlamydia psittaci* in tissue sections, *J. Clin. Pathol.*, 38, 712, 1985.
90. Szeredi, L. et al., Intestinal chlamydia in finishing pigs, *Vet. Pathol.*, 33, 369, 1996.
91. Juvonen, J. et al., Detection of *Chlamydia pneumoniae* in human nonrheumatic stenotic aortic valves, *J. Am. Coll. Cardiol.*, 29, 1054, 1997.
92. Buxton, D. et al., Ovine chlamydial abortion: Characterization of the inflammatory immune response in placental tissues, *J. Comp. Pathol.*, 127, 133, 2002.
93. Hotzel, H. et al., Occurrence of *Chlamydiaceae* spp. in a wild boar (*Sus scrofa* L.) population in Thuringia (Germany), *Vet. Microbiol.*, 103, 121, 2004.
94. Wood, M.M., and Timms, P., Comparison of nine antigen detection kits for diagnosis of urogenital infections due to *Chlamydia psittaci* in koalas, *J. Clin. Microbiol.*, 30, 3200, 1992.
95. Vanrompay, D. et al., Evaluation of five immunoassays for detection of *Chlamydia psittaci* in cloacal and conjunctival specimens from turkeys, *J. Clin. Microbiol.*, 32, 1470, 1994.
96. Sachse, K. et al., Detection of *Chlamydia suis* from clinical specimens: Comparison of PCR, antigen ELISA, and culture, *J. Microbiol. Methods*, 54, 233, 2003.
97. Wang, S.-P., and Grayston, J.T., Human serology in *Chlamydia trachomatis* infection with microimmunofluorescence, *J. Infect. Dis.*, 130, 388, 1974.
98. Verminnen, K. et al., Evaluation of a *Chlamydophila psittaci* diagnostic platform for zoonotic risk assessment, *J. Clin. Microbiol.*, 46, 281, 2008.
99. Messmer, T.O. et al., Application of a nested, multiplex PCR to psittacosis outbreaks, *J. Clin. Microbiol.*, 35, 2043, 1997.
100. Kaltenboeck, B., Schmeer, N., and Schneider, R., Evidence of numerous *omp*1 alleles of porcine *Chlamydia trachomatis* and novel chlamydial species obtained by PCR, *J. Clin. Microbiol.*, 35, 1835, 1997.
101. Yoshida, H. et al., Differentiation of *Chlamydia* species by combined use of polymerase chain reaction and restriction endonuclease analysis, *Microbiol. Immunol.*, 42, 411, 1998.
102. Sachse, K., and Hotzel, H., Detection and differentiation of chlamydiae by nested PCR, *Methods Mol. Biol.*, 216, 123, 2003.
103. Madico, G. et al., Touchdown enzyme time release-PCR for detection and identification of *Chlamydia trachomatis*, *C. pneumoniae*, and *C. psittaci* using the 16S and 16S-23S spacer rRNA genes, *J. Clin. Microbiol.*, 38, 1085, 2000.
104. Apfalter, P., Reischl, U., and Hammerschlag, M.R., In-house nucleic acid amplification assays in research: How much quality control is needed before one can rely upon the results? *J. Clin. Microbiol.*, 43, 5835, 2005.
105. Livak, K.J. et al., Oligonucleotides with fluorescent dyes at opposite ends provide a quenched probe system useful for detecting PCR product and nucleic acid hybridization, *PCR Methods Appl.*, 4, 357, 1995.
106. Heid, C.A. et al., Real-time quantitative PCR, *Genome Res.*, 6, 968, 1996.
107. Wittwer, C.T. et al., Continuous fluorescence monitoring of rapid cycle DNA amplification, *Biotechniques*, 22, 130, 134, 1997.
108. Ehricht, R. et al., Optimized DNA microarray assay allows detection and genotyping of single PCR-amplifiable target copies, *Mol. Cell. Probes*, 20, 60, 2006.
109. DeGraves, F.J., Gao, D., and Kaltenboeck, B., High-sensitivity quantitative PCR platform, *Biotechniques*, 34, 106, 2003.

110. Geens, T. et al., Development of a *Chlamydophila psittaci* species-specific and genotype-specific real-time PCR, *Vet. Res.*, 36, 787, 2005.
111. Pantchev, A. et al., New real-time PCR tests for species-specific detection of *Chlamydophila psittaci* and *Chlamydophila abortus* from tissue samples, *Vet. J.*, 181, 145, 2009.
112. Menard, A. et al., Development of a real-time PCR for the detection of *Chlamydia psittaci*, *J. Med. Microbiol.*, 55, 471, 2006.
113. Helps, C. et al., Use of real-time quantitative PCR to detect *Chlamydophila felis* infection, *J. Clin. Microbiol.*, 39, 2675, 2001.
114. Pantchev, A. et al., Detection of all *Chlamydophila* and *Chlamydia* spp. of veterinary interest using species-specific real-time PCR assays, *Comp. Immunol. Microbiol. Infect. Dis.*, 2009 (in press).
115. DeGraves, F.J. et al., Quantitative detection of *Chlamydia psittaci* and *C. pecorum* by high-sensitivity real-time PCR reveals high prevalence of vaginal infection in cattle, *J. Clin. Microbiol.*, 41, 1726, 2003.
116. Hoffmann, B. et al., A universal heterologous internal control system for duplex real-time RT-PCR assays used in a detection system for pestiviruses, *J. Virol. Methods*, 136, 200, 2006.
117. Sachse, K. et al., DNA microarray-based detection and identification of *Chlamydia* and *Chlamydophila* spp., *Mol. Cell. Probes*, 19, 41, 2005.
118. Borel, N. et al., Direct identification of chlamydiae from clinical samples using a DNA microarray assay—a validation study, *Mol. Cell. Probes*, 22, 55, 2008.
119. Reinhold, P. et al., Impact of latent infections with *Chlamydophila* species in young cattle, *Vet. J.*, 175, 202, 2009.
120. Bodetti, T.J. et al., Wide range of *Chlamydiales* types detected in native Australian mammals, *Vet. Microbiol.*, 96, 177, 2003.
121. Borel, N. et al., Seroprevalences for ovine enzootic abortion in Switzerland, *Prevent. Vet. Med.*, 65, 205, 2004.
122. Campbell, L.A., and Kuo, C.C., *Chlamydia pneumoniae*—an infectious risk factor for atherosclerosis? *Nat. Rev. Microbiol.*, 2, 23, 2004.
123. Kalman, S. et al., Comparative genomes of *Chlamydia pneumoniae* and *C. trachomatis*, *Nat. Genet.*, 21, 385, 1999.
124. Read, T.D. et al., Genome sequence of *Chlamydophila caviae* (*Chlamydia psittaci* GPIC): Examining the role of niche-specific genes in the evolution of the *Chlamydiaceae*, *Nucleic Acids Res.*, 31, 2134, 2003.
125. Thomson, N.R. et al., The *Chlamydophila abortus* genome sequence reveals an array of variable proteins that contribute to interspecies variation, *Genome Res.*, 15, 629, 2005.
126. Sachse, K. et al., Recent developments in the laboratory diagnosis of chlamydial infections, *Vet. Microbiol.*, 135, 2, 2009.
127. Verkooyen, R.P. et al., Evaluation of PCR, culture, and serology for diagnosis of *Chlamydia pneumoniae* respiratory infections, *J. Clin. Microbiol.*, 36, 2301, 1998.
128. Jee, J. et al., High prevalence of natural *Chlamydophila* species infection in calves, *J. Clin. Microbiol.*, 42, 5664, 2004.
129. Magnino, S. et al., Chlamydial infections in feral pigeons in Europe: Review of data and focus on public health implications, *Vet. Microbiol.*, 135, 54, 2009.
130. Kaltenboeck, B., Hehnen, H.R., and Vaglenov, A., Bovine *Chlamydophila* spp. infection: Do we underestimate the impact on fertility? *Vet. Res. Commun.*, 29, 1, 2005.
131. Biesenkamp-Uhe, C. et al., Therapeutic *Chlamydophila abortus* and *C. pecorum* vaccination transiently reduces bovine mastitis associated with *Chlamydophila* infection, *Infect. Immun.*, 75, 870, 2007.
132. DeGraves, F.J. et al., Reinfection with *Chlamydophila abortus* by uterine and indirect cohort routes reduces fertility in cattle preexposed to *Chlamydophila*, *Infect. Immun.*, 72, 2538, 2004.
133. Aydemir, S. et al., The effect of *Helicobacter pylori* on insulin resistance, *Dig. Dis. Sci.*, 50, 2090, 2005.
134. Mealey, B.L., and Rose, L.F., Diabetes mellitus and inflammatory periodontal diseases, *Curr. Opin. Endocrinol. Diabetes Obes.*, 15, 135, 2008.
135. Aboud, M. et al., Insulin resistance and HIV infection: A review, *Int. J. Clin. Pract.*, 61, 463, 2007.
136. Tong, C.Y., and Sillis, M., Detection of *Chlamydia pneumoniae* and *Chlamydia psittaci* in sputum samples by PCR, *J. Clin. Pathol.*, 46, 313, 1993.

47 *Elizabethkingia*, *Chryseobacterium*, and *Bergeyella*

Dongyou Liu

CONTENTS

47.1 INTRODUCTION

47.1.1 CLASSIFICATION

The genera *Elizabethkingia*, *Chryseobacterium*, and *Bergeyella* are classified in the family Flavobacteriaceae, order Flavobacteriales, class Flavobacteria, phylum Bacteroidetes. The phylum Bacteroidetes [formerly the *Cytophaga–Flavobacterium–Bacteroides* (CFB) group] belonging to the rRNA superfamily V is recognized as a separate line of descent within the domain Bacteria. The family Flavobacteriaceae in the CFB group consists of about 80 genera including *Elizabethkingia*, *Chryseobacterium*, *Bergeyella*, and *Riemerella*; and the genera *Bergeyella*, *Chryseobacterium*, and *Riemerella* appear to form a separate branch on the basis of rRNA cistron similarity and phenotypic characteristics.[1–3] Members of the *Bergeyella–Chryseobacterium–Riemerella* branch are characterized by their lack flagellation, low G + C content (30–38 mol%), production of menaquinones as the predominant respiratory quinones, possession of large percentages of branched-chain fatty acids, absence of carbohydrate fermentation, and similarity in hydrolytic enzyme patterns.

Currently, the genus *Chryseobacterium* (*chruseos*, "golden"; *bacterium*, a "small rod": *Chryseobacterium*, "a yellow rod") is composed of at least 45 valid species (i.e., *Chryseobacterium antarcticum*, *Chryseobacterium anthropi*, *Chryseobacterium aquaticum*, *Chryseobacterium aquifrigidense*, *Chryseobacterium arothri*, *Chryseobacterium balustinum*, *Chryseobacterium bovis*, *Chryseobacterium caeni*, *Chryseobacterium daecheongense*, *Chryseobacterium daeguense*, *Chryseobacterium defluvii*, *Chryseobacterium flavum*, *Chryseobacterium formosense*, *Chryseobacterium gambrini*, *Chryseobacterium gleum*, *Chryseobacterium gregarium*, *Chryseobacterium haifense*, *Chryseobacterium hispanicum*, *Chryseobacterium hominis*, *Chryseobacterium humi*, *Chryseobacterium hungaricum*, *Chryseobacterium indologenes*, *Chryseobacterium indoltheticum*, *Chryseobacterium jejuense*, *Chryseobacterium jeonii*, *Chryseobacterium joostei*, *Chryseobacterium koreense*, *Chryseobacterium luteum*, *Chryseobacterium marinum*, *Chryseobacterium molle*, *Chryseobacterium oranimense*, *Chryseobacterium pallidum*, *Chryseobacterium palustre*, *Chryseobacterium piscicola*, *Chryseobacterium piscium*, *Chryseobacterium scophthalmum*, *Chryseobacterium shigense*, *Chryseobacterium soldanellicola*, *Chryseobacterium soli*, *Chryseobacterium taeanense*, *Chryseobacterium taichungense*, *Chryseobacterium taiwanense*, *Chryseobacterium ureilyticum*, *Chryseobacterium vrystaatense*, and *Chryseobacterium wanjuense*).[4–22] While most *Chryseobacterium* species occur in aquatic environments or food products, some are pathogenic to humans and animals. For example, *Chryseobacterium gleum* and *C. indologenes* may cause nosocomial infections in humans, particularly in neonates or immunocompromised patients,[23–25] and *C. balustinum* and *C. scophthalmum* have been isolated from diseased fish.[26]

Until recently, *C. meningosepticum* (formerly *Flavobacterium meningosepticum* or CDCII-a) and *C. miricola* were allocated to the genus *Chryseobacterium*. However, comparative analysis of their 16S rRNA gene sequences in relation to *Chryseobacterium* species led to their transfer to the new genus *Elizabethkingia* (named in honor of Elizabeth O. King) as *Elizabethkingia meningoseptica* and *E. miricola*.

Elizabethkingia meningoseptica is considered the most serious human pathogen in the group and is known to cause meningitis in premature and newborn infants, and pneumonia, endocarditis, postoperative bacteremia, and meningitis in adults, usually associated with a severe underlying illness. The species has been also isolated from diseased birds, frogs, turtles, and cats, whereas *E. miricola* was isolated from condensation water in the space station Mir.

Members of the genus *Chryseobacterium* are strictly aerobic, nonmotile, nonspore-forming, gram-negative rods (0.5–0.7 × 1.1–1.3 μm) that are covered by fimbriae. On LB agar, *Chryseobacterium* colonies measure 1.0–1.5 mm in diameter, and are circular, convex. Some species are mucoid, while others are shiny, bright yellow–colored and smooth, with a ropy consistency. They show oxidase and catalase activities, grow at 30°C and 37°C under aerobic conditions, produce acid from glucose, maltose, and ethylene glycol, produce indole, hydrolyze esculin, starch, and gelatin, display alkaline phosphatase, trypsin (benzylarginine arylamidase), and pyrrolidonyl–aminopeptidase activities. They also produce flexirubin-type pigments (except for *C. hominis*) and have a DNA G + C content of about 29–39 mol%. *Chryseobacterium* spp. occur in a variety of environments, including soil, fresh- and seawater, raw milk and chicken, diseased fish, bioreactor sludge, and clinical samples.[7,12–14,18,19,27]

Chryseobacterium indologenes is oxidase-positive, glucose-nonfermenting gram-negative rod that forms smooth, circular, yellow-pigmented colonies of 1–2 mm on 5% sheep blood agar.

Chryseobacterium hominis (*hominis*, of human being) cells are nonmotile, gram-negative rods of 1–3 μm × 1.0–1.5 μm. They grow aerobically at 20°C, 30°C, and 37°C on standard media (e.g., tryptic soy agar and blood agar), with optimal growth at 30°C, but not on MacConkey agar, cetrimide agar, or 3% NaCl agar. Colonies are circular and mucoid; some are also sticky. Some strains exhibit a pale yellow or tan pigmentation, but do not generate flexirubin pigments. They produce acid oxidatively from glucose, maltose, and ethylene glycol, but are negative for urease, lysine decarboxylase, ornithine decarboxylase, and arginine dihydrolase activities. They also produce indole, and alkalize acetate but not citrate. In addition, they show alkaline phosphatase and trypsin (benzylarginine arylamidase) activities but do not hydrolyze esculin and gelatin.[19]

The genus *Elizabethkingia* is made up of two species: *Elizabethkingia meningoseptica* (formerly *Chryseobacterium meningosepticum*, or *Flavobacterium meningosepticum* or CDCII-a) and *Elizabethkingia miricola* (formerly *Chryseobacterium* miricola). *Elizabethkingia* cells are gram-negative, nonmotile, nonspore-forming rods (0.5 × 1.0–2.5 μm). They grow well on TSA and nutrient agar at 28°C–37°C, but not at 5°C or 42°C. Colonies are white–yellow, nonpigmented, semitranslucent, circular, and shiny, with entire edges. Being catalase-, oxidase-, phosphatase-, and *β*-galactosidase-positive, they hydrolyze casein, esculin, and gelatin, but not starch. They do not utilize malonate, nor reduce nitrate. They produce acid from D-fructose, D-glucose, lactose, D-maltose, D-mannitol, and trehalose, but not from L-arabinose, D-cellobiose, raffinose, sucrose, salicin, or D-xylose. Menaquinone MK-6 is the predominant quinone. The G + C content of *Elizabethkingia* DNA is 35.0–38.2 mol%.

Elizabethkingia meningoseptica (*meninx*, *meningos*, meninges, the membrane covering the brain; *septikos* putrefactive; *meningoseptica*, referring to association of the bacterium with both meningitis and septicemia, but not septic meningitis) was previously known as *Flavobacterium meningosepticum* (King 1959). *E. meningoseptica* cells are gram-negative, nonmotile, nonspore-forming, slender, slightly curved rods (0.5 × 1.0–2.0 μm). They show catalase, oxidase, and indole activities, but do not hydrolyze urea. The G + C content of *E. meningoseptica* DNA is 37.2 ± 0.6 mol% (37.1 mol% for the type strain).

Elizabethkingia miricola (*mirum*, *mir*, peace, the name of Russian space station; *-cola* from *incola*, inhabitant; *miricola*, inhabitant of the Mir space station) was previously known as *Chryseobacterium miricola* (Li et al. 2004). *E. miricola* cells are gram-negative, nonmotile, nonspore-forming rods (0.5 × 1.0–2.5 μm). They grow well on MacConkey agar, producing very sticky colonies on solid medium. They hydrolyze urea. The G + C content of *E. miricola* DNA is 35.3 ± 0.3 mol% (35.0 mol% for the type strain).

The genus *Bergeyella* consists of a single species *Bergeyella zoohelcum* (formerly *Weeksella zoohelcum*). *B. zoohelcum* is a nonfermentative, nonspore-forming nonmotile, aerobic, gram-negative rod that grows well on blood agar but not on MacConkey's agar. Colonies are circular, shiny, and smooth with entire edges. Biochemically, the bacterium is catalase, oxidase, and indole positive, and is nonsaccharolytic. *B. zoohelcum* is susceptible to penicillin, a feature that aids its differentiation from other similar bacteria (e.g., *Flavobacterium* and *Sphingobacterium* spp.). It is found as part of the normal oral flora of dogs, cats, and some other animals. Most clinical isolates come from infected animal bite wounds.[5,28–32]

47.1.2 Clinical Features

Elizabethkingia meningoseptica is a gram-negative rod that is widely distributed in the environment. As a former member of the genus *Chryseobacterium* (*Chryseobacterium meningosepticum*), *E. meningoseptica* is known to cause meningitis and bacteremia, predominantly in premature newborns and infants. Among neonates, meningitis predominates the cases of infection with a 57% mortality rate, followed by sepsis and pneumonia.[33] Among the postneonatal group, pneumonia is the most frequent infection, followed by sepsis, meningitis, endocarditis, cellulitis, abdominal infections, eye infections, sinusitis, bronchitis, and epididymitis.[24,34–39] *E. meningoseptica* outbreaks in neonatal intensive care and hemodialysis units and its transmission in long-term acute care hospitals with mechanically ventilated patients have been reported.[40–42] Occasionally, *E. meningoseptica* is responsible for bacterial meningitis in adults and adolescents, and cellulitis in immunocompromised patients.[39,43] Being resistant to multiple antibiotics that are prescribed for treating gram-negative bacterial infections, including extended-spectrum β-lactam agents and aminoglycosides, the bacterium represents a significant clinical concern.

Ozkalay et al.[43] reported a case of community-acquired *E. meningoseptica* meningitis and sepsis in a 17-year-old boy with thalassemia major who underwent splenectomy. *E. meningosepticum* was isolated in the cerebrospinal fluid (CSF) and blood samples of this patient, who recovered completely after receiving vancomycin therapy for 21 days. Bomb et al.[38] documented an unusual case of native valve endocarditis in a 58-year-old man due to *E. meningoseptica.* Tuon et al.[39] described a case of *E. meningoseptica* cellulitis with severe sepsis and hepatitis involving a 36-year-old male patient. The patient had a 7-day history of myalgia, fever, nausea, and vomiting before admission to hospital, and presented a right-shoulder cellulitis, tachycardia, tachypnoea, and hepatomegaly on physical examination. Laboratory tests revealed leukocytosis, renal failure, hepatitis, and metabolic acidosis. An *E. meningoseptica* strain was later identified in blood sample by a semiautomated system – Walkway (Dade-Behring). The patient recovered after undergoing ciprofloxacin treatment.

Chryseobacterium spp. are gram-negative, aerobic, motile rods that exist in soil, plants, foodstuffs, and water sources. As an opportunistic human pathogen, *C. indologenes* can cause bacteremia, wound sepsis, and ventilator-associated pneumonia in immunocompromised patients and patients that have received long-term broad-spectrum antibiotics via contaminated fluid-containing apparatuses and indwelling devices.[44,45] Cascio et al.[46] reported a *C. indologenes* bacteremia in a 2-year-old boy with type 1 diabetes mellitus. Two blood cultures on a standard blood culturing system (BACTEC 9120; Becton Dickinson) yielded *C. indologenes* isolates at day 5 after inoculation. Both isolates failed to produce acid from D-xylose and L-arabinose and were negative for esculin hydrolysis within 4 h of incubation but positive after 24 h of incubation. The biochemical profiles produced by the API 20NE system (BioMérieux) and Vitek GNI card using the Vitek-2 system (BioMérieux) confirmed the organism as *C. indologenes.* After treatment with ceftriaxone, the patient became afebrile after 48 h, and his general condition improved within 36 h.

Bergeyella zoohelcum is commonly isolated from the upper respiratory tract of dogs, cats, and other mammals. In one study, the organism was identified in 38%–90% of nasal and oral fluids and gingival scrapings of dogs. The bacterium is an occasional zoonotic pathogen typically associated with animal bites, and is known to cause cellulitis, leg abscess, tenosynovitis, septicemia, pneumonia, and meningitis in humans.[47] Shukla et al.[48] reported the isolation and characterization of a fastidious *Bergeyella zoohelcum* strain from acute cellulitis in the upper extremity of a 60-year-old woman. The isolate was characterized by PCR amplification and sequencing of the 16S rRNA gene with broad-range eubacterial primers. Recently, Lin et al.[49] described a 73-year-old man with liver cirrhosis who had no history of dog bite but had dog exposure. The patient developed cellulitis of the left lower leg, and *B. zoohelcum* was isolated from blood culture. This patient was treated with cefazolin and gentamicin with a good outcome.

47.1.3 Diagnosis

Phenotypic Identification. Members of the genera *Elizabethkingia, Chryseobacterium,* and *Bergeyella* are cultured on 5% sheep blood, eosin-methylene blue, and chocolate agars at 35°C in 5%–10% CO_2 for 24–48 h. Initial identification is based on Gram stain, and cell and colony morphology.[3] Specifically, cell morphologies are observed under a light microscope (1000× magnification) with cells grown in Anacker and Ordal's broth for 1–3 days. Gram reaction may be based on a nonstaining method (3% KOH).[50,51] Motility is examined by phase-contrast microscopy. Cellular morphology may be observed by transmission electron microscopy (TEM) with bacterial cells grown for 18 h on LB agar. The presence of flexirubin-type pigments is tested by observing the color shift when a small mass of bacterial cells collected on agar and deposited on a glass slide is flooded with a 20% (w/v) KOH aqueous solution.[19]

Hemolysis is recorded on trypticase soy agar supplemented with 5% (v/v) sheep blood (bioMérieux). Trypticase soy agar is also used to test the temperature tolerance at 5°C, 26°C, 37°C, and 42°C. Bacterial growth is tested on MacConkey no. 3 (Oxoid), thiosulfate-citrate-bile-sucrose (TCBS; Difco), brain heart infusion (Difco), Mueller Hinton (Difco), marine 2216 (Difco), and cetrimide. Oxidase activity is tested on oxidase test discs (bioMérieux Bactident) or oxidase test strips (Merck); and catalase activity, with 3% H_2O_2. The type of respiratory metabolism is determined by inoculating deep tubes of meat-liver medium containing 6% agar (BIO-RAD) Acid production is assessed using 1% glucose, H_2S production using peptone iron agar (Difco 289100) and starch hydrolysis using starch agar (Difco 272100). Hydrolysis of sodium alginate, deoxyribonucleic acid, gelatin, and Tween 80 are tested both on plain nutrient agar (containing 0.5% NaCl) and on the same medium enriched to 1.5% NaCl; production of indole and reduction of nitrate are also evaluated on nutrient agar. Further biochemical identification is performed by using API 20NE and API ID32 GN microtest systems (bioMerieux). Susceptibilities to antimicrobial agents are determined by E-test strips (AB Biodisk).[19,51–53]

Infections are deemed community acquired if the patient admitted with an acute illness and initial cultures at the time of presentation yield a positive result. Infections are considered nosocomial if cultures are negative at the time of admission or symptomatic infections develop after the first 72 h of hospitalization.

Genotypic Identification. DNA–DNA hybridization and 16S rRNA gene sequence analysis have proven useful for identification of members of the genera *Elizabethkingia, Chryseobacterium,* and *Bergeyella.*[54,55] Broad-range eubacterial primers (e.g., FD1: 5′-AGA GTT TGA TCC TGG CTC AG-3′ and RD1: 5′-AAG GAG GTG ATC CAG CC-3′) are used to amplify an ~1500-bp product from the 16S rRNA gene. The resulting PCR fragment is column purified and then sequenced from both sense and antisense directions by cycle sequencing.[48]

For genotyping purpose, pulsed-field gel electrophoresis (PFGE) and infrequent-restriction-site PCR (IRS-PCR) may be performed. PFGE and IRS-PCR demonstrate a similar efficiency in the discrimination of *Elizabethkingia meningoseptica*, leading to the identification of 11 distinct genotypes among the 11 clinical isolates in one study.[49] Ribotyping is also useful for epidemiological studies concerning *E. meningoseptica*.[56]

47.2 METHODS

47.2.1 Sample Preparation

Blood and CSF specimens are cultured on 5% sheep blood, eosin-methylene blue and chocolate agars at 35°C in 5%–10% CO_2 for 24–48 h. Cell and colony morphology and biochemical properties are examined using phenotypical procedures.[43]

The template DNA for 16S rRNA gene analysis may be prepared by boiling a few isolated colonies in 200 μL of sterile water for 10 min. The sample is centrifuged at 15,300 × *g* for 2 min to pellet the cell debris. The supernatant is collected and used directly as the template for PCR amplification. Alternatively, DNA is extracted with the DNeasy tissue kit (Qiagen).

47.2.2 Detection Procedures

(i) 16S rRNA Gene Analysis Protocol of Kim et al.[12] Kim et al.[12] utilized primers 9F (5′-GAGTTTGATCCTGGCT CAG-3′; positions 9–27 on *Escherichia coli* 16S rRNA numbering) and 1512R (5′-ACGGCTACCTTGTTACGA CTT-3′; positions 1512–1492) to amplify the complete 16S rRNA gene. They then used primers 9F (5′-GA GTTTGATCCTGGCTCAG-3′; 9–27), 341F (5′-CCTA CGGGAGGCAGCAG-3′; positions 341–357), 519F (5′-CAGCAGCCGCGGTAATAC-3′; positions 519–536), 907F (5′-AAACTCAAAKGAATTGACGG-3′; positions 907–926), 536R (5′-GTATTACCGCGGCTGCTG-3′; positions 536–519), 1100R (5′-GGGTTGCGCTCGTTG-3′; positions 1114–1100), and 1512R (5′-ACGGCTACCTT GTTACGACTT-3′; positions 1512–1492) to analyze the nucleotide sequence.

Procedure

1. PCR mixture (100 μL) consists of 1 μM each of primers 9F and 1512R, 0.1 μM each dNTP, 10× reaction buffer, 2.5 U *Taq* DNA polymerase, and 100 ng extracted DNA.
2. PCR amplification is conducted with 35 cycles of 94°C for 1 min, 60°C for 1 min, and 72°C for 2 min, followed by a final extension at 72°C for 10 min.
3. The amplified fragment is then sequenced using primers 9F, 519F, 907F, 536R, 1100R, and 1512R.
4. The full 16S rRNA gene sequences are compiled using SeqMan software and sequences of the test strains are edited using the BioEdit program and aligned using CLUSTAL X. The distance matrix is calculated by the BioEdit program after deleting regions containing ambiguous nucleotides. The phylogenetic tree is constructed by the neighbor-joining method using the Mega2 program.

(ii) 16S rRNA Gene Analysis Protocol of Han et al.[32] Han et al.[32] employed universal 16S rRNA gene primers A17F and 1512R for PCR amplification followed by DNA sequencing to identify *Elizabethkingia*, *Chryseobacterium*, and *Bergeyella* bacteria (Table 47.1).

Procedure

1. PCR mixture (25-μL) is made up of GoTaq DNA polymerase (Promega), and 2 μL of DNA. Amplification is performed with an initial denaturation at 94°C for 3 min; 28–32 cycles of 94°C for 1 min, 50°C for 1 min, and 72°C for 2 min; and a final extension at 72°C for 10 min, in an Applied Biosystems 2720 thermal cycler.
2. The PCR products are visualized by 1% agarose gel electrophoresis. The PCR products are then cloned into the pCR11 vector, and a total of 10 clones with positive 16S rRNA gene inserts are selected for species identification.
3. Plasmids from these 10 clones are extracted by using a WizardPlus SV Minipreps DNA purification system (Promega), and their inserts are sequenced by using primers GW1 and GW.
4. The sequences are assembled and aligned by using the VectorNTI program (Invitrogen). The NCBI BLAST nucleotide sequence database is searched for preliminary species identification. The 16S rRNA gene sequences are aligned by using the ClustalX1.8 program. A tree for phylogenetic analysis is constructed by using Molecular Evolutionary Genetics Analysis (version 4) software by calculation of a distance matrix and tree reconstruction by the neighbor-joining method. The statistical robustness of the analysis is estimated by bootstrapping with 1000 replicates.

TABLE 47.1
Primers for Amplification and Sequencing Analysis of *Elizabethkingia*, *Chryseobacterium*, and *Bergeyella* rRNA Gene

Primer	Sequence (5′→3′)	Purpose
A17F	GTTTGATCCTGGCTCAG	PCR
1512R	TACCTTGTTACGACTT	PCR
GW1	GTTGCAACAAATTGATGAGCAATGC	DNA sequencing
GW2	GTTGCAACAAATTGATGAGCAATTA	DNA sequencing

Sequences are considered monophylic when they have a connecting node within a genetic distance of 0.05. This analysis allows the presumptive identification of the bacteria by inferring the evolutionary relationships of homologous sequences. The similarities between the 16S rRNA genes within a species are expected to be >97%, and those between different species within a genus are between 93.3% and 99.9%.

47.3 CONCLUSION

The genera *Elizabethkingia*, *Chryseobacterium*, and *Bergeyella* in the family Flavobacteriaceae encompass a number of nonmotile, catalase-positive, oxidase-positive, indole-positive, nonglucose-fermenting, gram-negative bacilli that are distributed in a variety of environments. Some of these bacteria (e.g., *Elizabethkingia meningoseptica*, formerly *Flavobacterium meningosepticum* or *Chryseobacterium meningosepticum*; *Chryseobacterium indologenes*, and *Bergeyella zoohelcum*, formerly *Weeksella zoohelcum*) are associated with clinical diseases in humans, especially in immunosuppressed individuals and patients with underlying disease. The clinical symptoms of *E. meningoseptica* range from meningitis, bacteremia, pneumonia, endocarditis, cellulitis, abdominal infections, eye infections, sinusitis, bronchitis, and epididymitis.[33–49,57] Considering that *E. meningoseptica* is largely tolerant of antibiotics that are used presently to treat microbial infections (e.g., β-lactam agents and aminoglycosides), it is important to correctly and rapidly identify the bacterium for selecting appropriate antibiotic therapy. Overcoming the time consuming and variable nature of phenotypic procedure, nucleic acid–amplification technologies such as PCR provide a highly sensitive and precise alternative for laboratory identification and detection of *Elizabethkingia*, *Chryseobacterium*, and *Bergeyella* organisms. Further application of these state-of-art techniques will no doubt contribute to a better understanding of the roles of these bacteria in human diseases as well as to the designing of more effective control measures against these pathogens of increasing importance.

REFERENCES

1. Woese, C.R. et al., The *Flexibacter-Flavobacterium* connection, *Syst. Appl. Microbiol.*, 13, 161, 1990.
2. Hugo, C.J. et al., A polyphasic taxonomic study of *Chryseobacterium* strains isolated from dairy sources, *Syst. Appl. Microbiol.*, 22, 586, 1999.
3. Jooste, P.J., and Hugo, C.J., The taxonomy, ecology and cultivation of bacterial genera belonging to the family *Flavobacteriaceae*, *Int., J., Food Microbiol.*, 53, 81, 1999.
4. Yabuuchi, E. et al., *Sphingobacterium* gen. nov., *Sphingobacterium spiritivorum* comb. nov., *Sphingobacterium multivorum* comb. nov., *Sphingobacterium mizutae* sp. nov., and *Flavobacterium indologenes* sp. nov. glucose-nonfermenting gram-negative rods in CDC groups IIK-2 and IIb, *Int. J. Syst. Bacteriol.*, 33, 580, 1983.
5. Vandamme, P. et al., New perspectives in the classification of the flavobacteria: Description of *Chryseobacterium* gen. nov., *Bergeyella* gen. nov., and *Empedobacter* nom. rev., *Int. J. Syst. Bacteriol.*, 44, 827, 1994.
6. Bernardet, J.-F. et al., Cutting a Gordian knot: emended classification and description of the genus *Flavobacterium*, emended description of the family *Flavobacteriaceae*, and proposal of *Flavobacterium hydatis* nom. nov. (basonym, *Cytophaga aquatilis* Strohl and Tait 1978), *Int. J. Syst. Bacteriol.*, 46, 128, 1996.
7. Bernardet, J.-F., Nakagawa, Y., and Holmes, B., Proposed minimal standards for describing new taxa of the family *Flavobacteriaceae* and emended description of the family, *Int. J. Syst. Evol. Microbiol.*, 52, 1049, 2002.
8. Hugo, C.J. et al., *Chryseobacterium joostei* sp. nov., isolated from the dairy environment, *Int. J. Syst. Evol. Microbiol.*, 53, 771, 2003.
9. Kampfer, P. et al., *Chryseobacterium defluvii* sp. nov., isolated from wastewater, *Int. J. Syst. Evol. Microbiol.*, 53, 93, 2003.
10. Kampfer, P. et al., Description of *Chryseobacterium anthropi* sp. nov. to accommodate clinical isolates biochemically similar to *Kaistella koreensis* and *Chryseobacterium haifense*, proposal to reclassify *Kaistella koreensis* as *Chryseobacterium koreense* comb. nov. and emended description of the genus *Chryseobacterium*, *Int. J. Syst. Evol. Microbiol.*, 59, 2421, 2009.
11. Li, Y. et al., *Chryseobacterium miricola* sp. nov., a novel species isolated from condensation water of space station Mir, *Syst. Appl. Microbiol.*, 26, 523, 2003.
12. Kim, M.K. et al., *Kaistella koreensis* gen. nov., sp. nov., a novel member of the *Chryseobacterium–Bergeyella–Riemerella* branch, *Int. J. Syst. Evol. Microbiol.*, 54, 2319, 2004.
13. Kim, K.K. et al., *Chryseobacterium daecheongense* sp. nov., isolated from freshwater lake sediment, *Int. J. Syst. Evol. Microbiol.*, 55, 133, 2005.
14. Kim, K.K. et al., Transfer of *Chryseobacterium meningosepticum* and *Chryseobacterium miricola* to *Elizabethkingia* gen. nov. as *Elizabethkingia meningoseptica* comb. nov. and *Elizabethkingia miricola* comb. nov., *Int. J. Syst. Evol. Microbiol.*, 55, 1287, 2005.
15. de Beer, H. et al., *Chryseobacterium vrystaatense* sp. nov., isolated from raw chicken in a chicken-processing plant, *Int. J. Syst. Evol. Microbiol.*, 55, 2149, 2005.
16. de Beer, H. et al., *Chryseobacterium piscium* sp. nov., isolated from fish of the South Atlantic Ocean off South Africa, *Int. J. Syst. Evol. Microbiol.*, 56, 1317, 2006.
17. Shimomura, K., Kaji, S., and Hiraishi, A., *Chryseobacterium shigense* sp. nov., a yellow-pigmented, aerobic bacterium isolated from a lactic acid beverage, *Int. J. Syst. Evol. Microbiol.*, 55, 1903, 2005.
18. Quan, Z.-X. et al., *Chryseobacterium caeni* sp. nov., isolated from bioreactor sludge, *Int. J. Syst. Evol. Microbiol.*, 57, 141, 2007.
19. Vaneechoutte, M. et al., *Chryseobacterium hominis* sp. nov., to accommodate clinical isolates biochemically similar to CDC groups II-h and II-c, *Int. J. Syst. Evol. Microbiol.*, 57, 2623, 2007.
20. Hantsis-Zacharov, E. et al., *Chryseobacterium oranimense* sp. nov., a psychrotolerant, proteolytic and lipolytic bacterium isolated from raw cow's milk, *Int. J. Syst. Evol. Microbiol.*, 58, 2635, 2008.
21. Yassin, A.F. et al., *Chryseobacterium treverense* sp. nov., isolated from human clinical source, *Int. J. Syst. Evol. Microbiol.*, 60, 1993, 2010.

22 Pires, C. et al., *Chryseobacterium palustre* sp. nov. and *Chryseobacterium humi* sp. nov., isolated from industrially contaminated sediments, *Int. J. Syst. Evol. Microbiol.*, 60, 402, 2010.
23. Holmes, B. et al., *Flavobacterium gleum*, a new species found in human clinical specimens, *Int. J. Syst. Bacteriol.*, 34, 21, 1984.
24. Bloch, K.C., Nadarajah, R., and Jacobs, R., *Chryseobacterium meningosepticum*: An emerging pathogen among immunocompromised adults: Report of 6 cases and literature review, *Medicine*, 76, 30, 1997.
25. Al-Tatari, H., Asmar, B.I., and Ang, J.Y., Lumboperitonial shunt infection due to Chryseobacterium indologenes, *Pediatr. Infect. Dis. J.*, 26, 657, 2007.
26. Bernardet, et al., Polyphasic study of *Chryseobacterium* strains isolated from diseased aquatic animals, *Syst. Appl. Microbiol.*, 28, 640, 2005.
27. Hugo, C.J., and Jooste, P.J., Preliminary differentiation of food strains of *Chryseobacterium* and *Empedobacter* using multilocus enzyme electrophoresis, *Food Microbiol.*, 14, 133, 1997.
28. Holmes, B. et al., *Weeksella zoohelcum* sp. nov. (formerly group IIj) from human clinical specimens, *Syst. Appl. Microbiol.*, 8, 191, 1986.
29. Kivinen, P.K. et al., *Bergeyella zoohelcum* septicaemia of a patient suffering from severe skin infection, *Acta Derm. Venereol.*, 83, 74, 2003.
30. Beltran, A. et al., A case of *Bergeyella zoohelcum* bacteremia after ingestion of a dish prepared with goat blood, *Clin. Infect. Dis.*, 42, 891, 2006.
31. Han, Y.W. et al., Transmission of an uncultivated *Bergeyella* strain from the oral cavity to amniotic fluid in a case of preterm birth, *J. Clin. Microbiol.*, 44, 1475, 2006.
32. Han, Y.W. et al., Uncultivated bacteria as etiologic agents of intra-amniotic inflammation leading to preterm birth, *J. Clin. Microbiol.*, 47, 38, 2009.
33. Güngör, S. et al., A *Chryseobacterium meningosepticum* outbreak in a neonatal ward, *Infect. Control Hosp. Epidemiol.*, 24, 613, 2003.
34. Cone, L.A. et al., Osteomyelitis due to *Chryseobacterium (Flavobacterium) meningosepticum*, *Antimicrob. Infect. Dis. Newsl.*, 17, 60, 1998.
35. Chiu, C.H. et al., Atypical *Chryseobacterium meningosepticum* and meningitis and sepsis in newborns and the immunocompromised, Taiwan, *Emerg. Infect. Dis.*, 6, 481, 2000.
36. Lee, C.C. et al., Fatal case of community-acquired bacteremia and necrotizing fasciitis caused by *Chryseobacterium meningosepticum*: Case report and review of the literature, *J. Clin. Microbiol.*, 44, 1181, 2006.
37. Lin, P.Y. et al., Clinical and microbiological analysis of bloodstream infections caused by *Chryseobacterium meningosepticum* in nonneonatal patients, *J. Clin. Microbiol.*, 42, 3353, 2004.
38. Bomb, K., Arora, A., and Trehan, N., Endocarditis due to *Chryseobacterium meningosepticum,Indian J. Med. Microbiol.*, 25, 161, 2007.
39. Tuon, F.F. et al., *Chryseobacterium meningosepticum* as a cause of cellulitis and sepsis in an immunocompetent patient, *J. Med. Microbiol.*, 56, 1116, 2007.
40. Sztajnbok, J., and Troster, E.J., Community-acquired *Chryseobacterium meningosepticum* pneumonia and sepsis in a previously healthy child, *J. Infect.*, 37, 310, 1998.
41. Lim, L.C., Low, J.A., and Chan, K.M., *Chryseobacterium meningosepticum (Flavobacterium meningosepticum)*—a report of five cases in a local hospital, *Ann. Acad. Med. Singapore*, 28, 858, 1999.
42. Weaver, K.N. et al., Acute emergence of *Elizabethkingia meningoseptica* infection among mechanically ventilated patients in a long-term acute care facility, *Infect. Control Hosp. Epidemiol.*, 31, 54, 2010.
43. Ozkalay, N. et al., Community-acquired meningitis and sepsis caused by *Chryseobacterium meningosepticum* in a patient diagnosed with thalassemia major, *J. Clin. Microbiol.*, 44, 3037, 2006.
44. Christakis, G.B. et al., *Chryseobacterium indologenes* non-catheter-related bacteremia in a patient with a solid tumor, *J. Clin. Microbiol.*, 43, 2021, 2005.
45. Douvoyiannis, M. et al., *Chryseobacterium indologenes* bacteremia in an infant, *Int. J. Infect. Dis.*, 2009 September 1.
46. Cascio, A. et al., *Chryseobacterium indologenes* bacteraemia in a diabetic child, *J. Med. Microbiol.*, 54, 677, 2005.
47. Montejo, M. et al., Bergeyella zoohelcum bacteremia after a dog bite, *Clin. Infect. Dis.*, 33, 1608, 2001.
48. Shukla, S.K. et al., Isolation of a fastidious *Bergeyella* species associated with cellulitis after a cat bite and a phylogenetic comparison with *Bergeyella zoohelcum* strains, *J. Clin. Microbiol.*, 42, 290, 2004.
49. Lin, W.R., Chen, Y.S., and Liu, Y.C., Cellulitis and bacteremia caused by *Bergeyella zoohelcum*, *J. Formos. Med. Assoc.*, 106, 573, 2007.
50. Buck, J.D., Nonstaining (KOH) method for determination of Gram reactions of marine bacteria, *Appl. Environ. Microbiol.*, 44, 992, 1982.
51. Dees, S.B. et al., Chemical and phenotypic characteristics of *Flavobacterium thalpophilum* compared with those of other *Flavobacterium* and *Sphingobacterium* species, *Int. J. Syst. Bacteriol.*, 35, 16, 1985.
52. Fraser, S.L., and Jorgensen, J.H., Reappraisal of antimicrobial susceptibilities of *Chryseobacterium* and *Flavobacterium* species and method for reliable susceptibility testing, *Antimicrob. Agents Chemother.*, 41, 2738, 1997.
53. Vessillier, S. et al., Overproduction and biochemical characterization of the *Chryseobacterium meningosepticum* BlaB metallo-β-lactamase, *Antimicrob. Agents Chemother.*, 46,1921, 2002.
54. Ursing, J., and Bruun, B., Genetic heterogeneity of *Flavobacterium meningosepticum* demonstrated by DNA-DNA hybridization, *Acta Pathol. Microbiol. Immunol. Scand. Sect. B*, 95:33, 1987.
55. Drancourt, M., Berger, P., and Raoult, D., Systematic 16S rRNA gene sequencing of atypical clinical isolates identified 27 new bacterial species associated with humans, *J. Clin. Microbiol.*, 42, 2197, 2004.
56. Colding, H. et al., Ribotyping for differentiating *Flavobacterium meningosepticum* isolates from clinical and environmental sources, *J. Clin. Microbiol.*, 32, 501, 1994.
57. Lu, C.H. et al., An adult case of *Chryseobacterium meningosepticum* meningitis, *Jpn. J. Infect. Dis.*, 57, 214, 2004.

48 *Fusobacterium*

Dongyou Liu and Xiaoming Dong

CONTENTS

48.1 INTRODUCTION

48.1.1 Classification, Morphology, and Biology

The genus *Fusobacterium* covers a large number of gram-positive, nonspore-forming, nonmotile, anaerobic, rod-shaped bacterial species that come under the family Fusobacteriaceae, class Fusobacteria, phylum Fusobacteria. Currently, at least 25 species/subspecies are recognized within the genus: *Fusobacterium alocis*, *Fusobacterium canifelinum*, *Fusobacterium equinum*, *Fusobacterium gonidiaformans*, *Fusobacterium mortiferum*, *Fusobacterium naviforme*, *Fusobacterium necrogenes*, *Fusobacterium necrophorum* subsp. *funduliforme*, *Fusobacterium necrophorum* subsp. *necrophorum*, *Fusobacterium nucleatum* subsp. *animalis*, *Fusobacterium nucleatum* subsp. *fusiforme*, *Fusobacterium nucleatum* subsp. *nucleatum*, *Fusobacterium nucleatum* subsp. *polymorphum*, *Fusobacterium nucleatum* subsp. *vincentii*, *Fusobacterium perfoetens*, *Fusobacterium periodonticum*, *Fusobacterium plautii*, *Fusobacterium polysaccharolyticum*, *Fusobacterium prausnitzii*, *Fusobacterium pseudonecrophorum*, *Fusobacterium russii*, *Fusobacterium simiae*, *Fusobacterium sulci*, *Fusobacterium ulcerans*, and *Fusobacterium varium*.

While members of the genus *Fusobacterium* form part of commensal microflora on the mucosal surfaces (e.g., mouth, upper respiratory tract, gastrointestinal tract, female genital tract) of humans and animals,[1–4] some *Fusobacterium* spp. (e.g., *F. necrophorum*, *F. nucleatum*, *F. mortiferum*, and *F. varium*) have the capacity to invade host tissues, especially after accidental trauma, surgery, and edema, causing various purulent or gangrenous infections in man, especially young and elderly patients.[5–12] In this chapter, we focus on two most important and best characterized species of the genus *Fusobacterium*, that is, *F. necrophorum* and *F. nucleatum*.

Fusobacterium necrophorum (fusus, a spindle; fusobacterium, spindle-shaped bacterium) cells are gram-negative, obligately anaerobic, nonmotile, beta-hemolytic, nonspore forming, rods with rounded ends. *F. necrophorum* cells are often arranged singly or in pairs. Besides production of gas and indole, they generate butyric acid from peptone and glucose and propionic acid from lactate and threonine. However, they are unable to hydrolyze esculin and starch, reduce nitrate, produce acid from carbohydrates, and grow in 20% bile, in addition to being catralase-negative. Being a pleomorphic species, *F. necrophorum* is distinguished into two subspecies: *F. necrophorum* subsp. *necrophorum* (biovar A) and *F. necrophorum* subsp. *funduliforme* (biovar B).[13] Whereas the *necrophorum* subspecies is associated mostly with infections in domestic mammals (e.g., calf diphtheria, labial necrosis of rabbits, necrotic rhinitis of pigs, and foot rot of cattle, sheep, and goats), the *funduliforme* subspecies infects mainly humans, causing Lemierre's syndrome, a rare and once life-threatening disease that is characterized by acute pharyngitis, thrombophlebitis, and abscessation of the internal jugular vein. In addition, the *funduliforme* subspecies is also responsible for peritonsillar abscess and tonsillitis.

Fusobacterium necrophorum subsp. *necrophorum* cells measure 0.5–0.6 μm by 3–70 μm. Colonies on GAM agar supplemented with 5% defibrinated horse blood are about 2 mm in diameter and appear circular, with grayish to whitish color, and a flat and bumpy surface. The bacterium can liquefy gelatin and ferment maltose, and has a G + C content of 28–31 mol%. The type strain is strain VPI 2891 (= JCM 3718 = ATCC 25286).[13]

Fusobacterium necrophorum subsp. *funduliforme* (*fundula*, "sausage"; *forma*, "shape": *funduliforme*, "sausage shaped") cells measure 0.4–0.8 μm by 1–8 μm, with shorter cells (1–4 μm) being predominant. Colonies on GAM agar are about 1 mm in diameter, and appear circular with entire margins, grayish, translucent, and raised with smooth surfaces. The bacterium can liquefy gelatin, and ferment glucose weakly, and has a G + C content of 27–31 mol%. The type strain is strain Fn524 (= JCM 3724).[13]

Fusobacterium nucleatum is divided into five subspecies: *nucleatum*, *polymorphum*, *fusiforme*, *vincentii*, and *animalis*.[14–17] Being inhabitants of the natural cavities of man and animals, *F. nucleatum* generates butyric acid, metabolic end products, and irritates the fibroblast of the gum, leading to necrotic lesions, abscesses, periodontal disease and bacteremia.[18]

48.1.2 Clinical Features and Pathogenesis

A variety of clinical disease symptoms have been attributed fully or partially to *F. necrophorum*.[20,21] These include (i) nonstreptococcal tonsillitis, particularly, recurrent sore throat. In a survey of 251 human patients with *F. necrophorum* infection, the most frequent site of origin was throat (179), then ear (32), and then sinusitis (8), dental (4), esophageal (1), posttraumatic/postoperative (3), and unknown (24); (ii) peritonsillar abscess, which is a precursor to Lemierre's syndrome; (iii) deep neck space infections such as parapharyngeal or retropharyngeal abscess, in which *F. necrophorum* plays an accessory role; (iv) mediastinitis, which may evolve into empyema in the absence of underlying lung lesions; (v) mastoiditis, with meningitis being the most common complication accompanied by brain abscess, epidural abscess, sigmoid sinus thrombosis, subdural empyema, perisinus abscess, and transverse and cavernous sinus thrombosis; (vi) noma, which is an extreme form of necrotizing ulcerative gingivitis, with malnutrition and viral infection being the possible triggering factors; (vii) sinusitis, with some cases presenting intracranial complications such as meningitis, cavernous sinus thrombosis, and cerebral infarction; (viii) sinus thrombosis, which is a recognized complication of *F. necrophorum* infection, with the postanginal cases displaying internal jugular vein thrombophlebitis; (ix) cerebral abscess, meningitis, which can be aggressive, often with purulent CSF; (x) anaerobic bacteremia, which often occurs in patients with serious underlying conditions, including cancer, surgery, and chronic organ dysfunction; and (xi) septicemia, which may lead to death within 12 days in the preantibiotic era in the classic cases, or a distressing decline to death after 40–60 days in subacute cases, or a marked spike of fever following a throat infection in late metastatic manifestations, lasting weeks or months resulting in a lung abscess, septic arthritis, and liver abscess, or bacteremia without metastases.[21]

Metastatic complications comprise (i) pleuropulmonary lesions such as septic pulmonary emboli, which are the most common metastatic manifestation and archetypal radiologically identifiable lesions of Lemierre's syndrome; (ii) bone and joint manifestations such as osteomyelitis; (iii) pyomyositis and abscesses in muscle; (iv) skin and soft-tissue lesions, such as cutaneous pustules and subcutaneous abscesses; (v) intraabdominal sepsis, such as liver abscesses; and (vi) endocarditis and pericarditis.[21]

Lemierre's syndrome (or postanginal sepsis) is an exogenously acquired, monomicrobial, septicemic illness with metastatic abscesses secondary to a septic thrombophlebitis of the internal jugular vein infection with *F. necrophorum* subsp. *Funduliforme*.[22,23] Lemierre's syndrome is characterized by a history of recent oropharyngeal infection (acute sore throat or tonsillitis, followed by a neck mass and neck pain), clinical or radiological evidence of internal jugular vein thrombosis (predisposing to septic thrombophlebitis), a septicemia with septic emboli in lungs and other sites, and isolation of anaerobic pathogens, mainly *Fusobacterium necrophorum*. Lemierre's syndrome was relatively common in the preantibiotic era, affecting previously healthy young adults (with a median age of 19 years, of which 55%–65% were male), who lacked any identified risk factor. However, it has become rare these days (affecting about 1 per million persons per year in the 1990s) due to empirical antibiotic therapy of acute pharyngitis.[8,23,24] The mortality rate *F. necrophorum* infection ranges from 4.9% to 6.7% in patients. *F. necrophorum* subsp. *funduliforme* is likely transmitted human-to-human through close contact between carriers and susceptible individuals or enters the alimentary tract from animal sources, as the bacterium is present in the guts of herbivores such as cattle, pigs, and dogs.

Among the nonspore-forming anaerobes, *Fusobacterium necrophorum* is unique for its strong link with clinically distinctive necrobacillosis, postanginal sepsis, or Lemierre's syndrome. A number of virulence factors are produced by *F. necrophorum* to facilitate its invasion of host cells. These include: (i) leukotoxin, which is a large, secreted protein encoded by the leukotoxin operon (*lktBAC*) with neutrophil cytotoxicity[25–33]; (ii) endotoxin, which is a lipopolysaccharide known to be an important virulence factor in many gram-negative organisms, including *F. necrophorum*; (iii) hemolysin, which contributes to abscess formation in *F. necrophorum* infection; (iv) hemagglutinin, which is one of the components of the cell surfaces of *F. necrophorum* subsp. *necrophorum*; (v) adhesins, which allow *F. necrophorum* attachment to host cells; (vi) platelet aggregation, which is used by *F. necrophorum* subsp. *necrophorum* to enhance its pathogenicity; (vii) synergy, in which *F. necrophorum* is protected by an anaerobic environment created by facultative bacteria (e.g., group C streptococci) that utilize oxygen and lower redox. In addition, Epstein–Barr virus induces immunosuppression, with a transient decrease in T-cell-mediated immunity that may predispose to bacterial infection such as *F. necrophorum*.[21]

Besides the roles of *F. necrophorum* virulence factors in the disease process, several external factors (e.g., host age and immune status) may also exert influence on the disease outcome. For example, *F. necrophorum* is known to target young adults (late teens and early 20s), producing not only Lemierre's syndrome but also tonsillitis and peritonsillar abscess. The absence of immunity, concomitant viral infection (e.g., EBV) and genetic abnormality (e.g., single-nucleotide polymorphisms in Toll-like receptors that determine the innate immune response to bacteria) may also predispose the host to thrombophilia, then tonsillar and internal jugular vein thrombophlebitis, metastatic lesions, and severe sepsis.

F. nucleatum often aggregates with most oral bacteria and plays a critical part in the formation of dental plaque as a bridge between early colonizer (gram-positive bacteria) and late colonizer (gram-negative bacteria).[34–42] By scavenging oxygen and oxidative free radicals from dental plaque, *F. nucleatum* helps to maintain and support the conditions for major anaerobic periodontal pathogens.[43,44] In addition, *F. nucleatum* is capable of invading the mucosal keratinocytes and inducing proinflammatory cytokines and elastase.[45,46]

48.1.3 Diagnosis

Clinical Diagnosis. The occurrence of chills/rigors at around 4–5 days after the onset of the sore throat is a feature of Lemierre's syndrome or necrobacillosis, indicating that the organisms are gaining access to the circulation and in the process of initiating septicemia. Other symptoms include neck pain and sometimes stiffness, tender (normally unilateral) swelling at the angle of the jaw and the sternomastoid muscle (the cord sign), reflecting the development of internal jugular venous thrombophlebitis. The markedly raised level of C-reactive protein present in Lemierre's syndrome helps eliminate uncomplicated viral pharyngitis (e.g., EBV) as a possible diagnosis. Lemierre's syndrome is distinguished from acute bacterial pneumonia (e.g., *Staphylococcus aureus*), Legionnaires' disease, or aspiration pneumonia by the fact that lung involvement in Lemierre's syndrome is preceded by sore throat and may be accompanied by internal jugular vein thrombosis and that initially lung lesions consist of multiple diffusely disseminated nodular lesions which rapidly cavitate. Cavitation, septic pulmonary emboli, and jugular venous thrombosis are visible on chest X-ray, CT scanning or ultrasound scanning.[21] *F. necrophorum* may also cause false-positive results in *Mycoplasma pneumoniae* 16S rRNA PCR tests.[47]

Phenotypic Identification. *F. necrophorum* has a characteristic pleomorphic morphology, with filaments, short rods, and coccoid elements, which are clearly different from *F. nucleatum* and other gram-negative bacteria. *F. necrophorum* displays varied morphological features in tissue and pus (predominantly coccobacillary and often showing bipolar staining) in comparison with those in culture (polymorphic with irregularly stained rods, filamentous forms, and vesicular forms).[48,49]

For isolation of *F. necrophorum* from cerebral abscess pus, sample is inoculated immediately into semisolid medium that has been gassed out. Blood agar media supplemented with vitamin K, hemin, menadione, and a reducing agent such as cysteine hydrochloride provide more suitable media for cultivating *Fusobacterium* spp. Fastidious anaerobe agar produces the best growth of all media for *Fusobacterium* spp. On a suitable medium, *F. necrophorum* forms characteristic colonies of cream-yellow in color, and smooth, round, and entire in shape, with an odor redolent of cabbages. On horse blood agar most colonies show a narrow zone of complete hemolysis surrounding the colonies. Under long-wave UV light, colonies on fastidious anaerobe agar fluoresce with a vivid greenish yellow color.[21]

Biochemically, *F. necrophorum* produces indole, which is detectable directly from colonies on an agar plate by using the spot indole reagent *p*-dimethylcinnamaldehyde. In addition, *F. necrophorum* produces lipase on egg yolk agar. Use of commercial kits allows convenient biochemical characterization of *Fusobacterium* spp. (on the basis of indole-positive and alkaline phosphatase–negative activity). These include API rapid ID 32A (BioMérieux), AN-IDENT identification system (BioMérieux), Vitek ANI card (BioMérieux), and RapID-ANA II (Innovative Diagnostic Systems, Inc., Atlanta, GA). Analysis for the presence of volatile fatty acid end products by gas-liquid chromatography is also valuable. *Fusobacterium* spp. show a volatile fatty acid profile containing a single major peak of butyric acid, and *F. necrophorum* is the only *Fusobacterium* sp. that ferments lactate to propionate.[21]

Molecular Identification. Given the time consuming nature of culture-based testing for *Fusobacterium* spp., molecular methods have been increasingly utilized for species- and subspecies-specific determination.[47,50–59] The target genes often include 16S rRNA,[60–63] *rpoB* gene (RNA polymerase β-subunit),[64,65] hemagglutinin gene,[64] and the leukotoxin operon promoter-containing intergenic region.[32] While the *rpoB* gene is specific for *F. necrophorum*,[64] hemagglutinin gene is absent in *F. necrophorum* subsp. *Funduliforme*.[54] Furthermore, *F. necrophorum* subsp. *necrophorum* and *F. necrophorum* subsp. *funduliforme* can be differentiated by 16S-23S rRNA intergenic spacer region sequence analysis, random amplified polymorphic DNA–PCR, and ribotyping analysis.[66–69]

Aliyu et al. described a real-time PCR targeting the *rpoB* gene (RNA polymerase β-subunit) sequence for detection of *F. necrophorum* in throat swabs of patients with sore throat. D'Ercole et al.[70] showed that *F. nucleatum* is most frequently detected in oral specimens by using a multiplex PCR with primers from 16S rRNA gene. Shin et al.[71] developed strain-specific PCR for *F. nucleatum* subsp. *fusiforme* ATCC 51190(T) and *F. nucleatum* subsp. *vincentii* ATCC 49256(T), with two pairs of primers Fs17-F14/Fs17-R14 and pFv35-F1/Fv35-R1. More recently, real-time PCR has been also employed for improved determination and quantitation of *Fusobacterium* spp.[56,72]

48.2 METHODS

48.2.1 Sample Preparation

Sample from the disinfected tooth surface is collected with sterile paper points, transferred to fluid thioglycolate medium or Schaedler broth (Difco), and incubated at 37°C for 14 days in an anaerobic chamber (10% H^2, 5% CO^2, and 80% N^2). The cut files and paper points are placed in cryotubes containing 1 mL of Tris-EDTA buffer. After vigorous vortexing for 1 min, the suspension is stored at −20°C for subsequent DNA extraction.

Throat swabs are transported in Stuart's Transport Medium within 24 h of sampling. They are then inoculated on Columbia agar plates containing 5% sheep blood (Becton Dickinson) and incubated overnight at 35°C in 5% CO^2. Alternatively, the swabs are inoculated on plates containing 5% horse blood (SSI) v/v and incubated at 35°C in anaerobic jars containing an atmosphere of 5% CO^2, 10% H^2, and 85% N^2. Plates are examined after incubation for 1, 2, and 4 days. Throat swabs are stored for up to 2 days at 4°C until extraction of bacterial DNA.[47]

Initial identification of *F. necrophorum* is based on colony morphology, an odor of butyric acid, the presence of a greenish fluorescence from colonies when irradiated with UV light, and a characteristic gram-negative pleomorphic morphology under the microscope, supplemented by susceptibility to kanamycin and metronidazole on primary inoculated agar plates. Subsequent β-hemolysis on agar containing 5% horse blood, resistance to vancomycin, and susceptibility to polymyxin confirm the identification of *F. necrophorum*.[47]

For DNA extraction, *Fusobacterium* cells from overnight anaerobic BHI broth culture are pelleted by centrifugation at 5000 × *g* for 10 min at 4°C. The pelleted cells are washed and resuspended in 100 mM Tris/HCl (pH 8.0), 10 mM EDTA. The cell wall is digested with lysozyme (1 mg/mL), and the resulting lysate is treated with 1% sarkosyl followed by RNase A (20 μg/mL) and pronase (50 μg/mL). The sample is incubated on ice for 20 min and then sequentially extracted twice with phenol and chloroform. The DNA from the extract is precipitated in 0.1 volume of 3 M sodium acetate pH 5.2 and an equal volume of 2-propanol. The DNA is resuspended in 10 mM Tris/HCl pH 8.0, 1 mM EDTA. The concentration of extracted DNA is determined spectrophotometrically (Nanodrop Technologies). *Fusobacterium* nucleic acids can be also prepared directly from clinical samples by using commercial extraction kits.

48.2.2 Detection Procedures

48.2.2.1 PCR Identification of *F. necrophorum*

(i) PCR for F. necrophorum Bennett et al.[4] reported a PCR for *F. necrophorum* with primers targeting the leukotoxin-coding *lktA* gene, which is unique to *F. necrophorum* (including *F. necrophorum* subsp. *necrophorum* and *F. necrophorum* subspecies *funduliforme*) and not present in other *Fusobacterium* species/subspecies.[31] The *F. necrophorum*-specific primers *lktA*-up (5′-ACAATCGGAGTAGTAGGTTC-3′) and *lktA*-dn (5′-ATTTGGTAACTGCCACTGC-3′) amplify a 402 bp amplicon.

The PCR thermal profile consists of an initial denaturation step at 94°C for 5 min, followed by 35 cycles of 94°C for 30 s, 59°C for 30 s, and 72°C for 30 s, and a final extension at 72°C for 5 min. The PCR mixture includes 400 ng/μL bovine serum albumin (BSA) (New England Biolabs) and an extra 2.5 mM of $MgCl^2$ to reduce effects of inhibitors in the samples.

PCR products from both the *lktA* gene is then separated electrophoretically in 1.5% agarose gels containing 0.5 μg/mL of ethidium bromide, and visualized under UV transillumination.

(ii) Real-time PCR for F. necrophorum and F. necrophorum subsp. necrophorum Aliyu et al.[64] developed a quantitative real-time PCR with primers targeting the rpoB gene for detection of *F. necrophorum* in throat swabs collected from patients with pharyngitis. The samples positive for *F. necrophorum* are further analyzed with a hemagglutinin-related protein gene-specific PCR for *F. necrophorum* subspecies *necrophorum* (Table 48.1).

Procedure

1. All swabs are held in Amies transport media with charcoal, refrigerated for up to 2 days at 4°C. Bacteria are harvested from the swabs by soaking in 2 mL PBS for 10 min, vortexing for 30 s and then

TABLE 48.1
Primers and Probes for Real-Time PCR Identification of *F. necrophorum* and *F. necrophorum* Subspecies *funduliforme*

Gene	Primer	Sequence (5′→3′)	Position	Specificity
rpoB	RPO forward	TCTCTACGTATGCCTCACGGATC	171–193	
	RPO reverse	CGATATTCATCCGAGAGGGTCTCC	447–424	*F. necrophorum*
	RPO probe 1	GAAGACATGCCTTTCTTAGAGGAC–Fluo		
	RPO probe 2	LCRed-640–GAACTCATTTGGACGTTGTGTTA–Pho		
Haem	HAEMF	CATTGGGTTGGATAACGACTCCTAC	1786–1810	*F. necrophorum*
	HAEMR	CAATTCTTTGTCTAAGATGGAAGCGG	2096–2071	subspecies *necrophorum*

collecting the PBS from the swab by wringing out against the wall of the container. One mL of the processed PBS is transferred to a 1.5 mL Eppendorf tube and centrifuged for 10 min at 6000 × *g*. The supernatant is carefully decanted and the pellet resuspended in 300 μL PBS for extraction.

2. Nucleic acid is extracted from 200 μL of the PBS supernatant with a Total Nucleic Acid Isolation Kit (Roche Molecular Biochemicals) using an automated MagNA Pure Extraction System (Roche Molecular Biochemicals). The DNA is eluted in 60 μL and stored at −20°C until use.
3. PCR mixture (20 μL) is made up of 4 μL DNA extract, 2 μL LightCycler FastStart DNA master hybridization mixture (Roche Diagnostics), 3.2 μL 25 mM $MgCl^2$ (5 mM), 0.5 μL each of 20 pmol/μL RPO primers, 0.2 μL each of the 10 pmol/μL PRO probes and 9.4 μL RNase free water. Positive control (*F. necrophorum* subsp. *necrophorum* ATCC 25286^T DNA), a contamination control (sterile water), and extraction control (sterile swab processed as above) are included in each run. PCR quantification standards (10^{-4} down to 10^{-9}) are also included in every run of 32 samples to facilitate quantification of the bacterial load in positive samples.
4. PCR amplification is carried out in a LightCycler as follows: one cycle of denaturation at 95°C for 10 min followed by 50 cycles of 95°C for 0 s, 60°C for 2 s, and 72°C for 15 s, with a single transition rate of 20°C per s and a single fluorescence acquisition at 60°C. On completion of amplification, a single melt cycle is produced by holding the reaction at 95°C for 0 s, then at 45°C for 30 s, followed by slow heating at a transition rate of 0.1°C per s to 85°C and continuous fluorescence acquisition.
5. For specific identification of *F. necrophorum* subsp. *necrophorum*, a real-time PCR is performed on the LightCycler using FastStart DNA Master SYBR green 1 kit. PCR mixture (20 μL) is composed of 4 μL DNA extract, 2 μL of the LightCycler SYBR green kit, 2.4 μL 25 mM $MgCl^2$ (4 mM), 0.5 μL each of 20 pmol/μL HAEM primers, and 10.6 μL RNase free water.
6. PCR amplification is conducted in a LightCycler with one cycle of denaturation at 95°C for 10 min, 50 cycles of 95°C for 0 s, 65°C for 2 s and 72°C for 15 s with a single transition rate of 20°C per s and a single fluorescence acquisition at 72°C. On completion of amplification, a single melt cycle is produced by holding the reaction at 95°C for 0 s, then at 65°C for 30 s, followed by slow heating at a transition rate of 0.1°C per s to 95°C and continuous fluorescence acquisition.

Note: Jensen et al.[47] adapted this *F. necrophorum* species-specific real-time PCR assay for TaqMan chemistry. While the primers RPO-forward and RPO-reverse remain the same, only one TaqMan probe (RPO probe, 5′-TTGCCGGCGGA AGACATGCCTTTCTTA-3′, positions 167–193) is needed. The probe is labeled at the 5′ end with 6-carboxyfluorescein and at the 3′ end with Black Hole Quencher 1.

(iii) Real-Time PCR for F. necrophorum subsp. necrophorum and subsp. funduliforme Jensen et al.[47] described a TaqMan-based real-time quantitative PCR assay, targeting the *gyrB* gene, to detect and discriminate between the two subspecies of *F. necrophorum* in the throats of healthy volunteers, and to examine the possible role of *F. necrophorum* in nonstreptococcal tonsillitis (NST). A primer pair (*gyrB*-forward and *gyrB*-reverse) amplifies a 306 bp fragment from both *F. necrophorum* subsp. *necrophorum* and subsp. *funduliforme* subspecies, which are detected with two subspecies-specific TaqMan probes (probe 1 and probe 2). The subsp. *necrophorum*-specific probe 1 is labeled at the 5′ end with 6-carboxyfluorescein, and at the 3′ end with Black Hole Quencher 1, whereas the subsp. *funduliforme*-specific probe 2 is labeled at the 5′ end with hexachlorofluorescein and at the 3′ end with Black Hole Quencher 2 (Table 48.2). The assay is carried out in two separate PCR.

Procedure

1. Bacterial DNA is isolated by vortexing the swabs in 1 mL of physiological saline for 1 min and then extracting the nucleic acid from 200 μL of saline using the Kingfisher mL magnetic particle processor (Thermo Electron Corporation) and a Biosprint 15 DNA Blood Kit (Qiagen). Nucleic

TABLE 48.2
Primers and Probes for Real-Time PCR Identification of *F. necrophorum* subsp. *necrophorum* and subsp. *funduliforme*

Gene	Primer	Sequence (5′→3′)	Position	Specificity
gyrB	*gyrB*-forward	AGGATTGCATGGAGTAGGAA	27–46	*F. necrophorum*
	gyrB-reverse	CCTATTTCATTTCGACAATCCA	332–311	
	Probe 1	FAM-TCTACTTTGGAGGTTGGAGAAACAAC-BHQ	160–185	*F. necrophorum* subspecies *necrophorum*
	Probe 2	HEX-TCCGCTTTAGAGGCTGGAGAAACGAC-BHQ	160–185	*F. necrophorum* subspecies *funduliforme*

acids are eluted in 200 μL of the elution buffer supplied with the extraction kit, and stored at 4°C until use.

2. PCR for both assays is performed in 25-μL mixtures containing 12.5 μL of 2 × Brilliant QPCR Master Mix (Stratagene), 2.5 μL of 100 nM (final concentration) TaqMan probe, 2.5 μL of 300 or 400 nM (final concentrations) forward and reverse primer, for the *rpoB* and *gyrB* assays, respectively, and 5 μL of template DNA. DNA extracted from *F. necrophorum* subsp. *necrophorum* ATCC 25286 and subsp. *funduliforme* ATCC 51357 is included as positive controls and sterile water as negative control in each rune.
3. Thermal cycling on an Mx3000P Real-time PCR System (Stratagene) comprises 10 min at 95°C, followed by 55 cycles of 30 s at 95°C and 60 s at 60°C for both assays. All samples are run in duplicate. Results are analyzed using the Mx3000P software package (Stratagene).

Note: The minimum detection limit of the assays for both subspecies is between 1.5×10^2 and 1.5×10^3 CFU/swab (0.75 and 7.5 CFU/reaction).

48.2.2.2 PCR Identification of *F. nucleatum*

(i) Nested PCR for F. nucleatum Kulekci et al.[38] presented a nested PCR for detection of *F. nucleatum* in children with persistent middle-ear effusion (MEE) as a consequence of acute otitis media (AOM) and otitis media with effusion (OME). The first-round PCR relies on *E. coli* 16S ribosomal RNA (rRNA) universal primers common to all bacterial species (i.e., pHF: 5′-AAGGAGGTGATCCAGCCGCA-3′ and pAF: 5′-GAGTTTGATCATGGCTCAG-3′). MEE samples showing about 1400 bp amplicon with the universal primers are further assessed by using *F. nucleatum* specific primers from the 16S rRNA gene sequences (i.e., FnF: 5′-CTAAAT ACGTGCCAGCAGCC-3′, positions 487–506 and FnR: 5′-CGACCCCCAACACCTAGTAA-3′, positions 821–802).

Procedure

1. The external ear canal is disinfected and desquamated by placing 70% ethanol in the ear canal for 1 min. Immediately after myringotomy, the effusion is removed from the middle-ear cavity with an electronic suction device into a sterile Eppendorf tube and stored at −20°C until use.
2. A 100 μL aliquot of sample is diluted in 100 μL of 50 mM Tris-HCL pH 8.5, 50 mM EDTA, 2.5% SDS and digested overnight at 50°C after addition of 25 μL proteinase K (20 mg/mL). Sodium perchlorate is then added to a final concentration of 1 M, and the mixture is incubated at 50°C for 1 h. After addition of 200 μL of STE (150 mM NaCl, 10 mM Tris-HCl and 1 mM EDTA pH 8.5), the DNA is extracted with 400 μL phenol-chloroform (1:1), once with phenol-chloroform-isoamyl alcohol (25:24:1) and precipitated with ethanol, dried under vacuum, dissolved in 100 μL of sterile distilled water and stored at −20°C.
3. The PCR mixture (50 μL) is composed of 1 μM of each primer, 1.5 mM $MgCl^2$, 200 μM of each deoxynucleoside triphosphate, 1× Taq polymerase buffer (Boehringer Mannheim), 1.25 U of Taq polymerase, and 10 μL of template DNA. *F. nucleatum* ATCC 49256 is used as positive control and sterile distilled as a negative control for each reaction series.
4. PCR amplification is conducted in a DNA thermal cycler (GeneAmp 9600 thermal cycler; Perkin Elmer) with an initial denaturation at 94°C for 3 min; 36 cycles of 94°C for 45 s, 55°C for 30 s, 72°C for 45 s; and a final extension at 72°C for 7 min.
5. After amplification, 10 μL of PCR products is analyzed by electrophoresis on a 2% agarose gel, stained with 0.5 μg/mL ethidium bromide and visualized by ultraviolet light illumination.

Note: The size of the PCR products is estimated using f X174 DNA-Hae III Digest as a molecular weight marker. A band of 316 bp is expected for products from the second round PCR.

(ii) Multiplex PCR for F. nucleatum subsp. fusiforme and subsp. vincentii Shin et al.[71] designed strain-specific PCR primers for *F. nucleatum* subsp. *fusiforme* ATCC 51190T and *F. nucleatum* subsp. *vincentii* ATCC 49256T from the nucleotide sequence of the Fs17 (GenBank accession no., AY185204) and Fv35 (GenBank accession no., AY171472) DNA probes, respectively (Table 48.3).

Procedure

1. PCR mix (20 μL) is composed of 5 nmoles each of deoxynucleoside triphosphate, 0.8 mmoles of KCl, 0.2 mmoles of Tris-HCl pH 9.0, 0.03 mmoles of $MgCl^2$ and 1 U of Taq DNA polymerase, 20 pmoles of each primer and bacterial genomic DNA.
2. PCR is run for 32 cycles of 94°C for 1 min, 55°C or 50°C or 30 s, and 72°C for 1 min with a final 72°C for 10 min.
3. A 2 μL aliquot of the reaction mixture is then analyzed by 1.5% agarose gel electrophoresis in a

TABLE 48.3
Primers for Multiplex PCR Identification of *F. nucleatum* subsp. *fusiforme* and subsp. *vincentii*

Primer	Sequence (5′→3′)	Product (bp)	Annealing Temperature (°C)
Fs17-F14	GAT GAG GAT GAA AAG AAA CAA AGT A	393	55
Fs17-R14	CCA TTG AGA AGG GCT ATT GAC		
Fv35-F1	ATA ATG TGG GTG AAA TAA	208	50
Fv35-R1	CCC AAG GAA AAT ACT AA		

Tris-acetate buffer (0.04 M Tris-acetate, 0.001 M EDTA, pH 8.0) at 100 V for 30 min. The amplification products are stained with ethidium bromide, and visualized by UV transillumination.

(iii) Real-Time PCR for F. nucleatum Periasamy and Kolenbrander[18] designed a real-time PCR for quantitative detection of *F. nucleatum* in biofilm with primers from the 16S rRNA gene: forward primer 5′-CTTAGGA ATGAGACAGAGATG-3′ and reverse primer 5′-TGATGG TAACATACGAAAGG-3′.

Procedure

1. DNA is extracted from biofilms by immersing biofilm-covered pegs in 40 μL of sterile ultrapure water plus 160 μL of 0.05 M sodium hydroxide. After incubation at 60°C for 60 min, 18.4 μL of 1 M Tris-HCl pH 7.0 is added to neutralize the pH. The extract is used as the template DNA for the qPCR analyses. The bacterial genomic DNA used for standard curves is extracted from overnight cultures of *F. nucleatum* ATCC 10953, with a DNA extraction kit (Qiagen). Genomic DNA is stored at −20°C.
2. Quantification of *F. nucleatum* in the biofilms is performed by qPCR analysis using the SYBR green dye. The qPCR mixture (25 μL) is made up of 3 μL of template, 3.5 μL of diethyl pyrocarbonate-treated ultrapure water, 12.5 μL of Power SYBR green PCR master mixture (Applied Biosystems), and 3 μL each of the forward and reverse primers (375 nM).
3. qPCR is carried out with an MX3005P thermocycler (Stratagene) using the following thermal cycle recommended for the Power SYBR green PCR master mixture: 95°C for 10 min and then 40 cycles of 30 s at 95°C and 1 min at 56°C. Dissociation curves are generated by incubating reaction products at 95°C for 1 min and at 56°C for 30 s and then incrementally increasing the temperature to 95°C. Fluorescence data are collected at the end of the 56°C primer-annealing step for 40 amplification cycles and throughout the dissociation curve analysis.
4. Analysis of the melting curves with both primer sets reveals a single sharp peak. DNA concentrations (ng/mL) are calculated based on standard curves obtained by using tenfold serial dilutions of bacterial DNA isolated with a DNA extraction kit (Qiagen) and quantified with the PicoGreen fluorescence assay (Invitrogen). To convert ng DNA to numbers of cells, the following weights and genome sizes are used: 2.41 fg/genome and 2.4 Mb for fusobacteria for *F. nucleatum*.

48.2.2.3 PCR Identification of *Fusobacterium* spp.

(i) Real-Time PCR for Fusobacterium spp. Boutaga et al.[72] developed a real-time PCR with the 16S rRNA gene primers and probe for detection of *Fusobacterium* spp. in subgingival plaque samples from adult patients with periodontitis (Table 48.4). The probe is labeled with fluorescent dyes 6-carboxyfluorescein (FAM) at the 5′ end and 6-carboxytetramethylrhodamine (TAMRA) at the 3′ end.

Procedure

1. PCR mixture (25 μL) is made up of 12.5 μL of 2× TaqMan universal PCR master mixture (PCR buffer, deoxynucleoside triphosphates, AmpliTaq Gold, an internal reference signal [6-carboxy-X-rhodamine], uracil *N*-glycosylase, MgCl²; Applied Biosystems), 300 nM forwar primer, 300 nM reverse primer, 300 nM probe and 5 μL of purified DNA from plaque samples. The negative control is 5 μL of sterile H²O.
2. The mixture is subjected to an initial amplification cycle of 50°C for 2 min and 95°C for 10 min, followed by 45 cycles at 95°C for 15 s and 60°C for 1 min. The data are analyzed with ABI 7000 Sequence Detection System software.
3. The analytical sensitivity of each of the five real-time PCR sets is determined in triplicate with DNA isolated from *Fusobacterium* strain. Serial tenfold dilutions of homologous DNA are used as standard curves.

Note: The C^t parameter is defined as the fractional cycle number at which the fluorescence of the reporter dye generated by cleavage of the probe crosses an arbitrarily defined threshold within the logarithmic phase. This parameter can be used to compare different amplification reactions.

(ii) Real-Time PCR for Fusobacterium spp. Martin et al.[73] reported a real-time PCR for *Fusobacterium* spp. with primers and probe from a conserved region in the 16S rRNA gene (rDNA) sequence (Table 48.5). The probe is labeled with fluorescent dyes 6-carboxyfluorescein (FAM) at the 5′ end and 6-carboxytetramethylrhodamine (TAMRA) at the 3′ end.

Procedure

1. Bacteria are cultured to the late exponential phase, harvested by centrifugation (14,000 × *g* at 18 to 20°C for 2 min), washed, and resuspended in 10 mM phosphate buffer pH 6.7 containing 1 mg of lysozyme/mL, 1 mg of mutanolysin/mL, and 5 mM ZnCl². After incubation at 60°C for 30 min, DNA is

TABLE 48.4
Primers and Fluorogenic Probes for the Specific Detection of *Fusobacterium* spp.

Primer or Probe	Sequence (5′→3′)	Product (bp)
Forward	GGATTTATTGGGCGTAAAGC	162
Reverse	GGCATTCCTACAAATATCTACGAA	
Probe	FAM-CTCTACACTTGTAGTTCCG-TAMRA	

TABLE 48.5
Sequences of Oligonucleotide Primers and Probe for *Fusobacterium*

Primer or Probe[a]	Sequence (5′ → 3′)	T_m (°C)
Forward	AAGCGCGTCTAGGTGGTTATGT	58.8
Reverse	TGTAGTTCCGCTTACCTCTCCAG	58.6
Probe	FAM-CAACGCAATACAGAGTTGAGCCCTG CATT-TAMRA	69.9

[a] The *Fusobacterium*-specific primer and probe set also detects *F. periodonticum*, *F. alocis*, and *F. simiae* if present.

extracted and purified with a QIAamp DNA Mini Kit (QIAGEN). The DNA concentration (A^{260}) and purity (A^{260}/A^{280}) are measured.
2. To extract DNA from the anaerobic bacteria present in homogenized carious dentine, frozen suspensions are thawed on ice, and 80-μL samples are combined with 100 μL of ATL buffer (QIAGEN) and 400 μg of proteinase K (QIAGEN). The samples are vortexed for 10 s prior to incubation at 56°C for 40 min, with vortexing every 10 min to lyze the cells. Following the addition of 200 μg of RNase (Sigma), the samples are incubated for a further 10 min at 37°C before the DNA is purified with a QIAamp DNA Mini Kit.
3. Real-time PCR mixture (25-μL) is prepared with the TaqMan Universal PCR Master Mix containing 100 nM each forward primer, reverse primer, and probe and between 10 and 100 pg of template DNA/μL.
4. The real-time PCR amplification is conducted at 50°C for 2 min and 95°C for 10 min, followed by 40 cycles at 95°C for 15 s and 60°C for 1 min in an ABI PRISM 7700 Sequence Detection System (Applied Biosystems). Data analysis is performed with Sequence Detection software (version 1.6.3, Applied Biosystems).
5. The sensitivity of real-time PCR in detecting DNA is determined in duplicate with DNA extracted from *Fusobacterium* strain by using the appropriate homologous DNA as a standard.

(iii) Real-Time PCR for Fusobacterium spp. Suzuki et al.[55] described a quantitative real-time PCR targeting the 16S rRNA gene for detection of *Fusobacterium* spp. in subgingival plaque samples (Table 48.6).

Procedure

1. Real-time PCR mixture (20-μL) is composed of 1 μL of lyzed cells, 1× qPCR Master Mix (Eurogentec), sense and antisense primer at 200 nM each, and the TaqMan probe at 250 nM.
2. Amplification is conducted at 50°C for 2 min and 95°C for 10 min, followed by 60 cycles of 95°C for

TABLE 48.6
Primers and Probe for the Real-Time PCR Detection of *Fusobacterium* spp.

Primer or Probe	Sequence (5′→3′)	Product (bp)
Fs619F	CGCAGAAGGTGAAAGTCCTGTAT	101
Fs719R	TGGTCCTCACTGATTCACACAGA	
Probe Fs663T	FAM-ACTTTGCTCCCAAGTAACATGG AACACGAG-TAMRA	

15 s and 58°C for 1 min with the ABI PRISM 7700 Sequence Detection System (Applied Biosystems).
3. Standard curve is plotted with the *Ct* values obtained by amplifying successive tenfold dilutions of a known concentration of DNA extracted from bacterial cells containing 1.2×10^6 CFU (230 ng/μL) for *F. nucleatum* ATCC 10953.

(iv) Sequencing Analysis of Fusobacterium 16S rRNA Gene Cahill et al.[69] utilized the conserved DNA sequences flanking the regions of DNA variation in the 16S rRNA gene to design degenerate primers 16S2F (5′-CCTACGGRSGCAGCAG-3′, positions 299–320) and 16S2R (5′-GGACTACCMGGGNTATCTAATCCKG-3′, positions 741–764) for identification of *Fusobacterium* spp. from fetal membranes (amnion, chorion, and adherent decidua).

Procedure

1. Fetal membranes (amnion, chorion, and adherent decidua) are obtained within 30 min of delivery and placed in phosphate-buffered saline (PBS) supplemented with penicillin-streptomycin to limit further bacterial growth. Tissue samples (~200 mg) are frozen in liquid nitrogen and kept at −80°C until further study. DNA is extracted from the membrane tissue using the DNeasy Tissue Extraction Kit (Qiagen), resuspended in 200 μL of double distilled H^2O and stored at −20°C.
2. PCR mixture (50 μL) is composed of 1.5 mM $MgCl^2$, 50 mM KCl, 10 mM Tris-HCl pH 8.3, 200 μM of each deoxynucleotide triphosphate, 2 U of Amplitaq DNA polymerase (PE Applied Biosystems) and extracted DNA.
3. Amplification is performed in an ABgene® thermocycler with 95°C for 1 min, 35 cycles of 95°C for 30 s, 50°C for 30 s, 72°C for 1 min, and a final 72°C for 5 min. Negative control is included in each run with double distilled H^2O replacing DNA.
4. PCR products are resolved by agarose gel electrophoresis and the amplicon of required size is extracted using the Qiaex II gel extraction kit (Qiagen). The resulting DNA is ligated into the PCR®-TOPO® TA vector (Invitrogen) and transformed into One Shot® Top 10 chemically competent *E. coli* (Invitrogen).

Twelve clones per sample are sequenced. All sequencing results obtained are analyzed using the BLAST program and the resulting alignments analyzed.

Note: The integrity of the human DNA from these extractions may be confirmed by amplification of β-actin using specific primers Actin F and Actin R.[69] *F. nucleatum* is found to be the most commonly detected organism, being present in 9 of the 15 samples (60%) that contained bacterial DNA.[69]

48.3 CONCLUSION

Members of the genus *Fusobacterium* are gram-positive, nonspore-forming, nonmotile, anaerobic, rods that are present on the mucosal surfaces (e.g., mouth, upper respiratory tract, gastrointestinal tract, female genital tract) of humans and animals as commensal microflora.[21,74] Several *Fusobacterium* species/subspecies such as *F. necrophorum* and *F. nucleatum* have the capacity to cause various purulent or gangrenous infections in humans, particularly after accidental trauma, surgery, and edema.[75] While phenotypic identification of *Fusobacterium* spp. using cell and colony morphology as well as biochemical properties is possible, it is far too time consuming and complex to be of value for guiding the treatment and management of the infection.[21,76–78] The development of nucleic acid–amplification technologies such as PCR has helped streamline the identification and detection of *Fusobacterium* species/subspecies. Use of PCR assays has not only facilitated rapid, sensitive, and specific differentiation of *Fusobacterium* bacteria causing diseases in humans and animals, but also revealed insights in the molecular evolution and epidemiology of these organisms, leading to implementation of improved control and prevention measures against infections due to *Fusobacterium* spp.

REFERENCES

1. Dorsch, M., Love, D.N., and Bailey, G.D. *Fusobacterium equinum* sp. nov., from the oral cavity of horses, *Int. J. Syst. Evol. Microbiol.*, 51, 1959, 2001.
2. Batty, A., Wren, M.W., and Gal, M., *Fusobacterium necrophorum* as the cause of recurrent sore throat: Comparison of isolates from persistent sore throat syndrome and Lemierre's disease, *J. Infect.*, 51, 299, 2004.
3. Nagaraja, T.G. et al., *Fusobacterium necrophorum* infections in animals: Pathogenesis and pathogenic mechanisms, *Anaerobe*, 11, 239, 2005.
4. Bennett, G. et al., Detection of *Fusobacterium necrophorum* and *Dichelobacter nodosus* in lame cattle on dairy farms in New Zealand, *Res. Vet. Sci.*, 87, 413, 2009.
5. Altshuler, G., and Hyde, S., Clinicopathologic considerations of fusobacteria chorioamnionitis, *Acta Obstet. Gynecol. Scand.*, 67, 513, 1988.
6. Albandar, J.M., Brown, L.J., and Loe, H., Putative periodontal pathogens in subgingival plaque of young adults with and without early-onset periodontitis, *J. Periodontol.*, 68, 973, 1997.
7. Bourgault, A.-M., et al., *Fusobacterium* bacteremia: Clinical experience with 40 cases, *Clin. Infect. Dis.*, 25, S181, 1997.
8. Brazier, J.S., *Fusobacterium necrophorum* infections in man, *Rev. Med. Microbiol.*, 13, 141, 2002.
9. Berger, A. et al., Microbial invasion of the amniotic cavity at birth is associated with adverse short-term outcome of preterm infants, *J. Perinat. Med.*, 31, 115, 2003.
10. Carta, G. et al., Periodontal disease and poor obstetrical outcome, *Clin. Exp. Obstet. Gynecol.*, 31, 47, 2004.
11. Gardella, C. et al., Identification and sequencing of bacterial rDNAs in culture-negative amniotic fluid from women in premature labor, *Am. J. Perinatol.*, 21, 319, 2004.
12. Goepfert, A.R. et al., Periodontal disease and upper genital tract inflammation in early spontaneous preterm birth, *Obstet. Gynecol.*, 104, 777, 2004.
13. Shinjo, T., Fujisawa, T., and Mitsuoka, T., Proposal of two subspecies of *Fusobacterium necrophorum* (Flügge) Moore and Holdeman: *Fusobacterium necrophorum* subsp. *necrophorum* subsp. nov., nom. rev. (ex Flügge 1886), and *Fusobacterium necrophorum* subsp. *funduliforme* subsp. nov., nom. rev. (ex Halle 1898), *Int. J. Syst. Bacteriol.*, 41, 395, 1991.
14. Dzink, J.L., Sheenan, M.T., and Socransky, S.S., Proposal of three subspecies of *Fusobacterium nucleatum* Knorr 1922: *Fusobacterium nucleatum* subsp. *nucleatum* subsp. nov., comb. nov.; *Fusobacterium nucleatum* subsp. *polymorphum* subsp. nov., nom. rev., comb. nov.; and *Fusobacterium nucleatum* subsp. *vincentii* subsp. nov., nom. rev., comb. nov., *Int. J. Syst. Bacteriol.*, 40, 74, 1990.
15. Gharbia, S.E., and Shah, H.N., *Fusobacterium nucleatum* subsp. fusiforme subsp. nov. and *Fusobacterium nucleatum* subsp. animalis subsp. nov. as additional subspecies within *Fusobacterium nucleatum*, *Int. J. Syst. Bacteriol.*, 42, 296, 1992.
16. Kapatral, V. et al., Genome sequence and analysis of the oral bacterium *Fusobacterium nucleatum* strain ATCC 25586, *J. Bacteriol.*, 184, 2005, 2002.
17. Karpathy, S.E. et al., Genome sequence of *Fusobacterium nucleatum* subspecies *polymorphum*—a genetically tractable fusobacterium, *PLoS One*, 2, e659, 2007.
18. Periasamy, S., and Kolenbrander, P.E., *Aggregatibacter actinomycetemcomitans* builds mutualistic biofilm communities with *Fusobacterium nucleatum* and *Veillonella* species in saliva, *Infect. Immun.*, 77, 3542, 2009.
19. Roberts, G.L., Fusobacterial infections: An underestimated threat, *Br. J. Biomed. Sci.*, 57, 156, 2000.
20. Ludlam, H. et al., Epidemiology of pharyngeal carriage of *Fusobacterium necrophorum*, *J. Med. Microbiol.*, 58, 1264, 2009.
21. Riordan, T., Human infection with *Fusobacterium necrophorum* (necrobacillosis), with a focus on Lemierre's syndrome, *Clin. Microbiol. Rev.*, 20, 622, 2007.
22. Hagelskjaer, L.H. et al., Incidence and clinical epidemiology of necrobacillosis, including Lemierre's syndrome, in Denmark 1990–1995, *Eur. J. Clin. Microbiol. Infect. Dis.*, 17, 561, 1998.
23. Hagelskjaer Kristensen, L., and Prag, J., Human necrobacillosis, with emphasis on Lemierre's syndrome, *Clin. Infect. Dis.*, 31, 524, 2000.
24. Karkos, P.D. et al., Lemierre's syndrome: A systematic review, *Laryngoscope*, 119, 1552, 2009.
25. Narayanan, S.K. et al., Cloning, sequencing, and expression of the leukotoxin gene from *Fusobacterium necrophorum*, *Infect. Immun.*, 69, 5447, 2001.

26. Narayanan, S.K. et al., Leukotoxins of gram-negative bacteria, *Vet. Microbiol.*, 84, 337, 2002.
27. Narayanan, S.K. et al., *Fusobacterium necrophorum* leukotoxin induces activation and apoptosis of bovine leukocytes, *Infect. Immun.*, 70, 4609, 2002.
28. Tan, Z.L., Nagaraja, T.G., and Chengappa, M.M., Factors affecting the leukotoxin activity of *Fusobacterium necrophorum*, *Vet. Microbiol.*, 32, 15, 1992.
29. Tan, Z.L., Nagaraja, T.G., and Chengappa, M.M., *Fusobacterium necrophorum* infections: Virulence factors, pathogenic mechanism and control measures, *Vet. Res. Commun.*, 20, 113, 1996.
30. Ludlam, H.A. et al., lktA-encoded leukotoxin is not a universal virulence factor in invasive *Fusobacterium necrophorum* infections in animals and man, *J. Med. Microbiol.*, 58, 529, 2009.
31. Oelke, A.M. et al., The leukotoxin operon of *Fusobacterium necrophorum* is not present in other species of *Fusobacterium*, *Anaerobe*, 11, 123, 2005.
32. Zhang, F. et al., The two major subspecies of *Fusobacterium necrophorum* have distinct leukotoxin operon promoter regions, *Vet. Microbiol.*, 112, 73, 2006.
33. Tadepalli, S. et al., Human *Fusobacterium necrophorum* strains have a leukotoxin gene and exhibit leukotoxic activity, *J. Med. Microbiol.*, 57, 225, 2008.
34. Feuille, F.L. et al., Synergistic tissue destruction induced by *Porphyromonas gingivalis* and *Fusobacterium nucleatum*, *J. Dent. Res.*, 73, 159, 1994.
35. Bradshaw, D.J. et al., Role of *Fusobacterium nucleatum* and coaggregation in anaerobe survival in planktonic and biofilm oral microbial communities during aeration, *Infect. Immun.*, 66, 4729, 1998.
36. Han, Y.W. et al., Interactions between periodontal bacteria and human oral epithelial cells: *Fusobacterium nucleatum* adheres to and invades epithelial cells, *Infect. Immun.*, 68, 3140, 2000.
37. Han, Y.W. et al., *Fusobacterium nucleatum* induces premature and term stillbirths in pregnant mice: Implication of oral bacteria in preterm birth, *Infect. Immun.*, 72, 2272, 2004.
38. Kulekci, G. et al., PCR analysis of *Actinobacillus actinomycetemcomitans*, *Porphyromonas gingivalis*, *Treponema denticola* and *Fusobacterium nucleatum* in middle ear effusion, *Anaerobe*, 7, 241, 2001.
39. Moraes, S.R. et al., Clonality of *Fusobacterium nucleatum* in root canal infections, *Oral Microbiol. Immunol.*, 17, 394, 2002.
40. Nyfors, S. et al., Emergence of penicillin resistance among *Fusobacterium nucleatum* populations of commensal oral flora during early childhood, *J. Antimicrob. Chemother.*, 51, 107, 2003.
41. Haraldsson, G., Holbrook, W.P., and Könönen, E., Clonal similarity of salivary and nasopharyngeal *Fusobacterium nucleatum* in infants with acute otitis media experience, *J. Med. Microbiol.*, 53, 161, 2004.
42. Haraldsson, G., Holbrook, W.P., and Könönen, E., Clonal persistence of oral *Fusobacterium nucleatum* in infancy, *J. Dent. Res.*, 83, 500, 2004.
43. Thurnheer, T. et al., Infinite serovar and ribotype heterogeneity among oral *Fusobacterium nucleatum* strains? *Anaerobe*, 5, 79, 1999.
44. Diaz, P.I., Zilm, P.S., and Rogers, A.H., *Fusobacterium nucleatum* supports the growth of *Porphyromonas gingivalis* in oxygenated and carbon-dioxide-depleted environments, *Microbiology*, 148, 467, 2002.
45. Tew, J.G., Thomas, S.S., and Ranney, R.R., *Fusobacterium nucleatum*-mediated immunomodulation of the in vitro secondary antibody response to tetanus toxoid and *Actinobacillus actinomycetemcomitans*, *J. Periodont. Res.*, 22, 506, 1987.
46. Gaetti-Jardim Junior, E., and Avila-Campos, M.J., Haemagglutination and haemolysis by oral *Fusobacterium nucleatum*, *New Microbiol.*, 22, 63, 1999.
47. Jensen, A., Hagelskjaer Kristensen, L., and Prag, J., Detection of *Fusobacterium necrophorum* subsp. *funduliforme* in tonsillitis in young adults by real-time PCR, *Clin. Microbiol. Infect.*, 13, 695, 2007.
48. Hall, V. et al., A comparative study of *Fusobacterium necrophorum* strains from human and animal sources by phenotypic reactions, pyrolysis mass spectrometry and SDS-PAGE, *J. Med. Microbiol.*, 46, 865, 1997.
49. Brown, R., Lough, H.G., and Poxton, I.R., Phenotypic characteristics and lipopolysaccharides of human and animal isolates of *Fusobacterium necrophorum*, *J. Med. Microbiol.*, 46, 873, 1997.
50. Ashimoto, A. et al., Polymerase chain reaction detection of 8 putative periodontal pathogens in subgingival plaque of gingivitis and advanced periodontitis lesions, *Oral Microbiol. Immunol.*, 11, 266, 1996.
51. George, K.S., Reynolds, M.A., and Falkler, W.A. Jr., Arbitrarily primed polymerase chain reaction fingerprinting and clonal analysis of oral *Fusobacterium nucleatum* isolates, *Oral Microbiol. Immunol.*, 12, 219, 1997.
52. Suchett-Kaye, G., Décoret, D., and Barsotti, O., Clonal analysis by ribotyping of *Fusobacterium nucleatum* isolates obtained from healthy young adults with optimal plaque control, *J. Periodont. Res.*, 33, 179, 1998.
53. Jin, J. et al., Characterization of the 16S–23S rRNA intergenic spacer regions among strains of the *Fusobacterium necrophorum* cluster, *J. Vet. Med. Sci.*, 64, 273, 2002.
54. Narongwanichgarn, W. et al., Specific detection and differentiation of two subspecies of *Fusobacterium necrophorum* by PCR, *Vet. Microbiol.*, 91, 183, 2003.
55. Suzuki, N. et al., Quantitative microbiological study of subgingival plaque by real-time PCR shows correlation between levels of *Tannerella forsythensis* and *Fusobacterium* spp., *J. Clin. Microbiol.*, 42, 2255, 2004.
56. Jervøe-Storm, P.M. et al., Comparison of culture and real-time PCR for detection and quantification of five putative periodontopathogenic bacteria in subgingival plaque samples, *J. Clin. Periodontol.*, 32, 778, 2005.
57. Kim, H.S. et al., Development of strain-specific PCR primers based on a DNA probe Fu12 for the identification of *Fusobacterium nucleatum* subsp. *nucleatum* ATCC 25586T, *J. Microbiol.*, 43, 331, 2005.
58. Cogulu, D. et al., PCR-based identification of selected pathogens associated with endodontic infections in deciduous and permanent teeth, *Oral Surg. Oral Med. Oral Pathol. Oral Radiol. Endod.*, 106, 443, 2008.
59. Rôças, I.N., and Siqueira, J.F. Jr., Identification of bacteria enduring endodontic treatment procedures by a combined reverse transcriptase–polymerase chain reaction and reverse-capture checkerboard approach, *J. Endodontics*, 36, 45, 2010.
60. Greisen, K. et al., PCR primers and probes for the 16S rRNA gene of most species of pathogenic bacteria, including bacteria found in cerebrospinal fluid, *J. Clin. Microbiol.*, 32, 335, 1994.
61. Jalava, J. et al., Bacterial 16S rDNA polymerase chain reaction in the detection of intra-amniotic infection, *Br. J. Obstet. Gynaecol.*, 103,664, 1996.

62. Hitti, J. et al., Broad-spectrum bacterial rDNA polymerase chain reaction assay for detecting amniotic fluid infection among women in premature labor, *Clin. Infect. Dis.*, 24, 1228, 1997.
63. Jin, J. et al., Comparison of the 16S-23S rRNA intergenic spacer regions between *Fusobacterium varium* and "*Fusobacterium pseudonecrophorum*" strains, *J. Vet. Med. Sci.*, 64, 285, 2002.
64. Aliyu, S.H. et al., Real-time PCR investigation into the importance of *Fusobacterium necrophorum* as a cause of acute pharyngitis in general practice, *J. Med. Microbiol.*, 53, 1029, 2004.
65. Jin, J. et al., Phylogenetic analysis of *Fusobacterium necrophorum*, *Fusobacterium varium* and *Fusobacterium nucleatum* based on *gyrB* gene sequences, *J. Vet. Med. Sci.*, 66, 1243, 2004.
66. Avila-Campos, M.J. et al., Arbitrarily primed-polymerase chain reaction for identification and epidemiologic subtyping of oral isolates of *Fusobacterium nucleatum*, *J. Periodontol.*, 70, 1202, 1999.
67. Narongwanichgarn, W. et al., Differentiation of *Fusobacterium necrophorum* subspecies from bovine pathological lesions by RAPD-PCR, *Vet. Microbiol.*, 82, 383, 2001.
68. Rinttila, T. et al., Development of an extensive set of 16S rDNA-targeted primers for quantification of pathogenic and indigenous bacteria in faecal samples by real-time PCR, *J. Appl. Microbiol.*, 97, 1166, 2004.
69. Cahill, R.J. et al., Universal DNA primers amplify bacterial DNA from human fetal membranes and link *Fusobacterium nucleatum* with prolonged preterm membrane rupture, *Mol. Hum. Reprod.*, 11, 761, 2005.
70. D'Ercole, S. et al., Comparison of culture methods and multiplex PCR for the detection of periodontopathogenic bacteria in biofilm associated with severe forms of periodontitis, *New Microbiol.*, 31, 383, 2008.
71. Shin, H.S. et al., Development of strain-specific PCR primers for the identification of *Fusobacterium nucleatum* subsp. *fusiforme* ATCC 51190(T) and subsp. *vincentii* ATCC 49256(T), *Anaerobe*, 16, 43, 2010.
72. Boutaga, K. et al., Periodontal pathogens: A quantitative comparison of anaerobic culture and real-time PCR, *FEMS Immunol. Med. Microbiol.*, 45, 191, 2005.
73. Martin, F.E. et al., Quantitative microbiological study of human carious dentine by culture and real-time PCR: Association of anaerobes with histopathological changes in chronic pulpitis, *J. Clin. Microbiol.*, 40, 1698, 2002.
74. Batty, A., and Wren, M.W., Prevalence of *Fusobacterium necrophorum* and other upper respiratory tract pathogens isolated from throat swabs, *Br. J. Biomed. Sci.*, 62, 66, 2005.
75. Smith, G.R., and Thornton, E.A., Pathogenicity of *Fusobacterium necrophorum* strains from man and animals, *Epidemiol. Infect.*, 110, 499, 1993.
76. Hill, G.B., Preterm birth: Associations with genital and possibly oral microflora, *Ann. Periodontol.*, 3, 222, 1998.
77. Tan, Z.L. et al., Biological and biochemical characterization of *Fusobacterium necrophorum* leukotoxin, *Am. J. Vet. Res.*, 55, 515, 1994.
78. Narayanan, S. et al., Ribotyping to compare *Fusobacterium necrophorum* isolates from bovine liver abscesses, ruminal walls, and ruminal contents, *Appl. Environ. Microbiol.*, 63, 4671, 1997.

49 *Leptotrichia* and *Leptotrichia*-Like Organisms

Emenike Ribs K. Eribe and Ingar Olsen

CONTENTS

49.1 INTRODUCTION

49.1.1 Classification and Morphology

Classification. Trevisan[1] established the genus *Leptotrichia* in 1879 to distinguish filamentous microorganisms in the oral cavity from free-living filamentous *Leptothrix* species. The major metabolic end product of *Leptotrichia*,[2] lactic acid, separates it from other filamentous bacteria, *Fusobacterium* and *Bacteroides*.[3]

In the draft taxonomic outline[4] based on comparative analysis of 16S rRNA sequences, *Leptotrichia* is proposed as the first genus in the family Leptotrichiaceae, phylum Fusobacteria. Other proposed members of this family are the genera *Sebaldella*, *Streptobacillus*, and *Sneathia*.[4]

Molecular microbiological studies have revealed a wide biodiversity in *Leptotrichia*.[5,6] Most of these organisms cannot be cultured. A large variety of new 16S rRNA phylotypes has been detected in *Leptotrichia* from humans[5,7,8] and animals[9,10] as well as from environmental samples (http://www.meg.boun.edu.tr/MKC.pdf). The genus *Leptotrichia* contains six validated species: *L. buccalis*, which is the type species; *L. goodfellowii*, *L. hofstadii*, *L. shahii*, *L. trevisanii*,[3,11,12] and *L. wadei*.[5,11] "*L. amnionii*"[13] has not been formally validated but should probably be transferred to the genus *Sneathia*.[5]

The G + C content of DNA in *Leptotrichia* is 25 mol%[14] or 24–29.7 mol%.[12,15] DNA–DNA hybridization showed more than 70% homology between all species of *Leptotrichia*. With *L. buccalis*, *L. goodfellowii* showed 3.8%–5.5% DNA–DNA relatedness; *L. shahii* showed 24.5%–34.1% relatedness; *L. hofstadii* exhibited 27.3%–36.3% relatedness; and *L. wadei* 24.1%–35.9% relatedness, respectively.[5]

Leptotrichia is generally considered negative with respect to catalase, oxidase, and indole production. Some species are catalase-positive[5] (Table 49.1). *L. buccalis* hydrolyses aesculin.[2,5,16] Classification of *Leptotrichia* involves not only genetic and biochemical analyses but also fatty acid and protein profiles (see later).

Morphology. *Leptotrichia* are straight or curved rods, 0.8–1.5 μm wide and 5–15 μm long, nonsporing and nonmotile[5,15] with one or both ends pointed or rounded. Some features of *Leptotrichia* species are also shown in Table 49.1. Cellular morphology of the novel species *L. goodfellowii*, *L. hofstadii*, *L. shahii* and *L. wadei* was studied by examining cells grown anaerobically at 37°C on Columbia or brain-heart infusion (BHI) blood agar plates.[5] Light microscopy (LM), transmission electron microscopy (TEM) and scanning electron microscopy (SEM) were used for characterization of cell morphology and ultrastructure. *Leptotrichia* species had several common features. Cells are arranged in pairs, some slightly curved, others in chains joined by flattened ends. Electron micrographs of ultrathin sections revealed elongated cells with a cell-wall structure typical of gram-negative bacteria. *Leptotrichia* possess a double plasma membrane layer, a single electron-dense (intermediate) layer and a double outer layer with scale-like protrusions on the surface referred to as "fish scales" (Figure 49.1).[5,17–19] Fimbriae are not present. Septum is formed from the cytoplasmic membrane and the intermediate layer.[5,18] The outer

TABLE 49.1
Characteristics of *Leptotrichia* and *Leptotrichia*-Like Bacteria

Character	*L. buccalis*	*L. goodfellowii*	*L. hofstadii*	*L. shahii*	*L. wadei*	*L. trevisanii*	*F. nucleatum*	*Capnocytophaga*
Cellular morphology	Gram-negative[a]	Gram-negative	Gram-negative	Gram-negative	Gram-negative	Gram-negative	Gram-negative	Gram-negative
Cells on blood[b] agar	5–15 μm long, thick fusiform rods	2–4 μm short rods	7.5–15 μm long rods	7.5–17.5 μm long rods	2.5–10 μm short rods	6–13 μm long fusiform rod	3–10 μm long, slender fusiform rods	3–6 μm long, slender fusiform rods
Colony morphology on blood agar	Smooth, rhizoid, or convoluted	Speckled, convex, irregular pink periphery, grayish light brown glistening surface, opaque, dry	Glistening, granular surface, opaque, dry and circular, convex, entire (some irregular and lobate), grayish with a dark central spot	Very filamentous to rhizoid or convoluted, pale speckled, grayish with a dark central spot in old colonies, opaque, semidry in consistency	Convex, sparsely filamentous to irregular, old colonies grayish brown with a dark central spot, glistening, smooth with a rough edge, opaque, dry in consistency	Smooth, grayish, low convex, erose edged, dark central spot with transillumination, slightly convoluted	Smooth, convex, "flecked"	Flat, rough, spreading margin,[c] adherent or low-convex, smooth, pigmented
Growth in air and 5%–10% CO_2 on blood agar	+	+	+	+	+	+	–	+[c]
Indole	+	–	–	–	–	–	+	–
Catalase	–	+	+	+	+	+	+	–
Aesculin	+	+	+	+	+	UK[d]	–	+
Hemolysis on human blood agar	Nonhemolytic	β-hemolytic	β-hemolytic	Nonhemolytic	β-hemolytic	UK	UK	UK
Main fatty acids from glucose-peptone	Lactic	Lactic	Lactic	Lactic	Lactic	Lactic	Butyric	Acetic, succinic

[a] Occasionally gram-positive in young cultures.
[b] Human blood.
[c] Gliding motility.
[d] Unknown.

membrane ridge is firmly attached to the peptidoglycan sacculus, which may be the point of origin of the structure. When analyzed by SDS-PAGE, the macromolecules forming the ridge revealed a major component of 210-kDa polypeptide and a 15-kDa minor component. It has been suggested that the 210-kDa molecule is likely an adhesion component, but this is not proven.[18] A cross-section of the cell showed storage granules.

Leptotrichia are usually anaerobic on first isolation with some strains growing aerobically in presence of CO_2. The colonies on blood agar plates tend to be smooth, rhizoid, or convoluted and can also be sparsely filamentous to irregular and grayish brown in color, with a dark central spot in old colonies.[3] Occasionally, colonies are opaque and dry in consistency.[5]

49.1.2 Biology and Epidemiology

Leptotrichia are part of the normal flora in the oral cavity,[20] intestine,[15,21–29] and human female genitalia.[22,26,30] Prior to tooth eruption, 17% of children harbor *Leptotrichia*, and it continues to be present in up to 71% after eruption.[20,31] *Leptotrichia* can infect humans and animals. It has been reported to participate in oral diseases and diseases elsewhere in the body,[3] for example, in the intestine[15,21–29,32] and female genitalia[22,26,30] of healthy persons. *Leptotrichia* has also been recovered from patients with gingivitis, necrotizing ulcerative gingivitis, adult/juvenile periodontitis, "refractory" periodontitis,[33] noma,[3,7,22,34–38] aortic aneurysms,[8] cellulitis,[39] phagedenic chancroid,[40] salpingitis,[41] acute appendicitis,[42] and bacterial vaginosis.[43] Children with gingivitis harbor three

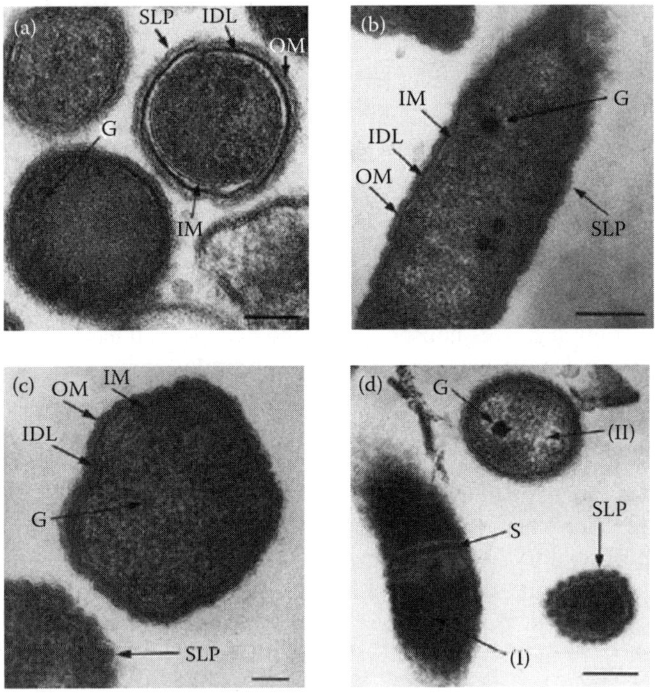

FIGURE 49.1 TEM (a, b, c, and d) of cells of novel *Leptotrichia* species after anaerobic culture on Columbia blood agar at 37°C for 48 h. (a), (b), (c), and (d) Cross-sectional views of cells of *L. goodfellowii* LB 57^T (a), *L. hofstadii* LB 23^T (b), *L. shahii* LB 37^T (c), and *L. wadei* LB 16^T (d). In (d) cells are visible in both a longitudinal (I) and a cross-sectional (II) view. G, Granule; IDL, intermediate dense layer; IM, inner cytoplasmic membrane; OM, outer membrane; S, septum; SLP, scale-like protrusion; CM, cytoplasmic membrane. Bars, 0.1 (a, c) and 0.25 (b, d) μm. (Modified from Eribe et al., *Int. J. Syst. Evol. Microbiol.*, 54, 583, 2004. With permission from the Society for Microbiology.)

times as many *Leptotrichia* species in subgingival plaque as adults with gingivitis.[44] *Leptotrichia* has also been reported to colonize permucosal implants of edentulous patients.[45] *L. buccalis* is highly saccharolytic, producing lactic acid, a property that may implicate its participation in tooth decay.[46–49] The organism has also been recovered from peritoneal fluid[50] and from blood of immunocompromised patients with neutropenia, HIV, leukemia, and endocarditis.[34,36,39,46,50–60] Furthermore, *Leptotrichia* has been isolated from infected material of domestic animals and after dog bites.[9,10]

Most *Leptotrichia* species have been isolated from blood and characterized as *L. buccalis*, *L. trevisanii*, *L. wadei*, and *L. goodfellowii*, whereas organisms such as *L. hofstadii*, *L. buccalis*, and *L. wadei* have been recovered from saliva. Human *Leptotrichia* infections have been reported to occur mostly in immunocompromised patients with predisposing factors, but there have also been incidences in healthy humans. There is no report of disease transmission between humans.

Leptotrichia buccalis. *L. buccalis* is the type species of the genus *Leptotrichia*. The type strain is ATCC 14201^T=NCTC 10249^T. It is a common member of the oral microflora in humans and has been isolated from dental plaque and saliva.[20,61] It is generally considered catalase-, oxidase- and indole-negative, and it produces lactate, but catalase production can occur. *L. buccalis* hydrolyses aesculin.[2,5,16] The DNA base content is 25 mol% G + C.[14] *L. buccalis* has previously been called *Leptothrix buccalis*,[62] *Fusobacterium plautivincentii* or *Fusobacterium fusiforme*.[21] *Leptotrichia* was first cultivated by Wherry and Oliver,[63] who named it *Leptotrichia innominata*. Thjøtta et al.[64] and Böe and Thjötta[65] gave the first adequate descriptions of *L. buccalis* based on the fact that it had many features in common with *Fusobacterium* and classified it as a gram-negative anaerobic bacterium. Hofstad and Selvig[17] used TEM to study its cell wall and concluded that *L. buccalis* belongs to gram-negative bacteria.

Leptotrichia goodfellowii. The type strain of *L. goodfellowii* is CCUG 32286^T=CIP 107915^T [5] and was originally isolated from the blood of an endocarditis patients.[56] It is a gram-negative, nonspore-forming, nonmotile short rod, varying in length, with one end tapered. Cells are arranged in pairs, some slightly curved; others, in chains joined by flattened ends. Colonies grow best anaerobically and sparsely aerobically at 37°C but not at 25°C and 42°C. Colonies are speckled, convex, irregular, pink in the periphery, and grayish light brown in color. They have a glistening surface appearance, are opaque, dry, and show β-hemolysis. *L. goodfellowii* is aesculin- and catalase-positive. Oxidase and indole are not produced. Arginine dihydrolase, β-galactosidase, β-glucosidase, β-*N*-acetyl glucosaminidase, alkaline phosphatase, arginine arylamidase, leucine arylamidase, and histidine arylamidase are produced.

Leptotrichia hofstadii. The type strain of *L. hofstadii* is CCUG 47504^T=CIP 107917^T,[5] and it was originally isolated from the saliva of a healthy person. It is a gram-negative, nonspore-forming, nonmotile, long rod, varying in length, with one end tapered. Cells are arranged in pairs, some slightly curved; others, in chains joined by flattened ends. Colonies grow best anaerobically and sparsely aerobically at 37°C but not at 25°C and 42°C. Colonies are circular, convex, entire (some are irregular and lobate), and grayish in color with a dark central spot in old colonies. They have a glistening and granular surface appearance, are opaque, dry, and nonhemolytic. *L. hofstadii* is aesculin- and catalase-positive. Oxidase and indole are not produced. β-Galactosidase-6-phosphate, α-glucosidase, β-glucosidase, and alkaline phosphatase are generated. Mannose is fermented.

Leptotrichia shahii. *L. shahii* has as type strain CCUG 47503^T=CIP 107916^T [5] and was originally isolated from a patient with gingivitis. It is a gram-negative, nonspore-forming, nonmotile long rod, varying in length, with one end tapered. Cells are arranged in pairs, some slightly curved; others in chains joined by flattened ends. Colonies grow best anaerobically and sparsely aerobically at 25°C and 37°C but not at 42°C. Colonies are very filamentous to rhizoid or convoluted, pale speckled, and grayish in color with a dark central spot in old colonies. They are opaque, semidry in consistency and nonhemolytic. *L. shahii* is aesculin- and catalase-positive. Oxidase and indole are not produced. α-Glucosidase and α-arabinosidase are generated.

Leptotrichia wadei. The type strain of *L. wadei* is CCUG 47505^T=CIP 107918^T [5] and was originally isolated from the saliva of a healthy person. It is a gram-negative, nonspore-forming, nonmotile, short rod, varying in length, with one end tapered. Cells, some slightly curved, are arranged in pairs; others, in chains joined by flattened ends. *L. wadei* grows sparsely aerobically at 37°C but not at 25°C and 42°C. Colonies are convex, sparsely filamentous to irregular, and grayish brown in color, with a dark central spot in old colonies. The surface appearance is glistening and smooth with a rough edge. Colonies are opaque, dry in consistency, and β-hemolytic. The organism is aesculin- and catalase-positive. Oxidase and indole are not produced. However, α-glucosidase and β-glucosidase are generated.

Leptotrichia trevisanii. *L. trevisanii* is a thin, filamentous, nonmotile, aerotolerant, gram-negative bacterium isolated from the blood of a man with acute myeloid leukemia.[12] The type strain of *L. trevisanii* is ATCC 700907^T [12]=DSM 22070^T. The deposition in a second culture collection in a different country (DSM) should now make *L. trevisanii* a validly published organism.[3,11] The organism is catalase-positive but indole-, oxidase-, and urease-negative. It is saccharolytic and produces lactate. The DNA base composition is 29.7 mol% G + C. The *16S rRNA* gene sequences of *L. trevisanii* showed 96% identity with those of the type strain of *L. buccalis*,[12] and 97% sequence identity with those of "*Leptotrichia*" species.[53]

"*Leptotrichia amnionii.*" "*L. amnionii*"[13] has pleomorphic coccobacillary, long, nonmotile, fusiform cells. Some cells are joined end to end in a filamentous form. This bacterium was isolated from the amniotic fluid of a pregnant woman with intrauterine fetal demise. "*L. amnionii*" grows anaerobically on blood agar. The isolate was characterized only by its *16S rRNA* gene sequences, which showed 96% relationship to *Sneathia sanguinegens*. "*L. amnionii*" is not yet validly published or recognized in IJSEM. Therefore, the taxonomic status of this species is uncertain, but it should probably be transferred to the genus *Sneathia*.[5]

49.1.3 Clinical Features and Pathogenesis

Clinical Features. *Leptotrichia* has been isolated from a variety of clinical lesions in man. The oral lesions comprise gingival hyperplasia,[43] gingivitis, necrotizing ulcerative gingivitis,[66] adult/juvenile periodontitis, "refractory" periodontitis,[33] noma,[3,7] root caries,[49] infected root canals,[67–69] and malodor.[70] It has also been found on permucosal implants of edentulous patients.[42] Besides the mouth, other major sites of primary *Leptotrichia* colonization include the genitourinary tract, where these organisms have been found in bacterial vaginosis,[71–75] abortion,[76,77] genitourinary tract infection,[78] salpingitis,[41] preterm labor,[79] preterm birth,[80] and amniotic fluid infection.[13]

Leptotrichia can also spread from primary sites of colonization, causing acute appendicitis,[42] cellulitis,[39] arthritis,[81] phagedenic chancroid,[40] and Lemierre syndrome.[82] It can further disseminate to the bloodstream of bone marrow transplants[52] and immunocompromised patients with and without neutropenia.[36,39,46,52,53,55–59,82–84] *Leptotrichia* has also been isolated from delivery and postpartum bacteremia.[84] Further, the organisms have been recovered from peritonitis[50] and from a hepatic abscess[54] and blood cultures from immunocompetent patients.[54,85] It has also been isolated from intestinal obstruction,[50] aortic valve disease,[34,40,58] aortic aneurysms,[8] endocarditis,[8,34,51,56,59,83] and esoinophilic pustular folliculitis, called "Ofuji's disease" (www.emedicine.com/derm/byname/Eosinophilic-Pustular-Folliculitis.htm). Furthermore, *Leptotrichia* has been detected in dog-bite lesions.[9,10]

Pathogenesis. *Leptotrichia* has protruding structures on the cell surface[18] (Figure 49.1). These ridges are probably adhesion structures.[19] Buccal epithelial cells, red blood cells, HeLa, and embryonic kidney cells bound to the hemagglutination fragments of *L. buccalis*, while saliva displayed hemagglutination-inhibition activity in a manner suggesting a binding-site interaction.[86] The receptor sites on human erythrocytes for *L. buccalis* may be sugar in nature.[87] These authors suggested that a salivary aggregating component may possess the same specific group with which *L. buccalis* reacts on human erythrocytes, and that the enhanced adherence of *L. buccalis* cells to saliva-coated enamel powder is due to a greater affinity of the organism for this component that adsorbs to human enamel powder. As a gram-negative bacterium *Leptotrichia* possesses a potent lipopolysaccharide (LPS) (endotoxin) that displays an O-antigen linked to lipid A. The latter may cause hemorrhage, fever, tumor necrosis, fatal shock, septicemia, and even abortion. These features have been associated with cases of *Leptotrichia* infections. Endotoxin obtained from *L. buccalis* had a potent biological activity[88] in New Zealand white female rabbits. The parameters studied were LD_{50}, febrile, and leukocytic responses, and the dermal Shwartzman reaction. The LD_{50} of *L. buccalis* was 93.15 µg/kg with 95% confidence limits of 66.44

and 131.00 μg/kg. Formal probit analysis revealed that the *L. buccalis* endotoxin, compared to that of *Escherichia coli*, had a relative potency of 19.45%, with 95% confidence limits of 13.50% and 28.01%. *L. buccalis* endotoxin was pyrogenic when given intravenously in a quantity of 0.039 μg/kg, whereas *E. coli* endotoxin was pyrogenic at 0.009 μg/kg. Parallel line bioassay methods applied to the febrile dose-response relationship of the toxins showed that, compared to *E. coli*, *L. buccalis* had a potency of 11.47% with 95% confidence limits of 9.69% and 13.63%. The quantity of endotoxin necessary to elicit a fever index of 40 cm^2 was estimated as 0.285 and 2.099 μg/kg for *E. coli* and *L. buccalis* endotoxins, respectively. A significant correlation was found between febrile response and fever index.

LPS isolated by phenol-water extraction from *L. buccalis* were endotoxic by their ability to alter dermal reactivity to epinephrine, and serologically specific by hemagglutination and hemagglutination-inhibition tests.[89] Furthermore, *L. buccalis* exerted a blastogenic response of lymphocytes in periodontitis patients, who were significantly more reactive than in dentulous and normal subjects.[90] This suggested that *L. buccalis* and other gram-negative bacteria tested, and the host response they evoke, are associated with advanced periodontal destruction. Systemic release of endotoxin and proinflammatory mediators from infected host tissue may contribute directly or indirectly to the sepsis syndrome associated with *Leptotrichia*. *L. buccalis* is highly saccharolytic, producing lactic acid, a property that may implicate participation in tooth decay.[47–49,91]

There is evidence supporting that *Leptotrichia* species are emerging human pathogens.[76] Opportunistic pathogens tend to cause disease when local or general predisposing factors are present. This also relates to *Leptotrichia*. Localized *Leptotrichia* infections are inclined to occur when immunological defense mechanisms are compromised, as has been seen in cases with neutropenia,[39,46,52,53,55–59] cancer,[55,92] diabetes,[81,93] pneumonia,[94] peritoneal dialysis,[50] and Lemierre syndrome.[82] *Leptotrichia* infections may be serious in hospitalized patients, particularly those with cancer, as observed with *L. shahii* and *L. wadei*,[92] and with bone marrow transplantation.[46,52,55]

Leptotrichia is apt to cause disease in the immunocompetent host.[54] Localized *Leptotrichia* infections can lead to generalized infections and, occasionally, to fatal bacteremia and endocardial arrest, as reported with *L. goodfellowii*.[51,56]

A recent study showed that *L. buccalis* can occur together with other fusiform bacteria, such as fusobacteria, *Tannerella forsythia*, and *Capnocytophaga* species, in gingivitis and acute necrotizing ulcerative gingivitis, with high prevalence.[66] In other cases of multiple organism infection, *Leptotrichia* species were recovered together with *Bacteroides*,[42] *Mycobacterium bovis*, *Candida albicans*, and cytomegalovirus and herpes virus.[50] Superinfection with the anaerobes *L. buccalis*, *Treponema vincentii*, and *Fusobacterium nucleatum*, responsible for the massive ulceration, has been associated with phagedenic chancroid patients. These patients present large, rapidly spreading necrotic ulcerations, which may result in extensive destruction or formation of a urethral fistula.[40] "*L. amnionii*" was associated with clinical characteristics that were consistent with bacterial vaginosis and bacterial vaginosis, defined by Nugent's and Amsel's criteria. This organism together with other fastidious bacteria was suspected of causing unrecognized infection, since none were associated with abnormal vaginal discharge.[95]

The protective function of the indigenous bacterial flora can be disrupted by broad-spectrum antibiotic therapy,[9,12,23,44,53,55–57,82,96] resulting in *Leptotrichia* infection. Enhanced *Leptotrichia* proliferation and tissue invasion can culminate into bloodstream invasion and dissemination.[36,54,85] This particularly occurs when the patient's immune system is compromised.

49.1.4 Diagnosis

Leptotrichia are gram-negative, non-sporing, anaerobic, saccharolytic rods. Fresh cells can occasionally stain gram-positive. Usually, *Leptotrichia* are anaerobic on the first isolation, but some strains grow aerobically in presence of CO_2[15] (Table 49.1). Some strains are aerotolerant.

The colonies on blood agar plates are 0.5–3.0 mm in diameter. They tend to be smooth, rhizoid, or convoluted but can also be sparsely filamentous to irregular and grayish brown in color, with a dark central spot in old colonies.[3] Occasionally, colonies are opaque and dry in consistency.[5] For correct diagnosis of *Leptotrichia*, adequate facilities for anaerobic cultivation and qualified expertise in anaerobic microbiology and taxonomy are needed.[5,6,97–100]

L. buccalis is isolated from clinical specimens after anaerobic culture on blood agar such as Brucella blood agar or FFA supplemented with sheep or horse blood. To isolate *L. buccalis* from the normal flora, addition of josamycin, vancomycin, and norfloxacin to the blood agar medium has been recommended.[15] Others have used *Fusobacterium* egg yolk agar for recovery of *Leptotrichia*.[101] A selective medium for culture of fusobacteria and leptotrichia has been established.[102] These authors found that Blood Agar Base (Difco), containing 5% defibrinated sheep blood in the natural or "chocolate" state with the addition of vancomycin (7.5 μg/mL of medium) and streptomycin (20 μg/mL), was an excellent medium for isolation of these bacteria.

A reliable diagnosis of *L. buccalis* can be made from cellular and colony morphology and a few biochemical tests. Biochemical reactions, cellular fatty acid (CFA) composition, and lactic acid as the major end product of glucose metabolism were used to distinguish them from fastidious gram-negative rods thought to be *Capnocytophaga* species.[103]

Diagnosis of *Leptotrichia* should include both phenotypic and genetic methods.[5,6,97] It is no doubt that *Leptotrichia* has been underdiagnosed in the past due to problems with distinguishing it from closely related organisms such as *Fusobacterium*, *Capnocytophaga*, and *Lactobacillus*, and the failure to cultivate many *Leptotrichia* strains.[5]

Useful in the presumptive identification of *Leptotrichia* may be their susceptibility to antibiotics. *L. buccalis* can be susceptible to β-lactamantibiotics, carbapenems, clindamycin, chloramphenicol, tetracycline, and metronidazole, but resistant to vancomycin and aminoglycosides.[21,57] Most strains are moderately resistant or resistant to erythromycin.

It is difficult to diagnose *Leptotrichia* infection based on clinical features alone. Fever can be an indication of LPS release associated with *Leptotrichia* infection. Also, clinical diagnosis of infection in the compromised host should cause consideration of *Leptotrichia* infection. It is important to realize that a number of *Leptotrichia* isolates have not yet been cultivated.[6,104] Molecular techniques should therefore be used to reveal the complete spectrum of *Leptotrichia* infections in man.[3]

49.2 METHODS

49.2.1 Sample Preparation

(i) Bacterial Culture and Phenotypic Characterization The present methods, described by Eribe et al.[5,6,97] have been used successfully for phenotypic and genetic characterization of *Leptotrichia* strains and for establishing new *Leptotrichia* species. For additional details and references, see these publications. Other methods may also be applicable, but the reported ones are working best in our hands.

Cultivation of Bacterial Strains. Suspected *Leptotrichia* strains are seeded onto Trypticase soy agar plates supplemented with 5% human blood, 50 mg/mL of hemin, and 5 mg/mL of menadione. They are cultured anaerobically (90% N_2, 5% H_2, and 5% CO_2) at 37°C for 2–5 days in evacuation jars (Anoxomat System, WS9000, Mart, The Netherlands). After replating, single colonies are picked and subcultured anaerobically on blood agar plates at 37°C for 2–3 days. After the culturing and harvesting of bacteria, cells are generally identified by gram staining, which is then followed by conventional phenotypic or genotypic procedures, or both. The isolated *Leptotrichia* strains may be identified by a commercial identification kit (rapid ID 32 A, bioMérieux), and CFA-, and protein-analyses. Alternatively or as a supplement, DNA is extracted from the isolates for RAPD, specific PCR, and *16S rRNA* gene sequence analysis. A polyphasic approach secures good taxonomy.

Enzymatic/Biochemical Testing. For enzymatic/biochemical testing the bacteria are cultured anaerobically at 37°C for 2–3 days on Columbia base blood agar supplemented with 5% human blood, 50 mg/mL of hemin, and 5 mg/mL of menadione. A bacterial suspension with a turbidity equivalent to a McFarland number 4 standard is made from the culture. Suspensions are transferred to the rapid ID 32 A kit (API), which is based on 29 standardized and miniaturized enzymatic and biochemical tests. The kits are filled using a robot inoculator and incubated anaerobically at 37°C for 4 h. Reading of the kits occurs automatically in an ATB reader. The results of the reactions are transferred into a numerical code and treated automatically in a database for *Leptotrichia* identification (API Plus). The experiments are done in triplicate, starting from the growth of the cultures through the analyses.

Cellular Fatty Acid Extraction. For CFA analysis the *Leptotrichia* cultures are transferred from secondary blood agar plates to liquid glass tubes containing 10 mL of prereduced anaerobically sterilized (PRAS) peptone-yeast-glucose (PYG) broth under anaerobic conditions, using a Virginia Polytechnic Institute and State University (VPI) (Blacksburg, VA) anaerobic culture system (Bellco, Inc., Vineland, NJ). The glass tubes are cultured anaerobically at 37°C for 2–5 days. After culturing, the bacterial cells are harvested by centrifugation, washed once in 0.85% NaCl and once in deionized distilled water, pelleted by centrifugation, and stored at −70°C until analyzed.

Pelleted cells (1–5 mg) kept in 8-mL glass tubes with Teflon-lined screw caps, are thawed prior to extraction. Saponification, methylation, and extraction of the cellular fatty acids are performed as recommended in the operation manual of the Microbial Identification System (MIS) software package (Microbial Identification System, ID). The cells are lyzed and saponified with 1 M saponification reagent [sodium hydroxide, methanol (certificate grade), distilled water]. The tubes are closed and their contents mixed for 10 s using a whirly mixer, and then heated in a boiling water bath for 5 min. After remixing the contents, the tubes are heated in the water bath for additional 25 min. They are then cooled for 30–60 s in a tray with cold tap water. For methylation, HCl-methanol (6.0 N HCl, methanol and sulfuric acid–methanol [50% H_2SO_4 (certificate grade), methanol] are added and mixed for 10 s. The tubes are heated at 80°C for 10 min and then again cooled for 30–60 s in a tray with cold tap water. The methylated components are extracted by adding hexane-ether [hexane and methyl-*tert*-butyl ether (HPLC grade)]. During extraction the tubes are placed in a test tube rotator, and mixing of their contents occurs by rotating the tubes end-over-end for 10 min. After the contents have separated in two phases, the lower aqueous phase is collected and discarded, whereas the upper (organic) phase is retained. Each extract is washed with NaOH–NaCl (sodium hydroxide in deionized distilled water saturated with sodium chloride). The tubes are again rotated end-over-end for 5 min, and the contents allowed to settle down. By using a Pasteur pipette, two-thirds of the top phase are transferred to a 2.0-mL gas chromatographic sample vial [Hewlett-Packard (H-P)]. The vials are closed using 11 mm caps with a red silicone septum lined with Teflon (H-P) and labeled. The experiments are done in triplicate, including cultures, extraction, and gas chromatography.

Cellular Fatty Acid Analysis by Gas Chromatography. CFA profiles are obtained using a model H-P-5980 series II gas chromatograph (H-P) equipped with an H-P U2 cross-linked 5% phenyl-methyl silicone fused silica capillary column (25 m × 0.2 m I.D.), a flame ionization detector, a model 3392A integrator, a model 7673A automatic sampler, and a computer 486DX with 16 MB of RAM. Hydrogen is used as carrier gas at 0.5 mL/min. The gas flow rates for the detector are approximately 40 mL/min for air, 30 mL/min for hydrogen, and 30 mL/min for nitrogen. The temperatures used are 250°C for the injection port and 300°C for the detector. After injection of 2 µL fatty acid extract, the oven temperature is increased from 170°C to 250°C at a rate of 5°C/min and then from 250°C to 300°C at a rate of 30°C /min, held at 300°C for 2 min and then returned to 170°C before the next sample is injected. Ninety-eight vials (49 bacterial strains in duplicate) are loaded onto a sample tray and analyzed. A calibration mixture (Calibration Standard Kit, Microbial ID) containing known fatty acids (straight-chain saturated $C_{9:0}$-$C_{20:0}$ fatty acid methyl esters and five hydroxy fatty acids (2-OH-$C_{10:0}$, 3-OH-$C_{10:0}$, 2-OH-$C_{14:0}$, 3-OH-$C_{16:0}$, and 2-OH-$C_{16:0}$), are included for quantitative assessment of the fatty acids, to compensate for peak area discrimination between low- and high-boiling-point fatty acids, and to provide accurate retention times for the straight-chain saturated fatty acids. The calibration mixture is analyzed before the fatty acid extracts and is automatically reanalyzed after every tenth injection. The quantitative data obtained from the fatty acid profiles are used as a basis for numerical analysis performed with assistance of the MIS software package (Microbial ID). The latter can identify the peaks (by retention time) and determine the area to height, the equivalent chain length (ECL), the total area, and the total area for named or listed compounds. The software package is used to calculate the percentage area for each named or listed compound compared to the total area of the compounds detected. These compounds are identified by using the Moore Broth Library, Version 3.9 (Microbial ID).

Numerical Analysis of Fatty Acid Data. Numerical analysis of the quantitative fatty acid data obtained from the FAME profiles is made using the Library Generation Software, Version 1.0 (Microbial ID). Peak area values for each fatty acid are calculated as percentages of the total peak area to eliminate the effect of variation in inoculum size. Similarities are calculated and the coefficient based on the Euclidean distance between pairs of bacteria determined. Clustering of strains is achieved by the unweighted pair group method for arithmetic average with a program provided in the H-P Library Generation Software, resulting in a dendrogram.

Culture for SDS-PAGE of Whole-Cell Proteins. For SDS-PAGE analysis of whole-cell protein, cultures are transferred from secondary plates to glass tubes containing 10 mL PRAS BHI broth (Biokar Diagnostics, Beauvais, France) supplemented with 0.1% calcium carbonate under anaerobic conditions (90% N_2, 5% H_2, 5% CO_2). The organisms are cultured anaerobically at 37°C for 2–5 days in evacuation jars. After culturing, the bacterial cells are harvested by centrifugation at 13,000 × *g* for 4 min, washed once with 0.85% NaCl and deionized distilled water, and stored at −20°C or −86°C until analyzed.

Whole-Cell Protein Extraction. Prior to the extraction of whole-cell proteins, the bacterial cells are thawed at room temperature and resuspended in TE buffer or in deionized distilled water. The cells are dispersed by sonication (Branson Ultrasonic Corporation) for 1 min at low amplification in an ice cooling bath. Acid-washed 106 µm glass beads are added to the tubes. The bacteria are vortexed twice, each time for 1 min, while kept on ice. From the vortexed bacterial suspension, 20 µL are transferred to Eppendorf tubes and 20 µL of treatment buffer consisting of stacking gel buffer (0.5 M Tris-HCl, pH 6.8, 10% SDS (w/v); 10% glycerol (w/v); 5% 2-mercaptoethanol (w/v); and 5 µL of loading dye (5 mg dissolved and filtered phenol red) are added. The mixture is boiled for 5 min, centrifuged at 13, 000 × *g* for 5 min and then kept on ice or at −86°C until gel electrophoresis can be performed.

SDS-PAGE of Soluble Whole-Cell Proteins. For protein electrophoresis 15 µL of the centrifuged supernatant from each strain are applied directly onto the surface of an 0.5 mm thin precast polyacrylamide 8%–18% (w/v) gradient gel (ExcelGel SDS, Amersham Pharmacia). The proteins are separated using a horizontal Multiphor II electrophoresis apparatus with a power supply (MultiDrive XL) at 600 V, 50 mA, and 30 W at a constant cooling temperature of 15°C (MultTemp II) for 90 min following the manufacturer's instructions. A molecular weight protein ladder standard is included in each electrophoretic run. Each strain is electrophoresed in triplicate, involving the entire experimental procedure from culture to extraction.

After electrophoresis, the SDS gel is examined with silver nitrate under constant shaking. The gels are fixed overnight in fixative (50% methanol, 37% formaldehyde) solution, followed by treatment in dithiothreitol (DTT) solution (1 mL DTT stock) (100 mg DTT/20 mL ddH_2O)/1000 mL for 45 min. The gels are stained with silver nitrate for 60 min and washed for 30 s in deionized distilled water. Thereafter, the gels are washed twice with 100 mL developer solution (0.28 M sodium carbonate, 37% formaldehyde solution), each time for 30 s, and suspended in developer solution for additional 5 min or until the bands are viable. To prevent the gel from overstaining, citric acid is added directly to the developer solution and allowed to settle down for 10 min. Finally, the gel is washed with deionized distilled water, photographed, and stored.

Band Pattern Analysis. The strain differences in band patterns generated by SDS-PAGE are first established visually by recording the presence or absence of bands without considering intensity.

SDS-PAGE electrophoresis bands on gels from different isolates are compared using the Dendron (Solltech Inc,) computer-assisted program. If the similarity coefficients (S_{AB}) for the patterns of two strains are identical, the S_{AB} is equal to 1.00. If one or more bands differ in size and intensity, there is a gradation of S_{AB} from 0.99 to 0.01, with decreasing similarity. An S_{AB} of 0.00 indicates that no bands in two patterns are of the same size. Dendrograms based on S_{AB} are generated using the Dendron computer program with the unweighted pair group method. In dendrogram construction, the program first searches for the two strains with the highest S_{AB} and groups them into a unit with a branching point corresponding to their S_{AB}. The program then searches again for the strain-strain or strain-unit pair with the highest S_{AB} and connects it. The process continues until all isolates are connected. A branching point connecting a unit and a strain is determined by the average S_{AB} between the member of the unit and another strain or unit.

(ii) Extraction of DNA for Genetic Analysis For genetic detection and analysis, DNA is extracted using the guanidium procedure.[5,6] *Leptotrichia* pellets are resuspended in TE buffer (10 mM Tris-HCl–1 mM EDTA, pH 8.0) and incubated with 10% SDS, lysozyme, and proteinase K (20 mg/mL) for 30 min at 37°C. The cells are lyzed with GES reagent (5 M guanidium thiocyanate, 100 mM EDTA and 0.5% sarkosyl). After cooling the sample on ice for 5 min, 7.5 M cold ammonium acetate is added. The phases are mixed gently by shaking and are kept on ice for additional 10 min. Chloroform-2-pentanol (24:1) is then added. Again, the phases are mixed by vigorous shaking, and centrifuged for 5 min at 13,000 × *g*. The aqueous phases are carefully transferred to a 1.5 mL Eppendorf tube. This procedure is repeated. Cold isopropanol (0.54 volumes) is added to a new tube containing the aqueous phase. The tube is inverted until the DNA precipitates, and centrifugation occurs at 13,000 × *g* for 4 min. The aqueous solution is aspirated, and the pellet is washed with 70% ethanol. A vacuum concentrator is used to dry the DNA for 15 min. The pellet is dissolved in TE buffer and treated with DNase-free RNase for 1 h at 37°C. Chloroform-2-pentanol is added and mixed by vigorous shaking, followed by centrifugation for 5 min at 13,000 × *g*. The aqueous phase is overlaid with 0.54 volume of cooled isopropanol (–20°C). The tube is inverted until the DNA precipitates and centrifugation occurs at 13,000 × *g* for 4 min. The aqueous solution is discarded, and the pellet washed with 70% ethanol. A vacuum concentrator is used to dry the DNA for 15 min. The DNA pellet is dissolved in TE buffer or deionized distilled water and stored at 4°C.

To check the quality of the extracted DNA, a total volume of 6 μL is made up by mixing 1 μL of DNA, 3 μL of 10× loading buffer plus ddH_2O. The mixture is loaded into 0.7% agarose gel and electrophoresed for 1 h at 150 V using a horizontal gel electrophoresis apparatus. The gel is stained with ethidium bromide for 30 min. The stain is removed with ddH_2O, and the gel is inspected under UV illumination. Only one band of chromosomal DNA is observed. No smears are seen. To insure good quality DNA for RAPD analysis, each DNA preparation is measured at ODs 260/280 and the absorbance ratio is recorded. Samples with ratios below or above 1.8–1.9 are reextracted.

49.2.2 Detection Procedures

RAPD Analysis. Amplification reactions for *Leptotrichia* are performed in GeneAmp PCR reaction tubes,[6] each containing a final volume of 20 μL, consisting of (i) 11.2 μL sterile double-distilled water; (ii) 1× buffer [10× PCR buffer (10 mM Tris-HCl, pH 8.3, 50 mM KCl)]; (iii) 1 mM dNTPs (10 mM); (iv) 3U (10 U/μL AmpliTaq DNA Polymerase Stoffel fragment (5 U/μL); (v) 2 μM primer (20 μM) (Operon); (vi) 1.5 mM $MgCl_2$ (25 mM); and (vii) 1.0 μL of 1:5 diluted DNA template (1–10 ng/μL). A master mix of the above reaction mixture is made in the order (i–vi) mixed together, and aliquots are made into individual MicroAmp reaction tubes. Component (vii) is added directly into each tube. All these steps are performed on ice. A negative control without DNA is included in each experiment together with a DNA molecular weight marker: 1 kb Ladder Plus.

DNA is amplified in a Perkin Elmer GeneAmp PCR System 2400. The thermal cycler temperature profile consists of denaturation at 96°C for 5 min, followed by 35 cycles of 96°C for 1 min, 48°C for 5 min, and 74°C for 1 min. Five μL of amplification products are analyzed by electrophoresis in 1.5% agarose gel on a horizontal gel electrophoresis apparatus. Gels are stained with ethidium bromide, and bands are visualized and photographed by using a UV transilluminator and a Kodak digital science™ electrophoresis gel documentation system 120.

Amplification of 16S rRNA and Purification of PCR Products. The 16S rRNA genes from the examined *Leptotrichia* strains are amplified under standardized conditions with the primers 9F and 1541R.[5] PCR is performed in thin-walled tubes with a Perkin-Elmer 9700 thermo cycler. One microliter of the diluted (1:5) DNA template is added to a reaction mixture (50 μL final volume) containing 1× Taq 2000 reaction buffer, 2.5 mM $MgCl_2$, 0.8 μM dNTPs, 400 nM of each primer, and 1 U Taq 2000 polymerase (Stratagene) in buffer containing Taqstart antibody (Sigma). In a hot-start protocol, samples are preheated at 94°C for 4 min, followed by amplification with 30 cycles of 94°C for 45 s, 60°C for 45 s, and 72°C for 1.5 min, with an additional 1 s for each cycle; and a final step at 72°C for 15 min. The PCR products are examined by electrophoresis in a 1% agarose gel. DNA is stained with ethidium bromide and visualized under short-wavelength UV light. Amplified *16S rRNA* genes are purified with the Sequenase kit for partial sequencing and on Sephadex G-50 resin columns before full-length sequencing.

16S rRNA Sequencing. Purified PCR is sequenced using an ABI Prism cycle-sequencing kit (BigDye Terminator cycle-sequencing kit). Quarter-dye chemistry is applied with 3.2 μM primers and 1.5 μL PCR products in a final volume of 20 μL. Cycle sequencing is done with an ABI 9700 PCR machine with 25 cycles of denaturation at 96°C for 10 s and annealing and extension at 60°C for 4 min. Purified DNA products are dried in a speed vac centrifuge for 75 min, resuspended in 3.2 μL loading dye, and denatured at 98°C for 2 min. Sequencing reactions are run on an ABI Prism 377 DNA sequencer.

Data Analyses of 16S rRNA Sequences. To determine the identity or approximate phylogenetic position of the *Leptotrichia* strains,[5] 410 bases are first sequenced. Using six additional sequencing primers,[5] full sequences (about 1500 bases) are obtained for representative strains. For identification of the closest relatives, the sequences of the unrecognized *Leptotrichia* strains are compared with the *16S rRNA* gene sequences of over 9000 bacteria in Paster and Dewhirst's database and 76,000 sequences in the Ribosomal Data Project.[105] Similarity matrices are corrected for multiple base changes at single positions by the method of Jukes and Cantor.[106] Phylogenetic trees are constructed by the neighbor-joining method of Saitou and Nei.[107] Sequences are aligned using the MegAlign program (DNASTAR) and imported into

TREECON, a software package for the Microsoft Windows environment, which is used for the construction and drawing of evolutionary trees.[108]

DNA–DNA Relatedness Assays. The S_1-nuclease procedure for the free-solution reassociation for DNA similarity assays is used for pair wise reactions[5,109] with selected strains of *Leptotrichia*. All procedures, including DNA isolation, French pressure cell fragmentation, hybridization, and S1-nuclease assays have been detailed elsewhere.[109] However, rather than labeling the probe DNA chemically with ^{125}I, the random primers method is used (Random Primers DNA Labeling System) to label DNA with (α-^{33}P)dCTP (Perkin Elmer Life Sciences). The probe and target DNAs are reassociated at 66.2°C ± 0.5°C for 24 h for *Leptotrichia* DNA, with a mean G + C content of 29.7 mol%.[55] Values for both homologous and heterologous reassociations are corrected for nonspecific heteroduplex formation by control reactions with salmon sperm single-stranded DNA (0% DNA relatedness) and are less than 10%. Each reaction is repeated at least three times. The mean of all reactions is reported as percentage DNA–DNA relatedness. DNA from *Erwinia amylovora* ATCC 29780 (EA 110) is used as outgroup.

49.3 CONCLUSION AND FUTURE PERSPECTIVES

There is no doubt that *Leptotrichia* are agents of emerging infections, although they are considered members of the normal flora in the oral cavity and genital tract. A particular reason for this is that improvements in medical treatment make people live longer. This causes a number of compromised hosts that are susceptible to indigenous *Leptotrichia*, particularly after breakage of mucosal barriers. Such breakage may serve as a portal of entry for *Leptotrichia*. However, *Leptotrichia* can also cause infections of the noncompromised host. This makes *Leptotrichia* more than opportunists. With such a wide pathogenic potential *Leptotrichia* infection should be expected relatively frequently. Up to now, this has not been so. A reason for that is that *Leptotrichia* can easily be confused with other bacteria, such as *Lactobacillus*, *Fusobacterium*, and *Capnocytophaga*. Furthermore, *Leptotrichia* infections cannot be said to have distinct clinical characteristics, although certain reactions such as fever may be ascribed to their lipopolysaccharide.

The isolation of *Leptotrichia* can be a challenge to both microbiologist and clinician. Recognition of a characteristic morphology together with a typical carbohydrate fermentation pattern with lactic acid as major end product after culture in PYG broth in an appropriate clinical setting should raise the suspicion of *Leptotrichia* infection. *Leptotrichia* are often considered contaminants if isolated from clinical infection, which may be unfair, particularly if the host is compromised. Recovery of *Leptotrichia* requires adherence to strict anaerobic techniques during clinical sampling, transport, and culture. Handling in the microbiology laboratory should also be completely anaerobic although a few strains may be aerotolerant and become used to aerobicity during repeated culture. Adequate experience in microbial diagnostics, taxonomy, and equipment are required to bring forward a correct *Leptotrichia* diagnosis. Unfortunately, not all laboratories are equipped for this purpose.

The taxonomic situation of *Leptotrichia* has been far from clear. Relatively few *Leptotrichia* species have been properly established. Species such as "*L. amnionii*" is still pending valid publication. A number of *Leptotrichia* strains have been listed in databases but cannot be cultured. We think that more efforts should be made to culture not-yet-cultivable *Leptotrichia*. This is important with regard to learning more about their pathogenic role. We also think that methods should be developed to test the possible significance of not-yet-cultivated strains in clinical settings. Little is known about the pathogenicity of *Leptotrichia*. Aside from a few studies on the toxicity of lipopolysaccharide, which is high compared to that of other organisms in the oral flora, practically nothing has been done. *Leptotrichia* often occurs together with other species. We do not know about synergistic mechanisms, although they may play a role. Nor do we know much about the host interactions. Little is also known of their immunological reactions, although antibodies are raised to the lipopolysaccharide of *Leptotrichia*. Because *Leptotrichia* can cause emerging infections, we should indeed pay more attention to the clinical significance and microbiology of these bacteria.

ACKNOWLEDGMENTS

This work was supported in part by the Faculty of Dentistry, University of Oslo and the Norwegian Geotechnical Institute, Norway. The support is highly appreciated and acknowledged.

REFERENCES

1. Trevisan, V., Prime linee d´introduzione allo studio dei Batterj italiani. Rendiconti dell´instituto Lombardo, *di Scienze et letter. Series* 2, 12, 133, 1879.
2. Hamilton, R.D., and Zahler, S.A., A study of *Leptotrichia buccalis*, *J. Bacteriol.*, 73, 386, 1957.
3. Eribe, E.R.K., and Olsen, I., *Leptotrichia* species in human infections, *Anaerobe*, 14, 131, 2008.
4. Ludwig, W., Euzéby, J., and Whitman, W.B., Draft taxonomic outline of the *Bacteroidetes*, *Planctomycetes*, *Chlamydiae*, *Spirochaetes*, *Fibrobacteres*, *Fusobacteria*, *Acidobacteria*, *Verrucomicrobia*, *Dictyoglomi*, and *Gemmatimonadetes*, www.bergeys.org/outlines/bergeys_vol 4-outline.pdf, last accessed November 3, 2010.
5. Eribe, E.R.K. et al., Genetic diversity of *Leptotrichia* and description of *Leptotrichia goodfellowii* sp. nov., *Leptotrichia hofstadii* sp. nov., *Leptotrichia shahii* sp. nov., and *Leptotrichia wadei* sp. nov., *Int. J. Syst. Evol. Microbiol.*, 54, 583, 2004.
6. Eribe, E.R., and Olsen, I., SDS-PAGE of whole-cell proteins and random amplified polymorphic DNA (RAPD) analyses of *Leptotrichia* isolates, *Microbial Ecol. Health Dis.*, 14, 193, 2002.
7. Paster, B.J. et al., Prevalent bacterial species and novel phylotypes in advanced noma lesions, *J. Clin. Microbiol.*, 40, 2187, 2002.

8. Marques da Silva, R. et al., Bacterial diversity in aortic aneurysms determined by 16S ribosomal RNA gene analysis, *J. Vasc. Surg.*, 44, 1055, 2006.
9. Goldstein, E.C. et al., Bacteriology of human and animal bite wounds, *J. Clin. Microbiol.*, 8, 667, 1978.
10. Hirsch, D.C., Biberstein, E.L., and Jang, S.S., Obligate anaerobes in clinical veterinary practice, *J. Clin. Microbiol.*, 10, 188, 1979.
11. Tindall, B.J., Taxonomic note. Rule 15 of the international code of nomenclature of bacteria: A current source of confusion, *Int. J. Syst. Evol. Microbiol.*, 58, 1775, 2008.
12. Tee, W. et al., Bacteremia due to *Leptotrichia trevisanii* sp. nov., *Eur. J. Clin. Microbiol. Infect. Dis.*, 20, 765, 2001.
13. Shukla, S.K. et al., *Leptotrichia amnionii* sp. nov., a novel bacterium isolated from the amniotic fluid of a woman after intrauterine fatal demise, *J. Clin. Microbiol.*, 40, 3346, 2002.
14. Page, L.R., and Krywolap, G.N., Determination of the deoxyribonucleic acid composition and deoxyribonucleic acid-deoxyribonucleic acid hybridization of *Fusobacterium fusiforme*, *Fusobacterium polymorphum*, and *Leptotrichia buccalis*: taxonomic considerations, *Int. J. Syst. Bacteriol.*, 26, 301, 1976.
15. Hofstad, T., and Olsen, I., *Fusobacterium* and *Leptotrichia*. In Borrelier, S.P. (Editor), *Topley and Wilson's Microbiology and Microbial Infections, Bacteriology*, 10th ed., vol. 2, p. 1945, Arnold, London, 2005.
16. Bennett, K.W., and Duerden, B.I., Identification of fusobacteria in a routine diagnostic laboratory, *J. Appl. Bacteriol.*, 59, 171, 1985.
17. Hofstad, T., and Selvig, K.A., Ultrastructure of *Leptotrichia buccalis*, *J. Gen. Microbiol.*, 56, 23, 1969.
18. Listgarten, M.A., and Lai, C.H., Unusual cell wall ultrastructure of *Leptotrichia buccalis*, *J. Bacteriol.*, 123, 747, 1975.
19. Smith, S.H., Murray, R.G., and Hall, M., The surface structure of *Leptotrichia buccalis*, *Can. J. Microbiol.*, 40, 90, 1994.
20. Könönen, E. et al., The oral gram-negative anaerobic microflora in young children: Longitudinal changes from edentulous to dentate mouth, *Oral Microbiol. Immunol.*, 9, 136, 1994.
21. George, W.L. et al., Gram-negative anaerobic bacilli: Their role in infection and patterns of susceptibility to antimicrobial agents. II. Little-known *Fusobacterium* species and miscellaneous genera, *Rev. Infect. Dis.*, 3, 599, 1981.
22. Söderberg, G., Lindberg, A.A., and Nord, C.E., *Bacteroides fragilis* in acute salpingitis, *Infection*, 7, 226, 1979.
23. Newman, M.G., et al., Predominant microbiota associated with periodontal health in the aged, *J. Periodontol.*, 49, 553, 1978.
24. Russell, C., and Melville, T.H., Bacteria in the human mouth, *J. Appl. Bacteriol.*, 44, 163, 1978.
25. Holdeman, L.V., Cato, E.P., and Moore, W.E.C., *Anaerobe Laboratory Manual*, 4th ed., Virginia Polytechnic Institute and State University, Blacksburg, 1977.
26. Moore, W.E.C., Johnson, J.L., and Holdeman, L.V., Emendation of Bacteroidaceae and *Butyrivibrio* and descriptions of *Desulfomonas* gen. nov. and ten new species in the genera *Desulfomonas*, *Butyrivibrio*, *Eubacterium*, *Clostridium* and *Ruminococcus*, *Int. J. Syst. Bacteriol.*, 26, 238, 1976.
27. Williams, B.L., Pantalone, R.M., and Sherris, J.C., Subgingival microflora and periodontitis, *J. Peridontol. Res.*, 11, 1, 1976.
28. Drasar, B.S., and Hill, M.J., *Human Intestinal Flora*, Academic Press, London, 1974, 263.
29. Hardie, J.M., and Bowden, G.H., The normal microbial flora of the mouth. In Skinner, F.A., and Carr, J.G. (Editors), *The Normal Microbial Flora of Man*, p. 47, Academic Press, London, 1974.
30. Evaldson, G. et al., Microbiological findings in pregnant women with premature rupture of the membranes, *Med. Microbiol. Immunol.*, 168, 283, 1980.
31. Könönen, E., Asikainen, S., and Jousimes-Somer, H., The early colonization of gram-negative anaerobic bacteria in edentulous infants, *Oral Microbiol. Immunol.*, 7, 28, 1992.
32. Rosebury, T., *Microorganisms Indigenous to Man*, p. 435, McGraw-Hill, New York, 1962.
33. Colombo, A.P. et al., Clinical and microbiological features of refractory periodontitis subjects, *J. Clin. Periodontol.*, 25, 169, 1998.
34. Duperval, R., Béland, S., and Marcoux, J.A., Infective endocarditis due to *Leptotrichia buccalis*: A case report, *Can. Med. Assoc. J.*, 130, 422, 1984.
35. Kohler, J.L. et al., Septicémie a *Leptotrichia buccalis* chez un immunodéprimé, *Pathol. Biol.*, 30, 161, 1982.
36. Morgenstein, A.A. et al., Serious infection with *Leptotrichia buccalis*. Report of a case and review of the literature, *Am. J. Med.*, 69, 782, 1980.
37. Thadepalli, H., and Rao, B., Clinical evaluation of mezlocillin, *Antimicrob. Agents Chemother.*, 16, 605, 1979.
38. Krywolap, C.N., and Page, L.R., Oral *Fusobacterium*, *Leptotrichia* and *Bacterionema*: II. Pathogenicity: A review of the literature, *J. Baltimore Coll. Dent. Surg.*, 32, 26, 1977.
39. Gupta, H., and Matlow, A., *Leptotrichia buccalis* causing cellulitis and bacteremia in a neutropenic child: A case report, *Clin. Microbiol. Newslett.*, 20, 195, 1998.
40. Benson, P.M., Sexually transmitted diseases, in Military Dermatol. http://www.heart-intl.net/HEART/110105/Sexually Transmitted.pdf.
41. Hebb, J.K. et al., Detection of novel organisms associated with salpingitis, by use of 16S rDNA polymerase chain reaction, *J. Infect. Dis.*, 190, 2109, 2004.
42. Pieper, R. et al., Acute appendicitis and *Bacteroides fragilis*, *Scand. J. Infect. Dis.* Suppl., 19, 92, 1979.
43. Fredricks, D.N., Schubert, M.M., and Myerson, D., Molecular identification of an invasive gingival bacterial community, *Clin. Infect. Dis.*, 41, e1, 2005.
44. Moore, W.E.C., et al., Bacteriology of experimental gingivitis in children, *Infect. Immun.*, 46, 1, 1984.
45. Nakou, M. et al., Early microbial colonization of permucosal implants in edentulous patients, *J. Dent. Res.*, 66, 1654, 1987.
46. Vernelen, K. et al., Bacteraemia with *Leptotrichia buccalis*: Report of a case and review of the literature, *Acta Clin. Belg.*, 51, 265, 1996.
47. van Houte, J., Role of micro-organisms in caries etiology, *J. Dent. Res.*, 73, 672, 1994.
48. Hofstad, T., Pathogenicity of anaerobic gram-negative rods: Possible mechanisms, *Rev. Infect. Dis.*, 6, 189, 1984.
49. Preza, D. et al., Bacterial profiles of root caries in elderly patients, *J. Clin. Microbiol.*, 46, 2015, 2008.
50. Grollier, G. et al., *Leptotrichia buccalis* isolated from peritoneal fluid of two immunocompromised patients, *Clin. Microbiol. Newslett.*, 12, 62, 1990.
51. Caram, L.B. et al., *Leptotrichia* endocarditis: Report of two cases from the International Collaboration on Endocarditis (ICE) database and review of previous cases, *Eur. J. Clin. Microbiol. Infect. Dis.*, 27, 139, 2008.
52. Lark, R.L. et al., Risk factors for anaerobic bloodstream infections in bone marrow transplant recipients, *J. Clin. Infect. Dis.*, 33, 338, 2001.
53. Patel, J.B. et al., Bacteremia caused by a novel isolate resembling *Leptotrichia* species in a neutropenic patient, *J. Clin. Microbiol.*, 37, 2064, 1999.

54. Messiaen, T., Lefebvre, C., and Geubel, A., Hepatic abscess likely related to *Leptotrichia buccalis* in an immunocompetent patient, *Liver*, 16, 342, 1996.
55. Schwartz, D.N. et al., *Leptotrichia buccalis* bacteremia in patients treated in a single bone marrow transplant unit, *Clin. Infect. Dis.*, 20, 762, 1995.
56. Hamman, R. et al., Endocarditis caused by a *Leptotrichia buccalis*-like bacterium in a patient with a prosthetic aortic valve, *Eur. J. Clin. Microbiol. Infect. Dis.*, 12, 280, 1993.
57. Weinberger, M. et al., *Leptotrichia buccalis* bacteremia in patients with cancer: Report of four cases and review, *Rev. Infect. Dis.*, 13, 201, 1991.
58. Baquero, F. et al., Capnophilic and anaerobic bacteria in a neutropenic patient: An oral source, *Rev. Infect. Dis.*, 12, 157, 1990.
59. Reig, M. et al., *Leptotrichia buccalis* bacteremia in neutropenic children, *J. Clin. Microbiol.*, 22, 320, 1985.
60. Prévot, A.R., Septicémie a "*Leptotrichia vaginalis*," *Sem. Hop.*, 30, 2040, 1968.
61. Sutter, V.L., Anaerobes as normal oral flora, *Rev. Infect. Dis.*, 6, Suppl 1, S62, 1984.
62. Robin, C., *Histoire Naturelle des Végétaux Parasites qui Croissent sur l'Homme et sur les Animaux vivants*, XVI, Baillière, Paris, 1853, 702.
63. Wherry, W.B., and Oliver, W.W., *Leptothrix innominata* (Miller), *J. Infect. Dis.*, 19, 299, 1916.
64. Thjøtta, T., Hartmann, O., and Bøe, J., A study of the *Leptotrichia trevisan*. History, morphology, biological and serological characteristics. *Skr. Nor. Vidensk.- Akad. Oslo. I. Mat.-Naturvidensk. Kl.*, 5, 1, 1939.
65. Böe, J., and Thjötta, T., The position of *Fusobacterium* and *Leptotrichia* in the bacteriological system, *Acta Pathol. Microbiol. Scand.*, 21, 441, 1944.
66. Gmür, R., et al., Gingival crevice microbiota from Chinese patients with gingivitis or necrotizing ulcerative gingivitis, *Eur. J. Oral. Sci.*, 112, 33, 2004.
67. Ruviére, D.B. et al., Assessment of the microbiota in root canals of human primary teeth by checkerboard DNA–DNA hybridization, *J. Dent. Child. (Chic)*, 74, 118, 2007.
68. Sassone. L.M. et al., Microbiological evaluation of primary endodontic infections in teeth with and without sinus tract, *Int. Endod. J.*, 41, 508, 2008.
69. Sassone, L. et al., Evaluation of the microbiota of primary endodontic infections using checkerboard DNA–DNA hybridization, *Oral Microbiol. Immunol.*, 22, 390, 2007.
70. Tyrrell, K.L. et al., Anaerobic bacteria cultured from the tongue dorsum of subjects with malodour, *Anaerobe*, 9, 243, 2003.
71. Fredricks, D.N. et al., Changes in vaginal bacterial concentrations with metronidazole therapy for bacterial vaginosis as assessed by quantitative PCR, *J. Clin. Microbiol.*, 2009, January 14. (Epub ahead of print).
72. Fredricks, D.N. et al., Targeted PCR for detection of vaginal bacteria associated with bacterial vaginosis, *J. Clin. Microbiol.*, 45, 3270, 2007.
73. Tamrakar, R. et al., Association between *Lactobacillus* species and bacterial vaginosis-related bacteria, and bacterial vaginosis scores in pregnant Japanese women, *BMC Infect. Dis.*, 7, 128, 2007.
74. Fredricks, D.N., Fiedler, T.L., and Marazzo, J.M., Molecular identification of bacteria associated with bacterial vaginosis, *N. Engl. J. Med.*, 353, 1899, 2005.
75. Fredricks D.N., and Marrazzo, J.M., Molceular methodology in determining vaginal flora in health and disease: Its time has come, *Curr. Infect. Dis. Rep.*, 7, 463, 2005.
76. Thilesen, C.M. et al., *Leptotrichia amnionii*, an emerging pathogen of the female urogenital tract, *J. Clin. Microbiol.*, 45, 2347, 2007.
77. Boennelycke, M. et al., *Leptotrichia amnionii* found in septic abortion in Denmark, *Scand. J. Infect. Dis.*, 39, 382, 2007.
78. Domann, E. et al., Culture-independent identification of pathogenic bacteria and polymicrobial infections in the genitourinary tract of renal transplant recipients, *J. Clin. Microbiol.*, 41, 5500, 2003.
79. DiGiulio, D.B. et al., Microbial prevalence, diversity and abundance in amniotic fluid during preterm labor: A molecular and culture-based investigation, *PLoS ONE*, 3, e 3056, 2008.
80. Han, Y.W. et al., Uncultivable bacteria as etiologic agents of intra-amniotic inflammation leading to preterm birth, *J. Clin. Microbiol.*, 47, 38, 2009.
81. Goto, M., Hitomi, S., and Ishii, T., Bacterial arthritis caused by *Leptotrichia amnionii*, *J. Clin. Microbiol.*, 45, 2082, 2007.
82. Hot, A. et al., Lemierre syndrome caused by *Leptotrichia buccalis* in a neutropenic patient, *Int. J. Infect. Dis.*, 12, 339, 2008.
83. Bhally, H.S. et al., *Leptotrichia buccalis* bacteremia in two patients with acute myelogenous leukemia, *Anaerobe*, 11, 350, 2005.
84. De Martino, S.J. et al., Peripartum bacteremias due to *Leptotrichia amnionii* and *Sneathia sanguinegens*, rare causes of fever during and after delivery, *J. Clin. Microbiol.*, 42, 5940, 2004.
85. Ulstrup, A.K., and Hartzen, S.H., *Leptotrichia buccalis*: A rare cause of bacteraemia in non-neutropenic patients, *Scand. J. Infect. Dis.*, 38, 712, 2006.
86. Falkler, W.A., Jr., Hawley, C.E., and Mongiello, J.R., *Leptotrichia buccalis* hemagglutination in cell binding and salivary inhibition studies, *J. Periodontal Res.*, 13, 425, 1978.
87. Kondo, W., Sato, M., and Ozawa, H., Haemagglutinating activity of *Leptotrichia buccalis* cells and their adherence to saliva-coated enamel powder, *Arch. Oral Biol.*, 21, 263, 1976.
88. Gustafson, R.L. et al., The biological activity of *Leptotrichia buccalis* endotoxin, *Arch. Oral Biol.*, 1, 1149, 1966.
89. Dearaujo, W.C., Varah, E., and Mergenhagen, S.E., Immunochemical analysis of human oral strains of *Fusobacterium* and *Leptotrichia*, *J. Bacteriol.*, 86, 837, 1963.
90. Patters, M.R. et al., Blastogenic response of human lymphocytes to oral bacterial antigens: Comparison of individuals with periodontal disease to normal and edentulous subjects, *Infect. Immun.*, 14, 1213, 1976.
91. Schüpbach, P., Osterwalder, V., and Guggenheim, B., Human root caries: Microbiota in plaque covering sound, carious and arrested caries root surfaces, *Caries Res.*, 29, 382, 1995.
92. Hooper, S.J. et al., A molecular analysis of the bacteria present within oral squamous cell carcinoma, *J. Med. Microbiol.*, 56, 1651, 2007.
93. McNamara, T.F. et al., The development of an altered gingival crevicular microflora in the alloxan-diabetic rat, *Arch. Oral Biol.*, 27, 217, 1982.
94. Kawanami, T. et al., Severe pneumonia with *Leptotrichia* sp. detected predominantly in bronchoalveolar lavage fluid by use of 16S rRNA gene sequencing analysis, *J. Clin. Microbiol.*, 47, 498, 2009.
95. Haggerty, C.L. et al., Clinical characteristics of bacterial vaginosis among women testing positive for fastidious bacteria, *Sex. Transm. Infect.*, 2008 November 12. (Epub ahead of print).

96. Gundi, V.A.K.B., Desbriere, R., and La Sola, B., *Leptotrichia amnionii* and the female reproductive tract, *Emerg. Infect. Dis.*, 10, 2056, 2004.
97. Eribe, E.R., Hofstad, T., and Olsen, I., Enzymatic/biochemical and cellular fatty acid analyses of *Leptotrichia* isolates, *Microbial Ecol. Health Dis.*, 14, 137, 2002.
98. Hofstad, T., *Fusobacterium* and *Leptotrichia*. In Collier, L., Balows, A., and Sussman, M. (Editors), *Topley and Wilson's Microbiology and Microbial Infections*, 9th ed., vol. 3,, 1356, Arnold, London, 1998.
99. Shah, H.N. et al., Approaches to the study of the systematics of anaerobic, gram negative, non spore-forming rods; current status and perspectives, *Anaerobe*, 2009, doi:10.1016/j.anaerobe.2009.08.003.
100. Olsen, I., and Shah, H.N., International Committee on Systematics of Prokaryotes. Subcommittee on the taxonomy of Gram-negative anaerobic rods. Minutes of the open and closed meeting, 2–3 September 2007, Edinburgh, UK, *Int. J. Syst. Evol. Microbiol.*, 58, 1051, 2008.
101. Murray, P.R., and Citron, D.M., Anaerobic Gram-negative bacilli and cocci. In Balows, A. (Editor), *Manual of Clinical Microbiology*, 5th ed., p. 488, ASM, Washington, DC, 1991.
102. McCarthy, C., and Snyder, M.L., Selective medium for fusobacteria and leptotrichia, *J. Bacteriol.*, 86, 158, 1963.
103. Bernard, K. et al., Use of chemotaxonomy as an aid to differentiate among *Capnocytophaga* species, CDC group DF-3, and aerotolerant strains of *Leptotrichia buccalis*, *J. Clin. Microbiol.*, 29, 2263, 1991.
104. Olsen, I. et al., Cultivated and not-yet-cultivated bacteria in oral biofilms, *Microbial Ecol. Health Dis.*, 21, 1, 2009.
105. Cole, J.R., et al., The Ribosomal Database Project (RDP-II): Previewing a new autoaligner that allows regular updates and the new prokaryotic taxonomy, *Nucleic Acids Res.*, 31, 442, 2003.
106. Jukes, T.H., and Cantor, C.R., Evolution of protein molecules. In Munro, H.N. (Editor), *Mammalian Protein Metabolism*, p. 21, Academic Press, New York, 1969.
107. Saitou, N., and Nei, M., The neighbor-joining method: A new method for reconstructing phylogenetic trees, *Mol. Biol. Evol.*, 4, 406, 1987.
108. Van de Peer, Y. et al., Database on the structure of small ribosomal subunit RNA, *Nucleic Acids Res.*, 24, 86, 1996.
109. Johnson, J.L., Similarity analysis of DNAs. In Gerhardt, P., Murray, R. G.E., Wood, W.A., and Krieg, N.R. (Editors), *Methods for General and Molecular Bacteriology*, p. 655, American Society for Microbiology, Washington, DC, 1994.

50 Porphyromonas

Stefan Rupf, Wolfgang Pfister, and Klaus Eschrich

CONTENTS

50.1 INTRODUCTION

50.1.1 Classification, Morphology, and Epidemiology

Classification. The genus *Porphyromonas* is classified to the family of Bacteroidaceae. So far, 17 species have been described (Table 50.1). Although *Porphyromonas* is closely related phylogenetically to the genera *Bacteroides* and *Prevotella*, Shah and Collins suggested a new taxon after reclassification of the *Bacteroides* spp. *B. asaccharolyticus*, *B. gingivalis*, and *B. endodontalis* on the basis of their biochemical and chemical properties.[1,3,6,27–29] Then 16S rRNA gene sequencing confirmed the independent taxon *Porphyromonas*. The phylogenetic differentiation of the members of the genus *Porphyromonas* was also evidenced by 16S rRNA gene sequence analysis and confirmed by DNA–DNA relatedness studies and by morphological, chemical, and biochemical criteria.[30–32] The discrimination of the species *P. gingivalis* and *P. gulae* was enabled by comparing 16S-23S rRNA gene internal transcribed spacer sequences.[8,21]

Members of the *Porphyromonas* genus have been isolated from a number of mammals. The species *P. asaccharolytica*,[1] *P. bennonis*,[4] *P. catoniae*,[5] *P. endodontalis*,[1] *P. gingivalis*,[1] *P. somerae*,[7] and *P. uenonis*[9] have been found worldwide in humans. Several *Porphyromonas* species, *P. macacae*,[26] *P. cansulci*,[10] *P. canoris*,[12] *P. cangingivalis*,[10] and *P. circumdentaria*,[15] were isolated from animals, where they are mostly associated with oral inflammatory processes. These species were frequently found in humans in infected dog or cat bite wounds.[33] *Porphyromonas* spp. are generally associated with inflammation and purulence, but some species primarily occur in gingival and periodontal disease processes, occasionally then causing accompanying systemic inflammative processes. *P. gingivalis* is closely associated with the progression of chronic periodontitis.

The genus *Porphyromonas* is subject to modifications. A number of strains isolated from human oral samples are genetically heterogeneous with regard to their 16S rRNA genes. These strains—among them *Porphyromonas* P3[34]—have not yet been comprehensively characterized. Numerous other strains primarily of oral origin are listed on the species level in the NCBI taxonomy database.[35,36] *P. macacae* used to be referred to as *P. salivosa*.[15] A species that was originally described as "*Porphyromonas denticanis*"[37] was reclassified as *Odoribacter denticanis*.

Morphology. The members of the genus *Porphyromonas* have been described as strictly anaerobic, nonmotile, nonspore-forming, gram-negative rods or coccoid-shaped. Depending on their stage of growth, the size of the pleomorphic *Porphyromonas* differs from 0.4 to 0.8 by 0.8–6 μm. They are equipped with the enzymes glutamate dehydrogenase and superoxide dismutase, but lack glucose-6-phosphatedehydrogenase and 6-phosphogluconatedehydrogenase. Their metabolism is proteolytic. Energy is produced by fermentation of amino acids. Carbohydrates are not fermented. The optimal pH of the culture is 7.5. *Porphyromonas* bacteria produce metabolic

TABLE 50.1
The Genus *Porphyromonas*: Classification, Origin, and Association with Diseases of the Members of the Genus (in Alphabetic Order)

Species	Source, Associated with	Cultural and Biochemical Characteristics
Human Origin		
P. asaccharolytica[1] (*B. melaninogenicus subsp. asaccharolyticus*[2]; *B. asaccharolyticus*[3])	*Abscesses* from different regions, nonoral mucosal surfaces, periodontal pocket	P: +; I: +; C: –; Gf: –; α-F: +; β-NAG –
P. bennonis[4]	*Wound infections and abscesses*	P: weakly +; I: –; C: different; Gf: –; α-F: –; β-NAG +
P. catoniae (*Oribaculum catoniae*[5])	Healthy and diseased human gingival, shallow pockets	P: –; I: –; C: –; Gf: +; α-F: +; β-NAG +
P. endodontalis[1] (*B. endodontalis*)[6]	*Apical periodontitis*, apical root canal, periodontitis, periodontal pocket, carious dentine	P: +; I: +; C: –; Gf: –; α-F: –; β-NAG –
P. gingivalis[1]	*Periodontitis*, oral cavity, periodontal pocket, apical periodontitis, root canal, carious dentine	P: +; I: +; C: –; Gf: –; α-F: –; β-NAG +
P. somerae (P. levii-like organisms)[7]	Skin and soft-tissue *abscesses*	P: +; I: –; C: –; Gf: weakly +; α-F: –; β-NAG +
P. uenonis (P. endodontalis-like organisms)[8,9]	Gastrointestinal tract	P: +; I: +; C: –; Gf: –; α-F: –; β-NAG –
Animal Origin		
P. asaccharolytica[1]	*Footrot*, goat	P: +; I: +; C: –; Gf: –; α-F: +; β-NAG –
P. cangingivalis[10,11]	Dental plaque, periodontal pocket, dog	P: +; I: +; C: +; Gf: most strains –; α-F: –; β-NAG different
P. canoris[12,13]	Plaque, periodontal pocket, dog	P: +; I: +; C: +; Gf: most strains –; α-F: –; β-NAG different
P. cansulci[10]	Plaque, periodontal pocket, dog	P: +; I: +; C: +; Gf: most strains –; α-F: –; β-NAG different
P. catoniae[5,13,14]	Gastrointestinal tract, piglet	P: –; I: –; C: –; Gf: +; α-F: +; β-NAG +
P. circumdentaria[15–17]	Plaque, gingival margins, dog, cats	P: +; I: +; C: +; Gf: most strains –; α-F: –; β-NAG different
P. crevioricanis[11,19,20]	Plaque, dog	P: +; I: +; C: +; Gf: most strains –; α-F: –; β-NAG different
P. endodontalis[16]	Plaque, dog	P: +; I: +; C: –; Gf: –; α-F: –; β-NAG –
P. gingivalis[11,21,22,23] (differentiation) into *P. gulae* (animal origin) and *P. gingivalis* (human origin)	Plaque, monkey, dog	P: +; I: +; C: –; Gf: –; α-F: –; β-NAG +
P. gingivicanis[11,19,20]	Plaque, dog	P: +; I: +; C: +; Gf: most strains –; α-F: –; β-NAG different
P. gulae[23]	Plaque, dog	P: +; I: +; C: +; Gf: most strains –; α-F: –; β-NAG different
P. levii[7,24,25]	Plaque, bear, cat, coyote, dog, wolf, monkey, bovine	P: +; I: –; C: –; Gf: weakly +; α-F: –; β-NAG +
P. macacae (*Syn: P. salivosa*)[26]	Plaque, infected cat bite wounds, primates	P: +; I: +; C: +; Gf: most strains –; α-F: –; β-NAG different

P, pigment production; I, Indole; C, Catalase; Gf, Glucose fermentation; F, α-Fucosidase; β-NAG, *N*-acetyl-β-glucosaminidase

products such as acetic acid, propionic acid, isobutyric acid, and isovaleric acid, as well as traces of succinic acid.

Originally, members of the *Porphyromonas* genus were described as asaccharolytic and further characterized by their production of porphyrins, heme-derived brown to black pigments, developing within 48 h of cultivation. Since *Porphyromonas catoniae* was included in the genus, the description was amended to include both nonpigmented and weakly saccharolytic species. *P. bennonis*, *P. somerae*, and *P. uenonis* are other species which are weakly pigmented only after prolonged cultivation on blood agar. Based on morphological, chemical, and biochemical criteria differentiation of the members of the genus was enabled by 16S rRNA gene sequencing and DNA–DNA relatedness studies. Table 50.1 lists characteristic properties of all species of the *Porphyromonas* genus described so far.[1,4,5,7–10,12,15,20,21,26,31]

Epidemiology. Epidemiological data for *Porphyromonas* spp. are increasingly available. *Porphyromonas* spp. are considered to be part of flora of the normal oral and gastrointestinal tract as well as of the vagina.[38,39] *Porphyromonas* spp. were found to be prevalent in abscesses, mostly of the head and neck; chronic otitis media, peritonitis, aspiration pneumonia, sinusitis, and periodontitis and were often isolated from mixed infections with other anaerobes and aerobic or facultative species.[40] *Porphyromonas* and/or *Prevotella* spp. can be isolated from 20% to 30% of all anaerobic infections in children, and 10% of anaerobic bacteraemia.[41–43] Studies on the composition of the subgingival microbiota in periodontitis patients have been performed in North American, European, Japanese, and Latin American populations.[14,44–50] Few studies are available addressing differences of the occurrence of *P. gingivalis* and *P. endodontalis* in different countries or different geographic regions.[46–55] These studies suggest that there may be substantial differences in prevalence and levels of subgingival or endodontal species. Mean proportions of *P. gingivalis* ranged from 2% to 34% of the microbiota in Chilean, Colombian, Spanish, United States, Brazilian, and Swedish patients with chronic periodontitis.[46,47] Using DNA–DNA checkerboard hybridization technique the detection frequency within Mexican subjects with chronic periodontitis reached 100% of all subjects with untreated chronic or aggressive periodontitis for *P. gingivalis* and 90% for *P. endodontalis*.[45] *P. endodontalis* has been isolated from samples of primary endodontic infections or endodontal abscesses of Brazilian, American, Italian and, Korean provenance in comparative investigations.[49,50,52–54] It has been concluded from these studies that differences of prevalence and proportion of *P. gingivalis* and *P. endodontalis* in different populations can occur. However, the small number of subjects as well as different subject-inclusion criteria may confound final statements.

50.1.2 Pathogenesis and Clinical Features

This section describes the pathogenesis and clinical features for the seven *Porphyromonas* species that were isolated from human sources. All of them are associated with inflammatory processes according to the current state of knowledge. However, their virulence differs significantly on both the species and subspecies level. *Porphyromonas* have several virulence factors. Some of them enable the bacteria to establish in tissues by suppressing the immune defense of the host; others are responsible for provision of nutrients. The most important virulence factors are membrane-associated cysteine proteases (gingipains) together with further proteases. Hemagglutinins and hemolysins are responsible for the iron supply. The docking to host tissues is enabled by fimbria, which themselves carry important enzymes. A hydrophilic capsule provides protection. Capsular vesicles contain antigens, for example, heat shock proteins. Compared to other bacteria, the lipopolysaccharides (LPS) of *Porphyromonas*, especially of *Porphyromonas gingivalis*, have a low biological activity. Metabolic products, such as ammonia, organic acids, amines, and hydrogen sulfide are toxic for the host's cells. *Porphyromonas* coaggregate with a large number of anaerobe and aerobe species.

The *Porphyromonas* species most often isolated from human samples were *P. asaccharolytica*, *P. endodontalis*, and *P. gingivalis*. The pathogenicity of *P. gingivalis* has been investigated most extensively, and this species might be the most important pathogen of the genus. Little information is available on other species isolated from human sources, such as *P. bennonis*, *P. catoniae*, *P. somerae*, and *P. uenonis*. *Porphyromonas* is known to inhabit the oral cavity, colon, and small intestine as well as the urogenital system. Most often, however, they have been isolated from periodontal pockets and infected root canals. The share of anaerobic bacteria in periodontal pockets can reach up to 90%. *Porphyromonas* and *Fusobacteria* can dominate these habitats. Members of the genus have been found in numerous abscesses in various body regions. *Porphyromonas* spp. also appear in actinomycoses. There are indications that members of the *Porphyromonas* genus occur as commensal, opportunistic, as well as pathogenic bacteria. Infection can be passed horizontally, for example, between siblings and between parents, as well as vertically, from parents to children, and probably via smear infections and from anaerobe habitats of the environment.[56]

P. asaccharolytica.[1] *Pathogenesis:* The physiology of *P. asaccharolytica* is very similar to *P. endodontalis*. The enzyme alpha-fucosidase distinguishes *P. asaccharolytica* from *P. endodontalis*. Together with *P. gingivalis* and *P. endodontalis*, this species is among those on which the original description of the genus *Porphyromonas* was based. Little is known about the virulence factors of *P. asaccharolytica*. The species is able to utilize human immunoglobulin G (IgG) as a substrate for growth. In contrast to *P. gingivalis*, only partial breakdown of IgG was observed.[57] *P. asaccharolytica* shows activity of glycylprolyl protease, elastase-like activity, and phospholipase C activity. It has no trypsin-like activity and cannot hydrolyze type I collagen. The lectin WGA reacts strongly with the cell surface of *P. asaccharolytica*.[6] OmpA proteins from *P. asaccharolytica* (OMP-PA), which are monomeric porins, have been found to induce release and expression of the proinflammatory cytokines IL-1-alpha, tumor necrosis factor (TNF) alpha, IFN-gamma, IL-6, and IL-10 in a murine model.[58] *P. asaccharolytica* can contain plasmids[59] and is susceptible to the nonoxidative killing mechanisms of human neutrophils.[60]

Clinical Features: P. asaccharolytica is a common medical pathogen. It has been isolated from numerous mucous membranes as well as from mostly intraabdominal abscesses with a rate of 41%.[61] *P. asaccharolytica* is part of the healthy and infected vaginal flora, with an increased incidence in women with infections or other gynecological problems.[62] This species is also associated with preterm deliveries[63] and is often isolated from men with genital ulcers.[64] It has also been shown to be associated with diversion colitis.[65] Some case reports associate *P. asaccharolytica* with Lemierre's disease.[66] *Fusobacterium* spp. are known to cause Lemierre's syndrome with *Porphyromonas* spp. as a possible accompanying species. *P. asaccharolytica* was identified as predominant bacterium in bone tissues, brain, liver, retroperitoneal, intraabdominal, chest wall, dental, and other abscesses in numerous further case reports.[40,61,67] *P. asaccharolytica* has been described to trigger bacteremia via intravenous catheters.[68] The pathogen has been isolated from patients with chronically inflamed maxillary sinuses[67] in a mixed flora of aerobe and anaerobe species. *P. asaccharolytica* can contribute to diseases of the cardiovascular system, as in the case of an infected left atrial myxoma.[69] It has also been isolated from the oral cavity. Its contribution to the pathogenesis of periodontitis is discussed controversially. According to Tran and coworkers, *P. asaccharolytica* does not seem to be part of the periodontopathic microbiota in humans.[70] It was isolated from only 10% of patients suffering from juvenile periodontitis.[71] In contrast, Rawlinson et al. found it to be the species most frequently identified in chronic periodontitis.[72] Together with *Fusobacterium nucleatum*, *Eubacterium lentum*, and *Peptostreptococcus micros*, it was isolated from infected root canals.[73] It is also frequently present in infected root dentin.[74] *P. asaccharolytica* can be effectively eradicated from abscesses and mucous membranes by standard antibiotic therapy. Antibiotic resistances do not seem to be very common. Beta-lactamase activity has been observed only in some strains.[61] In the treatment of periodontitis, root planning helps to reduce the frequency with which *P. asaccharolytica* can be isolated.[75]

P. bennonis.[4] This is the most recently described species of the genus *Porphyromonas*. It was first observed in 2009 in samples harvested from abscesses in various parts of the human body where it cohabitated with anaerobes and aerobes. It was most frequently isolated in chronic infections of skin and soft tissues in the perirectal/buttock/groin regions. Fifteen percent of the strains produced β-lactamase.

P. catoniae.[5] The species was identified in samples of gingival crevices of patients with gingivitis, active periodontal site, juvenile periodontitis as well as from humans with healthy gums.[76–78] Most frequently it is found in sites unaffected by destructive processes.[34] It was detected in 70% of the infants aged 12 months, but only in 2% of 2-month-old babies, making it obviously an early colonizer of the oral cavity.[14]

P. endodontalis.[1,6] *Pathogenesis: P. endodontalis* is a bacterium frequently found in the endodontium and in inflamed periapical tissues. As a strict anaerobe, it is adjusted to the conditions in the root canal and other anaerobe habitats. In order to survive in a community of organisms, *P. endodontalis* requires the porphyrin ring, preferably supplied as hemoglobin, as a growth supplement.[79] The end products n-butyric acid, acetic acid, hydrogen sulfide, and methyl mercaptan are cytotoxic,[80] for example, butyrate, a metabolic lipid byproduct causes G2/M phase arrests and inhibits protein synthesis.[81] The bacterium is frequently encountered in mixed infections together with example, *Fusobacterium nucleatum*, *P. gingivalis*, *Tannerella forsythia*, *Prevotella* sp., *Parvimonas micra* (*Peptostreptococcus micros*), *Selenomonas sputigena*, or *Campylobacter rectus*.[82,83] *P. endodontalis* binds hemoglobin that appears to serve as a sole carbon, iron, and nitrogen source.[79,84] *P. endodontalis* produces proteases and peptidases. The proteases are able to degrade fibrinogen, fibronectin, and collagen type IV, and also cleave angiotensin. A proline aminopeptidase degrades bradykinin and vasopressin.[85] The LPS from *P. endodontalis* was found to activate the complement system. *P. endodontalis* is able to degrade the complement factor C3.[86,87] Degradation or inactivation of IgG is thought to play an important role in polymicrobial infections.[57] LPS of *P. endodontalis* can rapidly induce the expression of the NF-κB-dependent gene interleukin-8 (IL-8) in dental-pulp stem cells (DPSCs), pulp fibroblasts, and osteoblasts.[88,89] It can upregulate the receptor activator of NF-κB ligand (RANKL) production in osteoblasts and is thus involved in developing apical periodontitis through the stimulation of RANKL production and activation of osteoclasts.[90] When human dental-pulp (HDP) cells were stimulated by LPS from *P. endodontalis*, the production of IL-6 always preceded that of IL-1beta. Since the IL-6 production was observed even in the presence of an IL-1beta receptor antagonist, it was concluded that IL-6 production was independent of the IL-1beta molecule in LPS-stimulated HDP cells.[91] LPS released from the infected root canal triggers the synthesis of IL-1 alpha and TNF-alpha from macrophages and is capable of stimulating PMNs.[92] These proinflammatory cytokines upregulate the production of MMP-1 by macrophages to promote periapical bone resorption.[93] *P. endodontalis* also was found to elevate MMP-2 production in both human pulp and PDL cell cultures.[94]

Via IL-8, a potent inducer of neutrophil chemotaxis and activation, which also has a proangiogenic effect, *P. endodontalis* stimulated the vascular network coincident to progression of the inflammation through the stimulation of vascular endothelial growth factor (VEGF).[95] In addition, proinflammatory cytokines and black-pigmented *Bacteroides* may be involved in developing pulpal inflammation through the stimulation of IL-6 production.[89] *P. endodontalis* also was found to be able to induce the antiinflammatory cytokine IL-10 but not TNF-alpha, IL-12, or IFN-gamma.[96] The activation of COX-2 may contribute to host degradative pathways in the pathogenesis of microbial-induced pulpal/periapical inflammation.[97]

In a rat model, the early stage of pulpal inflammation induced by *P. endodontalis* was accompanied by a shift of the CD4+:CD8+ ratio toward CD8+. The increase in IL-2 and IFN-gamma suggests a Th1 reaction.[55] *P. endodontalis* enhances tissue plasminogen activator (t-PA) production in human pulp, osteoblasts, and gingival cells.[98,99] Human coronary artery endothelial cells (HCAEC) and coronary artery smooth muscle cells (CASMC) can be invaded by *P. endodontalis*.[100]

Clinical Features: P. endodontalis has been isolated frequently in high counts from primarily and secondarily infected dental root canals, chronic periapical lesion, and submucous abscesses of endodontal origin.[101–105] The prevalence of the pathogen ranges between 10% and 70%,[106,107] depending on the methodology applied. Investigations comparing molecular detection methods with cultivation have shown higher prevalences for the former.[108] *P. endodontalis* can be present in symptomatic and asymptomatic cases, with the prevalence being higher in acute apical inflammations.[101,108] High levels of *P. endodontalis* were detected in carious dentine and in pulpal inflammation stages, from minimally inflammatory change to inflammatory degenerative change.[109] It is occasionally found on healthy oral mucosa and, more frequently, in gingivitis.[110] The species has been associated with periodontitis.[111] Differences in virulence between strains have been related to capsular material. On the basis of different types of capsules, three serotypes have been described.[112]

P. endodontalis is sensitive to a wide range of antibiotics, including penicillin, tetracycline, and metronidazole.[112]

P. gingivalis.[1] *Pathogenesis: P. gingivalis* is the most frequently found species of the genus and is the one most comprehensively investigated. This bacterium has its habitat in the human oral cavity and is associated with initiation and progression of a group of chronic inflammatory diseases of the gingival and supporting periodontal tissues. *P. gingivalis* possesses a wide range of virulence factors, helping it to settle in the oral ecosystem. Fimbriae are effective for adhesion on tissues, bacterial coaggregation,[113–117] and for initiating hemagglutination reactions.[118,119] The hemoglobin released by destruction of erythrocytes by hemolysins is made accessible to the nutritional supply.[120,121] Cell-surface anionic polysaccharides provide resistance to complement-mediated killing.[122] The LPS of *P. gingivalis* may contribute to inhibition of the innate immune response.[123,124] Different proteinases, particularly cysteine proteinases (gingipains), are considered the prominent virulence factor of the pathogen. These proteinases are responsible for the high proteolytic activity of the bacterium. *P. gingivalis* can survive inside as well as outside epithelial or endothelial cells.[125–127] It seems to be able to undermine immune clearance via exploitation of the complement receptor-/CR3 in monocytes/macrophages.[128] In biofilms, coaggregations of *P. gingivalis*, *T. forsythensis*, and *T. denticola*, or of *P. gingivalis* and *A. actinomycetemcomitans*, or *F. nucleatum*, are positively correlated with a progressive course of disease.[126,129,130]

The potential key virulence factors of *P. gingivalis* are briefly listed in the following (Figure 50.1):

(i) Fimbriae: Fimbriae mediate bacterial adherence to and entry into periodontal cells and contribute to coaggregation with species within the oral ecosystem.[131,132] *P. gingivalis* produces fimbriae of between 0.5 and 1.6 μm in length and approximately 5 nm in width.[133,134] Most strains of *P. gingivalis* produce fimbriae consisting of fimbrillin subunits of 41–49 kDa.[133] The gene encoding the *P. gingivalis* fimbriae (*fimA*) is resident on the chromosome as a single copy. All strains so far examined contain this gene.[135,136] Nucleic acid sequencing of fimA genes revealed five distinct clusters among *P. gingivalis* strains.[137] Fimbriae have an antigenic effect; proteases (gingipains) placed on them release cryptic ligands and promote hemagglutination.[138] They mediate the binding to a number of molecules and structures, such as saliva molecules, epithelial cells, other bacteria, fibrinogen, fibronectin, and lactoferrin.[139–144] In addition to adhesion, mediation fimbriae are supposed to possess chemotactic properties and to discharge cytokines. Intact fimbriae, fimbrillin-specific peptides, or synthetic fimbrial peptides can stimulate the production of fibroblast-derived thymocyte-activating factor from human gingival fibroblasts, interleukin (IL)-1, IL-6, IL-8, neutrophil chemotactic factor KC, and tumor necrosis factor (TNF)-alpha from mouse peritoneal macrophages or in human peripheral blood monocytes.[136]

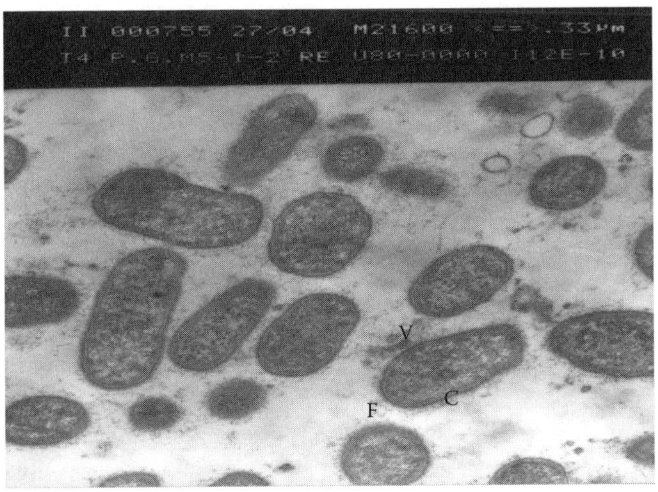

FIGURE 50.1 Transmission electron image of *Porphyromonas gingivalis* stained with ruthenium red. V, extracellular vesicles; C, capsule; F, fimbriae. Bar = 330 nm.

(ii) Minor fimbriae: A second type of fimbriae was isolated and characterized, which was found to have a higher molecular weight of the fimbrillin, 67 kDa. These fimbriae seem to play an important role in host bone loss.[145]

(iii) Membrane-associated proteinases: Thiol-activated trypsin-like cysteine proteinases (gingipains) have been recognized as major proteolytic enzymes of this organism. *Rgp*A and *Rgp*B (gingipain R: encoding genes: *rgp*A and *rgp*B) are arginine-specific proteinases, *Kgp* (gingipain K, *kgp* gene) is lysine-specific. *Rgp*A and *Kgp* contain separate catalytic and adhesion/hemagglutinin domains, while *Rgp*B has only a catalytic domain. The most important tasks of *Rgp*A and *Rgp*B are activation of the kallikrein-kinin system, inducing enhanced vascular permeability, and activation of the blood coagulation system. The activation of the kinin system is involved in breaking the vascular barrier that permits dissemination of *P. gingivalis*.[146] *Kgp* is the most potent fibrinogen/fibrin degrading enzyme and degrades also hemoglobin to heme. *Rgp*A activates coagulation factors and degrades fibrinogen/fibrin. Gingipains degrade macrophage CD14, thus inhibiting activation of leukocytes through the lipopolysaccharide (LPS) receptor.[147] The proteinases degrade complement factor C3, preventing the deposition of C3b on the bacterial cell surface. The degradation of C5 leads to generation of a chemotactic C5a-like fragment.[148] This contributes to significant leukocyte infiltration while simultaneously inactivating their C5a and FMLP receptors. This may lead to *P. gingivalis* being protected from phagocytosis by PMN.[149] C5a receptor cross-talks with TLR signaling pathways and inhibits IL-12.

This led to the assumption that *P. gingivalis* uses different complement-related mechanisms for escaping IL-12–mediated clearance.[88,128,150] Gingipains also have collagenolytic activity and degrade or inactivate inflammatory cytokines IL-6, IL-8, TNF-alpha, and IFN-gamma. The activation of host derived collagenases causes further tissue damage. Two other proteinases (Trp, PrtT) have MMP-activating or hemagglutinating properties. Dipeptidyl aminopeptidase IV (DPPIV) may act as a virulence factor by contributing to the degradation of connective tissue.[151]

(iv) Hemagglutinins and hemolysins: *P. gingivalis* has essential requirements for both iron and protoporphyrin IX, which it preferentially obtains from heme. Several distinct hemagglutinins (*Hag*A–*Hag*C and further) have been identified.[137,152,153] The proteinase Kgp specifically binds and cleaves hemoglobin.[154] Outer membrane receptors (HmuR, HmuY) are responsible for heme utilization.[155–157]

(v) Lipopolysaccharide: *P. gingivalis* LPS differs from the LPS of other bacteria. It induces in vivo endothelial cell expression of E-selectin at significantly lower levels compared to enterobacterial LPS, and inhibits E-selectin upregulation by other bacteria in vitro.[123,158] This may result in diminished neutrophil adhesion to endothelial cells. *P. gingivalis* expresses heterogeneous LPS, including atypical lipid A structures, that trigger TLR2 signaling, and weakly stimulate or antagonize TLR4 activation.[159,160] This may suppress TLR-mediated immunity.[159] *P. gingivalis* LPS upregulates the IL-1R–associated kinase (IRAK)-M and relatively poorly induce IL-1beta and TNF-alpha compared with *E. coli* LPS.[125,161,162] The binding of *P. gingivalis* LPS to LPS binding protein (LBP) is weak compared to *E. coli* LPS.[163] In mice, the LPS induces apoptosis before a specific immune response to eliminate early nonspecific activated lymphocytes. This suggests a direction of the periodontal inflammation infiltrate toward a Th2 reaction.[164]

(vi) Capsule: Most *P. gingivalis* strains are able to produce capsular polysaccharides. This results in reduced binding of *P. gingivalis* to PMN and inhibited phagocytosis.[165,166] Six capsular serotypes of *P. gingivalis* have been described, and a number of nontypeable capsule serotypes appear to exist. In a mouse model, the capsular serotypes were more virulent than nonencapsulated strains.[167–169]

(vii) Outer membrane vesicles (MV): *P. gingivalis* secretes outer membrane vesicles that contain major virulence factors, including the proteases Arg-gingipain (Rgp) and Lys-gingipain (Kgp). In vitro, MVs enter human epithelial cells via a lipid raft-dependent endocytic pathway and survive within the endosomes for an extended period. MV-associated gingipains degrade cellular functional molecules such as TfR and paxillin/FAK, resulting in cellular damage, indicating that *P. gingivalis* MVs are potent vehicles for the transmission of virulence factors into host cells and are involved in the etiology of periodontitis.[170,171]

(viii) Toxic metabolites: Organic acids, amines, ammonia, hydrogen sulfide, or butyric acid are inducers for apoptosis in B cells and T cells.[172,173]

Clinical Features: P. gingivalis has been strongly associated with the pathology of chronic periodontitis.[104] Periodontitis is a multibacterial infection that affects the tooth-supporting tissues by destroying connective tissues and alveolar bone. A cyclical course is characteristic for this disease. The presence of specific bacteria at high levels in the subgingival plaque as well as the individual susceptibility to periodontal disease influence the progression of periodontitis. Important influence factors are cigarette smoking and poorly controlled diabetes mellitus. Genetic determinants and other common diseases are discussed to be modifiers for the onset and severity of periodontitis.[174] The human oral cavity is composed of multiple epithelial and mucosal surfaces, as well as the calcified hard-substance enamel and the hard-tissue dentin. All oral surfaces are constantly moistened by saliva. Periodontopathogenic bacteria use saliva as a transmission medium. Various studies have shown that an organism is colonized by only one or at least one dominating *P. gingivalis* genotype. A large number of different strains could be isolated from different patients, supporting the hypothesis that *P. gingivalis* is to be considered an opportunistic pathogen, whose pathogenicity, however, cannot be assigned to a particular genotype. Precondition for the survival of the bacterium is its adhesion to epithelia or hard surfaces. The prevalence of *P. gingivalis* is lower in persons with healthy periodonts than in patients with periodontal diseases.[175–181] *P. gingivalis* can be isolated from periodontal pockets, supragingival plaque, root canals, saliva, tongue, and the buccal mucosa, but also from the tonsils.[161,180,182–184] Most frequently, the bacterium is found in deep periodontal pockets, where it also yields the largest counts. It has also been isolated from abscesses of the periodontium and endodontium. Its DNA was found in artheromatous plaques together with the DNA of other periodontal pathogens,[185] corneal scrapings,[186] in positive transient bacteremia subjects,[187] cavernous sinus thrombophlebitis,[185] from a severe brain abscess,[188] and a tubal-ovarian abscess.[189] *P. gingivalis* has been found in reduced prevalence in edentulous elderly.[176]

There is increasing evidence from epidemiological studies, case studies, and scientific observations that periodontal disease represents a significant risk factor for cardiovascular and cerebrovascular disease. Potentially periodontal diseases may be important for preterm delivery as well as low birth weight of newborns. Recent studies suggest that there is a bidirectional relationship between diabetes mellitus and periodontal disease.[190] Periodontal disease can compromise blood sugar regulation in diabetes on the one hand. On the other hand, diabetes has been identified as a significant risk factor for periodontal disease.[191] The occurrence of *P. gingivalis* has been found to be associated with stroke and arteriosclerosis, myocardial infarction, coronary heart disease, preterm births, diabetes mellitus, and respiratory diseases.[125,128,192,193]

***P. somerae* (*P. levii*-Like Organisms [PLLO]).**[7] *P. somerae* is associated with chronic skin and soft-tissue or bone infections, especially in the lower extremity. PLLO, now designated as *P. somerae*, have been isolated from diverse human clinical samples,[194] for example, from the vagina, from patients with chronic otitis media, and one case of bacterial vaginosis. A systemic predisposition such as diabetes mellitus, neuropathy or peripheral vascular disease seems to facilitate infections in which *P. somerae* is the predominant bacterium.[7] The species seems always to cohabitate with other anaerobes, such as anaerobic cocci, *Bacteroides fragilis* group, or *Prevotella* spp. About 20% of the strains are β-lactamase producers, and some strains have been found to be resistant to clindamycin.

***P. uenonis* (*P. endodontalis*-Like Organism [PELO]).**[8,9] *P. uenonis* appears most often in samples of intestinal origin, suggesting the human gut to be its preferred habitat. When isolated from infectious material, it was always accompanied by other anaerobes or aerobes. Some strains have been found to be resistant to penicillin G, ampicillin, and trimethoprim. *P. uenonis* seems to have a low virulence as it is exclusively found in mixed culture but not in blood cultures. It is not associated with serious infections.

50.1.3 Diagnostics

50.1.3.1 Conventional Techniques

Conventional techniques for the detection of *Porphyromonas* often have a lower sensitivity compared with molecular methods. Primarily, this relates to the limited viability of *Porphyromonas* outside anaerobic habitats. The transport of the samples from patient to a microbiological laboratory is therefore a particular challenge.[195,196] Once the bacteria have been successfully isolated, antimicrobial resistances can be determined on the basis of cultivation.[197]

Cultivation Techniques and Biochemical Identification. Cultivation still plays an important role in the detection of *Porphyromonas*. Samples are mostly mixed populations of aerobic and anaerobic microorganisms. Cultivation is performed by an anaerobic technique on blood agar[198]; identification, by the detection of specific enzyme patterns.[199] If bacteria have been successfully isolated, further differentiations on the species and subspecies level can be performed on the basis of cultivation using different methods of characterization.

Immunofluorescence. Specific poly- or monoclonal antibodies can be used for the direct detection and quantification of *P. gingivalis* in plaque samples as well as for its intracellular detection in gingival epithelial or endothelial cells.[200–203] Fluorescent dyes coupled to the antibodies make the targeted bacteria directly visible.

Matrix-Assisted Laser Desorption/Ionization Time-of-Flight Mass Spectrometry (MALDI-TOF-MS). MALDI-TOF-MS is a soft ionization method, which allows the phenotypic identification of different cultured bacteria.[204–206] Ions are separated and detected according to their masses and charges. Each mass peak corresponds to a molecular fragment released from the cell during laser desorption. Bacteria, including *Porphyromonas*, can be identified by comparing their mass spectrum with those obtained from known reference strains using methods of pattern analysis.[198,205,207]

Other Methods. The *BANA chair-side* test was a diagnostic method for the detection of bacterial enzymes. Trypsin-like proteinases, produced among others by *P. gingivalis* were detected by blue coloration of a test stripe. However, the test had an insufficient specificity.[208] *Serological diagnostics* detected antibody reactivity in the serum or sulcus fluid against specific bacterial antigens.[209] Restriction endonuclease analysis (REA) and *ribotyping* were successfully applied for the typing of *P. gingivalis*.[210] Analysis of electrophoretic mobilities of 16 metabolic enzymes of 100 *P. gingivalis* strains from humans and animals revealed 78 different multilocus enzyme electrophoresis (MLEE) types.[182]

50.1.3.2 Molecular Techniques

Nucleic acid–based methods are indispensable for the detection of *Porphyromonas*, particularly of *P. gingivalis*, as their introduction has enabled a fast and inexpensive diagnostic. They are clinically used in cases of severe or aggressive periodontitis for diagnosis, reevaluation after a treatment phase, and during recall.

The most important molecular methods for the detection and quantification of *P. gingivalis* include hybridization using species-specific DNA probes, conventional end-point PCR as well as quantitative end-point PCR, or quantitative real-time PCR using DNA intercalating dyes or fluorogenic DNA probes. The majority of current methods for detection and enumeration of *P. gingivalis* are based on 16S ribosomal RNA operon sequences.[211–213] Usage of the rRNA operon offers the advantage of every bacterial cell harboring several template copies. The *P. gingivalis* strain W83 is known to have 4 rRNA operons (The Ribosomal RNA Operon Copy Number Database, http://ribosome.mmg.msu.edu/rrndb/index.php). Single-copy gene assays based on species-specific virulence factor genes have been developed for *P. gingivalis* using sequences of the *Arg*-gingipain,[214] the *waaA* (3-deoxy-D-manno-oct-2-ulosonic acid transferase) gene, or the *fimA* genes.[215]

PCR Techniques. PCR is an established method for the detection of *P. gingivalis*.[216,217] Once its specificity has been tested, it has low demands on sampling and sample transportation, since neither living nor cultivable bacteria is required. Bacterial species and subspecies can be detected or quantified by means of different PCR techniques. Quantitative PCR also enables quantification of the entire number of copies of bacterial DNA contained in a sample.[218] Ribosomal and single-copy genes can be used as target sequences. The genes of 16S rRNA, *OMP*, *Fim*A, *rgp*B, or *Prt*C are frequently used for *P. gingivalis*.[216,219] Classic nonquantitative PCR, having a detection level of less than 100 cells, is often used in studies on the prevalence of this bacterium.[220] Another technique used was multiplex PCR using three 16S rRNA species-specific forward primers and a conserved reverse primer for the simultaneous detection of *A. actinomycetemcomitans*, *T. forsythensis*, and *P. gingivalis*.[221] While no quantitative information is obtained by means of classic PCR techniques, the number of bacterial cells can be determined by means of quantitative PCR. Competitive and real-time PCR techniques were developed for *Porphyromonas*.[213,215,219] In competitive PCR, inhibitors of amplification are easily recognized by preventing the formation of the competitor product. In contrast, in classic as well as in real-time PCRs, inhibitors of *Taq* polymerase and the presence of other bacterial or human DNA may lead to false-negative results. Real-time PCR techniques, which are routinely used for quantification of *Porphyromonas*, offer a fast quantification and monitoring of the amplification through DNA intercalating dyes (SYBR green), as well as control of the specificity by dye-labeled specific probes. The comparison of PCR with subsequent reverse hybridization and checkerboard DNA–DNA hybridization for the identification of bacterial species in subgingival plaque samples showed that both tests could distinguish samples from healthy and periodontitis subjects.[222] Different PCR based tests for detection and quantification of *P. gingivalis* are commercially available.[222,223] The clonal diversity of the taxon *P. gingivalis* was studied using molecular fingerprint techniques, like arbitrarily primed (AP) PCR, repetitive extragenic palindromic (REP),[210] and random amplified polymorphic DNA (RAPD)-PCR.[224]

RNA Hybridization. Identification and quantification of bacteria can be performed by direct hybridization of specific probes with the ribosomal RNA. Since many 16S rRNA copies exist per cell, no amplification of the target nucleic acids is necessary. However, since bacterial RNA is rapidly degraded after cell death, only living bacteria are detected. Hybridization of RNA requires a buffered transport system that protects bacteria from death but, at the same time, inhibits bacterial growth. The best-known hybridization test currently available on the market is IAI PadoTest 4.5.[225,226]

DNA Hybridization. The "checkerboard" DNA–DNA technique uses DNA probes on a single support membrane for hybridizing large numbers of DNA samples. Denatured DNA from up to 43 samples can be fixed in separate lanes on a single membrane mounted in a Miniblotter 45. The membrane is then rotated by 90° in the same device, which enables simultaneous hybridization with 43 different DNA probes. Hybridizations are performed with either digoxigenin-labeled whole genomic probes or 16S rRNA-based oligonucleotide probes directly conjugated to alkaline phosphatase. The method permits the simultaneous determination of the presence of multiple bacterial species in single or multiple dental plaque samples, thus suggesting its usefulness for a range of clinical or environmental samples.[227]

A modification of the method, the reverse-capture checkerboard hybridization, uses RT–PCR as a first step to amplify 16S rRNA genes. After that, hybridization with oligonucleotide probes allows to differentiate between closely related species or even subspecies.[228]

Other Molecular Techniques. PCR is often the basic technology for more complex applications. For example, there is a multilocus sequence typing (MLST) scheme for *P. gingivalis*. By a combination of PCR and DNA-sequencing, the heterogeneity of a set of target genes is used for a strain differentiation on the subspecies level. The following target genes are used for *P. gingivalis*: *gdpxJ*, *pga*, *hagB*, *mcmA*, *recA*, *ftsQ*, and *pepO*.[229] PCR is also the basic technology for the human oral microbe identification microarray (HOMIM) developed by The Forsyth Institute in Boston. HOMIM represents a high throughput technology to examine complex oral microbial diversity in a single hybridization. About 300 oral bacterial species (http://mim.forsyth.org/bacteria.html), including the "uncultivables" and *Porphyromonas* spp. are simultaneously detected.[230]

A DNA-chip has been developed[231] allowing a semiquantitative detection of periodontal bacteria, including *P. gingivalis*. Bacterial DNA has to be extracted, and part of the 16S rRNA gene is amplified by PCR using highly conserved primers labeled with a fluorophore (Cy5). Without further treatment, samples are hybridized to the DNA-chip, which is spotted with strain-specific DNA probes.

Terminal restriction fragment length polymorphism (T-RFLP) and denaturating gradient gel electrophoresis (DGGE) are alternative molecular approaches that allow the assessment of diversity of oral microbiota and rapid comparison of community structures. These techniques can contribute to the identification of yet unknown *Porphyromonas* spp.[232,233]

50.2 METHODS

50.2.1 Sample Preparation

Sample taking and preparation is not critical for cultured *Porphyromonas*. Clinical samples of *Porphyromonas*, mostly of oral origin, are very complex and may contain several hundred other bacteria species. In addition, oral samples are often contaminated by saliva, pus, and above all, blood. As hemoglobin strongly inhibits PCR and other enzyme-catalyzed reactions, an efficient DNA preparation plays a decisive role.

Sampling Methods.[184,234–237] The sampling strategy depends on the aim of the investigation and the method of analysis applied. In DNA based molecular techniques, cotton swabs or paper points can be shortly stored in 1.5 mL tubes at 4°C or at –20°C for longer periods of time. Aspirates from abscesses, plaque samples or whole saliva should be stored frozen. Large samples have to be centrifuged, and the pellets have to be resuspended in suitable smaller volumes, for example, Tris-EDTA buffer, before further processing. Smaller sample volumes can be suspended directly in 100–200 μL Tris-EDTA buffer. Sampling techniques routinely used for *Porphyromonas* are described in the following:

Cultures: One colony is to be scraped from agar plates and suspended in 200 μL Tris-EDTA buffer.

Mucosal (Skin, Abscess) Swabs: Streak the surface of the mucosa vigorously back and forth with a sterile swab. Resuspend the sample in 1 mL Tris-EDTA buffer.

Saliva: Salivary flow is stimulated by paraffin chewing for 2 min. Whole saliva samples are collected by expectoration. Saliva samples are centrifuged at 10,000 × *g* for 5 min, the supernatant has to be discarded, and the samples are resolved in 200 μL Tris-EDTA buffer.

Oral Lavage: Rinse with 10 mL sterile saline for 30 s and collect samples by expectoration. Then follow the procedure described for saliva.

Subgingival Plaque (Paper Points; Figure 50.2): Isolate sample sites with cotton rolls. Supragingival plaque has to be removed, and a sterile paper point (iso 40) is inserted into the pocket until meeting with resistance. After 10 s, the point is removed and stored in a 1.5 mL tube at –20°C until further investigation. There are site-specific and pooled-sample strategies. Pooled samples from the (minimally three) deepest active periodontal pockets provide representative information about the subgingival microflora of the whole oral cavity which is relevant for adjunctive systemic antibiotic therapy.[184]

Subgingival Plaque (Curette): Sample sites are isolated with cotton rolls. Supragingival plaque has to be removed and discarded. A Gracey curette is inserted to the base of the pocket or sulcus, and as much plaque as possible is removed with a single stroke. The curette plus sample has to be shaken in a 1.5 mL tube containing 200 μL Tris-EDTA until no visible plaque remains on the curette.

Supragingival Plaque (Curette): Sample sites are isolated with cotton rolls. A Gracey curette stroke is performed on the surface of the tooth without inserting the curette into the periodontal pocket.

DNA Preparation. The method of DNA preparation depends on sample characteristics and the detection method used. Inflammation exudate, blood, saliva, or human DNA in the sample can also influence the results. Many different DNA preparation protocols have been described that are subject to constant modification and improvement. There are two basic sample preparation strategies:

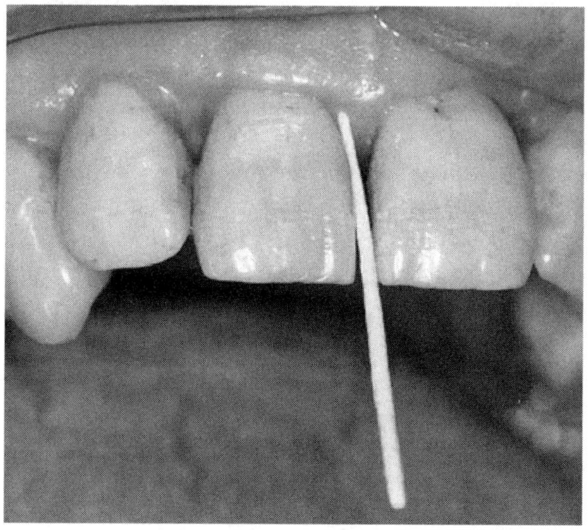

FIGURE 50.2 Sampling of subgingival plaque with a paper point at the mesio-buccal aspect of a right maxillary central incisor.

Simple preparation for cultured bacteria: Suspend a single bacterial colony in 50 μL of 0.7% (w/v) sodium chloride solution, and dilute this suspension subsequently 1:100 again in 0.7% sodium chloride; use 5 μL of this dilution as template for PCR.[229]

Complex preparation of clinical samples: Different kits for DNA preparation are available and commonly used in clinical and laboratory studies. A detailed DNA-isolation protocol is recommended including sophisticated lysis by lysozyme and Proteinase K and DNA precipitation by isopropanol/ethanol by The Forsyth Institute: (http://mim.forsyth.org/samplepreparation.html).

For quantitative competitive PCR of bacteria in plaque or saliva, the following protocol is used[238,239]: Add 50 μL of glass beads with diameters of 100–250 μm to 200 μL aliquots of the sample and vortex for 4 × 30 s at 30 Hz. Add 100 μL of a buffer containing 0.01 M Tris, 0.005 EDTA and 0.5% SDS and 1 μL lysozyme (50 mg/mL) to 15-μL aliquots of the supernatant. Add appropriate amounts of DNA competitor. Incubate the mix for 15 min at 4°C. Add 5 μL of proteinase K (19 mg/mL), and then incubate for 1 h at 50°C. For DNA precipitation, add 10 μL of sodium acetate and 500 μL of ethanol (96%). Incubate at –80°C for 30 min then centrifuge for 30 min at 4°C and 10,000 *g*. Next, 500 μL of 75% ethanol has to be followed by centrifugation for 15 min at 4°C and 10,000 × *g*. Redissolve the pellet in 15 μL of nanopure water.

50.2.2 Detection Procedures

50.2.2.1 Species-Specific PCR[213]

Primer Sequences (5′–3′): Target gene: 16S rRNA gene, forward (pg1): 5′-GGG ATT GAA ATG TAG ATG ACT GAT G-3′, reverse (pg2): 5′-CCT TCA GGT ACC CCC GAC T-3′, 488 bp.

Performed in a 50 μL reaction volume in 0.2 mL tubes, 2 μL of sample (template) supplemented with 40 pmol of each primer, 0.1 μmol $MgCl_2$, 2.5 nmol each of deoxynucleoside triphosphate, 1.25 U Thermus aquaticus polymerase, 5 μL 10× Taq buffer, and water to 50 μL. PCR amplification, performed in a standard DNA thermal cycler, initial denaturation step at 95°C for 10 min followed by 40 cycles of denaturation at 95°C, primer annealing at 55°C and extension at 72°C, for 1 min each, and a final step at 72°C for 10 min. The amplified material was stored at –20°C.

50.2.2.2 Quantitative Real-Time PCR

Primer Sequences (5′–3′): See 50.2.2.1.

LightCycler assay with kinetic analysis (fluorescence threshold method) (238, 239): PCR is performed with 20 μL reaction mixture containing 2 μL LightCycler-DNA Master SYBR Green I (Roche Molecular Biochemicals), 6 mM $MgCl_2$, 2–4 μL DNA template, and 0.5 mM of both forward and reverse primers. For amplification, 45 cycles of 94°C for 0 s, 55°C for 10 s, and 72°C for 20 s (temperature transition of 20°C/s), preceded by denaturation at 94°C for 3 min, are used, and the fluorescence reading is taken at 72°C. The melting curve analysis is performed with continuous fluorescence reading between 65°C and 95°C with a transition rate of 0.1°C/s. For the kinetic analysis, the second derivative maximum method is used.

50.2.2.3 Quantitative Competitive PCR (QC–PCR)[215,218,240]

Primer Sequences (5′–3′): target gene: 16S rRNA gene *P. gingivalis*, 450 bp PCR product, primer see Section 50.2.2.1 and pg-hy: 5′-GGGATTGAAATGTAGATGACTGATG TCAGCTCGTGCCGTGAG-3′.

This technique is an end-point PCR. Quantification of bacteria by PCR is carried out by coamplification of a homologous competitor. In order to facilitate the separation of the PCR products obtained from the sample and the homologous competitor, the latter is designed shorter than the sample template. Quantitative competitive PCR can be performed in a conventional block thermal cycler without using intercalating DNA dyes. This technique has been described for *P. gingivalis* and other bacteria and uses the genes of the 16S rRNA[218,240,241] or the *fimA* gene.[215]

The sample (template) (2–5 μL) is supplemented with aliquots of competitors corresponding to 10, 100, or 1000 bacteria, respectively; 40 pmol of each primer; 0.1 μmol $MgCl_2$; 2.5 nmol of each deoxynucleoside triphosphate; 1.25 U *Thermus aquaticus* (*Taq*) polymerase; 5 μL 10× *Taq* buffer; and water to 50 μL. PCR amplification, performed in a standard DNA thermal cycler, included an initial denaturation step at 95°C for 10 min, followed by 40 cycles of denaturation at 95°C; primer annealing at 55°C and extension at 72°C, for 1 min each; and a final step at 72°C for 10 min. The amplified material is stored at −20°C. Amplicons are detected by electrophoresis of 10 μL of PCR product on a 2% agarose gel. After staining with ethidium bromide DNA is quantified videodensitometrically. Counts of bacteria are calculated considering the amounts of 16S rRNA gene copies for the bacterial species investigated.

Competitive template construction: The competitor (407 bp) is obtained by PCR using the reverse primer (pg2) together with a hybrid primer (hy-pg) that binds with its 3′ part inside the *P. gingivalis* 16S rRNA and contains the sequence of the reverse primer at its 5′-terminus (Figure 50.3).

Alternatively, competitive templates can be constructed by cloning appropriate DNA fragments into plasmids constructing a heterologous *P. gingivalis* competitive template.[215]

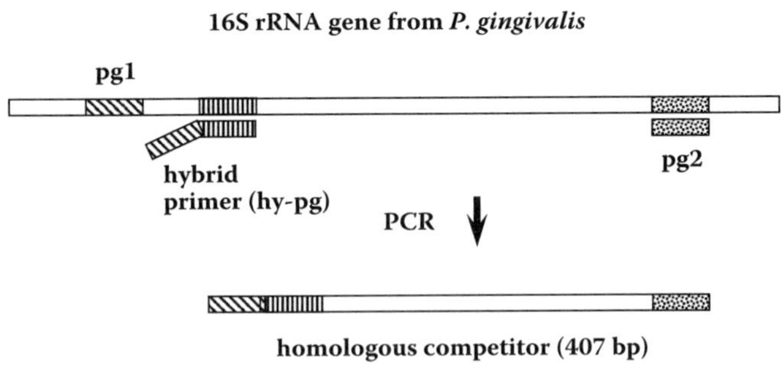

FIGURE 50.3 Principle of synthesis of the homologous competitor. With *P. gingivalis* 16S rRNA as template, the competitor DNA was obtained by PCR using the forward primer (pg1) together with the hybrid primer (hy-pg) that binds with its 3′-part inside the template and contains the sequence of the reverse primer (pg2) at its 5′-terminus.

50.3 CONCLUSIONS AND PERSPECTIVES

The fast development of molecular methods has brought about a specification of the taxonomic position of the genus *Porphyromonas*. Molecular approaches independent of cultivation have revealed the diversity, for example, of human oral microbiota and the existence of a large number of as-yet-to-be-cultured *Porphyromonas* spp. on the basis of 16S rRNA genes.[242] In addition, molecular methods enable the culture-independent detection of *Porphyromonas* spp. on the species as well as on the subspecies level and their association with a number of diseases. The species *P. asaccharolytica*, *P. endodontalis*, and especially, *P. gingivalis* have been the most thoroughly investigated so far. Surely, further species will be detected and classified as *Porphyromonas*. Molecular techniques have also given important insights into the interactions between bacterium and host of the genus *Porphyromonas*, particularly *P. gingivalis*. Numerous virulence factors are known whose functions in vitro have been largely elucidated. *P. gingivalis* may be a key species in the oral microbial community. It may be able to modulate innate host defense for surviving in tissues intra- as well as extracellularly.[243] The behavior of this organism in periodontal pockets may be unobtrusive or more aggressive. It still has to be finally clarified whether individual strains embody both characters or if different subspecies exist that are particularly virulent.[125] A random distribution of potential virulence factors of *P. gingivalis* detected in strains of different geographical origin led to the conclusion, that *P. gingivalis* may have a nonclonal population structure characterized by frequent recombination.[244] However, it is becoming increasingly clear that specific host factors play a decisive role in the clinical outcome.

Hybridization techniques and PCR methods are widely used in clinical science and practice. However, the use of molecular methods for diagnosis and therapy planning is still limited. The different molecular techniques have different limits of detection. PCR, in particular nested PCR, has a very low detection limit. Combination of PCR and hybridization techniques can detect 10^2–10^4 bacteria. The measured prevalence of *Porphyromonas* may differ according to the applied method. This makes the correlation of molecular detection results with clinical diagnoses complicated. Above all, the amount of *P. gingivalis* present in the periodontal pocket seems to be decisive for the progression of oral periodontal diseases,[245] which may also be true for other species of the genus *Porphyromonas*. For clinical application, the detection of *P. gingivalis* is a useful tool for motivating the patient by visualizing the infection. As *P. gingivalis* cannot be completely eradicated by mechanical periodontal therapy, the detection of this pathogen in high counts or percentages of the periodontal flora can support the decision to use systemic antibiotics adjunctively.[246] The application of quantitative molecular techniques offers advantages in comparison to nonquantitative detection. No molecular technique can yet predict, however, the pathogenic potential of an indigenous *P. gingivalis* population or a bacterial community for its host.

REFERENCES

1. Shah, H.N., and Collins, M.D., Proposal for reclassification of *Bacteroides asaccharolyticus*, *Bacteroides gingivalis*, and *Bacteroides endodontalis* in a new genus, *Porphyromonas*, *Int. J. Syst. Bacteriol.*, 38, 128, 1988.
2. Holdeman, L.V.R., Kelly, R.W., and Moore W.E.C. Genus I. *Bacteroides* Castellani and Chalmers 1919. In Krieg, N.R., and Holt, J.G. (Editors), *Bergey's Manual of Systematic Bacteriology*, Vol. I, pp. 604–631, The Williams & Wilkins Co., Baltimore, MD, 1984.
3. Finegold, S.M., and Barnes, E.M., Report of the ICSB taxonomic subcommittee on Gram-negative anaerobic rods, *Int. J. Syst. Bacteriol.*, 27, 388, 1977.
4. Summanen, P.H., Lawson, P.A., and Finegold, S.M., *Porphyromonas bennonis* sp. nov., isolated from human clinical specimens, *Int. J. Syst. Evol. Microbiol.*, 59, 1727, 2009.
5. Willems, A., and Collins, M.D., Reclassification of *Oribaculum catoniae* (Moore and Moore 1994) as *Porphyromonas catoniae* comb. nov. and emendation of the genus *Porphyromonas*, *Int. J. Syst. Bacteriol.*, 45, 578, 1995.
6. Van Steenbergen, T.J.M. et al., *Bacteroides endodontalis* sp. nov., an asaccharolytic black-pigmented *Bacteroides* species from infected dental root canals, *Int. J. Syst. Bacteriol.*, 34, 118, 1984.
7. Summanen, P.H. et al., *Porphyromonas somerae* sp. nov., a pathogen isolated from humans and distinct from *Porphyromonas levii*, *J. Clin. Microbiol.*, 43, 4455, 2005.
8. Conrads, G. et al., 16S-23S rRNA gene internal transcribed spacer sequences for analysis of the phylogenetic relationships among species of the genus *Porphyromonas*, *Int. J. Syst. Evol. Microbiol.*, 55, 607, 2005.
9. Finegold, S.M. et al., *Porphyromonas uenonis* sp. nov., a pathogen for humans distinct from *P. asaccharolytica* and *P. endodontalis*, *J. Clin. Microbiol.*, 42, 5298, 2004.
10. Collins, M.D. et al., Phylogenetic analysis of members of the genus *Porphyromonas* and description of *Porphyromonas cangingivalis* sp. nov. and *Porphyromonas cansulci* sp. nov., *Int. J. Syst. Bacteriol.*, 44, 674, 1994.
11. Allaker, R.P. et al., Dental plaque flora of the dog with reference to fastidious and anaerobic bacteria associated with bites, *J. Vet. Dent.*, 14, 127, 1997.
12. Love, D.N. et al., *Porphyromonas canoris* sp. nov., an asaccharolytic, black-pigmented species from the gingival sulcus of dogs, *Int. J. Syst. Bacteriol.*, 44, 204, 1994.
13. Su, Y. et al., Changes in abundance of *Lactobacillus* spp. and *Streptococcus suis* in the stomach, jejunum and ileum of piglets after weaning, *FEMS Microbiol. Ecol.*, 66, 546, 2008.
14. Kononen, E. et al., Establishment of oral anaerobes during the first year of life, *J. Dent. Res.*, 78, 1634, 1999.
15. Love, D.N. et al., Description of *Porphyromonas circumdentaria* sp. nov. and reassignment of *Bacteroides salivosus* (Love, Johnson, Jones, and Calverley 1987) as *Porphyromonas* (Shah and Collins 1988) salivosa comb. nov., *Int. J. Syst. Bacteriol.*, 42, 434, 1992.
16. Isogai, H. et al., Ecology of genus *Porphyromonas* in canine periodontal disease, *Zentralbl. Veterinarmed. B*, 46, 467, 1999.
17. Norris, J.M., and Love, D.N., Associations amongst three feline *Porphyromonas* species from the gingival margin of cats during periodontal health and disease, *Vet. Microbiol.*, 65, 195, 1999.
18. Norris, J.M., and Love, D.N., The isolation and enumeration of three feline oral *Porphyromonas* species from subcutaneous abscesses in cats, *Vet. Microbiol.*, 65, 115, 1999.
19. Allaker, R.P. et al., Prevalence of *Porphyromonas* and *Prevotella* species in the dental plaque of dogs, *Vet. Rec.*, 140, 147, 1997.
20. Hirasawa, M., and Takada, K., *Porphyromonas gingivicanis* sp. nov. and *Porphyromonas crevioricanis* sp. nov., isolated from beagles, *Int. J. Syst. Bacteriol.*, 44, 637, 1994.
21. Fournier, D. et al., *Porphyromonas gulae* sp. nov., an anaerobic, Gram-negative coccobacillus from the gingival sulcus of various animal hosts, *Int. J. Syst. Evol. Microbiol.*, 51, 1179, 2001.
22. Mikkelsen, D. et al., Phylogenetic analysis of *Porphyromonas* species isolated from the oral cavity of Australian marsupials, *Environ. Microbiol.*, 10, 2425, 2008.
23. Collins, J.G., Offenbacher, S., and Arnold, R.R., Effects of a combination therapy to eliminate *Porphyromonas gingivalis* in refractory periodontitis, *J. Periodontol.*, 64, 998, 1993.
24. Blum, S. et al., Isolation of *Porphyromonas levii* from vaginal samples from cows in herds negative for bovine necrotic vulvovaginitis, *Vet. Rec.*, 163, 745, 2008.
25. Finegold, S.M., and Jousimies-Somer, H., Recently described clinically important anaerobic bacteria: Medical aspects, *Clin. Infect. Dis.*, 25 Suppl 2, 88, 1997.
26. Love, D.N., *Porphyromonas macacae* comb. nov., a consequence of *Bacteroides macacae* being a senior synonym of *Polphyromonas salivosa*, *Int. J. Syst. Bacteriol.*, 45, 90, 1995.
27. Coykendall, A.L., Kaczmarek, F.S., and Slots, J., Genetic heterogeneity in *Bacteroides asaccharolyticus* (Holdeman and Moore, 1970; Finegold and Barnes, 1977; Approved Lists, 1980) and proposal of *Bacteroides gingivalis* sp. nov. and *Bacteroides macacae* (Slots and Genco) comb. nov., *Int. J. Syst. Bacteriol.*, 30, 559, 1980.
28. Holt, J.G. et al., Group 6. Gram-negative, anaerobic, straight, curved, andhelical bacteria. In *Bergey's Manual of Determinative Bacteriology*, 9th ed., pp. 291–334, The Williams & Wilkins Co., Baltimore, MD, 1994.
29. Mayrand, D., and Holt, S.C., Biology of asaccharolytic black-pigmented *Bacteroides* species, *Microbiol. Rev.*, 52, 134, 1988.
30. Shah, H.N., Biochemical and chemical properties of *Porphyromonas gingivalis*. In Shah, H.N. (Editor), *Biology of the Species Porphyromonas gingivalis*, pp. 59–82, CRC Press, Boca Raton, 1993.
31. Shah, H.N. et al., Reclassification of *Bacteroides levii* (Holdeman, Cato & Moore) in the genus *Porphyromonas*, as *Porphyromonas levii* comb. nov., *Int. J. Syst. Bacteriol.*, 45, 586, 1995.
32. Paster, B.J. et al., Phylogeny of *Bacteroides*, *Prevotella* and *Porphyromonas* spp. and related bacteria, *J. Bacteriol.*, 176, 725, 1994.
33. Talan, D.A. et al., Bacteriologic analysis of infected dog and cat bites. Emergency Medicine Animal Bite Infection Study Group, *N. Engl. J. Med.*, 340, 85, 1999.
34. de Lillo, A. et al., Culture-independent identification of periodontitis-associated *Porphyromonas* and *Tannerella* populations by targeted molecular analysis, *J. Clin. Microbiol.*, 42, 5523, 2004.

35. Sayers, E.W. et al., Database resources of the National Center for Biotechnology Information, *Nucleic Acids Res.*, 37, 5, 2009.
36. Benson, D.A. et al., GenBank, *Nucleic Acids Res.*, 37, 26, 2009.
37. Hardham, J.M. et al., Transfer of *Bacteroides splanchnicus* to *Odoribacter* gen. nov. as *Odoribacter splanchnicus* comb. nov., and description of *Odoribacter denticanis* sp. nov., isolated from the crevicular spaces of canine periodontitis patients, *Int. J. Syst. Evol. Microbiol.*, 58, 103, 2008.
38. Socransky, S.S., and Manganiello, S.D., The oral microbiota of man from birth to senility, *J. Periodontol.*, 42, 485, 1971.
39. Salako, N.O. et al., The bacteriology of the supragingival plaque of child dental patients in Kuwait, *Med. Princ. Pract.*, 13, 191, 2004.
40. Brook, I., *Prevotella* and *Porphyromonas* infections in children, *J. Med. Microbiol.*, 42, 340, 1995.
41. Finegold, S.M., *Anaerobic Bacteria in Human Disease*, Academic Press, New York, 1977.
42. Dunkle, L.M., Brotherton, T.J., and Feigin, R.D., Anaerobic infections in children: A prospective study, *Pediatrics*, 57, 311, 1976.
43. Thirumoorthi, M.C., Keen, B.M., and Dajani, A.S., Anaerobic infections in children: A prospective survey, *J. Clin. Microbiol.*, 3, 318, 1976.
44. Lopez, N.J., Occurrence of *Actinobacillus actinomycetemcomitans*, *Porphyromonas gingivalis*, and *Prevotella intermedia* in progressive adult periodontitis, *J. Periodontol.*, 71, 948, 2000.
45. Ximenez-Fyvie, L.A. et al., Subgingival microbiota of periodontally untreated Mexican subjects with generalized aggressive periodontitis, *J. Clin. Periodontol.*, 33, 869, 2006.
46. Herrera, D. et al., Subgingival microbial profiles in chronic periodontitis patients from Chile, Colombia and Spain, *J. Clin. Periodontol.*, 35, 106, 2008.
47. Haffajee, A.D. et al., Subgingival microbiota of chronic periodontitis subjects from different geographic locations, *J. Clin. Periodontol.*, 31, 996, 2004.
48. Sanz, M. et al., Differences in the composition of the subgingival microbiota of two periodontitis populations of different geographical origin. A comparison between Spain and The Netherlands, *Eur. J. Oral Sci.*, 108, 383, 2000.
49. Rocas, I.N. et al., Prevalence of selected bacterial named species and uncultivated phylotypes in endodontic abscesses from two geographic locations, *J. Endod.*, 32, 1135, 2006.
50. Foschi, F. et al., Detection of bacteria in endodontic samples by polymerase chain reaction assays and association with defined clinical signs in Italian patients, *Oral Microbiol. Immunol.*, 20, 289, 2005.
51. Lopez, N.J. et al., Subgingival microbiota of chilean patients with chronic periodontitis, *J. Periodontol.*, 75, 717, 2004.
52. Machado de Oliveira, J.C. et al., Bacterial community profiles of endodontic abscesses from Brazilian and USA subjects as compared by denaturing gradient gel electrophoresis analysis, *Oral Microbiol. Immunol.*, 22, 14, 2007.
53. Baumgartner, J.C. et al., Geographical differences in bacteria detected in endodontic infections using polymerase chain reaction, *J. Endod.*, 30, 141, 2004.
54. Siqueira, J.F. et al., Differences in prevalence of selected bacterial species in primary endodontic infections from two distinct geographic locations, *Oral Surg. Oral Med. Oral Pathol. Oral Radiol. Endod.*, 99, 641, 2005.
55. Kim, S.A., and Lim, S.S., T lymphocyte subpopulations and interleukin-2, interferon-gamma, and interleukin-4 in rat pulpitis experimentally induced by specific bacteria, *J. Endod.*, 28, 202, 2002.
56. Lemke, J.L. et al., Description of freshwater bacterial assemblages from the upper Paraná river floodpulse system, Brazil, *Microb. Ecol.*, 57, 94, 2009.
57. Jansen, H.J. et al., Degradation of immunoglobulin G by periodontal bacteria, *Oral Microbiol. Immunol.*, 9, 345, 1994.
58. Magalashvili, L. et al., Cytokine release and expression induced by OmpA proteins from the Gram-negative anaerobes, *Porphyromonas asaccharolytica* and *Bacteroides fragilis*, *FEMS Immunol. Med. Microbiol.*, 53, 252, 2008.
59. Hohne, C. et al., Occurrence of plasmids in black-pigmented Gram-negative anaerobes, *FEMS Immunol. Med. Microbiol.*, 6, 103, 1993.
60. Odell, E.W., and Wu, P.J., Susceptibility of *Porphyromonas gingivalis* and *P. asaccharolytica* to the nonoxidative killing mechanisms of human neutrophils, *Arch. Oral Biol.*, 37, 597, 1992.
61. Bahar, H. et al., Antimicrobial resistance and beta-lactamase production of clinical isolates of *Prevotella* and *Porphyromonas* species, *Chemotherapy*, 51, 9, 2005.
62. Egwari, L. et al., A study of the anaerobic bacterial flora of the female genital tract in health and disease, *Cent. Afr. J. Med.*, 41, 391, 1995.
63. Holst, E., Goffeng, A.R., and Andersch, B., Bacterial vaginosis and vaginal microorganisms in idiopathic premature labor and association with pregnancy outcome, *J. Clin. Microbiol.*, 32, 176, 1994.
64. Duerden, B.I., Black-pigmented Gram-negative anaerobes in genito-urinary tract and pelvic infections, *FEMS Immunol. Med. Microbiol.*, 6, 223, 1993.
65. Neut, C. et al., Treatment of diversion colitis with short-chain fatty acids. Bacteriological study, *Gastroenterol. Clin. Biol.*, 19, 871, 1995.
66. Sharma, D., and Rouphael, N., Case report: Atypical Lemierre's disease secondary to *Porphyromonas asaccharolytica*, *Curr. Opin. Pediatr.*, 21, 397, 2009.
67. Brook, I., and Frazier, E.H., Bacteriology of chronic maxillary sinusitis associated with nasal polyposis, *J. Med. Microbiol.*, 54, 595, 2005.
68. Castagnola, E. et al., Broviac catheter-related bacteraemias due to unusual pathogens in children with cancer: Case reports with literature review, *J. Infect.*, 34, 215, 1997.
69. Revankar, S.G., and Clark, R.A., Infected cardiac myxoma. Case report and literature review, *Medicine (Baltimore)*, 77, 337, 1998.
70. Tran, T. et al., Absence of *Porphyromonas asaccharolytica*, *Bacteroides fragilis* and *Chlamydia pneumoniae* in human subgingival plaque, *Oral Microbiol. Immunol.*, 12, 377, 1997.
71. Hall, E.R. et al., The gingival immune response to *Actinobacillus actinomycetemcomitans* in juvenile periodontitis, *J. Periodontol.*, 62, 792, 1991.
72. Rawlinson, A., Duerden, B.I., and Goodwin, L., New findings on the microbial flora associated with adult periodontitis, *J. Dent.*, 21, 179, 1993.
73. Ueno, K. et al., Cytotoxicity of anaerobic bacteria isolated from infected root canal, *Kokubyo Gakkai Zasshi*, 60, 408, 1993.
74. Peters, L.B. et al., Viable bacteria in root dentinal tubules of teeth with apical periodontitis, *J. Endod.*, 27, 76, 2001.
75. Rawlinson, A., Duerden, B.I., and Goodwin, L., Effects of root planing on the distribution of microorganisms adult periodontitis sites, *Eur. J. Prosthodont. Restor. Dent*, 1, 103, 1993.
76. Moore, L.V., and Moore, W.E., *Oribaculum catoniae* gen. nov., sp. nov.; *Catonella morbi* gen. nov., sp. nov.; *Hallella seregens* gen. nov., sp. nov.; *Johnsonella ignava* gen. nov., sp.

nov.; and *Dialister pneumosintes* gen. nov., comb. nov., nom. rev., Anaerobic Gram-negative bacilli from the human gingival crevice, *Int. J. Syst. Bacteriol.*, 44, 187, 1994.
77. Moore, W.E. et al., Comparative bacteriology of juvenile periodontitis, *Infect. Immun.*, 48, 507, 1985.
78. Moore, W.E. et al., The microflora of periodontal sites showing active destructive progression, *J. Clin. Periodontol.*, 18, 729, 1991.
79. Zerr, M.A. et al., Growth stimulation of *Porphyromonas endodontalis* by hemoglobin and protoporphyrin IX, *Oral Microbiol. Immunol.*, 15, 365, 2000.
80. Persson, S. et al., The formation of hydrogen sulfide and methyl mercaptan by oral bacteria, *Oral Microbiol. Immunol.*, 5, 195, 1990.
81. Ho, Y.C., and Chang, Y.C., Effects of a bacterial lipid byproduct on human pulp fibroblasts in vitro, *J. Endod.*, 33, 437, 2007.
82. Rocas, I.N., and Siqueira, J.F.Jr., Root canal microbiota of teeth with chronic apical periodontitis, *J. Clin. Microbiol.*, 46, 3599, 2008.
83. Sundqvist, G., Associations between microbial species in dental root canal infections, *Oral Microbiol. Immunol.*, 7, 257, 1992.
84. Zerr, M. et al., *Porphyromonas endodontalis* binds, reduces and grows on human hemoglobin, *Oral Microbiol. Immunol.*, 16, 229, 2001.
85. Rosen, G. et al., The purification and characterization of an 88-kDa *Porphyromonas endodontalis* 35406 protease, *Oral Microbiol. Immunol.*, 16, 326, 2001.
86. Horiba, N. et al., Complement activation by lipopolysaccharides purified from Gram-negative bacteria isolated from infected root canals, *Oral Surg. Oral Med. Oral Pathol.*, 74, 648, 1992.
87. Sundqvist, G. et al., Degradation of human immunoglobulins G and M and complement factors C3 and C5 by black-pigmented *Bacteroides*, *J. Med. Microbiol.*, 19, 85, 1985.
88. Chang, J. et al., NF-kappaB activation in human dental pulp stem cells by TNF and LPS, *J. Dent. Res.*, 84, 994, 2005.
89. Yang, L.C. et al., Induction of interleukin-8 gene expression by black-pigmented *Bacteroides* in human pulp fibroblasts and osteoblasts, *Int. Endod. J.*, 36, 774, 2003.
90. Chen, S.C. et al., The upregulation of receptor activator NF-kappaB ligand expression by interleukin-1alpha and *Porphyromonas endodontalis* in human osteoblastic cells, *Int. Endod. J.*, 42, 375, 2009.
91. Hosoya, S. et al., Stimulation of interleukin-1beta-independent interleukin-6 production in human dental pulp cells by lipopolysaccharide, *Biochem. Mol. Med.*, 59, 138, 1996.
92. Ko, H.J., and Lim, S.S., Production of macrophage inflammatory protein (MIP)-1alpha and MIP-1beta by human polymorphonuclear neutrophils stimulated with *Porphyromonas endodontalis* lipopolysaccharide, *J. Endod.*, 28, 754, 2002.
93. Hong, C.Y. et al., The role of lipopolysaccharide in infectious bone resorption of periapical lesion, *J. Oral Pathol. Med.*, 33, 162, 2004.
94. Chang, Y.C. et al., Stimulation of matrix metalloproteinases by black-pigmented *Bacteroides* in human pulp and periodontal ligament cell cultures, *J. Endod.*, 28, 90, 2002.
95. Yang, L.C. et al., Induction of vascular endothelial growth factor expression in human pulp fibroblasts stimulated with black-pigmented *Bacteroides*, *Int. Endod. J.*, 37, 588, 2004.
96. Jiang, Y., and Schilder, H., An optimal host response to a bacterium may require the interaction of leukocytes and resident host cells, *J. Endod.*, 28, 279, 2002.
97. Chang, Y.C. et al., Induction of cyclooxygenase-2 mRNA and protein expression in human pulp cells stimulated with black-pigmented *Bacteroides*, *J. Endod.*, 29, 240, 2003.
98. Huang, F.M. et al., Examination of the signal transduction pathways leading to upregulation of tissue type plasminogen activator by *Porphyromonas endodontalis* in human pulp cells, *Int. Endod. J.*, 38, 860, 2005.
99. Yang, S.F. et al., Effect of black-pigmented bacteria on the plasminogen-plasmin system in human pulp and osteoblastic cells, *Oral Surg. Oral Med. Oral Pathol. Oral Radiol. Endod.*, 95, 621, 2003.
100. Dorn, B.R. et al., Invasion of vascular cells in vitro by *Porphyromonas endodontalis*, *Int. Endod. J.*, 35, 366, 2002.
101. Siqueira, J.F. Jr. et al., Molecular detection of black-pigmented bacteria in infections of endodontic origin, *J. Endod.*, 27, 563, 2001.
102. Blome, B. et al., Molecular identification and quantification of bacteria from endodontic infections using real-time polymerase chain reaction, *Oral Microbiol. Immunol.*, 23, 384, 2008.
103. Sunde, P.T. et al., Assessment of periradicular microbiota by DNA–DNA hybridization, *Endod. Dent. Traumatol.*, 16, 191, 2000.
104. Bogen, G., and Slots, J., Black-pigmented anaerobic rods in closed periapical lesions, *Int. Endod. J.*, 32, 204, 1999.
105. Vickerman, M.M. et al., Phylogenetic analysis of bacterial and archaeal species in symptomatic and asymptomatic endodontic infections, *J. Med. Microbiol.*, 56, 110, 2007.
106. Tomazinho, L.F., and Avila-Campos, M.J., Detection of *Porphyromonas gingivalis*, *Porphyromonas endodontalis*, *Prevotella intermedia*, and *Prevotella nigrescens* in chronic endodontic infection, *Oral Surg. Oral Med. Oral Pathol. Oral Radiol. Endod.*, 103, 285, 2007.
107. Siqueira, J.F. Jr. et al., Bacteria in the apical root canal of teeth with primary apical periodontitis, *Oral Surg. Oral Med. Oral Pathol. Oral Radiol Endod*, 107, 721, 2009.
108. Gomes, B.P. et al., *Porphyromonas gingivalis*, *Porphyromonas endodontalis*, *Prevotella intermedia* and *Prevotella nigrescens* in endodontic lesions detected by culture and by PCR, *Oral Microbiol. Immunol.*, 20, 211, 2005.
109. Martin, F.E. et al., Quantitative microbiological study of human carious dentine by culture and real-time PCR: Association of anaerobes with histopathological changes in chronic pulpitis, *J. Clin. Microbiol.*, 40, 1698, 2002.
110. Persson, G.R. et al., *Tannerella forsythia* and *Pseudomonas aeruginosa* in subgingival bacterial samples from parous women, *J. Periodontol.*, 79, 508, 2008.
111. Kumar, P.S. et al., New bacterial species associated with chronic periodontitis, *J. Dent. Res.*, 82, 338, 2003.
112. van Winkelhoff, A.J., van Steenbergen, T.J., and de Graaff, J., *Porphyromonas* (*Bacteroides*) endodontalis: Its role in endodontal infections, *J. Endod.*, 18, 431, 1992.
113. Enersen, M. et al., *fimA* genotypes and multilocus sequence types of *Porphyromonas gingivalis* from patients with periodontitis, *J. Clin. Microbiol.*, 46, 31, 2008.
114. Hamada, S. et al., The importance of fimbriae in the virulence and ecology of some oral bacteria, *Oral Microbiol. Immunol.*, 13, 129, 1998.
115. Suzuki, N., Nakano, Y., and Kiyoura, Y., Characterizing the specific coaggregation between *Actinobacillus actinomycetemcomitans* serotype c strains and *Porphyromonas gingivalis* ATCC 33277, *Oral Microbiol. Immunol.*, 21, 385, 2006.
116. Yamada, M., Ikegami, A., and Kuramitsu, H.K., Synergistic biofilm formation by Treponema denticola and *Porphyromonas gingivalis*, *FEMS Microbiol. Lett.*, 250, 271, 2005.

117. Maeda, K. et al., Oral streptococcal glyceraldehyde-3-phosphate dehydrogenase mediates interaction with *Porphyromonas gingivalis* fimbriae, *Microbes. Infect.*, 6, 1163, 2004.
118. Hongo, H., Takano, H., and Morita, M., Dense fimbrial meshwork enhances *Porphyromonas gingivalis* adhesiveness: A scanning electron microscopic study, *J. Periodontal. Res.*, 42, 114, 2007.
119. Ogawa, T., and Hamada, S., Hemagglutinating and chemotactic properties of synthetic peptide segments of fimbrial protein from *Porphyromonas gingivalis*, *Infect. Immun.*, 62, 3305, 1994.
120. Lewis, J.P. et al., Hemoglobinase activity of the lysine gingipain protease (Kgp) of *Porphyromonas gingivalis* W83, *J. Bacteriol.*, 181, 4905, 1999.
121. Lewis, J.P., and Macrina, F.L., Localization of HArep-containing genes on the chromosome of *Porphyromonas gingivalis* W83, *Infect. Immun.*, 67, 2619, 1999.
122. Slaney, J.M. et al., Mechanisms of resistance of *Porphyromonas gingivalis* to killing by serum complement, *Infect. Immun.*, 74, 5352, 2006.
123. Liu, R. et al., *P. gingivalis* and *E. coli* lipopolysaccharides exhibit different systemic but similar local induction of inflammatory markers, *J. Periodontol.* 79, 1241, 2008.
124. Domon, H. et al., Up-regulation of the endoplasmic reticulum stress-response in periodontal disease, *Clin. Chim. Acta.*, 401, 134, 2009.
125. Hajishengallis, G., *Porphyromonas gingivalis*-host interactions: Open war or intelligent guerilla tactics? *Microbes. Infect.*, 11, 637, 2009.
126. Ebersole, J.L. et al., Comparative virulence of periodontopathogens in a mouse abscess model, *Oral Dis.*, 1, 115, 1995.
127. Yilmaz, O., The chronicles of *Porphyromonas gingivalis*: The microbium, the human oral epithelium and their interplay, *Microbiology*, 154, 2897, 2008.
128. Hajishengallis, G. et al., Subversion of innate immunity by periodontopathic bacteria via exploitation of complement receptor-3, *Adv. Exp. Med. Biol.*, 632, 203, 2008.
129. Yoneda, M. et al., Mixed infection of *Porphyromonas gingivalis* and *Bacteroides forsythus* in a murine abscess model: Involvement of gingipains in a synergistic effect, *J. Periodontal. Res.*, 36, 237, 2001.
130. Diaz, P.I. et al., Molecular characterization of subject-specific oral microflora during initial colonization of enamel, *Appl. Environ. Microbiol.*, 72, 2837, 2006.
131. Amano, A. et al., Variations of *Porphyromonas gingivalis* fimbriae in relation to microbial pathogenesis, *J. Periodontal. Res.*, 39, 136, 2004.
132. Njoroge, T. et al., A role for fimbriae in *Porphyromonas gingivalis* invasion of oral epithelial cells, *Infect. Immun.*, 65, 1980, 1997.
133. Okuda, K., Slots, J., and Genco, J., *Bacteroides gingivalis*, *Bacteroides asaccharolyticus* and *Bacteroides melanogenicus* subspecies: Cell surface morphology and adherence to erythrocytes and human buccal epithelial cells, *Curr. Microbiol.*, 6, 7, 1981.
134. Yoshimura, F. et al., Fimbriae from the oral anaerobe *Bacteroides gingivalis*: Physical, chemical, and immunological properties, *J. Bacteriol.*, 163, 730, 1985.
135. Takahashi, Y. et al., Detection of fimbrilin gene (fimA) in *Porphyromonas* (*Bacteroides*) *gingivalis* by Southern blot analysis, *J. Periodontal. Res.*, 27, 599, 1992.
136. Gibson, F.C., and Genco, C.A., The Genus *Porphyromonas*. In Dworkin, M., Falkow, S., Rosenberg, E., Schleifer, K.H., and Stackebrandt, E. (Editors), *The Prokaryotes: A Handbook on the Biology of Bacteria*, pp. 428–454, Springer, Singapore, 2006.
137. Olsen, I., Shah, H.N., and Gharbia, S.E., Taxonomy and biochemical characteristics of *Actinobacillus actinomycetemcomitans* and *Porphyromonas gingivalis*, *Periodontol. 2000*, 20, 14, 1999.
138. Kontani, M. et al., Adherence of *Porphyromonas gingivalis* to matrix proteins via a fimbrial cryptic receptor exposed by its own arginine-specific protease, *Mol. Microbiol.*, 24, 1179, 1997.
139. Amano, A. et al., Structural domains of *Porphyromonas gingivalis* recombinant fimbrillin that mediate binding to salivary proline-rich protein and statherin, *Infect. Immun.*, 64, 1631, 1996.
140. Amano, A. et al., Binding of *Porphyromonas gingivalis* fimbriae to proline-rich glycoproteins in parotid saliva via a domain shared by major salivary components, *Infect. Immun.*, 66, 2072, 1998.
141. Goulbourne, P.A., and Ellen, R.P., Evidence that *Porphyromonas* (*Bacteroides*) *gingivalis* fimbriae function in adhesion to Actinomyces viscosus, *J. Bacteriol.*, 173, 5266, 1991.
142. Lamont, R.J. et al., Involvement of *Porphyromonas gingivalis* fimbriae in adherence to Streptococcus gordonii, *Oral Microbiol. Immunol.*, 8, 272, 1993.
143. Murakami, Y. et al., *Porphyromonas gingivalis* fimbrillin is one of the fibronectin-binding proteins, *Infect. Immun.*, 64, 2571, 1996.
144. Sojar, H.T., Hamada, N., and Genco, R.J., Structures involved in the interaction of *Porphyromonas gingivalis* fimbriae and human lactoferrin, *FEBS Lett.*, 422, 205, 1998.
145. Umemoto, T., and Hamada, N., Characterization of biologically active cell surface components of a periodontal pathogen. The roles of major and minor fimbriae of *Porphyromonas gingivalis*, *J. Periodontol.*, 74, 119, 2003.
146. Hu, S.W. et al., Transvascular dissemination of *Porphyromonas gingivalis* from a sequestered site is dependent upon activation of the kallikrein/kinin pathway, *J. Periodontal. Res.*, 41, 200, 2006.
147. Imamura, T., The role of gingipains in the pathogenesis of periodontal disease, *J. Periodontol.*, 74, 111, 2003.
148. Wingrove, J.A. et al., Activation of complement components C3 and C5 by a cysteine proteinase (gingipain-1) from *Porphyromonas* (*Bacteroides*) *gingivalis*, *J. Biol. Chem.*, 267, 18902, 1992.
149. Cutler, C.W., Arnold, R.R., and Schenkein, H.A., Inhibition of C3 and IgG proteolysis enhances phagocytosis of *Porphyromonas gingivalis*, *J. Immunol.*, 151, 7016, 1993.
150. Hawlisch, H. et al., C5a negatively regulates toll-like receptor 4-induced immune responses, *Immunity*, 22, 415, 2005.
151. Kumagai, Y. et al., Molecular mechanism for connective tissue destruction by dipeptidyl aminopeptidase IV produced by the periodontal pathogen *Porphyromonas gingivalis*, *Infect. Immun.*, 73, 2655, 2005.
152. Holt, S.C. et al., Virulence factors of *Porphyromonas gingivalis*, *Periodontol. 2000*, 20, 168, 1999.
153. Potempa, J., Banbula, A., and Travis, J., Role of bacterial proteinases in matrix destruction and modulation of host responses, *Periodontol. 2000*, 24, 153, 2000.
154. Sroka, A. et al., Degradation of host heme proteins by lysine- and arginine-specific cysteine proteinases (gingipains) of *Porphyromonas gingivalis*, *J. Bacteriol.*, 183, 5609, 2001.
155. Liu, X. et al., Coordinate expression of the *Porphyromonas gingivalis* lysine-specific gingipain proteinase, Kgp,

arginine-specific gingipain proteinase, RgpA, and the heme/hemoglobin receptor, HmuR, *Biol. Chem.*, 385, 1049, 2004.
156. Wojtowicz, H. et al., Unique structure and stability of HmuY, a novel heme-binding protein of *Porphyromonas gingivalis*, *PLoS Pathog.*, 5, e1000419, 2009.
157. Simpson, W., Olczak, T., and Genco, C.A., Characterization and expression of HmuR, a TonB-dependent hemoglobin receptor of *Porphyromonas gingivalis*, *J. Bacteriol.*, 182, 5737, 2000.
158. Darveau, R.P. et al., Ability of bacteria associated with chronic inflammatory disease to stimulate E-selectin expression and promote neutrophil adhesion, *Infect. Immun.*, 63, 1311, 1995.
159. Darveau, R.P. et al., *Porphyromonas gingivalis* lipopolysaccharide contains multiple lipid A species that functionally interact with both toll-like receptors 2 and 4, *Infect. Immun.*, 72, 5041, 204.
160. Al-Qutub, M.N. et al., Hemin-dependent modulation of the lipid A structure of *Porphyromonas gingivalis* lipopolysaccharide, *Infect. Immun.*, 74, 4474, 2006.
161. Lamont, R.J., and Jenkinson, H.F., Life below the gum line: Pathogenic mechanisms of *Porphyromonas gingivalis*, *Microbiol. Mol. Biol. Rev.*, 62, 1244, 1998.
162. Domon, H. et al., Early and preferential induction of IL-1 receptor-associated kinase-M in THP-1 cells by LPS derived from *Porphyromonas gingivalis*, *J. Leukoc. Biol.*, 83, 672, 2008.
163. Cunningham, M.D. et al., Helicobacter pylori and *Porphyromonas gingivalis* lipopolysaccharides are poorly transferred to recombinant soluble CD14, *Infect. Immun.*, 64, 3601, 1996.
164. Isogai, E. et al., In vivo induction of apoptosis and immune responses in mice by administration of lipopolysaccharide from *Porphyromonas gingivalis*, *Infect. Immun.*, 64, 1461, 1996.
165. Haapasalo, M., Kerosuo, E., and Lounatmaa, K., Hydrophobicities of human polymorphonuclear leukocytes and oral *Bacteroides* and *Porphyromonas* spp., *Wolinella recta*, and *Eubacterium yurii* with special reference to bacterial surface structures, *Scand. J. Dent. Res.*, 98, 472, 1990.
166. Ingham, H.R. et al., Inhibition of phagocytosis in vitro by obligate anaerobes, *Lancet*, 2, 1252, 1977.
167. Laine, M.L., and van Winkelhoff, A.J., Virulence of six capsular serotypes of *Porphyromonas gingivalis* in a mouse model, *Oral Microbiol. Immunol.*, 13, 322, 1998.
168. Laine, M.L., Appelmelk, B.J., and van Winkelhoff, A.J., Prevalence and distribution of six capsular serotypes of *Porphyromonas gingivalis* in periodontitis patients, *J. Dent. Res.*, 76, 1840, 1997.
169. Laine, M.L., Appelmelk, B.J., and van Winkelhoff, A.J., Novel polysaccharide capsular serotypes in *Porphyromonas gingivalis*, *J. Periodontal. Res.*, 31, 278, 1996.
170. Furuta, N., Takeuchi, H., and Amano, A., Entry of *Porphyromonas gingivalis* outer membrane vesicles into epithelial cells causes cellular functional impairment, *Infect. Immun.*, 77, 4761, 2009.
171. Furuta, N. et al., *Porphyromonas gingivalis* outer membrane vesicles enter human epithelial cells via an endocytic pathway and are sorted to lysosomal compartments, *Infect. Immun.*, 77, 4187, 2009.
172. Kurita-Ochiai, T., Fukushima, K., and Ochiai, K., Butyric acid-induced apoptosis of murine thymocytes, splenic T cells, and human Jurkat T cells, *Infect. Immun.*, 65, 35, 1997.
173. Kurita-Ochiai, T., Ochiai, K., and Fukushima, K., Volatile fatty acid, metabolic by-product of periodontopathic bacteria, induces apoptosis in WEHI 231 and RAJI B lymphoma cells and splenic B cells, *Infect. Immun.*, 66, 2587, 1998.
174. Borrell, L.N., and Papapanou, P.N., Analytical epidemiology of periodontitis, *J. Clin. Periodontol.*, 32, 132, 2005.
175. Tanner, A. et al., Microbiota of health, gingivitis, and initial periodontitis, *J. Clin. Periodontol.*, 25, 85, 1998.
176. Cortelli, J.R. et al., Detection of periodontal pathogens in oral mucous membranes of edentulous individuals, *J. Periodontol.*, 79, 1962, 2008.
177. Cortelli, J.R. et al., Etiological analysis of initial colonization of periodontal pathogens in oral cavity, *J. Clin. Microbiol.*, 46, 1322, 2008.
178. Ximenez-Fyvie, L.A., Haffajee, A.D., and Socransky, S.S., Comparison of the microbiota of supra- and subgingival plaque in health and periodontitis, *J. Clin. Periodontol.*, 27, 648, 2000.
179. Haffajee, A.D. et al., Subgingival microbiota in healthy, well-maintained elder and periodontitis subjects, *J. Clin. Periodontol.*, 25, 346, 1998.
180. Socransky, S.S. et al., Microbial complexes in subgingival plaque, *J. Clin. Periodontol.*, 25, 134, 1998.
181. Colombo, A.P. et al., Clinical and microbiological features of refractory periodontitis subjects, *J. Clin. Periodontol.*, 25, 169, 1998.
182. Loos, B.G. et al., Genetic structure of populations of *Porphyromonas gingivalis* associated with periodontitis and other oral infections, *Infect. Immun.*, 61, 204, 1993.
183. Mager, D.L. et al., Distribution of selected bacterial species on intraoral surfaces, *J. Clin. Periodontol.*, 30, 644, 2003.
184. Beikler, T. et al., Sampling strategy for intraoral detection of periodontal pathogens before and following periodontal therapy, *J. Periodontol.*, 77, 1323, 2006.
185. Gaetti-Jardim, E. Jr. et al., Quantitative detection of periodontopathic bacteria in atherosclerotic plaques from coronary arteries, *J. Med. Microbiol.*, 58, 1568, 2009.
186. Rudolph, T. et al., 16S rDNA PCR analysis of infectious keratitis: A case series, *Acta. Ophthalmol. Scand.*, 82, 463, 2004.
187. Perez-Chaparro, P.J. et al., Genotypic characterization of *Porphyromonas gingivalis* isolated from subgingival plaque and blood sample in positive bacteremia subjects with periodontitis, *J. Clin. Periodontol.*, 35, 748, 2008.
188. Iida, Y. et al., Brain abscess in which *Porphyromonas gingivalis* was detected in cerebrospinal fluid, *Br. J. Oral Maxillofac. Surg.*, 42, 180, 2004.
189. Hirata, R. Jr. et al., Isolation of *Porphyromonas gingivalis* strain from tubal-ovarian abscess, *J. Clin. Microbiol.*, 33, 1925, 1995.
190. Soskolne, W.A., and Klinger, A., The relationship between periodontal diseases and diabetes: An overview, *Ann. Periodontol.*, 6, 91, 2001.
191. Matthews, D.C., The relationship between diabetes and periodontal disease, *J. Can. Dent. Assoc.*, 68, 161, 2002.
192. Seymour, G.J. et al., Infection or inflammation: The link between periodontal and cardiovascular diseases, *Future Cardiol.*, 5, 5, 2009.
193. Seymour, G.J. et al., Relationship between periodontal infections and systemic disease, *Clin. Microbiol. Infect.*, 13 Suppl 4, 3, 2007.
194. Jousimies-Somer, H.R., Update on the taxonomy and the clinical and laboratory characteristics of pigmented anaerobic Gram-negative rods, *Clin. Infect. Dis.*, 20 Suppl 2, 187, 1995.
195. Stoner, K.A. et al., Quantitative survival of aerobic and anaerobic microorganisms in Port-A-Cul and Copan transport systems, *J. Clin. Microbiol.*, 46, 2739, 2008.

196. van Steenbergen, T.J. et al., Survival in transport media of *Actinobacillus actinomycetemcomitans*, *Porphyromonas gingivalis* and *Prevotella intermedia* in human subgingival samples, *Oral Microbiol. Immunol.*, 8, 370, 1993.
197. Falagas, M.E., and Siakavellas, E., *Bacteroides*, *Prevotella*, and *Porphyromonas* species: A review of antibiotic resistance and therapeutic options, *Int. J. Antimicrob. Agents*, 15, 1, 2000.
198. Stingu, C.S. et al., Rapid identification of oral anaerobic bacteria cultivated from subgingival biofilm by MALDI-TOF-MS, *Oral Microbiol. Immunol.*, 23, 372, 2008.
199. Kitch, T.T., and Appelbaum, P.C., Accuracy and reproducibility of the 4-hour ATB 32A method for anaerobe identification, *J. Clin. Microbiol.*, 27, 2509, 1989.
200. Assmus, B. et al., Direct examination of subgingival plaque from a diseased periodontal site using confocal laser scanning microscopy (CLSM), *New Microbiol.*, 20, 155, 1997.
201. Wolff, L.F. et al., Fluorescence immunoassay for detecting periodontal bacterial pathogens in plaque, *J. Clin. Microbiol.*, 29, 1645, 1991.
202. van der Ploeg, J.R. et al., Quantitative detection of *Porphyromonas gingivalis* fimA genotypes in dental plaque, *FEMS Microbiol. Lett.*, 232, 31, 2004.
203. Pischon, N. et al., Effects of *Porphyromonas gingivalis* on cell cycle progression and apoptosis of primary human chondrocytes, *Ann. Rheum. Dis.*, 68, 1902, 2009.
204. Friedrichs, C. et al., Rapid identification of viridans streptococci by mass spectrometric discrimination, *J. Clin. Microbiol.*, 45, 2392, 2007.
205. Ruelle, V. et al., Rapid identification of environmental bacterial strains by matrix-assisted laser desorption/ionization time-of-flight mass spectrometry, *Rapid Commun. Mass Spectrom.*, 18, 2013, 2004.
206. Fagerquist, C.K. et al., Sub-speciating Campylobacter jejuni by proteomic analysis of its protein biomarkers and their post-translational modifications, *J. Proteome Res.*, 5, 2527, 2006.
207. Rupf, S. et al., Differentiation of mutans streptococci by intact cell matrix-assisted laser desorption/ionization time-of-flight mass spectrometry, *Oral Microbiol. Immunol.*, 20, 267, 2005.
208. Loesche, W.J. et al., Development of a diagnostic test for anaerobic periodontal infections based on plaque hydrolysis of benzoyl-DL-arginine-naphthylamide, *J. Clin. Microbiol.*, 28, 1551, 1990.
209. Gemmell, E. et al., Antibody responses of *Porphyromonas gingivalis* infected gingivitis and periodontitis subjects, *Oral Dis.*, 1, 63, 1995.
210. Teanpaisan, R., and Douglas, C.W., Molecular fingerprinting of *Porphyromonas gingivalis* by PCR of repetitive extragenic palindromic (REP) sequences and comparison with other fingerprinting methods, *J. Med. Microbiol.*, 48, 741, 1999.
211. Boutaga, K. et al., Comparison of real-time PCR and culture for detection of *Porphyromonas gingivalis* in subgingival plaque samples, *J. Clin. Microbiol.*, 41, 4950, 2003.
212. Yoshida, A. et al., Development of a 5′ fluorogenic nuclease-based real-time PCR assay for quantitative detection of *Actinobacillus actinomycetemcomitans* and *Porphyromonas gingivalis*, *J. Clin. Microbiol.*, 41, 863, 2003.
213. Rupf, S. et al., Comparison of profiles of key periodontal pathogens in periodontium and endodontium, *Endod. Dent. Traumatol.*, 16, 269, 2000.
214. Morillo, J.M. et al., Quantitative real-time polymerase chain reaction based on single copy gene sequence for detection of periodontal pathogens, *J. Clin. Periodontol.*, 31, 1054, 2004.
215. Doungudomdacha, S., Rawlinson, A., and Douglas, C.W., Enumeration of *Porphyromonas gingivalis*, *Prevotella intermedia* and *Actinobacillus actinomycetemcomitans* in subgingival plaque samples by a quantitative-competitive PCR method, *J. Med. Microbiol.*, 49, 861, 2000.
216. Sanz, M. et al., Methods of detection of *Actinobacillus actinomycetemcomitans*, *Porphyromonas gingivalis* and *Tannerella forsythensis* in periodontal microbiology, with special emphasis on advanced molecular techniques: A review, *J. Clin. Periodontol.*, 31, 1034, 2004.
217. Sakamoto, M., Umeda, M., and Benno, Y., Molecular analysis of human oral microbiota, *J. Periodontal. Res.*, 40, 277, 2005.
218. Rupf, S., Merte, K., and Eschrich, K., Quantification of bacteria in oral samples by competitive polymerase chain reaction, *J. Dent. Res.*, 78, 850, 1999.
219. Saito, D. et al., Real-time polymerase chain reaction quantification of *Porphyromonas gingivalis* and *Tannerella forsythia* in primary endodontic infections, *J. Endod.*, 35, 1518, 2009.
220. Slots, J. et al., Detection of putative periodontal pathogens in subgingival specimens by 16S ribosomal DNA amplification with the polymerase chain reaction, *Clin. Infect. Dis.*, 20 Suppl 2, 304, 1995.
221. Tran, S.D., and Rudney, J.D., Improved multiplex PCR using conserved and species-specific 16S rRNA gene primers for simultaneous detection of *Actinobacillus actinomycetemcomitans*, *Bacteroides forsythus*, and *Porphyromonas gingivalis*, *J. Clin. Microbiol.*, 37, 3504, 1999.
222. Haffajee, A.D. et al., Comparison between polymerase chain reaction-based and checkerboard DNA hybridization techniques for microbial assessment of subgingival plaque samples, *J. Clin. Periodontol.*, 36, 642, 2009.
223. Verner, C. et al., Carpegen real-time polymerase chain reaction vs. anaerobic culture for periodontal pathogen identification, *Oral Microbiol. Immunol.*, 21, 341, 2006.
224. Menard, C., and Mouton, C., Clonal diversity of the taxon *Porphyromonas gingivalis* assessed by random amplified polymorphic DNA fingerprinting, *Infect. Immun.*, 63, 2522, 1995.
225. Luterbacher, S. et al., Diagnostic characteristics of clinical and microbiological tests for monitoring periodontal and peri-implant mucosal tissue conditions during supportive periodontal therapy (SPT), *Clin. Oral Implants Res.*, 11, 521, 2000.
226. Eguchi, T. et al., Microbial changes in patients with acute periodontal abscess after treatment detected by PadoTest, *Oral Dis.*, 14, 180, 2008.
227. Socransky, S.S. et al., "Checkerboard" DNA–DNA hybridization, *Biotechniques*, 17, 788, 1994.
228. Paster, B.J., Bartoszyk, I.M., and Dewhirst, F.E., Identification of oral streptococci using PCR-based, reverse-capture, checkerboard hybridization, *Methods Cell Sci.*, 20, 223, 1998.
229. Koehler, A. et al., Multilocus sequence analysis of *Porphyromonas gingivalis* indicates frequent recombination, *Microbiology*, 149, 2407, 2003.
230. Colombo, A.P. et al., Comparisons of subgingival microbial profiles of refractory periodontitis, severe periodontitis, and periodontal health using the human oral microbe identification microarray, *J. Periodontol.*, 80, 1421, 2009.
231. Eberhard, J. et al., The stage of native biofilm formation determines the gene expression of human beta-defensin-2, psoriasin, ribonuclease 7 and inflammatory mediators: A novel approach for stimulation of keratinocytes with in situ formed biofilms, *Oral Microbiol. Immunol.*, 23, 21, 2008.
232. Zijnge, V. et al., Denaturing gradient gel electrophoresis as a diagnostic tool in periodontal microbiology, *J. Clin. Microbiol.*, 44, 3628, 2006.

233. Takeshita, T. et al., The ecological proportion of indigenous bacterial populations in saliva is correlated with oral health status, *ISME J.*, 3, 65, 2009.
234. Boutaga, K. et al., Comparison of subgingival bacterial sampling with oral lavage for detection and quantification of periodontal pathogens by real-time polymerase chain reaction, *J. Periodontol.*, 78, 79, 2007.
235. Krigar, D.M. et al., Two subgingival plaque-sampling strategies used with RNA probes, *J. Periodontol.*, 78, 72, 2007.
236. Teles, F.R., Haffajee, A.D., and Socransky, S.S., The reproducibility of curet sampling of subgingival biofilms, *J. Periodontol.*, 79, 705, 2008.
237. Jervoe-Storm, P.M. et al., Comparison of curet and paper point sampling of subgingival bacteria as analyzed by real-time polymerase chain reaction, *J. Periodontol.*, 78, 909, 2007.
238. Al-Robaiy, S., Rupf, S., and Eschrich, K., Rapid competitive PCR using melting curve analysis for DNA quantification, *Biotechniques*, 31, 1382, 1388, 2001.
239. Rupf, S. et al., Comparison of different techniques of quantitative PCR for determination of *Streptococcus mutans* counts in saliva samples, *Oral Microbiol. Immunol.*, 18, 50, 2003.
240. Rupf, S. et al., In vitro, clinical, and microbiological evaluation of a linear oscillating device for scaling and root planing, *J. Periodontol.*, 76, 1942, 2005.
241. Rupf, S. et al., Quantitative determination of *Streptococcus mutans* by using competitive polymerase chain reaction, *Eur. J. Oral Sci.*, 107, 75, 1999.
242. Paster, B.J. et al., The breadth of bacterial diversity in the human periodontal pocket and other oral sites, *Periodontol. 2000*, 42, 80, 2006.
243. Darveau, R.P., The oral microbial consortium's interaction with the periodontal innate defense system, *DNA Cell Biol.*, 28, 389, 2009.
244. Frandsen, E.V. et al., Evidence of recombination in *Porphyromonas gingivalis* and random distribution of putative virulence markers, *Infect. Immun.*, 69, 4479, 2001.
245. Byrne, S.J. et al., Progression of chronic periodontitis can be predicted by the levels of *Porphyromonas gingivalis* and *Treponema denticola* in subgingival plaque, *Oral Microbiol. Immuno.l*, 24, 469, 2009.
246. Walter, C. et al., Critical assessment of microbiological diagnostics in periodontal diseases with special focus on *Porphyromonas gingivalis*, *Schweiz. Monatsschr. Zahnmed.*, 115, 415, 2005.

51 *Prevotella*

Mario J. Avila-Campos, Maria R.L. Simionato, and Elerson Gaetti-Jardim Jr.

CONTENTS

51.1 INTRODUCTION

Anaerobic bacteria constitute members of the indigenous microbiota in humans and animals. Although these organisms do not usually cause disease, they have the capacity to take advantage of a weakened host's immune status (e.g., due to injury and immune-suppressing therapy) and can be involved in either mixed or monomicrobial infections.

Species of the genus *Prevotella* form the indigenous microbiota of the oral cavity and body surfaces in humans and animals. The *Prevotella* genus is a member of the family Bacteroidaceae and phylum Bacteroidetes. However, some species of *Prevotella* may cause local infections if the equilibrium of the habitat or the host-bacteria relationship is disrupted. In the last two decades, the taxonomy of the genus *Prevotella* has progressed slowly. The high degree of heterogeneity observed among these indigenous bacteria, and the similarity with other related species, have resulted in a problematic identification. Moreover, several new species have been partially characterized in recent years by using molecular methods, without previous knowledge about their virulence and ecological or epidemiological aspects, to support a significant role in the pathogenesis of infectious diseases or in metabolic traits of this resident microbiota.

51.1.1 Classification, Morphology, and Biology

The taxonomy of the gram-negative anaerobes was relatively cumbersome until the 1990s, when the *Bacteroides* genus was rearranged, and the genus *Prevotella* was formed.[1] A heterogeneous group of strictly anaerobic bacteria was described as *Bacterium melaninogenicum* that represented the core of the genus *Prevotella* and has undergone deep taxonomical rearrangements. This genus includes gram-negative, strictly anaerobic, nonspore-forming, nonmotile, saccharolytic or moderate, and pleomorphic rods. Colonies on blood agar plates vary from 0.5 to 1.0 mm in diameter and are usually circular, entire, convex, shiny, and smooth. They can be gray, light brown, or black-pigmented colonies, and particularly, coccobacilli that fluoresce red are in the pigmented *Prevotella* spp.–*Porphyromonas* spp. group.

The minimum medium setup includes nonselective, enriched, brucella base sheep blood agar plate supplemented with vitamin K_1 and hemin (BAP), and a kanamycin-vancomycin sheep blood agar (KVLB) for selection of *Bacteroides* and *Prevotella* spp. KVLB allows growth and rapid pigmentation of most *Prevotella* spp., but the concentration of vancomycin (7.5 µg/mL) can inhibit most *Porphyromonas* spp.

The hemolytic activity is variable; strains are able to growth from 25°C to 45°C, but the optimal growth is achieved at 37°C. Species of *Prevotella* are relatively saccharolytic, and most of them use peptides, free amino acids, nitrogen, and carbon as source of energy. In addition, hemin and menadione are the major requirements for growth, and the main fermentation products are acetic and succinic acids, as well as isobutyric, isovaleric, or lactic acids.

Historically, oral *Bacteroides* were separated in two groups, black-pigmented and nonpigmented rods. Those producing black- or brown-pigmented colonies on blood agar were called *B. melaninogenicus*; and the nonpigmented colonies, *B. oralis*. However, the taxonomic importance of the pigment production was overestimated, since the colonial pigmentation was dependent on the composition of the medium. The ability of most clinically relevant species in producing brown-black pigment in presence of hemin and blood is presented in Figure 51.1.

Nowadays, the genus *Prevotella* is composed of 43 species, occupying several ecological niches in resident microbiota from humans and other animals. Figures 51.2 and 51.3 present the most relevant biochemical tests to identify the medically relevant nonpigmented and pigmented *Prevotella*. However, various new species have been described within genus *Prevotella*,[2–18] and others have been transferred to different genera.[19] Therefore, the taxonomic analysis of these bacteria is effervescent.

The indole-positive and moderately saccharolytic *P. intermedia* group harbors *P. intermedia*, *P. falsenii*, *P. nigrescens*, and *P. pallens*.[5,18] *Prevotella intermedia*, *P. falsenii*, and *P. nigrescens* are phenotypically identical,[2] whereas *P. pallens* is lipase-negative and faintly pigment producing.[5] *Prevotella disiens* is similar, but produces neither pigment nor indole. New groups of *Prevotella* species have been described in recent years, such as *P. bergensis*, which forms a cluster with *P. buccae*, *P. dentalis*, and *P. baroniae*.[20] Both *P. enoeca* and *P. pleuritidis* are very similar[21]; *P. copri*, *P. veroralis*, *P. shahii*, and *P. stercorea* are genetically closely related, while the recently described *P. amnii* is closely related to *P. bivia*.[16]

Species of the genus *Prevotella* are part of the indigenous microbiota of mucous membranes, especially from the oral cavity,[8,9,15,21] and the respiratory, gastrointestinal, and genitourinary tracts of humans[12] and other mammals.[4,22] Members of the *P. melaninogenica* group, as well as *P. intermedia* and *P. nigrescens*, are among the first anaerobes to colonize the oral cavities of infants.[23] In addition, a new species from rice-plant residues (*P. paludivivens*) was recently characterized.

Moreover, genus *Prevotella* represents the most numerous members of the ruminal microbiota (42%–60%), however, typical ruminal species such as *P. bryantii*, *P. ruminicola*, and *P. brevis* represent 2%–4% of the total microbiota, and it is suggested that species of the resident microbiota in animals remain without a correct speciation.[4] Ruminal *Prevotella* species represent a reclassification of strains formerly classified as "*Bacteroides ruminicola*," made up of a group of genetically related species, such as *P. albensis*, *P. ruminicola*, *P. brevis*, and *P. bryantii*, and the hallmark of these four species is the nutritional versatility to use carbohydrates and amino acids.[22]

The indole-negative and lactose-fermenting *P. melaninogenica* group includes the phenotypically similar species, such as *P. melaninogenica*, *P. loescheii*, and *P. denticola*.[1] Closely related to these species are *P. oralis*, *P. buccae*, *P. dentalis*, *P. veroralis*, *P. bergensis*, *P. salivae*, *P. multiformis*, *P. baroniae*, *P. oulorum*, *P. maculosa*, *P. marshii*, *P. micans*, *P. multisaccharivorax*, *P. nanceiensis*, *P. shahii*, *P. paludivivens*, and *P. timonensis*.[1,6–8,10,11,14,17,20,21] In addition, oral *Prevotella* are inhibited by 20% bile and 6.5% NaCl, whereas species belonging to the intestinal microbiota are resistant to both substances. Most of strains grow properly in pH values

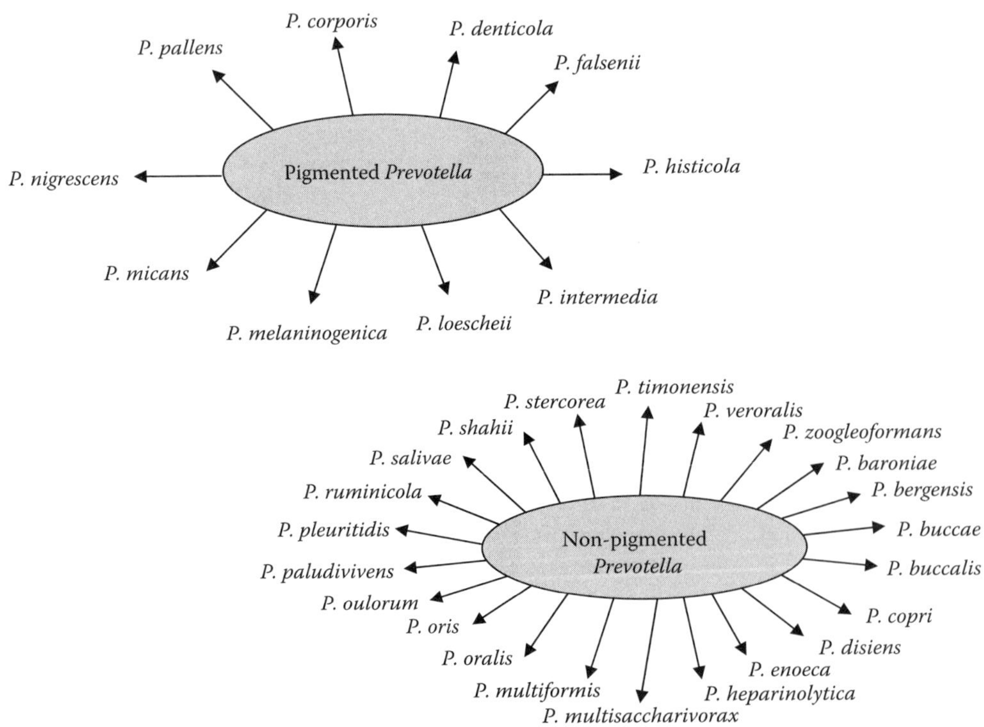

FIGURE 51.1 Major pigmented and nonpigmented species of genus *Prevotella*. *P. corporis* and *P. histicola* are variable in producing black-pigmented colonies.

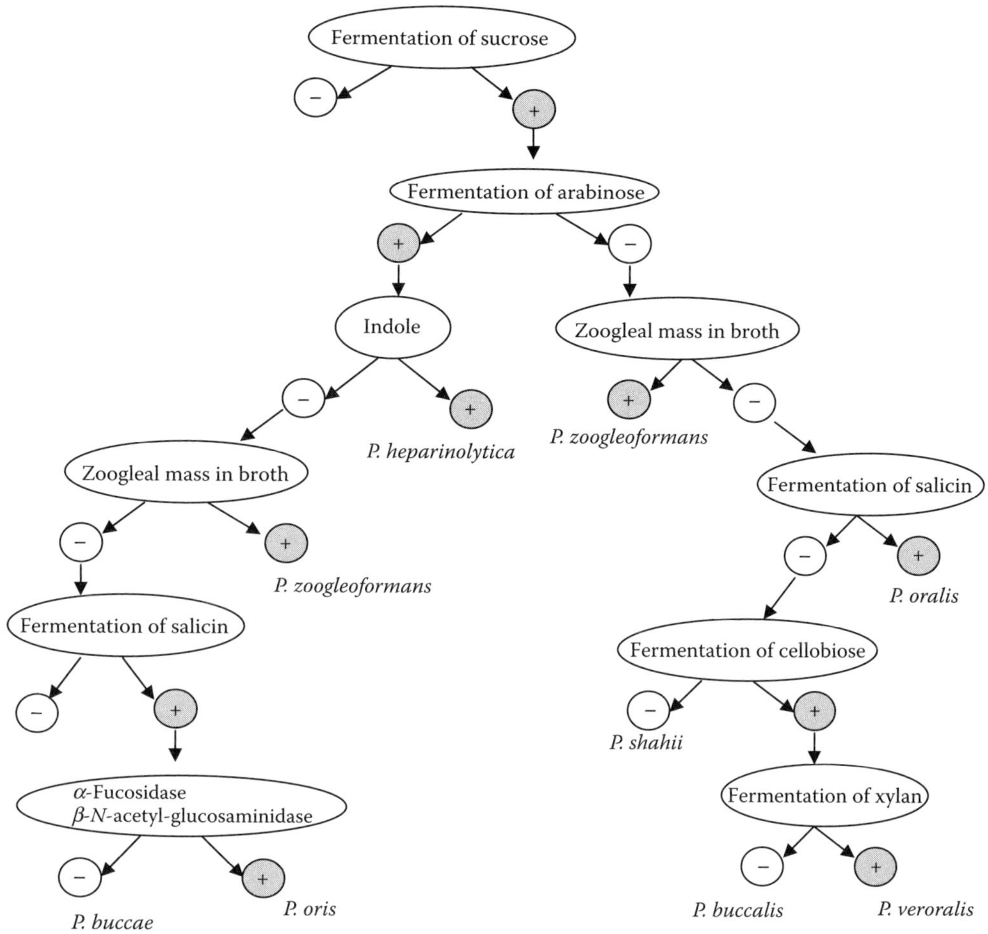

FIGURE 51.2 Identification of clinically relevant nonpigmented *Prevotella*.

varying from 5.7 to 6.7. This genus harbors a heterogeneous group of species, and the different habitats of *Prevotella* species are presented in Table 51.1. Its members are associated with opportunistic anaerobic infections in oral cavity and in other body sites, and they have been implicated as causative agents of several infectious processes. The infections most commonly associated with genus *Prevotella* are listed in Table 51.2.

51.1.2 Clinical Features

In general, infections caused by anaerobes are endogenous, and these organisms are particularly important in the ensuing phase, characterized by suppuration and abscess formation. The infections involving *Prevotella* species are similar to other infectious processes associated with gram-negative anaerobes. Usually, the infections involving *P. intermedia* and *P. melaninogenica* display acute clinical symptoms, especially in endodontic and periapical infections.[37]

Most of infectious processes associated with *Prevotella* spp. are observed in the oral cavity, especially in odontogenic infections. These infections are usually acute and suppurative, showing predominance of *Peptostreptococcus*, *Fusobacterium*, and both black-pigmented *Prevotella* and *Porphyromonas* species.[27] Moreover, the frequency of some uncultured *Prevotella* strains from head and neck infections suggests that these anaerobes can play a key role in the pathogenesis of these processes.

The presence and intensity of pain, especially in dental pulpitis, is associated with the presence of *P. intermedia* and other gram-negative anaerobes. The alogenic metabolites from anaerobic gram-negative bacteria in deep caries may partially explain why *P. intermedia* and the amount of lipopolysaccharide (LPS) in caries are positively related to heat sensitivity or pain. In addition, LPS activates the Hageman factor, leading to bradykinin production, a potent pain inducer.

A remarkable characteristic of *Prevotella* infections is the presence of malodor, and this feature seems to be associated with the production of volatile sulfur compounds by the aminoacid fermentation process of these anaerobes. Among the alogenic molecules produced by the metabolism, ammonia is the most potent pain inducer, followed by urea and indole. Many *Prevotella* species produce indole from tryptophan. Moreover, *Prevotella* species, especially *P. intermedia*, have a significant ability to induce abscess formation, and this characteristic seems to strengthen the virulence of other anaerobes, such as *Parvimonas micra*.[38] Thus, the presence of pus is a hallmark of the infectious diseases in which large numbers of these gram-negative anaerobes are found.[28,39]

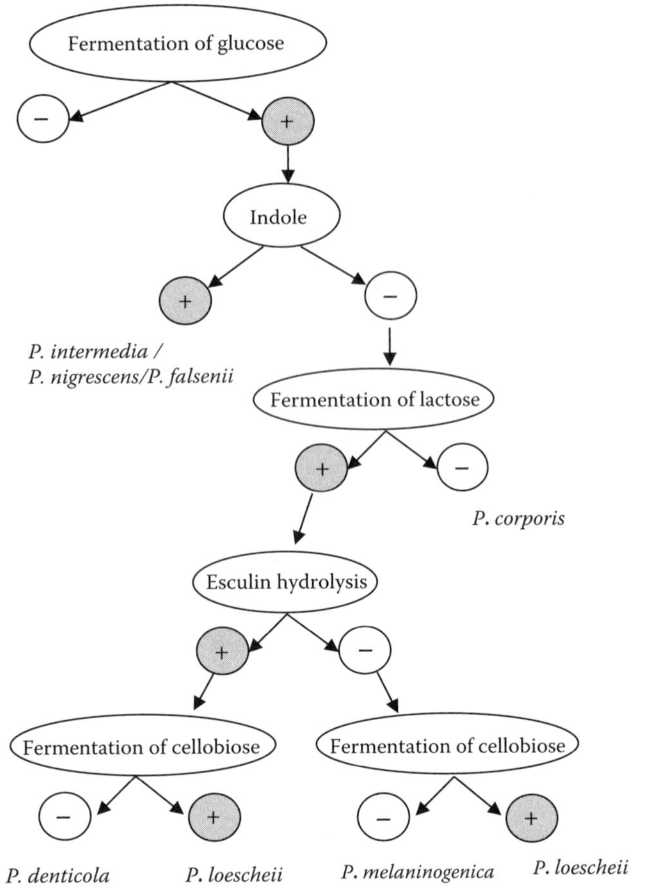

FIGURE 51.3 Identification of clinically relevant pigmented *Prevotella*.

In oral infections, especially periodontal infections, such as adult and advanced periodontitis, *P. intermedia* and *P. nigrescens* play an important role as pathogens and in early periodontal disease without deep periodontal pocket, where *Porphyromonas gingivalis*, an important periodontopathic bacteria is absent, since *Prevotella* species are more oxygen tolerant and can provide environmental conditions suitable to supporting other anaerobes.

It is well established that periodontal diseases are associated with a complex microbiota, and several bacterial species associated with periodontal diseases have been isolated and characterized,[32,40] mainly gram-negative anaerobic microorganisms, including *P. gingivalis*, *Tannerella forsythia*, *Treponema denticola*, *P. intermedia*, and *Fusobacterium nucleatum*.[41]

Several virulence factors in *P. intermedia*, such as enzyme production, including aminopeptidase-, chymotrypsin-, elastase-, and trypsine-like dipetidylpeptidase and alkaline phosphatase activities have been reported, and these can be harmful to the host.[42–46] Oral species of *P. intermedia* show the ability to adhere and to invade epithelial cells. Leung et al.[47] have suggested that the difference in adherence properties might be due to different types of fimbriae present in *P. intermedia* strains. Dorn et al.[48] demonstrated that *P. intermedia* invade KB cells and can be internalized within vacuoles. However, the invasion process can be

TABLE 51.1
Ecology of Genus *Prevotella*

Prevotella Species	Major Habitat	Reference
P. albensis	Rumen	4
P. amnii	Unknown[a]	16
P. baroniae	Oral cavity	7
P. bergensis	Unknown[b]	20
P. bivia	Human mucosa, especially oral cavity and genital tract	24
P. brevis	Rumen	4
P. bryantii	Rumen	4
P. buccae	Oral cavity	1
P. buccalis	Oral cavity	1
P. copri	Intestines	12
P. corporis	Human mucosa	1
P. dentalis	Human mucosa, especially oral cavity	19
P. denticola	Human mucosa, especially oral cavity	19
P. disiens	Human mucosa, especially oral cavity and genital tract	25
P. enoeca	Oral cavity	3
P. falsenii	Oral cavity from nonhuman primates	18
P. heparinolytica	Oral cavity	1
P. histicola	Oral cavity	16
P. intermedia	Human mucosa, especially oral cavity	1
P. loescheii	Human mucosa, especially oral cavity	19
P. maculosa	Oral cavity	21
P. marshii	Oral cavity	7
P. massiliensis	Unknown[c]	26
P. melaninogenica	Human mucosa, especially oral cavity	1
P. micans	Oral cavity	17
P. multiformis	Oral cavity	8
P. multisaccharivorax	Oral cavity	9
P. nanceiensis	Human mucosa	10
P. nigrescens	Human mucosa, especially oral cavity	2
P. oralis	Human mucosa, especially oral cavity	1
P. oris	Human mucosa, especially oral cavity	1
P. oulorum	Oral cavity	1
P. pallens	Oral cavity	5
P. paludivivens	Plant residues in the soil	14
P. pleuritidis	Unknown[d]	13
P. ruminicola	Rumen	4
Prevotella salivae	Oral cavity	6
Prevotella shahii	Oral cavity	6
P. stercorea	Intestines	12
P. tannerae	Oral cavity	3
P. timonensis	Unknown[e]	11
P. veroralis	Human mucosa, especially oral cavity	19
P. zoogleoformans	Oral cavity	3

[a] Isolated from human amniotic fluid.
[b] Isolated from skin and soft tissues infections in humans.
[c] Isolated from bacteremias in humans.
[d] Isolated from human suppurative pleuritis.
[e] Isolated from human breast abscess.

TABLE 51.2
Most Common Infectious Processes Associated with *Prevotella* Species

Microorganism	Infectious Disease or Clinical Condition	Reference
P. intermedia, *P. nigrescens*	Dental abscesses	27,28
Prevotella spp.	Breast abscesses	11
Prevotella spp.	Skin abscesses	20
Prevotella spp.	Bacteremias and septicemias	26
Prevotella spp.	Amniotic infections	16
Prevotella spp.	Bone infections	25,29
Prevotella spp.	Pleural infections	13
Prevotella spp.	Bite wound	29
Pigmented *Prevotella*	Otitis	30
Prevotella spp., particularly *P. intermedia* and *P. nigrescens*	Endodontic and periapical infections	27,28,31
P. bivia	Gynecology and obstetric diseases in patients evidencing preterm delivery	24
P. intermedia	Periodontal infections: chronic periodontitis, necrotizing gingivitis, pregnancy gingivitis	32–35
P. intermedia	Oral malodor	36
Prevotella spp.	Upper respiratory tract infections	29

inhibited by cycloheximide (an inhibitor of protein synthesis in mammalian cells), by cytochalasin D (an inhibitor of actin polymerization), and by monodansylcadaverine (an inhibitor of receptor-mediated endocytosis). In addition, the type C fimbriae of *P. intermedia* might be a key factor for the invasion since anti-type-C fimbriae antibodies effectively inhibit the *P. intermedia* invasion.[47] Studies in vivo have shown that *P. intermedia* as well as other periodontal pathogens can be detected in atherosclerotic arteries.[49–51]

51.1.3 Virulence and Pathogenesis

Anaerobes implicated in infectious diseases are usually in association with facultative and other anaerobic bacteria, and these infections depend on the adhesion to the host's cells. After these initial steps, tissue invasion may occur. However, the bacteria in the front of the invasion need to survive against the host's defenses and, at the same time, acquire nutrients to stay alive, and this process frequently ends with necrosis. In this sense, *Prevotella* species have a significant range of virulence factors that allow them to adhere to and invade the host's tissues, destroy cellular structures and matrix, and stimulate a powerful and persistent inflammatory process. Figure 51.4 summarizes most of the virulence mechanisms of *Prevotella* species involved in human infections.

51.1.3.1 Adhesion to Host's Cells and Intercellular Matrix

Bacterial adherence to host's cells is the initial step in colonization. Besides the adhesion process, the bacterial

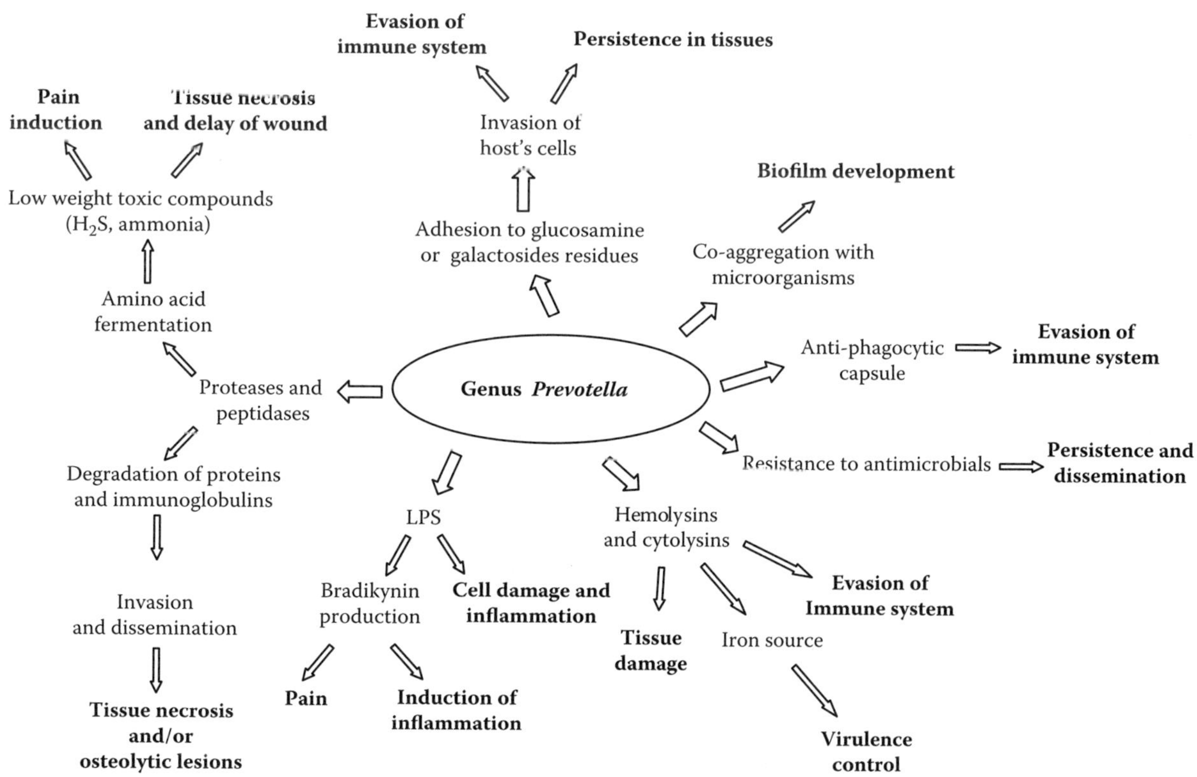

FIGURE 51.4 Overview of the virulence of *Prevotella* species.

coaggregation is extraordinarily important in the development of microbial communities involved in mixed anaerobic infections. Studies on the adhesion of *Prevotella* are considered for oral species, particularly *P. intermedia*, *P. loescheii*, and *P. nigrescens*. Many *Prevotella* species possess fimbriae, adhesins, and hemagglutinins.[52,53] The adhesins and hemagglutinins of *P. loescheii* and *P. melaninogenica* have affinity to galactose-containing carbohydrates residues on the host's cells; whereas, in *P. intermedia* and *P. nigrescens*, the adhesion molecules interact with glucosamine-containing carbohydrates. In addition, the hemagglutinating activity of *Prevotella* species is not mediated by proteolytic activities, unlike that *P. gingivalis*.

In addition, *P. intermedia* and *P. nigrescens* are able to adhere to large varieties of cells and proteins, such as fibroblasts and reconstituted membranes, epithelial cells, and erythrocytes,[54] as well as coaggregate with other microorganisms as *Actinomyces*, *Fusobacterium*, *Porphyromonas*, and *Streptococcus*.[55] *Prevotella intermedia* is able to coaggregate with several other early colonizers, such as *Streptococcus salivarius*,[56] and creating conditions for the establishment of some late colonizers, as *P. gingivalis*.[57] The coaggregation with *P. gingivalis* may be inhibited by addition of L-arginine and L-lysine.

Prevotella loescheii produces a galactoside-specific adhesin that mediates the hemagglutination of neuraminidase-treated human erythrocytes and lactose-inhibitable coaggregations with oral streptococci. The adhesin is a 75-kDa protein that is expressed on the cell surface in a maximum of 400 molecules per cell, and this adhesin is associated with the distal portion of the fimbriae. A second adhesin of 45 kDa that mediates lactose-noninhibitable coaggregation with *A. israelii* is also expressed by *P. loescheii*.[55]

The coaggregation partners of oral *Prevotella* species appear to be primarily the actinomycetes. *Prevotella* produces a coaggregation by adhesins and actinomyces by carbohydrate-containing receptors on its cell surface. In addition, a protease synthesized by *P. loescheii* is able to degrade a 75-kDa adhesin, suggesting that proteases may actually aid in detachment of this bacterium from the coaggregation partner cell, allowing finding a different niche in oral biofilm.[55]

51.1.3.2 Evasion of Immune System and Damage of Host's Tissues

In order to produce infections, a microorganism must be able to adhere to hosts' cells and colonize them, as well as produce harmful compounds and toxins to induce damage or be capable of invading tissues. In addition, evasion of immune defenses represents major strategy for microbial dissemination and persistence in the hosts, and several mechanisms may be involved in this phenomenon.

Capsule. Capsule is believed to help bacteria avoid phagocytosis and killing by polymorphonuclear leukocytes (PMNs) by preventing the deposition of opsonins on the bacterial surface or by mimicking mammalian cell-surface determinants.[58] Encapsulated anaerobes generally induce abscess, but nonencapsulated organisms are not able to induce it, as also described to *P. melaninogenica*, *P. oralis*, and *P. brevis*, and the ability to produce abscess is not associated with the presence of viable cells. Moreover, some species, such as *P. intermedia*, *P. oralis*, and *P. melaninogenica*, may be profusely capsulated.

Low-Weight Metabolic End Products, Lipopolysaccharides, Toxins, and Induction of Inflammation. These anaerobes are able to produce volatile sulfur compounds (hydrogen sulfide and methylmercaptans) associated with malodor, which may reduce the cellular proliferation and repair. Volatile sulfur compounds are produced by anaerobic bacteria derived from amino acids desulfuration containing sulfhydryl groups. For instance, hydrogen sulfide is derived from desulfuration of cysteine; and methylmercaptan, from methionine desulfuration. Thus, both *P. intermedia* and *P. nigrescens* are potent producers of these compounds.[59]

Species of *Prevotella* produce succinate, acetate, and isovalerate with a small amount of isobutyrate in the absence of glucose, but produce a large amount of ammonia, which has a significant immunosuppressive effect and delays the wound repair. These short-chain fatty acids are reported to inhibit the function and/or proliferation of neutrophils, T-lymphocytes, phagocytes, and fibroblasts.[60]

The presence of glucose represses the intracellular enzymes formation involved in the nitrogen-source metabolism. Thus, in the oral cavity around supragingival biofilm, where saliva and diet residues are available for the microbial metabolism, the fermentation of carbohydrates leads to environmental acidification, and the oxygen can reach the bacteria around the gingival margin. However, *P. intermedia* and *P. nigrescens* can survive and even grow under these conditions because they are saccharolytic, acid-tolerant, and oxygen-tolerant[61]; but in the bottom of the periodontal pockets and in deep internal infections, where carbohydrate supply is limited, the pathogenicity of *P. intermedia*, and *P. nigrescens* may be constantly elevated.[60]

Markers of systemic inflammation, such as the presence of C-reactive protein in plasma are indicative of myocardial infarction or ischemic stroke risk. Although several factors may increase plasmatic concentrations of C-reactive protein, this presence of *P. intermedia* and other anaerobes in subgingival samples was positively associated with elevated levels of C-reactive protein.[62] Since one of the hallmarks of periodontal disease is inflammation, mainly in chronic periodontal inflammation, it is possible that periodontal pathogens may be involved with the increase in C-reactive levels.

Lipopolysaccharide is a major component of the outer membrane of gram-negative bacteria, including *Prevotella* species. It has the ability to trigger a number of host cells, especially mononuclear phagocytes, to produce a wide variety of immunologically active mediators, including TNF-α, interleukins, and nitric oxide.[63] Moreover, LPS produced by *P. intermedia* may participate in the periodontal destruction and alveolar bone loss through the osteoclastogenesis stimulation,[64] as well as reduce bone formation. However, endotoxins of *Prevotella* spp. are less potent than endotoxins extracted from Enterobacteriaceae and are less likely to produce classic manifestations of toxic damage.[58]

Tissue Invasion. The ability to invade host tissues may be an important virulence factor in bacteria. *P. intermedia* can invade oral epithelial cells, but the type C fimbriae and a cytoskeleton rearrangement are required for invasion.[48] Moreover, *P. bivia* is also capable of adhering to and invading human cervix epithelial cells, and of inducing proinflammatory interleukins (IL-6 and IL-8), but this phenomenon is not related to the presence of fimbriae.[65] In addition, DNA of *P. intermedia* may be present in atherosclerotic plaques, and this anaerobe may play a role in development and progression of atherosclerosis, leading to a possible coronary vascular disease or other clinical sequel, suggesting a possible adhesion and invasion to the coronary cells.[66]

Some evidence suggests that the absence of hemin may limit the ability of *P. intermedia* to invade dentinal tubules.[67] However, the mechanism by which the iron interferes with invasion is not yet known. In theory, bacterial attachment to dentine and tubule invasion requires of adhesins, and it is known that stressed cells produce a different protein profile, particularly, to the cell survival.

Hemolysis. Another virulence factor is a cytolytic molecule with hemolytic activity that seems to contribute to the pathogenicity of these microorganisms by favoring the iron acquisition that is essential to their metabolism, to survival, and to act upon other host cells.[68] *Prevotella intermedia* lack the ability to synthesize heme, which is an important growth factor.[69] *P. intermedia* can use hemoglobin as a sole source of iron, and inorganic iron and iron-binding proteins, such as transferrin and lactoferrin, could not support the bacterial growth, suggesting that hemoglobin might be the main iron source. However, the mechanisms to the iron assimilation from hemoglobin have not been yet elucidated.

The whole cells of *P. intermedia* have a weak hemolytic activity on human erythrocytes, but when these microorganisms are lyzed and hemolysin-free and are treated by trypsin-like proteases, the lytic activity is significantly increased. In contrast, no increase in the lytic activity of *P. nigrescens* was observed after rupturing the bacteria, although proteolytically treated hemolysin produce a very high intensity of hemolysis, higher than *P. intermedia* hemolysin.[68]

Proteolytic Activities. *Prevotella* species produce a wide variety of enzymes associated with evasion of immunological responses or tissue damage. In addition, cysteine protease (linked to hemoglobin degradation) and other cysteine and serine proteases have been described.[70,71] All these proteases have been identified as secreted virulence factors producing nutrient generation, evasion of the immune response through the immunoglobulin inactivation, and by releasing bacterial proteins from cell surface.

Immunoglobulin A_1 protease, secreted by most oral *Prevotella* species, particularly *P. melaninogenica*, *P. intermedia*, and *P. nigrescens*, specifically cleaves IgA_1 at the same peptide bond in the hinge region, and this phenomenon occurs in vivo, especially in periodontitis patients. However, human serum from these patients present high titles of neutralizing antibodies able to inhibit partially the proteolytic activity of IgA_1 proteases produced by *Prevotella* species, such as *P. buccae*, *P. oris*, *P. loescheii*, *P. buccalis*, and *P. melaninogenica*.[72]

These enzymes may improve the pathogen dissemination, as well as provide amino acids used in the catabolic reactions of these anaerobes. Some evidences suggest that *P. loescheii*, *P. melaninogenica*, *P. intermedia*, and *P. oralis* recovered from dentoalveolar abscesses are more proteolytic than those obtained from healthy periodontal sites. In addition, *P. intermedia* and *P. nigrescens* are the most proteolytic among the oral *Prevotella* species.[73]

Moreover, serine proteases and exopeptidases produced by *P. loescheii* have been described recently, which help trypsin- and collagenase-like protease in the type I collagen degradation, but they are strongly inhibited by Cd^{2+}, Zn^{2+}, Hg^{2+}, Co^{2+}, and Cu^{2+}. Whole-cell extracts of *P. nigrescens* activate the host's metalloproteinases responsible for the matrix proteins' depolymerization and the dissemination of these anaerobes in connective tissues.[74,75]

Most of studies on proteolytic activity have been performed with oral *Prevotella* and ruminal species, particularly, *P. ruminicola*, *P. brevis*, *P. albensis*, and *P. bryantii*, that have also evidenced strong proteolytic abilities, mainly on casein, and several proteases such as serine, cysteine, and low molecular weight metalloproteases were detected.[76]

51.1.3.3 Resistance to Antimicrobial Drugs

Bacteria of the genus *Prevotella* are often resistant to antimicrobials, such as tetracycline,[77] clarithromycin, quinolones,[78] β-lactam antibiotics,[28,39,79] and clindamycin.[77] The β-lactam resistant *Prevotella* strains are β-lactamase producers, especially *P. bivia*, *P. intermedia*, *P. nigrescens*, *P. melaninogenica*, *P. loescheii*, *P. oralis*, *P. buccae*, and *P. oris*.[79,80]

β-lactamases enzymes produced by these anaerobes display properties of the class A-group 2e β-lactamases, which hydrolyze most penicillins and broad-spectrum cephalosporins but are inactive against penicillin plus clavulanic acid and carbapenems.[79] Moreover, it has been evidenced that oxygen exposition may affect the susceptibility of *P. intermedia* to antimicrobial drugs, probably due to upregulation of protective enzymes by oxidative stress.[61]

51.1.4 Microbial Diagnosis

The purpose of the microbial identification is to match an isolate with a previously recognized taxonomic group, using a number of phenotypic or genetic characters that can be evaluated. The identification systems must be reliable, convenient, rapid, easily performed, flexible, and cheaper.

Conventional Techniques. Usually, the conventional techniques for the diagnosis of anaerobic infections depend, at least to some extent, on culture. However, variations of environmental parameters, including redox potential, osmolarity, pH, temperature, and nutrients can interfere with detection of a pathogen. In microbiological diagnostics, the bacterial culture has been the gold standard for many years, although limitations with respect to detecting nonviable bacteria and the inability of some species to grow on selective media, as well as the high cost, are known. Despite their effectiveness and sensitivity, these procedures are extremely labor intensive and time consuming. Moreover, classification and identification based on phenotypic traits do not always provide clear-cut results and are sometimes unreliable.

The definitive identification of anaerobes is based on the use of biochemical tests in combination with gas–liquid chromatography, but this method is rather labor intensive and expensive. Thus, the development of automated methods for anaerobes identification has received interest. Some systems employ computerized high-resolution gas–liquid chromatography to determine the fatty acid composition from bacteria, which is then compared with a database. However, the performance of these systems is precarious for *Prevotella* species.[81]

Methods used to identify *Prevotella* strains includes colony morphology, pigmentation, Gram staining, carbohydrate fermentation, esculin hydrolysis, gelatin liquefaction, nitrate reduction, and indole production in PRAS media, and the detection of preformed enzymes by using RapID ANA II, RapiID 32, and API ZYM kits. Pure cultures must be stored in 20% skim milk at −80°C or liquid nitrogen. Using filter-paper spot tests for indole production, sialidase, α-glucosidase, β-glucosidase, α-fucosidase, and trypsin-like enzyme activities, the presumptive identification of oral *Prevotella* may be achieved. The bacterial pigmentation is largely media dependent, and the pigment production is used for distinguishing colonies in early stages of identification.

The screening using rapid tests is enough for identifying the most common oral *Prevotella* species, such as *P. intermedia*, *P. nigrescens*, and *P. melaninogenica*. In addition, the indole test would increase the accuracy of a simple identification scheme by recognizing occasional β-galactosidase (MUG)-positive *P. intermedia/P. nigrescens* strains, otherwise misidentified as belonging to the *P. melaninogenica* group.

Glucose, sucrose, and lactose fermentation, pigment, indole, and lipase production, as well as α-fucosidase and *N*-acetyl-β-glucosaminidase assays are the core test for identification of *P. intermedia/P. nigrescens* group, allowing the partial identification of *P. pallens*, *P. intermedia*, and *P. nigrescens*, *P. corporis*, and *P. disiens*.[5] Other phenotypical tests useful in the identification of *Prevotella* spp. include gelatin and esculin hydrolysis.[6]

On the other hand, ambiguous results are often obtained due to the emergence of phenotypically variable strains.[5,8] Moreover, the speciation of some species, such as *P. intermedia*, *P. nigrescens*, and *P. tannerae*, needs additional molecular tests, particularly, 16S rDNA-based polymerase chain reaction (PCR), since some black-pigmented *Prevotella* have been previously identified as *P. intermedia* by using rapid biochemical identification kits, and then as *P. nigrescens* by using SDS-PAGE. Thus these species were identified as *P. tannerae* when PCR and restriction enzyme analysis were performed.

Molecular Techniques. Molecular detection methodologies are powerful tools used for microbial identification, to study host-parasite relationships, and to clarify taxonomic positions of targeted microorganisms. Most of molecular techniques used in the microbial diagnosis are based on PCR, which can unequivocally detect phenotypically variable isolates and can be considered as a more specific and reliable method for bacterial identification; otherwise, it is a rapid and relatively inexpensive identification method.

Nowadays, the 16S rRNA gene PCR is considered as an accurate and reproducible method for separation among *P. nigrescens* and *P. intermedia*, as well as to identify the *Prevotella* species directly from clinical samples.[82–84] Moreover, other genes have been targeted with species-specific PCR, such as genes *phoC* (*P. intermedia* acid phosphatase), *phyA* (*P. melaninogenica* hemolysin), and *plaA* (*P. loescheii* adhesin precursor).[85]

The ribosomal RNA genes and elongation factor G have provided valuable phylogenetic information for anaerobes and other microorganisms. The 16S rRNA gene has been widely used for phylogenic studies, and has produced phylogeny comparable to the 23S rRNA genes. The phylogenetic analysis of *Prevotella* species have indicated some variations in the 16S rDNA sequences within the genus, and it is enough to make the 16S rRNA gene analysis a good tool for the species-specific PCR identification, both from cultures and directly from clinical samples.[82,86]

In addition, the 16S rDNA sequences have been the target for simultaneous detection of different species by multiplex PCR, including *P. nigrescens*, although other genes have been detected by using multiplex PCR in *P. intermedia*, *P. melaninogenica*, and *P. loescheii*.[85] Other molecular methods have been used to separate clinically relevant *Prevotella* species, particularly, *P. nigrescens* from *P. intermedia*, such as multilocus enzyme electrophoreses,[87] restriction enzyme analysis of total DNA, ribotyping,[88] SDS-PAGE protein electrophoresis,[87] 16S rRNA gene PCR-restriction fragment length polymorphism,[89] and 16S rRNA gene cloning and sequencing.[36,39]

The real-time PCR assays for *Prevotella* species allow an easy, rapid, and quantitative detection of these microorganisms directly from clinical samples. The TaqMan system is the most frequently used and confers high sensitivity and specificity to DNA amplification. Primers and probes sets for *Bacteroides-Prevotella* group,[90] *P. ruminicola* group,[22] *P. intermedia*, and *P. nigrescens*,[84] *P. melaninogenica, and P. loescheii*,[91] *P. melaninogenica*, and *Prevotella* sp.[22,83] have been described. In addition, protocols for quantitative detection of genes resistant to antimicrobials are also available.[79] Real-time PCR combines traditional end-point-detection PCR with fluorescent detection technologies to record the accumulation of amplicons in real time during each cycle of the PCR amplification. The detection of amplicons during the early exponential phase of the PCR enables the quantification of gene (or transcript) numbers when these are proportional to the starting template concentration.

This method has been shown to be a robust, highly reproducible, and sensitive method of quantifying functional genes in many conditions. The success of the application of this methodology is based on the specificity of real-time PCR assay, which is determined by the design of the primers (or internal probes). Ribosomal RNA or ribosomal 16S (16S rRNA) (specific prokaryote rRNA) is the most widely used macromolecule in phylogenetic and taxonomic studies. 16S rRNA analysis has been widely used to establish phylogenetic relationships in the world of prokaryotes, exerting a profound impact on our view of evolution and, as a result, on the classification and identification of bacteria.

One of the first techniques used in molecular biology to detect target microorganisms in clinical samples was the nucleic acid probe. A probe is formed by a short DNA fragment, containing only a few nucleotides, complementary to the DNA belonging to the target, and should be labeled with a molecule that is easily detected once hybridization has occurred. The main problem with the technique is the risk of cross-reaction. The checkerboard DNA–DNA hybridization technique allows the enumeration of large numbers of species in very large numbers of samples, and 16S rRNA gene clone library analysis can be used to survey the diversity of human oral microbiota and the existence of as-yet-to-be-cultured organisms that are presumed pathogens. In addition, terminal restriction fragment length polymorphism (T-RFLP) analysis has been also applied for assessment of diversity of human oral microbiota.

51.2 METHODS

51.2.1 Sample Preparation

The clinical samples commonly used to detect species of *Prevotella* are pus,[31,39,80] necrotic tissues,[80] biopsied tissues,[25] vaginal discharge,[24] and dental biofilm.[85] Bone infection clinical samples, especially of the skull bones, must avoid contact with external contaminants, and only five types of clinical specimens are acceptable for microbial diagnosis: bone biopsy, bone sequestra, bone marrow, granulation tissue, and aspirated pus. In order to avoid the artificial introduction of oral microorganisms into specimens, it now required that clinical specimens must be taken during surgical procedures, avoiding the contact with soft tissues, the oral environment, and sinus tracts. Whenever possible, bone sequestra or granulation tissue samples must also be obtained for the histopathological examination. The best strategy for collecting pus is needle aspiration and transportation inside a prereduced medium. The use of transport medium is unnecessary for short periods of time.[25,39] In vaginal infections, the use of cotton wool swabs is suitable when vaginal discharge is present.[24] For isolation of *Prevotella* species from rumen liquids and contents, samples[22] must be transported in an environment containing only CO_2 to avoid contact with oxygen or in prereduced transport media.

The most often used prereduced transport media for transportation of specimens containing anaerobic bacteria, including samples with *Prevotella* species, are Ringer-PRAS, RTF,[83,92] and VMGA III.[37,92] In addition, since preparation, storage, and handling of VMGA III medium are reliable, easily performed, and support the survival of anaerobic bacteria for 1 or 2 days, this medium is widely used as a standard in clinical laboratories. In oral samples, it is important to note when antiseptics were used to avoid sample collection if required, because the antimicrobial action can interfere with the bacterial recovery. Thus, use of sterile 5% sodium thiosulfate is recommended to eliminate residual antimicrobials.

51.2.2 Detection Procedures

The use of conventional techniques for the analysis of infections involving *Prevotella* species has shown limitations due to their susceptibility of the molecular oxygen. Strictly anaerobic bacteria are not equally susceptible to oxygen, and even different strains may display different levels of sensibility, which is reduced by continuous subcultures and by the production of protective enzymes acting against the toxic products derived from oxygen reduction. The ability to adapt to the oxygen changes is a remarkable physiologic trait of pathogenic *Prevotella*.[61] However, the prime isolation of these microorganisms is Eh-negative dependent, since fresh isolates are more susceptible to the oxygen's producing a reduction of 90%–99% of the microbial number after 15 h of exposition to molecular O_2.

Traditionally, culture for anaerobes is a time-consuming and expensive procedure. Most clinical and research laboratories use the so-called strictly anaerobic techniques or conventional anaerobic techniques for isolation of *Prevotella* species. These techniques differ from those considered "conventional" because the specimen is never exposed to air during the entire process of preparation and culturing, or at least the contact with molecular oxygen is minimized by using PRAS transport and dilution solutions, and the media used are also prereduced.

All pathogenic *Prevotella* species grow on nutritionally rich culture media, such as brucella agar, Columbia agar, brain heart infusion agar, fastidious anaerobe agar, or even trypticase soy agar, supplemented with hemin (0.05 mg/dL), menadione (0.01 mg/dL), and yeast extract (0.1%–5%). However, since most of *Prevotella* species are members of human and animal resident microbiota, their isolation requires the use of selective culture media, such as KVLB agar, in which kanamycin and vancomycin exert a selective role, especially for oral *Prevotella*. The association of 5% sheep or horse blood with 0.05 mg/dL hemin, 0.01 mg/dL vitamin K_1, 0.001% w/v nalidixic acid, and 2.5 mg/L vancomycin using Wilkins-Chalgren agar is also adequate for isolation of these anaerobes. Although, some *Prevotella* species are able to produce visible growth on agar plates after 48 h, the incubation must be performed during 7–15 days, to allow the development of the most fastidious species and the brown-pigment characteristic observed in the most virulent species.

The detection procedures of the molecular methods do not have specific peculiarities in relation to facultative anaerobes and other anaerobes. However, DNA extraction by heat (96°C, for 10 min) has been used with *Prevotella* spp.,[79] and in some circumstances DNA extraction and amplification may be affected by the presence of cellular debris and iron because they may affect the DNA amplification by chelating magnesium or interfering with DNA polymerase. To minimize this problem, the DNA extraction must be performed by using other methods, providing an additional clearance, such as phenol-chloroform or commercial kits.

In addition, due to the nature of PCR-based protocols with high levels of sensitivity, the main shortcomings in applying PCR assays include false-positive results due to background DNA contaminations. These concerns are particularly relevant in the microbial diagnosis of members of the human and animal resident microbiota. In addition, the presence of one microorganism is not necessarily an indication that it plays a pathogenic role in the infection development, and this is true in mixed anaerobic infections, where different microorganisms share the same habitat.

Bacterial anaerobic culture is based on living cells. This technique enables us to search for all the microorganisms present in a nonspecific way. Since it is not directed toward precise pathogenic targets, it remains the most objective technique (gold standard), but requires at least 10^3 pathogens for detection. Thus, many unexpected pathogen colonies appear during culture, whereas routine probes cannot detect them. Also, the cultures have the important advantage of allowing an antibiotic sensitivity. If we consider the interindividual or the intraindividual variations in the susceptibility of clinical isolates to antibiotics, in particular for *P. intermedia*, there is a good reason for to perform the conventional bacterial culture and antibiogram from each patient. On the other hand, the detection of pathogens by the real-time PCR is obtained even when they are at low levels, showing high sensitivity, and because the technique detects the DNA of both living and dead bacteria.

Most of studies on detection of *Prevotella* species were carried out using oral specimens, particularly from periodontal and endodontic/periapical infections, where *P. intermedia*, *P. nigrescens*, *P. melaninogenica*, and *P. loescheii* are common.

Based on the four *Prevotella* species-specific genes, Yoshida et al.[85] described one multiplex PCR focusing on *P. intermedia*, *P. nigrescens*, *P. melaninogenica*, and *P. loescheii* from oral cavity (Table 51.3).

Procedure. The oligonucleotide primers and their DNA sequences are listed in Table 51.3, and the specificity of the primer pairs is confirmed by sequencing. Multiplex PCR for black-pigmented oral *Prevotella* may be performed as follows: 95°C for 15 min, followed by 35 cycles of 94°C for 30 s, 54°C for 90 s, and 72°C for 90 s. A 50-μL PCR mixture for multiplex PCR consists of 25 μL 2× Qiagen Multiplex PCR Master Mix (Qiagen), 2 μM of each primer, and 1 μL of template DNA, and 1–10 ng of genomic DNA as template.

After completion of PCR cycles, add 4 μL of 10× DNA loading buffer into each tube; load 20 μL of PCR product on 1.0% agarose gel containing 0.5 μg/mL ethidium bromide. Include a DNA molecular marker in a spare lane. Run at 5 V per cm gel for about 30 min. Visualize under UV light transilluminator and photograph.

51.3 CONCLUSION AND FUTURE PERSPECTIVES

Almost all available knowledge about the physiology of these anaerobes is based on data obtained from oral *Prevotella* species, particularly *P. intermedia*, but this microbial genus is extremely heterogeneous, not only from the phenotypical level, but also mainly at the molecular level. Thus, it is entirely possible that new interactions between the newly described species and theirs hosts may be characterized in the coming years. These ecological interactions may explain the mechanisms involved in the pathogenicity of these anaerobes and may collaborate in the development of more ecological therapeutic approaches in the treatment endogenous infections associated with these microorganisms, and may also collaborate for a better understanding of the role played by these microorganisms as part of the resident microbiota of the oral cavity and other mucous membranes.

The characterization of *Prevotella paludivivens*, a hemicellulose-decomposing bacterium isolated from plant residue and rice roots in irrigated rice-field soil, opens the possibility of the participation of this strictly anaerobic genus in complex microbial communities not related to infections. The ability of these environmental microorganisms also represents an

TABLE 51.3
Multiplex PCR for Black-Pigmented *Prevotella* from Oral Cavity

Prevotella Species	Sequence of Oligonucleotide Primers	Gene	Amplicon Size (bp)
P. intermedia	5′-GAC TTT TGC ACA GAA TGC A-3′	*phoC*	718
	5′-CTT GGC AAC CTT GCC TTC-3′		
P. loescheii	5′-TGC CAA CTC CCG ATT TC-3′	*plaA*	637
	5′-TAC ACC AAG GTT TTC CCC-3′		
P. melaninogenica	5′-CGT CAT GAA GGA GAT TGG-3′	*phyA*	389
	5′-ATA GAA CCG TCA ACG CTC-3′		
P. nigrescens	5′-TTT CAT TGA CGG CAT CCG-3′	16S rRNA	825
	5′-CAC GTC TCT GTG GGC AG-3′		

Source: Yoshida, A. et al., *Oral Microbiol. Immunol.*, 20, 43, 2005.

exceptional opportunity to study the similarities and differences between species frequently isolated from infectious processes and those recovered from symbiotic communities in the external environment.

In addition, approximately three-quarters of all species of the genus *Prevotella* were identified and characterized during studies with clinical specimens from the human oral cavity and from infections, and almost nothing is known about the anaerobic microbiota from other animal species, even the nonhuman primates that are often used as animal models in studies involving medicine, veterinary medicine, and dentistry. Thus, it is likely that in the future, the number of species of this genus from animal origin will increase significantly, just as the knowledge available on the potential virulence of these microorganisms has.

Nowadays, several markers of resistance to antibiotics characteristic of other anaerobic species has been identified in species of the genus *Prevotella*, but the distribution of major genes associated with resistance to antimicrobials, especially those associated with resistance to β-lactams, tetracyclins, macrolides, and nitroimidazoles among species of the genus *Prevotella*, particularly *P. intermedia* and *P. nigrescens*, is not known, and the possibility of vertical and horizontal transmission of these markers to the other microorganisms inside the dental biofilm and intestinal microbiota remains unexplored.

Due to the development of molecular methods, it has become possible to evaluate the factors capable of controlling the expression of genes related to primary and/or secondary metabolism in opportunistic pathogens, enabling a greater understanding of how environmental factors or hosts' factors may exacerbate or depress the microbial virulence. This aspect seems particularly relevant to the oral species of the genus *Prevotella*, which are associated with gingival changes modified by hormonal changes. It is possible that this knowledge may enable the development of preventive strategies for numerous diseases associated with hormonal changes.

REFERENCES

1. Shah, H.N., and Collins, D.M., *Prevotella*, a new genus to include *Bacteroides melaninogenicus* and related species formerly classified in the genus *Bacteroides*, *Int. J. Syst. Bacteriol.*, 40, 205, 1990.
2. Shah, H.N., and Gharbia, S.E., Biochemical and chemical studies on strains designated *Prevotella intermedia* and proposal of a new pigmented species, *Prevotella nigrescens* sp. nov., *Int. J. Syst. Bacteriol.*, 42, 542, 1992.
3. Moore, L.V., Johnson, J.L., and Moore, W.E., Descriptions of *Prevotella tannerae* sp. nov. and *Prevotella enoeca* sp. nov. from the human gingival crevice and emendation of the description of *Prevotella zoogleoformans*, *Int. J. Syst. Bacteriol.*, 44, 599, 1994.
4. Avgustin, G., Wallace, R.J., and Flint, H.J., Phenotypic diversity among ruminal isolates of *Prevotella ruminicola*: Proposal of *Prevotella brevis* sp. nov., *Prevotella bryantii* sp. nov., and *Prevotella albensis* sp. nov. and redefinition of *Prevotella ruminicola*, *Int. J. Syst. Bacteriol.*, 47, 284, 1997.
5. Könönen, E. et al., Phylogenetic characterization and proposal of a new pigmented species to the genus *Prevotella*: *Prevotella pallens* sp. nov., *Int. J. Syst. Bacteriol.*, 48, 47, 1998.
6. Sakamoto, M. et al., *Prevotella shahii* sp. nov. and *Prevotella salivae* sp. nov., isolated from the human oral cavity, *Int. J. Syst. Evol. Microbiol.*, 54, 877, 2004.
7. Downes, J. et al., *Prevotella marshii* sp. nov. and *Prevotella baroniae* sp. nov. isolated from the human oral cavity, *Int. J. Syst. Evol. Microbiol.*, 55, 1551, 2005.
8. Sakamoto, M. et al., *Prevotella multiformis* sp. nov., isolated from human subgingival plaque, *Int. J. Syst. Evol. Microbiol.*, 55, 815, 2005.
9. Sakamoto, M. et al., *Prevotella multisaccharivorax* sp. nov., isolated from human subgingival plaque, *Int. J. Syst. Evol. Microbiol.*, 55, 1839, 2005.
10. Alauzet, C. et al., *Prevotella nanceiensis* sp. nov., isolated from human clinical samples, *Int. Syst. Evol. Microbiol.*, 57, 2216, 2007.
11. Glazunova, O. et al., *Prevotella timonensis* sp. nov., isolated from a human breast abscess, *Int. J. Syst. Evol. Microbiol.*, 57, 883, 2007.
12. Hayashi, H. et al., *Prevotella copri* sp. nov. and *Prevotella stercorea* sp. nov., isolated from human faeces, *Int. Syst. Evol. Microbiol.*, 57, 941, 2007.
13. Sakamoto, M. et al., *Prevotella pleuritidis* sp. nov. isolated from pleural fluid, *Int. Syst. Evol. Microbiol.*, 57, 1725, 2007.
14. Ueki, A. et al., *Prevotella paludivivens* sp. nov., a novel strictly anaerobic, gram-negative, hemicellulose-decomposing bacterium isolated from plant residue and rice roots in irrigated rice-field soil, *Int. J. Syst. Evol. Microbiol.*, 57, 1803, 2007.
15. Downes, J. et al., *Prevotella histicola* sp., nov., isolated from the human oral cavity, *Int. J. Syst. Evol. Microbiol.*, 58, 1788, 2008.
16. Lawson, P.A., Moore, E., and Falsen, E., *Prevotella amnii* sp. nov., isolated from human amniotic fluid, *Int. J. Syst. Evol. Microbiol.*, 58, 89, 2008.
17. Downes, J. et al., *Prevotella micans* sp. nov., isolated from the human oral cavity, *Int. J. Syst. Evol. Microbiol.*, 59, 771, 2009.
18. Sakamoto, M. et al., *Prevotella falsenii* sp. nov., a *Prevotella intermedia* like organism isolated from monkey dental plaque, *Int. J. Syst. Evol. Microbiol.*, 59, 319, 2009.
19. Willems, A., and Collins, M.D., 16S rRNA gene similarities indicate that *Hallella seregens* (Moore and Moore) and *Mitsuokella dentalis* (Haapasalo *et al.*) are genealogically highly related and are members of the genus *Prevotella*: Emended description of the genus *Prevotella* (Shah and Collins) and description of *Prevotella dentalis* comb. nov., *Int. J. Syst. Bacteriol.*, 45, 832, 1995.
20. Downes, J. et al., *Prevotella bergensis* sp. nov., isolated from human infections, *Int. J. Syst. Evol. Microbiol.*, 56, 609, 2006.
21. Downes, J. et al., *Prevotella maculosa* sp. nov., isolated from the human oral cavity, *Int. J. Syst. Evol. Microbiol.*, 57, 2936, 2007.
22. Stevenson, D.M., and Weimer, P.J., Dominance of *Prevotella* and low abundance of classical bacterial species in the bovine rumen revealed by relative quantification real-time PCR, *Appl. Microbiol. Biotechnol.*, 75, 165, 2007.
23. Kobayashi, N. et al., Colonization pattern of periodontal bacteria in Japanese children and their mothers, *J. Periodontol. Res.*, 43, 156, 2008.
24. Mikamo, H. et al., Bacterial flora in the uterine endometrial cavity of healthy subjects on the first and fifth days of the puerperium, *J. Infect. Chemother.*, 4, 12, 1998.

25. Fukuoka, M. et al., Pyogenic vertebral osteomyelitis caused by *Prevotella intermedia*, *Infect. Chemother.*, 8, 182, 2002.
26. Berger, P. et al., *Prevotella massiliensis* sp. nov. isolated from human blood, *Res. Microbiol.*, 156, 967, 2005.
27. Gomes, B.P.F.A. et al., Microbiological examination of infected dental root canals, *Oral Microbiol. Immunol.*, 19, 71, 2004.
28. Robertson, D., and Smith, A.J., The microbiology of the acute dental abscess, *J. Med. Microbiol.*, 58, 155, 2009.
29. Brook, I., Microbiology and management of joint and bone infections due to anaerobic bacteria, *J. Orthop. Sci.*, 13, 160, 2008.
30. Brook, I., The role of anaerobic bacteria in chronic suppurative otitis media in children: Implications for medical therapy, *Anaerobe*, 14, 297, 2008.
31. Al-Nawas, B., and Maeurer, M., Severe versus local odontogenic bacterial infections: Comparison of microbial isolates, *Eur. Surg. Res.*, 40, 220, 2008.
32. Moore, W.E.C., and Moore, L.V.H., The bacteria of periodontal diseases, *Periodontology*, 5, 66, 1994.
33. Botero, J.E. et al., Occurrence of periodontopathic and superinfecting bacteria in chronic and aggressive periodontitis subjects in a Colombian population, *J. Periodontol.*, 78, 696, 2007.
34. Lafaurie, G.I. et al., Demographic, clinical and microbial aspects of chronic and aggressive periodontitis in Colombia: A multicenter study, *J. Periodontol.*, 78, 629, 2007.
35. Herrera, D. et al., Subgingival microbial profiles in chronic periodontitis patients from Chile, Colombia and Spain, *J. Clin. Periodontol.*, 35, 106, 2008.
36. Haraszthy, V. et al., Identification of oral bacterial species associated with halitosis, *J. Am. Dent. Assoc.*, 138, 1113, 2007.
37. Cogulu, D. et al., PCR-based identification of selected pathogens associated with endodontic infections in deciduous and permanent teeth, *Oral Surg. Oral Med. Oral Pathol. Oral Radiol. Endodontol.*, 106, 443, 2008.
38. Araki, H. et al., The microbial synergy of *Peptostreptococcus micros* and *Prevotella intermedia* in a murine abscess model, *Oral Microbiol. Immunol.*, 19, 177, 2004.
39. Riggio, M.P. et al., Identification of bacteria associated with spreading odontogenic infections by 16S rRNA gene sequencing, *Oral Surg. Oral Med. Oral Pathol. Oral Radiol. Endodontol.*, 103, 610, 2007.
40. Socransky, S.S., and Haffajee, A.D., The bacterial etiology of destructive periodontal disease: Current concepts, *J. Periodontol.*, 63, 322, 1992.
41. Darby, I., and Curtis, M., Microbiology of periodontal disease in children and young adults, *Periodontology*, 26, 33, 2001.
42. Slots, J., Enzymatic characterization of some oral and nonoral gram-negative bacteria with the API ZYM system, *J. Clin. Microbiol.*, 14, 288, 1981.
43. Suido, H. et al., Arylaminopeptidase activities of oral bacteria, *J. Dent. Res.*, 65, 1335, 1986.
44. Suido, H. et al., Identification of periodontopathic bacteria based upon their peptidase activities, *Adv. Dent. Res.*, 2, 304, 1988.
45. Gazi, M.I. et al., Comparison of host tissue and bacterial dipeptidylpeptidases in human gingival crevicular fluid by analytical isoelectric focusing, *Arch. Oral Biol.*, 40, 731, 1995.
46. Gazi, M.I. et al., Characterization of protease activities in *Capnocytophaga* spp., *Porphyromonas gingivalis*, *Prevotella* spp., *Treponema denticola* and *Actinobacillus actinomycetemcomitans*, *Oral Microbiol. Immunol.*, 12, 240, 1997.
47. Leung, K.P. et al., Surface appendages, hemagglutination, and adherence to human epithelial cells of *Bacteroides intermedius*, *Oral Microbiol. Immunol.*, 4, 204, 1989.
48. Dorn, B.R, Leung, K.L., and Progulske-Fox, A., Invasion of human oral epithelial cells by *Prevotella intermedia*, *Infect. Immun.* 66, 6054, 1998.
49. Taylor-Robinson, D. et al., Oro-dental bacteria in various atherosclerotic arteries, *Eur. J. Clin. Microbiol. Infect. Dis.*, 21,755, 2002.
50. Fiehn, N.E. et al., Identification of periodontal pathogens in atherosclerotic vessels, *J. Periodontol.*, 75, 731, 2005.
51. Gaetti-Jardim Jr., E. et al., Quantitative detection of periodontopathic bacteria in atherosclerotic plaques from coronary arteries, *J. Med. Microbiol.*, 58, 1568, 2009.
52. Okamoto, M. et al., Hemolytic and hemagglutinating activities of *Prevotella intermedia* and *Prevotella nigrescens*, *FEMS Microbiol. Lett.*, 178, 299, 1999.
53. Haraldsson, G. et al., Properties of hemagglutination by *Prevotella melaninogenica*, *Anaerobe*, 11, 285, 2005.
54. Tamura, M. et al., Effects of zinc and copper on adhesion and hemagglutination of *Prevotella intermedia* and *Prevotella nigrescens*, *Oral Microbiol. Immunol.*, 20, 339, 2005.
55. Whittaker, C.J., Klier, C.M., and Kolenbrander, P.E., Mechanisms of adhesion by oral bacteria, *Annu. Rev. Microbiol.*, 50, 513, 1996.
56. Lévesque, C., Lamothe, J., and Frenette, M., Coaggregation of *Streptococcus salivarius* with periodontopathogens: Evidence for involvement of fimbriae in the interaction with *Prevotella intermedia*, *Oral Microbiol. Immunol.*, 18, 333, 2003.
57. Kamaguchi, A. et al., Adhesins encoded by the gingipain genes of *Porphyromonas gingivalis* are responsible for coaggregation with *Prevotella intermedia*, *Microbiology*, 149, 1257, 2003.
58. Stefanopoulos, P.K., and Kolokotronis, A.E., The clinical significance of anaerobic bacteria in acute orofacial odontogenic infections, *Oral Surg. Oral Med. Oral Pathol. Oral Radiol. Endodontol.*, 98, 398, 2004.
59. Siqueira, Jr., J.F., and Rôças, I.N., Bacterial pathogenesis and mediators in apical periodontitis, *Braz. Dent. J.*, 18, 267, 2007.
60. Saito, K. et al., Effects of glucose on formation of cytotoxic end-products and proteolytic activity of *Prevotella intermedia*, *Prevotella nigrescens* and *Porphyromonas gingivalis*, *J. Periodontol. Res.*, 36, 355, 2001.
61. Santos, S.G. et al., The influence of molecular oxygen exposure on the biology of *Prevotella intermedia*, with emphasis on its antibiotic susceptibility, *J. Appl. Microbiol.*, 103, 882, 2007.
62. Noack, B. et al., Periodontal infections contribute to elevated systemic C-reactive protein level, *J. Periodontol.*, 72, 1221, 2001.
63. Kim, S-J. et al., *Prevotella intermedia* lipopolysaccharide stimulates release of tumor necrosis factor-α through mitogen-activated protein kinase signaling pathways in monocyte-derived macrophages, *FEMS Immunol. Med. Microbiol.*, 51, 407, 2007.
64. Chung, Y-H. et al., Lipopolysaccharide from *Prevotella nigrescens* stimulates osteoclastogenesis in cocultures of bone marrow mononuclear cells and primary osteoblasts, *J. Periodontol. Res.*, 41, 288, 2006.
65. Strömbeck, L. et al., *Prevotella bivia* can invade human cervix epithelial (HeLa) cells, *APMIS*, 115, 241, 2007.
66. Pucar, A. et al., Correlation between atherosclerotic and periodontal putative pathogenic bacterial infections in coronary and internal mammary arteries, *J. Periodontol.*, 78, 677, 2007.
67. Love, R.M., Hemin nutritional stress inhibits bacterial invasion of radicular dentine by two endodontic anaerobes, *Int. Endodontol. J.*, 40, 94, 2007.

68. Silva, T.A et al., In vitro activation of the hemolysin in *Prevotella nigrescens* ATCC 33563 and *Prevotella intermedia* ATCC 25611, *Res. Microbiol.*, 155, 31, 2004.
69. Guan, S-M. et al., Purification and characterization of a hemoglobin-binding outer membrane protein of *Prevotella intermedia*, *FEMS Microbiol. Lett.*, 235, 333, 2004.
70. Guan, S.M. et al., Degradation of human hemoglobin by *Prevotella intermedia*, *Anaerobe*, 12, 279, 2006.
71. Mallorquí-Fernández, N. et al., A new autocatalytic activation mechanism for cysteine proteases revealed by *Prevotella intermedia* interpain A, *J. Biol. Chem.* 283, 2871, 2008.
72. Frandsen, E.V., Kjeldsen, M., and Kilian, M., Inhibition of *Prevotella* and *Capnocytophaga* immunoglobulin A1 proteases by human serum, *Clin. Diagn. Lab. Immunol.*, 4, 458, 1997.
73. Yanagisawa, M. et al., Proteinase activity of *Prevotella* species associated with oral purulent infection, *Curr. Microbiol.*, 52, 375, 2006.
74. Guan, S-M. et al., *Prevotella intermedia* induces metalloproteinase-9 expression in human periodontal ligament cells, *FEMS Microbiol. Lett.*, 283, 47, 2008.
75. Itoh, T. et al., The activation of matrix metalloproteinases by a whole-cell extract from *Prevotella nigrescens*, *J. Endodontol.*, 35, 55, 2009.
76. Griswold, K.E., White, B.A., and Mackie, R.I., Diversity of extracellular proteolytic activities among *Prevotella* species from the rumen, *Curr. Microbiol.*, 39, 187, 1999.
77. Hecht, D.W., Anaerobes: Antibiotic resistant, clinical significance, and the role of susceptibility testing, *Anaerobe*, 12, 115, 2006.
78. Mosca, A. et al., Antimicrobial profiles of *Prevotella* spp. and *Fusobacterium nucleatum* isolated from periodontal infections in a selected area of southern Italy, *Int. J. Antimicrob. Agents*, 30, 521, 2007.
79. Iwahara, K. et al., Detection of *cfxA* and *cfxA2*, the β-lactamase genes of *Prevotella* spp., in clinical samples from dentoalveolar infection by real-time PCR, *J. Clin. Microbiol.*, 44, 172, 2006.
80. Huits, R.M.H.G. et al., *Prevotella bivia* necrobacillosis following infectious mononucleosis, *J. Infect.*, 53, e53, 2006.
81. Stoakes, L. et al., Gas-liquid chromatographic analysis of cellular fatty acids for identification of Gram-negative anaerobic bacilli, *J. Clin. Microbiol.*, 29, 2636, 1991.
82. Ashimoto, A. et al., Polymerase chain reaction detection of 8 putative periodontal pathogens in subgingival plaque of gingivitis and advanced periodontitis lesions, *Oral Microbiol. Immunol.*, 11, 266, 1996.
83. Martin, F.E. et al., Quantitative microbiological study of human carious dentine by culture and real-time PCR: Association of anaerobes with histopathological changes in chronic pulpitis, *J. Clin. Microbiol.*, 40, 1698, 2002.
84. Kuboniwa, M. et al., Quantitative detection of periodontal pathogens using real-time polymerase chain reaction with TaqMan probes, *Oral Microbiol. Immunol.*, 19, 168, 2004.
85. Yoshida, A. et al., Multiplex polymerase chain reaction assay for simultaneous detection of black-pigmented *Prevotella* species in oral specimens, *Oral Microbiol. Immunol.*, 20, 43, 2005.
86. Riggio, M.P., and Lennon, A., Development of a novel PCR assay for detection of *Prevotella oris* in clinical specimens, *FEMS Microbiol. Lett.*, 276, 123, 2007.
87. Frandsen, E.V., Poulsen, K., and Kilian, M., Confirmation of the species *Prevotella intermedia* and *Prevotella nigrescens*, *Int. J. Syst. Bacteriol.*, 45, 429, 1995.
88. Teanpaisan, R. et al., Clonality of *Porphyromonas gingivalis*, *Prevotella intermedia* and *Prevotella nigrescens* isolated from periodontally diseased and healthy sites, *J. Periodontol. Res.*, 31, 423, 1996.
89. Milsom, S.E. et al., Rapid differentiation of *Prevotella intermedia* and *P. nigrescens* by 16S rDNA PCR-RFLP, *J. Med. Microbiol.*, 44, 41, 1996.
90. Okabe, S. et al., Quantification of host-specific *Bacteroides-Prevotella* 16S rRNA genetic markers for assessment of fecal pollution in freshwater, *Appl. Microbiol. Biotechnol.*, 74, 890, 2007.
91. Nagashima, S. et al., Use of the genomic subtractive hybridization technique to develop a real-time PCR assay for quantitative detection of *Prevotella* spp. in oral biofilm samples, *J. Clin. Microbiol.*, 43, 2948, 2005.
92. Dahlén, G. et al., A comparison of two transport media for saliva and subgingival samples, *Oral Microbiol. Immunol.*, 8, 375, 1993.

Section IV

Proteobacteria

Alphaproteobacteria

52 *Anaplasma*

Marina E. Eremeeva and Gregory A. Dasch

CONTENTS

52.1 INTRODUCTION

52.1.1 Classification and Morphology

The genus *Anaplasma* encompasses several obligately intracellular bacterial species (e.g., *Anaplasma phagocytophilum, A. marginale, A. centrale, A. bovis* and *A. platys*) that are classified in the family Anaplasmataceae, order Rickettsiales. While *Anaplasma phagocytophilum* infects the neutrophils of ruminants (previously referred to as *Ehrlichia phagocytophila*), equines (*Ehrlichia equi*), and humans (HGE agent) (Figure 52.1);[2] *A. marginale, A. centrale, A. bovis* and *A. platys* are essentially animal pathogens, which are beyond the scope of this chapter. The type strain of *A. phagocytophilum* is strain Webster. Different isolates of *A. phagocytophilum* are causative agents of tick-borne fever in sheep, cattle, and goats; equine granulocytic anaplasmosis; canine granulocytic anaplasmosis; and human granulocytic anaplasmosis. They exhibit a variety of specific features and host adaptations, which have been summarized in several reviews.[2–7] For the purpose of the current review, the scope of this chapter is limited to description of the agent of human granulocytic anaplasmosis (HGA) and the clinical characteristics of this disease.

Human granulocytic anaplasmosis was discovered in 1994 in the United States and was originally described as "human granulocytic ehrlichiosis."[8] Since then, numerous reviews and original articles have been published on this disease, which may be consulted for additional details.[9–16]

Anaplasma phagocytophilum grows as a cluster of small cocci or pleomorphic coccobacilli in the cytoplasm of neutrophils. Although they are gram-negative, the bacteria do not stain well with the Gram stain and are better visualized using Wright or Giemsa staining. Within infected cells, *A. phagocytophilum* is found inside intracytoplasmic vacuoles, where they form macrocolonies of 1.5–6 μm diameter called morulae.[17,18] Two morphological types of individual bacteria are present in morulae, a larger reticulate form and a smaller electron dense form (Figure 52.2); both types of cells divide by binary fission. The bacterium has a typical gram-negative cell wall ultrastructure, containing membrane proteins; however, the cell wall does not contain significant amount of peptidoglycan, and are lacking the lipopolysacharides, pili, and capsule.[17,18]

52.1.2 Biology, Habitat, Ecology, and Epidemiology

Anaplasma phagocytophilum has unusual capacity for infecting and surviving within the vacuole of neutrophils, which they enter via receptor-mediated endocytosis.[19–21] Binding of *A. phagocytophilum* to human neutrophils is mediated at least partially by the adherence of its 44-kDa major surface prtotein-2 (Msp2) to P-selectin glycoprotein ligand,[19,22] whereas interaction with murine neutrophils requires

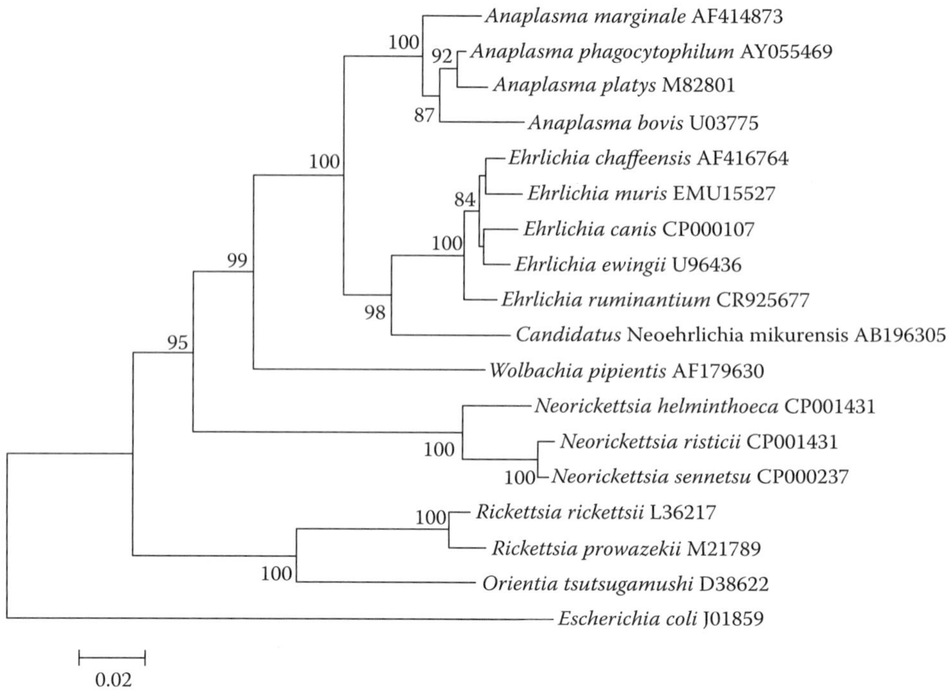

FIGURE 52.1 Reconstruction of the phylogenetic position of *Anaplasma phagocytophilum*. Phylogenetic tree constructed using nucleotide sequences from 16S rRNA gene of *Ehrlichia* and *Anaplasma* species and related bacteria, a homologous sequence of *E. coli* was used as an outgroup. Sequences were ClustalW aligned and tree was drawn using the MEGA4.0 software.[1] Distance matrix was calculated using Kimura-2 parameters and tree was obtained using the Neighbor-Joining method based on comparison of 1320 sites after complete deletion of gaps. The scale represents 0.02% divergence. Numbers at nodes are the proportions of 1000 bootstrap resamplings that support the topology shown. Only bootstrap values greater than 80 are shown. The sequences of the isolates used in this analysis are from the NCBI GenBank and the accession numbers are shown next to the sequence name.

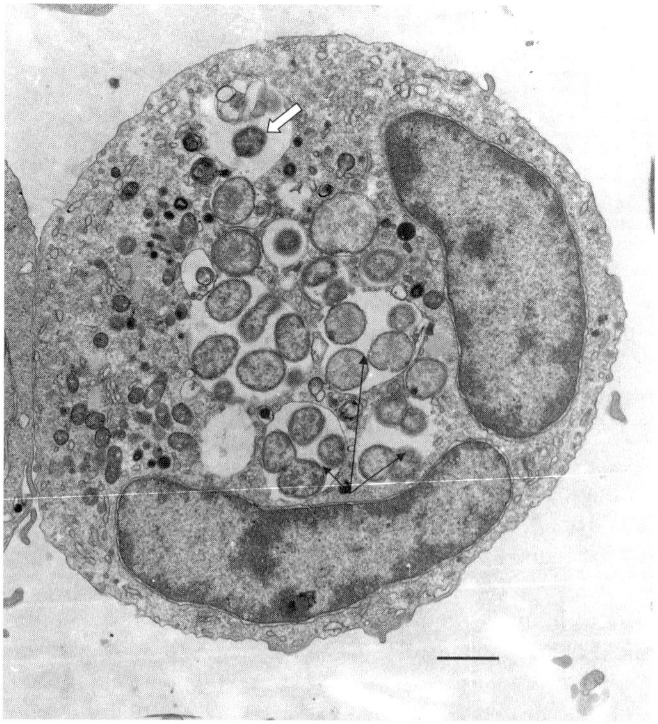

FIGURE 52.2 Ultrastructure of *A. phagocytophilum* in HL-60 cells. Scale bar = 1 μm. The cytoplasm contains several morulae of different size; morulae are shown with black arrows, morula with single reticulate body is shown with an open arrow. (Courtesy of V. Popov. With permission.)

expression of 1–3-fucosyltransferase.[23] Internalized *A. phagocytophilum* delays apoptosis of infected neutrophils,[24] and survives this superoxide-rich environment by scavenging O_2^-. Once infected by *A. phagocytophilum*, the modified endosome fails to fuse with lysosomal vesicles and form mature acidified phagolysosomes. *A. phagocytophilum* exploits host cell nutrients and divides until the cell is lysed, so it can infect other neutrophils. Beside alteration of phagocytic pathways, *A. phagocytophilum* reduces neutrophil adhesion to endothelial cells; upregulates production of chemokines, mediating transmigration of lymphocytes; and disrupts their motility, degranulation, and respiratory burst.[20,25]

Anaplasma phagocytophilum is a tick-borne agent that is naturally maintained in a zoonotic cycle between the ticks of the *Ixodes persulcatus* complex and mammalian reservoir hosts.[4,6,13,26] The ecology of *A. phagocytophilum* is complex. In the United States, it is found in *Ixodes scapularis* in the eastern and midwestern states, and in *I. pacificus* and *I. spinipalpis* in western coastal and mountain areas, respectively.[16] The prevalence of *A. phagocytophilum* in *I. scapularis* ranges from 4.9% to >50%.[27–30] *Ixodes persulcatus*, *I. ovatus*, and *I. ricinus* are vectors transmitting *A. phagocytophilum* in Asia and Europe.[26,31–36] DNA of *A. phagocytophilum* has been detected in 2%–45% of *I. ricinus* ticks (reviewed by Ref. 26); its presence varies from 2.1% to 4.5% in *I. persulcatus* from different parts of Russia[37] to 4%–4.3% in *I. persulcatus* from China and Korea.[32,36,38] In Japan, different genotypes of A. *phagocytophilum* were detected in 9.6% of *I. persulcatus* and 6.3% of *I. ovatus*.[35] Many species of domestic and wild animals are important reservoirs of *A. phagocytophilum* in European and Asian countries. Beside horses, cattle, sheep, and dogs, potential reservoirs include wood mouse (*Apodemus sylvaticus*), bank vole (*Clethrionymus graleolus*), shrew (*Sorex aranus*), roe deer (*Capreolus capreolus*) and other cervids; foxes (*Vulpes vulpes*), wild boar (*Sus scrofa*), bear, and rabbits.[33,39–42] The bacterium has also been found in *I. trianguliceps* from the UK and *I. ventalloi* from Portugal[43–46]; the role of these ticks in the natural cycle and transmission of the HGA agent needs further evaluation, but they are not known to feed on humans.

In the United States, the white-footed mouse, *Peromyscus leucopus*, is a primary mammalian reservoir for *A. phagocytophilum*; however, other small rodents from the genera *Neotoma*, *Apodemus*, *Microtus*, and *Clethrionomys* are likely reservoirs for *A. phagocytophilum* and hosts for its tick vector.[13,47] Other potential reservoirs in the United States include white-tailed deer (*Odocoileus virginianus*), raccoons (*Procyon lotor*), opossums (*Didelphis virginiana*), squirrels (*Sciurus* spp.), striped skunks (*Mephitis mephitis*), chipmunks (*Tamias*), rabbits, and lizards.[48–54] Some of the strains of *A. phagocytophilum* found in association with different mammalian species are believed to be nonpathogenic for humans, such as Ap-1 variant, which circulates in white-tailed deer.[55–57]

It is commonly accepted that immature and adult ticks acquire *A. phagocytophilum* while feeding on bacteremic animals and efficiently pass it transtadially to the next life stage. In contrast, transovarial transmission from an adult female tick to its progeny is considered rare.[13] However, recent data indicate that transovarial transmission of an *A. phagocytophilum* variant may occur in *Dermacentor albipictus*.[58] These observations suggest that some types of *A. phagocytophilum* do not exclusively depend on mammalian reservoirs and horizontal transmission during every generation of ticks.

Humans acquire infection as a result of exposures to infected ticks in a variety of habitats. Most infections occur during the spring to summer months, correlating with the biological cycles of feeding of overwintering, infected adult ticks and nymphs infected as larvae. Human anaplasmosis is reportable in the United States[59,60]; passively collected surveillance data show a steady, increasing trend in the number of annual cases, which reached 1026 in 2008. It is believed that the actual incidence and distribution are significantly greater than is reported.[10] Active surveillance estimated from 24 to 58 cases per 100,000 of population (0.02%–0.06%) in Connecticut[61] and the Upper Midwest.[62] The background seroprevalence rate was approximately 14.9% among healthy individuals from northwestern Wisconsin who had no known history of recent tick bite.[63]

HGA is not formally reportable outside the United States; therefore, extrapolations regarding the incidence and prevalence depend upon the number of laboratory-confirmed cases reported in the literature. These appear to be infrequent, or the disease is often undetected, since less than 100 laboratory-confirmed or probable cases of HGA have been reported from Europe and Russia.[26,64,65] The temporal distribution and demographic characteristics of the populations affected are similar in the United States and Europe. It has been noticed that a significant discrepancy exists between the high levels of seropositivity to *A. phagocytophilum* observed in populations from some European countries (up to 17.1%) and the relatively low level of *A. phagocytophilum* DNA detected in *I. ricinus* (0.25%–11.5%) in some regions. This suggests that people are frequently exposed to infected ticks and that they may exhibit persistent antibody responses to nonpathogenic strains of *A. phagocytophilum* found in *I. ricinus*.[66,67] Seroprevalence to *A. phagocytophilum* ranges from 8.8% to 20% in different regions of China[68,69]; antibodies to *A. phagocytophilum* were detected by IFA (geometric mean titer 1/747) in 1.8% ($n = 491$) of febrile patients from Korea.[70]

HGA can also be acquired rarely through exposure to *A. phagocytophilum* by blood transfusion, organ transplantation, nosocomial and perinatal transmission, or inhalation of aerosolized blood.[71–75]

52.1.3 Genetic and Genomic Features

The complete genome sequence has been determined for *A. phagocytophilum* strain HZ, isolated in 1995 from a patient from New York, USA.[76] The genome consists of a single circular chromosome of 1,471,282 bp (NCBI GenBank accession number CP000235) with 41% G + C content. The published annotation predicts 1264 proteins and 42 RNA genes. The gene order of *A. phagocytophilum* has significant synteny with the genomes of *Ehrlichia* species. The genome of *A. phagocytophilum* contains 462 open reading frames

(ORFs) or paralog clusters that are unique; the majority of these unique genes encode hypothetical, conserved hypothetical, and conserved domain proteins, as well as uncharacterized membrane proteins and lipoproteins. Other *A. phagocytophilum*-specific genes include those encoding the P44 outer membrane proteins and the HGE-14 and HGE-2 antigenic proteins. The tick-borne Anaplasmataceae (*Ehrlichia* spp. and *Anaplasma* spp.) may be the only Rickettsiales that are not transmitted transovarially in the invertebrate host. One ortholog cluster containing a class II aldolase/adducing domain protein (NSE_0849, RC0678, RP493, RT0479, WD0208) is absent only from *Ehrlichia* spp. and *Anaplasma* spp. It has been suggested that a lack of this aldolase/adducing domain protein may prevent transovarial transmission of these agents in the arthropod vector.[76] The superoxide dismutase gene, *sod*B is cotranscribed with components of the type IV secretion system in *A. phagocytophilum* suggesting that it has an important role in pathogenesis beyond elimination of superoxide. The *A. phagocytophilum* genome has three *omp*-1, one *msp*2, two *msp*2 homologs, one *msp*4, and 113 *p44* loci belonging to the OMP-1/MSP2/P44 superfamily. Microarray-based comparative genomic hybridization reveals that expansion of the *p44* family is a common feature in *A. phagocytophilum* strains.[76] The genome of *A. phagocytophilum* has numerous genes encoding HGE-14 protein that may be an excreted effector molecule of its type IV secretion system. *A. phagocytophilum* has genetic capacity to synthesize all nucleotides, most vitamins, and cofactors. A complete pyruvate dehydrogenase, tricarboxylic acid cycle, F0F1-ATPase, and electron transport chain were found. It appears to use host-derived carboxylates and amino acids but does not obtain carbon or energy from fatty acids or actively carry out glycolysis.

Differential gene transcription profiles were demonstrated for isolate HZ grown in HL-60, human microvascular endothelial cells (HMEC-1), and a tick cell line.[77] In particular, it was shown that seven *vir*B2 paralogs of the type IV secretion system exhibited human or tick cell dependent transcription. Unique cell-type dependent expression of previously unrecognized genes and coding sequences were identified, including expression of *p44/msp2* paralogs (of 114 total, only two are detected in HL-60, three in HMEC-1, and none in tick cells) suggesting variable synthesis of these proteins during different stages of the *A. phagocytophilum* life cycle.

52.1.4 Clinical Features and Pathogenesis

HGA is an acute illness that occurs 5–21 days after the bite of an infected tick[14]; its severity ranges from asymptomatic seroconversion to a mild or severe febrile illness. Most patients experience fever >101°F, myalgia, headache, and malaise.[14,62] Other symptoms occur less frequently and include nausea, abdominal pain, vomiting, diarrhea, cough, and arthralgias. Involvement of the skin or central nervous system is very rare during the course of HGA. Basic laboratory findings included thrombocytopenia (<100 × 10^9 platelets/L), leukopenia (<2.5 × 10^9 leukocytes/L), elevation of C-reactive protein concentration and moderately increased concentration of hepatic transaminases (alanine and aspartate aminotransferases) and lactic dehydrogenase without increase of bilirubin.[73,78,79] According to several reports, the likelihood of finding blood-cell counts below normal depends on the time of testing relative to onset of illness.[10,11,73,79] The lowest values of total white blood cells and platelets are observed around the seventh day of illness, after which cell counts return to the normal range, although clinical symptoms may persist. During the second week of illness, neutrophil counts gradually return to normal and are accompanied by a relative and absolute lymphocytosis, sometimes with atypical lymphocytes.[79] Over 77% of HGA patients have thrombocytopenia for one or more days during the first two weeks of illness[79]; most often thrombocytopenia is detected at least once between days 2 and 9 of the illness.

Based on limited data, as many as half of HGA patients required hospitalization, and about one-third of those needed treatment at an intensive care unit.[10,14] National surveillance data estimated a 33% hospitalization rate among HGA patients; 7% of hospitalized patients developed life-threatening complications.[59] Complications associated with HGA are infrequent; however, some patients may develop septic- or toxic-shock–like syndrome, respiratory insufficiency, pancarditis, acute renal failure, hemorrhage, and peripheral neuropathies that may persist for weeks or months.[10,13] Fatalities due to *A. phagocytophilum* infections are rare (0.7%) and typically arise from complicating opportunistic viral or fungal infections, advanced age, immunosuppression, or severe preexisting medical conditions.[10,13] The clinical features of HGA outside the United States are harder to assess because of the limited number of cases reported. However, the impression persists that the clinical manifestations and laboratory findings for HGA cases from central Europe are milder and resolve sooner, sometimes even without antibiotic treatment, compared to HGA cases reported in the United States.[26,64]

Systemic inflammation constitutes the main presentation of HGA pathogenesis; however, only limited pathologic and histologic data have been reported due to the paucity of fatal cases attributed to HGA.[62,71,80–83] Three of the eight fatal cases described died from opportunistic infections, and two other deaths were due to myocarditis or generalized lymphadenopathy and mononuclear phagocyte system activation. The accumulated pathologic findings include detection of perivascular lymphohistiocytic infiltrates in internal organs, hepatitis with signs of apoptosis, normocellular, or hypercellular bone marrow; mild lymphoid depletion; mononuclear phagocyte hyperplasia in spleen and lymph nodes; and some splenic necrosis.[13] Hemophagocytosis is seen in bone marrow, liver, and spleen. Vasculitis has not been detected, and *A. phagocytophilum* is rarely detected in lesions by immunohistochemical tests.[84] Furthermore, the lack of correlation between bacterial burden and severity of lesions, suggests that *A. phagocytophilum* just triggers the inflammatory process through interactions of bacterial effector molecules with host signaling pathways and transcription system.[20,25,85] The complete sequence of events is not completely understood;

however, involvement of Toll-like receptor-2 is clearly demonstrated, and a number of findings indicate an important role for fixed macrophages in the immunogenesis of HGA.[14,86]

52.1.5 Diagnosis

Diagnosis of HGA should be attempted during the acute stage of illness to reduce possible complications.[11,12,78] Definitive etiological diagnosis during the active stage of disease depends upon one or more of the following findings: detection of *A. phagocytophilum* morulae within the cytoplasmic vacuoles in peripheral blood neutrophils, immunohistochemical demonstration of *A. phagocytophilum* in neutrophils in tissues or peripheral blood, or demonstration of the presence of *A. phagocytophilum* nucleic acids in peripheral blood. Culture of *A. phagocytophilum* from the peripheral blood in a suitable cell-culture system, or serologic detection of immunoglobulin G (IgG) and IgM antibodies reactive with *A. phagocytophilum* primarily allows retrospective diagnosis of exposure because of the time required. Each method has its benefits and disadvantages, which have been discussed in detail by several authors.[12,15,78,83] A combination of results obtained using different acute tests increases the probability of accurate diagnosis of HGA infection, thus benefiting patient's proper management and timely recovery, while retrospective confirmation is essential for improved recognition in acute differential clinical diagnosis. To achieve the best diagnostic results for tests performed during the acute stage of illness, samples should be collected prior to the initiation of antibiotic therapy and no later than 24–48 h after treatment has begun, especially when appropriate therapy was initiated. Current requirements for laboratory-confirmed and probable cases of HGA, and their applicability to surveillance of HGA, are summarized by the Council of State and Territorial Epidemiologists (http://www.cste.org/ps2009/09-ID-15.pdf).

52.1.5.1 Examination of Peripheral Blood Smears

Careful scrutiny of peripheral blood smears stained with a Romanowsky method (Wright–Giemsa or Diff–Quick) during differential leukocyte counts for the presence of the intravacuolar bacterial inclusions provides the quickest provisional diagnosis of HGA.[8,10] The specific bacterial inclusions, called "morulae" (Latin for *mulberry*), are basophilic and stain blue to purple and can be detected in 25%–80% of acutely ill HGA patients; however, prolonged examination and counting 800–1000 granulocytes may be required because, if seen at all, *A. phagocytophilum* morulae are found in less than 1% of neutrophils.[80,87] Examination of buffy coat preparations may increase the likelihood of detecting morulae in patients presenting with leukopenia.[78] When used, buffy coat smears should be prepared within 24 h of blood collection because potential lysis of leukocytes during blood storage may complicate the examination of the smears. Examination of the smears is most sensitive during the first week of illness, and blood samples must be taken prior to doxycycline treatment since neutrophils with morulae will clear from the blood within 24–72 h after antibiotic treatment is initiated. Morphological identification of morulae may be difficult and inherently lacks specificity, so the cytology technicians need training to distinguish common intracellular inclusions and artefacts found within blood smears, including stain crystals, bacterial contamination of the stain solutions, Dohle bodies, Auer rods, toxic granulation, platelets overlying leukocytes, or phagocytosed materials from morulae.

52.1.5.2 Immunohistochemistry

There is little published information about application of immunochemistry to detection of *A. phagocytophilum* in tissue and blood smears.[73,80,81,83,84,88] This is both due to the lack of specific antisera to *A. phagocytophilum* and the specialized nature of the detection procedure. However, the great advantage of this technique is that use of specific antibodies permits definitive specific etiological confirmation of the morphologic findings and, probably, more accurate and specific detection of the morulae. However, specific anti-*A. phagocytophilum* antibodies without unwanted cross-reactivities are not commercialized, preventing broad application of this useful method.

52.1.5.3 Amplification of *A. phagocytophilum* DNA

Molecular diagnosis based on PCR amplification of *A. phagocytophilum* DNA has proven to be a very useful tool for the early confirmation of HGA infections based on compatible clinical symptoms (Table 52.1). The concentration of *A. phagocytophilum* DNA is greatest during the first week of infection, after which the bacterial load progressively decreases, providing medium-to-high relative sensitivity of the PCR test, primarily during the first week of illness.[11–13,62,78,89,90]

The success of the PCR assays used for diagnosis of HGA greatly depends on the availability and handling of appropriate test samples. Since *A. phagocytophilum* grows in neutrophils and their precursor cells, peripheral whole blood is the preferred test specimen.[8,67,91] Blood is collected in EDTA or other anticoagulants besides heparin; it can be used directly or fractionated to obtain leukocytes.[12] Early administration of antibiotics may be a critical factor in test failure since appropriate antibiotic treatment decreases agent abundance, and thus, PCR sensitivity. While different PCR assays may vary significantly in their sensitivity, single-step PCR assays targeting 16S rRNA or Msp2 gene can detect as low as one-fourth of an infected cell,[92] with a predicted theoretical sensitivity between 1 and 10 bacterial cells per assay.[91]

Various specific gene targets have been tested, and they have been found to be valuable for the specific detection and confirmation of *A. phagocytophilum* in clinical specimens, including 16S rRNA gene (*rrs*),[8,80,92]; *ank* (previously referred to as *epank*1 and *ank*A) encoding a 160-kDa ankyrin protein[93]; or *msp*2 (*p44* or *hge44*) encoding for the 44–49 kDa major immunodominant surface protein of *A. phagocytophilum*.[61] Other possible useful targets include *msp*4,[94] *gro*EL,[95] *rpo*B,[96] *glt*A,[97] and *fts*Z;[98] however, although they have been successfully used with veterinary or arthropod samples[94,96,97,99] or evaluated with culture-grown organisms,[94,96,98] their performance for laboratory diagnosis of HGA is unknown.

The sensitivity of PCR methods used for HGA apparently varies significantly. Dumler and Brouqui[12] summarized the results of 12 publications that used various sequence within the 16S rRNA gene, which had been applied to 1–152 patient samples (median = 12 patients per study), with a reported sensitivity ranging from 100% (one case analyzed) to 7% (28 patients, sensitivity median = 62.5%). Many publications describe PCR assays with apparent high analytical sensitivity and specificity that were typically evaluated using cell culture–derived materials, but they are not always supported by positive detection of *A. phagocytophilum* DNA in validated clinical specimens from patients with compatible clinical presentations and subsequent seroconversion to the HGA agent. It is difficult but valuable to use culture confirmed HGA cases in evaluating PCR sensitivity.[100] On average, the sensitivity of validated PCR assays used for diagnosis of HGA infection approaches 86%–87%, as demonstrated in studies with large cohorts of well-confirmed HGA patients, evaluated using different diagnostic tools.[11,62,100,101] Positive PCR results are less frequently obtained with European patients, a phenomenon possibly explained by the fact that HGA in Europe is typically considered to be milder, and where morulae are less frequently detected in blood smears.[26,102–104]

The best diagnostic assays require setting up nested reactions to ensure sufficient sensitivity or use of fluorescence PCR methods (quantitative PCR or so-called real-time PCR). In principle, the use of a multicopy gene, such as *msp*2, may provide enhanced sensitivity compared to single-copy gene targets, such as *rrs* or *gro*EL.[61,93]

Although false-positive results using molecular tests can occur, several procedural measures and additional controls can be applied to decrease their occurrence. In addition to the proper organization of the workflow in the laboratory to eliminate spurious amplicons, reduced use of homologous positive controls or use of engineered controls is recommended so that sample cross-contamination from this source can be easily recognized. Ideally, each positive PCR result should be confirmed by the sequencing of the amplicon or the use of hybridization probes. It has been suggested that a single positive PCR assay alone may be insufficient evidence of definite infection with *A. phagocytophilum*,[11,12,82] so amplification of a second target can be performed. Molecular data should always be interpreted in the full context of clinical symptoms, serologic data, culture, and blood smear detection of morulae.

New PCR assays have been developed that are slowly replacing standard gel-based PCR assays and nested PCR assays. The new detection platforms offer enhanced sensitivity, probe, or amplicon melt curve confirmation of product identity, single-tube assay, high throughput, and automation that eliminates the need to handle PCR amplicons in the lab, and thus, lowers the probability for amplicon contamination of laboratory and other samples. Several diagnostic assays have been developed and optimized for detection of *A. phagocytophilum* and other tick-borne pathogens. Fluorescence (e.g., SYBR-Green) PCR assays based on detection of the binding of a double-strand DNA intercalating dye to PCR amplicons[105] was originally developed for basic research but later was optimized to detect the presence of *Anaplasmataceae* in a variety of arthropod samples.[106,107] Subsequently, it was applied to detection of *A. phagocytophilum* DNA in human blood samples.[72] The *rrs* based primers are from a conserved portion of the 5′-end of the gene and will amplify DNA of most species in this family; therefore, sequence identification of the DNA detected is necessary. TaqMan probe hybridization assays for detection of *msp*2[108] and *rrs*[109] have also been developed. The latter report also demonstrated the advantages of using reverse transcription PCR, detecting abundant 16S rRNA compared to single-copy *rrs* DNA to enhance the sensitivity of the assay.[109] This assay also uses a novel enrichment procedure for target RNA which employs capture by biotin-labeled primer and streptavidin-coated magnetic beads.[109] Finally, a FRET-based detection assay was developed to detect *A. phagocytophilum* and *Ehrlichia* DNA using a conserved portion of the GroESL operon allowing sensitive and differential detection of several related tick-borne organisms by melting curve analysis.[110]

Due to the very nonspecific clinical manifestations of HGA and significant geographic overlap of regions associated with different tick-borne agents, patients with a compatible clinical picture should be evaluated for exposure to different agents, and for possible mixed infections. A new generation of multitarget PCR assays is likely to become available that permit the differentiation of specific amplicons based on nucleotide sequencing, analysis of melting curves, detection of single nucleotide polymorphisms, or probe hybridization methods.[108–110] For example, a reverse line blot hybridization assay was developed for identifying different variants of *A. phagocytophilum* in sheep blood, but this has not yet been applied to human samples.[111] Availability of the complete genome sequence of the *A. phagocytophilum* provides more opportunities to explore specific and unique targets for improved molecular diagnosis of HGA.

52.1.5.4 In Vitro Cultivation of *A. phagocytophilum*

Cultivation of *A. phagocytophilum* from a clinical sample is the gold standard for confirmatory diagnosis of HGA. However, isolation is also essential to understand the genetic diversity of *A. phagocytophilum* associated with human disease and for improvements in antimicrobial therapy or other adjunctive therapies and vaccine development. To recover *A. phagocytophilum*, the human HL-60 promyelocytic leukemia cells grown in RPMI-1640 medium supplemented with L-glutamine and fetal bovine serum are inoculated with whole EDTA-treated blood or leukocyte fraction and are incubated for several weeks at 37°C in 5% CO_2 atmosphere. Every 2–3 days, infected cultures are monitored by sampling infected cells, cytocentrifugation, and by staining them with a Wright–Giemsa or Diff–Quick method followed by microscopic evaluation and visualization of morulae. Microscopic detection of morulae can be confirmed by PCR amplification and sequencing of *A. phagocytophilum*-specific genes.

Culturing *A. phagocytophilum* is a simple and straightforward procedure[91,112]; however, because it requires the use of antibiotic-free cell cultures that are not typically maintained by clinical diagnostic laboratories, it is infrequently performed except by laboratories performing research with *A. phagocytophilum*. Blood samples collected for culture isolation should be obtained during the acute stage of illness prior to antibiotic treatment, which diminishes the probability of culture success.[14] Since *A. phagocytophilum* can survive refrigeration conditions in blood for almost two weeks,[113] an anticoagulated patient's blood can be refrigerated and transported to a reference diagnostic laboratory that performs culture isolation. If immediate isolation is not possible, samples can also be frozen at –70°C and transferred to liquid nitrogen for long-term storage; however, this freeze-thaw procedure may decrease culture success. It is believed that the sensitivity of culture isolation approaches the sensitivity of the PCR assay and, in principle, could exceed it since more sample can be used; however, positive culture results are usually not obtained before 1 week after inoculation, and they can even remain presumptively negative for over a 2-week period, precluding any value for confirmatory diagnosis during the acute stage of illness when treatment should be initiated.[11,79,100] Similarly, clinical samples that were not properly stored or that were accidently contaminated with other agents are not suitable for culture; however, they may still be PCR positive.

Anaplasma phagocytophilum can be also grown in HL-60 differentiated to granulocytic lineage by retinoic acid or dimethyl sulfoxide, KG-1 promyelocytic leukemia cells, THP-1 myelomonocytic cells, primary bone marrow cells, microvascular endothelial cells, and the tick embryo cell line IDE8 (ATCC #CRL11973), isolated from *I. scapularis*.[13,112,114–116] HL-60 cells differentiated into monocytic lineage do not support growth of *A. phagocytophilum*.[13]

52.1.5.5 Serologic Diagnosis

Serology is the most frequently used method for diagnosis of HGA[59,60,117]; however, it only provides a retrospective tool for confirming clinical diagnosis. Indirect fluorescent antibody test (IFA) is most commonly used to demonstrate a serologic reaction to *A. phagocytophilum* antigens and convalescent seroconversion in HGA. In this assay, serum antibodies are reacted with intact *A. phagocytophilum* in infected-tissue culture cells or, more rarely, with purified bacteria fixed to glass slides, followed by incubation with fluorescein isothiocyanate-conjugated antihuman anti-IgG or IgM immunoglobulin to detect the antibody–antigen complex. It has been estimated that the median incubation period for HGA is 8 days, and since most HGA patients do not seek medical care for another several days, it is expected that only 3%–30% of the patients suffering from HGA will be seropositive at initial presentation. The sensitivity of the IFA test is believed to be between 82% and 100% for IgG and 27%–37% for IgM antibody detection, and the specificity of the test is estimated to be 83%–100%,[10,118,119] but peak seropositivity is estimated at 30 days following the first doctor visit. Consequently, serologic diagnosis of HGA is based on comparison of acute and convalescent serum antibody titers and detection of seroconversion constituting a fourfold or greater increase in antibody titer, since acute serology alone is insensitive. Both IgG and IgM antibodies may be detected during the acute and early convalescent period of illness; however, IgM class antibodies in HGA do not usually persist beyond 45–60 days post onset of illness, while IgG antibodies may persist for several years.[10,118,119]

Several factors should be considered that may potentially affect the interpretation of positive serologic test results for HGA, particularly if only a single serum sample is available for testing. This includes long-term persistence of IgG class antibodies for months to years[117]; high background seropositivity rates have been shown in some endemic regions even among healthy individuals not known to be exposed to ticks.[63,117,118,120] Indeed, serum positivity can also be attributed to shared antigenic cross-reactivity between *Ehrlichia chaffeensis* and *A. phagocytophilum*[119,121] or even other species of *Anaplasmataceae*, or other conditions of infectious origin or autoimmune in nature.[14,122] It is not recommended to interpret a single positive serologic reaction in a patient lacking typical clinical manifestation as representing an active or persistent, or chronic or dormant infection.[117] Similarly, reductions in antibody titers cannot be used as indicators of effective treatment.[10]

Cross-reactivity detected by IFA when serum is tested against *A. phagocytophilum* and *E. chaffeensis* antigens may also occur, in part due to widely shared common bacterial antigens, such as the heat-shock proteins, including GroEL. When cross-reactivity is observed, the IFA titer to the infecting agent is usually higher than to the heterologous organisms.[67,87,123] Western blotting can also be used to characterize cross-reactive antisera and to demonstrate whether typical reactions to specific immunodominant antigens are present.[121,124] The 44-kDa surface protein (Msp2) is the antigen of *A. phagocytophilum* most often detected with sera from patients suffering from HGA.[125–128] Immunoassays using recombinant antigens expressed and purified from *E. coli* offer an alternative and more standardized approach for serologic diagnostic that does not depend on batch-to-batch variations of culture-grown *A. phagocytophilum*.[128,129]

52.2 METHODS

52.2.1 Sample Preparation

Because of the obligate intracellular growth of *A. phagocytophilum*, the bacterium has to be purified from host cell components before it is enriched enough for physiological or biochemical experiments, particularly, protein and antigen analyses. In contrast, some PCR-based tools can be applied with specimens containing variable amounts of the eukaryotic cell genetic material as long as it does not affect the efficiency of the target amplification and interpretation of results. However, whole genome sequencing would require highly purified *A. phagocytophilum* to be used for DNA preparation[76] to reduce the costs associated with sequencing host nuclear and mitochondrial DNA.

Preparation of *A. phagocytophilum* purified from host cell debris is described in great detail by Carlyon.[112] Heavily infected HL-60 cells are lysed by passing them through a syringe, and *A. phagocytophilum* partially separated from host cell membranes is recovered by differential centrifugation. Suspensions of semipurified *A. phagocytophilum* are suitable for in vitro infection assays, animal or tick inoculation, and IFA antigen preparation, or may be processed further to obtain bacteria that are highly purified from host cell material via ultracentrifugation through a discontinuous Renografin density gradient.[18,112] Purified *A. phagocytophilum* is concentrated in the interface of 30% and 42% Renografin on discontinuous gradient and collected with a pipette, residual Renografin is removed by washing the cells by centrifugation, and bacteria are aliquoted in sucrose glutamate buffer (pH 7.0) for long-term storage at −80°C or in liquid nitrogen. Suspensions of *A. phagocytophilum* purified through Renografin gradient density are not completely free from host cell components since they may be contaminated with variable amount of mitochondria because of their similar size and buoyant density, or trapped in residual phagosome-derived membranes. The purified preparations of *A. phagocytophilum* can be used for the isolation of nucleic acids, membrane fractions, and soluble protein fractions.

DNA of *A. phagocytophilum* is usually extracted using conventional methods combining treatments with lyzozyme, proteinase K, and sodium dodecyl sulfate or guanidine isothiocyanate. Several commercial kits with column binding or extraction and precipitation steps can be used for extraction and purification of *A. phagocytophilum* DNA, either from purified suspension or crude samples including human or animal blood, tissues or ticks; however, the resulting DNA preparation always includes eukaryotic DNA. Two hundred to 500 μL of EDTA whole blood can be used to extract high-quality DNA from a clinical specimen. Purified DNA is eluted using the manufacturer's recommended elution buffer, 10 mM Tris-HCl, 0.005 M EDTA (pH 8.0), or sterile distilled water, and kept refrigerated or frozen at −20°C for long-term storage. One to 5 μL of extracted DNA is typically used in a 20–25 μL PCR assay. If a larger amount of DNA is needed for testing, further PCR optimization may be required.

52.2.2 Detection Procedures

Molecular identification of *A. phagocytophilum*, particularly for confirmatory diagnostic purposes, is most often accomplished by PCR amplification and sequencing of 16S rRNA gene.[12] In our laboratory, we use a set of nested primers specific for *A. phagocytophilum* (Table 52.1); however, other sets of primers can be also applied. This diagnostic PCR assay is a nested reaction that utilizes a set of external primers (GE3a and GE2) targeting a 932-bp fragment and nested primers (GE9F and GE10R) amplifying a 546-bp fragment.[130] Typically, the PCR mixture is set up in 20–25 μL volume and contains an appropriate reaction buffer, 2.5–3 mM $MgCl_2$, 200 μM of each dNTP, 800 pMol of each forward and reverse primers and 2–4 μL of extracted DNA. Primary amplification is carried out for 40 cycles of 94°C for 1 min, 55°C for 30 s, and 72°C for 1 min. One μL of primary PCR product is used as a template for amplification with internal primers that is carried out for 30 cycles under the same conditions. The result of the amplification is assessed by examining 5 μL of nested PCR product on 1%–1.5% agarose gel in 1× Tris-acetate-EDTA buffer or 0.5× Tris-borate-EDTA buffer and visualization and detection of 546-bp fragment following staining with ethidium bromide or other less toxic dyes. When required, amplification of the *Gro*EL gene, *msp*2, or *ank* can be performed to further confirm detection of *A. phagocytophilum*. Examples of useful primer sequences and corresponding amplification conditions are listed in Table 52.1.

The identity of the PCR amplicon is then confirmed by DNA sequencing, which is performed on both the forward and reverse strands. The assembled nucleotide sequence that is generated is compared to reference sequences of *A. phagocytophilum* (U02521) deposited in the Genbank using the nucleotide-nucleotide BLAST tool (Basic Local Alignment Search Tool; http://www.ncbi.nlm.nih.gov/BLAST/).[133] Numerous isolates of *A. phagocytophilum* share at least 99.1% nucleotide sequence similarity in their 16S rRNA gene.[2] Several single nucleotide polymorphism (SNP) sites have been identified within the 16S rRNA gene that differentiate isolates of *A. phagocytophilum* depending on their geographic origin and mammalian hosts;[40,41,134–136] some of these SNPs are located in the 5′-end of the gene as is shown in Table 52.2. The *gro*EL sequences of different isolates of *A. phagocytophilum* share at least 98% nucleotide sequence similarity.[2,49,137] In contrast, *msp*2 sequences are characterized by significantly more diversity, ranging from 24.4% to 100% similarities to 56.3%–97.6% similarities among hypervariable regions of *msp*2 in strains of human, animals, or ticks in different geographic areas,[138,139] the feature reflecting continuous recombination among the multigene *msp*2 family.[140] Subsequently, direct sequencing of a *msp*2 amplicon is not possible; it requires cloning and sequencing individual clones. Particular attention needs to be paid to the quality of each sequence to be analyzed, since a decline in the quality of recently deposited *A. phagocytophilum* sequences has been demonstrated,[134] and this makes differentiation of isolates problematic.

52.3 CONCLUSION AND FUTURE PERSPECTIVES

The incidence of human infection due to *A. phagocytophilum* has been increasing in the United States, Europe, and Asia, and this trend has coincided with an increased incidence of other tick-borne diseases and the recognition of new, emerging, tick-borne etiological agents.[141–143] The clinical manifestations of HGA are nonspecific, so accurate clinical diagnosis is difficult without specific laboratory methods. However, the most

TABLE 52.1
PCR Assays and Primers Used for Detection and Identification of *Anaplasma phagocytophilum* in Diagnostic Laboratories

Gene Target	Assay	Primer Name	Designation	Primer Sequence (5′-3′)	Amplicon size (bp)	Assay Specificity	PCR Cycling Conditions: Cycle	Denaturation	Annealing	Extension	Reference
16S rRNA gene	Nested PCR	GE3A	Primary PCR	CACATGCAAGTCGAACGGAT	932	yes	40	95°C, 30″	55°C, 30″	72°C, 60″	129
		GE10R		TTCCGTTAAGAAGGATCTAATCTCC							
		GE9F	Nested PCR	AACGGATTATTCTTTATAGCTTGCT	546	yes	30	95°C, 30″	55°C, 30″	72°C, 60″	
		GE2		GGCAGTATTAAAAGCAGCTCCAGG							
	Conventional one-step PCR	EHR-521	Primary PCR	TGTAGGCGGTTCGGTAAGTTAAAG	270	yes	40	95°C, 30″	55°C, 30″	72°C, 60″	142
		HER-790		CTTAACGCGTTAGCTACAACACAG							
		PER1	Primary PCR	TTTATCGCTATTAGATGAGCCTATG	451	no	40	94°C, 11″	45°C, 10″	75°C, 15″	90
		PER2		CTCTACACTAGGAATTCCGCTAT							
	SYBR-Green PCR[a]	EC-SYBRF	Forward primer	AACACATGCAAGTCGAACGG	151	no	50	95°C, 60″	55°C, 60″		104, 105
		EC-SYBRR	Reverse primer	CCCCCGCAGGCATTATACA							
	TaqMan[b]	*Ehrlichia/Anaplasma* F	TaqMan primer	CTCAGAACGAACGCTGG	150	yes	45	95°C, 15″	50°C, 30″	60°C, 60″	108
		Anaplasma TaqMan R	TaqMan primer	CATTTCTAGTGGCTATCCC							
		A. phagocytophilum probe	TaqMan probe	TET-TTGCTATAAAGAATAATTA GTGGCAGACG-DABCYL							
*msp*2	Conventional one-step PCR	hge-396	Primary PCR	TCAAGACCAAGGGTATTAGAGATAG	526	yes	30–35	94°C, 60″	55°C, 60″	72°C, 60″	60, 143
		hge-921		GCCACTATGGTTTTTTCTTCGGG							
		msp3f	Primary PCR	CCAGCGTTTAGCAAGATAAGAG	334	yes	40	94°C, 30″	55°C, 30″	72°C, 60″	52
		msp3r		GCCCAGTAACATCATAAGC							
	TaqMan[c]	ApMSP2f	TaqMan primer	ATGGAAGGTAGTGTTGGTTATGGTATT	77	yes	40	95°C, 15″	60°C, 60″		107
		ApMSP2r	TaqMan primer	TTGGTCTTGAAGCGCTCGTA							
		ApMSP2p-HEX	TaqMan probe	HEX-TGGTGCCAGGGTTGAGCTTGA GATTG-TAMRA							
ank	Conventional one-step PCR	LA6	Primary PCR	GAGAGATGCTTATGGTAAGAC	444	yes	8	94°C, 30″	62–56°C, 30″	72°C, 30″	92
		LA1		CGTTCAGCCATCATTGTGAC			28	94°C, 30″	54°C, 30″	72°C, 30″	

continued

TABLE 52.1 (CONTINUED)
PCR Assays and Primers Used for Detection and Identification of *Anaplasma phagocytophilum* in Diagnostic Laboratories

Gene Target	Assay	Primer Name	Designation	Primer Sequence (5′-3′)	Amplicon size (bp)	Assay Specificity	PCR Cycling Conditions					Reference
							Cycle	Denaturation	Annealing	Extension		
*gro*ESL	Nested PCR	HS1	Primary PCR	TGGGCTGGTA(A/C)TGAAAT	1350	no	3	94°C, 60″	48°C, 120″	70°C, 90″		94
		HS6		CCICCIGGIACIA(C/T)ACCTTC			37	88°C, 60″	52°C, 120″	70°C, 90″		
		HS43	Nested PCR	AT(A/T)GC(A/T)AA(G/A) GAAGCATAGTC	480	no	30–35	94°C, 60″	55°C, 60″	70°C, 90″		
		HS45		ACTTCACG(C/T)(C/T)TCATAGAC								
	FRET PCR[4]	Esp-F	FRET primer	TACTCAGAGTGCTTCTCAATGT	359	yes	45	95°C, 10″	55°C, 15″	72°C, 12″		109
		Esp-R	FRET primer	GCATACCATCAGTTTTTTCAAC								
		Aph-FL	FRET probe	ATTTCAGCTAATGGAGATAAG AATATAG-FITC								

[a] PCR amplification is followed by melting curve analysis of the amplicon to enhance specificity for the identification of Anaplasmataceae. The final product should be sequenced because DNA of multiple representatives of Anaplasmataceae will be detected using these primers.[105,106]

[b] The assay can be set up as a reverse transcription PCR assay following an additional 48° for 30 min step to generate cDNA. This assay can be also used as a multiplex assay for detection of *A. phagocytophilum* and *A. platys* or with a different probe combination for detection of *E. chaffeensis*, *E. canis* and *E. ewingii*.[108] It has been tested with veterinary samples only.

[c] The original assay was designed as a multiplex assay for detection of *A. phagocytophilum* and *Borrelia burgdorferi*.[107]

[d] The original FRET PCR assay was designed as a multiplex assay for detecting and differentiating *A. phagocytophilum*, *Ehrlichia chaffeensis* and *E. ewingii*.[109] It has been recently demonstrated that an *Ehrlichia muris*-like agent can also be detected and differentiated based on its unique amplicon melting curve characteristics.

TABLE 52.2
Examples of Genetic Variations Detected in 16S rRNA Gene of *Anaplasma phagocytophilum* Isolates of Different Origin

				Nucleotide Position[1]															
No	Source of Isolation or Detection	Geographic Origin	Reference Accession Number	2	3	4	5	7	8	11	32	59	123	124	125	145	151	258	304
1	Human	USA	AY055469	T	A	A	A	A	A	G	A	G	T	T	A	A	T	G	T
2	Human	USA	U02521	T	A	A	A	A	A	G	A	G	T	T	A	A	T	G	T
3	Human	Sicily	DQ029028	T	A	A	A	A	A	G	A	G	T	T	A	A	T	G	T
4	Tick	Switzerland	AF084907	T	A	A	A	A	A	G	A	G	T	T	A	A	T	G	T
5	*Peromyscus* mouse	USA	AF189153	T	A	A	A	A	A	G	A	G	T	T	A	A	T	G	T
6	Tick	China	AF227954	T	A	A	A	A	A	G	**G**	G	T	T	A	**T**	T	**A**	T
7	Human	USA	CP000235	T	A	A	A	A	A	G	A	G	T	T	A	A	T	G	T
8	Rabbit	USA, Texas	DQ038128	T	A	A	A	A	A	G	A	G	T	T	A	A	T	G	T
9	Rabbit	USA, MA	AY144728	T	A	A	A	A	A	G	A	G	T	T	A	A	T	G	T
10	Human	USA, California	AF093789	T	A	A	A	A	A	**A**	A	G	T	T	A	A	T	G	T
11	Tick	South Korea	AF470701	T	A	A	A	A	A	**A**	A	G	T	T	A	A	T	G	T
12	Horse	USA, California	AF172165	T	A	A	A	A	A	**A**	A	G	T	T	A	A	T	G	T
13	Dog	South Africa	AY570538	T	A	A	A	A	A	**A**	A	G	T	T	A	A	T	G	T
14	Horse	USA, California	AF172164	T	A	A	A	A	A	**A**	A	G	T	T	A	A	**C**	G	T
15	Vole	Russia	AY094353	T	A	A	A	A	A	**A**	A	G	T	T	A	A	T	G	T
16	Roe deer	Swtizerland	AF384213	T	**G**	A	A	A	A	**A**	A	G	T	T	A	A	T	G	T
17	Deer, tickAp1 var	USA	AY193887	T	**G**	A	A	A	A	**A**	A	G	T	T	A	A	T	G	T
18	Tick	Sweden	AJ242784	T	**G**	A	A	A	A	**A**	A	G	T	T	A	A	T	G	T
19	Human or tick	Canada	AF311343	T	**G**	A	A	A	A	**A**	A	G	T	T	A	A	T	G	T
20	Tick	Sweden	AJ242783	T	**G**	A	A	A	A	G	A	G	T	T	A	A	T	G	T
21	Tick	Germany	AF136714	T	**G**	A	A	A	A	G	A	G	T	T	A	A	T	G	T
22	Tick	China	AY079425	T	A	A	A	A	**T**	G	A	G	T	T	A	A	T	G	T
23	Roe deer	Slovenia	AF481852	T	A	**G**	A	A	A	G	A	G	T	T	A	A	T	G	T
24	Tick	China	AF205140	T	**G**	**G**	A	**G**	A	**A**	A	G	T	T	A	A	T	G	T
25	Tick	Russia	AY587607	T	**G**	**G**	A	**G**	A	**A**	A	G	T	T	A	A	T	G	T
26	Rabbit	China	DQ458806	T	**G**	**G**	A	**G**	A	**A**	A	G	T	T	A	A	T	G	**C**
27	Rodent	China	DQ458808	T	**G**	**G**	A	**G**	A	**A**	A	G	T	T	A	A	T	G	T
28	Rabbit	USA, Georgia	DQ088129	T	A	A	**G**	A	A	G	A	G	T	T	A	A	T	G	T
29	Horse	USA, California	AF172166	T	A	A	**G**	A	A	**A**	A	G	T	T	A	A	T	G	T
30	Tick	Germany	AY281807	T	A	**G**	A	A	A	G	A	G	T	T	A	A	T	G	T
30	Deer, tick	Japan	AB195721	**C**	**G**	A	A	A	**0**	**A**	A	**A**	**C**	**A**	**T**	A	T	G	T
31	Horse	USA, California	AF172167	T	A	A	A	A	A	**A**	A	G	T	T	A	A	T	G	T
32	Horse	Switzerland	AF057707	T	A	A	A	A	A	G	A	G	T	T	A	A	T	G	T
33	Horse	Germany	AF482761	T	A	A	A	A	A	G	A	G	T	T	A	A	T	G	T
34	Horse	Sweden	AY527213	T	A	A	A	A	A	G	A	G	T	T	A	A	T	G	T

[1] Unique nucleotides are indicated with bold underlined letters; 0 indicates deletion.

efficient diagnostic methods available are useful either during a very short period of acute disease or retrospectively. This situation emphasizes the need for the development of more rapid and sensitive point-of-care diagnostic methods. Since several tick-borne agents may be present in the same tick, a high probability of mixed infections with unusual clinical manifestations requires access to multiplex assays allowing simultaneous detection of *Anaplasma*, *Ehrlichia*, *Rickettsia*, *Bartonella*, *Borrelia*, and *Babesia*, and perhaps other emerging disease agents. Although a number of valuable assays have been already developed and proven to have high analytical sensitivity and specificity, further development is needed to optimize their performance and turnaround time with clinical specimens. However, the biggest obstacle to improving the diagnosis of HGA remains the failure to collect and transport acute blood samples to specialty laboratories in a timely fashion and the inability to obtain and test convalescent samples by serology.

ACKNOWLEDGMENTS

We thank Robert Massung for critical review and helpful suggestions on this manuscript, Vsevolod Popov for electron microscopy image of *A. phagocytophilum*, Hubert Pan and Arianna Salazar for the literature search and resultant EndNote library. We apologize to our colleagues that many related publications could not be cited here due to space limitations.

The findings and conclusions in this report are those of the authors and do not necessarily represent the views of the Centers for Disease Control or the Department of Health and Human Services of the United States.

REFERENCES

1. Tamura, K. et al., MEGA4: Molecular Evolutionary Genetics Analysis (MEGA) software version 4.0, *Mol. Biol. Evol.*, 24, 1596, 2007.
2. Dumler, J.S. et al., Reorganization of genera in the families Rickettsiaceae and Anaplasmataceae in the order Rickettsiales: Unification of some species of *Ehrlichia* with *Anaplasma*, *Cowdria* with *Ehrlichia* and *Ehrlichia* with *Neorickettsia*, descriptions of six new species combinations and designation of *Ehrlichia equi* and "HGE agent" as subjective synonyms of *Ehrlichia phagocytophila*, *Int. J. Syst. Evol. Microbiol.*, 51, 2145, 2001.
3. Foley, J.E. et al, Ecology of *Anaplasma phagocytophilum* and *Borrelia burgdorferi* in the western United States, *J. Vector. Ecol.*, 29, 41, 2004.
4. Liz, J., Ehrlichiosis in *Ixodes ricinus* and wild mammals, *Int. J. Med. Microbiol.*, 291, Suppl 33, 104, 2002.
5. Woldehiwet, Z., *Anaplasma phagocytophilum* in ruminants in Europe, *Ann. N. Y. Acad. Sci.*, 1078, 446, 2006.
6. Woldehiwet Z., The natural history of *Anaplasma phagocytophilum*, *Vet. Parasitol.*, 167, 108, 2010.
7. Carrade, D.D. et al., Canine granulocytic anaplasmosis: A review, *J. Vet. Intern. Med.*, 23, 1129, 2009.
8. Chen, S.M. et al., Identification of a granulocytotropic *Ehrlichia* species as the etiologic agent of human disease, *J. Clin. Microbiol.*, 32, 589, 1994.
9. Bakken, J., and Dumler, J.S., Clinical diagnosis and treatment of human granulocytotropic anaplasmosis, *Ann. N. Y. Acad. Sci.*, 1078, 236, 2006.
10. Bakken, J., and Dumler, J.S., Human granulocytic anaplasmosis, *Infect. Dis. Clin. N. Am.*, 22, 433, 2008.
11. Bakken, J.S., and Dumler, J.S., Human granulocytic ehrlichiosis, *Clin. Infect. Dis.*, 31, 554, 2000.
12. Dumler, J.S., and Brouqui, P., Molecular diagnosis of human granulocytic anaplasmosis, *Expert. Rev. Mol. Diagn.*, 4, 559, 2004.
13. Dumler, J.S. et al., Human granulocytic anaplasmosis and *Anaplasma phagocytophilum*, *Emerg. Infect. Dis.*, 11, 1828, 2005.
14. Dumler, J.S. et al., Ehrlichioses in humans: Epidemiology, clinical presentation, diagnosis, and treatment, *Clin. Infect. Dis.*, 45, S45, 2007.
15. Thomas, R.J., Dumler, J.S., and Carlyon, J, A., Current management of human granulocytic anaplasmosis, human monocytic ehrlichiosis and *Ehrlichia ewingii* ehrlichiosis, *Expert. Rev. Anti-infect. Ther.*, 7, 709, 2009.
16. Goodman, J.L., Human granulocytic anaplasmosis. In Goodman, J.L., Dennis, D.T., and Sonenshine, D.E. (Editors), *Tick-Borne Diseases of Humans*, p. 218, ASM Press, Washington, DC, 2005.
17. Popov, V.L. et al., Ultrastructural differentiation of the genogroups in the genus *Ehrlichia*, *J. Med. Microbiol.*, 47, 235, 1998.
18. Rikihisa, Y. et al., Ultrastructural and antigenic characterization of a granulocytic ehrlichiosis agent directly isolated and stably cultivated from a patient in New York state, *J. Infect. Dis.*, 175, 210, 1997.
19. Herron, M.J. et al., Intracellular parasitism by the human granulocytic ehrlichiosis bacterium through the P-selectin ligand, PSGL-1, *Science*, 288, 1653, 2000.
20. Carlyon, J.A., and Fikrig, E., Invasion and survival strategies of *Anaplasma phagocytophilum*, *Cell. Microbiol.*, 5, 743, 2003.
21. Webster, P. et al., The agent of Human Granulocytic Ehrlichiosis resides in an endosomal compartment, *J. Clin. Invest.*, 101, 1932, 1998.
22. Park, J., Choi, K., and Dumler, J.S., Major surface protein 2 of *Anaplasma phagocytophilum* facilitates adherence to granulocytes, *Infect. Immun.*, 71, 4018, 2003.
23. Carlyon, J. et al., Murine neutrophils require alpha1,3-fucosylation but not PSGL-1 for productive infection with *Anaplasma phagocytophilum*, *Blood*, 102, 3387, 2003.
24. Ge, Y., and Rikihisa, Y., *Anaplasma phagocytophilum* delays spontaneous human neutrophil apoptosis by modulation of multiple apoptotic pathways, *Cell. Microbiol.*, 8, 1406, 2006.
25. Carlyon, J.A., and Fikrig, E., Mechanisms of evasion of neutrophil killing by *Anaplasma phagocytophilum*, *Curr. Opin. Hematol.*, 13, 28, 2006.
26. Blanco, J.R., and Oteo, J.A., Human granulocytic ehrlichiosis in Europe, *Clin. Microbiol. Infect.*, 8, 763, 2002.
27. Magnarelli, L.A. et al., Hemocytic rickettsia-like organisms in ticks: serologic reactivity with antisera to *Ehrlichiae* and detection of DNA of agent of human granulocytic ehrlichiosis by PCR, *J. Clin. Microbiol.*, 33, 2710, 1995.
28. Pancholi, P. et al., *Ixodes dammini* as a potential vector of human granulocytic ehrlichiosis, *J. Infect. Dis.*, 172, 1007, 1995.
29. Walk, S. et al., Correlation between tick density and pathogen endemicity, New Hampshire, *Emerg. Infect. Dis.*, 15, 585, 2009.

30. Massung, R. et al., Genetic variants of *Ehrlichia phagocytophila*, Rhode Island and Connecticut, *Emerg. Infect. Dis.*, 8, 467, 2002.
31. Cao, W.-C. et al., Granulocytic Ehrlichiae in *Ixodes persulcatus* ticks from an area in China where Lyme disease is endemic, *J. Clin. Microbiol.*, 38, 4208, 2000.
32. Cao, W.-C, et al., Natural *Anaplasma phagocytophilum* infection of ticks and rodents from a forest area of Jilin Province, China, *Am. J. Trop. Med. Hyg.*, 75, 664, 2006.
33. Liz, J.S. et al., PCR detection of granulocytic ehrlichiae in *Ixodes ricinus* ticks and wild small mammals in western Switzerland, *J. Clin. Microbiol.*, 38, 1002, 2000.
34. Morozova, O. et al., PCR detection of *Borrelia burgdorferi* sensu lato, tick-borne encephalitis virus, and the human granulocytic ehrlichiosis agent in *Ixodes persulcatus* ticks from western Siberia, Russia, *J. Clin. Microbiol.*, 40, 3802, 2002.
35. Ohashi, N. et al., *Anaplasma phagocytophilum*-infected ticks, Japan, *Emerg. Infect. Dis.*, 11, 1780, 2005.
36. Cao, W.-C. et al., Prevalence of *Anaplasma phagocytophila* and *Borrelia burgdorferi* in *Ixodes persulcatus* ticks from northeastern China, *Am. J. Trop. Med. Hyg.*, 68, 547, 2003.
37. Shpynov, S.N. et al., New evidence for the detection of *Ehrlichia* and *Anaplasma* in ixodes ticks in Russia and Kazakhstan, *Med. Parazitol. (Mosk)*, April–June, 10, 2004.
38. Chae, J. et al., Microbial pathogens in ticks, rodents and a shrew in northern Gyeonggi-do near the DMZ, Korea, *J. Vet. Sci.*, 9, 285, 2008.
39. Petrovec, M. et al., Infections of wild animals with *Anaplasma phagocytophila* in Austria and the Czech Republic, *Ann. N. Y. Acad. Sci.*, 990, 103, 2003.
40. Zhan, L. et al., *Anaplasma phagocytophilum* and *Borrelia burgdorferi* in rabbits from southeastern China, *Vet. Parasitol.*, 162, 354, 2009.
41. Rar, V.A. et al., The study of *Anaplasma* and *Ehrlichia* genetic variability in parasitic systems in the South-Western Siberia and Ural, *Bull. Sib. Med.*, 5, 116, 2006.
42. Brouqui, P., Ehrlichiosis in Europe. In Raoult, D., and Brouqui, P. (Editors), *Rickettsiae and Rickettsial Diseases at the Turn of the Third Millenium*, p. 220, Elsevier, Paris, France, 1999.
43. Bown, K.J. et al., Sympatric *Ixodes trianguliceps* and *Ixodes ricinus* ticks feeding on field voles (*Microtus agrestis*): potential for increased risk of *Anaplasma phagocytophilum* in the United Kingdom? *Vector Borne Zoonotic Dis.*, 6, 404, 2006.
44. Bown, K.J. et al., Relative importance of *Ixodes ricinus* and *Ixodes trianguliceps* as vectors for *Anaplasma phagocytophilum* and *Babesia microti* in field vole (*Microtus agrestis*) populations, *Appl. Environ. Microbiol.*, 74, 7118, 2008.
45. Santos, A.S. et al., Detection of *Anaplasma phagocytophilum* DNA in Ixodes ticks (Acari: Ixodidae) from Madeira Island and Setubal District, mainland Portugal, *Emerg. Infect. Dis.*, 10, 1643, 2004.
46. Santos, A.S. et al., PCR-based survey of *Anaplasma phagocytophilum* in Portuguese ticks (Acari: Ixodidae), *Vector Borne Zoonotic Dis.*, 9, 33, 2009.
47. Telford, S.R., 3rd et al., Perpetuation of the agent of human granulocytic ehrlichiosis in a deer tick-rodent cycle, *Proc. Natl. Acad. Sci. United States*, 93, 6209, 1996.
48. Belongia, E.A. et al., Prevalence of granulocytic *Ehrlichia* infection among white-tailed deer in Wisconsin, *J. Clin. Microbiol.*, 35, 1465, 1997.
49. Nieto, N.C., and Foley, J.E., Evaluation of squirrels (Rodentia: Sciuridae) as ecologically significant hosts for *Anaplasma phagocytophilum* in California, *J. Med. Entomol.*, 45, 763, 2008.
50. Nieto, N.C., and Foley, J.E., Reservoir competence of the redwood chipmunk (*Tamias ochrogenys*) for *Anaplasma phagocytophilum*, *Vector Borne Zoonotic Dis.*, 9, 573, 2009.
51. Foley, J.E., Clueit, S.B., and Brown, R.N., Differential exposure to *Anaplasma phagocytophilum* in rodent species in northern California, *Vector Borne Zoonotic Dis.*, 8, 49, 2008.
52. Foley, J.E. et al., Survey for zoonotic rickettsial pathogens in northern flying squirrels, *Glaucomys sabrinus*, in California, *J. Wildl. Dis.*, 43, 684, 2007.
53. Levin, M. et al., Comparison of the reservoir competence of medium-sized mammals and *Peromyscus leucopus* for *Anaplasma phagocytophilum* in Connecticut, *Vector Borne Zoonotic Dis.*, 2, 125, 2002.
54. Yabsley, M., Romines, J., and Nettles, V., Detection of *Babesia* and *Anaplasma* species in rabbits from Texas and Georgia, United States, *Vector Borne Zoonotic Dis.*, 6, 7, 2006.
55. Foley, J.E. et al., Possible differential host tropism in *Anaplasma phagocytophilum* strains in the Western United States, *Ann. N. Y. Acad. Sci.*, 1149, 94, 2008.
56. Massung, R.F. et al., Transmission efficiency of the AP-variant 1 strain of *Anaplasma phagocytophila*, *Ann. N. Y. Acad. Sci.*, 990, 75, 2003.
57. Reichard, M.V. et al., Inoculation of white-tailed deer (*Odocoileus virginianus*) with Ap-V1 or NY-18 strains of *Anaplasma phagocytophilum* and microscopic demonstration of Ap-V1 in *Ixodes scapularis* adults that acquired infection from deer as nymphs, *Vector Borne Zoonotic Dis.*, 9, 565, 2009.
58. Baldridge, G. et al., Transovarial transmission of *Francisella*-like endosymbionts and *Anaplasma phagocytophilum* variants in *Dermacentor albipictus* (Acari: Ixodidae), *J. Med. Entomol.*, 46, 625, 2009.
59. Demma. L.J. et al., Epidemiology of human ehrlichiosis and anaplasmosis in the United States, 2001–2002, *Am. J. Trop. Med. Hyg.*, 73, 400, 2005.
60. Chapman, A. et al., Diagnosis and management of tickborne rickettsial diseases: Rocky Mountain spotted fever, ehrlichioses, and anaplasmosis—United States: A practical guide for physicians and other health-care and public health professionals, *MMWR*, 55, 1, 2006.
61. IJdo, J.W. et al., The emergence of another tickborne infection in the 12-town area around Lyme, Connecticut: Human granulocytic ehrlichiosis, *J. Infect. Dis.*, 181, 1388, 2000.
62. Bakken, J.S. et al., Clinical and laboratory characteristics of human granulocytic ehrlichiosis, *JAMA*, 275, 199, 1996.
63. Bakken, J.S. et al., Seroprevalence of human granulocytic ehrlichiosis among permanent residents of northwestern Wisconsin, *Clin. Infect. Dis.*, 27, 1491, 1998.
64. Strle, F., Human granulocytic ehrlichiosis in Europe, *Int. J. Med. Microbiol.*, 293 Suppl 37, 27, 2004.
65. Sidel'nikov, I.N. et al., The first case of granulocytic ehrlichiosis in the Far East of the Russian Federation, *Klin. Med.* (Mosk), 81, 67, 2004.
66. Sreter, T. et al., *Anaplasma phagocytophilum*: An emerging tick-borne pathogen in Hungary and Central Eastern Europe, *Ann. Trop. Med. Parasitol.*, 98, 401, 2004.
67. Portillo, A. et al., Detection of a nonpathogenic variant of *Anaplasma phagocytophilum* in *Ixodes ricinus* from La Rioja, Spain, *Ann. N. Y. Acad. Sci.*, 1063, 333, 2005.
68. Zhang, L. et al., Rickettsial seroepidemiology among farm workers, Tianjin, People's Republic of China, *Emerg. Infect. Dis.* 14, 938, 2008.
69. Zhang, S. et al., Seroprevalence of human granulocytotropic anaplasmosis in central and southeastern China, *Am. J. Trop. Med. Hyg.*, 81, 293, 2009.

70. Heo, E.-J. et al., Serologic and molecular detection of *Ehrlichia chaffeensis* and *Anaplasma phagocytophila* (human granulocytic ehrlichiosis agent) in Korean patients, *J. Clin. Microbiol.*, 40, 3082, 2002.
71. Zhang, L. et al., Nosocomial transmission of human granulocytic anaplasmosis in China, *JAMA*, 300, 2263, 2008.
72. *Anaplasma phagocytophilum* transmitted through blood transfusion—Minnesota, 2007, *MMWR*, 57, 1145, 2008.
73. Bakken, J.S. et al., Exposure to deer blood may be a cause of human granulocytic ehrlichiosis, *Clin. Infect. Dis.*, 23, 198, 1996.
74. Dhand, A. et al., Human granulocytic anaplasmosis during pregnancy: Case series and literature review, *Clin. Infect. Dis.* 45, 589, 2007.
75. Adachi, J.A. et al., Human granulocytic ehrlichiosis in a renal transplant patient: Case report and review of the literature, *Transplantation*, 64, 1139, 1997.
76. Hotopp, J.C.D. et al., Comparative genomics of emerging human ehrlichiosis agents, *PLoS Genetics*, 2, e21, 2006.
77. Nelson, C. et al., Whole genome transcription profiling of *Anaplasma phagocytophilum* in human and tick host cells by tiling array analysis, *BMC Genomics*, 9, 364, 2008.
78. Aguero-Rosenfeld, M., Diagnosis of human granulocytic ehrlichiosis: State of the art, *Vector Borne Zoonotic Dis.*, 2, 233, 2002.
79. Bakken, J.S. et al., Serial measurements of hematologic counts during the active phase of human granulocytic ehrlichiosis, *Clin. Infect. Dis.*, 32, 862, 2001.
80. Bakken, J.S. et al., Human granulocytic ehrlichiosis in the upper Midwest United States. A new species emerging? *JAMA*, 272, 212, 1994.
81. Hardalo, C.J., Quagliarello, V., and Dumler, J.S., Human granulocytic ehrlichiosis in Connecticut: Report of a fatal case, *Clin. Infect. Dis.*, 21, 910, 1995.
82. Walker, D.H., Diagnosing human ehrlichioses: current status and recommendations, *ASM News*, 66, 287, 2000.
83. Walker, D.H., and Dumler, J.S., Emergence of the ehrlichioses as human health problems, *Emerg. Infect. Dis.*, 2, 18, 1996.
84. Lepidi, H. et al., Comparative pathology and immunohistology associated with clinical illness after *Ehrlichia phagocytophila*-group infections, *Am. J. Trop. Med. Hyg.*, 62, 29, 2000.
85. Choi, K.-S., Park, J., and Dumler, J.S., *Anaplasma phagocytophilum* delay of neutrophil apoptosis through the p38 mitogen-activated protein kinase signal pathway, *Infect. Immun.*, 73, 8209, 2005.
86. Choi, K.-S., Scorpio, D., and Dumler, J.S., *Anaplasma phagocytophilum* ligation to toll-like receptor (TLR) 2, but not to TLR4, activates macrophages for nuclear factor-kappa B nuclear translocation, *J. Infect. Dis.*, 189, 1921, 2004.
87. Aguero-Rosenfeld, M.E. et al., Human granulocytic ehrlichiosis: A case series from a medical center in New York State, *Ann. Intern. Med.*, 125, 904, 1996.
88. Jahangir, A. et al., Fatal pancarditis associated with human granulocytic ehrlichiosis in a 44-year-old man, *Clin. Infect. Dis.*, 27, 1424, 1998.
89. Dumler, J.S., and Bakken, J.S., Human granulocytic ehrlichiosis in Wisconsin and Minnesota: A frequent infection with the potential for persistence, *J. Infect. Dis.*, 173, 1027, 1996.
90. Dumler, J.S., and Bakken, J.S., Human ehrlichioses: Newly recognized infections transmitted by ticks, *Ann. Rev. Med.*, 49, 201, 1998.
91. Goodman, J.L., Direct cultivation of the causative agent of human granulocytic ehrlichiosis, *N. Engl. J. Med.*, 334, 209, 1996.
92. Massung, R., and Slater, K., Comparison of PCR assays for detection of the agent of human granulocytic ehrlichiosis, *Anaplasma phagocytophilum*, *J. Clin. Microbiol.*, 41, 717, 2003.
93. Walls, J.J. et al., Improved sensitivity of PCR for diagnosis of human granulocytic ehrlichiosis using *epank*1 genes of *Ehrlichia phagocytophila*-group ehrlichiae, *J. Clin. Microbiol.*, 38, 354, 2000.
94. de la Fuente, J. et al., Sequence analysis of the *msp*4 gene of *Anaplasma phagocytophilum* strains, *J. Clin. Microbiol.*, 43, 1309, 2005.
95. Sumner, J.W., Nicholson, W.L., and Massung, R.F., PCR amplification and comparison of nucleotide sequences from the *gro*ESL heat shock operon of *Ehrlichia* species, *J. Clin. Microbiol.*, 35, 2087, 1997.
96. Taillardat-Bisch, A.-V., Raoult, D., and Drancourt, M., RNA polymerase beta-subunit-based phylogeny of *Ehrlichia* spp., *Anaplasma* spp., *Neorickettsia* spp. and *Wolbachia pipientis*, *Int. J. Syst. Evol. Microbiol.*, 53, 455, 2003.
97. Inokuma, H. et al., Citrate synthase gene sequence: A new tool for phylogenetic analysis and identification of *Ehrlichia*, *J. Clin. Microbiol.*, 39, 3031, 2001.
98. Lee, K. et al., Characterization of the *ftsZ* gene from *Ehrlichia chaffeensis*, *Anaplasma phagocytophilum*, and *Rickettsia rickettsii*, and use as a differential PCR target, *DNA Cell Biol.*, 22, 179, 2003.
99. Petrovec, M. et al., Infection with *Anaplasma phagocytophila* in cervids from Slovenia: Evidence of two genotypic lineages, *Wien. Klin. Woch.*, 114, 641, 2002.
100. Horowitz, H.W. et al., Clinical and laboratory spectrum of culture-proven human granulocytic ehrlichiosis: Comparison with culture-negative cases, *Clin. Infect. Dis.*, 27, 1314, 1998.
101. Edelman, D.C., and Dumler, J.S., Evaluation of an improved PCR diagnostic assay for human granulocytic ehrlichiosis, *Mol. Diagn.*, 1, 41, 1996.
102. Petrovec, M. et al., Human disease in Europe caused by a granulocytic *Ehrlichia* species, *J. Clin. Microbiol.*, 35, 1556, 1997.
103. Lotric-Furlan, S. et al., Clinical and serological follow-up of patients with human granulocytic ehrlichiosis in Slovenia, *Clin. Diagn. Lab. Immunol.*, 8, 899, 2001.
104. Bjoersdorff, A. et al., Human granulocytic ehrlichiosis as a common cause of tick-associated fever in Southeast Sweden: Report from a prospective clinical study, *Scand. J. Infect. Dis.*, 34, 187, 2002.
105. Li, J.S. et al., Outer membrane protein-specific monoclonal antibodies protect SCID mice from fatal infection by the obligate intracellular bacterial pathogen *Ehrlichia chaffeensis*, *J. Immun.*, 166, 1855, 2001.
106. Eremeeva, M. et al., Detection and identification of bacterial agents in *Ixodes persulcatus* Schulze ticks from the north western region of Russia, *Vector Borne Zoonotic Dis.*, 7, 426, 2007.
107. Bermudez, S. et al., Detection and identification of rickettsial agents in ticks from domestic mammals in eastern Panama, *J. Med. Entomol.*, 46, 856, 2009.
108. Courtney, J. et al., Multiplex real-time PCR for detection of *Anaplasma phagocytophilum* and *Borrelia burgdorferi*, *J. Clin. Microbiol.*, 42, 3164, 2004.
109. Sirigireddy, K., and Ganta, R., Multiplex detection of *Ehrlichia* and *Anaplasma* species pathogens in peripheral blood by real-time reverse transcriptase-polymerase chain reaction, *J. Mol. Diagn.*, 7, 308, 2005.

110. Bell, C.A., and Patel, R., A real-time combined polymerase chain reaction assay for the rapid detection and differentiation of *Anaplasma phagocytophilum*, *Ehrlichia chaffeensis*, and *Ehrlichia ewingii*, *Diagn. Microbiol. Infect. Dis.*, 53, 301, 2005.
111. Stuen, S. et al., Identification of *Anaplasma phagocytophila* (formerly *Ehrlichia phagocytophila*) variants in blood from sheep in Norway, *J. Clin. Microbiol.*, 40, 3192, 2002.
112. Carlyon, J.A., Laboratory maintenance of *Anaplasma phagocytophilum*, *Curr. Protoc. Microbiol.*, Chapter 3: Unit 3A 2, 2005.
113. Kalantarpour, F. et al., Survival of the human granulocytic ehrlichiosis agent under refrigeration conditions, *J. Clin. Microbiol.*, 38, 2398, 2000.
114. Aguero-Rosenfeld, M.E. et al., Serology of culture-confirmed cases of human granulocytic ehrlichiosis, *J. Clin. Microbiol.*, 38, 635, 2000.
115. Munderloh, U. et al., Infection of endothelial cells with *Anaplasma marginale* and *A. phagocytophilum*, *Vet. Microbiol.*, 101, 53, 2004.
116. Munderloh, U.G. et al., Invasion and intracellular development of the human granulocytic ehrlichiosis agent in tick cell culture, *J. Clin. Microbiol.*, 37, 2518, 1999.
117. Bakken, J.S. et al., The serological response of patients infected with the agent of human granulocytic ehrlichiosis, *Clin. Infect. Dis.*, 34, 22, 2002.
118. Aguero-Rosenfeld, M. et al., Seroprevalence of antibodies that react with *Anaplasma phagocytophila*, the agent of human granulocytic ehrlichiosis, in different populations in Westchester County, New York, *J. Clin. Microbiol.*, 40, 2612, 2002.
119. Walls, J.J. et al., Inter and intralaboratory comparison of *Ehrlichia equi* and human granulocytic ehrlichiosis (HGE) agent strains for serodiagnosis of HGE by the immunofluorescent-antibody test, *J. Clin. Microbiol.*, 37, 2968, 1999.
120. Marshall, G. et al., *Ehrlichia chaffeensis* seroprevalence among children in the southeast and south-central regions of the United States, *Arch. Pediatr. Adoles. Med.*, 156, 166, 2002.
121. Wong, S.J., Brady, G.S., and Dumler, J.S., Serological responses to *Ehrlichia equi*, *Ehrlichia chaffeensis*, and *Borrelia burgdorferi* in patients from New York State, *J. Clin. Microbiol.*, 35, 2198, 1997.
122. IJdo, J.W. et al., Heat shock protein 70 of the agent of human granulocytic ehrlichiosis binds to *Borrelia burgdorferi* antibodies, *Clin. Diagn. Lab. Immunol.*, 5,118, 1998.
123. Anaplasmosis and ehrlichiosis—Maine, 2008. *MMWR*, 58, 1033, 2009.
124. Park, J.-H. et al., Detection of antibodies to *Anaplasma phagocytophilum* and *Ehrlichia chaffeensis* antigens in sera of Korean patients by western immunoblotting and indirect immunofluorescence assays, *Clin. Diagn. Lab. Immunol.*, 10, 1059, 2003.
125. Dumler, J.S., and Bakken, J.S., Ehrlichial diseases of humans: emerging tick-borne infections, *Clin. Infect. Dis.*, 20, 1102, 1995.
126. Ravyn, M.D. et al., Immunodiagnosis of human granulocytic ehrlichiosis by using culture-derived human isolates, *J. Clin. Microbiol.*, 36, 1480, 1998.
127. IJdo, J.W. et al., The early humoral response in human granulocytic ehrlichiosis, *J. Infect. Dis.*, 176, 687, 1997.
128. Zhi, N. et al., Cloning and expression of the 44-kilodalton major outer membrane protein gene of the human granulocytic ehrlichiosis agent and application of the recombinant protein to serodiagnosis, *J. Clin. Microbiol.*, 36, 1666, 1998.
129. Lodes, M.J. et al., Serodiagnosis of human granulocytic ehrlichiosis by using novel combinations of immunoreactive recombinant proteins, *J. Clin. Microbiol.*, 39, 2466, 2001.
130. Massung, R.F. et al., Nested PCR assay for detection of granulocytic ehrlichiae, *J. Clin. Microbiol.*, 36, 1090, 1998.
131. Kolbert, C., Detection of the agent of human granulocytic ehrlichiosis by PCR. In Persing, D. (Editor), *PCR Protocols for Emerging Infectious Diseases*, p. 106, ASM Press, Washington, DC, 1996.
132. Magnarelli, L.A. et al., Infections of granulocytic ehrlichiae and *Borrelia burgdorferi* in white-tailed deer in Connecticut, *J. Wildl. Dis.*, 35, 266, 1999.
133. Altschul, S.F. et al., Gapped BLAST and PSI-BLAST: A new generation of protein database search programs, *Nucl. Acids Res.*, 25, 3389, 1997.
134. Zeman, P., and Jahn, P., An entropy-optimized multilocus approach for characterizing the strains of *Anaplasma phagocytophilum* infecting horses in the Czech Republic, *J. Med. Microbiol.*, 58, 423, 2009.
135. Massung, R. et al., Isolation and propagation of the Ap-Variant 1 strain of *Anaplasma phagocytophilum* in a tick cell line, *J. Clin. Microbiol.*, 45, 2138, 2007.
136. Poitout, F. et al., Genetic variants of *Anaplasma phagocytophilum* infecting dogs in Western Washington State, *J. Clin. Microbiol.*, 43, 796, 2005.
137. Carpi, G. et al., *Anaplasma phagocytophilum groEL* gene heterogeneity in *Ixodes ricinus* larvae feeding on roe deer in Northeastern Italy, *Vector Borne Zoonotic Dis.*, 9, 179, 2009.
138. Morissette, E. et al., Diversity of *Anaplasma phagocytophilum* strains, United States, *Emerg. Infect. Dis.*, 15, 928, 2009.
139. Wuritu, F. et al., Characterization of *p44/msp2* multigene family of *Anaplasma phagocytophilum* from two different tick species, *Ixodes persulcatus* and *Ixodes ovatus*, in Japan, *Jap. J. Infect. Dis.*, 62, 142, 2009.
140. Wuritu, B. et al., Structural analysis of a *p44/msp2* expression site of *Anaplasma phagocytophilum* in naturally infected ticks inhabiting Japan, *J. Med. Microbiol.*, 58, 1638, 2009.
141. Doudier B, et al., Factors contributing to emergence of *Ehrlichia* and *Anaplasma* spp. as human pathogens, *Vet. Parasitol.*, 167, 149, 2010.
142. Beugnet, F., and Mari, J.-L., Emerging arthropod-borne diseases of companion animals in Europe, *Vet. Parasitol.*, 163, 298, 2009.
143. Parola, P., Paddock, C., and Raoult, D., Tick-borne rickettsioses around the world: Emerging diseases challenging old concepts, *Clin. Microbiol. Rev.*, 18, 719, 2005.

53 *Bartonella*

Alessandra Ciervo, Marco Cassone, Fabiola Mancini, and Lorenzo Ciceroni

CONTENTS

53.1 INTRODUCTION

The genus *Bartonella* was named after Alberto Leonardo Barton, a Peruvian physician who in 1909 discovered *Bartonella bacilliformis*, the causative agent of Carrion's disease. *B. bacilliformis* was the only member of the genus *Bartonella* until 1993. In that year, the genus *Bartonella* underwent a major taxonomic reorganization. Relying on 16S rRNA gene sequence analysis and DNA hybridization, the genus *Rochalimeae*, consisting of the four species *Rochalimeae quintana*, *Rochalimeae vinsonii*, *Rochalimeae henselae*, and *Rochalimmeae elizabethae*, was united with the genus *Bartonella*. At the same time, the genus *Bartonella* was removed from the order Rickettsiales, family Rickettsiaceae and transferred to the family Bartonellaceae, resulting in a single genus within the family Bartonellaceae.[1] In 1995, upon further 16S rRNA gene sequence analysis, DNA hybridization, G + C content, and phenotypic characterization studies, Birtles proposed the unification of the genera *Bartonella* and *Grahamella*, and the reclassification of the two species *Grahamella talpae* and *Grahamella peromysci* within the genus *Bartonella*.[2] Birtles also defined three new species: *Bartonella grahamii*, *Bartonella taylorii*, and *Bartonella doshiae*. In the subsequent years, the number of identified *Bartonella* spp. increased considerably, and new *Bartonella* species were identified as human and/or animal pathogens.

53.1.1 Classification, Morphology, and Biology

On the basis of 16S rRNA gene sequence comparisons, *Bartonella* is the only genus of the family Bartonellaceae, order Rhizobiales, class Alphaproteobacteria, phylum Proteobacteria.[3] The genus is currently composed of 24 validated species and three subspecies. At least 14 species of *Bartonella* have been reported to be pathogenic for humans. Among these, three species (*B. bacilliformis*, *B. quintana*, and *B. henselae*) are etiological agents of major emerging infectious diseases in humans. The remaining 11 species, including two subspecies, have also been associated with human disease or described as human pathogens in isolated cases reports (Table 53.1).

Members of the genus *Bartonella* are small (0.5–0.6 μm × 1.0 μm), gram-negative pleomorphic rods, with a distinct tendency to autoagglutinate. Cells are often curved and may show polar enlargement with granules at one or both ends. The cells also can be coccoid, beaded, filamentous, or in

TABLE 53.1
***Bartonella* Species Associated with Human Disease, Reservoirs, Vectors, and Distribution**

Bartonella sp	Reservoirs	Main Vectors	Main Diseases[a]	Distribution
B. bacilliformis	Humans	Sand flies	Carrion's disease	Andes
B. quintana	Humans	Human body lice	Trench fever, endocarditis, bacteremia, CSD, BA, CA	Worldwide
B. henselae	Cats	Cat fleas	CSD, BA, CA, endocarditis	Worldwide
B. elizabethae	Rats	Rat fleas, wild rodents fleas	Endocarditis, retinitis	Europe, USA, Asia
B. grahamii	Wild mice	Fleas	Retinitis	Europe, Canada, Asia
B. washoensis	Ground squirrels	Possibly ticks and fleas	Fever, myocarditis	USA
B. doshiae	Unknown	Possibly rat fleas	CSD	Europe
B. vinsoni subsp. berkhoffii	Coyotes, dogs	Possibly ticks	Endocarditis, myocarditis	Europe, USA
B. clarridgeiae	Cats	Cat fleas	CSD, endocarditis, bacteremia	Europe, USA, Asia
B. alsatica	Rabbits	Possibly fleas	Endocarditis	Europe
B. koehlere	Cats	Possibly fleas	Endocarditis, CSD	USA
B. vinsoni subsp. arupensis	White-footed mice	Possibly ticks and fleas	Bacteremia	Canada
B. rochalimaea	Possibly humans	Possibly fleas	Bacteremia, cutaneous lesions, splenomegaly	Peru
B. tamiae	Possibly humans	Unknown	Fever	Thailand
B. melophagi	Possibly humans	Unknown	Fever, splenomegaly	Peru

[a] CSD, cat scratch disease; BA, bacillary angiomatosis; CA, chronic adenopathy.

chains. *Bartonella* spp. are facultative intracellular bacteria that infect mainly mammalian erythrocytes and endothelial cells. If chronically infected, the host may serve as a reservoir. Human beings are the only known reservoir of *B. bacilliformis* and *B. quintana*, whereas for other *Bartonella* spp. the known reservoirs also include domestic and wild animals. The vectors that have been shown to play a role in the transmission of the different *Bartonella* species include several blood-sucking arthropods, such as sand flies, louse and biting flies, lice, and fleas[4] (Table 53.1). In the last few years, ticks, deer keds, and mites have been shown to harbor *Bartonella* species and are suspected to have a vector function.[5] The reservoir host and the vector are unknown, or have not been identified for several of the recently described *Bartonella* species (Table 53.1).

B. bacilliformis is the etiological agent of Carrion's disease, a vector-transmitted disease that has been known to occur in Peru since the pre-Inca times. Carrion's disease has been traditionally reported in the Andean valleys in Peru, Ecuador, and Colombia at altitudes between 500 and 3200 m above sea level,[6] where ecological conditions are favorable for sand flies of the genus *Lutzomyia*. Some of these species, especially *Lutzomyia verrucarum*, transmit the disease to humans through the bite of female flies. No reservoirs other than humans have been implicated in endemic areas. Carrion's disease is now a reemerging disease in Peru, Columbia, and Ecuador. During the last two decades, its epidemiology has changed, and it is affecting new areas that were previously free of the disease. Recent studies have shown genotypic diversity among *B. bacilliformis* isolates obtained from different areas of Peru, in different periods.[7] Genetic variability among *B. bacilliformis* isolates has also been suggested as a factor responsible for the diversity in the clinical progression of the disease and the mortality that was reported from endemic areas and the new foci of infections in Peru.[7]

B. quintana, transmitted by the body louse through abraded skin, was recognized in Europe as the causative agent of trench fever in 1915, during World War I[8] and is now recognized as a reemerging pathogen in large metropolitan areas among homeless populations as the cause of "urban trench fever." An outbreak of epidemic typhus associated with trench fever was also reported in refugee camps in Burundi in 1997.[9] Risk factors for urban trench fever include alcoholism, homelessness, poor living conditions, lack of hygiene, and exposure to body louse.

B. henselae was identified in 1992 as the etiological agent of cat scratch disease (CSD), a clinical entity first described by Debré et al. in France in 1950.[10] Domestic cats are the main reservoir of *B. henselae*. Several reports indicate that infection of domestic cats is worldwide, with the prevalence of antibodies and *B. henselae* bacteremia in cats being higher in warm and humid climates, where cat fleas are abundant.[11] Domestic cats can be bacteremic for a few months to years and are usually asymptomatic. However, uveitis, endocarditis, neurological signs, fever, lymphadenopathy, and reproductive disorders have been reported in naturally or experimentally infected cats.[12] Transmission of *B. henselae* among cats appears to require the cat flea, *Ctenocephalides felis*, as a vector. Humans mainly acquire infection through a cat scratch or bite.[4] There is strong evidence that transmission of *B. henselae* to humans occurs via infected flea feces.[4] *B. henselae* has also been found in dogs.[13] At least two genotypes have been identified in humans and cats and designed Houston-1 (also known as genotype I) and Marseille (also known as genotype type II).[14] Prevalence of *B. henselae* genotypes among cat populations varies worldwide.

B. vinsonii subsp. *berkhoffii*, isolated for the first time from a dog with endocarditis in 1993, was recognized as a human pathogen in 2000.[15] Current evidence indicates that Coyotes (*Canis latrans*), gray foxes (*Urocyon cinereoargenteus*), raccoons (*Procyon lotor*), and dogs may serve as reservoir hosts.[16] *B. clarridgeiae* is another possible but rarer cause of CSD. Like *B. henselae*, *B. clarridgeiae* may be transmitted to humans by cat bite. Cat infections with *B. clarridgeiae* have been documented in several parts of the world.[17] *B. koehlerae*, isolated for the first time from two cats in northern California, is also now recognized to be a human pathogen that causes culture-negative human endocarditis.[18] The mode of transmission of *B. koehlere* in cats has not been clarified; however, identification of *B. koehlerae* DNA in fleas strongly supports its flea-associated transmission. Among the rodent-borne bartonellae, five species have been documented as human pathogens: *B. elizabethae*, *B. grahamii*, *B. washoensis*, *B. vinsonii* subsp *arupensis*, *B. doshiae*.[19] Rats (*Rattus norvegicus*) are the main reservoir of *B. elizabethae*. Ground squirrels (*Spermophilus beecheyi*) have been identified as the main reservoir species of *B. washoensis* in the western United States. In 2006, Raoult et al.[20] reported the first documented case of endocarditis in a man infected with *B. alsatica*, which causes bacteremia in healthy wild rabbits. More recently, a human case of *B. alsatica* lymphadenitis has been reported.[21] *B. rochalimeae* was recently isolated from a case of splenomegaly in a patient who had travelled to South America.[22] In northern California, *B. rochalimeae* was found to be the predominant *Bartonella* spp. in both foxes (*Urocyon cinereoargenteus*) and the fleas (*Pulex* spp.) collected from them.[23] In 2008, Kosoy et al.[24] reported the isolation of *B. tamiae* from blood samples of patients during the screening of blood clot specimens from a prospective study performed to determine the etiology of febrile illness in Thailand. More recently, Maggi et al.[25] reported the isolation of candidatus *B. melophagi* from the blood cultures from two women, one of whom was coinfected with *B. henselae*. Sheep are the most likely reservoir hosts for this organism. Transmission among sheep is thought to occur via the sheep keds.

53.1.2 Pathogenesis and Clinical Features

Bartonella species share an elegant yet not completely characterized life cycle, with a series of main common stages and an array of species-specific aspects, which may account for the many different clinical manifestations observed in the final host. *Bartonella* have a tropism for erythrocytes, where they can satisfy their high requirement for hemin, and for endothelial cells; however, there is no definitive evidence pointing to endothelial cells, bone marrow, dendritic cells, or other sites as "sanctuary" or niche sites responsible for maintaining the chronic infection by repeated seeding of the bloodstream.[26] It is now apparent that the majority of clinical manifestations occur in a minority of humans when a high rate of hemolysis ensues after the invasion of erythrocytes (Carrion's disease), or when uncontrolled vascular endothelial proliferation follows localization to endothelial cells (bacillary angiomatosis and peliosis, verruga peruana). Moreover, an important contribution toward health care in several specialistic fields was offered by the discovery that bartonellae and especially *B. henselae* are the causative agents of CSD, as well as several ocular and neurological diseases that are characterized by neural and/or vascular infiltrates and exudates, and that were previously labeled as idiopathic (including Leber's idiopathic stellate neuroretinitis, Parinaud's oculoglandular syndrome, selected forms of aseptic meningitis, and others). Currently, the knowledge of the molecular factors involved in the exploitation of different cell types during the life cycle of *Bartonella* is rapidly improving, while clues on the specific factors responsible for the minority cases of overt disease are still scarce.

53.1.2.1 Pathogenesis

Bartonellae are successful parasites infecting most mammal species with a very high estimated prevalence. The low correlation between their evolutionary tree and that of the corresponding species is a sign of a relatively quick adaptability to new hosts,[27] confirmed by the usually asymptomatic course of the infection, unless they're inoculated in accidental hosts. Along with the human species *B. bacilliformis* and *B. quintana*, and with the human and feline species *B. henselae*, another well-studied species is *B. tribocorum*, which uses the rat as reservoir and has been used to develop animal models of infection.

Erythrocyte Invasion. *B. tribocorum* inoculated in the rat is promptly cleared from the bloodstream and takes hold in an unknown primary niche. After a synchronous, reproductive subcycle of about 5 days, it is seeded in the blood again and is now capable of erythrocyte adhesion. Bartonellae then multiply inside a membrane-bound compartment, and do not seem to affect the life of erythrocytes in most cases. The number of infected erythrocytes increases with recurring seeding at 5-day intervals from the primary niche; however, it generally stays well below 1%, causing asymptomatic bacteremia. Of note, in the human pathogens *B. henselae* and *B. Quintana*, the LPS has a different Lipid A composition and is much less likely to trigger toxic shock. On the contrary, it has been reported to have Toll-like receptor 4 inhibitory activity.[28] The role of humoral immunity in clearing the infection has been investigated; however, an arsenal of evasion mechanisms, including phase variation, can help the bacteria overcome it.[29]

Unlike *B. tribocorum* in rats, *B. quintana* in humans, and *B. henselae* in cats, *B. bacilliformis* infects a large number of erythrocytes, and causes severe hemolytic anemia, which accounts for the high mortality of infection from this species. In vitro, *B. bacilliformis* can adhere to erythrocytes much more efficiently than the other species, and a higher percentage of erythrocytes are infected at any given time. One reason may be found in the high motility, granted by multiple unipolar flagella. It has been demonstrated that impairing flagella activity by different methods invariably leads to much reduced binding to erythrocytes. This is also the case for nonflagellated mutants.[30] Regarding the unique hemolytic capabilities of *B. bacilliformis*, a protein with hemolytic activity has been localized on the outer membrane. Also, a complex of secreted proteins including deformin[31] have been implied, as well as the *ial* (invasion-associated) locus, which confers an invasive phenotype to nonenteroinvasive strains of *E. coli*. Interestingly enough, another major virulence factor, the type IV secretion system trw (trw–T4SS), is present in all human pathogenic bartonellae but not in *B. bacilliformis*. Trw–T4SS has a pivotal role in establishing adherence to erythrocytes, by binding to blood group antigens.[32] Indeed, *B. bacilliformis* occupies a different evolutionary branch than other bartonellae, with a limited ability to achieve subclinical persistence in the host.[33]

Endothelial Invasion and Proliferation. While it is well-known that endothelial cells have an important role in proinflammatory responses, the vasoproliferative lesions triggered by *Bartonella* rightly deserve to make the limited ranks of pathogen-induced tumors. In the lesion, it is possible to distinguish bartonellae around and inside the endothelial cells, while the infiltrating immune cells are prevalently macrophages and neutrophils. The endothelial cells are not lyzed, rather they undergo a proliferative process that only lasts as long as the bacteria are present, and is most likely necessary to satisfy the requirement for hemin by bartonellae when they are outside of the erythrocytes. Most in vitro data on the mechanisms of endothelial proliferation have been obtained in human umbilical vein endothelial cells (HUVEC).[34] The adhesion to HUVEC is mediated by several factors including the adhesin (*Bad*A in *B. henselae* and Vomp in *B. quintana*), a trimeric autotransporter adhesion homologous to the adhesin A of *Yersinia*, *Neisseria*, and others,[35] and the VirB-D4 T4-SS, a secretion system present in many gram-negative pathogens, capable of injecting virulence factors directly from the bacterial to the host cytoplasm.[36] The distinctive virB5 component is conserved across bartonellae, is immunogenic, and is possibly responsible for the peculiar tropism for endothelial cells of *Bartonella*.[37] The pivotal role of the VirB-D4 secretion system in the pathogenesis of bartonellosis is well documented, with several studies showing that VirB-D4 mutants fail to induce pathological alterations in HUVEC cells. Specific investigations are now focusing on the set of proteins known to be translocated in the host cytoplasm by the secretion system.[38] It should be noted that the VirB T4-SS, too, is absent in *B. bacilliformis*. After adhesion, bartonellae reorganize the cellular cytoskeleton, enter the cell, and take hold inside a membrane compartment. The proliferation of endothelial cells is then stimulated by the hypoxia inducible factor HIF-1,[39] which triggers a cascade of vasoactive cytokines including VEGF. Since VEGF is secreted, it may influence nonparasitated cells alike. Notably, the adhesin A also contributes to HIF-1 activation, besides its role in HUVEC binding. A role has been also advocated for other, intracellular mediators like CXCL8[40] and the chaperone protein GroEL.[41] Finally, *B. henselae* and *B. quintana* prevent the cell from entering the apoptotic pathway by inhibition of caspase activity and the late events that lead to chromatin lysis.[42]

53.1.2.2 Clinical Features

Bartonella infections can range from self-limited to life-threatening and may present different course and organ involvement due to a balance of bacterial and host factors that is often very difficult to gauge. A case report on CSD in a family showed that the same kitten transmitted the infection to four individuals, and each presented with different clinical characteristics.[43] More *Bartonella* species have been demonstrated to infect humans, and the literature concerning new clinical manifestations from previously known species, as well as newly identified species,[22] is getting richer every day. An intriguing aspect of the genera is its capacity to infect a very large range of mammalian hosts. It is believed that only a minority of species are nonpermissive. Advances in the knowledge of bartonellosis in domestic animals are of great importance toward preventing severe disease in pets (especially dogs), their owners, and in veterinarians. Common pets represent a main source of infection in most parts of the world, even though the importance of wild and exotic mammals like rats, gerbils, and squirrels should not be disregarded.[44] Some species, like *B. henselae* and *B. clarridgeiae*, are found in cats and dogs, while others are specific to either one of the two, probably due to host-restricted erythrocyte tropism. Dogs as opposed to cats tend to be accidental hosts and not the primary reservoir, and are actively studied because they manifest symptoms and clinical and pathologic findings similar to human infections,[45] including higher prevalence of infection among immunocompromised individuals (up to 27% in dogs). Three species are responsible for the vast majority of clinical bartonelloses in humans: *B. bacilliformis*, *B. quintana*, and *B. henselae*.

B. bacilliformis. It is the agent of Oroya fever (or Carrion's disease), a severe febrile illness accompanied by bacteremia possibly complicated by endocarditis, hemolysis causing severe anemia and jaundice and leading to circulatory collapse, or by encephalitis.[46] Alternatively, the course can be milder with fever and other "constitutional" symptoms such as nausea and headache. After this initial, acute, or subacute phase, a chronic phase called "verruga peruana" (Peruvian wart) ensues, which is characterized by diffuse skin tumors a few millimeters in diameter, due to local proliferation of new blood vessels, which is completely reversed by effective antimicrobial therapy. This disease was already known and illustrated thousands of years ago.

B. quintana. It is best known as the agent of trench fever. Trench fever is characterized by high relapsing fever, fatigue, headache, pain in the muscles of the back and the legs, and characteristically, in moving the eyeballs, as well as hyperesthesia of the shins. The most severe complication is endocarditis. While trench fever maintains the larger casistic (over 1 million cases reported, mostly during World War I), another important syndrome caused by *B. quintana* is chronic bacteremia, frequently associated with thrombocytopenia, in the homeless and in the alcoholics. The prevalence in the homeless is estimated to be relatively high. Like other bartonellae, *B. quintana* also has a second clinical face, bacillary angiomatosis (BA), attributable to involvement of the vascular endothelium, which proliferates, forming multiple tumors. While not malignant per se, these lesions can be found anywhere in the body and may cause fatal complications. Most cases of BA have been described in patients with AIDS.

B. henselae. It is known as the agent of CSD, an infection usually limited to the site of injury and presenting mild symptoms. However, in some cases granulomatous lesions can arise in internal organs, in both immonocompetent and immunocompromised individuals. The lesions can be life-threatening depending on the location, and pose problems of differential diagnosis with other granulomatous pathologies such as tuberculosis and sarcoidosis. *B. henselae* can also be the cause of BA, typically with liver involvement (peliosis), while subcutaneous and bone lytic lesions are frequent when the agent is *B. Quintana*.[47] Only recently, the high prevalence of bartonellosis is starting to be recognized. Antibodies to *B. quintana* or *B. henselae* have been found in up to 18% of all patients referred for fever of unknown origin. Up to 3% of the patients were culture-positive, despite the difficulties in isolating bartonellae from clinical specimens.

53.1.2.3 Prognosis and Treatment

Treatment options for *Bartonella* are somewhat limited by its intracellular localization, and might be even more limited in the future due to emerging resistance to quinolone and macrolide antibiotics.[48] Infections by *B. bacilliformis* are often lethal if not treated (mortality from Oroya fever can be as high as 88%).[49] Antibiotics with demonstrated efficacy are in this case chloramphenicol and, for adults, ciprofloxacin.

On the other hand, mortality is almost zero from uncomplicated CSD, and infections from *B. henselae* and *B. quintana* are usually only treated in case of complicated CSD (azithromycin), trench fever (doxicycline and gentamicin), retinitis (doxycycline with rifampin), and BA or peliosis (erythromycin or doxicycline).[50]

53.1.3 Diagnosis

Presumptive diagnosis of bartonellosis is clinically based and identified by symptoms, clinical examination, and from the patient's medical history. Laboratory findings are often limited to nonspecific infection markers, such as slightly elevated or diminished white blood cell counts, platelet counts, erythrocyte sedimentation rate, and C-reactive protein, while other laboratory parameters are within standard range. An important clue for the etiological diagnosis of bartonellosis is the patient's medical history, including contact with arthropods, scratches and bites from domestic or wild animals, occupational risks and travel in endemic areas in association with the reservoir hosts.[8] Accurate microbiological laboratory tests are essential for the clinical confirmation and therapeutic features. Moreover, the advancement of diagnostic techniques over the past years has led to the discovery of the role of *Bartonella* infection in an increasing number of pathological conditions as the number of confirmed cases continues to increase. Proper laboratory analyses to rule out other serious diseases with similar presentation are extremely important in atypical cases of bartonellosis with severe sequelae and may prevent invasive and expensive procedures. Diagnosis is problematic because the available commercial tests often show low sensitivity and specificity, still a drawback even for specialized laboratories due to the lack of standardization. The poor performance of those tests required the researchers to implement several in-house assays and further studies involving cultural methods, histological investigations, serology, and molecular biology procedures. The choice of procedures used to confirm *Bartonella* infections may be influenced by the facilities and equipment available in the laboratory and by the type of specimen collected.

53.1.3.1 Conventional Techniques

Imaging Studies. Medical imaging scans are very useful in localizing the pathology and may suggest organ or multiorgan involvement, which is very helpful for the diagnosis. Positive images can be important for both the evaluation of the infection and the direction of diagnostic workflow.

Histology and Microscopic Examination. Despite its invasive nature, histopathological diagnosis is frequently required. The pathology is not specific for *Bartonella* infection, and the morphology can be similar to other bacterial and fungal diseases and malignancies. Moreover, the magnitude of viable bacteria in tissues is low, and a mixed inflammatory infiltrate, including lymphocytes and neutrophils with areas of focal necrosis, is often present. The bacillus is difficult to detect by conventional staining methods, but bacterial clusters can be detected by silver Warthin-Starry stain or by electron microscopy. However, Warthin-Starry staining is difficult to interpret, and pitfalls caused by the intensity of stain development may be critical for the visualization of the bacteria. Bacilli can be more specifically visualized in biopsies or blood smears by direct immunofluorescence or immunohistochemistry.

Both intracellular and extracellular organisms can be observed by using house-made polyclonal rabbit or monoclonal mouse antibodies against the whole bacterium, recombinant *Bartonella* proteins, or specific epitope antigens.[51,52]

Culture. *Bartonella* spp. can be isolated from blood during the acute stage of infection and before the antibiotic treatment.[53] However, its microbiologic isolation onto solid/liquid media or in cell cultures is particularly problematic because of poor sensitivity, slow growth, low yield, and frequent contamination. Isolates can be obtained after 12–14 days of incubation, but sometimes up to 6 weeks of incubation are needed. This difference in time can be attributable to the type of sample, the stage of infection, the optimum growth conditions, and the laboratory procedures. *Bartonella* strain identification is a labor-intensive task due to divergence in similarity among isolates and the inter- or intraspecies heterogeneity.[54] Strain isolation by cultivation is essential for the characterization of biological properties and also for the full identification as an etiological agent of the disease. Solid medium is the most widely used method for the microbiological diagnosis of bartonellosis. Plates with a complex base medium are supplemented with blood or hemin and must be freshly prepared.[55,56] Automated systems may be also used for the isolation from samples of infected patients, but these have questionable sensitivity.[57] Over the last decade, new liquid media have been developed to support and improve the cultures from blood or other tissue samples, and other supplements, such as 10% fetal calf serum, carbohydrates, metabolites, amino acids, and horse or rabbit blood, have been reported to be helpful.[58–60] Cell cultures such as ECV304 human endothelial or bovine endothelial cell lines seem to be an effective tool for the isolation of this fastidious bacteria and sometimes cell co-cultivation is used as primary isolation before plating onto solid agar.[58]

Identification. Presumptive identification can be made by microscopic examination after Gram and Gimenez staining or, more specifically, by immunofluorescence. *Bartonella*, except for *B. bacilliformis* and *B. clarridgeiae*, have no flagella, and thus no motility can be detected. Species identification is obtained using techniques that range from the evaluation of the biochemical and phenotypic characteristics to more complex molecular genotyping analyses. Because of the inert nature of these organisms, conventional biochemical tests are poorly reliable and are only used for the identification to species level based on cellular fatty acid analysis.[55] Other approaches for *Bartonella* species differentiation can be supported by the employment of polyvalent antisera, or specific immunoblotting and protein profile.[61] Concerning the antimicrobial activity, in vitro studies have demonstrated that *Bartonella* is susceptible to many antibiotics. Despite this, most of them exhibit a low bactericidal power, which is probably attributable to the ineffective pharmacological bioavailability due to the intracellular localization of the bacterium.

Serology. Serological methods are the easiest and most practical laboratory procedures for the diagnosis of bartonellosis and involve Indirect Immunofluorescence Assay (IFA) (considered the gold standard), enzyme-linked immunosorbent assay (ELISA), and Western blot (WB). Notably, the primary immune response to *Bartonella* infection with a high diagnostic power is IgM antibody that appears early and is closely followed by IgG antibody response. Seroconversion may be established, but often, as in immunocompromised patients, a significant antibody response is not mounted. The production of specific antibodies decreases over time, and IgM and IgG titers may remain positive up to 3 months and 2 years, respectively.[62] Commercial and in-house kits have been developed; the performance of these methods may differ depending on the antigen used, working procedures, interpretation of the test (particularly for the IFA assay), the determination of the cut-off value, and the cross-reactivity between *Bartonella* species and other bacteria, particularly, with *C. burnetii* and *Chlamydia* species.[63–65] The IFA positivity is expressed as titers, while values for the ELISA test are assessed in optical density (OD) and can be adjusted to attain the sensitivity and specificity of the IFA cut-off titers.[65] Several WBs have been used for detection of *Bartonella* recombinant proteins such as rGroES, rGroEL, rSodB, rRplL, rBepA, and r17-kDa in different mixtures, but the sensitivity is low, and when available, quantitative ELISA is more effective than WB.[66]

53.1.3.2 Molecular Techniques

These fastidious, slow-growing pathogens are satisfactorily/adequately diagnosed by means of molecular biology. The most widely used procedure is the polymerase chain reaction (PCR), a technique that enables the amplification of specific sequences of nucleic acids. The PCR optimization is determined by various parameters such as quality and concentration of DNA template, design of primers, thermal cycling conditions, and also PCR buffer systems, including magnesium ions and the selection of DNA polymerase. The accuracy and reproducibility of these assays depend on variable sensitivity and technical accuracy. During the laboratory manipulations, the PCR specificity may be affected by contaminations from other samples or from previous amplification procedures (carryovers), causing a generation of false-positive results. In addition, false-negative findings may be due to the nucleic acid–extraction or amplification, given the presence of inhibitors in the specimen. To prevent those problems, laboratories should be equipped with separate rooms for the different steps of the procedures and rigorous quality-control systems. Another limitation of PCR DNA-based tests is the impossibility to distinguish viable from unviable organisms. For genetic *Bartonella* identification and typing, many in house-PCR methods have been proposed for various samples including body fluids and biopsies.

Molecular Detection and Identification. A broad-range of target genes have been successfully used to detect *Bartonella* spp. from clinical samples, with an excellent specificity (100%) and a sensitivity ranging from 43% to 76%.[61,67] Several oligonucleotide primers have been designed within highly conserved genes or in species-specific DNA sequences such as the16S rRNA, citrate synthase (*gltA*), 16S-23S rRNA intergenic spacer region, the *htrA*, the riboflavin synthase α chain (*ribC*), the β subunit of RNA polymerase (*rpoB*), genes encoding for PAP31, 35-kDa, the cell division (*ftsZ*) and the heat-shock (*groEL*) proteins.[68–74] For instance, the amplification of the *gltA* gene fragment and its enzymatic digestion or a nested PCR, allows discrimination of *Bartonella* species as well as the employment of degenerate primers and the hybridization of the product with an internal probe. Recently, the introduction of the real-time PCR has provided with a significant improvement in the PCR assays for a rapid, specific, and sensitive method for the detection and identification for *Bartonella*.[75–77] This technology allows the removal of postamplification detection procedures and monitoring the product formation since each PCR cycle can be measured as an increase in fluorescence. The *Bartonella* real-time PCRs described below utilize the LightCycler instrument and two different fluorogenic chemistries: the fluorescent dye (SYBR

green) and hybridization probes. The first application is based upon the detection of accumulated PCR product by the use of SYBR green, which intercalates double-stranded DNA. The second method is based on the principle of fluorescence resonance energy transfer (FRET), with two probes that hybridize in tandem on the template. This process occurs when two molecules and their fluorophores are in close proximity to each other (the distance between one and five nucleotides), the energy from an excited donor moiety is transferred to an acceptor moiety, and fluorescence is proportional to the amount of bound probes. The FRET chemistry supports the specificity of the PCR product and also the genotyping by the generation of the melting curve and the single-nucleotide polymorphism analysis (SNPs). A perfect match has a higher melting temperature than a mismatch. Mismatches in the target region will cause one or both probes to dissociate at a lower temperature than the identical target sequence. FRET probes can be also useful to detect a *Bartonella* coinfection with two or more different species.[14] Moreover, multiple amplicons of the same gene fragment with polymorphic sites can be identified in the same detection channel based on different melting peaks.[76]

Molecular Typing. Several typing methods for molecular characterization and discrimination among *Bartonella* species have been successfully employed, such as pulsed-field gel electrophoresis (PFGE), PCR-restriction fragment length polymorphism analysis (RFLP) on the 16S-23S rRNA intergenic spacer (ITS) region and amplicons digestion, repetitive extragenic palindromic (REP)–PCR, enterobacterial repetitive intergenic consensus (ERIC)–PCR, arbitrary primed (AP)–PCR, and more recently, also by multilocus sequence typing (MLST) and variable number tandem repeat (VNTR).[78–80]

PFGE is a molecular typing technique that uses restriction endonucleases to cut the DNA infrequently in the bacterial genomes producing a small number of very large DNA fragments resolved by the contour-clamped homogeneous electric field (CHEF). Criteria for interpreting differences in patterns from PFGE are related to genetic events (i.e., insertion, deletion, rearrangements), and differences in the positions of more than three bands probably indicate different strains. PFGE offers a high degree of reproducibility with an extraordinary discriminatory power and the prospect of interlaboratory standardization.

PCR–RFLP analysis of the 16S-23S rRNA ITS is conducted with PCR amplification followed by digestion of the products with different restriction endonucleases as *Hin*dIII, *Taq*I, *Hae*III, *Hin*fI, *Pvu*II, and *Alu*I, which generated fragments forming characteristic band patterns. REP and ERIC-PCR can amplify short repetitive DNA sequences, widely distributed in the genome. Specific primer sets are designed to match inverted repeat consensus sequences, and differences in band sizes represent polymorphisms in the distances between repetitive sequence elements. AP–PCR works through small nonspecific primers (typically 10–20 bases) and operate in nonstringent annealing conditions. MLST indexes the variation present in nucleotide sequences of internal fragments from housekeeping genes. This technique assigning alleles by DNA sequencing is ideally suited for the comparison of strains among different laboratories via internet. Approaches based on VNTR analysis are now increasingly used to characterize, in particular, recently emerged, highly monomorphic pathogens. This technique involves the amplification of multiple chromosomal VNTR loci and the size of amplicons can record allelic variants on the basis of length polymorphisms. Variation in the number of repeat units and the nucleotide changes within a VNTR locus, hence creating allelic variants.

To date, the most commonly applied methods for discrimination and the determination of genomic relationships between *Bartonella* species are PFGE, PCR–RFLP on 16S-23S rRNA ITS, ERIC–PCR, REP–PCR, and AP–PCR. Fingerprinting obtained by these techniques gives a genetic pattern profiling for *Bartonella* type or subtype.

53.2 METHODS

53.2.1 Sample Preparation

Immunohistochemistry. Samples from biopsy sections, aspirate, and blood sample are deparaffinized, if necessary, using xylene or a xylene substitute and adequately fixed (i.e., formaldehyde, glutaraldehyde, methanol, acetone). Fixed slides can be analyzed by unspecific Warthin-Starry silver impregnation stain or by immunofluorescence. Silver stain treatment can be observed with an ordinary light microscope where bacilli are visualized in black or dark brown in a yellow milieu. Permeabilization of the sample with a nonionic detergent is sometimes needed to allow the access of the antibodies. The primary antibodies include in-house monoclonal or polyclonal antibodies, anti-*Bartonella* antigens, generally diluted from 1:25 to 1:100, and a commercial labeled secondary antibody that react with the primary antibody. After the antigen-antibody reaction, highlighted bacteria can be analyzed under a fluorescence microscope fitted with the appropriate filters.

Culture. Heparinized blood, tissue, and biopsy materials are subject to the lysis of erythrocytes by freezing or by osmotic shock and to potter homogenization in a liquid medium to increase the yield of bartonellae. After a mild centrifugation, the supernatant is inoculated directly onto blood medium or cell culture. Different media can be used, such as blood agar, brain heart infusion blood medium, trypticase soy with 5% sheep blood, chocolate agar with 20% sheep blood, Columbia blood agar, Brucella blood medium, Shaedler's blood agar, and CDC agar. Microaerobic cultures are grown under an atmosphere of 5%–9% CO_2 at 30°C–35°C and examined weekly for growth verification. Supernatants from treated specimens can be cultured onto ECV 304 human endothelial cells in flasks or coverslip shell vials. Cultures and subcultures can also be performed in human erythrocytes. After inoculation, confluent monolayer and suspended cells are centrifuged at 600 × *g* at 22°C. One hour after infection, the medium is discarded, cells are washed two or three times with sterile phosphate-buffered saline (PBS), and then fresh complete medium (MEM or RPMI 1640 medium supplemented with 10% fetal calf serum, 1 mM L-glutamine 25 mM $NaHCO_3$ and 25 mM HEPES) is added and incubated at 37°C in 5% CO_2. Periodically, the medium is changed, and smears are made to determine the percentage of infection by stain with acridine orange or specific immunofluorescence.

Biochemical Identification. Biochemical tests can be performed by several identification systems such as Microscan rapid anaerobe panel, An-IDENT system, RapID ANA II, and Rapid ID 32 A. Standard biochemical tests do not include hemin for *Bartonella* growth, thus raising a potential problem. Generally, tests for catalase activity, indole, urease, acid or gas production from carbohydrates, nitrate reduction, Voges-Proskauer hippurate hydrolysis, alkaline phosphatase, tetrathionate reductase, pyrazinamidase, tributyrin, o-nitrophenyl-/3-D-galactoside, esculin hydrolysis, and arginine dihydrolase are all negative. The oxidase test is usually negative but may be variable for *B. quintana* and *B. vinsonii*. Motility test for flagella presence is positive for *B. bacilliformis* and *B. clarridgeiae*, while hemolysis is produced by *B. bacilliformis* and *B. elizabethae*. Phenylalanine aminopeptidase hydrolysis and succinate utilization has been demonstrated for *B. quintana*. Aminopeptidase activity for phenylalanine and trypsin is positive for *B. henselae*, *B. clarridgeiae*, and *B. elizabethae*. If hemin is added to the test media, positive results can be obtained using carbohydrates, Voges-Proskauer, pyrazinamidase, and hippurate hydrolysis, and it is possible to differentiate *B. henselae* from *B. quintana*.

Serology. Human antibodies recognizing *Bartonella* are measured using commercial or in-house IFA and ELISA tests. Several commercial serological kits have been developed for antibodies detection against *B. henselae* and *B. quintana*. Antigens preparation for serological methods consist of growth of microorganisms in blood agar plates, co-cultivation in culture cells, inactivation in 0.1% paraformaldehyde and then cells, suspended in PBS at an optimal protein concentration, are spotted onto slides or used to coat wells of ELISA microtiter plates. The IFA test is performed with twofold serum dilution added to appropriate slide wells in contact with the substrate, and incubated for 30 min in a humid chamber at 37°C. Subsequently, the slide is washed to remove unbound serum antibodies, the fluorescein-labeled antibody to IgM or IgG is added as the conjugate and incubated again. Following the incubation and washes, the slide is dried and examined using UV epifluorescence microscope. Sera giving a positive result ≥1:20 for IgM and ≥1:64 for IgG are then titrated to endpoint. Positive presence of IgM antibody is suggestive of recent infection. Positive IgG antibody titers 1:64–1:128 indicate current or past infection, while titers ≥1:256 strongly suggest infection within the last few months. The procedure for ELISA test includes a first step with bovine serum albumin to block unbound sites of the coated well and then serum samples, generally diluted 1:100 are incubated for 60 min at 37°C. Next, the plate is washed and the antihuman IgM or IgG alkaline phosphatase or horseradish peroxidase conjugates is placed, and subsequently the appropriate substrate for the enzyme is added. A positive sample results in the production of a colored end product that can be measured quantitatively in a spectrophotometer and is defined as one yielding an OD value that is at least two or three standard deviations (SD) above the mean value obtained from serum samples of the control group, generally blood donors.

DNA Extraction. DNA from biopsies, aspirates, paraffin-embedded formalin-fixed sections, blood, body fluids, and also from *Bartonella* isolates can be extracted by means of numerous and different commercially available kits following the manufacturer's instructions.

53.2.2 Detection Procedures

53.2.2.1 One-Step PCR Protocol Using a Single Set of Primers

Principle. Jensen et al.[71] described a single-step PCR assay used to amplify an internal portion of the 16S-23S rRNA ITS gene of bartonellae. Fragments of different sizes are amplified with the primers set 5′-(C/T)CTTCGTTTCTCTTTCTTCA-3′ and 5′-AACCAACTGAGCTACAAGCC-3′.

Procedure

1. Prepare the PCR mixture (50 μL) containing 100 ng of target DNA, 10 mM Tris (pH 8.3), 50 mM KCl, 3.5 mM $MgCl_2$, 200 μM each dNTPs, 1 μM each primer, and 2.5 U of Amplitaq Gold DNA polymerase (Applied Biosystems).
2. For PCR amplification use the following cycling conditions: an initial denaturation at 95°C for 5 min followed by 45 cycles of 95°C for 1 min, 60°C for 1 min, and 72°C for 30 s.
3. Visualize PCR amplification products by ethidium bromide fluorescence after electrophoresis in 3% agarose gels.

Note: The expected product sizes are: 211 bp for *B. bacilliformis*, 172 bp for *B. henselae*, and 157 bp for *B. quintana*.

53.2.2.2 PCR Protocol for Differentiating *Bartonella* Species Using Different Primers

Principle. Bereswill et al.[68] designed different sets of primers in the species-specific *ribC* gene for the differentiation of *Bartonella* species. Sets of primers are: PBH-L1 5′-GATATCGGTTGTGTTGAAGA-3′ and PBH-R1 5′-AATAAAAGGTATAAAACGCT-3′ for *B. henselae*; PBH-L1 5′-GATATCGGTTGTGTTGAAGA-3′ and PBQ-R1 5′-AAAGGGCGTGAATTTTG-3′ for *B. quintana*; and PBH-L1 5′-GATATCGGTTGTGTTGAAGA-3′ and PBB-R1 5′-AAAGGCGCTAACTGTTC-3′ *for B. bacilliformis*.

Procedure

1. Prepare the PCR mixture (50 μL) containing 1.5 mM $MgCl_2$, 200 μM each dNTPs, 25 pmol of each primer, 1 U of *Taq* DNA polymerase, and 100 ng of target DNA.

2. For PCR amplification use the following cycling conditions: 5 min of denaturation at 95°C, and then 30 cycles each of 1 min at 94°C, 2 min at 55°C, and 3 min at 72°C, followed by a terminal extension step of 10 min at 72°C.
3. Separate the PCR products electrophoretically on 1.2% or 1.6% agarose gels and stain with ethidium bromide.

Note: The expected product sizes are: 393 bp for *B. bacilliformis*, 386 bp for *B. henselae*, and 390 bp for *B. quintana*.

53.2.2.3 Real-Time PCR Protocol

Principle. The real-time PCR reported by Ciervo et al.[76] allows a single-step identification and differentiation of *Bartonella* species through detection of mutations in the 379-bp fragment of the *glt*A gene by melting curve analyses. For this assay, the following primers and FRET probes are used: primer BhCS.781p 5′-GGGGACCAGCTCATGGTGG-3′, primer BhCS.1137n 5′-AATGCAAAAAGAACAGTAAACA-3′, probe PAC1 5′-GCAAAAGATAAAAATGATTCTTTCCG-Fluorescin-3′ and probe PAC2 5′-LCRed640-CTTATGGG TTTTGGTCATCGAGT-3′ Phosphate.

Procedure

1. Prepare PCR mixture (20 μL) containing 10–50 ng of DNA, 1× LightCycler-DNA Fast Start Master hybridization probes, 4 mM $MgCl_2$, primers and probes at final concentrations of 1 and 0.2 μM, respectively.
2. Carry out real-time PCR in glass capillary tubes (Roche Diagnostics) using a LightCycler instrument (Roche Diagnostics). The conditions for thermal cycling are: initial denaturation at 95°C for 10 min, and then 45 cycles at 95°C for 10 s, 45°C for 10 s, and 72°C for 10 s, with the fluorescence measurement to the end of each annealing phase. The amplification is followed by a melting program, which started at 40°C for 23 s and then increased to 90°C at 0.1°C/s, with the fluorescence signal continuously monitored on-line.

Note: The specific melting peak temperatures for *B. henselae* and *B. quintana* are 59.8°C and 56.1°C, respectively.

53.2.2.4 Genotyping by PFGE

Principle. Sander et al.[78] described a genotyping method for *Bartonella* based on PFGE technique. This DNA fingerprinting system is an excellent tool for the molecular investigation on the genetic heterogeneity among different isolates.

Procedure

1. Harvest bacteria from chocolate agar plates after incubation at 37°C for 7 days; suspend in 0.9% NaCl and adjust to an optical density at 600 nm (OD_{600}) of 0.85; wash three times with 0.9% NaCl.
2. Prepare agarose blocks by adding 500 μL of cell suspension to 700 μL of 2% PFGE agarose (Sigma, Deisenhofen, Germany); incubate the solidified blocks in lysis buffer (0.5 mM EDTA, 1% lauroyl sarcosine, 1.8 mg of proteinase K per ml; pH 9.5) at 56°C overnight; after a thorough washing with TE buffer [10 mM Tris, 10 mM EDTA (pH 7.5)], incubate the blocks with 20 U of *Sma*I for 6 h under the conditions recommended by the enzyme manufacturer (New England BioLabs).
3. Perform PFGE with a CHEF DRIII electrophoresis unit (BioRad, Munich, Germany); prepare agarose gel (2%) in 0.5× TBE running buffer (90 mM Tris, 90 mM boric acid, 2 mM EDTA pH 8.5); electrophorese for 30 h at 5.9 V/cm with pulse times from 3 to 12 s at a constant temperature of 14°C; use low-range PFGE marker (New England BioLabs) as a molecular weight standard; stain agarose gel with ethidium bromide, and photograph under UV illumination.

Note: The fingerprints obtained are compared for similarity by visual inspection of band patterns. Sizes of DNA fragments amplified by PCR are determined by direct comparison with the DNA marker. Fingerprints are considered highly similar when all visible bands obtained have the same migration distance for each isolate. Variations in intensity and shape of bands among isolates are not considered differences. The presence or absence of one distinct band is considered a difference.

The fingerprints can be analyzed with the Windows version of GelCompar software version 4.0 (Applied Maths, Kortrijk, Belgium). The PFGE pattern is examined by applying the Dice coefficient to peaks. For clustering, the unweighted pair group method with arithmetic means (UPGMA method) is used. A tolerance in the band positions of 1.2% is applied for comparison of the fingerprint patterns. Fingerprint analysis and the methods and algorithms used are performed according to the instructions of the manufacturer.

53.3 CONCLUSION AND FUTURE PERSPECTIVES

During the last century, such factors as globalization, environmental alteration, human movement, and changed human lifestyle contributed to the emergence and reemergence of pathogens, including *Bartonella*. These mechanisms implicated and favored the spread of infectious diseases around the world. Bartonellae, in particular, are capable of producing infection on a variety of invertebrate and vertebrate hosts and several cell types. Molecular data on strains identification and on different vector-animals reservoirs, together with epidemiological information on cases, were and still are useful to discover new *Bartonella* species, some of which can infect humans. Clinicians faced with bartonellosis should be familiar with the laboratory findings and the clinical presentation of both bartonellosis and the other diseases involved in the differential diagnosis. In nonendemic regions, clinicians often fail to consider bartonellosis early in the diagnostic process, and as a consequence an accurate laboratory diagnosis is made when the infection is in advanced stage. Until now, no standardized procedures for the strain isolation, serology, or molecular biology have been successfully established, and often, several methods should be associated according to the illness presentation. Limitation of surveillance and the lack of extensive epidemiological studies are at present the two most important aspects concerning the modest data on *Bartonella* infections. An important perspective in the pathogenesis is the full understanding of the mechanisms used by *Bartonella* to escape the host immune system, a crucial factor for the identification of different disorders and for the implementation of novel strategies for the immuno/chemotherapy of bartonellosis. The convergence of epidemiology, new knowledge and insight in the development of molecular tools, biological assays, and recognition of intracellular pathways of pathogenesis and immunity will provide a comprehensive picture of the host-pathogen interaction and might contribute to the identification of new immunological strategies, therapeutic or prophylactic interventions to limit *Bartonella* virulence.

ACKNOWLEDGMENT

We are grateful to Giuseppina Mandarino for the language revision of the manuscript.

REFERENCES

1. Brenner, D.J. et al., Proposals to unify the genera *Bartonella* and *Rochalimaea*, with descriptions of *Bartonella quintana* comb. nov., *Bartonella vinsonii* comb. nov., *Bartonella henselae* comb. nov., and *Bartonella elizabethae* comb. nov., and to remove the family *Bartonellaceae* from the order *Rickettsiales*, *Int. J. Syst. Bacteriol.*, 43, 777, 1993.
2. Birtles, R.J. et al., Proposals to unify the genera *Grahamella* and *Bartonella*, with descriptions of *Bartonella talpae* comb. nov., *Bartonella peromysci* comb. nov., and three new species, *Bartonella grahamii* sp. nov., *Bartonella taylorii* sp. nov., and *Bartonella doshiae* sp. nov., *Int. J. Syst. Bacteriol.*, 45, 1, 1995.
3. O'Connor, S.P. et al., 16S rRNA sequences of *Bartonella bacilliformis* and cat scratch disease bacillus reveal phylogenetic relationships with the alpha-2 subgroup of the class Proteobacteria, *J. Clin. Microbiol.*, 29, 2144, 1991.
4. Chomel, B.B., Boulouis, H.J., and Breitschwerdt, E. B., Cat scratch disease and other zoonotic *Bartonella* infections, *J. Am. Vet. Med. Assoc.*, 224, 1270, 2004.
5. Saisongkorh, W. et al., Emerging Bartonella in humans and animals in Asia and in Australia, *J. Med. Assoc. Thai.*, 92, 707, 2009.
6. Alexander, B., A review of bartonellosis in Ecuador and Colombia, *Am. J. Trop. Med. Hyg.*, 52, 354, 1995.
7. Hambuch, T.M. et al., Population genetic analysis of *Bartonella bacilliformis* isolates from areas of Peru were Carrion's disease is endemic and epidemic, *J. Clin. Microbiol.*, 42, 3675, 2004.
8. Karem, K.L., Paddock, C.D., and Regnery, R.L., *Bartonella henselae*, *B. quintana*, and *B. bacilliformis*: Historical pathogens of emerging significance, *Microbes Infect.*, 2, 1193, 2000.
9. Raoult, D. et al., Outbreak of epidemic typhus associated with trench fever in Burundi, *Lancet*, 352, 353, 1998.
10. Debre, R. et al., Cat scratch disease, *Sem. Hop.*, 30, 1895, 1950.
11. Chomel, B.B. et al., *Bartonella henselae* prevalence in domestic cats in California: Risk factors and association between bacteremia and antibody titers, *J. Clin. Microbiol.*, 33, 2445, 1995.
12. Chomel, B.B. et al., *Bartonella* infection in domestic cats and wild felids, *Ann. N. Y. Acad. Sci.*, 1078, 410, 2006.
13. Gundi, V.A. et al., *Bartonella clarridgeiae* and *Bartonella henselae* in dogs, Gabon, *Emerg. Infect. Dis.*, 12, 389, 2006.
14. Gurfield, A.N. et al., Coinfection with *Bartonella clarridgeiae* and *Bartonella henselae* and with different *Bartonella henselae* strains in domestic cats, *J. Clin. Microbiol.*, 35, 2120, 1997.
15. Roux, V. et al., *Bartonella vinsonii* subsp. *bekhoffii* as an agent of febrile blood culture-negative endocarditis in a human, *J. Clin. Microbiol.*, 38, 1698, 2000.
16. Chung, C.Y. et al., *Bartonella* spp. DNA associated with biting flies from California, *Emerg. Infect. Dis.*, 10, 1311, 2004.
17. Marston, E.L. et al, Prevalence of *Bartonella henselae* and *Bartonella clarridgeiae* in an urban Indonesian cat population, *Clin. Diagn. Lab. Immunol.*, 6, 41, 1999.
18. Avidor, B. et al., *Bartonella koehlerae*, a new cat-associated agent of culture-negative human endocarditis, *J. Clin. Microbiol.*, 42, 3462, 2004.
19. Lamas, C. et al., Human bartonellosis: seroepidemiological and clinical features with an emphasis on data on Brasil. A review, *Mem. Inst. Osvaldo Cruz*, 103, 221, 2008.
20. Raoult, D. et al., First isolation of *Bartonella alsatica* from a valve of a patient with endocaditis, *J. Clin. Microbiol.*, 44, 278, 2006.
21. Angelakis, E. et al., Human case of *Bartonella alsatica* lymphadenitis, *Emerg. Infect. Dis.*, 14, 1951, 2008.
22. Eremeeva, M.E. et al., Bacteremia, fever, and splenomegaly caused by a newly recognized *Bartonella* species, *N. Engl. J. Med.*, 356, 2381, 2007.
23. Gabriel, M.W. et al., Zoonotic *Bartonella* species in Fleas collected on gray foxes (*Urocyon cinereoargenteus*), *Vector Borne Zoonotic Dis.*, 9, 597, 2009.
24. Kosoy, M. et al., *Bartonella tamiae* sp. nov., a newly recognized pathogen isolated from three human patients from Thailand, *J. Clin. Microbiol.*, 46, 772, 2008.
25. Maggi, R.G. et al., Isolation of candidatus *Bartonella melophagi* from human blood, *Emerg. Infect. Dis.*, 15, 66, 2009.
26. Greub, G., and Raoult, D., Bartonella: New explanations for old diseases, *J. Med. Microbiol.*, 51, 915, 2002.
27. Pretorius, A.M., Beati, L., and Birtles, R.J., Diversity of bartonellae associated with small mammals inhabiting free state province, South Africa, *Int. J. Syst. Evol. Microbiol.*, 54, 1959, 2004.
28. Popa, C. et al., *Bartonella quintana* lipopolysaccharide is a natural antagonist of Toll-like receptor 4, *Infect. Immun.*, 75, 4831, 2007.
29. Kyme, P., Dillon, B., and Iredell, J., Phase variation in *Bartonella henselae*, *Microbiology*, 49, 621, 2003.
30. Minnick, M.F., and Anderson, B.E., Bartonella interactions with host cells, *Subcell. Biochem.*, 33, 97, 2000.
31. Hendrix, L.R., and Kiss, K., Studies on the identification of deforming factor from *Bartonella bacilliformis*, *Ann. N. Y. Acad. Sci.*, 990, 596, 2003.
32. Seubert, A. et al., A bacterial conjugation machinery recruited for pathogenesis, *Mol. Microbiol.*, 49, 1253, 2003.

33. Chomel, B.B. et al., Ecological fitness and strategies of adaptation of *Bartonella* species to their hosts and vectors, *Vet. Res.*, 40, 29, 2009.
34. Maeno, N. et al., Live *Bartonella henselae* enhances endothelial cell proliferation without direct contact, *Microb. Pathog.*, 27, 419, 1999.
35. Zhang, P. et al., A family of variably expressed outer-membrane proteins (Vomp) mediates adhesion and autoaggregation in *Bartonella quintana*, *Proc. Natl. Acad. Sci. USA*, 101, 13630, 2004.
36. Cascales, E., and Christie, P.J., The versatile bacterial type IV secretion systems, *Nat. Rev. Microbiol.*, 1, 137, 2003.
37. Yeo, H.J. et al, Structural and functional characterization of the VirB5 protein from the type IV secretion system encoded by the conjugative plasmid pKM101, *Proc. Natl. Acad. Sci. USA*, 100, 15947, 2003.
38. Rhomberg, T.A. et al., Translocated protein of *Bartonella henselae* interferes with endocytic uptake of individual bacteria and triggers uptake of large bacterial aggregates via the invasome, *Cell. Microbiol.*, 11, 927, 2009.
39. Kempf, V.A. et al., Activation of hypoxia-inducible factor-1 in bacillary angiomatosis: Evidence for a role of hypoxia-inducible factor-1 in bacterial infections, *Circulation*, 111, 1054, 2005.
40. McCord, A.M., Cuevas, J., and Anderson, B.E., Bartonella-induced endothelial cell proliferation is mediated by release of calcium from intracellular stores, *DNA Cell. Biol.*, 26, 657, 2007.
41. Smitherman, L.S., and Minnick, M.F., *Bartonella bacilliformis* GroEL: Effect on growth of human vascular endothelial cells in infected cocultures, *Ann. N. Y. Acad. Sci.*, 1063, 286, 2005.
42. Kirby, J.E., and Nekorchuk, D.M., Bartonella-associated endothelial proliferation depends on inhibition of apoptosis, *Proc. Natl. Acad. Sci. U S A*, 99, 656, 2002.
43. Song, AT. et al., Familial occurrence of cat-scratch disease, with varying clinical expression, *Scand. J. Infect. Dis.*, 39, 728, 2007.
44. Inoue, K. et al., Exotic small mammals as potential reservoirs of zoonotic *Bartonella* spp., *Emerg. Infect. Dis.*, 15, 526, 2009.
45. Kelly, P. et al., *Bartonella quintana* endocarditis in dogs, *Emerg. Infect. Dis.*, 12, 1869, 2006.
46. Brouqui, P., and Raoult, D., New insight into the diagnosis of fastidious bacterial endocarditis, *FEMS Immunol. Med. Microbiol.*, 47, 1, 2006.
47. Koehler, J.E. et al., Molecular epidemiology of bartonella infections in patients with bacillary angiomatosis-peliosis, *N. Engl. J. Med.*, 337, 1876, 1997.
48. Biswas, S., Raoult, D., and Rolain, J.M., Molecular mechanisms of resistance to antibiotics in *Bartonella bacilliformis*, *J. Antimicrob. Chemother.*, 59, 1065, 2007.
49. Gray, G.C. et al., An epidemic of Oroya fever in the Peruvian Andes, *Am. J. Trop. Med. Hyg.* 42, 215, 1990.
50. Rolain, J.M. et al., Recommendations for treatment of human infections caused by *Bartonella* species, *Antimicrob. Agents Chemother.*, 48, 1921, 2004.
51. Liang, Z. et al., Production of *Bartonella* genus-specific monoclonal antibodies, *Clin. Diagn. Lab. Immunol.*, **8,** 847, 2001.
52. Rolain, J.M. et al., *Bartonella quintana* in human erythrocytes, *Lancet*, 360, 226, 2002.
53. Welch, D.F. et al., Bacteremia due to *Rochalimaea henselae* in a child: practical identification of isolates in the clinical laboratory, *J. Clin. Microbiol.*, 31, 2381, 1993.
54. Roux, V., and Raoult, D., Inter- and intraspecies identification of *Bartonella* (*Rochalimaea*) species, *J. Clin. Microbiol.*, 33, 1573,1995.
55. Clarridge, J.E. et al., Strategy to detect and identify *Bartonella* species in routine clinical laboratory yields *Bartonella henselae* from human immunodeficiency virus-positive patient and unique *Bartonella* strain from his cat, *J. Clin. Microbiol.*, 33, 2107, 1995.
56. Dougherty, M.J. et al., Evaluation of an extended blood culture protocol to isolate fastidious organisms from patients with AIDS, *J. Clin. Microbiol.*, 34, 2444, 1996.
57. Tierno, P.M., Inglima, K., and Parisi, M.T., Detection of *Bartonella (Rochalimaea) henselae* bacteremia using BacT/Alert blood culture system, *Am. J. Clin. Pathol.*, 104, 530, 1995.
58. La Scola, B., and Raoult, D., Culture of *Bartonella quintana* and *Bartonella henselae* from human samples: A 5-year experience (1993 to 1998), *J. Clin. Microbiol.*, 37, 1899, 1999.
59. Maggi, R.G., Duncan, A.W., and Breitschwerdt, E.B., Novel chemically modified liquid medium that will support the growth of seven Bartonella species, *J. Clin. Microbiol.,* 43, 2651, 2005.
60. Riess, T. et al., Analysis of a novel insect cell culture medium-based growth medium for *Bartonella* species, *Appl. Environ. Microbiol.,* 74, 5224, 2008.
61. Maurin, M. et al., Isolation and characterization by immunofluorescence, SDS polyacrylamide gel electrophoresis, Western blot, restriction-fragment length polymorphism PCR, 16S rRNA gene sequencing, and pulsed-field gel electrophoresis of *Rochalimaea quintana* from a patient with bacillary angiomatosis, *J. Clin. Microbiol.*, 32,1166, 1994.
62. Metzkor-Cotter, E. et al., Long-term serological analysis and clinical follow-up of patients with cat scratch disease, *Clin. Infect. Dis.*, 37, 1149, 2003.
63. La Scola, B., and Raoult, D., Serological cross-reactions between *Bartonella quintana*, *Bartonella henselae*, and *Coxiella burnetii*, *J Clin Microbiol.*, 34, 2270, 1996.
64. Sander, A. et al., Seroprevalence of antibodies to *Bartonella henselae* in patients with cat scratch disease and in healthy controls: Evaluation and comparison of two commercial serological tests, *Clin. Diagn. Lab. Immunol.*, 5, 486, 1998.
65. Vermeulen, M.J. et al., Serological testing for *Bartonella henselae* infections in The Netherlands: Clinical evaluation of immunofluorescence assay and ELISA, *Clin.. Microbiol. Infect.*, 13, 627–634, 2007.
66. McCool, T.L. et al., Discovery and analysis of *Bartonella henselae* antigens for use in clinical serologic assays, *Diagn. Microbiol. Infect. Dis.*, 60, 17, 2008.
67. Florin, T.A., Zaoutis, T.E., and. Zaoutis, L.B., Beyond cat scratch disease: Widening spectrum of *Bartonella henselae* infection, *Pediatrics*, 121, e1413, 2008.
68. Bereswill, S. et al., Molecular analysis of riboflavin synthesis genes in *Bartonella henselae* and use of the *ribC* gene for differentiation of *Bartonella* species by PCR, *J. Clin. Microbiol.*, 37, 3159, 1999.
69. Marston, E.L., Sumner, J.W., and Regnery, R.L., Evaluation of intraspecies genetic variation within the 60 kDa heat-shock protein gene (*groEL*) of *Bartonella* species, *Int. J. Syst. Bacteriol.*, 49, 1015, 1999.
70. Patel, R. et al., Use of polymerase chain reaction for citrate synthase gene to diagnose *Bartonella quintana* endocarditis, *Am. J. Clin. Pathol.*, 112, 36, 1999.
71. Jensen, W.A. et al., Rapid identification and differentiation of *Bartonella* species using a single step PCR assay, *J. Clin. Microbiol.*, 38, 1717, 2000.

72. Houpikian, P., and Raoult. D., 16S/23S rRNA intergenic spacer regions for phylogenetic analysis, identification, and subtyping of *Bartonella* species, *J. Clin. Microbiol.*, 39, 2768, 2001.
73. Renesto, P. et al., Use of *rpoB* gene analysis for detection and identification of *Bartonella* species, *J. Clin. Microbiol.*, 39, 430, 2001.
74. Zeaiter, Z., Liang, Z., and Raoult, D., Genetic classification and differentiation of *Bartonella* species based on comparison of partial *ftsZ* gene sequences, *J. Clin. Microbiol.*, 40, 3641, 2002.
75. Zeaiter, Z. et al., Diagnosis of *Bartonella* endocarditis by a real time nested-PCR assay using serum, *J. Clin. Microbiol.*, 41, 919, 2003.
76. Ciervo, A., and Ciceroni, L., Rapid detection and differentiation of *Bartonella* spp. by a single-run real-time PCR, *Mol. Cell. Probes*, 18, 307, 2004.
77. Ciervo, A. et al., Rapid identification of *Bartonella henselae* by real-time polymerase chain reaction in a patient with cat scratch disease, *Diagn. Microbiol. Infect. Dis.*, 53, 75, 2005.
78. Sander, A. et al., Comparison of different DNA fingerprinting techniques for molecular typing of *Bartonella henselae* isolates, *J. Clin. Microbiol.,* 36, 2973, 1998.
79. Arvand, M. et al., Multi-Locus Sequence Typing of *Bartonella henselae* ssolates from three continents reveals hypervirulent and feline-associated clones, *PLoS ONE*, 2, e1346, 2007.
80. Monteil, M. et al., Development of discriminatory multiple-locus variable number tandem repeat analysis for *Bartonella henselae, Microbiology*, 153, 1141, 2007.

54 *Brucella*

Sascha Al Dahouk, Heinrich Neubauer, and Herbert Tomaso

CONTENTS

54.1 INTRODUCTION

Brucellosis is a (re-)emerging disease in man and animals and the most common bacterial zoonosis worldwide.[1,2] Although many countries successfully established programs for eradication and control in animals, about 500,000 human cases are annually reported, mainly from endemic regions, including the Mediterranean Basin, the Arabian Peninsula, the Middle East, North Africa, Asia, Central America, and South America.

The disease-causing pathogen *Brucella* is transmitted from its animal reservoir to humans by direct contact with infected animals or, more often, through the consumption of raw food products of animal origin, for example, unpasteurized milk or cheese.

After September 11, 2001, further interest in *Brucella* arose from its status as a class B bioterrorist agent.[3] *Brucella* spp. were among the first biological agents weaponized in the 1950s because of their high contagiosity and due to the enormous economic impact of human and animal brucellosis.[4] Effective human vaccines are not available and countermeasures therefore have to rely on early diagnosis, adequate antibiotic treatment, and the control of livestock and products of animal origin that may pose a risk to consumers.

54.1.1 Taxonomy

Brucellae are gram-negative, nonmotile, nonspore-forming, and facultative intracellular coccobacilli.[5] The 16S rRNA gene sequence of the genus indicates that *Brucella* belongs to the family of *Rhizobiaceae* within the α-2 subdivision of the class *Proteobacteria* closely related to *Bartonella/Rochalimaea*, *Ochrobactrum*, and *Agrobacterium*.[6,7]

Historically, species and biovar (bv) classification of brucellae was based on natural host preference and phenotypic characteristics, that is, CO_2 requirement, H_2S production, urease activity, dye-sensitivity, oxidative metabolic profiles, lysis by *Brucella*-specific bacteriophages, and agglutination with monospecific antisera.[8,9] In accordance with these features, the genus *Brucella* (*B.*) currently comprises the six classical species *B. melitensis* bv 1–3 (primarily isolated from sheep and goats), *B. abortus* bv 1–6 and 9 (from cattle and other Bovidae), *B. suis* bv 1–3 (from pigs), bv 4 (from reindeer) and bv 5 (from small ruminants), *B. canis* (from dogs), *B. ovis* (from sheep), and *B. neotomae* (from desert wood rats).[10] Two novel species of marine origin, *B. pinnipedialis* (formerly *B. pinnipediae*, isolated from seals) and *B. ceti* (formerly *B. cetaceae*, isolated from dolphins and whales)[11,12]; *B. microti*, isolated from the common vole

Microtus arvalis,[13,14] from red foxes (*Vulpes vulpes*),[15] and also directly from soil,[16] have been proposed. Most recently *B. inopinata* sp. nov. isolated from a breast implant wound of a 71-year-old female patient has been described.[17,18] The animal reservoir of this species is still unknown.

Compared to the currently known *Brucella* species, *B. inopinata* sp. nov. is the first one that exhibits a unique 16S rRNA gene sequence. All the other *Brucella* species are highly related at the genetic level exhibiting identical 16S rRNA[19] and *recA* gene sequences,[20] as well as highly similar genomes, that is, sequence identity and gene synteny.[21–23] The majority of housekeeping genes only differ by single nucleotide polymorphisms among *Brucella* species.[24] Due to their high degree of DNA homology, exceeding 90% for all *Brucella* species, a monospecific genus has been proposed in which Abortus, Canis, Neotomae, Ovis, and Suis should be regarded as subspecies of *B. melitensis*.[25] However, in 2003 the Subcommittee on the Taxonomy of *Brucella* agreed to return to pre-1986 taxonomy and reapproved the six classical *Brucella* species with their recognized biovars.[26]

54.1.2 Pathogenicity and Biology

Although brucellae do not display classical virulence factors, such as exotoxins, cytolysins, proteases, fimbriae, flagellae, capsules, plasmids, or lysogenic phages, virulent strains are able to evade host defense mechanisms, penetrate host cells, alter intracellular trafficking, avoid killing in lysosomes, and modulate the intracellular environment.[27]

Especially in the early stage of infection, the nonendotoxic smooth lipopolysaccharide (sLPS) of *Brucella* helps to protect these bacteria from antimicrobial attacks of the host immune system.[28] *Brucella* LPS inhibits complement and cationic peptide-mediated killing and prevents the synthesis of immune mediators. Furthermore, the entry of *Brucella* in macrophages is partially dependent on the O-polysaccharide chain of LPS.[29] LPS also interferes with MHC II antigen presentation, impairing infected host cells from activating *Brucella* antigen-specific $CD4^+$ T cells.[30] However, cellular immunity is absolutely required for the control and resolution of *Brucella* infections in the host.[31]

Another key factor involved in the modulation of cell binding and penetration is the two-component regulatory system *Bvr*R/*Bvr*S, which controls the expression of various cell-surface proteins (Omp 3 family), allowing *Brucella* to escape from the lysosomal pathway. After internalization, virulent brucellae inhibit the fusion of their phagosomes with lysosomal compartments and modulate the phagosomal composition in order to evade intracellular degradation. The type IV secretion system encoded by the *Vir*B operon is also a major virulence factor and plays an important role in intracellular trafficking.[32] Effector molecules secreted by the type IV secretion machinery are responsible for maintaining the interactions of the *Brucella*-containing vacuole with the endoplasmic reticulum of the host cell until the replicative niche of *Brucella* is fully developed.[33] Within this compartment, also called "brucellosome," brucellae are able to survive and even replicate without restricting basic functions of the host cells. In the stationary phase, intracellular bacteria still have to withstand severe nutrient deprivation and microaerobic conditions.[34,35] Hence, a considerable degree of metabolic versatility is required for long-term survival and replication in this compartment.[36] The majority of proteins regulated at this stage of infection is involved in primary metabolism of *Brucella*, confirming a "stealthy" adaptation of the pathogen to the host cell environment.[37]

54.1.3 Brucellosis in Animals and Man

Although many domestic animals are susceptible to various *Brucella* species, cattle, sheep, goats, and pigs are primarily affected. Animal brucellosis is characterized by abortion, weak newborn, retained placenta, orchitis, and impaired fertility.[38] Hence, considerable economic damage is caused, not only through the loss of progeny, but also due to reduction in milk yield. Nonapparent infections have been described in all susceptible animal species, and the spread between herds usually occurs with the introduction of these asymptomatic chronically infected carriers.

Brucellae are disseminated within herds of cattle and flocks via contamination of sheds and pastures through abortion material or lochial discharge of infected animals after full-term parturition.[39] The infection is mainly transmitted orally by licking aborted fetuses and the genital discharges of an aborting animal or by feeding contaminated colostrum and milk. However, congenital (in utero) or perinatal infections may also occur. Especially in pigs, venereal transmission is a significant route of spread since brucellae are often localized in the nongravid uterus.

Four nomen species of the genus *Brucella* are pathogenic for humans, that is, according to their impact on public health, *B. melitensis*, *B. abortus*, *B. suis*, and rarely, *B. canis*.[1] Infrequent cases of brucellosis in humans caused by *B. pinnipedialis* or *B. ceti* have been described.[40–42] *B. inopinata* sp. nov. has been isolated from a woman only once.[17] Currently, only little is known about the pathogenicity of *B. microti*. Both, the source of infection and the endemic species, may vary according to the geographical region.

Brucella can be transmitted directly or indirectly from infected animals to humans and brucellosis usually is either an occupational or a foodborne infection. Direct transmission may involve the respiratory, conjunctival, and cutaneous route and requires close contact with infected animals. In endemic countries, brucellosis is recognized as an occupational disease and people engaged in the livestock industry, for example, shepherds, farmers, abattoir workers, and veterinarians, are often affected. Assistance in animal parturition is supposed to bear the highest risk of infection.[43]

The inhalation of aerosols containing viable brucellae is an important means of transmission to humans. The dose of infection by airborne transmission can be as low as 10 bacteria, and the attack rate of *Brucella* infections ranges between 30% and 100%. Therefore, brucellosis is the most frequent bacterial laboratory-acquired infection worldwide.[44]

However, indirect transmission remains the highest overall risk and mainly occurs through the consumption of unpasteurized milk or dairy products.[43]

Human brucellosis is a disease of protean manifestations. The clinical onset of disease is insidious, and the incubation period is extremely variable, ranging from weeks to months.

Acute brucellosis is usually accompanied by bacteremia and the spreading of the organism to various organ systems, mainly to reticuloendothelial tissues, that is, liver, spleen, skeletal, and hematopoietic system. The capability of *Brucella* spp. to survive and replicate in the mononuclear phagocytic system is responsible for the high frequency of prolonged disease, complications, and relapses. Clinical manifestations may vary widely including osteoarticular, dermal, gastrointestinal, respiratory, cardiovascular, and neurologic disorders. Human brucellosis may mimic many other infectious as well as noninfectious diseases.

Patients suffering from brucellosis usually present with nonspecific systemic signs and symptoms consistent with an influenza-like or septicemic illness, that is, fever, fatigue, malaise, weight loss, headaches, arthralgia, myalgia, chills, and sweats.

Brucella infections often develop as fever of unknown origin with an acute onset.[45] Fever >38.5°C is the leading symptom in the majority of patients. Because this fever can wax and wane, the disease is also called undulant fever.[46] Osteoarticular manifestations, for example, spondylitis, sacroiliitis, or arthritis are known to be the most frequent focal complications.[47,48]

Inadequate treatment may cause severe and debilitating chronic courses of infection and long-term sequelae. Patients can be plagued by serious life-threatening focal complications such as endocarditis and neurobrucellosis. *Brucella* endocarditis accounts for 80% of the fatal cases.[49] The overall case fatality rate reported is less than 1%.

Controlling acute illness and preventing complications and relapse are the most important tasks when treating human brucellosis. To reach this goal, prolonged chemotherapy using at least two synergistic antibiotic drugs is required. Therapeutic options are mainly based on four antibiotic substances, that is, doxycycline, rifampin, streptomycin (or other aminoglycosides), and trimethoprim-sulfamethoxazole (co-trimoxazole).[50–54] According to the former guidelines, the treatment of choice includes oral doxycycline 100 mg twice a day and rifampin 600–900 mg/day (15 mg/kg/day) in a single oral dose over a 6-week course.[51] Because the risk for a relapse is high under this antibiotic regimen (about 10% of patients), most authorities currently recommend a combination of streptomycin 1 g intramuscularly once daily for 2 weeks, instead of the administration of rifampin.[50,55] In clinical trials comparing the effectiveness of doxycycline plus rifampin with that of doxycycline plus streptomycin the latter combination showed lower relapse rates.[56–58] Especially in geographic regions with a high prevalence of tuberculosis the possible selection of resistant mycobacterial strains also makes the use of rifampin as the only drug for brucellosis treatment inadvisable.[59] Therefore, streptomycin and doxycycline are still the internationally favored antimicrobial agents for the treatment of brucellosis in endemic areas. Streptomycin may be substituted by other aminoglycosides, for example, gentamicin (5–6 mg/kg/day i.m. once daily for the first 7 days), which proved to be equally effective and showed fewer adverse effects.[60]

For the treatment of focal complications like *Brucella* endocarditis and neurobrucellosis triple or tetra, combinations of the already mentioned antimicrobial agents are necessary. The duration of therapy must be individualized in dependence of the clinical course but a longer treatment (>45 days) is definitively required.

54.1.4 Epidemiology of Brucellosis

As brucellosis is a zoonotic disease transmitted from an animal reservoir to man, the global number of human cases can only be reduced by effective surveillance and control of animal brucellosis. Bovine brucellosis (*B. abortus*) has been successfully reduced in the last decades, whereas the control of brucellosis in small ruminants (*B. melitensis*) has proven to be an intractable problem.

Test and slaughter programs as well as the vaccination of livestock helped to eradicate brucellosis. However, the spillover of brucellosis from wildlife to livestock and the globalization of animal trade promote the worldwide distribution of animal and human disease.[2,61] Hence, despite being controlled in many developed countries, brucellosis remains a major public health concern.

Bovine brucellosis (*B. abortus*) has been successfully eradicated in Canada, Japan, Northern Europe, and Australia. In the European Union, Sweden, Denmark, Finland, Germany, the United Kingdom (except for Northern Ireland), Austria, The Netherlands, Belgium, and Luxembourg are approved as "officially free from bovine and ovine/caprine brucellosis."[62] Norway and Switzerland are also considered to be brucellosis-free. In contrast, the situation is less favorable in southern European countries.[63]

Ovine/caprine brucellosis (*B. melitensis*) endemically occurs in countries surrounding the Mediterranean Sea, especially along its northern and eastern shores stretching through Central Asia. Furthermore, it is highly prevalent in countries around the Arabian Gulf. Parts of Latin America, especially Mexico, Peru, and northern Argentina are also seriously affected. Last but not least, *B. melitensis* infections in sheep have been reported in Africa and India. *B. melitensis* is not known to be enzootic in the United States, Canada, Northern Europe, Australia, New Zealand, or Southeast Asia, which is why only sporadic incursions have been notified in these regions.

Porcine brucellosis (*B. suis*) occurs in most areas in which pigs are kept. The human pathogenic biovars *B. suis* bv 1 and bv 3 are mainly isolated from wild boars, feral swine, and domestic pigs in Asia, South America, in the southeastern states of the United States, and in Queensland, Australia.[1,61] Available epidemiological evidence shows that *B. suis* bv 2 is the most common agent of porcine brucellosis in Europe, but the biovars 1 and 3 also occur.[64] *B. suis* bv 2 has only exceptionally been described as the causative agent of human brucellosis.[65,66]

The epidemiology of human brucellosis has significantly changed in the last decade for sanitary, socioeconomic, and political reasons. Currently, about a half million new cases are annually reported worldwide, and prevalence rates exceed 10 cases per 100,000 population in various endemic countries.[2] Taking into account that less than 10% of the *Brucella* infections are supposed to be recognized and notified because of the unspecific clinical symptoms of the disease, human brucellosis represents a significant and probably neglected health problem.[67,68]

Although ovine/caprine brucellosis has a limited geographic distribution, *B. melitensis* is by far the main cause of clinically apparent disease in humans worldwide.[69,70] In territories where ovine/caprine brucellosis is endemic and the population density of sheep and goats is high, there is a significant relationship between the number of infected flocks and the incidence of human cases.[71]

B. abortus and *B. suis* infections are often associated with certain occupational groups, for example, farm workers, veterinarians, and meatpacking employees, whereas human *B. melitensis* infections frequently occur without occupational exposure.[46] In the United States, the change of brucellosis from an occupational disease to a more common disease in the general population was accompanied by a shift from the *B. abortus* infections most frequently reported before 1960 to the *B. suis* infections predominating in the early 1970s and, currently, to an increased reporting of *B. melitensis* infections.[72]

In nonendemic countries, an association of human brucellosis with the immigrant population has been described.[73–75] Most cases have been imported, and the disease was acquired either through international travel or contaminated food products from endemic areas.[76] In the United States, 50%–60% of the brucellosis cases are notified in California and Texas, mainly among Hispanics.[67,77] Hispanic ethnicity, recent travel to endemic areas in Mexico and ingestion of nonpasteurized dairy products are the major risk factors for *Brucella* infections in the United States.[67,72,75,77]

54.1.5 Microbiological and Serological Methods in Diagnosis of Brucellosis

54.1.5.1 Isolation and Identification

Comparable to other bacterial infections, blood culture is still the gold standard in the diagnosis of human brucellosis. Although blood is the material most commonly examined, bone marrow cultures may yield a higher sensitivity, especially in patients with previous antibiotic treatment, and culture times may be faster.[78] Samples of any body fluid and of tissue collected during surgery or at necropsy can also be cultured. The isolation rate of brucellae from specimens with a competing microflora or from pus is low. In patients suffering from acute brucellosis the sensitivity of blood cultures may vary from 80% to 90% whereas in chronic disease direct bacteriologic confirmation is less successful ranging between 30% and 70% depending on the method of isolation used.[79]

Brucella grows on most standard media, for example, blood agar, chocolate agar, trypticase soy agar (TSA), or serum-dextrose agar (SDA). Two to five percent bovine or equine serum, which is needed for growth by various strains, is routinely added to the basal media. The inoculated agar plates should be incubated at 35°C–37°C in air supplemented with 5%–10% CO_2. Culturing the fastidious bacterium takes several days or even weeks. If samples of excreta or contaminated tissues are to be examined, selective media containing amphotericin B, bacitracin, cycloheximide/natamycin, D-cycloserine, nalidixic acid, polymyxin B, and vancomycin can be used.[80] As the number of *Brucella* organisms is likely to be low in tissue samples, enrichment culture using liquid media may be helpful. Castañeda's medium, a nonselective, biphasic medium, has been recommended for the isolation of *Brucella* from clinical samples.[81,82] However, weekly subcultures onto solid selective medium for up to 6 weeks are necessary. Therefore, this traditional method was replaced by automated culture systems, such as the lysis centrifugation method, which helped to reduce culture times and increased sensitivity.[83,84]

After a 2- to 3-day incubation period, punctate, nonpigmented, and nonhemolytic *Brucella* colonies are visible. Colonies of smooth (S) brucellae are raised, convex, circular, translucent, and 0.5–1 mm in diameter. After sub- or prolonged (>4 days) culture, morphology as well as virulence, antigenic properties, and phage-sensitivity tend to undergo variation. Smooth *Brucella* cultures dissociate to rough (R) forms growing in less convex and more opaque colonies with a dull, dry, yellowish-white granular appearance. Suspected culture colonies can be confirmed by the slide agglutination test using undiluted polyvalent *Brucella* antiserum (anti-S-serum) mixed with a saline suspension of colonies.

Brucella spp. are very small gram-negative coccobacilli, which microscopically appear only faintly stained and look like "fine sand." Oxidase- and urease-positive bacteria have to be suspected to be *Brucella* and have to be manipulated in a biological safety cabinet.

The identification of *Brucella* spp. and biovars is based on CO_2 requirement, H_2S production, urease activity, agglutination with monospecific sera (A and M), selective inhibition of growth on media containing dyes such as thionin or basic fuchsin, and phage typing.[8,9] These procedures are time consuming, hazardous, and subject to variable interpretation.

Using commercially available biochemical tests such as the API 20 NE™ (BioMerieux, Nürtingen, Germany), *Brucella* spp. may be misidentified, for example, as *Moraxella phenylpyruvica*.[85]

Since drug resistance does not significantly contribute to treatment failures and relapses, in vitro susceptibility testing of *Brucella* isolates is not generally recommended.[86] *Brucella* strains recovered from relapsed patients usually show antibiotic sensitivity profiles identical to the strains recovered during primary infection.[86,87]

54.1.5.2 Serological Tests

Because of the shortcomings of culture techniques, which are time consuming, hazardous, and not sensitive, most physicians rely on the indirect proof of *Brucella* infections. The detection of high or rising titers of specific antibodies in the serum allows a tentative diagnosis. IgM antibodies are predominant in the first week after inoculation with *Brucella*.[88] A switch to IgG isotype antibodies can be observed in the second week, followed by a continuous rise in both IgM and IgG levels, reaching a peak about four weeks after infection. Antibody titers usually decline with the beginning of an adequate antibiotic treatment, but significant antibody titers may persist for up to a year despite successful therapy.[88,89] IgG antibodies are related to active infection, and they are constantly detected in bacteriologically proven cases. Chronic courses of the disease are highly unlikely when IgG antibodies cannot be detected anymore.[88] However, antibody titers can peak a second time (usually only IgG), which might be a sign of relapse or an incomplete recovery.[90] In summary, the evaluation of *Brucella*-specific IgM and IgG antibodies helps to discriminate patients suffering from brucellosis at an early stage of disease or acute brucellosis from those with chronic brucellosis.[91]

The serological diagnosis of brucellosis is traditionally based on agglutination tests, for example, the serum agglutination test (SAT); the Rose Bengal test (RBT); and the antiglobulin, or Coombs', test (CT), which are all used for screening. The SAT developed by Wright and Smith in 1897 is still the most popular serological method in the diagnosis of human brucellosis.[92] In addition, SAT is frequently used as a reference to which other serological tests are compared.[93] Particularly the quality of the antigen preparations is essential for consistent and reproducible results in agglutination tests. Nevertheless, cross-reactions with various bacteria, for example, *Yersinia enterocolitica* O:9, *Escherichia coli* O:157, *Francisella tularensis*, *Salmonella urbana* group N, *Vibrio cholerae*, and *Stenotrophomonas maltophilia*, have been reported.[93] A confirmatory test is therefore always recommended to avoid false-positive results due to the low specificity of agglutination tests. Samples that reacted positive in a screening test are retested using the complement fixation test (CFT) or an enzyme-linked immunosorbent assay (ELISA), both of which are suitable for screening and confirmation. The sensitivity of various serological tests might also be low because they are based on secondary interactions such as the ability to agglutinate antigens or to fix complement. Especially in the early stage of infection, false-negative or only weak positive reactions may occur. Furthermore, the interpretation of test results is subjective reflecting the experience of the investigator.

Both, sensitivity and specificity of agglutination tests for brucellosis depend on the cut-off value used and on the background level of reactive antibodies in the population.[79] Therefore, cut-off values for agglutination titers considered to be confirmatory for active disease cannot be easily defined. In general, a titer ≥1:160 provides strong supportive evidence for active infection, and titers ≥1:80 are supposed to be suspicious, but even lower titers are not unusual in active brucellosis.[79,93] Especially in endemic areas, intermediate titers are difficult to interpret. In urban populations, a titer of 1:80 can be considered positive; whereas, in rural areas, where brucellosis is endemic, titers of ≥ 1:320 might be linked to disease.[94] Early in the course of disease or in persistent brucellosis, incomplete or blocking (nonagglutinating) antibodies may lead to negative serological test results even in patients with positive blood culture.[95]

Last but not least, the sensitivity of serological tests can be increased by testing paired serum samples. When testing a follow-up sample collected a few weeks after the initial diagnosis, seroconversion or a significant increase in antibody titers may prove active infection.

Conventional Serological Tests for Antibody Detection. *Serum Agglutination Test:* Several methods are suitable for agglutination tests, for example, the tube agglutination test (Wright test), slide or plate agglutination tests, and the card agglutination test. In the classical tube agglutination test, serial dilutions of the serum to be tested are mixed with the antigen, for example, a standardized phenol-inactivated suspension of a smooth culture of *B. abortus* organisms (strain 1119-3) adjusted to pH 7.2–7.4. The mixture is incubated in a water bath at 37°C for 48 h, and finally, the test results are recorded as grade of agglutination. The serum titer is defined as the dilution in which at least 50% of the bacterial cells are clumped. The tube agglutination test is labor intensive and time consuming.

Rose Bengal Test: The RBT was developed in the United States as a rapid screening test in the field surveillance of domestic cattle, especially in endemic areas. The RBT is a card test using a *B. abortus* strain 99 (Weybridge) or *B. abortus* strain 1119-3 (USDA) antigen suspension (8%), stained with Rose Bengal dye buffered to pH 3.65 ± 0.05.[96] Acidification loosens the bond between the antigen and nonspecific agglutinins, leading to an improved specificity and sensitivity of the RBT in comparison with SAT.[97,98]

Buffered plate antigen tests have proven to be inexpensive screening tests in humans with reduced nonspecific reactions in comparison with the SAT.[99] However, in endemic areas the use of the RBT as the sole technique for the diagnosis of brucellosis has to be considered very carefully, particularly in patients repeatedly exposed to *Brucella* or with a history of brucellosis.[100]

Complement Fixation Test: CFT is recommended as confirmatory test in the diagnosis of brucellosis because it is considered to be the most sensitive and specific conventional serological method. Especially during the incubation period and in the late chronic stage of disease, when SAT results are either negative or inconclusive, CFT has proven to be the diagnostic tool of choice.

Primary Binding Assays for Antibody Detection. *Enzyme-Linked Immunosorbent Assay (ELISA):* A great variety of antigens has been used for indirect enzyme-linked immunosorbent assay (iELISA), including whole cells, sLPS, native hapten polysaccharide, salt-extracted proteins, outer membrane proteins, and cytoplasmic proteins.[93] However, most ELISAs are based on bacterial extracts including high amounts of LPS, and false-positive reactions can occur after exposure to cross-reacting bacteria. Competitive enzyme immunoassays (cELISAs) are less prone to cross-reacting antibodies than conventional tests or iELISAs. Because of their additional high sensitivity, cELISAs can be used as confirmatory and screening tests.

ELISA is an excellent method for screening large populations for the presence of anti-*Brucella* antibodies.[101] In addition, ELISA techniques are helpful in classifying the stage of disease according to the detected antibody profile.[102] Patients suffering from acute brucellosis usually show very high levels of specific IgG and significant levels of IgM and IgA, whereas patients with chronic brucellosis show moderately high levels of IgG. Low levels of IgG in healthy persons may represent residual antibodies even a long time after complete recovery from brucellosis. Since the performance of ELISAs depends on the stage of disease, IgM- and IgG-ELISAs should generally be performed in parallel to increase sensitivity.[103–105]

Fluorescence Polarization Assay (FPA): The principle of fluorescence polarization is that light will be depolarized more intensively by a small, fast, rotating molecule than by a larger molecule. Since every molecule in a solution rotates randomly at a rate inversely proportional to its size, the rotation rate of a small antigenic molecule, measured using a fluorescent label and polarized light, will change after antibodies have been attached.[106]

FPA can be performed within a few minutes and requires minimal sample preparation. Under field conditions even whole blood samples can be analyzed.[107] For culture-confirmed human brucellosis cases the sensitivity of the FPA is 96%, and the specificity determined in healthy blood donors is 98%.[106]

Lateral Flow Assay (LFA): LFA can be used as a bedside test for the serodiagnosis of human brucellosis at all stages of disease.[108,109] For test performance, only a drop of blood obtained by a finger prick is needed. Furthermore, the LFA does not require a specific hands-on training, and test results are easy to interpret. All components, that is, the *B. abortus* antigen spotted onto a nitrocellulose strip and the monoclonal antihuman antibodies conjugated to colloidal dye particles for nonenzymatic detection, are stabilized and do not require cooling either during transport or for storage. Sensitivity and specificity of the LFA are high (>95%). Hence, LFA is a suitable rapid point-of-care assay ideal for field testing in an outbreak setting and for screening large populations.

54.1.6 Molecular Diagnostic Techniques

54.1.6.1 PCR

Numerous PCR techniques have been developed for the identification of *Brucella*.[110] PCR methods are increasingly used in the diagnosis of human brucellosis, since they can be more sensitive than blood culture and more specific than serological tests.[8] In addition, the work with DNA as opposed to highly infectious live cultures decreases the risk of laboratory-acquired infections. PCR turned out to be a suitable tool for the detection of *Brucella* in various matrices, including serum, whole blood and urine samples, different tissues, cerebrospinal fluid, synovial or pleural fluid, and pus.[111–115] Currently, the preferred clinical specimens used for molecular diagnosis of human brucellosis are whole blood and serum samples. Whether the serum fraction is superior to whole blood for molecular detection of the pathogen, or vice versa, is controversially discussed.[116,117] Although the concentration of PCR inhibitors is lower in serum samples, the small number of bacteria circulating after treatment may result in the absence of the target DNA in serum leading to false-negative results.

For rapid diagnosis of human brucellosis, the simple identification of *Brucella* by genus-specific PCR is adequate because antibiotic treatment is independent of the disease-causing species. A lot of target sequences have been used for detection of the genus *Brucella*, for example, the 16S rRNA,[118,119] the 16S-23S internal transcribed spacer region (ITS),[120] a gene coding for an outer membrane protein (*omp2*),[121] and the insertion element IS*711*.[122,123] However, the majority of genus-specific PCR assays target the *bcsp31* gene, which codes for a 31-kDa immunogenic outer membrane protein conserved among all *Brucella* spp.[124]

For species-specific surveillance programs, differential PCR assays are needed. The multiplex AMOS–PCR using a five primer cocktail was a first approach toward the differentiation of four *Brucella* species, that is, *B. abortus* (bvs 1, 2, and 4); *B. melitensis* (bvs 1, 2, and 3); *B. ovis*; and *B. suis* (bv 1).[125] One primer anneals to the insertion sequence IS*711*, which appears in several copies all over the genome, but distribution and number differ in *Brucella* species. The other four primers hybridize to species-specific regions downstream of this insertion element, leading to variable amplicon sizes. Recently, a conventional multiplex PCR assay suitable for the identification of all *Brucella* species (except for *B. inopinata*) and the vaccine strains, *B. abortus* RB51, *B. abortus* S19, and *B. melitensis* Rev1 has been developed.[126] The identification is based on the numbers and sizes of seven products amplified by the so-called Bruce-ladder PCR (Figure 54.1). In a worldwide multicenter study, the Bruce-ladder PCR has proven to be a useful tool for the rapid identification of *Brucella* strains in basic microbiology laboratories.[127] Another conventional multiplex PCR, based on primers targeting specific deletions and insertions in the different *Brucella* genomes, partially allows subtyping of *Brucella* species at the biovar level.[128]

Real-time PCR constitutes a further technological improvement for the molecular identification of the genus *Brucella* and the differentiation of its species. Especially in the context of biological warfare and agroterrorism, real-time PCR will allow a more rapid high-throughput screening of samples.[129,130] Test results are available within a few hours. The species-specific real-time PCRs established by Redkar et al.,[131] Newby et al.,[132] and Probert et al.[133] were described as rapid, sensitive, and specific diagnostic tools with a low risk of cross-contamination and the potential of automation. Single nucleotide polymorphism (SNP)-based differentiation of *Brucella* clades incorporated into real-time PCR assays may also provide a rapid method for the identification of *Brucella* species.[134]

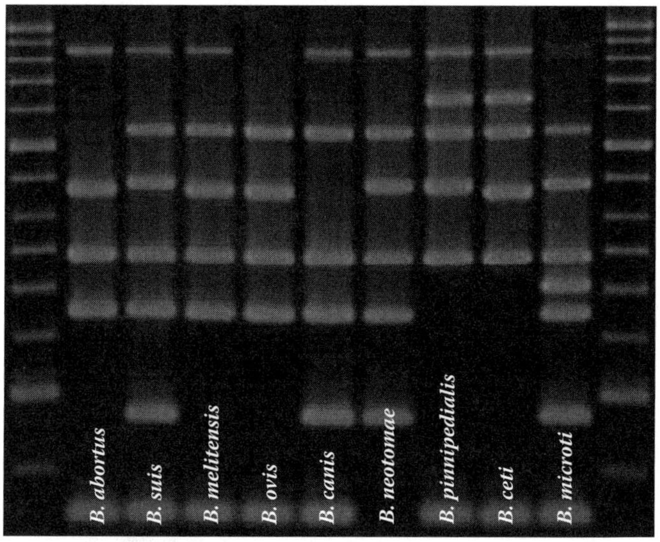

FIGURE 54.1 Identification and differentiation of *Brucella* species by the Bruce-ladder PCR.[126,127]

Various studies of genus-specific real-time PCR assays applied to human specimens have been published. A real-time PCR targeting the *bcsp31* gene appears to be suitable for screening whenever *Brucella* is suspected because false-negative results caused by rare species and biovars may be avoided by a genus-specific approach. Consecutively, a species-specific assay should be applied because the use of more than one marker-based PCR may increase sensitivity and specificity.[130]

Since PCR reactions may be inhibited by clinical samples,[135] an internal amplification control should be included.[130] Dilution of the samples may also reduce inhibitory effects of the sample matrix but the generally small amount of target DNA in environmental and clinical samples is challenging even for assays with a very low detection limit. Approximately five bacteria per reaction can be detected using well-established *Brucella* real-time PCR assays. Testing several replicates of the purified DNA in parallel may increase sensitivity.

Since the number of bacteria in the clinical samples of patients suffering from focal forms of brucellosis is usually small, the detection capacity of a valuable molecular assay has to be very high. Sensitivity of conventional and real-time PCR assays for the diagnosis of brucellosis can be further increased using the IS*711* insertion element of *Brucella* as a target sequence because it is found in multiple copies in *Brucella* chromosomes.[136,137]

Definite criteria to establish the success of treatment are still missing, and to date, none of the conventional microbiological methods has proven to be efficient to predict relapse. In the last years, more studies have been conducted to evaluate the usefulness of PCR methods during the incubation period and for posttreatment follow-up of *Brucella* infections. Quantitative real-time PCR appears to be a useful method to identify symptomatic nonfocal disease in patients for whom the classical microbiological methods fail.[138] In addition, real-time PCR using serum samples is a valuable tool, not only for the initial diagnosis of brucellosis, but also for the differentiation between active and past infection.[139] Using single-step conventional PCRs for posttreatment follow-up in whole blood samples, PCR results were negative in successfully treated patients, whereas test results were positive in relapsed patients.[116,140] In contrast, using real-time PCR techniques, *Brucella* DNA could be detected in the majority of brucellosis patients throughout treatment and follow-up, despite appropriate antibiotic therapy and apparent clinical recovery.[141,142] Although DNA load constantly decreases after the end of treatment, a significant number of patients exhibit bacterial DNA load even years after clinical cure and in the absence of any symptoms indicative of disease persistence or relapse.[142] However, *Brucella* may persist in human macrophages and replicate in such a low frequency that transient bacteremia can be controlled by the immune system. Furthermore, the diagnostic yield of modern real-time PCR assays is better as compared to conventional techniques leading to the detection of nonviable or phagocytosed microorganisms.

No difference in the evolution of DNA load can be observed between patients who relapse and those who do not.[143] Nevertheless, quantitative real-time PCR could be helpful in the diagnosis of chronic infection. *B. melitensis* DNA can be detected in asymptomatic subjects with a history of brucellosis, albeit in a smaller proportion than in the group of symptomatic patients suffering from chronic disease.[138] The clinical significance of persisting bacterial DNA load in asymptomatic patients long after successful completion of antibiotic therapy is still unclear. Further studies are urgently needed to clarify whether the persistence of *Brucella* DNA in human serum reflects the presence of active bacteria possibly leading to recurrent disease or just nonviable bacterial fragments.

Real-time PCR methods can also be useful in the differential diagnosis of brucellosis. For instance, in areas of high incidence of brucellosis and tuberculosis, differential diagnosis is often difficult because both diseases are characterized by clinical polymorphism and granoulomatous inflammation. In addition, *Brucella* spp. and *Mycobacterium tuberculosis* are slow-growing pathogens, and culture may be insensitive in extrapulmonary tuberculosis and focal manifestations of brucellosis. For rapid differential diagnosis a multiplex real-time PCR assay based on the intergenic region SenX3-ReX3 and *bcsp31* has recently been developed.[115]

54.1.6.2 Molecular Typing Methods

Molecular epidemiology significantly contributes to the analysis and understanding of infections caused by pathogenic bacteria.[144] In epidemiological studies, molecular typing methods can be used for trace-back analysis, which may help to identify the origin of an infection and the way it spreads. Fast and accurate typing procedures are therefore crucial for the eradication and control of brucellosis. However, molecular typing is difficult due to the high degree of genetic homology (>90%) among *Brucella* species, demonstrated by DNA–DNA hybridization,[25] due to identical 16S rRNA gene sequences among all *Brucella* species,[145] and because more than 90% of all genes share >98% sequence identity.[23] Nevertheless, various methods have been established for molecular subtyping of *Brucella* strains, for example, AP–PCR (arbitrary primed–PCR),[146] ERIC–PCR (enterobacterial repetitive intergenic consensus sequence–PCR), REP–PCR (repetitive intergenic palindromic sequence–PCR),[147,148] RAPD–PCR (random amplified polymorphic DNA–PCR),[149] PCR–RFLP (PCR–restriction fragment length polymorphism) of different genetic loci,[150] AFLP (amplified fragment length polymorphism),[151] SNP (single nucleotide polymorphism),[152,153] and MLST (multilocus sequence typing).[154] However, with the exception of SNP and MLST, these tests lack reproducibility, show a limited capability to differentiate single strains, and are not appropriate for routine typing.

Based on the genomes of *B. melitensis* 16M,[21] *B. suis* 1330,[22] and *B. abortus* strain 9-941[23] tandem repeat loci with a high degree of size polymorphism and variable number tandem repeats (VNTRs) could be identified. Tandem repeats are composed of perfect or imperfect copies of an elementary unit, and various alleles can be observed in different bacterial strains within a species. Tandem repeats are classified in satellites (megabases of DNA) present in many eukaryotic genomes, minisatellites (spanning hundreds of base pairs, with a repeat unit size of at least 9 bp), and microsatellites (spanning a few tens of nucleotides with a repeat unit size up to 8 bp).[155]

Fingerprints resulting from analysis of multiple loci can be highly discriminating or even unique. Multiple locus variable number tandem repeats analysis (MLVA) has proven to be discriminatory among unrelated *Brucella* isolates that could not be differentiated by classical microbiological methods. The hypervariability found by MLVA is in contrast to the well-known genetic homogeneity of the genus. However, tandem repeats generally mutate at different rates.[155] Some loci cluster *Brucella* isolates in accordance with the identified species, presumably because they mutate slowly and have a relatively low homoplasy rate. Others have a higher discriminatory index, and *Brucella* isolates originating from a restricted geographical area can still be discriminated, indicating the potential of VNTRs as an epidemiological tool.[156] The hypermutability of VNTR loci has been exploited for *Brucella* in large panels of animal isolates.[157–160] Various VNTR-based genotyping methods have been tested for subtyping human *Brucella* isolates both with a broad global presentation[156] and from small spatial scales, such as the Middle East,[161] Lebanon,[162] Italy,[163] Portugal,[164] Mexico,[165] and Peru.[166,167] Currently, the most frequently used molecular typing assay is the MLVA-16 scheme, which includes eight minisatellite loci (panel 1) and eight microsatellites (panel 2, which is subdivided into panels 2A and 2B) (Table 54.4).[156] The MLVA-16 data can be queried on the genotyping Web page at http://mlva.u-psud.fr/.

MLVA-16 has also proven to be useful for clinical applications. MLVA-16 allows identifying the source of laboratory-acquired *Brucella* infections[168] and helps to differentiate relapse from reinfection, which may have implications on therapy.[87]

Based on simple PCR techniques, *Brucella*-MLVA is accessible to a wide range of users. The alleles (PCR amplicons) can be analyzed by simple agarose gel electrophoresis (Figure 54.2) or automatic high-throughput procedures, that is, an automated capillary electrophoresis-based method[161] or a lab on a chip technology.[169] Standardization and quality control are easy to achieve. Especially in an outbreak setting including a great number of cases, high-throughput screening is possible without manipulation of the living agent. Last but not least, the MLVA data can be easily coded and exchanged by the repeat copy numbers for each locus and strain. Currently, MLVA is the most suitable molecular method for subtyping brucellae that fulfils all performance criteria recommended for a typing assay, that is, typeability, reproducibility, stability, discriminatory power, concordance with other typing techniques, and epidemiological concordance.[159]

54.2 METHODS

54.2.1 Sample Collection and Preparation

Blood, urine, cerebrospinal fluid, synovial fluid, and tissue can be valuable specimens for direct detection of *Brucella* DNA.[115] Due to the small number of brucellae in these clinical specimens, special methods for DNA preparation are required to reduce inhibitory effects caused by matrix components and to concentrate the DNA. The preparation of a bacterial DNA template from serum simply by boiling does not prevent the presence of PCR inhibitors.[135]

Commercial kits such as the QIAamp™ DNA Mini Kit (Qiagen Inc., Valencia, California) have been successfully used to extract *Brucella* DNA from whole blood, serum, and tissue samples.[142,143,170] Queipo-Ortuño et al. evaluated seven commercially available DNA extraction kits using human serum samples and obtained the best results with the UltraClean™ DNA BloodSpin Kit (MO BIO Laboratories Inc., Carlsbad, California).[171] However, comprehensive evaluation studies on DNA extraction methods in different human samples are still missing.

54.2.2 Detection Procedures

54.2.2.1 Multiplex PCR and MLVA

The previously described AMOS–PCR is highly effective for the differentiation of *B. abortus*, *B. melitensis*, *B. ovis*, and *B. suis*[125] (Table 54.1). This PCR employs one primer

TABLE 54.1
Primers for AMOS–PCR

Primers	Sequence (5′→3′)
BA	GAC GAA CGG AAT TTT TCC AAT CCC
BM	AAA TCG CGT CCT TGC TGG TCT GA
BS	GCG CGG TTT TCT GAA GGT TCA GG
BO	CGG GTT CTG GCA CCA TCG TCG
IS711	TGC CGA TCA CTT AAG GGC CTT CAT

Source: Bricker, B.J., and Halling, S.M., *J. Clin. Microbiol.*, 11, 2660, 1994.

(IS*711*) for binding with the insertion sequence IS*711*, which is present in all *Brucella* species, and four primers for recognizing the species-specific regions downstream of this insertion sequence, leading to formation of variable sizes of amplicons from these four *Brucella* species, and facilitating their identification.

To identify all *Brucella* species (except for *B. inopinata*) and the vaccine strains *B. abortus* RB51, *B. abortus* S19, and *B. melitensis* Rev1, the Bruce-ladder PCR can be applied.[126,127] Use of primers and PCR conditions shown in Tables 54.2 and 54.3, respectively, will result in the amplification of distinct bands from most *Brucella* species (Figure 54.1).

TABLE 54.2
Primers for Bruce-Ladder PCR

Primers	Sequence (5′→3′)	Amplicon Size (bp)
BMEI0998f	ATC CTA TTG CCC CGA TAA GG	1682
BMEI0997r	GCT TCG CAT TTT CAC TGT AGC	
BMEI0535f	GCG CAT TCT TCG GTT ATG AA	450 (1320 in *Brucella* strains isolated from marine mammals)
BMEI0536r	CGC AGG CGA AAA CAG CTA TAA	
BMEII0843f	TTT ACA CAG GCA ATC CAG CA	1071
BMEII0844r	GCG TCC AGT TGT TGT TGA TG	
BMEI1436f	ACG CAG ACG ACC TTC GGT AT	794
BMEI1435r	TTT ATC CAT CGC CCT GTC AC	
BMEII0428f	GCC GCT ATT ATG TGG ACT GG	587
BMEII0428r	AAT GAC TTC ACG GTC GTT CG	
BR0953f	GGA ACA CTA CGC CAC CTT GT	272
BR0953r	GAT GGA GCA AAC GCT GAA G	
BMEI0752f	CAG GCA AAC CCT CAG AAG C	218
BMEI0752r	GAT GTG GTA ACG CAC ACC AA	
BMEII0987f	CGC AGA CAG TGA CCA TCA AA	152
BMEII0987r	GTA TTC AGC CCC CGT TAC CT	
Bmispec f	AGA TAC TGG AAC ATA GCC CG	510
Bmispec r	ATA CTC AGG CAG GAT ACC GC	

BMEI and BMEII numbers designate loci in the *B. melitensis* genome; BR numbers designate loci in the *B. suis* genome; f, forward; r, reverse.
Source: García-Yoldi, D. et al., *Clin. Chem.*, 52, 779, 2006; López-Gõni, I. et al., *J. Clin. Microbiol.* 46, 3484, 2008.

TABLE 54.3
Bruce-Ladder PCR Cycling Program

	Temperature (°C)	Time	Cycles
Initial denaturation	95	15 min	1
Denaturation	94	30 s	25
Annealing	58	90 s	
Extension	72	120 s	
Last extension	72	10 min	1

The recently developed MLVA-16 assay[156] has proved to be highly discriminatory among unrelated human *Brucella* isolates that could not be differentiated by other PCR methods or by the classical microbiological techniques. The MLVA-16 is based on two panels of primers (Table 54.4), one comprising eight minisatellite markers (panel 1) useful for species identification, and a second group of eight complementary microsatellite markers (panel 2) showing a higher discriminatory power useful for molecular epidemiology. The panel 2 markers are split into two groups, panel 2A and 2B, composed of three and five markers, respectively. Panel 2B contains the most highly variable markers.

54.2.2.2 Real-Time PCR

Real-time PCR assays targeting *bcsp31* described by Probert et al. and Al Dahouk et al. can be recommended for the detection of *Brucella* for screening purposes (Tables 54.5 and 54.6).[130,133] Using IS*711* as target the sensitivity of real-time PCR technology can be further increased because this intergenic element occurs in multiple copies within the *Brucella* genome (Table 54.7).

Assays for the species-specific identification of *B. melitensis* described by Redkar et al. and Probert et al. have a low detection limit (Table 54.8).[130,131,133] Assays designed for the detection of *B. abortus* by Redkar et al., Probert et al., and Newby et al. only detected biovars 1, 2, and 4, because of the more distant relatedness of other biovars (Table 54.9).[131–133] A real-time PCR assay specific for *B. suis* bv 1 was also established by Redkar et al.[131]

Hybridization probes technology offers melting curve analysis which provides typical signals for specific amplification products. 5′-exonuclease assays are very useful for large-scale screening, as they can be performed on platforms including 96 or 384 tests in parallel. Since real-time PCR assays can be easily transferred from one technical platform to another,[172] diagnostic laboratories are able to choose between hybridization probes technology and 5′ exonuclease assays to meet their demands. Ready-to-use protocols for the detection of *Brucella* and the identification of the most relevant species by means of real-time PCR assays are given in Tables 54.5 through 54.9.

54.3 CONCLUSIONS AND FUTURE PERSPECTIVES

In the diagnosis of human brucellosis, time-consuming culture and phenotypic characterization of the isolate is still the "gold standard." The low yield of *Brucella* cultures, however, often results in diagnostic delay and the late initiation of appropriate antibiotic therapy. Serology is a more effective means of diagnostic, although cross-reactivity is still a major problem. PCR has proven to be a valuable tool when culture fails or serological results are inconclusive, and a test result can be available within a few hours. PCR assays targeting more than one gene of the *Brucella* genome may enhance the likelihood of detection. Genus-specific primers

TABLE 54.4
Primers for MLVA-16

VNTR Locus	Primer (5′→3′) Upper	Lower
Panel 1		
bruce06-RU1322_134bp_408bp_3u	ATGGGATGTGGTAGGGTAATCG	GCGTGACAATCGACTTTTTGTC
bruce08-BRU1134_18bp_348bp_4u	ATTATTCGCAGGCTCGTGATTC	ACAGAAGGTTTTCCAGCTCGTC
bruce11-BRU211_63bp_257bp_2u	CTGTTGATCTGACCTTGCAACC	CCAGACAACAACCTACGTCCTG
bruce12-BRU73_15bp_392bp_13u	CGGTAAATCAATTGTCCCATGA	GCCCAAGTTCAACAGGAGTTTC
bruce42-BRU424_125bp_539bp_4u	CATCGCCTCAACTATACCGTCA	ACCGCAAAATTTACGCATCG
bruce43-BRU379_12bp_182bp_2u	TCTCAAGCCCGATATGGAGAAT	TATTTTCCGCCTGCCCATAAAC
bruce45-BRU233_18bp_151bp_3u	ATCCTTGCCTCTCCCTACCAG	CGGGTAAATATCAATGGCTTGG
bruce55-BRU2066_40bp_273bp_3u	TCAGGCTGTTTCGTCATGTCTT	AATCTGGCGTTCGAGTTGTTCT
Panel 2A		
bruce18-BRU339_8bp_146bp_5u	TATGTTAGGGCAATAGGGCAGT	GATGGTTGAGAGCATTGTGAAG
bruce19-Bru324_6bp_163bp_18u	GACGACCCGGACCATGTCT	ACTTCACCGTAACGTCGTGGAT
bruce21-BRU329_8bp_148bp_6u	CTCATGCGCAACCAAAACA	GATCTCGTGGTCGATAATCTCATT
Panel 2B		
bruce04-BRU1543_8bp_152bp_2u	CTGACGAAGGGAAGGCAATAAG	CGATCTGGAGATTATCGGGAAG
bruce07-BRU1250_8bp_158bp_5u	GCTGACGGGGAAGAACATCTAT	ACCCTTTTTCAGTCAAGGCAAA
bruce09-BRU588_8bp_156bp_7u	GCGGATTCGTTCTTCAGTTATC	GGGAGTATGTTTTGGTTGTACATAG
bruce16-BRU548_8bp_152bp_3u	ACGGGAGTTTTTGTTGCTCAAT	GGCCATGTTTCCGTTGATTTAT
bruce30-BRU1505_8bp_151bp_6u	TGACCGCAAAACCATATCCTTC	TATGTGCAGAGCTTCATGTTCG

Source: Al Dahouk, S. et al., *J. Microbiol. Methods*, 69, 137, 2007.

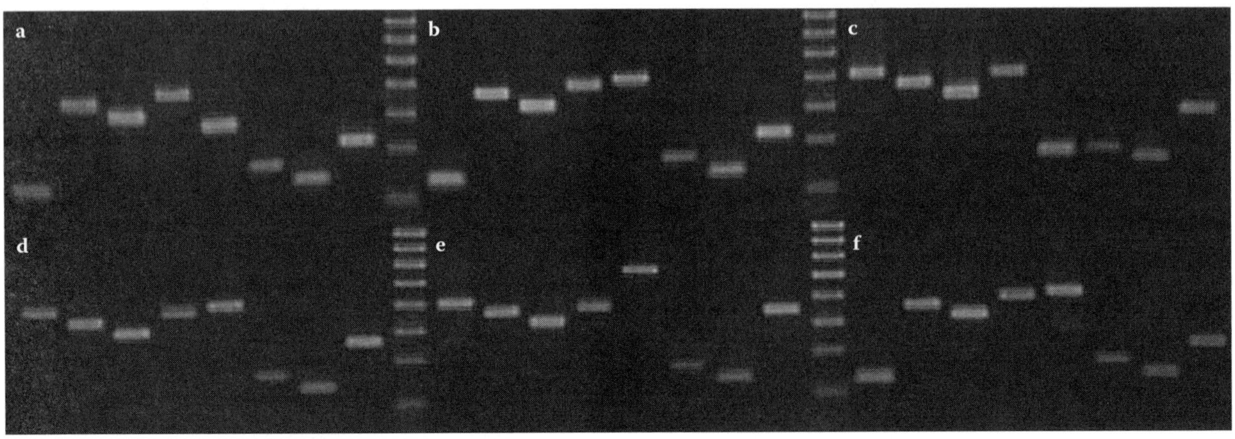

FIGURE 54.2 MLVA typing of *Brucella* field isolates of various geographic origins using panel 1 VNTR markers.[156] (a and b) Turkey, (c) Italy, (d) Pakistan, (e) United Arab Emirates, (f) Syria.

and probes should always be included in a diagnostic scheme to detect infrequently isolated *Brucella* species and atypical strains. A positive result in a genus-specific real-time PCR has to be specified by a species-specific assay. Serological tests should be performed in parallel to further consolidate the diagnosis, as the number of constantly seronegative patients is low.

There has been controversy about the preferred clinical specimen, the best method of DNA extraction, and the PCR assays with the lowest limit of detection. Most real-time PCR assays developed for the detection of *Brucella* can detect fewer than 10 bacteria per reaction, which is close to the technical limit of this method. In addition, a variety of DNA purification methods have been validated that eliminate

TABLE 54.5
Real-Time PCR Assay Targeting *bcsp31* for Specific Detection of Genus *Brucella*

A. PCR Mixture Makeup

Reagent	*Brucella* spp. (*bcsp31*)	Stock Concentration (pmol/μL = μM)	μL Per Reaction	Final Concentration (in 20 μL)
Water			9.8	
25 mM $MgCl_2$			1.2	2.5 mM
10× concentrated Reaction Mix	LightCycler FastStart DNA Master Hybridization Probes Cat. No. 2 239 272		2	
	Primers and Probes (5′→3′)			
B4	TGGCTCGGTTGCCAATATCAA	20 pmol/μL	0.5	0.5 μM
B5	CGCGCTTGCCTTTCAGGTCTG	20 pmol/μL	0.5	0.5 μM
Bru FL	AGGCAACGTCTGACTGCGTAAAGCC	8 μM	0.5	0.2 μM
Bru LC	LC Red 640-ACTCCAGAGCGCCCGACTTGATCG	8 μM	0.5	0.2 μM
Lambda F	ATGCCACGTAAGCGAAACA	20 pmol/μL	0.5	0.5 μM
Lambda R	GCATAAACGAAGCAGTCGAGT	20 pmol/μL	0.5	0.5 μM
Lambda FL	GGTGCCGTTCACTTCCCGAATAAC X	8 μM	0.5	0.15 μM
Lambda LC	LC Red 705-CGGATATTTTTGATCTGACCGAAGCG p	8 μM	0.5	0.15 μM
Phage Lambda DNA[a]	Plasmid DNA		1	
Sample DNA			2	

[a] Inclusion of an internal amplification control is especially useful for clinical samples.[130]

B. PCR Parameter

Parameter	Temperature (°C)	Time	Slope (°C/s)	Acquisition Mode
Activation of Fast Start Taq DNA polymerase	95	10:00 min	20	None
Amplification (45 cycles)	95	10 s	20	None
	55	10 s	20	Single
	72	12 s	20	None
Melting curve analysis	95	0	20	None
	55	30 s	20	None
	95	0	0.1	Continuous
Cooling	40	30 s	20	None

Note: The assay specifically detects all *brucellae* but no other organisms, with a detection limit of 16 fg DNA in buffer.

inhibitory components of the matrix. Remaining limitations are the small number of brucellae to be found in clinical specimens and, consequently, the even smaller amount of target DNA available for PCR. Despite of these biological and technical problems, especially for real-time PCR methods, various advantages compared to standard serological and bacteriological methods could have been demonstrated. However, PCR results are not always reproducible in different laboratories. PCR assays targeting different genes may help to exclude false-positive results. In order to determine inhibitory effects of the matrix, an internal amplification control has to be integrated into all PCR assays. Since practical experiences may vary between laboratories concerning optimal DNA preparation and PCR assays, results obtained with molecular methods generally have to be evaluated in the context of other diagnostic tests and of course the clinical signs and symptoms of the patient.

Molecular techniques may revolutionize the diagnosis of human brucellosis. Real-time PCR technology meets all requirements for a rapid diagnosis in clinical microbiology laboratories. The drastic reduction of diagnostic delay will have important prognostic implications in life-threatening complications of the disease, such as *Brucella* endocarditis or neurobrucellosis.

TABLE 54.6
Real-Time 5′-Exonuclease Assay Targeting *bcsp31* for Large-Scale Screening of Genus *Brucella*

A. PCR Mixture Makeup

Reagent	*Brucella* spp. (*bcsp31*)	Stock Concentration (pmol/μL = μM)	μL Per Reaction	Final Concentration (in 25μL)
Water			5.5	
Reaction buffer	TaqMan Universal MasterMix (uMM) Applied Biosystems Cat. No. 4304447	2×	12.5	1×
	Primers and Probes (5′→3′)			
Brucella spp fw	GCTCGGTTGCCAATATCAATGC	10 μM	0.75	0.3 μM
Brucella spp rev	GGGTAAAGCGTCGCCAGAAG	10 μM	0.75	0.3 μM
Brucella spp. Taq	6FAM-AAATCTTCCACCTTGCCCTTGCCATCA-DB	10 μM	0.5	0.2 μM
Sample DNA			5	

B. PCR Parameter

	Temperature (°C)	Time	Cycle
Decontamination	50	2 min	1
Initial denaturation	95	10 min	1
Denaturation	95	15 s	45
Annealing	57	1 min	

Note: The assay specifically detects all *brucellae* but no other organisms, with a detection limit of 16 fg DNA in buffer.
Source: Probert, W.S. et al., *J. Clin. Microbiol.*, 42, 1290, 2004.

TABLE 54.7
Real-Time PCR Assay Targeting Intergenic Element IS*711* for Specific and More Sensitive Detection of Genus *Brucella*

A. PCR Mixture Makeup

Reagent	*Brucella* spp. (IS*711*)	Stock Concentration (pmol/μL = μM)	μL Per Reaction	Final Concentration (in 20 μL)
Water			12.4	
25 mM $MgCl_2$			1.6	3 mM
10× concentrated Reaction Mix	LightCycler FastStart DNA Master Hybridization Probes Cat. No. 2 239 272		2	
	Primers and Probes (5′→3′)			
IS711_S	TTGTCGATGCTATCGGCCTAC	20 pmol/μL	0.5	0.5 μM
IS711_R	GGCAATGAAGGCCCTTAAGT	20 pmol/μL	0.5	0.5 μM
IS711_FL	GAAGCTTGCGGACAGTCACCATAAT--FL	8 μM	0.5	0.2 μM
IS711_LC	LC Red640-GCCGGGTGTTGGCTTTATTCG--PH	8 μM	0.5	0.2 μM
Sample DNA			2	

continued

TABLE 54.7 (continued)
Real-Time PCR Assay Targeting Intergenic Element IS*711* for Specific and More Sensitive Detection of Genus *Brucella*

B. PCR Parameter

Parameter	Temperature (°C)	Time	Slope (°C/s)	Acquisition Mode
Activation of Fast Start Taq DNA polymerase	95	10:00 min	20	None
Amplification (45 cycles)	95	10 s	20	None
	55	10 s	20	Single
	72	15 s	20	None
Melting curve analysis	95	0	20	None
	45	30 s	20	None
	95	0	0.1	Continuous
Cooling	40	30 s	20	None

TABLE 54.8
Real-Time PCR Assays Targeting BMEI1162/IS711 for Specific Detection of *B. melitensis* Designed as 5′-Exonuclease Assay for Large-Scale Screening

A. PCR Mixture Makeup

Reagent	*Brucella melitensis* BMEI1162/IS711	Stock Concentration (pmol/μL = μM)	μL Per Reaction	Final Concentration (in 25μL)
Water			5.5	
Reaction buffer	TaqMan Universal MasterMix (uMM) Applied Biosystems	2×	12.5	1×
	Primers and Probes (5′→3′)			
B. melitensis fw	AACAAGCGGCACCCCTAAAA	10 μM	0.75	0.3 μM
B. melitensis rev	CATGCGCTATGATCTGGTTACG	10 μM	0.75	0.3 μM
B. melitensis Taq	6FAM-CAGGAGTGTTTCGGCTCAGAATAAT CCACA-DB	10 μM	0.5	0.2 μM
Sample DNA			5	

B. PCR Parameter

	Temperature (°C)	Time	Cycle
Decontamination	50	2 min	1
Initial denaturation	95	10 min	1
Denaturation	95	15 s	45
Annealing	57	1 min	

Note: This assay specifically detects all *B. melitensis* isolates but no other orgnisms with a detection limit of 16fg DNA in buffer.
Source: Probert, W.S. et al., *J. Clin. Microbiol.*, 42, 1290, 2004.

TABLE 54.9
Real-Time PCR Assay Targeting *alkB*/IS711 for Specific Detection of *B. abortus* Designed as 5′-Exonuclease Assay for Large-Scale Screening

A. PCR Mixture Makeup

Reagent	*Brucella abortus alkB*/IS711	Stock Concentration (pmol/μL = μM)	μL Per Reaction	Final Concentration (in 25 μL)
Water			5.5	
Reaction buffer	TaqMan Universal MasterMix (uMM) Applied Biosystems	2×	12.5	1 x
	Primers and Probes (5′→3′)			
B. abortus fw	GCGGCTTTTCTATCACGGTATTC	10 μM	0.75	0.3 μM
B. abortus rev	CATGCGCTATGATCTGGTTACG	10 μM	0.75	0.3 μM
B. abortus Taq	6FAM-CGCTCATGCTCGCCAG ACTTCAATG-DB	10 μM	0.5	0.2 μM
Sample DNA			5	

B. PCR Parameter

	Temperature (°C)	Time	Cycle
Decontamination	50	2 min	1
Initial denaturation	95	10 min	1
Denaturation	95	15 s	45
Annealing	57	1 min	

Note: This assay specifically detects *B. abortus* bv 1, 2, and 4, but other biotypes cannot be detected reliably. The detection limit is 18 fg DNA in buffer.
Source: Probert, W.S. et al., *J. Clin. Microbiol.*, 42, 1290, 2004.

REFERENCES

1. Godfroid, J. et al., From the discovery of the Malta fever's agent to the discovery of a marine mammal reservoir, brucellosis has continuously been a re-emerging zoonosis, *Vet. Res.*, 36, 13, 2005.
2. Pappas, G. et al., The new global map of human brucellosis, *Lancet Infect. Dis.*, 6, 91, 2006.
3. Greenfield, R.A. et al., Bacterial pathogens as biological weapons and agents of bioterrorism, *Am. J. Med. Sci.*, 323, 299, 2002.
4. Christopher, G.W. et al., History of U.S. military contributions to the study of bacterial zoonoses, *Mil. Med.*, 170, 39, 2005.
5. Corbel, M.J., Microbiology of the genus *Brucella*. In Young, E.J., and Corbel, M.J. (Editors), *Brucellosis: Clinical and Laboratory Aspects*, p. 53, CRC-Press, Florida, 1989.
6. Jumas-Bilak, E. et al., Unconventional genomic organization in the alpha subgroup of the *Proteobacteria*, *J. Bacteriol.*, 180, 2749, 1998.
7. Moreno, E. et al., *Brucella abortus* 16S rRNA and lipid A reveal a phylogenetic relationship with members of the alpha-2 subdivision of the class *Proteobacteria*, *J. Bacteriol.*, 172, 3569, 1990.
8. Al Dahouk, S. et al., Laboratory-based diagnosis of brucellosis—a review of the literature. Part I: Techniques for direct detection and identification of *Brucella* spp., *Clin. Lab.*, 49, 487, 2003.
9. Alton, G.G. et al., *Techniques for the Brucellosis Laboratory*, Institut National de la Recherche Agronomique, Paris, 1988.
10. Corbel, M.J., and Brinley-Morgan, W.J., Genus *Brucella*. In Krieg, N.R., and Holt, J.G. (Editors), *Bergey's Manual of Systematic Bacteriology*, vol. 1, p. 370, Williams & Wilkins, Baltimore, 1984.
11. Cloeckaert, A. et al., Classification of *Brucella* spp. isolated from marine mammals by DNA polymorphism at the *omp*2 locus, *Microbes Infect.*, 3, 729, 2001.
12. Foster, G. et al., *Brucella ceti* sp. nov. and *Brucella pinnipedialis* sp. nov. for *Brucella* strains with cetaceans and seals as their preferred hosts, *Int. J. Syst. Evol. Microbiol.*, 57, 2688, 2007.
13. Hubalek, Z. et al., Brucellosis of the common vole (*Microtus arvalis*), *Vector Borne Zoonotic Dis.*, 7, 679, 2007.
14. Scholz, H.C. et al., *Brucella microti* sp. nov., isolated from the common vole *Microtus arvalis*, *Int. J. Syst. Evol. Microbiol.*, 58, 375, 2008.
15. Scholz, H.C. et al., Isolation of *Brucella microti* from mandibular lymph nodes of red foxes, *Vulpes vulpes*, in Lower Austria, *Vector Borne Zoonotic Dis.*, 9, 153, 2009.
16. Scholz, H.C. et al., Isolation of *Brucella microti* from soil, *Emerg. Infect. Dis.*, 14, 1316, 2008.
17. De, B.K. et al., Novel *Brucella* strain (BO1) associated with a prosthetic breast implant infection, *J. Clin. Microbiol.*, 46, 43, 2008.
18. Scholz, H.C. et al., *Brucella inopinata* sp. nov, isolated from a breast implant infection, *Int. J. Syst. Evol. Microbiol.*, 60, 801, 2010.

19. Gee, J.E. et al., Use of 16S rRNA gene sequencing for rapid confirmatory identification of *Brucella* isolates, *J. Clin. Microbiol.*, 42, 3649, 2004.
20. Scholz, H.C. et al., Genetic diversity and phylogenetic relationships of bacteria belonging to the *Ochrobactrum-Brucella* group by *recA* and 16S rRNA gene-based comparative sequence analysis, *Syst. Appl. Microbiol.*, 31, 1, 2008.
21. DelVecchio, V.G. et al., The genome sequence of the facultative intracellular pathogen *Brucella melitensis*, *Proc. Natl. Acad. Sci. USA*, 99, 443, 2002.
22. Paulsen, I.T. et al., The *Brucella suis* genome reveals fundamental similarities between animal and plant pathogens and symbionts, *Proc. Natl. Acad. Sci. USA*, 99, 13148, 2002.
23. Halling, S.M. et al., Completion of the genome sequence of *Brucella abortus* and comparison to the highly similar genomes of *Brucella melitensis* and *Brucella suis*, *J. Bacteriol.*, 187, 2715, 2005.
24. Whatmore, A.M., Perrett, L.L., and MacMillan, A.P., Characterisation of the genetic diversity of *Brucella* by multilocus sequencing, *BMC Microbiol.*, 7, 34, 2007.
25. Verger, J.M. et al., *Brucella*, a monospecific genus as shown by deoxyribonucleic acid hybridization, *Int. J. Syst. Bacteriol.*, 35, 292, 1985.
26. Osterman, B., and Moriyón, I., International Committee on Systematics of Prokaryotes Subcommittee on the Taxonomy of *Brucella*. Minutes of the meeting, 17 September 2003, Pamplona, Spain, *Int. J. Syst. Evol. Microbiol.*, 56, 1173, 2006.
27. Moreno, E., and Moriyón, I., *Brucella melitensis*: a nasty bug with hidden credentials for virulence, *Proc. Natl. Acad. Sci. USA*, 99, 1, 2002.
28. Lapaque, N. et al., *Brucella* lipopolysaccharide acts as a virulence factor, *Curr. Opin. Microbiol.*, 8, 60, 2005.
29. Porte, F. et al., Role of the *Brucella suis* lipopolysaccharide O antigen in phagosomal genesis and in inhibition of phagosome-lysosome fusion in murine macrophages, *Infect. Immun.*, 71, 1481, 2003.
30. Forestier, C. et al., *Brucella abortus* lipopolysaccharide in murine peritoneal macrophages acts as a down-regulator of T cell activation, *J. Immunol.*, 165, 5202, 2000.
31. Roop II, R.M. et al., Adaptation of the brucellae to their intracellular niche, *Mol. Microbiol.*, 52, 621, 2004.
32. Boschiroli, M.L. et al., The *Brucella suis VirB* operon is induced intracellularly in macrophages, *Proc. Natl. Acad. Sci. USA*, 99, 1544, 2002.
33. Celli, J., and Gorvel, J.P., Organelle robbery: *Brucella* interactions with the endoplasmic reticulum, *Curr. Opin. Microbiol.*, 7, 1, 2004.
34. Köhler, S. et al., The analysis of the intramacrophagic virulome of *Brucella suis* deciphers the environment encountered by the pathogen inside the macrophage host cell, *Proc. Natl. Acad. Sci. USA*, 99, 15711, 2002.
35. Köhler, S. et al., What is the nature of the replicative niche of a stealthy bug named *Brucella*? *Trends Microbiol.*, 11, 215, 2003.
36. Al Dahouk, S. et al., Proteomic analysis of *Brucella suis* under oxygen deficiency reveals flexibility in adaptive expression of various pathways, *Proteomics*, 9, 3011, 2009.
37. Al Dahouk, S. et al., Quantitative analysis of the intramacrophagic proteome of the pathogen *Brucella suis* reveals metabolic adaptation to the late stage of cellular infection, *Proteomics*, 8, 3862, 2008.
38. Miller, M.A., and Paige, J.C., Other food borne infections, *Vet. Clin. North Am. Food Anim. Pract.*, 14, 71, 1998.
39. Bercovich, Z., Maintenance of *Brucella abortus*-free herds: A review with emphasis on the epidemiology and the problems in diagnosing brucellosis in areas of low prevalence, *Vet. Quart.*, 20, 81, 1998.
40. Brew, S.D. et al., Human exposure to *Brucella* recovered from a sea mammal, *Vet. Rec.*, 144, 483, 1999.
41. McDonald, W.L. et al., Characterization of a *Brucella* sp. strain as marine-mammal type despite isolation from a patient with spinal osteomyelitis in New Zealand, *J. Clin. Microbiol.*, 44, 4363, 2006.
42. Sohn, A.H. et al., Human neurobrucellosis with intracerebral granuloma caused by a marine mammal *Brucella* sp., *Emerg. Infect. Dis.*, 9, 485, 2003.
43. Cooper, C.W., Risk factors in transmission of brucellosis from animals to humans in Saudi Arabia, *Trans. R. Soc. Trop. Med. Hyg.*, 86, 206, 1992.
44. Yagupsky, P., and Baron, E.J., Laboratory exposures to brucellae and implications for bioterrorism, *Emerg. Infect. Dis.*, 8, 1180, 2005.
45. Saltoglu, N. et al., Fever of unknown origin in Turkey: Evaluation of 87 cases during a nine-year-period of study, *J. Infect.*, 48, 81, 2004.
46. Young, E.J., An overview of human brucellosis, *Clin. Infect. Dis.*, 21, 283, 1995.
47. Colmenero, J.D. et al., Complications associated with *Brucella melitensis* infection: A study of 530 cases, *Medicine (Baltimore)*, 75, 195, 1996.
48. Gür, A. et al., Complications of brucellosis in different age groups: A study of 283 cases in southeastern Anatolia of Turkey, *Yonsei Med. J.*, 44, 33, 2003.
49. Peery, T.M., and Belter, L.F., Brucellosis and heart disease. II. Fatal brucellosis: A review of the literature and report of new cases, *Am. J. Pathol.*, 36, 673, 1960.
50. Food and Agriculture Organization-World Health Organization, Joint FAO/WHO Expert Committee on Brucellosis (fifth report), in *WHO Technical Report Series*, No. 464, p. 34, WHO, Geneva, 1971.
51. Food and Agriculture Organization-World Health Organization, Joint FAO/WHO Expert Committee on Brucellosis (sixth report), in *WHO Technical Report Series*, No. 740, p. 56, World Health Organization, Geneva, 1986.
52. Hall, W.H., Modern chemotherapy for brucellosis in humans, *Rev. Infect. Dis.*, 12, 1060, 1990.
53. Solera, J., Martinez-Alfaro, E., and Espinosa, A., Recognition and optimum treatment of brucellosis, *Drugs*, 53, 246, 1997.
54. Solera, J., Treatment of human brucellosis, *J. Med. Liban.*, 48, 255, 2000.
55. Corbel, M.J. et al., Treatment of human brucellosis. In Corbel, M.J., Elberg, S.S., and Cosivi, O. (Editors), *Brucellosis in Humans and Animals*, Chapter 5, World Health Organization, Geneva, 2006.
56. Ariza, J. et al., Comparative trial of rifampin-doxycycline versus tetracycline-streptomycin in the therapy of human brucellosis, *Antimicrob. Agents Chemother.*, 28, 548, 1985.
57. Ariza, J. et al., Treatment of human brucellosis with doxycycline plus rifampin or doxycycline plus streptomycin. A randomized, double-blind study, *Ann. Intern. Med.*, 117, 25, 1992.
58. Solera, J. et al., Doxycycline-rifampin versus doxycycline-streptomycin in treatment of human brucellosis due to *Brucella melitensis*. The GECMEI Group. Grupo de Estudio de Castilla-la Mancha de Enfermedades Infecciosas, *Antimicrob. Agents Chemother.*, 39, 2061, 1995.

59. Colmenero, J.D. et al., Comparative trial of doxycycline plus streptomycin versus doxycycline plus rifampin for the therapy of human brucellosis, *Chemotherapy*, 35, 146, 1989.
60. Solera, J. et al., Treatment of human brucellosis with doxycycline and gentamicin, *Antimicrob. Agents Chemother.*, 41, 80, 1997.
61. Godfroid, J., Brucellosis in wildlife, *Rev. Sci. Tech.*, 21, 277, 2002.
62. Godfroid, J., and Käsbohrer, A., Brucellosis in the European Union and Norway at the turn of the twenty-first century, *Vet. Microbiol.*, 90, 135, 2002.
63. Taleski, V. et al., An overview of the epidemiology and epizootology of brucellosis in selected countries of Central and Southeast Europe, *Vet. Microbiol.*, 90, 147, 2002.
64. European Food Safety Authority (EFSA), Scientific opinion of the panel on animal health and welfare (AHAW) on a request of the commission on porcine brucellosis (*Brucella suis*), *EFSA J.*, 1144, 1, 2009.
65. Paton, N.I. et al., Visceral abscesses due to *Brucella suis* infection in a retired pig farmer, *Clin. Infect. Dis.*, 32, 129, 2001.
66. Teyssou, R. et al., About a case of human brucellosis due to *Brucella suis* biovar 2, *Mèd. Mal. Infect.*, 19, 160, 1989.
67. Chomel, B.B. et al., Changing trends in the epidemiology of human brucellosis in California from 1973 to 1992: A shift toward foodborne transmission, *J. Infect. Dis.*, 170, 1216, 1994.
68. Wise, R.I., Brucellosis in the United States: past, present, and future, *JAMA*, 244, 2318, 1980.
69. Corbel, M.J., Brucellosis: an overview, *Emerg. Infect. Dis.*, 3, 213, 1997.
70. Pappas, G. et al., Medical Progress—Brucellosis, *N. Engl. J. Med.*, 352, 2325, 2005.
71. De Massis, F. et al., Correlation between animal and human brucellosis in Italy during the period 1997–2002, *Clin. Microbiol. Infect.*, 11, 632, 2005.
72. Fosgate, G.T. et al., Time-space clustering of human brucellosis, California, 1973–1992, *Emerg. Infect. Dis.*, 8, 672, 2002.
73. Al Dahouk, S. et al., Changing epidemiology of human brucellosis, Germany, 1962–2005, *Emerg. Infect. Dis.*, 13, 1895, 2007.
74. Eriksen, N. et al., Brucellosis in immigrants in Denmark, *Scand. J. Infect. Dis.*, 34, 540, 2002.
75. White, Jr., A.C., and Atmar, R.L., Infections in Hispanic immigrants, *Clin. Infect. Dis.*, 34, 1627, 2002.
76. Memish, Z.A., and Balkhy, H.H., Brucellosis and international travel, *J. Travel Med.*, 11, 49, 2004.
77. Troy, S.B., Rickman, L.S., and Davis, C.E., Brucellosis in San Diego. Epidemiology and species-related differences in acute clinical presentations, *Medicine*, 84, 174, 2005.
78. Gotuzzo, E. et al., An evaluation of diagnostic methods for brucellosis—the value of bone marrow culture, *J. Infect. Dis.*, 153, 122, 1986.
79. Franco, M.P. et al., Human brucellosis, *Lancet Infect. Dis.*, 7, 775, 2007.
80. Farrell, I.D., The development of a new selective medium for the isolation of *Brucella abortus* from contaminated sources, *Res. Vet. Sci.*, 16, 280, 1974.
81. Castañeda, M.R., A practical method for routine blood cultures in brucellosis, *Proc. Soc. Exp. Biol. Med.*, 64, 114, 1947.
82. Yagupsky, P., Detection of brucellae in blood cultures, *J. Clin. Microbiol.*, 37, 3437, 1999.
83. Bannatyne, R.M., Jackson, M.C., and Memish, Z., Rapid diagnosis of *Brucella* bacteremia by using the BACTEC 9240 system, *J. Clin. Microbiol.*, 35, 2673, 1997.
84. Mantur, B.G., and Mangalgi, S.S., Evaluation of conventional Castaneda and lysis centrifugation blood culture techniques for diagnosis of human brucellosis, *J. Clin. Microbiol.*, 42, 4327, 2004.
85. Barham, W.B. et al., Misidentification of *Brucella* species with use of rapid bacterial identification systems, *Clin. Infect. Dis.*, 17, 1068, 1993.
86. Ariza, J. et al., Relevance of in vitro antimicrobial susceptibility of *Brucella melitensis* to relapse rate in human brucellosis, *Antimicrob. Agents Chemother.* 30, 958, 1986.
87. Al Dahouk, S. et al., Failure of a short-term antibiotic therapy for human brucellosis using ciprofloxacin—a study on in vitro susceptibility of *Brucella* strains, *Chemotherapy*, 51, 352, 2005.
88. Reddin, J.L. et al., Significance of 7S and macroglobulin *Brucella* agglutinins in human brucellosis, *N. Engl. J. Med.*, 24, 1263, 1965.
89. Buchanan, T.M. et al., Brucellosis in the United States, 1960–1972: An abattoir-associated disease. II. Diagnostic aspects, *Medicine*, 53, 415, 1974.
90. Gazapo, E. et al., Changes in IgM and IgG antibody concentrations in brucellosis over time: Importance for diagnosis and follow-up, *J. Infect. Dis.*, 159, 219, 1989.
91. Araj, G.F., Human brucellosis: A classical infectious disease with persistent diagnostic challenges, *Clin. Lab. Sci.* 12, 207, 1999.
92. Wright, A.E., and Smith, F., On the application of the serum test to the differential diagnosis of typhoid fever and Malta fever, *Lancet*, 149, 656, 1897.
93. Al Dahouk, S. et al., Laboratory-based diagnosis of brucellosis—a review of the literature. Part II: Serological tests for brucellosis, *Clin. Lab.*, 49, 577, 2003.
94. Jumba, M.M., Mirza, N.B., and Mwaura, F.B., Agglutinins for brucellae antigens in blood sera of an urban and rural population in Kenya, *East Afr. Med. J.*, 73, 204, 1996.
95. Young, E.J., Human brucellosis, *Rev. Infect. Dis.* 5, 821, 1983.
96. Morgan, W.J.B. et al., The rose bengal plate agglutination test in the diagnosis of brucellosis, *Vet. Rec.*, 85, 636, 1969.
97. Patterson, J.M., Deyoe, B.L., and Stone, S.S., Identification of immunoglobulins associated with complement fixation, agglutination, and low pH buffered antigen tests for brucellosis, *Am. J. Vet. Res.*, 37, 319, 1976.
98. Rose, J.E., and Roepke, M.H., An acidified antigen for detection of nonspecific reactions in the plate-agglutination test for bovine brucellosis, *Am. J. Vet. Res.*, 18, 550, 1957.
99. Lucero, N.E., and Bolpe, J.E., Buffered plate antigen test as a screening test for diagnosis of human brucellosis, *J. Clin. Microbiol.*, 36, 1425, 1998.
100. Ruiz-Mesa, J.D. et al., Rose Bengal test: Diagnostic yield and use for the rapid diagnosis of human brucellosis in emergency departments in endemic areas, *Clin. Microbiol. Infect.*, 11, 221, 2005.
101. Caravano, R., Chabaud, F., and Oberti, J., Application of immunoenzymatic techniques for epidemiological surveys on brucellosis among human populations, *Ann. Inst. Pasteur Microbiol.*, 138, 79, 1987.
102. Magee, J.T., An enzyme-labeled immunosorbent assay for *Brucella abortus* antibodies, *J. Med. Microbiol.*, 13, 167, 1980.
103. Osoba, A.O. et al., Diagnostic value of *Brucella* ELISA IgG and IgM in bacteremic and nonbacteremic patients with brucellosis, *J. Chemother.*, 13, 54, 2001.
104. Cakan, G. et al., Assessment of diagnostic enzyme-linked immunosorbent assay kit and serological markers in human brucellosis, *Jpn. J. Infect. Dis.*, 61, 366, 2008.

105. Gómez, M.C., et al., Evaluation of seven tests for diagnosis of human brucellosis in an area where the disease is endemic, *Clin. Vaccine Immunol.*, 15, 1031, 2008.
106. Lucero, N.E. et al., Fluorescence polarization assay for diagnosis of human brucellosis, *J. Med. Microbiol.*, 52, 883, 2003.
107. Nielsen, K. et al., Fluorescence polarization assay for the diagnosis of bovine brucellosis: Adaptation to field use, *Vet. Microbiol.*, 80, 163, 2001.
108. Smits, H.L. et al., Development and evaluation of a rapid dipstick assay for serodiagnosis of acute human brucellosis, *J. Clin. Microbiol.*, 37, 4179, 1999.
109. Smits, H.L. et al., Immunochromatographic *Brucella*-specific immunoglobulin M and G lateral flow assays for rapid serodiagnosis of human brucellosis, *Clin. Diagn. Lab. Immunol.*, 10, 1141, 2003.
110. Bricker, B.J., PCR as a diagnostic tool for brucellosis, *Vet. Microbiol.*, 90, 435, 2002.
111. Colmenero, J.D. et al., Real-time polymerase chain reaction: a new powerful tool for the diagnosis of neurobrucellosis, *J. Neurol. Neurosurg. Psychiatry*, 76, 1025, 2005.
112. Debeaumont, C., Falconnet, P.A., and Maurin, M., Real-time PCR for detection of *Brucella* spp. DNA in human serum samples, *Eur. J. Clin. Microbiol. Infect. Dis.*, 24, 842, 2005.
113. Queipo-Ortuño, M.I. et al., Rapid diagnosis of human brucellosis by SYBR green I-based real-time PCR assay and melting curve analysis in serum samples, *Clin. Microbiol. Infect.*, 11, 713, 2005.
114. Queipo-Ortuño, M.I. et al., Rapid diagnosis of *Brucella* epididimo-orchitis by real-time polymerase chain reaction assay in urine samples, *J. Urol.*, 176, 2290, 2006.
115. Queipo-Ortuño, M.I. et al., Rapid differential diagnosis between extrapulmonary tuberculosis and focal complications of brucellosis using a multiplex real-time PCR assay, *PLoS One*, 4, e4526, 2009.
116. Mitka, S. et al., Evaluation of different PCR assays for early detection of acute and relapsing brucellosis in humans in comparison with conventional methods, *J. Clin. Microbiol.*, 45, 1211, 2007.
117. Zerva, L. et al., Serum is the preferred clinical specimen for diagnosis of human brucellosis by PCR, *J. Clin Microbiol.*, 39, 1661, 2001.
118. Herman, L., and De Ridder, H., Identification of *Brucella* spp. by using the polymerase chain reaction, *Appl. Environ. Microbiol.*, 6, 2099, 1992.
119. Romero, C. et al., Specific detection of *Brucella* DNA by PCR, *J. Clin. Microbiol.*, 3, 615, 1995.
120. Rijpens, N.P. et al., Direct detection of *Brucella* spp. in raw milk by PCR and reverse hybridization with 16S-23S rRNA spacer probes, *Appl. Environ. Microbiol.*, 5, 1683, 1996.
121. Leal-Klevezas, D.S. et al., Single-step PCR for detection of *Brucella* spp. from blood and milk of infected animals, *J. Clin. Microbiol.*, 12, 3087, 1995.
122. Ouahrani, S. et al., Identification and sequence analysis of IS6501, an insertion sequence in *Brucella* spp.: Relationship between genomic structure and the number of IS6501 copies, *J. Gen. Microbiol.*, 139, 3265, 1993.
123. Ouahrani-Bettache, S., Soubrier, M.P., and Liautard, J.P., IS6501-anchored PCR for detection and identification of *Brucella* species and strains, *J. Appl. Bacteriol.*, 81, 154, 1996.
124. Baily, G.G. et al., Detection of *Brucella melitensis* and *Brucella abortus* by DNA amplification, *J. Trop. Med. Hyg.*, 95, 271, 1992.
125. Bricker, B.J., and Halling, S.M., Differentiation of *Brucella abortus* bv. 1, 2, and 4, *Brucella melitensis*, *Brucella ovis*, and *Brucella suis* bv. 1 by PCR, *J. Clin. Microbiol.*, 11, 2660, 1994.
126. García-Yoldi, D. et al., Multiplex PCR assay for the identification of all *Brucella* species and the vaccine strains *Brucella abortus* S19 and RB51 and *Brucella melitensis* Rev1, *Clin. Chem.*, 52, 779, 2006.
127. López-Gõni, I. et al., Evaluation of a multiplex PCR assay (Bruce-ladder) for molecular typing of all *Brucella* species, including the vaccine strains, *J. Clin. Microbiol.* 46, 3484, 2008.
128. Huber, B. et al., Development of a PCR assay for typing and subtyping of *Brucella* species, *Int. J. Med. Microbiol.*, 299, 563, 2009.
129. Al Dahouk, S. et al., The detection of *Brucella* spp. using PCR-ELISA and real-time PCR assays, *Clin. Lab.*, 50, 387, 2004.
130. Al Dahouk, S. et al., Evaluation of genus-specific and species-specific real-time PCR assays for the identification of *Brucella* spp., *Clin. Chem. Lab. Med.*, 45, 1464, 2007.
131. Redkar, R. et al., Real-time detection of *Brucella abortus*, *Brucella melitensis* and *Brucella suis*, *Mol. Cell. Probes*, 15, 43, 2001.
132. Newby, D.T., Hadfield, T.L., and Roberto, F.F., Real-time PCR detection of *Brucella abortus*: A comparative study of SYBR green I, 5′-exonuclease, and hybridization probe assays, *Appl. Environ. Microbiol.*, 69, 4753, 2003.
133. Probert, W.S. et al., Real-time multiplex PCR assay for detection of *Brucella* spp., *B. abortus*, and *B. melitensis*, *J. Clin. Microbiol.*, 42, 1290, 2004.
134. Foster, J.T. et al., Real-time PCR assays of single-nucleotide polymorphisms defining the major *Brucella* clades, *J. Clin. Microbiol.*, 46, 296, 2008.
135. Queipo-Ortuño, M.I. et al., Preparation of bacterial DNA template by boiling and effect of immunoglobulin G as an inhibitor in real-time PCR for serum samples from patients with brucellosis, *Clin. Vaccine Immunol.*, 15, 293, 2008.
136. Bounaadja, L. et al., Real-time PCR for identification of *Brucella* spp.: A comparative study of IS*711*, *bcsp31* and *per* target genes, *Vet. Microbiol.*, 137, 156, 2009.
137. Hinic, V. et al., Novel identification and differentiation of *Brucella melitensis*, *B. abortus*, *B. suis*, *B. canis*, and *B. neotomae* suitable for both conventional and real-time PCR-systems, *J. Microbiol. Methods*, 75, 375, 2008.
138. Castaño, M.J., and Solera, J., Chronic brucellosis and persistence of *Brucella melitensis* DNA, *J. Clin. Microbiol.*, 47, 2084, 2009.
139. Queipo-Ortuño, M.I. et al., Usefulness of a quantitative real-time PCR assay using serum samples to discriminate between inactive, serologically positive and active human brucellosis, *Clin. Microbiol. Infect.*, 14, 1128, 2008.
140. Morata, P. et al., Posttreatment follow-up of brucellosis by PCR assay, *J. Clin. Microbiol.* 37, 4163, 1999.
141. Navarro, E. et al., Use of real-time quantitative polymerase chain reaction to monitor the evolution of *Brucella melitensis* DNA load during therapy and post-therapy follow-up in patients with brucellosis, *Clin. Infect. Dis.*, 42, 1266, 2006.
142. Vrioni, G. et al., An eternal microbe: *Brucella* DNA load persists for years after clinical cure, *Clin. Infect. Dis.*, 46, 131, 2008.
143. Maas, K.S. et al., Evaluation of brucellosis by PCR and persistence after treatment in patients returning to the hospital for follow-up, *Am. J. Trop. Med. Hyg.*, 76, 698, 2007.

144. van Belkum, A., High-throughput epidemiologic typing in clinical microbiology, *Clin. Microbiol. Infect.*, 9, 86, 2003.
145. Gee, J.E. et al., Use of 16S rRNA gene sequencing for rapid confirmatory identification of *Brucella* isolates, *J. Clin. Microbiol.*, 42, 3649, 2004.
146. Fekete, A. et al., Amplification fragment length polymorphism in *Brucella* strains by use of polymerase chain reaction with arbitrary primers, *J. Bacteriol.*, 23, 7778, 1992.
147. Mercier, E. et al., Polymorphism in *Brucella* strains detected by studying distribution of two short repetitive DNA elements, *J. Clin. Microbiol.*, 5, 1299, 1996.
148. Tscherneva, E. et al., Repetitive element sequence based polymerase chain reaction for typing of *Brucella* strains, *Vet. Microbiol.*, 51, 169, 1996.
149. Tscherneva, E. et al., Differentiation of *Brucella* species by random amplified polymorphic DNA analysis, *J. Appl. Microbiol.*, 88, 69, 2000.
150. Al Dahouk, S. et al., Identification of *Brucella* species and biotypes using polymerase chain reaction-restriction fragment length polymorphism (PCR-RFLP), *Crit. Rev. Microbiol.*, 31, 191, 2005.
151. Whatmore, A.M. et al., Use of amplified fragment length polymorphism to identify and type *Brucella* isolates of medical and veterinary interest, *J. Clin. Microbiol.*, 43, 761, 2005.
152. Marianelli, C. et al., Molecular characterization of the *rpoB* gene in *Brucella* species: New potential molecular markers for genotyping, *Microbes Infect.*, 8, 860, 2006.
153. Scott, J.C. et al., Multiplex assay based on single-nucleotide polymorphisms for rapid identification of *Brucella* isolates at the species level, *Appl. Environ. Microbiol.*, 73, 7331, 2007.
154. Whatmore, A.M., Perrett, L.L., and MacMillan, A.P., Characterisation of the genetic diversity of *Brucella* by multilocus sequencing, *BMC Microbiol.*, 7, 34, 2007.
155. Vergnaud, G., and Denoeud, F., Minisatellites: Mutability and genome architecture, *Genome Res.*, 10, 899, 2000.
156. Al Dahouk, S. et al., Evaluation of *Brucella* MLVA typing for human brucellosis, *J. Microbiol. Methods*, 69, 137, 2007.
157. Le Flèche, P. et al., Evaluation and selection of tandem repeat loci for a *Brucella* MLVA typing assay, *BMC Microbiol.*, 6, 9, 2006.
158. Whatmore, A.M. et al., Identification and characterization of variable-number of tandem-repeat markers for typing of *Brucella* spp., *J. Clin. Microbiol.*, 44, 1982, 2006.
159. Bricker, B.J., and Ewalt, D.R., Evaluation of the HOOF-print assay for typing *Brucella abortus* strains isolated from cattle in the United States: Results with four performance criteria, *BMC Microbiol.*, 5, 37, 2005.
160. Bricker, B.J., Ewalt, D.R., and Halling, S.M., *Brucella* "HOOF-prints": Strain typing by multi-locus analysis of variable number tandem repeats (VNTRs), *BMC Microbiol.*, 3, 15, 2003.
161. Tiller, R.V. et al., Comparison of two multiple-locus variable-number tandem-repeat analysis methods for molecular strain typing of human *Brucella melitensis* isolates from The Middle East, *J. Clin. Microbiol.*, 47, 2226, 2009.
162. Kattar, M.M. et al., Evaluation of a multilocus variable-number tandem-repeat analysis scheme for typing human *Brucella* isolates in a region of brucellosis endemicity, *J. Clin. Microbiol.*, 46, 3935, 2008.
163. Marianelli, C. et al., Molecular epidemiological characterization and antibiotic susceptibility of *Brucella* isolates from humans in Sicily, Italy, *J. Clin. Microbiol.*, 45, 2923, 2007.
164. Almendra, C. et al., "HOOF-Print" genotyping and haplotype inference discriminates among *Brucella* spp. isolates from a small spatial scale, *Infect. Genet. Evol.*, 9, 104, 2009.
165. Rees, R.K. et al., Single tube identification and strain typing of *Brucella melitensis* by multiplex PCR, *J. Microbiol. Methods*, 78, 66, 2009.
166. Smits, H.L. et al., MLVA genotyping of human *Brucella* isolates from Peru, *Trans. R. Soc. Trop. Med. Hyg.*, 103, 399, 2009.
167. Nöckler, K. et al., Molecular epidemiology of *Brucella* genotypes in patients at a major hospital in central Peru, *J. Clin. Microbiol.*, 47, 3147, 2009.
168. Marianelli, C. et al., Use of MLVA-16 to trace the source of a laboratory-acquired *Brucella* infection, *J. Hosp. Infect.*, 68, 274, 2008.
169. De Santis, R. et al., Lab on a chip genotyping for *Brucella* spp. based on 15-loci multi locus VNTR analysis, *BMC Microbiol.*, 9, 66, 2009.
170. Kattar, M.M. et al., Development and evaluation of real-time polymerase chain reaction assays on whole blood and paraffin-embedded tissues for rapid diagnosis of human brucellosis, *Diagn. Microbiol. Infect. Dis.*, 59, 23, 2007.
171. Queipo-Ortuno, M.I. et al., Comparison of seven commercial DNA extraction kits for the recovery of *Brucella* DNA from spiked human serum samples using real-time PCR, *Eur. J. Clin. Microbiol. Infect. Dis.*, 27, 109, 2008.
172. Christensen, D.R. et al., Detection of biological threat agents by real-time PCR: Comparison of assay performance on the R.A.P.I.D., the LightCycler, and the Smart Cycler platforms, *Clin. Chem.*, 52, 141, 2006.

55 *Ehrlichia*

Jere W. McBride, Juan P. Olano, and Nahed Ismail

CONTENTS

55.1 INTRODUCTION

55.1.1 Microbiology, Genetics, and Pathobiology

55.1.1.1 Taxonomy and Etiologic Agents

Many organisms in the order Rickettsiales, represented in the genera *Ehrlichia*, *Anaplasma*, *Cowdria*, and *Neorickettsia*, have recently been realigned based on genetic sequence information from two universally conserved gene sequences and the application of molecular phyologenetics. Previous criteria relied more heavily on morphologic attributes, cellular tropism, and antigenic similarity to define taxonomic position and classify various "rickettsia-like" organisms in the corresponding genera without the benefit of supporting genotypical information. A new taxonomy for the genus *Ehrlichia* was proposed in 2001 using data obtained from two highly conserved genes, *rrs* (16S ribosomal RNA genes) and *groESL*,[1] and completed genome sequences from *Ehrlichia* and *Anaplasma* spp. supported the taxonomic positions determined by individual gene sequences. The genus *Ehrlichia* is now part of a newly created family Anaplasmataceae, which also includes the genera *Anaplasma*, *Wolbachia*, and *Neorickettsia*, but it remains in the order Rickettsiales. The genus has five approved members (*E. canis*, *E. chaffeensis*, *E. muris*, *E. ruminantium*, and *E. ewingii*) that have at least 97.7% similarity in the 16S rRNA sequences, with the acquisition of one member from the genus *Cowdria* (*C. ruminantium*). Six previously recognized members have been reassigned to the genera, *Anaplasma* (*E. phagocytophila*, *E. equi*, *E. platys*, and *E. bovis*) and *Neorickettsia* (*E. sennetsu* and *E. risticii*). There are three recognized human pathogens in the genus, including *E. chaffeensis*, *E. ewingii*, and *E. canis* (Table 55.1), as well as a pathogen strictly of veterinary importance, *E. ruminantium*.

This chapter focuses on the ehrlichial agents associated with human disease; *E. chaffeensis*; the etiologic agent of human monocytotropic ehrichiosis (HME); *E. ewingii*, the agent of human ehrlichiosis ewingii (HEE); and *E. canis*, the primary etiologic agent of canine monocytic ehrichiosis (CME), and most recently associated with human infections in South America.[2]

E. chaffeensis. *E. chaffeensis* was isolated in 1991 and formally named after it caused the first human infection in 1986.[3,4] *E. chaffeensis* exhibits tropism for monocytes/macrophages and causes human monocytotropic ehrlichiosis (HME),

TABLE 55.1
***Ehrlichia* spp. That Cause Human Disease**

Agent	Target Cell	Natural Hosts	Disease
E. chaffeensis	Monocyte	White-tailed deer, canines and goats	Human monocytotropic ehrlichiosis (HME)
E. ewingii	Neutrophil	Canines and deer	Human ehrlichiosis ewingii (HEE)
E. canis	Monocyte	Canines	Canine monocytic ehrlichiosis (CME)

a life-threatening zoonosis that is also associated with mild to severe infections in dogs.[5,6] In humans, many (40%–60%) *E. chaffeensis* infections require hospitalization,[7,8] and there is a case fatality rate of 3% due to the difficulty in making an early and accurate diagnosis. *E. chaffeensis* is maintained in nature in a zoonotic cycle primarily in white-tailed deer, but dogs may also be a significant natural reservoir.[9] The primary vector is the lone star tick, *Amblyomma americanum*, which is distributed from west central Texas throughout the southeastern, south central and mid-Atlantic states.[9,10] *E. chaffeensis* DNA has been detected in other tick species, including *Dermacentor variabilis*, *Ripicephalus sanguineus*, *Ixodes pacificus*, *I. rincinus*, *A. testudinarium*, and *Haemaphysalis yeni*. Larval ticks become infected with *E. chaffeensis* after feeding on an infected vertebrate hosts and maintain a transstadial but not transovarial infection. The emergence of *E. chaffeensis* appears to coincide with changes in demographic and ecologic factors including increases in vector and mammalian host populations and human contact with natural foci, immunocompromised and aging human populations, and improved diagnosis and reporting.[9] *E. chaffeensis* can be cultivated in vitro in various mammalian and tick cell lines[11] but causes only transient subclinical infection in immunocompetent mice.[12]

E. ewingii. *E. ewingii* was described in 1971 as a pathogen of dogs[13] and has been associated commonly with two distinct clinical syndromes in canines, anemia, and polyarthritis.[14,15] *E. ewingii* is transmitted by the lone star tick, *A. americanum*, and exhibits host cell tropism for polymorphonuclear neutrophils.[16] The most common clinical features of *E. ewingii* infection in dogs are fever, lameness, and neurologic abnormalities.[17] *E. ewingii* as a human pathogen emerged in 1996, and four cases of human ehrlichosis ewingii (HEE) in immunocompromised patients were characterized between 1996 and 1998 that were clinically indistinguishable from disease caused by *E. chaffeensis*.[18] The natural host of *E. ewingii* is canines, and it appears to be an opportunistic pathogen in humans. Most cases of HEE are manifested in immunocompromised patients, such as those infected with the human immunodeficiency virus (HIV).[18,19] Many of the documented human cases reported contact with dogs before the onset of symptoms.[18] *E. chaffeensis* and *E. canis* appear to be more conserved antigenically and have substantial serologic cross reactivity, and *E. ewingii* appears to be more antigenically divergent. There are serologic cross-reactive antigens between *E. chaffeensis* and *E. ewingii*, associated with higher (>40 kDa) molecular weight proteins, but epitopes in the major outer membrane proteins (OMP-1s/p28s) appear to be antigenically distinct.[18,20] *E. ewingii* has not been cultivated in vitro; thus molecular characterization of this agent has been limited to a few genes that include the OMP-1/P28 gene families.[20]

55.1.1.2 Morphology

Ehrlichia species are gram-negative bacteria that stain dark blue with Romanovsky-type stains.[21] Ehrlichiae reside in an intracellular vacuole bounded by a host cell–derived membrane-forming inclusions (morulae) that contain variable numbers of bacteria.[9] By electron microscopy, individual bacteria within the vacuole appear coccoid to pleomorphic and vary in size from small (0.4 μm), medium (0.7 μm), large (1 μm), and occasionally very large (≤2 μm).[22,23] Two distinct morphological forms can be distinguished, dense-cored and reticulate cells, both of which have a gram-negative envelope, including a cytoplasmic membrane and an outer membrane (cell wall), separated by a periplasmic space. In contrast to free-living gram-negative bacteria, peptidoglycan and lipopolysaccharide have not been detected, which can be explained by the recent revelation that genes for the synthesis of these components are not present in the genome.[24] The dense-cored cells are usually smaller, and more consistently coccoid in appearance, with an electron-dense nucleoid that occupies most of the cytoplasm. By analogy to chlamydiae, these cells are also often referred to, particularly in the early literature, as elementary bodies. On the other hand, reticulate cells are larger, more pleomorphic, and have a dispersed filamentous nucleoid and ribosomes dispersed in the cytoplasm. Sometimes organisms of intermediate size and electron density are referred to as intermediate bodies.[22,23,25,26] Reticulate cells multiply by binary fission and intermediate bodies can also be observed dividing. Evidence suggests that dense-cored cells are the infectious form of the organism.[27,28] The morphology of *E. ruminantium* in experimentally infected ticks appears to be similar to that in mammalian cells, consisting of both dense-cored and reticulate forms.[29]

55.1.1.3 Genetic and Antigenic Features

The obligate intracellular lifestyle of *Ehrlichia* spp. has resulted in genomes that have evolved differently than free-living microbes. The genomes of *E. chaffeensis*, *E. canis*, and *E. ruminantium* are completed,[30–32] but the genome of *E. ewingii* has not yet been sequenced. Some of the common overall features of the completed genomes include a high degree of gene order, and all are relatively small (1.2, 1.3, and 1.5 Mbp, respectively) due to massive loss of biosynthetic genes through a process known as reductive evolution. Each genome has roughly 1000 protein-encoding genes and a consistently low average G + C content of only 27%–29%. These genomes have a substantially lower than average ratio of coding to noncoding sequence (62%–72% coding) and contain pseudogenes (17–32).[30,32] The genomes of these ehrlichiae exhibit a serine-threonine bias in proteins associated with host-pathogen interactions, proteins with tandem and/or ankyrin repeats, protein secretion systems, a large group of secreted proteins with transmembrane helicies suggestive of a complex membrane structure and paralogous protein families that may be involved in immune evasion.[30,32]

The evolution of *Ehrlichia* spp. as obligate intracellular organisms has resulted in loss of many metabolic pathway genes. Genes essential for the glycolytic pathway, ATP/ADP translocases are absent, thus they are unable to utilize glucose as a carbon or energy source. In contrast to *Rickettsia* spp. that are not confined to a vacuole, *Ehrlichia* spp. have complete pathways for the synthesis of purines and pyrimidines. Genes for enzymes involved in aerobic respiration, sets of genes involved in synthesis of lipids and phospholipids, and have tRNAs for all amino acids.[30,32] In addition, they lack genes for the biosynthesis of lipopolysaccharide and peptidoglycan.[24]

Molecular characterizations of some *Ehrlichia* spp. genes were independently identified and utilized for initial molecular characterizations and comparisons, including universally conserved genes, *rrs*, *rrl*, *rrf* (16S, 23S, and 5S rRNA, respectively),[33–35] *groEL*, and *groESL*,[36–38] quinolate synthetase A (*nadA*)[39] and citrate synthase (*gltA*).[40] Others involved in pathobiology have been examined for functional roles including thio-disulfide oxidoreductase (*dsb*),[41] ferric-ion binding protein (*fbp*),[42] a *virB* operon (type IV secretion machinery),[43] and two-component regulatory systems.[44,45]

The interaction of *Ehrlichia* with the host immune response has been examined using Western immunoblotting, and a small group of *E. chaffeensis*, *E. canis*, and *E. ewingii* proteins that are strongly recognized by antibody have been identified. Major immunoreactive *E. chaffeensis* proteins observed by immunoblotting include, 200-, 120-, 88-, 55-, 47-, 40-, 28-, and 23-kDa[46,47]; *E. canis*, 200-, 140-, 95-, 75-, 47-, 36-, 28-, and 19-kDa[48]; and *E. ruminantium*, 160-, 85-, 58-, 46-, 40-, 32-, and 21-kDa.[49] *E. ewingii* has not been cultured, and immunoreactive proteins are not well defined; however, sera from *E. ewingii* infected dogs consistently cross-react with high (>40 kDa) molecular mass *E. canis* and *E. chaffeensis* proteins but not with low molecular mass (<40 kDa) proteins including the OMP-1/P28/P30 paralogs/orthologs.[18,47]

The majority of the major immunoreactive proteins of *E. chaffeensis* and *E. canis* have been molecularly characterized.[50–60] Four of these contain tandem repeats (TR); one has ankyrin (Ank) repeats; and one represents a family of paralogous major outer membrane proteins. In *E. chaffeensis*, TRP120, TRP47, TRP32 (VLPT), Ank200, OMP-1/P28 (22 genes), and MAP2 are well defined, as well as the corresponding orthologs in *E. canis* [TRP140, TRP36, TRP19, Ank200, P30/P28s (25 genes), and MAP2, respectively]. Fewer of these orthologs have been molecularly identified and characterized in *E. ruminantium* [MAP1 (16 genes), MAP2, and mucin-like protein (clone hw26; TRP36/47 ortholog)],[49,61–63] but additional immunoreactive genes have been identified in *E. ruminantium*[63] that have not been described in *E. chaffeensis* or *E. canis*. Recently, the major outer membrane proteins representing the OMP-1 family have been characterized in *E. ewingii*.[20]

Many of the major immunoreactive *E. chaffeensis* and *E. canis* protein orthologs (Ank200s, TRP120/140, TRP47/36, OMP-1/P28/P30 paralogs) that have been identified and characterized, exhibit larger-than-predicted molecular mass, and three (TRP120/140, TRP47/36 and TRP32/19) have acidic serine-rich tandem repeats.[50,53,57,64] The TRP32 (VLPT) protein of *E. chaffeensis* also exhibits a mass slightly larger than predicted (25.8 kDa) and has characteristics (tandem repeats and serine-rich composition) found in other ehrlichial proteins.[54,64] Strain-dependent sequence polymorphisms have been demonstrated in the number (2–5) of serine/threonine-rich tandem repeats of TRP120[50,65,66] and TRP32 (3–6)[64] and in genes encoding the p28 multigene family.[67,68] The observed molecular masses of these proteins also vary depending on the number of repeat units (TRP120, 80–130 kDa; TRP32, 35–55 kDa).[64,66,69] Species-specific major antibody epitopes have been mapped in *E. chaffeensis* and *E. canis* the TRPs and Ank200s.[53–55,58,70]

55.1.1.4 Pathobiology

Ehrlichial infection is initiated by the binding E and L selectin[71] and internalization of dense-cored ehrlichiae to mononuclear phagocytes.[28] Entry involves transglutaminase activation, PLC-γ2 phosphorylation, tyrosine kinases, and calcium signaling,[72] suggesting that a receptor-mediated entry mechanism is involved. Caveolin-1 co-localizes with early and replicative inclusions and disruption of caveolae prevent *E. chaffeensis* internalization, indicating that the entry route involves caveolae.[73] Furthermore, removal of glycosylphosphatidylinositol-anchored proteins also reduces binding and internalization of *E. chaffeensis*, suggesting that the GPI-anchored proteins are receptors for the organism.[73] Following endocytosis, tyrosine phosphorylated proteins and PLC-γ2 co-localize with *E. chaffeensis* inclusions,[72,73] suggesting a role in bacterial survival and proliferation in the macrophage. *E. chaffeensis* inclusions have several early endosomal markers including an early endosomal antigen 1 (EEA1) and rab5, and also co-localize with vesicle-associated membrane protein 2 (VAMP2), major histocompatibility complex (MHC) class I and MHC class II molecules, β_2-microglobulin, and transferrin receptor.[74–76] After entry, *E. chaffeensis* do not fuse with lyosomes, and this inhibition appears to be related to the activity of ehrlichial two-component regulatory systems.[44,45] Host cell gene transcription is modulated during infection, affecting expression of various cytokines, Toll-like receptors (TLR), transcription factors, apoptosis inhibitors, cell cyclins, membrane trafficking proteins, and transferrin receptor gene expression.[75,77,78] The inhibition of p38 MAPK by *E. chaffeensis* has been linked to the downregulation of transcription factor PU.1, which can also regulate the expression of TLRs and other genes.[78] *E. chaffeensis* also blocks IFN-γ induced bacteriocidal mechanisms by inhibiting tyrosine phosphorylation of Janus kinase (Jak) and signal transducer and activator of transcription (Stat) signaling.[79] *E. chaffeensis* has eight genes encoding type IV secretion machinery that may be associated with intracellular survival and replication by delivering effector macromolecules to the host cell and are transcribed in the blood of acutely ill patients.[43] Some of these genes (*virB3*, *virB4*, and *virB6*) are downstream of superoxide dismutase B (*sodB*) and are co-transcribed with *sodB* through a *sodB* promoter.[43] TRP47 is an effector protein that is secreted by an unknown mechanism but interacts with multiple host cell targets involved in various host cell processes.[80] In addition, Ank200 is a nuclear translocated protein that interacts with Alu elements in a subset of host genes.[81] Iron plays a critical role in the survival of *E. chaffeensis* in the host cell,[75,82] and iron-acquisition mechanisms in *E. chaffeensis* appear to be evolutionarily divergent from those of other gram-negative bacteria.[42] A structurally and functionally conserved ferric-ion-binding protein (Fbp) has been identified in *E. chaffeensis* and *E. canis* that binds Fe (III), but not Fe (II), and Fbp is preferentially expressed on the surface of dense-cored ehrlichiae and is secreted extracellularly.[42] The role of the p28 multigene family in the pathogenesis of HME is unknown, but differential expression of p28 genes appears to occur in mammalian and tick hosts.[83–85] *E. chaffeensis* expresses 16 OMP-1/*p28* genes concurrently in vitro and in vivo.[83,86]

55.1.2 Epidemiology, Clinical Features, and Pathogenesis

55.1.2.1 Epidemiology and Public Health

All human ehrlichioses are nationally notifiable diseases. Health care providers notify local health departments, which in turn relay the information to state health departments. CDC notification is done through the National Electronic Telecommunications System for Surveillance (NETSS). HME and HEE are both tick-borne zoonotic infections maintained in nature by a cycle involving ticks and a mammalian host. Ehrlichiae are maintained in ticks transtadially, and no transovarial transmission has been documented. Humans are accidental hosts that play no role in maintaining Ehrlichiae in nature. Factors related to the emergence of HME include increased density of *A. americanum* ticks[87]; expanded vector geographic distribution[88] and vertebrate host populations of the tick vector *A. americanum*[88]; increase in reservoir host populations for *E. chaffeensis*[89]; increased human contact with natural foci of infection through recreational and occupational activities[90,91]; increased size, longevity, and immunocompromised status of the human population[19,92]; the availability of diagnostic reagents; and improved surveillance.[93,94] The most important factor in the emergence of *E. ewingii* appears to be increased immunocompromised populations because the infection has been observed primarily in HIV-infected individuals and patients on immunosuppressive therapies.[18,19]

55.1.2.2 Human Monocytotropic Ehrlichiosis

As of 2005, CDC had reported a total of 2396 cases (passive surveillance) of HME since the initial case in 1986. However, both incidence and prevalence are most likely underestimated since active surveillance studies performed in HME endemic areas in Missouri, Tennessee, and Georgia have revealed an incidence that is 10–100 times higher than the one reported by passive surveillance.[95] HME is a seasonal disease, with most reported cases occurring in the spring and summer coinciding with higher tick activity, although cases can occur in the fall in more southern latitudes. The geographic distribution of HME follows the distribution of its vector *A. americanum* (lone star tick), that begins in west central Texas and moves east along the Gulf Coast, north through Oklahoma and Missouri, eastward to the Atlantic coast, and proceeds northeast through New Jersey, encompassing all the south central, southeastern and mid-Atlantic states. The main zoonotic reservoir is the white-tail deer (*Odocoileus virginianus*), but other potentially important reservoirs are naturally infected with *E. chaffeensis* including goats, domestic dogs, and coyotes.[6,96,97] The states with highest incidence include Arkansas, North Carolina, Missouri, Oklahoma, and New Jersey.[93] Outside the United States, HME has been described in Cameroon, where confirmation of the diagnosis was based on PCR detection of *E. chaffeensis* DNA from ill patients and dogs.[98,99] *E. chaffeensis* has been found in 5%–15% of *A. americanum* ticks collected from endemic areas in the eastern United States,[100–102] and has been detected in ticks collected from at least 15 states. *E. chaffeensis* has not been isolated outside the United States, and the only proven tick vector is restricted to North America, but *E. chaffeensis* or a closely related *Ehrlichia* sp. has been detected in other tick species located in other regions of the world, including Russia, Korea, Thailand, and China.[103–107] Human infections with *E. chaffeensis* or antigenically related Ehrlichiae have been reported in Europe,[108] Asia,[109] South America,[110] and Africa.[111] Reports of HME from other countries are based on serological studies and therefore cannot be confirmed as HME cases.

55.1.2.3 Human Ehrlichiosis Ewingii

Ehrlichia ewingii was first described as the causative agent of canine granulocytic ehrlichiosis in 1971.[13] In humans, the first case was described in 1999 by PCR amplification of *E. ewingii* DNA from a blood sample from an ill patient.[18] The agent has not been cultivated, and therefore the diagnosis relies on nucleic acid amplification. To date, all cases have been described in immunocompromised patients, including those with HIV/AIDS.[18,19]

55.1.2.4 Clinical Features of Human Ehrlichosis

HME is more often diagnosed in male (>2:1) patients >40 years of age; the majority (>80%) report a tick bite,[8,95] and HME outbreaks are associated with recreational or occupational activities.[90,112] Tick attachment has to last for 24–48 h before disease can occur. HME presents as a more severe disease in patients older than 60 years of age and in immunocompromised patients, including HIV/AIDS patients in whom severe complications, such as adult respiratory distress syndrome, acute tubular necrosis, shock, and CNS involvement, can arise. Another risk factor for severe disease in

HIV/AIDS patients is sulfa intake, which has been reported to aggravate ehrlichiosis.

HME and HEE manifest as undifferentiated febrile illnesses 1–3 weeks after the bite of an infected tick. For HME, the most frequent clinical findings reported anytime during acute illness are fever, malaise, headache, dizziness, chills, and myalgias[7,8,95,113–116] (see Table 55.2 for a comprehensive description of clinical manifestations of HME). Patients with HEE present with a milder disease with few complications.[18] A vast majority of these patients are immunocompromised, further suggesting that *E. ewingii* is less pathogenic. Hematologic and biochemical abnormalities usually include leukopenia, thrombocytopenia, anemia, mildly elevated serum hepatic transaminase activities, and hyponatremia.[8,19,95] Lymphocytosis characterized by a predominance of $\gamma\delta$ T cells is often seen in patients during recovery.[117] A high proportion of immunocompetent (41%–62%) and immunocompromised patients (86%) require hospitalization,[8,19,95] and delays in antibiotic treatment are associated with more pulmonary complications, increased transfer to intensive care, and longer length of illness.[118] The case fatality rate for HME is estimated to be 3%,[93] and fatal disease is most often described in older patients and patients debilitated by underlying disease or immunodeficiency.[7,19] Immunocompromised patients (HIV-infected persons, transplant recipients, corticosteroid-treated patients) have a high risk of fatal infection associated with overwhelming infection not typically observed in immunocompetent patients.[19] No deaths have been reported as a result of infection with *E. ewingii*.[18,19]

55.1.2.5 Pathogenesis of Human Ehrlichiosis

The relatively low bacterial burden in the blood and tissues in nonimmunocompromised patients with HME suggests that the pathophysiology of ehrlichiosis may involve immunopathologic responses that are manifest as a toxic shock–like syndrome.[3,119] The first murine model of fatal human ehrlichiosis has been instrumental in understanding the mechanisms behind the toxic shock–like syndrome of severe and fatal human ehrlichiosis.[120] Mice inoculated with an ehrlichia (*Ixodes ovatus* ehrlichia; IOE) closely related to *E. chaffeensis* develop histopathology resembling that observed in HME patients, and a similar disease course is observed in the IOE murine model. Lethal infections with IOE are accompanied by extremely high levels of serum TNF-α,

TABLE 55.2
Clinical Manifestations in Published Series of HME Cases

Clinical Feature	Eng et al.	Fishbein et al.	Everett et al.	Schutze et al.[a]	Olano et al.	Olano et al.
Fever	100	97	100	100	95[b]	100
Malaise/weakness	61	84	30	NR	NR	38
Chills/Rigor	65	61	73	NR	NR	45
Cephalea	77	81	63	25	NR	72
Myalgia	53	68	43	58	NR	69
Arhtralgia	28	41	33	NR	NR	69
Anorexia	50	66	27	NR	NR	NR
Nausea	54	48	50	NR	NR	38
Vomiting	49	37	27	25	NR	38
Diarrhea	38	24	10	25	NR	10
Diaphoresis	21	53	NR	17	NR	NR
Abdominal pain	23	21	10	17	27	7
Cough	39	26	<10	17	NR	24
Pharyngitis	33	25	NR	17	NR	21
Rash	47	36	20	67	17	21
Confusion	29	20	20	8[c]	22	21
Photophobia	11	NR	13	NR	27	NR
Stupor	NR	NR	NR	NR	7	NR
Coma	NR	NR	NR	NR	2	NR
Seizures	NR	NR	NR	NR	2	NR
Lymphadenopathy	26	25	NR	NR	29	NR
Jaundice	15	NR	NR	NR	NR	NR
Dyspnea	23	NR	<10	NR	NR	NR
Hepatomegaly	15	NR	<10	25	2	NR
Splenomegaly	15	NR	NR	25	2	NR

[a] This series includes pediatric cases only (n = 12). Other reported signs and symptoms included puffy eyes, murmur, and dehydration.

[b] Malaise and headache usually accompanied with fever in this series.

[c] Described as irritable or combative instead of confusion.

Source: Modified from Olano, J.P., Human Ehrlichioses. In *Rickettsial Diseases* (Raoult, D., and Parola, P., eds.). Informa Healthcare, 2007.

a high frequency of TNF-α producing CD8+ splenic T cells, decreased *Ehrlichia*-specific CD4+ T lymphocyte proliferation, low IL-12 levels in the spleen, and a fortyfold decrease in the number of antigen specific IFN-γproducing CD4+ Th1 cells.[121] Furthermore, mice lacking TNF receptor I/II were more resistant to IOE-induced liver injury but exhibited higher bacterial burdens.[122]

55.1.3 Diagnosis of Human Ehrlichiosis

55.1.3.1 Conventional Techniques

HME is a serious and potentially fatal disease that is usually diagnosed empirically based on clinical findings and clues from medical history, such as history of tick bite and outdoor activities. However, specific laboratory testing is necessary to confirm the diagnosis.[123,124] Initial diagnosis of ehrlichiosis can be based on nonspecific biochemical and hematological findings.

Hematologic and Biochemical Abnormalities. Pancytopenias are a hallmark laboratory feature of HME early in the course of the illness. Anemia occurs within 2 weeks of illness and influences 50% of patients.[113] Mild to moderate leukopenia, and the largest decline in lymphocyte population is observed approximately in 60%–70% of patients during the first week of illness.[7,113] Atypical lymphocytes have been reported in patients with ehrlichiosis. Marked thrombocytopenia is also a common finding with HME, which is usually detected in 70%–90% of patients during their illness.[113,115] Mildly or moderately elevated hepatic transaminase levels are detected in ~90% of patients, associated with increased levels of alkaline phosphatase and bilirubin in some patients. Mild to moderate hyponatremia has been reported in as many as 50% of adult patients and 70% of pediatric patients.[115,116]

Other laboratory abnormalities that occur in severe disease are specific to the organ involved. Examples are increased serum creatinine, lactate dehydrogenase, creatine phosphokinase, amylase, and electrolyte abnormalities including hypocalcemia, hypomagnesemia, hypophosphatemia, prolonged prothrombin times, increased levels of fibrin degradation products, metabolic acidosis, profound hypotension, disseminated intravascular coagulopathy, hepatic and renal failure, adrenal insufficiency, and myocardial dysfunction.[7,113,115]

Serology. Serological testing by immunofluorescence (IFA) is at present considered to be the best choice for HME.[125] *E. ewingii* has not been cultivated in vitro, and serological diagnosis has been performed using *E. chaffeensis* antigens,[18] but the reliability of this procedure is not known. One of the major problems with IFA is cross-reactive antigens that are shared among ehrlichial pathogens as well as with conserved immunoreactive antigens shared by other bacteria. Molecular characterization and epitope mapping of major immunoreactive proteins of *E. chaffeensis* and *E. ewingii* have provided new opportunities to substantially improve the sensitivity and specificity of serologic tests for the HME and HEE. Peptides containing major antibody epitopes from *E. chaffeensis* TRP120 and TRP32 proteins can provide sensitive and specific serological diagnoses and can be utilized in solid phase assay formats allowing high-throughput screening.[54,70]

Paired sera collected during a 3- to 6-week interval represent the preferred specimens for serologic evaluation of HME. A single IgG antibody titer of at least 256; seroconversion from negative to positive antibody status (with a minimum titer of 64); and a fourfold rise in titer during convalescence is indicative of HME when acute- and convalescent-phase samples are compared.[94] Serological assays have limitations which, in the case of HME, can have a direct consequence on the outcome of the patient's illness due to delays in the administration of the appropriate therapy. Patients treated in the first week of illness, prior to the development of antibodies, have more favorable outcomes and less opportunity to develop severe disease manifestations.[7] Unfortunately, since most patients seek medical attention 4 days after onset of illness, a large majority (67%) of those with HME may be missed, since most patients do not have IgG antibodies during the first week, and IgM does not substantially improve clinical diagnostic sensitivity.[94] In addition, serologic testing has a high false-positive rate due to the presence of cross-reactive antibodies induced by cross-reactive antigens shared by different *Ehrlichia* species and *Anaplasma* (the causative agent of human anaplasmosis, previously called human granulocytic ehrlichiosis).[126] This serologic cross reactivity interferes with accurate diagnosis of the etiologic agent of the disease. Protein immunoblotting can be useful in classifying indeterminate cases,[18] but the procedure is time consuming and the lack of standardization remains a problem.

55.1.3.2 Molecular Techniques

Diagnosis of ehrlichial infection by PCR has become the test of choice for confirming serology indicating HME or other ehrlichial diseases due to its high specificity (60%–85%) and sensitivity (60%–85% for *E. chaffeensis*), as well as a rapid turnaround time. PCR detection is particularly important for detection of ehrlichial infection at early stages when antibody levels are very low or undetectable. PCR is also considered as the definitive diagnostic test for *E. ewingii* infection since the bacteria is uncultivable and serology is not specific, as mentioned above, although the sensitivity and specificity of this approach is unknown. Blood samples should be collected in EDTA or sodium citrate anticoagulants and obtained before or at the initiation of therapy to increase sensitivity. However, since doxycycline treatment is effective only at early stages of infection, treatment should start as soon as possible while waiting for laboratory result. PCR of CSF may be positive, but the sensitivity is lower than for whole blood, likely due to the significantly lower volume of infected cells. Several variations in PCR-based detection of single *Ehrlichia* spp., including nested PCR, reverse-transcriptase PCR, and real-time PCR and reverse transcription (RT–PCR), have been reported, targeting numerous genes including 16S rRNA, VLPT (TRP32), TRP120, dsb, and omp-1/p28.[127–134] Nested PCR, using the 16S rRNA gene, *rrs*, as the target, is the most widely used method and has the analytical sensitivity to detect low levels of circulating ehrlichial organisms.[18,19,94] Since ehrlichiosis can be caused by multiple agents, multiplex/multicolor methods of PCR testing to discriminate among *Ehrlichia* of medical and veterinary importance (*E. chaffeensis*, *E. ewingii*, and *E. canis*) in a single reaction have been developed with analytical sensitivity comparable to nested PCR, and this may translate to clinical improvement of diagnostic capability and improved surveillance to raise the level of diagnostic confirmation of HME and of ehrlichiosis caused by *E. ewingii*.[132,133] These new methods improve diagnostic capability and have excellent analytical sensitivity, but the clinical sensitivity and specificity of these assays has not been conclusively determined.

55.2 METHODS

55.2.1 Sample Preparation

PCR is useful and sensitive for detection of ehrlichial organisms in patients suspected of having active ehrlichiosis infections (within the first 2 weeks of infection), and prior to the administration of antibiotic therapy. Peripheral blood is the preferred sample for molecular detection of *Ehrlichia* spp. by PCR. Extraction of DNA from peripheral blood collected in EDTA is performed with the DNeasy kit (Qiagen) according to the manufacturer's protocol. DNA extractions should be performed in a physically separated laboratory.

55.2.2 Detection Procedures

The protocol outlined below provides step-by-step instructions for simultaneous real-time PCR detection of the *dsb* gene of *Ehrlichia* spp. (*E. chaffeensis*, *E. ewingii*, and *E. canis*) using a multicolor PCR assay according to the methods described by Doyle et al.[132] The *dsb* gene sequence is highly divergent among closely related genera, and is highly specific for the genus *Ehrlichia*. The protocol can be modified for detection of single or multiple *Ehrlichia* spp. by addition or subtraction of respective species-specific probes (Table 55.3).

Procedure

1. Prepare master mix according to the number of samples to be tested. Master mix consists of the following per reaction: 25 μL of iQ Supermix (supplied as 2× with 6 mM $MgCl_2$, BioRad cat. no. 170–8860); primers (*E. chaffeensis*, *E. ewingii*, and *E. canis* forward and reverse, 400 nM each); Taqman probes (1–3 probes [*E. chaffeensis*, *E. ewingii*, and *E. canis*]; 25 μM final concentration/probe); $MgCl_2$ (4 mM final concentration).
2. Aliquot into 0.2 mL thin-walled PCR tubes and add the following: DNA template (2 μL; 100–250 ng; add water for negative control), Nuclease-free water (to final volume of 50 μL).
3. Amplify samples with the following two-step thermal cycling protocol using an Eppendorf Realplex Mastercycler S: 1 cycle of 95°C for 5 min; 40 cycles of 95°C for 15 s, and 60°C for 1 min.

Note: A sample is considered positive if the fluorescence reaches a minimum threshold level above background fluorescence after base-line subtraction.

55.3 CONCLUSION AND FUTURE PERSPECTIVES

Ehrlichiosis is a tick-borne zoonosis caused by several *Ehrlichia* species, including *E. chaffeensis*, *E. ewingii*, and *E. canis*. Although serology remains the most sensitive diagnostic assay for human ehrlichioses, molecular detection of the organism is most sensitive during the earliest stages of infection, typically when patients present with symptoms and before antibodies are detectable.[94] It is therefore an indispensible diagnostic tool for early detection when therapeutic intervention is most effective, reducing severity and life-threatening complications.[135] New molecular detection technologies offer rapid species-specific detection of multiple ehrlichial agents capable of causing disease. The availability of genome sequence information has also enhanced the ability to rationally choose gene targets for molecularly based tests to detect genus and species-specific targets, such as *dsb*, which can be identified and utilized for future assay development. Increases in cases of human ehrlichiosis and recognition of multiple etiologic agents demonstrate a need for rapid multiplex diagnostic tests for detection of these

TABLE 55.3
Primers and Probes for Multicolor Real-Time PCR Detection of *Ehrlichia* spp.

Primers	Sequence (5′→3′)
E. chaffeensis, *E. ewingii*, and *E. canis*-forward primer	TTGCAAAATGATGTCTGAAGATATGAAACA
E. chaffeensis and *E. canis*-reverse primer	GCTGCTCCACCAATAAATGTATCYCCTA
E. ewingii-reverse primer	GCAGCTCCACCAATGAATGTATTTCCAA
Probes	
E. chaffeensis	Quasar 670-TGCTAGTGCTGCTTGAACAGCTTTCAGTGAT-BHQ-2
E. ewingii	TAMRA-AGCCAATGCTGCACGTACTGCTTTCAATGAT-BHQ-2
E. canis	FAM-AGCTAGTGCTGCTTGGGCAACTTTGAGTGAA-BHQ-1

Source: Doyle, C.K. et al., *J. Mol. Diagn.*, 10, 504, 2005.

agents. Further automation and standardization of molecular detection methods will improve diagnostic sensitivity and provide clinicians with the information necessary to implement appropriate and effective antimicrobial therapy in a timely manner.

REFERENCES

1. Dumler, J.S., et al., Reorganization of genera in the families *Rickettsiaceae* and *Anaplasmataceae* in the order *Rickettsiales*: unification of some species of *Ehrlichia* with *Anaplasma*, *Cowdria* with *Ehrlichia* and *Ehrlichia* with *Neorickettsia*, descriptions of six new species combinations and designation of *Ehrlichia equi* and "HGE agent" as subjective synonyms of *Ehrlichia phagocytophila*, *Int. J. Syst. Evol. Microbiol.*, 51, 2145, 2001.
2. Perez, M. et al., Human infection with *Ehrlichia canis* accompanied by clinical signs in Venezuela, *Ann. N. Y. Acad. Sci.*, 1078, 110, 2006.
3. Maeda, K. et al., Human infection with *Ehrlichia canis*, a leukocytic rickettsia, *N. Engl. J. Med.*, 316, 853, 1987.
4. Anderson, B.E. et al., Detection of the etiologic agent of human ehrlichiosis by polymerase chain reaction, *J. Clin. Microbiol.*, 30, 775, 1992.
5. Dawson, J.E., and Ewing, S.A., Susceptibility of dogs to infection with *Ehrlichia chaffeensis*, causative agent of human ehrlichiosis, *Am. J. Vet. Res.*, 53, 1322, 1992.
6. Breitschwerdt, E.B., Hegarty, B.C., and Hancock, S.I., Sequential evaluation of dogs naturally infected with *Ehrlichia canis*, *Ehrlichia chaffeensis*, *Ehrlichia equi*, *Ehrlichia ewingii*, or *Bartonella vinsonii*, *J. Clin. Microbiol.*, 36, 2645, 1998.
7. Eng, T.R. et al., Epidemiologic, clinical, and laboratory findings of human ehrlichiosis in the United States, 1988, *J. Am. Med. Assoc.*, 264, 2251, 1990.
8. Fishbein, D.B., Dawson, J.E., and Robinson, L.E., Human ehrlichiosis in the United States, 1985 to 1990, *Ann. Intern. Med.*, 120, 736, 1994.
9. Paddock, C.D., and Childs, J.E., *Ehrlichia chaffeensis*: A prototypical emerging pathogen, *Clin. Microbiol. Rev.*, 16, 37, 2003.
10. Childs, J.E., and Paddock, C.D., The ascendancy of *Amblyomma americanum* as a vector of pathogens affecting humans in the United States, *Annu. Rev. Entomol.*, 48, 307, 2003.
11. Dawson, J.E. et al., Isolation and characterization of an *Ehrlichia* sp. from a patient diagnosed with human ehrlichiosis, *J. Clin. Microbiol.*, 29, 2741, 1991.
12. Winslow, G.M. et al., Infection of the laboratory mouse with the intracellular pathogen *Ehrlichia chaffeensis*, *Infect. Immun.*, 66, 3892, 1998.
13. Ewing, S.A. et al., A new strain of *Ehrlichia canis*, *J. Am. Vet. Med. Assoc.*, 159, 1771, 1971.
14. Goldman, E.E. et al., Granulocytic ehrlichiosis in dogs from North Carolina and Virginia, *J Vet Intern Med*, 12, 61, 1998.
15. Cowell, R.L. et al., Ehrlichiosis and polyarthritis in three dogs, *J. Am. Vet. Med. Assoc.*, 192, 1093, 1988.
16. Anziani, O.S., Ewing, S.A., and Barker, R.W., Experimental transmission of a granulocytic form of the tribe Ehrlichieae by *Dermacentor variabilis* and *Amblyomma americanum* to dogs, *Am. J. Vet. Res.*, 51, 929, 1990.
17. Goodman, R.A. et al., Molecular identification of *Ehrlichia ewingii* infection in dogs: 15 cases (1997–2001), *J. Am. Vet. Med. Assoc.*, 222, 1102, 2003.
18. Buller, R.S. et al., *Ehrlichia ewingii*, a newly recognized agent of human ehrlichiosis, *N. Engl. J. Med.*, 341, 148, 1999.
19. Paddock, C.D. et al., Infections with *Ehrlichia chaffeensis* and *Ehrlichia ewingii* in persons coinfected with human immunodeficiency virus, *Clin. Infect. Dis.*, 33, 1586, 2001.
20. Zhang, C. et al., Identification of 19 polymorphic major outer membrane protein genes and their immunogenic peptides in *Ehrlichia ewingii* for use in a serodiagnostic assay, *Clin. Vaccine Immunol.*, 15, 402, 2008.
21. Rikihisa, Y., The tribe Ehrlichieae and ehrlichial diseases, *Clin. Microbiol. Rev.*, 4, 286, 1991.
22. Popov, V.L. et al., Ultrastructural differentiation of the genogroups in the genus *Ehrlichia*, *J. Med. Microbiol.*, 47, 235, 1998.
23. Popov, V.L. et al., Ultrastructural variation of cultured *Ehrlichia chaffeensis*, *J. Med. Microbiol.*, 43, 411, 1995.
24. Lin, M., and Rikihisa, Y., *Ehrlichia chaffeensis* and *Anaplasma phagocytophilum* lack genes for lipid A biosynthesis and incorporate cholesterol for their survival, *Infect. Immun.*, 71, 5324, 2003.
25. Pienaar, J.G., Electron microscopy of *Cowdria* (*Rickettsia*) *ruminantium* (Cowdry, 1926) in the endothelial cells of the vertebrate host, *Onderstepoort J. Vet. Res.*, 37, 67, 1970.
26. Prozesky, L., and du Plessis, J.L., Heartwater. The development and life cycle of *Cowdria ruminantium* in the vertebrate host, ticks and cultured endothelial cells, *Onderstepoort J. Vet. Res.*, 54, 193, 1987.
27. Prozesky, L., Bezuidenhout, J.D., and Paterson, C.L., Heartwater: An in vitro study of the ultrastructure of *Cowdria ruminatium*, *Onderstepoort J. Vet. Res.*, 53, 153, 1986.
28. Zhang, J.Z. et al., The developmental cycle of *Ehrlichia chaffeensis* in vertebrate cells, *Cell. Microbiol.*, 9, 610, 2007.
29. Kocan, K.M., and Bezuidenhout, J. D., Morphology and development of *Cowdria ruminantium* in *Amblyomma* ticks, *Onderstepoort J. Vet. Res.*, 54, 177, 1987.
30. Mavromatis, K. et al., The genome of the obligately intracellular bacterium *Ehrlichia canis* reveals themes of complex membrane structure and immune evasion strategies, *J. Bacteriol.*, 188, 4015, 2006.
31. Dunning Hotopp, J.C. et al., Comparative genomics of emerging human ehrlichiosis agents, *PLoS Genet.*, 2, e21, 2006.
32. Collins, N.E. et al., The genome of the heartwater agent *Ehrlichia ruminantium* contains multiple tandem repeats of actively variable copy number, *Proc. Natl. Acad. Sci. USA*, 102, 838, 2005.
33. Massung, R.F. et al., Characterization of the rRNA genes of *Ehrlichia chaffeensis* and *Anaplasma phagocytophila*, *DNA Cell Biol.*, 21, 587, 2002.
34. Anderson, B.E. et al., *Ehrlichia chaffeensis*, a new species associated with human ehrlichiosis, *J. Clin. Microbiol.*, 29, 2838, 1991.
35. Allsopp, M. et al., Different organisms associated with heartwater as shown by analysis of 16S ribosomal RNA gene sequences, *Vet. Parasitol.*, 71, 283, 1997.
36. Sumner, J.W., Nicholson, W.L., and Massung, R.F., PCR amplification and comparison of nucleotide sequences from the groESL heat shock operon of *Ehrlichia* species, *J. Clin. Microbiol.*, 35, 2087, 1997.
37. Sumner, J.W. et al., *Ehrlichia chaffeensis* expresses an immunoreactive protein homologous to the *Escherichia coli* GroEL protein, *Infect. Immun.*, 61, 3536, 1993.
38. Lally, N.C. et al., The *Cowdria ruminantium* groE operon, *Microbiology*, 141, 2091, 1995.
39. Yu, X.J., and Walker, D.H., Sequence and characterization of an *Ehrlichia chaffeensis* gene encoding 314 amino acids

highly homologous to the NAD A enzyme, *FEMS Microbiol. Lett.*, 154, 53, 1997.
40. Inokuma, H. et al., Citrate synthase gene sequence: A new tool for phylogenetic analysis and identification of *Ehrlichia*, *J. Clin. Microbiol.*, 39, 3031, 2001.
41. McBride, J.W. et al., Identification and functional analysis of an immunoreactive DsbA-like thio-disulfide oxidoreductase of *Ehrlichia* spp, *Infect. Immun.*, 70, 2700, 2002.
42. Doyle, C.K. et al., An immunoreactive 38-kilodalton protein of *Ehrlichia canis* shares structural homology and iron-binding capacity with the ferric ion-binding protein family, *Infect. Immun.*, 73, 62, 2005.
43. Ohashi, N. et al., Characterization and transcriptional analysis of gene clusters for a type IV secretion machinery in human granulocytic and monocytic ehrlichiosis agents, *Infect. Immun.*, 70, 2128, 2002.
44. Kumagai, Y. et al., Biochemical activities of three pairs of *Ehrlichia chaffeensis* two-component regulatory system proteins involved in inhibition of lysosomal fusion, *Infect. Immun.*, 74, 5014, 2006.
45. Cheng, Z. et al., Intra-leukocyte expression of two-component systems in *Ehrlichia chaffeensis* and *Anaplasma phagocytophilum* and effects of the histidine kinase inhibitor closantel, *Cell Microbiol.*, 8, 1241, 2006.
46. Chen, S.M. et al., Identification of the antigenic constituents of *Ehrlichia chaffeensis*, *Am. J. Trop. Med. Hyg.*, 50, 52, 1994.
47. Rikihisa, Y., Ewing, S.A., and Fox, J.C., Western immunoblot analysis of *Ehrlichia chaffeensis*, *E. canis*, or *E. ewingii* infections in dogs and humans, *J. Clin. Microbiol.*, 32, 2107, 1994.
48. McBride, J.W. et al., Kinetics of antibody response to *Ehrlichia canis* immunoreactive proteins, *Infect. Immun.*, 71, 2516, 2003.
49. Mahan, S.M. et al., Molecular cloning of a gene encoding the immunogenic 21 kDa protein of *Cowdria ruminantium*, *Microbiology*, 140, 2135, 1994.
50. Yu, X.J., Crocquet-Valdes, P., and Walker, D.H., Cloning and sequencing of the gene for a 120 kDa immunodominant protein of *Ehrlichia chaffeensis*, *Gene*, 184, 149, 1997.
51. Yu, X.J. et al., Molecular cloning and characterization of the 120-kilodalton protein gene of *Ehrlichia canis* and application of the recombinant 120-kilodalton protein for serodiagnosis of canine ehrlichiosis, *J. Clin. Microbiol.*, 38, 369, 2000.
52. Yu, X.J. et al., Characterization of the complete transcriptionally active *Ehrlichia chaffeensis* 28 kDa outer membrane protein multigene family, *Gene*, 248, 59, 2000.
53. Doyle, C.K. et al., Differentially expressed and secreted major immunoreactive protein orthologs of *Ehrlichia canis* and *E. chaffeensis* elicit early antibody responses to epitopes on glycosylated tandem repeats, *Infect. Immun.*, 74, 711, 2006.
54. Luo, T. et al., A variable-length PCR target protein of *Ehrlichia chaffeensis* contains major species-specific antibody epitopes in acidic serine-rich tandem repeats, *Infect. Immun.*, 76, 1572, 2008.
55. McBride, J.W. et al., Identification of a glycosylated *Ehrlichia canis* 19-kDa major immunoreactive protein with a species-specific serine-rich glycopeptide epitope, *Infect. Immun.*, 75, 74, 2006.
56. McBride, J.W. et al., Immunodiagnosis of *Ehrlichia canis* infection with recombinant proteins, *J. Clin. Microbiol.*, 39, 315, 2001.
57. McBride, J.W., Comer, J.E., and Walker, D.H., Novel immunoreactive glycoprotein orthologs of *Ehrlichia* spp., *Ann. N. Y. Acad. Sci.*, 990, 678, 2003.
58. Nethery, K.A. et al., *Ehrlichia canis* gp200 contains dominant species-specific antibody epitopes in terminal acidic domains, *Infect Immun.*, 75, 4900, 2007.
59. Ohashi, N. et al., Immunodominant major outer membrane proteins of *Ehrlichia chaffeensis* are encoded by a polymorphic multigene family, *Infect. Immun.*, 66, 132, 1998.
60. Ohashi, N. et al., Cloning and characterization of multigenes encoding the immunodominant 30-kilodalton major outer membrane proteins of *Ehrlichia canis* and application of the recombinant protein for serodiagnosis, *J. Clin. Microbiol.*, 36, 2671, 1998.
61. Sulsona, C.R., Mahan, S.M., and Barbet, A.F., The *map1* gene of *Cowdria ruminantium* is a member of a multigene family containing both conserved and variable genes, *Biochem. Biophys. Res. Commun.*, 257, 300, 1999.
62. Jongejan, F., and Thielemans, M.J., Identification of an immunodominant antigenically conserved 32-kilodalton protein from *Cowdria ruminantium*, *Infect. Immun.*, 57, 3243, 1989.
63. Barbet, A.F. et al., A subset of *Cowdria ruminantium* genes important for immune recognition and protection, *Gene*, 275, 287, 2001.
64. Sumner, J.W., Childs, J.E., and Paddock, C.D., Molecular cloning and characterization of the *Ehrlichia chaffeensis* variable-length PCR target: An antigen-expressing gene that exhibits interstrain variation, *J. Clin. Microbiol.*, 37, 1447, 1999.
65. Yabsley, M.J. et al., Molecular variation in the variable-length PCR target and 120-kilodalton antigen genes of *Ehrlichia chaffeensis* from white-tailed deer (*Odocoileus virginianus*), *J. Clin. Microbiol.*, 41, 5202, 2003.
66. Chen, S.M. et al., Genetic and antigenic diversity of *Ehrlichia chaffeensis*: Comparative analysis of a novel human strain from Oklahoma and previously isolated strains, *J. Infect. Dis.*, 175, 856, 1997.
67. Yu, X.J., McBride, J.W., and Walker, D.H., Genetic diversity of the 28-kilodalton outer membrane protein gene in human isolates of *Ehrlichia chaffeensis*, *J. Clin. Microbiol.*, 37, 1137, 1999.
68. Reddy, G.R., and Streck, C.P., Variability in the 28-kDa surface antigen protein multigene locus of isolates of the emerging disease agent *Ehrlichia chaffeensis* suggests that it plays a role in immune evasion, *Mol. Cell. Biol. Res. Commun.*, 1, 167, 2000; erratum, *Mol. Cell. Biol. Res. Commun.*, 3, 66, 2000.
69. Chen, S.M., Cullman, L.C., and Walker, D.H., Western immunoblotting analysis of the antibody responses of patients with human monocytotropic ehrlichiosis to different strains of *Ehrlichia chaffeensis* and *Ehrlichia canis*, *Clin. Diag. Lab. Immunol.*, 4, 731, 1997.
70. Luo, T., Zhang, X., and McBride, J.W., Major species-specific antibody epitopes of the *Ehrlichia chaffeensis* p120 and E. canis p140 orthologs in surface-exposed tandem repeat regions, *Clin. Vaccine Immunol.*, 16, 982, 2009.
71. Zhang, J.Z., McBride, J.W., and Yu, X.J., L-selectin and E-selectin expressed on monocytes mediating *Ehrlichia chaffeensis* attachment onto host cells, *FEMS Microbiol. Lett.*, 227, 303, 2003.
72. Lin, M., Zhu, M.X., and Rikihisa, Y., Rapid activation of protein tyrosine kinase and phospholipase C-gamma2 and increase in cytosolic free calcium are required by *Ehrlichia chaffeensis* for internalization and growth in THP-1 cells, *Infect. Immun.*, 70, 889, 2002.
73. Lin, M., and Rikihisa, Y., Obligatory intracellular parasitism by *Ehrlichia chaffeensis* and *Anaplasma phagocytophilum*

involves caveolae and glycosylphosphatidylinositol-anchored proteins, *Cell. Microbiol.*, 5, 809, 2003.
74. Barnewall, R.E., Rikihisa, Y., and Lee, E. H., *Ehrlichia chaffeensis* inclusions are early endosomes which selectively accumulate transferrin receptor, *Infect. Immun.*, 65, 1455, 1997.
75. Barnewall, R.E., Ohashi, N., and Rikihisa, Y., *Ehrlichia chaffeensis* and *E. sennetsu*, but not the human granulocytic erhlichiosis agent, colocalize with transferrin receptor and up-regulate transferrin receptor mRNA by activating iron-responsive protein 1, *Infect. Immun.*, 67, 2258, 1999.
76. Mott, J., Barnewall, R.E., and Rikihisa, Y., Human granulocytic ehrlichiosis agent and *Ehrlichia chaffeensis* reside in different cytoplasmic compartments in HL-60 cells, *Infect. Immun.*, 67, 1368, 1999.
77. Zhang, J.Z. et al., Survival strategy of obligately intracellular *Ehrlichia chaffeensis*: Novel modulation of immune response and host cell cycles, *Infect. Immun.*, 72, 498, 2004.
78. Lin, M., and Rikihisa, Y., *Ehrlichia chaffeensis* downregulates surface Toll-like receptors 2/4, CD14 and transcription factors PU.1 and inhibits lipopolysaccharide activation of NF-kappa B, ERK 1/2 and p38 MAPK in host monocytes, *Cell. Microbiol.*, 6, 175, 2004.
79. Lee, E.H., and Rikihisa, Y., Protein kinase A-mediated inhibition of gamma interferon-induced tyrosine phosphorylation of Janus kinases and latent cytoplasmic transcription factors in human monocytes by *Ehrlichia chaffeensis*, *Infect. Immun.*, 66, 2514, 1998.
80. Wakeel, A., Kuriakose, J.A., and McBride, J. W., An *Ehrlichia chaffeensis* tandem repeat protein interacts with multiple host targets involved in cell signaling, transcriptional regulation, and vesicle trafficking, *Infect. Immun.*, 77, 1734, 2009.
81. Zhu, B. et al., Nuclear translocated *Ehrlichia chaffeensis* ankyrin protein interacts with the mid A-stretch of host promoter and intronic *Alu* elements, *Infect. Immun.*, 2009.
82. Barnewall, R.E., and Rikihisa, Y., Abrogation of gamma interferon-induced inhibition of *Ehrlichia chaffeensis* infection in human monocytes with iron-transferrin, *Infect. Immun.*, 62, 4804, 1994.
83. Unver, A. et al., The omp-1 major outer membrane multigene family of *Ehrlichia chaffeensis* is differentially expressed in canine and tick hosts, *Infect. Immun.*, 70, 4701, 2002.
84. Singu, V. et al., *Ehrlichia chaffeensis* expresses macrophage- and tick cell-specific 28-kilodalton outer membrane proteins, *Infect. Immun.*, 73, 79, 2005.
85. Singu, V. et al., Unique macrophage and tick cell-specific protein expression from the p28/p30-outer membrane protein multigene locus in *Ehrlichia chaffeensis* and *Ehrlichia canis*, *Cell Microbiol.*, 8, 1475, 2006.
86. Long, S.W. et al., Antigenic variation of *Ehrlichia chaffeensis* resulting from differential expression of the 28-kilodalton protein gene family, *Infect. Immun.*, 70, 1824, 2002.
87. Ginsberg, H.S. et al., Increased population densities of *Amblyomma americanum* (Acari: Ixodidae) on Long Island, New York, *J. Parasitol.*, 77, 493, 1991.
88. Means, R.G., and White, D.J., New distribution records of *Amblyomma americanum* (Acari: Ixodidae) in New York State, *J. Vect. Ecol.*, 22, 133, 1997.
89. McCabe, T.R., and McCabe, R.E., Recounting whitetails past. In McShea, W.J., Underwood, H.B., and Rappole, J.H. (Editors), *The Science of Over Abundance: Deer Ecology and Population Management*, p. 11, Smithsonian Institution Press, Washington, DC, 1997.
90. Standaert, S.M. et al., Ehrlichiosis in a golf-oriented retirement community, *N. Engl. J. Med.*, 333, 420, 1995.
91. Tal, A., and Shannahan, D., Ehrlichiosis presenting as a life-threatening illness, *Am. J. Med.*, 98, 318, 1995.
92. Palella, F.J., Jr. et al., Declining morbidity and mortality among patients with advanced human immunodeficiency virus infection. HIV Outpatient Study Investigators, *N. Engl. J. Med.*, 338, 853, 1998.
93. McQuiston, J.H. et al., Human ehrlichioses in the United States, *Emerg. Infect. Dis.*, 5, 635, 1999.
94. Childs, J.E. et al., Outcome of diagnostic tests using samples from patients with culture-proven human monocytic ehrlichiosis: Implications for surveillance, *J. Clin. Microbiol.*, 37, 2997, 1999.
95. Olano, J.P. et al., Human monocytotropic ehrlichiosis, Missouri, *Emerg. Infect. Dis.*, 9, 1579, 2003.
96. Dugan, V.G. et al., Natural infection of domestic goats with *Ehrlichia chaffeensis*, *J. Clin. Microbiol.*, 38, 448, 2000.
97. Kocan, A.A. et al., Naturally occurring *Ehrlichia chaffeensis* infection in coyotes from Oklahoma, *Emerg. Infect. Dis.*, 6, 477, 2000.
98. Ndip, L.M. et al., Predominance of *Ehrlichia chaffeensis* in Rhipicephalus sanguineus ticks from kennel-confined dogs in Limbe, Cameroon, *Exp. Appl. Acarol.*, 50, 163, 2010.
99. Ndip, L.M. et al., Molecular and clinical evidence of *Ehrlichia chaffeensis* infection in Cameroonian patients with undifferentiated febrile illness, *Ann. Trop. Med. Parasit.*, 103, 719, 2009.
100. Ijdo, J.W. et al., Detection of *Ehrlichia chaffeensis* DNA in *Amblyomma americanum* ticks in Connecticut and Rhode Island, *J. Clin. Microbiol.*, 38, 4655, 2000.
101. Whitlock, J.E. et al., Prevalence of *Ehrlichia chaffeensis* (Rickettsiales: Rickettsiaceae) in *Amblyomma americanum* (Acari: Ixodidae) from the Georgia coast and barrier islands, *J. Med. Entomol.*, 37, 276, 2000.
102. Stromdahl, E.Y. et al., Prevalence of infection in ticks submitted to the human tick test kit program of the U.S. Army Center for Health Promotion and Preventive Medicine, *J. Med. Entomol.*, 38, 67, 2001.
103. Alekseev, A.N., et al. Evidence of ehrlichiosis agents found in ticks (Acari: Ixodidae) collected from migratory birds, *J. Med. Entomol.*, 38, 471, 2001.
104. Cao, W.C., et al. Identification of *Ehrlichia chaffeensis* by nested PCR in ticks from Southern China, *J. Clin. Microbiol.*, 38, 2778, 2000.
105. Ravyn, M.D. et al., Monocytic *Ehrlichia* in *Ixodes persulcatus* ticks from Perm, Russia, *Lancet*, 353, 722, 1999.
106. Kim, C.M. et al., Identification of *Ehrlichia chaffeensis*, *Anaplasma phagocytophilum*, and *A. bovis* in *Haemaphysalis longicornis* and *Ixodes persulcatus* ticks from Korea, *Vect. Borne Zoonotic Dis.*, 3, 17, 2003.
107. Parola, P. et al., Detection of *Ehrlichia* spp., *Anaplasma* spp., *Rickettsia* spp., and other eubacteria in ticks from the Thai-Myanmar border and Vietnam, *J. Clin. Microbiol.*, 41, 1600, 2003.
108. Nuti, M. et al., *Ehrlichia* infection in Italy, *Emerg. Infect. Dis.*, 4, 663, 1998.
109. Heppner, D.G. et al., Human ehrlichiosis in Thailand, *Lancet*, 350, 785, 1997.
110. Ripoll, C.M. et al., Evidence of rickettsial spotted fever and ehrlichial infections in a subtropical territory of Jujuy, Argentina, *Am. J. Trop. Med. Hyg.*, 61, 350, 1999.
111. Uhaa, I.J. et al., A case of human ehrlichiosis acquired in Mali: Clinical and laboratory findings, *Am. J. Trop. Med. Hyg.*, 46, 161, 1992.
112. Petersen, L.R. et al., An outbreak of ehrlichiosis in members of an Army Reserve unit exposed to ticks, *J. Infect. Dis.*, 159, 562, 1989.

113. Fishbein, D.B. et al., Human ehrlichiosis: Prospective active surveillance in febrile hospitalized patients, *J. Infect. Dis.*, 160, 803, 1989.
114. Everett, E.D. et al., Human ehrlichiosis in adults after tick exposure. Diagnosis using polymerase chain reaction, *Ann. Intern. Med.*, 120, 730, 1994.
115. Olano, J.P. et al., Clinical manifestations, epidemiology, and laboratory diagnosis of human monocytotropic ehrlichiosis in a commercial laboratory setting, *Clin. Diagn. Lab Immunol.*, 10, 891, 2003.
116. Schutze, G.E., and Jacobs, R.F., Human monocytic ehrlichiosis in children, *Pediatrics*, 100, E10, 1997.
117. Caldwell, C.W. et al., Lymphocytosis of gamma/delta T cells in human ehrlichiosis, *Am. J Clin. Pathol.*, 103, 761, 1995.
118. Hamburg, B.J. et al., The importance of early treatment with doxycycline in human ehrlichiosis, *Medicine (Baltimore)*, 87, 53, 2008.
119. Fichtenbaum, C.J., Peterson, L.R., and Weil, G.J., Ehrlichiosis presenting as a life-threatening illness with features of the toxic shock syndrome, *Am. J. Med.*, 95, 351, 1993.
120. Sotomayor, E. et al., Animal model of fatal human monocytotropic ehrlichiosis, *Am. J. Pathol.*, 158, 757, 2001.
121. Ismail, N. et al., Overproduction of TNF-alpha by CD8+ type 1 cells and down-regulation of IFN-gamma production by CD4+ Th1 cells contribute to toxic shock-like syndrome in an animal model of fatal monocytotropic ehrlichiosis, *J. Immunol.*, 172, 1786, 2004.
122. Ismail, N., Stevenson, H.L., and Walker, D.H., Role of tumor necrosis factor alpha (TNF-alpha) and interleukin-10 in the pathogenesis of severe murine monocytotropic ehrlichiosis: Increased resistance of TNF receptor p55- and p75-deficient mice to fatal ehrlichial infection, *Infect. Immun.*, 74, 1846, 2006.
123. Walker, D.H. et al., Ehrlichia chaffeensis: A prevalent, life-threatening, emerging pathogen, *Trans. Am. Clin. Climatol. Assoc.*, 115, 375, 2004.
124. Dumler, J.S. et al., Ehrlichioses in humans: Epidemiology, clinical presentation, diagnosis, and treatment, *Clin. Infect. Dis.*, 45, S45, 2007.
125. Walker, D.H., Diagnosing human ehrlichiosis: Current status and recommendations, *ASM News*, 66, 287, 2000.
126. Walker, D.H., Tick-transmitted infectious diseases in the United States, *Annu. Rev. Public Health*, 19, 237, 1998.
127. McBride, J.W. et al., PCR detection of acute *Ehrlichia canis* infection in dogs, *J. Vet. Diagn. Invest.*, 8, 441, 1996.
128. Sumner, J.W. et al., PCR amplification and phylogenetic analysis of groESL operon sequences from *Ehrlichia ewingii* and *Ehrlichia muris*, *J. Clin. Microbiol.*, 38, 2746, 2000.
129. Gusa, A.A. et al., Identification of a *p28* gene in *Ehrlichia ewingii*: Evaluation of gene for use as a target for a species-specific PCR diagnostic assay, *J. Clin. Microbiol.*, 39, 3871, 2001.
130. Stich, R.W. et al., Detection of *Ehrlichia canis* in canine carrier blood and in individual experimentally infected ticks with a p30-based PCR assay, *J. Clin. Microbiol.*, 40, 540, 2002.
131. Loftis, A.D., Massung, R.F., and Levin, M.L., Quantitative real-time PCR assay for detection of *Ehrlichia chaffeensis*, *J. Clin. Microbiol.*, 41, 3870, 2003.
132. Doyle, C.K. et al., Detection of medically important *Ehrlichia* spp. by quantitative multicolor Taqman real-time PCR of the *dsb* gene, *J. Mol. Diagn.*, 10, 504, 2005.
133. Sirigireddy, K.R., and Ganta, R.R., Multiplex detection of *Ehrlichia* and *Anaplasma* species pathogens in peripheral blood by real-time reverse transcriptase-polymerase chain reaction, *J. Mol. Diagn.*, 7, 308, 2005.
134. Felek, S. et al., Sensitive detection of *Ehrlichia chaffeensis* in cell culture, blood, and tick specimens by reverse transcription-PCR, *J. Clin. Microbiol.*, 39, 460, 2001.
135. Hamburg, B.J. et al., The importance of early treatment with doxycycline in human ehrlichiosis, *Medicine (Baltimore)*, 87, 53, 2008.

56 *Ochrobactrum*

Corinne Teyssier and Estelle Jumas-Bilak

CONTENT

56.1 INTRODUCTION

56.1.1 Classification, Morphology, and Biology

56.1.1.1 Classification

In 16S rRNA gene-based phylogeny, the genus *Ochrobactrum* belongs to the class *Alphaproteobacteria* and the family *Brucellaceae*, which also includes genera *Brucella*, *Mycoplana*,[1] *Pseudochrobactrum*,[2] *Daeguia*,[3] and *Paenochrobactrum*.[4] The genus *Ochrobactrum* was created in 1988 for the organisms formerly known as CDC group Vd.[5] *Ochrobactrum anthropi* was proposed as the type and sole species of the genus despite the genotypic and phenotypic heterogeneity of the strains included in the species. Further analyses led to the creation of a second species, *Ochrobactrum intermedium*,[6] with the reference strain of *O. anthropi* (LMG3301^T) transferred as the type strain of *O. intermedium*. Three other species, *Ochrobactrum tritici*, *Ochrobactrum grignonense*,[7] and *Ochrobactrum gallinifaecis*,[8] were then characterized mainly on the basis of 16S rRNA sequencing. Since 2005, the description of *Ochrobactrum* species underwent a tremendous change with the description of 10 novel species: *Ochrobactrum lupini*,[9] *Ochrobactrum oryzae*,[10] *Ochrobactrum cytisi*,[11] *Ochrobactrum pseudintermedium*,[12] *Ochrobactrum haematophilum*, *Ochrobactrum pseudogrignonense*,[13] *Ochrobactrum rhizosphaerae*, *Ochrobactrum thiophenivorans*,[14] *Ochrobactrum pituitosum*,[15] and finally *Ochrobactrum ciceri*.[16] Most of these species have been characterized on the basis of only one isolate.

56.1.1.2 Morphology

The *Ochrobactrum* strains are gram-negative short rods that are straight or slightly curved, with one end flame shaped. Cells are between approximately 0.6–1.2 and 2 μm in length. They are nonspore-forming, strictly aerobic, and nonfermentative. They display positive reaction to oxidase and do not produce indole. *Ochrobactrum* was first described as motile

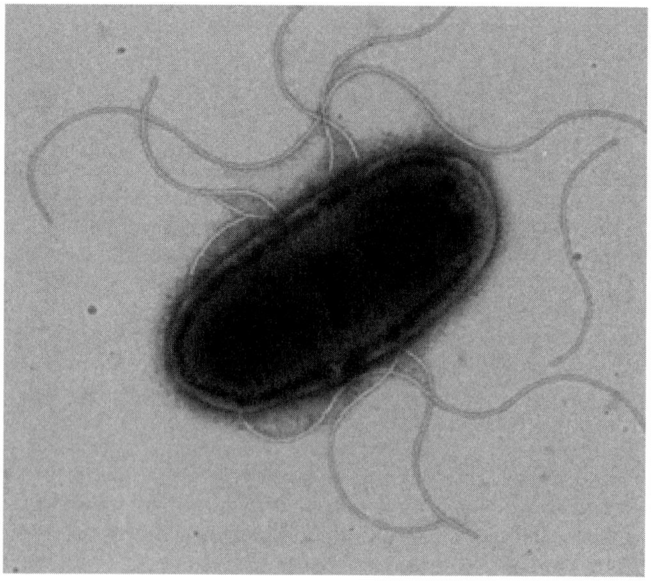

FIGURE 56.1 Electron microscopy of *Ochrobactrum pseudintermedium*, general morphology after negative staining.

bacteria,[5] but *O. gallinifaecis*, *O. lupini*, *O. haematophilum*, *O. pseudogrignonense*, and *O. thiophenivorans* cells were then found to be nonmotile. Concerning *O. rhizosphaerae*, the motility varies according to the growth curve since cells from the early exponential growth phase are strongly motile and become less motile or nonmotile at the stationary growth phase.[14] Electron microscopy after negative staining is an efficient method to show the presence of flagella, as presented in Figure 56.1 for *O. pseudintermedium*.

56.1.1.3 Genomic Features

The genomes of *O. anthropi*, *O. intermedium*, *O. tritici*, *O. grignonense*, and *O. gallinifaecis* have been shown to be complex with two independent circular chromosomes with two *rrn* in each chromosome.[17,18] The G + C content of the DNA is about 54.5–59 mol%.[5,12] The two chromosomes were frequently associated to circular or to linear plasmids and megaplasmids. The complete genomic sequence of *O. anthropi* ATCC49188^T and the unfinished sequence of *O. intermedium* LMG3301^T are available in the GenBank (www.ncbi.nlm.nih.gov). For *O. anthropi* chromosome 1, chromosome 2, plasmid 1, plasmid 2, plasmid 3, and plasmid 4, accession numbers are NC_009667 to NC_009672, respectively. For the *O. intermedium* unfinished sequence, accession numbers are NZ ACQA01000001, ACQA01000002, ACQA01000004, for chromosome 1, chromosome 2, and the plasmid, respectively.

Number and size of plasmids are highly variable among isolates of *O. anthropi* and *O. intermedium* leading to a large range of genome size even in a same species. For instance, the genome size of clinical isolates of *O. anthropi* varies from about 5 Mb to about 8 Mb.[18] A high level of genomic plasticity in a species has also been observed by restriction fragment length polymorphism (RFLP) in pulsed-field gel electrophoresis (PFGE),[19] amplified fragment length polymorphism (AFLP),[20] and PCR based on repeated sequences (REP–PCR).[7,21] Genomic fluidity related to a rapid microevolution has been proved for seven clonal clinical strains isolated from a same patient over a period of 1 year the strain encountered a large deletion of 150 kb including one *rrn* copy.[22] This suggested that the genomic polymorphism in *Ochrobactrum* could be related to the adaptation in various ecological niches and to the versatile lifestyles of these bacteria.

56.1.1.4 Ecological Niches and Lifestyles

Members of the genus *Ochrobactrum* are versatile bacteria isolated from a wide variety of ecological niches such as water, soil, plants, animals, and humans.[7,8,19,21,23,24] They display various lifestyles: free-living bacteria, members of the rhizosphere, nitrogen-fixing bacteria, xenobiotic-degrading bacteria, colonizers of nematods and insects, and opportunistic human pathogens. All these behaviors could be observed in the species *O. anthropi*.

They can establish with plants diverse relationships from members of the soil microbial communities to members of the rhizosphere, for instance the rhizosphere of wheat, potato and rice for *O. tritici*, *O. rhizosphaerae*, and *O. anthropi* respectively.[7,14,25] *Ochrobactrum* have been isolated from root nodules.[9-11,16,26] *O. lupini*, and *O. cytisi* establish a nitrogen-fixing symbiosis in the roots of *Lupinus honoratus* and *Cytisus scoparius*, respectively.[9,11]

Members of the genus *Ochrobactrum* also occupy polluted ecosystems,[14] such as agricultural soil, where *Ochrobactrum* are able to degrade organophosphorus and chlorinated pesticides; *s*-triazinic compounds, and atrazine.[27–29] Therefore, the use of *Ochrobactrum* may provide an efficient and economical method for pesticides bioremediation. *Ochrobactrum* is also a potential candidate for bioremediation of soils contaminated by xenobiotics such as nonylphenol,[30] methyl tert-butyl ether,[31] cyclic nitramine,[32] arsenic,[33] and chromate.[34] Some *Ochrobactrum* strains degrading polycyclic aromatic hydrocarbons are proposed for bioremediation in a crude oil contaminated environment.[35]

Members of the genus *Ochrobactrum* are also isolated from different animal samples. They are found in the digestive tract of invertebrates such as *Musca domestica*,[36] *Phlebotomus duboseqi*,[37] the termite *Zootermopsis angusticollis*,[38] the scorpion *Buthus martensii*,[39] and the screwworm *Cochliomyia macellaria*.[40] *O. anthropi* and *O. intermedium* were also recovered from the entomopathogenic nematodes belonging to the genus *Heterorhabditis*.[41] *Ochrobactrum* sp. and *O. gallinifaecis* type strain were collected from intestinal tract and feces of chicken.[8] Some *Ochrobactrum* strains seemed to be associated with pathologies in animals—gastritis of squirrel monkeys and spinal arthropathy in cane toads.[42,43]

56.1.2 Clinical Features and Antimicrobial Susceptibility

56.1.2.1 Clinical Features

O. anthropi and *O. intermedium* are the main *Ochrobactrum* species to be recovered from human clinical samples (respiratory tract, rectal swab, blood, gastric biopsy, etc.).[19,44–46]

Clinical isolates are present in monomicrobial or polymicrobial specimens and represent different relationships between human and bacteria: carriage, colonization without clinical symptoms, and infection. *O. anthropi* and *O. intermedium* are recognized as opportunistic human pathogens causing infections preferentially in hospitalized or immunocompromised patients.[44,45] Nosocomial infections due to *O. anthropi* have been increasingly reported during the last decade, particularly bacteremia and endocarditis in patients with indwelling central venous catheters[47–49] or others foreign bodies such as peritoneal dialysis catheter[50] or intraaortic body pump.[51] Other cases of infections and outbreaks have been described in patients in dialysis,[52] after surgery in ophthalmology,[53–56] in neurosurgery,[57] after transplantations[58] or after valve replacement.[59] Isolates of *O. anthropi* have been also recovered from the respiratory tract of a cystic fibrosis patient, but the clinical relevance of *O. anthropi* in cystic fibrosis should be investigated.[45] The virulence of *O. anthropi* is generally considered to be low, but reports suggested a high virulence for some strains involved in pyogenic monomicrobial infections,[60–62] life-threatening infections such as endocarditis,[59,63] and infections in immunocompetent hosts without underlying diseases.[63–65] The majority of these *Ochrobactrum* human infections have been imputed, to the species *O. anthropi*, except for 3 cases to *O. intermedium*: a liver abscess,[66] a pelvic abscess,[64] and a bacteremia in a patient with bladder cancer.[67] However, most of the clinical cases have been reported before the description of most of *Ochrobactrum* species. Three others *Ochrobactrum* species have been isolated from human samples: *O. pseudintermedium* (from rectal and axillary swabs), *O. pseudogrignonense* (from blood and ear swab), and *O. haematophilum* (from blood). A rapid survey of the sequences deposited in GenBank shows that only three strains of *O. pseudintermedium*, two strains of *O. pseudogrignonense*, and one strain of *O. haematophilum* were collected from human specimens. *Ochrobactrum* isolates are probably often misidentified due to the lack of cultural, biochemical, and physiological data defined among a large representative sampling, particularly for recently described species. Efficient identification procedures are necessary to evaluate the significance of each *Ochrobactrum* species in human infections as well as for epidemiological investigations.

56.1.2.2 Susceptibility to Antimicrobial Agents

Two methods are generally used to determine the antibiotic susceptibility profiles of *Ochrobactrum* strains: the disk diffusion assay or the minimum inhibitory concentration (MICs) assessment with Etest on Mueller Hinton agar. To date, all the *O. anthropi* clinical isolates tested using the disk diffusion assay have been highly resistant to all β-lactams, except imipenem. The disk diffusion assay showed no inhibition diameters for most of the β-lactams.[44] The MIC_{90} of 63 *O. anthropi* strains from clinical and environmental origin was higher than 256 µg/mL for 9 β-lactams except for imipenem, for which the MIC_{90} was equal to 4 µg/mL.[68] This resistance profile is consistent with the expression of the AmpC β-lactamase characterized in *O. anthropi*.[69–71] Identical susceptibility patterns were obtained for the strains of *O. intermedium* and *O. pseudintermedium* suggesting that a related β-lactamase or the same enzyme could be expressed in these three species.[12,44] In opposition with other members of the genus, the type strain *O. gallinifaecis* was sensible to all β-lactams except aztreonam. No results are available for *O. pseudogrignonense* and *O. haematophilum*. The results obtained using the Etest method are congruent with those obtained by the disk diffusion assay.[68] A general susceptibility to gentamicin, rifampicin, and fluoroquinolones was obtained using the disk diffusion assay. A constant susceptibility to trimethoprim-sulfamethoxazole was observed for the strains of *O. anthropi*, *O. intermedium*, and *O. tritici*, whereas the two strains of *O. grignonense* were resistant to this association. The *O. gallinifaecis* type strain was the sole strain found to be susceptible to chloramphenicol. Susceptibility to netilmicin and tobramycin varied according to the species. *O. anthropi* and *O. pseudintermedium* isolates were susceptible to netilmicin and tobramycin whereas *O. intermedium* strains resisted to these aminoglycosides.[12,44] In the paper describing *O. intermedium*, Velasco et al. showed that *O. intermedium* was resistant to colistin whereas *O. anthropi* were susceptible.[6] This characteristic was confirmed using the disk diffusion assay on a collection of 35 clinical strains,[44] and then on a larger scale since all *O. intermedium* clinical strains (n = 53) were resistant to colistin whereas only three *O. anthropi* clinical strains (n = 71) were resistant to this antibiotic (Teyssier et al., unpublished data). However, using Etest, Thoma et al. did not confirm the results obtained previously, since in a population of *O. anthropi* the MIC^{90} for colistin were >256 µg/mL.[68]

56.1.3 Diagnosis

56.1.3.1 Conventional Techniques

Bacteria of the genus *Ochrobactrum* are isolated from various human samples mainly, blood, expectoration, urine, and digestive samples. Phenotypic methods based on morphological, cultural and biochemical tests generally lead to a presumptive identification. The characteristics presented in this chapter correspond principally to those of the main *Ochrobactrum* species involved in human infections (i.e., *O. anthropi* and *O. intermedium*). For *O. pseudintermedium*, *O. haematophilum*, and *O. pseudogrignonense*, only a few clinical isolates are available; therefore, robust traits useful for the routine diagnosis cannot be proposed. However, when data are available, some characteristics will be given. The main differentiating characters between *Ochrobactrum* and related genera such as *Brucella* and *Pseudochrobactrum* will be added when necessary. The related genera *Mycoplana*, *Daeguia*, and *Paenochrobactrum* spp., never reported in humans, will not be considered in this chapter.

Morphology and Culture. *Ochrobactrum* cells are gram-negative, nonspore-forming, short, straight rods. *O. anthropi*, *O. intermedium*, and *O. pseudintermedium* cells appear to be motile under microscopic observation but *O. haematophilum* and *O. pseudogrignonense* do not.

Ochrobactrum spp. are strictly aerobic. *Ochrobactrum* isolated in clinical samples do not present exigent cultural requirements. Colonies appear after 24 h of incubation on tryptic-soy (TS) agar or nutrient agar. The optimal growth ranges from 30°C to 37°C. *O. intermedium* and *O. pseudintermedium* grow at 45°C on TS agar but *O. anthropi* does not. Colony size is about 1–2 mm in diameter. The brown pigment classically described for the genus *Ochrobactrum* is not visualized on TS agar and blood Columbia agar, even when a colony is harvested on a swab. Colony aspect at 37°C on TS agar clearly differs among species. *O. anthropi* develops circular, smooth, shining colonies whereas most of the colonies of *O. intermedium* and *O. pseudintermedium* are mucoid, opaque, and become quickly confluent. *O. haematophilum* and *O. pseudogrignonense* present beige, translucent, and shiny colonies. Hemolysis is never observed on blood Columbia agar. The colonies are lactose-negative for all species onto Drigalski and McConkey media. The positive culture on these two selective media allows to differentiate *Ochrobactrum* spp. from *Brucella* spp. Owing to their multiresistance to antibiotics, the isolation of *O. anthropi* and *O. intermedium* from polymicrobial samples can be performed using a selective medium such as McConkey agar or Drigalski agar supplemented with 4 mg ceftazidime/L. No growth was observed on cetrimide agar.

Biochemical Tests. *Ochrobactrum* displayed a positive reaction to oxidase test. Identification of the genus can be performed using commercially available biochemical identification systems such as API 20E and API 20NE systems (BioMérieux), or automated systems like VITEK 2 with ID-GNB card (BioMérieux) or BD Phoenix 100 system (Becton Dickinson). On the API 20NE and 20E strips, all the *Ochrobactrum* strains were negative for indole, H_2S, and acetoin production, carbohydrates fermentation and utilization of citrate. Moreover, arginine dihydrolase, lysine decarboxylase, ornithine decarboxylase, and gelatinase activities were not detected.[44,68] For the β-galactosidase and β-glucosidase activities, Thoma et al. showed that only five and four strains, respectively, among the 103 *Ochrobactrum* spp. tested present these enzymatic activities.[68] All the strains were positive for the presence of the cytochrome oxidase and for the assimilation of glucose, arabinose, mannose, *N*-acetylglucosamine, and malate.[44,68] 102 of the 103 *Ochrobactrum* strains tested by Thoma et al. were positive for assimilation of maltose.[68] Other characteristics varied among strains of a same species. However, the urease activity was a potential tool to distinguish *O. anthropi* and *O. intermedium* strains. Teyssier et al. showed that positive urease activity detected either by urea-indole medium (BioMérieux) or on API strips was observed for 20 out of the 21 *O. anthropi* clinical strains tested. In contrast, none of the *O. intermedium* nor the *O. pseudintermedium* strains displayed urease activity, whatever the method used to detect it.[12,44] This differential trait is confirmed on a larger collection since 69 out of 71 *O. anthropi* clinical strains present a urease activity whereas 51 out of 53 *O. intermedium* do not (Teyssier et al., unpublished data). The analyses performed by Thoma et al. using API 20NE strip showed that 23 of the 27 *O. anthropi* human strains of its collection presented an urease activity, whereas seven among the eight human *O. intermedium* isolates did not.[68] However, an inoculum heavier than that recommended by manufacturers led to positive reactions for most of the strains.[44] Thoma et al. showed a high variability of numerical profiles existing for *Ochrobactrum* using the API 20NE system. The mains numerical profiles (probably after 48 h incubation) for *O. anthropi* are 1247755, 1247775, and 1247777, whereas 1047755 is preferentially obtained for *O. intermedium* and *O. pseudintermedium* isolates.[68] In our experience, after 24 h incubation at 30°C, the main numerical profiles obtained for *O. anthropi* are 1243755 and 1247755, whereas 1043755 is preferentially obtained for *O. intermedium* and *O. pseudintermedium* isolates (Teyssier, unpublished data).

O. anthropi is the sole species of the genus *Ochrobactrum* included in the database of the API 20NE system. Therefore, the biochemical tests used in routine for *Ochrobactrum* identification are almost based on the characteristics of *O. anthropi*. Using the API 20NE system, nearly all the *Ochrobactrum* strains recovered from human samples are identified as *O. anthropi* after 24 h of incubation with the mentions "very good identification" or "excellent identification."[44] After 48 h, the same identification with the mentions "good identification" or "acceptable identification" was obtained. Therefore, reading the strips after 24 h gave the more confident identification. The absence of urease activity in *O. intermedium* and *O. pseudintermedium* strains did not alter the confidence level, since urease was presented as an inconstant trait for *O. anthropi* (84%) on the API 20NE system technical sheet. It is likely that the original calibration of the system used an *O. anthropi* population that was actually a mix of *O. anthropi* and *O. intermedium* strains. Finally, the API 20NE system gives satisfactory identification of the genus *Ochrobactrum* for the main species isolated from human beings.

The automated biochemical systems such as VITEK 2 with ID-GNB card or BD Phoenix 100 system allow *Ochrobactrum* spp. identification. For the VITEK 2, only a few of the 41 characteristics tested in the ID-GNB card were positive for *O. anthropi* or *O. intermedium*. All strains were positive for L-pyrrolidonyl arylamidase (PyrA) and negative for alkaline phosphatase (PHOS).[44] Further study showed that PyrA testing should be considered as a useful tool for the identification of nonfermentative gram-negative rods, and particularly to distinguish between *Ochrobactrum* and *Brucella*.[72] L-proline arylamidase, gamma glutamyl transferase, adonitol, L-lysine arylamidase, D-maltose, and palatinose were variable between strains without correlation with species identification.[44] No test appears to be species-specific. The VITEK 2 GNB card (bioMérieux) gave urease results that were generally consistent with the urea-indole test and the API systems. However, a few strains of *O. intermedium* present a urease activity. These discrepancies, as well as the inoculum effect underlined before, suggested that *O. intermedium* possessed a faint urease activity as previously suggested by Lebuhn et al.[7] This could also explain the discordances observed between urease results reported for *O. intermedium* in previous studies.[6,7,66]

Like for the API system, *O. anthropi* is the sole species included in the database of the GNB card. Since the VITEK 2 gives or suggests the identification of *O. anthropi* only for 25 strains among the 42 tested, this identification system appeared to be less powerful than the API 20NE system.[44] The results obtained with the BD Phoenix 100 systems are globally congruent with those obtained with the VITEK 2 but many misidentifications are noted, particularly with the genus *Pseudomonas*.[68] Sensitivity, specificity, and positive and negative predictive values of the Phoenix 100 system for the identification of *O. anthropi* ($n = 101$) were determined to be 84%, 32%, 67%, and 55%, respectively, whereas they were respectively 65%, 33%, 60%, and 37% for the API 20NE system.[68] Since *O. anthropi* is the only *Ochrobactrum* species included in the database of these identification systems, the BD Phoenix 100 system as well as VITEK 2 and API systems allow the identification at the genus level only. No study has been performed on *O. pseudintermedium*, *O. haematophilum*, and *O. pseudogrignonense* strains.

Identification in a Routine Practice of Medical Microbiology. In a routine practice, a presumptive diagnosis of *Ochrobactrum* genus must be considered for the isolation of a nonfastidious nonfermenting gram-negative short rod, which is oxidase-positive and resistant to all beta-lactams except for imipenem in disks diffusion assay. The bacteria belonging to the genus *Inquilinus* can also correspond to these characteristics.[73] In opposition to *Ochrobactrum*, *Inquilinus* strains do no grow on McConkey agar and display a remarkably huge inhibition diameter around an imipenem disk. Although *Brucella* spp. and *Ochrobactrum* spp. are genetically very closely related, they show clear differences in their phenotypes (i.e., most *Brucella* isolates are fastidious and slow-growing, gram-negative cocco-bacilli). However, misidentification could happen by using biochemical identification system such as API 20NE.[74] Particularly, *Brucella microti* was misidentified as *O. anthropi* on the basis of its rapid growth and biochemical traits.[75] Conventional methods do not permit to clearly differentiate between *Ochrobactrum* and *Pseudochrobactrum* genera, due in particular to the absence of robust differential characteristic between the two genera. In fact, each of the five *Pseudochrobactrum* species was described on the basis of only one isolate for each species.

Since *O. anthropi* is the only species included in the commercial identification systems' databases, the identification of *O. anthropi* ought to be considered genus identification. The use of these systems for species identification led generally to misidentification as *O. anthropi* of *O. intermedium* strains for example. This is the case for a report describing the involvement of an *O. anthropi* strain in a pelvic abscess due in fact to *O. intermedium*.[64] For routine identification at the species level, the urease test should be taken into account. Then, the aspect of the colonies, their growth at 45°C on TSA, and their susceptibility to tobramycin and netilmicin should be considered as additional characteristics allowing for the species identification. The susceptibility to colistin on disk diffusion assay is also recommended to distinguish between *O. anthropi* and *O. intermedium*. This routine protocol is efficient for *O. anthropi*, *O. intermedium*, and *O. pseudintermedium*. We cannot propose a robust methodology for the identification of *O. haematophilum* and *O. pseudogrignonense* due to the low number of isolates characterized for these species. Moreover, the antimicrobial susceptibility of these bacteria has never been analyzed.

56.1.3.2 Molecular Techniques

A diagnosis tool that permits accurate identification of *Ochrobactrum* clinical strains is essential for a better evaluation of the clinical and epidemiological significance of each *Ochrobactrum* species in human infections. Molecular methods based on genetic and genomic offer the opportunity to develop efficient diagnostic tools.

16S rRNA Gene Sequence Analysis. The 16S rRNA gene (16S rRNA) sequence analysis corresponds to a suitable tool for the affiliation of a bacterial strain to a species. Recently, a range of 98.7%–100% similarity level in the 16S rRNA gene sequence has been proposed for the bacterial species delineation.[76] The sequencing of 16S rRNA clearly differentiates members of the genus *Ochrobactrum* from related genera including *Pseudochrobactrum* spp., since the nucleotide sequence similarity was always below 95.5%, except for *Brucella* spp. and *Ochrobactrum* spp. which display a sequence similarity of over 97% for most of the strains.[2,44] In particular, *O. intermedium* shares 98.8% rRNA gene sequence similarity with *Brucella* spp.

The species of the genus *Ochrobactrum* could be discriminated by 16S rRNA sequencing except for *O. tritici*, *O. lupini*, and *O. cytisi*, which are nearly undistinguishable from *O. anthropi* (99.0%, 99.5%, and 100% identity in sequence with *O. anthropi* type strain, respectively).[7,9,11] The main *Ochrobactrum* species involved in clinical microbiology are discriminated by 16S rRNA sequencing. Similarity level between *O. anthropi* and *O. intermedium* ranged from 97.9% to 98.7% according to the strain.[44] However, identification could be impaired because many 16S rRNA sequences deposited in GenBank database correspond to misidentified strains. For example, 16S rRNA sequencing did not permit to identify the strain responsible for a published case of bacteremia since sequence similarities of 99% were obtained with both *O. intermedium* and *O. anthropi* sequences of misidentified strains.[67] Comparison should be done with type and reference strains and a particular care should be given to the results of the BLASTN program (www.ncbi.nlm.nih.gov/blast). The use of specialized databases such as Greengenes (http://greengenes.lbl.gov/) or RDPII (http://rdp.cme.msu.edu/index.jsp) should be recommended instead of GeneBank because the comparison with these databases gave the phylogenetic placement of the analyzed sequence. Alternatively, a phylogenetic tree including the type strains of the genus *Ochrobactrum* and the sequences for unknown strains should be reconstructed (Figure 56.2a). Moreover it is recommended to use complete 16S rRNA sequence instead of partial sequence. For instance, the complete 16S rRNA sequence permitted to affiliate to *O. intermedium* a clinical

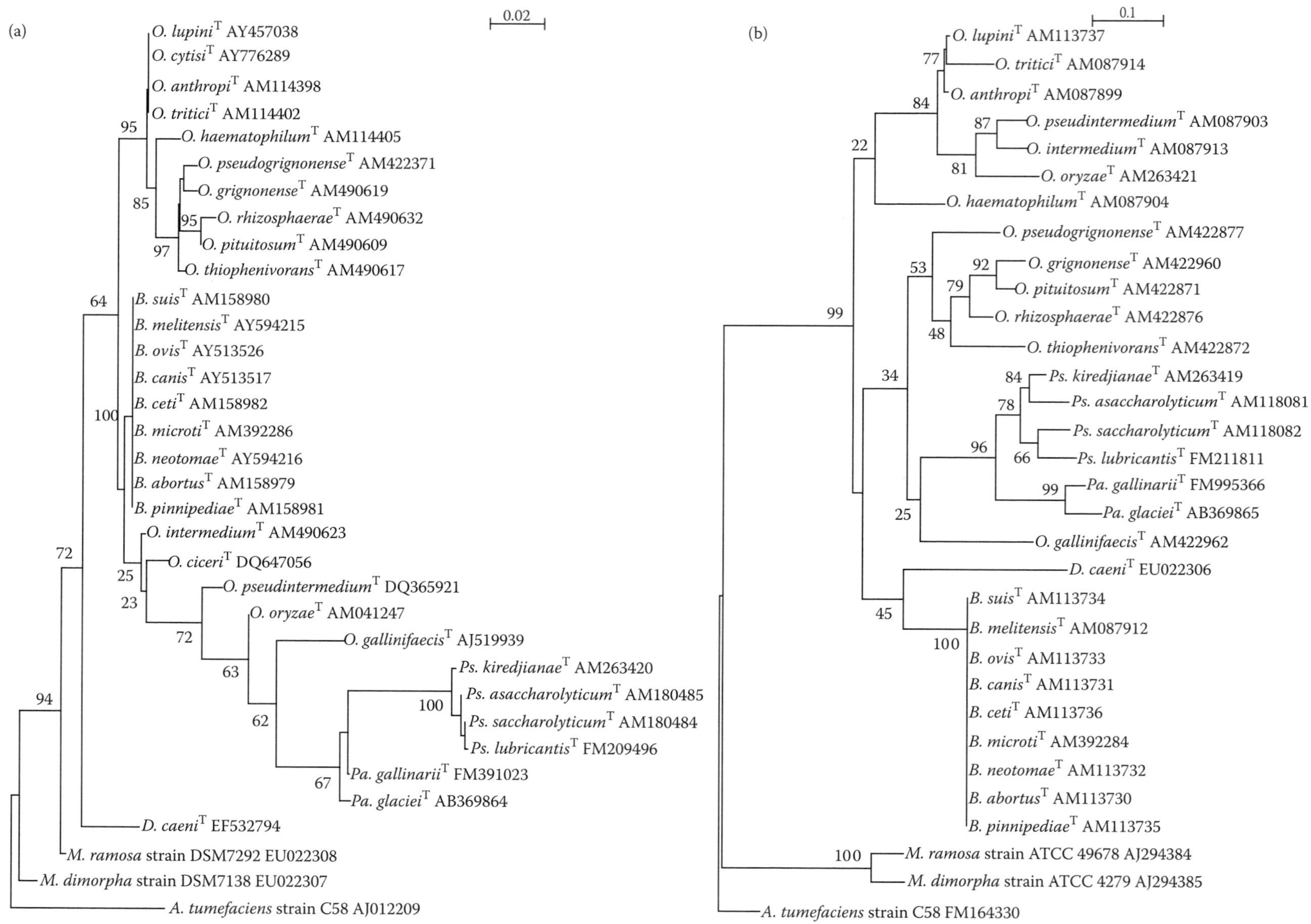

FIGURE 56.2 Phylogenetic structure of the genus *Ochrobactrum* among the *Brucellaceae* family. (a) The tree based on partial 16S rRNA gene sequences (1331 nt) obtained using maximum likelihood with GTR plus gamma distribution plus invariable sites as substitution model (phyML package). GenBank accession numbers are given with the species name. *Agrobacterium tumefaciens* was the outgroup organism. Numbers at nodes are bootstrap values based on 100 resamplings. Bar, 0.02 substitutions per site.

(b) The tree based on partial *recA* gene sequences (418 nt) obtained using maximum likelihood with GTR plus gamma distribution plus invariable sites as substitution model (phyML package). GenBank accession numbers are given with the species name. *Agrobacterium tumefaciens* was the outgroup organism. Numbers at nodes are bootstrap values based on 100 resamplings. Bar, 0.1 substitutions per site.

O., Ochrobactrum; B., Brucella; Ps. Pseudochrobactrum; P. Paenochrobactrum; D., Daeguia; M., Mycoplana; A., Agrobacterium.

strain previously identified as *O. anthropi* on the basis of a partial 16S rRNA sequence.[64,77] Kampfer et al. suggest that the BLASTN results should be carefully checked for their accuracy by the presence of *O. anthropi*- or *O. intermedium*-specific sequence motifs as indicated by Scholz et al.[77,78] The *O. intermedium*-specific sequence motifs are as follows: GGCTAAT and GGTTAGTGGAGACACTATCC, located at positions 487–493 and 905–924 of the 1390 bp 16S rRNA gene fragment. The latter motif is common with 16S rRNA sequence of *Brucella* spp. Instead of these motifs, *O. anthropi* presents GACTTTT and GGACACAGAGATGTGTCT at the same positions. A 16S rRNA amplification using primers F4 and R2 described as being specific for *Brucella* spp. and *O. intermedium* could allow differentiation of *O. anthropi* and *O. pseudintermedium* from *Brucella-O. intermedium*.[6,12,79] An insertion of 46 bp in position 187 was described to be present only in the 16S rRNA gene of *O. intermedium* strains. A PCR amplification anchored in this insertion could represent a supplementary tool contributing to *O. intermedium* identification.[22]

The identification to the species level in the genus *Ochrobactrum* by 16S rRNA sequencing could be questioned due to the low polymorphism of this gene in the *Brucellaceae*. Additional genetic markers should be analyzed.

Identification Based on Other Genetic Markers. The first candidate proposed for a better discrimination among *Ochrobactrum* species was the 16S/23S rRNA internal spacer region 1 (ITS).[80] However, the use of ITS for *Ochrobactrum* identification has not been evaluated. Then, the gene encoding the recombinase (*recA*) was used.[78,81] The *recA* gene was described to be useful for inter- and intraspecies-level differentiation of several bacteria, such as *Burkholderia stabilis*[82] and *Vibrio* spp.[83] Although no amplification was obtained for *O. grignonense*, *O. gallinifaecis*, and *O. pseudogrignonense*, *recA* gene was described to be more efficient than the 16S rRNA gene analysis.[13,78,81] Recently, the use of degenerated primers permits henceforth to amplify *recA* from *O. grignonense*, *O. gallinifaecis*, and *O. pseudogrignonense*.[81] *Brucella* was clearly differentiated from *O. anthropi* and *O. intermedium* by a *recA*-sequence identity of 86.2% and 87.5% in 1065 bp.[78] Moreover, *Pseudochrobactrum* and *Ochrobactrum* share only 80.7%–81.4% gene sequence similarities.[2] The sequence similarity between *O. anthropi* and *O. intermedium* was 91% only.[78] Thus, *recA*-sequence analysis represents an accurate and reliable tool for the identification at the species level of the clinically relevant *Ochrobactrum* strains (Figure 56.2b). The *recA* gene also discriminate all the known species in the genus *Ochrobactrum* (Figure 56.2b). In 2008, Scholz et al. developed a sensitive multiprimer single-target PCR (MP-ST-PCR) based on sequence variations in the *recA* gene that allows the accurate detection and correct identification of *O. anthropi*, *O. intermedium*, and *Brucella* spp. in a single reaction.[84] Alternatively, *dnaK* (encoding the 70 kDa heat-shock protein), *rpoB* (encoding the DNA-dependent RNA polymerase β-subunit), and *trpE* (encoding the anthranilate synthase) are markers useful for studying species diversity in the genus *Ochrobactrum*.[12,85] A MLST scheme for *O. anthropi* has been recently proposed. It is based on seven genes (*dnaK*, *recA*, *rpoB*, *trpE*, *aroC*, *omp25*, and *gap*), evolving mostly by neutral mutations. The MLST scheme applied to environmental and clinical *O. anthropi* strains revealed a human-associated clonal complex.[85]

Genomic Fingerprinting. Genomic fingerprinting could contribute to identification at the species level but also allows intraspecific diversity analysis, particularly in molecular epidemiology for the detection of nosocomial outbreaks. Methods such as PFGE–RFLP, amplification profiles produced by PCR anchored in repeated sequences (rep–PCR), and AFLP detect many classes of genomic variations mainly intragenomic recombination and horizontal gene transfer events. PFGE give a snapshot of these variations at the whole-genome level whereas PCR-based approaches focus on smaller amplified genomic regions.

RFLP–PFGE is considered as the gold standard method to study the bacterial intraspecific diversity. RFLP–PFGE has been performed for many years to compare the intraspecific diversity of both *O. anthropi* and *O. intermedium*.[19,66,86] At first, the macrorestriction of genomic *Ochrobactrum* DNA was carried out with the endonuclease *Xba*I.[66,86] *Xba*I generated about 30 DNA fragments from 20 to 300 kb but, among them, only 20 bands were suitable for patterns comparison. Genomic macrorestriction with the endonuclease *Spe*I gave PFGE patterns of about 25 well-separated bands ranging from 20 to 450 kb in size. The human clinical isolates appeared very diverse when analyzed by RFLP–PFGE.[19,22,85] For example, PFGE analysis performed after *Spe*I macrorestriction of 33 strains collected from different patients hospitalized at the University Hospital of Montpellier (France) showed highly variable genomic patterns.[85] Using RFLP–PFGE, rearrangements have been observed in the genomes of clonal clinical populations of *O. intermedium*. A loss of 150 kb was observed in the genome of a clinical strain of *O. intermedium* during its adaptation to the respiratory tract of a patient with chronic colonization.[22] Therefore, RFLP PFGE is a suitable tool to study the genomic polymorphism of clinical *O. anthropi* and *O. intermedium* strains. It currently appeared as the best methods for the detection of nosocomial outbreaks and for the study of microevolution related to adaptation of a particular strain.

BOX and REP–PCR are based on the amplification of highly variable genomic regions with primers matching repetitive extragenic palindromic DNA (REP) sequences.[87,88] Concerning *Ochrobactrum*, they allowed to handle the genetic diversity at the species level as well as microdiversity of *O. anthropi* isolated from soil and from the rhizoplane.[21] In this study, species-conform clustering was obtained in the REP analysis, but intraspecific subclusters were partly different from the BOX subclusters. BOX–PCR was described to be more discriminant than REP–PCR for analyzing *O. anthropi* soil isolates since it separated strains originating from soil and rhizoplane.[21] PCR based on enterobacterial repetitive intergenic consensus (ERIC) sequences was not a suitable tool in the genus *Ochrobactrum*, particularly for *O. anthropi* intraspecific diversity analysis, because the ERIC patterns contained too less amplification fragments (Teyssier, personal communication).

AFLP was used to analyze the genetic diversity among *Ochrobactrum* strains. AFLP patterns showed a genomic variability that separated the samples into distinct clusters corresponding to the species *O. anthropi* and *O. intermedium*. *O. intermedium* was found to be closely related to *Brucella* spp. AFLP also allowed to observe a genomic polymorphism among strains of the species *O. anthropi*.[20]

56.2 METHODS

56.2.1 Sample Preparation

Crude DNA Extract. Bacteria are grown on tryptic-soy agar (TSA) for 24 h at 37°C. One isolated colony is suspended in 50 μL sterile distilled water and the DNA is rapidly extracted by a method involving boiling and freezing method. DNA from *Ochrobactrum* spp. is also easily extracted by most of the commercialized kits.

Cleared DNA Extract.[81] One isolated colony is transferred from an agar plate to 200 μL 5× lysis buffer D (PCR optimation kit, Invitrogen, 1:5 diluted in Aqua dest), supplemented with 0.5% (v/v) Tween 20™ (ICI, American Limited, Merck), and 2 mg proteinase K/mL (Roche Diagnostics). After incubation at 56°C for 1 h and inactivation for 10 min at 95°C, the cell debris was centrifuged at 12,000 g for 3 min and 2 μL of the cleared lysate were used as template for the PCR amplification.

Pure DNA Extract.[44] Pure genomic DNA extract is prepared with an AquaPure genomic DNA isolation kit (Bio-Rad).

56.2.2 Detection Procedures

56.2.2.1 16S rRNA Gene Sequence Analysis[44]

The 16S rRNA gene is selectively amplified by PCR using universal primers 27f and 1492r (Table 56.1). The PCR is carried out in 50 μL of reaction mixture containing 200 nM (each) primer, 200 μM (each) deoxynucleoside triphosphate (dNTP), 1 U of *Taq* polymerase (Roche) in the appropriate reaction buffer, and 2 μL of crude DNA extract as the template. PCR conditions are 30 cycles of 1 min at 94°C, 1 min at 65°C, and 2 min at 72°C. PCR products of about 1400 bp are sequenced with 27f primer.

Primers F4 and R2 (Table 56.1) are often use to distinguish *O. anthropi* and *O. pseudintermedium* from *Brucella* and *O. intermedium*.[6,12,79] Primer anchored in the 46 bp insertion in the 16S rRNA gene is also listed in Table 56.1.

56.2.2.2 *recA* Gene Sequence Analysis[81]

The *recA* gene is selectively amplified by PCR using primers recA-BrucOchro-f binding to position 1–22 nucleotides (nt) and recA-BrucOchro-r binding to position 1046–1065nt (Table 56.1). The PCR is carried out in 50 μL ready-to-go MasterMix (Eppendorf) using 15 pmol of each primer and 2 μL of the cleared DNA extract. PCR conditions are 30 cycles of 30 s at 94°C, 30 s at 62°C, and 1 min at 72°C, followed by a final elongation of 7 min at 72°C. PCR products of 1065 bp are sequenced using primers recA-BrucOchro-f and recA-BrucOchro-r. Outward-directed internal primers recA-Ochro-seq1 and recA-Ochro-seq2 (Table 56.1) are used for sequencing the 3′- and 5′-terminal ends.

For the phylogenetically more distantly related species, degenerated primers recA-wob-f et recA-wob-r (Table 56.1)

TABLE 56.1
Primers Used for Gene Amplification

Target Gene or Sequence	Primers[a]	Primer Sequence 5′→3′	Reference
16S rRNA	27f	GTGCTGCAGAGAGTTTGATCCTGGCTCAG	89
	1492r	CACGGATCCTACGGGTACCTTGTTACGACTT	89
16S rRNA (46 bp insertion)	ins1	GCCCCCCTTTAAAATTTCAG	22
	1492r	CACGGATCCTACGGGTACCTTGTTACGACTT	89
16S rRNA	F4	TCGAGCGCCCGCAAGGG	79
	R2	AACCATAGTGTCTCCACTAA	79
recA	recA-BrucOchro-f	ATGTCTCAAAATTCATTGCGAC	78
	recA-BrucOchro-r	AGCATCTTCTTCCGGTCCGC	78
recA degenerated primers	recA-wob-f	TTYGGNAARGGNTCRATHATG	81
	recA-wob-r	NGCRTTYTCNCGNCCYTGNCC	81
rpoB	453f	ATCGTTTCGCAGATGCACCG	12
	1232r	CGCATGTTCATCTTCACGCGGCC	12
dnaK	289f	ATCGTCAAGGGCGACAATGGC	12
	1142r	CGTCCTTGACGTCGCCCTGCA	12
REP	REP1R-I	IIIICGICGICATCIGGC	87
	REP2-I	ICGICTTATCIGGCCTAC	87
BOX	BoxA1R	CTACGGCAAGGCGACGCTGACG	88

[a] f, forward; r, reverse.

are used in the same PCR conditions except for an annealing temperature of 50°C.

56.2.2.3 *dnaK* and *rpoB* Gene Sequencing[44]

Partial *dnaK* and *rpoB* gene amplifications are carried out on pure DNA extract with primers 289f/1142r and 453f/1232r, respectively (Table 56.1). The PCR is performed as described for 16S rRNA gene amplification. PCR products of about 850 bp for *dnaK* and 780 bp for *rpoB* are generated, respectively.

56.2.2.4 Tree Reconstruction

The sequences are compared with sequences deposited in databases using the program standard nucleotide-nucleotide basic local alignment search tool (BLAST) (http://www.ncbi.nlm.nih.gov/BLAST/). For 16S rRNA gene, sequences are also compared with Greengenes (http://greengenes.lbl.gov/) and RDPII (http://rdp.cme.msu.edu/index.jsp) databases. The genetic distance between strains is calculated using DNADIST program (option similarity table) in the PHYLIP package (evolution.genetics.washington.edu/phylip.html) after sequence alignment by CLUSTALW software (www.clustal.org). Neighbor-joining trees are reconstructed from the distance matrix (substitution model F84) using NEIGHBOR program in the PHYLIP package. Maximum likelihood trees are reconstructed using the program phyML (www.phylogeny.fr).

56.2.2.5 REP and BOX PCR[7,21,90]

BOX–PCR is carried out using primer BoxA1R. REP–PCR was carried out using primers REP1R-I and REP2-I. The PCR mix (25 μL) contains 6 mM of the primer, 1.25 mM (each) dNTP, 1.5 U of *Taq* polymerase (Eurogentec GoldstarTaq) in the PCR buffer additioning with 10% DMSO, 0.18% non-acetylated BSA, and 2 μL of bacterial cell suspension in H_2O as the template. PCR conditions are as follows: an initial denaturation step at 94°C for 5 min followed by 25 cycles of 45 s at 94°C, 1.5 min at 50°C, and 8 min at 65°C, and then a final extension step at 65°C for 16 min. Amplification products are separated by electrophoresis in 2% agarose in 1× TAE (Tris Acetate EDTA) buffer for 2.5 h at 100V.

REP–PCR conditions are as follows: an initial denaturation step at 95°C for 7 min followed by 30 cycles of 1 min at 94°C, 1 min at 44°C, and 8 min at 65°C and then a final extension step at 65°C for 15 min.

Cluster analyses of the fingerprinting profiles are performed using GelCompar II (Applied Maths). Pair wise profile similarities are calculated using the Pearson correlation coefficient and dendrograms were obtained applying the UPGMA clustering method.

56.2.2.6 Macrorestriction and PFGE[18,19,22]

Genomic DNA is prepared in agarose plugs for enzymatic digestion and pulsed-field gel electrophoresis (PFGE).

Bacterial cells grown on TSA plates are suspended in Tris-EDTA (TE) (10 mM Tris pH8, 0.1 mM EDTA) buffer to a turbidity of 1.25 at 650 nm, included in low-melting agarose (v/v) (SeaPlaque GTG, FMC BioProducts). Then agarose plugs are submitted to lysis by incubation for 48 h in 0.5 M EDTA, pH8, 1 mg of protease from *Streptomyces griseus* mL-1, and 1% (w/v) sodium dodecyl sulfate (SDS). Plugs are then washed two times in TE-PMSF (TE with 1 mM phenylmethylsulfonyl fluoride) and three times in TE before macrorestriction.

DNA is digested at 37°C with 40 U of *Spe*I (New England Biolabs). *Spe*I fragments are separated by PFGE using a contour-clamped homogeneous electric field apparatus (CHEF-DRII; Bio-Rad) in a 1% agarose gel in Tris-borate-EDTA buffer (TBE 0.5×). Pulse ramps were 5–35 s for 40 h at 150 V followed by 2–10 s for 10 h at 150 V. Molecular weight marker is a concatemer of phage λ (New England Biolabs).

Fingerprinting profiles generated by PFGE are standardized with PhotoCapt software (Vilbert Lourmat). The profiles are scored for the presence or absence of DNA bands. The extent of variability is determined using the RESTDIST program in the Phylip package v.3.66. Clustering is predicated by the unweighted pair group average method (UPGMA) using the SplitsTree v4.0.

56.3 CONCLUSION AND FUTURE PERSPECTIVES

Ochrobactrum spp. are described as widely distributed and versatile bacteria. A polyphasic approach consisting of MLST and multilocus phylogeny suggested an epidemic population structure for *O. anthropi*. *O. anthropi* could be considered as a human-specialized opportunistic pathogen.[85] Their role in human infections needs to be evaluated regarding their virulence as well as their exceptional natural resistance to most antimicrobial agents. The inability of *O. anthropi* type strain to replicate within professional and nonprofessional phagocytes[91] should be also verified. Molecular techniques for identification should be used preferably in order to give accurate identification at the species level. Indeed, the role of each species in human infections remains unclear probably due to identification errors.

The genomes of *O. anthropi* and *O. intermedium* are currently being sequenced. Total sequences analysis will bring a lot of information regarding virulence factors for these bacteria. A complete *virB* operon encoding the type IV secretory system is present on chromosome 1 of *O. anthropi* type strain. The maximum sequence similarity level is obtained with *Agrobacterium tumefaciens virB* operon. Comparative genomics, particularly with the genomes of *Brucella* spp. and other related alphaproteobacteria, will contribute to increasing our knowledge about the emergence of pathogens and about the relationships between versatile opportunistic bacteria and related strict pathogens.

ACKNOWLEDGMENT

The authors are very grateful to Bernard Gay for electron microscopy.

REFERENCES

1. Garrity, G.M., Bell J.A., and Lilburn T.G., Taxonomic outline of the prokaryotes. In *Bergey's Manual of Systematic Bacteriology*, 2nd ed., p. 55, Springer New York Berlin Heidelberg, 2004.
2. Kämpfer, P. et al., Description of *Pseudochrobactrum* gen. nov., with the two species *Pseudochrobactrum asaccharolyticum* sp. nov. and *Pseudochrobactrum saccharolyticum* sp. nov., *Int. J. Syst. Evol. Microbiol.*, 56, 1823, 2006.
3. Yoon, J.-H. et al., *Daeguia caeni* gen. nov., sp. nov., isolated from sludge of a textile dye works, *Int. J. Syst. Evol. Microbiol.*, 58, 168, 2008.
4. Kämpfer, P. et al., *Paenochrobactrum gallinarii* gen. nov., sp. nov., isolated from air of a duck barn and reclassification of *Pseudochrobactrum glaciei* as *Paenochrobactrum glaciei* comb. nov., *Int. J. Syst. Evol. Microbiol.*, 60, 1493, 2010.
5. Holmes, B. et al., *Ochrobactrum anthropi* gen. nov., sp. nov. from human clinical specimens and previously known as Group Vd, *Int. J. Syst. Bacteriol.*, 38, 408, 1988.
6. Velasco, J. et al., Evaluation of the relatedness of *Brucella* spp. and *Ochrobactrum anthropi* and description of *Ochrobactrum intermedium* sp. nov., a new species with a closer relationship to *Brucella* spp., *Int. J. Syst. Bacteriol.*, 48, 759, 1998.
7. Lebuhn, M. et al., Taxonomic characterization of *Ochrobactrum* sp. isolates from soil samples and wheat roots, and description of *Ochrobactrum tritici* sp. nov. and *Ochrobactrum grignonense* sp. nov., *Int. J. Syst. Evol. Microbiol.*, 50, 2207, 2000.
8. Kämpfer, P. et al., Towards a standardized format for the description of a novel species (of an established genus): *Ochrobactrum gallinifaecis* sp. nov., *Int. J. Syst. Evol. Microbiol.*, 53, 893, 2003.
9. Trujillo, M.E. et al., Nodulation of *Lupinus albus* by strains of *Ochrobactrum lupini* sp. nov., *Appl. Env. Microbiol.*, 71, 1318, 2005.
10. Tripathi, A.K. et al., *Ochrobactrum oryzae* sp. nov., an endophytic bacterial species isolated from deep-water rice in India, *Int. J. Syst. Evol. Microbiol.*, 56, 1677, 2006.
11. Zurdo-Piñero, J.L. et al., *Ochrobactrum cytisi* sp. nov. isolated from nodules of *Cytisus scoparius* in Spain, *Int. J. Syst. Evol. Microbiol.*, 57, 784, 2007.
12. Teyssier, C. et al., *Ochrobactrum pseudintermedium* sp. nov., a novel member of the family Brucellaceae, isolated from human clinical samples, *Int. J. Syst. Evol. Microbiol.*, 57, 1007, 2007.
13. Kämpfer, P. et al., *Ochrobactrum haematophilum* sp. nov. and *Ochrobactrum pseudogrignonense* sp. nov., isolated from human clinical specimens, *Int. J. Syst. Evol. Microbiol.*, 57, 2513, 2007.
14. Kämpfer, P. et al., *Ochrobactrum rhizosphaerae* sp. nov. and *Ochrobactrum thiophenivorans* sp. nov., isolated from the environment, *Int. J. Syst. Evol. Microbiol.*, 58, 1426, 2008.
15. Huber, B. et al., *Ochrobactrum pituitosum* sp. nov., isolated from an industrial environment, *Int. J. Syst. Evol. Microbiol.*, 60, 321, 2010.
16. Imran, A. et al., *Ochrobactrum ciceri* sp. nov., isolated from nodules of *Cicer arietinum Int. J. Syst. Evol. Microbiol.*, 60, 1548, 2010.
17. Jumas-Bilak, E. et al., Unconventional genome organisation in alpha proteobacteria, *J. Bacteriol.*, 180, 2749, 1998.
18. Teyssier, C. et al., Pulsed-Field Gel Electrophoresis to study the diversity of whole genome organization in the genus *Ochrobactrum*, *Electrophoresis*, 26, 2898, 2005.
19. Teyssier, C. et al., Species identification and molecular epidemiology of bacteria belonging to *Ochrobactrum* genus, *Pathol. Biol.*, 51, 5, 2003.
20. Leal-Klevezas, D.S. et al., Genotyping of *Ochrobactrum* spp. by AFLP analysis, *J Bacteriol.*, 187, 2537, 2005.
21. Bathe, S. et al., Genetic and phenotypic microdiversity of *Ochrobactrum* spp., *FEMS microbiol. Ecol.*, 56, 272, 2006.
22. Teyssier, C. et al., Atypical 16S rRNA gene copies in *Ochrobactrum intermedium* strains reveal a large genomic rearrangement by recombination between *rrn* copies, *J. Bacteriol.*, 185, 2901, 2003.
23. Berg, G., Eberl, L., and Hartmann, A., The rhizosphere as a reservoir for opportunistic human pathogenic bacteria, *Environ. Microbiol.*, 7, 1673, 2005.
24. Tadros, M.G., and Hughes, J.B., Degradation of polycyclic aromatic hydrocarbons (PAHs) by indigenous mixed and pure cultures isolated from coastal sediments, *Appl. Biochem. Biotechnol.*, 63, 865, 1997.
25. Tripathi, A.K., Verma, S.C., and Ron, E.Z., Molecular characterization of a salt-tolerant bacterial community in the rice rhizosphere, *Res. Microbiol.*, 153, 579, 2002.
26. N'Gom, A. et al., A novel symbiotic nitrogen-fixing member of the *Ochrobactrum* clade isolated from root nodules of *Acacia mangium*, *J. Gen. Appl. Microbiol.*, 50,17, 2004.
27. Zhang, H. et al., Functional assembly of a microbial consortium with autofluorescent and mineralizing activity for the biodegradation of organophosphates, *J. Agric. Food Chem.*, 56, 7897, 2008.
28. Khumar, M., Lakshmi, C.V., and Khanna, S., Biodegradation and bioremediation of endosulfan contaminated soil, *Bioresour. Technol.*, 99, 3116, 2007.
29. Mondragón-Parada, M.E. et al., Chemostat selection of a bacterial community able to degrade s-triaziniccompounds: Continuous simazine biodegradation in a multi-stage packed bed biofilm reactor, *J. Ind. Microbiol. Biotechnol.*, 35, 767, 2008.
30. Chang, B.V., Chiang, B.W., and Yuan, S.Y., Biodegradation of nonylphenol in soil, *Chemosphere*, 66, 1857, 2007.
31. Lin, C.W., Cheng, Y.W., and Tsai, S.L., Influences of metals on kinetics of methyl tert-butyl ether biodegradation by *Ochrobactrum cytisi*, *Chemosphere*, 69, 1485, 2007.
32. Panikov, N.S. et al., Biodegradation kinetics of the nitramine explosive CL-20 in soil and microbial cultures, *Biodegradation*, 18, 317, 2006.
33. Chopra, B.K. et al., The characteristics of rhizosphere microbes associated with plants in arsenic-contaminated soils from cattle dip sites, *Sci. Total Environ.*, 378, 331, 2007.
34. He, Z. et al., Isolation and characterization of a Cr(VI)-reduction *Ochrobactrum* sp. strain CSCr-3 from chromium landfill, *J. Hazard. Mater.*, 163, 869, 2008.
35. Arulazhagan, P., and Vasudevan, N., Role of a moderately halophilic bacterial consortium in the biodegradation of polyaromatic hydrocarbons, *Mar. Pollut. Bull.*, 58, 256, 2008.
36. Zurek, L., Schal, C., and Watson, D.W., Diversity and contribution of the intestinal bacterial community to the development of *Musca domestica* (*Diptera*: *Muscidae*) larvae, *J. Med. Entomol.*, 37, 924, 2000.
37. Volf, P., Kiewegová, A., and Nemec, A., Bacterial colonisation in the gut of *Phlebotomus duboseqi* (*Diptera*: *Psychodidae*): transtadial passage and the role of female diet, *Folia Parasitol. (Praha).*, 49, 73, 2002.
38. Wenzel, M. et al., Aerobic and facultatively anaerobic cellulolytic bacteria from the gut of the termite *Zootermopsis angusticollis*, *J. Appl. Microbiol.*, 92, 32, 2002.
39. Wang, B.J. et al., Microbial diversity in scorpion intestine (*Buthus martensii* Karsch), *Wei Sheng Wu Xue Bao*, 47, 888, 2007.

40. Ahmad, A., Broce, A., and Zurek, L., Evaluation of significance of bacteria in larval development of *Cochliomyia macellaria* (*Diptera*: *Calliphoridae*), *J. Med. Entomol.*, 43, 1129, 2006.
41. Babic, I. et al., Occurrence of natural dixenic association between the symbiont *Photorhabdus luminescens* and bacteria related to *Ochrobactrum* spp. in tropical entomopathogenic *Heterorhabditis* spp. (*Nematoda*, *Rhabditida*), *Microbiology*, 146, 709, 2000.
42. Khanolkar-Gaitonde, S.S. et al., Isolation of bacteria other than *Helicobacter pylori* from stomachs of squirrel monkeys (*Saimiri* spp.) with gastritis, *Dig Dis Sci.*, 45, 272, 2000.
43. Shilton, C.M. et al., Spinal arthropathy associated with *Ochrobactrum anthropi* in free-ranging cane toads (*Chaunus* [*Bufo*] *marinus*) in Australia, *Vet. Pathol.*, 45, 85, 2008.
44. Teyssier, C. et al., Molecular and phenotypic features for identification of the opportunistic pathogens *Ochrobactrum* spp., *J. Med. Microbiol*, 54, 1, 2005.
45. Menuet, M. et al., First isolation of two colistin-resistant emerging pathogens, *Brevundimonas diminuta* and *Ochrobactrum anthropi*, in a woman with cystic fibrosis: A case report, *J. Med. Case Reports.*, 2, 373, 2008.
46. Dharne, M.S. et al., Isolation of urease-positive *Ochrobactrum intermedium* in the stomach of a non-ulcer dyspeptic patient from north India, *J. Microbiol. Immunol. Infect.*, 41, 183, 2008.
47. Gill, M.V. et al., Intravenous line infection due to *Ochrobactrum anthropi* (CDC Group Vd) in a normal host, *Heart Lung*, 26, 335, 1997.
48. Mahmood, M.S. et al., Infective endocarditis and septic embolization with *Ochrobactrum anthropi*: Case report and review of literature, *J. Infect.*, 40, 287, 2000.
49. Stiakaki, E. et al., *Ochrobactrum anthropi* bacteremia in pediatric oncology patients, *Pediatr. Infect. Dis. J.*, 21, 72, 2002.
50. Duran, R. et al., *Ochrobactrum anthropi* bacteremia in a preterm infant with mecomium peritonitis, *Int. J. Infect. Dis.*, 13, Epub Oct7, 2008.
51. Arora, U., Kaur, S.. and Devi, P., *Ochrobactrum anthropi* septicaemia, *Indian J Med Microbiol.*, 26, 81, 2008.
52. Daxboeck, F. et al., *Ochrobactrum anthropi* bloodstream infection complaisant hemodialysis, *Am. J. Kidney Dis.*, 40, E17, 2002.
53. Berman, A.J., Del Priore, L.V., and Fischer, C.K., Endogenous *Ochrobactrum anthropi* endophthalmitis, *Am. J. Ophthalmol.*, 123, 560, 1997.
54. Greven, C.M., and Nelson, K.C., Chronic postoperative endophtalmitis secondary to *Ochrobactrum anthropi*, *Retina*, 21, 279, 2001.
55. Inoue, K. et al., *Ochrobactrum anthropi* endophthalmitis after vitreous surgery, *Br. J. Ophthalmol.*, 83, 502, 1999.
56. Song, S. et al., An epidemic of chronic pseudophakic endophtalmitis due to *Ochrobactrum anthropi*: Clinical findings and managements of nine consecutive cases, *Ocul. Immunol. Inflamm.*, 15, 429, 2007.
57. Christenson, J.C. et al., Meningitis due to *Ochrobactrum anthropi*: An emerging nosocomial pathogen. A report of 3 cases, *Pediatr. Neurosurg.*, 27, 218, 1997.
58. Ezzedine, H. et al., An outbreak of *Ochrobactrum anthropi bacteraemia* in five organ transplant patients, *J. Hosp. Infect.*, 27, 35, 1994.
59. Romero Gómez, M.P et al., Prosthetic mitral valve endocarditis due to *Ochrobactrum anthropi*: Case report, *J. Clin. Microbiol.*, 42, 3371, 2004.
60. Brivet, F. et al., Necrotizing fasciitis, bacteremia, and multiorgan failure caused by *Ochrobactrum anthropi*, *Clin. Infect. Dis.*, 17, 516, 1993.
61. Cieslak, T.J., Drabick, C.J., and Robb, M.L., Pyogenic infections due to *Ochrobactrum anthropi*, *Clin. Infect. Dis.*, 22, 845, 1996.
62. Wheen, L., Taylor, S., and Godfrey, K., Vertebral osteomylitis due to *Ochrobactrum anthropi*, *Intern. Med. J.*, 32, 426, 2002.
63. Ozdemir, D. et al., *Ochrobactrum anthropi* endocarditis and septic shock in a patient with no prosthetic valve or rheumatic heart disease: Case report and review of the literature, *Jpn. J. Infect. Dis.*, 59, 264, 2006.
64. Vaidya, S.A. et al., Pelvic abscess due to *Ochrobactrum intermedium* (corrected) in an immunocompetent host: Case report and review of the literature, *J. Clin. Microbiol.*, 44, 1184, 2006.
65. Battaglia, T.C., *Ochrobactrum anthropi* septic arthritis of the acromioclavicular joint in an immunocompetent 17 year old, *Orthopedics*, 31, 606, 2008.
66. Möller, L.V. et al., *Ochrobactrum intermedium* infection after liver transplantation, *J. Clin. Microbiol.*, 37, 241, 1999.
67. Apisarnthanarak, A., Kiratisin, P., and Mundy, L.M., Evaluation of *Ochrobactrum intermedium* bacteremia in a patient with bladder cancer, *Diagn. Microbiol. Infect. Dis.*, 53, 153, 2005.
68. Thoma, B. et al., Identification and antimicrobial susceptibilities of *Ochrobactrum* spp., *Int. J. Med. Microbiol.*, 299, 209, 2009.
69. Higgins, C.S. et al., Characterization, cloning and sequence analysis of the inducible *Ochrobactrum anthropi* AmpC β-lactamase, *J. Antimicrob. Chemother.*, 47, 745, 2001.
70. Higgins, C.S. et al. Resistance to antibiotics and biocides among nonfermenting Gram-negative bacteria, *Clin. Microbiol. Infect.*, 7, 308, 2001.
71. Nadjar, D. et al., Molecular characterization of chromosomal class C beta-lactamase and its regulatory gene in *Ochrobactrum anthropi*, *Antimicrob. Agents Chemother.*, 45, 2324, 2001.
72. Bombicino, K.A. et al., Evaluation of pyrrolidonyl arylamidase for the identification of nonfermenting Gram-negative rods, *Diagn. Microbiol. Infect. Dis.*, 57, 101, 2007.
73. Chiron, R. et al., Clinical and microbiological features of Inquilinus sp. isolates from five patients with cystic fibrosis, *J. Clin. Microbiol.*, 43, 3938, 2005.
74. Elsaghir, A.A., and James, E.A., Misidentification of *Brucella melitensis* as *Ochrobactrum anthropi* by API 20NE, *J. Med. Microbiol.*, 52, 441, 2003.
75. Scholz, H.C. et al., *Brucella microti* sp. nov., isolated from the common vole *Microtus arvalis*, *Int. J. Syst. Evol. Microbiol.*, 58, 375, 2008.
76. Stackebrandt, E., and Jonars, E., Taxonomic parameters revisited: Tarnished gold standards, *Microbiol. Today*, 33, 152, 2006.
77. Kampfer, P. et al., Difficulty in the identification and differentiation of clinically relevant Ochrobactrum species, *J. Med. Microbiol.*, 56, 1572, 2007.
78. Scholz, H.C. et al., Genotyping of *Ochrobacterium anthropi* by *recA* based comparative sequence, PCR-RFLP, and 16S rRNA gene analysis, *FEMS Microbiol. Lett.*, 257, 7, 2006.
79. Romero, C. et al., Specific detection of *Brucella* DNA by PCR, *J. Clin. Microbiol.*, 33, 615, 1995.
80. Lebuhn, M. et al., Comparative sequence analysis of the internal trancribed spacer 1 of *Ochrobactrum* species, *Syst. Appl. Microbiol.*, 29, 265, 2006.
81. Scholz, H.C. et al., Genetic diversity and phylogenetic relationships of bacteria belonging to the *Ochrobactrum-Brucella* group by *recA* and 16S rRNA gene-based comparative sequence analysis, *Syst. Appl. Microbiol.*, 31, 1, 2008.
82. Vandamme, P. et al., Identification and population structure of *Burkholderia stabilis* sp. nov. (formerly *Burkholderia cepacia* genomovar IV), *J. Clin. Microbiol.*, 38, 1042, 2000.

83. Thomson, C.C. et al., Use of *recA* as an alternative phylogenetic marker in the family *Vibrionaceae*, *Int. J. Syst. Evol. Microbiol.*, 54, 919, 2004.
84. Scholz, H.C. et al., Specific detection and differentiation of *Ochrobactrum anthropi*, *Ochrobactrum intermedium* and *Brucella* spp. by a multi-primer PCR that targets the *recA* gene, *J. Med. Microbiol.*, 57, 64, 2008.
85. Romano, S. et al., Multilocus sequence typing supports the hypothesis that *Ochrobactrum anthropi* displays a human-associated subpopulation, *BMC Microbiol.*, 18, 267, 2009.
86. Van Dijck, P. et al., Evaluation of pulsed-field gel electrophoresis and rep-PCR for the epidemiological analysis of *Ochrobactrum anthropi* strains, *Eur. J. Clin. Microbiol. Infect. Dis.*, 14, 1099, 1995.
87. Versalovic, J., Koeuth, T., and Lupski, J.R., Distribution of repetitive DNA sequences in eubacteria and application to fingerprinting of bacterial genomes, *Nucleic Acids Res.*, 19, 6823, 1991.
88. Rademaker, J.L.W., Louws, F.J., and De Bruijn, F.J., Characterization of the diversity of ecologically important microbes by rep-PCR genomic fingerprinting. In Akkermans, A.D.L., van Elsas, J.D., and De Bruijn, F.J. (Editors), *Molecular Microbial Ecology Manual*, vol. 3.4.3, PP-26, Kluwer Dordrecht, The Netherlands, 1998.
89. Weisburg, W.G. et al., 16S ribosomal DNA amplification for phylogenetic study, *J. Bacteriol.*, 173, 697, 1991.
90. Mercier, E. et al., Polymorphism in *Brucella* strains detected by studying distribution of two short repetitive DNA elements, *J. Clin. Microbiol.*, 34, 1299, 1996.
91. Barquero-Calvo, E. et al., The differential interaction of *Brucella* and *Ochrobactrum* with innate immunity reveals traits related to the evolution of stealthy pathogens, *PLOS*, 4, e5893, 2009.

57 Orientia

Harsh Vardhan Batra and Diprabhanu Bakshi

CONTENTS

57.1 INTRODUCTION

Orientia tsutsugamushi (formerly *Rickettsia tsutsugamushi*) represents one of the most covert bacterial pathogens of present times. It is the etiological agent of scrub typhus, also known as "tsutsugamushi disease, as well as many cases of "fevers of unknown origin" across most of the countries in Southeast Asia and the Asia-Pacific region. It is a single species that comprises of many antigenic and genotypic variants.[1] Scrub typhus is caused by the bite of the larvae of trombiculid mites harboring this etiological agent.

Orientia tsutsugamushi was originally implicated by Hayashi as the cause of tsutsugamushi fever in 1920, but it was then mistaken as a protozoan agent. *Orientia* was correctly characterized as bacterium in 1930 and given the name *Rickettsia orientalis* by Nagayo et al.[2] Subsequently, the name *Rickettsia tsutsugamushi* was commonly accepted in the literature (as per the 7th Edition of the *Bergey's Manual*) until 1995, when it was placed in the new genus *Orientia* based on numerous phenotypic and genetic differences that distinguish it from the species of *Rickettsia*.[3]

Members of the genera *Orientia* are obligate intracellular parasites that are morphologically and biochemically similar to other gram-negative bacteria. They are short, rod-shaped, or coccobacillary organisms, usually 0.8–2 μm long and 0.3–0.5 μm in diameter. *Orientia* are usually arthropod-associated bacteria that infect vertebrates, including humans, usually as accidental hosts. *Orientia* is mainly reported to be associated with the chigger mites *Leptotrombium* sp. Scrub typhus, caused by this obligate intracellular bacterium has found a place in world and military history, as one of the most significant rickettsioses affecting US troops during World War II.[4] Scrub typhus was responsible for numerous casualties, including deaths of American military personnel in the Asia-Pacific region. Before the war, the disease was little known outside Japan. Intensive exposure of scrub typhus to the Allied troops led to immense losses of man-days and caused thousands of deaths. Scrub typhus was the leading cause of fevers of unknown origin (FUO) among troops during the Vietnam War. Between 1966 and 1969, the incidences of scrub typhus ranked second only to sexually transmitted diseases.[4] FUOs were greater in combat troops who typically had intense jungle environment exposure than among support troops. Diseases like malaria, scrub typhus, and leptospirosis were more frequently contracted in heavily forested areas where anopheline mosquitoes and chigger vectors were common.[4]

57.1.1 Epidemiology, Classification, and Morphology

Epidemiology. Scrub typhus is known to be distributed throughout the Asia-Pacific region. This disease was first reported from Japan and subsequently from Sumatra, Taiwan, Philippines, India, Malaysia, Thailand, Australia, Burma, China, Nepal, Sri Lanka, Pakistan, and so forth.[2] More than 30 antigenically distinct serotypes of *Orientia tsutsugamushi* have been reported from the endemic regions.[5] The disease has widespread presence in Southeast Asia and Southwestern Pacific regions and has gained importance as one of the most rapidly reemerging infectious diseases of the present times.[6] The disease has been recently reported from several areas of south India and was the cause behind several deaths in Himachal Pradesh in 2003.[7–10] Scrub typhus is generally an incapacitating, febrile illness that could turn fatal due to serious multiorgan involvement caused by delayed or faulty diagnosis. When properly diagnosed, they are often easily treated but the unusual conditions created by war, military operations, or civil unrest often delay or prevent appropriate treatment.[4] This fact is compounded by the lack of awareness on this infectious agent in many developing Asian countries especially in poor rural areas where elementary healthcare systems are not well developed. In general, scrub typhus and rickettsial diseases is difficult to diagnose because they share symptoms with many other febrile diseases like leptospirosis, dengue, malaria, and typhoid with similar epidemiology. Despite the current focus of medical researchers on newly emerging diseases, scrub typhus, as well as other rickettsial diseases, has continued to evolve and emerge in unique guises without receiving much attention from the research fraternity. Unfortunately, the rickettsial diseases are not perceived to be emerging, even though they exhibit most of the characteristics of emerging and reemerging pathogens: underdiagnosed and unrecognized, present in new ecological settings, associated with increased clinical risk, reports of antibiotic resistance or new clinical presentations, more widespread geographically than previously believed, new genetic types of etiologic agents, and new arthropod vectors.[4] Also this listed biological warfare agent, especially its drug refractory strains, pose an increased threat in their use in bioterrorism activities.

Classification. Microorganisms in the genera *Rickettsia* and *Orientia* are obligately intracellular, gram-negative bacteria. Based on their 16S rRNA and *groEL* (60 kDda heat-shock chaperonin protein) genes, *Rickettsia* and *Orientia* represent very closely related evolutionary lineages of microorganisms in α-subdivision of proteobacteria. According to current concepts, *Rickettsia* and *Orientia* are the sole genera belonging to the family Rickettsiaceae, order Rickettsiales. *Orientia tsutsugamushi* is a single species that contains many antigenic and genotypic variants.[1] The genus *Rickettsia* had been classically separated into the typhus group (TG) and the spotted fever group (SFG) based on the presence of distinct group specific lipopolysaccharide (LPS) antigens. The TG contains two human pathogens namely *R. prowazekii* and *R. typhi*. The SFG contains more than 30 diverse genotypes, which cause at least 13 different pathological syndromes in humans.

Morphology. *Orientia tsutsugamushi* is an obligate intracellular parasite. It multiplies by binary fission in the host cell cytoplasm. It can be observed with a light microscope in specimens stained by a modification of the Gimenez procedure or by Giemsa stain. The fundamental morphology of *Orientia tsutsugamushi* is similar to that of gram-negative bacteria: the organism is surrounded by a cytoplasmic membrane and a cell wall showing a clear periplasmic space where amorphous materials are seen. In the cytoplasm, the electron-dense, ribosome-rich area and the less dense network area with DNA fibers are visible distinctively in electron microscopic studies, but in general, no other characteristic structures are seen.[2]

In *Orientia tsutsugamushi* cell wall, the outer leaflet is thicker than the inner one. It is surrounded by a very soft outer envelope, which is disrupted easily by osmotic shock.[11] In cell-free suspension, *Orientia* appeared to be round,[12] while they are rod or coccobacillary in the host cell cytoplasm.[13] By electron microscopic observation of antibody treated *Orientia tsutsugamushi*, the presence of a capsule-like layer as suggested by Silverman and Wisseman (1978).[14] *Orientia tsutsugamushi* is unique in family *Rickettsiales* as it lacks muramic acid, glucosamine, heptose, 3-deoxy D-mannooctulosonic acid (KDO), and hydroxy fatty acids, which are the general components of peptidoglycan or LPS.[15] In proteinase K-digested whole cells of *Orientia*, no LPS-like bands could be observed on PAGE analysis, leading to the conclusion that *Orientia tsutsugamushi* is deficient in both peptidoglycan and LPS in the cell wall.[15] This conclusion coincides well with the observation that *Orientia tsutsugamushi* is resistant to beta-lactam antibiotics, which disturbed the growth of other rickettsiae.[16–18] In addition, the finding that treatment of *Orientia tsutsugamushi* with periodate does not alter its antigenic properties in the neutralization test of infectivity with immune sera support the earlier assumptions about the lack of LPS in *Orientia*.

57.1.2 Antigenic and Genomic Structures

Major Antigens and Antigenic Variants in *Orientia*. SDS-PAGE analysis of purified *Orientia tsutsugamushi* lysates revealed as many as 30 polypeptide bands.[19] The major polypeptides migrated to 110, 80, 70, 60, 54–58, 47, 42, 35, 28, and 25 kDa, corresponding to molecular weight markers in SDS-PAGE PAGE. The 54–58 kDa polypeptide is present in the largest amounts. This polypeptide is the immunodominant antigen showing type-specific antigenicity and is easily digested by the treatment of intact *Orientia* with trypsin indicating that it is located on the cell surface.[19,20] It has been found that outer membrane protein antigens of *Orientia tsustugamushi* in infected humans and animals elicit vigorous antibody responses, which result in the production of IgM and IgG antibodies which are directed mostly against the 56 (54–58), 60, 47, 70, 22, and 110 kDa antigens. Antibodies to other surface antigens in low quantities can also be detected.[20,21] The 56 kDa polypeptide, also known as the type-specific antigen (TSA) of *Orientia tsutsugamushi*, which is present in the largest amount on the bacterial surface, reacts strongly with homologous antiserum but faintly or moderately with heterologous antisera, indicating that this polypeptide has strain or type-specific antigenicity. The 46–47 and 70 kDa polypeptides react not only with the homologous antiserum but also with heterologous antisera suggesting that these polypeptides have characteristics of group or species specific antigenicity. Most patient sera reacted with the 56 kDa polypeptide, suggesting that this polypeptide is the dominant antigen in the course of infection with *Orientia tsutsugamushi.* Many of the monoclonal antibodies raised against *Orientia tsutsugamushi* isolates have been found to react only with the corresponding homologous strain but not with heterologous strains indicating that this polypeptide has strain-specific epitopes. This property of strain-specific monoclonal antibodies directed against this polypeptide has been exploited by researchers to characterize antigenic variants of *Orientia tsutsugamushi* isolated from various sources across the world. However, many of the MAbs raised against the 56 kDa TSA also show cross reactivity with some or all heterologous strains indicating that this polypeptide consists of both strain-specific and common epitopes. A comprehensive mapping of the 56 kDa polypeptide of the Boryong strain isolated in South Korea was reported in 1997 by Seong and et al.[22] The Boryong 56 kDa *omp* (Bor56) was expressed in *E. coli* and a series of deletion mutants were generated as MalE fusion proteins. Antibody binding domains were characterized by using patient sera, mouse monoclonal antibodies, and Bor56 immunized mice sera. None of the antibodies bound to a fusion protein containing the carboxy terminal 140 amino acid stretch, suggesting that the carboxy terminal domain of Bor56 is not exposed on the cell surface. Human IgM predominantly bound to the antigenic domain I (AD I; amino acids 19–113) and AD III (amino acids 243–328). Human IgG showed preferential binding to AD I. The epitope recognized by strain-specific monoclonal antibodies was mapped to AD II (amino acids 142–203). These results suggested that AD I, AD II, and AD III are useful in the induction of humoral immunity against *Orientia tsutsugamushi.*

Amino acid sequences alignment of 56 kDa protein homologues from several antigenic variants of *Orientia tsutsugamushi* revealed the presence of four variable domains (VD I–VD IV). VD I was located to amino acids 105–132, VD II was located to amino acids 151–176, VD III to amino acids 202–246, and VD IV was located to amino acids 402–451.[2] These VDs harbored type-specific differences, although single or multiple base substitutions or deletions of contiguous amino acid residues are recognized throughout the antigen. The alignment studies highlighted both diversity and similarity among strains. Hydropathy analysis of these 56 kDa proteins showed alternate successions of hydrophilic and hydrophobic motifs, which is the characteristic of transmembrane proteins.[2] All the VDs were located in the hydrophilic regions. This finding supported the assumption that all or some of the domains are located on the surface of the pathogen and may serve as antigenic determinates.[2]

The 47 and 22 kDa antigens are the other important surface proteins of *Orientia tsutsugamushi.* The 47 kDa polypeptide react not only with homologous antiserum but also with heterologous antisera raised against several strains of *Orientia tsutsugamushi*, suggesting that these polypeptides have the characteristics of group or species specific antigenicity. The gene encoding the 47 kDa protein was cloned from the Karp strain of *Orientia tsutsugamushi.*[23] The 22 kDa polypeptide antigen showed subgroup specific properties.[2] The gene encoding this immunoreactive protein was cloned and sequenced in 1991 by Hickman et al.[24] Both these polypeptides were found to elicit a proliferative response in helper T cell line, suggesting the presence of stimulatory T cell epitopes. Consequently, both these molecules were predicted to be potential vaccine candidates for *Orientia tsutsugamushi* infections.[2]

The gene sequence of the 56 kDa TSA and the 47 kDa outer membrane protein have been used to develop sensitive conventional as well as real-time PCR-based diagnostic tests for the detection of *Orientia tsutsugamushi* from infected clinical materials, and have been extensively used and evaluated for the specific diagnosis of scrub typhus and also for phylogenetic analysis of *O. tsutsugamushi* isolates from diverse clinical and geographical sources.[5,25–32] The other polypeptides (i.e., the 60, 70, and 110 kDa antigens) were refractory in their reactivities against patient sera.[20]

The 56 kDa proteins elicit very early IgM and IgG serologic responses but convalescent sera also contain antibodies against the 22 and 47 kDa proteins of *O. tsutsugamushi.* Recombinant 56 kDa protein antigen has been found to be a suitable type-specific immunodiagnostic antigen and a vaccine candidate.[33,34] The recombinant 56 kDa TSA gene have been found to be a very suitable candidate for serodiagnosis of scrub typhus and have been used by many researchers in serodiagnostic assays for scrub typhus.[33,35–39]

Antigenic variation amongst strains of *Orientia tsutsugamushi* was analyzed and recognized by cross complement fixation, cross neutralization, and cross protection tests.[40–45] Early studies with cross CF tests established the Karp, Katō, and Gilliam antigen types of *Orientia* in Japan. With the advent of techniques like the indirect immunofluorescence assay (IFA) by Bozemann and Elisberg in 1963 and the increased use of monoclonal antibodies, several strains of antigenically diverse *Orientia* have been reported and continue to be reported from many countries.[2,46] In addition to the Karp, Katō, and Gilliam type strains, the other distinct strains, like the Shimokoshi, Boryong, Irie, Hirano, Kawasaki, Kuroki, and many more, were reported from several countries, like Korea, Japan, Australia, and Thailand.[47–60] Using data on the reactivity of human, rodent and arthropod vector isolates of *Orientia tsutsugamushi* to type-specific monoclonal antibodies (e.g., anti-Karp, anti-Gilliam, anti-Kato MAbs) and the gene sequence of the 56 kDa type-specific antigen (56 kDa TSA) gene, several isolates with low, medium, or high antigenic similarity with the prototype strains of *Orientia tsutsugamushi* has been described.[2] The 47 and 22 kDa outer membrane antigen genes and the GroEL heat-shock protein gene (*gro*EL), however, were found to be much more conserved.[2,61,62]

Genetic Structure. *O. tsutsugamushi* contains a chromosome of 2.0–2.7 Mbp, which is the largest in size in order

Rickettsiales.[3,63,64] It is a single circular chromosome containing significantly more noncoding or paralogous sequences and only about 45%–50% of the genome is estimated to have coding capacity. The average G + C contents determined for isolates so far varied between 28 and 30.5 mol%.[63–65] In the complete genome sequencing report available for the Boryong strain, there is a core of 512 genes that was shared with members of the other *Rickettsia* species.[63] The number of rRNA and tRNA genes was found to be more or less conserved with the *Rickettsia* group. Out of an average of 2000 ORFs detected in *O. tsutsugamushi* strains sequenced so far, 40%–45% has been found to represent fragmented genes. The majority of these fragmented genes coincided with repeated DNA regions distributed throughout the whole genome. Another important characteristic of the genome is that it has very poor colinearity to any of the previously sequenced *Rickettsia* genomes, which indicates that significant genome shuffling has taken place in the evolutionary course of *O. tsutsugamushi*.[64] In the Ikeda strain, 1196 out of the 1967 CDSs were found to be repeated genes constituting 85 families, where the members of each family exhibit more than 90% amino acid sequence identity. Many of these repeated genes were found to be pseudogenes. *O. tsutsugamushi* has been found to have 359 *tra* genes for conjugative type IV secretion systems (TFSS), which are arranged in fragmented repeat clusters with multiple *tra* genes per cluster. Many types of repetitive sequences characterize the genome. They have been extensively amplified and distributed throughout the genome. Many mobile genetic elements, transposable elements, and insertion sequences exclusive to *Orientia tsutsugamushi* have been observed. Each type of these elements has also been explosively amplified and their fragments and copies have been found scattered throughout this bacterial genome.[63,64]

The entire base sequence of the genes encoding the 56 kDa type-specific antigen, 22 and 47 kDa antigen genes and the heat-shock protein gene *gro*EL in several well characterized strains, have been clarified.[61,66–69] Based on the sequence, polymerase chain reaction (PCR) and nested PCR methods have been designed for the diagnosis of tsutsugamushi disease and for the typing of *O. tsutsugamushi* strains.[32,61,62,70–73]

57.1.3 Clinical Features and Pathogenesis

Clinical Features. Humans contract scrub typhus as accidental host when bitten by larvae of trombiculid mites (chiggers). The major mite species found to be associated with scrub typhus are *Leptotrombium pallidum*, *L. fletcheri*, *L. akamushi*, *L. arenicola*, *L. deliense*, and *L. akamushi*.[2] As clearly evident, certain occupational or recreational activities/habits tend to predispose humans to getting exposed to such mite-infested habitats mainly in forest areas. These mites have a four-stage lifecycle—namely egg, larva, nymph, and adult. The larval stage requires blood meal to develop into the nymph stage. Normally, wild rodents are the source of the blood meal for this stage of the mite. In this manner, *Orientia tsutsugamushi* is transmitted from the vertebrate reservoir to the invertebrate host. Humans accidentally bitten by this feeding larva infested with *Orientia tsutsugamushi* contracts this illness, which has an incubation period of 1–3 weeks with an average incubation period of 10–12 days. The onset of the illness is marked by fever, headache, myalgia, cough, and GI symptoms. During the first few days of illness, an accurate clinical diagnosis of rickettsial disease is very difficult and is strongly dependant on epidemiological clues.[74] A characteristic cutaneous eruption or eschar is reported in most cases.[75] The classical case scenario involves an eschar at the bite site, regional lymphadenopathy, and a macular or maculopapular rash that appears on the trunk and may extend to the arms and legs subsequently. However, the eschar develops only after several days of illness, generally at the site of the bite. It begins as a small papular eruption that increases in size and undergoes necrosis at the center, eventually acquiring a blackened crust with an erythematous halo that resembles a cigarette burn (Figure 57.1). The eschar marks the initial growth focus of *O. tsutugamushi* infection. However, the eschar may or may not be associated with many cases of scrub typhus, and it frequently occurs in difficult to find spots in the body and hence requires thorough examination by the physicians to detect one. Moreover, a history of arthropod exposure is not always elicited and the eschar is not always noticed or present. Physicians, therefore, often evaluate rather nonspecific clinical symptoms, including the

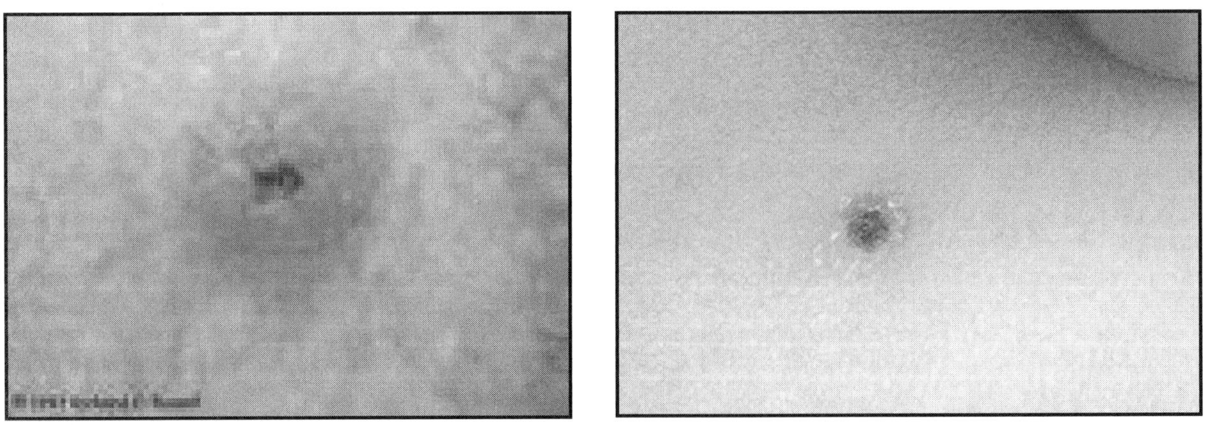

FIGURE 57.1 Development of eschar at the site of the bite.

onset of malaise, chills, myalgia, fever, and severe headache accompanied with suffused conjunctivae and face, drowsiness, apathy, pain in shins, and other muscles and more characteristically, generalized lymphadenopathy and hepatosplenomegaly.[76] Other symptoms may include nausea and vomiting, constipation, cough, epistaxis, tinnitus, and hyperacusis followed by deafness. In small proportion of cases, tremors, delirium, nervousness, slurred speech, and nuchal rigidity may develop in the second week of illness. ARDS as a complication of scrub typhus has also been reported.[77] Definitive diagnosis requires differentiation of the scrub typhus from various infectious diseases including leptospirosis, typhoid fever, dengue fever, malaria, and meningitis.

Pathogenesis. The pathogenesis of this disease is vasculitis caused by the proliferation of organisms in the endothelial lining of small arteries, veins, and capillaries. The pathogenic effect of *Orientia* has been attributed to their successful adherence, penetration, and intracellular multiplication in the cytoplasm of the eukaryotic cells. Each of these processes has been examined in great detail using electron microscopy by researchers.[78–80] The following steps of rickettsial entry can be differentiated by electron microscopy: (i) Adhesion of the rickettsiae to the host cell membrane; (ii) uptake by engulfment of the rickettsiae by host cell membrane extensions or by invagination of the host cell membrane and intracellular multiplication; (iii) release from the host cell.

(i) Penetration: The first step involves the attachment of the bacterium to the host cell surface. This attachment appears to be an interaction between a ligand on the bacterial surface and a receptor component on the host cell surface. Intact *Orientia tsutsugamushi* treated with trypsin before experimental infection failed to attach to host cell surface, indicating that trypsin inactivates or destroys these attachment components from the bacterial surface. Disappearance of the 21–25, 35, and the 54–56 kDa polypeptides upon trypsin treatment suggests the involvement of this polypeptide/s in adhesion to the host cells. The next stage in penetration is the formation of projections at the site of attachment indicating the initiation of phagocytosis. Subsequently, the cell surface is invaginated and engulfment of the bacteria takes place into a phagosome. In the cell cytoplasm, partial degradation of phagosomes and subsequent release of *Orienta* free of the phagosomal membrane is observed between 10 and 20 min after attachment. Both attachment and penetration may involve active participation of the host and the pathogen.[2]

(ii) Intracellular Multiplication: Orientia multiply in the host cell cytoplasm by binary fission. The cells are observed to form a constriction along the long axis during the process. Progeny bacterial cells in the host cell cytoplasm could be observed 48–72 h after entry of the pathogen. *O. tsutsugamushi* grows to high density in the cytoplasm of its infected cells and may also occasionally invade the nucleus.[80]

(iii) Release from Host Cell: The release mechanism for *Orientia* was observed to be similar to that of enveloped viruses. Microscopic observations reveal the bacterial progenies pushing at the host cytoplasmic membrane from inside out. The direction of the budding was not consistent as some cells budded perpendicularly and some horizontally to the long axis. Groups of budding bacteria can be clearly observed on the surface of the host cells by scanning electron microscopy. The budding rickettsiae adhere tenaciously to the host cell components and accumulate to high density at the surface of infected cells. *Orientia* may be released from the host covered with the host cell membrane, which may directly enter neighboring host cells or, alternatively, "slip-off" the membrane encapsulation to release the naked bacterium that invades other host cells by attachment as described earlier. Like the other rickettsiae, *O. tsutsugamushi* grows slowly in vertebrate cells at 37°C with a doubling time between 9 and 18 h.[81]

(iv) Cell Injury Due to Orientia Growth. Vascular injury is the major initial lesion observed in rickettsial diseases. The injury results from penetration and replication of the rickettsiae in endothelial cells. The damage inflicted on infected endothelial cells and resulting dissemination of rickettsiae into the surrounding and underlying tissues can be detected in histological study of severe cases. This damage results in enhanced permeability of tissue fluids into the infection foci, infiltration by monocytes and polymorphonuclear neutrophils, and inflammatory reactions and the formation of microthrombi.[82,83]

57.1.4 Laboratory Diagnosis of Scrub Typhus

(i) In vitro *and* In Vivo *Isolation* Isolation of *O. tsutsugamushi* from clinical specimens, mite vectors, and wild rodents has been accomplished, mainly by inoculation into laboratory mice with appropriately processed and stored test material. Direct isolation in tissue culture is also performed. Most of the *O. tsutsugamushi* strains isolated from clinical specimens have been found to highly pathogenic for mice. However, strains having low pathogenicity in mice have also been reported. Peripheral blood and buffy coat samples from human patients and liver and spleen homogenates from rodents and crushed insects are the main specimens used to isolate the pathogen.

Isolating *O. tsutsugamushi* by inoculation of blood-clot-derived extract directly to host cells may not be suitable as the host cells may get damaged due to the toxic materials in the hemolysate. Separation of the leukocytes from the heparinized (anticoagulated) blood and inoculation of the leukocyte homogenate is considered to be a better option for successful isolation of the pathogen.[2] Susceptible cell lines used commonly for the isolation of *O. tsutsugamushi* are L-929 mouse fibroblast cells, BS-C40 African Green Monkey Kidney cells, and Vero cells.

(ii) Serological Assays Detection of antibody against *O. tsutsugamushi* has been classically detected by the Weil Felix test, which is based on the reaction between an *O. tsutsugamushi*-specific antibody and some unknown component of *Proteus mirabilis* OXK that is common to *Orientia*. A complement fixation (CF) test employing yolk sac-derived antigens was also routinely used. However, the nonspecificity and low sensitivity of the Weil Felix test and the technical complexity

and difficulties due to anticomplementary activity in the CF test were major reasons of these diagnostic procedures to be replaced with indirect immunoflourescence (IIF) and indirect immunoperoxidase (IIP) methods. The indirect IF and IIP were subsequently recommended by the WHO task force meeting on the serodiagnosis of scrub typhus and considered the gold standard for the serological diagnosis of scrub typhus. The WF test is still in use in several rudimentary hospitals in developing countries in Southeast Asia, as these places lack the facilities and reagents required to undertake the indirect IF and IIP tests. However, proper sampling, follow-up, and analysis of the WF test could effectively aid in provisional diagnosis of scrub typhus. A fourfold or higher rise in antibody titers assessed in paired sera samples and determination of the correct cutoff values have been found to have greatly enhanced the positive predictive value the WF test.[84]

In the routinely used IIF and IIP tests too collection of paired sera samples 7–10 days apart is highly recommended. Another important consideration in the serological diagnosis of scrub typhus lies in the differentiation of IgM and IgG. It is also of high value to incorporate local type strains of *O. tsutsugamushi* along with the prototype strains in the antigen panel used for these assays.

(iii) Molecular Assays The inception of PCR as a diagnostic tool and further improvements in molecular detection technology by the development of real-time PCR techniques have vastly improved the molecular diagnostic potential for the specific and sensitive detection of pathogens in clinical samples. PCRs are capable of detecting *O. tsutsugamushi* during acute phase of infection, which occurs before specific antibody can be detected.[32,70,85] Moreover, the detection of the organism, its antigen(s) or its DNA seem more logical to antibody detection because it indicates active disease. This is particularly valid when used in an endemic area of infection, where the background antibodies often interfere with the interpretation of antibody assays. Nested, seminested, or reamplified PCR tests make it possible to detect *Orientia*[32,71–73,85–88] in clinical specimens more reliably than primary PCR, but at the expense of additional time and chance of false-positive results. PCR has been successfully used to follow the efficacy of antibiotic treatment in eradicating *Orientia* and *Rickettsia* from the blood of patients[89,90] and for the detection and/or eradication of orientiae from eschar, blood and buffy coats, and so forth.[25,29,30,89,91] The development of real-time PCR-based assays has lead to an increased sensitivity of assays for the detection of rickettsial infections from clinical samples and evaluation of antibiotic susceptibilities.[92] Quantitative and semiquantitative real-time PCRs specific for *O. tsutsugamushi* have been developed based on the 47 kDa, 56 kDa, and *groEL* genes.[1,27,31,62,91] The quantitative real-time PCR assay described by Jiang et al.[31] employed primers based on conserved regions of the 47 kDa *omp* antigen gene and could detect *O. tsutsugamushi* DNA from patient blood and sera samples at a concentration as few as 3–10 copies per reaction and has been reported to be even more sensitive in detecting the presence of *O. tsutsugamushi* than mouse inoculation.[93] As few as two copies of *O. tsutsugamushi* DNA could be detected in the *groEL* quantitative real-time PCR assay described recently.[62] A duplex PCR for the simultaneous detection of *O. tsutsugamushi* and *Leptospira* from wild rodents have been reported of late.[94] Development of molecular diagnostic tools and techniques has enabled simultaneous and differential diagnosis of these pathogens at the molecular level with high degree of simplicity and accuracy (Figure 57.2).

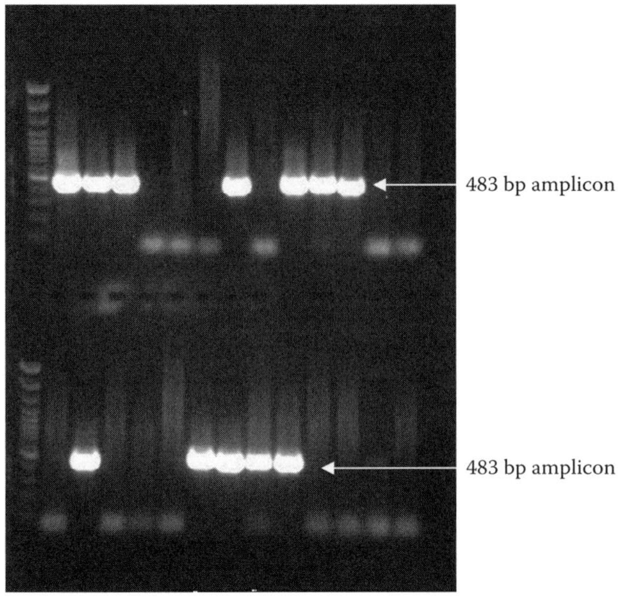

FIGURE 57.2 PCR detection of a specific 483 bp amplicon in *Orientia tsutsugamushi* using the protocol of Saisongkorh, W. et al., *Trans. R. Soc. Trop. Med. Hyg.*, 98, 360, 2004.

57.1.5 Treatment

Antimicrobial agents like chloramphenicol, rifampicin, the tetracyclines and the macrolides has clear-cut therapeutic effect on *O. tsutsugamushi*. Beta-lactam antibiotics are not effective against this pathogen. Tetracycline antibiotics (tetracycline, dimethyl-chlorotetracycline, doxycycline, and minocycline), chloramphenicol, kitasamysin (leukomysin, a macrolide antibiotic), and rifampicin all have strong inhibitory effects against *O. tsutsugamushi*. However, *O. tsutsugamushi* is reported to show higher resistance to some quinolone agents (e.g., norfloxacin, ciprofloxacin, and ofloxacin) than typhus and spotted fever group rickettsiae.[2]

Regarding the administration of effective antibiotics, there have been isolated reports that low-dose, long-term use is effective and prevents reoccurrence of the infections. For example, in classical scrub typhus, 50 mg tetracycline is given twice a day, the amount being halved after the fever subsides, with twice-daily administration to a total of 1.5 g. Some reports also strongly suggest that one dose of 200 mg of doxycycline-cured scrub typhus without resurgence that this regimen also served as a prophylactic against the disease. This trend in treatment is the one most followed in recent times for the treatment of scrub typhus, as well as for other rickettsioses.[95]

57.2 METHODS

57.2.1 Sample Preparation

57.2.1.1 Isolation of *Orientia* from Clinical Samples

Anticoagulated peripheral blood samples (obtained by addition of 10 U/mL Heparin or 0.135% EDTA) is centrifuged at 300 × *g* for 10 min and the thin layer of leukocytes or buffy coat is removed and resuspended in sterile sucrose phosphate buffered glutamic acid (SPG) solution (3 mM KH_2PO_4; 7.2 mM KH_2PO_4; 4 mM L-glutamic acid; 218 mM sucrose; pH 7.0) and used for mouse inoculation.

For the isolation of leukocytes, heparinized blood is mixed with an equal volume of the culture medium and 5 mL of the mixture is layered onto 3.5 mL of Ficoll-Conray mixture (density 1.114). The mixture is centrifuged at 500 rpm for 30 min and the leukocytes separated as a layer post centrifugation is collected and homogenized in a Dounce homogenizer in 0.3 mL of culture medium containing 2% calf serum. 0.1 mL of this homogenate is inoculated onto cell monolayers grown in Leighton tubes or shell vials. The inoculum is removed after 1–2 h of incubation at 37°C and the monolayer is replenished with culture media supplemented with 2% calf serum. The cells are maintained with media change given every 3–4 days until growth of *O. tsutsugamushi* is observed in cell smears stained with Giemsa or Gimenez staining or by IF.[96,97] If no infection is observed up to 6 weeks, the isolation is considered unsuccessful.

Alternately, fresh or stored (−80°C) blood clot from peripheral blood (drawn from suspected patient who has not been administered antibiotics) is homogenized with a mortar and pestle in SPG and the homogenate is used to inoculate hairy mice through the subcutaneous or intraperitoneal route. The mice are then injected i.p. with an immunosuppressant cyclophosphamide monohydrate at 0.2 mg/g body weight 1, 2, and 4 days after being injected with the test sample. If the mice were healthy, an additional injection of the immunosuppressant drug is given 7 days after the infection. After 14–17 days after infection, the mice are observed for the development of symptoms like ruffled fur, slowness in movement, weight loss, and so forth. If no symptoms are observed, the mice are sacrificed and the liver and spleen homogenates are prepared in SPG and injected i.p into other mice for blind passage. Infected mice should show signs of splenomegaly and heavy viscous exudates in the peritoneal cavity on autopsy as the first sign of successful isolation. The peritoneal smear could be observed by Giemsa or Gimenez staining, or by immunofluorescence microscopy. Blind passages are normally carried on until the third passage and sera samples stored at each step to monitor the presence or absence of antibodies. Blood samples drawn may also be analyzed by PCR techniques for the detection of the agent in peripheral blood.

57.2.1.2 Extraction of DNA from Clinical Specimens

For extraction of nucleic acid from blood clots and PBMC (Buffy Coat) samples, the clinical material is homogenized in distilled water and freshly prepared proteinase K and SDS was added at a final concentration of 0.2 mg/mL and 1% (wt/vol), respectively, in buffer containing 10 mM Tris-HCL and 1 mM EDTA, and pH 8.0 (TE). Digestion was performed at 55°C for 1 h and the DNA is extracted by phenol-chloroform-isoamyl alcohol (25:24:1) two to three times and once with chloroform-isoamyl alcohol (24:1) method and precipitated by adding one-tenth volume of 3 M sodium acetate and two volumes of 95% ethanol. The resultant DNA is washed once or twice with 70% ethanol and airdried. The DNA is resuspended in distilled water or TE buffer and 1–10 μL of the resuspended DNA is used for the PCRs.

For extraction of DNA from serum samples, Briefly, a volume of 100 μL of each sample is combined with the same volume of a solution containing (as final concentrations) 100 mM KCl, 20 mM Tris-HCl pH 8.3, 5 mM $MgCl_2$, 0.2 mg/mL gelatin, and 0.9% Tween 20 solution. Proteinase K is added to a final concentration of 60 μg/mL. The mixture is incubated at 55°C for 1 h. Proteinase K is then inactivated by heating the mixture at 95°C for 10 min followed by centrifugation at 10,000 rpm for 10 min in a microcentrifuge. A 10 μL aliquot of the supernatant is employed for PCR.

Alternatively, DNA could be extracted from the EDTA anticoagulated/heparinized blood samples using Blood Genomic DNA kit (Sigma) or Wizard SV Genomic DNA purification system (Promega), as per the manufacturer's instructions.

57.2.2 Detection Procedures

57.2.2.1 Nested PCR

This nested PCR methodology has been used most widely for the definitive diagnosis and prognosis of *O. tsutsugamushi* infection.[32] The nested PCR is performed with a reported set of *Orientia*-specific oligonucleotide primers p 34 (5′TC AAGCTTATTGCTAGTGCAATGTCTGC 3′) and p 55 (5′ AGGGATCCCTGCTCTGTGCTTGCTGCG3′)astheouterpair, and p 10 (5′ ATCAA GCTTCCTCAGCCTACTATAATGCC) and p 11 (5′CTAGGGATCCCGACAGATGCACTA TTAGGC 3′) as the inner pair.[32]

The first PCR mixture (25 μL) containing 1.5 mM $MgCl_2$ is prepared using Taq PCR core kit (Qiagen), as per the manufacturer's instructions. Alternately, the PCR mix may be prepared to constitute 50 mM KCl, 10 mM Tris-HCl (pH 8.3), 200 μM each of dATP, dTTP, dGTP, dCTP, 200 nM of primers and 0.625 U of Taq DNA polymerase (Qiagen). 5 μL of template DNA is added to the mixture.

The PCR amplification is optimized to comprise an initial denaturation at 94°C for 5 min followed by 30 cycles of 94°C for 1 min, 57°C for 1 min, and 72°C for 2 min in a thermal cycler (i-Cycler, BioRad). The products at the end of 30 cycles are subjected to a final extension step for 10 min at 72°C. 5 μL of amplified product from the first PCR is used as the template for the second, other conditions remaining the same. The PCR amplified products (5 μL) along with a 1 kb DNA ladder (Promega) are resolved in 1% agarose gel containing 0.5 μg/mL ethidium bromide in a submarine gel electrophoresis system. The bands were observed using Image Master VDS gel documentation system (Pharmacia Biotech, USA).

57.2.2.2 Serotype-Specific PCR

Serotype-specific amplification for samples positive by the nested PCR is performed using the product obtained from PCR with primers p34 and p55 as template. These serotype-specific primers are designed to amplify distinguishable fragments for the Karp, Kato, Gilliam, Kawasaki, and Kuroki strains. Individual primary PCR products are amplified using a combination of five primer sets comprising of a combination of primers p10, p11, KP (5′ ACAATATCGGATTTATAACC 3′), KT (5′ GAATATTTAATAGCACTGGA 3′), KW (5′ ATGCTGCTATTGATACAGGC3′),G(5′TTTATATCACTAT ATATCTT 3′), and KR (5′ CACCGGATTTACCATCATAT 3′) as per the protocol of Furuya et al.[32] The primers sets used for amplification of each serotype and the expected amplicon indicative of a particular serotype are as given below.

Primer Pairs	Serotype	Expected Amplicon Size
10 & G	Gilliam	407 bp
10 & KP	Karp	230 bp
10 & KT	Katō	242 bp
10 & KR	Kuroki	220 bp
11 & KW	Kawasaki	523 bp

The PCR conditions used was the same as that used in the nested PCR. 5 μL of PCR product from the first set of nested PCR (with p34 and p55) is combined with five individual sets of serotype-specific primers and the PCR mix is subjected to an initial denaturation at 94°C for 5 min, followed by 30 cycles of 94°C for 1 min, 57°C for 1 min, and 72°C for 2 min. The products at the end of 30 cycles are subjected to a final extension step for 10 min at 72°C.

57.2.2.3 Real-Time PCR

(i) Quantitative Real-Time PCR Based on 47 kDa Antigen Gene This procedure described by Jiang et al. as one of the earliest reports on the use of real-time PCR technique for detection of *O. tsutsugamushi* DNA.[31] This technique could reportedly detect as few as 3–10 copies of *O. tsutsugamushi* DNA from patient blood and sera samples at a concentration as few as 3–10 copies per reaction and has been reported to be even more sensitive in detecting the presence of *O. tsutsugamushi* than mouse inoculation.[93] The sequences of primers and probes were selected from the 47 kda outer membrane protein antigen gene based on the sequences of Karp, Kato, Gilliam, and Boryong strains with accession numbers L31934, L1697, L31933, and L319335, respectively.

Forward Primer (OtsuFP630): 5′ AACTGATTTT-ATTCAAACTAATGCTGCT 3′
Reverse Primer (OtsuRP747): 5′TATGCCTGAGTAA-GATACRTGAATRGAATT 3′

The primer pair is capable of producing an amplicon of 118 bp. A 31 bp fluorescent TaqMan probe (OtsuPR665), 5′-6FAM-TGGGTAGCTTTGGTGGACCGATGTTTAATCT-TAMRA-3′ is employed having 6-Carboxyfluorescien (FAM) reporter dye at the 5′end and 6-Carboxytetratmethyl rhodamine (TAMRA) quencher dye at the 3′end.

The qPCR assay is conducted in a total volume of 25 μL containing of 1 μL of DNA template, 2.5 μL of 10× PCR buffer with 50 mM $MgCl_2$, 2.5 μL of 2 mM dNTPs, 0.25 μL (5 U/μL) of platinum Taq DNA polymerase (Invitrogen), 0.25 μL (10 μM) of each primer, and 0.5 μL (10 μM) of the fluorescent probe. The qPCR mix is incubated at 94°C for 5 min followed by 45–50 cycles of two-step amplification at 94°C for 5 s and 60°C for 30 s on a Cepheid SmartCycler System (Sunnyvale, CA). Fluorescence is monitored during every thermal cycle at annealing step and data were analyzed with SmartCycler software (Version 1.2b). 10^5 copies/μL of a recombinant pWMC-Kt47 DNA is used as the positive control and monitored to return consistent cycle threshold values across reactions. This plasmid contains the open reading frame sequence of the Kato 47 kDa gene that is ligated into the plasmid VR1012 (Vical).

(ii) Real-Time PCR Based on 56 kDa Antigen Gene PCR was optimized with a set of oligonucleotide primers (Forward: 5′ TAATTGCTAGTGCAATGTCTGCGTT 3′, Reverse: 5′CCAAAGTCACGA TCAGCTATACT 3′) designed based on the sequence of the 56 kDa type-specific antigen (TSA) gene of *Orientia tsutsugamushi* (GenBank Accession nos: AY357216, AF050669, AF302994, M33004).[27] The primers are expected to amplify a 441 bp truncated portion of the 56 kDa TSA gene sequence. The PCR, when repeated with the same forward but a different reverse primer (5′GCATTATAGTAGGCTGAGGAGGTGT 3′) is expected to yield an amplicon of 411 bp. The primers are designed with the help of the software Gene Runner (Hastings Software Inc., USA). In this real-time PCR analysis, a SYBER-Green Real-Time PCR Premix (Genome, Dehra Dun, India) is used for all the reactions.

Reaction is performed in a 25 μL reaction volume comprising of 5 μL of extracted DNA, 10 pmol of each primer and 12.5 μL of the double strength real-time PCR premix in a Rotor-Gene 3000 (Corbett Research, Australia). The real-time PCR conditions are optimized to comprise an initial denaturation step of 10 min at 95°C, followed by 45 cycles of 95°C for 15 s, 60°C for 30 s, and 72°C for 20 s. A melt curve analysis is performed to observe melting characteristics of the amplicon to determine the presence of the specific product. The analysis immediately followed amplification at 95°C for 60 s, cooling to 55°C for 60 s, and a slow rise in temperature at a rate of 1°C/5 s with a continuous acquisition of fluorescence decline. The real-time amplified PCR products are also checked on agarose gel to demonstrate that under the specified real-time conditions, the amplicon of the desired size is produced.[27] Amplicons are randomly selected and purified from the gel using gel extraction kit (Qiagen) and confirmed by sequencing followed by alignment studies with reported sequences.

Real-Time PCR curves: Fluorescence curves based on threshold cycles (at which the fluorescence exceeds the background during the exponential phase of amplification) are constructed by Rotor-gene real-time analysis software 6.0 (Build 1).

(iii) Real-Time PCR Based on groEL Gene Recent studies have characterized the 60 kDa heat-shock protein GroEL of α-proteobacteria as a molecular indicator of various forms of cellular stress. The corresponding *groEL* gene has been exploited as a target for molecular diagnostics for detection of scrub typhus in a report published by D.H. Paris and coworkers.[62] A set of specific primers for the generation of a 160 bp amplicon of the *groEL* gene was designed using Primer Select version 6.1 software (DNA Star, USA). The sequences of the primers are as follows:

Forward: 5′TTGCAACRAATCGTGAAAAG 3′
Reverse: 5′TCTCCGTCTACATCATCAGCA 3′

The PCR mix comprised of primers (200 nM each), 2 µL of DNA template, 10 µL of master mix (QuantiMix Easy, Biotools, Spain) containing SYBER-Green, Taq polymerase, $MgCl_2$ (4 mM), dNTPs and distilled water in a final volume of 20 µL. Cycling is carried using Rotor-gene 3000 real-time thermocycler (Corbett Research, Mortlake, Australia). Cycling conditions are optimized to comprise of an initial denaturation at 95°C for 5 min followed by 40 cycles at 95°C for 15 s, 54°C for 15 s, and 72°C for 20 s, with fluorescence monitoring at the 54°C annealing step on a predetermined SYBR/FAM channel. Melting curve analysis is performed with increments of 1°C/step (72°C–95°C) to determine the change in peak fluorescence over time (dF/dT).

To determine limits of the assay, plasmids containing the amplified regions of *gro*EL were generated by ligation into pGEM-T Easy Vectors (Promega, USA). Subsequent quantitation of the plasmid is performed as per standard procedures as protocols and appropriate dilutions assessed by the real-time PCR to determine the detection limit of the assay. The assay returns a detection limit of less than 3 copies/µL of *O. tsutsugamushi* using serial dilutions of the reference plasmid as described earlier.

57.3 CONCLUSION AND FUTURE PERSPECTIVES

Orientia tsutsugamushi, formerly *Rickettsia tsutsugamushi*, is an obligate intracellular parasite that causes an incapacitating but potentially fatal disease known as scrub typhus when a susceptible individual is bitten by trombiculid mites harboring this pathogen. The history of this disease is now nearly 100 years old. It is one of those infectious agents, which changed the course of nations and played havoc with soldiers and civilian populations alike in all the major wartime as well as in peacetime history.[4] Increasing reports about the presence of this deadly pathogen in unexpected environments and reemergence of the diseases in places thought to have been freed of the menace have brought this agent under strict scrutiny once again.[6] When recognized, the disease is often easily treated, but the unusual conditions created by war, military operations or civil unrest may delay or prevent appropriate treatment. Scrub typhus is notoriously difficult to diagnose because they share symptoms with many other febrile diseases with similar epidemiology.[4]

Orientia tsutsugamushi is also regarded as a potential biological warfare agent and has been classified in the Risk Group 3+ category. Civilian and especially military history is replete with incidences of scrub typhus, a febrile zoonosis, which played havoc with susceptible population during the preantibiotic era. Although choice of treatment for scrub typhus has become available with the advent of antibiotics, it is notoriously difficult to diagnose because of its nonspecific symptoms. This listed biological warfare agent, especially its drug refractory strains, pose an increased threat in their use in bioterrorism activities.

O. tsutsugamushi infection elicits serological response against some of its outer membrane antigens, including the moieties which cross-react with other species of the bacterium and form the basis for the group, subgroup, and strain-specific range of protein antigens (e.g., the 56, 47, and 22 kDa outer membrane proteins). These outer membrane proteins and their genes have been extensively exploited and are still being researched for their potential in serological and molecular diagnosis.

It is important for researchers to understand the importance of rickettsial diseases as emerging, as they exhibit most of the characteristics of emerging and reemerging pathogens and are associated with increased clinical risk, which is compounded by reports of antibiotic resistance or new clinical presentations. Specific and sensitive serodiagnostic and molecular diagnosis tools are essential for preventive and prophylactic actions that aid in proper identification and eradication of this pathogen from the infected patients, as well as help in epidemiological research on this organism.

REFERENCES

1. Eremeeva, M.E., and Dasch, G.A., *Rickettsia* and *Orientia*. In Max Sussman (Editor), *Molecular Medical Microbiology*, Vol. II, pp. 2177–2217, Academic Press, Maryland Heights, MO, 2001.
2. Kawamura, A. Jr., Tsutusgamushi disease. In Kawamura, A., Tanaka, H., and Tamura, A. (Editors), *Tsutsugamushi Disease*, University of Tokyo Press, Tokyo, Japan, 1995.
3. Tamura, A. et al., Classification of *Rickettsia tsutsugamushi* in a new genus, *Orientia* gen. nov., as *Orientia tsutsugamushi* comb. nov., *Int. J. Syst. Bacteriol.*, 45, 589, 1995.
4. Kelly, D.J. et al., The past and present threat of rickettsial diseases to military medicine and international public health, *Clin. Infect. Dis.*, 34, S145, 2002.
5. Enatsu, T., Urakami, H., and Tamura, A., Phylogenetic analysis of *O. tsutsugamushi* strains based on the sequence homologies of 56-KDa type-specific antigen genes, *FEMS Microbiol. Lett.*, 180, 163, 1999.
6. Watt, G., and Parola, P., Scrub typhus and tropical rickettsiaoses, *Curr. Opin. Infect. Dis.*, 16, 429, 2003.
7. Sharma, A. et al., Investigation of an outbreak of scrub typhus in the Himalayan region of India, *Jpn. J. Infect. Dis*, 58, 208, 2005.

8. Sundhindra, B.K. et al., Rickettsial spotted fever in Kerala, *Nat. Med. J. Ind.*, 17, 51, 2004.
9. Mathai, E. et al., Outbreak of Scrub Typhus in Southern India during cooler months, *Ann. N. Y. Acad. Sci.*, 990, 359, 2003.
10. Mathai, E. et al., Serological evidence of the continued presence of human rickettsioses in southern India, *Ann. Trop. Med. Parasitol.*, 95, 395. 2001.
11. Takahashi, K., Urakami, H., and Tamura, A., Purification of cell envelopes of *Rickettsia tsutsugamushi*, *Microbiol. Immunol.*, 29, 475, 1985.
12. Tamura, A., Urakami, H., and Tsurahara, T., Purification of Rickettsia tsutsugamushi by Percoll density gradient centrifugation, *Microbiol. Immunol.*, 26, 321, 1982.
13. Urakami, H., Tsurahara, T., and Tamura, A., Electron microscopic studies on intracellular multiplication of *Rickettsia tsutsugamushi* in L cells, *Microbiol. Immunol.*, 28, 1191, 1984.
14. Silverman, D.J., and Wisseman, C.L. Jr., Comparative ultrastructural study on the cell envelopes of *Rickettsia prowazekii*, *Rickettsia rickettsii* and *Rickettsia tsutsugamushi*, *Infect. Immun.*, 21, 1020, 1978.
15. Amano, K., Tamura, A., and Ohashi, N., Deficiency of peptidoglycan and lipopolysaccharide components in *Rickettsia tsutsugamushi*, *Infect. Immun.*, 55, 2290, 1987.
16. Miyamura, S, Ohta, T., and Tamura, A., Comparison of *in-vitro* susceptibilities of *Rickettsia prowazekii*, *R. rickettsii*, *R. sibirica* and *R. tsutsugamushi* to antimicrobial agents, *Japan. J. Bacteriol*, 44, 717, 1989.
17. Wissemann, C.L. Jr., and Waddell, A., In-vitro susceptibility of *Rickettsia rickettsii* to doxycycline, *J. Infect. Dis.*, 145, 584, 1982.
18. Wissemann, C.L. Jr., Waddell, A.D., and Walsh, W.T., In-vitro studies of the action of antibiotics on *Rickettsia prowazekii* by two basic methods of cell culture, *J. Infect. Dis.* 130, 564, 1974.
19. Tamura, A. et al., Analysis of polypeptide composition and antigenic components of *Rickettsia tsutsugamushi* by polyacrylamide gel electrophoresis and immunoblotting, *Infect. Immun.*, 48, 671, 1985.
20. Ohashi, N., Tamura, A., and Suto, T., Immunoblotting analysis of antirickettsial antibodies produced in patients of tsutsugamushi disease, *Microbiol. Immunol.*, 32, 1085, 1988.
21. Oaks, E.V. et al., Antigenic and genetic relatedness of eight *Rickettsia tsutsugamushi* antigens, *Infect. Immun.*, 57, 3116, 1989.
22. Seong, S.Y. et al., Mapping of antigenic determinant regions of the Bor56 protein of *Orientia tsutsugamushi*, *Infect. Immun.*, 65, 5250, 1997.
23. Oaks, E.V., Stover, C.K., and Rice, R.M., Molecular cloning and expression of *Rickettsia tsutsugamushi* genes for two major protein antigens in *Escherichia coli*, *Infect. Immun.*, 55, 1156, 1987.
24. Hickman, C.J. et al., Molecular cloning and sequence analysis of a *Rickettsia tsutsugamushi* 22 kDa antigen containing B- and T-cell epitopes, *Microb. Pathog.*, 11, 19, 1991.
25. Piengchan, S. et al., Association of high *Orientia tsutsugamushi* DNA loads with disease of greater severity in adults with scrub typhus, *J. Clin. Microbiol.*, 47, 2, 430, 2009.
26. Kang, S.J. et al., Scrub typhus induced by peripheral blood stem cell transplantation in the immunocompromised patient: Diagnostic usefulness of nested polymerase chain reaction, *Transfusion*, 50, 467, 2010.
27. Bakshi, D. et al., Development of a Real Time PCR assay for the diagnosis of Scrub Typhus cases in India and evidence of the prevalence of new genotype of *Orientia tsutsugamushi*, *Acta Trop.*, 104, 63, 2007.
28. Kim, D.M. et al., Clinical usefulness of eschar polymerase chain reaction for the diagnosis of scrub typhus: A prospective study, *Clin. Infect. Dis.*, 43, 1296, 2006.
29. Lee, S.H. et al., Usefulness of eschar PCR for diagnosis of scrub typhus, *J. Clin. Microbiol.*, 44, 1169, 2006.
30. Watcheree, S., Mongol, C., and Khachornsakdi, S., Evaluation of nested PCR for the diagnosis of scrub typhus among patients with acute pyrexia of unknown origin, *Trans. Royal Soc. Trop. Med. Hyg.*, 98, 360, 2004.
31. Jiang, J. et al., Development of a quantitative real time polymerase chain reaction assay specific for *Orientia tsutsugamushi*, *Am. J. Trop. Med. Hyg.*, 71, 351, 2004.
32. Furuya, Y. et al., Serotype specific amplification of *Rickettsia tsutsugamushi* DNA by nested polymerase chain reaction, *J. Clin. Microbiol.*, 31, 1637, 1993.
33. Chen, W.J. et al., Recombinant 56-kilodalton major outer membrane protein antigen of *Orientia tsutsugamushi* Shanxi and its antigenicity, *Infect. Immun.* 71, 8, 4772, 2003.
34. Oaks, E.V., Stover, C.K., and Rice, R.M., Molecular cloning and expression of *Rickettsia tsutsugamushi* genes for two major protein antigens in *Escherichia coli*, *Infect. Immun.*, 55, 1156, 1987.
35. Jang, W.J. et al., Evaluation of an IgM capture ELISA assay for diagnosis of *Orientia tsutsugamushi* infection, *Clin. Diag. Lab. Immunol.*, 10, 394, 2003.
36. Tay, S.T. et al., Expression of recombinant proteins of *Orientia tsutsugamushi* and their applications in the serodiagnosis of scrub typhus, *Diagn. Microbiol. Infect. Dis.*, 44, 137, 2002.
37. Choi, M.S. et al., Homotypic and heterotypic antibody responses to a 56 kilodalton protein of *Orientia tsutsugamushi*, *Infect. Immun.*, 67, 6194, 1999.
38. Ching, W.M. et al., Expression and refolding of truncated recombinant major outer membrane protein antigen (r56) of *Orientia tsutsugamushi* and its use in enzyme linked immunosorbent assays, *Clin. Diag. Lab. Immunol.*, 5, 519, 1998.
39. Seong, S.Y. et al., Induction of homologous immune response to *Rickettsia tutsugamushi* Boryong with a partial 56-Kilodalton recombinant antigen fused with maltose-binding protein MBP-Bor56, *Infect. Immun.*, 65, 1541, 1997.
40. Bell, E.J., Benett, B.L., and Whitman, L., Antigenic differences between strains of scrub typhus ass demonstrated by cross-neutralization test, *Proc. Soc. Exp. Biol. Med.*, 62, 134, 1946.
41. Bengston, I.A., Apparent serological heterogeneity among strains of tsutsugamushi disease (scrub typhus), *Public Health Rep.*, 60, 1483, 1945.
42. Bennet, B.L., Smadel, J.E., and Gauld, R.L., Studies on scrub typhuss (Tsutsugamushi Disease), IV. Heterogeneity of strains of *R. tsutsugamushi* as demonstrated by cross-neutralization tests, *J. Immunol.*, 62, 453, 1949.
43. Fox, J.P., The neutralization technique I Tsutsugamushi disease (scrub typhus) and the antigenic differentiation of rickettsial strains, *J. Immunol.*, 62, 341, 1949.
44. Rights, F.L., and Smadel, J.E., Studies on scrub typhus (Tsutsugamushi disease), III. Heterogeneity of strains of *R. tsutsugamushi* as demonstrated by cross-vacccination studies, *J. Exp. Med.*, 87, 339, 1948.
45. Shishido, A. et al., The nature of immunity against scrub typhus in mice. II. The cross-protection test with mice for identification and differentiation of several strains of *Rickettsia orientalis* newly isolated in Japan, *Japan J. Med. Sci. Biol.* 12, 391, 1959.
46. Bozeman, F.M., and Elisberg, B.L., Serological diagnosis of scrub typhus by indirect immunofluorescence, *Proc. Soc. Exp. Biol. Med.*, 112, 568, 1963.

47. Philip, C.B., Observation on Tsutsugamushi disease (mite borne or scrub typhus) in Northwest Honshu Island, Japan in the fall of 1945. I. Epidemiological and ecological data, *Am. J. Hyg.* 46, 45, 1947.
48. Shishido, A., Identification and serological classification of the causative agents of scrub typhus in Japan, *Jpn. J. Med. Sci. Biol.*, 15, 330, 1962.
49. Shishido, A., Strain variation of *Rickettsia orientalis* in the complement fixation test, *J. Med. Sci. Biol.*, 17, 59, 1964.
50. Elisberg *et al.*, Physiographic distribution of scrub typhus in Thailand, *Acta Med. Biol.*, 15, S61, 1967.
51. Elisberg, B.L., Campbell, J.M., and Bozeman, F.M., Antigenic diversity of *Rickettsia tsutsugamushi*: Epidemiologic and ecologic significance, *J. Hyg. Epidemiol. Microbiol. Immunol.*, 12, 18, 1968.
52. Shirai, A. et al., *Rickettsia tsutsugamushi* strains found in chiggers collected in Thailand, *Southeast Asian J. Trop. Med. Pub. Health*, 12, 1, 1981.
53. Shirai, A. et al., Characterization of *Rickettsia tsutsugamushi* strains in two species of naturally infected laboratory reared chiggers, *Am. J. Trop. Med. Hyg.*, 31, 395, 1982.
54. Shirai, A. et al., Antigenic analysis by direct immunofluorescence of 114 isolates of *Rickettsia tsutsugamushi* recovered from febrile patients in rural Malaysia, *Jpn. J. Med. Sci. Biol.*, 32, 179, 1979.
55. Shirai, A. et al., Serological classification of *Rickettsia tsutsugamushi* organisms found in chiggers (Acarina: Trombiculidae) collected in Peninsular Malaysia, *Trans. Royal Soc. Trop. Med. Hyg.*, 75, 580, 1981.
56. Tamura, A. et al., Isolation of *Rickettsia tsutsugamushi* antigenically different from Kato, Karp and Gilliam strains from patients, *Microbiol. Immunol.*, 28, 873, 1984.
57. Kobayashi, Y. et al., Isolation of very low virulent strain of *Rickettsia tsutsugamushi* by use of cyclophosphamide treated mice. In Kazar, J., Ormsbee, R. A., and Tarasevich, I. N. (Editors), *Rickettsiae and Rickettsial Diseases*, pp. 181–188, VEDA Publ. House, Slovak Acad. Sci., Bratislava, 1995.
58. Chang, W-H. et al., Serological classification by monoclonal antibodies of *Rickettsia tsutsugamushi* isolated in Korea, *J. Clin. Microbiol.*, 28, 685, 1990.
59. Tange, Y., Kanemitsu, N., and Kobayashi, Y., Analysis of immunological characteristics of newly isolated strains of *Rickettsia tsutsugamushi* using monoclonal antibodies, *Am. J. Trop. Med. Hyg.*, 44, 371, 1991.
60. Yamamoto, S. et al., Immunological properties of *Rickettsia tsutsugamushi*, Kawasaki strain, isolated from a patient in Kyushu, *Microbiol. Immunol.*, 30, 611, 1986.
61. Ge, H. et al., Cloning and sequence analysis of the 22-kDa antigen genes of *Orientia tsutsugamushi* strains Kato, TA763, AFSC 7, 18-032460, TH1814 and MAK19, *Ann. N.Y. Acad. Sci.*, 1063, 231, 2005.
62. Paris, D.H. et al., A highly sensitive quantitative real-time PCR assay based on the *gro*EL gene of contemporary Thai strains of *Orientia tsutsugamushi*, *Clin. Microbiol. Infect.*, 15, 488, 2009.
63. Cho, N.H. et al., The *Orientia tsutsugamushi* genome reveals massive proliferation of the conjugative type IV secretion system and host-cell interaction genes, *Proc. Natl. Acad. Sci. USA*, 104, 7981, 2007.
64. Nakayama, K. et al., The whole genome sequencing of the obligate intracellular bacterium *Orientia tsutsugamushi* revealed massive gene amplification during reductive genome evolution, *DNA Res.*, 15, 185, 2008.
65. Kumura, K. et al., DNA base composition of *Rickettsia tsutsugamushi* determined by reversed-phase high-performance liquid chromatography, *Int. J. Syst. Bacteriol.*, 41, 247, 1991.
66. Kim, I.S. et al., High level expression of a 56 kilodalton protein gene (*bor56*) of *Rickettsia tsutsugamushi* Boryong and its application to enzyme-linked immunosorbent assays, *J. Clin. Microbiol.*, 31, 2057, 1993.
67. Ohashi, N. et al., Diversity of immunodominant 56 kDa type specific antigen (TSA) of *Rickettsia tsutsugamushi*. Sequence and comparative analyses of the genes encoding TSA homologous from four antigenic variants, *J. Biol. Chem.*, 267, 12728, 1992.
68. Ohashi, N. et al., Cloning and sequencing of the gene (*tsg56*) encoding a type specific antigen from *Rickettsia tsutsugamushi*, *Gene*, 91, 119, 1990.
69. Stover, C.K. et al., The 56 kilodalton major protein antigen of *Rickettsia tsutsugamushi*: Molecular cloning and sequence analysis of the *sta 56* gene and precise identification of a strain-specific epitope, *Infect. Immun.*, 58, 2076, 1990.
70. Furuya, Y., and Yoshida, Y., Detection of *Rickettsia tsutsugamushi* DNA by polymerase chain reaction, *Nippon Rinsho.*, 642, 422, 1992.
71. Furuya, Y. et al., Specific amplification of *Rickettsia tsutsugamushi* DNA from clinical samples by polymerase chain reaction, *J. Clin. Microbiol.*, 29, 2628, 1991.
72. Kawamori, F. et al., Two-step polymerase chain reaction for diagnosis of scrub typhus and identification of antigenic variants of *Rickettsia tsuutsugamushi*, *J. Vet. Med. Sci.*, 55, 749, 1993.
73. Murai, K. et al., Sensitivity of polymerase chain reaction assay for *Rickettsia tsutsugamushi* in patients' blood samples, *Microbiol. Immunol.*, 36, 1145, 1992.
74. Dumler, J.S., and Walker, D.H., Diagnostic tests for Rocky Mountain spotted fever and other rickettsial diseases, *Dermatol. Clin.*, 12, 25, 1994.
75. La Scola, B., and Raoult, D., Laboratory diagnosis of rickettsioses: Current approaches to diagnosis of both old and new rickettsial diseases, *J. Clin. Microbiol.*, 35, 2715, 1997.
76. Mahajan, S.K., and Bakshi, D., Acute reversible hearing loss in scrub typhus, *J. Assoc. Physicians Ind.*, 55, 512, 2007.
77. Wang, C.C. et al., Acute respiratory distress syndrome in scrub typhus, *Am. J. Trop. Med. Hyg.*, *76*, 1148, 2007.
78. Urakami, H., Tsurahara, T., and Tamura, A., Penetration of *Rickettsia tsutsugamushi* into cultured mouse fibroblasts (L cells): An electron microscopic observation, *Microbiol. Immunol.*, 27, 251, 1983.
79. Urakami, H., Tsurahara, T., and Tamura, A., Electron microscopic studies on intracellular multiplication of *Rickettsia tsutsugamushi* in L cells, *Microbiol. Immunol.*, 28, 1191, 1984.
80. Urakami, H., Tsuruhara, T., and Tamura, A., Intranuclear *Rickettsia tsutsugamushi* in cultured mouse fibroblasts (L cells), *Microbiol. Immunol.*, 26, 445, 1982.
81. Hanson, B., Improved plaque assay for *Rickettsia tsutsugamushi*, *Am. J. Trop. Med. Hyg.*, 36, 631, 1987.
82. Elghetany, M.T., and Walker, D.H., Hemostatic changes in Rocky Mountain spotted fever and Mediterranean spotted fever, *Am. J. Clin. Pathol.*, 112, 159, 1999.
83. Walker, D.H., The role of host factors in the severity of spotted fever and typhus rickettsioses, *Ann. N. Y. Acad. Sci.*, 590, 10, 1990.
84. Brown, G.W. et al., Diagnostic criteria for scrub typhus: probability values for immunofluorescent antibody and *Proteus* OXK agglutinin titers, *Am. J. Trop. Med. Hyg.*, 32, 1101, 1983.
85. Tay, S.T., Nazma, S., and Rohini, M.Y., Diagnosis of scrub typhus in Malaysian aborigines using nested polymerase chain reaction, *South East Asian J. Trop. Med. Pub. Health*, 27, 580, 1996.

86. Tay, S.T. et al., Isolation and PCR detection of rickettsiae from clinical and rodent samples in Malaysia, *Southeast Asian J. Trop. Med. Pub. Health*, 33, 772, 2002.
87. Horinouchi, H., Murai, K., and Okayama, A., Prevalence of genotypes of *Orientia tsutsugamushi* patients with scrub typhus in Miyazaki prefecture, *Microbiol. Immunol.*, 41, 503, 1997.
88. Ono, A. et al., Successful diagnosis using scab for PCR specimen in Tsutsugamushi disease, *Intern. Med.*, 41, 408, 2002.
89. Murai, K. et al., Eradication of *Rickettsia tsutsugamushi* from patients' blood by chemotherapy, as assessed by the polymerase chain reaction, *Am J. Trop. Med. Hyg.*, 52, 325, 1995.
90. Breitschwerdt, E.B., Papich, M.G., and Hegarty, B.C., Efficacy of Doxycycline, Azithromycin or Trovofloxacin for treatment of experimental Rocky Mountain spotted fever in dogs, *Antimicrob. Agents Chemother.*, 43, 813, 1999.
91. Kim, D.M., and Byun, J.N., Effects of antibiotic treatment on the results of nested PCR for scrub typhus, *J. Clin. Microbiol.*, 46, 3465, 2008.
92. Rolain, J.M. et al., Evaluation of antibiotic susceptibilities of three rickettsial species including *Rickettsia felis* by a quantitative PCR DNA assay, *Antimicrob. Agents Chemother.* 46, 2747, 2002.
93. Singhsilarak, T. et al., Detection of *Orientia tsutsugamushi* in clinical samples by a quantitative real-time polymerase chain reaction, *Am. J. Trop. Med. Hyg.*, 72, 640, 2005.
94. Piyada, W. et al., An application of Duplex PCR for the detection of *Leptospira* spp. and *Orientia tsutsugamushi* from wild rodents, *Jpn. J. Infect. Dis.*, 61, 407, 2008.
95. Brown, G.W. et al., Single dose doxycycline therapy for scrub typhus, *Trans. Roy. Soc. Trop. Med. Hyg.*, 72, 412, 1978.
96. Gimenez, D.F., Staining Rickettsiae in yolksack cultures, *Stain Technol.*, 39, 135, 1964.
97. NCCLS, Laboratory Diagnosis of Blood-Borne Parasitic Diseases. Approved Guideline M15-A, National Committee for Clinical Laboratory Standards, Villanova, PA, 2000.

58 *Rickettsia*

Marina E. Eremeeva and Gregory A. Dasch

CONTENTS

58.1 INTRODUCTION

The genus *Rickettsia* encompasses a diverse group of non-motile, gram-negative, nonspore-forming, highly pleomorphic bacteria that are obligate intracellular parasites of mammals, plants and insects. Among the 24 validated species within the genus, *Rickettsia rickettsii*, the etiologic agent of Rocky Mountain spotted fever (RMSF), is the prototypic pathogenic species. Being the first rickettsiosis described, RMSF was characterized at the end of the nineteenth and beginning of the twentieth century. However, despite the long history, our current knowledge about the mechanisms of infection, ecology, and epidemiology of the etiologic agent is still very fragmentary. Part of the problem is that a limited number of laboratories found primarily in the western hemisphere work with *R. rickettsii*, and the number of investigations has fluctuated greatly over time depending upon the priorities of funding agencies. Modern safety requirements and Select Agent regulations have contributed additional limitations for work with this organism. Moreover, different aspects of rickettsial biology were studied with a limited number of species; consequently, attempts to extrapolate observations from one species to the other should be taken very cautiously because the genus *Rickettsia* displays significant variability. This statement is well supported by comparisons of the complete genome sequences of different species; however, for the purpose of this review, we will refer to some findings obtained with different spotted fever group rickettsiae (SFGR) and comment about the values of such extrapolations to *R. rickettsii*.

58.1.1 Classification and Morphology

Rickettsia rickettsii belongs to the spotted fever group of rickettsiae in the genus *Rickettsia*, the family Rickettsiaceae, order Rickettsiales. Isolate Sheila Smith (ATCC VR-149) obtained from the blood of a patient from Montana in 1946 is the type strain of *R. rickettsii*. Strain Bitterroot (also erroneously called strain R) (ATCC VR-891) was isolated from *Dermacentor andersoni* and is often used as a substitute for isolate Sheila Smith.

Rickettsia rickettsii is an obligately intracellular, gram-negative bacterium and one of the best-studied species within the largest of the three evolutionary lineages of obligately intracellular microorganisms that comprise a major subdivision of the Alphaproteobacteria (Figure 58.1).[1–4] Beside *R. rickettsii*, the genus *Rickettsia* includes at least 24 validated species, several isolates with the status of *Candidatus*, and multiple unnamed isolates.[6] It is traditionally divided into typhus group and spotted fever group; however, recent publications have also proposed separating an ancestral group

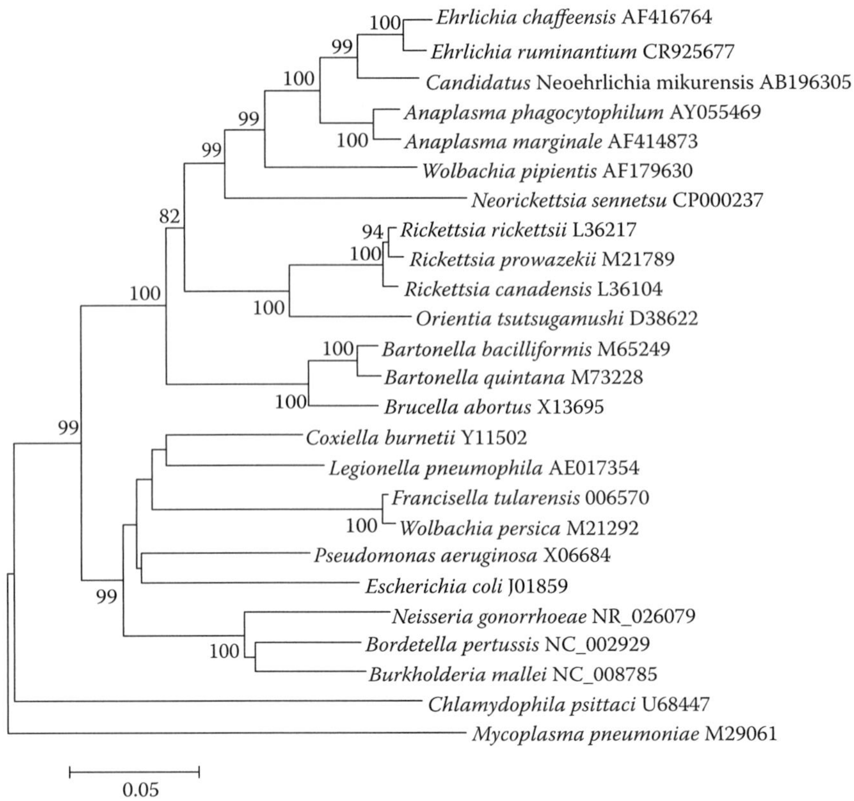

FIGURE 58.1 Analysis of the phylogenetic position of *Rickettsia rickettsii*. The phylogenetic tree was constructed using nucleotide sequences from 16S rRNA genes of *Rickettsia*, *Orientia*, *Ehrlichia*, and *Anaplasma* species and other bacteria belonging to different lineages of Alphaproteobacteria and other groups. Nucleotide sequences were aligned using ClustalW and the tree was drawn using the MEGA4.0 software.[5] Distance matrix was calculated using Kimura-2 parameters and tree was obtained using the Neighbor-Joining method based on comparison of 1169 sites after complete deletion of gaps. The scale represents 0.05% divergence. Numbers at nodes are the proportions of 1000 bootstrap resamplings that support the topology shown. Only bootstrap values greater than 80 are shown. The sequences of the isolates used in this analysis are from NCBI GenBank and the accession numbers are shown next to the species names.

consisting of *R. bellii* and *R. canadensis* and transitional group that includes *R. akari*, *R. australis*, and *R. felis*.[7,8]

Rickettsiia rickettsii is a rod-shaped bacterium, 0.3–0.5 μm in diameter and 0.8–2.0 μm in length. The cell wall of *R. rickettsii* resembles those of other gram-negative bacteria, with a bilayer inner membrane, a peptidoglycan layer, and a bilayer outer memebrane.[9] Rickettsial cell walls contain the constituents of peptidoglycan—namely sugars, amino acids, muramic acid, and diaminopimelic acid, but no teichoic acids. A microcapsular S-protein layer is associated with the outer leaflet of the outer membrane and is composed of repetitive electron-dense subunits.[10] External to the microcapsular layer surrounding individual rickettsia is a large, electron-translucent halo zone.[10] *Rickettsia* species, including *R. rickettsii*, do not manifest any true capsule or spore forms (Figure 58.2).

Species of *Rickettsia* do not stain well with the classical Gram stain; therefore, *R. rickettsii*, like the other species of *Rickettsia*, is stained most satisfactorily with Gimenez stain.[11] The method of Gimenez is based on the selective retention of carbol fuchsin by rickettsiae, which are seen as bright red, slender, coccobacillary forms against a pale greenish-blue background produced by the counter stain, typically malachite green. Alternatively, *Rickettsia* cells can be visualized

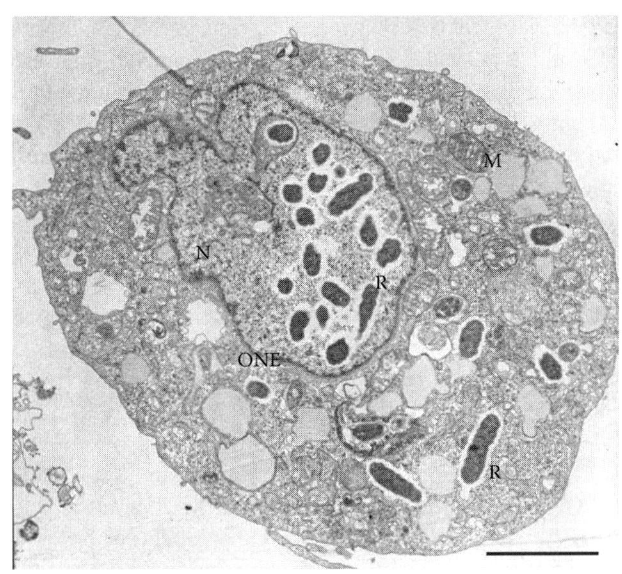

FIGURE 58.2 Transmission electron microphotograph of *Rickettsia rickettsii* grown in a primary culture of chicken embryo cells. Intracytoplasmic and intranuclear *R. rickettsii* are shown in a chicken embryo cell. *Abbreviations*: R, rickettsiae; M, mitochondria, N, nucleus; ONE, outer nuclear envelope. Scale bar, 5 μm. (Photograph is courtesy of David J. Silverman. With permission.)

by ultraviolet (fluorescent) microscopy following acridine orange staining. The dye binds preferentially to RNA and permits differentiation of physiologically active or dead microorganisms.[12] Viable microorganisms fluoresce a bright orange to red color, while dead microbes appear yellow-green, like the stained protein in the cytoplasm of their host cells.

58.1.2 Growth Characteristics and Plaque Formation

Members of the genus *Rickettsia* have not been grown axenically. *Rickettsia rickettsii* can be cultivated in the yolk sacs of embryonated chicken eggs. The optimal growth temperature for *R. rickettsii* is 32°C–34°C.[13] The embryo is killed early in the course of infection but the rickettsiae continue to multiply for the next 2–3 days; however, tissue culture cells are more suitable for the abundant growth and purification of SFG rickettsiae. *Rickettsia* species are rather promiscuous about their host cells as they can invade and grow in primary chicken fibroblasts and human umbilical vein endothelial cells (HUVEC), or in a variety of established permanent cell lines including green monkey kidney cells (Vero), L929 mouse fibroblasts, RAW264.6 mouse macrophage-like cells, and HMEC-1, ESV40, and EA.hy 926 human endothelial cells, as well as insect cell lines. Propagation of *R. rickettsii* in cultivated cells is typically done with a medium optimized for growth of the particular cell line, often RPMI-1640 or MEM supplemented with 2%–10% of fetal bovine serum, 2 mM of glutamine, and 5% CO_2 atmosphere. Typically, high concentration of fetal bovine serum are not beneficial for cultivation of *R. rickettsii*, particularly during primary isolation, because enriched medium may stimulate overgrowth of the eukaryotic cells while small inoculums of more slowly growing rickettsiae may be lost among rapidly dividing host cells. Treatment of host cells with emetine at 1 μg/mL may partially overcome this difficulty.[14]

While growing in monolayers of eukaryotic cells, *R. rickettsii* form so-called lytic plaques that are morphologically similar to viral lytic plaques and represent a useful taxonomic characteristic among representatives of the genus *Rickettsia*.[13–15] Among SFG, *R. rickettsii* forms clearly distinguishable plaques of 2–3 mm in diameter at 4–7 days after inoculation, although some variability in size and plaque morphology exists depending on the cell used, cell culture medium, incubation temperature, and the density of host cells.[14–18]

58.1.3 Epidemiology and Ecology

Rickettsia rickettsii is the etiologic agent of RMSF, a disease that is only known in the western hemisphere. *R. rickettsii* is transmitted by several species of ticks that flourish in different ecological settings. Those include *Dermacentor andersoni* in mountainous regions of the western United States, *D. variabilis* in the mid-Atlantic and Central states, and *Amblyomma cajennense* in southern Texas and throughout Central America to Brazil and Argentina, where it is also circulated by *A. aureolatum* and *A. oblongogutattum*.[19,20] Finally, endemic foci associated with the brown dog tick, *Rhipicephalus sanguineus* are found in the United States, Mexico and Brazil, possibly indicating a wide association of *R. rickettsii* with *Rh. sanguineus* and an important role of dogs in its ecology.[21–24] Other species of ticks such as *A. imitator* in Mexico may also be competent vectors for *R. rickettsii*.[25]

RMSF is endemic in Central and Southern America, with recent well-documented cases in Mexico, Costa Rica, Panama, Colombia, Brazil, and Argentina.[20] RMSF is a nationally notifiable disease in the United States and Brazil. According to US national surveillance records, the mean incidence rate for RMSF has increased in recent years.[26] Most cases are reported from the South Atlantic and West-South-Central US states, with North Carolina, followed by Arkansas, Oklahoma, and Tennessee reporting the highest incidence rates. Disease caused by *R. rickettsii* in Brazil is called Brazilian spotted fever (BSF); most cases of BSF are reported from Minas Gerais, São Paulo, Rio de Janeiro, Espirito Santo, and Santa Catarina.[27–29]

Since *R. rickettsii* is transmitted by tick bite, the peak of naturally acquired infection in the United States occurs during late spring through early fall, the time corresponding to feeding periods of adult and immature stages of *D. variabilis*, the primary vector. Humans are typically accidental hosts intersecting with the natural cycle of *R. rickettsii*. Similarly, they may be infected when encountering sites with dogs heavily infested with infected *Rh. sanguineus* that generally bite humans infrequently.

Rickettsia rickettsii is maintained in ticks by transovarial and transstadial transmission,[30,31] although, it can decrease tick fecundity and larvae and nymph survival.[32] It survives well in overwintering ticks and is excreted in tick feces.[31]

58.1.4 Genetic and Genomic Features

Complete genome sequences are available for strain Sheila Smith (AADJ01000001) and strain Iowa (NC_010263), which was attenuated during extensive laboratory passaging.[33] The genome of the virulent strain consists of single circular chromosome of 1,257,710 bp and G + C% content is 32%, while the Iowa isolate is 10,000 bp shorter; multiple INDEL events distinguish the two isolates. Genome sequences of six other isolates representing the six major genotypes of *R. rickettsii* have been completed.[34] The estimated coding capacity of *R. rickettsii* genome is approximately 75% and the number of putative open reading frames is between 1365 and 1527 depending on the algorithm used.[33,35] It encodes 33 tRNA genes, one copy of each ribosomal gene (5S, 16S and 23S), and three other RNA molecules, *ssr* (tmRNA), *rnp*B (RNA component of ribonuclease P), and *srp* (signal recognition particle 4.5S RNA). The ribosomal gene operon is split in *R. rickettsii*, like other *Rickettsia* where the 16S rRNA gene is widely separated from the 5S to 23S rRNA operon.[36] The genome of *R. rickettsii* has remarkable chromosome synteny and high gene homology to most well-identified functional proteins found in the genomes of other *Rickettsia*.[33,35,37,38] The genome of *R. rickettsii* encodes for several orthologs of surface protein antigens (Sca) and a complete set of type four secretion system proteins. Intergenic regions, sites consisting of tandem repeat elements, and INDEL sites identified between strains Sheila Smith and Iowa have been shown to have significant genetic intraspecies variability allowing sensitive genotyping of isolates of *R. rickettsii*.[20,33,39–41] Many of the INDELs identified in the attenuated Iowa isolate when compared to virulent Sheila Smith isolate are also present in isolates from fatal RMSF cases occurring west of the Mississippi and in most eastern isolates from the areas where *D. variabilis* is the primary vector of *R. rickettsii*[41]—therefore, most of these sites probably do not contribute to attenuation of the Iowa isolate. All three typing systems identified a unique genotype for Arizona isolates associated with *Rh. sanguineus* and distinguished among isolates from Brazil, Colombia, and Central America.[20,40,41] This analysis also demonstrated that isolate Hlp#2 is most accurately described as a nonpathogenic, genetically distinct subspecies of *R. rickettsii*.

58.1.5 Clinical Features and Pathogenesis

Clinical Features. Patients suffering from RMSF display a variety of nonspecific symptoms.[42,43] The disease begins to manifest following 1 or 2 weeks after the patient was bitten by an infected tick. The onset may be sudden or may be preceded by a few days of general malaise, after which the symptoms become more dramatic. The first sign is often a splitting headache (frequently described as the worst ever experienced by the patient) accompanied by pains in the back, joints, and legs. Patients are very sensitive to light, and their stiff neck sometimes leads physicians to suspect meningitis. The temperature rises rapidly to 102°F–104°F. The patient is usually restless, cannot sleep well, and frequently suffers from periods of delirium. The initial symptoms of RMSF may occasionally also mimic appendicitis or even the common cold, making the clinical diagnosis more difficult.

As the infection progresses, rickettsiae multiply within the endothelial cells lining the capillaries. Infected cells eventually swell—some burst—and by the third or fourth day of recognizable illness blood begins to seep through ruptures in the capillary walls. These hemorrhages cause the characteristic spots of the disease, which often look like the rash of measles and can be felt as slightly raised areas under the skin. They appear first on the wrists and ankles, spreading later to the limbs, the trunk and the face. The spots of RMSF often appear on the hands and soles of the feet and since only a few diseases produce this symptom, it is often considered a definitive clinical diagnostic sign for infection with *R. rickettsii*.[43]

If left untreated, most US patients will recover from their illness within 2 weeks. However, over 20% of patients, especially people over 40 years old, those with existing medical problems, and young children will die from infections. In different areas of Brazil, the mortality rate due to untreated RMSF may vary from 12.5% to 73.3%.[44,45] In a few cases, the disease is fulminant, killing its victim within a few days of onset, often before the characteristic rash has appeared. People with hereditary deficiency for glucose-6-phosphate dehydrogenase, particularly males of African American or Mediterranean origin, may exhibit increased susceptibility to severe *R. rickettsii* infections.[43,46]

Pathogenesis. The pathogenic effects of *Rickettsia* may largely be attributed to their entry and intracellular multiplication in the cytoplasm of microvascular endothelial cells, a process facilitated by inhibition of apoptosis in the infected cell and alteration of its transcriptional machinery.[47–50]

The uptake of rickettsiae by mammalian cells is sensitive to treatment with cytochalasins, *N*-ethylmaleimide, and NaF, requires metabolically active rickettsiae, and is probably calcium dependent and has been called induced phagocytosis.[51,52] Binding of viable rickettsia to the external cell surface appears to perturb the host cell plasma membrane in a way that results in their internalization at the site of adhesion.[51] The site of *R. rickettsii* adhesion and entry exhibits remarkable changes in its electron density.[48] As shown in an *R. conorii* model of infection, outer membrane protein OmpB binds specifically to Ku70, a component of the eukaryotic DNA-dependent protein kinase.[53–55] It has been speculated that this binding event is responsible for subsequent activation of membrane bound Ku70 and upregulation of a cascade of signaling events, including the small GTPase, Cdc42, phosphoinositidyl-3-kinase, Src-family tyrosine kinases, and the tyrosine phosphorylation of focal adhesion kinase. The orchestrated sequence of these events have been demonstrated or partially characterized for only a few species of *Rickettsia* including *R. rickettsii*.[49,56] In particular, *R. conorii* has been shown to stimulate the ubiquitination of Ku70 and recruitment of ubiquitin ligase c-Cbl at the entrance foci of rickettsiae.

Both internalization and escape of *R. rickettsii* from the phagocytic vacuole in HUVEC are very rapid processes since 10–30 min after internalization begins, many rickettsiae can be observed in the cytoplasm. Unipolar nucleation of actin tails and replication occur immediately after release of *R. rickettsii* from the phagocytic vacuole[48,57]; replicating rickettsiae spread to the adjacent cells causing accumulating damage at the site of infection and surrounding cells.

Formation of microvascular injury at the infection site and development of generalized microvasculitis is a hallmark of *R. rickettsii* infection. This damage results in enhanced permeability of tissue fluids into the infection foci, infiltration by monocytes and polynuclear neutrophils, an inflammatory reaction, and formation of microthrombi.[43] As shown with in vitro systems, the development of morphological injuries is associated with increased expression of eukaryotic cellular adhesion molecules (ICAM-1, VCAM-1, and E-selectin), induction of IL-8, IL-6, IL1-alpha, TNF-alpha, increased expression of tissue factor, activation of the fibrinolytic system, and secretion of prostaglandins.[47–49] Finally, it has been suggested that the intracellular presence of *R. rickettsii* induces production of reactive oxygen species, whose accumulation correlates with the intensity of necrotic changes in the infected cells and can be prevented by antioxidants.[49,58,59]

58.1.6 Diagnosis

RMSF is a tick-transmitted illness with an acute onset of fever and possible headache, malaise, myalgia, and nausea, vomiting, or neurologic signs. A macular or maculopapular rash occurs in many patients; it is frequently observed on the palms and soles. The clinical presentation is nonspecific as it is shared with other diseases. Laboratory confirmation requires a fourfold change in serum antibody titer against *R. rickettsii* antigen between paired serum samples by indirect fluorescent antibody tests (IFA), or demonstration of *R. rickettsii* antigen by immunohistochemical (IHC) methods, detection of *R. rickettsii* DNA by polymerase chain reaction (PCR) assay, or isolation and identification of *R. rickettsii* from a clinical specimen. A probable case is identified as a clinically compatible illness with serologic evidence of antibody reactive with *R. rickettsii* in a single serum. Clinically compatible cases without any laboratory data are interpreted as suspected cases.

Most patients seek medical care within the first 3 days of illness, a period when as few as 3% of patients present with the classic triad of fever, rash, and a history of tick exposure. This occurs typically prior to mounting the antibody response, which is routinely used for confirmatory diagnosis.[60] Consequently, conventional laboratory diagnosis of RMSF that relies on serology is largely retrospective. Below we describe advantages and shortfalls of conventional serologic analysis and more useful approaches for laboratory diagnosis of RMSF.

58.1.6.1 Clinical Chemistry of RMSF

Nonspecific host responses to rickettsial infection may be useful as presumptive diagnostic indicators of rickettsial infection.[61] Hyponatremia (less than 132 meq/L), hypoalbuminemia, anemia, and low platelet count develop as a result of massive vascular injury and increasing vascular permeability.[62] Low sodium levels or thrombocytopenia occurs in 52%–56% of the patients suffering from RMSF.[63] Of possible utility in distinguishing RMSF from bacterial infections such as *Neisseria meningitis* is the presence of a normal leukocyte count (less than 10,000/mkl) in 72%–78%, with more than 10% band-form neutrophils in 69%–89%.[63] Serum activity of two liver enzymes, alanine and aspartate aminotransferases, is typically elevated during the acute stage of RMSF; however, these changes alone are too nonspecific to ensure accurate diagnosis.[64]

58.1.6.2 Serologic Diagnosis

Microimmunofluorescence assays (MIF) also referred to as indirect fluorescent antibody tests (IFA) are presently the gold standard methods used for serologic diagnosis of rickettsioses including RMSF.[65] This is primarily because MIF is very sensitive and relatively rapid, it is suitable for use with small numbers of sera, and it requires exceedingly small amounts of antigen. The technique was originally described by Goldwasser and Shepard in 1959 for differential diagnosis of murine and epidemic typhus,[66] and later adapted by Philip et al. as the micromethod used nowadays.[65] To set up the test, killed rickettsial antigen is affixed as microdots to a microscope slide, then incubated with serial dilutions of serum and subsequently washed, incubated with fluorescein isothiocyanate (FITC)-labeled antibody (also referred to as the conjugate) against immunoglobulin in the sera being tested, washed again, mounted, and examined with a UV microscope equipped with appropriate filters to allow excitation and emission of specific visible apple-green light. The last dilution of serum producing a visible signal is recorded as the reciprocal endpoint titer with the antigen employed.

Because different antigen dots can be spotted on different zones of the same microwell, MIF is also convenient for rapidly screening multiple sera for reactivity against multiple agents. Unfortunately, reader fatigue and confusion about microspot orientation can occur. Moreover, determining endpoints and establishing clear cutoffs for significant positive reactions can be laborious. Lack of consistency in reagents and details of protocols between laboratories and operators can result in discrepant results. Moreover, some sera can be "sticky" and cause background of high fluorescence. High-quality antigens and conjugates, as well as proper serum processing and storage, are important for consistent assay accuracy and validity.

Although it is often assumed by analogy to the immune response to viral agents that the presence of IgM antibodies is an indication of acute rickettsial infection, determination of correct IgM antibody titers can be problematic as their assays may exhibit more background and low specificity [notably due in part to shared epitopes detected by *Proteus* Weil-Felix and *Legionella* lipopolysaccharide (LPS) cross-reactions].[67] The IgG immunoglobulins may interfere with binding of low affinity IgM antibodies. The IgM class antibody may also manifest prozone negative reactions at initial screening dilutions,[68] and rheumatoid factor may confuse the results. Consequently, MIF titers obtained with gamma-chain-specific IgG conjugates are used for primary confirmatory diagnostic testing at the CDC before determining IgM titers when they are of interest with IgG confirmed cases.

The sensitivity of the MIF assay depends substantially on when the sample is collected relative to the onset of disease. In most cases, initial samples obtained early in the infection will be negative or sufficiently close to the cutoff for positive assays that interpretation of the result has greater uncertainty and lower predictive value. MIF is estimated to be 94%–100% sensitive after 14 days, and confidence in the correct interpretation of low positive reactions is increased if paired samples are tested together.[64,69,70] Consequently, rapidly progressing infections like RMSF must often be treated without laboratory confirmation as mortality may result if treatment is delayed until laboratory testing is performed.[64] The duration of time that antibodies will persist after recovery from an infection is also variable and may depend upon the antibiotic therapy employed. For RMSF, IgG and IgM titers increase almost concurrently by the second week of illness, and IgM antibodies disappear after 3–4 months, whereas IgG titers may persist from 7 to 8 months to several years.[70,71] Unfortunately, the use of *R. rickettsii* (or *R. conorii* in Europe and Asia) and *R. typhi* antigen as surrogates for all spotted fever group and typhus group rickettsioses, respectively, is commonplace because other MIF antigens are not widely available. The degree to which different human sera recognize shared epitopes on the LPS and the major rickettsial surface protein antigens (OmpA and OmpB, also known as SPA surface layer protein antigens—two of the Sca family proteins) of different species of *Rickettsia* following infection with different rickettsial agents is poorly understood for either human or veterinary patients. These antibodies clearly differ in their affinity and specificity at different times after infection and between individuals. The extent to which the shared antigen epitopes are available for reaction in the MIF test also varies with the initial antigen preparation and its subsequent handling, including antigen inactivation technique and slide fixation protocol. Consequently, serological tests that detect antibodies reactive with *R. rickettsii* antigens may have resulted from past or present exposure to other SFGR species. Cross-reactivity of some sera from human rickettsioses to both typhus and spotted fever group antigens is also well documented.[65,72]

While other rickettsial serological procedures including complement fixation tests, latex agglutination, microagglutination, and hemagglutination assays all have a long history of effective use and are highly specific, they all have greater demands for purified quality antigens and assay standardization and most except latex agglutination provide less sensitivity than MIF so they have been largely superseded.[68,71,73,74] Indirect immunoassays with horseradish peroxidase (IIP) or alkaline phosphatase (IAP) conjugates instead of FITC conjugates have most of the advantages of MIF tests but do not require an expensive fluorescence microscope.[75,76] While the substrate step requires an additional incubation step to develop, the stain is stable so slides can be preserved while MIF slides cannot. The protocols are similar to those used in immunohistochemical assays but do not require the signal amplification step provided by biotin/streptavidin branches since highly antigenic rickettsiae are being detected rather than cells subject to harsh chemicals during formaldehyde fixation, paraffin embedding, and paraffin removal. High-quality enzyme conjugates and stains are available from many commercial sources and the same reagents can be used in Western blotting and dot-blotting protocols.

Enzyme-linked immunosorbent assays (ELISA) are also useful for diagnosis of rickettsioses, particularly in laboratories with access to purified antigens, and plate washers and readers.[70,77,78] The assay is particularly suitable for automating high throughput seroprevalence studies with multiple antigens or retrospective studies of paired sera. One reason ELISA has not supplanted MIF testing in most laboratories is that full-length rickettsial OmpA and OmpB antigens suitable for use in diagnosis of rickettsioses have proved difficult to produce by recombinant methods and epitopes recognized by human and animal sera are present in many sites on these large proteins.[79,80] However, high-quality OmpB proteins can be obtained relatively easily from purified typhus group rickettsiae.[81] Important immunoreactive epitopes are also heat labile, so fragments of the antigens exhibit inadequate seroreactivity to antisera preferentially recognizing conformational epitopes. Similarly, surrogates for the group-reactive LPS have not been developed although *Proteus* OX19 and OX2, and *Legionella bozemanii* and *L. micdadei* LPS contain some of the rickettsial LPS epitopes recognized by human serum following infection with both typhus and spotted fever rickettsiae.[67,80,82] This lack of LPS surrogates is unfortunate since it has limited the commercial availability of latex agglutination and hemagglutination assays that are rapid, reproducible and sensitive, and very flexible for use in point of care clinics with limited equipment as well as routine use of LPS in the ELISA.[77] Other limitations of ELISA are that enzyme reaction conditions with substrates can be difficult to standardize, have only a limited linear range relationship (saturation at high titers) of antibody titers relative to immunofluorescence, and the antigens may have additional domains that are not detected in MIF so the two assays are not strictly comparable.

Western blotting, chromatographic flow assays, and dot immunoassays are all sophisticated procedures that have been used in rickettsial serology.[80,83–85] They all require highly purified antigen and quality antibody detection reagents but their utility differs greatly. Chromatographic flow assays (similar to home pregnancy tests) are largely replacing dot immunoassays since they are easier to manufacture and read and they are particularly suitable for rapid point-of-care testing. Both assays are qualitative assays that are most valuable when positive results are obtained, but they lack sensitivity and specificity. On the other hand, Western blotting is very sensitive and the sole general method besides cross-absorption and blocking assays (by MIF or more efficiently by quantitative ELISA-blocking assays) which can identify the source of infections by the specificity of antibody reactions with the rickettsial OmpA and OmpB proteins. However, for some patient sera, and depending upon the time after infection a serum was collected, even Western blotting may require initial cross-adsorption of the sera to identify the agent causing infections (Figure 58.3).[84] In principle, capture assays can simplify detection of species-specific anti-*Rickettsia* antibodies, but appropriate antigens and monoclonal antibodies are not available for widespread use.[86]

58.1.6.3 Immunohistochemical Detection

Vector-transmission of a rickettsiosis generally involves initial infection through the skin at the site of the arthropod bite or abraded skin, although conjunctival or respiratory inoculation with arthropod juices or infected feces may also occur. Wolbach was the first to demonstrate rickettsial organisms while describing the cutaneous pathology of RMSF.[61] Dermal and cubcutaneous vascular lesions early in the course of disease are comprised of swollen, injured endothelial cells, which along with vascular smooth muscle cells contained pairs of rickettsial bacilli.[61,87] Later lesions exhibit perivascular infiltration by mononuclear and polymorphonuclear leucocytes and thrombosis and finally repair and recanalization of the thrombi. Since most, if not all, patients suffering from RMSF do not develop an eschar, the biopsy must be collected from rash sites, preferably petechial lesions. Unfortunately, this method requires that a rash is present to guide selection of the biopsy site, skill with the cryostat sectioning, and experience

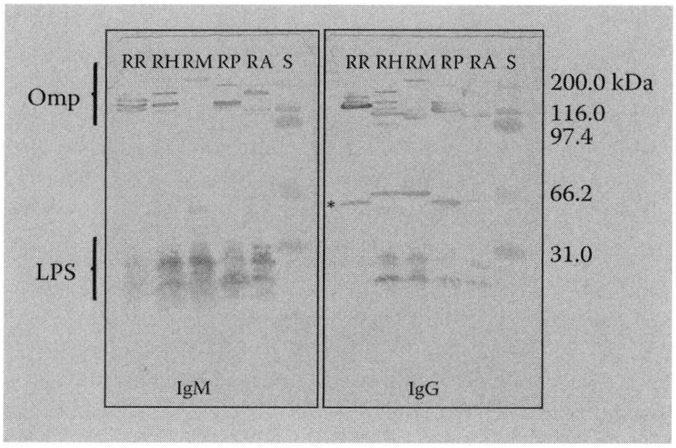

FIGURE 58.3 Cross-reactivity of a convalescent serum specimen from a patient confirmed to have Rocky Mountain spotted fever with other tick-borne spotted fever group rickettsiae found in the USA. The convalescent serum with IFA IgG titer 1/512 to *R. rickettsii* was collected at approximately one month after onset of illness and tested at 1:100 dilution. Acute serum of the same patient was IFA negative with *R. rickettsii* antigen (IgG /IgM ≤ 32). Renografin-purified rickettsiae were solubilized at room temperature in Laemmli sample buffer and 15 μg of total proteins per lane was loaded in the following order: *R. rickettsii* Bitterroot (RR); *R. rhipicephali* 3-7-♀6 (RH); *R. massiliae* AZT80 (RM); *R. parkeri* Portsmouth (RP); *R. amblyommii* GAT-30VS (RA). Numbers on the right are protein molecular weight markers. Reactivity of the surface protein antigens (Omp) and lipopolysaccharide (LPS) are marked with the brackets; the reactivity of other proteins is marked with a star (*). Western blotting experiment is courtesy of Mary E. Wikswo.

with histochemistry and immunostaining, and that requisites high affinity antibody reagents, conjugates, signal amplification reagents, precipitating stain, control tissues, and appropriate microscopes and cameras be available.

Immunostaining has allowed routine demonstration of *R. rickettsii* in skin and autopsy tissues.[88–91] The original application of this method used direct or indirect immunofluorescence detection of *R. rickettsii* in formalin-fixed, paraffin-embedded tissues after deparaffinization and trypsin digestion.[89,92] While IHC is often employed in examination of tissues autopsied from fatal cases,[93,94] in principle it can be applied to other organ biopsies. As IHC is performed currently, tissues are fixed in 10% buffered formalin and embedded in paraffin. Sections are deparaffinized and rickettsial antigen is detected by sequentially applying specific antibodies and biotinylated anti-immunoglobulin conjugates whose binding is detected with streptavidin-alkaline phosphatase conjugates and a precipitable alkaline phosphatase substrate. Polyclonal antibodies that are group reactive are generally employed, but species-specific monoclonal antibody reagents can also be employed. The sensitivity of the IHC is approximately 70% and its specificity is 100%.

As a complementary method, DNA can be extracted from parallel sections of blocks where rickettsial antigen has been detected and its characteristics determined by PCR and DNA sequencing.[95] Unfortunately, since only small fragments of specific genes can be amplified because of the fixation and embedding procedures employed in immunohistochemistry, it is preferable to obtain DNA from another portion of the biopsy that was not fixed, but this is frequently not available.

Unarguably, IHC provides effective and rapid diagnosis during the acute stage of RMSF when antibodies are still absent or nondiagnostic; however, this diagnostic test remains restricted to specialized research laboratories. Beside the limited availability of suitable antibody reagents, the drawback of this test is progressive loss of sensitivity following initiation of doxycycline treatment as the amount and integrity of rickettsial antigens decline.

Improvements in IHC analysis are dependent upon wider availability of stable monoclonal antibodies with increased specificity and affinity as well as development of multicolor assays that permit detection of mixed infections. Alternatively, it might be possible to detect activation of specific host response genes to rickettsial infection by detection of novel protein products associated with vascular injury and inflammatory cells. For sections with sparse rickettsial antigen loads, this may provide a more sensitive means of detecting these organisms early in the course of infection in skin biopsies. Isolation of specific cell types from biopsy samples may also provide means for enriching for less pathogenic or less-disseminated rickettsiae. PCR of fixed tissues may also be more efficient if laser microdissection of infected sections is performed so the ratio of rickettsial DNA to host DNA is increased and fewer inhibitory chemicals are present.

58.1.6.4 Diagnosis by Isolation

Isolation permits unequivocal demonstration of the etiology of disease caused by any of the rickettsial agents. Isolation remains more of an art than a science because of the variables involved in sample collection (tissue, time after disease onset, antibiotic treatment) and sample quality (amount, sterility, handling). In principle, only one viable organism in a sample is required for successful isolation in a susceptible animal host or the yolk sac of embryonated chicken eggs. However, most clinical laboratories do not have routine access to fertile eggs or animals, or ABSL-2 or ABSL-3 laboratories for housing the latter. Furthermore, selection of the appropriate animal host may be difficult if the patient history or presentation does not provide clues as to the specific species of *Rickettsia* involved. One myth that evolved prior to the antibiotic era for rickettsioses and modern biohazard containment hoods holds that rickettsiae are too dangerous to isolate.[61] In reality, laboratories with suitable facilities and properly trained personnel can routinely and safely undertake diagnosis by rickettsial isolation.[61,96]

Despite substantial inefficiencies (sterility, low titers), cell culture has become the primary means of isolation, particularly in laboratories that have experience with and are responsible for viral isolation and familiarity with suitable tissue culture cell lines. While the original inoculation and incubation may be conducted under BSL-2 conditions, in the United States, once primary isolation of *R. rickettsii* is confirmed, the isolate and infected patient materials must be transferred to a laboratory registered for Select Agent work (www.cdc.gov/od/sap). Further growth of *R. rickettsii* in larger quantities must then be done in a BSL-3 laboratory because of the greater potential for aerosol infection with this agent at higher titers; this level of containment and procedures is also highly recommended for other pathogenic rickettsiae.[97]

All of the same clinical samples (EDTA blood is only preferred over heparinized blood for PCR) and arthropod samples used for PCR confirmation of rickettsioses (see below for details) are suitable for isolation attempts.[21,64,96,98–101] However, while rickettsial DNA may be isolated from contaminated specimens as well as from same samples that were not properly handled to maintain rickettsial viability, these variables can adversely affect isolation success. If delays greater than 24–48 h in shipment to the isolation laboratory are anticipated, the samples should be frozen to −80°C after collection. Otherwise, the patient samples should be shipped and held at 4°C until processed. Maintaining adequate humidity and room temperature is important for live arthropods (damp but not wet plaster, cotton, or grass in small cryovials). Arthropods, skin/eschar biopsy specimens, and suboptimal late autopsy specimens present the biggest challenge to cell culture isolation because they may be contaminated with bacteria and fungi that can complicate cell culture isolation attempts. Arthropods are disinfected for 10 min in 10% bleach and 5 min in 70% ethanol and then rinsed before being bisected. One half can be extracted immediately for DNA isolation and PCR analysis and the other half frozen at −80°C in sucrose-phosphate-glutamate buffer (pH7.0) until PCR results are obtained. Biopsy specimens are commonly divided into several portions for immunohistochemistry, for DNA extraction for molecular analysis, and the remainder thoroughly minced with a scalpel and then homogenized for cell culture isolation. To reduce the amount of erythrocytes introduced into culture from blood specimens, buffy coat cells, Hypaque-purified lymphocytes, plasma, or washed pellets from hemolyzed blood may be employed.

Several cell cultures commonly used for isolation of *Rickettsia* include African green monkey kidney cells (Vero E6), irradiated or cycloheximide or emetine-treated mouse fibroblast L929 cells, and human embryonic lung (HEL) cells. Modified shell vial procedures first employed for virus isolation have been widely popularized for isolation of rickettsiae.[96] Two paired vials containing cover slips with adherent monolayers of selected cells are inoculated by centrifugation of the inoculum onto the monolayer. This has the advantage of requiring only small volumes of cells and inocula and facilitating rickettsial contact with the host cells. Inoculum debris and infectious contaminants are then removed by washing the infected monolayers three times after centrifugation. When there is concern about significant sample contamination, antibiotics (typically 0.2% of cell culture grade penicillin-streptomycin) and 1% fungizone (amphotericin B) can be added to the monolayers and removed at 48 h, and the monolayers washed with fresh medium. One monolayer can be stained at 48–72 h by immunofluorescence (MIF), immunoperoxidase, acridine orange, or Gimenez stains or debris in the supernatant analyzed by PCR or microscopy to detect rickettsiae. The other monolayer can be passaged to sustain the isolate or incubated longer for subsequent testing if the initial sample is negative. Alternatively, 25 cm² cell culture flasks or 6–96 well tissue culture plates can be used without cover slips and similar inoculation protocols. The larger flask area permits use of much larger inocula for samples with sparse organisms and repeated temporal sampling of the same flask by scraping cells. Both flasks and tissue culture plates are more convenient than shell vials for noninvasive examination of extracellular rickettsiae and development of cytopathic effects by inverted phase microscopy. The role of many factors that may stimulate initial rickettsial growth in culture are poorly understood (temperature, iron source, media, oxidation/reduction state, pH, host cell factors) but many isolates appear to undergo a long lag period of adaptation to the new host cell and often require blind passaging before sustained logarithmic growth in cell culture is obtained.

58.1.6.5 Molecular Techniques

The detection and characterization of specific fragments of rickettsial DNA amplified by polymerase chain reaction (PCR) is the most rapid method for indirect detection and accurate identification of small numbers of rickettsiae in clinical and environmental samples. Because complete genome sequences are now available for 18 species of *Rickettsia* and six major genotypes of *R. rickettsii*,[20,34,35,37,38] in principle, a large number of species-specific target regions can be selected. However, the gene targets that have been used most extensively were selected arbitrarily based on early rickettsial gene discovery efforts rather than on any of several possible ideal criteria.[102–105] Little attention has been directed toward development of protocols and rickettsial gene targets which have significantly enhanced amplification efficiency (a function of gene structure and primer properties as well as amplification conditions), the use of multicopy genes to increase sensitivity, selecting genetic traits unique for a particular species, or maximizing the degree of phylogenetic divergence of the specific gene regions being targeted for separating closely related rickettsiae. However, primers for three of the targets that have been widely used, the 17 kDa antigen gene, a fragment of the citrate synthase gene *glt*A, and the 5′ end of the major serotype surface antigen gene, *omp*A, were fortuitously located in sufficiently conserved regions that they have been useful for characterizing these gene regions for most species of rickettsiae (except the more divergent *R. bellii* and *R. canadensis*, which have not been shown to be pathogenic for humans).[102,103,106–108] For the most divergent environmental rickettsiae, PCR with conserved eubacterial 16S rDNA primers has been important for permitting rapid identification of novel agents which belong to the genus *Rickettsia*.[3,4,109] These four gene targets are still the mainstays of most direct and nested PCR assays that are used for molecular detection of rickettsiae in clinical specimens and other sources.[95,105,110–113] They have also been the primary targets for newer quantitative PCR assays based on TaqMan or SYBR Green methodologies.[114–117] However, these assays have also been augmented by development of primers for detecting and characterizing rickettsial genes like the 120 kDa cytoplasmic protein antigen (gene D, *sca*4), RNA polymerase subunit B (*rpo*B), *omp*B, the heat shock protein operon *gro*ESL, and the 5S–23S intergenic region.[118–122] However, there is even less experience with diagnostic assays using these genes than with the original gene targets. Like 16S rDNA, there is a large database for bacterial 60 kDa chaperonin GroEL genes, so rickettsial *gro*EL is a useful phylogenetic target for atypical *Rickettsia*-like agents.[123] However, the highly *Rickettsia*-conserved genes (*glt*A, *rrs*, *gro*EL) have disadvantages for distinguishing closely related species of *Rickettsia* and for assuring that other bacteria are not being detected. The presumptive identification of rickettsial species is often based on restriction fragment length polymorphism typing of variable *omp*A and *glt*A sites and more definitively by DNA sequencing of amplicons in laboratories with access to this technology.[100,102,106,124–126]

Although PCR of rickettsial 17 kDa antigen gene and 16S rDNA fragments from whole blood were first used to diagnose RMSF and typhus infections in 1989–1990,[104,105,110–112] these assays have been inefficient for diagnosing patients early in infection and required DNA probing or reamplification for confirmation of the source of the amplicon except in advanced or fatal cases. Typically low numbers of rickettsiae circulate in the blood in the absence of advanced disease or fulminant infection [$10^{0.7}$–$10^{1.2}$ $TCID_{50}$/mL (tissue culture infectious dose) of plasma from patients with mild to moderate diseases compared to 10^3 $TCID_{50}$/mL of plasma from fatally ill patients].[127] Although probe hybridization can replace DNA sequencing of amplicons produced from nested PCR reactions, further improvements in direct detection and identification of organisms present in blood samples will likely require development of creative methods to efficiently enrich rickettsial DNA from these samples,[128] or use of multicopy gene amplification methods like reverse transcription PCR of 16S rRNA to increase the amount of the template available for analysis.[129]

PCR appears to be more useful for detecting rickettsiae early after infection in eschar and skin or organ biopsies than in acute blood.[21,64] However, PCR testing of skin biopsies does not always offer ideal sensitivity because of limitations in biopsy sample volume, uncertainties about biopsy site selection in the absence of eschar, and because of the significant variation in distribution of rickettsiae in infected vessels that is apparent by immunohistochemistry. Laboratory confirmation of rickettsioses in the acute stage is improved when PCR is used in parallel with immunohistochemical staining to confirm that appropriate biopsy samples were obtained. If discordant results are obtained (positive IHC detection and negative blood PCR assay), PCR of DNA extracted from positive paraffin-embedded tissues permits confirmatory diagnosis of the rickettsial nature of the antigen.[95,130,131] IHC detection antibodies are also generally group cross-reactive, so it may only be possible to identify the specific rickettsial agent present by using molecular assays on the fixed tissue DNA.

Quantitative PCR (qPCR) assays offer significant advantages over nested assays as they are single tube assays that reduce the risk of amplicon contamination of work spaces and test samples.[114–116] Because qPCR assays, which incorporate probes, inherently have greater specificity (three recognition sites) than SYBR Green PCR assays, which just detect the generation of double stranded amplicons (end primer recognition only), they can provide greater assurance of the identity of the amplicon. However, small size amplicons are produced in quantitative assays; the use of probes that are too specific may simultaneously lead to false assumptions about the identity of the amplicon and produce false negative results for new agents. Multiplexed assays with conserved probes and human primer controls for DNA template quality and elimination of sample inhibitors can overcome these concerns. Coupling PCR detection with analysis of the melting curve (Tm) of the amplicon or detection of single nucleotide polymorphisms (SNPs) may add to accurate and specific detection of the amplicon. Assays intended for use with environmental samples may need to be substantially less specific than those targeted for specific well-established clinical diseases since numerous new rickettsial agents have been discovered in recent years and some of these agents appear to be emerging human pathogens.

58.2 METHODS

58.2.1 Sample Preparation

Clinical specimens suitable for PCR for *R. rickettsii* testing include whole blood, blood clots, buffy coat, skin biopsy, and necropsy specimens. Occasional positive results have been obtained with serum (most frequently in association with fatal cases) but these have not been evaluated systematically[21,132] and are less than optimal for early detection of rickettsial DNA. Fresh tissues or tissues frozen at −80°C are the best specimens for PCR, but formalin-fixed wet or paraffin-embedded tissues may also be used.[95,98,99,130] EDTA or sodium citrate anticoagulated blood specimens are better than heparinized samples for DNA preparation since heparin inhibits PCR assays. The addition of foreign carrier DNAs to samples low in DNA and volume (serum, cerebrospinal fluid, fixed tissues and paraffin sections) may be important for ensuring recovery of template DNA. While patient samples are important for case confirmation, arthropods collected from patients, pets, or the immediate environment of suspect cases can also be used for detection and isolation of rickettsiae to confirm the likely epidemiological sources responsible for disease transmission.

Rickettsial cells can be fractionated using a variety of ionic and cationic detergents, indicating ready susceptibility of rickettsial cells to lysis procedures and accessibility of its DNA.[81] Early molecular biology experiments used traditional phenol-chloroform methods of DNA extraction allowing isolation of high-quality DNA preparations suitable for DNA–DNA hybridization, cloning and library construction, gene and genome sequencing and biochemistry.[33,133–135] Crude extractions of the DNA prepared by boiling with Chelex-resin can be useful for quick assessments for the presence of rickettsial DNA[114]; however, this appears to work best for specimens with high template content of the rickettsial DNA, so samples prepared in such manner and containing less copious rickettsial DNA will likely be negative or suffer from inhibition by residual protein and other molecules.

Modern extraction procedures utilize a variety of commercial DNA extraction protocols including both column purification and liquid precipitation procedures applied to a single specimen or with rapid high throughput protocols developed for different robotic platforms.[136,137] Rickettsial DNA, including that of *R. rickettsii* is not infectious by itself; however, the original unfixed clinical or field materials suspected to be infected with *R. rickettsii*, should be processed in biological safety cabinets until samples are completely lysed and heat inactivated. When clinical samples are processed, each batch of the extracted DNA should include a no tissue extraction control sample to control potential cross-contamination of the samples processed and purity of the extraction reagents. As for common good laboratory practice, extraction of the test sample DNAs should be conducted in a site that is separate from spaces used for setting up PCR assays and analyzing the results.

58.2.2 Detection Procedures

As with other *Rickettsia*, the detection strategy for *R. rickettsii* DNA is primarily based on recognition of sequences within the genes encoding for 16S rRNA, 17 kDa protein, citrate synthase, and outer membrane proteins, OmpA, OmpB, and Sca4. Typically, the PCR reaction is set up in a final volume of 20–50 μL using 2–5 μL of the template DNA, 2.5 mM $MgCl_2$, 200 mM of each dNTPs, and 10–20 pmol each of forward and reverse primers. A variety of amplification reagents have been used successfully for detection of different genes of *R. rickettsii* in samples with different origins. Examples of primer sequences, cycling conditions and their application are provided in Table 58.1. It is usually advisable to test particular PCR conditions when new amplification reagents are evaluated and to optimize conditions when it is necessary. For example, we have noticed significant variable amplification efficiency of the *omp*A fragment when using reagents from different manufacturers (Figure 58.4). Optimization conditions typically concern changes in concentration of magnesium chloride in the 2–6 mM range, primer working concentrations (100–800 nM), the amount of the DNA template used and selection of temperatures for annealing and elongation steps, as well as number of amplification cycles needed. Controls for PCR inhibitors and DNA recovery based on actin or other human house-keeping genes are essential to include in order to ensure that quality DNA templates have been prepared.

As a rule, detection of *R. rickettsii* DNA in a clinical sample requires the sensitivities obtained with nested PCR reactions. For example, to amplify a fragment of 17 kDa protein gene, the primary reaction mixture is set up with 2–4 μL of extracted DNA and primers Rr17.122 and Rr17.500, and the reaction is carried out for 40 cycles of 94°C for 1 min, 55°C for 1 min, and 68°C for 1 min. One or 2 μL of primary PCR product is used as the template for nested PCR with primers TZ15 and TZ16 and carried out for 30 cycles under the same conditions. The result of the amplification is evaluated by running 5 μL of nested PCR product for detection of the 247-bp fragment on 1–1.5% agarose gel in 1× Tris-acetate-EDTA buffer or 0.5× Tris-borate-EDTA buffer and visualized following staining with ethidium bromide or any other fluorescent DNA dye. Amplification of the OmpA gene fragment is typically conducted at the same time to confirm positive detection of rickettsial DNA and to generate a fragment more useful for species identification by RFLP or sequencing. The *omp*A fragment is produced by amplification of the primary fragment with primers Rr190.70 and Rr190.701 followed by nested or seminested amplification using primers Rr190FN and Rr190RN or Rr190.70 and Rr190.602,[98,138,148] respectively, or another combination of internal primers.[114,116] Since the 70–602 nt *omp*A fragment of *R. rickettsii* Hlp#2 and 364D *Rickettsia* differ only by 3 nucleotides from all other isolates of *R. rickettsii*, amplification and sequencing of the intergenic regions should be performed to ensure accurate genotyping of the isolates.[40,138] Amplification and sequencing of *omp*B, *glt*A, *sca*4, or any other gene can provide additional information if analysis of *omp*A is not satisfactory or more complete characterization of the detected *Rickettsia* is desirable.

The identity of the PCR amplicon is then confirmed by sequencing performed on both forward and reverse strands. The assembled nucleotide sequence is compared to the reference sequences of *R. rickettsii* (16S rRNA gene L36217, 17 kDa antigen gene M16486, *glt*A U59729, *omp*A U43804, *omp*B X16353, and *sca*4 AF163000) available in GenBank using nucleotide-nucleotide BLAST tool (Basic Local Alignment Search Tool; http://www.ncbi.nlm.nih.gov/BLAST/).[140] The majority of the spotted fever group *Rickettsia* have 97.9–99.8% sequence similarity of their *rrs* and 93.3–99.9% of *glt*A, while *omp*A (55.4–98.8%), *omp*B (72.4–99.2%), and *sca*4 (85.4–99.3% are more variable.[141] The *R. rickettsii* strains we have analyzed do not exhibit nucleotide sequence variations in *omp*A fragment with the exception of Hlp#2[139]; in contrast, two genotypes are known for *glt*A (RpCS.877p-RpCS.1258n) fragment of *R. rickettsii* that differ by 3 nt INDEL.[102,124] The nucleotide sequence of 17 kDa protein gene is insufficient for identification of *R. rickettsii* because it is identical to that of several *Rickettsia*, including *R. conorii*, *R. parkeri*, *R. sibirica*, and several other less characterized *Rickettsia* species whose pathogenicity is not yet established.[142]

58.3 CONCLUSIONS AND FUTURE PERSPECTIVES

It is likely that IFA and IP assays will remain the primary diagnostic methods used for rickettsial serology because relatively few laboratories can produce the large quantities of highly purified antigens required for other serologic assays. Because the

TABLE 58.1
DNA Targets and PCR Assays Used for Detection and Identification of *Rickettsia rickettsii*. Primers and Probes Listed Are Not Specific for *R. rickettsii* and Will Amplify DNA of Other Species from the Genus *Rickettsia*; Sequencing of Amplicons Is Required to Complete Identification of Agent DNA.

Gene Target	Assay	Primer Name	Designation	Primer Sequence (5′→3′)	Amplicon Size (bp)	PCR Cycling Conditions				Reference
						Cycle	Denaturation	Annealing	Extension	
16S rRNA gene	Conventional one-step PCR	16S-WFD1T	Primary PCR	AGAGTTTGATCCTGGCTCAG	1073	40	95° 30″	55° 30″	72° 60″	109
		16S-ECOLI-1073R		ACGAGCTGACGACAGCCATG						
17 kDa protein gene	Nested PCR	Rr17–122	Primary PCR	CAGAGTGCTATGAACAAACAAGG	378	40	95° 30″	55° 30″	72° 60″	111,142
		R17–500		CTTGCCATTGCCCATCAGGTTG						
		TZ15	Nested PCR	TTCTCAATTCGGTAAGGGC	247	30	95° 30″	55° 30″	72° 60″	
		TZ16		ATATTGACCAGTGCTATTTC						
	TaqMan	R17K135F	TaqMan primer	ATGAATAAACAAGGKACNGGHACAC	115	50	94° 5″	60° 30"		115
		R17K249R	TaqMan primer	AAGTAATGCRCCTACACCTACTC						
		R17Kbprobe	TaqMan probe	FAM-CGCGACCCGAATTGAGAACCAAGTAAT GCGTCGCG-BHQ						
		R17K-C[a]	TaqMan probe	FAM-TTGGTTCTCAATTCGGTAAGGGTAAAGG-BHQ	115	50	95° 15″	57° 60″		146
gltA	Conventional one-step PCR	RPCS.877p	Primary PCR[b]	GGGGGCCTGCTCACGGCGG	381–384	35	95° 30″	55° 30″	68° 30″	102
		RPCS.1258n		ATTGCAAAAAGTACAGTGAACA						
	TaqMan	CS-5	TaqMan primer	GAGAGAAAATTATATATCCAAATGTTGAT	147	50	95° 15″	50° 30″	60° 30″	147
		CS-6	TaqMan primer	AGGGTCTTCGTGCATTTCTT						
		gltA probe	TaqMan probe	FAM-CATTGTGCCATCCAGCCTACGGT-BHQ1						
ompA	Nested or Seminested PCR	RR190-70	Forward primer[c]	ATGGCGAATATTTCTCCAAAA	601	40	95° 30″	57° 30″	68° 60″	102,107
		RR190–701	Reverse primer, primary PCR	GTTCCGTTAATGGCAGCATCT						
		RR190–602	Reverse primer, seminested PCR	AGTGCAGCATTCGCTCCCCCT	532	35	95° 30″	57° 30″	68° 30″	102
		190-FN1	Nested PCR	AAGCAATACAACAAGGTC	576	30	95° 30″	60° 30″	72° 60″	138, 148
		190-RN1		TGACAGTTATTATACCTC						
	SYBR Green PCR[d]	RR190–547	Forward primer	CCTGCCGATAATTATACAGGTTTA	155	50	95° 20″	57° 30″	65° 30″	114
		RR190–701	Reverse primer	GTTCCGTTAATGGCAGCATCT						
	SYBR Green PCR[e]	107F	Forward primer	GCTTTATTCACCACCTCAAC	193	50	95° 10″	58° 06″	72° 06″	116
		299R	Reverse primer	TRATCACCACCGTAAGTAAAT						

continued

TABLE 58.1 (Continued)
DNA Targets and PCR Assays Used for Detection and Identification of *Rickettsia rickettsii*. Primers and Probes Listed Are Not Specific for *R. rickettsii* and Will Amplify DNA of Other Species from the Genus *Rickettsia*; Sequencing of Amplicons Is Required to Complete Identification of Agent DNA.

Gene Target	Assay	Primer Name	Designation	Primer Sequence (5′→3′)	Amplicon Size (bp)	PCR Cycling Conditions: Cycle	Denaturation	Annealing	Extension	Reference
RR0155-*rpm*B intergenic region	Nested PCR	RR0155-PF	Primary PCR	GGATTTCATTTAAAGAGCGATTAGG	462	40	95° 30″	58° 30″	72° 60″	138
		RR0155-PR		GAAAAGATAGGCACTGATCACATTC						
		RR0155FN	Nested PCR	TTTCTAGCAGCGGTTGTTTTATC	290	35	95° 30″	50° 30″	72° 60″	40
		RR0155RN		TTAGCCCATGTTGACAGGTTTACT						
RR1240-*tlc*5 intergenic region	Nested PCR	RR1137-PF	Primary PCR	CTTTCTATGTCATTCCTGACTTCC	535	40	95° 30″	58° 30″	72° 60″	138
		RR1137-PR		CAAAACATTAAGGTACTCATCGG						
		RR1240 FN	Nested PCR	CGGGATAACGCCGAGTAATA	357	35	95° 30″	54° 30″	72° 60″	40
		RR1240RN		ATGCCGCTCTGAATTTGTTT						
*omp*B	Conventional one-step PCR	120-M59	Primary PCR	CCGCAGGGTTGGTAACTGC	764	40	95° 30″	50° 30″	68° 90″	118
		120–807		CCTTTTAGATTACCGCCTAA						
*sca*4	Conventional one-step PCR	D1f	Primary PCR	ATGAGTAAAGACGGTAACCT	928	40	95° 30″	50° 30″	68° 90″	120
		D928r		AAGCTATTGCGTCATCTCCG						

[a] Probe R17K-C was designed for use in TaqMan assay with R17K135F and R17K249R primers.[146]
[b] Original assay used a 48°C annealing temperature[102]; however, our data indicate that annealing temperature can be increased up to 55°C to enhance the specificity of this assay. Rolain et al.[117] used these primers for SYBR Green PCR assay and subsequent Tm analysis of the amplicon.
[c] Original assay used 48°C annealing temperature[102]; however, our data indicate that annealing temperature can be increased up to 57°C to enhance the specificity of this assay.
[d] PCR amplification is followed by melting curve analysis of the amplicon to enhance specificity for the identification of *Rickettsia*.[114,137,139]
[e] PCR amplification is followed by melting curve analysis of the amplicon to enhance specificity for the identification of *Rickettsia*. This primer set can be also used for conventional PCR amplification; PCR product can be sequenced to complete species identification.[116]

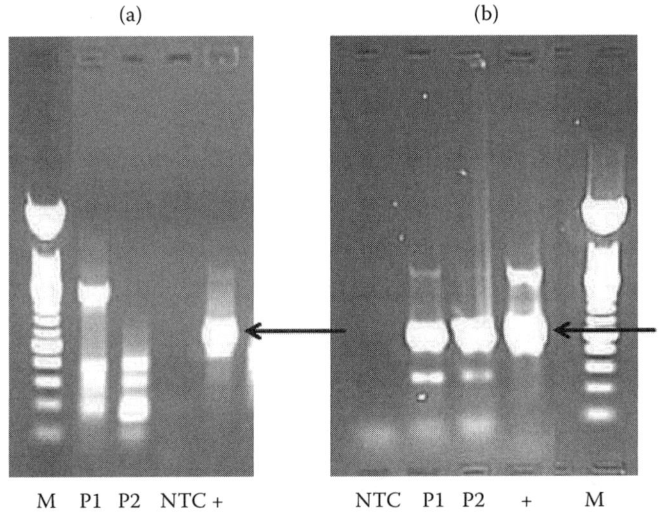

FIGURE 58.4 Demonstration of variable PCR amplification efficiency of *R. rickettsii omp*A fragment using different reagents. PCR amplification of *R. rickettsii omp*A fragment (70–602 nt) was performed using Amersham Ready-To-Go-Beads (a) and Qiagen Taq Master Mix (b) under the same conditions as described previously.[130] Arrows indicate migration position of the *R. rickettsii* positive PCR control. *Abbreviations*: M, DNA ladder; NTC, no template negative PCR control (+), *R. rickettsii* positive PCR control; P1 and P2, diagnostic specimens tested.

market for rickettsial diagnostic tests is relatively small, few commercial companies have invested in the development of alternative tests that require purified materials. Group-reactive IFA tests could be complemented by routine cross-adsorption assays in order to provide more accurate surveillance data on rickettsial diseases and earlier clinical identification of new rickettsioses, but this goal is again limited by the availability of rickettsial antigens. This lack of a differential or more specific serologic diagnostic assay led to recent changes in the US National surveillance case definition from RMSF to SFG rickettsiosis because of our present inability to differentiate serologically among *R. rickettsii*, *R. parkeri*, and other SFG rickettsioses in most human cases identified in the United States.[98,143] The present situation is unlikely to change until inexpensive methods for producing suitable rickettsial protein and lipopolysaccharide antigens using recombinant technology have been developed. However, even if this monovalent serological goal is not realized, it is likely that sophisticated rickettsial laboratories will be able to develop multiplex quantitative serological assays. These will probably be based primarily on Luminex bead technologies but more conventional ELISA capture assays and fluorescent microplate assays can be anticipated that will provide similar sensitivity and greater specificity compared to currently employed MIF or ELISA assays. However, even the identification of an increasing number of rickettsioses in the Americas and worldwide[5,98,138,144,147,148] is unlikely to provide a significant incentive to improve the specificity of the serological assays used since the clinical treatments still rely primarily on doxycycline for both the classic and the novel rickettsioses being discovered.

Molecular diagnostic assays will continue to be the field that will advance most rapidly, primarily because of the simplicity of the procedures involved and the tremendous advantages of agent DNA detection for rapid and specific diagnosis of rickettsial disease over serological methods. Increased sophistication of these assays will need to be accompanied by better coordination between laboratorians and physicians to assure that appropriate samples are collected promptly and submitted rapidly to the laboratory. It is likely that multiplexed DNA assays, either of conserved rickettsial gene targets combined with specific rickettsial targets or with gene targets from other tick-borne agents, will became the standard in the near future. Luminex bead assays coupled with increased use of the genomic information that is now available will provide robust approaches allowing substantial increases in the number of multiplexed assays that can be performed on small quantities of template DNA. One can anticipate increased use of whole genome amplification technologies to stretch precious clinical samples[145] that are often too small and may need to be shared by several laboratories.

Immunohistochemistry and rickettsial isolation will likely be used by more laboratories in the near future. The primary limiting factor for IHC is the need for increased production of good quality polyclonal and monoclonal antibodies against rickettsial antigens. Isolation will become more routine as the number of clinical laboratories with tissue culture experience and training in molecular identification of rickettsial DNAs become more commonplace.

Enhanced capabilities for diagnosis of rickettsioses in general and RMSF in particular should facilitate more rapid confirmation of classical clinical cases earlier in the course of infection and also ensure that diseases due to novel agents and unusual case presentations are more promptly recognized. The newly developed tests will need to satisfy the expectations of practicality, sensitivity and specificity and become widely available for point-of-care-testing. Continued improvements in diagnostic assays should enhance patient care and greatly improve the collection of accurate passive surveillance statistics to better characterize the prevalence and distribution of rickettsioses.

ACKNOWLEDGMENTS

We thank David J. Silverman for electron microscopy photograph and Perry Comegys for photographic work, Mary E. Wikswo for performing western blotting experiment, Arianna Salazar for assistance with literature search and creation of an EndNote library. Some of the materials used for this chapter were previously presented at 14 Congresso Brasileiro de Parasitologia Veterinaria and 2 Simposio Latino-Americano de Ricquetsioses, September 3-6, 2006, Centro de Convencoes, Ribeiro Preto, Brazil.

The findings and conclusions in this report are those of the authors and do not necessarily represent the views of the Centers for Disease Control or the Department of Health and Human Services of the United States.

REFERENCES

1. Ohashi, N. et al., Phylogenetic position of *Rickettsia tsutsugamushi* and the relationship among its antigenic variants by analyses of 16S rRNA gene sequences, *FEMS Microbiol. Let.*, 125, 299, 1995.
2. Viale, A.M. et al., Evolutionary relationships among eubacterial groups as inferred from GroEL (chaperonin) sequence comparisons, *Int. J. Syst. Bacteriol.*, 44, 527, 1994.
3. Stothard, D.R., and Fuerst, P.A., Evolutionary analysis of the spotted fever and typhus groups of *Rickettsia* using 16S rRNA gene sequences, *Syst. Appl. Microbiol.*, 18, 52, 1995.
4. Roux, V., and Raoult, D., Phylogenetic analysis of the genus *Rickettsia* by 16S rDNA sequencing, *Res. Microbiol.*, 146, 385, 1995.
5. Tamura, K. et al., MEGA4: Molecular Evolutionary Genetics Analysis (MEGA) software version 4.0, *Mol. Biol. Evol.*, 24, 1596, 2007.
6. Parola, P., Paddock, C., and Raoult, D., Tick-borne rickettsioses around the world: Emerging diseases challenging old concepts, *Clin. Microbiol. Rev.*, 18, 719, 2005.
7. Gillespie, J.J. et al., *Rickettsia* phylogenomics: Unwinding the intricacies of obligate intracellular life, *PLoS One*, 3, e2018, 2008.
8. Stothard, D.R., Clark, J.B., and Fuerst, P.A., Ancestral divergence of *Rickettsia bellii* from the spotted fever and typhus groups of *Rickettsia* and antiquity of the genus *Rickettsia*, *Int. J. Syst. Bacteriol.*, 44, 798, 1994.
9. Silverman, D.J., and Wisseman, C.L., Comparative ultrastructural study on the cell envelopes of *Rickettsia prowazekii*, *Rickettsia rickettsii*, and *Rickettsia tsutsugamushi*, *Infect. Immun.*, 21, 1020, 1978.
10. Silverman, D.J. et al., External layers of *Rickettsia prowazekii* and *Rickettsia rickettsii*: Occurrence of a slime layer, *Infect. Immun.*, 22, 233, 1978.
11. Gimenez, D.F., Staining rickettsiae in yolk-sac cultures, *Stain Technol.*, 39, 135, 1964.
12. Lauer, B.A., Reller, L.B., and Mirrett, S., Comparison of acridine orange and Gram stains for detection of microorganisms in cerebrospinal fluid and other clinical specimens, *J. Clin. Microbiol.*, 14, 201, 1981.
13. Weiss, E., and Moulder, J.W., Order I *Rickettsiales*, Gieszczkiewicz 1939, 25^{AL}. In Krieg, N.R., and Holt, J.G. (Editors), *Bergey's Manual of Systematic Bacteriology*, p. 687, The Williams & Wilkins, Baltimore, 1984.
14. Policastro, P.F., Peacock, M.G., and Hackstadt, T., Improved plaque assays for *Rickettsia prowazekii* in Vero 76 cells, *J. Clin. Microbiol.*, 34, 1944, 1996.
15. Wike, D.A. et al., Studies of the rickettsial plaque assay technique, *Infect. Immun.*, 5, 715, 1972.
16. Walker, D.H., and Cain, B.G., The rickettsial plaque. Evidence for direct cytopathic effect of *Rickettsia rickettsii*, *Lab. Invest.*, 43, 388, 1980.
17. Walker, D.H., Firth, W.T., and Edgell, C.J., Human endothelial cell culture plaques induced by *Rickettsia rickettsii*, *Infect. Immun.*, 37, 301, 1982.
18. Johnson, J.W., and Pedersen, C.E., Plaque formation by strains of spotted fever rickettsiae in monolayer cultures of various cell types, *J. Clin. Microbiol.*, 7, 389, 1978.
19. Labruna, M., Ecology of rickettsia in South America, *Ann. N. Y. Acad. Sci.*, 1166, 156, 2009.
20. Eremeeva, M.E., and Dasch, G.A., Closing the gaps between genotype and phenotype in *Rickettsia rickettsii*, *Ann. N. Y. Acad. Sci.*, 1166, 12, 2009.
21. Demma, L.J. et al., Rocky Mountain spotted fever from an unexpected tick vector in Arizona, *N. Engl. J. Med.*, 353, 587, 2005.
22. Labruna, M.B. et al., Prevalence of *Rickettsia* infection in dogs from the urban and rural areas of Monte Negro municipality, western Amazon, Brazil, *Vector Borne Zoonotic Dis.*, 7, 249, 2007.
23. Sanchez, R. et al., *Rhipicephalus sanguineus*-associated Rocky Mountain spotted fever in Mexicali, Mexico: Observations from an outbreak in 2008–2009. Abstract 75. The 23rd Meeting of the American Society for Rickettsiology. Hilton Head, South Carolina: Kansas State University, College of Veterinary Medicine, 2009.
24. Cunha, N.C. et al., First identification of natural infection of *Rickettsia rickettsii* in the *Rhipicephalus sanguineus* tick, in the State of Rio de Janeiro, *Pesq. Vet. Bras.*, 29, 105, 2009.
25. Oliveira, K.A., *Amblyomma imitator* ticks as vectors of *Rickettsia rickettsii*, Mexico. *Emerg. Infect. Dis.*, 16, 1282, 2010.
26. Adjemian, J. et al., Spatial clustering by disease severity among reported Rocky Mountain spotted fever cases in the United States, 2001–2005, *Am. J. Trop. Med. Hyg.*, 80, 72, 2009.
27. Angerami, R. et al., Brazilian spotted fever: A case series from an endemic area in southeastern Brazil: Clinical aspects, *Ann. N. Y. Acad. Sci.*, 1078, 252, 2006.
28. Moraes-Filho, J. et al., New epidemiological data on Brazilian spotted fever in an endemic area of the State of São Paulo, Brazil, *Vector Borne Zoonotic Dis.*, 9, 73, 2009.
29. Galvão, M.A. et al., Rickettsial diseases in Brazil and Portugal: Occurrence, distribution and diagnosis, *Rev. Saúde Públ.*, 39, 850, 2005.
30. Burgdorfer, W., and Brinton, L.P., Mechanisms of transovarial infection of spotted fever rickettsiae in ticks, *Ann. N. Y. Acad. Sci*,. 266, 61, 1975.
31. Burgdorfer, W., A review of Rocky Mountain spotted fever (tick-borne typhus), its agent, and its tick vectors in the United States, *J. Med. Entomol.*, 12, 269, 1975.
32. Niebylski, M.L., Peacock, M.G., and Schwan, T.G., Lethal effect of *Rickettsia rickettsii* on its tick vector (*Dermacentor andersoni*), *Appl. Env. Microbiol.*, 65, 773, 1999.
33. Ellison, D.W. et al., Genomic comparison of virulent *Rickettsia rickettsii* Sheila Smith and avirulent *Rickettsia rickettsii* Iowa, *Infect. Immun.*, 76, 542, 2008.
34. Dasch, G.A. et al., Genomic sequence variation in the complete genome sequences of multiple isolates of *Rickettsia rickettsii*, *R. prowazekii*, and *R. typhi*. Abstract #50. The 23rd Meeting of the American Society for Rickettsiology. Hilton Head, South Carolina: Kansas State University, College of Veterinary Medicine, 2009.
35. Eremeeva, M.E. et al., New perspectives on rickettsial evolution from new genome sequences of *Rickettsia*, particularly *R. canadensis*, and *Orientia tsutsugamushi*, *Ann. N. Y. Acad. Sci.*, 1063, 47, 2005.
36. Andersson, S.G. et al., Molecular phylogeny and rearrangement of rRNA genes in *Rickettsia* species, *Mol. Biol. Evol.*, 16, 987, 1999.
37. Ogata, H. et al., Mechanisms of evolution in *Rickettsia conorii* and *R. prowazekii*, *Science*, 293, 2093, 2001.
38. McLeod, M. et al., Complete genome sequence of *Rickettsia typhi* and comparison with sequences of other rickettsiae, *J. Bacteriol.*, 186, 5842, 2004.
39. Eremeeva, M.E. et al., Molecular typing of novel *Rickettsia rickettsii* isolates from Arizona, *Ann. N. Y. Acad. Sci.*, 1078, 573, 2006.

40. Karpathy, S.E., Dasch, G.A., and Eremeeva, M.E., Molecular typing of isolates of *Rickettsia rickettsii* by use of DNA sequencing of variable intergenic regions, *J. Clin. Microbiol.*, 45, 2545, 2007.
41. Kato, C.Y. et al., Insertion/Deletion (INDEL) typing of isolates of *Rickettsia rickettsii*. Abstract #35. The 23rd Meeting of the American Society for Rickettsiology. Hilton Head, South Carolina: Kansas State University, College of Veterinary Medicine, 2009.
42. Harden, V.A., The beginnings of scientific investigation. In Harden, V.A. (Editor), *Rocky Mountain Spotted Fever. History of a Twentieth-century Disease*, p. 23, The John Hopkins University Press, Baltimore and London, 1990.
43. Walker, D.H., Pathology and pathogenesis of vasculotropic ricketsioses. In Walker, D.H. (Editor), *Biology of Rickettsial Diseases*, p. 115, CRC Press, Inc., Boca Raton, FL, 1988.
44. Febre maculosa. www.cve.saude.sp.gov.br/htm/zoo/fm_i8503.htm. Accessed on November 16, 2010.
45. Surto de Febre Maculosa no municipio de Petropolis-RJ. Ministerio da Saude Secretaria de Vigilancia em Saude. http://portal.saude.gov.br/portal/arquivos/pdf/nota_maculosa_corrigida.pdf. Accessed on November 16, 2010.
46. Raoult, D. et al., Haemolysis with Mediterranean spotted fever and glucose-6-phosphate dehydrogenase deficiency, *Trans. Royal Soc. Trop. Med. Hyg.*, 80, 961, 1986.
47. Walker, D., and Ismail, N., Emerging and re-emerging rickettsioses: Endothelial cell infection and early disease events, *Nat. Rev. Microbiol.*, 6, 375, 2008.
48. Eremeeva, M.E., Dasch, G.A., and Silverman, D.J., Interaction of rickettsiae with eukaryotic cells. Adhesion, entry, intracellular growth, and host cell responses, *Sub-Cell. Biochem.*, 33, 479, 2000.
49. Sahni, S., and Rydkina, E., Host-cell interactions with pathogenic *Rickettsia* species, *Future Microbiol.*, 4, 323, 2009.
50. Martinez, J.J., and Cossart, P., Early signaling events involved in the entry of *Rickettsia conorii* into mammalian cells, *J. Cell Sci.*, 117, 5097, 2004.
51. Walker, T.S., and Winkler, H.H., Penetration of cultured mouse fibroblasts (L cells) by *Rickettsia prowazeki*, *Infect. Immun.*, 22, 200, 1978.
52. Walker, T.S., Rickettsial interactions with human endothelial cells in vitro: Adherence and entry, *Infect. Immun.*, 44, 205, 1984.
53. Cardwell, M.M., and Martinez, J.J., The Sca2 autotransporter protein from *Rickettsia conorii* is sufficient to mediate adherence to and invasion of cultured mammalian cells, *Infect. Immun.*, 77, 5272, 2009.
54. Chan, Y.G. et al., Rickettsial outer-membrane protein B (rOmpB) mediates bacterial invasion through Ku70 in an actin, c-Cbl, clathrin and caveolin 2-dependent manner, *Cell. Microbiol.*, 11, 629, 2009.
55. Martinez, J.J. et al., Ku70, a component of DNA-dependent protein kinase, is a mammalian receptor for *Rickettsia conorii*, *Cell*, 123, 1013, 2005.
56. Rydkina, E., Silverman, D.J., and Sahni, S., Similarities and differences in host cell signaling following infection with different *Rickettsia* species, *Ann. N. Y. Acad. Sci.*, 1063, 203, 2005.
57. Heinzen, R.A. et al., Directional actin polymerization associated with spotted fever group *Rickettsia* infection of Vero cells, *Infect. Immun.*, 61, 1926, 1993.
58. Silverman, D.J., and Santucci, L.A., Potential for free radical-induced lipid peroxidation as a cause of endothelial cell injury in Rocky Mountain spotted fever, *Infect. Immun.*, 56, 3110, 1988.
59. Eremeeva, M.E., and Silverman, D.J., Effects of the antioxidant alpha-lipoic acid on human umbilical vein endothelial cells infected with *Rickettsia rickettsii*, *Infect. Immun.*, 66, 2290, 1998.
60. Helmick, C.G., Bernard, K.W., and D'Angelo, L.J., Rocky Mountain spotted fever: Clinical, laboratory, and epidemiological features of 262 cases, *J. Infect. Dis.*, 150, 480, 1984.
61. Walker, D.H., and Peacock, M.G., Laboratory diagnosis of rickettsial diseases. In Walker, D.H. (Editor), *Biology of Rickettsial Diseases*, p. 135, CRC Press, Inc, Boca Raton, FL, 1988.
62. Elghetany, M.T., and Walker, D.H., Hemostatic changes in Rocky Mountain spotted fever and Mediterranean spotted fever, *Am. J. Clin. Pathol.*, 112, 159, 1999.
63. Kaplowitz, L.G., Fischer, J.J., and Sparling, P.F., Rocky Mountain spotted fever: A clinical dilemma, *Cur. Clin. Top. Infect. Dis.*, 2, 89, 1981.
64. Chapman, A.S. et al., Diagnosis and management of tickborne rickettsial diseases: Rocky Mountain spotted fever, ehrlichioses, and anaplasmosis—United States: A practical guide for physicians and other health-care and public health professionals, *MMWR Recomm. Rep.*, 55, 1, 2006.
65. Philip, R.N. et al., Microimmunofluorescence test for the serological study of Rocky Mountain spotted fever and typhus, *J. Clin. Microbiol.*, 3, 51, 1976.
66. Goldwasser, R.A., and Shepard, C.C., Fluorescent antibody methods in the differentiation of murine and epidemic typhus sera; specificity changes resulting from previous immunization, *J. Immun.*, 82, 373, 1959.
67. Raoult, D., and Dasch, G.A., Immunoblot cross-reactions among *Rickettsia*, *Proteus* spp. and *Legionella* spp. in patients with Mediterranean spotted fever, *FEMS Immun. Med. Microbiol.*, 11, 13, 1995.
68. Philip, R.N. et al., A comparison of serologic methods for diagnosis of Rocky Mountain spotted fever, *Am. J. Epidemiol.*, 105, 56, 1977.
69. Fournier, P.-E. et al., Kinetics of antibody responses in *Rickettsia africae* and *Rickettsia conorii* infections, *Clin. Diagn. Lab. Immunol.*, 9, 324, 2002.
70. Clements, M.L. et al., Serodiagnosis of Rocky Mountain spotted fever: comparison of IgM and IgG enzyme-linked immunosorbent assays and indirect fluorescent antibody test, *J. Infect. Dis.*, 148, 876, 1983.
71. Wilfert, C.M. et al., The incidence of Rocky Mountain spotted fever as described by prospective epidemiologic surveillance and the assessment of persistence of antibodies to *R. rickettsii* by indirect hemagglutination and microimmunofluorescence tests. In Burdorfer, W., and Anacker, R.L. (Editors), *Rickettsiae and Rickettsial Diseases*, p. 179, Academic Press, New York, 1981.
72. Hechemy, K.E. et al., Cross-reaction of immune sera from patients with rickettsial diseases, *J. Med. Microbiol.*, 29, 199, 1989.
73. Kleeman, K.T. et al., Early detection of antibody to *Rickettsia rickettsii*: Comparison of four serological methods: Indirect hemagglutination, indirect fluorescent antibody, latex agglutination, and complement fixation. In Burgdorfer, W., and Anacker, R.L. (Editors), *Rickettsiae and Rickettsial Diseases*, p. 171, Academic Press, New York, 1981.
74. Newhouse, V.F. et al., A comparison of the complement fixation, indirect fluorescent antibody, and microagglutination tests for the serological diagnosis of rickettsial diseases, *Am. J. Trop. Med. Hyg.*, 28, 387, 1979.
75. Raoult, D. et al., Serological diagnosis of Mediterranean spotted fever by the immunoperoxidase reaction. Eur. *J. Clin. Microbiol.*, 4, 441, 1985.

76. Tay, S.T., and Rohani, M.Y., The use of the indirect immunoperoxidase test for the serodiagnosis of rickettsial diseases in Malaysia, *Southeast Asian J. Trop. Med. Publ. Health*, 33, 314, 2002.
77. Jones, D. et al., Enzyme-linked immunosorbent assay for detection of human immunoglobulin G to lipopolysaccharide of spotted fever group rickettsiae, *J. Clin. Microbiol.*, 31,138, 1993.
78. Halle, S., Dasch, G.A., and Weiss, E., Sensitive enzyme-linked immunosorbent assay for detection of antibodies against typhus rickettsiae, *Rickettsia prowazekii* and *Rickettsia typhi*, *J. Clin. Microbiol.*, 6, 101, 1977.
79. Ching, W.-M. et al., Identification and characterization of epitopes on the 120-kilodalton surface protein antigen of *Rickettsia prowazekii* with synthetic peptides, *Infect. Immun.*, 64, 1413, 1996.
80. Eremeeva, M.E., Balayeva, N.M., and Raoult, D., Serological response of patients suffering from primary and recrudescent typhus: Comparison of complement fixation reaction, Weil-Felix test, microimmunofluorescence, and immunoblotting, *Clin. Diagn. Lab. Immun.*, 1, 318, 1994.
81. Dasch, G.A., Isolation of species-specific protein antigens of *Rickettsia typhi* and *Rickettsia prowazekii* for immunodiagnosis and immunoprophylaxis, *J. Clin. Microbiol.*, 14, 333, 1981.
82. Amano, K.I., Williams, J.C., and Dasch, G.A., Structural properties of lipopolysaccharides from *Rickettsia typhi* and *Rickettsia prowazekii* and their chemical similarity to the lipopolysaccharide from *Proteus vulgaris* OX19 used in the Weil-Felix test, *Infect. Immun.*, 66, 923, 1998.
83. Raoult, D., and Dasch, G.A., Line blot and western blot immunoassays for diagnosis of Mediterranean spotted fever, *J. Clin. Microbiol.*, 27, 2073, 1989.
84. La Scola, B. et al., Serological differentiation of murine typhus and epidemic typhus using cross-adsorption and Western blotting, *Clin. Diagn. Lab. Immun.*, 7, 612, 2000.
85. Broadhurst, L.E. et al., Laboratory evaluation of a dot-blot enzyme immunoassay for serologic confirmation of illness due to *Rickettsia conorii*, *Am. J. Trop. Med. Hyg.*, 58, 786, 1998.
86. Radulovic, S. et al., Antigen-capture enzyme immunoassay: a comparison with other methods for the detection of spotted fever group rickettsiae in ticks, *Am. J. Trop. Med. Hyg.*, 50, 359, 1994.
87. Walker, D.H., and Lane, T.W., Rocky Mountain spotted fever: Clinical signs, symptoms, and pathophysiology. In Walker, D.H. (Editor), *Biology of Rickettsial Diseases*, p. 63, CRC Press, Inc, Boca Raton, FL, 1988.
88. Walker, D.H., Cain, B.G., and Olmstead, P.M., Laboratory diagnosis of Rocky Mountain spotted fever by immunofluorescent demonstration of *Rickettsia* in cutaneous lesions, *Am. J. Clin. Pathol.*, 69, 619, 1978.
89. Walker, D.H., and Cain, B.G., A method for specific diagnosis of Rocky Mountain spotted fever on fixed, paraffin-embedded tissue by immunofluorescence, *J. Infect. Dis.*, 137, 206, 1978.
90. Woodward, T.E. et al., Prompt confirmation of Rocky Mountain spotted fever: Identification of rickettsiae in skin tissues, *J. Infect. Dis.*, 134, 297, 1976.
91. Green, W.R., Walker, D.H., and Cain, B.G., Fatal viscerotropic Rocky Mountain spotted fever. Report of a case diagnosed by immunofluorescence, *Am. J. Med.*, 64, 523, 1978.
92. Hall, W.C., and Bagley, L.R., Identification of *Rickettsia rickettsii* in formalin-fixed, paraffin-embedded tissues by immunofluorescence, *J. Clin. Microbiol.*, 8, 242, 1978.
93. Dumler, J.S. et al., Rapid immunoperoxidase demonstration of *Rickettsia rickettsii* in fixed cutaneous specimens from patients with Rocky Mountain spotted fever, *Am. J. Clin. Pathol.*, 93, 410, 1990.
94. Paddock, C.D. et al., Hidden mortality attributable to Rocky Mountain spotted fever: Immunohistochemical detection of fatal, serologically unconfirmed disease, *J. Infect. Dis.*, 179, 1469, 1999.
95. Rozental, T. et al., Fatal case of Brazilian spotted fever confirmed by immunohistochemical staining and sequencing methods on fixed tissues, *Ann. N. Y. Acad. Sci.*, 1078, 257, 2006.
96. La Scola, B., and Raoult, D., Diagnosis of Mediterranean spotted fever by cultivation of *Rickettsia conorii* from blood and skin samples using the centrifugation-shell vial technique and by detection of *R. conorii* in circulating endothelial cells: A 6-year follow-up, *J. Clin. Microbiol.*, 34, 2722, 1996.
97. Biosafety in Microbiological and Biomedical Laboratories. U.S. Department of Health and Human Services PHS, Centers for Disease Control and Prevention and National Institutes of Health, ed. Washington DC: U. S. Government Printing Office, 193–198, 2007. (http://www.cdc.gov/OD/OHS/biosfty/bmbl5/BMBL_5th_Edition.pdf)
98. Paddock, C.D. et al., *Rickettsia parkeri* rickettsiosis and its clinical distinction from Rocky Mountain spotted fever, *Clin. Infect. Dis.*, 47, 1188, 2008.
99. Paddock, C.D. et al., Isolation of *Rickettsia akari* from eschars of patients with rickettsialpox, *Am. J. Trop. Med. Hyg.*, 75, 732, 2006.
100. Eremeeva, M.E. et al., Isolation and identification of *Rickettsia massiliae* from *Rhipicephalus sanguineus* ticks collected in Arizona, *Appl. Environ. Microbiol.*, 72, 5569, 2006.
101. Pinter, A., and Labruna, M., Isolation of *Rickettsia rickettsii* and *Rickettsia bellii* in cell culture from the tick *Amblyomma aureolatum* in Brazil, *Ann. N. Y. Acad. Sci.*, 1078, 523, 2006.
102. Regnery, R.L., Spruill, C.L., and Plikaytis, B.D., Genotypic identification of rickettsiae and estimation of intraspecies sequence divergence for portions of two rickettsial genes, *J. Bacteriol.*, 173, 1576, 1991.
103. Anderson, B.E., and Tzianabos, T., Comparative sequence analysis of a genus-common rickettsial antigen gene, *J. Bacteriol.*, 171, 5199, 1989.
104. Carl, M. et al., Diagnosis of acute typhus infection using the polymerase chain reaction, *J. Infect. Dis.*, 161, 791, 1990.
105. Wilson, K.H., Blitchington, R.B., and Greene, R.C., Amplification of bacterial 16S ribosomal DNA with polymerase chain reaction, *J. Clin. Microbiol.*, 28: 1942, 1990.
106. Eremeeva, M., Yu, X., and Raoult, D., Differentiation among spotted fever group rickettsiae species by analysis of restriction fragment length polymorphism of PCR-amplified DNA, *J. Clin. Microbiol.*, 32, 803, 1994.
107. Roux, V., Fournier, P.-E., and Raoult, D., Differentiation of spotted fever group rickettsiae by sequencing and analysis of restriction fragment length polymorphism of PCR-amplified DNA of the gene encoding the protein rOmpA, *J. Clin. Microbiol.*, 34, 2058, 1996.
108. Roux, V. et al., Citrate synthase gene comparison, a new tool for phylogenetic analysis, and its application for the rickettsiae, *Int. J. Syst. Bacteriol.*, 47, 252, 1997.

109. Weisburg, W.G. et al., Phylogenetic diversity of the Rickettsiae, *J. Bacteriol.*, 171, 4202, 1989.
110. Wilson, K.H. et al., Probe directed at a segment of *Rickettsia rickettsii* rRNA amplified with polymerase chain reaction, *J. Clin. Microbiol.*, 27, 2692, 1989.
111. Tzianabos, T., Anderson, B.E., and McDade, J.E., Detection of *Rickettsia rickettsii* DNA in clinical specimens by using polymerase chain reaction technology, *J. Clin. Microbiol.*, 27, 2866, 1989.
112. Sexton, D.J. et al., The use of a polymerase chain reaction as a diagnostic test for Rocky Mountain spotted fever, *Am. J. Trop. Med. Hyg.*, 50, 59, 1994.
113. Ergas, D. et al., Early diagnosis of severe Mediterranean spotted fever cases by nested-PCR detecting spotted fever Rickettsiae 17-kD common antigen gene, *Scand. J. Infect. Dis.*, 40, 965, 2008.
114. Eremeeva, M.E., Dasch, G.A., and Silverman, D.J., Evaluation of a PCR assay for quantitation of *Rickettsia rickettsii* and closely related spotted fever group rickettsiae, *J. Clin. Microbiol.*, 41, 5466, 2003.
115. Jiang, J. et al., Development of a quantitative real-time polymerase chain reaction assay specific for *Orientia tsutsugamushi*, *Am. J. Trop. Med. Hyg.*, 70, 351, 2004.
116. Kidd, L. et al., Evaluation of conventional and real-time PCR assays for detection and differentiation of Spotted Fever group *Rickettsia* in dog blood, *Vet. Microbiol.*, 129, 294, 2008.
117. Rolain, J.-M. et al., Evaluation of antibiotic susceptibilities of three rickettsial species including *Rickettsia felis* by a quantitative PCR DNA assay, *Antimicrob. Agent Chemother.*, 46, 2747, 2002.
118. Roux, V., and Raoult, D., Phylogenetic analysis of members of the genus *Rickettsia* using the gene encoding the outer-membrane protein rOmpB (ompB), *Int. J. Syst. Evol. Microbiol.*, 50, 1449, 2000.
119. Vitorino, L. et al., rRNA intergenic spacer regions for phylogenetic analysis of *Rickettsia* species, *Ann. N. Y. Acad. Sci.*, 990, 726, 2003.
120. Sekeyova, Z., Roux, V., and Raoult, D., Phylogeny of *Rickettsia* spp. inferred by comparing sequences of "gene D," which encodes an intracytoplasmic protein, *Int. J. Syst. Evol. Microbiol.*, 51, 1353, 2001.
121. Lee, J.H. et al., Differentiation of rickettsiae by *groEL* gene analysis, *J. Clin. Microbiol.*, 41, 2952, 2003.
122. Drancourt, M., and Raoult, D., Characterization of mutations in the *rpo*B gene in naturally rifampin-resistant *Rickettsia* species, *Antimicrob. Agents Chemother.*, 43, 2400, 1999.
123. Emelyanov, V.V., Evolutionary relationship of Rickettsiae and mitochondria, *FEBS Lett.*, 501, 11, 2001.
124. Eremeeva, M.E. et al., Genetic analysis of isolates of *Rickettsia rickettsii* that differ in virulence, *Ann. N. Y. Acad. Sci.*, 990, 717, 2003.
125. Labruna, M. et al., *Rickettsia* species infecting *Amblyomma cooperi* ticks from an area in the state of São Paulo, Brazil, where Brazilian spotted fever is endemic, *J. Clin. Microbiol.*, 42, 90, 2004.
126. Kidd, L. et al., Molecular characterization of *Rickettsia rickettsii* infecting dogs and people in North Carolina, *Ann. N. Y. Acad. Sci.*, 1078, 400, 2006.
127. Kaplowitz, L.G. et al., Correlation of rickettsial titers, circulating endotoxin, and clinical features in Rocky Mountain spotted fever, *Arch. Intern. Med.*, 143, 1149, 1983.
128. Drancourt, M. et al., Diagnosis of Mediterranean spotted fever by indirect immunofluorescence of *Rickettsia conorii* in circulating endothelial cells isolated with monoclonal antibody-coated immunomagnetic beads, *J. Infect. Dis.*, 166, 660, 1992.
129. Sirigireddy, K., and Ganta, R., Multiplex detection of *Ehrlichia* and *Anaplasma* species pathogens in peripheral blood by real-time reverse transcriptase-polymerase chain reaction, *J. Mol. Diagn.*, 7, 308, 2005.
130. Estripeaut, D. et al., Rocky Mountain spotted fever, Panama, *Emerg. Infect. Dis.*, 13, 1763, 2007.
131. Zavala-Castro, J.E. et al., Fatal human infection with *Rickettsia rickettsii*, Yucatan, Mexico, *Emerg. Infect. Dis.*, 12, 672, 2006.
132. Mendes do Nascimento, E.M. et al., Evaluation of PCR-based assay in human serum samples for diagnosis of fatal cases of spotted fever group rickettsiosis, *Clin. Microbiol. Infect.*, 15, Suppl. 2, 232, 2010.
133. Tyeryar, F.J. et al., DNA base composition of rickettsiae, *Science*, 180, 415, 1973.
134. Myers, W.F., and Wisseman, C.L., Jr., The taxonomic relationship of *Rickettsia canada* to the typhus and spotted fever groups of the genus *Rickettsia*. In Burgdorfer, W., and Anacker, R.L. (Editors), *Conference on Rickettsiae and Rickettsial Diseases*, p. 313, Hamilton, Montana, September 3–5, 1980, Academic Press, 1981.
135. Schramek, S., Deoxyribonucleic acid base composition of members of the typhus group of rickettsiae, *Acta Virol.*, 16, 447, 1972.
136. Moriarity, J., Loftis, A., and Dasch, G., High-throughput molecular testing of ticks using a liquid-handling robot, *J. Med. Entomol.*, 42, 1063, 2005.
137. Bermudez, S. et al., Detection and identification of rickettsial agents in ticks from domestic mammals in eastern Panama, *J. Med. Entomol.*, 46, 856, 2009.
138. Shapiro, M.R. et al., *Rickettsia* 364D: A newly recognized cause of eschar-associated illness in California, *Clin. Infect. Dis.*, 50, 541, 2010.
139. Wikswo, M. et al., Detection and identification of spotted fever group rickettsiae in *Dermacentor* species from southern California, *J. Med. Entomol.*, 45, 509, 2008.
140. Altschul, S.F. et al., Gapped BLAST and PSI-BLAST: A new generation of protein database search programs, *Nucleic Acid Res.*, 25, 3389, 1997.
141. Fournier, P.-E. et al., Gene sequence-based criteria for identification of new rickettsia isolates and description of *Rickettsia heilongjiangensis* sp. nov., *J. Clin. Microbiol.*, 41: 5456, 2003.
142. Anderson, B.E. et al., Sequence analysis of the 17-kilodalton-antigen gene from *Rickettsia rickettsii*, *J. Bacteriol.*, 169, 2385, 1987.
143. Openshaw, J.J. et al., Rocky mountain spotted fever in the United States, 2000–2007: interpreting contemporary increases in incidence, *Am. J. Trop. Med. Hyg.*, 83, 174, 2010.
144. Parola, P., Labruna, M.B., and Raoult, D., Tick-borne rickettsioses in America: Unanswered questions and emerging diseases, *Curr. Infect. Dis. Rep.*, 11, 40, 2009.
145. Kato, C.Y. et al., Inexpensive and high throughput whole genome amplification for improved characterization of *Rickettsia* DNAs. Abstract #112. The 23rd Meeting of the American Society for Rickettsiology, Kansas State University,

College of Veterinary Medicine, Hilton Head, South Carolina, 2009.
146. Loftis, A. et al., Surveillance of Egyptian fleas for agents of public health significance: *Anaplasma*, *Bartonella*, *Coxiella*, *Ehrlichia*, *Rickettsia*, and *Yersinia pestis*, *Am. J. Trop. Med. Hyg.*, 75, 41, 2006.
147. Labruna, M.B. et al., *Rickettsia* species infecting *Amblyomma cooperi* ticks from an area in the state of Sao Paulo, Brazil, where Brazilian spotted fever is endemic, *J. Clin. Microbiol.*, 42, 90, 2004.
148. Paddock, C.D. et al., 2004. *Rickettsia parkeri*: a newly recognized cause of spotted fever rickettsiosis in the United States, *Clin. Infect. Dis.*, 38, 805, 2004.

Section IV

Proteobacteria

Betaproteobacteria

59 Achromobacter

Beth Mutai and Yi-Wei Tang

CONTENTS

59.1 INTRODUCTION

59.1.1 Classification and Biology

The genus *Achromobacter* is used to describe glucose-nonfermenting, gram-negative bacilli with peritrichous flagella rods that are oxidase-positive, attack carbohydrates aerobically, and do not produce 3-ketolactose from lactose.[1] *Achromobacter* species are straight rods 0.8–1.2 μm and 2.5–3.0 μm with rounded ends.[2] Originally classified as *Alcaligenes*, the *Achromobacter* was reclassified based on phenotypic characteristics, DNA base comparison, DNA–DNA similarity, and 16S rRNA sequence analysis.[3] The present *Achromobacter* genus contains *Achromobacter xylosoxidans* (with two subspecies, *denitrificans* and *xylosoxindans*), *Achromobacter ruhlandii*, and *Achromobacter piechaudii*. *A. xylosoxindans* subsp. *denitrificans* and *A. xylosoxindans* subsp. *xylosoxidans* were combined under one species because the rRNA cistrons were indistinguishable by hybridization technique despite DNA–DNA homology values of 30%–35%.[4] *A. ruhlandii* was added to the genus to accommodate the hydrogen oxidizing strains.[2] The third species *A. piechaudii* included isolates from human clinical material and strains originally named *Alcaligenes faecalis* and "*Achromobacter iophagus*"[4,5]; it was added to the genus as they were shown to be phenotypically and genotypically distinct from the other two species.[6]

Achromobacter species are distributed widely in nature and are frequently found in aqueous environments, mainly water and moist soil. *A. xylosoxidans* is an opportunistic human pathogen capable of causing a variety of infections and is widespread in oligotrophic aquatic niches.[6] *A. ruhlandii* is a commensal that has been isolated from soil[7] and not pathogenic to humans. *A. piechaudii* has been isolated from pharyngeal swabs, a nose wound, blood, human ear discharge, and soil.[4] *A. xylosoxidans* is frequently retrieved in the rhizosphere and many beneficial functions of this species have been reported in experimental assays. These functions includes control of some plant pathogens,[8–10] stimulation of ion transport to promote plant growth, and inhibition of afloxin production in *Aspergillus* sp.[11]

A. xylosoxidans subspecies *xylosoxidans* and *denitrificans* and *A. piechaudii* are the clinically relevant species of the genus, with *A. xylosoxidans* subsp. *xylosoxidans* being the most clinically important. *A. xylosoxidans* subsp. *xylosoxidans* probably is part of the endogenous flora of the ear and gastrointestinal tract and is a common contaminant of fluids. *A. xylosoxidans* is ideally suited to hospital environments in that it is resistant to many disinfectants and antibiotics. The organism has been isolated from swimming pools, dialysis fluid, distilled water, de-ionized and tap water, respirators and humidifiers, chlorhexidine solutions, and incubators.[12]

59.1.2 Clinical Manifestation

Clinical illness caused by *A. xylosoxidans* subsp. *xylosoxidans* has involved isolates from blood, peritoneal and pleural fluids, urine, respiratory secretions, and wound exudates. Bacteremia, often related to intravascular catheters, is the most commonly reported infection. Biliary tract sepsis, meningitis (sometimes with lymphocytic predominance in cerebrospinal fluid), pneumonia (nosocomial and community acquired), peritonitis, urinary tract infection, osteomyelitis, prosthetic knee infection, and prosthetic valve endocarditis have been reported. Patients often have an immunosuppressed state but this is not always the case, especially in nosocomial outbreaks. This organism can colonize the airways of

CF patients and has been associated with exacerbation of respiratory symptoms. Recovery in neonatal infection may result from perinatal transfer from the mother. The organism has been recovered from the eye, the ear, and the pharynx, but its pathogenic potential in these sites is unproved.

Strains of *A. xylosoxidans* subsp. *xylosoxidans* grow well on blood agar and McConkey's agar plates; they produce flat, spreading and rough colonies and have peritrichous flagellae, features that help distinguish them from pseudomonads. The organisms are oxidase positive, catalase positive, oxidize glucose to produce acid, and (as the species name indicates) oxidize xylose readily. An isolate of *A. xylosoxidans* subsp. *xylosoxidans* can easily be mistaken for a non-aeruginosa strain of *Pseudomonas*, but the unusual susceptibility pattern would suggest the correct identity.

Usually, strains of *A. xylosoxidans* subsp. *xylosoxidans* are susceptible to trimethoprim-sulfamethoxazole, ureidopenicillins, imipenem, ceftazidime, cefoperazone, and β-lactamase inhibitor combinations. Generally, they are resistant to narrow-spectrum penicillins, other cephalosporins (including cefotaxime and ceftriaxone), aztreonam, and aminoglycosides. Susceptibility to the fluoroquinolones is variable. Hyperproduction of β-lactamases has been implicated in resistance.

The clinical manifestation of infection caused by *A. xylosoxidans* is variable and includes primary bacteremia, pneumonia, meningitis, endocarditis, cholecystitis, peritonitis, pyelonephritis, osteomyelitis, lymphadenitis, and keratitis. Clinical manifestation in bacteremic patients infected with *Achromobacter xylosoxidans* includes fever, chills, hypotension, and altered consciousness.[13]

Treatment of infections caused by this organism is usually difficult as multidrug resistance is commonly encountered and as the optimal therapy is not yet determined. *Achromobacter* strains are frequently resistant to aminoglycosides, ampicillin, first- and second-generation cephalosporins, chloramphenicol, and fluoroquinolones, but are usually susceptible to anti-*Pseudomonas* third-generation cephalosporins, imipenem, and trimethoprim-sulfamethoxazole.[14] However, there are reports that empirical antibiotic therapy with piperacillin-tazobactam or a carbapenem would be a reasonable choice until results of susceptibility tests are available. Imipenem-resistant *A. xylosoxidans* carrying blaVIM-2-containing class 1 integron has been reported in Korea.[15]

59.1.3 Pathogenesis and Epidemiology

Strains of *A. xylosoxindans* subsp. *denitrification and A. piechaudii* are frequently isolated from clinical specimens. They are regarded as opportunistic pathogens.[4] Only *A. xylosoxindans* subtype *xylosoxindans* is considered to be of clinical significance. *A. xylosoxidans* is an uncommon but serious cause of nosocomial epidemics in hospitals for human beings. This species has been reported to be the etiologic agent of various human infections in immunocompromised and immunocompetent hosts,[16] especially patients with underlying diseases including endocarditis, meningitis, ventriculitis after neurosurgery, pneumonia, bacteremia osteomyelitis, arthritis, and peritonitis.[17–21]

Outbreaks of hospital-acquired infections have been reported in intensive care units and postsurgery recovery areas. The nosocomial infections have been attributed to *A. xylosoxindans* contaminating antiseptic solutions,[17] nonbacteriostatic solutions used in diagnostic tracer procedures,[22] dialysis fluids,[18] intravascular pressure transducer,[23] and diagnostic contrast solution.[24]

A. xylosoxidans is capable of persistence infection of the respiratory track of patients with cystic fibrosis (CF) but its role in contributing to pulmonary decline in this population is not clear. However it is important to remember that *A. xylosoxidans* infects approximately 9% of CF patients.[25] Persistent lymph node infection in a patient with hyperimmunoglobulin M syndrome was reported.[14]

In many instances this organism is considered as a colonizer and pathogenic role is generally dismissed and rarely recognized as a human pathogen. They have been isolated from clinical specimen including ear discharge, blood, cerebrospinal fluid, wound, peritoneal fluid, urine, stool, throat, sputum, and tracheal aspirate specimens. Despite its low intrinsic pathogenicity, it can cause serious infections in humans, leading to bacteremia, meningitis, pneumonia, endocarditis, peritonitis, osteomyelitis, urinary tract infection, and endophthalmitis. Most infections are nosocomially acquired, with neonates, burn victims, and other immunocompromised hosts being at greatest risk. Often clinical nosocomial outbreaks are attributed to disinfectant solution, dialysis fluids, saline solution, and de-ionized water.

Achromobacter is widely distributed in nature and is frequently found in aqueous environments mainly water and moist soil. The organism has been isolated from swimming pools, dialysis fluid, distilled water, de-ionized and tap water, respirators and humidifiers, chlorhexidine solutions, and incubators.[12]

A. xylosoxindans is capable of causing persistent infection of the respiratory tract of persons with cystic fibrosis, though its role in declining pulmonary is not clear. This species is very significant in CF patients, infecting 9% of CF patients and being frequently confused with species within the *Burkholderia cepacia* complex.[26,27]

Misidentification of *A. xylosoxindans* and related nonfermenting species seriously compromises infection control measures and cofounds efforts to more clearly understand the epidemiology and natural history of infection.[28]

The organism has been implicated in outbreaks of nosocomial infections associated with contaminated fluids, incubators and humidifiers, and contaminated soaps and disinfectants.[29,30] Contamination of water well used by patients was documented in a case of bacteremia.

59.1.4 Diagnosis

Accurate identification of *Achromobacter* is desirable for various reasons, as prerequisite for assessing the epidemiology and clinical significance of the species; *Achromobacter* species are commonly confused with other nonfermenters, like the members of the *Burkholderia* complexes, *Alcaligenes*, and *Pseudomonas*, which will have considerable impact on patient management. Most isolates have multidrug resistance,[31,32] and transmission of these isolates between close contacts seems to occur more frequently than was previously assumed.[33] A recent study used *P. aeruginosa*-specific duplex real-time PCR assays to further evaluate[26] *P. aeruginosa* sputum isolates from 561 cystic fibrosis patients identified by 17 Australian clinical microbiology laboratories. *A. xylosoxidans* (n = 21) was indicated as the species most commonly misidentified as *P. Aeruginosa*.[34]

Phenotypic Techniques. *Achromobacter* species are gram-negative rods of 0.8–1.2 μm by 2.5–3.0 μm, with rounded ends and 1–20 sheathed flagella arranged peritrichously.[3] Strains of *Achromobacter xylosoxidans* grow well on blood agar plates, produce flat spreading and rough colonies. *Achromobacter* grows on MacConkey Salmonella–Shigella agars and nutrient broth agar; they produce nonpigmented minute colonies at 25 h, progressing to moderate-sized and extremely mucoid colonies by 48 h,[35] oxidize positive and indole negative, produces urease and phenylalanine deaminase while metabolizing nitrate to nitrile and nitrogen gas. The predominant fatty acids are C16:0, C17:0 cyclo, 3-OH C 14:0, and C16:1007C.[36] *A. xylosoxidans* produces nonpigmented colonies on MacConkey agar and was oxidase positive and indole negative on spot biochemical test.[14]

Phenotypic identification of *Achromobacter* species has been done using API 20NE test (BioMèrieux). The API stripes are incubated for 48 h at 30°C under ambient air and the results are interpreted with the APILAB PLUS software version 3.3.3.37 API 20NE system was found to have better reliability than other phenotypic systems that gave good identification of *A. xylosoxidans*. Nevertheless, identification of *A. xylosoxidans* is still a challenge for the clinical microbiologist, and using biochemical and commercial kits is frequently confused with other nonfermentors—especially members of the *Burkholderia* complex.[37] Misidentification is worsened by the fact that the species are not in the database of commercial kits.[26,38] Also, identification with biochemical methods is difficult due to low metabolic activity and morphological and biochemical variability between isolates.

Unusual susceptibility patterns such as those generated by microdilution susceptibility test can improve identification. Antimicrobial minimum inhibitory concentrations are determined in cation-adjusted Mueller-Hinton broth (BBL; Becton Dickinson Microbiology Systems) on Sensititre microdilution plate (Trek Diagnostic System, Westlake, OH) by use of inoculums density of 3.8 × 105 CFU/mL in each 50 μL well. Plates are incubated in ambient air at 35°C for 24 h.[14]

Molecular Techniques. Several genotypic methods have been developed for the identification of nonfermenting gram-negative bacilli. Most of these methods are based on 16S rRNA gene polymorphism and involves PCR, restriction fragment length polymorphism analysis, single strand conformation polymorphism analysis, or partial sequencing.[38]

Fluorescence in situ hybridization (FISH) has been proven to be a rapid, easy to perform, and comparably cheap method for the identification of bacteria pathogens.[37,39,40] Wellinghausen et al.[37] utilized ARB program to design two probes (Ach-221 5′-CGC TCY AAT AGT GCA AGG TC-3′ and Alc-476 5′-CTG CAG ATA CCG TCA GCA GT-3′ bind to rRNA 221–240 and 476–495, respectively) for identification of *A. xylosoxidans*. These probes were synthesized with 5′ labeled with fluorescent dye Cy3. A universal bacteria probe (5′-GCT GCC TCC CGT AGG AGT-3′) that targets all bacteria was included as a positive control. The FISH assay was performed with bacterial suspension from an overnight culture on a glass slide followed by drying by air and fixing with methanol. *A. xylosoxidans* specific probes Ach-221 correctly identified all isolates of *A. xylosoxidans* showing a sensitivity of 100%. It showed a cross reactivity with an isolate of *Chryseobacterium* sp. giving a specificity of 99.2%. Due to the close relationship there was cross binding with *A. ruhlandii* but not the pathogenic *A. piechaudii*. The cross binding was concluded nonsignificant since *A. ruhlandii* has not been isolated from human specimens.[37] This technique was evaluated for the identification of *Achromobacter* and the closely related species *Alcaligenes feacalis* from cystic fibrosis patients using FISH probes. FISH was concluded to be an excellent method for specific identification of *A. xylosoxidans*, which is less time demanding, low cost, and simple to perform when compared to other molecular methods.[37]

Pulsed field gel electrophoresis (PFGE) is one of the molecular methods that can be used to type a broad array of bacteria species, it involves embedding organisms in agarose, lyzing the organism in situ and digesting the chromosomal DNA with restriction endonuclease that cleaves infrequently. Slices of agarose containing the DNA fragments are inserted into wells of an agarose gel, and the restriction fragments are resolved into a pattern of discrete bands. The DNA restriction patterns of the isolates are compared to determine the relatedness.[41] Using PFGE, several clusters of CF patients were revealed to carry genotypically identical *P. aeruginosa* and *A. xylosoxidans* strains in a rehabilitation center.[42]

16S rRNA gene sequencing is a powerful molecular tool that has been used for phylogenetic studies of bacteria from both the environment and human specimens. This technique allows accurate and rapid identification of bacteria to species level and it is used in clinical laboratories for identification of slow growing, unusual bacteria.[43,44] This method has been demonstrated to give good identification of *Achromobacter* isolates; however, public DNA databases contain faulty or incorrect assigned sequences because of quality control measures. Therefore, an experienced user is required to correctly interpret DNA sequencing data.[45] The technology was recently used in *Tunis* for identification of the bacterial pathogens, including *A. xylosoxidans* from the synovial fluid specimens of arthritic patients.[46]

Misidentification of *A. xylosoxidans* and related nonfermenting species seriously compromises infection control measures, and cofounds efforts to more clearly understand the epidemiology of *Achromobacter* species. Liu et al.[28] developed a PCR assay based on 16S rRNA gene sequences for identification of these species. The high degree of similarity among *Achromobacter* species did not allow designing of species specific primers in 16S rRNA gene. The primers were specific for *A. piechaudii*, *A. ruhlandii*, and *A. denitrificans* strains and negative for all the related bacteria including *Achromobacter*, *Alcaligenes*, *Bordetella*, *Burkholderia Pandoraea*, and *Ralstonia*. In another report, Nazarets et al.[47] described a ribosomal intergenic spacer analysis (RISA) coupled with high performance liquid chromatography (HPLC) for the detection and monitoring of CF patients' lung microbial colonizers including *Staphylococcus aureus*, *Haemophilus influenzae*, *P. aeruginosa*, the *Burkholderia cepacia* complex, *Stenotrophomonas maltophilia*, and *A. xylosoxidans*.

Matrix-assisted laser desorption ionization-time flight mass spectrometry (MALDI-TOF-MS) has been also used for bacterial identification by examining the profile of proteins detected directly from intact bacteria. This technique is based on relative molecular masses that allows desorption of peptides and proteins from whole different cultured microorganisms. Ions are separated and detected according to their molecular masses and charge with each bacteria strain producing a spectrum within minutes with a series of picks corresponding to m/z ratios of ions released from bacterial proteins during laser desorption.[48] Identification of the tested strains corresponds to the species of the reference strains with the best match in the database. This technique is used for the identification of nonfermenting, gram-negative bacilli isolated from cystic fibrosis patients including Achromobacter spp. and the identity of *A. xylosoxidans* is correctly determined. The technique is simple to run and requires a short time to run; it takes about 1 h to test 50 samples giving reliable identification.[49]

59.2 METHODS

59.2.1 Sample Preparation

Achromobacter can be grown on MacConkey Salmonella–Shigella agar, Mueller-Hinton agar, or nutrient broth agar 37°C for 24–48 h. DNA is extracted from isolated colonies using commercially available kits like QiaAmp DNA mini kit according to the instructions of the manufacturer. Alternatively, DNA is prepared by heating one or two colonies picked from an overnight grown plate at 95°C for 15 min in 20 μL of lysis buffer containing 0.25% (vol/vol) sodium dodecyl sulfate and 0.05 M NaOH. After lysis, 180 μL of sterile water is added to the lysis buffer and the DNA solution is stored frozen until used.[28]

59.2.2 Detection Procedures

(i) Ribosomal DNA–Directed PCR

Principle. Liu et al.[28] developed a PCR assay based on 16S rRNA gene sequences for identification of Achromobacter species. The primers (AX-F1: 5′-GCAGGAAAGAAACGTCGCGGGT-3′ and AX-B1: 5′-ATTTCACATCTTTCTTTCCG-3′) have been designed by aligning rDNA sequences of all Achromobacter, Alcaligenes, Bordetella, Burkholderia Pandoraea, and Ralstonia species available in the Genbank database. Use of these primers in PCR facilitates amplification of a 163 bp product from *A. xylosoxidans* strains.

Procedure

1. Prepare PCR mixture (25-μL) containing 2 μL of template, 1 U of Taq polymerase (Gibco-BRL), 250 mM concentrations of each deoxynucleotide triphosphate (Gibco-BRL), 1× PCR buffer (Gibco-BRL), 1.5 mM MgCl2 (Gibco-BRL), and a 1 μM concentration of each oligonucleotide primer (AX-F1 and AX-B1).
2. Amplify in a thermal cycler with 1 cycle of 95°C for 3 min; 30 cycles of 94°C for 1 min, 56°C for 1 min, and 72°C for 1 min 30 s; and a final 72°C for 10 min. Include all reaction mixture components except template DNA as negative control.
3. Separate the PCR product on 2% agarose gel in the presence of ethidium bromide, and visualize under UV light.

Note: The high degree of similarity among Achromobacter species did not allow designing of species specific primers in 16S rRNA gene. The primers AX-F1 and AX-B1 are specific for *A. piechaudii*, *A. ruhlandii*, and *A. denitrificans* strains and negative for all the related bacteria. The sensitivity and specificity of the PCR assay for *A. xylosoxidans* is 100% and 97%, respectively.[28]

(ii) 16S rRNA Gene Sequencing The nearly complete 16S rRNA gene (1490 bp) of *A. xylosoxidans* is amplified with conserved primers UFPL (5′-AGTTTGATCCTGGCTCAG-3′) and URPL (5′-GGTTACCTTGTTACGACTT0-3′)[50] and the resulting amplicon is sequenced using the primers 16SFor (5′-AGA GTT TGA TCC TGG CTC AG-3′) and 16SRev (5′-GGT TAC CTT GTT ACG ACT-3′).[28] Amplicons are purified by using the Promega Wizard PCR prep DNA purification kit (Promega). Sequence analysis is performed on a 310 Genetic analyser (ABI Prism Biosystems) using a Dye Terminator Cycle Sequencing Ready Reaction kit (ABI Prism). The sequences are then compared to sequences available in the GenBank database by using BLAST algorithm or EditSeq (DNAStar). In addition, the sequences can be compared to sequences of type strains of these species for instant, *Achromobacter xylosoxidans* subsp. *xylosoxidans* (ATCC 9220, GenBank accession number AF411021).

(iii) Genotyping by PFGE To identify *A. xylosoxidans* using PFGE, genomic DNA is extracted from logarithmic–phase cultures of *A. xylosoxidans* isolate grown in brain-heart broth (BBL; Becton Dickinson Microbiology Systems) prepared in low-melting point agarose plugs and digested with XbaI enzyme (New England Biolabs, Beverly MA) for 24 h[51] and bacteriophage lambda standard DNA size ladder is used (Bio-Rad Laboratories). Electrophoresis and gel photographing is performed.[14] Isolates are considered clonally related if there are more than three differences, according to criteria described.[42] PFGE of XbaI enzyme digested genomic DNA

is an established method for epidemiologic typing of *A. xylosoxidans*, Achromobacter field strains exhibit a wide diversity of restriction fragment length polymorphism.

59.3 CONCLUSION

A. xylosoxidans is a gram-negative oxidative organism that causes infections in patients with certain underlying illnesses. It is a rare but an important cause of bacteremia in immunocompromised individuals. Achromobacter isolates are often resistant to multiple antimicrobials. Empirical antibiotic therapy with piperacillin- tazobactam or a carbapenem is a reasonable choice until the results of susceptibility tests are available. Accurate identification of these organisms, which is mainly based on a series of phenotypic and genotypic methods, is important to enable better understanding of its epidemiology and clinical significance.

REFERENCES

1. Schreckenberger, P.C., Daneshver, M.I., and Hollis, D.G. *Acinetobacter, Achromobacter, Chryseobacterium, Moraxella*, and other nonfermentative Gram-negative rods. In Murray, P.R., et al. (Editors), *Manual of Clinical Microbiology*, 9th ed., pp. 770–802, American Society for Microbiology, Washington DC, 2007.
2. Yabuuchi, E., and Ohyama, A., Evaluation of 36 Minitek tests and a new approach for identification of nonfermentors, *J. Clin. Microbiol.*, 13, 572, 2007.
3. Yabuuchi, E., Kosako, Y., and Ezaki, T., Emendation of genus Achromobacter and *Achromobacter xylosoxidans* (Yabuuchi and Yano) and proposal of *Achromobacter ruhlandii* (Paker and Vishniac) comb. Nov *Achromobacter piechaudii* (Kiredjian et al.) *Achromobacter xylosoxidans* subsp. *denitrificants* (Ruger and Tan) comb. nov., *Microbiol. Immunol.*, 42, 429, 1998.
4. Kiredjian, M.B. et al., *Alcaligenes piechaudii*, a new species from human clinical specimens and the environments, *Int. J. System. Bacteriol.*, 36, 282, 1986.
5. Holmes, B., and Dawson, C.A., Numerical taxonomic studies on *Achromobacter* isolates from clinical materials. In Leclerc, H. (Editor), *Gram Negative Bacteria of Medical and Public Health Importance: Taxonomy, Identification, Applications*, pp. 331–341, Les Editions INSERMS, Paris, 1983.
6. Kersters, K., and De Ley, J., Genus *Alcaligenes* Cestellani and Chalmers. In Krieg, N.R., and Holt, J.G. (Editors), *Bergey's Manual of Systematic Bacteriology 1*, pp. 361–373, The Williams & Wilkins Co, Baltimore, 1984.
7. Packer, L. et al., Chemosynthetic fixation of carbon dioxide and characteristics of hydrogenase in resting cell suspensions of *Hydrogenomonas ruhlaindii* nov. spec., *J. Bacteriol.*, 70, 216, 1955.
8. Yuen G.Y. et al., Interactions of pseudomonas flourescens strains E6 with ornamental plants and its effect on the composition of root—colonizing microflora, *Phytopathogy*, 76, 176, 1986.
9. Benchimol, R.C. et al., Controle da fusariose em plantas de pimentado-reino com bacterias endofiticas, sobrevivencia e respostas morfofisiologicas, *Pesq. Agrop. Brasil*, 35, 1343, 2000.
10. Vaidya, R.J. et al., Production of chitinase and its optimization from a novel isolate *Alcaligenes xylosoxydans*: potential in antifungal biocontrol, *World J. Microbiol. Biotechnol.*, 17, 691, 2001.
11. Yan, P. et al., Cyclo(L-Leucyl-l-Prolyl) produced by Achromobacter xylosoxidans inhibits aflatoxins production by *Asoergillus parasiticus*, *Appl. Environ. Microbiol.*. 70, 7466, 2004.
12. Manjra, A.I., Moosa, A., and Bhamjee, A., Fatal neonatal meningitis and ventriculitis caused by multiresistant *Achromobacter xylosoxidans*, a case report, *South Afr. Med. J.*, 76, 571, 1989.
13. Shei, S.S. et al., Characteristics of *Achromobacter xylosoxidans* bacteremia in northern Taiwan, *J. Microbiol. Immunol. Infect. Dis.*, 38, 277, 2005.
14. Weitkamp, J. et al., Recurrent *Achromobacter xylosoxidans* bacteremia associated with persistent lymph node infection in a patient with hyper immunoglobin M syndrome, *Clin. Infect. Dis.*, 31, 1183, 2000.
15. Shin, K.S. et al., Imipenem-resistant A*chromobacter xylosoxidans* carrying blaVIM-2-containing class 1 integron, *Diagn. Microbiol. Infect. Dis.*, 53, 215, 2005.
16. Duggan, J.M. et al., *Achromobacter xylosoxidans* bacteremia: Report of four cases and review of the literature, *Clin. Infect. Dis.*, 23, 569, 1996.
17. Shigeta, S. et al., Cerebral ventriculitis associated with *Achromobacter xylosoxidans*, *J. Clin. Pathol.*, 31, 156, 1978
18. Reverdy, M.E. et al., Nosocomial colonization and infection by *Achromobacter xylosoxidans*, *J. Clin. Microbiol.*, 19, 140, 1984.
19. Legrand, C., and Anaissie, E., Bacteremia due to *Acliromobacter xylosoxidans* in patients with cancer, *Clin. Infect. Dis.*, 14, 479, 1992.
20. San Miguel, V.V. et al., *Achromobacter xylosoxidans* septic arthritis in a patient with systemic lupus erythematosus, *Arthritis Rheumatol.*, 34, 1484, 1991.
21. Walsh, R.D. et al., *Achromobacter xylosoxidans* osteomyelitis, *Clin. Infect. Dis.*, 16, 176, 1993.
22. McGuckin, M.B. et al., An outbreak of *Achromnobacter xylosoxidans* related to diagnostic tracer procedures, *Am. J. Epidemiol.*, 115, 785, 1982.
23. Gahrn-Hansen, B. et al., Outbreak of infection with *Achromobacter xylosoxidans* from contaminated intravascular pressure transducers, *J. Hosp. Infect.*, 12, 1, 1988.
24. Reina, J. et al., Nosocomial outbreak of *Achromobacter xylosoxidans* associated with a diagnostic contrast solution, *J. Clin. Pathol.*, 41, 920, 1988.
25. Burns, J.L. et al., Microbiology of sputum from patients at cystic fibrosis centers in the United States, *Clin. Infect. Dis.*, 27, 158, 1998.
26. Shelly, D.B. et al., Utility of commercial systems for identification of *Burkholderia cepacia* complex from cystic fibrosis sputum culture, *J. Clin. Microbiol.*, 38, 3112, 2000.
27. McMenamin, J.D. et al., Misidentification of *Burkholderia cepacia* in US cystic fibrosis treatment centers: an analysis of 1,051 recent sputum isolates, *Chest*, 117, 1661, 2000.
28. Liu, L. et al., Ribosomal DNA-directed PCR for identification of *Achromobacter (Alcaligenes) xylosoxidans* recovered from sputum samples from cystic fibrosis patients, *J. Clin. Microbiol.*, 40, 1210, 2002.
29. Sepkowitz, D.V., Bostic, D.E., and Maslow, M.J., *Achromobacter xylosoxidans* meningitis: Case report and review of the literature, *Clin. Pediatr.*, 26, 483, 1987.
30. Schoch, P.E., and Cunha, B.A., Nosocomial *Achromobacter xylosoxidans* infections, *Infect. Control Hosp. Epidemiol.*, 9, 84, 1988.

31. Klinger, J.D., and Thomassen, M.J., Occurrence and antimicrobial susceptibility of gram-negative nonfermentative bacilli in cystic fibrosis patients, *Diagn. Microbiol. Infect. Dis.*, 3, 149, 1985.
32. Lambiase, A. et al., Microbiology of airway disease in a cohort of patients with cystic fibrosis, *BMC Infect. Dis.*, 6, 4, 2006.
33. Krzewinski, J.W. et al., Use of random amplified polymorphic DNA PCR to examine epidemiology of *Stenotrophomonas maltophilia* and *Achromobacter (Alcaligenes) xylosoxidans* from patients with cystic fibrosis, *J. Clin. Microbiol.*, 39, 3597, 2001.
34. Kidd, T.J. et al., Low rates of *Pseudomonas aeruginosa* misidentification in isolates from cystic fibrosis patients, *J. Clin. Microbiol.*, 47, 1503, 2009.
35. Chester, B., and Cooper, L.H., *Achromobacter* species (CDC group Vd): Morphological and biochemical characterization, *J. Clin. Microbiol.*, 9, 425, 1979.
36. Coenye, T. et al., *Achromobacter insolitus* sp. nov. and *Achromobacter spanius* sp. nov. two novel species isolated from human clinical samples, *Int. J. System. Evol. Microbiol.*, 53, 1819, 2003.
37. Wellinghausen, N. et al., Fluorescence in situ hybridization for rapid identification of *Achromobacter xylosoxidans* and *Alcaligenes feacalis* recovered from cystic fibrosis patients, *J. Clin. Microbiol.*, 44, 3415, 2006.
38. Ferroni, A. et al., Use of 16S gene sequencing for identification of nonfermenting gram-negative bacilli recovered from patients attending a single cystic fibrosis center, *J. Clin. Microbiol.*, 40, 3793, 2002.
39. Manz, W.R. et al., Phylogenetic oligodeoxynucleotide probes for the major subclasses of proteobacteria: Problems and solutions, *System. Appl. Microbiol.*, 1, 593, 1992.
40. Amann, R.I. et al., Combination of 16S rRNA-targeted oligonucleotide probes with flow cytometry for analyzing mixed microbial populations, *Appl. Environ.Microbiol.*, 56, 1919, 1990.
41. Tenover, F.C. et al., Interpreting chromosomal DNA restriction patterns produced by pulsed—field gel electrophoresis: Criteria for bacterial strain typing, *J. Clin. Microbiol.*, 33, 2223, 1995.
42. Van Daele, S. et al., Shared genotypes of *Achromobacter xylosoxidans* strains isolated from patients at a cystic fibrosis rehabilitation center, *J. Clin. Microbiol.*, 43, 2998, 2005.
43. Mignard, S., and Flandrois, J.P., 16S rRNA sequencing in routine bacterial identification: A 30-month experiment, *J. Microbiol. Methods*, 67, 574, 2006.
44. Tang, Y.W. et al., Comparison of phenotypic and genotypic techniques for identification of unusual aerobic pathogenic gram-negative bacilli, *J. Clin. Microbiol.*, 36, 3674, 1998.
45. Wellinghausen, N. et al., Superiority of molecular techniques for identification of gram negative, oxidase-positive rods, including morphological nontyical *Pseudomonas aeruginosa*, from patients with cystic fibrosis, *J. Clin. Microbiol.*, 43, 4070, 2005.
46. Siala, M. et al., Broad-range PCR, cloning and sequencing of the full 16S rRNA gene for detection of bacterial DNA in synovial fluid samples of Tunisian patients with reactive and undifferentiated arthritis, *Arthritis Res. Ther.*, 11, 102, 2009.
47. Nazaret, S. et al., RISA-HPLC analysis of lung bacterial colonizers of cystic fibrosis children, *J. Microbiol. Methods*, 76, 58, 2009.
48. Demivev, P.A. et al., Microorganism identification by mass spectrometry and protein database searches, *Anal. Chem.*, 71, 1732, 1999.
49. Degand, N. et al., Matrix assisted laser desorption ionozation-time of flight mass spectrometry for identification of nonfermenting Gram-negative bacilli isolated from cystic fibrosis patients, *J. Clin. Microbiol.*, 46, 3361, 2008.
50. LiPuma, J.J. et al., Development of rRNA-based PCR assays for identification of *Burkholderia cepacia* complex isolates recovered from cystic fibrosis patients, *J. Clin. Microbiol.*, 37, 3167, 1999.
51. Cheron, M. et al., Investigation of hospital-acquired infections due to *Alcaligenes denitrificans* subsp. *xylosoxydans* by DNA restriction fragment length polymorphism, *J. Clin. Microbiol.*, 32, 1023, 1994.

60 *Bordetella*

Diego Omar Serra, Alejandra Bosch, and Osvaldo Miguel Yantorno

CONTENTS

60.1 INTRODUCTION

60.1.1 Incidence of Disease and Vaccination: A Historical Perspective

Whooping cough or pertussis is a highly contagious disease of the human respiratory tract caused by the strictly human pathogen *Bordetella pertussis*. It was described for the first time by Guillaume de Baillou after an epidemic in Paris in 1578.[1] As the disease became widely known it was given different names. The Italians spoke of the "dog bark," while in England it became known as the "chin cough" or "kin cough," later to be called whooping cough. The Chinese called it the "100 day cough" because of the protracted course of the disease. Thomas Sydenham first used the term "infantum pertussis" in 1670, "pertussis" meaning a violent cough of any kind.[2] Although the organism was detected microscopically in the sputum of a patient with whooping cough by 1900, Bordet and Gengou reported the isolation of *B. pertussis* for the first time in 1906.[3]

In the twentieth century, pertussis was one of the most common childhood diseases and a major cause of childhood mortality in the world. Before the introduction of the pertussis vaccine in the 1940s, more than 200,000 cases of pertussis were reported annually in the United States.[4] Between 1914 and 1923, Hess, Luttinger, and Madsen carried out the first anti-pertussis vaccine trials.[5,6] The first vaccine developed against pertussis contained whole-cell inactivated bacteria. Since widespread use of this pertussis vaccine began, incidence has decreased more than 80% compared with the prevaccine era. These first vaccines were developed when it was still unclear as to which specific antigens contributed to immunity. In addition, there were great differences among the approaches applied by the different manufacturers with respect to strains used, culture conditions, and inactivation procedures. As a result, vaccines with a wide range of efficacies were developed.[7]

For decades, the whole-cell pertussis vaccine has been formulated in combination with vaccines against diphtheria and tetanus. The combination is known as the DTP vaccine, triple vaccine, or DTwP, in which the other components are purified diphtheria and tetanus toxoids. Although these vaccines have contributed to decrease the morbidity and mortality due to pertussis, their use has been very controversial because of negative side effects attributed to *B. pertussis* whole cells. Among these effects, a possible association with neurological complications and deaths has led to decline the public confidence of whole-cell pertussis vaccines,[8] and therefore a subsequent reduction in immunization rates. In the mid-1970s, several countries, including Great Britain, Sweden, and Japan, discontinued the use of whole-cell pertussis vaccines due to safety concerns.[9,10] As a consequence, a resurgence of the disease occurred in these countries between 1978 and 1982. Coverage gradually recovered as public confidence was reestablished. In Great Britain, for instance, it was around 80% in 1990 and 90% in 1992.[11]

In the last decades, essentially due to the reactogenicity concerns of *B. pertussis* whole cells, several acellular pertussis vaccines have been formulated in developed countries, particularly for young children. These vaccines, known as DTaP, are also combinations of antigens of diphtheria and tetanus, and purified virulence factors of *B. pertussis*. The pertussis component always includes inactivated pertussis toxin (PT) as well as at least one other immunogen, such as filamentous hemagglutinin (FHA), pertactin (PRN), or fimbria (FIM) types 2 and 3.[12] These DTaP vaccines have been demonstrated to have relative good clinical efficacy and less reactogenicity than DTwP vaccines.[13–16] In spite of this, the current panorama shows that after several years of vaccination the disease still remains a considerable public health problem worldwide, being endemic with incidence peaks occurring every 3–5 years, even in countries with high vaccination coverage.[17–20] Due to this situation, pertussis is currently being considered a reemerging infectious disease.

Different reasons have been proposed to explain the resurgence of pertussis, including an increased awareness of the disease, the improvement in the microbiological diagnosis, low efficacy of particular batches of vaccine, waning immunity after vaccination or disease, and antigenic variations occurring in the bacteria populations,[21–23] which lead to antigenic divergence between circulating and vaccine strains.

Concomitantly with the resurgence of pertussis, a shift in its prevalence toward older age categories, as well as the appearance of atypical manifestations of the disease, were reported.[23] Historically, the occurrence of pertussis was associated with an acute infection and perceived as a childhood disease primarily in non- or underimmunized infants. However, currently it is becoming increasingly evident that a change in the incidence of the disease toward adolescents and adults with the appearance of persistent asymptomatic or mild infections.[23–25] Due to the absence of symptoms commonly manifested during the course of an acute infection, physicians often miss diagnosis of pertussis in adults and adolescents.[26] In this context, it has been estimated that 0.9–1.5 million cases of symptomatic pertussis occur annually in adults in the United States.[22]

Acellular pertussis boosters specifically formulated for adolescents and adults are now available for use in the United States.[27,28] These combinations, described as Tdap vaccines (commercially known as Adacel and Boostrix), include tetanus, diphtheria, and acellular pertussis, with reduced quantities of diphtheria and pertussis components compared with DTaP vaccines. It is expected that the expansion of their use in adolescents and adults could have a significant impact on the incidence of the disease, decreasing, probably, the circulation of *B. pertussis* in the population.

Some reports discuss about the efficacy of current pertussis vaccines arguing that a way to improve the vaccines could be to keep them updated, so that the strains used for vaccine production do not present differences in the most relevant antigens with respect to those strains circulating in the population.[29,30] In the last two decades clinical isolates recovered from pertussis-infected patients have been extensively analyzed in order to know the molecular evolution of the current circulating isolates.[19,31,32] Antigenic differences in isolates collected in different years and countries showed that pathogen adaptation has probably played an important role in the resurgence of the disease.[29,30,33] Therefore, the knowledge of the antigenic characteristic of clinical isolates in relation to the strains used in the vaccine production is increasingly being recognized as a major aspect to be considered in the design of novel strategy for controlling the disease.

60.1.2 Morphology, Biology, and Phylogeny

60.1.2.1 Morphology and Metabolism of *Bordetella*

The genus *Bordetella* includes a group of gram-negative, small coccobacilli with 0.2–0.5 μm in diameter and 0.5–2 μm in length, which are obligate aerobes (with the exception of *B. petrii*). *Bordetella* spp. are usually noninvasive parasites of the respiratory tracts of warm-blooded animals, including birds, with a predilection for the respiratory ciliated epithelium.[23,34] Most of the information available on the physiology of the *Bordetella* genus corresponds to the classical *Bordetella* species: *B. pertussis*, *B. parapertussis*, and *B. bronchiseptica*. These bacteria, but especially the human-restrict *B. pertussis*, are nutritionally fastidious; a reason why they are usually cultivated on rich media supplemented with blood. *B. pertussis* is a slow-growing organism, developing in 3–4 days. Clinical isolates of this species on Bordet–Gengou (BG) agar appear as smooth, shiny colonies with a high, domed profile. *B. pertussis* colonies on BG agar are β-hemolytic. On Regan–Lowe (RL) agar, colonies are small, domed, and shiny, with a white mother of pearl opalescence. *B. pertussis* has an unusual metabolism as it does not use sugars as carbon sources but appears to rely on amino acids for this.[35,36] At the metabolic level, *B. pertussis* can be distinguished from *B. parapertussis* in that *B. pertussis* is oxidase positive but urease negative, while *B. parapertussis* is oxidase negative and urease positive. *B. parapertussis* colonies grow faster, are more β-hemolytic on BG agar, and are grey or slightly brown in color. This is because this species produces a soluble brown pigment on agar media, while *B. pertussis* and *B. bronchiseptica* do not. *B. bronchiseptica* colonies grow within 24 h and are large, flatter, and have a dull rather than shiny appearance.

60.1.2.2 Genus *Bordetella*

Bordetella genus comprises nine species, namely *B. pertussis*, *B. bronchiseptica*, *B. parapertussis*$_{hu}$, *B. parapertussis*$_{ov}$, *B. avium*, *B. hinzii*, *B. holmesii*, *B. trematum*, and *B. petrii*.

Most members have adapted to live in close association with higher organisms, either as overt primary pathogens or in commensal associations that occasionally result in opportunistic diseases. Three species, *pertussis*, *parapertussis*, and *bronchiseptica* are the most associated.[37] They are exclusively respiratory pathogens of mammalian hosts and have been reclassified as "subspecies" of a single species with different host adaptations. *B. pertussis* is a pathogen for humans and possibly for higher primates, but no other reservoir is known. *B. parapertussis* consists of two distinct lineages: one found in humans and the other found in sheep (*B. parapertussis*$_{hu}$ and *B. parapertussis*$_{ov}$, respectively).[38] *B. parapertussis*$_{hu}$ produces infections with similar characteristics to pertussis but with milder clinical symptoms. *B. bronchiseptica*, which can live independently in the environment, causes infections that are typically chronic and often asymptomatic in a broad range of animals. This pathogen is associated with kennel cough in dogs, snuffles in rabbits, and atrophic rhinitis in pigs, but may also be present without symptoms in a majority of these and other animals.[23,39] It is only sporadically isolated from humans,[40] particularly from immunocompromised individuals, and human infections have been considered to be zoonotic.[41]

Phylogenetic analyses of the genus *Bordetella*, using pulsed-field gel electrophoresis, multilocus enzyme electrophoresis, and insertion sequence (IS) typing data, suggested that *B. pertussis*, *B. parapertussis*$_{hu}$, and *B. parapertussis*$_{ov}$ arose independently from *B. bronchiseptica*-like ancestors.[37] The genomes of a representative strain of each of the three *Bordetella* species (*B. pertussis* strain Tohama I; *B. parapertussis*$_{hu}$ strain 12822; and *B. bronchiseptica* strain RB50) were reported in 2003 by the Sanger Center.[42] The genomes of *B. pertussis* Tohama I and *B. parapertussis* 12822 are 4.09 and 4.77 Mb, respectively, while *B. bronchiseptica* RB50s is 5.34 Mb. Comparison of three genomes suggested that the time to the last common ancestor for *B. pertussis* and *B. parapertussis*$_{hu}$ and for *B. bronchiseptica* was 0.7–3.5 and 0.27–1.4 million years, respectively.[42]

The restriction in the host range, specifically the adaptation of *B. pertussis* and *B. parapertussis*$_{hu}$ to the human host, included a significant loss of genes, corresponding, perhaps, to a more rationalized genome. In comparison with *B. pertussis* Tohama I and *B. parapertussis* 12,822, a large portion of the extra DNA in *B. bronchiseptica* RB50 is attributed to prophage and prophage remnants.[42] Other aspect that distinguishes RB50 from Tohama I and 12,822 is that the last strains contain 358 and 200 pseudogenes, respectively, many of which have been inactivated by insertion of IS elements, in-frame stop codons, or frameshift mutations. Among the genes lost by *B. pertussis* Tohama I and *B. parapertussis*$_{hu}$ 12,822 are those involved in membrane transport, small-molecule metabolism, regulation of gene expression and synthesis of surface structures.[42] Remarkably, a few genes known or suspected to be involved in pathogenicity are missing in the genomes of human-adapted bordetellae, suggesting that virulence factors are critical for the strains to maintain their capabilities in the transmission and pathogenesis among the human populations.

60.1.3 Virulence and Pathogenesis

60.1.3.1 Virulence Factors and Their Regulation in *B. pertussis*

The expression of virulence factors by a pathogen is intimately associated with its ability to infect the host. *B. pertussis* presents a number of important virulence factors, which include adhesins, autotransporters, and toxins.[23] Among the factors that display adhesin functions are important to highlight: FHA (filamentous haemagglutinin), FIM (fimbriae), and PRN (pertactin), an autotransporter.[43] Other autotransporters, such as BrkA (Bordetella resistance killing antigen),[44] SphB1,[45] Vag8,[46] and TcfA (tracheal colonization factor),[47] also present in their sequences domains potentially involved in bacterial attachment. Among the toxins that *B. pertussis* produce, it is important to remark PT (pertussis toxin), AC (adenylate cyclase), DNT (dermonecrotic toxin), and TCT (tracheal cytotoxine).[23,48] This group also includes the type III secretion system, which translocates effector proteins directly into host target cells.[49] It is also important to include as virulence factors of *B. pertussis* the lipopolysaccharide (LPS),[50] which is a potent endotoxin, and the iron uptake systems.

An important aspect to distinguish among these virulence factors is the manner in which the bacterium regulates their expression. Like many other pathogens, *B. pertussis* controls the expression of most of these virulence factors in response to changes in the physicochemical conditions present in the surrounding microenvironment, particularly through a two-component signal transduction system called BvgAS (*Bordetella* *v*irulence *g*ene).[51] Given the central role of the BvgAS system, the virulence factors are usually classified according to whether or not they are under its control. FHA, FIM, PRN, PT, AC, DNT, and the type III secretion system are among the virulence factors whose expression is under the control of the BvgAS system, whereas the LPS, TCT, and iron uptake systems encompass some of the factors whose expression is regulated independently from the BvgAS.[23]

The BvgAS system consists of a transcriptional regulator, BvgA, and a sensor protein, BvgS, which is anchored to the external membrane. The activation of the BvgAS system involves a mechanism that consists of a series of phosphor-transfer reactions.[52] The factor/s that trigger the activation of the system in vivo still remain unknown. Originally, it was thought that the BvgAS system worked as an "on-off" switch, exerting control over two particular groups of genes, the *vag* (*v*irulence *a*ctivated *g*enes) that encode for the virulence factors described above, whose expression is activated by the system, and the *vgr* (*v*irulence *r*epressed *g*enes), whose functions are still unclear and whose expression is repressed by the BvgAS.[53] In vitro, under standard culture conditions, the BvgAS is turned "on," activating the expression of the *vag* genes and repressing the *vrg* genes. In this case, the pathogen displays a "virulent phase," also known as "Bvg$^+$ phase." At temperatures <26°C, or in presence of millimolar concentrations of modulating agents such as $MgSO_4$ (>4 0 mM) and nicotinic acid (>10 mM), the *vag* genes are repressed, whereas the *vrg* genes become active.[54] In this case the bacterium is in a "avirulent phase" or "Bvg⁻ phase." However, currently it is well known that the BvgAS system regulates, in a differential manner, the expression of the *vag* genes, which can be discriminated according to their temporal expression as *"early vags"* (i.e., *fhaB*, which encodes for the FHA adhesin) and *"late vags"* (i.e., cyaA, which encodes for the AC toxin).[55] It is also known that the BvgAS regulates a factor, BipA, which is specifically expressed in a third phase called "intermediate" or "Bvg^i phase,"[56] considered to be a transition between the Bvg^+ and Bvg^- phases. Therefore, the BvgAS functions more like a "rheostat," capable of controlling in a tight and coordinated manner the expression of several profiles of genes associated with the virulence of *Bordetella* in at least three phenotypic phases (Bvg^+, Bvg^i, and Bvg^-).[57] The Bvg^+ phase is characterized by the expression of the early and late *vags*, the Bvg^i phase by the expression of the early *vags* and BipA, whereas the Bvg^- phase is characterized by the expression of the *vrg* genes.[57] These sequential changes in the expression profiles of the virulence genes respond in vitro to subtle changes in the concentrations of the modulating agents ($MgSO_4$ or nicotinic acid) or in the temperature of cultivation.[54] This phase transition is known as "antigenic or phenotypic modulation." In spite of the proposal of models postulating that environmental and host factors may promote events of antigenic modulation in vivo that could favor the persistence of the pathogen as well as its transmission,[58] the exact role of this phenomenon in vivo still remains unclear.

Another important phenomenon associated with the virulence of the pathogen is "phase variation." This phenomenon is associated with reversible alterations in the genotype caused by mutations that change the reading frame and render the bacterium unable to synthesize toxins and others factors associated with the pathogenicity.[59] The emergence of natural phase variants during the infection process may be another mechanism displayed by *B. pertussis* to escape from the immune system of the host.

60.1.3.2 Stages of *B. pertussis* Pathogenesis

Pertussis consists of a localized infection of the respiratory tract, with secondary systemic effects resulting from the actions of several bacterial virulence factors and from the host response to those factors. At least four distinct events can be discerned in the infection process of *B. pertussis*: (i) adhesion, (ii) suppression and modulation of host immune response, (iii) local epithelial damage, and, eventually, (iv) persistence (Figure 60.1).

(i) The normal course of infection begins with the bacterium entering the host via the airways, contained within airborne droplets derived from the cough of an infected individual. The pathogen proceeds down the respiratory tract, adhering to ciliated epithelial cells in the nasopharynx and the trachea. A number of studies conducted by using human epithelial cell lines and animal infection models suggest that several protein factors (FHA, FIM, PT, TcfA, BrkA, Vag8, and PRN), some of them showing redundancy in their functions, participate at this stage.[43,44,46,47,60] As a consequence of this redundancy, the particular function of each single factor is often masked, thus becoming difficult to identify and design a protein as the major adhesin. In spite of this, most in vitro and in vivo studies still highlight FHA as the most important adhesin factor of *B. pertussis*.[61,62]

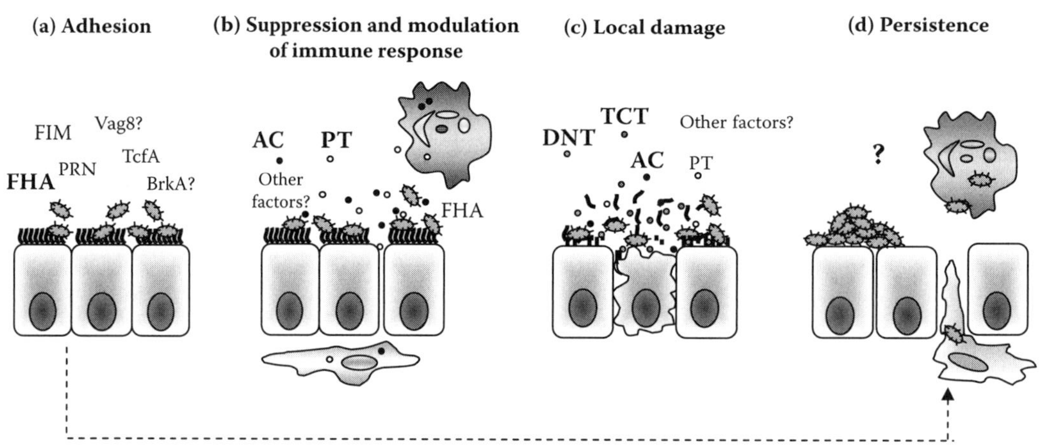

FIGURE 60.1 At least four distinct events can be discerned in the infection process of *B. pertussis*. (a) The infection process begins with the adhesion of *B. pertussis* to ciliated cells of the upper respiratory tract. Several protein factors are suggested to participate at this stage, among which FHA is the most relevant. (b) At this stage an interplay between virulence factors and immune mechanisms take place. Multiple virulence factors are devoted to suppress and modulate the host immune response. In several instances the effects of virulence factors are multiple and appear to be antagonistic to one another: cytotoxic, stimulatory, inhibitory, pro-, and anti-inflammatory. For instance, AC toxin is able to inhibit the phagocytic activity of polymorphonuclear leukocyte and macrophages. (c) The local damage in the ciliated epithelia is suggested to derive from the activity of the TCT and DNT toxins, but also from the AC and PT toxins. (d) Relatively little information is available regarding those events that may promote the persistence of the pathogen in the host. Two hypotheses are currently under investigation in our laboratory: the first hypothesis proposes the intracellular survival of *B. pertussis* within specific target cells as a mechanism of persistence, whereas second hypothesis proposes the biofilm lifestyle as the strategy adopted by the pathogen to survive within the human host.

(ii) Once attachment is initiated, the bacterium begins to replicate and colonizes adjacent areas. At this stage, interplay between virulence factors and immune mechanisms takes place. Accumulating recent evidence indicates that multiple virulence factors are devoted to suppress and modulate the host immune response.[63] In several instances, the effects of virulence factors are multiple and appear to be antagonistic to one another (cytotoxic, stimulatory, inhibitory, pro-, and anti-inflammatory). For example, it has been shown that FHA is both proinflammatory and able to induce apoptosis in macrophages.[64] Similarly, AC toxin, which is known for its inhibitory effects and cytotoxicity on neutrophils and other phagocytic cells,[65] also has proinflammatory effects, inducing IL-6 from respiratory epithelial cells[66] and cyclooxygenase-2 in cells expressing the integrin CD11b/CD18.[67] The activity of PT also interferes with the host defense mechanisms. This toxin has inhibitory effects on chemotaxis, oxidative responses, and lysosomal enzyme release in neutrophils and macrophages.[68] Several effects of PT are mediated by the ADP-ribosyl transferase activity of the catalytic domain.

(iii) The local damage in the ciliated epithelia, which affects the mucociliary clearance mechanisms, is suggested to derive from the activity of the TCT and DNT toxins[69,70] but also from the AC toxin. Likewise, it also indicated that the local damage in the respiratory tract is responsible for the paroxysmal cough, the most distinctive manifestation of the infection by *B. pertussis*. However, it is estimated that the duration of the cough episodes during a typical infectious process is longer than the time period in which the local damage occurs. It was also suggested that the paroxysmal cough may be caused by PT toxin.[68] Nevertheless, patients infected with *B. parapertussis*, which does not produce PT, also manifest a similar paroxysmal cough. All these evidences lead to speculation about the chance that another toxin, not yet identified, may contribute to the recurrent episodes of paroxysmal cough.

(iv) In spite of the efforts in elucidating the mechanisms of pathogenesis displayed by *B. pertussis* during the whole infectious cycle, particularly little information is available regarding those events that would promote the persistence of the pathogen in the host. Currently exist some lines of evidences supporting the idea that *B. pertussis* eventually persist within the human host. It is speculated that this may occur after the paroxysmal phase of the disease in a classic case, or in those atypical cases in which patients, most of them adolescents and adults, are infected but do not present symptoms or only manifest mild signs of the disease.[23] In this context, two hypotheses are currently under investigation. The first hypothesis proposes the intracellular survival of *B. pertussis* within specific target cells as a mechanism of persistence. Although it was thought that *B. pertussis* does not invade tissues, it was recently reported that the presence of this bacterium in mammalian cells, including human epithelial cell lines of respiratory origin (e.g., laryngeal Hep-2 cells, tracheal HTE cells, and lung carcinoma A549 cells) and professional phagocyte cells (macrophages and neutrophils).[71] The second hypothesis proposes the biofilm lifestyle as the strategy adopted by the pathogen to survive within the human host. *B. pertussis* was shown to adopt a distinctive physiology under this mode of life.[72] Recent in vivo studies utilizing a murine infection model demonstrated the presence of biofilm-like microcolonies formed by *Bordetella* in the nasal cavity, a primary host niche wherein the pathogen may escape immune defenses.[73]

60.1.4 Clinical Features

The clinical manifestations associated with pertussis disease may vary according to diverse factors such as the course of the infection, the age and the clinical conditions of the patient, previous history of immunization or infection, presence of passively acquired antibody, and the antibiotic therapy.[74]

Classic Manifestations. In a classic case the disease begins with a catarrhal phase, which also represents the period during which the risk of transmission is highest. This initial phase is indistinguishable from common upper respiratory infections with nasal congestion, rhinorrhea, and sneezing, variably accompanied by low-grade fever, tearing, and conjunctival suffusion.[74] After 7–10 days, the paroxysmal stage begins. This stage is characterized by repeated coughing fits with 5–10 forceful coughs during a single expiration (a paroxysm). At the end of a paroxysm, there is a massive inspiratory effort, during which the classic "whoop" occurs. Posttussive vomiting is common,[74] and cyanosis, bulging eyes, lacrimation, and distention of neck veins may occur. Fever is rarely associated with these symptoms. Following a group of paroxysmal episodes, the children are exhausted and may appear apathetic. Weight loss may occur because of frequent vomiting and the child's refusal to eat because he or she recognizes that eating often triggers paroxysms. The pneumonia may be a primary event in response to *B. pertussis* infection or may be due to a secondary infection with other pathogens.[75] Encephalopathy is most probably due to cerebral hypoxia related to severe paroxysms. The paroxysmal stage lasts for 1–6 weeks and sometimes longer. Then the convalescent stage starts. The transition is gradual and is associated with a decrease in the frequency of the paroxysms and in the severity of the events as well.[74] The convalescent stage usually lasts for 1–2 weeks but is occasionally prolonged. Most children with classic whooping cough associated to *B. pertussis* infection have an elevated white blood cell count with an absolute lymphocytosis.

Atypical Manifestations. Among immunized individuals, particularly adolescents and adults, the disease is often not severe and sometimes is confused with others causes of chronic cough that are common in this age group, such as asthma and postnasal drip.[76] Pertussis has been found to be the cause in 13%–20% of adults presenting with prolonged cough, although the nature of the referral population differs among studies.[75] The mean duration of cough in adults with pertussis is 36–48 days, and it may be only nocturnal.[77] According to a study analyzing pertussis in adults and infants,[78] scratchy throat and other pharyngeal symptoms occurred in approximately one-third of adults with pertussis, and episodes of sweating were reported by 40%–50% of patients >30 years old; paroxysmal cough in 70%–99%; whoop in 8%–82%; and posttussive vomiting in 17%–50%.

60.1.5 Epidemiology

In the prevaccine era (1922–1945), whooping cough was a major cause of morbidity and mortality for young children over the world. In the United States, the average yearly incidence of reported pertussis was 157 cases per 100,000 population.[79] Although mortality due to pertussis is markedly age related, the vast majority of deaths occur in infants. Data from England and Wales for 1945–1949 revealed the following percent distribution by age group: <1 year, 10.4%; 1–4 years, 56.9%; 5–9 years, 29.2%; 10–14 years, 2.0%; and ≥15 years, 1.4%.[23] With the introduction and widespread use of whole-cell pertussis vaccines between 1940 and the early 1950s, the rate of reported cases decreased approximately 150-fold.[2] From 1978 to 1981, in the United States the rate of pertussis was <1 case per 100,000 people.[79]

Despite vaccination, pertussis has persisted and, in the 1990s, resurged in a number of countries, even in those with highly vaccinated populations. In the United States, 25,616 and 15,632 probable or confirmed cases were reported in 2005 and 2006, respectively.[80] Pertussis also remains a substantial public health problem in other developed regions, such as the European Union and Canada, but especially in underdeveloped regions. According to reports of the World Health Organization (WHO), pertussis still ranks 5th in child mortality caused by vaccine-preventable disease.[81] Some reports indicate that there are approximately 48.5 million annual cases of pertussis worldwide, with 295,000 deaths.[23,82]

One of the hallmarks of the pertussis resurgence is a shift in disease prevalence toward older age categories. Over the last 20 years, it was evident that adults and adolescents, whose immunity decreases over time in absence of vaccine boosters, become susceptible of the disease,[24,25] and consequently the main source of contamination for vulnerable newborns too young to be vaccinated and for whom the consequences of the illness can be dramatic.[75] In Argentina, the number of cases of pertussis reported in 2002 was 267 but in 2005 this number increased to 2050.[83] Although the majority of these cases corresponded to children <1 year old, the higher increase in incidence was detected in adolescents. The waning immunity alone cannot explain the sudden and sharp incidence increase of pertussis cases over a wide age range in different countries. Recent works support the evidence that vaccination of young children with DTwP vaccine has led to the control of those isolates that are similar to the vaccine strains.[84] In this context, it is becoming increasingly clear that adaptation of circulating strains of the pathogen, driven by successive genetic changes, is a crucial aspect in the resurgence of pertussis.[29,30,33] Comparative analysis of clinical isolates collected in different countries using serotyping and different molecular genetic techniques, including DNA sequence analysis of virulence factors genes, comparative 16S rRNA analysis, multilocus enzyme electrophoresis and IS typing, revealed differences between vaccine strains and clinical isolates that might help to explain the pathogen adaptation.[19,31,32,85,86]

Antigenic divergence between clinical isolates and vaccine strains has been reported to be associated with polymorphism in the genes encoding for known virulence factors, particularly in PRN, FIM, and PT,[85] which are suggested to alter epitopes that might hinder recognition of the pathogen by the immune system of the host after vaccination. These virulence factors are present in the strains used for whole-cell vaccine production, but also are components included in acellular vaccines. In general, it was found that "non-vaccine-type" alleles of genes encoding for PT, PRN, and FIM have gradually replaced the "vaccine-type" alleles that were in circulation earlier. At least 13 alleles for the gene encoding for PRN, *prn*, have been described, making it one of the most polymorphic genes in *B. pertussis*.[87] Of them, *prn1*, *prn2*, and *prn3* are predominant. Most vaccine strains contain the *prn1* or *prn7* alleles.[29] Only one Swedish vaccine strain harbors *prn10*. In all countries investigated, variation in *prn* alleles has been observed. The genes for the two highly similar fimbriae *B. pertussis* produces are *fim2* and *fim3*. Two alleles for *fim2* were identified, *fim2-1* and *fim2-2*.[29] The *fim2-1* allele was found in most vaccine strains. Only one of the two strains used for the production of the Dutch whole-cell vaccine contained *fim2-2*.[29] In the case of the *fim3* gene, four alleles were found; however one allele, *fim3-1*, is present in most vaccine strains.[29] The *fim3-4* allele encode an identical protein to Fim3-1. PT is composed of five subunits. Of them, the PtxA subunit (also known as S1) is the most immunogenic and also the most polymorphic.[88] Eight *ptxA* alleles have been identified, four of which (*ptxA1*, *ptxA2*, *ptxA6*, and *ptxA7*) encode for an identical protein, PtxA1. Vaccine strains have been shown to contain *ptxA1*, *ptxA2*, and *ptxA4*.[29]

Mooi et al.[89] recently report that the dramatic increase in pertussis cases in Netherlands after 1995 could be associated with the emergence of *B. pertussis* strains carrying a novel allele for the pertussis toxin promoter (*ptxP3*). They alert that during 1989–2004 the allele *ptxP1* was completely replaced by *ptxP3*, which confers increased PT production (*ptxP3* strains produce 162 times more PT than *ptxP1* strains).[89] Consistent with the role of PT in the virulence of *B. pertussis*, the expansion of these strains resulted in augmented severity of human infections. Notoriously, in contrast to this situation, isolates not expressing PT or PRN were recently collected from infant infected with *B. pertussis* in France in 2007.[90] The isolates were found less pathogen in animal or cellular models. The appearance of these strains could be a consequence of the replacement of whole-cell vaccines by acellular vaccines, which contain only few bacterial antigens targeting the virulence of the bacterium. In this context, it is probable that the appearance of new strains expressing less, but also modified vaccine antigens, or exhibiting higher expression of nonvaccine antigens implicated in virulence, might occur.

In a recent work, Heikkinen et al.[91] reported that Finnish clinical strains isolated in 1950s and 1960s had only seven genes absent in relationship to Tohama I strain (isolated in 1954). A marked change associated with the deletion of 24 genes was noticed in strains isolated in 1970s. Then, the genes content of the strains isolated from the subsequent 20 years remained basically the same.[91] These studies confirm that *B. pertussis* population is dynamic and is continuously evolving, suggesting that the bacterium could use gene loss also as a strategy to adapt to highly immunized populations.

It is evident that in order to pertussis control, surveillance must continue involving standardized and specific diagnoses methods as well as extensive analysis of circulating isolates to rapid design appropriate vaccines.

60.1.6 Diagnosis

Determining who has pertussis and who does not is often difficult. The clinical case definition of pertussis includes (i) a person with cough illness lasting at least 2 weeks with one of the next characteristics: paroxysms, whooping and posttussive vomits[92]; (ii) a person with cough illness lasting 2 weeks or longer without other symptoms (applicable in outbreak settings). In general, the early clinical diagnosis of pertussis is rather difficult because many other infectious agents cause illnesses with a cough that can be confused with *Bordetella* infections.[76] Spasmodic attacks of coughing may be observed in children with bronchiolitis, bacterial pneumonia, cystic fibrosis, or tuberculosis. Classical symptoms of pertussis, including paroxysmal cough for more than 2 weeks, whooping and posttussive vomiting, are typical in nonvaccinated children.[75] These symptoms are much more variable and milder in neonates and previously vaccinated adults.

Clinical diagnosis must be confirmed by laboratory tests. In this context, rapid and reliable laboratory tests are required to guide treatment and to prevent transmission. Whenever possible, a nasopharyngeal swab or aspirate should be obtained from all persons with suspected cases. A properly obtained nasopharyngeal swab or aspirate is essential for optimal results in laboratory diagnosis. There are two types of approach to diagnosis, direct or indirect. Direct diagnosis consists of identifying the microorganism responsible for the disease by culture, polymerase chain reaction (PCR), or direct fluorescent antibody (DFA). Indirect diagnosis is essentially by serology and consists of detecting antibodies in the serum of an infected individual. The choice of the diagnosis technique usually depends on the age and the immune status of the suspected patients. According to the laboratory manual for the diagnosis of whooping cough caused by *B. pertussis*/*B. parapertussis* published by WHO,[93] culture is priority for infants, or PCR where culture is not available but serology is unreliable. For nonvaccinated children manifesting the first episodes of cough, culture and PCR are recommended. A serological test in children can be used if vaccination was not performed in the last 3 years. For adults serology is also recommended, but only if vaccination was not performed during the last 3 years.

60.1.6.1 Conventional Techniques

Culture. It is regarded as the "gold standard."[93–95] Definitive diagnosis of pertussis has traditionally been made by culture of the causative organism. It is also necessary for antimicrobial susceptibility testing and molecular typing. Culture of the organism is highly specific. A positive culture for *B. pertussis* confirms the diagnosis of whooping cough.[95] However, based on its growth requirements, *B. pertussis* is considered a fastidious bacterium; consequently, it is not easy to isolate. Additional factors such as the specimen collection and transportation, markedly influence the yield of culture.[94,95] Combination of these factors become the technique quite insensitive and slow, taking, sometimes, up to 2 weeks to obtain a clear result. Isolation rates are highest during the first 3–4 weeks of illness (catarrhal and early paroxysmal stages).[96] Cultures are less likely to be positive if performed later in the course of illness (more than 2 weeks after cough onset) or on specimens from persons who have received antibiotics or have been vaccinated.[96] In conclusion, bacterial culture is specific and confirms the diagnosis, however, it has low sensitivity, is time demanding, and may be difficult in practical terms.

Direct Fluorescent Antibody (DFA). The use of DFA testing for the rapid diagnosis of *B. pertussis* infections has been used for more than 40 years because it is rapid and inexpensive and can provide a positive result when cultures are negative due to the use of antibiotic. However, DFA tests lack sensitivity and specificity for *B. pertussis*.[97] False-positive results with polyclonal antibodies are due to several organisms of the oral and nasopharyngeal flora[98] including nonencapsulated *Haemophilus influenzae*, diphtherioids, and anaerobic or facultatively anaerobic bacterial species. Use of the mono clonal DFA test has improved the specificity,[97] but DFA still has a low sensitivity and should not be relied upon as a criterion for laboratory confirmation because of cross-reactions with members of the normal nasopharyngeal flora. DFA tests require care in all technical aspects and experienced personnel for their interpretation. Since it is not a confirmatory test, DFA should be used alongside culture or PCR.

Serology. It is regarded as an indirect approach to diagnosis.[93] Serology tests have been also used in epidemilogic investigations and in acellular pertussis vaccine trials.[23] Tests that have been used to measure serum antibodies to *B. pertussis* include enzyme-linked immunosorbent assay (ELISA), complement fixation, agglutination, and toxin neutralization. Prior to the development of ELISA techniques, the main serologic test for the diagnosis of pertussis was the demonstration of a fourfold increase in agglutinating-antibody titer.[23] Because of its convenience, ELISA using specific *B. pertussis* antigens (PT, FHA, PRN, FIM) is the most common method for serological diagnosis of the disease. Diagnosis of pertussis may be suspected by the use of ELISA showing an increase in IgG or IgA antibody titer to PT, FHA, PRN, or FIM. However, it is now clear that antibody responses to FHA and PRN also occur following other *Bordetella* infections, so that isolated increases in titers of antibody against these antigens are not specific for *B. pertussis* infection.[23] In this context, PT has advantage over the other virulence factors in that it is uniquely expressed by *B. pertussis* and can discriminate between infection by *B. pertussis* and the antigenically related but less virulent species, *B. parapertussis*. Currently, a generally accepted serologic criterion for diagnosis of pertussis is the use of ELISA to demonstrate a significant increase in IgG or IgA serum antibody concentrations (greater than or equal to twofold) against PT between acute and convalescent specimens.[96,99] Unfortunately this has a drawback: the ability to show seroconversion may be affected by the previous immunological priming of the patient (owing to vaccination or previous infection) and by when serum samples are collected. All these factors affect the reproducibility of the results.

For reasons of practicability and cost, routine diagnosis must be based on single-sample serology using a single or a continuous cutoff.[100] In adolescents and adults, single high values of IgG or IgA antibodies to PT also indicate infection. A serum IgG anti-pertussis toxin antibody level above 100–125 EU/mL has been considered a reasonable threshold for a positive test result.[78,101] Although IgA antibody to PT is more indicative of a recent antibody response, it is less consistent than a PT IgG response.

PT is useful for serodiagnosis in individuals who have not been vaccinated for pertussis. However, since PT is contained in most acellular pertussis vaccine formulations, its usefulness for the diagnosis in recently vaccinated individuals is questioned.[102] Given these limitations, Watanabe et al.[102] recently evaluated the use of antigens not included in acellular vaccine formulations, in parallel to the use of PT and FHA, for serologic diagnosis of pertussis by ELISA. The authors concluded that several of the nonvaccine antigens are promising candidates for the serodiagnosis of pertussis; however, they remarked that the PT IgG titer was by far the most sensitive and specific indicator of infection.[102] Likewise, authors indicate that combination of multiple ELISA targets is a good strategy to improve the sensitivity of the current assay for serological diagnosis.[102]

In conclusion, current serologic tests are more sensitive than culture; however, they only provide late or retrospective diagnosis. The variability of results, the usefulness of this method only late in the clinical course of the disease, and the lack of standardized commercial test kits make serologic testing difficult to use in common practice with reproducible results. Some European countries with statutory notification and laboratory confirmation accept serology as a proof of infection. In the United States, the Food and Drug Administration (FDA) has no cleared test available for measuring pertussis antibodies.

60.1.6.2 Molecular Techniques

PCR. PCR methods enhanced the probability of identification of *B. pertussis* since positive results may be obtained even when the organism is no longer culturable.[23,96,103] In addition, PCR has made the rapid diagnosis of pertussis possible. The first publication of *B. pertussis* PCR was in 1989.[104] Since then numerous protocols have been reported. A series of key factors and recommendations for the use of PCR in pertussis diagnosis have been previously reported.[103,105] Among them the following are important to highlight:

(i) Specimen collection and preparation. Collection methods for PCR are similar to those for culture, and often, although not recommended, the same sample can be used for both tests. However, calcium alginate swabs cannot be used to collect nasopharyngeal specimens for PCR.
(ii) Gene targets and primer selection. Primers derived from four regions of the genome have been used: (a) the promotor region of the PT gene, (b) repeated insertion sequences, (c) the AC toxin gene, and (d) a DNA region up stream of the porin gene. Common primers include IS*481*, which detects both *B. pertussis* and *B. holmesii*, and IS*1001*, which detects both *B. parapertussis* and *B. holmesii*. The other primers—except for the AC toxin primers, which do not distinguish between *B. pertussis* and *B. parapertussis*—are specific for *B. pertussis*. Conventional PCR methods amplify a single gene sequence, most often within the IS*481*, due to the presence of multiple copies. However, the use of protocols for multiple targets in separate assays in order to diminish further the risk of false-positive results due to sample-to-sample contamination is incrasingly recommended. For instance, a suggested PCR method consisted of a screening assay amplifying IS*481* and a confirmatory assay amplifying a sequence of the PT promoter region.[106] Positive results of both tests should be obtained to consider a specimen positive for *B. pertussis*.
(iii) Controls. Internal and external controls are necessary to minimize the risk of false-positive and false-negative results. Among the common causes of false-positive results are the contamination of the sample with DNA or whole organisms of laboratory strains of *B. pertussis* or related species and the carryover contaminations. Procedures to control contaminations include physical separation of sample-processing areas from amplification and detection areas and incorporation of both known negative samples and coded (blinded) negative samples. Some common causes of false-negative results includes insufficient and inadequate specimen collection, alteration of the specimen, very early or late sampling in the course of the disease, and the presence of inhibitors in the sample. Although it is not

possible to control for all of these factors, the use of an internal control for each sample facilitates the interpretation of the PCR results.

Although conventional PCR assays have greatly facilitated pertussis diagnosis, still problems with these assays persist: false-positive results occur with relative frequency, and differentiation between *Bordetella* species is not usually feasible when a single target is used.[103,107] In the last 10 years, concomitantly with use of conventional PCR assays, real-time PCR protocols have emerged. In contrast to conventional PCR methods, they are less prone to carryover contamination because postamplification handling is eliminated.[108] Product formation can be monitored in real time, making quantification of bacteria possible. Currently there exit sequence-specific formats that employ fluorescence resonance energy transfer (FRET) hybridization probes,[108] TaqMan probes,[109] or molecular beacons,[110] as well as nonsequence-specific formats using SYBR Green I.[111] Those PCR formats employing specific probes are particularly recommended, since they achieve higher sensitivity and specificity. Adapted versions of a real-time PCR format using LightCycler technology and targeting IS*481* are being used more often for *B. pertussis* diagnosis.[93,100] Basically, the LightCycler technology[112] works with two hybridization probes, which allow for sequence-specific detection by using fluorescence energy transfer between the fluorophores conjugated to the probes. In general, the Light Cycler PCR protocols represent an advance beyond conventional PCRs because they are faster and easier to perform than conventional assays. Their analytical sensitivity is high, ranging from 0.1 to 10 CFU/mL. In a study in which 113 specimens from patients with and without symptoms of pertussis were evaluated, LightCycler PCR increased the diagnostic sensitivity over that of culture by 2.3 fold.[108] However, the specificity in terms of *Bordetella* species still is relative. Therefore, it is recommended to monitor periodically the specificity of the target gene (IS*481*).[100] For practical purposes, a positive IS*481* PCR can be considered to be a *B. pertussis* infection, when the clinical symptoms are in accordance with this result. The specificity may be further substantiated by a positive PCR using the PT promoter region by reference laboratories.[100]

Among the drawbacks of the real-time PCR formats, which can be extended to conventional PCR assays, are the lack of standardization, the technical complexities, and the expense of equipment and consumables, which may influence their successful implementation in some clinical setting.

Pulsed-Field Gel Electrophoresis (PFGE). PFGE is a powerful tool for characterizing various strains at the DNA level. It is a technique using restriction nucleases to cut the chromosomal DNA of the bacteria at specific sites, thereby creating pieces of DNA of different sizes. The number of pieces and their sizes (a type of "DNA fingerprinting") are used to identify a profile that is useful as an epidemiologic marker. Thus, this technique is highly informative in epidemiological studies.[113] PFGE can be carried out on *B. pertussis* clinical isolates to assist track transmission and investigate the clonal relationships among them. PFGE results might show diversity within larger geographic areas such as cities, regions, and countries. This technique is not used for routine surveillance.

60.2 METHODS

60.2.1 Sample Preparation

Nasopharyngeal Specimen Collection. The laboratory manual for the diagnosis of whooping cough caused by *B. pertussis*/*B. parapertussis*,[93] as well as reports of the Centers for Disease Control and Prevention (CDC),[114] described a series of recommendations concerning the practical procedures to be considered by health care professionals previously and during collection of the specimen, which are focused on standardizing procedures, preventing sample contamination and optimizing results.

Collection of nasopharyngeal aspirates involves several steps. The first step consists of connecting a suction catheter with mucous trap to a vacuum pump or syringe with tubing that includes an inline filter. After immobilizing the head of the patient, the following steps are to insert the end of the catheter along the floor of the nasopharynx to the posterior pharynx, to apply suction by vacuum pump or syringe when the posterior pharynx is reached, and to maintain the catheter as it is slowly withdrawn to the middle of the nasal cavity. Once concluded, the suction is discontinued and the catheter is slowly removed. In the next step, the catheter is flushed out by aspirating 0.5–1 mL of Stainer–Scholte broth or 0.1% casamino acids solution through the catheter into the trap. After this some of the aspirated specimen is inoculated on a primary culture plate using a sterile swab or bacteriologic loop. Finally, to complete the protocol, it is appropriate to seal the trap and label the specimen and the primary culture plate accordingly.

Similar to the procedure described above, to collect the specimen by nasopharyngeal swabbing, the head of the patient has to be immobilized. In essence, the protocol consists of inserting the nasopharyngeal swab into a nostril until the posterior nares is reached and leaving it in place for some seconds. After this, the swab is slowly removed. At this step, it is recommended to use the swab to inoculate a primary culture plate, and then to insert it into the transport medium. To complete the protocol, it is appropriate to label the transport tube and the primary culture plate accordingly. As recommendation, swabs for culture or molecular diagnosis must be composed of Dacron with plastic shafts. Cotton-tipped swabs should not be used because some cotton fibers are inhibitory for cultures.[115] Calcium alginate and aluminum leached from metal swab shafts are also inhibitory for PCR-based assays for *B. pertussis* detection.[103]

Culture and Isolation of *B. pertussis*. Part of the nasopharyngeal aspirate or the material collected in the swab is used to isolate *B. pertussis* and other *Bordetella* species.[115] Optimally, material should be directly inoculated at the same time of specimen collection onto Regan–Lowe (RL) charcoal agar plates. This medium consists of charcoal agar as a basal medium[116] supplemented with cephalexin to inhibit indigenous bacteria of the nasopharynx and defibrinated horse blood (10% v/v) to support the growth of *Bordetella* species. In parallel, it is recommended to inoculate the specimens in RL charcoal agar plates without cephalexin, since a few strains (<10%) of *B. pertussis* could not grow on selective media.

If at the moment of collecting the specimen, the inoculation on selective isolation medium is not possible, then specimens can be inserted into semisolid RL transport medium (one-half strength charcoal agar supplemented with horse blood and cephalexin). Samples are transported at ambient temperature (15°C–30°C) to the laboratory within 24 h. Specimens transported in semisolid RL medium can be subcultured on receipt to RL isolation medium or BG medium with and without cephalexin. The transport medium is then incubated along with the primary plates and subculture on RL agar after 48 h of enrichment at 35°C in humidified ambient. Plates should be incubated for 7–12 days.[117]

60.2.2 Detection Procedures

60.2.2.1 Direct Fluorescent Antibody (DFA) Test for *B. pertussis*

A DFA test can be assayed directly by using nasopharyngeal swab specimens or material from a nasopharyngeal aspirate.[118] To accomplish the test, a small amount of the collected material is mixed with 2–3 drops of 1× PBS on a glass slide, then air dried, fixed in acetone for 10 min, and air dried again. In parallel, positive and negative control smears are prepared in other slide from PBS suspensions of stock of *B. pertussis* and *B. parapertussis* cultures grown on BG or RL medium (Optical density at 650 nm = 0.5). For staining, 3–5 drops of titered *B. pertussis* and *B. parapertussis* conjugates are added accordingly to the control and sample slides. Because of the high specificity of the conjugates, one is used as the nonspecific-staining/negative control for the other. Both polyclonal and monoclonal conjugates are commercially available. After addition of conjugates, the slides are incubated for 30 min at 37°C in a damp chamber (a plastic box lined with moistened absorbent paper). Then, slides are soaked in 1× PBS for 5 min, rinsed with distilled water, and air dried. Control and sample slides are finally examined by using a fluorescence microscopy. Smears of specimens are examined at ×400, and suspicious areas are examined under oil for confirmation. On DFA, the organisms should appear as small, coccobacilli with bright, apple-green peripheral fluorescence. They may appear singly, in pairs, or in clusters in smears prepared from highly tenacious specimens, the organisms may appear adherent to the mucous strands.

60.2.2.2 PCR Assay for *B. pertussis*

Approximately 100 different PCR protocols have been reported, including conventional and real-time formats. These vary by DNA purification techniques, PCR primers, reaction conditions, and product detection methods.[103] Laboratories must develop and validate their own PCR tests. However, it is imperative to do that following the guidelines and recommendations of the regulatory agencies.

In this section we briefly describe a simplex real-time PCR protocol using of LightCycler technology for amplification of the insertion element IS*481* of *B. pertussis*, recommended in the laboratory manual for the diagnosis of whooping cough caused by *B. pertussis*/*B. parapertussis* (WHO).[93] The LightCycler technology use two hybridization probes.[112] This allows sequence-specific detection by using fluorescence energy transfer between the fluorophores conjugated to the probes. The amount of fluorescence is directly proportional to the amount of target DNA generated during the PCR process. This protocol is very sensitive but lacks specificity in terms of *Bordetella* species.[93]

Material derived from swabs, nasopharyngeal aspirations, or sputum can be analyzed with this protocol. If necessary, specimens are firstly liquefied with fluidifiant preparations. Extraction of DNA from the primary specimen is necessary to limit inhibition of PCR. QIAamp DNA Mini Kit (Qiagen) or High Pure PCR Template Preparation Kit (Roche) is recommended to be used at this step.

The basic reagents required for PCR reactions are included in the LightCycler FastStart DNA Master PLUS Hybridization Probes kit provided by the manufacturer (Roche). The IS*481* primers and probes used with this protocol are BP-1 (forward) 5′-GATTCAATAGGTTGTATGCATGGTT-3′; BP-2 (reverse) 5′-TTCAGGCACACAAACTTGATGGGCG-3′; Probe 1 (BP-FLU) 5′-TCGCCAACCCCCCAGTTCACTCA-F-3′; and Probe 2 (PB-LCR) 5′-LC-Red 640-AGCCCGGCCGGATGAACACCC P-3′.

PCR is performed in glass capillaries with the LightCycler instrument. Reactions are prepared in a 20 μL volume containing 4 μL of FastStart Reaction Mix Hybridization Probes (component of the FastStart DNA Master Hybridization Probes kit), 5 μL of DNA extract; 1 μL of BP-1 and BP-2 primers (final concentration: 0.5 pmol/mL); 1 μL of BP-FLU and BP LCR probes (final concentration: 0.2 pmol/mL); 1 μL of uracil-DNA glycosylase; and 6 μL of H_2O PCR grade. Sample analysis has to be accompanied by internal amplification controls, positive and negative in-run controls as well as external quality controls.[100] PCR reactions are run using the following protocol: 10 min at 95°C, followed by 40 cycles of 10 s at 95°C, 10 s at 60°C, and 20 s at 72°C. Fluorescence increase (i.e., creation of specific product) is measured during the annealing step at 60°C. The conditions for the last amplification cycle are 95°C for 0 s, 40°C for 2 min, and 95°C for 0 s. Temperature change rates are 20°C/s for all steps. After the last cycle is completed, a melting curve analysis is performed. Samples are regarded as "positive," when the fluorescence signal increases and shows a typical amplification kinetic. Samples are regarded as "negative" when they do not fulfill the criteria mentioned above.

60.3 CONCLUSIONS AND FUTURE PERSPECTIVES

Despite high vaccination coverage, particularly in developed countries, pertussis still remains a substantial public health problem.[17–20] In the 1990s, the disease resurged, becoming the most prevalent vaccine-preventable disease in immunized populations. Concomitantly with the resurgence of the disease, a shift in its incidence toward older age groups was reported.[23–25] Adolescents and adults in whom the disease symptoms are often mild are now being recognized as a major source of infection for nonvaccinated children, increasing the level of circulation of *B. pertussis*.[23] The relatively short-lived immunity that results from vaccination in adolescents and adults has been considered a major reason for this shift.[24,25] However, it is true that other factors such as the greater awareness of pertussis in these age groups, which is related to the improvement in the microbiological diagnosis, also count as a cause. Indeed, in the present era, the extensive acellular pertussis vaccine efficacy trials and other epidemiological studies have made it possible to evaluate multiple methods for the diagnosis of *B. pertussis* infections. Currently, the sensitivity and specificity of the laboratory diagnosis of *B. pertussis* can equal those for many other bacterial infections, with the proper performance and management of culture, PCR and serologic tests.[103] In spite of this, diagnosis still requires to be improved, and the scientific community and government regulatory agencies agree about the necessity to advance toward a global standardization and harmonization of pertussis diagnosis tests, especially serological and PCR assays.[80]

The high circulation rate of *B. pertussis* has also been attributed to adaptation of the pathogen.[29,30] A number of studies demonstrated changes in the structure of *B. pertussis* populations after the introduction of vaccination[19,31,32] supporting this hypothesis. These changes have been associated with polymorphisms in the genes encoding for known virulence factors (PRN, PT, and FIM),[85] which are suggested to alter epitopes that might hinder recognition of the pathogen by the immune system of the host after vaccination. More recently, it was showed that these changes can also be associated with alterations in the level of expression of virulence genes,[89,90] which are speculated to become the pathogen less visible to the host immune system (in case of lowered expression[90]) or more efficient in colonizing the host (in case of increased expression[89]).

Despite the speculations about the advantage that antigenic divergence could provide to circulating *B. pertussis* strains to colonize and persist within the human host, no robust evidence regarding the mechanisms underlying these events is available. Mechanisms associated with intracellular survival and biofilm development have been proposed to explain long term bacterial colonization of the host.[71,72] In this context, we recently demonstrated that *B. pertussis* adopts a distinctive physiology under the biofilm mode of life.[72,119,120] Interestingly, the biosynthesis of a putative acidic-type polysaccharide polymer turned out to be the most distinctive feature of this lifestyle, but also a novel facet of *B. pertussis* physiology. Likewise, recent in vivo studies utilizing a murine infection model demonstrated the presence of biofilm-like microcolonies formed by *Bordetella* in the nasal cavity,[73] a primary host niche wherein the pathogen may escape host immune defenses. We believe that further investigations in this line may provide deeper insights into the mechanisms of pathogenesis of *B. pertussis*.

It is evident that current pertussis vaccines are effective at preventing the severe manifestations of the disease but are less effective at preventing bacterial colonization. The combination of the multiple factors described above make the prophylaxis programs unsuccessful in eradicating the pathogen, which remains circulating in the population. Pertussis is a multifactorial disease, therefore, it requires more than a simple approach to its recognition, treatment, and control. In this context, it is crucial to continue monitoring the evolution of these bacterial populations, to examine isolates for their antimicrobial susceptibility and to increase the knowledge about the mechanisms of pathogenesis that may facilitate the persistence of the pathogen in the host, in order to design novel strategies for controlling the disease.

ACKNOWLEDGMENT

We thank María Eugenia Rodríguez for helpful discussions.

REFERENCES

1. Still, G.F., *The History of Paediatrics*, Oxford University Press, London, 1931.
2. Cherry, J.D., Pertussis in the preantibiotic and prevaccine era, with emphasis on adult pertussis, *Clin. Infect. Dis.*, 28, S107, 1999.
3. Bordet, J., and Gengou, O., Le microbe de la coqueluche, *Ann. Inst. Pasteur Microbiol.*, 20, 48, 1906.
4. US Department of Commerce, Bureau of the Census. Historical Statistics of the United States, Colonial times to 1957, pp. 26–38, US Government Printing Office, Washington, DC, 1958.
5. Luttinger, P., Pertussis immunization, *NY Med. J.*, 101, 1043, 1915.
6. Madsen, T., Pertussis in Feroe Islands, Boston, *Med. Surg. J.*, 192, 50, 1925.
7. Njamkepo, E. et al., Thirty-five years' experience with the whole-cell pertussis vaccine in France: Vaccine strains analysis and immunogenicity, *Vaccine*, 20, 1290, 2002.
8. Cherry, J.D. et al., Report of the task force on pertussis and pertussis immunization, *Pediatrics*, 81, 939, 1988.
9. Romanus, V., Jonsell, R., and Bergquist, S.O., Pertussis in Sweden after the cessation of general immunization in 1979, *Ped. Infect. Dis. J.*, 6, 364, 1987.
10. Aoyama, T., Acellular pertussis vaccines developed in Japan and their application for disease control, *J. Infect. Dis.*, 174, S264, 1996.
11. Miller, E., Vurdien, J.E., and White, J.M., The epidemiology of pertussis in England and Wales, *Commun. Dis. Rep. CDR Rev.*, 2, R152, 1992.
12. American Academy of Pediatrics, Pertussis (whooping cough). In L. K. Pickering (Editor), *Red Book: 2006 Report of the Committee on Infectious Diseases*, 26th ed., pp. 498–520, American Academy of Pediatrics, Elk Grove Village, IL, 2007.
13. Rappuoli, R., Acellular Pertussis vaccines: A turning point in infant and adolescent vaccination, *J. Infect. Dis.*, 5, 21, 1996.
14. Gustafsson, L. et al., A controlled trial of a two-component acellular, a five-component acellular, and a whole-cell pertussis vaccine, *N. Engl. J. Med.*, 334, 349, 1996.

15. Olin, P. et al., Declining pertussis incidence in Sweden following the introduction of acellular pertussis vaccine, *Vaccine*, 21, 2015, 2003.
16. Pichichero, M.E. et al., A safety and immunogenicity comparison of 12 acellular pertussis vaccines and one whole-cell pertussis vaccine given as a fourth dose in 15- to 20-month-old children, *Pediatrics*, 100, 772, 1997.
17. Mathis, R.D., Shoaf, B., and Weiner T.I., Pertussis: The return of a bad penny, *Pediatr. Emerg. Care*, 9, 218, 1993.
18. Celentano, L.P. et al., Resurgence of Pertussis in Europe, *Pediatr. Infect. Dis. J.*, 24, 761, 2005.
19. Caro, V. et al., Pertussis in Argentina and France, *Vaccine*, 25, 4335, 2007.
20. Grgic-Vitek, M., Klavs, I., and Kraigher, A., Re-emergence of pertussis in Slovenia: Time to change immunization policy, *Vaccine*, 26, 1874, 2008.
21. Edwards, K.M., Overview of pertussis: Focus on epidemiology, sources of infection, and long term protection after infant vaccination, *Pediatr. Infect. Dis. J.*, 24, S104, 2005.
22. Cherry, J.D., The science and fiction of the "resurgence" of pertussis, *Pediatrics*, 112, 405, 2003.
23. Mattoo, S., and Cherry, J.D., Molecular pathogenesis, epidemiology, and clinical manifestations of respiratory infections due to *Bordetella pertussis* and other Bordetella subspecies, *Clin. Microbiol. Rev.*, 18, 326, 2005.
24. Guris, D. et al., Changing epidemiology of pertussis in the United States: Increasing reported incidence among adolescents and adults, 1990–1996, *Clin. Infect. Dis.*, 28, 1230, 1999.
25. Kapaskelis, A.M. et al., High prevalence of antibody titers against *Bordetella pertussis* in an adult population with prolonged cough, *Respir. Med.*, 102, 1586, 2008.
26. Deville, J.G. et al., Frequency of unrecognized *Bordetella pertussis* infections in adults, *Clin. Infect. Dis.*, 21, 639, 1995.
27. Li, W.C. et al., Boostrix: A reduced-dose acellular pertussis vaccine for use in adolescents and adults, *Expert. Rev. Vaccines*, 8, 1317, 2009.
28. Plosker, G.L., Combined, reduced-antigen content tetanus, diphtheria, and acellular pertussis vaccine (Boostrix): A review of its use as a single-dose booster immunization in individuals aged 10–64 years in the US, *BioDrugs*, 23, 253, 2009.
29. Mooi, F.R., *Bordetella pertussis* and vaccination: The persistence of a genetically monomorphic pathogen, *Infect. Genet. Evol.*, 10, 36, 2010.
30. Kallonen, T., and He, Q., *Bordetella pertussis* strain variation and evolution postvaccination, *Expert. Rev. Vaccines*, 8, 863, 2009.
31. van Amersfoorth, S.C. et al., Analysis of *Bordetella pertussis* populations in European countries with different vaccination policies, *J. Clin. Microbiol.*, 43, 2837, 2005.
32. Cassiday, P. et al., Polymorphism in *Bordetella pertussis* pertactin and pertussis toxin virulence factors in the United States, 1935–1999, *J. Infect. Dis.*, 182, 1402, 2000.
33. Litt, D.J., Neal, S.E., and Fry, N.K., Changes in genetic diversity of the *Bordetella pertussis* population in the United Kingdom between 1920 and 2006 reflect vaccination coverage and emergence of a single dominant clonal type, *J. Clin. Microbiol.*, 47, 680, 2009.
34. Parton, R., Review of the biology of *Bordetella pertussis*, *Biologicals*, 27, 71, 1999.
35. Thalen, M. et al., Rational medium design for *Bordetella pertussis*: Basic metabolism, *J. Biotechnol.*, 75, 147, 1999.
36. Thalen, M. et al., Fed-batch cultivation of *Bordetella pertussis*: Metabolism and Pertussis Toxin production, *Biologicals*, 34, 289, 2006.
37. Gerlach, G. et al., Evolutionary trends in the genus Bordetella, *Microbes Infect.*, 3, 61, 2001.
38. Porter, J.F., Connor, K., and Donachie, W., Isolation and characterization of *Bordetella parapertussis*-like bacteria from ovine lungs, *Microbiology*, 140, 255, 1994.
39. Goodnow, R.A., Biology of *Bordetella bronchiseptica*, *Microbiol. Rev.*, 44, 722, 1980.
40. Gueirard, P. et al., Human *Bordetella bronchiseptica* infection related to contact with infected animals: Persistence of bacteria in host, *J. Clin. Microbiol.*, 33, 2002, 1995.
41. Woolfrey, B.F., and Moody, J.A., Human infections associated with *Bordetella bronchiseptica*, *Clin. Microbiol. Rev.*, 4, 243, 1991.
42. Parkhill, J. et al., Comparative analysis of the genome sequences of *Bordetella pertussis*, *Bordetella parapertussis* and *Bordetella bronchiseptica*, *Nat. Genet.*, 35, 32, 2003.
43. Van Den Berg, B.M. et al., Role of *Bordetella pertussis* virulence factors in adherence to epithelial cell lines derived from the human respiratory tract, *Infect. Immun.*, 67, 3, 1056, 1999.
44. Fernandez, R.C., and Weiss, A.A., Cloning and sequencing of a *Bordetella pertussis* serum resistance locus, *Infect. Immun.*, 62, 4727, 1994.
45. Coutte, L. et al., Subtilisin-like autotransporter serves as maturation protease in a bacterial secretion pathway, *EMBO J.*, 20, 5040, 2001.
46. Finn, T.M., and Amsbaugh, D.F., Vag8, a *Bordetella pertussis* bvg regulated protein, *Infect. Immun.*, 66, 3985, 1998.
47. Finn, T.M., and Stevens, L.A., Tracheal colonization factor: A *Bordetella pertussis* secreted virulence determinant, *Mol. Microbiol.*, 16, 625, 1995.
48. Gentry-Weeks, C.R. et al., Dermonecrotic toxin and tracheal cytotoxin, putative virulence factors of *Bordetella avium*, *Infect. Immun*, 56, 1698, 1988.
49. Yuk, M.H., Harvill, E.T., and Miller, J.F., The BvgAS virulence control system regulates type III secretion in *Bordetella bronchiseptica*, *Mol. Microbiol.*, 28, 945, 1998.
50. Watanabe, M. et al., Biological properties of lipopolysaccharides from *Bordetella* species, *J. Gen. Microbiol.*, 136, 489, 1990.
51. Smith, A.M., Guzman, C.A., and Walker, M.J., The virulence factors of *Bordetella pertussis*: A matter of control, *FEMS Microbiol. Rev.*, 25, 309, 2001.
52. Uhl, M.A., and Miller, J.F., Autophosphorylation and phosphotransfer in the *Bordetella pertussis* BvgAS signal transduction cascade, *Proc. Nat. Acad. Sci. USA*, 91, 1163, 1994.
53. Coote, J.G., Antigenic switching and pathogenicity: Environmental effects on virulence gene expression in *Bordetella pertussis*, *J. Gen. Microbiol.*, 137, 2493, 1991.
54. Melton, A.R., and Weiss, A.A., Characterization of environmental regulators of *Bordetella pertussis*, *Infect. Immun.*, 61, 807, 1993.
55. Scarlato, V. et al., Sequential activation and environmental regulation of virulence genes in *Bordetella pertussis*, *EMBO J.*, 10, 3971, 1991.
56. Cotter, P.A., and Miller, J.F., A mutation in the *Bordetella bronchiseptica* bvgS gene results in reduced virulence and increased resistance to starvation, and identifies a new class of Bvg-regulated antigens, *Mol. Microbiol.*, 24, 671, 1997.
57. Deora, R. et al., Diversity in the Bordetella virulence regulon: Transcriptional control of a Bvg-intermediate phase gene, *Mol. Microbiol.*, 40, 669, 2001.
58. Vergara-Irigaray, N. et al., Evaluation of the role of the Bvg intermediate phase in *Bordetella pertussis* during experimental respiratory infection, *Infect. Immun.*, 73, 748, 2005.

59. Willems, R. et al., Fimbrial phase variation in *Bordetella pertussis*: A novel mechanism for transcriptional regulation, *EMBO J.*, 9, 2803, 1990.
60. Tuomanen, E. et al., Filamentous hemagglutinin and pertussis toxin promote adherence of *Bordetella pertussis* to cilia, *Dev. Biol. Stand.*, 61, 197, 1985.
61. Relman, D.A. et al., Filamentous hemagglutinin of *Bordetella pertussis*: Nucleotide sequence and crucial role in adherence, *Proc. Nat. Acad. Sci. USA*, 86, 2637, 1989.
62. van den Akker, W.M., The filamentous hemagglutinin of *Bordetella parapertussis* is the major adhesin in the phase-dependent interaction with NCI-H292 human lung epithelial cells, *Biochem. Biophys. Res. Commun.*, 252, 128, 1998.
63. Carbonetti, N.H., Immunomodulation in the pathogenesis of *Bordetella pertussis* infection and disease, *Curr. Opin. Pharmacol.*, 7, 272, 2007.
64. Abramson, T., Kedem, H., and Relman, D.A., Proinflammatory and proapoptotic activities associated with *Bordetella pertussis* filamentous hemagglutinin, *Infect. Immun.*, 69, 2650, 2001.
65. Confer, D.L., and Eaton, J.W., Phagocyte impotence caused by an invasive bacterial adenylate cyclase, *Science*, 217, 948, 1982.
66. Bassinet, L. et al., *Bordetella pertussis* adenylate cyclase-hemolysin induces interleukin-6 secretion by human tracheal epithelial cells, *Infect. Immun.*, 72, 5530, 2004.
67. Perkins, D.J. et al., *Bordetella pertussis* adenylate cyclase toxin (ACT) induces cyclooxygenase-2 (COX-2) in murine macrophages and is facilitated by ACT interaction with CD11b/CD18 (Mac-1), *Mol. Microbiol.*, 66, 1003, 2007.
68. Munoz, J.J., Action of pertussigen (pertussis toxin) on the host immune system in pathogenesis and immunity in pertussis. In Wardlaw, A.C., and Parton, R. (Editors), *Pathogenesis and Immunity in Pertussis*, pp. 173–192, John Wiley & Sons, Ltd., New York, 1988.
69. Luker, K.E. et al., Tracheal cytotoxin structural requirements for respiratory epithelial damage in pertussis, *Mol. Microbiol.*, 16, 733, 1995.
70. Fukui, A., and Horiguchi, Y., *Bordetella* dermonecrotic toxin exerting toxicity through activation of the small GTPase Rho, *J. Biochem.*, 136, 415, 2004.
71. Lamberti, Y.A. et al., Intracellular trafficking of *Bordetella pertussis* in human macrophages, *Infect. Immun.* 78, 907, 2010.
72. Serra, D.O. et al., Proteome approaches combined with Fourier transform infrared spectroscopy revealed a distinctive biofilm physiology in *Bordetella pertussis*, *Proteomics*, 8, 4995, 2008.
73. Sloan, G.P. et al., The Bordetella Bps polysaccharide is critical for biofilm development in the mouse respiratory tract, *J. Bacteriol.*, 189, 8270, 2007.
74. Wood, N., and McIntyre, P., Pertussis: Review of epidemiology, diagnosis, management and prevention, *Paediatr. Respir. Rev.*, 9, 201, 2008.
75. Crowcroft, N.S., and Pebody, R.G., Recent developments in pertussis, *Lancet*, 367, 1926, 2006.
76. Birkebaek, N.H. et al., *Bordetella pertussis* and chronic cough in adults, *Clin. Infect. Dis.*, 29, 1239, 1999.
77. Postels-Multani, S. et al., Symptoms and complications of pertussis in adults, *Infection*, 23, 139, 1995.
78. von Konig, C.H. et al., Pertussis of adults and infants, *Lancet Infect. Dis.*, 2, 744, 2002.
79. Cherry, J.D., The epidemiology of pertussis and pertussis immunization in the United Kingdom and the United States: A comparative study, *Curr. Probl. Paediatr.*, 14, 1, 1984.
80. Tondella, M.L. et al., International *Bordetella pertussis* assay standardization and harmonization meeting report. Centers for Disease Control and Prevention, Atlanta, Georgia, United States, 19–20 July 2007, *Vaccine*, 27, 803, 2009.
81. WHO, Challenges in Global Immunization and the Global Immunization Vision and Strategy 2006–2015, *Weekly Epidemiol. Rec.*, 81, 190, 2006.
82. Crowcroft, N.S. et al., How best to estimate the global burden of pertussis? *Lancet Infect. Dis.*, 3, 413, 2003.
83. Dirección Nacional de Epidemiología, Ministerio de Salud, Buenos Aires, Argentina, http://www.msal.gov.ar/htm/Site/sala_situacion/index.asp. Accessed October 2008.
84. Bouchez, V. et al., Genomic content of *Bordetella pertussis* clinical isolates circulating in areas of intensive children vaccination, *PLoS One*, 3, e2437, 2008.
85. Mastrantonio, P. et al., Antigenic variants in *Bordetella pertussis* strains isolated from vaccinated and unvaccinated children, *Microbiology*, 145, 2069, 1999.
86. Hallander, H.O. et al., Shifts of *Bordetella pertussis* variants in Sweden from 1970 to 2003, during three periods marked by different vaccination programs, *J. Clin. Microbiol.*, 43, 2856, 2005.
87. Mooi, F.R. et al., Epidemiological typing of *Bordetella pertussis* isolates: Recommendations for a standard methodology, *Eur. J. Clin. Microbiol. Infect. Dis.*, 19, 174, 2000.
88. Van Loo, I.H., and Mooi, F.R., Changes in the Dutch *Bordetella pertussis* population in the first 20 years after the introduction of whole-cell vaccines, *Microbiology*, 148, 2011, 2002.
89. Mooi, F.R. et al., *Bordetella pertussis* strains with increased toxin production associated with pertussis resurgence, *Emerg. Infect. Dis.*, 15, 1206, 2009.
90. Bouchez, V. et al., First report and detailed characterization of *Bordetella pertussis* isolates not expressing Pertussis Toxin or Pertactin, *Vaccine*, 27, 6034, 2009.
91. Heikkinen, E. et al., Comparative genomics of *Bordetella pertussis* reveals progressive gene loss in Finnish strains, *PLoS One*, 2, e904, 2007.
92. Council of State and Territorial Epidemiologists, CSTE Position Statement 1997-ID-9: Public health surveillance, control and prevention of pertussis, CSTE, 1997. http://www.cste.org/ps/1997/1997-id-09.htm. Accessed November 2010.
93. WHO, Department of Immunization, Vaccines and Biologicals, The laboratory manual for the diagnosis of whooping cough caused by *B. pertussis/B. parapertussis*, WHO Document Production Services, Geneva, Switzerland, WHO/IVB/04.14, 2007. http://www.who.int/vaccines-documents/DocsPDF04.www788.pdf. Accessed November 2010.
94. Wirsing von König, C.H. et al., Detection of *Bordetella pertussis* in clinical specimens. In Manclark (Editor), *Proceedings of the 6th International Symposium on Pertussis*, pp. 315–321, US FDA, Bethesda, MD, 1990.
95. Cimolai, N., Trombley, C., and O'Neill, D., Diagnosis of whooping cough: A new era with rapid molecular diagnostics, *Pediatr. Emerg. Care*, 12, 91, 1996.
96. Tozzi, A.E. et al., Diagnosis and management of pertussis, *CMAJ*, 172, 509, 2005.
97. McGowan, K.L, Diagnostic tests for pertussis: Culture vs. DFA vs. PCR, *Clin. Microbiol. Newsl.*, 24, 143, 2002.
98. Ewanowich, C.A. et al., Major outbreak of pertussis in northern Alberta, Canada: Analysis of discrepant direct fluorescent-antibody and culture results by using polymerase chain reaction methodology, *J. Clin. Microbiol.*, 31, 1715, 1993.

99. Simondon, F. et al., Evaluation of an immunoglobulin G enzyme-linked immunosorbent assay for pertussis toxin and filamentous hemagglutinin in diagnosis of pertussis in Senegal, *Clin. Diagn. Lab. Immunol.*, 5, 130, 1998.
100. Riffelmann, M. et al., Nucleic Acid amplification tests for diagnosis of Bordetella infections, *J. Clin. Microbiol.*, 43, 4925, 2005.
101. de Melker, H.E. et al., Specificity and sensitivity of high levels of immunoglobulin G antibodies against pertussis toxin in a single serum sample for diagnosis of infection with *Bordetella pertussis*, *J. Clin. Microbiol.*, 38, 800, 2000.
102. Watanabe, M., Connelly, B., and Weiss, A.A., Characterization of serological responses to pertussis, *Clin. Vaccine Immunol.*, 13, 341, 2006.
103. Müller, F.M., Hoppe, J.E., and Wirsing von Konig, C.H., Laboratory diagnosis of pertussis: State of the art in 1997, *J. Clin. Microbiol.*, 35, 2435, 1997.
104. Houard, S. et al., Specific identification of *Bordetella pertussis* by the polymerase chain reaction, *Res. Microbiol.*, 140, 477, 1989.
105. Meade, B.D., and Bollen, A., Recommendations for use of the polymerase chain reaction in the diagnosis of *Bordetella pertussis* infections, *J. Med. Microbiol.*, 41, 51, 1994.
106. Dragsted, D.M. et al., Comparison of culture and PCR for detection of *Bordetella pertussis* and *Bordetella parapertussis* under routine laboratory conditions, *J. Med. Microbiol.*, 53, 749, 2004.
107. Lievano, F.A. et al., Issues associated with and recommendations for using PCR to detect outbreaks of pertussis, *J. Clin. Microbiol.*, 40, 2801, 2002.
108. Kosters, K. et al., Real-time LightCycler PCR for detection and discrimination of *Bordetella pertussis* and *Bordetella parapertussis*, *J. Clin. Microbiol.*, 40, 1719, 2002.
109. Kösters, K., Riffelmann, M., and Wirsing von König, C.H., Evaluation of a real-time PCR assay for detection of *Bordetella pertussis* and *Bordetella parapertussis* in clinical samples, *J. Med. Microbiol.*, 50, 436, 2001.
110. Templeton, K.E. et al., Evaluation of real-time PCR for detection of and discrimination between Bordetella pertussis, *Bordetella parapertussis*, and *Bordetella holmesii* for clinical diagnosis, *J. Clin. Microbiol.*, 41, 4121, 2003.
111. Poddar, S.K., Rapid detection of *Bordetella pertussis* by real-time PCR using SYBR green I and a LightCycler instrument, *J. Clin. Lab. Anal.*, 18, 265, 2004.
112. Wittwer, C.T. et al., The LightCycler: A microvolume multisample fluorimeter with rapid temperature control, *BioTechniques*, 22, 176, 1997.
113. de Moissac, Y.R., Ronald, S.L., and Peppler, M.S., Use of pulsed-field gel electrophoresis for epidemiological study of *Bordetella pertussis* in a whooping cough outbreak, *J. Clin. Microbiol.*, 32, 398, 1994.
114. Murphy, T., Bisgard, K., and Sanden, G., Diagnosis and laboratory methods. In *Guidelines for the Control of Pertussis Outbreaks*, Chapter 2, National Immunization Program, Centers for Disease Control and Prevention, Atlanta, GA, 2000.
115. Hallander, H.O. et al., Comparison of nasopharyngeal aspirates with swabs for culture of Bordetella pertussis, *J. Clin. Microbiol.*, 31, 50, 1993.
116. Regan, J., and Lowe, F., Enrichment medium for the isolation of *Bordetella*, *J. Clin. Microbiol.*, 6, 303, 1977.
117. Katzko, G., Hofmeister, M., and Church, D., Extended incubation of culture plates improves recovery of Bordetella spp., *J. Clin. Microbiol.*, 34, 1563, 1996.
118. Koneman, E. et al., Miscellaneous fastidious Gram-negative bacilli. In *Color Atlas and Textbook of Diagnostic Microbiology*, 6th ed., pp. 518–519, Lippincott Williams & Wilkins, Philadelphia, US, 2006.
119. Bosch, A. et al., Characterization of *Bordetella pertussis* growing as biofilm by chemical analysis and FT-IR spectroscopy, *Appl. Microbiol. Biotechnol.*, 71, 736, 2006.
120. Serra, D.O., Continuous nondestructive monitoring of *Bordetella pertussis* biofilms by fourier transform infrared spectroscopy and other corroborative techniques, *Anal. Bioanal. Chem.*, 387, 1759, 2007.

61 *Burkholderia*

Karlene H. Lynch and Jonathan J. Dennis

CONTENTS

61.1 INTRODUCTION

Over 60 species of gram-negative β-proteobacteria are currently classified within the genus *Burkholderia*.[1] Little is known about the majority of these bacteria, with the exception of those that are clinically relevant, including *Burkholderia cepacia* complex (BCC) species, *Burkholderia pseudomallei*, *Burkholderia mallei*, and *Burkholderia gladioli*. This chapter will review the basic biology, pathogenesis, and diagnosis of these organisms (with an emphasis on the BCC) and will present a recently developed protocol for multilocus sequence typing (MLST) of *Burkholderia* spp.[2]

61.1.1 *Burkholderia cepacia* Complex

61.1.1.1 Classification, Morphology, Biology, and Epidemiology

The BCC is a subgroup of the *Burkholderia* genus comprising at least seventeen species. W. H. Burkholder was the first to describe these bacteria in 1950.[3] They were originally isolated from onions with sour skin rot, and so were named *Pseudomonas cepacia* (cepa meaning "onion" in Latin).[3] In 1992, the novel genus *Burkholderia* was created following assessment of molecular and phenotypic characteristics and *Pseudomonas cepacia* was renamed *Burkholderia cepacia*.[4]

The first description of the BCC as a group was published in 1997. Following DNA–DNA hybridization analysis, *B. cepacia* isolates could be divided into five distinct groups.[5] Each of these groups was designated as a genomic species or genomovar: I, II (identified as *Burkholderia multivorans*), III, IV, and V (previously named *Burkholderia vietnamiensis*).[5,6] In subsequent years, the genomovars were redesignated as species: I as *B. cepacia*, III as *Burkholderia cenocepacia*, and IV as *Burkholderia stabilis*.[7,8] In addition, new genomovars/species were added (VI/*Burkholderia dolosa*, VII/*Burkholderia ambifaria*, VIII/*Burkholderia anthina*, and IX/*Burkholderia pyrrocinia*) such that, until 2008, the complex consisted of nine different species.[9–12] Based on *recA* sequence analysis and other tests, an additional eight species were added to the BCC in 2008 and 2009: *Burkholderia latens*, *Burkholderia diffusa*, *Burkholderia arboris*, *Burkholderia seminalis*, *Burkholderia metallica*, *Burkholderia ubonensis*, *Burkholderia contaminans*, and *Burkholderia lata*.[13,14] As isolates have been found that occupy novel *recA* and MLST clusters not included in the current taxonomy, it is likely that the number of species present in the BCC will increase in the future.[14]

The description of BCC bacteria by W. H. Burkholder noted that the cells were gram-negative rods with an average diameter of 0.8 μm and length of 1.9 μm.[3] The cells have between one and three polar flagella and are motile.[3] Depending on the strain, colonies may appear smooth or rough and may have different morphologies on different carbon sources.[15–17] *B. cenocepacia* strains C1394 and K56-2, which both typically have a rough appearance, can convert to smooth variants which have different phenotypes in BCC infection models.[16,18] Isolates may produce a brown pigment similar to melanin or a yellow pigment more commonly found in environmental isolates than clinical isolates.[19,20] With regard to odor, colonies of *B. dolosa* are distinctive because they may smell like dirt or rotting potatoes.[21]

Although this chapter discusses the BCC in the context of clinical infection, it is important to note that these species have important roles in both environmental and industrial settings as well. These bacteria are commonly found associated with plant rhizospheres, including those of sugarcane, rice, corn, wheat, soybean, and alfalfa, as well as with bananas, mushrooms, and onions.[3,22–26] They can also be isolated from sewage, surface water sources, medical-grade water, and drinking water.[27–31] BCC bacteria can be found as contaminants in industrial processes and have caused significant clinical problems when present in medical products such as trypan blue, chlorhexidine, and heparin.[32–35]

BCC bacteria have three key characteristics that would make them promising candidates for both industrial and agricultural use. First, BCC species are able to degrade a wide variety of toxic organic compounds, including phenol, trichloroethylene, and diesel fuel.[36,37] Second, these bacteria produce antimicrobial, antifungal, and antiparasitic agents that have been found to be active against a wide variety of plant pathogens.[38,39] Finally, BCC strains have been shown to have nitrogenase activity and thus the ability to fix N_2.[6] However, due to concerns that environmental application of these bacteria may increase the risk of transmission to susceptible individuals, the Environmental Protection Agency (EPA) has restricted the commercial production and use of these strains.[40] These regulations are particularly important in light of studies that have shown that a *B. cenocepacia* epidemic strain can be found in soil and that over 20% of clinical isolates are of the same MLST sequence type (ST) (discussed below) as environmental isolates.[41,42]

In most cases, immunocompetent individuals are not susceptible to BCC infection. However, rare cases of community-acquired pneumonia, skin infections, endocarditis, and infections resulting from exposure to contaminated medical solutions have been reported for these individuals.[43–46] BCC infections are most commonly found in patients with cystic fibrosis (CF) and chronic granulomatous disease (CGD). CF is a genetic condition in which patients carry two mutated copies of the gene encoding the cystic fibrosis transmembrane conductance regulator (CFTR), a chloride channel.[47] In the absence of functional CFTR proteins, thick mucus builds up in the lungs, making these patients susceptible to *Pseudomonas*, *Staphylococcus*, *Haemophilus*, and *Burkholderia* infections.[48] CGD is a genetic condition caused by mutations in the genes encoding phagocytic NADPH oxidase components, which are crucial for oxidative burst activity.[49] Without this enzyme complex, patients become susceptible to a variety of different pathogens, including *Burkholderia*.[49]

Patients can be infected with BCC bacteria via person-to-person transmission, as shown for *B. contaminans*, *B. cenocepacia*, *B. multivorans*, *B. cepacia*, and *B. dolosa*, or from environmental sources.[14,50–52] Although all seventeen BCC species have been isolated clinically, the majority of infections are caused by *B. multivorans* and *B. cenocepacia*.[13,14,53,54] In the most recent BCC-specific epidemiological study from the United States, *B. multivorans* accounted for 38.7% of BCC infections in CF patients, while *B. cenocepacia* accounted for 45.6%.[55] While *B. cenocepacia* currently accounts for the greatest proportion of infections, the epidemiology of BCC infection has been changing in recent years. Between 1997 and 2004, the proportion of new infections caused by *B. cenocepacia* decreased by ~30% while the proportion caused by *B. multivorans* increased by ~20%.[55] It has been suggested that this shift is due to a decrease in person-to-person transmission of epidemic *B. cenocepacia* strains (a result of established infection-control protocols that segregate infected patients) and continued transmission of *B. multivorans* from environmental sources.[52,55,56] However, as epidemic *B. cenocepacia* strains can be found in the environment and *B. multivorans* strains can be transmitted from person-to-person, this answer may oversimplify the factors involved.[41,51]

61.1.1.2 Clinical Features and Pathogenesis

In 1984, Isles et al. published a key report detailing the outcomes of BCC infection.[57] This group analyzed patient records from 1970 to 1981 at a CF clinic in Toronto, Canada, and found that infections follow one of three distinct courses: chronic colonization of the lungs with few or no symptoms, chronic lung infection with pulmonary function worsening over time, and a serious condition that later became known as "cepacia syndrome."[57] In cepacia syndrome, which occurs in approximately 20% of BCC-infected patients, individuals who initially present with relatively mild symptoms rapidly develop pneumonia (with the occurrence of abscesses on the lungs) and septicemia, which is often fatal.[57,58] Both *B. multivorans* and *B. cenocepacia* infection can lead to the development of cepacia syndrome.[59]

When the two BCC species most commonly isolated from patients are compared, *B. cenocepacia* is considered to be more virulent than *B. multivorans*. An analysis of the patient records from a CF clinic in Manchester, United Kingdom showed that following infection by *B. cenocepacia*, almost all patients developed a chronic infection, while only half of those infected by *B. multivorans* did.[59] In addition, mortality of those chronically infected with *B. cenocepacia* was significantly higher than of those infected with *Pseudomonas aeruginosa*, while there was no significant difference in mortality following *B. multivorans* and *P. aeruginosa* infection.[59] While it is not known why *B. cenocepacia* is more pathogenic, it is thought that differences in gene distribution among members of the BCC are most likely important.[60]

Although BCC species express a wide range of putative virulence factors, there is no clear understanding as to which of these are the most important during infection. The characteristics that influence BCC pathogenicity have been grouped here into three categories: antibiotic resistance, surface structures, and secreted products. The following information discusses the BCC in general terms, and it should be noted both that this is not an exhaustive list and that not all of these characteristics are shared by all seventeen species.[61]

Antibiotic Resistance. From the first study of BCC epidemiology, it was understood that these bacteria are resistant to a wide range of antibiotics, although there are variations in sensitivity among both strains and species.[57,62,63] Some factors that contribute to this resistance are the expression of *β*-lactamases and low numbers of small porins (both causing *β*-lactam resistance), altered dihydrofolate reductase (causing trimethoprim resistance), efflux pumps (causing chloramphenicol, trimethoprim, and ciprofloxacin resistance), altered lipopolysaccharide (LPS) (causing aminoglycoside and polymyxin resistance), and biofilm formation.[64–70] As a result, it is difficult to treat infected patients effectively, even when using multiple drugs simultaneously.[71] In the three reported cases in which patients have developed cepacia syndrome and survived, high doses of multiple antibiotics needed to be administered intravenously over a short period of time (including ceftazidime, meropenem, ciprofloxacin, chloramphenicol, trimethoprim-sulphamethoxazole, tobramycin, and temocillin).[72]

Surface Structures. BCC bacteria encode multiple surface structures that allow them to carry out different functions. They may express type III, type IV, and type VI secretion systems, all of which have been implicated in virulence.[73–75] The LPS not only contributes to resistance, as mentioned above, but also promotes an inflammatory response.[69,76] Exopolysaccharide is thought to promote BCC persistence in the lungs.[18] *B. cenocepacia* may encode cable pili and an associated adhesin that can bind to both mucin and cytokeratin 13 on host cells and facilitate attachment.[77,78] Studies have suggested that BCC bacteria can invade and survive inside host cells, including epithelial cells and macrophages, thus avoiding the host immune response and allowing spread throughout the body.[79,80] Interestingly, *B. multivorans* and *B. cenocepacia* have been found to be more invasive than other BCC species.[81] Flagella have been shown to be important in this process.[82]

Extracellular Products. A variety of enzymes and other products are produced and released by BCC species. Three quorum-sensing systems exist in these bacteria, *cepIR*, *cciIR*, and *bviIR* (found in *B. vietnamiensis*).[83–85] The synthesis of several products, including protease, lipase, and ornibactin (a siderophore), is quorum sensing regulated in the BCC.[83] To protect cells from the oxidative bursts of phagocytes, these bacteria may produce a brown melanin-like pigment (mentioned above), catalase, catalase-peroxidase, and/or superoxide dismutase.[86,87] BCC bacteria also release hemolysin, but it may be produced by a minority of strains.[88]

61.1.1.3 Diagnosis

Conventional Techniques. In order to implement appropriate antibiotic therapy and infection-control procedures, BCC infections must be properly diagnosed.[89] However, BCC bacteria are often misidentified as other nonfermenting gram-negatives (and vice versa), with reported false-positive and false-negative rates of 11% and 36%, respectively.[89] Although molecular techniques for BCC detection are preferable to conventional techniques because of their specificity, using a combination of these methods can produce both fast and accurate results.[17] Tests such as growth on BCSA (discussed below) and at 42°C, lysine and ornithine decarboxylase activity, sucrose and adonitol oxidation, and hemolysis are acceptable preliminary tests for the identification of BCC organisms.[17]

Media: The three media generally used for propagation and selection of BCC bacteria are *Pseudomonas cepacia* agar (PCA), oxidation-fermentation polymyxin bacitracin lactose agar (OFPBL), and *Burkholderia cepacia* selective agar (BCSA).[90] These media contain high concentrations of polymyxin and other antibiotics to which BCC organisms are resistant, including ticarcillin, bacitracin, gentamicin, and vancomycin. BCSA is generally recommended over the other two media because it allows for better growth of BCC organisms, and it more effectively inhibits the growth of non-BCC organisms such as *Stenotrophomonas* and *Staphylococcus*.[90] However, it is not entirely selective as it can support the growth of similar organisms such as *P. aeruginosa* and *B. gladioli* (discussed below).[90]

Commercial Tests: Commercial systems commonly used for BCC identification include the Vitek GNI and GNI Plus, MicroScan Conventional Gram Neg Panel, API 20NE and RapID NF Plus.[91] These tests have been shown to be relatively inaccurate. An assessment of these and similar systems in 2000 found positive predictive values ranging from 71% to 98% and negative predictive values between 50% and 82%.[91] These statistics indicate that a sample that tests positive for BCC bacteria has a 2%–29% chance of being BCC-negative, whereas a sample that tests negative for BCC bacteria has an 18%–50% chance of being BCC-positive. As such, the use of commercial tests must be integrated with other diagnostic protocols in order to accurately identify these bacteria.[17]

Other Methods: Other conventional methods used to identify BCC bacteria that will not be discussed here include bacteriocin sensitivity assays, serotyping, mass spectrometry, protein and fatty acid profiling, and multilocus enzyme electrophoresis.

Molecular techniques. *rDNA-based methods:* Many rDNA sequence-based BCC identification protocols have been developed, including ribotyping, PCR ribotyping, fluorescent in situ hybridization (FISH), restriction fragment length polymorphism (RFLP) analysis or amplified ribosomal DNA restriction analysis (ARDRA), terminal restriction fragment length polymorphism (T-RFLP) analysis, length heterogeneity PCR (LH-PCR), species-specific PCR and sequencing.[92–98] Although each of these protocols has its own benefits and drawbacks, a problem common to all of these methods is that BCC species have very similar 16S rDNA sequences (>98%).[17] Because of this high homology, accurate identification to the species level is not possible using many of these techniques.[17]

Single-locus methods: Because of the problems associated with BCC rDNA-based diagnostics, other genes were identified that would allow for better differentiation of these bacteria. The most successful approach has been the use of the *recA* (recombinase A) gene. In a phylogenetic tree based on *recA* gene sequences, BCC strains form clusters that correlate with their species identification.[13,14,99,100] An exception is *B. cenocepacia*, which forms four different clusters designated as lineages IIIA, IIIB, IIIC, and IIID.[7] BCC identification protocols based on the *recA* gene sequence include sequencing, RFLP, and species-specific PCR.[99] Although this method has proven very successful overall, it has several disadvantages: it divides the most commonly isolated species into four divergent taxonomic groups; it has been shown to inaccurately identify strains in certain cases, and, with the exception of full gene sequencing, *recA*-based protocols have not been able to keep pace with the constantly changing BCC taxonomy.[7,17,100,101] Other single-locus methods, including those based on the *fur* (ferric uptake regulator), *gyrB* (DNA gyrase B), and *opcL* (peptidoglycan-associated outer-membrane lipoprotein) genes, are less commonly used and have not been updated to the current taxonomy.[102–104]

Multilocus sequence typing: Currently, the best molecular technique for the identification of BCC organisms is MLST. This method, first developed for the BCC in 2005, compares the sequences of seven different housekeeping genes: *atpD* (ATP synthase β chain), *gltB* (glutamate synthase large subunit), *lepA* (GTP binding protein), *phaC* (acetoacetyl-CoA reductase), *trpB* (tryptophan synthase subunit B), *recA*, and *gyrB* (both discussed above).[105] For all seven genes, each unique gene sequence is assigned an allele type. Each BCC strain has seven allele types (one for each gene sequenced) that together make up an allelic profile, which is then assigned a sequence type (ST). In the majority of cases, each strain has a different ST, so this protocol can be used to identify BCC isolates at both the species and the strain level.[105] This method has a number of advantages, as it allows more detailed analysis than previously developed techniques, it can be used to identify new taxonomic groups, it does not rely on qualitative observations, and so is easier to standardize, and, perhaps most importantly, it has been recently updated and is consistent with the current BCC taxonomy.[2,105]

Other methods: Other molecular protocols used for BCC identification that will not be discussed here include arbitrarily primed PCR (AP–PCR), enterobacterial repetitive intergenic consensus sequence PCR (ERIC–PCR), pulsed-field gel electrophoresis (PFGE), AFLP fingerprinting, randomly amplified polymorphic DNA (RAPD) fingerprinting, DNA–DNA hybridization, multilocus restriction typing (MLRT), and BOX–PCR.

61.1.2 *Burkholderia pseudomallei* and *Burkholderia mallei*

61.1.2.1 Classification, Morphology, Biology, and Epidemiology

Burkholderia pseudomallei and *Burkholderia mallei* are two closely related species, the latter a clone of the former.[106] Like the members of the BCC, *B. pseudomallei* and *B. mallei* were moved from the genus *Pseudomonas* to the genus *Burkholderia* in 1992.[4] In contrast to the BCC, however, the taxonomy of these species has remained relatively static. *B. pseudomallei* is a polar flagellated gram-negative rod with a diameter of 0.8 μm and a length of 1.5 μm.[107] *B. mallei* is a nonflagellated gram-negative rod with a diameter of 0.3–0.5 μm and a length that varies based on the age of the culture (between 0.7 and 5.0 μm).[107] Colonies of *B. pseudomallei* (which can grow at 42°C) grow quickly and may have a whitish or orange pigmentation, while colonies of *B. mallei* (which cannot grow at 42°C) tend to grow more slowly and have a yellow pigmentation that turns to brown over time.[107]

While *B. pseudomallei* can be found in a variety of environmental reservoirs, *B. mallei* is an obligate pathogen with limited ability to survive outside its host. *B. pseudomallei* is most commonly isolated from soils and water samples. A recent study in northeast Thailand showed that, out of 24 regions tested, 13 had soil that contained culturable *B. pseudomallei*, and five had soil that contained both *B. pseudomallei* and *B. thailandensis* (a closely related, generally nonpathogenic species).[108] *B. pseudomallei* can survive for long periods of time in a variety of soil conditions and remains culturable after two years at temperatures of 24°C and 32°C, at a pH between 5 and 8 or in soils with a water content of 40% or 80%.[109] *B. pseudomallei* can be isolated from both surface water and drinking water and remains in a culturable state after 3 years of incubation in distilled water.[30,110,111] It is thought that *B. pseudomallei* may be able to survive in the environment in a viable but nonculturable (VBNC) state or intracellularly in protozoa.[110] *B. mallei*, on the other hand, is only isolated from humans, animals such as horses, mules and donkeys, and animal-associated environments such as troughs.[112] It is thought that *B. mallei* cannot persist in the environment because it has lost genes for metabolic pathways that *B. pseudomallei* has maintained.[113] Because there are concerns that both *B. pseudomallei* and *B. mallei* could be propagated and released as bioweapons, both of these species are classified as Category B biological agents by the Centers for Disease Control and Prevention (CDC).[114]

Melioidosis, the disease caused by *B. pseudomallei*, is endemic in southeast Asia and northern Australia and may occasionally be found in other equatorial regions.[115] An epidemiological study from northern Australia in 2004 found that melioidosis had an incidence of 19.6 in 100,000 individuals, but was especially prevalent among males, aboriginal persons, and those over 45 years of age.[116] Risk factors include diabetes, alcoholism, and lung and kidney disease.[116] The disease is more common in the monsoon season than in the dry season, and it became especially prevalent following the tsunami in 2004.[117,118] *B. pseudomallei* is usually transmitted through broken skin, by inhalation or by ingestion.[119] Glanders and farcy, the diseases caused by *B. mallei*, have been eradicated in Western countries since 1949, with the exception of cases occurring due to accidental laboratory exposure.[120,121] These diseases are occasionally seen in South America and in the eastern hemisphere, including Africa, Asia, and the Middle East.[122] *B. mallei* is usually transmitted through the skin and mucosal surfaces following contact with infected animals.[123]

61.1.2.2 Clinical Features and Pathogenesis

B. pseudomallei infection may be asymptomatic or may present as one of four different conditions: acute localized infection, acute pulmonary infection, acute bloodstream infection, or disseminated infection.[119] The course of infection will be affected by both preexisting risk factors in the patient and the route of inoculation.[119] In the case of chronic infection, almost any organ in the body can be affected, including the skin, brain, lungs, liver, spleen, and bones.[119] Because of the variety of different symptoms associated with melioidosis, it is referred to as "the great mimicker."[124] *B. mallei* infections in humans are similar and present as one of the same conditions as *B. pseudomallei* infections.[123] The skin infection is referred to as "farcy," while the other presentations are collectively referred to as "glanders."[121]

As is the case for the BCC, it is not clear which virulence factors of *B. pseudomallei* and *B. mallei* are the most important for productive infection, particularly in the case of the less-studied *B. mallei*. Both species are resistant to a variety of antibiotics, including fluoroquinolones and cephalosporins, due to mechanisms such as efflux pumps and *gyrA* mutation.[125] Both species carry genes involved in secretion, iron uptake and the biosynthesis of LPS, exopolysaccharide, proteases, hemolysin and type IV pili.[126] Six hundred fifty different genes are in the *B. pseudomallei/B. mallei* virulome (consisting of the genes present in these two pathogenic species but absent in nonpathogenic *Burkholderia* strains), and, using in vivo models, several of these genes have already been shown to affect virulence.[127]

61.1.2.3 Diagnosis

Conventional Techniques. While a selective medium for *B. pseudomallei* (called Ashdown's agar) has been developed, no such medium exists for the isolation and propagation of *B. mallei*.[128] The best medium commercially available for these species is PCA (discussed above), on which most *B. pseudomallei* colonies have a white to yellow pigmentation and metallic sheen, and most *B. mallei* colonies are smooth and pinpoint with white to translucent pigmentation.[128] Identification using this medium is problematic, however, because not all colonies have this appearance, some *B. mallei* strains cannot grow on PCA and the growth of BCC bacteria is supported.[128] Commercial kits tend to be more effective in identifying *B. pseudomallei* than *B. mallei*. In one study using the API 20NE and Vitek 2 commercial systems, 86.7% and 78.3%, respectively, of *B. pseudomallei* isolates were correctly identified.[129] The remaining isolates were incorrectly identified as BCC, *Chromobacterium*, *Comamonas*, *Pseudomonas*, and *Shewanella*.[129] In a different study using the API 20NE and RapID NF Plus systems, 60% and 0%, respectively, of *B. pseudomallei* isolates were correctly identified, while neither system could identify *B. mallei*.[130] Other conventional techniques that will not be discussed here include bacteriophage plaque assays, latex agglutination, and indirect immunofluorescence assays.

Molecular Techniques. Similarly to the BCC, rDNA-based methods are relatively ineffective for identification of *B. pseudomallei* and *B. mallei*. These two species and *B. thailandensis* have 16S rDNA sequences that are over 99% similar, and so it is difficult to differentiate these species from one another using this method.[131] A single nucleotide polymorphism at position 75 in the 16S rDNA can be used to differentiate two of these species, as *B. pseudomallei* has a cytidine at this position, while *B. mallei* has a thymidine.[132] Similarly, at position 2143 in the 23S rDNA, *B. pseudomallei* has a cytidine, while *B. mallei* has a thymidine.[133] The most effective molecular technique for *B. pseudomallei* and *B. mallei* identification is MLST. The MLST scheme for these species, published in 2003, uses the genes *ace* (acetyl coenzyme A reductase), *gltB* (glutamate synthase), *gmhD* (ADP glycerol-mannoheptose epimerase), *lepA* (GTP-binding elongation factor), *lipA* (lipoic acid synthetase), *narK* (nitrite extrusion protein), and *ndh* (NADH dehydrogenase).[106] The most recent BCC MLST scheme (outlined below) also allows for the accurate identification of these species.[2] Other molecular techniques that will not be discussed here include DNA microarrays, real-time PCR (RT-PCR), single-locus PCR and sequencing, multiplex PCR, variable amplicon typing (VAT), multiple-locus variable-number tandem repeat analysis (MLVA), BOX–PCR, REP–PCR, RAPD fingerprinting, and PFGE.

61.1.3 *Burkholderia gladioli*

61.1.3.1 Classification, Morphology, Biology, and Epidemiology

The four pathovars of *B. gladioli* cause a spectrum of diseases in both plants and humans. As discussed earlier with respect to the BCC, *B. pseudomallei* and *B. mallei*, the species *Pseudomonas gladioli* was identified as a member of the *Burkholderia* genus in 1992.[4] Since 2003, this species has been divided into four pathovars: *gladioli*, *alliicola*, *agaricicola*, and *cocovenenans*.[134] *B. gladioli* cells are gram-negative rods, 0.5–0.8 μm in diameter and 1.5–2.5 μm in length, with one to three flagella.[135] Colonies of this species have a yellow pigmentation and are round and opaque.[136]

Different pathovars of *B. gladioli* have different environmental distributions. Pathovars *gladioli*, *alliicola*, and *agaricicola* are primarily phytopathogens. Pathovar *gladioli* has a global distribution and has been isolated from gladioli, crocuses, freesia, irises, ferns, and orchids.[137] Pathovar *alliicola* also has a global distribution, and, like the first characterized members of the BCC, causes skin rot in onions.[137] Pathovar *agaricicola* is a pathogen of fungi that infects mushrooms.[137] In contrast, pathovar *cocovenenans* is found in two traditional Asian food products: tempe bongkrek, an Indonesian fermented coconut cake, and fermented corn flour, used in China to make breads and noodles.[138,139]

Depending on the pathovar, *B. gladioli* can cause either acute or chronic infections or toxin-induced food poisoning. Nontoxigenic *B. gladioli* (considered pathovars *gladioli*, *alliicola*, and *agaricicola*) infects patients with CF, CGD, and conditions such as acquired immune deficiency syndrome (AIDS).[134,140,141] One of the reasons that nontoxigenic *B. gladioli* does not infect immunocompetent individuals is that the bacteria are killed by complement proteins in the serum.[140] Because epidemiological studies of *B. gladioli* infection have only identified isolates to the species level, it is not known if there is a difference in the clinical distribution of the three nontoxigenic pathovars.[142,143]

Most of the epidemiological data that has been published regarding these infections focuses on CF patients. *B. gladioli* infections occur most commonly in these individuals, and it was in this population that the organism was first recognized as an infectious human pathogen.[142–144] A study from a CF center in North Carolina found that, out of 769 patients admitted over a 6-year period, 33 (4.3%) became infected with *B. gladioli*.[142] Eighteen patients had an acute infection, twelve had a chronic infection, and there was insufficient data to assess the course of infection for three patients. An additional two patients with *B. gladioli* infection were identified who did not have CF: one had been placed on a respirator, and the other had abnormal cilia function. All of the patients had pulmonary infections and had not developed septicemia. There was no evidence of patient-to-patient spread. A second study on *B. gladioli* epidemiology from France was published in 2009.[143] The bacterium was isolated from 18 patients, 14 with CF and 4 without CF (three of whom had been intubated and one of whom had bronchiectasis). Eight patients had an acute infection, two had an intermittent infection, and three had a chronic infection (sufficient data was unavailable for the other patients). All 18 patients had pulmonary *B. gladioli* infections. One patient with CF who died due to *B. gladioli* infection had bacteria present at various time points in the lungs, in a mediastinal abscess and in the blood.

The epidemiology of pathovar *cocovenenans* is significantly different. In this case, the intoxication affects those who eat contaminated tempe bongkrek or fermented corn flour products. These intoxications are most common in the Banjumas province in Central Java, Indonesia, and in the provinces of Heilongjiang, Jilin, and Liaoning in northeastern China.[139,145] In Indonesia, 7216 cases and 850 deaths were reported between 1951 and 1975, while in China, 1842 cases and 703 deaths in 226 outbreaks were reported between 1953 and 1975.[139,146] The most important risk factor for intoxication is the quantity of contaminated food consumed.[139]

61.1.3.2 Clinical Features and Pathogenesis

In both acute and chronic infections, *B. gladioli* is most commonly isolated from the respiratory tract.[142–144] These infections are much less serious than BCC infections, and the effect on the patient is often minimal.[144] In a sample of seven *B. gladioli*-infected patients, lung function was worse for three individuals following infection, while function was no different or better for the other four.[142] Septicemia has been reported in those with CF, CGD, AIDS, and diabetes.[140,141,147,148] In rare cases, these bacteria may also cause abscesses (mentioned above), cervical adenitis, osteomyelitis, and keratitis/corneal ulcers.[140,143,149,150]

While nontoxigenic *B. gladioli* causes infections, pathovar *cocovenenans* causes foodborne intoxications. In an analogous fashion to *Clostridium botulinum*, the symptoms of pathovar *cocovenenans* poisoning are due to the toxins released by bacteria into the food, and not to the bacteria themselves. Within 1–10 h of eating contaminated food, individuals experience hyperglycemia, followed by hypoglycemia, spasms, loss of consciousness, and death.[139,145] Organ damage is often apparent following autopsy, especially in the liver, kidneys, and brain.[139]

Although studies have elucidated some of the *B. gladioli* virulence factors that cause disease in plants, it is not clear which of these factors are important for human infection. In comparison to other *Burkholderia* species, *B. gladioli* is relatively susceptible to antibiotics, including the β-lactams piperacillin and imipenem, the fluoroquinolone ciprofloxacin, and aminoglycosides.[143] However, clinical isolates may become resistant to multiple antibiotics, and these strains tend to be associated with chronic infections.[142] Graves et al. examined four *B. gladioli* clinical isolates and found that they produced siderophores but did not carry plasmids, did not have a hydrophobic surface and did not attach to eukaryotic cells (with one exception).[140] Clinical isolates of this species have been shown to both produce biofilms and survive and replicate in a macrophage cell line.[151] They have not been shown to secrete toxins, although a pathovar *gladioli* strain isolated from a gladiolus can produce toxoflavin (discussed below).[134,135]

Pathovar *cocovenenans* intoxication is caused by two toxins, toxoflavin, and bongkrekic acid. Toxoflavin is an electron carrier that becomes reduced and then transfers its electrons to oxygen, resulting in the formation of hydrogen peroxide.[152] Bongkrekic acid, thought to be the more dangerous of the two toxins, binds to the mitochondrial adenine nucleotide transporter, inhibiting oxidative phosphorylation.[153,154] Cells can no longer generate ATP, so glycogen is broken down during anaerobic glycolysis, causing hyperglycemia.[155,156] When no more glycogen is available, the affected individual becomes hypoglycemic and dies as a result of ATP depletion.[155]

61.1.3.3 Diagnosis

Conventional Techniques. As *B. gladioli* is an emerging pathogen in CF patients, it is important to be able to differentiate between this bacterium and other related CF-associated bacteria, such as those of the BCC. However, problems in identifying this species have been reported in the literature. A strain of *B. gladioli* was reported to have spread among six CF patients, providing the first supposed evidence of person-to-person *B. gladioli* transmission, but it was later shown that this strain was actually *B. cenocepacia*.[157,158] No selective medium has been developed for *B. gladioli*, but OFPBL has been shown to be the best medium for its isolation. *B. gladioli* can grow on OFPBL and PCA but poorly or not at all on BCSA.[143] OFPBL is recommended over PCA because the high concentration of ticarcillin in PCA could potentially inhibit *B. gladioli* growth.[143] With respect to commercial systems, the API 20NE is not effective as it misidentifies *B. gladioli* as *B. cepacia*.[143] The Vitek 2 GN is a better system for detection of this species, as it was able to correctly identify 15 out of 18 tested clinical isolates (the remaining three were misidentified as *Pseudomonas fluorescens* or *Pseudomonas oryzhihabitans*).[143] To differentiate between the pathovars of *B. gladioli*, a variety of phenotypic tests can be used (Table 61.1). For an extensive discussion of both conventional and molecular diagnosis of pathovar *cocovenenans*, please refer to Reference 159.

Molecular Techniques. The most commonly used molecular techniques to identify *B. gladioli* are rDNA-based methods. These include ARDRA, capillary electrophoresis-single-strand conformation polymorphism (CE-SSCP) analysis, multiplex PCR, and sequencing.[102,143,160,161] A downside of these techniques is that they cannot differentiate between the pathovars of *B. gladioli* because of their high homology. For example, the 16S rDNA sequences from the type strains of pathovar *gladioli*, and pathovar *cocovenenans* are 99.6% identical.[134] Furthermore, in a phylogenetic tree based on these sequences, not all strains cluster according to pathovar.[134] The same is true in a phylogenetic tree based on *B. gladioli gyrB* and *rpoD* sequences.[162] Although no MLST protocol has been developed specifically for *B. gladioli*, it can be accurately identified using the updated BCC protocol.[2]

61.2 METHODS

61.2.1 Sample Preparation

BCC. BCC bacteria should be cultured on BCSA. To make BCSA, combine 5.0 g NaCl, 10.0 g sucrose, 10.0 g lactose, 0.08 g phenol red, 0.002 g crystal violet, 10.0 g trypticase peptone, 1.5 g yeast extract, and 14.0 g agar in 1 L distilled water, and adjust the pH to 7.0 ± 0.1. After autoclaving at 15 lb/in^2 for 20 min, add 600,000 U polymyxin, 10 mg gentamicin, and 2.5 mg vancomycin.[20] To propagate BCC from sputum cultures, mix sputum with Sputasol (Oxoid, Basingstoke, Hampshire, UK) in a 1:1 dilution, incubate 30 min at 37°C, and streak out 10 μL of this solution onto BCSA. Incubate 48 h at 37°C.[163]

***B. pseudomallei* and *B. mallei*.** Both *B. pseudomallei* and *B. mallei* should be cultured on PCA. To make PCA, combine 1.0 g $(NH_4)_2SO_4$, 0.2 g $MgSO_4{\cdot}7H_2O$, 0.01 g $Fe(NH_4SO_4)_2{\cdot}6H_2O$, 0.02 g phenol red, 0.1 mL 1% crystal violet, 1.0 g proteose peptone no. 3, 1.5 g bile salts, and 15 g agar in 950 mL 0.04 M potassium phosphate buffer (pH 6.2). After autoclaving at 121°C for 15 min and cooling to 50°C, add 50 mL 10% pyruvic acid sodium salt, 100 mg ticarcillin sodium salt, and 300,000 U polymyxin B sulfate.[164] *B. pseudomallei* and *B. mallei* can be cultured from samples including blood, urine, and tissue.[165]

TABLE 61.1
Differential Characteristics among Pathovars of *B. gladioli*

	B. gladioli			
Characteristics	**Pathovar *gladioli* NCPPB 1891**	**Pathovar *agaricicola* NCPPB 3580**	**Pathovar *alliicola* NCPPB 947**	**Pathovar *cocovenenans* ATCC 33664 and 6 Strains**
Source	Gladiolus	Mushroom	Onion	Coconut, fermented corn
Bongkrekic acid production	−	−	−	100% (7/7)[a]
Growth at 41°C	+	−	+	100% (7/7)
Growth at 4°C	+	+	−	0% (0/7)
Nitrate reduction	−	−	+	100% (7/7)
Assimilation				
Xylitol	−	−	−	100% (7/7)
Gentiobiose	+	+	+	0% (0/7)
N-Acetyl-D-galactosamine	+	+	+	0% (0/7)
Acetic acid	+	−	+	86% (6/7)
L-Pyroglutamic acid	−	+	+	100% (7/7)
Maltose	−	+	−	0% (0/7)
D-Raffinose	−	−	+	0% (0/7)
Sucrose	−	−	+	0% (0/7)
Glycyl-L-aspartic acid	−	−	+	0% (0/7)

Source: Jiao, Z. et al., *Microbiol. Immunol.*, 47, 915, 2003. Table reproduced with permission from Wiley-Blackwell Publishing.
+, Positive or weak positive reaction; −, negative reaction.
[a] Percentage of positive strains.

B. gladioli. *B. gladioli* should be cultured on OFPBL. To make OFPBL, combine 9.4 g of OF basal medium (Difco Laboratories, Detroit, Michigan) and 15.0 g Bacto-Agar in 1 L distilled water, adjust the pH to 6.8 ± 0.1 and boil. After autoclaving at 15 lb/in^2 and 121°C for 15 min and cooling to 50°C, add 10.0 g lactose (in filter-sterilized solution), 300,000 U polymyxin B sulfate, and 200 U bacitracin.[166] *B. gladioli* can be cultured from samples including sputum, bronchoalveolar lavage, and tracheal aspirate.[142] For details on isolation, propagation, and toxin extraction from *B. gladioli* pv. *cocovenenans*, please refer to Reference 159.

61.2.2 Detection Procedures

The detection procedure for *Burkholderia* spp. described here is an updated version of the original BCC MLST scheme published by Baldwin et al. in 2005.[2,105] The 2005 protocol was able to effectively identify the nine BCC species that had been defined at the time, but it was not tested with non-BCC *Burkholderia* strains.[105] The updated protocol can identify all seventeen current BCC species, in addition to *B. pseudomallei*, *B. mallei*, *B. gladioli*, and others.[2]

MLST Protocol[2]

1. To isolate chromosomal DNA, resuspend one colony (following incubation on solid media overnight at 32°C) in 20 μL lysis buffer (0.25% (vol/vol) SDS and 0.05 N NaOH). Incubate at 95°C for 15 min. Add 180 μL high-performance liquid chromatography-grade water, and store at 4°C.[167]
2. Set up seven 25 μL PCR reactions containing 2 mM $MgCl_2$, 20 mM Tris-HCl, 50 mM KCl, 250 μM dNTP (each), 0.4 μM primer (each), 1 M betaine, 10% DMSO, 2 U of Taq polymerase, and 2 μL of the chromosomal DNA prepared in the previous step. Each reaction will contain one of the primer sets listed in Table 61.2.
3. Amplify the reactions in a DNA thermal cycler under the following conditions:
   Cycle 1: 95°C for 2 min
   Cycles 2–31: 94°C for 30 s, annealing temperature for 30 s, 72°C for 60 s
   Cycle 32: 72°C for 5 min
4. Purify and sequence the amplified DNA using the same primer sets.[168]
5. Compare the sequences from the isolate to the sequences deposited in the BCC MLST database (http://pubmlst.org/bcc) to determine the allele type for each gene. Compare the allelic profile of the isolate to the allelic profile of the strains in the database to assign a sequence type (and putative identity) for the isolate.

The authors tested these seven primer sets with 80 *Burkholderia* strains, and the primers amplified the expected product in each case. Sequences were obtained for all seven loci (from sequence databases or by using the protocol above) from strains of the BCC, *B. pseudomallei*, *B. mallei*, *B. gladioli*, *B. thailandensis*, *Burkholderia oklahomensis*, *Burkholderia glumae*, *Burkholderia plantarii*, *Burkholderia tropica*, *Burkholderia sacchari*, *Burkholderia mimosarum*, *Burkholderia xenovorans*, *Burkholderia phytofirmans*, *Burkholderia fungorum*, *Burkholderia terricola*, *Burkholderia caledonica*, *Burkholderia graminis*, *Burkholderia phenoliruptrix*, *Burkholderia tuberum*, *Burkholderia hospita*, *Burkholderia caribensis*, *Burkholderia*

TABLE 61.2
Oligonucleotide Primer Sequences and Annealing Temperatures for the Amplification and Sequencing of Seven MLST Loci for *Burkholderia* Species

Gene	Amplicon Size (bp)	Primer Sequence (5′→3′)	Annealing Temperature (°C)
atpD	756	ATGAGTACTRCTGCTTTGGTAGAAGG	56
		CGTGAAACGGTAGATGTTGTCG	
gltB	652	CTGCATCATGATGCGCAAGTG	58
		CTTGCCGCGGAARTCGTTGG	
gyrB	738	ACCGGTCTGCAYCACCTCGT	60
		YTCGTTGWARCTGTCGTTCCACTGC	
lepA	975	CTSATCATCGAYTCSTGGTTCG	55
		CGRTATTCCTTGAACTCGTARTCC	
phaC	525	GCACSAGYATYTGCCAGCG	58
		CCATSTCSGTRCCRATGTAGCC	
recA	704	AGGACGATTCATGGAAGAWAGC	58
		GACGCACYGAYGMRTAGAACTT	
trpB	787	CGCGYTTCGGVATGGARTG	58
		ACSGTRTGCATGTCCTTGTCG	

Source: From Spilker, T. et al., *J. Clin. Microbiol.*, 47, 2607, 2009. Table reproduced with permission from American Society for Microbiology.

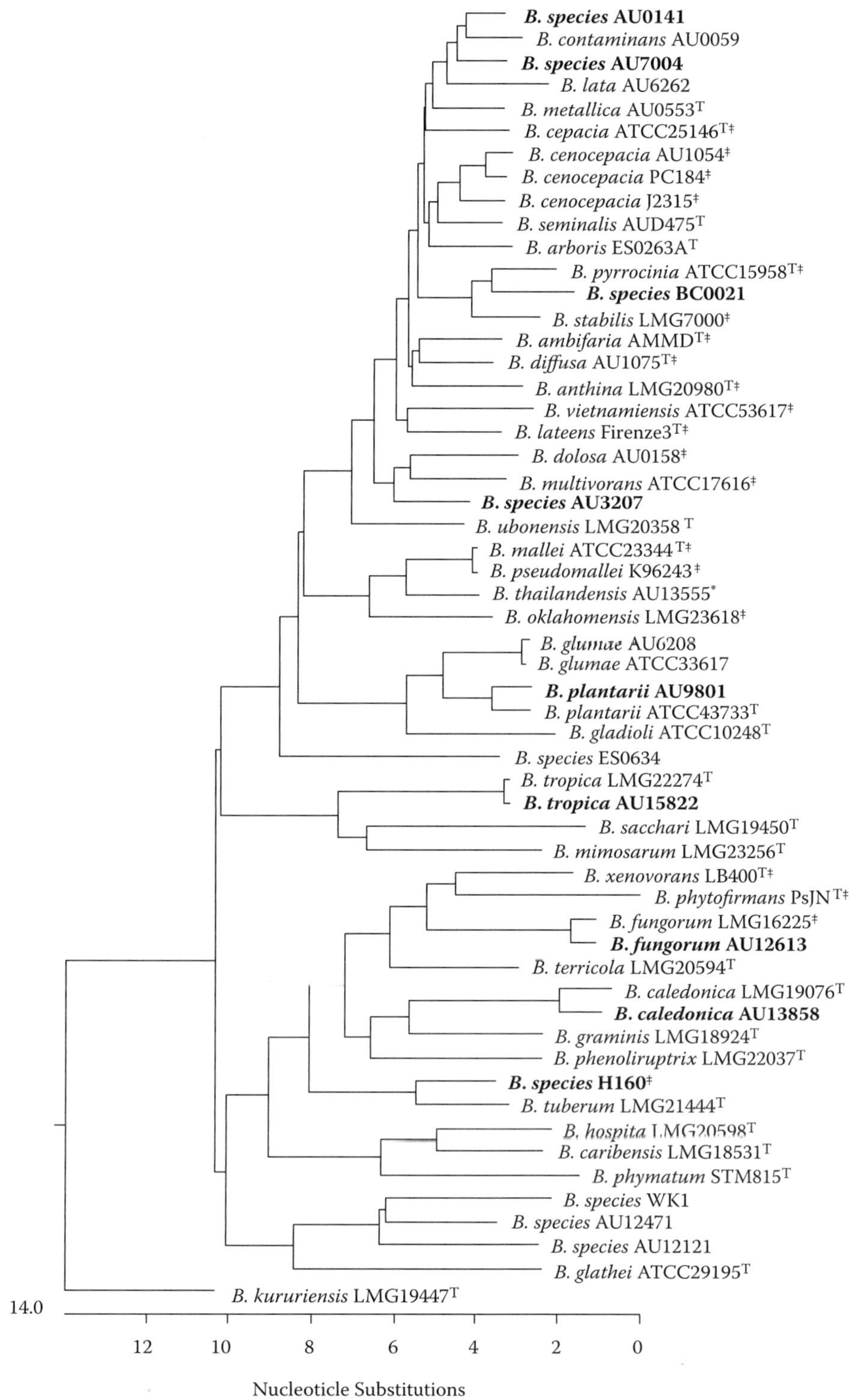

FIGURE 61.1 Phylogenetic tree of concatenated nucleotide sequences from the seven MLST loci, generated by using Clustal V. *Burkholderia* strains included are representatives of the strain set analyzed as part of the study by Spilker et al.[2] or strains for which genome sequence data were available in public databases (indicated by ‡ symbol). *Burkholderia* strains in bold type are discussed in the Spilker et al.[2] text. Strains belonging to various *Burkholderia* species and *Burkholderia* sp. strains are shown. (From Spilker, T. et al., *J. Clin. Microbiol.*, 47, 2607, 2009. Figure and legend reproduced and amended with permission from American Society for Microbiology.)

phymatum, *Burkholderia glathei*, and *Burkholderia kururiensis* (not shown).

When *Burkholderia* spp. are organized into a phylogenetic tree based on these sequences, they form clusters that correlate with their taxonomic designation (Figure 61.1). It is unknown if this protocol differentiates the pathovars of *B. gladioli*, as only two strains were included in the analysis. The inclusion of unclassified *Burkholderia* strains in this tree allows one to assess the relationships of these strains to others in this genus and, in some cases, suggests putative species assignments for these strains.[2]

61.3 CONCLUSION AND FUTURE PERSPECTIVES

A number of challenges exist with regard to the identification of *Burkholderia* species, including constantly changing taxonomies, the phenotypic similarity of these strains to other gram-negative nonfermenters, and the incredible diversity present within the genus. With the development of molecular techniques, diagnostics for these species have become much more defined, but problems remain. While rDNA-based methods are often a fundamental component of molecular diagnostics, they are relatively ineffective in identifying the members of this genus. With the development of MLST, a broadly applicable and readily updated technique, *Burkholderia* species can be accurately identified in both clinical and research laboratories.

REFERENCES

1. NCBI, Taxonomy browser (*Burkholderia*) [online], http://www.ncbi.nlm.nih.gov/Taxonomy/Browser/wwwtax.cgi?id=32008, 2009. Accessed November 4, 2010.
2. Spilker, T. et al., Expanded multilocus sequence typing for *Burkholderia* species, *J. Clin. Microbiol.*, 47, 2607, 2009.
3. Burkholder, W.H., Sour skin, a bacterial rot of onion bulbs, *Phytopathology*, 40, 115, 1950.
4. Yabuuchi, E. et al., Proposal of *Burkholderia* gen. nov. and transfer of seven species of the genus *Pseudomonas* homology group II to the new genus, with the type species *Burkholderia cepacia* (Palleroni and Holmes 1981) comb. nov., *Microbiol. Immunol.*, 36, 1251, 1992.
5. Vandamme, P. et al., Occurrence of multiple genomovars of *Burkholderia cepacia* in cystic fibrosis patients and proposal of *Burkholderia multivorans* sp. nov., *Int. J. Syst. Bacteriol.*, 47, 1188, 1997.
6. Gillis, M. et al., Polyphasic taxonomy in the genus *Burkholderia* leading to an emended description of the genus and proposition of *Burkholderia vietnamiensis* sp. nov. for N_2-fixing isolates from rice in Vietnam, *Int. J. Syst. Bacteriol.*, 45, 274, 1995.
7. Vandamme, P. et al., *Burkholderia cenocepacia* sp. nov. A new twist to an old story, *Res. Microbiol.*, 154, 91, 2003.
8. Vandamme, P. et al., Identification and population structure of *Burkholderia stabilis* sp. nov. (formerly *Burkholderia cepacia* genomovar IV), *J. Clin. Microbiol.*, 38, 1042, 2000.
9. Coenye, T. et al., *Burkholderia cepacia* genomovar VI, a new member of the *Burkholderia cepacia* complex isolated from cystic fibrosis patients, *Int. J. Syst. Evol. Microbiol.*, 51, 271, 2001.
10. Vermis, K. et al., Proposal to accommodate *Burkholderia cepacia* genomovar VI as *Burkholderia dolosa* sp. nov., *Int. J. Syst. Evol. Microbiol.*, 54, 689, 2004.
11. Coenye, T. et al., *Burkholderia ambifaria* sp. nov., a novel member of the *Burkholderia cepacia* complex including biocontrol and cystic fibrosis–related isolates, *Int. J. Syst. Evol. Microbiol.*, 51, 1481, 2001.
12. Vandamme, P. et al., *Burkholderia anthina* sp. nov. and *Burkholderia pyrrocinia*, two additional *Burkholderia cepacia* complex bacteria, may confound results of new molecular diagnostic tools, *FEMS Immunol. Med. Microbiol.*, 33, 143, 2002.
13. Vanlaere, E. et al., *Burkholderia latens* sp. nov., *Burkholderia diffusa* sp. nov., *Burkholderia arboris* sp. nov., *Burkholderia seminalis* sp. nov., and *Burkholderia metallica* sp. nov., novel species within the *Burkholderia cepacia* complex, *Int. J. Syst. Evol. Microbiol.*, 58, 1580, 2008.
14. Vanlaere, E. et al., Taxon K, a complex within the *Burkholderia cepacia* complex, comprises at least two novel species, *Burkholderia contaminans* sp. nov. and *Burkholderia lata* sp. nov., *Int. J. Syst. Evol. Microbiol.*, 59, 102, 2009.
15. Engelkirk, P.G., and Duben-Engelkirk, J., *Laboratory Diagnosis of Infectious Diseases: Essentials of Diagnostic Microbiology*, Wolters Kluwer Health, Baltimore, 2008.
16. Bernier, S.P., Nguyen, D.T., and Sokol, P.A., A LysR-type transcriptional regulator in *Burkholderia cenocepacia* influences colony morphology and virulence, *Infect. Immun.*, 76, 38, 2008.
17. Mahenthiralingam, E., Baldwin, A., and Dowson, C.G., *Burkholderia cepacia* complex bacteria: Opportunistic pathogens with important natural biology, *J. Appl. Microbiol.*, 104, 1539, 2008.
18. Chung, J.W. et al., Colonial morphology of *Burkholderia cepacia* complex genomovar III: Implications in exopolysaccharide production, pilus expression, and persistence in the mouse, *Infect. Immun.*, 71, 904, 2003.
19. Govan, J.R.W. et al., Evidence for transmission of *Pseudomonas cepacia* by social contact in cystic fibrosis, *Lancet*, 342, 15, 1993.
20. Henry, D.A. et al., Identification of *Burkholderia cepacia* isolates from patients with cystic fibrosis and use of a simple new selective medium, *J. Clin. Microbiol.*, 35, 614, 1997.
21. Henry, D.A. et al., Phenotypic methods for determining genomovar status of the *Burkholderia cepacia* complex, *J. Clin. Microbiol.*, 39, 1073, 2001.
22. Mendes, R. et al., Diversity of cultivated endophytic bacteria from sugarcane: Genetic and biochemical characterization of *Burkholderia cepacia* complex isolates, *Appl. Environ. Microbiol.*, 73, 7259, 2007.
23. Zhang, L., and Xie, G., Diversity and distribution of *Burkholderia cepacia* complex in the rhizosphere of rice and maize, *FEMS Microbiol. Lett.*, 266, 231, 2007.
24. Ramette, A., and Tiedje, J.M., Multiscale responses of microbial life to spatial distance and environmental heterogeneity in a patchy ecosystem, *Proc. Natl. Acad. Sci. USA*, 104, 2761, 2007.
25. Lee, Y.-A., and Chan, C.-W., Molecular typing and presence of genetic markers among strains of banana finger-tip rot pathogen, *Burkholderia cenocepacia*, in Taiwan, *Phytopathology*, 97, 195, 2007.
26. Yara, R. et al., A bacterium belonging to the *Burkholderia cepacia* complex associated with *Pleurotus ostreatus*, *J. Microbiol.*, 44, 263, 2006.
27. McNeely, D. et al., Isolation of *Burkholderia cenocepacia* and *Burkholderia vietnamiensis* from human sewage, *Int. J. Environ. Health Res.*, 19, 157, 2009.

28. Vermis, K. et al., Isolation of *Burkholderia cepacia* complex genomovars from waters, *Syst. Appl. Microbiol.*, 26, 595, 2003.
29. Douce, R.W. et al., Investigation of an outbreak of central venous catheter-associated bloodstream infection due to contaminated water, *Infect. Cont. Hosp. Ep.*, 29, 364, 2008.
30. Zanetti, F., De Luca, G., and Stampi, S., Recovery of *Burkholderia pseudomallei* and *B. cepacia* from drinking water, *Int. J. Food. Microbiol.*, 59, 67, 2000.
31. Pegues, D.A. et al., Acquisition of *Pseudomonas cepacia* at summer camps for patients with cystic fibrosis, *J. Pediatr.*, 124, 694, 1994.
32. Jimenez, L., and Smalls, S., Molecular detection of *Burkholderia cepacia* in toiletry, cosmetic, and pharmaceutical raw materials and finished products, *J. AOAC Int.*, 83, 963, 2000.
33. Sunenshine, R. et al., An outbreak of postoperative gram-negative bacterial endophthalmitis associated with contaminated trypan blue ophthalmic solution, *Clin. Infect. Dis.*, 48, 1580, 2009.
34. Heo, S.T. et al., Hospital outbreak of *Burkholderia stabilis* bacteraemia related to contaminated chlorhexidine in haematological malignancy patients with indwelling catheters, *J. Hosp. Infect.*, 70, 241, 2008.
35. Yang, C.-J. et al., Nosocomial outbreak of two strains of *Burkholderia cepacia* caused by contaminated heparin, *J. Hosp. Infect.*, 69, 398, 2008.
36. Folsom, B.R., Chapman, P.J., and Pritchard, P.H., Phenol and trichloroethylene degradation by *Pseudomonas cepacia* G4: Kinetics and interactions between substrates, *Appl. Environ. Microbiol.*, 56, 1279, 1990.
37. Mohanty, G., and Mukherji, S., Biodegradation rate of diesel range n-alkanes by bacterial cultures *Exiguobacterium aurantiacum* and *Burkholderia cepacia*, *Int. Biodeter. Biodegr.*, 61, 240, 2008.
38. El-Banna, N., and Winkelmann, G., Pyrrolnitrin from *Burkholderia cepacia*: antibiotic activity against fungi and novel activities against streptomycetes, *J. Appl. Microbiol.*, 85, 69, 1998.
39. Li, W. et al., Broad spectrum anti-biotic activity and disease suppression by the potential biocontrol agent *Burkholderia ambifaria* BC-F, *Crop Prot.*, 21, 129, 2002.
40. Environmental Protection Agency, *Burkholderia cepacia* complex; significant new use rule [online], http://www.epa.gov/fedrgstr/EPA-TOX/2003/June/Day-13/t15010.htm, 2003. Accessed November 4, 2010.
41. LiPuma, J.J. et al., An epidemic *Burkholderia cepacia* complex strain identified in soil, *Lancet*, 359, 2002, 2002.
42. Baldwin, A. et al., Environmental *Burkholderia cepacia* complex isolates in human infections, *Emerg. Infect. Dis.*, 13, 458, 2007.
43. Pujol, M. et al., Community-acquired bacteremic *Pseudomonas cepacia* pneumonia in an immunocompetent host, *Clin. Infect. Dis.* 15, 887, 1992.
44. Taplin, D., Bassett, D.C.J., and Mertz, P.M., Foot lesions associated with *Pseudomonas cepacia*, *Lancet*, 2, 568, 1971.
45. Hamilton, J. et al., Successful treatment of *Pseudomonas cepacia* endocarditis with trimethoprim-sulfamethoxazole, *Antimicrob. Agents Chemother.*, 4, 551, 1973.
46. Ghazal, S.S. et al., Outbreak of *Burkholderia cepacia* bacteremia in immunocompetent children caused by contaminated nebulized sulbutamol in Saudi Arabia, *Am. J. Infect. Control*, 34, 394, 2006.
47. Rich, D.P. et al., Expression of cystic fibrosis transmembrane conductance regulator corrects defective chloride channel regulation in cystic fibrosis airway epithelial cells, *Nature*, 347, 358, 1990.
48. Harrison, F., Microbial ecology of the cystic fibrosis lung, *Microbiology*, 153, 917, 2007.
49. Segal, B.H. et al., Genetic, biochemical, and clinical features of chronic granulomatous disease, *Medicine*, 79, 170, 2000.
50. LiPuma, J.J. et al., Disproportionate distribution of *Burkholderia cepacia* complex species and transmissibility markers in cystic fibrosis, *Am. J. Respir. Crit. Care Med.*, 164, 92, 2001.
51. Biddick, R. et al., Evidence of transmission of *Burkholderia cepacia*, *Burkholderia multivorans* and *Burkholderia dolosa* among persons with cystic fibrosis, *FEMS Microbiol. Lett.*, 228, 57, 2003.
52. Baldwin, A. et al., Elucidating global epidemiology of *Burkholderia multivorans* in cases of cystic fibrosis by multilocus sequence typing, *J. Clin. Microbiol.*, 46, 290, 2008.
53. Mahenthiralingam, E. et al., Diagnostically and experimentally useful panel of strains from the *Burkholderia cepacia* complex, *J. Clin. Microbiol.*, 38, 910, 2000.
54. Coenye, T. et al., Updated version of the *Burkholderia cepacia* complex experimental strain panel, *J. Clin. Microbiol.*, 41, 2797, 2003.
55. Reik, R., Spilker, T., and LiPuma, J.J., Distribution of *Burkholderia cepacia* complex species among isolates recovered from persons with or without cystic fibrosis, *J. Clin. Microbiol.*, 43, 2926, 2005.
56. Festini, F. et al., Isolation measures for prevention of infection with respiratory pathogens in cystic fibrosis: a systematic review, *J. Hosp. Infect.*, 64, 1, 2006.
57. Isles, A., Maclusky, I., and Corey, M., *Pseudomonas cepacia* infection in cystic fibrosis: An emerging problem, *J. Pediatr.*, 104, 206, 1984.
58. Coenye, T. et al., Taxonomy and identification of the *Burkholderia cepacia* complex, *J. Clin. Microbiol.*, 39, 3427, 2001.
59. Jones, A.M. et al., *Burkholderia cenocepacia* and *Burkholderia multivorans*: Influence on survival in cystic fibrosis, *Thorax*, 59, 948, 2004.
60. Bernier, S.P., and Sokol, P.A., Use of suppression-subtractive hybridization to identify genes in the *Burkholderia cepacia* complex that are unique to *Burkholderia cenocepacia*, *J. Bacteriol.*, 187, 5278, 2005.
61. Mahenthiralingam, E., Urban, T.A., and Goldberg, J.B., The multifarious, multireplicon *Burkholderia cepacia* complex, *Nat. Rev. Microbiol.*, 3, 144, 2005.
62. Nzula, S., Vandamme, P., and Govan, J.R.W., Infuence of taxonomic status on the in vitro antimicrobial susceptibility of the *Burkholderia cepacia* complex, *J. Antimicrob. Chemother.*, 50, 265, 2002.
63. Zhou, J. et al., Antimicrobial susceptibility and synergy studies of *Burkholderia cepacia* complex isolated from patients with cystic fibrosis, *Antimicrob. Agents Chemother.*, 51, 1085, 2007.
64. Hirai, K. et al., Purification and properties of a new β-lactamase from *Pseudomonas cepacia*, *Antimicrob. Agents Chemother.*, 17, 355, 1980.
65. Aronoff, S.C., Outer membrane permeability in *Pseudomonas cepacia*: Diminished porin content in a β-lactam-resistant mutant and in resistant cystic fibrosis isolates, *Antimicrob. Agents Chemother.*, 32, 1636, 1988.
66. Parr, Jr., T.R. et al., Role of porins in intrinsic antibiotic resistance of *Pseudomonas cepacia*, *Antimicrob. Agents Chemother.*, 31, 121, 1987.
67. Burns, J.L., Lien, D.M., and Hedin, L.A., Isolation and characterization of dihydrofolate reductase from trimethoprim-susceptible and trimethoprim-resistant *Pseudomonas cepacia*, *Antimicrob. Agents Chemother.*, 33, 1247, 1989.

68. Burns, J.L. et al., Invasion of respiratory epithelial cells by *Burkholderia (Pseudomonas) cepacia*, *Infect. Immun.*, 64, 4054, 1996.
69. Moore, R.A., and Hancock, R.E.W., Involvement of outer membrane of *Pseudomonas cepacia* in aminoglycoside and polymyxin resistance, *Antimicrob. Agents Chemother.*, 30, 923, 1986.
70. Caraher, E. et al., Comparison of antibiotic susceptibility of *Burkholderia cepacia* complex organisms when grown planktonically or as biofilm in vitro, *Eur. J. Clin. Microbiol.*, 26, 213, 2006.
71. Aaron, S.D. et al., Multiple combination bactericidal antibiotic testing for patients with cystic fibrosis infected with *Burkholderia cepacia*, *Am. J. Respir. Crit. Care Med.*, 161, 1206, 2000.
72. Grimwood, K., Kidd, T.J., and Tweed, M., Successful treatment of cepacia syndrome, *J. Cyst. Fibros.*, 8, 291, 2009.
73. Tomich, M. et al., Attenuated virulence of a *Burkholderia cepacia* type III secretion mutant in a murine model of infection, *Infect. Immun.*, 71, 1405, 2003.
74. Sajjan, S.U. et al., A type IV secretion system contributes to intracellular survival and replication of *Burkholderia cenocepacia*, *Infect. Immun.*, 76, 5447, 2008.
75. Aubert, D.F., Flannagan, R.S., and Valvano, M.A., A novel sensor kinase-response regulator hybrid controls biofilm formation and type VI secretion system activity in *Burkholderia cenocepacia*, *Infect. Immun.*, 76, 1979, 2008.
76. Silipo, A. et al., The complete structure and pro-inflammatory activity of the lipooligosaccharide of the highly epidemic and virulent gram-negative bacterium *Burkholderia cenocepacia* ET-12 (strain J2315), *Chem. Eur. J.*, 13, 3501, 2007.
77. Sajjan, S.U., and Forstner, J.F., Identification of the mucin-binding adhesin of *Pseudomonas cepacia* isolated from patients with cystic fibrosis, *Infect. Immun.*, 60, 1434, 1992.
78. Sajjan, U.S., Sylvester, F.A., and Forstner, J.F., Cable-piliated *Burkholderia cepacia* binds to cytokeratin 13 of epithelial cells, *Infect. Immun.*, 68, 1787, 2000.
79. Burns, J.L. et al., Invasion of respiratory epithelial cells by *Burkholderia (Pseudomonas) cepacia*, *Infect. Immun.*, 64, 4054, 1996.
80. Saini, L.S. et al., Intracellular survival of *Burkholderia cepacia* complex isolates in the presence of macrophage cell activation, *Microbiology*, 145, 3465, 1999.
81. Keig, P.M. et al., Differential invasion of respiratory epithelial cells by members of the *Burkholderia cepacia* complex, *Clin. Microbiol. Infect.*, 8, 47, 2002.
82. Tomich, M. et al., Role of flagella in host cell invasion by *Burkholderia cepacia*, *Infect. Immun.*, 70, 1799, 2002.
83. Lewenza, S. et al., Quorum sensing in *Burkholderia cepacia*: Identification of the LuxRI homologs CepRI, *J. Bacteriol.*, 181, 748, 1999.
84. Malott, R.J. et al., Characterization of the *cciIR* quorum-sensing system in *Burkholderia cenocepacia*, *Infect. Immun.*, 73, 4982, 2005.
85. Lutter, E. et al., Distribution of quorum-sensing genes in the *Burkholderia cepacia* complex, *Infect. Immun.*, 69, 4661, 2001.
86. Zughaier, S.M., Ryley, H.C., and Jackson, S.K., A melanin pigment purified from an epidemic strain of *Burkholderia cepacia* attenuates monocyte respiratory burst activity by scavenging superoxide anion, *Infect. Immun.*, 67, 908, 1999.
87. Lefebre, M.D., and Valvano, M.A., In vitro resistance of *Burkholderia cepacia* complex isolates to reactive oxygen species in relation to catalase and superoxide dismutase production, *Microbiology*, 147, 97, 2001.
88. Nakazawa, T., Yamada, Y., and Ishibashi, M., Characterization of hemolysin in extracellular products of *Pseudomonas cepacia*, *J. Clin. Microbiol.*, 25, 195, 1987.
89. McMenamin, J.D. et al., Misidentification of *Burkholderia cepacia* in US cystic fibrosis treatment centers: An analysis of 1,051 recent sputum isolates, *Chest*, 117, 1661, 2000.
90. Henry, D. et al., Comparison of isolation media for recovery of *Burkholderia cepacia* complex from respiratory secretions of patients with cystic fibrosis, *J. Clin. Microbiol.*, 37, 1004, 1999.
91. Shelly, D.B. et al., Utility of commercial systems for identification of *Burkholderia cepacia* complex from cystic fibrosis sputum culture, *J. Clin. Microbiol.*, 38, 3112, 2000.
92. LiPuma, J.J. et al., Person-to-person transmission of *Pseudomonas cepacia* between patients with cystic fibrosis, *Lancet*, 336, 1094, 1990.
93. Kostman, J.R. et al., Molecular epidemiology of *Pseudomonas cepacia* determined by polymerase chain reaction ribotyping, *J. Clin. Microbiol.*, 30, 2084, 1992.
94. Hogardt, M. et al., Specific and rapid detection by fluorescent in situ hybridization of bacteria in clinical samples obtained from cystic fibrosis patients, *J. Clin. Microbiol.*, 38, 818, 2000.
95. Segonds, C. et al., Differentiation of *Burkholderia* species by PCR-restriction fragment length polymorphism analysis of the 16S rRNA gene and application to cystic fibrosis isolates, *J. Clin. Microbiol.*, 37, 2201, 1999.
96. Rogers, G.B. et al., Bacterial diversity in cases of lung infection in cystic fibrosis patients: 16S ribosomal DNA (rDNA) length heterogeneity PCR and 16S rDNA terminal restriction fragment length polymorphism profiling, *J. Clin. Microbiol.*, 41, 3548, 2003.
97. Whitby, P.W. et al., Identification of members of the *Burkholderia cepacia* complex by species-specific PCR, *J. Clin. Microbiol.*, 38, 2962, 2000.
98. Coenye, T., and Vandamme, P., Diversity and significance of *Burkholderia* species occupying diverse ecological niches, *Environ. Microbiol.*, 5, 719, 2003.
99. Mahenthiralingam, E. et al., DNA-based diagnostic approaches for identification of *Burkholderia cepacia* complex, *Burkholderia vietnamiensis*, *Burkholderia multivorans*, *Burkholderia stabilis*, and *Burkholderia cepacia* genomovars I and III, *J. Clin.Microbiol.*, 38, 3165, 2000.
100. Moore, J.E. et al., Misidentification of a genomovar of *Burkholderia cepacia* by *recA* restriction fragment length polymorphism, *J. Clin. Pathol.*, 55, 309, 2002.
101. Vermis, K. et al., Evaluation of species-specific *recA*-based PCR tests for genomovar level identification within the *Burkholderia cepacia* complex, *J. Med. Microbiol.*, 51, 937, 2002.
102. Lynch, K.H., and Dennis, J.J., Development of a species-specific *fur* gene-based method for identification of the *Burkholderia cepacia* complex, *J. Clin. Microbiol.*, 46, 447, 2008.
103. Tabacchioni, S. et al., Use of the *gyrB* gene to discriminate among species of the *Burkholderia cepacia* complex, *FEMS Microbiol. Lett.*, 281, 175, 2008.
104. Plesa, M. et al., Conservation of the *opcL* gene encoding the peptidoglycan-associated outer-membrane lipoprotein among representatives of the *Burkholderia cepacia* complex, *J. Med. Microbiol.*, 53, 389, 2004.
105. Baldwin, A. et al., Multilocus sequence typing scheme that provides both species and strain differentiation for the

Burkholderia cepacia complex, *J. Clin. Microbiol.*, 43, 4665, 2005.
106. Godoy, D. et al., Multilocus sequence typing and evolutionary relationships among the causative agents of melioidosis and glanders, *Burkholderia pseudomallei* and *Burkholderia mallei*, *J. Clin. Microbiol.*, 41, 2068, 2003.
107. Hagan, W.A., Bruner, D.W., and Timoney, J.F., *Hagan and Bruner's Microbiology and Infectious Diseases of Domestic Animals*, 8th ed., Comstock Publishing Associates, Ithaca, 1988.
108. Palasatien, S. et al., Soil physicochemical properties related to the presence of *Burkholderia pseudomallei*, *Trans. R. Soc. Trop. Med. Hyg.*, 102, S5, 2008.
109. Tong, S. et al., Laboratory investigation of ecological factors influencing the environmental presence of *Burkholderia pseudomallei*, *Microbiol. Immunol.*, 40, 451, 1996.
110. Inglis, T.J.J. et al., Preliminary report on the northern Australian melioidosis environmental surveillance project, *Epidemiol. Infect.*, 132, 813, 2004.
111. Wuthiekanun, V., Smith, M.D., and White, N.J., Survival of Burkholderia *pseudomallei* in the absence of nutrients, *Trans. R. Soc. Trop. Med. Hyg.*, 89, 491, 1995.
112. Scholz, H.C. et al., Detection of the reemerging agent *Burkholderia mallei* in a recent outbreak of glanders in the United Arab Emirates by a newly developed *fliP*-based polymerase chain reaction assay, *Diagn. Microbiol. Infect. Dis.*, 54, 241, 2006.
113. Nierman, W.C. et al., Structural flexibility in the *Burkholderia mallei* genome, *Proc. Natl. Acad. Sci. USA*, 101, 14246, 2004.
114. Rotz, L.D. et al., Public health assessment of potential biological terrorism agents, *Emerg. Infect. Dis.*, 8, 225, 2002.
115. Currie, B.J., Dance, D.A.B., and Cheng, A.C., The global distribution of *Burkholderia pseudomallei* and melioidosis: An update, *Trans. R. Soc. Trop. Med. Hyg.*, 102, S1, 2008.
116. Currie, B.J. et al., Melioidosis epidemiology and risk factors from a prospective whole-population study in northern Australia, *Trop. Med. Int. Health*, 9, 1167, 2004.
117. Currie, B.J., Advances and remaining uncertainties in the epidemiology of *Burkholderia pseudomallei* and melioidosis, *Trans. R. Soc. Trop. Med. Hyg.*, 102, 225, 2008.
118. Athan, E. et al., Melioidosis in tsunami survivors, *Emerg. Infect. Dis.*, 11, 1638, 2005.
119. Centers for Disease Control and Prevention, Melioidosis [online], http://www.cdc.gov/nczved/divisions/dfbmd/diseases/melioidosis, 2010. Accessed November 4, 2010.
120. Womack, C.R., and Wells, E.B., Co-existent chronic glanders and multiple cystic osseous tuberculosis treated with streptomycin, *Am. J. Med.*, 6, 267, 1949.
121. Srinivasan, A. et al., Glanders in a military research microbiologist, *N. Engl. J. Med.*, 345, 256, 2001.
122. Ulrich, R.L. et al., Development of a polymerase chain reaction assay for the specific identification of *Burkholderia mallei* and differentiation from *Burkholderia pseudomallei* and other closely related Burkholderiaceae, *Diagn. Microbiol. Infect. Dis.*, 55, 37, 2006.
123. Centers for Disease Control and Prevention, Glanders [online], http://www.cdc.gov/nczved/divisions/dfbmd/diseases/glanders, 2010. Accessed November 4, 2010.
124. Wiersinga, W.J. et al., Melioidosis: Insights into the pathogenicity of *Burkholderia pseudomallei*, *Nat. Rev. Microbiol.*, 4, 272, 2006.
125. Viktorov, D.V. et al., High-level resistance to fluoroquinolones and cephalosporins in *Burkholderia pseudomallei* and closely related species, *Trans. R. Soc. Trop. Med. Hyg.*, 102, S103, 2008.
126. Kim, H.S. et al., Bacterial genome adaptation to niches: Divergence of the potential virulence genes in three *Burkholderia* species of different survival strategies, *BMC Genomics*, 6, 174, 2005.
127. Schell, M.A., Lipscomb, L., and DeShazer, D., Comparative genomics and an insect model rapidly identify novel virulence genes of *Burkholderia mallei*, *J. Bacteriol.*, 190, 2306, 2008.
128. Glass, M.B. et al., Comparison of four selective media for the isolation of *Burkholderia mallei* and *Burkholderia pseudomallei*, *Am. J. Trop. Med. Hyg.*, 80, 1023, 2009.
129. Deepak, R.N., Crawley, B., and Phang, E., *Burkholderia pseudomallei* identification: A comparison between the API 20NE and VITEK 2 GN systems, *Trans. R. Soc. Trop. Med. Hyg.*, 102, S42, 2008.
130. Glass, M.B., and Popovic, T., Preliminary evaluation of the API 20NE and RapID NF plus systems for rapid identification of *Burkholderia pseudomallei* and *B. mallei*, *J. Clin. Microbiol.*, 43, 479, 2005.
131. Gevers, D. et al., Re-evaluating prokaryotic species, *Nat. Rev. Microbiol.*, 3, 733, 2005.
132. Gee, J.E. et al., Use of 16S rRNA gene sequencing for rapid identification and differentiation of *Burkholderia pseudomallei* and *B. mallei*, *J. Clin. Microbiol.*, 41, 4647, 2003.
133. Bauernfeind, A. et al., Molecular procedure for rapid detection of *Burkholderia mallei* and *Burkholderia pseudomallei*, *J. Clin. Microbiol.*, 36, 2737, 1998.
134. Jiao, Z. et al., Need to differentiate lethal toxin-producing strains of *Burkholderia gladioli*, which cause severe food poisoning: Description of *B. gladioli* pathovar *cocovenenans* and an emended description of *B. gladioli*, *Microbiol. Immunol.*, 47, 915, 2003.
135. Ura, H. et al., *Burkholderia gladioli* associated with symptoms of bacterial grain rot and leaf-sheath browning of rice plants, *J. Gen. Plant Pathol.*, 72, 98, 2006.
136. Keith, L.M., Sewake, K.T., and Zee, F.T., Isolation and characterization of *Burkholderia gladioli* from orchids in Hawaii, *Plant Dis.*, 89, 1273, 2005.
137. Gitaitis, R., and Nischwitz, C., *Burkholderia*. In Gnanamanickam, S.S. (Editor), *Plant-Associated Bacteria*, pp. 645–670, Springer, Dordrecht, 2006.
138. van Veen, A.G., and Mertens, W.K., On the isolation of a toxic bacterial pigment, *Proc. Acad. Sci. Amsterdam*, 36, 666, 1933.
139. Meng, Z. et al., Studies on fermented corn flour poisoning in rural areas of China. I. Epidemiology, clinical manifestations, and pathology, *Biomed. Environ. Sci.*, 1, 101, 1988.
140. Graves, M. et al., Four additional cases of *Burkholderia gladioli* infection with microbiological correlates and review, *Clin. Infect. Dis.*, 25, 838, 1997.
141. Ross, J.P. et al., Severe *Burkholderia* (*Pseudomonas*) *gladioli* infection in chronic granulomatous disease: Report of two successfully treated cases, *Clin. Infect. Dis.*, 21, 1291, 1995.
142. Kennedy, M.P. et al., *Burkholderia gladioli*: Five year experience in a cystic fibrosis and lung transplantation center, *J. Cyst. Fibros.*, 6, 267, 2007.
143. Segonds, C. et al., Microbiological and epidemiological features of clinical respiratory isolates of *Burkholderia gladioli*, *J. Clin. Microbiol.*, 47, 1510, 2009.
144. Christenson, J.C. et al., Recovery of *Pseudomonas gladioli* from respiratory tract specimens of patients with cystic fibrosis, *J. Clin. Microbiol.*, 27, 270, 1989.
145. van Veen, A.G., The bongkrek toxins. In Mateles, R.I., and Wogan, G.N. (Editors), *Biochemistry of Some Foodborne Microbial Toxins*, pp. 43–50, MIT Press, Cambridge, 1966.

146. Mujimanto, Memeragi Bongkrek, *Topic*, 75, 24, 1975. *Cited in* Arbianto, P., Bongkrek Food Poisoning in Java, *Proc. GIAM-V*, 371, 1979.
147. Khan, S.U. et al., Empyema and bloodstream infection caused by *Burkholderia gladioli* in a patient with cystic fibrosis after lung transplantation, *Pediatr. Infect. Dis. J.*, 15, 637, 1996.
148. Shin, J.H. et al., Bacteremia due to *Burkholderia gladioli*: Case report, *Clin. Infect. Dis.*, 25, 1264, 1997.
149. Boyanton Jr., B.L. et al., *Burkholderia gladioli* osteomyelitis in association with chronic granulomatous disease: Case report and review, *Pediatr. Infect. Dis. J.*, 24, 837, 2005.
150. Lestin, F., Kraak, R., and Podbielski, A., Two cases of keratitis and corneal ulcers caused by *Burkholderia gladioli*, *J. Clin. Microbiol.*, 46, 2445, 2008.
151. Savoia, D., and Zucca, M., Clinical and environmental *Burkholderia* Strains: Biofilm production and intracellular survival, *Curr. Microbiol.*, 54, 440, 2007.
152. Latuasan, H.E., and Berends, W., On the origin of the toxicity of toxoflavin, *Biochim. Biophys. Acta*, 52, 502, 1961.
153. Buckle, K.A., and Kartadarma, E., Inhibition of bongkrek acid and toxoflavin production in tempe bongkrek containing *Pseudomonas cocovenenans*, *J. Appl. Bacteriol.*, 68, 571, 1990.
154. Henderson, P.J., and Lardy, H.A., Bongkrekic acid: An inhibitor of the adenine nucleotide translocase of mitochondria, *J. Biol. Chem.*, 245, 1319, 1970.
155. Welling, W., Cohen, J.A., and Berends, W., Disturbance of oxidative phosphorylation by an antibioticum produced by *Pseudomonas cocovenenans*, *Biochem. Pharmacol.*, 3, 122, 1960.
156. Schwerdt, G. et al., Inhibition of mitochondria prevents cell death in kidney epithelial cells by intra- and extracellular acidification, *Kidney Int.*, 63, 1725, 2003.
157. Wilsher, M.L. et al., Nosocomial acquisition of *Burkholderia gladioli* in patients with cystic fibrosis, *Am. J. Respir. Crit. Care Med.*, 155, 1436, 1997.
158. Clode, F.E. et al., Nosocomial acquisition of *Burkholderia gladioli* in patients with cystic fibrosis (multiple letters), *Am. J. Respir. Crit. Care Med*, 160, 374, 1999.
159. Lynch, K.H., and Dennis, J.J., *Burkholderia*. In Liu, D. (Editor), *Molecular Detection of Foodborne Pathogens*, pp. 331–343, CRC Press, Boca Raton, 2009.
160. Ghozzi, R. et al., Capillary electrophoresis-single-strand conformation polymorphism analysis for rapid identification of *Pseudomonas aeruginosa* and other gram-negative nonfermenting bacilli recovered from patients with cystic fibrosis, *J. Clin. Microbiol.*, 37, 3374, 1999.
161. Clode, F.E. et al., Evaluation of three oligonucleotide primer sets in PCR for the identification of *Burkholderia cepacia* and their differentiation from *Burkholderia gladioli*, *J. Clin. Pathol.*, 52, 173–176, 1999.
162. Maeda, Y. et al., Phylogenetic study and multiplex PCR-based detection of *Burkholderia plantarii*, *Burkholderia glumae* and *Burkholderia gladioli* using gyrB and rpoD sequences, *Int. J. Syst. Evol. Microbiol.*, 56, 1031, 2006.
163. Wright, R.M. et al., Improved cultural detection of *Burkholderia cepacia* from sputum in patients with cystic fibrosis, *J. Clin. Pathol.*, 54, 803, 2001.
164. Gilligan, P.H., Gage, P.A., and Bradshaw, L.M., Isolation medium for the recovery of *Pseudomonas cepacia* from respiratory secretions of patients with cystic fibrosis, *J. Clin. Microbiol.*, 22, 5, 1985.
165. New York State Department of Health—Wadsworth Center, *Burkholderia mallei/Burkholderia pseudomallei* [online], http://hawaii.gov/health/laboratories/sld-forms/card-burk.pdf, 2007. Accessed November 4, 2010.
166. Welch, D.F., Muszynski, M.J., and Pai, C.H., Selective and differential medium for recovery of *Pseudomonas cepacia* from the respiratory tracts of patients with cystic fibrosis, *J. Clin. Microbiol.*, 25, 1730, 1987.
167. Coenye, T., and LiPuma, J.J., Multilocus restriction typing: A novel tool for studying global epidemiology of *Burkholderia cepacia* complex infection in cystic fibrosis, *J. Infect. Dis.*, 185, 1454, 2002.
168. Spilker, T. et al., PCR-Based assay for differentiation of *Pseudomonas aeruginosa* from other *Pseudomonas* species recovered from cystic fibrosis patients, *J. Clin. Microbiol.*, 42, 2074, 2004.

62 Eikenella

Javier Enrique Botero, Adriana Jaramillo, and Adolfo Contreras

CONTENTS

62.1 INTRODUCTION

62.1.1 Classification, Morphology, and Biology

Eikenella corrodens, a facultative anaerobic gram-negative asaccharolytic rod, is the unique species within the genus *Eikenella*, family Neisseriaceae. *E. corrodens* was first described in 1948 as anaerobic slow-growing bacteria. It is part of the HACEK[1] group of organisms (*Hemophilus aphrophilus*, *Aggregatibacter actinomycetemcomitans*, *Cardiobacterium hominis*, *Eikenella corrodens*, and *Kingella kingae*) (Figure 62.1), which are part of the normal oropharyngeal microbiota in humans and which cause infective endocarditis, bacteremias, and abscesses.

E. corrodens cells are small, short, straight, unbranched, and slender rods, approximately 0.5 μm wide by 1.5–4 μm long.[2] A distinguishing feature of this organism is the ability to pit or corrode the agar in plated culture; however, corroding and noncorroding colonies may be present in individual strains on the same agar plate simultaneously. The colonies grow in the little grooves, and for this reason it is called "corroding bacillus." It was classified at first as *Bacteroides corrodens*. *E. corrodens* must be incubated at 35°C–37°C for 5–7 days before the colonies grow to a size sufficient for counting. When plated, the organism is dry, flat, and has a yellow-pigmented colony (Figure 62.2). The colony growth has three zones. There is a clear and moist center, a highly visible ring that appears like droplets, and an outer growth ring. The organism can produce either a musty or a bleach smell. It may be confused with *Bacteroides urealyticus*, which also exhibit pitting, but unlike *E. corrodens*, this is a urease-positive, obligate anaerobe.

E. corrodens is a nonsporeforming, nonencapsulated, nonhemolytic, and nonmotile rod. In cultures, it grows well in blood agar supplemented with hemin and menadione, under anaerobic atmosphere. The biochemical reactions used to identify it include oxidase, nitratase, and decarboxylase activity.[3] *E. corrodens* requires hemin when incubated aerobically; however, it does not need hemin in anaerobiosis. Plate growth may be stimulated within 5%–10% CO_2 environment, even though CO_2 is not essential. *E. corrodens* grows so slowly that it is sometimes hidden by other faster-growing bacteria. Adding 5 μg/mL of clindamycin to media increases recovery. The species is asaccharolytic and utilizes amino acids as energy source. The G + C content of *E. corrodens* genome is 56–58 mol%.

E. corrodens is a colonizer of the oral cavity and the upper respiratory tract,[4] and it has been implicated in etiology of mixed-oral and extraoral infections. No animal or environmental reservoirs exist for the microorganism. Common infections include periodontitis[5] and periodontal abscesses[6] and respiratory tract infections.[7] Other less frequent infections caused by *E. corrodens* are necrotizing fasciitis,[8] maternal and neonatal infections,[9] and genital abscesses.[10]

62.1.2 Clinical Features and Pathogenesis

Eikenella corrodens is a frequent habitant of the oral cavity, and it has been associated with the development of periodontitis and periodontal abscesses.[6,11,12] It can colonize the space around teeth (gingival sulcus/pocket), the tongue, and oral mucosa, and the upper respiratory tract.[13] While there is no distinct clinical feature for the infection of *E. corrodens*, lesions in which the microorganism is implicated are of an inflammatory nature. For example, periodontitis is an infectious-inflammatory disease that involves a biofilm formation on the tooth surface. With time, the infection

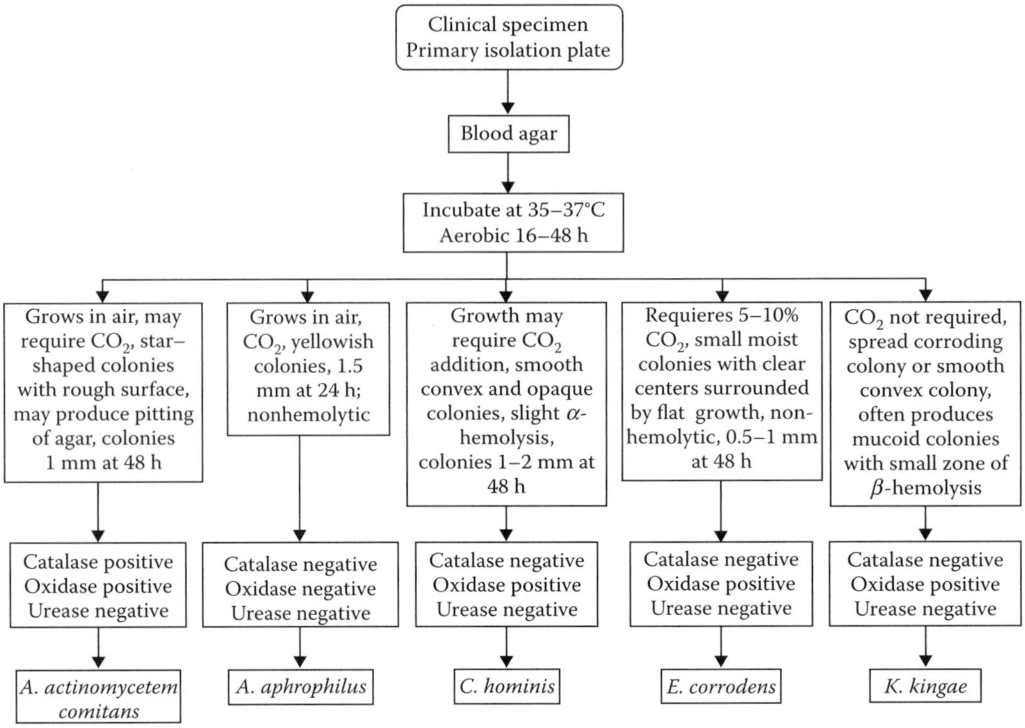

FIGURE 62.1 Flowchart for identification of HACEK microorganisms. (Reprinted with permission from Health Protection Agency (2009). Identification of Haemophilus species and the HACEK group of organisms National Standard Method BSOP ID Issue 2. http://www.hpa-standardmethods.org.uk/pdf_sops.asp)

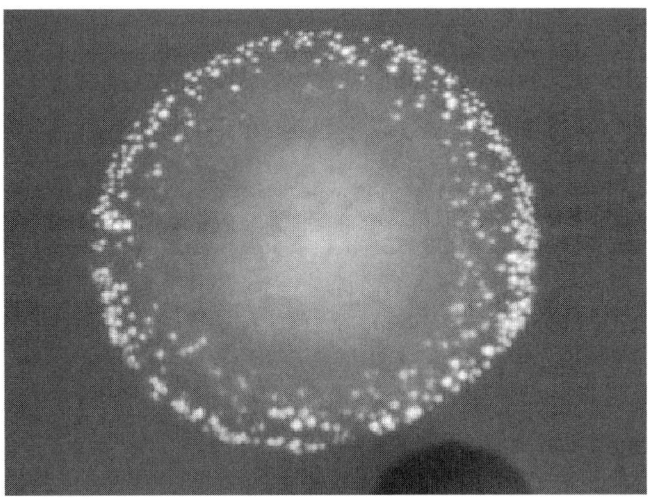

FIGURE 62.2 Aspect of a colony of *Eikenella corrodens* after 7 days anaerobic growth on blood agar supplemented with hemin/menadione. Colony is round with three distinct zones: a yellowish flat center surrounded by a circumferential clear halo encircled by spreading multiple-small microcolonies corroding the agar surface.

becomes chronic, and the inflammatory process causes the resorption of periodontal bone and degradation of gingival connective tissue, leading to dental loss. When a periodontal lesion closes and bacteria are trapped inside the pocket, a periodontal abscess develops. Signs of periodontal disease include gingival inflammation, redness, pocket formation, loss of gingival attachment, radiographic bone loss, tooth mobility, suppuration, and bleeding on probing. Pain is not frequently reported, but it can be reported in abscesses of the mouth.[14] Although *E. corrodens* is present in high numbers in periodontal lesions, it is not considered as the single causative agent, and therefore the association with other microorganisms in the periodontal biofilm is accepted in the pathogenesis of periodontitis.[15] The oral cavity, principally the periodontal pocket, represents a reservoir for *E. corrodens*, and it can be transferred to different parts of the body. The most common way of transfection is via saliva, but it can also enter the blood during bacteremia in a patient with periodontitis and colonize other sites.

It has been isolated from gynecologic infections such as Bartholin's abscess and chorioamnionitis.[16] The mode of transmission is not known, but it has been suggested that women with periodontitis are at greater risk of having the microorganisms transported to placental tissues by bacteremia. Several studies have found high levels of *E. corrodens* associated with poor oral health and complications during pregnancy, such as preeclampsia and preterm birth.[9,17,18] A recent study reported a chorioamnionitis case where oral-genital sex was a frequent routine,[19] and Jeppson et al.[20] suggested that these sexual preferences could be among the contributing factors for gynecologic infections.

Periodontitis represents an important risk factor for cardiovascular diseases. A strong association has been found for the presence of *E. corrodens* in coinfection with other periodontal pathogens in atheromatous plaques.[21] Furthermore, there are reports of patients with odontogenic infections who had suffered from brain abscess, and the microbiota found was composed of *E. corrodens* and *Prevotella* sp.[22]

In view of the fact that *E. corrodens* is a common colonizer of the surrounding tissues in the mouth, especially around teeth, the upper respiratory tract, and the gastrointestinal system, it is likely that saliva mainly contributes to the spread of the microorganism. The periodontal pocket, a deepened pathologic space around teeth, serves as an adequate environment for the growth of thousands of microorganism species, and hence allowing for the translocation to other body parts and/or transmission among individuals. Various ways of transmission/translocation have been suggested, but none has been demonstrated. Microorganisms from the periodontal pocket or in saliva could colonize wounds or exposed bone,[23] or gain access to the blood circulation during periodontal disease and establish in cardiac valves, atheromatous plaque, placenta, and fetus. It is well accepted that *Porphyromonas gingivalis* has the capability of degrading the oral epithelium and thus creating a way of infection for other bacteria,[24] which could enter the blood circulation or even during a dental procedure, chewing, flossing, or tooth-brushing.[25] However, *E. corrodens*-infected sites other than the mouth could be considered as opportunistic infections if the conditions for growth are present.

Multiple components of the bacterial surface of *E. corrodens* have been studied as potential virulence factors. First, KB cells exposed to soluble components from *E. corrodens* 1073, results in the production of IL-6, IL-8, and prostaglandin E2, indicating that all the components from the bacteria are possible virulence factors.[26] Lipopolysaccharide (LPS) from *E. corrodens* differs from gram-negative enteric microorganisms in that it lacks 2-keto-3 deoxyoctulosonic acid (KDO), β hydroxy myristic acid, and heptose. It has been indicated that there is heterogeneity in the structure and composition of LPS among strains of *E. corrodens* as determined by SDS-polyacrylamide gel electrophoresis (SDS-PAGE), but the results have shown that the pattern is mainly formed by bands of low molecular weight. Despite differences from enteric bacteria, *E. corrodens* LPS elicits potent immunologic reactions, such as the release of TNF and interleukin 1 from human monocytes as well as hemaglutination. In large quantities, *E. corrodens* LPS can induce Shwartzman reaction in experimental animals, periodontitis, and periapical bone lesions.[27–30] The outer membrane of *E. corrodens* is mainly composed of a principal outer membrane protein (POMP) accompanied by other several proteins of different molecular weights. At low concentrations, it induces the release of lysosomal enzymes from mice macrophages, while at increasing concentrations, it has a cytotoxic activity and reduction in macrophage phagocytosis.[31] In contrast to capsulated bacteria, *E. corrodens* appears to express a loosely bound exopolysaccharide layer like a slime layer. Although the antiphagocytic activity of the layer remains unknown, it has been observed that it can reduce the immune response in experimental animals.[32] The adhesion of a microorganism to epithelial surfaces is important for its pathogenicity. In addition to the hemaglutination activity exerted by *E. corrodens*, the expression of adhesions allows this microorganism to colonize gingival epithelial cells and even other bacteria and erythrocytes. These adhesins are of the type of lectin-like proteins (EcLS) and galactose-like receptors, since adhesion can be inhibited by treatment with trypsin and D-galactose and *N*-acetyl-D-galactosamine.[33,34] The expression of adhesins enhances the formation of biofilms,[35] and furthermore, the adhesion of *E. corrodens* to gingival epithelial cells results in an increase in the mRNA of ICAM-1 suggesting an important stimulant for the recruitment of neutrophils to the site.[36] Products from *E. corrodens* cause not only the stimulation of the immune system but also the inhibition of cell growth. A protein (p80) with decarboxylase activity has been identified with the capability of the removing lysine from the medium and thus, inhibiting HL60 cell growth after incubation.[37] As reviewed, all these virulence factors are responsible for the initiation of inflammatory lesions, colonization, and infection at different sites in the human body; however, further studies in this area are needed.

62.1.3 Diagnosis

Conventional Techniques. The gold standard for the detection of *E. corrodens* is by culture techniques as recommended by the Health Protection Agency,[1] but the use of molecular methods nowadays enables improved result availability and increased detection limit.

E. corrodens can be cultured from various clinical samples in different media and atmosphere conditions. A fresh clinical sample in an appropriate transport media is necessary to ensure growth of the microorganism. Goldstein et al.[38] compared growth of 10 isolates of *E. corrodens* in different media and atmospheres of incubation, and found that chocolate agar was the best medium to support growth in an aerobic or anaerobic or 5% CO_2 atmosphere. They also obtained growth of the bacterium on brucella agar supplemented with 5% blood and on Trypticase soy agar supplemented with 5% blood in both aerobic and 5% CO_2 atmospheres, and on Mueller-Hinton agar with 5% blood in a 5% CO_2 atmosphere of incubation. They did obtain growth when they tested nonsupplemented media under anaerobic conditions, but not under aerobic or 5% CO_2 conditions. Recently, a study conducted in Italy evaluated the accuracy to detect *A. actinomycetemcomitans*, *C. rectus*, *E. corrodens*, and *P. gingivalis* by conventional culture methods and multiplex PCR.[39] The findings of this study showed that both of these methods are able to accurately determine *E. corrodens*, confirming the utility of the multiplex PCR method for *E. corrodens* identification.

However, the main disadvantages of culture techniques for isolation and identification of *E. corrodens* are the variability of colony characteristics, and the fact that because the microorganism is associated with polymicrobial infections, its growth may be disturbed by faster-growing bacteria and difficult identification. Media with clindamycin may improve *E. corrodens* recovery in polymicrobial specimens. Nonetheless, isolation in primary cultures is useful to establish microbiological diagnosis and to facilitate in vitro antibiotic susceptibility testing, such as disk-agar diffusion, agar dilution, and E-test. In addition, cultured colonies could be frozen for future analysis and strain reference.

Penetration into the agar media by *E. corrodens* during translocation may account for the corroding colonial features. When growth onto chocolate agar, the characteristic three zones are less apparent, but an odor similar to hypochlorite or a musty odor is produced. In liquid media, the growth rates of individual strains may vary, but growth is improved by adding 0.1%–0.2% agar or 10 μg/mL of cholesterol to the broth medium. In thioglycolate broth supplemented with yeast extract, a band of growth develops a few centimeters below the media surface after 3 incubation days.

The biochemical characteristics of *E. corrodens* are depicted in Table 62.1. Generally, the organism is biochemically inactive, lacking fermentative characteristics and failing to produce urease, indole, or hydrogen sulfide. All isolates are oxidase-positive, and most isolates are able to reduce nitrate to nitrite. Catalase is usually negative. Tests to identify *E. corrodens* from the other HACEK organisms are shown in Table 62.1 and Figure 62.1.

Molecular Techniques. Development and application of molecular techniques vastly improve the sensibility, specificity, and speed of the laboratory diagnosis of infectious agents. These techniques can be used with many biological samples, and the extracted DNA can be preserved for long periods of time and used in the future. As opposed to culture, phenotypic viability of bacteria is not an issue, and small quantities of the microbial DNA can be detected by molecular methods. However, it is critical to maintain stringent specimen-handling procedures in order to avoid false-negatives or false-positives results.

Depending on nucleic acid integrity and the extraction method, the detection limit of molecular techniques can be as low as one bacterial cell. Fresh samples would provide a better starting material for nucleic acid extraction than poorly preserved samples or delayed processing. While end-point detection gives results that can be interpreted as positive or negative irrespective of the bacterial load, quantitative PCR (qPCR) can estimate the bacterial load based on a standard curve produced with known quantities (colony forming units CFU) of bacteria or DNA. There is considerable variation in the chemistry used for real-time PCR, such as SYBR green and Taqman probes. This should be thought out before conducting molecular testing, as results may vary between techniques.

TABLE 62.1
Biochemical Reactions of *Eikenella corrodens*

Catalase	–
Oxidase	+
Indole	–
Arginine dihydrolase	–
Nitrate to nitrite	+
Esculin hydrolysis	–
Ornithine decarboxylase	+
Acid from glucose, lactose, sucrose, xylose, maltose, and mannitol	–

62.2 METHODS

For molecular detection procedures, the following recommendations need to be considered first: (i) Clinical samples must be manipulated with care and under strict biohazard conditions. DNA isolation should be performed in a laminar flow cabinet and if DNA is not to be used immediately, it should be aliquoted and frozen at –70°C; (ii) molecular reagents must be used according to manufacturer instructions at all times and stored appropriately. Preparation and mixture of reagents should be performed in a clean room (different from the clinical sample manipulation room) in order to avoid contamination; (iii) careful pipetting is mandatory in order to avoid creating aerosols that could cause false-positives.

62.2.1 Sample Preparation

Sample Collection. It is important to emphasize that a fresh sample would yield better results, but the storing of samples is also possible. Samples can be collected from biological fluids and solid tissues. First, we describe the most common forms to obtain samples from fluids and later for solid tissues.

Oral cavity samples comprise whole saliva, pus from abscesses, and gingival crevicular fluid (GCF) from periodontal pockets. For abscesses and periodontal pockets, the site is carefully cleaned with a sterile gauze and isolated. A sterile absorbent paper point (#30) is inserted to the bottom of the pocket and kept in place for 30 s. Then, the paper point is collected in an eppendorf vial (1.5 mL) and stored at –70°C until detection. Swabs containing biological fluids can be collected in eppendorf vials (1.5 mL) by cutting the tip of the swab into the vial and storing it at –70°C. If fluids are collected in large quantities (saliva, pus, suppuration, amniotic fluid), the sample should be centrifuged (10,000 × *g* for 2 min), the pellet washed twice with 500 μL of TE buffer (10 mM Tris, 1 mM EDTA, pH 8.0), resuspended in 100 μL of TE buffer, and stored at –70°C until detection.

Another way to collect samples from biological fluids, blood, and buccal scrapes is to store the material directly onto FTA[40] (Flinders Technology Associates Rockville, MD, USA) matrix cards developed by Whatman. The cards are impregnated with a combination of buffers, free radical trap and protein denaturants that lyze cell membranes on contact, physically entraps DNA, and stabilizes and protects DNA from nuclease, oxidation, and UV damage and from both microbial and fungal degradation. One advantage of these matrix cards is that they can be stored at room temperature without any degrading of the genomic material. For detection, a Harris micropunch is used to create a small disc into a 1.5 mL vial. Then the disc is washed twice with 200 μL of FTA purification reagent (Whatman, WB120204) and rinsed once with 200 μL TE buffer. The TE buffer is removed and centrifuged at 3,000 × *g*. The remaining buffer is removed with a pipette, and the disc is dried in a heating block at 56°C for 15 min. Dry discs are transferred to 0.2-mL PCR tubes for amplification.

Solid tissue samples collected during biopsy or postmortem analysis should be handled under aseptic conditions in a laminar flow cabinet. Preservation of macromolecules (DNA, proteins) from tissues is generally achieved by immediately freezing the sample. A special storage of the sample in a tissue stabilization reagent (Allprotect Tissue Reagent, Qiagen) would be necessary, and processing of the tissue using DNA extraction kits should be used according to the manufacturer instructions (QIAmp Tissue Kit, Qiagen Hilden, Germany). For storage, stabilized tissues can be kept at –20°C or –80°C.

Culture. Since the primary ecology habitat of *E. corrodens* seems to be the subgingival biofilm in the oral cavity, oral and periodontal microbiologists cultivated and identified the organism.[41–43] However, in subjects with periodontitis, the bacteria is also frequently detected on the tongue, tonsils, cheek mucosa, and saliva[44] and it may be a normal inhabitant of the intestinal and genital tract.[29] Biological fluids can also be cultured.

After sampling, plating should be performed immediately for better recovery, but a maximum of 48 h after sampling to preserve the bacterial viability is acceptable. *E. corrodens* can be selectively cultivated on tryptic soy agar with 5% sheep blood, 5 mg/L hemin, and 0.5 mg/L menadione, containing 1 mg/L clindamycin and anaerobically incubated for 5–7 days at 35°C–37°C. By using a stereo microscope with a magnification of 50× and oblique illumination, the typical colony is described: colony is small (0.5–1 mm), round, and agar pitting or dry, flat, and radially spreading, with an irregular periphery and a moist central core. Noncorroding, translucent colonies of *E. corrodens* can also be present at the same plates with corroding colonies. Colony PCR could be performed to confirm the bacteria. A strong oxidase test and negative catalase and urease tests serve to presumptively identify *E. corrodens*.

DNA Extraction. For DNA preparation, the sample (swabs, paper points, pellets) is resuspended in 200 μL of TE buffer (10 mM Tris, 1 mM EDTA, pH 8.0) and vortexed vigorously for 5 min. Then the DNA can be isolated by conventional use of commercial kits or boiling. For DNA isolation kits, it is strongly recommended that the instructions from the manufacturer be followed.

To extract DNA by the boiling method, place the vial in boiling water (100°C) for 10 min, chill immediately on ice and centrifuge at 10,000 × *g* for 2 min. The supernatant is collected in a new tube and stored at –70°C for detection. When the vial is placed in boiling water, the top should be sealed, and care should be taken to avoid contamination. Only the part containing the sample should be submerged in the water.

62.2.2 Detection Procedures

(i) Single-Round PCR Ashimoto et al.[45] used a PCR technique to identify *E. corrodens* in clinical samples from periodontal pockets. A reaction mixture (volume 45 μL) is prepared with 5 μL 10× PCR buffer, 1.0 μM of each primer, 1.25 U Taq DNA polymerase, 0.2 mM of deoxyribonucleotides, and 1.5 mM $MgCl^2$ and the rest of the volume with PCR grade water; 5 μL of sample DNA is then added and amplified in a thermal cycler (Table 62.2). A 688 bp amplicon is observed in a 1.5% agarose gel stained with 0.5 μg/mL ethidium bromide under UV light (300 nm).

(ii) Nested PCR Rocas and Siqueira[46] developed a nested PCR to detect *E. corrodens* in endodontic clinical samples by using in the first amplification universal 16S rRNA gene primers (27F and 1.492R) and doing a second round according to Ashimoto et al.[45] A 25-μL reaction mixture containing 0.2 μM of each universal primers, 2.5 μL of 10× PCR buffer, 2 mM $MgCl^2$, 1.25 U of *Tth* DNA polymerase, and 25 μM of each deoxynucleoside triphosphate; 5 μL of sample DNA is added and amplified in a thermal cycler according to Table 62.2. For the second round, 1 μL of the universal reaction is used as template using primers and conditions by Ashimoto et al.[45] (see Table 62.2).

In comparison to single-step PCR, a nested PCR is more sensitive, and this would increase detection of low numbers of *E. corrodens* or when the clinical sample is small. However, a nested PCR should be performed carefully since cross-contamination could occur during the two amplification rounds, resulting in a false-positive. In some cases, it is advisable to run at least three serial dilutions to compare the results because inhibition of the reaction from a high concentrated DNA sample in the second round could occur and result in a negative amplification. Appropriate controls should be used at all times.

(iii) Real-Time PCR Real-time PCR allows for the quantification of any suspected pathogen within a sample by the detection of fluorescence intensity by an optical analyzer. One technique uses SYBR green as double-stranded DNA binding dye, as the product accumulates during the cycles. Nonetheless, SYBR green binds to nonspecific double-stranded DNA, thus, a well-optimized reaction is therefore essential for accurate results. After DNA preparation, a SYBR Green PCR Master Mix (Applied Biosystems) of all components except target DNA is for a total volume of 23 μL for each reaction. The concentration of the primers should be first optimized by titration to establish the best concentration without the amplification of nonspecific double-stranded DNA. The mix is then placed in a 96 well plate in triplicates and 2 μL of the extracted DNA is added.[21] The plate is sealed with the adhesive sheet and analyzed in an ABI Prism 7700 Sequence Detection System (Applied Biosystems). A plasmid containing the target gene is synthesized with a known concentration in order to perform a standard curve for the calculations. The ABI 7700 software calculates a Ct value that

TABLE 62.2
Description of Molecular Techniques for Detection of *E. corrodens*

Technique	Primers	Target Gene	Amplification Conditions	Amplicon (bp)	Reference
PCR	Forward 5′-CTA ATA CCG CAT ACG TCC TAA G-3′ Reverse 5′-CTA CTA AGC AAT CAA GTT C-3′	16S rRNA	1 cycle of 95°C for 2 min, 36 cycles of 95°C for 30 s, 60°C for 1 min, 72°C for 1 min 1 cycle of 72°C for 2 min	688	Ashimoto et al.[45]
Nested PCR	Universal (27F) forward 5′-AGA GTT TGA TCC TGG CTC AG-3′ Universal (1.492R) reverse 5′-ACG GCT ACC TTG TTA CGA CTT-3′ Second round primers as published by Ashimoto et al.[45]	16S rRNA	1 cycle of 97°C for 1 min 26 cycles of 97°C for 45 s, 55°C for 45 s, 72°C for 1 min; 1 cycle of 72°C for 4 min. Second round using primers and conditions as in Ashimoto et al.[45]	688	Rocas and Siqueira[46]
Real-time PCR-SYBR green	Forward 5′-GGG AAG AAA AGG GAA GTG CT-3′ Reverse 5′-TCT TCA GGT ACC GTC AGC AAA A-3′	16S rRNA	1 cycle of 94°C for 10 min, 40 cycles of 94°C for 15 s, 60°C for 1 min		Kozarov et al.[21]
Real-time PCR-TaqMan	Forward 5′-AGG CGA CGA AGG ACG TGT AA-3′ Reverse 5′-ATC ACC GGA TCA AAG CTC TAT TG-3′ Probe 5′-FAM-CCT GCG AAA AGC-MGB-3′	Intergenic Spacer Region (ISR)	1 cycle of 95°C for 10 min 40–45 cycles of 95°C for 15 s and 60°C for 60 s		Price et al.[47]
LAMP	FIP 5′-GGT AGG CCT TTA CCC CAC CAA CCA AGA CCT CGC GTT ATT CGA-3′ BIP 5′-ACG ATC AGT AGC GGG TCT GAG AAA TTC CCC ACT GCT GC TC-3′ F3 5′-CTA ATA CCG CAT ACG TCC TA-3′ B3 5′-GTTGCCCCCATTGTCCAA-3′ LB 5′-TGAGACACGGCCGCTCCTA-3′	16S rRNA	62°C–66°C for 60 min, 80°C for 2 min		Miyagawa et al.[48]

FIP, forward inner primer; BIP, backward inner primer; F3, forward outer primer; B3, backward outer primer; LB, two loop primer.

is proportional to the number of gene copies in the original sample and relative to the standard curve values.

Using TaqMan probe technology, the load of *E. corrodens* can be quantified in the sample in a more simple way. Using a set of specific primers coding for the Intergenic Spacer Region (ISR) located between the 16S and 23S rRNA genes and a specific probe, the increase in the fluorescence detectable above background and inversely proportional to the logarithm of the initial number of template molecules results in the quantity of the microorganism.[47] An appropriate positive control, mainly a synthesized plasmid containing the target sequence with a known concentration allows for the preparation of the standard curve. Triplicate preparations of each sample are preferred with a total volume of 20 μL containing: primers 300 nM, probe 100 nM, 10 μL 2× Taqman Universal PCR Master Mix (Applied Biosystems), 2 μL H_2O, and 2 μL of template or water (negative control). Amplification is performed in an ABI Prism 7700 Sequence Detection System (Applied Biosystems) with the conditions listed in Table 62.2.

(iv) LAMP An isothermal molecular amplification method was developed and termed LAMP (loop-mediated isothermal amplification).[49] This amplification procedure is carried on at a single temperature from 60°C to 66°C for 60 min using a set of four primers. A longer amplification time under a constant temperature allows for the amplification of large quantities of DNA target. It can also be visualized by the naked eye by adding of SYBR green I to the final mixture. The procedure can be accelerated by the addition of one or two loop primers (LF, LB). The reaction is simple and only requires one type of enzyme; no special apparatus is necessary.

The LAMP reaction is carried in a 25 μL mixture of 40 pmol of each FIB and BIP primers, 5 pmol of each F3 and B3 primers, 2 μL of template, 1 μL of *Bst* DNA polymerase

(8 U) and 12.5 mL of reaction mixture prepared in the commercial kit (Loopamp DNA amplification kit, Eiken Chemical Co. Ltd, Tochigi, Japan). To accelerate the reaction, 20 pmol of loop primer (LB) is added to the mixture and incubated at 60°C to 66°C for 60 min. The reaction is terminated by raising the temperature to 80°C for 2 min. The amplification can be detected by the naked eye by adding 1 μL of 10^{-1} diluted SYBR green I. For electrophoresis, 2 μL of the amplification product can be loaded in a 2% agarose gel stained with ethidium bromide and visualized under UV light. Although this is a simple reaction and procedure, it requires more research and validation for use with clinical samples in every laboratory.[48]

The detection of *E. corrodens* and specificity and sensitivity of any test performed is dependent on the quantity of the bacteria and quality of the DNA material. Appropriate storage and DNA processing is advisable.

62.3 CONCLUSION

Eikenella corrodens is an important microorganism that could be associated with infections in different human sites. It is frequently associated to other microorganisms that create polymicrobial infections. The ecological niche for *E. corrodens* seems to be the mouth and the periodontal pocket; however, the microorganism could disseminate from oral sites to other body sites, causing symptomatic infections. Collection of biological samples must be conducted under strict biohazard precautions.

For culture, a fresh viable sample is mandatory, whereas for molecular analysis, DNA isolation and storage is possible. Culture techniques are useful for the detection of the microorganism but are time consuming and require a fresh and viable sample. On the other hand, molecular techniques offer a simple and fast way to improve the detection of the microorganism. However, culture techniques should not be discarded; instead use them in combination with molecular methods as confirmatory and or for more detailed analysis, such as antimicrobial testing and strain analysis.

REFERENCES

1. Health Protection Agency, Identification *Haemophilus* species and the HACEK group of organisms. http://www.hpa-standardmethods.org.uk/pdf_sops.asp. *National Standard Method BSOP*, ID 12, issue 2, 2009. Accessed November 4, 2010.
2. Tanner, A., Lai, C-H., and Maiden, M., Characteristics of Oral gram-negative species. In Slots, J., and Taubman, M.A. (Editors), *Contemporary Oral Microbiology and Immunology*, p. 299, St Louis, Mosby, 1992.
3. Graevenitz, A.V., Zbinden, R., and Mutters, R., *Actinobacillus*, *Captnocytophaga*, *Eikenella*, *Kingella*, *Pasteurella* and other fastidious or ralely encountered Gram-negative rods. In Murray, P.R. (Editor), *Manual of Clinical Microbiology*, 9th ed., p. 621, ASM Press, Washington, DC, 2007.
4. Selby, T., Allaker, R.P., and Dymock, D., Characterization and expression of adjacent proline iminopeptidase and aspartase genes from *Eikenella corrodens*, *Oral. Microbiol. Immunol.*, 18, 256, 2003.
5. Slee, A.M., and Tanzer, J.M., Selective medium for isolation of *Eikenella corrodens* from periodontal lesions, *J. Clin. Microbiol.*, 8, 459, 1978.
6. Jaramillo, A. et al., Clinical and microbiological characterization of periodontal abscesses, *J. Clin. Periodontol.*, 32, 1213, 2005.
7. Wong, K.S., and Huang, Y.C., Bronchopleural cutaneous fistula due to *Eikenella corrodens*, *J. Pediatr.*, 81, 265, 2005.
8. Miller, A.T. et al., *Eikenella corrodens* causing necrotizing fasciitis after an elective inguinal hernia repair in an adult: A case report and literature review, *Am. Surg.*, 73, 876, 2007.
9. Garnier, F. et al., Maternofetal infections due to *Eikenella corrodens*, *J. Med. Microbiol.*, 58, 273, 2009.
10. Oztoprak, N. et al., *Eikenella Corrodens*, cause of a vulvar abscess in a diabetic adult, *Dis. Obstet. Gynecol.*, 63, 565, 2007.
11. Botero, J.E. et al., Occurrence of periodontopathic and superinfecting bacteria in chronic and aggressive periodontitis subjects in a Colombian population, *J. Periodontol.*, 78, 696, 2007.
12. Gajardo, M. et al., Prevalence of periodontopathic bacteria in aggressive periodontitis patients in a Chilean population, *J. Periodontol.*, 76, 289, 2005.
13. Udaka, T. et al., *Eikenella corrodens* in head and neck infections, *J. Infect.*, 54, 343, 2007.
14. Nagy, R.J., and Novak, M.J., Chronic periodontitis. In Newman, M.G., Takei, H., and Carranza, F. (Editors), *Carranza's Clinical Periodontology*, 9th ed., p. 398, W.B. Saunders, US, 2002.
15. Socransky, S.S., and Haffajee, A.D., Periodontal microbial ecology, *Periodontology*, 38, 135, 2005.
16. Kostadinov, S., and Pinar, H., Amniotic fluid infection syndrome and neonatal mortality caused by *Eikenella corrodens*, *Pediatr. Dev. Pathol.*, 8, 489, 2005.
17. Contreras, A. et al., Periodontitis is associated with pre-eclampsia in pregnant women, *J. Periodontol.*, 77, 182, 2006.
18. Herrera, J.A. et al., Periodontal disease severity is related to high levels of C-reactive protein in pre-eclampsia, *J. Hypertens.*, 25, 1459, 2007.
19. Jadhav, A.R., Belfort, M.A., and Dildy, G.A 3rd., *Eikenella corrodens* chorioamnionitis: Modes of infection? *Am. J. Obstet. Gynecol.*, 200, e4-5, 2009.
20. Jeppson, K.G., and Reimer, L.G., *Eikenella corrodens* chorioamnionitis, *Obstet. Gynecol.*, 78, 503, 1991.
21. Kozarov, E. et al., Detection of bacterial DNA in atheromatous plaques by quantitative PCR, *Microbes. Infect.*, 8, 687, 2006.
22. Ozüm, U. et al., A case of brain abscess due to *Entamoeba* species, *Eikenella corrodens* and *Prevotella* species, *Br. J. Neurosurg.*, 22, 596, 2008.
23. Støre, G., Eribe, E.R., and Olsen, I., DNA–DNA hybridization demonstrates multiple bacteria in osteoradionecrosis, *Int. J. Oral. Maxillofac. Surg.*, 34, 193, 2005.
24. Madianos, P.N. et al., *Porphyromonas gingivalis* FDC381 multiplies and persists within human oral epithelial cells in vitro, *Infect. Immun.*, 64, 660, 1996.
25. Silver, J.G., Martin, A.W., and McBride, B.C., Experimental transient bacteraemias in human subjects with varying degrees of plaque accumulation and gingival inflammation, *J. Clin. Periodontol.*, 4,92, 1977.
26. Yumoto, H. et al., Soluble products from *Eikenella corrodens* stimulate oral epithelial cells to induce inflammatory mediators, *Oral. Microbiol. Immunol.*, 16, 296, 2001.
27. Mashimo, J. et al., Fatty acid composition and Shwartzman activity of lipopolysaccharides from oral bacteria, *Microbiol. Immunol.*, 29, 395, 1985.

28. Mattison, G.D. et al., The effect of *Eikenella corrodens* endotoxin on periapical bone, *J. Endod.*, 13, 559, 2005.
29. Chen, C.K., and Wilson, M.E., *Eikenella corrodens* in human oral and nonoral infections: A review, *J. Periodontol.*, 63, 941, 1992.
30. Reddi, K. et al., Comparison of the pro-inflammatory cytokine-stimulating activity of the surface-associated proteins of periodontopathic bacteria, *J. Periodontal. Res.*, 31, 120, 1996.
31. Tufano, M.A. et al., Comparative study of some biological activities of porins extracted from various microorganisms, *Microbiologica*. 9, 431, 1986.
32. Behling, U.H., Pham, P.H., and Nowotny, A., Biological activity of the slime and endotoxin of the periodontopathic organism *Eikenella corrodens*, *Infect. Immun.*, 26, 580, 1979.
33. Yamazaki, Y., Ebisu, S., and Okada, H., *Eikenella corrodens* adherence to human buccal epithelial cells, *Infect. Immun.*, 31, 21, 1981.
34. Yamazaki, Y., Ebisu, S., and Okada, H., Partial purification of a bacterial lectin-like substance from *Eikenella corrodens*, *Infect. Immun.*, 56, 191, 1988.
35. Azakami, H. et al., Involvement of N-acetyl-D-galactosamine-specific lectin in biofilm formation by the periodontopathogenic bacterium, *Eikenella corrodens*, *Biosci. Biotechnol. Biochem.*, 70, 441, 2006.
36. Yamada, M. et al., N-acetyl-D-galactosamine specific lectin of *Eikenella corrodens* induces intercellular adhesion molecule-1 (ICAM-1) production by human oral epithelial cells, *J. Med. Microbiol.*, 51, 1080, 2002.
37. Levine, M. et al., Identification of lysine decarboxylase as a mammalian cell growth inhibitor in *Eikenella corrodens*: Possible role in periodontal disease, *Microb. Pathog.*, 30, 179, 2001.
38. Goldstein, E.J., Agyare, E.O., and Silletti, R., Comparative growth of *Eikenella corrodens* on 15 media in three atmospheres of incubation, *J. Clin. Microbiol.*, 13, 951, 1981.
39. D'Ercole, S. et al., Comparison of culture methods and multiplex PCR for the detection of periodontopathogenic bacteria in biofilm associated with severe forms of periodontitis, *N. Microbiol.*, 31, 383, 2008.
40. Rajendram, D. et al., Long-term storage and safe retrieval of DNA from microorganisms for molecular analysis using FTA matrix cards, *J. Microbiol. Methods*, 67, 582, 2006.
41. Tanner, A.C. et al., A study of the bacteria associated with advancing periodontitis in man, *J. Clin. Periodontol.*, 6, 278, 1979.
42. Savitt, E.D., and Socransky, S.S., Distribution of certain subgingival species in selected periodontal conditions, *J. Periodont. Res.*, 19, 111, 1984.
43. Mandell, R.L., A longitudinal microbiological investigation of *Actinobacillus actinomycetemcomitans* and *Eikenella corrodens* in juvenile periodontitis, *Infect. Immun.*, 45, 778, 1984.
44. Chen, C.K. et al., *Eikenella corrodens* in the human oral cavity, *J. Periodontol.*, 60, 611, 1989.
45. Ashimoto, A. et al., Polymerase chain reaction detection of 8 putative periodontal pathogens in subgingival plaque of gingivitis and advanced periodontitis lesions, *Oral. Microbiol. Immunol.*, 11, 266, 1996.
46. Rocas, I.N., and Siqueira, J.F. Jr., Culture-independent detection of *Eikenella corrodens* and *Veillonella parvula* in primary endodontic infections, *J. Endod.*, 32, 509, 2006.
47. Price, R.R. et al., Targeted profiling of oral bacteria in human saliva and in vitro biofilms with quantitative real-time PCR, *Biofouling*, 23, 203, 2007.
48. Miyagawa, J. et al., Rapid and simple detection of eight major periodontal pathogens by the loop-mediated isothermal amplification method, *FEMS. Immunol. Med. Microbiol.*, 53, 314, 2008.
49. Notomi, T. et al., Loop-mediated isothermal amplification of DNA, *Nucleic. Acids. Res.*, 28, E63, 2000.

63 *Kingella*

Dongyou Liu

CONTENTS

63.1 INTRODUCTION

63.1.1 Classification, Morphology, and Biochemical Properties

The genus *Kingella* comprises gram-negative, nonmotile, aerobic, coccobacilli belonging to the family Neisseriaceae, order Neisseriales, class Betaproteobacteria. The Neisseriaceae family is made up of 30 phylogenetically heterogeneous genera (i.e., *Alysiella*, *Andreprevotia*, *Aquaspirillum*, *Aquitalea*, *Bergeriella*, *Chitinibacter*, *Chitinilyticum*, *Chitiniphilus*, *Chromobacterium*, *Conchiformibius*, *Deefgea*, *Eikenella*, *Formivibrio*, *Gulbenkiania*, *Iodobacter*, *Kingella*, *Laribacter*, *Leeia*, *Microvirgula*, *Morococcus*, *Neisseria*, *Paludibacterium*, *Prolinoborus*, *Pseudogulbenkiania*, *Silvimonas*, *Simonsiella*, *Stenoxybacter*, *Uruburuella*, *Vitreoscilla*, and *Vogesella*), some of which (e.g., *Eikenella*, *Kingella*, *Laribacter*, and *Neisseria*) are associated with human diseases.[1] Currently, there are four species: *Kingella denitrificans*, *Kingella kingae*, *Kingella oralis*, and *Kingella potus* in the genus *Kingella*,[1] with a further species (*Kingella indologenes*) being transferred to the genus *Suttonella* as *Suttonella indologenes*.[2,3] Measuring 0.6–1 μm in diameter and 1–3 μm in length, *Kingella* cells are indole- and urease-negative, and produce acid from glucose and a limited number of other carbohydrates. The G + C content of *Kingella* DNA is 47–58 mol%.

Kingella kingae (*Kingella*, named after its discover Elizabeth O. King) was first described in 1960. The bacterium is oxidase-positive and catalase-negative. Besides producing a selective acid from glucose and maltose, *K. kingae* is hemolytic on blood agar. As a normal inhabitant of the upper respiratory tract that is at some time present in 75% of children over 6 months of age, *K. kingae* mainly causes bacteremia, arthritis, and spondylodiscitis in children.

Kingella denitrificans (formerly CDC group TM-1) differs from *K. kingae* by its inability to produce acid from maltose, in addition to other distinct phylogenetic characteristics.[4] The bacterium has been associated with endocarditis in patients with prosthetic valves, bacteremia, and empyema.[5–12]

Kingella oralis (*orale*, "oral, pertaining to the mouth") was identified from periodontal samples in 1993.[3] The bacterium is a gram-negative, aerobic or facultatively anaerobic, agar-corroding rod of 0.6–0.7 μm × 1–3 μm with rounded ends and occurs in pairs or chains. It is nonmotile by means of flagella, but its cells may form spreading colonies on agar. The bacterium is positive for oxidase, but negative for catalase, urease, indole, and esculin. *K. oralis* grows well on 5% sheep blood agar supplemented with 5 mg of hemin per liter and 0.5 pg of menadione per mL, but it does not grow on MacConkey agar. It is non–beta-hemolytic on blood agar and unable to produce acid from maltose or reduce nitrate or nitrite. The G + C content of *K. oralis* DNA is 56–58 mol%.[3] Human dental plaque appears to be the habitat of *K. orale*, and the prevalence of *K. oralis* in dental plaque ranges from 23% to 59% in different subject groups. The organism has been identified as a significant species in a few periodontitis sites.[13,14]

Kingella potus was isolated from a kinkajou bite wound in 2005.[15] Kinkajou (*Potus flavus*) is an arboreal mammal found in the rain forests of Central and South America. The bacterium is a gram-negative, nonspore-forming, nonmotile rod, that forms circular, low convex, yellow-pigmented, smooth, entire colonies of 1.5–2 mm in diameter and is friable on Columbia blood agar after 48 h of incubation at 37°C. Being nonhemolytic, it is oxidase-, casein-, and DNase-positive, but catalase-negative. In addition, it does not reduce nitrate and nitrite, hydrolyze esculin and urea, or produce indole. It is also unable to produce acid from glucose, fructose, mannose, mannitol, maltose, lactose, and sucrose. The G + C content of *K. potus* DNA is 58.4 mol%.[15]

63.1.2 Clinical Features and Pathogenesis

Kingella kingae is a slow-growing, small gram-negative bacterium that colonizes the oral cavity and the upper respiratory tract. Similar to other members of the HACEK group (including *Haemophilus influenzae*, *H. parainfluenzae*, *Cardiobacterium hominis*, *Aggregatibacter actinomycetemcomitans*, *A. aphrophilus*, *Eikenella corrodens*, and *Kingella kingae*) that are known to cause endocarditis on rare occasions, *Kingella kingae* may induce insidious clinical syndromes.[16] *K. kingae* has been most commonly reported to cause osteoarticular infections, occult bacteremia, lower respiratory tract infections and endocarditis in humans, particularly in immunocompromised hosts suffering from AIDS, lupus, or other vasculitis.[14,17–38] More often, the bacterium is responsible for septic arthritis, bacteremia, intervertebral disc infections (discitis), osteomyelitis, and central nervous system infections in children younger than 5 years old.[39–48] In one study, *K. kingae* accounted for 5%–29% of culture-positive osteoarticular infections (OAI) and for up to 48% of cases of septic arthritis in children under 2 years of age. During intervertebral disc infections, the disc space narrows after 2–4 weeks. This is followed by destruction of the adjacent cartilaginous vertebral end plates, with subsequent disc herniation into the vertebral body or vertebral body collapse.[49] Nontuberculous spondylodiscitis (NTS) in young patients may show limping, lumbar pain, back stiffness, refusal to sit or walk, increased irritability or crying, and neurological or abdominal symptoms.[50,51] Disc fibrosis or vertebral fusion represents some of the natural outcomes of NTS. The condition may be examined by magnetic resonance imaging (MRI), and a disc biopsy or aspirate can be performed for culture confirmation using blood culture bottles. In addition, *K. kingae* may cause concomitant upper respiratory tract infections, herpetic gingivostomatitis, or stomatitis in children with invasive viral infections.[49,52–54] Occasionally, *K. kingae* is associated with meningitis, corneal ulcers, endophthalmitis, pericarditis, and soft-tissue infections in immunocompetent hosts.[35,52] Apart from endocarditis, *K. kingae* infections are usually mild and easily treated with antibiotics, such as β-lactam drugs, with no long-term sequelae.

Kingella kingae is present in the pharynx of nearly 75% in children over 6 months of age and is often transmitted among daycare attendees.[52,55,56] The fading of vertically transmitted maternal immunity, and concomitant viral infections and stomatitis may damage the respiratory mucosa and allow the hematogenous spread of *K. kingae* into the bloodstream (causing occult bacteremia) and to distant organs, especially joints and bones, intervertebral disks, endocardium, and other remotes sites. The bacterium is known to produce a potent cytotoxin (RTX) that is toxic to a variety of human cell types, which may assist its colonization of the respiratory tract, invasion of the bloodstream, and damage to the joints.[57]

Similarly, *Kingella denitrificans* has been shown to cause infective endocarditis, septicemia, empyema, chorioamnionitis, and granulomatous disease on a number of occasions.[5–12]

63.1.3 Diagnosis

Phenotypic Identification. *Kingella* spp. are fastidious, gram-negative-staining microorganisms. They can be isolated from clinical samples (e.g., blood, synovial fluid exudate, joint, or disc biopsy) on chocolatized Columbia blood agar base (Oxoid) supplemented with 5% horse blood (or chocolatized trypticase-soy agar with 5% sheep blood, or blood culture bottle).[58–62] Plates are incubated at 37°C under an aerobic atmosphere with 5% CO_2. Observations on cellular and colonial morphology are based on a 2-day incubation.[63] Biochemical tests are carried out using API NH (BioMérieux), fermentation tests using phenol red broth base sugars (BBL Microbiological System), and other tests. Long-chain cellular fatty acids are extracted from cultured isolates and analyzed by gas chromatography (MIDI Sherlock, Newark, N.J.). The mol% G + C content of DNA may be determined by high-performance liquid chromatography. Antibiotic sensitivity testing may be carried out by using the E-test system (AB Biodisk, Sweden) with Diagnostic Sensitivity Test agar (Oxoid) supplemented with 5% saponin-lysed horse blood.

Molecular Identification. Broad-range PCR based on the universal 16S ribosomal RNA followed by sequencing has proven useful for identification and phylogenetic analysis of *Kingella* spp. A number of culture-negative arthritis and infant OAI due to *K. kingae* have been confirmed by this approach.[32,46,64–73]

K. kingae species-specific PCR assays targeting the RTX toxin gene, *cpn60*, and other genes have also been described.[69,74] The *K. kingae* RTX locus consists of five genes (*rtxA*, *rtxC*, *rtxD*, *rtxB*, and *tolC*), and its encoded product (RTX toxin) is a potent cytotoxin that is implicated in disease processes, including colonization of the respiratory tract, invasion of the bloodstream, and damage to the joints.[57]

Further development of real-time PCR assays renders rapid and sensitive detection of *K. kingae* possible.[69,75] For instance, Cherkaoui et al.[75] developed two qPCR assays targeting *rtxA* and *rtxB* genes for identification of *K. kingae*. Compared to seminested broad-range 16S rRNA gene PCR[69] with an analytical sensitivity of 300 CFU, the qPCR assays enable detection of 30 CFU based on serially diluted *K. kingae* genomic DNA.

For subtyping purpose, pulsed-field gel electrophoresis (PFGE) may be performed using a modified PulseNet protocol. Modifications include no proteinase K in plug preparation, *Eag*I used as the restriction endonuclease, the initial switch time of 2.2 s, and final switch time of 37.3 s. PFGE patterns are compared visually and using BioNumerics Software (Applied Maths, Kortrijk, Belgium) with the Dice coefficient.

63.2 METHODS

63.2.1 Sample Preparation

Throat swabs are brought in transport medium (liquid Stuart medium; Fisherbrand Transport Swabs, Houston, TX) and cultured within 4 h. Swabs are plated onto one plate each of trypticase-soy agar + 5% sheep blood with and without 2 μg/mL vancomycin (to suppress the growth of other oropharyngeal bacteria), columbia agar + 5% sheep blood, chocolate agar + isovitalex + bacitracin and/or McConkey agar (Becton Dickinson). Culture plates are incubated at 37°C in CO_2 and observed after 48 h of incubation. Cultures are screened visually for colonies compatible with *K kingae*. Suspect colonies are isolated, and conventional metabolic and morphologic tests are performed for identification.[76]

For DNA extraction, cotton swabs are mixed with 300 μL DNA extraction buffer I (150 mM Na_2EDTA, 225 mM NaCl; pH 8.5), and 45 μL lysozyme (50 mg/mL). After incubation at 37°C for 30 min, 9 μL 25% SDS, and 9 μL proteinase K (20 mg/mL) are added, and incubated for an additional 60 min at 37°C with agitation. Then, 150 μL of prewarmed (90°C) DNA extraction buffer II [100 mM Na_2EDTA, 400 mM Tris-HCl, 400 mM Na_2phosphate buffer pH 8.0, 5.55 M NaCl, 4% CTAB (hexadecyltrimethyl ammonium bromide); pH 8.0] and 27 μL 25% SDS are added. After incubation at 65°C for 60 min the samples are subjected to three cycles of freezing (–80°C) and thawing (65°C); 140 μL is subsequently used for purification of DNA with the QIAamp Viral RNA Mini Kit (Qiagen). DNA is eluted with 60 μL distilled water.[76]

63.2.2 Detection Procedures

(i) Real-Time PCR Protocol of Ilharreborde et al.[74] Ilharreborde et al.[74] developed a real-time PCR method for specific detection of *Kingella kingae* in joint fluid samples as well as blood samples collected before surgery. Two pairs of primers are designed from the sequence of the nonribosomal gene *cpn60*. The first primer pair Ksm1 and Ksm2 amplify a specific 175-bp fragment, and facilitate real-time PCR detection of *Kingella kingae* in combination with Kingprobe (Table 63.1). The second primer pair KingF and KingR amplify a 169-bp fragment from *Kingella* spp., including *Kingella kingae*, *Kingella* denitrificans, and *Kingella oralis* (Table 63.1).

Procedure

1. Blood sample (before surgery) is inoculated into an aerobic blood culture vial, and incubated in a continually monitored instrument (BacT/Alert 3D; BioMérieux). Joint fluid (during surgery) is immediately inoculated into aerobic blood culture bottles. A proportion of joint fluid sample is examined by Gram staining, cell count, and immediately inoculated onto Columbia blood agar (incubated in anaerobic conditions), chocolate agar (incubated in CO_2-enriched air), and brain heart broth. Blood culture bottles and the other media are incubated for, respectively, 5 and 10 days.
2. Aliquots (100–200 μL) of joint fluid are stored at −80°C for DNA extraction. 100–200 μL of plasma is separated from blood samples obtained for WBC or clotting tests and is stored at −80°C for DNA extraction.
3. DNA is extracted from specimens with the BioRobot EZ1 workstation using the EZ1 DNA tissue kit (Qiagen). Part of the DNA extract is immediately amplified, and the remainder is stored at −80°C. A negative control is included in each extraction run, consisting of all the reagents used for DNA extraction minus the sample.
4. Real-time PCR mixture (50 μL) is made up of 5 μL of DNA, 1 μL of each primer (Ksm1 and Ksm2) at 10 μM, 1 μL of probe (Kingprobe) at 10 μM, and 25 μL of IQ qPCR Mastermix (Eurogentec). A positive control consisting of DNA extracted from *K. kingae* CIP 80.16 and negative control (water) are included in each PCR run.
5. PCR amplification is carried out in an iCycler (Bio-Rad), with an initial denaturation step of 15 min at 95°C, followed by 45 cycles of 15 s at 95°C, 30 s at 55°C, and 30 s at 72°C, and a final extension step of 10 min at 72°C.

Note: The detection limit of the real-time PCR method is 200 copies of *K. kingae* genomes. The real-time PCR allows identification of *K. kingae* in culture-negative cases. The fact that *K. kingae* is detectable in joint fluid samples, but not blood, by real-time PCR enhances the etiological diagnosis of septic

TABLE 63.1
Primers and Probe for PCR Identification of *Kingella* spp. and *Kingella kingae*

Primer/Probe	Sequence (5′→3′)	Product (bp)	Specificity
Ksm1	GCAAGAAGTCGGCAAAGAG	175	*Kingella kingae*
Ksm2	GTCAAACAACAACACAAATGGG		
Kingprobe	FAM-CGCGATCGCGACAAGTAGCCAC GGTCAAGATCGCG-BHQ1		
KingF	TGTTGGCGCAAGCGATTGTTGCTG	169	*Kingella* spp.
KingR	CGCCCACTTGAGCGATTTGCTCG		

TABLE 63.2
Primers and Probes Targeting the *rtxA* and *rtxB* Genes in *K. kingae*

Gene	Primer/Probe	Primer Sequence (5′→3′)
rtxA	rtxA-F	TGCCAAAGTAAAACCAGCTGAA
	rtxA-R	AACTTACCTAATTTTGGCAAAGCAA
	rtxA P[a]	TGACAACAAACCGCTAATCATTTCTAAGGCC
rtxB	rtxB-F	CAACATAAGCCGCCAGTTGA
	rtxB-R	ACAATTAAAGCAATGGCAGTTGAG
	rtxB-P[a]	ATCCCAACGGCGCGTCATTTGT

[a] Fluorophore and quencher: 5′,6-carboxyfluorescein; 3′,6-carboxytetramethylrhodamine.

arthritis in children. Retrospective diagnosis is feasible for up to 6 days after antibiotic treatment initiation.

(ii) Real-Time PCR Protocol of Cherkaoui et al.[75] Cherkaoui et al.[75] described a real-time PCR assay for *Kingella kingae* with primers and probes from two independent genes *rtxA* and *rtxB* in the *K. kingae* RTX toxin locus (Table 63.2).

Procedure

1. *K. kingae* is grown on Columbia blood agar and chocolate agar at 37°C with 5% CO_2. *K. kingae* is stored at −80°C in skim milk with 15% glycerol for long-term storage.
2. For DNA extraction, *K. kingae* strains, osteoarticular fluids, and synovial biopsies are incubated for 1 h at 55°C with proteinase K and lysis buffer, and then purified with a MagNAPure LC instrument using the MagNAPure LC DNA Isolation Kit II (Roche Molecular Biochemicals), and eluted into 100 μL.
3. Real-time PCR mixture (25 μL) is prepared by using the qPCR TaqMan Universal PCR Master Mix with AmpErase UNG (Applied Biosystems) containing 0.5 μM of each primer, 0.25 μM of the probe, 5 μL input DNA, and nuclease-free water based on the parameters specified in the product insert. Each PCR analysis is performed in duplicate.

Note: The real-time PCR assay has a turnaround time of 4 h, including 1 h for DNA extraction/purification and 3 h for PCR amplification and analysis. A previously published seminested broad-range 16S rRNA gene PCR typically requires at least 2 days using primers BAK11w, BAK2, and BAK533r.[64,68] In addition, the real-time PCR assay exhibits a sensitivity of 30 CFU, which is tenfold more sensitive than the seminested broad-range 16S rRNA gene PCR.[64,68]

(iii) 16S rRNA Gene Sequencing Schabereiter-Gurtner et al.[76] utilized 16S rRNA gene primers for identification and phylogenetic analysis of *Kingella* spp. The primers 27f (5′-AGA GTT TGA TCC TGG CTC AG-3′) and 907r (5′-CCC CGT CAA TTC ATT TGA GTT T-3′) generate a PCR product corresponding to nucleotide positions 8–926 of the *Escherichia coli* 16S rDNA sequence.

Procedure

1. PCR mixture (25 μL) is made up of 12.5 pmol of each primer, 200 μM of each deoxyribonucleoside triphosphate, 2.5 μL of 10× PCR buffer (100 mM Tris-HCl, 15 mM $MgCl_2$, 500 mM KCl; pH 8.3), 0.5 U of *Taq* DNA polymerase (Roche Diagnostics), and 3 μL of the extracted DNA. For each run, two negative controls (distilled water instead of DNA template) are included.
2. PCR amplification is conducted in a RoboCcycler (Stratagene) with an initial denaturation at 95°C for 5 min, followed by 30 cycles of 95°C, 55°C, and 72°C for 1 min each, and a final extension at 72°C for 5 min.
3. 10 μL of PCR products is electrophoresed in 2% (wt/vol) agarose gel, stained with ethidium bromide (0.5 μg/mL) and visualized under UV light.
4. 5 μL of the PCR products is cloned with the pGEM-T Vector System (Promega). The ligation products are subsequently transformed into *E. coli* XLI-Blue, which allows blue-white screening. To screen for positive clones, clone inserts are amplified with the vector-specific primers SP6 (5′-ATT TAG GTG ACA CTA TAG AAT AC-3′) and T7 (5′-TAA TAC GAC TCA CTA TAG GG-3′).
5. For sequencing of clone inserts, fragments are amplified with primers SP6 and T7. 100 μL of PCR products is purified with a QIAquick PCR Purification Kit (Qiagen) and sequenced with a LI-COR DNA Sequencer Long Read 4200. Sequencing reactions are carried out by cycle sequencing with the SequiTherm system (EPICENTRE) with 2 pmol fluorescently labeled primers and 5 U SequiTherm thermostable DNA polymerase.
6. DNA sequences are edited and analyzed using Sequencher (Gene Codes, Ann Arbor, MI) and Vector NTI Suite (InforMax, Bethesda, Maryland), respectively, and compared with archived sequences for strain identification by conducting searches of rDNA databases or GenBank/EMBL database searches using the Fasta program. The sequences are retrieved from GenBank/EMBL and aligned with the newly determined sequence by using the SEQtools program. A phylogenetic tree is reconstructed according to the neighbor-joining method with the SEQtools and TREEVIEW programs, and the stability of the groupings is estimated by bootstrap analysis (1000 replications) using the same programs.

Note: Alternatively, a 1500 bp fragment of the 16S ribosomal RNA (rRNA) gene may be amplified using primers fD1 and rP1 for phylogenetic analysis of *Kingella* spp.

63.3 CONCLUSION

The genus *Kingella* consists of four gram-negative, nonmotile, aerobic, coccobacillary species: *Kingella denitrificans*, *Kingella kingae*, *Kingella oralis*, and Kingella *potus*. These bacteria form part of normal microflora in the upper respiratory tract, and have the capacity to take advantage of impaired host defense and injury to cause disease. In particular, *Kingella kingae* is responsible for bacteremia, arthritis, spondylodiscitis, and other diseases in children, and infective endocarditis in immunocompromised adults. Because phenotype-based procedures such as culture and biochemical tests are slow and costly, molecular techniques have been developed and increasingly applied for enhanced identification of *Kingella* spp. The ability of PCR assays to specifically detect *Kingella* in culture-negative samples highlights the diagnostic power of nucleic acid amplification technology, and provides further evidence on clinical importance of *Kingella* spp. in human diseases.

REFERENCES

1. Enright, M.C. et al., Phylogenetic relationships between some members of the genera *Neisseria*, *Acinetobacter*, *Moraxella*, and *Kingella* based on partial 16S ribosomal DNA sequence analysis, *Int. J. Syst. Bacteriol.*, 44, 387, 1994.
2. Dewhirst, F.E. et al., Transfer of *Kingella indologenes* (Snell and Lapage 1976) to the genus *Suttonella* gen. nov. as *Suttonella indologenes* comb. nov.; transfer of *Bacteroides nodosus* (Beveridge 1941) to the genus *Dichelobacter* gen. nov. as *Dichelobacter nodosus* comb. nov.; and assignment of the genera *Cardiobacterium*, *Dichelobacter*, and *Suttonella* to *Cardiobacteriaceae* fam. nov. in the gamma division of Proteobacteria on the basis of 16S rRNA sequence comparisons, *Int. J. Syst. Bacteriol.*, 40, 426, 1990.
3. Dewhirst, F.E. et al., Phylogeny of species in the family *Neisseriaceae* isolated from human dental plaque and description of *Kingella oralis* sp. nov., *Int. J. Syst. Bacteriol.*, 43, 490, 1993.
4. Hollis, D.G., Weaver, R.E., and Riley, P.S., Emended description of *Kingella denitrificans* (Snell and Lapage 1976): Correction of the maltose reaction, *J. Clin. Microbiol.*, 18, 1174, 1983.
5. Goldman, I.S. et al., Infective endocarditis due to *Kingella denitrificans*, *Ann. Intern. Med.*, 93, 152, 1980.
6. Khan, J.A. et al., *Kingella denitrificans* prosthetic endocarditis, *Am. J. Med. Sci.*, 291, 187, 1986.
7. Swann, R.A., and Holmes, B., Infective endocarditis caused by *Kingella denitrificans*, *J. Clin. Pathol.*, 37, 1384, 1984.
8. Brown, A.M. et al., Septicaemia with probable endocarditis caused by *Kingella denitrificans*, *J. Infect.*, 15, 225, 1987.
9. Molina, R. et al., Empyema caused by *Kingella denitrificans* and *Peptostreptococcus* spp. in a patient with bronchogenic carcinoma, *Eur. Respir. J.*, 1, 870, 1988.
10. Maccato, M. et al., Isolation of Kingella denitrificans from amniotic fluid in a woman with chorioamnionitis. A case report, *J. Reprod. Med.*, 36, 685, 1991.
11. Minamoto, G.Y., and Sordillo, E.M., *Kingella denitrificans* as a cause of granulomatous disease in a patient with AIDS, *Clin. Infect. Dis.*, 15, 1052, 1992.
12. Hassan, I.J., and Hayek, L., Endocarditis caused by *Kingella denitrificans*, *J. Infect.*, 27, 291, 1993.
13. Chen, C., Distribution of a newly described species, *Kingella oralis*, in the human oral cavity, *Oral Microbiol. Immunol.*, 11, 425, 1996.
14. Preza, D. et al., Bacterial profiles of root caries in elderly patients, *J. Clin. Microbiol.*, 46, 2015, 2008.
15. Lawson, P.A. et al., Description of *Kingella potus* sp. nov., an organism isolated from a wound caused by an animal bite, *J. Clin. Microbiol.*, 43, 3526, 2005.
16. Yagupsky, P., and Dagan, R., *Kingella kingae*: An emerging cause of invasive infections in young children, *Clin. Infect. Dis.*, 24, 860, 1997.
17. Ravdin, J.I. et al., Endocarditis resulting from *Kingella kingae*, presenting initially as culture-negative bacterial endocarditis, *Heart Lung*, 11, 552, 1982.
18. Verbruggen, A.M. et al., Infections caused by *Kingella kingae*: Report of four cases and review, *J. Infect.*, 13, 133, 1986.
19. De Groot, R. et al., Bone and joint infections caused by *Kingella kingae*: Six cases and a review of the literature, *Rev. Infect. Dis.*, 10, 998, 1988.
20. Kerlikowske, K., and Chambers, H.F., *Kingella kingae* endocarditis in a patient with the acquired immunodeficiency syndrome, *West. J. Med.*, 151, 558, 1989.
21. Amir, J., and Shockelford, P.G., *Kingella kingae* intervertebral disc infection, *J. Clin. Microbiol.*, 29, 1083, 1991.
22. Yagupsky, P. et al., High prevalence of *Kingella kingae* in joint fluid from children with septic arthritis revealed by the BACTEC blood culture system, *J. Clin. Microbiol.*, 30, 1278, 1992.
23. Sordillo, E.M. et al., Septicemia due to beta-lactamase-positive *Kingella kingae*, *Clin. Infect. Dis.*, 17, 818, 1993.
24. La Scola, B., Iorgulescu, I., and Bollini, G., Five cases of *Kingella kingae* skeletal infection in a French hospital, *Eur. J. Clin. Microbiol. Infect. Dis.*, 17, 512, 1998.
25. Sarda, H. et al., Infection multifocale invasive à *Kingella kingae*, *Arch. Pédiatr.*, 5, 159, 1998.
26. Luhmann, J.D., and Luhmann, S.J., Etiology of septic arthritis in children: an update for the 1990s, *Pediatr. Emerg. Care*, 15, 40, 1999.
27. Reekmans, A. et al., A rare manifestation of *Kingella kingae* infection, *Eur. J. Intern. Med.*, 11, 343, 2000.
28. Roiz, M.P., Peralta, F.G., and Arjona, R., *Kingella kingae* bacteremia in an immunocompetent adult host, *J. Clin. Microbiol.*, 35, 1916, 1997.
29. Wolak, T. et al., *Kingella* endocarditis and meningitis in a patient with SLE and associated antiphospholipid syndrome, *Lupus*, 9, 393, 2000.
30. Wolff, A.H. et al., *Kingella kingae* endocarditis: Report of a case and review of the literature, *Heart Lung*, 16, 579, 1987.
31. Wells, L., Rutter, N., and Donald, F., *Kingella kingae* endocarditis in a sixteen-month-old-child, *Pediatr. Infect. Dis. J.*, 20, 454, 2001.
32. Floyed, R.L., and Steele, R.W., Culture-negative osteomyelitis, *Pediatr. Infect. Dis. J.*, 22, 731, 2003.
33. Ross, J.J. et al., Pneumococcal septic arthritis: Review of 190 cases, *Clin. Infect. Dis.*, 36, 319, 2003.
34. Slonim, A. et al., Person-to-person transmission of *Kingella kingae* among day care center attendees, *J. Infect. Dis*,. 178, 1843, 1998.
35. Berkun, Y. et al., *Kingella kingae* endocarditis and sepsis in an infant, *Eur. J. Pediatr.*, 163, 687, 2004.
36. Yagupsky, P., *Kingella kingae*: From medical rarity to an emerging paediatric pathogen, *Lancet Infect. Dis.*, 4, 358, 2004.
37. Elyès, B. et al., *Kingella kingae* septic arthritis with endocarditis in an adult, *Joint Bone Spine*, 73, 472, 2006.

38. Dubnov-Raz, G. et al., Invasive *Kingella kingae* infections in children: Clinical and laboratory characteristics, *Pediatrics*, 122, 1305, 2008.
39. Lundy, D.W., and Kehl, D.K., Increasing prevalence of *Kingella kingae* in osteoarticular infections in young children, *J. Pediatr. Orthop.*, 18, 262, 1998.
40. Goutzmanis, J.J., Gonis, G., and Gilbert, G.L., *Kingella kingae* infection in children: ten cases and a review of the literature, *Pediatr. Infect. Dis. J.*, 10, 677, 1991.
41. Lacour, M. et al., Osteoarticular infections due to *Kingella kingae* in children, *Eur. J. Pediatr.*, 150, 612, 1991.
42. Lyon, R.M., and Evanich, J.D., Culture-negative septic arthritis in children, *J. Pediatr. Orthop.*, 19, 655, 1999.
43. Dodman, T., Robson, J., and Pincus, D., *Kingella kingae* infections in children, *J. Paediatr. Child Health*, 36, 87, 2000.
44. Moylett, E.H. et al., Importance of *Kingella kingae* as a pediatric pathogen in the United States, *Pediatr. Infect. Dis. J.*, 19, 263, 2000.
45. Costers, M. et al., Three cases of *Kingella kingae* infection in young children, *Eur. J. Pediatr.*, 162, 530, 2003.
46. Moumile, K. et al., Osteoarticular infections caused by *Kingella kingae* in children: Contribution of polymerase chain reaction to the microbiologic diagnosis, *Pediatr. Infect. Dis. J.*, 22, 837, 2003.
47. Wang, C.L. et al., Septic arthritis in children: Relationship of causative pathogens, complications, and outcome, *J. Microbiol. Immunol. Infect.*, 36, 41, 2003.
48. Claesson, B., Falsen, E., and Kjellman, B., *Kingella kingae* infections: A review and presentation of data from 10 Swedish cases, *Scand. J. Infect. Dis.*, 17, 233, 1985.
49. Amir, J., and Yagupsky, P., Invasive *Kingella kingae* infection associated with stomatitis in children, *Pediatr. Infect. Dis. J.*, 17, 757, 1998.
50. Bining, H.J. et al., *Kingella kingae* spondylodiscitis in a child, *Br. J. Radiol.*, 79, e181, 2006.
51. Yagupsky, P. et al., Outbreak of *Kingella kingae* skeletal system infections in children in daycare, *Pediatr. Infect. Dis. J.*, 25, 526, 2006.
52. Kiang, K.M. et al., Outbreak of osteomyelitis/septic arthritis caused by *Kingella kingae* among child care center attendees, *Pediatrics*, 116, e206, 2005.
53. Lebel, E. et al., *Kingella kingae* infections in children, *J. Pediatr. Orthop.*, 15, 289, 2006.
54. Yagupsky, P., and Dagan, R., *Kingella kingae* bacteremia in children, *Pediatr. Infect. Dis. J.*, 18, 1148, 1994.
55. Slonim, A., Steiner, M., and Yagupsky, P., Immune response to invasive *Kingella kingae* infections, age-related incidence of disease, and levels of antibody to outer-membrane proteins, *Clin. Infect. Dis.*, 37, 521, 2003.
56. Yagupsky, P. et al., Respiratory carriage of *Kingella kingae* among healthy children, *Pediatr. Infect. Dis. J.*, 14, 673, 1995.
57. Kehl-Fie, T.E., and St Geme III, J.W., Identification and characterization of an RTX toxin in the emerging pathogen *Kingella kingae*, *J. Bacteriol.*, 189, 430, 2007.
58. Yagupsky, P. et al., Epidemiology, etiology, and clinical features of septic arthritis in children younger than 24 months, *Arch. Pediatr. Adolesc. Med.*, 149, 537, 1995.
59. Lebjkowicz, F. et al., Recovery of *Kingella kingae* from blood and synovial fluid of two pediatric patients by using the BacT/Alert system, *J. Clin. Microbiol.*, 37, 878, 1999.
60. Yagupsky, P., Use of blood culture systems for isolation of *Kingella kingae* from synovial fluid, *J. Clin. Microbiol.*, 37, 3785, 1999.
61. Host, B. et al., Isolation of *Kingella kingae* from synovial fluids using four commercial blood culture bottles, *Eur. J. Clin. Microbiol. Infect. Dis.*, 19, 608, 2000.
62. Hughes, J.G. et al., Culture with BACTEC Peds Plus/F bottle compared with conventional methods for detection of bacteria in synovial fluid, *J. Clin. Microbiol.*, 39, 4468, 2001.
63. Gene, A. et al., Enhanced culture detection of *Kingella kingae*, a pathogen of increasing clinical importance in pediatrics, *Pediatr. Infect. Dis. J.*, 23, 886, 2004.
64. Goldenberger, D. et al., Molecular diagnosis of bacterial endocarditis by broad-range PCR amplification and direct sequencing, *J. Clin. Microbiol.*, 35, 2733, 1997.
65. Stahelin, J. et al., Polymerase chain reaction diagnosis of *Kingella kingae* arthritis in a young child, *Clin. Infect. Dis.*, 27, 1328, 1998.
66. van der Heijden, I.M. et al., Detection of bacterial DNA in serial synovial samples obtained during antibiotic treatment from patients with septic arthritis, *Arthritis Rheum.*, 42, 2198, 1999.
67. Yagupsky, P., Diagnosis of *Kingella kingae* arthritis by polymerase chain reaction analysis, *Clin. Infect. Dis.*, 29, 704, 1999.
68. Bosshard, P.P. et al., 16S rRNA gene sequencing versus the API 20 NE system and the VITEK 2 ID-GNB card for identification of nonfermenting gram-negative bacteria in the clinical laboratory, *J. Clin. Microbiol.*, 44, 1359, 2006.
69. Chometon, S. et al., Specific real-time polymerase chain reaction places *Kingella kingae* as the most common cause of osteoarticular infections in young children, *Pediatr. Infect. Dis. J.*, 26, 377, 2007.
70. Matta, M. et al., Molecular diagnosis of *Kingella kingae* pericarditis by amplification and sequencing of the 16S rRNA gene, *J. Clin. Microbiol.*, 45, 3133, 2007.
71. Rosey, A.L. et al., Development of a broad-range 16S rDNA real-time PCR for the diagnosis of septic arthritis in children, *J. Microbiol. Methods*, 68, 88, 2007.
72. Fuursted, K. et al., Broad-range PCR as a supplement to culture for detection of bacterial pathogens in patients with a clinically diagnosed spinal infection, *Scand. J. Infect. Dis.*, 40, 772, 2008.
73. Verdier, I. et al., Contribution of a broad range polymerase chain reaction to the diagnosis of osteoarticular infections caused by *Kingella kingae*: Description of twenty-four recent pediatric diagnoses, *Pediatr. Infect. Dis. J.*, 24, 692, 2005.
74. Ilharreborde, B. et al., New real-time PCR-based method for *Kingella kingae* DNA detection: application to samples collected from 89 children with acute arthritis, *J. Clin. Microbiol.*, 47, 1837, 2009.
75. Cherkaoui, A. et al., Molecular diagnosis of *Kingella kingae* osteoarticular infections by specific real-time PCR assay, *J. Med. Microbiol.*, 58, 65, 2009.
76. Schabereiter-Gurtner, C. et al., 16S rDNA-Based identification of bacteria from conjunctival swabs by PCR and DGGE fingerprinting, *Invest. Ophthalmol. Visual Sci.*, 42, 1164, 2001.

64 *Laribacter*

Patrick C.Y. Woo and Susanna K.P. Lau

CONTENTS

64.1 INTRODUCTION

64.1.1 Taxonomy

Laribacter hongkongensis is a facultative anaerobic gram-negative bacillus first described in 2001 as a novel genus and species.[1] Based on 16S ribosomal RNA (rRNA) gene sequence analysis, *L. hongkongensis* belongs to the Neisseriaceae family of β-subclass of the Proteobacteria (Figure 64.1).[2] It is most closely related to *Microvirgula aerodenitrificans*, a novel bacterium described in 1998, recovered from an upflow denitrifying filter inoculated with an activated sludge. Currently, the Neisseriaceae family contains 28 genera. Other important bacterial pathogens in the Neisseriaceae family include *Neisseria gonorrhoeae*, *Neisseria meningitidis*, *Chromobacterium violaceum*, *Eikenella corrodens*, and *Kingella kingae*. Among these, complete genome sequences are available for *N. gonorrhoeae*, *N. meningitidis*, and *C. violaceum*.

64.1.2 Microbiology and Molecular Detection

L. hongkongensis is a gram-negative, facultative anaerobic, seagull- or S-shaped, motile, bacillus. The name "*Laribacter*" represents "seagull-shaped bacterium." Bacterial cells appeared as S-shaped rods with flagellae inserted at both ends on scanning electron microscopy.[3] The bacterium grows in standard aerobic blood culture bottles and on sheep blood agar as nonhemolytic colonies at 37°C in both aerobic and anaerobic environment. Growth is also observed from 20°C to 42°C, in 0%–2% but not ≥3% NaCl, and on MacConkey agar. It is oxidase-, catalase-, urease-, and arginine-positive and reduces nitrate. It does not metabolize any sugar that has been tested in our laboratory. Cefoperaone MacConkey agar, containing 32 µg/mL of cefoperazone in MacConkey agar, is the selective medium for primary isolation of *L. hongkongensis* (Figure 64.2).[4] Its performance in quantitative recovery and in a clinical evaluation of 4741 human diarrheal stool specimens was superior to that of charcoal cefoperazone deoxycholate agar.[4]

The complete genome sequence of *L. hongkongensis* was published in 2009.[5] The genome size is 3,169,339 bp with a G + C content of 62.35% and seven rDNA operons. In line with its asaccharolytic nature, the *L. hongkongensis* genome lacks a complete set of enzymes for glycolysis, with orthologues of glucokinase, 6-phosphofructokinase, and pyruvate kinase being absent, and it also lacks a complete phosphotransferase system.[5] The genome contains complete sets of enzymes for gluconeogenesis, the pentose phosphate pathway, and the glyoxylate cycle.[5] The genome encodes a number of proteases for degradation of proteins into amino acids, amino acid transporters for transport of amino acids from the extracellular environment to the cytoplasm, and enzymes for the intracellular degradation of the amino acids.[5] These amino acids are probably used as the carbon source for *L. hongkongensis*, with the products of the reactions entering the central metabolic pathways, such as the tricarboxylate acid cycle and the gluconeogenesis pathway.[5] The genome contains enzymes for biosynthesis of all the 21 amino acids that are encoded genetically, including selenocysteine.[5] The genome contains enzymes for biosynthesis of both saturated and unsaturated fatty acids, as well as the complete set of enzymes for the β-oxidation pathway of saturated fatty acids.[5] Three terminal cytochrome oxidases, type aa_3, type cbb_3, and type bd oxidases, are present, which allow

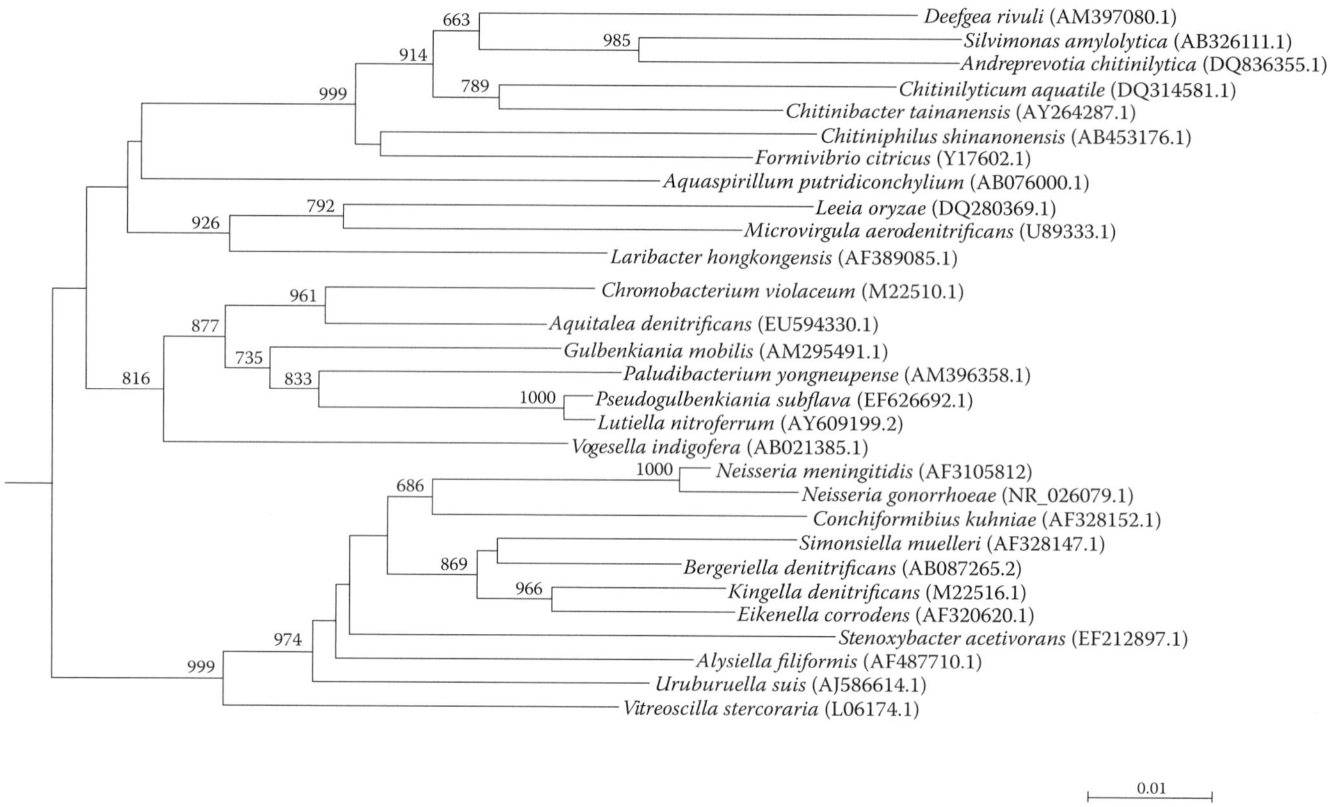

FIGURE 64.1 Phylogenetic tree showing the relationship of *L. hongkongensis* with representative species of the other 27 genera of the *Neisseriaceae* family. The tree was inferred from 16S rRNA data (1397 nucleotide positions) by the neighbor-joining method and was rooted using *Escherichia coli* (NC_000913). Numbers at nodes indicated levels of bootstrap support calculated from 1000 trees. The scale bar indicated the estimated number of substitutions for each 100 bases. Names and accession numbers are given as cited in the GenBank database.

L. hongkongensis to carry out respiration using oxygen as the electron acceptor under both aerobic conditions and conditions with reduced oxygen tension. The genome also encodes a number of reductases, including fumarate reductase, nitrate reductase, dimethylsulfoxide reductase, and tetrathionate reductase. These enable *L. hongkongensis* to carry out respiration with alternative electron acceptors to oxygen, including fumarate, nitrate, dimethylsulfoxide, and tetrathionate,

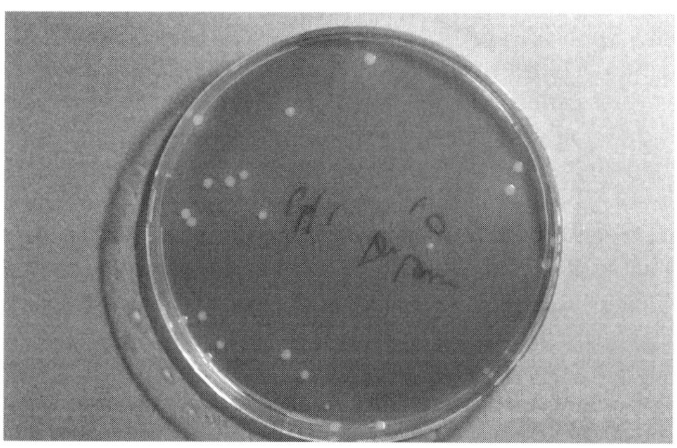

FIGURE 64.2 Colony morphology of *Laribacter hongkongensis* on cefoperazone MacConkey agar after 48 h incubation in aerobic environment.

under anaerobic conditions. The *L. hongkongensis* genome contains a complete urease gene cluster, with three structural genes (*ureA*, *ureB*, and *ureC*) and five accessory genes (*ureE*, *ureF*, *ureG*, *ureD*, and *ureI*). At least 38 genes, distributed in 6 gene clusters, are involved in the biosynthesis of flagella.

Among 21 human strains of *L. hongkongensis*, small plasmids were observed in 4 strains, and large ones in 6 strains.[6] Three of the small plasmids have been sequenced and characterized.[6–8] The smallest, 3264-bp plasmid, pHLHK19, has only one ORF that encodes a putative replication initiator protein and a predicted origin of replication (*ori*) with a DNAA box, three 18-bp direct repeats, and five pairs of inverted repeats.[6] Another plasmid, pHLHK8, consists of 8266 bp, with G + C content 53.93%.[7] The plasmid has four ORFs with all the genes encoded by the same DNA strand. The predicted amino acid sequences of the four ORFs are homologous to those of DNA invertase, type II restriction enzyme *Bsu*BI, ATPase involved in chromosome partitioning, and plasmid replication initiator protein. In addition, there is a predicted origin of replication that consists of a DNAA box and five 16-bp direct repeats.[7] The third plasmid, pHLHK26, consists of 8700 bp, with G + C content 51.3%.[8] There is a predicted origin of replication that consists of a DNAA box and five 22-bp direct repeats.[8] It has four ORFs with two genes encoded in the sense direction and the other two in the antisense direction.[8] These four ORFs encode a putative plasmid partitioning protein of the ParA family, a putative protein that contains putative ADP-ribose 1″-phosphatase activity belonging to the Appr-1-p processing enzyme family, a putative recombinase (TniR) of the resolvase/invertase family, and a putative replication protein, respectively.[8] Probably, pHLHK26 is a theta, possibly Class A, replicative plasmid, as it contains an origin of replication with AT-rich region, a number of iterons and a DNAA box, and a gene that encodes a replicative protein most homologous to those of other theta replicative plasmids, and it shares eight of the nine positions of the consensus sequence, TTAT(C/A)CA(C/A)A (TTTTCCACA in pHLHK26), in the DNAA boxes observed in other classical examples of Class A plasmids of this group.[8]

Since most microbiology laboratories are still not familiar with identification of *L. hongkongensis* by phenotypic characteristics, 16S rRNA gene sequencing remains the most accurate method for definitive identification.[9] When an oxidase-, catalase-, urease-, and arginine-positive, motile, facultative anaerobic asaccharolytic curved or S-shaped gram-negative bacillus is encountered, the possibility of *L. hongkongensis* should be considered and the isolate subjected to 16S rRNA gene PCR and sequencing. In addition, pulsed-field gel electrophoresis (PFGE) of *Spe*I digested chromosomal DNA and multilocus sequence typing (MLST) may be utilized for genotyping *L. hongkongensis*.[10–12]

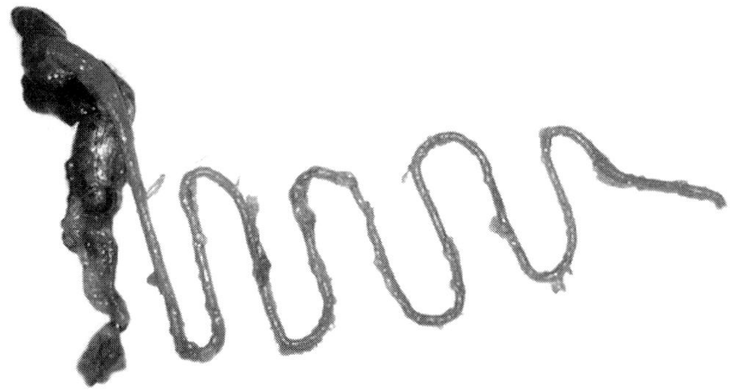

FIGURE 64.3 A grass carp and its dissected intestine.

FIGURE 64.4 Chinese tiger frogs caged in a retailed food market in Hong Kong.

64.1.3 Epidemiology and Clinical Manifestations

L. hongkongensis was first discovered in the blood culture and empyema pus of an alcoholic cirrhotic patient with empyema thoracis.[1] Subsequently, *L. hongkongensis* was documented to be associated with community acquired gastroenteritis and traveler's diarrhea, using cefoperazone MacConkey agar as the selective medium,[3,10] and has been included as a newly identified cause of diarrhea in multiple reviews.[13–15] *L. hongkongensis* gastroenteritis is present globally.[3,10,16] Cases have been found in patients who resided in or had recent travel histories to Asia, Europe, America, or Africa.[3,10,16] It is more prevalent in spring and summer in our locality (unpublished data). Most patients with *L. hongkongensis* gastroenteritis are immunocompetent. Eighty percent of patients with *L. hongkongensis* gastroenteritis had watery diarrhea, and 20% had bloody diarrhea. About 75% had abdominal pain, one-third had vomiting, and one-third had systemic symptoms, such as fever, chills, headache, myalgia, and arthralgia.[10]

L. hongkongensis has been recovered from up to 60% of the intestines of freshwater fish, mostly of the carp family, including grass carp (*Ctenoharyngodon idellus*) (Figure 64.3), bighead carp (*Aristichthys nobilis*), and mud carp (*Cirrhina molitorella*).[10,11] *L. hongkongensis* was commonly found in the gills, stomachs, and intestines in both grass carps and bighead carps, and on the skin surface of one fish, but not in other organs.[17] *L. hongkongensis* has been recovered from 80% of Chinese tiger frogs (*Hoplobatrachus chinensis*) (Figure 64.4), a widespread frog species commonly consumed in China and Southeast Asia.[18] In addition to freshwater fish and frogs, *L. hongkongensis* has also been recovered from six out of ten freshwater drinking water reservoirs in Hong Kong.[19] Similar to the data in patients, the recovery rate of *L. hongkongensis* from freshwater fish and drinking water reservoirs is also higher in spring and summer, when the environmental temperature is higher.[18,19] There was positive correlation between temperature and the isolation rates.[18,19] When *L. hongkongensis* was cultured in vitro at different temperatures, shorter lag time and higher growth rate were observed at higher temperatures, with 37°C being optimal among the tested temperatures.[18]

64.1.4 Pathogenicity and Therapy

After transmission via the oral route, the urease and enzymes of the arginine deiminase pathway, encoded by two *arc* gene clusters, enable *L. hongkongensis* to increase the local pH and survive through the hostile acidic environment of the stomach before reaching the small intestine.[5] After reaching the small intestine, efflux pumps, encoded by *acrAB-tolC*, *emrAB-tolC*, *acrAD-tolC*, *mdtABC-tolC*, and *ydgFE/mdtJI*, enable the bacterium to resist again bile.[5] *L. hongkongensis* possess a putative adhesin of the autotransporter family, type V protein secretion system of gram-negative bacteria, homologous to those of diffusely adherent *Escherichia coli* and enterotoxigenic *E. coli*.[5] To protect from the active oxygen species released from phagocytic cells, the genome of *L. hongkongensis* encodes superoxide dismutase and catalases, in line with its catalase-positive phenotype. For lipopolysaccharide synthesis, the set of genes that encode enzymes in the biosynthetic pathways of lipid A, the two Kdo units and the heptose units in *E. coli* are also present in the *L. hongkongensis* genome. Nine additional genes which encode putative enzymes for biosynthesis of the polysaccharide side chains are present in the *L. hongkongensis* genome. The *L. hongkongensis* genome contains a number of ORFs that encode putative cytotoxins, such as RTX toxin, hemolysins, and patatin-like proteins. Other enzymes of virulence potential, such as outer membrane phospholipase A, are also present. Using the available genetic manipulation systems of *L. hongkongensis*, the various putative virulence factors could be characterized.[5]

L. hongkongensis is resistant to ampicillin and to first, second, and third generation cephalosporins as a result of an *ampC/ampR* gene cluster in the *L. hongkongensis* genome, which is the first class C β-lactamase gene cloned in the β-subclass of the Proteobacteria.[20] Southern hybridization suggested that the *ampC* gene of *L. hongkongensis* was chromosomally encoded.[20] Subcloning experiments showed that the expression of the *ampC* gene of *L. hongkongensis* was regulated by its *ampR* gene, which acts as a repressor.[20] *L. hongkongensis* is susceptible to amoxicillin-clavulanate, piperacillin-tazobactam, the carbapenems, the fluoroquinolones, and aminoglycosides. Eight percent of *L. hongkongensis* strains are resistant to tetracycline, mediated by a 3566 bp gene cluster, which contains *tetR* and *tetA* genes.[21] The *tetR/tetA* gene cluster can be present on plasmids or chromosome.[21] Our recent study showed that Etest and disc diffusion appear to be reliable for evaluation of the susceptibilities of *L. hongkongensis* to imipenem, ciprofloxacin, and tetracycline. However, these methods may underestimate resistance to other β-lactams.[22]

Similar to *Salmonella* or *Campylobacter* gastroenteritis, no antibiotics would be required for immunocompetent patients with mild *L. hongkongensis* gastroenteritis. For patients with fever or severe diarrhea, a fluoroquinolone for adults and amoxicillin-clavulanate for children are recommended.[2,10]

64.2 METHODS

64.2.1 Sample Preparation

Laribacter hongkongensis is cultivated on blood agar at 37°C in ambient air for 24–48 h. A rapid procedure is used to prepare DNA from *L. hongkongensis* isolates. Namely, to 20 μL of bacterial cells suspended in distilled water, 80 μL of 50 mM NaOH is added; the mixture is incubated at 60°C for 45 min; then 6 μL of 1 M Tris/HCl (pH 7.0) is added to the mix; after diluting the mix by 100× in distilled water, 5 μL of the DNA extract is used in PCR (total volume 50 μL).[1]

64.2.2 Detection Procedures

(i) 16S rRNA Gene PCR and Sequencing Use of 16S rRNA gene PCR and sequencing provides an effective way to identify *L. hongkongensis*. Bacterial DNA extract is first amplified with 16S rRNA gene primers (LPW264 5′-GAGTTTGATCMTGGCTCAG-3′ and LPW265 5′-GNTACCTTGTTACGACTT-3′). The amplified product is then sequenced with PCR primers as well as additional primers (LPW266 5′-TCCCAGTGTGGCAGATCAT-3′, LPW267 5′-GAAAGGGAGCGGTAACGCA-3′).

Procedure

1. Prepare PCR mixture (50 μL) containing 5 μL of bacterial DNA, PCR buffer (10 mM Tris-HCl pH 8.3, 50 mM KCl, 2 mM $MgCl_2$, and 0.01% gelatin), 200 μM of each dNTPs and 1.0 U *Taq* polymerase, and 0.5 μM of each primer (LPW264 and LPW265).
2. Carry out amplification with 40 cycles of 94°C for 1 min, 55°C for 1 min, and 72°C for 2 min; and a final extension at 72°C for 10 min.
3. Sequence both strands of the PCR products should be sequenced twice using the PCR primers (LPW264 and LPW265) and additional sequencing primers (LPW266 and LPW267).

(i) Genotyping Two highly discriminative molecular typing methods are available for *L. hongkongensis*, pulsed-field gel electrophoresis of *Spe*I digested chromosomal DNA and multilocus sequence typing (MLST).[10–12] MLST of *L. hongkongensis* involves the amplification and sequencing of seven housekeeping gene loci, *rho*, *acnB*, *ftsH*, *trpE*, *ilvC*, *thiC*, and *eno*, which range from 362 to 504 bp.[12] A total of 3069 bp are sequenced, with an overall discriminatory power of 0.9861.[12] A Web site for *L. hongkongensis* MLST was set up and can be accessed at http://mlstdb.hku.hk:14206/MLST_index.html. Both methods showed that *L. hongkongensis* recovered from fish and those from human fell into separate clusters, suggesting that some clones could be more virulent than others.[11,12]

Pulsed-Field Gel Electrophoresis The bacterial suspension (9×10^8 cells/mL) of each strain of *L. hongkongensis* is mixed with an equal volume of a 2% low-melting-point agarose and poured into the block molds. The plugs for each isolate are equilibrated in *Spe*I buffer at room temperature for 3 h, and are then covered with 100 μL of *Spe*I reaction buffer containing 16 U *Spe*I. The reaction tubes are incubated at 37°C for 20 h.

TABLE 64.1
Primers for MLST Scheme in *L. hongkongensis*

	Primers		
Genes	Forward	Reverse	Amplicon Size (bp)
rho	5′-ATCGTCCTNNTGATTGACGAG-3′	5′-GATGTTGATGGCCNGGAA-3′	468
acnB	5′-GGCANGCCTTCGATTTCG-3′	5′-GCGNCCGGGNACNTACTG-3′	561
ftsH	5′-CGGTCGAAACGGCCNGGG-3′	5′-CGGNTGNGACGAAGCCAA-3′	480
trpE	5′-ACGGNGACATCATGCAGG-3′	5′-NACNGCNGTGCGGATGGC-3′	626
ilvC	5′-GCNGCCAACGGCGGCACCAA-3′	5′-AAGGCATCATNGCNCGCAG-3′	473
thiC	5′-ATCATGGCCAANTGGTGTCT-3′	5′-GCCTTGNNNAGCGCGTTGTC-3′	557
eno	5′-CGCTGCGGGGCGGAAATCTT-3′	5′-CCAGATCAACGGCCTTGAGC-3′	517

PFGE is performed using 1.5% (w/v) agarose gel in 0.5× TBE buffer (0.045 M Tris-borate, 0.001 M EDTA, pH 8.0) and the CHEF Mapper XA System (Bio-Rad Laboratories). The running temperature is 14°C. The voltage for the run is 6 V per cm. The included angle is 120°, and the ramp factor is linear. The total running time is 25 h, with switch time of 2.2–54.2 s. After PFGE, the gel is stained with ethidium bromide (1 μg/mL), and the patterns of the genomic DNA digest are visualized with a UV transilluminator. Digital images are stored electronically as TIFF files and analyzed visually and with GelCompar II (version 3.0; Applied Maths, Kortrijk, Belgium), using the Dice coefficient, and represented by UPGMA with 1% tolerance and 0.5% optimization settings.

MLST Extracted DNA from *L. hongkongensis* is used as the template for amplification of seven housekeeping genes [transcription termination factor Rho (*rho*); aconitate hydratase (*acnB*); cell division protein (*ftsH*); anthranilate synthase component I (*trpE*); ketol-acid reductoisomerase (*ilvC*); thiamin biosynthesis protein ThiC (*thiC*); enolase (*eno*)], using primers listed in Table 64.1.

The PCR mixture (100 μL) contains denatured *L. hongkongensis* DNA, PCR buffer (10 mM Tris-HCl pH 8.3 and 50 mM KCl), 2 mM $MgCl_2$, 200 μM of each deoxynucleoside triphosphates, and 2.5 U Ampli*Taq* Gold DNA polymerase (Applied Biosystems). For *rho*, *trpE*, *ilvC*, *thiC*, and *eno*, the sample is amplified in 40 cycles of 94°C for 1 min, 55°C for 1.5 min and 72°C for 2 min, and with a final extension at 72°C for 10 min in an automated thermal cycler. For *acnB* and *ftsH*, the sample is amplified using a reannealing temperature of 60°C. Twenty microliters of each amplified product is electrophoresed in 2% (w/v) agarose gel, with a molecular size marker (GeneRuler™ 50 bp DNA ladder, MBI Fermentas). Electrophoresis in Tris-borate-EDTA buffer is performed at 120 volts for 40 min. The gel is stained with ethidium bromide (0.5 μg/mL) for 15 min, rinsed, and photographed under ultraviolet light illumination.

The PCR product is gel-purified using the QIAquick PCR purification kit (QIAgen). Both strands of the PCR product are sequenced with an ABI Prism 3700 DNA Analyzer according to manufacturers' instructions (Applied Biosystems), using the PCR primers.

64.3 CONCLUSION

Laribacter hongkongensis is a recently discovered bacterium associated with community-acquired gastroenteritis and traveler's diarrhea. Freshwater fish of the carp family and Chinese tiger frogs are major reservoirs of *L. hongkongensis*. At the moment, two lines of evidence support a pathogenic role of *L. hongkongensis*. First, in all our patients, a moderate to heavy growth of the bacterium was usually found at presentation, but the bacterium could not be recovered from their stool samples after the diarrhea has subsided. Second, both PFGE and MLST showed that most patient isolates were clustered, suggesting the presence of clones that are more virulent than others. Further studies, including the setting up of animal and tissue culture models and characterization of potential virulence factors, should be performed.

The discovery of novel causes of "unexplained infectious disease syndromes" relies quite heavily on the description of novel microbes. For *L. hongkongensis*, when the bacterium was first discovered, only one strain was obtained. The description of the bacterium and the wide availability of molecular techniques, sophisticated database, and bioinformatics tools have made discovery of additional strains from other countries and rapid sharing of information possible. This led to a rapid confirmation of its association with gastroenteritis and determination of its reservoir. Recently, the completion of the genome sequence of *L. hongkongensis* has enabled us to begin to understand its potential strategies for adaptation to different environments as well as its virulence factors. As in other genome projects, more questions and hypotheses have come up than have been answered.

REFERENCES

1. Yuen, K.Y. et al., *Laribacter hongkongensis* gen. nov., sp. nov., a novel gram-negative bacterium isolated from a cirrhotic patient with bacteremia and empyema, *J. Clin. Microbiol.*, 39, 4227, 2001.
2. Woo, P.C.Y. et al., Current status and future directions of *Laribacter hongkongensis*, a novel bacterium associated with gastroenteritis and traveller's diarrhoea, *Curr. Opin. Infect. Dis.*, 18, 413, 2005.

3. Woo, P.C.Y. et al., *Laribacter hongkongensis*: A potential cause of infectious diarrhea, *Diagn. Microbiol. Infect. Dis.*, 47, 551, 2003.
4. Lau, S.K.P. et al., Use of cefoperazone MacConkey agar for selective isolation of *Laribacter hongkongensis*, *J. Clin. Microbiol.*, 41, 4839, 2003.
5. Woo, P.C.Y. et al., The complete genome and proteome of *Laribacter hongkongensis* reveal potential mechanisms for adaptations to different temperatures and habitats, *PLoS Genet.*, 5, e1000416, 2009.
6. Woo, P.C.Y. et al., Plasmid profile and construction of a small shuttle vector in *Laribacter hongkongensis*, *Biotechnol. Lett.*, 29, 1575, 2007.
7. Woo, P.C.Y. et al., Construction of an inducible expression shuttle vector for *Laribacter hongkongensis*, a novel bacterium associated with gastroenteritis, *FEMS Microbiol. Lett.*, 252, 57, 2005.
8. Woo, P.C.Y. et al., Characterization of a novel cryptic plasmid, pHLHK26, in *Laribacter hongkongensis*, *New Microbiol.*, 30, 139, 2007.
9. Woo, P.C.Y. et al., Then and now: Use of 16S rDNA gene sequencing for bacterial identification and discovery of novel bacteria in clinical microbiology laboratories, *Clin. Microbiol. Infect.*, 14, 908, 2008.
10. Woo, P.C.Y. et al., Association of *Laribacter hongkongensis* in community-acquired gastroenteritis with travel and eating fish: A multicentre case-control study, *Lancet*, 363, 1941, 2004.
11. Teng, J.L.L. et al., Ecoepidemiology of *Laribacter hongkongensis*, a novel bacterium associated with gastroenteritis, *J. Clin. Microbiol.*, 43, 919, 2005.
12. Woo, P.C.Y. et al., Development of a multi-locus sequence typing scheme for *Laribacter hongkongensis*, a novel bacterium associated with freshwater fish-borne gastroenteritis and traveler's diarrhea, *BMC Microbiol.*, 9, 21, 2009.
13. Schlenker, C., and Surawicz, C.M., Emerging infections of the gastrointestinal tract, *Best Pract. Res. Clin. Gastroenterol.*, 23, 89, 2009.
14. Marcos, L.A., and DuPont, H.L., Advances in defining etiology and new therapeutic approaches in acute diarrhea, *J. Infect.*, 55, 385, 2007.
15. Murray, P.R. et al., *Manual of Clinical Microbiology*, 9th ed., American Society for Microbiology, Washington, DC, 2007.
16. Ni, X.P. et al., *Laribacter hongkongensis* isolated from a patient with community-acquired gastroenteritis in Hangzhou City, *J. Clin. Microbiol.*, 45, 255, 2007.
17. Lau, S.K.P. et al., Seasonal and tissue distribution of *Laribacter hongkongensis*, a novel bacterium associated with gastroenteritis, in retail freshwater fish in Hong Kong, *Int. J. Food Microbiol.*, 113, 62, 2007.
18. Lau, S.K.P. et al., Isolation of *Laribacter hongkongensis*, a novel bacterium associated with gastroenteritis, from Chinese tiger frog, *Int. J. Food Microbiol.*, 129, 78, 2009.
19. Lau, S.K.P. et al., Isolation of *Laribacter hongkongensis*, a novel bacterium associated with gastroenteritis, from drinking water reservoirs in Hong Kong, *J. Appl. Microbiol.*, 103, 507, 2007.
20. Lau, S.K.P. et al., Cloning and characterization of a chromosomal class C β-lactamase and its regulatory gene in *Laribacter hongkongensis*, *Antimicrob. Agents Chemother.*, 49, 1957, 2005.
21. Lau, S.K.P. et al., Distribution and molecular characterization of tetracycline resistance in *Laribacter hongkongensis*, *J. Antimicrob. Chemother.*, 61, 488, 2008.
22. Lau, S.K.P. et al., Susceptibility patterns of clinical and fish isolates of *Laribacter hongkongensis*: comparison of the E-test, disc diffusion and broth microdilution methods, *J. Antimicrob. Chemother.*, 63, 704, 2009.

65 *Neisseria*

Paola Stefanelli

CONTENTS

65.1 INTRODUCTION

65.1.1 Classification, Morphology, and Biology

The classic discovery that various diseases are caused by microorganisms occurred in the twilight of the nineteenth century and symbolized a milestone in the history of infectious diseases. In 1879, Louis Pasteur demonstrated the link between microbiological organisms and puerperal sepsis.

The family Neisseriaceae covers gram-negative aerobic bacteria that are subdivided into fourteen genera,[1] including *Neisseria*, *Chromobacterium*, *Kingella*, and *Aquaspirillum*. The genus *Neisseria*, of 0.6–1.0 μm in diameter and usually seen in pairs, contains two important human pathogens, *Neisseria meningitidis* and *Neisseria gonorrhoeae*. Similar to other members of the genus, these organisms are gram-negative, aerobic, and nonmotile; harbor oxidase enzyme; and produce catalase. Their growth is stimulated by CO_2 and humidity, and they produce acid from carbohydrates oxidatively. *N. meningitidis*, the meningococcus, is a major cause of meningitis and septicemia worldwide, while *N. gonorrhoeae*, the gonococcus, is responsible for one of the most widespread sexually transmitted diseases.

Neisseria meningitidis. In 1887, an equally novel discovery was made by the Austrian pathologist Anton Weichselbaum, who clarified the etiological nature of meningococcal disease by showing, for the first time, that there was a connection between *N. meningitidis* (then known as *Diplococcus intracellularis meningitidis*) and "epidemic cerebrospinal meningitis." Reports of illness resembling meningococcal disease date back to the sixteenth century. In Europe, meningococcal disease was first reported by Vieusseux in the city of Geneva in 1805. Vieusseux provided a striking clinical picture of meningococcal sepsis and baptized the malady "*fièvre cérébrale maligne non-contagieuse*" (noncontagious malignant cerebral fever). Interestingly, during the epidemic of cerebrospinal fever in Geneva, Vieusseux hypothesized that meningococcal infection, rather than being transmitted through direct person-to-person contact, was spread by "bad air." Across the Atlantic, the first reported cases of this "singular and very mortal disease" occurred in Medfield, Massachusetts, USA.

N. meningitidis is divided into 13 serogroups defined by specific polysaccharides designated A, B, C, H, I, K, L, M, X, Y, Z, 29E, and W135, and most infections are caused by organisms belonging to serogroups A, B, C, Y, and W-135. Few case isolates from normally sterile sites [cerebrospinal fluid (CSF), blood, joint fluids] are nonserogroupable using current reagents, although meningococci isolated from nonsterile sites associated with carriage often do not express capsule and are, therefore, nonserogroupable. In some cases, down-regulation of capsule production can occur. Determination of the capsule serogroup in meningococcal invasive disease is important for case contact management, and especially in outbreaks where the use of a serogroup-specific vaccine may prevent further cases.

Meningococci are further classified into 20 serotypes (on the basis of class 2 or 3 OMP antigens), 10 subtypes (identifying class 1 OMP antigens), and 13 immunotypes (on the basis of lipooligosaccharide antigens). Molecular typing techniques for *N. meningitidis* are used for epidemiological purposes to investigate outbreaks and the spread of organisms, and to examine the population genetic structure of the organism to better understand its variation and evolution. Many investigators have employed molecular typing methods and shown that meningococcal disease is associated with a variety of different epidemiological patterns. The choice of a typing method depends on the epidemiological questions to be answered and on the population genetics of the organism under investigation.

Meningococci are genetically heterogeneous organisms and are usually commensal bacteria present in nasopharynx in humans.[2] Only a minority of the nasopharyngeal isolates cause invasive disease. Meningococci associated with invasive disease elaborate a capsule, which provides protection from desiccation during transmission and helps to evade host immune mechanisms. Adhesins, such as pili, one of the main adhesins that may target the CD46 receptor, the membrane cofactor protein, and the opacity-associated proteins, Opa and Opc, bind to CD66 and heparan sulfate proteoglycan receptors, respectively. Specific nutrient-acquisition factors, especially mechanisms for acquiring iron from human lactoferrin, transferrin, and hemoglobin, enhance the pathogenic potential of meningococci. A major factor in the virulence of the organism is the release of outer membrane vesicles that consist of lipooligosaccharide (endotoxin), outer membrane proteins, phospholipids, and capsular polysaccharides. Lipid A molecules of lipooligosaccharide (LOS) act as endotoxins, and their effects are due to interaction with innate immune receptors. The endotoxin of *N. meningitidis* is structurally distinct from the lipooligosaccharide of enteric gram-negative bacteria. Finally, the expression of pili, class 5 OMPs, capsule, and LOS is highly variable and subject to phase switching (on/off) and antigenic variation. These mechanisms can be used by the bacterium to circumvent host immunity. Special conditions are required to cause an invasive disease depending not only on bacterial properties, but also on environmental and social conditions, preceding or concomitant viral infections, and the immune status of the patient.[3] Noncapsulate meningococci are avirulent, and serogroups A, B, C are more invasive than other serogroups.

In sepsis, skin hemorrhages are the hallmark of invasive meningococcal disease. Microscopically, these lesions are characterized by endothelial damage and hemorrhages and microthrombi in small vessels, consistent with a generalized Sanarelli-Shwartzman reaction. Adrenal hemorrhages, diagnosed postmortem as Waterhouse-Friderichsen syndrome, may lead to transitory adrenal insufficiency.

In meningitis, the bacterium binds to the cerebral capillary endothelium that undergo fimbrial phase variation in order to cross the blood brain barrier.[4] Bacteria may also gain access to subarachnoid space within host phagocytic cells. Once in the subarachnoid space, where the principal humoral and cellular host defense mechanisms are absent, there is uncontrolled proliferation of meningococci. The evolving endotoxin liberation elicits compartmentalized (i.e., confined to the subarachnoid space) activation of proinflammatory cytokines. The subsequent release of neutrophil products contributes to the development of clinically overt meningitis.

Meningococcal disease occurs worldwide in both endemic and epidemic forms. Humans are the only natural reservoir of meningococcus. As many as 10% of adolescents and adults are asymptomatic transient carriers of *N. meningitidis*, most strains of which are not pathogenic. The primary mode of transmission is by respiratory droplet spread or by direct contact.

Meningococcal disease occurs throughout the year. Epidemic rates of meningococcal disease varies from <1 to 3/100,000 in many developed nations to 10–25/100,000 in some developing countries. The notification rate reported in Europe varies widely between countries, ranging from 0.25 to 4.4 per 100,000. These numbers reflect real differences in incidence, but also differences between surveillance systems and are certainly due in part to the variation in the methods used for confirming suspected cases (data available from: http://www.euibis.org/documents/2006_meningo.pdf). For example, in 2006, Malta and Ireland reported the highest notification rates, with 4.4 per 100,000 and 4.1 per 100,000, respectively. Children under five years old followed by the 15- to 24-year-olds were the age groups in which the disease was more frequent. In temperate regions the number of cases increased in the late winter and early spring. The highest notification rate was observed in the first 3 months of the year. Serogroups B and C together account for a large majority of cases in Europe and the Americas in children less than 5 years old and young adults (15–24 years old).

In studies of household contacts, only 3%–4% had secondary cases. This difference in attack rates reflects the difference in pathogenic properties of prevalent *N. meningitidis* strains and differences in socioeconomic and environmental conditions. The largest and most frequently recurring outbreaks have been in the semiarid area of sub-Saharan Africa. In the so-called "meningitis belt," a region of savannah that extends from Ethiopia in the east to Senegal in the west, serogroup A meningococcal disease has posed a recurrent threat to public health for at least 100 years.

Epidemiological studies by modern molecular methods have disclosed a complex picture of a few pathogenic meningococcal clones spreading worldwide. It appears that the occurrence of invasive meningococcal disease is not solely determined by the introduction of a new, virulent bacterial strain but also by other factors that enhance transmission.

Neisseria gonorrhoeae. Citations to the contagious nature of gonococcal infection dates back to biblical times (Lev. 15:1-15–19) and in ancient Chinese writings,[5] making gonorrhea one of the oldest recorded human diseases. Hippocrates referred to acute gonorrhea as "strangury" obtained from the "pleasures of Venus" in the fourth and the fifth centuries BC. The first usage of the term "gonorrhoea," by Galen in the second century, implied a "flow of seed." For centuries thereafter, gonorrhea and syphilis were confused, resulting from the fact that the two diseases were often present together in infected individuals. Paracelsus (1530) thought that gonorrhea was an early symptom of syphilis. Neisser described the causative agent of gonorrhea in 1879, and in 1882 Leistikow and Löffler finally cultivated the gonococcus. Since the discovery of gonorrhea in 1879 by Albert Neisser, this disease has become the second most common sexually transmitted bacterial infection. Neisser learned and adopted the techniques of his peers to visualize bacteria using the newest microscopic methods. By staining smears with methylene blue and using oil-immersion techniques, he could observe bacteria using ×1000 magnification. Originally, the organisms he saw were given the name "micrococcus"; however, it was a colleague, Paul Ehrlich, who first coined the name gonococcus. Since then, the *Neisseria* group of organisms, so named after Neisser himself, has been added to the list of known bacteria and comprehensively characterized. Trends in the prevalence of gonococci infection are apparent. Rises in infection have been noticed following wars, when as a result of their promiscuity while away from home, returning soldiers passed infections to their wives. During the 1950s, the advent of penicillin saw a decrease in infections; however, the rapid emergence of resistant strains in the 1960s and 1970s saw the infection rate increase once again. The 1980s brought the new danger of human immunodeficiency virus (HIV).

Numerous attempts have been made to subclassify *N. gonorrhoeae*. Classification based on the nutritional requirements, that is, auxotyping, or on the reactivity with a defined set of monoclonal antibodies (MAbs) against epitopes on protein I, the major outer membrane protein, that is, serovar determination, are the most widely used systems. A panel of 12 monoclonal antibodies to gonococcal outer membrane protein IA (PrIA) and IB (PrIB) are available. A-reactivity corresponds to the polyclonally defined serogroup WI and B-reactivity, to serogroup WII and WIII.[6,7]

Phenotypic characterization of *N. gonorrhoeae* by auxotype-serovar (A/S) classification is currently the most widely established method for differentiation of gonococci. However, the A/S classification system has some major limitations, such as the nonreproducibility of serotyping, especially when different batches of monoclonal antibody typing reagents are used, and the inability to differentiate strains present within major serovars. The tools now exist for a better understanding of gonococcal epidemiology. To overcome some of the limitations of the A/S classification system, newer molecular typing approaches such as restriction endonuclease analysis (REA), pulsed-field gel electrophoresis (PFGE), restriction fragment length polymorphisms of rRNA genes, multilocus enzyme electrophoretic analysis, *opa*-typing, direct sequencing of PIB gene, and several PCR-based forms of analysis have been developed. Most of the molecular typing methods have been shown to have greater discriminatory power than serotyping.

N. gonorrhoeae is an exclusive human pathogen highly adapted to mucosa tissue and transmitted by sexual contact. One of the most striking features of *N. gonorrhoeae* is the marked degree of variation between isolates. The source of this variation lies not only in spontaneous mutation inside the genes but also in the genetic exchange of DNA with other *Neisseria* species.

Strains of *N. gonorrhoeae* are naturally competent for DNA uptake and can also exchange genetic material by conjugation. The high degree of random recombination between chromosomal genetic loci, probably mediated by transformation, has to be frequent in nature, and this makes *N. gonorrhoeae* an example of pathogenesis with a fully sexual, that is, panmictic or nonclonal, population structure.[8–10] This causes genotypic and phenotypic variability, which is important for evasion or adaptation to the immune response of the host, and for the development of antibiotic resistance spread.

Although the urethra and the uterine cervix serve as the initial sites for gonococcal infection in men and women, respectively, conjunctiva can also be infected, and infection can damage the cornea, leading to blindness.[11,12] In addition, susceptibility to chronic complications associated with *N. gonorrhoeae* infection particularly affects women, and, as a consequence, their fertility and their newborns, and increases the risk of acquisition of HIV in both sexes. In particular, gonococcal infection increases fivefold the risk of HIV-1 transmission.[13] Rarely, *N. gonorrhoeae* invades the bloodstream, causing disseminated gonococcal infection (DGI).

To persist in the male urethra and avoid being washed away by the passage of urine, *N. gonorrhoeae* has developed effective methods to adhere to epithelial cells, and can also be found intracellularly.[14] During gonococcus infection, PMNs (polymorphonuclear cells) migrate into the urethral lumen and attach to the cells and extracellular matrix present in the surrounding environment. During the extravasion from the blood and attachment to surfaces, PMNs are activated and exhibit increased phagocytosis activity. Most gonococci taken up by the PMNs are killed, but up to 2% can survive because of their resistance to oxidative killing. The interaction of *N. gonorrhoeae* with PMNs is an important aspect of gonococcal disease that remains unclear. Several studies have suggested *N. gonorrhoeae* survives within PMN[15–18] and may replicate in PMN phagosomes with time. In fact, prolonged survival of even a low percentage of internalized bacteria might play a crucial role in the outcome of infection. The persistence of gonococci in resident macrophages, which represent a reservoir of infection, could provide a means to overcome host immune response and promote the infection. *N. gonorrhoeae* replicating within PMNs represents an attractive model to explain the chain of persistence of the infection and its possible role in transmission.

Successful colonization by gonococcus can result in a broad spectrum of clinical manifestations of the disease that are, in part, determined by either the specific strain initiating the infection or by the anatomical site infected. An example is the male urethra and the upper female genital tract, where the gonococci interact with epithelial cells, trigger cytokine release, and promote the influx of PMNs.

The incidence of gonorrhea declined from a peak during the Second World War, largely due to the advent of penicillin, and then increased during the sixties and early seventies.[19,20] This increase was thought to be caused by a number of different factors, such as increasing treatment failures due to the antibiotic resistance, the introduction of oral contraceptives, the liberation of sexual mores, and a growing travel and tourist industry.[19] Gonorrhea is a continuing threat to public health worldwide, especially in developing countries and within some lower socioeconomic groups in developed countries. Today, gonococcal infection remains a major global health problem, with increasing infection rates worldwide (http://www.cdc.gov/std/GISP2007).

The World Health Organization (WHO) estimates more than 60 million new cases of gonorrhea per year making it an important public health issue. It has been suggested that the above approximations are underestimated by almost half because of inadequate reporting measures in the regions where the disease is concentrated as well as the prevalence of asymptomatic infections.[21] Mainly teenagers and young adults, aged less than 20 or 25 years, are at risk for infection. High-risk groups for gonorrhea are also commercial sex workers and sexually active homosexual men; moreover, a previous episode of gonorrhea or a more general history of sexually transmitted diseases (STDs) is also documented to be a risk factor for gonorrhea infection.[22]

In Europe during 2006, gonorrhea was more commonly reported in men, who accounted for 74% of all cases reported. Over half of the cases were reported in people older than 25 years. Infected women tended to be younger; 67% of the women were under 25 years old, compared to 36% of the men. The proportion of reported gonorrhea cases among men who have sex with men has increased steadily over the last 10 years. As gonorrhea affects sexually active people, it is no surprise that the main age groups affected are 15–24 years and 25–44 years, for both sexes. Two-thirds (66%) of females were reported in the age category 15–24 years. The highest incidence rate was observed in the United Kingdom (33.98 per 100,000), followed by Latvia (30.09 per 100,000) and the lowest in Luxembourg (0.22 per 100,000), followed by Spain and Portugal (both with 0.42 per 100,000). However, different surveillance systems operate in the countries, making direct comparison difficult (http://ecdc.europa.eu/en/Health_Topics/Gonorrhoea/aer_07.aspx).

Gonorrhea can be effectively treated with antibiotics. However, as with most bacteria under antibiotic pressure, gonococcal antibiotic resistance has emerged. In fact, surveillance of antimicrobial resistance in several countries showed a progressive increase in resistance over time (Gonococcal Isolate Surveillance Program [USA] [GISP], Gonococcal Resistance to Antimicrobials Programme [Canada and United Kingdom] [GRASP], and Australian Gonococcal Surveillance Program [AGSP]). In particular, the emergence and spread of gonococcal resistance to quinolones in Northern Europe,[23] the Western Pacific Region,[24] and also in Australia[25] has dramatically increased. In 2002, a European network was established to identify reference centers for the European Gonococcal Susceptibility Programme (EuroGASP).

Increasing antibiotic resistance in *N. gonorrhoeae* as well as the importance of the disease as a public health issue suggest the need for an integrated approach in which provision of effective treatment is essential for the individual patient in order to interrupt the transmission chain and to reduce overall burden. Accurate surveillance data on the distribution of determinants of resistance is also essential for national and international treatment guidelines.

65.1.2 Clinical Features and Pathogenesis

Neisseria meningitidis. Infection with *N. meningitidis* has two presentations, acute bacterial meningitis and sepsis (meningococcemia), characterized by skin lesions (Figure 65.1, Table 65.1). The fulminant form of disease (with or without meningitis) is characterized by multisystem involvement and

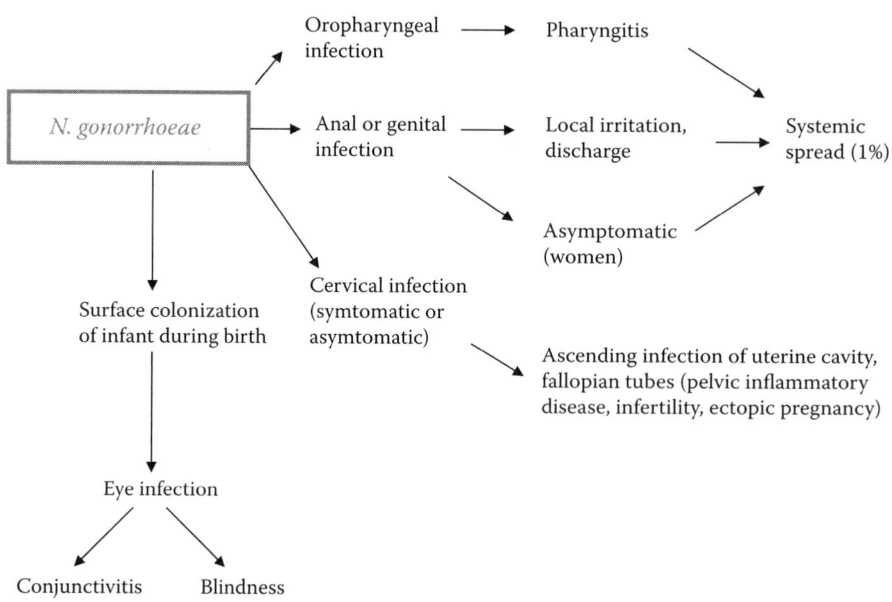

FIGURE 65.1 Transmission and clinical manifestations of *N. gonorrhoeae* infections.

TABLE 65.1
Two Types of Meningococcal Diseases

Disease Type	Manifestations
Meningitis	Bacteria attack the meningis
	Early symptom is a stiff, sore neck
	Meningis and brain swell, putting pressure on nerves
	Survivors often left deaf or with brain damage
Sepsis	Infection occurs through the bloodstream affecting the entire body
	Bacterial toxins rupture blood vessels and can rapidly shut down vital organs
	Survivors often left with permanent injuries and/or require amputations

high mortality. Infection is by aspiration of infective bacteria, which attach to epithelial cells of the nasopharyngeal and oropharyngeal mucosa, cross the mucosal barrier, and enter the bloodstream. It is not clear whether blood-borne bacteria may enter the central nervous system and cause meningitis.

The mildest form of disease is a transient bacteremia illness characterized by a fever and malaise; symptoms resolve spontaneously in 1–2 days. The most serious form is the fulminant form of disease complicated by meningitis. Chills, fever, malaise, and headache are the usual manifestations of infection. Signs of meningeal inflammation are also present.

The incubation period of meningococcal disease is 3–4 days, with a range of 2–10 days. Meningeal infection is similar to other forms of acute purulent meningitis, due to *Streptococcus pneumoniae*, *Haemophilus influenzae*, and *Escherichia coli*, with sudden onset of fever, headache, and stiff neck, often accompanied by other symptoms, such as nausea, vomiting, photophobia (eye sensitivity to light), and altered mental status. Meningococci can be isolated from the blood in up to 75% of persons with meningitis. Meningococcal sepsis (bloodstream infection or meningococcemia) occurs without meningitis in 5%–20% of invasive meningococcal infect ions. This condition is characterized by the abrupt onset of fever and a petechial or purpuric rash, often associated with hypotension, shock, acute adrenal hemorrhage, and multiorgan failure. Meningococci may spread from the nasopharynx to adjacent areas, occasionally causing less common presentations of meningococcal disease including pneumonia (5%–15% of cases), arthritis (2%), otitis media (1%), and epiglottitis (less than 1%). As many as 20% of survivors have permanent sequelae, such as hearing loss, neurologic damage, or loss of a limb. Family members of an infected person are at increased risk for meningococcal disease. Antecedent upper respiratory tract infection, household crowding, and both active and passive smoking also are also associated with increased risk.

A capsular polysaccharide, which protects the bacterium during the invasion process, is the outermost antigenic structure on the meningococcal surface and the primary target for mucosal and humoral immunity. The capsule is also an important vaccine component, with polysaccharide vaccines available against serogroups A, C, Y, and W-135, but not serogroup B. In contrast, in the main pool of the meningococcal population, which is represented by the isolates from healthy carriers, approximately 50% of the strains do not express a capsule and are therefore serologically not serogroupable.[26]

Neisseria gonorrhoeae. *N. gonorrhoeae* infections are acquired by sexual contact and usually affect the mucous membranes of the urethra in males and the endocervix and urethra in females, although the infection may disseminate to a variety of tissues. The pathogenic mechanism involves the attachment of the bacterium to nonciliated epithelial cells via pili (fimbriae) and the production of lipopolysaccharide endotoxin. Among the large number of virulence factors involved in the pathogenesis of gonorrhoeae infections, certainly Neisserial LOS is one of the most important. During infection, LOS and peptidoglycan are released by autolysis of cells. Both activate the host alternative complement pathway, while LOS also stimulates the production of tumor necrosis factor (TNF), which causes cell damage. Gonococcal LOS produces mucosal damage in fallopian tube organ cultures and brings about the release of enzymes that may be important in pathogenesis, such as proteases and phospholipases,. Thus, gonococcal LOS appears to have an indirect role in mediating tissue damage. It is also involved in the resistance of *N. gonorrhoeae* to the bactericidal activity of normal human serum. Specific LOS oligosaccharide types are known to be associated with serum-resistant phenotypes of *N. gonorrhoeae*.

The symptoms and signs of gonorrhea depend on the site of infection. Symptoms of gonorrhea infection may appear 1–14 days after exposure, although it is possible to be infected with gonorrhea and have no symptoms. Men are far more likely to notice symptoms as they are more apparent in men. The most common presenting symptom in males is a painful purulent urethral discharge. If left untreated, complications may include epididymitis, prostatitis, and urethral stricture. Anorectal infection is more common in homosexual males and is usually asymptomatic. It may cause pruritis, tenesmus, and discharge. Pharyngeal infection is usually asymptomatic. It is estimated that nearly half of the women who become infected with gonorrhea either experience no symptoms or have nonspecific symptoms such as a bladder infection. In females, an initial urethritis or cervicitis occurs a few days after exposure. It is frequently mild and passes unnoticed. Females may have abnormal vaginal discharge and postcoital bleeding. Later, pelvic inflammatory disease may develop causing ectopic pregnancy, infertility, or chronic pelvic pain. *N. gonorrhoeae* may also cause conjunctivitis, mostly in neonates, defined as *Ophtalmia neonatorum*.

Septicemia and septic arthritis, DGI, are rare complications may occur in up to 0.5%–3% of infected individuals and is most often associated with untreated asymptomatic urogenital infection.[27] Moreover, people deficient in late-acting complement factors: C7, C8, and C9 seem to be more likely to develop DGI.

The biological basis in HIV transmission is related to the increased load of HIV in the semen and vaginal fluid in persons with a concurrent gonorrhea infection.[28]

65.1.3 Diagnosis

The rapid and precise detection of a pathogen is essential for prompt and effective clinical and public health interventions. Diagnostic procedures can increase the number of microbiologically confirmed cases, thus improving the surveillance systems.

65.1.3.1 *Neisseria meningitidis*

Conventional Techniques. The surveillance of meningococcal diseases has improved in recent years due to refinement in the detection of the presence of the pathogen in clinical samples by molecular assays. Meningococcal disease should be suspected in the face of suggestive clinical features, and a prompt treatment with antibiotics is generally commenced prior to laboratory confirmation. The case definition for confirmed meningococcal disease requires isolation of *N. meningitidis* from a normal sterile site. For detailed case classification, refer to: http://www.cdc.gov/ncphi/disss/nndss/casedef/meningococcalcurrent.htm.

Suspect. (i) Clinical purpura fulminans in the absence of a positive blood culture; (ii) a clinically compatible case with gram-negative diplococci from a normally sterile site (e.g., blood or CSF).

Probable. A clinically compatible case that has either (i) evidence of *N. meningitidis* DNA using a validated polymerase chain reaction (PCR), obtained from a normally sterile site (e.g., blood or CSF); or (ii) evidence of *N. meningitidis* antigen by immunohistochemistry (IHC) on formalin-fixed tissue or latex agglutination of CSF.

Confirmed. A clinically compatible case and isolation of *N. meningitidis* from a normally sterile site (e.g., blood or CSF or, less commonly, synovial, pleural, or pericardial fluid), or skin scrapings of purpuric lesions.

However, the sensitivity of culture may be low, especially when performed after the initiation of antibiotic treatment. Typically the isolate comes from blood or CSF, but it can also be from joint, pleural, or pericardial fluid. The medium used to grow the organism is nonselective chocolate agar and sheep blood agar. Plates should be incubated in 5%–7% CO_2 at 35°C. Suspected *N. meningitidis* colonies are gray color, about 1 mm in diameter, low convex, with a smooth, moist, entire edge and glistening surface. Finally, carbohydrate utilization tests should be used for confirmation of isolates as meningococcus.

Gram staining is commonly used to identify the bacterium and continues to be a reliable and rapid method for presumptive identification. Intracellular gram-negative diplococci in CSF can be considered meningococci until proven otherwise. Negative direct Gram stain or antigen detection does not rule out meningococcal disease.

Direct tests of detection of meningococcal capsular polysaccharide in CSF, serum, or urine can also be done using commercially available kits based on antibody coated latex agglutination to detect capsular antigen of meningococcal serogroups A, B, C, Y, and W-135. This method is rapid, specific, and can provide a serogroup-specific diagnosis, but false-negative results are common.

Molecular Techniques. The use of molecular approaches is essential for rapid management of primary meningococcal disease case and contacts. They are particularly useful in patients who have received antimicrobial therapy prior to culture being taken. Various DNA-based approaches have been developed and applied to the characterization of meningococcal strains or in the culture-negative samples. These methods were those most used during the 1980s and 1990s: restriction analysis,[29] ribotyping,[30] insertion sequences analysis,[31] and pulsed-field gel electrophoresis.[32] Nowadays, the first-line nonculture method that can provide information equivalent to culture is PCR. Molecular diagnosis using PCR offers the advantage of identifying, at the same time, the species and the serogroup. Even when the organism is nonviable, PCR can still detect *N. meningitidis* DNA directly on nonviable bacteria. PCR based on various primers encoding *ctrA*, *porA*, *crgA*, 16S rRNA, *saiD*, and *nspA* have been developed.[33] Real-time PCR based on the capsule transfer gene (*ctrA*) is a significant aid in the diagnosis of meningococcal infection but fails to detect nongroupable strains generally associated with nasopharyngeal carriage. A *porA*-based TaqMan assay has been developed and reportedly found to be a highly specific method for detecting meningococcal DNA that is more sensitive than the *ctrA* assay for detecting meningococcal carriage and is particularly suitable for carriage studies where nongroupable strains and other *Neisseria* are present. Multiplex PCR assay for the contemporary detection of the three main pathogens responsible for meningitis: *N. meningitidis*, *S. pneumoniae*, and *H. influenzae* type b has been developed.[34] It is based on the *ctrA*, *ply*, and *bex* targets, respectively, enabling detection of 5–10 pg DNA. The sensitivity and specificity for the three targets was found to be very high (>90%), and more recently, they have been used in real-time PCR format.

Further molecular techniques to determine the genotype of meningococci should be applied for epidemiological purposes. Variable number tandem repeats analysis has high discriminatory power and may differentiate isolates that have been identified as similar using other molecular typing techniques. It is especially useful for the fine typing of meningococci belonging to the same clonal complex and helpful for short-term epidemiology during an outbreak.[35] The sequencing of individual genes permitted the development of a method for large-scale analysis of *N. meningitidis* and now also directly on DNA in culture-negative samples. MLST (multilocus sequence typing) is on the way to providing a really complete global picture of *N. meningitidis* epidemiology. MLST is a direct adaptation of multilocus enzyme electrophoresis in which genetic variants at multiple loci are identified based on their nucleotide sequence and are arbitrarily assigned an allele number.[36] The combination of allele numbers over the multiple loci is designated as a sequence type (ST). Seven housekeeping gene fragments of about 400–500 bp are sequenced, and the results are deposited in the public database (www.neisseria.org). MLST was developed to monitor the long-term epidemiology of circulating meningococci. For future development, two main goals need to be reached in molecular detection of meningococci infections: the use of point-of-care testing and the establishment of a multiple format for the complete characterization of meningococcus directly on clinical samples. Some recently documented experiences in this context[37] report on the use of rapid diagnostic tests to identify *N. meningitidis* serogroups A, C, W135, and Y, operated by nonspecialized staff, to be useful in extreme situations such as the African epidemic. The easy-to-use dipstick rapid diagnostic tests as a new POCT showed a satisfactory performance compared to PCR. Moreover, the authors underlined the use of the dipstick as a promising test for guiding individual treatment decisions as well as public health actions, including selection of which risk groups to vaccinate. Protein microarray technology has been applied to the investigation of genetic phase variation in meningococcus as a rapid screening approach of patient's sera. The identification of disease-specific proteins is a significant target in biomedical research because such proteins may have medical, diagnostic, and commercial potential as disease markers.[38]

65.1.3.2 *Neisseria gonorrhoeae*

The diagnosis of gonorrhea infection by laboratory confirmation is necessary since the clinical manifestations, whether symptomatic or asymptomatic, genital or extragenital are often similar to those of other sexually transmitted infections (STIs). In addition, strain isolation by culture permits the discovery of resistance to antimicrobials, which is considered a "warning" in many countries.

Conventional Methods. Presumptive diagnosis is made by microscopic examination of the urethral or endocervical discharge. However, the Gram staining showed limitation for extragenital specimens due to the presence of other *Neisseriae* in, for instance, the nasopharynx.

Culture confirmation of suspected colonies with typical morphology, positive oxidase reaction, and detection of gram-negative diplococci by microscopy has also to be confirmed by carbohydrate oxidation assays, biochemical and rapid enzyme substrate tests. Poor specimen collection, transport, and storage may contribute to false-negative results. Inoculated plates need to be immediately incubated between 35°C and 36.5°C in moist air containing 5% CO_2. GC agar base or Columbia agar supplemented with lysed or chocolate horse blood or a nonblood based supplement such as IsoVitalex (Becton-Dickinson) or Vitox (Oxoid) are the media of choice. When a single medium is used this should contain antimicrobial agents to select from normal flora. The choice of selected drugs depends on the sites of isolation. The transport swab should be stored in refrigerator at +4°C and transported to the lab as soon as possible and within 48 h. Plates should be incubated for a minimum of 48 h before being discarded. Plates are examined after 18–24 h of incubation and again after 48 h of incubation. After 18–24 h of incubation, typical colonies will be 0.5–1 mm in diameter, smooth, gray, translucent, and slightly raised. On further incubation they may reach a diameter of 1 cm and become less smooth. Colonial variation may occur especially if the nutrient quality of the medium is suboptimal. It is preferable to work with 18- to 24-h cultures which are easier to handle than older 48-h cultures.

Even under optimal laboratory conditions, the sensitivity of gonococcal cultures ranges from 85% to 95% for acute infection and falls to approximately 50% for females with chronic infections.

Definitive identification may be performed by a number of methods. The carbohydrate utilization test is rapid, economical, and reliable. Colonies which are oxidase-positive and which microscopically resemble gram-negative diplococci, are subcultured (to ensure a pure culture) onto a lysed blood agar (LBA) medium that contains 0.5% glucose to enhance the performance of carbohydrate utilization tests. These cultures are incubated overnight at 37°C in air containing 5% CO_2. Quality control program should be implemented on media, reagents, and equipment.

For a rapid identification a variety of commercial kits are now available. The immunological kits MicroTrak *N. gonorrhoeae* culture confirmation test (an Immunofluorescence test) the Phadebact Monoclonal GC test (a coagglutination test) and GonoGen II (a membrane immunoassay) are all based on the reaction of monoclonal antibodies against epitopes of PIA and PIB outer membrane proteins. Alexander and Ison[39] reported a greater sensitivity (>99%) of kits based on an immunological reaction compared to those based on biochemical reactions. The authors also recommended the use of two different methods, preferably biochemical and immunological, to avoid misclassification with other non-pathogenic *Neisseria*.

Molecular Tests. Confirmatory tests include biochemical, chromogenic enzyme substrate, immunoassays, and nucleic acid tests. The latter include either amplification-based (NAATs) or nonamplification-based methods (hybridization tests), and are increasingly used in clinical laboratories directly on clinical sample or when a viable culture cannot be obtained. Hybridization assays use a specific oligonucleotide probe to hybridize directly to *N. gonorrhoeae* nuclei acid present within a specimen. Diagnosis by DNA/RNA-hybridization based methods are rapid, can use noninvasive samples such as urine, do not require viable organisms, and can use potentially samples that can be collected by the patients themselves making pooling of specimens cost-effective. The sensitivity is lower than that of culture[40–42] especially for oropharyngeal infections. Disadvantages of the diagnostic DNA/RNA-based methods are poor specificity, the lack of approved protocol for some specimens (pharyngeal and rectal samples, for example), and that no viable bacteria are isolated for strain characterization or antibiograms for *N. gonorrhoeae*.

Table 65.2 shows the available molecular assays, some of which allow the contemporary detection of *C. thachomatis* and *N. gonorrhoeae*. In particular, the PACE 2NG test uses a single nucleic acid probe to detect the target rRNA of *N. gonorrhoeae* in endocervical specimens, while the Pace 2C test uses combination probes to detect both pathogens. The strand displacement amplification (SDA) system

TABLE 65.2
Commercial Nucleic Acid–Based Assays for *Neisseria gonorrhoeae*

Manufacturer	Gene Target	Technology
GenProbe PACE II	16S rRNA	Hybridization
Digene Hybrid Capture II assays	Specific chromosome DNA + cryptic plasmid	Hybridization
Roche Amplicor	Cytosine DNA methyltransferase	Polymerase chain reaction
ProbeTecSDA	Multicopy pilin gene-inverting protein homologue	Strand displacement amplification
Abbott LCx	Opacity protein	Ligase chain reaction
Gen-probe APTIMA	16S rRNA	Transcription-mediated amplification

is an isothermal amplification system; the ligase chain reaction (LCR) uses probes that target the *opa* and *pil* genes to detect gonorrhoeae, especially in endocervical, urine, and vaginal specimens. The amplification includes an elevated temperature cycle for the denaturation of target DNA to a single strand, a lower temperature for the annealing of adjacent probes to the target, gap-filling with DNA polymerase in the presence of deoxyguanosine triphosphate, and ligation of probes with ligase. The ligated amplicons are captured and detected by a fluorescence immunoassay. Transcription-mediated amplification (TMA) is a transcription-dependent amplification based on promoter primers that recognize specific target sequences and enable RNA amplicon synthesis.

The current recommendation from the Centers for Disease Control and Prevention (CDC) is that only bacterial culture should be used to test for pharyngeal and rectal gonorrhoea.[43] It has been suggested that any *N. gonorrhoeae* nucleic acid amplification tests (NAATs) positive result from an extragenital sites should be confirmed by a NAATs assay using a different genetic sequence. However, it is generally accepted that any nucleic acid detection methods need confirmation with another amplification method or gene target. Good laboratory practice and quality control should also be included to ensure true positive and negative results, and to rule out nucleic acid contamination or inhibition. The debate on the use of NAATs stems from the sensitivity of NAATs in relation to culture, and both approaches are influenced by the sample collected and/or transport medium used. The use of molecular approaches, especially as part of screening programs, is controversial because of the likelihood of false-positive results. At present, quality assurance programs for nucleic acid testing are more difficult to implement, and there is no interlaboratory standardization program in place for these tests. A panel of controls comprising low copy samples, high copy samples, DNA without target sequence, and a control without DNA are included in each test.

Moreover, several in-house PCR assays for the detection of *N. gonorrhoeae* have also been described. A conventional PCR assay using as a target gene the *gyrA* described by Airell et al.[44] showed a lack of specificity and sensitivity. Real-time PCR assays targeting the 16S rRNA gene have been described to date, but these assays were not internally controlled and also allowed the amplification of *Neisseria* spp. other than *N. gonorrhoeae*. Evaluation of *cppB* gene on the 2.6-MDa cryptic plasmid,[45] the *porA* pseudogene,[46] *opa*-based real-time PCR,[47] and the *orf1* gene[48] have also been described. Studies with PCR assays targeting the *porA* pseudogene and the *opa* gene have revealed high specificity, whereas assays targeting the *gyrA* gene have shown cross-reactivity with *N. subflava*.[49] Assays targeting the *cppB* gene have revealed its presence in some commensal *Neisseria* spp., but it is absent in around 6% of *N. gonorrhoeae* isolates. Assays targeting the 16S rRNA gene have also shown varying results. In particular, the 16S rRNA gene was chosen as the target of choice for the confirmatory assay in this study because it is a universal bacterial gene found in all *N. gonorrhoeae* strains. Titration studies with DNA from other *Neisseria* spp. revealed that concentrations greater than tenfold of *N. gonorrhoeae* could result in competitive amplification and false-negative results when performing real-time PCR with primers 100% specific for the genus *Neisseria*. This is of importance, as the presence of other *Neisseria* spp. has been shown to be the reason for false-positive results, which obviously has a social impact that needs to be considered. At the same time, to exclude the possibility of false-negative results because of the missing PCR target in certain *N. gonorrhoeae* subtypes, the 16S rRNA gene is preferable since it was found in all *N. gonorrhoeae* strains and for this reason accepted as a confirmatory target gene. Therefore, only the species-specific system of primers and probes in the 16S rRNA in combination with stringent amplification conditions, ensured detection of *N. gonorrhoeae* strains in the presence of large numbers of other *Neisseria* spp.

65.2 METHODS

65.2.1 Sample Preparation

Neisseria meningitidis. Cerebrospinal fluid, blood, or others should be collected early and at least prior to the antibiotic treatment when possible. In fact, the sensitivity of molecular diagnostics also decreases after antibiotic use. As documented by others,[50,51] samples for molecular investigation need to be collected within 24 h after the start of treatment. In a review, the importance of maintaining the sample at +4°C until the analysis was also underlined, and it is preferable to keep the samples frozen during transportation.[52] Purification of DNA with commercial kits is suggested, especially for blood samples, to avoid inhibition of the amplification reaction.

Neisseria gonorrhoeae. Swabs are taken from the infected area. The cells are stained with dye (Grain stain) and the bacteria can be identified under a microscope. This quick test is the most effective way of confirming the presence of gonorrhea in men. However, it is reliable in only about 50% of women with gonorrhea. It is also not effective for testing for gonorrhea in the throat or the anus. Swabs from the cervix, urethra, rectum, or throat require culture. The sample is placed in a culture media and allowed to grow for at least 48 h. Culture tests are more accurate than some other tests, but take more time. DNA probe test uses a discharge sample from the infected area and tests for genetic material of the gonorrhea bacteria. It is less accurate for samples collected from the throat. The DNA probe test may also identify the *Chlamydia* that frequently occurs with gonorrhea.

65.2.2 Detection Procedures

65.2.2.1 *Neisseria meningitidis*

PCR protocols are increasingly used in laboratories worldwide for the diagnosis and confirmation of invasive meningococcal infection. Negative and positive controls, such as internal controls in each run, are also required. In particular, a clinical sample spiked with a known amount of meningococcal DNA should be included. The external contamination needs to be avoided by following the standard procedures for PCR laboratory and PCR amplifications also by using commercial urail-N-glycosilate (UNG) and dUTP instead of dTTP that permit to eliminate carry-over products.

The first step is the identification of meningococcal DNA and the identification of serogroup. Several loci have been used: primers based on the gene encoding 16S rRNA,[53] the *ctr*A, which encodes an outer membrane protein involved in capsule transport; and the *crag*, which encodes a transcriptional regulator belonging to LysR family; and the *sia*D gene encoding the polysialtransferase of sialic acid–containing polysaccarides are the most frequently chosen.[52] The *sac*C gene, for serogroup A meningococci, is located inside the gene cassette required for biosynthesis of group A capsule.[52,54] A variety of techniques have been published for this purpose, including PCR–ELISA[55] and PCR. Multiplex PCR or fluorescence-based PCR methods were developed to confirm meningococcal disease, improving results from gel-based methods.[34] Despite the great number of publications, a general guideline is still lacking. Taha et al.[52] performed a comparison of different techniques performed in Reference European Laboratories. Collaboration within the European Monitoring Group on Meningococci (EMGM) will be a major step toward the widespread use of standardized protocols and extraction methods for DNA isolation from clinical samples.

Principle. A seminested PCR for the detection of *N. meningitidis* in clinical samples is based on the amplification of the 16S rRNA gene in the first step, and in a species-specific region in the second.[53] The sequences of 16S rRNA gene primers are U3: 5′-AACT(C/A)CGTGCCAGCAGCCAGCCGCGGTAA-3′ and ru8: 5′-AAGGAGGTGATCCA(G/A)CCGCA(G/C (G/C) TTC-3′; and *N. meningitidis*-specific primer is 5′-TGTTGG GCAACCTGATTG-3′.

Procedure

1. Add 5 μL of purified DNA to the first PCR mixture (25 μL) containing 0.2 mM of dNTP, 0.25 μM of primers U3 and ru8, 1 U of Taq DNA polymerase.
2. Carry out the first PCR amplification with 1 cycle of 94°C for 10 min; 51 cycles of 94°C for 30 s, 64–54°C for 30 s, 72°C for 1 min, for 51 cycles.
3. Dilute the first PCR amplicon (1/500 in water), and add 5 μL to the second PCR mixture (25 μL) prepared as for the first PCR, except that 0.25 μM of *N. meningitidis*-specific primer is used in place of primers U3 and ru8.
4. Perform the second PCR amplification with 25 cycles of 94°C for 30 s, 54°C for 30 s, and 72°C for 1 min.
5. Confirm the second PCR amplicon (about 0.71 kb) belonging to *N. meningitidis* in the sample through 1.5% agarose gel electrophoresis.

Note: This assay permits the identification of other bacterial DNA present in the sample, since sequence analysis of the first PCR amplicon with the same primers used for the amplification reveals its identity.

65.2.2.2 *Neisseria gonorrhoeae*

Due to their cost, the use of NAATs is limited especially in less-developed countries where disease rates are higher. For each NAATs tests are performed according to instructions included in the package insert. All specimens are collected in either a universal transport medium (M4 Remel Lenexa KS) or NAAT-specific transport medium. False GC-positive NAATs results can result from amplicon contamination, cross-reactions with other *Neisseria* species, laboratory errors, or specimen contamination in the clinic. False-negative results can be caused by gonococcus sequence variation, with some gonococci lacking certain target sequences. Currently no consensus emerges from the literature, and various approaches for the gonococci NATTs evaluations have been adopted. The combination with bacterial culture and other NATTs assays is probably considered the gold standard to confirm the NAATs results. For all these reasons, the drawbacks of NAATs include the increased cost, the inability to determine antimicrobial susceptibility, the possibility of contamination, the false-negative results, especially in extragenital infections, due to the occurrence of cross-reactivity with other *Neisseria*. The most frequent target genes of in-house methods are: *orf*1, *omp*III, *opa*, *cpp* genes, and the *por*A pseudogene for genital and extragenital infections.

Principle. One of the in-house PCRs to identify *N. gonorrhoeae* DNA used the *cpp*B gene located on the cryptic plasmid of *N. gonorrhoeae* and the chromosomal cytosine DNA methyltransferase; only positive results for both of the reactions confirm the presence of *N. gonorrhoeae* DNA in the sample.

Primer	Sequence (5′→3′)
NGA1	GCTACGCATACCCGCTTGCTTTGCT
NGA2	TTGGCGAAGACCTTCGAGCAGACATC
NGA3	GTGAACGTGTGCTGATTGTCGGA
NGA4	TCTGGCGACGTACTTCAGTTGTTG

Procedure

1. Prepare a PCR mixture (25 μL) containing 1 U of Taq DNA polymerase, 250 μM of dNTPs, 0.2 mM of dNTP, 0.25 μM each of primers (NGA1, NGA2, NGA3, and NGA4) and 10 μL of purified DNA.
2. Carry out the PCR with 1 cycle of 95°C for 5 min and 42 cycles of 95°C for 10 s, 65°C for 10 s, and 72°C for 10 s.
3. Detect the PCR amplicons with ethidium bromide-stained agarose gel.

Note: To validate the positive results obtained from the PCR above, analysis of the DNA sequences of a different gene target (e.g., 16S rRNA gene or the *por*A pseudogene) is necessary.

65.3 CONCLUSION AND FUTURE PERSPECTIVES

In clinical practice, early identification of infection causing microbes is the crucial requisite for a fast and optimal targeted treatment. For decades conventional phenotypic methods for bacterial detection and identification have depended on cultivation of microbial cultures in liquid or plated media and on biochemical testing. In contrast to conventional diagnostic methods lasting at least 24 h due to microbial growth requirements, DNA-based methods meet the need for a fast, reliable, and thereby life-saving diagnosis. However, without playing down the importance of molecular diagnostic systems in improving diagnostic rate, it is important to isolate the organism, not only to confirm an etiology of the infection, but also to perform antibiotic susceptibility testing for case management, epidemiological purposes, and development of vaccination strategies.

Now we know that the ability to control a disease is strictly correlated to the portion of cases undiagnosed or diagnosed later. In this respect, the genetic diversity of *Neisseria* species poses a formidable challenge in the speedy and accurate design of molecular assays for detection of these pathogens.

For meningococcal disease, early clinical recognition and treatment are recognized as important to reduce individual morbidity and mortality, but effective control will require the greater widespread use of broadly immunogenic, broadly protective, and long-lasting meningococcal vaccines. Molecular techniques have solved many of these problems, but a general consensus is lacking in the use of molecular approaches for diagnosis.

The management of invasive meningococcal disease and case contacts needs standardized methods for nonculture confirmed methods. The European Monitoring Group on meningococci (a consortium of microbiologists and epidemiologists working in reference laboratories in Europe) is working to develop and standardize the best methods for use by the references laboratories in the different EU countries in addition to the conventional bacteriological methods. It is also essential that all tests undergo rigorous quality assurance prior to use. Moreover, nonculture diagnosis of meningococcal diseases should not be considered a replacement for culture since viable bacteria are essential for the study on pathogenesis of host-pathogen interaction, especially if hypervirulent meningococci clones are involved. Milestones in the knowledge of the biology of meningococcal infection are still lacking and represent a challenge for the future development of new approaches for therapy and prophylaxis.

Gonorrhea prevalence and incidence data are subject to biases and limitations due to the nature of the disease and its social and behavioral implications. Owing to the variety of clinical presentations, not all of them due to gonococcus, the accuracy of the diagnosis is of paramount importance. The disease rates, not only in primary infections, but also in sequelae and complications in women and in neonates, are a major public health issue as far as controlling the spread of this sexual infection is concerned. Molecular techniques could solve the problem, but the antibiotic susceptibility pattern of the pathogen is also required.

The nucleic acid–based tests are divided in "probe tests" (hybridization tests) and "amplification tests." The latter are more sensitive and easier to interpret. However, there is an ongoing debate on the effective applications of these tests especially for extragenital infections and for the lack of information concerning antibiotic susceptibility. Without minimizing the utility of molecular approaches in enhancing diagnosis the next goal will be the choice of the better targets for molecular tests correlated to cost efficiency.

The generation of point-of-care testing and inexpensive diagnostic tests is very significant for developing countries. Future developments will include integration of amplification and signal detection technologies in single instruments, and the miniaturization of technologies, including applications of microarrays and microfluidics. Future point-of-care testing could include bedside or in-clinic use of DNA chips for the rapid assessment of *Neisseria* pathogens.

REFERENCES

1. Guerrero, R., Bergey's manuals and the classification of prokaryotes, *Int. Microbiol.*, 4, 103, 2000.
2. Stephens, D.S., Biology and pathogenesis of the evolutionarily successful, obligate human bacterium *Neisseria meningitidis*, *Vaccine*, 27 Suppl. 2, B71, 2009.
3. Tzeng, Y.L., and Stephens, D.S., Epidemiology and pathogenesis of *Neisseria meningitidis*, *Microbes Infect.*, 2, 687, 2000.
4. Manchanda, V., Gupta, S., and Bhalla, P., Meningococcal disease: History, epidemiology, pathogenesis, clinical manifestations, diagnosis, antimicrobial susceptibility and prevention, *Indian J. Med. Microbiol.*, 24, 7, 2006.
5. Handsfield, H.H., Old enemies. Combating syphilis and gonorrhea in the 1990s, *JAMA*, 19, 264, 1451, 1990.
6. Knapp, J.S. et al., Serological classification of *Neisseria gonorrhoeae* with use of monoclonal antibodies to gonococcal outer membrane protein I, *J. Infect. Dis.*, 150, 44, 1984.
7. Knapp, J.S. et al., III. Epidemiology of gonorrhea: Distribution and temporal changes in auxotype/serovar classes of *Neisseria gonorrhoeae*, *Sex Transm. Dis.*, 14, 26, 1987.
8. O'Rourke, M., and Stevens, E., Genetic structure of *Neisseria gonorrhoeae* populations: A nonclonal pathogen, *J. Gen. Microbiol.*, 139, 2603, 1993.

9. Viscidi, R.P., and Demma, J.C., Genetic diversity of *Neisseria gonorrhoeae* housekeeping genes, *J. Clin. Microbiol.*, 41, 197, 2003.
10. Sarafian, S.K., and Knapp, J.S., Molecular epidemiology of gonorrhoea, *Clin. Microbiol. Rev.*, Suppl 2, S49, 1989.
11. Harry, T.C., The outcome of oropharyngeal gonorrhoea treatment with different regimens, *Int. J. STD AIDS*, 16, 520, 2005.
12. MacDonald, N., Mailman, T., and Desai, S., Gonococcal infections in newborns and in adolescents, *Adv. Exp. Med. Biol.*, 609,108, 2008.
13. Klotman, M.E. et al., *Neisseria gonorrhoeae*-induced human defensins 5 and 6 increase HIV infectivity: Role in enhanced transmission, *J. Immunol.*, 180, 6176, 2008.
14. Apicella, M.A. et al., The pathogenesis of gonococcal urethritis in men: Confocal and immunoelectron microscopic analysis of urethral exudates from men infected with *Neisseria gonorrhoeae*, *J. Infect. Dis.*, 173, 636, 1996.
15. Shafer, W.M., and Rest, R.F., Interactions of gonococci with phagocytic cells, *Annu. Rev. Microbiol.*, 43, 121, 1989.
16. Wu, H.J. et al., Azurin of pathogenic *Neisseria* spp. is involved in defense against hydrogen peroxide and survival within cervical epithelial cells, *Infect. Immun.*, 73, 8444, 2005.
17. Seib, K.L. et al., Investigation of oxidative stress defenses of *Neisseria gonorrhoeae* by using a human polymorphonuclear leukocyte survival assay, *Infect. Immun.*, 73, 5269, 2005.
18. Simons, M.P., Nauseef, W.M., and Apicella, M.A., Interactions of *Neisseria gonorrhoeae* with adherent polymorphonuclear leukocytes, *Infect. Immun.*, 73, 1971, 2005.
19. De, S.A., and Meheus, A., Epidemiology of sexually transmitted diseases: The global picture, *Bull. WHO*, 68, 639, 1990.
20. Danielsson, D., Gonorrhea: Diagnostic and therapeutic strategies in the era of resistance to antibiotics, *Semin. Dermatol.*, 9, 109, 1990.
21. Tapsall, J.W., Limnios, E.A., and Murphy, D., Analysis of trends in antimicrobial resistance in *Neisseria gonorrhoeae* isolated in Australia, 1997–2006, *J. Antimicrob. Chemother.*, 61, 150, 2008.
22. Giuliani, M., and Suligoi, B., The STD Surveillance Working Group. Sentinel surveillance of sexually transmitted diseases in Italy, *Euro Surveill.*, 3, 55, 1998.
23. Martin, I.M., Hoffmann, S., and Ison, C.A., European Surveillance of Sexually Transmitted Infections (ESSTI): The first combined antimicrobial susceptibility data for *Neisseria gonorrhoeae* in Western Europe, *J. Antimicrob. Chemother.*, 58, 587, 2006.
24. Workowski, K.A., Berman, S.M., and Douglas, J.M.Jr., Emerging antimicrobial resistance in *Neisseria gonorrhoeae*: Urgent need to strengthen prevention strategies, *Ann. Intern. Med.*, 148, 606, 2008.
25. Samaan, G. et al., Evaluation of the Australian Gonococcal Surveillance Programme, *Commun. Dis. Intell.*, 29, 143, 2005.
26. Caugant, D.A., Tzanakaki, G., and Kriz, P., Lessons from meningococcal carriage studies, *FEMS Microbiol. Rev.*, 31, 52, 2007.
27. Dal Conte, I. et al., Disseminated gonococcal infection in an immunocompetent patient caused by an imported *Neisseria gonorrhoeae* multidrug-resistant strain, *J. Clin. Microbiol.*, 44, 3833, 2006.
28. Anzala, A.O. et al., Acute sexually transmitted infections increase human immunodeficiency virus type 1 plasma viremia, increase plasma type 2 cytokines, and decrease CD4 cell counts, *J. Infect. Dis.*, 182, 459, 2000.
29. Bjorvatn, B. et al., Applications of restriction endonuclease fingerprinting of chromosomal DNA of *Neisseria meningitidis*, *J. Clin. Microbiol.*, 19, 763, 1984.
30. Woods, T.C. et al., Characterization of *Neisseria meningitidis* serogroup C by multilocus enzyme electrophoresis and ribosomal DNA restriction profiles (ribotyping), *J. Clin. Microbiol.*, 30, 132, 1992.
31. Fox, A.J. et al., An epidemiologically valuable typing method for *Neisseria meningitidis* by analysis of restriction fragment length polymorphisms, *J. Med. Microbiol.*, 34, 265, 1991.
32. Bygraves, J.A., and Maiden, M.C., Analysis of the clonal relationships between strains of *Neisseria meningitidis* by pulsed field gel electrophoresis, *J. Gen. Microbiol.*, 138, 523, 1992.
33. Manchanda, V., Gupta, S., and Bhalla, P., Meningococcal disease: History, epidemiology, pathogenesis, clinical manifestations, diagnosis, antimicrobial susceptibility and prevention, *Indian J. Med. Microbiol.*, 24, 7, 2006.
34. Corless, C.E. et al., Simultaneous detection of *Neisseria meningitidis, Haemophilus influenzae, and Streptococcus pneumoniae* in suspected cases of meningitis and septicemia using real-time PCR, *J. Clin. Microbiol.*, 39, 1553, 2001.
35. Yazdankhah, S.P., Lindstedt, B.A., and Caugant, D.A., Use of variable-number tandem repeats to examine genetic diversity of *Neisseria meningitidis*, *J. Clin. Microbiol.*, 43, 1699, 2005.
36. Maiden, M.C. et al., Multilocus sequence typing: A portable approach to the identification of clones within populations of pathogenic microorganisms, *Proc. Natl. Acad. Sci. USA*, 95, 3140, 1998.
37. Boisier, P. et al., Field evaluation of rapid diagnostic tests for meningococcal meningitis in Niger, *Trop. Med. Int. Health*, 14, 111, 2009.
38. Steller, S. et al., Bacterial protein microarrays for identification of new potential diagnostic markers for *Neisseria meningitidis* infections, *Proteomics*, 5, 2048, 2005.
39. Alexander, S., and Ison, C., Evaluation of commercial kits for the identification of *Neisseria gonorrhoeae*, *J. Med. Microbiol.*, 54, 827, 2005.
40. Koumans, E.H., et al. Laboratory testing for *Neisseria gonorrhoeae* by recently introduced nonculture tests: A performance review with clinical and public health considerations, *Clin. Infect. Dis.*, 27, 1171, 1998.
41. Beltrami, J.F. et al., Martin DH. Evaluation of the Gen-Probe PACE 2 assay for the detection of asymptomatic *Chlamydia trachomatis* and *Neisseria gonorrhoeae* infections in male arrestees, *Sex Transmit. Dis.*, 25, 501, 1998.
42. Schachter, J. et al., Ability of the digene hybrid capture II test to identify *Chlamydia trachomatis* and *Neisseria gonorrhoeae* in cervical specimens, *J. Clin. Microbiol.*, 37, 3668, 1999.
43. Johnson, R.E. et al., Screening tests to detect *Chlamydia trachomatis* and *Neisseria gonorrhoeae* infections—2002, *MMWR Recomm. Rep.*, 51 (RR-15), 1, 2002.
44. Airell, A. et al., Verification of clinical samples, positive in AMPLICOR *Neisseria gonorrhoeae* polymerase chain reaction, by 16S rRNA and *gyr*A compared with culture, *Int. J. STD AIDS*, 16, 415, 2005.
45. Bruisten, S.M. et al., Multicenter validation of the *cpp*B gene as a PCR target for detection of *Neisseria gonorrhoeae*, *J. Clin. Microbiol.*, 42, 4332, 2004.
46. Hjelmevoll, S.O. et al., A fast real-time polymerase chain reaction method for sensitive and specific detection of the *Neisseria gonorrhoeae por*A pseudogene, *J. Mol. Diagn.*, 8, 574, 2006.

47. Whiley, D.M. et al., A real-time PCR assay for the detection of *Neisseria gonorrhoeae* in genital and extragenital specimens, *Diagn. Microbiol. Infect. Dis.*, 52, 1, 2005.
48. Chaudhry, U., and Saluja, D., Detection of *Neisseria gonorrhoeae* by PCR using orf1 gene as target, *Sex Transmit. Infect.*, 78, 72, 2002.
49. Whiley, D.M., Tapsall, J.W., and Sloots, T.P., Nucleic acid amplification testing for *Neisseria gonorrhoeae*: An ongoing challenge, *J. Mol. Diagn.*, 8, 3, 2006.
50. Bronska, E. et al., Dynamics of PCR-based diagnosis in patients with invasive meningococcal disease, *Clin. Microbiol. Infect.*, 12, 137, 2006.
51. Taha, M.K., and Fox, A., Quality assessed nonculture techniques for detection and typing of meningococci, *FEMS Microbiol. Rev.*, 31, 37, 2007.
52. Taha, M.K. et al., Interlaboratory comparison of PCR-based identification and genogrouping of *Neisseria meningitidis*, *J. Clin. Microbiol.*, 43, 144, 2005.
53. Backman, A. et al., Evaluation of an extended diagnostic PCR assay for detection and verification of the common causes of bacterial meningitis in CSF and other biological samples, *Mol. Cell. Probes*, 13, 49, 1999.
54. Swartley, J.S. et al., Characterization of the gene cassette required for biosynthesis of the (alpha1→6)-linked N-acetyl-D-mannosamine-1-phosphate capsule of serogroup A *Neisseria meningitidis*, *J. Bacteriol.*, 180, 1533, 1998.
55. Borrow, R. et al., Nonculture diagnosis and serogroup determination of meningococcal B and C infection by a sialyltransferase (*sia*D) PCR ELISA, *Epidemiol. Infect.*, 118, 111, 1997.

66 *Ralstonia*

Michael P. Ryan, J. Tony Pembroke, and Catherine C. Adley

CONTENTS

66.1 INTRODUCTION

The *Ralstonia* genus contains species of primarily environmental origin. The type species *Ralstonia pickettii* has been implicated in human pathogenicity and is a contaminant of high purity water (HPW) used in the manufacturing of medical injectibles and reagents. Another key member, *Ralstonia solanacearum*, is a major plant pathogen with a wide host range.

66.1.1 Taxonomy of *Ralstonia* Genus

The genus *Ralstonia* was created in 1995 to accommodate bacteria from diverse ecological niches previously classified as *Burkholderia* and *Alcaligenes* (*Burkholderia pickettii*, *Burkholderia solanacearum* and *Alcaligenes eutrophus*).[1] *Ralstonia pickettii* (ATCC27511, formerly called *Pseudomonas pickettii* or *Burkholderia pickettii*) is the type strain of this genus. Originally, the genus *Pseudomonas* was classified into five rRNA homology groups, with *Ralstonia pickettii* belonging to *Pseudomonas* homology group II. This classification was subsequently altered with four of the five homology groups being reclassified into separate genera.[2] One such reorganization occurred in 1992, with the proposal of the genera *Burkholderia* with seven species from *Pseudomonas* rRNA homology group II. This was based on the results of cellular lipid and fatty acid analysis and analysis of the 16*S* rRNA of the seven species.[3] It was observed that two of the species within the genus were different to the other *Burkholderia* species but were similar to each other. These were *B. pickettii* and *B. solanacearum*.[4]

In 1995, a series of tests demonstrated a close phenotypic and phylogenetic relationship between *Burkholderia pickettii*, *Burkholderia solanacearum*, and *Alcaligenes eutrophus*. Morphological, physiological, biochemical characterization of cellular lipid and fatty acids, phylogenetic analysis of 16*S* rRNA nucleotide sequences, and rRNA–DNA hybridization were carried out to elucidate the taxonomic position of these three species,[4] which resulted in their classification in a new genus, *Ralstonia*. As a result, new species names were introduced—namely, *Ralstonia pickettii*, *Ralstonia solanacearum*, and *Ralstonia eutropha*, respectively.[1] Since 1995, the taxonomy of the genus had been expanded to include several new species, reaching 13 species by 2004. In 2004 it was proposed to split this genus into two, *Ralstonia* and *Wausteria*, with *R. pickettii* remaining in the *Ralstonia* genus,[5] but splitting the genus into two separate groups—one the *Ralstonia pickettii* lineage and the other the *Ralstonia* (*Wautersia*) *eutropha* lineage. Sequence analysis of the 16*S* rRNA genes of theses bacteria (Figure 66.1), based on data available in August 2008, indicated that two distinct sublineages, with sequence dissimilarity of >4%, supported by a bootstrap value of 100%, are present within the genus *Ralstonia sensu lato*.

During the course of a long-term study of the biodiversity of various *Burkholderia cepacia*-like bacteria, it was discovered that a nearly complete 16*S* rRNA gene sequence that was deposited for *Cupriavidus necator* in the National Centre for Biotechnology Information (NCBI) database under the accession number AF191737 was very similar to that of *Wautersia eutropha* isolates.[6] The similarity level between the 16*S*

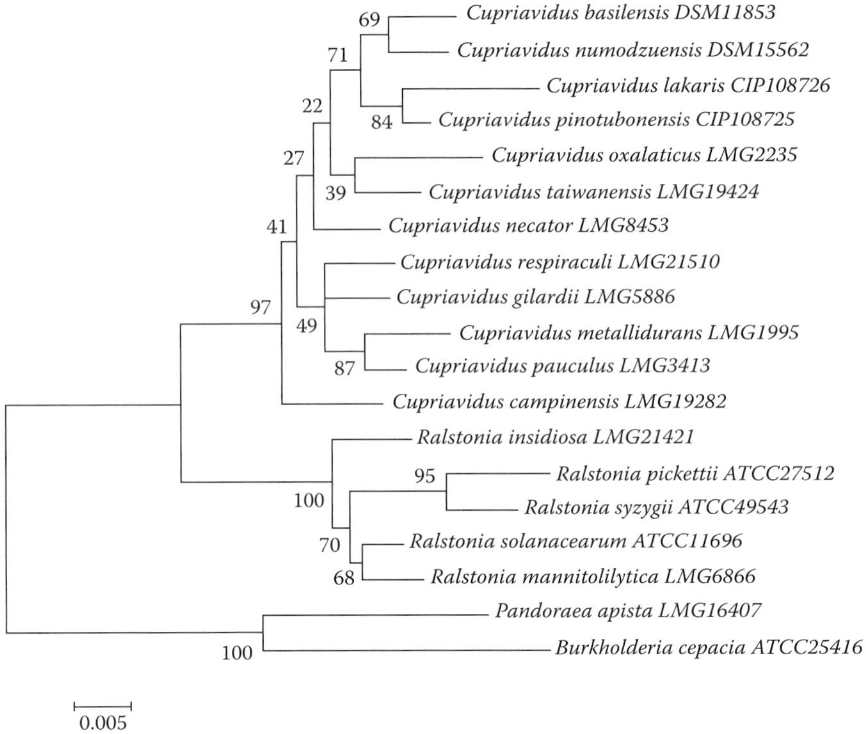

FIGURE 66.1 Phylogenetic tree based on 16S rRNA gene sequence homology for species of the genera *Ralstonia* and *Cupriavidus* and representative species of the β-Proteobacteria. Cluster analysis was based upon the neighbor-joining method. Numbers at branch-points are percentages of 1000 bootstrap resamplings that support the topology of the tree. The scale bar represents 0.005 substitutions per nucleotide position.

rRNA gene sequences of strains *C. necator* LMG8453 and *W. eutropha* LMG1199 was 99.7%.[6] *Cupriavidus necator* was described in 1987 to accommodate a nonobligate bacterial predator of various gram-negative and gram-positive soil bacteria and fungi.[7,8]

Subsequently, whole-cell protein electrophoretic profiles and DNA–DNA hybridization of *C. necator* and *W. eutropha* isolates were also carried out and demonstrated DNA–DNA binding values between 79% and 92% for the *W. eutropha* strains and *C. necator* LMG8453. This data indicated and supported the proposal that these bacteria were the same species.[9–12] The results of the extensive biochemical characterization of *C. necator* generally correlated well with those provided by Yabuuchi et al. and De Baere et al. for *W. eutropha*.[1,13] Rule 23a of the International Code of Nomenclature of Bacteria[14] specifies that each taxon above a species can allow only one correct name—that is, the earliest that is in accordance with the rules of the code, therefore the species called *Wautersia* should be called *Cupriavidus*.

The genus *Ralstonia* now is made up of the species *Ralstonia solanacearum*, *R. mannitolilytica*, *R. syzygii*, and the type strain *Ralstonia pickettii*. The genus *Cupriavidus* is made up of the species *Cupriavidus basilensis*, *Cupriavidus campinensis*, *Cupriavidus gilardii*, *Cupriavidus metallidurans*, *Cupriavidus oxalaticus*, *Cupriavidus pauculus*, *Cupriavidus respiraculi*, *Cupriavidus taiwanensis*, and the type strain *Cupriavidus necator* (*Wautersia eutropha*). Figure 66.1 shows a phylogenetic tree of the 16*S* rRNA gene sequence of *C. necator* LMG8453 with those of strains representing *Ralstonia*, *Cupriavidus,* and other β-Proteobacteria.

66.1.2 Phenotypic Characteristics of Genus *Ralstonia*

Members of the *Ralstonia* genus are gram-negative straight rods that are 1–5 µm in length and 0.5–1.0 µm in width.[2,15] *Ralstonia* spp. are nonspore-forming, motile, oxidase-positive bacteria that are nonencapsulated. Organisms of this genus grow under aerobic conditions at an optimum growth temperature of between 30°C and 37°C and they appear as nonfermenters on MacConkey agar.[2,15]

Ralstonia spp. have been isolated worldwide from a diverse range of ecological niches including plants, soils, and soils that have been contaminated with heavy metals.[16] Members of this genus are primarily environmental organisms found in water, soils, and on plant surfaces—particularly on both fruits and vegetables, where *R. solanacearum* is a significant plant pathogen.[17,18] *R. solanacearum* is the cause of wilting disease in over 200 plant species worldwide. Its hosts include potato, tobacco, and tomato.[19] Other members of the *Ralstonia* genus have been identified as clinical pathogens of varying significance and include *Ralstonia insidiosa*,[20] *R. mannitolilytica*[21] with *R. pickettii* being the only species of this genus that is of high clinical importance.[13] *Ralstonia* spp. are also capable of surviving in aqueous environments and can also form biofilms.[18]

Ralstonia pickettii. *Ralstonia pickettii* was first described in 1964 among a group of nonfermentative gram-negative rods. However, at the time, it was not recognized as a different species due to it similarly to both *Pseudomonas aeruginosa* and *P. cepacia*.[22] In 1973, it was recognized as a separate and distinct species due to its biochemical utilization properties and its DNA base composition and hybridization. The organism, originally isolated from hospital specimens, was at first misidentified as *P. pseudoalcaligenes*. Further studies revealed over forty distinguishing properties when compared to both *P. pseudoalcaligenes* and *P. alcaligenes*. DNA–DNA hybridization studies later revealed the close homology of the microbe with *P. solanacearum* and the name *Pseudomonas pickettii* was proposed for the newly identified species.[23] This new species also turned out to include strains of Centre for Disease Control (CDC) group Va-2[24,25] and possibly other strains where groups Va-1 and Va-2 were regarded as two different biovars of *Pseudomonas pickettii*.[22] Two groups could also be distinguished, based on both 16*S* rDNA sequencing and DNA–DNA hybridization for the CDC groups Va-2 and Va-1.[13] The DNA groups differed at positions 256 and 266 (*Escherichia coli* numbering),[26] with strains of DNA group 1 having the nucleotides A and T at those positions, whereas strains of DNA group 2 have G and C. By using restriction enzyme analysis of the amplified 16*S* rRNA gene with *Hae*III, these DNA groups could be distinguished quickly via restriction polyphorphisms. Although it had been previously reported that genotypic groups and biovars corresponded to each other, these findings were not supported by further research.[5] Extensive attempts to confirm the differential acidification of glucose and maltose in isolates of *R. pickettii* biovars Va-1 and Va-2 failed and turned out to be quantitative rather than qualitative. It proved impossible to delineate the biovars unambiguously on the basis of lactose and maltose acidification, due to the high variability observed and the absence of a clear gap between slow and rapid reactivity. Despite further extensive screening of biochemical characteristics, no clear-cut, unambiguous differentiation was possible between the two DNA groups. Therefore, it was suggested that the biovar designations should no longer be used and no association existed between DNA groups and the former biovar designations, in contrast to the previous results.[13]

Ralstonia pickettii itself can be described as an aerobic, nonfermentative, gram-negative, rod-shaped, oxidase-positive, nonspore-forming microorganism. The individual cells are usually 0.5–0.6 μm in width and 1.5–3.0 μm in length. Colonies are nonpigmented and soluble pigments are not produced. This species is an obligate aerobe with an optimum growth temperature of 35°C. It can grow at 41°C but not at 4°C.[23] *R. pickettii* was shown to grow at a wide range of temperatures (15°C–42°C), as well as in saline solution[27] and with a 64% GC content.[13] *R. pickettii* is a motile organism with polar monotrichous flagellum[3] and is distinguished from all other member of the genus *Ralstonia* and other nonfermentative, gram-negative rods by resistance to polymyxin B, acid production from glucose but not from ethanol, sucrose, or mannitol, alkali production from urea and malonate, and gas production from nitrate between 20°C and 30°C.[22]

Microscopically, it may appear as single or paired rods or occasionally in short chains.[28] Incubation of *R. pickettii* on blood agar plates reveals colonies that are opaque, circular, convex, approximately 1 mm in diameter, and nonhemolytic.[25] The organism is slow growing and barely visible after 24 h of incubation.[29]

Ralstonia insidiosa. *R. insidiosa* is the most closely related species to *R. pickettii* and is a gram-negative, nonsporulating, aerobic, nonfermentative, motile rod, with one to three polar flagella.[20] Aerobic growth is observed at 28°C, 32°C, and 37°C. Growth is also detected on *Burkholderia cepacia* selective agar. Catalase, oxidase, and lipase activities are present, but it cannot reduce nitrate[20] and the bacterium is resistant to colistin and desferrioxamine. Two biovars can be distinguished, one group (biovar 1, which includes the type strain of *Ralstonia insidiosa* LMG21421) are negative for assimilation of *N*-acetylglucosamine, gelatin hydrolysis, and alkalinization of mucate on Simmons' base agar, whereas strains of the other group (biovar 2 which includes strain CCUG46389) are positive for these characteristics. The first group is adipate-positive on bioMérieux API 20NE, whereas the second group is negative for this characteristic. DNA–DNA hybridization values for strains LMG21421 (the type strain was isolated from the sputum of a patient with acute lymphoblastic leukemia in the United States in 2001) and CCUG46389 are 87.9% and 87.4%, indicating that both of these biovars belong to the same species. The %GC content is between 63.9% and 64.3%.[5]

R. insidiosa has been isolated from the respiratory tracts of cystic fibrosis patients,[20,30] river and pond water, soil, activated sludge,[20] and has also been detected in water distribution systems[31] and laboratory purified water systems.[28] It has also been the cause of two cases of serious hospital infection in two immunocompromised individuals.[32]

Ralstonia mannitolilytica. *R. mannitolilytica* are gram-negative, nonsporulating rods that are motile by means of one polar flagellum (motility however was not observed in *R. mannitolilytica* Type strain, LMG 6866). Aerobic growth is observed at 30°C, 37°C, and 42°C, but strains are viable for less than 6 days on Tryptone soy agar (TSA) at room temperature and cultures are catalase and oxidase positive. *R. mannitolilytica* shows similar biochemical properties to *R. pickettii*,[5] but nitrate reduction (negative in *R. mannitolilytica*) and assimilation of D-arabitol and mannitol (both negative in *R. pickettii*) have been reported to enable differentiation.[13] The GC% content of *R. mannitolilytica* strains tested has been shown to be around 66.2%, which is slightly higher than the values for *R. pickettii* at 64.0%–64.3%. In addition to these above phenotypic differences,[5,13] *R. mannitolilytica* strains also differ from *R. pickettii* and *R. solanacearum* by their resistance toward desferrioxamine. *R. mannitolilytica* has also been shown to pass through 0.20 μm filters and to form biofilm.[33]

Ralstonia solanacearum. *R. solanacearum* strains can be differentiated from other *Ralstonia* species by acidification/assimilation of sucrose, lack of pyrrolidonyl arylamidase and assimilation of caprate, malonate, propionate, suberate, acetate, and lactate. *R. solanacearum* has been described as nonmotile[9,13]; however, the presence of one to four polar flagella[34] and the presence of flagellin genes in addition to observations of high motility in culture[34] suggest that at least some strains are motile. Indeed, it has been shown that only 1%–10% of the cell population may be flagellated at times, which may explain the variable reports of motility.[17] *R. solanacearum* is a soil-borne phytopathogen that causes lethal wilting disease in more than 200 plants worldwide,[17] with a wide host range covering both monocots and dicots, extending from annual plants to trees and shrubs.[17] It enters plant roots, invades the xylem vessels, and spreads rapidly through the plants vascular system. The disruption caused by this colonization causes wilting symptoms and eventually plant death. The importance of the disease lies in the pathogen's wide geographic distribution in warm and tropical climates but also in more temperate countries in Europe and North America as the result of adaption to cooler environmental conditions.[17]

Ralstonia syzygii. *R. syzygii* has been recognized as another genuine *Ralstonia* species rather than being a member of the genus *Pseudomonas*.[35] 16S rRNA gene sequences obtained for *R. syzygii* clustered with the *R. solanacearum* GenBank sequence (accession number X67036) with >99% similarity.[13] DNA–DNA hybridization studies have confirmed that *R. solanacearum* and *R. syzygii* are indeed two separate species. Based on this combined data, it was proposed that *Pseudomonas syzygii* should be renamed as *Ralstonia syzygii* comb. nov.[5]

66.1.3 Clinical Features and Pathogenesis

The *Ralstonia* genus is not considered to be of major human pathogenic significance and their virulence was thought to be low; thus, they are not routinely examined in routine hospital analysis.[18] However, a wide range of *Ralstonia* infections have been reported in the literature, indicating that these organisms may be of more pathogenic significance than was previously thought and the types of infections that they are associated with are more invasive and severe.

Types of Infections. Three of the five *Ralstonia* species have been found to cause disease in humans: *R. mannitolilytica*, *R. insidiosa*, and *R. pickettii*.[36] The types of infections caused by these bacteria are summarized in Table 66.1. *R. solanacearum* and *R. syzygii*, on the other hand, to date have not been found to cause disease in humans, and both are prevalent plant pathogens.[18] Many of the cases of infection with *Ralstonia* sp. are due to contaminated medical associated solutions,[37] especially with respect to *R. pickettii*. These sources of infection can include water for injection (WFI),[38] saline solutions formulated with purified water,[39] or sterile drug solutions.[40] These can be administered intravenously,[38] as a drip solution,[41] used to clean wounds, or for endotracheal suctioning.[39] These contaminated solutions have lead to both respiratory infections such as pneumonias and bloodstream infections such as bacteremias.[36] In many of these cases, contamination of

TABLE 66.1
Breakdown of *Ralstonia* sp. Infections Found in Literature

Condition	Number of Incidents
R. pickettii	
Bacteremia/Septicemia	27
Pneumonia	7
Respiratory Infection	5
Meningitis	2
Osteomyelitis	2
Spondylitis	1
Spinal Osteitis	1
Nonneutrocytic bacterascites (peritonitis)	1
Septic arthritis	1
Seminal infection	1
Endocarditis	1
Bacteremia + meningitis	1
Infection	3
Pseudobacteremia	6
Asymptomatic	5
Crohn's disease	1
Prostatitis	1
R. mannitolilytica	
Meningitis	1
Hemoperitoneum Infection	1
Post Renal Transplant Infection	1
Bacteremia	1
Respiratory Infection	1
R. insidiosa	
Infection	2

the solution usually happens at the manufacturing stage, involving poor good manufacturing practice (GMP). One of the most important may be the reported ability of *R. pickettii* to pass through both 0.45 and 0.22 μm filters used for the terminal sterilization of several medicinal products (e.g., saline solution and heat liable solutions).[27] Investigations in our laboratory have also found that *R. pickettii* can flow through 0.2 μm filters.[28]

An early example of how contamination occurs at the manufacturing stage was observed when five infants became infected with *R. pickettii* associated with a contaminated respiratory therapy solution.[42] The organism gained entry into the solution because an 82°C holding tank was bypassed during distilled water manufacture. This contaminated solution was then used for endotracheal suctioning allowing colonization of the patients. Other examples of contamination have resulted from the ability of *R. pickettii* to survive in hospital disinfectants, including chlorhexidine[43] and ethacridine lactate (acrinol).[44] Several patients have been infected with *R. pickettii* that has caused septicemia; the source of the contamination was traced back to the bidistilled water used to make up the 0.05% aqueous solution of chlorhexidine,[45] used for skin antisepsis before the insertion of a venous catheter. *R. pickettii* infections have also been found to be associated with permanent indwelling intravenous devices (PIIDs)[46] where four patients became infected with *R. pickettii*, the ultimate source of the bacteria however in this case was not determined. Contaminated sterile distilled water was implicated in a series of Hickman catheter-associated *R. pickettii* infections, when seven patients developed septicemias in a paediatric oncology unit.[47] *R. pickettii* has also been discovered as the cause of both unusual and severe invasive infections, including osteomyelitis,[48,49] a seminal infection,[50] a prostate infection,[51] septic arthritis in a drug user,[52] invasive infections in drug users[53] and meningitis,[54] and a case of a 38-year-old female who contracted both bacteremia and meningitis due to *R. pickettii* infection.[55] In many of these cases, the etiology was unknown or unreported. *R. pickettii* has also been isolated from cystic fibrosis patients[30] and Crohn disease sufferers.[56]

R. pickettii has also been linked with pseudo outbreaks,[55] which can cause unnecessary treatments to be given to patients, and waste time and resources in hospital laboratories. Pseudo outbreaks may be due to many different causes, which include contaminated distilled water used during the bacterial testing procedures,[29] phlebotomist error,[57] or contamination at the manufacturing stage of materials used in laboratory testing.[58] Verschraegen et al. reported that *R. pickettii* was the cause of pseudobacteremia in 19 patients in a surgical ward.[29] The patients did not demonstrate any signs of bacteremia, even though the organism was isolated from blood samples. The contamination was traced back to distilled water that was used in the testing procedures and a 0.5% chlorhexidine solution prepared using this distilled water and used as a bench wipe in the hospital pharmacy.[29]

R. mannitolilytica is closely related to *R. pickettii*, and has been described in a limited number of cases of hospital outbreaks with "*Pseudomonas thomasii*" and *R. pickettii* biovar 3/"*thomasii*" isolates being reported in the literature.[59–62] "*Pseudomonas thomasii*" is a synonym for *R. mannitolilytica*.[13] Early reports[63] dealt with bacteremia and bacteriuria in 20 patients due to parenteral fluids prepared with deionized water contaminated with "*P. thomasii*." In 1992, it was reported that 23 of 39 *R. pickettii* isolates of an epidemic involving 24 patients, caused by contaminated saline solution (prepared by the hospital pharmacy), belonged to "*P. thomasii*."[61] Reports have described it as causing meningitis and hemoperitoneum infection,[64] renal transplant infection,[65] and bacteremia.[66] *R. mannitolilytica* is the most prevalent species of the *R. pickettii* linage to be found in cystic fibrosis suffers,[30] and a pseudo outbreak has also been described.[67] To date, *R. insidiosa* has only been found in one infectious outbreak with two cases of infection, both in elderly patients due to contaminated catheters.[32] *R. insidiosa* is also on occasion found in the lung sputum of cystic fibrosis suffers.[30] However *R. insidiosa* is rarely distinguished from *R. pickettii* in standard clinical laboratories.

Underlying Causes. Many of the patients in these reported instances of *Ralstonia* infection had an underlying disease or condition that immunocompromised the patient, allowing the organism to infect. Such reports include a 53-year-old man who had suffered a myocardial infarction and contracted *R. pickettii*-related bacteraemia[68]; a 32-year old man, suffering from hepatitis C-related liver cirrhosis who acquired *R. pickettii*-related nonneutrocytic bacterascites (peritonitis)[69]; a 41-year-old male who contracted *R. pickettii* pneumonia while suffering from diabetes mellitus.[70] Wertheim and Markovitz reported a 71-year-old male patient who had chronic renal failure, diabetes mellitus, hypertension, and alcoholic cirrhosis, and had contracted *R. pickettii*-related osteomyelitis.[71] In 2006, five blood cancer patients were diagnosed with a catheter related *R. mannitolilytica* infection,[33] while a *R. insidiosa* infection of two older men—one with leukemia and another with chronic pulmonary disease—has also been reported.[32]

Morbidity Related to *Ralstonia* spp. Four instances of death have been recorded in cases linked to *R. pickettii* infection. The first recorded instance of *R. pickettii*-related death was reported in 1968, where a 33-year-old male drug user died of group IVd-related endocarditis,[72] later identified as *R. pickettii*.[73] The second was identified in 1987 where two patients, 71- and 74-year-old diabetics, died of *R. pickettii*-related septicemia. The source of the contamination was the ion exchange resins used to purify water for hospital use.[74] The third instance occurred in 1995 where a premature infant died due to *R. pickettii*-related pneumonia.[75] Finally, the fourth instance was recorded in 2005—in this case the deaths of two premature babies—were found to be associated with *R. pickettii*. Case 1 became septicemic, deteriorated rapidly, and died the following day; *B. cepacia* complex and *R. pickettii* isolates were found on antemortem blood cultures. Case 2 had clinical sepsis diagnosed, had a documented episode of *R. pickettii* bacteremia, and died soon after. The infection of *R. pickettii* was shown to be due to contaminated vials of WFI.[37] One death has been shown to be attributable to *R. mannitolilytica* in the case of a premature baby due a contaminated oxygen delivery device.[76] No instances of death have yet been reported for *R. insidiosa*.

Pathogenesis. There is limited information about the virulence of *Ralstonia* sp. and what information there is centered on *R. pickettii*. Major pathogenicity determinants appear to be lacking and studies have shown that *R. pickettii* and *R. insidiosa* have no extracellular proteolytic activity, and no elastase activity. However, all strains in a test group[58] displayed α-hemolysis activity (Ryan, M.P. et al., Unpublished data) and siderophore (an iron-complexing compound) activity has been discovered in some strains of *R. pickettii*.[77]

Ralstonia sp. have been shown to possess a wide array of antibiotic resistances, including resistance to gentamicin, chloramphenicol, colistin, tobramycin, polymyxin B, and many others (Ryan, M.P. et al., Unpublished data). The mechanisms of most of these resistances are unknown. *R. pickettii* has been shown to produce a wide range of β-lactamases, including OXA-60, a chromosomal, inducible, imipenem-hydrolysing class D β-lactamase,[78] and OXA-22, which in susceptibility tests indicated that *R. pickettii* isolates were resistant, or had decreased susceptibility, against amino and ureido penicillins, restricted spectrum cephalosporins, ceftazidime, and aztreonam resistances.[79] The enzymatic or molecular basis of the other resistance determinants has not been reported.

66.1.4 Diagnosis

66.1.4.1 Conventional Techniques

Conventional biochemical and culture techniques for the identification of *Ralstonia* sp. can be carried out as described by Chapin and Lauderdale[80] and Vaneechoutte et al.[81] All pathogenic members of this genus can be grown on *Burkholderia cepacia* selective agar (BCSA) supplemented with polymyxin B, which can be used as a preliminary selection for potential *Ralstonia* species.

Commercial identification kits such as the bioMérieux API 20NE, the BioMérieux ID 32 GN Vitek GNI+, GNB and NFC cards, the Uni-N/F-Tek Plate, the Remel RapID NF Plus, and the Biolog GN2 Microplate can all be used to identify *R. pickettii*, however, none have been updated to identify either *R. mannitolilytica* or *R. insidiosa*. The Remel RapID NF Plus has been shown to identify *R. pickettii* better than the BioMérieux API 20NE (Ryan, M.P. et al., Unpublished data).[81] This means that it is necessary to carry out additional tests to verify the identity of *Ralstonia* isolates. *R. mannitolilytica* does not assimilate mannitol and phenylacetate, *R. insidiosa* does not assimilate arabinose or *N*-acetylglucosamine, *R. mannitolilytica*, and *R. insidiosa* are nitrate reduction negative whereas *Ralstonia pickettii* is positive and *R. mannitolilytica* and *R. insidiosa* are resistant to desferrioxamine whereas *Ralstonia pickettii* is susceptible. Assimilation tests can also be carried out using the API 20NE; however, the results are not always reliable.[82,83]

Many identification tests have been shown to be non conclusive, for example tests carried out to determine nitrate reduction using nitrate reduction broth showed that out of 14 strains of *R. insidiosa* three gave positive results and the rest were negative and 11 out of 44 *R. pickettii* isolates gave negative results. This variability is also observed on occasion with the susceptibility to desferrioxamine. In one study of 44 isolates of *R. pickettii*, 11 were resistant to desferrioxamine and 33 were susceptible (Ryan, M.P. et al., Unpublished data).

Protein electrophoresis has been carried out to differentiate between different *Ralstonia* species including the three major pathogenic species[20,32,84]; however, this method can be time consuming and hard to interpret even with the use of specialist software such as BioNumerics. New techniques that are based on bacterial phenotypes such as matrix-assisted laser desorption ionization-time of flight mass spectrometry (MALDI-TOF-MS) have also been used to identify *R. pickettii* and *R. mannitolilytica*.[85] This technique works by detecting the profile of proteins directly from intact bacteria is based on relative molecular masses. It is a soft ionization method allowing desorption of peptides and proteins from whole cultured microorganisms. Ions are separated and detected according to their molecular mass and charge. For a given bacterial strain, this approach yields a reproducible spectrum within minutes, consisting of a series of peaks corresponding to *m/z* ratios of ions released from bacterial proteins during laser desorption.[85] Fourier transform infrared spectroscopy (FTIR) has also been used to identify bacteria including *R. pickettii*, from the lungs of cystic fibrosis suffers. No experiments have taken place to see if FTIR can distinguish between individual *Ralstonia* sp., however.[86] Experience in the laboratory does not demonstrate an easy or user-friendly experience (Ryan, M.P. et al., Unpublished data).

Bacteria initially isolated from blood plates should undergo Gram testing and fermentation studies using Durham tubes to determine if they are fermentative or nonfermentative. If the bacteria are nonfermentative and gram-negative, the isolates should be grown on *Burkholderia cepacia* media to determine if they are potential members of the general *Burkholderiaceae* family or relatives thereof. Following this, a biochemical identification-testing regime can be decided upon such as traditional or kit-based identification. If the kit-based methods are used and an identification of *R. pickettii* is confirmed, then further biochemical tests as outlined above should be carried out to determine if the isolate is *R. pickettii*, *R. insidiosa*, or *R. mannitolilytica*. The biochemical identification route can however be time consuming as growth of *Ralstonia* sp. can take up to 24 h on most media and incubation times for identification kits can vary from 4 to 48 h.

66.1.4.2 Molecular Techniques

Species-Specific PCR. In 2002, a species-specific PCR primer set was designed to identify two members of the *Ralstonia* genus, *R. pickettii* and *R. mannitolilytica* to reduce their misidentification as *B. cepacia* complex isolates[16] and another, designed in 2003, to identify *R. insidiosa*. This was designed to work in conjunction with a *R. pickettii* PCR primer set.[20] A general primer set that could identify all *Ralstonia* sp. was published in 2005. These were primarily developed in to help the identification of unusual bacteria from the lungs of cystic fibrosis patients.[30]

Single Gene Typing. PCR amplification, restriction digestion and/or sequencing of various genes including the *fliC* gene, the 16*S* rDNA gene and the 16S-23S interspacial region can be used to identify and distinguish between *Ralstonia* sp. Sequence differences between the *fliC* gene of *R. pickettii* and *R. solanacearum*[87] and the 16*S*-23*S* rDNA ISR of the same species have been reported which have been utilized in its identification (Ryan, M.P. et al., Unpublished data). Gene information has also recently become available for other single gene targets such as *gyrB* and *rpoB* that could be potentially used to identify *Ralstonia* sp. and utilized for phylogenetic analysis.[88]

Pulsed Field Gel Electrophoresis (PFGE). PFGE has been preformed on *R. pickettii* using *Sma*I, *Dra*I, *Ssp*I, and *Spe*I and shown to be highly discriminatory and reproducible[41,71,89]; it is, however, laborious and time consuming as a diagnostic technique. The possibility of limited genetic diversity within this species was suggested by PFGE studies.[41] PFGE studies have also been carried out on *R. mannitolilytica* with *Spe*I,[33,76] which indicated that *R. mannitolilytica* is a more diverse species than *R. pickettii*. To date no major studies have been carried out comparing *Ralstonia* sp. or with other members, such as *R. insidiosa*.

66.2 METHODS

66.2.1 Sample Preparation

Burkholderia cepacia **Selective Agar.** This media selects for *Burkholderia cepacia* and closely related bacteria such as the *Ralstonia* spp.[90] The following ingredients are mixed to make BCSA (per liter of distilled water): 5 g of sodium chloride, 10 g of lactose, 10 g of sucrose, 0.08 g of phenol red, 0.002 g of crystal violet, 10 g of trypticase peptone, 1.5 g of yeast extract, and 14 g of agar. The phenol red and crystal violet are prepared as aqueous solutions, and 10 mL of each are added per liter. The pH of the medium should be 7.0. After autoclaving for 20 min at 15 lb/in, 600,000 U of polymyxin (filter sterilized) and 10 mg of gentamicin are added.

DNA Preparation. One colony (picked from a culture grown overnight) is heated to 95°C for 20 min in 25 μL of a lysis buffer containing 0.25% (vol/vol) sodium dodecyl sulfate and 0.05 M NaOH. Following the heating step, 180 μL of sterile purified water is added to the lysis buffer\cell mixture. The DNA solutions can be stored at –20°C.[91]

66.2.2 Detection Procedures

(i) Species-Specific PCR for Identification of Ralstonia Species PCR assays are performed in 25 μL reaction mixtures, containing 1 μL DNA solution (as above), 1 U *Taq* polymerase, 250 μM (each) deoxynucleotide triphosphate, 1.5 mM $MgCl_2$, 1× PCR buffer (Bioline), and 20 pmol of Oligonucleotide primers (Table 66.2). For the *Ralstonia* species, primers denaturation at 95°C for 30 s and 30 amplification cycles are carried out, each consisting of 20 s at 94°C, 20 s at 58°C, and 40 s at 72°C. A final extension of 1 min at 72°C is applied. For the species-specific primers initial denaturation is carried out for 2 min at 94°C, 30 amplification cycles are then completed, each consisting of 1 min at 94°C, 1 min at 55°C (for identifying *R. pickettii* and *R. insidiosa*), 57°C (for identifying *R. mannitolilytica*) or 58°C, and 1 min 30 s at 72°C. A final extension of 10 min at 72°C is then applied. The PCR products should be analyzed by electrophoresis in a 1.5% (w/v) agarose gel for 1 h (100 V) and ethidium bromide (or SYBR®Safe) stained in the TBE buffer and photographed under the UV light with a DNA ladder included on all gels to allow standardization and sizing.

(ii) PCR–Ribotyping Polymorphisms can be sought in the 16*S*-23*S*-spacer region of the rRNA genes following DNA amplification using the primers complimentary to the conserved regions of the 16*S* and 23*S* bacterial rRNA genes. The primers used included 16SF 5′-TTGTACACACCGCCCGT CA-3′ and 23SR 5′-GGTACCTTAGATGTTTCAGTTC-3′.[92] Amplification is carried out with a 10× PCR buffer (in a total reaction of 100 μL) containing 200 μM deoxynucleoside triphosphates, 2.5 mM magnesium chloride, 100 pmol of each primer, 4 μL of genomic template DNA, and 1.25 U of *Taq* DNA polymerase. Initial denaturation at 95°C for 5 min followed by 30 cycles of PCR consisting of denaturation at 94°C for 1 min, annealing at 55°C for 1 min, and extension at 72°C for 1 min; in the last cycle, with an extension time of 10 min. Restriction digests can then be performed on the 16*S*-23*S* ISR PCR product in 20 μL reaction mixtures, containing 5 U restriction enzyme, 1× appropriate buffer, 0.2 μL bovine serum albumin at a concentration of 10 mg/mL and 15 μL of the PCR product. Restriction digests are carried out at the temperature recommended by the manufacturer for 4 h. Restriction enzymes can include *Cfo*I and *Hae*III (Roche, UK). Restriction fragments can be separated in 2% (w/v) agarose gels (Sigma) in 0.5× TBE buffer at 100 V for 75 min.

66.3 CONCLUSION

Ralstonia species are of growing importance in both clinical and industrial situations. Cases such as Wills et al. demonstrate that *R. pickettii* is becoming more and more prevalent in hospital environments.[93] Many reports implicate *Ralstonia* as an opportunistic pathogen particularly associated with medical fluids. Misdiagnosis of *Ralstonia* sp. is still a problem with many potential *Ralstonia* sp. being misidentified as *Burkholderia cepacia* and *Pseudomonas fluorescens*. Two whole genome sequences for *R. pickettii* (12J [CP001068, 5.5 Mbp] and 12D [CP001644, 5 Mbp]) have recently become available and will increase the knowledge available about this genus. These sequenced isolates originated from contaminated lake sediments and thus they might not prove relevant to aspects of *R. pickettii*'s virulence.

TABLE 66.2
PCR Primers Used in Identification of *Ralstonia* spp.

Primer	Oligonucleotide Sequence 5′→3′	Product Size	Reference
Rp-F1	ATGATCTAGCTTGCTAGATTGAT	210 bp	16
Rp-R1	ACTGATCGTCGCCTTGGTG		
R38R1	CACACCTAATATTAGTAAGTGCG	403 bp	20
Rm-F1	GGGAAAGCTTGCTTTCCTGCC	398 bp	16
Rm-R1	TCCGGGTATTAACCAGAGCCAT		
RalGS-F	CTGGGGTCGATGACGGTA	546 bp	30
RalGS-R	ATCTCTGCTTCGTTAGTGGC		

REFERENCES

1. Yabuuchi, E. et al., Transfer of two *Burkholderia* and an *Alcaligenes* species to *Ralstonia* gen. nov.: Proposal of *Ralstonia pickettii* (Ralston, Palleroni and Doudoroff 1973) comb. nov., *Ralstonia solanacearum* (Smith 1896) comb. nov. and *Ralstonia eutropha* (Davis 1969) comb. nov., *Microbiol. Immunol.*, 39, 897, 1995.
2. Gilligan, P.H., and Whittier, S., *Burkholderia, Stenotrophomonas, Ralstonia, Brevundimonas, Comamonas*, and *Acidovorax*. In Murray, P.R. et al. (Editors), *Manual of Clinical Microbiology*, 7th ed., p. 526, ASM Press, Washington, DC, 1999.
3. Yabuuchi, E. et al., Proposal of *Burkholderia* gen. nov. and transfer of seven species of the genus *Pseudomonas* homology group II to the new genus, with the type species *Burkholderia cepacia* (Palleroni and Holmes 1981) comb. nov., *Microbiol. Immunol.*, 36, 1251, 1992.
4. Gillis, M. et al., Polyphasic Taxonomy in the genus *Burkholderia* leading to an emended description of the genus and proposition of *Burkholderia vietnamiensis* sp. nov. for N_2-fixing isolates from rice in Vietnam, *Int. J. Syst. Evol. Microbiol.*, 45, 274, 1995.
5. Vaneechoutte, M. et al., *Wautersia* gen. nov., a novel genus accommodating the phylogenetic lineage including *Ralstonia eutropha* and related species, and proposal of *Ralstonia* [*Pseudomonas*] *syzygii* (Roberts et al.,. 1990) comb. nov., *Int. J. Syst. Evol. Microbiol.*, 54, 317, 2004.
6. Vandamme, P., and Coenye, T., Taxonomy of the genus *Cupriavidus*: A tale of lost and found, *Int. J. Syst. Evol. Microbiol.*, 54, 2285, 2004.
7. Makkar, N.S., and Casida, Jr. L.E., *Cupriavidus necator* gen. nov., sp. nov.: A nonobligate bacterial predator of bacteria in soil, *Int. J. Sys. Bacteriol.*, 37, 323, 1987.
8. Zeph, L.R., and Casida, L.E., Gram-negative versus Gram-positive (actinomycete) bacterial predators of bacteria in soil, *Appl. Environ. Microbiol.*, 52, 819, 1986.
9. Coenye, T. et al., Classification of *Alcaligenes faecalis*-like isolates from the environment and human clinical samples as *Ralstonia gilardii* sp. nov., *Int. J. Syst. Evol. Microbiol.*, 49, 405, 1999.
10. Coenye, T. et al., *Ralstonia respiraculi* sp. Nov., isolated from the respiratory tract of cystic fibrosis patients, *Int. J. Syst. Evol. Microbiol.*, 53, 1339, 2003.
11. Vandamme, P. et al., Assignment of centers for disease control group IVc-2 to the genus *Ralstonia* as *Ralstonia paucula* sp. nov., *Int. J. Syst. Evol. Microbiol.*, 49, 663, 1999.
12. Goris, J. et al., Classification of metal-resistant bacteria from industrial biotopes as *Ralstonia campinensis* sp. nov., *Ralstonia metallidurans* sp. nov. and *Ralstonia basilensis* Steinle *et al.*, 1998 emend, *Int. J. Syst. Evol. Microbiol.*, 51, 1773, 2001.
13. De Baere, T. et al., Classification of *Ralstonia pickettii* biovar 3/"thomasii" strains (Pickett 1994) and of new isolates related to nosocomial recurrent meningitis as *Ralstonia mannitolytica* sp. nov., *Int. J. Syst. Evol. Microbiol.*, 51, 547, 2001.
14. Lapage, S.P. et al. (Editors)., *International Code of Nomenclature of Bacteria (1990 Revision)*. Bacteriological Code. American Society for Microbiology, Washington, DC, 1992.
15. MacFaddin, J.F., Gram-negative bacteria. In MacFaddin, J.F. (Editor), *Biochemical Tests for Identification of Medical Bacteria*, 3rd ed., p. 640, Lippincott, Williams and Wilkins, Philadelphia, 2000.
16. Coenye, T., Vandamme, P., and LiPuma, J.J., Infection by *Ralstonia* species in cystic fibrosis patients: identification of *R. pickettii* and *R. mannitolilytica* by polymerase chain reaction, *Emerg. Infect. Dis.*, 8, 692, 2002.
17. Hayward, A.C., Biology and epidemiology of bacterial wilt caused by *Pseudomonas solanacearum*, *Annu. Rev. Phytopathol.*, 29, 65, 1991.
18. Gilligan, P.H. et al., *Burkholderia, Stenotrophomonas, Ralstonia, Brevundimonas, Comamonas, Delftia, Pandoraea* and *Acidovorax*. In Murray, P.R. et al. (Editors), *Manual of Clinical Microbiology*, 8th ed., p. 729, ASM Press, Washington, DC, 2003.
19. Flavier, A.B. et al., Hierarchical autoinduction in *Ralstonia solanacearum*: Control of acyl-homoserine lactone production by a novel autoregulatory system responsive to 3-hydroxypalmitic acid methyl ester, *J. Bacteriol.*, 179, 7089, 1997.
20. Coenye, T. et al., Classification of *Ralstonia pickettii*-like isolates from the environment and clinical samples as *Ralstonia insidiosa* sp. nov., *Int. J. Syst. Evol. Microbiol.*, 53, 1075, 2003.
21. Vaneechoutte, M. et al., One case each of recurrent meningitis and hemoperitoneum infection with *Ralstonia mannitolilytica*, *J. Clin. Microbiol.*, 39, 4588, 2001.
22. Pickett, M.J., and Greenwood, J.R., A study of the Va-1 group of *Pseudomonas* and its relationship to *Pseudomonas pickettii*, *J. Gen. Microbiol.*, 120, 439, 1980.
23. Ralston, E. et al., *Pseudomonas pickettii*, a new species of clinical origin related to *Pseudomonas solanacearum*, *Int. J. Syst. Bacteriol.*, 23, 15, 1973.
24. Tatum, H.W. et al., Miscellaneous Gram-negative bacteria. In Lennette, E.H., Spaulding, E.H., and Truant, J.P. (Editors), *Manual of Clinical Microbiology*, 2nd ed., p. 270, ASM Press, Washington, DC, 1974.
25. Riley, P.S., and Weaver, R.E., Recognition of *Pseudomonas pickettii* in the clinical laboratory: Biochemical characterization of 62 strains, *J. Clin. Microbiol.*, 1, 61, 1975.
26. Woese, C.R. et al., Detailed analysis of the higher-order structure of 16*S*-like ribosomal ribonucleic acids, *Microbiol. Rev.*, 47, 621, 1983.
27. Anderson, R.L. et al., Factors associated with *Pseudomonas pickettii* intrinsic contamination of commercial respiratory therapy solutions marketed as sterile, *Appl. Environ. Microbiol.*, 50, 1343, 1985.
28. Adley, C.C. et al., *Ralstonia pickettii*: Biofilm formation in high-purity water. In McBain, A. et al. (Editors), *Biofilms: Persistence and Ubiquity*, p. 261, Cardiff, Biofilm Club, 2005.
29. Verschraegen, G. et al., *Pseudomonas pickettii* as a cause of pseudobacteremia, *J. Clin. Microbiol.*, 21, 278, 1985.
30. Coenye, T. et al., Use of PCR analyses to define the distribution of *Ralstonia* species recovered from patients with cystic fibrosis, *J. Clin. Microbiol.*, 43, 3463, 2005.
31. Hoefel, D. et al., Profiling bacterial survival through a water treatment process and subsequent distribution system, *J. Appl. Microbiol.*, 99, 175, 2005.
32. Van der Beek, D. et al., Infection with *Ralstonia insidiosa* in two patients, *Clin. Microbiol. Newsl.*, 27, 159, 2005.
33. Gröbner, S. et al., Monoclonal outbreak of catheter-related bacteraemia by *Ralstonia mannitolilytica* on two haemato-oncology wards, *J. Infect.*, 55, 539, 2007.
34. Tans-Kersten, J. et al., *Ralstonia solanacearum* needs motility for invasive virulence on tomato, *J. Bacteriol.*, 183, 3597, 2001.
35. Seal, S.E. et al., Differentiation of *Pseudomonas solanacearum*, *Pseudomonas syzygii*, *Pseudomonas pickettii* and

the blood disease bacterium by partial 16S rRNA sequencing: Construction of oligonucleotide primers for sensitive detection by polymerase chain reaction, *J. Gen. Microbiol.*, 139, 1587, 1993.
36. Ryan, M.P. et al., *Ralstonia pickettii*: A persistent gram-negative nosocomial infectious organism, *J. Hosp. Infect.*, 62, 278, 2006.
37. Moreira, B.M. et al., *Ralstonia pickettii* and *Burkholderia cepacia* complex bloodstream infections related to infusion of contaminated water for injection, *J. Hosp. Infect.*, 60, 51, 2005.
38. Roberts, L.A. et al., An Australia-wide epidemic of *Pseudomonas pickettii* bacteraemia due to contaminated "sterile" water for injection, *Med. J. Aust.*, 152, 652, 1990.
39. McNeil, M.M. et al., Nosocomial *Pseudomonas pickettii* colonization associated with a contaminated respiratory therapy solution in a special care nursery, *J. Clin. Microbiol.*, 2, 903, 1985.
40. Fernandez, C. et al., Nosocomial outbreak of *Burkholderia pickettii* infection due to a manufactured intravenous product used in three hospitals, *Clin. Infect. Dis.*, 22, 1092, 1993.
41. Chetoui, H. et al., Comparison of biotyping, ribotyping, and pulsed-field gel electrophoresis for investigation of a common-source outbreak of *Burkholderia pickettii* bacteremia, *J. Clin. Microbiol.*, 35, 1398, 1997.
42. MMWR, *Pseudomonas pickettii* colonization associated with a contaminated respiratory therapy solution—Illinois, *MMWR*, 32, 495, 1983.
43. Verschraegen, G. et al., *Pseudomonas pickettii* as a cause of pseudobacteremia, *J. Clin. Microbiol.*, 21, 278, 1985.
44. Oie, S., and Kamiya, A., Bacterial contamination of commercially available ethacridine lactate (acrinol) products, *J. Hosp. Infect.*, 34, 51, 1996.
45. Kahan, A. et al., Nosocomial infections by chlorhexidine solution contaminated with *Pseudomonas pickettii* (Biovar VA-I), *J. Infect.*, 7, 256, 1983.
46. Raveh, D. et al., Infections caused by *Pseudomonas pickettii* in association with permanent indwelling intravenous devices: Four cases and a review, *Clin. Infect. Dis.*, 17, 877, 1987.
47. Lacey, S., and Want, S.V., *Pseudomonas pickettii* infections in a paediatric oncology unit, *J. Hosp. Infect.*, 17, 45, 1991.
48. Wertheim, W.A., and Markovitz, D.M., Osteomyelitis and intervertebral discitis caused by *Pseudomonas pickettii*, *J. Clin. Microbiol.*, 30, 2506, 1992.
49. Degeorges, R. et al., *Ralstonia pickettii* osteomyelitis of the trapezium, *Chirurgie de la Main*, 24, 174, 2005.
50. Carrell, D.T., Emery, B.R., and Hamilton, B., Seminal infection with *Ralstonia pickettii* and cytolysosomal spermophagy in a previously fertile man, *Fertil. Steri.l*, 79, 1665, 2003.
51. Ruano, O.D. et al., Acute bacterial prostatitis by *Ralstonia pickettii*: Clinical and epidemiological considerations of exceptional observation, *Med. Clin. (Barc)*, 133, 277, 2009.
52. Zellweger, C. et al., Failure of ceftriaxone in an intravenous drug user with invasive infection due to *Ralstonia pickettii*, *Infection*, 32, 246, 2004.
53. Maki, D.G. et al., Nosocomial *Pseudomonas pickettii* bacteremias traced to narcotic tampering. A case for selective drug screening of health care personnel, *JAMA*, 265, 981, 1991.
54. Fass, R.J., and Barnishan, J., Acute meningitis due to a *Pseudomonas*-like Group Va-1 bacillus, *Ann. Intern. Med.*, 84, 51, 1976.
55. T'Sjoen, G. et al., Avoidable "fever of unknown origin" due to *Ralstonia pickettii* bacteremia, *Acta Clin. Belg.*, 56, 51, 2001.
56. Parent, K., and Mitchell, P., Cell wall-defective variants of *Pseudomonas*-like (group Va) bacteria in Crohn's disease, *Gastroenterology*, 75, 368, 1978.
57. Luk, W.K., An outbreak of pseudobacteremia caused by *Burkholderia pickettii*: The critical role of an epidemiological link, *J. Hosp. Infect.*, 34, 59, 1996.
58. Boutros, N. et al., *Ralstonia pickettii* traced in blood culture bottles, *J. Clin. Microbiol.*, 40, 2666, 2002.
59. Baird, R.M., Elhag, K.M., and Shaw, E.J., *Pseudomonas thomasii* in a hospital distilled-water supply, *J. Med. Microbiol.*, 9, 493, 1976.
60. Dowsett, E., Hospital infections caused by contaminated fluid, *Lancet*, i, 1338, 1972.
61. Pan, H.J. et al., Identification and typing of *Pseudomonas pickettii* during an episode of nosocomial outbreak, *Chin. J. Microbiol. Immunol.*, 25, 115, 1992 (in Chinese).
62. Phillips, I., and Eykyn, S., Contaminated drip fluids, *Brit. Med. J.*, i, 746, 1972.
63. Phillips, I. et al., Outbreak of hospital infection caused by contaminated autoclaved fluids, *Lancet*, 2, 1258, 1972.
64. Vaneechoutte, M. et al., One case each of recurrent meningitis and hemoperitoneum infection with *Ralstonia mannitolilytica*, *J. Clin. Microbiol.*, 39, 4588, 2001.
65. Mukhopadhyay, C., Bhargava, A., and Ayyagari, A., *Ralstonia mannitolilytica* infection in renal transplant recipient: First report, *Indian J. Med. Microbiol.*, 21, 284, 2003.
66. Daxboeck, F. et al., Characterization of clinically isolated *Ralstonia mannitolilytica* strains using random amplification of polymorphic DNA (RAPD) typing and antimicrobial sensitivity, and comparison of the classification efficacy of phenotypic and genotypic assays, *J. Med. Microbiol.*, 54, 55, 2005.
67. Costas, M. et al., Investigation of a pseudo-outbreak of "*Pseudomonas thomasii*" in a special-care baby unit by numerical analysis of SDS-PAGE protein patterns, *Epidemiol. Infect.*, 105, 127, 1990.
68. Fujita, S., Yoshida, T., and Matsubara, F., *Pseudomonas pickettii* bacteremia, *J. Clin. Microbiol.*, 13, 781, 1981.
69. Yuen, K.H. et al., Monomicrobial nonneutrocytic bacterascites due to *Burkholderia pickettii*, *Am. J. Gastroenterol.*, 93, 2308, 1998.
70. Ahkee, S. et al., *Pseudomonas pickettii* pneumonia in a diabetic patient, *J. Ky. Med. Assoc.*, 93, 511, 1995.
71. Wertheim, W.A., and Markovitz, D.M., Osteomyelitis and intervertebral discitis caused by *Pseudomonas pickettii*, *J. Clin. Microbiol.*, 30, 2506, 1992.
72. Graber, C.D. et al., Endocarditis due to a lanthanic, unclassified Gram-negative bacterium (group IV d), *Am. J. Clin. Pathol.*, 49, 220, 1968.
73. Dimech, W.J. et al., Typing of strains from a single-source outbreak of *Pseudomonas pickettii*, *J. Clin. Microbiol.*, 31, 3001, 1993.
74. Poty, F. et al., Nosocomial *Pseudomonas pickettii* infection. Danger of the use of ion-exchange resins, *Presse Med.*, 16, 1185, 1987.
75. Timm, W.D. et al., 1995. An outbreak of multi-drug resistant *Pseudomonas pickettii* pneumonia in a neonatal intensive care unit, 35th ICAAC, San Francisco, Abstract J49.
76. Jhung, M.A. et al., A national outbreak of *Ralstonia mannitolilytica* associated with use of a contaminated oxygen-delivery device among pediatric patients, *Paediatrics*, 119, 1061, 2007.
77. Münzinger, M. et al., *S*, *S*-Rhizoferrin (enantio-rhizoferrin)-a siderophore of *Ralstonia* (*Pseudomonas*) *pickettii* DSM 6297—the optical antipode of *R*, *R*-rhizoferrin isolated from fungi, *Biometals*, 12, 189, 1999.
78. Girlich, D. et al., OXA-60, a chromosomal, inducible, and imipenem-hydrolyzing class D beta-lactamase from *Ralstonia pickettii*, *Antimicrob. Agents Chemother.*, 48, 4217, 2004.

79. Nordmann, P. et al., Biochemical-genetic characterization and distribution of OXA-22, a chromosomal and inducible class D beta-lactamase from *Ralstonia (Pseudomonas) pickettii*, *Antimicrob. Agents Chemother.*, 44, 2201, 2000.
80. Chapin, K.C., and Lauderdale, T. Reagents, stains and media: Bacteriology. In Murray, P.R., Baron, E.J., Jorgensen, J.H., Pfaller, M.A., and Yolken, R.H. (Editors), *Manual of Clinical Microbiology*, 8th ed., pp. 729–748, ASM Press, Washington, DC, 2003.
81. Adley, C.C., and Saieb, F.M., Comparison of BioMérieux API 20NE and Remel RapID NF Plus, identification systems of type strains of *Ralstonia pickettii*, *Lett. Appl. Microbiol.*, 4, 136, 2005.
82. Kiska, D.L. et al., Accuracy of four commercial systems for identification of *Burkholderia cepacia* and other gram-negative nonfermenting bacilli recovered from patients with cystic fibrosis, *J. Clin. Microbiol.*, 34, 886, 1996.
83. Van Pelt, C. *et al.*, Identification of *Burkholderia* spp. in the clinical microbiology laboratory: Conventional and molecular methods, *J. Clin. Microbiol.*, 37, 2158, 1999.
84. Lacey, S., and Want, S.V., *Pseudomonas pickettii* infections in a paediatric oncology unit, *J. Hosp. Infect.*, 17, 45, 1991.
85. Degand, N. et al., Matrix-assisted laser desorption ionization-time of flight mass spectrometry for identification of nonfermenting gram-negative bacilli isolated from cystic fibrosis patients. *J. Clin. Microbiol.*, 46, 3361, 2008.
86. Bosch, A. et al., Fourier transform infrared spectroscopy for rapid identification of nonfermenting gram-negative bacteria isolated from sputum samples from cystic fibrosis patients, *J. Clin. Microbiol.*, 46, 2535, 2008.
87. Schönfeld, J. et al., Specific and sensitive detection of *Ralstonia solanacearum* in soil on the basis of PCR amplification of *fliC* fragments, *Appl. Environ. Microbiol.*, 69, 7248, 2003.
88. Tayeb, L.A. et al., Comparative phylogenies of *Burkholderia*, *Ralstonia*, *Comamonas*, *Brevundimonas* and related organisms derived from *rpoB*, *gyrB* and *rrs* gene sequences, *Res. Microbiol.*, 159, 169, 2008.
89. Labarca, J.A. et al., A multistate nosocomial outbreak of *Ralstonia pickettii* colonization associated with an intrinsically contaminated respiratory care solution, *Clin. Infect. Dis.*, 29, 1281, 1999.
90. Henry, D.A. et al., Identification of *Burkholderia cepacia* isolates from patients with cystic fibrosis and use of a simple new selective medium, *J. Clin. Microbiol.*, 35, 614 1999.
91. Coenye, T. et al., Identification of *Pandoraea* species by 16*S* ribosomal DNA-based PCR assays, *J. Clin. Microbiol.*, 39, 4452, 2001.
92. Kostman, J.R. et al., Molecular epidemiology of *Pseudomonas cepacia* determined by polymerase chain reaction ribotyping, *J. Clin. Microbiol.*, 30, 2084, 1992.
93. Wills, T. et al., Empyema caused by *Ralstonia pickettii* in a hemodialysis patient, *Clin. Microbiol. Newsl.*, 29, 55, 2007.

Section IV

Proteobacteria

Gammaproteobacteria

67 Acinetobacter

Dongyou Liu

CONTENTS

67.1 INTRODUCTION

67.1.1 Classification

The genus *Acinetobacter* encompasses a diverse group of strictly aerobic, gram-negative cocobacillary rods that are classified taxonomically under the family Moraxellaceae, order Pseudomonadales, class Gammaproteobacteria, phylum Proteobacteria. Along with *Alkanindiges*, *Branhamella*, *Enhydrobacter*, *Moraxella*, *Perlucidibaca*, and *Psychrobacter*, the genus *Acinetobacter* makes up the seven-member family Moraxellaceae. Currently, there are at least 23 recognized species within the genus *Acinetobacter*—namely, *Acinetobacter baumannii* (formerly genospecies 2), *Acinetobacter baylyi*, *Acinetobacter beijerinckii*, *Acinetobacter bereziniae* (formerly genospecies 10), *Acinetobacter bouvetii*, *Acinetobacter calcoaceticus* (formerly genospecies 1), *Acinetobacter gerneri*, *Acinetobacter grimontii*, *Acinetobacter guillouiae* (formerly genospecies 1i), *Acinetobacter gyllenbergii*, *Acinetobacter haemolyticus* (formerly genospecies 4), *Acinetobacter johnsonii* (formerly genospecies 7), *Acinetobacter junii* (formerly genospecies 5), *Acinetobacter lwoffii* (formerly genospecies 8/9), *Acinetobacter parvus*, *Acinetobacter radioresistens*, *Acinetobacter schindleri*, *Acinetobacter soli*, *Acinetobacter tandoii*, *Acinetobacter tjernbergiae*, *Acinetobacter towneri*, *Acinetobacter ursingii*, and *Acinetobacter venetianus*.[1–10] In addition, there exist 12 *Acinetobacter* genospecies (i.e., 3, 6, 13TU, 13BJ, 14TU, 14BJ, 15BJ, 15TU, 16, 17, between 1 and 13, and close to 13U) in the genus that are yet to be given species names.[11,12]

Members of the genus *Acinetobacter* (*acinetus*, unable to move; *bacter*, a rod; *Acinetobacter*, nonmotile rod) are gram-negative, strictly aerobic, nonfermenting, nonfastidious, nonmotile, catalase-positive, oxidase-negative, rod-shaped bacteria. The G + C content of *Acinetobacter* DNA ranges from 39% to 47%. *Acinetobacter* spp. are widely distributed in environments that include soil, water, vegetables, and are frequently isolated from animals and humans.[13,14] Of the 45 species and genospecies in the genus, only a few, including *Acinetobacter baumannii* and *Acinetobacter* genospecies 3 and 13TU, cause nosocomial infections in humans. Together with the soil organism *Acinetobacter calcoaceticus*, *Acinetobacter baumannii*, and *Acinetobacter* genospecies 3 and 13TU are known as the *A. calcoaceticus–A. baumannii* (Acb) complex because of their phenotypic and genotypic similarities. For simplicity, *Acinetobacter baumannii* and *Acinetobacter* genospecies 3 and 13TU in the *A. calcoaceticus–A. baumannii* (Acb) complex are referred to as *Acinetobacter baumannii* (or *A. baumannii*) in this chapter.

Acinetobacter baumannii is renowned for its ability to cause both community and healthcare-associated infections (HAI). Given its ingenuity to accumulate mechanisms of pan-antimicrobial resistance, *A. baumannii* is often responsible for large, multicentered HAI outbreaks, with clinical presentations varying from pneumonia, urinary tract infections, bacteremia/septicemia, to surgical wound infections. Invasive procedures (e.g., mechanical ventilation, central venous, or urinary catheters) and use of broad-spectrum antimicrobials represent major risk factors for *A. baumannii* infections.

Acinetobacter beijerinckii is identified in human and animal specimens as well as various environmental sources, whereas *Acinetobacter gyllenbergii* is isolated exclusively in human clinical specimens. *A. beijerinckii* is differentiated from *A. gyllenbergii* by the former's inability to grow on l-arginine and the latter's ability to grow on azelate.[5]

Acinetobacter bereziniae (*Acinetobacter* genospecies 10) and *Acinetobacter guillouiae* (*Acinetobacter* genospecies 11) represent two phenetically and phylogenetically distinct groups within the genus *Acinetobacter.* While *A. bereziniae* strains are originated mainly from human clinical specimens, *A. guillouiae* strains are isolated from different environmental sources, as well as from human specimens. The ability of *A. bereziniae* to oxidize d-glucose and to grow at 38°C differentiates it from *A. guillouiae*, which is negative in these tests.[6]

Acinetobacter ursingii (phenon 1) and *Acinetobacter schindleri* (phenon 2) are distinguished from each other and from all known *Acinetobacter* (genomic) species by their specific phenotypic features and amplified rDNA restriction analysis patterns. *A. ursingii* appears to have the capacity to cause blood stream infections in hospitalized patients.[3]

Acinetobacter parvus is a glucose-nonacidifying, nonproteolytic *Acinetobacter* species that forms small colonies on agar media and is often isolated in human specimens. It is distinguishable from other *Acinetobacter* spp. by its ability to grow on ethanol and acetate as sole sources of carbon.[4]

67.1.2 Epidemiology and Clinical Features

Acinetobacter spp. are ubiquitous organisms that can be recovered from soil or surface water after enrichment culture, although not all species of the genus *Acinetobacter* use the environment as their natural habitat.[12] *A. baumannii,* and *Acinetobacter* genospecies 3, 11, and 13, are found in vegetables and soil.[14,15] *A. calcoaceticus, A. johnsonii, A. haemolyticus*, and *Acinetobacter* genospecies 11 are detected in soil and surface water, and *A. baylyi, A. bouvetii, A. grimontii, A. tjernbergiae, A. towneri*, and *A. tandoii* are isolated in activated sludge but not in humans.[16–18]

Acinetobacter spp. (e.g., *A. lwoffii, A. johnsonii, A. junii, A. radioresistens, Acinetobacter* genospecies 3 and 15BJ) have been shown to form part of communal microflora on human skin and mucous membranes. Certain *Acinetobacter* spp. are often present in one geographical location but not in other. For example, in studies carried out in Europe, *A. johnsonii* and *Acinetobacter* genospecies 11 are identified in fecal samples of 25% healthy individuals, and *A. baumannii* is rarely found on human skin (0.5%–3%) and in human feces (0.8%). A report from Hong Kong showed that *Acinetobacter* genospecies 3, 13TU, 15TU, and *A. baumannii* are frequently recovered, while *A. lwoffii, A. johnsonii*, and *A. junii* are only rarely seen in local population.[19] *A. schindleri* and *A. ursingii* have been identified only in human specimens, and *A. parvus* is isolated from humans and dog. Interestingly, *A. baumannii* is recovered in body lice from homeless people in one study.

A. baumannii is capable of persisting on hospital environmental surfaces for long periods and has been isolated from air, tap water faucets, bedside urinals, ventilators, gloves, and angiography catheters.[20,21] Thus, hospital equipment may function as a reservoir for *A. baumannii*, and contaminated hospital food may serve a potential source for *A. baumannii* acquisition.[15,22,23] Use of indwelling devices (e.g., endotracheal tubes, intravascular and urinary catheters), traumatic wounds, and exposure to broad-spectrum antibiotics represents important risk factors for *A. baumannii* infection.[24,25]

A. baumannii and *Acinetobacter* genospecies 3 and 13TU are the most clinically relevant species. Infections with *A. baumannii* often involve organ systems that have high fluid contents, such as the respiratory tract, peritoneal fluid, and the urinary tract. In hospitalized patients, *A. baumannii* frequently colonizes the skin, upper respiratory tract, and digestive tract. It has been also isolated in sputum, blood, urine, and feces from these patients.[26–30]

Over 90% of the reported cases of *A. baumannii* infections are hospital acquired, and only 4% are community acquired. The most common clinical presentations of *A. baumannii* infections include respiratory tract infections (pneumonia), exudates, abscesses, urinary tract infections, and bloodstream infection (bacteremia). *A. baumannii*-related meningitis (after neurosurgery with an external ventricular drain), and endophthalmitis or keratitis (contact lens use or following eye surgery) may also occur. The mortality rates of *A. baumannii* infections range from 5% in general wards to 54% in intensive care units.[31]

Although the incidence of *A. baumannii* nosocomial infection is relatively low, which occurs mostly in critically ill hospitalized patients, its danger to broad community lies in its capability to acquire antimicrobial-resistance genes extremely rapidly, leading to multidrug resistance. This is especially evident following the widespread use of broad-spectrum antibiotics cephalosporins and quinolones in the 1970s.[32,33]

One of the common ways for *A. baumannii* to build multidrug resistance is through the acquisition of mobile genetic elements (e.g., plasmids, transposon, integron, and natural transformation).[34–36] These genetic elements encode enzymes to degrade antimicrobials, efflux pumps to reduce cellular intake of antibiotics, and proteins to alter the antibiotic targets. For example, acquired by contact with *Pseudomonas aeruginosa*, class I integron carrying the [bla]VEB-1 extended-spectrum β-lactamase and six other antimicrobial-resistance genes enables *A. baumannii* to resist destruction by multiple antibiotics (including β-lactams such as carbapenems, aminoglycosides, quinolones, tetracyclines, and glycylcyclines).[37]

In addition, *A. baumannii* carries intrinsic carbapenem-hydrolysing class D β-lactamases (OXA-51/69 variants), and further acquisition of class B and D carbapenem-hydrolysing β-lactamases strengthens its resistance to carbapenems that are used to effectively treat multiresistant *Acinetobacter* infections. IMP, VIM, and SIM-1 enzymes represent some frequently identified class B metallo-β-lactamases, and OXA-23-like, OXA-24-like, and OXA-58 enzymes are common class D β-lactamases found in *A. baumannii*. Currently, polymyxins, tigecycline, and colistin represent effective options for the treatment of multiresistant *Acinetobacter* spp.[38–44]

Moreover, *Acinetobacter* spp. are able to evolve rapidly through natural transformation. The ability of *A. baumannii* to survive in a diversity of environments, to tolerate desiccation, and to build heavy-metal resistance further enhances its potential to cause infections in humans.

67.1.3 Diagnosis

Phenotypic Techniques. *Acinetobacter* spp. are gram-negative (or slightly gram-positive), nonmotile, catalase-positive, oxidase-negative, nonfermenting coccoid rods. *Acinetobacter* spp. are first isolated on solid media, such as sheep blood agar or tryptic soy agar, at 37°C. *A. calcoaceticus–A. baumannii* complex often form smooth, sometimes mucoid, grayish white colonies that are 1.5–3 mm after overnight culture, resembling to those of *Enterobacteriaceae*. Other *Acinetobacter* species may produce smaller and more translucent colonies. However, some *Acinetobacter* species outside *A. calcoaceticus–A. baumannii* complex do not grow on MacConkey agar. In addition, *A. haemolyticus* and *Acinetobacter* genospecies 6, 13BJ, 14BJ, 15BJ, 16, and 17 may show hemolysis on sheep blood agar, whereas *A. calcoaceticus–A. baumannii* complex do not. CHROMagar *Acinetobacter* (CHROMagar, Paris, France)-containing agents that inhibit the growth of most gram-positive organisms and carbapenem-susceptible gram-negative bacilli may be used for the rapid identification of multidrug resistant *A. baumannii* (MDRAB).[45]

Further phenotypical characterization is carried out with API 20NE (bioMèrieux) or other commercial systems such as Vitek 2, Phoenix, and MicroScan WalkAway. Nevertheless, phenotypic techniques do not allow unambiguous identification of *Acinetobacter* species, especially the genetically closely related *A. baumannii* and *A. calcoaceticus* in the *A. baumannii–A. calcoaceticus* (Acb) complex, as well as *Acinetobacter* genomic species 3 and 13TU.[46,47]

Genotypic Techniques. A number of molecular techniques have been developed for more precise identification of *Acinetobacter* species. These include DNA–DNA hybridization, amplified 16S rRNA gene restriction analysis (ARDRA), amplified fragment length polymorphism (AFLP), ribotyping, tRNA spacer fingerprinting, restriction and sequence analysis of the 16S-23S rRNA intergenic spacer sequences, and sequencing of the *rpoB* (RNA polymerase *β*-subunit) gene and its flanking spacers (Table 67.1).[48–60] Currently, ARDRA and AFLP are the most widely utilized methods for determination of *Acinetobacter* species, although ARDRA alone is not sufficient for the identification of all species.[61]

Although the *efp* gene is not an ideal target for specific detection of *A. baumannii*, it is useful for discriminating *Acinetobacter* strains at the species level.[62] A new PCR-based method, targeting the *gyrB* gene sequence, is shown to provide effective discrimination between *A. baumannii* and *Acinetobacter* genospecies 13TU.[58] The PCR technique has also been employed for the characterization of antibiotic resistant strains of *A. baumannii*.[63–71]

Typing. To assist epidemiological investigation and to control the spread of *Acinetobacter* organisms, various phenotypic and DNA-based typing methods have also been applied to identify the bacteria to subspecies level. Phenotypic typing methods include bacteriophage typing, serotyping, biotyping (biochemical profiles), antibiogram typing (antibiotic susceptibility patterns) and protein profiles. DNA-based typing methods include plasmid typing, ribotyping (using restriction enzymes *Eco*RI, *Hin*dIII, and *Hin*cII and *E. coli* rRNA-derived probe), arbitrary primed PCR (AP–PCR), randomly amplified polymorphism DNA (RAPD), repetitive extragenic palindromic sequence-based PCR (REP–PCR), pulsed-field gel electrophoresis (PFGE, using restriction enzyme *Apa*I), AFLP (using *Eco*RI + A and *Mse*I + C primers), multilocus sequence typing (MLST, based on *gltA*, *gyrB*, *gdhB*, *recA*, *cpn60*, *gpi*, and *rpoD* genes), PCR-ESI-MS, and DNA sequencing.[62,72–94] Use of typing techniques facilitates the identification of pan-European clone types I, II, and III, as well as Southeast England clones that have become endemic in hospital settings.

67.2 METHODS

67.2.1 Sample Preparation

Blood samples are divided and transferred into two bottles, one containing medium for aerobic growth and the other for anaerobic growth. MacConkey is also useful for seeding the gram-negative bacteria. The cultures are incubated at 35.5°C for 24 h. Blood cultures showing bacterial growth are regarded as positive and are further processed for identification by conventional microbial methods including Gram staining, seeding in various growth media, and an API test

TABLE 67.1
Primers for Amplification and Sequencing of Zones 1 and 2 of the *rpoB* Gene and Flanking Spacers of *Acinetobacter*

Primer	Sequence (5′→3′)	Position[a]	Temperature (°C)
Ac696F	TAYCGYAAAGAYTTGAAAGAAG	+2916	60
Ac1093R	CMACACCYTTGTTMCCRTGA	+3267	60
Ac1055F	GTGATAARATGGCBGGTCGT	+3263	60
Ac1598R	CGBGCRTGCATYTTGTCRT	+3773	58
AcintLBF	GAAGARCTTAAGAMDAARCTTG	−361	60
AcintLBR	CGTTTCTTTTCGGTATATGAGT	+29	60
AcintBCF	GTTCTTTAGGTATCAACATTGAA	+4048	60
AcintBCR	GACGCAAGACCAATACGRAT	+4207	59

[a] Positions are according to the *rpoB* gene sequence of *A. baumannii*.

system (BioMèrieux). Aliquots from the positive blood samples are processed for PCR identification. Samples showing no bacterial growth are regarded as negative after 7 days of incubation and discarded.

DNA from clinical isolates may be extracted on the ABI 6100 Nucleic Acid Prep Station with the NucPrep DNA isolation kit. Briefly, 10 μL of a 50-mg/mL stock solution of lysozyme and 10 μL of lysostaphin is added to a 96-well deep-well plate containing bacterial colonies suspended in 100 μL of Tris-EDTA buffer, and the plate is incubated at 37°C for 60 min. 20 μL of 20-mg/mL proteinase K is then added, followed by incubation at 70°C for 30 min. This is followed by the addition of 500 μL of NucPrep DNA purification solution, and the samples are further processed according to the manufacturer's instructions. The bacterial DNA is eluted in a total volume of 200 μL. Alternatively, 100 μL of 5 M guanidinium HCl is added to each well of a 96-well deep-well plate containing bacterial colonies suspended in 100 μL of Tris-EDTA buffer, and the plate is vortexed briefly. The plate is then sonicated in an ultrasonicator bath for 15 min to lyse the bacteria. This is followed by the addition of 1 ml of NucPrep DNA purification solution and brief vortexing. The samples are further processed according to the manufacturer's instructions. The bacterial DNA is eluted in a total volume of 200 μL.

To extract DNA from blood culture, 100 μL of a 12-mg/mL suspension of activated charcoal in 5 M guanidinium HCl is added to each well of a 96-well deep-well plate containing the 100 μL of seeded blood cultures. The plate is sonicated in an ultrasonicator bath for 15 min to lyse the bacteria. The resulting solution is filtered through a Tissue Prefilter tray II and collected in a 96-well microtiter plate. This plate is centrifuged at 1800 × *g* for 5 min, and the supernatant is transferred to a fresh 96-well deep-well plate. 100 μL of ABI NucPrep DNA purification solution is added to the wells, and the plate is briefly vortexed and then processed according to the manufacturer's instructions.

67.2.2 Detection Procedures

(i) PCR for Identification of Acinetobacter baumannii Pechorsky et al.[95] developed a PCR method for the identification of *Acinetobacter baumannii* in blood cultures, with primers (Forward: 5′-CCAGCGTTGTCTGAGTTGTATC-3′ and reverse 5′-CCACTTCCTGACTGGAAATCA-3′) from *gyrB* gene, yielding a 160 bp product.

Procedure

1. A 500 μL sample is taken directly from the BacT/ALERT blood culture bottle and added with 1 mL sterile water and centrifuged at 7000 × *g* for 5 min. The supernatant is removed and the pellet resuspended in 1.0 mL water. This washing step is repeated, the pellet is finally resuspended in 0.1 mL distilled water, and 3 μL aliquots are transferred to PCR mixture.
2. PCR mixture (20 μL) is made up of 3 μL of extracted DNA, 0.25 μM of each primer and 1× master mix.
3. The thermocycling conditions are 94°C for 5 min, followed by 40 cycles of 94°C for 15 s, 61°C for 15 s, and 72°C for 10 s, with a final extension at 72°C for 7 min in a GeneAmp PCR System 2400 (Roche).
4. After amplification, 10 μL of each reaction mixture is loaded onto a 2% gel agarose in Tris-acetate (TAE) buffer, stained with ethidium bromide and visualized under UV light. A 100 bp DNA ladder is used to estimate the size of the product.

(ii) Sequencing Analysis of 16S rRNA Gene Chang et al.[54] targeted the 16S-23S rRNA gene intergenic spacer (ITS) sequence for the differentiation of members in the *A. calcoaceticus–A. baumannii* complex as well as genospecies 3 and 13TU. The bacterium-specific universal primers 1512F (5′-GTCGTAACAAGGTAGCCGTA-3′) and 6R (5′-GGGTTYCCCCRTTCRGAAAT-3′) (where Y is C or T and R is A or G) are used to amplify a DNA fragment that encompassed a small fragment of the 16S rRNA gene region, the ITS, and a small fragment of the 23S rRNA gene region. The 5′ end of primer 1512F is located at position 1493 of the 16S rRNA gene, and the 5′ end of primer 6R is located at position 108 downstream of the 5′ end of the 23S rRNA gene (*E. coli* numbering).

Procedure

1. One colony of a pure culture is suspended in 50 μL of sterilized water and heated at 100°C for 15 min. After centrifugation in a microcentrifuge (6000 × *g* for 3 min), the supernatant is stored at −20°C.
2. PCR mixture (50 μL) is prepared with 5 μL (1–5 ng) of template DNA, 10 mM Tris-HCl pH 8.8, 50 mM KCl, 1.5 mM $MgCl_2$, 0.8 mM deoxyribonucleoside triphosphates (0.2 mM each), 1 μM (each) primer, and 1 U of *Taq* DNA polymerase.
3. The PCR cycling program consists of an initial denaturation at 94°C for 2 min; 35 cycles of 94°C for 1 min, 62°C for 1 min, and 72°C for 1 min; and a final extension step at 72°C for 7 min.
4. PCR products are purified with a PCR-M Clean Up kit and sequenced on a model 377 sequencing system (Applied Biosystems) with a BigDye Terminator cycle sequencing kit (version 3.1; Applied Biosystems).
5. The portions of the 16S and 23S rRNA gene regions (ACGAAAGATT at the 5′ end and GGGGTTGTAT at the 3′ end) are removed from the sequence data to obtain the exact ITS sequences. The ITS sequence of a strain is compared with that of the reference strain of the same species, and the PileUp command of the Wisconsin Genetics Computer Group package (version 10.3; Accelrys Inc., San Diego, CA) is used to obtain the ITS similarity scores.

(iii) Sequencing Analysis of rpoB Gene La Scola et al.[56] described an RNA polymerase β-subunit gene (*rpoB*)-based identification scheme for the delineation of species within the genus *Acinetobacter*, since the 16S rRNA gene is not polymorphic enough to clearly distinguish all *Acinetobacter* species. By analyzing the complete *rpoB* gene (4089 bp), all *Acinetobacter* species are separated into different clusters. Four pairs of primers are used to amplify and sequence two highly polymorphic partial sequences (350 and 450 bp) bordered by highly conserved regions: zone 1, between positions 2900 and 3250, and zone 2, between positions (3250 and 3700) of the *rpoB* gene.

Procedure

1. All strains are grown on Columbia agar plates with 5% sheep blood (Trypticase soy agar). The culture plates are incubated for 48 h at 37°C under aerobic conditions.
2. PCR mixture consists of 2.5×10^{-2} U of *Taq* polymerase per µL; 1× *Taq* buffer; 1.8 mM $MgCl_2$ (Gibco BRL, Life Technologies); 200 µM each of dATP, dCTP, dTTP, and dGTP; and 0.2 µM of each primer.
3. The PCR mixture is subjected to a denaturation step at 95°C for 2 min; 35 cycles of 94°C for 30 s, 48°C for 30 s, and 72°C for 1 min; and a final elongation step of 72°C for 10 min. The resulting PCR products are purified and sequenced.
4. The nucleotide sequences of the target *rpoB* gene fragments are processed by using standard software (Sequence Analysis Software; Applied Biosystems). The partially overlapping sequences encompassing the complete *rpoB* gene are combined into a single consensus sequence using sequence assembler software (Applied Biosystems). Multiple *rpoB* gene sequence alignments are made using the Clustal W program on the EMBL-EBI web server (http://www.ebi.ac.uk/clustalw/).
5. The phylogenetic relationship between complete sequences of the *rpoB* gene and two spacers is estimated by the neighbor-joining method. Phylogenetic trees are constructed, and their node reliabilities are estimated by performing bootstrap replicates. Bootstrap values are obtained from 1000 trees generated randomly using SEQBOOT (PHYLIP software package). Proportional similarities of partial and complete *rpoB* sequences and spacers (*rplL-rpoB* and *rpoB-rpoC*) among species are computed using the MEGA 2.1 program.

Note: For the routine identification of *Acinetobacter*, either of two partial sequences and spacers can be used as they have high discriminative power and short lengths. However, such an identification scheme may not be sufficient to distinguish *A. grimontii* from *A. junii* and *A. baylyi* from *Acinetobacter* genomic species 9. One way of addressing this problem is to incorporate another short variable sequence into the identification scheme. In order to identify an *Acinetobacter* isolate, zone 1 should be determined first as it is the shortest and allows clear identification of 20 of 24 species. If the sequences of zone 1 in *A. grimontii* and *A. junii* are similar, then the sequences of zone 2 should be examined to discriminate between these two species. On the other hand, if the sequences of zone 1 in *A. baylyi* and genomic species 9 are similar, then it will be useful to determine the sequence of the *rpoB-rpoC* spacer as it may make a clear distinction between these two species.[60]

67.3 CONCLUSION

Acinetobacter spp. are nonfermentative, strictly aerobic, gram-negative bacteria that are ubiquitous in nature and constitute part of the normal skin, throat, and rectal flora of humans and animals. Of the 45 species and genospecies in the *Acinetobacter* genus, several (e.g., *Acinetobacter baumannii* and *Acinetobacter* genospecies 3 and 13TU) are capable of causing nosocomial infections in humans, with clinical presentations ranging from pneumonia, exudates, abscesses, bacteremia, meningitis to endophthalmitis or keratitis. Given its ability to survive on dry surfaces for prolonged periods of time and to rapidly acquire multiple antibiotic resistance, *A. baumannii* has become an increasingly important pathogen in hospital setting, from which over 90% of clinical cases are reported.

Conventional phenotype-based procedures such as culture isolation and antibiotic susceptibility testing have contributed greatly to the characterization and identification of *A. baumannii*. More recently, molecular techniques have been developed and applied for the detection, typing, and drug resistance testing of *A. baumannii*. This has resulted in the significant improvement in the accuracy, sensitivity, and speed of *A. baumannii* diagnosis, and played a critical role in shortening the hospital care and preventing the spread of multidrug resistant *A. baumannii* to the wide community. Further streamlining in the clinical sample processing will render molecular techniques a more versatile tool for routine laboratory diagnosis of *A. baumannii* infections.

REFERENCES

1. Nishimura, Y., Ino, T., and Iizuka, H., *Acinetobacter radioresistens* sp. nov. isolated from cotton and soil, *Int. J. Syst. Bacteriol.*, 38, 209, 1988.
2. Bouvet, P.J.M., and Grimont, P.A.D., Taxonomy of the genus *Acinetobacter* with the recognition of *Acinetobacter baumannii* sp. nov., *Acinetobacter haemolyticus* sp. nov., *Acinetobacter johnsonii* sp. nov., and *Acinetobacter junii* sp. nov., and emended descriptions of *Acinetobacter calcoaceticus* and *Acinetobacter lwoffii*, *Int. J. Syst. Bacteriol.*, 36:228, 1986.
3. Nemec, A. et al., *Acinetobacter ursingii* sp. nov. and *Acinetobacter schindleri* sp. nov., isolated from human clinical specimens, *Int. J. Syst. Bacteriol.*, 51, 1891, 2001.
4. Nemec, A. et al., *Acinetobacter parvus* sp. nov., a small-colony-forming species isolated from human clinical specimens, *Int. J. Syst. Evol. Microbiol.*, 53, 1563, 2003.

5. Nemec, A. et al., *Acinetobacter beijerinckii* sp. nov. and *Acinetobacter gyllenbergii* sp. nov., haemolytic organisms isolated from humans, *Int. J. Syst. Evol. Microbiol.*, 59, 118, 2009.
6. Nemec, A. et al., *Acinetobacter berezinae* sp. nov. and *Acinetobacter guillouiae* sp. nov., to accommodate, respectively, *Acinetobacter* genomic species 10 and *Acinetobacter* genomic species 11, *Int. J. Syst. Evol. Microbiol.*, 60, 896, 2010.
7. Carr, E., Seviour, R.J., and Gürtler, V., Genomic fingerprinting of the 16S-23S gene spacer region suggests that novel *Acinetobacter* isolates are present in activated sludge, *Water Sci. Technol.*, 46, 449, 2002.
8. Vaneechoutte, M. et al., Oil-degrading *Acinetobacter* strain RAG-1 and strains described as "*Acinetobacter venetianus* sp. nov." belong to the same genomic species, *Res. Microbiol.*, 150, 69, 1999.
9. Vaneechoutte, M. et al., Naturally transformable *Acinetobacter* sp. strain ADP1 belongs to the newly described species *Acinetobacter baylyi*, *Appl. Environ. Microbiol.*, 72, 932, 2006.
10. Vaneechoutte, M. et al., Description of *Acinetobacter venetianus* ex Di Cello et al. 1997 sp. nov., *Int. J. Syst. Evol. Microbiol.*, 59, 1376, 2009.
11. Kämpfer, P., Tjernberg, I., and Ursing, J., Numerical classification and identification of *Acinetobacter* genospecies, *J. Appl. Microbiol.*, 75, 259, 1993.
12. Peleg, A.Y., Seifert, H., and Paterson, D.L., *Acinetobacter baumannii*: Emergence of a successful pathogen, *Clin. Microbiol. Rev.*, 21, 538, 2008.
13. Berlau, J. et al., Distribution of *Acinetobacter* species on skin of healthy humans, *Eur. J. Clin. Microbiol. Infect. Dis.*, 18, 179, 1999.
14. Berlau, J. et al., Isolation of *Acinetobacter* spp. including *A. baumannii* from vegetables: Implications for hospital acquired infections, *J. Hosp. Infect.*, 42, 201, 1999.
15. Corbella, X. et al., Relevance of digestive tract colonization in the epidemiology of nosocomial infections due to multiresistant *Acinetobacter baumannii*, *Clin. Infect. Dis.*, 23, 329, 1996.
16. Bergogne-Bérézin, E., and Towner, K.J., *Acinetobacter* spp. as nosocomial pathogens: Microbiological, clinical, and epidemiological features, *Clin. Microbiol. Rev.*, 9, 148, 1996.
17. Bergogne-Berezin, E., The increasing role of *Acinetobacter* species as nosocomial pathogens, *Curr. Infect. Dis. Rep.*, 3, 440, 2001.
18. Maragakis, L.L., and Perl, T.M., *Acinetobacter baumannii*: Epidemiology, antimicrobial resistance, and treatment options, *Clin. Infect. Dis.*, 15, 1254, 2008.
19. Linde, H.J. et al., Septicemia due to *Acinetobacter junii*, *J. Clin. Microbiol.*, 40, 2696, 2002.
20. Wendt, C. et al., Survival of *Acinetobacter baumannii* on dry surfaces, *J. Clin. Microbiol.*, 35, 1394, 1997.
21. Denton, M. et al., Role of environmental cleaning in controlling an outbreak of *Acinetobacter baumannii* on a neurosurgical intensive care unit, *J. Hosp. Infect.*, 56, 106, 2004.
22. Fournier, P.E., and Richet, H., The epidemiology and control of *Acinetobacter baumannii* in health care facilities, *Clin. Infect. Dis.*, 42, 692, 2006.
23. Wilks, M. et al., Control of an outbreak of multidrug-resistant *Acinetobacter baumannii-calcoaceticus* colonization and infection in an intensive care unit (ICU) without closing the ICU or placing patients in isolation, *Infect. Control Hosp. Epidemiol.*, 27, 654, 2006.
24. Nemec, A., Dijkshoorn, L., and Jezek, P., Recognition of two novel phenons of the genus *Acinetobacter* among non-glucose-acidifying isolates from human specimens, *J. Clin. Microbiol.*, 38, 3937, 2000.
25. Joly-Guillon, M.L., Clinical impact and pathogenicity of *Acinetobacter*, *Clin. Microbiol. Infect.*, 11, 868, 2005.
26. Seifert, H., Strate, A., and Pulverer, G., Nosocomial bacteremia due to *Acinetobacter baumannii:* Clinical features, epidemiology, and predictors of mortality, *Medicine (Baltimore)*, 74, 340, 1995.
27. Webster, C.A. et al., Molecular and antibiogram relationships of *Acinetobacter* isolates from two contrasting hospitals in the United Kingdom and South Africa, *Eur. J. Clin. Microbiol. Infect. Dis.*, 18, 595, 1999.
28. Chastre, J., Infections due to *Acinetobacter baumannii* in the ICU, *Semin. Respir. Crit. Care Med.*, 24, 69, 2003.
29. van Dessel, H. et al., Outbreak of a susceptible strain of *Acinetobacter* species 13 (sensu Tjernberg and Ursing) in an adult neurosurgical intensive care unit, *J. Hosp. Infect.*, 51, 89, 2002.
30. Wareham, D.W. et al., Bloodstream infection due to *Acinetobacter* spp: Epidemiology, risk factors and impact of multi-drug resistance, *Eur. J. Clin. Microbiol. Infect. Dis.*, 27, 607, 2008.
31. Dijkshoorn, L. et al., Prevalence of *Acinetobacter baumannii* and other *Acinetobacter* spp. in faecal samples from non-hospitalised individuals, *Clin. Microbiol. Infect.*, 11, 329, 2005.
32. Da Silva, G.J., Leitao, G.J., and Peixe, L., Emergence of carbapenem-hydrolyzing enzymes in *Acinetobacter baumannii* clinical isolates, *J. Clin. Microbiol.*, 37, 2109, 1999.
33. Coelho, J.M. et al., Occurrence of carbapenem-resistant *Acinetobacter baumannii* clones at multiple hospitals in London and Southeast England, *J. Clin. Microbiol.*, 44, 3623, 2006.
34. Gonzalez, G. et al., Presence of integrons in isolates of different biotypes of *Acinetobacter baumannii* from Chilean hospitals, *FEMS Microbiol. Lett.*, 161, 125, 1998.
35. Heritier, C. et al., Contribution of acquired carbapenem-hydrolyzing oxacillinases to carbapenem resistance in *Acinetobacter baumannii*, *Antimicrob. Agents Chemother.*, 49, 3198, 2005.
36. Hujer, K.M., Analysis of antibiotic resistance genes in multidrug-resistant *Acinetobacter* sp. isolates from military and civilian patients treated at the Walter Reed Army Medical Center, *Antimicrob. Agents Chemother.*, 50, 4114, 2006.
37. Marques, M.B. et al., Comparative in vitro antimicrobial susceptibilities of nosocomial isolates of *Acinetobacter baumannii* and synergistic activities of nine antimicrobial combinations, *Antimicrob. Agents Chemother.*, 41, 881, 1997.
38. Seward, R.J., Lambert, T., and Towner, K.J., Molecular epidemiology of aminoglycoside resistance in *Acinetobacter* spp., *J. Med. Microbiol.*, 47, 455, 1998.
39. Seward, R.J., and Towner, K.J., Detection of integrons in worldwide nosocomial isolates of *Acinetobacter* spp., *Clin. Microbiol. Infect.*, 5, 308, 1999.
40. Murray, C.K., and Hospenthal, D.R., Treatment of multidrug resistant *Acinetobacter*, *Curr. Opin. Infect. Dis.*, 18, 502, 2005.
41. Wright, M.O., Multi-resistant gram-negative organisms in Maryland: A statewide survey of resistant *Acinetobacter baumannii*, *Am. J. Infect. Control*, 33, 419, 2005.
42. Poirel, L., and Nordmann, P., Carbapenem resistance in *Acinetobacter baumannii*: Mechanisms and epidemiology, *Clin. Microbiol. Infect.*, 12, 826, 2006.

43. Poirel, L. et al., *Acinetobacter radioresistens* as a silent source of carbapenem resistance for *Acinetobacter* spp., *Antimicrob. Agents Chemother.*, 52, 1252, 2008.
44. Qi, C. et al., Characterization of genetic diversity of carbapenem-resistant *Acinetobacter baumannii* clinical strains collected from 2004 to 2007, *J. Clin. Microbiol.*, 46, 1106, 2008.
45. Gordon, N.C., and Wareham, D.W., Evaluation of CHROMagar *Acinetobacter* for detection of enteric carriage of multidrug-resistant *Acinetobacter baumannii* in samples from critically ill patients, *J. Clin. Microbiol.*, 47, 2249, 2009.
46. Bouvet, P.J.M., and Grimont, P.A.D., Identification and biotyping of clinical isolates of *Acinetobacter*, *Ann. Inst. Pasteur Microbiol.*, 138, 569, 1987.
47. Bouvet, P.J.M., and Jeanjean, S., Delineation of new proteolytic genomic species in the genus *Acinetobacter*, *Res. Microbiol.*, 140, 291, 1989.
48. Dolzani, L. et al., Identification of *Acinetobacter* isolates in the *A. calcoaceticus-A. baumannii* complex by restriction analysis of the 16S-23S rRNA intergenic-spacer sequences, *J. Clin. Microbiol.*, 33, 1108, 1995.
49. Vaneechoutte, M. et al., Identification of *Acinetobacter* genomic species by amplified ribosomal DNA restriction analysis, *J. Clin. Microbiol.*, 33, 11, 1995.
50. García-Arata, M.I. et al., PCR-amplified 16S and 23S rDNA restriction analysis for the identification of *Acinetobacter* strains at the DNA group level, *Res. Microbiol.*, 148, 777, 1997.
51. Ibrahim, A., Gerner-Smidt, P., and Liesack, W., Phylogenetic relationship of the twenty-one DNA groups of the genus *Acinetobacter* as revealed by 16S ribosomal DNA sequence analysis, *Int. J. Syst. Bacteriol.*, 47, 837, 1997.
52. Yamamoto, S., Bouvet, P.J., and Harayama, S., Phylogenetic structures of the genus *Acinetobacter* based on *gyrB* sequences: Comparison with the grouping by DNA–DNA hybridization, *Int. J. Syst. Bacteriol.*, 491, 87, 1999.
53. Vanderbrouckr-Grauls, C.M.J.E., and Savelkoul, P.H.M., Identification of epidemic strains of *Acinetobacter baumannii* by integrase gene PCR, *J. Clin. Microbiol.*, 39, 8, 2001.
54. Chang, H.C. et al., Species-level identification of isolates of the *Acinetobacter calcoaceticus-Acinetobacter baumannii* complex by sequence analysis of the 16S-23S rRNA gene spacer region, *J. Clin. Microbiol.*, 43, 1632, 2005.
55. Misbah, S., et al., Genomic species identification of *Acinetobacter* of clinical isolates by 16S rDNA sequencing, *Singapore Med. J.*, 46, 461, 2005.
56. La Scola, B. et al., Sequencing of the rpoB gene and flanking spacers for molecular identification of *Acinetobacter* species, *J. Clin. Microbiol.*, 44, 827, 2006.
57. Chen, T.L. et al., Comparison of one-tube multiplex PCR, automated ribotyping and intergenic spacer (ITS) sequencing for rapid identification of *Acinetobacter baumannii*, *Clin. Microbiol. Infect.*, 13, 801, 2007.
58. Higgins, P.G. et al., A PCR-based method to differentiate between Acinetobacter baumannii and Acinetobacter genomic species 13TU, *Clin. Microbiol. Infect.*, 13, 1199, 2007.
59. Maslunka, C. et al., Unique organization of the 16S-23S intergenic spacer regions of strains of *Acinetobacter baylyi* provides a means for its identification from other *Acinetobacter* species, *J. Microbiol. Methods*, 73, 227, 2008.
60. Gundi, V.A. et al., Validation of partial rpoB gene sequence analysis for the identification of clinically important and emerging *Acinetobacter* species, *Microbiology*, 155, 2333, 2009.
61. Dijkshoorn, L. et al., Evaluation of amplified ribosomal DNA restriction analysis for identification of *Acinetobacter* genomic species, *Syst. Appl. Microbiol.*, 21, 33, 1998.
62. Ecker, J.A. et al., Identification of *Acinetobacter* species and genotyping of *Acinetobacter baumannii* by multilocus PCR and mass spectrometry, *J. Clin. Microbiol.*, 44, 2921, 2006.
63. Bou, G. et al., Characterization of a nosocomial outbreak caused by a multiresistant *Acinetobacter baumannii* strain with a carbapenem-hydrolyzing enzyme: High-level carbapenem resistance in *A. baumannii* is not due solely to the presence of beta-lactamases, *J. Clin. Microbiol.*, 38, 3299, 2000.
64. van Dessel, H. et al., Identification of a new geographically widespread multiresistant *Acinetobacter baumannii* clone from European hospitals, *Res. Microbiol.*, 155, 105, 2004.
65. Naas, T., Poirel, L., and Nordmann, P., Pyrosequencing for rapid identification of carbapenem-hydrolysing OXA-type beta-lactamases in *Acinetobacter baumannii*, *Clin. Microbiol. Infect.*, 12, 1236, 2006.
66. Woodford, N. et al., Multiplex PCR for genes encoding prevalent OXA carbapenemases in *Acinetobacter spp.*, *Int. J. Antimicrob. Agents*, 27, 351, 2006.
67. Aktaş, Z., and Kayacan, C.B., Investigation of metallo-beta-lactamase producing strains of *Pseudomonas aeruginosa* and *Acinetobacter baumannii* by E-test, disk synergy and PCR, *Scand. J. Infect. Dis.*, 40, 320, 2008.
68. Feizabadi, M.M. et al., Antimicrobial susceptibility patterns and distribution of blaOXA genes among *Acinetobacte*r spp. isolated from patients at Tehran hospitals, *Jpn. J. Infect. Dis.*, 61, 274, 2008.
69. Mostachio, A.K. et al., Multiplex PCR for rapid detection of genes encoding oxacillinases and metallo-beta-lactamases in carbapenem-resistant *Acinetobacter* spp., *J. Med. Microbiol.*, 58, 1522, 2009.
70. Srinivasan, V.B. et al., Genetic relatedness and molecular characterization of multidrug resistant *Acinetobacter baumannii* isolated in central Ohio, USA, *Ann. Clin. Microbiol. Antimicrob.*, 8, 21, 2009.
71. Uma Karthika, R. et al., Phenotypic and genotypic assays for detecting the prevalence of metallo-beta-lactamases in clinical isolates of *Acinetobacter baumannii* from a South Indian tertiary care hospital, *J. Med. Microbiol.*, 58, 430, 2009.
72. Hartstein, A.I. et al., Plasmid DNA fingerprinting of *Acinetobacter calcoaceticus* subspecies *anitratus* from intubated and mechanically ventilated patients, *Infect. Control Hosp. Epidemiol.*, 11, 531, 1990.
73. Gerner-Smidt, P., Tjernberg, I., and Ursing, J., Reliability of phenotypic tests for identification of *Acinetobacter* species, *J. Clin. Microbiol.*, 29, 277, 1991.
74. Gerner-Smidt, P., Ribotyping of the *Acinetobacter calcoaceticus-Acinetobacter baumannii* complex, *J. Clin. Microbiol.*, 30, 2680, 1992.
75. Gerner-Smidt, P., and Tjernberg, I., *Acinetobacter* in Denmark. II. Molecular studies of the *Acinetobacter calcoaceticus-Acinetobacter baumannii* complex, *APMIS*, 101, 826, 1993.
76. Dijkshoorn, L. et al., Correlation of typing methods for *Acinetobacter* isolates from hospital outbreaks, *J. Clin. Microbiol.*, 31, 702, 1993.
77. Dijkshoorn, L. et al., Comparison of outbreak and nonoutbreak *Acinetobacter baumannii* strains by genotypic and phenotypic methods, *J. Clin. Microbiol.*, 34, 1519, 1996.
78. Seltmann, G. et al., Comparative classification of *Acinetobacter baumannii* strains using seven different typing methods, *Zentbl. Bakteriol.*, 282, 372, 1995.
79. Nowak, A., and Kur, J., Differentiation of seventeen genospecies of *Acinetobacter* by multiplex polymerase chain reaction and restriction fragment length polymorphism analysis, *Mol. Cell. Probes*, 10, 405, 1996.

80. Koeleman, J.G.M. et al., Comparison of amplified ribosomal DNA restriction analysis, random amplified polymorphic DNA analysis, and amplified fragment length polymorphism fingerprinting for identification of *Acinetobacter* genomic species and typing of *Acinetobacter baumannii*, *J. Clin. Microbiol.*, 36, 2522, 1998.
81. Koeleman, J.G. et al., Identification of epidemic strains of Acinetobacter baumannii by integrase gene PCR, *J. Clin. Microbiol.*, 39, 8, 2001.
82. Bou, G. et al., PCR-based DNA fingerprinting (REP-PCR, AP-PCR) and pulsed-field gel electrophoresis characterization of a nosocomial outbreak caused by imipenem- and meropenem-resistant *Acinetobacter baumannii*, *Clin. Microbiol. Infect.*, 6, 635, 2000.
83. Spence, R.P. et al., Comparison of *Acinetobacter baumannii* isolates from United Kingdom hospitals with predominant Northern European genotypes by amplified-fragment length polymorphism analysis, *J. Clin. Microbiol.*, 42, 832, 2004.
84. Bartual, S.G. et al., Development of a multilocus sequence typing scheme for characterization of clinical isolates of *Acinetobacter baumannii*, *J. Clin. Microbiol.*, 43, 4382, 2005.
85. Huys, G. et al., Repetitive-DNA-element PCR fingerprinting and antibiotic resistance of pan-European multi-resistant *Acinetobacter baumannii* clone III strains, *J. Med. Microbiol.*, 54, 851, 2005.
86. Huys, G. et al., Sequence-based typing of *adeB* as a potential tool to identify intraspecific groups among clinical strains of multidrug-resistant *Acinetobacter baumannii*, *J. Clin. Microbiol.*, 43, 5327, 2005.
87. Ecker, D.J. et al., Rapid identification and strain-typing of respiratory pathogens for epidemic surveillance, *Proc. Natl. Acad. Sci. USA*, 102, 8012, 2005.
88. Ecker, D.J. et al., Molecular genotyping of microbes by multilocus PCR and mass spectrometry: A new tool for hospital infection control and public health surveillance, *Methods Mol. Biol.*, 551, 71, 2009.
89. Turton, J.F. et al., Detection and typing of integrons in epidemic strains of *Acinetobacter baumannii* found in the United Kingdom, *J. Clin. Microbiol.*, 43, 3074, 2005.
90. Turton, J.F. et al., Use of sequence-based typing and multiplex PCR to identify clonal lineages of outbreak strains of *Acinetobacter baumannii*, *Clin. Microbiol. Infect.*, 13, 807, 2007.
91. Alp, E. et al., Genotypic analysis of *Acinetobacter* bloodstream infection isolates in a Turkish university hospital, *Scand. J. Infect. Dis.*, 38, 335, 2006.
92. Lee, J.H. et al., Differences in phenotypic and genotypic traits against antimicrobial agents between *Acinetobacter baumannii* and *Acinetobacter* genomic species 13TU, *J. Antimicrob. Chemother.*, 59, 633, 2007.
94. Wisplinghoff, H. et al., Molecular epidemiology of clinical *Acinetobacter baumannii* and *Acinetobacter* genomic species 13TU isolates using a multilocus sequencing typing scheme, *Clin. Microbiol. Infect.*, 14, 708, 2008.
95. Fontana, C. et al., *Acinetobacter baumannii* in intensive care unit: A novel system to study clonal relationship among the isolates, *BMC Infect. Dis.*, 8, 79, 2008.
95. Pechorsky, A., Nitzan, Y., and Lazarovitch, T., Identification of pathogenic bacteria in blood cultures: Comparison between conventional and PCR methods, *J. Microbiol. Methods*, 78, 325, 2009.

68 Aeromonas

Germán Naharro, Pedro Rubio, and José Maria Luengo

CONTENTS

68.1 INTRODUCTION

Members of the genus *Aeromonas* are gram-negative bacteria that occur in an aquatic environment and may cause several types of infection in humans and animals.[1] In humans, *Aeromonas* causes septicemia, usually associated with immunosuppression resulting from hepatitis, cirrhosis, biliary disease, pancreatic disease, or malignancies; wound infections and cellulite in individuals exposed to water or soil; extraintestinal infections such as peritonitis, meningitis, endocarditis, otitis of the middle ear, and so forth.[2] Several reports have implicated *Aeromonas* species as the aetiological agents of acute diarrhea, but the evidence of enteropathogenicity is controversial because the experimentation undertaken to solve the issue has proven insufficient, and conclusive human volunteer trials and animal models are required.[3] *Aeromonas* is not considered to be a normal inhabitant of the human gastrointestinal tract, but it is present in the feces of healthy animals and humans, probably due to the ingestion of water and foods contaminated with these microorganisms.[4] *Aeromonas* may form a biofilm after the colonization of drinking water distribution systems, where it is resistant to common disinfectants. However, no outbreaks attributable to *Aeromonas* through contact with drinking water have been described.[5]

The taxonomy of the genus *Aeromonas* is complex and confusing because of a lack of congruity between the phenotypic and genotypic properties of the various species based on biochemical and DNA–DNA hybridization characteristics, respectively, and multiple methods are required for accurate classification, which exceeds the analytical capacity of most clinical laboratories. In the most recent edition of Bergey's Manual, Martin-Carnahan and Joseph[1] described 17 DNA–DNA hybridization groups (HG) and 14 phenospecies. Several *Aeromonas* species have recently been added to the list, including *A. simiae*, *A. molluscorum*, *A. culicicola*, although the latter may be synonymous of *A. veronii*. Some are synonymous of previously recognized species, such as *A. ichthiosmia* (*A. veronii* biovar *sobria*) and *A. enteropelogenes* (*A. trota*). Some researchers consider the continued use of hybridization groups to be controversial, since all phenospecies named can be identified phenotypically.

Aeromonas infection in humans was first reported in 1954 in a Jamaican woman with myositis.[6] In 1968, von Graevenitz and Mensch reviewed 30 cases of infections and established *Aeromonas* species as significant human pathogens, causing a variety of extraintestinal infections in various organ systems.[7] *Aeromonas* infections typically follow trauma in an aquatic environment, or dissemination from the gastrointestinal tract as a result of underlying diseases such as cancer. Human infections are known to occur in both healthy hosts and immunocompromised individuals.[8,9] *Aeromonas* infections in humans are mainly produced by *A. hydrophila* (HG1), *A. caviae* (HG4), *A. veronii* biovar *sobria* (HG8), *A. jandei* (HG9), *A. veronii* biovar *veronii* (HG10), *A. schuvertii* (HG12), and *A. trotta* (HG14). Probably, *A. trotta* (HG14) has been underestimated, since it is sensitive to ampicillin, an antibiotic included in most *Aeromonas* isolation media.[9] Fortunately, 85% of human clinical isolates are *A. hydrophila* (HG1), *A. caviae* (HG4), and *A. veronii* biovar *sobria* (HG8).

Aeromonas species produce virulence factors, including cell-associated virulence factors and extracellular virulence factors. Among the cell-associated virulence factors are surface structures such as pili, flagella, outer membrane proteins, lipopolysaccharide, and capsules. The major extracellular products are cytotonic, cytolytic, hemolytic, enterotoxic, lipolytic, and proteolytic proteins.[8,10,11]

Some *Aeromonas* species are pathogenic for fish, amphibians, and reptiles. In fish, they are the aetiological agents of hemorrhagic disease, ulcerative disease, furunculosis, and septicemia.[12] *Aeromonas* species have also been described as pathogenic for birds and domestic animals, causing abortion, pneumonia, peritonitis, and various other diseases.[12]

68.1.1 Classification, Biology, and Epidemiology

The genus *Aeromonas* consists of straight coccobacillary to bacillary gram-negative bacteria that are classified within the family Aeromonadaceae, order Aeromonadales, class Gammaproteobacteria, phylum Proteobacteria. *Aeromonas* cells measure 0.3–1.0 × 1.0–3.5 μm with round ends[1] and often appear singly, in pairs, and occasionally in short chains. Motile strains produce a single polar flagellum, but peritrichous or lateral flagella may be formed by some species in solid media. *Aeromonas* spp. are facultatively anaerobic, catalase-positive, oxidase-positive, chemoorganotrophic bacteria with a fermentative and oxidative carbohydrate metabolism. In addition, they reduce nitrate, ferment D-glucose, and are resistant to the vibriostatic agent O129. *Aeromonas* spp. produce a broad variety of extracellular hydrolytic enzymes such as arylamidases, amylase, deoxyribonuclease, esterases, peptidases, elastase, chitinase, and lipase.[13,14] *Aeromonas* spp. grow optimally within a temperature range between 22°C and 35°C, although some species are able to grow in the 0°C–45°C temperature range.[15] Some species, including most nonmotile *A. salmonicida* strains, do not grow at 35°C.[1] *Aeromonas* spp. tolerate a pH range from 4.5 to 9.0, but the optimum pH range is from 5.5 to 9.0,[16] and the optimum sodium chloride concentration range is from 0% to 4%.

Aeromonas species are natural inhabitants of aquatic environments, where they may be primary pathogens for poikilotherms (cool-blooded animals). They are found in soils, sediments, and water in freshwater, estuarine, and marine environments. They are present in foods, sewage, and drinking water distribution systems. *Aeromonas* are found in the intestines of animals and humans where they may or may not cause gastrointestinal disease. Some aeromonads are pathogenic for humans, and most human clinical isolates belong to HG1, HG4, HG8, HG9, HG10, HG12, or HG14.[9] The proportion of strains within these hybridization groups able to cause human disease is unknown. The hybridization groups HG2, HG3, HG5, HG6, HG7, HG11, HG15, HG16, or HG17 are isolated from the environment or from diseased animals, and they are not usually considered to be human pathogens. Some strains of *Aeromonas* produce enterotoxins responsible for gastroenteritis in humans; however, the isolation of aeromonads from feces does not prove pathogenicity since these bacteria are widely distributed in water and foods, especially in summer.[9]

Aeromonas lipopolysaccharide, the major surface antigen (O), as in other gram-negative bacteria, is the main parameter used in serotyping.[17]

Most motile aeromonads are generally resistant to penicillin, ampicillin, carbenicillin, and ticarcillin, and they are typically susceptible to second- and third-generation cephalosporins, aminoglycosides, carbapenems, chloramphenicol, tetracyclines, trimethoprim-sulfamethoxazole, and quinolones.[9] Most aeromonads produce inducible chromosomal β-lactamase[18] but *A. trota* and up to 30% of *A. caviae* are susceptible to ampicillin. Resistance to streptomycin, chloramphenicol, tetracycline, cephalexin, erythromycin, furazolidone, and sulfathiazole is mediated by plasmids.[19]

The taxonomy of *Aeromonas* has been extensively revised during the last 20 years and became complicated in the 1980s because few phenotypic markers were available to differentiate species for the newly recognized hybridization groups (HG). The first observation of *Aeromonas* is attributed to Zimmermann in 1890,[20] who considered these microorganisms as belonging to the genus *Bacillus* (*Bacillus punctatus*). In 1891, Sanarelli[21] isolated a strain from infected frog blood: it was named *Bacillus hydrophilus fuscus* and was considered to be an aeromonad. In 1936, Kluyver and van Neil[22] coined the name *Aeromonas*, from the Greek words *aer* (air or gas) and *monas* (unities), which means "gas-producing units."

According to the second edition of *Bergey's Manual of Systematic Bacteriology*[1] the genus *Aeromonas* includes 14 phenospecies and at least 17 hybridization groups (HG), although the list is not closed since new species are continuously added (Table 68.1). The genus *Aeromonas* belongs to the family *Aeromonadaceae*, order *Aeromonadales*, subclass *γ-Proteobacteria*, class *Proteobacteria*. Current taxonomy is based on DNA–DNA hybridization analysis and 16S rRNA sequences. Nevertheless, some discrepancies have been observed between 16S rRNA sequences and DNA–DNA hybridization results owing to the high degree of conservation of 16S rRNA sequences (97.8%–100%) and the existence of several copies of the gene with intragenomic gene polymorphism, especially in *A. media* and *A. veronii*.[23] Recently, some investigators have started to use sequences other than 16S rRNA to identify *Aeromonas* to the genomic species level, such as *rpoB*, *gyrB*,[24] *dnaJ*,[25] and *recA*.[26]

Aeromonas species are ubiquitous and find multiple opportunities for transmission to humans through food, water, animal contact, and direct human contact. Extraintestinal infections may be acquired after trauma in aquatic environments, and intestinal infections are acquired by ingestion of contaminated food or water. In addition, intestinal infections in immunocompromised patients may disseminate, resulting in septicemia with multiple organ involvement. *Aeromonas* species have been recognized as significant agents in foodborne illnesses, gastroenteritis being the most commonly reported infection that range from mild, self-limiting watery diarrhea to a more severe, invasive *Shigella*-like dysenteric

TABLE 68.1
Current Genomospecies and Phenospecies of the Genus *Aeromonas*

HG	Type Strain	Genomospecies	Phenospecies
1	ATCC 7966	*A. hydrophila*	*A. hydrophila*
2	ATCC 14715	*A. bestiarum*	*A. hydrophila*-like
3	ATCC 33658	*A. salmonicida*	*A. salmonicida* ssp *salmonicida*
4	ATCC 15468	*A. caviae*	*A. caviae*
5A	CDC 0862-83	*A. media*	*A. caviae*-like
5B	CDC 0435-84	*A. media*	*A. media*
6	ATCC 23309	*A. eucrenophila*	*A. eucrenophila*
7	CIP 7433	*A. sobria*	*A. sobria*
8X	CDC 0437-84	*A. veronii*	*A. sobria*
8Y	ATCC 9071	*A. veronii*	*A. veronii* biovar *sobria*
9	ATCC 49568	*A. jandei*	*A. jandei*
10	ATCC 35624	*A. veronii* biovar *veronii*	*A. veronii* biovar *veronii*
11	ATCC 35941	Unnamed	*Aeromonas* spp. (ornithine positive)
12	ATCC 35941	*A. schubertii*	*A. schubertii*
13	ATCC 43946	*Aeromonas* Group 501	*A. schubertii*-like
14	ATCC 49657	*A. trota*	*A. trota*
15	CECT 4199	*A. allosaccharophila*	*A. allosaccharophila*
16	CECT 4342	*A. encheleia*	*A. encheleia*
17	LMG 1754	*A. popoffii*	*A. popoffii*

form.[27] *Aeromonas* is frequently isolated from foods, such as green vegetables, raw milk, ice cream, meat, and seafood.[28] Von Graevenitz defined *Aeromonas*-associated diarrhea as loose stools associated with the isolation of aeromonads in the stools.[27] However, the evidence of outbreaks with aeromonads is controversial and almost nonexistent in spite of their occurrence in drinking water and in foods such as seafood, meat, and vegetables.[29] The problem arises because of differences in the diarrheagenic potential among *Aeromonas* strains. A general consensus appears to be growing that certain strains are likely to be human enteric pathogens.[9] Several typing methods have been used to link environmental and clinical strains of aeromonads; however, the diversity of strains has resulted in little success in all but a few cases. Ribotyping was used to demonstrate that the ingestion of shrimp resulted in gastroenteritis in the first report of foodborne illness attributed to *Aeromonas* species.[30] The same method was used to confirm that a patient with chronic diarrhea carried the same strain for years and to demonstrate person-to-person transmission of *Aeromonas* species between a foster child and a foster parent.[31]

Only four phenospecies of the genus *Aeromonas* have been isolated from human feces with significant frequency: *A. hydrophila* (HG1), *A. caviae* (HG4), *A. veronii* biovar *sobria*, (HG8), and *A. trota* (HG14).[1,2] *A. jandei* (HG9) and *A. schubertii* (HG12) have been rare fecal isolates. In any case, it is expected that if these species are indeed agents of diarrhea, they should be more frequently isolated from patients with *Aeromonas*-associated diarrhea than in controls. Several studies have demonstrated that the epidemiological behavior of *Aeromonas* is different from that of other enteropathogenic bacteria, and there are even differences in the geographical and seasonal distributions, and the predominant species involved. Some studies observed isolation peaks in warmer seasons,[32] while others found no seasonal preferences.[33] *A. caviae* was the predominant species in most studies, followed by *A. hydrophila* and *A. veronii* biovar *sobria*.[3,32,33] *A. hydrophila* predominates in Brazil and Thailand.[34,35] *A. caviae* initially predominated in Bangladesh, but later *A. trota* was held responsible.[36] In Finnish tourists visiting Morocco, diarrhea was attributed to *A. veronii* biovar *sobria*.[37] While some studies did find significant differences in the frequency of aeromonads isolated from diarrheic patients including travellers[37] as compared to controls,[33] others did not.[38] The recovery rates of aeromonads from children with diarrhea vary geographically: 0.62%–4% in Malaysia,[39] 0.75% in Nigeria,[40] 2% in Sweden,[38] 4.8% in Switzerland,[41] and 6.8% in Greece.[42] In Thailand, the difference was significant for western Peace Corps workers but not for indigenous people.[35] In a quantitative study, a considerable variation in the number of colony-forming units (CFU) of aeromonads during different sampling periods was observed between patients and controls.[43] In another study, the amount of growth on selective medium was not found to be associated with the onset of diarrhea.[3] The isolation rate for human fecal specimens varies widely, since the geographical area, patient population, food habits, level of sanitation, and culture methods influence the recovery rate.[44]

68.1.2 Clinical Features and Pathogenesis

Some *Aeromonas* species are opportunistic pathogens of human beings, causing a variety of extraintestinal infections as well as gastrointestinal disease. Von Graevenitz and Mensch published the first comprehensive review of human infections caused by *Aeromonas* species.[7] Since then, a vast number of reviews and papers has been published on *Aeromonas* species as agents of human disease,[2,8,27] such as cellulite, abscesses, wound infection, necrotizing fasciitis, myonecrosis, pneumonia, empyema, septicemia, septic arthritis, osteomyelitis, endocarditis, meningitis, gastroenteritis, appendicitis, peritonitis, acute suppurative cholangitis, and corneal ulcer. The clinical symptoms of *Aeromonas* infections depend on the site and severity of infection. Septicemia may follow wound infection, and dissemination may result in meningitis or endocarditis. Patients presenting gastroenteritis symptoms range from self-limiting diarrhea to acute, severe diarrhea with abdominal cramping, vomiting, and fever. Children under 12 are likely to have more acute and severe illnesses, whereas adults may have chronic diarrhea and abdominal cramps. *A. caviae* and *A. hydrophila* are more associated with chronic diarrhea.[45]

The virulence of aeromonads is multifactorial and incompletely understood despite years of intensive investigation. Many putative virulence factors have been described, and have usually been grouped into two classes: cell-associated virulence factors (outer membrane proteins, S or A-layer, lipopolysaccharide, capsule, flagella, and pili), and extracellular virulence factors (toxins, enterotoxins, proteases, hemolysins and lipases, etc).[46] In the early literature, only one virulence factor—the A-layer of *A. salmonicida*—was unequivocally linked to disease causation. However, much progress has been made in understanding the virulence mechanisms of these microorganisms since then.[47]

68.1.2.1 Cell-Associated Virulence Factors

Invasins. Intracellular *Aeromonas* spp. have been demonstrated by electron microscopy, although no gene or gene product has been identified specifically with invasion, their invasive ability proving to be similar to that of *Campylobacter*.[48]

Adhesins. Adherence ability at the *Aeromonas* surface has been reported, together with a correlation between high-level HEp-2 cell adherence and enteropathogenicity.[49]

Outer Membrane Proteins. Outer membrane proteins (OMP) are involved in the adherence of *Aeromonas* spp. to cells. A 43-kDa OMP from *A. caviae* promoting adhesion to Hep-2 cells has been reported.[50] This 43-kDa OMP expressed higher levels when *A. caviae* was grown at 22°C than when grown at 37°C. Similar adherence characteristics have been observed in *A. hydrophila*,[51] which have been characterized at molecular level.

S-Layer Proteins. The S-layer (also termed A-layer) has been extensively studied and has been linked to virulence in that it provides protection against bactericidal serum ability, protecting cells from proteolysis, facilitating association with macrophages, binding collagen, laminin, fibronectin, and playing a role in colonization. When these genes are insertionally inactivated, the S-layer is lost and virulence is markedly reduced. S-layers are thought to protect against bacterial parasites such as *Bdellovibrio bacteriovorus*, although they are ineffective against protozoa.[52]

Lipopolysaccharide. Lipopolysaccharide is one of major surface antigens (O-antigen) of the gram-negative bacterial cell wall that consist of three domains: Lipid A, core oligosaccharide, and the distalmost O-specific polysaccharide or O antigen. Lipopolysaccharide plays a role in adhesion to epithelial cells, resistance to nonimmune serum, and virulence. The serotyping of aeromonads is based on the distalmost O-antigen of LPS. LPS is thought to be a colonization factor since its presence facilitates adherence to HEp-2 cells. The loss of LPS reduces adhesion by 60%.[53]

Capsule. Most motile aeromonads have no capsule, so few studies have examined the role of capsule polysaccharide as a virulence factor. Capsules have been reported in *A. hydrophila* serotypes O:11 and O:34 when grown in a glucose-rich medium. The capsule may play a role in septicemia since nonencapsulated strains are less virulent. The capsule material has the ability to protect cells from complement-mediated serum, killing activity.[54]

Flagella. *Aeromonas* generally have monotrichous and polar flagella. However, some strains present lateral flagella, and some strains are nonflagellated. Peritrichous flagella are unsheathed and are associated with a swarming movement across the surfaces of solid media. Flagellar production is under the control of over 40 genes and allows *Aeromonas* to reach target cells, where they colonize. Lateral flagella facilitate biofilm formation and persistence during infection.[55] Polar sheathed flagella are expressed constitutively in liquid media and peritrichous lateral flagella are produced in matrices that do not permit motility by a single polar flagellum, such as growth on solid media.[56]

Pili (Fimbriae). *Aeromonas* attachment to host cells is mediated by surface appendages such as pili which have been described as potential colonization factors in *A. hydrophila* and *A. veronii* biovar *sobria*.[57] Two morphotypes of pilus have been observed and reported to be important adhesive factors for mucosal surface attachment in *Aeromonas* species, short rigid pili (S/R type) and long wavy flexible pili (L/W type).[57]

68.1.2.2 Extracellular Virulence Factors

Proteases. Most aeromonads produce a broad variety of extracellular enzymes, some of which may be involved in pathogenesis,[58] and virulence. It has been described that aeromonads differ in the elaboration of proteases according to differences in strain origin, incubation temperature, and the culture medium.[15] Most strains of *A. hydrophila* and *A. veronii* biovar *sobria* produce two types of proteases, a thermostable metalloprotease (TSMP) that is sensitive to EDTA, and a thermolabile serine protease (TLSP) that is sensitive to PMSF.[59] The two major extracellular proteolytic activities of *A. hydrophila* described in our lab, a 38-kDa TSMP[13] and a 68-kDa TLSP,[14] are present in most *A. hydrophila* culture supernatants. We have also demonstrated that the 38-kDa TSMP, with high elastolytic activity, is essential for virulence in rainbow trout while the 68-kDa TLSP is not.[10,11]

Lipases. Several lipase-encoding genes have been cloned for *Aeromonas*, and in some cases mutation of these genes reduces lethal effects in mice and fish[60]; however, we have constructed an *A. hydrophila* lipase mutant by allelic replacement and the lethal effects in fish were not reduced (unpublished results).

Glycerophospholipid Cholesterol Acyl-Transferase (GCAT). GCAT activity has been described as a major virulence factor in *A. salmonicida* resulting in furunculosis in fish. A similar enzyme has also been described in *A. hydrophila* that hydrolyzes phospholipids. However, the role of GCAT in human pathogenicity is unknown.[61]

Superoxide Dismutase. An *A. hydrophila* manganese superoxide dismutase has been described with an equivalent function to the Cooper/Zinc superoxide dismutase of *E. coli*. This enzyme is located in the periplasmic space of *A. hydrophila* and may confer a virulence advantage on the bacteria.[62]

Toxins. *Aeromonas* species produce several toxins considered as virulence factors, such as hemolysins, and cytotoxic and cytotonic enterotoxins. The first toxin purified to homogeneity from an *Aeromonas* was a heat-labile 49- to 52-kDa *β*-hemolysin, which that induced fluid secretion in an animal model.[63] This is the major hemolysin produced by *Aeromonas* and is called aerolysin (*Aer*A), although it is also known by other names (cytotoxin enterotoxin, Asao toxin, and cholera toxin cross-reactive cytolytic enterotoxin). Aerolysin is produced by some strains of *A. hydrophila*, *A. veronii* biovar *sobria*, and *A. caviae*. The aerolysin coding gene, *aerA*, is present in all *Aeromonas* strains but its expression depends on the prevailing environmental conditions.[64]

Furthermore, three cytotoxic and cytotonic enterotoxins have been described in *Aeromonas* species: *Aeromonas* cytotoxic enterotoxin (Act), *Aeromonas* heat-labile (56°C) cytotonic enterotoxin (Alt), and *Aeromonas* heat-stable cytotonic enterotoxin (Ast).[65] Act is a single-chain polypeptide of 52-kDa that is related to aerolysin and has hemolytic, cytotoxic, and enterotoxic activities. The role of Act in *Aeromonas* virulence has been clearly demonstrated by LD_{50} studies and by the inability of *act* isogenic mutants to elicit a fluid secretory response in mouse ligated ileal loops. Alt is a heat-labile cytotonic enterotoxin exhibiting a size of 44-kDa single polypeptide that causes elevation of cyclic AMP and prostaglandin levels in CHO and intestinal epithelial cells, resulting in fluid accumulation in rat ileal loops, while Ast is a heat-stable cytotonic enterotoxin with a similar molecular mass and properties.[64]

68.1.3 Diagnosis

68.1.3.1 Conventional Techniques

A variety of selective and differential media have been proposed to isolate *Aeromonas* species, and even broth enrichment methods are frequently used to recover aeromonads from heavily contaminated samples.[66] Of more than 17 genomospecies of *Aeromonas*, however, only 7 are associated with human diarrhea. These include *A. hydrophila*, *A. caviae*, *A. veronii* biovar sobria, *A. veronii* biovar veronii, *A. schubertii*, *A. jandaei*, and *A. trota*. Eighty-five percent of clinical isolates belong to *A. hydrophila* (HG1), *A. caviae* (HG4), and *A. veronii* biovar sobria (HG8).[1] Selectivity is based on the use of media supplemented with ampicillin and/or inhibitors used in culture media for *Enterobacteriaceae* (bile salts, brilliant green, and sodium lauryl sulfate). Differential media are most often based on the presence of amylase in the microorganisms to be isolated and the pattern of carbohydrate fermentation, but no single medium has yet received general acceptance. Several studies have been performed by comparing a series of selective media with or without enrichment in alkaline peptone water (APW). Some authors have concluded that ampicillin dextrin agar (ADA) produces the best overall results; however, this selective medium is not appropriate for the isolation of *A. trota* and certain strains of *A. caviae* because they are sensible to ampicillin.[67] ADA medium supplemented with vancomycin (ADA-V) has been validated by the USEPA for the detection and enumeration of *A. hydrophila* in drinking water.[68] One interesting study evaluated seven different selective media and two enrichment broths for the isolation of mesophilic *Aeromonas* from meat and fish. It was concluded that enrichment in APW, and consecutive plating in two selective media—ampicillin-sheep-blood agar supplemented with 30 μg/mL of ampicillin (ASBA 30) and bile salts-Irgasan-brilliant green agar (BIBG)—should be recommended for qualitative isolation of *Aeromonas* species from meat and fish. Quantitative examination of samples may be performed with the same media, using a direct plating method. An incubation temperature of 25°C–30°C is recommended for environmental and food samples.[67] In general, the optimum recovery of aeromonads requires more than one agar medium. The same procedure may be used to isolate aeromonads from heavily contaminated clinical specimens such as feces, although in this case a temperature of incubation of 35°C is recommended.

BIBG medium was originally designed for the selective isolation of aeromonads from faeces.[67] Bile salts and brilliant green inhibit the growth of gram-positive bacteria and Irgasan inhibits the growth of gram-negative bacteria, which possess a type A nitratase. Bacteria that survive the selective process can be differentiated by their ability to assimilate-ferment xylose. Noninoculated BIBG is a purple-red, transparent medium. If xylose is fermented, acid is produced and the medium retains its original, noninoculated color. Mesophilic aeromonads are not able to ferment xylose and basic compounds are formed due to the breakdown of proteins, the medium changing from purple to a green-yellow color. *Aeromonas* species appear as translucent colonies and have a white-green appearance. The BIBG medium has been modified (mBIBG) by increasing the pH up to 8.7 and replacing xylose by soluble starch as a carbon source, with very good results.[69]

Once *Aeromonas* species have been isolated and detected, they must be identified at species level. The greatest difficulties in the identification of *Aeromonas* strains at species level derive from the current number of recognized taxa (n = 14) and the lack of appropriate phenotypic schemes that allow each of these groups to be distinguished from the others.[46] Most clinical microbiological laboratories use the Aerokey II system as an accurate and reliable system for the phenotypic identification of currently recognized *Aeromonas* species isolated from clinical specimens and related to human diarrhea, namely *A. hydrophila*, *A. caviae*, *A. veronii* biovar sobria, *A. veronii* biovar veronii, *A. schubertii*, *A. jandaei*, and *A. trota*.[70] The Aerokey II system correctly identified 97% of the 60 coded clinical isolates from an independent laboratory and 100% of the reference strains at species level, and it may be considered as a very accurate *Aeromonas* identification system.[70] However, Aerokey II and many other biochemical schemes used in clinical laboratories for aeromonad identification predate the description of newer taxa and may not be applicable for aeromonad identification described more recently.

An interesting study has been performed on 193 strains of *Aeromonas*, representing the 14 species currently recognized for 63 phenotypic traits.[71] In this study, a schema was designed for the identification of aeromonads, first at group and then at species level within each group, and it recommended the tests to be performed in clinical and reference laboratories. After isolation of aeromonads on selective media, it is possible to identify clinical isolates of *Aeromonas* species in one of the three major complexes or groups of significance in human disease: the *A. hydrophila* group (*A. hydrophila*, *A. bestiarum*, and *A. salmonicida*); the *A. caviae* group (*A. caviae*, *A. media*, and *A. eucrenophila*), and the *A. sobria* group (*A. veronii* HG8, *A. jandei*, *A. schubertii*, and *A. trota*). To separate species within complexes, sets of biochemical reactions were described.[1,71]

68.1.3.2 Molecular Techniques

Molecular methods such as nucleic acid amplification with the polymerase chain reaction (PCR) or genetic probes have been used profusely over the past 15 years for the direct detection of aeromonads in a variety of matrices such as raw milk, water, flies, fish, feces, and others.[72] PCR methods are usually more sensitive than culturing for complex matrices that are heavily contaminated with other bacteria that interfere with culture detection. A PCR method targeting the aerolysin gene has been described for the direct detection of *Aeromonas* species in feces. The PCR method detected the aerolysin gene in 68% of fecal samples as compared to 64% of aeromonads detected by conventional selective media. The *Aeromonas* species detected by PCR aerolysin gene amplification included *A. hydrophila* (55.8%), *A. caviae* (17.6%), *A. veronii* (10.2%), *A. schubertii* (4.4%), *A. jandaei* (2.9%), and *A. trota* (8.8%).[73]

PCR gene amplification has also been used for the detection of pathogenic bacteria and to differentiate pathogenic strains from nonpathogenic ones.[74] A multiplex PCR was developed to detect hemolysin and aerolysin genes in *A. hydrophila* and *A. sobria* and used to screen many clinical and reference strains. Five genotypes were identified based on the presence or absence of *ahh1*, *asa1*, and *aerA* genes.[75] The presence and number of hemolysin genes corresponded to cytotoxin titres when the phenotypic expression of toxin was evaluated in Vero cell culture cytotoxicity assays.

PCR targeting the *lip* gene has been utilized successfully at our laboratory for the detection and identification of *A. hydrophila* HG1.[76] We have reported that the PCR using specific primers for the *aroA* gene can be applied for the detection and identification of all aeromonads at genus level.[77] When PCR amplification of the *aroA* gene was combined with restriction fragment length polymorphism (RFLP) using the endonuclease *Hae* II, a different RFLP was obtained for each of the 14 reference *Aeromonas* strains used in this study, representing 14 *Aeromonas* species. In addition, the *aroA* housekeeping gene shows high genetic diversity and can be used in phylogenetic studies for *Aeromonas* species and strains identification.

In the last 15 years, the molecular identification of aeromonads has been increasingly relied on the analysis of 16S rRNA sequences.[78] However, the use of this molecular marker induces a complication because ribosomal genes are organized in operons and bacteria may contain up to 15 copies of the ribosomal operon, and the evidence suggests that intragenomic heterogeneity exists, even in *Aeromonas*.[23] A pioneer study to reconstruct the phylogeny of *Aeromonas* species revealed that *A. trota* and *A. caviae* only show three nucleotide differences in their 16S rRNA gene sequences but the DNA–DNA hybridization values for these species are 30%, indicating that they are distantly related. In contrast, *A. veronii* and *A. sobria* differ by 12 nucleotides but have a DNA–DNA hybridization value of 60%–65%, indicating that they are closely related.[79] This lack of congruence may be explained in terms of lateral gene transfer and recombination. This is why in the genus *Aeromonas* the 16S rRNA gene sequence is only useful to define isolates at genus level, and it is necessary to use other single-copy and stable housekeeping genes for correct *Aeromonas* identification.

Housekeeping genes are considered better molecular markers than the 16S rRNA genes for the study of taxonomic and phylogenetic relationships at species level, and the current consensus is that an informative level of phylogenetic data should be obtained from the determination of nucleotide sequences of a minimum of five housekeeping genes (genes under stabilizing selection for encoded metabolic functions).[80] However, several phylogenetic studies addressing the genus *Aeromonas* are based on a smaller number of housekeeping genes. Thus, the *rpoD* gene, encoding the σ^{70} factor of DNA-dependent RNA polymerase, was evaluated in comparison to *gyrB* (which encodes the B-subunit of DNA gyrase) for phylogenetic and taxonomic studies of 70 *Aeromonas* strains representing all known species of the genus with excellent results, especially for closely related taxa.[80]

A novel phylogenetic marker—the *dnaJ* gene—complemented with DNA–DNA hybridization studies has been used to clarify some controversies within the genus *Aeromonas*.[25] Recently, the *recA* gene sequence has been evaluated for the classification of *Aeromonas* strains at genotype level, with similar results to those obtained when this gene was included in a multilocus sequence typing scheme (MLST).[1,26]

68.2 METHODS

68.2.1 Sample Preparation

Clinical specimens, where *Aeromonas* species are usually present in high numbers, should be collected and transported according to recommended clinical laboratory practice.[71]

Recovery of *Aeromonas* from clinical specimens such as stool samples should use a broth enrichment procedure. Stool specimens may be inoculated in APW and incubated overnight aerobically with shaking (100 rpm). One loopful of enriched sample may be plated onto xylose-deoxycholate-citrate agar (XDCA) and sheep-blood agar (5% sheep blood) supplemented with 30 μg/mL ampicillin (ASBA), followed by incubation at 37°C overnight. Colorless colonies that do not ferment xylose on XDCA and ampicillin-resistant hemolytic colonies on ASBA should be tested for catalase and oxidase positivity and the resistance to the O129 vibriostatic agent. Presumptive *Aeromonas* colonies are ready for a phenotypic identification scheme.[71] If a molecular identification method is used, presumptive *Aeromonas* colonies are then suspended in 100 μL of TE buffer, vortexed for 1 min, and incubated at 96°C for 10 min in a water bath. After centrifugation of the tube at 12,000 g, the supernatant is transferred to a clean tube and 3–5 μL samples can be used as a DNA template for PCR assays.

To recover *Aeromonas* from food samples, 25 g of each food sample are weighted and transferred aseptically to a sterile stomacher bag and homogenized using a Stomacher device for 1–2 min with 225 mL of APW. For *Aeromonas* quantitative (CFU/g) analysis 0.1 mL of tenfold serial dilutions are plated, before enrichment, onto two selective culture media: Ampicillin-Dextin-Agar (ADA) and modified Bile Salts Irgasan-Brilliant Green (m-BIBG).[69] Both media are incubated for 24 h at 30°C and presumptive *Aeromonas* colonies enumerated. *Aeromonas* strains typically produce yellow colonies on ADA medium and purple colonies on mBIBG medium. Presumptive *Aeromonas* are ready for phenotypic or molecular identification schemes as commented earlier.

68.2.2 Detection Procedures

Principle. Molecular detection of *Aeromonas* may be more sensitive than culture for complex matrices with dense background microbiota. However, it is recommended to use an enrichment medium followed by selective media as described earlier. Presumptive *Aeromonas* colonies on selective media can be confirmed as *Aeromonas* species by using a very simple and easy PCR assay for the *aroA* gene, which is being used in our lab with excellent results, although other molecular methods may provide good results as well,[25,26] followed by restriction fragment length polymorphism (RFLP) analysis using *Hae*II endonuclease. The *aroA* gene is involved in a universal biosynthetic pathway for the biosynthesis of folic acid and aromatic amino acids, which is functional in all bacteria. The *aroA* gene nucleotide sequences is very well conserved within gram-negative bacterial groups and may be used as a powerful and appropriate tool for the identification of bacteria to the genus level by the use of a simple and inexpensive procedure such as *aroA* PCR amplification.

The *aroA* gene primers consist of a 24-nucleotide forward primer PF1 (5′-TTTGGAACCCATTTCTCGTGTGGC-3′) corresponding to positions 15 to 38 of the *A. salmonicida aroA* gene nucleotide sequence (from the GenBank data base under accession number L05002), and a 25-nucleotide reverse primer PR1 (5′-TCGAAGTAGTCCGGGAAGGTCT TGG-3′), which corresponds to positions 1253 to 1229. The combination of primers (PF1 and PR1) from *A. salmonicida aroA* gene enables the synthesis of an expected 1236-bp DNA fragment, which represents the bulk of *aroA* gene sequence of all *Aeromonas* reference strains tested.

Aeromonas spp. *aroA* PCR amplification may be performed from both extracted nucleic acids and bacterial cells; however, to prevent contaminations and avoid an excessive DNA manipulation, all data presented here were obtained from presumptive *Aeromonas* colonies, which simplified the procedure. Sensitivity was very high; amplification which resulted in detectable levels of PCR product was achieved when a minimum of 8 CFU of any *Aeromonas* spp. were lysed, on the basis of an average of five repeated testing of viable cells and PCR assays.

Procedure

1. Prepare PCR mixture (50 μL) containing 1 μL of DNA-containing sample (extracted by the rapid boiling method above), 1.25 U of *Taq* DNA polymerase (Boehringer GmbH, Mannheim, Germany), 5 μL of 10× PCR amplification buffer [100 mM Tris-HCl, 20 mM $MgCl_2$, 500 mM KCl (pH 8.3)], 1 μM of each primer (PF1 and PR1), 0.5 mM deoxynucleosides triphosphate, and double-distilled water (to a final volume of 50 μL). To minimize evaporation, add 50 μL of mineral oil to the mixture.
2. Carry out the following cycling program on a DNA thermal cycler (Perkin-Elmer Cetus, Norwalk, Conn. or other machine): 1 cycle of 94°C for 2 min; 40 cycles of 92°C for 1 min, 50°C for 1 min, and 72°C for 1 min; and 1 cycle of 72°C for 10 min. Keep the reaction mixture at 4°C until analysis.
3. Run 15 μL of PCR products to confirm the expected 1236-bp DNA fragment on a 1% agarose gel (prepared in 0.5× TBE) containing 0.5 μg/mL ethidium bromide (Figure 68.1).
4. Digest 20 μL of PCR product with 5 U of *Hae*II restriction endonuclease for 2 h, and separate in 3% agarose gels (prepared in 0.5× TBE) containing 0.5 μg/mL ethidium bromide (Figure 20.2).

Comments The PCR primers PF1 and PR1 from *A. salmonicida aroA* gene produce an expected 1236-bp DNA fragment, which represents most of *aroA* gene sequence of 13 *Aeromonas* reference strains (Figure 68.1). However, no PCR amplification product is obtained when members of *Enterobacteriaceae*, *Vibrionaceae*, and *Pseudomonadaceae* families' cells are used as negative controls. A typical example of *aroA* gene PCR amplification of 13 *Aeromonas* strains is shown in Figure 68.1.

The utility of this procedure is further enhanced when combined with restriction fragment length polymorphism

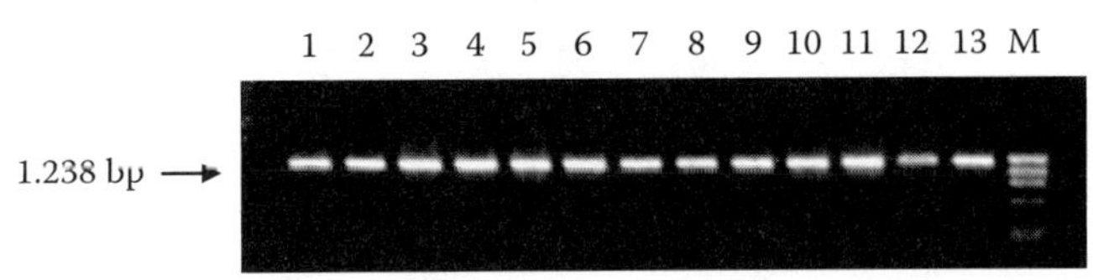

FIGURE 68.1 Agarose gel electrophoresis of PCR amplification products from *Aeromonas* spp. Lanes 1 to 13 *A. eucrenophila* CECT 4224, *A. jandaei* CECT 4228, *A. salmonicida* CECT 894, *A. hydrophila* CECT 839, *A. sobria* CECT 837, *A. enteropelogenes* CECT 4487, *A. schubertii* CECT 4241, *A. ichthiosmia* CECT 4486, *A. media* CECT 4232, *A. allosaccharophila* CECT 4199, *A. caviae* CECT 838, *A. encheleia* CECT 4341, and *A. trota* CECT 4255 respectively. M, standard DNA size marker ϕX174 *Hae*III digested.

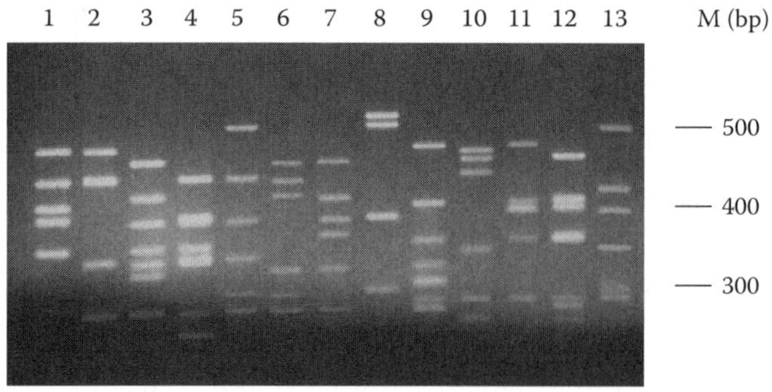

FIGURE 68.2 Agarose gel electrophoresis of fragments produced by *Hae*II digestion of 13 species of *Aeromonas aroA* genes amplified by PCR. Lanes: 1 to 13, reference *Aeromonas* strains ordered as in Figure 68.1. M, DNA molecular weight marker (100-bp ladder).

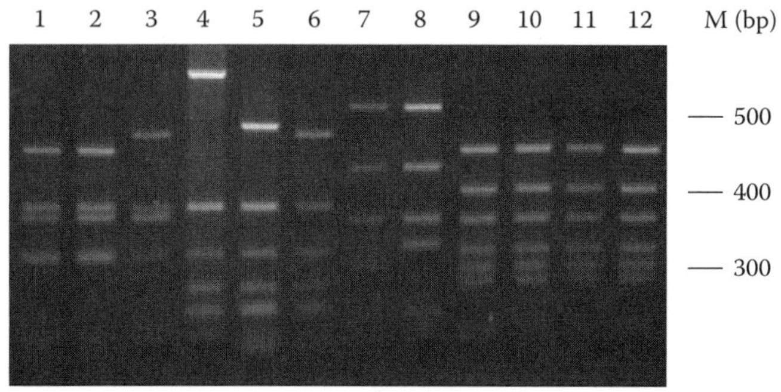

FIGURE 68.3 Agarose gel electrophoresis of fragments produced by *Hae*II digestion of *Aeromonas aroA* genes amplified by PCR. Lanes: 1 and 2, environmental strains phenotypically identified as *A. caviae*; 3, *A. caviae* CECT 838; 4 and 5, environmental strains phenotypically identified as *A. media*; 6, *A. media* CECT 4232; 7, *A. sobria* CECT 837; 8, environmental strain phenotypically identified as *A. sobria*; 9–11, clinical isolates identified as *A. salmonicida*; 12, *A. salmonicida* CECT 894.

(RFLP) of the amplified *aroA* gene, which results in specific patterns for each strain (Figure 68.2).

This molecular method (*aroA* gene PCR–RFLP) is also used for identification of *Aeromonas* environmental isolates (two strains of *A. caviae*, two strains of *A. media*, and one strain of *A. sobria*) or clinical isolates (three strains of *A. salmonicida* from diseased fish) that are previously identified by biochemical characteristics. We found good correlation between reference strain RFLPs and *Aeromonas* isolate RFLPs (Figure 68.3) as confirmed by phenotypic identification as well.

68.3 CONCLUSIONS AND FUTURE PERSPECTIVES

The genus *Aeromonas* consists of 14 recognized species of gram-negative, facultatively anaerobic, rod-shape bacteria measuring 0.3–1.0 μm × 1.0–3.5 μm[1] and often appearing singly, in pairs, and sometimes in short chains. Being widely distributed in aquatic environments (i.e., marine water and fresh water associated with aquatic animals and sediments), *Aeromonas* species are frequently isolated from foods and stools, and some are considered to be opportunistic pathogens of human beings and animals owing to the detection of suspected foodborne outbreaks.[1] *Aeromonas* spp. are motile through a polar flagellum, with the exception of *A. salmonicida* and *A. media*, which are nonmotile species. Most *Aeromonas* spp. are mesophilic, although *A. salmonicida* is a psychrophile. Most *Aeromonas* spp. also produce an array of extracellular enzymes: proteases, DNAses, RNAses, lipases, lecithinases, amylases, enterotoxins, cytotoxins, and hemolysins.[1]

Aeromonas has been frequently misidentified at species level owing to the current number of recognized taxa, with continuous incorporations of new species and a lack of a panel of phenotypic properties that allow species to be discriminated. The discrepancies of different sets of data using the same *Aeromonas* strains have contributed to the difficulties in the identification of these microorganisms, as well as the use of select biochemical characteristics for newly described species or different conditions of manipulation at different laboratories. However, recent studies have presented cumulative biochemical data for many *Aeromonas* strains representing all taxa described that allow accurate identification of current *Aeromonas* species with the depth required at laboratory level.[71] Recently, housekeeping gene sequence analysis (*gyrB*, *rpoB*, *rpoD*, *danJ*, or *recA*) complemented with DNA–DNA hybridization analysis and/or 16S rRNA sequence analysis have provided new tools for the elucidation of *Aeromonas* taxonomy and have contributed to delineating closely related species.[25,26,81]

Although aeromonads have been known for 100 years, the first report of *Aeromonas*-due human infection was in 1954[6] and the first strain isolated from feces occurred in 1961.[82] Since then, *Aeromonas* species have been associated with human diarrhea, although the evidence has been inconclusive because the criteria that could be utilized to solve this issue (Koch's postulates, an animal model reproducing human disease, and the failure of studies with human volunteers) have been insufficient and hence the debate remains open. Three *Aeromonas* species (*A. hydrophila* HG1, *A. caviae* HG4, and *A. veronii* biovar *sobria* HG8) have been isolated from human feces with significant frequency, representing 85% of isolates.[12] *A. trota* HG14, *A. jandei* HG9, and *A. schubertii* HG12 are rare fecal isolates.[12] If *Aeromonas* enteric infections are transmitted by food, drinking water, or person-to-person contact, and symptomatic infections are strain specific, it would be expected that the same strain would be isolated in patients and in foods, in drinking water or in an *Aeromonas*-transmitting person. However, this is not the general situation,[29] and only two cases have been reported that suggest transmission—one from food to a patient with *Aeromonas*-associated diarrhea[30] and another one from person to person.[31] In both studies, ribotyping was applied for identification purpose. Because of the abundance of *Aeromonas* species in nature and their potential involvement in human and animal disease, further investigation is needed to detect virulence mechanisms in sets of *Aeromonas* strains associated with human and animal disease. This solution would be of considerable importance for clinicians and for clinical microbiology laboratories with a view to investigating the presence of *Aeromonas* in clinical specimens. It would also be very important for public authorities to regulate specific microorganisms that may be found in drinking water, as well as food production, distribution, and consumption guidelines.

REFERENCES

1. Martin-Carnahan, A., and Joseph S.W., *Aeromonadaceae*. In Brenner, D.J. et al. (Editors), *The Proteobacteria, Part B, Bergey's Manual of Systematic Bacteriology*, 2nd ed., Vol. 2, pp. 556–578, Springer-Verlag, New York, 2005.
2. Figueras, M.J., Clinical relevance of Aeromonas, *Rev. Med. Microbiol.*, 16, 145, 2005.
3. Chu, Y.W. et al., Lack of association between presentation of diarrheal symptoms and faecal isolation of *Aeromonas* spp. amongst outpatients in Hong Kong, *J. Med. Microbiol.*, 55, 349, 2006.
4. Demarta, A. et al., Epidemiological relationships between *Aeromonas* strains isolated from symptomatic children and household environments as determined by Ribotyping, *Eur. J. Epidemiol.*, 16, 447, 2000.
5. Bomo, A.M. et al., Bacterial removal and protozoan grazing in biological sand filters, *J. Environ. Quality*, 33, 1041, 2004.
6. Hill, K.R., Caselitz, F.H., and Moody, L.M., A case of acute metastatic myositis caused by a new organism of the family *Pseudomonadaceae*, *West Indian Med. J.*, 3, 9, 1954.
7. von Graevenitz, A., and Mensch, A.H., The genus Aeromonas in human bacteriology. Report of 30 cases and review of the literature, *N. Engl. J. Med.*, 278, 245, 1968.
8. Janda, M.J., and Abbott, S.L., Human pathogens. In Austin, B. et al. (Editors), *The Genus Aeromonas*, p. 39, John Wiley and Sons, New York, 1996.
9. Janda, J.M., and Abbott, S.L., Evolving concepts regarding the genus *Aeromonas*: An expanding Panorama of species, disease presentations, and unanswered questions, *Clin. Infect. Dis.*, 27, 332, 1998.
10. Cascon, A. et al., Cloning, characterization, and insertional inactivation of a major extracellular serine protease gene with elastolytic activity from *Aeromonas hydrophila*, *J. Fish Dis.*, 23, 49, 2000.
11. Cascon, A. et al., A major secreted elastase is essential for pathogenicity of *Aeromonas hydrophila*, *Infect. Immun.*, 68, 3233, 2000.
12. Gosling, P., Aeromonas species in disease of animals. In Austin, B. et al. (Editors), *The Genus Aeromonas*, p. 39, John Wiley and Sons, New York, 1996.
13. Rivero, O. et al., Molecular cloning and characterization of an extracellular protease gene from *Aeromonas hydrophila*, *J. Bacteriol.*, 172, 3905, 1990.
14. Rivero, O. et al., Cloning and characterization of an extracellular temperature-labile serine protease gene from *Aeromonas hydrophila*, *FEMS Microbiol. Lett.*, 81, 1, 1991.
15. Mateos, D. et al., Influence of growth temperature on the production of extracellular virulence factors and pathogenicity of environmental and human strains of *Aeromonas hydrophila*, *J. Appl. Bacteriol.*, 74, 111, 1993.
16. Isonhood, J.H., and Drake, M., *Aeromonas* species in foods, *J. Food Prot.*, 65, 575, 2002.
17. Janda, J.M. et al., Further studies on biochemical characteristics and serologic properties of the genus *Aeromonas*, *J. Clin. Microbiol.*, 34, 1930, 1996.
18. Walsh, T.R. et al., Evaluation of a new E-test for detecting metallo-beta-lactamases in routine clinical testing, *J. Clin. Microbiol.*, 40, 2755, 2002.
19. Chaudhury, A. et al., Biochemical characterisation, enteropathogenicity and antimicrobial resistance plasmids of clinical and environmental *Aeromonas* isolates, *J. Med. Microbiol.*, 44, 434, 1996.
20. Zimmermann, O.E.R., Die bakteriem unserer trink-und Nutzwässer, insbesondere des wassers der chemnitzer wasserleitung, *Elfter. Bericht. Naturewiss. Gess. Chemnitz.*, 53, 154, 1890.
21. Sanarelli, G., Über einen neuen mikroorganismus des wassers welcher fur tiere mit veräenderlicher und konstanter temperatur pathogenist, *Zbl. Bakteriol. Parasitenk.*, 9, 222, 1891.
22. Kluyver, A.J., and Van Niel, C.B., Prospects for a natural system of classification of bacteria, *Zbl. Bakteriol. Parasitenkd. Infektionskr. Hyg.* Abt.II orig., 94, 369, 1936.
23. Morandi, A. et al., Evolutionary and diagnostic implications of intragenomic heterogeneity in the 16S rRNA gene in *Aeromonas* strains, *J. Bacteriol.*, 187, 6561, 2005.
24. Yáñez, M.A. et al., Phylogenetic analysis of members of the genus *Aeromonas* based on *gyrB* gene sequences, *Int. J. Syst. Evol. Microbiol.*, 53, 875, 2003.
25. Nhung P.H. et al., Use of the novel phylogenetic marker *dnaJ* and DNA–DNA hybridization to clarify interrelationships within the genus *Aeromonas*, *Int. J. Syst. Evol. Microbiol.*, 57, 1232, 2007.
26. Sepe, A. et al., Evaluation of *recA* sequencing for the classification of *Aeromonas* strains at the genotype level, *Lett. Appl. Microbiol.*, 46, 439, 2008.
27. von Graevenitz, A., The role of *Aeromonas* in diarrhea: A review, *Infection*, 35, 59, 2007.
28. Kirov, S.M., The public health significance of *Aeromonas* spp. in foods, *Int. J. Food Microbiol.*, 20, 179, 1993.
29. Palumbo, S.A., The *Aeromonas hydrophila* group in food. In Austin, B., Altwegg, M., Gosling, P., and Joseph, S. W. (Editors), *The Genus Aeromonas*, pp. 287–310, John Wiley and Sons, New York, 1996.
30. Altwegg, M. et al., *Aeromonas*-associated gastroenteritis after consumption of contaminated shrimp, *Eur. J. Clin. Microbiol. Infect. Dis.*, 10, 44, 1991.
31. Moyer, N.P. et al., Value of rRNA gene restriction patterns of *Aeromonas* spp. for epidemiological investigations, *Curr. Microbiol.*, 24, 15, 1992.
32. Moyer, N.P., Clinical significance of *Aeromonas* species isolated from patients with diarrhea, *J. Clin. Microbiol.*, 25, 2044, 1987.
33. Challapalli, M. et al., *Aeromonas*-associated diarrhea in children, *Pediatr. Infect. Dis. J.*, 7, 693, 1988.
34. Bauab, T.M. et al., Niche-specific association of *Aeromonas* ribotypes from human and environmental origin, *Microbiol. Immunol.*, 47, 7, 2003.
35. Pitarangsi, C. et al., Enteropathogenicity of *Aeromonas hydrophila* and *Plesiomonas shigelloides*: Prevalence among individuals with and without diarrhea in Thailand, *Infect. Immun.*, 35, 666, 1982.
36. Albert, M.J. et al., Prevalence of enterotoxin genes in *Aeromonas* spp. isolated from children with diarrhea, healthy controls, and the environment, *J. Clin. Microbiol.*, 38, 3785, 2000.
37. Hänninen, M.L. et al., Association of *Aeromonas* spp. with travelers' diarrhoea in Finland, *J. Med. Microbiol.*, 42, 26, 1995.

38. Svenungsson, B. et al., Enteropathogens in adult patients with diarrhea and healthy control subjects: A 1-year prospective study in a Swedish clinic for infectious diseases, *Clin. Infect. Dis.*, 30, 770, 2000.
39. Lee, W.S., and Puthucheary, S.D., Bacterial enteropathogens isolated in childhood diarrhoea in Kuala Lumpur—the changing trend, *Med. J. Malaysia*, 57, 24, 2002.
40. Kehinde, A.O. et al., Childhood gastroenteritis due to *Aeromonas hydrophila* in Ibadan, Nigeria, *Afr. J. Med. Sci.*, 30, 345, 2001.
41. Essers, B. et al., Acute community-acquired diarrhea requiring hospital admission in Swiss children, *Clin. Infect. Dis.*, 31, 192, 2000.
42. Maltezou, H.C. et al., Acute diarrhoea in children treated in an outpatient setting in Athens, Greece, *J. Infect.*, 43, 122, 2001.
43. George, W.L. et al., *Aeromonas*-related diarrhea in adults, *Arch. Intern. Med.*, 145, 2207, 1985.
44. Dumontet, S. et al., Incidence and characterisation of *Aeromonas* spp. in environmental and human samples in southern Italy, *New Microbiol.*, 26, 215, 2003.
45. del Val, A., Moles, J.R., and Garrigues, V., Very prolonged diarrhea associated with *Aeromonas hydrophila*, *Am. J. Gastroenterol.*, 85, 1535, 1990.
46. Janda, J.M., *Aeromonas* and *Plesiomonas*. In Sussman, M. (Editor), *Molecular Medical Microbiology*, p. 1237, Academic Press, San Diego, 2001.
47. Kay, W.W. et al., Purification and disposition of a surface protein associated with virulence of *Aeromonas salmonicida*, *J. Bacteriol.*, 147, 1077, 1981.
48. Shaw, J.G. et al., Invasion of tissue culture cells by *Aeromonas caviae*, *Med. Microbiol. Lett.*, 4, 324, 1995.
49. Kirov, S.M., and Sanderson, K., *Aeromonas* cell line adhesion, surface structures and in vivo models of intestinal colonization, *Med. Microbiol. Lett.*, 4, 305, 1995.
50. Rocha de Souza, C.M. et al., Identification of a 43-kDa outer-membrane protein as an adhesin in *Aeromonas caviae*, *J. Med. Microbiol.*, 50, 313, 2001.
51. Fang, H.M., Ge, R., and Sin, Y.M., Cloning, characterisation and expression of *Aeromonas hydrophila* major adhesin, *Fish Shellfish Immunol.*, 16, 645, 2004.
52. Sara, M., and Sleytr, U.B., S-layer proteins, *J. Bacteriol.*, 182, 859, 2000.
53. Aguilar, A. et al., Influence of osmolarity on lipopolysaccharides and virulence of *Aeromonas hydrophila* serotype O:34 strains grown at 37 degrees C, *Infect. Immun.*, 65, 1245, 1997.
54. Zhang, Y.L., Arakawa, E., and Leung, K.Y., Novel *Aeromonas hydrophila* PPD134/91 genes involved in O-antigen and capsule biosynthesis, *Infect. Immun.*, 70, 2326, 2002.
55. Merino, S., Shaw, J.G., and Tomás, J.M., Bacterial lateral flagella: an inducible flagella system, *FEMS Microbiol. Lett.*, 263, 127, 2006.
56. Canals, R. et al., Polar flagellum biogenesis in *Aeromonas hydrophila*, *J. Bacteriol.*, 188, 542, 2006.
57. Hokama., A., and Iwanaga, M., Purification and characterization of *Aeromonas sobria* pili, a possible colonization factor, *Infect. Immun.*, 59, 3478, 1991.
58. Pemberton, J.M., Kidd, S.P., and Schmidt, R., Secreted enzymes of *Aeromonas*, *FEMS Microbiol. Lett.*, 152, 1, 1997.
59. Leung, K.Y., and Stevenson, R.M.W., Characteristics and distribution of extracellular proteases from *Aeromonas hydrophila*, *J. Gen. Microbiol.*, 134, 151, 1988.
60. Merino, S. et al., Cloning, sequencing, and role in virulence of two phospholipases (A1 and C) from mesophilic *Aeromonas* sp. serogroup O:34, *Infect. Immun.*, 67, 4008, 1999.
61. Brumlik, M.J., and Buckley, J.T., Identification of the catalytic triad of the lipase/acyltransferase from *Aeromonas hydrophila*, *J. Bacteriol.*, 178, 2060, 1996.
62. Leclere, V., Bechet, M., and Blondeau, R., Functional significance of a periplasmic Mn-superoxide dismutase from *Aeromonas hydrophila*, *J. Appl. Microbiol.*, 96, 828, 2004.
63. Asao, T. et al., Purification and some properties of *Aeromonas hydrophila* hemolysin, *Infect. Immun.*, 46, 122, 1984.
64. Xu, X. et al., Role of a cytotoxic enterotoxin in *Aeromonas*-associated infections: Development of transposon and isogenic mutants, *Infect. Immun.*, 66, 3501, 1998.
65. Chopra, A.K. et al., The cytotoxic enterotoxin of *Aeromonas hydrophila* induces proinflammatory cytokine production and activates arachidonic acid metabolism in macrophages, *Infect. Immun.*, 68, 2808, 2000.
66. Moyer, N.P., Isolation and enumeration of aeromonads. In Austin, B. et al. (Editors), *The Genus Aeromonas*, p. 39, John Wiley and Sons, New York, 1996.
67. Gobat, P.F., and Jemmi, F., Comparison of seven selective media for the isolation of mesophilic *Aeromonas* species in fish and meat, *Int. J. Food. Microbiol.*, 24, 375, 1995.
68. USEPA Method 1605: *Aeromonas* in finished water by membrane filtration using ampicillin-dextrin agar with vancomycin (ADA-V). Washington, DC, 2001.
69. Neyts, K. et al., Modification of the bile salts-Irgasan-brilliant green agar for enumeration of *Aeromonas* species from food, *Int. J. Food Microbiol.*, 57, 211, 2000.
70. Carnahan, A.M., Behram, S., and Joseph, S.W., Aerokey II: A flexible key for identifying clinical *Aeromonas* species, *J. Clin. Microbiol.*, 29, 2843, 1991.
71. Abbott, S.L., Cheung, W.K.W., and Janda, J.M., The genus *Aeromonas*: Biochemical characteristics, atypical reactions, and phenotypic identification schemes, *J. Clin. Microbiol.*, 41, 2348, 2003.
72. Ji, N. et al., Universal primer PCR with DGGE for rapid detection of bacterial pathogens, *J. Microbiol. Methods*, 57, 409, 2004.
73. Kannan, S. et al., Direct detection of diarrheagenic *Aeromonas* from faeces by polymerase chain reaction (PCR) targeting aerolysin toxin gene, *Eur. Rev. Med. Pharm. Sci.*, 5, 91, 2001.
74. Sen, K., and Rodgers, M., Distribution of six virulence factors in *Aeromonas* species isolated from U.S. drinking water utilities: A PCR identification, *J. Appl. Microbiol.*, 97, 1077, 2004.
75. Wang, G. et al., Detection and characterization of the hemolysin genes in *Aeromonas hydrophila* and *Aeromonas sobria* by multiplex PCR, *J. Clin. Microbiol.*, 41, 1048, 2003.
76. Cascon, A. et al., Identification of *Aeromonas hydrophila* Hybridization Group 1 by PCR Assays, *Appl. Environ. Microbiol.*, 62, 1167, 1996.
77. Cascon, A. et al., RFLP-PCR análisis of the *aroA* gene as a taxonomic tool for the genus *Aeromonas*, *FEMS Microbiol. Lett.*, 156, 199, 1997.
78. Figueras, M.J. et al., Extended method for discrimination of *Aeromonas* spp. by 16S rDNA RFLP analysis, *Int. J. Syst. Evol. Microbiol.*, 50, 2069, 2000.

79. Martinez-Murcia, A.J., Benlloch, S., and Collins, M.D., Phylogenetic interrelationships of members of the genera *Aeromonas* and *Plesiomonas* as determined by 16S ribosomal DNA sequencing: lack of congruence with results of DNA-DNA hybridizations, *Int. J. Syst. Bacteriol.*, 42, 412, 1992.
80. Soler, L. et al., Phylogenetic analysis of the genus *Aeromonas* based on two housekeeping genes, *Int. J. Syst. Evol. Microbiol.*, 54, 1511, 2004.
81. Maiti, B. et al., Cloning and expression of an outer membrane protein OmpW of *Aeromonas hydrophila* and study of its distribution in *Aeromonas* spp., *J. Appl. Microbiol.*, 107, 1157, 2009.
82. Lautrop, H., Aeromonas hydrophila isolated from human feces and its possible pathological significance, *APMIS*, 51/S 144,299, 1961.

69 *Aggregatibacter*

Dongyou Liu

CONTENTS

69.1 INTRODUCTION

69.1.1 CLASSIFICATION, MORPHOLOGY, AND BIOCHEMICAL PROPERTIES

The genus *Aggregatibacter* is classified taxonomically in the family *Pasteurellaceae*, order *Pasteurellales*, class *Gammaproteobacteria*, phylum *Proteobacteria*. The family *Pasteurellaceae* currently includes 12 genera (*Pasteurella, Actinobacillus, Haemophilus, Lonepinella, Mannheimia, Phocoenobacter, Gallibacterium, Histophilus, Volucribacter, Avibacterium, Nicoletella, Aggregatibacter*) and 61 species. Members of the *Pasteurellaceae* family are characterized as gram-negative, nonmotile, aerobic, or facultatively anaerobic rods, coccoids, or filaments. Biochemically, due to their inability to synthesize nicotinamide adenine dinucleotide (NAD) de novo, they acquire this essential nutrient from their environment as either NAD or a limited number of precursors.[1] Within the class Gammaproteobacteria, the nonmotility and the positivity for oxidase and phosphatase separate members of the *Pasteurellaceae* from those of the *Enterobacteriaceae*, which are motile and oxidase and phosphatase negative. Similarly, the lack of motility and lack of tolerance to high salt concentrations separate members of the *Pasteurellaceae* from most members of the *Vibrionaceae*.[1]

The genus *Aggregatibacter* (*aggregare*, to come together; *bacter*, bacterial rod; *Aggregatibacter*, rod-shaped bacterium that aggregates with others), a recent addition to the family *Pasteurellaceae*, was established in 2006 to cover gram-negative, nonmotile, facultatively anaerobic rods or coccobacilli that were previously known as *Haemophilus* (*Actinobacillus or Bacterium*) *actinomycetemcomitans* (now *Aggregatibacter actinomycetemcomitans*), *Haemophilus aphrophilus* and *Haemophilus paraphrophilus* (which are now combined to form a single species *Aggregatibacter aphrophilus*), and *Haemophilus segnis* (now *Aggregatibacter segnis*).[2–4]

Compared to *Haemophilus* spp., which require blood factors hemin (factor X) and/or nicotinamide adenine dinucleotide (NAD, or factor V) for in vitro growth, *Aggregatibacter* spp. are mesophilic and capnophilic, being independent of factor X and variably dependent of factor V.[4,5] *Aggregatibacter* colonies are greyish white and nonhemolytic on sheep- and horse-blood agar. *Aggregatibacter* spp. are capable of producing acid from glucose, fructose, and maltose, but not from arabinose, cellobiose, melibiose, melezitose, salicin, and sorbitol. They show varied capacity to ferment galactose, lactose, mannitol, mannose, raffinose, sorbose, sucrose, trehalose, and xylose, which may be exploited for their speciation. They reduce nitrate and produce alkaline phosphatase, but are negative for indole, urease, ornithine, and lysine decarboxylases and arginine dihydrolase. In addition, they are oxidase negative or weakly positive, and catalase variable.[4] Forming part of the human oral flora, they are occasionally recovered from other body sites (including blood and brain) in association with endocarditis and abscesses.

Aggregatibacter actinomycetemcomitans (*actis*, a ray; *myces*, *mukes*, a fungus; *actinomyces*, ray fungus; *comitans*, accompanying; *actinomycetemcomitans*, accompanying an actinomycete) is the type species of the genus *Aggregatibacter*. The bacterium is a small rod which may appear as cocci in broth or actinomycotic lesions. *A. actinomycetemcomitans* cells may occur singly, in pairs, or in small clumps, and grow poorly in ambient air but well in 5% CO_2. Colonies are small, with a diameter of ≤0.5 mm after 24 h, and >1–2 mm after 48 h on chocolate agar. On primary isolation, colonies are rough textured and adherent (to glass) and show an internal, opaque, star-like pattern (or crossed cigars). The rough phenotype results from fimbriation and production of hexoseamine-containing exopolysaccharide. Successive in vitro subculturing of the rough phenotype on solid media may lead to its transformation into a smooth, nonadherent colony type with reduced ability to colonize the mouth of experimental animals. The smooth phenotype also displays a lower hemin binding activity and is less sensitive to proteinase K treatment than the rough phenotype. Furthermore, smooth phenotype tends to attach poorly and form weaker biofilms in comparison with rough phenotype.[6] *A. actinomycetemcomitans* is independent of both X and V factors for growth, and is catalase positive. The bacterium secretes a leukotoxin that belongs to the highly conserved repeat toxin (RTX) family. Based on the six structurally and antigenically distinct *O*-polysaccharide components of its respective lipopolysaccharide molecules, *A. actinomycetemcomitans* is divided into six serotypes (a–f).[7–12] Some nonserotypeable strains lacking serotype-specific polysaccharide antigen (S-PA) expression are identified as serotype a, b, c, or f by PCR-typing.[13] *A. actinomycetemcomitans* is distinguished from V factor-independent strains of *Aggregatibacter aphrophilus* by catalase and β-galactosidase (hydrolysis of ONPG) activity, as well as fermentation of lactose, sucrose, and trehalose. The G + C content of *A. actinomycetemcomitans* DNA is 42.7 mol% (T_m).[4,14]

Aggregatibacter aphrophilus (*aphros*, foam; *philos*, loving; *aphrophilus*, foam loving) is a short rod of 0.5 × 1.5–1.7 μm, with occasional filamentous forms. *A. aphrophilus* colonies are high-convex, granular, yellowish, and opaque, with a diameter of 1.0–1.5 mm within 24 h on chocolate agar in the presence of 5%–10% extra CO_2. Without extra CO_2, *A. aphrophilus* colonies appear very small with a few interspersed large colonies. In broth culture, *A. aphrophilus* adheres to the walls and sediments on the bottom of the tube. *A. aphrophilus* does not require X factor for growth, although its growth is enhanced in the presence of hemin. *A. aphrophilus* is able to hydrolyse ONPG. The G + C content of the *A. aphrophilus* DNA is 42 mol% (T_m).[4,15]

Aggregatibacter segnis (*segnis*, slow or sluggish) is a small, pleomorphic rod with a predominance of irregular, filamentous forms. *A. segnis* colonies are smooth or granular, convex, grayish white, and opaque, with a diameter of 0.5 mm after 48 h on chocolate agar. The bacterium depends on V factor but not X factor for growth, and ferments sucrose but not arabinose, cellobiose, lactose, mannitol, mannose, melibiose, melezitose, raffinose, salicin, sorbose, sorbitol, trehalose, and xylose. It is catalase and β-galactosidase (hydrolysis of ONPG) variable. The G + C content of *A. segnis* DNA is 43–44 mol% (T_m).[4]

69.1.2 Epidemiology, Clinical Features, and Pathogenesis

Epidemiology. *Aggregatibacter actinomycetemcomitans* constitutes a component of the normal flora in the oropharynx and oral cavity, with primary habitat being dental surfaces.[16,17] It is commonly found in human periodontal specimens, especially from juveniles with localized periodontitis, and is transmissible among individuals through saliva, inanimate objects (toothbrushes) and direct mucosal contact.[11,18–39] *A. actinomycetemcomitans* serotypes may be present in some geographic areas but not others. For example, while *A. actinomycetemcomitans* serotypes b, c, and a are frequently detected in German patients, serotypes c and d are found in Korean patients.[40] The distribution of *A. actinomycetemcomitans* serotypes in Indonesian young adults has shifted from predominantly serotype b to a more equal prevalence of serotypes b and c from 1994 to 2002.[41] *A. actinomycetemcomitans* clonal lineage (JP2) is isolated almost exclusively from adolescent periodontitis patients of West and Northwest African descent, including both Africans and Arabs. The JP2 clone is strongly associated with localized aggressive periodontitis as it harbors a unique 530-bp deletion (Δ530) in the promoter region of the *ltx* operon, resulting in enhanced production of leukotoxin, a key virulence factor.[41,42]

Apart from being responsible for periodontitis, *A. actinomycetemcomitans* is found in other pathological processes such as septic endocarditis, osteomyelitis, brain, lung, and subcutaneous abscesses. The bacterium was first isolated in 1912 from actinomycotic-like lesions, and its link to endocarditis was documented in 1964. As an unusual agent of infective endocarditis (IE), *A. actinomycetemcomitans* is often referred as a member of the HACEK group of slow growing, small gram-negative bacteria, consisting of *Haemophilus influenzae*, *H. parainfluenzae*, *Cardiobacterium hominis*, *Aggregatibacter actinomycetemcomitans, A. aphrophilus*, *Eikenella corrodens,* and *Kingella kingae* (or HACEK for short). The HACEK organisms are frequent colonisers of the oral cavity and are responsible for 3% of community-acquired cases of IE, with which *A. actinomycetemcomitans* is the most frequently associated. *A. actinomycetemcomitans* endocarditis is a subacute or chronic illness with a prolonged symptomatic period. Risk factors for *A. actinomycetemcomitans*-related endocarditis include underlying cardiopathy, previous valvular damage and surgery, and dental status.[30]

Aggregatibacter aphrophilus is also a member of the normal flora of the human oral cavity and pharynx. The bacterium has been isolated in brain abscess, infective endocarditis, and other infected sites such as peritoneum, pleura, wound, and bone.[15]

Aggregatibacter segnis is present in oral cavity, particularly in dental plaque, and the pharynx as part of commensal flora. The bacterium is occasionally associated with human infections, including infective endocarditis.[43,44]

Clinical Features. *A. actinomycetemcomitans* is one of the bacterial species that are commonly associated with periodontitis.[45] Over 500 different bacterial species in the oral cavity, high levels of *Porphyromonas gingivalis*, *Prevotella intermedia*, *A. actinomycetemcomitans*, *Tannerella forsynthia*, *Fusobacterium nucleatum*, *Selenomona* species, *Campylobacter rectus*, and *Eikenella corrodens* have been frequently identified in subgingival plaques, which are the initiators of the periodontal diseases such as localized juvenile periodontitis and severe adult periodontitis. Chronic periodontitis, a mixed bacterial infection, is characterized by a slow, gradual loss of tooth-supporting tissues, resulting in tooth loss in adults. Generalized aggressive periodontitis, frequently associated with *Aggregatibacter actinomycetemcomitans* and *Porphyromonas gingivalis*, often affects young adults under the age of 30, with pronounced episodes of periodontal attachment destruction. Localized aggressive periodontitis is characterized by specific infection with *A. actinomycetemcomitans*.

In *A. actinomycetemcomitans*-associated infective endocarditis, intermittent fever, weight loss, peripheral signs of endocarditis, anemia, and microscopic hematuria are noteworthy, with emboli, cardiac heart failure, and renal insufficiency being common complications. The main sites of infection include the aortic valve, mitral valve, and tricuspid valve.[30]

Besides their occasional association with infective endocarditis, *Aggregatibacter aphrophilus* may cause bacteremia,[46] and *Aggregatibacter segnis* is described in a case of acute pyelonephritis.[44]

Pathogenesis. *Aggregatibacter actinomycetemcomitans* is known to produce a heat-labile leukotoxin belonging to the repeat-in-toxin (RTX) family.[47,48] The RTX family members are expressed from an operon consisting of four genes (i.e., *ltxC*, *ltxA*, *ltxB*, and *ltxD*). The gene *ltx*A encodes a structural leukotoxin, the genes *ltx*B and *ltx*D encode proteins essential for its secretion, and the gene *ltx*C encodes an acyltransferase that is involved in the modification of prototoxin to the active toxin.[47,49,50] Being cell-specific, leukotoxin binds to neutrophils, monocytes, and a subset of lymphocytes, and is associated to evasion against the defense cells of the periodontal tissues.[51–55] The promoter regions of high leukotoxin producers such as JP2 clone possess a 530-bp deletion that represses leukotoxin expression.[56] Patients with aggressive periodontitis and localized juvenile periodontitis tend to have highly leukotoxic strains than chronic periodontitis individuals.[19,28,42,45,57,58] In addition, fimbriae, lipopolysaccharide (LPS), and extracellular polymeric substance (EPS) contribute to *A. actinomycetemcomitans* biofilm formation, which enhances its survival and pathogenic potential, as unattached cells are quickly removed by swallowing.[59,60]

69.1.3 Diagnosis

Phenotypic Identification. *Aggregatibacter* spp. are slow growers, and their growth is enhanced in enriched media and in the presence of carbon dioxide. Tryptic soy bacitracin vancomycin (TSBV) agar is widely used as a selective medium for growth of *A. actinomycetemcomitans*. In combination with a characteristic star-shaped morphology and a positive catalase test, a presumptive identification of *A. actinomycetemcomitans* may be achieved. Given its fastidious growth requirements, microbiological diagnosis of *A. actinomycetemcomitans* is typically made between days 6 and 30, with a mean duration of blood culture incubation before detection of growth at 7.1 days. For *A. actinomycetemcomitans* endocarditis, blood cultures remain a useful source material for diagnosis.[61,62]

Carbohydrate fermentation tests, oxidase (spot) test and assessment of hydrogen sulfide emission using lead acetate paper are useful.[63,64] V factor-dependence may be tested with tablets from Rosco Diagnostics placed on inoculated brain heart infusion agar plates. For detection of porphyrin synthesis from δ-aminolaevulinic acid, a small loopful (approximately 1 μL) of the test strain is suspended in 50 μL 2 mM δ-aminolaevulinic acid/80 μM $MgSO_4$/0.1 M phosphate buffer pH 6.9. After incubation for 4–24 h at 35°C, 15 μL is placed on plastic film on a UV light board and photographed.[65] Other phenotypic techniques for identification and detection of *Aggregatibacter* spp. include immunofluorescence, enzyme-linked immunosorbent assay, and multilocus enzyme electrophoresis.[7,66–70]

Genotypic Identification. A number of molecular techniques have been developed to characterize *Aggregatibacter* spp. These include DNA–DNA hybridization, restriction endonuclease analysis, ribotyping, and PCR.[37,71–81] 16S rRNA gene, *ltxC*, and *ltxA* and other genes are often targeted for genetic identification.[82–91] By using broad-range bacterial primers TPU1 (5′-AGA GTT TGA TCM TGG CTC AG-3′, corresponding to positions 8–27 in the *Escherichia coli* 16S rRNA gene) and RTU3 (5′-GWA TTA CCG CGG CKG CTG-3′, corresponding to complementary positions 519–536 in *E. coli* 16S rRNA) in PCR and an *A. actinomycetemcomitans*-specific 16S rRNA gene probe (5′-TCC ATA AGA CAG ATT C-3′) in dot blot hybridization, precise identification of *A. actinomycetemcomitans* is possible.[38]

Real-time PCR assays have been increasingly utilized for detection and quantitation of *A. actinomycetemcomitans* in the subgingival samples.[60,92–98] Morillo et al.[93] reported a TaqMan real-time PCR targeting a single copy gene *lktC* for quantitative detection of *A. actinomycetemcomitans* with forward primer 5′-ACG CAG ACG ATT GAC TGA ATT TAA-3′, reverse primer 5′-GAT CTT CAC AGC TAT ATG GCA GCT A-3′, and probe FAM-TCA CCC TTC TAC CGT TGC CAT GGG-TAMRA. Hyvärinen et al.[91] described a quantitative real-time PCR (qPCR) assay for *A. actinomycetemcomitans* using primer/probe set from conservative lipopolysaccharide-coding gene region. Kuboniwa et al.[92] developed a real-time PCR using 16S rRNA primers (Forward, 5′-CCC ATC GCT GGT TGG TTA-3′, Reverse, 5′-GGC ACG TAG GCG GAC C-3′) and TaqMan probe (Probe, FAM-CCT CTG TAT ACG CCA TTG TAG CAC GTG TGT-TAMRA) for quantitation of *A. actinomycetemcomitans* in periodontal specimens. Similarly, Periasamy and Kolenbrander[60] employed 16S rRNA gene primers (forward: 5′-GGACGGGTGAGTAATGCTTG-3′ and reverse: 5′-CCCTTACCCCACCAACTACC-3′) for specific real-time PCR detection of *A. actinomycetemcomitans* JP2 based on SYBR green dye chemistry.

An alternative platform for molecular detection of *A. actinomycetemcomitans* is loop-mediated isothermal amplification method (LAMP).[99] This technique uses four primers from six distinct regions of the *dam* gene (for *A. actinomycetemcomitans*) or the specific Δ530 deletion (for JP2 clone) together with additional loop primers to accelerate the reaction.

The assay is run at a constant temperature, usually around 65°C, with the help of *Bst* DNA polymerase large fragment, which has strand displacement activity. The resulting amplification products are a complex mixture of stem-loop DNA with inverted repeats and cauliflower-like structures that induce a white precipitate of magnesium pyrophosphate proportional to the amount of amplified DNA and visible to the naked eye.[100]

69.2 METHODS

69.2.1 Sample Preparation

Supragingival samples are obtained by using two sterile paper points inserted into the apical region of periodontal pockets or gingival crevices for 60 s and transported in VMGA III medium and ultrapure Milli Q water. Saliva is collected by using Salivette devices, which are centrifuged at 5000 × *g* for 2 min and the saliva was immediately diluted in VMG I and plated on selective and nonselective culture media.

Bacteria are grown on selective TSBV (trypticase soy bacitracin vancomycin) agar and nonselective Brucella agar enriched with 5% horse blood and supplemented with 5 μg/mL hemin and 1 μg/mL menadione. Plates were incubated in anaerobiosis (90% N_2 + 10% CO_2) at 37°C during 72 h and 14 days, respectively. Characteristic colonies of *A. actinomycetemcomitans* on TSBV agar were cultivated and identified by biochemical methods.[35,88]

Bacterial DNA is obtained from each sample collected in sterile ultrapure water by using a QIAamp DNA Mini Kit (Qiagen). The DNA concentration is determined at A_{260} nm.

69.2.2 Detection Procedures

(i) PCR–RFLP for Aggregatibacter spp. Riggio and Lennon[84] developed a PCR–RFLP method for rapid differentiation of *Aggregatibacter actinomycetemcomitans* and *Aggregatibacter aphrophilus* (*Haemophilus aphrophilus*). The method is based on PCR amplification of the 16S rRNA genes with primers 16S1 (5′-TCC TAC GGG AGG CAG CAG-3′; *Escherichia coli* positions 340–357) and 16S2 (5′-CCC GGG AAC GTA TTC ACC G-3′; *E. coli* positions 1387–1369), generating an amplicon of 1045 bp. Subsequent restriction enzyme digestion of the amplicon produces distinct band patterns for *A. actinomycetemcomitans* and *A. aphrophilus*.

Procedure

1. Samples are inoculated onto TSBV agar and chocolate agar plates and incubated in an atmosphere of 5% CO_2–95% air at 37°C for 3 days. For isolation of *A. actinomycetemcomitans* and *A. aphrophilus*, colonies from TSBV agar plates with characteristic colony morphology are subcultured to Columbia agar plates supplemented with 7.5% sterile defibrinated horse blood and are incubated as above. Gram-negative colonies that require carbon dioxide for growth, are catalase positive, and ferment glucose and maltose but not lactose, sucrose, salicin, or arabinose, are recorded as *A. actinomycetemcomitans*; colonies that are catalase negative are identified as *A. aphrophilus*.
2. DNA is prepared from each isolate by inoculation of two or three loopfuls of bacteria from the surface of an agar plate into 200 μL of sterile grade water, boiling for 10 min, removal of cell debris by centrifugation, and retention of the supernatant.
3. PCR mixture (100 μL) consists of 1× PCR buffer (10 mM Tris-HCl pH 8.8, 1.5 mM $MgCl_2$, 50 mM KCl, 0.1% Triton X-100), 2 U of Dynazyme I DNA polymerase (Flowgen Instruments Ltd), 0.2 mM each deoxynucleotide triphosphate, 0.2 μM each primer, and 5 μL of DNA extract. Negative control contains 5 μL of sterile water instead of template DNA.
4. PCR amplification is carried out with an initial 94°C for 5 min, then 30 cycles of 94°C for 1 min, 60°C for 1 min, and 72°C for 1.5 min, and finally 72°C for 10 min in an OmniGene thermal cycler (Hybaid Ltd).
5. Ten μL each of PCR product is electrophoresed on a 2% agarose gel, stained with ethidium bromide (0.5 μg/mL), and visualized under UV light. A 1045-bp fragment is expected.
6. PCR products are purified with the Wizard PCR Purification Kit (Promega). Approximately 0.5 μg of purified PCR product is digested in a total volume of 20 μL with 10 U (each) of the restriction enzymes *Hha*I and *Hin*fI (Life Technologies) at 37°C for 3 h.
7. Restriction fragments are separated by gel electrophoresis and visualized as above. The expected sizes of the restriction fragments by digestion of PCR products with *Hha*I are 735 and 279 bp for *A. actinomycetemcomitans*; and 634, 279, and 100 bp for *H. aphrophilus*. The sizes of expected restriction fragments with *Hin*fI digestion are 937 bp for *A. actinomycetemcomitans*; and 661 and 275 bp for *H. aphrophilus*.

Note: Although five *H. paraphrophilus* (*H. paraphrophilus*) were examined by the PCR–RFLP method in the original report, they may represent misidentified *Haemophilus parainfluenzae* isolates as shown recently.[4]

(ii) Nested PCR for A. actinomycetemcomitans Ready et al.[80] utilized a nested PCR method to detect *Aggregatibacter actinomycetemcomitans* in subgingival plaque. The first PCR is conducted with the universal primers from the 16S rRNA gene: 27F (5′-AGAGTTTGATCMTGGCTCAG-3′) and 1492R of 5′-TACGGYTACCTTGTTACGACTT-3′, yielding an expected amplicon of 1466 bp. The second PCR is performed with *A. actinomycetemcomitans*-specific forward primer (AaF) (5′-ATTGGGGTTTAGCCCTGGTG-3′) and the conserved reverse primer (ConR) (5′-ACGTCATCCCCACCTTCCTC-3′), yielding an expected product of 360 bp.

Procedure

1. The subgingival plaque is collected using a sterile curette and immediately placed in a sterile container with 1 mL of reduced transport fluid. The plaque samples are then dispersed in the reduced transport fluid by vortexing for 60 s. Genomic DNA is extracted from 500 μL of the sample using the Puregene DNA isolation kit (Gentra Systems), and the DNA is then stored at –20°C for further analysis.
2. The first PCR mixture (50 μL) comprises 1× PCR buffer II, 2.9 mM $MgCl_2$, 0.2 μM each of primers 27F and 1492R, 1 U of AmpliTaq gold (Applied Biosystems), 0.2 mM each of dATP, dCTP, and dGTP, and 600 μM dUTP, and 5 μl of genomic DNA. Sterile water and DNA (10 ng/μL) extracted from a pure culture of *Escherichia coli* are included as negative control and positive control, respectively.
3. The first PCR mixture is subjected to an initial denaturation at 95°C for 5 min, 22 cycles of 95°C for 1 min, 42°C for 2 min, and 72°C for 3 min, followed by an extension at 72°C for 10 min. The amplification product (of 1466 bp) is verified on a 1% agarose gel and compared with a 100-bp DNA molecular size marker (Promega).
4. The second PCR mixture (50 μL) is made up of 1× PCR buffer II, 2.9 mM $MgCl_2$, 0.15 μM primer AaF and 0.47 μM primer ConR, 1 U of AmpliTaq gold (Applied Biosystems), 0.2 mM each of dATP, dCTP, and dGTP, and 600 μM dUTP, and 2 μL of the first round PCR product.
5. The cycling parameters for the second PCR consist of an initial denaturation at 95°C for 10 min, 40 cycles of 95°C for 1 min, 61°C for 1 min, and 72°C for 5 min, and a final extension at 72°C for 10 min.
6. PCR products are separated on a 2.3% agarose gel, stained with ethidium bromide and visualized under UV light. The expected product from *A. actinomycetemcomitans* is 360 bp.

(iii) PCR for A. actinomycetemcomitans Vieira et al.[39] reported a PCR for specific detection of *A. actinomycetemcomitans* in clinical samples, with primer LKT-F (5′-GGA ATT CCT ATT TAT TGC GAA ACA ATT TGA TC-3′) and LKT-R (5′-GGA ATT CCT GAA ATT AAG CTG GTA ATC-3′).

Procedure

1. PCR mixture (25 μL) contains 1× PCR buffer, 0.2 mM of each dNTP, 0.5 U *Taq* DNA polymerase (Invitrogen), 0.4 μM of each primer pair and 10 ng of template.
2. Amplification is carried out in a GeneAmp PCR System 9700 (Perkin Elmer) with an initial 94°C for 5 min; 35 cycles at 94°C for 30 s, 55°C for 30 s, and 72°C for 1 min; and a final 72°C for 5 min.
3. PCR products are electrophoresed in 1% agarose gel (in 1× TBE buffer), stained with 0.5 μg/mL ethidium bromide and photographed on a UV light transilluminator.

(iv) PCR for Differentiation of A. actinomycetemcomitans and Clone JP2 Cortelli et al.[88] described a PCR assay to differentiate between highly and minimally leukotoxic *A. actinomycetemcomitans* strains. The primers are derived from the *A. actinomycetemcomitans* leukotoxin promoter region—forward: 5′-GCAGGATCCATATTAAATCTCC TTGT-3′ and reverse: 5′-GCGGTCGACAACCTGATAA CAGTATT-3′, generating either a 492 bp product characteristic of highly leukotoxic *A. actinomycetemcomitans* or a 1022 bp product characteristic of minimally leukotoxic *A. actinomycetemcomitans.*

Procedure

1. After the removal of supragingival plaque, subgingival plaque samples are taken with sterile paper points inserted to the depth of the pockets or sulci from the mesial aspect of five different teeth from each subject. The paper points are removed after 15 s, and immediately placed in phosphate-buffered saline (pH 7.4) on ice. The bacterial cells are pelleted at 12,000 × *g* for 1 min in a microfuge and DNA is isolated with Instagene Purification Matrix (BioRad).
2. PCR mixture (100 μL) is composed of 10 mM Tris-HCl pH 8.3, 50 mM KCl, 1.5 mM $MgCl_2$, 100 mM each of dNTPs, 0.001% (w/v) gelatin, 100 ng of each primer, 0.5 U *Taq* polymerase (Amersham), and 100 ng of genomic DNA.
3. PCR amplification is performed on a thermocycler (Perkin Elmer Cetus) with 25 cycles of 30 s at 95°C, 55°C, and 72°C, respectively.
4. The amplification products are analyzed by electrophoresis in 2% submarine agarose gels, stained with ethidium bromide at 0.5 μg/mL, visualized under 300 nm ultraviolet light and photographed.

(v) Real-Time PCR for A. actinomycetemcomitans Yoshida et al.[101] developed a TaqMan real-time PCR targeting *lktA* gene for detection of *A. actinomycetemcomitans* in subgingival plaque samples (Table 69.1).[102] The probe is labeled with fluorescent dyes 6-carboxyfluorescein (FAM) at the 5′ end and 6-carboxytetramethylrhodamine (TAMRA) at the 3′ end.

Procedure

1. Real-time PCR mixture (20-μL) is made up of 1× qPCR Master Mix (Eurogentec), sense and antisense primer at 200 nM each, the TaqMan probe at 250 nM and 1 μL of lysed cells.
2. Amplification is performed at 50°C for 2 min and 95°C for 10 min, followed by 60 cycles of 95°C for 15 s and 58°C for 1 min with the ABI PRISM 7700 Sequence Detection System (Applied Biosystems).

TABLE 69.1
Primers and Probe for Real-Time PCR Detection of *A. actinomycetemcomitans*

Primer	Sequence (5′→3′)
Forward	CAG CAT CTG CGA TCC CTG TA
Reverse	TCA GCC CTT TGT CTT TCC TAG GT
Probe	FAM-TCG AGT ATT CCT CAA GCA TTC TCG CAC G-TAMRA

69.3 CONCLUSION

The genus *Aggregatibacter* covers three gram-negative, nonmotile, facultatively anaerobic rod-shaped bacterial species (i.e., *A. actinomycetemcomitans*, *A. aphrophilus*, and *A. segnis*) in the family *Pasteurellaceae*. In comparison with members of the genus *Haemophilus*, which require blood factors hemin (factor X) and/or nicotinamide adenine dinucleotide (NAD, or factor V) for in vitro growth, *Aggregatibacter* spp. are independent of factor X and variably dependent of factor V.

Aggregatibacter spp., especially *A. actinomycetemcomitans*, are an important cause of localized juvenile periodontitis (an inflammation of the supporting tissues of the teeth that may lead to loss of teeth without medical intervention), and are occasionally involved in endocarditis, bacteremia, and other infections, particularly in patients undergoing surgery and with underlying diseases. Despite their close resemblance to *Haemophilus* spp. and other related bacteria, identification of *A. actinomycetemcomitans*, *A. aphrophilus*, and *A. segnis* by phenotypic techniques is generally reliable. The main benefits of using molecular techniques for detection and differentiation of *Aggregatibacter* spp. are their improved sensitivity, specificity, and speed. This is especially evident with recent development of real-time PCR, allowing automated detection and quantitation of *Aggregatibacter* spp. in clinical specimens. Application of these new methodologies will facilitate improved epidemiological investigation of infections due to *Aggregatibacter* bacteria, leading to effective intervention and control measures against these diseases.

REFERENCES

1. Martin, P.R., Shea, R.J., and Mulks, M.H., Identification of a plasmid-encoded gene from *Haemophilus ducreyi* which confers NAD independence, *J. Bacteriol.*, 183, 1168, 2001.
2. Tanner, A.C.R. et al., Classification and identification of *Actinobacillus actinomycetemcomitans* and *Haemophilus aphrophilus* by cluster analysis and deoxyribonucleic acid hybridizations, *J. Periodont. Res.*, 17, 585, 1982.
3. Potts, T.V., Zambon, J.J., and Genco, R.J., Reassignment of *Actinobacillus actinomycetemcomitans* to the genus *Haemophilus* as *Haemophilus actinomycetemcomitans* comb. nov., *Int. J. Syst. Bacteriol.*, 35, 337, 1985.
4. Nørskov-Lauritsen, N., and Kilian, M., Reclassification of *Actinobacillus actinomycetemcomitans*, *Haemophilus aphrophilus*, *Haemophilus paraphrophilus* and *Haemophilus segnis* as *Aggregatibacter actinomycetemcomitans* gen. nov., comb. nov., *Aggregatibacter aphrophilus* comb. nov. and *Aggregatibacter segnis* comb. nov., and emended description of Aggregatibacter aphrophilus to include V factor-dependent and V factor-independent isolates, *Int. J. Syst. Evol. Microbiol.*, 56, 2135, 2006.
5. Niven, D.F., and O'Reilly, T., Significance of V-factor dependency in the taxonomy of *Haemophilus* species and related organisms, *Int. J. Syst. Bacteriol.*, 40, 1, 1990.
6. Paju, S. et al., Heterogeneity of *Actinobacillus actinomycetemcomitans* strains in various human infections and relationships between serotype, genotype, and antimicrobial susceptibility, *J. Clin. Microbiol.*, 38, 79, 2000.
7. Page, R.C. et al., The immunodominant outer membrane antigen of *Actinobacillus actinomycetemcomitans* is located in the serotype-specific high-molecular-mass carbohydrate moiety of lipopolysaccharide, *Infect. Immun.*, 59, 3451, 1991.
8. Perry, M.B. et al., Structures of the antigenic O-polysaccharides of lipopolysaccharides produced by *Actinobacillus actinomycetemcomitans* serotypes a, c, d and e, *Eur. J. Biochem.*, 242, 682, 1996.
9. Perry, M B. et al., Characterization of the O-polysaccharide structure of lipopolysaccharide from *Actinobacillus actinomycetemcomitans* serotype b, *Infect. Immun.*, 64, 1215, 1996.
10. Kaplan, J.B. et al., Structural and genetic analyses of O polysaccharide from *Actinobacillus actinomycetemcomitans* serotype f, *Infect. Immun.*, 69, 5375, 2001.
11. Kilian, M. et al., The etiology of periodontal disease revisited by population genetic analysis, *Periodontol. 2000*, 42, 158, 2006.
12. Tinoco, E.M., Sivakumar, M., and Preus, H.R., The distribution and transmission of *Actinobacillus actinomycetemcomitans* in families with localized juvenile periodontitis, *J. Clin. Periodontol.*, 25, 99, 1998.
13. Kanasi, E. et al., Lack of serotype antigen in *A. actinomycetemcomitans*, *J. Dent. Res.*, 89, 292, 2010.
14. Nørskov-Lauritsen, N., Bruun, B., and Kilian, M., Multilocus sequence phylogenetic study of the genus *Haemophilus* with description of *Haemophilus pittmaniae* sp. nov., *Int. J. Syst. Evol. Microbiol.*, 55, 449, 2005.
15. Di Bonaventura, M.P. et al., *Complete genome sequence of Aggregatibacter (Haemophilus) aphrophilus* NJ8700, *J. Bacteriol.*, 191, 4693, 2009.
16. Poulsen, K. et al., Population structure of *Actinobacillus actinomycetemcomitans*: A framework for studies of disease-associated properties, *Microbiology*, 140, 2049, 1994.
17. Planet, P.J. et al., The widespread colonization island of *Actinobacillus actinomycetemcomitans*, *Nat. Genet.*, 34, 193, 2003.
18. Zambon, J.J., Christersson, L.A., and Slots, J., *Actinobacillus actinomycetemcomitans* in human periodontal disease. Prevalence in patient groups and distribution of biotypes and serotypes within families, *J. Periodontol.*, 54,707, 1983.
19. Zambon, J.J. et al., The microbiology of early-onset periodontitis: Association of highly toxic *Actinobacillus actinomycetemcomitans* strains with localized juvenile periodontitis, *J. Periodontol.*, 67, 282, 1996.
20. Alaluusua, S. et al., Ribotyping shows intrafamilial similarity in *Actinobacillus actinomycetemcomitans* isolates, *Oral Microbiol. Immunol.*, 8, 225, 1993.
21. Mombelli, A. et al., *Actinobacillus actinomycetemcomitans* in adult periodontitis. I. Topographic distribution before and after treatment, *J. Periodontol.*, 65, 820, 1994.
22. Preus, H.R. et al., The distribution and transmission of *Actinobacillus actinomycetemcomitans* in families with established adult periodontitis, *J. Periodontol.*, 65, 2, 1994.

23. Asikainen, S., Chen, C., and Slots, J., Likelihood of transmitting *Actinobacillus actinomycetemcomitans* and *Porphyromonas gingivalis* in families with periodontitis, *Oral Microbiol. Immunol.*, 11, 387, 1996.
24. Asikainen, S., and Chen, C., Oral ecology and person-to-person transmission of *Actinobacillus actinomycetemcomitans* and *Porphyromonas gingivalis*, *Periodontol. 2000*, 20, 65, 1999.
26. Slots, J., and Ting, M., *Actinobacillus actinomycetemcomitans* and *Porphyromonas gingivalis* in human periodontal disease: occurrence and treatment, *Periodontol. 2000*, 20, 82, 1999.
27. Lamell, C.W. et al., Acquisition and colonization stability of *Actinobacillus actinomycetemcomitans* and *Porphyromonas gingivalis* in children, *J. Clin. Microbiol.*, 38, 1196, 2000.
28. Kaplan, J.B. et al., Population structure and genetic diversity of *Actinobacillus actinomycetemcomitans* strains isolated from localized juvenile periodontitis patients, *J. Clin. Microbiol.*, 40, 1181, 2002.
29. Kaplan, J.B., Meyenhofer, M.F., and Fine, D.H., Biofilm growth and detachment of *Actinobacillus actinomycetemcomitans*, *J. Bacteriol.*, 185, 1399, 2003.
30. Paturel, L. et al., *Actinobacillus actinomycetemcomitans* endocarditis, *Clin. Microbiol. Infect.*, 10, 98, 2004.
31. Aas, J.A. et al., Defining the normal bacterial flora of the oral cavity, *J. Clin. Microbiol.*, 43, 5721, 2005.
32. Yang H-W. et al., Relationship of *Actinobacillus actinomycetemcomitans* serotypes to periodontal condition: Prevalence and proportions in subgingival plaque, *Eur. J. Oral Sci.*, 113, 28, 2005.
33. Fine, D.H., et al., *Aggregatibacter actinomycetemcomitans* and its relationship to initiation of localized aggressive periodontitis: Longitudinal cohort study of initially healthy adolescents, *J. Clin. Microbiol.*, 45, 3859, 2007.
34. Nakano, K. et al., Detection and serotype distribution of *Actinobacillus actinomycetemcomitans* in cardiovascular specimens from Japanese patients, *Oral Microbiol. Immunol.*, 22, 136, 2007.
35. Rêgo, R.O. et al., Transmission of *Aggregatibacter actinomycetemcomitans* between Brazilian women with severe chronic periodontitis and their children. *Braz. Dent. J.*, 18, 220, 2007.
36. Kulekci, G. et al., Salivary detection of periodontopathic bacteria in periodontally healthy children, *Anaerobe*, 14, 49, 2008.
37. Paju, S. et al., Detection of multiple pathogenic species in saliva is associated with periodontal infection in adults, *J. Clin. Microbiol.*, 47, 235, 2009.
38. Riep, B. et al., Are putative periodontal pathogens reliable diagnostic markers? *J. Clin. Microbiol.*, 47, 1705, 2009.
39. Vieira, E.M. et al., Occurrence of *Aggregatibacter actinomycetemcomitans* in Brazilian Indians from Umutina Reservation, Mato Grosso, Brazil, *J. Appl. Oral Sci.*, 17, 440, 2009.
40. Kim, T.S. et al., Serotypes of *Aggregatibacter actinomycetemcomitans* in patients with different ethnic backgrounds, *J. Periodontol.*, 80, 2020, 2009.
41. van der Reijden, W.A. et al., Java project on periodontal diseases: Serotype distribution of *Aggregatibacter actinomycetemcomitans* and serotype dynamics over an 8-year period, *J. Clin. Periodontol.*, 35, 487, 2008.
42. Haubek, D. et al., Risk of aggressive periodontitis in adolescent carriers of the JP2 clone of *Aggregatibacter (Actinobacillus) actinomycetemcomitans* in Morocco: A prospective longitudinal cohort study, *Lancet*, 371, 237, 2008.
43. Bangsborg, J.M., Tvede, M., and Skinhoj, P., *Haemophilus segnis* endocarditis, *J. Infect.*, 16, 81, 1988.
44. Lau, S.K. et al., Characterization of *Haemophilus segnis*, an important cause of bacteremia, by 16S rRNA gene sequencing, *J. Clin. Microbiol.*, 42, 877, 2004.
45. Haraszthy, V.I. et al., Evidence for the role of highly leucotoxic *Actinobacillus actinomycetemcomitans* in the pathogenesis of localized juvenile and other forms of early onset periodontitis, *J. Periodontol.*, 71, 912, 2000.
46. Querido, S.M.R. et al., Clinical and microbial evaluation of dental scaling associated with subgingival minocycline in chronic periodontitis subjects, *Braz. Oral Res.*, 18, 110, 2004.
47. Lally, E.T. et al., Structure/function aspects of *Actinobacillus actinomycetemcomitans* leukotoxin, *J. Periodontol.*, 67, 298, 1996.
48. Kraig, E., Dailey, T., and Kolodrubetz, D., Nucleotide sequence of the leukotoxin gene from *Actinobacillus actinomycetemcomitans*: homology to the alpha-hemolysin/leukotoxin gene family, *Infect. Immun.*, 58, 920, 1990.
49. Fives-Taylor, P.M. et al., Virulence factors of *Actinobacillus actinomycetemcomitans*, *Periodontol. 2000*, 20, 136, 1999.
50. Orru, G. et al., Usefulness of real time PCR for the differentiation and quantification of 652 and JP2 *Actinobacillus actinomycetemcomitans* genotypes in dental plaque and saliva, *BMC Infect. Dis.*, 6, 98, 2006.
51. Graber, K.R., Smoot, L.M., and Actis, L.A., Expression of iron binding proteins and hemin binding activity in the dental pathogen *Actinobacillus actinomycetemcomitans*, *FEMS Microbiol. Lett.*, 163, 135, 1998.
52. Zadeh, H.H., Nichols, F.C., and Miyasaki, K.T., The role of the cell-mediated immune response to *Actinobacilus actinomycetemcomitans* and *Porphyromonas gingivalis* in periodontitis, *Periodontol.*, 2000, 20, 239, 1999.
53. Johansson, A., Hänström, L., and Kalfas, S., Inhibition of *Actinobacillus actinomycetemcomitans* leukotoxicity by bacteria from the subgingival flora, *Oral Microbiol. Immunol.*, 15, 218, 2000.
54. Rhodes, E.R., et al., Iron acquisition in the dental pathogen *Actinobacillus actinomycetemcomitans*: What does it use as a source and how does it get this essential metal? *Biometals*, 20, 365, 2007.
55. Rhodes, E.R. et al., Genetic and functional analyses of the *Actinobacillus actinomycetemcomitans* AfeABCD siderophore-independent iron acquisition system, *Infect. Immun.*, 73, 3758, 2005.
56. Brogan, J.M. et al., Regulation of *Actinobacillus actinomycetemcomitans* leukotoxin expression: Analysis of the promoter regions of leukotoxic and minimally leukotoxic strains, *Infect. Immun.*, 62, 501, 1994.
57. Bueno, I.C., Mayer, M.P.A., and DiRienzo, J.M., Relationship between conversion of localized juvenile periodontitis-susceptible children from health to disease and *Actinobacillus actinomycetemcomitans* leukotoxin promoter structure, *J. Periodontol.*, 70, 998, 1998.
58. Haubek, D. et al., Stability of the JP2 clone of *Aggregatibacter actinomycetemcomitans*, *J. Dent. Res.*, 88, 856, 2009.
59. Amarasinghe, J.J., Scannapieco, F.A., and Haase, E.M., Transcriptional and translational analysis of biofilm determinants of *Aggregatibacter actinomycetemcomitans* in response to environmental perturbation, *Infect. Immun.*, 77, 2896, 2009.
60. Periasamy, S., and Kolenbrander, P.E., *Aggregatibacter actinomycetemcomitans* builds mutualistic biofilm communities with *Fusobacterium nucleatum* and *Veillonella* species in saliva, *Infect. Immun.*, 77, 3542, 2009.

61. Slots, J., Selective medium for *Actinobacillus actinomycetemcomitans*, *J. Clin. Microbiol.*, 15, 606, 1982.
62. Chen, C., and Slots. J., Microbiological tests for *Actinobacillus actinomycetemcomitans* and *Porphyromonas gingivalis*, *Periodontol. 2000*, 20, 53, 1999.
63. Kilian, M., A taxonomic study of the genus *Haemophilus*, with the proposal of a new species, *J. Gen. Microbiol.*, 93, 9, 1976.
64. Brondz, I., and Olsen, I., Multivariate analyses of carbohydrate data from lipopolysaccharides of *Actinobacillus* (*Haemophilus*) *actinomycetemcomitans*, *Haemophilus aphrophilus*, and *Haemophilus paraphrophilus*, *Int. J. Syst. Bacteriol.*, 40,405, 1990.
65. Slots, J., Salient biochemical characters of *Actinobacillus actinomycetemcomitans*, *Arch. Microbiol.*, 131, 60, 1982.
66. Bonta, Y. et al., Rapid identification of periodontal pathogens in subgingival plaque: comparison of indirect immunofluorescence microscopy with bacterial culture for detection of *Actinobacillus actinomycetemcomitans*, *J. Dent. Res.*, 64, 793, 1985.
67. Caugant, D.A., Selander, R.K., and Olsen, I., Differentiation between *Actinobacillus* (*Haemophilus*) *actinomycetemcomitans*, *Haemophilus aphrophilus* and *Haemophilus paraphrophilus* by multilocus enzyme electrophoresis, *J. Gen. Microbiol.*, 136, 2135, 1990.
68. Ebersole, J.L. et al., Serum antibody in *Actinobacillus actinomycetemcomitans*-infected patients with periodontal disease, *Infect. Immun.*, 59, 1795, 1991.
69. Myhrvold, V., Brondz, I., and Olsen, I., Application of multivariate analyses of enzymic data to classification of members of the *Actinobacillus-Haemophilus-Pasteurella* group, *Int. J. Syst. Bacteriol.*, 42, 12, 1992.
70. Nieminen, A., Kari, K., and Saxen, L., Specific antibodies against *Actinobacillus actinomycetemcomitans* in serum and saliva of patients with advanced periodontitis, *Scand. J. Dent. Res.*, 101, 196, 1993.
71. Potts, T.V. et al., Relationships among isolates of oral haemophili as determined by DNA–DNA hybridization, *Arch. Microbiol.*, 145, 136, 1986.
72. Tønjum, T., Bukholm, G., and Bøvre, K., Identification of *Haemophilus aphrophilus* and *Actinobacillus actinomycetemcomitans* by DNA–DNA hybridization and genetic transformation, *J. Clin. Microbiol.*, 28, 1994, 1990.
73. Preus, H.R. et al., Rapid identification of *Actinobacillus actinomycetemcomitans* based on analysis of 23S ribosomal RNA, *Oral Microbiol. Immunol.*, 7, 372, 1992.
74. Sedlacek, I. et al., Genetic relationship of strains of *Haemophilus aphrophilus*, *H. paraphrophilus*, and *Actinobacillus actinomycetemcomitans* studied by ribotyping, *Int. J. Med. Microbiol.*, 279, 51, 1993.
75. Ashimoto, A. et al., Polymerase chain reaction detection of 8 putative periodontal pathogens in subgingival plaque of gingivitis and advanced periodontitis lesions, *Oral Microbiol. Immunol.*, 11, 266, 1996.
76. Riggio, M.P. et al., Comparison of polymerase chain reaction and culture methods for detection of *Actinobacillus actinomycetemcomitans* and *Porphyromonas gingivalis* in subgingival plaque samples, *J. Periodont. Res.*, 31, 496, 1996.
77. Poulsen, K., Oum-Keltoum, E., and Haubek, D., Improved PCR for detection of the highly leukotoxic JP2 clone of *Actinobacillus actinomycetemcomitans* in subgingival plaque samples, *J. Clin. Microbiol.*, 41, 4829, 2003.
78. Perez, B.A. et al., Genetic analysis of the requirement for *flp-2*, *tadV*, and *rcpB* in *Actinobacillus actinomycetemcomitans* biofilm formation, *J. Bacteriol.*, 188, 6361, 2006.
79. Morikawa, M. et al., Comparative analysis of putative periodontopathic bacteria by multiplex polymerase chain reaction, *J. Periodont. Res.*, 43, 268, 2008.
80. Ready, D. et al., Disease severity associated with the presence in subgingival plaque of *Porphyromonas gingivalis*, *Aggregatibacter actinomycetemcomitans*, and *Tannerella forsythia*, singularly or in combination, as detected by nested multiplex PCR, *J. Clin. Microbiol.*, 46, 3380, 2008.
81. Fives-Taylor, P.M., Meyer, D.H., and Mintz, K.P., Virulence factors of the periodontopathogen *Actinobacillus actinomycetemcomitans*, *J. Periodontol.*, 67, 291, 1996.
82. Dewhirst, F.E. et al., Phylogeny of 54 representative strains of species in the family *Pasteurellaceae* as determined by comparison of 16S rRNA sequences, *J. Bacteriol.*, 174, 2002, 1992.
83. Tønjum, T., and Haas, R., Identification of *Actinobacillus actinomycetemcomitans* by leukotoxin gene-specific hybridization and polymerase chain reaction assays, *J. Clin. Microbiol.*, 31, 1856, 1993.
84. Riggio, M.P., and Lennon, A., Rapid identification of *Actinobacillus actinomycetemcomitans*, *Haemophilus aphrophilus*, and *Haemophilus paraphrophilus* by restriction enzyme analysis of PCR-amplified 16S rRNA genes, *J. Clin. Microbiol.*, 35, 1630, 1997.
85. Tran, S.D., and Rudney, J.D., Improved multiplex PCR using conserved and species-specific 16S rRNA gene primers for simultaneous detection of *Actinobacillus actinomycetemcomitans*, *Bacteroides forsythus*, and *Porphyromonas gingivalis*, *J. Clin. Microbiol.*, 37, 3504, 1999.
86. Contreras, A. et al., Frequency of 530-bp deletion in *Actinobacillus actinomycetemcomitans* leukotoxin promoter region, *Oral Microbiol. Immunol.*, 15, 338, 2000.
87. Hedegaard, J. et al., Phylogeny of the genus *Haemophilus* as determined by comparison of partial *infB* sequences, *Microbiology*, 147, 2599, 2001.
88. Cortelli, S.C. et al., Detection of highly and minimally leukotoxic *Actinobacillus actinomycetemcomitans* strains in patients with periodontal disease, *Pesqui Odontol. Bras.*, 17, 183, 2003.
89. Cortelli, S.C. et al., Detection of *Actinobacillus actinomycetemcomitans* in unstimulated saliva of patients with chronic periodontitis, *J. Periodontol.*, 76, 204, 2005.
90. Westling, K., and Vondracek, M., *Actinobacillus* (*Aggregatibacter*) *actinomycetemcomitans* (HACEK) identified by PCR/16S rRNA sequence analysis from the heart valve in a patient with blood culture negative endocarditis, *Scand. J. Infect. Dis.*, 40, 981, 2008.
91. Hyvärinen, K. et al., Detection and quantification of five major periodontal pathogens by single copy gene-based real-time PCR, *Innate Immun.*, 15, 195, 2009.
92. Kuboniwa, M. et al., Quantitative detection of periodontal pathogens using real-time polymerase chain reaction with TaqMan probes, *Oral Microbiol. Immunol.*, 19, 168, 2004.
93. Morillo, J.M. et al., Quantitative real-time polymerase chain reaction based on single copy gene sequence for detection of periodontal pathogens, *J. Clin. Periodontol.*, 31, 1054, 2004.
94. Boutaga, K. et al., Periodontal pathogens: A quantitative comparison of anaerobic culture and real-time PCR, *FEMS Immunol. Med. Microbiol.*, 45, 191, 2005.
95. Boutaga, K. et al., Comparison of subgingival bacterial sampling with oral lavage for detection and qualification of periodontal pathogens by real-time polymerase chain reaction, *J. Periodontol.*, 78, 79, 2007.
96. Atieh, M.A., Accuracy of real-time polymerase chain reaction versus anaerobic culture in detection of *Aggregatibacter*

actinomycetemcomitans and *Porphyromonas gingivalis*: A meta-analysis, *J. Periodontol.*, 79, 1620, 2008.
97. Gomes, S.C. et al., The effect of a supragingival plaque-control regimen on the subgingival microbiota in smokers and never-smokers: Evaluation by real-time polymerase chain reaction, *J. Periodontol.*, 79, 2297, 2008.
98. Gaetti-Jardim, E. Jr. et al., Quantitative detection of periodontopathic bacteria in atherosclerotic plaques from coronary arteries, *J. Med. Microbiol.*, 58, 1568, 2009.
99. Osawa, R. et al., Rapid detection of *Actinobacillus actinomycetemcomitans* using a loop-mediated isothermal amplification method, *Oral Microbiol. Immunol.*, 22, 252, 2007.
100. Seki, M. et al., Novel loop-mediated isothermal amplification method for detection of the JP2 clone of *Aggregatibacter actinomycetemcomitans* in subgingival plaque, *J. Clin. Microbiol.*, 46, 1113, 2008.
101. Yoshida, A. et al., Development of a 5′ fluorogenic nuclease-based real-time PCR assay for quantitative detection of *Actinobacillus actinomycetemcomitans* and *Porphyromonas gingivalis*, *J. Clin. Microbiol.*, 41, 863, 2003.
102. Suzuki, N., Yoshida, A., and Nakano, Y., Quantitative analysis of multi-species oral biofilms by TaqMan real-time PCR, *Clin. Med. Res.*, 3, 176, 2005.

70 Cardiobacterium

Efthimia Petinaki and George N. Dalekos

CONTENTS

70.1 INTRODUCTION

70.1.1 Classification, Morphology, and Biology

Cardiobacterium hominis is a small, catalase-negative, oxidase-positive, pleomorphic gram-negative coccobacillus, which is a member of the HACEK group of microorganisms (*Haemophilus* species, *Actinobacillus actinomycetemcomitans*, *Cardiobacterium hominis*, *Eikenella corrodens*, and *Kingella kingae*).[1,2] The microorganism is a part of the normal human nasal and oropharyngeal flora, but upon disturbance of the mucosal integrity it can become invasive and pathogenic.[1,3] Although it is present in the oral cavity of almost 70% of healthy individuals, it has also been detected in stool samples and genitourinary specimens by fluorescent antibody analysis.[4,5] The bacterium was first isolated in four patients with infective endocarditis in 1962.[6] At that time, it was classified as a *Pasteurella*-like organism and was designed group II ID. Slotnick and Dougherty subsequently proposed the name *Cardiobacterium hominis* in 1964.[7] Until 2004, *C. hominis* was the sole *Cardiobacterium* species known to cause human infections. In 2004, however, Han et al. identified a new *Cardiobacterium* species from a patient with endocarditis complicated by ruptured cerebral aneurysm; the new isolate called *Cardiobacterium valvarum*.[8]

Similar to other members of the HACEK group, *C. hominis* has historically been associated with culture-negative infective endocarditis due to its fastidious growth requirements, but recent improvements in blood culture systems and blood culture medium formulations have resulted in improved recovery of this pathogen, even though in most cases a prolonged period of incubation is needed.[9] Indeed, as a fastidious bacterium, *C. hominis* does not grow on selective enteric media such as MacConkey agar, while it grows well on standard enriched media (blood and chocolate agars).[10] Optimal growth of the microorganism occurs with supplemental CO_2 and increased humidity. It produces indole and ferments glucose, sorbitol, mannose, sucrose, and, in most cases, maltose and mannitol. Its capacity to produce indole differentiates it from *C. valvarum*, which has recently been reported as another rare cause of endocarditis.[8,11,12] *C. hominis* does not demonstrate urease, catalase, nitrate reductase, lysine decarboxylase, ornithine decarboxylase, and arginine dihydrolase activity. These characteristics help to distinguish *C. hominis* from other members of the HACEK group. Table 70.1 shows in detail the biochemical characteristics of *C. hominis* compared to other HACEK microorganisms.

70.1.2 Clinical Features and Pathogenesis

C. hominis is a low-virulence pathogen capable of causing infective endocarditis and, rarely, other focal infections under specific conditions.[1,9,13–16] In most of these cases, there is some form of preexisting cardiac disease and a history of dental manipulation, although in the majority of affected individuals no portal of entry can be identified.[1,9] Indeed, dental work, dental caries, and a history of gum bleeding have been proposed as risk factors leading to infection by *C. hominis*. In addition, previous history of aortic or mitral valve replacement, previous rheumatic heart disease, past history of infective endocarditis, known ventricular septal defect, known bicuspid aortic valve, congenital aortic valve disease, mitral valve prolapse with murmur and dilated cardiomyopathy have all been described as potential predisposing cardiac lesions in cases of *C. hominis* endocarditis.[1,9] Manipulations in the pharynx, such as upper gastrointestinal endoscopy, have also been rarely reported as a risk factor for infective endocarditis caused by *C. hominis* in bioprosthetic heart valve.[17] Neither intravenous drug use nor infection at

TABLE 70.1
Comparison of Biochemical Characteristics of *C. hominis* and Other HACEK Bacteria

Test	*Cardiobacterium hominis*	*Haemophilus aphrophilus*	*Actinobacillus actinomycetemcomicans*	*Eikenella corrodens*	*Kingella kingae*
Oxidase	+	v	–/w	+	+
Catalase	–	–	+	–	–
Nitrate	–	+	+	+	–
Indole reduction	+	–	–	–	+
Fermentation					
of glucose	+	+	+	–	+
of lactose	–	+	–	–	–
of maltose	+	+	+	–	+
of mannitol	+	–	v	–	–
of sucrose	+	+	–	–	–

+, positive; –, negative; v, variable; w, weak.

another site of the body has been described as risk factors for *C. hominis* endocarditis contrary to infective endocarditis caused by other pathogens (e.g., *Staphylococcus aureus*).

Since its first description in 1962, fewer than 75 cases of *C. hominis* endocarditis have been reported in the literature. Though unusual, this organism may cause focal infections outside the vascular system, namely bacterial meningitis, osteomyelitis, abdominal abscess, pericarditis, and septic arthritis.[13–16] *C. hominis* endocarditis has been reported in both men and women between the ages of 17 and 82 years, although it often affects the middle-aged population. The illness has a characteristically insidious onset, with the presence of low-grade fever but stigmata of endocarditis such as splenomegaly, anemia, and hematuria are usually already present during physical examination. Patients may present one or more of the following constitutional symptoms: fatigue, lethargy, night sweats, chills, weakness, myalgias, arthralgias, diffuse abdominal discomfort, anorexia, and weight loss. On the other hand, symptoms related to valves dysfunction or secondary to embolic events are rarely the only predominant clinical sign. Also, patients may be presented with typical clinical picture of subacute endocarditis, often feeling unwell for a period ranging from weeks to months before a diagnosis is achieved. The mean duration of the symptoms before diagnosis may be as long as of 4 months, which is significantly longer compared to the respective duration in cases of infective endocarditis caused, for example, by staphylococci and streptococci species.

C. hominis tends to form large, friable vegetation on endocardial surfaces that are associated with a significant risk of cerebral embolization (30%) or mycotic aneursm formation (10%), both of which appear to be more common in *C. hominis* endocarditis, than with endocarditis caused by other gram-negative bacilli.[1,18] However, small vegetations in the aortic valve without significant hemodynamic insufficiency have also been reported in infective endocarditis due to *C. hominis*.[19] Laboratory features of *C. hominis* endocarditis include mild to moderate anemia, and an increased white blood cell count, often less than 15×10^9/L in one-third of the affected cases, while the erythrocyte sedimentation rate is, typically, moderately raised. Severe thrombocytopenia caused by immune-mediated mechanisms has been reported.[20] Rheumatoid factors and C-reactive protein may also have been elevated but this determination lacks specificity. Rapid progressive glomerulonephriris treated with corticosteroids in combination with antibiotics has been described in cases with delayed diagnosis.[21]

70.1.3 Diagnosis

The diagnosis of *C. hominis* endocarditis requires a high concept of suspicion based on the clinical picture, the past history and the laboratory findings. Consequently, the confirmation of the clinical diagnosis should be done by the laboratory of microbiology; its role in the everyday clinical practice or in the academic venue is mandatory for the establishment of the diagnosis of endocarditis and its efficient management. It requires a rapid, reliable, and state-of-the-art identification of the etiological agent, followed by the susceptibility testing of the microorganism against various antimicrobial agents. These procedures help to accomplish a correct diagnosis, allow the appropriate approach toward the infection, and discourage the unwise utilization of antibiotics.[22,23] In this context, the diagnosis of *C. hominis* endocarditis poses a challenge giving rise to the development of conventional and molecular techniques for its detection in blood samples or heart valve tissues.[24–27]

Conventional Techniques. As in other cases of infective endocarditis, in patients with suspicion of *C. hominis* endocarditis, at least three sets of blood cultures over a 24-h period before starting antibiotics should be obtained. In case of valve replacement surgery, the heart valve tissues should be transferred immediately and sterile to the laboratory, since a positive tissue culture needs longer antibiotic treatment and seems to be negatively associated with the final outcome of the patient and the survival of the prosthetic valve.

The conventional approach for *C. hominis* isolation includes the culture of the clinical specimen, and if positive, the isolation and identification of the microorganism accompanied by the susceptibility testing. The improvements in blood culture medium formulations and the implementation of automated blood culture systems have resulted in increased recovery of this pathogen though in general, a prolonged period of incubation is needed.[28] Indeed, positive cultures for *C. hominis* have generally been reported after an incubation time ranging from 3 to 17 days. It seems rationale, however, that the sensitivity of the new blood culture systems for this pathogen depends on the number of bacterial cells that are circulating in the blood at the time of sample collection.[9,19,29] Therefore, in cases of suspected *C. hominis* endocarditis, we suggest a prolonged incubation period of blood cultures for at least 2 or even 3 weeks.

When a positive reading of blood culture is observed, subcultures on chocolate agar or on 5% sheep-blood agar, at 37°C in an atmosphere of 5% CO_2, should be performed. The colony of the microorganism after 4 days of incubation is usually tiny, opaque, and circular.[19] The initial identification of the microorganism is based on the biochemical characteristics, as described above and shown in Table 70.1. Recently, a new VITEK 2 NH card offers a useful method for the identification of diverse groups of fastidious organisms.[30] However, more work is needed in order to definitely evaluate the performance of this test with regard to *Cardiobacterium* species.[31] In the cases where the biochemical tests do not allow the identification of *C. hominis*, use of partial *16S rRNA* gene amplification followed by sequencing analysis provides an accurate identification of the microorganism to the species level.[32]

Molecular Techniques. Recently introduced in the clinical microbiological laboratory, molecular techniques have the ability to detect and identify the microorganisms directly in clinical samples such as blood, embolic material, and heart valve tissues.[23,33] In addition, these assays may help identify the causative agent in a wide range of biological samples such as serum, synovial fluid, cerebrospinal fluid, tonsillar tissues, and pleural fluid.[19,23,34–36] The basic principle of these techniques is based on the use of PCR primers that are targeted against the conserved regions of ribosomal (r) DNA. Thus, it is possible to design broad-range *16S rRNA* PCR that are capable of detecting DNA from almost any bacterial species as, indeed, bacterial rDNA is composed of highly conserved nucleotide sequences that are shared by all bacterial regions that are genus or species specific. The clinical specimen (blood, tissue, etc.) is sent to the clinical microbiological laboratory and total DNA is extracted (see Section 70.2). When a positive *16S rRNA* PCR result is obtained, a subsequent sequencing analysis of the PCR product is required in order to identify both novel and rare bacteria.[23] Results obtained by using this molecular approach are usually available in 2 days. This is clearly advantageous over conventional techniques that are often time consuming, particularly for fastidious microorganisms like *C. hominis*.[19] However, a major limitation of all molecular methods is the lack of the simultaneous provision of the expression of the antimicrobial susceptibility pattern. Therefore, it is a long way before the conventional methods are completely replaced by the molecular assays.

Susceptibility to Antimicrobial Agents. In the past, *C. hominis* was generally sensitive to penicillin.[1] Therefore, penicillin or ampicillin has been considered the drug of choice for treating infective endocarditis caused by this microorganism. However, recent reports have shown the emergence of beta-lactamase-producing *C. hominis* isolates[37,38] that are inhibited by clavulanic acid. These beta-lactamase-producing isolates can be readily detected by testing of colonies with nitrocefin. In addition, the Clinical and Laboratory Standard Institute (CLSI) recently published a new laboratory guideline for antimicrobial susceptibility testing of infrequently encountered or fastidious bacteria not covered in the previous CLSI publication. The emergence of *C. hominis* isolates that produce beta-lactamase emphasizes the need for susceptibility testing and provides further support to the notion that molecular techniques alone cannot be the solution for serious infections caused by fastidious bacteria.[39,40]

70.2 METHODS

70.2.1 Samples Preparation

Specimen Handling. Specimens obtained from the patient might be blood, explanted valve tissue or embolic material.

For the conventional methods, 10 mL of whole blood is injected into an aerobic blood culture bottle and 10 mL is also injected into an anaerobic bottle. Valve tissues or embolic materials are immediately transferred in sterile dry containers and shipped to the laboratory. By using a surgical sterile knife, the tissues are cut in very small pieces.[35] Then the pieces are put into blood cultures bottles, as mentioned above for blood samples.

For the molecular techniques, materials are sent to the laboratory (blood or tissue) in rubber-sealed, pyrogen-free tubes or containers for DNA extraction.

Culture and Chemical Testing. Cultures bottles are incubated in an appropriate instrument according to the procedures of the manufacturer. Where a positive reading of the bottles is observed, 10 μL fluid from the respective bottle is obtained and is cultured on chocolate agar at 37°C with 5% CO_2. After 3 or 4 days of incubation, the bacterium forms tiny, opaque, and circular colonies.[19] The absence of growth of the microorganism on MacConkey agar indicates the presence of a gram-negative fastidious organism. Gram stain from the colonies reveals the presence of straight gram-negative rods, located singly or in pairs, and of variable length depending on the medium. The first step of the identification includes testing with oxidase and catalase, followed by tryptophanase, glucose, fructose, maltose, sucrose, indole, lipase (weak), alkaline phosphatise, penicillinase, ornithine decarboxylase, urease, beta-galactosidase, proline arylamidase, and gamma-glutamyl transferase. For final identification, molecular tools and *16S rRNA* gene sequence analysis are indispensable (see Detection Procedures).

For susceptibility testing, the E-test could be used for the determination of MICs on blood Mueller-Hinton after 48 h of incubation in a CO_2 chamber.[11] Antimicrobials agents proposed are penicillin, amoxicillin, piperacillin, cefazolin, cefotiam, ceftriaxone, ceftazidime, imipenem, meropenem, gentamicin, tobramycin, amikacin, ciprofloxacin, moxifloxacin, erythromycin, tetracycline, chloramphenicol, vancomycin, and linezolid. Detection of beta-lactamase producing strains of *C. hominis* is carried out using a nitrocefin disk.

DNA Extraction. According to our experience, in each case of a serious infection, molecular methods must be applied in parallel with conventional cultures. DNA is extracted with the help of commercial kits. The extracted DNA is tested for the absence of PCR inhibitors after amplification using a pair of primers namely, RS42 (5′-GCTCACTCAGTGTGGCAAAG-3′) and Km (5′-GGTT GGCCAATCTACTCCCAGG-3′) targeting b2-globin.[19,23,34–36] At the same time, and in parallel with the DNA extraction, RNA is extracted directly from the clinical specimen and is stored at −80°C in order to test for the viability of the organism as the potential presence of bacterial DNA is not by definition a synonym of the viability of the microorganism.[36]

70.2.2 Detection Procedures

Principle. The presence of bacterial DNA is detected by PCR amplification of the extracted DNA, using a pair of universal primers targeting the *16S rRNA* gene: 5′-AGAGTTTGATCATGGCTCA-3′ (forward; positions 8–27) and 5′-ACGGCGACTGCTGCTGGCAC-3′ (reverse; positions 531–514).[19,41]

1. Prepare PCR mixture (50 µL) containing 1× PCR buffer, 200 µM of each dNTPs, 1 U Taq DNA polymerase, 50 pmol of each primer, and 5 µL purified template (corresponding to 200 ng of total DNA). Include 5 µL of template with a known concentration (50 pg) of *Escherichia coli* DNA in a separate tube as a positive control.
2. Carry out PCR amplification with an initial cycle of 95°C for 4 min; 35 cycles of 94°C for 30 s, 55°C for 30 s, and 72°C for 90 s; and a final extension at 72°C for 10 min.
3. Filter the total PCR volume through QIAquick spin column to remove primers and nucleotides; run 5 µL of the purified product on 1.5% agarose gel with ethidium bromide stain; visualize the product (a band of approximately 520 bp, corresponding to the specific amplification of the prokaryotic *16S rRNA* gene) under UV light.
4. Sequence the PCR amplicon in both directions using the Big-Dye Terminator, version 3.1, cycle sequencing kit (Perkin-Elmer), as described by the manufacturer.
5. Determine the identity of bacteria by comparing the obtained sequences with the existing sequences in the GenBank database using the BLAST program available at the National Centre for Biotechnology Information Web site.

Note: A sequence similarity less than 97% of the *16S rRNA* gene is the criterion used to define a potentially new bacterial species. The same protocol can be applied for colonies, after the extraction of DNA. Finally, in an attempt to assess the viability of the identified microorganism by the broad spectrum *16S rRNA* PCR and sequencing, a reverse transcription of the extracted RNA (see above) followed by PCR using the abovementioned primers is applied.[19,23,34–36] This test gives an additional advantage of the molecular-based assays as with it we are capable to assess whether the causative agent is alive or not.

70.3 CONCLUSION AND FUTURE PERSPECTIVES

In clinical practice, *C. hominis* is a rare cause of infective endocarditis, which in general is characterized by a more prolonged course, with symptoms presenting for weeks or months before an accurate diagnosis is established. As a fastidious microorganism, *C. hominis* is difficult to isolate and identify by using only conventional methods. Molecular-based assays have recently been used efficiently for a rapid and correct diagnosis in cases with *C. hominis* endocarditis and therefore, they may offer a great potential in the arena of *C. hominis* infections. In our opinion, molecular techniques should be used as an additional tool at least in patients highly suspected of infectious diseases by fastidious microorganisms like *C. hominis*. However, as mentioned in this chapter, a major limitation of all molecular assays is the lack of the simultaneous provision of the antimicrobial susceptibility pattern. Though this is not a major problem for *C. hominis* as so far, only a minority of the *C. hominis* strains are producing β-lactamase, one should keep in mind that molecular assays can not replace conventional methods at this stage. From the microbiological point of view and also from the clinical point of view, a combination of both conventional and molecular-based methods seems to be better than either alone for achieving an accurate diagnosis.

REFERENCES

1. Wormser, G.P., and Bottone, E.J., *Cardiobacterium hominis*: Review of microbiologic and clinical features, *Rev. Infect. Dis.*, 5, 680, 1983.
2. El Khizzi, N., Kasab, S.A., and Osoba A.O., HACEK group endocarditis at the Riyadh Armed Forces Hospital, *J. Infect.*, 34, 69, 1997.
3. Savage, D.D. et al., *Cardiobacterium hominis* endocarditis: Description of two patients and characterization of the organism, *J. Clin. Microbiol.*, 5, 75, 1977.
4. Slotnick, I.J., Mertz, J.A., and Dougherty, M., Fluorescent antibody detection of human occurrence of an unclassified bacterial group causing endocarditis, *J. Infect. Dis.*, 114, 503, 1964.
5. Slotnick, I.J., *Cardiobacterium* hominis in genitourinary specimens, *J. Bacteriol.*, 95, 1175, 1968.
6. Tucker, D.N. et al., Endocarditis caused by *Pasteurella*-like organism. Report of four cases, *N. Engl. J. Med.*, 267, 913, 1962.

7. Slotnick, I.J., and Dougherty, M., Further characterization of an unclassified group of bacteria causing endocarditis in man: *Cardiobacterium hominis* gen. et sp. n., *Antonie. Van. Leeuwenhoek*, 30, 261, 1964.
8. Han, X.Y. et al., Endocarditis with ruptured cerebral aneurysm caused by *Cardiobacterium valvarum* sp. nov, *J. Clin. Microbiol.*, 42, 1590, 2004.
9. Das, M. et al., Infective endocarditis caused by HACEK microorganisms, *Annu. Rev. Med.*, 48, 25, 1997.
10. Weaver, R, *Cardiobacterium*. In Krieg, N.R., and Holt, J.G., (Editors), *Bergeys Manual of Systematic Bacteriology*, 4th ed., pp. 583–585, Williams and Wilkins, Baltimore, 1984.
11. Han, X.Y., and Falsen, E., Characterization of oral strains of *Cardiobacterium valvarum* and emended description of the organism, *J. Clin. Microbiol.*, 43, 2370, 2005.
12. Hoover, S.E. et al., Endocarditis due to a novel *Cardiobacterium* species, *Ann. Intern. Med.*, 142, 229, 2005.
13. Apisarnthanarak, A.R. et al., *Cardiobacterium hominis* bioprosthetic mitral valve endocarditis presenting as septic arthritis, *Diagn. Microbiol. Infect. Dis.*, 42, 79, 2002.
14. Francioli, P.B., Roussianos, D., and Glauser, M.P., *Cardiobacterium hominis* endocarditis manifesting as bacterial meningitis, *Arch. Intern. Med.*, 143, 1483, 1983.
15. Kuzucu, C. et al., An unusual case of pericarditis caused by *Cardiobacterium hominis*, *J. Infect.*, 50, 346, 2005.
16. Nurnberger, M. et al., Pacemaker lead infection and vertebral osteomyelitis presumed due to *Cardiobacterium hominis*, *Clin. Infect. Dis.*, 27, 890, 1998.
17. Pritchard, T.M. et al., Prosthetic valve endocarditis due to *Cardiobacterium hominis* occurring after upper gastrointestinal endoscopy, *Am. J. Med.*, 90, 516, 1991.
18. Robinson, W.J., and Vitelli, A.S., Infective endocarditis caused by *Cardiobacterium hominis*, *South. Med. J.*, 78, 1020, 1985.
19. Gatselis, N. et al., Direct detection of *Cardiobacterium hominis* in serum from a patient with infective endocarditis by broad-range bacterial PCR, *J. Clin. Microbiol.*, 44, 669, 2006.
20. Arnold, D.M. et al., *Cardiobacterium hominis* endocarditis associated with very severe thrombocytopenia and platelet autoantibodies, *Am. J. Hematol.*, 76, 373, 2004.
21. Le Moing, V. et al., Use of corticosteroids in glomerulonephritis related to infective endocarditis: Three cases and review, *Clin. Infect. Dis.*, 28, 1057, 1999.
22. Mandell, G.L. et al., Update in infectious diseases, *Ann. Intern. Med.*, 135, 897, 2001.
23. Petinaki, E., and Dalekos, G.N., Molecular methods as tools for the diagnosis of bacterial infections: Useful or useless? *Res. Adv. Microbiol.*, 6, 23, 2006.
24. Bosshard, P.P. et al, Etiologic diagnosis of infective endocarditis by broad-range polymerase chain reaction: A 3-year experience, *Clin. Infect. Dis.*, 37, 167, 2003.
25. Brouqui, P., and Raoult D., Endocarditis due to rare and fastidious bacteria, *Clin. Microbiol. Rev.*, 14, 177, 2001.
26. Greub, G. et al., Diagnosis of infectious endocarditis in patients undergoing valve surgery, *Am. J. Med.*, 118, 230, 2005.
27. Podglajen, I. et al., Comparative molecular and microbiologic diagnosis of bacterial endocarditis, *Emerg. Infect. Dis.*, 9, 1543, 2003.
28. Petti, C. et al., Utility of extended blood culture incubation for isolation of *Haemophilus*, *Actinobacillus*, *Cardiobacterium*, *Eikenella and Kingella* organisms: A retrospective multicenter evaluation, *J. Clin. Microbiol.*, 44, 257, 2006.
29. Baron, E.J., Scott, J.D., and Tompkins, L.S., Prolonged incubation and extensive sub-culturing do not increase recovery of clinically significant microorganisms from standard automated blood cultures, *Clin. Infect. Dis.*, 41, 1677, 2005.
30. Robert, P. et al., Multicenter evaluation of the New Vitek 2 *Neisseria-Haemophilus* identification card, *J. Clin. Microbiol.*, 46, 2681, 2008.
31. Valenza, G. et al., Microbiological evaluation of the new VITEK 2 *Neisseria-Haemophilus* identification card, *J. Clin. Microbiol.*, 45, 3493, 2007.
32. Petti, C.A., Polage, C.R., and Schreckenberger, P., The role of 16S rRNA gene sequencing in identification of microorganisms misidentified by conventional methods, *J. Clin. Microbiol.*, 43, 6123, 2005.
33. Goldenberger, D. et al., Molecular diagnosis of bacterial endocarditis by broad-range PCR amplification and direct sequencing, *J. Clin. Microbiol.*, 35, 2733, 1997.
34. Foroulis, C.N. et al., Direct detection of *Clostridium sordellii* in pleural fluid of a patient with pneumonic empyema by a broad-range 16S rRNA PCR, *Scand. J. Infect. Dis.*, 39, 617, 2007.
35. Skoulakis, Ch. et al., Level of *Streptococcus pyogenes* in patients with recurrent tonsillitis and tonsillar hypertrophy, *Scand. J. Infect. Dis.*, 40, 899, 2008.
36. Makaritsis, K.P. et al., An immunocompetent patient presenting with severe septic arthritis due to *Ralstonia pickettii* identified by molecular-based assays: A case report, *Cases J.*, 2, 8125, 2009.
37. Lu, P.L. et al., Infective endocarditis complicated with progressive heart failure due to β-lactamase-producing *Cardiobacterium hominis*, *J. Clin. Microbiol.*, 38, 2015, 2000.
38. Le Quellec, A., Endocarditis due to beta-lactamase-producing *Cardiobacterium hominis*, *Clin. Infect. Dis.*, 19, 994, 1994.
39. Jorgensen, J.N., Need for susceptibility testing guidelines for fastidious or less fastidious isolated bacteria, *J. Clin. Microbiol.*, 42, 493, 2004.
40. Jorgensen, J., and Hindler, J., New consensus guidelines from the Clinical and Laboratory Standards Institute for Antimicrobial Susceptibility testing of infrequently isolated or fastidious bacteria, *Clin. Infect. Dis.*, 44, 280, 2007.
41. Rantakokko-Jalava, K. et al., Direct amplification of rRNA genes in diagnosis of bacterial infections, *J. Clin. Microbiol.*, 38, 32, 2000.

71 Cedecea

Maria Dalamaga and Georgia Vrioni

CONTENTS

71.1 INTRODUCTION

71.1.1 Classification, Morphology, and Biology

Bacterial members belonging to the family of Enterobacteriaceae continue to play an important role as causes of healthcare-associated infections.[1–3] Several new members of the family Enterobacteriaceae have been described over the past 30 years; however, most of these members constitute rare causes of human infections.[4] Despite the progress in bacterial taxonomy, unidentified bacterial strains are present in reference laboratories; these are often kept in a collection.

In 1977, laboratory workers at the Centers for Disease Control and Prevention (CDC) gave the vernacular name "Enteric Group 15" to a group of 17 unidentified Enterobacteriaceae strains that were lipase (corn oil)-positive.[5] The name *Cedecea*, an arbitrarily constructed name derived from the abbreviation CDC, was proposed in 1980 for this new genus, in the family of Enterobacteriaceae, formerly designated as CDC Enteric Group 15 by Grimont et al.[5] Originally, two named species, *Cedecea davisae* and *Cedecea lapagei,* were described together in 1981, and three unspecified, unnamed species were referred to as *Cedecea species* 3, 4, and 5 on the basis of their deoxyribonucleic acid (DNA) DNA hybridization and phenotypic characteristics.[4,5] *Cedecea davisae* was named in honor of Betty Davis, the American bacteriologist of the Enteric Bacteriology Laboratories, CDC, in Atlanta, Georgia, who made many contributions to the serological and biochemical identification of *Enterobacteriaceae* and *Vibrionaceae.*[5] *Cedecea lapagei* was named in honor of Stephen Lapage, a British bacteriologist who made many contributions to the *Enterobacteriaceae*, bacterial systematics, and in particular, as an editor of the Bacteriological Code.[4] DNA–DNA hybridization showed that three other groups were distinct and must be considered as three additional species of *Cedecea.*[4,5] The latter three groups were not given scientific names because only one strain was available for each. Therefore, the names *Cedecea* species 3, *Cedecea* species 4, and *Cedecea* species 5 were used. None of the original seventeen *Cedecea* strains was isolated from blood or spinal fluid, so their clinical significance remained unclear. Since the original description of *Cedecea* species 4, many additional strains have been studied; the scientific name *Cedecea neteri* was proposed in 1982 for species 4, leaving species 3 and 5 unnamed.[4–6] *Cedecea neteri* was named to honor Erwin Neter, an American physician and microbiologist who made many contributions to the clinical significance of *Enterobacteriaceae.*[6] Recently, unusual species, which appear different from the five described *Cedecea* species and which may represent additional species of the genus,[7] were referred to as *Cedecea* species 6.[8,9]

Cedecea strains are gram-negative straight rods, motile, peritrichous when they are motile, nonencapsulated, nonspore-forming, aerobic, and facultatively anaerobic. They grow on nutrient agar at 37°C, forming convex colonies about 1.5 mm in diameter.[4,5] They grow at pH 7–9 in peptone water containing 0%–4% NaCl and in tryptic soy broth at 15°–37°C.[5] *Cedecea* strains are catalase-positive, oxidase-negative, arabinose-negative, arginine-positive, reduce nitrates to nitrites, blacken esculin iron agar, and produce acid from D-arabitol, cellobiose, maltose, D-mannitol, D-mannose, salicin, and trehalose.[5] Isolates of *Cedecea* present the acetoin pathway of glucose catabolism, but are often negative in the standard (O'Meara) Voges-Proskauer test because of the small amount of produced 2,3-butanediol.[10] *Cedecea lapagei* is usually strongly positive for the Voges-Proskauer test with the O'Meara methodology, and the other *Cedecea* species appear negative for the test unless it is held for several hours after the reagents are added.[10]

Most *Cedecea* strains use the following substrates as sole carbon and energy sources: D-arabitol, D-cellobiose, D-fructose, citrate, fumarate, D-galactose, D-galacturonate, D-gluconate, D-glucosamine, D-glucose, D-glucuronate, glycerol, D-mannitol, D-mannose, D-ribose, L-malate, DL-glycerate, 2-ketogluconate, *N*-acetylglucosamine, DL-lactate, β-gentiobiose, pyruvate, salicin, and trehalose.[5] Based upon the first description of 17 *Cedecea* strains by Grimont P.A.D. et al.,[5] they could not utilize any of the following substrates as sole carbon and energy sources: adonitol, adipate, aconitate, acetamide, β-alanine, DL-3-aminobutyrate, L-ascorbate, L-arabinose, azelate, benzoate, butylamine, butyrate, L-cysteine, L-citrulline, DL-carnitine, dulcitol, ethanol, m-erythritol, D-fucose, L-fucose, glycine, glycolate, glutarate, hepatanoate, histamine, 2-hydroxybenzoate, 3-hydroxybenzoate, 4-hydroxybenzoate, isobutyrate, isobutanol, L-isoleucine, isovalerate, isopropanol, DL-kynurenine, lactose, L-lysine, L-leucine, maleate, D-malate, L-mandelate, D-mandelate, methanol, L-methionine, nicotinate, norvaline, L-norleucine, oxalate, pelargonate, pimelate, L-phenylalanine, propionate, quinate, raffinose, L-rhamnose, L-sorbose, sarcosine, suberate, D-tartrate, L-tartrate, trigonelline, tryptamine, L-threonine, L-tryptophane, D-tryptophane, L-tyrosine, L-valine, n-valerate, and D-xylitol.[5] Furthermore, those strains did not produce oxidase, pigments, indole from peptone water, H_2S from triple sugar iron agar, gelatinase (tube and plate method), deoxyribonuclease, chitinase, amylase, polygalacturonase, urease, phenylalanine deaminase, lysine decarboxylase, acid from glucose in the presence of iodoacetate or acid in peptone water containing adonitol, L-arabinose, erythritol, myo-inositol, mucate, dulcitol, or L-rhamnose. Finally, in the presence of chlorate there was any anaerobic growth and the Jordan-tartrate test was negative.[5]

The genus *Cedecea* is phenotypically distinct from other genera in the family of Enterobacteriaceae. In their biochemical reactions, they do not ferment arabinose or rhamnose nor do they produce extracellular deoxyribonuclease.[4,5] They do not resemble the genera *Escherichia, Citrobacter, Proteus, Providencia, Salmonella,* and *Shigella* because they are positive for the o-nitrophenyl-β-D-galactopyranoside and Voges-Proskauer test (especially *Cedecea lapagei*), mannitol, lipase, esculin and negative for H_2S, indole, urease, phenylalanine deaminase, and lysine decarboxylase.[5] They do not resemble the genera *Klebsiella, Erwinia, Enterobacter,* and *Serratia* because they are negative for gelatinase, deoxyribonuclease, chitinase, lysine decarboxylase, polygalacturonase, L-rhamnose, and L-arabinose.[4,5]

Table 71.1 depicts the biochemical characteristics that can differentiate *Cedecea* species in clinical bacteriology laboratories. *Cedecea lapagei* and *Cedecea davisae* can be distinguished by acid production from D-xylose and sucrose, and by ornithine decarboxylase and ascorbate tests (positive for *Cedecea davisae*), by production of β-xylosidase (positive for most *Cedecea davisae* strains) and the specific requirement in thiamine for growth of *Cedecea davisae*.[4,5] Unlike other *Cedecea* strains, *Cedecea lapagei* is positive for D-arabitol, malonate utilization, negative for sucrose and D-sorbitol, and growths in media without thiamine.[4,5,11] *Cedecea lapagei* is usually strong positive for the Voges-Proskauer test with the O'Meara methodology.[10] *Cedecea neteri* is positive for methyl red, Voges-Proskauer (very weak), arginine dihydrolase (Moeller's), malonate, nitrate reduction to nitrite, sucrose, D-sorbitol, D-xylose, and growths in media without thiamine.[6] The deoxyribonucleic acid–relatedness data (DNA–DNA hybridization) suggested that *Cedecea*

TABLE 71.1
Biochemical Tests That Differentiate *Cedecea* Species

	C. davisae	*C. lapagei*	*C. neteri*	*Cedecea* sp. 3	*Cedecea* sp. 5	*Cedecea* sp. 6
Test						
Acetate	–	V	–	V	V	+
Malonate	+	+	+	–	–	+
ODC	+	–	–	–	V	–
Acid from						
Lactose	V	V	V	–	–	–
Melibiose	–	–	–	+	+	ND
Raffinose	–	–	–	+	+	ND
Sucrose	+	–	+	V	+	+
D-Sorbitol	–	–	+	–	+	+
D-Xylose	+	–	+	+	+	–
Growth in media without thiamine	–	+	+	+	+	+

+, ≥90% positive; V, 10%–90% positive, and –, ≤10% positive; and ND, not determined.
ODC, ornithine decarboxylase.
Data from Janda, 2006,[9] Farmer et al., 1982,[6] and Farmer et al., 1985.[10]

constitutes a new genus in the family of Enterobacteriaceae comprising at least five species.[4,5] In particular, those studies indicated that *Cedecea* strains were 32%–100% related to each other and less than 23% related to other members of the Enterobacteriaceae.[5]

Regarding other properties, screening five genomic species in the Enterobacteriaceae family for D-glucose oxidation, *Cedecea,* and more specifically, *Cedecea lapagei* presented a different behavior.[12] *Cedecea lapagei* oxidized D-glucose without added pyrroloquinoline quinine (PQQ) like *Serratia marcescens, Yersinia frederiksenii,* and *Erwinia cypripedii.* Unlike *Cedecea lapagei, Escherichia coli* oxidized glucose only when PQQ was added. The extracellular D-glucose oxidation was determined in order to select conditions useful for taxonomic applications.[12] DEAE-cellulose paper chromatography was used to determine extracellular production of gluconate from ^{14}C-glucose by bacterial cells.[12]

Cedecea spp. strains can also be identified by their different fatty acid composition[13] from other members in the family of Enterobacteriaceae. The cellular fatty acids are extracted as fatty acid methyl esters and separated by gas chromatography.[13,14]

Based upon comparative enzymology, regulation of aromatic amino acid biosynthesis and sequencing of 16S rRNA, 5S rRNA, and tryptophan leader peptide, *Cedecea* belongs to the Enterocluster 3, which includes the genera *Kluyvera, Edwardsiella, Hafnia, Yersinia, Proteus, Providencia,* and *Morganella.* This enterocluster lacks the *trpG-trpD* gene fusion and doesn't present the three-step overflow pathway to L-phenylalanine.[15]

Another characteristic of *Cedecea davisae* is the production of potato-like odors when grown on agar media. Amid members of the family Enterobacteriaceae, most strains of *Serratia odorifera* and *Serratia ficaria* as well as a few strains of *Serratia rubidea* produce potato-like odor attributed to pyrazines, which are heterocyclic, nitrogen-containing compounds with unique organoleptic properties.[16] The major alkyl-methoxypyrazine produced by *Cedecea davisae* is 3-sec-butyl-2-methoxypyrazine and traces of 3-isopropyl-2-methoxy-5-methylpyrazine[16] identified by gas–liquid chromatography and mass spectrometry.[16] Alkyl-methoxypyrazines cause earthy, musty, bell-pepper or potato-like odors of many vegetables and other plants and are used in the food industry to improve flavor.[16,17] Pyrazines produced by *Cedecea, Serratia* spp., and *Pseudomonas* could be used for taxonomic and biotechnological purposes.

On the basis of the rapid catalase supplemental test for identification of members of the Enterobacteriaceae family, *Cedecea* spp. belongs to the immediate catalase group composed of *Cedecea, Serratia, Yersinia, Proteus, Morganella, Providencia,* and *Hafnia* spp.[18] Using this test, *Cedecea* spp. is distinguished from *Enterobacter,* which is a delayed catalase reactor.[18]

Finally, using the susceptibility profile to the antimicrobial agents, *Cedecea* cultures are resistant to colistin and cephalothin-like *Serratia* cultures.[5,19] *Cedecea* strains are usually susceptible to streptomycin, kanamycin, tobramycin, gentamicin, amikacin, carbenicillin, sulfonamide, trimethoprim, nalidixic acid, furantoin, chloramphenicol, tetracycline, and minocyclin.[5,6,20,21]

It is well established that members of the Enterobacteriaceae family are widespread throughout the environment.[4,11] They are found in water, soil, plants, and also in animals, ranging from insects to humans.[22] Most of them are inhabitants of the intestinal tract and therefore their presence in the environment reflects fecal excretion by animals and humans.[4,11] The natural habitat of *Cedecea* species remains unknown. A significant majority of *Cedecea* strains have been isolated from clinical specimens, mostly from blood in patients with bacteremia and septicemia (*Cedecea davisae, neteri, and lapagei*),[19–21,23] from sputum, BAL specimens and lung tissue,[5,24] from gall bladder (*Cedecea davisae*),[5] hand wounds,[5] burn wounds (*Cedecea lapagei*),[20] cutaneous ulcer,[7,21] eye swabs (*Cedecea davisae*),[5] stool,[5,25] urine,[5] throat cultures,[5] from peritoneal fluid (*Cedecea neteri* and *Cedecea lapagei*),[26,27] from scrotal abscess (*Cedecea davisae*),[28] and from veined catheters.[5] None of the original seventeen *Cedecea* strains were originated from blood or spinal fluid, suggesting that its clinical impact was insignificant.[5,6] The original isolate of *Cedecea* sp.3 was from sputum, and the original isolate of *Cedecea* sp.5 was from the toe of a Canadian patient.[10] Nowadays, the isolation of *Cedecea* species from clinical specimens in hospitals has been more and more frequent, so that their role in human disease as well as their ecologic niches will be better defined.[20,21]

Cedecea strains may be found in aqueous environment[21] and in agricultural dusts, together with other bacteria of the Enterobacteriaceae family.[26,29] Agricultural dust represents a potential hazard for workers because of the high concentrations of allergenic microorganisms and bacterial endotoxins.[26,29] The highest number of gram-negative organisms were recovered in association with maize and sorghum.[26,29]

Cedecea strains are also isolated, together with other members of the Enterobacteriaceae family (*Escherichia coli, Enterobacter* spp., *Citrobacter, Erwinia, Hafnia, Klebsiella, Kluyvera, Leclercia, Morganella, Proteus, Providencia, Rahnella, Salmonella,* and *Serratia*), from shell eggs collected during commercial processing, less frequently from fully processed eggs than from unwashed eggs, indicating that shell eggs are less contaminated with bacteria as a result of commercial washing procedures.[30]

Finally, strains identified as *Cedecea* spp. by their fatty acid composition were isolated from disinfecting footbaths containing TEGO 103G (amphoteric disinfectant) or TP-99 (alkyl amino acetate-based disinfectant) in dairy factories along with *Serratia marcescens* strains.[13,14]

71.1.2 Clinical Features and Pathogenesis

Many members of the Enterobacteriaceae family are opportunistic pathogens and are responsible for about 50% of nosocomial infections in the United States.[22,31,32]

Cedecea spp. constitute a rare pathogen.[5,10,20,21,23] *Cedecea* spp. are rarely isolated from clinical specimens of hospitalized patients.[10,20,21,23] Although *Cedecea* spp. strains are mostly recovered from clinical material, such as sputum, blood, wound swabs, and other specimens, their clinical significance remains unknown and their occurrence in clinical specimens has not been studied in depth.[20,21,23] It is unclear what role many *Cedecea* spp. could play in human infections; for some, the number of available strains is insufficient in order to determine clinical significance in humans or to establish a reservoir for infections.[5,10] In an unedited presentation from Brazil, the isolation of *Cedecea* spp. from hospitals has been more and more frequent, accounting for less than 1% of all registered infectious processes in Brazil.[32] According to this study, *Cedecea* spp. was recovered from 52 clinical specimens (mostly from urine, blood, veined catheters, and wound exudates).[32] *Cedecea lapagei* was identified more frequently than *Cedecea davisae* or other species.[32]

A literature search revealed few case reports of infections due to *Cedecea* spp.[6,7,10,19–21,23–28] Table 71.2 depicts the reported infectious cases due to *Cedecea* spp.

Cedecea davisae, C. lapagei, and *C. neteri* mostly caused bacteremia in patients presenting diabetes mellitus,[20,21] chronic heart disease[6,19,28] such as valvular heart disease, chronic obstructive pulmonary disease,[19,24] and systemic lupus erythematosus.[23] Only one case of bacteremia/septicemia in an immunosuppressed patients suffering from systemic lupus erythematosus was characterized as a nosocomial infection.[23]

For the other cases of bacteremia due to *Cedecea davisae* and *C. lapagei*, the findings supported a community-acquired origin for the infections because the patients were admitted with fever and other symptoms (or presented those symptoms soon after admission) and had no known exposure to hospitals in the preceding years.[6,19–21] In these cases, the bacterium was probably acquired from the patient's own dermatic flora and not from the hospital environment.[19–21] In one case, some of the evidence suggested that the contaminated central venous pressure catheter and Swan-Ganz lines may have been the foci of the *Cedecea davisae* bacteremia, since *Cedecea davisae* was isolated in large numbers from the tips and was recovered from blood during the acute phase of the illness.[19] Mangum and Radisch reported an isolate of *Cedecea* sp.3 from heart blood at autopsy but its clinical meaning was unclear because systemic infection with *Candida albicans* was also present, and *Cedecea* sp.3 was not isolated from blood cultures taken before death.[33] Laboratory findings of *Cedecea* bacteremia included leukocytosis with a left shift of the differential cell count, toxic granulation, a high erythrocyte sedimentation rate and a high CRP level.[6,19–21]

Cedecea davisae was the predominant microorganism of scrotal abscess in a patient with chronic heart and liver diseases.[28] *Cedecea lapagei* has also been linked to pneumonia in a 38-year-old male patient with chronic obstructive pulmonary disease who was admitted to hospital with subarachnoid hemorrhage, operated on, transferred to Intensive Care Unit of Reanimation, and underwent artificial ventilation.[24] In this case, *Cedecea lapagei* was isolated from bronchoalveolar lavage (BAL) fluid specimen, and it was considered that pneumonia originated from the aspiration of upper

TABLE 71.2
Documented Infectious Cases Due to *Cedecea* spp.

Number	*Cedecea* spp.	Infection	Isolation	Risk factors	Reference
1.	*C. lapagei*	Pneumonia	BAL specimen	Chronic obstructive pulmonary disease, subarachnoid hemorrhage, artificial ventilation	Yetkin et al.[24]
2.	*C. lapagei*	Bacteremia and burn wound infection	Blood, wound swabs	Diabetes mellitus, cement-related chemical burn injury	Dalamaga et al.[20]
3.	*C. lapagei*	Peritonitis	Peritoneal dialysis fluid	End-stage renal disease, liver transplantation secondary to cirrhosis related to hepatitis C	Davis and Wall[26]
4.	*C. davisae*	Bacteremia and leg ulcer infection	Blood, leg ulcer swabs and bullae material specimen	Diabetes mellitus	Dalamaga et al.[21]
5.	*C. davisae*	Bacteremia	Blood, central venous pressure catheter tip, Swan-Ganz tip	Chronic heart disease, chronic obstructive pulmonary disease	Perkins et al.[19]
6.	*C. davisae*	Scrotal abscess	Scrotal abscess specimen	Chronic heart and liver disease	Bae and Sureka[28]
7.	*C. neteri*	Bacteremia	Blood	Systemic lupus erythematosus	Aguilera et al.[23]
8.	*C. neteri*	Bacteremia	Blood	Valvular heart disease	Farmer et al.[6]
9.	*C. neteri* and *Escherichia vulneris*	Peritonitis	Peritoneal fluid	Agressive abdominal surgery	Anon et al.[27]
10.	Unusual *Cedecea* species (*Cedecea* sp. 6)	Cutaneous ulcer infection	Ulcer swab specimen	Diabetes mellitus, advanced age	Hansen and Glupczynski[7]

way secretion due to the patient's unconsciousness.[24] Two strains of *Cedecea davisae* from sputum were described by Bae et al.[28,34] in two elderly patients with heart disease who presented with pneumonia. In this case report, the isolates were referred to as "Enteric Group 15-Davis subgroup" classified later as *Cedecea davisae.* On a sputum culture in one patient's report, it was mentioned "scant growth of *Klebsiella oxytoca*." Therefore, *Cedecea davisae* was not the probable cause of pneumonia.[10,34]

Cedecea lapagei and *Cedecea neteri* were the causative agents of peritonitis in two cases.[26,27] In particular, *Cedecea lapagei*, isolated from peritoneal dialysis fluid, was the cause of continuous ambulatory peritoneal dialysis (CAPD)-related peritonitis in a 55-year-old male patient with a history of orthotopic liver transplantation secondary to cirrhosis related to hepatitis C, and end-stage renal disease secondary to glomerular disease and chronic tacrolimus toxicity.[26] In this case, *Cedecea lapagei* was considered as an environmental pathogen of peritonitis because breaks in technique resulted in contamination.[26] Usually, the major reservoirs for gram-negative organisms causing CAPD-related peritonitis are the gastrointestinal and urinary tracts.[26] *Cedecea neteri* as well as *Escherichia vulneris* were recovered from the peritoneal fluid of a patient with peritonitis after an aggressive abdominal surgical operation.[27]

In addition, *Cedecea lapagei, Cedecea davisae,* and *Cedecea* sp. 6 have been related to cutaneous infections in ulcers[7,21] and in burn wounds.[20] All three patients presented a history of diabetes mellitus, which is a classic predisposing factor for the evolution of infection. The alkaline environment due to the cement-related chemical burn injury facilitated the growth of *Cedecea lapagei* in a 47-year-old man suffering from bacteremia and wound infection after a cement-related chemical burn injury.[20] A possible way of bacterial transmission in this patient was the cross-infection and contact of the burned areas (main entrance of bacteria) with contaminated inanimate objects and through aqueous environment.[20]

From the reported cases, it can be concluded that the most prominent risk factors for *Cedecea* infections are diabetes mellitus, chronic obstructive pulmonary disease, chronic heart disease, and conditions or treatments that cause immunosuppression, such as systemic lupus erythematosus, liver transplantation, and end-stage renal disease.[6,10,19–28]

The treatment of these infections represents a challenge, since strains of *Cedecea* spp. are becoming resistant against a variety of antimicrobial agents, such as cephalothin, extended-spectrum cephalosporins, colistin, and several aminoglycosides.[20,21,32] In an unedited study from a Brazilian hospital regarding the incidence and resistance profile of *Cedecea* spp. isolated from clinical material, the susceptibility profile of *Cedecea* spp. isolates presents a clear tendency to multiresistance, with higher susceptibility to fluoroquinolones and carbapenems.[32] In a study that was undertaken in order to determine whether infants treated with third-generation cephalosporins develop fecal colonization with gram-negative bacteria that are resistant to these agents, it was shown that hospitalized children presented a high incidence of fecal colonization with gram-negative bacteria resistant to third-generation cephalosporins.[25] The commonest resistant strains belonged to *Enterobacter, Citrobacter, Serratia, Chromobacterium,* and *Cedecea.*[25] This colonization was not necessarily related to prolonged hospitalization or to the treatment with such antibiotics.[25] It can be suggested that there is a high prevalence of resistant strains within the hospital environment and in the community.[25] It would be important to reduce the spread of such resistant strains by an efficient infection control in the hospital environment and by the judicious use of antimicrobial agents.

71.1.3 Diagnosis

Conventional Techniques. *Cedecea* species grow well within 24 h after incubation at 35°C–37°C on most media, producing convex colonies about 1.5 mm in diameter.[4,5] Most *Cedecea* strains are lactose-negative and appear as nonfermenting, colorless colonies on lactose-containing differential media.[35] Isolation from blood had been achieved with both vented and unvented blood cultures incubated at 35°C in O_2 with either Columbia or tryptic soy broth and with Bactec bottles (Becton Dickinson Diagnostic Systems, Sparks, Maryland.). Cultures grow within 24 h of subculture to a variety of media, including MacConkey, chocolate, colistin-nalidixic acid, sheep blood, brain heart infusion, and phenyl alcohol agars.[35]

The biochemical tests that are useful in separating species within each genus are discussed in detail earlier. Table 71.1 depicts the biochemical tests for discriminating members of the genus *Cedecea*. Correct identification to species levels is increasingly important in recognizing strains that are high risk for carrying cephalosporinases, carbapenemases, and extended-spectrum β-lactamases.[11,36] *Cedecea* spp. are generally easily identified by commercial systems such as API 20E and by automated systems such as Microscan (Dade Behring), VITEK 2 (bioMérieux), Phoenix™ (Becton Dickinson), and so forth, which use a combined gram-negative identification panel and an automated microbroth dilution method for antibiotic susceptibility.[3,6,11,20,21] For a rapid discrimination of Enterobacteriaceae, a simple, semiquantitative slide catalase test can be performed[18] by using colonies from 18 to 24 h 5% sheep blood agar and MacConkey agar plates. The instant appearance of bubbles when a single colony is placed in a drop of 3% hydrogen peroxide on a glass slide using a bacteriological loop is called immediate.[18] The immediate catalase group consists of *Cedecea.* By using the rapid catalase test as a supplement test, *Cedecea* can be distinguished from *Enterobacter,* which belongs to the delayed catalase group.[18]

Resistance among members of the Enterobacteriaceae family is increasing, especially in bacteria harboring cephalosporinases, carbapenemases, and extended-spectrum

β-lactamases.[11,36] Most *Cedecea* strains are resistant to ampicillin, cephalothin, colistin, and polymyxins. In a Brazilian study, the susceptibility profile of *Cedecea* strains showed a tendency to multiresistance, with a higher susceptibility to fluoroquinolones and carbapenems.[32] Several methods can be used for testing the antibiotic susceptibility of Enterobacteriaceae, including *Cedecea* spp., but the most popular are disk diffusion on Mueller-Hinton agar and broth dilution.[11] When done on clinically significant isolates, the antibiotic susceptibility testing can be used as an epidemiologic marker by providing a useful tool for comparing isolates in epidemiologic studies. Also, the antibiogram of a *Cedecea* culture can be compared with antibiograms of known isolates of *Cedecea* in order to help identification.

Molecular Techniques. Conventional, culture-based, diagnostic assays use the phenotypic characteristics of *Cedecea* species. However, these identification criteria can be influenced by in vitro testing conditions and may lead to a misinterpretation and subsequent misidentification.[37,38]

Molecular techniques, in particular, nucleic acid–based detection and identification methods, take advantage of the use of stable genotypic characteristics.[38,39] The bacterial genome provides conserved regions for species identification and phylogenetic relationships, including 16S ribosomal RNA (rRNA) gene,[38] and protein-coding genes (e.g., *gyrB, hsp60, infB, rpoB*).[40] In particular, 16S rRNA sequences have been widely used to construct bacterial phylogenetic relationships or to detect pathogenic bacteria.[40,41] Bacterial analysis by 16S rRNA has become popular because these sections of RNA are conserved and easy to sequence. However, the classification of closely related enteric species of the family Enterobacteriaceae is difficult to achieve through this approach.[41–43] As an alternative to 16S rRNA analysis, protein-coding genes allowed the construction of reliable trees and established a well-accepted classification for the members of the family Enterobacteriaceae, but analyses of multiple genes are necessary to identify horizontal transfer of single genes. Moreover, different levels of selection in certain gene domains might cause constraints or distortions between the organisms being considered.[41–43]

Recently, other genes, such as the noncoding chromosomal replication origin *oriC* locus, *tuf*, and *atpD* genes, have been used for identification and phylogenetic analysis of enteric species belonging to the Enterobacteraceae family.[22,40] Especially, the use of *oriC* locus established a phylogenetic tree for the classification of different species of the genera *Escherichia*, *Shigella*, *Salmonella*, *Enterobacter*, *Citrobacter*, *Klebsiella*, *Raoultella*, *Kluyvera*, *Cedecea*, and *Buttiauxella*. Functional important protein-binding sites located in *oriC* are well conserved throughout the enteric group. Moreover, due to a high overall divergence value, phylogenetic trees were robust and well supported by bootstrap analysis. In comparison with 16S-rDNA analysis, the *oriC* sequences indicated a greater evolutionary divergence for bacteria.[40] Regarding the use of elongation factor Tu (*tuf* gene) and F-ATRase β-subunit (*atpD* gene), the phylogenetic trees generated from their partial sequences were similar overall, but they showed minor differences in branching. The *atpD* tree showed more monophyletic groups corresponding to species that belong to the same genus than did the *tuf* tree. According to *atpD* consensus tree, *Cedecea* is a monophyletic genus. Since *atpD* is more divergent than *tuf*, the former could allow better resolution for tree reconstruction.[22]

71.2 METHODS

71.2.1 Sample Preparation

Different samples may be received in the microbiology laboratory for *Cedecea* isolation, such as respiratory specimens, blood, abscess, cutaneous ulcer, and lung tissue.[8] Their preparation for conventional methods does not differ from those for isolation of other nonfastidious, facultative aerobes. *Cedecea* species grow well within 24 h after incubation at 35°C on most media. Isolation from blood had been achieved with both vented and unvented blood cultures incubated at 35°C in O_2 with either Columbia or tryptic soy broth and with Bactec bottles (Becton Dickinson Diagnostic Systems, Sparks, Maryland). Cultures grow within 24 h after subculture to a variety of media, including chocolate, colistin-nalidixic acid, sheep blood, MacConkey, brain heart infusion, and phenyl alcohol agars.[35] Most strains of *Cedecea* are lactose-negative and appear as nonfermenting, colorless colonies on lactose-containing differential media.[35]

Molecular procedures can vary depending on which manufacturer's reagents and equipment are used, but the basic steps remain the same. The first steps are growth and isolation of a potentially significant bacterium from a patient's specimen, followed by extraction of DNA from the isolate.[40] There are many options for extraction, including quick heat lyses, Chelex extraction, alkaline lysis, phenolchloroform extraction, mechanical disruption, and DNA isolation and purification kits.[40] To isolate chromosomal DNA with heat lyses method, one or two freshly grown colonies of bacteria are resuspended in sterile water. The bacterial suspension is then boiled (at 100°C for 10 min) to release the DNA.[41,43] Concerning DNA isolation kits, there are many different user-friendly commercial kits, and the extraction is made according the manufacturer's instructions.[13,22] All methods are equally effective for *Cedecea* species' DNA isolation.

71.2.2 Detection Procedures

Using conventional methods, *Cedecea* can be separated from other members of the Enterobacteriaceae family by negative biochemical tests for lysine decarboxylase (LDC), gelatinase, and deoxyribonuclease; their inability to ferment L-arabinose and L-rhamnose; and a positive arginine dihydrobase (ADH) test. The gelatinase and deoxyribonuclease reactions clearly separate *Cedecea* from *Serratia*. Reactions for ornithine, malonate, sucrose, D-sorbitol, xylose, and melibiose separate the five species of *Cedecea*. Differential tests used to identify

Cedecea species can be found in Table 71.1. In addition, some strains of *Cedecea* produce a pungent (potato-like) odor due to the production of alkyl-methoxypyrazines.[16]

For molecular methods, after sample preparation, the target is amplified by PCR. PCR primers depend on the genes that will be amplified, as well as PCR cycles and program.

For *oriC*-locus amplification, Roggenkamp[40] described five PCR assays based on five different sets of primers: the primer set Ori-85-f (5′-CAAAAGGAT CCKG ATAAAACATGGT-3′) and Ori-72-r (5′-GGSGA AACAC TGAAGATCAACAT-3′) for PCR 1; the primer set Ori-85-f and Ori-53-r (5′-AACATCCTYGAACATGARATT CCG-3′) for PCR 2; the primer set Ori-85-f and Ori-40-r (5′-CATGAGATTCCGGARGATCCG-3′) for PCR 3; the primer set Ori-50-f (5′-CGGGCCGTGGATTCTACTCA-3′) and Ori-40-r and for PCR 4; and the primer set EcOriC-73-f (5′-CACTGCCCTGTGGATAACAA-3′) and EcOriC-292-r (5′-TATACAGATCGTGCGATCTAC-3′) for PCR 5. Amplification conditions include a preheat at 95°C for 10 min; 35 cycles of 95°C for 30 s, annealing temperature (AT) (depending on the primer set) for 30 s, and 72°C for 1 min 30 s, and a final step at 72°C for 15 min. ATs are optimized by stepwise, increasing the AT by 2°C starting at 6°C below the lowest Tm of the primer sets used, until specific, single banded PCR products are obtained. The following AT must be chosen: 55°C for PCR 1, 56°C for PCR 2, and 57°C for PCR 3.[40]

Paradis et al.[41] designed one set of primers to amplify *tuf* gene and two sets to amplify *atpD* gene. A 884 bp fragment from the *tuf* region is amplified with the primer set T1: 5′-AAYATGATIACIGGIGCIGCICARATGGA-3′ and T2: 5′-CCIACIGTICKICCRCCYTCRCG-3′. For *atpD* gene amplification, the first primer set (A1: 5′-RTIATIGGIGCI GTIRTIGAYGT-3′ and A2: 5′-TCRTCIGCIGGIACRTAIA YIGCYTG-3′) generates a specific product of the same size as previously (e.g., 884 bp), while the second primer set (A3: 5′-TIRTIGAYGTCGARTTCCCTCARG-3′ and A2: 5′-TCR TCIGCIGGIACRTAIAYIGCYTG-3′) amplifies a product of 871 bp. The PCR mixtures are subjected to the following thermal cycling program: 95°C for 3 min; 35 cycles of 95°C for 1 min, 55°C (for *tuf*) or 50°C (for *atpD*) for 1 min, and 72°C for 1 min; a 72°C for 7 min.[41] At this point, it may be desirable to visualize PCR product in 2.0% agarose gels after electrophoresis and ethidium bromide staining to ensure that a single fragment of the appropriate size has been amplified.

For 16S-rDNA analysis, 16S rRNA gene is amplified with the primers KR1 (5′-TGGCTCAGATTGAACGCTGGCGGC -3′) and KR2 (5′-TACCTTGTTACGACTTCACCCCA-3′).[13] The PCR is initiated with a denaturation step at 94°C for 4 min. This is followed by 30 cycles of 95°C for 30 s, 55°C for 30 s, and 72°C for 90 s; and a final step at 72°C for 7 min.[13]

Cycle sequencing reactions are performed next using the ABI Prism BigDye Terminator Sequencing Ready Reaction kit (Perkin Elmer Applied Biosystems) on an automated DNA sequencer,[13,40,41] and the products of these reactions are analyzed by electrophoresis. If both strands of the amplicon are sequenced, the sequences must be assembled and edited for discrepancies. The final step is to compare the unknown sequence to a database. Depending on the type of sequencing, it may be necessary to purify the PCR amplicon before sequencing and to purify the sequencing products before electrophoresis.[22,40,41]

Bacterial identification by 16S rRNA gene sequence analysis has been greatly facilitated by the availability of commercial systems.[39] The finalized 16S rRNA sequence concerned is loaded onto one of the software packages that contains a database with 16S rRNA sequences of selected bacterial species, and the identity of the isolate is clarified. The most well known of these software packages include the Ribosomal Database Project (RDP-II), MicroSeq, Ribosomal Differentiation of Medical Microorganisms (RIDOM), and the SmartGene Intergrated Database Network System (SmartGene IDNS). The databases of RDP-II and SmartGene IDNS contain sequences downloaded from GenBank, whereas all sequences in the databases of RIDOM and MicroSeq were obtained by sequencing the 16S rRNA genes of bacterial strains of culture collections.[41,42]

71.3 CONCLUSION AND FUTURE PERSPECTIVES

The identification and reporting of Enterobacteriaceae have been debated and discussed for many years. The number of diagnostic assays in clinical bacteriology has increased steadily during the last decade. The following two main questions seem to always result: "How far should we go in identifying cultures?" and "How should we word reports?" Since the majority of clinical microbiology laboratories now use commercial and/or automated identification systems, the answer to the first question has been settled, since the test system usually has a fixed set of differential tests. The tests are run, and the resulting profile is checked in a code book or with a computer system. This final result is usually the name of an organism. As more data become available, these lists are revised with new and rare Enterobacteriaceae. Users of the commercial systems should have little difficulty in identification or reporting. Since most clinical laboratories use commercial systems, the vast majority of strains will be reported based on these methods. But, unlike a report of *Escherichia coli* or *Salmonella typhi*, a report of *Cedecea davisae* will probably have little meaning to a physician or allied health professional who might receive it. Thus, an explanation is desirable, if not essential, such as a comment on the report.[10]

Moreover, the detection and identification of unusual Enterobacteriaceae, such as *Cedecea* spp., can be simplified by the introduction of molecular methods. Such methods have provided new tools for rapid, specific, and sensitive detection and identification of this genus. However, conventional microbiology will not be replaced in the immediate future, but multiparameter identification of the most important pathogens using array-based detection technologies or rapid real-time PCR-based assays are becoming common in today's laboratories. Advancing automation and decreasing costs in molecular microbiology will bring these tools closer to the routine laboratory. The improvement of multiplexing strategies and analysis technologies might harbor the potential to cope with the evident problems of modern medicine. Combined with rapidly advancing innovations in microfluidics, the future trend clearly points toward complex multiparametric assays on miniaturized and automated lab-on-a-chip devices, also suitable for point-of-care diagnostics.

REFERENCES

1. Chastre, J., and Fagon, J.Y., Ventilator-associated pneumonia, *Am. J. Respir. Crit. Care Med.*, 165, 867, 2002.
2. Velasco, E. et al., Bloodstream infection surveillance in a cancer centre: A prospective look at clinical microbiology aspects, *Clin. Microbiol. Infect.*, 10, 542, 2004.
3. Stone, N.D. et al., Comparison of disk diffusion, VITEK 2, and broth microdilution antimicrobial susceptibility test results for unusual species of *Enterobacteriaceae*, *J. Clin. Microbiol.*, 45, 340, 2007.
4. Farmer, J.J. III., Boatwright, K.D., and Janda, M.J., *Enterobacteriaceae*: Introduction and identification. In Murray, P.R. et al. (Editors), *Manual of Clinical Microbiology*, 9th ed., pp. 649–669, ASM Press, Washington, DC, 2006.
5. Grimont, P.A.D. et al., *Cedecea davisae* gen.nov., sp. nov. and *Cedecea lapagei* sp. nov., new *Enterobacteriaceae* from clinical specimens, *Int. J. System. Bacteriol.*, 31, 317, 1981.
6. Farmer, J.J. III et al., Bacteremia due to *Cedecea neteri* sp. nov., *J. Clin. Microbiol.*, 16, 775, 1982.
7. Hansen, M.W., and Glupczynski, G.Y., Isolation of an unusual *Cedecea* species from a cutaneous ulcer, *Eur. J. Clin. Microbiol.*, 3, 152–153, 1984.
8. Winn, W. et al., *Koneman's Color Atlas and Textbook of Diagnostic Microbiology*, 6th ed., p. 211, Lippincott Williams & Wilkins, Baltimore, MD, 2006.
9. Janda, J.M., New members of the family *Enterobacteriaceae*. In Dworkin, M., Falkow, S., Rosenberg, E., Schleifer, K-H., and Stackebrandt, E. (Editors), *The Prokaryotes, A Handbook on the Biology of Bacteria, Proteobacteria: Gamma Subclass*, 3rd ed., vol. 6, pp. 5–40, Springer Science and Business Media, LLC, Singapore, 2006.
10. Farmer, J.J. III et al., Biochemical identification of new species and biogroups of *Enterobacteriaceae* isolated from clinical specimens, *J. Clin. Microbiol.*, 21, 46, 1985.
11. Abbott, S.L., *Klebsiella*, *Enterobacter*, *Citrobacter*, *Serratia*, *Plesiomonas*, and other *Enterobacteriaceae*. In Murray, P.R. et al. (Editors), *Manual of Clinical Microbiology*, 9th ed., pp. 698–715, ASM Press, Washington, DC, 2006.
12. Bouvet, O.M.M., and Grimont, P.A.D., Extracellular oxidation of D-Glucose by some members of the *Enterobacteriaceae*, *Ann. Inst. Pasteur Microbiol.*, 130, 59, 1988.
13. Langsrud, S., Møretrø, T., and Sundheim, G., Characterization of *Serratia marcescens* surviving in disinfecting footbaths, *J. Appl. Microbiol.*, 95, 186, 2003.
14. Sundheim, G., Sletten, A., and Dainty, R.H., Identification of pseudomonads from fresh and chill-stored chicken carcasses, *Int. J. Food Microbiol.*, 39, 185, 1998.
15. Ahmad, S., Weisburg, W.G., and Jensen, R.A., Evolution of aromatic amino acid biosynthesis and application to the fine-tuned phylogenetic positioning of enteric bacteria, *J. Bacteriol.*, 172, 1051, 1990.
16. Gallois, A., and Grimont, P.A.D., Pyrazines responsible for the potato-like odor produced by some *Serratia* and *Cedecea* strains, *Appl. Environ. Microbiol.*, 50, 1048, 1985.
17. Maga, J.A., Pyrazines in food: an update, *Crit. Rev. Food Sci. Nutr.*, 20, 1, 1982.
18. Chester, B., and Moskowitz, L.B., Rapid catalase supplemental test for identification of members of the family *Enterobacteriaceae*, *J. Clin. Microbiol.*, 25, 439, 1987.
19. Perkins, S. R., Beckett, T.A., and Bump, C. M., *Cedecea davisae* bacteremia, *J. Clin. Microbiol.*, 24, 675, 1986.
20. Dalamaga, M. et al., *Cedecea lapagei* bacteremia following cement-related chemical burn injury, *Burns*, 34, 1205, 2008.
21. Dalamaga, M. et al., Leg ulcer and bacteremia due to *Cedecea davisae*, *Eur. J. Dermatol.*, 18, 17, 2008.
22. Paradis, S. et al., Phylogeny of the *Enterobacteriaceae* based on genes encoding elongation factor Tu and F-ATPase β-subunit., *Int. J. Syst. Evol. Microbiol.*, 55, 2013, 2005.
23. Aguilera, A. et al., Bacteraemia with *Cedecea neteri* in patient with systemic lupus erythematosus, *Postgrad. Med. J.*, 71, 179, 1995.
24. Yetkin, G. et al., A pneumonia case caused by *Cedecea lapagei, Mikrobiyol. Bul.*, 42, 681, 2008.
25. Berkowitz, F.E., and Metchock, B., Third-generation cephalosporin-resistant Gram-negative bacilli in the feces of hospitalized children, *Pediatr. Infect. Dis. J.*, 14, 97, 1995.
26. Davis, O., and Wall, B.M., "Broom straw peritonitis" secondary to *Cedecea lapagei* in a liver transplant recipient, *Perit. Dial. Int.*, 26, 512, 2006.
27. Añón, M.T. et al., *Escherichia vulneris* infection. Report of 2 cases, *Enferm. Infect. Microbiol. Clin.*, 11, 559, 1993.
28. Bae, B.H., and Sureka, S.B., *Cedecea davisae* isolated from scrotal abscess, *J. Urol.*, 130, 148, 1983.
29. Pande, B.N. et al., Occupational biohazards in agricultural dusts from India, *Ann. Agric. Environ. Med.*, 7, 133, 2000.
30. Musgrove, M.T. et al., *Enterobacteriaceae* and related organisms isolated from shell eggs collected during commercial processing, *Poult. Sci.*, 87, 1211, 2008.
31. Brenner, D.J. et al., Attempts to classify Herbicola group-*Enterobacter agglomerans* strains by deoxyribonucleic acid hybridization and phenotypic tests, *Int. J. Syst. Bacteriol.*, 34, 45, 1984.
32. Antunes, M.M, Antunes, A.A., and Campos-Takaki, G.M., Incidence and resistance profile of *Cedecea* sp. isolated from hospital, powerpoint presentation in the internet, 2007 (*www.formatex.org/biomicroworld2007/virtual/ppt/865.ppt*)
33. Mangum, M.E., and Radisch, D., *Cedecea* species: Unusual clinical isolate, *Clin. Microbiol. Newslett.*, 117, 1982.
34. Bae, B.H.C., Sureka, S.B., and Ajamy, J.A., Enteric group 15 (*Enterobacteriaceae*) associated with pneumonia, *J. Clin. Microbiol.*, 14, 596, 1981.
35. Janda, J.M., and Abbott, S.L., *The Enterobacteria*, 2nd ed., pp. 357–376, ASM Press, Washington, DC, 2006.
36. Livermore, D.M., β-lactamases in laboratory and clinical resistance, *Clin. Microbiol. Rev.*, 8, 557, 1995.
37. Fahr, A.-M., and Hammann, R., Evaluation of BBL CHROMagar and CPS ID2 media for detection and presumptive identification of urinary tract pathogens, *J. Clin. Microbiol.*, 36, 1464, 1998.
38. Weile, J., and Knabbe, C., Current applications and future trends of molecular diagnostics in clinical bacteriology, *Anal. Bioanal. Chem.*, 394, 731, 2009.

39. Patel, J.B., 16S rRNA gene sequencing for bacterial pathogen identification in the clinical laboratory, *Mol. Diagn.*, 6, 313, 2001.
40. Roggenkamp, A., Phylogenetic analysis of enteric species of the family Enterobacteriaceae using the *oriC*-locus, *Syst. Appl. Microbiol.*, 30, 180, 2007.
41. Woo, P.C.Y. et al., Then and now: Use of 16S rDNA gene sequencing for bacterial identification and discovery of novel bacteria in clinical microbiology laboratories, *Clin. Microbiol. Infect.*, 14, 908, 2008.
42. Cilia, V., Lafay, B., and Christen, R., Sequence heterogeneities among 16S ribosomal RNA sequences, and their effect on phylogenetic analyses at the species level, *Mol. Biol. Evol.*, 13, 451, 1996.
43. Fukushima, M., Kakinuma, K., and Kawaguchi, R, Phylogenetic analysis of *Salmonella*, *Shigella*, and *Escherichia coli* strains on the basis of the *gyrB* gene sequence, *J. Clin. Microbiol.*, 40, 2779, 2002.

72 Citrobacter

Ignasi Roca and Jordi Vila

CONTENTS

72.1 INTRODUCTION

72.1.1 Classification and Morphology

The genus *Citrobacter* comprises a group of gram-negative bacilli belonging to the family Enterobacteriaceae, class Gammaproteobacteria, and phylum Proteobacteria. Description of this genera traces back to 1928, when an initial report described the isolation of a novel microorganism that was initially named *Bacterium freundii*[1] and then renamed as *Citrobacter freundii* in 1932 for its ability to use citrate as the sole source of carbon.[2] The genus *Citrobacter* was then enlarged with up to seven different species whose classification, however, underwent frequent modifications, and over the years they were further relocated within the genera *Colobactrum*, *Escherichia*, *Salmonella*, *Levinea*, and *Arizona*, among others. For several years, the names *Citrobacter diversus*, *Citrobacter koseri*, and *Levinea malonatica* were used to designate the same microorganism, and it was not until 1990 that *C. koseri* prevailed over the other two, which were eliminated as valid species names. In 1993, Brenner et al. showed that *Citrobacter* strains could be classified in 11 different species: *C. koseri*, *C. amalonaticus*, *C. farmeri*, *C. youngae*, *C. braakii*, *C. werkmanii*, *C. sedlakii*, *C. freundii*, and the unnamed genomospecies 9, 10, and 11, which were not named because only a very few strains had been isolated.[2a] Eventually, the isolation of additional strains led to their naming as *C. rodentium*, *C. gillenii*, and *C. muraliniae*, originating from the current classification for *Citrobacter*.[3]

Citrobacter species are typically straight rods (1 μm × 2–6 μm) found singly or in pairs, and they fall within the general definition of the family Enterobacteriaceae (gram-negative, oxidase-negative, not encapsulated, motile, facultative anaerobes, easily grown on ordinary media).

Organisms within this genus are typically straight, facultative anaerobic bacilli, having both a fermentative and respiratory type of metabolism and motile by means of peritrichous flagella, albeit some of them also bear class 1 fimbria.[4] *Citrobacter* species are usually nonencapsulated, despite some *C. freundii* strains producing Vi capsular antigen,[5] and all of them are eventual lactose fermenters. L-arabinose, maltose, L-rhamnose, threalose, D-xylose, D-mannitol, and D-sorbitol are also fermented by most *Citrobacter* strains, which produce acid and gas. *Citrobacter* spp. are catalase-positive, methyl-red positive, Voges-Proskauer–negative and lysine decarboxylase–negative.

72.1.2 Clinical Features

Citrobacter spp. can be found in the intestinal tract of both humans and animals and are also widely distributed in the environment. In the beginning, when *Citrobacter* strains were isolated from clinical specimens, they were considered contaminants or harmless colonizers, but they have gradually been recognized as true human pathogens. *Citrobacter* can cause a wide spectrum of infections in humans, such as urinary tract infections (UTIs), respiratory tract infections, wound infections, peritonitis, meningitis, and bacteremia.

Infection usually occurs in weakened/immunocompromised individuals of any age[6] or in neonates who develop sepsis and meningitis.[7] The most commonly isolated *Citrobacter* species from human samples are *C. freundii*, *C. koseri*, *C. braakii*, *C. youngae*, and *C. amalonaticus*,[8] although only *C. freundii*, *C. koseri*, and *C. braakii* seem to be related to human infections.[9–11] Among these infections, the UTIs are the most frequently found, followed by respiratory tract infections and skin and soft-tissue infections. Many *Citrobacter* infections are acquired in the nosocomial setting and in patients with underlying diseases.

Urinary Tract Infections. In a paper published by Mohanty and colleagues, *Citrobacter* was isolated as the cause of urinary tract infections (46.2%), respiratory tract infections (16.3%), bacteremia (15.8%), wound infections (12.1%), and infections of sterile body fluids in 9.3%.[11] Most of these patients had underlying diseases, such as neurosurgery (27.8%), abdominal surgery (13.6%), trauma (10.2%), and malignancies (11.7%). In a study by Hodges and colleagues,[12] 45% of the *Citrobacter* spp. caused UTIs. In addition, a recent retrospective study by Samonis et al.[13] over 78 patients with *Citrobacter* spp. isolates found that 52.6% of the patients had a UTI. *C. freundii* was isolated in 71.8% of the patients, whereas *C. koseri* was isolated from 18 patients (23%), and *C. braakii* in 3.8%. Among those patients with UTI, 30 had acute pyelonephritis; three had acute cystitis; and eight had bacteriuria. This data is in agreement with previous results that found that more than 15% of all UTI *Citrobacter* involved the upper urinary tract.[14] *Citrobacter* has also been shown to cause UTIs in the pediatric population, mainly in children with urinary tract infection, renal anomalies, or neurology impairment.[15]

Respiratory Tract Infections. The role of *Citrobacter* as a cause of respiratory tract infections sometimes is difficult to ascertain, since they are recovered in mixed cultures. In a recent paper Samonis et al. found that 40% of respiratory tract infections in which *Citrobacter* were isolated were polymicrobial.[13] However, in this study, this microorganism was not highly prevalent (6%) as a cause of respiratory tract infection in comparison with other previous reports in which the prevalence ranged from 16% to 25%.[12,14]

Intraabdominal Infections. Although peritonitis due to *Citrobacter* does not seem to be common, several cases have been described. Dervisoglu et al.[16] described a patient on continuous ambulatory peritoneal dialysis who developed peritonitis by *C. freundii*. Recently, in a study performed with 78 patients infected with *Citrobacter*, 14.1% of the patients showed an intraabdominal infection. In 30% of these patients, *Citrobacter* was detected, together with other microorganisms. Ten out of the 11 *Citrobacter* spp. isolated were *C. freundii*, whereas the remaining isolate was *C. koseri*. This infection may be of particular concern, as they have been frequently cited as the most common site of origin of *Citrobacter* bacteremia.[13]

Bacteremia. Bacteremia by *citrobacter* is uncommon, with an estimation of 1.2 cases per 10,000 patient discharges from the hospital,[17,18] although these microorganisms have been implicated as a serious cause of neonatal infections. The mortality rate of patients with bacteremia caused by *Citrobacter* spp. can reach between 18% and 56%.[10,17] This high mortality rate is most likely associated with the highly debilitated state of the host rather than intrinsic virulence of the microorganism. *Citrobacter* bacteremia is often nosocomially acquired, although one study found 61% of infections caused by this microorganism to be community acquired.[19] The most common portal of entry for *Citrobacter* bacteremia is the urinary tract; other sources are the gastrointestinal or biliary tracts.[17]

Meningitis. In addition to bacteremia, *Citrobacter* in neonates has also been shown as a cause of meningitis and brain abscess.[7,20,21] Seventy-seven percent of *C. koseri* meningitis patients develop brain abscesses.[20] It has also been observed experimentally that *C. koseri* has a propensity to cause abscess formation in brain in a mouse model.[20,21] Early-onset meningitis carries with it a higher mortality rate.[20] *C. koseri* recovered from patients with meningitis have been found to be more virulent than strains causing infections other than meningitis in an animal model and possess an outer membrane protein not seen in nonpathogenic bacteria.[22]

72.1.3 Pathogenesis

72.1.3.1 Virulence Determinants and Infection

Not being a "state-of-the-art" pathogen, there is very little information available on the molecular mechanisms and virulence factors that lead to *Citrobacter* infections in humans. There is a lot more knowledge concerning the mechanisms that lead to *Citrobacter* infections in mice, since *C. rodentium*, the etiologic agent of transmissible murine colonic hyperplasia (TMCH), is associated with subcellular alterations that are very similar to the *attaching and effacing* (A/E) lesions caused by two important human pathogens: enteropathogenic *E. coli* (EPEC) and enterohemorrhagic *E. coli* (EHEC): therefore, it is used as an infection model to investigate inflammatory bowel disease (IBD) and enteric intestinal infections. The hyperplasic state of the murine colon can lead to colon tumorigenesis which, in turn, is analogous to the increased risk for colorectal cancer in humans with ulcerative colitis.

The A/E lesions caused by EPEC and EHEC are characterized by the localized destruction (effacement) of the brush border microvilli, attachment of bacteria to the apical endothelium of the gut, and formation of a characteristic pedestal beneath the infecting bacteria very similar to that reported in TMCH. In addition, the bacterial outer membrane protein involved in attachment to the host cells, intimin, is homologous in *C. rodentium* and EPEC strains, and both infections trigger similar immune responses from the host.

During the initial stages of *C. rodentium* infection, bacteria are closely attached to the murine epithelium by means of adhesins and surface-associated filaments. An extensive repertoire of translocators and effectors are then injected into the host cell; these, in turn, activate several cell-transduction pathways, rearrange the host cell cytoskeleton, and cause the depolimerization of actin and the loss of microvilli. A more intimate contact ensues, and the accumulation of cytoskeletal elements at the site of bacterial attachment originates formation of the characteristic pedestal structure. Further delivery of additional bacterial effectors impairs host cell processes causing the loss of tight-junction integrity and eventually leading to cell death and the effacing phenotype.

The proteins involved in formation of A/E lesions in *C. rodentium* are located within a chromosomal locus termed LEE (*locus of enterocyte effacement*),[23] which comprises a 35-kb pathogenicity island containing the genes coding for: a general LEE transcriptional regulator (Ler), an outer membrane adhesin-intimin and its translocated receptor Tir, several secreted proteins (MAP, EspA, -D, -B, -F, and -G), additional regulators, and a number of undetermined open reading frames as well as a type III secretion system (TTSS).

TTSS are needle-like structures used to deliver bacterial effector proteins (usually virulence factors) inside the host cell. Overall, the effectors and the proteins involved in TTSS formation in *C. rodentium* are highly homologous to those of the TTSS systems found in EPEC and EHEC.

Deng et al.[24] showed that LEE-encoded proteins suffice for pedestal formation and that many of them are also involved in rearranging the host cytoskeleton, although LEE-encoded effectors fail to explain the full repertoire of symptoms caused by *C. rodentium*, suggesting that the LEE-encoded TTSS also translocates additional factors located somewhere else. Furthermore, a hierarchial expression of translocators and effectors has also been defined, with the former being secreted ahead of the later, which was somehow expected since translocators are used to translocate effectors inside the host cell.

Among the LEE-encoded genes, *espA*, *espB*, and *espD* are involved in formation of a translocation apparatus for delivery of effector molecules. EspB may also participate in disrupting the host cell cytoskeleton.[24] The *eae* gene encodes the outer membrane intimin, and *tir* produces the translocated intimin receptor (Tir) to which intimin binds. EspF is involved in disruption of the intestinal barrier function,[25,26] and EspG, EspH, and MAP participate in the modulation of the host cytoskeleton.[26,27]

As mentioned earlier, not all TTSS secreted effectors are encoded within the LEE locus, in fact, some of them are scattered through the bacterial chromosome, and several of them are carried on prophages or additional pathogenicity islands. EspI (also termed NleA for non-LEE-encoded effector A) has been shown to be essential for full virulence in *C. rodentium* infection, since *espI/nleA* mutants are highly attenuated and do not form colonic hyperplasia, but it does not seem to be required for A/E lesion formation.[24,26]

Additional non-LEE encoded secreted proteins have also been identified (*nleB, C, D, E, F, G*),[24] although their role in pathogenesis still needs further analysis. It has been shown that NleB is important for *C. rodentium* colonization of the intestine and for hyperplasia, but NleC or NleD do not seem to have any effect on host colonization or disease.[28] Strains that are deficient in *nleH* show a moderate competitive disadvantage over the wild-type strain in mixed infections and are slightly attenuated during the early stages of infection.[29,30] Similar results were also found for both *nleE* and *nleF*.[31,32]

Other virulence factors in *C. rodentium* not subjected to TTSS secretion and translocation include: *lifA*, encoding a lymphostatin proven to suppress lymphocyte activation in vivo and playing a role in colonization and hyperplasia through disruption of the intestinal barrier via modulation of Rho GTPases[33]; *rpoS*, the stationary phase sigma factor, which seems to have a role in the expression of LEE genes[34]; *regA*, a global transcriptional regulator presumably involved in the activation of the virulence factors needed for intestinal colonization[34]; the *pst-phoU* operon, involved in phosphate acquisition and metabolism that seems to have a role in adherence to host cells[35]; and *cfc*, encoding a CFC type IV pilus that might play a role in colonization, infection, transmission, or host specificity.[36]

Interestingly, *C. rodentium* seems to initiate colonization in the surface of the lymphoid tissue in the caecum that is analogous to the human Peyer's patches that are also preferred by EHEC. It is not clear what the advantages of colonizing this site are, since it contains a greater accumulation of immune cells and the immune system is likely to be triggered more promptly. It has been postulated that early modulation of the mucosal immune responses is a key step toward successful colonization, both in *C. rodentium* and EPEC. What we know for certain is that the adaptive immune system takes part in the clearance of *Citrobacter,* since T- and B-cell-deficient mice develop chronic infection, while normal mice clear all organisms after 21 days.[37,38]

In *C. koseri*, it has been shown that *fliP*, encoding an integral membrane protein associated with the basal body of the flagellar structure, is needed for macrophage persistence during neonatal meningitis, probably via the modulation of cytokine responses.[39]

72.1.3.2 Mechanisms of Resistance to Several Antimicrobial Agents

The main mechanism of resistance to beta-lactam antibiotics in gram-negative bacteria is the synthesis of beta-lactamases, either plasmid or chromosomally mediated. In this sense, *Citrobacter* has an inducible chromosomal cephalosporinase, and therefore the treatment with third-generation cephalosporins or aztreonam can select derepressed mutants resistant to all β-lactam antibiotics (also in combination with β-lactamase inhibitors) but carbapenems. In a surveillance program, *Citrobacter* spp. presented elevated rates of resistance to ceftazidime (14%) and aztreonam (12%), and 5.3% of resistance to piperacillin-tazobactam, indicating the prevalence of strains with stably derepressed production of chromosomal cephalosporinase.[40] In another study performed in France, 38.5% of *Citrobacter* strains presented derepressed AmpC beta-lactamases; 60% from *C. freundii* and 40% from *C. braaki*.[9] In addition both broad- and extended-spectrum β-lactamases have been described in this microorganism.

In a surveillance study performed in Korea analyzing the prevalence of ESBLs-producing *C. freundii,* 12.3% and 37% of the clinical isolates were resistant to cefotaxime and cefatzidime, respectively.[41] The three main types of ESBLs (SHV, TEM, and CTX-M) have been described in *Citrobacter* spp. SHV-12 was found in a *C. freundii* clinical isolate from a Korean study.[41] TEM-134 was found in a *C. koseri* clinical isolate during a national survey in Italy, and it was shown to be a natural derivative of TEM-2 with the following substitutions: E104K, R164H, and G238S.[42] This enzyme, however, hydrolyzed with variable efficiency penicillins, narrow-spectrum cephalosporins, cefepime, cefotaxime, ceftazidime, and aztreonam, which appeared to be the best substrate. The substitution R164H may be responsible for this narrow spectrum of activity.

On analyzing the mechanisms of resistance to third-generation cephalosporins in seven cefotaxime-resistant *C. freundii* isolates, ESBLs were four in three of them (2 SHV-12 and 1 CTX-M-27), and the genes encoding these enzymes were all plasmid mediated.[43] Other CTX-M β-lactamases, such as CTX-M-2 and CTX-M-30, have also been reported in *C. koseri* and *C. freundii.*[44,45] In a recent study, Liu and colleagues[46] undertook a survey of the occurrence of CTX-M and SHV extended-spectrum β-lactamases (ESBL) in Enterobacteriaceae from China. The overall rate of ESBL-positive isolates was 33.4%, with CTX-M found in 76.8% isolates, CTX-M-3 was found in *C. freundii,* and one isolate also carried SHV-11.

AmpC β-lactamases, both chromosomal and plasmid mediated, confer resistance to cefoxitin and cefotetan, and less frequently, to cefotaxime, ceftazidime, cefpodoxime, and ceftriaxone, and sometimes to aztreonam. It is well-known that most plasmid-mediated CMY-encoding genes are originated from *C. freundii* and that the insertion sequence element IS*Ecp1* is responsible for their mobilization.[47] Ahmed and Shimamoto[48] suggested that the chromosomally encoded CMY-37 found in *C. freundii*, which is a variant of the *C. freundii* chromosomal AmpC enzyme with seven amino acid substitutions, might be the origin of many plasmid-mediated AmpC β-lactamases.

The emergence and dissemination of Enterobacteriaceae isolates harboring carbapenemases represents a significant threat to the management of nosocomial infections, and in this sense, *Citrobacter* spp. is not an exception. Although the prevalence of carbapenem-resistant *Citrobacter* spp. is still low, a surveillance study revealed that 2 out of 1061 clinical isolates presented a slight increase in their MIC of imipenem associated with production of the KPC-3 β-lactamase.[40] Recently, a high-level carbapenem resistance in a *C. freundii* clinical isolate (MIC 256 mg/L) has been investigated, showing that this high level of resistance was due to the expression of a KPC-2 β-lactamase together with a decreased expression of two outer membrane porins with a molecular weight of 41 and 38 kDa, respectively.[49] The KPC-2 β-lactamase has been located in a plasmid, and the same plasmid carrying the bla_{KPC-2} gene was also found in a strain of *Klebsiella oxytoca* as well as in strains of *C. freundii*, thus demonstrating the potential for the horizontal transfer of carbapenemase-producing plasmids between clinically relevant gram-negative bacilli.[50]

Carbapenem resistance can also be associated with the production of class B metallo-β-lactamases (MBLs). Five types of MBLs (IMP, VIM, SPM, GIM, and AIM) enzymes have been reported so far, and they are mostly found as part of class 1 integrons. VIM-1 metallo-β-lactamases located within integron sequences have been reported in *C. freundii* clinical isolates in hospitals in Germany, Italy, and Greece.[51–53] Another metallo-β-lactamase, KHM-1, has also been found in a *C. freundii* clinical isolate, and its expression generates a low-level resistance to carbapenems.[54]

Fluoroquinolones are broad-spectrum antimicrobial agents inhibiting the activity of type II topoisomerases (DNA gyrase and topoisomerase IV). The DNA gyrase is a tetrameric enzyme with two A subunits encoded in the *gyrA* gene and two B subunits encoded in the *gyrB* gene. The topoisomerase IV is also a tetrameric enzyme, with two A subunits encoded in the *parC* gene and two B subunits encoded in the *parE* gene. Fluoroquinolone resistance in gram-negative bacteria is acquired by chromosomal mutations in specific genes as well as by the acquisition of plasmid carrying quinolone-resistant determinants. The chromosomal mutations can be distributed into two groups: (i) mutations in topoisomerase genes (*gyrA*, *gyrB*, *parC*, and *parE*) and (ii) mutations causing reduced drug accumulation, either by a decreased uptake or increased efflux.[55] Quinolone-resistance development because of mutations in the *gyrA* gene has been well established, which can be found clustered within the quinolone-resistance determining region (QRDR) located between amino acids Ala-67 and Gln-106 of the DNA gyrase A subunit. With a given breakpoint of 2 mg/L and 32 mg/L for ciprofloxacin and nalidixic acid (CLSI breakpoints), respectively, Navia et al.[56] showed that a point mutation resulting in the substitution of Thr-83 to Ile-83 produces a phenotype characterized by the acquisition of resistance to nalidixic acid (MIC of 1024 mg/L) and decreased susceptibility to ciprofloxacin (MIC of 0.25–0.5 mg/L). These isolates did not show further mutations in the *gyrB* or *parC* genes. Isolates with a nalidixic acid MIC >1024 mg/L, and ciprofloxacin MIC of 8–32 mg/L, however, showed concomitant mutations at codons 83 and 87 of the *gyrA* gene (Thr⇒Ile and Asp⇒Tyr, respectively) as well as a single mutation in codon 80 of the *parC* gene (Ser⇒Leu). Interestingly, two strains belonging to the same clone and with the same type of mutations in the *gyrA* and *parC* genes, showed different MICs of ciprofloxacin (8 and 32 mg/L, respectively). The analysis of the outer membrane protein profile between both strains did not show any significant differences, suggesting that a decreased permeability was not responsible for this observation. Later on, Sánchez-Céspedes and colleagues[57] showed that this difference in the MIC of ciprofloxacin was associated with the differential expression of the AcrAB efflux pump. AcrAB was found to be overexpressed in the strain with a ciprofloxacin MIC of 32 mg/L. In addition, the MIC of chloramphenicol, which is also a well-known substrate for AcrAB, was also different between both strains (4 and 32 mg/L, respectively).

Apart from chromosomally encoded fluoroquinolone resistance mechanisms, three quinolone-resistance plasmid-mediated determinants have also been described: (i) the Qnr protein, which protects both the DNA gyrase and topoisomerase IV from the action of quinolones; (ii) the *aac(6′)-Ib-cr* gene, encoding an aminoglycoside modifying enzyme that catalyzes the *N*-acetylation of the piperazinyl amine of some quinolones; and (iii) the QepA determinant, which is a plasmid-mediated efflux pump able to extrude some quinolones out of the cell.[58] Plasmid-mediated quinolone-resistance has also been described in *Citrobacter* spp. To date, 5 Qnr (QnrA, QnrB, QnrC, QnrD, and QnrS) determinants have been reported, although only QnrA, QnrB, and QnrS have been described in *Citrobacter.* Qnr was present in 23 (57%) out of 40 *C. freundii* clinical isolates. Six were QnrA1, thirteen were QnrB, and one was QnrS. The combinations QnrA + QnrB or QnrB + QnrS were found in just one strain each.[59] In a recent report by Dahmen et al. considering different Enterobacteriaceae, QnrB2 was found in two isolates, and QnrA6 and QnrB1, in one each.[60] Several variants of QnrB: QnrB4, QnrB6, QnrB8, QnrB12, QnrB15, and QnrB16[61–64] have been found in *Citrobacter freundii* and *Citrobacter werkmanii.* Whereas most of the Qnr determinants have been plasmid located, the QnrB determinant described in *Citrobacter* has mainly been found within the chromosome and inserted downstream of the *pspF* gene, which encodes a phage shock protein.[63] More recently, however, a plasmid-mediated *qnrB4* gene has been found in a large conjugative plasmid in *Klebsiella pneumoniae* together with the $bla_{\text{SHV-2a}}$, $bla_{\text{DHA-1}}$, *qnrB4*, and *aac(6′)-Ib-cr* genes.[65]

Although the *aac(6′)-Ib-cr* causes low-level ciprofloxacin resistance, it has been shown to act additively with other plasmid-mediated quinolone-resistance mechanisms, such as Qnr, and to favor the acquisition of higher levels of quinolone resistance. The *aac(6′)-Ib-cr* gene was found in 5 (3.7%) out of 134 *C. freundii* clinical isolates.[66] Another study reported the presence of *aac(6′)-Ib-cr* in 17% of the Enterobacteriaceae strains studied and among them in 26.7% of *C. freundii* strains.[59] Up to date, QepA has not been reported in *Citrobacter* spp. (Table 72.1).

72.1.4 Diagnosis and Treatment

Samples inspected for the presence of gram-negative rods are initially grown on blood agar plates as well as on differential media, such as MacConkey's agar. Lactose-positive *Citrobacter* colonies form pinkish colonies and can be further differentiated from *E. coli* by their ability to use citrate, as well as their growth in Salmonella–Shigella agar plates. Differentiation of lactose-negative *Citrobacter* from *Salmonella* is achieved by the biochemical testing of lysine decarboxylase (*Citrobacter*-negative, *Salmonella*-positive) and growth on KCN media (*Citrobacter*-KCN resistant, *Salmonella*-KCN sensitive). Diagnosis is made upon the identification of specific *Citrobacter* species by standard biochemical testing (Table 72.2). In circumstances where a more rapid diagnostic is needed, molecular techniques such as analysis of protein profiles in a MALDI-TOF mass spectrophotometer can also be used to obtain a species-specific diagnostic in a considerably shorter time.[67,68]

TABLE 72.1
Mechanisms of Resistance to β-Lactam Antibiotics and Quinolones

β-Lactam Antibiotics	
β-Lactamases	
	Chromosomal cephalosporinase AmpC
	Broad-spectrum β-lactamases
	Extended-spectrum β-lactamases
	Carbapenemases
Quinolones	
Chromosomal-mediated	
	Mutations in the *gyrA* and *parC* genes
	Overexpression of AcrAB
Plasmid-mediated	
	Qnr (QnrA1, QnrB4, QnrB6, QnrB8, QnrB15 and QnrB16, QnrS)
	Aac(6′)-Ib-cr

The different species of *Citrobacter* can show slight differences regarding the susceptibility to antimicrobial agents.[11,40,69,70] In addition, they can also display varying susceptibilities between countries. In Table 72.3, the percentages of resistance from several studies are shown. A high percentage of resistance (85%) to third-generation cephalosporins has been described in a study performed in a hospital in Northern India[11]; however, the authors did not mention whether this high level of resistance may be associated with a predominant clone, although this high resistance was found in both *C. freundii* and *C. koseri*. In an antibiotic susceptibility study of *Citrobacter* spp. causing skin and soft-tissue infections in the United States and Europe, the percentage of resistance to third-generation cephalosporins ranged from 17% to 38% to ceftazidime, and from 10% to 32% to cefotaxime.[69] In the ICU, the percentage of multiresistant isolates of *C. freundii* increased from 5.3% in 1993 to 11.1% in 2004.[70]

Resistance to ciprofloxacin was also very high (82%) for *Citrobacter* strains isolated from India[11]; however, in a multi-country study the range of resistance went from 5% to 21%,[69] and in a surveillance study performed with isolates from North America, Latin America, and Europe, the resistance was around 8%.[40] Carbapenems have the highest in vitro activities against *Citrobacter* spp.[71,72] However, Lockhart et al.[70] reported a steady increase in the resistance of *C. freundii* to ertapenem, tobramycin, and ciprofloxacin from 1993 to 2004.

The treatment of *Citrobacter* infections usually follows the principles for treatment of other Enterobacteriaceae infections. Based on the in vitro antimicrobial susceptibilities described earlier, carbapenems or fluoroquinolones would appear to be preferred therapeutic options for *Citrobacter* infections. However, other antimicrobial agents, such as third-generation cephalosporins and piperacillin-tazobactam, can also be used. As it has been mentioned earlier, carbapenem-resistant *Citrobacter* spp. strains can be isolated, and they are usually multiresistant, showing

TABLE 72.2
Biochemical Test to Differentiate *Citrobacter* Species

Characteristic	*C. freundii*	*C. amalonaticus*	*C. braakii*	*C. farmeri*	*C. gillenii*	*C. koseri*	*C. murliniae*	*C. rodentium*	*C. sedlakii*	*C. werkmanii*	*C. youngae*
Indole	d	+	d	+	–	+	+	–	+	–	d
Citrate-simmons	d	+	d	d	d	+	+	–	d	+	d
H_2S in iron agar	d	D	d	–	d	–	d	–	–	+	d
Ornithine	–	+	+	+	–	+	–	+	+	–	–
KCN growth	d	+	+	+	+	–	+	–	+	+	+
Malonate	d	d	–	–	+	+	–	+	+	+	–
Acid from											
Sucrose	+	d	–	+	d	d	d	–	–	–	d
Melibiose	+	–	d	+	d	–	d	–	+	–	–
Raffinose	d	–	–	+	–	–	d	–	–	–	–
Dulcitol	d	–	d	–	–	d	+	–	+	–	d
Adonitol	–	–	–	–	–	+	–	–	–	–	–
D-Arabitol	–	–	–	–	–	+	–	–	–	–	–

+, 90% or more of the strains are positive; –, 90% or more of the strains are negative; d, 11%–89% of the strains are positive.
Source: Adapted from Brenner et al. 1993.[2a]

TABLE 72.3
Resistance to Antimicrobial Agents of *C. freundii* and *C. koseri* Clinical Isolates

Antimicrobial Agent	% of Resistance				
	C. freundii[11,13]		*C. koseri*[11,13]		*Citrobacter* spp.[40]
Total of clinical isolates analyzed	20	29	185	6	1061
Amoxicillin/clavulanic acid	–	96	–	0	–
Piperacillin	65	–	82	–	–
Piperacillin/tazobactam	5	34.5	12	0	5
Ceftazidime	85	38	85	0	14
Cefotaxime	85	–	86	–	–
Ciprofloxacin	60	26	84	0	8
Amikacin	30	–	48	–	–
Imipenem	15	3.4	6.5	0	0.1
Nitrofurantoin	20	13.8	53	0	–

resistance to ciprofloxacin, aminoglycosides, and third-generation cephalosporins. Under these circumstances tigecycline can be a therapeutic alternative since different studies have shown that the tigecycline MIC_{50} for *Citrobacter* spp. was of 0.25 mg/L.[73]

72.2 METHODS

72.2.1 Sample Preparation

Carefully pick up a small amount of a freshly grown *Citrobacter* colony with the end of a tip and place it into an Eppendorf tube with 100 µL of 50 µL 1× TE buffer pH 8.0. Incubate the tubes at 96°C for 5 to 10 min. Centrifuge at 13,000 × *g* for 30 s, and chill on ice prior to analysis.

72.2.2 Detection Procedures

(i) Sequencing Analysis of gyrA and parC Genes from Citrobacter spp. A rapid PCR-based method is presented here for the amplification and sequencing of the QRDR determinants of both *gyrA* and *parC* genes from *Citrobacter* spp. using the primer pairs 5′-AAATCTGCCCGTGTCGTTGGT-3′ and 5′-GCCATACCTACGGCGATACC-3′ for *gyrA*; and 5′-AAACCTGTTCAGCGCCGCATT-3′ and 5′-CATCGCC GCGAACGA TTCGG-3′ for *parC*.

Procedure

1. Set up a 50-µL PCR reaction mix containing 50 mM Tris-HCl pH 9, 100 mM KCl, 1.5 mM $MgCl_2$, 0.2 mM dNTP mixture, 0.5 µM primers, 1.25 U Taq polymerase, and 5 µL of DNA extract.
2. Perform the PCR for 30 cycles of 96°C for 30 s, 55°C for 30 s, and 72°C for 60 s, followed by a final extension step at 72°C for 10 min.
3. Run the PCR products in a 2% (w/v) agarose gel and purify the amplification products.
4. Sequence the purified products using either one primer or the other for each sequencing reaction. Translate and compare the QRDR sequences with that of a reference strain.

(ii) Sequencing Analysis of the Citrobacter 16S rRNA Gene A 500 bp hypervariable region from the 16S rRNA gene is amplified by PCR with primers rRNA-F (5′-TCAG ATTTGAACGCTGGCGGCA-3′) and rRNA-R (5′-CGTAT TACCGCGGCTGCTGCCAC-3′). The resulting PCR products are sequenced for identification of *Citrobacter* spp.[74] DNA sequence analysis, alignment, and phylogenetic mapping are performed using the Lasergene (DNA-Star). For the phylogenetic analysis, an alignment is produced using the CLUSTAL W algorithm, applying a BLOSUM matrix with a gap-opening penalty of 10 and a gap-extension penalty of 0.1. The resultant alignment is analyzed using a maximal likelihood method with the standard parameters in Lasergene.

(iii) PCR Identification of Citrobacter koseri Targeting cko Gene The gene encoding the low-level constitutively expressed *β*-lactamase CK4, *cko*, is uniquely present in *C. koseri* isolates. Primers CKO-F (5′-ATG AGAAACGAGGAAGTCAT-3′) and CKO-R (5′-TTAA TCATAGACTGCGAGTG-3′) derived from this gene are useful for identification of *C. koseri*, yielding a specific 903 bp product.[74]

72.3 CONCLUSION

Like other Enterobacteriaceae, *Citrobacter* can cause a wide spectrum of infections in humans, such as urinary tract, respiratory, wound, peritoneum, and blood stream infections. Among the various sites of infection, the urinary tract is the most common, the majority of these infections being nosocomially acquired. Third-generation cephalosporins or aztreonam can select derepressed mutants resistant to all *β*-lactam antibiotics (also in combination with *β*-lactamase inhibitors) but carbapenems. In addition, the acquisition of extended-spectrum *β*-lactamases has also been reported. Usually, these ESBL-producing *Citrobacter* isolates are also resistant to fluoroquinolones, therefore a therapeutic alternative are carbapenems. However, carbapenem-resistant strains have been also shown. Therefore, the rapid identification of the clinical isolate would be advantageous to implement the appropriate therapy and, in this sense, the use of MALDI-TOF is an interesting tool for this purpose.

REFERENCES

1. Braak, H.R., *Onderzoekingen over Verfisting van Glycerine*, Delft, Netherlands, 1928.
2. Werkman, C.H., and Gillen, G.F., Bacteria producing trimethylene glycol, *J. Bacteriol.*, 23, 167, 1932.

2a. Brenner, D.J., et al., Classification of citrobacteria by DNA hybridization: designation of *Citrobacter farmeri* sp. nov., *Citrobacter youngae* sp. nov., *Citrobacter braakii* sp. nov., *Citrobacter werkmanii* sp. nov., *Citrobacter sedlakii* sp. nov., and three unnamed *Citrobacter* genomospecies, *Int. J. Syst. Bacteriol.*, 43, 645, 1993.

3. O'Hara, C.M., Roman, S.B., and Miller, J.M., Ability of commercial identification systems to identify newly recognized species of *Citrobacter*, *J. Clin. Microbiol.*, 33, 242, 1995.
4. Abraham, S.N. et al., Conservation of the D-mannose-adhesion protein among type 1 fimbriated members of the family *Enterobacteriaceae*, *Nature*, 336, 682, 1988.
5. Houng, H.S. et al., Expression of Vi antigen in *Escherichia coli* K-12: Characterization of ViaB from *Citrobacter freundii* and identity of ViaA with RcsB, *J. Bacteriol.*, 174, 5910, 1992.
6. Samonis, G. et al., Review of *Citrobacter* bacteremia in cancer patients over a sixteen-year period, *Eur. J. Clin. Microbiol. Infect. Dis.*, 10, 479, 1991.
7. Doran, T.I., The role of *Citrobacter* in clinical disease of children: Review, *Clin. Infect. Dis.*, 28, 384, 1999.
8. Janda, J.M. et al., Biochemical identification of citrobacteria in the clinical laboratory, *J. Clin. Microbiol.*, 32, 1850, 1994.
9. Lavigne, J.P. et al., Clinical and molecular epidemiology of multidrug-resistant *Citrobacter* spp. infections in a French university hospital, *Eur. J. Clin. Microbiol. Infect. Dis.*, 26, 439, 2007.
10. Lipsky, B.A., et al., *Citrobacter* infections in humans: Experience at the Seattle Veterans Administration Medical Center and a review of the literature, *Rev. Infect. Dis.*, 2, 746, 1980.
11. Mohanty, S., et al., *Citrobacter* infections in a tertiary care hospital in Northern India, *J. Infect.*, 54, 58, 2007.
12. Hodges, G.R., Degener, C.E., and Barnes, W.G., Clinical significance of *Citrobacter* isolates, *Am. J. Clin. Pathol.*, 70, 37, 1978.
13. Samonis, G., et al., *Citrobacter* infections in a general hospital: Characteristics and outcomes, *Eur. J. Clin. Microbiol. Infect. Dis.*, 28, 61, 2009.
14. Fields, B.N., et al., The so-called "Paracolon" bacteria. A bacteriologic and clinical reappraisal, *Am. J. Med.*, 42, 89, 1967.
15. Gill, M.A., and Schutze, G.E., *Citrobacter* urinary tract infections in children, *Pediatr. Infect. Dis. J.*, 18, 889, 1999.
16. Dervisoglu, E., Yumuk, Z., and Yegenaga, I., *Citrobacter freundii* peritonitis and tunnel infection in a patient on continuous ambulatory peritoneal dialysis, *J. Med. Microbiol.*, 57, 125, 2008.
17. Drelichman, V., and Band, J.D., Bacteremias due to *Citrobacter diversus* and *Citrobacter freundii*. Incidence, risk factors, and clinical outcome, *Arch. Intern. Med.*, 145, 1808, 1985.
18. Shih, C.C. et al., Bacteremia due to *Citrobacter* species: Significance of primary intraabdominal infection, *Clin. Infect. Dis.*, 23, 543, 1996.
19. Chen, Y.S. et al., Clinical features and antimicrobial susceptibility trends in *Citrobacter freundii* bacteremia, *J. Microbiol. Immunol. Infect.*, 35, 109, 2002.
20. Graham, D.R., and Band, J.D., *Citrobacter diversus* brain abscess and meningitis in neonates, *JAMA*, 245, 1923, 1981.
21. Soriano, A.L., et al., Pathophysiology of *Citrobacter diversus* neonatal meningitis: Comparative studies in an infant mouse model, *Infect. Immun.*, 59, 1352, 1991.
22. Kline, M.W., Mason, E.O. Jr., and Kaplan, S.L., Characterization of *Citrobacter diversus* strains causing neonatal meningitis, *J. Infect. Dis.*, 157, 101, 1988.
23. Deng, W., et al., Locus of enterocyte effacement from *Citrobacter rodentium*: Sequence analysis and evidence for horizontal transfer among attaching and effacing pathogens, *Infect. Immun.*, 69, 6323, 2001.
24. Deng, W., et al., Dissecting virulence: Systematic and functional analyses of a pathogenicity island, *Proc. Natl. Acad. Sci. USA*, 101, 3597, 2004.
25. McNamara, B.P., et al., Translocated EspF protein from enteropathogenic *Escherichia coli* disrupts host intestinal barrier function, *J. Clin. Invest.*, 107, 621, 2001.
26. Mundy, R., et al., Identification of a novel *Citrobacter rodentium* type III secreted protein, EspI, and roles of this and other secreted proteins in infection, *Infect. Immun.*, 72, 2288, 2004.
27. Tu, X., et al., EspH, a new cytoskeleton-modulating effector of enterohaemorrhagic and enteropathogenic *Escherichia coli*, *Mol. Microbiol.*, 47, 595, 2003.
28. Kelly, M., et al., Essential role of the type III secretion system effector NleB in colonization of mice by *Citrobacter rodentium*, *Infect. Immun.*, 74, 2328, 2006.
29. Hemrajani, C., et al., Role of NleH, a type III secreted effector from attaching and effacing pathogens, in colonization of the bovine, ovine, and murine gut, *Infect. Immun.*, 76, 4804, 2008.
30. Garcia-Angulo, V.A., et al., Regulation of expression and secretion of NleH, a new nonlocus of enterocyte effacement-encoded effector in *Citrobacter rodentium*, *J. Bacteriol.*, 190, 2388, 2008.
31. Echtenkamp, F., et al., Characterization of the NleF effector protein from attaching and effacing bacterial pathogens, *FEMS Microbiol. Lett.*, 281, 98, 2008.
32. Wickham, M.E., et al., *Citrobacter rodentium* virulence in mice associates with bacterial load and the type III effector NleE, *Microbes Infect.*, 9, 400, 2007.
33. Babbin, B.A., et al., The bacterial virulence factor lymphostatin compromises intestinal epithelial barrier function by modulating rho GTPases, *Am. J. Pathol.*, 174, 1347, 2009.
34. Dong, T., Coombes, B.K., and Schellhorn, H.E., Role of RpoS in the virulence of *Citrobacter rodentium*, *Infect. Immun.*, 77, 501, 2009.
35. Cheng, C., et al., Contribution of the *pst-phoU* operon to cell adherence by atypical enteropathogenic *Escherichia coli* and virulence of *Citrobacter rodentium*, *Infect. Immun.*, 77, 1936, 2009.
36. Mundy, R., et al., Identification of a novel type IV pilus gene cluster required for gastrointestinal colonization of *Citrobacter rodentium*, *Mol. Microbiol.*, 48, 795, 2003.
37. Vallance, B.A., et al., Modulation of inducible nitric oxide synthase expression by the attaching and effacing bacterial pathogen *Citrobacter rodentium* in infected mice, *Infect. Immun.*, 70, 6424, 2002.
38. Vallance, B.A., et al., Mice lacking T and B lymphocytes develop transient colitis and crypt hyperplasia yet suffer impaired bacterial clearance during *Citrobacter rodentium* infection, *Infect. Immun.*, 70, 2070, 2002.
39. Townsend, S.M., Gonzalez-Gomez, I., and Badger, J.L., *fliP* influences *Citrobacter koseri* macrophage uptake, cytokine expression and brain abscess formation in the neonatal rat, *J. Med. Microbiol.*, 55, 1631, 2006.

40. Deshpande, L.M., et al., Occurrence and characterization of carbapenemase-producing *Enterobacteriaceae*: Report from the SENTRY Antimicrobial Surveillance Program (2000–2004), *Microb. Drug Resist.*, 12, 223, 2006.
41. Choi, S.H., et al., Prevalence, microbiology, and clinical characteristics of extended-spectrum beta-lactamase-producing *Enterobacter* spp., *Serratia marcescens*, *Citrobacter freundii*, and *Morganella morganii* in Korea, *Eur. J. Clin. Microbiol. Infect. Dis.*, 26, 557, 2007.
42. Perilli, M., et al., Biochemical analysis of TEM-134, a new TEM-type extended-spectrum beta-lactamase variant produced in a *Citrobacter koseri* clinical isolate from an Italian hospital, *J. Antimicrob. Chemother.*, 60, 877, 2007.
43. Liu, C.P., et al., Cefotaxime-resistant *Citrobacter freundii* in isolates from blood in a tertiary teaching hospital in Northern Taiwan, *J. Infect.*, 55, 363, 2007.
44. Abdalhamid, B., et al., Community-onset disease caused by *Citrobacter freundii* producing a novel CTX-M beta-lactamase, CTX-M-30, in Canada, *Antimicrob. Agents Chemother.*, 48, 4435, 2004.
45. Muta, T., et al., A nosocomial outbreak due to novel CTX-M-2-producing strains of *Citrobacter koseri* in a hematological ward, *Jpn. J. Infect. Dis.*, 59, 69, 2006.
46. Liu, W., et al., Novel CTX-M {beta}-lactamase genotype distribution and spread into multiple species of *Enterobacteriaceae* in Changsha, Southern China, *J. Antimicrob. Chemother.*, 63, 895, 2009.
47. Philippon, A., Arlet, G., and Jacoby, G.A., Plasmid-determined AmpC-type beta-lactamases, *Antimicrob. Agents Chemother.*, 46, 1, 2002.
48. Ahmed, A.M., and Shimamoto, T., Emergence of a cefepime- and cefpirome-resistant *Citrobacter freundii* clinical isolate harbouring a novel chromosomally encoded AmpC beta-lactamase, CMY-37, *Int. J. Antimicrob. Agents*, 32, 256, 2008.
49. Zhang, R., et al., High-level carbapenem resistance in a *Citrobacter freundii* clinical isolate is due to a combination of KPC-2 production and decreased porin expression, *J. Med. Microbiol.*, 57, 332, 2008.
50. Rasheed, J.K., et al., Detection of the *Klebsiella pneumoniae* carbapenemase type 2 Carbapenem-hydrolyzing enzyme in clinical isolates of *Citrobacter freundii* and *K. oxytoca* carrying a common plasmid, *J. Clin. Microbiol.*, 46, 2066, 2008.
51. Aschbacher, R., et al., Linkage of acquired quinolone resistance (*qnrS1*) and metallo-beta-lactamase (*blaVIM-1*) genes in multiple species of *Enterobacteriaceae* from Bolzano, Italy, *J. Antimicrob. Chemother.*, 61, 515, 2008.
52. Protonotariou, E., et al., First identification of VIM-1-producing *Citrobacter freundii* in Greece, *Int. J. Antimicrob. Agents*, 32, 460, 2008.
53. Weile, J., et al., First detection of a VIM-1 metallo-beta-lactamase in a carbapenem-resistant *Citrobacter freundii* clinical isolate in an acute hospital in Germany, *Scand. J. Infect. Dis.*, 39, 264, 2007.
54. Sekiguchi, J., et al., KHM-1, a novel plasmid-mediated metallo-beta-lactamase from a *Citrobacter freundii* clinical isolate, *Antimicrob. Agents Chemother.*, 52, 4194, 2008.
55. Fabrega, A., et al., Mechanism of action of and resistance to quinolones, *Microb. Biotech.*, 2, 40, 2009.
56. Navia, M.M., et al., Analysis of the mechanisms of quinolone resistance in clinical isolates of *Citrobacter freundii*, *J. Antimicrob. Chemother.*, 45, 556, 2000.
57. Sanchez-Cespedes, J., and Vila, J., Partial characterisation of the *acrAB* locus in two *Citrobacter freundii* clinical isolates, *Int. J. Antimicrob. Agents*, 30, 259, 2007.
58. Strahilevitz, J., et al., Plasmid-mediated quinolone resistance: A multifaceted threat, *Clin. Microbiol. Rev.*, 22, 664, 2009.
59. Yang, H., et al., High prevalence of plasmid-mediated quinolone resistance genes *qnr* and *aac(6')-Ib-cr* in clinical isolates of *Enterobacteriaceae* from nine teaching hospitals in China, *Antimicrob. Agents Chemother.*, 52, 4268, 2008.
60. Dahmen, S., et al., Prevalence of plasmid-mediated quinolone resistance determinants in *Enterobacteriaceae* from Tunisia, *Clin. Microbiol. Infect.*, 16, 1019, 2009.
61. Kehrenberg, C., et al., Novel variant of the *qnrB* gene, *qnrB12*, in *Citrobacter werkmanii*, *Antimicrob. Agents Chemother.*, 52, 1206, 2008.
62. Minarini, L.A., et al., Plasmid-mediated quinolone resistance determinants among enterobacterial isolates from outpatients in Brazil, *J. Antimicrob. Chemother.*, 62, 474, 2008.
63. Sanchez-Cespedes, J., et al., Two chromosomally located *qnrB* variants, *qnrB6* and the new *qnrB16*, in *Citrobacter* spp. isolates causing bacteraemia, *Clin. Microbiol. Infect.*, 15, 1132, 2009.
64. Tamang, M.D., et al., Plasmid-mediated quinolone resistance determinants *qnrA*, *qnrB*, and *qnrS* among clinical isolates of *Enterobacteriaceae* in a Korean hospital, *Antimicrob. Agents Chemother.*, 52, 4159, 2008.
65. Corvec, S., et al., *Klebsiella pneumoniae* clinical isolate coproducing SHV-2a, DHA-1, QnrB4, and AAC(6')-Ib-cr determinants in France, *Diagn. Microbiol. Infect. Dis.*, 64, 462, 2009.
66. Kim, S.Y., et al., Prevalence and characteristics of *aac(6')-Ib-cr* in AmpC-producing *Enterobacter cloacae*, *Citrobacter freundii*, and *Serratia marcescens*: A multicenter study from Korea, *Diagn. Microbiol. Infect. Dis.*, 63, 314, 2009.
67. Sontakke, S., et al., Use of broad range16S rDNA PCR in clinical microbiology, *J. Microbiol. Methods*, 76, 217, 2009.
68. Klouche, M., and Schroder, U., Rapid methods for diagnosis of bloodstream infections, *Clin. Chem. Lab. Med.*, 46, 888, 2008.
69. Jones, M.E., et al., Epidemiology and antibiotic susceptibility of bacteria causing skin and soft tissue infections in the USA and Europe: A guide to appropriate antimicrobial therapy, *Int. J. Antimicrob. Agents*, 22, 406, 2003.
70. Lockhart, S.R., et al., Antimicrobial resistance among Gram-negative bacilli causing infections in intensive care unit patients in the United States between 1993 and 2004, *J. Clin. Microbiol.*, 45, 3352, 2007.
71. Hoban, D.J. et al., In vitro activity of three carbapenem antibiotics. Comparative studies with biapenem (L-627), imipenem, and meropenem against aerobic pathogens isolated worldwide, *Diagn. Microbiol. Infect. Dis.*, 17, 299, 1993.
72. Sader, H.S., and Jones, R.N., Antimicrobial activity of the new carbapenem biapenem compared to imipenem, meropenem and other broad-spectrum beta-lactam drugs, *Eur. J. Clin. Microbiol. Infect. Dis.*, 12, 384, 1993.
73. Brown, S.D., and Traczewski, M.M., Comparative in vitro antimicrobial activity of tigecycline, a new glycylcycline compound, in freshly prepared medium and quality control, *J. Clin. Microbiol.*, 45, 2173, 2007.
74. Underwood, S., and Avison, M.B., *Citrobacter koseri* and *Citrobacter amalonaticus* isolates carry highly divergent beta-lactamase genes despite having high levels of biochemical similarity and 16S rRNA sequence homology, *J. Antimicrob. Chemother.*, 53, 1076, 2004.

73 Coxiella

Claire Pelletier

CONTENTS

73.1 INTRODUCTION

Coxiella burnetii, currently the only species in the gram-negative genus *Coxiella*, is a strictly obligate intracellular bacterium that causes Q fever, a worldwide but rarely reported zoonotic disease.

Dr. E.H. Derrick, director of the laboratory section of the Queensland Health Department in Australia, first described Q fever in 1937, following outbreaks, which had been occurring since 1933, among a large number of workers in a meat plant in Brisbane.[1] The disease was first named "abattoir fever" or "Queensland rickettsia fever" but was finally called "Q fever" (the "Q" standing for "query") to highlight the uncertainties concerning its etiology and epidemiology and to avoid directing negative connotations at either the cattle industry or the state of Queensland.[2]

Coxiella burnetii is a biosafety level 3 pathogen and has been designated a category B bioterrorism agent.

73.1.1 Classification, Morphology, Genome, and Biology

Classification. *Coxiella burnetii* was first classified among the *Rickettsia* genus due to its morphological, tinctorial, and cultural characteristics. In 1939, E.H. Derrick proposed the name *Rickettsia burneti* in honor of F.M. Burnett, who demonstrated large numbers of *Rickettsia*-like bodies in the liver and the spleens of infected mice.[3]

At almost the same time, in 1938, G. Davis and H.R. Cox, working on the ecology of Rocky Mountain spotted fever (caused by *Rickettsia rickettsi*), isolated a filter-passing *Rickettsia*-like agent from ticks (*Dermacentor andersoni* species) collected near Nine Mile Creek, Montana, USA.[4] They suggested the name *Rickettsia diaporica* on account of its ability to pass through ordinary bacterial filters, unlike other *Rickettsia*.

In 1939, R.E. Dyer, director of the U.S. National Institutes of Health, established a definitive link between *Rickettsia burneti* and *Rickettsia diaporica*, showing that both strains gave cross immunity as well as cross protection in guinea pigs.[5] However, because Burnet's isolation was first, the name of *Rickettsia burneti* remained. In 1948, following further studies that showed that this organism is more resistant to physical and chemical agents than other *Rickettsiae*, C.B. Philip proposed that the Q-fever agent be classified as the type organism of a new genus named *Coxiella*, in honor of H.R. Cox. Hence, *Coxiella burnetii* superseded *Rickettsia burneti*.[6] Phylogenetics studies based on 16S ribosomal RNA (rRNA) analysis confirmed these initial phenotypical observations showing that genus *Coxiella* should be excluded from order Rickettsiales and class Alphaproteobacteria.[7] In the second edition of Bergey's *Manual of Systematic*

Bacteriology, *Coxiella burnetii*, the only species of the genus *Coxiella* to date, belongs to the phylum Proteobacteria (major group of bacteria, including a wide variety of pathogens, such as *Escherichia coli*, *Salmonella*, and *Vibrio,* and named for the Greek god Proteus, who could change his shape, because of the great diversity of forms found in this group), class Gammaproteobacteria, order Legionellales, and family Coxiellaceae.

Morphology. *Coxiella burnetii* is a short rod with a wall similar to that of gram-negative bacteria but is stained poorly by this technique. Effective staining methods are Gimenez (*Coxiella* burnetii are shown in red on a green background) and Stamp or Machiavello staining (*Coxiella burnetii* are shown in red on a blue background).[8,9]

Coxiella burnetii cannot be grown on axenic medium. It is cultured in embryonated eggs or, more commonly, in various cell lines (see Culture under Section 73.1.5.1). During in vitro culture, a phase change is observed that is comparable to the antigenic variation of "smooth-rough" Enterobacteria with modified outer membrane lipopolysaccharide (LPS).[10] The LPS on bacteria isolated from infected arthropods, animals, or humans are complete and characterize the phase I variant ("smooth"). After many passages in vitro, however, an incomplete form of LPS is present that has a different sugar composition and lacks some surface proteic determinants.[11] This characterizes the phase II variant or "rough" bacteria. These cells also have a large chromosomal deletion and lower infectivity.[12] Phase variation must exist in infected humans or animals because the humoral immune response is earlier and stronger towards phase II antigens, whereas the antibodies against phase I have an increased protective power.[9] Distinction between the two types of antibodies in the serum of infected patients allows differentiation of acute and chronic Q fever (see Serology under Section 73.1.5.1).

Two *Coxiella* variants can also be distinguished due to their extra- or intracellular location. Small-cell variants (SCVs) are extracellular rod shaped and 204 by 450 nm in size. The chromosome is condensed, and the periplasmic space is rich in proteins and accumulated peptidoglycans fragments. The large-cell variants (LCVs) are intracellular and can reach 2 μm in length. The chromosome is dispersed, and the periplasmic space is well individualized.[8]

Genome. The *Coxiella burnetii* genome consists of a circular chromosome of between 1.5 and 2.4 Megabases (Mb) (1,995,275 bp) for Nine Mile phase I RSA493 strain) and an optional plasmid of 33–42 kb.[13] The genome is predicted to contain 2134 coding sequences (CDS), of which 719 encode hypothetical proteins (i.e., without significant matches with known sequences), which is a high proportion compared to other Gammaproteobacteria.[13] Various intervening elements have been identified within physiologically vital genes. For example, the 23S rRNA gene is interrupted by both an intervening sequence and two group I self-splicing introns.[14,15] The presence of three selfish elements in the 23S rRNA gene is very unusual for an obligate intracellular bacterium. This suggests a recent shift to a new lifestyle from a previous niche with greater opportunities for gene transfer.[15] In the same way, *Coxiella burnetii* possesses various copies of insertion sequences (IS): *IS1111*, *IS30*, *ISAs1*, and three degenerate transposase genes of unknown lineage.[13] IS elements are dispersed around the chromosome, without any localized clustering. These IS-type elements played a role in generating *Coxiella burnetii* genetic diversity. Finally, genome analysis identified 83 pseudogenes (i.e., genes disrupted by various mechanisms such as frameshifts, point mutations, or truncations) suggesting that some genome reduction is underway.[13] Specific functions such as toxin production, resistance, transport, and binding are overrepresented among the pseudogenes, suggesting the elimination of functions no longer relevant to a new lifestyle and greater dependence on the host cell. Nevertheless, a high percentage of the genome is still coding (89.1% against only 76% for other species known to have undergone genome reduction, such as *Mycobacterium leprae*). This suggests that the reduction process is still at the beginning.[13]

DNA–DNA hybridization or 16S rRNA gene sequencing showed a low degree of genetic heterogeneity among *Coxiella burnetii* strains.[16] It was thus surprising that considerable heterogeneity was observed in genotyping studies using PCR-restriction fragment length polymorphism (RFLP) analysis (six genomic groups); multispacer sequence typing (30 genomic groups); infrequent restriction site PCR (IRS-PCR); and multiple loci variable number of tandem repeats analysis (MLVA) (36 genotypes were distinguished); repetitive element PCR-based differentiation on *IS1111* insertion sequences (five genomic groups); or microarray-based whole genome comparison to Nine Mile phase I reference isolate.[17–21]

Four types of optional plasmids are found in the *Coxiella burnetii* genome. The first plasmid described was the 36 kb QpH1 which is found at one to three copies per cell.[22] Then another type of plasmid was isolated from placental tissue of an aborted goat: QpRS, 39 kb.[23] A third plasmid was obtained from wild rodents: QpDG, 42 kb.[24] Finally, the last type of plasmid was found in a French strain isolated from an endocarditis patient: QpDV, 33 kb.[25]

The genome of plasmidless *Coxiella burnetii* isolates was found to contain some DNA sequences sharing homology to the QpRS plasmid.[26] These four types of plasmids have a common region of 30 kb, although polymorphisms specific to each type and to integrated plasmid-like sequences were described.[21–27] Nine predicted open reading frames (ORFs) are conserved among all plasmid types and integrated-plasmid sequences. Two of these encode proteins with similarity to phage proteins and may be remnants of a phage that integrated into the *Coxiella burnetii* plasmid.[21] A degenerate ORF flanks these phage genes, it is annotated as a putative toxin, with highest similarity to insecticidal and nematocidal toxins synthesized by the waterborne bacterium *Chromobacterium violaceum* and nematode symbionts.[28,29] This degenerated ORF was probably acquired by *Coxiella burnetii via* horizontal transfer mediated by phage integration and may have been beneficial for infecting an insect host range.[21]

Various studies attempted to correlate the genetic variability of *Coxiella burnetii* genomes and plasmids to isolate virulence.[23–30] However, later investigations suggested that predisposing host factors are more important than genetic-strain variation in the occurrence of acute or chronic Q-fever disease in human. Furthermore, observed genetic variations were apparently more closely connected to the geographical source of the isolate than to its clinical expression.[8]

Biology. *Coxiella burnetii* has developed an original strategy for entering its target cells (monocytes and macrophages). This strategy allows the bacteria to escape microbicidal activity and survive in the host cell.

Phase I virulent variants enter monocytes-macrophages through $\alpha v\beta 3$ integrin. They then interfere with the cosignal activity of integrin-associated protein (IAP) and the spatial distribution of Complement Receptor 3 (CR3) molecules, which remain excluded from pseudopodal protrusions implicated in bacterial endocytosis.[31,32] Activation of the IAP-CR3 pathway is critical for *Coxiella burnetii* uptake and microbicidal processes in monocytes-macrophages.[31,32] By contrast, the entrance of avirulent phase II variants is mediated by the $\alpha v\beta 3$-IAP complex, and CR3:CR3 activation leads to rapid phagocytosis and death of the bacterium.[31] The LPS is involved in the uptake of virulent phase I organisms through Toll-like receptor 4 (TLR-4) but not in that of avirulent phase II variants.[33]

After passive entry in monocytes-macrophages, *Coxiella burnetii* SCVs are internalized in phagocytic vesicles or phagosomes. Phagosomes acquire some characteristics of lysosomes, and these phagolysosome-like organelles fuse progressively to form a large unique vacuole.[34] The low pH (pH 4.7–5.2) present in the vacuole leads to the metabolic activation of SCVs and the formation of LCVs, the metabolically active intracellular form of *Coxiella burnetii*. The need for an acidic pH to activate metabolism and multiplication of microorganisms was also described for the amastigotes forms of the protozoan *Leishmania* sp. and for the bacterium *Francisella tularensis*.[9] LCVs divide actively in the vacuole by binary fission. During this multiplication, a sporulation-like process was observed. It is distinguished from the cell-division process because of the asymmetrical position of the septum, containing the dense core, close to one pole of the LCV.[35,36] The endogenous spore-like forms undergo further development to the metabolically inactive SCVs, which are then released from the infected host cell either by cell lysis or exocytosis. The physical or biochemical factors that induce the sporulation-like process in *Coxiella burnetii* are unknown; however the genome of the Nine Mile reference strain contains a sequence of 1741 bp with high homology to the gene *spoIIIE* involved in sporulation of *Bacillus subtilis*.[37] Nevertheless some authors dispute this sporulation-like process, suggesting instead that LCVs condense into SCVs.[9] A schematic representation of the intracellular cycle of *Coxiella burnetii* is shown in Figure 73.1.

The extracellular forms of *Coxiella burnetii* are resistant to harsh environmental conditions, such as desiccation, low or high pH, and pressure; to chemical products such as ammonium chloride, disinfectants (0.5% sodium chloride,

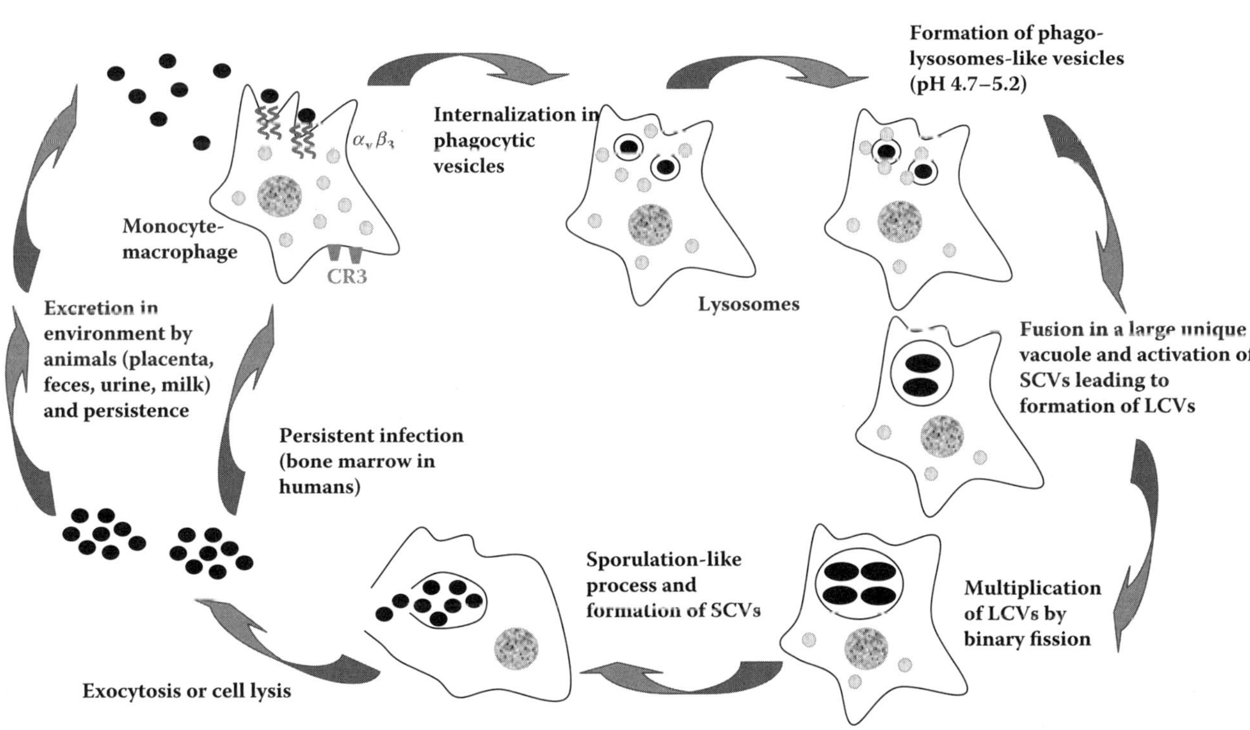

FIGURE 73.1 Schematic representation of the intracellular cycle of multiplication of *Coxiella burnetii* in monocyte-macrophages host cells. Phase I variants (or SCVs) are internalized in human monocytes via the $\alpha_v\beta_3$-IAP complex and survive in phagocytic vesicles. After fusion of phagosomes with lysosomes, the phago-lysosome-like organelles fuse progressively into large unique vacuole. The acid activation of SCVs leads to the formation of LCVs, the metabolically active intracellular form of *Coxiella burnetii*. LCVs multiplicate actively by binary fission and a sporulation-like process may occur, leading to the development of SCVs. The metabolically inactive forms of *Coxiella* are then released from infected cells either by lysis or exocytosis.

5% formol, 1% phenol); and UV radiation.[38] Only long-term exposure to ether, 5% chloroform, 70% ethanol, or 10% formol can destroy SCVs. These variants are also thermostables (they can resist one h at 60°C or 30 min at 63°C) but are inactivated after 10 min at 70°C.

Coxiella burnetii induces persistent infections in animals and humans.[39] It multiplies slowly inside cells, with a doubling time of around 20 h, similar to that of eukaryotic cells, which may partly explain the absence of damage to infected cells during prolonged infection. Roman et al.[40] proposed a model for persistent cell infection in which, when an infected cell divides, one daughter cell receives the unique *Coxiella burnetii* vacuole, whereas the other remains uninfected. The finding that uninfected cells are still present in cell cultures after several months of infection is compatible with this model.[8]

73.1.2 Epidemiology

Geographic Distribution. Q fever is a worldwide zoonotic disease, endemic in almost every country except New Zealand, as shown in a serological screening of sheepdogs and aborting cattle.[41,42] In most countries, Q fever is not included in the list of nationally notifiable diseases and is still rarely investigated and reported. Its epidemiology may only be extrapolated from investigations of human outbreaks, surveys conducted in humans and animals, or data obtained from public health or reference laboratories interested in *Coxiella burnetii*.[8]

Reservoirs. The reservoir for *Coxiella burnetii* is extremely large and includes many wild and domestic mammals, marsupials, arthropods such as ticks, and even fish, reptiles, and amphibians.[41]

Domestic ruminants represent the most frequent source of human *Coxiella burnetii* infections.[41] Infection in cattle, goats, and sheep is usually not harmful but animals are often chronically infected and shed bacteria intermittently in feces (the major route in sheep), urine, semen, vaginal mucus, and more continuously, in milk.[43,44] The main source of *Coxiella burnetii* shed by ruminants is birth products (placenta, birth fluids, fetus). Over 10^9 bacteria per g of placenta are released by infected females at the time of delivery.[41]

Parturient pets, namely dogs and cats, have also been reported as main sources of human infection.[45–47]

As mentioned above, the original Nine Mile reference strain of *Coxiella burnetii* was isolated from the tick species *Dermacentor andersoni*. Since this first description, over 40 tick species were found to be naturally infected with *Coxiella burnetii*, including, for example, *Rhipicephalus sanguineus,* found on dogs, or *Amblyomma triguttatum,* which infests kangaroos.[48,49] *Coxiella burnetii* multiplies in the cells of the middle gut or stomach of infected ticks. Heavy loads of *Coxiella burnetii* are expelled with feces onto the skin of the animal host at the time of feeding. Infection of tick ovaries was also described and may lead to germinative infection of the offspring.[41] Ticks may play a significant role in the transmission of coxiellosis among wild vertebrates, especially rodents, lagomorphs, and birds.[41–50] However, transmission of *Coxiella burnetii* to humans via a tick bite was seldom reported.[51] Thus, Q fever, as opposed to rickettsial diseases, is rarely an arthropod-borne disease.[8]

Coxiella burnetii was less frequently isolated from other domestic or wild mammals including horses, rabbits, swine, camels, water buffalo, rats, and mice.[41,50–52] Birds such as pigeons, chickens, ducks, geese, and turkeys may also be infected.[41] Finally, anti-*Coxiella burnetii* antibodies were found in snakes and tortoises in India, but the bacterium could not be isolated from these reptiles.[41]

Mode of Transmission. Q-fever epidemiology is characterized by two cycles of transmission from either wild or domestic animals. These cycles are illustrated in Figure 73.2.

Ticks are a major source of contamination of wild animals. As mentioned above, during feeding, ticks release *Coxiella burnetii* onto the skin of the host and infection can also take place during the transient bacteremia that occurs early after the tick attaches to the animal. Infected animals release *Coxiella burnetii* into the environment.

Transmission of coxiellosis among domestic animals mainly occurs by inhalation of contaminated aerosols. *Coxiella burnetii* aerosols may form directly from the parturient fluids of infected animals or be spread by the wind following environmental contamination.[53,54] *Coxiella burnetii* SCVs may survive for several weeks in areas where infected animals have been present or after spreading of contaminated manure. Furthermore, amoeba were shown to provide a potential protective niche for *Coxiella burnetii*, as demonstrated for *Legionella* species.[55,56]

Humans are "a dead end" in the epidemiological cycle of transmission of *Coxiella burnetii*. The primary route of human contamination is also the inhalation of contaminated aerosols. Ingestion (drinking unpasteurized milk, eating raw eggs) is probably a minor factor in the transmission of *Coxiella burnetii* and is a point of controversy.[57,58] Person-to-person transmission is also extremely rare. Sporadic Q-fever cases have occurred following contact with an infected parturient woman, via intradermal inoculation, blood transfusion, or transplacental transmission, which results in congenital infections.[8] Sexual transmission was demonstrated in infected mice.[59] However, this mode of transmission remains to be established in humans. Finally, as mentioned above, arthropod-borne transmission of Q fever is unlikely to be significant in humans.[8]

Factors Influencing the Incidence of Q Fever in Humans. In the south of France, the incidence of Q fever was noticed to vary seasonally, with acute cases more frequently reported in spring and early summer.[60] This is thought to be related to the outside lambing season, which leads to heavy contamination of the environment. In autumn, the majority of lambings occur inside and are not related to a higher incidence.[8]

Q fever is usually an occupational hazard. Persons at greatest risk are those in contact with livestock (farmers, abattoir workers, veterinarians). Laboratory personnel are also at risk of Q-fever infection when manipulating animal or human products containing Phase I variants.[8] Patients at risk of developing chronic Q fever are those with a previous

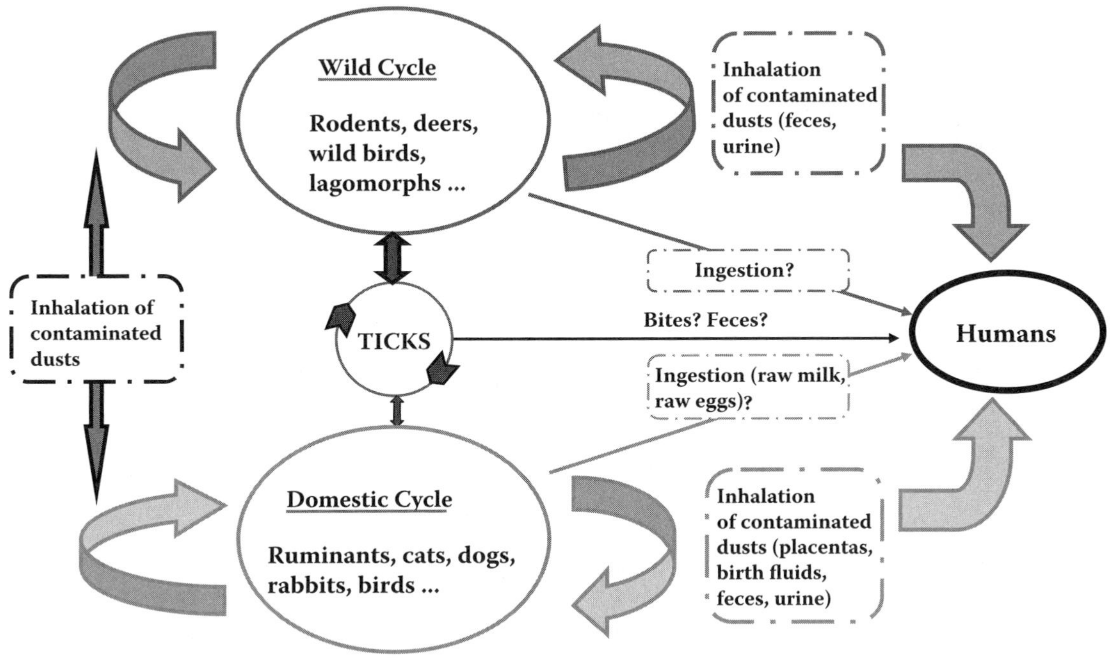

FIGURE 73.2 Cycles of transmission of *Coxiella burnetii* between domestic and wild animals and humans. Ticks represent the major source of contamination of wild animals, while transmission among domestic animals occurs by inhalation of contaminated aerosols. Humans are also mainly contaminated through inhalation and represent "a dead-end" in the epidemiological cycle of *Coxiella*.

valvulopathy, immunocompromised patients with cancer or AIDS, and pregnant women.[1,61–63]

The age and sex ratio of infection depends on the most predominant animal reservoir that is responsible for the contamination. The active population of men from 30 to 60 years old are more frequently contaminated by infected livestock, but a sex ratio of 1:1 is observed when parturient pets are at the origin of transmission.[8,64] Thus men and women are at equal risk of *Coxiella burnetii* contamination, although Q fever is more severe in men than in women.[65]

Biological Warfare and Significance for Bioterrorism. During the cold war, *Coxiella burnetii* was described as a potential biological weapon and was recently designated a category B bioterrorism agent, as an incapacitating organism.[66] Although it has a low case fatality rate, it meets criteria, such as stability in the environment, ability to become airborne and inhaled, and faculty of one single organism to cause disease in humans. This potential as a list B biothreat has revitalized research interest in *Coxiella burnetii.*

73.1.3 Pathogenesis and Pathology

Humans. Q fever is most commonly acquired by breathing infectious aerosols or dust contaminated with birth fluids or feces of infected domestic or wild animals.[41] Thus, alveolar macrophages are often the initial target host cell. The infectious dose for humans is estimated to be 10 microorganisms or even fewer.[67] In most infected patients, a transient bacteremia is observed, usually late in the incubation period which ranges from 1 to 3 weeks, depending on the inoculating dose.[68] Circulating monocytes provide a route for hematogenous spread of the pathogen via blood and the lymph system. Monocytes also allow its distribution in the macrophages of various organs including lungs, liver, spleen, bone marrow, and the female genital tract.[8] The majority of infections remain asymptomatic, but approximately 40% lead to mild to moderate flu-like symptoms with pneumonia and/or hepatitis.[69] Only 2% of patients need to be hospitalized. Approximately 10% of people convalescing from acute Q fever develop a fatigue syndrome lasting for longer than 6 months.[70] Patients with a previous valvulopathy, iatrogenic or other immunosuppression, or pregnant women may develop a chronic and often spontaneously fatal form of the disease.[8] Endocarditis is the most frequent clinical manifestation of chronic Q fever.

The pathological findings in acute pneumonia associated with Q fever are not specific to *Coxiella burnetii* infection: interstitial infiltrates of macrophages; lymphocytes and; to a lesser extent, polymorphonuclear leukocytes; alveolar exudates containing fibrin, erythrocytes; and mononuclear cells.[71,72]

Granulomatous hepatitis represents the most frequent indication of *Coxiella burnetii* infection of the liver. Portal triaditis, Kupffer-cell hyperplasia, and moderate fatty change are observed. Kupffer cells, which are the *Coxiella burnetii* target cell in the liver, may initiate local inflammation and the formation of granulomas.[8] Granulomas, which are often characterized by a central clear space and a fibrin ring, are referred to as "doughnut granuloma."[73] In chronic Q fever, the predominant histological observation is lymphocytic infiltration with spotty necrosis foci.

Bone marrow lesions usually correspond to doughnut granulomas similar to those found in the liver.[74]

Q-fever endocarditis involves the aortic, mitral, or prosthetic valves.[8] Histological findings in the infected cardiac valve are not specific. These include thrombi composed of fibrin and platelets, necrosis, and calcification foci.[75] Inflammatory cell infiltrates are mainly composed of lymphocytes and histiocytes, and foamy macrophages are frequently observed. The cardiac vegetation may cause arterial emboli and infarcts, especially in the spleen, kidneys and brain. Q-fever endocarditis may also induce circulating immune complex-related complications, like glomerulonephritis.[8]

The acute form of Q fever appears to stimulate an efficient immune response that eventually limits bacterial replication but fails, in many cases, to completely clear the bacteria. Thus, Marnion et al.[70] showed that *Coxiella burnetii* are still detected in the bone marrow of patients up to 12 years after infection. *Coxiella burnetii* evades and suppresses the host cell–mediated immune responses to survive the hostile environment of the pagolysosome-like vacuole and establish a persistent infection.[8,69]

One strategy to escape killing by the active oxygen and nitrogen intermediates produced during innate immune responses to infection is to avoid host cell activation.[69] Ghigo et al.[76] demonstrated that *Coxiella burnetii* survival in the THP-1 monocyte cell line involves impairment of phagosome maturation. *Coxiella burnetii* vacuoles acquire markers of late endosomes and late endosomes-early lysosomes but not the lysosomal hydrolase cathepsin D. Furthermore, the phase I LPS may interfere with dendritic cell and macrophage activation by masking ligands to Toll-like receptors (TLRs) and notably to TLR-2.[77,78] TLR-mediated recognition of microbial determinants is also critical for the further establishment of an acquired immune response that can contain and clear bacterial infections.[78]

The peripheral blood lymphocytes from patients suffering from endocarditis could no longer proliferate after in vitro exposure to *Coxiella burnetii* antigens, demonstrating immune suppression during chronic Q fever.[79] The suppression mechanism appears to involve the production of a high level of the prostaglandin E2 and TGFβ by peripheral blood mononuclear cells.[80] Patients with chronic Q fever also exhibited increased IL-10 secretion, leading to suppressed Th1-mediated cellular immunity and down-modulation of Th1 lymphokines (e.g., IL-2 and IFNγ).[80,81]

Naturally Infected Animals. During the acute phase of *Coxiella burnetii* infection, the presence of the bacterium was demonstrated in the blood, lungs, spleen, and liver, but most infected animals remain totally asymptomatic. *Coxiella burnetii* infection often becomes chronic. For example, in sheep, the agent persists in the liver, spleen, kidneys, lymph nodes, gut, and bone marrow, leading to persistent shedding in feces and urine.[8,9] During gestation, multiplication of *Coxiella burnetii* is reactivated, and a high level of bacteria is observed in the placenta and mammary glands.[41] The only pathological manifestations associated with chronic infection in animals are infertility, lower birth weight (mainly in cattle), and late endemic abortions (mainly in sheep and goat).[8,9] During subsequent gestations, infected females remain excretory, but in general no further abortions are noticed.[9] Some cases of pneumonia, metritis, conjunctivitis or arthritis were also described.[8,9]

73.1.4 Clinical Findings

Acute Q Fever. The most frequent clinical presentation of acute Q fever is prolonged fever, pneumonia, or hepatitis. These clinical expressions appear to vary depending on the geographical area of detection and the route of infection, but these three main symptoms can also coexist.[8]

Prolonged fever may reach 39°C–40°C and usually remains elevated all day. After 5–14 days, the temperature returns to normal. Twenty-five percent of acute Q fever patients experience a second phase of fever, usually at lower temperatures, and intermit for up to 19 days.[82]

Pneumonia is associated with fever, fatigue, chills, retro-orbital headaches, myalgia, and sweats.[83] Cough was often recorded. Nausea, vomiting, chest pain, diarrhea, and sore throat were reported less frequently.[8]

Q fever hepatitis is usually only revealed by an increase in hepatic enzyme levels such as alkaline-phosphatase, AST, and ALT (increase from two to three times).[84]

Besides these three major clinical findings, myocarditis, pericarditis, meningo-encephalitis, and skin rash have also been described.[8]

Chronic Q Fever. Chronic Q fever is characterized by clinical expression lasting more than six months and the presence of IgG or IgA directed against the phase I form of *Coxiella burnetii*.[8]

Endocarditis represents the major clinical presentation of chronic Q fever (60%–70% of all chronic Q fever cases). Endocarditis appears almost exclusively in patients with previous cardiac valve defects (over 90% of patients) but may also be related to chronic inflammation and cardiotoxic drugs administrated to organ transplant recipients and to patients suffering from cancer, chronic renal insufficiency, or AIDS. The male/female ratio is 75%, and most patients are older than 40 years.[8] Mortality is less than 10% when appropriate antibiotic therapy is administered, but Q-fever endocarditis can be spontaneously fatal when untreated.[8]

The clinical manifestations leading to diagnosis of Q fever endocarditis are not specific: chronic renal insufficiency, stroke, chronic hepatomegaly or splenomegaly, chronic eruptions, chronic elevation of liver enzyme levels (ALT, AST), necessity for early valve replacement after surgery, postoperative fever, and unexplained chronic inflammatory syndrome.[85] Thus Q-fever diagnosis is often established after several months or even years of evolution.[86] However, a delay in diagnosing Q-fever endocarditis does have a significant effect on the prognosis and may lead to complications. It explains the importance of recognizing Q-fever endocarditis as soon as possible.[87] Duke's criteria are used worldwide for calculating a diagnostic score for infective endocarditis.[88] These criteria are divided into major criteria (such as isolation of a typical microorganism from two separate blood cultures or cardiac vegetation on echocardiograms) and minor criteria (including fever, previous valvulopathy, vascular, or immunologic phenomena). Clinical criteria for infective endocarditis require the presence of two major criteria, or one major and three minor criteria, or five minor criteria. In the diagnosis of Q-fever endocarditis, isolation of *Coxiella burnetii* from blood and/or the presence of an antiphase I IgG minimal titer of 1 in 800 are also considered as major criteria.[89]

Other clinical manifestations of chronic Q fever are vascular infections, aneurysms or vascular grafts; osteoarticular infections (osteomyelitis, osteoarthritis) which are found more frequently in children or in immunocompromised adults with a joint prosthesis; chronic hepatitis; or pneumonia.[8]

Chronic Fatigue Syndrome. Chronic fatigue syndrome has been reported frequently in patients convalescing from acute Q-fever disease. Patients complained of fatigue, sweating, arthralgia, myalgia, muscle fasciculation, breathless on exertion, and blurred vision.[90] These symptoms can persist for many years in some patients.

Post–Q fever chronic syndrome should, and can, be distinguished from chronic Q fever by demonstrating an absence or low levels of antiphase I antibodies in the former patients.[8]

Q Fever During Pregnancy. By analogy with the pathogenesis observed in infected animals during gestation, and given the particular immune status associated with pregnancy, pregnant women are highly at risk of developing a chronic form of Q fever.

Q fever cases during pregnancy are rarely reported. Clinical findings are premature birth, spontaneous abortion, or death of fetus in utero.[91] *Coxiella burnetii* was recovered from placental and/or fetal tissues. Immune complexes generated during Q fever disease may lead to vasculitis or vascular thrombosis and, finally, to placental insufficiency.[8]

Pregnant women should thus avoid any contact with livestock and pets, especially females at the time of delivery.[8] Furthermore, Q fever should be added to the etiological agents of intrauterine infections associated with morbidity and mortality during pregnancy (TORCH) and investigated in pregnant women suffering from fever associated with pneumonia, hepatitis, and severe thrombocytopenia.[91]

73.1.5 Diagnosis

Coxiella burnetii is a highly infectious organism, mainly via inhalation, and laboratory workers have acquired cases of Q fever. Thus, only experienced personnel, working in biosafety level 3 laboratories, should be allowed to manipulate clinical samples and *Coxiella burnetii*–infected cell cultures or animals.

73.1.5.1 Phenotypic and Serological Techniques

Immunohistology. Immunohistology is especially informative for diagnosing Q-fever endocarditis and also during chronic hepatitis or vascular-graft infection.[8,92,93] Tissue biopsy samples are tested either fresh or after formalin fixation and paraffin embedding.[94] Immunodetection may be performed using the immunoperoxidase technique, capture enzyme-linked immunosorbent assay system or immunofluorescence with polyclonal or monoclonal antibodies.[94–96] Recently, an autoimmunohistochemistry method (immunohistochemical peroxidase-based method using patient's own serum) was developed to allow diagnosis of infective bacterial endocarditis in every laboratory because commercial antibodies are not available.[97]

On valve specimens, immunohistochemical techniques showed voluminous intracytoplasmic masses within infected mononuclear cells. Extracellular staining, as seen for most other etiological agents of infective endocarditis, is absent.[95]

Culture. Culturing of *Coxiella burnetii* as an isolation method should be performed in very few laboratories because of the risk of transmission (biosafety level 3 laboratories required).

Coxiella burnetii is a strict intracellular bacterium and therefore cannot be cultured for diagnostic purposes in axenic medium, even if complex *Coxiella* axenic medium was recently described to allow host cell–free growth.[98,99]

Coxiella burnetii has been successfully isolated in laboratory animals (mice, guinea pigs) or embryonated eggs. However, these techniques have been abandoned because they are more hazardous than in vitro cultures, and there is a risk of cross-contamination between infected and noninfected animals.[8]

Various cell lines can be used for in vitro cultures: mouse macrophage-like cells, fibroblasts, monkey kidney cells, or Vero cells.[8] Raoult and colleagues have developed a sensitive miniaturized shell-vial cell culture assay on human embryonic lung fibroblasts (HEL cell line).[100] Several human samples are suitable: heparinized blood, cardiac valves, lymph nodes, skin, lung or liver biopsies, vascular aneurysms, or grafts.[8–101] After 10–20 days of incubation, *Coxiella burnetii* is usually observed by microscopic examination with Gimenez or immunofluorescence staining.[101]

Serology. In general, Q-fever diagnosis relies on serological methods, which are quite easy to perform and, in some cases, commercially available. Antibodies are mostly detected from 2 to 3 weeks after the onset of infection. Depending on the target and the type of antibodies, serology allows the differentiation between acute and chronic Q fever.[8] However, Q fever is endemic to many countries, and therefore patients may have a positive serological titer unrelated to their current medical problems. For these reasons, demonstration of seroconversion or detection of IgM antibodies is particularly important to assess Q-fever diagnosis.

The immunofluorescence assay (IFA) is considered the reference method for the serodiagnosis of Q fever. It is the simplest, as well as one of the most accurate and commercially available serological techniques (e.g., *Coxiella burnetii* IFA slide from Vircell, Spain). This method can be used to determine antibodies to phases I and II in IgA, IgG, and IgM subclasses. Raoult and colleagues have proposed the following strategy for diagnosis of Q fever by microimmunofluorescence.[102] Sera are first screened at a 1:50 dilution with phase II antigens. Then, positive sera are serially diluted and tested with both phase II and phase I antigens for the presence of IgA, IgG, and IgM.[102] During acute Q fever, antibodies are detected by the third week after the onset of clinical symptoms in about 90% of cases. An IgG antiphase II titer of 1 in 200 and an IgM antiphase II titer of 1 in 50 are considered significant.[102] Chronic Q fever is characterized by the presence of antiphase I antibodies, and an IgG antiphase I titer of 1 in 800 is considered to be predictive of Q-fever endocarditis.[102]

However, the choice of negative cutoff titers may vary from one area to another depending on the amount of background antigen stimulation in the population studied.[64] Furthermore, manual IFA often lacks reproducibility, with results depending on the manipulator. Serological cross-reactions were also observed between *Coxiella burnetii* and other bacteria, such as *Bartonella quintana*, *Bartonella henselae,* or *Legionella micdadei*.[103,104] Nevertheless, the very high titers of specific antibodies against *Coxiella burnetii* observed during Q fever compared with the low titers of cross-reacting antibodies, mean that interpretation of results is usually straightforward.[8]

The multiplex protein microarray technique, a derivative of IFA, was recently developed for the serodiagnosis of blood culture–negative endocarditis and is commercially available (InoDiag, France).[105] This technique is rapid (1 h compared to 3 h for IFA) and the process is totally automated, from serum incubation to reading. Thus, interpretation is no longer "manipulator dependent."[105]

Enzyme-linked immunosorbent assay (ELISA) tests have been developed for epidemiologic screening and diagnosis of human Q fever. ELISA is a sensitive method. Both antiphase I and antiphase II IgG and IgM antibodies can be detected and commercial kits are available (PanBio, Brisbane, Australia; Vircell, Spain). However, cutoff values for ELISA tests may vary according to the method or the manufacturer and may not be fully adapted to the population studied. Although some studies have been reported, more comparative trials are needed to verify the performance of these assays.[106,107]

The complement fixation test was extensively used in the past. Sera were tested after heat inactivation of complement factors against either phase I or phase II antigens. The complement fixation technique is considered as mostly specific but is less sensitive than IFA or ELISA. False-negative results were obtained for chronic Q fever (prozone phenomenon) and false-positive results occurred due to cross-reactions with hen egg antigens. Furthermore, this technique is more time consuming than ELISA or IFA.[94]

Other techniques proposed for the serodiagnosis of Q fever include Western blotting, dot blotting, microagglutination, the indirect hemolysis test, or radio-immuno assays.[94]

73.1.5.2 Molecular Techniques

Diagnosis in Humans. Initially, radiolabeled DNA probes, targeting *Coxiella burnetii* 16S rRNA or plasmid sequences, were used to hybridize to nucleic acid in clinical samples such as blood, urine, or organs. Southern blotting in conjunction with polymerase chain reaction (PCR) was also developed; leading to improved detection.[108] These methods are sensitive and specific but could only be performed in specialized research laboratories.[94]

The development of *Coxiella burnetii* DNA amplification by PCR, firstly conventional or nested PCR and more recently real-time PCR, allowed "democratization" of the molecular diagnosis of Q fever.[109–112] Various PCR target sequences were proposed on single chromosomal genes (such as *com1*, *htpB*), plasmids (QpH1, QpRS), or the transposase gene insertion element *IS111* (see Table 73.1). Due to the multicopy number of the *IS1111* element, the corresponding PCR is very sensitive. However, a recent study showed that the number of *IS1111* elements varies considerably per genome depending on isolates (between 7 and 110 copies), so *Coxiella burnetii* cannot be quantified using this target.[113]

PCR can be performed for early diagnosis of acute Q fever in sera, during the first 4 weeks of the disease. It can also be carried out on target tissue (heart valves, placenta) during chronic Q fever or on patient sera during evolutive Q fever.[110–112,114,115] In addition, the recent identification of an outer membrane protein specific to *Coxiella burnetii* isolates associated with acute disease (AdaA) suggest that PCR targeting of the *adaA* gene will differentiate acute and chronic disease.[115] Finally, the antibiotic susceptibility of *Coxiella burnetii* isolates is assessed using real-time PCR.[116,117]

Diagnosis in Animal Hosts. Humans are especially prone to acquire Q fever from domestic ruminants, thus to control the spread of infection, reliable detection of *Coxiella burnetii* shedders is critical.[119] As in humans, the diagnosis often relies on serology involving a number of tests such as indirect IFA, ELISA, or a complement fixation test. Surprisingly, the later method remains a routine diagnostic assay for livestock, despite its lack of detectability. In general, currently commercially available tests use *Coxiella burnetii* phase II antigens isolated from the Nine Mile reference strain.[119] However, serological tests are difficult to interpret because of the need for kinetics at the individual level and of the fact that they only determine that *Coxiella burnetii* is probably circulating throughout the herd. Furthermore, these tests do not identify with certainty which animals are excreting bacteria: some seropositive animals do not shed bacteria, while some seronegative animals will be major shedders.[9]

The identification of the presence of the bacterium remains the only way to diagnose with certainty an infection to *Coxiella burnetti*. Samples are collected from aborted fetuses (lung, liver, or stomach), placenta, vaginal discharges, milk, colostrum and feces.[119] Staining, capture enzyme-linked immunosorbent assays, and immunochemistry can be used in case of an abortion. But these methods are still either both nonspecific and insensitive (staining), or not commercially available (immunohistochemistry).[119] PCR methods are currently used in veterinary diagnostic laboratories.[118–120] They can be performed on a large diversity of samples, and commercial kits are also available (AES-Chemunex, France; IDEXX, The Netherlands; LSI, France).

73.2 METHODS

73.2.1 Samples Preparation

Type of Samples and Shipping's Recommendations. (i) Blood (EDTA tube) or serum, at room temperature if shipped within 48 h; refrigerated at 5°C ± 3°C if shipped over 48 h; (ii) vaginal dry swabs, at room temperature if

TABLE 73.1
Primers and Probe Available for PCR Amplification of *Coxiella burnetii* Genes and Sequences

Target Gene or Sequence	Primers (Forward and Reverse) and Probe (P) Sequences
16S rRNA[94]	16S1 (5′-CTC CTG GCG GCG AGA GTG GC-3′)
	16S2 (5′-GTT AGC TTC GCT ACT AAG AAG GGA ACT TCC C-3′)
23S rRNA[94]	976F (5′-AGG TCC TGG TGG AAA GGA ACG-3′)
	1446R (5′-TCT CAT CTG CCG AAC CCA TTG C-3′)
16S-23S rRNA internal transcribed spacer[94]	16SF (5′-TTG TAC ACA CCG CCC GTC A-3′)
	23SR (5′-GGG T(CGT) CCC CAT TCG G-3′)
	16SS (5′-GAA GTC GTA ACA AGG TA-3′)
	23SS (5′-TCT CGA TGC CAA GGC ATC CAC C-3′)
Superoxide dismutase *sod*[109]	CB1 (5′-ACT CAA CGC ACT GGA ACG GC-3′)
	CB2 (5′-TAG CTG AAG CCA ATT CGC C-3′)
Plasmid QpRS[94]	QpRS01 (5′-CTC GTA CCC AAA GAC TAT GAA TAT ATC-3′)
	QpRS02 (5′-CAC ATT TTT CTT CCT CGA ATC TAT GAA T-3′)
Plasmid QpH1[94]	QpH11 (5′-TGA CAA ATA GAA TTT CTT CAT TTT GAT-3′)
	QpH12 (5′-GCT TAT TTT CTT CCT CGA ATC TAT GAA T-3′)
CbbE[94]	G4131 (5′-CTG ATG TGT CAA GTA ATG TCG G-3′)
	G4132 (5′-CTT CAT GGT TAT GAT TCT GCG-3′)
Com1[116]	FAF216 (5′-GCA CTA TTT TTA GCC GGA ACC TT-3′)
	RAF290 (5′-TTG AGG AGA AAA ACT GGA TTG AGA-3′)
DotA	dotA-F (5′-GCG CAA TAC GCT CAA TCA CA-3′)
	dotA-R (5′-CCA TGG CCC CAA TTC TCT T-3′)
	dotA-P (5′-6FAM-CCG GAG ATA CCG GCG GTG GG-TAMRA-3′)
IS30a	Cbis30aF (5′-AAT GTC TGC GGG AAA TAG GC-3′)
	Cbis30aR (5′-GAG GCC TTT TAC CGG AAT TC-3′)
	Cbis30aP (5′-6FAM-TCG AGA TCA TAG CGT CAT T-3′)
Icd[111]	Icd439F (5′-CGT TAT TTT ACG GGT GTG CCA-3′)
	Icd514R (5′-CAG AAT TTT CGC GGA AAA TCA-3′)
	Icd464TM (5′-6FAM-CAT ATT CAC CTT TTC AGG CGT TTT GAC CGT -TAMRA-3′)
IS-1111[94,112,114,118–120]	Trans-1 (5′-TAT GTA TCC ACC GTA GCC AGT C-3′)
	Trans-2 (5′-CCC AAC AAC ACC TCC TTA TTC-3′)
	Trans-3 (5′-CAA CTG TGT GGA ATT GAT GAG-3′)
	Trans-5 (5′-GCG CCA TGA ATC AAT AAC GT 3′)
	IS1111F (5′-GCG TCA TAA TGC GCC AAC ATA-3′)
	IS1111R (5′-CGC AGC CCA CCT TAA GAC TG-3′)
	IS1111P (5′-6FAM-TGC TCA GTA TGT ATC CAC CG-TAMRA-3′)
	CoxF (5′-GTC TTA AGG TGG GCT GCG TG-3′)
	CoxR (5′-CCC CGA ATC TCA TTG ATC AGC-3′)
	CoxTM (5′-AGC GAA CCA TTG GTA TCG GAC GTT TAT GG-TAMRA-3′)
	CoxbS (5′-GAT AGC CCG ATA AGC ATC AAC-3′)
	CoxbAs (5′-GCA TTC GTA TAT CCG GCA TC-3′)
	CoxbMGB (5′-6FAM -TCA TCA AGG CAC CAA T-MGB-3′)
	FCox (5′-AAC GGC GCT CTC GGT TTA T-3′)
	RCox (5′-GCG GAA CGA TGA AAT TGA TGA-3′)
	PCox (5′-6FAM-CGA GCG TGG GTG AC-MGB-3′)

shipped within 24 h or frozen below –20°C if shipped over 24 h; (iii) organs (biopsy of cardiac valves, liver, lungs), milk, and feces: refrigerated at 5°C ± 3°C if shipped within 24 h; frozen at –20°C or –80°C in dry ice over 24 h; (iv) placenta, at room temperature or frozen at –20°C.

Storage before Analysis. (i) Blood or serum, refrigerated at 5°C ± 3°C; (ii) vaginal dry swabs, at room temperature if DNA extraction is processed at reception, or, frozen below –20°C; (iii) organs (biopsy of cardiac valves, liver, lungs), milk and feces: frozen at reception below –20°C or –80°C; (iv) placenta: frozen at reception below –20°C.

DNA Extraction. For better reproducibility of DNA extraction yield, using commercial kits (silica columns or magnetic beads) is recommended. For safety reasons, it is mandatory to perform the steps preceding the final lysis and/or inactivation of *Coxiella burnetii* in biosafety level 2 laboratories, and doing open samples and microcentrifuge tubes under class II biological safety cabinets is mandatory.

The kits mentioned are cited as an examples; any equivalent product may be used unless validated.

(i) Blood or Serum with QIAamp DNA Blood Mini-Kit (Qiagen)

1. Pipet 20 μL proteinase K into the bottom of a 1.5-mL microcentrifuge tube.
2. Add 200 μL of whole blood or serum.
3. Follow manufacturer's spin protocol for blood and body fluids.
4. Elute in 50 μL of elution buffer.

(ii) Vaginal Dry Swabs with QIAamp DNA Mini-Kit (Qiagen)

1. Place vaginal swab in a 2-mL microcentrifuge tube, separate cotton from stick with scissors, add 200 μL of sterile PBS.
2. Inactivate *Coxiella burnetii* by incubation at 90°C for 10 min.
3. Incubate at 5°C ± 3°C for 5–10 min to cool before adding of proteinase.
4. Add 200 μL of lysis buffer ATL and 20 μL of proteinase K.
5. Incubate at 56°C for 30 min.
6. Add 400 μL of lysis buffer AL.
7. Incubate at 70°C for 10 min.
8. Add 400 μL of ethanol (96%–100%) and follow manufacturer's instruction for buccal swab spin protocol.

(iii) Organs (Biopsy of Cardiac Valves, Liver, Lungs) with QIAamp DNA Mini-Kit (Qiagen) or FastDNA Kit (Bio 101)

1. Depending on the commercial kit used, choose appropriate sample size and place the fragment in a microcentrifuge tube.
2. Inactivate *Coxiella burnetii* by incubation at 90°C for 10 min.
3. Incubate at 5°C ± 3°C for 5–10 min to cool before adding of proteinase.
4. Follow DNA extraction as described in manufacturer's protocol.

(iv) Milk with QIAamp DNA Mini-Kit (Qiagen)

1. Pipet 200 μL of milk in a 1.5-mL microcentrifuge tube.
2. Add 180 μL of lysis buffer (ATL) and 20 μL of proteinase K.
3. Incubate at 70°C for 1 h to perform inactivation and lysis of *Coxiella burnetii*.
4. Add 200 μL of AL buffer and incubate at 70°C for 10 min.
5. Follow DNA extraction as described in manufacturer's protocol.

(v) Feces with QIAamp DNA Stool Mini-Kit (Qiagen)

1. Mix 1 g of feces with 4 mL of DNase- and RNase-free water by vortexing for 30 s.
2. Pipet 400 μL in a 1.5-mL microcentrifuge tube and centrifuge at 6000 × *g* for 1 min.
3. Pipet 200 μL of supernatant into a 2-mL microcentrifuge tube.
4. Incubate at 90°C for 10 min to inactivate *Coxiella burnetii*.
5. Incubate at 5°C ± 3°C for 5–10 min to cool before adding of proteinase.
6. Follow the manufacturer's instructions.

(vi) Placenta with QIAamp DNA Mini-Kit (Qiagen)

1. Place the frozen placenta overnight at 5°C ± 3°C to allow thawing.
2. Pipet 200 μL of exsudate of thawing in a 1.5-mL microcentrifuge tube.
3. Incubate at 90°C for 10 min to inactivate *Coxiella burnetii*.
4. Incubate at 5°C ± 3°C for 5–10 min to allow cooling.
5. Add 200 μL of AL buffer and 20 μL of proteinase K.
6. Follow manufacturer's instructions for blood and body fluid spin protocol.
7. Elute in a final volume of 200 μL of elution buffer.

Storage of DNA Extracts before Analysis. If DNA extracts are processed for PCR within 8 h following DNA extraction, they can be stored at 5°C ± 3°C. For long-term storage, congelation below –20°C is recommended.

73.2.2 Detection Procedures

73.2.2.1 PCR Diagnosis of Q Fever

(i) Standard PCR Stein et al.[109] described a PCR for the diagnosis of Q fever, with primers from the superoxide dismutase (*sod*) gene: CB1 5′-ACT CAA CGC ACT GGA ACC GC-3′; CB2 5′-TAG CTG AAG CCA ATT CGC C-3′, resulting in a specific fragment of 257 bp.

Procedure

1. Prepare PCR mix (100 μL) containing 10 μL of DNA extracts, 1 μM of each primer, 200 μM of each dNTP, 50 mM of KCl, 10 mM Tris-HCl (pH 8.3), 2 mM $MgCl_2$, 0.01% gelatine, and 2.5 U of Taq DNA polymerase (Promega).
2. Subject the mix to 30 cycles of 95°C for 20 s, 50° C for 1 min, and 72°C for 1 min.
3. Electrophorese the product (257 bp) on 2% agarose gels, stain with ethidium bromide, and visualize under UV illumination.
4. Digest PCR products separately with AluI and TaqI endonucleases, separate the fragments on 20% polyacrylamide gels, stain with ethidium bromide, and view under UV illumination.

(ii) Real-Time Multiplex PCR Panning et al.[119] reported a real-time multiplex PCR for the diagnosis of Q fever with

primers and TaqMan®MGB probe for *Coxiella burnetii* detection targeting *IS1111* sequence: CoxbS 5′-GAT AGC CCG ATA AGC ATC AAC-3′; CoxbAs 5′-GCA TTC GTA TAT CCG GCA TC-3′; CoxbMGB 6FAM 5′-TCA TCA AGG CAC CAA T-3′. The assay includes an internal control plasmid, which is generated from the 86 bp amplicon of the *IS1111* sequence by introducing an alternative probe-binding site. The internal plasmid control is added during DNA extraction, in the lysis buffer. The corresponding 5′ nuclease probe is: prCoxmutant VIC5′-ATC GTT CGT TGA GCG ATT AGC AGT T-3′.

Procedure

1. Prepare PCR mix (25 μL) containing 5 μL of DNA extracts, 4 mM $MgCl_2$, 1× Platinum Taq polymerase reaction buffer (Invitrogen), 200 μM of each dNTP, 0.8 μM of each primer, 0.3 μM of CoxbMGB, 0.2 μM of prCoxmutant.
2. Amplify in ABI Prism 7000 (Applied Biosystems) with 1 cycle of 95°C for 2 min; 45 cycles of 95°C for 15 s, and 60°C for 30 s.
3. Analyze data with sequence detector software V2.1 (Applied Biosystems).

73.2.2.2 PCR Diagnosis of Ruminants' Abortions

(i) Standard PCR Berri et al.[120] developed a PCR assay for diagnosis of ruminants' abortions, with primers from the repetitive *IS1111* element: Trans-1 5′-TAT GTA TCC ACC GTA GCC AGT-3′ and Trans-2 5′-CCC AAC AAC ACC TCC TTA TTC-3′, resulting in a specific fragment of 687 bp.

Procedure

1. Prepare PCR mix (25 μL) containing 1× PCR buffer (Promega), 0.5 μM of each primer, 200 μM of each dNTP, 200 mM $MgCl_2$ and 0.5 U of Taq polymerase (Promega, France), and 5 μL of DNA extract; include positive DNA control and negative extraction control in each run.
2. Amplify in an automated thermal cycler (Eppendorf Mastercycler) with 1 cycle of 94°C for 10 min; 35 cycles of 94°C for 30 s, 64°C for 1 min, 72°C for 1 min, and a final 72°C for 10 min.
3. Analyze the PCR products on 1% agarose gel electrophoresis, stain with ethidium bromide, and visualize with ultraviolet transillumination.

(ii) Real-Time Multiplex PCR Pelletier et al.[118] described a real-time multiplex PCR for the diagnosis of ruminants' abortion. PCR primers and TaqMan®MGB probes were designed on the *IS1111* sequence for the detection of *Coxiella burnetii* and on *GAPDH* gene for the Internal Positive Control: FCox 5′-AAC GGC GCT CTC GGT TTA T-3′; RCox 5′-GCG GAA CGA TGA AAT TGA TGA-3′; PCox 6FAM 5′-CGA GCG TGG GTG AC-3′; FGAPDH 5′-AAG GCC ATC ACC ATC TTC CA-3′; RGAPDH 5′-CAG CAT CAC CCC ACT TGA TGT-3′; PGAPDH VIC 5′-AGC GAG ATC CTG CC-3′.

Procedure

1. Prepare PCR mix (25 μL) containing 1× QuantiTect Multiplex PCR Master Mix (Qiagen), 0.4 μM of each primer, 0.2 μM of each TaqMan®MGB probe, and 5 μL of DNAs extracts; include positive DNA control and negative extraction control in each run.
2. Amplify on ABI PRISM 7000 Real-Time Thermocycler (Applied Biosystems), with initial activation heating step for the HotStarTaq DNA polymerase at 95°C for 15 min; two-step cycling of denaturation at 94°C for 60 s, and annealing-extension-fluorescence data collection at 60°C for 60 s, repeated 40 times.
3. Analyze data with sequence detector software V2.1 (Applied Biosystems); determine baseline automatically, and select threshold values manually for each probe.

73.3 CONCLUSION AND FUTURE PERSPECTIVES

Q fever is a worldwide zoonotic disease, but it remains underdiagnosed, and many aspects of the disease are still unknown. Interest in research into *Coxiella burnetii* has been recently renewed following its classification as a category B bioterrorism agent. In this context, the development of methods for rapid and reliable identification of the microorganism in the environment, food, excretory animals, and infected people is necessary. Implementation of these methods should also be easy for sentinel public health and veterinary laboratories.

Currently, diagnosis of *Coxiella burnetii* infections in humans and animals relies mainly on serology. Serological methods are rapid, easy to perform, and commercial tests are available. However, some serological methods are not reproducible (IFA readings, for example, are manipulator dependent) and the cutoff levels of diagnostic thresholds must be adapted to the population studied. Some surveys found unacceptably low sensitivity using recommended diagnostic thresholds, while others used elevated thresholds to overcome specificity issues. In addition, although no comprehensive comparative study of commercially available diagnostics has yet been published, a poor correlation was found between assays that relied on antigens against different *Coxiella burnetii* strains. Furthermore, serological methods remain late diagnostic techniques (only 2–3 weeks after infection in acute Q fever) and especially in animals, serological titers often do not correlate with shedding and are a poor indicator of disease activity. Finally, these techniques cannot be used to monitor the environment and food for a risk of *Coxiella burnetii* contamination.

On the other hand, direct recovery of *Coxiella burnetii* in cultures is time consuming (e.g., 10–20 days for the miniaturized shell-vial cell culture assay) and hazardous because of the risk of personnel being contaminated (biosafety level 3 laboratories and experienced personnel required). *Coxiella burnetii* culture could also present a lack of detectability.

For all these reasons, implementing molecular methods for detecting *Coxiella burnetii* presents significant advantages, especially for diagnosis and surveillance of this agent.

First, molecular methods are very safe because samples can be pretreated by heating at 90°C, to inactivate *Coxiella burnetii,* and then tests are carried out on purified DNA extracts. Molecular analysis, therefore, does not necessarily require a biosafety level 3 laboratory. Second, molecular techniques are rapid (results in 1 day) and, depending on the target sequence and the primers and probe design, very sensitive (especially for the repetitive element *IS1111*) and specific.

Molecular methods can be applied on a large array of samples: blood, fresh or embedded tissues, vaginal swabs, feces, milk, eggs. Finally, real-time PCR was used to measure the antibiotic susceptibility of *Coxiella burnetii* and various molecular methods were described for isolate genotyping. These advantages explain why in France, for example, veterinary laboratories have been quick to use PCR techniques, especially since the existence of commercial kits, for diagnosing *Coxiella burnetii*–induced abortion in livestock.

Molecular techniques, however, do of course present some limitations. These include the disadvantages of all molecular techniques, such as the cost of equipment and reagents (especially for real-time PCR) and the risk associated with the use of intercalating agents in conventional PCR. Disadvantages specific to *Coxiella burnetii* detection by PCR are related to the particular nature of this agent and the pathogenesis of the disease.

Coxiella burnetii cells easily become airborne, and the bacterial load of some samples, such as ruminant placentas, is very high (up to 10^9 bacteria per g). DNA amplification is very sensitive, and artifactual results may be obtained due to intercontamination between samples, especially during nucleic acid extraction and purification. In our laboratory, we recommend that veterinarians perform on-site placental or vaginal swabs (preferred) after an abortion. We also take other precautions, such as opening samples in a type II cabinet, extracting DNA using methods involving the vacuum chamber, or centrifugation in "safe-lock" microtubes. Finally, we are particularly vigilant to decontaminate work areas and equipment to destroy both bacteria and their DNA, and we only include negative controls (to show that there has been no contamination from the extraction procedure) and no positive extraction controls in our PCR runs.

The samples (blood, placenta, feces) can also be rich in PCR inhibitors. In addition, because *Coxiella burnetii* is an intracellular bacterium, DNA extraction and purification methods must be adapted to the different matrices and validated on positive samples rather than on samples spiked with plasmids or genomic DNA. The use of an internal positive control (IPC), endogenous to the host cell, is particularly recommended to check both the process of extraction and the absence of inhibitors.

Lastly, a limitation of PCR for Q fever diagnosis is the pathogenesis of the disease. In humans, molecular techniques must be coupled with serological tests to distinguish between acute or chronic forms, assess reactivation of the disease, or monitor treatment.

Prospects for improved methods of molecular detection of *Coxiella burnetii* include: (i) validation using a common reference (in France, for example, the COFRAC document LAB GT04 for public health laboratories and the AFNOR standard in preparation for veterinary diagnostic laboratories); (ii) development of commercial kits for humans similar to those that exist in the veterinary field; (iii) comparative studies between molecular techniques and other methods to better define their use; (iv) development of multiplex PCR for the diagnosis of different pathogens responsible for pneumonia or endocarditis in humans, or to distinguish between acute or chronic forms of Q fever; and (v) the development of quantitative tests to determine the number of viable bacteria and estimate risks associated with animals shedders, manure spreading, or food.

On the research side, molecular techniques could provide interesting data for epidemiological typing of isolates and to study virulence factors and predictive elements of the evolution of the disease. Once genetic elements of interest have been identified, the development of DNA microarrays to simultaneously establish a diagnosis, type isolates, and identify virulence factors may be envisaged.

ACKNOWLEDGMENT

I sincerely thank Dr. Leigh Gebbie for reviewing the manuscript.

REFERENCES

1. Derrick, E.H., "Q" fever, a new fever entity: clinical features, diagnosis, and laboratory investigation, *M. J. Aust.*, 2, 281, 1937.
2. Dyer, R.E., Q fever: History and present status, *Am. J. Pub. Health*, 39, 471, 1949.
3. Burnet, F.M., and Freeman, M., Experimental studies on the virus of "Q" Fever, *M. J. Aust.*, 2, 299, 1937.
4. Davis, G.E. et al., A filter-passing agent isolated from ticks, *Pub. Health Rep.*, 53, 2259, 1938.
5. Dyer, R.E., Similarity of Australian "Q" fever and a disease caused by an infectious agent isolated from ticks in Montana, *Pub. Health Rep.*, 54, 1229, 1939.
6. Phillip, C.B., Comments on the name of the Q fever organism, *Pub. Health Rep.*, 63, 58, 1948.
7. Weisburg, W.G. et al., Phylogenetic diversity of the *Rickettsiae*, *J. Bacteriol.*, 171, 4202, 1989.
8. Maurin, M., and Raoult, D., Q fever, *Clin. Microbiol. Rev.*, 12, 518, 1999.
9. Euzeby, J.P., *Coxiella burnetii*, *Dictionnaire de Bactériologie Vétérinaire*, http://www.bacterio.cict.fr/bacdico/cc/coxiella.html, 2001, last accessed date 2010-11-05.
10. Hadckstadt, T.M. et al., Lipopolysaccharide variation in *Coxiella burnetii*: Intrastrain heterogeneity in structure and antigenicity, *Infect. Immun.*, 48, 359, 1985.
11. Schrameck, S., and Mayer, H., Different sugar composition of lipopolysaccharides isolated from phase I and pure phase II cells of *Coxiella burnetii*, *Infect. Immun.*, 38, 53, 1982.

12. Vodkin, M.H., and Williams, J.C., Overlapping deletion in two spontaneous phase variants of *Coxiella burnetii*, *J. Gen. Microbiol.*, 132, 2587, 1986.
13. Seshadri, R., Complete sequence of the Q fever pathogen *Coxiella burnetii*, *PNAS*, 100, 5455, 2003.
14. Afseth, G., Mo, Y.Y., and Mallavia, L.P., Characterization of the 23S and 5S rRNA genes of *Coxiella burnetii* and identification of an intervening sequence within 23S rRNA gene, *J. Bacteriol.*, 177, 2946, 1995.
15. Raghavan, R. et al., The unusual 23S rRNA gene of *Coxiella burnetii*: Two self-splicing group I introns flank a 34 base-pair exon and one element lacks the canonical ΩG, *J. Bacteriol.*, 189, 6572, 2007.
16. Vodkin, M.H., Williams, J.C., and Stephenson, E.H., Genetic heterogeneity among isolates of *Coxiella burnetii*, *J. Gen. Microbiol.*, 132, 455, 1986.
17. Willems, H., Jäger, C., and Balger, G., Physical and genetic map of the obligate intracellular bacterium *Coxiella burnetii*, *J. Bacteriol.*, 180, 3816, 1998.
18. Glazunova, O. et al., *Coxiella burnetii* genotyping, *Emerg. Infect. Dis.*, 11, 1211, 2005.
19. Aricau-Bouvery, N. et al., Molecular characterization of *Coxiella burnetii* isolates by infrequent restriction site-PCR and MLVA typing, *BMC Microbiol.*, 6,38, 2006.
20. Denison, A.M., Thompson, H.A., and Massung, R.F., IS-1111 insertion sequences of *Coxiella burnetii*: Characterization and use for repetitive element PCR-based differentiation of *Coxiella burnetii* isolates, *BMC Microbiol.*, 7, 91, 2007.
21. Beare, P.A et al., Genetic diversity of the Q fever agent, *Coxiella burnetii*, assessed by microarray-based whole-genome comparisons, *J. Bacteriol.*, 188, 2309, 2006.
22. Samuel, J.E. et al., Isolation and characterization of a plasmid from phase I *Coxiella burnetii*, *Infect. Immun.*, 41, 448, 1983.
23. Samuel, J.E, Frazier, M.E., and Mallavia, L.P., Correlation of plasmid type and disease caused by *Coxiella burnetii*, *Infect. Immun.*, 49, 775, 1985.
24. Mallavia, L.P., Genetic of *Rickettsiae*, *Eur. J. Epidemiol.*, 7, 213, 1991.
25. Valkova, D., and Kazar, J., A new plasmid (QpDV) common to *Coxiella burnetii* isolates associated with acute and chronic Q fever, *FEMS Microbiol. Lett.*, 125, 275, 1995.
26. Savinelli, E.A., and Mallavia, L.P., Comparison of *Coxiella burnetii* plasmids to homologous chromosomal sequences present in a plasmidless endocarditis-causing isolate, *Ann. N. Y. Acad. Sci.*, 590, 523, 1990.
27. Mallavia, L.P., Samuel, J.E., and Frazier, M.E., The genetics of *Coxiella burnetii*: etiologic agent of Q fever and chronic endocarditis. In Williams, J.C., and Thompson, H.A. (Editors), *Q Fever: The Biology of Coxiella burnetii*, p. 259, CRC Press, Inc., Boca Raton, FL, 1991.
28. Consortium BNGP, The complete genome sequence of *Chromobacterium violaceum* reveals remarkable and exploitable bacterial adaptability, *Proc. Natl. Acad. Sci. USA*, 102, 838, 2003.
29. Duchaud, E. et al., The genome sequence of the enthomopathogenis bacterium *Photorhabdus luminescens*, *Nat. Biotechnol.*, 21, 1307, 2003.
30. Hendrix, L.R., Samuel, J.E., and Mallavia, L.P., Differentiation of *Coxiella burnetii* isolates by analysis of restriction-endonuclease-digested DNA separated by SDS-PAGE, *J. Gen. Microbiol.*, 137, 269, 1991.
31. Capo, C. et al., Subversion of monocyte functions by *Coxiella burnetii*: Impairment of the cross-talk between $\alpha_v\beta_3$ integrin and CR3, *J. Immunol.*, 163, 6078, 1999.
32. Capo, C. et al., *Coxiella burnetii* avoids macrophages phagocytosis by interfering with spatial distribution of complement receptor 3, *J. Immunol.*, 170, 4217, 2003.
33. Honstettre, A. et al., Lipopolysaccharide from *Coxiella burnetii* is involved in bacterial phagocytosis, filamentous actin reorganization, and inflammatory responses through toll-like receptor 4, *J. Immunol.*, 172, 3695, 2004.
34. Hackstadt, T., and Williams, J.C., Biochemical stratagem for obligate parasitism of eukaryotic cells by *Coxiella burnetii*, *Proc. Natl. Acad. Sci. USA*, 78, 3240, 1981.
35. McCaul, T.F., The development cycle of *Coxiella burnetii*. In Williams, J.C., and Thompson, H.A. (Editors), *Q Fever: The Biology of Coxiella burnetii*, p. 223, pCRC Press, Inc., Boca Raton, FL, 1991.
36. McCaul, T.F., Williams, J.C., and Thompson, H.A., Electron microscopy of *Coxiella burnetii* in tissue culture. Induction of cell types as products of developmental cycle, *Acta. Virol.*, 35, 545, 1991.
37. Oswald, W., and Thiele, D., A sporulation gene in *Coxiella burnetii*?, *J. Vet. Med. B.*, 40, 366, 1993.
38. Scott, G.H., and Williams, J.C., Susceptibility of *Coxiella burnetii* to chemical desinfectants, *Ann. N. Y. Acad. Sci.*, 590, 291, 1990.
39. Baca, O.G., and Paretsky, D., Q fever and *Coxiella burnetii*: A model for host-parasite interactions, *Microbiol. Rev.*, 47, 127, 1983.
40. Roman, M.J., Coriz, P.D., and Baca, O.G., A proposed model to explain persistent infection of host cells with *Coxiella burnetii*, *J. Gen. Microbiol.*, 132, 1415, 1986.
41. Badudieri, B., Q fever: A zoonosis, *Adv. Vet. Sci.*, 5, 81, 1959.
42. Hilbink, F. et al., Q fever is absent from New Zealand, *Int. J. Epidemiol.*, 22, 945, 1993.
43. Rodolakis, A. et al., Comparison of *Coxiella burnetii* shedding in milk of dairy bovine, caprine and ovine herds, *J. Dairy Sci.*, 98, 5352, 2007.
44. Guatteo, R. et al., Shedding routes of *Coxiella burnetii* in dairy cows: Implications for detection and control, *Vet. Res.*, 37, 827, 2007.
45. Werth, D., The occurrence and significance of *Chlamydia psittaci* and *Coxiella burnetii* in dogs and cats. A study of literature, *Berl. Munch. Tierarztl. Wochenschr.*, 102, 156, 1989.
46. Buhariwalla, F., Carn, B., and Marrie, T.J., A dog-related outbreak of Q fever, *Clin. Infect. Dis.*, 23, 753, 1996.
47. Komiya, T. et al., Seroprevalence of *Coxiella burnetii* infections among cats in different living environments, *J. Vet. Med. Sci.*, 65, 1047, 2003.
48. Mantovani, A., and Benazzi, P., The isolation of *Coxiella burnetii* from *Rhipicephalus sanguineus* on naturally infected dogs, *J. Am. Vet. Med. Assoc.*, 122, 117, 1953.
49. Pope, J.H., Scott, W., and Dweyer, R., *Coxiella burnetii* in kangaroos and kangaroo-ticks in Western Queensland, *Aust. J. Exp. Biol.*, 38, 17, 1960.
50. Marrie, T.J. et al., Q fever pneumonia associated with exposure to wild rabbits, *Lancet*, 1, 427, 1986.
51. Eklund, C.M., Parker, R.R., and Lackman, D.B., A case of Q fever probably contracted by exposure to ticks in nature, *Pub. Health. Rep.*, 62, 1413, 1947.
52. Enright, J.B. et al., *Coxiella burnetii* in a wildlife-livestock environment. Distribution of Q fever in wild mammals, *Am. J. Epidemiol.*, 94, 79, 1971.
53. Tisso-Dupont, H. et al., A hyperendemic focus of Q fever related to sheep and wind, *Am. J. Epidemiol.*, 150, 67, 1999.
54. Tisso-Dupont, H. et al., Wind in November, Q fever in December, *Emerg. Infect. Dis.*, 10, 1264, 2004.

55. La Scola, B., and Raoult, D., Survival of *Coxiella burnetii* within free-living amoeba *Acanthamoeba castellanii*, *Clin. Microbiol. Infect.*, 7, 75, 2001.
56. Rowbotham, T.J., Preliminary report on the pathogenicity of *Legionella pneumophila* for fresh-water and soil amoebae, *J. Clin. Pathol.*, 33, 1179, 1980.
57. Benson, W.W., Brock, D.W., and Mather, J., Serologic analysis of a penitentiary group using raw milk from Q fever infected herd, *Pub. Health. Rep.*, 78, 707, 1963.
58. Fishbein, D.B., and Raoult, D., A cluster of *Coxiella burnetii* infections associated with exposure to vaccinated goats and their unpasteurized dairy products, *Am. J. Trop. Med. Hyg.*, 47, 35, 1992.
59. Tylewska-Wierbanowska, S.K., and Kruszewka, D., Q fever sexually transmitted infection?, *J. Infect. Dis.*, 161, 368, 1990.
60. Tissot-Dupont, H. et al., Epidemiologic features and clinical presentation of acute Q fever in hospitalized patients: 323 French cases, *Am. J. Med.*, 93, 427, 1992.
61. Raoult, D. et al., Acute and chronic Q fever in patients with cancer, *Clin. Infect. Dis.*, 14, 127, 1992.
62. Raoult, D. et al., Q fever and HIV infection, *AIDS*, 7, 81, 1993.
63. Raoult, D., and Stein, A., Q fever during pregnancy, a risk for women, fetuses and obstetricians, *N. Engl. J. Med.*, 330, 371, 1994.
64. Marrie, T.J., and Pollack, P.T., Seroepidemiology of Q fever in Nova Scotia: Evidence for age dependent cohorts and geographical distribution, *Eur. J. Epidemiol.*, 11, 47, 1995.
65. Leone, M. et al., Effect of sex on *Coxiella burnetii* infection: Protective role of 17beta-estradiol, *J. Infect. Dis.*, 189, 339, 2004.
66. Madriaga, M.G. et al., Q fever: A biological weapon in your backyard, *Lancet Infect. Dis.*, 3, 709, 2003.
67. Tigertt, W.D., Benenson, A.S., and Gouchenour, W.S., Airborne Q fever, *Bacteriol. Rev.*, 25, 285v, 1961.
68. Marrie, T.J., Acute Q fever. In Marrie, T.J. (Editor), *Q fever: The Disease*, p. 125, CRC Press, Inc., Boca Raton, FL, 1990.
69. Waag, D.M., *Coxiella burnetii*: Host and bacterial responses to infection, *Vaccine*, 25, 7288, 2007.
70. Marmion, B.P. et al., Long-term persistence of *Coxiella burnetii* after acute primary Q fever, *QJM*, 98, 7, 2005.
71. Perin, T.L., Histopathologic observations in a fatal case of Q fever, *Arch. Pathol.*, 47, 361, 1949.
72. Janigan, D.T., and Marrie, T.J., An inflammatory pseudotumor of the lung in Q fever pneumonia, *N. Engl. J. Med.*, 308, 86, 1983.
73. Pellegrin, M. et al., Granulomatous hepatitis in Q fever, *Hum. Pathol.*, 11, 51, 1980.
74. Silver, S.S., and McLeish, W.A., "Doughnut" granulomas in Q fever, *Can. Med. Assoc. J.*, 130, 102, 1984.
75. Brouqui, P.H. et al., Epidemiologic and clinical features of chronic Q fever: 92 cases from France (1982–1990), *Arch. Inter. Med.*, 153, 642, 1993.
76. Ghigo, E. et al., *Coxiella burnetii* survival in THP-1 monocytes involves the impairment of phagosome maturation: IFN-γmediates its restoration and bacterial killing, *J. Immun.*, 169, 4488, 2002.
77. Shannon, J.G., Howe, D., and Heinzen, R.A., Virulent *Coxiella burnetii* does not activate human dendritic cells: Role of lipopolysaccharide as shielding molecule, *Proc. Natl. Acad. Sci. USA*, 102, 8722, 2005.
78. Zamboni, D.S. et al., Stimulation of Toll-like Receptor 2 by *Coxiella burnetii* is required for macrophage production of pro-inflammatory cytokines and resistance to infection, *J. Biol. Chem.*, 279, 54405, 2004.
79. Koster, F.T., Williams, J.C., and Goodwin, J.S., Cellular immunity in Q fever: Specific lymphocyte unresponsiveness in Q fever endocarditis, *J. Infect. Dis.*, 152, 1283, 1985.
80. Capo, C. et al., Production of interleukin-10 and transforming growth factor β by peripheral blood mononuclear cells in Q fever endocarditis, *Infect. Immun.*, 64, 4143, 1996.
81. Mege, J.L. et al., *Coxiella burnetii*: The "query" fever bacterium. A model of immune subversion by a strictly intracellular microorganism, *FEMS Microbiol. Rev.*, 19, 209, 1997.
82. Derrick, E.H., The course of infection with *Coxiella burnetii*, *Med. J. Aust.*, 1, 1051, 1973.
83. Marrie, T.J., *Coxiella burnetii* (Q fever) pneumonia, *Clin. Infect. Dis.*, 21, 5253, 1994.
84. Marrie, T.J., Liver involvement in acute Q fever, *Chest*, 94, 896, 1988.
85. Raoult, D., Raza, A., and Marrie, T.J., Q fever endocarditis and other forms of chronic Q fever. In Marrie, T.J. (Editors), *Q Fever: The Disease*, p. 179, CRC Press, Inc., Boca Raton, FL, 1990.
86. Raoult, D. et al., Q fever endocarditis in the south of France, *J. Infect. Dis.*, 155, 570, 1987.
87. Raoult, D., Q fever: still a query after all these years, *J. Med. Microbiol.*, 44, 77, 1996.
88. Durack, D.T. et al., New criteria for diagnosis of infective endocarditis: Utilization of specific echocardiographic findings, *Am. J. Med.*, 96, 200, 1994.
89. Fournier, P.E. et al., Modification of the diagnostic criteria proposed by the Duke endocarditis service to permit improved diagnosis of Q fever endocarditis, *Am. J. Med.*, 100, 629, 1996.
90. Ayres, J.G. et al., Post-infection fatigue syndrome following Q fever, *Q. J. Med.*, 91, 105, 1998.
91. Stein, A., and Raoult, D., Q fever during pregnancy: A public health problem in Southern France, *Clin. Infect. Dis.*, 27, 592, 1998.
92. Lepidi, H., Gouriet, F., and Raoult, D., Immunohistochemical detection of *Coxiella burnetii* in chronic Q fever hepatitis, *Clin. Microbiol. Infect.*, 15, 169, 2009.
93. Lepidi, H. et al., Immunohistochemical detection of *Coxiella burnetii* in an aortic graft, *Clin. Microbiol. Infect.*, 15, 171, 2009.
94. Fournier, P.E., Marrie, T.J., and Raoult, D., Diagnosis of Q fever, *J. Clin. Microbiol.*, 36, 1823, 1998.
95. Brouqui, P., Dumler, J.S., and Raoult, D., Immunohistologic demonstration of *Coxiella burnetii* in the valves of patients with Q fever endocarditis, *Am. J. Med.*, 97, 451, 1994.
96. Thiele, D., Kano, M., and Krauss, H., Monoclonal antibody based capture ELISA/ELIFA for detection of *Coxiella burnetii* in clinical specimens, *Eur. J. Epidemiol.*, 8, 568, 1992.
97. Lepidi, H. et al., Autoimmunohistochemistry: A new method for the histological diagnosis of infective endocarditis, *J. Infect. Dis.*, 193, 1711, 2006.
98. Omsland, A. et al., Sustained axenic metabolic activity by the obligate intracellular bacterium *Coxiella burnetii*, *J. Bact.*, 190, 3203, 2008.
99. Omsland, A. et al., Host cell-free growth of the Q fever bacterium *Coxiella burnetii*, *Proc. Natl. Acad. Sci. USA*, 106, 4430, 2009.
100. Raoult, D. et al., Isolation of 16 strains of Coxiella burnetii from patients by using a sensitive centrifugation cell culture system and establishment of the strains in HEL cells, *J. Clin. Microbiol.*, 28, 2482, 1990.
101. Gouriet, F. et al., Use of shell-vial cell culture assay for isolation of bacteria from clinical specimens: 13 years of experience, *J. Clin. Microbiol.*, 43, 4993, 2005.

102. Tissot-Dupont, H.D., Thirian, H.X., and Raoult, D., Q fever serology: cut-off determination for microimmunofluorescence, *Clin. Diag. Lab. Immunol.*, 1, 189, 1994.
103. La Scola, B., and Raoult, D., Serological cross-reactions between *Bartonella quintana*, *Bartonella henselae* and *Coxiella burnetii*, *J. Clin. Microbiol.*, 34, 2270, 1996.
104. Musso, D., and Raoult, D., Serological cross-reactions between *Coxiella burnetii* and *Legionella micdadei*, *Clin. Diag. Lab. Immunol.*, 4, 2008, 1997.
105. Gouriet, F. et al., Multiplexed whole bacterial microarray, a new format for the automation of serodiagnosis: The culture-negative endocarditis paradigm, *Clin. Microbiol. Infect.*, 14, 1112, 2008.
106. Field, P.R. et al., Comparison of a commercial enzyme-linked immunosorbent assay with immunofluorescence and complement fixation test for detection of *Coxiella burnetii* (Q fever) Immunoglobulin M, *J. Clin. Microbiol.*, 38, 1645, 2000.
107. Field, P.R. et al., Evaluation of a novel commercial enzyme-linked immunosorbent assay detecting *Coxiella burnetii*-specific Immunoglobulin G for Q fever prevaccination screening and diagnosis, *J. Clin. Microbiol.,* 40, 3526, 2002.
108. Frazier, M.E. et al., DNA probes for the identification of *Coxiella burnetii* strains, *Ann. N. Y. Acad. Sci.*, 590, 445, 1990.
109. Stein, A., and Raoult, D., Detection of *Coxiella burnetii* by DNA amplification using polymerase chain reaction, *J. Clin. Microbiol.*, 30, 2462, 1992.
110. Zhang, G.Q. et al., Direct identification of *Coxiella burnetii* plasmids in human sera by nested PCR, *J. Clin. Microbiol.*, 36, 2210, 1998.
111. Fournier, P.E., and Raoult, D., Comparison of PCR and serology assays for early diagnosis of acute Q fever, *J. Clin. Microbiol.*, 41, 5094, 2003.
112. Klee, S.R. et al., Highly sensitive real-time PCR for specific detection and quantification of *Coxiella burnetii*, *BMC Microbiol.*, 6, 2, 2006.
113. Vaidya, V.M. et al., Comparison of PCR, Immunofluorescence assay, and pathogen isolation for diagnosis of Q fever in humans with spontaneous abortions, *J. Clin. Microbiol.*, 46, 2038, 2008.
114. Fenollar, F., Fournier, P.E., and Raoult, D., Molecular detection of *Coxiella burnetii* in the sera of patients with Q fever endocarditis on vascular infection, *J. Clin. Microbiol.*, 42, 4919, 2004.
115. Zhang, G. et al., Identification and characterization of an immunodominant 28-kilodalton *Coxiella burnetii* outer membrane protein specific to isolates associated with acute disease infection, *Infect. Immun.*, 73, 1561, 2005.
116. Brennan, R.E., and Samuel, J.E., Evaluation of *Coxiella burnetii* antibiotic susceptibilities by real-time PCR assay, *J. Clin. Microbiol.*, 41, 1869, 2003.
117. Boulos, A. et al., Measurement of the antibiotic susceptibility of *Coxiella burnetii* using real-time PCR, *Int. J. Antimicrob. Agents*, 23, 169, 2004.
118. Pelletier, C. et al., Validation of an internal method for the diagnosis of infections with *Chlamydophila abortus* and *Coxiella burnetii* by real-time multiplex PCR, *Dev. Biol.*, 126, 219, 2006.
119. Panning, M. et al., High throughput detection of *Coxiella burnetii* by real-time PCR with internal control system and automated DNA preparation, *BMC Microbiol.*, 8, 77, 2008.
120. Berri, M. et al., Simultaneous differential detection of *Chlamydophila abortus*, *Chlamydophila pecorum* and *Coxiella burnetii* from aborted ruminant's clinical samples using multiplex PCR, *BMC Microbiol.*, 9, 130, 2009.

74 Enterobacter

Angelika Lehner, Roger Stephan, Seamus Fanning, and Carol Iversen

CONTENTS

74.1 INTRODUCTION

74.1.1 Classification

The *Enterobacter* genus represents a large and heterogeneous group within the *Enterobacteriaceae* family. They are gram-negative, generally motile, oxidase-negative, nonspore forming, flagellated, rod-shaped, facultative anaerobes and they reduce nitrate.[1] While over the years a total of 22 *Enterobacter* species have been proposed, taxonomic studies have led to the transfer of three *Enterobacter* species to alternative genera (*E. intermedius* = *Kluyvera intermedia*, *E. agglomerans* = *Pantoea agglomerans*, *E. sakazakii* = *Cronobacter* spp.). In addition, *E. taylorae* has been recognized as a heterotypic synonym of *E. cancerogenus*, and *E. dissolvens* has been reassigned as *E. cloacae* subspecies *dissolvens* comb. nov. Thus, the genus *Enterobacter* currently contains 17 distinct species.[2–5] *E. dissolvens*, *E. asburiae*, *E. hormaechei*, *E. kobei*, *E. ludwigii*, and *E. nimipressuralis* are closely related to *E. cloacae*, having a DNA relatedness of over 60% and/or differing by one biochemical trait, and have been subsumed in the so-called *E. cloacae*-complex. Along with *E. aerogenes*, these are considered the main species that are frequently isolated from clinical samples.[6] Only one species—*Enterobacter sakazakii*—is recognized as a foodborne pathogen, particularly in relation to infections in infants after ingestion of reconstituted powdered nutritional products.[7]

Enterobacter aerogenes. *E. aerogenes* is closely related to *Klebsiella* and is a homotypic synonym of *K. mobilis*. It was transferred to the genus *Enterobacter* from *Aerobacter* in 1960, when the motile and ornithine decarboxylase (ODC)–positive strains were separated as *Enterobacter* from the nonmotile and ODC-negative *Klebsiella* strains.[8] *E. aerogenes* is widely distributed and occurs in fresh water, soil, sewage, vegetation, and animal feces. It is an opportunistic pathogen that can be isolated from all clinical sites, mainly causing burn, wound, and urinary tract infections (UTIs) as well as bacteremia and occasional cases of meningitis.

***Enterobacter cloacae* Complex.** The *E. cloacae* complex comprises 12 genetic clusters (I–XII) and one sequence crowd (XIII), which can be distinguished on the basis of *hsp60* sequence analysis. Species have been assigned to nine clusters (*E. asburiae* I, *E. kobei* II, *E. ludwigii* V, *E. hormaechei* VI–VIII, *E. nimipressuralis* X, *E. cloacae* subspecies *cloacae* XI, and *E. cloacae* subspecies *dissolvens* XII); as yet no species have been assigned to clusters III, IV, or IX. The *E. cloacae* complex has been described as two genetically distinct clades using *hsp*60- and *rpo*B-genotyping, multilocus sequence analysis (MLSA), and comparative genomic hybridization (CGH).[9] The younger clade 1 comprised *E. hormaechei* subspecies from hospitalized patients; these are the most frequently isolated *Enterobacter* in clinical environments, where there is high selective pressure from antibiotic use. Clade 2 consisted mainly of genetically heterogonous asymptomatic fecal isolates. Members of the *E. cloacae* complex are often not further classified beyond the genus level, as biochemical identification of these species and subspecies can be unreliable and irreproducible. "*E. cloacae*" are ubiquitous in foods, animals, and the environment and have been isolated from all clinical sites. They are considered an important cause of nosocomial infections.

Enterobacter sakazakii. *E. sakazakii*, previously known as "yellow-pigmented *Enterobacter cloacae*," was described as a new species in 1980 by Farmer et al. when the presence of certain biochemical traits, antibiotic susceptibilities, and DNA relatedness provided sufficient evidence to distinguish the species from *E. cloacae*.[10] However, from the beginning, members of this species were described as quite heterogeneous at the phenotypic as well as the molecular level.[10–14] Updating the original taxonomy of *E. sakazakii* by using a polyphasic approach has resulted in the recognition of six new species based on extensive geno- and phenotypic evaluations.[15,16] These species have been moved to a novel genus, *Cronobacter*, with the novel genus being synonymous with *E. sakazakii*.[17] A further study based on MLSA of several housekeeping genes present in *E. sakazakii* supports the classification of these organisms as a new genus.[18] *E. sakazakii* (*Cronobacter*) have been associated with infections in infants, including bacteremia, necrotizing enterocolitis (NEC), and meningitis. Infection in neonates has been epidemiologically linked to ingestion of contaminated infant formula.[7,19–26] Therefore, the International Commission for Microbiological Specification for Foods has ranked *E. sakazakii* (*Cronobacter*) as "severe hazard for restricted populations, life threatening or substantial chronic sequelae of long duration."[27] *E. sakazakii* (*Cronobacter*) infections rarely occur in adults although there have been reports of wound infection,[28] bacteremia,[29–32] and pneumonia,[33] mainly in elderly patients.

Other *Enterobacter* Species. *E. agglomerans* is commonly isolated from animal feces and the surfaces of plants. It is a rare opportunistic pathogen (usually of immunocompromised patients) and is mainly associated with penetrating wounds caused by vegetative material, catheter-related bacteremia, and contaminated intravenous products.[34] *E. amnigenus* is found in potable and surface water as well as unpolluted soils; two biogroups are recognized. It is a rare clinical isolate with similar clinical behavior and antimicrobial susceptibility to *E. cloacae*.[35] Originally isolated from a tree canker, *E. cancerogenus* is a heterotypic synonym of *E. taylorae*. Rare human wound infections and septicemia are usually associated with community-acquired trauma or crush injuries.[36] *E. cowanii* was formerly designated as NIH group 42 of *E. agglomerans*. Although infections are rarely encountered, the majority of *E. cowanii* strains have been isolated from clinical body fluids.[37] *E. gergoviae* can be isolated from wounds, urine, and blood, as well as water and cosmetics; its closest genotypic relative is *E. aerogenes*.[38] *E. helveticus, E. pulveris,* and *E. turicensis* were first described as isolates from fruit powders, and they can be found in other dried foods and manufacturing environments. No clinical cases have been attributed to these species although they are morphologically similar to *E. sakazakii* (*Cronobacter*) and are found in the same ecological niches, including infant formula.[3,4] *E. intermedius* (*K. intermedia*), isolated from water, soil, feces, wounds, blood, and the gall bladder, has been classified as an earlier heterotypic synonym of *Kluyvera cochleae*.[39] *E. oryzae* was originally isolated from the wild rice species *Oryza latifolia,* and there are no known clinical isolates.[5] *E. pyrinus*[40] and *E. radicincitans*[2] have both been isolated from blood and spinal fluid. However, they are mainly associated with brown leaf-spot lesions on pear trees and as a growth promoter in the phyllosphere of winter wheat, respectively.

74.1.2 Prevalence and Distribution

Environmental and Food Sources. Members of the genus *Enterobacter* can be found in the natural environment (soil, water, sewage, vegetation) and *E. cloacae*-complex species can routinely appear as commensal organisms of the animal and human gut.[1] *Enterobacter* can also be found in a wide range of foods, including fresh, frozen, dried, or powdered fruit and vegetables; meat, fish, eggs, and dairy products; animal feed (dried pellets); grains, nuts, and seeds; flour, pasta and confectionary; beverages and water. *Enterobacter* are not particularly thermotolerant, but have been found in pasteurized products, possibly due to postprocess contamination. *E. sakazakii* (*Cronobacter*) are the only species for which a link has been established between contaminated food and cases of infection. These organisms are capable of internalizing and surviving for a substantial time within plants, and this may be an important primary or intermediary habitat that enables them to enter the food chain.[41] *E. sakazakii* (*Cronobacter*) have attracted special attention as opportunistic neonatal pathogens associated with occasional contamination of powdered infant formula (PIF).[42–44] Presently, approved technology is not available to render PIFs commercially sterile. The *Enterobacter* species most frequently isolated from PIF are *E. cloacae, E. sakazakii, E. agglomerans, E. pulveris,* and *E. helveticus.* Entry of *Enterobacter* into the production environment may occur via dry raw materials,[45] inefficient air filtration,[46] or human carriage. However, there is no indication that raw milk and liquid ingredients are a potential hazard.[47] Material present in the environment of infant formula production facilities is thought to be a major source of (re-)contamination of the final products by these organisms. After contamination, *E. sakazakii* (*Cronobacter*) appear to exhibit greater resistance to desiccation stress and to persist longer in dried infant formula than most other *Enterobacter* species.[48] This may contribute to the fact that other *Enterobacter* have not been associated with foodborne infections in neonates.

Clinical Environments. *Enterobacter* spp. rarely causes disease in healthy individuals; however, they are becoming increasingly important as opportunistic pathogens. They can affect people of all ages, but infection occurs most often in neonates and in the elderly. At greatest risk are immunocompromised patients and patients with underlying trauma (such as severe wounds, crush injuries or burns, malignant tumors, liver disease, gastric ulcers, or in those who have undergone surgical procedures).[49] The *E. cloacae* complex and *E. aerogenes* account for the vast majority of *Enterobacter* infections.[1] However, *E. sakazakii* (*Cronobacter*) are also associated with severe infections in neonates.[50] Risk factors include extended hospitalization; invasive procedures (such as surgery, mechanical ventilation, or installation of catheters); serious underlying conditions (as listed above); and treatment with antimicrobial agents. The use of broad-spectrum cephalosporins or aminoglycosides in intensive care unit (ICU) settings is a specific risk factor for infection with multidrug-resistant *Enterobacter* strains, which may possess inducible β-lactamases and result in increased case fatality rates. Additional risk factors in neonates include premature birth and low birth weight.

Enterobacter have been isolated as common contaminants from hospital surfaces; blood products, total parenteral nutrition solutions, isotonic saline solutions, albumin, other medical and nutritional supplements; preparation and feeding equipment; catheters, endoscopes, blood pressure monitors, stethoscopes, digital thermometers, dialysis machines and other medical equipment; as well as the hands and clothing of medical staff. *Enterobacter* species are present in adult and neonatal ICUs throughout the world.

74.1.3 Clinical Symptoms and Epidemiology

Bloodstream Infections. Bacteremia is the presence of viable bacteria in the bloodstream, whether associated with active disease or not. The release of bacterial toxins into the circulation can elicit systemic inflammatory response syndrome, characterized by rapid breathing, low blood pressure, and fever. In as many as one-third of cases this can lead to multiple organ failure (septic shock). *E cloacae* complex, followed by *E. aerogenes,* are the most frequently isolated species in *Enterobacter* bacteremia, accounting for over 90% of all cases. However *E. sakazakii* (*Cronobacter*) are notable for their association with cases in neonates.[50] *E. cloacae, E. hormaechei,* and *E. gergoviae* have also been responsible for outbreaks in neonatal intensive care units (NICUs).[51–57] The route of entry into the bloodstream is often unknown, and mixed bacteremia with multiple isolates is common. Most cases of *Enterobacter* bacteremia are nosocomially acquired in the ICU, and case fatality rates for *Enterobacter* sepsis range from 6% to 50%. Molecular fingerprint epidemiology suggests that horizontal transmission from infected patients plays a significant role.

Necrotizing Enterocolitis. NEC is an inflammatory process of the small and large intestine and is one of the most common causes of death in NICUs. NEC has an incidence of 2%–5% in all preterm infants and up to 13% in those with low birth weight.[1] The reported case fatality rates range between 0% and 55% with low birth weight infants (<1500 g) at greater risk. NEC has no definitive known cause and research suggests the involvement of a combination of factors leading to intestinal damage, including intestinal mucosal immaturity/dysfunction, ischemia and/or reperfusion injury, the release of inflammatory mediators and bile acids, and downregulation of cellular growth factors. Empirical antibiotic use and intestinal bacteria are thought to be associated factors though no common infectious agent has been identified. In a study of 125 neonates with NEC, *Enterobacter* spp. were the most common organisms, being isolated from 29% of patients.[58] NEC is rarely seen in infants prior to initiation of oral feeding, and the use of infant formula as opposed to breast milk greatly increases the risk.[59] *E. cloacae* and *E. sakazakii* (*Cronobacter*) have been associated with outbreaks of NEC.[23,60]

Central Nervous System (CNS) Infections. Bacterial meningitis is an infection of the protective membranes covering the brain and spinal cord. Infections can increase inner cranial pressure, requiring aspiration of fluid and drainage to prevent cerebral damage. Meningitis is frequently fatal and can lead to long-term complications in survivors, including mental and motor disabilities, convulsive disorders, hydrocephalus, and deafness. *Enterobacter* spp. are estimated to cause 2.4%–4.5% of meningitis cases in adults and up to 10.4% of cases in children.[1] In adults, the main species implicated are *E. cloacae* and *E. aerogenes* with case fatality rates of 16%–20%. However, in neonates *E. sakazakii* (*Cronobacter*) CNS infections are reported with case fatality rates of up to 80%.[31,50] Risk factors in infants are low birth weight, prematurity, length of hospital stay, invasive procedure, and antibiotic use. Patients with *E. sakazakii* (*Cronobacter*) neonatal meningitis may suffer worse outcomes than those with meningitis attributed to *E. cloacae* and other gram-negative bacteria. *E. sakazakii* (*Cronobacter*) causes cystic changes, abscesses, fluid collection, dilated ventricles, and infarctions. Several cases of infections have occurred in NICUs, and a number of outbreaks have been traced back to the presence of these pathogens in reconstituted infant formula milk and/or their persistence in/on food-preparation equipment.[22–27] Infants of normal gestational age and birth weight who are infected during the first month postpartum are more likely to develop meningitis, compared to premature, low birth weight infants who usually develop later onset infections resulting in bacteremia with no progression to CNS disease.[50] A higher case fatality rate and incidence of adverse sequelae in survivors are associated with meningitis cases.

Other *Enterobacter* Infections. *Enterobacter* species cause up to 5% of nosocomial pneumonias being a significant cause of ventilator-associated and early post–lung transplant pneumonia. They are also associated with other lower respiratory tract infections including asymptomatic colonization, tracheobronchitis,[61] lung abscess, and empyema. Case fatality rates are particularly high in elderly patients, and risk factors include neurologic and chronic pulmonary obstructive disease, cancer, diabetes mellitus, alcohol abuse, and prior antimicrobial therapy. *Enterobacter* skin and soft-tissue infections include cellulitis, fasciitis, myositis, and abscesses as well as infection of wounds (surgical and community acquired), burns, and crush injuries. Prophylactic treatment with cephalosporin is associated with an increased risk of infection. Two-thirds of patients presenting with endocarditis due to *Enterobacter* are reported to have underlying cardiac disease, mainly mitral valve infection. *Enterobacter* causes up to 4% of nosocomial UTIs, including pyelonephritis, prostatitis, cystitis, and asymptomatic bacteriuria. *Enterobacter* UTI is usually associated with urinary catheters and/or prior antibiotic therapy. *Enterobacter* species that colonize the digestive tract can enter intraabdominal sites following intestinal perforation or surgery and are responsible for approximately 10% of postsurgical peritonitis cases. They are also occasionally associated with osteomyelitis, discitis, and septic arthritis in adults and children. *Enterobacter* bone and joint infections can be difficult to cure, and relapses require additional treatment.

74.1.4 Pathogenesis

Enterobacter have an outer membrane containing lipopolysaccharides from which lipid-A (endotoxin) plays a major role in sepsis. Lipid-A is the major stimulus for the release of cytokines, which mediate systemic inflammation and its complications. Nonlethal doses of endotoxin may also induce injury of the intestinal mucosal barrier, increase ileal permeability, and bacterial translocation from the gut to systemic organs via xanthine oxidase-generated oxygen-free radicals.[62,63] Several toxins have been reported from *Enterobacter* species. A Shiga-like toxin (SLT)–producing strain of *E. cloacae* was coisolated along with the SLT producing *E. coli* OR:H9 from an infant with hemolytic-uremic syndrome.[64] A heat stable hemolysin has been partially purified from a small percentage of clinical *E. cloacae* isolates,[65] and an *E. coli*-like α-hemolysin has been isolated from an *E. cloacae* strain.[66] In a study by Pagotto et al., it was reported that *E. sakazakii* (*Cronobacter*) strains were lethal to suckling mice at 10^8 CFU per mouse by intraperitoneal injection. Production of an enterotoxin-like compound was reported, but supporting data was not provided in this study.[67]

In order to cause an infection, a pathogen must be able to adhere to, colonize, and invade host surfaces.[68] *Enterobacter* species commonly produce type 1 or type 3 mannose-sensitive hemagglutinins (MSHA); *E. gergoviae* produces a mannose-resistant hemagglutinin. These are involved in cellular adhesion and contribute to colonization.[1,69] *E. sakazakii* (*Cronobacter*) exhibits clustered adhesion, a pattern that has also been associated with neonatal colitis–causing strains of *Klebsiella pneumoniae*.[70] Mange et al. demonstrated, using in vitro models, that *E. sakazakii* (*Cronobacter*) has adhesive capacities to a number of cell lines, including endothelial and transformed epithelial cell lines and proposed, that adhesion to epithelial cells is mainly nonfimbriae-based, suggesting the role of other virulence factors in binding.[70] Another tissue culture based study using human intestinal Caco-2 cells revealed the involvement of host actin filaments and microtubule structures in *E. sakazakii* (*Cronobacter*) entry and invasion and that disruption of the tight junction significantly enhanced in initial association as well as the efficiency of invasion.[71] The outer membrane protein OmpX has also been identified in *E. cloacae*. This protein shows high amino acid homology with the macrophage survival protein PagC and the serum resistance protein Rck, from *Salmonella* Typhimurium; as well as with the invasion protein Ail, from *Yersinia enterocolitica*. Expression of OmpX leads to decreased production of OmpF and OmpC and an increased antimicrobial resistance to β-lactam compounds. Overexpression of OmpX appears to increase the invasive potential of *E. cloacae*.[72] Similar to many other gram-negative bacteria, *E. sakazakii* (*Cronobacter*) expresses outer membrane protein A (OmpA). Recently, Singamsetty et al. provided evidence that the OmpA is crucial in the invasion of brain endothelial cells.[73] The pathogens also induce microtubule condensation at the sites of entry in the endothelial cells. This finding was supported by the study by Mittal et al., who used a newborn rat model to study the role of the OmpA protein in the onset of meningitis by *E. sakazakii* (*Cronobacter*).[74] It was shown for the first time that an orally administrated OmpA expressing the *E. sakazakii* (*Cronobacter*) strain traversed the intestinal barrier, multiplied in the blood, and subsequently penetrated the blood-brain barrier. The pathogens were present in high numbers in the brains of infected animals along with associated neutrophil infiltration, hemorrhage, and gliosis. The bacteria caused apoptosis of enterocytes in the intestinal segments of infected animals. OmpA expression was also involved in the resistance to blood and serum killing. In a similar study by Hunter et al., who focused on the onset of NEC in newborn rats, it was shown that *E. sakazakii* (*Cronobacter*) cells bind to enterocytes in rat pups at the tips of villi and to intestinal epithelial cells.[75] Exposure to the bacteria induced apoptosis and increased the production of interleukin-6 in intestinal epithelial cells (IEC-6) and in the animal model. It was suggested that *E. sakazakii* (*Cronobacter*) triggers intestinal disease by modulating enterocyte intracellular signaling pathways.[75]

To establish a systemic infection, pathogenic microbes need to circumvent the host immune system and also acquire important micronutrients, such as iron. Production of siderophores, such as enterobactin,[1,76] is common in *Enterobacter* species, and the production of aerobactin,[69] which is associated with microbial species able to cause invasive disease, has been reported in *E. cloacae*. Resistance to the lytic activity of complement has been reported for *E. cloacae* and *E. sakazakii* (*Cronobacter*). Monomeric and polymeric forms of an *E. cloacae* toxin from a strain exhibiting high hemolytic and leukotoxic activity was found to be able to generate oxidative stress and to provoke dose-dependent lysis of leukocytes.[77] Some strains of *E. sakazakii* (*Cronobacter*) have been reported to persist within macrophages.[78]

A significant factor in nosocomial *Enterobacter* infections appears to be prior exposure to antibiotics. Most species of *Enterobacter* have similar antibiotic resistance profiles, being intrinsically resistant to ampicillin, amoxicillin, amoxicillin-clavulanate, first-generation cephalosporins, and cephamycins, with variable resistance to expanded and broad-spectrum cephalosporins, tetracyclines, chloramphenicol, and antipseudomonal penicillins. The exceptions are *E. amnigenus*, *E. gergoviae*, *E. intermedius*, and *E. sakazakii* (*Cronobacter*), which are more susceptible to penicillins and cephalosporins than other *Enterobacters*.[79] Most *Enterobacter* isolates are susceptible to fluoroquinolones, trimethoprim-sulfamethoxazole, aminoglycosides, and carbapenems. Intrinsic cephalosporin activity is mediated through constitutively expressed chromosomal β-lactamase AmpC enzymes that upregulate β-lactamase expression in the presence of inducers, such as ampicillin, and cefoxitin. Derepression of these enzymes is increasing among clinical isolates, which then become resistant to third-generation cephalosporins, as well as ureido- and carboxypenicillins; carbapenems, and cefepime have a more stable β-lactam ring and retain reasonable activity against depressed strains. Many *Enterobacter* strains also carry plasmid-mediated β-lactamases, designated as SHV or TEM; plasmid-mediated extended spectrum beta lactamases (ESBL) activity is seen in *E. aerogenes, E. cloacae,* and, more rarely, *E. gergoviae*. An outbreak of aminoglycoside resistant *E. cloacae* infections displaying variable susceptibility to ciprofloxacin occurred in 2002. Fluoroquinolone susceptibility was associated with a plasmid-mediated *qnr*A1 gene present in 94% of outbreak isolates.[80] Low-level carbapenem resistance related to reduced expression of OmpC, a member of the subclass of enterobacterial porins, has been described in outbreak-related strains of *E. aerogenes*.[81]

As *Enterobacter* appear ubiquitous in the environment and can also be found as part of the normal human intestinal flora, it is likely that onset of infectious disease is a result of both predisposing host factors and bacterial virulence determinants. Currently ongoing alternative approaches such as whole genome analysis and comparisons of different *E. sakazakii* (*Cronobacter*) strains bear the potential for unraveling additional putative virulence factors and/or mechanisms of pathogenicity.

74.1.5 Diagnosis

74.1.5.1 Conventional Identification

Culture. There are no microbiological media specifically intended for the isolation of the *Enterobacter* genus. They grow on nonselective laboratory media aerobically and anaerobically at mesophilic incubation temperatures, and can be easily cultured from clinical samples. Most *Enterobacter* present as large, dull-gray, dry, or mucoid colonies on sheep blood agar. All *Enterobacter* species ferment glucose, and the most frequently encountered species will appear as typical Enterobacteriaceae colonies on violet-red bile, MacConkey, or Hektoen enteric agars. *Enterobacter* species contribute to the overall microbial quality of foods; however, only *E. sakazakii* (*Cronobacter*) are considered foodborne pathogens and as such, specific isolation methods have been developed for these organisms.

In 2002, the United States Food and Drug Administration (FDA) published a method for detection of *E. sakazakii* that included a preenrichment step in buffered peptone water (BPW), enrichment in Enterobacteriaceae enrichment (EE) broth, plating on violet-red bile glucose agar (VRBG), and picking of five typical colonies onto tryptone soy agar (TSA) plates.[82] After incubation at 25°C for 48–72 h, yellow-pigmented colonies on TSA plates are confirmed using the API 20E system. The main weak points of this procedure are the inability of some target strains to grow in the selective EE broth; the lack of discrimination between *Enterobacteriaceae* strains on VRBG; the variation in intensity of the pigmentation, with occasional observation of nonpigmented strains; and the reliability of the API 20E system.[17,83,84] Guillaume-Gentil et al. published an alternative procedure based on selective enrichment in a modified lauryl sulfate tryptose broth (mLST), incorporating 0.5 M NaCl and 10 mg/mL vancomycin hydrochloride.[85] This method was further improved by the replacement of VRBG with a chromogenic agar and forms the basis of the ISO Technical Specification for detection of *E. sakazakii* in milk-based infant formula, ISO/TS 22964:2006 "Milk and milk products—detection of *Enterobacter sakazakii*."[86]

Various chromogenic and fluorogenic agar media have been described in recent years for the detection of *E. sakazakii* (*Cronobacter*).[86–91] These are based mainly on the enzyme α-glucosidase, which is constitutively expressed in *E. sakazakii* (*Cronobacter*). However, other organisms also produce presumptive colonies on these agars, notably the newly described species *E. helveticus, E. turicensis, and E. pulveris*.[3,4] These species can be found in the same ecological niches as *E. sakazakii* (*Cronobacter*), such as dried food products and factory environments and present a challenge to culture-based as well as molecular isolation and identification methods.

It has been established that some isolates of *E. sakazakii* (*Cronobacter*) do not grow well in the enrichment broths currently proposed for isolation of these organisms.[83] In a recent study, a *Cronobacter* screening broth (CSB) has been developed to identify samples potentially contaminated with *Cronobacter* (*E. sakazakii*).[84] The broth is designed to circumvent the problems encountered with selective enrichment media for these organisms and to be complementary to current available chromogenic media in order to improve overall sensitivity and selectivity of *Cronobacter* (*E. sakazakii*) detection.

An alternative method to avoid the difficulties of selective broths is the MATRIX PSAK50 Method (Matrix MicroScience Ltd., UK), which uses cationic paramagnetic particle capture to concentrate a contaminating microorganism in a preenriched sample before plating directly onto isolation agar.[92] Other proposed rapid methods for the detection of *E. sakazakii* include two enzyme-linked immunoassays (EIAs): the Assurance for *Enterobacter sakazakii* (BioControl Systems, USA), and the TECRA HELIX *E. sakazakii* Method (TECRA International, Australia).

Phenotypic Identification. The *Enterobacter* genus is taxonomically complex, and phenotypic identification of individual *Enterobacter* species can be unreliable. A number of commercial identification galleries are available; however, their efficacy for determination of particular species is not always clear. To ensure the safety of infant formula and also to reduce unnecessary disposal of product, it is important to identify *E. sakazakii* (*Cronobacter*) species as accurately as possible. The recently developed fluoro- and chromogenic media are useful tools in the phenotypic selection of presumptive *E. sakazakii* (*Cronobacter*).[86–91] However *E. helveticus*, *E. pulveris*, *E. turicensis*, and *E. cloacae*-complex species may also give characteristic colonies on some of these media, and further confirmation is essential. The phenotypic identification of *E. sakazakii* (*Cronobacter*) with available commercial systems may be difficult.[17] Molecular methods revealed that several strains identified as *E. sakazakii* by commercial biochemical kits belonged to *E. cloacae*-complex species.[15,17]

74.1.5.2 Molecular Identification

Genus-Specific Detection and Identification. There are no specific molecular methods for identification of the genus *Enterobacter*. However, oligonucleotide probes have been designed to detect the 23S and 16S rRNA gene sequences of the family Enterobacteriaceae. Sequence analysis of the *recN*, *thdF*, *rpoA*, and/or *rpoB* genes can be used to estimate the genetic similarities between Enterobacteriaceae and facilitate identification and differentiate of species within this family, providing greater discrimination than 16S rRNA gene sequence data.[9,18]

A number of molecular approaches have been developed to specifically identify *E. sakazakii* (*Cronobacter*) strains. Conventional and real-time polymerase chain reaction (PCR) methods enable (quantitative), sensitive, specific, and rapid detection from infant formula, enrichment broths, and agar media. Targets for conventional PCR systems include the 16S rRNA gene,[92–95] the *omp*A gene,[96] the gene coding for the 1,6 α-glucosidase in *E. sakazakii*[97] and a gene encoding a zinc-containing metalloprotease.[98] Of the real-time assays that have been described so far, three of them target the 16S rRNA gene,[99–101] one is based on the region located between the 16S rRNA and the 23 rRNA genes,[102] another targets the region between the tRNA-glu and 23S rRNA genes,[103] and one targets the *dna*G gene.[104,105]

In a study by Lehner et al.,[97] the molecular basis of the α-glucosidase activity in *E. sakazakii* (*Cronobacter*) was determined, and the potential of this PCR system targeting the 1,6-α-glucosidase for the specific identification of *E. sakazakii* (*Cronobacter*) was evaluated in two studies.[17,93] The real-time PCR assay targeting the *dna*G gene, which is a component of the macromolecular synthesis operon, was originally developed by Seo and Brackett[104] and further modified by Drudy et al.[105] To date, these are the only two PCR-based methods that have been extensively evaluated on a broad panel of target as well as nontarget strains, and both of these proved to be 100% sensitive and specific for *E. sakazakii* (*Cronobacter*).[17] Both methods are described in detail below.

Genetic-based systems include the BAX® Assay for the Detection of *Enterobacter sakazakii* (Dupont Qualicon, USA) and the foodproof® *Enterobacteriaceae* plus *E. sakazakii* Detection System (BIOTECON Diagnostics, Germany). The latter qualitatively detects *Enterobacteriaceae* DNA while simultaneously identifying the presence of *Enterobacter sakazakii*. To avoid false-positive PCR results, the DNA of dead bacterial cells in the sample is eliminated using a light sensitive reagent that only penetrates the cell membranes of dead cells and covalently binds to DNA preventing amplification. The FDA has also developed a revised method involving an enrichment step, centrifugation, and a combination of two chromogenic agars and the *dna*G gene PCR for confirmation.[106]

The VIT® (vermicon identification technology, Munich, Germany) represents an alternative to the DNA-targeted PCR-based detection and identification systems for *E. sakazakii*. It is based on fluorescently labeled gene probes targeting specified regions on the ribosomal RNA of the bacteria; therefore, only live cells are detected by the system. The test is performed on 1 mL of overnight culture of enrichment broth or rich media broth (e.g., BHI, LB) after inoculation with presumptive colony material according to the instructions of the manufacturer. The method has successfully been used in a comparative study evaluating cultural and molecular identification systems for *E. sakazakii* (*Cronobacter*).[93] Another hybridization-based method using a peptide nucleic acid probe for the specific detection of *Cronobacter* genomospecies (*E. sakazakii*) has been published.[107]

Species-Specific Identification. As identification of *E. sakazakii* (*Cronobacter*) isolates to the species level is required for epidemiological studies, a PCR system for the differentiation of the six proposed species was developed in a recent study.[108] The *rpo*B gene was chosen as target to develop different conventional PCR systems which enable strains previously confirmed as belonging to the genus *Cronobacter* to be further discriminated to the species level. This method is described in detail below.

74.1.5.3 Subtyping

A number of methods have been developed that can be used to subtype strains of *Enterobacter* including antibiograms, biotyping, serogrouping, plasmid profiling, ribotyping, random amplification of polymorphic DNA (RAPD), arbitrarily primed PCR (AP–PCR), repetitive sequence primed PCR (REP–PCR), enterobacterial repetitive intergenic consensus (ERIC) PCR, amplified fragment length polymorphism (AFLP), and pulsed-field gel electrophoresis (PFGE). Techniques that generate a DNA fingerprint pattern are more discriminatory and reliable than classic techniques such as biotyping or antibiograms. During the course of a clinical outbreak the use of antibiotics can lead to acquired resistance altering the antibiograms of associated strains. *Enterobacter* strains frequently carry extrachromosomal elements; however, problems can arise during subtyping due to unstable or absent plasmids in some strains. PFGE is currently seen as the gold standard for molecular subtyping of foodborne pathogens and has been used to investigate outbreaks in NICUs involving *E. cloacae*, *E. aerogenes*, *E. gergoviae*, and *E. sakazakii* (*Cronobacter*). It has also been successfully used to trace the dissemination of *E. sakazakii* (*Cronobacter*) strains within manufacturing facilities.

The use of PFGE typing to investigate infection and colonization in hospitals has highlighted the major contribution of patient-to-patient transmission in the dissemination of *Enterobacter* strains and the importance of instilling effective infection control procedures among health care workers. Jalaluddin et al.[109] described the characterization of 45 nosocomial *E. aerogenes* strains in a Belgium Hospital using antibiograms, plasmid profiling, and PFGE. *E. aerogenes* was isolated from 19 respiratory specimens, 13 pus specimens, 7 blood specimens, 4 urinary specimens, 1 catheter specimen, and 1 heparin vial. The majority of the strains (82%) showed multiple resistances to commonly used antibiotics. Two major plasmid profiles were found, and PFGE with *Spe*I and/or *Xba*I restriction enzymes revealed that a single clone was responsible for causing 80% of infections or colonizations in different units throughout the year.[109]

PFGE of *Xba*I-digested chromosomal DNA from *Enterobacter gergoviae* strains in an NICU in Malaysia, where 11 babies were infected, revealed probable cross-contamination via a dextrose saline used for the dilution of parenteral antibiotics and the hands of a health care worker.[56]

Using PFGE, bacteremia and asymptomatic colonization due to an epidemic clone of *E. cloacae* were identified in one of two French neonatal units from December 2002 to March 2003. The nurse-to-patient ratio and hygiene practices were identified as key differences between the two units, and the outbreak was contained using improved hygiene practices combined with restriction of admission.[57] Clone relatedness of ESBL-producing *E. cloacae* in a Spanish cardiothoracic intensive care unit (CT-ICU) from July to September 2005 was determined by PFGE revealing two patterns. The outbreak was stopped by the implementation of barrier measures and cephalosporin restriction.[61] PFGE has also been used to investigate an outbreak of *E. cloacae* in the NICU of a hospital in Gauteng, South Africa. Three isolates from patient blood cultures, six from environmental sources, and one from the hands of a staff member belonged to the same genotypic cluster.[53] A surveillance program implementing PFGE to investigate the molecular epidemiology of *E. cloacae* in French neonatal units over a 2-year period revealed 90 different pulsotypes among 199 clinical and screening isolates from neonates, including three major epidemic clones. Mother-to-child transmission could be directly attributed to only 8.8% of *E. cloacae* intestinal colonizations.[57] Plasmid analysis and PFGE demonstrated similar discriminatory ability for analysis of systemic nosocomial infections caused by *E. cloacae* from 1995 to 1997 in the NICU of a Brazilian hospital. Three small, originally unrecognized, outbreaks occurred in different years, caused by strains with distinctive DNA profiles.[110] Fifty colonizing and infecting *E. cloacae* isolates from a Spanish NICU and four epidemiologically unrelated strains were characterized by PFGE, ribotyping, ERIC–PCR, and plasmid profiling. PFGE was the most discriminatory technique and identified 13 types compared with ERIC–PCR, plasmid profiling, and ribotyping, which identified 11, 11, and 7 types, respectively. Five different clones caused 15 cases of sepsis with disappearance of *E. cloacae* from the NICU after reinforcement of hygienic measures.[54] *To aid epidemiology, several isolates from each patient may need to be typed. E. cloacae* strains isolated from a variety of clinical specimens derived from 18 patients in a Dutch hospital produced 16 PFGE types. Several genotypes were encountered in multiple patients who had not been clustered in time or space, indicating the existence of a common infectious source. Five patients were colonized or infected by multiple types, with one patient colonized by at least four different PFGE genotypes.[111]

Smeets et al. used PFGE to confirm the epidemiological evidence that a contaminated dish brush used for cleaning bottles was the source of three cases of *E. sakazakii* (*Cronobacter*) infections in 1981.[112] In 1994, an outbreak occurred in France involving 17 neonates. *E. sakazakii* (*Cronobacter*) isolates were obtained from various anatomical sites, prepared feeds, and unused infant formula. PFGE analysis described four clusters, two of which contained isolates from neonates who ranged from asymptomatic colonized individuals to case fatalities. One other cluster contained isolates from the prepared feed, an asymptomatic neonate and a neonate with mild digestive problems. The fourth cluster comprised the isolates from the unused powdered formula. In this case, no link could be made between the powdered formula and the outbreak, and it is possible that the prepared feed became contaminated via an alternate source.[113] A further outbreak occurred in France in 2004 involving 13 infants; nine were asymptomatically colonized, two developed bacterial meningitis, one had conjunctivitis, and one suffered hemorrhagic colitis. The outbreak was investigated using automated ribotyping (using restriction endonucleases *Eco*RI, *Pst*I, and *Pvu*II) and by PFGE using the enzymes *Xba*I and *Spe*I. A total of nine isolates from eight neonates was found to have undistinguishable ribotypes and PFGE patterns from isolates obtained from four separate lots of infant formula, thus establishing a clear link between the outbreak and the product.[114] Recently, factory surveillance studies have employed a PFGE method for tracing *E. sakazakii* (*Cronobacter*) isolates within infant food-manufacturing facilities.[45,115] These studies were based on the protocol of Nazarowec-White and Farber[116] and used *Xba*I. Currently, in conjunction with the PulseNet Program at the Centers for Disease Control and Prevention (CDC) in the United States, a collaborative study is underway in several international laboratories to develop a standard protocol for *E. sakazakii* (*Cronobacter*) PFGE typing.

REP–PCR. Although PFGE is considered the gold standard for the detection of clonality in disease outbreaks, PCR-based methods are cheaper, easier to perform, and provide faster results. PFGE and ERIC- and REP–PCR were used to analyze 56 *E. cloacae* complex isolates belonging to genetic clusters III and VI at a hospital in Germany. In comparison to PFGE, a specificity of 14% was determined for ERIC–PCR, which appeared to differentiate between genovars rather than between strains. In contrast, REP–PCR yielded a specificity of 90% and differentiated better on the strain level.[117]

Restriction endonuclease analysis (REA) and REP–PCR were used to study 62 *E. aerogenes* isolates collected from 43 patients over 3 months in a U.S. tertiary-care hospital. REA with *Hind*III or *Eco*RI yielded complex banding patterns that differentiated community-acquired from some hospital-acquired organisms. Less complex fingerprints were obtained with REP–PCR, which also distinguished between epidemiologically unrelated strains. Two clusters of genomically distinct isolates were identified, suggesting nosocomial acquisition of *E. aerogenes*.[118]

More recently, the discriminative power of REP–PCR has been compared to PFGE for *E. sakazakii* (*Cronobacter*) isolates.[119] Using Simpson's index of diversity, values of 0.974 and 0.998 were calculated for REP–PCR and PFGE respectively at a similarity cutoff of 95%, demonstrating good correlation with a high degree of genetic heterogeneity among the isolates.

RAPD. RAPD typing has been used to analyze clonal relationships between *E. sakazakii* (*Cronobacter*) strains in a number of independent studies.[45,106,116,120–123] Nazarowec-White and Farber[116] developed protocols for the molecular subtyping of *E. sakazakii* (*Cronobacter*) by ribotyping, RAPD, and PFGE. The authors showed that RAPD and PFGE were the most discriminatory subtyping schemes for *E. sakazakii* (*Cronobacter*), followed by ribotyping and two microbiological typing methods—biotyping and antibiograms. In addition, Clementino et al. comparatively assessed the usefulness of the tRNA intergenic spacer, 16S-23S internal transcribed spacer, and RAPD for discrimination of *E. sakazakii* (*Cronobacter*) and *E. cloacae* isolates.[120]

Between 1994 and 1998, endemic *E. cloacae* strains responsible for gastrointestinal tract colonization and UTI of renal transplant patients in Poland were characterized using RAPD and PFGE. Increased resistance to aminoglycosides was recorded over the course of the study, and the majority of multiresistant *E. cloacae* belonged to two predominant clones.[124] RAPD has also revealed clonal identity between 21 highly fluoroquinolone-resistant *E. hormaechei* strains isolated from two hospitals in Marseille, France.[125]

E. aerogenes isolated from two ICUs in France during a 4-month period were compared using RAPD. In nine instances, isolates with indistinguishable RAPD patterns were detected in two-to-five patients over a 3- to 15-day period, suggesting patient-to-patient transmission. In one ICU, patients harbored 1–12 distinguishable isolates during their hospital stay, and it was concluded that successive patient-to-patient transmission of different *E. aerogenes* strains occurred. In contrast, in the second ICU isolates were indistinguishable by RAPD and probably involved the extensive spread of a single strain.[126] Also, RAPD and ERIC–PCR were used to determine the epidemiological relatedness of 123 *E. aerogenes* isolates collected from 23 hospital laboratories in France. A prevalent clone harboring the plasmid encoding for extended-spectrum beta-lactamase of the TEM-24 type was found in 21 of the laboratories. Dissemination appeared to be clonal and probably resulted from the general use of broad-spectrum cephalosporins and quinolones.[127]

Ribotyping. Ribotypes are generated by probing restriction fragments of genomic DNA for the highly conserved genes coding for the 16S and 23S rRNA. Small variations between strains occur in the less conserved, flanking genes and intergenic sections of the genome, resulting in fragments of unequal size. These are separated using gel electrophoresis, transferred to a membrane and the individual strain fingerprints revealed by hybridization of chemiluminescent probes. Riboprint patterns are analyzed based on the number, size, and signal intensity of the detected fragments. Comparison to existing entries in a riboprint database allows species- and subspecies-level identification.

The automated RiboPrinter™ Microbial Characterization System (Dupont Qualicon, USA) was used as part of a polyphasic taxonomic characterization of *E. sakazakii* (*Cronobacter*) strains.[15] Isolates were grown on TSA (18 h, 37°C) and prepared according to standard procedures[128] using the *Eco*RI restriction enzyme. The strains were divided into four distinct clusters; *C. dublinensis*, *C. turicensis*, and *C. muytjensii* occupied individual clusters, whereas *C. malonaticus* appeared as one subcluster among several subclusters of *C. sakazakii*.[15] Nazarowec-White and Farber[116] studying three *E. sakazakii* (*Cronobacter*) isolates obtained from one hospital over 11 years showed that they had indistinguishable ribotype patterns, indicating possible persistence in the environment.

Ceftazidime-resistant *E. aerogenes* isolated from blood cultures of three patients analysed using PFGE and ribotyping were found to be clonally similar to each other and to a strain isolated from a three-way stopcock connected to the indwelling catheter in one of the patients.[129]

In 1991, a comparison was conducted between ribotyping and a combination of serotyping, phage susceptibility, and biotype pattern for the subtyping of 45 *E. cloacae* isolates from 36 patients in nine UK hospitals. All isolates could be assigned to one of three biotypes, but many phage-sensitivity patterns were evident. Twenty-nine distinct strains were identified by combined typing. Ribotyping proved to be highly discriminating between strains producing 30 different profiles based on 28 bands. Overall, agreement between the ribotyping and combined typing methods was good (84.4%).[130] In December 1992, 41 strains of *E. cloacae*, including 12 from the oropharynx and respiratory tract of six ventilated neonates in a French NICU, were analyzed using plasmid profile analysis, AP–PCR, and ribotyping (with *Eco*RI or *Pvu*II endonuclease). Ribotyping and AP–PCR demonstrated that 12 epidemiologically unrelated isolates exhibited distinct genomic patterns, as did each of the strains isolated from neonates hospitalized before or after the epidemic peak. Two clones of outbreak strains were found, each comprising isolates from the ventilated neonates as well as isolates from children in other wards, with concordant results being obtained by the three typing methods.[52] *E. cloacae* strains with a derepressed cephalosporinase were studied in a French pediatric ward between February 1990 and January 1991. Fifty-two isolates were recovered from 200 neonates (stool, blood), and 14 strains were isolated from the neonatal environment. Typing of 36 *E. cloacae* isolates selected from 21 neonates, the neonatal ward environment and from other wards was carried out using antibiograms, biotyping, and ribotyping methods. No predominant biotyping pattern could be established, and 32 isolates had the same antibiotic resistance pattern, while ribotyping using two endonucleases (*Eco*RI and *Bam*HI) showed 28 isolates produced a single pattern.[51] In 1993, ribotyping of 109 *E. cloacae* isolates indistinguishable by biotyping, phagetyping, and O-serotyping, revealed some cases of cross-infection or common sources of infection. Problems arose with respect to biotyping and O-serotyping, as well as phagetyping; in particular, a number of isolates with the same ribotype showed differences in phage type, indicating that multiple typing schemes can be employed to increase discrimination between clonal strains.[131]

BOX–PCR. Proudy et al. examined the discriminative power of the 154 bp BOX element against the sequencing of the *fli*C gene and PFGE using 27 *E. sakazakii* (*Cronobacter*) strains from clinical and environmental sources.[132] The BOX–PCR results showed 92% agreement with PFGE results indicating the potential of this typing method for epidemiological investigation, whereas *fli*C gene sequencing was poorly discriminative. The application of BOX–PCR genotyping in a factory environment was demonstrated in a follow-up study.[133]

MLVA. Variable number tandem repeat (VNTR) motifs represent sources of genetic polymorphisms. These DNA sequence elements are often maintained within a bacterial species, with individual strains displaying different copy numbers. The length of a tandem repeat at a specific locus can vary as a consequence of DNA slippage during replication or of unequal crossover elements. These differences can be analyzed by amplification of the region and sizing of the resulting amplicons.[134] The high degree of polymorphism at these loci is particularly useful as a target for strain discrimination within bacterial species. Multilocus VNTR analysis (MLVA) is a subtyping method that involves amplification and fragment-size comparison of polymorphic VNTR regions and has successfully been used to type *Enterobacteriaceae*.[135–137] The availability of a complete *E. sakazakii* genome sequence (http://genome/wustl.edu/pub/organism/Microbes/Enteric_Bacteria/Enterobacter_sakazakii/assembly/Enterobacter_sakazakii-4.0/) enabled the identification of VNTR motifs within *C. sakazakii*. Subsequently, an MLVA subtyping scheme was developed and applied on a genotypically and phenotypically diverse collection of *E. sakazakii* (*Cronobacter*) isolates in a study by Mullane et al.[138]

AFLP and AP–PCR. The AFLP technique has been employed in plant and microbiological research to describe the molecular ecology of various niches and can be used to determine inter- and intraspecies relatedness.[139–141] Using AFLP analysis, Mougel et al. found that members of the same genomic species cluster consistently and suggested that future genomic delineation of bacterial species could be based on this approach.[142] This technique was included in a study by Iversen et al. to clarify the taxonomic relationship of over 200 strains previously identified as *E. sakazakii* and gave discriminatory results comparable to DNA–DNA hybridization.[15]

AFLP typing and AP–PCR have been used to investigate *E. cloacae* isolates from a hospital environment, neonates with line sepsis, and asymptomatic individuals. All the patient isolates and two isolates from thermometers were identical prompting the introduction of disposable thermometer covers, after which *E. cloacae* colonization slowly decreased.[55]

Van Acker et al. described 12 cases of NEC in neonates that occurred in 1998. Eleven strains of *E. sakazakii* (*Cronobacter*) were isolated from patient samples and 14 strains were isolated from infant milk preparations. The isolates produced three AP–PCR profiles, with the 14 milk isolates matching the profile from three patients. Four years earlier *E. sakazakii* (*Cronobacter*) had been isolated from a gastrostomy tube of a neonate fed the same type of milk. This original isolate was subsequently shown to have an AP–PCR profile almost identical to the 14 milk and patient isolates, suggesting a persistent contamination problem.[23]

Serotyping and Phagetyping. Traditionally serotyping of *E. cloacae* isolates is performed using antisera prepared in host animals.[143] Serotyping and phagetyping have been used to investigate contamination of total parenteral nutrition solutions as the origin of an outbreak of *E. cloacae* septicemia.[144] In a comparison of biotyping, phagetyping, and O-serotyping using 237 clinical isolates of *E. cloacae,* the typability of strains using the different methods was 100%, 83%, and 85%, respectively. Reproducibility of serotyping was 90%; and of phagetyping, 96%, allowing two major differences to determine distinguishable patterns, O-serotyping had the highest discriminatory power.[145] In an evaluation of the applicability of *E. cloacae* and *Klebsiella* typing reagents for classifying 75 clinical strains of *E. aerogenes*, none were agglutinated by *E. cloacae* O antisera or were sensitive to *E. cloacae* bacteriophages. In contrast, 70 *E. aerogenes* strains reacted with *Klebsiella* capsular antisera, and two-thirds of the strains were lysed by *Klebsiella* typing phages.[146]

Two groups[21,22] reported on four neonates with *E. sakazakii* (*Cronobacter*) infections in the state of Tennessee (USA). The organism was isolated from all four patients, a used can of infant formula milk and the blender (which had heavy growth of the organism). In these outbreak identical biotypes, antibiograms, and plasmid profiles were obtained for patients and environmental isolates.

Molecular-Based Serotyping. Another approach into the direction of subtyping of *E. sakazakii* (*Cronobacter*) was published recently by Mullane et al.[147] In this study, nucleotide polymorphism in the O-antigen coding locus *rfb* was determined and a PCR assay targeting the genes specific for the two prominent serotypes in *C. sakazakii* was developed, and it was reported that the major portion of isolates belonged to two serotype groups now designated O:1 and O:2.[147]

74.2 METHODS

74.2.1 Sample Preparation

Aliquots of matrices to be tested are preenriched normally in 9 × volume of BPW for 24 h at 37°C. A 100-fold dilution (0.1 mL in 10 mL) is transferred into *Cronobacter* screening broth [CSB: 18 g/L bromocressol purple broth, Sigma B2676, 10 g/L sucrose, and 10 mg/L vancomycin hydrochloride (Sigma V2002)][46] or alternative enrichment broths such as, for example, mLST.[13] Fermentation of the carbohydrate in CSB broth results in a color change from purple to yellow. After incubation at 41.5°C for up to 24 h, positive (yellow) broths are streaked onto chromogenic media, for example, chromogenic *Enterobacter sakazakii* agar (CM1055, Oxoid, UK), *Enterobacter sakazakii* Isolation Agar (ESIA, AES Cheminux, France), *Enterobacter sakazakii* Screening Plate (ESPM, bioMérieux, France), or equivalents.

For DNA extraction from broth cultures, commercially available kits, for example, the Qiagen Blood and Tissue Kit (Qiagen, Germany) are recommended. PCR and real-time PCR assays are set up using standard molecular handling procedures and by applying appropriate equipment and using standard instruments.

74.2.2 Detection Procedures

74.2.2.1 Genus-Specific Detection

(i) Standard PCR

Principle. The α-glucosidase (*gluA*) targeted amplification-based assay developed by Lehner et al.[97] employs the following primers: EsAgf: 5′-TGA AAG CAA TCG ACA AGA AG-3′ and EsAgr: 5′-ACT CAT TAC CCC TCC TGA TG-3′ generating a product of 1680 bp in size. The *glu*A_short fragment is amplified using primers EsAg5f: 5′-TAT CAG ATC TAC CCG CGC-3′ and EsAg5_5r: 5′-TTG ATG CCA AGC TGT TGC-3′ resulting in a 105 bp amplicon.

Procedure

1. Prepare a PCR mix (50 μL) containing 5 pmol of forward and reverse primers, 100 μM dNTPs, 1× *Taq* DNA polymerase buffer, and 2 U *Taq* DNA polymerase (Promega).
2. For PCR amplification use the following conditions: 94°C for 2 min; 29 cycles of 94°C for 30 s, 60°C for 60 s (short fragment: 62°C for 30 s), and 72°C for 90 s (short fragment: 72°C for 30 s); and a final elongation step at 72°C for 5 min.
3. Visualize PCR products in a 1% (short fragment: 2%) agarose gel and ethidium bromide staining.

(ii) Real-Time PCR

Principle. The real-time PCR assay targeting the *dna*G gene originally developed by Seo and Brackett[104] and modified by Drudy et al.[105] uses following primers for amplification: Fwd: 5′-GGG ATA TTG TCC CCT GAA ACA G-3′ and Rev: 5′-CAG GAA TAA GCC GCG ATT-3′, together with the TaqMan probe (6-FAM-ACA GAG TAG TAG TTG TAG AGG CCG TGC TTC C-TMR).

Procedure

1. Prepare a PCR mix (20 μL) using 200 μM deoxynucleoside triphosphates, 4.0 mM $MgCl_2$, 2.0 μL of 10× reaction buffer, 10 mM of each primer, 2.5 mM of probe, 1 U *Taq* DNA polymerase, and 2 μL DNA template (containing approximately 100 ng of DNA).
2. For PCR amplification use: 95°C for 15 min followed by 40 cycles of 95°C for 15 s and 60°C for 50 s.

Note: In the study by Drudy et al.[105] the RT–PCR was performed on a Corbett Research Rotor-Gene RG 3000 instrument.

74.2.2.2 Species-Specific Detection

Principle. The *rpo*B gene targeted assay for the identification of strains to the species level developed by Stoop et al.[108] employs following primers for the detection of *C. sakazakii*: Csakf 5′-ACG CCA AGC CTA TCT CCG CG-3′ and Csakr 5′-ACG GTT GGC GTC ATC GTG-3′; for *C. malonaticus*: Cmalf 5′-CGT CGT ATC TCT GCT CTC-3′ and Cmalr 5′-AGG TTG GTG TTC GCC TGA-3′; for *C. turicensis*: Cturf 5′-CGG TAA AAG AGT TCT TCG GC-3′ and Cturr 5′-GTA CCG CTC C-3′; for *C. dublinensis*: Cdublf 5′-GCA CAA GCG TCG TAT CTC C-3′ and Cdublr 5′-TTG GCG TCA TCG TGT TCC-3′; for *C. muytjensii*: Cmuyf 5′-TGT CCG TGT ATG CGC AGA CC-3′ and Cmuyr 5′-TGT TCG CAC CCA TCA ATG CG-3′; and for *C.* genomospecies 1: Cgenomf 5′-ACA AAC GTC GTA TCT CTG CG-3′ and Cgenomr 5′-AGC ACG TTC CAT ACC GGT C-3′. Following amplicon sizes are expected: 514 bp (*C. sakazakii*), 251 bp (*C. malonaticus*), 628 bp (*C. turicensis*), 418 bp (*C. dublinensis*), 289 bp (*C. muytjensii*), and 506 bp for *Cronobacter* genomospecies 1.

Procedure

1. Prepare a reaction mixture (total volume 50 μL) containing GoTaq Green Master Mix (Promega) with a final concentration of 1.5 mM $MgCl_2$, 200 μM dNTPs each, and primers at 10 pmol each.
2. For PCR use the following conditions: 94°C for 3 min; 30 cycles of denaturing at 94°C for 1 min, annealing at 67°C (*C. sakazakii*), 60°C (*C. malonaticus*), 61°C (*C. turicensis, C. muytjensii, C.* genomospecies 1), or 62°C (*C. dublinensis*) for 30 s, and elongation at 72°C for 1 min (*C. sakazakii, C. turicensis*), and 30 s for all other assays; and a final elongation step at 72°C for 5 min.
3. Visualize the reaction products on a 1.5% agarose gel followed by ethidium bromide staining and examination under UV light.

Note: As the *rpoB* gene sequences for *C. sakazakii* and *C. malonaticus* are very closely related, a two-step procedure in which the primer pairs Cmalf/Cmalr are used in a follow-up PCR on strains that have shown to be positive in the amplification with the *C. sakazakii* primers Csakf/Csakr is necessary in order to reliably discern *C. malonaticus* from *C. sakazakii*.

74.2.2.3 PFGE-Based Genotyping

Principle. PFGE is a nonamplified technique that separates long strands of DNA molecules (larger than 15–20 kb) previously digested with restriction enzyme(s). Periodical switching of the voltage in three directions results in the reorientation of DNA moving through the gel in a size dependent manner, which facilitates a finer resolution as it aligns and realigns to the applied electrical field. The most common restriction enzymes used for *Enterobacter* are *Xba*I, *Spe*I, *Not*I, and *Sma*I. The procedure described was recently used by Iversen et al.[45] for the subtyping of *Cronobacter* spp. isolates in a factory survey study.

Procedure

1. Inoculate 20 mL of brain heart infusion broth (BHI) with 200 μL of a BHI overnight culture of the respective strains, and incubate with shaking (200 rpm) for 3–4 h at 37°C until growth reached an OD of approximately 0.8 at 610 nm.
2. Harvest cells from 1 mL of broth by centrifugation (10,000 × *g* for 5 min), and wash once in wash buffer (0.85% NaCl).
3. Resuspend cells in 0.5 mL of wash buffer and mix with an equal volume of 1.2% pulsed-field certified agarose (BioRad) in water.
4. Dispense 100 μL aliquots of the bacteria agarose mixture into disposable plug moulds (BioRad), and allow the plugs to solidify at 4°C for 20 min.
5. Place plugs into tubes containing 1 mL lysis buffer (50 mM Tris, pH 8.0; and 50 mM EDTA, pH 8.0; 1% *N*-lauroylsarcosine, and 1 mg/mL freshly added proteinase K (Sigma) per plug and incubate at 55°C for 20–24 h.
6. Wash plugs in 1 mL per plug TE buffer (10 mM Tris pH 8.0, 1 mM EDTA pH8.0) four times for 1 h at room temperature and store in TE buffer at 4°C until restriction endonuclease digestion.
7. Cut agarose plugs into half and equilibrate twice in 300 μL of 1× restriction buffer H (Roche) per (half) plug.
8. Set up restriction digest of DNA containing agarose plugs in 300 μL fresh 1× restriction buffer H and 50 U of enzyme *Xba*I (Roche, Rotkreuz, Switzerland) for 4 h at 37°C.
9. Separate digested DNA in a 1% pulsed-field certified agarose (BioRad) gel in 0.5 TBE running buffer in a CHEF-DR III system (BioRad) instrument.
10. Use following conditions for the separation of the digested fragments: pulse time: 5–50 s linear ramping for 20 h at 14°C, 120° included angle.
11. Stain gel for 30 min with ethidium bromide (10 mg/L in water) and destain in water, and visualize gel on a UV transilluminator.
12. Use a CCD camera system (e.g., BioRad) to photograph the gel and save the patterns in TIFF file format for further analysis using a fingerprint analysis software (e.g., GelCompar tool of the BioNumerics 5.0 software, Applied Maths, Belgium).

Note: The running buffer may by supplemented with 50 mM Thiourea (Sigma), to overcome the problem of nontypable strains due to nucleolytic activity of a peracid derivative of Tris possibly formed at the anode during electrophoresis.[148]

74.3 FUTURE PERSPECTIVES

The heterogeneity of the genus *Enterobacter*, and further discovery of new *Enterobacter* species is likely to lead to ongoing taxonomic changes over the coming years. Important future work lies in the further characterization of clinical isolates to expand knowledge on the divergence of *E. cloacae* complex clades and the extent of virulence potential within these species. The impact of *Enterobacter* species in the clinical setting is likely to continue until advancements are made in regards to control of antibiotic resistance.

Management strategies to control both the spread of virulent *Enterobacter* strains in clinical settings, as well as the dissemination of contaminating *Enterobacter/Cronobacter* in manufacturing environments, can be aided by further development of appropriate subtyping schemata. Molecular methods such as MLVA may provide more convenient and economical alternatives to PFGE. Although the development of specific methods for other *Enterobacter* is not foreseen, improvements in methods for detection of *E. sakazakii* (*Cronobacter*) will continue with publication of EN ISO and AOAC horizontal standards and validation of alternative molecular-based methods. A reduction of the risk to children from *E. sakazakii* (*Cronobacter*) is being facilitated by continued improvements in infant formula hygiene standards, including recommendations to consumers and health care professionals concerning the handling of reconstituted infant formula.

Commonalities in some pathogenicity or virulence factors involved in bacterial infection of plants and humans have been demonstrated.[149–151] Furthermore, it has been suggested for some foodborne pathogens that intermediate residence in plants may modulate pathogenicity by initiation of expression of virulence associated factors.[152,153] Thus, whole genome expression studies of *E. sakazakii* (*Cronobacter*) internalized within plants may be an alternative model to identify/study the expression of relevant virulence factors in these organisms.

REFERENCES

1. Janda, J.M., and Abbott, S.L., *The Enterobacteria*, 2nd ed., ASM Press, Washington DC, 2005.
2. Kämpfer, P., Ruppel, S., and Remus, R., *Enterobacter radicinicitans* sp. nov., a plant growth promoting species of the

family *Enterobacetriaceae*, *Syst. Appl. Microbiol.*, 28, 213, 2005.
3. Stephan, R. et al., *Enterobacter turicensis* sp. nov. and *Enterobacter helveticus* sp. nov. isolated from fruit powder, *Int. J. Syst. Evol. Microbiol.*, 57, 820, 2007.
4. Stephan, R. et al., *Enterobacter pulveris* sp. nov. isolated from fruit powder, infant formula and infant formula production environment, *Int. J. Syst. Evol. Microbiol.*, 58, 237, 2008.
5. Peng, G. et al., *Enterobacter oryzae* sp. nov., a nitrogen-fixing bacterium isolated from the wild rice species Oryza latifolia, *Int. J. Syst. Evol. Microbiol.*, 59, 1650, 2009.
6. Hoffmann, H., and Roggenkamp, A., Population genetics of the nomenspecies *Enterobacter cloacae, Appl. Environ. Microbiol.*, 69, 5306, 2003.
7. Mullane, N. et al., *Enterobacter sakazakii* an emerging bacterial pathogen with implications for infant health, *Minerva Pediatr.*, 59, 137, 2007.
8. Hormaeche, E., and Edwards, P.R., A proposed genus *Enterobacter*, *Int. Bull. Bacteriol. Nomencl. Taxon.*, 10, 71, 1960.
9. Paauw, A. et al., Genomic diversity within the *Enterobacter cloacae* complex, *PLoS ONE*, 3, 8, e3018. doi:10.1371/journal.pone.0003018, 2008.
10. Farmer, J.J. et al., *Enterobacter sakazakii*: A new species of *Enterobacteriaceae* isolated from clinical species, *Int. J. Syst. Evol. Bacteriol.*, 30, 569, 1980.
11. Lehner, A., Tasara, T., and Stephan, R., 16S rRNA gene based analysis of *Enterobacter sakazakii* strains from different sources and development of a PCR assay for identification, *BMC Microbiol.*, 4, 43, 2004.
12. Iversen, C., Waddington, M., and Forsythe, S.J., Identification and phylogeny of *Enterobacter sakazakii* relative to *Enterobacter* and *Citrobacter* species, *J. Clin. Microbiol.*, 42, 5368, 2004.
13. Iversen, C. et al., The biochemical differentiation of *Enterobacter sakazakii* genotypes, *BMC Microbiol.,* 6, 94, 2006.
14. Iversen, C. et al., Identification of *Enterobacter sakazakii* from closely related species: The use of Artificial Neural Networks in the analysis of biochemical and 16S rDNA data, *BMC Microbiol.,* 6, 28, 2006.
15. Iversen, C. et al., The taxonomy of *Enterobacter sakazakii*: Proposal of a new genus *Cronobacter* gen. nov., and descriptions of *Cronobacter sakazakii* comb. nov., *Cronobacter sakazakii* subsp. *sakazakii*, comb. nov. *Cronobacter sakazakii* subsp. *malonaticus* subsp. nov., *Cronobacter turicensis* sp. nov., *Cronobacter muytjensii* sp. nov. *Cronobacter dublinensis* sp. nov. and *Cronobacter* genomospecies 1, *BMC Evol. Biol.*, 7, 64, 2007.
16. Iversen, C. et al., *Cronobacter* gen. nov., a new genus to accommodate the biogroups of *Enterobacter sakazakii*, and proposal of *Cronobacter sakazakii* gen. nov. comb. nov., *C. malonaticus* sp. nov., *C. turicensis* sp. nov., *C. muytjensii* sp. 3 nov., *C. dublinensis* sp. nov., *Cronobacter* genomospecies 1, and of three subspecies, *C. dublinensis* sp. nov. subsp. *dublinensis* subsp. nov., *C. dublinensis* sp. nov. subsp. *lausannensis* subsp. nov., and *C. dublinensis* sp. nov. subsp. *lactaridi* subsp. Nov., *Int. J. Syst. Evol. Microbiol.*, 58, 1442, 2008.
17. Iversen, C. et al., Identification of *Cronobacter* spp. (*Enterobacter sakazakii*), *J. Clin. Microbiol.*, 45, 3814, 2007.
18. Kuhnert, P. et al., Phylogeny and prediction of genetic similarity of *Cronobacter* and related taxa by multilocus sequence analysis (MLSA), *Int. J. Food Microbiol.*, 136, 152, 2009.
19. Anonymous, FAO-WHO, *Enterobacter sakazakii* and other microorganisms in powdered infant formula: Meeting report, MRA Series 6, WHO Geneva, 2004.
20. Anonymous, FAO-WHO, *Enterobacter sakazakii* and *Salmonella* in powdered infant formula, Second Risk Assessment Workshop, January 16–20, WHO Rome, Italy, 2006.
21. Simmons, B.P. et al., *Enterobacter sakazakii* infections in neonates associated with intrinsic contamination of powdered infant formula, *Infect. Control. Hosp. Epidemiol.*, 10, 398, 1989.
22. Clark, N.C. et al., Epidemiologic typing of *Enterobacter sakazakii* in two neonatal nosocomial outbreaks, *Diagn. Microbiol. Infect. Dis.*, 13, 467, 1990.
23. Van Acker, J. et al., Outbreak of necrotizing enterocolitis associated *with Enterobacter sakazakii* in powdered milk formula, *J. Clin. Microbiol.*, 39, 293, 2001.
24. Muytjens, H.L., and Kollee, L.A.A., *Enterobacter sakazakii* meningitis in neonates: Causative role of formula, *Pediatr. Infect. Dis. J.,* 9, 372, 1990.
25. Noriega, F.R. et al., Nosocomial bacteremia caused by *Enterobacter sakazakii* and *Leuconostoc mesenteroides* resulting from extrinsic contamination of infant formula, *Pediatr. Infect. Dis.,* 9, 447, 1990.
26. Biering, G. et al., Three cases of neonatal meningitis caused by *Enterobacter sakazakii* in powdered milk, *J. Clin. Microbiol.,* 27, 2054–2056, 1989.
27. ICMSF International Commission on Microbiological Specifications for Foods. Micro-organisms in Foods Number 7. Microbiological Testing in Food Safety Management. Kluwer Academic/Plenum Publishers, 2002.
28 Dennison, S.K., and Morris, J., Multiresistant *Enterobacter sakazakii* wound infection in an adult, *Infect. Med.,* 19, 533, 2002.
29. Jimenez, E., and Gimenez, C., Septic shock due to *Enterobacter sakazakii, Clin. Microbiol. Newsl.*, 4, 30, 1982.
30. Hawkins, R.E., Lissner, C.R., and Sanford, J.P., *Enterobacter sakazakii* bacteremia in an adult, *Southern Med. J.*, 84, 793, 1991.
31. Lai, K.K., *Enterobacter sakazakii* infections among neonates infants children and adults: Case reports and a review of the literature, *Med. Baltimore,* 80, 113, 2001.
32. See, K.C., Than, H.A., and Tang, T., *Enterobacter sakazakii* bacteraemia with multiple splenic abscesses in a 75-year-old woman: A case report, *Age Ageing*, 36, 595, 2007.
33. Gosney, M.A. et al., *Enterobacter sakazakii* in the mouths of stroke patients and its association with aspiration pneumonia, *Eur. J. Inter. Med.,* 17, 185, 2006.
34. Cruz, A.T., Cazacu, A.C., and Allen, C.H., *Pantoea agglomerans*, a plant pathogen causing human disease, *J. Clin. Microbiol.*, 45, 1989, 2007.
35. Izard, D. et al., Deoxyribonucleic acid relatedness between *Enterobacter cloacae* and *Enterobacter amnigenus* sp. nov., *Int. J. Syst. Evol. Microbiol.,* 31, 35, 1981.
36. Dickey, R.S., and Zumoff, C.H., Emended description of *Enterobacter cancerogenus comb.* nov. (Formerly *Erwinia cancerogena*), *Int. J. Syst. Bacteriol.,* 38, 371, 1988.
37. Kohaku Inoue, K. et al., *Enterobacter cowanii* sp. nov., a new species of the family *Enterobacteriaceae, Curr. Microbiol.*, 41, 417, 2000.
38. Brenner, D.J. et al., *Enterobacter gergoviae* sp. nov.: A new species of *Enterobacteriaceae* found in clinical specimens and the environment, *Int. J. Syst. Evol. Microbiol.*, 30, 1, 1980.

39. Pavan, M.E. et al., Phylogenetic relationships of the genus *Kluyvera*: Transfer of *Enterobacter intermedius* Izard et al. 1980 to the genus *Kluyvera* as *Kluyvera intermedia* comb. nov. and reclassification of *Kluyvera cochleae* as a later synonym of *Kluyvera intermedia*, *Int. J. Syst. Evol. Microbiol.*, 55, 437, 2005.
40. Young Ryun Chung, Y.R. et al., *Enterobacter pyrinus* sp. nov., an organism associated with brown leaf spot disease of pear trees, *Int. J. Syst. Evol. Microbiol.*, 43, 157, 1993.
41. Schmid, M. et al., Evidence for a plant associated natural habitat of *Cronobacter* spp, *Res. Microbiol.*, 160, 608, 2009.
42. Muytjens, H.L., Roelofs, W.H., and Jaspar, G.H.J., Quality of powdered substitutes for breast milk with regard to members of the family *Enterobacteriaceae*, *J. Clin. Microbiol.*, 26, 743, 1988.
43. Iversen, C., and Forsythe, S., Isolation of *Enterobacter sakazakii* and other *Enterobacteriaceae* from powdered infant formula milk and related products, *Food Microbiol.*, 21,771, 2004.
44. O'Brien, S. et al., *Cronobacter* species (*Enterobacter sakazakii*) in follow-on infant formula, *Lett. Appl. Microbiol.*, 48, 536, 2009.
45. Iversen, C. et al., Genotyping of *Cronobacter* (*Enterobacter sakazakii*) strains isolated from a infant formula processing plant, *Archives for Food Hygiene*, 60, 66, 2009.
46. Mullane, N. et al., Dissemination of *Cronobacter* spp. (*Enterobacter sakazakii*) in a powdered milk protein manufacturing facility, *Appl. Environ. Microbiol.*, 74, 5913, 2008.
47. Molloy, C. et al., Surveillance and characterisation by Pulsed-Field Gel Electrophoresis of *Cronobacter* (*Enterobacter sakazakii*) in farming and domestic environments, food production animals and retail foods, *Int. J. Food Microbiol.*, 136, 198, 2009.
48. Caubilla-Barron, J., and Forsythe, S., Dry stress and survival time of *Enterobacter sakazakii* and other *Enterobacteriaceae*, *J. Food Prot.*, 70, 2111, 2007.
49. Farmer, J.J. III, Boatwright, K.D., and Janda, J.M., *Enterobacteriaceae*. Introduction and identification. In Murray, P.R., Baron, E.J., Jorgensen, J., Pfaller, M.A., and Landry, M.L. (Editors), *Manual of Clinical Microbiology*, 9th ed., Chapter 42, ASM Press, Washington DC, 2007.
50. Bowen, A.B., and Braden, C.R., Invasive *Enterobacter sakazakii* disease in infants, *Emerg. Infect. Dis.*, 12, 1185, 2006.
51. Poilane, I., et al., *Enterobacter cloacae* cross-colonization in neonates demonstrated by ribotyping, *Eur. J. Clin. Microbiol. Infect. Dis.*, 12, 820, 1993.
52. Grattard, F. et al., Arbitrarily primed PCR, ribotyping, and plasmid pattern analysis applied to investigation of a nosocomial outbreak due to *Enterobacter cloacae* in a neonatal intensive care unit, *J. Clin Microbiol.*, 32, 596, 1994.
53. Van Nierop, W.H. et al., Molecular epidemiology of an outbreak of *Enterobacter cloacae* in the neonatal intensive care unit of a provincial hospital in Gauteng, South Africa, *J. Clin. Microbiol.*, 36, 3085, 1998.
54. Fernandez-Baca, V. et al., Molecular epidemiological typing of *Enterobacter cloacae* isolates from a neonatal intensive care unit: Three-year prospective study, *J. Hosp. Infect.*, 49, 173, 2001.
55. Dijk, Y. et al., Management of an outbreak of *Enterobacter cloacae* in a neonatal unit using simple preventive measures, *J. Hosp. Infect.*, 51, 21, 2002.
56. Ganeswire, R., Thongy, K.L., and Puthuchearyz, S.D., Nosocomial outbreak of *Enterobacter gergoviae* bacteraemia in a neonatal intensive care unit, *J. Hosp. Infect.*, 53, 292, 2003.
57. Gbaguidi-Haore, H. et al., Molecular epidemiology of *Enterobacter cloacae* in a neonatal department: A 2-year surveillance study, *Eur. J. Clin. Microbiol. Infect. Dis.*, 27, 643, 2008.
58. Chan, K.L. et al., A study of preantibiotic bacteriology in 125 patients with necrotizing enterocolitis, *Acta Paediatr. (Suppl.)*, 396, 45, 1994.
59. Lucas, A., and Cole, T.J., Breast milk and neonatal necrotizing enterocolitis, *Lancet*, 336, 1519–1523, 1990.
60. Powell, J. et al., Necrotizing enterocolitis. Epidemic following an outbreak of *Enterobacter cloacae* type 3305573 in a neonatal intensive care unit, *Am. J. Dis. Child*, 134, 1152, 1980.
61. Manzur, A. et al., Nosocomial outbreak due to extended-spectrum-beta-lactamase-producing *Enterobacter cloacae* in a cardiothoracic intensive care unit, *J. Clin. Microbiol.*, 45, 2365, 2007.
62. Deitch, E.A., Specian, R.D., and Berg, R.D., Endotoxin-induced bacterial translocation and mucosal permeability: Role of xanthine oxidase, complement activation, and macrophage products, *Crit. Care Med.*, 19, 785, 1991.
63. Townsend, S. et al., The presence of endotoxin in powdered infant formula milk and the influence of endotoxin and *Enterobacter sakazakii* on bacterial translocation in the infant rat, *Food Microbiol.*, 24, 67, 2007.
64. Paton, A.W., and Paton, J.C., *Enterobacter cloacae* producing a Shiga-like toxin II-related cytotoxin associated with a case of hemolytic-uremic syndrome, *J. Clin. Microbiol.*, 34, 463, 1996.
65. Simi, S., et al., A low molecular weight enterotoxic hemolysin from clinical *Enterobacter cloacae*, *Can. J. Microbiol.*, 49, 479, 2003.
66. Prada, J., and Beutin, L., Detection of *Escherichia coli* alpha-haemolysin genes and their expression in a human faecal strain of *Enterobacter cloacae*, *FEMS Microbiol. Lett.*, 63, 111, 1991.
67. Pagotto, F. et al., *Enterobacter sakazakii*: Infectivity and enterotoxin production in vitro and in vivo, *J. Food Prot.*, 66, 370, 2003.
68. Klemm, P., and Schembri, M.A., Bacterial adhesins: Function and structure, *Int J. Med. Microbiol.*, 290, 27, 2000.
69. Keller, R. et al., Occurrence of virulence-associated properties in *Enterobacter cloacae*, *Infect. Immun.*, 66, 645, 1998.
70. Mange, J.P. et al., Adhesive properties of *Enterobacter sakazakii* to human epithelial and brain microvascular endothelial cells, *BMC Microbiolol.*, 6, 58, 2006.
71. Kim, K.P., and Loessner, M.J., *Enterobacter sakazakii* invasion in human intestinal Caco-2 cells requires the host cell cytoskeleton and is enhanced by disruption of tight junctions, *Infect. Immun.*, 76, 562, 2008.
72. de Kort, G. et al., Invasion of rabbit ileal tissue by *Enterobacter cloacae* varies with the concentration of OmpX in the outer membrane, *Infect Immun.*, 62, 4722, 1994.
73. Singamsetty, V.K. et al., Outer membrane protein A expression in *Enterobacter sakazakii* is required to induce microtubule condensation in human brain microvascular endothelial cells for invasion, *Microb. Pathog.*, 42, 181, 2008.
74. Mittal, R. et al., Brain damage in newborn rat model of meningitis by *Enterobacter sakazakii*: A role for outer membrane protein A, *Lab. Invest.*, 89, 263, 2009.
75. Hunter, C.J. et al., *Enterobacter sakazakii* enhances epithelial cell injury by inducing apoptosis in a rat model of necrotizing enterocolitis, *J. Infect. Dis.*, 198, 586, 2008.
76. Raymond, K.N., Dertz, E.A., and Kim, S.S., Enterobactin: An archetype for microbial iron transport, *Proc. Nat. Acad. Sci. USA*, 100, 3584, 2003.

77. Paraje, M.G., Barnes, A.I., and Albesa, I., An *Enterobacter cloacae* toxin able to generate oxidative stress and to provoke dose-dependent lysis of leukocytes, *Int. J. Med. Microbiol.*, 295, 109, 2005.
78. Townsend, S.M. et al., *Enterobacter sakazakii* invades brain capillary endothelial cells, persists in human macrophages influencing cytokine secretion and induces severe brain pathology in the neonatal rat, *Microbiology,* 153, 3538, 2007.
79. Stock, I., and Wiedemann, B., Natural antibiotic susceptibility of *Enterobacter amnigenus*, *Enterobacter cancerogenous, Enterobacter gergoviae* and *Enterobacter sakazakii* strains, *Clin. Microbiol. Infect.,* 8, 564, 2001.
80. Paauw, A. et al., *Enterobacter cloacae* outbreak and emergence of quinolone resistance gene in dutch hospital, *Emerg. Infect. Dis.,* 12, 807, 2006.
81. Stürenburg, E. et al., Short communication—cloning and sequencing of *Enterobacter aerogenes* OmpC-type osmoporin linked to carbapenem resistance, *Int. J. Med. Microbiol.,* 291, 649, 2002.
82. Anonymous, U.S. Food and Drug Administration, Isolation and enumeration of *Enterobacter sakazakii* from dehydrated powdered infant formula, 2002. [Online] http://www.cfsan.fda.gov/~comm/mmesakaz.html, last accessed date 2010-07-13.
83. Iversen, C., and Forsythe, S., Comparison of media for the isolation of *Enterobacter sakazakii*, *Appl. Environ. Microbiol.*, 73, 48, 2007.
84. Iversen, C. et al., Development of a novel screening method for the isolation of *Cronobacter* spp. (*Enterobacter sakazakii*), *Appl. Environ. Microbiol.*, 74, 2550, 2008.
85. Guillaume-Gentil, O. et al., A simple and rapid cultural method for detection of *Enterobacter sakazakii* in environmental samples, *J. Food Prot.,* 68, 64, 2005.
86. Anonymous, Technical Specification ISO/TS 22964. Milk and milk products—detection of *Enterobacter sakazakii*. ISO/TS 22964:2006(E) and IDF/RM 210:2006(E).210, 2006(E), First Edition, 2006.
87. Iversen, C., Druggan, P., and Forsythe, S.J., A selective differential medium for *Enterobacter sakazakii*, *Int. J. Food Microbiol.*, 96, 133, 2004.
88. Leuschner, R.G.K., and Bew, J., A medium for the presumptive detection of *Enterobacter sakazakii* in infant formula: Interlaboratory study, *J. AOAC Int.*, 87, 604, 2004.
89. Oh, S.W., and Kang, D.H., Fluorogenic selective and differential medium for isolation of *Enterobacter sakazakii*, *Appl. Environ. Microbiol.*, 70, 5692, 2004.
90. Restaino, L. et al., A chromogenic plating medium for the isolation and identification of *Enterobacter sakazakii* from foods, food ingredients and environmental sources, *J. Food Prot.*, 69, 315, 2006.
91. Song, K.Y. et al., Evaluation of a chromogenic medium supplemented with glucose for detecting *Enterobacter sakazakii*, *Microbiol. Biotechnol.*, 18, 579, 2008.
92. Mullane, N.R. et al., Detection of *Enterobacter sakazakii* in dried infant milk formula by cationic magnetic bead capture, *Appl. Environ. Microbiol.*, 72, 6325, 2006.
93. Lehner, A. et al., Comparison of two chromogenic media and evaluation of two molecular based identification systems for *Enterobacter sakazakii* detection, *BMC Microbiol.*, 6, 15, 2006.
94. Witthuhn, R.C., Kemp, F., and Britz, T.J., Isolation and PCR detection of *Enterobacter sakazakii* in South African food products, specifically infant formula milks, *World J. Microbiol. Biotechnol.*, 23, 151, 2007.
95. Hassan, A.A. et al., Characterization of the gene encoding the 16S rRNA of *Enterobacter sakazakii* and development of a species-specific PCR method, *Int. J. Food Microbiol.*, 116, 214, 2007.
96. Nair Mohan, K.M., and Venkitanarayanan, K.S., Cloning and sequencing of the *omp*A gene of *Enterobacter sakazakii* and development of an *omp*A targeted PCR for rapid detection of *Enterobacter sakazakii* in infant formula, *Appl. Environ. Microbiol.*, 72, 2539, 2006.
97. Lehner, A. et al., Molecular characterization of the alpha glucosidase activity in *Enterobacter sakazakii* reveals the presence of a putative gene cluster for palatinose metabolism, *Syst. Appl. Microbiol.*, 29, 609, 2006.
98. Kothary, M.M. et al., Characterization of the zinc-containing metalloprotease encoded by *zpx* and development of a species-specific detection method for *Enterobacter sakazakii*, *Appl. Environ. Microbiol.*, 73, 4142, 2007.
99. Malorny, B., and Wagner, M., Detection of *Enterobacter sakazakii* strains by real-time PCR, *J. Food Prot.,* 68, 1623, 2005.
100. Lehmacher, A., Fiegen, M., and Hansen, B., Real-time PCR von *Enterobacter sakazakii* in Säuglingsanfangsnahrung, *J. Verbr. Lebensm.*, 2, 218, 2007.
101. Kang, S.E., Nam, Y.S., and Hong, K.W., Rapid detection of *Enterobacter sakazakii* using TaqMan real-time PCR assay, *Microbiol. Biotechnol.*, 17, 526, 2007.
102. Liu, Y. et al., Real time PCR using TaqMan and SYBR Green for detection of *Enterobacter sakazakii* in infant formula, *J. Microbiol. Methods*, 65, 21, 2006.
103. Derzelle, S., and Dilasser, F., A robotic DNA purification protocol and real-time PCR for the detection of *Enterobacter sakazakii* in powdered milk formulae, *BMC Microbiol.*, 6, 100, 2006.
104. Seo, K.H., and Brackett, R.E., Rapid, specific detection of *Enterobacter sakazakii* in infant formula using a real time PCR assay, *J. Food Prot.*, 68, 59, 2005.
105. Drudy, D. et al., Characterization of a collection of *Enterobacter sakazakii* isolates from environmental and food sources, *Int. J. Food Microbiol.*, 110, 127, 2006.
106. Lampel, K.A., and Chen, Y., Method for the isolation and detection of *Enterobacter sakazakii* (*Cronobacter*) from powdered infant formula, *Int. J. Food Microbiol.*, 136, 179, 2009.
107. Almeida, C. et al., Development and application of a novel peptide nucleic acid probe for the specific detection of *Cronobacter* genomospecies (*Enterobacter sakazakii*) in powdered infant formula, *Appl. Environ. Microbiol.*, 75, 2925, 2009.
108. Stoop, B. et al., Development and evaluation of *rpo*B based PCR systems to differentiate the six proposed species within the genus *Cronobacter*, *Int J. Food Microbiol.*, 136, 165, 2009.
109. Jalaluddin, S. et al., Molecular epidemiological study of nosocomial *Enterobacter aerogenes* isolates in a Belgian hospital, *J. Clin. Microbiol.*, 36, 1846, 1998.
110. Kuboyama, R.H., de Oliveira, H.B., and Moretti-Branchini, M.L., Molecular epidemiology of systemic infection caused by *Enterobacter cloacae* in a high-risk neonatal intensive care unit, *Infect. Cont. Hosp. Epidemiol.,* 24, 490, 2003.
111. van Belkum, A. et al., Rapid emergence of ciprofloxacin-resistant *Enterobacteriaceae* containing multiple gentamicin resistance-associated integrons in a Dutch hospital, *Emerg. Infect. Dis.*, 7, 2001.

112. Smeets, L.C. et al., Genetische karakterisatie van *Enterobacter sakazakii*-isolaten van Nederlandse patiënten met neonatale meningitis, *Ned. Tijdschr. Med. Microbiol.*, 6, 113, 1998.
113. Caubilla-Barron, J. et al., Genotypic and phenotypic analysis of *Enterobacter sakazakii* strains from an outbreak resulting in fatalities in a neonatal intensive care unit, *J. Clin. Microbiol.*, 45, 3979, 2007.
114. Insitut de Veille Sanitaire, Infections à *Enterobacter sakazakii* associées à la consommation d'une préparation en poudre pour nourrissons. France, octobre à décembre 2004, *Rapport d'invest.*, 2006.
115. Mullane, N.R. et al., Application of pulsed-field gel electrophoresis to characterise and trace the prevalence of *Enterobacter sakazakii* in an infant formula processing facility, *Int. J. Food Microbiol.*, 116, 73, 2007.
116. Nazarowec-White, M., and Farber, J.M., Phenotypic and genotypic typing of food and clinical isolates of *Enterobacter sakazakii*, *J. Med. Microbiol.*, 48, 559, 1999.
117. Stumpf, A.N., Roggenkamp, A., and Hoffmann, H., Specificity of enterobacterial repetitive intergenic consensus and repetitive extragenic palindromic polymerase chain reaction for the detection of clonality within the *Enterobacter cloacae* complex, *Diag. Microbiol. Infect. Dis.*, 53, 9, 2005.
118. Georghiou, P.R. et al., Molecular epidemiology of infectious due to *Enterobacter aerogenes*: Identification of hospital outbreak-associated strains by molecular techniques, *Clin. Infect. Dis.*, 20, 84, 1995.
119. Healy, B. et al., Evaluation of an automated rep-PCR system for subtyping *Enterobacter sakazakii*, *J. Food Prot.*, 71, 1372, 2008.
120. Clementino, M.M. et al., PCR analyses of tRNA intergenic spacer, 16S-23S internal transcribed spacer, and randomly amplified polymorphic DNA reveal inter and intraspecific relationships of *Enterobacter cloacae* strains, *J. Clin. Microbiol.*, 39, 3865, 2001.
121. Babalola, O.O. et al., Characterization of potential ethylene-producing rhizosphere bacteria of *Striga*-infested maize and sorghum, *Afr. J. Biotechnol.*, 1, 67, 2003.
122. Sanjaq, S. et al., Improvement and validation of RAPD in comparison to PFGE analysis of *Enterobacter sakazakii* strains, *Int. J. Med. Microbiol.*, 297, 97, 2007.
123. Kim, K. et al., Prevalence and genetic diversity of *Enterobacter sakazakii* in ingredients of infant foods, *Int. J. Food Microbiol.*, 122, 196, 2008.
124. Kamińska, W., Patzer, J., and Dzierżanowska, D., Urinary tract infections caused by endemic multi-resistant *Enterobacter cloacae* in a dialysis and transplantation unit, *J. Hosp. Infect.*, 51, 215, 2002.
125. Davin-Regli, A. et al., A nosocomial outbreak due to *Enterobacter cloacae* strains with the *E. hormaechei* genotype in patients treated with fluoroquinolones, *J. Clin. Microbiol.*, 35, 1008, 1997.
126. Davin-Regli, A. et al., Investigation of outbreaks of *Enterobacter aerogenes* colonisation and infection in intensive care units by random amplification of polymorphic DNA, *J. Med. Microbiol.*, 44, 89, 1996.
127. Bosi, C. et al., Most *Enterobacter aerogenes* strains in France belong to a prevalent clone, *J. Clin. Microbiol.*, 37, 2165, 1999.
128. Bruce, J., Automated system rapidly identifies and characterizes microorganisms in food, *Food Technol.*, 50, 77, 1996.
129. Goshi, S. et al., DNA analysis of nosocomial infection by *Enterobacter aerogenes* in three cases of septicaemia in Japan, *J. Hosp. Infect.*, 51, 221, 2002.
130. Garaizar, J., Kaufmann, M.E., and Pitt, T.L., Comparison of ribotyping with conventional methods for the type identification of *Enterobacter cloacae*, *J. Clin. Microbiol.*, 29, 1303, 1991.
131. Weischer, M., and Kolmos, H.J., Ribotyping of selected isolates of *Enterobacter cloacae* and clinical data related to biotype, phage type, O-serotype, and ribotype, *APMIS*, 101, 879, 1993.
132. Proudy, I. et al., Genotypic characterization of *Enterobacter sakazakii* isolates by PFGE, BOX-PCR and sequencing of the *fliC* gene, *Appl. Environ. Microbiol.*, 104, 26, 2008.
133. Proudy, I. et al., Tracing of *Enterobacter sakazakii* isolates in infant milk formula processing by BOX-PCR genotyping, *J. Appl. Microbiol.*, 105, 550, 2008.
134. Keim, P. et al., Multiple-locus variable-number tandem-repeat analysis reveals genetic relationship within *Bacillus anthracis*, *J. Bacteriol.*, 182, 2928, 2000.
135. Le Fleche, P. et al., A tandem repeats database for bacterial genomes: Application to the genotyping of *Yersinia pestis* and *Bacillus anthracis*, *BMC Microbiol.*, 1, 2, 2001.
136. Lindstedt, B.A. et al., DNA fingerprinting of Shiga-toxin producing *Escherichia coli* O157 based on multiple-locus variable-number tandem-repeats analysis (MLVA), *Ann. Clin. Microbiol. Antimicrob.*, 2, 12, 2003.
137. Lindstedt, B.A. et al., Multiple-locus variable-number tandem-repeats analysis of *Salmonella enterica* subsp. *enterica* serovar Typhimurium using PCR multiplexing and multicolor capillary electrophoresis, *J. Microbiol. Methods*, 59, 163, 2004.
138. Mullane, N.R. et al., Development of multiple-locus variable tandem repeat analysis for the molecular subtyping of *Enterobacter sakazakii*, *Appl. Environ. Microbiol.*, 74, 1223, 2008.
139. Vos, P. et al., AFLP: A new technique for DNA fingerprinting, *Nucleic Acids Res.*, 23, 4407, 1995.
140. Mueller, U.G., and Wolfenbarger, L.L., AFLP genotyping and fingerprinting, *Tree*, 14, 389, 1999.
141. Rademaker, J.L. et al., Comparison of AFLP and rep-PCR genomic fingerprinting with DNA–DNA homology studies: *Xanthomonas* as a model system, *Int J. Syst. Evol. Microbiol.*, 50, 665, 2000.
142. Mougel, B. et al., A mathematical method for determining genome divergence and species delineation using AFLP, *Int. J. Syst. Evol. Microbiol.*, 52, 573, 2002.
143. Gaston, M.A., Bucher, C., and Pitt, T.L., O serotyping scheme for *Enterobacter cloacae*, *J. Clin. Microbiol.*, 18, 1079, 1983.
144. Verschraegen, G. et al., Serotyping and phage typing to identify *Enterobacter cloacae* contaminating total parenteral nutrition, *Eur. J. Clin. Microbiol. Infect. Dis.*, 7, 306, 1988.
145. Weischer, M. et al., Biotyping, phage typing, and O-serotyping of clinical isolates of *Enterobacter cloacae*, *APMIS*, 101, 838, 1993.
146. Gaston, M.A. et al., Epidemiological typing of *Enterobacter aerogenes*, *J. Clin. Microbiol.*, 27, 564, 1989.
147. Mullane, N. et al., Molecular analysis of *Enterobacter sakazakii* O-antigen gene locus, *Appl. Environ. Microbiol.*, 74, 3783, 2008.
148. Ray, T., Mills, A., and Dyson, P., Tris-dependent oxidative DNA strand scission during electrophoresis, *Electrophoresis*, 16, 888–894, 1995.
149. Staskawicz, B.J. et al., Common and contrasting themes of plant and animal diseases, *Science*, 292, 2285, 2001.

150. Barak, J.D. et al., *Salmonella enterica* virulence genes are required for bacterial attachment to plant tissue, *Appl. Environ. Microbiol.,* 71, 5685, 2005.
151. Cao, H., Baldini, R., and Rhame, L.G., Common mechanisms for pathogens of plants and animals, *Annu. Rev. Phytopathol.*, 39, 259, 2001.
152. Brandl, M.T., Fitness of human enteric pathogens on plants and implications for food safety, *Annu. Rev. Phytopathol.*, 44, 367, 2006.
153. Bhagwat, A.A., Microbiological safety of fresh-cut produce: Where are we now? In Matthews, K.R. (Editor), *Microbiology of Fresh Produce*, pp. 121–165, ASM Press, Washington DC, 2006.

75 Escherichia

P. Elizaquível, G. Sánchez, and R. Aznar

CONTENTS

75.1 INTRODUCTION

75.1.1 Classification, Morphology, and Biology

The genus *Escherichia* is composed of six gram-negative, rod-shaped, and nonspore-forming bacterial species: *Escherichia adecarboxylata*, *Escherichia blattae*, *Escherichia coli*, *Escherichia fergusonii*, *Escherichia hermanii*, and *Escherichia vulneris*. Among these, *E. coli* is the most frequently isolated and the best understood and characterized living organism. *E. coli* cells measure about 2 μm in length and 0.5 μm in diameter. It is a facultative anaerobe, capable of reducing nitrates to nitrites and producing acid and gas when growing fermentatively on glucose or other carbohydrates. Most strains are oxidase-, citrate-, urease-, and hydrogen-sulfide-negative. They are positive for indol production and the methyl red test and can be differentiated from the closely related *Shigella* and *Salmonella* by their ability to ferment lactose.[1]

Members of *E. coli* are common inhabitants of the lower intestinal tract of humans and other vertebrates, where they often represent the most abundant facultative anaerobes; in fact, healthy humans typically carry more than a billion *E. coli* cells. They form a beneficial symbiotic relationship with mammal host, providing nutrients and immune regulation against other pathogens.[2] However, some *E. coli* strains have diverged to a more pathogenic nature, causing serious diseases both within the intestinal tract and elsewhere within the host. They can also be present in soil and water as the result of fecal contamination and therefore their presence has classically been used as an indicator of poor water and food quality.

High genetic heterogeneity of *E. coli* has been reported, with only half of the genome conserved among the different strains.[3] The virulence genes responsible of the clinical capabilities are carried on plasmids, lysogenic bacteriophages, or large chromosomal insertions known as "pathogenicity-associated islands" (PAIs).[4] As a result different disease-causing genotypes are recognized and referred to as pathotypes or categories. Typically, each pathotype has a different pathogenesis and is made up of a different combination of O (somatic), H (flagellar), and K (capsular) surface antigen profiles. So far, more than 700 antigenic types (serotypes) of *E. coli* are recognized based on these antigens.

Pathogenic *E. coli* strains are mainly responsible for three types of infections in humans: urinary tract infections (UTI), neonatal meningitis, and intestinal diseases (gastroenteritis), which, in some instances, can evolve to hemolytic-uremic syndrome (HUS). Currently, there are six pathotypes associated with intestinal diseases which are classified according to their unique interaction with eukaryotic cells as enterotoxigenic *E. coli* (ETEC), enteroinvasive *E. coli* (EIEC), enterohemorrhagic *E. coli* (EHEC), enteropathogenic *E. coli* (EPEC), enteroaggregative *E. coli* (EAEC), and diffusely adhering *E. coli* (DAEC).[5] Of these, only the first four pathotypes have been associated to food or waterborne illness. Moreover, two extraintestinal pathotypes, the uropathogenic *E. coli* (UPEC) and the meningitis-associated *E. coli* (MAEC), cause UTIs, ranging from asymptomatic bacteriuria to urosepsis, and neonatal septicemia/meningitis, respectively.

E. coli accounts for the majority of enteric disease cases and is the major cause of traveler's diarrhea. The most severe human enteric infections are due to EHEC strains. However, most of the reported human EHEC outbreaks and sporadic EHEC infections are associated with a minor number of O:H serotypes. Out of 100 serotypes that have been implicated in cases of human disease, the O157 serotype is the most frequently reported.[6,7] In addition, serotypes O26, O103, O111, and O145 have increasingly been isolated from clinical cases and outbreaks as well as from animals and environmental sources.[8,9]

Human infection may be acquired through the consumption of contaminated food or water, or by direct transmission from person to person or from infected animals to humans. EHEC (including serotype O157) have been isolated from many different animal species. Of them, the gastrointestinal tract of healthy ruminants seems to be the foremost important reservoir for EHEC. Foods of bovine and ovine origin are frequently reported as a source for human EHEC infections. Other important food sources include fecally contaminated vegetables and drinking water. The significance of many EHEC types that can be isolated from animals and foodstuffs for infections in humans is, however, not yet clear.

Data on the epidemiology of these organisms are very influenced by the ability to differentiate within the pathogenic groups and serotypes. On one hand, the enteric disease symptoms frequently derive from coinfection with ETEC and other enteric pathogens, which may lead to problems in the accurate diagnosis.[10] On the other hand, in contrast to serotype O157, only a few laboratories screen for the presence of non-O157 serotypes. Therefore, the relevance of non-O157 EHEC is still not well understood and might be underestimated.[8] Preliminary data reported for 2008 (FoodNet) on foodborne diseases in United States revealed that the incidence per 100,000 population was 1.12 for O157 serotype and 0.45 for non-O157 serotypes. (http://www.cdc.gov/mmwr/preview/mmwrhtml/mm5813a2.htm?s_cid=mm5813a2). In the EU incidence was 1.1 per 100,000 population according to the last EFSA report based on data from 2006.[7] EHEC O157 was also the most commonly identified EHEC serotype, but almost half of all cases caused by known serotypes were non-O157. Overall, more than one half of the reported EHEC cases occurs in 0- to 4-years-old children. Whilst most of the hospitalized persons, aged >50 years, the highest case fatality rate is among children aged <4 years.

A marked seasonality in human EHEC[7] and ETEC[11] cases is observed, with most cases being reported during the summer and autumn. This is linked, for instance, with the fact that fecal prevalence of *E. coli* O157:H7 in ruminants is the highest in summer, decreasing to low or undetectable levels during winter.[12] Seasonality for the different toxin phenotypes has also been suggested, with heat stable (ST)-producing ETEC strains being more common in summer whereas heat labile (LT)-producing ETEC strains are present all year round and do not show any seasonality.[10]

75.1.2 Clinical Features and Pathogenesis

With its large range of pathologies (Table 75.1), *E. coli* is a major cause of human morbidity and mortality around the world. Enteric disease is the most frequent syndrome caused by ETEC, while EHEC is the most commonly isolated pathotype.

75.1.2.1 Intestinal Pathotypes

ETEC is a frequent cause of watery diarrhea in humans and animals that can be mild in nature or in some instances can be a severe, cholera-like illness causing dehydration and even life threatening.[10] In endemic areas, ETEC is one of the most common causes of traveler's diarrhea. Virulence of ETEC is driven by the production of one or two enterotoxins, the heat-labile enterotoxin (LT), related to the *Vibrio cholerae* toxin, and the heat-stable enterotoxin (ST). ETEC also produces one or more of many defined colonization factors. All these virulence factors are encoded in transmissible plasmids. ETEC toxins are secreted in the terminal small intestine where the ETEC adhere.[10]

EPEC is defined as those diarrheagenic *E. coli* strains that cause attaching/effacing (A/E) lesions on intestinal epithelium but which lack Shiga toxins. EPEC strains are a significant cause of infant diarrhea in developing countries. They usually grow in the small intestine, causing watery diarrhea that may contain mucus but not blood. Vomiting, fever, malaise, and dehydration are also associated. Symptoms may last for several days, but chronic EPEC has also been described. The genes encoding for the A/E lesion are localized in a specific pathogenicity island, termed the "locus of enterocyte effacement" (LEE). In addition, most EPEC strains contain plasmids termed "EAF" ("EPEC adherence factor"), which are necessary to cause disease.[1] Two groups can be distinguished, typical EPEC (tEPEC) including the classical 12 O serogroups, and atypical EPEC (aEPEC) strains lacking the pEAF but causing diarrhea or containing additional virulence factors. Incidence of aEPEC has notably increased in the last years.[13]

EHEC, also known as Verotoxigenic *E. coli* (VTEC), shares the ability to cause A/E with EPEC but is characterized by the production of toxin(s) that are designated verotoxin(s), VT1 and VT2, which produce a cytopathic effect on Vero cells. It is also referred to as Shiga toxin-producing *E. coli* (STEC) because its cytotoxins closely resemble the Shiga toxin produced by *Shigella dysenteriae* 1 (Stx1 and Stx2). Shiga toxin genes are ubiquitous among *E. coli* strains due to their transmission as part of lambdoid phages. Human pathogenic EHEC usually harbor additional virulence factors that are important for the development of disease in humans. The O157:H7 strain was probably originated by the acquisition of Shiga toxin-encoding prophages and the large virulence plasmid pO157 common in EHEC strains.[14,15]

Symptoms vary from a simple watery diarrhea to a hemorrhagic colitis which consists bloody stools with ulcerations of the bowel. In most events, symptoms disappear in around five days, with a low rate of hospitalizations. However, in

TABLE 75.1
Characteristics Associated with the Different *E. coli* Pathotypes

	ETEC	EPEC	EHEC	EAEC	DAEC	EIEC	UPEC	MAEC
Clinical syndrome	Traveller's diarrhea	Infant diarrhea in developing countries	Hemorrhagic colitis, HUS	Childhood mucoid diarrhea	Diarrhea in older children	Watery diarrhea, dysentery	Urinary tract infections	Neonatal septicemia/ meningitis
Symptoms	Watery diarrhea	Watery diarrhea	Watery, very bloody diarrhea	Watery mucoid diarrhea	Watery diarrhea	Mucoid, bloody diarrhea	Cystitis, pyelonephritis, urosepsis	Meningitis
Damage location	Terminal small intestine	Small intestine	Colon	Terminial ileum colon	–	Colon, lower small intestine	Ureters, kidneys, bladder, prostate	Brain
Lessions on intestinal epithelium	–	Attaching/Effacing (A/E)	Attaching/Effacing (A/E)	–	–	–	–	–
Toxins	LT/ST toxins	–	Shiga toxins	EAST1	–	EIET	Cytotoxins	–
Virulence factors	Colonization factors	pEAF, BFP, Intimin	pO157	Mucinase AAF	F1845 fimbriae, adhesins	Invasion plasmid	Fimbriae, hemolysin	S-fimbriae, K1-capsule, 100-kb plasmid
Adherence to HEp-2	–	Localized	–	Aggregative	Diffuse	–	–	–
Serotypes	Various	O55, O127, O111, and others	O157:H7, O26, O111 and others	Various	O1,O2, O21, O75	Various	O1, O2, O4, O6, O7, O18, O75	Limited number (O18:K1:H7)
Infective dose	High (>10^6)	High (10^9)	Low (10–102)	High	Very high in adults	High (10^6)	–	–
Duration		Several days to chronic	Five days	More than 14 days	–	Self-limiting		

individuals at risk such as children and the elderly, EHEC infection can result in the HUS. HUS is characterized by acute renal failure, hemolytic anemia, and thrombocytopenia. It develops in up to 10% of patients infected with EHEC O157 and is the leading cause of acute renal failure in young children. The incidence of HUS in children who have EHEC infection is 8%–10%. Of these, 3%–5% will die, and 12%–30% will have severe sequelae, including renal impairment, hypertension, or central nervous system manifestations. The incubation period following infection is commonly 3–4 days (range 1–8 days). A low infection dose of 10–100 organisms are sufficient to cause disease, which indicates a high acid-tolerance ability of this microorganism compared to other enteric bacterial pathogens.[1]

EAEC does not posses LT enterotoxin or Shiga toxins but adheres to cultured Hep-2 cells in self-aggregates that are classically referred to as "stacked bricks." Disease is characterized by a watery diarrhea that in some cases occurs with abdominal cramps, but no fever.[5] There is no invasion of the bloodstream. There are no common serotypes of EAEC to aid in their recognition in the clinical laboratory. Regarding pathogenesis, several potential virulence factors are common to EAEC, such as the production of heat-stable enterotoxin EAST1, mucinase "Pic" and aggregative adherence fimbriae (AAF).

DAEC may cause diarrhea in very young children and is differentiated from other diarrheic *E. coli* by a distinct adhesion phenotype on Hep-2 cells. Little is known about its epidemiology and pathogenesis except for the synthesis of F1845 fimbriae, a sort of Dr adhesins that recognize and bind to the decay-accelerating factor (DAF), a cellular receptor. These adhesins are also found in some UPEC strains.[1]

EIEC is pathogenetically and phenotypically very close to *Shigella*, sharing the principal virulence genes, as well as the clinical symptoms. EIEC isolates are not motile and 70% nonlactose fermenters.[16] They share with other *E. coli* pathotypes the ability to produce gas from glucose and fermentation of xylose; and with *Shigella*, the inability to decarboxylate lysine. EIEC causes invasive inflammatory colitis and dysentery with the presence of blood and mucus in the stools, fever, and severe cramps. Like *Shigella* they invade intestinal epithelium, principally in the large intestine. Many of the genes necessary for cellular invasion and disease are carried on a large >200 kb plasmid found in both EIEC and *Shigella*.

75.1.2.2 Extraintestinal Pathotypes

Extraintestinal human diseases associated to *E. coli* are UTI and neonatal septicemia/meningitis caused by the UPEC and MAEC pathotypes, respectively. Some isolates have also been associated with both invasive neonatal disease and UTI.

UPEC is the principal cause of morbidity and mortality from either community- or hospital-acquired *E. coli* infections. UPEC strains have a number of putative virulence factors not found in common fecal *E. coli* strains, such as adhesins (e.g., Pap, Sfa, and Dra), hemolysin (Hly), cytotoxic necrotizing factor-1 (CNF-1), and the aerobactin (Aer) iron-sequestration systems. In addition, these strains share other factors critical for pathogenesis of extraintestinal disease that are common to all *E. coli*, such as the lipopolysaccharide, capsule production, and type 1 pili.[1]

MAEC group includes a limited number of serotypes accounting for the most common causes of neonatal meningitis together with Group B streptococci. Most of them have in common the presence of K1 capsule, a critical determinant in invasion across the blood-brain barrier. The newborn probably acquires the MAEC from its mother during passage through the birth canal. Neonatal meningitis is a severe disease with a high mortality rate and very likely long-term neurobiological problems in survivors.[17] In addition, apart from the several genes encoding for endothelial cell invasion, most of MAEC strains harbor a 100-kb plasmid that confers increased virulence.

75.1.3 Diagnosis

Pathogenic *E. coli* are identified based on its virulence factors; therefore, the analytical procedure for these pathogens generally requires the isolation and identification of the organisms as *E. coli* before testing for the specific virulence traits. Whereas identification of *E. coli* is easily performed by traditional culture tests, differentiation between pathotypes requires molecular-based techniques to detect the virulence factors or their genes, since there are limited examples where pathogenic *E. coli* strains differ from the commensals in their metabolic abilities. For instance, commensal *E. coli* strains generally use sorbitol, but *E. coli* O157:H7 does not. Most diarrheagenic strains cannot utilize D-serine as a carbon and nitrogen source, but uropathogenic and commensal fecal strains can use it.[1]

75.1.3.1 Conventional Techniques

Culture. *E. coli* detection is typically performed according to conventional recommended culture-based methods, that is, FDA's *Bacteriological Analytical Manual* (http://www.fda.gov/Food/ScienceResearch/LaboratoryMethods/BacteriologicalAnalyticalManualBAM/ucm070080.htm). *E. coli* can be recovered easily from clinical specimens on general or selective lactose-containing media, usually MacConkey or eosin methylene-blue agar, at 37°C under aerobic conditions. *E. coli* isolation from environmental or food samples normally requires previous steps such as filtration or preenrichment. A wide variety of selective and differential media are available to facilitate selection of presumptive colonies.

Biochemical Tests. Further identification is based on biochemical tests generally performed using commercially available strips, such as API tests, allowing quick and satisfactory results. Besides lactose fermentation, other specific biochemical traits have been applied to design rapid procedures for the isolation and identification of *E. coli*. For instance, it can be identified within 1 h by oxidase, indole, lactose, and beta-glucuronidase tests following growth on MacConkey agar. Moreover, presence of the enzyme beta-glucuronidase can be tested using broth or agar medium containing the substrate 4-methylumbelliferyl-β-D-glucuronide (MUG). This compound is colorless, but following hydrolysis by the glucuronidase, the cleaved moiety is fluorescent under long-wave ultraviolet light. The incorporation of MUG into MacConkey agar isolation plates allows direct determination of beta-glucuronidase activity, and identification can be completed in 5 min.

Discrimination of pathogenic *E. coli* from nonpathogenic strains is a key issue, which complicates the diagnosis because they have to be tested for the specific virulence factors. Serotyping has extensively been used to identify pathogenic strains. However, correlation between serotypes (combination of O and H antigens) and specific diarrheagenic *E. coli* strains is not frequent; an exception is the serotype O157:H7.

The recognized threat to human health of these pathogens led to the development of selective-differential media to aid in the detection, isolation, and presumptive identification of EHEC strains, particularly serotype O157:H7. Nearly all isolates of this serotype ferment D-sorbitol slowly, or not at all. This unique phenotypic feature is exploited by substituting the carbohydrate sorbitol for lactose in MacConkey agar to develop the Sorbitol–MacConkey (SMAC) agar. Sorbitol-negative colonies will appear colorless on SMAC. CHROMagar O157™ is also based on sorbitol fermentation, but using a highly specific chromogenic reaction, whereas *E. coli* O157 shows mauve colonies and *E. coli* non-O157 blue colonies. Cefixime and tellurite can be added for an increased selectivity (i.e., CT-SMAC, T-CHROMagar O157). Rainbow Agar O157 contains chromogenic substrates that are specific for two *E. coli*-associated enzymes: beta-galactosidase (a blue-black chromogenic substrate) and beta-glucuronidase (a red chromogenic substrate). The O157:H7 strains are typically glucuronidase negative and form black or gray colonies. Following, presumptive colonies that tested positive for agglutinating in O157 antiserum or O157 latex reagent should be identified biochemically as *E. coli*, since strains of several species cross-react with O157 antiserum. Confirmation of *E. coli* O157:H7 requires identification of the H7 flagellar antigen, which requires special expertise. In addition, isolates that are nonmotile or that are negative for the H7 antigen should be tested for production of the Shiga-like toxins to identify pathogenic strains. Specificity of selective-differential media can be increased, including a previous step by enriching the sample by immunocapture with magnetic beads coated with antibodies against *E. coli* O157.

In addition to *E. coli* O157:H7, atypical sorbitol-positive strains have increasingly been isolated from clinical cases, stressing the importance of screening for both sorbitol-positive and negative O157. Improved recovery of non-O157 EHEC serotypes (O26, O103, O111, and O145) has been approached by Possé et al.[9] using selective-differential media that combine the use of a chromogenic compound to signal beta-galactosidase activity and one or more fermentative carbon sources with a pH indicator and several inhibitory components.

Nevertheless, isolates have to be confirmed as toxigenic strains either by serotyping or testing for the presence of virulence factors using tissue culture methods. Alternatively, polymerase chain reaction (PCR)-based molecular methods, using serotype-specific PCR and PCR-based virulence typing (Table 75.2) have gained acceptance in the last decade.

Immunoassays. Traditionally, antibodies have been used for *E. coli* serotyping by mixing them with a culture of suspected *E. coli*. When antibodies react with the specific antigen, agglutination of antigens can be visualized on a slide. A number of serological tests (enzyme-linked immunosorbent assay [ELISA], radioimmunoassay, immunoblot and reversed-passive latex agglutination [RPLA]) have extensively been used for the detection of *E. coli* or *E. coli* toxins. Among them, ELISA test is the most popular format, where antibodies are fixed on a solid support, for example, wells of microtiter plate. Nowadays numerous of these immunological tests are commercially available, which is an advantage for routine laboratories. However, it is essential that an international and independent organization, for example, AOAC, validates or evaluates them prior to become an official method. In addition, some companies have completely automated the entire

TABLE 75.2
Most Frequent Target Genes for the Detection of *E. coli* Pathotypes by PCR or qPCR

E. coli Category	Target Gene	Virulence Factor	Reference
ETEC	*lt*	Heat-labile enterotoxin (LT)	5
	st	Heat-stable enterotoxin (ST)	
EPEC	*eae*	Intimin	18,19
	*bfp*A	Bundle forming pilus	
	EAF plasmid	EPEC adherence factor (EAF)	
EHEC	*eae*	Intimin	5,20,21
	*rfb*E	O157 antigen	
	*fli*C	H7 antigen	
	*stx*1	Shiga toxin 1	
	*stx*2	Shiga toxin 2	
	wzx	Flippase	
	*ihp*1	Inserted hypothetical protein 1	
	*uid*A	Glucuronidase	
	*hly*A	Hemolysin	
EIEC	*ial*	Invasive antigen locus	18,22
	*ipa*H	Invasive plasmid antigen	
	iuc	Aerobactin	
EAEC	*astA*	Enteroaggregative heat-stable enterotoxin	18,23,24
	*agg*R	Transcriptional activator	
	aap	Antiaggregation protein dispersin	
	*aat*A	ATP-binding cassette transporter system	
DAEC	*daa*D	Fimbrial adhesin	18

procedure. For instance, the VIDAS system (bioMérieux) is widely used for the detection of *E. coli* O157. This system consists of a reagents strip in which the enriched broth is added. Thereafter the strip is introduced in the multiparameter automated immunoanalyzer. The analytical module automatically performs all stages of the analysis, including printing out the full results. The immunoanalyzer contains five sections, each accepting six tests; thus, it can handle 30 tests per hour. This format is designed as a "sandwich" assay where a specific antibody is bound to a solid matrix which captures the target pathogen from enrichment cultures. Then, a second antibody conjugated to an enzyme fixes itself to the captured antigen. The detection takes place by measuring the intensity of the fluorescence emitted by an enzymatic reaction. Currently, there are two different VIDAS assays for the detection of *E. coli*: the VIDAS-ICE for the generic detection of *E. coli* and the VIDAS-ECO for the specific detection of serotype O157. In both cases, to confirm the presence of *E. coli* O157:H7 all presumptive positive results have to be confirmed culturally by streaking onto CT-SMAC and SMAC agar plates.

Besides the VIDAS system, the lateral flow technology is a simple, easy, and rapid method that allows taking the analysis out of the laboratory. Briefly, lateral flow tests are immunochromatographic assays, like the pregnancy tests, adapted to operate along a single axis to suit the test strip format. A number of variations of the technology have been developed into commercial products to detect *E. coli* O157, but they all operate according to the same basic principle. It is also a "sandwich" procedure but, instead of enzyme conjugates, the detection antibody is coupled to colored latex beads or to colloidal gold. In this way, the result is obtained within 10 min after cultural enrichment.

Another innovative development is the immunomagnetic separation system, where antibodies are coupled to magnetic beads that can easily be recovered. Thereafter, cells can be detected by direct plating, ELISA, PCR, or other procedures. This technique allows not only the detection, but also the concentration of the pathogen in the sample. It can be applied instead of the selective enrichment, thus preventing from the use of growth inhibitors that can cause cells stress-injury. Several systems based on magnetic beads have been developed for the isolation of *E. coli*, and some of them are currently commercially available. Moreover, fully automated immunomagnetic separation systems have also been developed for *E. coli* O157:H7 concentration,[25,26] reducing hands-on manipulations.

75.1.3.2 Molecular Techniques

Molecular methods are the most widely used techniques for differentiating pathogenic from nonpathogenic *E. coli* isolates. They mostly focus on the detection of EHEC by searching Shiga toxins production or associated genes. Among the ones that are currently in use, immunological and PCR-based methods are available in different formats. Besides, the next generation assays, such as biosensors and DNA chips, are an on going technology. Following, these molecular methods are presented.

PCR and Real-Time PCR (qPCR). Most of the methods currently used to identify *E. coli* in clinical or food samples involve PCR-based techniques. Primers used in PCR have successfully been applied to detect all six intestinal categories (Table 75.2). Moreover, in the last years real-time PCR (qPCR) assays have revolutionized nucleic acid detection by the high speed, sensitivity, reproducibility, and minimization of cross-contamination. Besides, it also offers the possibility to quantify the targeted pathogen in the sample. The qPCR technology is based on the detection of a fluorescent signal coming either from dsDNA specific dyes (e.g., SYBR Green I), or from specific probes which include hydrolysis probes (e.g., TaqMan or TaqMan-MGB) and hybridization probes (e.g., molecular beacons or Scorpion primers).

Theoretically, as few as three copies of the targeted gene per PCR can be detected.[27] However, due to the small sample volume (1–10 μL) analyzed, the reality is that for most PCRs the limit of detection is 1–1000 cells per reaction, which translates to 1000–1 × 10^6 cells per mL. Assuming that the pathogen might be present at lower levels, that is, water or food samples, initial cell numbers could be risen up, including an enrichment step or by concentrating the initial sample. Nevertheless the concentration step, other than immunoconcentration, may enhance the presence of PCR inhibitors. Therefore, inhibition of PCR or qPCR should be checked by dilution of the sample and using an internal amplification control (IAC).[27]

Presently, many qPCR assays for the detection of *E. coli* isolates have been described, most of them designed for serotype O157:H7 and targeting a single gen. For screening purposes, this represents a limitation since only a specific serotype is investigated. To overcome this issue, multiplex PCR assays allow simultaneous detection of various *E. coli* serotypes/pathotypes. Recently, Guion et al.[18] have developed a multiplex qPCR for the detection of all six intestinal *E. coli* categories. On the other hand, generic primers for the detection of *E. coli* have been described,[28] providing a rapid tool for *E. coli* monitoring.

Reverse-Transcription PCR (RT–PCR). The major obstacle with PCR assays is how to distinguish between DNA from dead and live cells. Intact DNA can be present although the organisms are dead. This is particularly relevant for qPCR quantification which may overestimate the number of live cells due to the relatively long persistence of DNA after cell death. The most commonly used strategy to avoid this overestimation relies on the detection of mRNA which is a direct indicator of the active bacterial metabolism. Thus, reverse transcription (RT)–PCR and nucleic acid sequence amplification (NASBA) have been applied for the detection of viable *E. coli* cells.[29–31] Other alternative assays consist on the use of DNA binding molecules like ethidium monoazide (EMA) and propidium monoazide (PMA) as a sample treatment previous to the PCR. Both dyes intercalate with DNA after a photolysis process using visible light inhibiting the PCR process. This treatment can be combined with the qPCR for the quantitative detection of viable cells, as recently shown for *E. coli* O157:H7 detection.[32]

Microarrays. Microarrays contain hundreds or thousands of micrometer-diameter spots, which can contain antibodies or probes to capture the target of interest. In the first case, for detection, a second labeled antibody is then used and the latter generally involves gene amplification followed by hybridization analysis. DNA arrays or DNA chips are available in formats ranging from high-density chips with over 100,000 gene probes, to low-density membranes with several genes only. Microarrays allow the screening of thousands samples in a single run, combined with high-throughput analysis. Microarrays based on antibodies and different types of probes have successfully been applied to detect *E. coli* O157:H7.[33–35] Moreover, PCR or multiplex PCR can be combined with a microarray hybridization step for the detection of different *E. coli* virulence genes. In addition, mRNA of *E. coli* has also been targeted, allowing the detection of the viable cells.[29]

Biosensors. Biosensors are devices that detect biological or chemical recognition complexes in the form of antigen-antibody, enzyme-substrate, or receptor-ligand, placed in proximity to a transducer that generates a signal. They are self-contained integrated devices capable of providing specific quantitative or semiquantitative analytical information. A biosensor should be clearly distinguished from a bioanalytical system, which requires additional processing steps, such as reagent addition.[36] Advantages of biosensors are the possibility of portability, miniaturization, and working on-site, and the ability to detect pathogens in complex matrices with minimal sample preparation. In addition to these advantages, biosensors are a promising technology for on-line monitoring of multiple pathogens in food. Different types of biosensors have been reported for *E. coli* detection, including optical, microgravimetric, and electrochemical devices.[37–39] Moreover, biosensors have also been assayed targeting mRNA, allowing the detection of *E. coli* viable cells.[40]

Pyrosequencing. Pyrosequencing technology is a real-time DNA sequencing method, based upon the release of pyrophosphate when any nucleotide triphosphate is successfully added to the growing polynucleotide chain, that is, it is sequencing by synthesis. This technology is particularly advantageous for the fast identification of short DNA sequences and detection of single nucleotide polymorphisms. This technique is simple to use, accurate, and allows a quantitative analysis of DNA sequences. This technology has successfully been applied to differentiate bacteria commonly associated with diarrhea by targeting a partial amplicon of the *gyr*B gene as well to differentiate O157 serotype from other *E. coli* serotypes.[41]

75.2 METHODS

Among the above described methods, PCR is undoubtedly the most rapid, sensitive, and reliable one to approach *E. coli* detection in different clinical, food, and environmental samples. Besides, it can be used as a supplement of conventional culture methods for the identification or molecular typing of isolates.

75.2.1 Sample Preparation

For most PCR detection applications, a previous DNA purification step is required especially when complex matrices, such as food, water, or clinical samples, have to be analyzed. Many factors affect the sensitivity of PCR methods. For instance, inhibitors can act in different points of the reaction: interfering in the cell lysis, degrading nucleic acids, or inhibiting polymerase activity.[42] Thus, DNA purification methods must provide enough removing of sample residues, additives, preservatives, and so forth to allow a proper amplification during the PCR reaction.

Conventional procedures for isolation of nucleic acids involve cell lysis with lysozyme, digestion of contaminating proteins with proteinase K, phenol/chloroform extraction of degraded proteins and cell debris, and a final ethanol precipitation of nucleic acids. These procedures are tedious and time consuming, and expose laboratory personnel to harmful chemicals. To overcome these problems, new techniques have been developed to improve and simplify nucleic acid preparation without depending on the use of hazardous reagents. Mechanical lysis by using solid beads or sonication can be employed to break up the bacterial cell. Silica oxide is also useful for separating lipopolysaccharides from DNA. Furthermore, DNA can be extracted by a chelating resin Chelex 100 without proteinase K digestion and phenol/chloroform extraction.[43,44] Chelex 100 is a resin that binds to cellular components after cell lysis, and nucleic acids (DNA and RNA) remain in the solution, which can be readily separated from other cell components bound to the resin by centrifugation. Heat treatment–based protocols are easier to handle and low in cost, but DNA extracts produced by these methods can be unstable, occasionally yielding turbid and brown DNA solutions.[45] Although generic extraction methods are available, the nature of the sample will drive the extraction procedure to be used.

Clinical Samples. Among clinical samples, stool is the one in which *E. coli* is more often searched. The nonuniformity of stool samples in terms of physical matter, target organisms, and associated background fecal microbiota makes DNA extraction from fecal specimens highly variable from specimen to specimen in terms of both the yield and the purity of the DNA.[46] In all cases, false-negative PCR results may be due to the presence of a small number of target organisms in the volume sample. Moreover, fecal samples carry variable amounts of inhibitory substances such as biliary salts, proteins, polysaccharide complexes, bilirrubins, and phenolic compounds.

Several extraction methods have been described to purify DNA from fecal samples, most of them based either on the isolation of microorganism prior to the lysis or direct cell lysis of the sample.[47] These methods are time consuming and depend on the use of toxic materials like phenol. Because of that, commercial kits have been developed and are being widely used for DNA purification, for example, FastDNA® Kit (Bio 101), Nucleospin® C + T Kit (Machery-Nagel), Quantum Prep® Aquapure Genomic DNA Isolation Kit

(Bio-Rad), QIAamp® DNA Stool Mini-Kit (Qiagen), and MoBio Fecal (MoBio). However, inhibitors, present on fecal samples are highly dependent on the diet, intestinal microbiota, and health conditions of the host, which makes essential the optimization of the DNA extraction method for each particular application.

Water Samples. The detection of *E. coli* in water is of a great interest as it is an indicator of fecal contamination. Regulations and guidelines for drinking water in different countries require absence of *E. coli* in 100 mL of water. *E. coli* is generally detected in water samples by using membrane filtration, selective broth enrichment, immunomagnetic separation (IMS), and isolation on selective agar culture plates, followed by confirmatory assays such as PCR or serological tests.[48] However, these methods have limitations regarding time of the procedure, interferences due to high levels of competing background microorganisms, and lack of specificity. Moreover, membrane filtration can also limit the volume processed for turbid surface waters due to filter clogging. Centrifugation is an alternative to membrane filtration and has an advantage of not being subject to potential sample volume processing constraints for turbid water samples, so the technique could potentially increase the sensitivity of detection.

Food Samples. Food samples are complex matrices and the presence of natural compounds such as polysaccharides, fat, carbohydrates, proteins, or salts may hamper DNA extraction.[49] Furthermore, these compounds may also affect PCR or qPCR efficiency. As a result, these inhibiting substances need to be removed during DNA extraction. A large number of commercial kits that fulfill these requirements are available and are being progressively introduced in diagnostic laboratories. Due to their ease of performance and reliability, filter column-based methods have increasingly been utilized in various protocols for improved DNA purification.[50] The filter columns selectively bind or retain nucleic acids. After removing other contaminating substances by washing, nucleic acids can be eluted with appropriate buffer. These highly reproducible techniques have constituted the basis of a number of commercial kits for rapid and reliable purification of DNA and RNA (e.g., the DNeasy® and RNeasy® kits from Qiagen). Therefore, they are very convenient to standardize routine procedures in food laboratories.

75.2.2 Detection Procedures

Principle. A qPCR assay for the specific detection of *E. coli* O157:H7, which has successfully been applied in food,[50–52] is described in detail as a representative molecular detection protocol. It is based on the use of primers and/or probe designed to target a highly conserved point mutation at position +93 of the β-glucuronidase gene (*uid*A). The uniqueness of this base change is shared by O157:H7 and nonmotile O157 strains. This mutation is included in primer PT2 for the SYBR Green system[53] and in a minor-groove binder (MGB) modified probe (uidAP266) for a TaqMan hydrolysis assay.[54] As shown in Table 75.3, for the specific detection of *E. coli* O157:H7 by TaqMan qPCR assay, primers uidAF and uidAR are used in combination with the TaqMan probe (uidAP); for the SYBR Green I approach, PT2 primer is used instead of *uid*AF.[55]

Procedure

1. Aseptically cut the sample (25 g) into small pieces, and add it to 225 mL of triptone soy broth (TSB) with novobiocin in a sterile filter stomacher bag and homogenize for 2 min. Use aliquots of the homogenized sample for DNA extraction prior enrichment (direct detection), or incubate to rise up the bacterial cells in the sample (enrichment). For the enrichment, incubate the homogenized sample at 41°C for 6 h. After that, dilute 1:10 in MacConkey broth with cefixime and potasic telurite and incubate 18 h at 35°C–37°C.
2. Extract DNA from 1 mL of either the initial homogenate or the enriched broth using the DNeasy Blood and Tissue Kit (Qiagen), which removes all PCR inhibitors from food samples.[50] After the extraction, elute DNA in a final volume of 100 μL bi-distilled water.
3. Prepare qPCR mixture (20 μL) containing 1× TaqMan PCR Buffer (TaqMan Core Reagents, Applied Biosystems) or 1× SYBR Green PCR

TABLE 75.3
qPCR Primers and Probe

Primer Designation	Sequence (5′→3′)	Amplicon Size (in bp)	Nucleotide Location	Reference
PT2	GCG AAA ACT GTG GAA TTG GG	132	230–250	55
*uid*AF	CAG TCT GGA TCG CGA AAA CTG	121	241–261	
*uid*AR	ACC AGA CGT TGC CCA CAT AAT T	132	383–362	54
*uid*AP*	NED-ATT GAG CAG CGT TGG-NFQ	121	266–286	

* MGB probe which is used in combination with primers *uid*AF and *uid*AR

Buffer (SYBR Green I Core Reagents, Applied Biosystems); 200 μM each dATP, dCTP, dGTP, and 400 μM dUTP; 250 nM of each primer; 25 nM of the uidA probe when using TaqMan approach, 3.5 mM of $MgCl_2$, 1 U of AmpErase uracil *N*-glycosidase; 1 U of AmpliTaq Gold DNA polymerase and 5 μL DNA template. The absence of PCR inhibitors can be assessed using 1 ng of purified DNA together with 5 μL of the extracted sample or by using an internal amplification control. Negative control of amplification is done by using 5 μL of water instead of DNA template. Reactions should be done at least in triplicate.

4. Amplify for 1 cycle of 50°C for 2 min, 1 cycle of 95°C for 10 min, 40 cycles of 95°C for 15 s, and 63°C for 1 min in a GeneAmp 5700 Sequence Detection System (PE Biosystems).

Note: qPCR results are given as the increase in the fluorescence signal of the reporter dye detected and visualized by the qPCR equipment. Threshold cycle (C_T) values represent the PCR cycle in which an increase in fluorescence, over a defined threshold, first occurred for each amplification plot.

Quantitative detection of *E. coli* can be approached on samples obtained prior to the enrichment, provided standard curves are generated. They can be constructed either by using: (i) tenfold serial dilutions of DNA extracted from an *E. coli* reference strain, covering at least a five log orders range, calculated based on the genome size of the pathogen,[56] or (ii) tenfold serial dilutions of a 18 h culture of the reference strain, covering at least a five log orders range (determined by plate count on TSA). Averaged C_T values (from triplicates) and standard deviations (SD) are calculated for each dilution. The standard curve is constructed by plotting the corresponding C_T values with respect to the log of DNA or cell concentration values. Linear relation between C_T values and the amount of target DNA or cells confirms qPCR validity.

75.3 CONCLUSION AND FUTURE PERSPECTIVES

E. coli constitutes a very versatile and diverse enterobacterial species whose genome is constantly subjected to rearrangements that frequently originate strains with new virulence traits. Therefore, despite the fact that both extraintestinal and intestinal pathotypes are well defined, the existence of atypical strains showing new combinations of pathogenic traits, as well as those showing new antigenic formula or being nontypeable, hampers detection and diagnosis.[13] In the last decade, the advent of the bioinformatics together with the large amount of sequence information generated in the era of "genomics" have lead to approach studies on the genomic diversity of *E. coli*. Derived from them, the pangenome calculations have recently revealed the existence of a reservoir of more than 13,000 genes within the *E. coli* genomes, many of which may be uncharacterized but important virulence factors.[3]

To date, genetic diversity and genome plasticity of *E. coli* has been underestimated probably because interest has been focused on strains of relevance in medical and veterinary fields. In addition, the amount of information on relevant serotypes monitoring, that is, EHEC in animals and food, provided by the reporting countries is relatively sparse. There are no harmonized recommendations for the monitoring of EHEC in animals and food, and many results are reported with no or limited serotype information.[6,7] A likely reason is that broad EHEC surveys are laborious and expensive to carry out. Besides, the lack of serotyping and characterization with regard to known virulence markers, means that the potential human health risk of the presence of EHEC in animals and food is very difficult to assess based on the available data. Therefore, detection and identification of the non-O157 EHEC should be approached to extend the data on the incidence of relevant pathogenic strains.

Molecular diagnostic techniques, and specially PCR-based methods are undoubtedly the best choice for rapid and reliable detection of pathogenic strains, even if they are atypical or nontypeable. However, the diagnostic technique is strongly influenced by the specificity of the primers used. Understanding the role of targeted genes in virulence as well as the significance of their presence in the isolate will reveal the right choice. In the meantime, both genotypic and phenotypic methods should be performed in order to reach a real picture on the incidence of pathogenic *E. coli*. Rapid and accurate identification of causative bacteria with a high degree of specificity and sensitivity is essential to outbreak surveillance and investigation. In addition, it would allow the implementation of proper clinical therapies as well as food ban regulations. Nowadays high-throughput systems are available for accurate identification by targeting several genes such as microarrays, an on going technology. In the same line, a promising technique is Pyrosequencing,™ a real-time DNA sequencing method, providing rapid and accurate genotypic identification that can also be applied in pathogen detection.[41]

Sensitivity is another goal for rapid detection methods. If the PCR procedure is designed for pathogen detection in water or food, it should ideally be able to detect as low as 1 CFU per g or mL without the enrichment step especially when a quantitative result is desired. To accomplish this requirement, several pre-PCR approaches are in progress, and will extend to next future, by concentrating the bacteria at the initial sample, that is, filtration[57] immunomagnetic separation[25]; by increasing the sensitivity of the detection system employed, that is, qPCR, biosensors[37–39]; or using a combination of both.

The application of quantitative detection of *E. coli* will depict the actual contamination levels of these organisms in naturally contaminated food and water samples. Moreover, improved procedures for the quantitative detection of viable cells will reveal the real risk that the presence of pathogenic *E. coli* strains in food and water pose to human health and will help in the definition of microbiological standards for quality regulations.

REFERENCES

1. Welch, R.A., The genus *Escherichia*. In Falkow, S., Rosenberg, E., Schleifer, K.-H., Stackebrandt, E., and Dworkin, M. (Editors), *The Prokaryotes*, Chapter 3, Springer, New York 2006.
2. Yan, F., and Polk, D.B., Commensal bacteria in the gut: Learning who our friends are, *Curr. Opin. Gastroenterol.*, 20, 565, 2004.
3. Rasko, D.A. et al., The pangenome structure of *Escherichia coli*: Comparative genomic analysis of *E. coli* commensal and pathogenic isolates, *J. Bacteriol.*, 190, 6881, 2008.
4. Hacker, J. et al., Pathogenicity islands of virulent bacteria: Structure, function and impact on microbial evolution, *Mol. Microbiol.*, 23, 1089, 1997.
5. Nataro, J.P., and Kaper, J.B., Diarrheagenic *Escherichia coli*, *Clin. Microbiol. Rev.*, 11, 142, 1998.
6. Anonymous, Surveillance for Foodborne Disease Outbreaks—United States, 2006. *Morbidity and Mortality Weekly Report (MMWR)* 58, 609, 2009. http://www.cdc.gov/mmwr/preview/mmwrhtml/mm5822a1.htm
7. European Food Safety Authority, The community summary report on trends and sources of zoonoses, zoonotic agents, antimicrobial resistance and foodborne outbreaks in the European Union in 2006, *EFSA J.*, 130, 3, 2007.
8. Bettelheim, K.A., The non-O157 shiga-toxigenic (verocytotoxigenic) *Escherichia coli*; under-rated pathogens, *Crit. Rev. Microbiol.*, 33, 67, 2007.
9. Posse, B. et al., Quantitative isolation efficiency of O26, O103, O111, O145 and O157 STEC serotypes from artificially contaminated food and cattle faeces samples using a new isolation protocol, *J. Appl. Microbiol.*, 105, 227, 2008.
10. Qadri, F. et al., Enterotoxigenic *Escherichia coli* in developing countries: Epidemiology, microbiology, clinical features, treatment, and prevention, *Clin. Microbiol. Rev.*, 18, 465, 2005.
11. Bettelheim, K.A., Non-O157 verotoxin-producing *Escherichia coli*: A problem, paradox, and paradigm, *Exp. Biol. Med.*, 228, 333, 2003.
12. Edrington, T.S. et al., Seasonal shedding of *Escherichia coli* O157:H7 in ruminants: A new hypothesis, *Foodborne. Pathog. Dis.*, 3, 413, 2006.
13. Hernandes, R.T. et al., An overview of atypical enteropathogenic *Escherichia coli*, *FEMS Microbiol. Lett.*, 297, 137, 2009.
14. Reid, S.D. et al., Parallel evolution of virulence in pathogenic *Escherichia coli*, *Nature*, 406, 64, 2000.
15. Lathem, W.W. et al., Acquisition of stcE, a C1 esterase inhibitor-specific metalloprotease, during the evolution of *Escherichia coli* O157:H7, *J. Infect. Dis.*, 187, 1907, 2003.
16. Silva, R.M. et al., Biochemical and cultural characteristics of invasive *Escherichia coli*, *J. Clin. Microbiol.*, 11, 441, 1980.
17. Stoll, B.J. et al., Changes in pathogens causing early-onset sepsis in very-low-birth-weight infants, *N. Engl. J. Med.*, 347, 240, 2002.
18. Guion, C.E. et al., Detection of diarrheagenic *Escherichia coli* by use of melting-curve analysis and real-time multiplex PCR, *J. Clin. Microbiol.*, 46, 1752, 2008.
19. Franke, J. et al., Nucleotide sequence analysis of enteropathogenic *Escherichia coli* (EPEC) adherence factor probe and development of PCR for rapid detection of EPEC harboring virulence plasmids, *J. Clin. Microbiol.*, 32, 2460, 1994.
20. Beutin, L. et al., Evaluation of the "GeneDisc" real-time PCR system for detection of enterohaemorrhagic *Escherichia coli* (EHEC) O26, O103, O111, O145 and O157 strains according to their virulence markers and their O- and H-antigen-associated genes, *J. Appl. Microbiol.*, 106, 1122, 2009.
21. Fu, Z. et al., Rapid detection of *Escherichia coli* O157:H7 by immunomagnetic separation and real-time PCR, *Int. J. Food Microbiol.*, 99, 47, 2005.
22. Kingombe, C.I. et al., Molecular strategies for the detection, identification, and differentiation between enteroinvasive *Escherichia coli* and *Shigella* spp, *J. Food Prot.*, 68, 239, 2005.
23. Cerna, J.F. et al., Multiplex PCR for detection of three plasmid-borne genes of enteroaggregative *Escherichia coli* strains, *J. Clin. Microbiol.*, 41, 2138, 2003.
24. Ruttler, M.E. et al., Evaluation of a multiplex PCR method to detect enteroaggregative *Escherichia coli*, *Biocell*, 30, 301, 2006.
25. Chandler, D.P. et al., Automated immunomagnetic separation and microarray detection of *E. coli* O157:H7 from poultry carcass rinse, *Int. J. Food Microbiol.*, 70, 143, 2001.
26. Prentice, N. et al., Rapid isolation and detection of *Escherichia coli* O157:H7 in fresh produce, *J. Rapid Methods Aut. Microbiol.*, 14, 299, 2007.
27. Bustin, S.A. et al., The MIQE guidelines: Minimum information for publication of quantitative real-time PCR experiments, *Clin. Chem.*, 55, 611, 2009.
28. Bernasconi, C. et al., Use of the tna operon as a new molecular target for *Escherichia coli* detection, *Appl. Environ. Microbiol.*, 73, 6321, 2007.
29. Liu, Y. et al., Electronic deoxyribonucleic acid (DNA) microarray detection of viable pathogenic *Escherichia coli, Vibrio cholerae,* and *Salmonella typhi*, *Anal. Chim. Acta*, 578, 75, 2006.
30. McIngvale, S.C. et al., Optimization of reverse transcriptase PCR to detect viable Shiga-toxin-producing *Escherichia coli*, *Appl. Environ. Microbiol.*, 68, 799, 2002.
31. Cook, N., The use of NASBA for the detection of microbial pathogens in food and environmental samples, *J. Microbiol. Methods*, 53, 165, 2003.
32. Wang, L. et al., Detection of viable *Escherichia coli* O157:H7 by ethidium monoazide real-time PCR, *J. Appl. Microbiol.*, 107, 1719, 2009. doi:10.1111/j.1365–2672.2009.04358.x.
33. Gehring, A.G. et al., Antibody microarray detection of *Escherichia coli* O157:H7: Quantification, assay limitations, and capture efficiency, *Anal. Chem.*, 78, 6601, 2006.
34. Kim, H. et al., A molecular beacon DNA microarray system for rapid detection of *E. coli* O157:H7 that eliminates the risk of a false negative signal, *Biosensors and Bioelectronics*, 22, 1041, 2007.
35. Willenbrock, H. et al., Design of a seven-genome *Escherichia coli* microarray for comparative genomic profiling, *J. Bacteriol.*, 188, 7713, 2006.
36. Rodriguez-Mozaz, S. et al., Biosensors as useful tools for environmental analysis and monitoring, *Anal. Bioanal. Chem.*, 386, 1025, 2006.
37. Wang, L. et al., A novel electrochemical biosensor based on dynamic polymerase-extending hybridization for *E. coli* O157:H7 DNA detection, *Talanta*, 78, 647, 2009.
38. Pyun, J.C. et al., Development of a biosensor for *E. coli* based on a flexural plate wave (FPW) transducer, *Biosens. and Bioelectron.*, 13, 839, 1998.
39. Zhu, P. et al., Detection of water-borne *E. coli* O157 using the integrating waveguide biosensor, *Biosens. Bioelectron.*, 21, 678, 2005.
40. Baeumner, A.J. et al., RNA biosensor for the rapid detection of viable *Escherichia coli* in drinking water, *Biosens. Bioelectron.*, 18, 405, 2003.
41. Hou, X.L. et al., Pyrosequencing analysis of the gyrB gene to differentiate bacteria responsible for diarrheal diseases, *Eur. J. Clin. Microbiol. Infect. Dis.*, 27, 587, 2008.
42. Wilson, I.G., Inhibition and facilitation of nucleic acid amplification, *Appl. Environ. Microbiol.*, 63, 3741, 1997.

43. Walsh, P.S. et al., Chelex 100 as a medium for simple extraction of DNA for PCR-based typing from forensic material, *Biotechniques*, 10, 506, 1991.
44. Rodriguez-Lazaro, D. et al., Rapid quantitative detection of, *Listeria monocytogenes* in salmon products: Evaluation of pre-real-time PCR strategies, *J. Food Prot.*, 68, 1467, 2005.
45. Merk, S. et al., Comparison of different methods for the isolation of *Burkholderia cepacia* DNA from pure cultures and waste water, *Int. J. Hyg. Environ. Health*, 204, 127, 2001.
46. Holland, J.L. et al., PCR detection of *Escherichia coli* O157:H7 directly from stools: Evaluation of commercial extraction methods for purifying fecal DNA, *J. Clin. Microbiol.*, 38, 4108, 2000.
47. Tang, J.N. et al., An effective method for isolation of DNA from pig faeces and comparison of five different methods, *J. Microbiol. Methods*, 75, 432, 2008.
48. Heijnen, L., and Medema, G., Quantitative detection of *E. coli*, *E. coli* O157 and other shiga toxin producing *E. coli* in water samples using a culture method combined with real-time PCR, *J. Water Health*, 4, 487, 2006.
49. Juste, A. et al., Recent advances in molecular techniques to study microbial communities in food-associated matrices and processes, *Food Microbiol.*, 25, 745, 2008.
50. Elizaquivel, P., and Aznar, R., Comparison of four commercial DNA extraction kits for PCR detection of *Listeria monocytogenes*, *Salmonella*, *Escherichia coli* O157:H7, and *Staphylococcus aureus* in fresh, minimally processed vegetables, *J. Food Prot.*, 71, 2110, 2008.
51. Elizaquivel, P., and Aznar, R., A multiplex RTi-PCR reaction for simultaneous detection of *Escherichia coli* O157:H7, *Salmonella* spp. and *Staphylococcus aureus* on fresh, minimally processed vegetables, *Food Microbiol.*, 25, 705, 2008.
52. Elizaquivel, P. et al., Quantitative detection of *Salmonella* spp., *Listeria monocytogenes* and *Escherichia coli* O157:H7 by real time PCR in naturally contaminated food products, *J. Food Prot.*, unpublished data.
53. Yoshitomi, K.J. et al., Optimization of a 3′-minor groove binder-DNA probe targeting the uidA gene for rapid identification of *Escherichia coli* O157:H7 using real-time PCR, *Mol. Cell. Probes*, 17, 275, 2003.
54. Jinneman, K.C. et al., Multiplex real-time PCR method to identify Shiga toxin genes stx1 and stx2 and *Escherichia coli* O157:H7/H- serotype, *Appl. Environ. Microbiol.*, 69, 6327, 2003.
55. Yoshitomi, K.J. et al., Detection of Shiga toxin genes stx1, stx2, and the +93 uidA mutation of *E. coli* O157:H7/H-using SYBR Green I in a real-time multiplex PCR, *Mol. Cell. Probes*, 20, 31, 2006.
56. Hayashi, T. et al., Complete genome sequence of enterohemorrhagic *Escherichia coli* O157:H7 and genomic comparison with a laboratory strain K-12, *DNA Res.*, 8, 11, 2001.
57. Brewster, J.D., Large-volume filtration for recovery and concentration of *Escherichia coli* O157:H7 from ground beef, *J. Rapid Methods Aut. Microbiol.*, 17, 242, 2009.

76 *Francisella tularensis*

Kiersten J. Kugeler and Jeannine M. Petersen

CONTENTS

76.1 INTRODUCTION

Tularemia is a zoonotic disease caused by the bacterium, *Francisella tularensis*. Though the organism was first isolated from ground squirrels in Tulare County, California, in 1912,[1] reports of a tularemia-like disease in both animals and humans in the United States, Japan, and Norway surfaced much earlier.[2] *F. tularensis* is found throughout the Northern Hemisphere and has been documented in hundreds of invertebrate and vertebrate species.[3,4] The organism is highly infectious: a dose as low as 10 cells can cause disease.[5,6] Humans are incidental hosts and clinical presentation and severity depend on the route of organism entry and strain type. Transmission occurs via arthropod bite, inhalation, ingestion of the organism or direct skin contact with infected tissues.[5,7,8] Because of its infectious nature and potential to cause significant morbidity, *F. tularensis* was part of offensive biological weapons programs in the mid-twentieth century, and is considered a high priority potential bioweapon.[9]

76.1.1 Classification and Morphology

F. tularensis is a gram-negative, aerobic, facultative intracellular organism.[10] Cells are small, nonmotile, occur singly, and are short rod or coccoid shaped, but pleomorphism is demonstrated in infected tissues.[10] The organism shows limited carbohydrate degradation resulting in acid but no gas formation, a growth requirement for cysteine and unique fatty acid composition.[10] Whole genome sequencing of *F. tularensis* subsp. *tularensis* (Schu S4 type strain) revealed a small genome (<2 MB) and a G + C content of roughly 33%.[11]

F. tularensis falls within the family *Francisellaceae*, a member of the gamma subclass of Proteobacteria. The family consists of a single genus, *Francisella*.[10] The closest relatives are *Piscirickettsiaceae*, and the nearest human pathogens are *Coxiella burnetii* and *Legionella pneumophila*, though both are quite distant relatives.[10,11] Taxonomy of the genus is constantly in flux as new organisms are discovered and new characterization methods are employed. There are three recognized species associated with human illness, *F. tularensis*, *F. novicida*, and *F. philomiragia*.[10,12,13] *F. tularensis*, the most widely recognized member of the genus, is the source of most knowledge on the genus, and as the causative agent of tularemia is responsible for nearly all human disease associated with *Francisellaceae*. *F. novicida* and *F. philomiragia* are less virulent and rarely cause human illness, primarily affecting immunocompromised individuals or near-drowning victims.[10,12,13] Additional *Francisellaceae* have been identified from a variety of sources including fish, ticks, soil, seawater, and humans.[14–19]

F. tularensis, *F. novicida*, and *F. philomiragia* show a high degree of DNA relatedness. DNA reassociation studies performed at 65°C showed 37%–43% relatedness between *F. tularensis* and *F. philomiragia*, 22%–51% between *F. philomiragia* and *F. novicida*, and 93%–95% between *F. tularensis* and *F. novicida*.[12] The 16S rRNA gene sequences of *F. tularensis* and *F. philomiragia* display 99.2% similarity and those of *F. tularensis* and *F. novicida* show 99.6% similarity.[20] *F. novicida* has been suggested as a subspecies of *F. tularensis* based on genetic similarity.[10,12,21] Whole genome comparisons provide evidence that *F. novicida* and *F. tularensis* represent distinct population structures with regard to recombination and nonsynonymous and synonymous site substitution rates.[22] In addition, illness caused by *F. novicida* displays little clinical similarity to tularemia caused by *F. tularensis*. Thus, it has been proposed that the species designation *F. novicida* be retained.[22]

F. tularensis is composed of three subspecies, subsp. *tularensis*, subsp. *holarctica*, and subsp. *mediasiatica*.[23] *F. tularensis* subsp. *tularensis* (type A) causes infection only in North America and is distinguished by its ability to ferment glycerol and the presence of a citrulline ureidase pathway; it is also considered the most virulent subspecies (Table 76.1).[10] *F. tularensis* subsp. *holarctica* (type B) is distributed throughout the Northern Hemisphere, does not ferment glycerol and produces mild to moderate disease in humans (Table 76.1). *F. tularensis* subsp. *mediasiatica,* found only in Central Asia, displays virulence in rabbits akin to type B strains and is rarely associated with human illness.[2,8,10,26] Among the three subspecies, 99.9% similarity is observed within the 16S rRNA gene.[20]

F. tularensis subsp. *tularensis* and *F. tularensis* subsp. *holarctica* have been further differentiated using molecular typing techniques. *F. tularensis* subsp. *tularensis* has been subdivided into clades, termed A1 (also known as A.I and A-east) and A2 (also known as A.II and A-west)[24,27] with clade A1 further separated into clades A1a and A1b (Table 76.1).[25,28] The geographic distribution of these clades differs, with A1 (A1a and A1b) infections predominating in the eastern half of the United States, though a few cases have occurred in the Western United States, whereas A2 infections appear restricted to the Western United States (Table 76.1).[24,25,29] Within *F. tularensis* subsp. *holarctica*, 10 separate clades have been identified.[27,30]

76.1.2 Ecology and Epidemiology

Most cases of tularemia are caused either by *F. tularensis* subsp. *tularensis* or *F. tularensis* subsp. *holarctica*. The ecology of these two subspecies is complex and poorly understood. *F. tularensis* subsp. *tularensis* has been linked to a terrestrial cycle involving lagomorphs and ticks whereas *F. tularensis* subsp. *holarctica* is associated with water systems and water-loving rodents including beavers and muskrats.[3,4] Epizootics are often noted or demonstrated in local animal populations before or concurrent to detection of human disease.[3,31] The reservoirs in nature are unknown; however, evidence has shown that *F. tularensis* can live in water courses.[4] Due to its growth requirement for cysteine, it is unlikely that the organism lives freely in the environment, and may live in association with amoebae.[32,33] *F. tularensis* has been demonstrated to persist in a viable but nonculturable state in water.[34]

Tularemia is a seasonal disease, with peak incidence in the spring and summer months when arthropods are most active. However, cases occur year-round, particularly among hunters.[35] Historically, tularemia has occurred most frequently in Scandinavia, North America, Japan, and countries of the former Soviet Union.[2] Annual cases are reported from Eastern Europe and less frequently from Western Europe, however, outbreaks of tularemia in Spain have occurred in recent years.[2,36,37] Outbreaks affecting hundreds of persons have also occurred in the last decade in Sweden and Kosovo.[38,39] True incidence worldwide is unknown.[8] In the United States, tularemia was more common in the early-to-mid twentieth century than in the present day. The highest annual occurrence was 2291 cases in 1939,[4] whereas about 125 cases from an average of 44 states are currently reported in the United States each year.[35]

Tularemia is transmitted to humans by a variety of mechanisms. In the United States, ticks (*Dermacentor* and *Amblyomma* spp.) are linked with transmission to humans.[3,4] Deer flies (*Chrysops* spp.) are vectors of disease in the Western United States.[40,41] In Sweden and Russia, mosquitoes (*Aedes* and *Ochlerotatus* spp.) have been implicated as vectors.[3,31,42,43] Most cases of tularemia resulting from direct skin contact are among hunters cleaning infected animals without gloves.[44,45] Human tularemia cases have also been attributed to bites from sick domestic cats.[25,46,47] Incidents of organism ingestion are due primarily to contamination of an untreated water supply, presumably by the waste or carcasses of infected animals.[4,48] Outbreaks of this nature have been documented in Turkey and Kosovo.[39,48–50] Inhalation of the organism occurs with generation of infective aerosols, for example, in the disruption of contaminated hay products or grass by lawn mowing or agricultural work.[51–54]

TABLE 76.1
***Francisella tularensis* Subspecies and Clades Causing Human Tularemia**

F. tularensis Subspecies or Clade	Geographic Distribution (Documented Cases)	Mortality[a] (US)
subsp. *tularensis* (type A)	North America (US and Canada)	9%
clade A2	Western US, Alaska	0%
clade A1a	Eastern US, California, Utah, Oregon	4%
clade A1b	Eastern US, California, Idaho, Oregon, Colorado, Alaska	24%
subsp. *holarctica* (type B)	Northern Hemisphere	7%
subsp. *mediasiatica*	Central Asia	N/A

[a] Mortality among culture-confirmed cases in the United States[24,25]

76.1.3 Pathogenesis and Clinical Disease

F. tularensis is an intracellular organism in vivo. *F. tularensis* can replicate within many cell types in vitro, although most infection studies have been conducted in macrophage cell lines.[2] Organs usually infected include lymph nodes, liver, spleen, kidneys, and lungs. Few virulence factors have been identified. The genetic basis for differences in virulence between strains is unknown, though some degree of genomic diversity exists not only between subspecies[55–57] but also between clades of the same subspecies.[25,27,28,30,58]

Tularemia can present as one of several different clinical manifestations depending on the route by which the organism enters the body, and after an incubation period usually of 3–5 days (range: 1–14 days).[9,59] General symptoms begin abruptly and can include fever, malaise, headache, fatigue, chills, and sore throat. Other symptoms of infection depend on the clinical form. The ulceroglandular form, the most common form of illness worldwide, is characterized by a tender ulcer at the point of organism entry (by arthropod bite or direct skin contact, usually a result of handling infected tissues).[5,8,59,60] In addition to the ulcer, regional lymphadenopathy is nearly always present. Glandular tularemia is similar, but lacks a noticeable ulcer.[8,59] Oropharyngeal tularemia is caused by ingestion of the organism and commonly develops as a pharyngitis or tonsillitis with cervical or retropharyngeal lymph node involvement.[8,59,61] Oculoglandular disease involves infection of the conjunctiva and is caused by direct inoculation, usually from contaminated hands.[8,59] The term "typhoidal" is commonly used to describe a severe systemic manifestation of tularemia without any obvious mode of acquisition, although reference to this as a distinct form of the disease has been questioned.[8] Pneumonic tularemia is not only a complication of other forms of tularemia but also results from direct inhalation of the organism (due to aerosolization in nature or laboratory). This is the most severe form of the disease and may lack specific pulmonary symptoms and display negative X-ray findings.[5,8,59]

Depending on the clinical manifestation and infecting strain type, the severity of infection can range from mild and self-limiting to fatal. Little to no tularemia-related mortality is reported in Europe and Asia, while mortality estimates in the United States range from 2% (those diagnosed by culture and serology) to 9% (culture-confirmed cases only).[24,45,60] Early in the study of *F. tularensis*, researchers noted differences in mortality rates in North America versus Eurasia and attributed this difference to the fact that infections with *F. tularensis* subsp. *tularensis* occur only in North America.[23] Molecular epidemiologic studies have refined this picture, and demonstrate that among culture-confirmed cases in the United States, human infections due to the A1b clade of *F. tularensis* subsp. *tularensis* are associated with significantly higher mortality (24%) than infections caused by clades A1a (4%) and A2 (0%), or *F. tularensis* subsp. *holarctica* (7%) (Table 76.1).[25]

Aminoglycosides are the antimicrobials of choice for treatment of tularemia.[8] Streptomycin is recommended, but due to issues involving side effects and supply, gentamicin is frequently and effectively used in its place.[8,59–62] Tetracyclines are also used, especially outside North America, although they are associated with a higher relapse rate than treatment with aminoglycosides.[7,8,59–61] Chloramphenicol is used in addition to aminoglycosides for cases of tularemic meningitis.[8,61] Fluoroquinolones are effective against *F. tularensis* in vitro, and recent reports have demonstrated promising results in clinical practice.[8,59,62–65] More clinical data are needed to determine if fluoroquinolones are an adequate alternative to aminoglycoside therapy for treatment of human tularemia. Delay in appropriate treatment can result in complications and lengthy recovery.[7] Full recovery can take months in some cases, and suppuration of lymph nodes requiring surgical drainage is the most common complication of infection.[61]

76.1.4 Diagnosis

Because of the rarity and sporadic nature of the disease as well as the varied clinical presentations, the diagnosis of tularemia can be quite complex. Prompt diagnosis often requires clinical suspicion based on exposure history. Tularemia, however, is commonly not in the differential diagnosis of physicians when submitting human specimens to clinical laboratories for testing.[66]

Ulceroglandular tularemia is the form of illness most easily diagnosed because it is the most common presentation and is characterized by the presence of recognizable ulcer. The most appropriate clinical specimens for diagnosis of tularemia are largely dependent on the disease manifestation: ulceroglandular, glandular, oculoglandular, oropharyngeal, pneumonic, or typhoidal tularemia.[8] Blood, though routinely collected for diagnosis of generalized systemic illness, is not very sensitive and may often be negative for all forms of illness. Lymph node aspirates and wound swabs are good specimens for presentations with lymphadenopathy or apparent ulcers.[8] Pharyngeal swabs and conjunctival swabs can be collected for oropharyngeal or oculoglandular presentations, respectively.[8] Bronchial washes, sputum, and pleural fluid are appropriate specimens for presumed respiratory infection.[8] Serum is a standard specimen usually taken for retrospective diagnosis of all forms of illness.

Due to the low infectious dose of *F. tularensis*, tularemia has been one of the most commonly reported laboratory-associated bacterial infections.[67,68] Culture manipulation presents the greatest risk to laboratory workers due to the high concentration of organisms. To reduce likelihood of exposure, biosafety level 3 (BSL-3) containment is recommended for all work with suspect *F. tularensis* cultures.[8,69]

76.1.4.1 Conventional Diagnostic Techniques

Culture. Culture of *F. tularensis* can be difficult because it is a slow-growing, fastidious organism. It is easily outcompeted by other organisms and requires cysteine/cystine supplementation for growth.[8,10] *F. tularensis* will grow on several media common to clinical laboratories including chocolate agar (CA), buffered charcoal-yeast extract agar (BCYE), and Thayer–Martin (TM) agar.[8,10] Cysteine heart agar supplemented with 9% chocolatized sheep blood (CHAB), is used for growth of *F. tularensis* in many reference laboratories.[8] The organism grows slowly, which may contribute to the relative insensitivity of culture as inoculated culture plates may only be held for 48 h in clinical laboratories, particularly if *F. tularensis* is not suspected in the diagnosis. Moreover, slow-growing cultures can lead to delayed organism identification and therefore delay of effective antibiotic treatment. Few studies have published on the sensitivity of culture for diagnosis of tularemia. In one study, when ulcer swabs were placed in Amies agar with charcoal and transported to the laboratory within 48 h, the sensitivity of culture was 62%.[70] Despite the challenges associated with culture of *F. tularensis*, this method provides conclusive diagnosis of tularemia, and allows for further isolate characterization [e.g., antimicrobial susceptibility profiles, identification of subspecies by biochemical methods (glycerol fermentation), and molecular epidemiologic analyses].

Serology. Serology is the most common method used for diagnosis of tularemia; however, detection of specific serum antibodies occurs 10–20 days after disease onset.[71] Thus, serologic diagnosis does not assist the clinician in making treatment decisions necessary to minimize complications of infection. The standard test for determining the presence of antibodies to *F. tularensis* is agglutination testing, either by tube agglutination (TA) or microagglutination (MA).[8,72,73] Formalin killed whole-cell antigen is commercially available for agglutination assays. Enzyme-linked immunoassays (ELISA) have been adopted for diagnostic use in parts of Europe.[8,74–76] The lipopolysaccharide (LPS) and/or outer membrane fraction are the primary antigens used in the ELISA format. Serologic tests do not differentiate between infections caused by *F. tularensis* subsp. *tularensis* and *F. tularensis* subsp. *holarctica*.

Antigen Detection Assays. Direct fluorescent-antibody (DFA) staining using a FITC-labeled hyperimmune rabbit polyclonal antibody directed against whole killed *F. tularensis* cells can be used to rapidly and specifically identify *F. tularensis* subsp. *tularensis* and *F. tularensis* subsp. *holarctica*. The test has a detection limit of 10^6 bacteria/mL and therefore is only useful in clinical specimens where the numbers of organisms are expected to be high (wound swabs, tissues, and aspirates).[8] As the reagent is not commercially available, this test is primarily performed in reference laboratories. A capture ELISA based on monoclonal antibodies specific for the LPS of *F. tularensis* subsp. *tularensis* and *F. tularensis* subsp. *holarctica* has also been developed, with a reported detection limit of 10^3–10^4 bacteria/mL.[77]

76.1.4.2 Molecular Diagnostic Techniques

Diagnostic PCR. Molecular diagnostics for detection of *F. tularensis* are a relatively recent advancement, arising primarily in response to the lack of sensitivity and rapidity in conventional diagnostics for detection of *F. tularensis* in clinical specimens.[78] Molecular methods allow for early diagnosis, thus having the potential to impact treatment. For the purpose of this chapter, we will focus on molecular diagnostic methods that have been evaluated with clinical samples. Because of the relative rarity of human tularemia, thorough evaluation of molecular diagnostics with clinical specimens has been challenging. PCR-based diagnostic methods have been used most commonly for diagnosis of ulceroglandular tularemia, the prevalent clinical form. The presence of a skin ulcer makes sampling relatively straightforward; it is also noninvasive as diagnostic material is obtained simply by swabbing the ulcer.

The most extensively evaluated molecular method for diagnosis of tularemia is a conventional PCR assay targeting the *lpnA* (*tul4*) gene (Table 76.2). The *lpnA* assay was developed for detection of *F. tularensis* DNA in wound samples from humans with ulceroglandular tularemia.[79] *lpnA* encodes for a highly conserved lipoprotein present in *F. tularensis* and *F. novicida* (99% identity); a lipoprotein with low similarity to *lpnA* has also been identified in *F. philomiragia*. The *lpnA* gene of *F. tularensis* shows no significant similarity to gene sequences in other bacteria. Standardized methods for transport and extraction of DNA from ulcer swabs have been developed for use with the *lpnA* PCR assay.[70,79] DNA is collected from ulcers using a cotton swab and transported to the laboratory in the presence of the chaotropic nuclease inhibitor guanidine isothiocyanate, shown to preserve *F. tularensis* DNA for 1 month. DNA is extracted from samples by binding of DNA to uniform glass beads in the presence of guanidine isothiocyanate. The detection limit of the conventional *lpnA* assay is 100 organisms.

When transport, purification and *lpnA* PCR methods were evaluated on wound specimens from patients with culture and/or serology-confirmed ulceroglandular tularemia, the *lpnA* PCR displayed 75% sensitivity as compared to 62% sensitivity for culture.[70] Additional studies have continued to show the utility of the *lpnA* assay for diagnosis of *F. tularensis* in wound samples, with results reported from over 200 cases of ulceroglandular tularemia (sensitivity of 75%–80%).[80,81] A number of findings have come from these studies. First, PCR results were shown to correlate with the appearance of the ulcer, with the presence of a crust seeming to provide a protective environment. When ulcer samples were obtained by lifting a crust, 35 of 36 (97%) specimens showed amplifiable DNA, as compared to 24 of 34 (71%) specimens from patients with nonencrusted ulcers.[80] In addition, the *lpnA*

TABLE 76.2
Diagnostic PCR-Based Assays for *F. tularensis*

PCR Assay[a]	Primers/Probes (5′→3′)
Standard	
*lpnA*_F	GCTGTATCATCATTTAATAACTGCTG
*lpnA*_R	TTGGGAAGCTTGTATCATGGCACT
Real time (TaqMan)	
*lpnA*_F	ATTACAATGGCAGGCTCCAGA
*lpnA*_R	TGCCCAAGTTTTATCGTTCTTCT
*lpnA*_P	TTCTAAGTGCCATGATACAAGCTTCCCAATTACTAAG
ISFtu2_F	TTGGTAGATCAGTTGGTGGGATAAC
ISFtu2_R	TGAGTTTTACCTTCTGACAACAATATTTC
ISFtu2_P	AAAATCCATGCTATGACTGATGCTTTAGGTAATCCA
*iglC*_F	TGAGATGATAACAAGACAACAGGTAACA
*iglC*_R	GGATGAGATCCTATACATGCAGTAGG
*iglC*_P	TCAGTTCTCACATGAATGGTCTCGCCA

[a] F, Forward primer; R, Reverse primer; P, TaqMan probe

PCR assay correctly diagnosed *F. tularensis* infections even after initiation of effective antibiotic treatment (up to 10 days post treatment).[80] Finally, for several cases with clinically compatible illness and negative culture and serology, amplifiable *F. tularensis* DNA was detectable by the *lpnA* PCR assay in wound specimens. In three of these cases, tularemia was confirmed by a cell-mediated immune response, suggesting that PCR may be used to diagnose some cases of tularemia undetectable by conventional diagnostic methods.[70] The *lpnA* conventional PCR assay is currently in routine use in several countries for diagnosis of ulceroglandular *F. tularensis* in wound specimens, permitting earlier diagnosis and initiation of effective treatment.

A limited number of samples from patients with serology or culture-confirmed oculoglandular tularemia have been tested with the conventional *lpnA* assay.[82] Biopsy specimens from cervical lymph nodes (two patients), a conjunctival swab (one patient), a surgically removed conjunctival granuloma (one patient), and blood samples (two patients) were tested, with DNA extracted using a standard phenol-chloroform procedure. In all six clinical specimens, the *lpnA* assay detected *F. tularensis* DNA, showing promising utility of this method for future clinical diagnosis of oculoglandular tularemia.

To increase the sensitivity of PCR-based diagnostics for *F. tularensis*, real-time PCR assays have been developed. The detection limit of real-time assays is typically 10 times better than standard PCR, in the range of 10 organisms; it is also more rapid, permitting amplification and simultaneous detection of nucleic acid products. Although a number of real-time assays have been developed for detection of *F. tularensis*, assays directed against ISFtu2, *iglC*, and *lpnA* have been utilized most commonly with clinical specimens (Table 76.2).[83–87] These assays were developed as 5′ nuclease TaqMan assays, but have also been modified for use with SYBR Green as the fluorescent detector.[50,83] ISFtu2 is a transposase present in multiple copies in the *F. tularensis* genome; the ISFtu2 transposase in *F. tularensis* displays 97% and 96% identity to transposases in *F. novicida* and *F. philomiragia*, respectively. The *iglC* gene encodes a protein (23 kDa) necessary for intracellular growth and lies within the pathogenicity island of *F. tularensis*; the *iglC* gene of *F. tularensis* shows 99% and 84% identity to genes in *F. novicida* and *F. philomiragia*, respectively. Of the three TaqMan real-time PCR assays (ISFtu2, *iglC*, and *lpnA*), only the *lpnA* assay discriminates *F. tularensis* from *F. philomiragia*.

As the TaqMan assays were not developed for diagnosis of a particular form of tularemia, no standardized method for transport or DNA purification was initially described.[83] Nonetheless, a growing number of specimens from a variety of clinical forms of tularemia (oropharyngeal, ulceroglandular, and pneumonic) have been tested using these assays. In two studies, 15 clinical specimens (lymph node aspirates, bronchial washes, pleural fluids, tissues, blood, and wound swabs) from a confirmed outbreak or from patients previously diagnosed by culture, serology, or DFA, were tested. DNA was extracted using the QIAamp DNA Mini Kit (Qiagen, Inc., Valencia, CA) and the manufacturer's associated DNA extraction protocols. All 15 (100%) clinical specimens were positive for *F. tularensis* by all three TaqMan assays (ISFtu2, *iglC*, and *lpnA*).[88,89]

In another set of studies, lymph node aspirates and tonsillar scrapings from 11 confirmed cases of oropharyngeal tularemia from two independent outbreaks were tested using the TaqMan *lpnA* assay.[90,91] Tonsillar swabs and lymph node aspirates were placed into guanidine isothiocyanate buffer for transport and DNA was purified using the glass bead procedure for the conventional *lpnA* assay. The TaqMan *lpnA* assay detected *F. tularensis* DNA in four of four tonsillar scrapings and nine of nine lymph node aspirates, with several lymph node samples originating from patients treated with antibiotics effective against *F. tularensis*.[90] In addition, identification of *F. tularensis* PCR-positive patients provided initial confirmation for one of the outbreaks, thereby allowing for subsequent identification of 129 additional cases. For follow-up testing during this outbreak, the ISFtu2, *iglC*, and *lpnA* assays were used with SYBR Green.[50] DNA was extracted from throat swabs (n = 25) and lymph node aspirates (n = 13) using the QIAamp DNA Mini Kit. The three real-time PCR assays were positive for 64% (16/25) of throat swabs and 62% (8/13) of lymph node aspirates.

A limitation of both the conventional *lpnA* and real-time TaqMan PCR (ISFtu2, *iglC*, *lpnA*) assays is the inability to discriminate *F. tularensis* from *F. novicida*. While not significant for patient management, or in endemic areas with common case presentations, correct species identification has public health value. It is important to rule out *F. novicida* as the cause of infection, particularly in areas where tularemia has not been previously reported. The subtyping assays described in the next section can be applied for this purpose. Although outside the clinical realm of this chapter, it is also important to note that a growing number of *Francisella*-like organisms have been identified in arthropods and environmental (water and soil) samples, which can cross-react with PCR assays for *F. tularensis* and lead to false positive results.[15,18,19,92,93]

16S rRNA PCR. For PCR detection of the 16S rRNA gene, *Francisella* genus specific primers (F5, F11), as well as primers that recognize both *F. tularensis* and *F. novicida* (FTS8, FTS12)—but not *F. philomiragia*—have been developed.[20] There are, however, only a limited number of reports utilizing these assays for direct detection of *F. tularensis* DNA in clinical specimens. In one case, the FTS8 and FTS12 genus specific primers were employed to detect *F. tularensis* DNA in a lymph node aspirate and lymph node biopsy from a culture-confirmed patient.[94] In another case, the F5–F11 PCR assay was evaluated on wound specimens from patients with culture and/or serology-confirmed ulceroglandular tularemia, with 16S rRNA PCR sensitivity only 50%, as compared to 73% sensitivity of the *lpnA* assay.[79]

Subtyping PCR. For diagnosis of *F. tularensis* in the clinical laboratory, identification by one of the diagnostic PCR methods described above is generally sufficient. There may be reasons, however, to differentiate between *F. tularensis* and *F. novicida* or to identify the infecting *F. tularensis* subspecies or clade. This is especially true in North America where both *F. tularensis* subsps. *tularensis* and *holarctica* are endemic, and infections caused by *F. novicida* also occur. The differences in virulence between subspecies (*tularensis* and *holarctica*) and species (*F. tularensis* and *F. novicida*) may have implications on therapy and potential disease outcomes. Differentiation of *F. novicida* from *F. tularensis* is also important outside North America, particularly in areas where tularemia has not been previously identified. A number of PCR subtyping assays (both conventional and real time) have been developed for identification of *F. tularensis* subspecies and clades, with many of these assays having the added advantage of differentiating *F. tularensis* from *F. novicida*. These PCR assays are used for supplemental characterization after *F. tularensis* has been identified by a primary PCR diagnostic method or on isolates recovered from human cases. PCR is advantageous for subtyping of isolates as these assays provide results more rapidly than conventional biotyping assays (i.e., glycerol fermentation).

Assays for identification of *F. tularensis* subsp. *holarctica* target Ft-M19, a 30 bp deletion specific to *F. tularensis* subsp. *holarctica* or the 3′ junction of the ISFtu2 transposase with an adjoining sequence showing similarity to transmembrane proteins (Genbank Accession no. AY062040) (Table 76.3).[57,88,95] *F. tularensis* subsp. *tularensis* assays target the *pdpD* gene, which is present in *F. tularensis* subsp. *tularensis* strains, almost entirely absent from *F. tularensis* subsp. *holarctica* strains and contains a 144 bp deletion in *F. novicida* strains (Table 76.3).[88,96] Real-time assays have also been developed for differentiation of *F. tularensis* subsp. *tularensis* clades A1 and A2 (Table 76.3).[89] The A2 assay targets a region encoding a hypothetical protein and a pseudogene for methyltransferase (Genbank Accession nos FTW_0430/FTW_0431); this region is present in A2 strains and absent from A1 strains. The A1 assay targets a region within a hypothetical protein (Genbank Accession nos. FTT1194c/

TABLE 76.3
Subtyping PCR Assays for *F. tularensis*

		Amplification Product or Size				
Assay[a]	Primers/Probes (5′→3′)	*F. t. tularensis;* clade A1	*F. t. tularensis;* clade A2	*F. t. holarctica*	*F. t. mediasiatica*	*F. novicida*
F. t. holarctica						
Ft-M19_F Ft-M19_R	AGGCGGAGATCTAGGAACCTTT AGCCCAAGCTGACTAAAATCTTT	250 bp	250 bp	220 bp	250 bp	250 bp
ISFtu2_F ISFtu2_R ISFtu2_P	CTTGTACTTTTATTTGGCTACTGAGAAACT CTTGCTTGGTTTGTAAATATAGTGGAA ACCTAGTTCAACCTCAAGACTTTTAGTA ATGGGAATGTCA	–	–	+	nt	–
F. t. tularensis						
*pdpD*_F *pdpD*_R *pdpD*_P	GAGACATCAATTAAAAGAAGCAATACCTT CCAAGAGTACTATTTCCGGTTGGT AAAATTCTGCTCAGCAGGATTTTG ATTTGGTT	+	+	–	nt	–
*pdpD*_F *pdpD*_R *pdpD*_FL *pdpD*_LC	TGGGTTATTCAATGGCTCAG TCTTGCACAGCTCCAAGAGT AGCAATACCTTATAATGAAGGTC TTCTAGAAAAT-FL 640-CTGCTCAGCAGGATTTTGATTTG	+	+	–	–	–
F. t. tularensis clade A1						
A1_F A1_R A1_P	CACTATCTTTACTTTAGCTTTGCCACAA TGATCCTTGGCATAAGAAAAAACA CTGGACAAAAATATTATCAAAGATTAAC AAGCCTACGC	+	–	–	nt	–
F. t. tularensis clade A2						
A2_F A2_R A2_P	GACAATCATTTAAGCAAAACGCTACT GCAGGTAATGTAGTTTTAGCAAATGC TTCTAGGATAATCATCTGCGATACCGTTGCC	–	+	–	nt	–

[a] F, Forward primer; R, Reverse primer; P, TaqMan probe; FL, Fluorescein; LC, LightCycler-Red 640

FTT1195c) present in A1 strains and absent from A2 strains. When tested against a large collection of isolates, all assays showed good specificity, consistently identifying only the *F. tularensis* subspecies or clade the assay was developed to detect (Table 76.3).[88,89,95,96]

16S rRNA Sequencing. Because of the infrequency of human tularemia, *F. tularensis* is often not easily identified, even when it is cultured. Traditional biochemical classification assays performed in clinical laboratories are of little value for *F. tularensis* isolates. Sequencing of the 16S rRNA gene has thus proven particularly useful for isolate identification.[36] 16S rRNA gene sequencing is also important for distinguishing *F. tularensis* from other *Francisellaceae*, with multiple examples of this methodology used for correct identification of clinical *F. novicida* and *F. philomiragia* isolates.[97–99]

Strain Typing. Although molecular typing of strains lacks direct clinical diagnostic value, it has enabled significant progress in epidemiological research and furthered our understanding of population structures and phylogenetic relationships. A number of different high resolution molecular subtyping methods, including multilocus variable number tandem repeat (MLVA), pulsed-field gel electrophoresis (PFGE), single nucleotide polymorphisms (SNPs), insertion-deletion markers, and multilocus sequencing, have been used to separate *F. tularensis* subsp. *tularensis* into clades A1 (also known as A.I and A-east) and A2 (also known as A.II and A-west).[24,25,27,100–102] PFGE and SNP typing further differentiate A1 into clades A1a and A1b.[25,28] MLVA and SNP typing show the best resolution for defining clades within *F. tularensis* subsp. *holarctica*.[27,30] For discrimination of individual *F. tularensis* strains, MLVA, based on 25 different repeats in the genome, provides the highest resolving power.[27,29]

76.2 METHODS

76.2.1 Collection, Transport, and Sample Preparation

For PCR testing of specimens, the best evaluated methods for collection and transport are for wound samples collected from patients with ulceroglandular tularemia.[70,79] Cotton swabs are recommended for collection of wound specimens. After sample collection, the cotton swab is placed in 0.7 mL of guanidine isothiocyanate-containing lysis buffer (L6).[8] L6 buffer is prepared by dissolving 30 g of guanidine isothiocyanate in 25 mL of 0.1 M Tris, pH 6.4. The guanidine isothiocyanate is dissolved in the Tris buffer by heating at 60°C in a waterbath. Subsequently, 5.5 mL of a 0.5 M EDTA solution adjusted with NaOH to pH 8.0 and 650 mg of Triton X-100 are added to the guanidine isothiocyanate/Tris buffer. The L6 buffer can be dispensed into 0.7 mL aliquots in 1.5 mL tubes and stored in the dark for 3 months. For transport of throat swabs and lymph node aspirates from patients with oropharyngeal tularemia, a method has been reported where swabs or 0.2 mL of aspirate are placed directly into tubes containing 0.5 mL of lysis buffer (5M guanidine isothiocyanate, 1/10 v/v sodium acetate and 0.5% sarcosil).[90,91]

For preparation of DNA from ulceroglandular swabs, the transport tube containing the cotton swab is subjected to freezing and thawing to lyse *F. tularensis*. After thawing, the tube is vigorously vortexed; 300 μL of L6 buffer and 5 μL of Glassmilk (MP Biomedicals, Irvine, CA) are added, samples vortexed for 5 s followed by incubation for 10 min at room temperature. The sample is then centrifuged at 10,000 × *g* for 15 s to pellet the Glassmilk with bound DNA. The supernatant is removed and the deposit is washed twice with 1.0 mL of L2 buffer, twice with 1.0 mL of ethanol and once with 1.0 mL of acetone. Between each step, samples are vortexed briefly and centrifuged at 10,000 × *g* for 15 s and supernatant discarded. L2 buffer is prepared by adding 60 g of guanidine isothiocyanate to 50 mL of 0.1 M Tris pH 6.4. After the final wash, the pellet is dried at 56°C for 10 min, with tube lids open. DNA is eluted from the Glassmilk by adding 30 μL TE, vortexing briefly and incubating at 56°C for 10 min. Tubes are vortexed briefly, spun at 12,000 × *g* for 2 min, and the supernatant containing the eluted purified DNA transferred to a new tube. This DNA preparation method has also been used for purification of DNA from throat swabs and lymph node aspirates.[90,91]

For preparation of DNA for testing by the real-time PCR assays, the QIAamp DNA Mini Kit has been used with a variety of clinical samples including lymph node aspirates, bronchial washes, pleural fluids, tissues, blood, wound swabs, and throat swabs. The kit provides the components necessary for DNA purification as well as protocols for purification from whole blood, body fluids, swabs, and tissue samples. Samples are lysed enzymatically, alcohol is added to samples and the lysates loaded onto the QIAamp spin column. DNA is absorbed onto the QIAamp silica membrane during a brief centrifugation step. Wash buffers are used to remove impurities and pure, ready-to-use DNA is then eluted in water or low-salt buffer.

76.2.2 Detection Procedures

(i) Standard lpnA Assay The primers for the conventional *lpnA* assay are listed in Table 76.2; a PCR amplification product of 386 bp is generated. A detailed protocol for the *lpnA* PCR assay is provided in the WHO Guidelines on Tularaemia[8] available online at http://www.cdc.gov/tularemia/resources/whotularemiamanual.pdf.

Briefly, the PCR mixture is composed of purified sample DNA (5 μL), 10× dNTP solution (2.5 μL) (Amersham Biosciences, Buckinghamshire, UK), 10× PCR buffer II (2.5 μL) (Applied Biosystems, Foster City, CA), 1.5 μL of a 10 μM stock of each primer, 3 μL of 25 mM $MgCl_2$, 0.4 μL of Ampli Taq Gold (5 U/μL) (Applied Biosystems), and distilled H_2O to a final volume of 25 μL. Thermal cycling conditions are 95°C for 10 min, followed by 30 cycles of 94°C for 30 s, 60°C for 1 min, and 72°C for 1 min. After amplification, 5 μL of each reaction mixture is subjected to electrophoresis in a 2% agarose gel and the amplified gene products visualized with UV light and ethidium bromide staining.

(ii) TaqMan Assays (ISFtu2, iglC, lpnA) The primer and probe sequences for the three TaqMan assays are listed in Table 76.2. The amplicon sizes are 97, 84, and 91 bp for the ISFtu2, *iglC,* and *lpnA* assays, respectively. The probe is labeled with a 6-carboxy fluorescein reporter molecule at the 5′ end and a Black Hole Quencher (BHQ) attached to the 3′ end.[83] The assays were developed with the LightCycler (Roche Diagnostics, Indianapolis, IN), but have also been optimized for other real-time instruments including an iCycler IQ (BioRad laboratories, Richmond, CA) and Quantica (Techne Inc., Cambridge, UK).[50,83,91] To maximize sensitivity, the three assays are set up independently.

For the LightCycler, the final concentrations of components per reaction are 1× LightCycler-FastStart Master Hybridization probe mix (Roche Diagnostics), 500 nM each primer, 100 nM probe, and 0.5 U uracil-DNA glycosylase. For the *iglC* and *lpnA* assays, the final concentration of $MgCl_2$ is 4 mM, and for the ISFtu2 assay the final concentration is 5 mM. Thermal cycling conditions are 50°C for 2 min, 95°C for 10 min, and 45 cycles at 95°C for 10 s and 60°C for 30 s, and then 45°C for 5 min. Crossing thresholds (Ct) are determined by performing automatic quantification using the second derivative maximum method with the y axis set to F1/F3. A positive result is observed by logarithmic amplification within 45 cycles.

(iii) 16S rRNA PCR For 16S rRNA genus specific PCR, the primer sequences are (F5: 5′-CCTTTTTGAGTTTCGCTCC-3′) and (F11: 5′-TACCAGTTGGAAACGACTGT-3′). For PCR-based recognition of *F. tularensis* and *F. novicida*, but not *F. philomiragia*, by the 16S rRNA gene, the primer sequences are (FTS8: 5′-CAAGGTTAATAGCCTTGGGGA-3′) and (FTS12: 5′-CCTTGTCAGCGGCAGTCTCA). The PCR mix and thermal cycling program for the F5–F11 primers is the same as used for the *lpnA* conventional assay. A detailed protocol for the F5–F11 PCR assay is available in the WHO Guidelines on Tularaemia.[8]

(iv) Subtyping Assays The PCR primers and probes for the *F. tularensis* subtyping assays are provided in Table 76.3. Detailed protocols for *F. tularensis* subsp. *holarctica* (Ft-M19 and ISFtu2 assays) and *F. tularensis* subsp. *tularensis* (*pdpD*) PCR assays are provided in the WHO Guidelines on Tularaemia.[8]

For the *F. tularensis* subsp. *tularensis* clade A1 and A2 assays, the final concentrations of reactants includes 1× LightCycler Fast Start master Hybridization probe mix (Roche Diagnostics), 750 nM each A1 primer or 500 nM each A2 primer, 200 nM probe, 5 mM $MgCl_2$, and 0.5 U uracil-DNA glycosylase. The thermal cycling conditions are 1 cycle of 50°C for 2 min, 1 cycle of 95°C for 10 min, 45 cycles at 95°C for 10 s, and 60°C for 30 s, and then 45°C for 5 min.

76.3 CONCLUSION AND FUTURE PERSPECTIVES

Molecular assays for diagnosis of tularemia are still in their infancy. Implementation of PCR in the clinical laboratory has been particularly successful in Sweden, where the ulceroglandular form dominates.[80,81] If a patient presents with an ulcer and a reported mosquito bite, clinicians are apt to suspect tularemia and send samples for rapid molecular testing. PCR diagnostics for tularemia have also been successfully incorporated into reference laboratories worldwide, particularly for use during outbreaks.[50,89–91]

Challenges remain, however, particularly with regard to integration of molecular diagnostics into clinical laboratories in areas with only sporadic cases, multiple disease manifestations, and lack of clinical suspicion for tularemia. As tularemia is a relatively rare disease in most areas of the Northern Hemisphere, clinicians have only limited or no experience with it. In the absence of clinical suspicion, the cost and time required to perform endless organism-specific molecular tests within the clinical laboratory would be prohibitive. Future development of low-cost multiplexed molecular assays for detection of multiple organisms associated with different specimen types, such as wound samples, respiratory samples, aspirate samples, and blood samples would be useful for clinical laboratories. This type of broad-scale approach is advantageous because it does not require a physician to request specific testing for *F. tularensis* but would still allow for early diagnosis. Standardized sampling, transport, and DNA purification protocols for tularemia specimens, other than wounds or ulcers, are also necessary for successful implementation of molecular diagnostics into clinical laboratories.

Use of molecular assays in clinical laboratories may allow for earlier diagnosis of tularemia, which in turn could lead to reductions in mortality and in complications associated with delay in diagnosis (i.e., suppurated lymph nodes requiring surgical drainage). Furthermore, early diagnosis should decrease medical costs associated with disease complications. Analysis of the impact of PCR-based diagnostics for tularemia, with respect to reduction in mortality, complications, and associated medical costs, will be important for demonstrating the utility of these assays in clinical laboratories.

Beyond clinical diagnosis of tularemia, molecular methods could also be used to inform future prevention and control strategies for tularemia. By comparing molecular fingerprints of clinical isolates with those of isolates recovered from presumed exposure sources, molecular typing of *F. tularensis* strains may identify previously unknown sources of exposure. With a better understanding of behaviors that put people at risk of infection, public health officials can work to implement education programs and effective control strategies. Molecular typing can also be used to identify geographic regions where infections with more virulent strains occur and educate physicians and health care workers in these areas accordingly. The coming years hold much promise for application of molecular methods to the diagnosis, prevention, and control of tularemia.

REFERENCES

1. McCoy, G., and Chapin, C., Further observations on a plague-like disease of rodents with a preliminary note on the causative agent, *Bacterium tularense*, *J. Infect. Dis.*, 10, 61, 1912.

2. Sjostedt, A., Tularemia: History, epidemiology, pathogen physiology, and clinical manifestations, *Ann. N. Y. Acad. Sci.*, 1105, 1, 2007.
3. Morner, T., The ecology of tularemia, *Rev, Sci, Tech.*, 11, 1123, 1992.
4. Jellison, W.L., *Tularemia in North America, 1930–1974*, University of Montana Missoula, 1974.
5. Ellis, J. et al., Tularemia, *Clin. Microbiol. Rev.*, 15, 631, 2002.
6. Saslaw, S. et al., Tularemia vaccine study. II. Respiratory challenge, *Arch. Intern. Med.*, 107, 702, 1961.
7. Eliasson, H. et al., Tularemia: current epidemiology and disease management, *Infect. Dis. Clin. North Am.*, 20, 289, 2006.
8. Tarnvik, A. ed., *WHO Guidelines on Tularaemia*, World Health Organization, Geneva, 2007.
9. Dennis, D.T. et al., Tularemia as a biological weapon: Medical and public health management, *JAMA*, 285, 2763, 2001.
10. Sjostedt, A., Family XVII. *Francisellaceae*, Genus I. *Francisella*. In Brenner, D.J. (Editor), *Bergey's Manual of Systematic Bacteriology*, pp. 200–210. Springer-Verlag, New York, NY, 2005.
11. Larsson, P. et al., The complete genome sequence of *Francisella tularensis*, the causative agent of tularemia, *Nat. Genet.*, 37, 153, 2005.
12. Hollis, D.G. et al., *Francisella philomiragia* comb. nov. (formerly *Yersinia philomiragia*) and *Francisella tularensis* biogroup *novicida* (formerly *Francisella novicida*) associated with human disease, *J. Clin. Microbiol.*, 27, 1601, 1989.
13. Wenger, J.D. et al., Infection caused by *Francisella philomiragia* (formerly *Yersinia philomiragia*), a newly recognized human pathogen, *Ann. Intern. Med.*, 110, 888, 1989.
14. Kugeler, K.J. et al., Isolation and characterization of a novel *Francisella* sp. from human cerebrospinal fluid and blood, *J. Clin. Microbiol.*, 46, 2428, 2008.
15. Barns, S.M. et al., Detection of diverse new *Francisella*-like bacteria in environmental samples, *Appl. Environ. Microbiol.*, 71, 5494, 2005.
16. Ottem, K.F. et al., Characterization of *Francisella* sp., GM2212, the first *Francisella* isolate from marine fish, Atlantic cod (*Gadus morhua*), *Arch. Microbiol.*, 187, 343, 2007.
17. Olsen, A.B. et al., A novel systemic granulomatous inflammatory disease in farmed Atlantic cod, *Gadus morhua* L., associated with a bacterium belonging to the genus *Francisella*, *J. Fish Dis.*, 29, 307, 2006.
18. Petersen, J.M. et al., Direct isolation of *Francisella* spp. from environmental samples, *Lett. Appl. Microbiol.*, 48, 663, 2009.
19. Scoles, G.A., Phylogenetic analysis of the *Francisella*-like endosymbionts of *Dermacentor* ticks, *J. Med. Entomol.*, 41, 277, 2004.
20. Forsman, M., Sandstrom, G., and Sjostedt, A., Analysis of 16S ribosomal DNA sequences of *Francisella* strains and utilization for determination of the phylogeny of the genus and for identification of strains by PCR, *Int. J. Syst. Bacteriol.*, 44, 38, 1994.
21. Euzeby, J.P., List of Prokaryotic names with Standing in Nomenclature http://www.bacterio.cict.fr/f/francisella.html, last accessed August 13, 2009.
22. Larsson, P. et al., Molecular evolutionary consequences of niche restriction in *Francisella tularensis*, a facultative intracellular pathogen, *PLoS Pathog.*, 5, e1000472, 2009.
23. Olsufjev, N.G., and Meshcheryakova, I.S., Infraspecific taxonomy of tularemia agent *Francisella tularensis* McCoy et Chapin, *J. Hyg. Epidemiol. Microbiol. Immunol.*, 26, 291, 1982.
24. Staples, J.E. et al., Epidemiologic and molecular analysis of human tularemia, United States, 1964–2004, *Emerg. Infect. Dis.*, 12, 1113, 2006.
25. Kugeler, K.J. et al., Molecular epidemiology of *Francisella tularensis* in the United States, *Clin. Infect. Dis.*, 48, 863, 2009.
26. Sandstrom, G. et al., Characterization and classification of strains of *Francisella tularensis* isolated in the central Asian focus of the Soviet Union and in Japan, *J. Clin. Microbiol.*, 30, 172, 1992.
27. Johansson, A. et al., Worldwide genetic relationships among *Francisella tularensis* isolates determined by multiple-locus variable-number tandem repeat analysis, *J. Bacteriol.*, 186, 5808, 2004.
28. Pandya, G. et al., Whole Genome Resequencing and SNP Genotyping of 40 *Francisella tularensis* strains [abstract 182c]. In *Program and Abstracts of ASM Biodefense and Emerging Diseases Research Meeting*, p. 55, American Society for Microbiology, Washington DC, 2007.
29. Farlow, J. et al., *Francisella tularensis* in the United States, *Emerg. Infect. Dis.*, 11, 1835, 2005.
30. Vogler, A.J. et al., Phylogeography of *Francisella tularensis*: global expansion of a highly fit clone, *J. Bacteriol.*, 191, 2474, 2009.
31. Tarnvik, A., Sandstrom, G., and Sjostedt, A., Epidemiological analysis of tularemia in Sweden 1931–1993, *FEMS Immunol. Med. Microbiol.*, 13, 201, 1996.
32. Berdal, B.P. et al., Field investigations of tularemia in Norway, *FEMS Immunol. Med. Microbiol.*, 13, 191, 1996.
33. Abd, H. et al., Survival and growth of *Francisella tularensis* in *Acanthamoeba castellanii*, *Appl. Environ. Microbiol.*, 69, 600, 2003.
34. Forsman, M. et al., *Francisella tularensis* does not manifest virulence in viable but non-culturable state, *FEMS Microbiol. Ecol.*, 31, 217, 2000.
35. Centers for Disease Control and Prevention. Tularemia—United States, 1990 2000, *Morb. Mortal. Wkly. Rep.*, 51, 181, 2002.
36. Anda, P. et al., Waterborne outbreak of tularemia associated with crayfish fishing, *Emerg. Infect. Dis.*, 7 (3 Suppl), 575, 2001.
37. Martin, C. et al., Outbreak of tularaemia in Castilla y Leon, Spain, *Euro Surveill.*, 12, E071108 1, 2007.
38. Payne, L., Endemic tularemia, Sweden, 2003, *Emerg. Infect. Dis.*, 11, 1440, 2005.
39. Reintjes, R. et al., Tularemia outbreak investigation in Kosovo: case control and environmental studies, *Emerg. Infect. Dis.*, 8, 69, 2002.
40. Klock, L.E., Olsen, P.F., and Fukushima, T., Tularemia epidemic associated with the deerfly, *JAMA*, 226, 149, 1973.
41. Petersen, J.M. et al., Multiple *Francisella tularensis* subspecies and clades, tularemia outbreak, Utah, *Emerg. Infect. Dis.*, 14, 1928, 2008.
42. Eliasson, H. et al., The 2000 tularemia outbreak: A case-control study of risk factors in disease-endemic and emergent areas, Sweden, *Emerg. Infect. Dis.*, 8, 956, 2002.
43. Petersen, J.M., Mead, P.S., and Schriefer, M. E., *Francisella tularensis*: An arthropod-borne pathogen, *Vet. Res.* 40, 7, 2009.
44. Stewart, S.J., Tularemia: association with hunting and farming, *FEMS Immunol. Med. Microbiol.*, 13, 197, 1996.
45. Taylor, J.P. et al., Epidemiologic characteristics of human tularemia in the southwest-central states, 1981–1987, *Am. J. Epidemiol.*, 133, 1032, 1991.

46. Arav-Boger, R., Cat-bite tularemia in a seventeen-year-old girl treated with ciprofloxacin, *Pediatr. Infect. Dis. J.*, 19, 583, 2000.
47. Capellan, J., and Fong, I.W., Tularemia from a cat bite: Case report and review of feline-associated tularemia, *Clin. Infect. Dis.*, 16, 472, 1993.
48. Leblebicioglu, H. et al., Outbreak of tularemia: A case-control study and environmental investigation in Turkey, *Int. J. Infect. Dis.*, 12, 265, 2008.
49. Helvaci, S. et al., Tularemia in Bursa, Turkey: 205 cases in ten years, *Eur. J. Epidemiol.*, 16, 271, 2000.
50. Willke, A. et al., An outbreak of oropharyngeal tularaemia linked to natural spring water, *J. Med. Microbiol.*, 58, 112, 2009.
51. Feldman, K.A. et al., An outbreak of primary pneumonic tularemia on Martha's Vineyard, *N. Engl. J. Med.*, 345, 1601, 2001.
52. Syrjala, H. et al., Airborne transmission of tularemia in farmers, *Scand. J. Infect. Dis.*, 17, 371, 1985.
53. Dahlstrand, S., Ringertz, O., and Zetterberg, B., Airborne tularemia in Sweden, *Scand. J. Infect. Dis.*, 3, 7, 1971.
54. McCarthy, V.P., and Murphy, M.D., Lawnmower tularemia, *Pediatr. Infect. Dis. J.*, 9, 298, 1990.
55. Thomas, R. et al., Discrimination of human pathogenic subspecies of *Francisella tularensis* by using restriction fragment length polymorphism, *J. Clin. Microbiol.*, 41, 50, 2003.
56. Broekhuijsen, M. et al., Genome-wide DNA microarray analysis of *Francisella tularensis* strains demonstrates extensive genetic conservation within the species but identifies regions that are unique to the highly virulent *F. tularensis* subsp. *tularensis*, *J. Clin. Microbiol.*, 41, 2924, 2003.
57. Johansson, A. et al., Evaluation of PCR-based methods for discrimination of *Francisella* species and subspecies and development of a specific PCR that distinguishes the two major subspecies of *Francisella tularensis*, *J. Clin. Microbiol.*, 38, 4180, 2000.
58. Molins-Schneekloth, C.R., Belisle, J.T., and Petersen, J.M., Genomic Markers for Differentiation of *Francisella tularensis* subsp. *tularensis* A.I and A.II, *Appl. Environ. Microbiol.*, 2007.
59. Tarnvik, A., and Chu, M.C., New approaches to diagnosis and therapy of tularemia, *Ann. N. Y. Acad. Sci.*, 1105, 378, 2007.
60. Evans, M.E. et al., Tularemia: A 30-year experience with 88 cases, *Medicine (Baltimore)*, 64, 251, 1985.
61. Penn, R.L. ed., *Francisella tularensis (Tularemia)*, Elsevier, Philadelphia, 2005.
62. Enderlin, G. et al., Streptomycin and alternative agents for the treatment of tularemia: Review of the literature, *Clin. Infect. Dis.*, 19, 42, 1994.
63. Syrjala, H., Schildt, R., and Raisainen, S., In vitro susceptibility of *Francisella tularensis* to fluoroquinolones and treatment of tularemia with norfloxacin and ciprofloxacin, *Eur. J. Clin. Microbiol. Infect. Dis.*, 10, 68, 1991.
64. Limaye, A.P., and Hooper, C.J., Treatment of tularemia with fluoroquinolones: Two cases and review, *Clin. Infect. Dis.*, 29, 922, 1999.
65. Johansson, A. et al., Ciprofloxacin for treatment of tularemia in children, *Pediatr. Infect. Dis. J.* 19, 449, 2000.
66. Tularemia—Missouri, 2000–2007, *Morb. Mortal. Wkly. Rep.*, 58, 744, 2009.
67. Pike, R.M., Laboratory-associated infections: Summary and analysis of 3921 cases, *Health Lab. Sci.*, 13, 105, 1976.
68. Overholt, E.L. et al., An analysis of forty-two cases of laboratory-acquired tularemia. Treatment with broad spectrum antibiotics, *Am. J. Med.*, 30, 785, 1961.
69. Department of Health and Human Services, P.H.S., *Biosafety in Microbiological and Biomedical Laboratories*, 5th ed., Department of Health and Human Services Washington, DC, 2007.
70. Johansson, A. et al., Comparative analysis of PCR versus culture for diagnosis of ulceroglandular tularemia, *J. Clin. Microbiol.*, 38, 22, 2000.
71. Koskela, P., and Salminen, A., Humoral immunity against *Francisella tularensis* after natural infection, *J. Clin. Microbiol.*, 22, 973, 1985.
72. Brown, S.L. et al., Evaluation of a safranin-O-stained antigen microagglutination test for *Francisella tularensis* antibodies, *J. Clin. Microbiol.*, 11, 146, 1980.
73. Bevanger, L., Maeland, J.A., and Naess, A.I., Agglutinins and antibodies to *Francisella tularensis* outer membrane antigens in the early diagnosis of disease during an outbreak of tularemia, *J. Clin. Microbiol.*, 26, 433, 1988.
74. Porsch-Ozcurumez, M. et al., Comparison of enzyme-linked immunosorbent assay, Western blotting, microagglutination, indirect immunofluorescence assay, and flow cytometry for serological diagnosis of tularemia, *Clin. Diagn. Lab. Immunol.*, 11, 1008, 2004.
75. Schmitt, P. et al., A novel screening ELISA and a confirmatory Western blot useful for diagnosis and epidemiological studies of tularemia, *Epidemiol. Infect.*, 133, 759, 2005.
76. Bevanger, L., Maeland, J.A., and Kvan, A.I., Comparative analysis of antibodies to *Francisella tularensis* antigens during the acute phase of tularemia and eight years later, *Clin. Diagn. Lab. Immunol.*, 1, 238, 1994.
77. Grunow, R. et al., Detection of *Francisella tularensis* in biological specimens using a capture enzyme-linked immunosorbent assay, an immunochromatographic handheld assay, and a PCR, *Clin. Diagn. Lab. Immunol.*, 7, 86, 2000.
78. Johansson, A., Forsman, M., and Sjostedt, A., The development of tools for diagnosis of tularemia and typing of *Francisella tularensis*, *Apmis*, 112, 898, 2004.
79. Sjostedt, A. et al., Detection of *Francisella tularensis* in ulcers of patients with tularemia by PCR, *J. Clin. Microbiol.*, 35, 1045, 1997.
80. Eliasson, H., Sjostedt, A., and Back, E., Clinical use of a diagnostic PCR for *Francisella tularensis* in patients with suspected ulceroglandular tularaemia, *Scand. J. Infect. Dis.*, 37, 833, 2005.
81. Eliasson, H., and Back, E., Tularaemia in an emergent area in Sweden: An analysis of 234 cases in five years, *Scand. J. Infect. Dis.*, 39, 880, 2007.
82. Kantardjiev, T., Padeshki, P., and Ivanov, I.N., Diagnostic approaches for oculoglandular tularemia: Advantages of PCR, *Br. J. Ophthalmol.*, 91, 1206, 2007.
83. Versage, J.L. et al., Development of a multitarget real-time TaqMan PCR assay for enhanced detection of *Francisella tularensis* in complex specimens, *J. Clin. Microbiol.*, 41, 5492, 2003.
84. Fujita, O. et al., Development of a real-time PCR assay for detection and quantification of *Francisella tularensis*, *Jpn. J. Infect. Dis.*, 59, 46, 2006.
85. McAvin, J.C. et al., Identification of *Francisella tularensis* using real-time fluorescence polymerase chain reaction, *Mil. Med.*, 169, 330, 2004.
86. Emanuel, P.A. et al., Detection of *Francisella tularensis* within infected mouse tissues by using a hand-held PCR thermocycler, *J. Clin. Microbiol.*, 41, 689, 2003.
87. Higgins, J.A. et al., Detection of *Francisella tularensis* in infected mammals and vectors using a probe-based polymerase chain reaction, *Am. J. Trop. Med. Hyg.*, 62, 310, 2000.

88. Kugeler, K.J. et al., Real-time PCR for *Francisella tularensis* types A and B, *Emerg. Infect. Dis.*, 12, 1799, 2006.
89. Molins, C.R. et al., Identification of *Francisella tularensis* subsp. *tularensis* A1 and A2 infections by real-time polymerase chain reaction, *Diagn. Microbiol. Infect. Dis.*, 64, 6, 2009.
90. Celebi, G. et al., Tularemia, a reemerging disease in northwest Turkey: Epidemiological investigation and evaluation of treatment responses, *Jpn. J. Infect. Dis.*, 59, 229, 2006.
91. Karadenizli, A. et al., Outbreak of tularaemia in Golcuk, Turkey in 2005: Report of 5 cases and an overview of the literature from Turkey, *Scand. J. Infect. Dis.*, 37, 712, 2005.
92. Sreter-Lancz, Z. et al., Detection of a novel *Francisella* in *Dermacentor reticulatus*: A need for careful evaluation of PCR-based identification of *Francisella tularensis* in Eurasian ticks, *Vector Borne Zoonotic Dis.*, 9, 123, 2009, Epub ahead of print.
93. Kugeler, K.J. et al., Discrimination between *Francisella tularensis* and *Francisella*-like endosymbionts when screening ticks by PCR, *Appl. Environ. Microbiol.*, 71, 7594, 2005.
94. Dolan, S.A., Dommaraju, C.B., and DeGuzman, G.B., Detection of *Francisella tularensis* in clinical specimens by use of polymerase chain reaction, *Clin. Infect. Dis.*, 26, 764, 1998.
95. Bystrom, M. et al., Tularemia in Denmark: identification of a *Francisella tularensis* subsp. *holarctica* strain by real-time PCR and high-resolution typing by multiple-locus variable-number tandem repeat analysis, *J. Clin. Microbiol.*, 43, 5355, 2005.
96. Tomaso, H. et al., Real-time PCR using hybridization probes for the rapid and specific identification of *Francisella tularensis* subspecies *tularensis*, *Mol. Cell. Probes*, 21, 12, 2007.
97. Leelaporn, A. et al., *Francisella novicida* bacteremia, Thailand, *Emerg. Infect. Dis.*, 14, 1935, 2008.
98. Mailman, T.L., and Schmidt, M.H., *Francisella philomiragia* adenitis and pulmonary nodules in a child with chronic granulomatous disease, *Can. J. Infect. Dis. Med. Microbiol.*, 16, 245, 2005.
99. Friis-Moller, A. et al., Problems in identification of *Francisella philomiragia* associated with fatal bacteremia in a patient with chronic granulomatous disease, *J. Clin. Microbiol.*, 42, 1840, 2004.
100. Svensson, K. et al., Evolution of subspecies of *Francisella tularensis*, *J. Bacteriol.*, 187, 3903, 2005.
101. Nubel, U. et al., Population structure of *Francisella tularensis*, *J. Bacteriol.*, 188, 5319, 2006.
102. Larsson, P. et al., Canonical insertion-deletion markers for rapid DNA typing of *Francisella tularensis*, *Emerg. Infect. Dis.*, 13, 1725, 2007.

77 Haemophilus

Dongyou Liu, Jianshun Chen, and Weihuan Fang

CONTENTS

77.1 INTRODUCTION

77.1.1 Classification and Biology

The genus *Haemophilus* is classified in the family Pasteurellaceae, order Pasteurellales, class Gammaproteobacteria, phylum Proteobacteria. It encompasses a large group of gram-negative, aerobic or facultatively anaerobic, pleomorphic coccobacilli, forming many distinct species (e.g., *H. aegyptius*, *H. avium*, *H. ducreyi*, *H. felis*, *H. haemolyticus*, *H. influenzae*, *H. parainfluenzae*, *H. paracuniculus*, *H. parahaemolyticus*, *H. pittmaniae*, *H. somnus*, etc.).[1–3]

The compositions of the *Haemophilus* genus (*haema*, blood; *philos*, friend, loving; *Haemophilus*, blood-lover) have undergone significant changes in recent years, following the application of DNA–DNA hybridization, nucleotide sequencing, and other molecular procedures. It has been showed that *H. aegyptius*, *H. parahaemolyticus*, and *H. paraphrohaemolyticus* do not merit specific status[1]; and *H. aegyptius* (causing conjunctivitis and Brazilian purpuric fever) is later proposed to be *H. influenzae* biogroup *aegyptius*.[4,5] Hemolytic V-factor-dependent strains from swine, formerly in *H. parahaemolyticus*, demonstrate clear difference from strains of human origin and are designated *H. pleuropneumoniae*. Only one of the several groups that were once assigned to *H. ducreyi* remains within the species. Covering one of V-factor-dependent hemophili and consisting mainly of strains from the human oral cavity, *H. segnis* is now known as *Aggregatibacter segnis*.[6] Similarly, *Actinobacillus (Haemophilus) actinomycetemcomitans* is reclassified as *Aggregatibacter actinomycetemcomitans*, and *Haemophilus aphrophilus* and *Haemophilus paraphrophilus* are combined to form a single species *Aggregatibacter aphrophilus* and emended an description of *Aggregatibacter aphrophilus* includes V-factor-dependent and V-factor-independent isolates.[6] *Haemophilus piscium* is found to be distinct from the *Haemophilus*; *H. suis* and *H. gallinarum* as well as *H. parasuis* and *H. paragallinarum* from swine and fowls are not hemin-dependent.[1,7,8] A novel species *Haemophilus pittmaniae* is established to cover several newly identified isolates.[9]

Haemophilus influenzae, a commensal bacillus present in the human upper respiratory tract, is associated with localized and invasive infections (e.g., bronchitis, otitis, pneumonia, meningitis, septicemia, and epiglottitis). Six structurally and serologically distinct capsular polysaccharides are expressed by capsulated *H. influenzae* isolates, which have been exploited in slide agglutination to differentiate the species into serotypes A to F.[10,11] Some *H. influenzae* strains do not contain capsule and are thus nontypeable (NTHi) by the technique. Prior to the implementation of *H. influenzae* type b (Hib) vaccination in the 1980s, over 95% of *H. influenzae* disease was attributable to *H. influenzae* type B.[12] The application of Hib conjugate vaccine has resulted in a significant reduction of the incidence of *H. influenzae* type B disease. However, infections with other capsular types (A and C to F) and noncapsulate (NC) or nontypeable (NTHi) *H. influenzae* strains have since become more common. In a recent survey of 52 *H. influenzae* isolates from patients with invasive disease, 26 are of serotype A, 20 of nontypeable, three of serotype B, and one isolate each of serotypes C, D, and F.[13]

Haemophilus parainfluenzae is commensal species that may induce pneumonia and endocarditis in immunologically weakened hosts. *H. parainfluenzae* is also subdivided into a number of biotypes (e.g., biotype 1 and biotype 2).[14] Comparison of several housekeeping genes in *H. parainfluenzae* with those in *H. influenzae* supports previous DNA–DNA hybridization data for the genus rather than the phylogeny inferred from 16S rRNA gene sequence comparison, indicating the benefit of preserving the former species in the genus *Haemophilus*.[1]

Haemophilus ducreyi is a fastidious coccobacillus causing the sexually transmitted disease chancroid, a major cause of genital ulceration that is characterized by painful sores on the genitalia and dark or light green shears in excrement.[15,16]

Haemophilus somnus (*Histophilus somni*) consists of commensal strains that colonize and reside relatively harmlessly in bovine mucosal surfaces (e.g., the upper respiratory and reproductive tracts) and of pathogenic strains that may spread systemically and cause opportunistic infections, leading to pneumonia, thrombotic meningoencephalitis, myocarditis, septicemia, arthritis, and abortion. Analysis of the genomes of *Haemonphilus somnus* 129Pt (a commensal preputial isolate), *Haemophilus influenzae* Rd (an avirulent laboratory strain), and *Haemophilus ducreyi* 35000HP (a pathogenic strain causing the human sexually transmitted disease chancroid) uncovers many genes that are common among these three organisms. However, the existence of distinct genomic features in *H. somnus* 129Pt, *H. influenzae*, and *H. ducreyi* underscores their ability to colonize mucosal niches in different hosts.[17–19]

77.1.2 Clinical Features and Pathogenesis

Haemophilus influenzae is a pleomorphic gram-negative bacillus that exists as a commensal in the human upper respiratory tract.[20–26] A recent study indicates the carriage rates of *H. influenzae* in children range from 56.4% in the winter to 38.6% in the summer (with an average of 47.4%).[27] The bacterium has the capacity to take advantage of host's injury and impaired immune defense, causing localized and invasive infections.[28]

Specifically, *H. influenzae* type b (Hib) strains colonize the nasopharynx, and can induce bacteremia by penetrating the epithelium and capillary endothelium, and meningitis by spreading via lymphatic drainage or hematogenous spread, leading to six "classical" disease entities: meningitis, bacteremic pneumonia, epiglottitis, septicemia, cellulitis, and osteoarticular infections (e.g., septic arthritis and osteomyelitis). In fact, meningitis often accounts for over 50% of the entire Hib disease spectrum and approximately 85% of Hib disease occur in patient group aged 0–4 years. It is estimated that Hib was responsible for about 8.13 million serious illnesses worldwide in 2000, with 371,000 deaths in children aged 1–59 months, especially in regions where Hib conjugate vaccine has not been applied.[29] Hib disease often strikes suddenly, with death or neurological sequelae as common outcome. Given that Hib infections are preventable, an expanded vaccination campaign together with an improved surveillance network will reduce childhood pneumonia and meningitis, and decrease child mortality.

Nontypeable *H. influenzae* strains are less invasive than Hib, and are also found in the nasopharynx as well as the trachea and bronchi and can infect mucosa that is damaged by viral disease or cigarette smoking, resulting in otitias media, sinusitis, tracheobronchitis, and pneumonia. While Hib causes fewer infections nowadays due to the success of Hib conjugate vaccine, nontypeable (NT) *H. influenzae* or *H. influenzae* of capsule types A, C, D, E, and especially F are increasingly incriminated as the culprits for *H. influenzae* invasive and other diseases.[25,30–38] Due to its proximity to the densely colonized upper respiratory tract, the lung is more readily invaded by less invasive organisms such as nontype B *H. influenzae*. In one recent report involving 166 *H. influenzae* isolates (identified between the years of 1996–2001), 58 (35%) belonged to Hib, 89 (54%) to non-B serotypes, while the other 19 (11%) were not serotyped. The non-B serotypes included 25 Hia (28%), 4 Hid (4%), 2 Hie (2%), 11 Hif (12%), and 47 (53%) were nontypeable isolates. Whereas patients with Hib and Hia infection tended to display meningitis, those infected with nontypeable serotypes often presented pneumonia.[39] In another study, two nonencapsulated (nontypeable) *H. influenzae* strains of biotypes III and VI were identified from a case of relapsing meningitis in a boy who had been successfully vaccinated against Hib.[40] In addition, Slater et al.[41] described five cases of adults with *H. influenzae* serotype F-associated bacteremia, secondary to respiratory tract infections (three cases of pneumonia and one of epiglottitis) and dysgammaglobulinaemia (one case). Similarly, Bruun et al.[36] documented 13 cases of invasive *H. influenzae* serotype F infection in adults, mostly occurring as a consequence of pneumonia, infective exacerbation of chronic obstructive pulmonary disease, meningitis, epiglottitis, osteomyelitis, cholangitis, and neutropenic bacteremia. Furthermore, besides Hib, *H. influenzae* serotype F has been implicated in necrotizing fasciitis (previously known as "hospital gangrene"), which is a rare but rapidly progressive infection affecting any of the layers within the soft tissue compartment (dermis, subcutaneous tissue, superficial fascia, deep fascia, or muscle) with thrombosis of the microcirculation.[42–46]

Other opportunistic pathogens within the genus *Haemophilus* include *H. influenzae* biogroup *aegyptius*, *H. ducreyi*, and *H. parainfluenzae*. *H. influenzae* biogroup *aegyptius* (formerly *H. aegyptius*) is known to cause purulent conjunctivitis in humans, and its role in Brazilian purpuric fever (BPF), a fulminant pediatric disease, has only been recognized recently. Usually occurring after a resolved purulent conjunctivitis, BPF is characterized by fever, with rapid progression to purpura, hypotensive shock, and death.[4,47] *H. ducreyi* infects its host by breaking the skin or epidermis, where pyogenic inflammation attracts lymphocytes, macrophages, and granulocytes, causing regional lymphadenitis in the sexually transmitted bacillus chancroid. From an erythematous papular lesion, chancroid develops into a painful bleeding ulcer with a necrotic base and ragged edge.

77.1.3 Diagnosis

77.1.3.1 Phenotypic Techniques

In Vitro Culture. *Haemophilus* spp. are gram-negative, nonmotile, facultatively anaerobic bacteria that require blood factors hemin (factor X) and/or nicotinamide adenine dinucleotide (NAD, or factor V) for growth, and are typically cultivated on blood agar plates. As chocolate agar allows for increased accessibility to these factors, it represents an excellent growth medium for *Haemophilus* spp. *H. ducreyi* also grows well on Mueller-Hinton agar with 5% sheep blood in a CO_2-enriched atmosphere. Sometimes, *Haemophilus* is inoculated together with *Staphylococcus* on a single blood agar plate (the so-called Staph streak technique), in which *Staphylococcus* produces the necessary blood factor byproducts essential for *Haemophilus* growth, with smaller "satellite" colonies of *Haemophilus* surrounding larger *Staphylococcus* colonies.

Biochemical Tests. The inability of *Haemophilus* spp. to synthesize NAD (V-factor dependence) and porphyrin/hemin (X-factor dependence) provides a presumptive identification. V-factor dependence may be assessed with Rosco diagnostic tablets placed on inoculated brain heart infusion agar plates; hemolysis is evaluated on agar plates containing 5% horse blood; and synthesis of porphyrins from δ-aminolevulinic acid is assessed by a modification of the porphyrin test. Further metabolic characteristics of *H. influenzae* include fermentation of D-xylose, D-ribose, and D-galactose, but not sucrose or D-mannose, which can be assessed with API 20E biochemical test strips (bioMèrieux). The oxidase (spot) test and assessment of hydrogen sulfide emission using lead acetate paper are also useful. *H. influenzae* is distinguished primarily from *H. haemolyticus* by latter's hemolytic action on erythrocytes. The biotypes of *H. influenzae* strains are typically determined on their ornithine decarboxylase, urease, indol production, and D-xylose fermentation,[1] resulting in the separation of eight biotypes (I–VIII).[48] In addition, immunoglobulin A1 (IgA1) protease activity may be detected by sodium dodecyl sulfate (SDS)-polyacrylamide gel electrophoresis of human myeloma IgA1 solution after overnight incubation with bacteria.

Serotyping. Traditionally, *H. influenzae* serotypes are ascertained in slide agglutination test, using six antisera (A–F) that are specific for each of the six polysaccharides.[49] Serotyping for capsular polysaccharides provides additional confirmation on *Haemophilus* identity, with *H. influenzae* isolates containing these polysaccharides forming six antigenically distinct groups (A–F), and isolates lacking these polysaccharides being called nontypeable.[50] Other phenotypic methods such as outer membrane protein analysis, lipopolysaccharide profiling, and multilocus enzyme electrophoresis may be also employed for the identification and epidemiological study of *Haemophilus* spp.

77.1.3.2 Genotypic Techniques

Species Identification. A number of PCR assays targeting 16S rRNA, *bexA*, *pal* and other genes have been described for species-specific determination of *H. influenzae*.[51–60] When targeting 16S rRNA gene for *H. influenzae* identification, primers complementary to universal regions (e.g., U3 and U8) are often used in the PCR, and then specific primers or probe are utilized for subsequent amplification/detection.[57] The *bexA* gene encodes the capsulation-associated BexA protein that is responsible for transporting capsular material. The presence or absence of the *bexA* gene determines whether an isolate is encapsulated or unencapsulated and therefore NT. Use of the *bexA* gene primers therefore enables amplification of DNA from all six *H. influenzae* types.[51] On the other hand, the *pal* gene encodes the outer membrane protein P6, which exists in both capsulated and noncapsulated *H. influenzae* and is thus useful for detection of any spontaneous capsule-deficient mutants of serotype B (B–) or noncapsulate (NC) strains of *H. influenzae*. However, primers from gene do not distinguish between B and NC strains.[51] In recent years, real-time PCR-based assays have been increasingly described for the detection of *H. influenzae* strains, using the *cap* locus, the capsule-producing gene (*bexA*), the 16S rRNA gene, the insertion-like sequence (IS*1016*), and the outer membrane protein D gene (*glpQ*) as targets. Due to their ease of use and reduced possibility of cross contamination, real-time PCR has become a method of choice in clinical laboratories for molecular detection of *H. influenzae* and other *Haemophilus* spp.[60] Apart from PCR, loop-mediated isothermal amplification (LAMP) of the outer membrane protein P6 gene has been described for detection of *H. influenzae*.[61]

Genotyping and Phylogenetic Analysis. Molecular techniques have proven valuable for characterization and typing of *Haemophilus* spp. A variety of technological platforms (e.g., PFGE, PCR-ribotyping, RAPD–PCR, real-time PCR, and DNA sequencing) targeting 16S rRNA gene, housekeeping genes, and other species and type-specific genes may be applied.[8,62–75]

Falla et al.[11] designed a PCR method for the unequivocal assignment of *H. influenzae* capsular type (types A–F), with primers from capsule type-specific DNA sequences from the capsular gene cluster of each of the six capsular types. Within three functionally defined regions (1–3) in the *cap* locus (encoding polysaccharide capsules for all encapsulated *H. influenzae* serotypes), regions 1 (*bexABCD*) and 3 (*hcsAB*), containing genes necessary for the processing and transport of the capsular material, are common to all six capsule types. Region 2, containing genes for capsule biosynthesis, is unique to each of the six capsule types.[76–82] By using multiple primer sets that detect sequences in the *bexA* gene in region 1 and capsule-specific (types A–F) genes in region 2, the capsule type of *H. influenzae* can be identified. Many phenotypically and serologically nontypeable strains lack the critical elements of the gene cluster, such as the capsule export gene *bexA* (region 1) or transport genes *hcsA* and *hcsB* (region 3), although their capsule-specific genes (region 2) are intact. When used in conjunction with PCR primers derived from the capsular gene bexA, capsulate, noncapsulate, and capsule-deficient type B mutant strains can be differentiated. PCR capsular typing obviates the shortcomings of the conventional slide agglutination serotyping procedure including its spurious cross-reaction and autoagglutination. Further optimization by Maaroufi et al.[83] led to the development of a two-step real-time PCR assay for discrimination of all six *H. influenzae* (i.e., serotypes A–F).[81]

Separately, genotyping based on *pgi* was shown as a surrogate for serotyping of encapsulated *H. influenzae*[84]; nonencapsulated *Haemophilus* strains are typeable by repetitive-element sequence-based PCR using intergenic dyad sequences[85]; presence of copper- and zinc-containing superoxide dismutase in *H. haemolyticus* isolates provides another useful marker to discriminate from nontypeable *H. influenzae*.[86] *H. influenzae* "cryptic genospecies biotype IV" strains (negative for tryptophanase/indole production and positive for urease and ornithine decarboxylase) are distinguishable from typical strains of *H. influenzae* by 16S rRNA gene sequence and DNA hybridization. In addition, *Haemophilus sodA* gene encoding manganese-dependent superoxide dismutase displays a markedly higher divergence than 16S rRNA gene, providing a valuable target for phylogenetic dissection of the genus *Haemophilus*.[87] Specifically, a pair of degenerate primers d1 (5′-CCITAYICITAYGAYG CIYTIGARCC-3′) and d2 (5′-ARRTARTAIGCRTGYTCC CAIACRTC-3′) is used to amplify a 449–458 bp fragment from the *sodA* gene. Subsequent sequencing analysis allows accurate identification of *Haemophilus* isolates from human clinical samples.[87]

Using RFLP to amplify and digest a specific 16S rRNA region of *H. influenzae* with restriction enzymes, and then a DNA probe of the 16S rRNA region to hybridize with the restricted genomic DNA, *H. influenzae* strain variability can be assessed.[88] Analysis of housekeeping genes in the X-factor-dependent species *H. influenzae*, *H. influenzae* biogroup *aegyptius*, and *H. haemolyticus* has uncovered a close relationship among these species, which appear to be more distant from another commensal species, *Haemophilus parainfluenzae*.[89]

Similarly, PCR assays targeting 16S rRNA gene, the *rrs* (16S)-*rrl* (23S) ribosomal intergenic spacer region, and the *gro*EL gene (encoding the *H ducreyi* GroEL heat shock protein) have been developed for specific and sensitive identification of *Haemophilus ducreyi*.[16,90]

77.2 METHODS

77.2.1 Sample Preparation

Culture. *Haemophilus* samples (e.g., nasopharyngeal swabs, BAL fluid, sputum, blood, cerebrospinal or joint fluid) are cultured on 5% horse blood agar and chocolate agar with 5% carbon dioxide at 35°C for 24–48 h. *Haemonphilus* spp. are identified tentatively by standard microbiological methods according to the following criteria: (i) colony morphology, (ii) negativity by Gram staining, (iii) catalase and oxidase production, and (iv) a requirement for V (hemin) and X (NAD) factors. The isolates may be subcultured in brain heart infusion (BHI) broth or BHI agar supplemented with 10 μg/mL hemin and 2 μg/mL NAD (Becton Dickinson).

DNA Extraction. A single colony of *Haemophilus* from agar plate is suspended in 30 μL of lysis solution (1 M Tris pH 8.9 containing 4.5% Nonidet P-40, 4.5% Tween 20, and 10 mg/mL proteinase K) in a microcentrifuge tube. The tube is heated at 60°C for 10 min and at 94°C for 5 min in a thermal cycler to lyze the bacterial cells. Bacterial lysate (2 μL) is used for PCR. Alternatively, a single colony is suspended in 50 μL of sterile water and lysed at 95°C for 5 min. The lysate is used immediately for PCR analysis or frozen at −20°C. Another rapid DNA preparation technique involves suspending 15–20 colonies from each culture in 100 μL of 10 mM Tris-HCl pH 8.0, boiling for 10 min, centrifuging at 2000 × *g* for 2 min, and collecting 80 μL of supernatant for PCR or storage at −20°C.

77.2.2 Detection Procedures

77.2.2.1 PCR Identification of *Haemophilus influenzae*

(i) Multiplex PCR for H. influenzae Billal et al.[91] developed a multiplex PCR for rapid identification of nontypeable and serotype b *Haemophilus influenzae* from nasopharyngeal secretions. Primers specific for the *bexA* gene-encoding capsular polysaccharides (H1 and H2), *p6* gene encoding a common outer membrane protein P6 (P6-S and P6-R), *β*-lactamase gene (TEM-S and TEM-R), and *cpsb* gene-encoding type b capsular polysaccharide (typeB-S and typeB-R) are used for the multiplex PCR to identify nontypeable, type B and *β*-lactamase producing *H. influenzae* strains, respectively (Table 77.1).

Procedure

1. 100 μL of nasopharyngeal secretions (NPS) are cultured on chocolate agar plate and incubated at 37°C for 24–48 h. *H. influenzae* strains are identified by colony morphology on the chocolate agar plates, no growth in blood agar plate, Gram staining, catalase test, and requirement of X and V factors. Productions of *β*-lactamase are examined by nitrocefinase disc (Nippon Becton Dickinson). Serotypes of *H. influenzae* are determined by the slide agglutination test with six specific antisera against each type of capsular polysaccharides (Denka Seiken Co., Ltd., Tokyo).
2. For extraction of genomic DNA, the NPS samples are diluted 1:3 with sterilized saline and centrifuged at 12,000 rpm for 30 min at 4°C. The pellets are collected and digested with 200 μL of lysis solution (1 M Tris pH 8.9, 4.5 v/v, nonident P-40, 4.5 v/v, Tween 20, 10 mg/mL Proteinase K) for 1 h at 60°C. After centrifugation at 12,000 rpm for 20 min at 4°C, the supernatants are mixed with 100 μL of 3 M sodium acetate buffer and then with 1 mL cold ethanol. Total genomic DNA is precipitated. Total

TABLE 77.1
Identity and Sequence of Multiplex PCR Primers

Primer Name	Target	Sequence (5′→3′)	Primer Length	Position	PCR Product (bp)
P6-S	*P6*	ACGATGCTGCAGGCAATGGT	20	141–160	198
P6-R		CATCAGTATTACCTTCTACTAAT	23	316–338	
TEM-S	*TEM-1*	TAAGAGAATTATGCAGTGCTGCC	23	350–372	458
TEM-R		TCCATAGTTGCCTGACTCCCC	21	787–807	
HI-1	*bexA*	CGTTTGTATGATGTTGATCCAGACT	25	3552–3577	343
HI-2		TGTCCATGTCTTTCAAAATGATG	22	3873–3895	
Typeb-S	*cpsb*	AGATACCTTTGGTCGTCTGC	20	5483–5502	224
Typeb-R		CTTACGCTTCTATCTCGGTG	20	5706–5725	

genomic DNA is also purified from *H. influenzae* isolates by lysis of a single colony from a chocolate agar plate in 30 μL of lysis solution for 10 min at 60°C and for 5 min at 94°C in the programmable thermal cycler (Gene Amp PCR System 9700, Perkin-Elmer).

3. PCR mixture (25-μL) is composed of 0.5 μL of each primers (final concentrations are 1 μM of H1 and H2, 1 μM of TEM-S and TEM-R, 1 μM of typeB-S and typeB-R, 3 μM of P6-S and P6-R), 3 μL template DNA, 12.5 μL Qiagen master mixture (Qiagen), 2 μL of 25 mM $MgCl_2$, and 4.5 μL distilled water.
4. Amplification is conducted with an initial denaturation at 94°C for 10 min, followed by 30 cycles of denaturation at 94°C for 30 s, annealing at 55°C for 30 s, extension at 72°C for 30 s, and further extension at 72°C for 10 min.
5. The PCR products are analyzed by 2% agarose gel electrophoresis with a molecular weight standard (100 bp). The expected sizes of DNA fragments are P6 gene (198 bp), Hib (224 bp), and *bexA* (343 bp), TEM-1 (458 bp).

Note: The multiplex PCR has a detection limit of 2×10^{-3} ng genomic DNA for *H. influenzae* strain.

(ii) Real-Time PCR for H. influenzae Morozumi et al.[54] described a real-time PCR for detection of *H. influenzae* in clinical samples, with species-specific primers and molecular beacon (MB) probe derived from 16S rRNA gene. The MB probe is labeled with a fluorescent reporter, 6-carboxyfluorescein (FAM), at the 5′ end and a black hole quencher 1 (BHQ-1) at the 3′ end. The reporter and quencher are connected to stem oligonucleotides with short oligonucleotides.

Primer/Probe	Sequence (5′→3′)	Amplicon Size (bp)
Forward primer	TTGACATCCTAAGAAGAGCTC	167
Reverse primer	TCTCCTTTGAGTTCCCGACCG	
Probe	FAM-CGCGATCCTGACGACAG CCATGCAGCACGATCGCG-BHQ1	

Procedure

1. Clinical samples are suspended in 1.5 mL of PPLO broth (pleuropneumonia-like organisms broth) (Difco). The sputum samples are first homogenized with 50 μL of SPUTAZYME solution (Kyokuto Pharmaceutical Industries Co., Ltd., Tokyo, Japan) and the suspensions are diluted to 1.5 mL with PPLO. Aliquots of 5 μL of are inoculated onto sheep blood agar plates and chocolate agar II plates (Becton Dickinson) and cultured at 37°C in an atmosphere containing 5% CO_2. After 20 h of incubation, the colonies on the plates are counted and bacteria are identified by standard methods.
2. For DNA extraction, a 1.0-mL aliquot of the PPLO broth in an Eppendorf tube is centrifuged at 2000 × *g* for 5 min at 4°C to collect bacterial cells, as well as epithelial and polymorphonuclear leukocyte cells. After discarding the supernatant, the pellet is resuspended in 150 μL of DNase- and RNase-free H_2O, and DNA is extracted using EXTRAGEN II (TOSOH Co., Tokyo, Japan).
3. Real-time PCR mix (50 μL) is composed of 25 μL real-time PCR Master Mix (TOYOBO Co., Osaka, Japan) and 0.15 μL of primers and MB probe solution (all with final concentrations at 300 nM) and 2 μL of sample DNA.
4. Amplification is conducted for 30 sec at 95°C, followed by 40 cycles of 95°C for 15 s, 50°C for 30 s, and 75°C for 30 s in a Stratagene Mx3000P.
5. The C_T value for a positive result is defined as the point at which the horizontal threshold line is crossed.

Note: The real-time PCR has a limit of detection of 10 copies for *H. influenzae*. The whole procedure (from DNA extraction to completion of PCR) takes less than 2 h. The sensitivity and specificity of the real-time PCR study relative to those of conventional cultures are 95.8% and 95.4%, respectively

(iii) Real-Time PCR for H. influenzae Serotype b Marty et al.[68] developed a real-time PCR assay targeting the capsulation locus of *H. influenzae* type B (Hib). The primers and the fluorogenic minor groove binding probe for region II of the capsulation locus of Hib are: forward

primer, 5′-CAAGATACCTTTGGTCGTCTGCTA-3′ (positions 5481 to 5504); reverse primer, 5′-TAGGCTCGAAGA ATGAGAAGTTTTG-3′ (positions 5631 to 5607). The probe is 5′-ATGATATGGGTACATCTGTT-3′ (positions 5563 to 5582), with the fluorescent reporter dye 6-carboxyfluorescein (FAM) at the 5′ end and the quencher 6-carboxy-*N,N,N′,N′*-tetramethylrhodamine (TAMRA) at the 3′ end.

Procedure

1. PCR mix (25-μL) is prepared with 2× TaqMan universal master mix (Perkin-Elmer Biosystems), 400 nM concentration of each primer, 200 nM probe, and 1 μL of DNA extract.
2. Amplification is performed with duplicate samples on an ABI PRISM 7700 sequence detection system for 2 min at 50°C and 10 min at 95°C, followed by 40 cycles of 15 s at 95°C and 1 min at 60°C.
3. Real-time data are analyzed with Sequence Detection Systems software, version 1.7.

Note: The real-time PCR assay shows a total agreement with capsular genotyping for Hib.

(iv) Real-Time Quantitative PCR for H. influenzae Serotype b Fukasawa et al.[92] presented a quantitative real-time PCR for *H. influenzae* serotype b with Hib-specific primers from *cap* gene and SYBR green dye.

Primer/Probe	Sequence (5′→3′)	Amplicon Size (bp)
Hib cap-F	GATACCTTAATTCGCCGCTCGTCAT	152
Hib cap-R	ATATCTCGCACATCGTGTTCAGCAC	

Procedure

1. DNA is extracted with the MORA-EXTRACT kit (Kyokuto).
2. Real-time PCR mix is prepared with Ex *Taq* DNA Hot Start polymerase (TaKaRa Bio) and SYBR Green dye.
3. The amplification is carried out in an iCycler iQ system (Bio Rad), with an initial incubation of 5 min at 95°C, followed by 40 cycles of 30 s at 95°C, 30 s at 55°C, and 30 s at 72°C.
4. For the preparation of external standard DNA, *H. influenzae* ATCC 10211 (Hib) is suspended in Tris/EDTA buffer to a density of 0.5 McFarland units. DNA is extracted with the MORA-EXTRACT kit. Six dilutions from 10^8 to 10^3 organisms/mL are prepared and run in duplicate as external standards in parallel with the test DNA. The threshold cycle (Ct) of the standard dilutions is plotted against the organism cell numbers to generate a standard curve.

77.2.2.2 PCR Identification of *Haemophilus ducreyi*

Suntoke et al.[93] reported a real-time PCR assay for reliable and accurate detection of *Haemophilus ducreyi* with primer and probe from 16S rRNA gene.[94] The probe contains FAM at 5′ end and MGB quencher at 3′ end that enhances probe-target annealing.

Primer/Probe	Sequence (5′→3′)	Position
Forward primer KO7A	CAAGTCGAACGGTAGCACGAAG	56–78
Reverse primer KO8A	TTCTGTGACTAACGTCAATCAATTTTG	495–468
Probe KO15	FAM- CCGAAGGTCCCACCCTTTAAT CCGA- MGB	211–186

Procedure

1. DNA is extracted from swab samples using the QIAamp DNA Blood Mini Kit (Qiagen).
2. The PCR mix (50 μL) is made up of ABI 2× Universal Master Mix, 250 nM of each primers (K07A and K08A) and 100 nM of probe (K015), and 10 μL of extracted sample or positive control DNA (in duplicate).
3. Assay is performed in 96-well plate format using an ABI 7900 HT Fast real-time PCR instrument with the following thermocycle parameters: 50°C for 2 min, 95°C for 10 min, and 40 cycles of 95°C for 15 s, 60°C for 1 min.
4. The threshold is selected based on positive and negative controls and used to determine a threshold-cycle value (C_T, the PCR cycle at which amplification signal crosses the threshold). Samples are counted as weakly positive if they had a C_T value greater than 36 and a fluorescence signal increase in the multicomponent view of at least 1000 units.

77.2.2.3 Genotyping and Phylogenetic Analysis of *Haemophilus* spp.

(i) Real-Time PCR Typing Maaroufi et al.[83] developed a two-step real-time PCR assay targeting all six capsulation loci of *H. influenzae* (i.e., serotypes A–F). Primers and probes are derived from the variable region 2 of the encapsulation *cap* locus. In addition, primer and probes are designed from the most conserved regions of *bexA* gene, as well as the gene coding for outer membrane lipoprotein P2 (*ompP2*), present in both encapsulated and noncapsulated (NC) *Haemophilus* strains, to confirm the *H. influenzae* species (Table 77.2).

Five conventional and three minor groove binder (MGB) fluorogenic probes, with a reporter dye on the 5′ end and a quencher [nonfluorescent quencher (NFQ) or black hole quencher (BHQ)] on the 3′ end, are utilized: *bexA*-Cy3 (where Cy3 is indocarbocyanin), Hia-FAM (where FAM is 6-carboxyfluorescein), Hib-JOE (where JOE is 6-carboxy-4′,5′-dichloro-2′,7′-dimethoxyfluorescein), Hic-VIC, Hid-NED (Applied Biosystems), Hie-FAM, Hif-Cy3, and *ompP2*-FAM, used for the detection of *bexA*, Hia, Hib, Hic, Hid, Hie, Hif, and *ompP2* DNA, respectively (Table 77.1).

Strains are first tested for the presence or absence of the *bexA* gene, region 2 of the *cap* locus of Hib, and the *ompP2*

TABLE 77.2
Primers and Probes for Real-Time PCR Detection of Capsule Type-Specific *H. influenzae*

Capsular Type	Primer/Probe	Sequence (5′→3′)[a]	Probe Label 5′	Probe Label 3′	T_m (°C)	Concentration (nM)	Position	Amplicon (bp)
Hia	Forward primer	GCAACCATCTTACAACTTAGCGAATAC			60	900	2300–2326	83
	Probe	AAGTGAAGCATGTCGCCATTCGTCCA	FAM	BHQ1	70	250	2331–2356	
	Reverse primer	GGTCTGCGGTGTCCTGTGTT			60	900	2382–2363	
Hib	Forward primer	TGTTCGCCATAACTTCATCTTAGC			59	900	5579–5602	147
	Probe	CACAAAACTTCTCATTCTTCGAGCCTA	JOE	BHQ1	70	250	5605–5631	
	Reverse primer	CTTACGCTTCTATCTCGGTGATTAATAA			59	900	5725–5698	
Hic	Forward primer	TCTGTGTAGATGATGGTTCAGTAG			55	900	34–57	67
	Probe	TGCAGCTAAGATTATT	VIC	MGBNFQ	67	125	59–74	
	Reverse primer	TTAGGATATTTACGCTGCCATT			57	900	100–79	
Hid	Forward primer	TATTGATGACCGATACAACCTGTTTAAA			60	900	4–31	120
	Probe	AATGGTTGTAAAACTCTTCT	NED	MGBNFQ	69	125	38–57	
	Reverse primer	CCAGAAATTATTTCTCCGTTATGTTGA			60	300	123–97	
Hie	Forward primer	GTTGAAAACAAACCGCACTTT			58	900	1–21	73
	Probe	AACGAATGTAGTGGTAGTTAGA	FAM	MGBNFQ	70	250	25–46	
	Reverse primer	ATCTTTAATTACCAGATCCCTTTCAT			59	900	73–48	
Hif	Forward primer	GGATAATCAAATACCACATTGGCTTA			58	900	8121–8146	125
	Probe	TCATCGTGAGATCATTGATCACGAT	CY3	BHQ2	70	250	8178–8202	
	Reverse primer	GTAGATTAGCCTCAATAACATGTGAATTAA			58	900	8245–8216	
bexA	Forward primer	CTGAATTRGGYGATTATCTTTATGA			59	300	358–382	101
	Probe	AGGGATGAAAGCYCGRCTTGCAT	CY3	BHQ2	69	500	404–426	
	Reverse primer	ACAATCAAAYTCAACHGAAAGHGA			58	300	458–435	
ompP2	Forward primer	GGTGCATTCGCAGCTTCAG			59	300	215217–215235	151
	Probe	TTGTTTATAACAACGAAGGGACTAACGT	FAM	BHQ1	68	50	215251–215278	
	Reverse primer	GATTGCGTAATGCACCGTGTT			59	900	215367–215347	

[a] R, A or G; Y, C or T; H, A, T, or C.

gene by a multiplex real-time PCR-based assay. The *ompP2* gene is used as an internal control for DNA extraction and amplification. Non-b capsulated *H. influenzae* strains are then checked in four real-time PCR-based assays, including a duplex assay for detection of types A and F and three uniplex assays for detection of types C, D, and E.

Procedure

1. PCR mixture (25 μL) is prepared with 12.5 μL TaqMan universal master mix (Applied Biosystems), 1 μL template DNA, and primers and probes at specified amounts and the combination outlined earlier (Table 77.1).
2. For multiplex amplification covering *bexA*-Hib-*ompP2*, the cycling parameters include 50°C for 2 min and 95°C for 10 min, followed by 40 cycles of 95°C for 15 s and 58°C for 1 min using an ABI Prism 7500 sequence detection system (Applied Biosystems). For duplex PCR covering Hia and Hif, the same conditions are used, except for an annealing temperature of 60°C. For uniplex amplifications, annealing temperatures are 58°C (types c and d) and 56°C (type e).

Note: The sensitivities of the PCR are 10 CFI for types A, B, and C; 10^3 CFU for type E; and 10^4 CFU for types D and F. The primer/probe system specific for the *bexA* gene recognize only encapsulated strains, but not NC *H. influenzae* strains. Compared to gel-based capsular PCR, this assay includes an internal control for extraction and amplification (*ompP2*), with a shorter turnaround time (1.5 h) instead of 5–6 h.

(ii) 16S rRNA PCR and Sequencing Bahrani-Mougeot et al.[95] utilized PCR to amplify 16S rRNA gene followed by cloning into plasmid and sequencing for phylogenetic analysis of bacterial species including *Haemopnilus* in clinical samples (oral swab from the dorsal tongue and bronchoalveolar lavage fluid). The following bacterial universal primers are used: 9F forward primer, 5′-GAGTTTGATYMTGGCTCAG-3′, positions 9 to 27; 1541R reverse primer, 5′-AAGGAGGTGWTCCARCC-3′, positions 1541 to 1525.

Procedure

1. Oral swab samples are rotated for 4–5 h at room temperature to release bacteria into the solution. Bacterial DNA is isolated with a modification of the QIAamp DNA mini kit (QIAGEN). BAL samples are centrifuged at high speed (18,000 × *g* for 15 min), and DNA is isolated from the cell pellet with the QIAamp DNA mini kit. DNA is stored at −20°C until analysis.
2. PCR mix (50 μL) is prepared with *Taq* polymerase (Invitrogen) in thin-walled tubes, and subjected to 94°C for 4 min; 30 cycles of 94°C for 45 s, 60°C for 45 s, and 72°C for 90 s; 72°C for 15 min; and 4°C soaking on Gene Amp PCR System 9700 (PE Applied Biosystems).
3. PCR products (5 μL) are analyzed by electrophoresis on a 1% agarose gel, stained by ethidium bromide and visualized with a UV transilluminator.
4. The remaining 45 μL of PCR product is dried, resuspended in 5 μL of water, and run on a 0.8% low-melting-point agarose gel in Tris-borate-EDTA buffer. The 16S rRNA band of the expected size is excised from the gel and purified by using spin columns (QIAquick gel extraction kit; QIAGEN). The purified PCR product (3–4 μL) is cloned with a TOPO TA cloning kit (Invitrogen). The insert sizes of 50 clones are verified by PCR by using M13 universal primers. Clones with the correct-size inserts (1500 bp) are analyzed by sequencing.
5. Purified PCR products are sequenced with an ABI Prism cycle sequencing kit using the BigDye Terminator cycle sequencing ready reaction kit (PE Applied Biosystems). The primers used are 9F and 533R (5′-TTACCGCGGCTGCTG-3′). Sequencing reactions are run on an ABI model 3100 DNA sequencer.
6. Approximately 500 bases are obtained first to determine identity or approximate phylogenetic position. Full sequences (about 1500 bases obtained with five or six additional sequencing primers) are analyzed for novel species. For identification of closest relatives, sequences of the unrecognized inserts are compared to the 16S rRNA gene sequences in the Ribosomal Database Project (http://wdcm.nig.ac.jp/RDP/html/index.html) and GenBank databases by BLAST (http://www.ncbi.nlm.nih.gov/).

Note: This method allows identification of known and putative respiratory bacterial pathogens.

77.3 CONCLUSION

The genus *Haemophilus* consists of multiple pleomorphic gram-negative bacterial species that are found as commensals in human mucosal surfaces. While *H. influenzae* is responsible for a range of localized and invasive infections such as bronchitis, otitis, pneumonia, meningitis, septicemia, and epiglottitis, *H. ducreyi* mainly causes the sexually transmitted disease chancroid.[16,90,96,97] Of the six serotypes (A–F) within the species *H. influenzae*, type b (Hib) is highly invasive, and has accounted for a majority of *Haemophilus* infections prior to the application of conjugate vaccine in the 1980s. Today, the incidence of Hib disease has dropped significantly, while other *H. influenzae* serotypes (A, C–F) and nontypeable strains have been isolated from human infections with regularity.[12,98,99]

In the post-Hib vaccine era, it is therefore critical to have rapid, sensitive and specific methods at our disposal to accurately identify and monitor an increasing diversity of *H. influenzae* serotypes and nontypeable strains that are isolated from human patients.[96,100] Due to their obvious advantages over phenotypic techniques such as culture, biochemical, and serological characterization, PCR assays, in particular real-time PCR, have been widely adopted for diagnosis of *H. influenzae* and other *Haemophilus* diseases.[64] It is envisaged that further automation in the sample preparation and continuing refinement in other aspects of molecular procedures will greatly enhance our capacity to detect and tract *Haemophilus* organisms involved in human epidemic outbreaks.

REFERENCES

1. Kilian, M., A taxonomic study of the genus *Haemophilus*, with the proposal of a new species, *J. Gen. Microbiol.*, 93, 9, 1976.
2. Murphy, T.F. et al., *Haemophilus haemolyticus*: A human respiratory tract commensal to be distinguished from *Haemophilus influenzae*, *J. Infect. Dis.*, 195, 81, 2007.
3. McCrea, K.W. et al., Relationships of nontypeable *Haemophilus influenzae* strains to hemolytic and nonhemolytic *Haemophilus haemolyticus* strains, *J. Clin. Microbiol.*, 46, 406, 2008.
4. Brenner, D.J. et al., Biochemical, genetic, and epidemiologic characterization of *Haemophilus influenzae* biogroup aegyptius (*Haemophilus aegyptius*) strains associated with Brazilian purpuric fever, *J. Clin. Microbiol.*, 26, 1524, 1988.
5. Kilian, M., Poulsen, K., and Lomholt, H., Evolution of the paralogous *hap* and *iga* genes in *Haemophilus influenzae*: Evidence for a conserved hap pseudogene associated with

microcolony formation in the recently diverged *Haemophilus aegyptius* and *H. influenzae* biogroup aegyptius, *Mol. Microbiol.*, 46, 1367, 2002.

6. Nørskov-Lauritsen, N., and Kilian, M., Reclassification of *Actinobacillus actinomycetemcomitans*, *Haemophilus aphrophilus*, *Haemophilus paraphrophilus* and *Haemophilus segnis* as *Aggregatibacter actinomycetemcomitans* gen. nov., comb. nov., *Aggregatibacter aphrophilus* comb. nov. and *Aggregatibacter segnis* comb. nov., and emended description of *Aggregatibacter aphrophilus* to include V factor-dependent and V factor-independent isolates, *Int. J. Syst. Evol. Microbiol.*, 56, 2135, 2006.
7. Niven, D.F., and O'Reilly, T., Significance of V-factor dependency in the taxonomy of *Haemophilus* species and related organisms, *Int. J. Syst. Bcateriol.*, 40, 1, 1990.
8. Hedegaard, J. et al., Phylogeny of the genus *Haemophilus* as determined by comparison of partial *infB* sequences, *Microbiology*, 147, 2599, 2001.
9. Nørskov-Lauritsen, N., Bruun, B., and Kilian, M., Multilocus sequence phylogenetic study of the genus *Haemophilus* with description of *Haemophilus pittmaniae* sp. nov., *Int. J. Syst. Evol. Microbiol.*, 55, 449, 2005.
10. Musser, J. et al., Genetic relationships of serologically nontypable and serotype b strains of *Haemophilus influenzae*, *Infect. Immun.*, 52, 183, 1986.
11. Falla T.J. et al., PCR for capsular typing of *Haemophilus influenzae*, *J. Clin. Microbiol.*, 32, 2382, 1994.
12. Peltola, H., Worldwide *Haemophilus influenzae* type b disease at the beginning of the 21st century: Global analysis of the disease burden 25 years after the use of the polysaccharide vaccine and a decade after the advent of conjugates, *Clin. Microbiol. Rev.*, 13, 302, 2000.
13. Tsang, R.S.W. et al., Invasive *Haemophilus influenzae* in Manitoba, Canada, in the postvaccination era, *J. Clin. Microbiol.*, 44, 1530, 2006.
14. Nørskov-Lauritsen, N., Overballe, M.D., and Kilian, M., Delineation of the species *Haemophilus influenzae* by phenotype, multilocus sequence phylogeny, and detection of marker genes, *J. Bacteriol.*, 191, 822, 2009.
15. Treees, D.L., and Morse, S.A., Chancroid and *Haemophilus ducreyi*: An update, *Clin. Microbiol. Rev.*, 8, 357, 1995.
16. Lewis, D.A., Chancroid: clinical manifestations, diagnosis, and management, *Sex. Transm. Infect.*, 79, 68, 2003.
17. Fleischmann, R.D. et al., Whole-genome random sequencing and assembly of *Haemophilus influenzae* Rd., *Science*, 269, 496, 1995.
18. Hobbs, M.M. et al, Organization of the *Haemophilus ducreyi* 35000 chromosome, *Microbiology*, 142, 2587, 1996.
19. Challacombe, J.F. et al., Complete genome sequence of *Haemophilus somnus* (*Histophilus somni*) strain 129Pt and comparison to *Haemophilus ducreyi* 35000HP and Haemophilus influenzae Rd., *J. Bacteriol.*, 189, 1890, 2007.
20. Mohle-Boetani, J.C. et al., Carriage of *Haemophilus influenzae* type b in children after widespread vaccination with conjugate *Haemophilus influenzae* type b vaccines, *Pediatr. Infect. Dis. J.*, 12, 589, 1993.
21. Saito, M. et al., Clonal spread of an invasive strain of *Haemophilus influenzae* type b among nursery contacts accompanied by a high carriage rate of non-disease-associated strains, *J. Med. Microbiol.*, 49, 845, 2000.
22. Peerbooms, P.G.H. et al., Nasopharyngeal carriage of potential bacterial pathogens related to day care attendance, with special reference to the molecular epidemiology of *Haemophilus influenzae*, *J. Clin. Microbiol.*, 40, 2832, 2002.
23. Vitovski, S. et al., Nontypeable *Haemophilus influenzae* in carriage and disease, *JAMA*, 287, 1699, 2002.
24. Farjo, R.S. et al., Diversity and sharing of *Haemophilus influenzae* strains colonizing healthy children attending day-care centers, *Pediatr. Infect. Dis. J.*, 2341, 2004.
25. Hammitt, L.L. et al., Assessment of carriage of *Haemophilus influenzae* type a after a case of invasive disease, *Clin. Infect. Dis.*, 43, 386, 2006.
26. Mukundan, D. et al., Pharyngeal colonization dynamics of *Haemophilus influenzae* and *Haemophilus haemolyticus* in healthy adult carriers, *J. Clin. Microbiol.*, 45, 3207, 2007.
27. Hashida, K. et al., Nasopharyngeal *Haemophilus influenzae* carriage in Japanese children attending day-care centers, *J. Clin. Microbiol.*, 46, 876, 2008.
28. Moxon, E.R., and Kroll, J.S., Type b capsular polysaccharide as a virulence factor of *Haemophilus influenzae*, *Vaccine*, 6, 113–115, 1988.
29. Watt, J.P. et al., Burden of disease caused by *Haemophilus influenzae* type b in children younger than 5 years: Global estimates, *Lancet*, 374, 903, 2009.
30. Falla, T.J. et al., Population-based study of non-typable *Haemophilus influenzae* invasive disease in children and neonates, *Lancet*, 341, 851, 1993.
31. Waggoner-Fountain, L.A. et al., The emergence of *Haemophilus influenzae* types e and f as significant pathogens, *Clin. Infect. Dis.*, 21, 1322, 1995.
32. Urwin, G. et al., Invasive disease due to *Haemophilus influenzae* serotype f: Clinical and epidemiologic characteristics in the *H. influenzae* serotype b vaccine era, *Clin. Infect. Dis.*, 22, 1069, 1996.
33. Adderson, E.E. et al., Invasive serotype a *Haemophilus influenzae* infections with a virulence genotype resembling *Haemophilus influenzae* type b: Emerging pathogen in the vaccine era? *Pediatrics*, 108, 1, 2001.
34. Campos, J. et al., Infections due to *Haemophilus influenzae* serotype E: Microbiological, clinical, and epidemiological features, *Clin. Infect. Dis.*, 37, 841, 2003.
35. Campos, J. et al., Analysis of invasive *Haemophilus influenzae* infections after extensive vaccination against *H. influenzae* type b, *J. Clin. Microbiol.*, 42, 524, 2004.
36. Bruun, B. et al., Clonal relationship of recent invasive *Haemophilus influenzae* serotype f isolates from Denmark and the United States, *J. Med. Microbiol.*, 53, 1161, 2004.
37. Xie, J. et al., Identification of new genetic regions more prevalent in nontypeable *Haemophilus influenzae* otitis media strains than in throat strains, *J. Clin. Microbiol.*, 44, 4316, 2006.
38. Dworkin, M.S., Park, L., and Borchardt, S.M., The changing epidemiology of invasive *Haemophilus influenzae* disease, especially in persons ≥65 years old, *Clin. Infect. Dis.*, 44, 810, 2007.
39. McConnell, A. et al., Invasive infections caused by *Haemophilus influenzae* serotypes in twelve Canadian IMPACT centers, 1996–2001, *Pediatr. Infect. Dis. J.*, 26, 1025, 2007.
40. Kunze, W. et al., Recurrent posttraumatic meningitis due to nontypable Haemophilus influenzae: Case report and review of the literature, *Infection*, 36, 74, 2008.
41. Slater, L.N. et al., Bacteraemic disease due to *Haemophilus influenzae* capsular type f in adults: Report of five cases and review, *Rev. Infect. Dis.*, 12, 628, 1990.
42. Collette, C.J., Southerland, D., and Corall, C.J., Necrotizing fasciitis associated with *Haemophilus influenzae* type b, *Am. J. Dis. Child.*, 141, 1146, 1987.

43. Stumvoll, M., and Fritsche, A., Necrotizing fasciitis caused by unencapsulated *Haemophilus influenzae*, *Clin. Infect. Dis.*, 25, 327, 1997.
44. Gilsdorf, J.R., Marrs, C.F., and Foxman, B., *Haemophilus influenzae*: Genetic variability and natural selection to identify virulence factors, *Infect. Immun.*, 72, 2457, 2004.
45. McLellan, E., Suvarna, K., and Townsend, R., Fatal necrotizing fasciitis caused by *Haemophilus influenzae* serotype f, *J. Med. Microbiol.*, 57, 249, 2008.
46. Saito, T. et al., Necrotizing fasciitis caused by *Haemophilus influenzae* type b in an elderly patient, *J. Clin. Microbiol.*, 47, 852, 2009.
47. Harrison, L.H., Simonsen, V., and Waldman, E.A., Emergence and disappearance of a virulent clone of *Haemophilus influenzae* biogroup *aegyptius*, cause of Brazilian purpuric fever, *Clin. Microbiol. Rev.*, 21, 594, 2008.
48. Wallace, R.J. Jr., et al., *Haemophilus influenzae* infections in adults: characterization of strains by serotypes, biotypes, and β-lactamase production, *J. Infect. Dis.*, 144, 101, 1981.
49. Shively, R.G. et al., Typing of *Haemophilus influenzae* by coagglutination and conventional slide agglutination, *J. Clin. Microbiol.*, 14, 706, 1981.
50. LaClaire, L. et al., Identification of *Haemophilus influenzae* serotypes by standard slide agglutination serotyping and PCR-based capsule typing, *J. Clin. Microbiol.*, 41, 393, 2003.
51. van Ketel, R.J., de Wever, B., and van Alphen, L., Detection of *Haemophilus influenzae* in cerebrospinal fluids by polymerase chain reaction DNA amplification, *J. Med. Microbiol.*, 33, 271, 1990.
52. Radstrom, P. et al., Detection of bacterial DNA in cerebrospinal fluid by an assay for simultaneous detection of *Neisseria meningitidis*, *Haemophilus influenzae*, and streptococci using a seminested PCR strategy, *J. Clin. Microbiol.*, 32, 2738, 1994.
53. Hassan-King, M. et al., Detection of *Haemophilus influenzae* and *Streptococcus pneumoniae* DNA in blood culture by a single PCR assay, *J. Clin. Microbiol.*, 34, 2030, 1996.
54. Hassan-King, M. et al., A polymerase chain reaction for the diagnosis of *Haemophilus influenzae* type b disease in children and its evaluation during a vaccine trial, *Pediatr. Infect. Dis. J.*, 17, 309, 1998.
55. Quentin, R. et al., Genetic identification of cryptic genospecies of *Haemophilus* causing urogenital and neonatal infections by PCR using specific primers targeting genes coding for 16S rRNA, *J. Clin. Microbiol.*, 34, 1380, 1996.
56. Corless, C. et al., Simultaneous detection of *Neisseria meningitidis*, *Haemophilus influenzae*, and *Streptococcus pneumoniae* in suspected cases of meningitis and septicemia using real-time PCR, *J. Clin. Microbiol.*, 39, 1553, 2001.
57. Failace, L. et al., Simultaneous detection of *Neisseria meningitidis*, *Haemophilus influenzae* and *Streptococcus* sp. by polymerase chain reaction for the diagnosis of bacterial meningitis, *Arq. Neuropsiquiatr.*, 63, 920, 2005.
58. Strålin, K. et al., Design of a multiplex PCR for *Streptococcus pneumoniae*, *Haemophilus influenzae*, *Mycoplasma pneumoniae* and *Chlamydophila pneumoniae* to be used on sputum samples, *APMIS*, 113, 99, 2005.
59. Morozumi, M. et al., Simultaneous detection of pathogens in clinical samples from patients with community-acquired pneumonia by real-time PCR with pathogen-specific molecular beacon probes, *J. Clin. Microbiol.*, 44, 1440, 2006.
60. Tuyama, M. et al., The utility of the polymerase chain reaction assay for aetiologic definition of unspecified bacterial meningitis cases, *Mem. Inst. Oswaldo Cruz.*, 103, 138, 2008.
61. Torigoe, H. et al., Detection of *Haemophilus influenzae* by loop-mediated isothermal amplification (LAMP) of the outer membrane protein P6 gene, *Jpn. J. Infect. Dis.*, 60, 55, 2007.
62. Smith-Vaughan, H.C. et al., Long PCR-ribotyping of nontypeable *Haemophilus influenzae*, *J. Clin. Microbiol.*, 33, 1192, 1995.
63. Saito, M., Umeda, A., and Yoshida, S., Subtyping of *Haemophilus influenzae* strains by pulsed-field gel electrophoresis, *J. Clin. Microbiol.*, 37, 2142, 1999.
64. Bokermann, S. et al., Evaluation of methodology for serotyping invasive and nasopharyngeal isolates of *Haemophilus influenzae* in the ongoing surveillance in Brazil, *J. Clin. Microbiol.*, 41, 5546, 2003.
65. Gonin, P., Lorange, M., and Delage, G., Performance of a multiplex PCR for the determination of *Haemophilus influenzae* capsular types in the clinical microbiology laboratory, *Diagn. Microbiol. Infect. Dis.*, 37, 1, 2000.
66. Yadav, M.C. et al., Rapid detection of *Haemophilus influenzae* by hel gene polymerase chain reaction, *Lett. Appl. Microbiol.*, 37, 190, 2003.
67. Hotomi, M. et al., Genotyping of *Streptococcus pneumoniae* and *Haemophilus influenzae* isolated from paired middle ear fluid and nasopharynx by pulsed-field gel electrophoresis, *ORL J. Otorhinolaryngol. Relat. Spec.*, 6, 233, 2004.
68. Marty, A. et al., Detection of *Haemophilus influenzae* type b by real-time PCR, *J. Clin. Microbiol.*, 42, 3813, 2004.
69. Erwin, A.L. et al., Characterization of genetic and phenotypic diversity of invasive nontypeable *Haemophilus influenzae*, *Infect. Immun.*, 73, 5853, 2005.
70. Erwin, A.L. et al., Analysis of genetic relatedness of *Haemophilus influenzae* isolates by multilocus sequence typing, *J. Bacteriol.*, 190, 1473, 2008.
71. Ohkusu, K., Nash, K.A., and Inderlied, C.B., Molecular characterisation of *Haemophilus influenzae* type a and untypeable strains isolated simultaneously from cerebrospinal fluid and blood: Novel use of quantitative real-time PCR based on the cap copy number to determine virulence, *Clin. Microbiol. Infect.*, 11, 637, 2005.
72. Sacchi, C.T. et al., High level of sequence diversity in the 16S rRNA genes of *Haemophilus influenzae* isolates is useful for molecular subtyping, *J. Clin. Microbiol.*, 43, 3734, 2005.
73. Karlsson, E., and Melhus, A., Nontypeable *Haemophilus influenzae* strains with the capsule-associated insertion element IS*1016* may mimic encapsulated strains, *APMIS*, 114, 633, 2006.
74. Abdeldaim, G.M.K. et al., Detection of *Haemophilus influenzae* in respiratory secretions from pneumonia patients by quantitative real-time polymerase chain reaction, *Diagn. Microbiol. Infect. Dis.*, 64, 366, 2009.
75. Xirogianni, A. et al., Development of a single-tube polymerase chain reaction assay for the simultaneous detection of *Haemophilus influenzae*, *Pseudomonas aeruginosa*, *Staphylococcus aureus*, and *Streptococcus* spp. directly in clinical samples, *Diagn. Microbiol. Infect. Dis.*, 63, 121, 2009.
76. Kroll, J.S. et al., Common organization of chromosomal loci for production of different capsular polysaccharides in *Haemophilus influenzae*, *J. Bacteriol.*, 171, 3343, 1989.
77. Kroll, J.S. et al., The *bex* locus in encapsulated *Haemophilus influenzae*: A chromosomal region involved in capsule polysaccharide export, *Mol. Microbiol.*, 4, 1853, 1990.
78. St. Geme, J.W. III. et al., Evidence for capsule gene sequences among pharyngeal isolates of nontypeable *Haemophilus influenzae*, *J. Infect. Dis.*, 169, 337, 1994.

79. van Eldere, J. et al., Region II of the *Haemophilus influenzae* type b capsulation locus is involved in serotype-specific polysaccharide synthesis, *Mol. Microbiol.*, 15, 107, 1995.
80. Ogilvie, C. et al., Capsulation loci of non-serotype b encapsulated *Haemophilus influenzae*, *J. Infect. Dis.*, 184, 144, 2001.
81. Satola, S.W. et al., Capsule gene analysis of invasive *Haemophilus influenzae*: Accuracy of serotyping and prevalence of IS1016 among nontypeable isolates, *J. Clin. Microbiol.*, 45, 3230, 2007.
82. Sukupolvi-Petty, S., Grass, S., and St. Geme III, J.W., The *Haemophilus influenzae* type b *hcsA* and *hcsB* gene products facilitate transport of capsular polysaccharide across the outer membrane and are essential for virulence, *J. Bacteriol.*, 188, 3870, 2006.
83. Maaroufi, Y. et al. Real-time PCR for determining capsular serotypes of *Haemophilus influenzae*, *J. Clin. Microbiol.*, 45, 2305, 2007.
84. Anyanwu, J.N. et al., *pgi* genotyping is a surrogate for serotyping of encapsulated *Haemophilus influenzae*, *J. Clin. Microbiol.*, 41, 2080, 2003.
85. Bruant, G. et al., Typing of nonencapsulated *Haemophilus* strains by repetitive-element sequence-based PCR using intergenic dyad sequences, *J. Clin. Microbiol.*, 41, 3473, 2003.
86. Fung, W.W. et al., Presence of copper- and zinc-containing superoxide dismutase in commensal *Haemophilus haemolyticus* isolates can be used as a marker to discriminate them from nontypeable *H. influenzae*, *J. Clin. Microbiol.*, 44, 4222, 2006.
87. Cattoir, V. et al., The sodA gene as a target for phylogenetic dissection of the genus *Haemophilus* and accurate identification of human clinical isolates, *Int. J. Med. Microbiol.*, 296, 531, 2006.
88. Lancellotti, M. et al., Ribotyping, biotyping and capsular typing of *Haemophilus influenzae* strains isolated from patients in Campinas, southeast Brazil, *Braz. J. Infect. Dis.*, 12, 430, 2008.
89. Meats, E. et al., Characterization of encapsulated and noncapsulated *Haemophilus influenzae* and determination of phylogenetic relationships by multilocus sequence typing, *J. Clin. Microbiol.*, 41, 1623, 2003.
90. Lewis, D.A., Diagnostic tests for chancroid, *Sex. Transm. Infect.*, 76, 137, 2000.
91. Billal, D.S. et al., Rapid identification of nontypeable and serotype b *Haemophilus influenzae* from nasopharyngeal secretions by the multiplex PCR, *Int. J. Pediatr. Otorhinolaryngol.*, 71, 269, 2007.
92. Fukasawa, C. et al., A mixed bacterial infection of a bronchogenic lung cyst diagnosed by PCR, *J. Med. Microbiol.*, 55, 791, 2006.
93. Suntoke, T.R. et al. Evaluation of multiplex real-time PCR for detection of *Haemophilus ducreyi*, *Treponema pallidum*, herpes simplex virus type 1 and 2 in the diagnosis of genital ulcer disease in the Rakai District, Uganda, *Sex. Transm. Infect.*, 85, 97, 2009.
94. Orle, K.A. et al., Simultaneous PCR detection of *Haemophilus ducreyi*, *Treponema pallidum*, and herpes simplex virus types 1 and 2 from genital ulcers, *J. Clin. Microbiol.*, 34, 49, 1996.
95. Bahrani-Mougeot, F.K. et al., Molecular analysis of oral and respiratory bacterial species associated with ventilator-associated pneumonia, *J. Clin. Microbiol.*, 45, 1588, 2007.
96. Galil, K. et al., Reemergence of invasive *Haemophilus influenzae* type b disease in a well-vaccinated population in remote Alaska, *J. Infect. Dis.*, 179, 101, 1999.
97. Hammitt, L.L. et al., Outbreak of invasive *Haemophilus influenzae* serotype a disease, *Pediatr. Infect. Dis. J.*, 24, 453, 2005.
98. Satola, S.W., Schirmer, P.L., and Farley, M.M., Genetic analysis of the capsule locus of *Haemophilus influenzae* serotype f, *Infect. Immun.*, 71, 7202, 2003.
99. Satola, S.W., Schirmer, P.L., and Farley, M.M., Complete sequence of the *cap* locus of *Haemophilus influenzae* serotype b and nonencapsulated b capsule-negative variants, *Infect. Immun.*, 71, 3639, 2003.
100. Heath, P.T. et al., Non-type b *Haemophilus influenzae* disease: Clinical and epidemiologic characteristics in the *Haemophilus influenzae* type b vaccine era, *Pediatr. Infect. Dis. J.*, 20, 300, 2001.

78 *Klebsiella*

Beatriz Meurer Moreira, Ana Cristina Gales, Maria Silvana Alves, and Rubens Clayton da Silva Dias

CONTENTS

78.1 INTRODUCTION

78.1.1 Classification, Morphology, and Biology

Historically, the medical importance of the genus *Klebsiella* has led to the recognition of three species: *Klebsiella pneumoniae*, responsible for pneumonia; *Klebsiella ozaenae*, the cause of ozaenia (chronic atrophic rhinitis); and *Klebsiella rhinoscleromatis*, the cause of rhinoscleroma (chronic granulomatous disease), as described in extensive reviews.[1,2] In the latest edition (1984) of Bergey's Manual of Systematic Bacteriology, based on phenotypic characteristics and data derived from DNA–DNA hybridization studies, the genus *Klebsiella* was classified into five species: *K. pneumoniae*, *Klebsiella oxytoca*, *Klebsiella terrigena*, *Klebsiella ornithinolytica*, and *Klebsiella planticola*. *K. pneumoniae* comprised three subspecies: *K. pneumoniae* subsp. *pneumoniae*, *K. pneumoniae* subsp. *ozaenae*, and *K. pneumoniae* subsp. *rhinoscleromatis*.[3] With increasing use of DNA–DNA hybridization techniques and additional molecular studies, new species have been described and others reclassified or moved to the new genus *Raoultella*. In the current text, the term *Klebsiella* refers to *Klebsiella* and *Raoultella* species, unless specified differently. Thus, by sequencing of 16S rRNA and *phoE* genes, *Calymmatobacterium granulomatis*, the supposed agent of donovanosis, was renamed *Klebsiella granulomatis*; based on 16S rRNA and *rpoB* sequence analysis, the new genus *Raoultella* was proposed to accommodate the three species previously called *K. planticola*, *K. terrigena*, and *K. ornithinolytica*, while *K. oxytoca* was shown to form a distinct genogroup; two new *Klebsiella* species were described: *Klebsiella variicola*, from analysis of *rpoB*, *gyrA*, *mdh*, *infB*, *phoE*, and *nifH* sequences, and *Klebsiella singaporensis*, based on 16S rRNA genes and *rpoB* gene sequences; and finally, 16S rRNA gene sequences for two proposed species, *Klebsiella milletis* and *Klebsiella senegalensis*, and 16S rRNA and *rpoB* sequences for the proposed species *Klebsiella alba* were deposited in the GenBank (http://www.ncbi.nlm.nih.gov) in 2003 and 2007, respectively.[4–7] However, no data has been published describing these three last species to date. A summary of the current classification of *Klebsiella* is presented in Table 78.1.

Subgroups within species were also determined. Based on nucleotide variations of the *gyrA*, *parC*, and *rpoB* genes, clinical isolates of *K. pneumoniae* fall into four phylogenetic groups named KpI, KpII-A, KpII-B, and KpIII.[8,11,13] The species *K. variicola* corresponds to group KpIII.[6] In addition, *K. oxytoca* has been shown to include five distinct phylogenetic lines (KoI, KoII, KoIII, KoIV, and KoVI) identified by 16S rRNA, *rpoB*, *gyrA*, *gapDH*, and bla_{OXY} gene sequencing.[15]

Klebsiella is ubiquitous in nature, and has been found in surface water, sewage, soil, and plants, and colonize mucosal

TABLE 78.1
Classification of *Klebsiella*

Species	Subspecies	Reference
K. pneumoniae	*K. pneumoniae* subsp. *pneumoniae*	3,5,8–11
	K. pneumoniae subsp. *Ozaenae*;	3,5,8–10
	K. pneumoniae subsp. *Rhinoscleromatis*	
K. granulomatis	–[a]	4,5,9,12
K. variicola	–	6,8,10,13
K. singaporensis	–	7
Genomospecies *K. oxytoca*	–	3,5,8–11,14–16
R. planticola	–	5
R. terrigena	–	5
R. ornithinolytica	–	5

[a] It has been suggested *K. granulomatis* belongs to *K. pneumoniae*, or to a new cluster.[4,8,12]

surfaces of mammals such as humans, horses, and swine.[1] The investigation of natural surface water (freshwater, salt water, and brackish water) in Germany led to the isolation of *Klebsiella* species from 110 samples from streams, lakes, and the Baltic Sea in various geographic areas; *K. pneumoniae* was detected in 52%, *K. oxytoca* in 27%, and *Raoultella planticola* in 22% of the samples.[17] Fecal shedding of *K. pneumoniae* by cows contributes to the presence of a variety of strains in dairy herds, which may in turn be a source of colonization for humans.[18]

K. pneumoniae causes serious infections in humans.[19] It is reported also as an animal pathogen, and recognized as a main cause of clinical mastitis in cows, resulting in high milk losses and mortality of the affected animal.[18] The presence of virulence traits in *K. pneumoniae* from water and contemporary human clinical isolates has been shown to be comparable, indicating environmental isolates are potential sources of significant pathogens for humans.[17,20]

78.1.2 Clinical Features and Pathogenesis

78.1.2.1 Community-Acquired *Klebsiella* Infections

K. pneumoniae has been recognized as cause of community-acquired pneumonia (CAP) since its first description more than 120 years ago. The clinical presentation of *K. pneumoniae* CAP is dramatic, with sudden onset high fever, hemoptysis, and toxemia. Chest radiograph abnormalities such as bulging interlobar fissure and cavitary abscesses are prominent.[21] Early initiation of an adequate antibiotic therapy has been shown to improve the prognosis; however, CAP still remains one important cause of death, especially in the elder and those with underlying diseases.[22]

An emerging and invasive presentation of *K. pneumoniae* infection is primary pyogenic liver abscess (PLA).[21] *K. pneumoniae* PLA differs from traditional liver abscess because it is community-acquired, patients have no history of hepatobiliary diseases, and may present other septic metastatic lesions.[21,23] Most reports originate from Taiwan and Southeast Asia, but this disease has now been described also in Europe, North America, and Australia.[24–26] Apparently, there is a geographical preponderance of a syndrome consisting of liver abscess, meningitis, and endophthalmitis in Taiwan.[26] Although the reasons for the geographical diversity are unknown, genetic susceptibility may be important.[27] The disease has been caused by K1 capsular serotype hypermucoviscous (HMV*Kp*) strains. A main virulence factor in HMV*Kp* strains is *magA*, a gene encoding an iron transport protein (originally described in *Magnetospirillum* sp.), responsible for the hypermucoviscous phenotype.[25] Genotyping of K1 *K. pneumoniae* isolates from three different continents by pulsed field gel electrophoresis (PFGE) and multilocus sequence typing (MLST) revealed that a major clone causes the disease; isolates belonging to a main PFGE cluster, with one exception, belonged to ST23.[26]

K. pneumoniae is also an important cause of community-acquired central nervous system infection in East Asia, despite the presence of a predisposing condition. Meningitis and brain abscess were reported from patients in Singapore, some of them with diabetes mellitus.[28] A strong association between community-acquired bacterial meningitis due to *K. pneumoniae* and liver cirrhosis in adult patients was reported.[29]

Other serious *Klebsiella* community-acquired infections are rhinoscleroma, caused by *K. pneumoniae* subsp. *rhinoscleromatis*, and ozena, a primary atrophic rhinitis caused by *K. pneumoniae* subsp. *ozaenae*. Rhinoscleroma is a chronic granulomatous disease of the respiratory tract. The nasal cavity is most often affected, but lesions may also involve the larynx, oral cavity, paranasal sinuses, soft tissue of the lips and nose, trachea, bronchi and the orbit or middle ear.[30] Endemic regions are tropical Africa, India, Southeast Asia, Central and South America, and Central Europe.[31] The incidence of the disease in the US population may be on the rise with migration and travel abroad. The diagnosis in nonendemic areas is often delayed due to unfamiliarity with the disease, stage-dependent clinical and histological findings, and the fact that about 40% of cultures are negative for *K. rhinoscleromatis*. Although this disease is rarely lethal, late diagnosis may lead to nasal and airway obstruction and nasal deformity from erosive processes.[32] Ozena is characterized by mucosal atrophy together with bone resorption and a fetid thick endonasal crust.[33] Osseous wall thickening of the maxillary and ethmoid sinuses has also been evidenced.[34]

Klebsiella is also an accidental cause of diarrhea. A study of a gastroenteritis outbreak revealed that an LT-enterotoxin producing *K. pneumoniae* capsular serotype K15 caused the diarrhea, and the outbreak source was contaminated turkey, having been kept at improper temperatures for a long time.[35] In addition, *K. oxytoca* is an emerging cause of antibiotic-associated diarrhea. Hemorrhagic colitis following administration of amoxicillin with or without clavulanate, metronidazole, or clarithromycin was reported in five young healthy outpatients.[36] Spontaneous resolution of symptoms was observed in all cases 3–7 days after stopping the antimicrobial therapy. *K. oxytoca* stool carriage may be an important factor in the development of antibiotic-associated diarrhea.

78.1.2.2 Healthcare-Associated Infections Caused by *Klebsiella*

Between 1997 and 2002, *Klebsiella* was responsible for 7%–10% of all healthcare-associated bloodstream infections in Europe, Latin America, and North America, as reported by the SENTRY Antimicrobial Surveillance Program.[37] Immunocompromised patients, especially the elderly and infants, are the population at most risk for infections.[38] *K. pneumoniae* may cause a variety of healthcare-associated infections, including bloodstream infections, ventilator-associated pneumonia, meningitis, soft tissue, and urinary tract infections.[1] Endemic infections and outbreaks have plagued hospitalized patients, usually in intensive care units, where the neonate is especially susceptible.[39] *K. oxytoca* is the second most frequent *Klebsiella* isolate recovered in hospitals. *R. planticola* is seen as an aquatic, botanic, and soil bacterium, but has been rarely isolated from clinical specimens, and may cause bloodstream infection and pancreatitis.[40]

Colonization of the gastrointestinal tract of patients usually precedes infection.[41–47] In fact, the gastrointestinal tract constitutes a main reservoir for the spread of these organisms on hands and medical equipment to uncolonized patients admitted to hospitals. *Klebsiella* isolates have been known to be well adapted to the hospital environment. These organisms can survive longer than other enteric bacteria on hands and hospital environmental surfaces, facilitating cross infection within hospitals.[39,48,49] Poor hand hygiene by healthcare personnel is probably a main issue in the spread of these nosocomial pathogens between patients. Other environmental sources are important; for example, contamination of enteral feeding by *K. oxytoca* has been shown to be the source of throat and stool colonization of premature babies.[45] The strategy to decrease the prevalence of healthcare-associated infections caused by *Klebsiella* isolates should include restriction of antimicrobial use, particularly third-generation cephalosporins, and the reinforcement of the correct employment of barrier precautions.[42,45,48] The effectiveness of infection control measures is enhanced by the early detection of patients whose gastrointestinal tracts have been colonized.[47,50]

Antimicrobial resistance represents a serious problem in this bacterial species. The prevalence of isolates that produce extended-spectrum β-lactamase (ESBL), which hydrolyses broad-spectrum cephalosporins, monobactams, and penicillins, is increasing worldwide due to the spread of mobile genetic elements and the dispersion of epidemic clones.[51] ESBL-producing strains are often resistant to aminoglycosides, sulfonamides, and fluoroquinolones.[52] The escape of *K. pneumoniae* strains from hospitals to the community poses a serious challenge to the treatment of community-acquired infections.[23] The frequency of ESBL-production community-acquired *K. pneumoniae* isolates varies in different countries; however, these strains already emerged in Europe, Asia, and South and North America.[23,53–55] Carbapenem-resistant *K. pneumoniae* is an additional serious problem observed in hospitals, but there is little doubt that these other resistant strains will appear in the community.[23] Carbapenem resistance may involve a combination of several mechanisms: decreased outer membrane permeability, up-regulation of efflux systems, and the increased production of *Amp*C type β-lactamases, ESBLs, and, especially, carbapenemases.[56–58] *K. pneumoniae* carbapenemase (KPC) emerged in New York, but is currently globally present; prevalence of KPC production is especially high in *K. pneumoniae*.[56] Moreover, the emergence of *K. pneumoniae* simultaneously producing several β-lactamases has been increasingly reported. For instance, KPC-2 producing *K. pneumoniae*, which produce also the metallo- β-lactamase VIM-1, was recently described in Greece.[59]

Nosocomial infections are caused by highly diverse *K. pneumoniae* strains that may be considered as opportunistic pathogens. Nevertheless, with the rise in the number resistant infections, the mortality of *K. pneumoniae* has increased and treatment has become more complicated.[39,60]

78.1.2.3 Virulence of *K. pneumoniae*

Virulence factors of *K. pneumoniae* include the capsular serotype, the lipopolysaccharide, aerobactin production, fimbrial and nonfimbrial adhesions, and a mucoid phenotype.[61,62] The polysaccharidic capsule is the most important virulence determinant, and protects against the bactericidal action of serum and impairs fagocytosis. During colonization of mucosal surfaces, the capsular polysaccharide obstructs bacterial binding and internalization to the mucosal membrane, inducing a defective immunological host response.[63] A total of 77 capsular types (K) have been described, but types K1–K5 are more often associated with severe infections.[61,62] Nevertheless, the full characterization of 235 *K. pneumoniae* isolates of diverse environmental and clinical origins study of several virulent lineages revealed that individual capsular types were not always associated to specific clones.[61] This finding is explained by the horizontal transfer of the *cps* operon, responsible for the synthesis of the capsular polysaccharide. But, indeed, the main virulent clones were clearly associated with specific infections. Two other main findings of this study were that (i) two capsular type K1 clones (named CC23K1 and CC82K1) were strongly associated with PLA and respiratory infection, and (ii) strains associated with ozena and rhinoscleroma corresponded to one monomorphic clone and, therefore, *K. pneumoniae* subsp. *ozaenae* and *K. pneumoniae* subsp. *rhinoscleromatis* should be regarded as specific virulent clones derived from K. *pneumoniae*.

78.1.3 Diagnosis

Klebsiella is frequently isolated in the clinical laboratory, where these organisms can be obtained from blood, urine, respiratory secretions, and other specimens. Standard laboratory media are suitable for the isolation of this organism.[18]

MacConkey agar is an excellent media, because it allows for the clear observation of the typical appearance of large, mucoid, or rough colonies. Biochemical identification must include several tests, usually with the help of media that combine some of them. Culture media containing citrate media may be used to facilitate isolation of *Klebsiella* strains because these organisms can use citrate as a sole carbon source.[3]

Several different methods may be used for characterization of *Klebsiella* spp.: biochemical and molecular tests[10]; serotyping with specific antisera by Quellung test[64] or countercurrent immunoelectrophoresis[65]; in situ hybridization to identify *Klebsiella* in phagocytes from blood specimens[66]; phenotypic or genotypic methods to determine antimicrobial susceptibility[67–70]; and DNA fingerprinting by ribotyping, analysis of chromosomal DNA restriction patterns by PFGE and PCR; and single nucleotide polymorphism (SNP) analysis.[8,71–73] Currently, a complete identification of all *Klebsiella* species, the determination of a definitive antimicrobial susceptibility profile, and a meaningful strain typing analysis are attained only with molecular tests. For example, *K. pneumoniae* and *K. variicola* cannot be confidently separated by biochemical methods, and *K. granulomatis* is not cultivable in artificial laboratory media.[4,10] In addition, some of the tests necessary for identification of species either are not available or are not standardized in most laboratories.

78.1.3.1 Conventional Techniques

The correct identification of *Klebsiella* species is not easily accomplished because several species share a similar biochemical profile. The following profile has been suggested for isolates belonging to the most common *Klebsiella* and *Raoultella* species: nonmotile, gram-negative bacilli, glucose fermenters, and form typical red colonies on MacConkey agar; negative results for cytochrome oxidase activity, hydrogen sulfide production in triple sugar iron agar, arginine and ornithine (except *R. ornithinolytica*) decarboxylation, phenylalanine deamination, and citrate utilization.[10] Isolates which are negative for indole production, unable to assimilate histamine and D-melezitose and to grow at 10°C are identified as *K. pneumoniae* or *K. variicola*. *R. ornithinolytica* usually shows a positive ornithine test, however, isolates with a negative result have been described.[74] Possible *K. oxytoca* isolates are positive for indole production, lysine decarboxylation, Voges-Proskauer test, and growth at 10°C and negative for histamine assimilation. *R. planticola* are positive for indole production, histamine assimilation, and growth at 10°C and negative for ornithine and D-melezitose. A summary of useful key tests to differentiate *Klebsiella* and *Raoultella* clinical isolates is show in Table 78.2.

Although a scheme with essential phenotypic tests for differentiation of *Klebsiella* and *Raoultella* species may be used, these tests are not always able to classify all isolates into one species. In addition, either some of the tests necessary for identification of species are not available or are not standardized in most laboratories. Thereby, as mentioned previously, molecular tests are necessary for a complete identification of all *Klebsiella* species.

The laboratory detection of *Klebsiella* usually includes determination of susceptibility to antimicrobial agents. Unfortunately, antimicrobial resistance has become a serious problem in this bacterial group. Standardized susceptibility tests include disk diffusion method for most antimicrobial agents, but determination of minimum inhibitory concentrations (MIC) is mandatory in many situations, as recommended by the European Committee on Antimicrobial Susceptibility Testing (EUCAST, http://www.eucast.org) and the Clinical and Laboratory Standards Institutes.[75] Yet, the susceptibility test is complete only with molecular tests in several situations. For example, a recent challenge emerging worldwide is *Klebsiella* resistance to carbapenems, one of the most useful drug classes for the therapy of severe infections. Unfortunately, detection of carbapenem resistance may be complex. Detection of KPC-producing bacteria based only on susceptibility testing is particularly difficult due to heterogeneous expression of β-lactam resistance.[58] For example, isolates with decreased disk zone diameters to carbapenems are suspect of carbapenemase production despite having MICs or disk zone diameters within the susceptibility range. In fact, several KPC-producing bacteria have been reported as susceptible to carbapenems, a feature that might have been related to the fast and initially unnoticed widespread of KPC-producing bacteria.

A modified Hodge test has been recommended to detect KPC-producing *Klebsiella*, which consists of the spread of

TABLE 78.2
Useful Phenotypic Tests to Differentiate *Klebsiella* and *Raoultella* Clinical Isolates

	Species[b]				
Test[a]	*K. pneumoniae*	*K. oxytoca*	*R. ornithinolytica*	*R. planticola*	*R. terrigena*
Indole	–	+	–	V	–
Growth at 10°C	–	+	+	+	+
Histamine	–	–	+	+	+
D-Melezitose	–	v	–	–	+
Ornithine	–	–	V	–	–

[a] –, negative result; +, positive result; v, variable result.

[b] Phenotypic tests do not differentiate *K. pneumoniae* from *K. variicola*, and ornithine-negative *R. ornithinolytica* from indole-negative *R. planticola*.[10,74]

an indicator strain over a plate to investigate the presence of synergy zones around carbapenem disks.[67] The test is time consuming, and interpretation is subject to individual judgment. Molecular tests are easier to perform and interpret, and may detect the presence of several resistance mechanisms, especially β-lactamases. Detection of specific resistance markers is important not only for therapy, but also for infection control purposes. The increasing use of molecular methods disclosed the rapid international spread of each several new resistance mechanisms, sometimes by unique epidemic lineages, sometimes by the spread of the resistance determinants.[56,75]

78.1.3.2 Molecular Techniques

The study of bacterial taxonomy and phylogeny evolved robustly by comparison of rRNA genes or spacer sequences, and several protocols for molecular identification of pathogens were developed based on analysis of these genes. Thus, the rRNA gene region appeared as the most prominent target in microbial detection. The popularity of this region is surely due to a flexible mix of highly conserved segments and moderately to highly variable segments that the region contains.

The 23S rRNA gene, which encodes the large ribosomal RNA unit, has been used less frequently for diagnostic purposes, perhaps because of its greater size. Thereby, the number of complete bacterial 23S rRNA gene sequences available in databases is still very small compared to that of 16S rRNA sequences. Indeed, the manageable size of the 16S rRNA gene (~1500 bp) made it the most characterized region of the rRNA operon in bacteria with thousands and thousands of sequences available on the GenBank database (http://www.ncbi.nlm.nih.gov). rRNA genes became ideal candidates for identification and evolutionary studies because of their relative conservation within bacterial species.[76] However, the analysis of 16S rRNA sequences may have limitations: for example, the presence of several and variable copies of ribosomal coding genes have been described in *K. variicola*[6] and the variable regions may be not sensitive enough to allow for a clear differentiation of closely related microorganisms.[77]

The 16S-23S rDNA intergenic spacer region also called internal transcribed spacer, located between the two major ribosomal rRNA genes, can be an attractive alternative target for bacterial identification. In addition to sequence variation, the presence of size variation may render this segment suitable for identification and differentiation. Since the 16S-23S rDNA intergenic spacer is not subject to the same selective pressure as the rRNA genes, this region appears to be able to overcome the apparent limitation of rRNA genes.[78,79] In fact, several studies have used this spacer region as a tool for bacterial species and subspecies identification,[80–82] evolutionary studies,[83,84] and strain typing.[82,85] PCR methods based on the 16S-23S rDNA intergenic spacer region have been developed for detection and identification of *K. pneumoniae*.[72,86,87] For example, the technique can be employed to identify *K. pneumoniae* by using one of two pairs of specific primers: Pf/Prl or Pf/Pr2. The primer sequences are: Pf, 5'-ATTTGAAGAGGTTGCAAACGAT-3'; Prl, 5'-TTCACTCTGAAGTTTTCTTGTGTT C-3'; and Pr2, 5'-CCGAAGATGTTTCACTTCTGATT-3'.[86]

Sequence analysis of several additional housekeeping genes have been used for identification of *Klebsiella*, including the *rpoB* gene (RNA polymerase β-subunit), *gyrA* (DNA gyrase A subunit), *mdh* (malate dehydrogenase), *infB* (translation initiation factor 2), *phoE* (phosphoporine E), and *nifH* (nitrogenase reductase).[4–7] The comparison of the first *rpoB* partial sequences of *Enterobacteriaceae* species revealed that divergence levels of sequences were clearly superior to those of 16S rRNA genes.[5,88] In addition, phylogenetic trees of *Enterobacteriaceae* isolates obtained with *rpoB* fit better with the classification of these organisms than 16S rRNA trees. Therefore, *rpoB* sequencing has been frequently used as a means of identification and phylogenetic analysis of enterobacteria.

On the other hand, the study of the *rpoB* gene has not been straightforward. For example, five different primers were used when analysis of this gene was proposed for bacterial identification: two for *rpoB* amplification and three for sequencing.[88] In addition, utilization of the proposed CM_7 and CM_{31b} primer pair can result in more than one band, making the purification of PCR product for sequencing more difficult.[10,88] The primers rubF (5'-CTCTGGGCGATCTGGATA-3') and rubR (5'-CATGTTCGCACCCATCAA-3'), proposed for PCR amplification and sequencing, result in only one band, and make this approach easier to perform.[89] However, the initial steps of bacterial isolation and biochemical identification are still necessary, at least at the *Enterobacteriaceae* level.

Sequence analysis of nonhousekeeping genes, such as the chromosomal *bla* gene, is a promising alternative for identification of *Klebsiella* and *Raoultella* species because of a greater allelic diversity.[90] Identification of *Raoultella* species can be further approached by *bla* PCR-restriction fragment length polymorphism (RFLP) analysis.[74] In this technique, the PCR product from the *bla* gene of *R. ornithinolytica*, but not of *R. planticola*, may be digested by the restriction enzyme *Not*I, generating two restriction fragments. Thereby, *bla* PCR-RFLP analysis, with no sequencing, is able to separate ornithine-negative *R. ornithinolytica* isolates from indole-negative *R. planticola*.

Traditional sequencing methods are based on separation of DNA fragments of different sizes in gels. A promising automated, fast, and nongel-based DNA sequencing method named pyrosequencing has been developed. This technique is built on a four-enzyme real-time monitoring of DNA synthesis by bioluminescence. The detection system is based on the pyrophosphate released when a nucleotide is introduced in the DNA-strand. Since pyrosequencing assays can be incorporated into 24- or 96-well plates, multiple sequence targets can be analyzed simultaneously. Nevertheless, sequence analysis is limited to less than 500 bp. Applications of this technology are highly diverse and include detection, identification, strain typing, and recognition of drug-resistance mutations in virus and bacteria.[91–98]

Other new promising approaches have been recently described for direct detection of *Klebsiella* in blood culture samples.[99–101] One technique involves fluorescence in situ hybridization (FISH) with peptide nucleic acid (PNA) probes targeting rRNA genes. The PNA probe targets the *K. pneumoniae* 23S rRNA gene, and preliminary data suggest it provides an accurate diagnosis within 3 h.

Real-Time PCR. A sensible and specific technique is real-time PCR.[102] This is a valuable quantitative PCR-based tool for the diagnoses of bacterial pathogens in clinical microbiology laboratories. Real-time PCR detects the increasing DNA amounts as they are amplified, and is faster than conventional PCR because the detection steps are performed together with amplification in the same closed vessel. Sensitivity and specificity may be better or comparable to conventional PCR. There are distinct real-time PCR instruments that vary regarding to the nucleic acid probe formats supported, excitation and detection wavelengths, maximum number of samples per run, reaction volumes, and relative thermocycling times.

One detection method for nucleic acid detection with real-time PCR uses double strand intercalating dyes such as SYBR Green, SYTO9, and LC Green[103] to detect the accumulation of any double-stranded DNA product. SYBR Green may be used with instruments that perform a melting curve analysis, which permits detection of different amplification products based upon the %G + C content and length of the amplification product. This dye reveals the presence of amplified DNA, but although sensitive, it is nonspecific. Specific detection is possible with real-time PCR by using novel fluorescent probe technology probes. Three other types of nucleic acid detection methods have been used most frequently with real-time PCR testing platforms in clinical microbiology: 5 nuclease (TaqMan probes), molecular beacons, and fluorescence resonance energy transfer (FRET) hybridization probes. The real-time PCR instruments allow for the use of all or some of the dyes used for TaqMan probes and molecular beacons. However, only the LightCycler supports FRET hybridization probe detection with melting curve analysis.

TaqMan is a hydrolysis probe that increases the specificity of real-time PCR assays.[104] It relies on the 5'–3' nuclease activity of *Taq* polymerase to cleave a dual-labeled probe during hybridization to the complementary target sequence and fluorophore-based nucleic acid detection. Therefore, as in other real-time PCR methods, the resulting fluorescence signal permits quantitative measurements of the product accumulation during the exponential stages of the PCR. Similar to TaqMan, molecular beacons are oligonucleotide hybridization probes with the two dye molecules attached to a single probe.[105] The first is a fluorescent dye, and the second can be either a quencher dye or another fluorescent dye which can absorb fluorescent light transferred from the first dye and reemit light at a different wavelength. The proximity of the two dyes in the probe is determined by the hairpin shaped probe structure.

Contrasting with TaqMan or molecular beacons probes, FRET hybridization probes (also referred to as LightCycler) have the dyes attached separately to two probes designed to anneal next to each other in a head-to-tail configuration on target nucleic acid DNA.[106] The upstream probe has a fluorescent dye on the 3' end and the downstream probe has an acceptor dye on the 5' end. If both probes anneal to the target PCR product, fluorescence from the 3' dye is absorbed by the adjacent acceptor dye on the 5' end of the second probe. The second dye is excited and emits light at a third wavelength and this third wavelength is detected. For all types of FRET probes, as the distance between adjacent dye molecules increases, FRET decreases. In addition, the mechanisms to achieve a fluorescent signal with the TaqMan, molecular beacon, or FRET hybridization probe format are different.

In one study, a LightCycler real-time PCR assay (Roche Molecular Biochemicals) to detect *K. pneumoniae* 16S rRNA, developed for the rapid identification of *K. pneumoniae*, was 100% sensitive and specific for positive blood cultures, which included *Pseudomonas aeruginosa*, *E. coli* and *K. oxytoca* as negative controls.[100] A quantitative estimate is possible with this assay, which was able to detect as few as 10 bacterial cells. The assay was complete in 2 h. A second study, using a new commercial LightCycler real-time PCR assay (LightCycler SeptiFast Test M Grade, Roche Molecular Systems) described the test performed directly from blood samples.[101] The assay targets specific rDNA internal transcribed spacer regions. The test was complete in 6 h, and detects 20 different pathogens or groups of pathogens, but *K. pneumoniae* is not differentiated from *K. oxytoca*. In the study described, the PCR test detected more pathogens than the conventional blood culture system; however, no samples were positive for *Klebsiella* (*pneumoniae/oxytoca*) in the patient population studied.[101]

Concerning antimicrobial resistance determinants, the presence and/or the expression of β-lactamases (bla_{SHV} and bla_{KPC}), outer membrane proteins (*ompK-35* and *ompK-36*) encoding genes and/or hyperexpression of AcrAB efflux systems have been also evaluated by real-time PCR in *K. pneumoniae*.[68–70,107] In the present chapter, we have chosen to describe in detail LightCycler real-time PCR assay as a molecular approach for detection and quantification of antimicrobial resistance gene expression of *K. pneumoniae*. However, the initial steps of bacterial isolation are still necessary for gene expression analysis.

Molecular Typing. Molecular typing of pathogens has become an important tool for the epidemiologic investigation of infectious diseases. Different molecular typing methods have enhanced our understanding of the epidemiology of several bacterial infectious diseases, by helping to determine modes of transmission of pathogens, the source and the risk factors for infections. Strain typing methods are used to address two different kinds of problems—to determine if isolates recovered from a localized outbreak of disease belong to a single or to different strains (short-term or local epidemiology), or to determine if strains causing disease in one geographic area are related to others obtained worldwide (long term or global epidemiology).

Molecular strain typing methods together with classic epidemiology studies have been essential for the understanding of the transmission dynamics of infections caused by

Klebsiella.[108,109] For instance, a molecular epidemiology study led to the identification of a contaminated ultrasonography coupling gel container of an emergency room as the source of a hospital outbreak.[110] Strain typing by PFGE revealed that patient isolates and an isolate from the ultrasonography gel belonged to a single strain. This investigation disclosed two modes of *K. pneumoniae* transmission: ultrasound scanning by transvaginal ultrasonography, and peripartum colonization of neonates, and led to the complete control of the outbreak.

Analysis of chromosomal DNA restriction patterns by PFGE is an appropriate and discriminatory technique for typing different microorganisms.[108] However, PCR-based techniques such as random amplified polymorphic DNA–PCR (RAPD–PCR), enterobacterial repetitive intergenic consensus-PCR (ERIC–PCR), and repetitive extragenic palindromic–PCR (REP–PCR) analysis are faster and easier to perform[111–113] and provide useful discriminatory powers. For typing *K. pneumoniae*, up to two band differences is a functional cutoff, with ERIC–PCR being the most reliable of the methods.[113]

PFGE and PCR-based methods have the disadvantage of relying on comparison of banding patterns generated by gel electrophoresis, which is difficult to standardize. Therefore, comparability of patterns within and between laboratories requires strict adherence to established protocols, even though reproducibility is not always enough and normalization of patterns is complex. To overcome the poor portability problems of electrophoresis banding techniques, an appealing, sequence-based alternative was developed—MLST, based on the nucleotide sequence variability of internal regions of housekeeping loci. Small sequence variations allow for the assignment of alleles, and the combination of alleles for several loci provides an allele profile that determined a sequence type (ST). Subsequently, STs are grouped into clonal complexes based on similarity to a central allelic profile. MLST has been proved useful for local and global epidemiologic studies. In 2005, Diancourt and colleagues[114] developed an MLST scheme for *K. pneumoniae*, available at the http://www.pasteur.fr/mlst Web site. The scheme uses internal fragments of seven housekeeping genes: *rpoB*, *gapA* (glyceraldehyde 3-phosphate dehydrogenase), *mdh*, *pgi* (phosphoglucose isomerase), *phoE*, *infB*, and *tonB* (periplasmic energy transducer). The analysis of nosocomial isolates with this protocol revealed a high (96%) discriminatory power, compared to ribotyping.

Interesting data about *Klebsiella* infections have been provided by studies using MLST. Paulin-Curlee and colleagues[115] evaluated the genetic diversity of *K. pneumoniae* isolated from clinical mastitis cases, to define genotypes most commonly associated with the disease. Authors used the same MLST protocol proposed by Diancourt and colleagues[114] in addition to PFGE and REP–PCR. A significant genetic diversity among the isolates was observed regardless of the fingerprinting method used: Simpson's diversity index was 97%, 96%, and 94%, when strains were analyzed by MLST, PFGE, and PCR-base DNA fingerprinting, respectively. Noteworthy was that three isolates from mastitis in cows belonged to the same MLST types previously observed among human clinical isolates. This finding indicates that *K. pneumoniae* isolates of bovine origin may be involved in human disease, and illustrates how the increasing use of portable typing data will impact our knowledge about the molecular epidemiology of infections.

The patterns of spread of KPC-producing *K. pneumoniae* over the United States were investigated by PFGE and MLST.[56] Surprisingly, a predominant strain, which belongs to ST 258, was found in 10 of the 19 states investigated, and accounted for 70% of the isolates sent to the Centers for Diseases Control and Prevention for reference testing from 1996 to 2008. This same strain is also present in Israel. Interestingly, ST 258 is a single locus variant of ST 11, a *K. pneumoniae* lineage described in Hungary, which produces an epidemic ESBL-type, CTX-M-15.

78.2 METHODS

We present below the PCR LightCycler real-time PCR assay to detect *K. pneumoniae* and quantify antimicrobial resistance gene expression in this species. Primers and fluorescent probes used for identification of *K. pneumoniae* and detection/quantification of expression of antimicrobial resistance genes by real-time PCR are presented in Table 78.3.

78.2.1 Sample Preparation

Extraction of DNA for PCR Detection. Obtaining DNA is a crucial step for all genotypic tests, independent of their purpose or the technique chosen, and can be accomplished either by manual or automated methods. Extraction methods must be optimized taking into consideration the pathogen and clinical specimen where it will be isolated from. Usually the gram-negative cell wall is more easily broken than that of gram-positive bacteria. Substances such as heme in blood or bile in stool must be removed during this process since they might inhibit DNA amplification by PCR. In addition, DNA must be concentrated into a small volume and maintained in an aqueous solution to protect its degradation.

DNA can be obtained from pure colonies or directly from clinical specimens. To extract DNA from bacterial colonies, cells must be lysed prior to PCR amplification. Bacterial DNA can be extracted by distinct methods. For many isolates, a single boiling step is sufficient to break the bacterial cell and obtain DNA. When possible, this method is preferred because is little time consuming and requires no special reagents or equipments.[89] Two to five bacterial colonies are suspended in 200 μL of sterile double, distillated water and incubated at 95°C–105°C for 10 min. Then, the samples are centrifuged at 20,800 × *g* for 5 min and the supernatant containing the target DNA is utilized for PCR reaction.

For DNA isolation from blood and body fluids, a commercially available kit such as QIAamp blood and tissue kit (QIAGEN, Valencia, California) may be used following strictly manufacturer's instructions. This kit can be also used

TABLE 78.3
Primers and Fluorescent Probes Used for Identification of *K. pneumoniae* and Antimicrobial Resistance Gene Detection/Quantification of Expression by Real-Time PCR

Target DNA	Primers	Probes	Real-Time PCR	Reference
		Detection of *K. pneumoniae* from blood cultures		
126-bp target sequence of the 16S rRNA gene, positions 44–170 of *K. pneumoniae* strain EMBL accession no. X93214	K16SF 5'-AGCACAGAGAGCTTG-3' K16SR 5'-ACTTTGGTCTTGCGAC-3'	Fluorescence-labeled probes to hybridize to the species-specific region of *K. pneumoniae*. The probes are labeled with LightCycler Red 640 at the 5' end and phosphorylated at the 3' end (19-mer; melting temperature [*Tm*] = 62.7°C. The anchor probe is labeled with isothiocyanate at the 3' end (30-mer, *Tm* = 69.9°C)	Sensitivity: 10 copies of target DNA Specificity C_T value of ≤13 was achieved for *K. pneumoniae* T° melting: 66°C	100
		Antimicrobial resistance gene detection/quantification of expression		
ompK35	F 5-TCCCTGCCCTGCTGGTAG-3 R 5-CTGGTGTCGCCATTGGTGG-3	One-step QuantiFast SYBR Green RT-PCR Kit (Qiagen)		69
ompK36	F 5-GCGACCAGACCTACATGCGT-3 R 5-AGTCGAAAGAGCCCGCGTC-3			
ompK37	F 5-GCCCTCGTTATTCCGGCTT-3 R 5-CCGTCTTTTTTCGAGTCGCTG-3			
all bla_{KPC}	F 5-GATACCACGTTCCGTCTGG-3 R 5-GCAGGTTCCGGTTTTGTCTC-3	6-carboxyfluorescein–5'-AGCGGCAGCAGTTTGTTGATTG-3'–6-carboxytetramethylrhodamine	2 min at 50°C, 10 min at 95°C, and 50 cycles of 15 s at 95°C and 1 min at 60°C. Sensitivity: 1 CFU Specificity C_T value of 40	116
bla_{KPC2-3}	F 5-TCTGGACCGCTGGGAGCTGG-3 R 5-TGCCCGTTGACGCCCAATCC-3	6-carboxyfluorescein–5'-CGCGCGCCGTGACGGAAAGC-3'–6-carboxytetramethylrhodamine	2 min at 95°C, and 35 cycles of 2 s at 94°C, 10 s at 62°C, and 15 s at 72°C	68
acrA	F 5-GTCCGGCACGCTGGA-3 R 5-ATAGCGCGTAGCGTGATTGA-3	6-carboxyfluorescein–5'-CGGATGTCACCGTCGATCAGACCAC-3'–6-carboxytetramethylrhodamine		107
ramA	F 5-GCATCAACCGCTGCGTATT-3 R 5- CGTTGCAGATGCCATTTCG -3	6-carboxyfluorescein–5'-TCGCTCGCCATGCCGGGTAT-3'–6-carboxytetramethylrhodamine		

for extracting DNA directly from bacterial colonies, when DNA extraction cannot be accomplished by boiling the bacterial colonies. Other manual methods of nucleic acid extraction and purification for rapid real-time PCR assays are also commercially available, such as High Pure (Roche Applied Science; www.roche-applied-science.com) and IsoQuick (Orca Research; www.bioexpress.com). All of these kits allow DNA extraction from plasma, whole blood, stool, respiratory tract specimens, sterile body fluids, and swabs.

Automated methods using kits/instruments of distinct manufacturers have been used for isolation of bacterial DNA. Many authors employ the MagNA Pure LC DNA isolation kit III with MagNA Pure LC extractor (www.Roche-Applied-Science.com; www.magnapure.com) and the NucliSENS easyMAG System (www.biomerieux.com). The extraction of DNA by these systems has been mainly employed for fast detection of antibiotic resistant genes by real-time PCR[116] or identification of pathogens from blood or body fluids by automated systems such as LightCycler SeptiFast MGRADE test (Roche Diagnostics).[117] Although both systems have shown an equivalent performance, it was reported that the NucliSENS easyMAG System (100% sensitivity) was slightly more efficient than the MagNA Pure LC extractor (98% sensitivity) for obtaining DNA from rectal swab specimens for bla_{KPC} detection.[116] This difference was probably related to the maximum sample volume allowed by the extraction method (100 μL vs. 200 μL for the MagNA Pure LC and NucliSENS easyMAG System, respectively). In addition, the MagNA Pure LC (20–30 min) required longer sample manipulation than NucliSENS easyMAG System (10 min).[116]

Extraction of RNA for Antimicrobial Resistance Gene Expression Measurements

1. Grow the bacterial cells on 10 mL of Luria Bertani broth at 37°C on a rotary shaker to reach an optical density of 1.5–2.0 at 600 nm wave length.

2. Total RNA can be extracted using the kit RNeasy Mini Kit (Qiagen) and treated with Rnase-free DNase Set (Quiagen) according to manufacturer's instructions.
3. Quantify the total amount of RNA extracted using a spectrophotometer at 260 and 280 nm wave lengths. An $OD_{260/280} \geq 2.0$ is desirable for the RT-PCR.
4. A gel electrophoresis (1.2% agarose, 6.7% formaldehyde, 200 mM MOPS, 50 mM sodium acetate, 10 mM EDTA, and DEPC water) is performed to confirm the integrity of RNA.
5. In addition, a PCR reaction using primers targeting ribosomal DNA (K16SF and K16SR) is carried out to confirm the absence of DNA. Any amplification signal is suggestive of DNA contamination, and new treatment with Rnase-free DNase Set must be done.

78.2.2 Detection Procedures

78.2.2.1 Real-Time PCR Identification

1. Prepare the master mixture using: 2 μL of 10× LightCycler-FastStart DNA master hybridization probes (Roche Diagnostics), 3 mM $MgCl_2$ (final concentration), 0.5 μM final concentration of *K. pneumoniae*-specific primers, 0.2 μM (final concentration) of hybridization probes, and double distilled water to reach a final volume of 18 μL per reaction.
2. To complete the PCR mixtures, add 2 μL of the template DNA to 18 μL of master mixture on the LightCycler glass capillaries.
3. After a short centrifugation (700 × *g* for 10 s), place the sealed capillaries into the LightCycler.
4. Set the thermocycling conditions as one cycle of denaturation at 95°C for 10 min, followed by 30 amplification cycles (temperature transition rate of 20°C/s), each comprising denaturation (95°C for 10 s), annealing (55°C for 15 s), and extension (74°C for 10 s).
5. The PCR step is followed by melting curve analysis as illustrated in Figure 78.1. Melting curve analysis takes place after all thermocycling is complete. The temperature of the reaction vessel is lowered below the annealing temperature established for the assay. This allows annealing of the probes to the target DNA of the intended sequence. The temperature is then slowly increased. Probes that are hybridized to these DNA molecules will separate from complementary DNA, which results in a loss of FRET signals. These probes will separate from the target DNA (melt) at lower temperatures if there is a mismatch between the complementary sequences of the probes and the DNA to which they are annealed. Comparisons of negative derivatives of fluorescence over time with temperature generate melting curves, which are unique for a specific complementary DNA sequence.

78.2.2.2 Detection of Antimicrobial Resistance Gene Expression Measurements

Before a gene expression measurement can be performed by real-time PCR, the mRNA in the sample must be copied to cDNA by reverse transcription (RT) (i.e., in an RT reaction RNA molecules are transcribed to DNA copies by reverse transcriptase).

Synthesis of cDNA. cDNA is synthesized by using "High Capacity cDNA Archive kit" (Applied Byosistems) according to manufacturer's instructions using approximately 5 μg/mL of RNA as template.

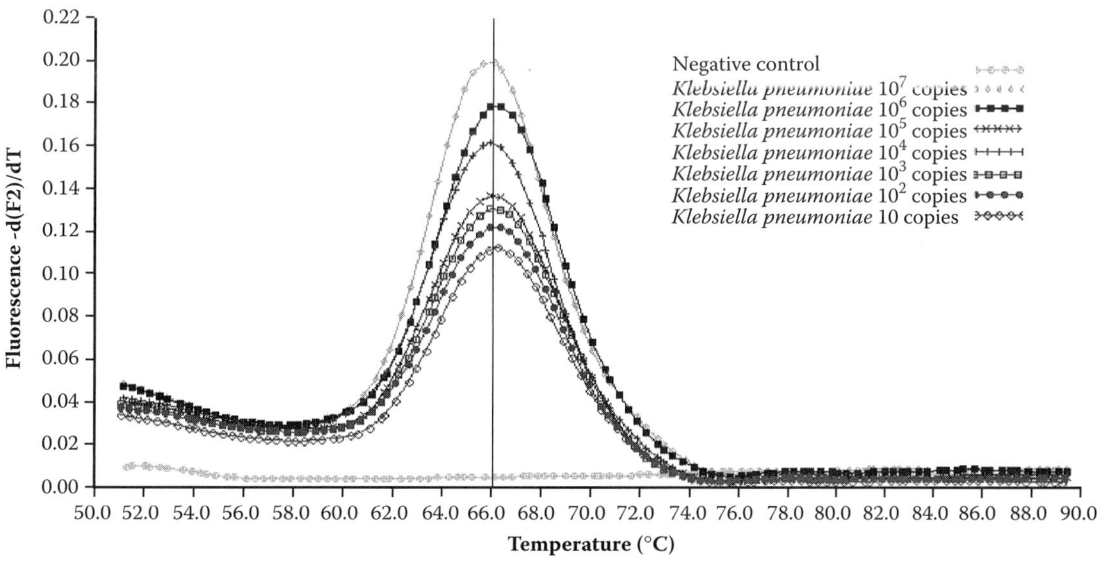

FIGURE 78.1 Melting curve for the LightCycler KPS and KPA FRET hybridization probes. The specific melting peak can be observed with the quantified number of copies of *K. pneumoniae* DNA. (From Kurupati, P. et al., *J. Clin. Microbiol.*, 42, 1337, 2004. With permission from American society of Microbiology.)

Real-Time PCR. (i) Perform the real-time PCR, in triplicate, in a LightCycler (Roche, www.roche.com) using the one-step QuantiTect SYBR Green PCR master mix (QIAGEN, Valencia, California) according to the manufacturers' protocol. Briefly, 2 μL of cDNA must be added to 12.5 μL of master mixture containing the SYBR Green, 1 μL of each 10 μM primer solution and double, distilled water to complete a final volume of 25 μL. (ii) Adjust the thermocycling conditions according to the tested primers.

Analysis of mRNA by Real-Time PCR Using Comparative C_T ($\Delta\Delta C_T$). (i) The mRNA of a housekeeping gene such as *rpoB* or*16S RNA* and of a reference strain for the studied gene must be chosen as the endogenous reference RNA for relative quantification. (ii) Relative quantities of mRNA from each gene of interest is determined by the comparative threshold cycle method (C_T), where the C_T value is defined as the number of cycles needed for the fluorescence, which reflects the quantity of amplified RNA, to exceed a value of 0. (iii) The magnitude of expression (ΔC_T) of the gene of interest is obtained by normalization against the housekeeping gene, to correct for variation in RNA content and amplification efficiency between samples. The equation 2^{-C_T}, where $\Delta C_T = C_{T\text{-gene of interest}} - C_{T\text{-housekeeping gene}}$, and $\Delta\Delta C_T = \Delta C_{T\text{-gene of interest}} - \Delta C_{T\text{-gene of sample reference}}$, is calculated to allow the relative quantification of differences in gene expression levels between two strains. Another method to quantify relative expression of a gene of interest is the evaluation of standard curve. In this case, the efficiency of the reaction is not necessary.[118]

78.3 CONCLUSIONS AND FUTURE PERSPECTIVES

Molecular tests are a promising and necessary alternative for the identification of *Klebsiella* because several biochemical tests are currently needed to separate species, many species share a similar biochemical profile, and the presence of atypical strains, such as fastidious, slow growing or even pleiotropic, makes the identification even more difficult. Real-time PCR has been shown to be a useful tool for identification and direct detection of *Klebsiella* in clinical specimens.

Klebsiella is a repository of antimicrobial resistance genes.[119] The more resistant strains are usually hospital-acquired, but multidrug-resistance is being already seen in the community. Persistent carries are an important source of resistance dissemination. Recently, an editorial called the attention for six microorganisms that have escaped the lethal action of antimicrobial agents in hospitals. *K. pneumoniae* is part of that list, together with *Enterococcus faecium*, *Staphylococcus aureus*, *Acinetobacter baumannii*, *Pseudomonas aeruginosa*, and *Enterobacter*.[120] Clearly, a more devoted and coordinated effort from research agencies and heath authorities must be put forward to face the problems these pathogens raise. Fast detection and quantification of antimicrobial gene expression have an important role in this scenario.

Finally, an aspect to be solved is the existence of reservoirs as a source of more virulent *K. pneumoniae* strains, such as the hypermucoviscous serotype K1, and antibiotic associated-*K. oxytoca* colitis. A rapid molecular test would be useful to detect and investigate the epidemiologic importance of such strains in different environments.

The tireless searching for new technologies grows in compass with the need for speed in the diagnosis of infectious diseases. Pyrosequencing, a DNA sequencing technology based on the sequencing-by-synthesis principle, seems to be a promising state-of-the-art technology for the next generation of automated DNA sequencing apparatus. This analysis instrument put together the speed and the "gold standard" result of molecular analysis (the sequence). Pyrosequencing delivers qualitative sequencing data, quantitative allele frequency information, and allows for rapid and automated comparisons of results against local and global databases.

ACKNOWLEDGMENTS

Work presented by Authors was supported by Conselho Nacional de Desenvolvimento Científico e Tecnológico (CNPq) and Fundação de Amparo à Pesquisa do Estado do Rio de Janeiro (FAPERJ) of Brazil, and the Fogarty International Program in Global Infectious Diseases (D43 TW006563–07) of the National Institute of Health.

REFERENCES

1. Podschun, R., and Ullmann, U., *Klebsiella* spp. as nosocomial pathogens: Epidemiology, taxonomy, typing methods, and pathogenicity factors, *Clin. Microbiol. Rev.*, 11, 589, 1998.
2. Martinez, J. et al., How are gene sequence analyses modifying bacterial taxonomy? The case of *Klebsiella*, *Int. Microbiol.*, 7, 261, 2004.
3. Ørskov, I., *Bergey's Manual of Systematic Bacteriology*, 1, p. 461, Krieg and Holt, Baltimore, 1984.
4. Carter, J.S. et al., Phylogenetic evidence for reclassification of *Calymmatobacterium granulomatis* as *Klebsiella granulomatis* comb. nov., *Int. J. Syst. Bacteriol.*, 49, 1695, 1999.
5. Drancourt, M. et al., Phylogenetic analyses of *Klebsiella* species delineate *Klebsiella* and *Raoultella* gen. nov., with description of *Raoultella ornithinolytica* comb. nov., *Raoultella terrigena* comb. nov. and *Raoultella planticola* comb. nov., *Int. J. Syst. Evol. Microbiol.*, 51, 925, 2001.
6. Rosenblueth, M. et al., *Klebsiella variicola*, a novel species with clinical and plant-associated isolates, *Syst. Appl. Microbiol.*, 27, 27, 2004.
7. Li, X. et al., *Klebsiella singaporensis* sp. nov., a novel isomaltulose-producing bacterium, *Int. J. Syst. Evol. Microbiol.*, 54, 2131, 2004.
8. Brisse, S., and Verhoef, J., Phylogenetic diversity of *Klebsiella pneumoniae* and *Klebsiella oxytoca* clinical isolates revealed by randomly amplified polymorphic DNA, *gyrA* and *parC* genes sequencing and automated ribotyping, *Int. J. Syst. Evol. Microbiol.*, 51, 915, 2001.
9. Boye, K., and Hansen, D.S., Sequencing of 16S rDNA of *Klebsiella*: Taxonomic relations within the genus and to other *Enterobacteriaceae*, *Int. J. Med. Microbiol.*, 292, 495, 2003.
10. Alves, M.S. et al., Identification of clinical isolates of indole-positive and indole-negative *Klebsiella* spp., *J. Clin. Microbiol.*, 44, 3640, 2006.

11. Fevre, C. et al., Variants of the *Klebsiella pneumoniae* OKP chromosomal beta-lactamase are divided into two main groups, OKP-A and OKP-B, *Antimicrob. Agents Chemother.*, 49, 5149, 2005.
12. Kharsany, A.B. et al., Phylogenetic analysis of *Calymmatobacterium granulomatis* based on 16S rRNA gene sequences, *J. Med. Microbiol.*, 48, 841, 1999.
13. Brisse, S. et al., Development of a rapid identification method for *Klebsiella pneumoniae* phylogenetic groups and analysis of 420 clinical isolates, *Clin. Microbiol. Infect.*, 10, 942, 2004.
14. Fournier, B. et al., Chromosomal β-lactamase genes of *Klebsiella oxytoca* are divided into two main groups, bla_{OXY-1} and bla_{OXY-2}, *Antimicrob. Agents Chemother.*, 40, 454, 1996.
15. Fevre, C. et al., Six groups of the OXY β-lactamase evolved over millions of years in *Klebsiella oxytoca*, *Antimicrob. Agents Chemother.*, 49, 3453, 2005.
16. Granier, S.A. et al., New *Klebsiella oxytoca* β-lactamase genes bla_{OXY-3} and bla_{OXY-4} and a third genetic group of *K. oxytoca* based on bla_{OXY-3}, *Antimicrob. Agents Chemother.*, 47, 2922, 2003.
17. Podschun, R. et al., Incidence of *Klebsiella* species in surface waters and their expression of virulence factors, *Appl. Environ. Microbiol.*, 67, 3325, 2001.
18. Munoz, M.A. et al., Molecular epidemiology of two *Klebsiella pneumoniae* mastitis outbreaks on a dairy farm in New York State, *J. Clin. Microbiol.*, 45, 3964, 2007.
19. Abbott, S.L., *Manual of Clinical Microbiology*, 9th ed., 698pp., ASM press, Washington, DC, 2007.
20. Struve, C., and Krogfelt, K.A., Pathogenic potential of environmental *Klebsiella pneumoniae* isolates, *Environ. Microbiol.*, 6, 584, 2004.
21. Ko, W-C. et al., Community-acquired *Klebsiella pneumoniae* bacteremia: Global differences in clinical patterns, *Emerg. Infect. Dis.*, 8, 160, 2002.
22. Mandell, L.A. et al., Infectious disease society of America/American thoracic society consensus guidelines on the management of community-acquired pneumonia in adults, *Clin. Infect. Dis.*, 44, S27, 2007.
23. Keynan, Y., and Rubinstein, E., The changing face of *Klebsiella pneumoniae* infections in the community, *Intern. J. Antimicrobial Agents*, 30, 386, 2007.
24. Ni, Y-H. et al., Community-acquired brain abscess in Taiwan: Etiology and probably source of infection, *J. Microbiol. Immunol. Infect.*, 37, 231, 2004.
25. Fang, C-T. et al., *Klebsiella pneumoniae* genotype K1: An emerging pathogen that causes septic ocular or central nervous system complications from pyogenic liver abscess, *Clin. Infect. Dis.*, 45, 284, 2007.
26. Turton, J.F. et al., Genetic similar isolates of *Klebsiella pneumoniae* serotype K1 causing liver abscess in three continents, *J. Med. Microbiol.*, 56, 593, 2007.
27. Cheng, D.L. et al., Septic metastatic lesions of pyogenic liver abscess. Their association with *Klebsiella pneumoniae* bacteremia in diabetes patients, *Arch. Intern. Med.*, 151, 1557, 1991.
28. Habib, A.G., and Tambyah, P.A., Community-acquired *Klebsiella pneumoniae* central nervous system infections in adults in Singapore, *Eur. J. Clin. Microbiol. Infect. Dis.*, 22, 486, 2003.
29. Chang, W.N. et al., Community-acquired spontaneous *Klebsiella pneumoniae* meningitis in adult cirrhotic patients with and without diabetes, *Eur. J. Clin. Microbiol. Infect. Dis.*, 22, 271, 2003.
30. Andraca, R., Edson, R.S., and Kern, E.B., Rhinoscleroma: A growing concern in the United States? Mayo Clinic experience, *Mayo Clin. Proc.*, 68, 1151, 1993.
31. De Champs, C. et al., Laryngeal scleroma associated with *Klebsiella pneumoniae* subsp. *ozaenae*, *J. Clin. Microbiol.*, 43, 5811, 2005.
32. Chan, T.V., and Spiegel, J.H., *Klebsiella rhinoscleromatis* of the membranous nasal septum, *J. Laryngol. Otol.*, 121, 998, 2007.
33. Medina, L.M. et al., Clinical, genetic and immunologic analysis of a family affected by ozena, *Eur. Arch. Otorhinolaryngol.*, 260, 390, 2003.
34. Talmi, Y.P. et al., Computed tomography study of sinus involvement in ozena, *Am. J. Rhinol.*, 9, 281, 1995.
35. Rennie, R.P. et al. *Klebsiella pneumoniae* gastroenteritis masked by *Clostridium perfringens*, *J. Clin. Microbiol.*, 28, 216, 1990.
36. Högenauer, C. et al., *Klebsiella oxytoca* as a causative organism of antibiotic-associated hemorrhagic colitis, *N. Engl. J. Med.*, 355, 2418, 2006.
37. Biedenbach, D.J., Moet, G.J., and Jones R.N., Occurrence and antimicrobial resistance pattern comparisons among bloodstream infection isolates from SENTRY Antimicrobial Surveillance Program (1997–2002), *Diagn. Microbiol. Infect. Dis.*, 50, 59, 2004.
38. Podschun, R. et al., Isolation of *Klebsiella planticola* from newborns in a neonatal ward, *J. Clin. Microbiol.*, 36, 2331, 1998.
39. Pessoa-Silva, C.L. et al., Extended-spectrum β-lactamase-producing *Klebsiella pneumoniae* in a neonatal intensive care unit: risk factors for infection and colonization, *J. Hosp. Infect.*, 53, 198, 2003.
40. Alves, M.S. et al., A case of severe pancreatitis complicated by *Raoultella planticola* infection, *J. Med. Microbiol.*, 56, 696, 2007.
41. De Champs, C. et al., Prospective survey of colonization and infection caused by expanded spectrum β lactamase-producing members of the family *Enterobacteriaceae* in an intensive care unit, *J. Clin. Microbiol.*, 27, 2887, 1989.
42. Peña, C. et al., Epidemiology and successful control of a large outbreak due to *Klebsiella pneumoniae* producing extended-spectrum β-lactamases, *Antimicrob. Agents Chemother.*, 42, 53, 1998.
43. Asensio, A. et al., Outbreak of a multiresistant *Klebsiella pneumoniae* strain in an intensive care unit: antibiotic use as risk factor for colonization and infection, *Clin. Infect. Dis.*, 30, 55, 2000.
44. Moustaoui, N. et al., Digestive tract colonization with extended spectrum beta-lactamase producing *Enterobacteriaceae* in a surgical intensive care unit in Casablanca, *J. Hosp. Infect.*, 46, 238, 2000.
45. Berthelot, P. et al., Nosocomial colonization of premature babies with *Klebsiella oxytoca*: Probable role of enteral feeding procedure in transmission and control of the outbreak with the use of gloves, *Infect. Control. Hosp. Epidemiol.*, 22, 148, 2001.
46. Moustaoui, N. et al., Molecular biology of extended-spectrum β-lactamase-producing *Enterobacteriaceae* responsible for digestive tract colonization, *J. Hosp. Infect.*, 57, 202, 2004.
47. De Almeida, V.C. et al., Genetic relatedness among extended-spectrum β-lactamase-producing *Klebsiella pneumoniae* outbreak isolates associated with colonization and invasive disease in a neonatal intensive care unit, *Microb. Drug Resist.*, 11, 21, 2005.

48. Casewell, M.W., and Phillips, I., Aspects of plasmid-mediated antibiotic resistance and epidemiology of *Klebsiella* species, *Am. J. Med.*, 70, 459, 1981.
49. Kristóf, K. et al., Extended-spectrum beta-lactamase-producing *Klebsiella* spp. in a neonatal intensive care unit: Risk factors for the infection and the dynamics of the molecular epidemiology, *Eur. J. Clin. Microbiol. Infect. Dis.*, 26, 563, 2007.
50. Lucet, J.C. et al., Control of a prolonged outbreak of extended-spectrum β-lactamase-producing *Enterobacteriaceae* in a university hospital, *Clin. Infect. Dis.*, 29, 1411, 1999.
51. Cantón, R. et al., Prevalence and spread of extended-spectrum β-lactamase-producing Enterobacteriaceae in Europe, *Clin. Microbiol. Infect.*, 14, 144, 2008.
52. Du, B. et al., Extended-spectrum beta-lactamase-producing *Escherichia coli* and *Klebsiella pneumoniae* bloodstream infections: risk factors and clinical outcome, *Intensive Care Med.*, 28, 1718, 2002.
53. Bou, G. et al., Identification and broad dissemination of the CTX-M-14 β-lactamase in different *Escherichia coli* strains in northwest area of Spain, *J. Clin. Microbiol.*, 40, 4030, 2002.
54. Colodner, R., Risk factors for the development of extended-spectrum beta-lactamase-producing bacteria in nonhospitalized patients, *Eur. J. Clin. Microbiol. Infect. Dis.*, 23, 163, 2004.
55. Pitout, J.D. et al., Emergence of *Enterobacteriaceae* producing extended-spectrum β-lactamases (ESBLs) in the community, *J. Antimicrob. Chemother.*, 56, 52, 2005.
56. Kitchel, B. et al., Molecular epidemiology of KPC-producing *Klebsiella pneumoniae* isolates in the United States: Clonal expansion of multilocus sequence type 258, *Antimicrob. Agents Chemother.*, 53, 3365, 2009.
57. Leavitt, A. et al., Ertapenem resistance among extended-spectrum β-lactamase-producing *Klebsiella pneumoniae* isolates, *J. Clin. Microbiol.*, 47, 969, 2009.
58. Nordmann, P., Cuzon, G., and Naas, T. The real threat of *Klebsiella pneumoniae* carbapenemase producing bacteria, *Lancet Infect. Dis.*, 9, 228, 2009.
59. Giakkoupi, P. et al., Emerging *Klebsiella pneumoniae* isolates coproducing KPC-2 and VIM-1 carbapenemases, *Antimicrob. Agents Chemother.*, 53, 4048, 2009.
60. Blomberg, B. et al., High rate of fatal cases of pediatric septicemia caused by gram-negative bacteria with extended-spectrum beta-lactamases in Dar es Salaam, Tanzania, *J. Clin. Microbiol.*, 43, 745, 2005.
61. Brisse, S. et al., Virulent clones of *Klebsiella pneumoniae*: Identification and evolutionary scenario based on genomic and phenotypic characterization, *PLoS One*, e4982, 2009.
62. Yu, V.L. et al., Virulence characteristics of *Klebsiella* and clinical manifestations of *K. pneumoniae* bloodstream infections, *Emerg. Infect. Dis.*, 13, 986, 2007.
63. Evrard, B. et al., Roles of capsule and lipopolysaccharide O antigen in interactions of human monocyte-derived dendritic cells and *Klebsiella pneumoniae*, *Infect. Immun.*, 78, 210, 2010.
64. Mizuta, K. et al., Virulence for mice of *Klebsiella* strains belonging to the O1 group: Relationship to their capsular (K) types, *Infect. Immun.*, 40, 56, 1983.
65. Palfreyman, J.M., *Klebsiella* serotyping by counter-current immunoelectrophoresis, *J. Hyg.*, 81, 219, 1978.
66. Matsuhisa, A. et al., Detection of bacteria in phagocyte-smears from septicemia-suspected blood by in situ hybridization using biotinylated probes, *Microbiol. Immunol.*, 38, 511, 1994.
67. Clinical and Laboratory Standards Institute, Performance standards for antimicrobial susceptibility testing; Nineteenth informational supplement. CLSI document M100-S19. CLSI, Wayne, PA, 2009.
68. Cole, J.M. et al., Development and evaluation of a real-time PCR assay for detection of *Klebsiella pneumoniae* carbapenemase genes, *J. Clin. Microbiol.*, 47, 322, 2009.
69. Doumith, M. et al., Molecular mechanisms disrupting porin expression in ertapenem-resistant *Klebsiella* and *Enterobacter* spp. clinical isolates from the UK, *J. Antimicrob. Chemother.*, 63, 659, 2009.
70. Randegger, C.C., and Hächler, H., Real-time PCR and melting curve analysis for reliable and rapid detection of SHV extended-spectrum beta-lactamases, *Antimicrob. Agents Chemother.*, 45, 1730, 2001.
71. Haeggman, S. et al., Diversity and evolution of the class A chromosomal beta-lactamase gene in *Klebsiella pneumoniae*, *Antimicrob. Agents Chemother.*, 48, 2400, 2004.
72. Lopes, A.C.S. et al., Application of PCR ribotyping and tDNA-PCR for *Klebsiella pneumoniae* identification, *Mem. Inst. Oswaldo Cruz*, 102, 827, 2007.
73. Vimont, S. et al., Comparison of PFGE and multilocus sequence typing for analysis of *Klebsiella pneumoniae* isolates, *J. Med. Microbiol.*, 57, 1308, 2008.
74. Walckenaer, E. et al., How to identify *Raoultella* spp. including *R. ornithinolytica* isolates negative for ornithine decarboxylase? The reliability of the chromosomal *bla* gene, *J. Microbiol. Methods*, 75, 405, 2008.
75. Woodford, N. et al., Arrival of *Klebsiella pneumoniae* producing KPC carbapenemase in the United Kingdom, *J. Antimicrob. Chemother.*, 62, 1261, 2008.
76. Gutell, R.R., Larsen, N., and Woese, C.R., Lessons from an evolving rRNA: 16S and 23S rRNA structures from a comparative perspective, *Microbiol. Rev.*, 58, 10, 1994.
77. Palys, T., Nakamura, L.K., and Cohan, F.M., Discovery and classification of ecological diversity in the bacterial world: The role of DNA sequence data, *Int. J. Syst. Bacteriol.*, 47, 1145, 1997.
78. Barry, T. et al., The 16S/23S ribosomal spacer region as a target for DNA probes to identify eubacteria, *PCR Methods Appl.*, 1, 51, 1991.
79. Gürtler, V., and Stanisich, V.A., New approaches to typing and identification of bacteria using the 16S-23S rDNA spacer region, *Microbiol.*, 142, 3, 1996.
80. Mora, D. et al., 16S-23S rRNA intergenic spacer region sequence variation in *Streptococcus thermophilus* and related dairy streptococci and development of a multiplex ITS-SSCP analysis for their identification, *Microbiol.*, 149, 807, 2003.
81. Nakanishi, S. et al., Rapid species identification and partial strain differentiation of *Clostridium butyricum* by PCR using 16S-23S rDNA intergenic spacer regions, *Microbiol. Immunol.*, 49, 613, 2005.
82. Regassa, L.B. et al, Differentiation of group VIII *Spiroplasma* strains with sequences of the 16S-23S rDNA intergenic spacer region, *Can. J. Microbiol.*, 50, 1061, 2004.
83. Antón, A.I., Martinez-Murcia, A.J., and Rodriguez-Valera, F., Sequence diversity in the 16S-23S intergenic spacer region (ISR) of the rRNA operons in representatives of the *Escherichia coli* ECOR collection, *J. Mol. Evol.*, 47, 62, 1998.
84. Pérez Luz, S. et al., Variation of the ribosomal operon 16S-23S gene spacer region in representatives of *Salmonella enterica* subspecies, *J. Bacteriol.*, 180, 2144, 1998.

85. Maggi, R.G. et al., A *Bartonella vinsonii berkhoffii* typing scheme based upon 16S-23S ITS and Pap31 sequences from dog, coyote, gray fox, and human isolates, *Mol. Cell Probes*, 20, 128, 2006.
86. Liu, Y. et al., PCR detection of *Klebsiella pneumoniae* in infant formula based on 16S-23S internal transcribed spacer, *Int. J. Food Microbiol.*, 125, 230, 2008.
87. Wang, M. et al., Analysis of the 16S-23S rRNA gene internal transcribed spacer region in *Klebsiella* species, *J. Clin. Microbiol.*, 46, 3555, 2008.
88. Mollet, C., Drancourt, M., and Raoult, D., *rpoB* sequence analysis as a novel bases for bacterial identification, *Mol. Microbiol.*, 26,1005, 1997.
89. Moreira, B.M. et al., *Klebsiella*. In Dongyou Liu (Editor), *Molecular Detection of Foodborne Pathogens*, Chapter 28, Taylor & Francis CRC Press, FL, 2009.
90. Walckenaer, E. et al., Genetic and biochemical characterization of the chromosomal class A β-lactamases of *Raoultella* (formerly *Klebsiella*) *planticola* and *Raoultella ornithinolytica*, *Antimicrob. Agents Chemother.*, 48, 305, 2004.
91. Ahmadian, A., Ehn, M., and Hober, S., Pyrosequencing: history, biochemistry and future, *Clin. Chim. Acta*, 363, 83, 2006.
92. Clarke, S.C., Pyrosequencing: Nucleotide sequencing technology with bacterial genotyping applications, *Expert Rev. Mol. Diagn.*, 5, 947, 2005.
93. Cristea-Fernstrom, M. et al., Pyrosequencing of a short hypervariable 16S rDNA fragment for the identification of nontuberculous mycobacteria-a comparison with conventional 16S rDNA sequencing and phenotyping, *APMIS*, 115, 1252, 2007.
94. Deyde, V.M. et al., Pyrosequencing as a tool to detect molecular markers of resistance to neuraminidase inhibitors in seasonal influenza A viruses, *Antiviral Res.*, 81, 16, 2009.
95. Grahn, N. et al., Identification of mixed bacterial DNA contamination in broad-range PCR amplification of 16S rDNA V1 and V3 variable regions by pyrosequencing of cloned amplicons, *FEMS Microbiol. Lett.*, 219, 87, 2003.
96. Haanpera, et al., Typing of SHV extended-spectrum β-lactamases by pyrosequencing in *Klebsiella pneumoniae* strains with chromosomal SHV β-lactamase, *Antimicrob. Agents Chemother.*, 52, 2632, 2008.
97. Jonasson, J., Olofsson, M., and Monstein, H.J., Classification, identification and subtyping of bacteria based on pyrosequencing and signature matching of 16S rDNA fragments, *APMIS*, 110, 263, 2002.
98. Unemo, M. et al., Molecular characterisation of *Neisseria gonorrhoeae* identifies transmission and resistance of one ciprofloxacin resistant strain, *APMIS*, 115, 231, 2007.
99. Søgaard, M. et al., Peptide nucleic acid fluorescence in situ hybridization for rapid detection of *Klebsiella pneumoniae* from positive blood cultures, *J. Med. Microbiol.*, 56, 914, 2007.
100. Kurupati, P. et al., Rapid detection of *Klebsiella pneumoniae* from blood culture bottles by real-time PCR, *J. Clin. Microbiol.*, 42, 1337, 2004.
101. Mancini, N. et al., Molecular diagnosis of sepsis in neutropenic patients with haematological malignancies, *J. Med. Microbiol.*, 57, 601, 2008.
102. Espy, M.J. et al., Real-time PCR in clinical microbiology: applications for routine laboratory testing, *Clin. Microbiol. Rev.*, 19, 165, 2006.
103. Monis, P.T., Giglio, S., and Saint, C.P., Comparison of SYTO9 and SYBR Green I for real-time polymerase chain reaction and investigation of the effect of dye concentration on ampliWcation and DNA melting curve analysis, *Anal. Biochem.*, 340, 24, 2005.
104. Heid, C.A. et al., Real time quantitative PCR, *Gen. Res.*, 6, 986, 1996.
105. Piatek, A.S. et al., Molecular beacon sequence analysis for detecting drug resistance in *Mycobacterium tuberculosis*, *Nat. Biotechnol.*, 16, 359, 1998.
106. Chen, X., and Kwok, P.Y., Homogeneous genotyping assays for single nucleotide polymorphisms with fluorescence resonance energy transfer detection, *Genet. Anal.*, 14, 157, 1999.
107. Ruzin, A., Immermann, F.W., and Bradford, P.A., Real-time PCR and statistical analyses of *acrAB* and *ramA* expression in clinical isolates of *Klebsiella pneumoniae*, *Antimicrob. Agents Chemother.*, 52, 3430, 2008.
108. Tenover, F.C. et al., Interpreting chromosomal DNA restriction patterns produced by pulsed-field gel electrophoresis: Criteria for bacterial strain typing, *J. Clin. Microbiol.*, 33, 2233, 1995.
109. Tenover, F.C., Arbeit, R.D., and Goering, R.V., How to select and interpret molecular strain typing methods for epidemiological studies of bacterial infections: a review for healthcare epidemiologists. Molecular Typing Working Group of the Society for Healthcare Epidemiology of America, *Infect. Control. Hosp. Epidemiol.*, 18, 426, 1997.
110. Gaillot, O. et al., Nosocomial outbreak of *Klebsiella pneumoniae* producing SHV-5 extended-spectrum β-lactamase, originating from a contaminated ultrasonography coupling gel, *J. Clin. Microbiol.*, 36, 1357, 1998.
111. Gori, A. et al., Comparison of pulsed-field gel electrophoresis and randomly amplified DNA polymorphism analysis for typing extended-spectrum-β-lactamase-producing *Klebsiella pneumoniae*, *J. Clin. Microbiol.*, 34, 2448, 1996.
112. Gazouli, M. et al., Study of an outbreak of cefoxitin-resistant *Klebsiella pneumoniae* in a general hospital, *J. Clin. Microbiol.*, 35, 508, 1997.
113. Cartelle, M. et al., Risk factors for colonization and infection in a hospital outbreak caused by a strain of *Klebsiella pneumoniae* with reduced susceptibility to expanded-spectrum cephalosporins, *J. Clin. Microbiol.*, 42, 4242, 2004.
114. Diancourt, L. et al., Multilocus sequence typing of *Klebsiella pneumoniae* nosocomial isolates, *J. Clin. Microbiol.*, 43, 4178, 2005.
115. Paulin-Curlee, G.G. et al., Genetic diversity of mastitis-associated *Klebsiella pneumoniae* in dairy cows, *J. Dairy Sci.*, 90, 3681, 2007.
116. Hindiyeh, M. et al., Rapid detection of bla_{KPC} carbapenemase genes by real-time PCR, *J. Clin. Microbiol.*, 46, 2879, 2008.
117. Westh, H. et al., Multiplex real-time PCR and blood culture for identification of bloodstream pathogens in patients with suspected sepsis, *Clin. Microbiol. Infect.*, 15, 544, 2009.
118. Bustin, S.A., Absolute quantification of mRNA using real-time reverse transcription polymerase chain reaction assays, *J. Mol. Endocrinol.*, 25, 169, 2000.
119. Bistué, A.J.C.S. et al., *Klebsiella pneumoniae* multiresistance plasmid pMET1: Similarity with the *Yersinia pestis* plasmid pCRY and integrative conjugative elements, *Proc. Natl. Acad. Sci.*, 3, e1800, 2008.
120. Rice, L.B., Federal funding for the study of antimicrobial resistance in nosocomial pathogens: No ESKAPE, *J. Infect. Dis.*, 197, 1079, 2008.

79 *Legionella*

Dianna J. Bopp, Kimberlee A. Musser, and Elizabeth J. Nazarian

CONTENTS

79.1 INTRODUCTION

Legionella species were first described in 1943 and 1947 using procedures developed for the isolation of *Ricksettsia*.[1] However, it wasn't until 1976 that this bacteria was described as a significant pathogen in humans following an outbreak of respiratory illness at an American Legion Conference in a Philadelphia hotel,[2,3] although *Legionella* was not determined to be the causative agent until almost a year later. More recently, there is evidence that cases of legionellosis are on the rise[4] and an estimated 8000–18,000 people are hospitalized resulting in a 5%–30% fatality rate.[5] To date, a significant amount of research, new diagnostic methods, and surveillance activities have been described that have impacted the current investigation of *Legionella* disease. This chapter describes all the diagnostic approaches, from culture methods to the advanced molecular methods and protocols used for both clinical specimens and environmental samples.

79.1.1 *Legionella* Genus

The family *Legionellaceae* is composed of a single genus, *Legionella*. These bacteria are characterized as gram-negative rods that are obligate aerobes and have nutritionally fastidious growth requirements including the need for L-cysteine. Currently, there are 52 validated species and 71 serogroups within the genus *Legionella* (Table 79.1) that have been characterized between 1979 and 2007 (http://www.bacterio.cict.fr/l/legionella.html). Of these recognized species, *Legionella pneumophila* is believed to cause 90% of disease,[6] however as many diagnostic tests are specific to this species, the contribution of other *Legionella* species may be underestimated and underreported. Currently, 24 *Legionella* species are described as occasional human pathogens, including: *L. anisa*, *L. birminghamensis*, *L. bozemanae*, *L. cincinnatiensis*, *L. dumoffii* *L. feeleii*. *L. gormanii*, *L. hackeliae*, *L. jordanis*, *L. lansingensis*, *L. longbeachae*, *L. lytica*, *L. maceachernii*, *L. micdadei*, *L. oakridgensis*, *L. parisiensis*, *L. quinlivanii*, *L. rubrilucens*, *L. sainthelensi*, *L. tucsonensis*, *L. wadsworthii*, *L. waltersii*, and *L. worsleiensis*.[7–19]

79.1.2 Human Disease

While the bacteria within the genus *Legionella* are responsible for an acute, self-limiting, non-pneumonic illness known as Pontiac fever, a far more significant public-health issue is the potentially fatal pneumonic form of legionellosis, Legionnaires' disease. Symptoms of legionellosis may include fever, nonproductive cough, headache, myalgias, rigors, dyspnea, diarrhea, and delirium.[20] In its clinical manifestations, a pneumonia caused by *Legionella* is not always distinguishable from pneumonias of other bacterial etiologies[21]; this lack of specific symptoms can contribute to delay in isolation and identification of the causal agent. Diagnosis is highly dependent on performing the appropriate testing when the patient may be at risk for this disease.[6]

Mortality rates associated with Legionnaires' disease are influenced by immunosuppression, underlying disease, severity of pneumonia, whether antibiotic therapy is administered and how quickly, and possibly whether the disease is community acquired or hospital acquired.[22] For example, community-acquired *Legionella* has a mortality rate of about 10% but may be as high as 27% if antibiotics are not received.[23,24]

TABLE 79.1
Fifty-Five *Legionella* Species and Subspecies with Standing in Nomenclature

L. adelaidensis	*L. lytica*
L. anisa	*L. maceachernii*
L. beliardensis	*L. micdadei*
L. birminghamensis	*L. moravica*
L. bozemanae	*L. nautarum*
L. brunensis	*L. oakridgensis*
L. busanensis	*L. parisiensis*
L. cherrii	*L. pittsburghensi*
L. cincinnatiensis	*L. pneumophila*
L. drancourtii	*L. pneumophila subsp. Fraseri*
L. drozanskii	*L. pneumophila subsp. pascullei*
L. dumoffii	*L. pneumophila subsp. pneumophila*
L. erythra	*L. quateirensis*
L. fairfieldensis	*L. quinlivanii*
L. fallonii	*L. rowbothamii*
L. feeleii	*L. rubrilucens*
L. geestiana	*L. sainthelensi*
L. gormanii	*L. santicrucis*
L. gratiana	*L. shakespearei*
L. gresilensis	*L. spiritensis*
L. hackeliae	*L. steigerwaltii*
L. impletisoli	*L. taurinensis*
L. israelensis	*L. tucsonensis*
L. jamestowniensis	*L. wadsworthii*
L. jordanis	*L. waltersii*
L. lansingensis	*L. worsleiensis*
L. londiniensis	*L. yabuuchiae*
L. longbeachae	

However, the highest mortality rates (80%) are found in immunocompromised patients receiving no appropriate antibiotic therapy and those untreated with pneumonia complicated by adult respiratory distress syndrome (>80%).[22]

79.1.3 Microbial Ecology

As natural inhabitants of aquatic environments, legionellae are ubiquitous.[25] These bacteria have been isolated frequently from freshwater and saltwater all over the world, and potting soils have also been implicated.[26] *Legionella* can live in fresh water environments as a parasite of amoebae[27] and within biofilms.[28] It has also been demonstrated that *Legionella* aerosolized from a manmade water system can enter the human lung and cause severe pneumonia because of its ability to replicate within alveolar macrophages.[29] Engineered water systems are particularly problematic because they can serve as an amplifier for *L. pneumophila* by providing suitable conditions for growth and multiplication.[30] Some recent studies have shown that as many as 30%–60% of the cooling towers tested contained *L. pneumophila*.[31,32]

Many cases of legionellosis are eventually traced to these manmade aquatic environments where the water temperature is higher than ambient temperature. Thermally altered aquatic environments can shift the balance between protozoa and bacteria, resulting in rapid multiplication of legionellae, translating into human disease. Legionellosis is a disease that has emerged in the last half of the twentieth century because of human alteration of the environment. It is believed that if legionellae remained only in their natural state, they would be an extremely rare cause of human disease, as natural freshwater environments have not been implicated as reservoirs of outbreaks.[6]

Their presence can be particularly problematic when found in the water distribution systems of health care facilities. Of particular concern is the risk of nosocomial and community-acquired infection via aerosolization of water from contaminated water supplies, which could potentially infect the immunocompromised, the elderly, and individuals with underlying respiratory conditions.

79.1.4 Diagnosis

(i) Culture Isolation Culture is still considered the gold standard for diagnosis of Legionnaire's Disease, even though the lack of sensitivity and timeliness are recognized weaknesses. The techniques for isolation of *Legionella* have not changed significantly in recent years, but the addition of molecular tests has been helpful in obtaining a rapid diagnosis and in directing culture efforts toward those specimens with a higher probability of yielding a positive result.[33] *Legionella* do not colonize humans and, therefore, detection of *Legionella* in respiratory tract specimens is always significant and the specificity of culture is 100%. Culture methods can detect all cultivable species of *Legionella*, although it is important to use both selective and nonselective media because some species are susceptible to the antibiotics in some media formulations.[34] Expectorated sputum and lower respiratory tract specimens are most suitable for culture; however, *Legionella* have been isolated from a variety of other specimens. The yield is often best if areas of purulence have been selected, although specimens should not be rejected if they contain few polymorphonuclear leukocytes.[35] If overgrowth is a problem, specimens can be pretreated with acid (HCl–KCl, pH 2.2 for 4 min)[36] or heat (50°C for 30 min)[37] to lower competing bacterial flora. The standard media for the culture of *Legionella* is buffered charcoal yeast extract with α-ketoglutarate (BCYEα) and BCYEα with added antibiotics (polymyxin B, anisomycin, and cefamandole; or polymyxin B, anisomycin, and vancomycin). Culture plates are incubated in a humidified atmosphere at temperatures of 35°C–37°C and observed for at least 10 days. *Legionella* colonies do not appear for 2–3 days, and this characteristic is useful in separating *Legionella* from more rapidly growing bacteria. *Legionella* colonies are best observed using a dissecting microscope, have a grainy appearance, and frequently have a bluish or greenish tint. Suspect colonies are subcultured to BCYEα and BCYEα without L-cysteine (blood agar plates may be used for this purpose) to determine the growth requirement for L-cysteine. These colonies are Gram

stained to confirm that they are slender, gram-negative rods. Further identification following colony and cell morphology and growth requirement for L-cysteine often involves the use of specific antisera [direct fluorescent antibody (DFA) testing or agglutination]. Commercial assays have been reported to be useful for the detection of *L. pneumophila* and isolates of other *Legionella* species[38] and may be useful for detection in water, although the sensitivity may not be high enough for this purpose.[39]

Investigation of cases of legionellosis often requires the testing of environmental samples (potable water, cooling towers, biofilms) if a potential source is suspected. A variety of concentration (filtration, centrifugation) and pretreatment methods (acid, heat) should be employed to optimize isolation. More selective media with added bromocresol purple, bromothymol blue, glycine, vancomycin, and polymyxin B is useful for environmental samples in addition to BCYEα.

(ii) Immunological Detection **DFA.** Detection of *Legionella* in respiratory tract specimens by DFA was one of the first diagnostic tests that offered a rapid result as part of *Legionella* testing. However, it requires expertise to obtain optimal results and has been shown to have low sensitivity.[40–42] Cross-reactivity with non-*Legionella* have also been described, and for these reasons it is not recommended for routine use on respiratory tract specimens.[42,43]

Serology. Serological tests were among the first diagnostic tests for *Legionella* and the first to be standardized.[44] Although serology can be useful for epidemiological studies or cases that do not have an established etiology, it does not usually provide information early enough to aid in treatment. Disadvantages of this method include that seroconversion can take a number of weeks[45,46]; not all serogroups are detected,[47] not all cases seroconvert,[40] and a significant percentage of well persons demonstrate antibodies to *Legionella*.[48–50]

Urinary Antigen. The detection of antigen in urine has become a commonly used laboratory test[51–53] and offers an early diagnosis that may be important in ensuring that patients receive appropriate antimicrobial therapy. It has the obvious advantage of ease of collection as well as rapid results. In addition, it's use with other specimen types such as serum and respiratory specimens has been assessed.[54] Both enzyme immunoassay (EIA) and, more recently, immunochromatography (ICT) tests have been shown to be sensitive and specific. ICT offers a simpler methodology and more rapid results. Several ICT assays have demonstrated comparable results to EIA (Rapid U),[55,56] (SAS),[57] although not all have proven adequate for routine use (SD Bioline).[58] A shortcoming of the urinary antigen test is that it detects primarily *L. pneumophila* serogroup 1 with some potential for detecting the other species and serogroups less well. In addition, the sensitivity is lower both for milder infections, due to the lower antigen burden,[52] and the for nosocomial infections, due to the subtypes usually found in these patients.[59] The sensitivity for subtypes that are more often the cause of travel and community-acquired cases is higher; however, these cases may be milder and therefore be underrecognized.[52] Thus, reliance on the urinary antigen assay alone for diagnosis will underestimate the true burden of illness and the diversity of the causative agents.

(iii) Nucleic Acid Detection Nucleic acid amplification tests (NAAT) for *Legionella* spp. are being used in conjunction with or replacing phenotypic identification methods. These methods are used to detect *Legionella* DNA from a variety of sources: respiratory secretions, tissue, urine, serum, and environmental water. The first molecular methods employed used PCR with visualization of DNA amplicons by ethidium bromide staining in agarose gels, reverse dot blot hybridization, or Southern blotting with a digoxigenin-labeled probe.[29,40] The potential for rapid results, the ability of legionellae to exist in a viable but unculturable state, the ability to identify species other than *Legionella pneumophila*, the known serological cross-reactivity, and the limited availability of antibodies to newer strains[60] have all contributed to the probability of nucleic acid detection as a promising choice for the identification of *Legionella* spp.

Conventional PCR. With conventional PCR, oligonucleotide primers are designed to conserved regions within a gene or other DNA specific target for the organism to be identified. PCR amplification is visualized on an agarose gel resulting in a presumptive identification. Confirmatory testing can involve southern blotting or nucleic acid sequence analysis for a definitive identification. Several genes are described in the literature for *Legionella* identification. The 16S rRNA and *mip* genes[29] are widely used for the identification of *Legionella* species and *Legionella pneumophila* specifically. Success has also been found with the 5S rRNA gene.[40,61] These targets have been shown to have an inability to allow species determination of legionellae without further nucleic acid sequence analysis.

Real-Time PCR. In the forefront of NAAT is detection by real-time PCR. Real-time PCR chemistries allow for the detection of PCR amplification during the early phases of a reaction in a closed tube system. This provides a distinct advantage over traditional PCR detection, which uses agarose gels for the detection of PCR amplification at the endpoint of the PCR reaction. Real-time PCR minimizes the postamplification processing required for a definitive identification in conventional PCR, and therefore reduces the turnaround time, required training, and the possibility of laboratory contamination with another PCR reaction. Real-time instrumentation has also improved to provide higher levels of multiplexing, which allows for the amplification and detection of two or more targets in a single reaction.[62] Several multiplex real-time PCR assays have been published for the simultaneous detection of the *Legionella* genus and specifically *Legionella pneumophila*.[33,41,63–65] Real-time PCR assays have also been described that are specific for *Legionella pneumophila*.[66–68]

A disadvantage of real-time PCR with respect to *Legionella* identification is the inability to differentiate the species of *Legionella* or serogroup of *Legionella pneumophila* present. Most assays only provide identification of the genus with the exception of assays specific for *Legionella pneumophila*. Species identification is typically accomplished by nucleic acid sequencing of the 16S rRNA, *mip*, *dnaJ*, or *rpoB* genes.[69] The ability to serogroup *Legionella pneumophila* is also a challenge with molecular testing due to the homology of the targeted genes among the different serogroups. The *dnaJ* gene has been proposed as a potential gene to differentiate between *Legionella* subspecies as well as between *Legionella pneumophila* serogroups.[69] Other methods, such as randomly amplified polymorphic DNA (RAPD) analysis and PCR-based "serotyping," have been investigated for this purpose.[70,71]

Targets. Several genes have been described to identify the *Legionella* genus and specifically *Legionella pneumophila* by both conventional and real-time PCR. These genes include the 16S rRNA,[41,64,66,72–75] 23S rRNA,[33] 5S rRNA,[76] the 23S-5S rRNA gene spacer region,[65] and the *mip* gene.[33,41,61,64,66–68,73,77,78] The 16S, 5S, and 23S rRNA genes are essential for the survival of all bacteria and are usually present in multiple copies. The genes encode ribosomal RNA transcribed from a single operon. These genes contain highly conserved regions and unique areas of DNA sequence (hypervariable regions). The conserved regions allow primers to be designed to large subsets of bacteria and when the nucleic acid of the unique areas undergoes nucleic acid sequencing, allows the species identification of the bacteria. These genes are typically used for *Legionella* genus identification or species determination if sequencing methods are used. The macrophage infectivity potentiator, *mip* gene, encodes a 240 kDa surface protein (Mip) that is required for efficient infection of phagocytic cells[6] is often used for specific detection of *Legionella pneumophila*. Table 79.2 provides a list of PCR assays utilized for *Legionella* detection that have been published to date.

Inhibition. The assessment of PCR inhibition is essential to determine that a negative sample is truly negative and not negative due to an inhibitory substance in the specimen itself or introduced from the reagents or materials used for DNA isolation. Inhibitory components of specimens include heme from blood, and rust or dirt in environmental water samples, among others.[33] Inhibitory components that may be introduced through the DNA isolation procedures include excess KCl, NaCl, ionic detergents, ethanol, isopropanol, and phenol. The direct binding of these agents to single-stranded or double-stranded DNA can prevent amplification and facilitate copurification of the inhibitor and DNA.[79] It is important that DNA isolation procedures are utilized that have efficacy for removing inhibitors.

Inhibition of PCR with respect to *Legionella* has been assessed by different methods. Spiking a known low concentration of *Legionella pneumophila* cells directly to the mastermix[33] or into negative samples[41] can be an effective way to assess inhibition. The addition of an internal control to each sample that coamplifies with the target of interest has also been used effectively.[64,66]

Subtyping. Subtyping of isolates is essential to establish the genetic relatedness between outbreak associated patient isolates and to determine the environmental source of the outbreak. *Legionella* is often found in aquatic environments, but it is only a subset of these that is commonly found to cause infection in humans.[80] Therefore, the presence of *Legionella* in an aquatic environment is not sufficient to implicate a source as the cause of an outbreak; a specific subtype found in both patients with legionellosis and the potential source is necessary. Subtyping has also been used to demonstrate the persistence of strains of *Legionella* over time and to demonstrate the variety of strains that can be present in aquatic environments that can serve as sources of exposure.[81–83] Many techniques have been employed for this purpose: monoclonal antibody subtyping, ribotyping, pulsed-field gel electrophoresis (PFGE), amplified fragment length polymorphism (AFLP), consensus sequence based typing (SBT), and variable number tandem repeats analysis (VNTR-MLVA). PFGE is highly discriminatory[81,83] but requires standardization in order to compare subtypes, and is complex to perform, although improvements continue to be described.[84] Recently, sequence-based typing has been standardized by the European Working Group for *Legionella* Infections (EWGLI) and interlaboratory reproducibility assessment has been evaluated.[85] MLVA is another technique that is fast, easy, and has the potential for automation and comparison between laboratories.[86,87] Optimally, the subtyping technique or combination of techniques employed should show a high degree of discrimination, easy and economical to perform, and show a high degree of standardization to allow comparison of isolates over time and among different facilities.

New Technologies. In the last several years, additional diagnostics have been reported and suggested for a possible role in *Legionella* investigations. One such report has suggested the role of analyzing the host genes as a way to determine susceptibility to particular pathogens. This approach suggests evaluating several important genes to determine if a particular host will be permissive to *Legionella* infection.[88] A newer method growing in utility over the past decade is the microarray. This method shows promise in increasing the detection of *Legionella* infection, often not assessed by culture. A novel DNA microarray designed to detect bacterial pathogens in patients with chronic obstructive pulmonary disease directly from sputum has been recently described. This study found the microarray to offer the advantage of

TABLE 79.2
Molecular Assays Developed for *Legionella* Detection

Molecular Method	Targeted Gene	Reference
Real-time PCR	23S rRNA	33
	16S rRNA	41,65,67,74,75
	Mip	33,41,65,67,68,78
	23S–5S rRNA	64,66
Conventional PCR	16S rRNA	76
	5S rRNA	77
	Mip	62,79
DNA sequence analysis	16S rRNA	33,71
	Mip	71,74
	dnaJ	71

both unlimited target detection and simultaneous detection of organisms in clinical specimens with complex flora.[89] Future work of interest will include the addition of new targets and a larger clinical evaluation. In addition, reports of the use of multiplex reverse transcription-PCR assay with manual (enzyme hybridization) or automated (electronic microarray) detection,[90] as well as biochip technology using such technology as surface-enhanced Raman scattering (SERS) have also been described.[91]

79.2 METHODS

79.2.1 Sample Preparation

Clinical Specimens. *Legionella* DNA has been recovered from a variety of specimen types. Acceptable specimens for the isolation of *Legionella* DNA are expectorated sputum and other lower respiratory specimens (tracheal aspirates and bronchoalveolar lavages), lung biopsies, urine, and serum. The quality of the specimen is important for efficient isolation of DNA. Some specimens require additional sample preparation procedures before DNA isolation can be performed. Due to the mucolytic nature of some respiratory specimens, pretreatment with a liquefying agent is necessary. This can involve treatment by enzymatic (proteinase K), chemical (dithiothreitol, Sputolysin),[33,64] or mechanical (MagNA Lyser, Roche)[77,92] methods. Bronchoalveolar lavage and urine specimens are typically concentrated by centrifugation with the resulting pellet resuspended in a volume of PCR buffer or tris-EDTA.[75,93] After the appropriate sample preparation procedures are performed, DNA may be isolated.

Environmental Specimens. Environmental water samples are another widely tested specimen for *Legionella* for both surveillance of public water systems, and for outbreak and remediation investigations. Specimen preparation prior to DNA isolation involves concentration by centrifugation and filtration methods. In one method, a representative volume of water is concentrated by centrifugation and the resulting pellet is resuspended in PCR buffer or tris-EDTA.[33] In filtration preparations, specimens are filtered through polycarbonate filters. The filters are placed in a volume of sterile water and sonicated.[94] Commercially available kits have also been developed utilizing immunomagnetic separation technology. These incorporate the use of a magnet and high-affinity antibody-coated beads to capture and concentrate *Legionella* from environmental water samples.[95]

DNA Extraction. Efficient isolation of *Legionella* DNA is essential for its detection with molecular techniques. Early extraction methods often employed the use of organic methods such as extraction by phenol-chloroform-isoamyl alcohol and precipitation with absolute alcohol.[61,93] Commercially available nucleic acid isolation and purification kits are now widely used for the extraction of legionellae from clinical specimens and environmental samples.[33,40,41,64,66,75,93] Several manufacturer's (Qiagen and Roche Applied Sciences) provide kits that use spin columns where nucleic acids bind to silica particles while contaminants pass through. Wash steps are added to remove PCR inhibitors and proteins leaving behind purified nucleic acid. A salt-precipitation protocol for total nucleic acid extraction is also available (Epicentre Biotechnologies) that can be used for the isolation of legionellae DNA. Automated systems are also available for total nucleic acid extraction and include steps that bind DNA to magnetic particles and then wash the DNA from the particles leaving them behind (NucliSENS® easyMAG® bioMerieux; MagNA Pure LC, Roche Applied Science). As legionellae are ubiquitous, extreme care must be taken particularly when using commercially available DNA isolation kits. Reports have been published describing contamination of spin columns with *Legionella* DNA.[64,96]

79.2.2 Detection Procedures

(i) Real-Time PCR Detection Nazarian et al.[33] described a real-time multiplex TaqMan® PCR for the detection of *Legionella pnemophila* and differentiation from other *Legionella* spp. in clinical and environmental samples. Primers and probe targeting the 23S rRNA gene were used for the specific detection of the *Legionella* genus: forward primer Leg23SF (5′-ccc atg aag aaa gtt gaa-3′); reverse primer Leg23SR (5′-aca atc agc caa tta gta cga gtt agc-3′); probe Lsp23SP VIC®-TAMRA (5′-tcc ac acct cgc cta tca acg tcg tag t-3′); and primers and probe targeting the *mip* gene were used for the specific detection of *Legionella pneumophila*: forward primer LmipF (5′-aaa ggc atg caa gac gct atg-3′); reverse primer LmipR (5′-gaa act tgt taa gaa cgt ctt tca ttt g-3′); and probe LmipP FAM-TAMRA (5′-tgg cgc tca att ggc ttt aac cga-3′).

1. Perform nucleic acid extraction using the MasterPure™ Complete DNA Purification Kit (Epicentre Biotechnologies) according to the manufacturer's instructions for both clinical and environmental samples.
2. Prepare PCR reaction mix (25 μL) containing 5 μL of DNA extract, 1× LightCycler FastStart DNA Master Hybridization Probe Mix (Roche Applied Science), 4.0 mM of $MgCl_2$, 900 nM of each primer, and 250 nM of each probe.
3. Include with each run a no template control (NTC), which consists of reaction mix and water, a negative extraction control, and a positive amplification control.
4. To assess for inhibition of PCR, prepare an additional PCR reaction mix as described earlier with the addition of 5 μL of the positive control for each sample tested. Negative samples with an inhibition test threshold cycle (Ct) value within 3 Cts of the control would not be considered inhibitory.
5. Perform thermal cycling, fluorescent data collection, and data analysis on ABI 7500 Real-time PCR Instrument (Applied Biosystems) using SDS 2.1 sequence analysis software according to the manufacturer's instructions.

6. The thermal cycling profile includes an initial incubation at 95°C for 10 min, followed by 40 cycles of 95°C for 15 s and 60°C for 1 min.

(ii) Conventional PCR Detection Conventional PCR can be utilized in order to determine the species of *Legionella*. Primers cited by Nazarian et al.[33] are used for the detection of *Legionella* species using PCR of the 16S rRNA gene with sequencing confirmation: forward primer LEG 225 (5′-aag att agc ctg cgt ccg at-3′) and reverse primer LEG 858 (5′-gtc aac tta tcg cgt ttg ct-3′), resulting in a 634 bp product.

1. Prepare a 1 McFarland suspension of a *Legionella* that has been streaked for purity and grown for 24 h on BCYEα at 37°C.
2. Heat treat the suspension in a screw top microcentrifuge tube at 95°C for 20 min in a heat block.
3. Prepare PCR reaction mix (50 μL) containing 10 μL of heat treated bacterial suspension, 1 μM of each primer, 200 μM of each dNTP, 3.0 mM of $MgCl_2$, 1× PCR Buffer II, and 2.5 U of AmpliTaq Gold DNA polymerase (Applied Biosystems).
4. Subject the mix to one hold at 95°C for 9 min followed by 40 cycles of 95°C for 30 s, 62°C for 1 min, and 72°C for 1 min, followed by a hold at 72°C for 7 min.
5. When the cycling is completed, electrophorese the product (634 bp) on a 2.0% agarose gel stained with ethidium bromide and visualize under UV illumination.
6. If product is present, prepare for sequencing: combine 15 μL PCR product to 5 μL ExoSAP-IT (USB) and incubate at 37°C for 15 min, followed by an inactivation at 80°C for 15 min.
7. Perform sequencing with primer Leg 225 (5′-aag att agc ctg cgt ccg at-3′) using BigDye Chemistry on a Sequencer (Applied Biosystems). Sequence analysis can be performed by a BLAST search of the GenBank database or other proprietary software such as MicroSeq (Applied Biosystems).

(iii) Genotyping The PFGE technique separates large DNA fragments and results in banding patterns that allow for discrimination between strains.

Procedure

1. Prepare DNA from *Legionella* cells grown for 24 h on BCYEα at 37°C.
2. Prepare a suspension in cell suspension buffer (100 mM Tris, 100 mM EDTA pH 8.0) to 0.5 on a Dade Microscan turbidity meter.
3. Add 20 μL of proteinase K to 400 μL of the suspension and add 400 μL of 1% SeaKem Gold in TE Buffer (10 mM Tris, 1 mM EDTA, 1% sodium dodecyl sulfate) at 50°C, gently mix, and dispense into a plug mold.
4. Remove solidified plug to 5 mL lysis solution (50 mM Tris, 50 mM EDTA pH 8.0, 1% Sarcosyl) with 25 μL proteinase K added and incubate in a water bath with rotation for 2 h at 50°C.
5. Wash plugs 4× with 10–15 mL sterile distilled water and 4× with 10–15 mL of TE Buffer (10 mM Tris, 1 mM EDTA pH8.0) at 50°C.
6. Restrict a small plug slice with endonuclease *Sfi* I (New England Biolabs) and incubate for 2–4 h at 50°C; include size controls in lanes to bracket isolates.
7. Separate the fragments in a 1.0% agarose gel on a clamped homogeneous electric field apparatus (CHEF Mapper, BioRad Laboratories) with the initial pulse time of 7.0 s increased linearly to 70 s over 21 h.
8. Stain with ethidium bromide, destain in distilled water, visualize, and capture using a GelDoc XR (BioRad Laboratories).
9. Analyze using Bionumerics software (Applied Maths, Belgium).

79.3 CONCLUSION AND FUTURE PERSPECTIVE

Public Health. A rapid, sensitive, and specific method for the detection of *Legionella* is clearly important to identify infected individuals and to expedite cleanup of contaminated water systems so as to prevent additional cases of infection. State reference laboratories are often able to perform *Legionella* testing for numerous hospitals and facilities in their states each year in the context of outbreak and remediation investigations. As one such example, in New York State, a real-time PCR assay is utilized to detect both the *Legionella* genus and, specifically, *Legionella pneumophila* serogroups 1–16 in a single reaction, from both environmental and clinical specimens.[33] In addition, a testing algorithm that combines this assay with culture and 16S rRNA gene sequence analysis is available (Figure 79.1).

This public-health strategy for testing *Legionella* samples includes both culture and real-time PCR analysis. Real-time PCR analysis and culture can be set up on day 1. PCR results are analyzed, and Ct values are determined for each sample tested. Culture is then performed to determine whether viable *Legionella* is present. If an isolate is confirmed as *Legionella* by culture and DFA, and the patient and environmental isolates both exist from the same outbreak, PFGE is performed to determine molecular epidemiology links. If an isolate is not obtained from a given sample, even after repeat culture attempts, PCR and sequence analysis can be performed in order to identify the *Legionella* species or potentially serogroup present.

It has been suggested that as many as 35% of *Legionella pneumophila*-positive samples are only detected by PCR.[33] Although PCR is not the accepted standard diagnostic tool, PCR detection of these samples was noteworthy as culture was not ideal for these samples. When available, isolates from culture are important for outbreak investigation, so that subtyping techniques like PFGE can be performed. While culture has historically been considered the standard test for

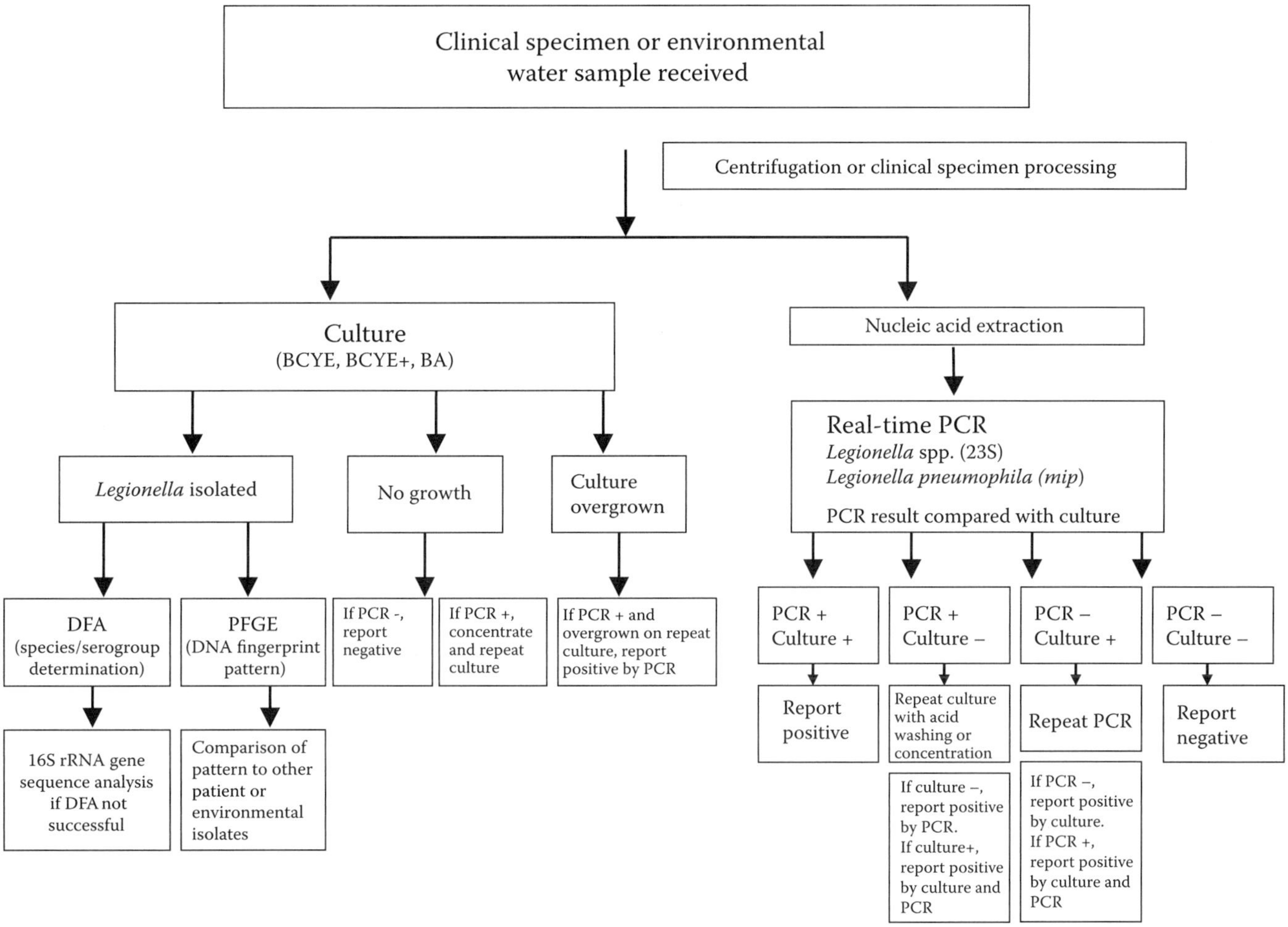

FIGURE 79.1 Algorithm for *Legionella* testing of clinical specimens and environmental water. (From Nazarian E.J. et al., *Diag. Microbiol. Infect. Dis.*, 62, 125, 2008. With permission from ELS.)

defining cases of legionellosis, substantial technical challenges exist for the laboratorian, including low viability due to antibiotic treatment of the patient, availability of specimens and degradation due to poor transport conditions. In addition, delays in specimen transport time can occur, given that *Legionella* is not always the initial diagnosis especially if patients are urinary antigen negative. Any of these factors can reduce analytical productivity and can delay assignment of genus and species to a sample. Respiratory specimens are particularly susceptible to these problems, and frequently culture is not even possible; this is the case for lung tissue specimens that are paraffin embedded or formalin fixed. Under these circumstances, real-time PCR becomes an indispensable tool. The combination of real-time PCR with culture methods, DFA, and PFGE in a vigorously tested algorithm provides the best approach to the testing of both clinical specimens and environmental samples for a public-health laboratory.

Clinical Microbiology. In clinical microbiology, routine culture for *Legionella* may not be cost effective due to small numbers of test requests and the difficulty in keeping specialized media in date and passing quality control. However, urinary antigen testing is advised for the rapid diagnosis of the most common legionellosis disease. However, as noted, this may only detect a percentage of the cases for which *L. pneumophila* serogroup 1 is the causative agent. Currently, no FDA-approved molecular test is available, so laboratories that would like to provide molecular analysis such as real-time PCR need to evaluate or develop assays and validate these assays for use in their laboratory. It has been proposed that the urinary antigen test, PCR, and culture were conducive to the diagnosis of *Legionella* infection, and ordering of all three tests was recommended to ensure a definite and rapid diagnosis of *Legionella*.[97]

Future of the Field. In the next decade it would be anticipated that more cases of *Legionella* will continue to be reported due to the potential increases in diagnostic testing methods, and as the number of immunocompromised and elderly in the United States rises. These testing methods are more specific and sensitive for *Legionella* and include newer diagnostics for *Legionella* species in panels of microorganisms in one test. *Legionella* also poses a difficult infection control issue because it can exist for prolonged periods in water and biofilms and can resist chlorination and complete disinfection. In addition, standards and guidelines for the sampling, diagnostic testing, and interpretation are important to consider for the future and these continue to become available in the clinical and public-health settings. Finally, infection with *Legionella* may decrease in the future if resources are directed at improved disinfection and design of water systems.

ACKNOWLEDGMENTS

The authors would like to acknowledge Nellie Dumas for a critical reading of this material.

REFERENCES

1. McDade, J.E. et al., Legionnaires' disease: Isolation of a bacterium and demonstration of its role in other respiratory disease, *N. Engl. J. Med.*, 297, 1197, 1977.
2. Brenner, D.J., Steigerwalt, A.G., and McDade J.E., Classification of the Legionnaires' disease bacterium: *Legionella pneumophila*, genus novum, species nova, of the family *Legionellaceae*, familia nova, *Ann. Intern. Med.*, 90, 656, 1979.
3. Fraser, D.W. et al., Legionnaires' disease: description of an epidemic of pneumonia, *N. Engl. J. Med.*, 297, 1189, 1977.
4. Carrarala, J., and Garcia-Vidal, C., An update on *Legionella*, *Curr. Opin. Infect. Dis.*, 23, 2010.
5. Smith, P. et al., Surveillance for travel-associated Legionnaires disease—United States, 2005–2006, *MMWR*, 56, 1261, 2007.
6. Fields, B.S., Benson, R.F., and Besser, R.E., *Legionella* and Legionnaires' Disease: 25 years of investigation, *Clin. Microbiol. Rev.*, 15, 506, 2002.
7. Berger, P. et al., Amoeba-associated microorganisms and diagnosis of nosocomial pneumonia, *Emerg. Infect. Dis.*, 12, 248, 2006.
8. Wilkinson, H.W., Cruce, D.D., and Broome, C.V., Validation of *Legionella pneumophila* indirect immunofluorescence assay with epidemic sera, *J. Clin. Microbiol.*, 13, 139, 1981.
9. Miller, M.L., Hayden, R., and Gaur, A. *Legionella bozemanii* pulmonary abscess in a pediatric allogeneic stem cell transplant recipient, *Pediatr. Infect. Dis. J.*, 26, 760, 2007.
10. Thacker, W.L. et al., *Legionella jordanis* isolated from a patient with fatal pneumonia, *J. Clin. Microbiol.*, 26, 1400, 1988.
11. Yu, V.L. et al., Distribution of *Legionella* species and serogroups isolated by culture in patients with sporadic community-acquired legionellosis: an international collaborative survey, *J. Infect. Dis.*, 186, 127, 2002.
12. David, C. et al., Legionellosis due to *Legionella gormanii* fortuitously found in a man with chronic lymphocytic leukemia: A case study and literature review, *Med. Mal. Infect.*, 36, 172, 2006.
13. Thacker, W.L. et al., *Legionella lansingensis* sp. nov. isolated from a patient with pneumonia and underlying chronic lymphocytic leukemia, *J. Clin. Microbiol.*, 30, 2398,1992.
14. De Gheldre, Y. et al., Rapid identification of clinically relevant *Legionella* spp. by analysis of transfer DNA intergenic spacer length polymorphism, *J. Clin. Microbiol.*, 39, 162, 2001.
15. Thacker, W.L. et al., *Legionella tucsonensis* sp. nov. isolated from a renal transplant recipient, *J. Clin. Microbiol.*, 27, 1831, 1989.
16. Chee, C.E., and Baddour, L.M., *Legionella maceachernii* soft tissue infection, *Am. J. Med. Sci.*, 334, 410, 2007.
17. Korvick, J.A., Yu, V.L., and Fang, G.D., *Legionella* species as hospital-acquired respiratory pathogens, *Semin. Respir. Infect.*, 2, 34, 1987.
18. Benson, R.F. et al., *Legionella sainthelensi* serogroup 2 isolated from patients with pneumonia, *Res. Microbiol.*, 141, 453, 1990.
19. Edelstein, P.H. et al., *Legionella wadsworthii* species nova: A cause of human pneumonia, *Ann. Intern. Med.*, 97, 809, 1982.
20. Tsai, T. F. et al., Legionnaires' disease: Clinical features of the epidemic in Philadelphia, *Ann. Intern. Med.*, 90, 509, 1979.
21. Mulazimoglu, L., and Yu, V., Can Legionnaires disease be diagnosed by clinical criteria? *Chest*, 120, 1049, 2001.
22. Edelstein, P., Antimicrobial chemotherapy for Legionnaires' disease: A review, *Clin. Infect. Dis.*, 21, S265, 1995.
23. Falco, V. et al., A cause of severe community-acquired pneumonia, *Chest*, 100, 1007, 1991.
24. Fiumefreddo, R. et al., Clinical predictors for *Legionella* in patients presenting with community-acquired pneumonia to the emergency department, *BMC Pulm. Med.*, 9, 4, 2009.
25. Fliermans, C.B. et al., Ecological distribution of *Legionella pneumophila*, *Appl. Environ. Microbiol.*, 41, 9, 1981.
26. Velonakis, E.N. et al., First isolation of *Legionella* species, including *L. pneumophila* serogroup 1, in Greek potting soils: Possible importance for public health, *Clin. Microbiol. Infect.*, 16, 763, 2010, Epub ahead of print, 2009.
27. Rowbotham, T.J., Preliminary report on the pathogenicity of *Legionella pneumophila* for freshwater and soil amoebae, *J. Clin. Pathol.*, 33, 1179, 1980.
28. Murga, R. et al., The role of biofilms in the survival of *Legionella pneumophila* in a model potable water system, *Microbiol.*, 147, 3121, 2001.
29. Fields, B.S., The molecular ecology of legionellae, *Trends Microbiol.*, 4, 286, 1996.
30. Atlas, R.M., *Legionella*: From environmental habitats to disease pathology, detection and control, *Environ. Microbiol.*, 1, 283, 1999.
31. Bentham, R.H., Routine sampling and the control of *Legionella* spp. in cooling tower water systems, *Curr. Microbiol.* 41, 271, 2000.
32. Garbe, P.L. et al., Nosocomial Legionnaires' disease: Epidemiologic demonstration of cooling towers as a source, *JAMA*, 254, 521, 1985.
33. Nazarian, E.J. et al., Design and implementation of a protocol for the detection of *Legionella* in clinical and environmental samples, *Diag. Microbiol. Infect. Dis.*, 62, 125, 2008.
34. Lee, T.C. et al., Growth of 28 *Legionella species* on selective culture media: A comparative study, *J. Clin. Microbiol.*, 31, 2764, 1993.
35. Ingram, J.G., and Plouffe, J.F., Danger of sputum purulence screens in culture of *Legionella* species, *J. Clin. Microbiol.*, 32, 209, 1994.
36. Buesching, W.J., Brust, R.A., and Ayers, L.W., Enhanced primary isolation of *Legionella pneumophila* from clinical specimens by low-pH treatment. J. *Clin. Microbiol.*, 17, 1153, 1983.
37. Edelstein, P.H., Snitzer, J.B., and Brudge, J.A., Enhancement of recovery of *Legionella pneumohila* from contaminated respiratory tract specimens by heat, *J. Clin. Microbiol.*, 16, 1061, 1982.
38. Helbig, J.H. et al., Evaluation of the Duopath *Legionella* lateral flow assay for identification of *Legionella pneumophila* and *Legionella* species culture isolates, *Appl. Environ. Microbiol.*, 72, 4489, 2006.

39. Koide, M., et al., Comparative evaluation of Duopath *Legionella* lateral flow assay against the conventional culture method using *Legionella pneumophila* and *Legionella anisa* strains, *Jpn. J. Infect. Dis.*, 60(4), 214, 2007.
40. Lindsay, D.S.J. et al., Laboratory diagnosis of Legionnaires' disease due to *Legionella pneumophila* serogroup 1: Comparison of phenotypic and genotypic methods, *J. Med. Microbiol.*, 53, 183, 2004.
41. Hayden, R.T. et al., Direct detection of *Legionellla* species from bronchoalveolar lavage and open lung biopsy specimens: comparison of light cycler PCR, in situ hybridization, direct fluorescence antigen detection, and culture, *J. Clin. Micobiol.*, 39, 2618, 2001.
42. She, R.C. et al., Limited applicability of direct fluorescent-antibody testing for *Bordetella* sp. and *Legionella* sp. specimens for the clinical microbiology laboratory, *J. Clin. Microbiol.*, 45, 2212, 2007.
43. Murdoch, D.R., Diagnosis of *Legionella* infection, *Clin. Infect. Dis.*, 36, 64, 2003.
44. Wilkinson, H.W., Cruce, D.D., and Broome, C.V., Validation of *Legionella pneumophila* indirect immunofluorescence assay with epidemic sera, *J. Clin. Microbiol.*, 13, 139, 1981.
45. Den Boer, J.W., and Yzerman, E.P.F., Diagnosis of *Legionella* infection in Legionnaires' disease, *Eur. J. Clin., Microbiol. Infect. Dis.*, 23, 871, 2004.
46. Plouffe, J.F. et al., Reevaluation of the definition of Legionnaire's disease: Use of the urinary antigen assay, *Clin. Infect. Dis.*, 20, 1286, 1995.
47. Bernander, S. et al., Serologic study of an outbreak of Legionnaires' disease: Variation of sensitivity associated with the subgroup of *Legionella pneumophila* sg1 antigen used and evidence of concurrent reactivity to other atypical pneumonia agents. In Cianciotto, N.P., et al., *Legionella: State of the Art 30 Years After Its Recognition*, pp. 63–67. ASM Press, Washington, DC, 2006.
48. Lee, H.K. et al., Prevalence of antibodies in response to *Legionella* species, analysis of a healthy population from Jeollanam-do province, *Korean J. Microbiol.*, 46, 160, 2008.
49. Borella, P. et al., Prevalence of anti-*legionella* antibodies among Italian hospital workers, *J. Hosp. Infection*, 69:148, 2008.
50. Rudbeck, M., Molbak, K., and Uldum, S., High prevalence of antibodies to *Legionella* spp. in Danish blood donors. A study in areas with high and average incidence of Legionnaires' disease, *Epidemiol. Infect.*, 136, 257, 2008.
51. Benin, A.L., Benson, R.F., and Besser, R.E., Trends in legionnaires disease, 1980–1998: declining mortality and new patterns of diagnosis, *Clin. Infect. Dis.*, 35, 1039, 2002.
52. von Baum, H. et al., Community-acquired *Legionella* pneumonia: New insights from the German Competence Network for Community Acquired Pneumonia, *Clin. Infect. Dis.*, 46, 1356, 2008.
53. Joseph, C.A., and Ricketts, K.D., Legionnaires' Disease in Europe 1995–2004: A ten-year review. In Cianciotto, N.P., et al., *Legionella: State of the Art 30 Years After Its Recognition*, p. 89, SM Press, Washington, DC, 2006.
54. Koide, M. et al., Detection of *Legionella* species in clinical samples: Comparison of polymerase chain reaction and urinary antigen detection kits, *Infection*, 5, 264, 2006.
55. Blanco, S. et al., Detection of *Legionella* antigen in non-concentrated and concentrated urine samples by a new immunochromatographic assay, *Eur. J. Clin. Microbiol. Infect. Dis.*, 27, 1249, 2008.
56. Diederen, B.M.W., and Peeters, M.F., Evaluation of rapid U *Legionella* plus test, a new immunochromatographic assay for detection of *Legionella pneumophila* serogroup 1 antigen in urine, *Eur. J. Clin. Microbiol. Infect. Dis.*, 25, 733, 2006.
57. Diederen, B.M.W., and Peeters, M.F., Evaluation of the SAS *Legionella* test, a new immunochrommatograhpic assay for detection of *Legionella pneumophila* serogroup 1 antigen in urine, *Clin. Microbiol. Infect.*, 13, 86, 2006.
58. Diederen, B.M.W., and Peeters, M.F., Evaluation of two new immunochromatographic assays (Rapid U *Legionella* Antigen Test and SD Bioline *Legionella* Antigen Test) for detection of *Legionella pneumophila* serogroup 1 antigen in urine, *J. Clin. Microbiol.*, 44, 2991, 2006.
59. Helbig, J. et al., Clinical utility of urinary antigen detection for diagnosis of community-acquired, travel-associated, and nosocomial Legionnaires' disease, *J. Clin. Microbiol.*, 41, 838, 2003.
60. Edelstein, P.H., *Legionella*. In Murray, P.R., Baron, E.J., Jorgensen, J.H., Landry, M.L., and Pfaller, M.A. (Editors), *Manual of Clinical Microbiology*, 9th ed., pp. 835–849, ASM Press, Washington, DC, 2007.
61. Lindsay, D.S.J., Abraham, W.H., and Fallon, R.J., Detection of *mip* gene by PCR for diagnosis of Legionnaires' disease, *J. Clin. Microbiol.*, 32, 3068, 1994.
62. Cirino, N.M., Musser, K.A., and Egan, C., Multiplex diagnostic platforms for detection of biothreat agents, *Expert Rev. Mol. Diagn.*, 4, 841, 2004.
63. Herpers, B.L. et al., Real-time PCR assay targets the 23S-5S spacer for direct detection and differentiation of *Legionella* spp. and *Legionella pneumophila*, *J. Clin. Microbiol.*, 41, 4815, 2003.
64. Templeton, K.E. et al., Development and clinical evaluation of an internally controlled, single-tube multiplex real-time PCR assay for detection of *Legionella pneumophila* and other *Legionella* species, *J. Clin. Microbiol.*, 41, 4016, 2003.
65. Yang, G. et al., Dual detection of *Legionella pneumophila* and *Legionella* species by real-time PCR targeting the 23S 5S rRNA gene spacer region, *Clin. Microbiol. Infect.*, 16, 255, 2010, [Epub ahead of print].
66. Wellinghausen, N., Frost, C., and Marre, R., Detection of legionellae in hospital water samples by quantitative real-time lightcycler PCR, *Appl. Environ. Microbiol.*, 67, 3985, 2001.
67. Ballard, A.L. et al., Detection of *Legionella pneumophila* using a real-time PCR hybridization assay, *J. Clin. Microbiol.*, 38, 4215, 2000.
68. Diederen, B.M.W. et al., Detection and quantification of *Legionella pneumophila* DNA in serum: Case reports and review of the literature, *J. Med. Microbiol.*, 55, 639, 2006.
69. Yong, S.F, Goh, F.N., and Ngeow, Y.F., *Legionella* species and serogroups in Malaysian water cooling towers: Identification by latex agglutination and PCR-DNA sequencing of isolates. *J. Water Health*, 8, 92, 2010.
70. Liu, H. et al., Use of the *dnaJ* gene for the detection and identification of all *Legionella pneumophila* serogroups and description of the primers used to detect 16S rDNA gene sequences of major members of the genus *Legionella*, *Microbiol. Immunol.*, 47, 859, 2003.
71. Perola, O. et al., Persistent *Legionella pneumophila* colonization of a hospital water supply: Efficacy of control methods and a molecular epidemiological analysis, *APMIS*. 113, 45, 2005.
72. Thürmer, A. et al., PCR-based "serotyping" of *Legionella pneumophila*, *J. Med. Microbiol.*, 58, 588, 2009.

73. Stolhaug, A., and Bergh, K., Identification and differentiation of *Legionella pneumophila* and *Legionella* spp. with real-time pcr targeting the 16S rRNA gene and species identification by *mip* sequencing, *Appl. Environ. Microbiol.*, 72, 6394, 2006.
74. Reischl, U. et al., Direct detection and differentiation of *Legionella* spp. and *Legionella pneumophila* in clinical specimens by dual-color real-time PCR and melting curve analysis, *J. Clin. Microbiol.*, 40, 3814, 2002.
75. Cloud, J.L. et al., Detection of *Legionella* species in respiratory specimens using PCR with sequencing confirmation, *J. Clin. Microbiol.*, 38, 1709, 2000.
76. Murdoch, D.R. et al., Use of the polymerase chain reaction to detect *Legionella* DNA in urine and serum samples from patients with pneumonia, *Clin. Infect. Dis.*, 23, 475, 1996.
77. Diederen, B.M.W. et al., Utility of real-time PCR for diagnosis of Legionnaires' disease in routine clinical practice, *J. Clin. Microbiol.*, 46, 671, 2008.
78. Ratcliff, R.M. et al., Sequence-based classification scheme for the genus *Legionella* targeting the mip gene, *J. Clin. Microbiol.*, 36, 1560, 1998.
79. Bessetti, J., *An Introduction to PCR Inhibitors, Profiles in DNA*, Promega Corporation, 2007.
80. Harrison, T.G. et al., Distribution of *Legionella pneumophila* serogroups, monoclonal antibody subgroups and DNA sequence types in recent clinical and environmental isolates from England and Wales (2000–2008), *Eur. J. Clin. Microbiol. Infect. Dis.*, 28, 781, 2009.
81. Oberdorfer, K., Mussigbrodt, G., and Wendt, C., Genetic diversity of *Legionella pneumophila* in hospital water systems, *Int. J. Hyg. Environ. Health*, 211, 172, 2008.
82. Garcia-Nunez, M. et al., Persistence of *Legionella* in hospital water supplies and nosocomial Legionnaires' disease, *FEMS Immunol. Med. Microbiol.*, 52, 202, 2008.
83. Sanchez, I. et al., Genotypic variability and persistence of *Legionella pneumophila* PFGE patterns in 34 cooling towers from two different areas, *Environ, Microbiol.*, 10, 395, 2008.
84. Chang, B., Anemura-Maekawa, J., and Watanabe, H., An improved protocol for the preparation and restriction enzyme digestion of pulsed-field gel electrophoresis agarose plugs for the analysis of *Legionella* isolates, *Jpn. J. Infect. Dis.*, 62, 54, 2009.
85. Afshar, B. et al., External quality assessment of a DNA sequence-based scheme for epidemiological typing of *Legionella pneumophila* by an international network of laboratories, *J. Clin. Microbiol.*, 45, 3251, 2007.
86. Nederbragt, A.J., et al., Multiple-locus variable-number tandem repeat analysis of *Legionella pneumophila* using multi-colored capillary electrophoresis, *J. Mircobiol. Methods*, 73, 111, 2008.
87. Pourcel, C. et al., Identification of variable number tandem-repeat (VNTR) sequences in *Legionella pneumophila* and development of an optimized multiple-locus VNTR analysis typing scheme, *J. Clin. Micobiol.*, 45, 1190, 2007.
88. Tuite, A., and Gros, P., The impact of genomics on the analysis of host resistance to infectious disease, *Microbes Infect.*, 8, 1647, 2006.
89. Curran, T. et al., Development of a novel DNA microarray to detect bacterial pathogens in patients with chronic obstructive pulmonary disease (COPD), *J. Microbiol. Methods*, 80, 257, 2010, January 14. [Epub ahead of print], 2010.
90. Kumar, S. et al., Detection of 11 common viral and bacterial pathogens causing community-acquired pneumonia or sepsis in asymptomatic patients by using a multiplex reverse transcription-PCR assay with manual (enzyme hybridization) or automated (electronic microarray) detection, *J. Clin. Microbiol.*, 46, 3063–3072, 2008.
91. Grow, A.E. et al., New biochip technology for label-free detection of pathogens and their toxins, *J. Microbiol. Methods*, 53, 221, 2003.
92. Bencini, M.A. et al., A multicenter comparison of molecular methods for the detection of *Legionella* in sputum samples, *J. Clin. Microbiol.*, 10, 505, 2007.
93. Rantakokko-Jalava, K., and Jalava, J., Development of conventional and real-time PCR assays for detection of *Legionella* DNA in respiratory specimens, *J. Clin. Microbiol.*, 39, 2904, 2001.
94. Joly, P. et al., Quantitative real-time *Legionella* PCR for environmental water samples: Data interpretation, *Appl. Environ. Microbiol.*, 72, 2801, 2006.
95. Invitrogen, Speed detection of *Legionella pneumophila* in every water sample: Dynabeads MAX *Legionella. Mag. Sep. Tech.*, 2008.
96. van der Zee, A. et al., Qiagen DNA extraction kits for sample preparation for *Legionella* PCR are not suitable for diagnostic purposes, *J. Clin. Microbiol.*, 40, 1126, 2002.
97. Jespersen, S. et al., The relationship between diagnostic tests and case characteristics in Legionnaires' disease, *Scand. J. Infect. Dis.*, 41, 425, 2009.

80 Moraxella

Suzanne J.C. Verhaegh and John P. Hays

CONTENTS

80.1 INTRODUCTION

The genus *Moraxella* belongs to the family *Moraxellaceae*, which includes the closely related genera *Acinetobacter* and *Psychrobacter*. The *Moraxella* genus itself currently contains 15 different species, including *M. lacunata*, *M. atlantae*, *M. boevrei*, *M. bovis*, *M. canis*, *M. caprae*, *M. catarrhalis*, *M. caviae*, *M. cuniculi*, *M. equi*, *M. lincolnii*, *M. nonliquefaciens*, *M. osloensis*, *M. ovis*, and *M. saccharolytica*.[1] Recently, *M. pluranimalium* has been described as a new species of the genus *Moraxella*, after its isolation from sheep and pigs.[2] Although all of these 15 species have been classified within the genus *Moraxella*, the classification of the *Moraxellaceae* family is still not definitive but continuously evolving. For example, molecular systematic studies (of which 16S ribosomal RNA (rRNA) gene sequencing has long been the "gold standard") has led to the reclassification of *Riemerella anatipestifer*, *Psychrobacter phenylpyruvicus*, and *Oligella urethralis*, which were formally known as *Moraxella anatipestifer*, *Moraxella phenylpyruvica*, and *Moraxella urethralis*, respectively.[1,3]

Of the 15 *Moraxella* species, *M. caprae*, *M. caviae*, *M. cuniculi*, *M. equi*, *M. lincolnii*, *M. ovis*, *M. saccharolytica*, *M. boevrei*, and *M. bovis* are only associated with diseases in animals, and are thus clinically not relevant. Occasionally, however, some *Moraxella* species are associated with human disease. Specifically, *M. lacunata* has been associated with chronic conjunctivitis, arthritis, and endocarditis in humans,[4–8] while *M. nonliquefaciens* has been known to cause bacterial infection (botryomycosis) and inflammation (endophthalmitis).[9–11] *M. atlantae* is an unusual and only rarely isolated species that has been associated with bacteremia in humans,[12,13] while *M. osloensis* rarely causes infections but has been associated with bacteremia and endophthalmitis.[14–16] Interestingly, *M. canis* is an upper-airway commensal bacterium in cats and dogs, and has been considered nonpathogenic for humans even though it has been isolated from humans in a few cases.[17,18] The single most important exception to this "*Moraxella*-animal disease" relationship is *M. catarrhalis*, an organism that is a strictly human pathogen and that is frequently associated with a range of human disease states.

80.1.1 Classification, Morphology, Biology, and Epidemiology

Classification. *M. catarrhalis* (formally known as *Branhamella catarrhalis*) has undergone several name changes in the past 100 years.[19,20] It was first described at the end of the nineteenth century, when it was named *Micrococcus catarrhalis*, which was later changed to *Neisseria catarrhalis* because of its similarity in phenotype and ecological niche to species of *Neisseria*.[21] In 1970, the bacterium was transferred to a new genus, *Branhamella*, based on differences between *N. catarrhalis* and the type species of the genus *Neisseria* and six other species of this genus.[19] In 1984, *Branhamella catarrhalis* was reassigned to the genus *Moraxella* as *Moraxella (Branhamella) catarrhalis*.[22] *M. catarrhalis* is now the most widely accepted and currently preferred name for this bacterial species.

Morphology. *M. catarrhalis* is an oxidase-positive, Gram-negative diplococcus that grows well on both blood and chocolate agars, as well as several selective agar media. Optimal growth is achieved at a temperature of 35°C and may be enhanced in an atmosphere of 5% carbon dioxide. Colonies are generally nonhemolytic, gray to white, opaque and smooth, and can be pushed along the agar surface without losing colonial integrity, the so-called hockey-puck sign[23,24] (Figure 80.1).

Selective agar media incorporating vancomycin, trimethoprim, amphotericin B, and acetazolamide have been described for the selective culture of *M. catarrhalis* from mixtures of commensal bacteria present in the human respiratory tract.[25] Biochemically, *M. catarrhalis* is asaccharolytic, meaning that it is unable to produce acid from glucose, maltose, sucrose, lactose, and fructose. Although the species fails to ferment carbohydrates, it is able to reduce nitrate and nitrite.[23] It is possible to distinguish *M. catarrhalis* from other closely related and commensal, oxidase-positive, Gram-negative cocci (i.e., *Neisseria* species) by its ability to hydrolyze tributyrin due to the production of butyrate esterase.[26] This is a useful and simple differential biochemical test that is relatively easy to set up in the routine microbiological laboratory. However, accurate *M. catarrhalis* species identification from clinical samples should ideally rely on a combination of biochemical and molecular testing protocols.

Biology and Epidemiology. *M. catarrhalis*, an exclusive human pathogen, is an important colonizer of the respiratory tract. Until recently, *M. catarrhalis* was thought to be a harmless commensal. However, it is now recognized as one of the more important bacterial pathogens associated with both upper and lower respiratory tract infections in humans. Colonization with *M. catarrhalis* begins very early in life with a high percentage of infants being colonized, but a much lower percentage of adults being colonized (1%–5%). In fact, an *M. catarrhalis* colonization peak is observed at 2 years of age, with both infants and children tending to acquire and eliminate a number of different strains.[27,28] Factors that have been particularly associated with an increased likelihood of *M. catarrhalis* colonization include the presence of siblings and day care center attendance.[28a] Further, studies have shown that frequent colonization with *M. catarrhalis* increases the risk of otitis media (OM) and that patterns of colonization are important determinants of otitis media–related disease.[24,29,30]

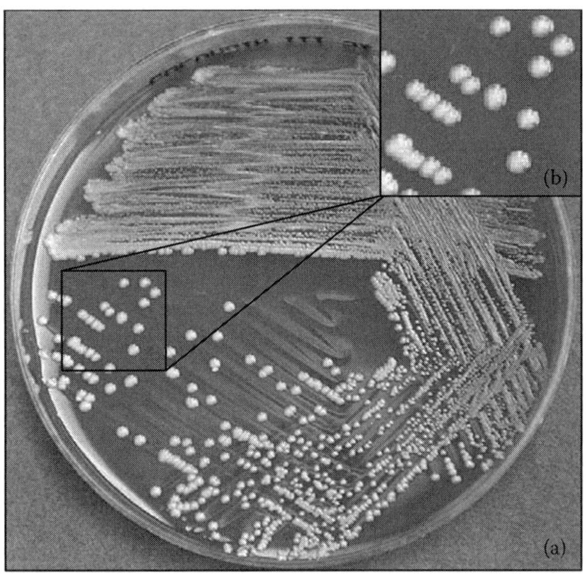

FIGURE 80.1 (a) Colonies of *M. catarrhalis* growing on blood agar plates containing 5% sheep blood and incubated at a temperature of 35°C in 5% CO_2 for 48 h. (b) Close-up view of individual *M. catarrhalis* colonies. These colonies may be pushed along the surface of the agar without losing integrity (hockey-puck sign).

Interestingly, a considerable variation in the colonization rate of healthy children attending day care centers has been observed between different geographic regions and time periods. For example, a study conducted in Cuba in the period of January to May 2000, showed that 65% of children up to 5 years old were colonized with *M. catarrhalis*, with the colonization rate reaching 87% in the 2–4-year age group.[31] In contrast, in the Czech Republic during the winter of 2004–2005 (December 2004–April 2005), 22% of children 3–6 years of age were colonized with *M. catarrhalis*, and in a Japanese study conducted in 1999, children aged 1 month to 5 years, 35% were found to be colonized.[32,33] In The Netherlands, a comparative study of 3–36-month-old children from 4 January through 9 March 1999, indicated a carriage rate for children attending day care centers versus controls (children not attending day care centers) of 80% and 48%, respectively.[34]

In contrast to colonization rates in infants and children, the rate of *M. catarrhalis* colonization in adults is relatively low. However, *M. catarrhalis* is relatively frequently isolated from adults with exacerbations of chronic obstructive pulmonary disease (COPD), suggesting a role for this bacterium in lower airway infections.[24]

80.1.2 Clinical Features

M. catarrhalis is associated with a range of human diseases, including acute, chronic, and chronic suppurative otitis media and exacerbations of COPD.

Acute Otitis Media (AOM). AOM involves inflammation of the middle ear, including the eardrum, the inner ear, and the Eustachian tube. Although AOM is generally considered a bacterial infection, a crucial role in the development of AOM for respiratory viruses has been reported in several epidemiologic studies.[35] Essentially, a respiratory viral infection may result in a prolonged dysfunction of the Eustachian tube, congestion of the respiratory mucosa, and alteration in middle ear pressure, leading to the accumulation of fluid within the middle ear. These abnormalities may then facilitate the invasion and proliferation of colonizing microbial pathogens into the middle ear. Despite the possible interaction between viruses and bacteria in the etiology of AOM, there is no indication that respiratory infections caused by a given viral species favor the growth of a certain bacterial pathogen in the middle ear fluid (MEF) more than another.[36]

Bacteria may be isolated from the MEF in approximately 70% of children with AOM and, depending on the geographic region, *S. pneumoniae* is generally the predominant species cultured, followed by *H. influenzae* and *M. catarrhalis*. However, this ranking is subject to some debate as there are still questions regarding which bacterial species plays the most important role in middle ear infection and therefore AOM.[37]

Chronic Otitis Media with Effusion (OME). OME represents a condition in which fluid within the middle ear space collects over time as a result of a negative pressure that builds up due to dysfunctioning of the Eustachian tube. There are several other predisposing etiologic factors that relate to the development of OME, including environmental factors such as climate, pollution, hygiene, nutrition, and standard of living. Another important factor involves allergies or an immune deficiency.[38] Further, it is well established that pathogenic bacteria may also contribute to OME. Bacteria have been cultured from 30% to 50% of chronic serous and mucoid middle ear effusions, where *H. influenzae* predominates, and to a lesser extent, *S. pneumoniae* and *M. catarrhalis*.[37,39]

Chronic Suppurative Otitis Media (CSOM). CSOM is one of the most common chronic diseases of childhood in many underdeveloped countries (especially among the poor) and in certain populations of the developed nations. CSOM is a stage of ear disease in which there is chronic infection of the middle ear cleft and in which a nonintact tympanic membrane (e.g., perforation or tympanostomy tube) are present, as well as the presence of pus draining out of the ear (otorrhea). CSOM is an active bacterial infection within the middle ear space that may last for several weeks or more. In contrast to AOM and OME, *Pseudomonas aeruginosa* is the most predominant species isolated, with a minor role for *M. catarrhalis*.[40,41]

Chronic Obstructive Pulmonary Disease (COPD). In contrast to the upper respiratory tract infections (URTI) associated with *M. catarrhalis* in children, *M. catarrhalis* in adults is mostly associated with lower respiratory tract infections (LRTI). COPD is a major LRTI and a considerable public health problem. It is predicted to become the fifth leading cause of death worldwide in 2020 and the prevalence of, and mortality from, this disease are currently increasing.[42] The definition of COPD given by the global initiative for chronic obstructive lung disease (GOLD; http://www.goldcopd.com) is now rapidly gaining general acceptance (i.e., "COPD is a disease state characterized by airflow limitation that is not fully reversible. The airflow limitation is usually both progressive and associated with an abnormal inflammatory response of the lungs to noxious particles or gases").

Several lines of evidence developed in the last decade have established that *M. catarrhalis* is associated with exacerbations of COPD. For example, nontypeable *H. influenzae*, *M. catarrhalis*, and *S. pneumoniae* are the three predominant bacterial species isolated from sputum in 40%–60% of acute exacerbations of COPD, with *M. catarrhalis* being isolated in 4%–21% of sputa.[43] The high prevalence of bacterial colonization of the airways observed in COPD is generally related to the degree of airflow obstruction and cigarette smoking.[44] However, more research is needed to test the hypothesis that bacterial colonization accelerates the progressive decline in lung function seen in COPD (the vicious circle hypothesis), although neutrophilic airway inflammation and systemic inflammation are more intense in the presence of bacteria.[43,45]

Other Respiratory Diseases. Another important URTI in children is sinusitis. In acute sinusitis, viral upper respiratory tract infections frequently precede bacterial super infection by *S. pneumoniae*, *H. influenzae*, and *M. catarrhalis*.[46] In approximately 29% of episodes of the common cold, one or more of the bacterial pathogens *S. pneumoniae*, *H. influenzae*, and *M. catarrhalis* are involved, and other sporadic cases of infections with *M. catarrhalis* include meningitis, urinary tract infection, and conjunctivitis.[47,48]

Treatment and Management. *M. catarrhalis* is known to produce a β-lactamase (BRO), which is an enzyme responsible for the resistance of *M. catarrhalis* to β-lactam antibiotics. Moreover, in the last 30 years major changes have occurred in the incidence of β-lactamase production in *M. catarrhalis* isolates. The first β-lactamase-producing isolate was detected in clinical specimens reported by Malmvall et al. in 1977, with a prevalence rate of 3.8%.[49] However, by 1980 75% of *M. catarrhalis* isolates from the United States produced β-lactamase.[50] Recent studies have shown that 30 years after the first discovery of β-lactamase producing strains, over 95% of the clinical isolates worldwide produce β-lactamase.[51]

All strains of *M. catarrhalis* are resistant to amoxicillin due to the production of BRO β-lactamase, but the discovery of the β-lactamase inhibitor clavulanic acid in the year 1972 and its coadministration with amoxicillin as amoxicillin/clavulanate still allows the continued use of β-lactam antibiotics to treat infections caused by *M. catarrhalis*. The efficacy of this combination has been demonstrated in a range of clinical trials, and, since 1981, amoxicillin/clavulanate has been registered for the treatment of URTIs and other diseases.[52] Although amoxicillin is recommended as the initial antibacterial agent of choice for most children, it is shown that recurrent OM occurs more often in children originally treated with amoxicillin.[53] Frequent use of antibiotics as therapy for AOM tends to result in a rise in antibiotic resistance and more evidence is becoming available regarding the limited clinical efficacy of antibiotics in AOM.[54]

The American Academy of Pediatrics published a guideline in 2004 that addressed the diagnosis and treatment of AOM. This guideline recommended the use of observation as a potential strategy for the treatment of AOM, although rates of antibiotic prescription for AOM still vary to a great extent internationally.[54,55]

An alternative to the use of antibiotics is vaccination. However, there is currently no licensed vaccine against

M. catarrhalis, and to date none of the putative vaccine candidates so far described (Table 80.1) have progressed to clinical trials. The challenge in identifying potential vaccine candidates for *M. catarrhalis* lies in identifying antigens that (i) are able to generate an appropriate immune response to prevent colonization/infection/disease and (ii) are conserved among global strains. A further major obstacle in vaccine development has been the absence of a satisfactory animal model for *M. catarrhalis*-mediated diseases, and consequently, problems associated with the identification and verification of correlates of protection against *M. catarrhalis*.[24]

On a further note, effective vaccines against viruses and bacteria involved in AOM could have a dramatic impact on the overall incidence of this disease. However, the potential use of vaccines against for example nontypeable *H. influenzae* and *S. pneumoniae* (Prevnar, Wyeth) may lead to a reduction in the incidence of these two pathogens in the etiology of AOM, possibly allowing *M. catarrhalis* to capitalize

TABLE 80.1
***Moraxella catarrhalis* Genes Used in PCR-Mediated *M. catarrhalis* Research Laboratories and Possibly Useful for Routine Diagnostic Detection**

Target Gene	Protein Function	Comment	Primer Sequence [5′–3′]	Reference
uspA1	Adhesin, binding of complement regulators, autotransporter	Laminin, fibronectin and CEACAM-binding domain. Phase-variable expression	RTF1–8 [cgttatgcactaaaagagcaggtc] RTB1–8 [gcatctgaccagcttagaccaatc]	56,57,58,59
uspA2	Adhesin, mediates serum resistance	Laminin, fibronectin, vitronectin and C3-binding domain. Phase-variable expression	RTF2–10 [gcatctgcggataccaagtttg] RTB2–10 [ttgagccatagccaccaagtgc]	56,59–61
uspA2H	Adhesin	Hybrid protein that most closely resembles UspA1 at its N-terminal and UspA2 at its C-terminal region	uspA2H-f [ggcagttttgggcagtttattg] uspA2H-r [ataaggctggaatagacc]	62
mid/hag	Adhesin, IgD binding	Phase-variable expression	mid-f [atgtcaacgatggcaatcaagagcc] mid-r [ccccaagcttaaagtgaaaacctgcaccaactgctgcc]	63–65
mhaC	Adhesin, transporter	–	P1 [agcgattcatggtggactgtcctt] P2 [actttggcagaaagaaaggcggtg]	66
mhaB1	Adhesin	–	P3 [aacgcatgggtttgtcctgtcctt] P4 [ttgggcgttgataggaatgcgttg]	66
mhaB2	Adhesin	–	P5 [ttaggctcgcattgcaca] P6 [acggtatcggtgtgggttagcttt]	66
mcaP	Adhesin, lipolytic activity (phospholipase B) autotransporter	Highly conserved at the amino acid level (98%–100% identity)	mcaP-f [cgcaataaagatcaccatgcttg] mcaP-r [cgggatcccgctgacacattgcattgataaa]	67,68
ompCD	Adhesin, mucin-binding activity	–	Mcat.ompCD.f [acgcactggcaagaagctaga] Mcat.ompCD.r [gacctgcaccaaccaagacat]	69,70
ompJ	May function as a porin	Found in all worldwide isolates screened so far (>200)	19kDres.f [ctaacgctgccatcagctat] 19kDres.r [gttgcattacggctggtaac]	71
copB	Involved in iron acquisition	–	McCOPB2F [ggcgtgcgtgttgaccgttttg] McCOPB2R [gtttggcaggcgataggcgacat]	69,72
copB	–	Real-time PCR for quantitative detection of *M. catarrhalis*. The fluorescent reporter dye at the 5′ end of the probe is hexachloro-6-carboxyfluorescein (Biosearch Technologies Inc.); the quencher at the 3′ end is a dark quencher called QSY-7 (Biosearch Technologies Inc.)	copB-f [gtgagtgccgctttacaacc] copB-r [tgtatcgcctgccaagacaa] Probe [tgcttttgcagctgttagccagcctaa]	73,74
LOS types A, B, and C	Invasion of human epithelial cells	Detoxified form is potential vaccine antigen	Pr 406 [caaaagaagacaaacaagcagc] Pr 408 [catcaaaaaccccccctacc] Pr 649 [atcctgctccaactgactttc]	75,76
16S rRNA	Ribosomal RNA	–	16S-8F [agagtttgatcctggttcag] 16S-556R [ctttacgcccatttaatccg]	69,77
16S rRNA	Ribosomal RNA	Multiplex PCR for detection of *H. influenzae* (Hi), *S. pneumoniae* (Sp), *M. catarrhalis* (Mc) and *A. otitidis* (Ao). A Cy-5-labeled primer is used as a common reverse primer (UR)	Hi [cgtattatcggaagatgaaagtgc] Sp [aaggtgcacttgcatcactac] Mc [cccataagccctgacgttac] Ao [ggggaagaacacggatagga] UR [ctacgcatttcaccgctacac]	78,79

on the available niche and become the main bacterial AOM pathogen of the future.

80.1.3 Pathogenesis of Infection

The recent recognition of *M. catarrhalis* as an important pathogen in both the upper and lower respiratory tract has resulted in an increased interest in both the bacterium's interaction with the human host and its antigenic composition.

Two distinct genetic lineages related to 16S rRNA type have been identified for *M. catarrhalis*, which differ phenotypically in their ability to resist the killing effect of human serum (seroresistant versus serosensitive) and in their ability to adhere to human epithelial cells.[77,80] Further, the seroresistant lineage of *M. catarrhalis* (16S rRNA type 1 lineage) is more virulent than 16S rRNA types 2 and 3 (lineage 2 isolates).[80]

There is increasing information regarding the mechanisms facilitating *M. catarrhalis* colonization and infection of the respiratory tract. In particular, bacterial adherence to the respiratory mucosa is an essential first step towards colonization of the human respiratory tract epithelium and the development of disease. Important adhesins responsible for the attachment of *M. catarrhalis* to host cells include the family of ubiquitous surface protein A (UspA) proteins UspA1 and UspA2, which have been extensively studied in *M. catarrhalis* and were initially described approximately 15 years ago.[81] The closely related UspA2H protein appears to be a hybrid protein that most closely resembles UspA1 at its N-terminal and UspA2 at its C-terminal region.[62] UspA1 and UspA2 acquire their adherence characteristics through multifunctional binding sites, which include domains that have the ability to attach to epithelial cells via cell-associated fibronectin, laminin, and vitronectin.[56] Further, UspA1 includes a critical binding site for carcinoembryonic antigen-related cell adhesion molecules (CEACAMs), which are expressed in various human tissues including respiratory epithelial cells.[82] The CEACAM-1 receptor-binding domain is located in the stalk region of UspA1 and consists of an extended, rod-like left-handed trimeric coiled-coil. The stalk region is capable of changing conformation after CEACAM-1 binding to allow a closer approach between *M. catarrhalis* and the host cell membrane.[83] The stalk region of UspA1 and UspA2 share multiple amino acid repeats and distinct sequence "motifs," a finding that suggests that the make-up of the UspA proteins might be interchangeable possibly under the influence of immune pressure. This extensive sharing of sequences between UspA1 and UspA2 does not include the N- and C-terminal regions of these proteins.[84,85] The UspA stalk region contains one or two copies of the NINNY motif [which contains the epitope for the protective monoclonal antibody (MAb) 17C7], and the CEACAM-binding motif (which contains the epitope for a separate anti-UspA1 antibody that also inhibits CEACAM binding).[85,86] The exchange of repeats and motifs in UspA proteins may allow *M. catarrhalis* populations to exhibit various combinations of binding properties, which could help define the virulence of individual strains.[85]

Another outer membrane protein (OMP) that is important for bacterial attachment is the *Moraxella* immunoglobulin D-binding protein (MID), also referred to as *hag* (human erythrocyte agglutinin), due to the ability of *M. catarrhalis* to agglutinate human erythrocytes and to bind immunoglobulin in a nonimmune-dependant manner.[63,64] The adhesive domain of MID is located in the 149 amino acid fragment $MID^{764-913}$ of *M. catarrhalis* strain Bc5 and promotes the attachment of bacteria to epithelial cells, whereas the IgD-binding domain is located within the 238 amino acid fragment $MID^{962-1200}$ of the same protein.[63,87] Further, antibodies directed against $MID^{764-913}$ effectively inhibit attachment of the high MID-expressing isolates to alveolar epithelial cells, and animals immunized with $MID^{764-913}$ clear *M. catarrhalis* much more efficiently when compared to mice immunized with bovine serum albumin (BSA).[88]

M. catarrhalis has developed several mechanisms to evade the effect of human complement, allowing it to evade the host innate immune response. Three biochemical pathways activate the complement system: the classical pathway, the alternative pathway, and the mannose-binding lectin (MBL) pathway. The classical pathway of the complement system is initiated by the binding of the complement component (C)-1 complex to antibodies bound to an antigen on the surface of a bacterial cell, whereas the alternative pathway is initiated by the covalent binding of a small amount of C3b to the surface of a microbe and is activated by cleavage of C3 in plasma. Activation of these systems leads to a cascade of protein deposition on the bacterial surface, resulting in formation of the membrane attack complex (MAC) and opsonization of the pathogen, followed by phagocytosis.[89,90]

M. catarrhalis has evolved mechanisms to interfere with complement pathway activation, including the binding of complement inhibitor C4-binding protein (C4BP) (classical pathway) and by noncovalently binding C3 (alternative pathway). The *M. catarrhalis* OMPs UspA1 and UspA2 bind C4BP and C3, and UspA2 interferes with the proper formation of the MAC porin complex that inserts into bacterial outer membranes, due to the stable binding of vitronectin by UspA2.[91–94]

Another example of the ability of *M. catarrhalis* to evade the host immune system is phase variation and antigenic variation.[95] Phase variation in general refers to a reversible switch between an on–off expressing phase, resulting in variation in the level of expression of one or more OMPs between individual bacteria of a clonal population. Antigenic variation refers to the expression of functionally conserved OMPs within a clonal population of bacteria that are antigenically distinct.[96] As OMPs are found on the bacterial cell surface and may be exposed to the immune system, antigenic variation influences the efficacy of antibody-mediated defense mechanisms against the respective OMPs and hence against the bacterium per se. Currently, *M. catarrhalis* OMPs UspA1, UspA2, and MID/Hag are known to undergo phase-variation, with antigenic variation reported in the target region of MAb 17C7 (a conserved UspA1 and UspA2 binding site), although this variation involves only minor amino acid changes.[57,60,65,97,98]

Whether antigenic variation is actually a virulence mechanism for *M. catarrhalis* is a subject of debate, as Meier et al. demonstrated that reduced or absent expression of UspA1, rather than antigenic variation, was responsible for nonreactivity with the MAb 17C7.[97]

Interestingly, *M. catarrhalis* releases outer membrane vesicles (OMVs), or "blebs" from the surface of the cell during growth in various environments, including liquid culture, solid culture, and during biofilm formation. Blebs are thought to contain some of the underlying periplasm, together with OMPs and lipopolysaccharides (LPS) of the outer membrane layer.[99] The OMVs that are secreted by *M. catarrhalis* carry UspA1 and UspA2, and have been shown to interfere with successful complement activation against *H. influenzae* in mixed infections. Thus, OMVs from *M. catarrhalis* may contribute to the pathogenicity of pure and mixed *M. catarrhalis* populations by binding and depleting complement component C3 in their immediate environment.[100]

Another important component of the innate immune system is the transmembrane toll-like receptors (TLRs). Toll-like receptors are pathogen recognition receptors (PRRs) that recognize pathogen-associated molecular patterns (PAMPs), small conserved molecular motifs found in microbes, for example, bacterial LPS. The specific inflammatory responses induced by *M. catarrhalis* may be mediated by activation of these PRRs, with the inflammatory immune response to *M. catarrhalis* in respiratory epithelial cells to being mainly dependent on TLR2.[101] This TLR2 inflammatory immune response is mediated by the cytokine interleukin (IL)-8, which has a critical role in regulating neutrophil and monocyte chemotaxis toward sites of infection. Further, the activation of protein kinase C (PKC) through nuclear factor (NF)-κB signaling pathways, which induces IL-8 release, has been found to be dependent on expression of UspA2.[102] In addition, IL-8 release is inhibited by UspA1-dependent interactions with CEACAM-1.[103]

A single highly homologous 32–amino acid sequence of *M. catarrhalis* UspA1 and UspA2 is also responsible for binding and neutralizing α1-antichymotrypsin. *M. catarrhalis* antichymotrypsin neutralization is a novel microbial virulence mechanism that may induce excessive inflammation, resulting in a more exposed extracellular matrix that may be beneficial for bacterial colonization.[104]

It is known that *M. catarrhalis* exists in biofilms and that it is able to prolong its survival through biofilm formation on mucosal surfaces in the nasopharynx. Biofilms are usually defined as microbial communities, surrounded by an extracellular polymeric substance (EPS) matrix, which adhere to biological or nonbiological surfaces.[105] Biofilm formation has been demonstrated for numerous pathogens and is clearly an important microbial survival strategy.[106] Several other pathogens associated with infections, including *H. influenzae* and *S. pneumoniae* in chronic otitis media, are also linked to biofilm formation. These bacteria benefit from this microbial community thanks to, for instance, the phenomenon of indirect pathogenicity where *S. pneumoniae* is protected from the action of certain β-lactam antibiotics in the presence of BRO β-lactamase positive *M. catarrhalis*, and where *M. catarrhalis* protects *H. influenzae* due to binding and depleting C3 by release of OMVs.[100,107,108]

Studies have shown that several *M. catarrhalis* genes are associated with biofilm formation—for example, *uspA1*, *uspA2*, *uspA2H*, and *hag*—although the exact role of each of the resultant proteins in biofilm formation is still open to question. Further, Wang et al. showed that growth of *M. catarrhalis* in a biofilm results in increased expression of genes encoding a nitrate reductase, a nitrite reductase, and a nitric oxide reductase, which can function not only in energy generation but also in resisting the innate immune response.[109]

80.1.4 Diagnosis

The diagnosis of *M. catarrhalis* is based on both conventional and molecular detection methods. The conventional approach of identifying *M. catarrhalis* is based on the use of selective culture media and biochemical tests, in addition to Gram staining (for colony morphology) (Figure 80.2). In contrast, the molecular approach is based on specific nucleic acid amplification methods or specific gene detection. In addition, several typing techniques are available for genotyping *M. catarrhalis* and for determining the presence of clonal relationships between *M. catarrhalis* isolates.

80.1.4.1 Phenotypic Identification

Selective Culture Media. A variety of selective media have been developed to differentiate *M. catarrhalis* from other respiratory tract pathogens. A semiselective agar that

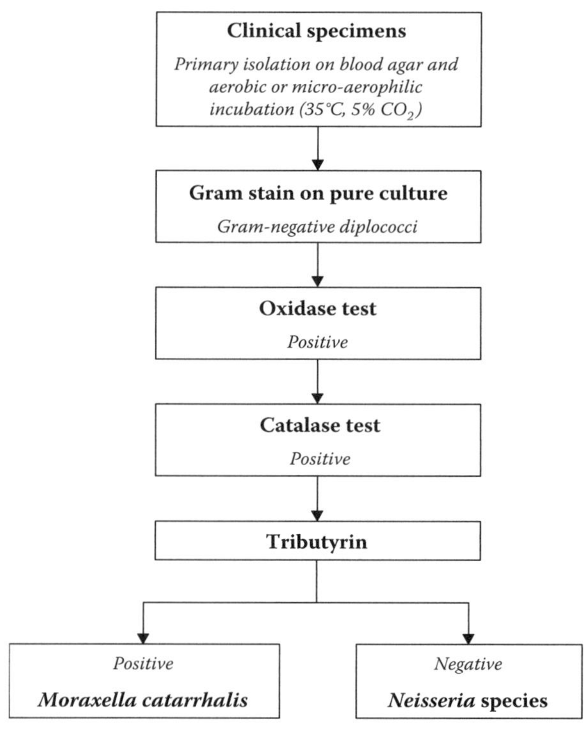

FIGURE 80.2 Flowchart showing the conventional techniques, used in the majority of clinical laboratories, for the identification of *Moraxella catarrhalis*.

contains vancomycin, trimethoprim, and amphotericin B has been reported to inhibit the growth of respiratory tract flora other than *M. catarrhalis*.[110] However, these antibiotics are known not to inhibit commensal *Neisseria* species, which share many phenotypic characteristics with *M. catarrhalis* and may also be found occupying the same human niche. Therefore, further differentiation of cultured colonies is necessary when using this agar. Another selective agar media incorporating acetazolamide (a synthetic sulfonamide carbonic anhydrase inhibitor) has been described for the selective culture of *M. catarrhalis* from mixtures of commensal bacteria present in the human respiratory tract. The addition of acetazolamide to the medium (which also includes the antibiotics vancomycin, trimethoprim and amphotericin B), increases the selectivity of the medium and inhibits the growth of *Neisseria* species.[25]

Another selective approach for *M. catarrhalis* is to incorporate vancomycin, trimethoprim, and amphotericin B to deoxyribonuclease (DNase) test agar. DNase test agar (with toluidine blue O) is used for the detection of deoxyribonuclease activity, a characteristic useful in aiding the identification of bacteria isolated from clinical specimens. DNase is an enzyme that breaks down DNA and may be used as one of the biochemical test for the differentiation of *M. catarrhalis* from other members of the families *Neisseriaceae* and *Moraxellaceae*. Only *Kingella denitrificans*, *M. caviae*, and *M. catarrhalis* are able to produce DNase within these bacterial families, whereas *K. denitrificans* produces a negative catalase reaction, in contrast with *M. caviae* and *M. catarrhalis*. Further, *M. caviae* is only associated with animal disease and is not clinically relevant, leaving *M. catarrhalis* as the only relevant species isolated from human clinical specimens to produce DNase from the *Neisseriaceae* and *Moraxellaceae* families.[111]

Chromogenic and Fluorogenic Enzyme-Substrate Tests. Commercial systems—for example, the indoxyl butyrate strip test (Butylase; Lab M Limited)—are available for the identification of *M. catarrhalis*. These systems use biochemical substrates that, on hydrolysis by the enzyme butyrate esterase, yield a colored end product that is either detected directly or via the addition of a detection reagent. Usually, the activity of the butyrate esterase enzyme is detected using either the chromogenic substrates tributyrin or indoxyl butyrate, or the fluorogenic substrate 4-methylumbelliferyl butyrate. The tributyrin disc test (ROSCO Diagnostica) distinguishes *M. catarrhalis* from other closely related and commensal, oxidase-positive, Gram-negative cocci, (i.e., *Neisseria* species), by the ability of *M. catarrhalis* to hydrolyze tributyrin via butyrate esterase.[26]

The possibility of rapid detection of bacterial enzymes (β-D-glucosidase, β-D-galactosidase, β-D-glucuronidase, *N*-acetyl-β-D-glucosaminidase, α-D-mannosidase) by 4-methylumbelliferyl (MUB) substrates was first reported in 1975.[112] This enzyme activity can be detected because the 4-methylumbelliferone moiety, released from the hydrolyzed substrate, is fluorescent when viewed at 366 nm using an intensive UV source. Vaneechoutte et al. described a test that used MUB (Sigma–Aldrich) as a substrate in the identification of *M. catarrhalis*.[113]

When the different test set-ups are compared, the indoxyl butyrate strip test appears to give results similar to those of the MUB test. However, tests based on the hydrolysis of indoxyl butyrate are quicker and do not rely on the availability of a UV light source. Both the indoxyl butyrate strip test and the MUB test only require one or two colonies of an overnight agar plate culture, which is in contrast to the tributyrin disc test, where a heavier inoculum of an overnight plate culture is required.[114] Further, the tributyrin disc test requires a longer incubation time of 4 h (up to 24 h), and this biochemical test, though popular, tends to be more expensive than the other tests.[26,115]

These three rapid biochemical detection methods are useful tools helping differentiate *M. catarrhalis* from *Neisseria* species.[26,113] However, other *Moraxella* species [e.g., *M. lacunata*, *M. nonliquefaciens,* and *M. osloensis* (occasionally associated with human disease)] may also generate positive reactions with indoxyl butyrate, MUB, and tributyrin, meaning that the interpretation of a positive indoxyl butyrate, MUB or tributyrin test as definitive proof of *M. catarrhalis* identification should be made only after reference to 2 other biological parameters, such as colony morphology and Gram stain (*M. catarrhalis* are Gram-negative diplococci, while *M. lacunata*, *M. nonliquefaciens,* and *M. osloensis* are Gram-negative bacilli), as well as other carbohydrate utilization tests.

Nowadays, integrated chromogenic enzyme-substrate detection systems have been introduced into many laboratories. An example of such a system is the VITEK® 2 *Neisseria/Haemophilus* (NH) Identification card (bioMérieux), which has replaced many of the standard conventional biochemical tests that were previously performed by hand.[116]

80.1.4.2 Molecular Identification

The ultimate goal of diagnostic microbiology is the rapid and accurate identification of bacteria in their "natural" environments. Molecular techniques such as PCR and 16S rRNA gene sequencing are helping revolutionize the rapid and sensitive detection, and specific identification, of many bacterial species, including *M. catarrhalis*.

PCR-Based Bacterial Detection. Molecular techniques, such as the PCR, detect bacterial DNA and do not rely on bacterial biochemistry. The advantage of these techniques, therefore, is their ability to detect metabolically inactive bacteria that may be present in culture-negative clinical specimens (e.g., in antibiotic treated middle-ear effusions). Research has shown that the use of PCR technology increases detection rates for one or more of the respiratory pathogens *S. pneumoniae*, *H. influenzae*, and *M. catarrhalis* from approximately 25% to nearly 80% in clinical specimens obtained from children suffering from OME.[117,118]

Currently, many research groups use standard singleplex PCR screening techniques to detect *M. catarrhalis*-specific genes in clinical specimens. These PCRs are usually targeted to genes coding for the OMPs of *M. catarrhalis*.[73] As an example, OMP J is an OMP of *M. catarrhalis* that exists in two major forms, OMP J1 and OMP J2, respectively.[71] These proteins are coded by the genes *ompJ1* and *ompJ2* (Table 80.1). Research studies using OMP J-specific PCR primers have shown that one of these two genes is present in all *M. catarrhalis* isolates tested to date (>200 worldwide isolates). Further, this PCR distinguishes between the two major 16S rRNA lineages of *M. catarrhalis* (lineage 1 = OMP J2, lineages 2 and 3 = OMP J1), which can be useful in assessing the virulence of any *M. catarrhalis* strains detected. However, despite being widely applied in the field of research, no *M. catarrhalis* PCR (including the *ompJ* PCR) has yet been accepted or accredited for use in the microbiological diagnostic laboratory.

Another OMP of *M. catarrhalis* (CopB) is frequently used in real-time PCR-based bacterial detection studies. This *M. catarrhalis*-specific gene has been used in studies to both detect and to quantify *M. catarrhalis* in clinical specimens. Both real-time PCR and quantitative real-time PCR (Q-PCR) are laboratory techniques based on PCR to detect and quantify a specific sequence in a DNA sample. Real-time PCR technologies are currently used in many research laboratories as investigational tools. However, real-time detection and quantitative PCR protocols that are specific for *M. catarrhalis* have not yet been introduced into the diagnostic clinical laboratory, although *M. catarrhalis*-based real-time PCR protocols have been occasionally used in clinical studies. For example, Greiner et al. and Kais et al. used quantitative real-time PCR to detect *M. catarrhalis* in nasopharyngeal secretions and lower respiratory tract samples.[73,74] They concluded that real-time PCR is faster and more precise than conventional culture and identification procedures, and therefore may provide a potentially reliable tool for the diagnosis of *M. catarrhalis* infection.

Another PCR-based bacterial detection method is multiplex PCR. Hendolin et al. developed a protocol that allowed the simultaneous detection of several bacterial species at the same time, including *M. catarrhalis*. The researchers targeted the bacterial 16S rRNA gene, and designed primers for detection of the pathogens most commonly encountered by culture in the middle ear (i.e., *H. influenzae*, *S. pneumoniae*, and *M. catarrhalis*), as well as the slowly growing bacterium *Alloiococcus otitidis*.[78] Based on the results of this study, and other studies that adapted this method, the multiplex PCR method improved the detection rate of these bacteria significantly and was a more reliable method in comparison to the conventional culture methods for use in routine laboratories.[78,79,119,120]

The advantage of using singleplex PCR, multiplex PCR, real-time PCR, and Q-PCR is that bacterial culture becomes unnecessary as a first step in the detection procedure. Instead, extraction of DNA directly from clinical specimens, such as nasopharyngeal secretions and sputum, is performed without the need for culture.

16S rRNA Gene PCR, Sequencing, and PCR–RFLP. Information useful for the identification of bacteria may be obtained by sequence analysis of amplified DNA and comparison with sequence databases.[3,121] For bacterial phylogenetic studies, 16S rRNA gene sequencing is frequently used due to the fact that this gene is highly conserved between isolates. Essentially, universally conserved PCR primers are initially used to amplify a portion of the bacterial 16S rRNA gene. In addition to highly conserved primer binding sites, 16S rRNA gene sequences also contain hypervariable regions that provide species-specific signature sequences useful for bacterial identification. These sequences can be compared to a database of 16S rRNA gene sequences obtained from different species of bacteria—for example using the basic local alignment search tool (BLAST; http://blast.ncbi.nlm.nih.gov/Blast.cgi)—in order to determine the actual bacterial species present. As a result, 16S rRNA gene sequencing is often applied in medical microbiology research laboratories as an accurate method of bacterial species identification. However, as yet this technique has not been widely adopted by diagnostic routine medical microbiology laboratories.

The phylogeny of the family *Moraxellaceae* has been studied by 16S rRNA sequence analysis.[3] Essentially, two distinct genetic lineages, related to 16S rRNA type, have been identified for *M. catarrhalis*, namely a seroresistant lineage (16S rRNA type 1) and a serosensitive lineage (16S rRNA types 2 and 3).[80] Using this knowledge (and the accompanying sequence data) has allowed the development of a PCR–RFLP test, a relatively simple and rapid method for *M. catarrhalis* lineage identification. In this method, PCR–RFLP restriction enzymes *FspB*I and *Hha*I generate lineage specific fragments from PCR products amplified from the 16S rRNA gene of 16S rRNA types 1, 2, and 3 isolates.[69,77]

Fluorescence In Situ Hybridization (FISH). FISH is a cytogenetic technique that may be used to detect and localize specific DNA sequences of bacteria within clinical samples. In microbiology, the most commonly used target molecule for FISH is 16S rRNA because of its genetic stability, its domain structure (which comprises both conserved and variable regions), and its high copy number.[122] FISH uses 16S rRNA fluorescent probes that are able to hybridize to these variable regions, and depending on the probe, FISH can be used to detect bacteria of different phylogenetic levels. FISH hybridized bacterial cells may be detected by either fluorescent microscopy or flow cytometry.[123,124]

During the last decade, FISH has become a particularly important tool in research laboratories as a means of investigating the microflora present in bacterial biofilms, with a number of oligonucleotide probes having been designed and evaluated for the major bacterial species frequently found in biofilms.[123,125] Hall-Stoodley et al. used FISH to detect *M. catarrhalis* in bacterial biofilms on the middle-ear mucosa (MEM) of children with chronic otitis media. They reported that *M. catarrhalis* could be detected in MEM biopsy specimens using 16S rRNA *M. catarrhalis* specific probes.[123] Further, Heiniger et al. used FISH to show that colonization with *M. catarrhalis* is more frequent than determined by routine culture and concluded that colonization and invasion by *M. catarrhalis* is indeed frequent, due to the organism residing both within and beneath the epithelium of host cells.[126]

80.1.4.3 *Moraxella catarrhalis* Typing

Many typing methods have been developed and widely used in several major fields of microbiological research.[127] Several of these methods have been used to characterize strains of *M. catarrhalis* in order to compare the genotypic relatedness of geographically diverse and outbreak-related isolates. The most frequently used typing methods for *M. catarrhalis* are currently pulsed-field gel electrophoresis (PFGE) and multilocus sequence typing (MLST). Further, ribotyping, amplification fragment length polymorphism (AFLP) and randomly amplified polymorphic DNA (RAPD) analysis have also been used in several epidemiological studies.[77,127,128]

Pulsed-Field Gel Electrophoresis (PFGE). PFGE is used to resolve (large) DNA molecules by continuous reorientation. PFGE is able to separate much higher molecular weight DNA molecules than standard DNA gel electrophoresis (i.e., fragments >50 kb in size), and it separates them with greater resolution than conventional electrophoresis.[129]

Many studies have been published using PFGE as a tool in the epidemiological genotyping of *M. catarrhalis*. One of the earliest studies using PFGE for genotyping of *M. catarrhalis* isolates was published by Roberts et al. in 1991, who characterized tetracycline-resistant *M. catarrhalis* isolates using PFGE.[130] Currently, epidemiological studies using PFGE as a tool have revealed the heterogeneity and genetic diversity of *M. catarrhalis* strains isolated from identical locations, as well as in different geographical regions and in nosocomial outbreaks.[69,128,131]

While PFGE has proven to be an important molecular tool for epidemiological studies involving *M. catarrhalis*, it suffers from a lack of interlaboratory assessment in relation to its performance and reproducibility, due to the different methodologies used and the difficulty in accurately mapping band separation lengths.[132] On the other hand, MLST allows easy interlaboratorial comparison to be made.

Multilocus Sequence Typing (MLST). MLST is a technique used in the field of molecular biology for typing multiple loci. In the MLST scheme developed for *M. catarrhalis*, isolates are characterized using the DNA sequences of internal fragments of eight housekeeping genes, namely *glyRS* (glycyl-tRNA synthetase beta subunit), *ppa* (pyrophosphate phospho-hydrolase), *efp* (elongation factor P), *fumC* (fumarate hydratase), *trpE* (anthranilate synthase component I), *mutY* (adenine glycosylase), *adk* (adenylate kinase), and *abcZ* (ATP-binding protein). Approximately 450–500 bp internal fragments of each housekeeping gene are PCR amplified and sequenced. For every gene, a specific number is assigned that represents each distinct sequence allele present. The isolate is then assigned a unique MLST number ("ST-type") dependent on the combination of these 8 allele numbers. For further information, the reader is referred to http://mlst.ucc.ie/.

The great advantage of MLST is that sequence data are unambiguous and that allelic profiles of isolates can easily be compared to those in large central databases via the Internet. This allows accurate and rapid interlaboratory comparisons to be made. Unfortunately, however, MLST typing is not as sensitive as PFGE genotyping.

Matrix-Assisted Laser Desorption/Ionization Time-of-Flight Mass Spectrometry (MALDI–TOF MS). MALDI–TOF MS can also be used to detect and type bacteria, but using proteins instead of DNA. Hsieh et al. showed that mass spectrometry techniques might be an alternative for highly efficient and accurate bacterial identification; as yet, however, this technique has not been applied to the detection of *M. catarrhalis*.[133] However, Schaller et al. have developed a MALDI–TOF MS method (based on the expression profiles of *M. catarrhalis* OMPs) to differentiate between the two major 16S rRNA *M. catarrhalis* lineages. This method was reported to be a rapid and robust tool for *M. catarrhalis* strain typing that would help to establish the role of *M. catarrhalis* subpopulations in colonization and disease manifestations.[134]

80.2 METHODS

80.2.1 Sample Preparation

Singleplex PCR, PCR–RFLP, and MLST. Relatively pure DNA is required for singleplex PCR, 16S rRNA gene PCR, PCR–RFLP, and MLST. (i) Inoculate a *M. catarrhalis* isolate onto a blood agar plate for overnight incubation at 35°C in an atmosphere of 5% CO_2 in air. (ii) Extract DNA from bacterial cells either manually (e.g., Wizard® Genomic DNA Purification Kit, Promega) or by automation (e.g., the MagNA Pure LC 2.0 System, Roche Applied Science).

Multiplex PCR. Hendolin et al. described a modified QIAamp (Qiagen) DNA extraction protocol for middle ear effusion (MEE) specimens.[79] (i) Mix 100 μL of the MEE with 50 μL of a freshly prepared mixture of 75 mM NaOH and 1.5% sodium dodecyl sulfate (SDS). (ii) Incubate the suspension at 95°C for 5 min, and then cool the suspension on ice. (iii) Neutralize the suspension with 50 μL of 75 mM Tris-HCl (pH 7.0), and add 200 μL of buffer AL (QIAamp kit). (iv) Incubate the suspension again at 95°C for 5 min and cool on ice. (v) Add 210 μL of 99% ethanol. (vi) Follow the protocol described in the instructions for the QIAamp kit (Qiagen). (vii) Elute the DNA from the columns twice using 90 and 60 μL of sterile distilled water (Amersham Pharmacia Biotech) at 70°C.

Real-Time PCR. DNA can be extracted from nasopharyngeal secretions using an adapted protocol of Greiner et al.[73] (i) Centrifuge 1 ml of nasopharyngeal secretion at 12,000 × *g* for 10 min. (ii) Resuspend the pellet in 200 μL of digestion buffer [50 mM Tris-HCl (pH 8.5), 1 mM EDTA, 0.5% SDS, 200 mg of proteinase K/mL]. (iii) Incubate with shaking for 1 h at 55°C. (iv) Purify the DNA using the QIAamp tissue kit (Qiagen).

Pulsed-Field Gel Electrophoresis (PFGE). A DNA extraction procedure of Verduin et al.[135] may be used. (i) Inoculate an *M. catarrhalis* isolate onto a blood agar plate for overnight incubation at 35°C in an atmosphere of

5% CO_2 in air. (ii) Suspend a few colonies of the overnight *M. catarrhalis* culture in a 100 mM EDTA, 10 mM EGTA, and 10 mM Tris (pH 8.0) buffer (EET-buffer), and mix with an equal volume of molten 1.4% InCert agarose (BMA) to form a bacterial/agar mix. (iii) Transfer the mix into PFGE plug molds (Bio-Rad) and allow to set. (iv) Lyse *M. catarrhalis* within the agarose plug by immersing the plug in 1 mL of EET-lysis buffer (EET buffer containing 5 mg of lysozyme/mL). (v) Incubate the plug at 37°C for 3–4 h. (vi) Transfer the agarose plug into 1 mL of deproteinization buffer [0.5 mM Tris-NaCl (pH 8.0), 1% SDS, 3 mg of proteinase K/mL]. (vii) Incubate the plug overnight at 37°C. (viii) Wash the plug six times in a 10 mM Tris (pH 8.0), 1 mM EDTA storage buffer ($T_{10}E_1$-buffer).

Fluorescence In Situ Hybridization (FISH). FISH protocols do not require access to relatively pure DNA per se, but require "permeabilization" of bacterial cells in order to facilitate DNA:DNA hybridization. The sample preparation protocol presented here for 16S rRNA FISH has been adapted from Hall-Stoodley et al.[123] (i) Fix MEM biopsy specimens with fresh 4% paraformaldehyde in phosphate-buffered saline (PBS). (ii) Wash the specimens in PBS and PBS-ethanol (1:1), and subsequently incubate for 3 min in 80% and 100% ethanol. (iii) Incubate the specimens with 10 mg/mL of lysozyme (Sigma–Aldrich) in 0.1 M Tris (hydroxymethyl) aminomethane hydrochloride and 0.05 M Na_2EDTA at 37°C. (iv) Wash the specimens with sterile distilled water. Note: the lysozyme step may be omitted for some MEM specimens to optimize conditions for Gram-negative bacteria.

80.2.2 Detection Procedures

80.2.2.1 Identification

(i) Singleplex PCR Various PCR mixes and primer combinations have been described in the literature for the singleplex detection of *M. catarrhalis* genes (Table 80.1). For the detection of the *ompJ1*/*ompJ2* genes, PCR screening is performed as mentioned below:

Procedure

1. Prepare a PCR mixture (25 μL) containing 0.2 mM dNTP, 2 mM $MgCl_2$, 1× *Taq* buffer with KCl, 12.5 pmol/μL of each primer (i.e., 19kDres.f: 5′-ctaacgctgccatcagctat-3′ and 19kDres.r: 5′-gttgcatt acggctggtaac-3′), 0.25 U *Taq* DNA Polymerase (Fermentas) and 1 μL pure DNA.
2. Run the following touchdown thermocycling program in a GeneAmp® PCR System 9700 (Applied Biosystems): initial denaturation at 94°C for 4 min; 15 cycles of 94°C for 30 s, 70°C for 1 min (which is reduced by 1°C per cycle to a final annealing temperature of 56°C), and 72°C for 2 min; 20 cycles of 94°C for 30 s, 55°C for 1 min, and 72°C for 2 min; and a final extension at 72°C for 10 min.
3. Load 8 μL of PCR product on a 1% agarose gel alongside a 100 bp DNA ladder (Invitrogen).
4. Visualize the expected PCR product sizes of approximately 363 bp (*ompJ1*) and 333 bp (*ompJ2*) under UV light and photograph.

(ii) Multiplex PCR of Hendolin et al.[78] For multiplex PCR, the protocol of Hendolin et al. has been used.[78] This protocol uses an internal control (IC) and a dUTP-uracil-*N*-glycosylase (UNG) system (Boehringer Mannheim). The reaction mixture is optimized for Mg^{2+}, primer, and AmpliTaq Gold concentrations.

Procedure

1. Prepare a PCR mixture containing 3.6 μM *A. otitidis*-specific primer (5′-ggggaagaacacggatagga-3′), 1.8 μM *H. influenzae*-specific primer (5′-cgtattatcggaaga tgaaagtgc-3′), 0.6 μM *M. catarrhalis*-specific primer (5′-cccataagccctgacgttac-3′), 0.2 μM *S. pneumoniae*-specific primer (5′-aaggtgcacttgcatcactac-3′), 0.8 μM Cy-5-labeled common reverse primer (5′-ctacgcatttca ccgctacac-3′), 200 μM each of dNTP, 400 μM dUTP, 0.2% Tween 20, 1 attogram (ag) of IC, 1× GeneAmp PCR buffer II, 2.8 mM $MgCl_2$, 2.5 U of AmpliTaq Gold, and 0.3 U of thermolabile UNG.
2. Run the following thermocycling program: initial denaturation at 94°C for 9.5 min; 38 cycles of 95°C for 30 s, 66°C for 45 s, and 72°C for 1 min; final extension at 72°C for 10 min; and 55°C for 30 min.

(iii) Real-Time PCR of Greiner et al.[73]

Procedure

1. Prepare a 25 μL PCR mixture containing 2× Platinum Quantitative PCR SuperMix-UDG (Life Technologies Inc.), which includes dUTP and uracil-*N*-glycosylase and to which Blue 636 (Invitrogen) is added as passive reference dye to a final concentration of 0.1 pmol/μL, 333 nM of each primer, 200 nM fluorescence-labeled probe and 1 μL of DNA extract.
2. Run the following thermocycling program in an ABI PRISM 7700 sequence detection system (Perkin-Elmer, Applied Biosystems): 50°C for 2 min and 95°C for 10 min, followed by 40 cycles of 95°C for 15 s and 60°C for 1 min.
3. Analyze real-time data by using for example Sequence Detection Systems software version 1.7 (Applied Biosystems).[73]

(iv) Fluorescence In Situ *Hybridization (FISH)* Fluorescence in situ hybridization may be performed as described by Hall-Stoodley et al. and further adapted by Amann et al., Kempf et al., and Hogardt et al.[123,124,136,137]

Procedure

1. Label an oligonucleotide probe that hybridizes to the 16S rRNA of *M. catarrhalis* [5′-ccgccacuaaguaucaga-3′], as well as a universal eubacterial probe EUB338 [5′-gctgcctcccgtaggagt-3′]

with Cy3, Cy5, or 6-FAM dyes (Integrated DNA Technologies).

2. Hybridize fixed MEM biopsy specimens by application of 10 μL of hybridization buffer [0.9 M NaCl, 20 mM Tris-HCl (pH 7.2), 0.1% SDS, 30% formamide] containing 5 ng of each specific oligonucleotide probe for 30 min at 45°C.
3. Remove any nonspecific binding of the probe by incubating the specimens for 15–30 min at 48°C in wash buffer [0.9 M NaCl, 20 mM Tris-HCl (pH 7.2), 0.1% SDS].
4. Rinse the specimens with phosphate-buffered saline (PBS).
5. Evaluate the hybridized specimens with confocal laser scanning microscopic (CLSM) imaging, using for example a Leica DM RXE microscope attached to a TCS SP2 AOBS confocal system and high-resolution objectives (Leica Microsystems).

80.2.2.2 Typing

(i) 16S rRNA PCR–RFLP of Verhaegh et al.[69]

Procedure

1. Prepare a PCR mixture (25 μL) containing 0.2 mM dNTP, 2 mM $MgCl_2$, 1× *Taq* buffer with KCl, 12.5 pmol/μL of each primer (i.e., 19kDres.f: 5′-ctaacgctgccatcagctat-3′ and 19kDres.r: 5′-gttgcatta cggctggtaac-3′), 0.25 U *Taq* DNA Polymerase (Fermentas) and 1 μL extracted DNA.
2. Run the following thermal cycling reaction in a GeneAmp® PCR System 9700 (Applied Biosystems): initial denaturation at 94°C for 4 min; 40 cycles of 94°C for 1 min, 55°C for 1 min, and 72°C for 1 min.
3. Digest 8 μL of the 16S rRNA gene PCR product overnight at 37°C in a total reaction volume of 20 μL using 10 U of *Fsp*BI or *Hha*I (Fermentas), 1× Tango buffer (Fermentas) and sterile water.
4. Run the PCR-digest on a 2% agarose gel alongside a 100 bp DNA ladder (Invitrogen).
5. Visualize the gel under UV light and photograph. The expected band sizes and 16S rRNA types that may be differentiated are listed in Table 80.2.

TABLE 80.2
Differentiation of 16S rRNA Types Using 16S rRNA PCR–RFLP

	Restriction Enzyme Digestion	
	***Fsp*BI**	***Hha*I**
	Expected Fragment Size	**Expected Fragment Size**
16S rRNA type 1	416/127 bp	No restriction
16S rRNA type 2	309/127/107 bp	No restriction
16S rRNA type 3	309/127/106 bp	347/195 bp

Note: The enzymes *Fsp*BI and *Hha*I are chosen based on the 16S rRNA gene sequences reported in the study of Bootsma et al., and are available under GenBank accession numbers AF192339 (type 1), AF192340 (type 2), and AF192341 (type 3).[77]

(ii) Pulsed-Field Gel Electrophoresis (PFGE) The PFGE procedure detailed below has been described by Verduin et al. and Verhaegh et al.[69,135]

Procedure

1. Prestabilize the PFGE plugs stored in $T_{10}E_1$-buffer in a 10 mM Tris (pH 8.0)/0.1 M EDTA buffer ($T_{10}E_{0.1}$-buffer).
2. Digest the plugs overnight at 37°C, using the restriction enzyme *Spe*I (20 U) (Fermentas).
3. Embed the plugs in a 1% agarose gel (Bio-Rad).
4. Separate the DNA fragments within the plugs by electrophoresis in a CHEF Mapper system (Bio-Rad) using the following protocol: a constant voltage of 6 V cm^{-1}, in which the pulse time increases from 3.5 to 25 s during the first 12 h, followed by a block of 8 h, where the pulse time increases linearly from 1 to 5 s.
5. Stain the gel in 2 L of sterile water containing 100 μL of ethidium bromide (10 mg/mL).
6. Visualize the gel under UV light and photograph.
7. Analyze the PFGE patterns using comparison software [e.g., BioNumerics (Applied Maths) with gel lanes normalized against a lambda DNA ladder (Bio-Rad)]. Band tolerance is usually set to 1.5%.

(iii) Multilocus Sequence Typing (MLST) PCR is performed to detect and sequence the *glyRS*, *ppa*, *efp*, *fumC*, *trpE*, *mutY*, *adk*, and *abcZ* genes as described in the guidelines available at the multilocus sequence typing Web site (http://mlst.ucc.ie/).

Procedure

1. For all genes, a standard PCR protocol comprising an annealing temperature of 52°C is used, except for *glyRS* (58°C) and *adk* (54°C). Thermal cycling reaction conditions consist of an initial denaturation at 94°C for 4 min; 25 cycles of denaturation at 94°C for 30 s, annealing for 1 min at gene dependent temperature, extension at 72°C for 2 min; and a final extension at 72°C for 10 min.
2. PCR primer pairs (at 12.5 pmol/μL) are used for amplicon sequencing at a concentration of 1.25 pmol/μL, except for the genes *glyRS*, *fumC*, and *mutY*, which require separate sequencing primers, using the BigDye® Terminator v3.1 Cycle Sequencing Kit (Applied Biosystems) and 1 μL of PCR product.
3. Determine the sequences of PCR products, obtained from the eight housekeeping loci in the

   M. catarrhalis MLST scheme, using (for example) the Applied Biosystems 3130*xl* Genetic Analyzer.
4. Assign the alleles at each locus a specific number by referring to the database available at http://mlst.ucc.ie/.
5. Define the allelic profile, also known as sequence type (ST).
6. Compare the allelic profiles of the isolates from different geographic regions that are entered into the *M. catarrhalis* database at the Internet address given above.

80.3 CONCLUSIONS

M. catarrhalis is part of the normal bacterial flora of the human respiratory tract which relatively recently has been acknowledged as a bacterial pathogen and not simply being a commensal. Recognition of its pathogenic status means that the detection of *M. catarrhalis* in clinical specimens is now an important topic in disease management and diagnosis. In the routine clinical laboratory, the detection of *M. catarrhalis* tends to be based on conventional methods only, although a variety of molecular detection techniques are being used within research laboratories to investigate *M. catarrhalis* specific genes and virulence. Despite being somewhat old fashioned, time consuming, and unable to detect unculturable isolates, the conventional approaches for identifying *M. catarrhalis* (including selective culture media, Gram stain morphology, and biochemical testing, for example the tributyrin disc test) are still extremely useful.

Modern molecular detection techniques for *M. catarrhalis* are currently commonly applied in research microbiology laboratories. These include PCR and DNA hybridization-based bacterial detection methods, which are usually reliant on the specific detection and amplification of genes encoding for *M. catarrhalis* outer membrane proteins. For both research and diagnostic purposes, further details concerning the relatedness of *M. catarrhalis* isolates may be obtained by using one of several *M. catarrhalis* typing techniques. These epidemiological investigation techniques include PFGE and MLST as the most popular, as well as electrophoretic profiling of outer membrane proteins, iso-electric focusing of β-lactamase enzymes and serological typing of lipopolysaccharide.[138–140] More recent developments include a *M. catarrhalis* microarray assay that has been developed to study gene expression of *M. catarrhalis* under varying environmental conditions.[109]

Culture and biochemical testing remains the "gold standard" in *M. catarrhalis* diagnosis. The use of PCR tests directed against *M. catarrhalis* genes in the routine clinical microbiology laboratories is hindered by the lack of sensitivity, specificity, and reproducibility data, as well as the lack of an accepted library of quality control *M. catarrhalis* isolates to test. Issues related to cost per unit specimen, as well as the fact that bacterial culture tends to be a "catch-all" technique (rather than the more specific molecular techniques), also play a role in current decisions related to the implementation of molecular tests for *M. catarrhalis* in the routine clinical microbiology laboratory.

REFERENCES

1. Garrity, G.M., Bell, J.A., and Lilburn, T.G., *Taxonomic Outline of the Prokaryotes Bergey's Manual of Systematic Bacteriology*, 2nd ed., Springer, New York, 2004.
2. Vela, A.I. et al., *Moraxella pluranimalium* sp. nov., isolated from animal specimens, *Int. J. Syst. Evol. Microbiol.*, 59, 671, 2009.
3. Pettersson, B. et al., Phylogeny of the family *Moraxellaceae* by 16S rDNA sequence analysis, with special emphasis on differentiation of *Moraxella* species, *Int. J. Syst. Bacteriol.*, 48, 75, 1998.
4. Juvin, P. et al., *Moraxella lacunata* infectious arthritis, *J. R. Soc. Med.*, 84, 629, 1991.
5. Kutlesa, M. et al., *Moraxella lacunata*: A rare cause of infective endocarditis, *Scand. J. Infect. Dis.*, 39, 917, 2007.
6. Nagano, N. et al., Presumed endocarditis caused by BRO β-lactamase-producing *Moraxella lacunata* in an infant with Fallot's tetrad, *J. Clin. Microbiol.*, 41, 5310, 2003.
7. Noman, F. et al., "*Moraxella lacunata*" endocarditis treated with penicillin, *J. Pak. Med. Assoc.*, 58, 336, 2008.
8. Ringvold, A., Vik, E., and Bevanger, L.S., *Moraxella lacunata* isolated from epidemic conjunctivitis among teen-aged females, *Acta Ophthalmol. (Copenh.)*, 63, 427, 1985.
9. Feldman, S.R., and Petersen, M.J., Botryomycosis caused by *Moraxella nonliquefaciens*, *Cutis*, 43, 140, 1989.
10. Laukeland, H., Bergh, K., and Bevanger, L., Posttrabeculectomy endophthalmitis caused by *Moraxella nonliquefaciens*, *J. Clin. Microbiol.*, 40, 2668, 2002.
11. Sherman, M.D. et al., Endophthalmitis caused by β-lactamase-positive *Moraxella nonliquefaciens*, *Am. J. Ophthalmol.*, 115, 674, 1993.
12. Buchman, A.L., and Pickett, M.J., *Moraxella atlantae* bacteraemia in a patient with systemic lupus erythematosis, *J. Infect.*, 23, 197, 1991.
13. De Baere, T. et al., Bacteremia due to *Moraxella atlantae* in a cancer patient, *J. Clin. Microbiol.*, 40, 2693, 2002.
14. Berrocal, A.M. et al., Endophthalmitis caused by *Moraxella osloensis*, *Graefes Arch. Clin. Exp. Ophthalmol.*, 240, 329, 2002.
15. Shah, S.S., Ruth, A., and Coffin, S.E., Infection due to *Moraxella osloensis*: Case report and review of the literature, *Clin. Infect. Dis.*, 30, 179, 2000.
16. Sifri, C.D. et al., *Moraxella osloensis* bacteremia in a kidney transplant recipient, *Transpl. Int.*, 21, 1011, 2008.
17. Christensen, J.J. et al., A case of *Moraxella canis*-associated wound infection, *Scand. J. Infect. Dis.*, 33, 155, 2001.
18. Vaneechoutte, M. et al., Isolation of *Moraxella canis* from an ulcerated metastatic lymph node, *J. Clin. Microbiol.*, 38, 3870, 2000.
19. Catlin, B.W., Transfer of the organism named *Neisseria catarrhalis* to *Branhamella* gen. nov., *Int. J. Syst. Bacteriol.*, 20, 155, 1970.
20. Henriksen, S.D., and Bovre, K., The taxonomy of the genera *Moraxella* and *Neisseria*, *J. Gen. Microbiol.*, 51, 387, 1968.
21. Frosch, P., and Kolle, W., Die mikrokokken. In Flugge, C. (Editor), *Die Mikroorganismen*, Vol. 2, p. 154, Verlag von Vogel, Leipzig, 1896.
22. Bovre, K., The genus *Moraxella*. In Krieg, N.R., and Holt, J.G. (Editors), *Bergey's Manual of Systematic Bacteriology*, Vol. 1, p. 296, The Williams & Wilkins Co., Baltimore, 1984.

23. Doern, G.V., *Branhamella catarrhalis*: Phenotypic characteristics, *Am. J. Med.*, 88, 33S, 1990.
24. Murphy, T.F., and Parameswaran, G.I., *Moraxella catarrhalis*, a human respiratory tract pathogen, *Clin. Infect. Dis.*, 49, 124, 2009.
25. Vaneechoutte, M. et al., Selective medium for *Branhamella catarrhalis* with acetazolamide as a specific inhibitor of *Neisseria* spp, *J. Clin. Microbiol.*, 26, 2544, 1988.
26. Speeleveld, E. et al., Comparison of three rapid methods, tributyrine, 4-methylumbelliferyl butyrate, and indoxyl acetate, for rapid identification of *Moraxella catarrhalis*, *J. Clin. Microbiol.*, 32, 1362, 1994.
27. Faden, H., Harabuchi, Y., and Hong, J.J., Epidemiology of *Moraxella catarrhalis* in children during the first 2 years of life: relationship to otitis media, *J. Infect. Dis.*, 169, 1312, 1994.
28. Revai, K., Mamidi, D., and Chonmaitree, T., Association of nasopharyngeal bacterial colonization during upper respiratory tract infection and the development of acute otitis media, *Clin. Infect. Dis.*, 46, e34, 2008.
28a. Verhaegh, S.J. et al., Determinants of *Moraxella catarrhalis* colonization in healthy Dutch children during the first 14 months of life, *Clin. Microbiol. Infect.*, 16, 992, 2010.
29. Faden, H. et al., Relationship between nasopharyngeal colonization and the development of otitis media in children. Tonawanda/Williamsville Pediatrics, *J. Infect. Dis.*, 175, 1440, 1997.
30. Faden, H. et al., Nasopharyngeal flora in the first three years of life in normal and otitis-prone children, *Ann. Otol. Rhinol. Laryngol.*, 100, 612, 1991.
31. Quinones, D. et al., Nasopharyngeal colonization by *Moraxella catarrhalis* and study of antimicrobial susceptibility in healthy children from Cuban day-care centers, *Arch. Med. Res.*, 36, 80, 2005.
32. Zemlickova, H. et al., Characteristics of *Streptococcus pneumoniae*, *Haemophilus influenzae*, *Moraxella catarrhalis* and *Staphylococcus aureus* isolated from the nasopharynx of healthy children attending day-care centres in the Czech Republic, *Epidemiol. Infect.*, 134, 1179, 2006.
33. Masuda, K. et al., Incidences of nasopharyngeal colonization of respiratory bacterial pathogens in Japanese children attending day-care centers, *Pediatr. Int.*, 44, 376, 2002.
34. Peerbooms, P.G. et al., Nasopharyngeal carriage of potential bacterial pathogens related to day care attendance, with special reference to the molecular epidemiology of *Haemophilus influenzae*, *J. Clin. Microbiol.*, 40, 2832, 2002.
35. Ruuskanen, O. et al., Viruses in acute otitis media: increasing evidence for clinical significance, *Pediatr. Infect. Dis. J.*, 10, 425, 1991.
36. Kleemola, M. et al., Is there any specific association between respiratory viruses and bacteria in acute otitis media of young children? *J. Infect.*, 52, 181, 2006.
37. Ruuskanen, O., and Heikkinen, T., Otitis media: Etiology and diagnosis, *Pediatr. Infect. Dis. J.*, 13 (Suppl 1), S23, 1994.
38. Portmann, M., The etiology and pathogenesis of otitis media with effusion. A review, *Acta Otolaryngol. Suppl*, 414, 41, 1984.
39. Bluestone, C.D., Stephenson, J.S., and Martin, L.M., Ten-year review of otitis media pathogens, *Pediatr. Infect. Dis. J.*, 11 (Suppl), S7, 1992.
40. Brook, I., The role of anaerobic bacteria in chronic suppurative otitis media in children: Implications for medical therapy, *Anaerobe*, 14, 297, 2008.
41. Saunders, J.E. et al., Current bacteriology of suppurative otitis: Resistant patterns and outcomes analysis, *Otol Neurotol.*, 30, 339, 2009.
42. Murray, C.J., and Lopez, A.D., Evidence-based health policy—lessons from the Global Burden of Disease Study, *Science*, 274, 740, 1996.
43. Sethi, S., and Murphy, T.F., Bacterial infection in chronic obstructive pulmonary disease in 2000: A state-of-the-art review, *Clin. Microbiol. Rev.*, 14, 336, 2001.
44. Zalacain, R. et al., Predisposing factors to bacterial colonization in chronic obstructive pulmonary disease, *Eur. Respir. J.*, 13, 343, 1999.
45. Sethi, S. et al., Inflammatory profile of new bacterial strain exacerbations of chronic obstructive pulmonary disease, *Am. J. Respir. Crit. Care Med.*, 177, 491, 2008.
46. Wald, E.R., Microbiology of acute and chronic sinusitis in children and adults, *Am. J. Med. Sci.*, 316, 13, 1998.
47. Lutwick, L., and Fernandes, L., The other siblings: Respiratory infections caused by *Moraxella catarrhalis* and *Haemophilus influenzae*, *Curr. Infect. Dis. Rep.*, 8, 215, 2006.
48. Pappas, D.E. et al., Symptom profile of common colds in school-aged children, *Pediatr. Infect. Dis. J.*, 27, 8, 2008.
49. Malmvall, B.E., Brorsson, J.E., and Johnsson, J., In vitro sensitivity to penicillin V and β-lactamase production of *Branhamella catarrhalis*, *J. Antimicrob. Chemother.*, 3, 374, 1977.
50. Shurin, P.A. et al., Emergence of β-lactamase-producing strains of *Branhamella catarrhalis* as important agents of acute otitis media, *Pediatr. Infect. Dis.*, 2, 34, 1983.
51. Johnson, D.M. et al., Susceptibility trends of *Haemophilus influenzae* and *Moraxella catarrhalis* against orally administered antimicrobial agents: Five-year report from the SENTRY Antimicrobial Surveillance Program, *Diagn. Microbiol. Infect. Dis.*, 47, 373, 2003.
52. Ball, P., The clinical development and launch of amoxicillin/clavulanate for the treatment of a range of community-acquired infections, *Int. J. Antimicrob. Agents*, 30 (Suppl 2), S113, 2007.
53. Bezakova, N. et al., Recurrence up to 3.5 years after antibiotic treatment of acute otitis media in very young Dutch children: Survey of trial participants, *BMJ*, 338, b2525, 2009.
54. Schilder, A.G., Lok, W., and Rovers, M.M., International perspectives on management of acute otitis media: A qualitative review, *Int. J. Pediatr. Otorhinolaryngol.*, 68, 29, 2004.
55. American Academy of Pediatrics Subcommittee on Management of Acute Otitis Media. Diagnosis and management of acute otitis media, *Pediatrics*, 113, 1451, 2004.
56. Tan, T.T., Forsgren, A., and Riesbeck, K., The respiratory pathogen *Moraxella catarrhalis* binds to laminin via ubiquitous surface proteins A1 and A2, *J. Infect. Dis.*, 194, 493, 2006.
57. Lafontaine, E.R., Wagner, N.J., and Hansen, E.J., Expression of the *Moraxella catarrhalis* UspA1 protein undergoes phase variation and is regulated at the transcriptional level, *J. Bacteriol.*, 183, 1540, 2001.
58. Aebi, C. et al., A protective epitope of *Moraxella catarrhalis* is encoded by two different genes, *Infect. Immun.*, 65, 4367, 1997.
59. Meier, P.S. et al., The outer membrane proteins UspA1 and UspA2 of *Moraxella catarrhalis* are highly conserved in nasopharyngeal isolates from young children, *Vaccine*, 20, 1754, 2002.
60. Attia, A.S., and Hansen, E.J., A conserved tetranucleotide repeat is necessary for wild-type expression of the *Moraxella catarrhalis* UspA2 protein, *J. Bacteriol.*, 188, 7840, 2006.
61. Aebi, C. et al., Phenotypic effect of isogenic *uspA1* and *uspA2* mutations on *Moraxella catarrhalis* 035E, *Infect. Immun.*, 66, 3113, 1998.

62. Lafontaine, E.R. et al., The UspA1 protein and a second type of UspA2 protein mediate adherence of *Moraxella catarrhalis* to human epithelial cells in vitro, *J. Bacteriol.*, 182, 1364, 2000.
63. Forsgren, A. et al., The immunoglobulin D-binding protein MID from *Moraxella catarrhalis* is also an adhesin, *Infect. Immun.*, 71, 3302, 2003.
64. Forsgren, A. et al., Isolation and characterization of a novel IgD-binding protein from *Moraxella catarrhalis*, *J. Immunol.*, 167, 2112, 2001.
65. Mollenkvist, A. et al., The *Moraxella catarrhalis* immunoglobulin D-binding protein MID has conserved sequences and is regulated by a mechanism corresponding to phase variation, *J. Bacteriol.*, 185, 2285, 2003.
66. Balder, R. et al., *Moraxella catarrhalis* strain O35E expresses two filamentous hemagglutinin-like proteins that mediate adherence to human epithelial cells, *Infect. Immun.*, 75, 2765, 2007.
67. Timpe, J.M. et al., Identification of a *Moraxella catarrhalis* outer membrane protein exhibiting both adhesin and lipolytic activities, *Infect. Immun.*, 71, 4341, 2003.
68. Lipski, S.L. et al., The *Moraxella catarrhalis* autotransporter McaP is a conserved surface protein that mediates adherence to human epithelial cells through its N-terminal passenger domain, *Infect. Immun.*, 75, 314, 2007.
69. Verhaegh, S.J. et al., Age-related genotypic and phenotypic differences in *Moraxella catarrhalis* isolates from children and adults presenting with respiratory disease in 2001–2002, *Microbiology*, 154, 1178, 2008.
70. Holm, M.M. et al., The *Moraxella catarrhalis* porin-like outer membrane protein CD is an adhesin for human lung cells, *Infect. Immun.*, 72, 1906, 2004.
71. Hays, J.P. et al., Identification and characterization of a novel outer membrane protein (OMP J) of *Moraxella catarrhalis* that exists in two major forms, *J. Bacteriol.*, 187, 7977, 2005.
72. Aebi, C. et al., Expression of the CopB outer membrane protein by *Moraxella catarrhalis* is regulated by iron and affects iron acquisition from transferrin and lactoferrin, *Infect. Immun.*, 64, 2024, 1996.
73. Greiner, O. et al., Quantitative detection of *Moraxella catarrhalis* in nasopharyngeal secretions by real-time PCR, *J. Clin. Microbiol.*, 41, 1386, 2003.
74. Kais, M. et al., Quantitative detection of *Streptococcus pneumoniae*, *Haemophilus influenzae*, and *Moraxella catarrhalis* in lower respiratory tract samples by real-time PCR, *Diagn. Microbiol. Infect. Dis.*, 55, 169, 2006.
75. Edwards, K.J. et al., Multiplex PCR assay that identifies the major lipooligosaccharide serotype expressed by *Moraxella catarrhalis* clinical isolates, *J. Clin. Microbiol.*, 43, 6139, 2005.
76. Spaniol, V. et al., Outer membrane protein UspA1 and lipooligosaccharide are involved in invasion of human epithelial cells by *Moraxella catarrhalis*, *Microbes Infect.*, 10, 3, 2008.
77. Bootsma, H.J. et al., Analysis of *Moraxella catarrhalis* by DNA typing: Evidence for a distinct subpopulation associated with virulence traits, *J. Infect. Dis.*, 181, 1376, 2000.
78. Hendolin, P.H. et al., Use of multiplex PCR for simultaneous detection of four bacterial species in middle ear effusions, *J. Clin. Microbiol.*, 35, 2854, 1997.
79. Hendolin, P.H., Paulin, L., and Ylikoski, J., Clinically applicable multiplex PCR for four middle ear pathogens, *J. Clin. Microbiol.*, 38, 125, 2000.
80. Wirth, T. et al., The rise and spread of a new pathogen: Seroresistant *Moraxella catarrhalis*, *Genome Res.*, 17, 1647, 2007.
81. Helminen, M.E. et al., A large, antigenically conserved protein on the surface of *Moraxella catarrhalis* is a target for protective antibodies, *J. Infect. Dis.*, 170, 867, 1994.
82. Hill, D.J., and Virji, M., A novel cell-binding mechanism of *Moraxella catarrhalis* ubiquitous surface protein UspA: Specific targeting of the N-domain of carcinoembryonic antigen-related cell adhesion molecules by UspA1, *Mol. Microbiol.*, 48, 117, 2003.
83. Conners, R. et al., The *Moraxella* adhesin UspA1 binds to its human CEACAM1 receptor by a deformable trimeric coiled-coil, *Embo. J.*, 27, 1779, 2008.
84. Brooks, M.J. et al., Modular arrangement of allelic variants explains the divergence in *Moraxella catarrhalis* UspA protein function, *Infect. Immun.*, 76, 5330, 2008.
85. Cope, L.D. et al., Characterization of the *Moraxella catarrhalis uspA1* and *uspA2* genes and their encoded products, *J. Bacteriol.*, 181, 4026, 1999.
86. Hill, D.J. et al., Carcinoembryonic antigen-related cell adhesion molecule (CEACAM)-binding recombinant polypeptide confers protection against infection by respiratory and urogenital pathogens, *Mol. Microbiol.*, 55, 1515, 2005.
87. Nordstrom, T., Forsgren, A., and Riesbeck, K., The immunoglobulin D-binding part of the outer membrane protein MID from *Moraxella catarrhalis* comprises 238 amino acids and a tetrameric structure, *J. Biol. Chem.*, 277, 34692, 2002.
88. Forsgren, A., Brant, M., and Riesbeck, K., Immunization with the truncated adhesin *Moraxella catarrhalis* immunoglobulin D-binding protein (MID764–913) is protective against *M. catarrhalis* in a mouse model of pulmonary clearance, *J. Infect. Dis.*, 190, 352, 2004.
89. Walport, M.J., Complement. Second of two parts, *N. Engl. J. Med.*, 344, 1140, 2001.
90. Walport, M.J., Complement. First of two parts, *N. Engl. J. Med.*, 344, 1058, 2001.
91. Nordstrom, T. et al., The emerging pathogen *Moraxella catarrhalis* interacts with complement inhibitor C4b binding protein through ubiquitous surface proteins A1 and A2, *J. Immunol.*, 173, 4598, 2004.
92. Attia, A.S. et al., Binding of vitronectin by the *Moraxella catarrhalis* UspA2 protein interferes with late stages of the complement cascade, *Infect. Immun.*, 74, 1597, 2006.
93. Verduin, C.M. et al., Differences in complement activation between complement-resistant and complement-sensitive *Moraxella* (*Branhamella*) *catarrhalis* strains occur at the level of membrane attack complex formation, *Infect. Immun.*, 62, 589, 1994.
94. Nordstrom, T. et al., Ionic binding of C3 to the human pathogen *Moraxella catarrhalis* is a unique mechanism for combating innate immunity, *J. Immunol.*, 175, 3628, 2005.
95. Lukacova, M., Barak, I., and Kazar, J., Role of structural variations of polysaccharide antigens in the pathogenicity of Gram-negative bacteria, *Clin. Microbiol. Infect.*, 14, 200, 2008.
96. van der Woude, M.W., and Baumler, A.J., Phase and antigenic variation in bacteria, *Clin. Microbiol. Rev.*, 17, 581, 2004.
97. Meier, P.S. et al., *Moraxella catarrhalis* strains with reduced expression of the UspA outer membrane proteins belong to a distinct subpopulation, *Vaccine*, 23, 2000, 2005.
98. Hays, J.P. et al., Total genome polymorphism and low frequency of intra-genomic variation in the *uspA1* and *uspA2* genes of *Moraxella catarrhalis* in otitis prone and non-prone children up to 2 years of age. Consequences for vaccine design? *Vaccine*, 21, 1118, 2003.
99. Kuehn, M.J., and Kesty, N.C., Bacterial outer membrane vesicles and the host-pathogen interaction, *Genes. Dev.*, 19, 2645, 2005.

100. Tan, T.T. et al., *Haemophilus influenzae* survival during complement-mediated attacks is promoted by *Moraxella catarrhalis* outer membrane vesicles, *J. Infect. Dis.*, 195, 1661, 2007.
101. Slevogt, H. et al., *Moraxella catarrhalis* is internalized in respiratory epithelial cells by a trigger-like mechanism and initiates a TLR2- and partly NOD1-dependent inflammatory immune response, *Cell. Microbiol.*, 9, 694, 2007.
102. Slevogt, H. et al., Differential regulation of *Moraxella catarrhalis*-induced interleukin-8 response by protein kinase C isoforms, *Eur. Respir. J.*, 31, 725, 2008.
103. Slevogt, H. et al., CEACAM1 inhibits Toll-like receptor 2-triggered antibacterial responses of human pulmonary epithelial cells, *Nat. Immunol.*, 9, 1270, 2008.
104. Manolov, T. et al., *Moraxella*-dependent alpha 1-antichymotrypsin neutralization: A unique virulence mechanism, *Am. J. Respir. Cell. Mol. Biol.*, 38, 609, 2008.
105. Hall-Stoodley, L., Costerton, J.W., and Toodley, P., Bacterial biofilms: From the natural environment to infectious diseases, *Nat. Rev. Microbiol.*, 2, 95, 2004.
106. Hall-Stoodley, L., and Stoodley, P., Evolving concepts in biofilm infections, *Cell. Microbiol.*, 11, 1034, 2009.
107. Budhani, R.K., and Struthers, J.K., Interaction of *Streptococcus pneumoniae* and *Moraxella catarrhalis*: Investigation of the indirect pathogenic role of β-lactamase-producing *Moraxellae* by use of a continuous-culture biofilm system, *Antimicrob. Agents Chemother.*, 42, 2521, 1998.
108. Hol, C. et al., Experimental evidence for *Moraxella*-induced penicillin neutralization in pneumococcal pneumonia, *J. Infect. Dis.*, 170, 1613, 1994.
109. Wang, W. et al., Metabolic analysis of *Moraxella catarrhalis* and the effect of selected in vitro growth conditions on global gene expression, *Infect. Immun.*, 75, 4959, 2007.
110. Corkill, J.E., and Makin, T., A selective medium for non-pathogenic aerobic Gram-negative cocci from the respiratory tract: With particular reference to *Branhamella catarrhalis*, *Med. Lab. Sci.*, 39, 3, 1982.
111. Soto-Hernandez, J.L. et al., Selective medium with DNase test agar and a modified toluidine blue O technique for primary isolation of *Branhamella catarrhalis* in sputum, *J. Clin. Microbiol.*, 26, 405, 1988.
112. Maddocks, J.L., and Greenan, M.J., A rapid method for identifying bacterial enzymes, *J. Clin. Pathol.*, 28, 686, 1975.
113. Vaneechoutte, M. et al., Rapid identification of *Branhamella catarrhalis* with 4-methylumbelliferyl butyrate, *J. Clin. Microbiol.*, 26, 1227, 1988.
114. Dealler, S.F. et al., Identification of *Branhamella catarrhalis* in 2.5 min with an indoxyl butyrate strip test, *J. Clin. Microbiol.*, 27, 1390, 1989.
115. Louie, M., Ongsansoy, E.G., and Forward, K.R., Rapid identification of *Branhamella catarrhalis*. A comparison of five rapid methods, *Diagn. Microbiol. Infect. Dis.*, 13, 205, 1990.
116. Valenza, G. et al., Microbiological evaluation of the new VITEK 2 *Neisseria-Haemophilus* identification card, *J. Clin. Microbiol.*, 45, 3493, 2007.
117. Post, J.C. et al., Molecular analysis of bacterial pathogens in otitis media with effusion, *JAMA*, 273, 1598, 1995.
118. Rayner, M.G. et al., Evidence of bacterial metabolic activity in culture-negative otitis media with effusion, *JAMA*, 279, 296, 1998.
119. Eser, O.K. et al., Efficacy of nasopharyngeal culture in identification of pathogen in middle ear fluid in chronic otitis media with effusion, *Indian J. Med. Microbiol.*, 27, 237, 2009.
120. Leskinen, K. et al., *Alloiococcus otitidis* in acute otitis media, *Int. J. Pediatr. Otorhinolaryngol.*, 68, 51, 2004.
121. Chen, K. et al., Broad range DNA probes for detecting and amplifying eubacterial nucleic acids, *FEMS Microbiol. Lett.*, 48, 19, 1989.
122. Woese, C.R., Bacterial evolution, *Microbiol. Rev.*, 51, 221, 1987.
123. Hall-Stoodley, L. et al., Direct detection of bacterial biofilms on the middle-ear mucosa of children with chronic otitis media, *JAMA*, 296, 202, 2006.
124. Amann, R.I. et al., Combination of 16S rRNA-targeted oligonucleotide probes with flow cytometry for analyzing mixed microbial populations, *Appl. Environ. Microbiol.*, 56, 1919, 1990.
125. Gok, U. et al., Bacteriological and PCR analysis of clinical material aspirated from otitis media with effusions, *Int. J. Pediatr. Otorhinolaryngol.*, 60, 49, 2001.
126. Heiniger, N. et al., A reservoir of *Moraxella catarrhalis* in human pharyngeal lymphoid tissue, *J. Infect. Dis.*, 196, 1080, 2007.
127. van Belkum, A. et al., Role of genomic typing in taxonomy, evolutionary genetics, and microbial epidemiology, *Clin. Microbiol. Rev.*, 14, 547, 2001.
128. Vu-Thien, H. et al., Comparison of randomly amplified polymorphic DNA analysis and pulsed-field gel electrophoresis for typing of *Moraxella catarrhalis* strains, *J. Clin. Microbiol.*, 37, 450, 1999.
129. Den Dunnen, J.T., and van Ommen, G.J., Methods for pulsed-field gel electrophoresis, *Appl. Biochem. Biotechnol.*, 38, 161, 1993.
130. Roberts, M.C. et al., Tetracycline resistance in *Moraxella (Branhamella) catarrhalis*: Demonstration of two clonal outbreaks by using pulsed-field gel electrophoresis, *Antimicrob. Agents Chemother.*, 35, 2453, 1991.
131. Martinez, G. et al., DNA restriction patterns produced by pulsed-field gel electrophoresis in *Moraxella catarrhalis* isolated from different geographical areas, *Epidemiol. Infect.*, 122, 417, 1999.
132. Pingault, N.M. et al., A comparison of molecular typing methods for *Moraxella catarrhalis*, *J. Appl. Microbiol.*, 103, 2489, 2007.
133. Hsieh, S.Y. et al., Highly efficient classification and identification of human pathogenic bacteria by MALDI-TOF MS, *Mol. Cell. Proteomics*, 7, 448, 2008.
134. Schaller, A. et al., Rapid typing of *Moraxella catarrhalis* subpopulations based on outer membrane proteins using mass spectrometry, *Proteomics*, 6, 172, 2006.
135. Verduin, C.M. et al., Complement-resistant *Moraxella catarrhalis* forms a genetically distinct lineage within the species, *FEMS Microbiol. Lett.*, 184, 1, 2000.
136. Kempf, V.A., Trebesius, K., and Autenrieth, I.B., Fluorescent In situ hybridization allows rapid identification of microorganisms in blood cultures, *J. Clin. Microbiol.*, 38, 830, 2000.
137. Hogardt, M. et al., Specific and rapid detection by fluorescent in situ hybridization of bacteria in clinical samples obtained from cystic fibrosis patients, *J. Clin. Microbiol.*, 38, 818, 2000.
138. Bartos, L.C., and Murphy, T.F., Comparison of the outer membrane proteins of 50 strains of *Branhamella catarrhalis*, *J. Infect. Dis.*, 158, 761, 1988.
139. Nash, D.R. et al., Isoelectric focusing of β-lactamases from sputum and middle ear isolates of *Branhamella catarrhalis* recovered in the United States, *Drugs*, 31 (Suppl 3), 48, 1986.
140. Vaneechoutte, M. et al., Serological typing of *Branhamella catarrhalis* strains on the basis of lipopolysaccharide antigens, *J. Clin. Microbiol.*, 28, 182, 1990.

81 *Pasteurella*

Francis Dziva and Henrik Christensen

CONTENTS

81.1 INTRODUCTION

The genus *Pasteurella* Trevisan 1887 is one of the 14 genera included in the family Pasteurellaceae, Pohl 1981, which also contains the prominent genera *Actinobacillus* and *Haemophilus*. Members of the genus *Pasteurella sensu stricto* are gram-negative, nonmotile, nonspore-forming rods, coccobacilli, or filaments with acrobic, microaerophilic, or facultatively anaerobic respiration and chemoorganotrophic metabolism, sharing these properties with other members of the Pasteurellaceae family. The genus *Pasteurella sensu stricto* is separated from other Pasteurellaceae by an array of biochemical tests, including a negative urease test, positive indole and porphyrin reactions, a positive alpha-glucosidase (PNPG) test, and acid formation from D-fructose, D-galactose, D-mannose, and sucrose.[25,44] Species of the genus *Pasteurella* are predominantly obligate parasites or commensals that colonize mucosal surfaces of vertebrate hosts. The majority cause economically important infections in livestock, companion, and wild animals. Human infections, though infrequent, are invariably associated with direct or indirect contact with such animals; hence, pasteurellosis is a recognized zoonosis. This chapter briefly updates taxonomic aspects of the genus *Pasteurella*, highlights significant infections in humans and describes molecular detection techniques for this group of pathogens in order to inform about control strategies.

81.1.1 Taxonomic Aspects of Genus *Pasteurella*

The genus *Pasteurella* was first proposed by Trevisan in 1887 in recognition of Pasteur's work on these bacteria.[69] At the turn of the twentieth century, the genus *Pasteurella* initially catered for a multitude of unrelated species based on a few common phenotypic traits, thus seriously creating confusion. Throughout the twentieth century, dramatic changes were observed, for example, both *Yersinia pestis* and *Franciscella tulariensis* were temporarily classified under the genus *Pasteurella*. The *Approved Lists of Bacterial Names*[82] excluded most of the problematic names, and only seven species were assigned under the genus *Pasteurella*. The recent application of molecular techniques has significantly improved the taxonomy of this group of organisms. The current taxonomy includes a group of *Pasteurella sensu stricto*, and species excluded from this genus are placed in brackets (Table 81.1). *Pasteurella sensu stricto* is defined on the basis of DNA–DNA hybridization studies, nucleotide sequence comparisons of 16S rRNA and other conserved genes, and some selected important phenotypic characteristics. Strictly speaking, the genus *Pasteurella* comprises only four species: *Pasteurella multocida*, *Pasteurella canis*, *Pasteurella stomatis*, *Pasteurella dagmatis,* and the unnamed *Pasteurella* species B.[22,31,69] Any other species including [*Pasteurella*] *aerogenes*, [*Pasteurella*] *pneumotropica*, [*Pasteurella*] *caballi*, [*Pasteurella*] *mairii*, [*Pasteurella*] *bettyae,* and [*Pasteurella*] *testudinis* are unrelated to *Pasteurella sensu stricto* and are therefore excluded.[70,71] The species *P. multocida* was named by Rosenbusch and Merchant[77] and is the type species of the genus.[82] Based on DNA–DNA hybridization studies and comparison of phenotypic characters, *P. multocida* is subdivided into three subspecies: *P. multocida* subsp. *multocida*; *P. multocida* subsp. *Gallicida*; and *P. multocida* subsp. *septica*.[69] Relationships

TABLE 81.1
Some Clinical Conditions Associated with *Pasteurella* Species in Humans

Species	Predisposing Factor(s)/ Underlying Condition(s)	Transmission	Disease(s)	Reference
P. multocida	Diabetes	Contact with dog	Septicemia	36
	Unknown*	Cat scratch	Septic arthritis and shock	37
	Immune-compromised and renal transplant operation	Licking of mucosal surfaces	Invasive sinusitis	79
	Old age, acute renal failure, and hypercalcemia	Cat bite	Meningitis and septic arthritis	58
	Neurosurgery	Contact with pets	Meningitis	46
	Unknown*	No history of contact with animals	Epiglottitis	43
	Unknown*	Cat bites, scratches and licking, dog bites and licking	Wound infections, meningitis, peritonitis	51
	Urologic abnormality and subsequent surgical correction	Contact with household pet	Urinary tract infection	63
Taxon 45 and 46 of Bisgaard	Unknown*	Bites from large cats	Bite wound infections	23
P. dagmatis	Diabetes	Contact with dog	Septicemia	36
	Cirrhosis	Dog scratch	Peritonitis and septicemia	4
	Unknown*	Dog bite	Cellulitis, abscess, and wound infections	51
P. stomatitis	Unknown*	Dog and cat bites	Bite wound infections, abscesses	51
P. canis	Bites	Dog bite	Osteomyelitis and cutaneous abscess	49 51
	Unknown*	Dog bite	Bite wound infection	
[*P.*] *caballi*	Unknown*	Horse bite	Bite wounds	35
[*P.*] *aerogenes*	Unknown*	Pig bite	Bite wound infections	6,62
[*P.*] *bettyae*	Human immunodeficiency virus	Not defined	Pleuropneumonia, peripartum bacteremia	67,81

[*P.*] = indicates that the species should not be classified with *Pasteurella sensu stricto*. * = patient presumed healthy.

between the type strains of *P. multocida* subspecies can be shown by multilocus enzyme electrophoresis, ribotyping, 16S rRNA sequence, and partial *rpoB* and *atpD* sequence comparisons. Strikingly, subspecies *multocida* is closely related to *gallicida*, whereas subsp. *septica* is distantly related to these taxa.[9,54,57,73]

P. canis, which is mainly associated with dogs, is separated from *P. multocida* on the basis of its inability to ferment (-)-D-mannitol. On the other hand, *P. stomatis*, also associated with dogs and cats, is ornithine-decarboxylase–negative, thus differentiating it from *P. multocida* and *P. canis,* which both give a positive reaction in this character. *P. dagmatis,* also isolated from dogs and cats, is positive in the urease reaction, and produces acid from maltose and dextrin, thus distinguishing itself from three other species: *P. multocida, P. canis,* and *P. stomatis.*[18]

Taxon 45 of Bisgaard represents a group of pathogens that cannot be separated by phenotypic characteristics from *P. multocida* but have a distinct genotype (*Pasteurella* genomospecies). Members of this group are carried in the oral cavity of large cats such as lion and tigers and can be isolated from septic bite wounds inflicted by such animals.[23] Recently, a new genus, *Avibacterium*, comprising the species *Avibacterium gallinarum*, *Avibacterium avium*, *Avibacterium volantium,* and *Avibacterium paragallinarum,* includes the reclassified *Pasteurella gallinarum*, *Pasteurella avium*, *Pasteurella volantium,* and [*Haemophilus*] *paragallinarum*, respectively.[10] While members of this group were believed to be restricted to birds, there are incidences where some, especially *Av. gallinarum* (formerly, *P. gallinarum*) have been isolated from human infections. Intriguingly, there is increasing speculation that these so-called human isolates of *Av. gallinarum* could possibly have been misidentified.[41] Indeed, *Av. gallinarum* shares many phenotypic characteristics with other Pasteurellaceae, especially members of the *Haemophilus* group, hence the increased risk for misidentification.[19]

Interestingly, there are reported incidences of human cases associated with a number of species now removed from the genus *Pasteurella,* and some of these are potential human pathogens (http://www.hpa.org.uk). *Bibersteinia trehalosi* is now the validly published name that relates to the reclassified [*Pasteurella*] *trehalosi.*[11] This species causes an economically important disease in sheep. Similarly, [*Pasteurella*] *anatis,* which is mainly associated with birds, has been reclassified and is now under the genus *Gallibacterium.*[7]

Pasteurella langaaensis, also of avian origin, is unrelated to *Pasteurella sensu stricto*.[17] The species [*P.*] *aerogenes* and [*P.*] *mairii* have also been removed from *Pasteurella sensu stricto*.[72] Descriptions of these two species have recently been revised[24]; hence, it is worthwhile to consult this work for characterization and identification purposes. Though predominantly isolated from cases of porcine abortion, other hosts for [*P.*] *aerogenes* include rabbit, dog, horse, and human infections related to pig bites.[24,62] Unlike [*P.*] *aerogenes*, there appears to be a lack of reports associating [*P.*] *mairii* with human infections.

Taxonomic revision of microorganisms is a continual process hence it is imperative to adopt new genera and species designations introduced by recent changes in taxonomy. These validly published names appear on the "List of Prokaryotic Names with Standing in Nomenclature" (see http://www.bacterio.cict.fr/p/pasteurella.html) alongside their "old" names, which were once validly published. Such information is freely available on the Web site or on the homepage of the International Committee on Systematics of Prokaryotes (ICSP) Subcommittee on the Taxonomy of *Pasteurellaceace* (http://www.the-icsp.org/taxa/Pasteurellaceaelist.htm). The recent taxonomy of Pasteurellaceae has grouped previously misclassified species, including those exclusively associated with animal diseases. Significantly, some species have been removed from the genus *Pasteurella* and thus have become irrelevant as causes of pasteurellosis in humans. While identification is optimally performed with the most recent information, the change in terminology causes problems when it comes to the interpretation of investigations based on reports in the older literature. Consequently, it is recommended to trust information where pheno- or genotypic properties are clearly described for the organism under consideration.

81.1.2 Epidemiology, Pathogenesis, and Clinical Features

Consistent with the published literature, this section describes human infections caused by members of the genus *Pasteurella sensu stricto* and some excluded species. Species of the genus *Pasteurella*, including some previously misclassified strains, form part of the normal flora of many animals, including domestic dogs and cats. Typically, *P. multocida* is found in the oropharynx of 70%–90% of domestic cats, 25%–50% of dogs, 15% of pigs, 14% of rats, and many other animals.[52,86] Clinically observable diseases in these animal species are often regarded as opportunistic infections. In recent years, human infections caused or complicated by *Pasteurella sensu stricto* acquired from animals are steadily increasing. This is thought to be partly due to a canine population that has grown four times faster than the human population[45] or to an increase in an immune-compromised human population. The risk for being bitten by animals varies with age, being higher in infants and children, who may tease pets, disregarding their temperament. The World Health Organisation (http://www.who.int/zoonoses/vph_intro/en/) lists pasteurellosis as a significant zoonotic disease among other diseases like salmonellosis and *E. coli* infections. A significant proportion of human infections are attributed to *P. multocida*, with the remainder shared among other members of the genus. In 2005 and 2006, *P. multocida* accounted for the majority of 920 laboratory confirmed cases of pasteurellosis reported in the UK (http://www.hpa.org.uk). Well-recognized predisposing factors for bite wound infections and their associated sequelae include (i) immature immune system of infants, which increases the risk of invasive infections like septicemia, meningitis, and osteomyelitis; (ii) concurrent immunosuppressive viral infections; (iii) stress-related immunosuppression; and (iv) exposure to animals, and commonly, to domestic pets.[45] The mode of transmission of *Pasteurella* infections in humans is predominantly through dog and cat bites, scratches, licking of injured skin or mucosal surfaces, and indirect exposure to oral secretions. In some cases, human infections have been related to bite wounds inflicted by larger carnivores.[23,51,63] Though *P. multocida* may colonize the respiratory tract mucosa of humans, with or without discernible clinical symptoms, it is still unclear whether person-to-person transmission occurs. Respiratory tract infections have been described among veterinarians and workers who are often in contact with animals: farmers, milkmen, and persons employed in areas where animal tissues are processed.[52]

A wide spectrum of virulence factors, including capsular polysaccharide and lipopolysaccharide, have established roles in pathogenesis of *P. multocida* infections in animals, but it is unknown whether these play similar roles in human infections. A principal virulence factor for atrophic rhinitis in pigs, termed the osteolytic or dermonecrotic toxin, (i) induces increased DNA synthesis in osteoblasts, leading to accelerated osteoblast growth with reduced bone mineralization and collagen formation; (ii) suppresses host immune responses; and (iii) is highly mitogenic on cultured cells.[59,60,68,78] The name refers to a skin test for the toxin in guinea pigs. The toxin is also simply referred to as *P. multocida* toxin (PMT). This toxin has been demonstrated in *P. multocida* strains from human patients with respiratory infections but not in strains from animal bite wounds.[32] In addition, there is a single report of the toxin in a *P. canis* strain isolated from a human bitten by a dog.[51] The role of the toxin in humans is unknown but is thought to be linked to the development of tumors because of its mitogenic effect.[60] Whereas virulence factors of *P. multocida* strains–associated animal infections are well described, little is known of those factors mediating disease in humans. Clearly, more research is seriously needed to understand how *Pasteurella* species cause disease(s) in humans. Human and bacterial factors influencing mucosal colonization have remained poorly characterized. An RTX toxin, PaxA, which mediates hemolytic activity on sheep blood agar and gives a positive CAMP reaction, has been described in [*P.*] *aerogenes* strains,[56] but again, it has no known role in the pathogenesis of human pasteurellosis. This paucity of information is further observed in the rest of species within *Pasteurella sensu stricto*, where comparatively very little research has been undertaken. Consequently, the pathogenesis of human pasteurellosis is not fully understood, and the global epidemiology of human pasteurellosis has also remained poorly defined.

Commonly encountered human infections can be grouped as systemic or localized wound or arthritic infections,[86] and examples are given in Table 81.1. Often, these diseases are associated with well-defined predisposing factors, and some of these are described later. Invasive sinusitis caused by *P. multocida* was reported in an immune-compromised renal transplant patient that had been licked on mucosal surfaces by a pet dog.[79] Central nervous system (CNS) infections caused by *P. multocida* were reported to be prevalent in neonates and the elderly.[27] CNS infections caused by *P. multocida* arose following neurosurgery[46] and also secondary to a cat bite.[58] Acute osteomyelitis and cutaneous abscesses due to *P. canis* were linked to a domestic dog bite.[49] Fatal peritonitis and septicemia caused by *P. dagmatis* was reported in a patient with underlying cirrhosis who had been scratched by a pet dog.[4] In addition to *P. canis* and *P. dagmatis*, *P. stomatis* is another important cause of dog or cat bite wound infections and abscesses in humans.[51,75] In some instances, concurrent or subsequent infection with *Pasteurella* species may exacerbate preexisting nonlife-threatening conditions. Life-threatening complications may also arise from routine surgical or medical conditions. A patient preinfected with human immunodeficiency virus succumbed to pleuropneumonia exacerbated by [*P.*] *bettyae*.[67] This infrequently reported species has also been isolated from a case of peripartum bacteremia.[81] While the majority of *Pasteurella* infections can be linked to some form of predisposing factor, in some instances, this has not been possible. Severe septic arthritis, later degenerating into septic shock, ensued after a cat scratch in an otherwise healthy woman.[37] *P. multocida* septicemia was diagnosed in a 4-week-old girl with no history of scratches or direct contact with animals.[28] Acute epiglottitis caused by *P. multocida* was reported in an adult patient without any history of exposure to animals.[43]

Sporadic infections caused by species now excluded from *Pasteurella sensu stricto* have also been observed, and some of these have been associated with animal bites (see Table 81.1). [*P.*] *caballi* infection was reported in humans following horse bites.[35] [*P.*] *aerogenes* was isolated from human wounds following a pig bite.[6] Interestingly, human isolates obtained from infections related to pig bites seem to be related to the emended [*P.*] *aerogenes* rather than the misclassified biovar 17 of [*P.*] *aerogenes*.[61,62] Biovar 17 of [*P.*] *aerogenes* is related to taxon 10 of Bisgaard and has been isolated from human infections after a horse bite.[53] In other instances, a well-defined predisposing factor and/or a history of contact with animals may be lacking. Specifically, *Av. gallinarum* (formerly *P. gallinarum*) has been linked to human infections, with no apparent risk factor or source of infection. Intriguingly, *Av. gallinarum* was isolated from three unrelated cases: (i) a patient with infective endocarditis that ensued following surgical operation of the truncus arteriosus[1]; (ii) from the blood of a patient with symptoms of acute gastroenteritis[2]; and (iii) from fatal septicemia in a neonatal baby.[3] Diagnosis in all these instances was confirmed with phenotypic characteristics, which for *Av. gallinarum* are highly variable, thus increasing the risk of misidentification. This possible misidentification has previously been highlighted,[41] and it is clear that reliable identification of such doubtful isolates should be undertaken with molecular techniques such as DNA sequencing.

81.1.3 Diagnosis of *Pasteurella* Infections

81.1.3.1 Clinical Diagnosis

In general, clinical syndromes caused by members of the genus *Pasteurella* in humans are not pathognomonic. A presumptive diagnosis of *Pasteurella* species is often derived from a history of (i) contact with domestic pets; (ii) having been bitten or scratched; and (iii) some pertinent predisposing factors (mentioned in Table 81.3). In the absence of such underpinning history, symptoms alone are often not indicative since many other bacterial pathogens may give a similar clinical presentation. A definitive diagnosis of members of *Pasteurella* is always obtained from one or more laboratory tests. Although this may not influence subsequent clinical management of the condition, it helps to determine the prevalence and antibiotic sensitivity, and more importantly, it provides the foundation for a further assessment of the virulence of the species. The type of clinical specimen for laboratory diagnosis depends on the disease condition; for example, sputum and blood are recommended for suspected respiratory and septicemic *Pasteurella* infections, respectively. Swabs with or without transport media are usually appropriate for localized septic wounds. In cases of meningitis and arthritis, the preferred samples are aseptically collected taps of cerebrospinal and synovial fluids, respectively.

81.1.3.2 Conventional Identification Methods

Direct Microscopy and Isolation Techniques. Conventional laboratory processing of bacterial specimen involves direct smear preparation at the time of culturing or enrichment. Smears are fixed and stained preferably with Gram's stain or any other suitably available stain. Microscopic examination of Gram's stained smears provides a clue for the likely presence of *Pasteurella*, which assumes a gram-negative rod or coccobacillus morphology. Depending on the species or serotype, cells may exhibit bipolar staining, which is a common characteristic of recently recovered *P. multocida*. This property is often subsequently lost through serial subculture on laboratory media. In addition, direct smear examination allows numbers of predominant bacterial cells within the sample to be roughly gauged. This will indicate possible reasons for the failure to culture, which for *Pasteurella* species may be due to high levels of bacteriostatic agents in a clinical sample. While a presumptive diagnosis might be obtained from direct smear examination, morphological appearance is insufficient for species identification. Final identification relies on further phenotypic (including biochemical tests) and/or genotypic tests after primary isolation. Primary isolation is routinely undertaken on 5%–10% sheep blood agar and incubation at 37°C under capnophilic conditions often improves recovery of *Pasteurella* species. Heavily contaminated samples may be subjected to an initial enrichment step or are directly plated on selective media. A wide spectrum of selective media are available for recovering mainly *P. multocida*, but these are confounded with variable degrees of success. The most successful enrichment appears to be injecting into the peritoneum of mice and harvesting the spleen after 24 h. Mice, in general, are very susceptible to all serotypes of *P. multocida* and would die within 24 h. Homogenates from spleen are plated on 10% sheep blood agar or may be stored in liquid nitrogen for future reference. However, this technique carries serious ethical and animal welfare implications, hence the need to develop highly specific and sensitive molecular detection techniques. Though isolated from cases of pig pneumonia,[55] it is unclear whether V-factor dependent strains of *P. multocida* subsp. *multocida* cause human infections. Perhaps these are excluded because media supplying NAD[8] are not routinely used for the isolation of *P. multocida* in most standard microbiology laboratories.

Culture-Based Identification. The ease of making a presumptive identification of *Pasteurella* based on colonial morphology varies with the species involved. For example, *P. multocida* can be identified by examining for a characteristic shiny colonial morphology, a sweetish smell of indole and an absence of hemolysis on blood agar. However, other members of genus exhibit variable phenotypic characteristics that may be confused with other *Pasteurellaceae*. Consequently, colonial characteristics are of very limited value in species identification. Following successful isolation, suspected colonies are subjected to a variety of phenotypic tests that primarily comprise a panel of biochemical tests. Typically, phenotypic identification of *P. multocida* is based on characteristics listed in Christensen and Bisgaard[18] that also cater for divergent strains.[22] *P. multocida* is separated from other members of the genus *Pasteurella* mainly by (i) positive reactions in ornithine decarboxylase, indole, and acid production from (-)-D-mannitol; (ii) lack of acid formation from maltose and dextrin; and (iii) a negative urease reaction. Subspecies of *P. multocida* are separated on the basis of differences in acid production from (-)-D-sorbitol and dulcitol. *P. multocida* subsp. *gallicida* is the only subspecies that produces acid from dulcitol while *P. multocida* subsp. *septica* is the only subspecies negative in the sorbitol reaction.[69,71] Major variations in phenotypes reported for *P. multocida* and its subspecies include reactions of ornithine decarboxylase, indole, mannitol, sucrose, and sorbitol.[22,23,57] These phenotypic variations have led to misclassification in some instances. Classic examples are those of biovar 2 of *P. canis* and *P. avium*, which have subsequently been shown by genotypic methods to be *P. multocida*.[21]

After successfully identifying *P. multocida* isolates, it is a common practice to undertake serotyping for establishing relationships between host and disease. Unfortunately, serotyping is insufficient in separating isolates from different disease outbreaks and frequently, is unable to separate different subspecies of *P. multocida* in relation to diseases and hosts.[18] *P. multocida* is serologically classified into five capsular types[13,76] and 16 somatic serotypes.[50] Capsular typing is undertaken by passive hemagglutination[13] and/or by a multiplex PCR,[84] whereas gel-diffusion precipitin assays are for assigning somatic serotypes.[50] Reference serotype strains for *P. multocida* are also given in Table 81.2, though these may not be relevant with regards to human infections.

Semiautomated Systems. Automated laboratory systems using phenotypic tests generally improve the turnaround time for obtaining a definitive diagnosis. Significantly, the BACTEC 9000 series (Becton Dickinson) and the BacT/Alert (BioMerieux) monitor the growth of blood-borne bacteria by measuring the production of CO_2 using fluorescent and colorimetric sensors, respectively. However, false-positive results may be obtained, thus necessitating the use of molecular methods to confirm the presence of microbial DNA in clinical specimens. Common reasons for false-positives include (i) antibiotic usage by the patient; (ii) elevated patient leukocyte counts; and (iii) an increase in background carbon dioxide concentration. Advanced systems that also produce accurate identification with an added advantage of simultaneously providing antimicrobial sensitivities are the VITEK 2 (BioMerieux) and Phoenix (Becton Dickinson). These systems are generally expensive and thus are not routinely used in most standard laboratories. However, a similarly semiautomated system based on an analytical profile index (API) is preferentially cheaper, but a possible misidentification has been highlighted for *Pasteurella* spp.[48] Another key limitation of the API system is the lack of sporadically encountered species of *Pasteurella* from the database. These species occasionally do not fit into any known phenotypic pattern of the genus or species. Consequently, such isolates are often misidentified; hence, these commercial identification systems may need to be supplemented with specific phenotypic tests or molecular techniques.

81.1.3.3 Molecular Identification and Typing Techniques

Molecular detection techniques provide a faster alternative to obtaining a diagnosis and are increasingly gaining popularity in modern diagnostic and research laboratories. This requirement becomes even more important for septicemic infections, such as some forms of pasteurellosis, where a rapid diagnosis is required to indicate appropriate and prompt chemotherapeutic treatment. Over the years, a variety of genotype-based techniques have been developed and validated for the diagnosis of specific, important animal diseases caused by *P. multocida*; for a comprehensive review, see Reference 33. Indeed, these genus- and species-specific techniques described for diagnosis of *P. multocida* infections in animals will equally be applicable for the diagnosis of human infections caused by this species. One of the key requirements for successfully achieving molecular detection is the inclusion of positive controls in the form of reference strains. Standard reference strains can be obtained from well-recognized culture collections like the American Type Culture Collection (ATCC), National Collection of Type Cultures (NCTC), or Culture

TABLE 81.2
List of Reference Strains of *Pasteurella* Available from Culture Collections with Public Access

Species	Strain[a]	Alternative Strain Number(s)	Reference Strain for
P. multocida	CCUG 25971	X-73, ATCC 11039	Somatic type 1
P. multocida	CCUG 25973	P1059, ATCC 15742	Somatic type 3
P. multocida	CCUG 25974	P1662	Somatic type 4
P. multocida	CCUG 25975	P1702	Somatic type 5
P. multocida	CCUG 25976	P2192	Somatic type 6
P. multocida	CCUG 25977	P1997	Somatic type 7
P. multocida	CCUG 25978	P1581	Somatic type 8
P. multocida	CCUG 25979	P2095	Somatic type 9
P. multocida	CCUG 25980	P2100	Somatic type 10
P. multocida	CCUG 25981		Somatic type 11
P. multocida	CCUG 25982		Somatic type 12
P. multocida	CCUG 25983		Somatic type 13
P. multocida	CCUG 25984	P2235	Somatic type 14
P. multocida	CCUG 25985	P2237	Somatic type 15
P. multocida	CCUG 25986		Somatic type 16
P. multocida subsp. *multocida*	ATCC 43017	NCTC10323	Capsular type B
P. multocida subsp. *multocida*	ATCC 43019	NCTC10325	Capsular type D
P. multocida subsp. *multocida*	ATCC 43020	NCTC10326	Capsular type E
P. multocida subsp. *multocida*	CCUG 17976^T	W9217^T NCTC10322^T	Type strain of species Capsular type A
P. multocida subsp. *gallicida*	CCUG 17977^T	NCTC10204^T	Type strain of subspecies gallicida
P. multocida subsp. *septica*	CCUG 17978^T	NCTC11995^T	Type strain of subspecies septica
P. multocida	CCUG 26990	NCTC12177	Toxinogenic Capsular type A
P. multocida	CCUG 26985	NCTC12178	Toxinogenic Capsular type D
P. multocida	ATCC 6530		HS positive
Pasteurella canis	ATCC 43326T	NCTC 11621^T, CCUG 12400^T	
Pasteurella dagmatis	ATCC 43325^T	NCTC 11617^T, CCUG 12397^T	
Pasteurella stomatis	ATCC 43327^T	NCTC 11623^T, CCUG 17979^T	
[*Pasteurella*] *bettyae*	CCUG 2042^T	NCTC 10535^T, ATCC 23273^T	
Bisgaard taxon 45	ATCC BAA-600		
[*Pasteurella.*] *aerogenes*	ATCC 27883^T	CCUG 27904^T, CIP 80.14^T, DSM 10153^T	
[*Pasteurella*] *mairii*	ATCC 33393^T	CCUG 27189^T	
[*Pasteurella*] *caballi*	ATCC 49197^T	CIP 103915^T CCUG 28833^T	
[*Pasteurella*] *pneumotropica*	NCTC8141^T	ATCC35149^T CIP6616^T CCUG12398^T	

[a] CCUG, Culture Collection, University of Goteborg (http://www.ccug.se); ATCC, the American Type Culture Collection, Rockville, USA (http://www.lgcpromochem-atcc.com/common/catalog/bacteria/bacteriaIndex.cfm); NCTC, National Collection of Type Cultures, London (http://www.hpacultures.org.uk/collections/nctc.jsp); CIP, Collection de l'Institut Pasteur, Institut Pasteur, Paris (http://www.crbip.pasteur.fr/index2.jsp?lang=ang).

Collection of the University of Goteborg (CCUG) (see Table 81.2). Described below are examples of molecular techniques that have been employed for the successful detection of *Pasteurella* species in clinical or experimental settings.

Conventional, Nested, and Multiplex PCR. Polymerase chain reaction (PCR)–based methods are the cornerstone of modern-day molecular diagnostic assays. The basic principle involves specific amplification of a single-copy target gene, often unique to a bacterial pathogen or species under consideration, using small, complimentary oligonucleotide sequence (usually 17–21 bases) designated primers. This form of technique is referred to as conventional PCR. A standard PCR-reaction mixture comprises (i) a DNA template; (ii) a mixture of deoxynucleoside triphosphates (A, C, G, and T) for incorporation during the amplification process; (iii) specific primers binding the target gene to initiate amplification;

(iv) a polymerase enzyme, usually *Thermus aquaticus* (Taq); and (v) a conducive environment provided by buffers and an enzyme cofactor, usually magnesium. A range of PCR tests are available for *P. multocida*[20,33] but not so for other members of *Pasteurella sensu stricto.* The PCR test of Miflin and Blackall.[66] targeting the 23S rRNA gene is highly recommended for the detection of *Pasteurella* species, though it can also detect some representative strains of taxon 45 of Bisgaard.[23] Species-specific and type-specific PCR assays have been developed for *P. multocida*,[83] and a genus-specific PCR test targeting *tRNA*-intergenic spacer produces specific patterns for members of the *Pasteurella sensu stricto.*[14] These assays are promising to rapidly detect *Pasteurella* species, but evidence for their practical application is lacking. More importantly, the performance of these tests needs to be further assessed with a globally representative strain collection of *P. multocida.*

Members of the genus *Pasteurella* often colonize mucosal surfaces as mixed microflora, hence prior enrichment and/or isolation may be required before a PCR test. To increase the sensitivity, genomic DNA purification may be undertaken, but frequently, this step is unwarranted. The enzyme buffer contains a potent detergent that lyses bacterial cells to avail DNA template strands in the reaction tube. However, in some instances, a crude DNA template may be obtained by briefly boiling a small colony for 5–10 min in minimal volume of sterile distilled water and separating cell debris by centrifugation. The supernatant is used as the template for PCR reaction. While template DNA may be provided in this way, the use of purified genomic DNA should not be precluded.

Targeting a higher resolution with decreased costs, multiparametric PCR tests have been evaluated in *P. multocida*, mainly for epidemiological and phylogenetic studies. A multiplex PCR for simultaneous capsular typing of *P. multocida* now exists.[84] Other multiplex virulotyping PCR techniques continue to develop based on conserved housekeeping or virulence genes. A multiplex PCR assay targeting the genes *pfhA*, *hgbB*, *tbpA*, and *toxA* in *P. multocida* has recently been described.[5] The advantage of multiplexing is seen in the reduction of time and reagents, thus providing a significant cost benefit. To improve sensitivity, a nested PCR has been developed for detecting the *toxA* gene in demonecrotic toxin-producing strains of *P. multocida.*[16] Furthermore, nonspecific PCR tests, primarily for molecular epidemiology, in some instances, have also been useful in confirming *P. multocida* isolates. Important ones include single-primer PCR, enterobacterial repetitive intergenic consensus (ERIC), and repetitive extragenic palindromic (REP) PCR.[33]

Real-Time PCR. Real-time PCR (RT–PCR) is increasingly preferred to conventional PCR in clinical laboratories due to the need for less postamplification handling and the ability to estimate the pathogen load in a specimen.[47] The completion time of the assay is less than 5 h, thus significantly reducing throughput period. Furthermore, specific detection can be obtained from tissues harboring as little as 10^3 colony forming units (CFU) per gram. In *P. multocida*, a species-specific 5′ *Taq* nuclease assay offers rapid diagnosis and has been validated in fowl cholera strains.[29] This is a real-time PCR-based technique that utilizes minor groove binder dual-labeled probe technology. This technology is promising to provide a shorter turnaround time for the detection of *Pasteurella* species.

Nucleotide Sequence Comparison. Sequence analysis of target conserved genes may facilitate accurate identification of some species under the genus *Pasteurella.* A classic example is the 16S rRNA gene. While analysis of 16S rRNA sequences has provided the most significant improvement in phylogeny of the Pasteurellaceae family, this has also been exploited for the definitive identification of some *Pasteurella* species associated with human disease.[40] The 16S rRNA of the query strain is compared with that of the reference strain stored in the database. A commercially available 16S rRNA bacterial sequencing kit (MicroSeq500; Applied Biosystems) caters for the identification of bacterial isolates with ambiguous biochemical profiles.[39] The MicroSeq system includes software and a database containing full-length 16S rRNA gene sequences for more than 1400 different bacteria. However, commercially available 16S rRNA bacterial sequencing kits may not always be updated with the newest bacterial species and sequences. A freely accessible server, Eztaxon[26] compares the query sequence to all type strains of bacteria with the newest validly published names and in this way provides a better and more accurate identification at species level. The alternative is to compare sequences directly against those in the nucleotide databases (GenBank/EMBL/DDBJ) by BLAST search. A key limitation of this approach is that sequences of reference strains cannot easily be separated from those of less characterized isolates in the databases. However, there are specific examples where nucleotide sequence comparison enabled accurate identification of *Pasteurella* species from human infections. 16S rRNA nucleotide comparison confirmed the diagnosis of septicemia due to [*P.*] *pneumotropica* in a 72-year-old woman with no underlying disease[40] and also that of *P. dagmatis*-induced peritonitis in a patient undergoing dialysis.[85] Although 16S rRNA nucleotide sequence comparison was successfully implemented in the confirmation of *Pasteurella* species; in some instances, and as earlier mentioned, new taxa have been realized. A new taxon, Bisgaard taxon 46, related to *P. dagmatis* was proposed based on 16S rRNA sequence analysis.[23] Indeed, the power of 16S rRNA sequence analysis is further evidenced by its ability to unravel genetic diversity among *P. multocida* strains at subspecies level.[57,73]

Another recently targeted conserved gene for the accurate identification of *Pasteurella* isolates of human origin is *sodA*. The *sodA* gene encodes for a manganese-dependent superoxide dismutase and is widely used for the detection of members of the Pasteurellaceae family and other human bacterial pathogens like *Haemophilus* and *Streptococcus.* Sequence analysis of an amplified internal fragment of *sodA* provides a rapid and accurate identification of all *Pasteurella* and related species.[42] And with the increasing recognition of the power of nucleotide sequence comparison, novel gene targets are being evaluated for phylogenetic studies and diagnostic exploitation. Recently, conserved housekeeping genes, including *rpoB* (that encodes for a B-subunit of RNA polymerase), *infB* (encodes for translation initiation factor 2), and *atpD* (encodes for a B-subunit of ATP synthase) have been assessed.[21] Phylogenetic relationships within the Pasteurellaceae were observed based on partial sequence analysis of these genes, and some form of association of some species with animals was revealed.[21,54]

Fluorescent In Situ Hybridization (FISH). Hybridization techniques rely on the complementary nature of building blocks of DNA (nucleosides), that is, adenine binds to thymine, and cytosine binds to guanine. Based on this principle, techniques for the specific detection of conserved genes have been developed for diagnostic purposes. These techniques are based on use of the labeled, single-stranded, part or entire conserved gene that binds on denatured double-stranded target DNA or rRNA in the tissues. The reporter system either emits fluorescence upon excitation or breaks down a chromogenic substrate. This direct detection of bacterial DNA from infected tissues with a labeled probe eliminates the need for culturing and phenotypic identification. Using a specific Cy3 or flourescein-labeled 16S rRNA oligonucleotide probe, *P. multocida* was successfully detected in formalin-fixed paraffin-embedded lung tissues from deliberately infected pig lungs and experimentally challenged birds.[65] The diagnostic exploitation of this technique in other members of the genus is feasible, but this does not appear to have been pursued.

Molecular Typing. Though primarily used for epidemiological studies, molecular typing techniques can crudely be adapted for the identification of bacterial species by examining for a 100% similarity match of DNA fingerprint profiles with a reference strain. For *P. multocida*, pulse-field gel electrophoresis (PFGE) perfectly matched a strain from human urinary tract infection with that of *P. multocida* from a domestic cat.[63] In some instances, molecular typing tools may reveal a relationship between subgroups of strains and clinical presentation. A noteworthy correlation between taxonomic subgroups of *P. multocida* and clinical presentations was shown by 16S rRNA sequence analysis and REP–PCR.[15] Multiple locus sequence types (MLST) assays may similarly be adapted for definitive identification of bacterial species. A well-recognized scheme for *P. multocida* is based on the genes: *adk*, *aroA*, *deoD*, *gdhA*, *g6pd*, *mdh*, and *pgi*,[30] but this is primarily used for epidemiological phylogenetic studies.

81.1.3.4 Protein- and Antibody-Based Assays

To our knowledge, currently there are no protein-based detection techniques for *Pasteurella* species. Matrix-assisted laser dissection ionization–time-of-light (MALDI-TOF) mass spectrometry has been employed for research purposes, particularly in the identification of in vivo-induced proteins of *P. multocida*,[12] but its application in the molecular detection of pasteurellae is yet to be seen. However, there are selected antibody-based assays targeting specific proteins of *P. multocida*. Important examples include (i) PMT ELISA targeting dermonecrotic toxin-producing strains[38]; (ii) monoclonal antibodies targeting outer membrane proteins H and W of porcine strains[64]; and (iii) ELISA assays targeting a 37 kDa protein[74] and NanH sialidase.[80] A widely used serotyping scheme for *P. multocida* is based on capsular polysaccharide and lipopolysaccharide.

81.2 METHODS

81.2.1 Sample Preparation

DNA extraction is a prerequisite for the majority of molecular detection techniques. Extraction of bacterial DNA may be achieved by (i) simple boiling, (ii) phenol:chloroform extraction, or (iii) use of commercial kits, for example, the BioSprint 96 One-For-All Vet Kit (Qiagen). Commercial kits are particularly useful when extracting DNA directly from tissue samples. Here, we give a brief description of the standard phenol:chloroform extraction method commonly used in most ordinary research laboratories. An overnight broth culture is pelleted by centrifugation. The pellet is resuspended in Tris-EDTA buffer pH8.0. Lysates of bacterial cells are digested with proteinase K for 1–2 h at 37°C. Cell debris is precipitated with ammonium sulphate or cetyltrimethylammonium bromide (CTAB) in the presence of NaCl. Following three phenol:chloroform:isoamyl alcohol extractions, DNA is precipitated by addition of 0.7 volumes of isopropanol. Pelleted DNA is washed in 70% ethanol, air-dried, and resuspended in sterile distilled water or Tris bufffer, and then stored at −20°C or used in any of the downstream molecular detection techniques.

81.2.2 Detection Procedures

Genotypic detection relies on the resolution and subsequent staining of DNA on agarose gels. Traditionally, DNA in agarose gels is detected by staining with ethidium bromide, which fluoresces upon exposure to an ultraviolet transilluminator. However, ethidium bromide is highly toxic hence newer nontoxic DNA staining dyes are slowly entering the market. Offering nontoxic, environmentally safe, and ultrasensitive staining of DNA and surpassing SYBR® Safe (Molecular Probes Inc.) are GelRed™ and GelGreen™ (http://www.biotium.com). These nontoxic staining dyes are likely to exclusively replace ethidium bromide in the near future. Summarized detection procedures for members of the genus *Pasteurella* causing zoonotic infections are described in Table 81.3.

(i) Standard PCR Detection

Principle. A targeted conserved gene is amplified through a series of steps using a DNA polymerase enzyme. Successfully targeted genes of *Pasteurella* spp. include 23S rRNA,[66] *toxA*,[16] adenylate cyclase,[34] and capsular genes.[84] The amplification steps comprise an initial denaturation single step of a minimum of 25 cycles, consisting of denaturation for separating DNA template strands, annealing for primer binding, and extension for incorporating free nucleotides; a final single extension step; and an eventual holding temperature of 4°C.

TABLE 81.3
Selected Molecular Detection Techniques for *Pasteurella* Species

Protocol	Detection Procedures	Examples of Target Genes	Reference
1. PCR	Amplify according to the specific cycling parameters outlined in original reports. Resolve amplification products on agarose gel: stain with ethidium bromide or GelRed and fluorescence viewed under UV transilluminator: check the size of the amplified product against that given by the positive reference strain.	(i) 23S rRNA for *P. multocida* (ii) *toxA* for toxinogenic *P. multocida* (iii) tRNA intergenic spacer for genus-specific detection (iv) Adenylate cyclase gene for *Pasteurella* spp. (v) Capsular genes of *P. multocida*	66 16 14 34 84
2. RT–PCR 5′ *Taq* nuclease assay	Amplify with 50 cycles of 95°C for 15 s and 60°C for 1 min and a threshold value of 0.1 in a Rotor-Gene 3000 (Corbett Research) or similar equipment. Analyze results with an appropriate computer software. The amount of PCR product should correspond to the fluorescence emitted by reporter molecule incorporated during the amplification process.	(i) 16S rRNA for *P. multocida*	29
3. Nucleotide sequence comparison	Amplify with one cycle at 94°C for 2 min, 35 cycles of 94°C for 30 s, 55°C for 1 min, 72°C for 2 min, and one cycle at 72°C for 7 min. Analyze sequence reactions on Automatic DNA Sequencer. Compare sequences obtained using AlignX software with sequences from databases like GenBank. Microseq 500 16S rDNA Bacterial Sequencing Kit (Applied Biosystems, Foster City, CA) is available for 16S rDNA-based identification. Alternatively, sequencing may be outsourced.	(i) 16S rRNA (ii) *rpoB* gene (iii) *sodA* gene	15,19,85 21,54 42
4. Fluorescent in situ hybridization	Prehybridize slides, and denature flourescein- or Cy3-labeled 16S rRNA-based oligonucleotide probe. Hybridize in a humidified chamber overnight at 37°C; remove unbound probe by washing in appropriate buffers. View Vectashield-preserved slides with a fluorescent microscope	(i) 16S rRNA	65
5. Molecular typing tools, for example, PFGE	Digest purified genomic DNA with restriction enzyme *Sal*1 overnight for 37°C. Separate digested fragments on agarose gel at 150 V, with pulse times of 20 s for 12 h and then 5–17 s for 17 h. Stain DNA and view on UV transilluminator for similarities in patterns of fragments.	None	63

Procedure

1. In a single PCR tube, mix 2.5 μL of 10× enzyme buffer, 2 μL of 25 mM $MgCl_2$, 2 μL of 10× dNTPs, DNA template (or colony), 2 μL each of 10 pmol forward and reverse primers.
2. Adjust the volume to 24.5 μL with distilled sterile water, and add 0.5 μL of *Taq* polymerase.
3. Set the tube in a PCR cycling machine with a heated lid and start the reaction.
4. Resolve and detect the amplicon by agarose gel electrophoresis as described above.

Note: The extension time in each cycle is dependent on the size of the target gene product, usually 1 min for every 1 kb. The annealing temperature is usually set at 2°C–3°C below the lower melting temperature (Tm) of the two primers.

(ii) Real-Time PCR Detection

Principle. This type of PCR requires a pair of primers just as regular PCR and an additional fluorogenic oligonucleotide probe conjugated to both a reporter fluorescent dye and a quencher dye. The amount of PCR product corresponds to the fluorescence emitted by reporter molecule that is incorporated during amplification. The example from Corney et al.[29] targeting part of the 16S rRNA gene utilizes the primers: forward primer, 5′-ATAACTGTGGGAAACTGCAGCTAA-3′; reverse primer, 5′-GGTCCCACCCTTT(A/C)CTCCTC-3′; and MGB probe, 5′-6FAM-CCGCGTA(A/T)TCTCT-MGBNFQ-3′.

Procedure

1. Prepare reaction mixture in sterile distilled water to a final volume of 25 μL, containing: 2 μL of template, RealMasterMix™ Probe mixture, 0.2 μM per primer, and 0.3 μM the probe.
2. Amplification cycles: 95°C for 15 min, followed by 50 cycles of 95°C for 15 s and 60°C for 1 min, and a threshold value of 0.1 in a Rotor-Gene 3000 (Corbett Research) or similar equipment.

Note: Prior optimization of probe and primer concentrations that give the minimum cycle threshold (CT) values with maximum normalized fluorescence is critical. The technique can detect as little as 10^3 colony forming units (CFU) of *P. multocida* per gram of tissue.

(iii) Genotyping

Principle. Simple genotyping involves a visual comparison of the DNA fragment patterns of a reference strain and an isolate in question. These DNA fragment profiles are generated by digesting purified genomic DNA with a chosen restriction enzyme, for example, SalI for PFGE.[63]

Procedure

1. Prepare high-quality genomic DNA from the strains under consideration in 1% low-melting agarose.
2. Digest genomic DNA in agarose gel with restriction enzyme *Sal*I overnight at 37°C. Separate digested fragments by electrophoresis using 150 Volts at 10°C, with pulse times of 20 s for 12 h, and then 5–17 s for 17 h.
3. Stain DNA in agarose gel with ethidium bromide or GelRed™, view on UV transilluminator, and check for similarities or differences in profiles.

Note: The choice of a perfect RE is important since each RE potentially generates a different profile. Alternatively, commercial kits like the Genepath Reagent kit (Bio-Rad Laboratories) may be employed according to the manufacturer's protocol. And analysis of DNA fingerprints can also be undertaken by computer software, for example, Molecular Analyst Fingerprinting (Bio-Rad Laboratories).

81.3 CONCLUSION AND FUTURE PERSPECTIVES

Although the global incidence of pasteurellosis and the extent of cross-species transmission are poorly understood, there is undisputed evidence that different species of *Pasteurella* pose a significant zoonotic threat. *Pasteurella* infections in humans may resemble other bacterial infections, and if not linked to a historical contact with pets or other animals, it is frequently not considered as a differential diagnosis and hence may be overlooked. Evidently, research on the epidemiological significance, particularly highlighting the zoonotic importance of members of genus *Pasteurella,* is seriously needed. An early diagnosis is vital for initiating a prompt treatment, and in this respect, molecular detection techniques provide a quicker diagnostic platform. However, for the majority of *Pasteurella* species, these techniques are currently unavailable, most probably owing to the lack of genome sequences of the species associated with human infections. With decreased costs of nucleotide sequencing, it is expected that availability of such genomic sequences will facilitate the design of genus-specific detection techniques. The development of multiplex-PCR– and microarray-based technologies is increasingly enabling individual tests to be combined, thus providing a significant cost benefit, but such benefits are yet to be fully realized for members of the genus *Pasteurella*. Another hurdle in the development of detection techniques is the lack of a clear description of some strains sharing pasteurella-like characteristics that are encountered in human infections. Classic examples are provided by some unclassified strains under Bisgaard taxon 45 and 46 that have been isolated from bite wounds of large wild cats. Genome sequencing of such strains will pave a way for a proper classification and also enable development of genus-specific molecular detection tools. Although the ultimate value of these detection tests may not lie in the treatment of respective infections, this may be realized in the identification of latent carriers, mostly for instituting prophylactic treatment before the development of serious complications. The importance of this will only come to light when the role of humans in the epidemiology of pasteurellosis is clearly defined.

The field of molecular diagnostics is fast developing in routine clinical microbiology laboratories, thereby slowly replacing or complementing conventional culture-based and immunological assays. These gains are yet to be realized in *Pasteurella* species associated with human infections. While considerable progress has been witnessed in the field of taxonomy, studies on molecular mechanisms of virulence of *Pasteurella* species have, comparatively, received very little attention. Probably, a relatively low global incidence of *Pasteurella* infections in humans and a generally good response to chemotherapy have contributed to the very little interest received so far. Truly, it is baffling that in these modern days, when sequencing costs have been dramatically reduced, only one complete genome of a *P. multocida* strain associated with fowl cholera exists. Evidently, more research on *Pasteurella* species is seriously needed to address fundamental questions on molecular determinants of virulence. To aid this, whole genome sequencing of more *Pasteurella* strains of defined virulence becomes a prerequisite. In silico analysis will facilitate the identification of, among potential virulence genes, a wide range of conserved genes suitable for the design of better, more sensitive, and genus-specific diagnostic tools. The number of commercial molecular diagnostic assays in clinical laboratories is steadily increasing and with further availability of more whole genome sequences, this growing trend will also be expected for *Pasteurella* species. Already, rapid real-time PCR techniques and array-based technologies are proving very popular in other pathogens with sequenced genomes. Indeed, whole genome sequencing is the gateway to newer molecular detection technologies; hence, the future promises to witness a plethora of complex multiparametric automated assays. Currently, the concept of "Lab-on-the-Chip" is fast changing the world of microbial detection, both under field conditions and in clinical laboratories.

Recent developments in the postgenomic era indicate that MALDI-TOF mass spectrometry may provide a high-throughput, economical, simple, and accurate alternative to rapid microbial identification. Whether this will be applicable to *Pasteurella* species in the future remains to be proven. Although these new-age molecular detection tools are likely to provide a faster alternative to diagnostic confirmation, conventional techniques are expected to remain the hub of clinical microbiology laboratories because they provide the foundation to unrivaled services leading to epidemiological investigations, developing, and testing intervention strategies. This is even expected to continue for *Pasteurella* species, where little is known on virulence mechanisms, genomic structure, and host-specific factors mediating the disease process.

REFERENCES

1. Christensen, H. et al., Proposed minimal standards for the description of genera, species and subspecies of the *Pasteurellaceae, Int. J. Syst. Evol. Microbiol.*, 57, 166, 2007.
2. Gregersen, R.H. et al., Comparative studies on [*Pasteurella*] *testudinis* and [*Pasteurella*] *testudinis*-like bacteria and proposal of *Chelonobacter oris* gen. nov., sp. nov. as a new member of the *Pasteurellaceae, Int. J. Syst. Evol. Microbiol.*, 59, 1583, 2009.
3. Mutters, R. et al., Reclassification of the genus *Pasteurella* Trevisan 1887 on the basis of DNA homology with proposals for the new species *Pasteurella dagmatis*, *Pasteurella canis*, *Pasteurella stomatis*, *Pasteurella anatis*, and *Pasteurella langaa*, *Int. J. Syst. Bacteriol.*, 35, 309, 1985.
4. Skerman, V.B. et al., Approved Lists of Bacterial Names. *Int. J. Syst. Bacteriol.*, 30, 225, 1980.
5. Fajfar-Whetstone, C.J.T. et al., *Pasteurella multocida* septicemia and subsequent *Pasteurella dagmatis* septicemia in a diabetic patient, *J. Clin. Microbiol.*, 33, 202, 1995.
6. Fernández-Valencia, J.A., García, S., and Prat, S., *Pasteurella multocida* septic shock after a cat scratch in an elderly otherwise healthy woman: A case report, *Am. J. Emerg. Med.*, 26, 380.e1, 2008.
7. Schmulewitz, L. et al., Invasive *Pasteurella multocida* sinusitis in a renal transplant patient, *Transpl. Infect. Dis.*, 10, 206, 2008.
8. Layton, C.T., *Pasteurella multocida* meningitis and septic arthritis secondary to a cat bit, *J. Emerg. Med.*, 17, 445, 1999.
9. Guennoc, X. et al., *Pasteurella multocida* meningitis following neurosurgery, *Med. Mal. Infect.*, 36, 223, 2006.
10. Glickman, M., and Klein, R.S., Acute epiglottitis due to *Pasteurella multocida* in an adult without animal exposure, *Emerg. Infect. Dis.*, 3, 408, 1997.
11. Holst, E. et al., Characterization and distribution of *Pasteurella* species recovered from infected humans, *J. Clin. Microbiol.*, 30, 2984, 1992.
12. Liu, W. et al., *Pasteurella multocida* urinary tract infection with molecular evidence of zoonotic transmission, *Clin. Infect. Dis.*, 36, e58, 2003.
13. Christensen, H. et al., Characterization of sucrose negative variants of *Pasteurella multocida* including isolates from large cat bite-wounds, *J. Clin. Microbiol.*, 43, 259, 2005.
14. Ashley, B.D. et al., Fatal *Pasteurella dagmatis* peritonitis and septicaemia in a patient with cirrhosis: A case report and review of the literature, *J. Clin. Pathol.*, 57, 210, 2004.
15. Hara, H. et al., *Pasteurella canis* osteomyelitis and cutaneous abscess after a domestic dog bite, *J. Am. Acad. Dermatol.*, 46 (Suppl.), S151, 2002.
16. Escande, F., Vallee, E., and Aubart, F., *Pasteurella caballi* infection following a horse bite, *Zentrabl. Bakteriol.*, 285, 440, 1997.
17. Barnham, M., Pig bite injuries and infection: report of seven human cases, *Epidemiol. Infect.*, 101, 641, 1988.
18. Lindberg, J. et al., Problems of identification in clinical microbiology exemplified by pig bite wound infections, *Zentralbl. Bakteriol.*, 288, 491, 1998.
19. Moritz, F. et al., Fatal *Pasteurella bettyae* pleuropneumonia in a patient infected with human immunodeficiency virus, *Clin. Infect. Dis.*, 22, 591, 1996.
20. Shapiro, D.S. et al., Peripartum bacteremia with CDC group HB-5 (*Pasteurella bettyae*), *Clin. Infect. Dis.*, 22, 1125, 1996.
21. Christensen, H. et al., Genotypical characterization of atypical isolates of *Pasteurella multocida* and revised description of *P. multocida* to include phenotypic variants previously classified as biovar 2 of *P. canis* and *P. avium, Microbiol.*, 150, 1757, 2004.
22. Dewhirst, F.E. et al., Phylogeny of the *Pasteurellaceae* as determined by comparison of 16S ribosomal ribonucleic acid sequences, *Zentralbl. Bakteriol.*, 279, 35, 1993.
23. Mutters, R., Mannheim, W., and Bisgaard, M., Taxonomy of the group. In Adlam, C., and Rutter, J.M. (Editors), *Pasteurella and Pasteurellosis*, p. 3, Academic Press, London, 1989.
24. Mutters, R., Christensen, H., and Bisgaard, M., Genus *Pasteurella* Trevisan 1887, 94, AL Nom. Cons. Opin. 13, Jud. Comm. 1954 In Brenner, D.J., Krieg, N.R., and Staley, J.T. (Editors), *Bergey's Manual of Systematic Bacteriology*, 2nd ed., Springer-Verlag, New York, 2005.
25. Rosenbusch, C.T., and Merchant, I.A., A study of the hemorrhagic septicemia *Pasteurellae*, *J. Bacteriol.*, 37, 69, 1939.
26. Blackall, P.J. et al., Population structure and diversity of avian isolates of *Pasteurella multocida* from Australia, *Microbiology*, 144, 279–289, 1998.
27. Korczak, B.M., and Kuhnert, P., Phylogeny of *Pasteurellaceae*. In Kuhnert, P., and Christensen, H. (Editors), *Pasteurellaceae: Biology, Genomics and Molecular Aspects*, p. 27, Caister Academic Press, Norfolk, UK, 2008.
28. Kuhnert, P. et al., Phylogenetic analysis of *Pasteurella multocida* subspecies and molecular identification of feline *P. multocida* subsp. *septica* by 16S rRNA gene sequencing, *Int. J. Med. Microbiol.*, 290, 599, 2000.
29. Petersen, K.D. et al., Genetic diversity of *Pasteurella multocida* fowl cholera isolates as demonstrated by ribotyping and 16S rRNA and partial *atpD* sequence comparisons, *Microbiol.*, 147, 2739, 2001.
30. Christensen, H., and Bisgaard, M., The genus *Pasteurella*. In Dworkin, M., and Lyons, C. (Editors), *The Prokaryotes. An Evolving Electronic Resource for the Microbiological Community*, release 3.13, Springer-Verlag, New York, 2003.
31. Blackall, P.J. et al., Reclassification of *Pasteurella gallinarum*, [*Haemophilus*] *paragallinarum*, *Pasteurella avium* and *Pasteurella volantium* as *Avibacterium gallinarum* gen. nov., comb. nov., *Avibacterium paragallinarum* comb. nov., *Avibacterium avium* comb. nov. and *Avibacterium volantium* comb. nov., *Int. J. Syst. Evol. Microbiol.*, 55, 353, 2005.
32. Frederiksen, W., and Tonning, B., Possible misidentification of *Haemophilus aphrophilus* as *Pasteurella gallinarum, Clin. Infect. Dis.*, 32, 987, 2001.
33. Christensen, H. et al., Genotypical heterogeneity of *Pasteurella gallinarum* as shown by ribotyping and 16S rRNA sequencing, *Avian Pathol.*, 31, 603, 2002.
34. Blackall, P.J. et al., Reclassification of [*Pasteurella*] *trehalosi* as *Bibersteinia trehalosi* gen. nov., comb. nov., *Int. J. Syst. Evol. Microbiol.* 57, 666, 2007.
35. Bisgaard, M. et al., Classification of the taxon 2 and taxon 3 complex of Bisgaard within *Gallibacterium* and description of *Gallibacterium melopsittaci* sp. nov., *Gallibacterium trehalosifermentans* sp. nov. and *Gallibacterium salpingitidis* sp.

nov., *Int. J. Syst. Evol. Microbiol.*, 59, 735, 2009.
36. Christensen, H., and Bisgaard M., Taxonomy and biodiversity of members of *Pasteurellaceae*. In Kuhnert, P., and Christensen, H. (Editors), *Pasteurellaceae: Biology, Genomics and Molecular Aspects*, p. 1, Caister Academic Press, Norfolk, UK, 2008.
37. Olsen, I. et al., Family *Pasteurellaceae*. In Brenner, D.J., Krieg, N.R., and Staley, J.T. (Editors), *Bergey's Manual of Systematic Bacteriology*, 2nd ed., p. 851, Springer-Verlag, New York, 2005.
38. Christensen, H. et al., Emended description of porcine [*Pasteurella*] *aerogenes*, [*Pasteurella*] *mairii* and [*Actinobacillus*] *rossii*, *Int. J. System. Evol. Microbiol.*, 55, 209, 2005.
39. Kristinsson, G., *Pasteurella multocida* infections, *Pediatr. Rev.*, 28, 472, 2007.
40. Weber, D.J. et al., *Pasteurella multocida* infections: Report of 34 cases and review of the literature, *Medicine (Baltimore)*, 63, 133, 1984.
41. Griego, R.D. et al., Dog, cat, and human bites: a review, *J. Am. Acad. Dermatol.*, 33, 1019, 1995.
42. Lax, A.J., and Grigoriadis, A.E., *Pasteurella multocida* toxin: The mitogenic toxin that stimulates signalling cascades to regulate growth and differentiation, *Int. J. Med. Microbiol.*, 291, 261, 2001.
43. Lax, A.J. et al., The *Pasteurella multocida* toxin interacts with signalling pathways to perturb cell growth and differentiation, *Int. J. Med. Microbiol.*, 293, 505, 2004.
44. Mullan, P.B., and Lax, A.J., *Pasteurella multocida* toxin is a mitogen for bone cells in primary culture, *Infect. Immun.*, 64, 959, 1996.
45. Rozengurt, E. et al., *Pasteurella multocida* toxin: Potent mitogen for cultured fibroblasts, *Proc. Natl. Acad. Sci. USA*, 87, 123, 1990.
46. Donnio, P.Y. et al., Dermonecrotic toxin production by strains of *Pasteurella multocida* isolated from man, *J. Med. Microbiol.*, 34, 333, 1991.
47. Kuhnert, P. et al., Characterization of PaxA and its operon: A cohemolytic RTX toxin determinant from pathogenic *Pasteurella aerogenes*, *Infect. Immun.*, 68, 6, 2000.
48. Clapp, D.W. et al., *Pasteurella multocida* meningitis in infancy: an avoidable infection, *Am. J. Dis. Child.*, 140, 444, 1986.
49. Pouedras, P. et al., *Pasteurella stomatitis* infection following a dog bite, *Eur. J. Clin. Microbiol. Infect. Dis.*, 12, 65, 1993.
50. Cohen-Adam, D. et al., *Pasteurella multocida* septicemia in a newborn without scratches, licks or bites, *Isr. Med. Assoc. J.*, 8, 657, 2006.
51. Lester, A. et al., Phenotypical characters and ribotyping of *Pasteurella aerogenes* from different sources, *Zentralbl. Bakteriol.*, 279, 75, 1993.
52. Kjældgaard, P. et al., Taxon 10, a potential pathogen associated with infected horse bite wounds. Poster abstract. International *Pasteurellaceae* Society Conference, Sorento, 2008.
53. Ahmed, K. et al., *Pasteurella gallinarum* neonatal meningitis, *Clin. Microbiol. Infect.*, 8, 55, 2002.
54. Al Fadeh Saleh, M. et al., First case of human infection caused by *Pasteurella gallinarum* causing infective endocarditis in an adolescent 10 years after surgical correction for truncus arteriosus, *Paediatrics*, 95, 944, 1995.
55. Arashima, Y. et al., First case of *Pasteurella gallinarum* isolation from blood of a patient with symptoms of acute gastroenteritis in Japan, *Clin. Infect. Dis.*, 29, 698, 1999.
56. Krause, T.H. et al., V-factor dependent strains of *Pasteurella multocida* subsp *multocida*, *Zentralbl. Bakteriol. Hyg. A*, 266, 255, 1987.
57. Blackall, P., and Nørskov-Lauritsen, N., *Pasteurellaceae*—the view from the diagnostic laboratory. In Kuhnert, P. and Christensen, H. (Editors), *Pasteurellaceae: Biology, Genomics and Molecular Aspects*, p. 227, Caister Academic Press, Norfolk, UK, 2008.
58. Christensen, H. et al., Comparative phylogenies of the housekeeping genes *atpD, inf*B and *rpo*B and the 16S rRNA gene within *Pasteurellaceae*, *Int. J. Syst. Evol. Microbiol.*, 54, 1601, 2004.
59. Carter, G.R., Studies on *Pasteurella multocida*. I. A haemagglutination test for the identification of serological types, *Am. J. Vet Res.*, 16, 481, 1955.
60. Rimler, R.B., and Rhoades, K.R., Serogroup F, a new capsule serogroup of *Pasteurella multocida*, *J. Clin. Microbiol.*, 25, 615, 1987.
61. Heddleston, K.L., Gallagher, J.E., and Rebers, P.A., Fowl cholera: Gel diffusion precipitin test for serotyping *Pasteurella multocida* from avian species, *Avian Dis.*, 16, 925, 1972.
62. Townsend, K.M. et al., Genetic organization of *Pasteurella multocida* cap loci and development of a multiplex capsular PCR typing system, *J. Clin. Microbiol.*, 39, 924, 2001.
63. Hamilton-Miller, J.M.T., A possible pitfall in the identification of *Pasteurella* spp. with the API system, *J. Med. Microbiol.*, 39, 78, 1993.
64. Dziva, F. et al., Diagnostic and typing options for investigating diseases associated with *Pasteurella multocida*, *Vet. Microbiol.*, 128, 1, 2008.
65. Christensen, H. et al., PCR detection of [*Haemophilus*] *paragallinarum*, [*H.*] *somnus*, *Mannheimia* (*Pasteurella*) *haemolytica*, *Mannheimia* spp., [*P.*] *trehalosi*, and *Pasteurella multocida*. In Sachse, K., and Frey, J. (Editors), *PCR Detection of Microbial Pathogens: Methods and Protocols*, p. 257, The Humana Press Inc., Towota, New Jersey, 2003.
66. Miflin, J.K., and Blackall, P.J., Development of a 23 S rRNA-based PCR assay for the identification of *Pasteurella multocida*, *Lett. Appl. Microbiol.*, 33, 216, 2001.
67. Townsend, K.M. et al., Development of PCR assays for species and type-specific identification of *Pasteurella multocida* isolates, *J. Clin Microbiol.*, 36, 1096, 1998.
68. Catry, B. et al., *tRNA*-intergenic spacer PCR for the identification of *Pasteurella* and *Mannheimia* spp., *Vet. Microbiol.*, 98, 251, 2004.
69. Atashpaz, S., Shayegh, J., and Hejazi, M.S., Rapid virulence typing of *Pasteurella multocida* by multiplex PCR, *Res. Vet. Sci.*, 87, 355, 2009.
70. Choi, C., and Chae, C., Enhanced detection of toxigenic *Pasteurella multocida* directly from nasal swabs using a nested polymerase chain reaction, *Vet J.*, 162, 255, 2001.
71. Guenther, S. et al., Real-time PCR assay for the detection of species of the genus *Mannheimia*, *J. Microbiol. Methods*, 75, 75, 2008.
72. Corney, B.G. et al., *Pasteurella multocida* detection by 5′ *Taq* nuclease assay: A new tool for use in diagnosing fowl cholera, *J. Microbiol. Methods,* 69, 376, 2007.
73. Frebourg, N.B. et al., Septicemia due to *Pasteurella pneumotropica*: 16S rRNA sequencing for diagnosis confirmation, *J. Clin. Microbiol.*, 40, 687, 2002.
74. Fontana, M.C. et al., Use of the MicroSeq 500 16S rRNA gene-based sequencing for identification of bacterial isolates that commercial automated systems failed to identify correctly, *J. Clin. Microbiol.*, 43, 615, 2005.

75. Chun, J. et al., EzTaxon: a web-based tool for the identification of prokaryotes based on 16S ribosomal RNA gene sequences, *Int. J. Syst. Evol. Microbiol.*, 57, 2259, 2007.
76. Wallet, F. et al., Molecular identification of *Pasteurella dagmatis* peritonitis in a patient undergoing peritoneal dialysis, *J. Clin. Microbiol.*, 38, 4681, 2000.
77. Gautier, A-L. et al., Rapid and accurate identification of human isolates of *Pasteurella* and related species by sequencing the *sodA* gene, *J. Clin. Microbiol.*, 43, 2307, 2005.
78. Mbuthia, P.G. et al., Specific detection of *Pasteurella multocida* in chickens with fowl cholera and in pig lung tissues using fluorescent rRNA in situ hybridization, *J. Clin. Microbiol.*, 39, 2627, 2001.
79. Chen, H.I., Hulten, K., and Clarridge III, J.E., Taxonomic subgroups of *Pasteurella multocida* correlate with clinical presentation, *J. Clin. Microbiol.*, 40, 3438, 2002.
80. Davies, R.L., MacCorquodale, R., and Reilly, S., Characterisation of bovine strains of *Pasteurella multocida* and comparison with isolates of avian, ovine and porcine origin, *Vet. Microbiol.*, 99, 145, 2004.
81. Boyce, J.D. et al. Analysis of the *Pasteurella multocida* outer membrane subproteome and its repsonse to the in vivo environment of the natural host, *Proteomics*, 6, 870, 2006.
82. Foged, N.T., Nielsen, J.P., and Pedersen, K.B., Differentiation of toxigenic from nontoxigenic isolates of *Pasteurella multocida* by enzyme-linked immunosorbent assay, *J. Clin. Microbiol.*, 26, 1419, 1988.
83. Marandi, M.V., and Mittal, K.R., Identification and characterization of outer membrane proteins of *Pasteurella multocida* serotype D by using monoclonal antibodies, *J. Clin. Microbiol.*, 33, 952, 1995.
84. Peterson, R.R., Deeb, B.J., and DiGiacomo, R.F., Detection of antibodies to *Pasteurella multocida* by capture enzyme immunoassay using a monoclonal antibody against P37 antigen, *J. Clin. Microbiol.*, 35, 208, 1997.
85. Sanchez, S. et al., Serological response to *Pasteurella multocida* NanH sialidase in persistently colonized rabbits, *Clin. Diagn. Lab. Immunol.*, 11, 825, 2004.
86. Escande, F., and Crasnier, M., Detection of adenylate cyclase gene in *Pasteurella* species, *Zentralbl. Bakteriol.*, 279, 45, 1993.

82 Photobacterium

Carlos R. Osorio and Manuel L. Lemos

CONTENTS

82.1 INTRODUCTION

82.1.1 Classification, Morphology, and Epidemiology

Classification. The genus *Photobacterium* is included in the family Vibrionaceae and belongs to the class Gammaproteobacteria. First proposed by Beijerinck in 1889,[1] the genus is currently composed of the following seventeen species: *Photobacterium phosphoreum*,[1,2] *P. leiognathi*,[3] *P. fischeri*,[1,2] *P. angustum*,[4] *P. damselae* (subsp. *damselae* and subsp. *piscicida*),[5,6] *P. iliopiscarium*,[7] *P. profundum*,[8] *P. indicum*,[9] *P. rosenbergii*,[10] *P. lipolyticum*,[11] *P. frigidiphilum*,[12] *P. aplysiae*,[13] *P. ganghwense*,[14] *P. halotolerans*,[15] *P. lutimaris*,[16] *P. kishitanii*,[17] and *P. aquimaris*.[18]

Currently, *P. damselae* includes strains classified into two distinct subspecies, *damselae* and *piscicida*,[6] which include the species formerly denominated as *Vibrio damsela* and *Pasteurella piscicida*, respectively. These two current subspecies were originally placed into two different families (Vibrionaceae and Pasteurellaceae, respectively) showing a wide range of phenotypic differences. However, a study that included the analysis of 16S rRNA gene sequences and DNA–DNA hybridization values demonstrated that the formerly named *Vibrio damsela* and *Pasteurella piscicida* were indeed members of a same species, *Photobacterium damselae*.[6]

P. damselae subsp. *damselae* was first identified as "an unnamed marine *Vibrio*," after being isolated as the causative agent of a human infectious case in 1971.[19] Later, this organism was isolated from skin ulcers of damselfish (*Chromis punctipinnis*) and therefore named *Vibrio damsela*.[20] In 1985, it was assigned to the genus *Listonella* as *Listonella damsela*, based on its 5S rRNA sequence. However, further genetic and phenotypic studies suggested that this organism was closely related to species of the genus *Photobacterium*, and it was named *Photobacterium damsela*.[5] In 1995, the 16S rRNA sequence analysis of a fish pathogen formerly named *Pasteurella piscicida*,[21] together with DNA–DNA hybridization data, suggested that this species was closely related to *P. damselae*, and both organisms were assigned to the same species epithet, *P. damselae*, with category of subspecies.[6] Thus, *P. damselae* subsp. *piscicida* comprises the strains that fall within the description of the formerly named *Pasteurella piscicida*, whereas the strains having the characteristics of *Vibrio damsela* have been since then included in *P. damselae* subsp. *damselae*.

Sequence analysis of the 16S rRNA gene of a large collection of strains of these two subspecies showed that they share exactly the same nucleotide sequence.[22] A high similarity between the strains of these two subspecies has also been reported for other genetic regions, in addition to the 16S rRNA gene, as is the case with the 23S and 5S rRNA genes and their respective intergenic spacer regions,[23,24] the virulence regulator gene *toxR*,[25] the iron regulator gene *fur*, and genes for heme transport.[26,27] However, despite this similarity at the nucleotide sequence level in many genes and the high percentage of DNA–DNA relatedness, strains of these two subspecies are clearly distinguished by a number of phenotypical traits (Table 82.1).[28,29] Of special relevance is the ability of strains of the subspecies *damselae* to grow at 37°C (a temperature otherwise inhibitory for subsp. *piscicida* strains), a trait that is essential for a bacterium to colonize and establish an infection in a homeotherm animal. Other phenotypical traits that are positive only for subsp. *damselae* isolates, and of special interest in subspecies discrimination for identification purposes, are motility, nitrate reduction and hemolysis on sheep blood agar, as well as urease activity (although urease-negative strains have been described).

Morphology. Morphologically, the members of the genus *Photobacterium* are straight rods of 0.8–1.3 × 1.8–2.4 μm, although *P. damselae* and *P. leiognathi* appear as coccobacilli, with *P. damselae* subsp. *piscicida* displaying a typical bipolar stain. Except for *P. damselae* subsp. *piscicida*, all *Photobacterium* species are motile by one to three polar flagella. These flagella, unlike those of *Vibrio*, are unsheathed. The species are facultative anaerobic, chemoorganotrophic, having both a respiratory and a fermentative metabolism. *Photobacterium* strains require sodium ions for growth, achieving optimal growth with 160–280 mM Na^+. The optimal growth temperature range is 18°C–25°C, with the notable exception of *P. profundum*, a psychrophilic species that cannot grow at temperatures over 20°C. *Photobacterium* strains accumulate poly-β-hydroxybutyrate (PHB) under certain conditions of cultivation, but they cannot use the exogenous monomer β-hydroxybutyrate.

Epidemiology. *Photobacterium* species are widespread in marine environments and have been isolated from seawater, sediment, the specialized light organs of fish, the surface and intestines of healthy marine animals as well as from the organs of diseased marine animals, and from humans. *P. damselae* subsp. *damselae* and *P. damselae* subsp. *piscicida* are pathogenic for several marine species. Among all the species, *P. damselae* subsp. *damselae* is the only one with pathogenic potential for homeotherms including humans and will constitute the main subject of this chapter.

P. damselae subsp. *damselae* is an autochthonous member of aquatic ecosystems. Strains of this subspecies have been isolated from aquatic environments and from a variety of aquatic animals as well as from humans. It is considered as a primary pathogen causing wound infections and hemorrhagic septicemia in different species of wild fish (damselfish, shark, stingray, catfish, and so on), as well as in reared fish that are of economical importance in aquaculture.[30] Some of the species affected include sea bream (*Sparus aurata*), sea bass (*Dicentrarchus labrax*), turbot (*Psetta maxima*), rainbow trout (*Oncorhynchus mykiss*), and yellowtail (*Seriola quinqueradiata*) among others. Most isolates that are virulent for fish show a wide host range, including a large diversity of economically important farmed fish and also mice. Since its first description as a fish pathogen, it has been recognized as a pathogen also for reptiles, such as the leatherback sea turtle (*Dermochelys coriacea*),[31] mollusks (*Octopus joubini*),[32] and crustacean species.[33,34] There are also reports on the isolation of *P. damselae* subsp. *damselae* from different species of dolphins (*Tursiops truncatus* and *Delphinus delphis*)[35,36] and from Bryde's whale (*Balaenoptera edeni*).[36]

It is known that virulent strains of this pathogen can infect new hosts (fish) through water, and the spread of the disease depends largely on the temperature and salinity of the water.[37] Similarly, virulent *P. damselae* subsp. *damselae* strains can survive in seawater microcosms at 14°C–22°C as culturable bacteria for long periods of time and maintain their infectivity for fish.[38] *P. damselae* subsp. *damselae* can be recovered from seawater and seaweeds, as well as from uninfected marine animals.[39] Since the aquatic ecosystem can be considered as a natural reservoir for fish virulent strains, and the infection is known to be transmitted through saline water and warm temperatures, the presence of this subspecies in aquatic environments represents a threat not only for marine animals but also for humans. It is suggested that the disease caused by this bacterium in marine animals might follow a seasonal distribution, since warmer water temperatures and an increase of salinity in estuarine environments are factors that contribute to an increase in the number of infections caused by *P. damselae* subsp. *damselae*.

Most of the reported human infections due to *P. damselae* subsp. *damselae* have their origin in wounds inflicted during the handling of fish or after exposure to seawater, marine animals, or the marine environment in general. Table 82.2 shows a summary of cases in which *P. damselae* subsp. *damselae* was found to be the cause of disease in humans. The majority of the cases occurred in coastal areas of the United States and Japan, mostly associated with wounds suffered while handling either fish or fishing tools.

TABLE 82.1
Summary of Main Phenotypical Traits That Differentiate Two Subspecies of *Photobacterium damselae*

Test	subsp. *damselae*	subsp. *piscicida*
Motility	+	−
Growth in TCBS agar	+	−
Urease	+	−
Gelatinase	V	−
Amylase	V	−
Nitrate reduction	+	−
Gas from glucose	+	−
Growth at 37°C	+	−
Growth at 5% NaCl	+	−
Hosts	Fish, marine invertebrates, reptiles, mammals, and humans	Fish
Hemolysis	+	−

TABLE 82.2
A Summary of Reports of Infections in Humans Caused by *P. damselae* subsp. *damselae*

Date and Reference	Location	Patient	Way of Infection	Underlying Disease	Outcome
1982[19]	N/A	Six cases	Wounds exposed to brackish water	No (healthy persons)	N/A
1985[40]	USA	61-year-old man	Handling of catfish	Pancreatitis and diabetes	Fatal
1986[41]	USA	N/A	N/A	N/A	Necrotizing infection
1989[42]	Australia	N/A	Wound infections	N/A	N/A
1993[43]	USA	70-year-old man	Fish handling	No (healthy person)	Fatal fulminant septicemia
1993[44]	Southeast Asia	N/A	Injury by rabbit fish	No (healthy person)	Fatal necrotizing fasciitis
1996[45]	N/A	63-year-old man	N/A	Diabetes and alcoholic liver disease	N/A
1997[46]	USA	64-year-old man	Injury by fish handling	Heart disease	Fatal
1999[47]	N/A	N/A	N/A	N/A	Necrotizing fasciitis
2000[48]	N/A	N/A	A man lashed by a stingray	N/A	Necrotizing fasciitis
2004[49]	USA	69-year-old man	N/A	N/A	Fatal necrotizing fasciitis
2004[50]	Japan	58-year-old man	Fishing	Diabetes mellitus	Fatal necrotizing fasciitis
2004[51]	Japan	76-year-old man	Fishing	No	Fatal necrotizing fasciitis
2005[52]	Jamaica	Child	N/A	Sickle-cell disease	Bacteremia
2006[53]	USA	22-year-old pregnant woman	N/A	No	Urinary tract infection
2008[54]	Japan	68-year-old man	Harpoon injury	Diabetes mellitus	Necrotizing fasciitis
2009[55]	N/A	14-year-old man	Wound infection	N/A	N/A

N/A, not available.

82.1.2 Pathogenesis

Virulence factors employed by *P. damselae* subsp. *damselae* to cause disease have been studied mainly in fish isolates. Infected fish show hemorrhaged areas on the body surface and ulcerative lesions as typical signs of the disease. The ulcers typically occur in the region of the pectoral fin and caudal peduncle and may reach 5–20 mm in diameter, as has been described for damselfish.[20] In turbot, the most remarkable symptoms are extensive hemorrhage in the eyes and mouth, including the palate and jaw.[55] In strains virulent for fish, several putative virulence factors have been proposed, which include (i) resistance to fish serum, (ii) iron-uptake systems, and (iii) exotoxins, such as phospholipase and others.

Resistance to Fish Serum. It is known that the resistance to the bactericidal effect of fish serum complement is correlated with pathogenicity in *P. damselae* subsp. *damselae*. In addition, only virulent strains survive and multiply in serum, and their growth is enhanced by the addition of iron in the form of hemoglobin.[56] The ability to obtain iron from host tissues is crucial for most bacterial pathogens in order to establish an infection.[57] In this regard, *P. damselae* subsp. *damselae* is not an exception. Bacterial pathogens usually obtain iron by two main mechanisms, one consisting of utilizing heme-containing proteins, and a second main mechanism is dependent on the production of siderophores, small iron-chelating molecules with high affinity for iron that are further transported to the bacterial cell cytoplasm via specific outer membrane receptors. It has been reported that virulent *P. damselae* subsp. *damselae* strains (isolated from fish and also from human isolates) are capable of obtaining iron from both artificial and natural iron-restricted environments (strains resist the bacteriostatic effect of serum transferrins).[56,58]

Iron-Uptake Systems. Pathogenicity assays conducted with iron-overloaded animals demonstrated that the availability of iron increases host susceptibility to infection by virulent *P. damselae* subsp. *damselae* strains, showing a significant reduction in the LD_{50}. *P. damselae* subsp. *damselae* can also utilize heme and hemoglobin as iron sources. The production of hemolysins is believed to play a role in the increase of free heme. Recently, a set of genes encoding a system for the uptake and utilization of heme as source of iron was identified in a human isolate of *P. damselae* subsp. *damselae*.[27] In addition to the use of heme as iron source, this bacterium is known to produce a hydroxamate-type siderophore as well as a series of high molecular weight outer membrane proteins that are induced under conditions of iron limitation. One of these proteins is believed to constitute the specific outer membrane receptor for the iron-siderophore complex.[56]

Exotoxins. Several enzymatic activities have been detected in the extracellular products (ECPs) of *P. damselae* subsp. *damselae*, which include phospholipase and hemolysin activities.[59] The experimental in vivo assays carried out with fish showed that the infection by this subspecies causes hemorrhages and necrosis in tissues. In addition, the in vitro assays with different cell lines indicated existence of cytotoxic and cytolytic effects.[59] The histological analysis of internal organs from experimentally infected turbot, indicated that the exotoxins and cells of virulent strains of this pathogen cause similar tissue damage. This includes structural changes, necrosis, and destruction of cells, as well as the presence of numerous blood cells in interstitial tissue, evidencing internal hemorrhages. A positive correlation was found between virulence and the toxicity of ECPs, since the most toxic ECPs were found to derive from the most virulent strains.[55]

The main exotoxin so far identified in *P. damselae* subsp. *damselae* strains is a potent cytolysin named damselysin. In 1984,[60] the correlation observed between the ability of 19 isolates of *P. damselae* subsp. *damselae* to cause disease in mice and to produce in vitro large amounts of a heat-labile cytolytic toxin was reported. In further studies, this toxin was purified, its molecular weight estimated in ca. 69,000, and its activity against the erythrocytes of a variety of animals demonstrated.[61] Later, this toxin was defined as a phospholipase-D, active against sphingomyelin with hemolytic activity.[62] The gene encoding the damselysin toxin, named *dly*, was cloned and sequenced.[63] Although the genomic context of this gene remains uncharacterized, it has been proposed that it could be carried on an element of mobile DNA, since the loss of *dly* gene, together with extensive flanking sequences, has been detected in *P. damselae* subsp. *damselae* strains that spontaneously showed a markedly reduced hemolytic phenotype.[63] These same authors reported the existence of hemolytic strains of this pathogen that tested negative for presence of the *dly* gene, an observation suggesting that those strains might encode a different cytolysin. In a further work, it was reported that presence of *dly* gene was not a prerequisite for the hemolytic activity and for the pathogenicity for mice or fish.[64] In this case, only 6 out of 17 strains isolated from humans, marine animals, and seawater tested positive for the *dly* gene, while certain strains negative for this gene showed hemolytic activities against sheep and rabbit erythrocytes and virulence degrees for mice and fish comparable to some of the *dly*-positive isolates. Thus, the results not only demonstrate that *dly* gene is not an indicator of the pathogenicity of this bacterium for mice, but also point at the existence of alternative genes encoding proteins with hemolytic activity different from damselysin in *P. damselae* subsp. *damselae* strains. The nature of this protein(s) remain unknown.

Comparison of Clinical and Environmental Isolates. In a recent study,[65] the phenotypic and genotypic differences between two deadly clinical isolates of *P. damselae* subsp. *damselae* and other isolates of this subspecies from the environment, as well as their pathogenicity for mice, were investigated. The sequence analysis of *gyrB*, *toxR*, and *ompU* genes revealed that the two clinical isolates, together with one environmental isolate, clustered together distinctly from other environmental isolates. However, using whole genome typing techniques as ribotyping, AFLP, and PFGE analysis, no significant differences between clinical and environmental strains were evidenced. The two clinical strains showed profiles distinct from the environmental strains in a series of biochemical properties tested. These differences included enzymatic activities that were present in clinical isolates and absent from virtually all the environmental strains (namely, gelatinase production), and the ability to ferment a collection of substrates (esculin, salicin, and some sugars, among others). The two clinical isolates along with four environmental strains showed large hemolytic haloes on sheep blood agar, whereas 9 out of 13 environmental strains showed small haloes. The two clinical strains showed to be the most strongly pathogenic for mice, their LD_{50} values being one to two orders of magnitude lower than those of environmental strains. Other evidence drawn from that study was that the pathogenicity of *P. damselae* subsp. *damselae* for mice and hemolytic activity are usually positively correlated, in such a way that those strains with large hemolytic haloes ranked among the most pathogenic strains for mice. This correlation was also demonstrated in other studies.[60,64] Takahashi et al.[65] analyzed plasmid content of *P. damselae* subsp. *damselae* strains and found that plasmids were observed in all the strongly hemolytic isolates but not in the weakly hemolytic ones (with the exception of one isolate). This might suggest a contribution of plasmid-encoded genes to hemolytic activities, although this hypothesis would need further investigation.

Thus, it could be said that clinical and environmental isolates of this subspecies are closely and genetically related and not clearly distinguishable from environmental isolates, although phenotypic typing shows distinct profiles for a few biochemical traits that, broadly, do not show an evident link with virulence (although gelatinase production by clinical strains might be related to the ability to invade host tissues). The most distinctive characteristic of clinical isolates is thus their pathogenicity profile for mice.

82.1.3 Clinical Features

As it can be deduced from the cases listed in Table 82.2, most infections caused by *P. damselae* subsp. *damselae* have their primary origin in wounds that are exposed to salt or brackish water or by handling of fish.[19,40–54] In a majority of the reported clinical cases, *P. damselae* subsp. *damselae* is responsible for an extreme variant of necrotizing fasciitis, highly severe, and that advances along contiguous tissue following a remarkably aggressive course. The potential of this pathogen to cause fulminant systemic effects has been reported in many cases. In fact, some authors emphasize the importance of surgically debriding and amputating without hesitation at a very early point of infection by this pathogen, since this rapid action might be the only intervention capable of saving the lives of patients.[49] In one of the very first cases reported, in 1984, of fatal infections due to this pathogen, the patient injured his hand while handling a catfish, and in less than 24 h, he had a marked edema of the forearm and bullae formation on the hand.[40] Although extensive debridement of the affected area, which extended to the shoulder and the chest, was performed, the patient died after a series of complications which included disseminated intravascular coagulation, acute renal failure, and hypercalcemia. Tissue from the hand and bullae were taken for culture. *P. damselae* subsp. *damselae* was recovered in high numbers from the tissue sample but only in very small numbers from the bullae fluid.[40]

Some patients infected by this organism develop multiple organ failure within a few hours from the onset of initial symptoms, despite intensive chemotherapy and surgical treatments. This was the outcome of a fatal case in a 64-year-old man caused by *P. damselae* subsp. *damselae* after he was injured while handling fish.[46] The pathogen was isolated in pure culture from wound specimens but could not be isolated in cultures from blood specimens. These authors suggested that a virulence factor or systemic toxin released by this bacterium contributed to this outcome, but septicemia did not. In other cases of infection by *P. damselae* subsp. *damselae*, the organism was recovered from blood.[43,45] In two fatal cases reported in Japan, this organism caused a sepsis that was proven by blood culture.[60] The patients developed multiple organ failure within 20–36 h from the onset of the initial symptoms despite chemotherapy and surgical treatments.

To date, a unique case of urinary tract infection by *P. damselae* subsp. *damselae* has been described, affecting a pregnant woman.[52] According to the reports, necrotizing fasciitis due to *P. damselae* subsp. *damselae* tended to demonstrate more serious complications and a higher mortality rate than those caused by *Vibrio vulnificus*. One of the main differences between the cases due to either of these two pathogens resides in the fact that necrotizing fasciitis by *V. vulnificus* usually affects persons with underlying diseases (mainly chronic liver disease and diabetes mellitus), whereas necrotizing fasciitis by *P. damselae* subsp. *damselae* sometimes occurs in healthy hosts (Table 82.2). Also, in many cases, antibiotic administration proved unable to control the progression of fatal infections due to this pathogen.[40,46,50]

82.1.4 Diagnosis

82.1.4.1 Phenotypic Techniques

Conventional plate and tube biochemical tests, together with microscopy observation can lead to a good identification of this pathogen once it has been isolated in pure culture. *P. damselae* subsp. *damselae* includes strains able to grow at 37°C that form round to slightly irregular, off-white, opaque, flat, shiny colonies. Gram-negative rods, motile but nonswarming, capable of growing in TCBS as green colonies, catalase- and oxidase-positive, and sensitive to the vibriostatic agent O/129. They are positive for arginine dihydrolase (ADH) but negative for ornithine decarboxylase (ODC) and lysine decarboxylase (LDC); positive for Voges-Proskauer, production of gas from glucose, and production of lipase and urease (although urease-negative isolates have been identified[29]). Able to grow at 1%–6% NaCl, with an optimum of 3% NaCl, it produces phosphatase and DNase, but not gelatinase (although some strains were found to be gelatinase producers). Most of strains produce β-hemolysis on agar supplemented with sheep erythrocytes.

The API 20E multigallery (Biomerieux, France) currently includes the profile for *P. damselae* subsp. *damselae* and allows a good identification of this subspecies. One typical characteristic of *P. damselae* subsp. *damselae* strains is that they are rather homogeneous in their biochemical and physiological properties, irrespective of their origin. However, some variation occurs for certain phenotypical traits as cited above. In this regard, the isolation of different *P. damselae* subsp. *damselae* phenotypes from the same host at the same time is noteworthy. As an example, two different hemolytic phenotypes were isolated from the tissue of a patient with a fatal wound infection, and the two isolates differed in their lecithinase, lipase, and hemolytic activities.[40]

There is a high heterogeneity in the molecular composition of the outer membrane. Thus, up to seven different profiles of lipopolysaccharides and seven different patterns of outer membrane proteins have been reported.[30] Antigenic variation has been also reported in *P. damselae* subsp. *damselae*. At least four O-antigen serogroups have been identified,[30] although serogroups do not correlate with the source of isolation. Rapid diagnosis of the different serogroups can be achieved by dot blot assay. Nevertheless, during the last years, isolates of *P. damselae* subsp. *damselae* belonging to new serogroups have been recovered from diseased fish species in Europe, which suggests that the serotyping system of this subspecies should be revised.

82.1.4.2 Molecular Techniques

The 16S rRNA molecule constitutes a crucial tool in molecular bacteriology phylogeny studies, and the determination and comparative analyses of 16S sequences made possible the revealing of the evolutionary relationships between species. This ribosomal gene can be amplified by PCR and subjected to direct DNA sequencing following a very simple methodology. For its PCR amplification, several pairs of universal primers that target conserved positions at both 5′ and 3′ ends of the 16S rRNA gene are available in the literature.[66,67] The 16S gene was sequenced in 18 strains of subsp. *piscicida* and eight strains of subsp. *damselae*, and it was evidenced that the 16S rRNA consensus sequence of *P. damselae* subsp. *damselae* and subsp. *piscicida* are identical.[22] A comparative analysis of these 26 *P. damselae* 16S sequences with those of 16S rRNA genes from a collection of *Vibrio* and *Photobacterium* species, allowed the selection of variable regions within this gene that were specific for *P. damselae*, and were used to design three primers to be employed in a nested-PCR approach for the detection of both *P. damselae* subsp. *piscicida* and subsp. *damselae*.[22]

Despite the fatal cases reported after infections with *P. damselae* subsp. *damselae*, this subspecies has received little attention from microbiologists in terms of designing molecular detection protocols. In addition, since the subsp.

damselae does not have the economic importance of the subsp. *piscicida* in sea-farming industry, the 16S gene–based protocol reported for the PCR detection of *P. damselae* subsp. *damselae* is indeed an indirect consequence of the design of a molecular detection method for subspecies *piscicida*.[68] The high degree of overall genomic similarity between subspecies *damselae* and subspecies *piscicida* has indeed hampered the design of a PCR-based procedure that specifically detects subsp. *damselae* without giving cross-reactions with subsp. *piscicida*. However, a multiplex PCR method has been reported to discriminate between the two subspecies of *P. damselae* by using a combination of primers targeted to the 16S gene and the *ureC* gene.[69] One of the phenotypical traits characteristic of *P. damselae* subsp. *damselae* isolates as opposed to subsp. *piscicida* is the ability of the first to hydrolize urea. Although rarely occurring within the family Vibrionaceae, urease genes are present in several species of *Vibrio* and *Photobacterium*. The combination of 16S gene and *ureC* gene primers in the same PCR reaction allowed the identification of *P. damselae* subsp. *damselae* isolates from pure cultures.[69] This multiplex PCR approach has since been successfully utilized by other authors for diagnosis of *P. damselae* subsp. *damselae* in crustaceans.[34]

As we have described earlier, the hemolytic activities of *P. damselae* subsp. *damselae* strains is not always correlated to the presence of the *dly* gene encoding the toxin damselysin.[64] Screening of *dly* gene in a collection of 17 *P. damselae* subsp. *damselae* strains isolated from marine animals, seawater, and humans revealed that six strains were positive for this gene (including an human isolate), whereas 11 strains tested negative. Since no *dly*-related sequences have been encountered so far in other *Vibrio* or *Photobacterium* species, the PCR detection of *dly* gene in a isolate suspected to belong to *P. damselae* subsp. *damselae* can be of utility.

82.2 METHODS

82.2.1 Sample Preparation

To date, PCR detection of *P. damselae* subsp. *damselae* has only been evaluated using bacterial colonies grown on agar plates. Simple and fast DNA extraction procedures from single bacterial colonies, like Instagene matrix from Bio-Rad Laboratories, have proven useful for PCR detection of this pathogen.[22] An alternative efficient simple protocol for DNA extraction from bacterial pure cultures is described below.

DNA Extraction

1. Transfer several *P. damselae* subsp. *damselae* fresh colonies from LB agar or sheep blood agar into 5 mL of LB broth, and incubate at 37°C for about 18 h with shaking at 200 rpm.
2. Pellet the cells by centrifugation at 13,000 rpm (in an Eppendorf microcentrifuge) for 5 min, and re-suspend bacterial cells in 500 μL of TES buffer (10 mM Tris-HCl pH 8.0, 1 mM EDTA, 0.1 M NaCl); add 5 μL of freshly prepared lysozyme (10 mg/mL), and incubate at 37°C for 30 min.
3. Add 5 μL each of proteinase K (Sigma) (10 mg/mL) and RNase (Sigma) (10 mg/mL), and incubate the mixture for 1 h at 65°C in a water bath.
4. Add 50 μL of 20% SDS, and return tubes immediately to the 65°C water bath for a further 10 min. The lysate becomes clear. Let the tubes cool down to room temperature.
5. Add 1 volume (500 μL) of a mixture of phenol/chloroform (1:1) (Amresco) equilibrated at pH 8.0 and mix by inversion.
6. Centrifuge at 6000 rpm (Eppendorf microcentrifuge) for 10 min and transfer the upper aqueous phase to a new 1.5 mL Eppendorf tube. If this phase is not clear, complete the volume to 500 μL with TES buffer, and repeat steps 4 and 5.
7. Precipitate DNA with 2.5 volumes of cold absolute ethanol (–20°C), mix gently by inversion, and pellet the DNA by centrifugation at 13,000 rpm for 5 min.
8. Discard supernatant, invert the tube on a paper towel, and air dry for 10 min.
9. Dissolve the DNA pellet in 50–100 μL sterile water or TE buffer and store at –20°C.
10. DNA quality and concentration can be checked by loading a 5 μL sample on a 1% agarose gel and, alternatively, by spectrophotometrical measure at 260 nm.

For clinical specimens (blood samples, tissue samples), fast DNA extraction methods can also be of utility. In addition, commercially available DNA extraction kits specific for tissue and blood samples might be used. However, there are no data available about the effective detection of this pathogen directly by PCR from clinical samples.

82.2.2 Detection Procedures

82.2.2.1 PCR Detection Using 16S rRNA Gene

Principle. The 16S rRNA gene of *P. damselae* subsp. *damselae* can be specifically detected from a DNA sample by using the primer pair Car1/Car2 (Table 82.3) and an annealing temperature of 65°C.[22] At this annealing temperature, genetically closely related species of *Vibrio* and *Photobacterium* do not yield amplification products of the expected size. In order to increase both the sensitivity and the specificity of this test, the amplification products can be used as template for a second round of PCR (a nested-PCR), using the primer pair Car1 and Nestcar1 (Table 82.3).[22] Due to the 100% similarity between the 16S genes of *P. damselae* subsp. *damselae* and subsp. *piscicida*, this test would also amplify subsp. *piscicida* DNA if present. However, subspecies *piscicida* has never been encountered in samples from homeotherm animals and does not survive

TABLE 82.3
Primers Used for Identification of *P. damselae* subsp. *damselae* by PCR Amplification

Target Gene	Primers	Primer Sequences (5′→3′)	PCR Product Size (bp)
16S rRNA			
	Car1	GCTTGAAGAGATTCGAGT	267
	Car2	CACCTCGCGGTCTTGCTG	
	Nested PCR		
	Car1	GCTTGAAGAGATTCGAGT	259
	Nestcar1	GGTCTTGCTGCCCTCTG	
ureC (encoding urease)			
	Ure-5′	TCCGGAATAGGTAAAGCGGG	448
	Ure-3′	CTTGAATATCCATCTCATCTGC	
dly (encoding damselysin)			
	Dly-1	GCTTAACCTGCCAGATAGCG	2615
	Dly-2	GGATGGTGAATCTCTATCGG	
	Dly-5′	CCTATGGGACATGAATGG	567
	Dly-3′	GCTCTAGGCTAAATGAATC	

at 37°C, thus ruling out the possibility of false-positives due to subsp. *piscicida* cells during the detection of subsp. *damselae* from clinical specimens.

Procedure

1. Prepare a first-round PCR reaction mixture (100 μL) containing 1 μL each of primers Car1 and Car2 (stocks at 10 μM), 2 U of *Taq* polymerase (PerkinElmer), 10 μL of 10× *Taq* polymerase buffer (PerkinElmer), 4 μL of a 50 mM $MgCl_2$ solution, each dNTP at a concentration of 200 μM, and 100 ng of template DNA. Include a tube containing PCR mixture without DNA template as a negative control.
2. Run the following cycling program: one cycle of 95°C for 4 min, followed by 30 cycles consisting of 95°C for 1 min, 65°C for 1 min, and 72°C for 20 s. Perform a final extension step consisting of 5 min at 72°C.
3. After completion of this first round PCR, take a 0.5 μL portion of the amplification product and use it as a DNA template for a second round (heminested PCR) using primers Car1 and Nestcar1 (Table 82.3) using the same amplification conditions as above.
4. Amplification products can be analyzed on 1% (wt/vol) agarose gels with TAE (0.04 M Tris-acetate, 1 mM EDTA) electrophoresis buffer and visualized with a UV transilluminator after staining with ethidium bromide. A 1-kb DNA ladder (Bio-Rad) must be included as a molecular weight marker. After the two PCR rounds, a specific DNA band of 259 bp must be visualized in the amplified product.

82.2.2.2 Multiplex PCR 16S rRNA-*ureC*

Principle. A multiplex PCR method has been reported to discriminate between the two subspecies of *P. damselae*, by using a combination of primers targeted to the 16S rRNA

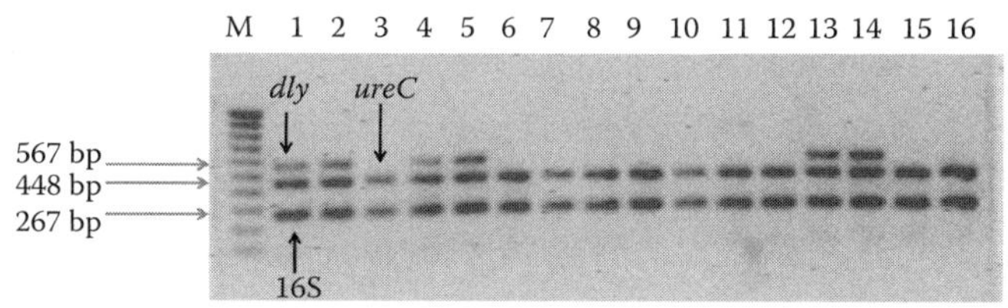

FIGURE 82.1 Agarose gel showing the results of applying a multiplex PCR assay to 16 strains of *P. damselae* subsp. *damselae*. Three primer pairs for three genes were used in conjunction: *dly* gene (567 bp amplicon), *ureC* gene (448 bp amplicon) and 16S rRNA gene (267 bp amplicon).

gene and the *ureC* gene[69] using primers Ure-5′ and Ure-3′ (Table 82.3). This primer pair flanks a 448 bp internal fragment of *ureC*, and can be combined in the same PCR reaction with primers Car1 and Car2 described above, using an annealing temperature of 60°C. This multiplex PCR procedure was tested with 16 *P. damselae* subsp. *damselae*, and in all the cases it allowed the specific identification of *P. damselae* subsp. *damselae* isolates from pure cultures.[69] With this protocol, isolates of *P. damselae* subsp. *damselae* yield two amplification fragments corresponding to 16S rRNA (267 bp) and *ureC* (448 bp) (Figure 82.1).

Procedure

1. Prepare a PCR reaction mixture (100 μL) containing 1 μL each of primers Car1, Car2, Ure-5′, and Ure-3′ (stocks at 10 μM); 2 U of *Taq* polymerase (PerkinElmer); 10 μL of 10× *Taq* polymerase buffer (PerkinElmer); 2 mM $MgCl_2$; 200 μM of each dNTP; and 100 ng of template DNA. Include a tube containing PCR mixture without DNA template as a negative control.
2. Run the following cycling program: one cycle of 95°C for 4 min, followed by 30 cycles consisting of

95°C for 1 min, 60°C for 1 min, and 72°C for 40 s. Perform a final extension step consisting of 5 min at 72°C.
3. Amplification products can be analyzed on 1% (wt/vol) agarose gels with TAE (0.04 M Tris-acetate, 1 mM EDTA) electrophoresis buffer and visualized with a UV transilluminator after staining with ethidium bromide. A 1-kb DNA ladder (Bio-Rad) must be included as a molecular weight marker. After the amplification two specific DNA bands of 267 and 448 bp must be visualized in the amplified product (Figure 82.1).

82.2.2.3 PCR Detection of Damselysin (*dly*) Gene

Principle. The nucleotide sequence of *dly* gene was used to design two primer sets (Table 82.3) for PCR detection of this gene.[64] The primer pair Dly-1/Dly-2 amplifies the almost complete *dly* gene, yielding an amplicon of a 2615 bp, whereas primer pair Dly-5′/Dly-3′ flanks a 567 bp internal fragment of *dly* gene. However, as mentioned above, this gene is not present in all pathogenic strains of *P. damselae* subsp *damselae*.[64] Thus, the PCR detection of this gene alone cannot be used as a *P. damselae* identification method.

Procedure

1. Prepare a PCR reaction mixture (100 μL) containing 1 μL each of primers Dly-1/Dly-2 (stocks at 10 μM), 2 U of *Taq* polymerase (PerkinElmer), 10 μL of 10× *Taq* polymerase buffer (PerkinElmer), 2 mM $MgCl_2$, 200 μM of each dNTP, and 100 ng of template DNA. Include a tube containing PCR mixture without DNA template as a negative control.
2. Run the following cycling program: one cycle of 95°C for 4 min, followed by 30 cycles consisting of 95°C for 1 min, 55°C for 1 min, and 72°C for 3 min. Perform a final extension step consisting of 10 min at 72°C.
3. Amplification products can be analyzed on 1% (wt/vol) agarose gels with TAE (0.04 M Tris-acetate, 1 mM EDTA) electrophoresis buffer and visualized with a UV transilluminator after staining with ethidium bromide. A 1-kb DNA ladder (Bio-Rad) must be included as a molecular weight marker. After the amplification a specific DNA band of 2615 bp must be visualized in the amplified product (Figure 82.1).

The primer pair Dly-5′/Dly-3′ can be used in combination with 16S (Car1/Car2) and *ureC* (Ure-5′ and Ure-3′) primers in a multiplex PCR following the protocol described in Section 82.2.2.2. This multiplex primer combination allows the identification of *P. damselae* subsp. *damselae* isolates, as well as the simultaneous screening of presence of the damselysin gene *dly* (Figure 82.1). After the amplification three specific DNA bands of 267, 448, and 567 bp must be visualized in the amplified product.

82.3 CONCLUSIONS AND FUTURE PERSPECTIVES

The genus *Photobacterium* currently comprises 17 species. The species *Photobacterium damselae* includes two subspecies, subsp. *damselae* (formerly *Vibrio damsela*) and subsp. *piscicida* (formerly *Pasteurella piscicida*). *P. damselae* subsp. *damselae* is the only member of this genus that has been isolated as the causative agent of disease in humans. This pathogen was initially recovered from ulcers of the damselfish (*Chromis punctipinnis*), and thereafter has been associated with infections in humans, causing not only wound infections but also primary septicemia in healthy individuals. Little is known about the molecular basis of the pathogenesis of this subspecies, although a cytolysin with hemolytic activity named damselysin has been identified as a likely contributor of virulence in some strains. Despite being the cause of a number of fatal infections, this pathogen has not received much attention from clinical microbiologists. Actually, most of our knowledge comes from studies carried out with fish isolates since this subspecies is also the causative agent of disease in a variety of marine animals which include fish species of economical importance in aquaculture. A multiplex PCR method, based on the amplification of sequences of the 16S rRNA gene and the *ureC* gene has been reported for the presumptive identification of *P. damselae* subsp. *damselae* from plate cultures.

The seriousness of *P. damselae* subsp. *damselae* as a human pathogen cannot be overstated. Although the number of cases of disease attributed to this organism is rather low, the severity of the disease, with a fatal outcome in more than half of the cases reported so far, should increase the attention of microbiologists for this bacterium. Very little is known about the virulence factors that enable this aquatic bacterium, usually related to diseases in fish, to cause such a fatal necrotizing fasciitis in humans. This pathogen is closely related to *P. damselae* subsp. *piscicida*, a fish pathogen unable to cause infection in homeotherms. There is a need for additional genetic studies aimed at decipherating the differential gene contents between strains of the two subspecies. The comparative genomic analysis of the two subspecies of *P. damselae* will shed some light in understanding the genetic determinants that enable *P. damselae* subsp. *damselae* to invade, colonize and cause disease in humans. The identification of novel virulence factors in this subspecies will also allow the future design of subspecies-specific molecular procedures for the fast, sensitive, and specific detection of this pathogen from clinical specimens.

REFERENCES

1. Beijerinck, M.W., Le *Photobacterium luminosum*. Bacterie luminosum de la Mer Nord, *Arch. Neer. Sci.*, 23, 401, 1889.
2. Reichelt, J.L., and Baumann, P., Taxonomy of the marine, luminous bacteria, *Arch. Mikrobiol.*, 94, 283, 1973.
3. Boisvert, H., Chatelain, R., and Bassot, J.M., Study on a *Photobacterium* isolated from the light organ of the *Leiognathidae* fish, *Ann. Inst. Pasteur (Paris)*, 112, 521, 1967.

4. Reichelt, J.L., Baumann, P., and Baumann, L., Study of genetic relationships among marine species of the genera *Beneckea* and *Photobacterium* by means of in vitro DNA/DNA hybridization, *Arch. Microbiol.*, 110, 101, 1976.
5. Smith, S.K. et al., Evaluation of the genus *Listonella* and reassignment of *Listonella damsela* (Love et al.) MacDonell and Colwell to the genus *Photobacterium* as *Photobacterium damsela* comb. nov. with an emended description, *Int. J. Syst. Bacteriol.*, 41, 529, 1991.
6. Gauthier, G. et al., Small-subunit ribosomal-RNA sequences and whole DNA relatedness concur for the reassignment of *Pasteurella piscicida* (Snieszko et al.) Janssen and Surgalla to the genus *Photobacterium* as *Photobacterium damsela* subsp. *piscicida* comb. nov., *Int. J. Syst. Bacteriol.*, 45, 139, 1995.
7. Urakawa, H., Kita-Tsukamoto, K., and Ohwada, K., Reassessment of the taxonomic position of *Vibrio iliopiscarius* (Onarheim et al. 1994) and proposal for *Photobacterium iliopiscarium* comb. nov., *Int. J. Syst. Bacteriol.*, 49, 257, 1999.
8. Nogi, Y., Masui, N., and Kato, C., *Photobacterium profundum* sp. nov., a new, moderately barophilic bacterial species isolated from a deep-sea sediment, *Extremophiles*, 2, 1, 1998.
9. Xie, C.H., and Yokota, A., Transfer of *Hyphomicrobium indicum* to the genus *Photobacterium* as *Photobacterium indicum* comb. nov., *Int. J. Syst. Evol. Microbiol.*, 54, 2113, 2004.
10. Thompson, F.L. et al., *Photobacterium rosenbergii* sp. nov. and *Enterovibrio coralii* sp. nov., vibrios associated with coral bleaching, *Int. J. Syst. Evol. Microbiol.*, 55, 913, 2005.
11. Yoon, J.H. et al., *Photobacterium lipolyticum* sp. nov., a bacterium with lipolytic activity isolated from the Yellow Sea in Korea, *Int. J. Syst. Evol. Microbiol.*, 55, 335, 2005.
12. Seo, H.J. et al., *Photobacterium frigidiphilum* sp. nov., a psychrophilic, lipolytic bacterium isolated from deep-sea sediments of Edison Seamount, *Int. J. Syst. Evol. Microbiol.*, 55, 1661, 2005.
13. Seo, H.J. et al., *Photobacterium aplysiae* sp. nov., a lipolytic marine bacterium isolated from eggs of the sea hare *Aplysia kurodai*, *Int. J. Syst. Evol. Microbiol.*, 55, 2293, 2005.
14. Park, Y.D. et al., *Photobacterium ganghwense* sp. nov., a halophilic bacterium isolated from sea water, *Int. J. Syst. Evol. Microbiol.*, 56, 745, 2006.
15. Rivas, R. et al., *Photobacterium halotolerans* sp. nov., isolated from Lake Martel in Spain, *Int. J. Syst. Evol. Microbiol.*, 56, 1067, 2006.
16. Jung, S.Y. et al., *Photobacterium lutimaris* sp. nov., isolated from a tidal flat sediment in Korea, *Int. J. Syst. Evol. Microbiol.*, 57, 332, 2007.
17. Ast, J.C. et al., *Photobacterium kishitanii* sp. nov., a luminous marine bacterium symbiotic with deep-sea fishes, *Int. J. Syst. Evol. Microbiol.*, 57, 2073, 2007.
18. Yoshizawa, S. et al., *Photobacterium aquimaris* sp. nov., a luminous marine bacterium isolated from seawater, *Int. J. Syst. Evol. Microbiol.*, 59, 1438, 2009.
19. Morris, J.G. Jr. et al., Illness caused by *Vibrio damsela* and *Vibrio hollisae*, *Lancet*, 319, 1294, 1982.
20. Love, M. et al., *Vibrio damsela*, a marine bacterium, causes skin ulcers on the damselfish *Chromis punctipinnis*, *Science*, 214, 1139, 1981.
21. Janssen, W.A., and Surgalla, M.J., Morphology, physiology, and serology of a *Pasteurella* species pathogenic for white perch (*Roccus americanus*), *J. Bacteriol.*, 96, 1606, 1968.
22. Osorio, C.R. et al., 16S rRNA gene sequence analysis of *Photobacterium damselae* and nested PCR method for rapid detection of the causative agent of fish pasteurellosis, *Appl. Environ. Microbiol.*, 65, 2942, 1999.
23. Osorio, C.R. et al., Characterization of the 23S and 5S rRNA genes and 23S-5S intergenic spacer region (ITS-2) of *Photobacterium damselae*, *Dis. Aquat. Organ.*, 61, 33, 2004.
24. Osorio, C.R. et al., Variation in 16S-23S rRNA intergenic spacer regions in *Photobacterium damselae*: A mosaic-like structure, *Appl. Environ. Microbiol.*, 71, 636, 2005.
25. Osorio, C.R., and Klose, K.E., A region of the transmembrane regulatory protein ToxR that tethers the transcriptional activation domain to the cytoplasmic membrane displays wide divergence among *Vibrio* species, *J. Bacteriol.*, 182, 526, 2000.
26. Juiz-Rio, S., Osorio, C.R., and Lemos, M.L., Identification and characterisation of the fur genes in *Photobacterium damselae* ssp. *piscicida* and ssp. *damselae*, *Dis. Aquat. Organ.*, 58, 151, 2004.
27. Rio, S.J., Osorio, C.R., and Lemos, M.L., Heme uptake genes in human and fish isolates of *Photobacterium damselae*: existence of *hutA* pseudogenes, *Arch. Microbiol.*, 183, 347, 2005.
28. Thyssen, A. et al., Phenotypic characterization of the marine pathogen *Photobacterium damselae* subsp. *piscicida*, *Int. J. Syst. Bacteriol.*, 48, 1145, 1998.
29. Botella, S. et al., Amplified fragment length polymorphism (AFLP) and biochemical typing of *Photobacterium damselae* subsp *damselae*, *J. Appl. Microbiol.*, 93, 681, 2002.
30. Fouz, B. et al., Characterization of *Vibrio damsela* strains isolated from turbot *Scophthalmus maximus* in Spain, *Dis. Aquat. Organ.*, 12, 155, 1992.
31. Obendorf, D.L., Carson, J., and Mcmanus, T.J., *Vibrio damsela* infection in a stranded leatherback turtle (*Dermochelys coriacea*), *J. Wildl. Dis.*, 23, 666, 1987.
32. Hanlon, R.T. et al., Fatal penetrating skin ulcers in laboratory-reared octopuses, *J. Invertebr. Pathol.*, 44, 67, 1984.
33. Song, Y.L., Cheng, W., and Wang, C.H., Isolation and characterization of *Vibrio damsela* infectious for cultured shrimp in Taiwan, *J. Invertebr. Pathol.*, 61, 24, 1993.
34. Vaseeharan, B. et al., *Photobacterium damselae* ssp *damselae* associated with diseased black tiger shrimp *Penaeus monodon* Fabricius in India, *Lett. Appl. Microbiol.*, 45, 82, 2007.
35. Fujioka, R.S. et al., *Vibrio damsela* from wounds in bottlenose dolphins *Tursiops truncatus*, *Dis. Aquat. Organ.*, 4, 1, 1988.
36. Buck, J.D. et al., Bacteria associated with stranded cetaceans from the Northeast USA and Southwest Florida Gulf coasts, *Dis. Aquat. Organ.*, 10, 147, 1991.
37. Fouz, B. et al., Evidence that water transmits the disease caused by the fish pathogen *Photobacterium damselae* subsp *damselae*, *J. Appl. Microbiol.*, 88, 531, 2000.
38. Fouz, B. et al., Survival of fish-virulent strains of *Photobacterium damselae* subsp. *damselae* in seawater under starvation conditions, *FEMS Microbiol. Lett.*, 168, 181, 1998.
39. Buck, J.D. et al., Aerobic microorganisms associated with free-ranging bottlenose dolphins in coastal Gulf of Mexico and Atlantic Ocean waters, *J. Wildl. Dis.*, 42, 536, 2006.
40. Clarridge, J.E., and Zighelboimdaum, S., Isolation and characterization of 2 hemolytic phenotypes of *Vibrio damsela* associated with a fatal wound infection, *J. Clin. Microbiol.*, 21, 302, 1985.
41. Coffey, J.A. et al., *Vibrio damsela*—another potentially virulent marine vibrio, *J. Infect. Dis.*, 153, 800, 1986.
42. Dryden, M. et al., *Vibrio damsela* wound infections in Australia, *Med. J. Aust.*, 151, 540, 1989.

43. Perez-Tirse, J., Levine, J.F., and Mecca, M., *Vibrio damsela*—a cause of fulminant septicemia, *Arch. Intern. Med.*, 153, 1838, 1993.
44. Yuen, K.Y. et al., Fatal necrotizing fasciitis due to *Vibrio damsela*, *Scand. J. Infect. Dis.*, 25, 659, 1993.
45. Shin, J.H. et al., Primary *Vibrio damsela* septicemia, *Clin. Infect. Dis.*, 22, 856, 1996.
46. Fraser, S.L. et al., Rapidly fatal infection due to *Photobacterium (Vibrio) damsela*, *Clin. Infect. Dis.*, 25, 935, 1997.
47. Tang, W.M., and Wong, J.W.K., Necrotizing fasciitis caused by *Vibrio damsela*, *Orthopedics*, 22, 443, 1999.
48. Barber, G.R., and Swygert, J.S., Necrotizing fasciitis due to *Photobacterium damsela* in a man lashed by a stingray, *N. Engl. J. Med.*, 342, 824, 2000.
49. Goodell, K.H. et al., Rapidly advancing necrotizing fasciitis caused by *Photobacterium (Vibrio) damsela*: A hyperaggressive variant, *Crit. Care Med.*, 32, 278, 2004.
50. Yamane, K. et al., Two cases of fatal necrotizing fasciitis caused by *Photobacterium damsela* in Japan, *J. Clin. Microbiol.*, 42, 1370, 2004.
51. Knight-Madden, J.F.M. et al., *Photobacterium damsela* bacteremia in a child with sickle-cell disease, *Pediatr. Infect. Dis. J.*, 24, 654, 2005.
52. Alvarez, J.R. et al., An unusual case of urinary tract infection in a pregnant woman with *Photobacterium damsela*, *Infect. Dis. Obstet. Gynecol.*, 2006, 80682, 2006.
53. Nakamura, Y. et al., Necrotizing fasciitis of the leg due to *Photobacterium damsela*, *J. Dermatol.*, 35, 44, 2008.
54. Aigbivhalu, L., and Maraqa, N., *Photobacterium damsela* wound infection in a 14-year-old surfer, *South. Med. J.*, 102, 425, 2009.
55. Fouz, B. et al., Histopathological Lesions Caused by *Vibrio damsela* in cultured turbot, *Scophthalmus maximus* (L)—inoculations with live cells and extracellular products, *J. Fish Dis.*, 18, 357, 1995.
56. Fouz, B. et al., Role of iron in the pathogenicity of *Vibrio damsela* for fish and mammals, *FEMS Microbiol. Lett.*, 121, 181, 1994.
57. Ratledge, C., and Dover, L.G., Iron metabolism in pathogenic bacteria, *Annu. Rev. Microbiol.*, 54, 881, 2000.
58. Fouz, B., Biosca, E.G., and Amaro, C., High affinity iron-uptake systems in *Vibrio damsela*: Role in the acquisition of iron from transferrin, *J. Appl. Microbiol.*, 82, 157, 1997.
59. Fouz, B. et al., Toxicity of the extracellular products of *Vibrio damsela* isolated from diseased fish, *Curr. Microbiol.*, 27, 341, 1993.
60. Kreger, A.S., Cytolytic activity and virulence of *Vibrio damsela*, *Infect. Immun.*, 44, 326, 1984.
61. Kothary, M.H., and Kreger, A.S., Purification and characterization of an extracellular cytolysin produced by *Vibrio damsela*, *Infect. Immun.*, 49, 25, 1985.
62. Kreger, A.S. et al., Phospholipase D activity of *Vibrio damsela* cytolysin and its interaction with sheep erythrocytes, *Infect. Immun.*, 55, 3209, 1987.
63. Cutter, D.L., and Kreger, A.S., Cloning and expression of the damselysin gene from *Vibrio damsela*, *Infect. Immun.*, 58, 266, 1990.
64. Osorio, C.R. et al., Presence of phospholipase-D (*dly*) gene coding for damselysin production is not a pre-requisite for pathogenicity in *Photobacterium damselae* subsp. *damselae*, *Microb. Pathog.*, 28, 119, 2000.
65. Takahashi, H. et al., Difference of genotypic and phenotypic characteristics and pathogenicity potential of *Photobacterium damselae* subsp *damselae* between clinical and environmental isolates from Japan, *Microb. Pathog.*, 45, 150, 2008.
66. Alm, E.W. et al., The oligonucleotide probe database, *Appl. Environ. Microbiol.*, 62, 3557, 1996.
67. Marchesi, J.R. et al., Design and evaluation of useful bacterium-specific PCR primers that amplify genes coding for bacterial 16S rRNA, *Appl. Environ. Microbiol.*, 64, 2333, 1998.
68. Osorio, C.R., Juiz-Rio, S., and Lemos, M.L., A siderophore biosynthesis gene cluster from the fish pathogen *Photobacterium damselae* subsp. *piscicida* is structurally and functionally related to the *Yersinia* high-pathogenicity island, *Microbiology*, 152, 3327, 2006.
69. Osorio, C.R. et al., Multiplex PCR assay for ureC and 16S rRNA genes clearly discriminates between both subspecies of *Photobacterium damselae, Dis. Aquat. Organ.*, 40, 177, 2000.

83 *Plesiomonas*

Jesús A. Santos, Andrés Otero, and María-Luisa García-López

CONTENTS

83.1 INTRODUCTION

The genus *Plesiomonas* currently comprises a single gram-negative bacterial species, *Plesiomonas shigelloides*, that is recognized as a cause of extraintestinal diseases and, increasingly, as an agent of gastrointestinal illness. This introduction provides a general overview of the biology, taxonomy, and pathogenicity of *P. shigelloides*, with a particular focus on its role as a foodborne pathogen.

83.1.1 History and Current Taxonomic Status

The genus *Plesiomonas* (*plesios*, "neighbor"; *monas*, "unit," for "neighbor to *Aeromonas*") consists of only one homogeneous species with a single DNA hybridization group (*P. shigelloides*). The organisms within the genus are motile, facultative anaerobic, oxidase- and catalase-positive, glucose-fermenting, gram-negative, rod-shaped bacteria. The complex history of the taxonomy of this genus has been recently reviewed by González-Rey[1] and by Janda and Abbot.[2] This bacterium was first described in 1947 after being isolated from the feces of a patient; it received the name "Strain C27" and was thought to be a member of the family Enterobacteriaceae.[3] In 1954, the C27 organism was moved to the genus *Pseudomonas* with the species name of *shigelloides*, in consideration of its *Shigella*-like characteristics.[4] Ewing and Johnson[5] proposed its transfer to the genus *Aeromonas* (within the family Vibrionaceae) as *A. shigelloides*, based on the cytochrome oxidase activity, flagella morphology, and other biochemical properties, although they were aware that it differed from the rest of the described species of *Aeromonas*. The transfer of C27 strain to the generic name *Plesiomonas* was proposed in 1962 since the organism did not exhibit some important characteristics, such as the enzymatic activity, traditionally linked to the genus *Aeromonas*.

The bacterium has remained within the family Vibrionaceae, together with the genera *Vibrio* and *Aeromonas*, mainly because the genera included in this family shared common features, as oxidase-positive activity, polar flagellation, and aquatic habitat, rather than an evolutionary relationship among them.[6] Since the early 1980s, with the advent of techniques to measure evolutionary sequences, like DNA hybridization and small-scale DNA sequencing, it became clear that the genera *Aeromonas* and *Plesiomonas* had to be removed from the Vibrionaceae.[7,8] The molecular data obtained by Martinez-Murcia et al.[9] and Ruimy et al.[10] demonstrated the relationship of the genus *Plesiomonas* with the family Enterobacteriaceae and thus was included in the last edition of the taxonomic outline of the *Bergey's Manual of Systematic Bacteriology*,[11] being the only oxidase-positive member of this family. Ruimy et al.[10] had suggested that additional data could support the allocation of the genus *Plesiomonas* in a new family, the Plesiomonadaceae, and it has to be considered that the family Enterobacteriaceae is the closest neighbor, though with a relatively low level of 16S rRNA homology (93%–95%),[2] but the results obtained by Salerno et al.[12] using multilocus sequence typing (MLST) of 77 isolates from different sources and of different geographical origin, showed that *P. shigelloides* constitutes a branch nested deep within the family Enterobacteriaceae, with no significant intraspecies structures. They attributed, without further explanation, the oxidase activity to a derived characteristic that evolved secondarily from an oxidase-negative ancestor.[12]

83.1.2 Morphology and Phenotypic Characteristics

Cells of *P. shigelloides* are long and straight rods, motile by two to eight polar flagella and may also produce peritrichous flagella.[13] Polyphosphates inclusion bodies has been detected after seven hours of incubation in BHI broth and do not disappear during the stationary phase of growth, indicating that these bodies could be used as an energy source to initiate cell division under unfavorable environmental conditions.[14,15] *P. shigelloides* is a fermentative and facultative anaerobic microorganism. It can ferment glucose without gas-production, and all the strains ferment inositol and decarboxylate lysine and ornithine. They lack extracellular hydrolytic activities on starch, gelatine, and DNA, but they can degrade elastine[16] and chitin.[17] The phenotypic profile is very homogeneous, with only minor differences in carbohydrate fermentation.[18,19] Serology of *P. shigelloides* has many points in common with the Enterobacteriaceae; many strains cross-react with different serovars of *Shigella dysenteriae*, *S. sonnei*, *S. boydii*, and *S. flexneri*,[3,20] and it also produces the enterobacterial common antigen.[21]

83.1.3 Habitat, Ecology, and Epidemiology

P. shigelloides is widely distributed in nature and has been isolated worldwide, being considered an aquatic organism, preferably found in fresh water (rivers, lakes, streams, ponds, sediments, recreational), but can also be isolated from estuarine and marine environments.[2,22–25] Its minimum growth temperature is reported to be 8°C,[26] and this implies that it is more frequently isolated from tropical and subtropical climates, and also temperate climates in warmer periods,[2] but it is not restricted to these climates, as it has been isolated from such extreme aquatic environments as a lake located north of the polar circle.[27] In addition to temperature, the availability of nutrients and the level of pollution are important factors that influence the presence of *P. shigelloides* in surface waters; Medema and Schets[25] have reported that *Plesiomonas* density was significantly correlated with index parameters for the trophic state (Secchi depth, a measurement of water clarity, and chlorophyll A) and for fecal pollution (*Escherichia coli*).

It seems unlikely that the presence of *P. shigelloides* in aquatic environments is always related to pollution or sewage contamination, suggesting that fish, reptiles, amphibians, and other water-related animals may serve as a reservoir for this bacterium.[2,28–30] The organism has been isolated from a wide range of animals, both warm and cold-blooded, as well as from wild and domestic species. Fish and shellfish in the natural habitats of *P. shigelloides* appear to act as reservoirs for the bacterium, which has been isolated from crabs, bivalve molluscs, and from the intestines of 59% of freshwater fish in Zaire; 10.2% of freshwater fish in Japan; and 23% of marine fish in Spain.[28,29,31,32] Wild birds have also been found to carry *P. shigelloides*, and most bird species from which the bacterium has been isolated live in aquatic ecosystems and/or feed on fish. Isolation of *P. shigelloides* has also been reported from marine mammals, reptiles, and amphibians, monkeys, food animals (pigs, cattle, poultry, sheep, and goats), and pet animals (cats and dogs). The organism has been associated with disease in cats, freshwater fish, reptiles, and turtles.[28]

P. shigelloides is not a part of the normal gut flora of man, and the rate of carriage among healthy people is generally low (0.2%–3.2%) although it varies considerably with location, the highest rates corresponding to tropical and subtropical countries.

Water has been implicated as a vehicle of *P. shigelloides* infection. Deficient drinking-water supplies are considered responsible for a number of large outbreaks of gastrointestinal disease with high attack rates,[33,34] and some cases seems to be associated with recreational use of water[35] and water from aquaria.[36]

Fish and shellfish appear to be the most common cause of foodborne infections. Outbreaks of gastroenteritis have been attributed to oysters, salt mackerel, cuttlefish salad, crab, shrimps, scallops, and sushi consumption.[2,19,37,38] The usual route of transmission is by ingestion of contaminated water or raw or undercooked fish and shellfish, though there is at least one report of an outbreak of diarrheal disease with chicken as suspected vehicle of transmission,[39] and the microorganism has been isolated from samples of fresh vegetables non related to cases of disease.[40]

The information on the factors affecting the growth and survival of *P. shigelloides* in foods is very limited. The effects of intrinsic and extrinsic factors have been reviewed by Varnam and Evans[41] and by the International Commission on Microbiological Specifications for Foods.[26]

The minimum growth temperature is widely accepted to be 8°C although at least one strain has been reported to grow at 0°C[42] and Herrera[43] found that of 13 isolates from fish and clinical specimens none grew at 6°C or 8°C, but most did at 10°C requiring up to 9 days for visible growth to occur. Optimum growth appears to occur in the range 37°C–39°C, and the maximum growth temperature is around 45°C. Strains of this organism can be recovered from frozen foods that have been stored for years.

Miller and Koburger[44] characterized 40 *P. shigelloides* strains from a variety of sources (water, sediment, fish, bivalve molluscs, and stools) and concluded that the organism is able to withstand up to 5% NaCl in trypticase soy broth but only up to 3% in a less nutritious medium such as tryptone broth. All cultures grew in the pH range 4.5–8.5, with 60% growing at pH 4, and 85% at pH 9. The salt tolerance of the isolates tested in trypticase soy broth by Herrera[43] was dependent on strain origin; thus, strains from clinical specimens tolerated less salt (3.5%) than those from marine environments (5.5%). For all isolates, tolerance to acid and alkaline conditions was similar to that reported by Miller and Koburger.[44] Even fewer studies on *P. shigelloides* growth under reduced water activity (a_w) conditions have been reported. Minimum a_w for growth of *P. shigelloides*, which varied with strain and the type of humectant used, ranged from 0.92 (glycerol) to 0.97 (NaCl and sucrose).[43]

Knowledge of the effect of food processing on *P. shigelloides* is also extremely limited. Studies to date indicate that pasteurization at 60°C for 30 min is an effective means of destroying *P. shigelloides* cells.[44] Ingham[45] evaluated the effects of storage in air and under vacuum and a modified atmosphere, consisting of 80% CO_2, on the growth of *P. shigelloides* spiked on cooked crayfish tails. The organism did not grow at 8°C under any packaging system. At 11°C, the organism grew well in air but was strongly inhibited by the modified atmosphere and, to a lesser extent, by vacuum, and at 14°C, only the modified atmosphere inhibited *P. shigelloides* growth. As for other gram-negative, mesophilic, facultative, anaerobic pathogenic bacteria, low temperatures and CO_2 are effective means of preventing *P. shigelloides* growth in raw foods packaged under oxygen-reduced conditions.

83.1.4 Clinical Features

Since its first isolation, *P. shigelloides* has been regarded as a human pathogen, but it has received little attention in the medical literature, because it was considered as a sporadic opportunist pathogen, because of its apparently low virulence, or because it was simply overlooked by diagnostic laboratories.[46] Human infections caused by *P. shigelloides* can be grouped into two categories: gastrointestinal infections and extraintestinal diseases; it is also recognized as an agent of animal infections.

Gastroenteritis. Gastroenteritis is the most common syndrome associated whit *P. shigelloides* infection and the most controversial as well. There are a number of epidemic outbreaks of diarrhea attributed to *P. shigelloides* as the causative agent and a series of case reports predominantly from tropical and subtropical countries. In cold and temperate climates, *Plesiomonas* cases of gastroenteritis are often associated with foreign travel. Traveler's diarrhea is seen year-round, whereas locally acquired infections show a marked seasonal trend related to environmental contamination of freshwater, with the peak occurring during the warmer seasons.[47–49] Approximately 70% of people who present diarrhea linked to this organism have either an underlying disease (cancer, Crohn's disease, achlorhydria, diverticulosis, cirrhosis, etc.) or an identifiable risk factor (e.g., tropical travel and/or seafood consumption).[2,50] Several authors[46,48] have reported that, in contrast with other pathogenic foodborne bacteria that have a predilection for children such as *Aeromonas* or *Yersinia enterocolitica*, the organism affects all age groups, although most studies indicate a higher incidence in adults. Both males and females are equally affected.

At least three major clinical presentations of *P. shigelloides* gastroenteritis occur: (i) a secretory (watery) type of diarrhea; (ii) a more invasive disease resembling shigellosis; and (iii) a subacute or chronic disease lasting between two weeks and 2–3 months.[46,51] On rare occasions, the organism, in association with other enteric pathogens (i.e., *Aeromonas sobria*), has been reported to cause a cholera-like disease.[52] Brenden et al.[46] concluded that *Plesiomonas* diarrhea is most often secretory in nature, but findings from Holmberg et al.[50] and Kain and Kelly[48] suggest that the dysenteric form (enteroinvasiveness) rather than the secretory (enterotoxin production) diarrhea is the most common presentation of *Plesiomonas* gastroenteritis.

The infective dose is unknown but is presumed to be very high.[2] On average, symptoms may begin between 24 and 50 h after consumption of contaminated food or water although shorter incubation times of 1–1.5 h have been reported.[53] Diarrhea is the predominant symptom, occurring in 94% of cases. Accompanying symptoms vary, but severe abdominal pain or cramping, nausea and/or vomiting, and low-grade fever are most common. Less frequent symptoms include chills, headache, and some degree of dehydration.[2,37,48,50]

Patients with stool positive for *P. shigelloides* have either watery diarrhea or diarrhea with blood and mucus. The secretory form varies in severity from a mild illness of short duration to severe diarrhea. It is usually reported to last between 1 and 7 days but can persist as long as 21 days, with up to 30 stools per day at the peak of the disease. The dysenteric form is usually severe, being characterized primarily by abdominal pain with the presence of macroscopic blood and mucus in the stools, which are often greenish tinged and slimy.[46] In most cases of *P. shigelloides* gastroenteritis, the episode resolves spontaneously in a few days, but in a significant proportion, infection becomes subacute or chronic and does not resolve until the patient is placed on the appropriate antimicrobial therapy. There have also been reports of fatal cases subsequent to primary enteric illness.[2,52,54]

For *P. shigelloides* diarrhea, the drugs of choice have been tetracycline and trimethoprim-sulfamethoxazole. It should be noted that *P. shigelloides* strains present natural resistance to penicillins, roxithromycin, clarithromycin, lincosamides, streptogramins, glycopeptides, and fusidic acid.[43,55]

Extraintestinal Disease. In addition to gastroenteritis, *P. shigelloides* has been implicated in a number of cases of extraintestinal infections, the majority of them occurring in immunocompromised hosts or patients with an underlying condition. There are reports of septicemia, meningitis, intraabdominal infections (cholecystitis, pseudoappendicitis, peritonitis), ocular diseases, wound infections, cellulitis, osteomyelitis, arthritis, proctitis, pyosalpingitis, and pneumonia.

Septicemia occurs mainly in adults, and in most of the cases, *P. shigelloides* is the only bacterium isolated, though in some instances it appeared in association with other bacteria (*Clostridium difficile*, *Escherichia coli*, and *Edwardsiella tarda*).[2,56,57] The systemic infection is usually preceded by gastrointestinal illness and, although predisposing factors are poorly understood, high iron levels in blood (due to an underlying liver disease or transfusions) are thought to increase the chance of *Plesiomonas* infection.[58,59] More than 60% of the reported cases require hospitalization because of the severity and prolonged course of the illness. The overall mortality rate is about 50%, but the incidence of mortality is decreasing for reasons like better medical care, and the use of more active antimicrobials.[2,46]

Meningitis is another category of systemic infection associated, to date, with neonates probably due to vertical transmission during birth. At least in one case, *P. shigelloides* was isolated from the stool of the mother, though she had been asymptomatic; in other cases the women reported recent seafood consumption or symptoms suggesting infection such as chills and fever.[2,37] The mortality rate approached 80%, but treatment with cefotaxime was effective, and the last case reported was in 1999.[2,60]

Other extraintestinal infections are observed rarely. Intraabdominal infections seem to be foodborne, but not always causing gastrointestinal disease. If surgery is needed, *P. shigelloides* can develop postoperative septic events.[61,62] Ocular and wound infections are occasionally reported, with *P. shigelloides* isolated together with other bacteria, and contact with contaminated water or consumption of shellfish were the most probable source of the microorganisms.[63–65]

Disease in Animals. *P. shigelloides* has been isolated from a wide range of animal species, both poikilotherms and homeotherms, though not always related to disease. Several surveys were carried out to determine the prevalence of *P. shigelloides* in animals, and all showed a high incidence of this bacterium in cats, which can be a consequence of the supplementation of the diet with prebiotics.[66] Symptoms of infection in animals vary from enteritis to sepsis in mammals and enteritis, skin and liver pathologies, or tumors in other animals such as fish, reptiles, and birds. An extensive review of *P. shigelloides* from the veterinary perspective is undertaken by Jagger.[28]

83.1.5 Pathogenicity and Virulence Factors

Although there has been interest in the pathogenicity of *Plesiomonas*, many aspects remain unknown. One of the problems is the lack of animal models for conclusive identification of virulence determinants and the negative result obtained in one human volunteer study.[67] More recently, Vitovec et al.,[68] studying coinfection with *P. shigelloides*, *Aeromonas* spp., and *Cryptosporidium parvum* indicated that experimental infection of neonatal BALB/c mice, may serve as an initial experimental animal model for studying the initial steps of gastrointestinal colonization and the diarrheal disease syndrome caused by these bacterial and protozoan pathogens and that such a model can serve as a step leading toward improved models for understanding gastrointestinal colonization and the diarrheal disease syndrome.

Current evidence indicates that the exact mechanism of *P. shigelloides* pathogenicity is not fully elucidated and that more than one virulence factor is required to cause diarrhea. Experience with other enteric pathogens (e.g., mesophilic *Aeromonas*, *Yersinia enterocolitica*, or *Escherichia coli*) suggests that perhaps only some strains can cause disease in certain host populations. A number of putative virulence factors have been identified, but none of them is widely accepted as being important for *Plesiomonas*-associated infections or is tested for routinely. In addition, very few have been investigated in any detail.

Toxins. Whether *P. shigelloides* produce toxins or not is a controversial issue. Sanyal et al.[69] screened six *P. shigelloides* isolates for enterotoxigenicity in the adult rabbit ligated ileal loop assay. Very few positive (4 out of 29) results were observed, with some of the strains producing only a small amount of fluid accumulation within the intestinal lumen. Ljungh et al.[70] reported a positive Y-1 adrenal cell assay but a negative rabbit ileal loop assay with two strains of *P. shigelloides*. Gurwith and Williams[71] assayed two strains of *P. shigelloides* by means of the Y-1 adrenal cell assay, and both strains elicited a positive reaction for enterotoxin. Penn et al.[72] reported that heat-labile and heat-stable enterotoxins were negative on the *P. shigelloides* strains isolated from a patient. Negative results for the production of enterotoxin and cytotoxin were also reported by Johnson and Lior[73] and by Pitarangsi et al.[74] Saraswhati et al.[75] described the production of heat-stable and heat-labile enterotoxins from all 70 *Plesiomonas* strains tested, irrespective of origin or serotype. Partial purification and characterization of the heat-labile and heat-stable enterotoxins by Manorama et al.[76] revealed that the former was active in the rabbit ileal loop and the latter in both the rabbit ileal loop and suckling mice. Gardner et al.[77] grew 28 strains of *P. shigelloides* and a type strain in an iron-depleted medium. Filtrates of 24 of the 29 strains produced elongation of CHO cells. These changes could be prevented by heating or by preincubation of the filtrate culture with cholera antitoxin, the authors concluding that *P. shigelloides* elaborates a cholera-like toxin. Later on, they demonstrated complete inhibition of production of this CHO cell elongation factor(s) by growth in an iron-containing medium.[78] Herrington et al.[67] showed that genetic probes for heat-stable enterotoxins related to those of enterotoxigenic *Escherichia coli* were negative and also that heat-labile enterotoxins were not detected when a modified GM_1-enzyme linked immunosorbent assay was used. According to Matthews et al.,[79] the key to detecting *P.* enterotoxins may be serial passage of isolates in vivo, also suggesting that the enterotoxin of *P. shigelloides* appears to be novel. The fact that enterotoxic activity appears to be rapidly lost upon in vitro subculture could explain the failure of many early studies to consistently identify such a factor.[51] Falcón et al.[80] reported that two out of five *P. shigelloides* strains from water were positive for enterotoxin production when tested in the suckling mouse model. They also observed cytotoxic activity in CHO, HeLa, and Vero cells, which was lost when the cell-free supernatants were heated at 56°C for 15 min and partially inactivated by antiserum to the cytotoxin of *Aeromonas hydrophila*. Okawa et al.[81] found and isolated a new heat-stable cytotoxin [anticholera toxin-reactive protein-lypopolysaccharide complex (ACRP-LPS) complex] from the culture filtrate of one clinical *P. shigelloides* strain, which also gave a positive reaction in the suckling mouse model and suggested that both cytotoxicity and enterotoxigenicity by ACPR in the cytotoxin was the most important virulence factor in the pathogenesis of their *P. shigelloides* strain.

Invasiveness. Information on *P. shigelloides* mechanisms involved in attachment, chemotaxis, and penetration of the gastrointestinal epithelium, and its associated mucous layer is scarce. Adherence to host cells is a fundamental step in bacterial infection, and many enteropathogens possess surface structures that facilitate adherence to host-cell epithelial surfaces.[82] For *P. shigelloides*, it is not known if flagella have a role in adherence. A glycocalyx was not detected on any *P. shigelloides* isolates by Theodoropoulos et al.,[83] although there is one report of a glycocalyx detected on the outer surface of this bacterium.[84] Tsugawa et al.[85] demonstrated that GroEL, known to be a chaperone or heat-shock protein, has a positive influence on the attachment of *P. shigelloides* to Caco-2 cells.

The Sereny test has always yielded uniformly negative results.[46,84] Studies investigating the invasive capability of *P. shigelloides* by using cell cultures (namely, Y-1 mouse adrenal tumor cells, Hep-2 derived from larynx human carcinoma, and HeLa derived from human cervical carcinoma) have been inconclusive, since some of them obtained positive results[86] while others did not find invasive potential.[67,87,88] By means of transmission electron microscope, Theodoropoulos et al.[83] demonstrated that isolates of *P. shigelloides* derived from clinical samples and one type strain were capable of adhering to and entering the human colon carcinoma cell line Caco-2. Their observations suggested that initial uptake may occur through a phagocytic-like process and also that the noted free *P. shigelloides* in the cytosol of Caco-2 cells was due to escape from cytoplasmic vacuoles. The inconclusive data from previous studies were attributed to the fact that no enteric cell lines chosen, which are not representative of the target cells in vivo. The authors also indicated the possibility that, like *E. coli*, *P. shigelloides* comprises different pathogenic phenotypes. Based on an invasion assay and flow cytometry, the mechanisms of invasion of the *P. shigelloides* clinical isolate strain P-1 into Caco-2 cells have been studied by Tsugawa et al.[89] The authors suggested that the P-1 strain invades through signal transduction followed by the rearrangement of the cytoskeletal proteins, which is important for the endocytosis of bacteria by the epithelial cells. They also conclude that the invasion of *P. shigelloides* induces apoptotic cell death. Finally, studying the interplay between this bacterium and host defense mechanisms, Pavlova et al.[90] presented an evidence of the inhibitory activities of *P. shigelloides* toward papain-like proteinases and proposed that cathepsin inhibition is important to the survival and spread of this pathogen in a mammalian host.

Hemolytic Activity. Efficient mechanisms for iron acquisition from the host during an infection are considered essential for virulence. A mechanism for iron acquisition is the production of hemolysins, which release iron from intracellular heme and hemoglobin. For *P. shigelloides*, the detection of hemolytic activity has also been questionable, but using appropriate methods, a number of workers have provided evidence that most isolates from different sources and geographic areas have the ability to display cytolytic activity against erythrocytes.[16,80,91–94] In addition, the genes encoding the heme iron utilization system of *P. shigelloides*, which is similar to that of *Vibrio cholerae*, have been isolated and characterized. One of them, the *hug*A gene, encodes an outer membrane receptor, HugA, required for *P. shigelloides* heme iron utilization.[95] Detection of the *hug*A gene is common in clinical and environmental *P. shigelloides* isolates.[32]

The presence of hemolytic molecules and the HugA outer membrane receptor may represent an important pathway for iron acquisition. Furthermore, vacuolating activity associated to hemolysin production has been reported for some strains of *P. shigelloides*.[80] The hemolysin(s) roles in the ability to escape from the intravacuolar compartment and enterotoxigenicity have also been suggested.[80,83] In conclusion, the above studies indicate that *P. shigelloides* could produce, at least, two hemolytic factors, their expression and detection being influenced considerably by environmental growth conditions and testing procedures.[80,91–93] For this bacterium, the overlay assay appears to be the best routine procedure for detecting hemolytic activity.

Plasmids. Many *P. shigelloides* strains contain plasmids. A very large plasmid (>150 MDa, *ca.* 230 kb) was found by Holmberg et al.[50] in 12 of 27 clinical isolates, but it was not the same as the large virulence plasmid described for *Shigella* species and enteroinvasive *Escherichia coli*. The five clinical isolates studied by Herrington et al.[67] also possessed in common a very large plasmid (between 118 and 312 MDa), which in gnotobiotic piglets appeared to facilitate the uptake of *P. shigelloides* into the mucosa of the distal ileum. Furthermore, when one gnotobiotic piglet was infected with a cured strain, the animal remained healthy. It is possible that unstable virulence plasmids may be involved in *P. shigelloides* pathogenicity.

Another plasmid-mediated property is antimicrobials resistance. Marshall et al.[96] examined five *P. shigelloides* strains isolated from Louisiana blue crabs (*Callinectes sapidus*) for resistance to selected antibiotics and presence of plasmids. They found that each isolate carried three plasmids of approximately 2.5 kb, 3.8 kb, and 5.3 kb. Plasmid curing linked the streptomycin resistance determinant with the 3.8 kb and/or 5.3 kb plasmids. However, others workers have failed in demonstrating any association between plasmids and antibiotic resistance.[97]

Other Putative Virulence Factors. *P. shigelloides* strains show elastolytic activity, which may be involved in connective tissue degradation. Santos et al.[16] reported that elastin degradation was cell associated, enhanced when the strains where grown in an iron-depleted medium, and lost after thermal treatment at 100°C for 10 min. They also found that the enzymatic activity was inactivated by phenyl methyl sulfonyl fluoride that is an inhibitor of serine proteases.

Another enzymatic activity displayed by *P. shigelloides* strains is chitinase. Chitin is an abundant polymer in the marine environment, and chitinases occur in several species of aquatic bacteria, members of the genus *Vibrio* and *P. shigelloides* among them.[17,98] The detection of the chitinase gene in strains of *P. shigelloides* is of interest, not only from the ecological point of view, but also for its possible relationship with the virulence, as was suggested for other bacterial pathogen, as *Vibrio cholerae*.[99,100]

Tetrodotoxin (TTX) is a potent marine neurotoxin, named after the order of fish from which it is most commonly associated, the Tetraodontiformes. The toxin can be produced by different bacterial species, one of them being *P. shigelloides*.[101,102] This bacterium has also been identified as a histamine producer in scombroid fish species by López-Sabater et al.[103] and González-Rey.[1] The former authors suggested that *P. shigelloides* could play an important role within fish histidine decarboxylating bacteria because of its association with aquatic environments. Data from our laboratory support this suggestion, as we have seen that strains isolated from fish samples can produce more than 300 ppm of histamine in culture broth (González-Huerta and Santos, unpublished results).

83.1.6 Laboratory Diagnosis of *P. shigelloides*

Conventional Diagnosis. *P. shigelloides* can be found on several kinds of samples (clinical and environmental, including water, fishes, and foods), and the purpose of the analysis can be very different in each situation. The selection of a concrete procedure of analysis will be influenced by (i) the required selectivity (mainly, the ability to suppress background microbiota); (ii) the desired diagnostic feature (mainly, the ability to differentiate between *P. shigelloides* and other morphologically or biochemically similar bacteria); (iii) the presumed presence of injured cells; and (iv) the need for quantifying the *Plesiomonas* population.

The classical procedure for detection of *P. shigelloides* on clinical and environmental samples includes the isolation of more or less typical colonies on a selective and differential agar and the confirmation of the identity of isolates by a set of morphological and biochemical tests. For the isolation, different solid media have been used, particularly enteric agars (MacConkey agar, Salmonella–Shigella agar, xylose lysine deoxycholate agar, Hektoen enteric agar, deoxycholate lactose agar, and Endo agar, among others). However, the morphology of *Plesiomonas* colonies on such media is very similar to that of colonies of other growing bacteria (*Aeromonas*, Enterobacteriaceae).

Two differential media, formulated specifically for the isolation and enumeration of *P. shigelloides* are inositol brilliant green bile salts agar (IBB)[104] and *Plesiomonas* agar (PL).[37] Besides the selective components (brilliant green and bile salts), IBB agar includes inositol as carbon source, which can be used only for a few competing bacteria. At the same time, *Plesiomonas* ferments inositol, giving red to pinkish colonies on IBB, while nonfermeting inositol strains, such as *Aeromonas* strains, appear colorless (in overcrowded plates, *Plesiomonas* colonies can appear small and white),[18] and coliforms form greenish or pink colonies.[105]

On PL agar lysine decarboxylase-positive and nonfermenting organisms such as *P. shigelloides* form pink colonies surrounded by a red zone.[106] However, because of the reduced concentration of bile salts, PL agar is less inhibitory of the competitive microbiota than IBB agar, the latter being generally preferred for the isolation of *Plesiomonas* from environmental samples. On the other hand, PL agar shows a better performance in recovering heat- and cold-stressed cells of *Plesiomonas*.[106]

For routine analysis of environmental samples, incubation of plates at 35°C for 24–48 h is generally recommended,[107] although 44°C was found to be the optimal incubation temperature to easily differentiate colonies of *Plesiomonas* and *Aeromonas* in 24 h when *Plesiomonas* differential agar is used.[108]

Enrichment of samples before plating is a controversial practice. Four media have been proposed: alkaline peptone water (APW), bile peptone broth, and tetrathionate broth with or without iodine. As an example of the contradictory results, Freund et al.[109] found that enrichment with tetrathionate broth without iodine gave a high recovery rate for *P. shigelloides* from fresh water, whereas Van Damme and Vandepitte[29] reported no increase in the isolation rate of *P. shigelloides* from freshwater fish using tetrathionate with or without iodine. If tetrathionate broth (with or without iodine) is selected as enrichment media, incubation at 40°C for 24 h seems to provide the highest recovery of *P. shigelloides*.[109] In order to increase the recovery of *P. shigelloides*, it is a common practice to combine the direct plating on two selective media (usually IBB agar and PL agar) and an enrichment on tetrathionate broth without iodine followed by streaking on the two selective media.[110]

The identification of suspect isolates presents little difficulties, and it can be done by inoculation both in triple sugar iron (TSI) slants and inositol gelatine deeps, with incubation at 35°C for 24 h. An oxidase test and Gram stain are also necessary. *P. shigelloides* is a gram-negative rod, oxidase-positive, alkaline over acid with no gas or hydrogen sulfide in TSI, ferments inositol, and fails to hydrolyze gelatine. A complete list of useful characteristics of *P. shigelloides* can be found elsewhere.[110] Alternatively to the classical biochemical identification, miniaturized systems, like API 20E, BBL Crystal E/NF, Phoenix 100 ID/AST, and NID Panel, offer reliable results. Common bionumbers for *P. shigelloides* are 7144204 and 7144244 (API 20E system) and 2203200257 and 2223200257 (BBL Crystal E/NF).[32,41,43,111]

Molecular Diagnosis. Due to the low pathogenic potential of *P. shigelloides*, limited investigations have been conducted to facilitate the detection of the microorganism. The molecular methods reported for *P. shigelloides* are focused on PCR. A summary of the characteristics of the methods, including target, primers, visualization procedures, and application is provided in Table 83.1. The first published procedure was directed toward specific sequences of the 23S rRNA,[112] amplifying a fragment of 284 bp. This method was checked against a number of isolates of *P. shigelloides* from aquatic environments, clinical, and animal origin, but it was applied only to pure cultures, and a procedure for the testing of clinical and environmental samples was not provided.

TABLE 83.1
Characteristics of PCR Assays Developed for *P. shigelloides*

Target	Primers and Probes (5′→3′)	Visualization	Application	Reference
23S rRNA	PS23FW3: CTCCGAATACCGTAGAGTGCTATCC PS23RV3: CTCCCCTAGCCCAATAACACCTAAA	Agarose gel (284 bp), quantification by image analysis	Pure cultures, clam and oyster tissue	González-Rey et al.[112] and Gu and Levin[113,114]
23S rRNA	Forward: AGCGCCTCGGACGAACACCTA Reverse: GTGTCTCCCGGATAGCAC Probe: LCRed640-GGTAGAGCACTGTTAAGGCTAGGGGGTCATC-P	Real-Time FRET PCR	Stool samples	Loh and Yap[115]
hugA	Forward: GCGAGCGGGAAGGGAAGAACC Reverse: GTCGCCCCAAACGCTAACTCATCA	Agarose gel (435 bp)	Fish samples	Herrera et al.[32]

Based on the PCR procedure of González-Rey et al.[112] Gu and Levin developed two modified methods that allowed for quantification of *P. shigelloides*.[113,114] The first method was carried out in homogenates of shellfish tissue (clams and oysters) that were then seeded with known numbers of a collection strain of *P. shigelloides*.[113] The bacteria were recovered by differential centrifugation and treated with formaldehyde. DNA was extracted by lysing the cells with a mixture of Triton X-100 and sodium azide and purified with a commercial clean-up system. After PCR amplification, the PCR products were resolved by electrophoresis in controlled conditions and stained with GelStar™ (Cambrex, Rockland, MN, USA). The gel was photographed, and the fluorescent intensities of the DNA band were analyzed with the public domain Image software (http://rsb.info.nih.gov/ij/index.html). The relative fluorescent intensity of the PCR products was plotted against the log of CFU/g. The detection level was 60 CFU/g for clams and 200 CFU/g for oysters, but they were improved to 4 and 40 CFU/g, respectively, with the introduction of a nonselective enrichment step. The second modification was developed to differentiate live and dead cells by using a DNA intercalating dye [ethidium bromide monoazide (EMA)] unable to penetrate into live cells.[114] The dye would be incorporated mainly by dead cells with damaged membranes, inhibiting PCR amplification of the target DNA. At the same time, quantification of the viable cells is achieved by analyzing the relative fluorescent intensities of DNA bands obtained after a controlled PCR amplification.

A quantitative assay was developed by Loh and Yap,[115] based on the amplification of 23S rRNA, with the specific primers presented in Table 83.1. An LCRed640 labeled probe was also designed to specifically hybridizate to the amplicon. The reaction mixture included SYBR Green I, which acts as a generic donor in fluorescent resonance energy transfer (FRET) to excite LCRed640, and this is only possible if amplification occurs. The intensity of the measured fluorescence is proportional to the amount of DNA generated during the amplification process and the time of detection is related to the initial amount of DNA. The sensitivity of the assay was tested using serially diluted *P. shigelloides* DNA, and the specificity was determined against 27 other bacterial species implicated in gastrointestinal disease. The PCR procedure was applied on stool samples from patients with diarrhea, using the QIAamp® DNA Stool Mini Kit (Qiagen, Hilden, Germany) for DNA extraction.

A different PCR assay was developed by Herrera et al.[32] for the detection of *P. shigelloides* in fish samples. The PCR was directed toward the *hugA* gene, which encodes an outer membrane receptor required for heme iron utilization[95] and is probably implicated in the virulence mechanisms of *Plesiomonas*. The assay has been applied both to pure cultures of bacteria and to food samples. The selectivity was tested against strains of bacteria commonly found in aquatic environments or of relevance as foodborne pathogens. This procedure will be further detailed as a working protocol.

Typing. Serotyping has been an important tool for differentiating among strains of *P. shigelloides*, since biotyping is of limited value due to the phenotypic homogeneity of this species. Two major schemes were initially developed and later unified in an international antigenic scheme.[116] Serotyping has been used for epidemiological studies with clinical and environmental strains[117,118] but nowadays it is on the decline and likely to be replaced by molecular methods. Even though many isolates of *Plesiomonas* carry plasmids (see Section 83.1.5), plasmid profiling does not seem to be a suitable procedure for epidemiological studies because of their heterogenicity. Molecular typing, such as RAPD, PFGE, and more recently, MLST, has been used to investigate the diversity and relationships of isolates from different origins.[12,27,47,119] González-Rey et al.[27] showed that RAPD and PFGE were able to discriminate among strains belonging to different serotypes but not between strains from the same serotype. All the methods tested had good performance; considering that RAPD is fast, simple, and inexpensive, it could be a promising method for routine typing of *P. shigelloides* until a standardized procedure is available.

83.2 METHODS

83.2.1 Sample Collection and Preparation

P. shigelloides is considered a fish-borne pathogen, and the procedure that we have developed has been used successfully to test fish and seafood samples,[32] but it may work with other kind of food samples as well. For analysis of stool samples, the procedure developed by Loh and Yap[115] may be utilized for the extraction of bacterial DNA, which can then be incorporated to the PCR assay.

1. Blend fish samples (10 g) with 90 mL of TSBYE and incubated at 35°C ± 2°C for 24 h with constant shaking (150 rev min^{-1}).
2. Withdraw 1 mL of the upper phase of the culture to a microcentrifuge tube and centrifuge for 5 min at 10,000 × *g*. Carefully remove the supernatant and process the pellet for DNA extraction. The pellets from the enrichment culture may be stored in TSB plus 40% glycerol or −40°C for DNA extraction at a later date; before testing, transfer 20 µL of the frozen culture to 5 mL of fresh TSBYE and incubate overnight at 37°C. Centrifuge 1 mL of the culture as described earlier.
3. Wash the pellet three times with PBS and resuspend in 500 µL of a 1% Triton X-100 solution. Boil the suspension for 5 min and immediately cool on ice. Centrifuge the boiled suspension at 10,000 × *g* for 5 min at 4°C and transfer the supernatant to a fresh tube and freeze at −20°C for long-term storage. Other procedures can be used for DNA extraction from the washed pellet. In our laboratory we have checked commercial kits from different suppliers (GE Healthcare, http://www.gehealthcare.com; QIAGEN, http://www.qiagen.com; Bio-Rad, http://www.biorad.com) with no differences in the performance of the PCR.
4. When the sample material is a pure culture of a microorganism, prepare a suspension from a single colony in 200 µL of PBS and process as described in step 3.

A similar enrichment procedure (TSB supplemented with yeast extract) is useful for the detection of *P. shigelloides* in clam and oyster.[113] The addition of formaldehyde at a final concentration of 4% decreases the inhibition of the PCR by enzymes present in the shellfish tissue. Moreover, the presence of 0.1% BSA to the PCR reaction mixtures improves the DNA amplification.

83.2.2 PCR Detection

1. For each sample to be tested, make up a mix (50 µL) containing 5 µL of 10× PCR buffer, 25 pmol of each primer (Forward 5′-GCG AGC GGG AAG GGA AGA ACC-3′ and Reverse 5′-GTC GCC CCA AAC GCT AAC TCA TCA-3′), 0.1 mM of dNTPs, 0.8 µL (0.8 U) of DNA polymerase, and 5 µL (*ca.* 50 ng) of template DNA. A master mix can be prepared for all the amplification reactions of a PCR run. In our laboratory we routinely use enzymes and chemicals from Biotools, S.A (http://www.biotools.net) and from 5 Prime (http://www.5prime.com) with good results.
2. Prepare a positive control with DNA from a reference strain of *P. shigelloides* (type strain ATCC 14029, supplied by the Spanish Type Culture Collection, CECT 4262, is the choice in our laboratory) and a negative control with 5 µL of double-distilled sterile water instead of DNA template. The controls are freshly prepared for every set of PCR reactions.

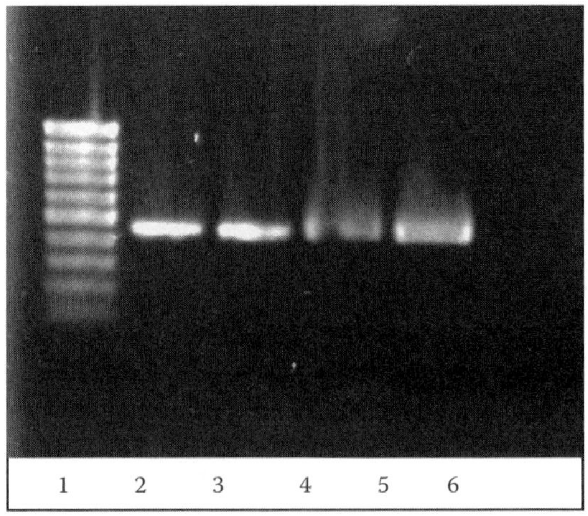

FIGURE 83.1 Results of amplification of a 435 bp fragment of *hug*A gene from *P. shigelloides*. Lane 1, Molecular size marker 100 bp ladder; lane 2, *P. shigelloides* ATCC 14029 (type strain); lane 3, *P. shigelloides* P-1 (isolated from grouper); lane 4, *P. shigelloides* P2 (isolated from halibut); lane 5, *P. shigelloides* C1 (clinical strain isolated from a patient with diarrhea); lane 6, negative control.

3. Spin briefly the tubes to be sure that the reaction mix is at the bottom of the tube.
4. Place tubes in a thermal cycler fitted with a heated lid to prevent evaporation (we use a MasterCycler®, from Eppendorf, http://www.eppendorf.com), and run a program consisting of an initial step at 94°C for 3 min; 30 cycles of 92°C for 30 s, 63°C for 30 s, and 72°C for 1.5 min; and a final step at 72°C for 3 min.
5. Load 5 µL of the PCR products into a 1% (w/v) agarose gel and perform the electrophoresis until a good resolution of the bands is achieved (about 1 h at 6 V/cm).
6. Stain the gel with ethidium bromide and visualize under UV light.

Note: Results obtained with different strains of *P. shigelloides* are shown in Figure 83.1. The presence of a band of 435 bp is considered a positive result.

83.3 CONCLUSION AND FURTHER PERSPECTIVES

Isolation of *P. shigelloides* from clinical and environmental samples is not well defined, and most laboratories do not use special procedures to isolate it. Several enrichment procedures and isolation media have been proposed, but there is evidence that none of them is particularly effective, and a combination of enrichment broth incubated at 40°C, and isolation in two selective plating media is advisable to enhance the recovery of this bacterium.[109,110] On the other hand, identification of the isolates presents little difficulty, either using conventional biochemical techniques or modern commercial miniaturized systems, like API 20E and BBL Crystal E/NF.[32,41]

Thus, the introduction of molecular methods is of great importance as they contribute to the detection and isolation of *P. shigelloides* as well as the identification of isolates. All the methods revised in this chapter work well for pure cultures; two of them have been used to test food samples and one to test stool samples (Table 83.1). The alleged limitations of the procedures are related to the presence of PCR inhibitors in the samples or in the enrichment broths, and these limitations can be overcome by using nonselective enrichment broths, such as TSB or TSBYE or commercial kits.[32,113,115]

The selectivity of the PCR assays for *P. shigelloides* has been checked against a number of strains of *Plesiomonas* isolated from different origins as well as strains of closely related bacteria, as members of the genera *Vibrio*, *Aeromonas*, or *Escherichia*. Herrera et al.[32] indicated a nonspecific amplification both with primers directed toward 23S rRNA and *hug*A gene with a particular strain of *E. coli* harboring the *stx*2 gene, but it seems improbable that this situation would occur in the routine analysis.

Identification of suitable targets can be a promising way of developing new procedures for molecular detection of plesiomonads. In our laboratory, we are exploring the potential use of the histidine decarboxilase gene, implicated in the production of histamine and characteristic of some gram-negative bacteria, and the *chi*A gene, which codified for a chitinase presented in different species of marine bacteria. Also, the formulation of new culture media and the definition of new procedures to enhance the recovery of *P. shigelloides* from clinical and environmental samples will be of help.

Further molecular investigations are needed to elucidate the role of *P. shigelloides* in human and animal disease as well as to identify the pathogenic mechanisms and to facilitate the detection, identification, and characterization of this microorganism.

ACKNOWLEDGMENTS

The authors are grateful to the Spanish Ministry of Education and Science (grants AGL2002-762, AGL2004–4672/ALI and CONSOLIDER - Ingenio 2010: CARNISENUSA (CSD2007-00016)), the Junta de Castilla y León (grant LEO34A06 and research group GR155), and the University of León (grant ULE2001-12) for the financial support to study different aspects of gram-negative foodborne pathogens.

REFERENCES

1. González-Rey, C., Studies on *Plesiomonas shigelloides* isolated from different environments, PhD thesis, Dept. of Veterinary Microbiology, Swedish Univ. of Agricultural Sciences, 2003.
2. Janda, J.M., and Abbott, S.L., The genus *Plesiomonas*. In *The Enterobacteria*, 2nd ed., 335–356, ASM Press, Washington, DC, 2006.
3. Ferguson, W.W., and Henderson, N.D., Description of strain C27: A motile organism with the major antigen of *Shigella sonnei* phase I, *J. Bacteriol.*, 54, 179, 1947.
4. Bader, R.E., Über die herstellung eines agglutinierenden serums gegen die rundform von *Shigella sonnei* mit einem stamm der gattung *Pseudomonas*, *Z. Hyg. Infektionskr.*, 140, 450, 1954.
5. Ewing, W.H., and Johnson, J.G., The differentiation of *Aeromonas* and C27 cultures from *Enterobacteriaceae*, *Int. Bull. Bacteriol. Nomencl. Taxon.*, 10, 223, 1960.
6. Baumann, P., and Schubert, R.H.W., Family II. *Vibrionaceae* In Krieg, N.R., and Holt, J.G. (Editors), *Bergey's Manual of Systematic Bacteriology*, Vol. 1, pp. 516–517, Williams & Wilkins, Baltimore, 1984.
7. Colwell, R.R., MacDonell, M.T., and DeLey, J., Proposal to recognize the family *Aeromonadaceae* fam. nov., *Int. J. Syst. Bacteriol.*, 36, 473, 1986.
8. MacDonell, M.T., and Colwell, R.R., Phylogeny of the *Vibrionaceae*, and recommendation for two new genera, *Listonella* and *Shewanella*, *Syst. Appl. Microbiol.*, 6, 171, 1985.
9. Martinez-Murcia, A.J., Benlloch, S., and Collins, M.D., Phylogenetic interrelationships of members of the genera *Aeromonas* and *Plesiomonas* as determined by 16S ribosomal DNA sequencing: lack of congruence with results of DNA-DNA hybridizations, *Int. J. Syst. Bacteriol.*, 42, 412, 1992.
10. Ruimy, R. et al., Phylogenetic analysis and assessment of the genera *Vibrio*, *Photobacterium*, *Aeromonas*, and *Plesiomonas* deduced from small-subunit rRNA sequences, *Int. J. Syst. Bacteriol.*, 44, 416, 1994.
11. Garrity, G.M., Bell, J.A., and Lilburn, T.G. Taxonomic outline of the prokaryotes. In Garrity, G.M. et al. (Editors), *Bergey's Manual of Systematic Bacteriology*, 2nd ed., p. 120, Springer-Verlag, New York, 2004.
12. Salerno, A. et al., Recombining population structure of *Plesiomonas shigelloides* (*Enterobacteriaceae*) revealed by multilocus sequence typing, *J. Bacteriol.*, 189, 7808, 2007.
13. Inoue, K. et al., Peritrichous flagellation in *Plesiomonas shigelloides* strains, *Jpn. J. Med. Sci. Biol.*, 44, 141, 1991.
14. Pastian, M.R., and Bromel, M.C., Inclusion bodies in *Plesiomonas shigelloides*, *Appl. Environ. Microbiol.*, 47, 216, 1984.
15. Ogawa, J., and Amano, Y., Electron microprobe X-ray analysis of polyphosphate granules in *Plesiomonas shigelloides*, *Microbiol. Immunol.*, 31, 1121, 1987.
16. Santos, J.A. et al., Hemolytic and elastolytic activities influenced by iron in *Plesiomonas shigelloides*, *J. Food Prot.*, 62, 1475, 1999.
17. Ramaiah, N. et al., Use of a *chi*A probe for detection of chitinase genes in bacteria from the Chesapeake Bay, *FEMS Microbiol. Ecol.*, 34, 63, 2000.
18. Schubert, R.H.W., Genus IV. *Plesiomonas*. In Krieg, N.R., and Holt, J.G. (Editors), *Bergey's Manual of Systematic Bacteriology*, Vol. 1, pp. 548–550, Williams & Wilkins, Baltimore, 1984.
19. Koburger, J.A., *Plesiomonas shigelloides*. In Doyle, M.P. (Editor), *Foodborne Bacterial Pathogens*, pp. 311–325, Marcel Dekker, New York, 1989.
20. Albert, M.J. et al., Characterisation of *Plesiomonas shigelloides* strains that share type-specific antigen with *Shigella flexneri* 6 and common group 1 antigen with *Shigella flexneri* spp. and *Shigella dysenteriae* 1, *J. Med. Microbiol.*, 39, 211, 1993.
21. Whang, H.Y., Heller, M.E., and Neter, E., Production by *Aeromonas* of common enterobacterial antigen and its possible taxonomic significance, *J. Bacteriol.*, 110, 161, 1972.

22. Krovacek, K. et al., Isolation, biochemical and serological characterisation of *Plesiomonas shigelloides* from freshwater in Northern Europe, *Comp. Immunol. Microbiol. Infect. Dis.*, 23, 45, 2000.
23. Gibotti, A. et al., Prevalence and virulence properties of *Vibrio cholerae* non-O1, *Aeromonas* spp. and *Plesiomonas shigelloides* isolated from Cambe Stream (State of Parana, Brazil), *J. Appl. Microbiol.*, 89, 70, 2000.
24. Islam, M.S., Alam, M.J., and Khan, S.I., Distribution of *Plesiomonas shigelloides* in various components of pond ecosystems in Dhaka, Bangladesh, *Microbiol. Immunol.*, 35, 927, 1991.
25. Medema, G., and Schets, C., Occurrence of *Plesiomonas shigelloides* in surface water: Relationship with faecal pollution and trophic state, *Zentralbl Hyg Umweltmed*, 194, 398, 1993.
26. ICMSF, *Microorganisms in Foods 5: Characteristics of Microbial Pathogens*, Blackie Academic & Professional, London, 1996.
27. González-Rey, C. et al., Unexpected finding of the "tropical" bacterial pathogen *Plesiomonas shigelloides* from lake water north of the Polar Circle, *Polar Biol.*, 26, 495, 2003.
28. Jagger, T.D., *Plesiomonas shigelloides*: A veterinary perspective, *Infect. Dis. Rev.*, 2, 199, 2000.
29. Van Damme, L.R., and Vandepitte, J., Frequent isolation of *Edwardsiella tarda* and *Plesiomonas shigelloides* from healthy Zairese freshwater fish: A possible source of sporadic diarrhea in the tropics, *Appl. Environ. Microbiol.*, 39, 475, 1980.
30. Sakata, T., and Todaka, K., Isolation of *Plesiomonas shigelloides* and its distribution in fresh water environments, *J. Gen. Appl. Microbiol.*, 33, 497, 1987.
31. Arai, T. et al., A survey of *Plesiomonas shigelloides* from aquatic environments, domestic animals, pets and humans, *J. Hyg. (Lond)*, 84, 203, 1980.
32. Herrera, F.C. et al., Occurrence of *Plesiomonas shigelloides* in displayed portions of saltwater fish determined by a PCR assay based on the *hug*A gene, *Int. J. Food Microbiol.*, 108, 233, 2006.
33. Tsukamoto, T. et al., Two epidemics of diarrhoeal disease possibly caused by *Plesiomonas shigelloides*, *J. Hyg. (Lond)*, 80, 275, 1978.
34. Centers for Disease Control and Prevention, *Plesiomonas shigelloides* and *Salmonella* serotype Hartford infections associated with a contaminated water supply—Livingston County, New York, 1996, *MMWR*, 47, 394, 1998.
35. Dziuban, E.J. et al., Surveillance for waterborne disease and outbreaks associated with recreational water-United States, 2003–2004, *MMWR Surveill. Sum.*, 55, 1, 2006.
36. Centers for Disease Control and Prevention, Aquarium-associated *Plesiomonas shigelloides* infection - Missouri, *MMWR*, 38, 617, 1989.
37. Miller, M.A., and Koburger, J.A., *Plesiomonas shigelloides*: An opportunistic food and waterborne pathogen, *J. Food Prot.*, 48, 449, 1985.
38. Kirov, S.M., *Aeromonas* and *Plesiomonas* species. In Doyle, M.P., Beuchat, L.R., and Montville, T.J. (Editors), *Food Microbiology: Fundamentals and Frontiers*, 2nd ed., pp. 301–327, ASM Press, Washington, DC, 2001.
39. Newsom, R., and Gallois, C., Diarrheal disease caused by *Plesiomonas shigelloides*, *Clin. Microbiol. Newsl.*, 4, 158, 1982.
40. Monge, R., Arias-Echandi, M.L., and Utzinger, D., Presence of cytotoxic *Aeromonas* and *Plesiomonas shigelloides* in fresh vegetables, *Rev. Biomed.*, 9, 176, 1998.
41. Varnam, A.H., and Evans, M.G., *Foodborne Pathogens: An Illustrated Text*, Wolfe, London, 1991.
42. Rouf, M.A., and Rigney, M.M., Growth temperatures and temperature characteristics of *Aeromonas*, *Appl. Microbiol.*, 22, 503, 1971.
43. Herrera, F.C., Microbiology of marine fish: Indicator, spoilage and pathogenic microorganisms, with reference to *Plesiomonas shigelloides*, PhD thesis, University of León, 2004.
44. Miller, M.A., and Koburger, J.A., Tolerance of *Plesiomonas shigelloides* to pH, sodium chloride and temperature, *J. Food Prot.*, 49, 877, 1986.
45. Ingham, S.C., Growth of *Aeromonas hydrophila* and *Plesiomonas shigelloides* on cooked crayfish tails during cold storage under air, vacuum and a modified atmosphere, *J. Food Prot.*, 53, 665, 1990.
46. Brenden, R.A., Miller, M.A., and Janda, J.M., Clinical disease spectrum and pathogenic factors associated with *Plesiomonas shigelloides* infections in humans, *Rev. Infect. Dis.*, 10, 303, 1988.
47. Shigematsu, M. et al., An epidemiological study of *Plesiomonas shigelloides* diarrhoea among Japanese travellers, *Epidemiol. Infect.*, 125, 523, 2000.
48. Kain, K.C., and Kelly, M.T., Antimicrobial susceptibility of *Plesiomonas shigelloides* from patients with diarrhea, *Antimicrob. Agents Chemother.*, 33, 1609, 1989.
49. Tseng, H.K. et al., Characteristics of *Plesiomonas shigelloides* infection in Taiwan, *J. Microbiol. Immunol. Infect.*, 35, 47, 2002.
50. Holmberg, S.D. et al., *Plesiomonas* enteric infections in the United States, *Ann. Intern. Med.*, 105, 690, 1986.
51. Clark, R.B., and Janda, J.M., *Plesiomonas* and human disease, *Clin. Microbiol. Newslett.*, 13, 49, 1991.
52. Sawle, G. et al., Fatal infection with *Aeromonas sobria* and *Plesiomonas shigelloides*, *Br. Med. J.*, 292, 525, 1986.
53. Wouafo, M. et al., An acute foodborne outbreak due to *Plesiomonas shigelloides* in Yaounde, Cameroon, *Foodborne Pathog. Dis.*, 3, 2006.
54. Nolte, F.S. et al., Proctitis and fatal septicemia caused by *Plesiomonas shigelloides* in a bisexual man, *J. Clin. Microbiol.*, 26, 388, 1988.
55. Stock, I., and Wiedemann, B., Natural antimicrobial susceptibilities of *Plesiomonas shigelloides* strains, *J. Antimicrob. Chemother.*, 48, 803, 2001.
56. Körner, R.J., MacGowan, A.P., and Warner, B., The isolation of *Plesiomonas shigelloides* in polymicrobial septicaemia originating from the biliary tree, *Int. J. Med. Microbiol.*, 277, 334, 1992.
57. Leung, M.J., *Plesiomonas shigelloides* and sucrose-positive *Edwardsiella tarda* bacteremia in a man with obstructive jaundice, *Pathology*, 28, 68, 1996.
58. Humphreys, H., Keogh, B., and Keane, C.T., Septicaemia and pleural effusion due to *Plesiomonas shigelloides*, *Postgrad. Med. J.*, 62, 663, 1986.
59. Tzanetea, R. et al., *Plesiomonas shigelloides* sepsis in a thalassemia intermedia patient, *Scand. J. Infect. Dis.*, 34, 687, 2002.
60. Bravo, L. et al., Fatal *Plesiomonas shigelloides* in a newborn, *Mem. Inst. Oswaldo Cruz*, 94, 661, 1999.
61. Claesson, B.E. et al., *Plesiomonas shigelloides* in acute cholecystitis: A case report, *J. Clin. Microbiol.*, 20, 985, 1984.
62. Kennedy, C.A., Goetz, M.B., and Mathisen, G.E., Postoperative pancreatic abscess due to *Plesiomonas shigelloides*, *Rev. Infect. Dis.*, 12, 813, 1990.

63. Khardori, N., and Fainstein, V., *Aeromonas* and *Plesiomonas* as etiological agents, *Annu. Rev. Microbiol.*, 42, 395, 1988.
64. Butt, A.A., Figueroa, J., and Martin, D.H., Ocular infection caused by three unusual marine organisms, *Clinical infectious diseases*, 24, 740, 1997.
65. Marshman, W.E., and Lyons, C.J., Congenital endophthalmitis following maternal shellfish ingestion, *Clin. Exp. Ophthalmol.*, 26, 161, 1998.
66. Sparkes, A.H. et al., Effect of dietary supplementation with fructo-oligosaccharides on fecal flora of healthy cats, *Am. J. Vet. Res.*, 59, 436, 1998.
67. Herrington, D.A. et al., In vitro and in vivo pathogenicity of *Plesiomonas shigelloides*, *Infect. Immun.*, 55, 979, 1987.
68. Vitovec, J. et al., Enteropathogenicity of *Plesiomonas shigelloides* and *Aeromonas* spp. in experimental mono- and coinfection with *Cryptosporidium parvum* in the intestine of neonatal BALB/c mice, *Comp. Immunol. Microbiol. Infect. Dis.*, 24, 39, 2001.
69. Sanyal, S.C., Singh, S.J., and Sen, P.C., Enteropathogenicity of *Aeromonas hydrophila* and *Plesiomonas shigelloides*, *J. Med. Microbiol.*, 8, 195, 1975.
70. Ljungh, A., Popoff, M., and Wadstrom, T., *Aeromonas hydrophila* in acute diarrheal disease: detection of enterotoxin and biotyping of strains, *J. Clin. Microbiol.*, 6, 96, 1977.
71. Gurwith, M.J., and Williams, T.W., Gastroenteritis in children: A two-year review in Manitoba. I. Etiology, *J. Infect. Dis.*, 136, 239, 1977.
72. Penn, R.G. et al., *Plesiomonas shigelloides* overgrowth in the small intestine, *J. Clin. Microbiol.*, 15, 869, 1982.
73. Johnson, W.M., and Lior, H., Cytotoxicity and suckling mouse reactivity of *Aeromonas hydrophila* isolated from human sources, *Can. J. Microbiol.*, 27, 1019, 1981.
74. Pitarangsi, C. et al., Enteropathogenicity of *Aeromonas hydrophila* and *Plesiomonas shigelloides*: Prevalence among individuals with and without diarrhea in Thailand, *Infect. Immun.*, 35, 666, 1982.
75. Saraswathi, B., Agarwal, R.K., and Sanyal, S.C., Further studies on enteropathogenicity of *Plesiomonas shigelloides*, *Indian J. Med. Res.*, 78, 12, 1983.
76. Manorama, T.V., Agarwal, R., and Sanyal, S., Enterotoxins of *Plesiomonas shigelloides*: Partial purification and characterization, *Toxicon Suppl*, 3, 269, 1983.
77. Gardner, S.E., Fowlston, S.E., and George, W.L., In vitro production of cholera toxin-like activity by *Plesiomonas shigelloides*, *J. Infect. Dis.*, 156, 720, 1987.
78. Gardner, S.E., Fowlston, S.E., and George, W.L., Effect of iron on production of a possible virulence factor by *Plesiomonas shigelloides*, *J. Clin. Microbiol.*, 28, 811, 1990.
79. Matthews, B.G., Douglas, H., and Guiney, D.G., Production of a heat stable enterotoxin by *Plesiomonas shigelloides*, *Microb. Pathog.*, 5, 207, 1988.
80. Falcón, R. et al., Intracellular vacuolation induced by culture filtrates of *Plesiomonas shigelloides* isolated from environmental sources, *J. Appl. Microbiol.*, 95, 273, 2003.
81. Okawa, Y. et al., Isolation and characterization of a cytotoxin produced by *Plesiomonas shigelloides* P-1 strain, *FEMS Microbiol. Lett.*, 239, 125, 2004.
82. Sansonetti, P.J., Bacterial pathogens, from adherence to invasion: Comparative strategies, *Med. Microbiol. Immunol.*, 182, 223, 1993.
83. Theodoropoulos, C. et al., *Plesiomonas shigelloides* enters polarized human intestinal Caco-2 cells in an in vitro model system, *Infect. Immun.*, 69, 2260, 2001.
84. Kirov, S.M., *Aeromonas* and *Plesiomonas* species. In Doyle, M.P., Beuchat, L.R., and Montville, T.J. (Editors), *Food Microbiology: Fundamentals and Frontiers*, pp. 265–287, ASM Press, Washington, DC, 1997.
85. Tsugawa, H. et al., Cell adherence-promoted activity of *Plesiomonas shigelloides* GroEL, *J. Med. Microbiol.*, 56, 23, 2007.
86. Binns, M.M. et al., Invasive ability of *Plesiomonas shigelloides*, *Zbl. Bakt. Hyg. A*, 257, 343, 1984.
87. Sanyal, S.C., Saraswathi, B., and Sharma, P., Enteropathogenicity of *Plesiomonas shigelloides*, *J. Med. Microbiol.*, 13, 401, 1980.
88. Abbott, S.L., Kokka, R.P., and Janda, J.M., Laboratory investigations on the low pathogenic potential of *Plesiomonas shigelloides*, *J. Clin. Microbiol.*, 29, 148, 1991.
89. Tsugawa, H. et al., Invasive phenotype and apoptosis induction of *Plesiomonas shigelloides* P-1 strain to Caco-2 cells, *J. Appl. Microbiol.*, 99, 1435, 2005.
90. Pavlova, A. et al., Inhibition of mammalian cathepsins by *Plesiomonas shigelloides*, *Folia Microbiol. (Praha)*, 51, 393, 2006.
91. Daskaleros, P.A., Stoebner, J.A., and Payne, S.M., Iron uptake in *Plesiomonas shigelloides*: Cloning of the genes for the heme-iron uptake system, *Infect. Immun.*, 59, 2706, 1991.
92. Janda, J.M., and Abbott, S.L., Expression of hemolytic activity by *Plesiomonas shigelloides*, *J. Clin. Microbiol.*, 31, 1206, 1993.
93. Baratéla, K.C. et al., Effects of medium composition, calcium, iron and oxygen on haemolysin production by *Plesiomonas shigelloides* isolated from water, *J. Appl. Microbiol.*, 90, 482, 2001.
94. González-Rodríguez, M.N. et al., Cell-associated hemolytic activity in environmental strains of *Plesiomonas shigelloides* expressing cell-free, iron-influenced extracellular hemolysin, *J. Food Prot.*, 70, 885, 2007.
95. Henderson, D.P. et al., Characterization of the *Plesiomonas shigelloides* genes encoding the heme iron utilization system, *J. Bacteriol.*, 183, 2715, 2001.
96. Marshall, D.L., Kim, J.J., and Donnelly, S.P., Antimicrobial susceptibility and plasmid-mediated streptomycin resistance of *Plesiomonas shigelloides* isolated from blue crab, *J. Appl. Bacteriol.*, 81, 195, 1996.
97. Kelly, M.T., and Kain, K.C., Biochemical characteristics and plasmids of clinical and environmental *Plesiomonas shigelloides*, *Experientia*, 47, 439, 1991.
98. Martínez, O. et al., Detection of chitinase gene in strains of *Plesiomonas* from environmental and clinical origin, *9th International Symposium on Aeromonas and Plesiomonas*, p. 84, 2009.
99. Bartlett, D.H., and Azam, F., Chitin, cholera, and competence, *Science*, 310, 1775, 2005.
100. Pruzzo, C., Vezzulli, L., and Colwell, R.R., Global impact of *Vibrio cholerae* interactions with chitin, *Environ. Microbiol.*, 10, 1400, 2008.
101. Simidu, U. et al., Marine bacteria which produce tetrodotoxin, *Appl. Environ. Microbiol.*, 53, 1714, 1987.
102. Cheng, C.A. et al., Microflora and tetrodotoxin-producing bacteria in a gastropod, *Niotha clathrata*, *Food Chem. Toxicol.*, 33, 929, 1995.
103. Lopez-Sabater, E.I. et al., Incidence of histamine-forming bacteria and histamine content in scombroid fish species from retail markets in the Barcelona area, *Int. J. Food Microbiol.*, 28, 411, 1996.

104. Schubert, R.H.W., Ueber den Nachweis von *Plesiomonas shigelloides* Habs and Schubert, 1962, und ein Elektivmedium, den Inositol-Brillantgrun-Gallesalz-Agar, *Ernst Rodenwaldt-Arch.*, 4, 97, 1977.
105. von Graevenitz, A., and Bucher, C., Evaluation of differential and selective media for isolation of *Aeromonas* and *Plesiomonas* spp. from human feces, *J. Clin. Microbiol.*, 17, 16, 1983.
106. Jeppesen, C., Media for *Aeromonas* spp., *Plesiomonas shigelloides* and *Pseudomonas* spp. from food and environment, *Int. J. Food Microbiol.*, 26, 25, 1995.
107. Miller, M.A., and Koburger, J.A., Evaluation of Inositol Brilliant Green Bile Salts and Plesiomonas Agars for recovery of *Plesiomonas shigelloides* from aquatic samples in a seasonal survey of the Suwanee river estuary, *J. Food Prot.*, 49, 274, 1985.
108. Huq, A. et al., Optimal growth temperature for the isolation of *Plesiomonas shigelloides*, using various selective and differential agars, *Can. J. Microbiol.*, 37, 800, 1991.
109. Freund, S.M., Koburger, J.A., and Wei, C.I., Enhanced recovery of *Plesiomonas shigelloides* following an enrichment technique, *J. Food Prot.*, 51, 110, 1988.
110. Palumbo, S.A. et al. *Aeromonas*, *Arcobacter*, and *Plesiomonas*. In Downes, F.P., and Ito, K. (Editors), *Compendium of Methods for the Microbiological Examination of Foods*, 4th ed., pp. 283–300, APHA, Washington, DC, 2001.
111. O'Hara, C.M., Evaluation of the Phoenix 100 ID/AST system and NID panel for identification of *Enterobacteriaceae*, *Vibrionaceae*, and commonly isolated nonenteric gram-negative bacilli, *J. Clin. Microbiol.*, 44, 928, 2006.
112. González-Rey, C. et al., Specific detection of *Plesiomonas shigelloides* isolated from aquatic environments, animals and human diarrhoeal cases by PCR based on 23S rRNA gene, *FEMS Immunol. Med. Microbiol.*, 29, 107, 2000.
113. Gu, W.M., and Levin, R.E., Quantitative detection of *Plesiomonas shigelloides* in clam and oyster tissue by PCR, *Int. J. Food Microbiol.*, 111, 81, 2006.
114. Gu, W., and Levin, R.E., Quantification of viable *Plesiomonas shigelloides* in a mixture of viable and dead cells using ethidium bromide monoazide and conventional PCR, *Food Biotechnol.*, 21, 145, 2007.
115. Loh, J.P., and Yap, E.P.H., Rapid and specific detection of *Plesiomonas shigelloides* directly from stool by LightCycler PCR. In Reischl, U., Wittwer, C., and Cockerill, F. (Editors), *Rapid Cycle Real Time PCR. Methods and Applications. Microbiology and Food Analysis*, pp. 161–169, Springer-Verlag, Berlin, 2002.
116. Nair, G.B., and Holmes, B., International Committee on Systematic Bacteriology Subcommittee on the Taxonomy of Vibrionaceae. Minutes of the closed meeting, 19 May 1998, Atlanta, GA, USA, *Int. J. Syst. Bacteriol.*, 49, 1945, 1999.
117. González-Rey, C. et al., Serotypes and anti-microbial susceptibility of *Plesiomonas shigelloides* isolates from humans, animals and aquatic environments in different countries, *Comp. Immunol. Microbiol. Infect. Dis.*, 27, 129, 2004.
118. Ciznar, I. et al., Potential virulence-associated properties of *Plesiomonas shigelloides* strains, *Folia Microbiol. (Praha)*, 49, 543, 2004.
119. Gu, W. et al., Genetic variability among isolates of *Plesiomonas shigelloides* from fish, human clinical sources and fresh water, determined by RAPD typing, *Food Biotechnol.*, 20, 1, 2006.

84 Proteus

Antoni Różalski and Paweł Stączek

CONTENTS

84.1 INTRODUCTION

84.1.1 CLASSIFICATION AND BIOLOGY

Classification. *Proteus* rods were described for the first time by Hauser in 1885.[1] *Proteus* belongs to the Enterobacteriaceae family, and formerly this genus was placed in the tribe *Proteeae* with *Morganella* and *Providencia* rods.[2] Apart from *Proteus mirabilis* and *Proteus vulgaris*, the two first species described, the genus *Proteus* contains also biochemically distinct bacteria, named *P. myxofaciens*. This species is the only *Proteus* species without any significance in the pathogenicity of human beings.[3] *P. myxofaciens* produces slime rods, isolated from larvae of the gypsy moth.[4] The biochemical as well as DNA–DNA hybridization analyses showed the homogeneity of *P. mirabilis* and heterogeneity of *P. vulgaris* species, which was subdivided into three biogroups based on fermentation of salicin, esculin hydrolysis, and indole production. Results of further studies made it possible to name biogroup 1 (genomospecies 1) *Proteus penneri*. These bacteria are indole-, esculin-, and salicin-negative.[5] The name *P. vulgaris* was reassigned to biogroup 2 (genomospecies 2), as it turned out to be homogenous and represented a single species.[6] The biogroup 3, consisting of four genomospecies, were studied using molecular and biochemical methods. Results of this study allowed O'Hara et al.[7] to propose the separation of genomospecies 3 (DNase-, lipase-, and tartrate-negative) from another genomospecies, which was named *Proteus hauseri*. The *Proteus* genomospecies 4, 5, and 6 remained unnamed until now. Currently, the genus *Proteus* consists of five named species: *P. mirabilis* (type strain ATCC 29906), *P. vulgaris* (ATCC 29905), *P. penneri* (ATCC 33519), *P. hauseri* (ATCC 700826 and ATCC 13315), and *P. myxofaciens* (ATCC 19692), as well as *Proteus* genomospecies 4 (ATCC 51469), 5 (ATCC 51470), and 6 (ATCC 51471).[3,6]

Environmental Distribution and Pathogenicity. *Proteus* rods are a group of gram negative bacteria that are widespread in the natural environment. Because of their proteolytic activity and their ability to hydrolyze urea to ammonia and carbon dioxide, as well as the oxidative deamination of amino acids, these bacteria play an important role in decomposing organic matter of animal origin. Therefore, they can most often be found in polluted water, soil, and manure. *Proteus* bacteria are also a part of natural flora of the human and animal gastrointestinal tracts.[2] They are frequently isolated from human and animal feces, as well as from associated sources such as sewage and decomposing meat. *P. mirabilis* is a common inhabitant of dogs, cows, and birds. *P. vulgaris* is often recovered from cows and birds; however, in contrast to *P. mirabilis*, this species is more frequently isolated from pigs than from dogs.[8] Human feces were found as a source of nosocomial infections caused by *P. mirabilis*.[9] *P. vulgaris* and *P. penneri* were isolated from stools with a wide range of frequencies: 0.45%–6% and 0.4%–1.2%, respectively.[3] There is no evidence supporting the classification of *Proteus* rods as enteropathogens.

Proteus rods are opportunistic pathogens, which cause urinary tract infections (UTIs), wounds infections, meningitis in neonates or infants, and rheumatoid arthritis. These bacteria are also associated with nosocomial infections.[3,10–12]

Proteus plays a particularly important role in UTI. It can cause two types of UTI—hematogenous infections and ascending infections; however, the latter are more common to these bacteria. They are frequently responsible for UTI in patients with a urinary catheter in place or with structural and/or functional abnormalities within the urinary tract, as well as after surgical intervention in the urogenital system. *P. mirabilis* causes UTI with the highest frequency among all *Proteus* species. It was found that *P. mirabilis* is a common cause of catheter-associated UTIs. This species is the third most common cause of complicated UTI, with a frequency of 12%, and the second most common cause of catheter-associated bacteriuria in patients who have been catheterized for a long time.[13,14]

Swarming Phenomenon. *Proteus* are dimorphic bacteria. While growing in liquid media they are motile, peritrichously flagellated short rods (1.0–2.0 μm in length with 6–10 flagella) called swimmer cells; however, when transferred onto solid media these bacteria change themselves into elongated (20–80 μm in length), hyperflagellated, multinucleated, nonseptated swarmer cells. These differentiated swarmer cells migrate on solid surfaces in a coordinated manner that is dependent upon cell-to-cell signaling and multicellular interactions.[15–17] Swarmer cells' translocation is facilitated by an extracellular acidic polysaccharide designated Cmf (Colony migration factor). It acts as a lubricant, reducing surface friction.[18] The migration of swarmer cells from the original inoculation site continues until the number of bacteria in the population is reduced by the loss of individual cells, or when bacterial mass changes the direction of moving. The next step of the swarming phenomenon, called "consolidation," is accompanied by the swarmer cells' disintegration into short rods.[15]

Differentiation of short rods into swarmer cells is induced by the contact of bacteria with a solid surface and the inhibition of flagellar rotation.[19] The molecular mechanism of swarming phenomenon has not been explained until now. It is a metabolically complicated process, genetically coordinated by over 50 genes.[14,16] The most important for swarmer cell differentiation is 30-fold increase of expression of *flhDC* master operon.[16] Several factors involved in swarming regulation have been identified, including the global regulator **l**eucine responsive **r**egulatory **p**rotein (Lrp); RsbA protein (**r**egulation of **s**warming **b**ehavior) homologous to membrane sensor—histidine kinase; 4 Umo (**u**pregulated of **m**aster **o**peron) proteins, localized in cell membrane and outer membrane; CcmA proteins (**c**urved **c**ell **m**orphology) maintaining linearity of elongated swarmer cells; WosA (**w**ild type **o**nset with **s**uperswarming)—a possible regulator of *flhDC* expression; CheW, which facilitates the transfer of information between the methyl accepting chemotaxis proteins (MCPs) and CheA kinase; enzymes GalU and RfaD important in lipopolysaccharides (LPSs) biosynthesis; PepQ—proline dipeptidase, possibly required for signaling of swarm cell differentiation; DsbA—an oxidoreductase important for disulfide bonds formation in periplasmic proteins.[16,20] Identified signaling molecules in *P. mirabilis* comprise autoinducer 2 (AI-2), cyclic dipeptide, glutamine, and putrescine.[14–17,20] The latter seems to play an important role in swarmer cell differentiation. It is interesting that this amino acid can be found in a complex with galacturonic acid residue present in *Proteus* LPSs and Cmf.[21] It is postulated that putrescine complexed with cell-surface polysaccharides is a signal molecule for differentiation of bacteria.[20] *Proteus* swarming is inhibited by saturated fatty acids: lacuric, myristic, and palmitic acids.[16]

The role of swarmer cell differentiation in the virulence of *Proteus* rods is still not clear. Swarmer cells demonstrate a higher expression of flagellin synthesis, production of urease, HpmA hemolysin, and IgA metaloprotease ZapA as compared to swimmer cells.[14,20] A study by Allison et al.[22] showed in vitro that swarmers could invade ueroepithelial cells, whereas nonflagellated and nonswarming mutant cells were noninvasive. However, these findings were not confirmed by other authors[23,24] who found that short vegetative cells, whose number increased due to the reversion of long swarmer bacteria in liquid medium, were internalized. Jansen et al.[25] showed in mouse model of ascending UTI that the predominant cell types are short swimmer cells but not elongated swarmer cells. Consequently, it is possible that swarming phenomenon may play a role during the first stages of infection, when bacteria migrate from uretral meatus to the bladder, and it is not required for crystalline biofilm formation in this part of urinary tract.[14]

84.1.2 Virulence Factors of *Proteus* Bacteria

Proteus has evolved multiple virulence factors including fimbriae, flagella, Pst phosphate transporter, enzymes (urease, proteases, deaminases), hemolysins, endotoxin (lipopolysaccharide, LPS), and invasiveness.[26–31]

Proteus bacilli are capable of producing different types of adherence factors which enable them to cause complicated and uncomplicated UTI. Roberts et al.[32] showed that, out of all gram-negative bacteria, *P. mirabilis* has the greatest ability to attach to catheters. This bacterium can attach and swarm across the surface of urinary catheters and then colonize the bladder and upper part of urinary tract. To counteract the flushing motion of urine flow, bacteria adhere to uroepithelial cells using fimbrial and afimbrial adhesins. These surface structures play also a role in the colonization of kidneys as well as in crossing the renal barrier into the bloodstream. Six types of adhesions have been shown for *Proteus* strains: MR/P (**m**annose **r**esistant ***Proteus*** like fimbriae), MR/K (**m**annose **r**esistant ***Klebsiella*** hemagglutinins), PMF (***Proteus mirabilis*** **f**imbriae), NAF (**nonag**glutinating **f**imbriae), ATF (**a**mbient **t**emperature **f**imbriae), and *P. mirabilis* P-like fimbriae (fimbriae, which are similar to *E. coli P. pili*).[14,26]

MR/P fimbriae are encoded by *mrp* operon containing nine genes and located on the chromosome. MrpA, MrpG, MrpH, and MrpD proteins serve as a major structural subunit, minor subunit, tip adhesion, and chaperon protein, respectively. Expression of these fimbriae is regulated by phase variation phenomenon.[14] It was demonstrated that oxygen-limiting conditions enrich fimbriated bacteria.[33] MR/P fimbriae elicite a strong immune response directed against MrpA, facilitate colonization of the upper urinary tract, contribute to biofilm formation, and are more often found on strains which cause pyelonephritis.[14,27] MR/K hemagglutinins are associated more frequently with *P. penneri* strains and most probably play a role in the adherence to surface of the urinary catheters.[14] PMF are encoded by *pmf* operon consisting of five genes.[26] There are conflicting results concerning the importance of these fimbriae for pathogenesis. Massad et al.[34] showed that they were important in colonization of the bladder, but not of the kidneys. Later Zuninio et al.[35] demonstrated their role in colonization of both the bladder and kidneys. UCA/NAF (uroepithelial cell adhesion/nonagglutinating fimbriae) were proved to adhere to uroepithelial cells, and it has been suggested that they may be involved in the colonization of intestines due to their homology to F17 *E. coli* pilin.[14] UCA/NAF binds a receptor containing GalNAc-(β1,4)-Gal disaccharide present in asialogangliogIycolipids.[36] ATF as well as *P. mirabilis* P-like fimbriae are not important in the pathogenicity.[14,28,29] Analysis of the genome sequence of the *P. mirabilis* strain HI 4320 revealed 12 additional chaperon-usher fimbrial operons.[37] Thus, the total described and postulated fimbriae encoded by *P. mirabilis* amounts to 17 types. Several genes involved in the assembly of type IV pili were also noticed, which suggests that *P. mirabilis* may produce also this type of fimbriae.[37] *P. mirabilis* strains could repress the motility during fimbrial expression due to MrpJ and additional paralogues of this protein, which is a transcriptional regulator reported as the repressor of migration. Therefore, these bacteria alternate between motile and adherent forms.[38]

Flagella play an important role in dissemination of bacteria from the initial site of infection. It was found in the mouse model of ascending UTI that flagella-negative *P. mirabilis* mutant is 100-fold less virulent compared to the wild type, which indicates the importance of flagella in the virulence.[39] Flagella as H-antigen are strongly immunogenic. The antigenic variation via flagellin gene rearrangement allows *P. mirabilis* rods to avoid host immune response. Particularly, this mechanism permits bacteria to evade secretory IgA directed toward flagella during the colonization of urinary tract.[40]

Proteus spp. can also avoid the sIgA action since these bacteria have proteolytic activity. They produce metalloproteases which reveal highest activity in alkaline conditions (pH 8), which exist in urinary tract due to the urease activity (see below). Compared to other proteolytic bacteria, *Proteus* proteases showed expanded substrate specificity including: subclasses of IgA (sIgA and IgA2), as well as IgG antibodies, proteins of the complement system, cell matrix proteins, cytoskeletal proteins and antimicrobial defense components of human innate immune system peptides hBD1, and cathelicidin LL-37.[14,41] Two proteases—designated ZapA and ZapE—were described. These enzymes belong to the serralysin family—zinc homologous metalloproteases, produced by *Serratia marcescens*, *Pseudomonas aeruginosa*, and *Erwinia chrysanthemi*.[42] They are encoded by *zap* operon containing five genes, including *zapB*, *zapC*, and *zapD* coding ABC transporter, membrane-spanning fusion protein, and polypeptide involved in outer membrane transport, respectively.[43] ZapA is important for UTI since *zapA*$^-$ mutants were found to fail for IgA degradation activity in mouse as well as rat model of UTI, and they were less virulent, compared to the wild type.[44]

Proteus ssp. also produces amino acids deaminases encoded by *aad* gene that enable these bacteria to produce α-keto acids, serving them as siderophores.[45] This type of mechanism of iron acquisition is characteristic for *Proteus*, *Providencia*, and *Morganella* species. The genomic sequence of *P. mirabilis* HI 4320 led to identification of at least eight loci involved in iron acquisition.[37]

Jacobsen et al.[30] postulated the high-affinity phosphate transporter Pst described in *P. mirabilis* as a virulence factor for this bacterium. It is suggested that Pst transport system plays a role in biofilm formation, and therefore, it is important for the establishing of chronic infections.

Urease, an enzyme digesting urea to ammonia and carbon dioxide, was demonstrated as the most important virulence factor of *Proteus* strains.[14,26–29] It was found that urease-negative mutants were less virulent than the parental strains.[46] Urease is urea-inducible multimeric nickel metalloenzyme encoded by *ure* operon containing seven genes. Urease mediates urea hydrolysis to ammonia and carbon dioxide that elevates the pH of local environment, which results in the precipitation of magnesium and calcium ions and the formation of urinary stones containing struvite and carbonate apatite.[28] The accumulation of toxic levels of ammonia also results in damages to the urinary tract epithelium. Particularly, the toxic NH_3 damages bladder surface glycoaminoglycans involved in the prevention of bacterial adherence to uroepithelium.[14]

Proteus rods synthesize cell-associated, calcium-independent HpmA hemolysin, which is 166 kDa protein.[47] It belongs to the **t**wo-**p**artner **s**ecretion (TPS) pathway family in different organisms, including bacteria. The component A (HmpA) of hemolysin capable of lysing mammalian cells is secreted by the Sec-dependent transport system to the periplasm. The extracellular secretion and activation of HpmA occurs due to the activity of the component B (HpmB).[48] This toxin seems to be not so important in pathogenesis of this genus, which was shown by the use of HpmA$^-$ mutant.[49] HpmA hemolysin was found to be cytotoxic against a number of cell-culture lines.[14] Three species *P. vulgaris*,[50] *P. penneri*,[51] and *P. hauseri* (Kwil, I. et al., unpublished data, 2006) also produce cell-free, calcium-dependent HlyA hemolysin, classified into RTX proteins (**r**epeated in **t**o**x**in). HlyA belongs to the *Morganella*, *E. coli*, *Proteus*, and *Pasteurella* family of hemolysins.[50] Both *Proteus* HpmA and HlyA hemolysins are members of pore forming toxins family, which are able not only to act against erythrocytes but also to destroy other types of cells.[28]

Alamuri and Mobley[31] described an **a**uto**t**ransporter (AT) in *P. mirabilis* that is both cytotoxin and agglutinin. This bifunctional bacterial product was designated Pta (***P**roteus* **t**oxic **a**gglutinin). Pta is an inducible subtilisin-like alkaline serine protease, optimally expressed in the exponential growth phase at pH 8.5–9.5 and in the presence of Ca^{2+}. This enzyme exhibits both cytotoxic and autoagglutinin activities. Pta is able to lyse cultured bladder epithelial cells and to mediate the autoaggregation of bacterial cells. During autoaggregation, Pta most probably cooperates with MR/P fimbriae, since they were shown to contribute to this phenomenon.[52] Alamuri et al.,[53] by the use of *pta* and *hmpA*-negative mutants, reported the important role of both agglutinin and hemolysin in infection. Moreover, the vaccination of mice with PtA provides protection against *P. mirabilis* UTI. Most likely Pta is not the only autotransporter of *P. mirabilis*. The genome sequence of strain HI 4320 predicts six putative ATs in this microorganism.[37]

Several authors have shown the ability of *P. mirabilis*, *P. vulgaris*, and *P. penneri* to penetrate different cell lines, including Vero (the African green monkey kidney cells), HeLa (cancer cells of the uterus neck), L929 mouse fibroblast, human bladder cell lines (T24, EJ/28, and 5637), Hu 609 (human ureteric ueroepithelial cell lines), HRPTEC (human renal proximal tubular epithelial cells), Int 407 (embryonic intestinal epithelial cells), and HCT-8 (ileocecal epithelial cells).[3,14] It was found that *P. mirabilis* strains—highly active in production of HpmA hemolysin—are the most invasive.[23,54,55] The penetration of cell lines by *P. vulgaris* and *P. penneri* strains producing HlyA hemolysin is accompanied by the cytotoxic effect of this cytolysin.[51] The mechanism of *Proteus* invasiveness is nonelucidated. It was demonstrated that internalized bacteria in HRPTEC were present within membrane—bound vacuoles and that microfilament formation is not involved in the internalization process because of the actin polymerization.[56] Some authors[56] found that the bacteria can replicate within cell lines, whereas other authors did not confirm this.[23,51]

Proteus as all gram-negative bacteria contains lipopolysaccharide (LPS), which, when released from bacteria, is a biologically active endotoxin. LPS causes a broad spectrum of pathophysiological effects, particularly septic shock.[57] *Proteus* LPS contains three parts: O-specific polysaccharide (O-antigen), core region, and lipid A—an endotoxin biological center.[58]

The O-specific polysaccharide (OPS) chain is exposed outside bacteria and, together with capsular polysaccharides, is involved in glycocalyx formation. Most likely, O-specific polysaccharide is also important for the swarming phenomenon since the R mutants producing LPS without this part were unable to swarm or expressed only a limited ability to migrate on the solid medium.[58] The acidic *Proteus* O-antigens may play an important role in stones formation in the urinary tract. Torzewska et al.[59] showed that stone formation due to the urease activity may be modified by bacterial OPSs. The sugar composition of *Proteus* LPS may either enhance or inhibit the crystallization of struvite and apatite, depending on its chemical structure and ability to bind cations. The O-specific part of LPS of *Proteus* spp. S-forms contributes to their resistance against serum's bactericidal action. It was demonstrated that hydrophobic MAC (**m**embrane **a**ttack **c**omplex) of complement cannot pass the hydrophilic barrier of a long chain of OPSs to gain outer and inner membranes and forms pores leading to bacterial lyses. Indeed, *P. mirabilis* R-mutants without OPS are sensitive to the action of serum, whereas most of the *Proteus* S-forms are resistant.[58] The OPS turned out to be important in swarming growth of *Proteus* spp., since the above mentioned rough mutants were unable to swarm or showed limited ability to migrate on agar media, while most of *Proteus* S-forms could swarm intensively.[58] Most likely, OPS as well as capsular polysaccharides (CPS) facilitate translocation of swarmer cells by a reduction of cell friction.[18] The differences in the structure of O-antigens serve as the basis for serological classification of *Proteus* strains[28] (see below).

The second antigen, also exposed on *Proteus* surfaces, is ECA (**e**nterobacterial **c**ommon **a**ntigen). It is a lipoglycan present in all bacteria of Enterobacteriaceae family. In *E. coli* it occurs in three structurally different forms: $ECA_{PG,}$ which is anchored in the outer membrane (OM) via glycerophospholipid; $ECA_{LPS,}$ wherein ECA is linked to the core oligosaccharide of LPS; and $ECA_{CYC,}$ located in the periplasm in a cyclic form (trisaccharide repeating units containing Fuc4NAc, ManNAcA, and GlcNAc) without lipid carrier.[60,61] Recently, it has been shown that, similarly to *E. coli*, the full core region of *P. mirabilis* LPS is required as an acceptor of ECA. This common *P. mirabilis* antigen turned out to be immunogenic in both forms as ECA_{LPS} and ECA_{PG}; the latter, when combined with proteins of serum of immunized host.[62]

84.1.3 The Formation of *Proteus* Biofilm

Bacterial biofilm is defined as matrix-enclosed bacterial population adherent to surfaces. Bacteria located in glycocalyx capsule are protected against the action of antibodies and other antibacterial substances as well as immune mechanisms.[63] They also differ in the expression of particular virulence factors and in metabolic activity in comparison to bacteria growing in liquid media.[64] *Proteus mirabilis* forms a crystalline biofilm, particularly on catheters, including Foley catheters, which are usually used in clinics. The development of this type of biofilm involves several stages. Usually, implanted catheter is rapidly coated by host proteins from the surrounding body fluids. These proteins serve as receptors for bacterial fimbrial and afimbrial adhesins. *P. mirabilis* can also bind directly to the catheter surface.[65,66] The most important factor in the development of crystalline biofilm is urease. It was found that the *ure*⁻ mutant was not able to form this type of biofilm.[67] Acidic OPS and capsular polysaccharide (CPS) could accelerate the formation of struvite crystals.[59,68] There are several physical and chemical factors involved in the development of *Proteus* biofilm, and catheter and bacterial surfaces are the most important.[65] Hydrophobic surfaces are more likely colonized than hydrophilic ones. The colonization is increased in the alkaline pH of urine.[67] From several biocids investigated for antimicrobial activity, triclosan showed the highest activity against

P. mirabilis. It was also found that triclosan and nalidic acid were able to significantly inhibit encrustation process.[69] The reduction of catheter encrustation and blockage by crystalline biofilm was observed in vitro where EDTA (sodium ethylenediaminotetracetic acid) was used.[70] Broomfield et al.[71] established that the most effective way to prevent catheter encrustation is to dilute urine and increase its citrate concentration. Morgan et al.[66] suggested that the antibacterials incorporated into catheters to prevent encrustation must diffuse out from their surfaces to reduce the number of viable urease-producing bacteria. Dispersion of a biofilm formed by *P. mirabilis* can be induced by cis-2-decanoic acid. This substance is active in nanomolar concentration; it is functionally and structurally related to short-chain fatty acids described as signaling molecules in the cell-to-cell communication of bacteria.[72]

84.1.4 Laboratory Diagnosis

84.1.4.1 Phenotypic Identification

The ability of *Proteus* ssp. to swarm makes recognition of these bacteria on agar media surface easier than with other bacteria from the Enterobcteriaceae family; however, the swarming phenomenon makes difficult to isolate non-*Proteus* strains growing on the same media. Microscopically, the swarming phenomenon appears as concentric rings of growth starting from the colony or inoculum site.[3,15,73] Several substances are used to inhibit swarming, including phenol, sodium azide, sodium salicylate, ρ-nitrophenyl glycerol, bile salts, and a high concentration of agar.[3] Since almost all *Proteus* strains typically swarm, they are visible on most common agar media used in Enterobacteriaceae diagnosis, such as MacConkey, Hektoen enteric, S–S (Salmonella–Shigella), and xylose-lysine-deoxycholate.[3,74]

A number of selective media were proposed for *Proteeae* isolation. Hawkey et al.[75] described one of them called PIM (*Proteeae* isolation medium), which is a selective and differential medium containing clindamycin and colistin as selective agents and tryptophan/tyrosine as differential ones. PIM is useful to differentiate rods of the *Proteeae* tribe from contaminating bacteria in clinical specimens. Bacteria of this tribe degrade tyrosine and produce a reddish-brown pigment containing melanin because of the action of the enzyme tyrosine phenol-lyase. The selective medium called "SPM" (selective *Proteeae* medium) was described by Urbanova.[76] Identification of *Proteeae* on SPM is based on the detection of the phosphatase activity, which is produced by bacteria. This medium contains a tryptose phosphate with phenylphtalein monophosphate as the substrate, bile salts and polymyxin B as inhibitors of enteric bacteria, and methyl green as the indicator. *P. mirabilis* and *P. vulgaris* grow on SPM as light green colonies.

Bacteria from the tribe *Proteeae* can be differentiated by use of biochemical tests. Those belonging to the genus *Proteus* are lactose -negative, urease-, and phenylalanine-deaminase-positive. They also produce hydrogen sulfide and oxidatively deaminated tryptophan.[3,73,74] Properties used to identify the *Proteus* species isolated from clinical materials and for identification of non pathogenic *P. myxofaciens* are presented in Table 84.1. Six tests—indole production, ornithine decarboxylase, maltose, salicin, and xylose fermentation as well as esculin hydrolysis—are enough to identify species of the genus *Proteus*.[77] It is worth stressing that both *P. mirabilis* and *P. penneri* are indole-negative; however, *P. penneri* and 25% of *P. mirabilis* strains produce green color when tested by Kovacs' test.[78] In addition, susceptibility to chloramphenicol is also suggested to be useful in identification of the *Proteus* species.[79]

Several rapid, automated, and manual systems of identification of *Proteus* bacteria to the species level are available. Most of them, including Crystal ID-E/NF (Becton Dickinson Microbiology Systems), GNI, GNI+ and API 20E (BioMerieux Inc), Rapid Neg ID3 (Dade Behring Inc.), Vitek GNI card and Vitek GNI+ card (BioMerieux Vitek), are highly accurate (>90%) in identification of three clinically significant species: *P. mirabilis*, *P. vulgaris*, and *P. penneri*, however, the latter is difficult to recognize. Misidentification within *Proteus* sp. as well as identification of *Proteus*

TABLE 84.1
Differentiation within Genus *Proteus*[a]

Biochemical Test/Property	*P. mirabilis*	*P. vulgaris*	*P. penneri*	*P. hauseri*	*P. myxofaciens*
Indole production	−	+	−	+	−
Ornithine decarboxylase	+	−	−	−	−
Maltose fermentation	−	+	+	+	+
Salicin fermentation	−	+	−	−	−
D-Xylose fermentation	+	+	+	+	−
Esculin hydrolysis	−	+	−	−	−
Chloramphenicol susceptibility	S	V	R	S	S
Importance in human pathogenicity	I	I	I	I	NI

+, 90%–100% positive; −, 0%–9.9% positive; S, susceptible; R, resistant; V, variable susceptibility; I, important; NI, not important.

[a] References 3, 28, 74, 75, and 78.

bacteria as *Morganella morganii* or *Providencia rettgeri* were reported.[3,7]

One of the oldest assay applied to epidemiological typing is the Dienes test.[3,12,75] When two different strains of *Proteus* ssp. are inoculated onto an agar plate and allowed to swarm toward each other, the demarcation line of growth appears, when the strains are incompatible. Such reaction does not occur if the strains are identical. The boundary region is called the Dienes line. The exact basis for the Dienes phenomenon has not been explained. It is believed that it is related to production and susceptibility to bacteriocins (proticins) by *Proteus* strains. It was found that the Dienes line is associated with the formation of rounded cells, which originate exclusively from one of the two colliding swarming cells, and have decreased viability. Thus, from the two swarming bacterial strains one can be designated as the dominant over the other one. Budding et al.[80] suggested that both swarming strains play a role in the Dienes line formation. The round cells occurred as a result of a short-range killing mechanism demonstrated by the dominant strain having a competitive advantage, an advantage that is highly specific for the swarming state. The Dienes test is a method easily applied to epidemiological investigations and yields highly discriminatory results.[12]

Bacteriophage and bacteriocin typing are not routinely used in clinical laboratory work. They can be used in epidemiological investigations. There are three schemes of phage typing described: first for *Proteus* and other members of *Proteeae*; second, for *P. mirabilis* and *P. vulgaris*; and third, only for *P. mirabilis* strains.[12] Hickman and Farmer[81] differentiated 200 *P. mirabilis* strains into 113 types using 23 lytic bacteriophages. Senior[82] examined *Proteus* strains for production of proticins (P) as well as proticins sensitivity (S) and found that by use of this assay 250 strains can de differentiated into 90 P/S types. The P/S type of *Proteus* isolate is independent of O serogrouping.[83]

Serotyping can be also recommended in epidemiological investigations. The differences between the O-antigens and H-antigens (flagella) are the molecular basis for serological classification of *Proteus* rods. The Kaufman and Perch scheme includes 49 different *P. mirabilis* and *P. vulgaris* O-serogroups and 19 serologically distinct H-antigens.[84] This classification was later supplemented by Larson,[85] as well as by Penner and Hennesy[86] with additional serogroups. Chemical and serological studies conducted subsequently allowed to establish additional O-serogroups, which were created to classify *P. penneri* and *P. hauseri* strains as well as *P. myxofaciens*, some *P. mirabilis*, and *P. vulgaris* strains unclassified by the above mentioned schemes.[87,88] The serological classification currently consist of 78 serogroups.[58,89,90] The most frequently isolated *Proteus* strains in clinical infections were classified into serogroups O3, O6, O10, O26, O27, O28, and O30.[12,84]

The serological specificity of *Proteus* O-antigens was studied using the polyclonal rabbit anti-O sera specific to particular serogroups.[28,58,88–90] Polyclonal anti-O sera contain antibodies of different specificity types. Usually, the major antibodies fraction recognizes the main epitope, which defines the group specificity, whereas the minor fractions can bind other epitopes in O-antigen or in core region of LPS. Several tests are recommended in immunodiagnosis of *Proteus* sp., including the agglutination test, enzyme immunosorbent assay (EIA) as well as a passive hemolysis test (PHT).[91–93] The agglutination test for determination of O-antigens should be performed with cultures that have been boiled for 1 h to destroy the H-antigen.[12] EIA and PHT with isolated LPS representing particular serogroups as antigens are used to study the antisera specificity.

Uronic acids and hexosamines, the characteristic compounds of *Proteus* OPSs, play an important role in the serological specificity of these antigens.[28,29,58] α-D-GlcA and β-D-GlcA present as a branch of O-specific polysaccharide chain, as well as in linear O-polysaccharides usually serve as immunodominant sugars in specific epitopes.[91,94,95] The most common epitopes found in *Proteus* O-antigens were uronic acids substituted by amino acids. The importance of α-D-GalA(-L-Lys) in the specificity of *P. mirabilis* O3,[96] O26,[97] and O28[98] as well as α-D-GalA(-L-Thr) in the specificity of *P. mirabilis* O11[99] was described.

O-specific antibodies may cross react with LPS of strains belonging to the same species or genus but classified into other serogroups. Such cross-reactivity occurs even with the LPSs of taxonomically different bacteria. One-way or two-way cross-reactivity is common in *Proteus* due to the phenomenon of sharing minor epitopes in O-antigens or epitopes in the core region. For example, thanks to the common α-L-FucNAc-(1→3)-α/β-D-GlcNAc disaccharide, cross-reactivity was observed in antigen-antibodies systems of LPS and heterologous antisera of *P. vulgaris* O8, O12, O39, and *P. mirabilis* O6.[58] Cross-reactivity was also found in the heterologous antigen-antibody system of *P. vulgaris* O45 and *P. penneri* O17 because of the presence of D-Fuc3N (3-amino-3,6-dideoxygalactose) in the O-antigen of the former, and in the core region of LPS of the latter.[100]

Proteus O-antigens also show a similarity with OPSs from the bacteria of other genera. For example, cross-reactivity was found between the anti-*P. mirabilis* O13 serum, as well as the anti-*P. myxofaciens* serum and the LPS from *Providencia alcalifaciens* O14 and O23. Serological studies revealed the important role of common compound alanino-lysine (AlaLys) for the specificity of these O-antigens.[101,102] The common D-Qui4N residue carrying N-linked *N*-acetyl-aspartic acid is responsible for antigenic relationship of O-antigens of *P. mirabilis* O38, as well as *P. alcalifaciens* O4 and O33.[103,104] Significant antigenic similarities were found for *P. vulgaris* O21, *P. mirabilis* O48, and for *Hafnia alvei* 744 and PCM 1194.[105] The knowledge of the chemical *Proteus* O-antigens structure and of the the specificity of O-antisera helps avoid mistakes in immunodiagnosis of these bacteria.

P. mirabilis OXK (serogroup O3) showed cross-reactivity with *Orientia tsutsugamushi* antisera. Similar phenomenon was shown for *P. vulgaris* OX19 (serogroup O1) and *P. vulgaris* OX2 (serogroup O2) with antisera against *Rickettsia* sp. *Proteus* strains belonging to these serogroups are used in

Weil–Felix test, which makes possible diagnostics of rickettsial infections. It was shown that common epitopes responsible for such cross-reactivity reside on the *Proteus* OPS; however, their exact structure, as well as the structures of the corresponding Orientia and Rickettsia antigens, remain unknown.[28]

84.1.4.2 Molecular Identification

In comparison with the methods based on the phenotypic features of bacteria, nucleic acid–based detection and differentiation assays allow fast, reliable, and reproducible testing of the samples. As in the case of other pathogens that may be present in very low concentrations in specimens containing mixed populations, the biggest problem in proper molecular identification of *Proteus* may be posed by the presence of accompanying bacterial flora. Many molecular techniques, especially these based on the amplification of the variable regions with the use of primers anchored in the conservative flanking sequences, may yield false-positive products originating even from low amounts of cross-contaminating bacterial flora DNA. In order to be used as routine methods, they require very strict regime of sample handling, a proper set of controls, and careful examination of the products. On the other hand, the sensitivity of the nucleic acids–based assays decrease significantly when they are applied directly to clinical samples because of the presence of reaction inhibitors, an abundance of host cells/DNA, or uneven pathogen distribution. Thus, to date most of the proposed methods for *Proteus* show high specificity only when pure cultures are raised prior to molecular detection and identification and are rarely applied directly to the clinical specimens. Because of the specific features of the *Proteus* rods that allow their relatively easy identification, at least to the genus level, when grown on microbiological media, no specific standardized protocol has been proposed for their routine detection and identification at the molecular level.

PCR. In recent years, many PCR-based methods have been developed for the rapid detection of human pathogens, among them those belonging to the genus *Proteus*. These assays use specific primers for the amplification of a single gene/region or several target sequences (multiplex PCR) generating genus- or species-specific products. Extensively studied rRNA operons are widely used target region for pathogen detection as well as for elucidating phylogenetic relationships among bacteria. Their relatively high copy number per cell increases the sensitivity of tests, which is especially useful when the specimens contain low levels of bacterial cells. The commonly utilized gene within this region is 16S rRNA, the sequences of which are currently available for most of human pathogens. Due to the presence of variable regions located between highly conservative sequences, it is possible to design pairs of universal primers suitable for amplification of the DNA from different bacteria. A relatively low level of amplicon polymorphisms requires detailed additional analysis, of which sequencing is the most reliable; however, still a time-consuming method. Even then, in case of very closely related species, such as *P. vulgaris* and *P. penneri*, differentiation based on the analysis of 16S rRNA may bring unsatisfactory results.[106]

Dempsey at al.[107] sequenced the cloned 16S rRNA amplicons that were obtained with a pair of universal primers 27f: 5′-AGA GTT TGA TCC/A TGG CTC AG-3′ and 1387r: 5′-GGG CGG A/TGT GTA CAA GGC-3′, which correspond to *E. coli* nucleotides 8–27, and 1387–1404, respectively, in order to identify bacterial species colonizing 10 prosthetic hip joints removed due to clinical infection or the aseptic loosening of the prosthesis. Using a culture-dependent as well as culture-independent approach, several species have been identified in the mixed populations colonizing the surface of the hip joints, among which *P. mirabilis* accounted for 3.5% of the clones analyzed.

A pair of universal primers 27f: 5′-AGA GTT TGA TCC TGG CTC AG-3′, and 907r: 5′-CCC CGT CAATTC ATT TGA GTT T-3′ for amplification of ~920 bp region of 16S rRNA was used by Schabereiter-Gurtner et al.[108] to identify pathogens in monomicrobial and polymicrobial ocular infections. DNA extraction was performed directly from the conjunctival swabs, and the PCR products were reamplified in the nested-PCR reaction with a pair of primers: 341fGC 5′-CCT ACG GGA GGC AGC AG-3′ containing 40 nt GC-clamp attached to the 5′ end (5′-CGC CCG CCG CGC GCG GCG GGC GGG GCG GGG GCA CGG GGG G-3′), and 518r 5′-ATT ACC GCG GCT GCT GG-3′. Resulting fragments, corresponding to nucleotide positions 341–534 of *E. coli* 16S rRNA gene, were subjected to differentiation by denaturing gradient gel electrophoresis (DGGE). In this technique, migration of amplicons toward an increasing concentration of denaturing agent in the gel leads to the sequence-dependent, partial disruption of hydrogen bonds and specifically reduces the mobility of the molecules. Among several DGGE pro files obtained, the one specific for *Proteus* spp. has been found. Sequencing of the PCR product identified this strain as *P. mirabilis*.

Another pair of 16S rRNA universal primers (U1: 5′-CCA GCA GCC GCG GTA ATA CG-3′; U2: 5′-ATC GG(C/T) TAC CTT GTT ACG ACT TC-3′) was designed by Lu et al.[109] for detection of pathogens in cerebrospinal fluid (CSF). Amplifications performed on the set of lysates of 18 different reference strains originating from American Type Culture Collection (ATCC) generated products of the same size (996 bp), which differed in restriction fragments length polymorphism (RFLP) analysis with HaeIII endonuclesase. In order to determine the reproducibility of the method, over 230 clinical isolates, including 3 *P. mirabilis* strains, were tested, yielding identical RFLP patterns as they reference counterparts. This strategy, when applied to the DNAs isolated directly from 150 CSF specimens obtained from patients with meningitis, revealed two cases of *P. mirabilis* infection, among other pathogens. The sensitivity of the assay, determined on the basis of serial dilutions of *E. coli* and *S. aureus* cells in CSF, was estimated for 10 gram-negative and 250 gram-positive bacteria. Such levels seem to be adequate for pathogen detection in CSF, as during infection the amount of cells usually reaches $>10^3$ CFU mL[110];

however, it may not be sufficient in septicemia where the number of pathogen cells often does not exceed 1 CFU/mL.[111] This method is relatively fast, requiring 8–10 h to be accomplished.

The strategy that significantly improved sensitivity and time of pathogen detection is real-time PCR. It quantitates the product formation during amplification cycles and estimates its specificity while it is still in the test tube, which additionally reduces the risk of cross-contamination. Bacterial suspensions obtained from single colonies as well as DNAs isolated from blood cultures were amplified by Vliegen et al.[112] in real-time reaction with a pair of primers: 16S-1: 5′-TGG AGA GTT TGA TCC TGG CTC AG-3′ and 16S-2: 5′-TAC CGC GGC TGC TGG CAC-3′, spanning the 16S rRNA region, which corresponds to *E. coli* nucleotide position 4–532. A control over product concentration and specificity allowed to omit electrophoresis step and move immediately to sequencing of the amplicons in order to identify pathogen species. Among 94 species used in the study, *P. vulgaris* and *P. mirabilis* were precisely identified using this method in about 6.5–7 h, depending on the origin of samples.

Several approaches have been undertaken to increase the discriminatory power of 16S rRNA-based PCR methods. Multiplex PCR utilizing a set of three pairs of primers anchored in the conservative regions of 16S rRNA (ER10: 5′-GGCGGACGGGTGAGTAA-3′ and ER11G: 5′-GGACTGCTGCCTCCCGTAG-3′, ER14: 5′-GCTAACTCCGTGCCAGCA-3′ and ER15: 5′-GCGTGGACTACCAGGGTATC-3′, P11P: 5′-GAGGAAGGTGGGGATGACGT-3′ and P13P: **5′**-AGGC CCGGGAA CGTATTCAC-3′) was designed[113] and generated fragments corresponding to *E. coli* nucleotide positions 103–357, 506–810, and 1173–1389, respectively. These amplicons were differentiated by a single-stranded DNA conformation polymorphism (SSCP). In this technique. DNA fragments are heat denatured in order to obtain single strands prior to the nondenaturing polyacrylamide gel electrophoresis. The unique electrophoretic mobility of particular single-stranded DNA fragments is due to their nucleotide sequence within the variable regions, as they adopt specific secondary structures upon renaturation. In addition, labeling the primers with fluorescent dyes allowed detection of the products during resolution in an automatic sequencing device. Species-specific, laser-induced fluorescence signal patterns received by the sensor at the bottom of the electrophoretic path were analyzed and stored in the computer. This approach avoided gel staining procedures, thus making the method more sensitive and rapid. Moreover, automatic sequencer precisely controls temperature of a gel, and allows the addition of internal markers, which significantly improves the reproducibility of the results. Sensitivity of the multiplex PCR followed by the automated SSCP was estimated at >10^1 CFU of bacteria in the samples consisting of pure cultures, and the whole procedure lasted only eight hours. Using 178 ATCC strains and clinical isolates representing 51 species belonging to 21 genera, specific SSCP patterns were obtained, among which signals common for 4 *P. mirabilis* and for seven *P. vulgaris* were distinguished. Analysis of particular primer pairs revealed that P11P-P13P set generated pattern specific for a genus *Proteus*, while ER14-ER15 was capable of distinguishing *P. mirabilis* from *P. vulgaris*.

As noted, some reports proved that the discriminatory power of 16S rRNA analyzes be too low for distinguishing closely related species and even genera. For that reason, Cao et al.[114] focused their attention on 16S-23S rRNA internal transcribed spacer region (ITS) in order to distinguish three most common pathogens of the *Proteus* genus, namely: *P. vulgaris*, *P. mirabilis*, *P. penneri*, as well as *P. myxofaciens*, which is nonpathogenic for humans. The ITS region is considered to be subject to a lower evolutionary pressure than 16S rRNA and thus more prone to accumulate mutations, which leads to its polymorphic composition. Universal primers wl-5793: 5′-TGT ACA CAC CGC CCG TC-3′ and wl-5794: 5′-GGT ACT TAG ATG TTT CAG TTC-3′ designed on the basis of available *Proteus* rRNA sequences were anchored in the conserved end of 16S rRNA and the proximal region of 23S rRNA, respectively. Electrophoretic analysis of the amplification products generated on DNA templates from 46 isolates belonging to those four species revealed multiple bands of between 700 and 1000 bp. All 10 *P. penneri* and two *P. myxofaciens* strains as well as two *P. mirabilis* and four *P. vulgaris* showed two band patterns, whereas in the remaining twenty *P. mirabilis* and nine *P. vulgaris* strains, three band patterns were observed. Cloning and sequencing of these bands allowed to identify tRNA gene classes as well as conservative and variable regions scattered along the ITS region, forming mosaic-like structure in all *Proteus* strains tested. One variable region designated v2, located at the 3′ end of the ITSs and flanked by two conservative regions c2 and c3, has been found useful for the classification of each of the four species, which is especially important in terms of closely related *P. vulgaris* and *P. penneri* differentiation.

Besides great usefulness of rRNA genes in identification of bacterial species, other targets also focus the attention of investigators in order to develop new, specific techniques for detection of particular pathogens. *P. mirabilis* presence was confirmed by Takeuchi et al.[115] in urinary calculi using *ure*C1 and *ure*C2 complementary to the urease gene. Among 87 stones, from which DNA was extracted by boiling pulverized pieces, *P. mirabilis* was detected in struvite stones (calcium oxalate, calcium phosphate, uric acid, cystine), and was not detectable in metabolic calculi. The sensitivity of this method was found to be higher than for the stone and urine cultivation techniques.

A search for the species-specific regions resulted in the cloning of an undefined 3.5-kb fragment of *P. mirabilis* genome in which terminal regions were sequenced and served to design a pair of primers.[116] These two primers: MMKAP1: 5′-ACC TAG TCC CAG AAA AGA ACC CTC-3′ and MMKAP2: 5′TGG TGT TAA TCA CAT GCT TGA TGG-3′ gave rise to a 3.5-kb amplicon only on the template DNAs from 18 different *P. mirabilis* isolates, while no products were detected in case of *P. vulgaris*, *P. penneri*, *P. myxofaciens*, *Morganella morganii*, different species of *Providencia* genus and nearly 30 ther unrelated pathogens. The sensitivity of the assay tested with the DNA isolated from pure cultures was estimated for 10 fg, which corresponds to ~17 cells.

Oligonucleotide Arrays. The development of microarrays for the analysis of gene expression created new opportunities in the field of microbial diagnostics. This technology utilizes hybridization between short, specific, oligonucleotide probes attached to a solid support, usually a glass-slide (the so-called DNA chip), and nucleic acid targets isolated directly from the cells or preamplified by PCR methods. Advances in microfabrication allowed the preparation of very dense DNA chips, with almost unlimited probe capacity, thus making it possible to detect thousands of different targets at the same time. However, the hardware required for such dense chip printing and signal analysis is rather expensive, requires trained personnel to handle, and so far, may neither be affordable nor practicable in most clinical laboratories. Several approaches to avoid these limitations and simplify the technology are currently being tested in order to obtain cheap, portable, easy-to-handle biosensors that can be used in everyday routine situations, similar to common glucometers.

Microarray designed for detection of UTI pathogens was designed and validated by Liao et al.[117] In this approach, 16S rRNA was used as a target molecule, which decreased time of detection as no amplification step was required to set up the procedure. Bacteria pelleted from 1-mL aliquotes of broth cultures, inoculated urine as well as urine clinical samples were disrupted by alkaline method, and the raw lysates were subjected to only 10 min hybridization at 65°C with a set of fluorescein-labeled detector oligonucleotides. These probes targeted specific rRNA regions of most common uropathogen species and groups: *E. coli*, *P. mirabilis* (5′-GGT CCG TAG ACA TTA TGC GGT ATT AGC CAC CGT TT-3′), *Pseudomonas aeruginosa*, *Enterococcus* spp., *Klebsiella*, and *Enterobacter* spp., Enterobacteriaceae. When applied onto the surface of sensor arrays that contained the complementary set of capture probes (*P. mirabilis* 5′-GGG TTC ATC CGA TAG TGC AAG GTC CGA AGA GCC CC-3′) attached, a consecutive 15 min hybridization between the rRNA-detector probe complex and the capture probe could occur, thus linking the fluorescein label to the specific sensor. A chip placed in the chip reader chamber was then developed using antifluorescein horseradish peroxidase-coupled antibodies, followed by the covering of the sensor area with 3,3′5,5′-tetramethybenzidine (TMB)-H_2O_2 solution, and then amperometric analysis of peroxidase-catalyzed red-ox reaction was performed. Despite the contact with bacteria originating directly from the urine, which may pose a threat to this detection strategy due to the samples' varying levels of pH, high protein concentrations, and contamination with white and red blood cells, a high signal-to-noise ratio was maintained throughout the tests. It has to be noted that because single strains are involved in most UTIs, clinical validation of the UTI chip was performed on samples containing only one uropathogen. Under this condition the assay was able to detect *P. mirabilis* and other uropathogens at the concentration >4 × 10^4 CFU/mL, which is high enough to detect clinically relevant bacteria in urine, and the procedure took less than 1 h. Despite the great speed of detection and the good sensitivity level, the major drawback of the UTI chip, which limited its application in the portable devices, was the need to provide high temperatures during subsequent hybridization steps, which required high voltage and made the device rather bulky. In the follow-up studies, the same group of investigators performed a detailed analysis of the target region covering the 16S rRNA helix 18 bulge to optimize capture and detection probes in order to reduce a gap between their hybridization sites, relocated the fluorescein marker within the detector probe,[118] and ultimately shortened the capture and detector probes to the range of 8–21 nucleotides, depending on the sensor's target species (the *P. mirabilis* sensor contained capture probe 5′-CAT CCG ATA GTG CAA GGT CCG-3′ and hybridized to 16S rRNA linked to the labeled detector probe 5′-AAG AGC CCC TGC TTT GGT CCG-3′).[119] These refinements increased greatly a sensitivity of the UTI biosensor, allowing it to detect ~300 cells; moreover, it made it possible to conduct the whole assay at ambient temperatures of 20°C–25°C. Target sequences differing by only single nucleotide, such as the ones originating from *P. mirabilis* and *K. pneumoniae*, could be differentiated unambiguously despite the low-level cross-reactivity. According to the authors, further improvements to the UTI chip sensitivity and the discriminatory range of gram-negative and gram-positive uropathogens should bring us closer to the moment when it can be used in a handheld device for rapid uropathogen detection at the point-of-care.

A microchip intended to simultaneously detect and identify multipathogenic intestinal infection was proposed by Jin et al.[120] Due to the large variety of potential candidate species, two regions, one within 16S rRNA (corresponding to *E. coli* region 34–528 nt) and another one within 23S rDNA (nt 440–1081) were chosen for the preparation of the oligonucleotide probes and target DNA amplification (primers 16S-F1: 5′-CGC TGG CGG CAG GCC TAA CAC ATG C-3′, 16S-R1: 5′-CGC GGC TGC TGG CAC GGA GTT AGC C-3′ and 23S-F3: 5′-ACC GAT AGT GAA CCA GTA CCG TGA G-3′ 23S-R4: 5′-TTA AAT GAT GGC TGC TTC TAA GCC-3′). *Proteus* spp. 24 nt probe 5′-GGT ACG CAG TCA CAG GAC AAA GCC-3′ complementary to 23S rDNA region was found successful in detection of this genus and its discrimination among 15 other pathogens of different genera/species. Analysis of stool samples yielded sensitivity level of 10^3 CFU/1 mL when a single pathogen was present and decreased down to 10^5 CFU/1 mL when the sample was contaminated with six different pathogens.

A different strategy of signal-to-noise increase was applied to microarray technology by Kostic et al.[121] in a case when the relative abundance of the target pathogen is low within the microbial community of the analyzed specimen. A set of 35 nt capture probes corresponding to the *gyrB* species- or genus-specific fragment, and their reverse complementary (RC) oligonucleotide counterparts lacking the 3′ terminal cytosine residue were designed. Fragments of *gyrB* gene were amplified from 20 pathogen species, including *P. mirabilis*, dephosphorylated, and then served as the templates for end labeling of the RC oligos with the fluorescently labeled ddCTP. The labeling was possible only when hybridization between the available PCR product and the RC oligo had occurred. Liberated, end-labeled RC oligos were allowed to interact with the capture probes deposited at the surface of the microarray, and the position of positive signals helped to identify a pathogen. This strategy significantly reduced the concentration of labeled nucleic acids in the reaction, which allowed a decrease in the level of nonspecific signals and improved the sensitivity of the assay. Pathogens were detected by this approach even if they constituted 0.1% of the total microbial community analyzed.

As mentioned earlier, until portable biosensors are invented, microchip analysis requires rather sophisticated hardware that may not be commonly available to the clinical laboratories. Therefore, attempts are undertaken to simplify the technology while keeping all advantages of the method. Hong et al.[122] developed simple, nylon membrane supported microarray to differentiate among intestinal pathogens based on the 23S rDNA polymorphism. A pair of universal primers P1: 5′-ACC AGG ATT TTG GCT TAG AAG-3′ and 5′ digoxigenin end-labeled P2 5′-CAC TTA CCC CGA CAA GGA AT-3′ was used to amplify 888 bp variable region corresponding to *E. coli* 1051–1938 nucleotide sequence. Specific products obtained from 14 species of pathogens were then hybridized to the array containing a set of 21 species/genus specific probes (*P. vulgaris* probe sequence: 5′-GTA GGC AGA GTG ATT AGG CAA A-3′). Detection was performed using anti-digoxygenin antibodies coupled with alkaline phosphatase, which catalyzed the formation of the color product from the NBT/BCIP substrate. *P. vulgaris* was detected at the 10^1 CFU level in the presence of two accompanying pathogens. This method, although more time consuming (~8 h) than other microarray strategies, has a great advantage, as the analysis of the results can be evaluated by a naked eye, and it can be used in moderately equipped laboratories.

Yet a completely different type of microarray support was used by Hou et al.[123] to improve the array designed for rapid identification of bacterial species responsible for bacteremia.[124] Because only up to ~50 pathogen species may be involved in this type of infection,[124] it creates opportunity to abandon densely printed, costly, glass-slide chips in favor of bead-bound probes in aqueous suspension. Briefly, specific 23S rDNA probes[124] (*P. mirabilis* probe sequence 5′-ATA GCC CCG TAT CTG AAG ATG CT-3′) tailed with 18 dTTP linker were coupled to the beads of characteristic color code due to the different ratios of red fluorophore. Variable regions of 23S rDNA amplified on pure DNA samples from pathogenic species using a pair of universal primers[124] 6F: 5′-GCG ATT TCC/T GAA T/CGG GGA/G AAC CC-3′ and biotin 5′ end-labeled 10R: 5′-TTC GCC TTT CCC TCA CGG TAC. The attractiveness of the method is that each pathogen-derived PCR product may be hybridized to a color-coded bead-probe set in an ordinary thermal cycler, which significantly reduces reaction volume and simplifies handling. However, deciphering the results by identification of the color code being a conjunction of target fluorescence with specific fluorescence of particular bead-probe complex, requires more specialized hardware. All 34 clinical specimens proved by cultivation methods to be pathogen contaminated gave positive signals, and 2 *P. mirabilis* strains were identified among them. The sensitivity of the assay was quite low as it reached 1×10^7 value for the starting template of *Listeria monocytogenes* used as a test strain, probably because of the authors' intention to apply a very simple DNA extraction method by cell boiling in 0.5% SDS, a method which can be commonly used by any routine laboratory. Nevertheless, this array was not designed for differentiation within *Proteus* genus, as the probes used in the original method have failed to distinguish between *P. vulgaris* and *P. mirabilis*.[124]

Typing. One of the most common typing methods is ribotyping. Chromosmal DNA from the members of *Proteeae* tribe digested with *Eco*RV or *Hinc*II restriction endonucleases and subjected to hybridization with a set of oligonucleotides targeted to rRNA genes revealed good discriminatory capacity of this method.[125] On the other hand, the Dienes test, described earlier, is an extremely simple, low-cost, and easy-to-interpret assay, and thus may be performed routinely in any laboratory. Using a collection of 63 *P. mirabilis* clinical isolates, Pfaller et al.[126] compared this test with the automated ribotyping, which requires application of specialized hardware. Both tests identified 40 independent clones, while the remaining 23 strains were assigned to 13 Dienes types and 12 ribotypes, thus reaching almost identical discriminatory power. Other investigators found Dienes test unreliable while testing isolates randomly,[127,128] thus several other fingerprinting techniques allowing inter- and intraspecies differentiation have been developed. Bingen et al.[129] investigated *P. mirabilis* cross-infection in hospital environment by the application of both ribotyping (using 16S and 23S rRNA probes hybridization to *Hind*III/*Eco*RI digested chromosomal DNAs) and RAPD. Application of the single 10 nucleotide primer (5′-TCACGATGCA-3′) allowed thorough characterization of the isolates from 18 pregnant patients and excluded the common source of infection, despite such suggestions from the prior phenotypic analysis. Michelim et al.[130] tested several other short primers for RAPD analysis to find OPZ20 (5′-ACT TTG GCG G-3′), OPX13 (5′-ACG GGA GCA A-3′), and OPA11 (5′-CAA TCG CCG T-3′) to have the highest discriminatory indices when applied to *P. mirabilis* typing. RAPD typing procedure was also successfully applied to test genetic diversity of *P. mirabilis* isolates from urinary tract of arthritis patients.[131] JWOPA12: 5′-TCG GCG ATA G-3′ and JWP1: 5′-GCA CTG AAG TGA CCA AGC GG-3′ primers were found to generate diagnostically useful patterns, which differentiated 93 isolates into 14 types.

Another strategy of random amplification was proposed by Kwil et al.[132] Despite the fact, that primers hpm1: 5′-TAT CCC CAT CTA TAG CCA GTT C-3′ and hpm2: 5′-TTG GCA CTA GGG CGG TGC TAT C-3′ targeted a specific internal region of *P. mirabilis HpmA* hemolysin gene, the amplification carried out under low-stringency conditions generated nonspecific random products creating multiple band patterns that are useful, not only for a strain typing, but also for identification of *P. penneri* strains among other *Proteeae* and different unrelated pathogens. Aside from clone-specific patterns obtained, four major PCR products of ~150, 250, 500, and 750 bp were found to be characteristic for all 69 *P. penneri* clinical isolates and clearly distinguished that species from its relatives.

Repetitive elements such as ERIC, REP, or BOX sequences, which are interspersed in the genomes of *Eubacteria*,[133] suit very well the purpose of bacterial fingerprinting. Primers recognizing these elements (ERIC1R: 5′-ATG TAA GCT CCT GGG GAT TCA C-3′ and ERIC2 5′-AAG TAA GTG ACT GGG GTG AGC G-3′, Rep1R-Dt: 5′-III NCG NCG NCA TCN GGC-3′ and Rep2-Dt: 5′-NCG NCT TAT CNG GCC TAC-3′, as well as BoxA1R: 5′-CTA CGG CAA GGC GAC GCT GAC G-3′)[133] helped generate products of different sizes, due to the genus-, species-, or even clone-specific distribution and orientation of these elements. Discriminatory values were above 0.95 of Simpson's index in case of ERIC1/2 and Box typing of *P. mirabilis* isolate collection, which indicates very good differentiation abilities, and much lower for Rep1/2 pair (0.621).[130] Using the same primers and additionally BoxA2R (5′-ACG TGG TTT GAA GAG ATT TTC G-3′),[133] Serwecińska et al.[134] fingerprinted chromosomal DNAs isolated from pure *P. mirabilis*, *P. vulgaris*, and *P. penneri* cultures. Besides the intraspecies differentiation, bands specific for particular species were identified. In *P. vulgaris* Rep amplification patterns, two specific bands, 1880 and 260 bp were distinguishable, whereas for *P. mirabilis* 560 bp specific band was observed. The most complex set of bands was found for *P. penneri*, where seven unique fragments ranging from 240 to 3650 bp were identified. ERIC amplification yielded three bands specific for *P. vulgaris*—330, 830, and 1690 bp; one band for *P. mirabilis*—740 bp, three bands for *P. penneri*—700, 960, and 1350 bp), BoxA2 (610 bp band for *P. vulgaris*, 1050 bp for *P. mirabilis*), and BoxA1 (one specific band for *P. vulgaris*, *P. mirabilis*, and *P. penneri* was identified of 610, 530, and 2490 bp, respectively). Recently, BoxA1 primer and a pair of Rep-Dt primers were used to confirm a case of *P. mirabilis* strain transmission from intestines to the urinary tract of hospitalized patient.[135]

Proteus strains were also fingerprinted based on the microsatellite sequences. Typing 40 *P. mirabilis* laboratory strains, and 42 clinical isolates with two microsatellite sequences $(GACA)_4$ and $(CAAT)_4$ revealed a high discriminatory power for both primers (0.992 and 0.940, respectively),[136] which places them within the range of Dienes test (0.98), ribotyping (0.92), and PFGE (0.979) established by Pfaller et al.[126] Other primers toward microsatellites, such as $(AC)_8T$ and $(AG)_8A$, were also found to be useful in *P. mirabilis* differentiation and showed good discrimination indices.[130]

84.2 METHODS

84.2.1 Sample Preparation

Clinical specimens may frequently contain concomitant bacterial flora and themselves pose a risk to the enzymatic reactions due to the presence of several inhibitors. Hence, the sensitivity and specificity of molecular methods for the direct detection and identification of pathogens may decrease significantly. Therefore, the amount of expected pathogens should be estimated initially in the analyzed specimen along with the evaluation of bacterial population complexity. Only then should a decision be made about whether the sensitivity range of the proposed detection technique is appropriate to the expected amount of material, and if not, preliminary stages of sample concentration should be considered. It is a good standard that full microbiological diagnostics should be performed, especially since in the case of *Proteus* bacilli it gives decent results, at least at the genus level. In such cases, molecular methods should serve as a reliable confirmation of classical techniques or should decide in a hard-to-diagnose case. As a method of choice, individually these techniques should be used only in life threatening cases, for example, sepsis, when the prompt identification of pathogen may be crucial to a patient's prognosis.

The most common clinical material from which *Proteus* rods are isolated is urine. If low amount of bacteria is present, samples may be concentrated by centrifugation at 2500 × *g* for 20 min at 4°C. In other types of specimens such as wound exudates, swabs, and so forth, cultivation methods may be required in order to obtain pure cultures if the amount of pathogen is low or there is a problem with the large number of concomitant bacteria. *Proteus* can be isolated from the samples after cultivation on selective and differential media (described in part 84.2) within 24 h at 37°C. The swarming ability of *Proteus* rods can be inhibited by increasing concentration of agar in the medium or by use of one of a number of substances specified above. Well-isolated colony should be used for inoculation of liquid medium such as Luria–Bertani broth when nucleic acid isolation is required prior to molecular analysis.

Isolation of the DNA suitable for any molecular test may be performed using commercially available DNA extraction kits. If such a kit is not available, highly pure DNA may be extracted by placing 1 ml of pure overnight culture in a sterile Eppendorf tube and centrifuged at 14,000 × *g* for 1 min. The pellet is then resuspended in 250 μL of Tris-EDTA (TE) buffer; 50 μL of lysozyme in TE (10 mg/mL) is added, followed by an incubation for 15 min at 37°C. After addition of 60 μL of STEP buffer (50 μL Tris-Cl pH-8, 50 μL 10% SDS, 160 μL 0.5 M EDTA pH-8, 1 mg proteinase K, H_2O ad 1 mL), the sample requires further incubation for 2 h at 50°C, and DNA is extracted two to three times with an equal volume of phenol:chloroform solution, each time collecting aqueous phase into new tube. A final extraction step with an equal volume of chloroform:isoamyl alcohol solution (24:1) helps to remove traces of phenol. Then, aqueous phase should be collected in a new tube; 0.1 volume of 3 M sodium acetate is added, and the sample is mixed and precipitated with 2.5 volumes of 96% cold ethanol, followed by centrifugation at 14,000 × *g* for 10 min at 4°C. The pellet should be dried, resuspended in 100 μL of TE buffer and subjected to treatment with 5 μL of RNaseA (5 mg/mL in TE) for 30 min at 37°C. Single-step extraction with chloroform: isoamyl alcohol solution removes RNase traces, and a sample needs to be precipitated with ethanol, the pellet should be washed with 1 mL of 70% ethanol, centrifuged, and dried. DNA is then resuspended in 100 μL of TE buffer, and its concentration should be determined spectrophotometrically. It may be stored for a long time at −20°C.

If the speed of the procedure is a major concern, DNA may be prepared by washing ~10^5 bacterial cells in a 1 mL of lysis buffer containing 1% Triton X-100, 10 mM Tris (pH 8.0), and 1 mM EDTA and pelleting them by centrifugation at 13,000 × *g* for 5 min. A pellet is then resuspended in 100 μL of the same buffer and boiled in a water bath for 30 min. A cell debris is then centrifuged at 13,000 × *g* for 5 min and a supernatant may serve as a source of DNA for molecular detection.[109] In our experience though, this method gives irreproducible results, especially when the DNA is subjected to fingerprinting.

84.2.2 Detection Procedures

As noted, no particular method has been officially recommended for molecular detection of *Proteus* bacilli. Among different protocols the ones focused on the amplification of 16S rDNA region seem to be the most universal and relatively simple to perform. There are several primer pairs in use that are anchored in the 16S rDNA conservative region, which allows for the amplification of the variable regions of this gene. We propose a pair of universal primers harbored in 3′ end of the gene, designed originally by Lu et al.[109] (U1: 5′-CCA GCA GCC GCG GTA ATA CG-3′; U2: 5′-ATC GG(C/T) TAC CTT GTT ACG ACT TC-3′) and in the case of *Proteus*, yields the 997 bp fragment.

For 50 μL PCR mixture, 5 μL of 10× polymerase buffer is used along with 200 μM of each dNTP, 0.2 μM of each primer, 20–50 ng of genomic DNA (a negative control in which deionized water should be used instead of DNA), and 2.5 U of *Taq* DNA polymerase. The cycling parameters include an initial 94°C for 10 min; 35 cycles of 94°C for 1 min, 55°C for 1 min, and 72°C for 2 min; a final 72°C for 10 min. The size and the specificity of the PCR product may be checked by running 5 μL on the 1% agarose gel in TAE buffer, along with the negative control and the linear DNA ladder. For result confirmation, the amplicon is sequenced and analyzed using AlignX software with the sequences from the GenBank database (accession numbers: *P. mirabilis*—AJ301682, *P. vulgaris* —AJ233425, *P. penneri*—AJ634474). Alternatively, PCR products may be analyzed using restriction fragment length polymorphism (RFLP). *Proteus* amplicons digested with *Hae*III, and *Mnl*I endonucleases[79] give genus-specific pattern in case of *Hae*III (34, 126, 161, 180, 217, and 279 bp); whereas *Mnl*I yields species-specific patterns (*P. mirabilis*—14, 47, 65, 104, 119, 126, 157, 176, 189 bp; *P. vulgaris*—14, 14, 47, 65, 104, 119, 126, 157, 175, 176 bp). Similarly, *Hpa*II endonuclease digestion results in legible species-specific patterns (*P. mirabilis*—11, 81, 130, 130, 202, 443 bp; *P. vulgaris*—11, 81, 130, 130, 137, 202, 306 bp). For digestion, 5 μL of PCR product should be used, 2 μL of digestion buffer supplied by the enzyme manufacturer, 5–10 U of appropriate restriction endonuclease, filled up to 20 μL with deionized water. After 1 h incubation at 37°C, a sample should be precipitated with 96% ethanol, dried, and resuspended in 10 μL of water. Then, it may be resolved by running through 7% polyacrylamide gel in TBE buffer, along with the linear DNA ladder of the required range, and stained with ethidium bromide in order to visualize bands under UV light.

84.3 CONCLUSIONS

Genus *Proteus* is a group of opportunistic pathogens whose clinical relevance is constantly increasing, especially regarding *P. mirabilis*.[6,137] Several different community and nosocomial infections of the respiratory tract, wounds, burns, blood, empyemas, meninges, and most often, the urinary tract, are associated with the presence of these bacteria, where they appear individually or as an opportunistic pathogen in polymicrobial infection. In clinical practice, diagnosis of *Proteus* is usually based on the classical microbiological methods, aimed at the detection of a very typical motile growth on agar media, followed by the analysis of specific biochemical properties. At present, molecular techniques are used rather for confirmation of traditional identification methods and for epidemiological purposes, that is, typing of clinical strains in order to monitor clone dissemination within hospital environment. However, under favorable conditions, especially within the group of patients with chronically infected urinary tract, a systemic spread of these pathogens may occur that requires immediate prompt identification and determination of drug-resistance profile of the contaminating agent. Like other members of Enterobacteriaceae family, *Proteus* spp. frequently harbor plasmids, which may contain antimicrobial resistance determinants.[138,139] Because of the mobile nature of these resistance cassettes, they can be easily spread horizontally among bacterial population and confer new drug-resistance to the previously sensitive strains. Tsakris et al.[140] described for the first time *P. mirabilis* isolate producing VIM-1 class B metallo-β-lactamase responsible for carbapenem resistance. A strain of *P. mirabilis*, isolated from diabetic patient, and showing imipenem, meropenem, and ertapenem resistance was found for the first time to acquire transmissible plasmid-mediated carbapenemase$bla_{\text{KPC-2}}$ gene,[141] while in another *P. mirabilis* isolate a presence *Salmonella* genomic island 1 (SGI1) variant and *Tn1826*-derived class 2 integron was reported.[142] *P. mirabilis* SGI1 carried five antibiotic resistance genes: *dfrA15*, *sul1*, *floR*, *tetG*, and *blaP1* which confer resistance to trimetoprim, sulphonamides, chloramphenicol, tetracyclines, and β-lactams, respectively, while class 2 integron contained *sat2* and *aadA2* genes, conferring resistance to streptothricin and streptomycin/spectinomycin, respectively.[143] These discoveries point out that, as in case of many pathogens, increasing multidrug resistance may limit in near future therapeutic options available to clinicians, and development of fast, reliable tests for detection of these determinants should be immediately considered to complement molecular assays for detection and differentiation of pathogens. Recently, Palka-Santini et al.[144] described medium-density DNA chip for identification of pathogens involved in sepsis that is able to analyze hundreds of specific gene segments, including drug-resistance determinants, amplified simultaneously by a large-scale multiplex PCR.

It seems that DNA microarray technology is the future in the field of fast diagnostics and typing of *Proteus* and other pathogens originating from biological specimens. The increasing popularization of the automated microchip readers, along with their decreasing cost; miniaturization; and the development of commercially available standardized microchips containing variety of probes against wide range of genera, species, strains, or genes of special interest will create a chance to utilize this method even in smaller laboratories. Currently, the major drawback of DNA microchip technology in case of *Proteus* is incomplete knowledge about the genome sequence, which limits the development of specific probes and primers. Only recently, the first complete sequence of *P. mirabilis* genome was published.[37] *P. penneri* genome sequence is currently under investigation as a part of the Human Microbiome Project (http://nihroadmap.nih.gov/hmp/), while *P. vulgaris* and the remaining species of this genus, so far, have not been taken under consideration. Development of new, fast, and relatively inexpensive methods of whole-genome sequencing,[145] among which pyrosequencing[146] becomes most popular, creates an opportunity for promising changes in this field.

REFERENCES

1. Hauser, G., Über Fäulnisbakterien und deren Beziehung zur Septicämia, p. 107, Vogel, Leipzig, 1885.
2. Penner, J.L., The genera *Proteus*, *Providencia* and *Morganella*. In Balows, A., Trüper, H.G., Harder, W., and Schleifer, K.H. (Editors), *The Prokaryotes*, vol. III, p. 2849, Springer Verlag, Berlin, 1992.
3. Janda, J.M., and Abbot, S.L., The genus *Proteus*. In *The Enterobacteriaae*, 2nd ed., Chapter 14, ASM Press, Washington, DC, 2006.
4. Cosenza, B.J., and Podgwaite, J.D., A new species of *Proteus* isolated from the larvae of the gypsy moth *Pothetria dispar* (L), *Antonie Leeuwenhoek*, 32, 187, 1966.
5. Hickman, F.W. et al., Identification of *Proteus penneri* sp. nov., formerly known as *Proteus vulgaris* indole negative or as *Proteus vulgaris* biogroup 1, *J. Clin. Microbiol.*, 15, 1097, 1982.
6. O' Hara, M.C., Brenner, F.W., and Miller, J.M., Classification, identification, and clinical significance of *Proteus*. *Providencia*, and *Morganella*, *Clin. Microbiol Rev.*, 13, 534, 2000.
7. O' Hara, M.C. Classification of *Proteus vulgaris* biogroup 3 with recognition of *Proteus hauseri* sp. nov. nom. rev. and unnamed *Proteus* genomospecies 4, 5 and 6, *Int. J. Syst. Evol. Mirobiol.*, 50, 1869, 2000.
8. Philips, J.E., In vitro studies on *Proteus* organisms of animal origin, *J. Hyg. Camb.*, 53, 26, 1955.
9. Burke, J.P. et al., *Proteus mirabilis* infections in hospital nursery traced to human carrier, *N. Engl. J. Med.*, 284, 115, 1971.
10. Rashid, T., and Ebringer, A., Rheumatoid arthritis is linked to *Proteus*—the evidence, *Clin. Rheumatol.*, 26, 1036, 2007.
11. Ebringer, A., and Rashid, T., Rheumatoid arhithis is caused by *Proteus*: The molecular mimicry theory and Karl Popper, *Front Biosci.*, 1, 577, 2009.
12. Senior, B.W., *Proteus*, *Morganella* and *Providencia*. In Colier, L. (Editor), *Topley and Wilson's Microbiology and Microbial Infections*, P. 1035, Arnold. London, 1998.
13. Warren, J.W., Clinical presentations and epidemiology of urinary tract infections. In Mobley, H.L.T., and Warren, J.W. (Editors), *Urinary Tract Infections, Molecular Pathogenesis and Clinical Management*, p. 2, ASM Press, Washington DC, 1996.
14. Jacobsen, S.M. et al., Complicated catheter-associated urinary tract unfections due to *Escherichia coli* and *Proteus mirabilis*, *Clin. Microbiol. Rev.*, 21, 26, 2008.
15. Belas, R., The swarming phenomenon of *Proteus mirabilis*, *ASM News*, 58, 15, 1992.
16. Verstraeten, N. et al., Living on a surface: swarming and biofilm formation, *Trends Microbiol.*, 16, 496, 2008.
17. Belas, R., *Proteus mirabilis* swarmer differentiation and urinary tract infection. In Mobley, H.L.T., and Warren, J.W. (Editors), *Urinary Tract Infections, Molecular Pathogenesis and Clinical Management*, p. 271, ASM Press, Washington DC, 1996.
18. Gygi, D. et al., A cell surface polysaccharide that facilitates rapid population migration by differentiated swarm cell of *Proteus mirabilis*, *Mol. Microbiol.*, 17, 1167, 1995.
19. Belas, R., and Suvanasuthi, R., The ability of *Proteus mirabilis* to sense surfaces and regulate virulence gen expression involves FliL, a flagellar basal body protein, *J. Bacteriol.*, 187, 6789, 2005.
20. Rather, P.N., Swarmer cell differentiation in *Proteus mirabilis*, *Environ. Microbiol.*, 7, 1065, 2005.
21. Vinogradov, E., Sidorczyk, Z., and Knirel, Y.A., Structure of the lipopolysaccharide core region of bacteria of the genus *Proteus*, *Aus. J. Chem.*, 55, 61, 2002.
22. Allison, C. et al., Ability of *Proteus mirabilis* to invade urothelial cells is coupled to motility and swarming differentiation, *Infect. Immun.*, 60, 4740, 1992.
23. Różalski, A., Długońska, H., and Kotełko, K. Cell invasiveness ot *Proteus mirabilis* and *Proteus vulgaris* strains, *Arch. Immunol. Ther. Exp.*, 34, 505, 1986.
24. Chippendale, G.R. et al., Internalisation of *Proteus mirabilis* by human renal epithelial cells, *Infect. Immun.*, 62, 3115, 1994.
25. Jansen, A.M. et al., Visualization of *Proteus mirabilis* morphotypes in the urinary tract: The elongated swarm cell is rarely observed in ascending urinary tract infections, *Infect. Immun.*, 71, 3607, 2003.
26. Coker, C. et al., Pathogenesis of *Proteus mirabilis* urinary tract infections, *Microbes Infect.*, 2, 1497, 2000.
27. Mobley, H.L.T., Virulence of *Proteus mirabilis*. In Mobley, H.L.T., and Warren, J.W. (Editors), *Urinary Tract Infections, Molecular Pathogenesis and Clinical Management*, p. 245, ASM Press, Washington DC, 1996.
28. Różalski, A., Sidorczyk, Z., and Kotełko, K., Potential virulence factors of *Proteus* bacilli, *Microbiol. Mol. Biol. Rev.*, 61, 65, 1997.
29. Różalski, A., Molecular basis of the pathogenicity of *Proteus* bacteria, *Adv. Clin. Exp. Med.*, 11, 3, 2002.
30. Jacobsen, S.M. et al., The high-affinity phosphate transporter Pst is a virulence factor for *Proteus mirabilis* during complicated urinary tract infection, *FEMS Immunol. Med. Microbiol.*, 52, 180, 2008.
31. Alamuri, P., and Mobley, H.L.T., A novel autotransporter of uropathogenic *Proteus mirabilis* is both a cytotoxin and an agglutinin, *Mol. Microbiol.*, 68, 9978, 2008.
32. Roberts, J.A., Fussel, E.N., and Kaack, M.B., Bacterial adherence to the urethral catheters, *J. Urol.*, 144, 264, 1990.

33. Lan, M.C. et al., Oxygen-limiting conditions enrich for fimbriate cells of uropathogenic *Proteus mirabilis* and *Escherichia coli*, *J. Bacteriol.*, 191, 1382, 2009.
34. Massad, G. et al., *Proteus mirabilis* fimbriae: Construction of an isogenic *pmfA* mutant and analysis of virulence in a CBA mouse model of ascending urinary tract infection, *Infect. Immun.*, 62, 536, 1994.
35. Zunino, P. et al., *Proteus mirabilis* fimbriae (PMF) are important for both bladder and kidney colonization in mice, *Microbiology*, 149, 3231, 2003.
36. Lee, K.K. et al., The binding of *Proteus mirabilis* nonagglutinating fimbriae to ganglio-series asialoglycolipids and lactosyl ceramide, *Can. J. Microbiol.*, 46, 961, 2000.
37. Pearson, M.M. et al., Complete genome sequence of uropathogenic *Proteus mirabilis*, a master of both adherence and motility, *J. Bacteriol.*, 190, 4027, 2008.
38. Pearson, M.M., and Mobley, H.L.T., Repression motility during fimbrial expression; identification of 14 *mrpJ* gene paralogues in *Proteus mirabilis*, *Mol. Microbiol.*, 69, 548, 2008.
39. Mobley, H.L.T. et al., Construction of a flagellum-negative-mutant of *Proteus mirabilis*: Effect and internalisation by human renal epithelial cells and virulence in a mouse model of ascending urinary tract infection, *Infect. Immun.*, 64, 5332, 1996.
40. Belas, R., Expression of multiple flagella-encoding genes of *Proteus mirabilis*, *J. Bacteriol.*, 176, 7169, 1994.
41. Belas, R., Manos, J., and Suvanasuthi, R., *Proteus mirabilis* ZapA metalloprotease degrades a broad spectrum of substrates, including antimicrobial peptides, *Infect. Immun.*, 72, 5159, 2004.
42. Walker, K.E. et al., ZapA, the IgA-degrading metelloprotease of *Proteus mirabilis*, is a virulence factor expressed specifically in swarmer cells, *Mol. Microbiol.*, 32, 825, 1999.
43. Wassif, C., Cheek, D., and Belas, R., Molecular analysis of metalloprotease from *Proteus mirabilis*, *J. Bacteriol.*, 177, 5790, 1995.
44. Phan, V. et al., ZapA, a virulence factor in a rat model of *Proteus mirabilis*–induced acute and chronic prostatitis, *Infect. Immun.*, 76, 4859, 2008.
45. Drechsel, H.A. et al., α-Keto acids are novel siderophores in the genera *Proteus*, *Providencia* and *Morganella* and are produced by amino acid deaminases, *J. Bacteriol.*, 175, 2727, 1993.
46. Jones, B.D. et al., Construction of urease-negative mutant of *Proteus mirabilis*: Analysis of virulence in mouse model of ascending urinary tract infection, *Infect. Immun.*, 58, 1120, 1990.
47. Welch, R.A., Identification of two different hemolysin determinants in uropathogenic *Proteus* isolates, *Infect. Immun.*, 55, 2183, 1987.
48. Uphoff, T.S., and Welch, R.A., Nucleotide sequencing of the *Proteus mirabilis* calcium independent hemolysin genes (*hpmA* and *hpmB*) reveals sequence similarity with *Serratia marcescens* hemolysin genes (*shlA* and *shlB*), *J. Bacteriol.*, 172, 1206, 1990.
49. Swihart, K.G., and Welch, R.A., Cytotoxic activity of *Proteus* hemolysin HpmA, *Infect. Immun.*, 58, 1861, 1990.
50. Koronakis, V. et al., The secreted hemolysins of *Proteus mirabilis*, *Proteus vulgaris*, and *Morganella morganii* are genetically related to each other and to the alpha-hemolysin of *Escherichia coli*, *J. Bacteriol.*, 169, 1509, 1987.
51. Różalski, A., and Kotełko, K., Hemolytic activity and invasiveness in strains of *Proteus penneri*, *J. Clin. Microbiol.*, 25, 1094, 1987.
52. Rocha, S.P. et al., Aggregative adherence of uropathogenic *Proteus mirabilis* to cultured epithelial cells, *FEMS Immunol. Med. Microbiol.*, 51, 319, 2007.
53. Alamuri, P. et al., Vaccination with *Proteus* toxic agglutinin, a hemolysis-independent cytotoxin in vivo, protects against *Proteus mirabilis* urinary tract infections, *Infect. Immun.*, 77, 632, 2009.
54. Peerbooms, P.G., Verweij, A.M.J.J., and MacLaren, D.M., Vero cell invasiveness of *Proteus mirabilis*, *Infect. Immun.*, 4, 1068, 1984.
55. Peerbooms, P.G.M., Verweij, A.M.J.J., and MacLaren, D.M., Uropathogenic properties of *Proteus mirabilis* and *Proteus vulgaris*, *J. Med. Microbiol.*, 19, 55, 1985.
56. Chippendale, G.R. et al., Internalisation of *Proteus mirabilis* by human renal epithelial cells, *Infect. Immun.*, 62, 3115, 1994.
57. Raetz, C.R., and Whitefield, C., Lipopolysaccharide endotoxin, *Ann. Rev. Biochem.*, 71, 635, 2002.
58. Różalski, A., Lipopolysaccharide (LPS, endotoxin) of *Proteus* bacteria—chemical structure, serological specificity and the role in the pathogenicity, *Folia Biol. Oecol.*, 4, 5, 2008.
59. Torzewska, A., Stączek, P., and Różalski, A., The crystallization of urine mineral components may depend on the chemical nature of *Proteus* endotoxin polysaccharides, *J. Med. Microbiol.*, 52, 471, 2003.
60. Kuhn, H.M., Meier-Dieter, U., and Mayer, H., ECA, the enterobacterial common antigen, *FEMS Microbiol. Rev.*, 4, 195, 1988.
61. Lugowski, C., Romanowska, E., and Kenne, L., Identification of trisaccharide repeating unit in the enterobacterial common antigen, *Carbohydr. Res.*, 118, 173, 1983.
62. Duda, K.A. et al., ECA-immunogenicity of *Proteus mirabilis* strains, *Archiv. Immunol. Ther. Exp.*, 57, 147, 2009.
63. Donlan, R.M., and Costerton, J.W.E., Biofilms: Survival mechanism of clinically relevant microorganisms, *Clin. Microbiol. Rev.*, 15, 167, 2002.
64. Lewis, K., Persister cells and the riddle of biofilm survival, *Biochemistry (Mosc.)*, 70, 267, 2005.
65. Stickler, D.J., Bacterial biofilms in patients with indwelling urinary catheters, *Nat. Clin. Pract. Urol.*, 5, 598, 2008.
66. Morgan, S.D., Rigby, D., and Stickler, D., A study of the structure of the crystalline bacterial biofilms that can encrust and block silver Foley catheters, *Urol. Res.*, 31, 89, 2009.
67. Stickler, D.J., Observation on the adherence of *Proteus mirabilis* onto polymer surfaces, *J. Appl. Microbiol.*, 100, 1028, 2006.
68. Dumanski, A.J. et al., Unique abbility of *Proteus mirabilis* capsule to enhance mineral growth in infectious urinary calculi, *Infect. Immun.*, 62, 2998, 1994.
69. Williams, G.J., and Stickler, D.J., Effect of triclosan on the formation of crystalline biofilms by mixed communities of urinary tract pathogens on urinary catheters, *J. Med. Microbiol.*, 57, 1135, 2008.
70. Percival, S.L. et al., The effect of EDTA instillations on the rate of development of encrustation and biofilms in Foley catheters, *Urol. Res.*, 37, 205, 2009.
71. Broomfield, R.J. et al., Crystalline bacterial biofilm formation on urinary catheters by urease producing urinary tract pathogens: A simple method of control, *J. Med. Microbiol.*, 58, 1367, 2009.
72. Davies, D.G., and Marques, C.N., A fatty acid messenger is responsible for inducing dispersion in microbial biofilms, *J. Bacteriol.*, 191, 1393, 2009.

73. Hawkey, P.M., *Proteus*, *Providencia* and *Morganella* spp., In Gillespie, S.H., and Hawkey, P.M. (Editors), *Principles and Practice in Clinical Bacteriology*, 2nd ed., Chapter 32, John Wiley & Sons, Ltd. England, 2006.
74. Weissfeld, A.S. et al., Enterobacteriaceae. In Howard, B.J. et al. (Editors), *Clinical and Pathogenic Microbiology*, 2nd ed., Chapter 16, Mosby, St. Louis, USA, 1994.
75. Hawkey, P.M., Mc Cormick, A., and Simpson, R.A., Selective and differential medium for primery isolation of members of the *Proteeae*, *J. Clin. Microbiol.*, 23, 600, 1986.
76. Urbanova, E., Selective medium for primery isolation of members of the tribe *Proteeae*, *Folia Microbiol.*, 44, 629, 1999.
77. Senior, B.W., Media and tests to simplify the recognition and identification of members of the *Proteeae*, *J. Med. Microbiol.*, 46, 39, 1997.
78. Müller, H.E., *Proteus penneri* showing a green colour reaction with Kovacs' indole reagent, *Zentbl. Bacteriol. Hug. A*, 261, 198, 1986.
79. Farmer, III, J.J. et al., Biochemical identification of new species and biogroups of Enterobacteriaceae isolated from clinical specimens, *J. Clin. Microbiol.*, 21, 46, 1985.
80. Budding, A.E. et al., The Dienes phenomenon: Competition and territoriality in swarming *Proteus mirabilis*, *J. Bacteriol.*, 191, 3892, 2009.
81. Hickman, F.W., and Farmer III, J.J., Differentiation of *Proteus mirabilis* by bacteriophage typing and the Dines reaction, *J. Clinical. Microbiol.*, 3, 350, 1976.
82. Senior, B.W., Typing of *Proteus* strains by proticine production and sensitivity, *J. Med. Microbiol.*, 10, 7, 1977.
83. Senior, B.W., and Larsson, P., A highly discriminatory multityping scheme for *Proteus mirabilis* and *Proteus vulgaris*, *J. Med. Microbiol.*, 16, 193, 1983.
84. Kauffmann, F., *The Bacteriology of Enterobacteriaceae*, 333 pp., The Williams & Wilkins, Co. Baltimore, 1966.
85. Larsson, P., Serology of *Proteus mirabilis* and *Proteus vulgaris*, *Methods Microbiol.*, 14, 187, 1984.
86. Penner, J.L., and Hennessy, J.N., Separate O-grouping schemes for serotyping clinical isolates of *Proteus vulgaris* and *Proteus mirabilis*, *J. Clin. Microbiol.*, 12, 77, 1980.
87. Knirel, Y.A. et al., Structural study of O-specific polysaccharide of *Proteus*, *J. Carbohydr. Chem.*, 12, 379, 1993.
88. Knirel, Y.A. et al., Structure of O-antigenic polysaccharides of *Proteus* bacteria, *Pol. J. Chem.*, 73, 859, 1999.
89. Drzewiecka, D., Zych, K., and Sidorczyk, Z., Characterization and serological classification of a collection of *Proteus penneri* clinical strains, *Arch. Immunol. Ther. Exp.*, 52, 121, 2004.
90. Drzewiecka, D. et al., Structural and serological studies of the O-polysaccharides of strains from newly created *Proteus* O78 serogroup prevalent in Polish patients, *FEMS Immunol. Med. Microbiol.*, 58, 269, 2010.
91. Bartodziejska, B. et al., Structural and serological studies on a new acidic O-specific polysaccharide of *Proteus vulgaris* O32, *Eur. J. Biochem.*, 256, 488, 1998.
92. Torzewska, A. et al., Structure of the O-specific polysaccharide from *Proteus vulgaris* O37 and close serological relatedness of the lipopolysaccharides of *P. vulgaris* O37 and *P. vulgaris* O46, *FEMS Immunol. Med. Microbiol.*, 31, 227, 2001.
93. Torzewska, A. et al., Structure and serology of the O-antigen of *Proteus* strains classified into serogroup O17 and former serogroup O35, *Arch. Immunol. Therap. Exp.*, 52, 277, 2006.
94. Cedzynski, M. et al., Structural and immunochemical studies of two cross-reactive *Proteus mirabilis* O-antigens, O6 and O23,containing β-1$\rightarrow$3-linked 2-acetamido-2-deoxy-D-glucopyranose residues, *Microbiol. Immunol.*, 42, 7, 1998.
95. Toukach, F.V. et al., Structure of a new acidic O-antigen of *Proteus vulgaris* O22 containing O-acetylated 3-acetamido-3,6-dideoxy-D-glucose, *Carbohydr. Res.*, 318, 146, 1999.
96. Kaca, W. et al., Structure of the O-specific polysaccharide of *Proteus mirabilis* S 1959, *Arch. Immunol. Ther. Exp.*, 35, 431, 1987.
97. Shashkov, A.S. et al., Structures of new acidic O-specific polysaccharides of the bacterium *Proteus mirabilis* serogroups O26 and O30, *FEBS Lett.*, 386, 247, 1996.
98. Radziejewska-Lebrecht, J. et al., Structure and epitope characteristic of O-specific polysaccharide of *Proteus mirabilis* O28 containing amides of D-galacturonic acid with L-serine and L-lysine, *Eur. J. Biochem.*, 230, 705, 1995.
99. Arbatsky, N.P. et al., Structure of the O-specific polysaccharide of *Proteus mirabilis* O11, another *Proteus* O-antigen containing an amide of D-galacturonic acid with L-threonine, *Carbohydr. Res.*, 323, 81, 2000.
100. Perepelov, A.V. et al., Structure of the O-specific polysaccharide of *Proteus vulgaris* O45 containing 3-acetamido-3,6-dideoxy-D-galactose, *Carbohydr. Res.*, 338, 327, 2003.
101. Kocharova, N.A. et al., Structure of the O-specific polysaccharide of *Providencia rustigianii* O14 containing N$^{\boxtimes}$-[(S)-1-carboxyethyl]-Nα-(D-galacturonoyl)-L-lysine, *Carbohydr. Res.*, 338, 1009, 2003.
102. Torzewska, A. et al., Serological characterization of the O-specific polysaccharide of *Providencia alcalifaciens* O23, *Arch. Immunol. Therap. Exp.*, 52, 43, 2004.
103. Kocharova, N.A. et al., Structure of the O-polysaccharide of *Providencia stuartii* O4 containing 4-(N-acetyl-L-aspart-4-yl) amino-4,6-dideoxy-D-glucose, *Carbohydr. Res.*, 339, 195, 2004.
104. Torzewska, A. et al., Structure of the O-polysaccharide of *Providencia stuartii* O33 containing 4-(N-acetyl-D-aspart-4-yl)amino-4,6-dideoxy-D-glucose, *FEMS Immunol. Med. Microbiol.*, 41, 133, 2004.
105. Bartodziejska, B. et al., Structural and serological studies of two related O-specific polysaccharides of *Proteus vulgaris* O21 and *Proteus mirabilis* O48 having oligosaccharide-phosphate repeating units, *Eur. J. Biochem.*, 267, 6888, 2000.
106. Manos, J., and Belas, R., The genera *Proteus*, *Providencia*, and *Morganella*, In Dworkin, M., Falkow, S., Rosenberg, E., Schleifer, K.-H., and Stackebrandt, E. (Editors), *The Prokaryotes*, Vol. 6, pp. 245–269, Springer Verlag, Berlin, 2006.
107. Dempsey, K.E. et al., Identification of bacteria on the surface of clinically infected and non-infected prosthetic hip joints removed during revision arthroplasties by 16S rRNA gene sequencing and by microbial culture, *Arthritis Res. Ther.*, 9, R46, 2007.
108. Schabereiter-Gurtner, C. et al., 16S rDNA-based identification of bacteria from conjunctival swabs by PCR and DGGE fingerprinting, *Invest. Ophthalmol. Vis. Sci.*, 42, 1164, 2001.
109. Lu, J.J. et al., Use of PCR with universal primers and restriction endonuclease digestions for detection and identification of common bacterial pathogens in cerebrospinal fluid, *J. Clin. Microbiol.*, 38, 2076, 2000.
110. La Scolea, L. Jr., and Dryja, D., Quantitation of bacteria in cerebrospinal fluid and blood of children with meningitis and its diagnostic significance, *J. Clin. Microbiol.*, 19, 187, 1984.
111. Henry, N.K. et al., Microbiological and clinical evaluation of the ISOLATOR lysis-centrifugation blood culture tube, *J. Clin. Microbiol.*, 17, 864, 1983.

112. Vliegen, I. et al., Rapid identification of bacteria by real-time amplification and sequencing of the 16S rRNA gene, *J. Microbiol. Methods*, 66, 156, 2006.
113. Widjojoatmodjo, M.N., Fluit, A.C., and Verhoef, J., Molecular identification of bacteria by fluorescence-based PCR-single-strand conformation polymorphism analysis of the 16S rRNA gene, *J. Clin. Microbiol.*, 33, 2601, 1995.
114. Cao, B. et al., 16S-23S rDNA internal transcribed spacer regions in four *Proteus* species, *J. Microbiol. Methods*, 77, 109, 2009.
115. Takeuchi, H. et al., Detection of *Proteus mirabilis* urease gene in urinary calculi by polymerase chain reaction, *Int. J. Urol.*, 3, 202, 1996.
116. Mansy, M.S. et al., Amplification of *Proteus mirabilis* chromosomal DNA using the polymerase chain reaction, *Mol. Cell. Probes*, 13, 133, 1999.
117. Liao, J.C. et al., Use of electrochemical DNA biosensors for rapid molecular identification of uropathogens in clinical urine specimens, *J. Clin. Microbiol.*, 44, 561–570, 2006.
118. Liao, J.C. et al., Development of an advanced electrochemical DNA biosensor for bacterial pathogen detection, *J. Mol. Diagn.*, 9, 158, 2007.
119. Mastali, M. et al., Optimal probe length and target location for electrochemical detection of selected uropathogens at ambient temperature, *J. Clin. Microbiol.*, 46, 2707, 2008.
120. Jin, L.Q. et al., Detection and identification of intestinal pathogenic bacteria by hybridization to oligonucleotide microarrays, *World J. Gastroenterol.*, 11, 7615, 2005.
121. Kostic, T. et al., A microbial diagnostic microarray technique for the sensitive detection and identification of pathogenic bacteria in a background of nonpathogens, *Annal. Biochem.*, 360, 244, 2007.
122. Hong, B.X. et al., Application of oligonucleotide array technology for the rapid detection of pathogenic bacteria of foodborne infections. *J. Microbiol. Methods*, 58, 403, 2004.
123. Hou, X.-L., et al., Using oligonucleotide suspension arrays for laboratory identification of bacteria responsible for bacteremia, *J. Zhejiang Univ. Sci. B*, 9, 291, 2008.
124. Anthony, R.M., Brown, T.J., and French, G.L., Rapid diagnosis of bacteremia by universal amplification of 23S ribosomal DNA followed by hybridization to an oligonucleotide array, *J. Clin. Microbiol.*, 38, 781, 2000.
125. Pignato, S. et al., Molecular characterization of the genera *Proteus*, *Morganella*, and *Providencia* by ribotyping. *J. Clin. Microbiol.*, 37, 2840, 1999.
126. Pfaller, M.A. et al., Evaluation of the discriminatory powers of the Dienes test and ribotyping as typing methods for *Proteus mirabilis*, *J. Clin. Microbiol.*, 38, 1077, 2000.
127. DeLouvois, J., Serotyping and the Dienes reaction on *Proteus mirabilis* from hospital infections, *J. Clin. Pathol.*, 22, 263, 1969.
128. Hickman, F. W., and Farmer, J.J., III, Differentiation of *Proteus mirabilis* by bacteriophage typing and the Dienes reaction, *J. Clin. Microbiol.* 3, 350, 1976.
129. Bingen, E. et al., Arbitrarily primed polymerase chain reaction provides rapid differentiation of *Proteus mirabilis* isolates from a pediatric hospital, *J. Clin. Microbiol.*, 31, 1055, 1993.
130. Michelim, L. et al., Comparison of PCR-based molecular markers for the characterization of *Proteus mirabilis* clinical isolates, *Bras. J. Infect. Dis.*, 12, 423, 2008.
131. Whiteford, J.R. et al., Genetic diversity in *Proteus mirabilis* isolates found in the urinary tract in reumathoid arthritis patients, *J. Infect.*, 41, 245, 2000.
132. Kwil, I. et al., Use of randomly amplified polymorphic DNA (RAPD) analysis for identification of *Proteus penneri*, *Adv. Exp. Med. Biol.*, 485, 321, 2000.
133. Versalovic, J., Koeuth, T., and Lupski, J.R., Distribution of repetitive DNA sequences in eubacteria and application to fingerprinting of bacterial genomes, *Nucleic Acids Res.*, 19, 68231, 1991.
134. Serwecińska, L. et al., Genomic fingerprinting of *Proteus* species using repetitive sequence based PCR (rep-PCR), *Acta Microbiol. Pol.*, 47, 313, 1998.
135. Drzewiecka, D. et al., Structure and serological properties of the O-antigen of two clinical *Proteus mirabilis* strains classified to a newly created *Proteus* O77 serogroup, *Immunol. Med. Microbiol.*, 45, 185, 2008.
136. Cieślikowski, T. et al., Tandem tetramer-based microsatellite fingerprinting for typing of *Proteus mirabilis* strains, *J. Clin. Microbiol.*, 41, 1673, 2003.
137. Liu, P.Y.F. et al., Survey of the prevalence of β-lactamases amongst 1000 Gram-negative bacilli isolated consecutively at the Royal London Hospital, *J. Antimicrob. Chemother.*, 30, 429, 1992.
138. Hall, R.M., and Collis, C.M., Antibiotic resistance in gram-negative bacteria: The role of gene cassettes and integrons, *Drug Resist. Updates*, 1, 109, 1998.
139. Giske, C.G. et al., Clinical and economic impact of common multidrug-resistant gram-negative bacilli, *Antimicrob. Agents Chemother.*, 52, 813, 2008.
140. Tsakiris, A. et al., Transmission in the community of clonal *Proteus mirabilis* carrying VIM-1 metallo-β-lactamase, *J. Antimicrob. Chemother.*, 60, 136, 2007.
141. Tibbetts, R. et al., Detection of KPC-2 in a clinical isolate of *Proteus mirabilis* and first reported description of carbapenemase resistance caused by KPC β-lactamase in *P. mirabilis*, *J. Clin. Microbiol.*, 46, 3080, 2008.
142. Ahmed, A.M., Hussein, A.I.A., and Shimamoto, T., *Proteus mirabils* clinical isolate harbouring a new variant of *Salmonella* genomic variant island 1 containing the multiple antibiotic resistance region, *J. Antiicrob. Chemother.*, 59, 184, 2007.
143. Sundstrom, L., Roy, P.H., and Skold, O., Site-specific insertion of three structural gene cassettes in transpozon *Tn7*, *J. Bacteriol.*, 173, 3025, 1991.
144. Palka-Santini, M. et al., Large scale multiplex PCR improves pathogen detection by DNA microarrays, *BMC Microbiol.*, 9, 1, 2009.
145. Bentley, D.R., Whole-genome resequencing, *Curr. Opin. Genet. Dev.*, 16, 545, 2006.
146. Ronaghi, M., Uhlen, M., and Nyren, P., A sequencing method based on real-time pyrophosphate, *Science*, 281, 363, 1998.

85 *Providencia*

Brunella Posteraro, Maurizio Sanguinetti, and Patrizia Posteraro

CONTENTS

85.1 INTRODUCTION

85.1.1 Classification, Morphology, and Biology/Epidemiology

The family Enterobacteriaceae covers a large group of gram negative bacteria, which were previously classified into 11 genera and 26 species on the basis of biochemical and antigenic analyses. Recent application of various molecular techniques, such as polymerase chain reaction (PCR) and 16S ribosomal RNA (rRNA) gene sequencing, has resulted in an exponential increase in the number of new genera and species within the family. To date, a total of 47 genera, 247 species, and 56 subspecies are recognized in the Enterobacteriaceae family.[1] An updated list of Enterobacteriaceae species can be found at the Web site http://www.bacterio.cict.fr.

The genus *Providencia*, along with the genera *Proteus* and *Morganella*, belongs to the tribe *Proteeae* of the family Enterobacteriaceae.[2,3] Presently the genus *Providencia* contains six species: *Providencia alcalifaciens*, *Providencia heimbachae*, *Providencia rettgeri*, *Providencia rustigianii*, *Providencia stuartii*, and *Providencia vermicola*. Figure 85.1 summarizes the major taxonomic changes within the genus since early 1918, when Hadley et al.[4] designated as *Bacterium rettgeri* (later renamed *Bacillus rettgeri*) an organism previously isolated by Rettger from chickens during an epidemic resembling fowl cholera. Several decades later, *P. rustigianii* (1983) and *P. heimbachae* (1986) were identified by Hickman-Brenner et al.[5] and Müller et al.,[6] respectively. Subsequently, Somvanshi et al.[7] (2006) described the isolation of three strains from infective juveniles of the entomopathogenic nematode *Steinernema thermophilum*, which were distinguishable from the already recognized *Providencia* species based on a combination of phylogenetic, genomic, and metabolic properties. One of these strains was designated as *P. vermicola*, a novel species in the genus *Providencia*.

Members of the genus *Providencia* are facultative anaerobic rods whose motility is assisted by their peritrichous flagella. They are able to ferment D-glucose and other sugars with or without gas production, are catalase-positive and oxidase negative, reduce nitrate to nitrite, and grow well on MacConkey agar. As shown in Table 85.1, shared phenotypic characteristics (e.g., the ability to deaminate phenylalanine) are used to differentiate the tribe *Proteeae* from other Enterobacteriaceae. In addition, *Providencia* can be distinguished from *Proteus* by the production of acid from a variety of sugars, and *Proteus* is further characterized by the hydrolysis of gelatin and the production of lipase and hydrogen sulfide. Apart form *P. rettgeri* and some *P. stuartii* strains, other *Providencia* species are not typical urease producers.[8] According to the classification established in 1972, Ewing et al.[9] identified two species (*P. alcalifaciens* and *P. stuartii*) and six biogroups (4 for *P. alcalifaciens* and two for *P. stuartii*), based on gas production from D-glucose and acid production from adonitol and *i*(*myo*)-inositol.

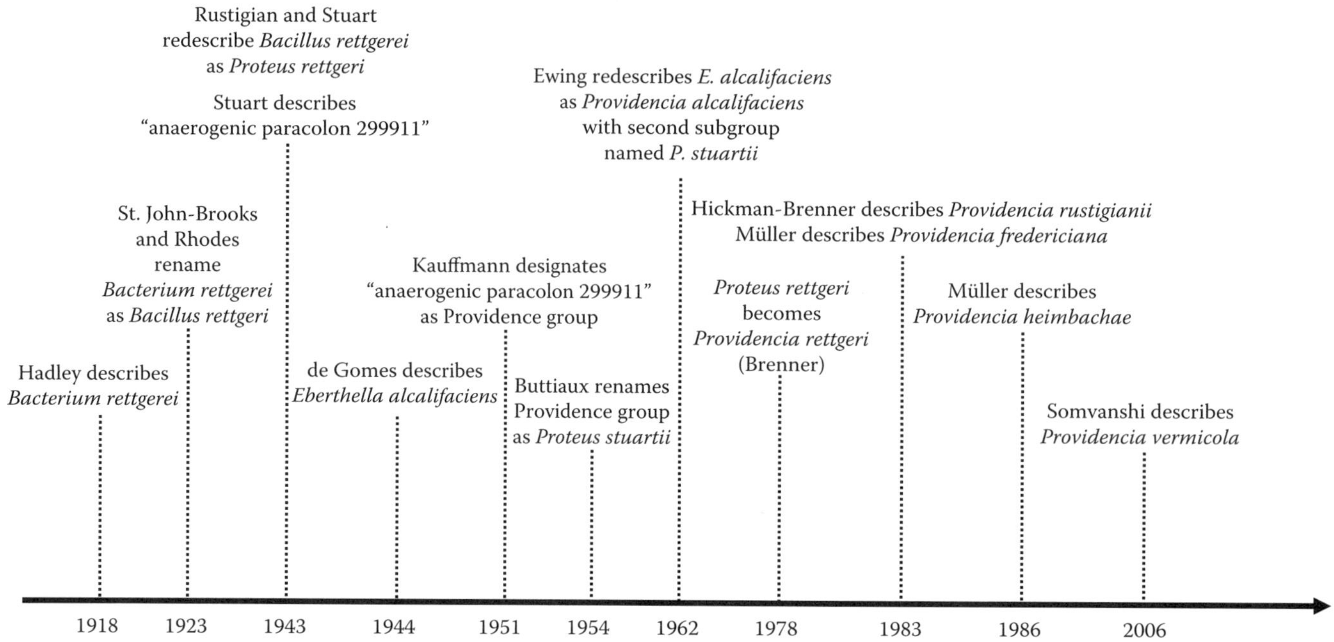

FIGURE 85.1 Timeline of the genus *Providencia*.

TABLE 85.1
Principal Characteristics Common to the Genera *Proteus, Providencia,* and *Morganella*

	Biochemical Test								
Genus	Arginine Dihydrolase	Lysine Decarboxylase	Phenylalanine Deaminase	Growth on KCN	D-Glucose from Acid	Acid from Melibiose	Nitrite from Nitrate	Oxidase Production	ONPG Production
Proteus	–	–	+	+	+	–	+	–	–
Providencia	–	–	+	+	+	–	+	–	–
Morganella	–	–	+	+	+	–	+	–	–

+, present; –, absent; KCN, potassium cyanide; and ONPG, *o*-nitrophenyl-*β*-D-galactopyranoside.

Source: Modified from Manos, J., and Belas, R., *The Prokaryotes. A Handbook on the Biology of Bacteria: Ecophysiology, Isolation, Identification, Applications*, Springer, New York, 2006, Chapter 3.3.12.

With the help of DNA–DNA hybridization techniques, several species within the *Proteeae* tribe have been reassigned to separate genera.[8] For example, *P. alcalifaciens* biogroup 3 has been reclassified as a distinct species and named *P. rustigianii*[5] to honor Robert Rustigian, known for his early studies on the *Proteus* group (see also Figure 85.1). In addition, analysis of the G + C content of the type strains of *Providencia* indicated that the genus has a narrow and homogeneous range of 39–43 mol%.[3] Further subdivision of the *Providencia* species could arise from the use of molecular typing by 16S rRNA PCR-restriction fragment length polymorphism (ribotyping), as suggested by Pignato et al.[10] The authors showed that distinct ribogroups were detected within clinical isolates of *P. alcalifaciens* (two with *Eco*RV and *Hinc*II restriction enzymes), *P. rettgeri* (four with *Eco*RV and five with *Hinc*II), *P. stuartii* (two with *Eco*RV), and *P. rustigianii* (two with *Hinc*II) (see also Section 1.3.2). Recently, the 16S rRNA genes of *P. alcalifaciens* strain DSM 30120 (GenBank file name AJ301684), *P. heimbachae* strain DSM 3591 (GenBank file name AM040490), *P. rettgeri* strain DSM 4542 (GenBank file name AM040492), *P. rustigianii* strain DSM 4541 (GenBank file name AM040489), and *P. stuartii* strain DSM 4539 (GenBank file name AM040491) have been completely sequenced. A phylogenetic evaluation of the 16S rRNA sequences from selected *Proteeae* demonstrated a close relationship between the two *Providencia* species studied, *P. stuartii* and *P. rettgeri*.[11] Nonetheless, analysis of the 16S rRNA sequences has proven to be not so fruitful for solving taxonomic problems among the closely related species of the family Enterobacteriaceae, due to their very high degree of conservation.[12] Further, as 16S rRNA gene exists in several copies within the genome, which may show sequence divergence among different gene copies, direct sequencing of PCR products is seldom suitable for achieving a representative sequence. Thus, other genes, such as the *gyrB* gene[13] and the genes encoding elongation factor Tu and F-ATPase β-subunit (i.e., *tuf* and *atpD*)[14] have been used to resolve the phylogeny of enterobacterial. By comparing these sequences from 96 strains, representing 78 clinically relevant enterobacterial species including *Providencia* species with the 16S rRNA gene trees generated using sequence data available in public databases, it was demonstrated that *tuf* and *atpD* genes provided a higher discriminating power for distinguishing between species, thereby offering the possibility of designing gene-specific primers and molecular probes for identification purposes.[14]

Along with *Proteus* and *Morganella*, *Providencia* species are widespread in the environment and are normal inhabitants of the gut. Specifically, *P. alcalifaciens* and *P. rustigianii* have been isolated from the colon and feces of humans; *P. stuartii* from wounds, urinary tract, and respiratory tract of humans; *P. rettgeri* from urinary tract of humans, poultry, feces of reptiles, and other environments; *P. heimbachae* from feces of penguins and, very rarely, humans (*P. heimbachae*). Unlike the more pathogenic members of the Enterobacteriaceae (e.g., *Salmonella*, *Shigella*, and *Escherichia coli*) but in common with *Proteus* and *Morganella*, the genus *Providencia* encompasses species that are usually considered opportunistic pathogens.[3] However, these organisms are capable of causing a range of human infections of different severities, being isolated frequently in both the community and hospital settings.[15] *P. stuartii* strains are frequent causes of urinary tract infections (UTIs), mainly associated with prolonged catheterization and urinary surgery. In fact, *P. stuartii* has long been recognized as the most frequent pathogen for nursing home patients with long-term urinary catheters, and also implicated in bacteremia with symptoms typical of other septicemias, usually fatal due to antibiotic resistance of the infecting organisms.[16] More recently, Woods and Watanakunakorn[17] found a mortality rate of 25% in a review of 49 cases of *P. stuartii* bacteremia; however, given that 51% of these cases were polymicrobial bacteremia, the mortality rate for *P. stuartii* bacteremia alone was likely lower. Although *P. stuartii* and *P. rettgeri* are quite different at biochemical level, with the latter species metabolizing the sugars D-arabitol, adonitol, and erythritol (see also Section 85.1.3.1), most *P. rettgeri* strains exhibit pathogenic properties similar to those of *P. stuartii*, being these strains responsible for UTIs in catheterized and otherwise compromised patients. These infections are also difficult to treat owing to multiple antibiotic resistance of *P. rettgeri* strains, even though the overall resistance is less marked in this species than in *P. stuartii*.[18] As with *P. stuartii*, bacteremia caused by *P. rettgeri* was found in a high proportion of nursing home patients with UTIs, the majority of whom had long-term indwelling urinary catheters.[17] Recently, Kim et al.[19] reviewed 132 episodes of bacteremia due to the tribe *Proteeae* during 1991–2000 years at a tertiary-care centre in Korea; among infrequent cases of bacteremia caused by *Providencia* species (*P. stuartii* followed by *P. rettgeri*), three were primary bacteremia, and the remaining five cases were associated with biliary (2) and urinary (2) tract infections and peritonitis (1). In contrast to *P. stuartii* and *P. rettgeri*, *P. alcalifaciens* is mainly an invasive enteric pathogen and a cause of diarrheal disease in travelers and children in developing countries.[20,21] However, recent findings by Yoh et al.[22] showed that *P. rettgeri* can also be considered an important cause of gastroenteritis in adults. Sporadic ocular infections owing to *P. rettgeri* have also caused keratitis, dacryocystitis, conjunctivitis, and endophthalmitis.[23] In the five study patients, concurrent UTIs and immunosuppression were identified as important risk factors for *P. rettgeri* infection.

While antimicrobial resistance surveys have investigated specific subsets of infections caused by these organisms, there were relatively few studies focusing on the epidemiology of infections caused by the *Proteeae* as a whole. Laupland et al.[15] conducted a population-based laboratory surveillance for all *Proteeae* isolates in a large Canadian health region, with the aim of defining their incidence and associated demographical risk factors. They found that the overall annual population incidence of *Proteeae* isolation was 75.9/100,000 population, whereas the incidence rate for *Providencia* species was 3.4/100,000. Among isolates from community-onset infections, *Providencia* species were less likely to be isolated from outpatients and more likely to be isolated from nursing home residents than other *Proteeae*. Furthermore, *P. stuartii* (149 cases) was the most common incident species in the genus, followed by *P. rettgeri* (72) and un speciated *Providencia* (7). It was notable that 68 (46%) of 149 *P. stuartii* isolates were obtained from acute-care hospitals and other community settings. There was also a significant relationship between age, gender, and the incidence of *Proteeae* isolation, with the rates increasing dramatically above the 60-years age, probably reflecting the high rate of bacteriuria seen in the elderly; females had twice the risk of *Proteeae* isolation compared to males. The rate of bacteremia was also increased significantly with increased age, supporting the hypothesis that age is a true risk factor for development of *Proteeae* infections.

In order to assess the burden of disease due to *Providencia* infection at our Institution (a large University Hospital in Rome, Italy), we estimated the species distribution over a 5-year period (2004 through 2008), for a total of 353 isolates collected from urine, wound, respiratory tract, blood, and genital specimens. *P. stuartii* and *P. rettgeri* were the most prevalent species, with annual frequencies of occurrence of 85% and 9%, respectively. Increased isolation rate (from 3% to 9%) of *P. rustigianii* was observed between 2005 and 2008, whereas *P. alcalifaciens* was isolated with 1%–2% frequency only in 2006 and 2008. Despite the fact that *P. stuartii* remained the most common *Providencia* species causing infection in our patients, this organism did not show a consistent increase or decrease in isolation rate over time, thereby emphasizing its clinical relevance. On the other hand, isolation over the same time period of relatively rare species like *P. rustigianii* may pose a threat of emergence in certain settings.

85.1.2 Clinical Features and Pathogenesis

Although the genus *Providencia* contains six species, only three (i.e., *P. alcalifaciens*, *P. rettgeri*, and *P. stuartii*) have been clearly identified as human pathogens.[8] As specified in Figure 85.1, de Gomes first described as *Eberthella* (now *Providencia*) *alcalifaciens* a strain isolated from an 11-month-old child with dysentery.[24] While *P. alcalifaciens* has usually been considered a commensal of the gastrointestinal tract, several reports[20,21] have documented that the organism is an invasive intestinal pathogen and a recognized cause of gastroenteritis. Prior studies by Albert et al.[25] on *P. alcalifaciens* stool isolates indicated that the organism may cause diarrhea by an invasive mechanism, as isolates have been shown to invade in vitro cultured HEp-2 cells and to induce diarrhea in rabbits by the removable intestinal tie adult rabbit diarrhea (RITARD) model. Further evidence supporting the role of *P. alcalifaciens* in bacterial gastroenteritis was garnered by Janda et al.[26] who reported that some *P. alcalifaciens* isolates recovered from people with diarrhea were able to invade HEp-2 and other eukaryotic cell lines, but *P. stuartii* and *P. rettgeri* isolates were not. However, several years later, Yoh et al.[22] obtained from fecal specimens, through the use of an efficacious selective medium, the polymyxin-mannitol-xylitol medium for *Providencia* (PMXMP) (see also Section 85.2.2), a large number of *P. rettgeri* isolates, most of them were found to invade Caco-2 cells.

In an attempt to elucidate the *P. alcalifaciens* pathogenicity, Sobreira et al.[27] analyzed by PCR 36 Brazilian isolates from Recife city for the presence of DNA sequences homologous to the virulence genes inv_{ent} and *ail* of *Yersinia enterocolitica*, inv_{pstb} of *Yersinia pseudotuberculosis*, *senA* of *E. coli, and Shigella* and *psa* of *Yersinia pestis* and *Y. pseudotuberculosis*, as well as for their ability to invade HeLa cells. No homologous sequences of virulence genes were noted with any of the isolates studied, whereas only 10 of 36 isolates were shown to invade HeLa cells. Thus, these authors did not find any genetic or virulence-associated characteristics that could potentially explain the strain-related differences in cell invasiveness. Finally, the work by Khashe et al.[28] suggested that the strain variation in the ability of *P. alcalifaciens* to invade HEp-2 cells observed previously may be linked to expression of key adhesion(s) on the cell surface of invasive isolates. Therefore, the inability of *P. alcalifaciens* to adhere to eukaryotic cells would preclude cellular invasion and, consequently, its disease-producing potential. Recently, the *P. alcalifaciens* pathogenic mechanisms were further investigated by Chen et al.[29] who first identified in this species a gene encoding the manganese superoxide dismutase Mn-SOD (*sodA*), which is known to be a major defense cellular mechanism against the molecular oxygen and hydrogen peroxide oxidative stress. These authors investigated the role of this gene in the *P. alcalifaciens* virulence by constructing two strains, a *sodA* deletion mutant and its complemented strain. When the strains were tested in the J774 macrophages, the *sodA* strain was found to be more susceptible to the macrophage killing than the wild-type and its complemented strain. Moreover, when DDY strain mice were injected intraperitoneally with the same strains, the *sodA* deletion mutant was significantly less virulent, while the complemented strain recovered the virulence compared to the wild-type strain. These results suggested that Mn-SOD may be involved in the cell invasive process of *P. alcalifaciens* by conferring it the ability to survive inside intestinal and macrophage cells, as *Salmonella* or *Yersinia* does.[29]

While *P. stuartii* seems to be not particularly invasive, UTIs due to this organism tend to persist longer than those due to other gram-negative bacteria. *P. stuartii* has long been recognized as a common cause of antibiotic-resistant bacteriuria in patients with long-term indwelling catheters.[16] On these devices, *P. stuartii* may form a biofilm which provides specific adherence characteristics allowing for its persistence in the catheterized urinary tract.[30] Another possible explanation of these persistent infections may be the ability of *P. stuartii* to increasingly adhere to uroepithelial cells by the production of hemagglutinin (HA). When a large number of strains of *Morganella*, *Proteus*, and *Providencia* were examined for their ability to produce hemagglutinins, two main kinds of hemagglutinins, namely mannose-sensitive (MS) and mannose-resistant/*Klebsiella*-like (MR/K), have been detected.[31] Interestingly, *P. stuartii* isolates from patients experiencing bacteriuria of long duration were found to express significantly more MR/K hemagglutinin than those from patients experiencing bacteriuria of short duration. Also, a significant proportion of isolates expressing MR/K hemagglutinin were shown to bind to catheter material. Investigation of molecular mechanisms involved in the persistence of *P. stuartii* in catheterized UTIs has led to the identification of MR/K as an important factor enabling *P. stuartii* to persist and adhere to the catheter. Conversely, MS hemagglutinin binds to Tamm-Horsfall protein and may prevent persistence of *P. stuartii* in the catheterized urinary tract.[31] There is evidence that urease-producing *P. stuartii* strains are able to form extensive biofilm communities, generating ammonia from urea, and elevating the pH of the urine and biofilm. Under these alkaline conditions, crystals of magnesium ammonium phosphate (struvite) precipitate in the renal pelvis and bladder, and encrust on urinary catheters with subsequent development of catheter blockages.[32] Struvite is a component of urinary stones that impede the clearing of the infecting organism from the urinary tract by interfering with its voiding. In addition to stone formation, the basic urine damages the uroepithelium, inhibits the action of the complement, and reduces the efficiency of certain antibiotics.[33] For these reasons, urease importantly contributes to the virulence of *P. stuartii* in the urinary tract. The plasmid-encoded urease gene cluster contains seven tandem genes encoding urease subunits (*ureABC*) and accessory polypeptides (*ureDEFG*), whereas an eighth gene termed *ureR*, that is found 414 bp upstream and divergently transcribed from the *ureDABCEFG* cluster, is a member of the AraC family of transcriptional activators. In the infected urinary tract, urea freely diffuses into the bacterium *P. stuartii* where it binds to UreR. This interaction both activates

UreR, which promotes transcription of the urease genes, and alerts the organism of its immediate environment.[33]

In long-term catheterized patients, the urine and catheter biofilms are more commonly colonized by polymicrobial communities, frequently containing *Proteus mirabilis* along with other uropathogens, such as *P. stuartii* and *Klebsiella pneumoniae*. Although a certain antagonism between *P. mirabilis* and some of the other urinary tract organisms was observed, only temporary were the effects exerted by the other species on the ability of catheter encrustation and blockage by *P. mirabilis*.[30]

85.1.3 Diagnosis

85.1.3.1 Conventional Techniques

Conventional diagnosis of *Providencia* species infections is usually dependent on cultivation and identification of the microorganism(s) from clinical specimens (e.g., urine, stool, or blood). After their isolation from primary culture plates (Columbia sheep blood agar and/or MacConkey agar), microorganisms are checked for purity and used to inoculate commercially available identification systems. Table 85.2 shows some of biochemical characteristics used to differentiate the *Providencia* species. Briefly, all species are positive in tests for phenylalanine deaminase, but negative in tests for lysine decarboxylase, ornithine decarboxylase and arginine dihydrolase, and they produce acid from D-mannose.[3]

A myriad of systems, both manual (e.g., API 20E, RapID onE) and automated (e.g., bioMérieux Vitek and BD Phoenix 100) have been developed over the last years for the accurate identification of Enterobacteriaceae, that include all clinically important *Providencia* species. In a recently published review, O'Hara[34] gave details relevant to each product or system. All identification methods are based on one of five different technologies or a combination thereof. These include pH-based reactions that require 15–24 h of incubation, enzyme-based reactions that require 2–4 h, utilization of carbon sources, visual detection of bacterial growth or detection of volatile or nonvolatile fatty acids via gas chromatography. The last technology is more complex and scarcely utilized.

Because of its wide acceptance by the clinical microbiology laboratory market, the manual API 20E, owned since 1986 by bioMérieux, Inc. (Durham, N.C.), became somewhat of a "gold standard" among commercial systems to which many other identification systems have been compared. With the continued expansion of the database (103 gram-negative taxa in 2009), the strip has been subjected to various reevaluations for its accuracy in the identification of Enterobacteriaceae, although early studies had shown that this method was able to identify by 24 h the 78.7% of organisms less routinely isolated like, for example, *P. stuartii*.[35] Another manual system, the RapID onE, marketed by Remel (Lenexa, KS), employs conventional and chromogenic substrates for the identification of Enterobacteriaceae. Its current database contains 28 genera, 70 species, and several biogroups within species. By this method, Kitch et al.[36] had analyzed 364 strains of Enterobacteriaceae and found an accuracy rate of 97.6% without use of additional tests.

Within the last two decades, fully automated identification systems have been introduced, as methods alternative to manual ones, in many clinical microbiology laboratories worldwide. These systems, such as Vitek (bioMérieux, Marcy l'Etoile, France), MicroScan (Dade, West Sacramento, CA), and Phoenix (Becton Dickinson, Sparks, MD) enable clinical microbiologists to identify more rapidly and accurately medically relevant bacteria, thereby contributing to better management of infected patients. The Vitek 2 card (NGNC; bioMérieux), created recently for identification of gram-negative bacilli, contains 47 colorimetric tests that measure utilization of carbon sources, enzymatic activities, and resistance inhibitory substances, leading to the identification of 159 gram-negative taxa. Test reactions are read automatically

TABLE 85.2
Key Biochemical Reactions Used for Differentiation of *Providencia* Species

	% Positive at 48 h for				
Test	***P. alcalifaciens***	***P. heimbachae***	***P. rettgeri***	***P. rustigianii***	***P. stuartii***
Indole production	99	0	99	98	98
Citrate, Simmons'	98	0	95	15	93
Urea, Christensen	0	0	98	0	100
Growth in KCN	100	8	97	100	100
Acid production from					
D-Adonitol	98	92	100	0	5
D-Arabitol	0	92	100	0	0
Erythritol	0	0	75	0	0
i(*myo*)-Inositol	1	46	90	0	95
D-Mannitol	2	0	100	0	10
L-Rhamnose	0	100	70	0	0
Trehalose	2	0	0	0	98

Source: Modified from O'Hara, C.M. et al., *Clin. Microbiol. Rev.*, 13, 534, 2000.

by the Vitek 2 instrument every 15 min, then a specific kinetic algorithm calculation allows for a final identification result after 2–10 h of incubation. A recent evaluation of the Vitek 2 NGNC card was carried out by Funke and Funke-Kissling[37] on 35 clinical isolates of *Providencia* species (4 *P. alcalifaciens*, 18 *P. rettgeri*, 7 *P. rustigianii*, and 6 *P. stuartii*). All of the isolates were correctly identified between 5 and 6 h. Interestingly, for 33% of *P. stuartii* isolates, identification was achieved in less of 4 h; when two of *P. rettgeri* isolates were tested directly from primary isolation plates, both the isolates were identified with the same level of accuracy. Designed and marketed by Becton Dickinson, the BD Phoenix 100 automatic system has been introduced in 2003 for the rapid identification of gram-negative and gram-positive bacteria of human origin. The NID panel contains 45 substrates (including 16 enzymatic substrates, 23 carbon source substrates, and 5 utilization and growth inhibition tests) and 2 fluorescent positive control wells. The panels are available as an identification panel only or as part of a combination identification-antimicrobial susceptibility testing panel. The current database is version 3.34 and contains 60 genera, 155 species, and 5 CDC enteric groups. In 2006, O'Hara[38] evaluated the performance of the Phoenix 100 ID/AST System and NID Panel by testing 507 isolates of the family Enterobacteriaceae, that included 7 isolates of *P. alcalifaciens*, 8 isolates of *P. rettgeri*, 2 isolates of *P. rustigianii*, and 14 isolates of *P. stuartii*. Of the 507 isolates of the Enterobacteriaceae, 456 (89.9%) were correctly identified to the genus and species levels. Misidentifications of the genus and species levels occurred for 29 (5.7%) isolates of the Enterobacteriaceae. Among the *Providencia* isolates tested, two isolates were not correctly identified as follows: one isolate of *P. rettgeri* was correctly identified to the genus level (misidentified as *P. stuartii*) and one isolate of *P. stuartii* (urease-positive) was erroneously identified as *P. mirabilis*.

Although *P. heimbachae* is a rarely encountered pathogen, it should be noted that all the identification systems mentioned above fail to identify this organism because of its exclusion from the respective databases. Thus, in the case of biochemical test results that might lead to suspecting the presence of this organism in a stool specimen (e.g., reactions negative for indole and citrate, and positive for adonitol, D-arabitol, D-galactose or L-rhamnose fermentation) (see Table 85.2), hospital laboratories would not expect to identify this organism without using a reference laboratory.[39]

In order to reduce the turnaround time for laboratory diagnosis of bacteremia due to gram-negative pathogens, the Vitek 2 and Phoenix 100 systems allow the direct inoculation of positive blood cultures into the identification cards or panels, as mentioned above.[40,41]

Serotyping methods rely on differences in antigenic determinants of organisms belonging to the same species or subspecies that are recognized by specific polyclonal or monoclonal antibodies. *P. stuartii* and *P. rettgeri* can be subtyped serologically into serogroups based on the organism's somatic antigen (O polysaccharide), and into serotypes based on its flagellar protein (H antigen). Such serotyping had been used in the past for the identification of specific strains endemic in different hospitals.[42,43]

P. rettgeri and *P. stuartii* are generally resistant to gentamicin and tobramycin, but susceptible to amikacin.[3] Urine isolates are also susceptible to broad- and expanded-spectrum cephalosporins, ciprofloxacin, amoxicillin-clavulanic acid, imipenem, and trimethoprim-sulfamethoxazole.[44] However, Fass et al.[45] reported that the susceptibility of *P. stuartii* to ciprofloxacin decreased from 100% to 46% over a 6-year period in their institution, indicating the potential for emerging resistance in this group. *P. heimbachae*, although infrequently seen in humans, is resistant to tetracycline, but susceptible to cephalosporins, aminoglycosides, ciprofloxacin, and trimethoprim-sulfamethoxazole.[39] In addition, all *Providencia* isolates are virtually capable of producing β-lactamases that hydrolyze primary and extended-spectrum penicillins and cephalosporins. It is well known that β-lactam and fluoroquinolone antibiotics are important drug classes used to treat infections caused by Enterobacteriaceae. Nonetheless, emerging resistance mechanisms against these agents have recently been described in these organisms and include the production of newer β-lactamases and plasmid-mediated quinolone resistance.[46] For these reasons, drug susceptibility testing of clinical isolates of *Providencia* species needs to be routinely performed.

85.1.3.2 Molecular Techniques

Whereas classic culture-based identification methods have dominated clinical microbiology in the past, new molecular methods are slowly but inexorably displacing the traditional methods, not only in research laboratories but also in clinical laboratories. Thus, genotypic identification methods have become widely used, and most of them are based on the polymorphism of the 16S (also called the small subunit) rRNA genes. Species-specific PCR, restriction patterns and, more recently, partial DNA sequencing are used for species identification. In particular, bacteria can be identified rapidly (i.e., within 24 h) and accurately by using primers to amplify by PCR conserved regions of 16S rRNA genes, thereby obtaining a partial sequence to be compared with the ever-growing sequence databases. This approach can improve clinical microbiology by better identification of poorly described, rarely isolated, or biochemically aberrant strains. Also, sequencing of the 16S rRNA genes is a standard tool of bacteriologists for elucidating phylogenetic relatedness.[47] As stated above, the 16S rRNA gene is not sufficiently discriminative for Enterobacteriaceae because of the closely relatedness between the species, and variation may result from slight differences between multiple gene copies within a strain.[13] Thus, Delmas et al.[48] developed a real-time PCR and sequencing assay targeting a fragment internal to *gyrB* (*gyrBint*) of the Enterobacteriaceae. By this method, 240 isolates belonging to *Enterobacteriaceae* species commonly and uncommonly encountered in the clinical laboratory were identified in a reasonable time (<8 h), suggesting that direct sequencing of *gyrBint* could be useful as a complementary tool for the identification of clinical isolates.

Strain-typing methods based on PCR assays can be essentially divided in two major groups: those that rely on random sequences in the whole genome, such as arbitrarily primed (AP)–PCR and randomly amplified polymorphic DNA (RAPD)–PCR, and those based on the heterogeneity within endonuclease restriction sites, such as PCR-restriction enzyme analysis (REA) and PCR-restriction fragment length polymorphism (RFLP). In the first group, a random mixture of nucleotide hexamers that anneal to many sites on a bacterial chromosome is used as primers in RAPD–PCR, thus producing a mixture of differently sized amplicons, which appear as multiple bands after electrophoretic separation and staining. Although RAPD–PCR can easily distinguish between different strains of a bacterial species, a main drawback of this technique is that the banding pattern can vary from laboratory to laboratory and even from experiment to experiment in the same laboratory. In the second group, techniques require that part of the target sequence to be amplified is known and that such targets are heterogeneous between strains belonging to the same species. The number of bands generated by use of a specific restriction enzyme can be predicted according to the number of nucleotide changes that occur at a restriction site in the PCR amplicon, and thus the bands reflect the genetic relationships of the strains being analyzed. An example of these techniques is the ribosomal DNA (rDNA) fingerprinting (ribotyping), based on RFLP in the chromosomal DNA-containing rRNA genes.[10] When used for characterization of clinical isolates of different *Providencia* species, this method appeared to have a high discriminatory power, because not only distinct rRNA gene restriction patterns were generated from different species, but also different ribogroups and ribotypes were found in the majority of the species. Among the genome-based typing methods, pulsed field gel electrophoresis (PFGE) uses endonucleases that recognize rare restriction site sequences, thus producing large pieces of DNA, which generate electrophoretic patterns with resolution and number of bands amenable for analysis.

Finally, matrix-assisted laser desorption ionization–time-of-flight mass spectrometry (MALDI–TOF MS), which can be used to analyze the protein composition of a bacterial cell, has emerged as a new technology for species identification. By measuring the exact sizes of peptides and small proteins, which are assumed to be characteristic for each bacterial species, it is possible to determine the species within a few min when the analysis is started with whole cells, cell lysates, or crude bacterial extracts. However, because of the limited availability of reference data sets, MALDI–TOF MS has not been widely used for species identification of *Enterobacteriaceae*, although it should be possible to build a complete database for all species belonging to this family, including those of the genus *Providencia*.[49]

Molecular techniques can be also used in the practice of hospital epidemiology for the characterization of drug-resistant *Providencia* strains. While *P. stuartii* is naturally resistant to aminopenicillins and narrow-spectrum cephalosporins due to a chromosomally expressed Ambler class C cephalosporinase (AmpC), as with other members of the family Enterobacteriaceae, resistance to expanded-spectrum cephalosporins in *Providencia* species results not only from overexpression of AmpC but also from acquisition of extended-spectrum β-lactamases (ESBLs). These enzymes are predominantly plasmid mediated and are derived from broad-spectrum β-lactamase TEM-1, TEM-2, or SHV-1 by a limited number of mutations.[50] By the use of PCR and sequence analysis, Arpin et al.[51] reported that 1 of 24 (4.2%) TEM-24 producing isolates belonged to *P. stuartii*, whereas Tumbarello et al.[52] reported that, among 160 of 223 (52%) ESBL-producing *P. stuartii* isolates, TEM-52 and TEM-72 were identified in 87% and 13% of isolates, respectively. Other newly emerging non-TEM, non-SHV ESBLs such as enzymes of CTX-M, PER, VEB, and GES families can also be encountered in these organisms.[53,54] Shiroto et al.[55] described the characterization of eight *P. rettgeri* isolates with reduced susceptibility to imipenem and classified as IMP-1 metallo-β-lactamase producers. These isolates harbored a conjugative plasmid containing an integron on which $bla_{\text{IMP-1}}$ was encoded. Interestingly, the eight $bla_{\text{IMP-1}}$-positive strains were isolated from two hospitals and showed two different PFGE patterns and two different integron structures that corresponded to the two hospitals. As first identified in *P. stuartii* isolates from Italian hospital patients,[54] the $bla_{\text{PER-1}}$ gene has also been detected in a clinical *P. stuartii* isolate originating in Kosovo,[56] whereas *P. stuartii* and *P. rettgeri* isolates simultaneously carrying $bla_{\text{VIM-2}}$ and $bla_{\text{PER-1}}$ have been found in a Korean tertiary-care hospital.[57] In the last cases, patients from whom *Providencia* has been isolated were hospitalized in the same hospital room and were catheterized with urinary catheter.

85.2 METHODS

85.2.1 Sample Preparation

Clinical specimens from which *Providencia* species are usually isolated include urine, blood, stool, throat, perineum, axilla, and wound. These specimens are processed following standard microbiological procedures that are different depending on their clinical source. Stool cultures sent to the laboratory with a request to isolate and identify the cause of a possible intestinal infection, must be processed rapidly, and, if it is not possible, a small portion of feces or a swab coated with feces should be placed in a transport medium such as Cary-Blair or buffered glycerol saline. Among extraintestinal specimens, urine is screened by a commercially available automated system, then positive samples are usually submitted to quantitative bacterial culture by inoculating appropriate media (i.e., blood agar and MacConkey agar) with a measured amount of urine, most commonly via a plastic calibrated loop. All other types of specimens (e.g., throat, wound, and abscess), properly collected, are directly plated onto the media. With regard to blood specimens, conventional broth culture systems using nutritionally enriched liquid media are useful to recover most of bacteria including *Providencia* species. After inoculation, bottles are examined daily for evidence of growth, by employing fully automated monitoring methods, such as BacT/Alert (bioMérieux) and BACTEC 9000 (Becton Dickinson) systems.

Providencia species are routinely grown by cultivating clinical specimens on the nutrient media used commonly for general identification of Enterobacteriaceae. As alternatives, specific media have been developed to enhance primary isolation of *Providencia* species from clinical specimens. In 1992, Thaller et al.[58] used a modified MacConkey agar containing methyl green phosphatase (MCP) that greatly improved identification of *P. stuartii* strains. The peculiar phosphatase activity demonstrated by this species allowed for its differentiation from other gram-negative bacteria, except for *Morganella morganii*, on MCP medium. All of 103 *P. stuartii* strains from a total of 1278 seeded urine samples from elderly patients were correctly identified within 48 h by coupling MCP medium and the API 20E system, whereas 16%–18% of strains were misidentified by standard procedures (MacConkey agar and API 20E strips). Thus, strains of *P. stuartii* and *M. morganii* that were phosphatase-positive on MCP could be easily distinguished from each another by using a simple test like ornithine decarboxylase or Simmons' citrate. The MCP properties make it of great utility for nursing home settings, where the incidence of *P. stuartii* in patients with long-term catheters is reported to be very high. Another medium, the CHROMagar Orientation, has been developed for direct detection of gram-negative bacilli from mixed cultures of clinical specimens (i.e., stool, wound, urine, and other fluids). Presumptive identification of the organism(s) is based on the colony color produced by enzymatic action on a chromogenic substrate present in the medium, where *Providencia* isolates appear as diffuse brown colonies owing to tryptophan deaminase production. Using CHROMagar Orientation in combination with indole, lysine, and ornithine decarboxylase tests, and serological testing, Ohkusu[59] correctly identified 98.7% of clinical isolates including those of *Providencia* species. Senior[60] described a *P. alcalifaciens* medium (PAM), containing mannitol, xylose, and galactose, that enabled highly specific and sensitive isolation of *P. alcalifaciens* from feces. Red colonies of *P. alcalifaciens* appeared quite distinct from the lemon-yellow acid-forming colonies from all of the other bacteria fermenting on one or more sugars. As specified earlier, Yoh et al.[22] established a selective medium for *Providencia*, termed PMXMP, that was obtained by adding sugars (maltose and xylose) and polyximine B sulphate to the PAM medium of Senior.[60] The suspected colonies, red to pink colored, were mostly *Providencia* species or *M. morganii*, and, occasionally, *Serratia liquefaciens* and *Pseudomonas* species. Once again, the ornithine decarboxylase test distinguished *Providencia species* from *M. morganii* (and *S. liquefaciens*), and the oxydase test distinguished *Providencia* species from *Pseudomonas* species.

The starting material for all nucleic acid–based identification and strain-typing procedures is, of course, the bacterial genomic DNA. This is extracted from whole cells that are grown aerobically in Luria-Bertani broth, by using standard procedures or a commercial system (e.g., QIAmp DNA minikit, Qiagen, Hilden, Germany). In the standard procedures, the bacterial cell wall and outer and inner membranes are removed by sequential addition to overnight culture of EDTA (ethylendiaminotetraacetic acid), lysozyme, detergents, and finally, proteinase. At alkaline pH, double-stranded chromosomal and plasmid DNA separate into single-stranded molecules that are soluble in aqueous solution. When the DNA-containing solution's pH is lowered to approximately 4.5–4.8 (by acid acetic added to DNA-containing solution), high molecular weight DNA precipitate, while the plasmid DNA remains in solution. This allows separation of plasmids from the chromosome. Alternatively, several commercial kits can be used to obtain exhaustive and reproducible DNA preparations for molecular analyses. Before the extracted and purified DNA materials are used for molecular studies, their quantity and quality need to be carefully assessed.

85.2.2 Detection Procedures

85.2.2.1 PCR Identification

Principle. For molecularly based identification, the DNA is used as a template for PCR to amplify a segment of about 500 or 1500 bp of the 16S rRNA gene sequence, using broad-range or universal primers complementary to conserved regions as described.[47] The resulting PCR products are purified to remove excess primers and nucleotides; to this end, several good commercial kits, such as the QIAquick PCR purification kit (Qiagen), are available. For full-sequence 16S rRNA gene sequencing, the PCR products are automatically sequenced using the PCR primers and additional primers designed from the first round of sequencing results. The generated DNA sequences are assembled by aligning the forward and reverse sequences, then compared with a database library by using analysis software. Alternatively, through the use of MicroSeq 500 16S rDNA-based bacterial identification system (Perkin-Elmer Applied Biosystems Division, Foster City, CA), the first 527-bp fragment of the 16S rRNA gene of the bacterial strain is amplified, subjected to automated sequencing and analyzed using the database of the system (see below). At present, while PCR amplification, purification of PCR products, DNA sequencing, and sequence editing have to be performed manually; the only step that can be automated is input of the 16S rDNA sequence of the bacterial isolate for identification into one of the software packages that will generate the result of the identity of the isolate on the basis of its sequence database. The most well known of these software packages include Ribosomal Database Project (RDP), MicroSeq microbial identification system, Ribosomal Differentiation of Medical Microorganisms (RIDOM), and the SmartGene Integrated Database Network System (SmartGene IDNS). The databases of RDP-II and SmartGene IDNS contain sequences downloaded from GenBank, whereas all sequences in the databases of RIDOM and MicroSeq were obtained by sequencing the 16S rRNA genes of bacterial strains from culture collection. The MicroSeq microbial identification system includes two databases, MicroSeq ID 16S rDNA 500 Library version 2.0, which contains sequences from the 5′-end 527 bp of 16S rRNA genes, and MicroSeq ID 16S rDNA Full Gene Library version 2.0, which contains full 16S rRNA gene sequences. However, interpretation of the identification results is often difficult for inexperienced users, so the use of 16S rRNA gene sequencing in clinical microbiology remains largely limited to the identification of bacterial strains with ambiguous biochemical profiles, whereas additional/supplementary methods are necessary for bacterial species that cannot be identified confidently by 16S rRNA gene sequencing alone.[61]

Procedure

1. Extract genomic DNA from cultured bacteria by using EZ1 DNA tissue kit (Qiagen) followed by a BioRobot EZ1 automated DNA extraction system (Qiagen).
2. Amplify and sequence a portion of the 16S rDNA as specified by the manufacturer for the MicroSeq 500 16S rDNA bacterial sequencing kit. Briefly, 2 μL of bacterial lysate is added to a 48-μL volume consisting of 23 μL of sterile water and 25 μL of 2× master mix in a 0.2-mL microcentrifuge tube. The mixture is then subjected to thermal cycling conditions as follows: 95°C for 10 min, followed by 30 cycles of 95°C for 30 s, 60°C for 30 s, and 72°C for 45 s, followed by a final incubation at 4°C. Thus, agarose gel electrophoresis is performed to check PCR reactions for the presence of a single amplification product approximately 500 bp in length (Figure 85.2).
3. Prepare sequencing reactions for the sense and antisense strands by adding 8 μL of the PCR product, obtained above and purified using the Minielute PCR purification kit (Qiagen), to 13 μL of strand-specific sequencing master mix from the MicroSeq kit. The mixtures are then thermal cycled according to the following parameters: 95°C for 10 min, followed by 25 cycles of 96°C for 10 s, 50°C for 5 s, and 60°C for 4 min, followed by a final incubation at 4°C.
4. Prior to electrophoresis, the sequencing products are purified by ethanol precipitation and reconstitute in 12 μL of HiDi formamide. Thus, samples are run on an ABI Prism 3100 (50-cm capillary) instrument, and the generated sequence electropherograms are examined for signal strength and base-calling artifacts. After alignment of forward and reverse strands, nucleotide ambiguities are resolved, and the resulting consensus sequence is used as a query within the appropriate DNA sequence database.

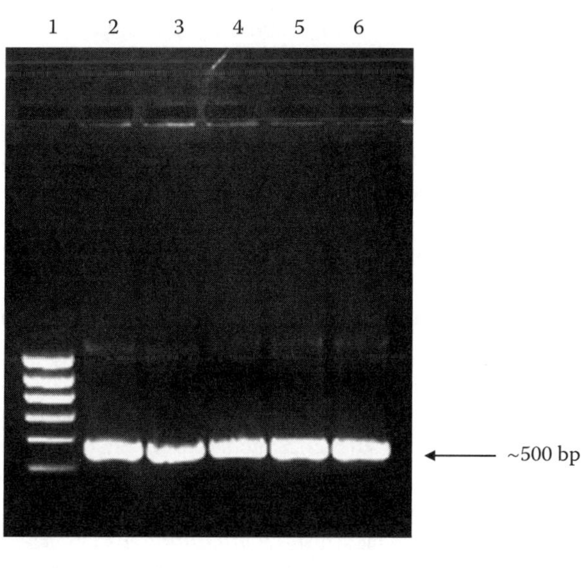

FIGURE 85.2 Agarose gel electrophoresis analysis of 16S rDNA PCR products from five bacterial DNA extracts (lanes 2–6). The exact size of the amplicons is estimated by comparison with DNA bands from appropriate molecular-weight marker (lane 1).

Note: Although 16S rDNA sequencing and sequence analysis has become practical for implementation in diagnostic settings, the major limitation of this method is that there exist no ideal databases and no universally accepted guidelines for interpretation of 16S rDNA sequence data, especially when close homology among recently diverged species limits resolution capabilities. However, it can be considered a valid adjunct to routine methods (e.g., commercial identification methods) for identification of bacteria, such as those of the genus *Providencia*, in the clinical microbiology laboratory.

85.2.2.2 Ribotyping Analysis

Principle. Molecular typing by ribosomal DNA (rDNA) fingerprints (ribotyping) on the basis of restriction fragment length polymorphism in the chromosomal DNA containing rRNA genes is a useful technique for taxonomic and epidemiological studies of microorganisms. Pignato et al.[10] utilized two enzymes (*Eco*RV and *Hinc*II) to generate distinct rRNA gene restriction patterns from the *Proteus-Providencia* group, leading to the confirmation of *P. mirabilis*, *P. penneri*, *M. morganii*, *P. rustigianii*, and *P. heimbachae* as discrete and homogeneous species.

Procedure

1. Extract genomic DNA from cultured bacteria by lysis in sodium dodecyl sulfate and pronase solutions, followed by an Autogen 540 automated DNA extraction system (AutoGen Instruments, Beverly, Massachusetts).
2. Digest the purified bacterial DNA (5 μg in 10 mM Tris-1 mM EDTA) with *Eco*RV or *Hinc*II in respective buffer for 4 h at 37°C.
3. Electrophorese the digested DNA at 50 V for 16 h in a horizontal 0.8% (wt/vol) agarose gel in TBE buffer [89 mM Tris base, 89 mM boric acid, 2.5 mM EDTA (pH 8)], using *Mlu*I-digested genomic DNA of *Citrobacter koseri* CIP 105177 as a reference to determine the sizes of DNA fragments.
4. Denature the gel with alkaline followed by neutralization, and transfer the DNA onto a nylon filter

(Hybond N; Amersham) with a VacuGene apparatus (Pharmacia LKB Biotechnology, Uppsala, Sweden).
5. Hybridize the nylon filter with digoxigenin (DIG)-labeled rRNA OligoMix5 probes and perform immunoenzymatic detection of hybridizing fragments by using the DIG Easy Hyb kit (Boehringer, Mannheim, Germany) and the DIG nucleic acid detection kit (Boehringer).
6. Scan the banding patterns with One-Scanner (Apple Computer, Cupertino, California) and calculate the fragment sizes by comparing migration values to the average of the two nearest standard lanes, weighted on the basis of lane proximity.

Note: *Eco*RV- or *Hinc*II-digested DNA from the type strains of the 10 different species of the genera *Proteus*, *Morganella*, and *Providencia* showed remarkably different patterns for each strain. The distinct number and position of the restriction fragments in the patterns permitted calculation of large genetic distances between the type strains, and differentiation of distinct species, ribogroups, and ribotypes.[10] The variability of rDNA restriction patterns in almost all of the species studied indicates that ribotyping can be a very sensitive method for epidemiological investigations of *Proteus*, *Morganella*, and *Providencia* bacteria.

85.2.2.3 PFGE Analysis

Principle. For PFGE analysis, a contour-clumped homogeneous electric field (CHEF) method is used. Although a variety of electrophoretic methods other than CHEF [e.g., field inversion gel electrophoresis (FIGE)] with different hardware configurations have been described, they are all based on the common principle of an electrophoretic current "pulsed" in different directions over a gradient of time intervals (i.e., ramping). By these methods, a range of DNA molecules, including those of megabase size, can be easily resolved. As CHEF is used by a great majority of laboratories, it has become essentially synonymous with the term "PFGE" in common parlance. Thus, agarose plugs containing genomic DNA are digested with *Sfi*I[55] or *Sma*I[62,63] restriction enzyme. Fragments are separated by agarose gel electrophoresis on a CHEF Mapper system (Bio-Rad Laboratories, Hercules, CA). The banding patterns are evaluated by using fingerprinting II DST software (Bio-Rad) with Dice and UPGMA coefficients, as described.[55] According to dendrogram results, isolates with a genetic similarity of >80% are considered to be clonal.

Procedure

1. Wash bacterial cells, obtained by culturing overnight a single colony of the isolate to be tested, and adjust their optical density at a wavelength of 610 nm to 1.40.
2. Mix 0.5 mL of cell suspension with 0.5 mL of chromosomal-grade agarose (Bio-Rad), and pipet the mixture into plug molds.
3. Incubate the plugs with lysis buffer (50 mM Tris, pH 8.0; 50 mM EDTA, pH 8.0; 1% sarcosine; 1 mg of proteinase K per mL) overnight in a 55°C water bath.
4. Inactivate proteinase K and wash samples.
5. Digest overnight with restriction enzyme as indicated above.
6. Electrophorese the digested samples on the CHEF Mapper system at the following conditions: 0.8%–1.0% agarose; 0.5× TBE buffer; initial switch time, 2.16 s; final switch time, 54.17 s; run time, 22 h; angle, 120°; gradient, 6 V/cm; temperature, 14°C, ramping factor, linear.
7. Stain gels with ethidium bromide (10 mg/mL) and observe PFGE results.

Note: Genomic DNA patterns generated by PFGE are highly specific for strains from a variety of microorganisms and have significant value in epidemiological investigations of outbreaks caused by most clinically relevant bacterial pathogens, including the *Providencia* species. This is because PFGE, although not every genetic change is noted and not every restriction fragment is visualized, provides an opportunity to detect multiple variations throughout the chromosome as they influence rare restriction sites and the distances between them.

85.3 CONCLUSION AND FUTURE PERSPECTIVES

Human infections due to the *Providencia* species remain relatively uncommon in clinical practice, although they represent an important cause of nosocomial and community-acquired illnesses, particularly in immunocompromised or elderly patients. However, our understanding of the pathogenic role of the *Providencia* species is restricted to only three of them, *P. stuartii*, *P. rettgeri*, and *P. alcalifaciens*, whereas rudimentary remains that of *P. rustigianii* and, even more, that of *P. heimbachae*, as the latter is a very rarely encountered species, probably because of difficulty of identifying it with the current methodologies. In this context, it is now essential that laboratories consider use of newer broad-spectrum molecular methods to aid in the species identification of clinical isolates. Much work remains to be done in investigating the role of urease in *P. rettgeri* pathogenicity. In this regard, efforts to construct isogenic urease-negative *P. rettgeri* mutants should be encouraged.

Another fundamental question in the study of bacterial pathogens is to identify the environmental signals triggering the expression of virulence genes, but sensing and signaling mechanisms responsible for interaction between host cells and pathogenic organisms are not exactly known. At the time of writing, the type strains *P. alcalifaciens* DSM 30120, *P. rettgeri* DSM 1131, *P. rustigianii* DSM 4541, and *P. stuartii* ATCC 25827, all isolated from human feces, are part of an ongoing comprehensive, sequence-based survey of members of the intestinal microbiome, the mutually interacting system comprising the host cells and the residing microbial community. The human gut microbiome project was launched as a joint research effort of the Washington University Genome Sequencing Center and the Center for Genome Sciences at Washington University School of Medicine, to provide the general scientific community with a broad view of the gene content of 100 representatives of the major divisions represented in the intestine's microbial community. This information will provide a frame of reference for analyzing metagenomic studies of the human gut microbiome that will allow individualization of new strategies for the maintenance of human health.

Drug-resistant strains of *Providencia* species appear to be increasing in frequency, leading to a shift in the antibiotics of choice or an increased patient mortality when antibiotics inactive against such organisms are used. Antibiotic choice is seriously reduced for infections with ESBL producers, because either the organisms are able to hydrolyze many β-lactam antibiotics or the plasmids bearing the genes encoding ESBLs also carry genes encoding resistance to aminoglycosides and trimethoprim/sulphamethoxazole. Therefore, infection control measures and antibiotic management interventions have become necessary to curtail the spread of ESBL producers. In this context, clinical microbiology laboratories have several reliable tests to detect ESBL production, which is particularly important in organisms from samples, such as urine (e.g., *P. stuartii*), that may represent the potential for transfer of such organisms to other patients. It is also pivotal to be able to determine whether nosocomial *Providencia* infections represent monoclonal or oligoclonal outbreaks, and to this end, several molecular techniques can help to assess the genetic relationships between nosocomially acquired organisms. If these techniques become widely exploited in clinical microbiology laboratories, it is possible that subdivision of members of the *Providencia* species into one or more subspecies may be considered appropriate in the future.

REFERENCES

1. Janda, J.M., New members of the family Enterobacteriaceae. In Balows, A., Trüper, H.G., Dworkin, M., Harder, W., and Schleifer, K.H. (Editors), *The Prokaryotes. A Handbook on the Biology of Bacteria: Ecophysiology, Isolation, Identification, Applications*, Chapter 3.3.1, Springer, New York, 2006.
2. Farmer, J.J. III, Enterobacteriaceae: Introduction and identification. In Murray, P.R, Baron, E.J., Jorgensen, J.H., Pfaller, M.A., and Yolken, R.H. (Editors), *Manual of Clinical Microbiology*, Chapter 41, ASM Press, Washington DC, 2003.
3. O'Hara, C.M., Brenner, F.W., and Miller, J.M., Classification, identification, and clinical significance of *Proteus*, *Providencia*, and *Morganella*, *Clin. Microbiol. Rev.*, 13, 534, 2000.
4. Hadley, P.B., Elkins, M.W., and Caldwell, D.W., The colon-typhoid intermediates as causative agent of disease in birds. I. The paratyphoid bacteria, *R. I. Agric. Exp. Stn. Bull.*, 174, 180, 1918.
5. Hickman-Brenner, F.W. et al., *Providencia rustigianii*: A new species in the family Enterobacteriaceae formerly known as *Providencia alcalifaciens* biogroup 3, *J. Clin. Microbiol.*, 17, 1057, 1983.
6. Müller, H.E. et al., *Providencia heimbachae*, a new species of Enterobacteriaceae isolated from animals, *Int. J. Syst. Bacteriol.*, 36, 252, 1986.
7. Somvanshi, V.S. et al., *Providencia vermicola* sp. nov., isolated from infective juveniles of the entomopathogenic nematode *Steinernema thermophilum*, *Int. J. Syst. Evol. Microbiol.*, 56, 629, 2006.
8. Manos, J., and Belas, R., The genera *Proteus*, *Providencia*, and *Morganella*. In Balows, A., Trüper, H.G., Dworkin, M., Harder, W., and Schleifer, K.H. (Editors), *The Prokaryotes. A Handbook on the Biology of Bacteria: Ecophysiology, Isolation, Identification, Applications*, Chapter 3.3.12, Springer, New York, 2006.
9. Ewing, W.H., Davis, B.R., and Sikes, S.V., Biochemical characterization of *Providencia*, *Public Health Lab.*, 30, 25, 1972.
10. Pignato, S. et al., Molecular characterization of the genera *Proteus*, *Morganella*, and *Providencia* by ribotyping, *J. Clin. Microbiol.*, 37, 2840, 1999.
11. Thompson, J.D., Higgins, D.G., and Gibson, T.J., CLUSTAL W: Improving the sensitivity of progressive multiple sequence alignment through sequence weighting, position-specific gap penalties and weight matrix choice, *Nucleic Acids Res.*, 22, 4673, 1994.
12. Spröer, C. et al., The phylogenetic position of *Serratia*, *Buttiauxella* and some other genera of the family Enterobacteriaceae, *Int. J. Syst. Bacteriol.*, 49, 1433, 1999.
13. Dauga, C., Evolution of the *gyr*B gene and the molecular phylogeny of Enterobacteriaceae: A model molecule for molecular systematic studies, *Int. J. Syst. Evol. Microbiol.*, 52, 531, 2002.
14. Paradis, S. et al., Phylogeny of the Enterobacteriaceae based on genes encoding elongation factor Tu and F-ATPase beta-subunit, *Int. J. Syst. Evol. Microbiol.*, 55, 2013, 2005.
15. Laupland, K.B. et al., Population-based laboratory surveillance for tribe *Proteeae* isolates in a large Canadian health region, *Clin. Microbiol. Infect.*, 13, 683, 2007.
16. Warren, J.W., *Providencia stuartii*: A common cause of antibiotic-resistant bacteriuria in patients with long-term indwelling catheters, *Rev. Infect. Dis.*, 8, 61, 1986.
17. Woods, T.D., and Watanakunakorn, C., Bacteremia due to *Providencia stuartii*: Review of 49 episodes, *South Med. J.*, 89, 221, 1996.
18. Piccolomini, R.L. et al., Comparative in vitro activities of 13 antimicrobial agents against *Morganella-Proteus-Providencia* group bacteria from urinary tract infections, *Antimicrob. Agents Chemother.*, 31, 1644, 1987.
19. Kim, B.N. et al., Bacteraemia due to tribe *Proteeae*: A review of 132 cases during a decade (1991–2000), *Scand. J. Infect. Dis.*, 35, 98, 2003.
20. Haynes, J., and Hawkey, P.M., *Providencia alcalifaciens* and travellers' diarrhoea, *BMJ*, 299, 92, 1989.
21. Albert, M.J., Faruque, A.S., and Mahalanabis, D., Association of *Providencia alcalifaciens* with diarrhea in children, *J. Clin. Microbiol.*, 36, 1433, 1998.
22. Yoh, M. et al., Importance of *Providencia* species as a major cause of travellers' diarrhoea, *J. Med. Microbiol.*, 54, 1077, 2005.
23. Koreishi, A.F., Schechter, B.A., and Karp, C.L., Ocular infections caused by *Providencia rettgeri*, *Ophthalmology*, 113, 1463, 2006.
24. de Gomes, L.S., Sobre una nova specie do genero *Eberthella* Buchanan, isolada de fezes patologica de crianca, *Rev. Inst. Lutz*, 4, 183, 1944.

25. Albert, M.J. et al., Characteristics of invasion of HEp-2 cells by *Providencia alcalifaciens*, *J. Med. Microbiol.*, 42, 186, 1995.
26. Janda, J.M. et al., Invasion of HEp-2 and other eukaryotic cell lines by *Providenciae*: Further evidence supporting the role of *Providencia alcalifaciens* in bacterial gastroenteritis, *Curr. Microbiol.*, 37, 159, 1998.
27. Sobreira, M. et al., Molecular analysis of clinical isolates of *Providencia alcalifaciens*, *J. Med. Microbiol.*, 50, 29, 2001.
28. Khashe, S. et al., Non-invasive *Providencia alcalifaciens* strains fail to attach to HEp-2 cells, *Curr. Microbiol.*, 43, 414, 2001.
29. Chen, X. et al., Demonstration and characterization of manganese superoxide dismutase of *Providencia alcalifaciens*, *Microbiol. Immunol.*, 51, 951, 2007.
30. Macleod, S.M., and Stickler, D.J., Species interactions in mixed-community crystalline biofilms on urinary catheters, *J. Med. Microbiol.*, 56, 1549, 2007.
31. Mobley, H.L. et al., MR/K hemagglutination of *Providencia stuartii* correlates with adherence to catheters and with persistence in catheter-associated bacteriuria, *J. Infect. Dis.*, 157, 264, 1988.
32. Kunin, C.M., Blockage of urinary catheters: Role of microorganisms and constituents of the urine on formation of encrustations, *J. Clin. Epidemiol.*, 42, 835, 1989.
33. Gendlina, I. et al., Urea-dependent signal transduction by the virulence regulator UreR, *J. Biol. Chem.*, 277, 40, 37349, 2002.
34. O'Hara, C.M., Manual and automated instrumentation for identification of Enterobacteriaceae and other aerobic gram-negative bacilli, *Clin. Microbiol. Rev.*, 18, 147, 2005.
35. O'Hara, C.M., Rhoden, D.L., and Miller, J.M., Reevaluation of the API 20E identification system versus conventional biochemicals for identification of members of the family Enterobacteriaceae: A new look at an old product, *J. Clin. Microbiol.*, 30, 123, 1992.
36. Kitch, T.T., Jacobs, M.R., and Appelbaum, P.C., Evaluation of RapID onE system for identification of 379 strains in the family Enterobacteriaceae and oxidase-negative, gram-negative nonfermenters, *J. Clin. Microbiol.*, 32, 931, 1994.
37. Funke, G., and Funke-Kissling, P., Evaluation of the new VITEK 2 card for identification of clinically relevant gram-negative rods, *J. Clin. Microbiol.*, 42, 4067, 2004.
38. O'Hara, C.M., Evaluation of the Phoenix 100 ID/AST system and NID panel for identification of Enterobacteriaceae, *Vibrionaceae*, and commonly isolated nonenteric gram-negative bacilli, *J. Clin. Microbiol.*, 44, 928, 2006.
39. O'Hara, C.M., Steigerwalt, A.G., and Green, D., Isolation of *Providencia heimbachae* from human feces, *J. Clin. Microbiol.*, 37, 3048, 1999.
40. Chen, J.R. et al., Rapid identification and susceptibility testing using the VITEK 2 system using culture fluids from positive BacT/ALERT blood cultures, *J. Microbiol. Immunol. Infect.*, 41, 259, 2008.
41. Funke, G., and Funke-Kissling, P., Use of the BD Phoenix automated microbiology system for direct identification and susceptibility testing of gram-negative rods from positive blood cultures in a three-phase trial, *J. Clin. Microbiol.*, 42, 1466, 2004.
42. Penner, J.L. et al., O serotyping of *Providencia stuartii* isolates collected from twelve hospitals, *J. Clin. Microbiol.*, 9, 11, 1979.
43. Penner, J.L., and Hennessy, J.N., Application of O-serotyping in a study of *Providencia rettgeri* (*Proteus rettgeri*) isolated from human and nonhuman sources, *J. Clin. Microbiol.*, 10, 834, 1979.
44. Abbott, S.L., *Klebsiella*, *Enterobacter*, *Citrobacter*, *Serratia*, *Plesiomonas*, and other Enterobacteriaceae. In Murray, P.R, Baron, E.J., Jorgensen, J.H., Pfaller, M.A., and Yolken, R.H. (Editors), *Manual of Clinical Microbiology*, Chapter 44, ASM Press, Washington DC, 2003.
45. Fass, R.J., Barnishan, J., and Ayers, L.W., Emergence of bacterial resistance to imipenem and ciprofloxacin in a university hospital, *J. Antimicrob. Chemother.*, 36, 343, 1995.
46. Pitout, J.D., Multiresistant Enterobacteriaceae: New threat of an old problem, *Expert. Rev. Anti Infect. Ther.*, 6, 657, 2008.
47. Clarridge, J.E. III, Impact of 16S rRNA gene sequence analysis for identification of bacteria on clinical microbiology and infectious diseases, *Clin. Microbiol. Rev.*, 17, 840, 2004.
48. Delmas, J. et al., Rapid identification of Enterobacteriaceae by sequencing DNA gyrase subunit B encoding gene, *Diagn. Microbiol. Infect. Dis.*, 55, 263, 2006.
49. Degand, N. et al., Matrix-assisted laser, desorption ionization–time of flight mass spectrometry for identification of nonfermenting gram-negative bacilli isolated from cystic fibrosis patients, *J. Clin. Microbiol.*, 46, 3361, 2008.
50. Paterson, D.L., and Bonomo, R.A., Extended-spectrum beta-lactamases: A clinical update, *Clin. Microbiol. Rev.*, 18, 657, 2005.
51. Arpin, C. et al., Extended-spectrum β-lactamase-producing Enterobacteriaceae in community and private health care centers, *Antimicrob. Agents Chemother.*, 47, 3506, 2003.
52. Tumbarello, M. et al., ESBL-producing multidrug-resistant *Providencia stuartii* infections in a university hospital, *J. Antimicrob. Chemother.*, 53, 277, 2004.
53. Aubert, D. et al., Novel genetic structure associated with an extended-spectrum beta-lactamase bla_{VEB} gene in a *Providencia stuartii* clinical isolate from Algeria, *Antimicrob. Agents Chemother.*, 49, 3590, 2005.
54. Luzzaro, F. et al., Trends in productions of extended- spectrum β-lactamases among enterobacteria of medical interest: report of the second Italian nationwide survey, *J. Clin. Microbiol.*, 44, 1659, 2006.
55. Shiroto, K. et al., Metallo-beta-lactamase IMP-1 in *Providencia rettgeri* from two different hospitals in Japan, *J. Med. Microbiol.*, 54, 1065, 2005.
56. Poirel, L. et al., Multidrug-resistant *Providencia stuartii* expressing extended-spectrum beta-lactamase PER-1, originating in Kosovo, *J. Antimicrob. Agents*, 61, 1392, 2008.
57. Lee, H.W. et al., Multidrug-resistant *Providencia* isolates carrying bla_{PER-1}, bla_{VIM-2}, and *armA*, *J. Clin. Microbiol.*, 45, 272, 2007.
58. Thaller, M.C. et al., Modified MacConkey medium which allows simple and reliable identification of *Providencia stuartii*, *J. Clin. Microbiol.*, 30, 2054, 1992.
59. Ohkusu, K., Cost-effective and rapid presumptive identification of gram-negative bacilli in routine urine, pus, and stool cultures: evaluation of the use of CHROMagar orientation medium in conjunction with simple biochemical tests, *J. Clin. Microbiol.*, 38, 4586, 2000.
60. Senior, B.W., Media for the detection and recognition of the enteropathogen *Providencia alcalifaciens* in faeces, *J. Med. Microbiol.*, 46, 524, 1997.
61. Woo, P.C.Y. et al., Then and now: Use of 16S rDNA gene sequencing for bacterial identification and discovery of novel bacteria in clinical microbiology laboratories, *Clin. Microbiol. Rev.*, 14, 908, 2008.
62. Murata, T. et al., A large oubreak of foodborne infection attribute to *Providencia alcalifaciens*, *J. Infect. Dis.*, 184, 1050, 2001.
63. Saida, N.B. et al., Clonality of *Providencia stuartii* isolates involved in outbreak that occurred in a burn unit, *Burns*, 34, 829, 2008.

86 *Pseudomonas*

Timothy J. Kidd, David M. Whiley, Scott C. Bell, and Keith Grimwood

CONTENTS

86.1 INTRODUCTION

Pseudomonas sensu stricto is an extensive genus of gram-negative bacteria comprising several species of medical importance. Most human infections are opportunistic and associated with impaired host susceptibility. *Pseudomonas aeruginosa* represents the major pathogenic species of this group and causes a variety of acute and chronic illnesses. Infections with other species are rarely encountered, but can also be serious.

Pseudomonas spp. are traditionally identified in clinical samples by microscopy, culture, and phenotypic techniques. Whilst culture is important for determining antimicrobial susceptibility, detection can be hindered by prior antimicrobial treatment, multiple antibiotic resistance, atypical phenotypic features, inadequate commercial identification systems, and the presence of other nonfermenting gram-negative bacteria. To overcome these difficulties, several molecular-based detection and identification strategies have recently been devised.

86.1.1 CLASSIFICATION AND BIOLOGY

First proposed as a genus in 1894, original taxonomic descriptions loosely defined *Pseudomonas* as aerobic nonsporulating gram-negative bacilli demonstrating motility via polar flagella. The advent of ribosomal RNA–DNA hybridization techniques allowed pseudomonads to be divided into five distinct homological groups: RNA group I (*Pseudomonas* sensu stricto), RNA group II (*Burkholderia* and *Ralstonia* spp.), RNA group III (*Comomonas, Acidovorax, Delftia* and *Hydrogenophaga* spp.), RNA group IV (*Brevundimonas* spp.), and RNA group V (*Stenotrophomonas and Xanthomonas* spp.). Subsequently, *Pseudomonas* was reduced to include only those species belonging to RNA group I.[1] Of the 160 currently listed species, 12 are of clinical relevance to humans.[2]

The genus *Pseudomonas* belongs to the Pseudomonadaceae family and the γ subclass of Proteobacteria. Genus members are typically straight or slightly curved gram-negative non-spore-forming bacilli, 0.5–1.0 × 1.5–5.0 μm in size. Most have polar flagella and utilize glucose and other carbohydrates via oxidative pathways. Although categorized as strict aerobes, *Pseudomonas* spp. can survive in microaerophilic environments by using nitrate or arginine as terminal electron acceptors.[2,3] All clinically relevant species, except *Pseudomonas luteola* and *Pseudomonas orzyihabitans*, have cytochrome oxidase activity.[2]

Each of the 12 clinically relevant species can be divided into two distinct groups based on their ability to produce fluorescent pigments. The six species belonging to the fluorescent group (*P. aeruginosa, Pseudomonas fluorescens, Pseudomonas putida, Pseudomonas veronii, Pseudomonas monteilii,* and *Pseudomonas mosselii*) produce pyoverdin, a water-soluble pigment that fluoresces white to blue-green when exposed to ultraviolet light. Many *P. aeruginosa* strains also produce the blue pigment pyocyanin, which, with pyoverdin, creates typical bright green colonies. This characteristic separates *P. aeruginosa* from the other members of the fluorescent group. *P. aeruginosa* strains can also produce other diffusible pigments, such as pyomelanin (brown/black) and pyorubin (red). The corresponding six nonfluorescent species are *Pseudomonas stutzeri, Pseudomonas mendocina, Pseudomonas alcaligenes, Pseudomonas pseudoalcaligenes, P. luteola,* and *P. oryzihabitans*.[2]

Pseudomonas spp. are extremely versatile, being able to survive in a broad range of habitats.[1] *P. aeruginosa* flourishes in moist environments, and it reaches large numbers, even with minimal nutrients. In health care settings, it can be found in sinks, showers, and respiratory equipment.[4] Other *Pseudomonas* spp. can metabolize various aromatic hydrocarbons and environmental pollutants.[5–7]

The first complete *Pseudomonas* genome sequence (*P. aeruginosa* strain PA01) was annotated in 2000.[8] It revealed a complex and highly regulated genome of 6.3 Mbp (G + C content 66.6%), containing numerous paralogous gene families encoding important adaptive functions. These functions include the capacity to import and metabolize various organic substrates, synthesize virulence factors, export antibiotics using efflux systems, scavenge iron, and form biofilms via quorum sensing pathways. Genomic analyzes of other clinical and environmental *P. aeruginosa* strains confirm that the core genome is highly regulated and conserved.[9–11] *P. fluorescens* and *P. putida* also have genetic determinants that provide an exceptional platform for metabolic versatility.[6,7] This remarkable level of genetic conservation and functional adaptability allow *Pseudomonas* spp. to grow in a diverse range of habitats.

86.1.2 Epidemiology

Pseudomonas spp. act as opportunistic pathogens. Of these, *P. aeruginosa* is the most clinically significant. It is an important cause of lower respiratory tract, bloodstream, and urinary tract infections amongst patients in intensive care units or with compromised immunity.[12,13] *P. aeruginosa* is often isolated from moist environmental reservoirs within health care settings, such as taps, toilets, showers, respiratory equipment, and cleaning solutions.[4] Skin, throat, and fecal carriage amongst healthy individuals is also reported, and while most healthcare-associated infections are acquired from contaminated inanimate sources, infections arising from either endogenous carriage or person-to-person transmission can also occur.[4,14,15]

P. aeruginosa is also an important pathogen in chronic suppurative lung diseases, most notably in cystic fibrosis (CF), where, by early adulthood, approximately 80% of those with CF are infected.[16] Most CF patients possess unique strains of *P. aeruginosa*. These strains seem to be acquired from environmental sources, although their origins and acquisition pathways are unknown.[17] However, siblings with CF can have indistinguishable *P. aeruginosa* strains, suggesting cross-infection may also occur.[18] Indeed, CF clinics in the United Kingdom, Europe, and Australia report patients sharing one or more clonal strains of *P. aeruginosa*.[19–22] Some clonal strains are found in CF clinics that are widely separated from one another, while others are associated with increased symptoms and treatment costs and can cause infections in patients with non-CF pulmonary disorders.[19,22–26] One strain in particular, Clone C, is widely dispersed in the physical environment and has been isolated from patients in CF clinics and intensive care units in Europe, the United Kingdom, and Canada.[21,22,27] It seems that, while many with chronic pulmonary disorders can acquire *P. aeruginosa* from environmental sources, others may become infected following person-to-person transmission.

86.1.3 Clinical Features

P. aeruginosa. *P. aeruginosa* was first recognized as a pathogen in 1882, when it was isolated from bluish-green pus from two patients with superficial wounds. Since then *P. aeruginosa* has been recognized as causing diseases ranging from superficial infection to systemic sepsis.[4] Acute infection in otherwise healthy individuals includes folliculitis and otitis externa ("swimmer's ear") following exposure to contaminated swimming pools, hot tubs, spa baths, or sponges; os calcis osteochondritis after suffering penetrating heel injuries when wearing sneakers, and contact lens–associated keratitis.[4,28] Except for CF, most other *P. aeruginosa* infections are in critically ill patients in intensive care units, burn patients, and the immunocompromised. These infections include ventilator-associated pneumonia, catheter-related bloodstream or urinary tract infections, other infections from medical devices or penetrating trauma, sepsis in neutropenic cancer and transplant patients, skin and soft-tissue infections in burn and diabetic patients, postoperative endophthalmitis and necrotizing otitis externa in diabetics.[4,29,30]

In CF, the initial acquisition of *P. aeruginosa* within the respiratory tract is typically transient, and the bacteria are cleared by either the host immune response or antibiotics. Eventually, a single clone becomes established, following which onset of chronic *P. aeruginosa* CF lung infection is marked by mucoid variants that are virtually impossible to eradicate, even with aggressive antibiotic treatment.[31] Subsequently, there is an accelerated decline in pulmonary function, an increase in treatment requirements, and life expectancy is reduced by about 10 years.[32–34]

Other *Pseudomonas* spp. Infections with species other than *P. aeruginosa* are uncommon. Table 86.1 shows that these species are mostly limited to the immunocompromised (e.g., intensive care, oncology, and postoperative patients).[2] These species lack recognized virulence determinants and are often acquired from contaminated solutions or equipment, including water, detergents, intravenous catheters, blood products, and ventilators. Ocular infections occur postoperatively or after penetration injury. *P. fluorescens* pseudobacteremia is a well-recognized entity where positive blood cultures are detected in patients not showing signs of infection. Such instances are usually attributed to contaminated blood sampling and intravenous catheter devices and, if unrecognized, can lead to unnecessary treatment.

86.1.4 Pathogenesis

The primary determinant for establishing acute infection is a breakdown of anatomical or functional host defences. *P. aeruginosa* possesses numerous virulence factors, which contribute to its pathogenesis. Pili and flagella permit host cell adherence, motility, and biofilm formation. Resistance to innate host defences and antibiotics involve lipopolysaccharide and protease production, sequestration of iron from siderophore uptake systems, protein and efflux systems, and biofilm formation. Exotoxin A, phospholipase, leukocidin, type III secretion systems, and hemolysin each contribute to cell and tissue destruction, releasing additional nutrients and further impairing host defences.[4]

The mechanisms of initial *P. aeruginosa* infection in CF patients are incompletely understood, but may involve impaired mucociliary clearance secondary to low airway surface liquid volumes.[47] Early infecting *P. aeruginosa* strains are motile, possess smooth lipopolysaccharides with long side chains, are susceptible to antipseudomonal antibiotics, and form nonmucoid colonies.[17] These motile strains infiltrate mucous plugs within the airways, where under conditions of oxidative stress and nutrient deprivation, they undergo genetic adaptation that selects for strains with loss of function mutations in regulatory genes. Adaptation to the microenvironment of the CF lung may be accelerated by isolates labeled as "hyper-mutators," that have defects in DNA mismatch repair systems.[48,49] These adaptive changes result in loss of motility, biofilm formation, upregulation of drug efflux pumps, and conversion to mucoid variants, the latter following downregulation of the *mucA* gene and overexpression of the *algD* promoter gene.[50–52] Consequently, *P. aeruginosa* undergoes metabolic adaptation that ensures survival within the microaerophilic environment of CF airway mucous plugs, while at the same time reducing susceptibility to antibiotic therapy and increasing resistance to clearance by the host immune system.[53]

86.1.5 Diagnosis

86.1.5.1 Conventional Techniques

Traditional techniques for detecting *Pseudomonas* spp. involve microscopy and culture. After primary culture and selection, isolates deemed clinically significant are identified

TABLE 86.1
Non-*P. aeruginosa Pseudomonas* spp. Isolated from Humans

Organism	Clinical Presentations	Reference
P. fluorescens	CF; ventilator-associated pneumonia; transfusion and catheter-related sepsis; posttraumatic endophthalmitis; UTI; and pseudobacteremia	2,35–39
P. putida	CF; meningitis; transfusion and catheter-related sepsis; bacteremia, septic arthritis, UTI and pneumonia amongst immunocompromised patients; dialysis associated peritonitis; keratitis; cellulitis; post traumatic wound infection	2,35,40
P. stutzeri	CF; meningitis and ventriculitis; haemodialysis-related sepsis, bacteremia, bone infections and pneumonia amongst immunocompromised patients; community acquired pneumonia and empyema; postoperative endophthalmitis and panophthalmitis; UTI	2,41
P. luteola	Dialysis associated peritonitis; cellulitis; sepsis; prosthetic valve endocarditis; and postoperative endophthalmitis	2,42
P. oryzihabitans	Dialysis associated peritonitis; sepsis; catheter-related bacteremia in immunocompromised patients; and postoperative endophthalmitis	2,43
P. mendocina	Rare instances of spondylodiscitis and endocarditis	44
P. alcaligenes	One reported instance of catheter-related endocarditis in a bone marrow transplant recipient	2
P. pseudoalcaligenes	One case of infantile meningitis	2
P. veronii	Associated with intestinal inflammatory pseudotumors	2
P. mosselii and *P. monteilii*	Isolated from a variety of human specimens but clinical significance remains unknown	45,46

CF, cystic fibrosis; UTI, urinary tract infection.

by their colonial morphology, growth, and biochemical characteristics. Commercial identification systems, such as the API 20NE (bioMérieux Vitek), VITEK (bioMérieux Vitek), Phoenix (BD), Crystal Enteric/Nonfermenter (BD), and RapidID NF PLUS (Remel) are also widely used by clinical microbiology laboratories.[2]

Microscopy, Culture, and Phenotypic Identification. Pseudomonads appear as straight or slightly curved gram-negative rods, 0.5–0.8 by 1.5–3.0 μm in size. They are usually easily differentiated from Enterobacteriaceae by their lack of a bipolar staining appearance. Isolates exhibiting the typical CF mucoid phenotype are often visualized as clusters of rods encapsulated in a thick pink staining alginate coat.[2]

Pseudomonas spp. grow on several different nonselective agar and broth preparations, including nutrient, 5% blood, chocolate, and MacConkey agars. For isolation from clinical specimens, MacConkey agar allows easy selection of the nonlactose fermenting colonies produced. Most species grow satisfactorily in aerobic conditions at 37°C and within 24 h form visible colonies. Yet some species other than *P. aeruginosa* may have improved growth at 28°C–30°C. Isolates from CF patients may also require at least 48 h incubation before an analysis is undertaken.[2,54]

P. aeruginosa. *P. aeruginosa* is characteristically oxidase positive, grows at 42°C, is unable to ferment lactose, but hydrolyzes acetamide (Table 86.2). Colonial morphology is variable (particularly in CF specimens), ranging from the typical spreading, flat colonies with serrated edges and metallic sheen to smooth, gelatinous, small colony or mucoid forms. *P. aeruginosa* also produces several different diffusible pigments, including the typical bright green (combined pyoverdin and phenazine pigments), brown to brown-black (pyomelanin), and red (pyorubin) pigments. Notably, some strains are nonpigmented and often produce mucoid colonies. Most *P. aeruginosa* isolates are susceptible to colistin sulphate, which helps to differentiate them from members of the *Burkholderia cepacia* complex and other important related pathogens.[2]

Whilst most *P. aeruginosa* isolates are easy to identify, those from CF patients can sometimes present taxonomic challenges. Phenotypic identification of CF-related isolates can be hampered by the absence of pigments, auxotrophic metabolic activity, antibiotic resistance (including colistin sulfate), atypical colonial morphology, and alginate production. Commercial identification platforms are unreliable.[2,54] Furthermore, CF patients may harbor other nonfermenting gram-negative bacilli, such as *Burkholderia*, *Achromobacter*, *Stenotrophomonas*, and *Inquilinus* spp., which can impede identification.[55–57] Most clinical microbiology laboratories still identify *P. aeruginosa* by traditional phenotypic techniques, and low rates of *P. aeruginosa* misidentification are reported from microbiology laboratories relying solely upon phenotypic procedures to identify CF clinical isolates.[57]

***P. fluorescens* and *P. putida*.** *P. fluorescens* and *P. putida* are arginine dihydrolase (ADH) and oxidase positive, but cannot grow at 42°C or reduce nitrate, and they lack distinctive colonial features. *P. fluorescens* is differentiated from *P. putida* by producing gelatinase and growth at 4°C. *P. fluorescens* and *P. putida* are identified by most commercial identification systems, but additional phenotypic tests such as growth at 4°C may be required to further differentiate these species.[2]

***P. stutzeri*.** *P. stutzeri* produces characteristic adherent, reddish brown, wrinkled colonies that can pit the agar surface. They are oxidase-positive, ADH-negative, hydrolyze starch, and produce acid from maltose. Most commercial identification platforms correctly identify *P. stutzeri*.[41]

***P. luteola* and *P. oryzihabitans*.** Each is oxidase negative and produces a nondiffusible yellow pigment. *P. luteola* is differentiated from *P. oryzihabitans* by its ability to hydrolyze esculin and grow at 42°C. Both are identified by several commercial identification systems.[2]

***P. mendocina*.** Colonies of *P. mendocina* appear flat and smooth with slightly yellow to brown nondiffusible pigmentation. They are oxidase- and ADH-positive, reduce nitrate, grow at 42°C, and demonstrate a negative acetamide hydrolysis reaction.[2]

***P. alcaligenes* and *P. pseudoalcaligenes*.** Both are difficult to identify given their lack of distinctive colonial features and biochemical reactivity. Only some strains grow at 42°C, and they are incapable of hydrolyzing gelatin, acetamide, starch, or esculin. *P. pseudoalcaligenes* is differentiated from *P. alcaligenes* by its ability to produce acid from fructose. Commercial identification platforms are unreliable.[2]

***P. veronii*, *P. mosselii*, and *P. monteilii*.** Each is rarely encountered in clinical specimens. *P. veronii* is ADH- and oxidase-positive and appears as smooth, nonpigmented, nonhemolytic colonies, which grow at 4°C–36°C. They can reduce nitrates and are differentiated from *P. aeruginosa* by their inability to hydrolyze acetamide. *P. mosselii and P. monteilii* grow at 10°C–36°C and are ADH- and oxidase-positive, but they are unable to reduce nitrates. Most *P. mosselii* strains produce gelatinase. In contrast, *P. monteilii* lacks gelatinase, cannot ferment xylose, and fails to grow in 6% NaCl. None of these species is reliably identified by commercial identification systems.[2]

86.1.5.2 Molecular Techniques: Detection and Identification

Nucleic acid–based techniques offer routine clinical microbiology laboratories the opportunity to provide rapid and definitive results for patients with complex polymicrobial and life-threatening illnesses. They provide a valuable means of genus and species differentiation in the reference and research laboratory setting, as is often the case with isolates from CF patients. Robotic liquid handling and DNA extraction instruments further facilitate molecular detection of *Pseudomonas* spp.

Polymerase Chain Reaction (PCR). Early amplification targets for *P. aeruginosa* (*toxA*, *oprL*, *algD*, and 16S rRNA gene assays) suffered from suboptimal sensitivity or

TABLE 86.2
Phenotypic Features of Clinically Relevant *Pseudomonas* spp.[a]

	Oxidase	Growth at 42°C	Growth at 4°C	Growth in 6% NaCl	Pyoverdin	Yellow Pigment	Arginine Dihydrolase	Gelatin Hydrolysis	Esculin Hydrolysis	Acetamide Hydrolysis	Starch Hydrolysis	Nitrate Reduction	Glucose	Lactose	Maltose	Xylose	Fructose
P. aeruginosa	+[b]	+	–	V	+	–	+	V	–	+	–	+	+	–	–	+	+
P. fluorescens	+	–	+	V	+	–	+	+	–	–	–	V	+	V	–	+	V
P. putida	+	–	+	+	+	–	+	–	–	–	–	–	+	V	V	+	V
P. stutzeri	+	V	V	V	–	–	–	–	–	–	+	+	+	–	+	+	V
P. luteola	–	+	NP	V	–	+	+	V	+	NP	–	V	+	–	+	+	+
P. oryzihabitans	–	V	NP	V	–	+	V	V	–	NP	–	–	+	V	+	+	+
P. mendocina	+	+	NP	+	–	–	+	–	–	–	–	+	+	–	–	V	+
P. alcaligenes	+	V	NP	V	–	–	V	–	–	NP	–	V	–	–	–	–	–
P. pseudoalcaligenes	+	+	NP	V	–	–	V	–	–	NP	–	+	–	–	–	V	V
P. veronii	+	–	+	NP	+	–	+	V	NP	–	NP	+	+	NP	NP	+	+
P. mosselii	+	–	–	+	+	–	+	V	–	NP	–	–	+	–	V	–	+
P. monteilii	+	–	–	–	+	–	+	–	–	–	–	–	+	–	–	–	+

[a] Data collated from References 2, 3, 41, 45, 46, 102, and 103.
[b] +, 90% or more of the strains are positive; –, 90% or more of the strains are negative; V, 11%–89% of the strains are positive; NP, no data available.

specificity.[55,58,59] Recently, DNA gyrase subunit B (*gyrB*), 16S-23S rRNA internal transcribed spacer (ITS), and the extracellular-cytoplasmic function sigma factor determinant (*ecfX*) target sequences have shown excellent sensitivity and specificity for identifying *P. aeruginosa*.[58–60] Simultaneous targeting of *ecfX* and *gyrB* genes in a Taqman probe–based real-time PCR assay provides a highly suitable platform for the definitive identification of *P. aeruginosa* from clinical and environmental sources.[55,57] A multiplex approach reduces potential false-positives from assay cross-reactivity and false-negatives from sequence variation.[61]

Few PCR assays exist for identifying non–*P. aeruginosa* species. Conserved *oprI* gene sequences can be used to screen for *Pseudomonas* spp.[62] Similarly, another assay amplifies a fragment of the *groE* heat-shock protein gene from several *Pseudomonas* spp.[35] A conventional gel-based detection assay using primers for 16S rRNA gene sequences helps to differentiate between *P. stutzeri* and other *Pseudomonas* spp.[63] The performance of these assays when applied to a wider range of clinical isolates requires further investigation.

Direct detection of *P. aeruginosa* DNA from extracted respiratory samples has been limited to the *toxA*, *oprL*, and 16S rRNA gene targets in relatively small investigations.[64,65] Further analysis of improved genomic targets, involving a broader range of samples, is required to fully investigate the validity of this molecular detection approach.

16S rRNA PCR–Restriction Fragment Length Polymorphism (RFLP) Analysis. RFLP analysis of amplified 16S rRNA gene PCR products can be successfully applied to the identification of several different *Pseudomonas* spp.[57,66–70] It relies on culturing the organism of interest and is particularly useful at identifying species other than *P. aeruginosa* when no specific PCR assays are available. Computer-assisted restriction-site mapping and fragment analysis have further improved this technique.[67] Limitations include a slightly higher cost and turnaround time than PCR, and the need for a defined set of species control strain RFLP patterns that can be readily compared to isolates of interest.

DNA Sequencing. 16S rRNA gene sequence analysis is a widely utilized molecular technique to identify most bacterial genera. Genomic DNA extracted from bacterial cultures becomes the PCR template for amplifying a 500–1500 bp segment of the highly conserved 16S rRNA gene sequence. Purified PCR products undergo terminally labeled cycle sequencing, where forward and reverse fragments are generated in separate reactions with forward and reverse primers. Sequence determination of purified fragments is performed with automated capillary or gel electrophoresis. After editing, the complementary sequences are compared to a database sequence library.[69] Sequence databases for *Pseudomonas* spp. include GenBank, EMBL Nucleotide Sequence Database (EMBL-Bank), Ribosomal Database Project (RDP-II), DNA Data Bank of Japan (DDBJ), and MicroSEQ® ID 16S rRNA 500 Library v2.1. The benefits of using 16S rRNA gene sequence analysis for identifying *Pseudomonas* spp. isolated from CF respiratory secretions are well documented.[56,57]

Other than identifying cultivable organisms, 16S rRNA gene sequence analysis is also an extremely powerful tool for diagnosing culture-negative infections. In this approach, bacterial DNA extracted directly from clinical samples is used as the template for 16S rRNA gene PCR amplification. Sequence analysis of the purified PCR products allows the direct detection and identification of bacterial DNA within clinical samples. Its application is limited to samples from sterile sites where only single bacterial species are expected. Samples containing more than one species, such as sputum, produce mixed and uninterpretable sequence data.[69] Instead, this method is more suitable for detecting *Pseudomonas* spp. in patients, for example, with meningitis who have previously received antibiotics.[70]

Although 16S rRNA gene sequence analysis involves high costs from DNA-sequencing equipment and requires staff with enhanced technical skills the recent availability of cheaper, simpler sequencing kits, and their introduction into many diagnostic laboratories makes this approach of bacterial identification a viable alternative in many clinical settings.

Terminal Restriction Fragment Length Polymorphism (T-RFLP). In T-RFLP, bacterial nucleic acids extracted directly from clinical samples are used as templates for PCR amplification of 16S rRNA gene or 16S-23S ITS sequences. PCR primers bind to both conserved and variable sequences between forward and reverse primer sites, which serve to generate unique amplicons for each species present within a sample. Following amplification, each PCR product is digested with a restriction enzyme to produce a discrete series of terminal restriction fragments generated by species-characteristic lengths of the first endonuclease cut position. Each fragment is measured by high resolution polyacrylamide gel electrophoresis, and bacterial species are predicted by comparison with in-silco sequence database predictions. Automation from using 5′ fluorescently labeled primers coupled with DNA sequence analysis is also possible.[71] Compared with traditional culture-based methodologies, T-RFLP provides rapid and accurate information on bacterial diversity in clinical samples. Sample extraction, amplification, restriction enzyme digestion, and fragment sequence analysis can be achieved within 12 h. T-RFLP analyses of respiratory samples from patients with CF and chronic obstructive pulmonary disease have identified numerous bacterial genera and species, including various *Pseudomonas* spp.[72,73] Moreover, T-RFLP allows discovery of novel or unexpected pathogens that would otherwise go undetected if specific assays such as PCR are used. Disadvantages include a technically demanding methodology and the need for sophisticated and costly gel analysis software and dedicated DNA-sequencing equipment. Some bacterial species produce indistinguishable terminal restriction fragments, which require further resolution using additional restriction enzymes.[72,73] For these reasons T-RFLP analysis of routine clinical samples is not currently recommended.

Fluorescence In Situ Hybridization (FISH). FISH is a nonamplified molecular method, which allows rapid and direct detection of bacterial species in clinical samples. It involves developing fluorescently labeled DNA or peptide nucleic acid probes targeted at species-specific 16S and 23S rRNA regions. Samples are typically fixed onto glass slides, hybridized with a species-specific probe and then screened using fluorescence microscopy. Several FISH assays are available for detecting *P. aeruginosa* in respiratory secretions and blood cultures. While they remain an alternative to routine culture, they have lower sensitivity and specificity than standard culture and phenotypic identification techniques.[75–77] Most specificity issues associated with FISH assays arise from nonspecific probe interactions with other gram-negative bacilli, including other *Pseudomonas* spp., *Achromobacter* spp., *Rhizobium radiobacter*, *S. maltophilia*, *B. cepacia* complex, and *Escherchia coli*.[74–76,78] Despite these limitations, FISH probes with confocal laser scanning microscopy provide an invaluable tool for studying the *in-vivo* ecology and spatial organization of *P. aeruginosa* biofilms in polymicrobial infections such as those encountered in chronic wounds and the CF lung.[79,80]

86.1.5.3 Molecular Techniques: Strain Analysis

Bacterial strain molecular typing is fundamental to epidemiological investigations and control of infection outbreaks. Recent reports of clonal *P. aeruginosa* strains in CF patients highlight the role these techniques may play in controlling person-to-person transmission in CF clinics.

Pulsed Field Gel Electrophoresis (PFGE). PFGE is considered the "gold standard" DNA-fingerprinting technique and specific analysis criteria are available.[81] It involves separation of restriction enzyme digested whole-cell DNA by electrophoresis, leading to unique patterns of DNA fragments (or "genomic fingerprint"). PFGE has high discriminatory power but suffers from being a costly, technically demanding, and time-consuming procedure, and it is incapable of assessing the overall relatedness of isolates with unrelated fingerprints. In addition, PFGE provides no direct measure of allelic diversity, and so insights into evolutionary processes shaping *P. aeruginosa* populations are unable to be obtained.[82]

Computer-Assisted *Sal*I Fingerprinting. Another fingerprinting-based *P. aeruginosa* genotyping strategy involves analysis of high molecular weight *Sal*I digested DNA fragments. DNA is first extracted from pure cultures, digested for 1 h and then separated by standard horizontal agarose gel electrophoresis. The numbers of restriction fragments corresponding to particular molecular weight ranges are then used to calculate the overall genetic distances between isolates expressed by unweighted pair-group method with arithmetic mean (UPGMA) trees. When compared to PFGE, additional benefits from this simple and rapid categorical typing method include high levels of reproducibility, equitable discriminatory power, portability, and low cost.[83,84]

PCR-Based DNA Fingerprinting. PCR-based typing methods have simplified the technical complexity of typing bacteria, and they are also less expensive than PFGE. Two repetitive extragenic palindromic (rep)-PCR assays, enterobacterial repetitive intergenic consensus (ERIC)–PCR and BOX–PCR, and random amplified polymorphic DNA (RAPD)–PCR are rapid and inexpensive tests that have been successfully applied to the epidemiological analysis of *P. aeruginosa* isolates.[85–87] However, because of their lower levels of discrimination, they are more useful as screening tools.

A limitation of PFGE and PCR-based typing techniques is that they are best applied to small sets of isolates collected during localized outbreak investigations over a brief time frame.[88] This is because epidemiological markers with continuous values can generate inconsistent data from a lack of precision in their measurement.[89] Application of these techniques to longitudinally collected or geographically distinct isolates for which no epidemiological information is available, can produce misleading or uninterpretable results.[84] Since the genetic organization of *P. aeruginosa* is known to be hypervariable secondary to the influence of environmental selection and mobile genetic elements, genomic fingerprinting techniques may provide ambiguous results suggesting that isolates are unrelated, when they may in fact share a very close evolutionary relationship.[90,91]

Multilocus Sequence Typing (MLST). MLST schemes have been developed recently for several clinically important bacterial species, including *P. aeruginosa*.[91,92] This involves PCR amplification and sequence analysis of seven housekeeping genes (*acsA*, *aroE*, *guaA*, *mutL*, *nuoD*, *ppsA*, and *trpE*), which were selected because of their relationships to vital cellular functions and slow rates of mutagenesis.[82] Following amplification and cycle sequencing, the different sequences at each locus are allocated unique allele numbers. Each strain is defined by the alleles at the seven loci, and each distinctive allelic profile is assigned a sequence type (ST). Each isolate's ST is readily comparable with other STs. A particular advantage of this method is that isolates collected longitudinally or from epidemiologically unrelated patients or sites can be meaningfully compared with each other and with those previously published on the PubMLST Web site (www.pubmlst.org/paeruginosa).[91] Another advantage of MLST typing and the eBURST algorithm is being able to organize each ST based upon its relatedness to other STs into "clonal complexes," which contain isolates of the same ancestral origin. Clonal complex analysis provides useful information on bacterial population structures, helping to determine geographical relationships and elucidate the rates and patterns of evolutionary descent.[93] This is of particular relevance to understanding the origins, acquisition, transmission pathways, and genetic changes shown by clonal *P. aeruginosa* strains in CF patients.

Variable Number of Tandem Repeats (VNTR) and Multiple-Locus Variable Number Tandem-Repeat (MVLA) Analyses. Recent whole-genome sequence analyses of several bacterial species have revealed the presence of short repeat DNA motifs called variable number of tandem repeats (VNTR). The number of repeats is often strain dependant, whereby variation can occur during chromosomal replication. As with MLST and computer-assisted *Sal*I fingerprinting techniques, VNTR typing represents another categorical genotyping tool, which expresses an allelic profile that corresponds to the number of repeat sequences present within each VNTR loci. The number of repeat units is determined by measuring the size of the amplified PCR products generated when primers that flank the repeat regions are used. Multiple-locus VNTR analysis (MVLA) schemes are described for several bacterial species and, where applicable, offer a rapid, discriminatory, portable, and reproducible means of assessing epidemiological relationships between bacterial strains.[94] Preliminary investigations of the *P. aeruginosa* genome have identified fifteen VNTR markers that can be used for genotyping.[95,96] However, at present, additional studies comparing a broader range of isolates and genotyping techniques are needed to fully investigate the suitability of this technique for typing *P. aeruginosa* isolates.

86.2 METHODS

In light of the currently available molecular techniques, and taking into consideration cost, time, and resource implications, the following section outlines a molecular approach for sample preparation, detection, identification, and strain typing of *Pseudomonas* spp. from clinical samples. The methods described should provide medium to large clinical reference and research laboratories with a relatively easy and achievable framework for implementing these techniques. Figure 86.1 summarizes the methods proposed.

86.2.1 Sample Preparation

86.2.1.1 Sample Collection, Culture, and Basic Phenotypic Screening

Pseudomonas can cause single or polymicrobial infections and are reliably grown from most clinical samples, including blood, cerebrospinal fluid, urine, purulent material from any site, sputum, and bronchoalveolar lavage fluid. Based on this premise and the need to perform antimicrobial susceptibility testing, in most instances presumptive *Pseudomonas* isolates are isolated by standard bacterial culture and interpretation techniques. Most clinical samples can be cultured and incubated on a minimum of MacConkey, blood, and/or chocolate-based agar media for 24–48 h. Respiratory samples derived from patients with CF and other chronic pulmonary disorders should be incubated for at least 48 h to ensure sufficient growth of all *Pseudomonas* spp. present.

Presumptive isolate screening includes testing for oxidase activity (excluding *P. luteola* and *P. orzyihabitans*), and temperature growth characteristics, such as growth at 42°C. Other important characteristics that should be noted are colistin sulfate susceptibility, presence of diffusible and nondiffusible pigments, and distinctive colonial features, as is the case with *P. stutzeri* and mucoid *P. aeruginosa*.

86.2.1.2 DNA Extraction

Principle. DNA extraction can be easily achieved using a range of standard heat-denatured, commercial, and in-house DNA extraction and preparation techniques. Heat-denatured protocols are the cheapest and easiest to perform and can be undertaken as follows:

Procedure

1. Using a sterile loop, emulsify a well-isolated colony in 1 mL of nuclease-free water (or saline).
2. Heat the suspension at 95°C for 10 min, and then centrifuge at high speed (10,000–14,000 rpm) for 3 min.
3. The supernatant can be used directly as a template for PCR amplification.

Note: In our experience, heat-denatured samples may produce suboptimal results for applications such as PCR fingerprinting assays. The implementation of any of these extraction techniques should therefore be fully tested and validated for each intended application before routine clinical use. Most other standard manual and automated bacterial

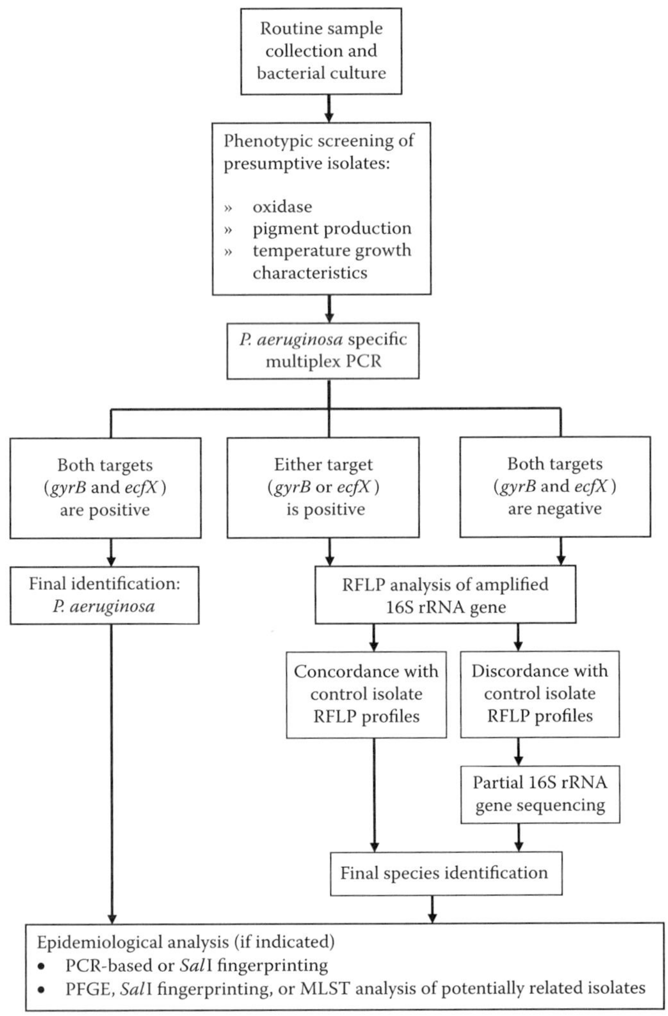

FIGURE 86.1 Proposed molecular identification and strain typing workflow. PCR, polymerase chain reaction; *gyrB*, DNA gyrase subunit B gene sequence; *ecfX*, extracellular-cytoplasmic function sigma factor determinant gene sequence; RFLP, restriction fragment length polymorphism; PFGE, pulsed field gel electrophoresis; MLST, multilocus sequence typing.

DNA extraction protocols are well suited for generating genomic DNA from *Pseudomonas* spp. DNA yield can be further enhanced by lysozyme treatment of suspended bacterial cells before DNA extraction. To ensure assay reproducibility, extraction of a standard bacterial cell concentration that is easily determined and adjusted (if required) according to McFarland standards should be undertaken.

86.2.2 Detection Procedures

86.2.2.1 PCR Identification of *P. aeruginosa* Isolates

Principle. Following presumptive phenotypic screening, suspected *P. aeruginosa* isolates are confirmed by a multiplex PCR assay targeting species-specific *gyrB* and *ecfX* regions.[55]

Procedure

1. Prepare a 25 μL PCR mixture with 12.5 μL 2× QuantiTect Probe PCR Kit (Qiagen), 10 pmol of each primer, 4 pmol FAM labelled *gyrB*-TM probe, 4 pmol Yakima Yellow or JOE labeled *ecfX* probe, 2 μL of DNA, and nuclease-free water (to 25 μL). Ensure appropriate positive, negative, and reagent controls are included in each run.
2. Cycling can be performed on an ABI Prism 7500 Detection System (Applied Biosystems). The first cycle is at 95°C for 15 min, followed by another 40 cycles at 95°C for 15 s and 60°C for 1 min.
3. Isolates producing amplicons for both PCR targets can be reliably identified as *P. aeruginosa*. Those exhibiting negative or discordant PCR results require further investigation with other nucleic acid–based methods, such as 16S rRNA gene PCR–RFLP or sequence analysis.

Note: If further discrimination of nonaeruginosa *Pseudomonas* spp. from *P. aeruginosa* and other nonfermenting gram-negative bacteria is required at this level, an *oprI* or *groE* based *Pseudomonas* spp. PCR assay can be undertaken.[35,62]

86.2.2.2 16S rRNA Gene PCR–RFLP Identification of *Pseudomonas* spp.

Principle. 16S rRNA PCR–RFLP analysis allows the universal examination of all *Pseudomonas* spp. and related genera that may be encountered in clinical samples. After broad-range 16S rRNA gene amplification, each PCR product is digested by at least two suitable restriction endonucleases in separate reactions. Restricted fragments are separated and analyzed using horizontal electrophoresis of ethidium bromide–stained agarose gels. Each isolate's RFLP profile is compared with those produced by related nonfermenting gram-negative control strains.

Procedure

1. Prepare a 25 μL PCR mixture containing 1 U Taq DNA polymerase (Qiagen), 1× PCR Buffer, 250 μM dNTPs, 20 pmol each of primers fD1 and rD1,[68] 5 μL of DNA, and nuclease-free water (to 25 μL).
2. PCR amplification can be performed in a standard thermal cycler using the following conditions: one cycle at 94°C for 3 min; 40 cycles of 94°C for 1 min, 55°C for 1 min, and 72°C for 2 min; and one final extension step at 72°C for 10 min.
3. Prepare two separate restriction enzyme digestion mixtures containing 5 U of *Dde*I and *Msp*I (Roche Diagnostics), 2.5 μL of the appropriate SuRE/Cut Buffer, 5 μL of amplified PCR product, and nuclease-free water (up to 25 μL). Incubate each tube at 37°C for a minimum of 1 h.
4. After restriction digestion, add 2 μL of 6× DNA loading buffer to 10 μL of digested PCR product; separate the fragments in a 3% MS agarose gel (Roche Diagnostics) containing 5 μL of 10 mg/mL ethidium bromide at 100 V for 75 min.
5. Visualize and photograph under UV light. In most cases, isolates that exhibit RFLP profiles identical to those of the control strains can be identified as that particular species. Isolates that exhibit atypical RFLP profiles should be further analyzed by complete or partial 16S rRNA gene sequencing.

86.2.2.3 16S rRNA Gene Sequence Analysis

Principle. DNA sequence analysis of amplified 16S rRNA gene sequences can be performed on isolates that cannot be reliably identified by species-specific PCR or by 16S rRNA PCR–RFLP analysis. PCR amplification and sequencing is performed following in-house or commercial protocols.

Procedure

1. Initial 16S rRNA gene PCR primers, reagents, and conditions are identical to those described in Section 86.2.2.2.
2. Purify each PCR product using the QIAquick PCR Purification Kit (Qiagen) or equivalent.
3. Sequencing PCR (i.e., forward and reverse) should be undertaken using 10 ng of purified PCR product, 3.2 pmol of the primers fD1 and rD1,[68] and the BigDye Terminator v3.1 Cycle Sequencing Kit (Applied Biosystems) as per the manufacturers' instructions.
4. Sequencing PCR amplification can be performed in a standard thermal cycler using the following conditions: one cycle at 96°C for 1 min; 25 cycles of 96°C for 10 s, 50°C for 5 s, and 60°C for 4 min.
5. Postreaction cleanup can be undertaken using the Agencourt CleanSEQ® sequencing purification system (Beckman Coulter) or equivalent according to the manufacturers' instructions. Capillary separation can be achieved using an AB3730xl or equivalent genetic analyzer (Applied Biosystems).
6. Edited sequences should be compared with at least two database sequence libraries, such as GenBank (http://blast.ncbi.nlm.nih.gov/Blast.cgi), EMBL-Bank

(http://www.ebi.ac.uk/embl/), RDP-II (http://rdp.cme.msu.edu/), and DDBJ (http://www.ddbj.nig.ac.jp/), with a minimum of 500 bp providing sufficient sequence data for reliable discrimination of most *Pseudomonas* spp.

86.2.2.4 Direct Molecular Detection of *P. aeruginosa* from Clinical Specimens

Samples for culture-independent molecular detection procedures are those obtained from normally sterile sites, such as cerebrospinal fluid, blood, and joint aspirates. As with cultured isolates, a multiplex real-time PCR-based approach is likely to represent the best strategy for rapidly and reliably detecting *P. aeruginosa* in clinical samples. Such assays require in-house development and validation as no published studies using *gyrB*, *ecfX,* or 16S-23S rRNA ITS targets have reported the direct detection of *P. aeruginosa* DNA in clinical samples.

Sequence analysis of amplified 16S rRNA gene sequences is another approach; it allows detection and identification of bacterial DNA within clinical samples. Its application is more expensive and time consuming, however, and direct detection strategy is limited to samples from normally sterile sites, where only individual bacterial species are expected. Following standard tissue DNA extraction, real-time or conventional broad-range 16S rRNA gene PCR assays help determine whether bacterial DNA is present. Amplicon sequence analysis of positive samples allows the culture-independent identification of the bacterial species present. Before it can be introduced into clinical laboratories, preliminary assay development and validation for each sample type is required as is the routine use of appropriate positive, negative, and sample inhibition controls.

86.2.3 Strain Typing

The epidemiological analysis of clinical *P. aeruginosa* isolates is reliably performed by several different genotyping methods. When rapid and less expensive genotyping methods are required, either PCR-based (e.g., ERIC or RAPD) tests as preliminary screening tools or computer-assisted *Sal*I fingerprinting for assessing strain relatedness can be employed. To ensure that PCR-based assays are performing reliably, a representative selection of isolates displaying a high degree of relatedness are also analyzed by either PFGE or MLST. Investigation of longitudinally collected or geographically distinct isolates for which no epidemiological information is available are best approached using categorical genotyping methodologies, such as MLST or computer-assisted *Sal*I fingerprinting techniques.

86.3 CONCLUSION AND FUTURE PERSPECTIVES

Advances in molecular technology have seen methods such as PCR become invaluable tools in *Pseudomonas* research. In addition, molecular detection techniques are being introduced into microbiology diagnostic laboratories to replace traditional phenotypic methods when these fail to provide reliable results. The success of PCR is largely attributed to its robust, but adaptable platforms, which have been enhanced further by real-time PCR technology. Real-time PCR is cheaper and easier to use than conventional PCR and provides results within a few hours. The multiplex PCR identification procedure offers a suitable alternative to phenotypic identification of *P. aeruginosa,* with the possibility of developing similar assays for other *Pseudomonas* spp.

Further research is required to investigate the emergence of clonal *P. aeruginosa* in patients with CF, which highlights the need for typing methods suitable for clinical laboratory settings. The available typing methods for *Pseudomonas* spp. are many, and the one used will depend largely upon the research or clinical question. Overall, it appears that sequence-based typing technologies, such as MLST, will be the future preferred method for the molecular typing of *Pseudomonas,* since they combine assay objectivity and robustness with the ability to investigate both temporally and geographically diverse isolates. These techniques will be facilitated by advances in whole-genome comparative sequencing technologies. DNA microarray techniques, capable of simultaneous detection of thousands of targets within a single reaction, have already been successfully applied to *Pseudomonas* spp. for both identification and typing purposes.[97–99] However, these techniques are currently very labor intensive and expensive to perform. Newer systems using matrix-assisted laser desorption/ionization time-of-flight mass spectrometry (MALDI-TOF MS) are now available in a form suitable for commercial systems.[100,101] Such advances may see multitarget systems developed to identify, type, and detect resistance and virulence markers for multiple pathogens, as well as their relative abundance within a single clinical sample.

Even though molecular techniques are being used increasingly, we do not currently see the routine isolation of *Pseudomonas* spp. by bacterial-culture techniques being replaced by nonculture molecular methods. From a diagnostic laboratory perspective, the convenience of a rapid molecular result obtained directly from a clinical sample will continue to be overshadowed by the ease at which this group of organisms is cultured from biological specimens and the need for antimicrobial susceptibility testing, which currently cannot be confidently predicted by molecular means. Similarly, the polymicrobial nature of many clinical isolates and extensive sequence homology shared between *Pseudomonas* spp. and *Enterobacteriaceae* will continue to pose challenges for the direct application of many molecular typing strategies to clinical samples. For the foreseeable future, bacterial isolation will remain an integral part of *Pseudomonas* clinical laboratory and research investigation. Molecular testing will help to identify *Pseudomonas* spp., including those present in low numbers associated with polymicrobial infections and following antibiotic administration. Finally, molecular techniques will continue to genotype *P. aeruginosa* strains as part of outbreak investigations and when there is a suspected breakdown of infection control.

ACKNOWLEDGMENTS

The authors received support from Queensland Health, the National Health & Medical Research Council, the Royal Children's Hospital Foundation, and the Australian Cystic Fibrosis Research Trust.

REFERENCES

1. Palleroni, N.J., The road to the taxonomy of Pseudomonas. In Cornelis, P. (Editor), *Pseudomonas: Genomic and Molecular Biology,* p. 1, Caister Academic Press, Norfolk, 2008.
2. Blondel-Hill, E., Henry D.A., and Speert D.P., *Pseudomonas*. In Murray, P.R. et al. (Editors), *Manual of Clinical Microbiology*, 9th ed., p. 734, ASM Press, Washington DC, 2007.
3. Palleroni, N.J., *Pseudomonas* classification. In Garrity, G.M. (Editor), *Bergey's Manual of Systemic Bacteriology. Part B. The Gammaproteobacteria*, 2nd ed, p. 323, Springer, New York, 2005.
4. Pier, G.B., and Ramphal R., *Pseudomonas aeruginosa*. In Mandell, G.L., Bennett, J.E., and Dolin, R. (Editors), *Mandell, Douglas, and Bennett's Principles and Practice of Infectious Diseases*, 6th ed., p. 2587, Elsevier Churchill Livingstone, Philadelphia, 2005.
5. Habe, H., and Omori, T., Genetics of polycyclic aromatic hydrocarbon metabolism in diverse aerobic bacteria, *Biosci. Biotechnol. Biochem.*, 67, 225, 2003.
6. Nelson, K.E. et al., Complete genome sequence and comparative analysis of the metabolically versatile *Pseudomonas putida* KT2440, *Environ. Microbiol.*, 4, 799, 2002.
7. Paulsen, I.T. et al., Complete genome sequence of the plant commensal Pseudomonas fluorescens Pf-5, *Nat. Biotechnol.*, 23, 873, 2005.
8. Stover, C.K. et al., Complete genome sequence of *Pseudomonas aeruginosa* PA01, an opportunistic pathogen, *Nature*, 406, 959, 2000.
9. Lee, D.G. et al., Genomic analysis reveals that *Pseudomonas aeruginosa* virulence is combinatorial, *Genome Biol.*, 7, R90, 2006.
10. Wolfgang, M.C. et al., Conservation of genome content and virulence determinants among clinical and environmental isolates of *Pseudomonas aeruginosa*, *Proc. Natl. Acad. Sci. USA*, 100, 8484, 2003.
11. Mathee, K. et al., Dynamics of *Pseudomonas aeruginosa* genome evolution, *Proc. Natl. Acad. Sci. USA*, 105, 3100, 2008.
12. Kohlenberg, A. et al., Time-trends for Gram-negative and multidrug-resistant Gram-positive bacteria associated with nosocomial infections in German intensive care units between 2000 and 2005, *Clin. Microbiol. Infect.*, 14, 93, 2008.
13. Richards, M.J. et al., Nosocomial infections in combined medical-surgical intensive care units in the United States, *Infect. Control Hosp. Epidemiol.*, 21, 510, 2000.
14. Morrison, A.J. Jr., and Wenzel, R.P., Epidemiology of infections due to *Pseudomonas aeruginosa*, *Rev. Infect. Dis.*, 6 Suppl 3, S627, 1984.
15. Trautmann, M., Lepper, P.M., and Haller, M., Ecology of P*seudomonas aeruginosa* in the intensive care unit and the evolving role of water outlets as a reservoir of the organism, *Am. J. Infect. Control*, 33, S41, 2005.
16. Cystic Fibrosis Foundation, Cystic Fibrosis Foundation Patient Patient Registry 2007 Annual Data Report, Bethesda, MD.
17. Gibson, R.L., Burns, J.L., and Ramsey, B.W., Pathophysiology and management of pulmonary infections in cystic fibrosis, *Am. J. Respir. Crit. Care Med.*, 168, 918, 2003.
18. Speert, D.P. et al., Epidemiology of *Pseudomonas aeruginosa* in cystic fibrosis in British Columbia, Canada, *Am. J. Respir. Crit. Care Med.*, 166, 988, 2002.
19. Armstrong, D.S. et al., Detection of a widespread clone of *Pseudomonas aeruginosa* in a pediatric cystic fibrosis clinic, *Am. J. Respir. Crit. Care Med.*, 166, 983, 2002.
20. O'Carroll, M.R. et al., Clonal strains of *Pseudomonas aeruginosa* in paediatric and adult cystic fibrosis units, *Eur. Respir. J.*, 24, 101, 2004.
21. Romling, U. et al., A major *Pseudomonas aeruginosa* clone common to patients and aquatic habitats, *Appl. Environ. Microbiol.*, 60, 1734, 1994.
22. Scott, F.W., and Pitt, T.L., Identification and characterization of transmissible *Pseudomonas aeruginosa* strains in cystic fibrosis patients in England and Wales, *J. Med. Microbiol.*, 53, 609, 2004.
23. Al-Aloul, M. et al., Increased morbidity associated with chronic infection by an epidemic *Pseudomonas aeruginosa* strain in CF patients, *Thorax*, 59, 334, 2004.
24. Armstrong, D. et al., Evidence for spread of a clonal strain of *Pseudomonas aeruginosa* among cystic fibrosis clinics, *J. Clin. Microbiol.*, 41, 2266, 2003.
25. Jones, A.M. et al., Increased treatment requirements of patients with cystic fibrosis who harbour a highly transmissible strain of *Pseudomonas aeruginosa*, *Thorax*, 57, 924, 2002.
26. Robinson, P. et al., *Pseudomonas* cross-infection from cystic fibrosis patients to non-cystic fibrosis patients: implications for inpatient care of respiratory patients, *J. Clin. Microbiol.*, 41, 5741, 2003.
27. Romling, U. et al., Worldwide distribution of *Pseudomonas aeruginosa* clone C strains in the aquatic environment and cystic fibrosis patients, *Environ. Microbiol.*, 7, 1029, 2005.
28. Bottone, E.J., and Perez, A.A. 2nd., *Pseudomonas aeruginosa* folliculitis acquired through use of a contaminated loofah sponge: An unrecognized potential public health problem, *J. Clin. Microbiol.*, 31, 480, 1993.
29. Gould, I.M., and Wise, R., *Pseudomonas aeruginosa*: Clinical manifestations and management, *Lancet*, 2, 1224, 1985.
30. Lyczak, J.B., Cannon, C.L., and Pier, G.B., Establishment of *Pseudomonas aeruginosa* infection: lessons from a versatile opportunist, *Microbes Infect.*, 2, 1051, 2000.
31. Rosenfeld, M., Ramsey, B.W., and Gibson, R.L., *Pseudomonas* acquisition in young patients with cystic fibrosis: Pathophysiology, diagnosis, and management, *Curr. Opin. Pulm. Med.*, 9, 492, 2003.
32. FitzSimmons, S.C., The changing epidemiology of cystic fibrosis, *J. Pediatr.*, 122, 1, 1993.
33. Kosorok, M.R. et al., Acceleration of lung disease in children with cystic fibrosis after Pseudomonas aeruginosa acquisition, *Pediatr. Pulmonol.*, 32, 277, 2001.
34. Nixon, G.M. et al., Clinical outcome after early *Pseudomonas aeruginosa* infection in cystic fibrosis, *J. Pediatr.*, 138, 699, 2001.
35. Clarke, L. et al., Development of a diagnostic PCR assay that targets a heat-shock protein gene (groES) for detection of *Pseudomonas* spp. in cystic fibrosis patients, *J. Med. Microbiol.*, 52, 759, 2003.
36. Bahrani-Mougeot, F.K. et al., Molecular analysis of oral and respiratory bacterial species associated with ventilator-associated pneumonia, *J. Clin. Microbiol.*, 45, 1588, 2007.

37. Franzco, R.W.E., Charles, P.G.P., and Franzco, P.J.A., Three cases of post-traumatic endophthalmitis caused by unusual bacteria, *Clini. Experiment. Ophthalmol.*, 32, 435, 2004.
38. Gershman, M.D. et al., Multistate outbreak of *Pseudomonas fluorescens* bloodstream infection after exposure to contaminated heparinized saline flush prepared by a compounding pharmacy, *Clin. Infect. Dis.*, 47, 1372, 2008.
39. Murray, A.E. et al., Blood transfusion-associated *Pseudomonas fluorescens* septicaemia: Is this an increasing problem? *J. Hosp. Infect.*, 9, 243, 1987.
40. Carpenter, R.J. et al., *Pseudomonas putida* war wound infection in a US Marine: A case report and review of the literature, *J. Infect.*, 56, 234, 2008.
41. Lalucat, J. et al., Biology of *Pseudomonas stutzeri*, *Microbiol. Mol. Biol. Rev.*, 70, 510, 2006.
42. Uy, H.S. et al., Chronic, postoperative *Pseudomonas luteola* endophthalmitis, *Ocul. Immunol. Inflamm.*, 15, 359, 2007.
43. Yu, E.N., and Foster, C.S., Chronic postoperative endophthalmitis due to pseudomonas oryzihabitans, *Am. J. Ophthalmol.*, 134, 613, 2002.
44. Chi, C.Y. et al., *Pseudomonas mendocina* spondylodiscitis: A case report and literature review, *Scand. J. Infect. Dis.*, 37, 950, 2005.
45. Dabboussi, F. et al., *Pseudomonas mosselii* sp. nov., a novel species isolated from clinical specimens, *Int. J. Syst. Evol. Microbiol.*, 52, 363, 2002.
46. Elomari, M. et al., *Pseudomonas monteilii* sp. nov., isolated from clinical specimens, *Int. J. Syst. Bacteriol.*, 47, 846, 1997.
47. Murray, T.S., Egan, M., and Kazmierczak, B.I., *Pseudomonas aeruginosa* chronic colonization in cystic fibrosis patients, *Curr. Opin. Pediatr.*, 19, 83, 2007.
48. Mena, A. et al., Genetic adaptation of *Pseudomonas aeruginosa* to the airways of cystic fibrosis patients is catalyzed by hypermutation, *J. Bacteriol.*, 190, 7910, 2008.
49. Oliver, A. et al., High frequency of hypermutable *Pseudomonas aeruginosa* in cystic fibrosis lung infection, *Science*, 288, 1251, 2000.
50. Martin, D.W. et al., Mechanism of conversion to mucoidy in *Pseudomonas aeruginosa* infecting cystic fibrosis patients, *Proc. Natl. Acad. Sci. USA*, 90, 8377, 1993.
51. Smith, E.E. et al., Genetic adaptation by *Pseudomonas aeruginosa* to the airways of cystic fibrosis patients, *Proc. Natl. Acad. Sci. USA*, 103, 8487, 2006.
52. Wagner, V.E., and Iglewski, B.H., *P. aeruginosa* biofilms in CF infection, *Clin. Rev. Allergy Immunol.*, 35, 124, 2008.
53. Hoboth, C. et al., Dynamics of adaptive microevolution of hypermutable *Pseudomonas aeruginosa* during chronic pulmonary infection in patients with cystic fibrosis, *J. Infect. Dis.*, 200, 118, 2009.
54. Gilligan, P.H., Kiska, D.L., and Appleman, M.D., Cumitech 43, Cystic fibrosis microbiology. In Appleman, M.D. (Editor), *ASM Cumitech*, p. 1, ASM Press, Washington DC, 2006.
55. Anuj, S.N. et al., Identification of Pseudomonas aeruginosa by a duplex real-time polymerase chain reaction assay targeting the ecfX and the gyrB genes, *Diagn. Microbiol. Infect. Dis.*, 63, 127, 2009.
56. Bosshard, P.P. et al., 16S rRNA gene sequencing versus the API 20 NE system and the VITEK 2 ID-GNB card for identification of nonfermenting Gram-negative bacteria in the clinical laboratory, *J. Clin. Microbiol.*, 44, 1359, 2006.
57. Kidd, T.J. et al., Low rates of Pseudomonas aeruginosa misidentification in isolates from cystic fibrosis patients, *J. Clin. Microbiol.*, 47, 1503, 2009.
58. Lavenir, R. et al., Improved reliability of *Pseudomonas aeruginosa* PCR detection by the use of the species-specific ecfX gene target, *J. Microbiol. Methods*, 70, 20, 2007.
59. Qin, X. et al., Use of real-time PCR with multiple targets to identify *Pseudomonas aeruginosa* and other nonfermenting gram-negative bacilli from patients with cystic fibrosis, *J. Clin. Microbiol.*, 41, 4312, 2003.
60. Tyler, S.D. et al., Oligonucleotide primers designed to differentiate pathogenic pseudomonads on the basis of the sequencing of genes coding for 16S-23S rRNA internal transcribed spacers, *Clin. Diagn. Lab. Immunol.*, 2, 448, 1995.
61. Whiley, D.M. et al., False-negative results in nucleic acid amplification tests-do we need to routinely use two genetic targets in all assays to overcome problems caused by sequence variation? *Crit. Rev. Microbiol.*, 34, 71, 2008.
62. De Vos, D. et al., Detection of the outer membrane lipoprotein I and its gene in fluorescent and non-fluorescent pseudomonads: implications for taxonomy and diagnosis, *J. Gen. Microbiol.*, 139, 2215, 1993.
63. Bennasar, A., Guasp, C., and Lalucat, J., Molecular methods for the detection and identification of *Pseudomonas stutzeri* in pure culture and environmental samples, *Microb. Ecol.*, 35, 22, 1998.
64. Karpati, F., and Jonasson, J., Polymerase chain reaction for the detection of *Pseudomonas aeruginosa, Stenotrophomonas maltophilia* and *Burkholderia cepacia* in sputum of patients with cystic fibrosis, *Mol. Cell. Probes*, 10, 397, 1996.
65. Xu, J. et al., Early detection of Pseudomonas aeruginosa: Comparison of conventional versus molecular (PCR) detection directly from adult patients with cystic fibrosis (CF), *Ann. Clin. Microbiol. Antimicrob.*, 3, 21, 2004.
66. Laguerre, G., Rigottier-Gois, L., and Lemanceau, P., Fluorescent *Pseudomonas* species categorized by using polymerase chain reaction (PCR)/restriction fragment analysis of 16S rDNA, *Mol. Ecol.*, 3, 479, 1994.
67. Porteous, L.A., Widmer, F., and Seidler, R.J., Multiple enzyme restriction fragment length polymorphism analysis for high resolution distinction of *Pseudomonas* (sensu stricto) 16S rRNA genes, *J. Microbiol. Methods*, 51, 337, 2002.
68. van Pelt, C. et al., Identification of Burkholderia spp. in the clinical microbiology laboratory: Comparison of conventional and molecular methods, *J. Clin. Microbiol.*, 37, 2158, 1999.
69. Clarridge, J.E. 3rd., Impact of 16S rRNA gene sequence analysis for identification of bacteria on clinical microbiology and infectious diseases, *Clin. Microbiol. Rev.*, 17, 840, 2004.
70. Xu, J. et al., Employment of broad range 16S rDNA PCR and sequencing in the detection of aetiological agents of meningitis, *New Microbiol.*, 28, 135, 2005.
71. Liu, W.T. et al., Characterization of microbial diversity by determining terminal restriction fragment length polymorphisms of genes encoding 16S rRNA, *Appl. Environ. Microbiol.*, 63, 4516, 1997.
72. Rogers, G.B. et al., characterization of bacterial community diversity in cystic fibrosis lung infections by use of 16s ribosomal DNA terminal restriction fragment length polymorphism profiling, *J. Clin. Microbiol.*, 42, 5176, 2004.
73. Spasenovski, T. et al., Molecular analysis of diversity within the genus Pseudomonas in the lungs of cystic fibrosis patients, *Diagn. Microbiol. Infect. Dis.*, 63, 261, 2009.
74. Hogardt, M. et al., Specific and rapid detection by fluorescent in situ hybridization of bacteria in clinical samples obtained from cystic fibrosis patients, *J. Clin. Microbiol.*, 38, 818, 2000.

75. Peleg, A.Y. et al., Utility of peptide nucleic acid fluorescence in situ hybridization for rapid detection of *Acinetobacter* spp. and *Pseudomonas aeruginosa*, *J. Clin. Microbiol.*, 47, 830, 2009.
76. Sogaard, M., Stender, H., and Schonheyder, H.C., Direct identification of major blood culture pathogens, including *Pseudomonas aeruginosa* and *Escherichia coli*, by a panel of fluorescence in situ hybridization assays using peptide nucleic acid probes, *J. Clin. Microbiol.*, 43, 1947, 2005.
77. Wellinghausen, N. et al., Superiority of molecular techniques for identification of gram-negative, oxidase-positive rods, including morphologically nontypical *Pseudomonas aeruginosa*, from patients with cystic fibrosis, *J. Clin. Microbiol.*, 43, 4070, 2005.
78. Gunasekera, T.S. et al., Specific detection of Pseudomonas spp. in milk by fluorescence in situ hybridization using ribosomal RNA directed probes, *J. Appl. Microbiol.*, 94, 936, 2003.
79. Bjarnsholt, T. et al., *Pseudomonas aeruginosa* biofilms in the respiratory tract of cystic fibrosis patients, *Pediatr. Pulmonol.*, 44, 547, 2009.
80. Kirketerp-Moller, K. et al., Distribution, organization, and ecology of bacteria in chronic wounds, *J. Clin. Microbiol.*, 46, 2717, 2008.
81. Tenover, F.C. et al., Interpreting chromosomal DNA restriction patterns produced by pulsed-field gel electrophoresis: Criteria for bacterial strain typing, *J. Clin. Microbiol.*, 33, 2233, 1995.
82. Urwin, R., and Maiden, M.C., Multilocus sequence typing: A tool for global epidemiology, *Trends Microbiol.*, 11, 479, 2003.
83. Al-Samarrai, T.H. et al., Simple and inexpensive but highly discriminating method for computer-assisted DNA fingerprinting of *Pseudomonas aeruginosa*, *J. Clin. Microbiol.*, 38, 4445, 2000.
84. Schmid, J. et al., *Pseudomonas aeruginosa* transmission is infrequent in New Zealand cystic fibrosis clinics, *Eur. Respir. J.*, 32, 1583, 2008.
85. Mahenthiralingam, E. et al., Random amplified polymorphic DNA typing of *Pseudomonas aeruginosa* isolates recovered from patients with cystic fibrosis, *J. Clin. Microbiol.*, 34, 1129, 1996.
86. Syrmis, M.W. et al., Rapid genotyping of *Pseudomonas aeruginosa* isolates harboured by adult and paediatric patients with cystic fibrosis using repetitive-element-based PCR assays, *J. Med. Microbiol.*, 53, 1089, 2004.
87. Van Daele, S. et al., Survey of *Pseudomonas aeruginosa* genotypes in colonised cystic fibrosis patients, *Eur. Respir. J.*, 28, 740, 2006.
88. Maiden, M.C. et al., Multilocus sequence typing: A portable approach to the identification of clones within populations of pathogenic microorganisms, *Proc. Natl. Acad. Sci. USA*, 95, 3140, 1998.
89. Blanc, D.S., The use of molecular typing for epidemiological surveillance and investigation of endemic nosocomial infections, *Infect. Genet. Evol.*, 4, 193, 2004.
90. Pitt, T.L., Epidemiological typing of *Pseudomonas aeruginosa*, *Eur. J. Clin. Microbiol. Infect. Dis.*, 7, 238, 1988.
91. Curran, B. et al., Development of a multilocus sequence typing scheme for the opportunistic pathogen *Pseudomonas aeruginosa*, *J. Clin. Microbiol.*, 42, 5644, 2004.
92. Feil, E.J., and Enright, M.C., Analyses of clonality and the evolution of bacterial pathogens, *Curr. Opin. Microbiol.*, 7, 308, 2004.
93. Feil, E.J. et al., eBURST: Inferring patterns of evolutionary descent among clusters of related bacterial genotypes from multilocus sequence typing data, *J. Bacteriol.*, 186, 1518, 2004.
94. van Belkum, A., Tracing isolates of bacterial species by multilocus variable number of tandem repeat analysis (MLVA), *FEMS Immunol. Med. Microbiol.*, 49, 22, 2007.
95. Onteniente, L. et al., Evaluation of the polymorphisms associated with tandem repeats for *Pseudomonas aeruginosa* strain typing, *J. Clin. Microbiol.*, 41, 4991, 2003.
96. Vu-Thien, H. et al., Multiple-locus variable-number tandem-repeat analysis for longitudinal survey of sources of *Pseudomonas aeruginosa* infection in cystic fibrosis patients, *J. Clin. Microbiol.*, 45, 3175, 2007.
97. Wu, L. et al., Microarray-based whole-genome hybridization as a tool for determining procaryotic species relatedness, *Isme J.*, 2, 642, 2008.
98. Zhu, L.X. et al., Development of a base stacking hybridization-based microarray method for rapid identification of clinical isolates, *Diagn. Microbiol. Infect. Dis.*, 59, 149, 2007.
99. Wiehlmann, L. et al., Population structure of Pseudomonas aeruginosa, *Proc. Natl. Acad. Sci. USA*, 104, 8101, 2007.
100. Baker, S. et al., High-throughput genotyping of *Salmonella enterica* serovar Typhi allowing geographical assignment of haplotypes and pathotypes within an urban District of Jakarta, Indonesia, *J. Clin. Microbiol.*, 46, 1741, 2008.
101. Honisch, C. et al., Automated comparative sequence analysis by base-specific cleavage and mass spectrometry for nucleic acid-based microbial typing, *Proc. Natl. Acad. Sci. USA*, 104, 10649, 2007.
102. Pitt, T.L., and Simpson, J.H., *Pseudomonas* and *Burkholderia* spp. In Gillespie, S.H., and Hawkey, P.M. (Editors), *Principles and Practice of Clinical Bacteriology*, 2nd ed., p. 427, John Wiley & Sons, Chichester, 2005.
103. Elomari, M. et al., DNA relatedness among *Pseudomonas* strains isolated from natural mineral waters and proposal of Pseudomonas veronii sp. Nov., *Int. J. Syst. Bacteriol.*, 46, 1138, 1996.

87 *Salmonella*

Mathilde H. Josefsen, Charlotta Löfström, Katharina E.P. Olsen, Kåre Mølbak, and Jeffrey Hoorfar

CONTENTS

87.1 INTRODUCTION

87.1.1 Biology

In 1885, the bacterium causing "hog cholera" was isolated from infected pigs by two American veterinarians, Salmon and Smith. The bacterium was named *Salmonella choleraesuis*, and thereby the *Salmonella* genus was founded.[1,2] *Salmonella* are gram-negative facultative anaerobic rods belonging to the Enterobacteriaceae family. The size of the rods ranges from 0.7 to 1.5 μm by 2.2–5.0 μm, producing colonies of approximately 2–4 mm in diameter. They are nonspore-forming, oxidase-negative and catalase-positive. Members of the genus are predominantly motile by petrichous flagella, chemoorganotrophic, able to reduce nitrate to nitrite, able to grow on citrate as sole carbon source, capable of producing acid and gas from glucose, able to produce hydrogen sulfide on triple sugar iron and decarboxylate lysine and ornithine, and incapable of hydrolyzing indole and urea. However, deviations from these characteristics are occasionally observed.[1,3]

Salmonella are known to be quite resilient toward environmental influences, and prolonged persistence in hostile surroundings is reported.[4–6] This enhanced survival supports the transmission of *Salmonella* from the environment to a new host.[7] The optimal growth temperature for *Salmonella* is 37°C, but the temperature range of growth is quite broad (7°C–45°C), and the preconditioning of cells can result in growth at extreme temperatures at both ends of the scale. Growth of *Salmonella* in various foods has been observed down to 2°C–4°C. The normal boundaries for growth regarding water activity and pH can also be forced in either direction as a function of preconditioning.[8–10] The optimum pH for growth is 6.5–7.5, but salmonellae can proliferate in the pH 4.5–9.5 range. Growth of *Salmonella* has not been reported in foods with an a_w of less than 0.93.[9] The combined effects of environmental actions on *Salmonella* growth in different surroundings can be predicted by modeling. Modeling of *Salmonella* can be performed in a database, such as ComBase, containing data sets describing the growth, survival, and inactivation of bacteria under diverse environments related to food-processing operations.[11]

87.1.2 Classification

The taxonomic classification and nomenclature of the *Salmonella* genus have undergone a number of rearrangements over the years, as a natural consequence of the improvement in methods for determining the true relatedness of this group of bacteria. Currently, the *Salmonella* genus consists of the species *S. enterica* and *S. bongori*, two lineages suspected to have diverged early in evolution. *S. enterica* is further divided into the following six subspecies: *S. enterica* subsp. *enterica*; *S. enterica* subsp. *salamae*; *S. enterica* subsp. *arizonae*; *S. enterica* subsp. *diarizonae*; *S. enterica* subsp. *houtenae*; and *S. enterica* subsp. *indica*. Subspecies are separated into serovars according to the antigenic White-Kauffmann-Le Minor scheme,[12] based on somatic (O), flagellar (H), and capsular (Vi) antigens. A notation after the serovar contains the antigenic description in which the major antigenic groups are separated by colons (O antigens:Vi antigens:phase 1 antigens:phase 2 antigens). *Salmonella* isolates are currently denoted by genus, species, subspecies, and serovar, for example, *Salmonella enterica* subsp. *enterica* serovar Typhimurium. Today, a total of 2579 serovars are recognized, the vast majority (>99.5%) of which belong to the *S. enterica* species.[12] Of these, more than half belong to *S. enterica* subsp. *enterica* and are mainly associated with warm-blooded animals; the remaining subspecies are predominantly found in cold-blooded animals and in the environment.[13] Some serovars are known to exhibit host specificity, whereas others are ubiquitous, not having adapted to any specific niche. As proposed by Uzzau et al.,[13] the host-specific serovars are divided into those that are *host restricted*, associated with one particular host, for example, *S. enterica* subsp. *enterica* serovar Typhi in humans and *S. enterica* subsp. *enterica* serovar Gallinarum in fowl, and *host adapted*, prevalent in one particular host but also able to cause disease in other hosts, for example, *S. enterica* subsp. *enterica* serovar Dublin, and *S. enterica* subsp. *enterica* serovar Choleraesuis.[13]

87.1.3 Epidemiology

Salmonella continues to be an important zoonotic pathogen, globally affecting the health of both humans and animals and thus having wide socioeconomic impact. Human *Salmonella* infection manifests itself as gastroenteritis, enteric (typhoid) fever, or in rare cases, systemic infections. The invasive systemic disease typhoid fever is caused by the host-restricted serovars *S. enterica* subsp. *enterica* serovar Typhi and serovar Paratyphi A, B, and C. The remaining serovars usually cause self-limiting gastroenteritis; however, in rare cases, often associated with immunocompromized hosts at extremes of age, nontyphoidal *Salmonella* can cause invasive systemic infections.[3,14,15]

Globally, *Salmonella* has been estimated to be responsible for 112 million human infections annually, and numbers from the World Health Organization (WHO) point to the fact that typhoid fever accounts for 21.7 million of these cases, resulting in more than 200,000 deaths each year.[16] Currently, in the United States 40,000 human *Salmonella* infections are reported annually; however, the true number is estimated to be 1.4 million cases a year, resulting in 600 deaths,[17] with an estimated cost of $3 billion.[18] In the European Union, 151,995 confirmed cases of human *Salmonella* infection were reported to the European Food Safety Agency (EFSA) in 2007.[19] From 2000 to 2002, a survey of the global distribution of *Salmonella* serovars proved *S. enterica* subsp. *enterica* serovar Enteritidis to be by far the most common serovar isolated from humans, accounting for 65% of all isolates. This tendency is even more pronounced in Europe, where *S. enterica* subsp. *enterica* serovar Enteritidis was associated with 85% of all human cases.[20] The second most frequently isolated serovar from human infections is *S. enterica* subsp. *enterica* serovar Typhimurium (12%), followed by serovar Newport (4%).[20] The dominance of these serovars could be attributable to their invasive behavior in important food animals.[21] Since the majority of human isolates have similar serovars, efficient source tracing in foodborne *Salmonella* outbreaks has to rely on further characterization of isolates. Additional discriminatory typing methods like phage typing, antimicrobial susceptibility testing and pulsed-field gel electrophoresis are applied for this purpose.

The principal habitat of *Salmonella* is the intestinal tract of warm-blooded animals. The natural excretion into the surrounding environment, where the bacteria are able to persist, explains the ubiquity of *Salmonella*. The intensive production of food animals in the agricultural industry, and the transmission of *Salmonella* to this production through contaminated feed and animal supply flocks, wildlife reservoirs, and the environment sustain this existence.[3]

Commonly, *Salmonella* will colonize the intestinal tract of food animals without causing symptoms of illness, and during slaughter contamination of meat and meat products by intestinal contents can occur. The establishment of *Salmonella* combined with the agricultural industry's intensive therapeutic and growth-promoting use of antimicrobial compounds has resulted in problems with antibiotic resistance. The antibiotics used in the primary production of food animals selects for strains that confer resistance toward them, and during subsequent human infection with these strains, treatment options will exclude these antibiotics. Treatment failure is currently a problem in both developed and developing countries.[14]

The majority of human *Salmonella* infections are contracted through consumption of food of animal origin. Meat, poultry, eggs, and milk are reported as vehicles of transmission.[18] However, a wide variety of foods are known to be associated with *Salmonella* infections, including fresh and dry fruits and vegetables, where contamination can occur during fertilization, irrigation, harvesting, or handling under nonoptimal hygienic conditions.[2] The transmission of *Salmonella* is cyclic between humans, animals, food, and environmental sources, and the basic routes are depicted in Figure 87.1.

Human *Salmonella* infections are mainly sporadic; however foodborne outbreaks do occur regularly.[18] Foods commonly associated with outbreaks of *Salmonella* are the following: milk-based products (chocolate, cheese, salads, ice cream, etc.), sliced cold meat, vegetables, and fruits.[3] Pasteurized milk contaminated with *S. enterica* subsp. *enterica* serovar Typhimurium was responsible for one of the largest outbreaks, with 16,284 confirmed cases in the United States in 1985[22] and, recently, peanut butter and peanut butter–containing products contaminated with *S. enterica* subsp. *enterica* serovar Typhimurium have been responsible for 741 human infection in United States through April 2009.[23] In Denmark, an ongoing outbreak with an unidentified source has so far caused more than 1300 human infections with *S. enterica* subsp. *enterica* serovar Typhimurium.[24]

The transmission of enteric fevers occurs via the fecal-oral route, often under poor hygienic conditions, with water or food as vehicles. The disease is seen infrequently in the developed part of the world but constitutes a major problem in many underdeveloped countries.[14,25]

87.1.4 Clinical Features

The main reservoir of *S. enterica* subsp. *enterica* serovar Typhi, and serovar Paratyphi are humans, and clinical features of infection are primarily enteric fever (typhoid and paratyphoid fever). Human infections with nontyphoid *Salmonella* serovars present commonly as acute gastrointestinal illness and, less frequently, septicemia.

87.1.4.1 Clinical Features of Nontyphoid *Salmonella* Infections

Exposure of humans to nontyphoid *Salmonella* from contaminated foods or from environmental sources occurs frequently and may lead to clinical or subclinical infections. The serovar, number of organisms ingested, vehicle of infection, microbial environment of the gut, and several other host-associated factors are important in determining the outcome of exposure. In both nontyphoidal *Salmonella* gastroenteritis and *S. enterica* subsp. *enterica* serovar Typhi infection, low bacterial inocula may cause disease, particularly in association with deficiencies in host defenses or a food vehicle that protects the bacteria from the gastric acid.[26]

Nontyphoid *Salmonella* infection presents often as an acute gastrointestinal illness, indistinguishable from that due to many other gastrointestinal pathogens. Within 6–48 h after ingestion of contaminated food, diarrhea occurs. Most patients develop symptoms between 24 and 48 h after exposure, but with ingestion of a high dose the incubation period

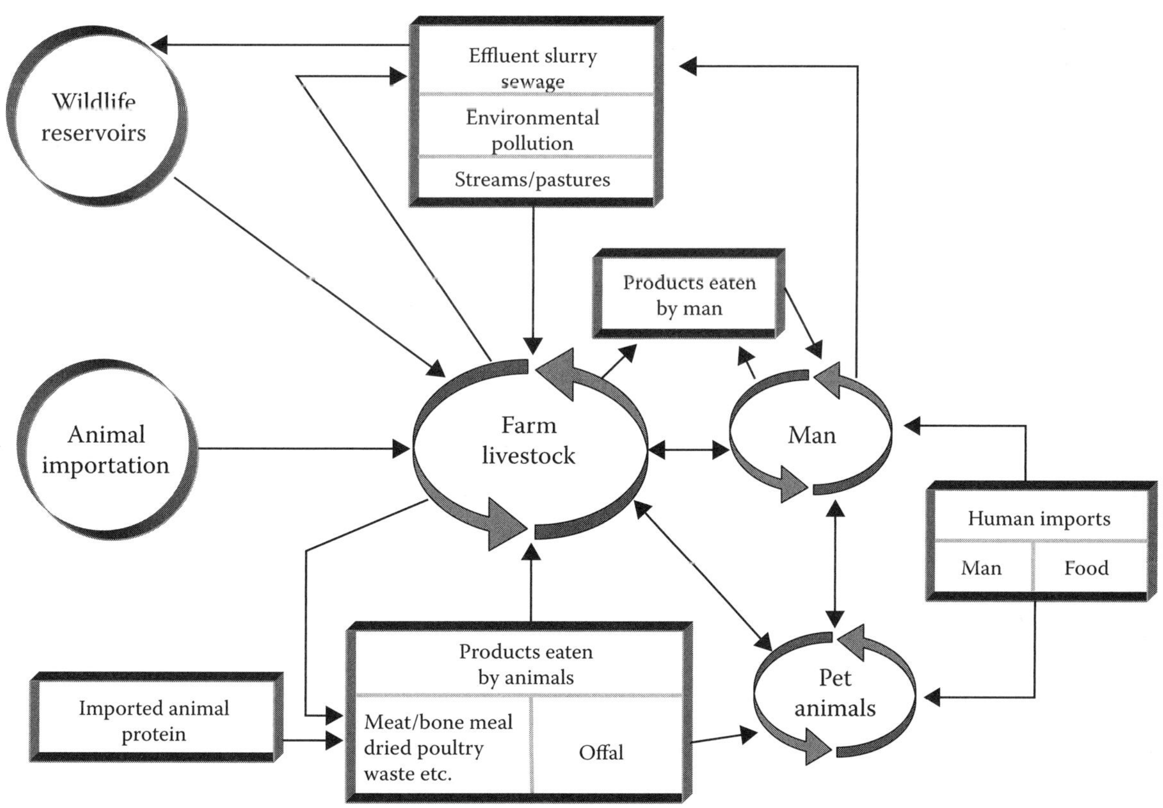

FIGURE 87.1 Basic transmission routes of *Salmonella*. (Adapted with permission from Linton, A.H., *Guidelines on prevention and control of Salmonellosis*, World Health Organisation (WHO), Geneva, Switzerland, 1983.)

may be as short as a few hours. In other situations, patients may initially be subclinically colonized, and symptoms develop more than a week after exposure. In addition to diarrhea, other common symptoms include abdominal pain or cramps, fever, chills, nausea, vomiting, pain in joints, headache, myalgia, general malaise, and weight loss. Diarrhea varies in volume and frequency and may contain blood. In uncomplicated cases, fever usually resolves within 48–72 h, and the average time to recovery is 1–2 weeks.

Between 3% and 7% of immunocompetent persons infected with *S. enterica* subsp. *enterica* serovar Typhimurium or serovar Enteritidis have positive blood cultures.[27,28] The risk of bloodstream infection is higher for less common serovars such as Dublin, Choleraesuis, Oranienburg, and Virchow.[29] The proportion of blood isolates is higher in patients of extreme ages, particularly the elderly.

Immunosuppresion and chronic underlying illness, including inflammatory bowel disease, organ transplantation, malignancy, and malnutrition are risk factors for severe gastroenteritis, as well as for bloodstream infection. Subsets of patients develop a primary septicemia without prominent gastrointestinal symptoms.

Without spread to the bloodstream, most infections are self-limiting. However, in cases of underlying illnesses or severe symptoms, the clinician will often prescribe antibiotics. Antimicrobial drug resistance against clinically important drugs may lead to reduced efficacy of early treatment and limitation of therapeutic choice. Resistance also leads to increased transmission and mortality.[30–32] Persistence of bowel symptoms is common and is responsible for considerable morbidity and health care costs.[33] More severe complications are less common. Among 3328 Danish patients, extraintestinal disease was present in 4% and complications in 7%, including 27 unnecessary appendectomies.[27] Complications include endocarditis or arterial infections, intestinal perforation, and abdominal, soft-tissue, and urogenital infections.[34] Between 2% and 15% of *Salmonella* gastroenteritis episodes are followed by reactive arthritis, Reiter's syndrome, or erythema nodosa (all correlated with HLA-B27 antigen). Arthritis occurs at an average of 10 days after the onset of diarrhea. Between 0.2% and 0.6% of patients with nontyphoid *Salmonella* infections develop carriage, but this carriage will not be persistent, as it is for *S. enterica* subsp. *enterica* serovar Typhi.

Acute-phase mortality has been estimated to be 1.3% in the United States[35] and 1.2% in Denmark.[27] In a registry-based study, one-year mortality risk was 3.1%, 2.9 times higher than a matched sample of the Danish population, suggesting that the mortality after *Salmonella* infection may be underestimated.[30]

87.1.4.2 Enteric Fever Caused by *S. enterica* subsp. *enterica* Serovar Typhi and Serovar Paratyphi

Enteric fever, which continues to be common in developing countries, is a severe systemic illness characterized by fever and abdominal symptoms. In developed countries, enteric fever is often associated with foreign travel or immigration, in particular, Southeast Asia, the Indian subcontinent, South America, and Africa.

Typhoid fever has an insidious onset, and the clinical picture may be mild or atypical. The incubation period ranges from 3 days to one month with a usual range of 8–14 days. After an incubation period of 1–10 days, *S. enterica* subsp. *enterica* serovar Paratyphi may cause gastroenteritis resembling nontyphoid *Salmonella*.

Some typhoid fever patients have a prodromal phase with mild diarrhea and abdominal pains. Symptoms in the first week include headache, malaise, rising fever, constipation, or nonproductive cough. In the second week of illness, fever is a classic sign but is not always present. Some patients may look dazed and apathetic and have sustained fever. Other symptoms may include abdominal pain, diarrhea, chills, anorexia, a slightly distended abdomen, and splenomegaly. In Caucasian patients, up to 50% have rose spots of 2–4 mm in diameter on the trunk. These spots are caused by bacterial embolization and may also be found in invasive nontyphoid *Salmonella* infections. Bradycardia is common in adults during the first 2 weeks of disease. In the third week, the patients become more dazed and ill. Without treatment, the high fever persists and a delirious confused state sets in (typhoid state). Abdominal distension is pronounced, and abdominal sounds are scanty. Diarrhea is common, with liquid, foul, green-yellow stools (pea-soup diarrhea). Pulmonary symptoms are also often present. Death may occur from septicemia, myocarditis, and intestinal hemorrhage or perforation. Patients who survive into the *fourth* week slowly improve, but intestinal complications may still occur. Paratyphoid fever is milder and of shorter duration than typhoid fever.

In the preantibiotic era the case-fatality rate of typhoid fever was high (10%–20%), and mortality continues to be high in some areas of Africa and Asia. The fatality may be reduced to less than 1% with prompt and relevant antibiotic therapy.

Between 10% and 20% of patients treated with antimicrobial drugs suffer a relapse after initial recovery. The relapse is usually milder and shorter than the initial illness. Common complications of enteric fever are intestinal hemorrhage and perforation, and the recognition of these complications may be difficult. The liver, gall bladder, and pancreas may also be affected. Other complications include focal infections, an often fatal toxic myocarditis, and degeneration of skeletal muscle fibers. A number of neurological or psychological symptoms may develop during convalescence. Subclinical disseminated intravascular coagulation occurs commonly in typhoid fever.

Around 10% of untreated typhoid patients discharge bacteria for three months after onset of symptoms, and between 1% and 4% of patients with *S. enterica* subsp. *enterica* serovar Typhi infection develop chronic carriage. Considerably fewer patients with *S. enterica* subsp. *enterica* serovar Paratyphi become chronic carriers. The frequency of chronic carriage is higher in women and in persons with biliary abnormalities. *S. enterica* subsp. *enterica* serovar Typhi carriers working as food handlers are important sources of infection. Gall bladder stones are thought to be the locus of the carrier state. Fluoroquinolones are excreted in the bile and are excellent for the management of chronic carriers, even in the presence of biliary pathology. A recent decline in carrier rates may be associated with the shift in therapeutic choice from chloramphenicol or co-trimoxazole to fluoroquinolones. Carriers of *S. enterica* subsp. *enterica* serovar Typhi have been reported to have an increased risk of hepatobiliary cancer. Gallstones increase the risk of gallbladder cancer and are almost always present in typhoid carriers, so the causal route of this condition is debatable.[36,37]

87.1.5 Pathogenesis

The pathogenesis of *Salmonella enterica* infections is complex, and there are still many unknown or disputed angles in our understanding of the role and significance of different virulence factors. However, it is well established that important differences between the pathogenesis of gastrointestinal disease caused by nontyphoid *Salmonella* serovars and enteric fever caused by *S. enterica* subsp. *enterica* serovar Typhi and serovar Paratyphi exist and still evolve.

To understand the diverse clinical manifestations and pathology of *Salmonella* infections, several virulence factors have to be considered, for example, adhesins, type III secretion systems (T3SS) translocating effector proteins into eukaryotic host cells, enterotoxins, and iron uptake systems, and so forth. The various *Salmonella* virulence factors are required for the colonization, intracellular replication, and neutralization of host defense mechanisms. These virulence factors are encoded either by pathogenicity islands (PAIs), plasmids, or prophages, enabling horizontal gene transfer in the evolution of virulence.

87.1.5.1 Nontyphoid *Salmonella*

Five major PAIs have been encountered in *S. enterica* subsp. *enterica* serovar Typhimurium (*Salmonella* pathogenicity islands, SPI). The two distinct T3SSs, encoded by pathogenicity islands 1 (SPI-1 T3SS) and 2 (SPI-2 T3SS), cover more than 30 effector proteins involved in signal transduction, membrane trafficking, and proinflammatory immune responses and are considered to be of major importance in virulence.[38] The SPI-1 T3SS is a prerequisite for invasion of nonphagocytic cells mediating the translocation of effector proteins into eukaryotic host cells, whereas the SPI-2 is necessary for intracellular replication.[39,40]

In contrast to luminal pathogens causing secretory diarrhea (e.g., *Vibrio cholerae*), highly invasive pathogens such as nontyphoid *Salmonella* cause inflammatory diarrhea ascribed to the presence of the type III secretion system required for invasion of epithelial cells.[38,41] The pathological changes in tissue indicating inflammatory diarrhea are increased vascular permeability; neutrophil influx in the intestinal mucosa, preventing bacterial dissemination; and the predominance of neutrophils in stool samples.[42] Bacterial invasion into the lamina propria is believed to be followed by detection of pathogen-associated molecular patterns (PAMPs), through pattern-recognition receptors such as Toll-like receptors (TLRs) expressed by epithelial and mononuclear (e.g., macrophages) cells. The pattern recognition leads to activation of the innate immune system by the production of CXC chemokines, such as interleukin 8 (IL-8) attracting neutrophils.[43,44] *S. enterica* subsp. *enterica* serovar Typhimurium strains translocate into the lamina propria within 1–4 h after infecting the intestinal mucosa, followed by intracellular location within macrophages/dendritic cells and neutrophils.[45] These phagocytic cells produce toxic oxygen compounds (e.g., hydrogen peroxide, superoxide anions, hydroxyl radicals, etc.) to eliminate ingested bacterial cells. *S. enterica* subsp. *enterica* serovar Typhimurium strains harbor a lambdoid prophage encoded virulence factor (Gifsy-2): superoxide dismutase, which enables it to scavenge the superoxide anions and persist in the infected host.[46] The *spv* operon, encoded by a virulence plasmid (90 kb), is likewise involved in intramacrophage survival.[47]

Fever is associated with both inflammatory diarrhea and enteric fever mediated by the innate immune system via pyrogenic cytokines (IL-1β, IL-6, tumor necrosis factor alpha (TNF-α), and interferon) that enter the circulation and subsequently the brain (i.e., hypothalamus) followed by an increase in the body temperature.[48] However, serum levels of pyrogenic cytokines (TNF-α and IL-1β) are lower in patients with enteric fever compared with patients with septic shock due to gram-negative bacteria.[49,50]

87.1.5.2 Typhoid *Salmonella*

The typhoid *Salmonella*, in contrast to the nontyphoid *Salmonella*, does not cause a classical antibacterial response characterized by local neutrophil recruitment, tissue exudates, neutrophilia (increased amount of neutrophils in the blood), and during sepsis, septic shock. Enteric fever is characterized by interstitial inflammation recruiting mononuclear cells as macrophages and dendritic cells into inflammatory infiltrates (Table 87.1).[51] Only a third of patients with enteric fever develop diarrhea, and in these patients the fecal leukocyte content is dominated by mononuclear cells.[52]

How does typhoid *Salmonella* avoid the innate immune recognition in the intestine? Purified lipopolysaccharide (LPS) and flagellin of nontyphoid and typhoid *Salmonella* stimulate to the same degree TLR4 and TLR5, respectively, and therefore cannot explain the differences in host response for the two groups of *Salmonella*.[53,54] Salmonella pathogenicity island 7 (SPI7) is present in *S. enterica* subsp. *enterica* serovar Typhi but absent in nontyphoid *Salmonella*.[55] The SPI7 harbors the *viaB* locus, which is involved in preventing recognition by TLR4 and TLR5. The *viaB* locus contains genes that are involved in the regulation, biosynthesis and export of the Vi-capsule.[56] Expression of the Vi-capsule reduces the TLR4-mediated production of TNF-α in vivo, probably by masking the LPS (TLR4 agonist) on the surface of the bacteria.[57] TviA, a regulator protein encoded by the *viaB* locus, is involved in both the positive regulation of the Vi-capsule and in repressing the transcription of the flagellin gene *fliC* encoding the flagellin (TLR5 agonist).[58] Thus,

TABLE 87.1
Features of Nontyphoid and Typhoid *Salmonella* Infection

	Clinical Syndrome	Invasive	Pyrogenic Cytokines	Type III Secretion Systems	*Salmonella* Pathogenicity Island 7 (SPI7)	Inflammation	Leukocytes in Stool Samples
Nontyphoid *Salmonella*	Inflammatory diarrhea	Yes	High level	Yes	No	Exudative	Neutrophils
Salmonella Typhi	Enteric fever	Yes	Low level	No	Yes	Interstitial	Macrophages and dendritic cells

TviA might be the missing link to explain the host-derived differences in the pathology between nontyphoid and typhoid *Salmonella*.

87.1.6 Diagnosis

87.1.6.1 Conventional Techniques

The detection and isolation procedures for *Salmonella* are principally alike in the standard methods recommended by the U.S. Food and Drug Administration's (FDA's) Bacteriological Analytical Manual (BAM), Association of Official Analytical Chemists International (AOACI), and the International Organization for Standardization (ISO); however, the media used in the different steps differ.[59–61]

Food, animal feed, and environmental samples usually contain low numbers of potentially stressed *Salmonella* bacteria, and an enrichment procedure is therefore needed. The conventional routine analysis methods for isolating *Salmonella* from food, animal feed, and environmental samples rely on four steps, the first being preenrichment in nonselective media to aid the recovery of sublethally injured *Salmonella*.[62] Even though *Salmonella* is viable and able to cause disease under the right conditions, the organisms can easily be killed if they are grown under selective pressure, such as high temperature or in the presence of chemical additives.[63] It has, however, been suggested that preenrichment in nonselective enrichment can mask the presence of *Salmonella* by allowing growth of competing microflora. Buffered peptone water (BPW) and lactose broth are two of the most widely used media for preenrichment.

The second step is enrichment in selective media to increase the levels of *Salmonella* enough to detect them on selective agar plates. This is accomplished by inhibiting the growth of accompanying microflora by the addition of different selective agents in the media. The most commonly used media for selective enrichment are Rappaport–Vassiliades soy (RVS) broth, selenite cysteine broth, and tetrathionate broth. Modified semisolid Rappaport–Vassiliades is another selective media that is particularly useful for detecting salmonellae in feces and environmental samples.[64] It is based on the ability of salmonellae to migrate through the selective medium ahead of competing motile organisms, thus producing opaque halos of growth.

The third step in the analysis method is isolation on selective agar plates by streaking out the selective enrichment broths to the surface of selective solid media to obtain isolated colonies. The growth of other bacteria is selectively repressed by the addition of inhibitory compounds, and salmonellae are differentiated from other bacteria, for example, by the production of H_2S or acid from sugars. It is recommended that at least two different media be employed because the agars exert different selective pressures. Commonly employed media include brilliant green, xylose lysine Tergitol-4, bismuth sulphide, Hektoen enteric, and xylose lysine deoxycholate agars. Recently, several chromogenic agars have been compared favorably with these, being more sensitive and specific than the common agars.[65–70] An added advantage to using chromogenic agar is that it reduces the need for follow-up confirmation of isolates by subculture and biochemical tests because identification from the primary medium is possible. Chromogenic compounds added in the agar act as substrates for specific enzymes and typically result in color change.[71]

Finally, confirmation of presumptive *Salmonella* isolates is performed by biochemical tests and serotyping. Presumptive colonies are inoculated onto nonselective plates in order to get well-isolated colonies that can be used for further tests. Several biochemical tests are then employed, that is, triple sugar iron agar, mannitol, urea, ornithin decarboxylase, and lysine decarboxylase.[72] A serological verification by determining the antigenic composition is performed. The antigens are classified as somatic (O) and flagellar (H) and are determined by an agglutination test using somatic and flagellar polyvalent antisera. Miniaturized tests such as API 20 E (bioMérieux) and BBL™ Enterotube™ II (BD Diagnostics) have been developed for confirmation and they present an efficient and labor-saving alternative for identification of suspected *Salmonella* isolates.[73]

Human gastroenteritis caused by *Salmonella* is diagnosed by culturing of fecal samples. Normally, isolation is accomplished by direct plating on selective plates from fecal samples containing high numbers of target bacteria.[74] An enrichment step in selenite broth has been reported to increase the sensitivity of the culture method, but this has to be weighed against the additional 24 h until a diagnostic answer can be given. Confirmation of presumptive *Salmonella* is performed by biochemical and serological testing as described earlier.

Typhoid fever is conventionally diagnosed by culture from mainly blood but also from stool, bone marrow, bowel fluid, or specific anatomical lesions. In some cases, a probable diagnosis is based on positive serodiagnosis or antigen detection tests. Ox bile medium is used as a selective blood culture medium for *S. enterica* subsp. *enterica* serovar Typhi and

Paratyphi, but the nonselective broths, such as brain heart infusion or tryptic soy, are often applied if the symptoms are less unambiguous.[75] From stool samples, typhoid fever is diagnosed by enrichment in selenite F broth. Postenrichment isolation and confirmation is performed, as described earlier.

Conventional methods for detection of *Salmonella* are time consuming and labor intensive. Furthermore, these methods suffer from poor specificity due to, for instance, difficulties in recovering sublethally injured cells, problems in the identification of atypical colonies and a high degree of false-positive results. An important advantage of conventional culture methods is that they produce an isolate that can be characterized further.

87.1.6.2 Molecular Techniques

Since *Salmonella* worldwide remain a source of human infection, causing serious illness and socioeconomic problems, a lot of effort has been devoted to the development and improvement of detection methods in clinical and environmental samples, but especially in food samples. The new methods must as a minimum correspond to the conventional techniques regarding sensitivity and selectivity, and furthermore, factors such as rapidity, cost, ease of handling, and possibility of automation are important.[61] As for conventional detection techniques, the detection limit of molecular techniques has to comply with demands of international legislation that there be an absence of *Salmonella* in 25 g of food sample,[60] and for this reason, a preliminary enrichment step is often necessary.

The following sections present a more detailed description of some of the nucleic acid–based molecular techniques commonly applied for the detection of *Salmonella*.

Nucleic acid–based techniques rely on detection of DNA or RNA specific to the organism of analysis. PCR is the most widely used nucleic acid–based technique today. In PCR, repeated cycles of DNA synthesis are applied in order to replicate a target DNA sequence. In a few hours, PCR amplification can produce millions of copies of a target DNA, in theory from a single copy, thus enabling the detection of DNA initially present in very low amounts.[76]

Numerous PCR assays for the detection of *Salmonella* in a variety of sample matrixes have been described in the literature. The majority of them deal with detection in food, reflecting the epidemiology of this human pathogen. The PCR-based detection methods in food commonly employ a preenrichment step combined with subsequent PCR detection. The preenrichment times reported vary from 6 to 24 h, depending on the artificial inoculation level of *Salmonella* in the experimental design of the studies.[77–83] The studies taking into consideration the limit of detection of 1 cfu/25 g in food samples report a minimum of preenrichment of 8–10 h.[81,83–85] Several of these PCR assays amplify a part of the *invA* gene, encoding a protein involved in the invasion of epithelial cells, however, it has been shown that *invA* is lacking in some strains.[86,87] Malorny et al. recently designed a PCR assay for amplification of a part of the *ttrRSBCA* locus encoding proteins used for tetrathionate respiration.[80] The inclusivity and exclusivity of this assay was tested applying 110 *Salmonella* strains and 87 non*Salmonella* strains, and was found to be 100% in both cases. The assay has furthermore been validated in collaborative trials and found to perform well.[88]

PCR detection methods have been described for human clinical samples as well. A number of them are designed to target specific serovars and have little use as diagnostic screening tools, but broader methods for *Salmonella enterica* have compared favorably with conventional culture-based methods in several studies. The fecal samples from a 3-year case-control study in the United Kingdom was reanalyzed by PCR, and the overall number of *Salmonella*-positive samples increased by 23% compared with the analysis results previously obtained by culture.[89] An attempt to apply real-time PCR as a routine screening method for detection of *Salmonella* was successfully conducted in a Dutch clinical microbiological laboratory. The limits of detection and specificity with the real-time PCR method was comparable to the conventional culture-based method, and was found to be significantly more sensitive in analyzing more than 2000 stool samples. Furthermore, the time of analysis was shortened substantially employing the real-time PCR method.[90]

Also for detection of typhoid fever, molecular techniques have been developed. In endemic areas, laboratory and culture facilities are rarely available, and for this reason, efforts have been devoted to developing direct molecular methods based on the detection of specific antibodies present in blood serum. Formerly serodiagnosis was performed by the Widal test, but a lack of sensitivity and specificity, together with problems of interpreting the test results, prompted the development of improved immunoassays to replace it.[91] Commercial rapid test kits for typhoid fever such as TUBEX™ (IDL Biotech, Bromma, Sweden), Typhidot-M® (Malaysian Biodiagnostic Research SDN BHD, Singapore, Malaysia), and Multi-Test Dip-S-Ticks (PANBIO INDX Inc., Baltimore, Maryland) are also available and generally reported to be superior to the Widal test in both sensitivity and specificity.[75,92] PCR has also been shown to be an indispensable tool in the diagnosis of complicated cases of enteric fever caused by *S. enterica* subsp. *enterica* serovar Paratyphi A with unusual clinical manifestations.[93]

Commercial systems based on the PCR detection principle are available for *Salmonella*. One of the most widely used is the BAX® (DuPont Qualicon), an integrated system where preenrichment of 22–26 h, followed by 3 h of regrowth depending on the sample matrix, is followed by DNA extraction and automated PCR.[94–96] The BAX® system has been validated against traditional culture methods and is currently approved by several standardization organizations for *Salmonella* testing in a variety of foods.[97] Two examples of commonly used kits for PCR-based detection of salmonellae are the TaqMan™ *Salmonella* kit (Applied Biosystems), and the iQ-Check™ *Salmonella* kit (Bio-Rad Laboratories).[84,98,99] In both kits DNA is extracted from a sample preenriched in BPW for 16 h and analyzed by PCR.

DNA microarrays for detection and characterization of pathogens such as *Salmonella* are being developed more frequently. The principle behind microarray is hybridization of labeled DNA to numerous oligonucleotides immobilized on a glass slide. The oligonucleotides are designed to recognize a variety of genes; for *Salmonella*, this regards genes involved in virulence, antibiotic resistance, or serovar-specific genes, and information on presence and expression can be obtained.[100–103] DNA microarrays targeting and partly characterizing multiple pathogens, including *Salmonella*, have been described in the literature as diagnostic tools for rapid and simultaneous identification of pathogens involved in clinical diarrhea.[104]

GeneQuence™ (Neogen, Lansing, Michigan) is a commercial assay employing microwell-format DNA hybridization technology to detect *Salmonella* in selected food matrices. Following 24 h of preenrichment, the DNA hybridizes to labeled rRNA specific probes, and an enzyme-mediated color change will occur if *Salmonella* is present.[105]

87.2 METHODS

87.2.1 Real-Time PCR Detection of *Salmonella* in Meat and Environmental Samples

The following section presents a protocol for the real-time PCR detection of *Salmonella* in raw meat. The method has been validated and approved by the Nordic Organization for Validation of Alternative Methods (NordVal).[106] The method is based on shortened preenrichment in buffered peptone water (BPW) followed by DNA extraction and real-time PCR amplification of a region within the *ttr*RSBCA locus required for tetrathionate respiration in *Salmonella*.[80] An internal amplification control is included in the method. The time of analysis is 14 h, and the limit of detection is 1–10 cfu per 25 g.

87.2.1.1 Sample Preparation

1. Weigh 25 g of sample in a stomacher bag and preheat the sample to 37°C ± 1.0°C. Add 225 mL 37 ± 1.0°C BPW. Homogenize manually for 5 s, and incubate at 37°C ± 1.0°C for 12 ± 2 h. Withdraw 5 mL for DNA extraction. Centrifuge the sample at 3000 × *g* for 5 min, and discard the supernatant. Keep the sample pellet at –20.0°C ± 2°C until performing DNA extraction.
2. Extract DNA from the sample pellet with the MagneSil KF Genomic System (Promega) on a KingFisher instrument (Thermo Fisher Scientific) or equivalent equipment as previously described.[83] Briefly, resuspend the sample pellet in 200 µL of lysis buffer, and prepare the plates for the automated DNA extraction (75 µL of paramagnetic beads is used). Include the following controls in the DNA extraction procedure: (1) one process blank consisting of only DNA extraction reagents and water, and (2) one positive DNA extraction control consisting of approx. 2000 CFU *Salmonella*. Place the prepared plates in the automated DNA extraction platform and run the program.

87.2.1.2 Detection Procedures

1. Prepare the PCR mixture according to Table 87.2. Always prepare a batch of PCR mixture large enough to analyze all samples to avoid intraanalysis variations, and calculate a small excess of mixture to ensure a sufficient amount for all samples. Include the following controls in the real-time PCR run: (i) one PCR tube containing PCR-grade water instead of DNA template as a nontemplate control (NTC); (ii) one negative DNA control (for instance, *Escherichia coli* DNA in a concentration of 5 ng); and (iii) one positive DNA control (*Salmonella* DNA in a concentration of 5 ng). Also include the DNA extraction controls in the PCR run.
2. Run the following thermal profile on the Mx3005p real-time PCR thermal cycler (Stratagene, La Jolla, California) or equivalent equipment: 95°C for 3 min, followed by 40 cycles of 95°C for 30 s, 65°C for 60 s, and 72°C for 30 s. Collect fluorescence data from the FAM and HEX channel in the annealing step. Assign the threshold using the software option: background-based threshold, that is, the standard deviation of all amplifications is determined from cycle 5 to 9 (or a similar appropriate interval), and this value is multiplied by a background sigma multiplier of 10.
3. A real-time PCR result should be interpreted as follows:
   Positive when a threshold cycle (C_T) value <36 is obtained.
   Negative when no C_T value or a C_T value >36 is obtained and a C_T value for the IAC is <40. In case of inhibition the analysis should be repeated to exclude the possibility of a false-negative result.

87.2.2 Real-Time PCR Detection of *Salmonella* in Human Clinical Samples

The following section presents a protocol for the real-time PCR detection of *Salmonella* in human clinical samples described by Schuurman et al.[90] based on amplification of 119 bp of the invasion (*invA*) gene (GenBank accession no. M90846) described by Hoorfar et al.[107] The method is based on direct DNA extraction from a fecal suspension followed by real-time PCR analysis. The method has an improved detection rate and sensitivity compared with the conventional culture-based detection method, and a comparable limit of detection of approximately 18 cfu/g of feces. The time of analysis for a negative sample is 1 day for the real-time PCR method, compared with 3–4 days for the conventional method. A preliminary positive diagnosis is also available

TABLE 87.2
Reagents for PCR Mixture to the Real-Time PCR Method for Detection of *Salmonella* in Meat Samples

Reagent	Volume/ Sample (µL)	Concentration of Stock Solution	Sequence or Other Relevant Information
PCR H_2O	0.25	–	
PCR buffer	2.5	10×	10× buffer: 100 mM Tris-HCl. pH 8.9 (25°C), 1 M KCl, 15 mM $MgCl_2$, 500 µg/mL BSA, 0.5 % Tween 20 (v/v)
dUTP mix	1.0	12.5 mM	2.5 mM each of dATP, dCTP, dGTP and 5.0 mM of dUTP
$MgCl_2$	4.0	25 mM	
Glycerol	2.0	87%	
Forward primer	1.0	10 pmol/µL	5′ CTC ACC AGG AGA TTA CAA CAT GG 3′
Reverse primer	1.0	10 pmol/µL	5′ AGC TCA GAC CAA AAG TGA CCA TC 3′
BSA	1.25	20 mg/mL	Dissolved in 50 mM Tris-HCl, 100 mM NaCl, 0.25 mM EDTA, 1 mM 2-mercaptoethanol, 50% glycerol (v/v), pH 7.5
Tth DNA polymerase	0.3	5 U/µL	
DMSO	0.5		Dimethyl sulfoxide, PCR Reagent, SIGMA D9170–1VL or equivalent
Salmonella LNA probe	0.6	10 pmol/µL	5′(6FAM) CG+ACGGCG+AG+ACCG (BHQ1)3′ (+ indicates LNA substitution)
IAC probe	0.6	10 pmol/µL	5′(JOE) CAC ACG GCG ACG CGA ACG CTT T (BHQ1) 3′
IAC target	1.0	–[a]	5′-GACTCACCAGGAGATTACAACATGGCTCTTGCTGTGCATC ATCGCAGAACATCAAAGCGTTCGCGTCGCCGTGTGGGATGGTCACTT TTGGTCTGAGCTAC-3′
Template DNA	9.0 µL	–	
Total volume/sample	25.0 µL	–	

[a] The concentration used should result in a C_T value in the HEX channel of 28–32.

within 1 day, however a subsequent culture confirmation additionally requires 3–4 days.

87.2.2.1 Sample Preparation

1. Prepare a fecal suspension of 33%–50% [wt/vol].
2. Extract DNA from the fecal suspension with NucliSens magnetic extraction reagents on a NucliSens miniMAG instrument (bioMérieux) or equivalent equipment according to the manufacturer's instructions. Include the following controls in the DNA extraction procedure: (i) one process blank consisting of only DNA extraction reagents and water, and (ii) one positive DNA extraction control consisting of approx. 2,000 cfu *Salmonella*.
3. Add approximately 6000 copies of phocine herpesvirus 1 (PhHV) to the extracted DNA as an internal amplification control (IAC).

87.2.2.2 Detection Procedures

1. Prepare the PCR mixture (25 µL) containing 1 × TaqMan Universal PCR master mix (Applied Biosystems), 2.5 µg bovine serum albumin (BSA, Roche Diagnostics), 5 µL DNA template, 300 nM of both target primers (forward: 5′-TCG TCA TTC CAT TAC CTA CC-3′, reverse: 5′-AAA CGT TGA AAA ACT GAG GA-3′), 400 nM target probe (5′ FAM-TCT GGT TGA TTT CCT GAT CGC A-TAMRA), 300 nM IAC primers (forward: 5′-GGG CGA ATC ACA GAT TGA ATC-3′, reverse: 5′-GCG GTT CCA AAC GTA CCA A-3′), and 100 nM IAC probe (5′-NED-CGC CAC CAT CTG GAT-NFQ-MGB-3′). Always prepare a batch of PCR mixture large enough to analyze all samples to avoid intraanalysis variations, and calculate a small excess of mixture to ensure a sufficient amount for all samples. Include the following controls in the real-time PCR run: (i) one PCR tube containing PCR-grade water instead of DNA template as a nontemplate control (NTC), (ii) one negative DNA control (for instance, *Escherichia coli* DNA in a concentration of 5 ng), (iii) one positive DNA control (*Salmonella* DNA in a concentration of 5 ng). Also include the DNA extraction controls in the PCR run.
2. Run the following thermal profile on the real-time PCR thermal cycler: 50°C for 2 min, 95°C for 10 min, followed by 40 cycles of 95°C for 15 s, 50°C for 15 s, and 60°C for 1 min. Collect fluorescence data from the FAM and NED channel in the annealing step.
3. A real-time PCR result should be interpreted as follows:

   Positive when a threshold cycle (C_T) value <40 is obtained.

   Negative when no C_T value is obtained and the signal from the IAC does not exceed the mean C_T value for uninhibited samples ± 2 standard deviations. In case of inhibition, the analysis should be repeated to exclude the possibility of a false-negative result.

87.3 CONCLUSIONS AND FUTURE PERSPECTIVES

While the worldwide burden of human infections caused by *Salmonella* and the socioeconomic consequences of these remain significant, efforts are devoted to innovate and improve methods of detection. The conventional culture-based techniques are still widely employed in routine laboratories analyzing clinical, environmental, and food samples for the presence of *Salmonella*. However, novel molecular techniques offer several advantages in comparison, regarding rapidity, sensitivity, specificity, cost, ease of handling, and possibility of automation.

For detection of *Salmonella*, the use of PCR-based methods is being increasingly employed. The improvement of the detection part of the PCR technique that came with the development of real-time PCR rendered it more rapid and sensitive as well as less labor intensive. The closed-tube system of real-time PCR also renders the exposure of PCR products to the surrounding environment superfluous, thereby eliminating the risk of carry-over contamination, which otherwise is a drawback of the PCR technique. Furthermore, the need for quantitative detection in order to perform meaningful risk assessment to support modeling and decision making regarding prevention and control measures can be fulfilled with the real-time PCR technique.

Despite their obvious advantages, PCR-based detection and quantification methods for *Salmonella* have not been widely implemented in routine laboratories. There are several explanations for this, but the main hindrance can be found in the lack of standardization. Method validation is necessary; the performance of novel molecular-based detection methods has to be systematically compared with the performance of a reference method.[108] If molecular-based detection of *Salmonella* is to replace conventional culture, officially approved methods are needed.

Another major challenge is sample preparation, where the following three aspects have to be considered: (i) the target pathogen is only present in low levels; (ii) from a large sample volume (25 g or mL) only <10 μL is added to the PCR; and (iii) substances from the sample matrix can inhibit the PCR. In many sample types, *Salmonella* is only present in low numbers, and therefore a preenrichment step is needed before PCR amplification. Even though preenrichment is time consuming, this step can be reduced substantially compared with the conventional culture-based methods. This preenrichment step will circumvent the problems arising from an uneven distribution of *Salmonella* cells in a sample as well as the problems of reducing the large sample volume, but in nonenriched samples this can still present a problem. Finally, clinical, environmental, and food samples are complex biological matrices containing a variety of PCR inhibitory substances, and for this reason, a large number of different sample-preparation procedures have been developed. This lack of a universal sample-preparation procedure is an obstacle for implementation of PCR-based *Salmonella* detection in routine laboratories.

In conclusion, the diagnostic possibilities produced by molecular-based detection of *Salmonella* present an interesting alternative to conventional culture-based detection. The advantages of using real-time PCR as a screening tool in routine diagnosis are manifold, especially for negative samples. Improved sample-preparation methods ensuring the capture, separation, and concentration of *Salmonella* from the embedded matrix will accelerate the dissemination of real-time PCR in routine laboratories, and hopefully circumvent the need for the preenrichment of samples containing low numbers of bacteria. Recent initiatives combining the detection and characterization step in one assay would also present routine laboratories with a better incitement to replace culture-based detection with molecular.

Finally, improved detection, quantification, and characterization methods will provide superior information to develop models founding the basis for the risk assessment of *Salmonella*, and eventually enable improved prevention and control of this important human pathogen.

ACKNOWLEDGMENTS

This work was supported financially by the European Union–funded Integrated Project BIOTRACER (FOOD-2006-CT-036272) under the 6th RTD Framework.

REFERENCES

1. Le Minor, L., Genus III. *Salmonella*. In Krieg, N.R., Bergey, D.H., and Holt, J.G. (Editors), *Bergey's Manual of Systemic Bacteriology*, p. 427, Williams and Wilkins, Baltimore, 1984.
2. Humphrey, T., Public health aspects of *Salmonella enterica* in food production. In Mastroeni, P., and Maskell, D. (Editors), *Salmonella Infections. Clinical, Immunological and Molecular Aspects*, p. 89, Cambridge University Press, New York, 2006.
3. D'Aoust, J.Y., *Salmonella* species. In Doyle, M.P., Beuchat, L.R., and Montville, T.J. (Editors), *Food Microbiology. Fundamentals and Frontiers*, p. 129, ASM Press, Washington DC, 1997.
4. D'Aoust, J.Y., Psychrotrophy and foodborne *Salmonella*, *Int. J. Food Microbiol.*, 13, 207, 1991.
5. Tietjen, M., and Fung, D.Y., Salmonellae and food safety, *Crit. Rev. Microbiol.*, 21, 53, 1995.
6. Jensen, A.N. et al., Survival and transmission of *Salmonella* enterica serovar typhimurium in an outdoor organic pig farming environment, *Appl. Environ. Microbiol.*, 72, 1833, 2006.
7. Winfield, M.D., and Groisman, E.A., Role of nonhost environments in the lifestyles of *Salmonella* and *Escherichia coli*, *Appl. Environ. Microbiol.*, 69, 3687, 2003.
8. Ferreira, M.A.S.S., and Lund, B.M., The influence of pH and temperature on initiation of growth of *Salmonella* spp., *Lett. Appl. Microbiol.*, 5, 67, 1987.
9. D'Aoust, J.Y., *Salmonella*. In Doyle, M.P. (Editor), *Foodborne Bacterial Pathogens*, p. 327, Marcel Dekker Inc., New York, 1989.
10. Thomas, L.V., Wimpenny, J.W., and Peters, A.C., Testing multiple variables on the growth of a mixed inoculum of *Salmonella* strains using gradient plates, *Int. J. Food Microbiol.*, 15, 165, 1992.

11. Anonymous, Combase, From, http://wyndmoor.arserrc.gov/combase/, 2009. Accessed November 4, 2010.
12. Grimont, A.D., and Weill, F.X., *Antigenic Formulae of the Salmonella Serovars*, 9th ed., 1, 2007.
13. Uzzau, S. et al., Host adapted serotypes of *Salmonella enterica*, *Epidemiol. Infect.*, 125, 229, 2000.
14. Cooke, F.J., Threlfall, E.J., and Wain, J., Current trends in the spread and occurrence of human salmonellosis: Molecular typing and emerging antibiotic resistance. In Rhen M., et al. (Editors), *Salmonella. Molecular Biology and Pathogenesis*, p. 1, Horizon Bioscience, Norfolk, 2007.
15. D'Aoust, J.Y., Pathogenicity of foodborne *Salmonella*, *Int. J. Food Microbiol.*, 12, 17, 1991.
16. Crump, J.A., Luby, S.P., and Mintz, E.D., The global burden of typhoid fever, *Bull. World Health Organ.*, 82, 346, 2004.
17. Anonymous, Salmonellosis, From, http://www.cdc.gov/nczved/dfbmd/disease_listing/salmonellosis_gi.html#8, 2009. Accessed November 4, 2010.
18. Anonymous, Drug-resistant *Salmonella,* From http://www.who. int/mediacentre/factsheets/fs139/en/, 2009. Accessed November 4, 2010.
19. Anonymous, The community summary report on trends and sources of zoonoses and zoonotic agents in the European Union in 2007, *EFSA J.*, 223, 21–108, 2009.
20. Galanis, E. et al., Web-based surveillance and global *Salmonella* distribution, 2000–2002, *Emerg. Infect. Dis.*, 12, 381, 2006.
21. Humphrey, T., *Salmonella*, stress responses and food safety, *Nat. Rev. Microbiol.*, 2, 504, 2004.
22. Lecos, C., Of microbes and milk: Probing Americas worst *Salmonella* outbreak, *Dairy Food Sanit.*, 6, 136, 1986.
23. Centers for disease control and prevention. *Salmonella*. Investigation update: Outbreak of *Salmonella* Typhimurium infections 2008–2009, From, http://www.cdc.gov/salmonella/typhimurium/update.html, 2009.
24. Statens Serum Institut, Nyheder. 3. februar 2009—Status på det landsdækkende Salmonellaudbrud uge 5, From, http://www.ssi.dk/sw174.asp?PAGE=1&ArtNo=3647090, 2009. Accessed November 4, 2010.
25. Coburn, B., Grassl, G.A., and Finlay, B.B. *Salmonella*, the host and disease: A brief review. *Immunol. Cell Biol.*, 85, 112, 2007. Accessed July 2, 2009.
26. Molbak, K., Olsen, J.E., and Wegener, H.C., *Salmonella* infections. In Riemann, H.P., and Cliver, D.O. (Editors), *Foodborne Infections and Intoxications*, p. 57, Elsevier Academic Press, 2006.
27. Fisker, N. et al., Clinical review of nontyphoid *Salmonella* infections from 1991 to 1999 in a Danish county, *Clin. Infect. Dis.*, 37, e47, 2003.
28. Saphra, I., and Winter, J.W., Clinical manifestations of salmonellosis in man; an evaluation of 7779 human infections identified at the New York *Salmonella* Center, *N. Engl. J. Med.*, 256, 1128, 1957.
29. Schonheyder, H.C., and Ejlertsen, T., Survey of extraintestinal nontyphoid *Salmonella* infections in a Danish region. Inverse relation of invasiveness to frequency of isolation, *APMIS,* 103, 686, 1995.
30. Helms, M. et al., Short and long term mortality associated with foodborne bacterial gastrointestinal infections: Registry based study, *BMJ,* 326, 357, 2003.
31. Travers, K., and Barza, M., Morbidity of infections caused by antimicrobial-resistant bacteria, *Clin. Infect. Dis.*, 34 Suppl 3, S131, 2002.
32. Molbak, K., Human health consequences of antimicrobial drug-resistant *Salmonella* and other foodborne pathogens, *Clin. Infect. Dis.*, 41, 1613, 2005.
33. Neal, K.R., Hebden, J., and Spiller, R., Prevalence of gastrointestinal symptoms six months after bacterial gastroenteritis and risk factors for development of the irritable bowel syndrome: Postal survey of patients, *BMJ,* 314, 779, 1997.
34. Miller, S.I., Hohmann, E.L., and Pegues, D.A., *Salmonella* (including *Salmonella* Typhi). In Mandell, G.L., Bennett, J.E., and Dolin, R. (Editors), *Principles and Practice of Infectious Diseases*, p. 2103, Churchill Livingstone, New York, 1995.
35. Cohen, M.L., and Tauxe, R.V., Drug-resistant *Salmonella* in the United States: An epidemiologic perspective, *Science,* 234, 964, 1986.
36. Mellemgaard, A., and Gaarslev, K., Risk of hepatobiliary cancer in carriers of *Salmonella* typhi, *J. Natl. Cancer Inst.*, 80, 288, 1988.
37. Dutta, U. et al., Typhoid carriers among patients with gallstones are at increased risk for carcinoma of the gallbladder, *Am. J. Gastroenterol.*, 95, 784, 2000.
38. Schmidt, H., and Hensel, M., Pathogenicity islands in bacterial pathogenesis, *Clin. Microbiol. Rev.*, 17, 14, 2004.
39. Ehrbar, K. et al., Characterization of effector proteins translocated via the SPI1 type III secretion system of *Salmonella* typhimurium, *Int. J. Med. Microbiol.*, 291, 479, 2002.
40. Waterman, S.R., and Holden, D.W., Functions and effectors of the *Salmonella* pathogenicity island 2 type III secretion system, *Cell. Microbiol.*, 5, 501, 2003.
41. Tsolis, R.M. et al., Contribution of *Salmonella* typhimurium virulence factors to diarrheal disease in calves, *Infect. Immun.*, 67, 4879, 1999.
42. Santos, R.L. et al., Animal models of *Salmonella* infections: Enteritis versus typhoid fever, *Microbes Infect.*, 3, 1335, 2001.
43. Hobbie, S. et al., Involvement of mitogen-activated protein kinase pathways in the nuclear responses and cytokine production induced by *Salmonella* typhimurium in cultured intestinal epithelial cells, *J. Immunol.*, 159, 5550, 1997.
44. Iwasaki, A., and Medzhitov, R., Toll-like receptor control of the adaptive immune responses, *Nat. Immunol.*, 5, 987, 2004.
45. Santos, R.L. et al., Morphologic and molecular characterization of *Salmonella* typhimurium infection in neonatal calves, *Vet. Pathol.*, 39, 200, 2002.
46. Figueroa-Bossi, N. et al., Variable assortment of prophages provides a transferable repertoire of pathogenic determinants in *Salmonella*, *Mol. Microbiol.*, 39, 260, 2001.
47. Rotger, R., and Casadesus, J., The virulence plasmids of Salmonella, *Int. Microbiol.*, 2, 177, 1999.
48. Rothwell, N.J., Mechanisms of the pyrogenic actions of cytokines, *Eur. Cytokine Netw.*, 1, 211, 1990.
49. Girardin, E. et al., Tumor necrosis factor and interleukin-1 in the serum of children with severe infectious purpura, *N. Engl. J. Med.*, 319, 397, 1988.
50. Keuter, M. et al., Patterns of proinflammatory cytokines and inhibitors during typhoid fever, *J. Infect. Dis.*, 169, 1306, 1994.
51. Kraus, M.D., Amatya, B., and Kimula, Y., Histopathology of typhoid enteritis: Morphologic and immunophenotypic findings, *Mod. Pathol.*, 12, 949, 1999.
52. Harris, J.C., Dupont, H.L., and Hornick, R.B., Fecal leukocytes in diarrheal illness, *Ann. Intern. Med.*, 76, 697, 1972.

53. Bignold, L.P. et al., Inhibition of chemotaxis of neutrophil leukocytes to interleukin-8 by endotoxins of various bacteria, *Infect. Immun.*, 59, 4255, 1991.
54. Wyant, T.L., Tanner, M.K., and Sztein, M.B., *Salmonella* typhi flagella are potent inducers of proinflammatory cytokine secretion by human monocytes, *Infect. Immun.*, 67, 3619, 1999.
55. Raffatellu, M. et al., Capsule-mediated immune evasion: A new hypothesis explaining aspects of typhoid fever pathogenesis, *Infect. Immun.*, 74, 19, 2006.
56. Virlogeux, I. et al., Role of the viaB locus in synthesis, transport and expression of Salmonella typhi Vi antigen, *Microbiology,* 141, 3039, 1995.
57. Wilson, R.P. et al., The Vi-capsule prevents Toll-like receptor 4 recognition of *Salmonella*, *Cell. Microbiol.,* 10, 876, 2008.
58. Winter, S.E. et al., The *Salmonella* enterica serotype Typhi regulator TviA reduces interleukin-8 production in intestinal epithelial cells by repressing flagellin secretion, *Cell. Microbiol.,* 10, 247, 2008.
59. Andrews, W.H., and Hammack, T.S., Bacteriological analytical manual. In *Salmonella*, 8th Ed., Chapter 5, from, U.S. Food and Drug Administration (FDA) http://www.cfsan.fda.gov/~ebam/bam-mm.html, 2009.
60. Anonymous, Microbiology of food and animal feeding stuffs—Horizontal method for the detection of *Salmonella* spp., ISO 6579:2002, International Organisation for Standardisation (ISO), Geneva, Switzerland, 2002.
61. van der Zee, H., and Huis in't Veld, J.H.J., Methods for the rapid detection of *Salmonella*. In Wray, C., and Wray, A. (Editors), *Salmonella in Domestic Animals*, p. 373, Cabi Publishing, Wallingford, 2000.
62. Fleet, G.H., Microorganisms in food ecosystems, *Int. J. Food Microbiol.,* 50, 101, 1999.
63. Waltman, W.D., Methods for cultural isolation of *Salmonella*. In Wray, C., and Wray, A. (Editors), *Salmonella in Domestic Animals*, p. 355, Cabi Publishing, Wallingford, 2000.
64. De Smedt, J.M. et al., Rapid *Salmonella* detection in foods by motility enrichment on a modified semisolid Rappaport–Vassiliadis medium, *J. Food Prot.,* 49, 510, 1986.
65. Gaillot, O. et al., Comparison of CHROMagar *Salmonella* medium and hektoen enteric agar for isolation of salmonellae from stool samples, *J. Clin. Microbiol.,* 37, 762, 1999.
66. Eigner, U. et al., Evaluation of a new chromogenic medium for the isolation and presumptive identification of *Salmonella* species from stool specimens, *Eur. J. Clin. Microbiol. Infect. Dis.,* 20, 558, 2001.
67. Maddocks, S., Olma, T., and Chen, S., Comparison of CHROMagar *Salmonella* medium and xylose-lysine-desoxycholate and *Salmonella-Shigella* agars for isolation of *Salmonella* strains from stool samples, *J. Clin. Microbiol.,* 40, 2999, 2002.
68. Cassar, R., and Cuschieri, P., Comparison of *Salmonella* chromogenic medium with DCLS agar for isolation of *Salmonella* species from stool specimens, *J. Clin. Microbiol.,* 41, 3229, 2003.
69. Perez, J.M. et al., Comparison of four chromogenic media and Hektoen agar for detection and presumptive identification of *Salmonella* strains in human stools, *J. Clin. Microbiol.,* 41, 1130, 2003.
70. van Dijk, S., Bruins, M.J., and Ruijs, G.J., Evaluation and implementation of a chromogenic agar medium for *Salmonella* detection in stool in routine laboratory diagnostics, *J. Clin. Microbiol.,* 47, 456, 2009.
71. Manafi, M., Kneifel, W., and Bascomb, S., Fluorogenic and chromogenic substrates used in bacterial diagnostics, *Microbiol. Rev.,* 55, 335, 1991.
72. Anonymous, *Salmonella*. Detection in food, Method no. 71, 5th ed., Nordic Committee on Food Analysis, Åbo, Findland, 1999.
73. Hayek, L.J., and Willis, G.W., A comparison of two commercial methods for the identification of the Enterobacteriaceae—API 20E and the Enterotube—with conventional methods, *J. Clin. Pathol.,* 29, 158, 1976.
74. Gray, L.D., *Escherichia, Salmonella, Shigella* and *Yersinia*. In Murray, P.R., et al. (Editors), *Manual of Clinical Microbiology*, p. 450, ASM press, Washington DC, 1995.
75. Anonymous, Background document: The diagnosis, treatment and prevention of typhoid fever. World Health Organisation, Department of vaccines and biologicals, Geneva, Switzerland, 7, 2003.
76. Persing, D.H., In vitro nucleic acid amplification techniques. In Persing, D.H., et al. (Editors), *Diagnostic Molecular Microbiology. Principles and Applications*, pp. 51–88, Mayo Foundation, US, 1993.
77. Agarwal, A., Makker, A., and Goel, S.K., Application of the PCR technique for a rapid, specific and sensitive detection of *Salmonella* spp. in foods, *Mol. Cell. Probes,* 16, 243, 2002.
78. Eyigor, A., Carli, K.T., and Unal, C.B., Implementation of real-time PCR to tetrathionate broth enrichment step of *Salmonella* detection in poultry, *Lett. Appl. Microbiol.,* 34, 37, 2002.
79. Ellingson, J.L. et al., Twelve hour real-time PCR technique for the sensitive and specific detection of *Salmonella* in raw and ready-to-eat meat products, *Mol. Cell. Probes,* 18, 51, 2004.
80. Malorny, B. et al., Diagnostic real-time PCR for detection of *Salmonella* in food, *Appl. Environ. Microbiol.,* 70, 7046, 2004.
81. Hein, I. et al., Real-time PCR for the detection of *Salmonella* spp. in food: An alternative approach to a conventional PCR system suggested by the FOOD-PCR project, *J. Microbiol. Methods,* 66, 538, 2006.
82. Myint, M.S. et al., The effect of pre-enrichment protocol on the sensitivity and specificity of PCR for detection of naturally contaminated *Salmonella* in raw poultry compared to conventional culture, *Food Microbiol.,* 23, 599, 2006.
83. Josefsen, M.H. et al., Optimization of a 12-hour TaqMan PCR-based method for detection of Salmonella bacteria in meat, *Appl. Environ. Microbiol.,* 73, 3040, 2007.
84. Chen, S. et al., The evaluation of a fluorogenic polymerase chain reaction assay for the detection of *Salmonella* species in food commodities, *Int. J. Food Microbiol.,* 35, 239, 1997.
85. Gado, I. et al., Rapid combined assay for *Salmonella* detection in food samples, *Acta Microbiol. Immunol. Hung.,* 47, 445, 2000.
86. Rahn, K. et al., Amplification of an invA gene sequence of *Salmonella* typhimurium by polymerase chain reaction as a specific method of detection of *Salmonella*, *Mol. Cell. Probes,* 6, 271, 1992.
87. Galan, J.E., and Curtiss, R., 3rd., Distribution of the invA, -B, -C, and -D genes of *Salmonella* typhimurium among other *Salmonella* serovars: invA mutants of *Salmonella* typhi are deficient for entry into mammalian cells, *Infect. Immun.,* 59, 2901, 1991.
88. Malorny, B. et al., Multicenter validation study of two blockcycler- and one capillary-based real-time PCR methods for the detection of *Salmonella* in milk powder, *Int. J. Food Microbiol.,* 117, 211, 2007.

89. Amar, C.F. et al., Detection by PCR of eight groups of enteric pathogens in 4,627 faecal samples: re-examination of the English case-control Infectious Intestinal Disease Study (1993–1996), *Eur. J. Clin. Microbiol. Infect. Dis.*, 26, 311, 2007.
90. Schuurman, T. et al., Feasibility of a molecular screening method for detection of *Salmonella enterica* and *Campylobacter jejuni* in a routine community-based clinical microbiology laboratory, *J. Clin. Microbiol.*, 45, 3692–3700, 2007.
91. Chart, H. et al., Serodiagnosis of *Salmonella* enterica serovar Typhi and *S. enterica* serovars Paratyphi A, B and C human infections, *J. Med. Microbiol.*, 56, 1161, 2007.
92. Olsen, S.J. et al., Evaluation of rapid diagnostic tests for typhoid fever, *J. Clin. Microbiol.*, 42, 1885, 2004.
93. Chaudhry, R. et al., Utility of PCR in diagnosing complicated cases of unusual clinical manifestations of *Salmonella* enterica var. paratyphi A, *Am. J. Med.*, 118, 799, 2005.
94. Bailey, J.S., Detection of *Salmonella* cells within 24 to 26 hours in poultry samples with the polymerase chain reaction BAX system, *J. Food Prot.*, 61, 792, 1998.
95. Bennett, A.R. et al., Rapid and definitive detection of *Salmonella* in foods by PCR, *Lett. Appl. Microbiol.*, 26, 437, 1998.
96. D'Aoust, J.Y. et al., Evaluation of the BAX gel and fluorometric systems for the detection of foodborne *Salmonella*, *J. Food Prot.*, 70, 835, 2007.
97. DuPont Qualicon, Bax Sytem PCR assay *Salmonella*, From, http://www2.dupont.com/Qualicon/en_US/products/BAX_System/bax_salmonella.html, 2009. Accessed November 4, 2010.
98. Uyttendaele, M., Vanwildemeersch, K., and Debevere, J., Evaluation of real-time PCR vs automated ELISA and a conventional culture method using a semi-solid medium for detection of *Salmonella*, *Lett. Appl. Microbiol.*, 37, 386, 2003.
99. Fakhr, M.K. et al., Adding a selective enrichment step to the iQ-Check real-time PCR improves the detection of *Salmonella* in naturally contaminated retail turkey meat products, *Lett. Appl. Microbiol.*, 43, 78, 2006.
100. Malorny, B. et al., Molecular characterisation of *Salmonella* strains by an oligonucleotide multiprobe microarray, *Mol. Cell. Probes*, 21, 56, 2007.
101. Wattiau, P. et al., Evaluation of the Premi Test *Salmonella*, a commercial low-density DNA microarray system intended for routine identification and typing of *Salmonella enterica*, *Int. J. Food Microbiol.*, 123, 293, 2008.
102. van Hoek, A.H. et al., Detection of antibiotic resistance genes in different *Salmonella* serovars by oligonucleotide microarray analysis, *J. Microbiol. Methods*, 62, 13, 2005.
103. Scaria, J. et al., Microarray for molecular typing of *Salmonella enterica* serovars, *Mol. Cell. Probes*, 22, 238, 2008.
104. You, Y. et al., A novel DNA microarray for rapid diagnosis of enteropathogenic bacteria in stool specimens of patients with diarrhea, *J. Microbiol. Methods*, 75, 566, 2008.
105. Mozola, M.A., Peng, X., and Wendorf, M., Evaluation of the GeneQuence DNA hybridization method in conjunction with 24-hour enrichment protocols for detection of *Salmonella* spp. in select foods: Collaborative study, *J. AOAC Int.*, 90, 738, 2007.
106. Löfström, C. et al., Validation of a same-day real-time PCR method for screening of meat and carcass swabs for *Salmonella*, *BMC Microbiol.*, 9, 85, 2009.
107. Hoorfar, J., Ahrens, P., and Radstrom, P., Automated 5′ nuclease PCR assay for identification of *Salmonella* enterica, *J. Clin. Microbiol.*, 38, 3429, 2000.
108. Malorny, B. et al., Standardization of diagnostic PCR for the detection of foodborne pathogens, *Int. J. Food Microbiol.*, 83, 39, 2003.

88 *Serratia*

Kevin B. Laupland and Deirdre L. Church

CONTENTS

88.1 INTRODUCTION

88.1.1 Classification and Epidemiology

Classification. *Serratia* species are classified in the tribe Klebsielliae within the family Enterobacteriaceae along with several other genera, including *Klebsiella*, *Enterobacter*, *Hafnia*, and *Pantoea*. *Serratia marcescens* is the most important cause of human disease amongst all species recovered from clinical specimens. The *S. liquefaciens* group (consisting of *S. liquefaciens sensu stricto*, *S. proteamaculans*, and *S. grimesii*) and *S. rubidaea* are less common but well-documented causes of human disease. Other *Serratia* species that may occasionally cause human disease include *S. fonticola* and *S. odorifera*, whereas *S. ureilytica*, *S. quinivorans*, *S. plymuthica*, *S. entomophila*, and *S. ficaria* are predominantly environmental organisms and are rare causes of human disease.[1,2]

Epidemiology. *Serratia* spp. may be found broadly in the environment. They are commonly present in water sources and may also be found colonizing plants, animals, and insects. Although widely recognized as an important contemporary cause of human disease, *Serratia* spp. have only been considered to be significant pathogens for the past half century.[3,4] Outbreaks of nosocomial urinary tract infections received attention in the 1950s, and clinical case reports and series of invasive infections with *Serratia* spp. were reported with increasing frequency over the subsequent decades.[3,4] At present, *Serratia* spp., most commonly *S. marcescens*, are widely recognized agents causing serious hospital-acquired infections, including pneumonia, wound infection, bloodstream infection, and urinary tract infection. Contemporary, hospital-based series and surveillance surveys indicate that *Serratia* spp. cause approximately 1%–2% of hospital-acquired infections overall, with a higher proportion observed in critically ill patients.[5–8]

The ability of *Serratia* spp., and in particular *S. marcescens*, to cause large hospital- and healthcare-associated outbreaks is widely recognized. Most other members of the Enterobacteriaceae family are frequently found in the normal intestinal flora, but this does not appear to be the case with *Serratia* spp., where most infections are believed to be exogenously acquired. *Serratia* spp. have a notable ability to contaminate fluids and devices, and represent the most commonly reported cause of outbreaks due to contaminated blood products.[9,10] Numerous large hospital outbreaks with *Serratia* spp. have been reported, with sources related to contaminated parenteral infusions and injections, particularly due to multiple use of single-dose vials[11–17]; environmental sources, such as contaminated faucets, air conditioners, and respiratory equipment[18–22]; and non-antimicrobial-containing soaps.[23,24] Patient-to-patient transmission is believed to occur primarily by healthcare workers as a result of inadequate hygiene practices.[25,26]

But exposure to contaminated substances and inadequate hygiene practices aside, patient-related risk factors for nosocomial infections due to *Serratia* spp. are multifold and largely similar to those for other major nosocomial pathogens. These are not limited to prolonged stays, admission to intensive care units (ICU), broad-spectrum antimicrobial use, and the use of devices and catheters.[27] It has been recognized that *Serratia* spp. have a propensity to cause infections in patients admitted to neonatal ICUs, with the associated major risk factors of low birth weight, prematurity, and mechanical ventilation.[28] In adult ICUs, neurosurgical and burn patients appear to be at increased risk.

Reports of large outbreaks, hospital-based surveillance reports, anecdotal case reports of severe cases of disease, and clinical experience all attest to *Serratia* spp. being an important cause of nosocomial infection. However, basing the overall epidemiology of an infectious disease on such data may be misleading. In particular, the outbreak situation represents a special circumstance in time and location and often is not reflective of the epidemiology of disease occurring at baseline. In addition, outbreak investigations during unusual circumstances, or surveillance reviews conducted in selected tertiary-care referral hospitals may not be widely generalized elsewhere. Finally, by limiting evaluation to hospitalized patients, the importance of an organism as a cause of community-onset disease may be grossly underestimated. To best define the epidemiology of an infectious disease, population-based studies that study all episodes of disease occurring in all hospitalized and nonhospitalized residents within a defined population are optimal. Two recent studies, one from Canada[29] and the other from Australia,[30] have investigated the incidence and determinants of *Serratia* spp. infections at the population level and provided important new perspectives on their epidemiology.

Laupland and colleagues conducted population-based laboratory surveillance from 2000 to 2005 in Calgary, Canada (population 1.1 million), to define the incidence and demographic risk factors for isolating *Serratia* spp. from clinical specimens.[29] A total of 715 incident *Serratia* spp. isolates were identified for an annual incidence of 10.8 per 100,000 residents. Bacteremic disease occurred in 0.9 per 100,000 residents annually. Most of the incident isolates were *Serratia marcescens* (659; 92%), with the remainder being the *Serratia liquifaciens* group (27; 4%), *Serratia odorifera* (7; 1%), *Serratia rubidaea* (6; 1%), *Serratia fonticola* (2), *Serratia plymuthica* (1), and *Serratia* not speciated (13; 2%). The majority (65%) of incident *Serratia* spp. isolates were of community onset. The 249 hospital-onset cases were isolated from a total of 62 different wards in the four area hospitals studied. A total of 10 clusters, all *Serratia marcescens*, were observed, involving a total of 40 patients on 8 different wards. Several medical and surgical wards, an adult ICU, and a rehabilitation ward had clusters involving three to five patients per cluster. Although a substantial number of nosocomial diseases occurred, this study demonstrated that *Serratia* spp. are most commonly isolated as community-onset pathogens, and that when hospital-onset disease occurred, it was typically in non-ICU areas.

Engel et al. reported on the epidemiology of *Serratia* spp. bacteremias in Canberra, Australia, during a 10-year period.[30] These investigators identified 38 episodes of *Serratia* spp. bacteremia in area residents for an annual incidence of 1.03 per 100,000. *Serratia marcescens* was the predominant isolate (82%), followed by *S. liquifaciens* (13%), *S. rubidaea* (3%), and *S. fonticola* (3%). Eighteen (47%) cases had onset in the community and of these eleven (29%) episodes were classified as community associated. The remaining seven (18%) community-onset cases had complications of an indwelling device, medical procedure, or instrumentation, or were associated with cytotoxic-induced neutropenia and deemed to have healthcare-associated community-onset disease. Twenty (53%) were deemed to have inpatient healthcare-associated (nosocomial) bacteremia. This study demonstrated the importance and highlighted the underappreciation of *Serratia* spp. as significant agents of community-onset disease.

88.1.2 Clinical Features and Pathogenesis

Clinical Features. Although they are mostly recognized as an agent of urinary tract, respiratory tract, wound, and bloodstream infections, *Serratia* spp. have been reported to cause a wide range of clinical disease. There are numerous published anecdotal reports of bone and joint infections, including osteomyelitis and septic arthritis; contact lens–associated keratitis and endopthalmitis; peritonitis; otitis media; endocarditis; and meningitis and other central nervous system infections.[1,11,31] Recent years have also witnessed an increasing number of reports of *Serratia* spp. infections associated with high severity. These include severe soft-tissue infections associated with necrotizing fasciitis[32] and purpura fuliminans.[33] However, while the epidemiological context may suggest that *Serratia* spp. may be potential pathogens, there are no particular or distinctive clinical features or syndromes that allow a clinician to diagnose these infections, and an etiologic diagnosis is dependent on laboratory confirmation. Case-fatality rates for *Serratia* spp. infections are dependent on age, resistance, comorbidities, severity of illness, and focus of infection.[17] In their population-based study from Australia, Engel et al. identified 7-day, 1-month, 3-month, and 6-month case-fatality rates of 5%, 18%, 29%, and 37%, respectively, for *Serratia* spp. bacteremia.[30]

Pathogenesis. The ability of *Serratia* spp. to cause important human infections involves a number of potential virulence factors, but its ability to resist many antimicrobial agents is also an important determinant. As with all gram-negative organisms, lipopolysaccharide (LPS) in the outer membrane is believed to be an important virulence determinant because it resists serum killing and stimulates a potentially maladaptive host inflammatory response. *Serratia* spp. may produce a number of exoproteins that may contribute to virulence, such as chitinase, proteases, nuclease, lipase, lecithinase, and hemolysin.[34] Kurz et al. investigated potential virulence factors using a *Caenorhabditis elegans* gut infection model,[35] and implicated virulence genes as those that function in lipopolysaccharide (LPS) biosynthesis, iron uptake, and hemolysin production.[35]

Another important aspect of the pathogenesis of *Serratia* spp. infections is the organism's ability to attach to and move along surfaces and form biofilms. Strains may possess mannose-resistant and mannose-sensitive hemagglutinins, and fimbriae and pili may allow adherence to uroepithelial cells, which facilitates the development of and damage associated with urinary tract infections. The formation of a biofilm increases the chances of survival of the organisms in the setting of oxidative stress, antibiotics, and host response.[36–38] Biofilms are complex communities of organisms attached to a surface embedded within self-produced exopolymeric substances.[37,38] Formation of a surface biofilm involves a cycle of attachment of organisms, surface motility, maturation into a biofilm community structure, and then detachment of planktonic cells.[36] A number of determinants are potentially involved in this process, including the surfactant serrawettin, prodigiosin, and flagellae that enable swarming and swimming motility.[37] The cell-cell communication via quorum sensing is believed to be mediated primarily by *N*-acyl-L-homoserine lactones and to a lesser extent by Autoinducer-2 (AI-2).[37]

Clinicians recognize that the intrinsic and acquired resistance of *Serratia* spp. to antimicrobial agents is a major determinant of the acquisition, severity, and outcome of infections caused by this genus. Like a number of other Enterobacteriaceae, including *Enterobacter* spp., *Citrobacter freundii*, and *Morganella morganii*, *Serratia* spp. naturally encode an inducible chromosomal *Amp*C-type β-lactamase that hydrolyzes a broad spectrum of β-lactam and cephalosporin antibiotics.[39,40] The main mechanisms of β-lactamase resistance in *S. marcescens* arises from the de-repression of the chromosomal *Amp*C cephalosporinase within the SRT families, including SST-1, ART-1, SRT-2, and SRT-3, and the recently described S4 gene.[41,42] *Serratia* spp. will therefore typically demonstrate in vitro susceptibility to most β-lactam and cephalosporin antibiotics, but they may develop resistance to these antibiotics during therapy, resulting in treatment failure.[43,44]

Serratia also demonstrated a major ability to acquire other types of antimicrobial resistance through the acquisition of plasmids. Although extended-spectrum β-lactamases (ESBL) are most commonly present among Enterobacteriaceae in *E. coli* and *K. pneumoniae*, they may also be frequently observed in clinical isolates of *S. marcescens*.[39,40,45,46] These ESBLs-producing strains of Enterobacteriaceae are typically multiresistant to other antimicrobial agents, such as the aminoglycosides, trimethoprim, sulfonamides, tetracyclines, and chloramphenicol, because the large plasmids that encode the β-lactamase genes also carry other resistance genes.[47] Most ESBLs identified in clinical isolates have been of the SHV and TEM types that result in resistance to penicillins and first-generation cephalosporins because of point mutation around the enzyme's active site.[39,40,48] The newer CTX-M β-lactamase enzymes are the most recently described types of ESBLs that originated from the environmental bacterium *Kluyvera* spp. These CTX-M-type ESBLs cause resistance to cefotaxime but may remain sensitive to ceftazidime (e.g., CTX-M-15 hydrolyzes ceftazidime efficiently) and are divided into five groups based on their amino acid identities: the CTX-M-1 group (includes CTX-M-1, -3, -10, -12, -15, -22, -23, -28, -29, -30, -32, -33, -36, -54, and UOE-1); the CTX-M-2 group (includes CTX-M-2, -4, -5, -6, -7, -20, -31, -44, FEC-1); the CTX-M-8 group (includes CTX-M-8 and -40); the CTX-M-9 group (includes CTX-M-9, -13, -14, -16, -17, -19, -21, -24, -27, -45, -46, -47, -48, -49, and -50); and the CTX-M-25 group (includes CTX-M-25, -26, -39, and -41).[39,40,47]

Carbapenem-hydrolyzing β-lactamases are emerging in gram-negative pathogens worldwide and are a real threat to effective antibiotic therapy because carbapenems often remain the only therapeutic option for treating multidrug-resistant bacteria. Resistance to carbapenems has occasionally been described in *S. marcescens*, either due to the production of plasmid-mediated Ambler class B metallo-β-lactamases (MBLs), such as IMP-1, IMP-6, and VIM-2,[49–51] or to chromosomally encoded SME-type Ambler class A β-lactamases.[49,52–55] Three-point mutant variants (SME-1, SME-2, and SME-3) have been isolated worldwide.[55,56] These enzymes significantly hydrolyze penicillins, cephalosporins, and carbapenems and are inhibited by clavulanic acid. Carbapenem-hydrolyzing KPC β-lactamases have also recently been described in *S. marcescens* isolates, and these enzymes belong to Bush group 2f, molecular class A β-lactamases that are capable of hydrolyzing carbapenems, penicillins, cephalosporins, and aztreonam and are inhibited by clavulanic acid and tazobactam.[49,55–57]

Resistance to quinolones is usually mediated by chromosomal mutations in genes coding for the target enzymes DNA gyrase and/or topoisomerase IV or genes regulating expression of the efflux pump.[58–60] Plasmid-mediated quinolone resistance (PMQR) has been increasingly described in different species of *Enterobacteriaceae* due to the presence of *qnr*A, *qnr*B, and *qnr*S determinants. More recently, PMQR has been described due to *aac(6′)-Ib-cr* (i.e., one of approximately 30 variants of *aac(6′)-Ib* that has two amino acid changes; Trp102Arg and Asp179Tyr), which acetylate quinolones such as ciprofloxacin and norfloxacin that have unsubstituted piperazinyl nitrogen.[60–64] The increasing prevalence of *aac(6′)-Ib-cr* in *Enterobacteriaceae,* including *Amp*C-producing strains of *Enterobacter cloacae*, *Citrobacter freundii,* and *S. marcescens,* has recently been reported.[65]

88.1.3 Laboratory Diagnosis

88.1.3.1 Phenotypic Techniques

Clinical microbiology laboratories continue to rely primarily on standard biochemical methods to identify and perform antibiotic susceptibility testing on *Serratia* spp. isolates. Microscopy typically demonstrates gram-negative bacilli resembling enteric flora. The *Serratia* spp. are facultative anaerobes and readily grow on a variety of nonselective (e.g., blood agar) and selective (e.g., MacConkey agar) biochemical media at temperatures ranging from 25°C to 37°C. A number of culture characteristics have been recognized that may suggest that the species *S. odorifera* produces alkylmethoxypyrazines that have a potato-like odor. *S. rubidaea* colonies typically demonstrate a red, orange, or pink color due to the production of prodigiosin pigment, and these colors may also be produced by some strains of *S. marcescens* and *S. plymuthica*.[66]

Phenotypic tests can readily differentiate clinically important species within the genus. *Serratia* are slow or weak lactose fermentors and may appear as colorless colonies on MacConkey agar after 24 h of aerobic incubation at 35°C to 37°C, demonstrating a slightly pink color only after 24–48 h.[67] Clinical strains of *Serratia* are typically motile and oxidase- and methyl-red-negative, and are phenotypically unique amongst the Enterobacteriaceae because isolates within the genera produce the hydrolytic enzymes lipase, gelatinase, and DNase.[68] Commercial automated systems (i.e., Vitek 2 GN, bioMerieux, Hazelwood, Missouri; NEG ID Type 2 and Rapid NEG ID Type 3, Microscan, Dade Behring, Siemens, Phoenix, Arizona; NID, BD Diagnostic Systems; and Sensititre GNID, Trek Diagnostic Systems, Inc.) readily identify clinically important *Serratia* spp. except for the *S. liquefaciens* group, which requires carbon assimilation and enzymatic tests that can be performed using a combination of API 50 CH and API ZYM panels, respectively (bioMerieux, Hazelwood, Missouri).[67] The *S. liquefaciens* group is isolated infrequently from clinical sources and *S. marcescens* can be differentiated from this group by its ability to ferment L-arbinose.[68]

Antibiotic susceptibility testing should be performed on all *Serratia* isolates cultured from clinical specimens using standard methods.[69] Although phenotypic methods are used initially to identify resistance, there are currently no standard phenotypic methods for confirmation of the presence of *Amp*C, ESBL, or carbapenem-hydrolyzing enzymes in *Serratia* spp. The phenotypic double antibiotic disk methods that rely on inhibition by clavulanate used to detect the presence of ESBLs in *Escherichia coli*, *Klebsiella pneumoniae*, or *Proteus* spp. are technically demanding and difficult to interpret in organisms, such as *Enterobacter*, *Citrobacter*, and *Serratia*, where other types of β-lactamases are also commonly present (i.e., *Amp*C-type enzyme).[70,71] Molecular methods are therefore increasingly being used to detect the type of β-lactamase present and to characterize antibiotic resistance gene mechanism(s) present in a given pathogenic strain.

88.1.3.2 Molecular Techniques

Identification. PCR assays targeting species-specific genes are valuable techniques for identification of *Serratia* spp. Zhu et al.[72] developed a single-round PCR assay for detection of *Serratia* spp. with primers from *pfs* and *luxS* genes, which are involved in the AI-2-dependent quorum sensing system. The biosynthetic pathway leading to AI-2 production has been previously described, and the enzymes *pfs* and *luxS* are key in this pathway.[73]

In addition, considering that multiple copies of the highly conserved ribosomal RNA genes are present in all bacterial genomes, amplification and sequencing analysis of the initial part of the 16S rRNA gene with universal primers provides a reliable identification of many pathogens.[74] Several broad-range 16S rRNA gene primer sets have been published and applied successfully to the identification of bacterial pathogens.[75,76] However, because some species may share 99%–100% 16S rRNA sequence identities, alternative DNA gene targets need to be exploited to reliably identify these microorganisms. For example, separation of genus and species within the Enterobacteriaceae family can be difficult due to minimal sequence variation (i.e., *E. coli* and *Shigella* spp.), or it may be difficult to separate the distinct genogroups within a given genus such as *Serratia*. Despite these limitations, 16S rRNA sequencing provides an excellent identification of *S. marcescens*.[74] To help differentiate closely related species, published guidelines document which alternative targets may be useful for most human pathogens: targets such as *rec*A, *rpo*B, *tuf*, *gyr*A, *gyr*B, and the *cpn*60 family of proteins have functionally conserved regions with flanking regions of variability.[74]

Furthermore, broad-range PCR using the 16S rRNA gene with restriction fragment length polymorphisms (RFLP) analysis of DNA products has been explored to clinically detect and identify bacterial species causing endophthalmitis that might not be detected by conventional cultivation.[77,78] Amplification of the 16S rRNA gene followed by RFLP analysis of the 16S rRNA PCR products using a cocktail of restriction endonucleases to generate a fragment pattern that is specific to a given pathogen allows identification of the organisms present in the clinical sample. This type of molecular analysis has been done on aqueous and vitreous intraocular samples. Also, broad-range PCR targeting the 16S rRNA gene with denaturing gradient gel electrophoresis (DGGE) analysis of products is invaluable for the detection and identification of bacterial species causing polymicrobial eye infections.[79] DGGE analysis allows the sequence-specific separation of amplicons of the same length amplified from the 16S rRNA gene. During gel electrophoresis, short 16S rRNA amplicons migrate toward increasing denaturing concentrations, leading to a partial melting of the DNA helix and to a decrease and subsequent ending of electrophoretic migration. As a consequence, a band pattern is produced in which each band theoretically represents a bacterial taxon. Band excision followed by 16S rRNA nucleic acid sequencing enables identification of the pathogen(s) of interest.

Microarrays can simultaneously detect tens to hundreds of thousands of unique DNA molecules (probes) of known sequence in a given assay. Diagnostic assays typically use custom or commercial arrays where the presynthesized probes (oligonucleotides or PCR products) are attached to a rigid surface (nylon membrane or chip). *Serratia marcescens* may be detected in polymicrobial infections using microarray technology. A recent study developed a selective diagnostic method for assessing live bacteria by combining sample treatment with the DNA-intercalating dye propidium monoazide with molecular detection using microarray technology.[80] The microarray was designed to detect multiple pathogens as may be encountered in mixed infections and included microarray probes for the following prototype organisms: *Escherichia coli* O157:H7, *Listeria monocytogenes*, *Pseudomonas aeruginosa*, *S. marcescens*, and *Salmonella typhimurium*. PCR products obtained from amplification of chaperonin 60 genes (*cpn*60) were hybridized to a custom-designed microarray containing strain-specific *cpn*60-based 35-mer oligonucleotide probes.[80] Comparison of the microarray assay results with those of quantitative PCR showed that propidium monoazide could inhibit amplification of DNA from killed cells in the mixtures.

Antibiotic Resistance Gene Detection. Molecular detection of ESBLs plays an essential role in the clinical laboratory for screening, tracking, and monitoring the spread of Enterobacteriaceae-producing CTX-M enzymes, particularly in the nosocomial setting. CTX-M enzymes are detected by PCR amplification of the $bla_{\text{CTX-M}}$ gene, leading to the identification of a specific CTX-M phenotype.[81,82] In addition to the detection of the CTX-M-3 genotype in *S. marcescens*,[42] several recent publications provide alternative oligonucleotide primers for amplifying a wide range of CTX-M genes in Enterobacteriaceae.[45,46,82] The presence of an *Amp*C β-lactamase can also be confirmed using oligonucleotide primers specific for the SRT family genes.[41,42] The molecular method commonly used to detect TEM and SHV-type ESBLs is PCR amplification with oligonucleotide primers followed by sequencing. Plasmid-mediated quinolone resistance due to *aac(6′)-Ib-cr* can be also detected using PCR.[60–65] Further, molecular analysis can determine which carbapenem-hydrolyzing genes are present in clinical isolates of *S. marcescens* demonstrating phenotypic resistance to carbapenems.

Serratia and other gram-negative bacteria obtain new resistance genes by acquisition of mobile elements, not only plasmids, but also integrons that encode multiple types of antibiotic resistance genes. An integron is defined as having conserved features such as the ability to incorporate gene cassettes such as the *int*I gene (a specialized site-specific recombination enzyme, integrase), an *att*I site (a recombination site), and a promoter that drives the transcription of genes within the variable region.[50,83–85] Classes 1 and 2 integrons are the most frequently occurring types amongst gram-negative bacteria. Both types of integrons have been found in *Serratia marcescens* using molecular methods.[50,84]

Typing. Because of the frequency of nosocomial outbreaks caused by *S. marcescens*, clinical microbiology laboratories must be able to differentiate between individual isolates to pinpoint potential sources of contamination and to track transmission between patients. Genotyping of *S. marcescens* has been performed using various molecular methods, particularly PFGE.[86–91] PFGE separates different strains by analyzing the chromosomal DNA restriction patterns. PFGE is performed by embedding the organism into agarose, lysing the organisms in situ, and digesting the chromosomal DNA with restriction endonucleases that cleave at infrequent restriction sites.[92] Agarose slices that contain the chromosomal DNA fragments are inserted into the wells of an agarose gel. The restriction fragments are resolved into a pattern of discrete bands in the gel by a CHEF Mapper instrument that changes the direction of current according to a predetermined pattern. The DNA restriction patterns of isolates are then compared; strains are considered related if their PFGE patterns differ by ≤3 bands according to the criteria set out by Tenover et al.[92] After obtaining a pure growth of the organism, standard PFGE for most organisms including *S. marcescens* typically takes 3 days to complete. More rapid modifications of the standard protocol steps prior to preparing the plug mold have been reported that may provide better band resolution.[86,88] The GenePath system (Bio-Rad) has been used to perform PFGE electrophoresis on *S. marcescens* isolates causing outbreaks in neonatal intensive care units.[90] DNA is digested with *spe*I and separated by PFGE with a CHEF-DRII apparatus (Bio-Rad Laboratories) using a 1.1% agarose gel using run parameters of 5–35 s for 24 h at 200 V.[90,91] Gels were analyzed with Molecular Analyst Fingerprinting Plus software (Bio-Rad) to determine strain relatedness.

However, PCR-based typing methods are increasingly being used for the rapid genotyping of *S. marcescens* isolates because of the labor intensity and technical constraints of PFGE. *Serratia marcescens* can be genotyped using PCR-based methods including randomly amplified polymorphic DNA (RAPD), repetitive element (RE)–PCR, and amplification of DNA fragments surrounding rare restriction sites (ADSRRS-fingerprinting).[89,93–95] Several different primer pairs have been described for RAPD analysis, while RE primers based on extragenic palindromic (REP) elements, enterobacterial repetitive intergenic consensus (ERIC) sequences and polymorphic GC-rich repetitive sequences (PGRS) have also been utilized. Although these methods have had limited comparison to the "gold standard" PFGE method, PCR-based methods using specific primer pairs are comparable with respect to the discrimination and reproducibility for epidemiological studies of *S. marcescens* strains in nosocomial outbreaks. One study that compared the use of different primers for genotyping 62 *S. marcescens* clinical isolates showed that PGRS–PCR and RAPD–PCR using the 1254 primer pairs were the most discriminatory, whereas the ERIC–PCR provided poor separation.[94] In addition, PCR-fingerprinting of *S. marcescens* using arbitrary primers (i.e., core sequence of phage M13) by the method previously described was able to discriminate all 15 strains shown by PFGE.[91]

88.2 METHODS

88.2.1 Sample Preparation

Bacterial cultures should be inactivated prior to processing for DNA extraction as follows: Using a sterile loop, transfer bacterial colonies from agar medium into a 1.5 mL screw-capped reaction tube containing 500 μL of phosphate buffered saline, and strongly vortex to create a homogenous suspension. Place the reaction tube into a Thermomixer set at 80°C, 500-rpm mixing speed, for 10 min prior to centrifuging at maximum speed (14,000 rpm) for 10 min and saving the sediment.

Serratia spp. may also be detected using molecular methods from a wide variety of clinical samples including; blood, other sterile fluids (i.e., bronchoalveloar lavage fluid, cerebrospinal fluid, dialysate and other peritoneal fluid, pleural fluid, and synovial fluid), tissues, superficial and deep wound specimens, and urines. In addition, *Serratia* spp. may be the primary pathogen recovered from prosthetic medical devices (i.e., joints, heart valves, bone grafts, gortex grafts) and peripheral and central venous catheters.

In order to get optimal DNA yields for downstream molecular testing, initial sample preparation is required for individual specimen types as follows:

(i) Sterile fluids (nonbloody)—aliquot a minimum of 1 mL of fluid into a sterile 1.5 mL screw-capped reaction tube and centrifuge at maximum speed (14,000 rpm) for 10 min; save the sediment.
(ii) Tissues—mince small biopsies or pieces of tissue with a sterile scalpel in a petri-dish containing 1 mL of thioglycollate broth and distilled water before transferring the processed sample to a sterile 1.5 mL screw-capped reaction tube and centrifuging for 3 min at maximum speed (14,000 rpm).
(iii) Prosthetic medical devices/catheters—place the sample in a plastic bag and immerse in 2 mL sterile phosphate buffered saline prior to ultrasound disruption at 45 kh for 10 min.
(iv) Urine—aliquot a minimum of 1 mL of urine into a 1.5 mL screw-capped reaction tube and centrifuge at maximum speed (14,000 rpm) for 10 min; save 100 μ of the sediment.
(v) Swabs in transport medium—place the swab in 2 mL of sterile phosphate buffered saline and strongly vortex for 3 min. Transfer 1 mL of the suspension into a 1.5 mL screw-capped reaction tube and centrifuge at maximum speed (14,000 rpm) for 10 min; save the sediment.
(vi) Blood and bloody sterile fluids—transfer a 1 mL aliquot aspirated from the blood culture bottle or specimen container using a sterile syringe and needle into a 1.5 mL screw-capped reaction tube.

DNA extraction techniques involve initial cellular rupture using conventional procedures that include the use of lysis buffers containing lysozyme, proteinase K, and sodium dodecyl sulfate, as previously described.[86,93] Released DNA is subsequently purified using conventional phenol/chloroform extraction, precipitation in alcohol, and redissolving the DNA in TE Buffer.[96,97] However, commercialized kits may also be used for genomic DNA extraction, such as the Qiagen DNeasy Blood & Tissue Kit of QIAamp Mini-Blood Kit (Qiagen), AquaPure DNA Isolation Kit (Bio-Rad Laboratories), Wizard® SV Genomic DNA Purification Kit (Promega), and PureLink™ Genomic DNA Kit (Invitrogen, Life Technologies). The purity of extracted DNA should be checked using scanning ultraviolet spectrophotometry from 200 to 300 nm. A ratio of 260:280 of approximately 1.8 indicates that the sample is free of protein contamination that absorbs strongly at 280 nm.[96,97]

88.2.2 Detection Procedures

88.2.2.1 Identification

(i) Single-Round PCR Detection Zhu et al.[72] utilized primers from *pfs* and *luxS* genes involved in the AI-2-dependent quorum sensing system for PCR identification of *Serratia* spp. Specifically, a 193-bp fragment is amplified from the *pfs* gene with the primer set Fpfs1: (5′-CCG GCA TCG GCA AAG TCT-3′) and Rpfs2: (5′-ATC TGG CCC GGC TCG TAG CC-3′); and a 102-bp fragment is amplified from the *lux*S gene with the primer set FluxS1: (5′-GCT GGA ACA CCT GTT CGC-3′) and FluxS2: (5′-ATG TAG AAA CCG GTG CGG-3′).

Procedure

1. PCR mixture (50-μL) is made up of 2 μL of bacterial genome DNA extract and 48 μL of reaction mix that includes: 15 pmol of each primer, 200 μM each of the four dNTPs, 5 μL of 10× amplification buffer, and 2.5 U of *Taq* DNA polymerase.
2. PCR amplification is conducted with one cycle of 94°C for 5 min; 30 cycles of 94°C for 45 s, 55°C for 30 s and 72°C for 15 s; and a final extension at 72°C for 10 min.
3. PCR products are resolved by electrophoresis in 1.5% (w/v) agarose gels and visualized under UV after ethidium bromide staining.

Note: The expected product size after single-round PCR of the *pfs* gene is 193-bp and the *luxS* gene is 102 bp. A positive and DNA template negative control should be included in each PCR run.

(ii) 16S rRNA PCR and Sequencing Useful primer targets within the first 1000 bp of the 16S rRNA gene include: (i) forward sequences: 4F (5′-TTG GAG AGT TTG ATC CTG GCTC-3′) and 27F (5′-AGA GTT TGA TCM TGG CTC AG-3′) and (ii) reverse sequences: 534R (5′-TAC CGC GGC TGC TGG CAC-3′) and 801R (5′-GGC GTG GAC TTC CAG GGT ATCT-3′).[74] The sequence information obtained from the first 500 bp of the 16S rRNA gene often provides adequate differentiation for identification purposes, and most sequence information provided in public databases corresponds to this region of the gene.

Most microbial identification protocols utilize PCR amplification of the 16S rRNA gene, amplicon detection by gel electrophoresis, purification of product followed by cycle sequencing. Although many DNA amplification and sequencing protocols have been described in the literature, clinical laboratories should use the guidelines outlined in CLSI/NCCLS document MM9 and CLSI document MM10 for verifying optimal performance of amplification and sequencing chemistry.[74]

Bidirectional sequencing of the target regions, that is, generated sequence data from both strands of the target molecule is recommended, and can facilitate differentiation between closely related species when sequence variation within the target sequence is minimal. Laboratories should routinely review the sequencing electropherogram for quality score metrics, as outlined in published CLSI guidelines.[74]

Clinical laboratories should use the cutoff values for percent identity scores to identify microorganisms in a consistent manner as follows: (i) ≥99.0% identify for species identification; (ii) ≥97.0% identity for genus identification; and (iii) ≥95% identity cannot be definitively identified by 16S rRNA target sequencing.

(iii) 16S rRNA-Based PCR-RFLP Amplification of the 16S rRNA gene followed by RFLP analysis of the 16S rRNA PCR products with restriction endonucleases generates fragment patterns that are specific to a given pathogen, permitting identification of the organisms present in the clinical sample. Application of this type of molecular analysis to aqueous and vitreous intraocular samples has been described. Multiple 16S rRNA primer sets have been used for this nested-PCR procedure. Okbravi et al.[77,78] used the primer sets 16SF (5′-TTG GAG AGT TTG ATC CTG GCT C-3′) (positions 4–25) and 16SR (5′-ACG TCA TCC CCA CCT TCC TC-3′) (positions 1174–1194) and the primer set NF (5′-GGC GGC AKG CCT AAY ACA TGC AAG T-3′) (positions 42–66) and NR (5′-GAC GAC AGC CAT GCA SCA CCT GT-3′) (positions 1044–1067) to amplify a nested PCR product corresponding to these nucleotide positions of the *E. coli* 16S rRNA sequence.

Procedure

1. PCR mixture (25 μL) is composed of 20 mM Tris-HCl (pH 8.3), 100 mM KCl, 0.01% Tween 20, 0.1% NP-40, 100 μM of each dNTP, 0.3 μM of each primer, 1.5 mM $MgCl_2$, and 3 U of *Taq* DNA polymerase.
2. The first-round PCR (using primers 16SF and 16SR) is performed with one cycle of 95°C for 5 min; 30 cycles of 94°C for 10 s, 56.2°C for 10 s, and 72°C for 15 s; the nested PCR (using primers NF and NR) is conducted with one cycle of 95°C for 5 min; 30 cycles of 94°C for 7 s, 64°C for 7 s, and 72°C for 10 s.
3. PCR products are resolved by electrophoresis in 1.5% (w/v) agarose gels and visualized under UV after ethidium bromide staining. The expected product size after initial-round and nested-PCR amplification of the 16S rRNA gene is ~1170-bp and 1036 bp respectively. PCR product is excised from the gel and recovered into solution using a commercial kit.
4. Restriction enzyme analysis is then carried out on the purified DNA PCR products as follows: The restriction enzyme cocktail (5 μL each enzyme) is added to approximately 1 μg of DNA/PCR product in PCR buffer that has been adjusted to contain 100 mM NaCl, 1 mM dithiothreitol and a final concentration of 7 mM $MgCl_2$. A restriction enzyme cocktail containing the following nine restriction endonucleases is used to achieve speciation: *Afl*III, *Bss*HII, *Cla*I, *Dra*I, *Dra*III, *Hpa*I, *Nde*I, *Nsi*I, and *Sal*I. Restriction digests are performed at 37°C for 18 h, and the reaction is halted by freezing. The restriction fragments are analyzed on 10% TBE/polyacrylamide gels.

Note: *E. coli* and *S. marcescens* have identical RFLP patterns using this cocktail of restriction endonucleases, reflecting the similarity of the 16S rRNA gene sequences for these two species (96.7% identity).

(iv) 16S rRNA-Based PCR–DGGE Broad-range PCR amplification of the 16S rRNA gene followed by denaturing gradient gel electrophoresis (DGGE) analysis of the products has been proven useful for the detection and identification of bacterial species causing polymicrobial eye infections.[79] Multiple primer sets have been described for this nested-PCR procedure. Schabereiter-Gurtner et al.[79] used the initial-round PCR primer set 27f (5′-AGA GTT TGA TCC TGG CTC AG-3′) and 907r (5′-CCC CGT CAA TTC ATT TGA GTT T-3′) generating a PCR product corresponding to nucleotide positions 8–926 of the *E. coli* 16S rRNA sequence, and the nested-PCR primer set 341fGC (5′-CCT ACG GGA GGC AGC AG-3′) to which a 40-base GC clamp was added to the 5′ end (5′-CGC CCG CCG CGC GCG GCG GGC GGG GCG GGG GCA CGG GGG G-3′), and 518r (5′-ATT ACC GCG GCT GCT GG-3′) generating a PCR product corresponding to nucleotide positions 341–534 in the *E. coli* 16S rRNA sequence.

Procedure

1. The first-round PCR mixture (25-μL) contains 2.5 μL of 10× PCR buffer (100 mM Tris-HCl, 100 mM KCl, 15 mM $MgCl_2$, 500 mM KCl; pH 8.3), 200 μM of each dNTP, 12.5 pmol of each primer, 0.5 U of *Taq* DNA polymerase, and 2 μL of extracted bacterial DNA. The nested-PCR mixture (100 μL) is similar except that 4 μL of DNA product from the first-round PCR is used as template.
2. The cycling conditions for the first-round (primers 27f and 907r) and nested-PCR (primers 341f and 518r) cycles include one cycle of 95°C for 5 min; 30 cycles of 95°C for 1 min, 55°C for 1 min, and 72°C for 1 min; and a final extension at 72°C for 5 min.
3. PCR products are resolved by electrophoresis in 2.0% (w/v) agarose gels and visualized under UV with ethidium bromide staining before DGGE analysis. The expected product size of the 16S rRNA gene using the 341f/581r nested-PCR primer set is 193-bp.

4. Ninety μL of nested-PCR product is precipitated in 96% alcohol, resuspended in 15 μL ddH$_2$0 and analyzed by DGGE as previously described in a linear denaturant gradient from 25% to 60% in a D GENE-System (Bio-Rad). After completion of electrophoresis, gels are stained in an ethidium bromide solution and documented with a UVP documentation system.

88.2.2.2 Antibiotic Resistance Gene Detection

(i) AmpC-Type β-Lactamase and ESBLs PCR amplification with oligonucleotide primers followed by sequencing is commonly used to detect TEM and SHV-type ESBLs. While targeting the *bla*$_{CTX-M}$ gene allows identification of a specific CTX-M phenotype,[81,82] confirmation of the SRT family genes reveals the presence of an *Amp*C β-lactamase.[41,42] The following oligonucleotide primers have been used to amply the SHV-, TEM-, and CTX-M-related gene from the plasmid DNA of clinical Enterobacteriaceae isolates with the ESBL phenotype.[42] TEM-forward (5′-ATA AAA TTC TTG AAG ACG AAA-3′), TEM-reverse (5′-GAC AGT TAC CAA TGC TTA ATC-3′) are used to amplify *bla*$_{TEM}$; SHV forward (5′-TGG TTA TGC GTT ATA TTC GCC-3′), SHV-reverse (5′-GGT TAG CGT TGC CAG TGCT-3′) are used to amplify *bla*$_{SHV}$; and CTX-M forward (5′-TGT TGT TAG GAA GTG TGC CGC-3′) and CTX-M-reverse (5′-TCG TTG GTG GTG CCA TAG TC-3′) are used to amplify the *bla*$_{CTX-M-3}$.[42] The complete coding sequences of the *bla*$_{CTX-M-3}$ genes are amplified with the primers CTX-M-3-forward (5′-GGA TCC ATG GTT AAA AAA TCA CTG CG-3′) and CTX-M-3-reverse (5′-AAG CTT TTA CAA ACC GTC GGT GAC-3′) as described previously.[42,98] The *Amp*C genes are detected using a pair of specific primers, SRT-forward (5′-GCC GAT ACC CTG CAA CCT AAG-3′) and SRT-reverse (5′-CGC CTG GAC GAT GTG GTA AG-3′) to amplify the *bla*SRT-1 gene.[41,42]

Procedure

1. Plasmid DNA from clinical isolates is extracted by previously published methods. Genomic DNA of *S. marcescens* is extracted using the QIAamp DNA Mini Kit (Qiagen).
2. PCR mixture is prepared.
3. The cycling conditions for all these genes include one cycle of 94°C for 3 min; 35 cycles of 94°C for 1 min, 55°C for 1 min, and 72°C for 2 min; and a final extension at 72°C for 7 min.
4. Amplicons are detected by electrophoresis on a 1.5% agarose gel with staining by ethidium bromide and visualization by ultraviolet (UV) light.
5. Amplicons are purified using QIAquick PCR Purification Kit (Qiagen).
6. DNA sequence analysis is performed using Big DYE terminator cycle sequencing chemistry on an automated sequencer (ABI PRISM 3730, Applied Biosystems, Foster City, California).
7. The sequencers are compared with the reported sequences in GenBank.

(ii) Aac(6′)-Ib-c The following oligonucleotide primers have been used to detect the presence of *aac(6′)-Ib*[61,65]: forward (5′-TTG CGA TGC TCT ATG GGC TA-3′) and reverse (5′-CTC GAA TGC CTG GCG TGT TT-3′) to produce a 482-bp PCR fragment. These primers recognize all known *aac(6′)-Ib* variants.

Procedure

1. PCR mixture is prepared as reported.[61,65] Strains positive and negative for *aac(6′)-Ib* are included as controls.
2. The PCR is subjected to 34 cycles of 94°C for 45 s, 55°C for 45 s, and 72°C for 2 min; and a final extension at 72°C for 7 min.
3. All *aac-(6′)-Ib* positives are further analyzed by digestion with the restriction enzyme *Bst*F51 (New England Biolabs) and/or direct DNA sequencing of the PCR products using primer (5′-CGT CAC TCC ATA CAT TGC AA-3′) to verify the presence of *aac(6′)-Ib-cr*, which lacks that *Bst*F51 restriction site present in the wild-type gene.

(iii) Carbapenem-Hydrolyzing Enzymes The following oligonucleotide primer pairs are used to detect the presence of *bla* genes by PCR analysis[99]: for *bla*$_{IMP}$-forward (1241F 5′-CTA CCG CAG AGT CTT TG-3′) and *bla*$_{IMP}$-reverse (1828R 5′-AAC CAG TTT TGC CTT ACC AT-3′) that amplify a 587-bp product for *bla*$_{TEM}$-forward (90F 5′-TCG GGG AAA TGT GCG CG-3′) and for *bla*$_{TME}$-reverse (1062R 5′-TGC TTA ATC AGT GAG GCA CC-3′) that amplify a 971-bp product, for *bla*$_{SHV}$-forward (103F 5′-CAC TCA AGG ATG TAT TGT G-3′) and *bla*$_{SHV}$-reverse (988R 5′-TTA GCG TTG CCA GTG CTC G-3′) that amplify a 885-bp product, and for *bla*$_{SME}$-forward (151F 5′-AAC GGC TTC ATT TTT GTT TAG-3′) and *bla*$_{SME}$-reverse (981 R 5′-GCT TCC GCA ATA GTT TTA TCA-3′) that amplify a 831-bp product. All the primer pairs are designed as previously reported.[54,83] Alternative primers for the PCR identification of plasmid- and integron-borne *bla*$_{IMP}$-1 and *bla*$_{IMP}$-10 in clinical isolates of *S. marcescens* have also recently been reported.[50] The following oligonucleotide primer pairs are used to detect the presence of *bla*$_{KPC}$-α genes by PCR analysis: KPC-1 forward (5′-TGT CAC TGT ATC GCC GTC-3′) and KPC-1 reverse (5′-CTC AGT GCT CTA CAG AAA ACC-3′).[100]

Procedure

1. PCR is performed for the various primer pairs according to previously published reaction mixtures and cycling conditions for *bla*$_{IMP}$, *bla*$_{TEM}$, *bla*$_{SHV}$, and *bla*$_{SME}$.
2. The PCR mixture (100 μL) for detection of *bla*$_{KPC}$ contain 0.5 μM of each primer, 250 μM of each dNTPs, 2 mM $MgCl_2$, and 2.5 U of *Taq* DNA polymerase prepared in a 1× reaction buffer supplied by the manufacturer (Applied Biosystems).
3. Cycling parameters for the *bla*$_{KPC}$ PCR are as follows: one cycle of 95°C for 5 min; 35 cycles of 95°C

for 1 min, 58°C for 30 s, and 72°C for 1 min; a final cycle at 72°C for 1 min.

4. Amplicons are detected by electrophoresis on a 1.5% agarose gel with ethidium bromide and visualized under ultraviolet (UV) light.
5. The amplicons from *bla* gene PCR reactions are purified using QIAquick PCR Purification Kit (Qiagen, Germany). The purified PCR products are then cloned and the recombinant plasmid is transformed into *E. coli* DH5α prior to sequencing the *bla* gene insert as previously described.[100]
6. Additional primer pairs are used to perform DNA sequence analysis of the bla_{KPC} plasmid insert as previously described.[100] DNA sequence analysis is performed using Big DYE terminator cycle ABI PRISM 3730 (Applied Biosystems).
7. The sequences are compared with the reported sequences in GenBank.

(iv) Integrons Bacteria possessing integron are capable of incorporating gene cassettes, such as the *int*I gene (a specialized site-specific recombination enzyme, integrase), an *att*I site (a recombination site), and a promoter that drives the transcription of genes within the variable region.[50,83–85] These features enhance bacterial resistance to antibiotics. Class 1 and 2 integrons are the most frequently found amongst gram-negative bacteria, including *Serratia marcescens*.[50,84] The following oligonucleotide primer pairs are used to perform integron analysis by PCR: Int-1F (5′CS) (5′-GGC ATC CAA GCA GCA AGC-3′), and Int-1R (3′CS) (5′-AAG CAG ACT TGA CCT GAT-3′); qacEΔ1 F (5′-ATC GCA ATA GTT GGC GAA GT-3′) and qacEΔ1 B (5′-CAA GCT TTT GCC CAT GAA GC-3′) that amplify a 225 bp product; *Sul*1 F (5′-CTT CGA TGA GAG CCG GCG GC-3′) and *Sul*1 B (5′-GCA AGG CGG AAA CCC GCG CC-3′) that amplify a 436-bp product; hep74 (5′CS) (5′-CGG GAT CCC GGA CGG CAT GCA CGA TTT GTA-3′) and hep51 (3′CS) (5′-GAT GCC ATC GCA AGT ACG AG-3′); Int1A (5′CS) (5′-AAA ACC GCC ACT GCG CCG TTA-3′), and Int1B (3′CS) (5′-GAA GAC GGC TGC ACT GAA CG-3′) that amplify a 1200-bp product; Int2A (5′CS) (5′-ATG TCT AAC AGT CCA TTT TTA AAT TCT A-3′) and Int2B (3′CS) (5′-AAA TCT TTA ACC CGC AAA CGC-3′) that amplify a 443-bp product; and RW3A (5′-TCG CTC AAA ACA ACG ACA CC-3′). All the primer pairs are designed as previously reported.[101–106] Other primer pairs have also been reported for detection of some integron genes (*int*I, qacEΔ1, and *Sul*1 in *Serratia*.[50]

Procedure

1. The PCR mixture (50 μL) for detection of integrons contains 25 pmol of each primer, 200 μM/L dNTPs, 1.5 mM/L $MgCl_2$, and 2.5μ of *Taq* DNA polymerase prepared in a 10× reaction buffer (10 mMol/L Tris-HCl, pH 8.0, 100 mMol/L NaCl, 0.1 mMol/L dithiothreitol, 50% glycerol, and 1% triton X-100). For each run, a negative control is included.
2. Cycling parameters for PCR are as follows: one cycle of 94°C for 5 min; 35 cycles of 94°C for 1 min, 54°C for 1 min, and 72°C for 2 min; and a final extension cycle at 72°C for 1 min.
3. Amplicons are detected by electrophoresis on a 1.5% agarose gel with ethidium bromide and visualized under ultraviolet (UV) light.
4. Amplified DNA products are purified using a QIAquick PCR Purification Kit (Qiagen).
5. The purified amplicons are sequenced by automated methods. Briefly, class 2 integron sequencing is performed using a forward and reverse read of the amplicon generated by the primer pairs hep74 and hep51 which result in two sequences of ~800 bp. Another primer pair is designed to allow amplification of the central portion of the 2143-bp amplicon to complete the sequencing.
6. The sequencers are compared with the reported sequences in GenBank.

88.2.2.3 Genotyping

Genotyping *Serratia marcescens* is often achieved by using randomly amplified polymorphic DNA (RAPD), repetitive element (RE)–PCR, and amplification of DNA fragments surrounding rare restriction sites (ADSRRS-fingerprinting).[89,93–95] The following oligonucleotide primers have shown the best discrimination of *S. marcescens* strains using PCR-based methods: (i) RAPD–PCR: 1254 (5′-CCG CAG CCAA-3′); (b) PGRS–PCR: (5′-CCG CG TTG GCC GTT GCC GCCG-3′); and (c) phage M13 (5′-GAG GGT GGC GGT TTCT-3′).[91]

Procedure

1. PCR mixtures for RAPD–PCR and PGRS–PCR are prepared as previously reported.[94] The PCR mixture (50 μL) for detection of phage M13 contains 25 pmol of M13 primer, 200 μM of each dNTPs, 1.5 mM $MgCl_2$, and 2.5 U of *Taq* DNA polymerase prepared in a 10× reaction buffer (10 mM Tris-HCl, pH 8.3, 50 mM KCl). For each run, a negative control is included.
2. Cycling parameters for phage M13-PCR reactions are as follows: 35–40 cycles of 93°C for 1 min, 50°C for 1 min, and 72°C for 20 s with a final extension cycle at 72°C for 6 min.
3. Amplicons are detected by electrophoresis on a 1.2% agarose gel for 6 h at 3 V/cm and staining with ethidium bromide and visualized under ultraviolet (UV) light.

88.3 CONCLUSION AND FUTURE PERSPECTIVES

During the past half-century, the *Serratia* spp., especially *S. marcescens*, have emerged as major causes of human disease. While mostly recognized as agents of nosocomial infections, particularly in critically ill patients, *Serratia* spp. are also a significant and underappreciated causes of community-onset disease. The *Serratia* spp. cause a wide range of clinical infections which may range in severity from uncomplicated urinary tract infections to severe life-threatening invasive infections. While the pathogenesis of *Serratia* spp. infections has not been investigated in as much detail as for many of the other Enterobacteriaceae, a number of potential virulence factors have been identified, and the factors associated with biofilm formation have been studied. The *Serratia* spp. are characterized by their inducible chromosomal resistance to β-lactam antibiotics and propensity to acquire resistance to many antimicrobial agents via mobile genetic elements. With increasingly complex and aging patient populations and the associated use of medical devices, *Serratia* spp. will inevitably increase in importance as human pathogens. Future studies directed to defining pathogenesis of these infections will be important to identify targets for prevention and for improving therapies, especially in the setting of biofilm formation.

The clinically important species within the genus and their antimicrobial susceptibilities are generally readily identified using standard phenotypic techniques. Molecular methods are however increasing in importance in the clinical laboratory for detection, defining mechanisms of resistance, and for epidemiologic analyses. The development and implementation into routine clinical practice of rapid molecular techniques for the early detection and prediction of antimicrobial activity will likely be a necessary step in improving adequate early therapy and resulting outcomes of serious infections due to this genus.

REFERENCES

1. Badenoch, P.R., Thom, A.L., and Coster, D.J., *Serratia ficaria* endophthalmitis, *J. Clin. Microbiol.*, 40, 1563, 2002.
2. Traub, W.H., Antibiotic susceptibility of *Serratia marcescens* and *Serratia liquefaciens*, *Chemotherapy*, 46, 315, 2000.
3. Yu, V.L., *Serratia marcescens*: Historical perspective and clinical review, *N. Engl. J. Med.*, 300, 887, 1979.
4. Wheat, R.P., Zuckerman, A., and Rank, L.A., Infection due to *Chromobacteria*: Report of eleven cases, *Arch. Intern. Med.*, 88, 461, 1951.
5. Kollef, M.H. et al., Epidemiology and outcomes of health-care-associated pneumonia: Results from a large US database of culture-positive pneumonia, *Chest*, 128, 3854, 2005.
6. Ksycki, M.F., and Namias, N., Nosocomial urinary tract infection, *Surg. Clin. North Am.*, 89, 475, 2009.
7. Gill, A.W., Analysis of neonatal nosocomial infection rates across the Australian and New Zealand Neonatal Network, *J. Hosp. Infect.*, 72, 155, 2009.
8. Gaynes, R., and Edwards, J.R., Overview of nosocomial infections caused by gram-negative bacilli, *Clin. Infect. Dis.*, 41, 848, 2005.
9. Vonberg, R.P., and Gastmeier, P., Hospital-acquired infections related to contaminated substances, *J. Hosp. Infect*, 65, 15, 2007.
10. Ramirez-Arcos, S. et al., Fatal septic shock associated with transfusion-transmitted *Serratia marcescens*, *Transfusion*, 46, 679, 2006.
11. Civen, R. et al., Outbreak of *Serratia marcescens* infections following injection of betamethasone compounded at a community pharmacy, *Clin. Infect. Dis.*, 43, 831, 2006.
12. Su, J.R. et al., Epidemiologic investigation of a 2007 outbreak of *Serratia marcescens* bloodstream infection in Texas caused by contamination of syringes prefilled with heparin and saline, *Infect. Control Hosp. Epidemiol.*, 30, 593, 2009.
13. Sunenshine, R.H. et al., A multistate outbreak of *Serratia marcescens* bloodstream infection associated with contaminated intravenous magnesium sulfate from a compounding pharmacy, *Clin. Infect. Dis.*, 45, 527, 2007.
14. Pan, A. et al., An outbreak of *Serratia marcescens* bloodstream infections associated with misuse of drug vials in a surgical ward, *Infect. Control Hosp. Epidemiol.*, 27, 79, 2006.
15. Henry, B., Plante-Jenkins, C., and Ostrowska, K., An outbreak of *Serratia marcescens* associated with the anesthetic agent propofol, *Am. J. Infect. Control*, 29, 312, 2001.
16. Yoon, H.J. et al., Outbreaks of *Serratia marcescens* bacteriuria in a neurosurgical intensive care unit of a tertiary care teaching hospital: A clinical, epidemiologic, and laboratory perspective, *Am. J. Infect. Control*, 33, 595, 2005.
17. Shih, H.I. et al., *Serratia marcescens* bacteremia at a medical center in southern Taiwan: High prevalence of cefotaxime resistance, *J. Microbiol. Immunol. Infect.*, 38, 350, 2005.
18. Horcajada, J.P. et al., Acquisition of multidrug-resistant *Serratia marcescens* by critically ill patients who consumed tap water during receipt of oral medication, *Infect. Control Hosp. Epidemiol.*, 27, 774, 2006.
19. Takahashi, H. et al., Nosocomial *Serratia marcescens* outbreak in Osaka, Japan, from 1999 to 2000, *Infect. Control Hosp. Epidemiol.*, 25, 156, 2004.
20. Cullen, M.M. et al., *Serratia marcescens* outbreak in a neonatal intensive care unit prompting review of decontamination of laryngoscopes, *J. Hosp. Infect.*, 59, 68, 2005.
21. Ostrowsky, B.E. et al., *Serratia marcescens* bacteremia traced to an infused narcotic, *N. Engl. J. Med.*, 346, 1529, 2002.
22. Harnett, S.J., Allen, K.D., and Macmillan, R.R., Critical care unit outbreak of *Serratia liquefaciens* from contaminated pressure monitoring equipment, *J. Hosp. Infect.*, 47, 301, 2001.
23. Rabier, V. et al., Hand washing soap as a source of neonatal *Serratia marcescens* outbreak, *Acta Paediatr.*, 97, 1381, 2008.
24. Buffet-Bataillon, S. et al., Outbreak of *Serratia marcescens* in a neonatal intensive care unit: Contaminated unmedicated liquid soap and risk factors, *J. Hosp. Infect.*, 72, 17, 2009.
25. Ivanova, D. et al., Extended-spectrum beta-lactamase-producing *Serratia marcescens* outbreak in a Bulgarian hospital, *J. Hosp. Infect.*, 70, 60, 2008.
26. de Vries, J.J. et al., Outbreak of *Serratia marcescens* colonization and infection traced to a healthcare worker with long-term carriage on the hands, *Infect. Control Hosp. Epidemiol.*, 27, 1153, 2007.
27. Vincent, J.L., Nosocomial infections in adult intensive-care units, *Lancet*, 361, 2068, 2003.
28. Al Jarousha, A.M. et al., An outbreak of *Serratia marcescens* septicaemia in neonatal intensive care unit in Gaza City, Palestine, *J. Hosp. Infect.*, 70, 119, 2008.
29. Laupland, K.B. et al., Population-based laboratory surveillance for *Serratia* species isolates in a large Canadian health region, *Eur. J. Clin. Microbiol. Infect. Dis.*, 27, 89, 2008.
30. Engel, H.J. et al., *Serratia* sp. bacteremia in Canberra, Australia: A population-based study over 10 years, *Eur. J. Clin. Microbiol. Infect. Dis.*, 28, 821, 2009.

31. Parkins, M.D., and Gregson, D.B., Community-acquired *Serratia marcescens* spinal epidural abscess in a patient without risk factors: Case report and review, *Can. J. Infect. Dis. Med. Microbiol.*, 19, 250, 2008.
32. Liangpunsakul, S., and Pursell, K., Community-acquired necrotizing fasciitis caused by Serratia marcescens: Case report and review, *Eur. J. Clin. Microbiol. Infect. Dis.*, 20, 509, 2001.
33. Teertstra, T.K. et al., *Serratia marcescens* septicemia presenting as purpura fulminans in a premature newborn, *Eur. J. Clin. Microbiol. Infect. Dis.*, 24, 846, 2005.
34. Kida, Y. et al., *Serratia marcescens* serralysin induces inflammatory responses through protease-activated receptor 2, *Infect. Immun.*, 75, 164, 2007.
35. Kurz, C.L. et al., Virulence factors of the human opportunistic pathogen *Serratia marcescens* identified by in vivo screening, *Embo J.*, 22, 1451, 2003.
36. Labbate, M. et al., Quorum-sensing regulation of adhesion in *Serratia marcescens* MG1 is surface dependent, *J. Bacteriol.*, 189, 2702, 2007.
37. Van Houdt, R., Givskov, M., and Michiels, C.W., Quorum sensing in *Serratia*, *FEMS Microbiol. Rev.*, 31, 407, 2007.
38. Costerton, J.W., Stewart, P.S., and Greenberg, E.P., Bacterial biofilms: A common cause of persistent infections, *Science*, 284, 1318, 1999.
39. Paterson, D.L., and Bonomo, R.A., Extended-spectrum beta-lactamases: A clinical update, *Clin. Microbiol. Rev.*, 18,657, 2005.
40. Pitout, J.D., Multiresistant Enterobacteriaceae: New threat of an old problem, *Expert Rev. Anti. Infect. Ther.*, 6, 657, 2008.
41. Yu, W.L. et al., Institutional spread of clonally related *Serratia marcescens* isolates with a novel AmpC cephalosporinase (S4): A 4-year experience in Taiwan, *Diagn. Microbiol. Infect. Dis.*, 61, 460, 2008.
42. Wu, L.T. et al., Survey of CTX-M-3 extended-spectrum beta-lactamase (ESBL) among cefotaxime-resistant *Serratia marcescens* at a medical center in middle Taiwan, *Diagn. Microbiol. Infect. Dis.*, 49, 125, 2004.
43. Choi, S.H. et al., Emergence of antibiotic resistance during therapy for infections caused by Enterobacteriaceae producing AmpC beta-lactamase: Implications for antibiotic use, *Antimicrob. Agents Chemother.*, 52, 995, 2008.
44. Kim, B.N. et al., Outcome of antibiotic therapy for third-generation cephalosporin-resistant Gram-negative bacteraemia: An analysis of 249 cases caused by *Citrobacter*, *Enterobacter* and *Serratia* species, *Int. J. Antimicrob. Agents*, 22, 106, 2003.
45. Pitout, J.D. et al., Molecular epidemiology of CTX-M-producing *Escherichia coli* in the Calgary Health Region: Emergence of CTX-M-15-producing isolates, *Antimicrob. Agents Chemother.*, 51, 1281, 2007.
46. Pitout, J.D. et al., Molecular characteristics of travel-related extended-spectrum-beta-lactamase-producing *Escherichia coli* isolates from the Calgary Health Region, *Antimicrob. Agents Chemother.*, 53, 2539, 2009.
47. Pitout, J.D., and Laupland, K.B., Extended-spectrum beta-lactamase-producing Enterobacteriaceae: An emerging public-health concern, *Lancet Infect. Dis.*, 8, 159, 2008.
48. Szabo, D. et al., SHV-type extended-spectrum beta-lactamase production is associated with reduced cefepime susceptibility in *Enterobacter cloacae*, *J. Clin. Microbiol.*, 43, 5058, 2005.
49. Deshpande, L.M. et al., Occurrence and characterization of carbapenemase-producing Enterobacteriaceae: Report from the SENTRY Antimicrobial Surveillance Program (2000–2004), *Microb. Drug Resist.*, 12, 223, 2006.
50. Hu, Z., and Zhao, W.H., Identification of plasmid- and integron-borne blaIMP-1 and blaIMP-10 in clinical isolates of *Serratia marcescens*, *J. Med. Microbiol.*, 58, 217, 2009.
51. Ito, H. et al., Plasmid-mediated dissemination of the metallo-beta-lactamase gene blaIMP among clinically isolated strains of *Serratia marcescens*, *Antimicrob. Agents Chemother.*, 39, 824, 1995.
52. Carrer, A. et al., Occurrence of an SME-2-producing *Serratia marcescens* isolate in Canada, *Int. J. Antimicrob. Agents*, 31, 181, 2008.
53. Gales, A.C. et al., Carbapenem-resistant *Serratia marcescens* isolates producing Bush group 2f beta-lactamase (SME-1) in the United States: Results from the MYSTIC Programme, *Diagn. Microbiol. Infect. Dis.*, 39, 125, 2001.
54. Naas, T. et al., Cloning and sequence analysis of the gene for a carbapenem-hydrolyzing class A beta-lactamase, Sme-1, from *Serratia marcescens* S6, *Antimicrob. Agents Chemother.*, 38, 1262, 1994.
55. Poirel, L., Pitout, J.D., and Nordmann, P., Carbapenemases: Molecular diversity and clinical consequences, *Future Microbiol.*, 2, 501, 2007.
56. Walther-Rasmussen, J., and Hoiby, N., Class A carbapenemases, *J. Antimicrob. Chemother.*, 60, 470, 2007.
57. Cai, J.C. et al., Emergence of *Serratia marcescens*, *Klebsiella pneumoniae*, and *Escherichia coli* isolates possessing the plasmid-mediated carbapenem-hydrolyzing beta-lactamase KPC-2 in intensive care units of a Chinese hospital, *Antimicrob. Agents Chemother.*, 52, 2014, 2008.
58. Mammeri, H. et al., Emergence of plasmid-mediated quinolone resistance in *Escherichia coli* in Europe, *Antimicrob. Agents Chemother.*, 49, 71, 2005.
59. Pitout, J.D. et al., Emergence of Enterobacteriaceae producing extended-spectrum beta-lactamases (ESBLs) in the community, *J. Antimicrob. Chemother.*, 56, 52, 2005.
60. Nordmann, P., and Poirel, L., Emergence of plasmid-mediated resistance to quinolones in Enterobacteriaceae, *J. Antimicrob. Chemother.*, 56, 463, 2005.
61. Park, C.H. et al., Prevalence in the United States of aac(6′)-Ib-cr encoding a ciprofloxacin-modifying enzyme, *Antimicrob. Agents Chemother.*, 50, 3953, 2006.
62. Park, Y.J. et al., Spread of *Serratia marcescens* coharboring aac(6′)-Ib-cr, bla CTX-M, armA, and bla OXA-1 carried by conjugative IncL/M type plasmid in Korean hospitals, *Microb. Drug Resist.*, *15*, 97, 2009.
63. Robicsek, A. et al., Fluoroquinolone-modifying enzyme: A new adaptation of a common aminoglycoside acetyltransferase, *Nat. Med.*, 12, 83, 2006.
64. Pitout, J.D. et al., Surveillance for plasmid-mediated quinolone resistance determinants in Enterobacteriaceae within the Calgary Health Region, Canada: The emergence of aac(6′)-Ib-cr, *J. Antimicrob. Chemother.*, 61, 999, 2008.
65. Kim, S.Y. et al., Prevalence and characteristics of aac(6′)-Ib-cr in AmpC-producing *Enterobacter cloacae*, *Citrobacter freundii*, and *Serratia marcescens*: A multicenter study from Korea, *Diagn. Microbiol. Infect. Dis.*, 63, 314, 2009.
66. Williams, R.P., Green, J.A., and Rappo-Port, D.A., Studies on pigmentation of *Serratia marcescens*. I. Spectral and paper chromatographic properties of prodigiosin, *J. Bacteriol.*, 71, 115, 1956.
67. Abbott, S.L., *Klebsiella, Enterobacter, Citrobacter, Serratia, Plesiomonas,* and Other Enterobacteriaceae. In Murray, P.R., et al. (Editors), *Manual of Clinical Microbiology*, 9th ed., vol. 1, pp. 698–715, American Society for Microbiology, Washington, DC, 2007.

68. Janda, J.M., and Abbott, S.L., *The Enterobacteria*, 2nd ed., ASM Press, Washington, DC, 2006.
69. CLSI, Performance Standards for Antimicrobial Susceptibility Testing; Nineteenth Informational Supplement, 2009.
70. Drieux, L. et al., Phenotypic detection of extended-spectrum beta-lactamase production in Enterobacteriaceae: Review and bench guide, *Clin. Microbiol. Infect.*, 14, 90, 2008.
71. Pitout, J.D. et al., Modification of the double-disk test for detection of enterobacteriaceae producing extended-spectrum and AmpC beta-lactamases, *J. Clin. Microbiol.*, 41, 3933, 2003.
72. Zhu, H., Sun, S.J., and Dang, H.Y., PCR detection of *Serratia* spp. using primers targeting pfs and luxS genes involved in AI-2-dependent quorum sensing, *Curr. Microbiol.*, 57, 326, 2008.
73. Schauder, S. et al., The LuxS family of bacterial autoinducers: Biosynthesis of a novel quorum-sensing signal molecule, *Mol. Microbiol.*, 41, 463, 2001.
74. CLSI, MM18-A, Interpretive criteria for identification of bacteria and fungi by DNA target sequencing, Approved guideline, 2008.
75. Weisburg, W.G. et al., 16S ribosomal DNA amplification for phylogenetic study, *J. Bacteriol.*, 173, 697, 1991.
76. Clarridge, J.E., Impact of 16S rRNA gene sequence analysis for identification of bacteria on clinical microbiology and infectious diseases, *Clin. Microbiol. Rev.*, 17, 840, 2004.
77. Okhravi, N., Adamson, P., and Lightman, S., Use of PCR in endophthalmitis, *Ocul. Immunol. Inflamm.*, 8, 189, 2000.
78. Okhravi, N. et al., PCR-RFLP-mediated detection and speciation of bacterial species causing endophthalmitis, *Invest. Ophthalmol. Vis. Sci.*, 41, 1438, 2000.
79. Schabereiter-Gurtner, C. et al., 16S rDNA-based identification of bacteria from conjunctival swabs by PCR and DGGE fingerprinting, *Invest. Ophthalmol. Vis. Sci.*, 42, 1164, 2001.
80. Nocker, A. et al., Selective detection of live bacteria combining propidium monoazide sample treatment with microarray technology, *J. Microbiol. Methods*, 76, 253, 2009.
81. Pitout, J.D. et al., Population-based laboratory surveillance for *Escherichia coli*-producing extended-spectrum beta-lactamases: Importance of community isolates with blaCTX-M genes, *Clin. Infect. Dis.*, 38, 1736, 2004.
82. Pitout, J.D., Hossain, A., and Hanson, N.D., Phenotypic and molecular detection of CTX-M-beta-lactamases produced by *Escherichia coli* and *Klebsiella* spp., *J. Clin. Microbiol.*, 42, 5715, 2004.
83. Arakawa, Y. et al., A novel integron-like element carrying the metallo-beta-lactamase gene blaIMP, *Antimicrob. Agents Chemother.*, 39, 1612, 1995.
84. Crowley, D., Cryan, B., and Lucey, B., First detection of a class 2 integron among clinical isolates of Serratia marcescens, *Br. J. Biomed. Sci.*, 65, 86, 2008.
85. Poirel, L. et al., Integron mobilization unit as a source of mobility of antibiotic resistance genes, *Antimicrob. Agents Chemother.*, 53, 2492, 2009.
86. Alaidan, A. et al., Rapid PFGE method for fingerprinting of *Serratia marcescens* isolates, *J. Microbiol. Methods*, 78, 238, 2009.
87. Alfizah, H., Nordiah, A.J., and Rozaidi, W.S., Using pulsed-field gel electrophoresis in the molecular investigation of an outbreak of *Serratia marcescens* infection in an intensive care unit, *Singapore Med. J.*, 45, 214, 2004.
88. Ishii, Y. et al., Rapid pulsed-field gel electrophoresis technique for determination of genetic diversity of *Serratia marcescens*, *J. Infect. Chemother.*, 8, 368, 2002.
89. Krawczyk, B. et al., Evaluation and comparison of random amplification of polymorphic DNA, pulsed-field gel electrophoresis and ADSRRS-fingerprinting for typing *Serratia marcescens* outbreaks, *FEMS Immunol. Med. Microbiol.*, 38, 241, 2003.
90. Maragakis, L.L. et al., Outbreak of multidrug-resistant *Serratia marcescens* infection in a neonatal intensive care unit, *Infect. Control Hosp. Epidemiol.*, 29, 418, 2008.
91. Steppberger, K. et al., Nosocomial neonatal outbreak of *Serratia marcescens*-analysis of pathogens by pulsed field gel electrophoresis and polymerase chain reaction, *Infection*, 30, 277, 2002.
92. Tenover, F.C. et al., Interpreting chromosomal DNA restriction patterns produced by pulsed-field gel electrophoresis: Criteria for bacterial strain typing, *J. Clin. Microbiol.*, 33, 2233, 1995.
93. Enciso-Moreno, J.A. et al., Identification of *Serratia marcescens* populations of nosocomial origin by RAPD-PCR, *Arch. Med. Res.*, 35, 12, 2004.
94. Patton, T.G. et al., Genotyping of clinical *Serratia marcescens* isolates: A comparison of PCR-based methods, *FEMS Microbiol. Lett.*, 194, 19, 2001.
95. Versalovic, J., Koeuth, T., and Lupski, J.R., Distribution of repetitive DNA sequences in eubacteria and application to fingerprinting of bacterial genomes, *Nucleic Acids Res.*, 19, 6823, 1991.
96. Heptinstall, J., Spectrophotometric analysis of nucleic acids. In Rapley, R. (Editor), *The Nucleic Acid Protocols Handbook*, pp. 57–60, Human Press, Totowa, NJ, 2000.
97. Corkill, G., and Rapley, R., The manipulation of nucleic acids. In Walker, J. M., and Rapley, R. (Editors), *Molecular Biomethods Handbook*, 2nd ed., pp. 3–15, Humana Press, Totowa, NJ, 2008.
98. Cheng, K.C. et al., Clinical experiences of the infections caused by extended-spectrum beta-lactamase-producing *Serratia marcescens* at a medical center in Taiwan, *Jpn. J. Infect. Dis.*, 59, 147, 2006.
99. Zhao, W.H. et al., Characterization of imipenem-resistant *Serratia marcescens* producing IMP-type and TEM-type beta-lactamases encoded on a single plasmid, *Microbiol. Res.*, 162, 46, 2007.
100. Yigit, H. et al., Novel carbapenem-hydrolyzing beta-lactamase, KPC-1, from a carbapenem-resistant strain of *Klebsiella pneumoniae*, *Antimicrob. Agents Chemother.*, 45, 1151, 2001.
101. Bissonnette, L., and Roy, P.H., Characterization of In0 of *Pseudomonas aeruginosa* plasmid pVS1, an ancestor of integrons of multiresistance plasmids and transposons of gram-negative bacteria, *J. Bacteriol.*, 174, 1248, 1992.
102. Del Vecchio, V.G. et al., Molecular genotyping of methicillin-resistant *Staphylococcus aureus* via fluorophore-enhanced repetitive-sequence PCR, *J. Clin. Microbiol.*, 33, 2141, 1995.
103. Falbo, V. et al., Antibiotic resistance conferred by a conjugative plasmid and a class I integron in Vibrio cholerae O1 El Tor strains isolated in Albania and Italy, *Antimicrob. Agents Chemother.*, 43, 693, 1999.
104. Sandvang, D., Aarestrup, F.M., and Jensen, L.B., Characterisation of integrons and antibiotic resistance genes in Danish multiresistant *Salmonella enterica* Typhimurium DT104, *FEMS Microbiol. Lett.*, 160, 37, 1998.
105. Stokes, H.W., and Hall, R.M., A novel family of potentially mobile DNA elements encoding site-specific gene-integration functions: Integrons. *Mol. Microbiol.*, 3, 1669, 1989.
106. White, P.A., McIver, C.J., and Rawlinson, W.D., Integrons and gene cassettes in the enterobacteriaceae, *Antimicrob. Agents Chemother.*, 45, 2658, 2001.

89 Shigella

K.R. Schneider, L.K. Strawn, K.A. Lampel, and B.R. Warren

CONTENTS

89.1 INTRODUCTION

89.1.1 Classification, Morphology, and Biology

Shigella spp. are the causative agent of shigellosis or "bacillary dysentery," and has been implicated in many worldwide foodborne outbreaks annually. According to the Centers for Disease Control and Prevention's (CDC's) Emerging Infections Program (EIP), Foodborne Diseases Active Surveillance Network (FoodNet), *Shigella* was the third most reported foodborne bacterial pathogen in the United States in 2002. Humans, and possibly higher primates, have been the only identified hosts of this pathogen, and transmission occurs via the fecal-oral route. A common route of contamination with *Shigella* in foods is by an infected food handler who practices poor personal hygiene. Shigellosis outbreaks usually involve large numbers of individuals, and the outbreaks can linger due to the fact that *Shigella* can be carried asymptomatically.[1] *Shigella* is acid resistant, salt tolerant, and can survive at infective levels in many types of foods, such as fruits and vegetables, low pH foods, prepared foods, and foods held in modified atmosphere packaging or vacuum packaging. Survival is often increased when food is held at refrigerated temperatures. Detection methods for *Shigella* include conventional culture methods, immunological methods, and molecular-based methods. Conventional culture of *Shigella* in foods is often problematic due to the lack of appropriate selective media. Immunological methods for *Shigella* have been researched, yet there is only one commercially available test kit. Molecular microbiological methods such as PCR, oligonucleotide microarrays, and rep-PCR have also been developed for the detection and identification of *Shigella*. This chapter reviews *Shigella* classification, morphology, epidemiology, clinical features, pathogenesis characteristics, prevalence, growth and survival, and methods for the detection of *Shigella* in food.

Shigella was first discovered over 100 years ago by Kiyoshi Shiga, a Japanese scientist.[2] Shigellae, members of the family Enterobacteriaceae, are nearly identical genetically to *Escherichia coli* (*E. coli*) and are closely related to *Salmonella* and *Citrobacter* spp.[3] Shigellae are gram-negative, facultatively anaerobic, nonspore-forming, nonmotile rods that typically do not ferment lactose. In addition, they are lysine-decarboxylase-, acetate-, and mucate-negative and do not produce gas from glucose (some *S. flexneri* 6 serotypes may produce gas).[4,5]

There are four serogroups of *Shigella*: *S. dysenteriae* (serogroup A) serotypes 1–15, *S. flexneri* (serogroup B) serotypes 1–8 (9 subtypes), *S. boydii* (serogroup C) serotypes 1–19, and *S. sonnei* (serogroup D) serotype 1. The four serogroups of *Shigella* differ in epidemiology.[6] *S. dysenteriae* is primarily associated with epidemics,[6] with serotype 1 associated with the highest fatality rate (5%–15%).[7] *S. flexneri* predominates in areas of endemic infection, while *S. sonnei* has been implicated in source outbreaks in developed countries.[8] *S. boydii* is associated with source outbreaks in Central America and South America and is rarely isolated in North America. In particular, *S. sonnei* accounts for over two-thirds of shigellosis cases in the United States.[7] Table 89.1 shows the percentage of the four *Shigella* serogroups and total number of isolates reported annually from 1995 to 2005.[9] The average number of *Shigella* isolates associated with each serogroup remains relatively unchanged throughout the 10-year span as do the total number of isolates (Table 89.1).[9]

TABLE 89.1
Percentage of *Shigella* Isolates in the United States Reported by PHLIS from 1995 to 2005

Serogroup	1995	1996	1997	1998	1999	2000	2001	2002	2003	2004	2005
S. sonnei	76.6%	72.9%	71.5%	75.2	73.0%	79.6%	77.3%	83.5%	80.2%	68.9%	74.4%
S. flexneri	15.6%	19.2%	20.9%	17.7%	20.1%	13.6%	15.7%	12.2%	14.4%	17.2%	13.6%
S. boydii	1.2%	2.0%	2.1%	1.7%	1.6%	1.4%	1.2%	0.8%	1.1%	1.8%	1.2%
S. dysenteriae	0.5%	0.7%	0.6%	0.7%	0.5%	0.4%	0.5%	0.3%	0.4%	0.4%	0.5%
Ungrouped	6.1%	5.2%	4.9%	4.8%	4.8%	5.0%	5.3%	3.2%	3.9%	11.7%	10.3%
Total isolates	19,330	14,071	12,314	12,485	10,084	12,732	10,598	12,992	11,552	9343	10,484

Source: Reference 9.

Shigella has been classically characterized as a waterborne pathogen,[10] and outbreaks have occurred from contaminated community water sources that are un- or underchlorinated.[11–13] Foodborne outbreaks of *Shigella* are also common, especially with foods that are subjected to processing or preparation by hand; are exposed to a limited heat treatment; or are served/delivered raw to the consumer.[14] Several examples of food products from which *Shigella* has been isolated are potato salad, ground beef, bean dip, raw oysters, fish, and raw vegetables.

89.1.2 Clinical Features and Epidemiology

There are an estimated 400,000 cases of shigellosis each year, but only approximately 20,000 cases are actually reported.[15] Most cases are due to poor personal hygiene; however, shigellosis can be associated with contaminated food and water and international travel.[7] Each year approximately 20% of Americans who travel to developing countries become infected with Shigella.[15] Recently, male homosexual activity has been associated with an increase of *Shigella* infection. Outbreaks of shigellosis, primarily serogroups *flexneri* and *sonnei*, have been reported within the gay male population in North America, Europe, and Australia.[1]

The infective dose for *Shigella* is very low: 10 cells of *S. dysenteriae* to 500 cells of *S. sonnei*.[16] At-risk populations, such as children aged under 5 years, the elderly, or persons with decreased immune function, may be more susceptible to infection. Due to the low infective dose of *Shigella*, person-to-person transmission is common, especially in day-care settings (*S. sonnei*) where toddlers commonly practice poor personal hygiene. Typical symptoms of infection include bloody diarrhea, abdominal pain, fever, and malaise. Although the mechanism is unknown, seizures have been reported in 5.4% of shigellosis cases involving children.[17] Chronic sequelae from *S. dysenteriae* serotype 1 infections can include hemolytic uremic syndrome (HUS), whereas *S. flexneri* infections are associated with later development of reactive arthritis, especially in persons with the genetic marker HLA-B27.[18] Reactive arthritis is characterized by joint pain, eye irritation, and painful urination.[18]

Foodborne Outbreaks Involving *Shigella*. A summary of selected recent foodborne shigellosis outbreaks is given in Table 89.2. Characteristic of foodborne shigellosis in several recent outbreaks have been associated with foods consumed raw[18,22,26,28,29,31,32,45,46] and processed or prepared by hand.[18,23–25,33,34,40–42,47] In 2000, a multistate outbreak of shigellosis was traced to a commercially prepared five-layer bean dip.[36] This outbreak involved 406 ill persons (14 hospitalizations, 0 deaths) across 10 states. After extensive epidemiological investigation, numerous problems were identified in the manufacturing process, and investigators determined that the source of the outbreak was most likely an infected food handler.[36]

Prevalence of *Shigella*: Food and Food Handlers. As noted earlier, according to the CDC's EIP FoodNet, *Shigella* was the third most reported foodborne bacterial pathogen in 2002.[48] Of 16,580 laboratory-diagnosed cases, *Shigella* accounted for 3,875 cases (23.4%) behind only *Salmonella* (6,028 cases) and *Campylobacter* (5,006 cases).[48] From 2001 to 2002, the national incidence of *Shigella* in the United States increased from 3.8 to 4.5 cases per 100,000.[9,49] The incidence of each serogroup in the United States was similar to previous years, with serogroup D (*S. sonnei*) accounting for the majority of all isolates (83.5%), followed by serogroup C (*S. flexneri*, 12.2%), serogroup B (*S. boydii*, 0.8%), and serogroup A (*S. dysenteriae*, 0.3%).[9]

Despite the high incidence of shigellosis, there have been some studies indicating the prevalence of *Shigella* among food handlers or on food products. For instance, a study investigating the presence of enteropathogens among food handlers in Irbid, Jordan, showed that *Shigella* was isolated from the stools of four out of 283 examined food handlers.[50] Mensah et al. evaluated 511 food items from the streets of Accra, Ghana, from which *S. sonnei* was isolated from one sample of macaroni. It was noted that the macaroni was served using bare hands instead of clean utensils, which may have led to the *S. sonnei* contamination.[51] Wood et al. examined foods from Mexican homes; commercial sources in Guadalajara, Mexico; and from restaurants in Houston, Texas, for contamination with bacterial enteropathogens.[52] While no *Shigella* was isolated from the foods sampled from 12 Houston restaurants or from food commercially prepared in Guadalajara, Mexico, four isolates were obtained from meals prepared in Mexican homes. These studies demonstrate the importance of proper food handling and the role of food handlers in the transmission of *Shigella*.

TABLE 89.2
Selected Foodborne Outbreaks Involving *Shigella*

Year	Serogroup	Food Product(s) Implicated	Reference
1983	*S. sonnei*	Tossed salad	Martin et al., 1986[19]
1986	*S. sonnei*	Shredded lettuce	Davis et al., 1988[20]
1986	*S. sonnei*	Raw oysters	Reeve et al., 1989[21]
1987	*S. sonnei*	Watermelon	Fredlund et al., 1987[22]
1988	*S. sonnei*	Uncooked tofu salad	Yagupsky et al., 1991[23]; Lee et al., 1991[24]
1989	*S. flexneri* 4a	German potato salad	Lew et al., 1991[25]
1992	*S. flexneri* 2	Tossed salad	Dunn et al., 1995[26]
1994	*S. sonnei*	Iceberg lettuce	Long et al. 2002[27]; Kapperud et al., 1995[28]; Frost et al., 1995[29]
1995–1996	*S. sonnei*	Fresh pasteurized milk cheese	García-Fulgueiras et al., 2001[30]
1996	*S. flexneri*	Salad vegetables	PHLS, 1997[31]
1998	*S. sonnei*	Uncooked, chopped curly parsley	CDC, 1999[32]
1998	*S. flexneri*	Restaurant-associated, source unknown	Trevejo et al., 1999[33]
1999	*S. boydii* 18	Bean salad (parsley or cilantro)	CDPH, 1999[34]
2000	*S. sonnei*	Five-layer bean dip	CDC, 2000[35]; Kimura et al., 2004[36]
2001	*S. sonnei*	Raw oysters	Terajima et al., 2004[37]
2002	*Shigella* spp.	Greek-style pasta salad	TPH, 2002[38]
2004	*S. sonnei*	Dip, cheese	CSPI, 2006[39]
2004	*S. sonnei*	Raw carrots	Gaynor et al., 2009[40]
2004	*S. flexneri*	Macaroni salad, coleslaw, potato salad	CSPI, 2006[39]
2004	*S. flexneri*	Tamale	CSPI, 2006[39]
2005	*S. sonnei*	Chicken dishes	CSPI, 2006[39]
2005	*S. flexneri*, serotype 2a	Beef, other	CSPI, 2006[39]
2006	*S. sonnei*	Beef, picadilla	CDC, 2006[18]
2006	*S. sonnei*	Beans, unspecified	CDC, 2006[18]
2006	*S. sonnei*	Salad, unspecified	CDC, 2006[18]
2006	*S. sonnei*	Salads, lettuce based	CDC, 2006[18]
2007	*S. sonnei*	Raw baby corn	Lewis et al., 2009[41]
2008	*S. sonnei*	Salad	Hung-Wei et al., 2009[42]
2009	*S. sonnei*	Sugar peas	Heier et al., 2009[43]; Muller et al., 2009[44]

In response to President Clinton's National Food Safety Initiative (January 1997) and Produce & Imported Foods Safety Initiative (October 1997), the U.S. Food and Drug Administration (FDA) has investigated the presence of foodborne pathogens, including *Shigella*, on imported and domestic produce from 1999 to the present.[53,54] An FDA survey of imported broccoli, cantaloupe, celery, parsley, scallions, loose-leaf lettuce, and tomatoes found *Shigella* contamination in nine of 671 total samples: three of 151 cantaloupe samples, two of 84 celery samples, one of 116 lettuce samples, one of 84 parsley samples, and two of 180 scallion samples.[53] Another FDA survey of domestically grown fresh cantaloupe, celery, scallions, parsley, and tomatoes found *Shigella* contamination in five of 665 total samples: one of 164 cantaloupe samples, three of 93 scallion samples, and one of 90 parsley samples.[54] An additional survey of imported produce was conducted; however, at the time of this publication results were not publicly available.[54]

89.1.3 Pathogenesis

Pathogenesis of *Shigella*. The invasion of colonic (large intestine) epithelial cells by *Shigella* involves four steps: entry into epithelial cells, intracellular multiplication, intra- and intercellular spreading, and killing of the host cell.[55] The genetic machinery responsible for invasion resides on a large (180–220 kDa) virulence plasmid. The virulence plasmid contains invasion plasmid antigen (*ipa*) genes, which encode four highly immunogenic polypeptides; IpaA, IpaB, IpaC, and IpaD. Studies in which Tn5 insertions affecting the expression of the *ipa* genes reveal that expression of *ipaB*, *ipaC*, and *ipaD* is strongly associated with entry, whereas *ipaA* is not.[55]

These effector cells, along with approximately 25 additional proteins, are secreted from the pathogen into host cells via the type III secretory apparatus (see Reference 56 for review). The invasion of *Shigella* into the host is carried out by this type III secretory apparatus. *Shigella* can enter the host intestinal epithelial cells by secreting virulence factors directly into the epithelial host cells using type III secretion.[57] This secretion system is composed of nearly 50 proteins, along with the chaperones IpgA, IpgC, IpgE, and Spa15, and several transcription activators as well as the effector proteins, which are encoded on the large virulence plasmid harbored by all four pathogenic strains. The complex mechanism that involves the assembly of the type III secretory needle and the subsequent translocation of the effector protein molecules is controlled by a cascade of regulatory activators and partially activated when the pathogen is in contact with the host cell.

Invasion is also mediated by the *virF* gene, located on the virulence plasmid, and the *virR* gene, which is located on the chromosome. The *virF* gene encodes a 30-kDa protein that positively regulates the expression of the *ipa* genes and *icsA* (also known as *virG*), another virulence plasmid gene that encodes intra- and intercellular spread. Environmental factors that affect the expression of *virF* are not known. The *virR* gene is a repressor of the plasmid invasion genes in a temperature-dependant manner.[55] When shigellae are grown at 30°C, they do not express the Ipa polypeptides and are therefore, noninvasive at this temperature. However, shigellae grown at 37°C are fully invasive, and all virulence plasmid polypeptides are encoded.[55]

Once ingested, shigellae transit through the gastrointestinal tract to the colon where they translocate the epithelial barrier via M cells that overlay the solitary lymphoid nodules.[58] Upon reaching the basolateral side of the M cells, *Shigella* infect macrophages and induce cell apoptosis.[58] Infected macrophages release interleukin-1β, which elicits a strong inflammatory response.[59] Once *Shigella* is released from the macrophage, it enters neighboring epithelial cells (enterocytes) through a process known as macropinocytosis. This involves the host cell extending its membrane into formations known as pseudopodia, which engulf large volumes, close around them, and form a vacuole. Macropinocytosis requires actin polymerization and the presence of myosin, an actin-binding protein.[60] Common stimuli that induce macropinocytosis include cytokines and bacterial antigens. In response to invasion, epithelial cells produce proinflammatory cytokines that contribute to inflammation of the colon.[58]

Once engulfed by epithelial cells, shigellae immediately disrupt the vacuole releasing them to the host-cell cytoplasm, where they multiply rapidly. Sansonetti et al. observed the generation time of *S. flexneri* in HeLa cells to be approximately 40 min.[61] Efficient intracellular growth requires the acquisition of host-cell nutrients. Since little free iron exists within mammalian host cells, *Shigella* spp. must also express high-affinity iron acquisition systems. In order to obtain iron, *Shigella* synthesize and transport the siderophores aerobactin and enterobactin[62] and utilize a receptor/transport system in which iron is obtained from heme. Siderophores are low molecular weight, iron-binding compounds that remove iron from host proteins. Enterobactin is produced by some but not all *Shigella* spp.,[63,64] while aerobactin is synthesized by *S. flexneri*, *S. boydii*,[65] and some *S. sonnei*.[66] Headley et al. demonstrated that aerobactin systems, although active in extracellular environments, are not expressed intracellularly.[67] This suggests that siderophore-independent iron acquisition systems can provide essential iron during intracellular multiplication.[67]

The capacity for *Shigella* to spread intracellularly and infect adjacent epithelial cells is critical in the infection process.[55] Intra- and intercellular spreading is controlled by the *icsA* (*virG*) gene located on the virulence plasmid. The *icsA* gene encodes the protein IcsA, which enables actin-based motility[68] and intercellular spread.[69] IcsA is a surface-exposed outer membrane protein consisting of three distinctive domains: a 52 amino acid N-terminal signal sequence, a 706 amino acid α-domain, and a 344 amino acid C-terminal β-core.[70–72] The α-domain is the exposed portion while the β-core is embedded in the outer membrane.[72]

IcsA is distributed at only one pole of the outer membrane surface. This asymmetrical distribution allows the polar formation of actin tails and thus directional movement of *Shigella* within host-cell cytoplasm. The polar localization of IcsA is primarily affected by two events: (i) the rate of diffusion of outer membrane IcsA[73–76] and (ii) the specific cleavage of IcsA by the protease IcsP (SopA).[77–79] The rate of diffusion of outer membrane IcsA is directly affected by the O-side chains of the membrane lipopolysaccharide (LPS). Sandlin et al. demonstrated this relationship with a *S. flexneri* LPS mutant, which did not produce O-antigen. Without O-side chains, the LPS mutant polymerized actin in a nonpolar fashion.[74] In addition to O-side chains, the correct composition of the IcsA C-terminus α domain is also required for polar localization and polar movement of *S. flexneri*.[80] A *S. flexneri* mutant, in that a segment of the IcsA C-terminus α domain had been deleted, was unable to polymerize actin in a polar fashion or move unidirectionally.[80]

The *icsP* gene encodes the outer membrane protease IcsP (also called SopA) that cleaves laterally diffused IcsA, thus promoting polar localization.[78] An *E. coli* K-12 strain, engineered to express the *icsA* gene, was shown to diffuse IcsA along its outer membrane.[81] When the same *E. coli* K-12 strain was engineered to express the *icsP* gene with the *icsA* gene, the number of bacteria that polymerized actin at one pole increased.[81] The reader is directed to the cited references for information on molecular mechanisms of actin assembly in *Shigella*.[58,70,82–84]

Intercellular spreading is dependent upon an actin-based motility mechanism.[81] *Shigella* cells first form a membrane-bound protrusion into an adjacent cell. This protrusion must distend two membranes: one from the donor cell, and another from the recipient cell.[85] As the protrusion pushes further into the recipient cell, it is taken up by the recipient cell resulting in the bacteria enclosed in a double-membrane vacuole.[81] Intercellular spread is completed when *Shigella* rapidly escapes from the double-membrane vacuole, releasing it into the cytosol of the secondary cell. Monack and Theriot observed intercellular spread of an *E. coli* K-12 strain expressing the *icsA* and *icsP* genes in HeLa cells.[81] After invading adjacent cells, the *E. coli* K-12 strain was observed both enclosed in a double-membrane vacuole and free in the cytosol of host cells.

The early killing of host cells by *Shigella* requires invasion and is mediated by the genes found on the virulence plasmid. Sansonetti demonstrated the inability of noninvasive *S. flexneri* to kill host cells, whereas the invasive phenotype killed efficiently and rapidly.[55] Early killing of host cells involves metabolic events that rapidly drop the intracellular concentration of ATP, increase pyruvate, and arrest lactate production.[86] Shiga toxin (STX), a potent toxin produced by *S. dysenteriae* serotype 1, does not play a role in the early killing of host cells. Fontaine et al. constructed a Tox⁻ mutant strain of *S. dysenteriae serotype 1* and found that the mutant killed as efficiently as the wild-type strain.[87]

Toxins Produced by *Shigella* spp. Only *S. dysenteriae* serotype 1 produces the potent toxin STX. Although STX is not necessary to sustain an infection, its expression increases the severity of disease. Three biological activities associated with STX are cytotoxicity, enterotoxicity, and neurotoxicity, while the one known biochemical effect is the inhibition of protein synthesis.[88] STX is considered the prototype to a family of toxins known as Shiga-like toxins (SLT), which are similar in structure and function, and share the same receptor sites. Perhaps the most widely known human pathogen that produces SLTs is *E. coli* O157:H7, which has two toxin variants (SLT I and SLT II).

STX is composed of two polypeptides: an A subunit (32,225 kDa) and five B subunits (7691 kDa each).[88] The B subunits mediate binding to cell surface receptors, which have been identified as glycolipids containing terminal galactose-α(1–4)galactose disaccharides, such as galabiosylceramide and globotriasylceramide (Gb3).[89,90] The A subunit, once inside the host-cell cytoplasm, acts enzymatically to cleave the *N*-glycosidic bond of adenine at nucleotide position 4324 in the 28S rRNA of the 60S ribosomal unit.[88]

Recently, two additional enterotoxins, shigella enterotoxin 1 (ShET 1) and shigella enterotoxin 2 (ShET 2), have been characterized and are believed to play a role in the clinical manifestation of shigellosis.[91] ShET 1, which is chromosomally encoded, was only prevalent in isolates of *S. flexneri* 2a. ShET 2, however, is encoded on the virulence plasmid, and was detected in all *Shigella* isolates tested, except isolates that had been cured of the virulence plasmid.

Environmental Factors on Survival of *Shigella*. *Shigella* spp. are heat sensitive, acid resistant, salt tolerant bacteria that can withstand low levels of organic acids.[92–94] Zaika studied the survival of *S. flexneri* strain 5348 in brain heart infusion (BHI) broth as a function of pH (2–5) and temperature (4°C–37°C). When inoculated into BHI broth adjusted to pH 5, *S. flexneri* demonstrated growth when held at 19°C, 28°C, and 37°C, whereas counts declined over time at temperature of 12°C or lower. When inoculated into BHI broth adjusted to pH 2, 3, or 4, inoculated *S. flexneri* counts declined over time at all temperatures tested. *S. flexneri* in BHI broth adjusted to pH 2 reached undetectable levels in 1–3 days when held at temperature of 19°C or lower. In general, *S. flexneri* survival was greater in BHI broth incubated at lower temperatures and adjusted to higher pH. This study demonstrates that acidic foods may support the survival of *Shigella* over a long period of time.[92]

Zaika studied the survival of *S. flexneri* strain 5348 in BHI broth (pH 4–6) containing 0.5%–8% NaCl. In BHI adjusted to pH 6, *S. flexneri* grew in the presence of ≤6% NaCl when held at 19°C and 37°C, and in the presence of ≤7% NaCl when held at 28°C. Growth of *S. flexneri* was also observed in BHI broth adjusted to pH 5 containing ≤2%, ≤4%, ≤4%, and ≤0.5% NaCl when held at 37°C, 28°C, 19°C, and 12°C, respectively.[93] *S. flexneri* populations gradually declined in BHI adjusted to pH 4 at all incubation temperatures and all levels of NaCl tested.[93] Results from this study suggest that *S. flexneri* is salt tolerant and may survive for extended periods of time in salty foods such as pickled vegetables, caviar, pickled herring, dry cured ham, and certain cheeses.[93] *S. flexneri* survival was also studied in BHI broth supplemented with organic acids commonly found in fruits and vegetables (citric, malic, and tartaric acid) or fermentation acids commonly used as preservatives (acetic and lactic acids) at 0.04M and adjusted to pH 4 with HCl or NaOH.[94] Fermentation acids (acetic and lactic acid) had a greater effect on survival than citric, malic, and tartaric acids.[94] When incubated at 37°C, *S. flexneri* survived for 1–2 days in the presence of each organic acid tested. At 4°C, *S. flexneri* survived in the presence of all the organic acids tested for longer than 55 days.[94] This study suggests that organic acids may aid in the inactivation of *Shigella*; however, foods with low levels of acids stored at low temperatures may support the survival of the bacterium for extended periods of time.[94]

Temperature is an important factor in survival of shigellae. Freezing (–20°C) and refrigeration (4°C) temperatures support survival, but not growth, of *Shigella*. When studied in nutrient broth, the observed temperature ranges that permitted growth of *S. sonnei* and *S. flexneri* were 6.1°C–47.1°C and 7.9°C–45.2°C, respectively (cited in Reference 5); however, *Shigella* can survive for extended periods of time when held at –20°C or at 4°C. Elevated temperatures are less permissive for *Shigella* survival, and traditional pasteurization and cooking temperatures are sufficient for inactivation. Evans et al. calculated the decimal reduction time (D-value) of *S. dysenteriae* in pasteurized whole milk to be 0.0008 s at 82.2°C. When studied in nutrient broth, most strains of *S. sonnei* and *S. flexneri* were inactivated within 5 min at 63°C (cited in Reference 5). Sublethal heat exposure can sensitize *Shigella* to selective components of microbiological media. Tollison and Johnson demonstrated that *S. flexneri* sublethally heat-stressed by exposure to 50.0°C for 30 min in phosphate buffer became sensitive to 0.85% bile salts and 0.50% sodium desoxycholate. Since these compounds are ingredients in several enrichment and isolation media that are used for detection of *Shigella*, the thermal history of the food sample to be analyzed should be known.[95]

Survival of *Shigella* on Fomites. Inanimate objects, or fomites, can serve as vectors for transmission of *Shigella* and there have been several reports on the survival of *Shigella* on various surfaces.[96–98] Spicer studied the survival of *S. sonnei* dried on cotton threads at room temperature and under refrigeration at various levels of relative humidity. In general, survival was better at refrigerated temperatures and at high (84%) and low (0%) relative humidity (RH). *S. sonnei* remained detectable on the cotton threads after 12 days at 5°C–10°C (84% RH).[96] Similar results were observed using 10 different strains of *S. sonnei* inoculated on cotton, glass, wood, paper, and metal at various temperatures

(−20°C–45°C).[97] When held at −20°C, most of the strains survived for more than 14 days on each surface, however surfaces held at 45°C did not support survival of most strains.[97] Investigations on the survival of *S. dysenteriae* serotype 1 on cloth, wood, plastic, aluminum, and glass objects suggest that 1.5–4 h postinoculation, *S. dysenteriae* serotype 1 enters a viable but nonculturable (VBNC) state.[98] While no *S. dysenteriae* serotype 1 could be recovered after 5 days by conventional culture methods, viable cells could be observed using fluorescent antibody techniques.[98] Whether or not *Shigella* is able to achieve a true VBNC state or whether these results demonstrate the inadequacy of plating media in recovering viable *Shigella* has not been fully investigated. Nevertheless, these studies demonstrate that fomites can sustain viable *Shigella* for an extended time and serve as vehicles in the transmission of the pathogen.

Survival of *Shigella* in Water and Food. *Shigella* can survive in water with little decline in population. Rafii and Lunsford inoculated distilled water with *S. flexneri* at 2.8×10^8 colony forming units (CFU)/mL and held the samples at 4°C. After 26 days, 9.2×10^7 CFU/mL of *S. flexneri* survived. This high survival rate of *S. flexneri* in water supports the historical association of shigellosis outbreaks and water sources.[99]

Fruits and vegetables can support the growth or survival of *Shigella*. Escartin et al. artificially contaminated fresh cut papaya, jicama, and watermelon with *S. sonnei*, *S. flexneri*, or *S. dysenteriae,* and within 6 h at room temperature, growth was observed. Rafii and Lunsford inoculated raw cabbage, onion, and green pepper with *S. flexneri* and although counts decreased slightly at 4°C, survival was observed after 12 days on onion and green pepper, at which time sampling was terminated due to spoilage.[99] *S. flexneri* survival was observed on the cabbage for 26 days. Wu et al. studied survival of *S. sonnei* on whole and chopped parsley leaves. At 21°C, growth on chopped parsley was observed at a similar rate to that in nutrient broth.[14] At 4°C, populations declined on both chopped and whole parsley throughout the 14-day storage period, however *S. sonnei* survived regardless of initial population.[14] These studies demonstrate *Shigella* survival on refrigerated produce for periods of time that exceed expected shelf life.

Low pH foods can support survival of *Shigella* when held at refrigerated temperatures. Bagamboula et al. demonstrated *S. sonnei* and *S. flexneri* survival in apple juice (pH 3.3–3.4) and tomato juice (pH 3.9–4.1) held at 7°C for 14 days. No reduction was observed in the tomato juice, while a 1.2–3.1 $\log_{10}$ CFU reduction was observed in apple juice over the 14-day study.[100] Rafii and Lunsford observed *S. flexneri* survival in carrot salad (pH 2.7–2.9), potato salad (pH 3.3–4.4), coleslaw (pH 4.1–4.2), and crab salad (pH 4.4–4.5) held at 4°C. Sampling was terminated at day 11 for the carrot and the potato salads, at which time *S. flexneri* counts decreased from 4.3×10^6 to 4.2×10^2 CFU/g and from 1.32×10^6 to 8.5×10^2 CFU/g, respectively. Sampling of the coleslaw and the crab salads ceased due to product spoilage on days 13 and 20, respectively, however *S. flexneri* survived at levels of 2.16×10^4 and 2.4×10^5 CFU/g, respectively. These studies indicate that *Shigella* survived at refrigerated temperatures despite the presence of background microflora and low pH.[99]

Prepared foods can also support the survival of *Shigella*. Islam et al. investigated the growth and survival of *S. flexneri* in boiled rice, lentil soup, milk, cooked beef, cooked fish, mashed potato, mashed brinjal, and raw cucumber. All food samples, except raw cucumber, were autoclaved prior to inoculation. Ten-gram or 10-mL samples of each food were inoculated with 10^5 cells of *S. flexneri*, incubated at 5°C, 25°C, or 37°C and sampled over 72 h. All of the foods tested supported growth up to 10^8–10^{10} cells per g or mL within 6–18 h after inoculation at 25°C and 37°C.[101] Initial inoculum levels were maintained throughout the 72 h holding period for all foods, except rice and milk. *S. flexneri* counts in rice decreased by approximately 1 log unit after 72 h at 5°C, whereas counts in milk dropped after 48 h but then returned to the initial inoculum level by 72 h. These results demonstrate the ability of *Shigella* to grow and survive in a variety of prepared foods that may be contaminated by an infected food handler.

Shigella is able to survive on produce packaged under vacuum or modified atmosphere. Satchell et al. investigated the survival of *S. sonnei* in shredded cabbage packaged under vacuum or in a modified atmosphere of nitrogen and carbon dioxide when stored at room temperature or under refrigeration. When test samples were stored room temperature, counts of inoculated *S. sonnei* remained level for up to 3 days, at which time populations began to drop. When test samples were stored under refrigeration, *S. sonnei* counts did not drop after 7 days. In the latter study, refrigerated samples maintained a constant pH throughout the study, while samples stored at room temperature had a drop in pH that may have contributed to the decline in *S. sonnei* cell numbers.[102]

89.2 DIAGNOSIS AND DETECTION

89.2.1 Conventional Techniques

Traditional Microbiological Media for Enrichment and Isolation of *Shigella*. Traditional microbiological techniques make use of selective and differential media for the enrichment and isolation of *Shigella*. Many variants of enrichment and plating media have been investigated for optimal recovery, often with conflicting results. Although *Shigella* is readily isolated from clinical samples, food samples are more problematic. Isolation of *Shigella* from food samples can be inhibited by indigenous microflora, especially the coliform bacteria and *Proteus* spp.[5] The addition of the antibiotic novobiocin to liquid and solid media has been shown to improve the isolation of *S. flexneri* and *S. sonnei* from investigated foods (cited in Reference 5). Contamination of food products with *Shigella* results primarily through contact with a food handler with poor personal hygiene; therefore, the concentration of *Shigella* may be very low compared to that of the indigenous microflora.[103] Currently, selective media are not available that will adequately suppress the growth of background microflora; therefore *Shigella* is often overgrown by competing microorganisms.[103] More research is needed to determine more appropriate selective media and enrichment conditions for the isolation of *Shigella* from food samples.

Two enrichment broths initially used for the isolation of *Shigella* were Selenite-F (SF) and Tetrathionate (TT) broth. These broths were originally designed for the isolation of salmonellae, but due to the lack of specific enrichment media for shigellae, they were used as all-purpose enteric enrichment broths.[104] Sodium selenite, although selective for salmonellae, is toxic to *Shigella* (and most enterics); therefore, it is no longer used in enrichment procedures for *Shigella*. TT is a peptone-based broth with bile salts and sodium thiosulfate that inhibits growth of most gram-positives and Enterobacteriaceae. While TT is routinely used for the enrichment of *Salmonella,* it is rarely used for *Shigella*. Gram-negative (GN) broth is a peptone-based broth with glucose and mannitol. The concentration of mannitol in GN broth is higher than in glucose to promote mannitol fermentors, like *Shigella*. Both TT broth and GN broth contain bile salts, which can be inhibitory to stressed cultures.[95] Furthermore, GN broth contains sodium desoxycholate, which has been shown to inhibit heat-stressed shigellae.[105] Still, GN broth is listed as an alternate enrichment medium for the detection of *Shigella* from food by some standard methods.[3]

Current enrichment procedures[3,106,107] use a low carbohydrate medium, *Shigella* broth (SB) with addition of novobiocin, for the detection/isolation of *Shigella*. One study reported that acids produced by other Enterobacteriaceae during the fermentation of carbohydrates were toxic to shigellae[108]; however, other studies have shown that *Shigella* can grow at pH 4.5–4.75[100] and survive at pH 4.0.[93] Since SB contains very little carbohydrate, the effect of a low pH environment during enrichment is limited. SB is also less stringent than TT broth and GN broth for the enrichment of *Shigella* since it contains neither bile salts nor sodium desoxycholate. A recent study investigated SB, GN broth, tryptic soy broth, and Enterobacteriaceae Enrichment (EE) broth with the addition of novobiocin for enrichment/detection of shigellae.[105] When incubated in GN broth, *S. sonnei* was unable to grow to comparable levels as observed in SB and EE broths. EE broth, however, has been reported inhibitory to *S. boydii*.[109,110]

Multiple plating media with differing selectivity can be used to increase the chances of *Shigella* isolation. The most common low-selectivity medium used for plating *Shigella* is MacConkey agar (MAC). Eosin methylene blue (EMB) or Tergitol-7 (T7) agars can also be used. Since differentiation on MAC is solely based on lactose fermentation, *Shigella* colonies look similar to those of many lactose negative competitors.[105] On MAC, *Shigella* colonies are translucent or slightly pink, with or without rough edges. *Shigella* produce colorless colonies on EMB and bluish colonies on the yellowish-green T7 agar.[3]

Intermediate selectivity media useful in isolating *Shigella* are desoxycholate citrate agar (DCA) and xylose lysine desoxycholate agar (XLD). *Shigella* produce colorless colonies on both DCA and XLD. Bhat and Rajan (1975) reported XLD superior to DCA for the isolation of *Shigella* since DCA required a 48-h incubation to show clear colony morphology as opposed to overnight incubation for XLD. Unfortunately, D-xylose, which serves as a differentiating agent on XLD agar, is fermented by some strains of *S. boydii* while most *Shigella* cannot ferment xylose.[111] Thus, some strains of *Shigella* will be missed if XLD is used as the sole plating medium.

Highly selective media for *Shigella* spp. include *Salmonella–Shigella* agar (SSA) and Hektoen Enteric agar (HEA). A problem associated with SSA and HEA is that they may be too stringent for some strains of *Shigella*, especially if the culture is stressed.[3,105] *Shigella* produce colorless, translucent colonies on SSA and green colonies on HEA.

A newly developed plating medium, Chromogenic *Shigella* spp. Plating Medium (CSPM) (R&F Laboratories, West Chicago, Illinois) offers medium selectivity (bile salts, antibiotic supplementation) with an alternative to differentiation based on lactose fermentation. Instead, differentiation on CSPM is based on proprietary components consisting of select carbohydrates, pH indicators, and chromogens (L. Restaino, personal communication). *Shigella* spp. produce white-to-clear colonies on CSPM while competitors produce colored colonies. CSPM has been compared to MAC and SSA for the recovery of *S. boydii* and *S. sonnei* from tomato surfaces with no significant differences in recovery observed.[109,110] Further evaluation of CSPM against other strains of *S. boydii* and *S. sonnei* as well as the other serogroups of *Shigella* is needed.

In the current FDA import and domestic produce assignments, select produce commodities are analyzed using a PCR-based assay and bacteriological plating on chromogenic agar. Samples are rinsed and then processed immediately (0 h) by PCR and concurrently added to 2× SB supplemented with 0.5 μg/mL novobiocin. At 6 h and overnight enrichment, aliquots are removed for both PCR analysis and plating on *Shigella/Aeromonas* Rainbow Agar (Biolog Inc., Haywood, California). This approach affords an opportune means to screen food samples and also isolate any *Shigella* spp. present.

The BAM Culture Method for Detection of *Shigella* in Foods. The FDA Bacteriological Analytical Manual (BAM) culture method for the isolation and detection of *Shigella* spp. from food utilizes a combination of low carbohydrate enrichment, anaerobic conditions, and elevated temperature.[106] Briefly, a 25 g sample of the food product is transferred to 225 mL of SB to which novobiocin (0.5 μg/mL for *S. sonnei*; 3.0 μg/mL for other *Shigella* spp.) has been added. Samples are held at room temperature for 10 min and periodically shaken. Sample supernatants are transferred to an Erlenmeyer flask and the pH adjusted to 7.0 ± 0.2 with sterile 1 N NaOH or 1 N HCl. Flasks are incubated anaerobically for 20 h (44°C for *S. sonnei*; 42°C for all other *Shigella* spp.) and the enrichments are streaked on MAC. Confirmation of suspicious colonies involves tests for motility, H_2S, gas formation, lysine decarboxylase, and fermentation of sucrose or lactose. Isolates negative for all confirmatory tests are further tested using biochemical reactions including adonitol, inositol, lactose, potassium cyanide, malonate, citrate, salicin, and methyl red. Shigellae are negative for all except methyl red.

Antisera agglutination is then used to identify any culture displaying typical *Shigella* characteristics.

June et al. evaluated the effectiveness of the BAM culture method for *Shigella*. Two strains of *S. sonnei*, strains 9290 and 25931, were inoculated on potato salad, chicken salad, cooked shrimp salad, lettuce, raw ground beef, and raw oysters. Using either unstressed or chilled stressed cells, acceptable recovery was achieved for both strains from the potato salad, chicken salad, cooked shrimp salad, and lettuce samples, but not from the ground beef and raw oyster samples. Approximately 4 log unit differences in recovery from ground beef samples were observed between the two strains suggesting high strain variability. Given the low infective dose of *Shigella* (as low as 10 cells), the BAM was considered ineffective for the evaluation of raw ground beef and raw oysters.[112]

In 2002, the BAM culture method for *Shigella* was evaluated using two strains of unstressed, chill-stressed, or freeze-stressed *S. sonnei* (strains 357 and 20143) on selected types of produce.[113] Acceptable recovery of unstressed cells ($<1.0 \times 10^1$ CFU/25 g) and chill-stressed or freeze-stressed cells ($<5.2 \times 10^1$ CFU/25 g) were observed for all produce types tested.[113] In our laboratory, similar tests of a modified protocol showed unacceptable recovery of unstressed *S. sonnei* (patient isolate from an outbreak involving an unknown source) and *S. boydii* (patient isolate from an outbreak involving bean salad) on tomatoes at 1.9×10^2 CFU/tomato and $>5.3 \times 10^5$ CFU/tomato, respectively.[114] These results support the observation that significant variation exists among strains of *S. sonnei* and demonstrate the importance of including the other serogroups of *Shigella* when evaluating culture methods for detection in food.

Other Culture Methods for the Detection of *Shigella* in Foods. Alternate culture methods for the detection of *Shigella* in foods can be found in Health Canada's Compendium of Analytical Methods[107] and the American Public Health Association's Compendium of Methods for the Microbiological Examination of Foods (CMMEF).[3] The Health Canada method is similar to the BAM culture method with a few modifications; most notably enrichments for all *Shigella* spp. (including *S. sonnei*) are supplemented with 0.5 μg/mL novobiocin and incubated anaerobically at 42°C. The CMMEF culture method is also similar to the BAM culture method, however the level of novobiocin in *S. sonnei* enrichment media is lower (0.3 μg/mL) and enrichment conditions are aerobic at 37°C. The CMMEF further suggests that two to three plates of various selective media be used to streak the enriched cultures: MAC for low selectivity, XLD for intermediate selectivity, and HEA for high selectivity. Confirmation of suspicious colonies is similar to methods described above for the BAM culture method.

Recently, the CMMEF culture method and an enrichment procedure involving EE broth have been compared to the BAM culture method for the detection of *S. boydii* and *S. sonnei* on tomato surfaces.[109] Natural tomato microflora was found to have a great impact on recovery of *S. sonnei* and completely inhibited recovery of *S. boydii* in all three culture methods. No significant differences ($P > 0.05$) were observed among the culture methods for detection of *S. sonnei*, or between the BAM and CMMEF culture methods for the detection of *S. boydii*. EE broth was found to be inhibitory to *S. boydii*. These results suggest the need for more selective enrichment protocols for the evaluation of *Shigella* spp. in food.

89.2.2 Molecular Detection Procedures

PCR Detection of *Shigella* in Foods. Polymerase chain reaction (PCR) methods for detection of *Shigella* in food have previously demonstrated higher sensitivity than comparable culture methods.[109] PCR assays for *Shigella* spp. have targeted the invasion associated locus (*ial*),[101,114] the *virA* gene,[115,116] or the *ipaH* gene[109,117–120] for amplification. The same PCR primers are used to detect each of the serogroups of *Shigella* and enteroinvasive *E. coli*. The *ial* (now referred to as the *mxi/spa* genes) and *virA* genes are located on the large virulence plasmid (sometimes referred to as the invasion plasmid); however, sequencing of the *S. flexneri* genome[121] revealed the *ipaH* gene to be encoded multiple times on both the chromosome and the large virulence plasmid. Thus, detection of the *ipaH* gene is possible in the event the large virulence plasmid has been lost, which has been shown to occur when food samples are stored for a prolonged periods of time prior to analysis or in the case of *S. sonnei*, on continuous passage.[120] The discussion below will be limited to PCR detection of *Shigella* in food samples; however, several reports of detection in fecal samples appear in the literature. These methods along with other rapid methods are summarized in Table 89.3.

Vantarakis et al. developed a multiplex PCR method to detect both *Shigella* spp. and *Salmonella* spp. in mussels. Multiplex PCRs involve the use of two or more sets of primers in the same PCR such that multiple targets can be amplified in the same reaction. Artificially inoculated *S. dysenteriae* and *S.* Typhimurium were recovered by homogenizing mussel meat with peptone water. DNA from an aliquot of the homogenate was purified using a guanidine isothionate method and concentrated via ethanol precipitation. The *virA* gene of *Shigella* spp. and the *invA* genes of *Salmonella* spp. were amplified. Sample homogenates were inoculated at various concentrations to establish the lowest detection limit for the method. When samples were not preenriched prior to analysis, the multiplex PCR method was able to detect *S. dysenteriae* at 1.0×10^3 CFU/mL in the homogenate. Following 22 h incubation in buffered peptone water the multiplex PCR method was able to detect *S. dysenteriae* levels as low as 1.0×10^1–1.0×10^2 CFU/mL in the homogenate. Similar results were observed for *S. typhimurium*.[116]

Villalobo and Torres investigated PCR for the detection of *S. dysenteriae* serotype 1 in mayonnaise. Samples were homogenized in buffered peptone water and artificially contaminated with various concentrations of *S. dysenteriae* serotype 1. Bacterial cells were lysed with detergent, the DNA extracted with phenol-chloroform and precipitated with

TABLE 89.3
Selected Rapid Methods for Detection of *Shigella*

Serogroup	Target	Matrix	Sample Preparation	Detection	Detection Limit	Reference
S. sonnei *S. boydii*	*ipaH*	Tomatoes	FTA filtration	Nested PCR	<10^1 CFU/tomato	Warren, 2003[109]
S. flexneri	*ipaH*	Water	SE, boiled extract	Seminested PCR	≥1.1×10^1 CFU/mL	Theron et al., 2001[119]
S. dysenteriae 1	*virA*	Mayonnaise	CEP	PCR	10^2–10^3 CFU/mL	Villalobo and Torres, 1998[115]
S. dysenteriae	*virA*	Mussels	PE, CEP	PCR	10^1–10^2 CFU/mL	Vantarakis et al., 2000[116]
S. flexneri	plasmid DNA	Lettuce	Alkaline denaturation	PCR	1.0×10^4 CFU/g	Lampel et al., 1990[122]
S. flexneri	*ial*	Various foods	SE, BDC	Nested PCR	1.0×10^1 CFU/g	Lindqvist, 1999[114]
S. sonnei form I	LPS, *ial*	Feces	IMS	PCR	1.0–1.5×10^1 CFU/mL	Achi-Berglund and Lindberg, 1996[123]
S. dysenteriae *S. flexneri* *S. boydii* *S. sonnei*	LPS	Rectal swabs	PE, DPSM	EIA	ND	Sonjai et al., 2001[124]
S. dysenteriae	LPS	Various foods	None	Biosensor	≥4.9×10^4 CFU/mL	Sapsford et al., 2004[125]
S. dysenteriae 1 *S. flexneri* *S. sonnei*	LPS, *ial*	Feces	IMS	PCR	ND	Achi et al., 1996[126]
S. dysenteriae 1 *S. flexneri*	LPS, *ial*	Feces	IMS	PCR	1.0×10^1 CFU/g	Islam and Lindberg, 1992[127]
S. dysenteriae *S. flexneri* *S. boydii* *S. sonnei*	LPS, 16S rRNA	Sewage	Immunocapture	UPPCR	5.0×10^1 CFU/mL	Peng et al., 2002[128]
S. dysenteriae 1 *S. flexneri*	LPS	Feces	IMS	EIA	1.0×10^3 CFU	Islam et al., 1993[101]

SE, selective enrichment; PE, preenrichment; ND, not determined; LPS, lipopolysaccharide; EIA, enzyme immunoassay; BDC, buoyant density centrifugation; UPPCR, universal primer PCR; DPSM, direct plate on selective media; CEP, chemical extraction/ethanol precipitation.

ethanol. A multiplex PCR was then used to amplify regions from the *virA* gene and the 16S rRNA gene. The lowest level of inoculation detected by this method was 1.0×10^2–1.0×10^3 CFU/mL in the homogenate.[115]

In a study by Lindqvist, nested PCR was investigated for the detection of the *ial* locus of *Shigella* spp. from spiked lettuce, shrimp, milk, and blue cheese samples. Nested PCR involves the use of two sequential PCRs in which the target sequence in the second PCR lies within the amplified sequence in the first PCR. Nested PCRs can be used to achieve extremely low sensitivities. Inoculated food samples were homogenized in physiological saline and bacteria were isolated by buoyant density centrifugation (separation of components based on density). Single PCR, using only the external primer sets, was able to detect *S. flexneri* in aqueous solution at 0.5–1.0×10^5 CFU/mL. Nested PCR was more sensitive and was able to detect 1.0×10^3 CFU/mL. The nested PCR assay in combination with buoyant density centrifugation was able to detect *S. flexneri* inoculated onto all four foods at 1.0×10^1 CFU/g.[114]

Theron et al. investigated a seminested PCR for the detection of the *ipaH* gene of *S. flexneri* in spiked environmental water samples. The detection limits in the various environmental water samples were 2.0×10^3 CFU/mL for well water, 1.4×10^1 CFU/mL for lake water, 5.8×10^2 CFU/mL for river water, 6.1×10^2 CFU/mL for treated sewage water, and 1.1×10^1 CFU/mL for tap water. Variability in results among the water samples was attributed to the presence of humic substances that inhibited PCR. Humic substances are transformed products from soil organic matter that do not belong to the known classes of biochemistry. These include humic acids, fulvic acids, and humins. Preenrichment in GN broth served to dilute humic substances while allowing the *S. flexneri* to multiply, thereby increasing the concentration of target DNA.[119]

Improved PCR Detection of *Shigella* by FTA® Filtration. FTA® cards (Whatman, Clifton, New Jersey) have been used in template preparation for PCR detection of pathogenic microorganisms.[109,111,120] Moisture in the sample activates chemicals in the FTA® card that lyse cells, denature enzymes, inactivate pathogens, and immobilize genomic DNA.[129] Chemicals in the FTA® cards also protect DNA and RNA from light, free radicals, enzymes, and pathogens during dry, room-temperature storage. FTA® filtration as sample preparation for PCR has been developed for detection of *Cyclospora* and *Cryptosporidium* in water.[106] Recently, this sample preparation technique has been investigated for detection of *S. boydii* and *S. sonnei* from tomato rinses by nested

PCR.[109,111] Briefly, tomato rinses were passed through a two-stage filter where the first stage contained filter paper for size exclusion and the second stage contained FTA® filter paper. After purification and washing, 6 mm punches were taken from the FTA® filters and used directly as template in the first step of the nested PCR. The FTA® filtration/nested PCR (FTA–PCR) assay detected *S. boydii* and *S. sonnei* at 6.2×10^0 CFU/tomato and 7.4×10^0 CFU/tomato, respectively.[109,111]

Although FTA–PCR had excellent sensitivity when used to analyze tomato rinses, which are relatively clean, similar results were not observed when recovery of inoculated *S. boydii* and *S. sonnei* was attempted from cantaloupes, strawberries, or retail Valencia oranges (Dr. K.R. Schneider, unpublished data). FTA–PCR was unable to detect inoculated *Shigella* from strawberries, suggesting that the purification and washing protocol used on the FTA® filters may not be adequate to remove PCR inhibitors.[109] Despite several size exclusion filtering techniques applied prior to FTA® filtration, the FTA® filters were routinely clogged by cantaloupe fibers or the wax coating from retail oranges, which entered the buffer rinse during recovery procedures.[109] The advantage of the FTA–PCR technique is that no enrichment is necessary to obtain low detection limits due to the large volume of sample analyzed by the filters. The technique warrants further investigation for the analysis of foods for *Shigella* contamination as improvements in prefiltration techniques could improve the sensitivity.

Additional Molecular Microbiological Techniques for Detection of *Shigella*. An interesting new method, immunocapture universal primer PCR (iUPPCR) has been recently reported.[128] Universal primer PCR employs primers designed against highly conserved regions, such as 16S rRNA genes, thus the resulting PCR can be used for amplification of almost any bacteria. Typically, the sequence of the resulting amplicon would be further analyzed for identification of the bacteria. In iUPPCR however, immunocapture is employed for specificity and universal primer PCR is used for detection of captured bacteria. Monoclonal antibodies specific for individual serotypes of *S. dysenteriae*, *S. flexneri*, *S. boydii*, or *S. sonnei* were coated into the wells of a 96-well polystyrene plate. Cross-reactivity tests demonstrated the specificity of the monoclonal antibodies to the strain level. Wells were challenged with bacteria and captured bacteria were detected using universal primer PCR, that is, PCR with primers specific for 16S rRNA sequences conserved among closely related bacteria. The detection limit for shigellae in broth was 5.0×10^1 CFU/mL.[128] Presumably, detection of up to 96 pathogens sharing the conserved 16S rRNA sequence could be accomplished in the same plate using specific monoclonal antibodies for bacterial capture (differentiation of cell types) and the universal primer PCR for detection. Further investigation of this method with respect to food samples as opposed to bacteria in broth are required.

Advances in oligonucleotide microarrays have enabled the simultaneous detection and identification of bacterial foodborne pathogens. Oligonucleotide microarrays involve the ordered immobilization of specific probes to a solid surface followed by hybridization with labeled sample DNA. After hybridization, development of the label can allow identification/enumeration of complementary sample DNA sequences. In a recent report, Kakinuma et al. describes an oligonucleotide microarray using probes specific for the *gyrB* gene to detect and differentiate *E. coli*, *Shigella*, and *Salmonella*. PCR amplified regions of the *gyrB* gene were fluorescently labeled and hybridized to detection probes immobilized on glass slides. Based on reaction patterns, three species of *Shigella*, seven serovars of *Salmonella*, and one strain of *E. coli* were correctly identified at the species level. Identification at the subspecies level of *Salmonella* was problematic when multiple serovars were present in the same sample due to the overlap of microarray patterns. Assay formats such as this could be expanded to potentially detect and identify a wide range of enteric pathogens.[130]

Finally, repetitive-sequence-based PCR (rep-PCR) has been demonstrated for the identification and molecular typing of members of the family Enterobacteriaceae, including *Shigella*.[131,132] In rep-PCR, primers are designed complementary to bacterial interspersed repetitive sequences and the regions lying outward of the primers are amplified resulting in DNA fragments of varying lengths. These fragments can then be separated by electrophoresis to form a bacterial fingerprint unique to individual bacterial strains. The DiversiLab™ System (Spectral Genomics, Inc., Houston, Texas) generates such fragments by rep-PCR that are then analyzed using microfuidics lab-on-a-chip and the Agilent 2100 Bioanalyzer (Agilent Technologies, Inc. Palo Alto, California). Results can be visualized in several formats including dendograms, electropherograms, gel-like image, or scatter plots.[132] Rep-PCR technology has been shown to be reproducible and can allow the differentiation of species, subspecies, and strains in as little as 4 h. Some *Shigella* serogroups show >95% similarity with the DiversiLab™ Enteric Kit Beta version, however the DiversiLab™ *Shigella* kit offers greater differentiation among serogroups and strains.[132] Strains of *E. coli* (nonpathogenic, enterohemorragic, and enterotoxigenic) were effectively differentiated from *Shigella* spp. using rep-PCR[131,132]; however, no enteroinvasive *E. coli* strains were included in the studies.

89.3 CONCLUSION

Shigella spp. continues to be a significant agent of foodborne illness in the United States. While *Shigella* can survive in many types of food, there is some information indicating that the prevalence of *Shigella* on food or among food handlers remains a concern in regard to food safety. Despite much research, an appropriate conventional culture protocol for the detection of *Shigella* in foods is still needed. As of this publication, few immunological-based detection kits for *Shigella* are commercially available. Several molecular and microbiological methods, such as PCR-based assays, microarrays, and rep-PCR, have been developed; however, a more robust and selective protocol is needed. While the development of rapid test methods will facilitate more timely and cost-effective testing, a more profound reduction of foodborne shigellosis may depend on the education of employees and consumers as to proper personal hygiene.

ACKNOWLEDGMENTS

A portion of the material presented in this chapter has been previously published by Taylor and Francis (www.informaworld.com) in: Warren, B.R., Parish, M.E., and Schneider, K.R., *Shigella* as a foodborne pathogen and current methods for detection, *Crit. Rev. Food. Sci. Nutr.,* 46, 551–567, 2006.

The authors would also like to acknowledge the editorial work performed by Ms. A. Balaguero. Her work on this chapter was greatly appreciated.

REFERENCES

1. Centers for Disease Control and Prevention (CDC), *Shigella flexneri* Serotype 3 infections among men who have sex with men—Chicago, Illinois, 2003—2004, *Morb. Mortal. Wkly. Rep.* 54, 820, 2005.
2. Anonymous, About-*Shigella*. Available at: www.about-shigella.com. 2002. Accessed April 8, 2009.
3. Lampel, K.A., *Shigella*. In Downes, F.P., and Ito, K. (Editors), *Compendium of Methods for the Microbiological Examination of Foods*, 4th ed., pp. 381–385, American Public Health Association, Washington, DC, 2001.
4. Echeverria, P., Sethabutr, O., and Pitarangsi, C., Microbiology and diagnosis of infections with *Shigella* and enteroinvasive *Escherichia coli*, *Rev. Infect. Dis.*, 13, S220, 1991.
5. International Commission on Microbiological Specifications for Foods (ICMSF), *Shigella*. In Roberts, T.A., Baird-Parker, A.C., and Tompkins, R.B. (Editors), *Microorganisms in Foods*, 5, pp. 280–298, Blackie Academic & Professional, Chapman & Hall, London, 1996.
6. Ingersoll, M., Groisman, E.A., and Zychlinsky, A., Pathogenicity islands of *Shigella*, *Curr. Topics Microbiol. Immunol.,* 264, 49, 2002.
7. Centers for Disease Control and Prevention (CDC), Disease information: Shigellosis. Available at: http://www.cdc.gov/nczved/dfbmd/disease_listing/shigellosis_gi.html. 2008. Accessed July 23, 2008.
8. Hale, T., Genetic basis of virulence in *Shigella* species, *Microbiol. Rev.,* 55, 206, 1991.
9. Centers for Disease Control and Prevention (CDC), PHLIS surveillance data. *Shigella*: Annual summary, 1995 through 2005. Available at: http://www.cdc.gov/ncidod/dbmd/phlisdata/shigella.htm. 2003. Accessed April 6, 2009.
10. Smith, J.L., *Shigella* as a foodborne pathogen, *J. Food Prot.,* 50, 788, 1987.
11. Blostein, J., Shigellosis from swimming in a park pond in Michigan, *Public Health Rep*., 106, 317, 1991.
12. Fleming, C.A. et al., An outbreak of *Shigella sonnei* associated with a recreational spray fountain, *Am. J. Public Health*, 90, 1641, 2000.
13. Centers for Disease Control and Prevention (CDC), Shigellosis outbreak associated with an unchlorinated fill-and-drain wading pool—Iowa, 2001, *Morb. Mortal. Wkly. Rep.,* 50, 797, 2001.
14. Wu, F.M. et al., Fate of *Shigella sonnei* on parsley and methods of disinfection, *J. Food Prot.,* 63, 568, 2000.
15. Mead, P.S. et al., Food-related illness and death in the United States, *Emerg. Infect. Dis.*, 5, 607, 1999.
16. Kothary, M.H., and Babu, U.S., Infective dose of foodborne pathogens in volunteers: A review, *J. Food Safety*, 21, 49, 2001.
17. Galanakis, E. et al., Rate of seizures in children with shigellosis, *Acta Paediatr.,* 91, 1001, 2002.
18. Centers for Disease Control and Prevention (CDC), Foodborne outbreaks due to confirmed bacterial etiologies. Available at: http://www.cdc.gov/foodborneoutbreaks/documents/2006_line_list/2006_line_list.pdf. 2006. Accessed April 6, 2009.
19. Martin, D.L., et al., Contaminated produce—a common source for two outbreaks of *Shigella* gastroenteritis. *Am. J. Epidemiol.*, 124, 299, 1986.
20. Davis, H.J., et al., A shigellosis outbreak traced to commercially distributed lettuce. *Am. J. Epidemiol.*, 128, 1312, 1988.
21. Reeve, G., et al., An outbreak of shigellosis associated with the consumption of raw oysters. *N. Engl. J. Med.*, 321, 224, 1989.
22. Fredlund, H. et al., Water-melon as a vehicle of transmission of shigellosis. *Scand. H. Infect. Dis.,* 19, 219, 1987.
23. Yagupsky, P. et al., Use of multiple markers for investigation of an epidemic of *Shigella sonnei* infections in Monroe County, New York, *J. Clin. Microbiol,* 29, 2850, 1991.
24. Lee, L.A. et al., An outbreak of shigellosis at an outdoor music festival, *Am. J. Epidemiol.,* 133, 608, 1991.
25. Lew, J.F. et al., An outbreak of shigellosis aboard a cruise ship caused by a multiple-antibiotic-resistant strain of *Shigella flexneri*, *Am. J. Epidemiol,.* 134, 413, 1991.
26. Dunn, R.A. et al., Outbreak of *Shigella flexneri* linked to salad prepared at a central commissary in Michigan, *Public Health Rep.,* 110, 580, 1995.
27. Long, S.M., et al., General outbreaks of infectious intestinal disease linked with salad vegetables and fruit, England and Wales, 1992–2000. *Common Dis. Public Health.,* 5, 101, 2002.
28. Kapperud, G. et al., Outbreak of *Shigella sonnei* infection traced to imported iceberg lettuce, *J. Clin. Microbiol.* 33, 609, 1995.
29. Frost, J.A. et al., An outbreak of *Shigella sonnei* infection associated with consumption of iceberg lettuce, *Emerg. Infect. Dis.,* 1, 26, 1995.
30. García-Fulgueiras, A., et al., A large outbreak of *Shigella sonnei* gastroenteritis associated with consumption of fresh pasteurised milk cheese. *Eur. J. Epidemiol.*, 17, 533, 2001.
31. Public Health Laboratory Service (PHLS), Outbreaks of foodborne illness in humans, England and Wales: Quarterly Report, *CDR Wkly.,* 7, 207, 1997.
32. Centers for Disease Control and Prevention (CDC), Outbreaks of *Shigella sonnei* infection associated with eating fresh parsley—United States and Canada, July–August 1998, *Morb. Mortal. Wkly. Rep,* 48, 285, 1999.
33. Trevejo, R.T. et al., An untypeable *Shigella flexneri* strain associated with an outbreak in California, *J. Clin. Microbiol.,* 37, 2352, 1999.
34. Chicago Department of Public Health (CDPH), Small cluster of a rare *Shigella* serotype. Available at: http://egov.cityofchicago.org/webportal/COCWebPortal/COC_EDITORIAL/shig_arbo_may_99.pdf. 1999. Accessed November 5, 2004.
35. Centers for Disease Control and Prevention (CDC). Public health dispatch: outbreak of *Shigella sonnei* infections associated with eating a nationally distributed dip—California, Oregon, and Washington, January 2000. *Morb. Mortal. Wkly. Rep.*, 49, 60, 2000.

36. Kimura, A.C. et al., Multistate shigellosis outbreak and commercially prepared food, United States, *Emerg. Infect. Dis.*, 10, 1147, 2004.
37. Terajima, J., et al., A multi-prefectural outbreak of *Shigella sonnei* infections associated with eating oysters in Japan. *Microbiol. Immunol.*, 48, 49, 2004.
38. Toronto Public Health. 2002. Health advisory: shigellosis food poisoning investigation. Available at: http://www.city.toronto.on.ca/health/advisory_shigella.htm. Accessed April 8, 2009.
39. Center for Science in the Public Interest (CSPI). 2006. *Shigella* Outbreaks. Available at http://www.cspinet.org/foodsafety/outbreak/outbreaks.php?column=pathogenGroup&colval=Shigella. Accessed April 6, 2009.
40. Gaynor, K. et al., International foodborne outbreak of *Shigella sonnei* infection in airline passengers, *Epidemiol. Infect.*, 137, 335, 2009.
41. Lewis, H.C. et al., Outbreaks of *Shigella sonnei* infections in Denmark and Australia linked to consumption of imported raw baby corn, *Epidemiol. Infect.*, 137, 326, 2009.
42. Hung-Wei, K. et al., A food-borne outbreak of *Shigella sonnei* gastroenteritis, Austria, 2008, *Middle Eur. J. Med.*, 121, 157, 2009.
43. Heier, B.T., et al., *Shigella sonnei* infections in Norway associated with sugar peas, May–June 2009. Eurosurveillance, 14. www.Eurosurveillance.org. 2009.
44. Muller, L., et al., Imported fresh sugar peas as suspected source of an outbreak of *Shigella sonnei* in Denmark, April–May 2009. Eurosurveillance, 14. www.Eurosurveillance.org. 2009.
45. Davis, H.J. et al., A shigellosis outbreak traced to commercially distributed lettuce, *Am. J. Epidemiol.*, 128, 1312, 1998.
46. Cook, K.A. et al., Scallions and shigellosis: A multistate outbreak traced to imported green onions. [abstract], *Proc. 44th Ann. Conf. Epid. Intel. Serv.*, 36. March 27–31, 36, Atlanta, GA.
47. Toronto Public Health. Health advisory: Shigellosis food poisoning investigation. Available at: http://www.city.toronto.on.ca/health/advisory_shigella.htm. 2002. Accessed April 8, 2009.
48. Centers for Disease Control and Prevention (CDC), Preliminary FoodNet data on the incidence of foodborne illnesses—selected sites, United States, 2002, *Morb. Mortal. Wkly. Rep.*, 52, 340, 2003.
49. Centers for Disease Control and Prevention (CDC), Preliminary FoodNet data on the incidence of foodborne illnesses—selected sites, United States, 2001, *Morb. Mortal. Wkly. Rep.*, 51, 325, 2002.
50. al-Lahham, A.B., Abu-Saud, M., and Shehabi, A.A., Prevalence of *Salmonella*, *Shigella* and intestinal parasites in food handlers in Irbid, Jordan, *J. Diarrhoeal. Dis. Res.*, 8, 160, 1990.
51 Mensah, P. et al., Street foods in Accra, Ghana: How safe are they? *Bull. World Health Organ.*, 80, 546, 2002.
52. Wood, L.V. et al., Incidence of bacterial enteropathogens in foods from Mexico, *Appl. Environ. Microbiol.*, 46, 328, 1983.
53. Food and Drug Administration (FDA), FDA survey of imported fresh produce, FY 1999.
54. Food and Drug Administration (FDA), FDA survey of imported fresh produce: Imported produce assignment, FY, 2001.
55. Sansonetti, P.J., Genetic and molecular basis of epithelial cell invasion by *Shigella* species, *Rev. Infect. Dis.*, 13, S285, 1991.
56. Parsot, C., *Shigella* type III secretion effectors: How, where, when, for what purposes? *Curr. Opin. Microbiol.*, 12, 110, 2009.
57. Tamano, K. et al., Supramolecular structure of the *Shigella* type III secretion machinery: The needle part is changeable in length and essential for delivery of effectors, *EMBO J.*, 19, 3876, 2002.
58. Suzuki, T., and Sasakawa, C., Molecular basis of the intracellular spreading of *Shigella*, *Infect. Immun.*, 69, 5959, 2001.
59. Zychlinsky, A. et al., Interleukin 1 is released by murine macrophages during apoptosis induced by *Shigella flexneri*, *J. Clin. Invest.*, 94, 1328, 1994.
60. Stendahl, O.I. et al., Distribution of actin-binding protein and myosin in macrophages during spreading and phagocytosis, *J. Cell. Biol.*, 84, 215, 1980.
61. Sansonetti, P.J. et al., Multiplication of *Shigella flexneri* within HeLa cells: Lysis of the phagocytic vacoule and plasmid-mediated contact hemolysis, *Infect. Immun.*, 54, 461, 1986.
62. Vokes, S.A. et al., The aerobactin iron transport system genes in *Shigella flexneri* are present within a pathogenicity island, *Mol. Microbiol.*, 33, 63, 1999.
63. Perry, R.D., and San Clemente, C.L., Siderophore synthesis in *Klebsiella pneumoniae* and *Shigella sonnei* during iron deficiency, *J. Bacteriol.*, 140, 1129, 1979.
64. Payne, S.M. et al., Expression of hydroxamate and phenolate siderophores by *Shigella flexneri*, *J. Bacteriol.*, 155, 949, 1983.
65. Lawlor, K.M., and Payne, S.M., Aerobactin genes in *Shigella* spp., *J. Bacteriol.*, 160, 266, 1984.
66. Payne, S.M., Iron and virulence in the family Enterobacteriaceae, *CRC Crit. Rev. Microbiol.*, 16, 81, 1988.
67. Headley, V. et al., Expression of aerobactin genes by *Shigella flexneri* during extracellular and intracellular growth, *Infect. Immun.*, 65, 818, 1997.
68. Bernardini, M.L. et al., Identification of icsA, a plasmid locus of *Shigella flexneri* that governs bacterial intra- and intercellular spread though interaction with F-actin, *Proc. Natl. Acad. Sci. USA*, 86, 3867, 1989.
69. Makino, S. et al., A genetic determinate required for continuous reinfection of adjacent cells on a large plasmid of *Shigella flexneri* 2a, *Cell*, 46, 551, 1986.
70. Goldberg, M.B. et al., Unipolar localization and ATPase activity of IcsA, a *Shigella flexneri* protein involved in intracellular movement, *J. Bacteriol.*, 175, 2189, 1993.
71. Lett, M.-C. et al., *virG*, a plasmid-coded virulence gene of *Shigella flexneri*: Identification of the *virG* protein and determination of the complete coding sequence, *J. Bacteriol.*, 171, 353, 1989.
72. Suzuki, T., Lett, M.-C., and Sasakawa, C., Extracellular transport of VirG protein in *Shigella*, *J. Biol. Chem.*, 270, 30874, 1995.
73. Sandlin, R.C. et al., Avirulence of rough mutants of *Shigella flexneri*: Requirement of O antigen for correct unipolar localization of IcsA in the bacterial outer membrane, *Infect. Immun.*, 63, 229, 1995.
74. Sandlin, R.C., Goldberg, M.B., and Maurelli, A.T., Effect of O side-chain length and composition on the virulence of *Shigella flexneri* 2a, *Mol. Microbiol.*, 22, 63, 1996.
75. Sandlin, R.C., and Maurelli, A.T., Establishment of unipolar localization of IcsA in *Shigella flexneri* 2a is not dependent on virulence plasmid determinants, *Infect. Immun.*, 67, 350, 1999.
76. Robbins, J.R. et al., The making of a gradient: IcsA (VirG) polarity in *Shigella flexneri*, *Mol. Microbiol.*, 41, 861, 2001.

77. d'Hauteville, H. et al., Lack of cleavage of IcsA in *Shigella flexneri* causes aberrant movement and allows demonstration of a cross-reactive eukaryotic protein, *Infect. Immun.*, 64, 511, 1996.
78. Egile, C. et al., SopA, the outer membrane protease responsible for polar localization of IcsA in *Shigella flexneri*, *Mol. Microbiol.*, 23, 1063, 1997.
79. Steinhauer, J. et al., The unipolar *Shigella* surface protein IcsA is targeted directly to the bacterial old pole: IcsP cleavage of IcsA occurs over the entire bacterial surface, *Mol. Microbiol.*, 17, 945, 1999.
80. Suzuki, T., Saga, S., and Sasakawa, C., Functional analysis of *Shigella* VirG domains essential for interaction with vinculin and actin-based motility, *J. Biol. Chem.*, 271, 21878, 1996.
81. Monack, D.M., and Theriot, J.A., Actin-based motility is sufficient for bacterial membrane protrusion formation and host cell uptake, *Cell. Microbiol.,* 3, 633, 2001.
82. Miki, H., Miura, K., and Takenawa, T., N-WASP, a novel actin-depolymerizing protein, regulates the cortical cytoskeletal rearrangement in a PIP2-dependent manner downstream of tyrosine kinases, *EMBO J.*, 15, 5326, 1996.
83. Loisel, T.P. et al., Reconstitution of actin-based motility of *Listeria* and *Shigella* using pure proteins, *Nature*, 401, 613, 1999.
84. Shibata, T. et al., Cdc42 facilitates invasion but not the actin-based motility of *Shigella*, *Curr. Biol.*, 12, 341, 2002.
85. Parsot, C., and Sansonetti, P.J., Invasion and the pathogenesis of *Shigella* infections, *Curr. Top. Microbiol. Immunol.*, 209,.25, 1996.
86. Sansonetti, P.J., and Mounier, J., Metabolic events mediating early killing of host cells infected by *Shigella flexneri*, *Microb. Pathog.*, 3, 53, 1987.
87. Fontaine, A., Arondale, J., and Sansonetti, P.J., Role of shiga toxin in the pathogenesis of bacillary dysentery studied using a tox- mutant of *Shigella dysenteriae* 1, *Infect. Immun.*, 56, 3099, 1988.
88. Donohue-Rolfe, A., Acheson, D.W.K., and Keusch, G.T., Shiga toxin: Purification, structure, and function, *Rev. Infect. Dis.*, 13, S293, 1991.
89. Brown, J.E., Echeverria, P., and Lindberg, A.A., Digalactosyl-containing glycolipids as cell surface receptors for shiga toxin of *Shigella dysenteriae* 1 and related cytotoxins of *Escherichia coli*, *Rev. Infect. Dis.*, 13, S298, 1991.
90. Keusch, G.T. et al., Shiga toxin: intestinal cell receptors and pathophysiology of enterotoxic effects, *Rev. Infect. Dis.*, 13, S304, 1991.
91. Yavzori, M., Cohen, D., and Orr, N., Prevalence of the genes for *Shigella* enterotoxins 1 and 2 among clinical isolates of *Shigella* in Israel, *Epidemiol. Infect.*, 128, 533, 2002.
92. Zaika, L., The effect of temperature and low pH on survival of *Shigella flexneri* in broth, *J. Food Prot.*, 64, 1162, 2001.
93. Zaika, L., The effect of NaCl on survival of *Shigella flexneri* in broth as affected by temperature and pH, *J. Food Prot.*, 65, 774, 2002.
94. Zaika, L., Effect of organic acids and temperature on survival of *Shigella flexneri* in broth at pH 4, *J. Food Prot.*, 65, 1417, 2002.
95. Tollison, S.B., and Johnson, M.G., Sensitivity to bile salts of *Shigella flexneri* sublethally heat stressed in buffer or broth, *Appl. Environ. Microbiol.*, 50, 337, 1985.
96. Spicer, C.C., The survival of *Shigella sonnei* on cotton threads, *J. Hyg.*, 57, 210, 1959.
97. Nakamura, M., The survival of *Shigella sonnei* on cotton, glass, wool, paper and metal at various temperatures, *J. Hyg.*, 60, 35, 1962.
98. Islam, M.S. et al., Survival of *Shigella dysenteriae* type 1 on fomites, *J. Health Popul. Nutr.*, 19, 177, 2001.
99. Rafii, F., and Lunsford, P., Survival and detection of *Shigella flexneri* in vegetables and commercially prepared salads, *J. AOAC Int.*, 80, 1191, 1997.
100. Bagamboula, C., Uyttendaele, M., and Debevere, J., Acid tolerance of *Shigella sonnei* and *Shigella flexneri*, *J. Appl. Microbiol.*, 93, 479, 2002.
101. Islam, D. et al., Rapid detection of *Shigella dysenteriae* and *Shigella flexneri* in faeces by an immunomagnetic assay with monoclonal antibodies, *Eur. J. Clin. Mibrobiol. Infect. Dis.*, 12, 25, 1993.
102. Satchell, F.B. et al., The survival of *Shigella sonnei* in shedded cabbage, *J. Food Prot.*, 53, 558, 1990.
103. Lampel, K.A., and Maurelli, A.T., *Shigella* species. In Doyle, M.P., Beuchat, L.R., and Montville, T.J. (Editors), *Food Microbiology. Fundamentals and Frontiers*, 2nd ed., Chapter 12, pp. 216–227, American Society of Microbiology Press, Washington, DC, 2001.
104. Taylor, W.I., and Schelhart, D., Isolation of Shigellae VII. Comparison of gram-negative broth with Rappaport's enrichment broth, *Appl. Microbiol.*, 18, 393, 1969.
105. Uyttendaele, M. et al., Evaluation of culture media for enrichment and isolation of *Shigella sonnei* and *S. flexneri*, *Int. J. Food Microbiol.*, 70, 255, 2001.
106. Food and Drug Administration (FDA), *Bacteriological Analytical Manual*, *Shigella*, 8th ed., AOAC International, Arlington, VA, 1998.
107. Health Canada, Compendium of analytical methods. MFLP-25: Isolation and identification of *Shigella* spp. from foods. Available at: http://www.hc-sc.gc.ca/fn-an/res-rech/analy-meth/microbio/volume3/mflp25-eng.php. Accessed on November 5, 2010.
108. Mehlman, I.J., Romero, A., and Wentz, B.A., Improved enrichment for recovery of *Shigella sonnei* from foods, *J. AOAC*, 68, 552, 1985.
109. Warren, B.R., Comparison of conventional culture methods and polymerase chain reaction for the detection of *Shigella* spp. on tomato surfaces, Masters thesis, University of Florida, Gainesville, Fl, 2003.
110. Warren, B.R., Parish, M.E., and Schneider, K.R., Comparison of chromogenic *Shigella* spp. plating medium with standard media for the recovery of *Shigella boydii* and *Shigella sonnei* from tomato surfaces, *J. Food Prot.*, 68, 621, 2005.
111. Bhat, P., and Rajan, D., Comparative evaluation of desoxycholate citrate medium and xylose lysine desoxycholate medium in the isolation of Shigellae, *Am. J. Clin. Pathol.*, 64, 399, 1975.
112. June, G.A. et al., Evaluation of the Bacteriological Analytical Manual culture method for the recovery of *Shigella sonnei* from selected foods, *J. AOAC Int.*, 76, 1240, 1993.
113. Jacobson, A.P. et al., Evaluation of the Bacteriological Analytical Manual (BAM) culture method for the detection of *Shigella sonnei* in selected types of produce. [poster] *FDA Science Forum*, February 20–21, Washington, DC 2002.
114. Lindqvist, R., Detection of *Shigella* spp. in food with a nested PCR method—sensitivity and performance compared with a conventional culture method, *J. Appl. Microbiol.*, 86, 971, 1999.
115. Villalobo, E., and Torres, A., PCR for the detection of *Shigella* spp. in mayonnaise, *Appl. Environ. Microbiol.*, 64, 1242, 1998.

116. Vantarakis, A. et al., Development of a multiplex PCR detection of *Salmonella* spp. and *Shigella* spp. in mussels, *Lett. Appl. Microbiol.*, 31, 105, 2000.
117. Sethabutr, O. et al., Detection of Shigellae and enteroinvasive *Escherichia coli* by amplification of the invasion plasmid antigen H DNA sequence in patients with dysentery, *J. Infect. Dis.*, 167, 458, 1993.
118. Sethabutr, O. et al., Detection of PCR products of the *ipaH* gene from *Shigella* and enteroinvasive *Escherichia coli* by enzyme linked immunosorbent assay, *Diagn. Microbiol. Infect. Dis.*, 37, 11, 2000.
119. Theron, J. et al., A sensitive seminested PCR method for the detection of *Shigella* in spiked environmental water samples, *Water Res.*, 35, 869, 2001.
120. Lampel, K.A., and Orlandi, P.A., Polymerase chain reaction detection of invasive *Shigella* and *Salmonella enterica* in food, *Methods Mol. Biol.*, 179, 235, 2002.
121. Jin, Q. et al., Genome sequence of *Shigella flexneri* 2a: Insights into pathogenicity though comparison with genomes of *Escherichia coli* K12 and O157, *Nucleic Acids Res.*, 30, 4432, 2002.
122. Lampel, K.A., J et al., Polymerase chain reaction for detection of invasive *Shigella flexneri* in food. *Appl. Environ. Microbiol.*, 56, 1536, 1990.
123. Achi-Berglund, R., and Lindberg, A.A. Rapid and sensitive detection of *Shigella sonnei* in feces by the use of an O-antigen-specific monoclonal antibody in a combined immunomagnetic separation-polymerase chain reaction assay. *Clin. Microbiol. Infect.*, 2, 55, 1996.
124. Sonjai, K., et al., Validation of salmonellosis and shigellosis diagnostic test kits at a provincial hospital in Thailand. *Asian Pac. J. Allergy Immunol.*, 19, 115, 2001.
125. Sapsford, K.E., et al., Detection of *Campylobacter* and *Shigella* species in food samples using an array biosensor. *Anal. Chem.*,76, 433, 2004.
126. Achi, R., et al., Immunomagnetic separation and PCR detection show Shigellae to be common faecal agents in children from urban marginal communities of Costa Rica. *J. Infect.*, 32, 211, 1996.
127. Islam, D., and Lindberg, A.A. Detection of *Shigella dysenteriae* type 1 and *Shigella flexneri* in feces by immunomagnetic isolation and polymerase chain reaction. *J. Clin. Microbiol.*, 30, 2801, 1992.
128. Peng, X. et al., Rapid detection of *Shigella* species in environmental sewage by an immunocapture PCR with universal primers, *Appl. Environ. Microbiol.*, 68, 2580, 2002.
129. Whatman website. BioScience FAQs. Available at: http://www.whatman.com/support/?pageID=3.6.124.112#fa112. Accessed November 12, 2004.
130. Kakinuma, K., Fukushima, M., and Kawaguchi, R., Detection and identification of *Escherichia coli*, *Shigella*, and *Salmonella* by microarrays using the *gyrB* gene, *Biotech. Bioeng.*, 83, 721, 2003.
131. Raza, S. et al., Identification and strain differentiation of food and waterborne bacterial theat agents using automated rep-PCR. [poster]. *BioDefense Research Meeting*, *Am. Soc. Microbiol.*, March 9–12, Baltimore, MD, 2003.
132. Lising, M. et al., Identification of Gram-negative bacteria using the DiversiLab system. [poster]. *104th General Meeting*, *Am. Soc. Microbiol.*, March 23–27, New Orleans, LA, 2004.

90 Stenotrophomonas

Martina Adamek and Stephan Bathe

CONTENTS

90.1 INTRODUCTION

90.1.1 Genus *Stenotrophomonas*

Stenotrophomonas maltophilia was first described as *Pseudomonas maltophilia* in 1961.[1] This name was revived again in 1981 by Hugh.[2] Shortly thereafter, the organism was renamed to *Xanthomonas maltophilia* because of its closer relationship to the xanthomonads.[3] In 1993, a separate genus, *Stenotrophomonas*, was finally proposed with the single species *S. maltophilia*.[4]

In the recent years, additional *Stenotrophomonas* species, *S. nitritireducens*,[5] *S. acidaminiphila*,[6] *S. rhizophila*,[7] *S. koreensis*,[8] *S. terrae* and *S. humi*,[9] *S. chelatiphaga*,[10] and *S. ginsengisoli*[11] have been described. The species *S. africana*[12] was later found to be synonymous to *S. maltophilia*.[13] The recently described species *S. dokdonensis*[14] has meanwhile been moved to the genus *Pseudoxanthomonas*.[15]

Altogether, the genus *Stenotrophomonas* presently contains nine phylogenetically distinct species. Figure 90.1 shows the 16S rRNA diversity within the genus.

In a phylogenetic context, the genetic diversity of the genus was first extensively investigated in 1999 by Hauben et al.[16] In this study, more than 100 strains of *Stenotrophomonas* were investigated by amplified fragment length polymorphism (AFLP) genotyping, 16S rRNA sequence analysis, and DNA–DNA hybridization. The AFLP study revealed 10 distinct genogroups, and sequences of representatives of these groups have been included in Figure 90.1. At least seven of these genogroups still belong to *S. maltophilia*. A later study partly confirms the existence of these groups by restriction fragment length polymorphism (RFLP) analysis of the *gyrB* gene.[17] Recently, an extensive investigation of the genetic diversity within *S. maltophilia* based on multilocus sequence typing (MLST) of the genes *atpD, gapA, guaA, mutM, nuoD, ppsA*, and *recA* has been published.[18] This study provides support for the classification of the Hauben genogroups 1–7 and 9 (16S rRNA gene sequences of a group 8 strain assign this isolate to *S. rhizophila*) as genetically distinct lineages within *S. maltophilia* and additionally describes five new groups (A–E). Our own studies, based on rep-PCR and *gyrB* gene sequencing, again confirm the existence of several of these groups (unpublished data). It should be noted here that strains of AFLP group 9 within *S. maltophilia* form a distinct 16S rRNA lineage that differs by approximately 1% from the remaining *S. maltophilia* sequences. Strains of this group have to date only been isolated from environmental

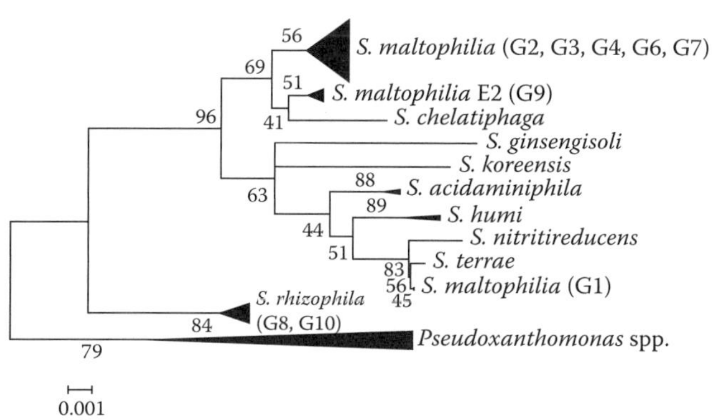

FIGURE 90.1 Neighbor-joining tree (1000 bootstrap repetitions) of 16S rRNA gene sequences of the genus *Stenotrophomonas*. Clades containing sequences belonging to the genomic groups indicated by Hauben et al.[16] are indicated with a G and the group number.

sources. This lineage was also recognized by other researchers as group E2, and it has been speculated that it may form a separate species within *Stenotrophomonas*.[19]

Thus, the genus *Stenotrophomonas* presently consists of nine species, with *S. maltophilia* as the type species of the genus. *S. maltophilia* itself displays a vast genetic diversity, and several genogroups can be reproducibly recognized using a variety of genetic typing methods. To date, it seems that *S. maltophilia* is the only opportunistic pathogenic species within the genus *Stenotrophomonas*.

90.1.2 Morphology, Biology, and Epidemiology

S. maltophilia is a gram-negative nonfermentative bacillus that forms rod-shaped cells of 0.5–1.5 μm in length, which are straight to slightly curved. They are motile and carry several polar flagella. It grows obligately aerobic at temperatures between 5°C and 37°C on a wide variety of media [nutrient agar, lysogeny broth (LB), MacConkey] and forms white-to-yellow, smooth colonies. Some strains show dark discoloration of the medium on clear agar plates. Most strains are methionine- and/or cysteine-auxotrophic.

S. maltophilia has a ubiqitous distribution regarding global geography as well as habitat, apparently with a preference for moist and aqueous habitats.[20] It has been isolated from fountain drains,[21] tap water,[22] drinking water treatment and distribution systems,[23] and bottled water.[24] It occurs in wastewater environments[25–27] and natural water sources.[28] There are also some reports of detection of the organism in association with animals[29–34] and plants.[35–38] It also has been found in soils, mainly in rhizosphere soils.[9–43] Several reviews have addressed parallels in rhizospheric symbiotic interactions and the possible pathogenic properties of opportunistic pathogens using *S. maltophilia* as an example, which might explain a possible reservoir function of the plant rhizosphere for these organisms.[44,45]

With respect to the hospital environment, numerous sources have been reported as reviewed by Denton and Kerr,[20] including water reservoirs, nebulizer equipment, aerators, and disinfectant reservoirs.[22,46–50]

Epidemiological studies concerning *S. maltophilia* strains isolated from patients show a general trend: strains belonging to a number of genogroups indicated in previous phylogenetic studies seem to display a stronger tendency to cause infection or at least to occur as commensal organisms. This dominantly seems to be the case for respiratory and cystic fibrosis (CF) isolates. Coenye et al.[17] noted that 36 out of 40 CF isolates tested in their study clustered in two genogroups as defined by RFLP analysis of the *gyrB* gene. The recent study by Kaiser et al.[18] showed an accumulation of CF isolates (12 of 19) in genogroup 6. Among other groups, this group falls into one of the clusters described in the *gyrB* RFLP analysis. On the other hand, some genogroups analyzed in these and in our own studies (unpublished) clearly show clades that exclusively consist of environmental isolates, which indicates that a different environmental niche is occupied by these organisms. This is the case for strains associated with genogroups 5 and 9.

In contrast to the MLST and AFLP studies mentioned above, which aim to identify the phylogenetic relatedness of strains, purely epidemiological studies with the main purpose of identifying outbreaks or the clonal identity of strains isolated in one or several hospitals have found that *S. maltophilia* is predominantly acquired from the environment. A patient-to-patient spread has only been detected in a few cases. In a study of 25 *S. maltophilia* bacteremia cases, the isolates were typed by PFGE, and as many different PFGE patterns as investigated patients were found.[51] A Brazilian study of 39 strains revealed 33 different RAPD patterns among these isolates and indicated three pairs of patients with identical patterns. It was observed that each pair was situated on the same ward, making a double infection with the same strain in each case likely.[52] An AP-PCR study of 52 strains from Turkish hospitals indicated two small outbreaks of five and three strains isolated from a neurology and a pediatrics clinic, respectively.[53] A further study has investigated the genetic relatedness of 139 strains isolated from a Spanish hospital using PFGE.[54] Three episodes of the cross-transmission of three different respiratory strains between six, three, and two patients as well as two strains occurring in two pairs of bacteremic patients. However, this study and a further investigation[55] identified both medical devices and worker-related transmission of strains in two outbreak situations.

In short, *S. maltophilia* presents itself as an opportunistic pathogen consisting of at least 14 genogroups. Some of these genogroups contain exclusively environmental strains, indicating a low pathogenicity and possibly different niche differentiation of these isolates. Both clinical and environmental isolates are distributed among the other genogroups, with a prevalence of respiratory isolates in two to three genogroups, whereas strains causing other disease manifestations do not show a clear association with distinct genogroups.

It should be noted that, besides its pathogenic properties, *S. maltophilia* strains show a high versatility with respect to other biological properties: strains of the organism have been shown to produce antifungal substances[56] and to possess antifungal properties,[19] which may also lead to the plant-growth-promoting properties observed in some isolates.[57] Also, the production of indole-acetic acid (IAA), a plant hormone, has been observed.[57] A further characteristic supporting the growth of plants and influencing soil properties is the ability to fix nitrogen from air.[35] In addition, extremely heavy-metal tolerant strains have been described,[58,59] and there are a number of publications indicating the involvement of *S. maltophilia* in the geochemical cycling of selenium.[41,60–62] Also, the degradation of xenobiotic compounds such as herbicides[63] and aromatic hydrocarbons[64] has been noticed.

90.1.3 Clinical Features and Pathogenesis

The disease manifestations *S. maltophilia* has been associated with are as diverse as the organism itself. As an opportunistic pathogen that does not seem to possess specific toxins or a tissue/organ specificity, the organism establishes itself at its point of entry into the human body and may occasionally spread from this place to cause a systemic disease. There is a large grey area between colonization and true infection caused by this organism.

Numerous types of infections have been described, as previously noted by Denton and Kerr.[20] A recent overview of *S. maltophilia* as an emerging nosocomial pathogen, including epidemiology, infections, and treatment options, is given by Looney et al.[65] They identify pneumonia as the most common type of infection caused by the bacterium, followed by bloodstream, wound, and urinary tract infections. Apart from this, there are numerous reports of rare infection types, such as infections of the skin, bone and joints, the eye, the gastrointestinal tract, and the nervous system.[20,65]

Especially in intensive care units (ICUs), the pathogen is gaining importance.[66] The severity of the disease before *S. maltophilia* infection is an important risk factor and was shown to be connected with the mortality rate.[67] Most patients had a previous therapy with broad-spectrum antibiotics, such as carbapenems, ceftadicime, fluorquinolones, and glycopeptides. A lengthy antibiotic therapy promotes the infection, thus the intrinsic antibiotic multiresistance of *S. maltophilia* is an advantage in colonization.[68–70] The duration of antibiotic donation was also shown to have a significant influence on the outcome of the disease.[71] Cystic fibrosis (CF) patients are generally more susceptible to bacterial infections and therefore are often treated with antibiotics over extended periods of time, which increases their susceptibility to the antibiotic multiresistant *S. maltophilia*.[72] Not only CF but also other respiratory tract infections, such as chronic obstructive pulmonary disease (COPD), predispose patients for an infection. In case of mechanical ventilation or several kinds of other indwelling devices, as intravascular catheters or tracheostomy tubes, patients show an increased susceptibility. Due to the adhesion of the organism to medical surfaces and the possibility of biofilm formation, these devices are considered to be the source of infection. In this context, it is called ventilator associated pneumonia (VAP). The duration of the mechanical ventilation plays a significant role, so the infection can manifest as late-onset VAP. In bloodstream infection, *S. maltophilia* often occurs in communities with other bacteria, for example, *Pseudomonas* spp. The risk for acquiring bacteremia is higher in the presence of a central venuous catheter. In such cases, quick removal of the device leads to the patient's fast recovery, whereas failure to remove the catheter increases the risk of relapse.[73,74] For hematologic patients with neutropenia and hypertension, *S. maltophilia*–associated bacteremia can be a serious complication. Neutropenia could be an accessory symptom induced by cancer chemotherapy.[75] Other diseases contracted by cancer patients, immunocompromised because of chemotherapy, are mucositis and diarrhea, which were shown to be significant risk factors associated with *S. maltophilia* bacteremia.[68] Therefore, it can be said that risk factors for acquisition of an infection with *S. maltophilia* are prior antibiotic therapy, catheterization, chemotherapy, prolonged hospital stay, ICU admission, and mechanical ventilation as well as underlying diseases and malignancies.[20,65] In addition, ICU admission and underlying hematological disease are named as connected risk factors in the case of cancer patients, as well as shock and thrombocytopenia as risk factors for patients with bloodstream infection and pneumonia.

S. maltophilia is well known for its intrinsic antibiotic multiresistance. Determination of antibiotic resistances in the laboratory is, however, quite variable and to a high degree dependent on the technique used. Tatman-Otkun et al.[53] found using a comparison of E-test, disc diffusion, and agar dilution methods and showed a limited correlation of these methods. Currently, the agar dilution method is recommended by National Committee for Clinical Laboratory Standards (NCCLS) for the detection of antimicrobial susceptibility of *S. maltophilia*. Bacterial resistances to broad-spectrum β-lactam antibiotics, such as carbapenems and aminoglycosides except gentamicin, which are typically used for prior antibiotic therapy, have been described. Furthermore, *S. maltophilia* shows resistance to macrolides, tetracycline, chloramphenicol, and older quinolones.[76] It is presumed that life in the plant rhizosphere contributes to natural formation of antibiotic resistances. The rich content of nutrients leads to high competition among bacteria colonizing the rhizosphere. On the one hand, a high number of different natural antibiotics occurs in this environment; on the other hand, horizontal gene transfer is enhanced. Both effects contribute to increasing antibiotic resistances.[44]

Several mechanisms leading to antibiotic resistance have been elucidated. The recent complete genome sequencing of *S. maltophilia* strain K279a revealed nine different resistance nodulation division (RND)–type efflux pumps, which are a basis for the intrinsic multidrug resistance of the bacterium.[76] Of the previously described RND tripartite efflux pumps, SmeDEF is known to be responsible for resistance to fluorquinolones, chloramphenicol, and tetracycline.[77] *S. maltophilia* is also resistant against all clinically available aminoglycosides, which is referred to genes *aph(3′)-IIc* carrying an aminoglycoside phosphotransferase enzyme and *aac(6′)-Iz* encoding an aminoglycoside 6′-*N*-acetyltransferase.[78,79] In an early study, Wheat et al.[80] observed the temperature-dependent variation to antibiotic resistance against aminoglycosides and polymyxinB. At a temperature of 30°C, the *S. maltophilia* isolates were most resistant and at 37°C, most susceptible. This was later explained in a couple of studies by Rahmanti-Bahram et al.,[81–83] who found that conformation change of the lipopolysaccharides (LPS) in the outer membrane at a 30°C leads to a lowered binding affinity and therefore to reduced aminoglycoside uptake. Furthermore a phosphoglucomutase (PGM) was described to be associated with LPS production. In *S. maltophilia* the *spgM* gene was discovered, which encodes an enzyme with PGM and phosphomannomutase activities. Mutants lacking *spgM* were shown to produce less LPS and therefore to display an increase in susceptibility for several antibiotics.[84] A common treatment for *S. maltophilia* infections is the application of trimethoprim/sulfamethoxazole (TMP/SMX), because of the so-far promising in vitro susceptibility rates.[85] A problem is an increasing number of TMP/SMX resistant *S. maltophilia* strains. Recent studies revealed the presence of class 1 integron carrying *sul1* gene and insertion sequence common region (ISCR) elements carrying *sul2* on several *S. maltophilia* isolates. *sul* genes were described to be linked to TMP/SXT resistance.[86,87] A serious threat would be a mobilization of these genes caused by an increased use of TMP/SXT.

Although *S. maltophilia* does not possess specific toxins, there are several factors involved in virulence, such as the ability to colonize medical devices, production of proteases, lipases, and other extracellular enzymes. In several studies, the ability of biofilm formation on different surfaces is described as the first step of colonization. In this context, a severe problem is colonization of medical devices, for example, catheters or endotracheal tubes, form which the bacteria have access to the different patient's tissues. A study by Jucker et al.[88] describes adhesion to negatively charged glass and Teflon, which is uncommon for bacteria because usually at physiological pH bacterial surfaces are negatively charged as well. *S. maltophilia* was found to have only traces of LPS associated proteins on the surface and lacks of negatively charged groups. This leads to a slightly positive charge of the surface caused by the few outer membrane proteins. Another main aspect of adherence to different surfaces is production of fimbriae and flagellae. Electron microscopy scans by de Oliveria-Garcia et al.[89] showed two kinds of fibers, long, thick filaments of 40–50 nm, probably flagellae, and short fine fimbriae (5–7 nm) distributed around the bacteria. Furthermore, the fimbrin (SMF-1) was purified and shown to be involved in biofilm formation, adherence to cultured epithelial cells, and causing agglutination of mouse and hen erythrocytes.[90] Another factor proposed to be involved in *S. maltophilia* virulence is an alkaline serine protease encoded by *StmPr1* gene. This protease has already been shown to degrade several human proteins from serum and connective tissue.[91] The immunostimulatory properties of *S. maltophilia* were determined by comparison to a well-characterized *Pseudomonas aeruginosa* strain. In this assay, stimulation of airway epithelial and macrophage cell lines leaded to significant interleukin-8 and tumor-necrosis-factor-α (TNFα) expression, both signs of inflammation responses.[92] In addition, a diffusible signal factor (DSF)–dependent cell-cell signaling system, similar to the DSF signaling system of the plant pathogen *Xanthomonas campestris pv. campestris*, was also shown to contribute to *S. maltophilia* virulence.[93]

As mentioned earlier, the genome sequence of the bacteremia causing isolate K279a was the first to be published.[76] The genome analysis of this organism has revealed a high number of putative drug resistance genes, many of which belong to the class of RND (resistance nodulation division) efflux pumps. These enzyme systems are efflux transporters with broad substrate specificities which have been described in many multidrug resistant nosocomial pathogens. These pumps mediate a proton-motive force driven efflux in conjunction with a membrane fusion protein spanning the cytoplasmic space and an outer membrane protein. Mutation analysis of the genes coding for the pumps have revealed effects on drug resistance in five of these systems. Furthermore, a number of other putative drug resistance genes have been found in the K279a genome sequence, including those for aminoglycoside, macrolide, and beta-lactam resistance.

Besides K279a, the genomes of two additional isolates have been sequenced. One is a strain that has been isolated as an endophyte of poplar trees, R551–3[37]; the other is a marine isolate, SKA14. Very limited information is available about these strains. An initial genomic comparison of K279a and R551–3 has recently been published.[94] There is a clear difference in genome size (4.85 Mbp of K279a vs. 4.57 Mbp of R551–3), but the genomes show the same GC content and share a common scaffold. The differences are mostly due to additional DNA sequences present in K279a, but R551–3 also shows some unique DNA segments. Genes present on those genomic islands that are unique to either one of the strains are metal resistance genes, open reading frames (ORFs) coding for secretion systems, hemaggluttinin genes, and LPS genes, with K279a showing the majority of additional ORFs. Prophages and insertion sequences have also been located on some of these genomic islands.

90.1.4 Diagnosis

Conventional Techniques. In the clinical laboratory, *S. maltophilia* is usually detected as a gram-negative, nonfermenting bacillus. It grows on a variety of media. including MacConkey and blood agar, is DNAse-positive, oxidase-negative, and shows a late or no oxidation of glucose. In contrast, it is positive for oxidation of maltose (hence the name *maltophilia*) and shows lysine decarboxylase activity. By now, *S. maltophilia* has been included as a target species in most commercial biochemical identification systems, such as the VITEK2 card, API 20NE, BIOLOG, and others.

A number of selective media for culturing of *S. maltophilia* from clinical and environmental sources have also been described:

The first one is Xanthomonas maltophilia selective medium (XMSM) developed by Juhnke et al. in 1989.[95] This is a tryptone-containing medium with maltose as an additional carbon source, including a bromothymol blue indicator to detect any acidification of the medium. The basis of the selection is the presence of six different antibiotics and two antimycotic compounds to suppress the growth of fungi from environmental sources.

Later, Bollet et al.[96] published a two-step isolation procedure. This is based on preenrichment of a sample in liquid nutrient broth amended with methionine. Bacteria grown in this medium are then plated on Müller–Hinton agar followed by placement of Imipenem-containing antibiotic discs on the agar surface. Colonies grown in the inhibition zones are then regarded as presumptive *S. maltophilia* strains.

Another direct selective agar is VIA agar developed by Kerr et al.[97] This is based on a mannitol agar containing a bromothymol blue indicator supplemented with vancomycin, imipenem, and amphotericin B as selective agents. Non-mannitol-fermenting colonies growing on this medium are being considered as *S. maltophilia*. This medium was recently modified by replacement of imipenem with meropenem because of the better commercial availability of the latter compound.[98]

Molecular Techniques. Several studies containing protocols for the detection of *S. maltophilia* by polymerase chain reaction (PCR) have been published. An early investigation by Karpati et al.[99] had the aim of detecting three major CF pathogens, *Pseudomonas aeruginosa*, *Burkholderia cepacia*, and *Stenotrophomonas maltophilia*, in the sputum of CF patients by using a multiplex PCR approach. However, this protocol does not give primers with a high species specificity for *S. maltophilia* and may not be suitable for general diagnostic purposes. Later, a PCR protocol aimed at species-specific detection of *S. maltophilia*, again with the main aim of analyzing the sputum of CF patients, was published.[100] The PCR primer pair presented was based on the 23S rRNA gene and produced negative results for various species of *Pseudomonas, Burkholderia, Klebsiella, Moraxella, Ralstonia*, and *Alcaligenes*, thus demonstrating a species specificity for organisms within clinical settings. A later investigation of phylogenetically more closely related organisms (*Stenotrophomonas* species other than *S. maltophilia*, *Xanthomonas* spp.) revealed that these PCR primers were not absolutely specific for *S. maltophilia*, as they also recognized other organisms that may occupy similar environmental niches.[101]

Similarly, several studies containing 16S rRNA specific oligonucleotide probes for use in fluorescence in situ hybridization (FISH) followed by epifluorescence microscopy or flow cytometry for detection and quantification of the organism have been published. The first of these studies had a clinical background of detecting and differentiating the main pathogenic bacteria in sputum samples of CF patients.[102] Five out of six culture-positive samples could by confirmed by FISH with subsequent microscopy. The same FISH probe was later used to identify pathogens in artificially spiked blood cultures by FISH with subsequent flow cytometric analysis.[103] However, investigation of the specificity of this FISH probe using the ProbeMatch tool of the Ribosomol Database Project II shows a low number of targeted database entries for *Stenotrophomonas*, although nontarget entries outside the genus are not targeted. In contrast, a FISH probe developed in a recent ecological study[28] shows a much higher number of targeted database entries without simultaneously detecting other *Stenotrophomonas* species. It should be noted here that some database entries outside the genus *Stenotrophomonas* were also targeted. Altogether, it seems that the newer FISH probe developed by Piccini et al.[28] is better suited to detect a range of *S. maltophilia* strains than the one designed in the study by Hogardt et al.[102]

90.2 METHODS

This section describes protocols for the detection and identification of *S. maltophilia* by PCR and by prokaryotic FISH. Both detection methods are aimed at ribosomal RNA sequences (DNA or RNA) as targets, so that a high phylogenetic specificity can be achieved here.

90.2.1 Sample Preparation

Preparation of DNA Extracts from Bacterial Cultures or Environmental Samples. DNA is extracted from biomass and soil samples using standard molecular biological protocols. A range of commercially available kits can also be used, dependent on the preferences of the worker and the procedures of the laboratory. The protocol described here produces a minipreparation of genomic DNA from bacteria,[104] which has proven to be suitable for pure cultures, activated sludge, and wastewater biomass samples, as well as for samples from sediments with low humic acid contents.

Approximately 50 mg of biomass pellet is suspended in 567 μL TE buffer. Thirty μL of 10% SDS and 3 μL of proteinase K (20 mg/mL) are added, and the sample is incubated for 1 h at 37°C. Then, 100 μL of 5 M NaCl and 80 μL of 10% CTAB/0.7 M NaCl are added; the sample is mixed thoroughly and incubated for 10 min at 65°C. The solution is then extracted with chloroform, phenol/chloroform, and chloroform again. The DNA is precipitated with 0.6 vol of isopropanol, pelleted by centrifugation for 15 min at 15,000 × *g*, washed with 1 mL of 70% ethanol, and finally resuspended in 100 μL of water.

Alternatively, crude DNA extracts from pure bacterial cultures are prepared as follows: Resuspend a suitable amount of biomass in 800 μL sterile 0.85% KCl (the suspension should be visibly turbid but not dense). Wash cells once more in 0.85% KCl by centrifugation and resuspension, centrifuge again, take off the supernatant, and finally resuspend in 400 μL of ultrapure water suitable for PCR. This cell suspension can directly be used as template in PCR. For cell-free extracts, lyse cells in the suspension by heating for 10 min at 95°C to release DNA into the solution, pellet cell debris by centrifugation, and collect 300 μL of the supernatant, which can be used again as the template in PCR.

Fixation of Biomass for FISH. Fixation of bacterial cells for FISH is carried out using 4% paraformaldehyde (PFA) in phosphate buffered saline (PBS).[105] Briefly, biomass is washed and finally resuspended using PBS. One volume of biomass suspension is then added to three volumes of 4% paraformaldehyde solution in PBS and incubated for 1–3 h (may be extended until overnight) at 4°C. The biomass is then centrifuged (4°C, 5000 rpm, 5 min) and washed again in PBS to remove residual PFA, and suspended in 1:1 mixture of PBS and ethanol. This suspension can be stored at −20°C for several months.

90.2.2 Detection Procedures

The following PCR primers have been used in our studies to confirm *S. maltophilia* strains by screening of colonies. The primers also yield products from environmental samples containing *S. maltophilia*, but the sensitivity and detection limits of this reaction are currently not known.

The forward primer SM1f is a modification of the FISH probe developed by Piccini et al.[28] and has the sequence 5′-GTTGGGAAAGAAATCCAGC-3′. This primer binds in the 16S rRNA gene. The reverse primer SM4 has been described in the study by Whitby et al.[100] and binds in the 23S rRNA gene. The sequence is 5′-TTAAGCTTGCCACGAACAG-3′. The PCR amplicon produced has a length of approximately 2.1 kb, depending on the size of the ITS region.

A PCR mix (in a total volume of 25 μL) contains 1 U of Taq DNA polymerase, PCR buffer, 2.5 mM $MgCl_2$, 0.25 mM of each dNTP, 0.2 μM of each primer, and 1 μL of template DNA. The thermal program consists of an initial denaturation at 95°C for 5 min followed by 35 cycles of 30 s at 95°C, 1 min at 57°C, 2 min at 72°C, and is concluded by a final extension of 5 min at 72°C.

PCR products are then separated and visualized using agarose gel electrophoresis and staining with ethidium bromide.

90.3 CONCLUSION AND FUTURE PERSPECTIVES

It is now clear that bacteria of the species *Stenotrophomonas maltophilia* are highly versatile, combining within the species opportunistic human pathogens as well as potentially useful strains for biotechnological or biocontrol purposes. An in-depth analysis of current and yet-to-be determined genome sequences combined with genomic and transcriptomic microarrays should further reveal the genetic basis for virulence, pathogenicity, and antibiotic resistance. On the other hand, these investigations should shed further light on the population structure of this organism and also should identify molecular markers which are suitable to differentiate pathogenic and nonpathogenic strains by linking evolutionary relationships with genetic and/or phenotypic properties.

In light of the high general degree of antibiotic resistance as well as upcoming new strains that show resistance against the commonly applied antibiotics such as trimethoprim/sulfamethoxazol, additional treatment options should be developed. This may be based on molecular targets revealed by genomic analyses. Further on, the continued development of molecular detection strategies is desirable. This should serve two purposes: the first is a satisfying coverage of the *S. maltophilia* population diversity in order to reduce the number of false-negatives. The second should lead to the development of faster and easier detection protocols. One option that could be further be exploited here is loop-mediated isothermal amplification,[106] a fast, reliable, and very sensitive detection method. Once established, this method requires only very little laboratory equipment and can be applied qualitatively as well as quantitatively.

In conclusion, *S. maltophilia* is an emerging multiresistant opportunistic pathogen that deserves a high degree of attention in terms of surveillance, epidemiology, and the development of detection and treatment methods. This organism may well acquire traits enhancing its pathogenicity and hence has the potential to develop into a serious pathogen.

REFERENCES

1. Hugh, R., and Ryschenkow, E., *Pseudomonas maltophilia*, an alcaligenes-like species, *J. Gen. Microbiol.*, 26, 123, 1961.
2. Hugh, R., *Pseudomonas maltophilia* sp. nov., nom. rev., *Int. J. Syst. Bacteriol.*, 31, 195, 1981.
3. Swings, J. et al., Transfer of *Pseudomonas maltophilia* Hugh 1981 to the genus *Xanthomonas* as *Xanthomonas maltophilia* (Hugh 1981) comb. nov., *Int. J. Syst. Bacteriol.*, 33, 409, 1983.
4. Palleroni, N.J., and Bradbury, J.F., *Stenotrophomonas*, a new bacterial genus for *Xanthomonas maltophilia* (Hugh 1980) Swings et al. 1983, *Int. J. Syst. Bacteriol.*, 43, 606, 1993.
5. Finkmann, W. et al., Characterization of N_2O-producing *Xanthomonas*-like isolates from biofilters as *Stenotrophomonas nitritireducens* sp. nov., *Luteimonas mephitis* gen. nov., sp. nov. and *Pseudoxanthomonas broegbernensis* gen. nov., sp. nov., *Int. J. Syst. Evol. Microbiol.*, 50, 273, 2000.
6. Assih, E.A. et al., *Stenotrophomonas acidaminiphila* sp. nov., a strictly aerobic bacterium isolated from an upflow anaerobic sludge blanket (UASB) reactor, *Int. J. Syst. Evol. Microbiol.*, 52, 559, 2002.
7. Wolf, A. et al., *Stenotrophomonas rhizophila* sp. nov., a novel plant-associated bacterium with antifungal properties, *Int. J. Syst. Evol. Microbiol.*, 52, 1937, 2002.
8. Yang, H.C. et al., *Stenotrophomonas koreensis* sp. nov., isolated from compost in South Korea, *Int. J. Syst. Evol. Microbiol.*, 56, 81, 2006.
9. Heylen, K. et al., *Stenotrophomonas terrae* sp. nov. and *Stenotrophomonas humi* sp. nov., two nitrate-reducing bacteria isolated from soil, *Int. J. Syst. Evol. Microbiol.*, 57, 2056, 2007.
10. Kaparullina, E. et al., *Stenotrophomonas chelatiphaga* sp. nov., a new aerobic EDTA-degrading bacterium, *Syst. Appl. Microbiol.*, 32, 157, 2009.
11. Kim, H.B. et al., *Stenotrophomonas ginsengisoli* sp. nov., a bacterium isolated from a ginseng field, *Int. J. Syst. Evol. Microbiol.*, 60, 1522, 2010.
12. Drancourt, M., Bollet, C., and Raoult, D., *Stenotrophomonas africana* sp. nov., an opportunistic human pathogen in Africa, *Int. J. Syst. Bacteriol.*, 47, 160, 1997.
13. Coenye, T. et al., *Stenotrophomonas africana* Drancourt et al. 1997 is a later synonym of *Stenotrophomonas maltophilia* (Hugh 1981) Palleroni and Bradbury 1993, *Int. J. Syst. Evol. Microbiol.*, 54, 1235, 2004.
14. Yoon, J.H. et al., *Stenotrophomonas dokdonensis* sp. nov., isolated from soil, *Int. J. Syst. Evol. Microbiol.*, 56, 1363, 2006.
15. Lee, D.S. et al., *Pseudoxanthomonas sacheonensis* sp. nov., isolated from BTEX-contaminated soil in Korea, transfer of *Stenotrophomonas dokdonensis* Yoon et al. 2006 to the genus *Pseudoxanthomonas* as *Pseudoxanthomonas dokdonensis* comb. nov. and emended description of the genus *Pseudoxanthomonas*, *Int. J. Syst. Evol. Microbiol.*, 58, 2235, 2008.
16. Hauben, L. et al., Genomic diversity of the genus *Stenotrophomonas*, *Int. J. Syst. Bacteriol.*, 49, 1749, 1999.
17. Coenye, T. et al., Identification of genomic groups in the genus *Stenotrophomonas* using *gyrB* RFLP analysis, *FEMS Immunol. Med. Microbiol.*, 40, 181, 2004.
18. Kaiser, S., Biehler, K., and Jonas, D., A *Stenotrophomonas maltophilia* multilocus sequence typing scheme for inferring population structure, *J. Bacteriol.*, 191, 2934, 2009.
19. Minkwitz, A., and Berg, G., Comparison of antifungal activities and 16S ribosomal DNA sequences of clinical and environmental isolates of *Stenotrophomonas maltophilia*, *J. Clin. Microbiol.*, 39, 139, 2001.

20. Denton, M., and Kerr, K.G., Microbiological and clinical aspects of infection associated with *Stenotrophomonas maltophilia*, *Clin. Microbiol. Rev.*, 11, 57, 1998.
21. Brooke, J.S. et al., Investigation of bacterial pathogens on 70 frequently used environmental surfaces in a large urban U.S. university, *J. Environ. Health*, 71, 17, 2009.
22. Cervia, J.S., Ortolano, G.A., and Canonica, F.P., Hospital tap water as a source of *Stenotrophomonas maltophilia* infection, *Clin. Infect. Dis.*, 46, 1485, 2008.
23. Hoefel, D. et al., Profiling bacterial survival through a water treatment process and subsequent distribution system, *J. Appl. Microbiol.*, 99, 175, 2005.
24. Wilkinson, F.H., and Kerr, K.G., Bottled water as a source of multi-resistant *Stenotrophomonas* and *Pseudomonas* species for neutropenic patients, *Eur. J. Cancer Care (Engl)*, 7, 12, 1998.
25. Huys, G. et al., Characterization of oxytetracycline-resistant heterotrophic bacteria originating from hospital and freshwater fish farm environments in England and Ireland, *Syst. Appl. Microbiol.*, 23, 599, 2000.
26. Prazmo, Z. et al., Exposure to bioaerosols in a municipal sewage treatment plant, *Ann. Agric. Environ. Med.*, 10, 241, 2003.
27. Mispagel, H., and Gray, J.T., Antibiotic resistance from wastewater oxidation ponds, *Water Environ. Res.*, 77, 2996, 2005.
28. Piccini, C. et al., Blooms of single bacterial species in a coastal lagoon of the southwestern Atlantic Ocean, *Appl. Environ. Microbiol.*, 72, 6560, 2006.
29. Furushita, M. et al., Isolation of multidrug-resistant *Stenotrophomonas maltophilia* from cultured yellowtail (*Seriola quinqueradiata*) from a marine fish farm, *Appl. Environ. Microbiol.*, 71, 5598, 2005.
30. Diaz, M.A. et al., Plasmid-mediated high-level gentamicin resistance among enteric bacteria isolated from pet turtles in Louisiana, *Appl. Environ. Microbiol.*, 72, 306, 2006.
31. Cloud-Hansen, K.A. et al., Thirteen-lined ground squirrels (Spermophilus tridecemlineatus) harbor multiantibiotic-resistant bacteria, *J. Am. Assoc. Lab. Anim. Sci.*, 46, 21, 2007.
32. Hejnar, P. et al., *Stenotrophomonas maltophilia* as a part of normal oral bacterial flora in captive snakes and its susceptibility to antibiotics, *Vet. Microbiol.*, 121, 357, 2007.
33. Matyar, F., Kaya, A., and Dincer, S., Antibacterial agents and heavy metal resistance in Gram-negative bacteria isolated from seawater, shrimp and sediment in Iskenderun Bay, Turkey, *Sci. Total Environ.*, 407, 279, 2008.
34. Romanenko, L.A. et al., Occurrence and antagonistic potential of *Stenotrophomonas* strains isolated from deep-sea invertebrates, *Arch. Microbiol.*, 189, 337, 2008.
35. Park, M. et al., Isolation and characterization of diazotrophic growth promoting bacteria from rhizosphere of agricultural crops of Korea, *Microbiol. Res.*, 160, 127, 2005.
36. Liba, C.M. et al., Nitrogen-fixing chemo-organotrophic bacteria isolated from cyanobacteria-deprived lichens and their ability to solubilize phosphate and to release amino acids and phytohormones, *J. Appl. Microbiol.*, 101, 1076, 2006.
37. Taghavi, S. et al., Genome survey and characterization of endophytic bacteria exhibiting a beneficial effect on growth and development of poplar trees, *Appl. Environ. Microbiol.*, 75, 748, 2009.
38. Garbeva, P. et al., Analysis of endophytic bacterial communities of potato by plating and denaturing gradient gel electrophoresis (DGGE) of 16S rDNA based PCR fragments, *Microb. Ecol.*, 41, 369, 2001.
39. Duarte, G.F. et al., Analysis of bacterial community structure in sulfurous-oil-containing soils and detection of species carrying dibenzothiophene desulfurization (dsz) genes, *Appl. Environ. Microbiol.*, 67, 1052, 2001.
40. Krsek, M., and Wellington, E.M., Assessment of chitin decomposer diversity within an upland grassland, *Antonie Van Leeuwenhoek*, 79, 261, 2001.
41. Di Gregorio, S., Lampis, S., and Vallini, G., Selenite precipitation by a rhizospheric strain of *Stenotrophomonas* sp. isolated from the root system of *Astragalus bisulcatus*: A biotechnological perspective, *Environ. Int.*, 31, 233, 2005.
42. Vallini, G., Di Gregorio, S., and Lampis, S., Rhizosphere-induced selenium precipitation for possible applications in phytoremediation of se polluted effluents, *Z. Naturforsch [C]*, 60, 349, 2005.
43. Hawle-Ambrosch, E. et al., Biodegradation of fuel oil hydrocarbons by a mixed bacterial consortium in sandy and loamy soils, *Biotechnol. J.*, 2, 1564, 2007.
44. Berg, G., Eberl, L., and Hartmann, A., The rhizosphere as a reservoir for opportunistic human pathogenic bacteria, *Environ. Microbiol.*, 7, 1673, 2005.
45. Berg, G., and Smalla, K., Plant species and soil type cooperatively shape the structure and function of microbial communities in the rhizosphere, *FEMS Microbiol. Ecol.*, 68, 1, 2009.
46. Weber, D.J. et al., Faucet aerators: A source of patient colonization with *Stenotrophomonas maltophilia*, *Am. J. Infect. Control*, 27, 59, 1999.
47. Arvanitidou, M. et al., Occurrence and antimicrobial resistance of Gram-negative bacteria isolated in haemodialysis water and dialysate of renal units: Results of a Greek multicentre study, *J. Appl. Microbiol.*, 95, 180, 2003.
48. Denton, M. et al., *Stenotrophomonas maltophilia* contamination of nebulizers used to deliver aerosolized therapy to inpatients with cystic fibrosis, *J. Hosp. Infect.*, 55, 180, 2003.
49. Siebor, E. et al., Presumed pseudobacteremia outbreak resulting from contamination of proportional disinfectant dispenser, *Eur. J. Clin. Microbiol. Infect. Dis.*, 26, 195, 2007.
50. Szymanska, J., Bacterial contamination of water in dental unit reservoirs, *Ann. Agric. Environ. Med.*, 14, 137, 2007.
51. Jumaa, P.A. et al., The molecular epidemiology of *Stenotrophomonas maltophilia* bacteraemia in a tertiary referral hospital in the United Arab Emirates 2000–2004, *Ann. Clin. Microbiol. Antimicrob.*, 5, 32, 2006.
52. Travassos, L.H. et al., Phenotypic properties, drug susceptibility and genetic relatedness of *Stenotrophomonas maltophilia* clinical strains from seven hospitals in Rio de Janeiro, Brazil, *J. Appl. Microbiol.*, 96, 1143, 2004.
53. Tatman-Otkun, M. et al., The antimicrobial susceptibility of *Stenotrophomonas maltophilia* isolates using three different methods and their genetic relatedness, *BMC Microbiol.*, 5, 24, 2005.
54. Valdezate, S. et al., High genetic diversity among *Stenotrophomonas maltophilia* strains despite their originating at a single hospital, *J. Clin. Microbiol.*, 42, 693–699, 2004.
55. Park, Y.S. et al., Pseudooutbreak of *Stenotrophomonas maltophilia* bacteremia in a general ward, *Am. J. Infect. Control*, 36, 29, 2008.
56. Jakobi, M. et al., Maltophilin: a new antifungal compound produced by *Stenotrophomonas maltophilia* R3089, *J. Antibiot. (Tokyo)*, 49, 1101, 1996.
57. Suckstorff, I., and Berg, G., Evidence for dose-dependent effects on plant growth by *Stenotrophomonas* strains from different origins, *J. Appl. Microbiol.*, 95, 656, 2003.

58. Pages, D. et al., Heavy metal tolerance in *Stenotrophomonas maltophilia*, *PLoS One*, 3, e1539, 2008.
59. Chien, C.C., Hung, C.W., and Han, C.T., Removal of cadmium ions during stationary growth phase by an extremely cadmium-resistant strain of *Stenotrophomonas* sp., *Environ. Toxicol. Chem.*, 26, 664, 2007.
60. Antonioli, P. et al., *Stenotrophomonas maltophilia* SeITE02, a new bacterial strain suitable for bioremediation of selenite-contaminated environmental matrices, *Appl. Environ. Microbiol.*, 73, 6854, 2007.
61. Siddique, T. et al., Bacterial reduction of selenium in coal mine tailings pond sediment, *J. Environ. Qual.*, 36, 621, 2007.
62. Dungan, R.S., Yates, S.R., and Frankenberger, W.T. Jr., Transformations of selenate and selenite by *Stenotrophomonas maltophilia* isolated from a seleniferous agricultural drainage pond sediment, *Environ. Microbiol.*, 5, 287, 2003.
63. Marecik, R. et al., Atrazine degradation by aerobic microorganisms isolated from the rhizosphere of sweet flag (*Acorus calamus L.*), *Biodegradation*, 19, 293, 2008.
64. Lee, E.Y. et al., Degradation characteristics of toluene, benzene, ethylbenzene, and xylene by *Stenotrophomonas maltophilia* T3-c, *J. Air Waste Manag. Assoc.*, 52, 400, 2002.
65. Looney, W.J., Narita, M., and Muhlemann, K., *Stenotrophomonas maltophilia*: An emerging opportunist human pathogen, *Lancet Infect. Dis.*, 9, 312, 2009.
66. Jones, R.N., Sader, H.S., and Beach, M.L., Contemporary in vitro spectrum of activity summary for antimicrobial agents tested against 18569 strains non-fermentative Gram-negative bacilli isolated in the SENTRY Antimicrobial Surveillance Program (1997–2001), *Int. J. Antimicrob. Agents*, 22, 551, 2003.
67. Paez, J.I., and Costa, S.F., Risk factors associated with mortality of infections caused by *Stenotrophomonas maltophilia*: A systematic review, *J. Hosp. Infect.*, 70, 101, 2008.
68. Apisarnthanarak, A. et al., Risk factors for *Stenotrophomonas maltophilia* bacteremia in oncology patients: A case-control study, *Infect. Control Hosp. Epidemiol.*, 24, 269, 2003.
69. Graff, G.R., and Burns, J.L., Factors affecting the incidence of *Stenotrophomonas maltophilia* isolation in cystic fibrosis, *Chest*, 121, 1754, 2002.
70. Meyer, E. et al., *Stenotrophomonas maltophilia* and antibiotic use in German intensive care units: data from Project SARI (Surveillance of Antimicrobial Use and Antimicrobial Resistance in German Intensive Care Units), *J. Hosp. Infect.*, 64, 238, 2006.
71. Nseir, S. et al., Intensive care unit-acquired *Stenotrophomonas maltophilia*: Incidence, risk factors, and outcome, *Crit. Care*, 10, R143, 2006.
72. Talmaciu, I. et al., Risk factors for emergence of *Stenotrophomonas maltophilia* in cystic fibrosis, *Pediatr. Pulmonol.*, 30, 10, 2000.
73. Boktour, M. et al., Central venous catheter and *Stenotrophomonas maltophilia* bacteremia in cancer patients, *Cancer*, 106, 1967, 2006.
74. Lai, C.H. et al., Central venous catheter-related *Stenotrophomonas maltophilia* bacteraemia and associated relapsing bacteraemia in haematology and oncology patients, *Clin. Microbiol. Infect.*, 12, 986, 2006.
75. Micozzi, A. et al., Bacteremia due to *Stenotrophomonas maltophilia* in patients with hematologic malignancies, *Clin. Infect. Dis.*, 31, 705, 2000.
76. Crossman, L.C. et al., The complete genome, comparative and functional analysis of *Stenotrophomonas maltophilia* reveals an organism heavily shielded by drug resistance determinants, *Genome Biol.*, 9, R74, 2008.
77. Alonso, A., and Martinez, J.L., Cloning and characterization of SmeDEF, a novel multidrug efflux pump from *Stenotrophomonas maltophilia*, *Antimicrob. Agents Chemother.*, 44, 3079, 2000.
78. Lambert, T. et al., Characterization of the chromosomal aac(6′)-Iz gene of *Stenotrophomonas maltophilia*, *Antimicrob. Agents Chemother.*, 43, 2366, 1999.
79. Okazaki, A., and Avison, M.B., Aph(3′)-IIc, an aminoglycoside resistance determinant from *Stenotrophomonas maltophilia*, *Antimicrob. Agents Chemother.*, 51, 359, 2007.
80. Wheat, P.F., Winstanley, T.G., and Spencer, R.C., Effect of temperature on antimicrobial susceptibilities of *Pseudomonas maltophilia*, *J. Clin. Pathol.*, 38, 1055, 1985.
81. Rahmati-Bahram, A., Magee, J.T., and Jackson, S.K., Growth temperature-dependent variation of cell envelope lipids and antibiotic susceptibility in *Stenotrophomonas (Xanthomonas) maltophilia*, *J. Antimicrob. Chemother.*, 36, 317, 1995.
82. Rahmati-Bahram, A., Magee, J.T., and Jackson, S.K., Temperature-dependent aminoglycoside resistance in *Stenotrophomonas (Xanthomonas) maltophilia*; alterations in protein and lipopolysaccharide with growth temperature, *J. Antimicrob. Chemother.*, 37, 665, 1996.
83. Rahmati-Bahram, A., Magee, J.T., and Jackson, S.K., Effect of temperature on aminoglycoside binding sites in *Stenotrophomonas maltophilia*, *J. Antimicrob. Chemother.*, 39, 19, 1997.
84. McKay, G.A. et al., Role of phosphoglucomutase of *Stenotrophomonas maltophilia* in lipopolysaccharide biosynthesis, virulence, and antibiotic resistance, *Infect. Immun.*, 71, 3068, 2003.
85. Zelenitsky, S.A. et al., Antibiotic combinations significantly more active than monotherapy in an in vitro infection model of *Stenotrophomonas maltophilia*, *Diagn. Microbiol. Infect. Dis.*, 51, 39, 2005.
86. Barbolla, R. et al., Class 1 integrons increase trimethoprim-sulfamethoxazole MICs against epidemiologically unrelated *Stenotrophomonas maltophilia* isolates, *Antimicrob. Agents Chemother.*, 48, 666, 2004.
87. Toleman, M.A. et al., Global emergence of trimethoprim/sulfamethoxazole resistance in *Stenotrophomonas maltophilia* mediated by acquisition of sul genes, *Emerg. Infect. Dis.*, 13, 559, 2007.
88. Jucker, B.A., Harms, H., and Zehnder, A.J., Adhesion of the positively charged bacterium *Stenotrophomonas (Xanthomonas) maltophilia* 70401 to glass and Teflon, *J. Bacteriol.*, 178, 5472, 1996.
89. de Oliveira-Garcia, D. et al., Characterization of flagella produced by clinical strains of *Stenotrophomonas maltophilia*, *Emerg. Infect. Dis.*, 8, 918, 2002.
90. de Oliveira-Garcia, D. et al., Fimbriae and adherence of *Stenotrophomonas maltophilia* to epithelial cells and to abiotic surfaces, *Cell. Microbiol.*, 5, 625, 2003.
91. Windhorst, S. et al., The major extracellular protease of the nosocomial pathogen *Stenotrophomonas maltophilia*: Characterization of the protein and molecular cloning of the gene, *J. Biol. Chem.*, 277, 11042, 2002.
92. Waters, V.J. et al., Immunostimulatory properties of the emerging pathogen *Stenotrophomonas maltophilia*, *Infect. Immun.*, 75, 1698, 2007.
93. Fouhy, Y. et al., Diffusible signal factor-dependent cell-cell signaling and virulence in the nosocomial pathogen *Stenotrophomonas maltophilia*, *J. Bacteriol.*, 189, 4964, 2007.
94. Rocco, F. et al., *Stenotrophomonas maltophilia* genomes: A start-up comparison, *Int. J. Med. Microbiol.*, 299, 535, 2009.

95. Juhnke, M.E., and des Jardin, E., Selective medium for isolation of *Xanthomonas maltophilia* from soil and rhizosphere environments, *Appl. Environ. Microbiol.*, 55, 747, 1989.
96. Bollet, C., Davin-Regli, A., and De Micco, P., A simple method for selective isolation of *Stenotrophomonas maltophilia* from environmental samples, *Appl. Environ. Microbiol.*, 61, 1653, 1995.
97. Kerr, K.G. et al., A new selective differential medium for isolation of *Stenotrophomonas maltophilia*, *Eur. J. Clin. Microbiol. Infect. Dis.*, 15, 607, 1996.
98. Foster, N.F., Chang, B.J., and Riley, T.V., Evaluation of a modified selective differential medium for the isolation of *Stenotrophomonas maltophilia*, *J. Microbiol. Methods*, 75, 153, 2008.
99. Karpati, F., and Jonasson, J., Polymerase chain reaction for the detection of *Pseudomonas aeruginosa, Stenotrophomonas maltophilia* and *Burkholderia cepacia* in sputum of patients with cystic fibrosis, *Mol. Cell. Probes*, 10, 397, 1996.
100. Whitby, P.W. et al., Identification and detection of *Stenotrophomonas maltophilia* by rRNA-directed PCR, *J. Clin. Microbiol.*, 38, 4305, 2000.
101. Foster, N.F. et al., Cross-reaction of *Stenotrophomonas* and *Xanthomonas* species in a 23S rRNA gene-directed PCR for detection of *S. maltophilia*, *J. Clin. Microbiol.*, 46, 4111, 2008.
102. Hogardt, M. et al., Specific and rapid detection by fluorescent in situ hybridization of bacteria in clinical samples obtained from cystic fibrosis patients, *J. Clin. Microbiol.*, 38, 818, 2000.
103. Kempf, V.A. et al., Rapid detection and identification of pathogens in blood cultures by fluorescence in situ hybridization and flow cytometry, *Int. J. Med. Microbiol.*, 295, 47, 2005.
104. Ausubel, F.M. et al., *Short Protocols in Molecular Biology*, Wiley and Sons, New York, 1999.
105. Manz, W. et al., Phylogenetic oligodeoxynucleotide probes for the major subclasses of proteobacteria: Problems and solutions, *Syst. Appl. Microbiol.*, 15, 593, 1992.
106. Mori, Y., and Notomi, T., Loop-mediated isothermal amplification (LAMP): A rapid, accurate, and cost-effective diagnostic method for infectious diseases, *J. Infect. Chemother.*, 15, 62, 2009.

91 *Vibrio*

Asim Bej

CONTENTS

91.1 INTRODUCTION

Vibrios are natural inhabitants of fresh, brackish, or marine waters. Typically rod-shaped, gram-negative, asporogenic-curved bacteria, many of them are pathogenic to humans and other marine invertebrates. *V. cholerae*, the causative agent of cholera disease in humans, was the first member of the Vibrionaceae family to be reported in 1854.[1] Since then, more than 45 species have been listed in the 2005 edition of Bergey's Manual.[1] The pathogenic vibrios cause mild to life-threatening diseases in humans, such as gastroenteritis, septicemia, wound infection, and ear infection. *Vibrio*-related fatality often occurs if the infection is untreated or the individual presents immunocompromised or other predisposing conditions. The primary modes of transmission of these pathogens could be (i) through potable water; (ii) consumption of raw or poorly cooked seafood; or (iii) exposure of wounds to warm seawater leading to infection. Environmental waters such as fresh potable, estuarine, and marine serve as the reservoir of this pathogen. However, the primary source of transmission of *V. cholerae* is consumption of contaminated drinking or potable waters polluted by large numbers of this pathogen from the feces of infected individuals in poor hygiene locations. Vibrios inhabiting coastal waters are primarily transmitted to humans through the consumption of raw or undercooked shellfish. Commonly, the disease manifestation following infection by most of the vibrios is gastroenteritis, which is often self-limiting. However, some cases of wound infection and septicemia could also occur, particularly to those susceptible individuals who have had liver disease or been otherwise immunocompromised, often leading to fatal consequences. Although most early studies on seafood-related vibrio infections were from the United States and Japan, over the last decade numerous cases have also been described from Asia, Europe, Australia, and New Zealand, suggesting that these pathogens are prevalent in warm coastal waters throughout the world. Although *V. cholerae*-related cholera disease has not been reported in the Unites States since 1982, frequent incidences and sporadic outbreaks in South America and in Southeast Asia have been reported. Such outbreaks are often widespread during the summer months and during occurrences of natural calamities, which result in the unavailability of clean potable waters and poor hygienic conditions. In the United States, the seafood-related vibrios are prevalent in warm waters of Gulf of Mexico, although they have also been isolated from Pacific Northwest and Eastern coastal waters. Seafood industry is a major contributor to the local economy to the U.S. Gulf of Mexico and to some extent the Northeast and Pacific Northwest States. Over the last 25 years, the United States has experienced several shellfish-related disease outbreaks, with only 4% being due to fecal contaminants (e.g., coliforms, *Salmonella* sp., Norwalk virus, Hepatitis A), whereas 20% of the diseases and 99% of the deaths were linked to *Vibrio* spp. present in seafood.[2] The microbial pathogens, primarily vibrios are indigenous to the Gulf of Mexico waters and are accumulated by the oysters in their tissues during filter-feeding.[3–5] Therefore, consumers inevitably ingest these and other harmful microorganisms when they consume untreated or poorly handled/treated raw oysters. The epidemiological data suggests that the pathogenic vibrios for the incidences and outbreaks of seafood-related diseases in the United States and throughout the world are *V. parahaemolyticus* and *V. vulnificus*. Many cases of *Vibrio*-associated gastroenteritis are substantially underrecognized because vibrios are not readily identified in routine stool cultures and may require time-consuming biochemical tests. Therefore the need for rapid and reliable molecular methodologies have been sought and frequently reported. Although the culture-based biochemical and microbiological methodologies are still in place in many laboratories, the merits of the molecular approaches are slowly but steadily being realized and gaining popularity in clinical laboratories, as well as in seafood industry. This chapter will elaborate biology of pathogenic vibrios and molecular approaches for their detection in food.

91.1.1 Classification of *Vibrio*

The family Vibrionaceae under the Order Vibrionales includes the members of the genus *Vibrio*. This genus is non-spore-forming, gram-negative bacteria, identified by their unique short rod or curved shape morphology. Also, some members may manifest straight (approximately 1.5–3.0 μm × 0.5 μm) or S-shaped cells when individual cells are joined end to end. Since morphological diversity exists, it is necessary to implement physiological and biochemical taxonomic tests for identification and phylogenetic analysis. Although *Vibrio* species is motile with a single polar flagellum, two or more polar or lateral flagella have also been demonstrated, especially when cultured on solid media.[1] *Vibrio* species are facultative anaerobes, manifest both a respiratory (oxygen-utilizing) and a fermentative metabolism. Most of them produce oxidase and catalase, and ferment glucose without producing gas.[6] Oxygen has been reported to be the universal electron acceptor. Normally, most vibrios inhabit in estuarine, salt, or fresh waters and can be isolated by culturing them on standard growth media such as nutrient agar/broth or media supplemented with required concentrations of NaCl (0.5%–3%) and/or a mixture of salts (preferably sea salts). Over the years, the classification of many of the individual vibrio isolates has gone through numerous taxonomic rearrangements. The DNA G + C composition has also played an important role in the taxonomic rearrangement of the *Vibrio* species. For example, the demarcation between the genera *Pseudomonas* and *Vibrio* has been reconsidered based on the physiological and nucleic acids characteristics. A distinct taxonomical separation among the members of the genus *Vibrio* and *Aeromonas*, *Plesiomonas*, *Photobacterium*, and *Lucibacteriu* could be somewhat controversial. This is because all of them manifest fermentation but do not produce gas (*Aeromonas* could vary) on media supplemented with carbohydrate; are oxidase positive (*Photobacterium* could vary); and none of them produce diffusible pigment. Based on these characteristics, Bauman and Schubert[6] have included the following genera: *Vibrio*, *Photobacterium*, *Aeromonas*, and *Plesiomonas* under the family Vibrionaceae, which was originally described by Veron.[7] However, the most recent edition of *Bergey's Manual*, Farmer and Janda[1] reclassified these genera. In their description, *Aeromonas* and *Plesiomonas* have been classified under different families. This reclassification is based upon the introduction of new methodologies, such as various aspects of proteomics and genomics for the identification of the organisms. In the current *Bergey's Manual*, the family Vibrionaceae comprises three genus: *Vibrio* (Genus I), *Photobacterium* (Genus II), and *Salinivibrio* (Genus III).[1]

Vibrio vulnificus and *V. parahaemolyticus* are natural inhabitants of estuarine and marine waters, respectively, and have been subjected to taxonomic reclassification for many members on this group.

91.1.2 Biology and Pathogenesis

After the first description of *Vibrio cholerae* O1 by Pacini (1854)[7a] as the causative agent for cholera in humans, numerous strains have been isolated and taxonomic positions have been assigned. Most vibrios inhabit in aquatic environment (freshwater, estuarine, and marine) as a free-living organism or associated with aquatic animals and plants. However, some vibrios are pathogenic and are isolated from humans when infected and disease symptoms manifested; yet some are pathogenic to aquatic (primarily marine) animals such as fish and eels. Some luminescence vibrios coexist with marine squid in a symbiotic manner.[8] In this chapter, we will focus on vibrios that are primarily pathogenic to humans. Most human pathogenic vibrios reside in freshwater, costal waters (estuarine), and marine environments. The pathogenic vibrios cause infections in humans with the following clinical syndromes: (i) gastroenteritis; (ii) wound infection; and (iii) septicemia. The mode of transmissions could be through the ingestion of raw or poorly cooked seafood; drinking of potable fresh water, causing gastroenteritis; or exposure of warm coastal waters to cut and bruises on the skin, leading to septicemia. Some pathogenic vibrios infect normal individuals while others infect immuncompromised persons, particularly those individuals with liver diseases. Most cases of gastroenteritis are self-limiting, while the septicemia could be fatal. The Center for Disease Control and Prevention (CDC), in association with the United States Food and Drug Administration (FDA) and Gulf Coast States (Alabama, Florida, Mississippi, and Texas), has established the *Cholera* and Other *Vibrio* Illnesses Surveillance System (COVIS) in 1988, which monitors and reports the *Vibrio*-related illnesses. A comprehensive summary of the human *Vibrio* isolates since 1998 can be obtained on the CDC Web site (http://www.cdc.gov/nationalsurveillance/cholera_vibrio_surveillance.html). CDC serotypes all *V. parahaemolyticus* isolates received from state health departments and screens for cholera toxin production in all *V. cholerae* isolates. Most states now voluntarily submit reports of clinical cases of *Vibrio* infections to this site. The primary purpose for the COVIS is to maintain a reliable database of reported cases of *Vibrio* infections and strain information, and to educate public about these pathogens, health risks of seafood and other source of contracting disease by these pathogens, and how to avoid them.

91.1.2.1 *V. cholerae*

A number of *V. cholerae* serotypes are potentially pathogenic to humans, whereas nonpathogenic strains have been also been reported. *V. cholerae* O1 is primarily considered to be causative agent for cholera disease in humans. *V. cholerae* O1 is subdivided into the most common biotype El Tor or classic; serotype Inaba, Ogawa, or Hikojima; and toxin producing toxigenic or nontoxigenic.[9] In 1992, toxigenic *V. cholerae* O139 was first discovered on the Indian subcontinent (Bangladesh) and recognized as another cause of cholera.[10] Incidences of *V. cholerae* O139 have been reported in 1993 in the United States.[11] Due to its similar epidemic potential, monitoring and surveillance is recommended.[11] *V. cholerae* O1 and O139 are considered to be the pandemic strains. The epidemic strains of *V. cholerae* are usually the El Tor biotype.

Infection with this organism could cause severe watery diarrhea often described as "rice water" stool, vomiting, and muscle cramps, which could lead to severe conditions such as the loss of body weight, loss of normal skin turgor, dry mucous membranes, sunken eyes, lethargy, anuria, weak pulse, altered consciousness, and circulatory collapse and possibly death if not.[12] Severe cholera disease causes devastating fluid imbalance leading to dehydration, profound hypokalemia (loss of potassium), metabolic acidosis (from bicarbonate loss), and renal failure.

Recently, the *V. cholera* strains that exhibit resemblance to the clinical strains in their biochemical characteristics but fail to agglutinate in either anti-O1 or anti-O139 sera have been grouped as non-O1/139 *V. cholera*. These strains of *V. cholerae* are found in environmental waters, primarily in coastal estuarine waters. These strains could cause cholera-like diarrhea that is less severe and are sporadic in nature but rarely been reported to cause outbreaks.[13,14]

As early as in 1817, *V. cholerae* seemed to have been causing cholera disease in Asian countries but spread in the estuarine waters around Africa, Europe, and the Americas, causing six pandemic outbreaks between 1832 and 1866. The seventh pandemic caused by the *V. cholera* El Tor biotype was recorded in 1961, spreading from Indonesia to other continents. In 1991, this strain was spread in South America believed to be through ship ballast waters and oceanic current and zooplanktons.[15] The outbreak caused by the *V. cholera* O139 biotype in early 1990s could be referred as the Eighth Cholera Pandemic.[16,17] The spectrum of illness in cholera includes asymptomatic infection (75%), mild illness (18%), moderate illness (5%), and severe illness (2%).[18]

91.1.2.2 *V. parahaemomlyticus*

V. parahaemomlyticus is a halophilic estuarine organism that inhabits the coastal waters of virtually all temperate regions, with occurrences primarily during the warmer months of the year. *V. parahaemolyticus* is the leading cause for gastroenteritis-related illness in humans (irrespective of the health condition). The clinical manifestations of *V. parahaemolyticus* are acute watery diarrhea and abdominal cramp associated with nausea typically lasting 1–3 days. However severe cases with bloody diarrhea may require hospitalization. The primary source of infection by this pathogen is the consumption of raw seafood, especially raw oysters. Fujino et al. (1953)[19] first reported *V. parahaemolyticus* as a cause for foodborne gastroenteritis in 1950 due to Shirasu food-poisoning in Japan. In 1971, the outbreak of *V. parahaemolyticus* was first reported in Maryland from the consumption of crabs.[20,21] Forty outbreaks of *V. parahaemolyticus* in the United States have been reported to CDC between 1973 and 1998.[22,23] These infections occurred primarily during the warmer months and through the consumption of shellfish. In 1997–1998, a relatively large *V. parahaemolyticus*-related outbreak was reported to affect more than 425 individuals in the United States. The affected individuals consumed raw oysters harvested from the warm waters of the Gulf of Mexico.[22–27]

All *V. parahaemolyticus* strains react with an H antigen. However, several O type and K (capsular) antigen-reactive strains have been reported, yet many strains are untypable.[28] Recently, a newly emerged *V. parahaemolyticus* O3:K6 serogroup was involved in outbreaks from contaminated oysters harvested from Oyster Bay, New York, and Galveston Bay, Texas, in 1998.[27] The *V. parahaemolyticus*-related disease outbreak in Galveston Bay was the largest reported in the United States (416 persons). In 1997, an outbreak caused by *V. parahaemolyticus* occurred in the Pacific Northwest, resulting in one death and 209 gatroenteritis-related illnesses. Each of these infected persons had consumed raw oysters.[27] Beside gastroenteritis, *V. parahaemolyticus* can cause wound infection and septicemia, and there have been cases of *V. parahaemolyticus*-related infections reported in the Unites States between 1988 and 1997, indicating 59% of patients presented with gastroenteritis, 34% with wound infections, 5% with primary septicemia, and 2% with other sites of infections from the consumption of raw shellfish.[22,23]

The pathogenic and nonpathogenic strains of *V. parahaemolyticus* are distinguished primarily based on their ability to produce beta-type hemolysin on Wagatsuma blood agar, which is known as Kanagawa phenomenon.[29,30,32,33] This hemolysin is referred as thermostable direct hemolysin, which is encoded by the *tdh* gene.[29,31] However, there have been cases of illness that were associated with strains that did not produce thermostable direct hemolysin. These strains are referred as Kanagawa negative, but produce a different hemolysin, the *trh*-encoded thermostable direct-related hemolysin.[22,23] It has been reported that there is a strong association between pathogenicity and the presence of either the *tdh* gene or the *trh* gene or both, indicating that in *V. parahaemolyticus* both of these genes are potentially virulence factors.[34] An increase in the incidence of *V. parahaemolyticus* around the world was observed 1996, which was caused by a unique clone of *V. parahaemolyticus* serotype O3:K6.[35,36] It was determined that many of the O3:K6 strains isolated since 1996 contain a filamentous phage, f237, and that this phage contains a unique open reading frame, ORF8. In addition, another group of *V. parahaemolyticus* strain produces a thermolabile hemolysin, which is encoded by the *tlh* gene.[37] The thermolabile hemolysin seems to not associate with the pathogenicity, but present in all *V. parahaemolyticus* isolates so far been tested.[38]

91.1.2.3 *V. vulnificus*

V. vulnificus is halophilic and inhabits in coastal warm waters (primarily estuarine water with low ppt salt contents) around the world. In the United States, *V. vulnificus* is prevalent in the Gulf of Mexico; however, they are also found in East and West Coast waters. During the warmer months of the year, *V. vulnificus* counts dramatically increase in the coastal waters. In the United States, *V. vulnificus* is considered to be the most invasive pathogen among all vibrios causing high fatality to infected individuals, especially who have had liver disease or otherwise immunocompromised. The fatality rate could reach to 60% in individuals with liver disease and/or have excess levels of iron in the blood serum.[39,40] In addition, open wounds exposed to *V. vulnificus* in marine waters have been documented as a route of entry for this pathogen, leading to skin breakdown and, in some cases, amputation of the infected body part.[40,41] In healthy individuals, infection with *V. vulnificus* can cause vomiting, diarrhea, and abdominal pains. It was first identified and described by the CDC in 1976.[42] Since then, it has become the leading cause of seafood-associated deaths in the United States.

The virulence mechanism for *V. vulnificus* is still not well understood. However, a number of putative virulence factors such as hemolysin, polysaccharide capsule, expression of surface pili, and cellular mechanisms for iron acquisition have been proposed.[39,43,44] Among these, it is notable that the capacity for iron acquisition seems to underlie the pathogenicity of many *V. vulnificus* infections.[45] A number of studies associate strain virulence with the ability to utilize transferrin-bound iron or with the production of significant amounts of catechol siderophore during infection.[46] Moreover, most of the clinical isolates have been identified with thick capsules. Therefore, it is proposed that encapsulated form of *V. vulnificus* is invasive type. Most of the environmental isolates are found to be without capsule and therefore considered nonpathogenic.[47] It has been proposed that *V. vulnificus* produces a number of enzymes—such as lipase, mucinase, hyaluronidase, DNAase, and protease—that may facilitate pathogenesis process.[18]

91.1.2.4 Other Vibrios

Besides *V. cholerae*, *V. parahaemolyticus*, and *V. vulnificus*, a few other marine vibrios, although not prevalent, have been reported to cause infections in humans. Some of the halophilic vibrios inhabit brackish waters that have been reported to cause human disease are *V. alginolyticus*, *V. fluvialis*, *V. furnissii*, *V. metchnikovii*, and *V. hollisae*. A short description of these vibrios and their possible pathogenic mechanisms are described below:

1. *V. alginolyticus* has been isolated from patients with wound infection, although it can cause ear infection too. However, unlike the other three primary vibrio pathogens, *V. alginolyticus* rarely causes gastroenteritis.[1,18] The virulence mechanism for this pathogen is yet to be described.
2. *V. fluvialis* has been known to cause diarrhea, often with bloody stools, fever, vomiting, dehydration, and severe abdominal pain.[18] This pathogen has been previously referred as "marine aeromonads," Group F is in England and Group EF6 is at CDC.[1] Gastroenteritis associated with this pathogen has often been described with "cholera-like" symptoms. A large outbreak of *V. fluvialis* related diarrhea was reported in Bangladesh in 1976–1977.[48] Reports of infection by this pathogen are often from the geographical regions where pathogens causing gastroenteritis are found in feces.[1] This pathogen rarely causes septicemia or wound infection.[18] The mechanism of pathogenicity has been inflicted upon "enterotoxin-like" molecules tested in rabbit ileal loop.[49,50]
3. *V. furnissii* is normally found in estuarine environment and sometimes been isolated from animal feces. Often *V. furnissii* were isolated from patients along with other pathogenic vibrios. Therefore the role of *V. furnisssii* as the actual cause for diarrhea remains unclear.[1] The pathogenic mechanism by this pathogen is yet to be described.
4. *V. metchnikovii* is normally isolated from brackish, marine or fresh water, but rarely from clinical samples. Normally *V. metchnikovii* is isolated from nonhuman sources such as river, fowl, and other seafood, which could potentially serve as the reservoir for this pathogen to be transmitted to humans.[1] It was isolated from the blood of a diabetic individual with acute gall bladder problem.[51] *V. metchnikovii* was isolated from children with diarrhea in Peru and no subsequent infection was reported.[52] The nature of the virulence factor or the mechanism of pathogenesis is not understood.
5. *V. hollisae* is a halophilic bacteria isolated from sporadic cases of patients with diarrhea or in the blood.[1] However, the actual role of this pathogen causing this disease has not been firmly established. It could cause septicemia and wound infections.[53,54]

91.1.3 Diagnosis of *Vibrio*

The conventional culture methods for the detection of vibrios often rely upon the enrichment of the sample and plating on vibrio-specific agar plates followed by a series of other biochemical tests for the confirmation of the results. As a proactive measure to prevent or reduce the vibrio-related incidences and outbreaks, monitoring these pathogens in relevant environmental and food samples by early detection has been recommended. For clinical situations where a patient has been suspected to be infected with a vibrio pathogen, rapid detection and initiation of treatments regimes are essential. This is particularly important for invasive vibrios such as *V. cholerae* and *V. vulnificus*, because these may rapidly progress to fatal consequences.[55,56] Many cases of *Vibrio*-associated gastroenteritis are underrecognized because most clinical laboratories do not routinely use the selective medium, thiosulfate citrate bile salts -sucrose (TCBS) agar, for processing of stool specimens unless they are specifically instructed to do so. Therefore, a heightened awareness and understanding of appropriate measures by implementation of postharvest treatment and rapid detection methods of the seafood products by the seafood industry, government agencies, clinicians, laboratory technicians, and epidemiologists are essential.[57]

Vibrio species grow in the presence of relatively high levels of bile salts. Since they are facultative anaerobic and are inhabitants of marine or brackish waters, alkaline conditions along with various concentrations of salts are preferable and essential for their optimum growth. For example, *V. parahaemolyticus* is strictly halophilic; therefore, a 2%–3% NaCl is necessary for optimum growth. However, for *V. vulnificus* and other vibrios inhabiting in the brackish waters, approximately 0.5%–1% NaCl is sufficient. Normally, the thiosulfate citrate bile salts sucrose (TCBS) agar, which was originally developed for the isolation of pathogenic vibrios is widely used for the isolation of *Vibrio* species.[58] TCBS utilizes alkaline pH (8.6), ox bile, and NaCl (1%) to suppress the growth of nontarget bacteria, including members of the family Enterobacteriaceae, the genus *Pseudomonas*, and gram-positive bacteria. The stool specimens can be cultured on blood and MacConkey agars, but isolation is enhanced by using TCBS agar.[59,60] *V. cholerae* appears as yellow colonies on TCBS agar. Serological test can be conducted for cholera diagnosis. The serologic conversion, which is vibriocidal antibody titer of greater than 1:640 suggests that recent infection or a fourfold rise in vibriocidal antibody titer is relatively reliable approach. An increase in the titers several weeks after exposure followed by a decrease could often be seen in patients who have had been exposed to *V. cholerae*.[18]

V. parahaemolyticus isolates can be maintained on nutrient-rich (Tryptone) agar supplemented with 3% NaCl. However, similar to other vibrios, *V. parahaemolyticus* may be overlooked if plated on nonselective media; therefore, they are plated on TCBS agar. On TCBS agar *V. parahaemolyticus* isolates appear as distinct green colonies. Virulent strains of *V. parahaemolyticus* can be determined by β-hemolysis of red blood cells using Wagatsuma blood agarm,[30] although newer methods use DNA gene probes and PCR have been described.[38]

The microbiological growth media and conditions for isolation and culture confirmation of *V. vulnificus* have been reviewed.[4] Normally, the enriched cultures of *V. vulnificus* in oyster or other seafood homogenate can be isolated on TCBS agar. The *V. vulnificus* appears as green colonies on TCBS agar. Although the TCBS agar is widely used for the isolation of *V. vulnificus*, *V. cholerae*, and *V. parahaemolyticus*, it has limitations.[60] The cellobiose polymyxin B colistin (CPC) and modified-CPC (mCPC) agar media were specifically formulated for the isolation of *V. vulnificus* in environmental samples.[61] These agar media have been shown over 99% efficient in isolating *V. vulnificus* and are commonly used most laboratories. Besides the mCPC agar, VVMc agar has shown promising results in the isolation of *V. vulnificus*.[62,63] This agar medium is the modification of VVM agar that does not contain polymyxin B.[64] A comparison of the efficiency of the recovery of *V. vulnificus* from environmental and clinical sources was compared on VVM, VVMc, CC, and TCBS agars exhibited highest on VVMc (53%) and lowest on TCBS (42%).[62,63] It is important to note that the samples from eels did not support any growth of *V. vulnificus* on VVMc or CC agar. This suggests that the *V. vulnificus* Biotype II, which is primarily found in eels, requires different media composition for their growth.

The *Vibrio* species can be identified and characterized using a commercially available API 20E Biolog system. Although the method itself could be useful, this method requires (i) isolation of individual colonies after at least 24 h if incubation on TCBS agar and (ii) selection of large number of colonies on TCBS agar for testing, as this agar medium itself is not confirmative. Therefore, this approach is not practical for screening a large number of samples. The application of conventional culture methods followed by a series of confirmative biochemical tests for the evaluation of a large number of food or clinical samples is not convenient.[65] The molecular techniques have been gaining popularity for the detection of pathogenic vibrios in coastal waters, seafood, and clinical samples. Besides plate cultures and confirmatory biochemical tests, immunoassays for the detection of *V. parahaemolyticus* or *V. vulnificus* are described. However, the specificity and the sensitivity of the detection of specific *Vibrio* species were not reliable. The genetic-based approaches are considerably reliable and often take much less time than culture-based biochemical testing. DNA-based molecular approaches offer time-efficient and specific detection of pathogenic *Vibrio* species without compromising the required sensitivity. The evolution of the DNA-based molecular approaches occurred in a relatively short period of time and notable achievements in the detection technology have been made.

For the molecular approach, a DNA–DNA colony hybridization using either a radiolabeled or a colorimetric tag gene-specific DNA probe was first reported. Yamamoto et al. described the use of a 3.2 kbp segment consisting of the *V. vulnificus* cytolysin gene (also referred as *cthA*, *hlyA*, or *vvhA*, Genbank accession no. M34670) labeled with radionucleotide as a DNA probe, the specificity of which was later confirmed for the detection of environmental and clinical isolates of *V. vulnificus*.[66] Later, an oligonucleotide probe labeled with alkaline phosphatase referred to as VVAP was developed and tested for its specificity for the colorimetric detection of *V. vulnificus* in environmental samples.[64,66] In addition to the detection of *V. vulnificus*, *tlh* and *tdh* gene-specific probes labeled with alkaline phosphatase have been developed for the detection of total and pathogenic stains of *V. parahaemolyticus*.[67–70] The use of the alkaline phosphatase-labeled probes in association with the Whatman 541 filter paper for colony lifts method of hybridization detection of targeted *Vibrio* pathogens are relatively inexpensive and takes 2 days to complete. However, this approach is dependent upon the growth of the targeted pathogen on selective agar media. Microbial pathogens in compromised physiological conditions often fail to grow on selective agar media or cluster when high CFU are present in a culture. This leads to either a "false negative" results or undetermined assay. The colorimetric colony hybridization using an alkaline-labeled probe is a FDA-recommended approach still practiced primarily by the FDA Seafood Laboratories for the detection of *Vibrio* pathogens in shellfish. In this context, it is appropriate to mention that the conventional qualitative method of plate culturing and the most-probable-number (MPN) technique requires 4–7 days to complete.[65]

The recent applications of the PCR technology for the detection of *Vibrio* species have significantly improved the time and specificity necessary for early assessment of the presence of these pathogens in seafood. After a series of reports of the use of the conventional PCR technology, the real-time PCR became more popular for such purposes and currently being proposed to be used as a standard detection method for *Vibrio* species in seafood. For *V. cholerae* O1 strain, the cholera toxin gene (*ctx*) has been targeted. However, many strains do not possess the *ctx* gene and *V. mimicus*, being closest to *V. cholerae*, also possesses this gene with over 90% similarity. Conventional PCR targeting a 302 bp DNA fragment of the *ctx* gene for the detection of pathogenic *V. cholerae* in shellfish has been reported.[71] In another study, the *V. cholerae* classical El Tor strains was detected targeting the hemolysin (*hlyA*) gene, whereas the rest of the other biotypes were identified targeting a 869 bp segment of the *ompU*, a 779 bp of the *toxR*, and a 0.862 bp region of the *tcpAI* genes.[72,73] The PCR amplifications by Brasher et al.[71] and Rivera et al.[72,73] were performed in a multiplexed format in which all targeted genes were amplified together in a single reaction. Multiplexed PCR is a cost-effective rapid detection method of microbial pathogens. Lyon[74] described real-time PCR detection of *V. cholerae* O1, O139, non-O1, and non-O139 strains using oligonucleotide primers and a Taqman probe targeting a 70 bp segment of the *hlyA* gene achieving <10 CFU level of sensitivity of detection. The detection of toxigenic *V. cholerae* by Taqman PCR targeting the *ctx* gene was validated against the conventional culture method in bottled water, shrimp, milk, and potato salad.[75] The detection level of an initial inoculum of 1–2 CFU/g was achieved in 8 h, including 6 h of enrichment. The toxigenic *V. cholerae* in oysters, sediments, and seawater from the Gulf of Mexico was detected with the minimum level of detection <10 CFU.[76] In this study, the quantitative analysis was determined using an internal amplification control DNA.

Since most incidences and outbreaks of *V. parahaemolyticus*-related gastroenteritis have been reported due to the consumption of raw or poorly cooked oysters, the detection of this pathogen has primarily been targeted to the seafood. PCR-based detection methodologies have targeted species-specific *tlh* gene, and *tdh* and *trh* genes for the detection of pathogenic *V. parahaemolyticus*. A conventional multiplexed PCR amplification targeting all three genes for the detection of total and pathogenic strains of *V. parahaemolyticus* in oyster tissue homogenate have been established.[37] Later, in 1998, the emergence of a new pandemic pathogenic *V. parahaemolyticus* serogroup O3:K6 prompted the development of a PCR-based detection of this pathogen. Using conventional PCR method, Myers et al.[77] targeted a 369 bp segment of the ORF8 for the detection of this newly emerged pandemic pathogen. A multiplexed real-time Taqman PCR detection targeting the species-specific *tlh* gene and the pandemic strain O3:K6-specific ORF8 DNA segments developed by Rizvi et al.[78] has been applied on oysters and waters from the Gulf of Mexico. The detection level was 1 CFU per ml of initial inoculum of *V. parahaemolyticus* O3:K6 strain in homogenized oyster shell stock or the Gulf of Mexico waters. Other molecular approaches such as the pulse field gel electrophoresis (PFGE) and arbitrarily primed polymerase chain reaction (AP–PCR) of the pandemic *V. parahaemolyticus* O3:K6 strains and newly emerged O4:K68, O1:K25, O1:K26, and O1:K untypeable strains (collectively referred to as the "pandemic group") targeting the *toxRS/new* or orf8 DNA segments exhibited a distinct genotypic cluster.[79] In their study, the strains belonged to the "pandemic group" and four other O3:K6 strains that did not possess the *tdh* gene but possessed the *toxRS*/new sequence. In addition, three O3:K6 strains that clearly belonged to the pandemic group by PFGE and AP–PCR did not possess the orf8 sequence. Since PCR primers targets only a small segment of the targeted DNA/gene, which is subjected to sequence polymorphisms, it would be necessary to perform a DNA–DNA hybridization experiment using the entire orf8 DNA segment as a probe and confirm that in fact these three O3:K6 pandemic strains are devoid of this marker gene. A multiplexed PCR was developed by Okura et al.[79] targeting a 263 bp segment of the *tdh* and 651 bp segment of the *toxRS/new* sequence that distinguished the pandemic strains from the rest of the *V. parahaemolyticus* strains. In another study, the multiplexed real-time PCR assay targeting the *tlh*, *tdh*, *trh*, and the *orf8* gene segments and four TaqMan probes labeled with four different fluorophores for the detection of species and pathogenic strains of *V. parahaemolyticus* in oysters and Gulf of Mexico waters was developed.[80] The sensitivity of the combined four-locus multiplexed TaqMan PCR was found to be an initial inoculum of 1 CFU *V. parahaemolyticus* per gram of oyster tissue homogenate following overnight enrichment. This assay was tested on oysters from the Gulf of Mexico exhiniting 17/33 samples that were positive for *tlh* and 4/33 samples that were positive for *tdh*. No sample was positive *trh* or the orf8 DNA segment.

In a separate study, Taqman quantitative PCR (qPCR) was developed targeting the beta subunit of the DNA gyrase (*gyrB*) gene and tested on 300 oysters collected from the East China Sea. The sensitivity of detection has been reported to be <10 CFU/g of oyster tissue homogenate. Out of the 300 oysters tested, 78 (26%) were positive for *V. parahaemolyticus* by the conventional culture method and 97 (32.3%) positive by the real-time PCR assay. All culture-positive samples were PCR-positive. However, 19 samples positive by PCR were culture negative. These results suggest that although real-time Taqman PCR could be more sensitive than the culture approach by targeting the "injured" or the "physiologically compromised" *V. parahaemolyticus* cells, one must be careful as "non-viable" or "dead" cells or "free DNA" from this species could be amplified, resulting in a false-positive detection.

A multiplexed real-time Taqman qPCR targeting the *tlh*, *tdh*, and *trh* gene segments with an internal amplification control (IAC) was recently described for the detection of species-specific and pathogenic strains of *Vibrio parahaemolyticus* in oysters.[81] The study tested 150 strains of *V. parahaemolyticus* and other *Vibrio* species for confirmation of the specificity of the selected primers. The *tdh* was positive for three strains of *V. hollisae*, suggesting that this assay may not be specific when targeted for this gene. The real-time PCR assay was combined with a most-probable-number format to estimate the quantitative values of the targeted pathogens. In some reactions, the inhibition of the IAC amplification by the oyster tissue homogenate was noticed. Nonetheless the IAC allows monitoring of *V. parahaemolyticus* by real-time PCR amplification and provides a quantitative estimation of the targeted pathogen and helps interpret any "false-negative" results due to the inhibition of the reaction by the oyster matrix.

For the detection of *V. vulnificus*, initially conventional PCR was developed in seeded oyster tissue homogenate by targeting the *vvhA* gene.[82] Then a different region of the same target was used by Brauns et al.[83] for the detection of viable but nonculturable cells. An alternative target of *V. vulnificus*-specific 23S rDNA segment was used for a nested PCR amplification with the sensitivity of detection of 120 CFU.[84] Multiplexed PCR amplification targeting *V. vulnificus* and other seafood-borne pathogens was also developed.[71,72,85] Due to the inhibition of the oyster tissue matrix, it was often necessary to perform two successive rounds of PCR to achieve the required level of sensitivity of detection. Following multiplexed PCR targeting *V. vulnificus* and other shellfish-borne microbial pathogens, a relatively simple confirmatory detection approach utilizing the Covalink™ microtiter-based colorimetric detection technology was developed by Lee et al.[86] Detection sensitivity of *V. vulnificus* in seeded oyster homogenates after 3 h enrichment was 100 CFU/g.

The conventional PCR led to the development of a rapid and quantitative detection of *V. vulnificus* in oysters using the real-time PCR using double-stranded DNA-binding SYBR Green I fluorescent dye or Taqman DNA probes.[87] The applications of the real-time PCR in various foods have been reviewed by Levin.[88] Several recent reviews detailing other real-time applications[88] and problems[89] are available.

Real-time PCR using SYBR Green I dye targeting a 205 bp segment of the *vvhA* gene was developed by Panicker et al.[90] The assay was shown to be specific and an initial inoculum of 1 CFU could be detected after 5 h of enrichment. Furthermore, the relationship between cell number and cycle threshold signal (Ct value) was linear for both enriched and unenriched samples. This indicates that this method can be used for quantitative analysis of *V. vulnificus* in oysters. Also, the conventional MPN culture method can be combined with the SYBR Green I-based real-time PCR to establish a rapid and cost-effective confirmatory quantitative test. This will also allow detection of viable cells in the samples and avoid the possibility of the detection of "non-viable" or "dead" cell in the samples.

In addition to the SYBR Green I-based real-time PCR assay, Taqman probe was also used for the detection of *V. vulnificus* in oysters targeting a 100 bp DNA fragment of the *vvhA* gene.[91] The sensitivity of detection was determined to be 100 CFU/g of oyster tissue homogenate. This study showed that a much higher cell counts (10^7 CFU/g) are necessary for the detection by the Taqman assay.

Three different primer sets and their internal Taqman probes targeting the *vvhA* gene have been compared by real-time PCR method.[92] Although all 3 primer sets exhibited positive amplification of all 81 *V. vulnificus* strains tested, a few other *Vibrio* species exhibited positive Ct values indicating false-positive detection by one of the three sets of primer. This is most likely due to the smaller amplicon size leading to the formation of PCR artifacts rather than the specificity of the targeted gene for *V. vulnificus*. Identification of clinical strains of *V. vulnificus* was developed using real-time PCR and SYBR Green I targeting the vulnibactin (*viuB*) gene, the product of which is involved in chelating iron from the outer environment and transport it into the cytoplasm.[90] In this study, a 504 bp segment of the *viuB* gene and 205 bp segment of the *vvhA* gene were used in a multiplexed format for the detection of total and clinical strains of *V. vulnificus* in oysters with a sensitivity level of an initial inoculum of 1 CFU/g following overnight enrichment. Application of this assay on oysters collected from the Gulf of Mexico exhibited 51% of the samples positive for *vvhA* and 15% of them were positive for *viuB*.

During the winter months, *V. vulnificus* has been reported to remain in a viable but nonculturable (VBNC) state.[93,94] Therefore, the conventional culture techniques may detect this pathogen when present in this state. However, the genetic-based technologies targeting species-specific genes have the potential to detect *V. vulnificus* and other *Vibrio* species during their existence in the VBNC state. Campbell and Wright[91] used the Taqman PCR targeting *vvhA* gene on a VBNC culture and were able to achieve positive detection at 10^7 CFU *V. vulnificus* per gram of oyster tissue homogenate. Detection of this high level of CFU coincides with the previous observation by Brauns et al.[83] Similarly, Ward and Bej[80] showed relatively high Ct values on cold-stressed VBNC *V. parahaemolyticus* cultures with low CFU per milliliter as compared to the cultures maintained in normal temperature, indicating that high cell counts are necessary for the detection of VBNC cultures by PCR methods.

Detection of viable cells could be better determined by targeting the mRNA of metabolically active signature genes of a pathogen. Studies for such detection have been documented for several *Vibrio* species. The application of the reverse transcriptase PCR (RT–PCR) on *vvhA* transcript in VBNC *V. vulnificus* showed sustained expression during this physiological state.[95] In a separate study, the RT–PCR on the pathogenic genes in VBNC *V. parahaemolyticus* cultures were shown to remain transcriptionally silent.[96,97] Therefore it may be wise to select an essential housekeeping gene to for PCR amplification detection of VBNC cultures. Sun et al.[98] reported that VBNC cultures of *V. harveyi* do not cause infection to the host (fish), but when resuscitated the cultures become infectious. This result suggests that although the virulence potential for the pathogenic *Vibrio* species is diminished during the VBNC state, they remain pathogenic when resuscitated. This observation is supported by a previous study, in which VBNC cultures of *V. alginolyticus* and *V. parahaemolyticus* were noninfectious to mice but become infectious when resuscitated.[99] A RT–qPCR analysis of the VBNC *V. cholerae* culture showed expression of an essential protein synthesis and a stress responsive gene—*relA* and *rpoS*, respectively.[100] This study indicates that careful selection of target genes for the detection of the viable *Vibrio* cells in environmental and food samples are essential.

Although the RT–PCR approach seems to be an effective method for the detection of VBNC *Vibrio* species, challenges related to the inhibition of the reaction by the sample matrix, effectiveness of the reverse transcriptase reaction, and possibility of the positive detection signal of the nonviable cells in the sample remain. In addition the issues relevant to the bacterial viability, "true" existence of the VBNC cells in the nature and in seafood remain controversial.

91.2 METHODS

91.2.1 Sample Preparation

Sample preparation remains the crucial step for successful detection of pathogenic *Vibrio* species in seafood samples by PCR amplification. The primary foci for the sample preparation for successful detection of *Vibrio* species in seafood are to avoid the inhibition of the PCR leading to a false-negative detection; achieve sensitivity of detection that complies with the Interstate Shellfish Sanitation Conference (ISSC) guidelines; and avoid detection of nonviable/dead cells, which could lead to a false-positive detection. Current general consensus of all procedures is to include a sample enrichment step prior to the PCR amplification, which helps to achieve the aforementioned objectives.

An enrichment growth medium suitably selective for *V. cholerae* has not been documented. Currently Alkaline Peptone Water (APW)[52] is used for enrichment, in which *V. cholerae* can grow to a sufficient density for PCR amplification detection within 6–8 h. Since APW is not selective, it supports other competing microorganisms to grow simultaneously affecting the growth rate for *V. cholerae*. However, Depaola and Hwang[68] have shown that overnight enrichment resulted into an effective isolation of *V. cholerae*. Longer enrichment may be necessary for better recovery of the *V. cholerae* and other pathogenic vibrios if the samples are subjected to postharvest treatments such as freezing and thawing, or high-pressure treatment resulting in injured cells. In addition, the incubation of raw oyster shellstock in APW at 42°C for 6–8 h has been shown to be effective for the isolation of *V. cholerae* and is currently recommended.[69]

The general enrichment protocol and conventional culture-based identification and enumeration methods for pathogenic *Vibrio* species currently recommended by the FDA Bacteriological Analytical Methods[101] (http://www.cfsan.fda.gov/~ebam/bam-9.html; http://www.cfsan.fda.gov/~ebam/bam-9.html) are as follows:

(i) V. cholerae

1. Dilute raw oyster shellstock to 1:100 ratio of oyster to APW. Transfer 25 g of sample into a tared jar (preferable capacity approximately 500 mL). Products such as seafood or vegetables may be blended or cut into small pieces with sterile scissors or knife.
2. Add 225 mL APW (pH 8.4) to jar. Thoroughly mix the sample or blend 2 min at high speed.
3. Incubate APW at 35 ± 2°C for 6–8 h. Reincubate the jar overnight if desired.
4. For analysis of raw oysters, it is recommended that a second tared flask with 25 g of product plus 2475 mL APW. This flask should be incubated 18–21 h at 42°C ± 0.2°C in a water bath.[65,68]

Comment: An enumeration technique by most probable number (MPN) may also be performed if desired. A sample from the MPN-positive tubes can be subjected to PCR amplification targeting the species-specific gene(s) for the detection and establishing a quantitative assessment of the targeted pathogen (see section below for PCR amplification detection).

5. For conventional culture-based detection, the cultures can be tenfold serially diluted in APW and applied onto TCBS agar or modified-CPC agar plates by spread plating or streaking the pellicle of the sample with a 1 mm sterile loop in a manner that will yield single colony isolation. TCBS agar is incubated overnight (18–24 h) at 35°C ± 2°C. Incubate mCPC and CC overnight at 39°C–40°C or at 35°C–37°C. *V. cholerae* on TCBS agar can be identified by the morphology and texture such as large (2–3 mm), smooth, yellow, and slightly flattened with opaque centers and translucent peripheries. Typical colonies of *V. cholerae* on mCPC or CC agar are small, smooth, opaque, and green to purple in color, with a purple background on extended incubation.
6. For biochemical identification, colonies from crowded plates must be streaked to a nonselective agar (T1N1, T1N3, or TSA-2% NaCl agar) for purity.

Incubate overnight at 35°C ± 2°C and proceed with identification using a single isolated colony.

7. Subculture three or more typical colonies from each plating medium to T1N1 agar slants or motility test medium stabs. Incubate slants or stabs overnight at 35°C ± 2°C.

(ii) V. vulnificus

1. For fish samples, the surface tissues and gills, and for shellfish samples, the shellstock (meat and liquor) from 12 to 15 animals of approximately 50 g of weight are analyzed. The samples are pooled and blended at high speed for 90–120 s (if necessary with pulses) until the homogenate is formed. For crustaceans such as shrimp or crayfish the entire animal is used. However if the animal is too large then the central portion, including gill and gut, is used. (This step is same for *V. parahaemolyticus*).
2. Tissue homogenates (preferably three tubes each with 1 mL) are inoculated in 100 mL APW and incubated at 35°C–37°C for 18–24 h in a rotary shaker set at 150–170 rpm.
3. For plate culture analysis, 3 mm loopful from the top 1 cm of APW tubes is streaked onto mCPC or CC selective agars. Incubate mCPC and CC agars overnight at 39°C–40°C (alternatively 35°C–37°C if 39°C–40°C not available). On either agar, colonies appear as round, flat, opaque, yellow, and 1–2 mm in diameter.
4. Along with the plate cultures, a three-tube MPN method as described above for the *V. cholerae* can be conducted. The MPN-positive tubes can be subjected to PCR amplification for the confirmative and quantitative detection of *V. vulnificus*.
5. Individual colonies can be picked by using a sterile toothpick and resuspended in 10 μL sterile water and boiled for 10 min to release DNA. The released DNA can be subjected to PCR amplication using *V. vulnificus* gene-specific primers (*vvh*A) for detection.

(iii) V. parahaemolyticus

1. The enrichment step is same as *V. vulnificus* described above. After enrichment, the culture is streaked on TCBS agar using the method described above and grown at 35°C–37°C overnight. *V. parahaemolyticus* colonies appear as round, opaque, green or bluish colonies, 2–3 mm in diameter on TCBS agar. *V. parahaemolyticus* colonies can be differentiated from *V. alginolyticus* colonies, which are large, opaque, and yellow. Most strains of *V. parahaemolyticus* will not grow on mCPC or CC agar. If growth occurs, colonies will be green-purple in color due to lack of cellobiose fermentation.
2. An aliquot of the culture can be subjected to three-tube MPN enrichment if desired. The method is described above.
3. Biochemically, *V. parahaemolyticus* and *V. vulnificus* are phenotypically similar, but can be differentiated by differences of the ONPG, salt-tolerance, cellobiose, and lactose reactions.
4. The urease positive for *V. parahaemolyticus* indicates the potentially pathogenic strain.
5. Either the enriched or the MPN enriched cultures could be subjected to PCR amplification detection of species and pathogenic strains of *V. parahaemolyticus*.

91.2.2 Detection Procedures

91.2.2.1 Standard PCR

Principle. Standard PCR assays based on agarose gel format have been developed for identification of pathogenic *V. cholerae*,[71,74] *V. vuliníficus*,[71] and *V. parahaemolyticus*[37] (Table 91.1). We present below stepwise PCR protocols for *Vibrio* detection with primers from these authors.

Procedure

1. Prepare PCR master mix containing 200 μM of each dNTP, 1 μM of each primer (see Table 91.1), 2.5 U of Taq DNA polymerase, 1× PCR buffer [50 mM Tris·Cl pH 8.9, 50 mM KCl, 2.5 mM $MgCl_2$ (referred as buffer C)], and sterile distilled water to 25 or 50 μL per reaction (depending upon the make and model of the thermal cycler). Transfer an appropriate aliquot into a PCR tube. Add 1–3 μL of boiled DNA sample per reaction. Include a tube with purified genomic DNA as positive control and a tube with master mix (plus distilled water) as negative control.
2. Perform PCR amplification in a thermocycler with 1 cycler of 94°C for 3 min; 25–30 cycles of 94°C for 0.5–1 min, 55°C–60°C for 0.5–1 min, and 72°C for 0.5–1 min; and 1 cycle of 72°C for 5–7 min followed by the soak cycle at 4°C–10°C until gel electrophoresis analysis.
3. Separate PCR product (5 μL) on a 1%–1.2% (w/v) agarose gel in 1× TAE (pH 8.2) buffer containing ethidium bromide (0.5 μg/mL) at 5 V/cm. Visualize the specific products under UV light.

Note: If further confirmation is necessary, Southern blot DNA–DNA hybridization can be performed using the standard method described in Ausubel et al.[102] The probes specific for the hybridization detection of the amplified DNA are described in Table 91.1.

91.2.2.2 Real-Time PCR

Principle. Several real-time PCR assays based on TaqMan have been reported, which facilitate improved detection and identification of pathogenic *V. cholerae*,[74,76] *V. vuliníficus*,[92] and *V. parahaemolyticus*[77–80] (Table 91.1).

TABLE 91.1
Description of Primers and Probes for PCR Amplification Detection of Pathogenic *Vibrio* Species

Microorganism	Targeted Gene	Primers/Probes	Tm (°C)	Amplicon Size (bp)	Reference
Conventional PCR					
V. cholerae	*ctxA* (Classical El Tor)	L-CTX: 5′-ctcagacgggatttgttaggcacg-3′	74	302	Brasher et al., 1998[71]
		R-CTX: 5′-tctatctctgtagccggtattacg-3′	70		
		P-CTX2: 5′-attccatactccccaaatata-3′			
	Nonclassical hlyA (El Tor, O139; non- El Tor, non-O139)	Forward 5′-tgcgttaaacacgaacgcat-3′	58	70	Lyon, 2001[74]
		Reverse 5′-aagtcttacattgtgcttgggtca-3′	59		
		Probe: 5′-tcaaccgatgcgattgcccaaga-3′			
V. vulnificus	*vvhA* (species specific	L-CTH: 5′-ttccaacttcaaaccgaactatgac-3′	70	205	Brasher et al., 1998[71]
		R-CTH: 5′-gctactttctagcattttctctgc-3′	68		
		P-CTH: 5′-gaagcgcccgtgtctgaaactggcgtaacg-3′			
V. parahaemolyticus	*tlh* (species specific)	L-TL:5′-aaagcggattatgcagaagcactg-3′	70	450	Bej et al., 1999[38]
		R-TL:5′-gctactttctagcattttctctgc-3′	68		
		P-TL: 5′-acggacgcaggtgcgaagaacttcatgttg-3′			
	tdh (pathogenic strain)	L-TDH:5′-gtaaaggtctctgacttttggac-3′	68	269	
		R-TDH:5′-tggaatagaaccttcatcttcacc-3′	68		
		P-TDH: 5′-gcggtgtctggctstssgsgcggtcssttct-3′			
	trh (pathogenic strain)	L-TRH:5′-ttggcttcgatattttcagtatct-3′	66	500	
		R-TRH:5′-B-cataacaaacatatgcccatttccg-3′	60		
		P-TRH: 5′-aaaactgaatcaccagttaacgcaatcgtt-3′			
	orf8 (pandemic O3:K6 serotype)	F-O3MM824 5′aggacgcagttacgcttgatg-3′	64	369	
		R-O3MM1192 5′-ctaacgcattgtccctttgtag-3′	64		
Real-time PCR					
V. cholerae	*ctxA* (Classical El Tor)	Forward: 5′-tttgttaggcacgatgatggat-3′		84	Blackstone et al., 2007
		Reverse: 5′-accagacaatatagtttgacccactaag-3′			
		Probe: FAM-5′-tgtttccacctcaattagtttgagaagtgccc-3′ BHQ			
	Nonclassical *hlyA* (El Tor, O139; non- El Tor, non-O139)	Forward 5′-tgcgttaaacacgaacgcat-3′	58	70	Lyon, 2001[74]
		Reverse 5′-aagtcttacattgtgcttgggtca-3′	59		
		TaqMan probe FAM 5′-tcaaccgatgcgattgcccaaga-3′ TAMRA			
V. vulnificus	*vvhA*	F-vvh785 ttccaacttcaaaccgaactatgac-3′	66	205	Panicker and Bej, 2005[92]
		R-vvh990 attccagtcgatgcgaatacgttg-3′	70		
		P-vvh874 FAM-5′-aactatcgtgcacgctttggtaccgt-3′BHQ-1			
V. parahaemolyticus	*tlh* (species specific)	L-TL:5′-aaagcggattatgcagaagcactg-3′		450	Ward and Bej, 2006[80]; Rizvi et al., 2006
		R-TLH:5′-gctactttctagcattttctctgc-3′			
		P-TL: 5′-TEXR-aagaacttcatgttgatgacact-BHQ2–3′			
	tdh (Pathogenic strain)	L-TDH:5′-gtaaaggtctctgacttttggac-3′	68	269	
		R-TDH:5′-tggaatagaaccttcatcttcacc-3′	68		
		P-TDH:5′-CY5-attttacgaacacagcagaatga-3-Iowa Black-RQ-3			
	trh (pathogenic strain)	L-TRH:5′-ttggcttcgatattttcagtatct-3′	66	500	
		R-TRH:5′-B-cataacaaacatatgcccatttccg-3′	60		
		P-TRH:5′-TET-tatttgtygttagaaatacaacaat-BHQ1–3′			
	orf8 (pandemic O3:K6 serotype)	F-O3MM824 5′aggacgcagttacgcttgatg-3′	64	369	Myers et al., 2003[77]
		R-O3MM1192 5′-ctaacgcattgtccctttgtag-3′	64		
		P-ORH8: 5′-6-FAM-aagccattaacagttgaaggcgttgact-BHQ1–3′			

Y = C or T; M = A or C; R = G or A; F, Forward primer; R, Reverse primer; P, Probe; TexR, sulforhodamine 101 (Texas Red) fluorescent dye; BHQ-2, Black Hole-2 quencher dye; 6-FAM, 6-fluorescein fluorescent dye; BHQ-1, Black Hole-1 quencher dye; Cy5, carbocyanine fluorescent dye; Iowa Black-RQ, Iowa Black quencher dye; TET, tetrachloro-6-carboxyfluorescein fluorescent dye; ROX, 6-carboxy-X-rhodamine; BHQ-2, Black Hole-2 quencher dye; FAM, 6-fluorescein; BHQ-1, Black Hole-1 quencher dye; Tm (°C) = 2 (A + T) + 4 (G + C); bp, base pair.

(i) V. cholerae [Nonclassical hlyA (El Tor, O139; Non-El Tor, Non-O139)]

1. Prepare PCR mix (50 μL) consisting of 1 μL TaqMan buffer A (Perkin Elmer Applied Biosystems); 5 mM $MgCl_2$; 200 μM (each) of dATP, dCTP, and dGTP; 400 μM dUTP; 0.02 μM Taqman probe specific for the nonclassical *hlyA* gene (Table 91.1); 0.3 μM (each) *V. cholerae* specific primers targeting the nonclassical *hlyA* gene (Table 91.1); 1 U of AmpErase uracil *N*-glycosylase; 2.5 U of AmpliTaq Gold DNA polymerase; and DNA sample (2.5 μL, 100 ng/μL) in 0.2 mL of MicroAmp Optical reaction tubes with MicroAmp Optical tube caps (PE Applied Biosystems).
2. Keep the PCR mixture at 50°C for 5 min and then denature at 95°C for 10 min. Carry out 40 cycles of 95°C for 20 s followed by 60°C for 1 min using the ABI Prism 7700 sequence detection system (PE Applied Biosystems) using an ABI Prism 7700 sequence detection system and SOS (version 1.7) application software (PE Applied Biosystems). Other real-time PCR thermocyclers may also be used.

(ii) V. cholerae (Classical El Tor)

1. Prepare PCR mix (25 μL) consisting of 1× PCR buffer (Invitrogen, Carlsbad, CA), 5 mM $MgCl_2$, 0.2 mM each of the dNTPs, 1.25 U of Platinum Taq polymerase (Invitrogen), 250 nM primers targeting the nonclassical *hlyA* gene (Table 91.1), 100 nM Taqman probe specific for the *ctx* gene (Table 91.1), and DNA sample (1.5 μL, 100 ng/μL).
2. Perform real-time PCR with 1 cycle of 94°C for 2 min; 45 cycles of 94°C for 10 s, 63°C for 30 s (with the instrument optics on) using a Cepehid Smart Cycler™ II with all default analysis parameters except that the manual threshold fluorescence units setting is adjusted to 10. Other real-time PCR thermocyclers may also be used.

(iii) V. vulnificus

1. Prepare PCR mix (25 μL) consisting of 2.5 μL of 10× PCR buffer [200 mM Tris-Cl (pH 8.4), 500 mM KCl], 5 μL of 25 mM $MgCl_2$, 200 μM concentrations of deoxynucleoside triphosphates, 0.24 μM Taqman probe specific for the *vvhA* gene, 0.4 μM of the primers targeting the *vvh*A gene, 1.5 U of *Taq* DNA polymerase, and DNA sample (1.5 μL, 100 ng/μL).
2. Perform real-time PCR with 1 cycle of 94°C for 2 min; 40 cycles of 94°C for 15 s, 58°C for 15 s, and 72°C for 20 s using a Cepehid Smart Cycler™ II with all default analysis parameters and the set cycle threshold set at 30.

(iv) V. parahaemolyticus

For amplification of *V. parahaemolyticus* species and pathogenic strains a multiplexed PCR is recommended, with four sets of primers targeting the *tlh*, *tdh*, *trh*, and *orf8* genes, and four gene-specific Taqman probes (Table 91.1).

1. Prepare PCR mix (25 μL) consisting of 1× PCR buffer [20 mM Tris-Cl (pH 8.4), 50 mM KCl], 5 mM $MgCl_2$, 400 μM each deoxynucleoside triphosphate, 0.4 μM of each of the eight gene-specific oligonucleotide primers, 0.2 μM of each of the four fluorescence-labeled Taqman probes, 100 ng bovine serum albumin, 2 U of Platinum *Taq* DNA polymerase (Invitrogen, Carlsbad, CA).
2. Perform real-time PCR with 1 cycle of 94°C for 2 min; 45 cycles of 94°C for 20 s, 56°C for 20 s, and 72°C for 30 s using a Cepehid Smart Cycler™ II. The increase in fluorescence for each channel is measured and recorded during the annealing step of each cycle. The cycle threshold (*Ct*) value is set at 15 fluorescence units.
3. Recently Nordstrom et al.[81] described the use of an internal amplification control (IAC) DNA, which is coamplified with the targeted gene(s). The IAC avoids the "false-negative" detection and helps determine the quantitative estimation of the initial copies of the targeted gene(s). The use of this approach is useful for samples that have not been enriched. Since the seafood or the environmental water samples for the detection of *Vibrio* species are often enriched, the quantitative estimation of the targeted pathogens cannot be determined accurately by this approach. Therefore, the use of the IAC for the detection of *Vibrio* species is restricted to confirm the integrity of the PCR amplification process.

91.2.2.3 Loop-Mediated Isothermal Amplification (LAMP)

The loop-mediated isothermal amplification is a relatively new method applied for the detection of some of the *Vibrio* pathogens. LAMP assay is faster and relatively easier to perform than conventional PCR assays and may not require an expensive thermal cycler.[103] The LAMP method has been applied for the detection of *V. alginolyticus; V. cholerae*, *V. nigripulchritudo*, *V. parahaemolyticus*, and *V. vulnificus* in environmental and clinical samples.[104–114] The LAMP assay yields a large amount of amplicons in a relatively short reaction time and can be detected by simple turbidity. Therefore this approach provides an excellent new platform for a simple, rapid, and cost-effective detection of *Vibrio* pathogens.

The LAMP method uses a *Bst* DNA polymerase large fragment for the detection of a specific DNA sequence with specific characteristics by autocycling strand displacement DNA synthesis. The method requires four or six complex and carefully selected primers that specifically recognize six or eight distinct regions of the targeted DNA. The reaction is conducted at an isothermal condition (60°C–65°C); therefore unlike real-time or conventional PCR, an expensive thermal cycler is necessary to execute the process (http://loopamp.eiken.co.jp/e/lamp/index.html). The amplification is achieved in 15–60 min and typically the targeted DNA is amplified 109–1010 times, yielding a large quantity of amplicon. Because the LAMP assay synthesizes a large amount of DNA, the products can be detected by simple turbidity visually or using a turbidimeter. The list of primers and optimized reaction conditions for the detection of pathogenic vibrios can be available in the literatures cited above and the reagents are commercially available at Eiken Genome Site (http://loopamp.eiken.co.jp/e/index.html) or other vendors. For LAMP real-time amplification, normally the primers and the template DNA along with the *Bst* DNA polymerase with reaction buffer and dNTPs are mixed and then incubated at 60°C–65°C for 60 min and then at 80°C for 2 min to terminate the reaction. The reaction can be performed in a Loopamp real-time turbidimeter (LA-320; Teramecs, Kyoto, Japan). LAMP amplification is detected as a value of turbidity at 650 nm using the turbidimeter (LA-320) in real time. The reaction is considered to be positive when the turbidity reached 0.1 within 60 min. Turbidity visible with the unaided eye can also be considered to indicate a successful LAMP reaction.[111–113]

91.3 CONCLUSIONS AND FUTURE PERSPECTIVES

Historically, *V. cholerae* within the *Vibrioneaceae* family received the most attention for the outbreaks, including several pandemic outbreaks of cholera disease throughout the world, and it continues be a problem primarily in South America, Southeast Asia, and Africa. However, since the last report of the identification of pathogenic strains of this pathogen in the Gulf of Mexico water, this pathogen has not been a problem in the United States. It is important to note that although the O1 serogroup has not been a problem in the United States, the on-O1 *V. cholerae*-related diarrhea has been reported. Therefore, it is necessary to continuously monitor *V. cholerae* in food and waters by using rapid detection technologies, such as real-time PCR or DNA-microarray to ensure the safety of public health. In contrast, *V. parahaemolyticus* and *V. vulnificus* have been a constant problem in the United States as well as other countries, such as Japan, Southeast Asia, South America, and some European countries. Therefore, effort has been devoted to explore rapid, reliable, sensitive, and cost-effective detection technologies by using molecular approaches. During the last two decades, the progress in such detection techniques has been phenomenal. The molecular technologies for the detection of pathogenic vibrios have started with the colorimetric colony blot hybridizations using species-specific DNA probes. The methods were relatively successful meeting the detection compliances and have been adopted by the FDA seafood laboratories. The second-generation molecular detections have to take advantage of the PCR amplification technology. During the first phase to this effort, the conventional PCR methodologies were successfully explored to the identification of the species-specific targeted genes and optimizations of the PCR parameters. The conventional PCR approach led to the second phase of PCR-based detection of the pathogenic vibrios, which is the use of the real-time PCR approach either using the Taqman probes or a double-stranded DNA-binding fluorescent dye such as SYBR Green I or the Eva Green. The Taqman PCR method is much faster than the conventional PCR, and often the amplification of the targeted genes can be monitored in real time as each cycle of amplification progresses. However, a SYBR Green I-based PCR approach, although rapid when compared with the conventional PCR, requires completion of sufficient number of amplification cycles, at which point a DNA melt-curve analysis is performed to confirm the amplification of the targeted gene segments. In that sense, unlike the Taqman PCR, the SYBR Green I-based PCR method of detection cannot be referred to as real-time PCR. The transition from the conventional PCR to either Taqman real-time PCR or the SYBR Green I-based PCR required some optimization of the reaction parameters, such as the length of the amplicons, reagents concentrations, and the shorter PCR temperature cycling parameters. This has led to many studies redesign the primer locations on the targeted genes that had been previously established by the conventional PCR methods.

Besides the PCR approach, a number of reports have taken the advantage of other molecular approaches, such as DNA microarray,[90] loop-mediated amplification, or the helicase-dependent isothermal amplification (Mojib and Bej, unpublished data). The DNA microarray offers detection of numerous microbial pathogens simultaneously on a single glass slide with immobilized gene arrays. However, the process is based on the principles of DNA–DNA hybridization requiring multiple steps of wash cycles followed by computer analysis of the positive hybridization results. Therefore, this approach does not offer a real advantage at this time until an automated method is developed. The isothermal amplification approach seems to be the next generation detection technology of the *Vibrio* species. A few reports have already shown the promise that the isothermal technology offers a cost-effective method of detection without the use of a relatively expensive real-time PCR thermocycler.[103–114] In addition, the isothermal LAMP amplification approach seems effective in the detection of several pathogenic vibrios in clinical, seafood, and environmental samples. However, the current format of the loop-mediated isothermal amplification technology requires rather complex and carefully selected primers and a rigorous optimization process. The other approach of the use of the thermostable helicase-dependent isothermal amplification technology (HDA assay) offers a relatively simple approach of primer selection and isothermal amplification approach. The HDA can be purchased in the kit format from BioHelix, Inc. (www.biohelix.com).

One of the challenging issues that yet to be solved is a "realistic" quantitative analysis of *Vibrio* species and the pathogenic strains in shellfish and other seafood without enrichment step. This is a rather complex because of the facts that (i) ISSC guidelines recommend the minimum level of specific *Vibrio* species as a standard for "safe/consumable" shellfish. This also applies to the postharvest treated oysters; (ii) due to the complex tissue matrix, the detection of the ISSC-recommended minimum level of *Vibrio* pathogens without enrichment is often difficult to achieve; (iii) without enrichment there is a possibility of the detection of dead cells or cell-free DNA from the targeted pathogens;

(iv) enrichment step may not accurately reflect the initial number of cells in the sample. However, enrichment ensures the detection of viable cells in the samples. To address these issues, a reverse transcriptase polymerase chain reaction (RT–PCR) was conducted in which successful amplification of *vvhA* mRNA allowed the investigators to confirm the detection of viable pathogens.[95] However, this approach can be better achieved when applied to the enriched samples and be technically challenging due to relatively short half-life of most prokaryotic mRNA.

A quantitative assessment of the pathogenic *Vibrio* species can be better addressed by combining the conventional most-probable-number (MPN) method with the real-time PCR approach.[115,116] The conventional MPN is specifically applicable to the PHP oysters, which often requires process validation that is often labor intensive and expensive. The detection of targeted *Vibrio* species in the MPN-positive tubes by real-time PCR is rapid, reliable, and saves valuable time for subsequent lengthy time-consuming analysis for confirmation. However, the MPN/real-time PCR approach requires handling numerous culture tubes and many real-time PCR and could be somewhat expensive.

The other issue with *Vibrio* species is that during the winter months, they seem to enter into a viable but nonculturable (VBNC) state, which often cannot be detected by standard microbiological culture methods.[93] Since the direct detection of low level of *Vibrio* species in seafood, especially in shellfish tissue without enrichment, is often unreliable and leads to "false-negative" detection, revival followed by enrichment of the VBNC cells may be necessary. It is a relatively difficult issue and much attention is needed to come up with a reliable approach to achieve detection of this pathogen when present in the VBNC state.

The next generation nucleic acid detection of the targeted genes from *Vibrio* species could be the applications of the biosensor technology, including the nanotechnology.[117] This approach is based on the principles of DNA–DNA hybridization on a monolayer of single-stranded DNA probes, which are layered on silicon microcantilever. The positive hybridization of the probe DNA with the targeted gene amplicons can be detected label-free by measuring the changes of the tension in the monolayer as a result of hydration. The tension is changed rapidly when the single-stranded probes on the monolayer interact with the complementary single-stranded amplicons forming hybrids by the nucleic acid base-pair rules. This is approach is very sensitive and reported to detect femtomole quantity of DNA. This approach is currently suggested for the detection of multiple gene targets in a microarray format. The nanotechnology approach has further been developed and tested for the detection of a single or multiple microbial pathogens in bacterial barcode or single-layered carbon nanofiber coated with aptamer format in near real-time format.[118,119]

Molecular detection of *Vibrio* species using real-time PCR and LAMP methods is rapid, specific, and could be cost effective. Also, the PCR and LAMP-based methods enable us to detect total and pathogenic (clinical) strains of *Vibrio* species simultaneously in a single reaction. The "up and coming" detection technology using nanotechnology, such as single layer carbon nanofibers and transistors, should revolutionize the single-cell *Vibrio* detection in clinical, environmental, and food samples. The nucleic acids and peptide-based approaches, along with the conventional culture methods of detection, have become widely accepted for the detection of vibrios in government and academic laboratories. The combined approaches have the major advantage of establishing a "fail-safe" detection of this pathogen. With significant improvement in the quality of the reagents and the instrumentation for these methodologies the reliability of the detection of *Vibrio* species in seafood and clinical samples has become more achievable. However, three aspects of detection—(i) a sample processing method that will yield sufficiently pure targeted genomic DNA for inhibition-free amplification of the targeted genes; (ii) a "true" sensitivity of detection without the sample enrichment; and (iii) detection of viable "live" cells in shellfish and other seafood samples—need further improvement. In addition, a consensus primer/probe combination for the detection of *Vibrio* species and the pathogenic strains should be evaluated and established by an independent research entity so that the application of a "gold standard" approach can be offered to the seafood industry and the clinical laboratories for a rapid and effective detection of *Vibrio* pathogens. Alternatively, the development of the nanotechnology approach may simply use the "whole-cell" detection without the complex processing of the samples such as enrichment, DNA purification, primer design, optimization of the reaction and internal control DNA, and so forth. Such development of the new state-of-the-art microbial diagnostic technology, when adopted for the detection of pathogenic *Vibrio* spp., is expected to help the seafood industry and processing facilities for routine monitoring of seafood, especially PHT oysters and other shellfish, thereby ensuring seafood safety, consumer confidence, and protection of public health. This would also enable clinicians to achieve rapid detection thereby establishing a timely treatment regimen for patients with *Vibrio* infection.

ACKNOWLEDGMENT

Some of the PCR and HDA isothermal work for vibrio detection in seafood by the author's lab, presented in this chapter, has been supported by the Mississippi Alabama Sea Grant Consortium (MASGC Project Nos. R/AT-9-PD and R/SP-8).

REFERENCES

1. Farmer, J.J., and Janda, M., Vibrionaceae. In Garrity, G.M. et al. (Editors), *Bergey's Manual of Systematic Bacteriology*, 2nd ed., Vol. 2 The *Proteobacteria*, p. 491, Springer-Verlag, New York, 2005.
2. Lipp, E.K., and Rose, J.B., The role of seafood in foodborne diseases in the United States of America, *Rev. Sci. Technol.*, 16, 620, 1997.
3. DePaola, A. et al., Molecular, serological, and virulence characteristics of *Vibrio parahaemolyticus* isolated from environmental, food, and clinical sources in North America and Asia, *Appl. Environ. Microbiol.*, 69, 3999, 2003.

4. Harwood, V.J. et al., Methods for isolation and confirmation of *Vibrio vulnificus* from oysters and environmental sources: A review, *J. Microbiol. Methods*, 59, 301, 2004.
5. Wallace, D.J. et al., Incidence of foodborne illnesses reported by the foodborne diseases active surveillance network (FoodNet)-1997. FoodNet Working Group, *J. Food. Prot.*, 63, 807, 2000.
6. Bauman, P., and Schubert, R.H.W., Family II. *Vibrionaceae*. In Kraig, N.R., and Holt, H.G. (Editors), *Bergey's Manual of Systematic Bacteriology*, 1st ed., Vol. 1, p. 516, The Williams and Wilkins Co., Baltimore, MD, 1984.
7. Veron, M.M., La position taxonomique des *Vibrio* et de certines bacteries comparables, *C. R. Acad. Sci. Paris.*, 261, 5243, 1965.
8. Nyholm, S.V., and McFall-Ngai, M.J., The winnowing: Establishing the squid-vibrio symbiosis, *Nat. Rev. Microbiol.*, 2, 632, 2004.
9. Greenough, W.B.III., *Vibrio cholerae* and cholera. In Mandell, G.L., Bennett, J.E., and Dolin, R. (Editors), *Principles and Practices of Infectious Diseases*, 4th ed., p. 1934, Churchill Livingstone, New York, 1995.
10. Ramamurthy, T. et al., Emergence of novel strain of *Vibrio cholerae* with epidemic potential in southern and eastern India, *Lancet*, 341, 703, 1993.
11. CDC, Imported cholera associated with a newly described toxigenic *Vibrio cholerae* O139 strain—California, *MMWR*, 42, 501, 1993.
12. Pierce, N.F., and Mondal, A., Clinical features of cholera. In Barua, D., and Burrows, W. (Editors), *Cholera*, p. 209, WB Saunders Company, Philadelphia, 1974.
13. Morris, J.G. Jr., *Vibrio cholerae* O139 Bengal: Emergence of a new epidemic strain of cholera, *Infect. Agents Dis.*, 4, 41, 1995.
14. Sharma, C. et al., Molecular analysis of non-O1, non-O139 *Vibrio cholerae* associated with an unusual upsurge in the incidence of cholera-like disease in Calcutta, India, *J. Clin. Microbiol.*, 36, 756, 1998.
15. Faruque, S.M. et al., Epidemiology, genetics, and ecology of toxigenic *Vibrio cholerae*, *Microbiol. Mol. Biol. Rev.*, 62, 1301, 1988.
16. Berche, P. et al., The novel epidemic strain O139 is closely related to the pandemic strain O1 of *Vibrio cholerae*, *J. Infect. Dis.*, 170, 701, 1994.
17. Swerdlow, D.L., and Ries, A.A., *Vibrio cholerae* non-O1—the eighth pandemic? *Lancet*, 342(8868), 382, 1993.
18. Daniels, N.A., and Shafaie, A., A review of pathogenic *Vibrio* infections for clinicians, *Infect. Med.*, 17, 665, 2000.
19. Fujino, T. et al., On the bacteriological examination of shirasu-food poisoning, *Med. J. Osaka Univ.*, 4, 299, 1953.
20. Dadisman, T.A. Jr. et al., *Vibrio parahaemolyticus* gastroenteritis in Maryland. I. Clinical and epidemiologic aspects, *Am. J. Epidemiol.*, 96, 414, 1972.
21. Dadisman, T.A. et al., *Vibrio parahaemolyticus* gastroenteritis in Maryland—clinical and epidemiologic aspects, *Am. J. Epidemiol.*, 96, 414, 1973.
22. Daniels, N.A. et al. *Vibrio parahaemolyticus* infections in the United States, 1973–1998, *J. Infect. Dis.*, 181, 1661, 2000.
23. Daniels, N.A. et al., Emergence of a new *Vibrio parahaemolyticus* serotype in raw oysters: Prevention quandary, *JAMA*, 284, 1541, 2000.
24. CDC, Gastroenteritis caused by *Vibrio parahaemolyticus* aboard a cruise ship, *MMWR*, 27, 67, 1978.
25. CDC, Foodborne diseases active surveillance network, 1996, *MMWR*, 46, 258, 1997.
26. CDC, Outbreak of *Vibrio parahaemolyticus* infections associated with eating raw oysters—Pacific Northwest, 1997, *MMWR*, 47, 457, 1998.
27. CDC, Outbreak of *Vibrio parahaemolyticus* infection associated with eating raw oysters or clams harvested from Long Island Sound—Connecticut, New Jersey, and New York, 1998, *MMWR*, 48, 48, 1999.
28. Miwatani, T. et al., Antibody titers against the thermostable direct hemolysin in the sera of patients with *Vibrio parahaemolyticus* infection, *Kansenshogaku Zasshi*, 50, 46, 1976.
29. Miyamoto, Y. et al., In vitro hemolytic characteristic of *Vibrio parahaemolyticus*: Its close correlation with human pathogenicity, *J. Bacteriol.*, 100, 1147, 1969.
30. Wagatsuma, S., A medium for the test of the hemolytic activity of *Vibrio parahaemolyticus*, *Media Circle*, 13, 159, 1968.
31. Honda, T., and Iida, T., The pathogenicity of *Vibrio parahaemolyticus* and the role of the thermostable direct haemolysin and related haemolysins, *Rev. Med. Microbiol.*, 4, 106, 1993.
32. Honda, T. et al., Evidence of immunologic cross-reactivity between hemolysins of *Vibrio hollisae* and *Vibrio parahaemolyticus* demonstrated by monoclonal antibodies, *J. Infect. Dis.*, 160, 1089, 1989.
33. Honda, T., Ni, Y., and Miwatani, T., Purification of a TDH-related hemolysin produced by a Kanagawa phenomenon-negative clinical isolate of *Vibrio parahaemolyticus* O6:K46, *FEMS Microbiol. Lett.*, 57, 241, 1989.
34. Shirai, H. et al., Molecular epidemiologic evidence for association of thermostable direct hemolysin (TDH) and TDH-related hemolysin of *Vibrio parahaemolyticus* with gastroenteritis, *Infect. Immun.*, 58, 3568, 1990.
35. Matsumoto, C. et al., Pandemic spread of an O3:K6 clone of *Vibrio parahaemolyticus* and emergence of related strains evidenced by arbitrarily primed PCR and toxRS sequence analyses, *J. Clin Microbiol.*, 38, 578, 2000.
36. WHO, *Vibrio parahaemolyticus*, Japan, 1996–1998, *Wkly. Epidemiol. Rec.*, 74, 357, 1999.
37. Taniguchi, H. et al., 1985. Cloning and expression in *Escherichia coli* of *Vibrio parahaemolyticus* thermostable direct hemolysin and thermolabile hemolysin genes, *J. Bacteriol.*, 162, 510, 1985.
38. Bej, A.K. et al., 1999. Detection of total and hemolysin-producing *Vibrio parahaemolyticus* in shellfish using multiplex PCR amplification of *tl*, *tdh* and *trh*, *J. Microbiol. Methods*, 36, 215, 1999.
39. Linkous, D.A., and Oliver, J.D., Pathogenesis of *Vibrio vulnificus*, *FEMS Microbiol. Lett.*, 174, 207, 1999.
40. Patel, V.J., Gardner, E., and Burton, C.S., *Vibrio vulnificus* septicemia and leg ulcer, *J. Am. Acad. Dermatol.*, 46, S144, 2002.
41. Ulusarac, O., and Carter, E., Varied clinical presentations of *Vibrio vulnificus* infections: A report of four unusual cases and review of the literature, *South Med. J.*, 97, 163, 2004.
42. Hollis, D.G. et al., Halophilic *Vibrio* species isolated from blood cultures, *J. Clin. Microbiol.*, 3, 425, 1976.
43. Paranjpye, R.N. et al., The type IV leader peptidase/N-methyltransferase of *Vibrio vulnificus* controls factors required for adherence to Hep-2 cells and virulence in iron-overloaded mice, *Infect. Immun.*, 66, 5659, 1998.
44. Strom, M.S., and Paranjpye, R.N., Epidemiology and pathogenesis of *Vibrio vulnificus*, *Microbes Infect.*, 2, 177, 2000.
45. Wright, A.C., Simpson, L.M., and Oliver, J.D., Role of iron in the pathogenesis of *Vibrio vulnificus* infections, *Infect. Immun.*, 34, 503, 1981.

46. Hor, L.I. et al., Mechanism of high susceptibility of iron-overloaded mouse to *Vibrio vulnificus* infection, *Microbiol. Immunol.*, 44, 871, 2000.
47. Hayat, U. et al., Capsular types of *Vibrio vulnificus*: An analysis of strains from clinical and environmental sources, *J. Infect. Dis.*, 168, 758, 1993.
48. Huq, M.I. et al., Isolation of *Vibrio*-like group, EF-6, from patients with diarrhea, *J. Clin. Microbiol.*, 11, 621, 1980.
49. Chikahira, M., and Hamada, K., Enterotoxigenic substance and other toxins produced by *Vibrio fluvialis* and *Vibrio furnissii*, *Nippon Juigaku Zasshi.*, 50, 865, 1988.
50. Nishibuchi, M., and Seidler, R.J., Medium-dependent production of extracellular enterotoxins by non-O-1 *Vibrio cholerae*, *Vibrio mimicus*, and *Vibrio fluvialis*, *Appl. Environ. Microbiol.*, 45, 228, 1983.
51. Jean-Jacques, W., Vibrio, In Murray, P.R. et al. (Editors), *Manual of Clinical Microbiology*, 6th ed., p. 465, ASM Press, Washington, DC, 1995.
52. Dalsgaard., A. et al., Clinical manifestations and molecular epidemiology of five cases of diarrhoea in children associated with *Vibrio metschnikovii* in Arequipa, Peru, *J. Med, Micriobiol.*, 45, 494, 1996.
53. Hlady, W.G., and Klontz, K.C., The epidemiology of *Vibrio* infections in Florida, 1981–1993, *J. Infect. Dis.*, 173, 1176, 1996.
54. Rank, E.L., Smith, I.B., and Langer, M., Bacteremia caused by *Vibrio hollisae*, *J. Clin. Microbiol.*, 26, 2, 1988.
55. Vollberg, C.M., and Herrara, J.L., *Vibrio vulnificus* infection: an important cause of septicemia in patients with cirrhosis, *South Med. J.*, 90, 1040, 1997.
56. Zide, N., Davis, J., and Ehrenkranz, N.J., Fulminating *Vibrio parahaemolyticus* septicemia, *Arch. Intern. Med.*, 133, 479, 1974.
57. Marano, N.N. et al. Stool culturing practices for *Vibrio* species at clinical laboratories in Gulf Coast states, *J. Clin. Microbiol.*, 38, 2267, 2000.
58. Kobayashi, T. et al., A new selective isolation medium for vibrio group on a modified Nakanishi's medium (TCBS agar medium), *Jpn. J. Bacteriol.*, 18, 387, 1963.
59. Atlas, R.M., *Handbook of Microbiological Media*, p. 529, CRC Press, Boca Raton, FL, 1993.
60. Lotz, M.J., Tamplin, M.L., and Rodrick, G.E., Thiosulphatecitrate-bile salts-sucrose agar and its selectivity for clinical and marine vibrio organisms, *Ann. Clin. Lab.*, 13, 45, 1983.
61. Tamplin, M.L. et al., Enzyme immuno assay for identification of *Vibrio vulnificus* in seawater, sediment, and oysters, *Appl. Environ. Microbiol.* 57, 1235, 1991.
62. Cerda-Cuellar, M., Jofre, J., and Blanch, A.R., A selective medium and a specific probe for detection of *Vibrio vulnificus*, *Appl. Environ. Microbiol.*, 66, 855, 2000.
63. Cerda-Cuellar, M. et al., Comparison of selective media for the detection of *Vibrio vulnificus* in environmental samples, *J. Appl. Microbiol.*, 91, 322, 2001.
64. Wright, A.C. et al., Rapid identification of *Vibrio vulnificus* on nonselective media with an alkaline phosphatase-labeled oligonucleotide probe, *Appl. Environ. Microbiol.*, 59, 541, 1993.
65. Elliot, E.L. et al., *Vibrio cholerae*, *V. parahaemolyticus*, *V. vulnificus* and other *Vibrio* spp, p. 9.01–9.27. In *FDA Bacteriological Analytical Manual*, 8th ed., AOAC International, Gaithersburg, MD, 1995.
66. Miceli, G.A. et al., Direct plating procedure for enumerating *Vibrio vulnificus* in oysters (*Crassostrea virginica*), *Appl. Environ. Microbiol.*, 59, 3519, 1993.
67. McCarthy, S.A. et al., Evaluation of alkaline phosphatase-and digoxigenin-labeled probes for detection of the thermolabile hemolysin (*tlh*) gene of *Vibrio parahaemolyticus*, *Lett. Appl. Microbiol.*, 28, 66, 1999.
68. DePaola, A., and Hwang, G.C., Effect of dilution, incubation time, and temperature of enrichment on cultural and PCR detection of *Vibrio cholerae* obtained from the oyster *Crassostrea virginica*, *Mol. Cell. Probes*, 9, 75, 1995.
69. DePaola, A., Motes, M.L., and McPhearson, R.M., Comparison of the APHA and elevated temperature enrichment methods for recovery of *Vibrio cholerae* from oysters: Collaborative study, *J. Assoc. Off. Anal. Chem.*, 71, 584, 1988.
70. Gooch, J.A. et al., Evaluation of two direct plating methods using nonradioactive probes for enumeration of *Vibrio parahaemolyticus* in oysters, *Appl. Environ. Microbiol.*, 67, 721, 2001.
71. Brasher, C.W. et al., Detection of microbial pathogens in shellfish with multiplex PCR, *Curr. Microbiol.*, 37, 101, 1998.
72. Rivera, I.N. et al., Genotypes associated with virulence in environmental isolates of *Vibrio cholerae*, *Appl. Environ. Microbiol.*, 67, 2421, 2001.
73. Rivera, I.N. et al., Method of DNA extraction and application of multiplex polymerase chain reaction to detect toxigenic *Vibrio cholerae* O1 and O139 from aquatic ecosystems, *Environ. Microbiol.*, 5, 599, 2003.
74. Lyon. W.J., TaqMan PCR for detection of *Vibrio cholerae* O1, O139, non-O1, and non-O139 in pure cultures, raw oysters, and synthetic seawater, *Appl. Environ. Microbiol.*, 67, 4685, 2001.
75. Fedio, W. et al., Rapid detection of the *Vibrio cholerae* ctx gene in food enrichments using real-time polymerase chain reaction, *J. AOAC Int.*, 90, 1278, 2007.
76. Blackstone, G.M. et al., Use of a real time PCR assay for detection of the *ctxA* gene of *Vibrio cholerae* in an environmental survey of Mobile Bay, *J. Microbiol. Methods*, 68, 254, 2007.
77. Myers, M.L., Panicker, G. and Bej, A.K., PCR detection of a newly emerged pandemic *Vibrio parahaemolyticus* O3:K6 pathogen in pure cultures and seeded waters from the Gulf of Mexico, *Appl. Environ. Microbiol.*, 69, 2194, 2003.
78. Rizvi, A.V. et al., Detection of pandemic *Vibrio parahaemolyticus* O3:K6 serovar in Gulf of Mexico water and shellfish using real-time PCR with Taqman fluorescent probes, *FEMS Microbiol. Lett.*, 262, 185, 2006.
79. Ocurra, M. et al., Genotypic analyses of *Vibrio parahaemolyticus* and development of a pandemic group-specific multiplex PCR assay, *J. Clin. Microbiol.*, 41, 4676, 2003.
80. Ward, L.N., and Bej, A.K., Detection of *Vibrio parahaemolyticus* in shellfish by use of multiplexed real-time PCR with TaqMan fluorescent probes, *Appl. Environ. Microbiol.*, 72, 2031, 2006.
81. Nordstrom, J.L. et al., Development of a multiplex real-time PCR assay with an internal amplification control for the detection of total and pathogenic *Vibrio parahaemolyticus* bacteria in oysters, *Appl. Environ. Microbiol.*, 73, 5840, 2007.
82. Hill, W.E. et al., Polymerase chain reaction identification of *Vibrio vulnificus* in artificially contaminated oysters, *Appl. Environ. Microbiol.*, 57, 707, 1991.
83. Brauns, L.A., Hudson, J.D., and Oliver, J.D., Use of the polymerase chain reaction in detection of culturable and nonculturable *Vibrio vulnificus* cells, *Appl. Environ. Microbiol.*, 57, 2651, 1991.
84. Arias, C.R., Garay, E., and Aznar, R., Nested PCR method for rapid and sensitive detection of *Vibrio vulnificus* in fish, sediments, and water, *Appl. Environ. Microbiol.*, 61, 3476, 1995.

85. Wang, H., Fadl, A.A., and Khan, M.I., Multiplex PCR for avian pathogenic mycoplasmas, *Mol. Cell. Probes*, 11, 3, 1997.
86. Lee, C.Y., Panicker, G., and Bej, A.K., Detection of pathogenic bacteria in shellfish using multiplex PCR followed by CovaLink NH microwell plate sandwich hybridization, *J. Microbiol. Methods*, 53, 199, 2003.
87. Higuchi, R. et al., Simultaneous amplification and detection of specific DNA sequences, *Biotechnology (NY)*, 10, 413, 1992.
88. Levin, R.E., The application of real-time PCR to food and agricultural systems: A review, *Food Biotechnol.*, 18, 97, 2004.
89. Jaykus, L., Challenges to developing real-time methods to detect pathogens in foods, *ASM News*, 69, 341, 2003.
90. Panicker, G. et al., Detection of pathogenic *Vibrio* spp. in shellfish by using multiplex PCR and DNA microarrays, *Appl. Environ. Microbiol.*, 70, 7436, 2004.
91. Campbell, M.S., and Wright, A.C., Real-time PCR analysis of *Vibrio vulnificus* in oysters, *Appl. Environ. Microbiol.*, 69, 7137, 2003.
92. Panicker, G., and Bej, A.K., Real-time PCR detection of *Vibrio vulnificus* in oysters: Comparison of oligonucleotide primers and probes targeting *vvhA*, *Appl. Environ. Microbiol.*, 71, 5702, 2005.
93. Oliver, J.D., The viable but nonculturable state in bacteria, *J. Microbiol.*, 43, 93, 2005.
94. Whitesides, M.D., and Oliver, J.D., Resuscitation of *Vibrio vulnificus* from the viable but nonculturable state, *Appl. Environ. Microbiol.*, 63, 1002, 1997.
95. Fischer-Le Saux, M. et al., Detection of cytotoxin-hemolysin mRNA in nonculturable populations of environmental and clinical *Vibrio vulnificus* strains in artificial seawater, *Appl. Environ. Microbiol.*, 68, 5641, 2002.
96. Coutard, F. et al., Real-time reverse transcription-PCR for transcriptional expression analysis of virulence and housekeeping genes in viable but nonculturable *Vibrio parahaemolyticus* after recovery of culturability, *Appl. Environ. Microbiol.*, 73, 5183, 2007.
97. Coutard, F. et al., mRNA detection by reverse transcription-PCR for monitoring viability and potential virulence in a pathogenic strain of *Vibrio parahaemolyticus* in viable but nonculturable state, *J. Appl. Microbiol.*, 98, 951, 2005.
98. Sun, F. et al., Characterization and virulence retention of viable but nonculturable *Vibrio harveyi*, *FEMS Microbiol. Ecol.*, 64, 37, 2008.
99. Baffone, W. et al., Retention of virulence in viable but nonculturable halophilic *Vibrio* spp., *Int. J. Food Microbiol.*, 89, 31, 2003.
100. González-Escalona, N. et al., Quantitative reverse transcription polymerase chain reaction analysis of *Vibrio cholerae* cells entering the viable but non-culturable state and starvation in response to cold shock, *Environ. Microbiol.*, 8, 658, 2006.
101. Kaysner, C.A., and DePaola, A., Jr., *Vibrio cholerae*, *V. parahaemolyticus*, *V. vulnificus*, and other *Vibrio* spp. In *Bacteriological Analytical Manual*, 8th ed., Revision A, 2004.
102. Ausubel, F.M. et al. (Editors), *Current Protocols in Molecular Biology*, pp. 2.10–2.11, John Wiley & Sons Inc., New York, 1987.
103. Notomi, T. et al., Loop-mediated isothermal amplification of DNA. *Nucleic Acids Res.*, 28, e63, 2000.
104. Ding, W.C., Establishment of loop-mediated isothermal amplification for detection of *Vibrio alginolyticus*. *Fen Zi Xi Bao Sheng Wu Xue Bao*, 42, 70, 2009.
105. Fall, J. et al., Establishment of loop-mediated isothermal amplification method (LAMP) for the detection of *Vibrio nigripulchritudo* in shrimp, *FEMS Microbiol. Lett.*, 288, 171, 2008.
106. Han, F., and Ge, B., Evaluation of a loop-mediated isothermal amplification assay for detecting *Vibrio vulnificus* in raw oysters, *Foodborne Pathog. Dis.*, 5, 311, 2008.
107. Nemoto, J. et al., Rapid detection of *Vibrio vulnificus* using a loop-mediated isothermal amplification method, *Kansenshogaku Zasshi*, 82, 407, 2008.
108. Nemoto, J. et al., Rapid and specific detection of the thermostable direct hemolysin gene in *Vibrio parahaemolyticus* by loop-mediated isothermal amplification, *J. Food Prot.*, 72, 748, 2009.
109. Okada, K. et al., A rapid, simple, and sensitive loop-mediated isothermal amplification method to detect toxigenic *Vibrio cholerae* in rectal swab samples, *Diagn. Microbiol. Infect. Dis.*, 66, 135, 2010.
110. Srisuk, C. et al., Rapid and sensitive detection of *Vibrio cholerae* by loop-mediated isothermal amplification targeted to the gene of outer membrane protein ompW, *Lett. Appl. Microbiol.*, 50, 36, 2010.
111. Yamazaki, W. et al., Development of a loop-mediated isothermal amplification assay for sensitive and rapid detection of *Vibrio parahaemolyticus*, *BMC Microbiol.*, 30, 163, 2008.
112. Yamazaki, W. et al., Sensitive and rapid detection of cholera toxin-producing *Vibrio cholerae* using a loop-mediated isothermal amplification, *BMC Microbiol.*, 8, 94, 2008.
113. Yamazaki, W. et al., Development of a loop-mediated isothermal amplification assay for sensitive and rapid detection of the *tdh* and *trh* genes in *Vibrio parahaemolyticus* and related Vibrio species, *Appl. Environ. Microbiol.*, 76, 820, 2010.
114. Ren, C.H. et al., Sensitive and rapid identification of *Vibrio vulnificus* by loop-mediated isothermal amplification, *Microbiol Res.*, 164, 514, 2009.
115. Luan, X. et al., Rapid Quantitative Detection of *Vibrio parahaemolyticus* in Seafood by MPN-PCR, *Curr. Microbiol.*, 57, 218, 2008.
116. Wright, A.C. et al., Evaluation of postharvest-processed oysters by using PCR-based most-probable-number enumeration of *Vibrio vulnificus* bacteria, *Appl. Environ. Microbiol.*, 73, 7477, 2007.
117. Mertens, J. et al., Label-free detection of DNA hybridization based on hydration-induced tension in nucleic acid films, *Nat. Nanotechnol.*, 3, 301, 2008.
118. Kaittanis, C., Santra, S., and Perez, J.M., Emerging nanotechnology-based strategies for the identification of microbial pathogenesis, *Adv. Drug. Deliv. Rev.*, 62, 408, 2010.
119. So, H.M. et al., Detection and titer estimation of *Escherichia coli* using aptamer-functionalized single-walled carbon-nanotube field-effect transistors. *Small*, 4, 197, 2008.

92 Yersinia

Mikael Skurnik, Peter Rådström, Rickard Knutsson, Bo Segerman, Saija Hallanvuo, Susanne Thisted Lambertz, Hannu Korkeala, and Maria Fredriksson-Ahomaa

CONTENTS

92.1 INTRODUCTION

92.1.1 TAXONOMICAL ASPECTS OF GENUS *YERSINIA*

The genus *Yersinia*—a member of family Enterobacteriaceae of the γ-purple bacteria[1]—now comprises 14 species, three of which contain human pathogenic strains: *Y. pestis*, *Y. pseudotuberculosis*, and *Y. enterocolitica*. *Y. pestis* is the causative agent of the feared bubonic plague that spreads from infected rodents to humans via a flea vector.[2] Pathogenic strains of *Y. pseudotuberculosis* and *Y. enterocolitica* cause yersiniosis, usually a mild diarrheal disease, the infection of which usually takes place by ingestion of contaminated foodstuffs. *Y. pseudotuberculosis* and *Y. enterocolitica* and related species (*Y. aldovae*, *Y. aleksiciae*, *Y. bercovieri*, *Y. frederiksenii*, *Y. intermedia*, *Y. kristensenii*, *Y. mollaretii*, *Y. rohdei*, *Y. massiliensis*, and *Y. similis*, the two latter of which were only recently described[3,4]) have several serovariants, whereas *Y. pestis* has three biovariants but is serologically of a single type. Five serotypes have been described for *Y. ruckeri* that causes the enteric red mouth disease of fish.[5]

During the last 5–10 years, genomic sequencing projects of strains representing 11 *Yersinia* species have been initiated and the available genomic characteristics are outlined in Table 92.1. (For updated information, see the list of complete genomes and genomes in progress at NCBI Web site, http://www.ncbi.nlm.nih.gov/genomes/lproks.cgi.) The whole-genome sequencing efforts have generated new knowledge about the evolution, the emergence of virulence in the genus *Yersinia*, and a broader understanding of the taxonomical aspects. The size of the genomes varies between 3.7 and 5.4 Mb and the GC content is in the range of 47%–49%. This far, however, only the genome sequences of *Y. pestis*, *Y. pseudotuberculosis*, and *Y. enterocolitica* have been completed and annotated. Table 92.1 demonstrates that a common feature for these pathogenic *Yersinia* is the presence of a variety of plasmids, transposons, IS-elements, and loci exhibiting foreign DNA. The genome projects of 4 *Y. pseudotuberculosis* and 20 *Y. pestis* strains have been initiated thus far and for historical and biosafety reasons the genome of *Y. pestis* has been the most thoroughly studied genome. The speciation of

TABLE 92.1
Major Characteristics of Draft and Complete Genome Sequences of *Yersinia* Species

Yersinia strain	Total Size (Mbp)	GC (%)	Protein Coding Genes	tRNA and rRNA Genes	Status of Genome	No. of Plasmids	Chromosomal Genes Related to: Plasmid Functions	Chromosomal Genes Related to: Phage	Chromosomal Genes Related to: Transposon or IS	*Y. pestis* Type (biovar)
Y. aldovae ATCC 35236	4.30	47.0	3967	66	Draft					
Y. bercovieri ATCC 43970	4.35	48.0	3880	67	Draft					
Y. enterocolitica 8081	4.68	47.2	4402	96	Complete	1[a]	2	62	167	
Y. fredriksenii ATCC 33641	4.88	46.0	4358	69	Draft					
Y. intermedia ATCC 29909	4.71	47.0	4122	73	Draft					
Y. kristensenii ATCC 33638	5.04	47.0	4542	69	Draft					
Y. mollaretii ATCC 43969	4.55	49.0	4002	62	Draft					
Y. pestis CO92	4.82	47.6	5034	86	Complete	3[b]	2	62	167	Epidemic (orientalis)
Y. pestis Nepal 516	4.65	47.6	4096	93	Complete	2[c]	2	35	177	Epidemic (antiqua)
Y. pestis Antiqua	4.88	47.7	4635	89	Complete	3[d]	8	28	310	Epidemic (antiqua)
Y. pestis Pestoides F	4.73	47.7	4158	87	Complete	2[e]	16	45	171	Endemic (antiqua)
Y. pestis 91001	5.42	47.7	4142	93	Complete	4[f]	3	33	149	Endemic (microtus)
Y. pestis KIM	4.95	47.7	4457	94	Complete	5[g]	6	88	172	Epidemic (mediaevalis)
Y. pestis Angola	4.69	47.6	4351	92	Complete	2[h]	11	41	276	Endemic (antiqua)
Y. pestis CA88–45	4.66	47.0	4023	112	Draft	2				Epidemic (orientalis)
Y. pestis FV-1	4.47	47.0	4815	6	Draft					Epidemic (orientalis)
Y. pestis MG05–1020	4.98	47.0	4396	104	Draft					Epidemic (orientalis)
Y. pestis UG05–0454	4.83	47.0	4188	89	Draft					Epidemic (antiqua)
Y. pestis B42003004	4.84	47.0	4256	84	Draft					Epidemic (antiqua)
Y. pestis K1973002	4.72	47.0	4141	102	Draft					Epidemic (mediaevalis)
Y. pestis E1979001	4.84	47.0	4225	81	Draft					Epidemic (antiqua)
Y. pestis F1991016	4.90	47.0	4254	82	Draft					Epidemic (orientalis)
Y. pestis IP275	5.31	47.0	4758	71	Draft					Epidemic (orientalis)
Y. pestis India 195	4.72	47.0	4679	106	Draft					Epidemic (orientalis)

Y. pestis PEXU2	4.77	47.0	4665	108	Draft					Epidemic (orientalis)
Y. pestis Pestoides A	4.75	47.0	4731	106	Draft					Endemic
Y. pseudotuberculosis IP32953	4.84	47.6	4116	106	Complete	2[i]	3	92	67	
Y. pseudotuberculosis IP 31758	4.93	47.2	4363	108	Complete	2[j]	42	35	42	
Y. pseudotuberculosis PB1/+	4.77	47.0	4237	97	Complete	1[k]				
Y. pseudotuberculosis YPIII	4.70	47.0	4192	98	Complete					
Y. rohdei ATCC 43380	4.32	46.0	3853	59	Draft					
Y. ruckeri ATCC 43969	3.74	47.0	3365	53	Draft					

[a] pPYV.
[b] pCD1, pMT1, pPCP1.
[c] pMT, pPCP.
[d] pMT, pPCP, pCD.
[e] pCD, pMT.
[f] pCD1, pCRY, pMT1, pPCP1
[g] pCD, pCD, pMT1, pMT1, pPCP1.
[h] pCD, pPCP.
[i] pYptb.
[j] 59 Kb plasmid, 153 Kb plasmid.
[k] pYPTS01.

the highly monomorphic *Y. pestis* has been a rapid evolutionary event and this is reflected by the large-scale genetic flux indicated by the presence of pseudogenes, insertion sequences, chromosomal rearrangements, and gene loss.[6] The extreme virulence of *Y. pestis* compared to the closely related *Y. pseudotuberculosis* is believed to be caused by the presence in *Y. pestis* of multiple loci of foreign DNA, the presence of ca. 300 inactivated genes (i.e., pseudogenes), and the two extra plasmids essential for pathogenesis; a large plasmid designated pMT or PFra and a small plasmid designated pPst or pPCP1 (see below).[6]

In Figure 92.1, the *Yersinia* strains for which draft and complete genomic sequences are available are used to illustrate the phylogenetic relationships between *Yersinia* species and in particular between individual *Y. pestis* strains. The close relationship between *Y. pestis* and *Y. pseudotuberculosis* can be seen in Figure 92.1a; *Y. pestis* is considered a recently (within the last 1500–40,000 years) diverged clone of *Y. pseudotuberculosis* serotype O:1b.[7–10] Figure 92.1b illustrates the relationships within the well conserved *Y. pestis* group.

Table 92.1 and Figure 92.1b demonstrate that the genome of *Y. pestis* is highly dynamic and 11 major groups of *Y. pestis* are defined globally.[7] The phenotypic and genotypic methods that have been used for classification and typing of *Y. pestis* strains were recently reviewed by Lindler[11] and he concludes, mainly based on recent genotyping efforts, that *Y. pestis* strains can be classified into two basic types: (i) the *epidemic strains* that are able to cause human plague transmitted by the flea vector, and (ii) the *endemic strains* are virulent for rodents but do not generally cause human infections by normal route of infection.[11]

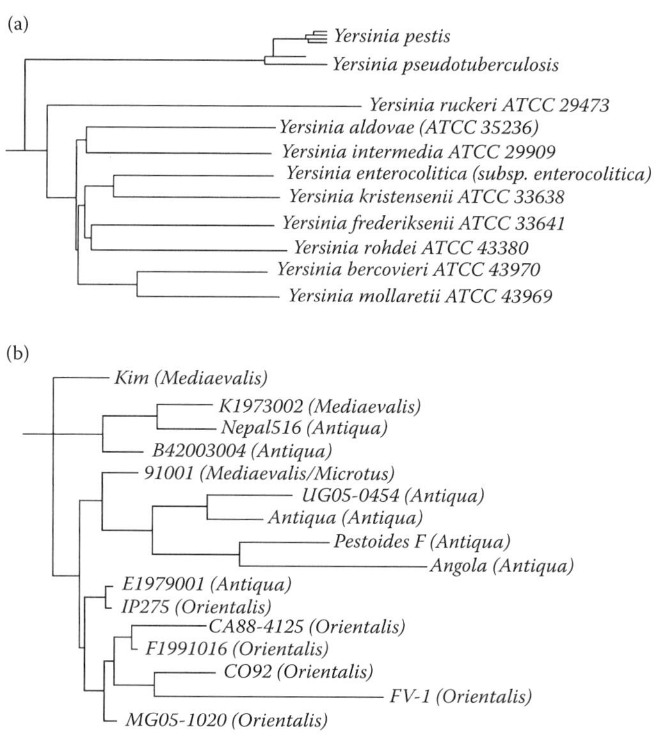

FIGURE 92.1 Whole-genome comparison of *Yersinia* strains whose genome sizes are approximately 4.5–4.8 Mb. The dendrograms are based on a distance matrix created from average blast score values of 5.4 million cross-blast comparisons spanning the entire genomes. (a) Relationships at genus level. (b) Relationships between *Y. pestis* genomes. Biovariant classifications are indicated. The biovariants might not be completely monophyletic, which has also been observed with multilocus variable number of tandem repeat analysis (MLVA).[7]

92.1.2 *Y. pestis*—Causative Agent of Plague

92.1.2.1 Biology and Pathogenesis

Among the all-pathogenic *Yersiniae*, *Y. pestis* causes the most fatal and devastating disease—bubonic plague, known during the Middle Ages as the Black Death. The number of people killed by this bacterium over the span of world history approaches 200 million. It has been responsible for three human pandemics—the Plague of Justinian in the fifth to seventh centuries; the Black Death in the thirteenth to fifteenth centuries, which is claimed to have killed one-third of the European population; and modern plague since 1870s. Although plague is no longer so serious threat it still occurs occasionally in developing countries.[2,12]

Y. pestis is a vector-borne pathogen, and has a complex life cycle involving a mammalian reservoir (primarily rodents) and a flea vector. The flea acquires the bacteria by taking a blood meal from an infected mammal. In the flea's gut, the bacterium replicates and forms a clumped mass, which blocks the flea intestine from receiving further blood meals. An infected insect is, in effect, starving and actively looking for a new host. When an infected flea bites a human, it regurgitates fragments of a mass containing microbial pathogen. The bacteria disseminate rapidly to the lymphatic system and locate in the nearby lymph nodes, where the massive inflammatory response causes painful swellings, the factual buboes that are the hallmark symptoms of bubonic plague. At this point, *Y. pestis* may enter the bloodstream and travel to the lungs, where it can cause pneumonic plague. Pneumonic plague can be transmitted by aerosols from one human to another. Ninety percent of untreated victims infected by this route die within few days due to the septic shock.[12] In addition, *Y. pestis* can be acquired orally; eating contaminated camel meat has caused plague cases.[13,14]

Wild type *Y. pestis* strains carry three virulence factor encoding plasmids: pPst, pFra, and pYV (also known as pPCP1, pMT1, and pCD1, respectively). While the latter, encoding the type III secretion system (T3SS) and the secreted effector toxins called Yops, is also present in the other pathogenic *Yersiniae*, the other two plasmids are unique to *Y. pestis*. The pPst plasmid carries the *pla* gene that encodes the plasminogen activator Pla, a very important protease indispensable for bacterial dissemination after subcutaneous injection into mammalian tissues.[15] Other *Y. pestis* unique factors, like capsular protein F1 and Ymt (*Yersinia* murine toxin), are encoded by the genes present on pFra plasmid. They appear to prevent phagocytosis and play a role in the colonization of the flea gut.[16]

92.1.2.2 Epidemiology

Serologically, *Y. pestis* strains are very homogenous because their lipopolysaccharide (LPS) is rough (i.e., lacking O-antigen).[17] *Y. pestis* has three biovariants—*mediaevalis*, *orientalis*, and *antiqua*—based on differences in fermentation of glycerol (+, −, +) and ability to reduce nitrate to nitrite (−, +, +), respectively.[18,19] Variety *orientalis* is endemic in India, Southwest Asia, South America, and western North America; *antiqua* is present in Southeast and northern Asia (China and Russia) and Africa; and *mediaevalis* is limited to Turkey and Iran.[19] For a recent discussion of *Y. pestis* typing see the review of Lindler.[11] During the relatively short time, *Y. pestis* evolved from the less-virulent *Y. pseudotuberculosis*, and the pre-*Y. pestis* bacterium experienced very significant genetic changes that collectively have resulted in the huge increase in virulence when compared to *Y. pseudotuberculosis*.

92.1.2.3 Laboratory Diagnosis

The samples from where *Y. pestis* are conventionally cultured are sputum or bronchial washes, blood, aspirates, or biopsies, but also from environmental or other nonclinical samples. The CDC Plague homepage (http://www.cdc.gov/ncidod/dvbid/plague/index.htm) informs that suspicion for the presence of *Y. pestis* is raised if gram-negative and/or bipolar-staining coccobacilli are seen on a smear taken from affected tissues after Gram or other differential staining. A presumptive identification of *Y. pestis* can be made if the capsule F1 antigen can be detected from the specimen by immunofluorescence. *Y. pestis* grows slowly and usually visible colonies appear within 48 h. The potential *Y. pestis* colonies can be identified as *Y. pestis* by biochemical tests and the use of specific antisera and bacteriophages. At present time, PCR and other sequence-based identification methods are used to confirm the identification (see below).

Due to increased funding to research biothreat agents, recent years have evidenced a burst of molecular methods designed for the detection of *Y. pestis*. The methods range from simple single-PCR assays via multiplex real-time PCR assays to microarray based protocols and also include other detection technologies (Table 92.2). Specific attention has been placed on the danger that the very close relative *Y. pseudotuberculosis* poses on the detection method as a false positive.

Handling of *Y. pestis* strains requires Biosafety Level 3 (BSL-3) capabilities for appropriate safety. The reason is that the most important route of *Y. pestis* exposure of laboratory workers is by inhalation. BSL-3 is used for work with indigenous agents when the potential for infection by aerosol is real and the disease may have serious or lethal consequences.[55] Special engineering design criteria for BSL-3 laboratories for laboratory work on *Y. pestis* involve the access zone, sealed penetrations, and directional airflow to aid in protecting the employee from respiratory exposure. In addition, appropriate cabinets, dress codes, risk assessments, and deactivation and emergency procedures are needed. For deactivation of *Y. pestis*, 10% sodium hypochlorite can be used before autoclaving, and formaldehyde fumigation can be used for cabinet and room decontamination. Consequently, the isolation and molecular detection methods of *Y. pestis* have to be implemented in order to fulfill the biosafety requirements.

92.1.3 Foodborne Yersiniosis

Y. enterocolitica and *Y. pseudotuberculosis* are important enteropathogenic bacteria causing infections in humans, especially in industrialized countries but also in developing countries.[56,57] They are zoonotic bacterial pathogens, capable of being transmitted from animals to humans.[58] *Y. enterocolitica* and *Y. pseudotuberculosis* infections in humans are usually acquired indirectly by ingestion of contaminated food products or water. Transmission can also occur directly from animal to human or from person to person, usually via the fecal-oral route but in rare cases through blood transfusion.[59]

Yersiniosis is the third most common bacterial gastroenteritis after campylobacteriosis and salmonellosis in Europe.[56] Due to the inconsistency of surveillance system in different countries, the comparison of incidences of yersinia infections between different countries is only suggestive. In 2007, the overall incidence of yersiniosis in Europe was 2.8 confirmed cases per 100,000 inhabitants. The highest incidences were reported in Lithuania (16.8), Finland (9.1), Sweden (6.2), and Germany (6.1).[56] The number of infections is clearly lower in United States, where an overall incidence of 0.36 laboratory confirmed cases per 100,000 population was reported in 10 states in 2007.[60] However, the incidence has shown to be very high among black infants (141.9).[61] In many countries, the actual number of infections is much higher than reported. The incidence is highest among children under 5 years of age.[60,62] The majority of yersinosis cases are domestically acquired and no seasonal distribution has been reported in Europe.[62]

Y. pseudotuberculosis and *Y. enterocolitica* can cause a variety of intestinal and extraintestinal illnesses ranging from mild enteritis to septicemia. The severity of the infection depends on the age and immunity of the infected person, the virulence of the strain, and the infection dose. Thus, the incidence and severity of infection is much higher in some groups of a population, such as children, the elderly, and immunocompromised people. Typical symptoms for enteric yersiniosis are gastroenteritis with diarrhea and fever.[58] Sometimes the diarrhea may be bloody, and the fever may be high, especially in infants.[63] Diarrhea is a more dominant symptom in *Y. enterocolitica* infections while fever and right lower quadrant abdominal pain due to mesenteric lymphadenitis, which is often clinically indistinguishable from acute appendicitis, are often the most frequent symptoms of *Y. pseudotuberculosis* infections.[64,65] The minimal infection dose for yersinosis in humans is unknown. Most likely many strain-dependent and host factors have an influence on the infection dose and the clinical outcome. The incubation period for gastrointestinal symptoms is about one week ranging from 1 to 11 days

TABLE 92.2
Molecular Methods for Detection and Identification of *Y. pestis*

Method	Target Chromosome or *Plasmid*	Samples, Detection Method, and Comments	Reference
Conventional single PCR assays	*pla*	Wild fleas; gel detection	20
	16S rRNA, *cafl*	Soil samples; gel detection and sequencing	21
	cafl, pla, lcrV, 16S rRNA	Suitable for identification of *Yersinia* species; gel detection	22
	41.7 kb insertion site	The target is specific for *Y. pestis;* gel detection	23
	cafl	Madagaskar field study; gel detection	24
	16S and 23S rRNA	Gel detection and a fluorescent probe for in situ hybridization	25
	pla	6th century teeth; gel detection and sequencing	26
	28 signature genes	Genes identified by microarray; PCR analysis by gel detection	27
	wzx-wbyIJ	Specific identification of *Y. pestis;* gel detection	28
Real-time PCR assays	*pla*	Spiked fleas, blood oropharyngeal swabs	29
	gyrA	Ciprofloxacin resistance	30,31
	pla	Sputum	32
	Chromosomal *YP48* gene	Field-deployable RAPID ("Ruggedized" advanced pathogen identification device)	33
	Not given	Differentiation between *Y. pestis* and *Y. pseudotuberculosis.* Detection by TaqMan minor groove binder probe	34
	pla, pim, caf1	Detection by dual-labeled fluorogenic probe	35
	pst	Detection by TaqMan fluorogenic probe	36
	pla	Simultaneous detection of other biothreat agents; detection by TaqMan minor groove binder probe	37
	16S rRNA	Detection by TaqMan probe	38
Multiplex PCR assays	*cafl, pla, lcrV, irp2*	Samples from 1986 Paraiba outbreak; gel detection	39
	cafl, pla, lcrV, ymt, 16S rRNA	Real-time PCR; detection using fluorescent probes	40
	cafl, pla, yopM, inv	Gel detection	41
	yihN, cafl, pla, lcrV	Detection by TaqMan probes	42
	Not given	Rapid simultaneos detection of select category A bacterial agents from blood; detection by SYBR Green and Tm	43
	entF3, cafl, pla, pCD1-region	Real-time multiplex PCR; detection by TaqMan probes	44
	pla, ypo2088, wzz	Real-time multiplex PCR; dual-labeled fluorogenic probes; plasmid and chromosomal targets. *wzz* detects both *Y. pestis* and *Y. pseudotuberculosis;* internal ampification control	45
Microarray-based assays	Multiple targets	Multiplex PCR microarray; simultaneous detection of bioterror agents in blood	46
	Multiple targets	Multiplexed high-density DNA array	47
	Not given	Multiplexed PCR-coupled liquid bead array for simultaneous detection of many biothreat agents	48
Other methods	*F1-antigen*	Rapid diagnostic test (RDT) to study a medieval burial place	49
	F1-antigen	Fiber optic biosensor for detection from blood, etc.	50
	F1-antigen	Magnetic-beads based magnetic biosensor for detection from serum and blood samples	51
	F1-antigen	Immunochromatography, Plague BioThreat Alert test strips, ABICAP columns, ELISA, flow cytometry, and IF-microscopy	52
	Not given	Immunoassay plus real-time PCR to analyze aerosols	53,54

with symptoms typically persisting from 5 to 14 days but occasionally lasting for several months.[64,66]

Postinfectious complications like joint pain (reactive arthritis, ReA) and skin rash (erythema nodosum) are relatively common especially among adults.[65–67] The incidence of ReA after a *Y. pseudotuberculosis* O:3 outbreak was shown to be high. However, the incidence of ReA appears to be lower in *Y. pseudotuberculosis* serotype O:1 outbreaks.[67] ReA is considered less common in children than in adults and there is an association between ReA and HLA-B27.[67] In East Asia (Japan, Korea, and the Russian Far East), the clinical picture has sometimes been more severe, including erythematous skin rash, skin desquamation, exanthema (scarlatinoid rash), and toxic shock syndrome.[68] This disease has been termed Far East scarlet-like fever or Izumi fever in Japan. It has also been linked to Kawasaki syndrome, an acute multisystem vasculitis of young children.

In most cases, the infection is self-limiting and no antimicrobial therapy is needed. However, in severe cases, like systemic infection and bacteremia, antimicrobials may be useful. Treatment of *Yersinia* enteritis should also be considered for patients who are immunocompromised and patients with iron overload.[69] In contrast to *Y. pseudotuberculosis*, which lacks β-lactamase activity and is hence susceptible to penicillins, *Y. enterocolitica* does possess β-lactamase activity and is therefore generally resistant to penicillins. However, a number of antimicrobial agents are active in vitro against *Yersinia* strains isolated from human and nonhuman sources. These include aminoglycosides, third generation cephalosporins, co-trimoxazole, tetracyclines, chloramphenicol, and fluoroquinolones.[69,70]

92.1.4 *Y. PSEUDOTUBERCULOSIS*

92.1.4.1 Biology and Pathogenesis

Y. pseudotuberculosis is an enteropathogen widely spread in the environment (soil, water, and vegetables). It was first described in 1884.[71] This bacterium is a primary pathogen of wild and domestic animals in all continents.[68] In humans, *Y. pseudotuberculosis* infections are not frequent, although outbreaks associated with the consumption of water or foodstuffs contaminated with animal feces have been reported.

Major virulence factors associated with *Y. pseudotuberculosis* are the pYV-encoded Yop proteins that enter host mammalian cells through T3SS.[72] Other virulence factors are adhesion molecules, invasin and YadA[73] and the superantigenic toxin designated YPM (*Y. pseudotuberculosis*-derived mitogen) which mediates activation of the immune system.[74] In addition, the LPS O-polysaccharide[75,76] and type IV pili[77] contribute to pathogenicity.

The current *Y. pseudotuberculosis* antigenic scheme lists 21 O-serotypes.[78,79] The division into different groups is based on different O-factors that reflect the different antigenic sugar residues present in the O-polysaccharide (O-antigen).[79] Some of the O-factors (specifically those carrying 3,6-dideoxyhexoses) are shared with the O-antigenic structures of *Salmonella* and *E. coli* and may cause cross-reactivity.[80,81]

92.1.4.2 Epidemiology

Y. pseudotuberculosis infections among humans have been reported on all continents, but the incidence appears to be less frequent than for *Y. enterocolitica*.[56,68] In France, a sudden increase of human *Y. pseudotuberculosis* infections was reported in 2004–2005.[82] The surveillance data of *Y. pseudotuberculosis* is very limited throughout the world. Many countries do not report *Y. pseudotuberculosis* separately from *Yersinia* spp.[60,62] Most of the infections are sporadic, but in recent years several outbreaks have been reported in Finland, Russia, and Japan (Table 92.3). The outbreaks in Finland have involved large-scale kitchens and mainly affected children in day-care centers and schools. One milk-related outbreak has been reported in households in Canada.[83] The majority of the outbreaks have been associated with contaminated vegetables (Table 92.3).

TABLE 92.3
Outbreaks of Human *Y. pseudotuberculosis* Infection

Country	Year	Serotype	No. of Cases	Source	Reference
Canada	1998	O:1	74	Milk	83
Finland	1997	O:3	35	Unknown	84
	1998	O:3	47	Iceberg lettuce	85
	2001	O:1, O:3	125	Iceberg lettuce	86
	2003	O:1	111	Carrots	66
	2004	O:1	125	Carrots	87
	2006	O:1	>400	Carrots	88
	2008	O:1	>30	Carrot	89
Japan	1982	O:4	140	Water	90
	1984	O:5	39	Barbecue	90
	1984	O:3	63	Unknown	90
	1984	O:4	11	Water	90
	1985	O:4	68	Unknown	90
	1986	O:4	549	Unknown	90
	1991	O:5	732	Unknown	91
Russia	2005	Not reported	33	Vegetables	ProMED (20050202.0359)[a]
	2005	Not reported	69	Vegetables	ProMED (20050427.1169)
	2005	Not reported	13	Vegetables	ProMED (20051031.3178)
	2005	Not reported	24	Vegetables	ProMED (20051216.3617)
	2007	Not reported	11	Unknown	ProMED (20070501.1412)
	2007	Not reported	16	Unknown	ProMED (20070803.2511)
	2007	Not reported	121	Vegetables	ProMED (20071001.3240)
	2008	Not reported	141	Unknown	ProMED (20080718.2184)

[a] ProMED (www.promedmail.org) archive number.

The epidemiology of *Y. pseudotuberculosis* is complex and still mostly unclear.[92] Most *Y. pseudotuberculosis* strains are pathogenic to humans. The majority of *Y. pseudotuberculosis* strains isolated in Europe from humans belong to serotypes O:1a, O:1b, or O:3. In Japan, serotypes O:3, O:4, and O:5 are the most prevalent.[68,90,93] The same serotypes have been isolated from a wide variety of animals and therefore the animals are suspected as reservoirs for *Y. pseudotuberculosis* and, hence, sources of human infections.

Y. pseudotuberculosis has been isolated on all continents from several animal species including farm and pet animals, zoo animals, rodents, and birds.[94–96] However, the isolation rates have been low.[97] The principal reservoir hosts are believed to be wild animals, especially birds, rodents, and their predators.[98] *Y. pseudotuberculosis* O:3 has frequently been isolated from asymptomatic pigs at slaughter in Finland[99–101] and Eastern Europe.[96] Animals are mostly healthy carriers but they may become ill and excrete the bacteria due to stress like lactation, parasite infection, transportation, cold and humid weather, or starvation.[102] In a recent study, fatal yersiniosis due to *Y. pseudotuberculosis* O:3 in farmed deer in Japan was reported.[103] Monkeys are especially sensitive to *Y. pseudotuberculosis*, and many fatal cases and outbreaks in breeding monkeys have been reported in Japan.[104]

Although *Y. pseudotuberculosis* is considered a foodborne bacterium, it has very rarely been isolated from foods.[97,105] This pathogen has sporadically been isolated from meat products, especially from pork.[90,105] An association between human infections and pork has been suggested.[106] Fresh products are uncommon vehicles of *Y. enterocolitica*; however, *Y. pseudotuberculosis* has been isolated form vegetables in Russia[68] and iceberg lettuce and raw carrots implicated in foodborne outbreaks in Finland[85,87,107] (Table 92.3).

Y. pseudotuberculosis can survive for long periods in the environment, like in water, soil, and plants. The environment can be contaminated by feces of infected animals—mainly wild animals like deer, rodents, and birds.[106] *Y. pseudotuberculosis* has been isolated in Japan and Korea from fresh water, such as river, well, and mountain stream water at a considerably high rate.[90,108]

The transmission routes of *Y. pseudotuberculosis* are still unclear. In humans, the infection most probably occurs through contaminated food or water. Wild animals harboring *Y. pseudotuberculosis* may contaminate the environment (soil and water). Thus, unchlorinated drinking water from wells, springs, or streams contaminated with the feces of wild animals is considered an important transmission route in mountain areas in Japan.[106] In Korea, untreated mountain spring water could be linked to human *Y. pseudotuberculosis* infection.[109] In Finland, iceberg lettuce was probably contaminated with irrigation water.[85] In a large point-source outbreak in Finland caused by raw carrots contaminated by *Y. pseudotuberculosis*, the same strain was also found from the soil and from production line at the farm of origin.[66] In a recent outbreak, the same *Y. pseudotuberculosis* strain that was found from human patients was also isolated from the carrots and from a shrew intestine from the farm, indicating that shrews were the primary infection source.[107] Furthermore, family pets such as dogs and cats may serve as an infection source for human *Y. pseudotuberculosis* infections. An epidemiological study has shown evidence that *Y. pseudotuberculosis* was transmitted by the feces from an infected cat to humans in a familial outbreak in Japan.[106]

92.1.4.3 Laboratory Diagnosis

The isolation of *Y. pseudotuberculosis* from naturally contaminated samples is demanding since *Y. pseudotuberculosis* grows slowly and the minute colonies are easily overgrown by other bacteria. In addition, only limited information is available on reliable protocols of *Y. pseudotuberculosis* recovery from clinical, food, and environmental samples. Cold enrichment for 2 weeks in slightly selective phosphate-buffered saline broth supplemented with 1% mannitol and 0.15% bile salts and an alkali treatment after cold enrichment just before plating on solid media was shown to improve the isolation rate of *Y. pseudotuberculosis*.[99]

Selective agar plates are needed for the isolation of *Y. pseudotuberculosis* strains. However, the growth of *Y. pseudotuberculosis* is often poor and delayed as many selective agars are designed for *Y. enterocolitica*. The *Yersinia*-selective CIN-agar is also the most commonly used agar for *Y. pseudotuberculosis*[110]; however, some *Y. pseudotuberculosis* strains may not grow and the use of a second agar such as MacConkey agar is advised despite its low selectivity.[111] *Y. pseudotuberculosis* can be identified using some biochemical key tests or with commercial identification kits. Among *Y. pseudotuberculosis* strains there is little variation in biochemical reactions, except for the sugars melibiose and raffinose, and citrate, which can be used to divide the species into four biotypes.[78] The commercially available anti-O:1-O:6 antisera have been used for serotyping of *Y. pseudotuberculosis*; however, no antisera are commercially available to other serotypes. A PCR-based O-genotyping scheme that covers all the 21 O-serotypes was recently published.[28,112] The majority of *Y. pseudotuberculosis* strains are pathogenic; however, a number of characteristics associated with the virulence are available to evaluate the pathogenic potential of *Y. pseudotuberculosis* isolates (e.g., PCR methods specific for virulence factor encoding genes).[68]

The isolation of pathogenic *Yersinia* from clinical, food, and environmental samples is always demanding. The low isolation rates of *Y. pseudotuberculosis* from naturally contaminated samples may be due to a limited sensitivity of culture methods. PCR methods (most designed to study pure cultures) to identify and characterize *Y. pseudotuberculosis* strains are listed in Table 92.4. A systematic application of these PCR assays directly to sample materials would probably provide better estimation of the occurrence of *Y. pseudotuberculosis* in humans, animal reservoirs, foods, and environments.

TABLE 92.4
PCR-Based Detection Methods for *Y. pseudotuberculosis*

Method	Target Chromosome or Plasmid	Comments on the Samples and Detection Method	Reference
Single PCR assays	*inv, virF(lcrF)*[a]	Pure culture and gel detection	113
	16S and 23S rRNA[b]	Pure culture, gel detection and in situ hybridization	25
	wzz[c]	Chromosomal target part of O-antigen gene cluster	28
	yadA	Stool samples, real-time detection using SYBRGreen.	108
	ail[c]	Enrichment, DNA extration; includes IAC; TaqMan probe; specific for most serotypes	114
Multiplex PCR assays	*inv, virF(lcrF)*[a]	Pure culture and gel detection	115
	inv, virF(lcrF)[a]	Food samples and gel detection	116
	inv, virF(lcrF)[a]	Nested multiplex PCR and gel detection	117
	ail, virF(lcrF)[a]	Pure culture and gel detection	118
	16S rRNA[b]	Blood samples, nested multiplex PCR and gel detection	46
	fcl, prt, manB, abe, wbyL, wbyH, ddhAB, wbyK, wzx	Pure culture and gel detection; useful for O-genotyping	28

[a] Both *Y. pseudotuberculosis* and *Y. enterocolitica* will be detected.
[b] Capable to differentiate between *Y. pestis*, *Y. pseudotuberculosis*, and *Y. enterocolitica*.
[c] Both *Y. pseudotuberculosis* and *Y. pestis* will be detected.

92.1.5 *Y. ENTEROCOLITICA*

92.1.5.1 Biology and Pathogenesis

Y. enterocolitica was first isolated in 1930s[119] and initially described in 1963 as *Pasteurella* X.[120,121] *Y. enterocolitica* are facultative anaerobic rods that do not possess capsule and are motile below 28°C. *Y. enterocolitica* are able to grow at temperatures ranging from 4°C to 43°C. However, optimal growth and metabolic activity for this species is 28°C–30°C. Ability to grow at refrigeration temperatures allows it to mul tiply in stored foods.[122]

Y. enterocolitica and related species comprise over 70 serotypes, which are mainly determined by the variability of O-antigen.[123] The human and animal pathogenic *Y. enterocolitica* strains carry the 70–75 kb virulence plasmid of *Yersinia*, pYV, and most commonly belong to serotypes O:3, O:5,27, O:8, and O:9.[124] Less frequently encountered pathogenic serotypes are O:1,2,3, O:4,32, O:13a,18, and O:21.

As a species, *Y. enterocolitica* comprises organisms that differ in biochemical and genetic characteristics and can be classified into six biotypes known as 1A, 1B, 2, 3, 4, and 5.[125] These, on the other hand, fall into three lineages with different pathogenic potential: (i) biotype 1A strains that are nonpathogenic and lack pYV, (ii) biotype 2–5 strains that carry pYV but are weakly pathogenic and do not kill mice, and (iii) biotype 1B strains that carry pYV and are mouse-lethal and highly pathogenic. The biotype 1A strains are usually called nonpathogenic or environmental strains; however, as they are frequently isolated from human and animal (stool) samples, their reliable differentiation from pathogenic strains is important.

Several chromosome and pYV-encoded virulence factors contribute to the virulence of *Y. enterocolitica* and many of them are common with *Y. pseudotuberculosis* and *Y. pestis*. The pYV encodes an outer membrane protein YadA (*Yersinia* adhesin A), the T3SS, and the effector Yops.[126–131] YadA is an outer membrane protein, which is an important virulence factor of *Y. enterocolitica* but seems to be of less significance for the virulence of *Y. pseudotuberculosis*,[129,132–135] and in *Y. pestis* YadA is not expressed as *yadA* is a pseudogene.[131,136] At least two chromosomal-encoded virulence factors, Inv (invasion) and Ail (attachment invasion locus), mediate cell invasion.[137–141] The chromosome encoded Yst (*Yersinia* stable toxin) of *Y. enterocolitica* resembles closely the heat-stable toxin of enterotoxigenic *Escherichia coli*; however, at present the role of Yst in the pathogenesis of *Y. enterocolitica* diarrhea in humans is uncertain.[142] Also similar to *Y. pseudotuberculosis*, LPS contributes to the pathogenicity of *Y. enterocolitica*.[134,143–149]

The major difference in virulence between the biotype 1B and other pathogenic strains is the presence of iron uptake system in biotype 1B strains. Thus, biotype 1B serotype O:8 strains kill mice while biotype 4 serotype O:3 strains do not unless iron is made available to bacteria by pretreatment of mice with desferroxamine or iron.[150]

92.1.5.2 Epidemiology

Y. enterocolitica is the most common *Yersinia* species reported in human yersiniosis. *Y. enterocolitica* infections have been reported from humans on all continents.[58] Most cases occur sporadically without an apparent source and outbreaks are rare. Only certain bioserotypes have been associated with human *Y. enterocolitica* infection.[58] Bioserotype 4/O:3 is the most common type found in humans through the world, followed by 2/O:9 and 2/O:5,27. Bioserotype 3/O:3 has been responsible for human yersinosis in Japan and China.[105] Bioserotype 1B/O:8, originally found in the United States, has also been sporadically isolated in Europe and Japan.[58,151] However, a dramatic increase of *Y. enterocolitica* O:8 infections in Poland has recently been reported.[152]

Pigs are considered the major reservoir for *Y. enterocolitica* infections in humans.[153] A close genetic relationship between pig and human *Y. enterocolitica* strains has been demonstrated by several DNA-based methods.[124] Fattening pigs are mostly asymptomatic carriers of human pathogenic strains, in particular strains of bioserotype 4/O:3, the most common human type. This bioserotype has frequently been isolated from the oral cavity, especially from the tonsils and the tongue but also from submaxillar lymph nodes and intestinal content of pigs at slaughter.[124,154,155] Other bioserotypes associated with human disease like 3/O:3, 2–3/O:5,27, 2–3/O:9, and 1B/O:8 have also been isolated from pigs but the isolation rates have been low. Human pathogenic *Y. enterocolitica* strains have only sporadically been isolated from other animal sources like pet animals, ruminants, birds, and wild animals.[58,95,154,156] In Japan, wild rodents may play an important role as source of human *Y. enterocolitica* O:8 infection.[151]

Pathogenic *Y. enterocolitica* has seldom been isolated from food samples even though food is expected to be the main infection source of human yersiniosis. Raw pork and pork products have been widely studied. Using PCR, pathogenic *Y. enterocolitica* has been detected frequently, especially on pig tongues but also on the surfaces of freshly slaughtered pig carcasses, in minced pork meat, and pork sausages.[154,157–159] Moreover, pathogenic *Y. enterocolitica* has sporadically been detected in chicken, lettuce, and water samples. Stream water especially has shown to be contaminated with serotype O:8 in Japan.[151] Most *Y. enterocolitica* strains recovered from environmental samples are nonpathogenic, differing both biochemically and serologically from human pathogenic strains. However, bioserotype 4/O:3 strains have occasionally been isolated from the environment of pig slaughterhouses and butchers, and sewage water.[124]

Consumption of pork products has been demonstrated to be associated with *Y. enterocolitica* infections.[63,160,161] In case-control studies, a correlation has been demonstrated between *Y. enterocolitica* infection and consumption of undercooked pork and sausage products and untreated water.[160–162] In addition, use of a baby's dummy and contact with domestic animals has shown to be important risk factors for domestic yersiniosis in Swedish children.[160] Recently, some small outbreaks due to consumption of Christmas brawn in Norway and semicooked cocktail sausages in New Zealand have been reported.[163,164]

The most common transmission route of pathogenic *Y. enterocolitica* is thought to be fecal-oral via contaminated food, especially pork and related products.[124] Dogs and cats may be an important transmission link between pigs and young children having close contact with pets. It has been shown that dogs, which were fed raw pork, excreted pathogenic *Y. enterocolitica* 4/O:3 in the feces.[165] One transmission link may be direct contact with pigs, a common risk for pig farmers and slaughterhouse workers.[166] Direct person-to-person contact has not been demonstrated, but Lee et al. reported *Y. enterocolitica* O:3 infections in infants who were probably exposed to infection by their caretakers during preparation of pork intestine (chitterlings).[167] This may happen when basic hygiene and hand-washing habits are inadequate.

92.1.5.3 Laboratory Diagnosis

Several different cultural methods for detection of *Y. enterocolitica* have been described.[157] So far no single method is satisfactory for the isolation of all bioserotypes. The *Y. enterocolitica* isolation method usually involves a direct culture on selective agar plates and/or an enrichment step in one or two enrichment broths followed by plating onto one or two selective agar plates, identification of typical colonies, biotyping, serotyping, and testing for virulence properties.

Cold enrichment in different non and low-selective culture media has been widely used for the detection of the pathogenic *Y. enterocolitica* from various sources, including human clinical samples. A benefit of this method is that all pathogenic *Y. enterocolitica* strains can be enriched; however, nonpathogenic *Yersinia* or other psychrotrophic bacteria contaminating the sample may also multiply during enrichment and challenge the detection of the pathogens. Another drawback is the long incubation period, typically 7–21 days. Potassium hydroxide (KOH) treatment of the enrichment culture can sometimes reduce the background flora, making selection of *Yersinia* colonies less laborious. Some selective broths for isolation of pathogenic *Y. enterocolitica* at higher temperatures have been developed to increase the selectivity and to shorten the incubation time.[157] The irgasan-ticarcillin-potassium chlorate (ITC) broth and modified Rappaport broth (MRB) are the most often used enrichment media that have been designed for recovery of strains of bioserotype 4/O:3. Both media, however, inhibit strains of bioserotypes 2/O:5,27, 2/O:9, and 1B/O:8.

Many different selective agar plating media are available for isolation of *Y. enterocolitica* from naturally contaminated samples. The most commonly used selective agar plates are cefsulodin-irgasan-novobiocin (CIN), *Salmonella–Shigella* deoxycholate calcium chloride (SSDC), and MacConkey (MAC) agar plates. CIN agar is the most frequently used agar because of the high confirmation rate of presumptive isolates and its relatively high selectivity for fecal samples. In addition, *Y. enterocolitica* also grow on solid media such as XLD or Hektoen agar, however, with similar appearance as many other gram-negative species.

Presumptive urea-positive *Yersinia* isolates can be identified using selected biochemical key tests or by commercial identification tests. The most widely used biochemical kit for identification of *Y. enterocolitica* and other *Yersinia* spp. is the API 20E system (BioMérieux, Inc.). It is a kit for species determination and does not differentiate between virulent and avirulent strains; therefore, additional biochemical or DNA-based tests are necessary. These tests may add up to 5 days' more work to the 2 days of species identification. Although previous reports diverge,[168] Neubauer et al. found that the API 20E system properly identified pathogenic strains of *Y. enterocolitica*, and that 95% of the nonpathogenic *Y. enterocolitica* strains investigated were identified.[169] Biotyping of *Y. enterocolitica* isolates is recommended because the biotype is a reliable marker of pathogenicity.[170] At least five different biochemical tests are needed to distinguish the six biotypes.[125] As already mentioned, isolates belonging to biotype 1A are regarded as nonpathogenic. Serotyping using commercial antisera to O:3, O:5, O:8, and O:9 has been used extensively; however, these antigens can sometimes be found also in nonpathogenic *Y. enterocolitica* strains and in other *Yersinia* species. Therefore, a strategy including both biotype and serotype determination of the isolate is highly recommended. Finally, it is very important to assess the pathogenic potential of the isolates because majority of strains recovered from asymptomatic carriers, food, and environmental samples are nonpathogenic and have no clinical significance. In addition to using virulence-factor-gene-specific PCR assays, the pyrazinamidaze test and a number of phenotypic characteristics associated with the virulence plasmid (calcium dependence, detected as growth restriction on magnesium oxalate agar, autoagglutination at 35°C–37°C, and uptake of Congo red and crystal violet) can be applied.[157]

Pathogenic *Y. enterocolitica* strains have been detected from different natural sources, such as animals, food, and environment both by culture and by PCR. The detection efficiency of pathogenic strains of *Y. enterocolitica* in nonhuman sources is clearly lower, with culture compared to PCR. Based on our experience of the use of PCR for identification and on results from others studies, it can be concluded that PCR can identify strains of pathogenic *Y. enterocolitica* and distinguish them from nonpathogenic strains within a couple of hours of work. Use of PCR for identification has the advantage of not being dependent on the phenotypic expression and therefore, biochemically atypical organisms can be identified as readily as, and in addition to, typical ones. For example, sucrose-negative isolates have been regarded as nonpathogenic and are therefore not proposed for further testing in existing reference methods. Recently, some sucrose-negative *Y. enterocolitica* isolates have been reported to harbor virulence characteristics. These strains were readily identified by use of a PCR method developed for detection of pathogenic *Y. enterocolitica*.[171] Likewise, Harnett et al. reported that the pyrazinamidaze test, previously believed to accurately differentiate pathogenic from nonpathogenic serotypes, gave for serotype O:1,2,3 strains varying results, while in a PCR assay these strains were identified unambiguously as virulent.[172]

It has been argued that a major limitation of the use of PCR for detection is that showing the presence of only one gene or target sequence does not adequately cover the pathogenic potential of a strain. Thoerner et al. used five different PCR assays to analyze *Yersinia* strains for the presence of five virulence-associated genes[173]; however, a multiplex PCR assay would be a more efficient system.

When using real-time PCR, there is generally no need for a postamplification analysis. In certain cases, sequencing the amplicon can be useful (e.g., to distinguish based on the amplified *ail*-gene sequences the mouse-lethal "American" *Y. enterocolitica* bioserotype 1B/O:8 strains from the less-virulent European bioserotype 4/O:3, 2/O:9, and 2/O:5,27 strains).[174] The amplified *ail*-gene fragments of bioserotype 4/O:3, O:9, and O:5,27 strains are 100% identical, while the 1B/O:8 sequences contain four differences.

Table 92.5 summarizes some PCR-based detection methods available for pathogenic *Y. enterocolitica*.

92.2 METHODS

PCR methods can be used for rapid detection of pathogens directly from clinical samples or from enrichment media to reach diagnosis without time-consuming conventional culture methods. Samples giving positive results could be further studied with culture method to get the strain for further characterization (e.g., serotyping, antimicrobial resistance testing, genotyping). PCR can also be used as an identification tool to characterize strains that already have been isolated by a conventional culture method. In this context PCR replaces conventional confirmation (i.e., the lengthy and complicated biochemical and/or serological identification procedures). PCR is in this case performed either on pure fluid cultures or on colonies. PCR inhibitory problems are therefore rarely encountered and PCR can be applied after only a short sample preparation step (e.g., a short DNA extraction procedure).

92.2.1 Sample Preparation

Yersinia isolation from human clinical samples is usually done by direct plating of the sample onto (*Yersinia*-selective) agar; only rarely is the enrichment procedure considered necessary. DNA may be extracted from pure liquid cultures or colonies using either commercial ready-to-use purification kits or simply by boiling of bacteria in 100 µL of water for 10 min. To obtain DNA of good quality for PCR analysis it is recommended to use one of the commercial ready-to-use kits for extraction.

1. Culture the sample onto Cefsulodin Irgasan Novobiocin-agar (CIN) plate. Incubate at 30°C for 21 ± 3 h (*Y. enterocolitica*) and 48 ± 4 h (*Y. pseudotuberculosis*).
2. Pick from the CIN-agar plate presumptive *Yersinia* colonies. Perform DNA extraction by transfer of the bacteria either to a tube containing 100 µL sterile distillated water and boil for 10 min or by following the instruction of the manufacturer if using a commercial ready-to-use kit. As an example the Instagene Matrix extraction kit (BioRad) protocol is used:
   - (i) Suspend 1–2 presumptive *Yersinia* colonies in 100–200 µL of Instagene solution.
   - (ii) Vortex the sample for 15–20 s.
   - (iii) Incubate for 30 min at 56°C in a heat block.
   - (iv) Vortex the sample for 15–20 s.
   - (v) Incubate for 15 min at 95°C in a heat block.
   - (vi) Vortex the sample for 15–20 s.
   - (vii) Centrifuge for 5 min at 13,200 rpm.

3. Use 5 μL of the DNA-lysate for PCR analysis. Store the rest for (possible) virulence plasmid investigation and/or for reanalysis. For longer storage keep the samples at −20°C and for temporary storage keep at 4°C. Centrifuge the stored samples at 13,200 rpm for 5 min prior to reanalysis.

92.2.2 Detection Procedures

In this section we briefly comment on DNA-based methods for the detection of the three pathogenic *Yersinia* spp. and give detailed procedures for the detection of pathogenic *Y. enterocolitica* and *Y. pseudotuberculosis* by using real-time PCR targeting the *ail* gene of both organisms.

Also, commercial ready-to-use PCR mastermixes are available, both for conventional PCR and for real-time PCR use. Special mastermixes are available for multiplexed PCR analyses, that is, if more than one target is intended for detection in the same PCR analysis. The commercial kits can be recommended because they contain optimized concentrations of all components and are easy to handle.

Y. pestis. For *Y. pestis*, the detection assays targeting the *pla* gene appear to be the most sensitive, and the assay of Loiez et al. is a good candidate (Table 92.2). The real-time PCR assay is based on TaqMan technology coupled with automated DNA extraction and uses a minor groove binder-conjugated small DNA probe for DNA detection. Details of the method are given in the original publication.[32] The only caveat in the assay is the location of the *pla* gene on the small pPst plasmid that can be lost during handling. Therefore, assays targeting also chromosomal *Y. pestis*-specific genes such as *YP48*, *ypo2088*, *yihN*, and *entF3*[34,42,44,45] should be considered (Table 92.2).

Y. pseudotuberculosis. The *inv* gene, located in the chromosome and the *virF/lcrF* gene on the plasmid of *Y. pseudotuberculosis*, has often been used as targets for PCR (Table 92.4). Both genes can be detected using multiplex PCR with primers that amplify 295 bp *inv* and 591 bp *virF/lcrF* fragments according to Nakajima et al.[195] Other specific target for detection of *Y. pseudotuberculosis* is the *wzz*-gene[28,45]; however, in positive cases, the presence of *Y. pestis* has to be excluded [e.g., by one of the *Y. pestis*–specific targets (see above)]. Below, however, we will describe an *ail* gene-specific real-time PCR application for the detection of *Y. pseudotuberculosis.*[114]

Y. enterocolitica. A number of methods for specific detection of pathogenic *Y. enterocolitica* have been described (Table 92.5). The methods are designed to target either chromosomal or virulence plasmid-located genes or both; for detection using the chromosome as target molecule either *ail*, *yst*, or 16S rDNA have been used and for target genes located on the virulence plasmid *yadA*, *virF (lcrF)*, or *yopN (lcrE)*.

TABLE 92.5
Molecular Detection Methods Available for *Y. enterocolitica*

Methodology	Target Chromosome or *Plasmid*	Sample Preparation and Comments	Reference
Single PCR	*yadA*	Enrichment; nested PCR; gel detection	174
	lcrE	DNA extraction; gel detection; detecting also *Y. pseudotuberculosis* and *Y. pestis*	175
	wbcT (rfbB)	Gel detection; specific for serotype O:3; enrichment; two nested single PCR	176
	ail, virF	Gel detection; specific for all pathogenic serotypes	177
Multiplex PCR assay	*bipA*, ERIC	DNA extraction; real-time analysis	178
	ail, virF	Enrichment/DNA extraction method evaluation; gel detection	179
	yadA, 16S rRNA	Enrichment/buoyant density gradient centrifugation and *Yersinia* PCR-compatible enrichment (YPCE) medium; gel detection	180,181
	ail, virF	Enrichment; most probable number-multiplex PCR for quantitation; gel detection	182
Real-Time PCR	*yadA*	Enrichment/DNA extraction; detection by SYBR Green	108
	per	DNA extraction; including IAC; detection by TaqMan probe; specific for serotype O:9	183
	16S rRNA	Flotation to enrich target bacteria; detection by SYBR Green	184
	ail	Enrichment; DNA extraction; including IAC; Taqman probe; specific for all pathogenic serotypes	173
	ail	Enrichment; DNA extraction; TaqMan based	185
	yst	Enrichment; DNA extraction. TaqMan-based	186
PCR-DNA Microarray	*virF, ail, yst, blaA*, 16S rRNA	DNA extraction; multiplex PCR; detection by microarray	187
	23S rRNA	Enrichment; analysis of PCR products by oligonucleotide array	188
	gyrB	DNA extraction; analysis of PCR products by oligonucleotide array	189
Other methods	*rpoB*	DNA extraction; PCR-DGGE (denaturing gradient gel electrophoresis)	190
	16S rRNA	RNA extraction; microfluidic chip-based system	191
	Surface antigens	Enrichment; multiplexed chemiluminescent immunoassay	192
	rRNA	Cell lysis; RNA hybridization-electrochemiluminescence	193
	O:3 LPS	Enrichment; surface adhesion immunofluorescence	194

To be fully virulent pathogenic *Y. enterocolitica* strains must harbor a virulence plasmid. Since the plasmid may easily be lost (e.g., during subculture in the laboratory or storage over time), the virulence plasmid is not considered as a suitable target for the detection of the bacterium. In studies aiming to examine the presence of the virulence plasmid, it is therefore important to take into account the identified conditions (temperature, pH, calcium dependence, etc.) that may retain the plasmid-bearing cells in the sample before PCR analysis.[196] For that reason, we recommend the use of a method for detection/identification of pathogenic *Y. enterocolitica* that targets the chromosomally located *ail*-gene[173] and thereafter in positive cases to perform an additional PCR to check for the presence of the virulence plasmid. Alternatively, a multiplex PCR that identifies both targets simultaneously may be used.

Furthermore, according to ISO 20838:2006,[197] if conventional PCR is used, the identity of the obtained PCR product should be confirmed by a method other than size determination, for example, by DNA sequencing, by hybridization with specific DNA probes, or by restriction digestion analysis of the PCR product.

Procedure. The same DNA extraction procedure and PCR mastermix can be used for identification of pathogenic *Y. enterocolitica*, *Y. pseudotuberculosis*, and *Y. pestis*. The precautions when working with *Y. pestis* have been mentioned earlier (see Section 92.1.2.3).

The real-time PCR conditions and primer sequences based on the *ail*-genes of *Y. enterocolitica*[173] and *Y. pseudotuberculosis*[114] are presented in Tables 92.6 and 92.7.

Quality Assurance. The implementation of quality assurance and control measures in laboratories using PCR is critical to ensure the reliability of PCR results. According to ISO standard 22174:2005, inclusion of controls (e.g., negative controls and amplification controls) to monitor the analysis performance is considered obligatory and must be added in each PCR run when analyzing for human pathogens.[199] A negative PCR control (typically sterile water in place of test sample) is used to detect any PCR-fragment carry-over contamination. An amplification control [either internal (IAC) or external amplification controls, typically consisting of nucleic acid from a known positive sample] serves to control the amplification and will monitor for false-negative PCR results. Thus, a negative PCR result from both the amplification control and the target bacteria is indicative of PCR inhibition. These samples can be rerun after dilution by 1:10 (or more). The concentration of the added control DNA fragment should be as low as possible in order to detect even small inhibitions and at the same time give a statistical reproducible positive result. Both internal (IAC) and external amplification controls exist.

92.3 CONCLUSION AND FURTHER PERSPECTIVES

This chapter reviews the taxonomical, biological, pathological, and epidemiological properties of the genus *Yersinia* and discusses the application of molecular methods for the detection, isolation, and characterization of *Yersinia* spp.

TABLE 92.6
TaqMan Real-Time PCR Reagents and Conditions for the Detection of *ail*-Gene

Component (Stock Conc.)	Volume/Reaction (Final Conc.)	Comments
10× TaqMan buffer A[a]	2.5 μL (1×)	Applied Biosystems, Foster City, CA, USA
$MgCl_2$ (25 mM)[a]	3.5 μL (3.5 mM)	
dNTPs (50 mM)[a]	0.1 μL (200 μM)	
AmpliTaqGold® (5 U/μL)[a]	0.25 μL (0.02 U/μL)	
Primer 1 (10 μM)	0.75 μL (300 nM)	For sequence see Table 92.7
Primer 2 (10 μM)	0.75 μL (300 nM)	For sequence see Table 92.7
Probe (20 μM)	0.25 μL (200 nM)	For sequence see Table 92.7
IAC[b]	Approximately 100 copies	
Test sample	5 μL	
Sterile distilled water	Add to 25 μL	
Real-time PCR cycling parameters (optimized for the 7500 ABI system)		
1 cycle at 95°C for 10 min		
45 cycles 95°C for 15 s and 60°C for 1 min		

[a] These components can be replaced by the 10× TaqMan® Universal PCR Master mix (Applied Biosystems) or by similar buffers from other suppliers. When naturally contaminated samples are tested it is strongly recommended to use the Master mix with the inclusion of dUTP instead of dTTP. (The final concentration of the Master mix components: 1× TaqMan Universal PCR mix, 900 nM of each primer, 200 nM of probe, 5 μL test sample; total volume 25 μL).

[b] Two internal amplification controls (here named IPC and IAC) are presented here: (i) A commercially available TaqMan® Exogenous Internal Positive Control (Applied Biosystems). The reagent kit includes primers, a Vic™probe, IPC target DNA, and blocking solution. The IPC target DNA needs to be diluted 10 times to achieve a copy number of approximately 100 per PCR reaction. The PCR product length is not given to the customer. (ii) An open formula pUC19-based internal amplification control IAC developed by Fricker et al.[198] can be used. Approximately 50–100 copies of target DNA (pUC19) are used per PCR. The size of this IAC is 119 bp.

TABLE 92.7
Primers and Probes Used for Real-Time PCR Detection of Pathogenic *Y. enterocolitica* and *Y. pseudotuberculosis*

Primer or Probe	Sequence (5′–3′)	Position[a]	Melting Temperature (°C)[b]	Amplicon Size
	Y. enterocolitica			
F-real 10A	ATGATAACTGGGGAGTAATAGGTTCG	2379–2404	63.0	
R-real 9A	CCCAGTAATCCATAAAGGCTAACATAT	2516–2541	61.5	163 bp
Ye probe[d]	FAM-TCTATGGCAGTAATAAGTTTGGTCACGGTGATCT-TAMRA[c]	2436–2469	68.1	
	***Y. pseudotuberculosis* (NOTE: These primers will also detect *Y. pestis*)**			
F- Yps 1	CGTCTGTTAATGTGTATGCCGAAG	1395870–1395893	59.6	
R- Yps 2	GAACCTATCACTCCCCAGTCATTATT	1396001–1396026	58.8	157 bp
Probe[d]	VIC-CGTGTCAAGGACGATGGGTACAAGTTGG-TAMRA[c]	1395929–1395956	69.5	

[a] Nucleotide positions correspond to the *Y. enterocolitica* O:3 *ail*-gene sequence (Genbank acc.no AJ605740) and to *Y. pseudotuberculosis* YPIII complete genome sequence (Genbank acc.no CP000950.1).
[b] Calculated with the Primer Express program, version 3.0 (Applied Biosystems).
[c] TAMRA™; FAM™; VIC® (Applied Biosystems).
[d] The *Y. enterocolitica* probe can be exchanged for the MGB probe: FAM-TGACCAAACTTATTACTGCCATA-MGB. The *Y. pseudotuberculosis* probe can be exchanged for the MGB probe: NED-ATGCTCAAAGTCGTGTCAA-MGB. MGB, minor groove binder. (Probes from Applied Biosystems).
Source: From references 114 and 173.

Due to their obvious advantages, nucleic acid amplification technologies, in particular PCR, have been widely adopted in research and clinical laboratories for identification of *Yersinia* bacteria. Sample preparation influences both the specificity and the sensitivity of the PCR protocol. To minimize PCR result variations it is necessary to thoroughly optimize the pre-PCR processing conditions. In response to the growing demand for rapid, robust, and simple PCR protocols, further development in the integration of DNA extraction and the PCR amplification process will be invaluable. From a quality point of view, as few steps as possible should be included in the process of diagnostic PCR. The use of unconventional DNA polymerases together with streamlined sampling conditions will in future reduce the amount of manual handling involved in the analysis of pathogenic bacteria. In addition, with the growing number of sequenced (bacterial) genomes, there is scope for identification of novel biomarkers for PCR and microarray assays. Along with the development of novel detection devices, this will allow the design of high-throughput systems for cost-efficient, simultaneous detection of major pathogens from different types of samples.

Real-time PCR is a useful extension of the basic PCR technique. In contrast to conventional PCR, real-time PCR facilitates automation, computerization, and quantification of nucleic acids. A significant improvement introduced by real-time PCR, in addition to earlier amplicon detection, is the increased throughput and reduced carry-over contamination. This is largely due to reduced cycle times, removal of separate post-PCR detection procedures, and the use of sensitive fluorescence detection equipment. Future automation of the entire real-time PCR process will not only enhance the usefulness of this method in laboratory diagnostics, epidemiological studies, and the food industry, but also provide a powerful tool for pathogen quantification in naturally contaminated samples and selective detection of viable cells only.

Moreover, high-throughput technologies such as DNA microarray, mass spectrometry-PCR, and whole-genome sequencing platforms will strengthen the portfolio of molecular based diagnostic tools for pathogenic *Yersinia* spp. These high-throughput-based technologies allow rapid molecular characterization of *Yersinia* spp. and determination of antibiotic resistance of the isolate; and they have been employed for the detection and characterization of *Y. pestis* in a number of bio preparedness programs. Following the completion of several *Y. pestis* genome sequences (Figure 92.1a), the challenge is to utilize this information in the high-throughput *Yersinia* detection methods that include optimized sampling and sample preparation and integrated workflows for bio security and food safety concerns.

ACKNOWLEDGMENTS

The work of MS has been supported by the Academy of Finland (projects 114075 and 104361) and the European Union Network of Excellence "EuroPathoGenomics" (contract LSHB-CT-2005–512061). The work of PR has been supported by the Swedish Research Council for Environment, Agricultural Sciences, and Spatial Planning (FORMAS).

REFERENCES

1. Woese, C.R., Bacterial evolution, *Microbiol. Rev.*, 51, 221, 1987.
2. Perry, R.D., and Fetherston, J.D., *Yersinia pestis*—etiologic agent of plague, *Clin. Microbiol. Rev.*, 10, 35, 1997.

3. Merhej, V. et al., *Yersinia massiliensis* sp. nov., isolated from fresh water, *Int. J. Sys. Evol. Microbiol.*, 58, 779, 2008.
4. Sprague, L.D. et al., *Yersinia similis* sp. nov, *Int. J. Sys. Evol. Microbiol.*, 58, 952, 2008.
5. Davies, R.L., O-serotyping of *Yersinia ruckeri* with special emphasis on European isolates, *Vet. Microbiol.*, 22, 299, 1990.
6. Chain, P.S. et al., Insights into the evolution of *Yersinia pestis* through whole-genome comparison with *Yersinia pseudotuberculosis*, *Proc. Natl. Acad. Sci. USA*, 101, 13826, 2004.
7. Achtman, M. et al., Microevolution and history of the plague bacillus, *Yersinia pestis*, *Proc. Natl. Acad. Sci. USA*, 101, 17837, 2004.
8. Achtman, M. et al., *Yersinia pestis*, the cause of plague, is a recently emerged clone of *Yersinia pseudotuberculosis*, *Proc. Natl. Acad. Sci. USA*, 96, 14043, 1999.
9. Skurnik, M., Peippo, A., and Ervelä, E., Characterization of the O-antigen gene clusters of *Yersinia pseudotuberculosis* and the cryptic O-antigen gene cluster of *Yersinia pestis* shows that the plague bacillus is most closely related to and has evolved from *Y. pseudotuberculosis* serotype O:1b, *Mol. Microbiol.*, 37, 316, 2000.
10. Achtman, M., Age, descent and genetic diversity within *Yersinia pestis*. In Carniel, E., and Hinnebusch, B.J. (Editors), *Yersinia. Molecular and Cellular Biology*, p. 17, Horizon Bioscience, Wymondham, UK, 2004.
11. Lindler, L.E., Typing methods for the plague pathogen, *Yersinia pestis*, *J. AOAC Int.*, 92, 1174, 2009.
12. Gage, K.L., and Kosoy, M.Y., Natural history of plague: Perspectives from more than a century of research, *Annu. Rev. Entomol.*, 50, 505, 2005.
13. Bin Saeed, A.A., Al-Hamdan, N.A., and Fontaine, R.E., Plague from eating raw camel liver, *Emerg. Infect. Dis.*, 11, 1456, 2005.
14. Arbaji, A. et al., A 12-case outbreak of pharyngeal plague following the consumption of camel meat, in north-eastern Jordan, *Ann. Trop. Med. Parasitol.*, 99, 789, 2005.
15. Sodeinde, O.A. et al., A surface protease and the invasive character of plague, *Science*, 258, 1004, 1992.
16. Wren, B.W., The yersiniae—a model genus to study the rapid evolution of bacterial pathogens, *Nat. Rev. Microbiol.*, 1, 55, 2003.
17. Brubaker, R.R., Factors promoting acute and chronic diseases caused by *Yersiniae*, *Clin. Microbiol. Rev.*, 4, 309, 1991.
18. Devignant, R., Variétés de l'éspèce *Pasteurella pestis*. Nouvelle hypothèse, *Bull. WHO*, 4, 247, 1951.
19. Brubaker, R.R., The genus *Yersinia*: Biochemistry and genetics of virulence, *Curr. Top. Microbiol. Immunol.*, 57, 111, 1972.
20. Engelthaler, D.M. et al., PCR detection of *Yersinia pestis* in fleas: Comparison with mouse inoculation, *J. Clin. Microbiol.*, 37, 1980, 1999.
21. Kuske, C.R. et al., Environmental survey for four pathogenic bacteria and closely related species using phylogenetic and functional genes, *J. Forensic Sci.*, 51, 548, 2006.
22. Neubauer, H. et al., A combination of different polymerase chain reaction (PCR) assays for the presumptive identification of *Yersinia pestis*, *J. Vet. Med. B*, 47, 573, 2000.
23. Radnedge, L. et al., Identification of nucleotide sequences for the specific and rapid detection of *Yersinia pestis*, *Appl. Environ. Microbiol.*, 67, 3759, 2001.
24. Rahalison, L. et al., Diagnosis of bubonic plague by PCR in Madagascar under field conditions, *J. Clin. Microbiol.*, 38, 260, 2000.
25. Trebesius, K. et al., Development of rRNA-targeted PCR and in situ hybridization with fluorescently labelled oligonucleotides for detection of *Yersinia* species, *J. Clin. Microbiol.*, 36, 2557, 1998.
26. Wiechmann, I., and Grupe, G., Detection of *Yersinia pestis* DNA in two early medieval skeletal finds from Aschheim (Upper Bavaria, 6th century A.D.), *Am. J. Phys. Anthropol.*, 126, 48, 2005.
27. Zhou, D. et al., Identification of signature genes for rapid and specific characterization of *Yersinia pestis*, *Microbiol. Immunol.*, 48, 263, 2004.
28. Bogdanovich, T. et al., Use of O-antigen gene cluster-specific PCRs for the identification and O-genotyping of *Yersinia pseudotuberculosis* and *Yersinia pestis*, *J. Clin. Microbiol.*, 41, 5103, 2003.
29. Higgins, J.A. et al., 5′ nuclease PCR assay to detect *Yersinia pestis*, *J. Clin. Microbiol.*, 36, 2284, 1998.
30. Lindler, L.E., Fan, W., and Jahan, N., Detection of ciprofloxacin-resistant *Yersinia pestis* by fluorogenic PCR using the LightCycler, *J. Clin. Microbiol.*, 39, 3649, 2001.
31. Lindler, L.E., and Fan, W., Development of a 5′ nuclease assay to detect ciprofloxacin resistant isolates of the biowarfare agent *Yersinia pestis*, *Mol. Cell. Probes*, 17, 41, 2003.
32. Loiez, C. et al., Detection of *Yersinia pestis* in sputum by real-time PCR, *J. Clin. Microbiol.*, 41, 4873, 2003.
33. McAvin, J.C. et al., A real-time fluorescence polymerase chain reaction assay for the identification of *Yersinia pestis* using a field-deployable thermocycler, *Mil. Med.*, 168, 852, 2003.
34. Chase, C.J. et al., Real-time PCR assays targeting a unique chromosomal sequence of *Yersinia pestis*, *Clin. Chem.*, 51, 1778, 2005.
35. Christensen, D.R. et al., Detection of biological threat agents by real-time PCR: Comparison of assay performance on the R.A.P.I.D., the LightCycler, and the Smart Cycler platforms, *Clin. Chem.*, 52, 141, 2006.
36. Iqbal, S.S. et al., Detection of *Yersinia pestis* by pesticin fluorogenic probe-coupled PCR, *Mol. Cell. Probes*, 14, 109, 2000.
37. Skottman, T. et al., Simultaneous real-time PCR detection of *Bacillus anthracis*, *Francisella tularensis* and *Yersinia pestis*, *Eur. J. Clin. Mircobiol. Infect. Dis.*, 26, 207, 2007.
38. Yang, S. et al., Rapid polymerase chain reaction-based screening assay for bacterial biothreat agents, *Acad. Emerg. Med.*, 15, 388, 2008.
39. Melo, A.C., Almeida, A.M., and Leal, N.C., Retrospective study of a plague outbreak by multiplex-PCR, *Lett. Appl. Microbiol.*, 37, 361, 2003.
40. Tomaso, H. et al., Rapid detection of *Yersinia pestis* with multiplex real-time PCR assays using fluorescent hybridisation probes, *FEMS Immunol. Med. Microbiol.*, 38, 117, 2003.
41. Tsukano, H. et al., Detection and identification of *Yersinia pestis* by polymerase chain reaction (PCR) using multiplex primers, *Microbiol. Immunol.*, 40, 773, 1996.
42. Stewart, A. et al., A quadruplex real-time PCR assay for the detection of *Yersinia pestis* and its plasmids, *J. Med. Microbiol.*, 57, 324, 2008.
43. Selvapandiyan, A. et al., A novel semiquantitative fluorescence-based multiplex polymerase chain reaction assay for rapid simultaneous detection of bacterial and parasitic pathogens from blood, *J. Mol. Diagn.*, 7, 268, 2005.
44. Woron, A.M. et al., Development and evaluation of a 4-target multiplex real-time polymerase chain reaction assay for the detection and characterization of *Yersinia pestis*, *Diagn. Microbiol. Infect. Dis.*, 56, 261, 2006.

45. Matero, P. et al., Real-time multiplex PCR assay for detection of *Yersinia pestis* and *Yersinia pseudotuberculosis*, *APMIS*, 117, 34, 2009.
46. Tomioka, K. et al., A multiplex polymerase chain reaction microarray assay to detect bioterror pathogens in blood, *J. Mol. Diagn.*, 7, 486, 2005.
47. Song, L., Ahn, S., and Walt, D.R., Fiber-optic microsphere-based arrays for multiplexed biological warfare agent detection, *Anal. Chem.*, 78, 1023, 2006.
48. Wilson, W.J. et al., A multiplexed PCR-coupled liquid bead array for the simultaneous detection of four biothreat agents, *Mol. Cell. Probes*, 19, 137, 2005.
49. Bianucci, R. et al., Technical note: A rapid diagnostic test detects plague in ancient human remains: An example of the interaction between archeological and biological approaches (southeastern France, 16th-18th centuries), *Am. J. Phys. Anthropol.*, 136, 361, 2008.
50. Cao, L.K. et al., Detection of Yersinia pestis fraction 1 antigen with a fiber optic biosensor, *J. Clin. Microbiol.*, 33, 336, 1995.
51. Meyer, M.H. et al., Magnetic biosensor for the detection of *Yersinia pestis*, *J. Microbiol. Methods*, 68, 218, 2007.
52. Tomaso, H. et al., Comparison of hand-held test kits, immunofluorescence microscopy, enzyme-linked immunosorbent assay, and flow cytometric analysis for rapid presumptive identification of *Yersinia pestis*, *J. Clin. Microbiol.*, 45, 3404, 2007.
53. Hindson, B.J. et al., Development of an automated sample preparation module for environmental monitoring of biowarfare agents, *Anal. Chem.*, 76, 3492, 2004.
54. Hindson, B.J. et al., Autonomous detection of aerosolized biological agents by multiplexed immunoassay with polymerase chain reaction confirmation, *Anal. Chem.*, 77, 284, 2005.
55. Fleming, D.O., and Hunt, D.L., *Biological Safety Principles and Practices*, 3rd ed., ASM Press, Washington DC, USA, 2000.
56. EFSA (European Food Safety Authority), The community summary report on trends and sources of zoonoses, zoonotic agents in the European Union in 2007, *EFSA J.*, 223, 223, 2009.
57. Okwori, A.E.J. et al., Pathogenic *Yersinia enterocolitica* 2/O:9 and *Yersinia pseudotuberculosis* 1/O:1 strains isolated from human and non-human sources in the Plateau State of Nigeria., *Food Microbiol.*, 26, 872, 2009.
58. Bottone, E.J., *Yersinia enterocolitica*: Overview and epidemiologic correlates, *Microbes Infect.*, 1, 323, 1999.
59. Leclercq, A. et al., Fatal *Yersinia enterocolitica* biotype 4 serovar O:3 sepsis after red blood cell transfusion, *Transfusion*, 45, 814, 2005.
60. CDC (Centers for Disease Control and Prevention), Preliminary FoodNet data on the incidence of infection with pathogens transmitted commonly through food—10 states, *MMWR*, 58, 333, 2009.
61. Ray, S.M. et al., Population-based surveillance for *Yersinia enterocolitica* infections in FoodNet sites, 1996–1999: Higher risk of disease in infants and minority populations, *Clin. Infect. Dis.*, 38 Suppl 3, S181, 2004.
62. European Centre for Disease Prevention and Control (ECDC), Annual Epidemiological report on Communicable Diseases 2008. http://ecdc.europa.eu/en/publications/Publications/ 0812_SUR_Annual_Epidemiological_Report_2008.pdf
63. Jones, T.F., From pig to pacifier: Chitterling-associated yersiniosis outbreak among black infants, *Emerg. Infect. Dis.*, 9, 1007, 2003.
64. Cover, T.L., and Aber, R.C., *Yersinia enterocolitica*, *N. Engl. J. Med.*, 321, 16, 1989.
65. Tertti, R. et al., Clinical manifestations of *Yersinia pseudotuberculosis* infection in children, *Eur. J. Clin. Mircobiol. Infect. Dis.*, 8, 587, 1989.
66. Jalava, K. et al., An outbreak of gastrointestinal illness and erythema nodosum from grated carrots contaminated with *Yersinia pseudotuberculosis*, *J. Infect. Dis.*, 194, 1209, 2006.
67. Hannu, T. et al., Reactive arthritis after an outbreak of *Yersinia pseudotuberculosis* serotype O:3 infection, *Ann. Rheum. Dis.*, 62, 866, 2003.
68. Fukushima, H. et al., Geographical heterogeneity between Far Eastern and Western countries in prevalence of the virulence plasmid, the superantigen *Yersinia pseudotuberculosis*-derived mitogen, and the high-pathogenicity island among *Yersinia pseudotuberculosis* strains, *J. Clin. Microbiol.*, 39, 3541, 2001.
69. Abdel-Haq, N.M. et al., Antibiotic susceptibilities of *Yersinia enterocolitica* recovered from children over a 12-year period, *Int. J. Antimicrob. Agents*, 27, 449, 2006.
70. Stock, I., and Wiedemann, B., An in-vitro study of the antimicrobial susceptibilities of *Yersinia enterocolitica* and the definition of a database, *J. Antimicrob. Chemother.*, 43, 37, 1999.
71. Malassez, L., and Vignal, W., Sur le micro-organisme de la tuberculose zoologique, *Arch. Physiol. Norm. Pathol.*, 3, 81, 1884.
72. Cornelis, G.R., The *Yersinia* Ysc-Yop virulence apparatus, *Int. J. Med. Microbiol.*, 291, 455, 2002.
73. Heesemann, J., Sing, A., and Trulzsch, K., *Yersinia's* stratagem: Targeting innate and adaptive immune defense, *Curr. Opin. Microbiol.*, 9, 55, 2006.
74. Abe, J., and Takeda, T., Characterization of a superantigen produced by *Yersinia pseudotuberculosis*, *Preparat. Biochem. Biotechnol.*, 27, 173, 1997.
75. Karlyshev, A.V. et al., Application of high-density array-based signature-tagged mutagenesis to discover novel *Yersinia* virulence-associated genes, *Infect. Immun.*, 69, 7810, 2001.
76. Mecsas, J., Bilis, I., and Falkow, S., Identification of attenuated *Yersinia pseudotuberculosis* strains and characterization of an orogastric infection in BALB/c mice on day 5 postinfection by signature-tagged mutagenesis, *Infect. Immun.*, 69, 2779, 2001.
77. Collyn, F. et al., *Yersinia pseudotuberculosis* harbors a type IV pilus gene cluster that contributes to pathogenicity, *Infect. Immun.*, 70, 6196, 2002.
78. Tsubokura, M., and Aleksic, S., A simplified antigenic scheme for serotyping of *Yersinia pseudotuberculosis*: Phenotypic characterization of reference strains and preparation of O and H factor sera, *Contrib. Microb. Immunol.*, 13, 99, 1995.
79. Skurnik, M., Lipopolysaccharides of *Yersinia*. In Carniel, E., and Hinnebusch, B.J. (Editors), *Yersinia: Molecular and Cellular Biology*, p. 215, Horizon Bioscience, Wymondham, UK, 2004.
80. Tsubokura, M. et al., Addition of new serogroups and improvement of the antigenic designs of *Yersinia pseudotuberculosis*, *Curr. Microbiol.*, 11, 89, 1984.
81. Tsubokura, M. et al., Characterization of *Yersinia pseudotuberculosis* serogroups O9, O10 and O11—subdivision of O1 serogroup into O1a, O1b, and O1c subgroups, *Int. J. Med. Microbiol.*, 278, 500, 1993.
82. Vincent, P. et al., Sudden onset of pseudotuberculosis in humans, France, 2004–05, *Emerg. Infect. Dis.*, 14, 1119, 2008.

83. Nowgesic, E. et al., Outbreak of *Yersinia pseudotuberculosis* in British Columbia—November 1998, *Can. Commun. Dis. Rep.*, 25, 97, 1999.
84. Pebody, R.G. et al., Outbreak of *Yersinia pseudotuberculosis* infection in central Finland, *Eurosurveillance*, 1, 1997. http://www.eurosurveillance.org/ViewArticle.aspx?ArticleId=1051
85. Nuorti, J. P. et al., A widespread outbreak of *Yersinia pseudotuberculosis* O:3 infection from iceberg lettuce, *J. Infect. Dis.*, 189, 766, 2004.
86. Jalava, K. et al., Multiple outbreaks of *Yersinia pseudotuberculosis* infections in Finland, *J. Clin. Microbiol.*, 42, 2789, 2004.
87. Takkinen, J. et al., *Yersinia pseudotuberculosis* infection traced to raw carrots in Finland, *Eurosurveillance*, 8, 2004. http://www.eurosurveillance.org/ViewArticle.aspx?ArticleId=2563
88. Fukushima, H., Tsunomori, Y., and Seki, R., Duplex real-time SYBR green PCR assays for detection of 17 species of food- or waterborne pathogens in stools, *J. Clin. Microbiol.*, 41, 5134, 2003.
89. Anonymous, Outbreak of *Y. pseudotuberculosis* in Kajaani-grated carrots suspected as a vehicle of infections, *Bull. Natl. Public Health Institute Finland*, 7, 22, 2008.
90. Tsubokura, M. et al., Special features of distribution of *Yersinia pseudotuberculosis* in Japan, *J. Clin. Microbiol.*, 27, 790, 1989.
91. Toyokawa, Y. et al., Large scale outbreak of *Yersinia pseudotuberculosis* serotype 5a infection at Noheji-machi in Aomori Prefecture, *Kansenshogaku Zasshi*, 67, 36, 1993.
92. Fredriksson-Ahomaa, M., Lindström, M., and Korkeala, H., *Yersinia enterocolitica* and *Yersinia pseudotuberculosis*. In Juneja, V.K., and Sofos, N.J. (Editors), *Pathogens and Toxins in Foods: Challenger and Interventions*, pp. 164–180, ASM Press., Washington, DC, 2010.
93. Aleksic, S., Bockemühl, J., and Wuthe, H.-H., Epidemiology of *Y. pseudotuberculosis* in Germany, 1983–1993, *Contrib. Microb. Immunol.*, 13, 55, 1995.
94. Fukushima, H., and Gomyoda, M., Intestinal carriage of *Yersinia pseudotuberculosis* by wild birds and mammals in Japan, *Appl. Environ. Microbiol.*, 57, 1152, 1991.
95. Niskanen, T. et al., *virF*-positive *Yersinia pseudotuberculosis* and *Yersinia enterocolitica* found in migratory birds in Sweden, *Appl. Environ. Microbiol.*, 69, 4670, 2003.
96. Ortiz Martìnez, P., et al., High prevalence of enteropathogenic *Yersinia* in Estonian, Latvian and Russian (Leningrad Region) pigs., *Foodborne Pathog. Dis.*, 6, 719, 2009.
97. Chiesa, C. et al., *Yersinia pseudotuberculosis* in Italy—attempted recovery from 37,666 samples, *Microbiol. Immunol.*, 37, 391, 1993.
98. Fukushima, H. et al., Putative origin of *Yersinia pseudotuberculosis* in western and eastern countries. A comparison of restriction endonuclease analysis of virulence plasmids, *Int. J. Med. Microbiol.*, 288, 93, 1998.
99. Niskanen, T., Fredriksson-Ahomaa, M., and Korkeala, H., *Yersinia pseudotuberculosis* with limited genetic diversity is a common finding in tonsils of fattening pigs, *J. Food Prot.*, 65, 540, 2002.
100. Niskanen, T. et al., Distribution of *virF/lcrF*-positive *Yersinia pseudotuberculosis* serotype O:3 at farm level, *Zoonoses Public Health*, 55, 214, 2008.
101. Laukkanen, R. et al., Transmission of *Yersinia pseudotuberculosis* in the pork production chain from farm to slaughterhouse, *Appl. Environ. Microbiol.*, 74, 5444, 2008.
102. Seimiya, Y.M. et al., Caprine enteritis associated with *Yersinia pseudotuberculosis* infection, *J. Vet. Med. Sci.*, 67, 887, 2005.
103. Zhang, S. et al., Fatal yersiniosis in farmed deer caused by *Yersinia pseudotuberculosis* serotype O:3 encoding a mannosyltransferase-like protein WbyK, *J. Vet. Diagn. Invest.*, 20, 356, 2008.
104. Iwata, T. et al., Virulence characteristics of *Yersinia pseudotuberculosis* isolated from breeding monkeys in Japan, *Vet. Microbiol.*, 129, 404, 2008.
105. Fukushima, H. et al., Introduction into Japan of pathogenic *Yersinia* through imported pork, beef and fowl, *Int. J. Food Microbiol.*, 35, 205, 1997.
106. Fukushima, H. et al., Restriction endonuclease analysis of virulence plasmids for molecular epidemiology of *Yersinia pseudotuberculosis* Infections, *J. Clin. Microbiol.*, 32, 1410, 1994.
107. Kangas, S. et al., *Yersinia pseudotuberculosis* O:1 traced to raw carrots, Finland, *Emerg. Infect. Dis.*, 14, 1959, 2008.
108. Rimhanen-Finne, R. et al., *Yersinia pseudotuberculosis* causing a large outbreak associated with carrots in Finland, 2006, *Epidemiol. Infect.*, 137, 324, 2009.
109. Han, T.H., Paik, I.K., and Kim, S.J., Molecular relatedness between isolates *Yersinia pseudotuberculosis* from a patient and an isolate from mountain spring water, *J. Korean Med. Sci.*, 18, 425, 2003.
110. Schiemann, D.A., Synthesis of a selective agar medium for *Yersinia enterocolitica*, *Can. J. Microbiol.*, 25, 1298, 1979.
111. FDA, *Yersinia enterocolitica* and *Yersinia pseudotuberculosis*. Bacteriological analytical manual online, Food and Drug Administration, USA, 2007.
112. Bogdanovich, T.M. et al., Genetic (sero)typing of *Yersinia pseudotuberculosis*. In Skurnik, M., Granfors, K., and Bengoechea, J.A. (Editors), *The Genus Yersinia: Entering the Functional Genomic Era*, p. 337, Kluwer Academic/Plenum Publishers, New York, 2003.
113. Thoerner, P. et al., PCR detection of virulence genes in *Yersinia enterocolitica* and *Yersinia pseudotuberculosis* and investigation of virulence gene distribution, *Appl. Environ. Microbiol.*, 69, 1810, 2003.
114. Lambertz, S.T., Nilsson, C., and Hallanvuo, S., TaqMan-based real-time PCR method for detection of *Yersinia pseudotuberculosis* in food, *Appl. Environ. Microbiol.*, 74, 6465, 2008.
115. Nakajima, H. et al., Degradation of a polymerase chain reaction (PCR) product by heat-stable deoxyribonuclease (DNAse) produced from *Yersinia enterocolitica*, *Microbiol. Immunol.*, 38, 153, 1994.
116. Kaneko, S., Ishizaki, N., and Kokubo, Y., Detection of pathogenic *Yersinia enterocolitica* and *Yersinia pseudotuberculosis* from pork using the polymerase chain reaction. In Ravagnan, G., and Chiesa, C. (Editors), *Yersiniosis: Present and Future*, p. 153, Karger, Basel/Switzerland, 1995.
117. Kageyama, T. et al., *Yersinia pseudotuberculosis* infection in breeding monkeys: Detection and analysis of strain diversity by PCR, *J. Med. Primatol.*, 31, 129, 2002.
118. Lambertz, S.T., and Danielsson-Tham, M.L., Identification and characterization of pathogenic *Yersinia enterocolitica* isolates by PCR and pulsed-field gel electrophoresis, *Appl. Environ. Microbiol.*, 71, 3674, 2005.
119. Schleifstein, J.I., and Coleman, M.B., An unidentified microorganism resembling *B. lignieri* and Past. pseudotuberculosis, and pathogenic for man, *New York State J. Med.*, 34, 1749, 1939.
120. Winblad, S., Nilehn, B., and Jonsson, M., Two further cases, bacteriologically verified, of human infection with "Pasteurella X" (syn Yersinia enterocolitica), *Acta Pathol. Microbiol. Scand.*, 67, 537, 1966.

121. Winblad, S., Nilehn, B., and Sternby, N.H., *Yersinia enterocolitica* (*Pasteurella* X) in human enteric infections, *Brit. Med. J.*, 2, 1363, 1966.
122. Goverde, R.L. et al., The psychrotrophic bacterium *Yersinia enterocolitica* requires expression of pnp, the gene for polynucleotide phosphorylase, for growth at low temperature (5 degrees C), *Mol. Microbiol.*, 28, 555, 1998.
123. Wauters, G. et al., Somatic and flagellar antigens of *Yersinia enterocolitica* and related species, *Contrib. Microb. Immunol.*, 12, 239, 1991.
124. Fredriksson-Ahomaa, M., Stolle, A., and Korkeala, H., Molecular epidemiology of *Yersinia enterocolitica* infections, *FEMS Immunol. Med. Microbiol.*, 47, 315, 2006.
125. Wauters, G., Kandolo, K., and Janssens, M., Revised biogrouping scheme of *Yersinia enterocolitica*, *Contrib. Microb. Immunol.*, 9, 14, 1987.
126. Cornelis, G.R., The *Yersinia* deadly kiss, *J. Bacteriol.*, 180, 5495, 1998.
127. Cornelis, G.R. et al., The virulence plasmid of *Yersinia*, an antihost genome, *Microbiol. Mol. Biol. Rev.*, 62, 1315, 1998.
128. Boland, A., and Cornelis, G.R., Interaction of Yersinia with host cells, *Subcell. Biochem.*, 33, 343, 2000.
129. El Tahir, Y., and Skurnik, M., YadA, the multifaceted *Yersinia* adhesin, *Int. J. Med. Microbiol.*, 291, 209, 2001.
130. Iriarte, M., and Cornelis, G.R., Molecular determinants of *Yersinia* pathogenesis, *Microbiologia*, 12, 267, 1996.
131. Skurnik, M., and Wolf-Watz, H., Analysis of the *yopA* gene encoding the Yop1 virulence determinants of *Yersinia* spp., *Mol. Microbiol.*, 3, 517, 1989.
132. Han, Y.W., and Miller, V.L., Reevaluation of the virulence phenotype of the *inv yada* double mutants of *Yersinia pseudotuberculosis*, *Infect. Immun.*, 65, 327, 1997.
133. Bölin, I., and Wolf-Watz, H., Molecular cloning of the temperature-inducible outer membrane protein 1 of *Yersinia pseudotuberculosis*, *Infect. Immun.*, 43, 72, 1984.
134. Biedzka-Sarek, M., Venho, R., and Skurnik, M., Role of YadA, Ail, and lipopolysaccharide in serum resistance of *Yersinia enterocolitica* serotype O:3, *Infect. Immun.*, 73, 2232, 2005.
135. Tamm, A. et al., Hydrophobic domains affect the collagen-binding specificity and surface polymerization as well as the virulence potential of the YadA protein of *Yersinia enterocolitica*, *Mol. Microbiol.*, 10, 995, 1993.
136. Rosqvist, R., Skurnik, M., and Wolf-Watz, H., Increased virulence of *Yersinia pseudotuberculosis* by two independent mutations, *Nature*, 334, 522, 1988.
137. Miller, V.L., and Falkow, S., Evidence for two genetic loci in *Yersinia enterocolitica* that can promote invasion of epithelial cells, *Infect. Immun.*, 56, 1242, 1988.
138. Miller, L., Finlay, B.B., and Falkow, S., Factors essential for the penetration of epithelial cells by *Yersinia*, *Curr. Topics Microbiol. Immunol.*, 138, 15, 1988.
139. Miller, V.L., *Yersinia* invasion genes and their products, *AMS News*, 58, 26, 1992.
140. Pepe, J.C., and Miller, V.L., *Yersinia enterocolitica* invasin: A primary role in the initiation of infection, *Proc. Natl. Acad. Sci. USA*, 90, 6473, 1993.
141. Pepe, J.C., and Miller, V.L., The biological role of invasin during a *Yersinia enterocolitica* infection, *Infect. Agents Dis. Rev. Issues Comment*, 2, 236, 1993.
142. Delor, I., and Cornelis, G.R., Role of *Yersinia-Enterocolitica* Yst toxin in experimental infection of young rabbits, *Infect. Immun.*, 60, 4269, 1992.
143. Zhang, L. et al., Molecular and chemical characterization of the lipopolysaccharide O-antigen and its role in the virulence of *Yersinia enterocolitica* serotype O:8., *Mol. Microbiol.*, 23, 63, 1997.
144. Al-Hendy, A., Toivanen, P., and Skurnik, M., Lipopolysaccharide O side chain of *Yersinia enterocolitica* O:3 is an essential virulence factor in an orally infected murine model, *Infect. Immun.*, 60, 870, 1992.
145. Skurnik, M. et al., The lipopolysaccharide outer core of *Yersinia enterocolitica* serotype O:3 is required for virulence and plays a role in outer membrane integrity, *Mol. Microbiol.*, 31, 1443, 1999.
146. Najdenski, H. et al., Proper expression of the O-antigen of lipopolysaccharide is essential for the virulence of *Yersinia enterocolitica* O:8 in experimental oral infection of rabbits, *FEMS Immunol. Med. Microbiol.*, 38, 97, 2003.
147. Bengoechea, J.A., Najdenski, H., and Skurnik, M., Lipopolysaccharide O antigen status of *Yersinia enterocolitica* O:8 is essential for virulence and absence of O antigen affects the expression of other *Yersinia* virulence factors, *Mol. Microbiol.*, 52, 451, 2004.
148. Perez-Gutierrez, C. et al., Expression of the *Yersinia enterocolitica* pYV-encoded type III secretion system is modulated by lipopolysaccharide O-antigen status, *Infect. Immun.*, 75, 1512, 2007.
149. Skurnik, M. et al., Characterization and biological role of the O-polysaccharide gene cluster of *Yersinia enterocolitica* serotype O:9, *J. Bacteriol.*, 189, 7244, 2007.
150. Robins-Browne, R.M., and Prpic, J.K., Effects of iron and desferrioxamine on infections with *Yersinia enterocolitica*, *Infect. Immun.*, 47, 774, 1985.
151. Hayashidani, H. et al., Potential sources of sporadic human infection with *Yersinia enterocolitica* serovar O:8 in Aomori Prefecture, Japan, *J. Clin. Microbiol.*, 33, 1253, 1995.
152. Rastawicki, W. et al., A dramatic increase of *Yersinia enterocolitica* serogroup O:8 infections in Poland, *Eur. J. Clin. Mircobiol. Infect. Dis.*, 28, 535, 2009.
153. Fredriksson-Ahomaa, M. et al., Sporadic human *Yersinia enterocolitica* infections caused by bioserotype 4/O:3 originate mainly from pigs, *J. Med. Microbiol.*, 55, 747, 2006.
154. Bucher, M. et al., Epidemiological data on pathogenic *Yersinia enterocolitica* in Southern Germany during 2000–2006, *Foodborne Pathog. Dis.*, 5, 273, 2008.
155. Wesley, I.V., Bhaduri, S., and Bush, E., Prevalence of *Yersinia enterocolitica* in market weight hogs in the United States, *J. Food Prot.*, 71, 1162, 2008.
156. Milnes, A.S. et al., Intestinal carriage of verocytotoxigenic *Escherichia coli* O157, *Salmonella*, thermophilic *Campylobacter* and *Yersinia enterocolitica*, in cattle, sheep and pigs at slaughter in Great Britain during 2003, *Epidemiol. Infect.*, 136, 739, 2008.
157. Fredriksson-Ahomaa, M., and Korkeala, H., Low occurrence of pathogenic *Yersinia enterocolitica* in clinical, food, and environmental samples: A methodological problem, *Clin. Microbiol. Rev.*, 16, 220, 2003.
158. Lambertz, S.T. et al., Evaluation of a combined culture and PCR method (NMKL-163A) for detection of presumptive pathogenic *Yersinia enterocolitica* in pork products, *J. Food Prot.*, 70, 335, 2007.
159. Fredriksson-Ahomaa, M. et al., High prevalence of *Yersinia enterocolitica* 4:O3 on pig offal in southern Germany: A slaughtering technique problem, *Syst. Appl. Microbiol.*, 24, 457, 2001.

160. Boqvist, S. et al., Sources of sporadic *Yersinia enterocolitica* infection in children in Sweden, 2004: A case-control study, *Epidemiol. Infect.*, 137, 897, 2009.
161. Tauxe, R.V. et al., *Yersinia enterocolitica* infections and pork: The missing link, *Lancet*, 1129, 1987.
162. Ostroff, S.M. et al., Sources of sporadic *Yersinia enterocolitica* infections in Norway—a prospective case-control study, *Epidemiol. Infect.*, 112, 133, 1994.
163. Anonymous, *Yersiniosis*, Sausages-New Zealand, 2007.
164. Grahek-Ogden, D. et al., Outbreak of *Yersinia enterocolitica* serogroup O:9 infection and processed pork, Norway, *Emerg. Infect. Dis.*, 13, 754, 2007.
165. Fredriksson-Ahomaa, M., Korte, T., and Korkeala, H., Transmission of *Yersinia enterocolitica* 4/O:3 to pets via contaminated pork, *Lett. Appl. Microbiol.*, 32, 375, 2001.
166. Seuri, M., and Granfors, K., Possible confounders of the relationship between occupational swine contact and *Yersinia enterocolitica* O:3-antibody and O:9-antibody, *Eur. J. Epidemiol.*, 8, 532, 1992.
167. Lee, L. et al., *Yersinia enterocolitica* O:3 infections in infants and children, associated with the household preparation of chitterlings, *N. Engl. J. Med.*, 322, 984, 1990.
168. Archer, J.R. et al., Identification of *Yersinia* spp. with the API 20E system, *J. Clin. Microbiol.*, 25, 2398, 1987.
169. Neubauer, H. et al., Comparison of systems for identification and differentiation of species within the genus *Yersinia*, *J. Clin. Microbiol.*, 36, 3366, 1998.
170. EFSA (European Food Safety Agency), The community summary report on trends and sources of zoonoses, zoonotic agents, antimicrobial resistance and foodborne outbreaks in the European Union in 2006., *EFSA J.*, 130, 206, 2007.
171. Fredriksson-Ahomaa, M. et al., Characterisation of sucrose-negative Yersinia enterocolitica 4/O:3 isolates recovered from pig tonsils, *Int. J. Food Microbiol.*, 75, 19, 2002.
172. Harnett, N., Lin, Y.P., and Krishnan, C., Detection of pathogenic *Yersinia enterocolitica* using the multiplex polymerase chain reaction, *Epidemiol. Infect.*, 117, 59, 1996.
173. Lambertz, S.T. et al., Real-time PCR method for detection of pathogenic *Yersinia enterocolitica* in food, *Appl. Environ. Microbiol.*, 74, 6060, 2008.
174. Waage, A.S. et al., Detection of low numbers of pathogenic *Yersinia enterocolitica* in environmental water and sewage samples by nested polymerase chain reaction, *J. Appl. Microbiol.*, 87, 814, 1999.
175. Viitanen, A.-M. et al., Application of the polymerase chain reaction and immunofluorescence techniques to the detection of bacteria in *Yersinia*-triggered reactive arthritis, *Arthritis Rheum.*, 34, 89, 1991.
176. Weynants, V. et al., Detection of *Yersinia enterocolitica* serogroup O:3 by a PCR method, *J. Clin. Microbiol.*, 34, 1224, 1996.
177. Lambertz, S.T. et al., A combined culture and PCR method for detection of pathogenic *Yersinia enterocolitica* in food, *Int. J. Food Microbiol.*, 57, 63, 2000.
178. Aarts, H.J.M. et al., Rapid duplex PCR assay for the detection of pathogenic *Yersinia enterocolitica* strains, *J. Microbiol. Methods*, 47, 209, 2001.
179. Bhaduri, S., A comparison of sample preparation methods for PCR detection of pathogenic *Yersinia enterocolitica* from ground pork using swabbing and slurry homogenate techniques, *Mol. Cell. Probes*, 17, 99, 2003.
180. Lantz, P.G. et al., Detection of pathogenic *Yersinia enterocolitica* in enrichment media and pork by a multiplex PCR: A study of sample preparation and PCR-inhibitory components, *Int. J. Food Microbiol.*, 45, 93, 1998.
181. Knutsson, R. et al., Evaluation of selective enrichment PCR procedures for *Yersinia enterocolitica*, *Int. J. Food Microbiol.*, 73, 35, 2002.
182. Hudson, J.A. et al., Detection, isolation and enumeration of *Yersinia enterocolitica* from raw pork, *Int. J. Food Microbiol.*, 123, 25, 2008.
183. Jacobsen, N.R. et al., A real-time PCR assay for the specific identification of serotype O:9 of *Yersinia enterocolitica*, *J. Microbiol. Methods*,. 63, 151, 2005.
184. Wolffs, P. et al., Rapid quantification of *Yersinia enterocolitica* in pork samples by a novel sample preparation method, flotation, prior to real-time PCR, *J. Clin. Microbiol.*, 42, 1042, 2004.
185. Jourdan, A.D., Johnson, S.C., and Wesley, I.V., Development of a fluorogenic 5′ nuclease PCR assay for detection of the *ail* gene of pathogenic *Yersinia enterocolitica*, *Appl. Environ. Microbiol.*, 66, 3750, 2000.
186. Vishnubhatla, A. et al., Rapid 5′ nuclease (TaqMan) assay for detection of virulent strains of *Yersinia enterocolitica*, *Appl. Environ. Microbiol.*, 66, 4131, 2000.
187. Myers, K.M., Gaba, J., and Al-Khaldi, S.F., Molecular identification of *Yersinia enterocolitica* isolated from pasteurized whole milk using DNA microarray chip hybridization, *Mol. Cell. Probes*, 20, 71, 2006.
188. Hong, B.X. et al., Application of oligonucleotide array technology for the rapid detection of pathogenic bacteria of foodborne infections, *J. Microbiol. Methods*, 58, 403, 2004.
189. Kostic, T. et al., A microbial diagnostic microarray technique for the sensitive detection and identification of pathogenic bacteria in a background of nonpathogens, *Anal. Biochem.*, 360, 244, 2007.
190. Cocolin, L., and Comi, G., Use of a culture-independent molecular method to study the ecology of Yersinia spp. in food, *Int. J. Food Microbiol.*, 105, 71, 2005.
191. Ikeda, M. et al., Rapid and simple detection of food poisoning bacteria by bead assay with a microfluidic chip-based system, *J. Microbiol. Methods*, 67, 241, 2006.
192. Magliulo, M. et al., A rapid multiplexed chemiluminescent immunoassay for the detection of *Escherichia coli* O157:H7, *Yersinia enterocolitica*, *Salmonella typhimurium*, and *Listeria monocytogenes* pathogen bacteria, *J. Agric. Food Chem.*, 55, 4933, 2007.
193. Chaney, R., Rider, J., and Pamphilon, D., Direct detection of bacteria in cellular blood products using bacterial ribosomal RNA-directed probes coupled to electrochemiluminescence, *Transfus. Med.*, 9, 177, 1999.
194. Cloak, O.M. et al., Isolation and detection of *Listeria* spp, *Salmonella* spp and *Yersinia* spp using a simultaneous enrichment step followed by a surface adhesion immunofluorescent technique, *J. Microbiol. Methods*, 39, 33, 1999.
195. Nakajima, H. et al., Detection and identification of *Yersinia pseudotuberculosis* and pathogenic *Yersinia enterocolitica* by an improved polymerase chain reaction method, *J. Clin. Microbiol.*, 30, 2484, 1992.
196. Li, H., Bhaduri, S., and Magee, W.E., Maximizing plasmid stability and production of released proteins in *Yersinia enterocolitica*, *Appl. Environ. Microbiol.*, 64, 1812, 1998.

197. Anonymous, EN ISO 20838. International standard. Microbiology of food and animal feeding stuffs—Polymerase chain reaction (PCR) for the detection of foodborne pathogens—Requirements for detection and amplification for qualitative methods, *International Organisation for Standardisation*, Geneva, Switzerland, 2006.
198. Fricker, M. et al., Diagnostic real-time PCR assays for the detection of emetic *Bacillus cereus* strains in foods and recent food-borne outbreaks, *Appl. Environ. Microbiol.*, 73, 1892, 2007.
199. Anonymous, EN ISO 22174. International standard. Microbiology of food and animal feeding stuffs—Polymerase chain reaction (PCR) for the detection of foodborne pathogens—General requirements and definitions., *International Organisation for Standardisation*, Geneva, Switzerland, 2005.

Section IV

Proteobacteria

Epsilonproteobacteria

93 *Arcobacter*

Kurt Houf

CONTENTS

93.1 INTRODUCTION

Campylobacters were first isolated at the beginning of the twentieth century. In 1913, McFadyean and Stockman isolated *Vibrio*-like organisms from aborted ovine fetuses.[1] Five years later, Smith described the isolation of *Vibrio/Spirillum*-like organisms from aborted bovine fetuses, which he considered as the same species as described by McFadyean and Stockman.[2] Although the cell shape of the those organisms was characteristic of the family *Vibrionaceae*, they failed to ferment carbohydrates, and were therefore transferred into a new genus, *Campylobacter*.[3] Seventy years later, Ellis et al.,[4,5] and Higgins and Degre,[6] reported the isolation of aerotolerant, spiral-shaped *Campylobacter*-like organisms from the organs of porcine and bovine fetuses. Examination of all *Campylobacter* and *Campylobacter*-like isolates revealed two distinct biochemical groups: the group 1 isolates were identified as *Campylobacter* fetus, whereas the group 2 isolates were considered as the "aerotolerant campylobacters."[7,8] In a comprehensive study in 1985, Neill et al.[9] provided a thoroughly documented description for a new species, *Campylobacter cryaerophila*, to include these "aerotolerant campylobacters" but also emphasized their phenotypic heterogeneity. In 1988, Thompson et al.[10] showed by partial 16S rRNA sequence analysis that *C. cryaerophila* and *Campylobacter nitrofigilis*, an organism isolated from the roots of *Spartina alterniflora*, exhibited more homology with each other than with other campylobacters, suggesting that classification of both *C. cryaerophila* and *C. nitrofigilis* in another genus would be appropriate. In the early 1990s, Kiehlbauch et al.[11] found two subgroups among *C. cryaerophila* strains by phenotypic characterization and DNA–DNA hybridization. Catalase-positive strains that were able to grow under aerobic conditions at 30°C but not in a medium with glycine or on MacConkey agar were considered as *C. cryaerophila*. The name *Campylobacter butzleri* was proposed for the aerotolerant isolates that were negative to weakly catalase positive, and for which growth was observed in glycine minimal medium an on MacConkey agar. In 1991, after a polyphasic taxonomic study, Vandamme et al.[12] transferred *C. cryaerophila* and *C. nitrofigilis* into a new genus, *Arcobacter*. Subsequently, *C. butzleri* was reclassified as *Arcobacter butzleri* and *Arcobacter skirrowii* was proposed for yet another group of animal associated *Arcobacter* strains.[13]

93.1.1 Genus *Arcobacter*

Due to the close phenotypic and genotypic affiliation, the genus *Arcobacter* was classified together with the genus *Campylobacter* into the family *Campylobacteraceae*.[14] Together with the genera *Helicobacter*, *Wolinella*, and *Sulfurospirillum*, they constitute the most important representatives of a distinct group referred to as the rRNA superfamily VI, or as the ε-division of the class *Proteobacteria*.[12] Arcobacters are gram-negative, nonspore-forming, motile, spiral-shaped rods (0.2–0.8 × 0.5–5 μm) that are able to grow microaerobically or aerobically.[12] They have the ability to grow from 15°C to 37°C, with an optimal growth in microaerobic conditions at 30°C. The growth at 15°C and the aerotolerance are the features that differentiate *Arcobacter* from *Campylobacter* species.[12,15] In 2007, the complete genome sequence analysis of *A. butzleri* revealed, however, that the majority of its proteome is most similar to those of *Sulfuromonas denitrificans* and *Wolinella succinogenes*, and to those of the deep-sea vent Epsilonproteobacteria *Sulfurovum* and *Nitratiruptor*.[16] The presence of pathways and loci associated with nonhost-associated organisms, as well as genes associated with virulence, suggests that *A. butzleri* is a free-living, waterborne organism.[16]

To date, nine species have been described: *Arcobacter butzleri*, *Arcobacter cibarius*, *Arcobacter cryaerophilus*, *Arcobacter halophilus*, *Arcobacter marinus*, *Arcobacter mytili*, *Arcobacter nitrofigilis*, *Arcobacter skirrowii*, and *Arcobacter thereius*.[12,17–19] *A. butzleri* has been predominantly associated with human infection as it has been isolated from stool of patients with diarrhea,[20,21] but not from healthy humans.[22] Furthermore, the presence of virulence genes,[16] and its cytopathogenic effect on in vitro cell lines,[23,24] resulted in the classification of this species as an emerging pathogen.[25] Besides humans, *A. butzleri* has been isolated from healthy and ill livestock,[26] poultry,[27,28] nonhuman primates,[29–31] and diverse environmental matrices such as water and sludge.[32] *Arcobacter cryaerophilus* is a genotypically heterogeneous species, which was originally divided into two subgroups by fatty acid analysis[13] or by restriction fragment length polymorphism-DNA analysis.[33] As for *A. butzleri*, *A. cryaerophilus* strains have been isolated from a whole range of different matrices, including from stool of healthy humans,[22] as well as from patients with enteritis.[20] *Arcobacter skirrowii* has first been isolated from feces of lambs with diarrhea,[13] but later on also from preputial fluids of bulls, organs of bovine, porcine, and ovine aborted fetuses, animal feces, food, and water. Association of *A. skirrowii* with human disease has only rarely been reported.[34,35] In 2005, *A. cibarius* was described as a new *Arcobacter* species, with *A. cryaerophilus* as the closest phylogenetic neighbor.[18] The first representatives were isolated from broiler carcasses,[18] but association with pigs has been suggested as well.[36] Whereas *Arcobacter marinus* and *Arcobacter mytili* were identified in seawater and seafood (mussels), *Arcobacter thereius* has been isolated in the organs of aborted piglets and duck feces.[37,38]

Besides the animal and human related species, different free-living, environmental *Arcobacter* species have been described. *Arcobacter nitrofigilis* is a nitrogen-fixing bacterium isolated from the roots of the salt marsh plant *Spartina alterniflora*.[17] The strains prefer microaerobic growth conditions, but aerobic growth is possible in adapted culture media.[39] An obligate microaerophile marine sulfide-oxidizing autotrophic bacterium has been described that produced hydrophilic filamentous sulfur as a novel metabolic end product for which the name *Candidatus Arcobacter sulfidicus* sp. nov. has been proposed.[40] A phylogenetic affiliation of a gram-negative bacterium isolated from water collected at a hypersaline lagoon on the Hawaiian Islands to the genus *Arcobacter* was confirmed for which the name *A. halophilus* was introduced.[19] Cells are slightly curved, obligate halophile, and growth is observed in culture media containing 2%–4% salt or 0.1% potassium nitrate under aerobic and microaerobic conditions at room temperature and at 37°C, and under anaerobic conditions at 37°C. Finally, DNA from a number of *Arcobacter*-like organisms has been detected in diverse sources as saltwater lakes, coastal seawater, oil wells, bio-catalytic calcification reactor, sediments in the Black Sea, sludge, and production waters.[41–52]

93.1.2 Arcobacters in Humans

Interest for arcobacters in veterinary and human public health has enhanced since the first report of the isolation of arcobacters from food of animal origin.[53] Since then, studies worldwide have reported the occurrence of arcobacters on food and have highlighted the possible transmission to the human population. However, since the clarification of the taxonomic position of *Arcobacter*, only a few surveys have dealt with the clinical course of *Arcobacter* infection in humans. In 2000, Engberg et al.[54] isolated one *A. butzleri* and one *A. cryaerophilus* from a total of 3267 clinical stool specimens. In the same year, Lastovica et al.[55] reported an *A. butzleri* prevalence of 0.39% in a study realized on 19,535 diarrheal stool of pediatric patients. During an 8-year study period, Vandenberg et al.[20] reported *A. butzleri* as the fourth most common *Campylobacter*-like organism isolated from 67,599 stool specimens form patients with diarrhea. In the study, *A. butzleri* was more associated with a persistent and watery diarrhea—and less with bloody diarrhea—than compared with *C. jejuni* infection, and the organism has been recovered from patients of different ages (<1–90 years old). Similar results have been reported from a surveillance network in France in 2006.[21]

Most clinical *Arcobacter* infections are single cases with the source of infection rarely identified. The first association of *Arcobacter* with human infection was reported in 1987. From the feces of a 35-year-old man with acute diarrhea and abdominal pain, *A. cryaerophilus* was isolated without detection of other pathogens.[56] Since then, *A. cryaerophilus* and *A. butzleri* infections were reported several times from stool samples of patients with acute diarrhea.[11,35,54,57–60] In 2004, the first human isolation of *A. skirrowii* was reported by Wybo et al.[34] from a 73-year-old patient with chronic diarrhea. More recently, *A. skirrowii* has been detected in diarrheal stool samples in South Africa.[35]

Besides the reports of the single cases of *Arcobacter*-associated enteritis, two *Arcobacter* outbreaks have been described. The first outbreak occurred in an Italian nursery and primary school where children suffered from recurrent abdominal cramps without diarrhea. The isolates were identified as *A. butzleri* and the identical phenotypic characteristics and genotypic profiles of the isolates suggested an epidemiological relation for all cases.[60] Person-to-person transmission was assumed as the ongoing cause of infection. A second outbreak appeared during a scout camp, where 94 girls suffered from nausea, vomiting, abdominal cramps, and diarrhea.[61] The outbreak was assumed to be correlated with the breakdown of the automated water chlorinating system. No clear cause of infection could be identified by the examination of stool samples, but *A. butzleri* was the only pathogen detected in the drinking water.

Besides the correlation with gastroenteritis, *Arcobacter* has also been implicated in extra-intestinal invasive diseases. *Arcobacter butzleri* was isolated from the blood of a neonate and the clinical data indicated an in utero sepsis.[62] Yan et al.[63] reported an *A. butzleri* isolation from two blood cultures of a man with liver cirrhosis displaying high fever and esophageal variceal bleeding, and in another bacteremia report also mentioned, beside *Escherichia coli* and *Streptococcus milleri*, the isolation of *A. butzleri* from a patient with an acute gangrenous appendicitis.[59] Bacteremia due to *A. cryaerophilus* has been reported in a patient with hematogenous pneumonia,[64] and in a traffic accident victim.[65]

The significance of *Arcobacter* as a cause of human diarrhea is still largely unknown, since clinical samples are not routinely tested for *Arcobacter* species. The symptoms are similar to those of *Campylobacter* infections, and *Arcobacter* infections often have a spontaneous recovery.[66] This makes prevalence-determination difficult and mostly incorrect.[20,67]

93.1.3 Virulence Factors

To assess the pathogenicity of arcobacters for humans and animals, evaluation of potential virulence factors is required. However, up to now, little is known about the mechanisms of pathogenicity. A necessary state in the successful colonization, establishment, and ultimately production of disease by microbial pathogens is the ability to adhere to host surfaces such as mucous membranes, gastric and intestinal epithelial or endothelial tissue. Therefore, it is a common trait of microbial pathogens to express adherence factors responsible for recognizing and binding to specific receptor moieties of cells, thus enabling the bacteria to resist host strategies that would impede colonization.

Since campylobacterioses and *A. butzleri*-related illnesses have similar clinical features, it might be expected that some *C. jejuni* virulence factors would be present in *Arcobacter*. In the genomic sequence analysis of *A. butzleri* strain LMG 10828^T, homologs of the fibronectin binding proteins CadF and Cj1349, the invasin protein CiaB, the virulence factor MviN, the phospholipase PldA, and the TlyA hemolysin were detected.[16] Several other *Campylobacter* virulence-associated genes however were not present with most notably the genes encoding the cytolethal distending toxin (CDT), which correlated with the study by Johnson and Murano who were unable to detect CDT-genes in *Arcobacter* species by PCR.[68] Furthermore, three additional putative virulence determinants have been identified: irgA, which encodes an iron-regulated outer membrane protein; hecA, a member of the filamentous hemagglutinin family; and hecB, a related hemolysin activation protein.[16] The role and the functionality of these virulence determinants has not been determined yet.

The first in vivo study with *Arcobacter* strains intravenous or intraperitoneal inoculated in rodents was unsuccessful.[6] No clinical symptoms were present and lesions were not observed by autopsy. However, the invasion and colonization capacity has later been demonstrated by Wesley et al.[69] as *A. butzleri* was isolated from feces and different organs up to seven days after experimental inoculation of piglets. Also, the invasive capacity of *A. cryaerophilus* has been studied in an experimentally infected rainbow trout with death and clinical abnormalities reported.[70] Agglutination of *A. butzleri* strains with human, rabbit, and sheep erythrocytes revealed the presence of adhesion molecules in arcobacters.[71] No fimbriae or pili were observed by electron microscopic examination, but an immunogenic haemagglutinin of 20 kDa was characterized. The haemagglutinin consisted of a lectin-like molecule, sensitive to heat treatment and enzymatic proteolysis, which could interact with a D-galactose-containing erythrocyte receptor.

Several in vitro studies on the adhesion, invasion, and interleukine-8 and toxin-production capacity by *Arcobacter* strains of different origin on Vero-, Hep-2, INT407, Caco-2, IPI-2I, and HeLa-cells have been performed.[22–24,72,73] A strong cytopathogenic effect was observed in Vero cells comparable to the effect of VT-toxin of *E. coli* O157,[72,74] whereas Hep-2- and HeLa-cells showed weak cytopathogenic liason. Cell rounding and nuclear pyknosis was observed with *Arcobacter* isolates on HeLa and Intestine 407 cells and cell elongation on CHO-cells.[24] The presence of a vacuole-forming toxin, different from *Campylobacter* CDT, in *Arcobacter* strains has been demonstrated in a Vero cell culture setup.[75]

Fernandez et al.[73] and Carter[76] performed in situ studies about the existence of toxigenic and invasive capacities of arcobacters. In those studies, the toxigenic capacity of arcobacters from animal origin was determined in the rat and rabbit and pig ileal loop tests, respectively. In both studies, distention of the ileal loops with fluid accumulation and enhanced electrolyte concentrations was observed.

The information based on molecular and in vitro studies, presently available suggest that adhesion, invasion, and toxin production could be mechanisms by which *Arcobacter* species may cause disease. However, whereas colonization of the intestine and production of cytotoxins by bacteria is generally associated with bloody diarrhea, this symptom is rarely described in the human cases reported to date. Further studies are certainly necessary to elucidate the pathobiology of *Arcobacter* species.

93.1.4 Antimicrobial Susceptibility

Specific standardized procedures for susceptibility testing of *Campylobacteraceae* and resistance breakpoints have not been established. Consequently, a number of different testing methods, such as agar and broth (micro)dilution,[77–79] disc diffusion,[80–82] and the E-test [63,83] have been used for *Arcobacter* susceptibility testing in clinical, veterinary, and food microbiology. Furthermore, due to the fastidious nature and the microaerobic growth requirements of those microorganisms, the quality control limits given for nonfastidious organisms in aerobic atmosphere are not adequate.[84] It is known that an increased level of carbon dioxide does decrease the effect of certain antimicrobials such as macrolides and fluoroquinolones. This will certainly occur in the microaerobic atmosphere required for the growth of *Campylobacteraceae*, and should be taken into account when interpreting susceptibility patterns.[84] Comparison of the broth microdilution, the E-test, and the agar dilution method showed overall comparable results for *Campylobacter* susceptibility testing when performed under the same microaerobic conditions and incubation temperature. Especially the minimum inhibitory concentration (MIC) obtained by the three methods of ciprofloxacin and erythromycin were in accordance with each other.[85]

In 1992, Kiehlbauch et al.[79] applied the broth microdilution technique under aerobic atmosphere for susceptibility testing of the same panel of antimicrobials for *A. butzleri* and *A. cryaerophilus* isolates. The MICs for ciprofloxacin, erythromycin, doxycycline, and nalidixic acid differ by no more than one dilution from those obtained in the study of Houf et al.[86] However, with gentamicin, MICs for *A. butzleri* and *A. cryaerophilus* ranged from ≤0.12 to 0.5 μg/mL, whereas in the study of Houf et al.[86] The MICs ranged from 0.5 to 4 μg/mL and 0.25 to 2 μg/mL, respectively. It is unknown whether the higher carbon dioxide concentration may decrease the activity of aminoglycosides as well. In the susceptibility study by Fera et al.,[77] even higher MIC ranges for both ciprofloxacin and gentamicin were obtained, although the same incubation atmosphere and temperature were applied. Whether the different origin of the strains in the studies, water versus human stools and poultry carcasses, is the cause of those MIC shifts is not clear.

In contrast to thermophilic *Campylobacter* species, of which some strains demonstrate resistance to quinolones and cross-resistance between nalidixic acid and quinolones, most of the *Arcobacter* strains seem to be susceptible to both antimicrobial agents. Remarkable, however, is the lowered susceptibility and even resistance to ciprofloxacin of *A. butzleri* strains isolated from poultry.[86] The latter finding is also demonstrated by the concentrations required to inhibit growth of 50% of the strains (MIC_{50}): the MIC_{50} for *A. butzleri* isolated from poultry is 0.12 whereas the MIC_{50} for human strains is 0.03.[86] The use of fluoroquinolones for treatment of poultry may be the basis for this decreased susceptibility.

Since poultry products in particular are incriminated for the transmission of arcobacters to humans, the presence of antimicrobial-resistant *Arcobacter* species in fresh poultry products can have public health implications. Today, data indicate that some *Arcobacter* isolates from poultry products are resistant and that multidrug resistance occurs. Especially the resistance to erythromycin and the decreased susceptibility to ciprofloxacin may have human health implications, as the two antimicrobials are generally prescribed as first-line drugs for the treatment of infections with *Campylobacteraceae*.

93.1.5 *Arcobacters* in Animals

Apart from *A. nitrofigilis* and *A. halophilus*, *Arcobacter* species have been incriminated with various animal diseases including abortion, septicemia, mastitis, gastritis, and enteritis.[11,29,87–91] However, several studies reported the occurrence of arcobacters in healthy livestock and poultry, detected by different isolation methods and molecular techniques.[26,92–99] Beside the clinical relevance, the occurrence of arcobacters in healthy animals may act as significant reservoir and infection source to humans.

In cattle, arcobacters have been associated with pathologies such as mastitis and reproduction disorders, but have more commonly been isolated from feces of clinically healthy animals.[93,95] In 1977, arcobacters were isolated for the first time from the placenta and the internal organs of aborted bovine fetuses.[4] Although several studies have confirmed these observations,[6,100–102] no studies have clearly identified arcobacters as a causative agent of disease in cattle.

As in cattle, the first reports of arcobacters in pigs were associated with reproduction disorders. On farms with a history of *Arcobacter*-associated abortions, sows with reproduction problems expressed high antibody serum titers in a microscopic agglutination test.[103–106]

There are conflicting reports in literature whether or not *Arcobacter* is part of the poultry intestinal flora.[94,95,107–113] Besides the reports of *Arcobacter* in chickens and turkeys, the presence of arcobacters in ducks and geese has also been described.[37,114–131]

The occurrence of arcobacters in horses has been examined by Van Driessche et al.[26] The presence of *Arcobacter* spp. in raccoons (*Procyon lotor*) was reported for the first time in 2004.[117] Arcobacters have been isolated both from ill and healthy nonhuman primates. Several cases have been reported of the isolation of *A. butzleri* from healthy infant macaques (*Macaca nemestrina*) in a monkey nursery facility and from Rhesus macaques (*Macaca mulatta*) with and without diarrhea.[29–31]

Besides contaminated water, food of animal origin is another possible route of transmission of arcobacters to humans, possibly via manipulation of raw meat, the consumption of undercooked products, and cross-contamination. Recent studies have indicated that also arcobacters are common on broiler carcasses. Arcobacters have also been isolated from skin samples of commercially reared ducks and turkeys. Eggs seem not to be infected. Apart from chickens, arcobacters have been isolated from geese and ducks. The variations in recovery rates can be due to differences in country, farm management, and hygiene in slaughterhouses and processing companies.[118]

93.1.6 Identification and Characterization of *Arcobacters*

Members of the genus *Arcobacter* are gram-negative, nonspore-forming organisms. Cells are usually slender, curved rods, 0.2–0.9 μm wide and 0.5–3 μm long. S-shaped or helical cells are often present. Cells in old cultures may form spherical or coccoid bodies and loose spiral filaments up to 20 μm long.[132] Arcobacters are motile with a characteristic corkscrew-like motion by means of a single polar unsheathed flagellum at one or both ends of the cell.[13] Although the cell surface is critical in pathogenic processes such as the colonization of the host, resistance to host defense systems and invasion of cells, the cell surface characteristics of arcobacters are still largely unknown.[133] Research about the presence of filamentous appendages, or specific proteins with porin and adhesive properties, which have received detailed research because of their role in the pathogenesis in both *Campylobacter* and *Helicobacter* species, has not been performed. Neither specific proteins in the S-Layer, as described for *Campylobacter fetus*, nor polysaccharide components responsible for serotype specificity, are known to date.

Optimal growth occurs at 30°C under microaerobic conditions, with a respiratory type of metabolism. Hydrogen is not required. After primary isolation in a microaerobic environment, growth is possible in aerobic or anaerobic atmosphere. Growth can occur at 15°C–37°C and growth at 42°C is described for some *A. butzleri* and *A. skirrowii* strains.[134] Colonies of *A. butzleri* and *A. cryaerophilus* grown for 48 h under optimal conditions are respectively 2–4 mm and 1–3 mm in diameter and are convex with an entire edge. Growth of *A. butzleri* has a whitish appearance whereas colonies of *A. cryaerophilus* have mostly a dirty yellow pigment. Colonies of *A. skirrowii* grown for 48 h under optimal conditions are 1–3 mm in diameter and have a flat irregular shape. Growth has a grayish appearance and is usually not profuse.

Identification of arcobacters at species level by biochemical characteristics is difficult, as members of the genus display little metabolic activity.[13,134] Classical biochemical differentiation of the genus *Arcobacter* from the related genera *Campylobacter* and *Helicobacter*, is primarily achieved by the identification of the individual species. In general, arcobacters can be differentiated from campylobacters by their lower optimal growth temperatures (25°C–30°C compared to 30°C–42°C for campylobacters) and aerotolerance. However, in the identification of *Campylobacteraceae* at species level by the use of phenotypic tests, some fundamental problems can occur. First, many phenotypic tests used to differentiate other bacterial groups, such as the members of the family *Enterobacteriaceae*, have no discriminatory power for *Campylobacteraceae* because of their fastidious growth requirements and their relative metabolic inertness. For example, *Arcobacter* and *Campylobacter* species do not ferment nor oxidize carbohydrates. Second, there is a lack of standardization for the tests that are used. For example, the outcome of a given test may be influenced by the inoculum size, cultural age, and basal medium used. The third problem is the lack of objectivity in the schemes available. Most tables so far described are used by comparing the test results of the unknown with the phenotypic profiles of known taxa and more importance is often attached to a single test result that is considered as an essential character than to the remainder of the phenotype. In a comprehensive study in 1996, On et al.[135] documented an identification scheme for *Campylobacteraceae* and *Helicobacter* species, by which most *Campylobacteraceae* can be identified accurately and objectively with phenotypic tests when probabilistic methods of data assessment are employed.

In contrast to other organisms such as *Salmonella*, *Campylobacter*, and *Listeria*, serology is not used for *Arcobacter* identification as attempts to perform genus or species identification by specific antibody agglutination were not successful.[136] The high antigenic heterogeneity within the *Arcobacter* species may be on the basis of this failure. Due to the rather disappointing results, serological identification was not further extended, and is not further used to date.

Differentiation between *Arcobacter* and *Campylobacter* isolates is possible by the determination of the cellular fatty acid composition. *Arcobacter* species possess a unique combination of a tetradecenoic acid C14:1 and two isomers of C16:1,[137] later identified as C14:1ω7-*cis*, C16:1ω7-*cis* (common in most bacteria), and C16:1ω7-*trans* (considered unique for *Arcobacter*).[138,139] Within the genus *Arcobacter*, *A. nitrofigilis*, *A. skirrowii*, and the two *A. cryaerophilus* subgroups were differentiated by gas chromatography of the cellular fatty acids, but it was not able to differentiate *A. butzleri* from *A. cryaerophilus* subgroup 2.[13] To date, due to the rather complex analysis protocol and the availability of faster, less complex, and cheaper molecular-based methods, identification based on fatty acid profiles is not commonly applied, though its usefulness has been demonstrated by Jelinek et al.[140]

Identification based on whole-cell proteins profiles obtained by SDS–PAGE has been the gold standard method since the description of the genus *Arcobacter*. By SDS–PAGE, all known *Arcobacter* species can be identified including the differentiation between A. *cryaerophilus* subgroup 1 and 2. However, an enforced standardization of the protocol combined with a profile library of known and related species and genera is necessary. As this method is rather time consuming, it can hardly be applied in routine analysis.

(i) Molecular Identification Differentiating *Arcobacter* species by using phenotypic tests might give erroneous results because of the shortage of clear-cut differentiating tests, a phenomenon that has also been observed in the closely related genus *Campylobacter*. Therefore, several DNA-based assays were developed for the identification of arcobacters at the genus and species level. Recently, a microarray technique and a real-time fluorescence resonance energy transfer PCR for the detection and identification of *Arcobacter* spp. were reported, but are not yet routinely applied.[141,142]

Restriction Fragment Length Polymorphism (RFLP). Using whole-cell chromosomal digests by the restriction enzym *Pvu*II and hybridization with probes derived from the *E. coli* 16S and 23S rRNA genes, Kiehlbauch et al.[33,143] developed a DNA-based method to differentiate the genera

Campylobacter, Arcobacter, Helicobacter, and *Wolinella.* Although it is not able to distinguish *A. butzleri* from *A. skirrowii,* the method can be used for the differentiation of *A. cryaerophilus* from the other *Arcobacter* taxa and for the two subgroups of the latter species.[144] Sequence analysis of the conserved 16S rRNA gene of *Arcobacter* allowed Wesley et al.[144] to design a genus-specific nucleic acid probe and a species-specific DNA probe for *A. butzleri.* Southern blot hybridization of *Pvu*II digested DNA using the *Arcobacter* genus-specific probe or the *A. butzleri*-specific probe end-labeled with [γ^{32}P]ATP provided a reliable identification method for arcobacters at genus level and for *A. butzleri.*

PCR. Analysis of the ribosomal gene sequence has proven to be a valuable tool in the determination of phylogenetic relationships between prokaryotes.[10] A high (>94%) 16S rRNA gene sequence similarity was detected among the published *Arcobacter* species type strains. On the other hand, similarity to other members of the epsilon *Proteobacteria* was low (<90%).[18,144] One of the first described DNA-based identification techniques included a genus- and species-specific PCR reaction developed by Bastyns et al.,[145] with five primers targeting the 23S rRNA. Based on the sequence of the *Arcobacter* and *A. butzleri*-specific DNA probes described by Wesley et al.,[144] two primer pairs were developed that could be used in an *Arcobacter* genus- and an *A. butzleri*-specific PCR reaction with different annealing temperature.[146] When the species-specific primers were replaced by the species-specific primers of Bastyns et al.,[145] the first *Arcobacter* multiplex-PCR (m-PCR) was created: the genus *Arcobacter* (1223 bp) and the species *A. butzleri* (686 bp) were identified in one PCR amplification.[147] By means of five primers, a PCR product of 401 bp was generated for *A. butzleri,* 257 bp for *A. cryaerophilus,* and 641-bp for *A. skirrowii.* Those three species were also identified by the PCR assay developed by Kabeya et al.,[148] but *A. cryaerophilus* subgroup 1 and 2 were also differentiated from each other.

A PCR-hybridization protocol differentiated *A. butzleri* from the related *C. jejuni, C. coli, C. lari,* and *C. upsaliensis* strains.[149] The conserved *gly*A gene region of isolates was amplified during a PCR reaction followed by a hybridization reaction of the amplicons with species-specific oligodeoxyribonucleotide probes. Another example is a culture-PCR method, used to detect arcobacters on chicken meat.[150] After enrichment of the meat sample, identification of arcobacters in the medium was performed by a new developed genus-PCR assay that generated an amplicon of 181-bp for *Arcobacter* positive samples.

RFLP–PCR. Combined use of PCR with RFLP was first described by Cardarelli-Leite et al.[151] RFLP analysis of a PCR-amplified DNA fragment of the gene coding for 16S rRNA of *Campylobacter, Helicobacter, Arcobacter,* and *Wolinella succinogenes,* generated a 283 bp fragment from all species belonging to the examined genera. Initial restriction of the amplicon by *Dde*I delivered a unique pattern for *A. butzleri.* Performing additional digestion using *Hpa*II on the *Dde*I digested fragments, *A. cryaerophilus, A. skirrowii,* and *A. nitrofigilis* can be distinguished as a single group from the *Campylobacter* and *Helicobacter* species, although further differentiation at species level is not possible. Hurtado and Owen[152] performed a comparable study in which amplicons ranging from 2.6 to 3.0 kb are generated by a PCR assay using primers in the conserved region of the 23S rRNA gene of *Campylobacter* and *Arcobacter.* Digesting these amplicons with *Hae*III, *Cfo*I, *Hpa*II, and *Hin*fI, species-specific patterns for *A. butzleri* and *A. nitrofigilis* and identical patterns for *A. cryaerophilus* and *A. skirrowii* can be obtained. In 1999, Marshall et al.[153] combined a PCR assay with RFLP for the identification of arcobacters at the species level. By amplifying a 1004 bp fragment using primers targeting a conserved region of the 16S rRNA gene, followed by restriction endonuclease digestion with *Dde*I and *Taq*I, species-specific RFLP patterns were obtained for *A. butzleri, A. cryaerophilus,* and *A. skirrowii.*

AFLP. The genotyping AFLP technique may also be used for concurrent species identification of the family *Campylobacteraceae,* including the *Arcobacter* species.[38,154] The species *A. butzleri, A. cryaerophilus,* and *A. skirrowii* form well-distinguished clusters in the dendrogram obtained after numerical analysis of yielded patterns.

(ii) Molecular Characterization The determination of the relation of isolates below species level has become very important for the identification of transmission routes and biological reservoirs in epidemiological studies. Phenotyping methods such as biotyping,[155] serotyping,[156] or comparisons of whole-cell proteins[13] are of limited use due to their low discriminatory power and the instability and low reproducibility of the phenotypic characteristics.[134]

The suitability of pulsed-field gel electrophoresis (PFGE) and random amplification of polymorphic DNA (RAPD) as characterization techniques described by Lior and Wang[157] was confirmed for fecal or meat isolates in recent reports.[96,123,125,158,159] The AFLP technique proved its qualities as genotyping method for all members of the *Campylobacteraceae.*[15,38,154] Good distinguished clusters were obtained for *A. butzleri, A. cryaerophilus,* and *A. skirrowii,* with profiles reproducing the clonal relation from the isolates within each species. The AFLP-method is a robust method with high discriminatory power, but it is nevertheless a demanding technique and requires a large reference database.[66]

An enterobacterial repetitive intergenic consensus (ERIC) PCR was optimized for the characterization of *A. butzleri, A. cryaerophilus,* and *A. skirrowii* strains in combination with a rapid DNA-extraction method.[159] The technique is described in detail in Section 93.2.2.4. Fingerprints generated with ERIC–PCR are complex enough to differentiate at strain and substrain level and have a good reproducibility.

93.2 METHODS

93.2.1 Sample Preparation

Arcobacter Isolation. The isolation of arcobacters is generally performed by a selective enrichment step, followed by plating on a selective or nonselective agar plate and takes approximately 4–5 days. This selective isolation is achieved predominantly by the use of antimicrobial agents in the enrichment and plating media or by incubations below 30°C.

The first arcobacters were coincidentally isolated using the semisolid Ellinghausen–McCullough–Johnson–Harris (EMJH) medium containing 5-fluorouracil as selective supplement, originally developed for the isolation of *Leptospira*.[4] Since the description of the genus *Arcobacter*, different isolation methods and protocols have been used to examine the presence of arcobacters in different matrices. The first attempts were modifications of methods developed for thermophilic campylobacters or *Yersinia* species,[58,129] and applied commercially available selective supplements such as cefoperazone, amphotericin B, teicoplanin (CAT)[160]; cefoperazone, charcoal, deoxycholate agar (CCDA)[128]; or cefsulodin, irgasan, novobiocin (CIN).[129] Those methods are either time consuming, do not support the growth of all *Arcobacter* species, or do not sufficiently inhibit the accompanying flora.[78] The first specific selective *Arcobacter* isolation medium developed for *Arcobacter* isolation from meat and meat products contained cefoperazone, trimethoprim, piperacillin, and cycloheximide in the selective supplement.[53] The isolation protocol consists of an enrichment in *Arcobacter* selective broth followed by plating on a semisolid *Arcobacter* selective medium. After incubation, plates are examined for the presence of a motility zone. Since antibiotic sensitivity tests revealed that the piperacillin and cefoperazone concentrations in the medium inhibited the growth of *A. skirrowii*, this medium is not generally applied any longer. Another medium has been developed by Johnson and Murano in which the inhibition of the accompanying flora was obtained with a selective broth and agar containing cefoperazone and 5-fluorouracil as selective additives.[161]

A study on the minimal inhibitory concentrations (MICs) of selective agents on the growth of *Arcobacter* illustrated that most of the isolation media contained concentrations that inhibited the recovery of one or more *Arcobacter* species or did not sufficiently suppress the accompanying flora present in the sample.[78] Based on these findings, a qualitative and quantitative isolation medium was developed by Houf et al.[162] A broth and agar medium contained a combination of different antimicrobial agents as selective supplement: amphotericin B, cefoperazone, 5-fluorouracil, novobiocin, and trimethoprim. The qualitative isolation procedure included enrichment in broth followed by plating on selective agar. Enumeration of the arcobacters in the sample was performed by direct plating of 100 μL homogenate onto the selective agar plate. After microaerobic incubation, colonies were counted and the bacterial load could be determined. This protocol provides a fast and reliable method for the isolation and enumeration of all food-related *Arcobacter* species with a good suppression of the contaminating flora.

DNA Extraction. Arcobacters can be detected directly in the enrichment media of food samples, in blood or fecal material, or after isolation on selective or nonselective agar plate. Detection at enrichment phase or directly in the organic specimen is, however, often hindered by the presence of inhibiting components as blood, bile salts, and organic material, in particular fat. Furthermore, a minimum number of arcobacters must be present. The detection can, however, be enhanced by centrifugation of the enrichment broths or homogenates followed by several wash steps of the bacterial pellet to dilute the inhibiting components. Extraction of the DNA can then be performed by several commercially available DNA-extraction kits, often designed specifically for a particular clinical specimen, or by heating the washed bacterial pellet for 15 min at 95°C, or by a more extended DNA-extraction procedure as the guanidine thiocyanate extraction procedure according to Pitcher et al.,[163] described in detail later. The use of commercial extraction kits or the guanidine thiocyanate extraction procedure often results in more pure, nonfragmented DNA and, after standardization of the amount of dsDNA, is recommended for the molecular-based characterization of *Arcobacter* isolates.

DNA-Extraction Protocol

1. Suspend an *Arcobacter* colony in 500 μL RS buffer [0.1 M EDTA (ethylene diaminetetraactetic acid) and 0.15 M NaCl, pH 8.0];
2. Centrifuge for 2 min at 15,000 × *g* and remove the supernatant;
3. Wash the pellet in 500 μL RS buffer;
4. Centrifuge 2 min at 15,000 × *g* and remove the supernatant;
5. Suspend the pellet in 100 μL TE buffer (10 mM Tris-HCl pH 8, 1 mM EDTA);
6. Add 500 μL GES solution (Guanidium thiocyanate-EDTA-Sarkosyl; for 100 mL: 60 g Guanidiumthiocyanate, 20 mL 0.5 M EDTA, 20 mL Milli Q, 1 g *N*-Lauroylsarcosine), and mix until cell lysis occur;
7. Incubate for 10 min on ice;
8. Add 250 μL cold NH_4Ac (7.5 M) and mix;
9. Incubate for 10 min on ice;
10. Add 500 μL cold chloroform/isoamylalcohol (24:1) and vortex;
11. Centrifuge 20 min at 11,400 × *g* at 4°C;
12. Collect as much as possible the upper layer without disturbing the visible interface;
13. Add approximately 800 μL cold isopropanol and mix until strands of DNA are visible;
14. Centrifuge 10 min at 15,000 × *g* and discard the supernatant;
15. Wash the pellet three times with 100 μL 70 % ethanol;
16. Dry the pellet in air and dissolve in 100 μL TE buffer; store at 4°C or –20°C.

93.2.2 Detection Procedures

93.2.2.1 Genus-Specific PCR

For a fast and reliable identification of arcobacters at genus level, the PCR assay developed in 1996 by Harmon and Wesley[146] can be applied. The assay utilizes primers ARCO I and ARCO II targeted to the genes encoding 16S rRNA, and a specific 1223-bp fragment is yielded for all presently described *Arcobacter* species. DNA may be extracted by commercially available DNA-extraction kits, as well as by the guanidine thiocyanate-based method, or by simply boiling the isolates for 15 min at 95°C. After cooling and a short spin to pellet the cell debris, the whole-cell lysate is ready to use in the assay. Quantification and standardization of the DNA extract or lysate is not necessary but should not exceed 200 ng ds DNA/μL.

PCR Protocol

1. Prepare the PCR mixture (50 μL) containing:
   - 1× PCR buffer (10 mM Tris-HCl, 50 mM KCl)
   - 1.5 mM $MgCl_2$
   - Primers: 50 pmol of
     ARCO I: 5′-AGA GAT TAG CCT GTA TTG TAT-3′
     ARCO II: 5′-TAG CAT CCC CGC TTC GAA TGA-3′
   - 200 μM dATP, dTTP, dGTP, and dCTP
   - 1.25 U *Taq* DNA polymerase
   - 1–2 μL of DNA extract or lysate
2. Run the following cycling program on a thermal cycler: 1 cycle of 94°C for 4 min; 25 cycles of 94°C for 1 min, 56°C for 1 min, and 72°C for 1 min.
3. Separate the amplified products on 1.5% agarose gel (in 1× TAE or 0.5× TBE buffer) containing 0.5 μg/mL of ethidium bromide and visualize on an UV transilluminator.

93.2.2.2 Species-Specific PCR

A species-specific PCR assay for the identification of *A. butzleri*, *A. cryaerophilus* 1A and 1B and *A. skirrowii* has been developed by Kabeya et al.[148] The species-specific forward primers from the most variable region of the 23S rRNA gene are N.butz, N.cl.A, N.c.1B, and N.ski, and the reverse primer ARCO-U recognizes a conserved *Arcobacter* specific region. DNA may be extracted by commercially available DNA-extraction kits, as well as by the guanidium thiocyanate-based method, or by simply boiling the isolates for 15 min at 95°C. After cooling and a short spin to pellet the cell debris, the whole-cell lysate is ready to use in the assay. Authors recommend to adjust the DNA concentration to 10 ng DNA/μL. In the one-step PCR assay, a specific 728 bp fragment is generated for *A. cryaerophilus* 1A, a 692 bp fragment for *A. butzleri*, a 488 bp fragment for *A. skirrowii*, and a 152 bp fragment for *A. cryaerophilus* 1B.

PCR Protocol

1. Prepare PCR mixture (20 μL) containing:
   - 1× PCR buffer (10 mM Tris-HCl, 50 mM KCl)
   - 1.5 mM $MgCl_2$
   - Primers: 50 pmol of
     N.butz 5′-AGCGTTCTATTCAGCGTAGAAGATGT-3′
     N.c1A 5′-ACCGAAGCTTTAGATTCGAATTTATTCG-3′
     N.c1B 5′-GGACTTGCTCCAAAAAGCTGAAG-3′
     N.ski 5′-CGAGGTCACGGATGGAAGTG-3′
     ARCO-U 5′-TTCGCTTGCGCTGACATCAT-3′
   - 200 μM dATP, dTTP, dGTP, and dCTP
   - 1.0 U *Taq* DNA polymerase
   - 2 μL of DNA extract or lysate
2. Perform PCR amplification as follows: 1 cycle of 94°C for 3 min; 30 cycles of 94°C for 30 s, 62°C for 1 min, and 72°C for 1 min; 1 cycle of 72°C for 5 min.
3. Separate the amplified products on 2% agarose gel (in 1× TAE or 0.5× TBE) containing 0.5 μg/mL of ethidium bromide and visualize on an UV transilluminator.

93.2.2.3 Species-Specific PCR–RFLP

As none of the abovementioned methods are able to discriminate the six currently accepted species, a 16S rDNA–RFLP method for the identification of all *Arcobacter* species was developed by Figueras et al.[164] The assay comprises an initial amplification of a 1026 bp fragment targeted the 16S rRNA gene using a modified PCR assay of Marchall et al.[153] followed by an endonuclease digestion and electrophoresis of the fragments. DNA may be extracted by commercially available DNA-extraction kit, by a guanidine thiocyanate- or phenol/chloroform-based method, or by preparing a whole-cell lysate by boiling. The DNA concentration should be adjusted to 100 ng ds DNA/μL.

PCR Protocol

1. Prepare PCR mixture (50 μL) containing:
   - PCR buffer (10 mM Tris-HCL, 50 mM KCl)
   - 1.5 mM $MgCl_2$
   - Primers: 50 pmol of
     mCAH1a 5′-AAC ACA TGC AAG TCG AAC GA-3′
     CAH1b 5′-TTA ACC CAA CAT CTC ACG AC-3′
   - 200 μM dATP, dTTP, dGTP, and dCTP
   - 2.5 U *Taq* DNA polymerase
   - 5 μL whole-cell lysate
2. Perform PCR amplification using following cycling program: 1 cycle of 95°C for 2 min; 30 cycles of 94°C for 30 s, 52°C for 30 s, and 72°C for 90 s; and 1 cycle of 72°C for 5 min.
3. Digest 10 μL of the amplification product with 10 U of the enzyme *Mse*I (and 2 μL of the 10× buffer R) in a total volume of 20 μL at 65°C for 3 h.
4. Separate the restriction fragments on 15% polyacrylamide gel electrophoresis in 1× TBE buffer at 350 V for 5 h.
5. Stain the gel with ethidium bromide and photograph on a UV transilluminator.

93.2.2.4 Characterization by ERIC–PCR

Houf et al. optimized an enterobacterial repetitive intergenic consensus (ERIC)–PCR for the genetic characterization of arcobacters at strain level and assessed the performance and discriminatory abilities of this method.[159] In this and later studies, ERIC–PCR has shown to be a valuable, fast, and robust technique for characterizing *A. butzleri*, *A. cryaerophilus*, *A. skirrowii*, and *A. cibarius* isolates.[18,93,159]

PCR Protocol

1. Prepare PCR mixture (50 μL) containing:
   - PCR buffer (10 mM Tris-HCL, 50 mM KCl)
   - 4 mM $MgCl_2$
   - Primers: 25 pmol of
     ERIC1R: 5′-ATGTAAGCTCCTGGGGATTCC-3′
     ERIC2: 5′-AAGTAAGTGACTGGGGTGAGCG-3′
   - 200 μM dATP, dTTP, dGTP, and dCTP
   - 5 U *Taq* DNA polymerase
   - 1 μL of DNA extract or 1 μL whole-cell lysate
2. Perform PCR amplification as follows: 1 cycle of 94°C for 5 min; 40 cycles of 94°C for 1 min, 25°C for 1 min, and 72°C for 2 min.
3. Separate the PCR products (8 μL) on 2% agarose gel (in 0.5× TBE) containing 0.5 μg/mL ethidium bromide for 2.5 h at 100 V.

Note: The banding patterns used to determine the ERIC–PCR type comprise DNA fragments between 100 and 2072 bp. For the interpretation of the fingerprints, use computer-based normalization and interpolation of the DNA profiles, and numerical analysis using the Pearson product moment correlation coefficient, with 1% position tolerance. Construct dendrograms using the unweighted pair group linkage analysis method (UPGMA). DNA patterns that differ in one or more DNA fragments represent different types. Furthermore, groups of identical fingerprints, or fingerprints with slight differences in band intensity, are characterized by Pearson correlation coefficients of 90% or higher.

93.3 CONCLUSIONS AND FUTURE PERSPECTIVES

Arcobacters are commonly isolated from animals, food, and environmental samples worldwide, and cases of human *Arcobacter* infection have regularly been reported. However, often the infection source remains unclear. Several major aspects actually hamper the risk assessment for these bacteria. First, besides the fact that clinical samples are not routinely tested for *Arcobacter* species as is done for *Salmonella* or *Campylobacter*, most isolations are performed by methods designed for thermophilic *Campylobacter* species. The selectivity of those media is mostly achieved by the incorporation of antimicrobial agents, and though arcobacters have sporadically been isolated in this way, several studies about the *Arcobacter* susceptibility to the antimicrobials included showed that none of those supplements allowed the growth of all *Arcobacter* species or strains and at the same time sufficiently suppressed the accompanying flora present in biological samples. The low recovery rate reported today from human diarrheal specimens is therefore certainly an underestimation of the real prevalence. In order to determine the exposure assessment of arcobacters to humans, efficient, robust, and reliable methods for isolation, identification, and characterization of arcobacters are needed. Validation should be based on the determination of the parameters specificity, sensitivity, repeatability, reproducibility, and detection limit, both for classical bacteriological isolation as for molecular-based methods.

Correct identification of arcobacters is another challenge in microbiology, and too often arcobacters are misidentified as campylobacters, especially when phenotypical methods are applied. Due to the relatively metabolic inertness and the antigenic heterogeneity, biochemical or serological identification of arcobacters is not recommended. These are also time consuming and laborious, and clear-cut parameters are not always present in all strains of a certain species. For now, identification at genus level can reliable be performed by genus-specific PCR assays, but species identification is often limited to some of the well-established species. For the more recently described species, such as *A. cibarius*, and potential new species yet to be described, no (or not routinely) applicable methods are available. Furthermore, as most of the one-step PCR assays target the 16S or 23S rRNA genes, it can be expected that some new species can not be distinguished by this approach as only few differences exist between the several species at gene level. Though the PCR–RFLP approach has shown to be the only routinely applicable method to differentiate the six species at present, the occurrence of single mutations or the presence of the same cleaving place within two species cannot be excluded. In the future, genome analysis will probably identify more stable and discriminatory genes to create a more robust identification assay.

Direct detection of arcobacters by PCR, or even direct quantification by RT–PCR in a biological specimen, is not commonly applied at this moment. In contrast to more vulnerable bacteria, such as *Helicobacter* species, arcobacters are relatively easily to culture and there is no demand in human or veterinary medicine for a rapid analysis tool at the moment. Furthermore, as biological matrices are complex and often comprise inhibiting factors, preparation and cleanup of the sample is expensive and hamper their use in routine laboratories. Furthermore, many studies are focused on further examination of the isolates as microbiological susceptibility and typing for which the preservation of the isolates is needed. The use of those direct methods can however have a future in large-scale surveys.

The existence of considerable heterogeneity among arcobacters in one specimen is yet another difficulty in *Arcobacter* research. This phenomenon was already reported in early biotyping and serotyping studies. Possible explanations are multiple sources of contamination, the existence of multiple parent genotypes, and a high degree of genomic recombination among the progeny of parent genotypes, but none of these hypotheses has been demonstrated yet. Furthermore, in vitro cultivation of *Arcobacter* isolates over a period of time did not induce genomic rearrangements, as identical fingerprints were obtained in the reproducibility studies. The use of molecular typing has also shown that this heterogeneity is not linked to artefacts by phenotypic characterization, but truly exists in the *Arcobacter* population. As a result, typing *Arcobacter* isolates often yields an overwhelming jumble of different genotypes with no extra contribution to the identification of possible transmission routes. Also, a bias caused by the isolation methods has clearly been demonstrated. Therefore, although commonly applied in current food and medical molecular microbiology, typing at strain level for such heterogeneous bacteria should seriously be considered and aims clearly stated before outset. In the first place, typing should be reserved to determine the relation between a direct potential contamination source and the patient's isolate for which epidemiological-based evidence is also available. When typing is appropriate, several methods are available, though only molecular-based systems are currently applied. In practice, the choice often depends on the availability of equipment, knowledge, and experience in the lab. Several studies have shown that ERIC–PCR or RAPD are easy to perform and have shown to be reproducible in their outcome. For fast, cheap, and easy to perform characterization needs, ERIC–PCR is therefore recommended as first-line tool, but can be complemented by AFLP, MLST, or by a non-PCR based method as PFGE. However, the latter methods are far more demanding and do not necessarily have a larger discriminatory power. Standardization of the DNA amount analyzed in a PCR assay is often recommended, but the use of a DNA extraction by commercial kits, in-house extraction methods, or simply boiling of the isolates to prepare whole-cell extracts as such result in the same outcome of the genotypic relatedness of the isolates. However, as they have an influence of the fingerprint patterns, they cannot be combined within the same analysis.

Finally, to establish epidemiological links between arcobacters isolated from different to closely related sources or to elucidate the transmission routes, the clonally relatedness and the significance of a single fragment difference as often noticed in PCR-based fingerprints should be elucidated. The application of MLST and the setting up of a MLST-type bank, as established for the closely related campylobacters, may be helpful to accomplish these goals.

REFERENCES

1. McFadyean, J., and Stockman, S., Report of the Departmental Committee appointed by the Board of Agriculture and Fisheries to enquire into epizootic abortion. Appendix to Part II. Abortion in sheep, *His Majesty's Stationary Office, London*, 1, 1, 1913.
2. Smith, T., Isolation of *Spirillum*-like organisms from aborted bovine fetuses, *J. exp. Med.*, 28, 701, 1918.
3. Sebald, M., and Véron, M., Teneur en bases de l'ADN et classification des vibrions, *Ann. I. Pasteur Paris*, 105, 897, 1963.
4. Ellis, W.A. et al., Isolation of Spirillum-Vibrio-like organisms from bovine fetuses, *Vet. Rec.*, 100, 451, 1977.
5. Ellis, W.A. et al., Isolation of spirillum-like organisms from pig fetuses, *Vet. Rec.*, 102, 106, 1978.
6. Higgins, R., and Degre, R., Isolation of spirillum-like organisms from pig and bovine fetuses, *Vet. Rec.*, 104, 262, 1979.
7. Neill, S.D. et al., Biochemical characteristics of *Campylobacter*-like organisms from cattle and pigs, *Res. Vet. Sci.*, 25, 368, 1978.
8. Neill, S.D. et al., Designation of aerotolerant *Campylobacter*-like organisms from porcine and bovine abortions to the genus *Campylobacter*, *Res. Vet. Sci.*, 27, 180, 1979.
9. Neill, S.D. et al., Taxonomic position of *Campylobacter cryaerophila* sp. nov, *Int.J. Syst. Bacteriol.*, 35, 342, 1985.
10. Thompson, L.M. et al., Phylogenetic study of the genus *Campylobacter*, *Int.J. Syst. Bacteriol.*, 38, 190, 1988.
11. Kiehlbauch, J.A. et al., *Campylobacter butzleri* sp. nov. isolated from humans and animals with diarrheal illness, *J. Clin. Microbiol.*, 29, 376, 1991.
12. Vandamme, P. et al., Revision of *Campylobacter*, *Helicobacter*, and *Wolinella* taxonomy: Emendation of generic descriptions and proposal of *Arcobacter gen. nov*, *Int.J. Syst. Bacteriol.*, 41, 88, 1991.
13. Vandamme, P. et al., Polyphasic taxonomic study of the emended genus *Arcobacter* with *Arcobacter butzleri* comb-nov and *Arcobacter skirrowii* sp-nov, an aerotolerant bacterium isolated from veterinary specimens, *Int.J. Syst. Bacteriol.*, 42, 344, 1992.
14. Vandamme, P., and De Ley, J., Proposal for a new family, *Campylobacteraceae*, *Int. J. Syst. Bacteriol.*, 41, 451, 1991.
15. On, S.L.W., Taxonomy of *Campylobacter*, *Arcobacter*, *Helicobacter* and related bacteria: Current status, future prospects and immediate concerns, *J. Appl. Microbiol.*, 90, 1S, 2001.
16. Miller, W.G. et al., The complete genome sequence and analysis of the Epsilonproteobacterium *Arcobacter butzleri*, *PLoS ONE*, 2,e1358, 2007.
17. Mcclung, C.R. et al., *Campylobacter nitrofigilis* sp. nov, a nitrogen-fixing bacterium associated with roots of *Spartina alterniflora* Loisel, *Int.J. Syst. Bacteriol.*, 33, 605, 1983.
18. Houf, K. et al., *Arcobacter cibarius* sp nov., isolated from broiler carcasses, *Int. J. Syst. Evol. Microbiol.*, 55, 713, 2005.
19. Donachie, S.P. et al., *Arcobacter halophilus* sp nov., the first obligate halophile in the genus *Arcobacter*, *Int. J. Syst. Evol. Microbiol.*, 55, 1271, 2005.
20. Vandenberg, O. et al., *Arcobacter* species in humans, *Emerg. Infect. Dis.*, 10, 1863, 2004.
21. Prouzet-Mauleon, V. et al., *Arcobacter butzleri*: Underestimated enteropathogen, *Emerg. Infect. Dis.*, 12, 307, 2006.
22. Houf, K., and Stephan, R., Isolation and characterization of the emerging foodborn pathogen *Arcobacter* from human stool, *J. Microbiol. Methods*, 68, 408, 2007.
23. Ho, H.T.K. et al., Interaction of *Arcobacter* spp. with human and porcine intestinal epithelial cells, *FEMS Immun. Med. Microbiol.*, 50, 51, 2007.
24. Musmanno, R.A. et al., In vitro virulence factors of *Arcobacter butzleri* strains isolated from superficial water samples, *Microbiologica*, 20, 63, 1997.
25. International Commission on Microbiological Specifications for Foods (ICMSF), Microorganisms in foods? Microbiological testing in food safety management. Kluwer Academic/Plenum Publishers, New York, 171, 2002.
26. Van Driessche, E. et al., Isolation of *Arcobacter* species from animal feces, *FEMS Microbiol. Lett.*, 229, 243, 2003.

27. Houf, K. et al., Occurrence and distribution of *Arcobacter* species in poultry processing, *J. Food Prot.*, 65, 1233, 2002.
28. Andersen, M.M. et al., Prevalence of *Arcobacter* species in market-weight commercial turkeys, *Antonie Van Leeuwenhoek*, 92, 309, 2007.
29. Anderson, K.F. et al., *Arcobacter (Campylobacter) butzleri*-associated diarrheal illness in a nonhuman primate population, *Infect. Immun.*, 61, 2220, 1993.
30. Higgins, R. et al., *Arcobacter butzleri* isolated from a diarrhoeic non-human primate, *Lab. Anim.*, 33, 87, 1999.
31. Russell, R.G. et al., Uncommon *Campylobacter* species in infant *Macaca nemestrina* monkeys housed in a nursery, *J. Clin. Microbiol.*, 30, 3024, 1992.
32. Lehner, A. et al., Relevant aspects of *Arcobacter spp.* as potential foodborne pathogen, *Int. J. Food Microbiol.*, 102, 127, 2005.
33. Kiehlbauch, J.A. et al., Restriction fragment length polymorphisms in the ribosomal genes for species identification and subtyping of aerotolerant *Campylobacter* species, *J. Clin. Microbiol.*, 29, 1670, 1991.
34. Wybo, I. et al., Isolation of *Arcobacter skirrowii* from a patient with chronic diarrhea, *J. Clin. Microbiol.*, 42, 1851, 2004.
35. Samie, A. et al., Prevalence of *Campylobacter* species, *Helicobacter pylori* and *Arcobacter* species in stool samples from the Venda region, Limpopo, South Africa: Studies using molecular diagnostic methods, *J. Infect.*, 54, 558, 2007.
36. Chinivasagam, H.N. et al., Detection of *Arcobacter* spp. in piggery effluent and effluent-irrigated soils in southeast Queensland, *J. Appl. Microbiol.*, 103, 418, 2007.
37. Atabay, H.I. et al., Detection and diversity of various *Arcobacter* species in Danish poultry, *Int. J. Food Microbiol.*, 109, 139, 2006.
38. On, S.L.W. et al., Differentiation of *Arcobacter* species by numerical analysis of AFLP profiles and description of a novel *Arcobacter* from pig abortions and turkey faeces, *J. Appl. Microbiol.*, 95, 1096, 2003.
39. Park, Y.B., and Han, Y.H., The optimal conditions for enhancing cell growth of *Arcobacter nitrofigilis*, *Korean J. Microbiol.*, 32, 384, 1994.
40. Sievert, S.M. et al., Growth and mechanism of filamentous-sulfur formation by Candidatus *Arcobacter sulfidicus* in opposing oxygen-sulfide gradients, *Environ. Microbiol.*, 9, 271, 2007.
41. Voordouw, G. et al., Characterization of 16S rRNA genes from oil field microbial communities indicates the presence of a variety of sulfate-reducing, fermentative, and sulfide-oxidizing bacteria, *Appl. Environ. Microbiol.*, 62, 1623, 1996.
42. Teske, A. et al., Molecular identification of bacteria from a coculture by denaturing gradient gel electrophoresis of 16S ribosomal DNA fragments as a tool for isolation in pure cultures, *Appl. Environ. Microbiol.*, 62, 4210, 1996.
43. Snaidr, J. et al., Phylogenetic analysis and in situ identification of bacteria in activated sludge, *Appl.Environ. Microbiol.*, 63, 2884, 1997.
44. Llobet-Brossa, E. et al., Microbial community composition of Wadden Sea sediments as revealed by fluorescence in situ hybridization, *Appl. Environ. Microbiol.*, 64, 2691, 1998.
45. Gevertz, D. et al., Isolation and characterization of strains CVO and FWKOB, two novel nitrate-reducing, sulfide-oxidizing bacteria isolated from oil field brine, *Appl. Environ. Microbiol.*, 66, 2491, 2000.
46. Thamdrup, B. et al., Microbial manganese and sulfate reduction in Black Sea shelf sediments, *Appl. Environ. Microbiol.*, 66, 2888, 2000.
47. Watanabe, K. et al., Molecular characterization of bacterial populations in petroleum-contaminated groundwater discharged from underground crude oil storage cavities, *Appl. Environ. Microbiol.*, 66, 4803, 2000.
48. Frias-Lopez, J. et al., Partitioning of bacterial communities between seawater and healthy, black band diseased, and dead coral surfaces, *Appl. Environ. Microbiol.*, 68, 2214, 2002.
49. Romero, J. et al., Bacterial 16S rRNA gene analysis revealed that bacteria related to *Arcobacter* spp. constitute an abundant and common component of the oyster microbiota (*Tiostrea chilensis*), *Microbial. Ecol.*, 44, 365, 2002.
50. Wirsen, C.O. et al., Characterization of an autotrophic sulfide-oxidizing marine *Arcobacter* sp that produces filamentous sulfur, *Appl.Environ. Microbiol.*, 68, 316, 2002.
51. Hammes, F. et al., Molecular, biochemical and ecological characterisation of a bio-catalytic calcification reactor, *Appl. Environ. Microbiol.*, 62, 191, 2003.
52. Grabowski, A. et al., Microbial diversity in production waters of a low-temperature biodegraded oil reservoir, *FEMS Microbiol. Ecol.*, 54, 427, 2005.
53. de Boer, E. et al., A selective medium for the isolation of *Arcobacter* from meats, *Lett. Appl. Microbiol.*, 23, 64, 1996.
54. Engberg, J. et al., Prevalence of *Campylobacter*, *Arcobacter*, *Helicobacter*, and *Sutterella* spp. in human fecal samples as estimated by a reevaluation of isolation methods for campylobacters, *J. Clin. Microbiol.*, 38, 286, 2000.
55. Lastovica, A.J., and le Roux, E., Efficient isolation of campylobacteria from stools, *J. Clin. Microbiol.*, 38, 2798, 2000.
56. Tee, W. et al., *Campylobacter cryaerophila* isolated from a human, *J. Clin. Microbiol.*, 26, 2469, 1988.
57. Lerner, J. et al., Severe diarrhea associated with *Arcobacter butzleri*, *Eur. J. Clin. Microbiol. Infect. Dis.*, 13, 660, 1994.
58. Burnens, A.P. et al., Isolation of *Arcobacter butzleri* from a girl with gastroenteritis on *Yersinia* selective CIN agar, *Med. Microbiol. Lett.*, 251, 1992.
59. Lau, S.K.P. et al., Identification by 16S ribosomal RNA gene sequencing of *Arcobacter butzleri* bacteraemia in a patient with acute gangrenous appendicitis, *Mol. Pathol.*, 55, 182, 2002.
60. Vandamme, P. et al., Outbreak of recurrent abdominal cramps associated with *Arcobacter butzleri* in an Italian school, *J. Clin. Microbiol.*, 30, 2335, 1992.
61. Rice, E.W. et al., Isolation of *Arcobacter butzleri* from ground water, *Lett. Appl. Microbiol.*, 28, 31, 1999.
62. On, S.L.W. et al., Isolation of *Arcobacter butzleri* from a neonate with bacteremia, *J. Infect.*, 31, 225, 1995.
63. Yan, J.J. et al., *Arcobacter butzleri* bacteremia in a patient with liver cirrhosis, *J. Formos. Med. Assoc.*, 99, 166, 2000.
64. Hsueh, P.R. et al., Bacteremia caused by *Arcobacter cryaerophilus* 1B, *J. Clin. Microbiol.*, 35, 489, 1997.
65. Woo, P.C.Y. et al., Identification of *Arcobacter cryaerophilus* isolated from a traffic accident victim with bacteremia by 16S ribosomal RNA gene sequencing, *Diagn. Microbiol. Infect. Dis.*, 40, 125, 2001.
66. Phillips, C.A., *Arcobacter* spp in food: Isolation, identification and control, *Trends Food Sci. Technol.*, 12, 263, 2001.
67. Ho, H.T.K. et al., *Arcobacter*, what is known and unknown about a potential foodborne zoonotic agent! *Vet. Microbiol.*, 115, 1, 2006.
68. Johnson, L.G., and Murano, E.A., Lack of a cytolethal distending toxin among *Arcobacter* isolates from various-sources, *J. Food Prot.*, 65, 1789, 2002.

69. Wesley, I.V. et al., Infection of cesarean-derived colostrum-deprived 1-day-old piglets with *Arcobacter butzleri*, *Arcobacter cryaerophilus*, and *Arcobacter skirrowii*, *Infect. Immun.*, 64, 2295, 1996.
70. Yildiz, H., and Aydin, S., Pathological effects of *Arcobacter cryaerophilus* infection in rainbow trout (Oncorhynchus mykiss Walbaum), *Acta Veterinaria Hungarica*, 54, 191, 2006.
71. Tsang, R.S.W. et al., Immunochemical characterization of a haemagglutinating antigen of *Arcobacter* spp, *FEMS Microbiol. Lett.*, 136, 209, 1996.
72. Carbone, M. et al., Adherence of environmental *Arcobacter butzleri* and *Vibrio* spp. isolates to epithelial cells in vitro, *Food Microbiol.*, 20, 611, 2003.
73. Fernandez, H. et al., Toxigenic and invasive capacities: Possible pathogenic mechanisms in *Arcobacter cryaerophilus*, *Memorias do Instituto Oswaldo Cruz*, 90, 633, 1995.
74. Kalman, M., Czermann, B., and Szöllosy, E., *Arcobacter butzleri* strains in Csongrad county and their cytotoxin production in different cell lines, *Acta. Microbiol. Immunol. Hung.*, 43, 143, 1996.
75. Villarruel-Lopez, A. et al., Isolation of *Arcobacter* spp. from retail meats and cytotoxic effects of isolates against Vero cells, *J. Food Prot.*, 66, 1374, 2003.
76. Carter, E.R., Enteropathogenicity of *Arcobacter butzleri* in rabbit and pig ileal loops, Thesis Master of Science, Faculty of the Virginia Polytechnic Institute and State University, Blacksburg, Virginia, USA, 1996.
77. Fera, M.T. et al., In vitro susceptibility of *Arcobacter butzleri* and *Arcobacter cryaerophilus* to different antimicrobial agents, *Int. J. Antimicrob. Agents*, 21, 488, 2003.
78. Houf, K. et al., Susceptibility of *Arcobacter butzleri*, *Arcobacter cryaerophilus*, and *Arcobacter skirrowii* to antimicrobial agents used in selective media, *J. Clin. Microbiol.*, 39, 1654, 2001.
79. Kiehlbauch, J.A. et al., In vitro susceptibilities of aerotolerant *Campylobacter* isolates to 22 antimicrobial agents, *Antimicrob. Agents Chemother.*, 36, 717, 1992.
80. Kabeya, H. et al., Prevalence of *Arcobacter* species in retail meats and antimicrobial susceptibility of the isolates in Japan, *Int. J. Food Microbiol.*, 90, 303, 2004.
81. Atabay, H.I., and Aydin, F., Susceptibility of *Arcobacter butzleri* isolates to 23 antimicrobial agents, *Lett. Appl. Microbiol.*, 33, 430, 2001.
82. Harrass, B. et al., Identification and characterization of *Arcobacter* isolates from broilers by biochemical tests, antimicrobial resistance patterns and plasmid analysis, *Zentralbl. Veterinarmed. B*, 45, 87, 1998.
83. Vandenberg, O. et al., Antimicrobial susceptibility of clinical isolates of non-jejuni/coli campylobacters and arcobacters from Belgium, *J. Antimicrob. Chemother.*, 57, 908, 2006.
84. Hakanen, A. et al., Quality control strains used in susceptibility testing of *Campylobacter* spp, *J. Clin. Microbiol.*, 40, 2705, 2002.
85. Luber, P. et al., Comparison of broth microdilution, E test, and agar dilution methods for antibiotic susceptibility testing of *Campylobacter jejuni* and *Campylobacter coli*, *J. Clin. Microbiol.*, 41, 1062, 2003.
86. Houf, K. et al., Antimicrobial susceptibility patterns of *Arcobacter butzleri* and *Arcobacter cryaerophilus* strains isolated from humans and broilers, *Microb. Drug Resist.*, 10, 243, 2004.
87. Schroeder-Tucker, L. et al., Phenotypic and ribosomal RNA characterization of *Arcobacter* species isolated from porcine aborted fetuses, *J. Vet. Diagn. Invest.*, 8, 186, 1996.
88. On, S.L.W. et al., Prevalence and diversity of *Arcobacter* spp. isolated from the internal organs of spontaneous porcine abortions in Denmark, *Vet. Microbiol.*, 85, 159, 2002.
89. de Oliveira, S.J. et al., Classification of *Arcobacter* species isolated from aborted pig fetuses and sows with reproductive problems in Brazil, *Vet. Microbiol.*, 57, 347, 1997.
90. Logan, E.F. et al., Mastitis in dairy cows associated with an aerotolerant *Campylobacter*, *Vet. Rec.*, 110, 229, 1982.
91. Suarez, D.L. et al., Detection of *Arcobacter* species in gastric samples from swine, *Vet. Microbiol.*, 57, 325, 1997.
92. Van Driessche, E. et al., Occurrence and strain diversity of *Arcobacter* species isolated from healthy Belgian pigs, *Res. Microbiol.*, 155, 662, 2004.
93. Van Driessche, E. et al., Prevalence, enumeration and strain variation of *Arcobacter* species in the faeces of healthy cattle in Belgium, *Vet. Microbiol.*, 105, 149, 2005.
94. Van Driessche, E., and Houf, K., Discrepancy between the occurrence of *Arcobacter* in chickens and broiler carcass contamination, *Poult. Sci.*, 86, 744, 2007.
95. Wesley, I.V. et al., Fecal shedding of *Campylobacter* and *Arcobacter* spp. in dairy cattle, *Appl. Environ. Microbiol.*, 66, 1994, 2000.
96. Hume, M.E. et al., Genotypic variation among *Arcobacter* isolates from a farrow-to-finish swine facility, *J. Food Prot.*, 64, 645, 2001.
97. Golla, S.C. et al., Determination of the occurrence of *Arcobacter butzleri* in beef and dairy cattle from Texas by various isolation methods, *J. Food Prot.*, 65, 1849, 2002.
98. Kabeya, H. et al., Distribution of *Arcobacter* species among livestock in Japan, *Vet. Microbiol.*, 93, 153, 2003.
99. Ongor, H. et al., Investigation of arcobacters in meat and faecal samples of clinically healthy cattle in Turkey, *Lett. Appl. Microbiol.*, 38, 339, 2004.
100. Fernandez, H. et al., First isolation in Chile of *Arcobacter cryaerophilus* from a bovine abortion, *Archivos de Medicina Veterinaria*, 27, 111, 1995.
101. Parvanta, M.F., *Campylobacter cryaerophila* and *Campylobacter fetus* subspecies *venerealis* as a cause of serial abortions in two cattle herds in North-Rhine-Westfalia, *Tierarztliche Umschau*, 54, 364, 1999.
102. Gill, K.P., Aerotolerant *campylobacter* strain isolated from a bovine preputial sheath washing, *Vet. Rec.*, 112, 459, 1983.
103. de Oliveira, S.J. et al., Antigenic diversity among strains of *Arcobacter* spp. isolated from pigs in Rio Grande do Sul, Brazil and presence of agglutinin titers in serum samples of sows with reproductive problems, *Ciência Rural*, *Santa Maria*, 29, 705, 1999.
104. Jahn, B., *Campylobacter* im genitaltrakt des schweines. Inaugural dissertation, Tierärztliche Hochshule Hannover, Hannover, Germany, 1993.
105. Ho, T.K.H. et al., Potential routes of acquisition of *Arcobacter* species by piglets, *Vet. Microbiol.*, 114, 123, 2006.
106. Harvey, R.B. et al., Prevalence of *Campylobacter*, *Salmonella*, and *Arcobacter* species at slaughter in market age pigs, *Adv. Exp. Med. Biol.*, 473, 237, 1999.
107. Atabay, H.I. et al., Diversity and prevalence of *Arcobacter* spp. in broiler chickens, *J. Appl. Microbiol.*, 84, 1007, 1998.
108. Eifert, J.D. et al., Comparison of sampling techniques for detection of *Arcobacter butzleri* from chickens, *Poult. Sci.*, 82, 1898, 2003.
109. Gude, A. et al., Ecology of *Arcobacter* species in chicken rearing and processing, *Lett. Appl. Microbiol.*, 41, 82, 2005.

110. Talhouk, R.S. et al., Prevalence, antimicrobial susceptibility and molecular characterization of *Campylobacter* isolates recovered from humans and poultry in Lebanon, *J. Med. Liban.*, 46, 310, 1998.
111. Corry, J.E.L., and Atabay, H.I., Poultry as a source of *Campylobacter* and related organisms, *J. Appl. Microbiol.*, 90, 96S, 2001.
112. Wesley, I.V., and Baetz, A.L., Natural and experimental infections of *Arcobacter* in poultry, *Poult. Sci.*, 78, 536, 1999.
113. Houf, K. et al., Molecular characterization of *Arcobacter* isolates collected in a poultry slaughterhouse, *J. Food Prot.*, 66, 364, 2003.
114. Atabay, H.I. et al., Prevalence of Arcobacter species in domestic geese (Anser anser) in Turkey, *Zoonoses Public Hlth.*, 54, 73, 2007.
115. Ridsdale, J.A. et al., Prevalence of campylobacters and arcobacters in ducks at the abattoir, *J. Appl. Microbiol.*, 85, 567, 1998.
116. Ridsdale, J.A. et al., *Campylobacter* and *Arcobacter* spp. isolated from the carcasses and caeca of commercially reared ducks, *Anaerobe*, 5, 317, 1999.
117. Hamir, A.N. et al., *Campylobacter jejuni* and *Arcobacter* species associated with intussusception in a raccoon (Procyon lotor), *Vet. Rec.*, 155, 338, 2004.
118. Phillips, C.A., Arcobacters as emerging human foodborne pathogens, *Food Control*, 12, 1, 2001.
119. Manke, T.R. et al., Prevalence and genetic variability of *Arcobacter* species in mechanically separated turkey, *J. Food Prot.*, 61, 1623, 1998.
120. Johnson, L.G., and Murano, E.A., Comparison of three protocols for the isolation of *Arcobacter* from poultry, *J. Food Prot.*, 62, 610, 1999.
121. Atabay, H.I. et al., The prevalence of *Arcobacter* spp. on chicken carcasses sold in retail markets in Turkey, and identification of the isolates using SDS-PAGE, *Int. J. Food Microbiol.*, 81, 21, 2003.
122. Vytrasova, J. et al., Isolation of *Arcobacter butzleri* and *A. cryaerophilus* in samples of meats and from meat-processing plants by a culture technique and detection by PCR, *Folia Microbiologica*, 48, 227, 2003.
123. Rivas, L. et al., Isolation and characterisation of *Arcobacter butzleri* from meat, *Int. J. Food Microbiol.*, 91, 31, 2004.
124. Morita, Y. et al., Isolation and phylogenetic analysis of *Arcobacter* spp. in ground chicken meat and environmental water in Japan and Thailand, *Microbiol. Immun.*, 48, 527, 2004.
125. Festy, B., et al., Poultry meat and water as the possible sources of *Arcobacter butzleri* associated human desease in Paris, France, *Acta-enterogastrologica Belgica*, suppl. 6, 35, 1993.
126. Houf, K. et al., Development of a multiplex PCR assay for the simultaneous detection and identification of *Arcobacter butzleri*, *Arcobacter cryaerophilus* and *Arcobacter skirrowii*, *FEMS Microbiol. Lett.*, 193, 89, 2000.
127. Scullion, R. et al., Prevalence of *Arcobacter* spp. in raw milk and retail raw meats in northern Ireland, *J. Food Prot.*, 69, 1986, 2006.
128. Zanetti, F. et al., Prevalence of thermophilic *Campylobacter* and *Arcobacter butzleri* in food of animal origin, *Int. J. Food Microbiol.*, 33, 315, 1996.
129. Collins, C.I. et al., Detection of *Arcobacter* spp in ground pork by modified plating methods, *J. Food Prot.*, 59, 448, 1996.
130. Ohlendorf, D.S., and Murano, E.A., Prevalence of *Arcobacter* spp. in raw ground pork from several geographical regions according to various isolation methods, *J. Food Prot.*, 65, 1700, 2002.
131. Van Driessche, E., and Houf, K., Characterization of the *Arcobacter* contamination on Belgian pork carcasses and raw retail pork, *Int. J. Food Microbiol.*, 118, 20, 2007.
132. Mansfield, L.P., and Forsythe, S.J., *Arcobacter butzleri*, *A. skirrowii and A. cryaerophilus*—potential emerging human pathogens, *Rev. Med. Microbiol.*, 11, 161, 2000.
133. Penn, C.W., *Campylobacter*, *Arcobacter* and *Helicobacter*: Surface characteristics. In *Campylobacter*, *Helicobacter* and *Arcobacter*, Summer 2000 Conference, Society for Applied Microbiology, Strathclyde, UK, 3, 2000.
134. On, S.L.W., Identification methods for campylobacters, helicobacters, and related organisms, *Clin. Microbiol. Rev.*, 9, 405, 1996.
135. On, S.L.W. et al., A probability matrix for the identification of campylobacters, helicobacters and allied taxa, *J. Appl. Bacteriol.*, 81, 425, 1996.
136. Boudreau, M. et al., Biochemical and serological characterization of *Campylobacter cryaerophila*, *J. Clin. Microbiol.*, 29, 54, 1991.
137. Lambert, M.A. et al., Differentiation of *Campylobacter* and *Campylobacter*-like organisms by cellular fatty acid composition, *J. Clin. Microbiol.*, 25, 706, 1987.
138. Moss, C.W., and Lambert-Fair, M.A., Location of double bonds in monounsaturated fatty acids of *Campylobacter cryaerophila* with dimethyl disulfide derivatives and combined gas chromatography-mass spectrometry, *J. Clin. Microbiol.*, 27, 1467, 1989.
139. Moss, C.W. et al., Isoprenoid quinones of *Campylobacter cryaerophila*, *C. cinaedi*, *C. fennelliae*, *C. hyointestinalis*, *C. pylori*, and "*C. upsaliensis*," *J. Clin. Microbiol.*, 28, 395, 1990.
140. Jelinek, D. et al., Identification of *Arcobacter* species using phospholipid and total fatty acid profiles, *Folia Microbiologica*, 51, 329, 2006.
141. Abdelbaqi, K. et al., Development of a real-time fluorescence resonance energy transfer PCR to detect *Arcobacter* species, *J. Clin. Microbiol.*, 45, 3015, 2007.
142. Miller, W.G. et al., Novel multilocus sequence typing methods for multiple human pathogenic species of *Arcobacter* reveal substantial diversity within the genus. In: 107th General Meeting of the American Society for Microbiology, Metro Toronto Convention Centre, Toronto, Canada, C-305, 2007.
143. Kiehlbauch, J.A. et al., Evaluation of ribotyping techniques as applied to *Arcobacter*, *Campylobacter* and *Helicobacter*, *Mol. Cell. Probes*, 8, 109, 1994.
144. Wesley, I.V. et al., *Arcobacter*-specific and *Arcobacter butzleri*-specific 16S ribosomal-RNA-based DNA probes, *J. Clin. Microbiol.*, 33, 1691, 1995.
145. Bastyns, K. et al., A variable 23S rDNA region is a useful discriminating target for genus-specific and species-specific PCR amplification in *Arcobacter* species, *Syst. Appl. Microbiol.*, 18, 353, 1995.
146. Harmon, K.M., and Wesley, I.V., Identification of *Arcobacter* isolates by PCR, *Lett. Appl. Microbiol.*, 23, 241, 1996.
147. Harmon, K.M., and Wesley, I.V., Multiplex PCR for the identification of *Arcobacter* and differentiation of *Arcobacter butzleri* from other arcobacters, *Vet. Microbiol.*, 58, 215, 1997.
148. Kabeya, H. et al., One-step polymerase chain reaction-based typing of *Arcobacter* species, *Int. J. Food Microbiol.*, 81, 163, 2003.

149. Al Rashid, S.T. et al., Identification of *Campylobacter jejuni*, *C. coli*, *C. lari*, *C. upsaliensis*, *Arcobacter butzleri*, and *A. butzleri-like* species based on the glyA gene, *J. Clin. Microbiol.*, 38, 1488, 2000.
150. Gonzalez, I. et al., Development of a combined PCR-culture technique for the rapid detection of *Arcobacter* spp. in chicken meat, *Lett. Appl. Microbiol.*, 30, 207, 2000.
151. Cardarelli-Leite, P. et al., Rapid identification of *Campylobacter* species by restriction fragment length polymorphism analysis of a PCR-amplified fragment of the gene coding for 16S rRNA, *J. Clin. Microbiol.*, 34, 62, 1996.
152. Hurtado, A., and Owen, R.J., A molecular scheme based on 23S rRNA gene polymorphisms for rapid identification of *Campylobacter* and *Arcobacter* species, *J. Clin. Microbiol.*, 35, 2401, 1997.
153. Marshall, S.M. et al., Rapid identification of *Campylobacter*, *Arcobacter*, and *Helicobacter* isolates by PCR-restriction fragment length polymorphism analysis of the 16S rRNA gene, *J. Clin. Microbiol.*, 37, 4158, 1999.
154. On, S.L.W. et al., Genotyping and genetic diversity of *Arcobacter butzleri* by amplified fragment length polymorphism (AFLP) analysis, *Lett. Appl. Microbiol.*, 39, 347, 2004.
155. Lior, H., and Woodward, D.L., *Arcobacter butzleri*: A biotyping scheme, *Acta Gastro-enterologica Belgica* (Suppl), 6, 28, 1993.
156. Lior, H., and Woodward, D.L., *Arcobacter butzleri*: A serotyping sheme, *Acta Gastro-enterologica Belgica* (Suppl), 6, 29, 1993.
157. Lior, H., and Wang, G., Differentiation of *Arcobacter butzleri* by pulsed field gel electrophoresis (PFGE) and random amplified polymorphic DNA (RAPD), *Acta Gastro-enterologica Belgica* (Suppl), 6, 29, 1993.
158. Atabay, H.I. et al., Discrimination of *Arcobacter butzleri* isolates by polymerase chain reaction-mediated DNA fingerprinting, *Lett. Appl. Microbiol.*, 35, 141, 2002.
159. Houf, K. et al., Assessment of the genetic diversity among arcobacters isolated from poultry products by using two PCR-based typing methods, *Appl. Environ. Microbiol.*, 68, 2172, 2002.
160. Atabay, H.I., and Corry, J.E.L., The prevalence of campylobacters and arcobacters in broiler chickens, *J. Appl. Microbiol.*, 83, 619, 1997.
161. Johnson, L.G., and Murano, E.A., Development of a new medium for the isolation of *Arcobacter* spp, *J. Food Prot.*, 62, 456, 1999.
162. Houf, K. et al., Development of a new protocol for the isolation and quantification of *Arcobacter* species from poultry products, *Int. J. Food Microbiol.*, 71, 189, 2001.
163. Pitcher, D.G. et al., Rapid extraction of bacterial genomic DNA with guanidium thiocyanate, *Lett. Appl. Microbiol.*, 8, 151, 1989.
164. Figueras, M.J., Collado, L., and Guarro, J., A new 16S rDNA-RFLP method for the discrimination of the accepted species of *Arcobacter*, *Diagn. Microbiol. Infect. Dis.*, 73, 53, 2007.

94 *Campylobacter*

Rodrigo Alonso, Cecilia Girbau, and Aurora Fernández Astorga

CONTENTS

94.1 INTRODUCTION

Campylobacter species are causative agents of human enterocolitis, and one of the most commonly identified bacterial causes of acute gastroenteritis worldwide. As animal pathogens, they have been long related to diarrhea in cattle and septic abortion in both cattle and sheep[1]; however, it was not before the 1970s that they were successfully associated with human enteritis.[2,3] They were initially classified as *Vibrio*[4] because of its similarity to so-called *V. fetus* (current *Campylobacter fetus*) found in aborting cattle. (For a historical account, see Skirrow.[5])

Infection in humans is termed campylobacteriosis, and ranks from self-limiting gastroenteritis to more serious systemic infections (i.e., bacteremia or hemolytic uremic syndrome).[6,7] Postinfectious complications are rare but demyelinating disorders such as Guillain–Barré syndrome (GBS) and Miller-Fisher syndrome (MFS) are now recognized as sequelae of campylobacter infection.[8]

The epidemiology of *Campylobacter* infection is complex since the organism is widely distributed in the environment being considered as part of the normal intestinal biota of a wide range of domestic and wild animals. Fecal contamination of meat often occurs during slaughter, and human infection is usually acquired through the consumption of undercooked contaminated meat or other cross-contaminated food products. The risks to human health vary between the different animal species and also differ between countries, often due to variations in food preparation and consumption patterns.[6]

94.1.1 Classification, Morphology, and Biology

Classification. Members of the genus *Campylobacter* are classified into the epsilon division of the class *Proteobacteria*, also referred to as rRNA super-family VI. They share a close taxonomic framework and biological and clinical characteristics, which make species-specific identification complicated in some cases. The genus, first described by Véron and Chatelain[9] in 1973, comprises gram-negative, nonspore-forming bacilli that have a curved or spiral shape with tapering ends. Cells possess a polar flagellum at one or both ends of the cells. They are generally microaerophilic, requiring modified atmospheres for optimal growth conditions. Furthermore, they are neither able to ferment nor oxidize carbohydrates or degrade complex substances. Instead, they obtain energy from amino acids or tricarboxylic acid cycle intermediates.[10,11]

At present, the genus includes at least 16 species and six subspecies, of which *Campylobacter jejuni* subsp. *jejuni*, *C. jejuni* subsp. *doylei*, *Campylobacter coli*, *Campylobacter lari*, *Campylobacter upsaliensis*, and *Campylobacter helveticus* form a genetically closed group of species[12] classed as thermophilic, owing its optimal growth at 42°C. Thermophilic campylobacters infect both human and warm-blooded animals. Although primarily occurring as commensals in a wide range of domestic and wild birds and animals, thermophilic campylobacters (except for *C. jejuni* subsp. *doylei*) can also cause diseases and they are most commonly isolated from human and animal diarrhea.[6,12,13] Within the species *C. jejuni*, two subspecies, ssp. *doylei* and ssp. *jejuni*, can be distinguished on the basis of nitrate reduction, cephalothin susceptibility, and catalase activity. With the pathogenic role of the ssp. *doylei* yet not defined,[11] hereafter in the chapter *C. jejuni* will refer to *C. jejuni* ssp. *jejuni*.

Morphology and Genome Structure. *Campylobacter* cells have classic spiral morphology that is 0.2–0.8 μm wide and 0.5–5.0 μm long. In old cultures or under stressed conditions they trend to spherical morphology (coccoid forms) with a diameter of approximately 1 μm. Although transition to coccoid forms has been related to a nonculturable state,[14,15] later studies have suggested no correlation between culturability and cell morphology state[16,17]; *Campylobacter* are motile by means of unipolar or bipolar unsheathed flagella. The insertion of the flagellum in a cone-like depression allows its rotation in a characteristic corkscrew-like manner, providing an extremely rapid and spinning motion to the cell.[18] Due to this ability, *Campylobacter* can move in very viscous media and can easily colonize and pass through the mucous layer in the intestinal tract of human and animals.[19]

The genome of *C. jejuni* is 1.6–1.7 Mbp of adenine and thymine-rich DNA in size and contains a low G + C ratio—30 mol% on average.[20,21] This is a relatively small genome compared with those of enteropathogens as *Escherichia coli* (4.5 Mbp) and *Salmonella enterica* (4.9 Mbp). Some of the remarkable biochemical characteristics (i.e., their inability to ferment nor to oxidize carbohydrates and to degrade complex substances, as well as their requirement for complex media for growth) stem from the small size of the genome.[21] The small size of the genome has facilitated additional genome sequencing since the first NCTC 11168 *C. jejuni* genome sequence data were available in 2000.[22] At present, a total of five *C. jejuni* whole genomes are sequenced and eight are in progress, all of which are available on the Web site of the National Center for Biotechnology Information (NCBI, http://www.ncbi.nlm.nih.gov). The genomes of other *Campylobacter* are either sequenced (*C. concisus*, *C. curvus*, *C. fetus*, and *C. hominis*) or in progress (*C. coli*, *C. fetus* subsp. *venerealis*, *C. gracilis*, *C. rectus*, *C. showae*, and *C. upsaliensis*). Of note are the hipervariable regions found in the *C. jejuni* genome which might be important both in survival[23] and pathogenesis of the organisms.[24,25] The genome sequence, however, contains no transposons, phage remnants, or insertion sequence elements and very few repeat sequences.[22] That makes *C. jejuni* almost unique among sequenced bacterial pathogens.[26]

Biology and Epidemiology. Typically microaerophilic and moderately capnophilic, the organisms require lower O_2 and higher CO_2 concentrations than the ambient atmosphere and may be cultured in atmospheres with 5%–10% oxygen supplemented with 2%–10% carbon dioxide. However, *C. jejuni* can also grow fairly well in humidified CO_2 incubators ranging from 5% to 14% CO_2, as reported by Gaynor et al.[27] The range of growth temperatures is narrow from 30°C to 44°C, with an optimal temperature of 37°C and 42°C. Even at optimal conditions the growth is quite slow, and subsequently the times for primary isolation are long, 2–5 days.

Campylobacteriosis is considered to be a zoonotic disease. The vast reservoir in domestic and wild animals is probably the ultimate source for most enteric *Campylobacter* infections in humans.[28] They are commensal organisms routinely found in cattle, sheep, swine, goats, dogs, cats, rodents, and avian species.[29–31] The role of other vectors such as litter beetles and house flies have also been identified as potential transmission risks.[32] Other sources include untreated water, untreated milk, and sewage contamination.[13] Foods of animal origin, in particular poultry, have been identified as significant sources of *Campylobacter* as a result of infection and contamination at the preharvest and harvest level.[33] Fecal contamination of meat often occurs during slaughter, and human infection is usually acquired through the consumption of undercooked contaminated meat or other cross-contaminated food products.[34] Natural water surfaces are frequently contaminated with *C. jejuni*, *C. coli*, and *C. lari* throughout the year. Infection can be caused by drinking unchlorinated, contaminated water or consuming food prepared with untreated or improperly treated water.[28] Because the infectious dose of *Campylobacter* is quite high in comparison with that of other organisms (*Shigella*, *Giardia*), person-to-person transmission is unusual.[7]

The single most important source of *Campylobacter* infections in industrialized countries remains the consumption and handling of chicken.[7] Although accounting for a small fraction of *Campylobacter* infections in humans (most infections are sporadic), outbreaks have been traced to the consumption of raw milk and contaminated water, as well as contact with pets and farm animals.[29]

94.1.2 Clinical Features and Pathogenesis

Acute enteritis is the most common presentation of *Campylobacter* infection, with symptoms varying from one day to one week or longer. Clinically, *Campylobacter* enteritis cannot be distinguished from other acute bacterial enteritis such as salmonellosis or shigellosis.[35] However, *Campylobacter* are far less often recognized as a cause of outbreaks than *Salmonella* or *Shigella* is. Within the genus, *C. jejuni* and *C. coli* are predominant species associated with campylobacteriosis, which account for more than 95% of the isolates from diarrheal stool samples in developed countries[29,36,37] and cause clinically similar illnesses[35]; they are followed, to a lesser extent, by *C. upsaliensis*, *C. hyointestinalis*, and *C. lari*.[38]

Although some outbreaks have been reported,[39–41] most of the campylobacteriosis are normally sporadic cases affecting children and young adults in developed countries and mainly infants and young children in developing countries.[7] Owing the incidence, though to be an estimated 1% of the population per year in developed countries and even higher in developing countries, as well as the spectrum of disease with chronic sequelae and the food risks, it is not surprising that the health and social burden of campylobacteriosis could be as large as estimated.[42]

The epidemiological and symptomatic patterns of disease differ greatly between developed and developing countries. In developed countries, the infection shows a seasonal peak in the summer and early fall, and targets mostly young adults and young children, but is also found in older people. Infection most often manifests as inflammatory diarrhea with severe cramping in young adults and is usually self-limiting. In developing countries, the infection is mostly restricted to children, who might naturally vaccinate them against becoming infected as adults,[43] and does not show clear seasonality. Also, there is a high rate of asymptomatic carriage, and milder clinical symptoms of watery, noninflammatory diarrhea are usually seen in young children.[29,44]

Disease outcome is likely to be dependent on virulence of the infecting strain and host immune status.[45] Gastrointestinal symptoms are usually preceded by a prodromal stage of fever, headache, and malaise, which lasts about a day. The most common symptoms with *Campylobacter* infection are diarrhea, malaise, fever, and severe abdominal pain. The diarrhea produces stools that are often foul-smelling and can vary from being profuse and watery to bloody and dysenteric. In some patients, cramping abdominal pain, typically periumbilical, can be the predominant manifestation of illness.[46] Nausea is common, but vomiting is rare. In most patients, the illness is self-limited, and recovery occurs within 2–6 days, and usually lasts no more than 10 days, although there may be occasional relapses.[7,34] A recurrence of symptoms is experienced by 25% of patients, often characterized by abdominal pain and varying from a relatively mild gastroenteritis to an enterocolitis with bloody diarrhea and accompanying abdominal pain lasting for several weeks.[47] As a postinfectious manifestation, irritable bowel syndrome (IBS), which is characterized by abdominal pain sometimes associated with altered bowel habit, has been increasingly linked to enteric *C. jejuni* infection.[48]

The clinical symptoms of *Campylobacter* enteritis are difficult to distinguish from those of *Salmonella*, *Shigella*, or other causes of bacterial enteritis. However, abdominal pain tends to be more severe in *Campylobacter* infection and sometimes mimics acute appendicitis.

Asymptomatic carriage of *Campylobacter* is a common feature following infection and a patient may disseminate *Campylobacter* in the feces for over 4 months.

As with other intestinal pathogens, the clinical picture of *Campylobacter* infection varies from symptomless excretion to severe disease. Although it is normally a self-limiting disease, *Campylobacter* can cause extraintestinal infections, mainly found with *C. jejuni* infections. They are usually the result of systemic spread. Extraintestinal infections such as cholecystitis, peritonitis, pancreatitis, endocarditis, nephritis, and meningitis have been reported.[35]

C. jejuni enteritis can be associated with complications, although these are rare. Such complications result in the development of severe neurological and autoimmune illnesses such as Guillain–Barré,[49] Miller-Fisher, and Reiter's syndromes.[50,51] These are known to occur in about 1%–2% of individuals who have been infected with *C. jejuni*. The development of GBS is associated with lipooligosaccharides (LOS) that mimic human gangliosides.[43,52] Thus, antibodies to the LOS core structures result in an autoimmune response affecting peripheral nerves. It is estimated that as many as one out of a thousand gastroenteritis cases will develop some form of GBS.[25]

A reasonable understanding of the general clinical, microbiological, and epidemiological aspects of *Campylobacter* infection has been achieved. However, the molecular mechanisms involved in pathogenesis are still poorly understood. Despite its global importance and the progress that has been made in recent years, there are still large gaps in our knowledge on basic aspects of the mechanisms of *Campylobacter* pathogenicity and host responses to infection.[43] One of the major obstacles to solve these problems is the lack of a suitable small animal model of infection that mimics human disease as closely as possible.[53,54] We can find many references on literature about the use of animal models for *Campylobacter* infection over past years, but all of them have major disadvantages in terms of disease reproducibility or inadequate biological characterization. Nonhuman primates,[55] mice,[56,57] ferrets,[58] piglets,[59] hamsters,[60] rabbits,[61] and chickens[62] have been used as models for *C. jejuni* pathogenesis. Recently new and promising murine models have been developed using mice with limited enteric biota[63] or immunodeficient mice.[64–66] In spite of the many attempts, today there is no particular model that is suitable for the study of *Campylobacter* infection in mammalians. Polarized intestinal epithelial lines,[67] as well as human intestinal tissue,[68] are also used to investigate the pathogenic properties of *C. jejuni* and immune response of host.

Few of the virulence determinants involved in *Campylobacter* pathogenesis have or are known to have a proven role. The virulence factors are generally not well characterized and some are rather controversial. In association with food or water, campylobacters enter the host intestine via the stomach acid barrier and colonize the distal ileum and colon. After colonization of the mucus and adhesion to intestinal cell surfaces, campylobacters perturb the absorptive capacity of the intestine by damaging epithelial cell function.

Chemotaxis and Motility. Effective colonization requires chemotaxis. Genome sequence analysis has shown that the *C. jejuni* genome encodes most features of the *E. coli* chemotaxis system.[22,69] *C. jejuni* displays chemotactic motility toward mucin, L-serin, and L-fucose; amino acids that are found in the chick gastrointestinal tract; and components of mucus.[70] Mutants that lack either the chemotaxis receptors DocB or Cj0262c show decreased chicken colonization.[71] Several other components of the chemotaxis system of *C. jejuni* have been identified, including CheY, CheV, CetA, and CetB.[43]

Motility of *Campylobacter* necessitates the production of the flagellum, the best characterized virulence determinant of campylobacters. Flagella and flagellar motility are vital to host colonization, virulence in ferret models, secretion, and host-cell invasion.[43] The flagellum of *C. jejuni* consists of an unsheathed polymer of flagellin subunits, which are encoded by the adjacent *flaA* and *flaB* genes. Both genes are subjected to antigenic and phase variation. Mutants of *flaA*, the primary structural gene for flagella, are unable to colonize chicks and cannot invade human intestinal epithelial cells in vitro.[72] Adhesion and invasion are dependent on both motility and flagella expression, as *C. jejuni* mutants with reduced motility show reduced adherence and no invasion. This indicates that while flagella are involved in adherence, other adhesins are involved in subsequent internalization.[34]

Adhesion and Invasion. Adhesion by bacterial pathogens is often mediated by fimbrial structures. Genome annotations of several *C. jejuni* strains do not include obvious pilus or pilus-like ORFs. However, several proteins contribute to *C. jejuni* adherence to eukaryotic cells. CadF mediates adhesion by binding to the cell matrix protein fibronectin.[73] CadF is required for maximal binding and invasion by *C. jejuni* in vitro, and *cadF* mutants are unable to bind fibronectin and do not colonize chicks.[74] Another characterized adhesin, JlpA, is a surface-exposed lipoprotein that has been shown to bind to Hsp90α on Hep-2 cells. This binding resulted in activation of NF-κB and p38 mitogen-activated protein kinase, both of which contribute to inflammatory response.[75,76] The lipoprotein CapA, a putative autotransporter, has been described as a possible adhesin that plays a role in adhesion and invasion of Caco-2 cells. In addition, CapA-deficient mutants have decreased colonization and persistence in a chick model.[77] Another reported adhesin is Peb1, a periplasmic ABC binding protein that is required for adherence to HeLa cells and for intestinal colonization of mice.[78]

C. jejuni is generally considered to be invasive,[79] although the mechanism of uptake into epithelial cells remains vague. There are no specialized type III secretion systems similar to those of other enteric pathogens that mediate invasion into epithelial cells. Instead, the flagella appears to function as a type III secretion organelle that secretes a number of proteins, some of which are reported to affect invasion. The Cia proteins (for *Campylobacter* invasion antigens) are secreted through the flagella filament and affect invasion of some strains of *C. jejuni*.[80,81] Although CiaB is internalized into epithelial cells, the mechanism by which CiaB mediates invasion has not been detailed.[25] CiaB and other secreted Cia proteins (Cia A–H) require a functional flagellar export apparatus for their secretion.[81] As well as the Cia proteins, this flagellar export apparatus secretes FlaC, which is also required for invasion. Thus, the flagellar export apparatus is an important secretion mechanism in *C. jejuni* and is required for host-cell invasion.[43] The ability to cross the epithelial cell barrier either through epithelial cell invasion or via tight junctions allows the bacterium to move to the basolateral surface. At this point, it can reinvade the epithelial cell or be taken up into macrophages.[25] It has been shown that *C. jejuni* can replicate intracellularly in macrophages and induce apoptosis.[82] Interactions of *C. jejuni* with epithelial cells, dendritic cells, and macrophages can release chemokines and cytokines that contribute to both inflammatory diarrhea and clearance of infection.[25]

The exact mechanisms by which *C. jejuni* induces disease remain unknown; symptoms could be result of cytolethal distending toxin-induced host-cell death and the following inflammatory responses.[83]

Cytolethal Distending Toxin. The best-characterized virulence factor present in the genome of *C. jejuni* is cytolethal distending toxin (CDT). CDT consists of three subunits (CdtA, -B, and -C) encoded by a three-gene operon *cdtABC*[45] and isogenic *C. jejuni cdt* mutants lost all CDT activity.[84] The role of CDT in *C. jejuni* pathogenesis remains unclear, but its mechanism of action is becoming understood.[43] CDT is a tripartite toxin in which CdtB is the active subunit and CdtA and CdtC comprise the binding components. Once the toxin is bound to the cell, CdtB is transported to the nucleus where it acts as DNase and arrests the cell in the G2 phase of the cell cycle.[85] CDT activity causes certain cell types (such as HeLa cells and Caco-2 cells) to become slowly distended, which progresses into cell death.[34] Also, CDT induces IL-8 secretion from epithelial cells, which would contribute to inflammatory diarrhea.[86] Although all *C. jejuni* strains tested contain *cdt* genes, there is a profound variation in CDT titres.[87,88]

Glycosylation System. Recently, the long-accepted dogma that bacteria only express nonglycosylated protein has been disproved.[45,89] In 1989, Logan et al. provided clear evidence that *Campylobacter* flagellin was posttranslationally modified.[90] Today it is known that *Campylobacter* contains both a general *N*-linked protein glycosylation pathway (responsible for posttranslational modification on at least 30 proteins) and an *O*-linked system (responsible for flagellar glycosylation).[91] Both systems have been involved in *C. jejuni* virulence. Studies about the biological role for *N*-linked glycosylation clearly link it to bacterial virulence; a *pglH* mutant showed reduced ability to adhere/invade human epithelia and colonize chicks.[92] It has been shown that *O*-linked glycosylation is essential for successful flagellin assembly and motility hence influences adhesion, invasion, and virulence in vivo.[93]

Capsule and LOS. Genome sequences and DNA microarray data have demonstrated that *C. jejuni* genome contains about 22 variable regions on the chromosome. The most variable chromosomal regions are those involved in biosynthesis of surface carbohydrate structures, all of which play a role in virulence: the lipooligosaccharide (LOS) core, a polysaccharide capsule (CPS), and the locus for O-linked glycosylation.[25] The *C. jejuni* polysaccharide capsule is important for serum resistance, the adherence and invasion of epithelial cells, chick colonization, and virulence.[94] Another study utilizing a capsule-deficient mutant suggested that contribution of bacterial CPS to host-pathogen interactions may be species dependent.[95]

The *C. jejuni* LOS are highly variable due to differences in monosaccharide linkage and composition. Various LOS structures resemble human neuronal gangliosides. This molecular mimicry is thought to lead to autoimmune disorders, including Guillain–Barré and Miller-Fisher syndromes.[43] The *C. jejuni* LOS and flagella have been shown to be sialylated, which is thought to be responsible for the ganglioside mimicry leading to GBS.[34,96] Mutations in the various genes that are involved in LOS biosynthesis affect serum resistance, and adherence/invasion of INT 407 cells.[97]

94.1.3 Diagnosis

Campylobacter diagnostics and determination of antibiotic resistance are of interest not only for the treatment of infected individuals, but also for epidemiological surveillance. Clinical investigations and symptoms may orientate to a suspected diagnosis but a definitive diagnosis of *Campylobacter* enteritis can only be made bacteriologically.[35] Among the specimens, stool sample should be routinely submitted to the laboratory for isolation *Campylobacter* species from patients with diarrheal illness.[98] Transport of stools requires refrigeration at 4°C, mainly if it delays more than a day. In these cases, transport medium such as alkaline peptone water with thioglycolate and cystine,[99] modified Stuart medium,[100] or Cary-Blair medium[101] should be used.

Direct detection of *Campylobacter* in fresh stool samples is of interest as a rapid method for presumptive diagnostics during the acute stage of illness. Because of the high numbers the organisms can be visualized by performing a Gram stain in smears of fresh stools.[102]

94.1.3.1 Conventional Techniques

Culture techniques allow the recovery of *Campylobacter* from fecal samples by direct plating onto selective media. However, in some cases, selective enrichment broths as Preston broth[103] are also required to improve the recovery of *Campylobacter* from stool samples. That is, when low numbers of organisms are expected (i.e., after the acute stage, long delays in transport, or suspected complications as GBS).[104] Furthermore, filtration techniques combined with semisolid media can be used to improve the recovery of *Campylobacter* that are susceptible to antibiotics present in most selective media.[105]

All incubations must be under microaerophilic conditions using microaerobic atmosphere-generating systems. A microaerobic atmosphere comprising of approximately 5% O_2, 10% CO_2, and 85% N_2 is adequate for optimal recovery of most *Campylobacter* species.

Both solid and liquid selective media are normally achieved by using antibiotics. Typically, these are cefoperazone, cycloheximide, trimethoprim, rifampicin, vancomycin, polymyxin B, and amphotericin or nystatin combined in different manners. Most current media use a combination of cefoperazone and amphotericin B, active against grampositive bacteria and fungi respectively, together with vancomycin or trimethoprim. In addition, selective media usually contain sheep or horse sterile lysed blood to neutralize the toxic effects of compounds formed in presence of light and oxygen. Charcoal cefoperazone deoxycholate agar[106] and charcoal-based selective medium[107] are examples of blood-free media, and Skirrow medium[3] is blood-containing medium used for routine isolation procedures.

Instead or in addition to the lysed blood, charcoal, hematin and a combination of ferrous sulfate, sodium metabisulfite, and sodium pyruvate (FBP supplement) can be used as oxygen-quenching ingredients. Attempts to improve oxygen protection include lysed blood and FBP supplement in the same selective media.

There is little consensus on methods for isolating *Campylobacter* from stools, and a number of different isolation methods are used worldwide. However, it is likely that no single method is the ideal to ensure the recovery of all *Campylobacter* from fecal samples, and a combination of media could be more appropriate.

At primary isolation, confirmation of the suspected colonies can be achieved by minimal standards, which are colony morphology, Gram stain, motility, and an oxidase test.[108] Colonies that show oxidase-positive reactions along with typical gram-negative spiral rods can be presumptively reported as *Campylobacter* spp. Subsequent identification of *Campylobacter* species is based on biochemical tests such as nitrate/nitrite reduction, hippurate hydrolysis, and catalase and oxidase tests.[109] However, the genus includes closely related bacteria that share many biochemical characteristics, which makes species differentiation difficult. By way of an example, the hippurate hydrolysis is thought only positive in *C. jejuni*.[110] However, some *C. jejuni* isolates are hippuricase negative, making it impossible to differentiate *C. coli* from hippuricase-negative *C. jejuni* using purely biochemical tests.[72,111] Furthermore, false-positive test results for species other than *C. jejuni* are not rare.[112] To avoid some of these problems, an algorithm for identification of the thermophilic *Campylobacter* species was recently proposed.[113]

Therefore, with species identification relying on just a few biochemical tests, in many cases not conclusive, it is not surprising that these conventional methods may sometimes lead to equivocal results and true traps for diagnostics. All of that accounts especially for those organisms considered fastidious owing the slow growth, susceptibility to environmental factors, and restrictive culture conditions such as *Campylobacter*.

94.1.3.2 Molecular Techniques

The aforementioned problems inherent to the conventional methods for species identification of *Campylobacter* can be avoided by the application of DNA-based methods that use PCR, as the most versatile and widely used amplification technique. The products of PCR are normally separated by electrophoresis in agarose gel and detected by a DNA stain. Alternatively, it can be detected by hybridization with labeled probes[114] or nucleotide sequencing analysis.[115] The hazards of cross-contamination with the PCR products have promoted the development and wide use of the real-time PCR.[116] This technique combines PCR methodology with hybridization allowing the amplification and the detection of the products in a single reaction vessel. Detection is based either on fluorescent-labeled probes or fluorescent dyes as SYBR green, both binding to the PCR products as fast as they are synthesized by the DNA polymerase. This allows direct detection and quantification as the number of cycles required to reach the threshold value can be directly correlated with the initial numbers of cells in the sample. In comparison, real-time

PCR is faster than conventional PCR and may be carried out in automated devices that require less expertise.[117] Both techniques allow to use multiplex PCR by including various primer sets within a single reaction. Multiplex PCR may be useful on identification of the pathogens by combining genus- and species-specific genes as targets,[118] as well as in epidemiological studies where a wide range of items may be analysed.[119]

In the last years, a growing number of PCR assays for species-specific identification of *Campylobacter* have been described, several of which are based on a variety of genes. A resumed description of some of those is given in Table 94.1.

A consensus on major advantages of the PCR-based detection is the speed, together with high specificity and sensibility. Among the disadvantages, the presence of inhibitory substances in the fecal samples, which can affect primer binding or *Taq* activity then resulting in false negative assays, is highlighted. Purification of the DNA to be used in the amplification by in-house methods or commercial kits minimizes this limitation.[158]

The selection of the genetic target to be amplified is one of the most critical aspects on designing a PCR method. The ribosomal RNA presents highly conserved regions into the species of the genus, which is used for *Campylobacter* spp. detection. In addition, the high numbers of copies per cell

TABLE 94.1
Identification of *Campylobacter* Species by Molecular Methods

Target	Isolation Procedure[a]	PCR Product Detection	Reference
	Real-time PCR		
C. jejuni/ coli, C. lari, C. upsaliensis, C. fetus, C. helveticus, C. lanienae (16S rRNA)	Colony, stool	Melting peak analysis	90
C. jejuni (gyrA)	Stool	Melting peak analysis	120
C. jejuni (yphC, gyrA)	Stool	Taqman probe	121
C. jejuni (hipO), C. coli (glyA)	Colony, stool	Taqman probe	122
C. jejuni (mapA), C. coli (ceuE)	Stool	Taqman probe	123
C. jejuni (mapA), C. coli (ceuE)	Stool	Taqman probe	124
C. jejuni (mapA)	Stool and colony	Taqman probe	125
	Conventional PCR		
C. jejuni/coli (flaA)	Colony, stool	Gel electrophoresis	126
Thermophilic campylobacters *(23S rRNA), C. jejuni, C. coli, C. lari, C. upsaliensis (23S rRNA)*	Colony	Gel electrophoresis	127
Campylobacter spp. *(flaA)*	Stool	Gel electrophoresis	128
C. jejuni, C. coli (ceuE)	Colony	Gel electrophoresis	129
C. jejuni/coli (16S rRNA), C. jejuni (hipO), C. coli (asp)	Stool	Gel electrophoresis	118
C. jejuni, C. coli, C. lari, C. upsaliensis (23S rRNA)	Colony	RFLP (AluI, Tsp509I enzymes)	130
C. jejuni, C. coli, C. lari, C. upsaliensis, C. helveticus (16S rRNA)	Stool	Gel electrophoresis	131
C. jejuni, C. coli, C. lari, C. upsaliensis, C. fetus, C. hyointestinalis (16S rRNA)	Colony, stool	PCR-ELISA (8 capture probes)	132
C. jejuni, C. coli, C. lari, C. upsaliensis (GTPase)	Colony	Reverse hybridization LiPA (line probe assay)	133
C. jejuni, C. coli, C. lari, C. upsaliensis (glyA)	Colony	Southern hybridization (species-specific probes)	134
Campylobacter spp., *C. jejuni, C. coli, (16S/23S rRNA* intergenic spacer region)	Stool	PCR/DNA probe membrane-based colorimetric assay	135
C. jejuni, C. coli (ORF-C)	Colony	PCR ELISA DIG detection kit and five biotinylated capture probes	136
Campylobacter spp., *C. jejuni, C. coli, C. lari, C. upsaliensis (16S/23S rRNA* intergenic spacer region)	Stool	PCR/DNA probe membrane-based colorimetric assay	137
Campylobacter spp. *(gale)*	Colony	Gel electrophoresis	138
C. jejuni (hipO), C. coli (asp)	Stool	Gel electrophoresis	139 140
C. jejuni, C. coli, C. lari, C. upsaliensis, C. fetus, C. hyointestinalis, C. lanienae, C. helveticus, C. mucosalis (groEL)	Colony	RFLP (AluI, ApoI enzymes)	141
C. jejuni (mapA)	Appendix (paraffin blocks)	Gel electrophoresis	142

continued

TABLE 94.1 (Continued)
Identification of *Campylobacter* Species by Molecular Methods

Target	Isolation Procedure[a]	PCR Product Detection	Reference
C. jejuni, C. coli, C. lari, C. upsaliensis, C. fetus, C. hyointestinalis, C. helveticus, C. mucosalis, C. concisus, C. curvus, C. gracilis, C. rectus, C. showae, C. sputorum (cpn60)	Colony	Direct sequencing	143
C. jejuni/coli (16S rRNA)	Stool	Gel electrophoresis	144
Campylobacter spp. (*16S rRNA*)	Stool	Gel electrophoresis	123
	Multiplex PCR		
C. jejuni, C. coli, C. lari, C. upsaliensis (lpxA)	Colony	Gel electrophoresis	145
C. jejuni, C. coli (ceuE)	Colony, stool	Gel electrophoresis	146
C. jejuni (hipO), C. coli, C. lari, C. upsaliensis (glyA), C. fetus (sapB2)	Colony	Gel electrophoresis	147 148
C. jejuni (hipO), C. coli (asp)	Stool	Gel electrophoresis	149
Thermophilic campylobacters (*23S rRNA*), *C. jejuni (lpxA), C. coli (ceuE)*	Colony	Gel electrophoresis	150
C. jejuni (cj0414), C. coli (ask), C. lari (glyA), C. upsaliensis (lpxA), C. fetus (cstA), C. hyointestinalis (23S rRNA)	Colony	Gel electrophoresis	151
C. jejuni, C. coli, C. fetus (cdtA, cdtB, cdtC)	Colony	Gel electrophoresis	152
	DNA array		
C. jejuni, C. coli, C. lari, C. upsaliensis (fur, glyA, ceuB-C, cdtABC, fliY)	Colony	Oligonucleotide probes	153
C. jejuni (16S rRNA, 23S rRNA)	Colony	Oligonucleotide probes	154
C. jejuni (mapA, ceuE, cdt)	Stool	PCR probes	155
	DNA FISH		
Campylobacter spp. (*23S rRNA*), thermophilic campylobacters (*23S rRNA*), *C. jejuni (23S rRNA), C. lari (16S rRNA), C. upsaliensis (16S rRNA)*	Colony	Oligonucleotide probes	156
	LAMP (Loop-mediated isothermal amplification)		
C. jejuni (cj0414), C. coli (CCO0367)	Stool	Turbidity	157

[a] Stool: direct DNA extraction from human fecal samples; Colony: DNA extraction from bacterial colonies.

provide a naturally amplified target, hence increasing the analytic accuracy. The 16S rRNA genes are the most suitable for this purpose and so they are widely used for *Campylobacter* spp. detection. However, not all the PCRs targeting the 16S rRNA are directed to the same position into the gene sequence. Other genetic targets are selected amid genes exclusively found in a concrete species. Such genes are used for differentiating *C. jejuni* from the other species in the genus. An example of this is the hippurate gene, *hipO*, codifying for the hippurate hydrolase that is found exclusively in *C. jejuni* species. Since hippurate-negative strains were detected, this gene is no longer used in the identification of *C. jejuni*. Instead, the interest is turned on to the species-specific fragments into common genes suitable for discriminating between *C. jejuni* and *C. coli*, which are very closely related. As an example, the *ceuE* gene codifying for a virulence determinant is present in both species, but choosing adequate primers a specific fragment for each one may be obtained.[129]

A major disadvantage concerning the DNA-based methods is the question of whether or not they are detecting DNA from dead cells. Due to the long persistence of DNA in cells postdeath, the correlation between presence of DNA and viability is not clear.[159] However, this does not account for those methods using a preenrichment period.

The whole-genome-based methods have been developed from the microarray technology that offers the possibility of depositing a high density of different specific probes on a very small area. They are also known as biochip, DNA chip, DNA microarray, and gene array. The DNA microarray consists of a series of samples spotted on a solid support, most commonly glass or silicon. Each spot contains DNA fragments or oligonucleotides as probes.[160,161] Probes appear as spots in the final image, where each spot represents a unique probe sequence and spots are usually 100–200 Am in size and located within 200–500 Am of each other. The targets are labeled with fluorescent dyes before being applied to the microarray and the probe-target hybridization can be detected and quantified by fluorescence-based detection.[162] Depending on the objective, targets may be PCR products, genomic DNA, total RNA, mRNA, cDNA, plasmid DNA, or oligonucleotides.[163] The uses of the DNA arrays are likely unlimited as they can combine the amplification of nucleic acids with their massive screening capabilities. Simultaneous detection or level of expression of thousands of specific DNA can be achieved in a single DNA microarray, resulting in sensitivity, specificity, and high-throughput capacity.

Two main types of arrays are produced: genomic arrays and oligonucleotide arrays. The genomic microarrays comprise either whole genome, genomic fragments from a cDNA library, or open reading frames from a bacterial strain. These arrays are useful for comparative phylogenetic studies and genetic diversity by determining constant and variable genes among the isolates. The oligonucleotide arrays, containing probes from 18 to 70 nucleotides long,[164] are useful for genomic analysis because they can consist of multiple sequence variants of a target gene and can be adapted for pathogen detection in different environments.

94.2 METHODS

PCR is a powerful, rapid method for diagnosis of microbial infections. However, the usefulness of diagnostic PCR is limited in part by the presence of inhibitory substances in complex biological samples, which reduce or block the PCR amplification reaction. It is known that feces contain large amounts of phenolic and metabolic compounds and polysaccharides that are inhibitory for PCR. Common inhibitors are DNases, polysaccharides, and proteases.[165] Thus, selection of the nucleic acid extraction method is of critical importance. Strategies to neutralize these PCR inhibitors include the use of effective DNA purification protocols or PCR facilitators. Amplification facilitators are substances that can enhance the efficiency of PCR. For feces, it was shown that the addition of bovine serum albumin is most effective for overcoming PCR-inhibitory substances.[166] Another strategy to overcome PCR inhibition is to use a polymerase that is more resistant than *Taq* (i.e., t*Tth* or *Pwo* polymerases).[167] DNA purification methods can, in addition to removing inhibitors, concentrate total genomic DNA, especially from bacteria. Many commercially available extraction kits incorporate a silica matrix membrane (typically in column format) to trap the DNA. Several wash steps are required to remove proteins and other macromolecules, and the purified DNA is then eluted from the membrane. Silica membrane columns provide a convenient method for the purification of DNA, which is relatively free of inhibitors.[168]

Below we present procedures for the extraction of *Campylobacter* DNA from culture and directly from fecal samples, and their subsequent identification by molecular techniques. A multiplex PCR-based method for detection of *C. jejuni* and *C. coli* is described.

94.2.1 Sample Preparation

Extraction of DNA from Bacterial Cultures. DNA is extracted from bacterial culture using the NucleoSpin Tissue Kit according to the manufacturer's instructions (Macheray Nagel, Germany) and the purified DNA is quantified using a ND-1000 spectrophotometer (NanoDrop).

1. Grow culture of *C. jejuni* in liquid BHI media. Incubate overnight at 37°C under microaerobic conditions.
2. Centrifuge 1 mL culture for 5 min at 8000 × *g*. Remove supernatant.
3. Resuspend the pellet in 180 μL Buffer T1 by pipetting up and down. Add 25 μL Proteinase K solution. Vortex vigorously and incubate at 56°C until complete lysis is obtained (at least 1–3 h). Vortex occasionally or use a shaking incubator.
4. Add 200 μL Buffer B3, vortex vigorously and incubate at 70°C for 10 min.
5. Add 210 μL ethanol (96%–100%) to the sample and vortex vigorously.
6. Place one NucleoSpin® Tissue Column into a Collection Tube. Apply the sample to the column. Centrifuge for 1 min at 11,000 × *g*. Discard the flow-through and place the column back into the Collection Tube.
7. Add 500 μL Buffer BW. Centrifuge for 1 min at 11,000 × *g*. Discard flow-through and place the column back into the Collection Tube.
8. Add 600 μL Buffer B5 to the column and centrifuge for 1 min at 11,000 × *g*. Discard flow-through and place the column back into the Collection Tube.
9. Centrifuge the column for 1 min at 11,000 × *g*.
10. Place the NucleoSpin® Tissue Column into a 1.5 mL microcentrifuge tube and add 150 μL prewarmed Elution Buffer BE (70°C). Incubate at room temperature for 1 min. Centrifuge 1 min at 11,000 × *g*.
11. Determine DNA concentration in each sample using a spectrophotometer.
12. Take an aliquot of the DNA suspension and dilute to give a stock template DNA concentration of 20 ng/μL. DNA suspensions may be stored at 4°C or −20°C.

Alternatively, *Campylobacter* DNA can be extracted from colonies using the boiling method; namely, suspending a 10 μL loopful of growth in 500 μL of sterile distilled water, heated at 100°C for 10 min, and centrifuged at 13,000 rpm for 5 min.[141]

Extraction of DNA from Human Fecal Samples. DNA is extracted from approximately 0.2 g of the stool specimens using the QIAamp DNA stool mini kit (Qiagen) according to the manufacturer's instructions, selecting the option to incubate the specimens in the lysis buffer at 95°C.[120,122,135] DNA is kept at −20°C until use.

1. Weigh 180–220 mg stool in a 2 mL microcentrifuge tube and place the tube on ice.
2. Add 1.4 mL Buffer ASL to each stool sample. Vortex continuously for 1 min or until the stool sample is thoroughly homogenized.
3. Heat the suspension for 5 min at 95°C.
4. Vortex for 15 s and centrifuge sample at 20,000 × *g* for 1 min to pellet stool particles.
5. Pipette 1.2 mL of the supernatant into a new 2 mL tube and discard the pellet.

6. Add 1 InhibitEX Tablet to each sample and vortex immediately and continuously for 1 min or until the tablet is completely suspended. Incubate suspension for 1 min at room temperature.
7. Centrifuge sample at 20,000 × *g* for 3 min to pellet inhibitors bound to InhibitEX matrix.
8. Pipette all the supernatant into a new 1.5 mL tube and discard the pellet. Centrifuge the sample at 20,000 × *g* for 3 min.
9. Pipette 15 μL of proteinase K into a new 1.5 mL tube.
10. Pipette 200 μL supernatant from step 8 into the 1.5 mL tube containing proteinase K.
11. Add 200 μL Buffer AL and vortex for 15 s.
12. Incubate at 70°C for 10 min. Spin down the sample before next step.
13. Add 200 μL of ethanol (96%–100%) to the sample, and mix by vortexing. Spin down the sample before next step.
14. Label the lid of a new QIAamp spin column placed in a 2 mL collection tube. Carefully apply the complete lysate from step 13 to the QIAamp spin column without moistening the rim. Close the cap and centrifuge at 20,000 × *g* for 1 min. Place the QIAamp spin column in a new 2 mL collection tube, and discard the tube containing the filtrate.
15. Carefully open the QIAamp spin column and add 500 μL Buffer AW1. Centrifuge at 20,000 × *g* for 1 min. Place the QIAamp spin column in a new 2 mL collection tube, and discard the collection tube containing the filtrate.
16. Carefully open the QIAamp spin column and add 500 μL Buffer AW2. Centrifuge at 20,000 × *g* for 3 min. Discard the collection tube containing the filtrate.
17. Place the QIAamp spin column in a new 2 mL collection tube. Centrifuge at 20,000 × *g* for 1 min. Discard the collection tube containing the filtrate.
18. Transfer the QIAamp spin column into a new, labeled 1.5 mL microcentrifuge tube and pipet 200 μL Buffer AE directly onto the QIAamp membrane. Incubate for 1 min at room temperature, then centrifuge at 20,000 × *g* for 1 min to elute DNA.

94.2.2 Detection Procedures

The PCR conditions are those proposed by Denis et al.[112] with minor modifications affecting the dNTPs concentration and 16S rRNA gene primer concentration.[169]

Multiplex PCR amplification for specific detection of the 16S rRNA (*Campylobacter* spp.), *mapA* (*C. jejuni*), and *ceuE* (*C. coli*) genes is performed with the three primer sets described in Table 94.2.

1. PCR amplifications are performed in 30-μL volumes containing 1× PCR buffer; 1.5 mM $MgCl_2$; 0.2 mM dNTPs; 0.6 U *Taq* polymerase; 0.5 μM of MD16S1 and MD16S2 primers; and 0.42 μM of MDmapA1, MDmapA2, COL3, and MDCOL2 primers; and 100 ng of template DNA.
2. The PCR is run in a conventional thermal cycler using the following parameters: 95°C for 10 min; 35 cycles of 95°C for 30 s, 59°C for 1 min 30 s, 72°C for 1 min; and a final 72°C for 10 min.
3. Mix 10 μL of the PCR product with 2 μL of gel loading buffer and put them into wells of a 2% TBE-agarose gel. In a parallel well, dispense 5 μL of the HyperLadder IV DNA molecular weight marker.
4. Subject to electrophoresis at 90 V for approximately 1.5–2 h in 1× TBE buffer.
5. Remove gel and submerge in an ethidium bromide DNA staining solution (0.5 μg/mL) for 30 min.
6. Wash the gel with distilled water and visualize the DNA amplicons by exposure to a UV-transilluminator.
7. Expected product sizes are as follows: an 857 bp *Campylobacter* specific fragment of the 16S rRNA gene, and two species-specific fragments of 589 and 462 bp corresponding to *C. jejuni* and *C. coli*, respectively.

94.3 CONCLUSIONS AND FUTURE PERSPECTIVES

Campylobacter species are causative agents of human enterocolitis being one of the most commonly identified bacterial causes of acute gastroenteritis worldwide. They include highly variable organisms adapted to a wide range

TABLE 94.2
Oligonucleotide Primers and Probes for Identification of *Campylobacter* spp.

Target	Name	Sequence (5′→3′)	Size	Reference
16S rRNA	MD16S1	ATCTAATGGCTTAACCATTAAAC	857 bp	118
	MD16S2	GGACGGTAACTAGTTTAGTATT		
mapA	MDmapA1	CTATTTTATTTTTGAGTGCTTGTG	589 bp	170
	MDmapA2	GCTTTATTTGCCATTTGTTTTATT		
ceuE	COL3	AATTGAAAATTGCTCCAACTATG	462 bp	129
	MDCOL2	TGATTTTATTATTTGTAGCAGCG		

of ecological niches making the ubiquitous distribution of *C. coli* and *C. jejuni* in animals, foods, and the environment difficult for the clarification of the epidemiology of human campylobacteriosis.

In spite of its global importance, insights into *C. jejuni* are limited compared to other enteropathogens like *Salmonella* and *E. coli*. As a consequence, much effort has been directed toward an improved understanding of the *C. jejuni*, its epidemiology and pathogenic mechanisms; however, many questions are yet to be answered.

C. jejuni is typically microerophilic and moderately capnophilic, which largely reduces the risk of growth in environment and foods. However, human infections are usually acquired through the consumption of undercooked contaminated meat, water, or other cross-contaminated food products, being the foods of poultry origin the most common sources of infections. How these organisms in both the environment and the food chain survive is by now an open question. As they can enter a viable but nonculturable state under adverse conditions it has been suggested that this VBNC state can be a survival strategy in environments not supporting growth of *Campylobacter*. However, the significance of the VBNC-forms in transmission of disease is today uncertain, as the resuscitation back to the actively growing state is still controversial.

Another interesting open question is how *Campylobacter* elicits illness in humans but establishes commensal relationships with other animals. A variety of virulence factors have been identified, among them LOS structures resembling human neuronal gangliosides and cytolethal distending toxin. But there is general agreement that *C. jejuni* is unique in not having an identifiable enterotoxin as mechanisms to cause diarrhea, and the mechanisms to cause diarrhea remain unclear. Hence, there is still a real need to understand the mechanisms by which *Campylobacter* cause diarrheal illness in humans. This limited understanding on *C. jejuni* pathogenesis is partly because of the lack of a suitable small animal model of infection that mimics human disease as closely as possible.

There is a continued need for improvements in routine identification procedures for *C. coli* and *C. jejuni* to ensure accurate results of epidemiological studies. Because *Campylobacter* isolates are not routinely speciated by microbiology laboratories, the number of reported cases of each species cannot be determined directly from laboratory reports. However, the development of strategies for the successful control of this organism will depend on directed, species-specific research. Because they may sometimes lead to equivocal results and true traps for diagnostics the conventional diagnostic methods are not suitable for these purposes. The DNA-based methods, like PCR-based methods, have the potential to improve the clinical management and epidemiological tracking of *Campylobacter* enteritis. The uses of the DNA arrays are likely unlimited as it can combine the amplification of nucleic acids with its massive screening capability. Extending the future research to this area should allow the identification of improved epidemiological markers, which in turn should allow a rapid increase in the knowledge of the ecology, epidemiology, and pathogenesis of *Campylobacter*.

ACKNOWLEDGMENTS

Work in the authors' laboratory is supported by grants from the Spanish Government (AGL2008–03626) and the Basque Government (GIC07/52-IT-308–07).

REFERENCES

1. Smith, T., and Taylor, M.S., Some morphological and biochemical characters of the spirilla (*Vibrio fetus* n. sp.) associated with disease of the fetal membranes in cattle, *J. Exp. Med.*, 30, 299, 1919.
2. Butzler, J.P. et al., Related *vibrio* in stools, *J. Pediatr.*, 82, 493, 1973.
3. Skirrow, M.B., *Campylobacter* enteritis: A "new" disease, *Br. Med. J.*, 2, 9, 1977.
4. Jones, F.S., Orcutt, M., and Little, R.B., Vibrios (*V. jejuni* n. sp.) associated with intestinal disorders in cows and calves, *J. Exp. Med.*, 53, 853, 1931.
5. Skirrow, M.B., John McFadyean and the centenary of the first isolation of *Campylobacter* species, *Clin. Infect. Dis.*, 43, 1213, 2006.
6. Humphrey, T., O'Brien, S., and Madsen, M., *Campylobacters* as zoonotic pathogens: A food production perspective, *Int. J. Food Microbiol.*, 117, 237, 2007.
7. Allos, B.M., *Campylobacter jejuni* infections: Update on emerging issues and trends, *Clin. Infect. Dis.*, 32, 1201, 2001.
8. Nachamkin, I., Chronic effects of *Campylobacter* infection, *Microbes Infect.*, 4, 399, 2002.
9. Véron, M., and Chatelain, R., Taxonomic study of the genus *Campylobacter* (Sebald and Veron) and designation of the neotype strain for the type species, *Campylobacter fetus* (Smith and Taylor) Sebald and Veron, *Int. J. Syst. Bacteriol.*, 23, 122, 1973.
10. Griffiths, P.L., and Park, R.W.A., Campylobacters associated with human diarrhoeal disease, *J. Appl. Bacteriol.*, 69, 281, 1990.
11. Vandamme, P., Taxonomy of the family *Campylobacteraceae*. In Nachamkin, I., and Blaser, M.J. (Editors), *Campylobacter*, 2nd ed., p. 3, ASM Press, Washington, DC, 2000.
12. On, S.L.W., Taxonomy of *Campylobacter*, *Arcobacter*, *Helicobacter* and related bacteria: Current status, future prospects and immediate concerns, *J. Appl. Microbiol.*, 90, 1S, 2001.
13. Ketley, J.M., Pathogenesis of enteric infection by *Campylobacter*, *Microbiol.*, 143, 5, 1997.
14. Rollins, D.M., and Colwell, R.R., Viable but nonculturable stage of *Campylobacter* and its role in survival in the natural aquatic environment, *Appl. Environ. Microbiol.*, 52, 531, 1986.
15. Jones, D.M., Sutcliffe, E.M., and Curry, A., Recovery of viable but non-culturable *Campylobacter jejuni*, *J. Gen. Microbiol.*, 137, 2477, 1991.
16. Hazeleger, W.C. et al., Temperature-dependent membrane fatty acid and cell physiology changes in coccoid forms of *Campylobacter jejuni*, *Appl. Environ. Microbiol.*, 61, 2713, 1995.

17. Lázaro, B. et al., Viability and DNA maintenance in nonculturable spiral *Campylobacter jejuni* cells after long-term exposure to low temperatures, *Appl. Environ. Microbiol.*, 65, 4667, 1999.
18. Guerry, P. et al., Role of flagellin genes in *Campylobacter* mobility, *J. Bacteriol.*, 173, 4757, 1991.
19. Black, R. E. et al., Experimental *Campylobacter jejuni* infection in humans, *J. Infect. Dis.*, 157, 472, 1988.
20. Chang, N., and Taylor, D.E., Use of pulsed-field agarose gel electrophoresis to size genomes of *Campylobacter* species and to construct a *Sal*I map of *Campylobacter jejuni* UA580, *J. Bacteriol.*, 172, 5211, 1990.
21. Nuijten, P.J.M. et al., Size and physical map of the *Campylobacter jejuni* chromosome, *Nucleic Acids Res.*, 18, 6211, 1990.
22. Parkhill, J. et al., The genome sequence of the foodborne pathogen *Campylobacter jejuni* reveals hypervariable sequences, *Nature*, 403, 665, 2000.
23. On, S.L.W. et al., Numerical analysis of DNA microarray data of *Campylobacter jejuni* strains correlated with survival, cytolethal distending toxin and haemolysin analyses, *Int. J. Med. Microbiol.*, 296, 353, 2006.
24. Hofreuter, D. et al., Unique features of a highly pathogenic *Campylobacter jejuni* strain, *Infect. Immun.*, 74, 4694, 2006.
25. Poly, F., and Guerry, P., Pathogenesis of *Campylobacter*, *Curr. Opin. Gastroenterol.*, 24, 27, 2008.
26. Dorrell, N. et al., Whole genome comparison of *Campylobacter jejuni* human isolates using a low-cost microarray reveals extensive genetic diversity, *Genome Res.*, 11, 1706, 2001.
27. Gaynor, E.C. et al., The *Campylobacter jejuni* stringent response controls specific stress survival and virulence-associated phenotypes, *Mol. Microbiol.*, 56, 8, 2005.
28. Hu, L., and Kopecko, J., *Campylobacter* species. In Miliotis, M.D., and Bier, J.W. (Editors), *International Handbook of Foodborne Pathogens*, p. 181, Marcel Dekker Inc., New York, NY, 2003.
29. Friedman C.R. et al., Epidemiology of *Campylobacter jejuni* infections in the United States and other industrialized nations. In Nachamkin, I., and Blaser, M.J. (Editors), *Campylobacter*, 2nd ed., p. 139, ASM Press, Washington, DC, 2000.
30. Franco, D.A., and Williams, C.E., *Campylobacter jejuni*. In Hui, Y.H., Pierson, M.D., and Gorham, J.R. (Editors), *Foodborne Disease Handbook*, 2nd ed., p. 83, Marcel Dekker Inc., New York, NY, 2001.
31. Waldenström, J. et al., Prevalence of *Campylobacter jejuni*, *Campylobacter lari*, and *Campylobacter coli* in different ecological guilds and taxa of migrating birds, *Appl. Environ. Microbiol.*, 68, 5911, 2002.
32. Chuma, T., Hashimoto, S., and Okamoto, K., Detection of thermophilic *Campylobacter* from sparrows by multiplex PCR: The role of sparrows as a source of contamination of broilers with *Campylobacter*, *J. Vet. Med. Sci.*, 62, 1291, 2000.
33. Pezzotti, G. et al., Occurrence and resistance to antibiotics of *Campylobacter jejuni* and *Campylobacter coli* in animals and meat in northeastern Italy, *Int. J. Food Microbiol.*, 82, 281, 2003.
34. van Vliet, A.H.M., and Ketley, J.M., Pathogenesis of enteric *Campylobacter* infection, *J. Appl. Microbiol.*, 90, 45S, 2001.
35. Blaser, M.J., and Engberg, J., Clinical Aspects of *Campylobacter jejuni* and *Campylobacter coli* infections. In Nachamkin, I., Szymanski, C.M., and Blaser, M.J. (Editors), *Campylobacter*, 3rd ed., p. 99, ASM Press, Washington, DC, 2008.
36. World Health Organization. 2001. The increasing incidence of human campylobacteriosis. Report and proceedings of a WHO consultation of experts. Copenhagen, Denmark, 21 to 25 November 2000. World Health Organization, Geneva, Switzerland. http://whqlibdoc.who.int/hq/2001/who_cds_csr_aph_2001.7.pdf.
37. Matsuda, M., and Moore, J.E., Urease-positive thermophilic *Campylobacter* species, *Appl. Environ. Microbiol.*, 70, 4415, 2004.
38. Lastovica, A.J., and Allos, M., Clinical significance of *Campylobacter* and related species other than *Campylobacter jejuni* and *Campylobacter coli*. In Nachamkin, I., Szymanski, C.M., and Blaser, M.J. (Editors), *Campylobacter*, 3rd ed., p. 123, ASM Press, Washington, DC, 2008.
39. Evans, M.R. et al., *Campylobacter* outbreak associated with stir-fried food, *Epidemiol. Infect.*, 121, 275, 1998.
40. Lehner, A. et al., Epidemiologic application of pulsed-field gel electrophoresis to an outbreak of *Campylobacter jejuni* in an Austrian youth centre, *Epidemiol. Infect.*, 125. 13, 2000.
41. Llovo, J. et al., Molecular typing of *Campylobacter jejuni* isolates involved in a neonatal outbreak indicates nosocomial transmission, *J. Clin. Microbiol.*, 41, 3926, 2003.
42. Havelaar, A.H. et al., Health burden in the Netherlands due to infection with thermophilic *Campylobacter* spp., *Epidemiol. Infect.*, 125, 505, 2000.
43. Young, K.T., Davis, L.M., and Dirita, V.J., *Campylobacter jejuni*: Molecular biology and pathogenesis, *Nat. Rev. Microbiol.*, 5, 665, 2007.
44. Oberhelman, R.A., and Taylor, D.N., *Campylobacter* infection in developing countries. In Nachamkin, I., and Blaser, M.J. (Editors), *Campylobacter*, 2nd ed., p. 139, ASM Press, Washington, DC, 2000.
45. Zilbauer, M. et al., *Campylobacter jejuni*-mediated disease pathogenesis: An update, *Trans. R. Soc. Trop. Med. Hyg.*, 102, 123, 2008.
46. Blaser, M.J., *Campylobacter* species. In Mandell, G.L. et al. (Editors), *Principles and Practice of Infectious Diseases*, 2nd ed., p. 1221, John Wiley and Sons, New York, 1985.
47. Mandel, B.K., De Mol, P., and Butzler, J.P., Clinical aspects of *Campylobacter* infections in humans. In Butzler, J.P. (Editor), *Campylobacter Infection in Man and Animals*, p. 22, CRC Press, Boca Ratón, FL, 1984.
48. Spiller, R.C., Role of infection in irritable bowel syndrome, *J. Gastroenterol.*, 42, 41, 2007.
49. Allos, B.M., Association between *Campylobacter jejuni* infection and Guillain-Barré syndrome, *J. Infect. Dis.*, 2, S125, 1997.
50. Peterson, M.C., Rheumatic manifestations of *Campylobacter jejuni* and *C. fetus* infections in adults, *Scan. J. Rheumatol.*, 23, 167, 1994.
51. Ebright, J.R., and Ryan, L.M., Acute erosive reactive arthritis associated with *Campylobacter jejuni*-induced colitis, *Am. J. Med.*, 76, 321, 1984.
52. Bergman, M., et al. Development of joint disease is associated with Campylobacter jejuni strains harbouring lipo-oligosaccharide biosynthese gene cluster A, *Zoonosis Public Health*, 54, S19, 2007.
53. Young, V.B., Schauer, D.B., and Fox, J.G., Animal models of *Campylobacter* infection. In Nachamkin, I., and Blaser, M.J. (Editors), *Campylobacter*, 2nd ed., p. 287, ASM Press, Washington, DC, 2000.
54. Newell, D.G., Animal models of *Campylobacter jejuni* colonization and disease and the lessons to be learned from similar *Helicobacter pylori* models, *J. Appl. Microbiol.*, 90, 57S, 2001.

55. Russell, R.G. et al., Experimental *Campylobacter jejuni* infection in *Macaca nemestrina*, *Infect. Immun.*, 57, 1438, 1989.
56. Baqar, S., Applebee, L.A., and Bourgeois, A.L., Immunogenicity and protective efficacy of a prototype *Campylobacter* killed whole-cell vaccine in mice, *Infect. Immun.*, 63, 3731, 1995.
57. Hodgson, A.E. et al., Experimental campylobacter infection and diarrhoea in immunodeficient mice, *J. Med. Microbiol.*, 47, 799, 1998.
58. Bell, J.A., and Manning, D.D., A domestic ferret model of immunity to *Campylobacter jejuni*-induced enteric disease, *Infect. Immun.*, 58, 1848, 1990.
59. Babakhani, F.K., Bradley, G.A., and Joens, L.A., Newborn piglet model for campylobacteriosis, *Infect. Immun.*, 61, 3466, 1993.
60. Humphrey, C.D., Montag, D.M., and Pittman, F.E., Experimental infection of hamsters with *Campylobacter jejuni*, *J. Infect. Dis.*, 151, 485, 1985.
61. Caldwell, M.B., Simple adult rabbit model for *Campylobacter jejuni* enteritis, *Infect. Immun.*, 42, 1176, 1983.
62. Cawthraw, S.A. et al., Increased colonization potential of *Campylobacter jejuni* strain 81116 after passage through chickens and its implication on the rate of transmission within flocks, *Epidemiol. Infect.*, 117, 213, 1996.
63. Fox, J.G. et al., Gastroenteritis in NF-kappaB-deficient mice is produced with wild-type *Camplyobacter jejuni* but not with *C. jejuni* lacking cytolethal distending toxin despite persistent colonization with both strains, *Infect. Immun.*, 72, 1116, 2004.
64. Chang, C., and Miller, J.F., *Campylobacter jejuni* colonization of mice with limited enteric flora, *Infect. Immun.*, 74, 5261, 2006.
65. Watson, R.O. et al., A MyD88-deficient mouse model reveals a role for Nramp1 in *Campylobacter jejuni* infection, *Infect. Immun.*, 75, 1994, 2007.
66. Mansfield, L.S. et al., C57BL/6 and congenic interleukin-10-deficient mice can serve as models of *Campylobacter jejuni* colonization and enteritis, *Infect. Immun.*, 75, 1099, 2007.
67. Byrne, C.M., Clyne, M., and Bourke, B., *Campylobacter jejuni* adhere to and invade chicken intestinal epithelial cells in vitro, *Microbiol.*, 153, 561, 2007.
68. MacCallum, A.J. et al., *Campylobacter jejuni* infected human epithelial cell lines vary in their ability to secrete interleukin-8 compared to in vitro-infected primary human intestinal tissue, *Microbiology*, 152, 3661, 2006.
69. Marchant, J., Wren, B., and Ketley, J., Exploiting genome sequence: Predictions for mechanisms of *Campylobacter* chemotaxis, *Trends Microbiol.*, 10, 155, 2002.
70. Hugdahl, M.B., Beery, J.T., and Doyle, M.P., Chemotactic behavior of *Campylobacter jejuni*, *Infect. Immun.*, 56, 1560, 1988.
71. Hendrixson, D.R., and DiRita, V.J., Identification of *Campylobacter jejuni* genes involved in commensal colonization of the chick gastrointestinal tract, *Mol. Microbiol.*, 52, 471, 2004.
72. Fields, P.I., and Swerdlow, M.D., *Campylobacter jejuni*, *Clin. Lab. Med.*, 19, 489, 1999.
73. Konkel, M.E. et al., Identification and molecular cloning of a gene encoding a fibronectin-binding protein (CadF) from *Campylobacter jejuni*, *Mol. Microbiol.*, 24, 953, 1997.
74. Ziprin, R.L. et al., The absence of cecal colonization of chicks by a mutant of *Campylobacter jejuni* not expressing bacterial fibronectin-binding protein, *Avian Dis.*, 43, 586, 1999.
75. Jin, S. et al., JlpA, a novel surface-exposed lipoprotein specific to *Campylobacter jejuni*, mediates adherence to host epithelial cells, *Mol. Microbiol.*, 39, 1225, 2001.
76. Jin, S. et al., JlpA of *Campylobacter jejuni* interacts with surface-exposed heat shock protein 90α and triggers signalling pathways leading to the activation of NF-κB and p38 MAP kinase in epithelial cells, *Cell. Microbiol.*, 5, 165, 2003.
77. Ashgar, S.S. et al., CapA, an autotransporter protein of *Campylobacter jejuni*, mediates association with human epithelial cells and colonization of the chicken gut, *J. Bacteriol.*, 189, 1856, 2007.
78. Pei, Z. et al., Mutation in the *peb1A* locus of *Campylobacter jejuni* reduces interactions with epithelial cells and intestinal colonization of mice, *Infect. Immun.*, 66, 938, 1998.
79. Kopecko, D.J., Hu, L., and Zaal, K.J., *Campylobacter jejuni*-microtubule-dependent invasion, *Trends Microbiol.*, 9, 389, 2001.
80. Konkel, M.E. et al., Bacterial secreted proteins are required for the internalization of *Campylobacter jejuni* into cultured mammalian cells, *Mol. Microbiol.*, 32, 691, 1999.
81. Konkel, M.E. et al., Secretion of virulence proteins from *Campylobacter jejuni* is dependent on a functional flagellar export apparatus, *J. Bacteriol.*, 186, 3296, 2004.
82. Hickey, T.E., Majam, G., and Guerry, P., Intracellular survival of *Campylobacter jejuni* in human monocytic cells and induction of apoptotic death by cytholethal distending toxin, *Infect. Immun.*, 73, 5194, 2005.
83. Snelling, W.J. et al., *Campylobacter jejuni*, *Lett. Appl. Microbiol.*, 41, 297, 2005.
84. Purdy, D. et al., Characterisation of cytolethal distending toxin (CDT) mutants of *Campylobacter jejuni*, *J. Med. Microbiol.*, 49, 473, 2000.
85. Lara-Tejero, M., and Galan, J.E., Cytolethal distending toxin: Limited damage as a strategy to modulate cellular functions, *Trends Microbiol.*, 10, 147, 2002.
86. Hickey, T.E. et al., *Campylobacter jejuni* cytolethal distending toxin mediates release of interleukin-8 from intestinal epithelial cells, *Infect. Immun.*, 68, 6535, 2000.
87. Eyigor, A. et al., Cytolethal distending toxin genes in *Campylobacter jejuni* and *Campylobacter coli* isolates: detection and analysis by PCR, *J. Clin. Microbiol.*, 37, 1646, 1999.
88. Eyigor, A. et al., Detection of cytolethal distending toxin activity and *cdt* genes in *Campylobacter* spp. isolated from chicken carcasses, *Appl. Environ. Microbiol.*, 65, 1501, 1999.
89. Schmidt, M.A., Riley, L.W., and Benz, I., Sweet new world: Glycoproteins in bacterial pathogens, *Trends Microbiol.*, 11, 554, 2003.
90. Logan, S.M., Trust, T.J., and Guerry, P., Evidence for posttranslational modification and gene duplication of *Campylobacter* flagellin, *J. Bacteriol.*, 171, 3031, 1989.
91. Szymanski, C.M., and Wren, B.W., Protein glycosylation in bacterial mucosal pathogens, *Nat. Rev. Microbiol.*, 3, 225, 2005.
92. Karlyshev, A.V. et al., The *Campylobacter jejuni* general glycosylation system is important for attachment to human epithelial cells and in the colonization of chicks, *Microbiol.*, 150, 1957, 2004.
93. Guerry, P. et al., Changes in flagellin glycosylation affect *Campylobacter* autoagglutination and virulence, *Mol. Microbiol.*, 60, 299, 2006.
94. Bacon, D.J. et al., A phase-variable capsule is involved in virulence of *Campylobacter jejuni* 81–176, *Mol. Microbiol.*, 40, 769, 2001.

95. Bachtiar, B.M., Coloe, P.J., and Fry, B.N., Knockout mutagenesis of the *kpsE* gene of *Campylobacter jejuni* 81116 and its involvement in bacterium-host interactions, *FEMS Immunol. Med. Microbiol.*, 49, 149, 2007.
96. Nachamkin, I., Allos, B.M., and Ho, T., *Campylobacter* species and Guillain-Barré syndrome, *Clin. Microbiol. Rev.*, 11, 555, 1998.
97. Fry, B.N. et al., The *galE* gene of *Campylobacter jejuni* is involved in lipopolysaccharide synthesis and virulence, *Infect. Immun.*, 68, 2594, 2000.
98. Fitzgerald, C., Whichard, J., and Nachamkin, I., Diagnosis and antimicrobial susceptibility of *Campylobacter* species. In Nachamkin, I., Szymanski, C.M., and Blaser, M.J. (Editors), *Campylobacter*, 3rd ed., p. 227, ASM Press, Washington, DC, 2008.
99. Wang, W.L. et al., Evaluation of transport media for *Campylobacter jejuni* in human fecal specimens, *J. Clin. Microbiol.*, 18, 803, 1983.
100. Aho, M., Kauppi, M., and Hirn, J., The stability of small number of campylobacteria in four different transport media, *Acta Vet. Scand.*, 29, 437, 1988.
101. Cary, S.G., and Blair, E.B., New transport medium for shipment of clinical specimens. I. Fecal specimens, *J. Bacteriol.*, 88, 96, 1964.
102. Wang, H., and Murdoch, D.R., Detection of *Campylobacter* species in faecal samples by direct Gram stain microscopy, *Pathology*, 36, 343, 2004.
103. Bolton, F.J., and Robertson, L.A., Selective medium for isolating *Campylobacter jejuni/coli*, *J. Clin. Pathol.*, 35, 462, 1982.
104. Nachamkin, I., Microbiologic approaches for studying *Campylobacter* species in patients with Guillain-Barré syndrome, *J. Infect. Dis.*, 176, S106, 1997.
105. Goossens, H. et al., Semisolid blood-free selective-motility medium for the isolation of campylobacters from stool specimens, *J. Clin. Microbiol.*; 27, 1077, 1989.
106. Hutchinson, D.N., and Bolton, F.J., Improved blood free selective medium for the isolation of *Campylobacter jejuni* from faecal specimens, *J. Clin. Pathol.*, 37, 956, 1984.
107. Karmali, M.A. et al., Evaluation of a blood-free, charcoal-based, selective medium for the isolation of *Campylobacter* organisms from feces, *J. Clin. Microbiol.*, 23, 456, 1986.
108. Barrett, T.J., Patton, C.M., and Morris, G.K., Differentiation of *Campylobacter* species using phenotypic characterization, *Lab. Med.*, 19, 96, 1988.
109. On, S.L.W., Identification methods for campylobacters, helicobacters and related organisms, *Clin. Microbiol. Rev.* 9, 405, 1996.
110. Harvey, S.M., Hippurate hydrolysis by *Campylobacter fetus*, *J. Clin. Microbiol.*, 11, 435, 1980.
111. Rautelin, H., Jusufovic, J., and Hanninen, M.J., Identification of hippurate-negative thermophilic Campylobacters, *Diagn. Microbiol. Infect. Dis.*, 35, 9, 1999.
112. Denis, M. et al., Development of a m-PCR assay for simultaneous identification of *Campylobacter jejuni* and *C. coli*, *Lett. Appl. Microbiol.*, 29, 406, 1999.
113. Fitzgerald, C., and Nachamkin, I., *Campylobacter* and *Arcobacter*. In Murray, P.R., Baron, E.J., Jorgensen, J.H., Landry, M.L., and Pfaller, M.A. (Editors), *Manual of Clinical Microbiology*, p. 933, ASM Press, Washington, DC, 2007.
114. O'Sullivan, N.A. et al., Detection and differentiation of *Campylobacter jejuni* and *Campylobacter coli* in broiler chicken samples using a PCR/DNA probe membrane based colorimetric detection assay, *Mol. Cell. Probes*, 14, 7, 2000.
115. Nayak, R., Stewart, T.M., and Nawaz, M.S., PCR identification of *Campylobacter coli* and *Campylobacter jejuni* by partial sequencing of virulence genes, *Mol. Cell. Probes*, 19, 187, 2005.
116. Heid, C.A. et al., Real time quantitative PCR, *Genome Res.*, 6, 986, 1996.
117. Abubakar, I. et al., A systematic review of the clinical, public health and cost-effectiveness of rapid diagnostic tests for the detection and identification of bacterial intestinal pathogens in faeces and food, *Health Technol. Assess.*, 11, 36, 2007.
118. Linton, D. et al., PCR detection, identification to species level, and fingerprinting of *Campylobacter jejuni* and *Campylobacter coli* direct from diarrheic samples, *J. Clin. Microbiol.*, 35, 2568, 1997.
119. Martinez, I. et al., Detection of *cdtA*, *cdtB*, and *cdtC* genes in *Campylobacter jejuni* by multiplex PCR, *Int. J. Med. Microbiol.*, 296, 45, 2006.
120. Fukushima, H., Tsunomori, Y., and Seki, R., Duplex real-time SYBR green PCR assays for detection of 17 species of food- or waterborne pathogens in stools, *J. Clin. Microbiol.*, 41, 5134, 2003.
121. Iijima, Y. et al., Improvement in the detection rate of diarrhoeagenic bacteria in human stool specimens by a rapid real-time PCR assay, *J. Med. Microbiol.*, 53, 617, 2004.
122. LaGier, M. J. et al., A real-time multiplexed PCR assay for rapid detection and differentiation of *Campylobacter jejuni* and *Campylobacter coli*, *Mol. Cell. Probes*, 18, 275, 2004.
123. Amar, C. et al., Detection by PCR of eight groups of enteric pathogens in 4,627 faecal samples: Re-examination of the English case-control Infectious Intestinal Disease Study (1993–1996), *Eur. J. Clin. Microbiol. Infect. Dis.*, 26, 311, 2007.
124. Best, E.L. et al., Specific detection of *Campylobacter jejuni* from faeces using single nucleotide polymorphisms, *Epidemiol. Infect.*, 135, 839, 2007.
125. Schuurman, T. et al., Feasibility of a molecular screening method for detection of *Salmonella enterica* and *Campylobacter jejuni* in a routine community-based clinical microbiology laboratory, *J. Clin. Microbiol.*, 45, 3692, 2007.
126. Oyofo, B.A. et al., Specific detection of *Campylobacter jejuni* and *Campylobacter coli* by using polymerase chain reaction, *J. Clin. Microbiol.*, 30, 2613, 1992.
127. Eyers, M. et al., Discrimination among thermophilic *Campylobacter* species by polymerase chain reaction amplification of *23S rRNA* gene fragments, *J. Clin. Microbiol.*, 31, 3340, 1993.
128. Waegel, A., and Nachamkin, I., Detection and molecular typing of *Campylobacter jejuni* in faecal samples by polymerase chain reaction, *Mol. Cell. Probes*, 10, 75, 1996.
129. Gonzalez, I. et al., Specific identification of the enteropathogens *Campylobacter jejuni* and *Campylobacter coli* by using a PCR test based on the *ceuE* gene encoding a putative virulence determinant, *J. Clin. Microbiol.*, 35, 759, 1997.
130. Fermer, C., and Engvall, E.O., Specific PCR identification and differentiation of the thermophilic campylobacters, *Campylobacter jejuni*, *C. coli*, *C. lari*, and *C. upsaliensis*, *J. Clin. Microbiol.*, 37, 3370, 1999.
131. Lawson, A.J. et al., Large-scale survey of *Campylobacter* species in human gastroenteritis by PCR and PCR-enzyme-linked immunosorbent assay, *J. Clin. Microbiol.*, 37, 3860, 1999.
132. Metherell, L.A., Logan, J.M., and Stanley, J., PCR-enzyme-linked immunosorbent assay for detection and identification of *Campylobacter* species: Application to isolates and stool samples, *J. Clin. Microbiol.*, 37, 433, 1999.

133. van Doorn, L.J. et al., Rapid identification of thermotolerant *Campylobacter jejuni*, *Campylobacter coli*, *Campylobacter lari*, and *Campylobacter upsaliensis* from various geographic locations by a GTPase-based PCR-reverse hybridization assay, *J. Clin. Microbiol.*, 37, 1790, 1999.
134. Al Rashid, S.T. et al., Identification of *Campylobacter jejuni*, *C. coli*, *C. lari*, *C. upsaliensis*, *Arcobacter butzleri*, and *A. butzleri*-like species based on the *glyA* gene, *J. Clin. Microbiol.*, 38,1488, 2000.
135. Collins, E. et al., Evaluation of a PCR/DNA probe colorimetric membrane assay for identification of *Campylobacter* spp. in human stool specimens, *J. Clin. Microbiol.*, 39, 4163, 2001.
136. Sails, A.D. et al., Development of a PCR ELISA assay for the identification of *Campylobacter jejuni* and *Campylobacter coli*, *Mol. Cell. Probes*, 15, 291, 2001.
137. Maher, M. et al., Evaluation of culture methods and a DNA probe-based PCR assay for detection of *Campylobacter* species in clinical specimens of feces, *J. Clin. Microbiol.*, 41, 2980, 2003.
138. Nawaz, M.S. et al., Detection of *galE* gene by polymerase chain reaction in campylobacters associated with Guillain-Barre syndrome, *Mol. Cell. Probes*, 17, 313, 2003.
139. Sinha, S. et al., Detection of preceding *Campylobacter jejuni* infection by polymerase chain reaction in patients with Guillain-Barre syndrome, *Trans. R. Soc. Trop. Med. Hyg.*, 98, 342, 2004.
140. Samie, A. et al., Prevalence of *Campylobacter* species, *Helicobacter pylori* and *Arcobacter* species in stool samples from the Venda region, Limpopo, South Africa: Studies using molecular diagnostic methods, *J. Infect.*, 54, 558, 2007.
141. Kärenlampi, R.I., Tolvanen, T.P., and Hanninen, M.L., Phylogenetic analysis and PCR-restriction fragment length polymorphism identification of *Campylobacter* species based on partial *groEL* gene sequences, *J. Clin. Microbiol.*, 42, 5731, 2004.
142. Campbell, L. et al., Molecular detection of *Campylobacter jejuni* in archival cases of acute appendicitis, *Mod. Pathol.*, 19, 1042, 2006.
143. Hill, J.E. et al., Identification of *Campylobacter* spp. and discrimination from *Helicobacter* and *Arcobacter* spp. by direct sequencing of PCR-amplified *cpn60* sequences and comparison to cpnDB, a chaperonin reference sequence database, *J. Med. Microbiol.*, 55, 393, 2006.
144. Abu Elamreen, F.H., Abed, A.A., and Sharif, F.A., Detection and identification of bacterial enteropathogens by polymerase chain reaction and conventional techniques in childhood acute gastroenteritis in Gaza, Palestine, *Int. J. Infect. Dis.*, 11, 501, 2007.
145. Klena, J.D. et al., Differentiation of *Campylobacter coli*, *Campylobacter jejuni*, *Campylobacter lari*, and *Campylobacter upsaliensis* by a multiplex PCR developed from the nucleotide sequence of the lipid A gene *lpxA*, *J. Clin. Microbiol.*, 42, 5549, 2004.
146. Houng, H.S. et al., Development of a *ceuE*-based multiplex polymerase chain reaction (PCR) assay for direct detection and differentiation of *Campylobacter jejuni* and *Campylobacter coli* in Thailand, *Diagn. Microbiol. Infect. Dis.*, 40, 11, 2001.
147. Wang, G. et al., Colony multiplex PCR assay for identification and differentiation of *Campylobacter jejuni*, *C. coli*, *C. lari*, *C. upsaliensis*, and *C. fetus* subsp. *fetus*, *J. Clin. Microbiol.*, 40, 4744, 2002.
148. Nakari, U.M., Puhakka, A., and Siitonen, A., Correct identification and discrimination between *Campylobacter jejuni* and *C. coli* by a standardized hippurate test and species-specific polymerase chain reaction, *Eur. J. Clin. Microbiol. Infect. Dis.*, 27, 513, 2008.
149. Persson, S., and Olsen, K.E., Multiplex PCR for identification of *Campylobacter coli* and *Campylobacter jejuni* from pure cultures and directly on stool samples, *J. Med. Microbiol.*, 54, 1043, 2005.
150. Gilpin, B. et al., Application of pulsed-field gel electrophoresis to identify potential outbreaks of campylobacteriosis in New Zealand, *J. Clin. Microbiol.*, 44, 406, 2006.
151. Yamazaki-Matsune, W. et al., Development of a multiplex PCR assay for identification of *Campylobacter coli*, *Campylobacter fetus*, *Campylobacter hyointestinalis* subsp. *hyointestinalis*, *Campylobacter jejuni*, *Campylobacter lari* and *Campylobacter upsaliensis*, *J. Med. Microbiol.*, 56, 1467, 2007.
152. Asakura, M. et al., Development of a cytolethal distending toxin (*cdt*) gene-based species-specific multiplex PCR assay for the detection and identification of *Campylobacter jejuni*, *Campylobacter coli* and *Campylobacter fetus*, *FEMS Immunol. Med. Microbiol.*, 52, 260, 2008.
153. Volokhov, D. et al., Microarray-based identification of thermophilic *Campylobacter jejuni*, *C. coli*, *C. lari*, and *C. upsaliensis*, *J. Clin. Microbiol.*, 41, 4071, 2003.
154. Jin, D. Z. et al., Detection and identification of intestinal pathogens in clinical specimens using DNA microarrays, *Mol. Cell. Probes*, 20, 337, 2006.
155. You, Y. et al., A novel DNA microarray for rapid diagnosis of enteropathogenic bacteria in stool specimens of patients with diarrhea, *J. Microbiol. Methods*, 75, 566, 2008.
156. Poppert, S. et al., Identification of thermotolerant campylobacter species by fluorescence in situ hybridization, *J. Clin. Microbiol.*, 46, 2133, 2008.
157. Yamazaki, W. et al., Development and evaluation of a loop-mediated isothermal amplification assay for rapid and simple detection of *Campylobacter jejuni* and *Campylobacter coli*, *J. Med. Microbiol.*, 57, 444, 2008.
158. On, S.L.W., and Jordan, P.J., Evaluation of 11 PCR assays for species-level identification of *Campylobacter jejuni* and *Campylobacter coli*, *J. Clin. Microbiol.*, 41, 330, 2003.
159. Keer, J.T., and Birch, L., Molecular methods for the assessment of bacterial viability, *J. Microbiol. Methods*, 53, 175, 2003.
160. Al-Khaldi, S.F. et al., DNA microarray technology used for studying foodborne pathogens and microbial habitats: Minireview, *J. AOAC Int.*, 85, 906, 2002.
161. Ramsay, G., DNA chips: State-of-the art, *Nat. Biotechnol.*, 16, 40, 1998.
162. Vora, G.J. et al., Nucleic acid amplification strategies for DNA microarray-based pathogen detection, *Appl. Environ Microbiol.*, 70, 3047, 2004.
163. Call, D.R., Borucki, M.K., and Loge, F.J., Detection of bacterial pathogens in environmental samples using DNA microarrays, *J. Microbiol. Methods*, 53, 235, 2003.
164. Bryant, P.A. et al., Chips with everything: DNA microarrays in infectious diseases, *Lancet*, 4, 100, 2004.
165. Wilson, I.G., Inhibition and facilitation of nucleic acid amplification, *Appl. Environ. Microbiol.*, 63, 3741, 1997.
166. Abu Al-Soud, W., and Radström, P., Effects of amplification facilitators on diagnostic PCR in the presence of blood, feces, and meat, *J. Clin. Microbiol.*, 38, 4463, 2000.

167. Abu Al-Soud, W., and Radström, P., Capacity of nine thermostable DNA polymerases to mediate DNA amplification in the presence of PCR-inhibiting samples, *Appl. Environ. Microbiol.*, 64, 3748, 1998.
168. Malorny, B., and Hoorfar, J., Toward standardization of diagnostic PCR testing of fecal samples: Lessons from the detection of salmonellae in pigs, *J. Clin. Microbiol.*, 43, 3033, 2005.
169. Mateo, E. et al., Evaluation of a PCR assay for the detection and identification of *Campylobacter jejuni* and *Campylobacter coli* in retail poultry products, *Res. Microbiol.*, 156, 568, 2005.
170. Stucki, U. et al., Identification of *Campylobacter jejuni* on the basis of a species-specific gene that encodes a membrane protein, *J. Clin. Microbiol.*, 33, 855, 1995.

95 Helicobacter

Athanasios Makristathis and Alexander M. Hirschl

CONTENTS

95.1 INTRODUCTION

95.1.1 CLASSIFICATION, MORPHOLOGY, AND EPIDEMIOLOGY

In the new family *Helicobacteraceae*, the genus *Helicobacter* presently comprises more than 40 validly named species and four *Candidatus* species, a designation of provisional status adopted by the International Committee on Systematic Bacteriology for incompletely described procaryotes. *Helicobacter* species colonize the gastrointestinal and hepatobiliary tracts of humans and a variety of animal species. Helicobacters are nonspore-forming, gram negative bacteria with curved, spiral, or fusiform cellular morphology that may vary with the age, growth conditions, and species identity of the cells. Strains of all species are motile, with a rapid corkscrew-like or slower wave-like motion due to flagellar activity. Most species have bundles of multiple sheathed flagella with a polar or bipolar distribution. Oxidase activity is present in all species, whereas most species also produce catalase. Many species, in particular gastric helicobacters, produce urease. The species of primary importance for human medicine is *Helicobacter (H.) pylori*. Other *Helicobacter* species are primarily pathogenic in animals; human infections may be classified as zoonoses with a variety of animals such as dogs, cats, pigs, birds, rabbits, hamsters and other rodents, horses, and nonhuman primates as potential sources of infection. However, these infections are of minor importance as compared with infection with *H. pylori*. Therefore, this chapter will focus on the human pathogen *H. pylori*.

Detection of spiral-shaped bacteria in autopsy specimens obtained from the gastric mucosa dates back to the nineteenth century. However, the decisive advance was achieved by the Australian scientists Warren and Marshall, who not only rediscovered this organism, but were also able to identify and demonstrate its association with the development of gastritis and peptic ulcers.[1] The bacterium was originally named "*Campylobacter pyloridis*" and then "*Campylobacter pylori*" because it resembled the genus *Campylobacter* in several aspects.[2] However, early electron micrographs and subsequent 16S rRNA sequence analysis resulted in the classification of the spiral/helical-shaped bacteria as the genus *Helicobacter* with *H. pylori* being one of the first members of the new genus.[3]

Based on an assumption persisting for decades that ulcer disease is predominantly a result of stress, malnutrition, or genetic factors, the hypothesis of a bacterial cause of ulcer disease was originally met with considerable skepticism. Today, *H. pylori* is considered the primary etiological agent in not only gastritis but also in gastric and duodenal ulcer, as well as MALT lymphoma and gastric cancer.

Morphologically, the bacterial cells are curved to S-shaped rods, 0.5/2.5–5 µm. They are motile by means of 4–8 unipolar or bipolar, sheathed flagella. The bacteria grow as small, translucent colonies (diameter 1–2 mm); optimum growth is obtained under microaerobic conditions at 37°C with no growth at 42°C or under anaerobic conditions. For primary isolation, 3–7 days of incubation are required, which can be reduced to 2 days for subsequent subculture. All strains have been reported to be oxidase, catalase, and urease positive. The latter is produced in large amounts and represents up to 6% of the bacterial protein content. In contrast to other bacterial ureases, *H. pylori* urease is partly surface-associated. *H. pylori* is susceptible to cold, dehydration, and oxygen. Under adverse environmental conditions the rod shape may convert to a coccoid form. Although showing minor metabolic activity, this coccoid form has been unculturable to date.

Previous studies employing different molecular typing methods showed a high degree of genomic variability among strains.[4,5] This strain-specific genetic diversity was suggested to be responsible for the ability of the microorganism to cause different diseases or even be nonpathogenic for the infected host. In 1999, the comparison of the complete genome of two unrelated isolates revealed a high degree of similarity in overall genomic organization, gene order, and predicted proteins.[6] However, horizontal genetic exchange and free recombination were suggested to be much more frequent in *H. pylori* than in other species, which may be of particular importance considering that gastric colonization with more than one *H. pylori* strain is common.[7] A whole-genome microarray analysis of different *H. pylori* strains showed that 22% of the genes are dispensable in one or more strains, indicating extensive genomic plasticity.[8] While a minimal functional core of 1281 genes encode most metabolic and cellular processes, strain-specific genes encode restriction modification enzymes, transposases, and cell surface proteins, including genes unique to *H. pylori*, which may facilitate adaptation to different hosts or to the changing conditions within one and the same host during long-term infection.

The most significant difference among *H. pylori* strains concerns the presence or absence of a pathogenicity island (*cag*PAI). This is a region of 40 kb comprising about 30 genes that encode proteins assembling a type IV secretion system (T4SS) and an effector protein CagA.[9] Furthermore, there is considerable variation in several loci (s, i, and m) of the cytotoxin encoding gene *vacA*, which is present in all strains. This variation is associated with striking differences in the biological activity of the cytotoxin; s1i1m1 is the most active VacA genotype, whereas s2 type VacA lacks in vitro activity.[9] Strain-specific diversity has also been observed in the expression of different outer membrane proteins, which may be of particular importance in case of the blood group antigen-binding and sialic acid-binding adhesins, BabA, and SabA.[9]

Infection with *H. pylori* shows worldwide prevalence and may be traced back to prehistoric times. Genetic evaluations have shown clones with circumscribed geographic distribution reflecting the pathways of human migration originating in tribal populations in Africa and Asia.[10,11] It can be assumed that gastric *Helicobacter* has always been present in mammals and that such bacteria have accompanied human beings from the days of their origin to the present day.

H. pylori primarily colonizes the human gastric mucosa; it has also been described in zones of gastric metaplasias (esophagus, duodenum) but also on ectopic gastric epithelium (Meckel's diverticulum). In Western countries, the infection affects about 20% of persons aged under 50 years and 50% of those older than 60 years, while it is rather rare in children except for those with a migration background.[12,13] As a rule, the infection with *H. pylori* is acquired at childhood age (i.e., before the age of 10). Increase of prevalence with age is the result of a cohort effect (varying acquisition rates at childhood age). In addition, prevalence is inversely correlated with the socioeconomic status, especially at childhood, while genetic factors play only a subordinate role, if at all.[14]

In nonindustrialized countries, prevalence is extremely high, especially in the poorest segments of the population, and may reach ≥80%[12,13]; acquisition predominantly occurs between 2 and 8 years of age, but may already reach 50%–60% at the age of 2 years.[15] Poverty and the resultant limited household and living space, together with poor or completely lacking sanitary facilities, constitute the most relevant risk factors. Improvement of living conditions is invariably associated with a decrease of *H. pylori* infection.[12,16] An example is provided by a study in children in Saint Petersburg, Russia. In the year 1995, prevalence of *H. pylori* infection had been 44%, but had decreased to 13% only 10 years later. In children aged younger than 5, prevalence in 1995 accounted for 30%, while it was only 2% 10 years later.[17]

The human being itself is the only relevant source of infection with *H. pylori*. Gastro-oral (e.g., contact with vomit) as well as fecal-oral routes are considered the most important pathways of transmission, although indirect transmission via foodstuffs and water also appears possible.[18] In industrialized countries, transmission is intrafamilial in the majority of cases[19,20]; however, in developing countries the origin of infection and routes of transmission may be more complex. Thus, in a recent study using a noninvasive genotyping method in addition to clonal lineages, a considerable number of distinct *H. pylori* clones could be shown in African households.[21]

Genuine reinfection (defined as renewed detection ≥12 months following successful eradication therapy) is rare in industrialized countries with ≤1%, while substantially higher rates (10%–20%) have been reported from nonindustrialized countries.[22]

95.1.2 Clinical Features and Pathogenesis

Infection with *H. pylori* is one of the most common human infections and is invariably associated with gastritis. The most important complications may include ulcer disease, adenocarcinoma, and gastric MALT lymphoma. For the majority of those infected, the infection will remain without clinical signs and symptoms, but usually will persist for lifetime. However, up to 50% of those infected will develop atrophic gastritis, up to 10% peptic ulcers, and 1%–2% will develop gastric carcinoma.[23]

Enzymatic urease activity will induce the release of ammonia from the urea, thereby neutralizing gastric acid in the immediate neighborhood of the bacterium.[24] Thus, urease protects *H. pylori* against the bactericidal effect of acid. After passing the gastric lumen, the marked mobility of the organism and its spiral shape allow penetration into the highly viscous gastric mucus, ensuring its survival and growth under adequate protection against the immune response and at virtually neutral pH. Acute infection develops as a mild transitory disorder associated with epigastric pain and vomiting. However, as acute infection mostly develops at early childhood and involves uncharacteristic symptoms frequently encountered for this age group, the infection usually remains unrecognized. Although the causative pathogen will usually persist, symptoms and complaints will usually regress and reverse within 1 week. Self-tests have shown that the infection is primarily associated with neutrophilic gastritis followed by gradual infiltration with all kinds of inflammatory cells, especially lymphocytes.[25,26] It also involves a significant reduction of acid secretion as a result of the toxic effect of *H. pylori* and the inhibitory action of inflammatory cytokines such as interleukin-1 (IL-1). The infection-related reduction of the number of D-cells in the antrum induces a decrease of somatostatin production with the resultant hypergastrinemia stimulating acid secretion in turn so that most patients will experience normalization of gastric acid secretion.[27] As a rule, acid secretion will normalize within a few weeks to a few months. Regardless of this, spontaneous healing will virtually never be encountered with most cases showing development of chronic active gastritis.

Several factors allow *H. pylori* to persist in the stomach overcoming gastric peristalsis, antimicrobial factors (digestive enzymes, lactoferrin, and defensins), epithelial cell turnover, and humoral and cellular immune responses. Most bacteria are free-living in the mucus layer, which is the permanent reservoir of *H. pylori*. But bacteria that may invade gastric epithelial cells may also be important for persistence under unfavorable conditions. Adherence to gastric epithelial cells was shown to be inhibited by urease in vitro, which may hinder the interaction between the pathogen and the host.[28] Toll-like receptors (TLRs) on host cells recognize microbial components; TLR4 normally recognizes LPS and TLR5 flagellin, resulting in the induction of an initial proinflammatory response and priming of the adaptive immune response. However, TLR5 on gastric epithelial cells does not recognize *H. pylori* flagellin and also *H. pylori* LPS is only a weak agonist for TLR4.[29] Bacterial antigens such as arginase, γ-glutamyl transpeptidase, and VacA were even shown to inhibit activation-induced proliferation of T cells, which may also result in immunosuppression.[9]

Moreover, upregulation of the expression of growth factors by *H. pylori* may contribute to the persistent nature of the infection. Their release may result in tissue repair and the reconstitution of the gastric epithelial barrier thus affecting the host-pathogen interaction. Particularly, the growth factor TGF-β1 may induce *H. pylori*-specific CD4 + CD25 high regulatory T cells, which were shown to naturally occur during infection playing a role in suppressing the immune response to *H. pylori* in the gastric mucosa.[30,31]

Interactions of *H. pylori* with the epithelium result in the recruitment of neutrophils, macrophages, mast cells, dendritic cells, and lymphocytes and proinflammatory cytokine release in gastric tissue. Inflammation and tissue damage may be essential for chronic colonization due to the improvement of the nutritional supply for bacterial cells through disrupted cell adhesion junctions in the gastric epithelium.

H. pylori strains can be roughly categorized in those producing virulence factors such as *cag*PAI, active VacA, and BabA, which enhance the host-pathogen interactions (type I strains) and those lacking these factors. Particularly in Western populations, type I strains are associated with an increased risk of peptic ulcer disease, premalignant lesions, and gastric adenocarcinoma. During intimate contact, type I strains inject by the T4SS bacterial antigens such as the CagA protein into the host cell, CagA then becomes tyrosine phosphorylated. There is inter- and even intrastrain diversity in the number of tyrosine phosphorylation Glu-Pro-Ile-Tyr-Ala (EPIYA) motifs, which are characterized as EPIYA-A, -B, -C, and -D according to the surrounding amino acid sequence and arise from intragenomic recombination between direct DNA repeats in the 3′ portion of *cagA*. There is growing evidence that the number EPIYA motifs and, in particular, the number of EPIYA-C (Western CagA) and -D (East Asian CagA) motifs correlate with gastric cancer and precursor lesions.[9]

Phosphorylation-dependent and independent effects of CagA include, among others, actin cytoskeletal rearrangements and the disruption of cell adhesion junctions. Importantly, phosphorylated CagA activates SHP-2 phosphatase, which is associated with malignant disorders.[32] In a CagA-independent manner the *cag*PAI is also associated with increased inflammation due to induction of proinflammatory cytokines such as interleukin-8. Since *cag*PAI, active VacA, and BabA coexist in type I strains, though genetically unlinked, they are likely to serve a common function resulting in their selection, which may be the regulation of the inflammatory response. In this context, while under high inflammation conditions nutrients transit through the leaky epithelium into the gastric mucous, bacterial factors exhibiting immunosuppressive activities—amongst others VacA—enter into the submucosa inducing conditions of low inflammation.[33] Thus, inflammation is probably best thought of as a continuum that may cycle from one extreme to another, especially in infections with type I strains and may differ between anatomic regions of the stomach. Depending on the degree of inflammation and nutrient levels, antigenic or phase variation of adhesins such as BabA and the resulting alteration of the adhesion properties of type I strains has been suggested as a further possibility to control the inflammatory process in the gastric mucosa.[33] Thus, *H. pylori* has evolved several mechanisms of regulation of the innate and adaptive arms of the immune response in order to achieve an adequate degree of inflammation, which is crucial for persistent infection.

Essentially, two types of chronic gastritis may be differentiated, namely nonatrophic antral gastritis associated with ulcer disease and progressive multifocal atrophic gastritis, which has been associated with gastric carcinoma.[34] For many patients, antral gastritis represents the only relevant long-term pathological alteration caused by *H. pylori*. These patients will usually be fully asymptomatic or will only complain of nonspecific dyspeptic symptoms. Such symptoms may affect up to 25% of the general population in the course of their lifetime. The clinical picture is characterized by epigastric complaints and pain in the upper abdomen. In the absence of manifest endoscopic or ultrasonographic evidence this condition is termed functional dyspepsia.[35] The Maastricht III Consensus Conference recommends a so-called test and treat strategy in adult patients under the age of 45 years with persistent dyspeptic complaints and without alarming symptoms.[36] While this strategy is only 10% more effective than placebo with regard to symptom improvement,[37] it is cost efficient in patient populations with relatively high *H. pylori* prevalence (>20%). Moreover, it will reduce the risk of peptic ulcer, atrophic gastritis, and the development of gastric carcinoma. With a low *H. pylori* prevalence, empirical treatment with proton pump inhibitors (PPI) may be considered as equivalent option.[36]

Bacterial and host-specific factors as well as environmental effects will complicate the clinical course by development of different sequelae. Thus, some patients will experience increased acid secretion. In such patients, infection will only cause significant gastritis in antrum and cardia, while the acid-producing corpus mucosa remains protected from colonization and will therefore be affected to a much lesser extent. In response to the increased acid load islets of gastric metaplasia will develop in the mucosa of the bulbus duodeni, which may be colonized by *H. pylori*. From the resultant chronic inflammatory lesions, ulcers may ultimately develop under the influence of acid.[27] Thus, duodenal ulcers are virtually always associated with *H. pylori*, while gastric ulcers may also develop without *H. pylori* infection and are frequently causally related to treatment with nonsteroidal anti-inflammatory drugs (NSAIDs). For a *H. pylori*-associated peptic ulcer, eradication of the causative pathogen will produce a dramatic reduction of the risk of recurrence and will usually provide for prolonged remission of ulcer disease, usually for the full lifetime.[38] There is an additive or even synergistic effect between infection with *H. pylori* and NSAID for the development of ulcer and bleeding; therefore, eradication of the pathogen is indicated before starting therapy with an NSAID.[36,39]

However, in cases with low acid production, which may be due to malnutrition or recurrent bacterial infections in childhood (tonsillitis, diarrhea) and an associated cytokine release, the corpus mucosa will not be adequately protected against development of *H. pylori* gastritis and patients will develop pangastritis followed by chronic atrophic gastritis associated with a high cancer risk.[27]

Gastric carcinoma is the third most frequent type of malignant tumor in men and ranks fifth in frequency in women. Annually, an estimated 1 million new cases must be expected worldwide; thus, gastric carcinoma accounts for 10% of all new cancer cases per year. With about 700,000 deaths, gastric cancer is the most common type of lethal tumor with the exception of lung cancer. Despite a decrease of this type of cancer in industrialized countries, it remains a major worldwide problem, especially in developing countries. The 5-year survival rates with advanced gastric carcinoma have remained rather low, with less than 20% even in industrialized countries, while the extensive and intricate screening programs for early gastric carcinoma practiced in Japan has ensured a 5-year survival rate of about 70%.

Epidemiological studies have shown a definite association between *H. pylori* infection and distal gastric carcinoma, concluding that about 70% of cases with this type of carcinoma are the result of the infection.[40] On account of this association, the World Health Organization (WHO) had already classified *H. pylori* as a class I carcinogen in 1994 and has recently confirmed this classification.[41] Especially for the intestinal type of gastric carcinoma, the histologically defined cascade starting with *H. pylori* infection and development of chronic active gastritis followed by atrophy, metaplasia and dysplasia, and ultimately terminating in cancer, has been well confirmed by appropriate evidence.[42] Development of premalignant alterations of the gastric mucosa may be avoided by successful eradication therapy.[43,44] Thus, an Asian-Pacific consensus conference recommended screening in high-risk countries with subsequent eradication therapy as measure for reducing carcinoma incidence.[45] However, etiology of gastric carcinoma must be considered as being multifactorial, with host factors such as proinflammatory cytokine profile and positive family history playing a prominent role in addition to *H. pylori* and its virulence factors. Moreover, nutritional habits and socioeconomic factors will always be of importance.[43]

Formation of mucosa-associated lymphatic tissue (MALT) constituting the starting point for the development of low-grade MALT lymphoma is also clearly associated with *H. pylori* infection. Several studies have shown that *H. pylori* eradication will provide for complete remission of the lymphoma (within 12 months) in more than 50% of cases.[35,46] For this reason, MALT lymphoma is considered one of the undisputed indications for eradication therapy.

The association between infection with *H. pylori* and thrombocytopenic purpura and iron deficiency anemia of otherwise unclarified etiology is considered as clearly confirmed and is also reflected in treatment recommendations.[36] An inverse association between *H. pylori* and the occurrence of allergic asthma and atopic disorders has been proposed[9] though currently being without any clinical consequences.

The interaction between *H. pylori* infection and reflux disease has been a controversial topic. Studies have shown that prevalence of *H. pylori* in patients with reflux disease is low.[47] Moreover, the worldwide decrease of prevalence of *H. pylori* infection, which has been accompanied by an increase of reflux disease and the occurrence of esophageal adenocarcinoma, has been taken as an evidence of a protective effect of *H. pylori* for both these disorders.[9] However, eradication therapy is recommended for infected patients receiving long-term treatment with PPI for reflux disease because long-term PPI treatment involves the risk of pangastritis followed by atrophies.[36]

The combination of a PPI + clarithromycin + amoxicillin or metronidazole is still considered standard therapy for the treatment of *H. pylori* infection. However, increasing macrolide resistance requires alternative treatment approaches (e.g., quadruple therapy with PPI + bismuth + tetracycline + metronidazole or PPI + amoxicillin + levofloxacin) and suggests an urgent need to select therapy based on the type of resistance determined. This is especially true in populations with a high prevalence of resistant strains. Currently, no vaccine is yet available. Hygienic precautions are strictly recommended, especially in the context of endoscopy.

Only a small proportion of patients with dyspeptic disorders (0.01%–1%) will harbor gastric *Helicobacter* species other than *H. pylori* (in 10%–30% of cases infection with *H. suis*). Such infections are accompanied by active chronic gastritis with a severity being less marked than with *H. pylori* infection, and they have only rarely been associated with gastric erosions, duodenal ulcers, and MALT lymphoma.[48] Moreover, infection with enterohepatic *Helicobacter* has been discussed as potential risk factor for progression of inflammatory hepatic disease to cirrhosis and hepatocellular carcinoma, especially in patients chronically infected with hepatitis C virus.[49]

95.1.3 Diagnosis

Methods for diagnosis of *H. pylori* infection may be classified as invasive (culture, histology, urease, or DNA detection in biopsy) or noninvasive methods (breath tests, serology, stool antigen test, DNA detection in feces) (Table 95.1). Culture and tests utilizing enzymatic activity (urease production) of *H. pylori* are considered as reliable methods for detection of active infection. Noninvasive tests are also termed global tests, because in contrast to biopsy-based tests they involve assessment of the complete organ affected. Culture and histology are still considered as gold standards of diagnostics. The essential indications for noninvasive tests include the screening of young dyspeptic patients in the context of a "test and treat strategy" and posttreatment follow-up.[36]

TABLE 95.1
Routine Methods for the Detection of *H. pylori* Infection

Specimen	Method
Biopsy	Rapid urease test
	Histology
	Culture
	PCR, FISH
Exhaled breath	^{13}C- or ^{14}C- urea breath test
Stool	Antigen test
	PCR
Serum	Detection of specific antibodies

95.1.3.1 Conventional Techniques

For detection by culture, at least one biopsy from corpus and antrum is required; detection in gastric juice and mucus may also be possible under specific circumstances. Use of an appropriate transport medium and short transport times are of particular importance.[50] The material should be homogenized and applied on solid selective and nonselective growth media. Incubation should be at 37°C under microaerobic conditions for at least 7 days. Identification is based on the typical micro- and macromorphological criteria and a positive catalase, oxidase, and urease reaction.[51] Under optimized conditions, culture provides for more than 90% sensitivity with 100% specificity. The major advantages of culture include the unlimited opportunity of resistance testing, typing, and the detection of virulence factors. Drawbacks may include the long time until results are obtained and the limited availability. Isolation of other gastric *Helicobacter* species from the human stomach was only possible in isolated cases and isolates were subsequently identified as *Helicobacter bizzozeronii*.[52] For culture of gastroenteritis-causing *Helicobacter* species (e.g., *H. cinaedi*, *H. fennelliae*, *H. pullorum*) from stool samples selective growth media, prolonged incubation times, and a microaerobic atmosphere are required. In isolated cases, enterohepatic helicobacters were grown in blood cultures[53]

The urease test is the most rapid and cost-effective invasive method for the detection of *H. pylori* and it is primarily performed using antral biopsies.[50] Because of the marked urease activity of the bacterium and the negligible or lacking colonization of the stomach by other urease-producing bacteria, the specificity of this test is high. Sensitivity will depend on the extent of colonization with at least 10^5 organisms required for a positive reaction. As with other tests based on the enzymatic activity of the bacterium, recent antibiotic therapy, administration of PPIs, and minor colonization of the mucosa (in the presence of intestinal metaplasia and atrophy) will have an unfavorable effect on test sensitivity.

Apart from pathogen detection, histopathologic examination will also provide evidence of the status of the gastric mucosa. As a rule, several biopsies are collected from the antrum and corpus. Histology will also allow for detection of noncultivable *Helicobacter* species. A wide variety of different stains (modified Giemsa stain, Warthin Starry silver stain, acridine orange, etc.) will be available for detection. Histological detection methods are widely available; however, lacking standardization due to subjective evaluation constitutes a limitation of this test.

Today, noninvasive tests are an indispensable part of *H. pylori* diagnostics. Because of excellent sensitivity and specificity the ^{13}C- and ^{14}C-urea breath test represent the gold standard of noninvasive diagnostic assessment.[54,55] These tests are based on the principle that *H. pylori*—if present in the stomach—will break down orally ingested ^{13}C or ^{14}C labeled urea. Labeled CO_2 diffuses into blood and is exhaled via the lungs to be measured in the exhaled breath. As a rule, a baseline measurement is followed by second measurement 30 min after oral urea ingestion. The test is characterized by a simple test procedure and easy handling of reagent and sample material, although analytics for the ^{13}C-urea breath test is quite complex, involving use of mass spectrometry or infrared spectrometry. Other well-proven tests include the stool antigen test and, especially, laboratory-based quantitative EIA using monoclonal antibodies for antigen detection. Sensitivity and specificity of these tests is excellent with ≥90% in untreated patients, while variable results have been reported when used for therapeutic follow-up.[55,56] In obvious contrast to the breath test, rapid transfer, and processing of samples is required. Regardless of the test chosen, therapeutic follow-up should not be done any earlier than 4 weeks after discontinuation of treatment as transient reduction of bacterial counts may also be expected upon failed eradication.

As regards antibody detection, a variety of different tests (mostly based on EIA) are available; these tests include both quantitative and qualitative methods.[57] Best results are obtained with laboratory-based tests in serum, for which IgG detection is sufficient. Local validation is of primary importance. Rapid tests in serum and antibody tests in urine and mucus are generally not recommended.[50] As an essential drawback of serology antibody tests may still be positive even months to years after successful eradication. For this reason, serology is only of limited value for therapeutic follow-up. However, serologic evaluation provides for some specific applications including infections with low pathogen density (atrophy, extensive metaplasia, patients under PPI therapy), MALT lymphoma, ulcer bleeding and—for epidemiological reasons—also identification of past infection.[50]

95.1.3.2 Molecular Techniques

During the last 20 years, PCR has been the most widely used molecular technique for the detection of *H. pylori* in clinical specimens, susceptibility testing, and the analysis of determinants of pathogenicity.

Many of the PCR protocols aiming at the detection of the infection in clinical samples target genes of the urease operon such as *ureA* and *glmM* (formerly *ureC*) and the 16S rRNA gene.[58] Standard PCR can accurately be used for the detection of *H. pylori* infection in gastric biopsies.[58] The method has also been applied to gastric juice aspirates preferentially collected by the string method, which is considered to be a minimally invasive approach, showing satisfactory diagnostic accuracy.[59] However, a dramatic loss of sensitivity was observed when standard PCR was used for the detection of *H. pylori* DNA in stool specimens. In order to increase sensitivity, nested and seminested PCR protocols have been introduced also allowing for the detection of the infection in feces.[60] However, due to the considerable risk of airborne contamination by amplicons the use of nested PCR may not be generally recommended.

During recent years, real-time PCR having several advantages over conventional PCR such as short working time, high specificity, and low risk of contamination, proved to be a powerful diagnostic tool in *H. pylori* infection. Real-time PCR protocols using the hyprobe or *Taq*Man technology have been used occasionally for the quantitative detection of the pathogen in gastric biopsies.[58] Most importantly, protocols using the FRET technology followed by melting curve analysis either on the basis of biprobes[61] or hyprobes,[62] allowed for simultaneous detection of *H. pylori* and clarithromycin susceptibility testing. This was achieved by amplifying a specific region of the 23S rRNA gene, including the sites of possible point mutations responsible for resistance to clarithromycin. Real-time FRET PCR alike PCR–RFLP, which was the first technique described for detection of these point mutations, was applied on cultured isolates or gastric biopsies delivering accurate results. A particular challenge has been the development of a biprobe real-time PCR protocol for detection of the pathogen and clarithromycin susceptibility testing using stool specimens. The right choice of highly sensitive and specific primers has been of crucial importance considering that *H. pylori* as a nonintestinal pathogen is present only in low numbers in feces and the high sequence homology in the 23S rRNA region between different species and especially *Helicobacter* species. The clinical evaluation of this assay clearly demonstrated its suitability not only for biopsies but also for stool specimens.[63]

Determination of *H. pylori* resistance to other antibiotics such as tetracycline and fluoroquinolones has also been reported (e.g., by hyprobe real-time PCR assays allowing for the detection of relevant mutations in the 16S rRNA and *gyrA* gene, respectively).[64,65] As yet, no molecular technique allowing for the reliable detection of resistance to metronidazole is available. This is due to the considerable number of possible mutations in the *rdxA* gene, some of them being irrelevant for resistance. However, metronidazole susceptibility testing may be of limited interest due to the poor association with the clinical outcome.

PCR has been used also for the detection of pathogenic factors such as genes of the *cag*PAI and in particular *cagA*; polymorphisms in the s, i, and m loci of *vacA*; genes involved in adherence; and so forth.[66–68] While this analysis has limited implications on an individual basis (patients tested positive for *H. pylori* should undergo eradication treatment irrespective of the pathogenic features of the infecting strain and moreover, the great majority of Asian strains are already *cag*PAI positive), it is of interest for shedding light on the pathogenesis of gastric diseases and for epidemiological investigations. Protocols can be applied to both culture isolates and gastric biopsies; however, in the former case it is advisable to test multiple colonies in the event of an infection with more than one strain or intrastrain diversity.

Furthermore, techniques utilizing the principle of reverse hybridization such as DNA enzyme immunoassay (DEIA) and line probe assay (LiPA) have been used for the specific detection of amplification products and mutations associated with macrolide resistance.[69] Protocols may be applied not only to culture isolates but also to gastric biopsies. A PCR-LiPA test allowing for the detection of factors of pathogenicity has also been designed.[70]

The only PCR-independent molecular method allowing for the rapid and accurate detection and clarithromycin susceptibility testing of *H. pylori* in gastric biopsies (native or paraffin-embedded) is the FISH test.[71] Similar to real-time FRET PCR and PCR-LiPA, this method is convenient for routine use.

Molecular techniques are of particular importance for the detection of other gastric and enterohepatic *Helicobacters* because of major difficulties with conventional methods. The most practical approach for species differentiation and identification may be the application of a genus-specific PCR using an appropriate target gene followed by sequence analysis of the amplicon.[52,72]

Apart from diagnostic purposes, molecular techniques have been widely used for epidemiological investigations. Endoscopy is the usual approach, since most genotyping methods aiming at strain identification used so far (e.g., RAPD-PCR, MLST) may deliver accurate results only when applied on cultured *H. pylori* isolates.[5,73] Recently, a noninvasive protocol based on two biprobe real-time PCR assays and sequencing has been introduced allowing for the accurate detection and typing of *H. pylori* in stool.[21]

95.2 METHODS

95.2.1 Sample Preparation

Total bacterial genomic DNA can be extracted by commercial tests such as the DNeasy isolation kit (Qiagen) or the QIAamp DNA Mini Kit using the protocol for isolation of genomic DNA from bacterial cultures (Qiagen). Alternatively, standard protocols utilizing phenol-chloroform-isoamylalcohol extraction and ethanol precipitation or more simple methodology can also be used. From biopsy homogenates (100 μL), DNA can be extracted by the QIAamp DNA Mini Kit using the blood and body fluid protocol (Qiagen). From stool specimens (0.2 g), DNA can be extracted by the QIAamp DNA Stool Mini Kit (Qiagen) using the protocol for isolation of DNA for pathogen detection.

Preanalytic and laboratory practice are of importance, particularly when stool specimens are analyzed; *H. pylori* as a nonintestinal pathogen is present only in low numbers in feces and most of the bacteria may not be intact with their DNA being exposed to enzymatic or mechanical degradation. Prolonged storage/transportation of samples at room temperature, repeated freeze/thaw of stool/biopsy homogenates on different occasions, and freeze/thaw or prolonged storage at 4°C of the DNA extract prior to PCR should be avoided.

95.2.2 Detection Procedures

95.2.2.1 Detection and Clarithromycin Susceptibility Testing of *H. pylori*

In *H. pylori*, resistance to clarithromycin is due to point mutations in the 23S rRNA gene. These mutations mainly concern an adenine-to-guanine transition at positions 2142 and 2143 followed by the less prevalent adenine-to-cytosine transversion at position 2142. The following protocols allow for both detection of the infection and the mutations mentioned above. Furthermore, currently commercially available tests exist based on FISH, real-time FRET PCR, and PCR-LiPA, the latter also allowing for the detection of resistance to fluoroquinolones; all of these tests can be applied to culture isolates and gastric biopsies, real-time FRET PCR can be used for the examination of stool specimens as well.

(i) Real-Time PCR Based on a considerable number of published sequences, a real-time FRET PCR assay based on the biprobe technology was designed using the following primers and probe[63]:

23S-F: 5′-AGATGGGAGCTGTCTCAACCAG-3′
23S-R: 5′- TCCTGCGCATGATATTCCC-3′
23S-S: 5′-Cy5-AAGACGGAAAGACCCCGT-biotin-3′.

Primers for the 23S rRNA gene were selected from a region as close as possible to the mutation sites. Probe 23S-S shows 100% homology to the wild-type (clarithromycin sensitive) *H. pylori* genotype, but also a greater number of bacterial genera; therefore, the specificity of the assay is mainly provided by the primers.

Procedure

1. Prepare a 20-μL reaction mixture as follows: 2 μL of LightCycler-FastStart DNA Master SYBR Green I (Roche Molecular Biochemicals), 4 mM $MgCl_2$, 0.05 μM primer 23S-F, 0.5 μM primer 23S-R, 0.3 μM probe 23S-S, and 2 μL of DNA extract made up to the final volume with water.
2. Perform the amplification reaction with preliminary denaturation for 10 min at 95°C, followed by 70 amplification cycles (with a temperature transition rate of 20°C/s) of denaturation at 95°C for 5 s, annealing at 65°C for 10 s, and primer extension at 72°C for 6 s, with a single fluorescence acquisition step at the end of extension.
3. Perform melting analysis of the probe-PCR product duplex consisting of 95°C for 0 s, then cooling to 40°C for 60 s before the temperature is raised to 95°C at a rate of 0.2°C/s with continuous fluorescence acquisition, and a final cooling step at 40°C for 10 s. When using the LightCycler apparatus (Roche Diagnostics), light emission is monitored through the F1 channel of the instrument and Cp values are determined by the second derivative function. Melting curves are constructed automatically and analyzed in the F3 channel of the instrument. Samples were considered as *H. pylori* positive upon presence of a biprobe-specific melting curve.

The melting temperatures are 63°C for the wild-type genotype, 58°C for the A2142C mutant, and 54°C for both the A2142G and A2143G mutants.

Note: This protocol can be used for both biopsy and stool DNA extracts. In case of the latter it is of particular importance to run samples in duplicate; a sample is considered as positive if at least one of the reactions is positive. This procedure together with the relatively high number of amplification cycles resulted in an excellent diagnostic accuracy irrespective of the nature of the sample analyzed.

(ii) FISH One of the advantages of FISH as compared to PCR is the visualization of the bacteria including coccoid forms; a disadvantage may be the observer-dependence of results due to difficulty in reading, which may occur sometimes. A test was designed using essentially the following *H. pylori*-specific probes[71]:

Hpy-1: 5′-CACACCTGACTGACTATCCCG-3′ (16S)
ClaR1: 5′-CGGGGTCTTCCCGTCTT-3′ (23S, mutation A2142G)
ClaR2: 5′-CGGGGTCTCTCCGTCTT-3′ (23S, mutation A2143G)
ClaR3: 5′-CGGGGTCTTGCCGTCTT-3′ (23S, mutation A2142C)

Procedure

1. Cover deparaffinized gastric tissue sections with 50 μL of hybridization buffer (0.9 M NaCl; 0.02 M Tris/HCl, pH 8.0; 0.01% SDS) containing 30% formamide and 5 ng/μL of each fluorescence-labeled probe. The probes for resistance detection can be labeled with the fluorochrome Cy3 (red signal) and the probe for *H. pylori* identification with fluorescein (green signal). Use coverslips to minimize evaporation.
2. Perform hybridization at 46°C for 90 min in a humid chamber and stringent washing at 48°C in a buffer containing 0.112 M NaCl, 20 mM Tris/HCl (pH 8.0), and 0.01% SDS.
3. Rinse hybridized samples quickly with PBS, mount in Citifluor (Citifluor Ltd., London, UK) and examine with a fluorescent microscope equipped with filters for green and red fluorescence.

Note: Using this protocol, *H. pylori* cells sensitive to clarithromycin shine green whereas resistant bacteria shine yellow due to the simultaneous hybridization of green and red fluorescent probes.

95.2.2.2 Detection of Determinants of Pathogenicity

CagA, one of the products of the *cag*PAI, and an active VacA are specific virulence determinants, which have been associated with disease. Recently, the biological activity of CagA has been suggested to correlate with the number of tyrosine phosphorylation EPIYA-C (Western CagA) and -D (East Asian CagA) motifs. VacA activity depends on the allelic types (1 or 2) of three variable regions (s, i, and m) with the most active genotype being s1i1m1. Appropriate primer pairs for the specific detection of the *cag*A region encoding the EPIYA motifs and the determination of the *vac*A genotype are shown in Table 95.2.

Procedure

1. Prepare a 50-μL reaction mixture as follows: 10 mM Tris, 50 mM KCl, 1.5 mM $MgCl_2$ for *cagA* and 2 mM for *vacA* PCR, 0.01% gelatin, 1 U of *Taq* (Fermentas UAB), 200 μM of each dNTP, 0.5 μM each of the primers (cagA2530S and cagA3000AS for *cagA*, VA1-F and VA1-R for *vacA* s region, and VA4-F, VA4-R, VA7-F, and VA7-R for *vacA* m region), and 50 ng of microbial genomic DNA.
2. Perform the amplification reaction with preliminary denaturation at 94°C for 5 min, followed by 35 cycles as follows: *cagA*, 94°C for 30 s, 50°C for 45 s, and 72°C for 45 s; *vacA* s region, 94°C for 60 s, 52°C for 60 s, and 72°C for 60 s; *vacA* m region, 94°C for 30 s,

TABLE 95.2
Primers for Determination of the *cag*A Region Encoding the EPIYA Motifs and Allelic Types of the Variable Regions s, i, and m of *vacA*

Target	Primer	Sequence (5′→3′)	Reference
cagA	cagA2530S	GTTAARAATRGTGTRAAYGG, R = A/G, Y = T/C	66
	cagA3000AS	TTTAGCTTCTGATACCGC	
vacA s	VA1-F	ATGGAAATACAACAAACACAC	67
	VA1-R	CTGCTTGAATGCGCCAAAC	
vacA i	VacF1	GTTGGGATTGGGGGAATGCCG	68
	C1R	TTAATTTAACGCTGTTTGAAG	
	C2R	GATCAACGCTCTGATTTGA	
vacA m	VA4-F	GGAGCCCCAGGAAACATT	67
	VA4-R	CATAACTAGCGCCTTGCAC	
	VA7-F	GTAATGGTGGTTTCAACACC	
	VA7-R	TAATGAGATCTTGAGCGCT	

56°C for 60 s, and 72°C for 90 s. For each of the reactions perform a final extension at 72°C for 5 min.

3. Determine the molecular weight of amplicons on 1.5% agarose gel. For *cagA*, amplicons of 370, 470, 570, and 670 bp correspond to AB, ABC, ABCC, ABCCC, respectively. For *vacA*, amplicons of 259 and 286 bp correspond to s1 and s2, whereas those of 630, 352, and 705 bp correspond to m1, m2, and m1/m2 allelic types, respectively.

Note: It is advisable to use the empty site-positive PCR assay for the confirmation of *cagA*-negative results.[74] The *vacA* i region may be determined by a PCR protocol as described recently.[68] However, due to sequence similarity of the reverse primers some additional optimization may be required to avoid cross-reactivity. Alternatively, sequence analysis of the amplicon may be performed to identify the i allelic type.

95.3 CONCLUSIONS AND FUTURE PERSPECTIVES

H. pylori being associated with different digestive diseases such as gastritis, peptic ulcer, MALT lymphoma and gastric cancer is the most important human pathogen of the genus *Helicobacter.* This microaerophilic, gram-negative, spiral-shaped bacterium persistently colonizes the gastric mucosa of half of the world's population. The infection showing worldwide prevalence predominantly occurs in early childhood and is inversely correlated with socioeconomic status. During the chronic course of infection most individuals will be asymptomatic or complain of nonspecific dyspeptic symptoms, with antral gastritis being the only pathological alteration. However, due to bacterial but also host specific and environmental factors, some patients will develop different sequelae such as peptic ulcer and gastric cancer, which is one of the most frequent types of lethal malignant tumors. Therefore, eradication treatment is recommended for dyspeptic patients tested positive for *H. pylori* infection.

At present, several diagnostic assays for the detection of *H. pylori* infection are available. An invasive approach requiring gastric endoscopy followed by rapid urease testing, culture, and histology still represent the "gold standard" in *H. pylori* diagnostics. Molecular methods such as real-time FRET PCR, PCR-LiPA, and FISH have been shown to be useful for routine use allowing for accurate detection of the pathogen and susceptibility testing to antibiotics. PCR may also be used for the determination of pathogenic factors (e.g., within the scope of epidemiological investigations).

During recent years, noninvasive methods, particularly urea breath testing and fecal antigen detection, gained in importance; their use is suggested when endoscopy is either not required or unfeasible. However, no additional information (e.g., on resistance against antibiotics such as clarithromycin) can yet be obtained with these tests. Since clarithromycin is a widely used antimicrobial drug in industrialized countries, the prevalence of clarithromycin resistant *H. pylori* strains is increasing continually. In the case of resistance the use of a clarithromycin containing treatment regimen would result in eradication failure. Considering the limited availability of alternative treatment regimens, especially for children, pretreatment information on susceptibility to clarithromycin is strictly recommended in order not to abandon the use of this antibiotic in regions of high prevalence of resistant strains. As alternative to an invasive approach stool, real-time FRET PCR can currently be used for detection and clarithromycin susceptibility testing of *H. pylori.*

Current molecular techniques would also allow for a noninvasive or minimally invasive approach for susceptibility testing for other relevant antibiotics with an increasing prevalence of resistant strains (e.g., fluoroquinolones); however, appropriate protocols are not yet available but maybe in the near future. Furthermore, noninvasive *H. pylori* genotyping protocols may be suitable for large-scale epidemiological investigations, which should finally allow for the characterization of the common transmission patterns of the pathogen in different populations. This information is of great value because it may have an impact in terms of reduction of the risk of acquisition of the pathogen, which may be of particular importance for developing countries, where the prevalence of the infection is expected to remain at a very high level during the next decades.

REFERENCES

1. Marshall, B.J., and Warren, J.R., Unidentified curved bacilli in the stomach of patients with gastritis and peptic ulceration, *Lancet*, i, 1311, 1984.
2. Marshall, B.J., and Goodwin, C.S., Revised nomenclature of *C. pyloridis*, *Int. J. Syst. Bacteriol.*, 37, 68, 1987.
3. Goodwin, C.S. et al., Transfer of *Campylobacter pylori* and *Campylobacter mustelae* to *Helicobacter* gen. nov. as *Helicobacter pylori* comb. nov. and *Helicobacter mustelae* comb. nov., respectively, *Int. J. Syst. Bacteriol.*, 39, 397, 1989.
4. Taylor, D.E. et al., Construction of a *Helicobacter pylori* genome map and demonstration of diversity at the genome level, *J. Bacteriol.*, 174, 6800, 1992.
5. Akopyanz, N. et al., DNA diversity among clinical isolates of *Helicobacter pylori* detected by PCR-based RAPD fingerprinting, *Nucleic Acids Res.*, 20, 5137, 1992.
6. Alm, R.A. et al., Genomic-sequence comparison of two unrelated isolates of the human gastric pathogen *Helicobacter pylori*, *Nature*, 397, 176, 1999.
7. Suerbaum, S. et al., Free recombination within *Helicobacter pylori*, *Proc. Natl. Acad. Sci. USA*, 95, 12619, 1998.
8. Salama, N. et al., A whole-genome microarray reveals genetic diversity among *Helicobacter pylori* strains, *Proc. Natl. Acad. Sci. USA*, 97, 14668, 2000.
9. Cover, T.M., and Blaser, M.J., *Helicobacter pylori* in health and disease, *Gastroenterology*, 136, 1863, 2009.
10. Linz, B. et al., An African origin for the intimate association between humans and *Helicobacter pylori*, *Nature*, 445, 915, 2007.
11. Falush, D. et al., Traces of human migrations in *Helicobacter pylori* populations, *Science*, 299, 1582, 2003.
12. Malaty, H.M., Epidemiology of *Helicobacter pylori* infection, *Best Pract. Res. Clin. Gastroenterol.*, 21, 205, 2007.

13. Bruce, G.M., and Maaroos, H.I., Epidemiology of *Helicobacter pylori* infection, *Heliocbacter*, 13, S1, 2008.
14. Malaty, H.M. et al., A co-twin study of the effect of environment on *Helicobacter pylori* acquisition, *Am. J. Epidemiol.*, 148, 793, 1998.
15. Bhuiyan, T.R. et al., Infection by *Helicobacter pylori* in Bangladeshi children from birth to two years, relation to blood group, nutritional status, and seasonality, *Pediatr. Infect. Dis. J.*, 28, 79, 2009.
16. Huck, J.T., and Goh, K.L., Changing epidemiology of *Helicobacter pylori* in Asia, *J. Dig. Dis.*, 9, 186, 2008.
17. Tkachenko, M.A. et al., Dramatic changes in the prevalence of *Helicobacter pylori* infection during childhood: A 10-year follow-up study in Russia, *J. Pediatr. Gastroenterol. Nutr.*, 45, 428, 2007.
18. Dube, C., Tanih, N.F., and Ndip, R.N., *Helicobacter pylori* in water sources: A global environmental health concern, *Rev. Environ. Health*, 24, 1, 2009.
19. .Bamford, K.B. et al., *Helicobacter pylori*: Comparison of DNA fingerprints provides evidence for intrafamilial infection, *Gut*, 34, 1348, 1993.
20. Kivi, M. et al., Concordance of *Helicobacter pylori* strains within families, *J. Clin. Microbiol.*, 41, 5604, 2003.
21. Puz, S. et al., A novel noninvasive genotyping method of *Helicobacter pylori* using stool specimens, *Gastroenterology*, 135, 1543, 2008.
22. Zhang, Y.Y. et al., Review article: "True" re-infection of *Helicobacter pylori* after successful eradication—worldwide annual rates, risk factors and clinical implications, *Aliment. Pharmacol. Ther.*, 29, 145, 2009.
23. Kuipers, E.J., Exploring the link between Helicobacter pylori and gastric cancer, *Aliment. Pharmacol. Ther.*, 13, S3, 1999.
24. Marshall, B.J. et al., Urea protects *Helicobacter (Campylobacter) pylori* from the bactericidal effect of acid, *Gastroenterology*, 99, 697, 1990.
25. Marshall, B.J. et al., Attempt to fulfil Koch's postulates for pyloric campylobacter, *Med. J. Aust.*, 142, 436, 1985.
26. Morris, A.J. et al., Long-term follow-up of voluntary ingestion of *Helicobacter pylori*, *Ann. Intern. Med.*, 114, 662, 1991.
27. Blaser, M.J., and Atherton, J.C., *Helicobacter pylori* persistence: Biology and disease, *J. Clin. Invest.*, 113, 321, 2004.
28. Makristathis, A. et al., Urease prevents adherence of *Helicobacter pylori* to Kato III gastric epithelial cells, *J. Infect. Dis.*, 184, 439, 2001.
29. Algood, H.M., and Cover, T.L., *Helicobacter pylori* persistence: An overview of interactions between *H. pylori* and host immune defenses, *Clin. Microbiol. Rev.*, 19, 597, 2006.
30. Lundgren, A. et al., *Helicobacter pylori*-specific $CD4^+$ $CD25^{high}$ regulatory T cells suppress memory T-cell responses to *H. pylori* in infected individuals, *Infect. Immun.*, 71, 1755, 2003.
31. Kandulski, A. et al., Naturally occurring regulatory T cells (CD4+, CD25high, FOXP3+) in the antrum and cardia are associated with higher *H. pylori* colonization and increased gene expression of TGF-beta1, *Helicobacter*, 13, 295, 2008.
32. Hatakeyama, M., *Helicobacter pylori* and gastric carcinogenesis, *J. Gastroenterol.*, 44, 239, 2009.
33. Cooke, C.L., Huff, J.L., and Solnick, J.V., The role of genome diversity and immune evasion in persistent infection with *Helicobacter pylori*, *FEMS Immunol. Med. Microbiol.*, 45, 11, 2005.
34. Egan, B.J., *Helicobacter pylori* gastritis, the unifying concept for gastric diseases, *Helicobacter*, 12, 39, 2007.
35. Kandulski, A., Selgrad, M., and Malfertheiner, P., *Helicobacter pylori* infection: A clinical overview, *Dig. Liv. Dis.*, 40, 619, 2008.
36. Malfertheiner, P. et al., Current concepts in the management of *Helicobacter pylori* infection: The Maastricht III Consensus Report, *Gut*, 56, 772, 2007.
37. Moayyedi, P. et al., Eradication of *Helicobacter pylori* for non-ulcer dyspepsia, *Cochrane Database Syst. Rev.*, Issue 2, Art. No. CD002096, 2006.
38. Ford, A. et al., Eradication therapy for peptic ulcer disease in *Helicobacter pylori* positive patients, *Cochrane Database Syst. Rev.*, Issue 2, Art. No. CD003840, 2006.
39. Kiltz, U. et al., Use of NSAIDs and infection with *Helicobacter pylori*—what does the rheumatologist need to know, *Rheumatology*, 47, 1342, 2008.
40. Ekström, A.M. et al., *Helicobacter pylori* in gastric cancer established by CagA immunoblot as a marker of past infection, *Gastroenterology*, 121, 784, 2001.
41. Bouvard, V. et al., Special report: Policy: A review of human carcinogens-Part B: Biological agents, *Lancet Oncol.*, 10, 321, 2009.
42. Correa, P., and Houghton, J., Carcinogenesis of *Helicobacter pylori*, *Gastroenterology*, 133, 659, 2007.
43. Mbulaiteye, S.M., Hisada, M., and El-Omar, E.M., *Helicobacter pylori* associated global gastric cancer burden, *Front. Biosci.*, 14, 1490, 2009.
44. Ito, M. et al., Clinical prevention of gastric cancer by *Helicobacter pylori* eradication therapy: A systematic review, *J. Gastroenterol.*, 44, 365, 2009.
45. Fock, K.M. et al., Asia-Pacific consensus guidelines on gastric cancer prevention, *J. Gastroenterol. Hepatol.*, 23, 351, 2008.
46. Zullo, A. et al., Eradication therapy for *Helicobacter pylori* in patients with gastric MALT Lymphoma: A pooled data analysis, *Am. J. Gastroenterol.*, 104, 1932, 2009.
47. Raghunath, A. et al., Prevalence of *Helicobacter pylori* in patients with gastro-oesophageal reflux disease: Systematic review, *BMJ*, 326, 737, 2003.
48. Haesebrouck, F. et al., Gastric Helicobacters in domestic animals and nonhuman primates and their significance for human health, *Clin. Microbiol. Rev.*, 22, 202, 2009.
49. Pandey, M., and Shukla, M., *Helicobacter* species are associated with possible increase in risk of hepatobiliary tract cancers, *Surg. Oncol.*, 18, 51, 2009.
50. Megraud, F., and Lehours, Ph., *Helicobacter pylori* detection and antimicrobial susceptibility testing, *Clin. Microbiol. Rev.*, 20, 280, 2007.
51. Ndip, R. et al., Culturing Helicobacter pylori from clinical specimens: Reviews of microbiological methods, *J. Pediatr. Gastroenterol. Nutr.*, 36, 612, 2003.
52. Baele, M. et al., Non-helicobacter pylori helicobacters detected in the stomach of humans comprise several naturally occurring Helicobacter species in animals, *FEMS Immunol. Med. Microbiol.*, 55, 306, 2009.
53. Fox, J.G., and Megraud, F., *Manual of Clinical Microbiology*, 9th ed., Chapter 60, ASM press, Washington D.C., 2007.
54. Graham, D.Y. et al., *Campylobacter pylori* detected noninvasively by the ^{13}C-urea breath test, *Lancet*, i, 1174, 1987.
55. Hirschl, A.M., and Makristathis, A., Non-invasive *Helicobacter pylori* diagnosis: Stool or breath tests, *Dig. Liv. Dis.*, 37, 732, 2005.
56. Gisbert, J.P., del la Morena, F., and Abraira, V., Accuracy of monoclonal stool antigen test for the diagnosis of *H. pylori* infection: A systematic review and meta-analysis, *Am. J. Gastroenterol.*, 11, 1921, 2006.

57. Herbring, P., and van Doorn, L.P., Serological methods for diagnosis of *Helicobacter pylori* infection and monitoring of eradication therapy, *Eur. J. Clin. Microbiol. Infect. Dis.*, 19, 164, 2000.
58. Mégraud, F., and Lehours, P., *Helicobacter pylori* detection and antimicrobial susceptibility testing, *Clin. Microbiol. Rev.*, 20, 280, 2007.
59. Yoshida, H. et al., Use of a gastric juice-based PCR assay to detect *Helicobacter pylori* infection in culture-negative patients, *J. Clin. Microbiol.*, 36, 317, 1998.
60. Makristathis, A. et al., Detection of *Helicobacter pylori* in stool specimens by PCR and antigen enzyme immunoassay, *J. Clin. Microbiol.*, 36, 2772, 1998.
61. Chisholm, S.A. et al., PCR-Based diagnosis of *Helicobacter pylori* infection and real-time determination of clarithromycin resistance directly from human gastric biopsy samples, *J. Clin. Microbiol.*, 39, 1217, 2001.
62. Oleastro, M. et al., Real-time PCR assay for rapid and accurate detection of point mutations conferring resistance to Clarithromycin in *Helicobacter pylori*, *J. Clin. Microbiol.*, 41, 397, 2003.
63. Schabereiter-Gurtner, C. et al., Novel real-time PCR assay for detection of *Helicobacter pylori* infection and simultaneous clarithromycin susceptibility testing of stool and biopsy specimens, *J. Clin. Microbiol.*, 42, 4512, 2004.
64. Glocker, E. et al., Real-time PCR screening for 16S rRNA mutations associated with resistance to tetracycline in *Helicobacter pylori*, *Antimicrob. Agents Chemother.*, 49, 3166, 2005.
65. Glocker, E., and Kist, M., Rapid detection of point mutations in the *gyrA* gene of *Helicobacter pylori* conferring resistance to ciprofloxacin by a fluorescence resonance energy transfer-based real-time PCR approach, *J. Clin. Microbiol.*, 42, 2241, 2004.
66. Panayotopoulou, E.G. et al., Strategy to characterize the number and type of repeating EPIYA phosphorylation motifs in the carboxyl terminus of CagA protein in *Helicobacter pylori* clinical isolates, *J. Clin. Microbiol.*, 45, 488, 2007.
67. Atherton, J.C. et al., Simple and accurate PCR-based system for typing vacuolating cytotoxin alleles of *Helicobacter pylori*, *J. Clin. Microbiol.*, 37, 2979, 1999.
68. Rhead, J.L. et al., A new *Helicobacter pylori* vacuolating cytotoxin determinant, the intermediate region, is associated with gastric cancer, *Gastroenterology*, 133, 926, 2007.
69. Owen, R.J., Molecular testing for antibiotic resistance in *Helicobacter pylori*, *Gut*, 50, 285, 2002.
70. van Doorn, L.J. et al., Typing of *Helicobacter pylori vacA* gene and detection of *cagA* gene by PCR and reverse hybridization, *J. Clin. Microbiol.*, 36, 1271, 1998.
71. Trebesius, K. et al., Rapid and specific detection of *Helicobacter pylori* macrolide resistance in gastric tissue by fluorescent in situ hybridisation, *Gut*, 46, 608, 2000.
72. Man, S.M. et al., Detection of enterohepatic and gastric helicobacter species in fecal specimens of children with Crohn's disease, *Helicobacter*, 13, 234, 2008.
73. Schwarz, S. et al., Horizontal versus familial transmission of *Helicobacter pylori*, *PLoS Pathog.*, 4, e1000180, 2008.
74. Akopyants, N.S. et al., Analyses of the cag pathogenicity island of *Helicobacter pylori*, *Mol. Microbiol.*, 28, 37, 1998.

Section V

Spirochaetes

96 Borrelia

Nataliia Rudenko, Maryna Golovchenko, James H. Oliver Jr., and Libor Grubhoffer

CONTENTS

96.1 INTRODUCTION

96.1.1 Classification and Genome Organization

Lyme borreliosis is a multisystem disorder and the most frequent infectious arthropod-borne disease in Europe and the United States. The disease is induced by spirochetes from *Borrelia burgdorferi* sensu lato complex (s.l.) and transmitted by ixodid ticks.

B. burgdorferi sensu stricto (s.s.) was first isolated in 1982 by W. Burgdorfer[1] from the tick *Ixodes dammini*. Subsequently, *I. dammini* was shown to be conspecific to *I. scapularis* so the latter name is now considered the correct name.[2] Today, 17 *B. burgdorferi* s.l. species are well recognized around the world. Eleven of them were identified in and strictly associated with Eurasia (*B. afzelii*, *B. bavariensis*, *B. garinii*, *B. japonica*, *B. lusitaniae*, *B. sinica*, *B. spielmanii*, *B. tanukii*, *B. turdi*, *B. valaisiana* and *B. yangtze*), while another four (*B. americana*, *B. andersonii*, *B. californiensis* and *B. carolinensis*) were identified only in the USA. *B. burgdorferi* s.s. and *B. bissettii* share the distinction of being present in both the Old and the New World. In addition to the well-established populations of the above-named species another undescribed North American species was delineated from a group of atypical American strains and was proposed as genomospecies 2 (Table 96.1). Descriptions of new species and variants continue to be recognized, so the current number of described species is probably not final.

B. burgdorferi belongs to the phylum Spirochaetes, class Spirochaetes, order Spirochaetales, and genus *Borrelia*. It is a spiral, unicellular, gram-negative bacterium without a rigid cellular wall, 10–40 μm in length, and 0.2–0.5 μm in diameter. The most distinctive property is the presence of from 7 to 14 periplasmic flagella that are wound around the helical protoplasmic cylinder lengthwise and situated between the peptidoglycan layer and the outer membrane. Movement of the flagellum produces a screw-like motion that propels the organism. The flagella consist mainly of two types of flagellin proteins—minor FlaA (38 kDa) and major FlaB (41 kDa).[15] Targeted inactivation of the FlaB gene induces loss of motility and the rod-like form of these mutants.[16] Changing the culture conditions, cultivation in medium with nonoptimal pH, presence of antibiotics, absence of serum, aged cultures, or generally, stress conditions can affect the spirochete morphology.[17–19] The cytomorphologic features of *B. burgdorferi* show marked polymorphism, resulting in the appearance of variations in tissue sections.[20] These variations may reflect important differences in the spirochete tendency to penetrate or adhere to the tissues of the host or vector tick or can be correlated with it virulence, infectivity, or antigenicity.[21,22]

TABLE 96.1
Currently Known *Borrelia burgdorferi* Sensu Lato Species/Genomospecies

Borrelia Species	Vector	Hosts/Reservoirs	Geographical Distribution	Reference
Borrelia afzelii	*I. ricinus, I. persulcatus*	Rodents	Europe, Asia	3
Borrelia americana	*I. pasificus, I. minor*	Birds	United States	142
Borrelia andersonii	*I. dentatus*	Cotton tail rabbit	United States	4
Borrelia bavariensis	*I. ricinus*	Rodents	Europe	143
Borrelia bissettii	*I. ricinus, I. scapularis, I. pacificus, I. minor*	Rodents	Europe, United States	5
Borrelia burgdorferi sensu stricto	*I. ricinus, I. scapularis, I. pacificus*	Rodents, birds, lizards, big mammals	North America, Europe	6
Borrelia californiensis	*I. pacificus, I. jellisonii, I. spinipalpis*	Kangaroo rat, mule deer	United States	7
Borrelia carolinensis	*I. minor*	Rodents, birds	United States	8
Borrelia garinii	*I. ricinus, I. persulcatus, I. hexagonus, I. nipponensis*	Birds, lizards, rodents	Europe, Asia	6
Borrelia japonica	*I. ovatus*	Rodents	Japan	9
Borrelia lusitaniae	*I. ricinus*	Rodents, lizards	Europe, North Africa	10
Borrelia sinica	*I. ovatus*	Rodents	China	11
Borrelia tanukii	*I. tanuki*	Unknown (possibly dogs and cats)	Japan	12
Borrelia turdi	*I. turdus*	Birds	Japan	12
Borrelia spielmanii	*I. ricinus*	Rodents	Europe	13
Borrelia valaisiana	*I. ricinus, I. granulatus*	Birds, lizards	Europe, Asia	14
Borrelia yangtze	*H. longicornis, I. granulatus*	Rodents	China	141
Genomospecies 2	*I. pacificus*	Unknown	United States	7

At the molecular level, a unique feature of *Borrelia* is the unusual organization of its genome (1997,[23] updated 2000[24]). Upon sequencing and analysis of *Borrelia*, 21 extra chromosomal DNA elements were found. This is the largest number known for any bacterium. *Borrelia* has a segmented genome, consisting of an approximately 911 kb linear chromosome with an average G + C content of 28.6%, and numerous circular and linear plasmids[23] ranging in size from 9 to 32 kb and 5 to 56 kb, respectively.[24] The chromosome carries 853 predicted open reading frames (ORFs) from which the biological roles are assigned to 59% and 28% of ORFs, which represent new genes. Chromosomal genes code proteins necessary for replication, transcription, translation, energy metabolism, and transport of nutrients and ions, DNA repair, homologous recombination, heat-shock response, and chemotactic factors coding genes.[23,25] About 8% of *Borrelia* genome is devoted to producing lipoproteins exposed on the outer surface.[26]

The plasmids encode 535 genes, and 90% of these have no similarity to genes outside the *Borrelia* genus, suggesting their specialized functions related to the adaptation of the spirochete to different hosts.[26] Only 16% of plasmid encoding genes have assigned biological functions, whereas 58% represent genes of unknown function, suggesting that their encoded proteins may define the unique biologic and pathogenic aspects of *Borrelia*.[23] Some *Borrelia* plasmids carry genes essential for survival.[26–36] *Borrelia* plasmids vary in stability; some of them are unstable and lost during cultivation. The consequences of plasmid loss also vary; some plasmids encode functions that are critical for survival in natural host, whereas others are lost without impact on the infectious cycle. Genetic manipulation with *Borrelia* is difficult due to highly segmented, unstable genomes and the presence of plasmid encoded restriction modification systems that degrade foreign DNA.[37]

Studies of *Borrelia* isolates with known plasmid patterns[38] separate plasmids into three groups: one group includes plasmids lp25 and lp28–1, which are absolutely necessary for persistent infection of a mammalian host. Bacteria that have lost either of these two plasmids remain capable of in vitro growth, but they lose the ability to cause persistent infection. The second group includes plasmids not required for infectivity (cp9, cp32–3, lp21, lp28–2, lp28–4, lp56), and in the third group plasmids are present in all isolates (cp26, cp32–1, cp32–4, cp32–6, cp32–7, cp32–8, cp32–9, lp17, lp28–3, lp36, lp38, and lp54).[38]

Some linear plasmids have features similar to the linear chromosome and should be regarded as "mini chromosomes."[39] An unusual component of the *B. burgdorferi* genome is a family of closely related 32-kb circular plasmids. Nine distinct cp32 plasmids can coexist within the same bacteria. These have been found to be prophage genomes and are involved in the horizontal transfer of DNA among spirochetes that share a common geographical and ecological niche.[40,41]

The chromosomes of different *B. burgdorferi* isolates are in general very similar, but, some isolates are known to have an extra fragments of DNA (between 7 and 20 kb) added to the chromosomal 3′-end.[23,24] It was noted that an extra 7.2 kb of DNA from the 3′-end of the *B. burgdorferi* B31 chromosome is similar to plasmid sequences. Curiously, there is little similarity between the linear plasmids and the 5′-end of the B31 chromosome.

Neither chromosomes nor plasmids contain genes for cellular biosynthetic reactions. This is the reason for the limited metabolic capacity of *B. burgdorferi* s.l., which lacks the enzymes necessary for the biosynthesis of aminoacids, fatty acids, nucleotides, and cofactors.

96.1.2 Epidemiology

Lyme borreliosis is a global public health problem, and the number of annual reported cases continues to increase because of the actual spread of the disease or improved diagnostic methods, or both. It is an old disease that was recognized only at the end of the twentieth century. The symptoms of the disease were first described in Europe at the end of the nineteenth century, although the first descriptions of an illness similar to Lyme arthritis were registered earlier at the beginning of the eighteenth century in a document from Jura island (Hebrides). Nevertheless, the term Lyme arthritis was not coined until the mid-1970s.[42,43] Some other cases of early Lyme borreliosis were registered in Europe in the late nineteenth and beginning of the twentieth century: 1893, Dr. Buchwald, Germany—description of first case of acrodermatis chronica atroficans; 1909, Dr. Afzelius, Sweden—description of first case of erythema chronicum migrans; 1922, Dr. Garrin and Dr. Bujadoux, France—description of lymphocytic meningitis. A. Steere was the first American to recognize the importance of the disease, which he named Lyme arthritis in 1977.[44] One year later, Spielman described a tick species that he incorrectly thought was a new species (*Ixodes dammini*), which was later found to be the vector of the disease. Subsequently, *B. burgdorferi* s.s. was found to be the etiologic agent of Lyme disease.[1] An understanding of the different clinical manifestations, epidemiological characteristics, and microbiological features of this spirochete and the development of diagnostics clearly defined this as a different disease. Lyme borreliosis has now become a recognized problem of worldwide concern.

Of the total vector-borne illnesses found in the United States, Lyme borreliosis is the most prominent; it accounts for more than 90% of all reported cases of vector-borne illness (www.cdc.gov). Approximately 16,000–20,000 cases of Lyme borreliosis are reported annually, though it is believed to be vastly underreported. There have been several dominant tenets in the past concerning Lyme borreliosis in the United States.[45,46] Though Lyme borreliosis has been reported in all 50 states, only 12 states located along the northeastern and mid-Atlantic seaboards and in the upper north-central region of the United States are considered by Centers for Disease Control and Prevention (CDC) to be highly or moderately endemic. Contrary to dogma, recently published results confirm the presence and wide distribution of *B. burgdorferi* s.l. in many areas of the United States in which it was not previously thought to occur, in particular, the southeastern United States.[8,46,47]

Surveillance of Lyme borreliosis in Europe varies and does not allow direct comparison among countries. In most countries, evidence is mainly based on diagnostic laboratories' reporting of patients with positive tests, but these data are often not directly comparable because of different serological tests used to detect antibodies to *B. burgdorf*eri s.l.[48] Even considering the limitations of serological methods, it is clear that Lyme borreliosis shows a gradient of increasing occurrence from west to east with the highest incidences in Central and Eastern Europe. The highest incidences of Lyme borreliosis in Europe are found in the Baltic States and Sweden in the north, and in Austria, the Czech Republic, Germany, and Slovenia in Central Europe (EUCALB, 2009; www.eucalb.com).

Dr. Z. Hubalek[49] summarized Lyme borreliosis related data from 36 European countries, 12 states from the United States of America, Canada, nine Asian countries and four from Africa. He concluded that latitude and altitude, seasonal distribution, weather related factors and patient characteristics are the most significant factors determining the distribution of Lyme borreliosis. It was shown that the disease occurs between 35°N and 60°N in Europe and 30°N and 55°N in North America. Its incidence decreases rapidly outside of this range; tick-infection rates decrease with elevation; nevertheless, *I. ricinus* infected with *B. burgdorferi* s.l. was detected as high as 1350 m in Austria.[50] The disease is also season-related—the peak of its incidences occurs in May–July in most of the countries, but in northern countries could include August. Age distribution of the disease is bimodal in most countries, with the lower maximum occurring in children from 5 to 9 years old and maximum in adults aged 50–64 years; summer temperature and rainfall affect tick vector population. The duration of tick attachment is also important for transmission of *Borrelia*. Whereas authorities from the United States assured that *Borrelia* transmission may occur only 48 h after tick attachment, European scientists believe that spirochete transmission may occur as fast as 8 h after tick attachment.

The epidemiological data from many European countries and the US show a dramatic increase of the diagnosed cases of Lyme disease. The 2006 report of Lingren and Jaenson[51] stated that about 85,000 Lyme borreliosis cases were estimated in Europe alone, with additional 16,000–20,000 annual cases in the United States. Taking into consideration the significant number of underreported cases,[52] the total annual number of Lyme borreliosis cases in the world might be as many as 255,000. In addition to all the factors that could be responsible for the recent increase of reported cases, the definition of the disease has also changed. What was not considered Lyme borreliosis before might be considered a disease now. Development of new progressive diagnostic methods during the last decades contributed to the increasing number of diagnosed cases.[49]

96.1.3 Clinical Features and Pathogenesis

The clinical symptoms of Lyme borreliosis are heterogeneous and basically depend on several factors including the tropism of the infectious agent. The complete presentation of the disease, starting with skin lesion (after the tick bite), followed by nervous system and heart association and later on by arthritis with the involvement of joints, eyes, and skin, is extremely unusual.[53] Lyme borreliosis is medically described as a three-phase disease: stage 1 is early localized disease with skin inflammation; stage 2 is early disseminated disease with heart and nervous system involvement, including palsies and meningitis; and stage 3 is late disease featuring motor and sensory nerve damage and brain inflammation as well as arthritis. These distinctions are only valuable for informational reasons, as very often they are not in agreement with clinical findings, and each person may have a unique presentation and progression of symptoms. The protocol of antibiotic treatment of different stages varies due to unique pathophysiology of the disease, which involves not only bacterial infection, but also the host immune response.[54] Stage 3 is often referred to as "chronic Lyme borreliosis" by European physicians. This is unfortunate because that term has a different meaning to many U.S. physicians, who readily acknowledge that some patients suffer motor and sensory nerve damage, brain inflammation, and arthritis but do not consider these symptoms to be chronic Lyme disease because they may occur during active cases. Most U.S. physicians do not recognize a chronic long-term stage of Lyme borreliosis due to the lack of supportive laboratory data. The existence of chronic Lyme borreliosis is very controversial between patient advocacy organizations and many physicians.

At the early phase (stage 1), within days to weeks of the tick bite, the skin around the bite develops an expanding ring of redness called erythema migrans (EM). EM is the most common symptom, occurring in approximately 75% of patients. It is usually accompanied by other signs of the disease including fatigue, muscle and joint stiffness, swollen lymph nodes ("swollen glands"), headache, fever, and nuchal rigidity, resembling symptoms of a virus infection. Because of the close temporal proximity of tick bites and the appearance of EM, this manifestation has a seasonal occurrence.

Secondary or early disseminated Lyme borreliosis typically occurs weeks to months after the tick bite, when the infection spreads via the lymph system or bloodstream. It is also called "Bannwarth syndrome." At this stage, symptoms usually include blurred vision, fainting, fatigue, general discomfort, uneasiness, or ill feeling (malaise), headache, heart palpitations, lightheadedness, joint inflammation in the knees and other large joints, lethargy, muscle pains, and stiff neck. Meningitis is one of the most common symptoms of early disseminated disease. Neuropathy often affects the seventh cranial nerve and may cause facial palsies, especially in children, and may be bilateral. Arthritis occurs in about 20% of patients and usually involves the knee joint. About 15% of people develop neurologic problems; about 8%, heart problems.[55]

Stage 3 or late (or chronic) Lyme borreliosis can occur from months to years after the tick bite. Untreated or inadequately treated patients may develop severe and chronic symptoms that affect many parts of the body, including brain, nerves, eyes, joints, and heart. Infection of central nervous system causes more severe neurological symptoms. It is typically associated with intermittent or persistent arthralgia involving one or few large joints. It was originally believed that 50%–75% of patients with Lyme borrelosis had arthritis. However, recent analyses suggest that the incidence of arthritis in patients with late or chronic disease is closer to 25%.[56] In addition to arthritis, several purely neurological symptoms may occur, including headaches, aseptic meningitis, and rheumatoid arthritis, atrioventricular conduction abnormalities (AV heart block), facial nerve (Bell's) palsy, and encephalitis or encephalopathy that may be manifested by sleep disorder, mood changes, memory loss, and psychiatric symptoms such as panic, anxiety, or depression.[57] A small percent of infected patients develop a type of neurological dysfunction that can increase their risk of aggressiveness, resulting in significant suicidal and aggressive tendencies.[58] In some patients, Lyme borreliosis can be fairly easy to treat, whereas in others it seems like a never-ending battle. As noted above, chronic Lyme borreliosis is surrounded by much controversy. Whereas some physicians accept it as a real condition caused by a persistent presence of *B. burgdorferi* in the body that sometimes causes complete debilitation; other medical experts are convinced that it is a fictional disease resulting from such psychological factors as paranoia or hypochondria.

Borrelia spirochetes are transmitted to humans by all three developmental stages of ixodid ticks, but the nymphal stage appears to be the most important. Considering the human sensitivity to *B. burgdorferi* s.l. and taking into account results of the newest publications, the complex of 17 *Borrelia* species and two genomospesies can be divided into two groups:

- Ten species and two genomospecies that have never been detected in or isolated from humans. This group includes *B. americana*, *B. andersonii*, *B. bavariensis*, *B. californiensis*, *B. carolinensis*, *B. japonica*, *B. tanukii*, *B. turdi*, *B. sinica*, *B. yangtze* and genomospecies 2.
- Seven species with pathogenic potential. This group includes *B. afzelii*, *B. bissettii*, *B. burgdorferi* s.s., *B. garinii*, *B. lusitaniae*, *B. spielmanii*, and *B. valaisiana*. Until recently, only three species were considered to be the causative agents of Lyme borreliosis, that is, *B. burgdorferi* s.s., *B. afzelii*, and *B. garinii*. However, the DNA of *B. valaisiana*, *B. lusitaniae*, *B. spielmanii*, and *B. bissettii* has been detected in samples of human origin or the spirochetes themselves isolated from patients with Lyme borreliosis symptoms.[59–66]

It was suggested that different species possess different organotropisms and may preferentially cause distinct clinical manifestations of the disease. Lyme arthritis is the most common musculoskeletal symptom resulting from *B. burgdorferi* s.s. infection. About 60% of untreated patients with EM in America experienced brief attacks of arthritis.[67] In contrast, only 3%–15% of patients suffered from arthritis in Europe,[68] where *B. garinii* and *B. afzelii* are more frequently recovered than *B. burgdorferi* s.s.

Serotyping studies of isolates from Europe reveal a striking correlation between neuroborreliosis and infection with

B. garinii. *B. burgdorferi* s.s., and *B. afzelii* can also be associated with neurological manifestation but not at such a high rate.

B. afzelii in humans seems to have an organotropism for the skin since it preferentially causes lymphadenosis benigna cutis[69] and acrodermatitis chronica atrophicans (ACA). Molecular studies of isolates from patients in several European countries confirmed association of ACA with *B. afzelii* infection.[70] However, *B. afzelii* is the predominant but not the exclusive etiologic agent of ACA; *B. garinii* has also been detected. It is interesting to note that ACA has never been observed in Americans who have never left continental United States, confirming that endemic *B. burgdorferi* s.s. rarely, if ever, induces this form of Lyme borreliosis.

The pathogenic potential of *B. valaisiana* is suspected among patients with EM based on polymerase chain reaction (PCR).[60] Association of *B. valaisiana* with chronic clinical symptoms is also known.[63] Molecular analysis of spirochete isolates from a Portuguese patient identified as *B. lusitaniae* suggests a clinical pattern for *B. lusitaniae* that is different from the other *Borrelia* spp. examined so far.[64] The role of *B. spielmanii* as a causative agent of Lyme borrelosis was confirmed by reported cases of patients with EM in Netherlands, Germany, Hungary, and Slovenia.[71] Members of *B. bissettii* DN127 genomic group had been previously isolated from ixodid ticks and small reservoir hosts only, suggesting that this species may not be pathogenic to humans. But, about decade ago, Strle presented clinical findings that some patients from Slovenia had Lyme borreliosis that was caused by *Borrelia* species with genotypic and phenotypic similarity to strain 25015, a member of the DN127 genomic group of *B. bissettii*.[61] These results confirmed that *B. bissettii* shared with *B. burgdorferi* s.s. the distinction of being present in both the Old and New Worlds. Strain 25015 was previously found to be infectious but nonpathogenic using a mouse model.[72] A recent study with a different murine model showed that this strain is mildly arthritogenic.[73] Later, Schneider reported the potential pathogenesis of *B. bissettii* to cause human Lyme borreliosis in a mouse model.[74] It was proved that two Colorado isolates of *B. bissettii* (CO-Bb) induced lesions of the bladder, heart, and femorotibial joint 8 weeks after inoculation into mice.[74] Isolation of *B. bissettii* from patients in Slovenia,[59,61,75] detection of *B. bissettii* in cardiac valve tissue of the patient with endocarditis and aortic valve stenosis in the Czech Republic,[65] and the detection of *B. bissettii* in the serum samples of patients with symptomatic borreliosis or chronic borrelial infection[66] are all strong supporting evidence that this species is involved in Lyme borreliosis in Europe. However, the small number of patients infected with *B. bissettii* does not allow any conclusions to be made yet on the association of this species with organotropism or species-specific clinical manifestations.

The basic features of Lyme borreliosis are similar worldwide, but there are regional variations, primarily among the illness found in the United States, Europe, and Asia. EM affects all ages and both sexes, with slight predominance of females in Europe but not in the United States. The previous name, "erythema chronicum migrans," was shortened to "erythema migrans" because these lesions proved not to be chronic in the United States as they can be in Europe. In North America, EM is caused by *B. burgdorferi* s.s., whereas in Europe, EM is most often caused by *B. afzelii* (70%–90%), less frequently by *B. garinii* (10%–20%), rarely by *B. burgdorferi* s.s., and exceptionally by other species of *Borrelia*.[53] In comparison to Europeans, American patients had higher rates of systemic symptoms such as fatigue and malaise, headache, myalgia, and arthralgia. In European patients with EM, fever is recalled in fewer than 5%,[76] whereas in the United States it is recalled in one-third of all Lyme borreliosis patients.[77] A briefer duration of EM prior to presentation is observed in the United States. American patients have higher rates of seropositivity, whereas the Europeans with EM are mostly seronegative. Secondary EM lesions are also more common in the United States than in Europe.[76] Some of the skin manifestations of Lyme borreliosis such as borrelial lymphocytoma, ACA, and morphea-like lesions, occur almost exclusively in Europe. Such late neurologic feature as chronic encephalomyelitis characterized by spastic paraparesis, cranial neuropathy, or cognitive impairment with marked intrathecal antibody production is caused by *B. garinii* infection and is typical only of European patients.[62,78,79]

96.1.4 Diagnosis

Each disease has specific signs or symptoms. Due to diversity of clinical symptoms Lyme borelliosis is often referred to as the "great imitator." It may mimic myriad conditions. This is partly determined by the diversity of *B. burgdorferi* s.l. species causing Lyme borreliosis.[70,80,81] It makes diagnosing Lyme borreliosis very difficult and demands the use of the whole complex of the standardized laboratory methods, which are supposed to explain the diagnosis, not simply confirm it.

The diagnosis of Lyme borreliosis has recently been made by clinical, epidemiologic and serologic techniques; however, all such confirmations can be unreliable and problematic.[82] Antigen detection by monoclonal antibodies does not exceed the experimental level, has a low yield, and cannot be used as a routine test. Lyme borreliosis diagnosis by serological detection of *Borrelia*-specific analysis has low sensitivity during early infection.[83,84] Antibody titers are usually not detectable before 4–6 weeks postinfection. In contrast, positive Lyme serology may indicate past *Borrelia* infection without any proof of actual disease. Serologic tests are subject to false-positive results at a rate of 20%–40%.[85] A silver staining technique can identify other spirochetes, besides *B. burgdorferi*, increasing the number of false-positive results. Enzyme-linked immunosorbent assay (ELISA), currently used for diagnosis, appeared to yield some false-positive and false-negative results; moreover, it does not distinguish different spirochete genospecies involved in Lyme borreliosis. Western blotting improves the specificity of ELISA, but it still does not solve the problem, confirming only 30%–50% of suspected cases. Direct detection of spirochetes by cultivation or dark-field microscopy cannot be considered reliable because of a low pathogen concentration in body fluids.[83] Recently, PCR has proved to be a sensitive and fast method for the diagnosis of microorganisms that are difficult to culture.[86,87]

Primary, diagnosis of Lyme borreliosis is based on clinical information such as history, symptoms, and response to therapy. The only sign that enables a reliable clinical diagnosis of early Lyme borreliosis is EM lesion. The patient who presents with an EM lesion will likely be seronegative, since the lesion often appears prior to development of a diagnostic, adaptive immune response. A positive serology is not needed to make the diagnosis in this case. The latest knowledge reveals that this is only valid in Europe as in the United States skin lesion of STARI (Southern tick-associated rash illness) is very similar to EM.[88] The majority of the symptoms and signs have minimal diagnostic value. Laboratory confirmation of the infection is needed for all manifestations.

The diagnostic tools of Lyme borreliosis in clinical laboratories can be divided into microscope-based assays, in vitro cultivation, antigen detection assays, and nucleic acid–based detection of *B. burgdorferi*. The sensitivity and specificity of these methods depend on various factors and are also variable among laboratories.

96.1.4.1 Conventional Techniques

The liquid media currently used to grow *B. burgdorferi* were originally modified from Kelly medium.[89] Current versions of this medium [Barbour-Stoenner-Kelly II medium,[90] BSK-H,[91] and Kelly medium Preac-Mursic (MKP)[92]] are an improvement and better support spirochete recovery from low inocula, show shorter generation times of the spirochete, and maximal concentration of spirochetes in culture.[93] *B. burgdorferi* s.l. can also be grown on solid media with agarose under microaerophilic or anaerobic conditions,[90,94] producing individual colonies, which is a big advantage in analysis of cases with coinfection.

Cultivated borrelia is the best confirmation of the active infection. The majority of successful isolates was obtained from patients with early Lyme borreliosis who did not receive appropriate antibiotic treatment. Borrelia were isolated or identified in clinical specimens of different origin including: biopsy, specimens of EM skin lesions, biopsy specimens of ACA skin lesions, biopsy specimens of borrelial lymphocytoma skin lesions, cerebrospinal fluid (CSF) specimens, blood specimens, synovial fluid, iris (see Reference 82), cardiac tissue,[65] and serum.[66] According to some data, the success of cultivation ranged from 5% to 86%.[93,95] The sensitivity of culture for clinical specimens is undefined yet, but as low as one spirochete can be recovered from laboratory adapted and continuously propagated strains of *B. burgdorferi* s.s.[90] Cultivation of borrelia is more expensive, more labor intensive, and much slower, requiring up to 6 weeks of incubation, and is directly dependent on the frequency of microscopic examination. Moreover, such examination still has limited value in laboratory confirmation of Lyme borreliosis.[96–98] Borrelia can be directly seen in human body fluids under dark-field microscopy or after immunological staining on histological sections. However, this method is of very low sensitivity due to the sparseness of organisms in them and is unreliable and extremely time consuming.

Due to limitations in the direct detection of borrelia in clinical specimens, antibody detection methods have been the main laboratory approach used in the clinical diagnosis of Lyme borreliosis. The immunoserologic tests used to assess *B. burgdorferi* antibodies include immunofluorescence assays and a variation of this assay using antigens attached to a membrane (FIAX),[99] ELISA and enzyme-linked fluorescent assay (ELFA),[100] particle agglutination (PA) method and Western blotting. Among these, ELISA and immunofluorescence assays are the most popular. Methods less frequently used are assays that detect borreliacidal (functional) antibodies, antibodies bound to immune complexes (IC), and hemagglutinating antibodies.[101,102] The results of the tests are influenced by the isolate of *B. burgdorferi* s.l., concentration of serum antibody, cutoff points, and interpretative criteria. A number of reports have expressed concern about the sensitivity, specificity, and reproducibility of mentioned methods.[103–107] Although serological tests are used widely, a positive antibody assay in the absence of the corresponding clinical symptoms is not sufficient for the diagnosis of Lyme borreliosis, but it can provide very useful supporting evidence, especially in patients with disseminated or chronic infection.

The immune response to *B. burgdorferi* s.l. infection begins with the appearance of specific immunoglobulin M (IgM) antibodies, usually within the first several weeks after initial exposure. IgM response may persist for many months or even years despite effective antimicrobial therapy. That is why the presence of specific IgM antibodies cannot be used as the sole criterion to diagnose a recent infection. Most patients will have detectable IgG antibodies after 1 month of active infection. Like that of IgM, the IgG response can persist for years after Lyme borreliosis symptoms have resolved. Both IgG and IgM responses can be greatly diminished or absent in patients receiving antimicrobial therapy.

The *B. burgdorferi* IgG/IgM indirect fluorescent antibody (IFA) test is designed for the qualitative and semiquantitative presumptive detection of total antibodies to *B. burgdorferi* in human serum. This test should only be used for patients with signs and symptoms that are consistent with Lyme borreliosis. Limitations of this assay are the need for fluorescence microscopy, well-trained personnel, and subjectivity in reading and interpreting the results. That is why the IFA method has been replaced by ELISA. Different antigens can be used for this test: whole-cell lysates, partially purified antigens, or recombinant antigens. One of the limitations of ELISA for the detection of *B. burgdorferi* s.l. antibodies is the lack of standardization.

PA method is more specific than immunofluorescence assays, but it is also more expensive and labor intensive and thus is rarely used for screening.

A two-step testing strategy that includes ELISA or immunofluorescence assay, followed by Western blotting has been recommended both in the United States[108] and in Europe[109] for the serodiagnosis of Lyme borreliosis.

Western blot is believed to be much more accurate and specific than ELISA or IFA and is usually used when positive results were obtained on one of the above mentioned tests. According to criteria for interpretation of Western blot accepted by CDC,[109a] a positive IgM blot is defined by the presence of two of three particular immunoreactive bands (OspC, p41 or p39 kDa). The IgG criteria require the presence of at least 5 of 10 particular bands (p93, p66, p58, p45, p41, p39, p30, p28, p21, or p18).[110,111] The guidelines for Western blot interpretation state that IgM or IgG criteria can be used during the first 4 weeks of illness, but that only IgG criteria can be used after 4 weeks from onset of the disease. For Europe *de novo* criteria led to the following recommendations: for Western blot for IgG, at least two bands among p83/100, p58, p43, p39, p30, OspC, p21, p17, and p14 for *B. afzelii* (PKo), and at least one band among p83/100, p39, p30, OspC, p21, and p17 for *B. garinii* (PBi). For Western blot for IgM, at least one band among p39, OspC, and p17 or a strong p41 band for *B. afzelii* and at least one band among p39 and OspC or a strong p41 band for *B. garinii*. Western blot with *B. afzelii* (PKo) was the most sensitive, and this strain was recommended for use in Western blot for the serodiagnosis of Lyme borreliosis throughout Europe.[112] Because of the antigenic variability among *B. burgdorferi* s.l. species, so far not a single antigen has demonstrated sufficient sensitivity and specificity to warrant replacing the two-step testing procedure.

Antibodies to certain antigens of *B. burgdorferi* s.l., in particular to certain outer surface proteins, may have bactericidal activity.[113] Such antibodies have been used in the borreliacidal antibody assays of early and late Lyme borreliosis.[114,115] In this assay, live *B. burgdorferi* s.s. is incubated with patient serum plus an exogenous source of complement. Growth inhibition of *B. burgdorferi* s.s. can be determined by visual inspection of the percentage of nonmotile spirochetes, color changes by use of a pH indicator, or flow cytometry after staining with acridine orange.

Another antibody detection method is the detection of antibodies bound to circulating immune complexes (IC). It has been proposed that seronegativity in early Lyme borreliosis is mainly the result of the formation of specific antigen-antibody complexes that prevent the detection of free antibodies by conventional methods. Detection of antibodies bound to immune complexes involves the treatment of serum with polyethylene glycol to precipitate the IC, followed by dissociation through alkalinization of the complexes to release antibodies that can be detected by ELISA or Western blot. Some studies have suggested that immune complexes are found not only in serum but also in other body fluids such as cerebrospinal and synovial fluids.[116–118]

A new rather controversial testing method is the Bowen Q-RiBb (Quantitative Rapid Identification of *Borrelia burgdorferi*) test, which has been approved as U.S. patent (Test for Lyme disease, United States Patent 6,838,247). Developed by Whitaker, Fort, and Hamilton the Q-RiBb test is unique in its approach and potentially offers much greater accuracy than other conventional testing. The method uses a fluorescent antibody technique on whole blood. The fluorescent antibodies are added to the diluted whole blood. The solution containing the antigen of the whole *B. burgdorferi* is progressively diluted until a count of the antigen in that particular blood sample remains. So, the test can determine the extent of infection and may be able to distinguish the carriers from the patients with serious disease. Another advantage is the short time required to complete the test (within 24 h of receiving the specimen). Finally, the Q-RiBb is unaffected by whether the patient is currently (or recently has been) taking antibiotics.

Borrelia infections cause malfunctioning of the immune system. Coinfections only add to this problem. The antigenic heterogeneity of *B. burgdorferi* s.l. also influences the sensitivity and specificity of serological tests for borreliosis, especially on the European continent, where more pathogenic species of *B. burgdorferi* s.l. are recognized. The false-negative results occur if the infection goes into cyst form, which reduces the immune response, or into intracellular cell-wall deficient form, which can not be detected by antibodies. The antibodies can form the immune complexes, and so can not be detected by the test. The spirochetes can be encapsulated in host tissue (lymphocytic cell walls) or go deep into it, especially into tissue with a very poor blood supply, and thus hide from the immune system. Recent treatment with various drugs, including anti-inflammatories and antibiotics or other factors which cause immunosuppression can result in the false-negative serology.

96.1.4.2 Molecular Techniques

As serology has a number of limitations various molecular and biochemical approaches have been developed and used for detection and identification of Lyme borreliosis–related spirochetes in clinical samples. These methods include species-specific PCR and PCR-based restriction fragment length polymorphism (RFLP) analysis,[119] sequence analysis of different *Borrelia* genes, reverse-line blotting,[120] real-time PCR,[121] multilocus sequence analysis (MLSA),[7] multilocus sequence typing,[122] multilocus enzyme electrophoresis and DNA–DNA reassociation analysis,[123] rRNA gene restriction analysis (ribotyping),[123] pulsed-field gel electrophoresis and large restriction fragment length polymorphism (LRFLP),[124] plasmid fingerprinting,[125] randomly amplified polymorphic DNA fingerprinting analysis,[126] and others. Only a brief review of the PCR-based techniques is presented in this chapter.

PCR-based molecular techniques are employed for (i) confirmation of the clinical diagnosis of suspected Lyme borreliosis; (ii) identification and typing of spirochetes species in clinical specimens or in cultured isolates; and (iii) detection of coinfection within *B. burgdorferi* s.l. species as well as with other tick-borne pathogens. Under experimental conditions, the sensitivity of PCR is extremely high, and DNA of a single gene copy can be detected. Application of molecular typing methods for the classification of *B. burgdorferi* s.l. provides the framework for the systematic approach to characterizing the differences in infectivity as well as in pathogenicity between strains. Sensitive molecular typing techniques do not require large amounts of material or cultivation of spirochetes and thus play an increasingly important role in elucidation of the pathogenic potential of different *B. burgdorferi* genotypes.[82] Due to high heterogeneity of *B. burgdorferi*, the most important consideration in selecting a target for amplification is genetic stability. Loss or alteration of the target sequence may result in a loss of reactivity. PCR targets that have been recently utilized by laboratories as targets for PCR analysis to detect *B. burgdorferi* s.l. DNA include: *rrf-rrl* intergenic spacer, 16S rRNA, Osp A, Osp C, flagellin, p66, recA, rpoB, groEL, hbb, pepX, clpX, rplX, rplB, and many others (www.ncbi.nlm.nih.gov). Despite of the large number of target genes used to detect Lyme disease spirochetes in clinical specimens, the success of PCR is actually dependent on selecting the correct target for the specific clinical sample (tissue or body fluid) and disease stage. For example, the sensitivity of PCR targeting p66 gene in ACA patient samples is two times higher compared to 23S rRNA gene amplification.[127] The targeting of the OspA gene in biopsy samples from ACA patients is more sensitive than the targeting of the *rrf-rrl* intergenic spacer region.[60]

The sensitivity of PCR in clinical samples can be decreased by degradation of the *B. burgdorferi* s.l. DNA during sample transport, storage, and processing. It can be lowered by losses during DNA preparation, the presence of extraneous DNA (host DNA, in particular),[128] or by contamination with interfering inhibitory substances that may be present in various clinical samples (blood, urine, synovial fluid, and CSF). So the test sensitivity is dependent on technical details and the particular specimen used for analysis. The number of spirochetes in human body fluids is generally low, especially in late chronic disease.[129] The number of genomes in the urine or plasma of infected patients is generally less than 50 per mL and rarely exceeds 5000 per mL,[130] in CSF the number of organisms might even be lower.[131] PCR sensitivity is really high in detection of *B. burgdorferi* s.l. DNA in skin biopsy samples from patients with EM,[132] but such testing is rarely necessary, as a clinical diagnosis can be easily made if the characteristic skin lesion is present. In general, the sensitivity of PCR for detection of *B. burgdorferi* s.l DNA in blood, plasma, serum, or urine samples from patients with Lyme borreliosis was for a long time believed to be rather low. But many approaches have been established and optimized in recent years to use the PCR from body fluids as a diagnostic tool for Lyme borreliosis. The determination of the number of spirochetes present in infected patients by real-time quantitative PCR could contribute to an understanding the variability of disease manifestations,[133] although a lot of practical doctors consider this method to be more appropriate for laboratory use than for clinical diagnosis.

96.2 METHODS

96.2.1 Sample Preparation

There are no specific recommendations for sample preparation involving serological methods, except for some minor considerations. All samples should be fresh or freshly frozen. Repeated freezing and thawing of the samples should be avoided. Blood should be drawn using standard venipuncture techniques, and serum should be separated from the blood cells as soon as possible. Samples should be allowed to clot for 1 h at room temperature, centrifuged for 10 min at 4°C, and then the serum should be extracted. Plasma should be collected using sodium citrate, EDTA, or heparin as an anticoagulant. To ensure optimal recovery and minimal platelet contamination after collection there must be quick separation of plasma with less than 30 min on ice, followed by centrifugation for 10 min (4°C) to remove any particulates. Fresh serum or plasma samples, free from hemolysis should be used. Highly lipemic, icteric, or microbially contaminated sera or plasma samples or concentrated immunoglobulin preparations can lead to unreliable results.

Sample preparation significantly influences the subsequent PCR assay. Many methods have already been described, but almost all of them include a DNA purification step. Yet despite the positive features, any purification leads to a loss of DNA. Thus the method for purification of nucleic acids described by Boom[134] was adopted for daily use by the majority of laboratories. Since the target bacterial DNA present in clinical specimens occurs in rather low amounts, any loss of DNA during purification can bring undesirable complications. Numerous data have been published comparing purification techniques based on different strategies of DNA extraction and differing in cost, labor, and time consumption (see Reference 71). Briefly, the classical four-step process with the use of commercially available kits provided better results than the standard phenol–chloroform procedure. The spin-column technique is a three-step process that has the advantage of not requiring any chemical reagents for extraction. Microwave irradiation or resin-binding is an alternative approach that bypasses the need to perform the DNA extraction step. Using size-fractionated silica particles, this method allows small-scale DNA purification from different specimens. The anion-binding resin method (GeneFizz) is a single-tube, single-step process without any centrifugation; lysis and DNA extraction are accomplished directly in the amplification tube. An autoclave method preparing bacterial DNA for PCR eliminates the use of detergents, organic solvents and mechanical cellular disruption approaches, thereby significantly reducing processing time and cost.

All the above techniques represent relatively rapid, reliable, simple, cost- and time-saving methods that reduce the loss of material and the risk of cross contamination. All of them, however, have a similar limiting factor—the potential reduction of Taq-polymerase activity due to residual inhibitory substances that may interfere with enzyme reactions during PCR; this makes the methods less suitable for treating multiple clinical samples in routine detection of pathogen DNA.

Procedures that employ other-than-DNA purification steps have been developed; these include boiling the specimen,[135] centrifugation and boiling,[136] alkali lysis,[128] boiling and concentration by centrifugation and ultrafiltration,[137] or the use of a cation exchanger resin, Chelex 100.[138] The last method was optimized for use with clinical samples.[82] The sample preparation involved no DNA purification. The resin binds all possible inhibitors of the PCR reaction present in the samples. It was shown that 10 pg of pathogen DNA can be detected in the spiked human serum samples after double treatment with Chelex 100 resin.[82]

96.2.2 Detection Procedures

The most frequently used and reliable PCR-based procedures for *B. burgdorferi* detection are presented in this chapter.

96.2.2.1 Single-Round PCR Detection of *B. burgdorferi* Sensu Lato

Principle. Demaerschalck et al.[139] described a single-round PCR with oligonucleotide primers based on *B. burgdorferi* sl ospA gene sequences to type all or each genospecies involved in Lyme borreliosis. (SL primers: Forward: 5′-AATAGGTCTAATAATAGCCTTAATAGC-3′; Reverse: 5′-CTAGTGTTTTGCCATCT TCTTTGAAAA-3′) (GI for *B. burgdorferi* ss: Forward: 5′-AACAAAGACGGCAAGTACG ATCTAATT-3′; Reverse: 5′-TTACAGTAATTGTTAAAGTT GAAGTGCC-3′; GII for *B. garinii*: Forward: 5′-TGAT AAAAACAACGGTTCTGGAAC-3′; Reverse: 5′-GTAACT TTCAATGTTGTTTTGCCG-3′; GIII for *B. afzelii*: Forward: 5′-TAAAGACAAAACATCAACAGATGAAATG-3′; Reverse: 5′-TTCCAATGTTACTTTATCATTAGCTACTT-3′).

Procedure. PCR reaction is conducted according to protocol of Boehringer Mannheim Biochemica (Mannheim, Germany). A 100 ng of total genomic DNA is amplified for 35 cycles under the following conditions: 93°C for 1 min, 69°C for 1 min, and 72°C for 1 min. For the GIII primer set, the annealing step is performed at 68°C for 1 min. For the SL primer set, the annealing step is performed at 65°C for 1 min.

Note: The expected product size is 304 bp for *B. burgdorferi* sl, 544 bp for *B. burgdorferi* ss., 345 bp for *B. garinii*, and 190 bp for *B. afzelii*.

96.2.2.2 PCR–RFLP Analysis

Principle. Fukunaga et al.[119] designed the PCR primers for the amplification of an approximately 580-bp DNA fragment of borrelia *flagellin* gene. Restriction of amplicon with HapII, HhaI, CelII, HincII, or DdeI endonucleases results in different species-specific PCR-RFLP patterns. The primers used for amplification of partial *fla* gene are forward: 5′-GCAGTTCAATCAGGTAACGG-3′ and reverse: 5′-AGGTITTCAATAGCATACTC-3′.

Procedure. The PCR is conducted in a 50-μL (total volume) mixture containing 5 μL of DNA, 200 μM of each dNTP, 1 μM of each primer, 1.5 U of Taq polymerase (Ex Taq; Takara), and Ex Taq reaction buffer. The reaction mixtures are subjected to 30 thermal cycles consisting of 94°C for 40 s, 50°C for 40 s, and 72°C for 40 s. The amplified DNA is digested with *Hap*II, *Hha*I, *Hinc*II (Takara), *Cel*II (Pharmacia), or *Dde*I (TOYOBO, Japan) and electrophoresed in a 2.5% agarose gel (NuSieve GTG; FMC). All DNA bands are visualized by ethidium bromide staining. DNA ladders [100-bp ladder (100–1000 bp) and 20-bp ladder (20–1000 bp); GenSura Laboratories, United States] are used as size markers.

96.2.2.3 Identification of Genomic Groups of *B. burgdorferi* Sensu Lato by RLB

Principle. Rijpkema et al.[120] described a modification of the reverse dot blot for *B. burgdorferi*. After amplification of 5S and 23S rRNA intergenic spacer region by nested PCR the product is hybridized to a probe specific to all genomic groups of *B. burgdorferi* sl and to species-specific oligonucleotides. Primers for 1st PCR are: 23SN1: 5′-ACCATAGACTCTTATTACTTTGAC-3′ and 23SC1: 5′-TA AGCTGACTAATACTAATTACCC-3′. Primers for 2nd PCR are: 23SN2: 5′-ACCA TAGACTCT TATTACTTTGACCA-3′ and 5SCB: 5′-biotin-GAGAGTAG GTTATTGCCAGGG-3′.

Procedure. The first PCR is performed in a volume of 25 μL containing 0.625 U of Taq DNA polymerase, Saiki buffer supplemented with 2.5 mM $MgCl_2$, 200 μM of each dNTP, 5 pmol of each primers 23SN1 and 23SC, and 5 μL of the sample. The program for the first PCR consists of 1 round of 94°C for 1 min and 25 rounds of 94°C for 30 s, 52°C for 30 s, and 72°C for 1 min. For the second PCR, a 5-μL reaction mixture containing 0.625 U of Taq DNA polymerase, Saiki buffer supplemented with 2.5 mM $MgCl_2$, 1 mM (each) dNTP, and 5 pmol of biotin-labeled primer 5SCB and of primer 23SN2 is added to each tube. The PCR conditions include 1 round of 94°C for 1 min and 40 rounds of 94°C for 30 s, 55°C for 30 s, and 72°C for 1 min. The product of the first PCR has a size of approximately 380 bp, and DNA amplification by the second PCR results in amplicon of 225 bp. The probes (*B. burgdorferi* sl 5′-aCTTTGACCATATT TTTATCTTCCA-3′, *B. burgdorferi* ss 5′-aAACACCAATA TTTAAAAAACATAA-3′, *B. garinii* 5′-aAACATGAACA TCTAAAAACATAAA-3′, *B. afzelii* 5′-aAACATTTAAA AAATAAATTCAAGG-3′, and *B. valaisiana* 5′-aCATTA AAAAAATATAAAAAATAAATTTAAGG-3′) are covalently linked to an activated Bio-dyne C membrane by the 5′ aminolink group. Membrane is activated by incubation for 15 min in 16% 1-ethyl-3-(3-dimethyl-aminopropyl) carbodiimide (Merck), rinsed with water, and placed in a miniblotter system (Immunetics, Cambridge, Massachusetts). Five slots are filled with 150 μL of an oligonucleotide suspension, which consisted of 12.5–100 pmol of probe (Perkin-Elmer) in 500 mM NaHCO3 (pH 8.4). After 1 min of incubation, solution is aspirated, and the blot is removed from the blotter and inactivated by incubation with 100 mM NaOH for 10 min. After a rinse with water, the blot is incubated at 56°C in 2 × SSPE [360 mM NaCl, 20 mM NaH_2PO_4, 2 mM EDTA (pH 7.2)]—0.1% sodium dodecylsulfate (SDS) for 10 min, and rinsed in 2 × SSPE. The filter is again placed in the miniblotter and rotated 90 degrees to the previous position. The slots are filled with 150 μL of heat denatured amplicons (10 μL of the PCR mixture suspended in 140 μL of 2 × SSPE—0.1% SDS, boiled for 10 min, and chilled on ice) and incubated for 60 min at 45°C. The slots are aspirated; blot is removed from the miniblotter and washed twice for 10 min at 40°C with 2 × SSPE/0.5% SDS. Ten μL of streptavidin-peroxidase conjugate (Boehringer Mannheim GmbH) diluted 1:4000 in 2 × SSPE/0.5% SDS is added to the blot and incubated at 40°C for 30 min. Blot is rinsed twice with 2 × SSPE—0.5% SDS at 40°C and twice with 2 × SSPE. Bound streptavidin labeled PCR product is detected by chemiluminescence with the ECL detection system (Amersham), and visualized by exposure of the blot to an X-ray film (Hyperfilm; Amersham).

96.2.2.4 Real-Time PCR Detection

Principle. Pahl et al.[121] introduced real-time PCR assay for the quantification of *Borrelia burgdorferi*. This technology measures the release of fluorescent oligonucleotides during the PCR. *Flagellin* of *B. burgdorferi* was chosen as the target sequence. A linear quantitative detection range of 5 logs with a calculated detection limit of one to three spirochetes per assay reaction mixture is observed. Primers and probes were selected for the *flagellin* gene of *B. burgdorferi*: forward: 5′-TCTTTTCTCTGGTGAGGGAGCT-3′; reverse: 5′-TCCTT CCTGTTGAACACCCTCT-3′; internal probe: 5′-([FAM] AAACTGC[TAMRA]TCAGGCTG CACCGGTTC-3′). Probes were 5′ end labeled with 6-carboxy-fluorescein (FAM) and internally labeled with 6-carboxy-N,N,N9,N9-tetramethylrhodamine (TAMRA).

Procedure. The PCR mixture (25-μL total volume) consists of 300 nM primer, 200 nM probe, 200 μM deoxynucleoside triphosphates, 3.5 mM $MgCl_2$, 1 μL of DNA, 1 U of Ampli Taq Gold, and 1× PCR buffer [50 mM KCl, 10 μM EDTA, 10 mM Tris-HCl (pH 8.3)] (Perkin-Elmer, Germany). Amplification and detection are performed with an ABI 7700 system with the following profile: 95°C for 10 min and 45 cycles of 95°C for 15 s and 60°C for 1 min. The fluorescence is continuously monitored. Signals for each dye are extracted from the emission spectrum. The FAM signal is standardized to the passive reference ROX, which is included in the reaction buffer. Furthermore, the background fluorescence, which is calculated from cycles 3 to 10, is subtracted. Quantification is performed by determining the threshold cycle (C_t). This is defined as the cycle at which the FAM fluorescence exceeds 10 times the standard deviation of the mean baseline emission for cycles 3–10. Calculation of spirochete numbers is based on standard curves of threefold dilution series representing the plot of C_t values versus the log of copy numbers included in each PCR run.

96.3 CONCLUSION AND FUTURE PERSPECTIVES

In spite of considerable research progress on the biology of *B. burgdorferi* in the recent years, some features of this bacterium still remain unclear. A high level of awareness of the complex nature of Lyme borreliosis is desirable for its recognition, treatment, and prevention. But still the vast majority of publications focus on one species—*B. burgdorferi* sensu stricto. This may be due to the concentration of *Borrelia* research groups in the United States, where *B. burgdorferi* s.s. is still considered to be the only pathogenic species. That is why the creation of international networks of *Borrelia* researchers to combine their efforts on additional species is of great importance.

Lack of reliable standardized microbiologic or immunologic criteria for the diagnosis or cure of Lyme borreliosis hamper the development of optimal therapeutic modalities of this infection. There is a great need to develop additional simple, sensitive, and rapid procedures to correctly diagnose Lyme borreliosis and to distinguish those who are actively infected with *B. burgdorferi* from those who have recovered from a previous infection.

Since the genome of *B. burgdorferi* B31 has now been completely sequenced, and the sequencing of genomes of *B. afzelii* and *B. garinii* is almost completed, the application of this information by modern technologies is needed. For example, techniques such as microarrays and proteomics could improve diagnosis as well as provide new insights on the pathogenesis of Lyme borreliosis and pathogen-specific host response mechanisms. A combination of powerful experimental technologies and bioinformatics would help to identify from different *Borrelia* species common antigens that are pathogenic for humans and that would function as reliable markers for serodiagnosis.

Automation will be of great help in applying molecular methods of Lyme borreliosis diagnosis in clinical practice. Most of the procedures developed and used in research centers may not be easily adaptable to routine clinical work. The next generation of PCR equipment will probably include processors for automatic sample preparation, amplification, and detection. Several companies have already introduced such machines.[140] After standardization of sample preparation, selection of target genes, establishing the PCR and detection methods, physicians need to be trained to interpret results. No doubt the number of users of PCR tests will increase, and a new dimension in the diagnosis of *B. burgdorferi* infections will occur.

For many years, investigations on new anti-Lyme vaccines have been conducted. The first vaccine on the market (LYMErix) was based on recombinant OspA. Typically, the aim in making a vaccine is to recreate a normal immune response without exposing a patient to the pathological elements of a disease. LYMErix followed a completely different goal—to prevented transmission of the *B. burgdorferi* from arthropod vector to human host. Though it looked promising, in 2002, the manufacturer of the LYMErix pulled the vaccine off the market, blaming poor sales. However, the reasons for its withdrawal were much more complicated. Currently, research to develop a multivalent vaccine to increase vaccine disease-prevention effectiveness is being done. The availability of data on complete genome sequence from

B. burgdorferi now enables the search for vaccine candidates among the complete proteome of this pathogen. Development of vaccine against Lyme borreliosis using newly identified immunogenic proteins that are expressed during the infection and persistence of the pathogen in the host is useful goal. These genes could be ideal candidates for vaccine. Another generation of the vaccines—a vaccine against an arthropod vector—is now under investigation as well. As one vector may carry multiple different pathogens, vaccinating against the vector itself may prevent infection of multiple disease agents and be far more cost effective than producing vaccines for each of those pathogens. Though a lot of progress has been made in this field, there is still no reliable vaccine available as of 2009.

ACKNOWLEDGMENTS

This work was partially supported by the grants MSM 6007665801 and LC06009 from Ministry of Education, grant Z60220518 (NR, MG, LG) and grant 206/09/1782 of the Czech Science Foundation and also by grants CZB1-2963-CB-09 (MG) and CZB1-2966-CB-09 (NR) from Central Research and Development Foundation (CRDF).

REFERENCES

1. Burgdorfer, W. et al., Lyme disease-a tick-borne spirochetosis? *Science*, 216, 1317, 1982.
2. Oliver, J.H. Jr., et al., Conspecificity of the ticks *Ixodes scapularis* and *Ixodes dammini* (Acari: Ixodidae), *J. Med. Entomol.*, 30, 54, 1993.
3. Canica, M.M. et al., Monoclonal antibodies for identification of *Borrelia afzelii* sp. nov. associated with late cutaneous manifestations of Lyme borreliosis, *Scand. J. Infect. Dis.*, 25, 441, 1993.
4. Marconi, R.T., Liveris, D., and Schwartz, I., Identification of novel insertion elements, restriction fragment length polymorphism patterns, and discontinuous 23S rRNA in Lyme disease spirochetes: Phylogenetic analyses of rRNA genes and their intergenic spacers in *Borrelia japonica* sp. nov. and genomic group 21038 (*Borrelia andersonii* sp. nov.) isolates, *J. Clin. Microbiol.*, 33, 2427, 1995.
5. Postic, D. et al., Expanded diversity among Californian borrelia isolates and description of *Borrelia bissettii* sp. nov. (formerly Borrelia group DN127), *J. Clin. Microbiol.*, 36, 3497, 1998.
6. Baranton, G. et al., Delineation of *Borrelia burgdorferi* sensu stricto, *Borrelia garinii* sp. nov., and group VS461 associated with Lyme borreliosis, *Int. J. Syst. Bacteriol.*, 42, 378, 1992.
7. Postic, D., Garnier, M., and Baranton, G., Multilocus sequence analysis of atypical *Borrelia burgdorferi* sensu lato isolates-description of *Borrelia californiensis* sp. nov., and genomospecies 1 and 2, *Int. J. Med. Microbiol.*, 297, 263, 2007.
8. Rudenko, N. et al., *Borrelia carolinensis* sp.nov.—a new (14th) member of *Borrelia burgdorferi* sensu lato complex from the southeastern United States, *J. Clin. Microbiol.*, 47, 134, 2009.
9. Kawabata, H., Masuzawa, T., and Yanagihara, Y., Genomic analysis of *Borrelia japonica* sp. nov. isolated from *Ixodes ovatus* in Japan, *Microbiol. Immunol.*, 37, 843, 1993.
10. Le Fleche, A. et al., Characterization of *Borrelia lusitaniae* sp. nov. by 16S ribosomal DNA sequence analysis, *Int. J. Syst. Bacteriol.*, 47, 921, 1997.
11. Masuzawa, T. et al., *Borrelia sinica* sp. nov., a lyme disease-related Borrelia species isolated in China, *Int. J. Syst. Evol. Microbiol.*, 51, 1817, 2001.
12. Fukunaga, M. et al., *Borrelia tanukii* sp. nov. and *Borrelia turdae* sp. nov. found from ixodid ticks in Japan: rapid species identification by 16S rRNA gene-targeted PCR analysis, *Microbiol. Immunol.*, 40, 877, 1996.
13. Richter, D. et al., Delineation of *Borrelia burgdorferi* sensu lato species by multilocus sequence analysis and confirmation of the delineation of *Borrelia spielmanii* sp. nov., *Int. J. Syst. Evol. Microbiol.*, 56, 873, 2006.
14. Wang, G. et al., Genetic and phenotypic analysis of *Borrelia valaisiana* sp. nov. (Borrelia genomic groups VS116 and M19), *Int. J. Syst. Bacteriol.*, 47, 926, 1997.
15. Ge, Y. et al., Structure and expression of the FlaA periplasmic flagellar protein of *Borrelia burgdorferi*, *J. Bacteriol.*, 180, 2418, 1998.
16. Motaleb, M.A. et al., *Borrelia burgdorferi* periplasmic flagella have both skeletal and motility functions, *Proc. Natl. Acad. Sci. USA*, 97, 10899, 2000.
17. Barbour, A.G., and Hayes, S.F., Biology of borrelia species, *Microbiol. Rev.*, 50, 381, 1986.
18. MacDonald, A.B., Spirochetal cyst forms in neurodegenerative disorders, hiding in plain sight, *Med. Hypotheses*, 67, 819, 2006.
19. MacDonald, A.B., Plaques of Alzheimer's disease originate from cysts of *Borrelia burgdorferi*, the Lyme disease spirochete, *Med. Hypotheses*, 67, 592, 2006.
20. Aberer, E., and Duray, P.H., Morphology of *Borrelia burgdorferi*: Structural patterns of cultured Borreliae in relation to staining methods, *J. Clin. Microbiol.*, 29, 764, 1991.
21. Kurtti, T.J. et al., Colony formation and morphology in *Borrelia burgdorferi*, *J. Clin. Microbiol.*, 25, 2054, 1987.
22. Brorson, O., and Brorson, S.H., Transformation of cystic forms of *Borrelia burgdorferi* to normal, mobile spirochetes, *Infection*, 25, 240, 1997.
23. Fraser, C.M. et al., Genomic sequence of a Lyme disease spirochaete, *Borrelia burgdorferi*, *Nature*, 390, 580, 1997.
24. Casjens, S. et al., A bacterial genome in flux: The twelve linear and nine circular extrachromosomal DNAs in an infectious isolate of the Lyme disease spirochete *Borrelia burgdorferi*, *Mol. Microbiol.*, 35, 490, 2000.
25. Grubhoffer, L. et al., Lyme borreliosis: Insights into tick-/host-borrelia relations, *Folia Parasitol. (Praha)*, 52, 279, 2005.
26. de Silva, A.M., Tyson, K.R., and Pal, U., Molecular characterization of the tick-Borrelia interface, *Front. Biosci.*, 14, 3051, 2009.
27. Schwan, T.G. et al., Induction of an outer surface protein on *Borrelia burgdorferi* during tick feeding, *Proc. Natl. Acad. Sci. USA*, 92, 2909, 1995.
28. Stevenson, B., Schwan, T.G., and Rosa P.A., Temperature-related differential expression of antigens in the Lyme disease spirochete, *Borrelia burgdorferi*, *Infect. Immun.*, 63, 4535, 1995.
29. Indest, K.J. et al., Cell-density-dependent expression of *Borrelia burgdorferi* lipoproteins in vitro, *Infect. Immun.*, 65, 1165, 1997.
30. Carroll, J.A., Garon, C.F., and Schwan, T.G., Effects of environmental pH on membrane proteins in *Borrelia burgdorferi*, *Infect. Immun.*, 67, 3181, 1999.
31. Ojaimi, C. et al., *Borrelia burgdorferi* gene expression profiling with membrane-based arrays, *Methods Enzymol.*, 358, 165, 2002.
32. Revel, A.T., Talaat, A.M., and Norgard, M.V., DNA microarray analysis of differential gene expression in *Borrelia burgdorferi*, the Lyme disease spirochete, *Proc. Natl. Acad. Sci. USA*, 99, 1562, 2002.

33. Ojaimi, C. et al., Profiling of temperature-induced changes in *Borrelia burgdorferi* gene expression by using whole genome arrays, *Infect. Immun.*, 71, 1689, 2003.
34. Byram, R., Stewart, P.E., and Rosa P., The essential nature of the ubiquitous 26-kilobase circular replicon of *Borrelia burgdorferi*, *J. Bacteriol.*, 186, 3561, 2004.
35. Stewart, P.E. et al., Delineating the requirement for the *Borrelia burgdorferi* virulence factor OspC in the mammalian host, *Infect. Immun.*, 74, 3547, 2006.
36. Jayaram, M., Split target specificity of ResT: A design for protein delivery, site selectivity and regulation of enzyme activity? *Mol. Microbiol.*, 64, 575, 2007.
37. Rosa, P.A., Tilly, K., and Stewart, P.E., The burgeoning molecular genetics of the Lyme disease spirochaete, *Nat. Rev. Microbiol.*, 3, 129, 2005.
38. Purser, J.E., and Norris, S.J., Correlation between plasmid content and infectivity in *Borrelia burgdorferi*, *Proc. Natl. Acad. Sci. USA*, 97, 13865, 2000.
39. Actis, L.A., Tolmasky, M.E., and Crosa, J.H., Bacterial plasmids: Replication of extrachromosomal genetic elements encoding resistance to antimicrobial compounds, *Front Biosci.*, 4, D43, 1999.
40. Eggers, C.H., and Samuels, D.S., Molecular evidence for a new bacteriophage of *Borrelia burgdorferi*, *J. Bacteriol.*, 181, 7308, 1999.
41. Eggers, C.H. et al., Transduction by f BB-1, a bacteriophage of *Borrelia burgdorferi*, *J. Bacteriol.*, 183, 4771, 2001.
42. Weber, K., Aspects of Lyme borreliosis in Europe, *Eur. J. Clin. Microbiol. Infect. Dis.*, 20, 6, 2001.
43. Kristoferitsch, W., Lyme borreliosis in Europe, *Rheum. Dis. Clin. North. Am.*, 15, 767, 1989.
44. Steere, A.C. et al., Erythema chronicum migrans and Lyme arthritis. The enlarging clinical spectrum, *Ann. Intern. Med.*, 86, 685, 1977.
45. Lane, R.S., Piesman, J., and Burgdorfer, W., Lyme borreliosis: Relation of its causative agent to its vectors and hosts in North America and Europe, *Annu. Rev. Entomol.*, 36, 587, 1991.
46. Oliver, J.H. Jr., Lyme borreliosis in the southern United States: A review, *J. Parasitol.*, 82, 926, 1996.
47. Oliver, J.H. Jr. et al., An enzootic transmission cycle of Lyme borreliosis spirochetes in the southeastern United States, *Proc. Natl. Acad. Sci. USA*, 100, 11642, 2003.
48. Santino, I. et al., Detection of four *Borrelia burgdorferi* genospecies and first report of human granulocytic ehrlichiosis agent in *Ixodes ricinus* ticks collected in central Italy, *Epidemiol. Infect.*, 129, 93, 2002.
49. Hubálek, Z., Epidemiology of lyme borreliosis, *Curr. Probl. Dermatol.*, 37, 31, 2009.
50. Stünzner, D. et al., Prevalence of *Borrelia burgdorferi* sensu lato in the tick *Ixodes ricinus* in the Styrian mountains of Austria, *Wien Klin. Wochenschr.*, 118, 682, 2006.
51. Lindgren, E., and Jaenson, T.G.T., Lyme borreliosis in Europe: Influences of climate and climate change, epidemiology and adaptation measures. In Menne, B., and Ebi, K.L. (Editors), *Climate Change and Adaptation Strategies for Human Health*, pp. 157–188, Steinkopff, Darmstadt, 2006.
52. Campbell, G.L. et al., Estimation of the incidence of Lyme disease, *Am. J. Epidemiol.*, 148, 1018, 1998.
53. Strle, F., and Stanek, G., Clinical manifestations and diagnosis of Lyme borreliosis, *Curr. Probl. Dermatol.*, 37, 51, 2009.
54. Hansmann, Y., Treatment and prevention of Lyme disease, *Curr. Probl. Dermatol.*, 37, 111, 2009.
55. Wormser, G.P., Clinical practice. Early Lyme disease, *N. Engl. J. Med.*, 354, 2794, 2006.
56. Ziska, M.H., Donta, S.T., and Demarest, F.C., Physician preferences in the diagnosis and treatment of Lyme disease in the U.S., *Infection*, 23, 1, 1995.
57. Fallon, B., and Nields, J.A., Lyme disease: A neuropsychiatric illness, *Am. J. Psych.*, 141, 1571, 1994.
58. Chabria, S.B., and Lawrason, J., Altered mental status, an unusual manifestation of early disseminated Lyme disease: A case report, *J. Med. Case Rep.*, 1, 62, 2007.
59. Picken, R.N. et al., Patient isolates of *Borrelia burgdorferi* sensu lato with genotypic and phenotypic similarities of strain 25015, *J. Infect. Dis.*, 174, 1112, 1996.
60. Rijpkema, S.G. et al., Detection of *Borrelia afzelii*, *Borrelia burgdorferi* sensu stricto, *Borrelia garinii* and group VS116 by PCR in skin biopsies of patients with erythema migrans and acrodermatitis chronica atrophicans, *Clin. Microbiol. Infect.*, 3, 109, 1997.
61. Strle, F. et al., Clinical findings for patients with Lyme borreliosis causing by *Borrelia burgdorferi* sensu lato with genotypic and phenotypic similarities to strain 25015, *Clin. Infect. Dis.*, 25, 273, 1997.
62. Wang, G. et al., Molecular typing of *Borrelia burgdorferi* sensu lato: Taxonomic, epidemiological, and clinical implications, *Clin. Microbiol. Rev.*, 2, 633, 1999.
63. Ryffel, K. et al., Scored antibody reactivity determined by immunoblotting shows an association between clinical manifestations and presence of *Borrelia burgdorferi* sensu stricto, *B. garinii*, *B. afzelii*, and *B. valaisiana* in humans, *J. Clin. Microbiol.*, 37, 4086, 1999.
64. Collares-Pereira, M. et al., First isolation of *Borrelia lusitaniae* from a human patient, *J. Clin. Microbiol.*, 42, 1316, 2004.
65. Rudenko, N. et al., Detection of *Borrelia bissettii* in cardiac valve tissue of a patient with endocarditis and aortic valve stenosis in the Czech Republic, *J. Clin. Microbiol.*, 46, 35, 2008.
66. Rudenko, N. et al., Molecular detection of *Borrelia bissettii* DNA in serum samples from patients in the Czech Republic with suspected borreliosis, *FEMS Microbiol. Lett.*, 292, 274, 2009.
67. Steere, A.C., Lyme disease, *N. Engl. J. Med.*, 263, 201, 1989.
68. Oschmann, P. et al., Stages and syndromes of neuroborreliosis, *J. Neurol.*, 245, 262, 1998.
69. Grange, F. et al., *Borrelia burgdorferi*-associated lymphocytoma cutis simulating a primary cutaneous large B-cell lymphoma, *J. Am. Acad. Dermatol.*, 47, 530, 2002.
70. van Dam, A.P. et al., Different genospecies of *Borrelia burgdorferi* are associated with distinct clinical manifestations of Lyme borreliosis, *J. Clin. Infect. Dis.*, 17, 708, 1993.
71. Maraspin, V., Ruzic-Sabljic, E., and Strle, F., Lyme borreliosis and *Borrelia spielmanii*, *Emerg. Infect. Dis.*, 12, 1177, 2006.
72. Anderson, J.F., Barthold, S.W., and Magnarelli, L.A., Infectious but nonpathogenic isolate of *Borrelia burgdorferi*, *J. Clin. Microbiol.*, 28, 2693, 1990.
73. Fikrig, E. et al., *Borrelia burgdorferi* strain 25015: Characterization of outer surface protein A and vaccination against infection, *J. Immunol.*, 148, 2256, 1992.
74. Schriefer, B.S. et al., *Borrelia bissettii* isolates induce pathology in a murine model of disease, *Vector Borne Zoonotic Dis.*, 8, 623, 2008.
75. Strle, F., Lyme borreliosis in Slovenia, *Zentralbl Bakteriol.*, 289, 643, 1999.
76. Strle, F. et al., Clinical and epidemiological findings for patients with erythema migrans. Comparison of cohorts from the years 1993 and 2000, *Wien Klin. Wochenschr.*, 114, 493, 2002.
77. Dandache, P., and Nadelman, R.B., Erythema migrans, *Infect. Dis. Clin. North Am.*, 22, 235, 2008.

78. Nadelman, R.B., and Wormser, G.P., Lyme borreliosis, *Lancet*, 352, 557, 1998.
79. Steere, A.C., Lyme disease, *N. Engl. J. Med.*, 345, 115, 2001.
80. Hubalek, Z., and Halouzka, J., Distribution of *Borrelia burgdorferi* sensu lato genomic groups in Europe, a review, *Eur. J. Epidemiol.*, 13, 957, 1997.
81. Saint Girons, I. et al., Identification of *Borrelia burgdorferi* sensu lato species in Europe, *Zentbl. Bakteriol.*, 287, 190, 1998.
82. Rudenko, N. et al., Improved method of detection and molecular typing of *Borrelia burgdorferi* sensu lato in clinical samples by polymerase chain reaction without DNA purification, *Folia Microbiol.*, 50, 31, 2005.
83. Craft, J.E., Grodzicki, R.L., and Steere, A.C., Antibody response in Lyme disease: Evaluation of diagnostic tests, *J. Infect. Dis.*, 149, 789, 1984.
84. Kmety, E., Dynamics of antibodies in *Borrelia burgdorferi* sensu lato infections, *Bratisl Lek Listy.*, 101, 5, 2000.
85. Agger, W.A., and Case, K.L., Clinical comparison of borreliacidal-antibody test with indirect immunofluorescence and enzyme-linked immunosorbent assays for diagnosis of Lyme disease, *Mayo Clin. Proc.*, 72, 510, 1997.
86. Rosa, P.A., and Schwan, T.G., A specific and sensitive assay for the Lyme disease spirochete *Borrelia burgdorferi* using the polymerase chain reaction, *J. Infect. Dis.*, 160, 1018, 1989.
87. von Stedingk, L.V. et al., Polymerase chain reaction for detection of *Borrelia burgdorferi* DNA in skin lesions of early and late Lyme borreliosis, *Eur. J. Clin. Microbiol. Infect. Dis.*, 14, 1, 1995.
88. Masters, E.J., Grigery, C.N., and Masters, R.W., STARI, or Masters disease: Lone Star tick-vectored Lyme-like illness, *Infect. Dis. Clin. North Am.*, 22, 361, 2008.
89. Kelly, R., Cultivation of *Borrelia hermsii*, *Science*, 173, 443, 1971.
90. Barbour, A.G., Isolation and cultivation of Lyme disease spirochetes, *Yale J. Biol. Med.*, 57, 521, 1984.
91. Pollack, R.J., Telford, III S.R., and Spielman, A., Standardization of medium for culturing Lyme disease spirochetes, *J. Clin. Microbiol.*, 31, 1251, 1993.
92. Preac-Mursic, V., Wilske, B., and Schierz, G., European *Borrelia burgdorferi* isolated from humans and ticks culture conditions and antibiotic susceptibility, *Zentbl. Bakteriol. Mikrobiol. Hyg.*, 263, 112, 1986.
93. Aguero-Rosenfeld, M.E. et al., Diagnosis of Lyme Borreliosis, *Clin. Microbiol. Rev.*, 18, 484, 2005.
94. Preac-Mursic, V., Wilske, B., and Reinhardt, S., Culture of *Borrelia burgdorferi* on six solid media, *Eur. J. Clin. Microbiol. Infect. Dis.*, 10, 1076, 1991.
95. Berger, B.W. et al., Cultivation of *Borrelia burgdorferi* from erythema migrans lesions and perilesional skin, *J. Clin. Microbiol.*, 30, 359, 1992.
96. de Koning, J., Bosma, R.B., and Hoogkamp-Korstanje, J.A., Demonstration of spirochaetes in patients with Lyme disease with a modified silver stain, *J. Med. Microbiol.*, 23, 261, 1987.
97. Johnston, Y.E. et al., Lyme arthritis. Spirochetes found in synovial microangiopathic lesions, *Am. J. Pathol.*, 118, 26, 1985.
98. Snydman, D.R. et al., *Borrelia burgdorferi* in joint fluid in chronic Lyme arthritis, *Ann. Intern. Med.*, 104, 798, 1986.
99. Pennell, D.R., Wand, P.J., and Schell, R.F., Evaluation of a quantitative fluorescence immunoassay (FIAX) for detection of serum antibody to *Borrelia burgdorferi*, *J. Clin. Microbiol.*, 25, 2218, 1987.
100. Wormser, G.P., Aguero-Rosenfeld, M.E., and Nadelman, R.B., Lyme disease serology: Problems and opportunities, *JAMA*, 282, 79, 1999.
101. Pavia, C.S. et al., An indirect hemagglutination antibody test to detect antibodies to *Borrelia burgdorferi* in patients with Lyme disease, *J. Microbiol. Methods*, 40, 163, 2000.
102. Creson, J.R. et al., Detection of anti-*Borrelia burgdorferi* antibody responses with the borreliacidal antibody test, indirect fluorescent-antibody assay performed by flow cytometry and Western immunoblotting, *Clin. Diagn. Lab. Immunol.*, 3, 184, 1996.
103. Bakken, L.L. et al., Performance of 45 laboratories participating in a proficiency testing program for Lyme disease serology, *JAMA*, 268, 891, 1992.
104. Hedberg, C.W., and Osterholm, M.T., Serologic tests for antibody to *Borrelia burgdorferi*: Another Pandora's box for medicine? *Arch. Intern. Med.*, 150, 732, 1990.
105. Hedberg, C.W. et al., An interlaboratory study of antibody to *Borrelia burgdorferi*, *J. Infect. Dis.*, 155, 1325, 1987.
106. Luger, S.W., and Krauss, E., Serologic tests for Lyme disease: Interlaboratory variability, *Arch. Intern. Med.*, 150, 761, 1990.
107. Schwartz, B.S. et al., Antibody testing in Lyme disease: A comparison of results in four laboratories, *JAMA*, 262, 3431, 1989.
108. Wormser, G.P. et al., A limitation of 2-stage serological testing for Lyme disease: Enzyme immunoassay and immunoblot assay are not independent tests, *Clin. Infect. Dis.*, 30, 545, 2000.
109. Wilske, B. et al., MIQ 12 Lyme-Borreliose. In Mauch, H., and Lutticken, R. (Editors), *Qualitatsstandards in der Mikrobiologisch-Infektiologischen Diagnostik*, pp. 1–59, Urban and Fischer Verlag, Munich, Germany, 2000.
109a. Centers for Disease Control and Prevention (CDC). Recommendations for test performance and interpretation from the Second National Conference on Serologic Diagnosis of Lyme Disease. *MMWR Morb. Mortal Wkly Rep.*, 44, 590, 1995.
110. Engstrom, S.M., Shoop, E., and Johnson, R.C., Immunoblot interpretation criteria for serodiagnosis of early Lyme disease, *J. Clin. Microbiol.*, 33, 419, 1995.
111. Dressier, F. et al., Western blotting in the serodiagnosis of Lyme disease, *J. Infect. Dis.*, 167, 392, 1993.
112. Hauser, U., Lehnert, G., and Wilske, B., Validity of interpretation criteria for standardized Western blots (Immunoblots) for serodiagnosis of Lyme Borreliosis based on sera collected throughout Europe, *J. Clin. Microbiol.*, 37, 2241, 1999.
113. Callister, S.M. et al., Characterization of the borreliacidal antibody response to *Borrelia burgdorferi* in humans: A serodiagnostic test, *J. Infect. Dis.*, 167, 158, 1993.
114. Callister, S.M. et al., Detection of borreliacidal antibodies by flow cytometry. An accurate, highly specific serodiagnostic test for Lyme disease, *Arch. Intern. Med.*, 154, 1625, 1994.
115. Agger, W.A., and Case, K.L., Clinical comparison of borreliacidal-antibody test with indirect immunofluorescence and enzyme-linked immunosorbent assays for diagnosis of Lyme disease, *Mayo Clin. Proc.*, 72, 510, 1997.
116. Coyle, P.K. et al., Cerebrospinal fluid immune complexes in patients exposed to *Borrelia burgdorferi*: Detection of Borrelia-specific and -nonspecific complexes, *Ann. Neurol.*, 28, 739, 1990.
117. Schutzer, S.E. et al., Sequestration of antibody to *Borrelia burgdorferi* in immune complexes in seronegative Lyme disease, *Lancet*, 335, 312, 1990.

118. Zhong, W., Oschmann, P., and Wellensiek, H.J., Detection and preliminary characterization of circulating immune complexes in patients with Lyme disease, *Med. Microbiol. Immunol. (Berlin)*, 186, 153, 1997.
119. Fukunaga, M. et al., Phylogenetic analysis of Borrelia species based on flagellin gene sequences and its application for molecular typing of Lyme disease borreliae, *Int. J. Syst. Bacteriol.*, 46, 898, 1996.
120. Rijpkema, S.G. et al., Simultaneous detection and genotyping of three genomic groups of *Borrelia burgdorferi* sensu lato in Dutch *Ixodes ricinus* ticks by characterization of the amplified intergenic spacer region between 5S and 23S rRNA genes, *J. Clin. Microbiol.*, 33, 3091, 1995.
121. Pahl, A. et al., A.quantitative detection of *Borrelia burgdorferi* by real-time PCR, *J. Clin. Microbiol.*, 37, 1958, 1999.
122. Margos, G. et al., MLST of housekeeping genes captures geographic population structure and suggests a European origin of *Borrelia burgdorferi*, *Proc. Natl. Acad. Sci. USA*, 105, 8730, 2008.
123. Stackebrandt, E., and Goebel, B.M., Taxonomic note: A place for DNA–DNA reassociation and 16S rRNA sequence analysis in the present species definition in bacteriology, *Int. J. Syst. Bacteriol.*, 44, 846, 1994.
124. Belfaiza, J. et al., Genomic fingerprinting of *Borrelia burgdorferi* sensu lato by pulsed-field gel electrophoresis, *J. Clin. Microbiol.*, 31, 2873, 1993.
125. Rosa, P.A., Hogan, D., and Margolis, N., Molecular analysis of the major outer surface protein locus from a divergent *Borrelia burgdorferi* isolate from Europe. In Schutzer, S.E. (Editor), *Lyme Disease: Molecular and Immunologic Approaches*, pp. 95–110, Cold Spring Harbor Laboratory, Cold Spring Harbor, NY, 1992.
126. Williams, J.G.K. et al., DNA polymorphisms amplified by arbitrary primers are useful as genetic markers, *Nucleic Acids Res.*, 18, 6531, 1990.
127. Brettschneider, S. et al., Diagnostic value of PCR for detection of *Borrelia burgdorferi* in skin biopsy and urine samples from patients with skin borreliosis, *J. Clin. Microbiol.*, 36, 2658, 1998.
128. McGuire, B.S. et al., Detection of *Borrelia burgdorferi* in human blood and urine using the polymerase chain reaction, *Pathobiology*, 60, 163, 1992.
129. Schmidt, B.L., PCR in laboratory diagnosis of human *Borrelia burgdorferi* infections, *Clin. Microbiol. Rev.*, 10. 185, 1997.
130. Goodman, J.L. et al., Bloodstream invasion in early Lyme disease: results from a prospective, controlled, blinded study using the polymerase chain reaction, *Am. J. Med.*, 99, 6, 1995.
131. Lebech, A.M., and Hansen, K., Detection of *Borrelia burgdorferi* DNA in urine samples and cerebrospinal fluid samples from patients with early and late Lyme neuroborreliosis by polymerase chain reaction, *J. Clin. Microbiol.*, 30, 1646, 1992.
132. Dumler, J.S., Molecular diagnosis of Lyme disease: Review and meta-analysis, *Mol. Diagn.*, 6, 1, 2001.
133. Liveris, D. et al., Quantitative detection of *Borrelia burgdorferi* in 2-millimeter skin samples of erythema migrans lesions: Correlation of results with clinical and laboratory findings, *J. Clin. Microbiol.*, 40, 1249, 2002.
134. Boom, R. et al., Rapid and simple method for purification of nucleic acids, *J. Clin. Microbiol.*, 28, 495, 1990.
135. Krüger, W.H., and Pulz, M., Detection of *Borrelia burgdorferi* in cerebrospinal fluid by the polymerase chain reaction, *J. Med. Microbiol.*, 35, 98, 1991.
136. Debue, M. et al., Detection of *Borrelia burgdorferi* in biological samples using the polymerase chain reaction assay, *Res. Microbiol.*, 142, 565, 1991.
137. Pachner, A.R., and Delaney, E., The polymerase chain reaction in the diagnosis of Lyme neuroborreliosis, *Ann. Neurol.*, 34, 544, 1993.
138. Walsh, P.S., Metzger, D.A., and Higuchi, R., Chelex 100 as a medium for simple extraction of DNA for PCR-based typing from forensic material, *Biotechniques*, 10, 506, 1991.
139. Demaerschalck, I. et al., Simultaneous presence of different *Borrelia burgdorferi* genospecies in biological fluids of Lyme disease patients, *J. Clin. Microbiol.*, 33, 602, 1995.
140. Jungkind, D., Tech.Sight. Molecular testing for infectious disease, *Science*, 294, 1553, 2001.
141. Chu, C.Y. et al., Novel genospecies of *Borrelia burgdorferi* sensu lato from rodents and ticks in southwestern China, *J. Clin. Microbiol.*, 46, 3130, 2008.
142. Rudenko, N. et al., Delineation of a new species of the *Borrelia burgdorferi* sensu lato complex, *Borrelia americana* sp. nov., *J. Clin. Microbiol.*, 47, 3875, 2009.
143. Margos, G. et al., A new borrelia species defined by multilocus sequence analysis of housekeeping genes, *Appl. Environ. Microbiol.*, 75, 5410, 2009.

97 *Leptospira*

Rudy A. Hartskeerl

CONTENTS

97.1 INTRODUCTION

Leptospirosis is an emerging infectious disease of global importance, as illustrated by recent large outbreaks on virtually all continents. The disease is caused by pathogenic leptospires and characterized by a broad spectrum of clinical manifestations, varying from unapparent infection to fulminant, life-threatening disease. In its mild form, leptospirosis may present with fever, headache, and myalgias. Severe leptospirosis, characterized by jaundice, renal disorders, and hemorrhagic diathesis, is referred to as Weil's syndrome. Weil's syndrome is associated with case fatality rates (CFR) ranging from 5% to 20%. Either concomitant or not, severe pulmonary hemorrhage syndrome (SPHS) is worldwide emerging and, with a CFR exceeding 70%, presents the most severe form of the disease.[1] Because of the wide variety of clinical manifestations, mostly uncharacteristic, leptospirosis mimics many other diseases such as influenza; hepatitis; dengue; Hantavirus infections and other viral hemorrhagic fevers; yellow fever, malaria, typhoid fever and other enteric diseases; and pneumonia.[2–7] Therefore, it is frequently confused with any of these other diseases that often are endemic and epidemic under the same ecological and climatologic conditions. Confirmation of clinically suspected leptospirosis cases at the laboratory also has many bottlenecks. Standard tests, such as culturing and the microscopic agglutination test (MAT) are tedious, laborious, and require well-equipped laboratories with experienced staff. Therefore, performance of these tests is restricted to a few "expert" centers. Novel or adapted simplified diagnostic tests are badly needed. Several rapid tests for human use are currently available.[8] Meaningful comparisons to ascertain their relative merits as diagnostic tools are scarce,[9–11] and performance appears to vary in distinct regions.[12] Therefore, these tests are for screening purposes only, and results must be confirmed or locally evaluated by standard tests. In addition, serological tests are applicable only at a later stage of disease when effective treatment with antibiotics is less effective. The lack of awareness is mainly due to the uncharacteristic clinical manifestations and the technically demanding laboratory tests, making the disease difficult to diagnose. This makes leptospirosis one of the most overlooked and neglected infectious diseases. Worldwide surveys performed by the International Leptospirosis Society (ILS) revealed that from 300,000 to 500,000 recorded severe leptospirosis cases occurred annually.[13,14] Hantavirus infections and dengue are viral hemorrhagic fevers that are frequently confused with leptospirosis[6,7,15] and that benefit from more attention. Comparison of global numbers of severe cases and CFRs of leptospirosis with those of dengue and Hantavirus infections, shows that leptospirosis is at least equal to both as a public health hazard.[15,16] It should be noted that notification systems are lacking in most high-prevalence countries and that usually only hospitalized cases of Weil's syndrome are recognized. The unbalanced reporting to ILS by predominantly low-prevalence partners probably caused a severe underestimation of the actual annual number of severe global cases. Thus the true spread and increase of leptospirosis cases are as yet unknown.

Pathogenic leptospires live in the kidneys of natural hosts. A wide range of mammalian species are carriers. The leptospires are excreted with the urine into the environment where they can survive for several months, depending on favorable conditions (warm and humid climates). Infection of accidental hosts, including humans, occurs via direct contact with the carrier's urine or indirectly through a urine contaminated environment. Unlike natural hosts, accidental hosts often develop disease. The severity of the disease depends on a number of unknown and known factors, among which are the infecting serovar and the host's immune status. Humans are dead-end hosts. Human-to-human transmission is only rarely recorded.

The epidemiology of leptospirosis is complex and dynamic. Multiple serovars have been identified, each adapted to one or more hosts. Adaptation is a dynamic process changing the spatial and temporal distribution of serovars and clinical manifestations in accidental hosts. Also climatologic (global warming, El Niño) and ecologic changes (introduction of new animals and crops, expanding cities, deforestation) will affect the distribution of *Leptospira* serovars and consequently the prevalence and clinical features of leptospirosis cases, while human practices and animal management systems are likely to determine exposure and infection risks.

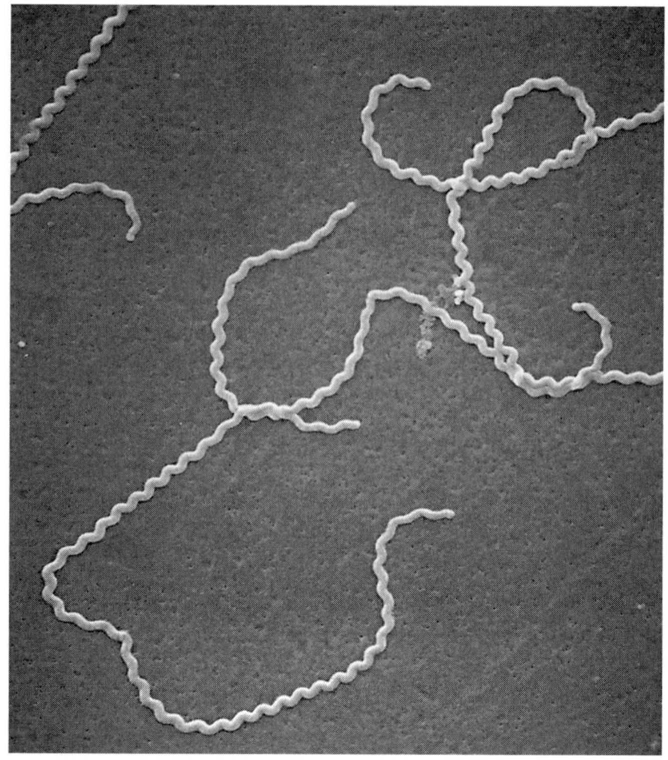

FIGURE 97.1 Electron microscopic image of *Leptospira interrogans.*

97.1.1 Classification, Morphology, and Genome Structure

The genus *Leptospira* belongs to the family Leptospiracea, order Spirochetales. Conventional classification separated the genus into two species, that is, *Leptospira interrogans* representing the pathogenic leptospires and *L. biflexa* comprising free-living (saprophytic) ones. Leptospires are 6–20 μm long and usually 0.1 μm thick (Figure 97.1). They stain poorly with Giemsa but can be seen microscopically by dark-field examination and after silver impregnation staining. They have a tightly corkscrewed form with one or two hooked ends; "interrogans" refers to the form of a question mark, and "biflexa" indicates two hooks. Leptospires cannot be differentiated on the basis of morphological features. Discrimination between pathogenic and saprophytic leptospires is possible by differences in growth characteristics established by applying distinct differential culture conditions[17,18]; by virulence testing in experimental animals, usually hamsters,[18] or by PCR.[19,20] This chapter focuses on pathogenic leptospires.

Currently there are two distinct classification systems that have a poor association (Table 97.1), that is, the conventional serological classification and the relatively new DNA-based classification.

In the serological classification, the serovar is the basic taxon. The serovar is determined by a strict serological criterion defined by a technique called the cross agglutinin absorption test (CAAT).[21] CAAT is a tedious and time-consuming approach that is confined to a few reference centers. To date, more than 250 pathogenic serovars are recognized.[21] Serologically related serovars have been placed into 26 serogroups, including an "undefined" group.[22] These groups do not have a formal taxonomic status but are supportive in the serological characterization. In addition, factor analysis,[23] a fine-tuning of the CAAT method, has been applied to identify several serovars. Monoclonal antibodies provide an easy and quick approach to type *Leptospira* serovars. However, the use of mAbs has its limitations as the method cannot be applied on all known serovars,[24] is subjective to a degree[2] and does not always give unambiguous results.[25]

DNA-based or genotypic classification measures DNA relatedness by the hydroxyapatite method[26,27] or by Southern hybridization[28] and clusters related strains in DNA hybridization groups, here referred to as "species." Currently, nine pathogenic species are known. These hybridization methods are complicated and use isotope-labeled DNA, thus undermining a broad applicability. Therefore, numerous alternative methods have been developed to assess DNA traits for speciation. These methods include bacterial restriction-endonuclease DNA analysis (BRENDA) either with[29,30] or without[31–33] a subsequent Southern hybridization step. Pulsed-field gel electrophoresis (PFGE) is an application of BRENDA in which DNA is digested by rare-cutting restriction endonucleases into large fragments that are separated on agarose gels by a special electrophoresis technique. Thus, relatively simple patterns are generated that largely coincide with established serovar states.[34,35] Gravekamp and coworkers[36]

TABLE 97.1
Species with Major Serogroups

Species	Major Serogroups Present
	***L. interrogans* sensu lato**
L. interrogans sensu stricto	Australis, Autumnalis, Canicola, Icterohemorrhagiae, Pomona, Pyrogenes, Sejroe
L. santarosai	Hebdomadis, Mini, Pyrogenes, Sejroe, Tarrassovi
L. borgpetersenii	Ballum, Javanica, Sejroe, Tarassovi
L. kirschneri	Autumnalis, Grippotyphosa
L. noguchi	Australis, Icterohemorrhagiae
L. weilii	Celledoni, Javanica, Tarassovi
L. alexanderi	Hebdomadis, Manhao
L. alstonii	Ranarum
L. fainei[a]	Hurstbridge
L. meyeri[a]	Javanica, Mini, Semaranga
L. inadai[a]	Lyme, Manhao
	***L. biflexa* sensu lato**
L. biflexa sensu stricto[b]	Andamana, Semaranga
L. wolbachii[b]	Codice
L. vanthielii[b]	Tentative serogroup Holland
L. terpstrae[b]	Icterohemorrhagiae
L. yanagawae[b]	Semaranga

[a] Species containing strains with a questionable pathogenic status (intermediate group) or with both pathogenic and saprophytic strains.

[b] Saprophytic species.

Note: *L. alstonii*, *L. vanthielii*, *L. terpstrae*, and *L. yanagawae* replace the denotations L. genomospecies 1, 3, 4, and 5, respectively according to the recommendations of the Taxonomic Subcommittee on Leptospira.[20a]

described a polymerase chain reaction (PCR) with one primer pair G1/G2 amplifying DNA from all pathogenic *Leptospira* species, except *L. kirschneri*. For this species, they designed a second specific primer set, B64I/B64II. As such, B64I/B64II represents the first described species-specific primer, followed only in 2006 by an *ompL1*-based set of PCR primers enabling the discrimination of the six major pathogenic species.[37] Various PCR-based methods have claimed discrimination of leptospires at the serovar level, including low-stringency (LS) PCR, known under several synonyms[38] and *Leptospira*-specific insertion element (IS)-derived PCR.[39,40] Evaluation of these approaches used too few strains to allow conclusions on the extent of their discriminative power. Other approaches used the properties of PCR products (amplicon), either generated for diagnostic purposes or not, analyzed by techniques such as differential electrophoresis in nondenaturing polyacrylamide gels, followed by silver staining,[41] single-strand conformation polymorphism analysis (SSCP) of PCR products[42] and low-stringency single specific primer (LSSP) PCR.[43] The claim that these molecular methods have the potential to identify leptospires at the serovar level should be considered with caution. The serovar status is predominantly determined by the lipopolysaccharide (LPS) exposed at the surface of the cell envelope (Figure 97.2). Therefore, it can be argued that only molecular approaches targeting LPS coding DNA would justify such a claim. Most, if not all molecular methods have the additional disadvantage of poor reproducibility or a need of viable cultures for extracting sufficient amounts of DNA. New molecular methods, largely evading such needs and markedly contributing to the field of molecular epidemiology and pathogen evolution, are fluorescent amplified fragment length polymorphisms (FAFLP),[44,45] multiple-locus variable number of tandem repeats analysis (MLVA),[45,46] and sequencing. Traditionally the *rrs* sequences coding for the 16SrRNA is the gene of choice used for phylogeny. However, for *Leptospira*, the potential of this gene is limited due to its conserved nature.[47,48] Moreover, using the sequence of only one locus for phylogenic determinations is particularly risky for *Leptospira* considering the high fluidity of its genome,[47,49] obscuring conclusions on single locus speciation. Therefore, multilocus sequence typing (MLST) is probably the most robust approach for phylogenetic assessments[48,50,51] (Figure 97.3). Sequencing and phylogeny have revealed an increasing group of intermediate species[52,53] isolated from animals and humans with disputable clinical relevance because virulence cannot be convincingly demonstrated.[52,53]

Leptospires are gram-negative obligate aerobic bacteria that have an optimal in vitro growth temperature of 28°C–30°C. Special culture media are used for growing leptospires. Vitamin B2 and B12 are essential growth factors, long-chain fatty acids present the sole carbon source, and ammonium salts are used as nitrogen source.[19,54–56] Leptospires grow slowly with generation times of 6–8 h, or sometimes 14–18 h.

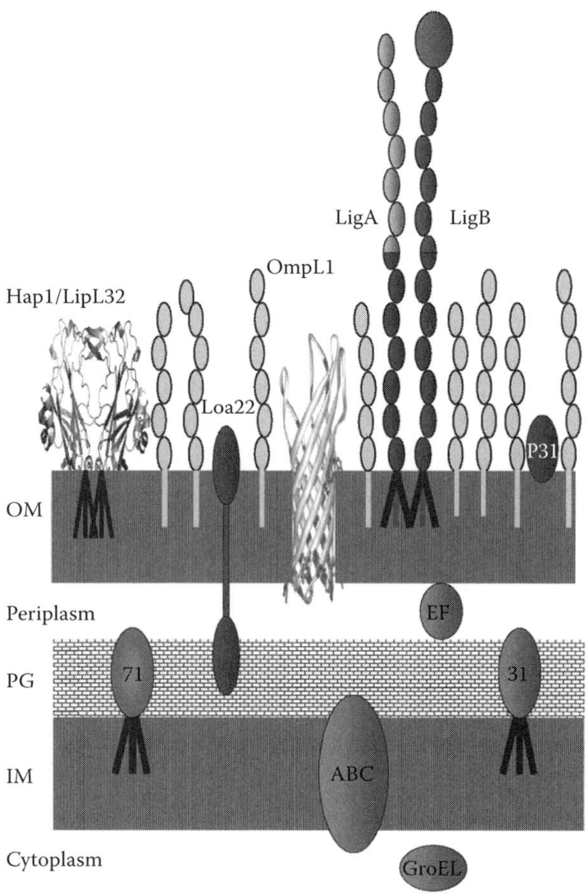

FIGURE 97.2 Cell wall structure of *Leptospira*. OM is outer membrane, IM is inner membrane, PG is peptidoglycan layer. The outer layer of the OM mainly consists of lipopolysaccharides (not detailed). Various proteins are indicted in the OM, IM, PG, and cytoplasm.

Like other spirochetes, they have a typical double-membrane structure separated by a periplasmic space (Figure 97.2). The cytoplasmic or inner membrane (IM) is closely associated to a peptidoglycan layer. The surface-exposed layer of the outer membrane (OM) contains LPS with, as far as is currently known, a (three-dimensional) structure that differs from the LPS of *Escherichia coli* and has a lower endotoxic capacity.[19] The OM has a relatively low protein content, which includes pore-forming proteins like OmpL1 and several lipoproteins such as LipL32 (Figure 97.2). A number of proteins, such as LipL32, LipL41, and LigB are restricted to pathogenic leptospires,[57–59] and differential expression between in vivo and in vitro cultured leptospires has been reported for several OM proteins[60] and LPS.[61] Motility is facilitated by two periplasmic flagella, consisting of axial filaments that have polar anchors in the IM.

The genome of pathogenic *Leptospira* consists of two circular replicons, denoted as CI and CII, with sizes varying from 3.6 to 4.3 × 10^3 kilo base pairs (kbp) for CI and approximately 0.3 × 10^3 kbp for CII, and a G + C content of 35% to over 40%.[49,58,59,62] *Leptospira* genomes are highly fluid, testifying for multiple rearrangement events during the evolution of the various species. Pathogenic genomes contain a high number of transposases known to mediate such rearrangements as well as multiple copies of genes involved in chemotaxis.[58] Genome fluidity and multiplication of "environmental signaling" genes might be associated with the unusually high adaptation potential of notably pathogenic leptospires to a variety of environmental stimuli, both within and outside the host. Interesting in this respect is the smaller size of CI in *L. borgpetersenii* including a reduction of the number of genes involved in environmental signaling, coinciding with a lower survival capacity outside the host.[49]

97.1.2 Clinical Features and Pathogenesis

Many infected persons remain asymptomatic.[63,64] Serologic evidence of a past unapparent infection is found in 15%–40% of persons who have been exposed but have not become ill.[3,4,63,64] More than 90% of symptomatic persons have the relatively mild and usually anicteric form of leptospirosis, with or without meningitis. The number of mild cases is unknown but probably forms the major burden of leptospirosis. Weil's syndrome develops in about 10% of symptomatic individuals. SPHS, manifesting spontaneously or triggered by intubation for mechanic ventilation, is a rapidly emerging complication.[1] In the past, fatal pulmonary hemorrhages have been almost exclusively attributed to serovar Lai, circulating in China and Southeast Asia. To date, several serovars have been associated with the manifestation in other parts of the world,[7,65,66] making it unlikely that the cause of the syndrome is serovar restricted. This is in line with the current notion that distinct clinical syndromes are not associated with specific serovars, although for some serovars there is a tendency to cause more severe disease than in others.

An acute leptospiremic phase is followed by an immune leptospiruric phase. The distinction between the first and second phase is not always clear. Uveitis may develop as a late sequel. Leptospirosis is either categorized as mild anicteric and severe icteric leptospirosis[67] or as (i) mild; (ii) Weil's syndrome; (iii) meningitis/meningoencephalitis; and (iv) pulmonary hemorrhage with respiratory failure.[2] Here, we use the categorizations of "mild" and "severe" leptospirosis. Mild leptospirosis is usually but not exclusively anicteric, and vice versa, severe leptospirosis often is icteric. For details please consult.[67]

Mild Leptospirosis. Leptospirosis may present as an acute illness with fever, chills, severe headache, nausea, vomiting, and myalgias. Muscle pain in the calves, back, and abdomen is an important feature of leptospiral infection. Sometimes photophobia and conjunctional suffusion, mild pulmonary disorders, a sore throat, rash, muscle tenderness, and mild jaundice develop.

Most patients become asymptomatic within 1 week, but the illness may recur after 2–3 days. This coincides with the immune phase, resulting in the clearance of the leptospires from all sites in the host except the eye, the proximal renal tubules, and perhaps the brain. Symptoms are more variable

FIGURE 97.3 Phylogenetic tree of strains belonging to the pathogenic *Leptospira* species, *L. interrogans*, *L. kirschneri*, *L. santarosai*, *L. borgpetersenii*, and *L. noguchii* rooted by the saprophytic *L. biflexa*.

than during the first phase. The fever is less pronounced, and the myalgias are less severe. Aseptic meningitis may develop and is more common among children than among adults. Mortality in mild leptospirosis is low.

Severe Leptospirosis. Weil's syndrome is characterized by jaundice, renal dysfunction, hemorrhagic diathesis, and often pulmonary and cardiac involvement. The mortality rate ranges from 5% to 20% but occasionally exceeds 50%.[68] Early manifestations are not different from those of mild leptospirosis; however, after 4–9 days, jaundice as well as renal and vascular dysfunction generally develop. A biphasic disease pattern is mostly lacking. Jaundice and kidney damage leading to oliguria or anuria often develop during the second week of illness. Dialysis is sometimes required.

Severe gastrointestinal bleeding rarely occurs, but pulmonary hemorrhage with respiratory failure is increasingly reported. SPHS accounts for high mortality rates[1] and thereby presents the most severe complication of the disease. Hemolysis, myocarditis, pericarditis, congestive heart failure, shock, SPHS, necrotizing pancreatitis, and multiorgan failure have all been described in case of severe, fatal leptospirosis. In general, organ damage is reversible and organs recover completely, although tissue scars might be seen after persistent or recurrent manifestations.

Pathogenesis. The pathogenesis of leptospirosis is incompletely understood. Leptospires enter the body of the host through broken skin or via the intact mucous membranes of eyes, nose, and mouth. This implies that infection can occur via consumption of contaminated food or water or by inhalation of contaminated aerosols.[16,24,49] The incubation period is usually 1–2 weeks but ranges from 2 to 30 days. After entry of the leptospires, leptospiremia develops, with a subsequent spread to all organs. All pathogenic leptospires can damage the wall of small blood vessels, leading to vasculitis and leakage. In the kidney, leptospires migrate to the interstitium, renal tubules, and tubular lumen, causing interstitial nephritis and tubular necrosis. Altered capillary permeability may contribute to the development of renal failure. Leptospirosis does not cause severe hepatocellular necrosis but centrilobular necrosis with proliferation of Kupffer cells may be observed. Pulmonary involvement is the result of hemorrhage and not of inflammation. Invasion of skeletal muscle by leptospires results in focal necrosis.

When antibodies are formed, leptospires are eliminated from the blood and most organs. The systemic immune response eliminating the organism may also produce symptomatic inflammatory reactions.

Treatment with antimicrobial agents should be given as early in the course of illness as possible. Penicillin G and doxycycline have been shown to reduce morbidity from leptospirosis.[69–71] Penicillin G (1.5–2.0 million units every 6 h) is recommended for the treatment of severe leptospirosis. Third-generation cephalosporins (ceftriaxone, 1 g IV once a day; cefotaxime, 1 g IV every 6 h) are an effective alternative. Doxycycline (100 mg every 12 h), ampicillin, or amoxicillin (500 mg every 6 h) can be used as oral regimens for the treatment of mild leptospirosis. After the start of antimicrobial treatment for leptospirosis, a Jarisch-Herxheimer reaction may develop, but it is a relatively rare event compared to the reaction occurring in other spirochetal diseases. Prompt triage of high-risk patients and supportive care are required for the treatment of hypotension, renal, and respiratory distress, and hemorrhage. Timely initiation of dialysis and mechanical ventilation is essential to preventing mortality.

97.1.3 Diagnosis

The clinical diagnosis of leptospirosis is difficult because of its protean manifestations, which can include one or more of a variety of signs and symptoms, such as fever, myalgia, severe headache, chills, diarrhea, nausea, and vomiting, oliguria/anuria, jaundice, conjunctival suffusion, aseptic lymphadenopathy, hepatomegaly, splenomegaly, and others.[3,16] Notably, in the early acute phase, signs and symptoms are uncharacteristic and may be confused with those of several other diseases.[2]

Also clinical laboratory data are mostly nonspecific.[2,3,16,67] Findings related to kidney involvement range from urinary sediment changes and mild proteinuria in anicteric leptospirosis to renal failure in severe disease. In severe leptospirosis, the erythrocyte sedimentation rate (ESR) is usually elevated, and leukocytosis is often marked. Unlike acute viral hepatitis, leptospirosis displays elevated serum levels of bilirubin and alkaline phosphatase. Levels of aminotransferases and creatine phosphokinase may be elevated. Abnormalities in the cerebrospinal fluid (CSF) are common in the second week of illness. In severe leptospirosis, development of pulmonary radiographic abnormalities 3–9 days after the onset of illness is not uncommon. Usually a patchy alveolar pattern that corresponds to scattered alveolar hemorrhage is observed.[72]

The nonspecific clinical laboratory findings may at best suggest a diagnosis of leptospirosis. For confirmation of the diagnosis, specific microbiological tests are necessary. A definite diagnosis of leptospirosis is based either on detection or isolation of the organism from the patient, or on seroconversion or a significant titer rise in the MAT. To choose an optimal approach, the various stages in leptospirosis must be known. During the incubation period, leptospires multiply and disseminate to all organs via the blood. Antibodies are being produced within the first week of illness and clear the blood of leptospires after about 1 week.[3] Generally IgM antibodies are produced slightly earlier than IgG antibodies, and IgM levels drop off within a few months. IgG antibodies may persist for several years. The presence of IgM antibodies may thus serve as an indication of a current or recent infection. Leptospires appear in CSF from day 2 on and remain for about 3 weeks. The shedding of leptospires in the urine starts during the first week after onset of the disease and may continue for a month in a convalescent shedder.[3] This implies that antigen is present in blood and various organs in the first week of illness, after which it quickly disappears. Antigen detection in CSF and urine may possible for a longer period. Serology will have a very low sensitivity during the first week of illness, increasing to optimal levels in the late acute phase. Although unsuitable for antibody detection, early acute-phase blood samples are valuable for establishing seroconversion or a significant titer rise.

97.1.3.1 Conventional Techniques

(i) Dark-Field Microscopy Leptospires are too thin to be observed under the ordinary light microscope. Staining has been applied to improve the detection by direct microscopy, including immunoperoxidase staining of leptospires in blood and urine[73] and (immuno)histopathological[60] and silver staining.[74]

For direct visualization of leptospires, dark-field microscopy (DFM) also referred to as dark-ground microscopy (DGM), is required. The dark-field microscope contains a special condenser that interrupts the central light. Only light scattered by objects present in the illumination path enters the objective lens. Leptospires appear as bright fibers in a dark background. It is essential to use a good quality dark-field microscope with an intensely bright point-light source. A dry-dark-field condenser is usually combined with a suitable ×10 or ×20 objective and a ×10 ocular.

DFM examination of body fluids, such as blood samples, CSF, and urine, has been used for the diagnosis of leptospirosis. A concentration of approximately10^4 leptospires/mL is considered to be the detection limit by DFM. A quantitative buffy coat method, originally developed to check blood for malaria parasites, increased the sensitivity of DFM detection of leptospires in blood to nearly 10^3 leptospires/mL.[75] DFM is prone to false-positive outcomes due to the confusion of artefacts with leptospires. Notably, blood of febrile patients contains fibrin or protein threads that, by virtue of Brownian motion, are virtually indistinguishable from live leptospires. Because of its lack of sensitivity and specificity, DFM is not recommended as a sole diagnostic tool.[2] It is important to note that only during the first few days of acute illness leptospires are present in blood in sufficient numbers to be detectable. DFM on urine samples is not suitable for early diagnosis as excretion is not always evident at that stage.

Thus, DFM is a very simple and fast technique for direct detecting leptospires in liquid clinical samples but has a low reliability. The mere value for DFM lays within the inspection of leptospires in cultures and in agglutination reactions.

(ii) Isolation of Leptospires A positive culture of leptospires from clinical material from a suspected patient provides evidence for leptospirosis. Leptospires are fastidious organisms, requiring experienced staff and special culture media to grow. A number of culture media have been described, including Korthof-Babudieri and Stuard's liquid medium and Fletcher's semisolid medium.[3,18,76] EMJH media described by Ellinghausen and McCullough as modified by Johnson and Harris[77,78] is the most successful medium to date. It is used in liquid and semisolid form and might be supplemented with 5-fluorouracil or cocktails of suitable antibiotics to suppress growth of contaminating microorganisms[18,24,76] or by animal serum, usually slightly hemolyzed rabbit serum, as enrichment.[24,76,79] Current culture media have been developed more than four decades ago. Significant improvements have not been achieved since then.

Because leptospires disseminate to virtually all organs via the blood, a variety of clinical samples are suitable for isolating leptospires by culture. Blood and urine are the most straightforward samples. In case of postmortem investigations, tissue samples of kidney, liver, lungs, and heart are often used. The success of isolation not only varies with the choice of the material combined with the stage of the disease, but the period spanning the collection of the sample and the inoculation into the culture medium also affects success. It should be noted that clinical specimens contain growth inhibitors. Big inoculums will suppress growth; "more" is not equal to "better."

Blood samples must be cultured as early in the illness as possible. Leptospiremia only occurs during the early acute phase of the disease. Besides, antibiotics kill the leptospires, and therefore, blood samples for culturing should be collected early and before the start of treatment. Venipunctured blood or capillary blood taken from the finger is ideally inoculated at the bedside into culture medium. One or two drops of anticoagulated blood in 5-mL culture medium is optimal. Serum might contain fewer leptospires but is also suitable for culturing. Usual culture success rates vary from 10% to 30 %, but success can be increased by applying "leptospires enrichment" steps prior to inoculation. Wuthiekanun et al.[80] reported a yield of 96.8% by using a sampling strategy in which culture is performed using whole blood and deposit from spun plasma.

CSF and urine could be cultured during the first week or from the start of the second week of symptomatic illness, respectively.[81] Fresh midstream urine is collected and inoculated by adding one drop of undiluted urine into a vial containing 5 mL of culture medium. Enrichment of leptospires by inserting a centrifugation step does not guarantee greater success. Medium is preferably supplemented with 5-fluorouracil, and inoculation of the first vial should be followed by two serial tenfold dilutions because urine samples are notorious for containing contaminating microorganisms and growth inhibitors. Leptospires are highly sensitive to the acid environment and detrimental components in urine. Urine samples should be inoculated immediately but at least within 2 h after voiding. The time to inoculation can be prolonged by alkalization,[18,76] but speed remains an issue.

For tissue-derived cultures, small segments of tissue materials (0.5 × 0.5 × 0.5 cm) are grounded in approximately 1 mL of phosphate-buffered saline or culture medium, and subsequently inoculated and diluted in culture medium as described for urine.[2,76] Leptospires do not survive long in dead tissue because of autolysis of the cells and a consequent decrease of pH. Autolysis, and thus the time interval to culture medium inoculation, can be delayed 2–4 days by storing the sample at +4°C. Longer periods of storage are possible but might reduce the success rate because pathogenic leptospires loose survival capacity at low temperatures.[3,17,18,24]

To optimize culturing from a variety of samples, culture media and supplements in various combinations are used. Because leptospires grow slowly,[3,17,18,24] maintaining cultures for 6 months before discarding them as negative is recommended. However, this poses logistic problems for most laboratories, and more feasible approaches with one type of medium and shorter incubation times are used. Because of the long time it takes for cultures to become positive, isolation does not contribute to early diagnosis. However, isolation and subsequent characterization of leptospires is highly valuable for epidemiological purposes and for optimizing panels of *Leptospira* serovars in the MAT (see below) that preferably include local isolates.

Especially in the early days of leptospirosis diagnosis, leptospires have been isolated from blood and urine by a passage through experimental animals.[82] Young guinea pigs and golden hamsters are intraperitoneally injected with about 0.5-mL sample, and after 3 days, peritoneal fluid is withdrawn from the lower quadrant of the abdomen by puncture with a capillary pipette. The peritoneal fluid is then examined by DFM, and if leptospires are observed, the animals are bled, under anaesthesia, by heart puncture for culturing. If the experimental animal dies, kidney or liver tissue can be used.

(iii) Serological Diagnostic Tests Most cases of leptospirosis are diagnosed by serological tests. The enzyme-linked immunosorbent assay (ELISA) and the MAT are widely used. The application of the indirect fluorescence test (IFAT) is becoming less, mainly because of the need of costly and demanding reagents and equipment. MAT is the cornerstone of the serodiagnosis of leptospirosis because of its high sensitivity and unsurpassed (antigen) specificity and is often

referred to as gold standard. ELISA is increasingly used because it is often much more practical to use than MAT.

The interpretation of serological data should rely on the examination of paired blood specimens; the first one taken within a minimum time period after the onset of symptoms, and the second one, collected 10–14 days later. A third convalescent serum sample may be required for confirmation and can be helpful for determination of the infecting serogroup by MAT. Serological tests do have a low sensitivity within the first 5–10 days after the onset of symptoms.[3,83] Moreover, patients with fulminant leptospirosis may die before sufficient antibodies have been produced. In those cases, confirmation by serology is not possible.

Microscopic Agglutination Test. The MAT is based on the agglutination-lysis test developed by Martin and Pettit.[84] This method has been adapted into a format that can be performed and eventually read in microtiter plates. The test is to some extent subjective. Visual interpretation introduces an intraobserver variation, usually limited to one titer step between experienced staff of one laboratory. Other variables that can influence the results of the tests are the composition of the *Leptospira* panel and the age and density of the antigen culture. Cutoff titers differ among areas with different incidences of leptospirosis and other potentially cross-reacting diseases. Therefore, the MAT is difficult to standardize into a global uniform platform. An international proficiency-testing scheme has been implemented to provide a means of external control on the performance of the MAT under a variety of situations.[85]

The method consists of mixing serial twofold dilution of test serum in PBS with an equal volume of a culture of leptospires and then examining the degree of agglutination by DFM. Starting dilutions are commonly 1:20 or 1:50, including the volume of the antigen. Incubation for allowing antibodies to react with the antigen is usually done 1–2 h at 30°C but can be prolonged when performed at lower temperatures. The degree of agglutination, which is a measure for the concentration of anti-*Leptospira* antibodies, is assessed by determining the end-point titer. This titer is defined as the highest dilution of serum at which 50% agglutination occurs, leaving 50% of the leptospires free compared to a control culture diluted 1:2 in PBS.[18,24,76] The MAT detects anti-*Leptospira* antibodies, irrespective of the host species, implicating a broad applicability of the test in human and veterinary settings. The humoral response is primarily directed against the infecting serovar and therefore displays a certain level of antigen specificity. For adequate sensitivity, a battery of representative strains that are known to be present locally must be employed. Care should be taken to also detect newly introduced serovars or infections acquired in other regions where different serovars are present. For this reason, a saprophytic strain (usually *L. biflexa* strain Patoc I) is included that cross-reacts with human antibodies generated by most pathogenic serovars.[2,24] "Freshly" isolated local strains may give higher titers than their reference laboratory counterparts. If knowledge of local strains is lacking, MAT can be started with a recommended battery of reference strains.[2]

Live cultures of all serovars required for use as antigens must be maintained by subculturing in liquid medium every week. MAT agglutination titers depend on the density and age of the culture. Recommendations indicate the use of suspensions with a density of about 2×10^8 leptospires per mL. The density can be determined by counting, by spectrophotometry, or by nephelometry.[18,24,76] Of note, recommendations do not have to be followed strictly. Most important is that the test procedure within one laboratory is standardized, and that the cutoff titer is deduced from actual, local findings, taking into account background and cross-reactive titers in healthy persons and patients with other defined diseases. Applying a well-defined cutoff titer is particularly important when only testing a single serum sample. Cutoff titers vary from ≥1:160 in low-prevalence regions in the Northern and Southern hemisphere to a single titer of ≥1:800[18] in symptomatic patients in most tropical countries.

Serial dilutions of sera are prepared in plastic microtiter plates with 96 flat-bottomed wells. No particular type of plate is specified if the results are read following the transfer of the contents of a small drop onto a glass slide. However, if the plate is observed directly under an inverted DFM, a plastic microtiter plate with good optical qualities is necessary.

To avoid incorrect conclusions on presumptive infecting serogroups, the use of well-characterized strains is essential.[2,18,24,76] The identity of strains should be checked at regular intervals by serological and/or molecular techniques.[24,86]

Views vary as to whether it is preferable to use live or killed antigen in the MAT. Antigens killed with 0.5% formalin overcome laboratory-acquired infections associated with the use of live antigens. Besides, formalized antigens can be stored for weeks at +4°C, providing batches of antigen with standard quality, at the same time evading the need for weekly subculturing. However, titers obtained with these antigens are one titer step lower, and more cross-reactions are detected.[24,76,87] In veterinary testing, presumptive information on infecting antigens is of relevance, and therefore, the use of formalized antigens is avoided.

The MAT is usually positive 10–12 days after the appearance of the first clinical symptoms. The seroconversion may sometimes occur as early as 5–7 days after the onset of the illness. Antibiotic therapy started before the test may interfere with the development of the antibody response.

Interpretation of the MAT is complicated by the high degree of cross-reactions that occurs between different serogroups, especially in acute-phase samples. In the first few weeks of the disease it is not uncommon that heterologous reactions are stronger than the homologous ones.[24,88] The homologous reaction may even be absent. In this case, this phenomenon is called "paradoxical reaction."[89]

The presence of agglutinating antibodies is evidence of infection, present or past. Acute infection is suggested by a single strongly elevated titer detected in association with an acute febrile illness. Titers following acute infection can take months or even years to fall to low levels. A weak but highly strain-specific agglutination supports a past infection.

Paired sera are required to confirm a diagnosis with certainty. Seroconversion or a fourfold or greater rise in titer between paired sera confirms the diagnosis regardless of the interval between samples. The interval between the first and second samples greatly depends on the delay between onset of symptoms and presentation of the patient. If symptoms of overt leptospirosis are present, an interval of 3–5 days may be adequate to detect rising titers. However, if the patient presents earlier in the course of the disease or if the date of onset is not known precisely, then an interval of 10–14 days between samples is more appropriate. Less often, seroconversion does not occur with such rapidity, and repeated sampling is necessary.

The test may give an indication of the serogroup to which the causative serovar belongs, but it only rarely identifies it.[90] MAT detects both IgM- and IgG-class antibodies with IgG predominantly present in convalescent samples, giving a more antigen-specific reaction.[91] If possible, it is important to examine several sera taken at intervals after the acute disease in order to determine the presumptive infecting serogroup. Some patients have serological evidence of previous infection with a different leptospiral serogroup. In these cases, serological diagnosis is complicated further by the "anamnestic response," in which the first rise in antibody titer is usually directed against the infecting serovar from the previous exposure. Only later, as the titer of specific antibody rises, does it become possible to identify the serovar or serogroup responsible for the current infection. Presumptive serogroup reactivity data should be used only to gain a broad idea of the serogroups present at the population level.[90]

Enzyme-Linked Immunosorbent Assay. ELISA uses polystyrene microtiter plates coated with *Leptospira* antigen. Test serum is added to the coated well, and antibodies are allowed to react with the antigen. Following several washing steps to eliminate excess antibodies, conjugated secondary antibodies are added and allowed to react with the host antibodies. Unbound conjugate is carefully removed by repeated washing, and substrate is subsequently added. Basically, the conjugate converts a colorless substrate into a colored form. The color change can be determined by measuring the optical density (OD) with a spectrophotometer equipped with a suitable filter. The color indicates that anti-*Leptospira* antibodies are present in the serum sample. The intensity of the color relates to the concentration of anti-*Leptospira* antibodies in the sample. This can de deduced from the decrease of the intensity of the color in a serial dilution of the serum or by measuring the OD intensity in a single well format. In both cases, well-defined positive and negative sera need to be included in the assay as internal control. ELISA that uses serial dilutions is more robust than single well-ELISA as the influence of interwell variations is reduced.

Sensitivity and specificity of the ELISA are similar to those of MAT.[11,83] In contrast to the MAT, the ELISA uses a single, broadly reactive antigen and, therefore, is not suitable for serogroup/serovar identification. Besides, the conjugate is host specific, which confines the applicability of a distinct ELISA to a single host species. On the other hand, the use of a single antigen makes the test relatively simple in comparison with the MAT. It can be used to concomitantly process many samples within a relatively short period. The conjugate is directed at either isotype IgM or IgG antibodies and hence is suitable to confirm a recent infection and to determine the seroprevalence of leptospirosis in epidemiological studies, respectively. Cutoff values depend on the level of persisting antibodies in the general population. Usually, IgM ELISA becomes positive one day earlier in the course of the disease than the MAT,[3] thus providing an important gain for early diagnosis. Anti-*Leptospira* IgM antibodies have been detected by ELISA in CSF from patients with icteric leptospirosis[92] and from patients with meningitis without a proven etiology.[93] Anti-*Leptospira* IgM is also present in saliva, but the sensitivity and specificity of ELISA on CSF and saliva is lower compared to serum.[94,95]

ELISA's have been described with a wide variety with different antigens and preparation protocols and with various reagents.[95,96] The selection of antigen affects the sensitivity of the ELISA assay. Although the test is genus specific, homologous serovars tend to give higher titers. Other variables that affect the result of the ELISA are, notably, the antigen concentration to coat the plates and the conjugate. Coating and antigen concentration must be standardized, while a checkerboard titration needs to be executed to determine the optimal concentration of conjugate in a new batch.

Sandwich ELISA has been applied for the detection of leptospiral antigens in clinical material and could detect 10^5 leptospires/mL. It offers greater specificity than DFM, but the potential for greater sensitivity[3] so far remains hypothetical.

Indirect Fluorescent Antibody Test. Immunofluorescence makes use of antibodies conjugated to a fluorescent dye as detection reagent.[24,76,97,98] The method can detect specific antibodies in body fluids or antigen in tissue sections. The IFAT for the detection of anti-*Leptospira* antibodies in fluids is broadly reactive. Leptospires are coated onto teflon-coated microscope slides. Serum dilutions are applied on the antigen spots, and bound anti-*Leptospira* antibody is detected with a fluorescein-conjugated class-specific antibody under the fluorescence microscope. The fluorescence is scored as ++++ for brilliant fluorescence to ± for weak fluorescence. The titer is defined as the highest serum dilution giving a + (visible fluorescence) signal. The sensitivity and specificity of the IFAT correspond to those of the ELISA.

Detection of leptospires in tissue can be done by specific antibodies conjugated to a fluorescent (direct immunofluorescence) or indirectly by using a second conjugated antibody as in body fluids.

The fluorescence of the preparation is subject to fading during illumination and might change the color. Illumination during microscopic examination should be minimized.

(iv) Conventional and Novel Screening Tests MAT and ELISA are relatively complicated and therefore not always available. Rapid serological screening assays contribute to the

diagnosis of leptospirosis in situations where well-equipped laboratories are absent. A wide variety of rapid tests have been reported that basically all need confirmation by MAT. Selection of an assay will mainly balance suboptimal performance, on one hand, with availability, ease of use, costs, and reagent stability avoiding the need of cooling during transport and storage, on the other hand. This section focuses on generally known tests and excludes those that have not been widely used.

Macroscopic Slide Agglutination Test (MSAT). MSAT uses dense suspensions of formalin or heat-killed *Leptospira* serovars.[99,100] The antigen should consist of all locally prevalent strains. The strains can be mixed into pools serving as screening batteries. A drop of serum is mixed on a glass plate with a drop of the antigen. After a brief incubation, the glass plate is placed over a light source and inspected by eye for agglutination. The sores are ++++ for 100% agglutination (all organisms appear clumped) to + for 25% agglutination. The (pooled) antigens should be prepared according to standardized procedures. Antigens are stable and can be stored at +4°C for at least 1 year. Glycerol might be added for stabilization and to prevent the drying of the suspensions during the test. Several modifications of this test have used a single-serovar antigen, usually serovar Patoc.[100–102] The MSAT is relatively insensitive for diagnosis but may be useful for epidemiological screening.[103]

Indirect Hemagglutination Assay (IHA). A number of methods using sensitized red blood cells have been described. The extraction of an erythrocyte-sensitizing substance (ESS) led to the development of both a hemolytic assay[104] and a genus-specific hemagglutination assay, IHA.[105,106] These assays detect both IgM and IgG antibodies and normally use a single broad reactive saprophytic strain such as Patoc I for ESS preparation. In the IHA, heat-inactivated serum (30 min at 56°C), which is safer to handle, is mixed in a well of a microtiter plate with sensitized red cells, usually from sheep. Red cells agglutinate if anti-*Leptospira* antibodies are present in the serum. Agglutination is judged by eye, following an incubation time up to 18 h at room temperature. Positive agglutination is judged ++++ if compact granular agglutination has occurred, to ++ if a smooth mat of cells is seen. In a doubtful case (+), small rings of cells occur at the bottom of the well. Quantization can be achieved by using dilutions of serum. The procedure of the lysis test is similar but incubation is done in the presence of complement. Details on the preparation of ESS and the coating of erythrocytes can be found in textbooks.[76] IHA is commercially available (MRL Diagnostics, California, USA). Recent estimates of the sensitivity of the IHA in populations in which leptospirosis is endemic have varied from good[107] to poor.[108,109] Differences might be explained by differences in case ascertainment and study design, including inclusion of population and the unavailability of prospective unbiased samples.[11,110]

Latex Agglutination Test (LAT). LAT is similar to the IHA but uses commercial latex particles sensitized with broad reactive *Leptospira* antigen.[106] It can be used on heat-inactivated serum samples. Several protocols of antigen preparation have been described. The Dri-Dot is a simple test that consists of sensitized colored latex particles dried onto an agglutination card. The latex particles are suspended in 10 μL human serum, and agglutination occurs within 30 s if anti-*Leptospira* antibodies are present.[8] The test is applicable on human serum only. The Dri-Dot has a satisfying performance,[111] and by its stable and robust format is a very useful point-of-care. Unfortunately, the test is currently unavailable.

Microcapsule Agglutination Test (MCAT). The MCAT uses a synthetic polymer in place of red blood cells.[112] The test has been evaluated at various centers[113] and shown a sensitivity similar to the MAT and IgM-ELISA in early acute-phase samples, but it failed to detect infections caused by some serovars.[114] The MCAT can be applied for both human and veterinary purposes.

Complement Fixation (CF) Test. The CF test is a generally used serological technique and can also be applied for antigen detection. The details of the test are common in principle to all methods and can be found in any textbook on microbiological and serological methods.[76] The test is based on the fact that the complement complexes cell-antibody bindings, leading to lysis of the cells. Usually, sheep erythrocytes sensitized with antibodies are used. In the test, antigen of the suspected infectious agent is added. Causative agent-specific antibodies bind to the antigen, and the resulting complex binds the complement, thus competing with the complement binding to the antibody-sensitized erythrocytes and inhibiting the hemolysis of the erythrocytes. Absence of hemolysis constitutes a positive test. For the diagnosis of leptospirosis, often strain Patoc I is used as the antigen. The test can be applied on heat-inactivated serum. It is essential to standardize the erythrocyte suspension and other variable reagents in the test and to include suitable positive and negative controls. The test can be applied on human and animal sera.

Lateral-Flow (LF) Assay. The LF assay is a second-generation format of the dipstick test.[10] Basically, it is an ELISA, but the leptospiral antigen is coated onto a nitrocellulose membrane, and the conjugate is bound to a dye instead of an enzyme. The dipstick is submerged into a mixture of conjugate and serum and incubated for 3 h at room temperature to allow antibodies to react with the antigen. The strips are then inspected by eye. The test includes an internal control band. A positive reaction, therefore, shows two bands, whereas a negative reaction only shows the control stain. The test is genus-specific, and the conjugate is directed at human IgM for detecting acute infection in humans only. The Dip-S-Tick is a commercial IgM dot-ELISA format test (PanBio Inc., Australia) that is quite similar to the dipstick.

In the LF test, the nitrocellulose strip is inserted into a lateral-flow device.[115] The conjugate is incorporated in the device as well. The serum sample and subsequently the running buffer are added to the sample well. By a lateral-flow motion, the serum passes and elutes the conjugate, then reaches the leptospiral antigen to bind anti-*Leptospira* antibodies, and subsequently passes the control antigen to bind other IgM antibodies. The 10-min reaction time of the LF test is markedly shorter than that of the dipstick. An additional advantage of the LF assay is that it can be applied on whole blood samples. The LF is commercially available. Both the dipstick and LF test allow semiquantification by the subjective visual estimation of the staining intensity. As can be expected, the performance characteristics of the dipstick and LF tests are similar to those of the ELISA.[10,115]

Due to its wide applicability, stability, rapidity, and ease of use, the LF assay is highly suitable as a point-of-care test for diagnosis and surveillance in "basic" situations such as occur during outbreaks in disaster areas

(v) Recombinant Antigen-Based Serological Tests Current ELISA and rapid formats all use whole-*Leptospira*-based antigen, often using the same strain for antigen preparation. Therefore, it is to be expected that these tests perform with similar characteristics and only differ in detail.[11] However, various evaluations revealed significant differences in sensitivity and specificity across geographical regions. Notably, in certain high-prevalence settings for leptospirosis and other infectious diseases, the specificity of tests appeared poor.[12,116] A major reason for this might be that the crude antigens comprise components that are also present in other infectious agents,[117] thus introducing false-positive scores.

To evade loss of specificity due to cross-reactions, the use of recombinant proteins as antigens in serodiagnostic tests has been advocated in recent years. Several *Leptospira* proteins have been explored as novel antigens, such as the outer membrane proteins LipL32[118] and LipL41,[119] the cytoplasmic chaperon protein GroEL,[119] the *Leptospira* immunoglobulin (Ig)-like (Lig) proteins.[120]

A probable drawback is that such proteins may lack homology between distinct *Leptospira* species, limiting their good performance to homologous infections. Hence, there arises a realistic risk that a higher specificity is reached at the expense of a lower overall sensitivity.

97.1.3.2 Molecular Techniques

The major limitation of serological tests is the low sensitivity at the first days of illness, when clinical decision making on the treatment is the most important. A variety of molecular techniques, including Southern blotting, dot-blotting,[121] in situ hybridization,[122] and nucleic acid (NA) amplification techniques have been applied to detect leptospiral DNA or RNA[123] as a marker for the presence of leptospires. When targeting patient's blood in which leptospires are present in the early acute phase of the illness, the molecular approach provides the badly needed early diagnostic confirmation. Reports on hybridization techniques are anecdotic. In contrast, the PCR has been extensively described as a promising, sensitive, and widely used molecular detection technique that can be applied on a variety of clinical samples.

Conventional PCR. The introduction of the PCR into the laboratory has provided a valuable tool for the rapid and early confirmation of the diagnosis of leptospirosis. With the technique, leptospiral DNA can be specifically amplified by repeated cycles of DNA denaturing, primer annealing, and DNA duplication by a heat-stable DNA polymerase enzyme. Theoretically, a single DNA target can be multiplied a millionfold. In contrast to the single copy, the multiplied fragment can easily be detected after electrophoresis in an agarose gel.

The first PCRs for the detection of leptospires in clinical samples have been described in the early 1990s, when the technique still was in its infancy.[36,124–126] With the current availability of high quality heat-stable DNA polymerases, NA extraction kits, and an abundance of available sequences for selecting suitable primers, PCR has become a rapid and solid method. To date, a wealth of PCR methods have been described.

It was recognized early that anomalous products (nonspecifically amplified DNA fragment) with the same electrophoretic mobility as the specific DNA amplicon contributed to false-positive results. To evade this type of false-positive score, a subsequent hybridization step with a specific internal probe was advocated. On the one hand, such a Southern hybridization presents an extra step, delaying the analysis. On the other hand, as an unintended but advantageous side effect, Southern blotting resulted in an almost tenfold increase in the sensitivity, thus enabling a detection threshold of 1–10 copies.[36,81] Another approach, increasing both the specificity and the sensitivity, was found in the use of nested primers.[124] Nested primers are positioned within the region amplified by the first primer pair and by specific annealing, mimic the function of a specific probe by increasing the specificity and detection potential. The product can be directly detected from the agarose gel by visual inspection; a subsequent hybridization step is not needed, and therefore, the use of nested primers makes the test less laborious. In spite of this advantage, nested PCR is only rarely used for the detection of leptospires.

Conventional PCR is particularly prone to contamination, notably by amplicons. Strict procedures for sample processing and internal control are mandatory. A false-positive outcome resulting from contamination is obvious in cases where negative internal controls give positive signals,[127] but an apparent better performance of PCR compared to the reference test in other studies is debatable.[128,129]

A major disadvantage of PCR is that, in spite of the multitude of methods that have been reported, only two clinical evaluations have been executed, albeit on relatively small and geographically restricted sets of samples.[128,130] The performance of PCR depended on the type of clinical sample, varying from a clinical sensitivity of 44%–62%,[109,128,130] while clinical specificity has not adequately been addressed. Because of the insufficient evaluation and partly disappointing and doubtful results, the value of the conventional PCR for clinical application has remained uncertain.

Several primer pairs for PCR detection of leptospires have been described, some based on defined gene targets, most frequently 16S or 23S rRNA genes[124,125,127] and repetitive elements,[39,131,132] while others have been constructed from genomic libraries.[36] Leptospiral ribosomal genes have low sequence variation,[47,48] often leading to amplification of DNA from both pathogenic and saprophytic strains. Since specific detection of pathogenic leptospires is essential for a diagnostic PCR, an additional hybridization with a pathogen-specific probe is an indispensable requirement.[124]

Another limitation of PCR assays for the diagnosis of leptospirosis is their inability to identify the infecting serovar. While this is not significant for individual patient management, the identity of the serovar has an important epidemiological and public health value.

The specificity of the PCR depends on the properties of the primers or probes, enabling detection at the level of the genus, pathogenic, and saprophytic species and subspecies. Various strategies have been designed to achieve a certain level of strain specificity. Discrimination between pathogenic and saprophytic species has been achieved by selecting specific primers[19,52,133] or probes,[124] respectively. A recent approach to specifically target pathogenic leptospires is to select genes, such as *lipL32*, that are restricted to pathogenic species. However, information on such genes has become available relatively recently, and their application in convectional PCR is rare. PCR with primer pairs that have the potential to identify one or more pathogenic *Leptospira* species have been described by Gravekamp et al.[36] and Reitstetter.[37] Various approaches have claimed discrimination of leptospires at the serovar level, such as arbitrarily amplified (AP)-PCR,[38] IS-driven PCR,[39] differential electrophoresis in nondenaturing polyacrylamide gels,[41] SSCP and RFLP analysis of PCR products[42,134] and LSSP–PCR,[43] and sequencing of PCR products.[48,50,51,135]

Real-Time PCR. A large variety of real-time PCRs for the detection of leptospires have been described, a number of them targeting genes that are restricted to pathogenic leptospires.[134,135,136–141] Real-time PCR has several strong advantages over conventional PCR: (i) it greatly reduces carry-over contamination risk because amplification and detection occurs in the same containment; (ii) it is not dependent of an end-of-process analysis; amplification can be followed during the reaction; (iii) the amplification reaction and analysis is done within 2 h, which is more rapid than the conventional PCR, often exceeding a working day; and (iv) it provides a straightforward tool for quantification of the number of copies of a specific sequence in a DNA sample.

The procedure follows the general principle of PCR. The key feature is that the amplified DNA is quantified as it accumulates in the reaction in real time after each amplification cycle. Two common methods of detection are the use of fluorescent dyes that intercalate with double-strand DNA and the use of fluorescent reporter probes.

Intercalating agents produce fluorescence under appropriate illumination when bound to double-stranded DNA.[142] An increase in DNA product during PCR therefore leads to an increase in fluorescence intensity and is measured at each cycle. Real-time PCR mostly uses SYBR Green as the fluorescent agent. The SYBR Green real-time PCR has the limitation that all double-stranded DNA, including nonspecific amplifications, produces a signal. Nonspecific binding of SYBR Green and premature termination of the dye during the reaction is avoided by using small targets, preferable not exceeding 200 bp. To avoid erroneous positive scores by anomalous amplification, the specificity of the accumulated DNA is checked by a subsequent determination of its melting temperature (Tm). This analysis is based on the fact that by increasing temperature, strands of a specific DNA fragment will separate at a specific temperature, designated Tm. Strand separation at Tm results in a sudden decrease of the fluorescence intensity that can be expressed in a melting curve. The specificity of the products is based on the specificity of the Tm, which is defined by the specific NA contents. Since *Leptospira* spp. are based on differences in DNA composition, amplification of DNA from distinct species might generate amplicons with different NA contents' and thus species-specific Tm. This property provides the SYBR Green real-time PCR with the tool to use Tm for concomitant species determination. In practice, this requires targets that are conserved within the species, such as *rrs*.[140] It does not hold for segments with variable NA composition within species.[141] SYBR Green real-time PCR has the advantage of being relatively easy and affordable. SYBR Green real-time PCR appears highly suitable for early diagnosis. Analytical sensitivities may range from 1 to 30 copies, depending on the species of the strain and the type of clinical sample, while diagnostic sensitivity and specificity range up to 100%.[138,141]

Fluorescent reporter probes, also referred to as molecular beacons, are RNA or DNA-based probes that specifically hybridize to the target sequence. In principle, they carry a fluorescent reporter at one end and a quencher of fluorescence at the opposite end. In bound form of the probe, the quencher is in close proximity to the reporter, inhibiting its fluorescence upon illumination at a suitable wave length. During the process of strand duplication, the $5' \rightarrow 3'$ exonuclease activity of the heat-stable DNA polymerase degrades the probe. The reporter becomes separated from the quencher and starts emitting light. An increase in the product targeted by the reporter probe at each PCR cycle therefore causes a proportional increase in fluorescence by the released reporter. Determination of a melting curve is not possible. But this lack is compensated by the sequence-specific annealing of the probe that ensures the specificity of the signal. A great advantage of this approach is the possibility of using different colored labels. Currently, up to five reporters with emissions at different wave lengths can be detected simultaneously. This enables the design of multiplex systems, for example, for targeting different genes in one reaction or for species- or strain-specific detection. Real-time PCR based on molecular beacons reaches a detection threshold of 1–10 copies. It is often advocated as being more accurate and reliable than the SYBR Green approach. However, the method is more expensive, and in practice both approaches reach a comparable diagnostic accuracy.[141,143] The choice for either approach may be driven by the need for cost-efficiency.

In diagnostic laboratories, detection of pathogenic leptospires by real-time PCR offers considerable benefits compared to conventional methods. It combines rapidity with a low contamination risk and displays a high diagnostic accuracy. There are two real-time PCRs, either on SYBR Green or on molecular beacon technology, which have been validated.[141,143]

A major bottleneck of PCR, conventional or real-time, lies in the NA extraction step, which requires the presence of 10–1000 leptospires per mL clinical fluids in order to be detectable, provided that extraction occurs with 100% efficiency.[138,141] This greatly obviates the need for improved NA extraction assays or nonenzymatic amplification methods that do not need a high degree of DNA purity.

97.2 METHODS

97.2.1 Sample Preparation

For culturing and serology, a variety of samples can be used, depending on the presence of leptospires or antibodies in a distinct sample at a certain stage of illness. Usually, samples that are appropriate for culturing are not useful for antibody detection, and vice versa. Precautions and sample preparation procedures of blood, either coagulated or not, are, according to standard methods, mentioned in text books. Heat inactivation of serum samples at 56°C for 30 min diminishes infection risks with leptospires and reduces risk of infection with other infectious agents, while leaving serological properties of the sample largely unchanged. Heat inactivation of noncoagulated blood samples leads to clotting and therefore is not possible.

PCR can be applied on all samples that are suitable for isolation by culturing, although PCR on blood samples might suffer less from (recently started) antibiotic treatment. Crossover contamination with amplicons, leading to false-positive results, presents a major concern for PCR. Contamination can be reduced by taking the proper precautions.

The most important precaution is the implementation of a strict management policy. This consists of the use of clean rooms for DNA extraction from clinical specimens and reaction-mixture preparation, which should be well separated from rooms used for amplification and detection. Entering rooms prone to amplicon contamination prior to entering the clean rooms should be prohibited, and the inclusion of a suitable and appropriate number of negative controls in the reaction run is mandatory.

The problem related to crossover of amplicons can greatly be reduced by application of the UNG-dUTP approach.[144] The enzyme uracil-*N*-glycosylase (UNG) initiates degradation of DNA containing uracil instead of thymine as a base. Most heat-stable DNA polymerases are capable of incorporating 2′-deoxyuradine. When adding dUTP to the PCR mix, the nucleoside will be inserted in the amplicon, making it sensitive to UNG. By including incubation with UNG prior to the amplification, eventual contaminating amplicons having 2′-deoxyuradine incorporated will be degraded. The approach is elegant but suffers from the disadvantage that the sensitivity of the PCR is reduced at various degrees depending on the ability of the polymerase to process dUTP. This can be compensated by mixing dTTP and dUTP to an optimal ratio,[81] but the balance is fragile and mixing does not fully restore the maximum sensitivity. Probably, therefore, the approach has not found a wide applicability.

Several NA extraction procedures have been applied. The silica gel method of Boom and coworkers[145] is most widely used. Several commercial silica-based NA extraction kits are available and potentially reduce the processing time of the samples prior to PCR. PCR is very sensitive and in principle allows the detection of a single copy. However, extraction methods are the main bottleneck, increasing the practical sensitivity to 10^2–10^3 leptospires per mL of clinical liquid.

Inhibiting substances notably remain when extracting DNA from tissue (kidney) samples and urine. To meet with this, a repeated extraction on the sample has been applied.[143] We use the QIA amp DNA Mini Kit (Qiagen, GmbH, D-40724 Hilden, Germany) for DNA extraction, following the specific recommendations for NA extraction from body fluids and tissue.[141] We achieved better results from kidney-extracted DNA by using a tenfold dilution in sterile, highly purified distilled water prior to amplification. It should be noted that both methods result in a lower DNA load due to loss of target material or dilution, but the overall balance is in favor of the extra steps.

97.2.2 Detection Procedures

97.2.2.1 Conventional PCR Identification

(i) secY-Deduced PCR for Pathogenic Leptospira spp. The PCR has been developed to amplify a secY segment from all pathogenic spp. comprising the 285 bp segment as amplified by primer set G1/G2 described by Gravekamp et al.[36] and enables subsequent sequencing and reliable phylogeny[50] (Figure 97.3). This PCR is also applicable for detection of leptospires in (clinical) materials. It uses primers SecYII (5′-GAATTTCTCTTTTGATCTTCG-3′) and SecYIV (5′-GAGTTAGAGCTCAAATCTAAG-3′). Gene *secY* sequencing and phylogeny can also be obtained with the primers SecYIVF and SecYIV described below but following the protocol here.[141]

Procedure. The reaction is performed in a final volume of 25 μL consisting of 2.5 μL 10× HotGoldstar DNA Polymerase buffer (Eurogentec, Belgium), 3 μL 25 mM $MgCl_2$, 0.25 μL 25 mM dNTPs, 0.2 μL 20 μM of each primer, 0.1 μL 5 U/μL Goldstar DNA polymerase (Eurogentec), 10 μL target DNA (or distilled water for the negative control), and 8.75 μL distilled water.

For primers SecYII and SecFIV the following cycling parameters are used: initial cycle of 95°C for 10 min; 40 cycles of 94°C for 1 min, 55°C for 1 min, and 72°C for 2 min followed by 72°C for 8 min and 18°C for 5 min.

For primers SecYIVF and SecYIV, the following cycling parameters are used: initial cycle of 95°C for 10 min; 40 cycles of 94°C for 1 min, 54°C for 30 s, 72°C for 1 min followed by 72°C for 8 min and 18°C for 5 min.

Note: With samples extracted from tissue materials, better results are obtained by a tenfold dilution of extracted DNA to evade PCR inhibition. For sequencing, the amplicon is applied to (1.5%) agarose gel electrophoreses. The DNA band is excized from the gel according to standard procedures[142] and purified with the MinElute PCR Purification Kit (Qiagen).

(ii) PCR for the Discrimination of Pathogenic and Saprophytic Leptospira spp. Murgia et al.[19] developed PCR assays for distinguishing saprophytic and pathogenic leptospires, with primers from *rrs:* (i) Sapro and Sapro2 for saprophytic leptospires; and (ii) Lepat1, Lepat2, L3, and L4 for pathogenic leptospires. The pathogen-specific primers can be used in various combinations, single round or nested[19] but usually Lepat1 and Lepat2 are applied.

Procedure. The reactions are carried out in 50 μL reaction mixtures containing 10 mM TRIS-HCl pH 8.3, 50 mM KCl, 1.5 mM $MgCl_2$, 0.001% gelatine, 200 μM each deoxynucleoside triphosphates, 1.25 U of *Taq* DNA polymerase (Promega), and 0.5 μM of each primers. Ten ng of purified DNA is used as template.

With primers Sapro1 and Sapro2 the following cycle conditions are used: 93°C for 3 min and then 35 cycles of 93°C for 1 min, 63°C for 1.5 min, and 72°C for 2 min. After the last cycle, extension is continued for a further 10 min. With Lepat 1-Lepat 2, the only modification is the annealing temperature at 48°C.

(iii) ompL-Derived PCR for Species Specific Detection of Pathogenic Leptospira PCR is performed with seven primer sets that amplify 6 major pathogenic *Leptospira* spp.[37]

Procedure. DNA is extracted from cultures (30°C for 6–8 days) with a Qiagen Blood and Cell Culture DNA Midi kit (Qiagen). The amount of isolated DNA is adjusted to 25 ng/μL with nuclease-free water (Ambion). The PCR mix (25 μL) consists of 10 μL Eppendorf MasterMix (final concentrations: 1.25 U *Taq* polymerase, 50 mM KCl, 30 mM Tris-HCl, 1.5 mM Mg^{2+}, 0.1% Igepal-CA630, 200 μM of each dNTP) (Eppendorf), 1.5 μL of each forward and reverse primer (2 pmol/μL stocks), 10 μL nuclease-free water and 2 μL DNA template (nuclease-free water for negative controls). The following cycling parameters are used: initial cycle of 94°C for 2 min; 30 cycles of 94°C for 30 s, 62°C for 30 s and 72°C for 30 s and 72°C for 7 min using a GeneAmp PCR System 9700 (Applied Biosystems) or Mastercycler ep-gradient (Eppendorf). For less stringent PCR conditions, the temperature gradient mode on the Master ep-gradient is set to an annealing temperature range between 39.9°C and 60.1°C.

97.2.2.2 Real-Time PCR Identification

Various real-time PCRs have been described, mostly without satisfying validation. Because of the importance of the availability of validated methods, we focus on two real-time PCRs that have been sufficiently validated.

(i) Validated secY-Derived Real-Time PCR (SYBR Green) for Detection of Pathogenic Leptospira spp. The primers SecYIVF (5′-GCGATTCAGTTTAATCCTGC-3′) and SecYIV (5′-GAGTTAGAGCTCAAATCTAAG-3′) specifically amplify a 203 bp fragment of the *secY* gene of pathogenic *Leptospira* species.[141] The PCR includes an internal amplification control (IAC) constructed by insertion of a similar *secY* segment of *Treponema pallidum*, flanked by the SecYIVF and SecYIV sequences into vector pGEM-T. The IAC amplification product can de distinguished from *Leptospira*-derived amplicons by their distinct Tm's.

Usage	Primer	Sequence (5′→3′)	Expected Product (bp)
Saprophytic leptospires	Sapro1	AGAAATTTGTGCTAATACCGAATGT	240
	Sapro2	GGCGTCGCTGCTTCAGGCTTTCG	
Pathogenic leptospires	Lepat1	GAGTCTGGGATAACTTT	308
	Lepat2	TCACATCG(CT)TGCTTATTTT	
Pathogenic leptospires	L3	GAT(CTG)T(GTA)(CT)(GA)GGTAAAGATT(CT)ATT	
	L4	TGAGGGTTAAAACCCCCAAC	

Species	Primer	Sequence (5′→3′)
L. interrogans	Intergroup A fwd	CTACTGGCGGCTTGATCAAC
	Intergroup A rev	CTGGATCTGTTCCGTCTGCGATC
L. interrogans	Intergroup B fwd	CTTGATAGAACCACTGGTGGTGCC
	Intergroup B rev	CTGGATCGGTTCCATCGCTCAG
L. borgpetersenii	Borgpeter fwd	CTTGATAGAACAACAGGCGGCATCATC
	Borgpeter rev	GCTAATAAGTTTGCAATGCTCGTAAC
L. kirschneri	Kirschner fwd	CGGTTTGATCAATGCGAGAAGCACC
	Kirschner rev	TTGGATCCGTTCCGTCTGCGATT
L. santarosai	Santarosai fwd	CTTATCAATGCAAGATCTACCAAAGGT
	Santarosai rev	GCGGATATGTTCCCGAGTAGTAATC
L. noguchii	Noguchii fwd	GCGGATTTATCAATGCAAGAAGTACA
	Noguchii rev	CCGGATCGGTTCCGTCTGCGATCAG
L. weilii	Weilii fwd	AGGCTGATATTGCAGGCTTC
	Weilii rev	CGGAATCGAATATGTTCACGAGTG

Procedure. The real-time PCR mix (25 μL) contains 12.5 μL 2× IQ SYBR Green Supermix (IQ™SYBR® Green Supermix, Bio-Rad, with 100 mM KCl, 40 mM Tris-HCl, pH 8.4, 0.4 mM of each dNTP, 50 U/mL iTaq DNA polymerase, 6 mM $MgCl_2$, 20 nM fluorescein, and stabilizers), 10 μL DNA template, and 0.5 μL IAC corresponding to 0.87.10 pg per reaction. Forward and reverse primers are added at a final concentration of 400 nM each.

PCR amplification is performed in iCycler iQ PCR plates (96 wells), sealed with iCycler iQ Optical Tape using an iQ™5 Multicolor Real-Time PCR Detection System (Bio-Rad Laboratories), with the following cycling parameters: initial cycle at 95°C for 10 min; 40 cycles of 95°C for 5 s, 54°C for 5 s, 72°C for 15 s. The reaction is then stopped at 95°C for 2 min, cooled at 20°C for 1 min, and the product is subsequently melted (70°C–94°C with plate readings every 0.5°C) to determine the Tm. Analysis is done with the Bio-Rad iQ5 2.0 Standard Edition Optical System Software (V2.0.148.060623).

(ii) Validated Real-Time PCR (TaqMan LightCycler) for Detection of Pathogenic Leptospira spp. This is a *rrs*-derived molecular beacon PCR with primers LeptoF (5′-CCC GCGTCCGATTAG-3′) and LeptoR (5′-TCCATTGTGCCGA/GACAC-3′) and probe 5′-[FAM]CTCACCAAGGCGACG ATCGGTAGC-3′[TAMRA].[143]

Procedure. The PCR mix (20 μL) contains 10 μL mastermix composed of 2 μL Fast-start Hybridization Probe Mix (Roche), 3.6 μL of 25 mM $MgCl_2$, 20 pM of forward and revered primers each, 4 pM 6-carboxyfluorescien-labeled (FAM) Black Hole Quencher (BHQ1) quenched Taqman probe, filled up with double distilled water (ddH_2O), and 10 μL DNA extract, loaded into a LightCycler capillary and then centrifuged to mix. Positive control contains DNA of *L. interrogans* serovar Australis strain Ballico (2 μL plus 8 μL ddH_2O), and negative controls contain DNA of *L. biflexa*, serovar Patoc strain Patoc I (2 μL plus 8 μL ddH_2O), and 10 μL ddH_2O, respectively.

The amplification is performed using the Roche LightCycler version 1 (other versions might be equally applicable), and fluorescent data acquisition is performed at the end of each annealing/extension cycle, in this case in Channel F1. Amplification cycles consist of 95°C for 8 min followed by 50 cycles of 95°C for 10 s and 60°C for 1 min.

(iii) Quantitative Real-Time PCR (SYBR Green-LightCycler) for Detection of Pathogenic Leptospira spp. This quantitative real-time PCR (qPCR) amplifies a 330 bp DNA segment in a LightCycler (Roche Applied Science) using the *rrs*-deduced primers LA (5′-GGCGGCGCGTCTTAAACATG-3′) and LB (5′-TTCCCCCCATTGAGCAAGATT-3′) for quantitation of leptospires in hamster tissue.[146] For quantification, signals are systematically normalized to the guinea pig housekeeping gene glyceraldehyde-3-phosphate dehydrogenase (GAPDH) using forward primer 5′-AATGGGAAGCTCACAGGTATGG-3′, and reverse primer 5′-ATGTCATCGTATTTGGCCGGT-3′. For application on tissue of other mammalian species other standards need to be developed.

Procedure. Reactions are performed with LightCycler FastStart DNA master SYBR Green I (Roche Applied Science) using a 20 μL volume in each reaction capillary. PCR conditions are as follows (ramp rates of 20°C/s): initial denaturation 95°C for 10 min, followed by 45 cycles of 95°C for 10 s, 57°C for 8 s and 72°C for 10 s, and fusion at 95°C for 6 min. Results are expressed as the number of *Leptospira* detected in 1 μg tissue DNA or in 1 μL serum deduced from a standard curve with DNA extracted from tenfold dilutions of known numbers of leptospires.

97.3 CONCLUSIONS AND FUTURE PERSPECTIVES

Leptospirosis presents in a spectrum ranging from a mild to severe, potentially fatal illness and has a wide variety of clinical manifestation, notably at an early stage. As such it mimics several other infectious diseases that are endemic and epidemic in the same environmental conditions. Therefore, leptospirosis is difficult to diagnose on clinical grounds, and support of the laboratory is indispensable.

Confirmation of the presumptive diagnosis at an early stage is relevant as antibiotic treatment is most effective when started early. Conventional diagnosis is based on the direct demonstration of infecting leptospires in clinical samples or indirectly, on the detection of anti-*Leptospira* antibodies. Direct detection of leptospires by DFM is notoriously unreliable. Isolation of leptospires by culturing provides evidence for leptospirosis but takes too long to be of value for timely diagnosis, while histological methods are mainly suitable for investigating post mortem materials.

Also serology has the drawback in proving late case confirmation. Serological tests reach optimal sensitivity when sufficient antibodies have been produced. Usually detection in blood becomes possible 5–10 days after onset of symptoms, which is too late for starting effective antibiotic treatment.

Thus, in spite of the wide variety of tests that are currently available, clinicians have to decide on treatment upon clinical suspicion and, when available, compatible epidemiological information. Confirmation usually is obtained afterwards. Yet, laboratory confirmation is important as this (i) raises alertness for the occurrence of leptospirosis and contributes to outbreak alert and investigation and (ii) might contribute to the awareness of clinicians to focus also on less-known forms of leptospirosis that are usually missed. Especially the design of novel, easy, rapid, and stable point-of-care tests presents a significant improvement in situations that are devoid of a proper laboratory infrastructure, which actually counts for most high-prevalence countries.

Probably mild and unknown or unexpected forms of leptospirosis present the major burden of the disease. Assessment of such other forms of leptospirosis can be achieved by active case detection on cases with undifferentiated fever or by cross-sectional epidemiological studies. Conventional tests are very useful tools for the execution of epidemiological surveys, revealing spread and occurrence of the disease and its causative serovars in animals and humans.

Nucleic acid amplification techniques have provided a significant step forwards to the badly needed early confirmation of leptospirosis. The PCR, in particular the real-time PCR, is very promising. In potential, the real-time PCR is highly sensitive, enabling the detection of as few as one copy of the genome. As far as satisfying evaluations have been presented, the test has a high diagnostic accuracy, notably at the early acute stage of the disease when leptospires are abundant in the blood. There is, however, a drawback. The heat-stable DNA polymerase used in the reaction is very sensitive to inhibitors and needs highly pure DNA to act effectively. Available DNA extraction methods or kits are inadequate in the sense that they are not able to result in highly concentrated DNA copies in the final extract. While the PCR is able to amplify one copy, this equals at least 50 copies/mL body fluid. Although promising, notably real-time PCR is too expensive for wide application in economically deprived countries where leptospirosis occurs most frequently. Isothermal amplification methods require less demanding equipment and might be a more feasible option for such situations. The isothermal Nucleic Acid Sequence Based Amplification (NASBA) technique has been presented as an alternative to PCR almost two decades ago[123] but failed to find wide applicabilty mainly because of technical complications and hence, disappointing performance in its practical application. Loop-mediated isothermal amplification (LAMP), which is much similar to NASBA, has been described as a follow-up only recently[147] but its usefulness as a diagnostic tool remains to be evaluated.

ACKNOWLEDGMENTS

I wish to thank Dr. Jarlath Nally, Veterinary Sciences Centre, University College, Dublin, and Dr. David Haake, Division of Infectious Diseases, UCLU School of Medicine, Los Angeles for their generous gifts of the electron microscopic image of leptospires and the graphical presentation of the leptospiral cell wall, respectively. The phylogenetic tree is kindly provided by A. Ahmed, WHO/FAO/OIE and National Collaborating Centre for Reference and Research on Leptospirosis.

REFERENCES

1. Gouveia, E.L. et al., Leptospirosis-associated severe pulmonary hemorrhage syndrome, Salvador, Brazil, *Emerg. Infect. Dis.*, 14, 505, 2008.
2. World Health Organization, *Human Leptospirosis: Guidance for Diagnosis, Surveillance and Control*, World Health Organization, Geneva, 2003.
3. Levett, P.N., Leptospirosis, *Clin. Microbiol. Rev.*, 14, 296, 2001.
4. Bharti, A.R. et al., Leptospirosis: A zoonotic disease of global importance, *Lancet Infect. Dis.*, 3, 757, 2003.
5. Zaki, R.S., Shieh, W.-J., and the Epidemic Working Group, Leptospirosis associated with outbreak of acute febrile illness and pulmonary hemorrhage, Nicaragua, 1995, *Lancet*, 347, 535, 1996.
6. Sanders, E.J. et al., Increase of leptospirosis in dengue-negative patients after a hurricane in Puerto Rica in 1996, *Am. J. Trop. Med. Hyg.*, 61, 935, 1999.
7. Ko, A.I. et al., Urban epidemic of severe leptospirosis in Brazil. Salvador, *Lancet*, 354, 820, 1999.
8. Smits, H.L. et al., Simple latex agglutination assay for rapid diagnosis of human leptospirosis, *J. Clin. Microbiol.*, 38, 1272, 2000.
9. Smits, H.L. et al., International multicenter evaluation of the clinical utility of a dipstick assay for the detection of *Leptospira*-specific immunoglobulin M antibodies in human serum specimens, *J. Clin. Microbiol.*, 37, 2904, 1999.
10. Smits, H.L., Hartskeerl, R.A., and Terpstra, W.J., An international multicentre evaluation of a dipstick assay for human leptospirosis, a quick and easy test for the serodiagnosis of acute human leptospirosis, *Trop. Med. Int. Health*, 5, 124, 2000.
11. McBride, A.J.A. et al., Evaluation of four whole-cell *Leptospira*-based serological tests for the diagnosis of urban leptospirosis, *Clin. Vac. Immunol.*, 14, 1245, 2007.
12. Blacksell, S.D. et al., Limited diagnostic capacities of two commercial assays for the detection of *Leptospira* immunoglobulin M antibodies in Laos, *Clin. Vac. Immunol.*, 13, 1166, 2006.
13. World Health Organization, Leptospirosis worldwide, 1999, *Wkly. Epidemiol. Rec.*, 74, 237, 1999.
14. LeptoNet: www.leptonet.net.
15. Hartskeerl, R.A., Leptospirosis: current status and future trends, *Indian J. Med. Microbiol.*, 24, 309, 2006.
16. Hartskeerl, R.A., Leptospirose: een onderschatte infectieziekte, *Ned. Tijdschft. Med. Microbiol.*, 4, 94, 2004.
17. Johnson, R.C., and Harris, V.G., Differentiation of pathogenic and saprophytic leptospires. I. Growth at low temperatures, *J. Bacteriol.*, 94, 27, 1967.
18. Faine, S. et al., *Leptospira and Leptospirosis*, 2nd ed., Medi Sci Writing, Melbourne, 1999.
19. Murgia, R. et al., Oligonucleotides specific for pathogenic and saprophytic *Leptospira* occurring in water, *FEMS Microbiol. Lett.*, 148, 27, 1997.
20. Kositanont, U. et al., Detection and differentiation between pathogenic and saprophytic *Leptospira* spp. by multiplex polymerase chain reaction, *Diagn. Microbiol. Infect. Dis.*, 57, 117, 2007.

20a. International Committee on Systematics of Prokaryotes: Subcommittee on the taxonomy of *Leptospiraceae*. Minutes of the closed meeting, 18 September 2007, Quito, Ecuador. *Int. J. Syst. Evol. Microbiol.*, 58, 1049, 2008.

21. Kmety, E., and Dikken, H., *Classification of the Species Leptospira interrogans and History of Its Serovars*, University Press Groningen, Groningen, 1993.
22. Cerqueira, G.M., and Picardeau, M., A century of Leptospira strain typing, *Infect. Gen. Evol.*, 9, 760, 2009.
23. Kmety, E., Factoranalyse von Leptospiren der Icterohaemorrhagiae und eigener verwanter serogruppen, *Slovak Academy of Sciences, Bratislava*, 13, 3, 1967.
24. Hartskeerl, R.A. et al., *Manual International Course on Laboratory Methods for the Diagnosis of Leptospirosis*, KIT, Amsterdam, 2006.
25. De los Valverde, M.A. et al., Arenal, a new *Leptospira* serovar of serogroup Javanica, isolated from a patient in Costa Rica, *Infect. Genet. Evol.*, 8, 529, 2008.
26. Yasuda, P. et al., Deoxyribonucleic acid relatedness between serogroups and serovars in the family Leptospiraceae with proposals for seven new *Leptospira* species, *Int. J. Syst. Bacteriol.*, 37, 407, 1987.

27. Brenner, D.J. et al., Further determination of DNA relatedness between serogroups and serovars in the family Leptospiraceae with a proposal for *Leptospira alexanderi* sp. nov. and four new *Leptospira* genomospecies, *Int. J. Syst. Bacteriol.*, 49, 839, 1999.
28. Ramadass, P. et al., Genetic characterization of pathogenic *Leptospira* species by DNA hybridization, *Int. J. Syst. Bacteriol.*, 42, 215, 1992.
29. Le Febre, R.B., and Thiermann, A.B., DNA homology studies of leptospires of serogroups Sejroe and Pomona from cattle and swine, *Am. J. Vet. Res.*, 47, 959, 1986.
30. Van Eys, G.J.J.M. et al., Characterization of serovars of the genus *Leptospira* by DNA hybridization with Hardjobovis and Icterohaemorrhagiae recombinant probes with special attention to serogroup Sejroe, *J. Clin. Microbiol.*, 29, 1042, 1991.
31. Marshall, R.B., Wilton, B.E., and Robinson, A.J., Identification of *Leptospira* serovars by restriction-endonuclease analysis, *J. Med. Microbiol.*, 14, 163, 1981.
32. Robinson, A.J. et al., Differentiation of subtypes within *Leptospira interrogans* serovars Hardjo, Balcanica and Tarassovi, by bacterial restriction-endonuclease DNA analysis (BRENDA), *J. Med. Microbiol.*, 15, 331, 1982.
33. Ellis, W.A., Montgomery, J.M., and Thiermann, A.B., Restriction endonuclease analysis as a taxonomic tool in the study of pig isolates belonging to the Australis serogroup of *Leptospira interrogans*, *J. Clin. Microbiol.*, 29, 957, 1991.
34. Hermann, J.L. et al., Pulsed-field gel electrophoresis of *Not*I digests of leptospiral DNA: A new rapid method of serovar identification, *J. Clin. Microbiol.*, 30, 1696, 1992.
35. Galloway, R.L., and Levett, P.N., Evaluation of a modified Pulsed-Field Gel Electrophoresis approach for the identification of *Leptospira* serovars, *Am. J. Trop. Med. Hyg.*, 78, 628, 2008.
36. Gravekamp, C. et al., Detection of seven species of pathogenic leptospires by PCR using two sets of primers, *J. Gen. Microbiol.*, 139, 1691, 1993.
37. Reitstetter, R.E., Development of species-specific PCR primer sets for the detection of *Leptospira*, *FEMS Microbiol. Lett.*, 264, 31, 2006.
38. Brown, P.D., and Levett, P.N., Differentiation of *Leptospira* species and serovars by PCR-restricted endonuclease analysis, arbitrarily promed PCR and low-stringency PCR, *J. Med. Microbiol.*, 46, 173, 1997.
39. Zuerner, R.L., Alt, D., and Bolin, C.A., IS*1533*-based PCR assay for the identification of *Leptospira interrogans* sensu lato serovars, *J. Clin. Microbiol.*, 33, 3284, 1995.
40. Barocchi, M.A. et al., Identification of new repetitive element in *Leptospira interrogans* serovar Copenhageni and its application to PCR-based differentiation of *Leptospira* serogroups, *J. Clin. Microbiol.*, 39, 191, 2000.
41. Oliveira, M.A.A. et al., The use of nondenaturing silver-stained polyacrylamide gel analysis of polymerase chain reaction amplification products for the differential diagnosis of *Leptospira interrogans* infection, *Diagn. Microbiol. Infect. Dis.*, 22, 343, 1995.
42. Matthias, M.A., and Levett, P.N., Identification of leptospira serovars by PCR-SSCP analysis, *West Indian Med. J.*, 49, 24, 2000.
43. Oliveira, M.A.A. et al., Low-stringency single specific primer PCR for identification of *Leptospira*, *J. Med. Microbiol.*, 52; 127, 2003.
44. Vijayachari, P. et al., Use of fluorescent amplified fragment length polymorphism for molecular epidemiology of leptospirosis in India, *J. Clin. Microbiol.*, 42, 3575, 2004.
45. Slack, A. et al., An improved multiple-locus variable number of tandem repeats analysis for *Leptospira interrogans* serovar Australia: A comparison with fluorescent amplified fragment length polymorphism analysis and its use to redefine the molecular epidemiology of this serovar in Queensland, Australia, *J. Med. Microbiol.*, 55, 1549, 2006.
46. Zuerner, R.L., and Alt, D.P., Variable Nucleotide Tandem-Repeat Analysis revealing a unique group of *Leptospira interrogans* serovar Pomona isolates associated with California sea lions, *J. Clin. Microbiol.*, 47, 1202, 2009.
47. Haake, D.A. et al., Molecular evolution and mosaicism of leptospiral outer membrane proteins involves horizontal DNA transfer, *J. Bacteriol.*, 186, 2818, 2004.
48. Ahmed, N. et al., Multilocus sequence typing method for identification and genotypic classification of pathogenic *Leptospira* species, *Ann. Clin. Microbiol. Antimicrob.*, 5, 28, 2007.
49. Bulach, D.M. et al., Genome reduction in *Leptospira borgpetersenii* reflects limited transmission potential, *Proc. Natl. Acad. Sci. USA*, 103, 14560, 2006.
50. Victoria, B. et al., Conservation of the *S10-spc-alpha* locus within otherwise highly plastic genomes provides phylogenetic insight into the Genus *Leptospira*, *PLoS One*, 3, e2752, 2008.
51. Thaipadungpanit, J. et al., A dominant clone of *Leptospira interrogans* associated with an outbreak of human leptospirosis in Thailand, *PloS. Negl. Trop. Dis.*, 1, e56, 2007.
52. Perolat, P. et al., *Leptospira fainei* sp. nov., isolated from pigs in Australia, *Int. J. Syst. Bacteriol.*, 48, 851, 1998.
53. Matthias, M.A. et al., Human leptospirosis caused by a new, antigenically unique *leptospira* associated with a rattus species reservoir in the peruvian Amazon, *PLoS Negl. Trop. Dis.*, 2, e213, 2008.
54. Ellinghausen, H.C., Some observations on cultural and biochemical characteristics of Leptospira Pomona, *J. Inf. Dis.*, 106, 237, 1960.
55. Vaneseltine, W.P., and Staples, S.A., Nutritional requirements of leptospirae. I. Studies on oleic acid as growth factors for a strain of *Leptospira Pomona*, *J. Inf. Dis.*, 108, 262, 1961.
56. Taneck, J.L., Henneberry, R.C., and Cox, C.D., Growth requirements of pathogenic leptospires, *Infect. Immun.*, 7, 886, 1973.
57. McBride, A.J. et al., Genetic diversity of the Leptospiral immunoglobulin-like (Lig) genes in pathogenic *Leptospira* spp., *Infect.Genet.Evol.*, 9, 196, 2009.
58. Nascimento, A.L.T.O. et al., Comparative genomics of two *Leptospira interrogans* serovars reveals novel insights into physiology and pathogenesis, *J. Bacteriol.*, 186, 2164, 2004.
59. Picardeau, M. et al., Genomic sequence of the saprophyte *Leptospira biflexa* provides insights into the evolution of *Leptospira* and the pathogenesis of leptosprosis, *PloS One*, 3, e1607, 2008.
60. Barnett, J.K. et al., Expression and distribution of leptospiral outer membrane components during renal infection of hamsters, *Infect. Immun.*, 67, 853, 1999.
61. Nally, J.E. et al., Changes in lipopolysaccharide O antigen distinguish acute versus chronic *Leptospira interrogans* infections, *Infect. Immun.*, 73, 3251, 2005.
62. Ren, S.X. et al., Unique physiological and pathogenic features of *Leptospira interrogans* revealed by whole-genome sequencing, *Nature*, 422, 888, 2003.
63. Vijayachari, P. et al., Leptospirosis among schoolchildren of the Andaman & Nicobar Islands, India: Low levels of morbidity and mortality among pre-exposed children during an epidemic, *Epidemiol. Infect.*, 132, 1115, 2004.

64. Ashford, D.A. et al., Asymptomatic infection and risk factors for leptospirosis in Nicaragua, *Am. J. Trop. Med. Hyg.*, 63, 2000, 249, 2000.
65. Vijayachari, P. et al., *Leptospira interrogans* serovar Valbuzzi: A cause of severe pulmonary haemorrhages in the Andaman Islands, *J. Med. Microbiol.*, 52, 1, 2003.
66. Trevejo, R.T. et al., Epidemic leptospirosis associated with pulmonary hemorrhage—Nicaragua, 1995, *J. Infect. Dis.*, 178, 1457, 1998.
67. Speelman, P., and Hartskeerl, R., Leptospirosis. In Fauci, A.S. et al. (Editors), *Harrison's Principles of Internal Medicine*, 17th ed., Chapter 164, McGraw Hill Medical, New York, 2008.
68. McBride, A.J. et al., 2005. Leptospirosis, *Curr. Opin. Infect. Dis.*, 18, 376, 2005.
69. Watt, G. et al., Placebo-controlled trial of intravenous penicillin for severe and late leptospirosis, *Lancet*, 331, 433, 1988.
70. Sehgal, S.C., Randomized controlled trial of doxycycline prophylaxis against leptospirosis in an endemic area, *Int. J. Antimicrob. Agents*, 13, 24, 2000.
71. Takafuji, E.T., An efficacy trial of doxycycline chemoprophylaxis against leptospirosis, *N. Eng. J. Med.*, 310, 497, 1984.
72. Wagenaar, J.F.P., De Vries, P.J., and Hartskeerl, R.A., Leptospirosis with pulmonary hemorrhage, caused by a new strain of serovar Lai: Langkawi, *J. Travel. Med.*, 11, 3879, 2004.
73. Terpsta, W.J., Jarboury-Postema, J., and Korver, H., Immunoperoxidase staining of leptospires in blood and urine, *Zbl. Bakt. Mikrobiol. Hyg.*, I, 254, 534, 1983.
74. Faine, S., Silver staining of spirochaetes in single tissue sections, *J. Clin. Pathol.*, 18, 381, 1965.
75. Kramer, K.J. et al., Evaluation of the quantitative buffy coat analysis (QBC) system for the detection of *Leptospira* in human blood, *South Asian J. Trop. Med. Pub. Health*, 25, 788, 1994.
76. Faine, S., *Guidelines for Leptospirosis Control*, WHO offset Publication, Geneva, 1982.
77. Ellinghausen, H.C., and McCullough, W.G., Nutrition of *Leptospira* Pomona and growth of 13 other Serotypes: Fractionation of oleic albumin complex and a medium of bovine albumin and polysorbate 80, *Am. J. Vet. Res.*, 26, 45, 1965.
78. Johnson, R.C., and Harris, V.G., Differentiation of pathogenic and saprophytic leptospires. I. Growth at low temperatures, *J. Bacteriol.*, 94, 27, 1967.
79. Hartskeerl, R.A. et al., Classification of *Leptospira* from the eyes of horses suffering from recurrent uveitis, *J. Vet. Med.*, B, 51, 110, 2004.
80. Wuthiekanun, V. et al., Optimization of culture of *Leptospira* from humans with leptospirosis, *J. Clin. Microbiol.*, 45, 1363, 2007.
81. Bal, A.E. et al., Detection of leptospires in urine with the polymerase chain reaction for early diagnosis of leptospirosis, *J. Clin. Microbiol.*, 32, 1894, 1994.
82. Wolff, J.W., *The Laboratory Diagnosis of Leptospirosis*, Charles C Thomas, Springfield, IL, 1954.
83. Gussenhoven, G.C. et al., LEPTO dipstick, a dipstick assay for detection of *Leptospira*-specific immunoglobulin M antibodies in human sera, *J. Clin. Microbiol.*, 35, 92, 1997.
84. Martin, L., and Petit, A., *Spirochetose icterohemorragique*, Manson et Cie, Paris, 1919.
85. Chappel, R.J. et al., Impact of proficiency testing on results of the microscopic agglutination test for the diagnosis of leptospirosis, *J. Clin. Microbiol.*, 42, 5484, 2004.
86. Cerqueira, G.M. et al., Monitoring *Leptospira* strain collections: The need for quality control, *Am. J. Trop. Med. Hyg.*, 82, 83, 2010.
87. Wolff, J.W., *The Laboratory Diagnosis of Leptospirosis*, Blackwell, Oxford, 1954.
88. Borg-Petersen, C., and Fagroeus, A., The influence of the antigen density and other factors on the serum titer in the agglutination-lysis-test for leptospirosis, *Acta Path. XXVI*, 4, 1949.
89. Kmety, E., Betrachtungen zum problem der paradoxen reaction und deren bedeutung in der serogiagnostik einiger leptospirosen, *Zbl. Bakt. Infekt. Hyg. I*, 170, 597, 1957.
90. Levett, P.N., Usefulness of serologic analysis as a predictor of the infecting serovar in patients with severe leptospirosis, *Clin. Infect. Dis.*, 36, 447, 2003.
91. Adler, B., and Faine, S., The antibodies involved in the human immune response to leptospiral infection, *J. Med. Microbiol.*, 11, 387, 1978.
92. Camargo, E.D. et al., ELISA-IgM applied to cerebrospinal fluid in human leptospirosis, *Serodiagn. Immunother. Infect. Dis.*, 7, 19, 1995.
93. Silva, M.V. et al., Application of anti-leptospira ELISA-IgM for the etiologic elucidation of meningitis, *Rev. Inst. Med. Trop. Sao Paulo*, 38, 153, 1996.
94. Silva, M.V. et al., Immunodiagnosis of human leptospirosis by dot-ELISA for the detection of IgM, IgG, and IgA antibodies, *Am. J. Trop. Med. Hyg.*, 56, 650, 1997.
95. McBride, A.J.A. et al., Evaluation of the EIE-IgM-Leptospirose assay for the serodiagnosis of leptospirosis, *Acta Trop.*, 102, 206, 2007.
96. Terpstra, W.J., Ligthart, G.S., and Schoone, G.J., ELISA for the detection of specific IgM and IgG in human leptospirosis, *J. Gen. Microbiol.*, 131, 377, 1985.
97. Cherry, W.B., Immunofluorescence techniques. In Lennette, E.H. et al. (Editors), *Manual of Clinical Microbiology*, 3rd ed., Chapter 11, American Society for Microbiology, Washington, 1980.
98. Torten, M., Shenberg, E., and Van der Hoeden, J., The use of immunofluorescence in the diagnosis of human leptospirosis by a genus-specific antigen, *J. Infect. Dis.*, 116, 537, 1966.
99. Galton, M.M. et al., A rapid microscopic slide screening test for the serogiagnosis of leptospirosis, *Am. J. Vet. Res.*, 19, 505, 1958.
100. Wolff, J.W., and Bohlander, H.J., Evaluation of Galton's macroscopic slide test for the serodiagnosisi of leptospirosis in human serum samples, *Ann. Soc. Belg. Med. Trop.*, 46, 123, 1966.
101. Mazzonelli, J., Dorta de Mazzonelli, G., and Mailloux, M., Possibilité de diagnostique sérologique des leptospires à l'aide d'un antigèn unique, *Méd. Malad. Infect.*, 4, 252, 1974.
102. Wanyanga, S.W. et al., Comparison of the DIFCO and Patoc I slide antigens in the screening of leptospirosis, *Comp. Immunol. Microbiol. Infect. Dis.*, 10, 155, 1987.
103. Marı´n-Le´on, I. et al., Prospective evaluation of the macroagglutination slide test for *Leptospira*, *Serodiagn. Immunother. Infect. Dis.*, 8, 191, 1997.
104. McComb, D.E. et al., The use of erythrocyte sensitizing substances in the diagnosis of leptospirosis. II. The sensitized erythrocyte lysis test, *Am. J. Trop. Med. Hyg.*, 6, 101, 1957.
105. McComb, D.E. et al., The use of erythrocyte sensitizing substances in the diagnosis of leptospirosis. I. The sensitized erythrocyte agglutination test, *Am. J. Trop. Med. Hyg.*, 6, 90, 1957.
106. Kelen, A.E., and Labzoffsky, N.A., Studies on latex agglutination test for leptospirosis, *Can. J. Microbiol.*, 6, 463, 1960
107. Levett, P.N., and Whittington, C.U., Evaluation of the indirect hemagglutination assay for diagnosis of acute leptospirosis, *J. Clin. Microbiol.*, 36, 11, 1998.

108. Effler, P.V. et al., Evaluation of the indirect hemagglutination assay for diagnosis of acute leptospirosis in Hawaii, *J. Clin. Microbiol.*, 38, 1081, 2000.
109. Yersin, C. et al., Field evaluation of a one-step dipstick assay for the diagnosis of human leptospirosis in the Seychelles, *Trop. Med. Int. Health*, 4, 38, 1999.
110. Hull-Jackson, C. et al., Evaluation of a commercial latex agglutination assay for the serological diagnosis of leptospirosis, *J. Clin. Microbiol.*, 44, 1853, 2006.
111. Smits, H.L. et al., Latex based, rapid and easy assay for human leptospirosis in a single test format, *Trop. Med. Int. Health*, 6, 114, 2001.
112. Arimitsu, Y. et al., Development of a simple serological method for diagnosing leptospirosis: A microcapsule agglutination test, *J. Clin. Microbiol.*, 15, 835, 1982.
113. Sehgal, S.C., Vijayachari, P., and Subramaniam, V., Evaluation of *Leptospira* Micro Capsular Agglutination Test (MCAT) for the sero-diagnosis of leptospirosis, *Indian J. Med. Res.*, 106, 504, 1997.
114. Arimitsu, Y. et al., Evaluation of the one-point microcapsule agglutination test for diagnosis of leptospirosis, *Bull. WHO*, 72, 395, 1994.
115. Smits, H.L. et al., Lateral-flow assay for the rapid serodiagnosis of human leptospirosis, *Clin. Diag. Lab. Immunol.*, 8, 166, 2001.
116. Wagenaar, J.F. et al., Rapid serological assays for leptospirosis are of limited value in southern Vietnam, *Ann. Trop. Med. Parasitol.*, 98, 843, 2004.
117. Matsuo, K., Isogai, E. and Araki, Y., Occurrence of [→3)-β-D-Manp (1–4)-β-D-Manp-(1→}n units in the antigenic polysaccharides from *Leptospira biflexa* serovar patoc, strain Patoc I, *Carbohydrate Res.*, 328, 517, 2000.
118. Boonyod, D. et al., LipL32 as outer membrane protein of *Leptospira*, as an antigen in a dipstick assay for diagnosis of leptospirosis, *Asian Pac. J. Allergy Immunol.*, 23, 133, 2005.
119. Flannery, B. et al., Evaluation of recombinant *Leptospira* antigen-based enzyme-linked immunosorbent assays for the serodiagnosis of leptospirosis, *J. Clin. Microbiol.*, 39, 3303, 2001.
120. Croda, J. et al., *Leptospira* immunoglobulin-like proteins as serodiagnostic marker for acute leptospirosis, *J. Clin. Micrbiol.*, 45, 1528, 2007.
121. Terpstra, W.J., Schoone, G.J., and Ter Schegget, J., Detection of leptospiral DNA by nucleic acid hybridization with ^{32}P- and biotin-labeled probes, *J. Med. Microbiol.*, 22, 23, 1986.
122. Terpstra, W.J. et al., Detection of *Leptospira interrogans* in clinical specimens by in situ hybridization using biotin-labeled DNA probes, *J. Gen. Microbiol.*, 133, 911, 1987.
123. Colebrander, A. et al., Detectie van pathogene leptospiren met behulp van NASBA, *Ned. Tijdschr. Geneeskd.*, 138, 436, 1994.
124. Merien, F. et al., Polymerase chain reaction for detection on *Leptospira* spp. in clinical samples, *J. Clin. Microbiol.*, 30, 2219, 1982.
125. Hookey, J.V., Detection of Leptospiaceae by amplification of 16 S ribosomal DNA, *FEMS Microbiol. Lett.*, 90, 267, 1982.
126. Van Eys, G.J.J.M. et al., Detection of leptospires in urine by polymerase chain reaction, *J. Clin. Microbiol.*, 27, 2258, 1989.
127. Zhang, Y., Li, S., and Dai, B., Amplified 23S rRNA gene of 52 strains of *Leptospira* and detection of leptospiral DNA in 55 patients by PCR, *Hua Xi Yi Ke Da Xue Xue Bao*, 24, 262, 1993.
128. Merien, F., Baranton, G., and Perolat, P., Comparison of polymerase chain reaction with microagglutination test and culture for diagnosis of leptospirosis, *J. Infect, Dis.*, 172, 2815, 1995.
129. Yersin, C. et al., Field evaluation of a one-step dipstick assay for the diagnosis of human leptospirosis in the Seychelles, *Trop. Med. Int. Health*, 4, 38, 1999.
130. Brown, P.D. et al., Evaluation of the polymerase chain reaction for early diagnosis of leptospirosis, *J. Med. Microbiol.*, 43, 110, 1995.
131. Pacciarini, M.L. et al., The search for improved methods for diagnosing leptospirosis: The approach of a laboratory in Brescia, Italy, *Rev. Sci. Tech.*, 12, 647, 1993.
132. Savio, M.L. et al., Detection and identification of *Leptospira interrogans* serovars by PCR coupled with restriction endonuclease analysis of amplified DNA, *J. Clin. Microbiol.*, 32, 935, 1994.
133. Woo, T.H. et al., Rapid distinction between *Leptospira interrogans* and *Leptospira biflexa* by PCR amplification of 23S ribosomal DNA, *FEMS Microbiol. Lett.*, 150, 9, 1997.
134. Kawabata, H. et al., *flaB*-polymerase chain reaction (*flaB*-PCR) and its restriction fragment length polymorphism (RFLP) analysis are an efficient tool for detection and identification of *Leptospira* spp., *Microbiol. Immunol.*, 45, 491, 2001.
135. Slack, A.T. et al., Identification of pathogenic *Leptospira* species by conventional or real-time PCR and sequencing of the DNA gyrase subunit B encoding gene, *BMC Microbiol.*, 6, 95, 2006.
136. Palaniappan, R.U. et al., Evaluation of *lig*-based conventional and real time PCR for the detection of pathogenic leptospires, *Mol. Cell. Probes*, 19, 111, 2005.
137. Smythe, L.D. et al., A quantitative PCR (TaqMan) assay for pathogenic *Leptospira* spp., *BMC Infect. Dis.*, 2, 13, 2002.
138. Levett, P.N. et al., Detection of pathogenic leptospires by real-time quantitative PCR, *J. Med. Microbiol.*, 54, 45, 2005.
139. Stoddard, R.A. et al., Detection of pathogenic *Leptospira* spp. through TaqMan polymerase chain reaction targeting the LipL32 gene, *Diagn. Microbiol. Infect. Dis.*, 64, 247, 2009.
140. Merien, F. et al., A rapid and quantitative method for the detection of *Leptospira* species in human leptospirosis, *FEMS Microbiol. Lett.*, 249, 139, 2005.
141. Ahmed, A. et al., Development and validation of a real-time PCR detection of pathogenic *Leptospira* species in clinical materials. *PLoS One*, 4, 1, 2009.
142. Maniatis, T., Fritsch, E.F., and Sambrook. J., *Molecular Cloning: A Laboratory Manual*, Cold Spring Harbor Laboratory, Cold Spring Harbor, New York, 1983.
143. Slack, A. et al., Evaluation of a modified Taqman assay detecting pathogenic Leptospira spp. against culture and *Leptospira*-specific IgM enzyme-linked immunosorbent assay in a clinical environment, *Diag. Microbiol. Infect. Dis.*, 57, 361, 2007.
144. Udaykumar, Epstein, J.S., and Hewlett, I.K., A novel method employing UNG to avoid carry-over contamination in RNA-PCR, *Nucleic Acids Res.*, 21, 3917, 1993.
145. Boom, R. et al., Rapid and simple method for purification of nucleic acids, *J. Clin. Microbiol.*, 28, 495, 1990.
146. Lourdault, K., Aviat, F., and Picardeau, M., Use of quantitative real-time PCR for studying the dissemination of *Leptospira interrogans* in the guinea pig infection model of leptospirosis, *J. Med. Microbiol.*, 58, 648, 2009.
147. Lin X, et al., Application of a loop-mediated isothermal amplification method for the detection of pathogenic *Leptospira*. *Diagn. Microbiol. Infect. Dis.*, 63, 237, 2009.

98 Treponema

Rita Castro and Filomena Martins Pereira

CONTENTS

98.1 INTRODUCTION

98.1.1 Classification, Morphology, and Biology

Members of the genus *Treponema* are spiral-shaped bacteria that are classified in the family Spirochaetaceae, order Spirochaetales. Along with those from the genus *Borrelia*, *Treponema* bacteria represent the most important human pathogens within the family. The genus *Treponema* has a great variety of natural hosts, including humans, other primates, rabbits, cattle, dogs, and termites. The best-known pathogen of this genus is *Treponema pallidum* subspecies *pallidum*, the etiological agent of syphilis. Other pathogenic treponema that cause disease in humans include *Treponema carateum*—the causative agent of pinta, and two subspecies of *Treponema pallidum*, *T. pallidum* subspecies *pertenue*—the causative agent of yaws, and *T. pallidum* subspecies *endemicum*—the etiological agent of endemic syphilis. These treponemes may be propagated in laboratory animals, with the exception of *Treponema carateum*, which can only be propagated in primates. General characteristics of pathogenic treponemes are summarized in Table 98.1.

Various other species of treponemes can be found in the mouth and in the gastro-intestinal and genital tracts. Most of them can be cultured and have a low pathogenicity, with the exception of *Treponema denticola* and *Treponema lecithinolyticum*, which have been associated with periodontal disease.

Treponema pallidum genome sequence became available in 1998.[1] This organism has one of the smallest bacterial genomes, a circular chromosome with approximately 1.14 million base pairs (Mb) and a limited metabolism, in accordance with its great host-dependent nature. The genome of *T. denticola*[2] has a larger size (2.48 Mb) and contains some genes for biosynthetic activities but is also host dependent, needing the gingival microenvironment for the acquisition of nutrients. The treponemal cell ranges between 5 and 20 μm in length and between 0.1 and 0.5 μm in diameter. These measures do not allow them to be seen in light microscopy. Dark-field or phase contrast microscopy is needed to identify treponemes microscopically. The treponemal cell may have a regular or irregular helical shape or flat-wave morphology.[3,4]

The organisms of the genus *Treponema* have a similar architecture to gram-negative bacteria, with an outer and cytoplasmatic membrane, a thin peptidoglycan cell wall, and a periplasmic space. In contrast to conventional gram-negative bacteria, the outer membrane of *T. pallidum* does not contain lipopolysacharide.[1,5] It also has a significantly lower protein content.[1,6] Consequently, it has a poorly antigenic nature, being called "stealth pathogen" because of its capacity to evade the host immune response and to remain in a latent stage. Studies

TABLE 98.1
Characteristics of Pathogenic Treponemes

	Treponema pallidum subsp.			
Characteristics	***pallidum***	***pertenue***	***endemicum***	***Treponema carateum***
Pathogenicity	Invasive Local and systemic infection	Moderately invasive	Moderately invasive	Without systemic involvement
Diseases	Syphilis	Yaws (framboesia, pian, buba)	Endemic syphilis (bejel)	Pinta
Geographical distribution	Worldwide	Tropical regions	Northern Africa, Middle East, Southeast Europe	Central and South America
Transmission	Sexual contact Congenital infection	Skin contact with lesions	Mucous membrane contact Rarely congenital infection	Skin contact with lesions
Age of onset	Adolescents and adults	Children	Children and adults	Children and adolescents
Known experimental hosts	Rabbit > primate > guinea-pig > hamster	Rabbit = hamster	Rabbit = hamster	Primate

Source: Adapted from Larsen et al., Syphilis and related treponematoses. In Hausler, W.J. Jr., and Sussman, M. (Editors), *Microbiology and Microbial Infections*, Vol. 3, pp. 641–668, Oxford University Press, Inc., New York, 1998.

using anti-*T. pallidum* antibodies have demonstrated that antibodies bind poorly to the surface of *T. pallidum*. Treponemes are mobile due to flagella, which are called periplasmatic flagella, or endoflagella, because they are positioned in the periplasmic space. Depending on the species, a high or low number of flagella are inserted subterminally at each end of the cell. In *T. pallidum,* three flagella are located at each end of the cell,[7] and they wind around the cell body toward the center of the bacterium, where they overlap. *Treponema* flagella filaments have a unique structure, characterized by an outer sheath protein (FlaA) that surrounds three core proteins (FlaB1, FlaB2, FlaB3), whose function is not clear.[8]

The metabolism of treponemes is best known in *T. pallidum*, *T. denticola,* and *T. phagedenis*, in the first two based on the knowledge obtained through the complete sequence of their genomes. The complete genomic sequence of *T. pallidum*[1] reveals that the bacteria contains all the genes necessary for the Embden–Meyerhoff–Parnas pathway, that the genes encoding for the fermentation enzymes are not present, and only two enzymes of the Entner–Doudoroff pathway exist. It also lacks heme proteins taking part in electron transport and all genes for the Krebs cycle; in addition, the genes for the synthesis of most amino acids, purines, pyrimidines, and fatty acids are absent.[9] Related to this poor biosynthetic capacity is the existence of 57 genes encoding for 18 putative transport systems for amino acids, carbohydrates, and cations. Genes for enzymes that can protect *T. pallidum* from the toxic effects of oxygen radicals are missing, thus making this bacterium highly sensitive to the air oxygen atmosphere. Some *T. pallidum* genes are associated with virulence, of which the most important seem to be the *tpr* genes, encoding hemolisines, regulating proteins, polysaccharides biosynthesis, and proteins. Twelve of these genes encode proteins similar to the *T. denticola* major membrane protein and are designed *Treponema pallidum* repeat A-L.[10] Although *T. pallidum* can grow in vitro, multiplication of this organism has not been achieved after subculture.

98.1.2 Epidemiology

Syphilis is a sexually transmitted disease with a worldwide distribution. The World Health Organization estimates that there are 12 million new cases of syphilis each year, of which 2 million are pregnant women, the great majority of these cases being seen in developing countries. The prevalence of this infection during pregnancy ranges from 4% to 15%.[11] The disease is more common among populations with limited resources that lack access to health care or live on the margins of society, and among those with many sexual partners.[12] Humans are the only reservoir of *T. pallidum,* and transmission is mainly through sexual contact, although vertical transmission remains a major concern. This last mode of transmission is responsible for congenital syphilis, resulting in stillbirth or spontaneous abortion, and if infants survive, they may suffer from abnormal growth and development and mental retardation.[13]

Untreated syphilis progresses slowly over several years with some degree of morbidity and mortality.[14] Morbidity with neurological and cardiovascular complications is often associated to HIV coinfection.[15] The rate of transmission from an infected individual from a single sexual contact is estimated to be 30%.[16] In pregnancy, transmission to the fetus may happen shortly after the onset of infection[17] and as early as the ninth week of pregnancy.[18] Pregnant women may infect their fetuses many years after acquiring infection, but this risk decreases with time. About 25% of infected pregnancies result in stillbirth or spontaneous abortion. Low birth weight or severe infection are complications in another 25% of the cases; both these situations are associated with a higher risk of perinatal death.

In the early 1990s, reported cases fell to their lowest levels in Western European countries and in the United States.[19,20] A national syphilis-elimination program aimed at African American heterosexuals was started in 1999 in the United States, but between 2000 and 2005, the number of syphilis cases increased, especially among men who have sex with men (MSM). It is estimated that syphilis has increased in the Eastern European countries[21] and in China,[22] with very high numbers in Southeast Asia and in Sub-Saharian Africa.[12]

98.1.3 Clinical Features

Natural History of Syphilis. Without treatment, syphilis is a chronic systemic infection that progresses with symptomatic stages (primary, secondary, and tertiary syphilis) and asymptomatic stages, over a period of many years. However, the period of time between primary and secondary syphilis is much shorter than the time from secondary to tertiary syphilis.

Treponemes may invade the host through normal mucous membranes or abrasions on the skin surface. They multiply at the site of inoculation, and shortly, some bacteria gain access to hematogenous and lymphatic circulation and disseminate all over the body, including the central nervous system (CNS). Individuals with primary- and secondary-stage syphilis, including secondary relapses of early latency (when moist, mucocutaneous lesions are present), are the most infectious.

Primary Syphilis. Primary syphilis is characterized by the appearance of an ulcerated, nontender, indurated, and nonpurulent lesion (chancre) at the local of inoculation, most commonly the anogenital area. The chancre is usually painless, and vaginal or anal lesions may not be noticed. In general, it is a single lesion with a clear base, with no exudate or with scanty serous exudate, unless it is secondary infected. However, multiple primary lesions may occur, predominantly in HIV-infected patients. This lesion appears 10–90 days after exposure. Atypically, the chancre may be painful and soft, and not uncommonly, the primary lesion may be absent. Primary lesions can be seen in extra-anogenital sites, such as lips, tongue, and fingers, where they can have an atypical appearance. Regional lymphadenopathy, frequently bilateral, with moderately enlarged rubbery and painless lymph nodes is associated with the chancre. The chancre heals within 3–6 weeks, even in the absence of treatment, but lymphadenopathy usually persists longer.

Secondary Syphilis. After the primary chancre is healed, untreated individuals may develop the secondary stage of infection in a period of weeks to several months. In rare cases, this phase of symptoms may be coincident with the primary chancre. This stage is related to hematogenous and lymphatic dissemination of *T. pallidum,* with multiorgan involvement. The most characteristic symptoms of this stage are skin lesions, but secondary syphilis can affect virtually any organ or system. A period of systemic illness with low-grade fever, malaise, sore throat, and headache can initiate this stage.

The initial skin lesion is generally a macular rash, after which a symmetric papular eruption appears in the trunk and extremities, with a characteristic distribution on the palms of the hands and soles of the feet. This rash usually resolves within 2–6 weeks, but may recur in untreated patients. *Condylomata lata*, a larger raised whitish or gray lesion can be observed, more commonly in moist areas. Diffuse nontender lymphadenopathy is present in the majority of patients.[23] A less common presentation of secondary syphilis includes hepatitis,[24] gastric involvement,[25] periostitis,[26] anterior uveitis,[27] meningitis,[28] cranial nerve palsies,[29] and glomerulonephritis.[30] Circulating immune complexes are present in this stage of infection, and their deposition can be the explanation for some of these complications, such as anterior uveitis or glomerulonephritis.

Although not usually accompanied by neurological signs, the early dissemination of *T. pallidum* to the CNS affects at least 25% of patients.[17,31] Secondary syphilis resolves spontaneously in 3–12 weeks, leading to the stage referred to as latency.[32]

Latent Syphilis. After resolution of secondary syphilis manifestations, untreated patients enter a stage without any symptoms or signs, which is called latent syphilis. This asymptomatic stage is divided into early latent syphilis (patients infected for less than 1 year for Centers for Disease Control and Prevention (CDC) and 2 years for WHO) and late latent syphilis (patients infected for more than 1 year for CDC or 2 years for WHO, or with unknown duration of infection). This difference is important because during early years of infection patients can have spirochetemia episodes, and they may therefore transmit the infection to others. The majority (90%) of relapses take place during the first year of infection, but they may occur as late as 5 years afterward.[32] In the natural evolution of the disease and with the improvement of immunological control, there is a reduction in spirochetemia[30] and an end of the infectiousness. However, this does not correspond to the total eradication of infection, since treponemes may persist, and the tertiary stage may occur years later.

Tertiary Syphilis. Tertiary syphilis is now rarely seen, which is probably related to the generalized use of antibiotic therapy. If left untreated, 30% of the patients develop clinical manifestations of tertiary syphilis.[33] These late manifestations involve many organs, such as bones, heart, CNS, and skin and are the main cause of syphilis morbidity and mortality in adults and can occur from 1 to 30 years after infection. These manifestations are usually divided into three major groups: late benign syphilis, cardiovascular syphilis, and neurosyphilis.

(i) Late benign syphilis, or gummas, results from a proliferative and destructive granulomatous process. Usually they affect the skin, bone, or mucous membranes, but they may also occur in the liver, heart, brain, stomach, and upper respiratory tract.[34] *T. pallidum* can be identified in gummatous lesions.[35]
(ii) Cardiovascular syphilis was a frequent complication prior to the advent of penicillin therapy. This

manifestation becomes clinically apparent after a period of 15–30 years from the onset of the infection, occurring in approximately 10% of individuals with untreated syphilis.[36] Aortic aneurysm, aortic insufficiency, coronary artery stenosis, and rarely, myocarditis are the symptoms of cardiovascular involvement and contribute to the difficulty of making a differential diagnosis with more common varieties of cardiac diseases.[37] Cardiovascular syphilis results from *T. pallidum* dissemination to the heart, where they stay in the aortic wall remaining dormant for years. They have a predilection for the vasa vasorum of the aorta, producing transmural inflammatory lesions and causing obliterative endarteritis, adventitia scarring, and media necrosis with destruction of elastic fibers.

(iii) Neurosyphilis is an infection of the CNS, and classically, it has been associated with tertiary syphilis. Presently, neurosyphilis is divided into early and late forms.[28]

As noted above, *T. pallidum* disseminates to the cerebrospinal fluid (CSF) and meninges very early in the infection.[28] *T. pallidum* is often clears spontaneously from the CSF, but in few cases symptomatic disease develops within the first weeks or months after initial infection. Early neurosyphilis occurs within weeks to few years (usually 5–12) after the onset of infection, affecting mainly the CNS, cerebral blood circulation, and meninges, but it may be asymptomatic. Symptomatic forms manifest as meningitis, with or without cranial nerves or eye involvement, and meningovascular disease.[25,34,38] Late neurosyphilis is extremely rare, appearing two to three decades after untreated infection and affecting the meninges and brain or spinal cord parenchyma. Its manifestations include general paresis; a demential syndrome with psychotic features; and tabes dorsalis, which presents as sensory ataxia; paresthesis; and bowel and bladder dysfunction.

HIV Infection and Syphilis. HIV coinfection with syphilis is common, the two infections affecting each other in many ways. Primary syphilis lesions increase HIV transmission,[39,40] while *T. pallidum* infection in HIV patients decreases CD4 and increases HIV viral loads.[41] The course of syphilis may be altered in patients infected with HIV, resulting in a more aggressive disease,[42] with more constitutional symptoms and a more rapid natural evolution.

Clinical presentation in HIV patients is generally similar to presentation in those who are not infected, although some differences may be present, such as larger painful multiple ulcers[43]; manifestations of secondary syphilis appear more frequently with concomitant primary ulcer[42] and a greater organ involvement, atypical florid rash and a more rapid evolution for neurosyphilis, and ophthalmic involvement.[44,45] The increased rate of early neurosyphilis among HIV-infected patients is probably related to an increase in the difficulty in clarifying the initial neuroinvasion by *T. pallidum*.[28]

98.1.4 Pathogenesis

T. pallidum attaches to a variety of cell types[46] and has specific proteins (TP0155, TP0483, and TP0751) that bind to laminin[47] and fibronectin, which are extracellular matrix components (ECM). This allows *T. pallidum* to adhere and disseminate. The motility of this bacterium also facilitates invasion and dissemination. *T. pallidum* lacks lipopolysaccharide (LPS), but the lipoproteins under the outer membrane are very immunogenic. However, because they are not on its surface, the immune system does not immediately detect this microorganism and thus it has time to multiply and disseminate.[48]

Symptoms and signs of infection with *T. pallidum* are consequences of inflammatory and immune responses rather than the direct influence of the bacterium on the cells.[7,48] Some studies have shown that endoflagellar proteins, TPR proteins, and other lipoproteins stimulate T-cell response.[7]

Humoral responses are also found in *T. pallidum* infection, since antibodies can be detected soon after the onset of primary syphilis, increasing with dissemination of the infection, although the role of these antibodies is not well known.[48]

Many factors have been attributed as cause for the persistence of *T. pallidum* infection, but the almost nonexistence of surface proteins is presently accepted as the main factor, together with antigenic variation in some proteins that allow them to pass unrecognized and persist to cause long-term infection.[34,49]

Untreated patients seem to have a certain degree of immunity, while treated patients are very often reinfected, as shown in experiments performed in both humans and rabbits.[32] However, the mechanisms of this protective response are still not clear.

98.1.5 Laboratory Diagnosis

98.1.5.1 Conventional Techniques

Direct Methods. In the presence of lesions, direct detection of the organism is the most specific and easiest method for the laboratory diagnosis of syphilis. It can be made using a dark-field microscopy or a direct fluorescent antibody technique. The proper specimen for dark-field microscopy is the serous fluid (the collection of blood must be avoided) from moist primary- or secondary-stage lesions or from the fluid aspirated from enlarged regional lymph nodes. Oral or rectal samples can not be observed by microscopy because *T. pallidum* cannot be differentiated from nonpathogenic treponemes colonizing those sites. This method has a great sensitivity if performed by an experienced microscopiste. Although *T. pallidum* subspecies cannot be differentiated from other pathogenic treponemes, they can be distinguished from other corkscrew-shaped organisms by a characteristic forward and backward rotation movement along the longitudinal axis. Dark-field examination must be done immediately after sampling to maintain the viability of treponemes.

The direct fluorescent antibody test (DFA *T. pallidum*) also has a good sensitivity (73%–100%),[50] and it is especially useful when smears can not be visualized immediately or in the presence of oral-rectal lesions, as the use of a specific conjugate for pathogenic treponemes makes this technique a more specific test.[51] This method should also be used in biopsies, given its greater sensitivity and specificity over silver staining.[52]

Culture. Continuous "in vitro" culture of *T. pallidum* is not possible,[53] which makes experimental infection of rabbits the preferred method for its propagation. The rabbit infectivity test (RIT) is a method for direct treponemal detection and is considered the gold standard test, presenting a high sensitivity and specificity. However, it is only used as a research tool because of the need for animals, the very long incubation time (several weeks to months), and variation in rabbit's susceptibility to infection.[54]

Serological Tests. Serological tests are currently the most commonly used methodology for syphilis diagnosis, and they are the only practical method that can be used to diagnose latent syphilis. Syphilis serological tests are divided into to two types: the nontreponemal or nonspecific tests and the treponemal or specific tests, based on the characteristics of the antigen used.

Nontreponemal Tests (NTTs). The NTTs detect antibodies referred to as phospholipid or cardiolipin or nonspecific antibodies and use cardiolipin, cholesterol, and lecitin as antigen. They are reactive in the presence of IgM and IgG produced by the host in response to lipid release from damaged host cells and possibly to cardiolipin released from treponemes, and can be detected by flocculation tests and by enzyme immunoassays. The flocculation tests include the veneral disease research laboratory (VDRL) slide test, the unheated serum reagin test (URS), the rapid plasma reagin test (RPR), and the toluidine red unheated serum test (TRUST).

VDRL and URS are microscopic tests, and results are read microscopically. They are both performed on a glass slide, but while the VDRL antigen needs to be prepared fresh daily, the URS antigen is stabilized, and therefore does not need to be prepared daily. RPR and TRUST are macroscopic tests with a stabilized antigenic suspension, to which charcoal particles (RPR) and azo pigment (TRUST) were added to aid in the visualization of reactive sera.

Reactive results must be qualitative and quantitative. The result of the quantitative test must be reported as the reciprocal of the dilution, the titer being the last dilution showing reactivity. This titer establishes a baseline of reactivity from which changes can be measured. New infection or relapse is demonstrated by a fourfold rise in titer, while successful treatment is reflected by, at least, a fourfold decrease in titer.

In the natural evolution of infection, NTTs reactive in 2–3 weeks and the titers of antibodies will peak during the secondary stage. After that, and even when the patient is not treated, the titers of these tests will decrease gradually.[31] In 30% of syphilis patients without treatment, the sera will became nonreactive for NTT during their lifetimes.[41]

NTTs may originate false-negative or false-positive results. Antilipoidal antibody reactivity can result from other treponemal infections, from nontreponemal diseases with tissue damage of an acute or chronic nature. They can be transitory (>6 months) and are generally associated with acute infection, immunizations, pregnancy (>6 months), chronic autoimmune disease, and the elderly.

False-negative results can occur in 1%–2% of patients with secondary syphilis due to a prozone reaction that takes place when antibody is in excess and blocks the normal antigen antibody reaction.

Treponemal Tests. Treponemal tests use *T. pallidum* or its components as the antigen and are confirmatory tests. Most patients who have syphilis will have reactive treponemal tests for the rest of their lives, regardless of treatment or disease activity, which make these tests inappropriate for monitoring treatment.

Treponemal tests become reactive in early primary syphilis and remain reactive for many years with or without treatment.[51] False-positive reactions associated with autoimmune diseases, toxoplasmosis, and intravenous drug users may occur with treponemal tests.[50]

Classically, NTTs, such as the RPR or the VDRL, are used to screen for syphilis, and reactive results are confirmed by a treponemal test. In Europe, where the prevalence of syphilis is supposed to be low, some investigators recommend the use of treponemal tests for the primary screening of samples, with another treponemal test to confirm a positive result and a quantitative NTT on the first day of therapy to provide a baseline for monitoring response to treatment.[55]

There are several tests used for the detection of specific treponemal antibodies. The first treponemal test to be used was the *T. pallidum* immobilization (TPI), based on immobilization of *T. pallidum* in the presence of antibodies and complement.[56] This test is no longer used because of its complexity, difficulty in reproducibility, cost, and requirement for animals. The most commonly used tests, based on indirect fluorescent antibody and in agglutination techniques, and other tests based in techniques such as EIA, immunochromatography, and Western blot will be described.

The fluorescent treponemal antibody test (FTA),[57] which become more specific with the addition of an absorption step (FTA-Abs), uses antihuman IgG or IgM, labeled with fluorescein, to detect antibodies IgG or IgM bound to *T. pallidum* organisms on microscopic slides. This test is the first test to become reactive during the course of syphilis.

The hemagglutination test is technically less complicated than the FTA-Abs and does not require the use of a fluorescent microscope. In this test, reactive antibodies are detected by the agglutination of red blood cells used as carriers for the *T. pallidum* antigens.[58] This test is being replaced in some laboratories by the *T. pallidum* particle agglutination assay,[59] which is based on the same principle but uses a lyophilized preparation of colored gelatin particles as the carrier of the *T. pallidum* antigen, instead of blood cells. The sensitivity and the specificity of the TPHA and the TPPA are similar, but both seem to be slightly less sensitive than FTA-Abs in primary syphilis, although the difference in sensitivity seems to be smaller for the TPPA than for the TPHA.[60]

There are various commercialized EIA tests to detect treponemal antibodies described in the literature, using *T. pallidum* antigen,[61,62] recombinant proteins,[63,64] or cloned components of *T. pallidum,* such as the 47-kDa antigen.[65] They have a similar sensitivity, depending of the stage of syphilis, and a high specificity. The great advantage of these tests is that they can be fully automatized, making them very useful for studying a great number of samples.

Western blot techniques have shown higher sensitivity and specificity when compared to other serological tests.[66] They are considered positive when at least three of the immunodeterminants, with molecular masses of 15.5, 17, 44.5, and 47 kDa are present.[67]

All previously described tests require trained personnel, electricity, and laboratory equipment. As those conditions are not available in some regions of the world, rapid tests were developed that can use whole blood from a finger prick as well as from serum or plasma. They can be stored at room temperature and do not require any type of equipment. In general, it seems that available rapid tests have a high specificity and adequate sensitivity, which is better in serum than with whole-blood samples.[68]

98.1.5.2 Molecular Techniques

Polymerase chain reaction (PCR) techniques using different gene targets to identify *T. pallidum* DNA have been developed by several investigators. The gene *tpp*47 (TP0574) that encodes a penicillin-binding protein with carboxypeptidase activity is the most commonly used target gene, alone or in multiplex PCR,[69–71] together with the *pol*A.[72,73]

The sensitivity and specificity of PCR techniques using the *tpp*47 and/or *pol*A to identify *T. pallidum* in different clinical material and in various stages of syphilis, has also been studied. The most commonly used samples were genital ulcers of primary syphilis,[70–72] and together with lesions of secondary syphilis, they yielded the most sensitive results.[74,75] Detection of *T. pallidum* in blood samples would be of great help in the diagnosis of latent syphilis, because diagnosis is only made on the basis of serological-test reactivity. However, PCR sensitivity in samples from patients with latent syphilis is very low (around 40%), although in general, plasma samples showed a slight higher sensitivity than blood or sera.[73,75–77] One of the most promising samples for the diagnosis of latent syphilis seems to be the ear lobe, as described by Castro et al.[73] An interesting finding was the detection of *T. pallidum* in the blood of some individuals with no symptoms or signs of syphilis and nonreactive serology, who had been exposed to patients with syphilis.[76,78]

A multiplex PCR using the *tpp*47 target gene can be performed in swabs from genital ulcers to differentiate between *T. pallidum*, *Haemophilus ducreyi*, and Herpes simplex virus types 1 and 2,[70,71] the most common agents of genital ulcerations in the same PCR reaction tube. This PCR assay seems to be more sensitive than standard diagnostic tests for the detection of those agents from genital ulcers.

In some of the PCR studies described here, specificity was also studied, using a variety of other pathogens, including other sexually transmitted microorganisms, some from the normal flora of the skin, and a variety of other treponemes, which were all negative. Studies on the value of PCR techniques for the diagnosis of neurosyphilis have also been done, but sensitivity has been very low.[79,80]

Real-time PCR is a fast, sensitive method for the detection of a variety of microorganisms. Target genes for the detection of *T. pallidum* have also been the *pol*A[81,82] and the *tpp*47 genes.[83,84] This technique has been assayed for the detection of *T. pallidum* in different biological specimens, such as blood[81,83] and lesion exudates[82–84] and was shown to have a specificity of 99%–100%[83,84] and a high sensitivity (80%) in primary syphilis lesions, as demonstrated by Gayet-Aeron.[83] However sensitivity was relatively low in other samples, such as blood, sera, lesion, and urine samples. The detection of *T. pallidum* in urine in 44% of patients with primary and secondary syphilis has some advantage, since urine is an easy-to-obtain noninvasive sample.

98.1.5.3 Interpretation of Syphilis Diagnostic Tests

T. pallidum infection will be transmitted to nearly 30% of the sexual contacts of untreated syphilis patients.[51] During the incubation period, no available test can detect *T. pallidum*, although Marfin et al.[76] reported the detection of *T. pallidum* DNA by PCR technique in four of eight patients considered to be in this period.

Direct detection of *T. pallidum* in genital or extragenital ulcers is definitive evidence of primary syphilis. As previously noted, this can be done using dark-field or direct fluorescence microscopy. Anti-*T. pallidum* antibodies generally are not detected until 2 or 3 weeks after infection, and at this time, serology is positive in only 30%–50% of cases, depending on the test used.[85] The first tests to be positive are those that can detect IgM, as is the case of EIA and FTA-Abs 19S IgM. In contrast to primary syphilis, in secondary syphilis all serological tests are uniformly reactive, and treponemes may be found in secondary lesions (condilomata lata, mucosal lesions, skin papules, lymph nodes aspirate) by direct microscopy.

Since in some studies, PCR techniques have shown to have a 100% sensitivity and specificity in ulcers and skin,[76] they could be useful in the case of atypical lesions, which are more frequent in HIV patients. When serological tests are not available, a suggestive diagnosis can be made, based on the presence of clinical manifestations and a report of sexual contact with an individual with syphilis within the previous 6 months.

The natural evolution of untreated syphilis is spontaneous resolution (latent syphilis), resulting in no clinical evidence of disease along with reactive serological tests. However, as the length of time between initial infection and serological testing increases, the likelihood of a reactive NTT decreases and may become negative. Definitive diagnosis of latent syphilis is not possible, because in this stage there are no specimens suitable for the direct detection of *T. pallidum*. This is a situation in which the PCR method could perhaps be of some help by demonstrating the presence of *T. pallidum*

DNA in the blood.[73,77] PCR is also useful in lesions that are coinfected with other causes of genital ulcers, and multiplex PCR has the advantage of performing a simultaneous differential diagnosis of Herpes simplex and *Haemophilus ducreyi* ulcers.[70,84]

Late syphilis occurs in approximately one-third of persons who where not treated, and because NTTs could be negative in 30% of these cases, both treponemal and NTTs must be done in a patient with suspected late syphilis.

The diagnosis of neurosyphilis depends on the combination of clinical observation, serological tests, and CSF examination (VDRL/RPR, white cell count, and proteins). A presumptive diagnosis can be made in the presence of clinical signs and symptoms associated with a reactive treponemal test in the sera, increased CSF cell count, and proteins. Reactive NTTs (VDRL/RPR) in CSF are considered diagnostic of neurosyphilis, but although they present a very high specificity, they have a low sensitivity.[86,87] The CSF–FTA-Abs test is highly sensitive, but has a low specificity. These characteristics led some specialists to say that neurosyphilis should be rolled out when a nonreactive result in the CSF is obtained.[86]

98.1.5.4 Subtyping *Treponema pallidum*

A molecular method for the subtyping of *T. pallidum* could be an important tool for a better understanding of the epidemiology of this infection. Pillay et al.[88] described a molecular subtyping method based on the combination of amplification of the *arp* gene (TP0433) encoding acid repeat protein and amplification of the *tpr* E, G, J genes (TP0313, TP0317, TP0621). This technique has allowed the identification of different subtypes of *T. pallidum* and showed to have a good reproducibility.[78,88,89]

98.2 METHODS

98.2.1 Sample Preparation

Lesion Exudates. Use a sterile cotton or Dacron-tipped swab to collect material from the genital ulcer or secondary lesions of syphilitic patients. Press and roll the swab twice along the base of the ulcer, or along the skin lesion. Suspend the swab in a tube containing 1 mL transport medium (PBS), roll it along the side of the tube to express the excess fluid, discard and mix well. If processed during the next 2 days, store at 4°C. Otherwise, aliquot and store at –70°C.

Blood. Collect 5–10 mL of blood into a tube with and without the anticoagulant EDTA to obtain total blood, plasma, and sera. Whole blood—if processed during the next 2 days—should be stored at 4°C. Otherwise, aliquot and store at –70°C. If the QIAamp Blood/Tissue Kit (Qiagen) is being used for extraction, and if the sample cannot be processed, add 200 μL buffer AL to 200 μL whole blood and store at –20°C until processed. Plasma—centrifuge the blood with EDTA to obtain plasma. If processed during the next 2 days, store at 4°C. Otherwise, aliquot and store at –70°C. Sera—centrifuge the tube with blood without EDTA to obtain sera. If processed during the next 2 days, store at 4°C. Otherwise, aliquot and store at –70°C.

CSF. It must be free of blood. If processed during the next 2 days, it should be stored at 4°C. Otherwise, aliquot and store at –70°C.

Ear Lobe Scrapings. To obtain this specimen, use a scalpel to scrape the ear lobe and collect the resulting blood with a sterile, dry dracon swab. After collection, agitate swabs vigorously in a tube containing 1 mL of sterile PBS for 30 s, press against the side of the tube to express the liquid, and discard swabs. If processed during the next 2 days it should be stored at 4°C. Otherwise, aliquot and store at –70°C.

Chromosomal DNA Extraction. All pipetting must be performed using plugged pipette tips or positive displacement pipettes. Samples must be prepared in an area free of amplified DNA. All specimens can be extracted with the Mini Kit–QIAamp Blood/Tissue (Qiagen), in accordance with the manufacturer's instructions, with the following modifications:

1. Use 400 μL from blood with EDTA, instead of the 200 μL recommended.
2. Before precipitation with ethanol, add 3 μL of a 1:10 dilution of Calf thymus DNA (10.5 μg/μL, Sigma) to all samples, with the exception of blood.
3. Increase incubation period to 3 h for total blood with EDTA, and to 1 h for plasma, sera, and ear lobe scrapings.
4. Perform step 9a in the Qiagen protocol (this will eliminate residual Buffer AW2 in the spin column).
5. Before the elution step, use a sterile fine-tipped swab or strip of Whatmann paper to absorb traces of Buffer AW on the inside of the Qiagen spin column.
6. Use 80–100 μL of Buffer AE to elute DNA from the spin column: incubate 5 min at room temperature.
7. Store DNA at –4°C for up 7 days. For longer periods of time, store at –20°C.

98.2.2 Detection Procedures

98.2.2.1 Diagnostic PCR (Table 98.2)

Procedure

1. Prepare PCR mix (25 μL) as shown in Table 98.3. In every run, include a negative control (distilled water; two or more depending on the number of samples to be amplified), and a positive control (e.g., purified DNA from *Treponema pallidum* Nichols strain). Assay all samples in duplicate, and include multiple negative and positive controls for each set of the 25 specimens. Add positive control after all clinical specimens have been pipetted.
2. Amplify with 1 cycle of 95°C for 5 min to activate polymerase and denature DNA; 40 cycles of 94°C for 20 s, 60°C for 20 s, and 72°C for 20 s (for 47-PCR); or 45 cycles of 94°C for 30 s, 62°C for 30 s, and

TABLE 98.2
Primers for Diagnostic PCR

Gene	Primer	Sequence (5′→3′)	Amplicon (bp)
tp47[70]	KO3A–F	AAGTTTGTCCCAGTTGCGGTT	260
	KO4–R	AGAGCCATCAGCCCTTTTCA	
polA[77]	polA–F	AAGTTTGTCCCAGTTGCGGTT	378
	polA–R	CACAGTGCTCAAAAACGCGTGCACG	

TABLE 98.3
Reaction Mixture Used for 47-PCR and *pol*A-PCR

47-PCR	Per Reaction	*pol*A-PCR	Per Reaction
Sterile distilled water	–	Sterile distilled water	–
10× PCR buffer	1×	10× PCR buffer	1×
dNTP	0.4 mM	dNTP	0.4 mM
$MgCl_2$	2 mM	$MgCl_2$	2.5 mM
KO3A	0.2 μM	polA–F	0.2 μM
KO4	0.2 μM	polA–R	0.2 μM
Taq polymerase (Imolase, Citomed)	0.5 U	*Taq* polymerase (Imolase, Citomed)	0.5 U
DNA	5 μL	DNA	5 μL
Total volume	25 μL	Total volume	25 μL

TABLE 98.4
Primers for the Nested PCR[88]

Gene	Primer	Sequence (5′→3′)	Amplicon (bp)
tpr	B1–F	ACTGGCTCTGCCACACTTGA	2186
	A2–R	CTACCAGGAGAGGGTGACGC	
	IP6–F	CAGGTTTTGCCGTTAAGC	1836
	IP7–R	AATCAAGGGAGAATACCGTC	

72°C for 30 s (for *pol*A-PCR); and a final extension at 72°C for 5 min (47-PCR) or 15 min (*pol*A-PCR). Store tubes at 4°C until analyzed.

3. Electrophorese the PCR products on a 1.5% agarose gel for 1 h at 100 V. Include a molecular weight marker into one well of each row in the agarose gel to position size of the amplicon products. Stain gel with ethidium bromide (0.5 μg/mL, Sigma) and visualize on a UV transilluminator.

Note: The specimens are scored as positive for 47-PCR if a band of 260 bp is visualized and for *pol*A-PCR if the band has 378 bp or negative when no band is visualized.

98.2.2.2 Subtyping *Treponema pallidum* Subspecies *pallidum*

This method is applied to positive samples obtained in the diagnostic PCR technique.

(i) Analysis of tpr Genes For the subtyping of *T. pallidum*, *tpr* genes are amplified by a nested PCR, and the final products analyzed by RFLP. The use of outer B1/A2 primers in the first PCR facilitates amplification of a 2186 bp fragment, whereas the use of inner IP6/IP7 primers in the second PCR produces an internal fragment of 1836 bp (Table 98.4).

Procedure

1. Prepare the first PCR mix (25 μL) containing outer B1/A2 primers (as shown in Table 98.5). In every run, include a negative control (distilled water) and a positive control (*Treponema pallidum* Nichols strain DNA).
2. Amplify with 1 cycle of 95°C for 5 min; 35 cycles of 94°C for 1 min, 59°C for 2 min, and 72°C for 2 min; and a final extension at 72°C for 15 min.
3. Prepare the second PCR mix (25 μL) containing inner primers IP6/IP7 (as shown in Table 98.5).

TABLE 98.5
Reaction Mixture Used for the Nested PCR

Outer	Per Reaction	Inner	Per Reaction
Sterile distilled water	–	Sterile distilled water	–
10× PCR buffer	1×	10× PCR buffer	1×
dNTP	1 mM	dNTP	1 mM
$MgCl_2$	2 mM	$MgCl_2$	3 mM
B1–F	0.6 μM	IP6–F	0.2 μM
A2–R	0.6 μM	IP6–R	0.2 μM
Taq polymerase (Imolase, Citomed)	1.25 U	*Taq* polymerase (Imolase, Citomed)	1.25 U
DNA	5 μL	DNA	1–2 μL
Total volume	25 μL	Total volume	25 μL

4. Amplify with 1 cycle of 95°C for 5 min; 40 cycles at 94°C for 1 min, 62°C for 1 min, and 72°C for 1 min; and a final extension at 72°C for 10 min.
5. Examine 4–6 μL of PCR products on a 1% agarose gel; then digest 10–13 μL of the products with restriction enzyme *Mse*I in a total volume of 20 μL (containing 10–13 μL of the PCR amplicon, 2 μL reaction 10× buffer, 5 U of restriction enzyme *Mse*I, Gibco/LifeTechonologies) at 37°C overnight.
6. Electrophorese restriction fragments on a 1.5% or 2% agarose gel for 1 h at 100 V. Compare RFLP patterns to the fragments of the *tpr* amplicon of *T. pallidum* Nichols strain, with the help of molecular weight markers.

(ii) Analysis of arp Gene The primers for the amplification of the 60 bp tandem repeat region within the *arp* gene are shown in Table 98.6.

TABLE 98.6
Primers Used to Amplify the *arp* Gene

Gene	Primer	Sequence (5′→3′)	Amplicon (bp)
arp	ARP 1-F	CAA GTCAGG ACG GAC TGT CC	Size of the fragments in relation to the number of repeats
	ARP 2-R	GGT ATC ACC TGG GGA TGC	

TABLE 98.7
Reaction Mixture Used to Amplify *arp* Gene

	Per Reaction
Sterile distilled water	–
10× PCR buffer	1×
dNTP	1 mM
$MgCl_2$	2 mM
ARP1-F	0.4 μM
ARP2-R	0.4 μM
Taq polymerase (Imolase, Citomed)	2 U
DNA	5 μ
Total volume	25 μL

Procedure

1. Prepare PCR mix (25 μL) as in Table 98.7. In every run, include a negative control (distilled water) and a positive control (consisting of *T. pallidum* Nichols strain DNA).
2. Amplify with 1 cycle of 95°C for 5 min; 40 cycles at 94°C for 1 min, 61°C for 1 min, and 72°C for 2 min and 30 s; a final extension at 72°C for 15 min.
3. Electrophorese the *arp* amplicons on a 1.5% or 2% agarose gel for 1 h at 100 V. Estimate the number of *arp* gene tandem repeats through comparison with the molecular weight marker fragments and the *arp* PCR products from the *T. pallidum* Nichols strain (14 repeats—1155 bp).

Note: The subtype is determined by combining the results of the number of the repeated sequences of 60 pb of the *arp* gene and the *tpr* RFLP patterns and designated with lower case letter, as described by Pillay et al.[88]

98.3 CONCLUSION AND FUTURE PERSPECTIVES

Syphilis has been known for many centuries, and although there have been many efforts to control this disease, it remains a very common infection worldwide. Coinfection with HIV has made this disease even more problematic.

The publication of *T. pallidum* genome resulted in increased knowledge of its structure, pathogenesis, and biology that was not possible before, due to the inability of *T. pallidum* to multiply in vitro. However, further studies to identify surface antigens and their respective functions, to find the means to induce protective immunity, and to develop an in vitro culture system where *T. pallidum* is able to grow and multiply are needed.

On the other hand, syphilis diagnosis is far from being ideal, especially in very early infection, in asymptomatic congenital syphilis, in neurosyphilis, and in differentiating treated from untreated infections. PCR techniques for the detection of *T. pallidum* are now available, but they lack sensitivity in latent syphilis and in CSF samples. Improving diagnosis for these stages is a great priority, as are improved tests for the follow-up of syphilis, particularly in HIV-infected patients. The results obtained in follow-up testing are often confusing, and patients are often treated repeatedly without needing it. However, access to *T. pallidum* genome sequence makes it easier to study and experiment with new targets for the development of new rapid tests and others.

To control this infection, much remains to be done; clinicians must be aware of the multiple aspects of the disease, namely, presentation, diagnosis, and treatment. Since a syphilis vaccine will not be available in the near future, the control of the disease will continue to focus on risk reduction, rapid identification and the tracing of contacts, and the early diagnosis and treatment of patients suffering from syphilis. Other control measures, such as antenatal screening, are routine procedures in many countries, but unfortunately not yet in many parts of the world.

Health education, condom distribution, syndromic management when laboratory diagnosis is not available, and early access to diagnosis and treatment are the existent tools that need to be applied and will help to control this infection. Probably, innovative structural interventions to alter the sociocultural and legal environment for risk would be of great help. Mass screening should be implemented in regions where there is a high prevalence of syphilis and for that, the implementation of rapid, sensitive, and specific tests is of great importance, since they have all the characteristics that make them ideal tests to be used in those regions.

Some years ago, syphilis elimination was thought to be possible within few years, but it seems that this infection will probably remain with us for the next century. Although the elimination of syphilis is feasible, since man is its only reservoir, and it can be easily and inexpensively diagnosed and treated with penicillin, without new approaches, we will never be able to achieve that goal or to substantially decrease the global number of syphilis cases.

REFERENCES

1. Fraser, C.M. et al., Complete genome sequence of *Treponema pallidum* the syphilis spirochete, *Science,* 281, 375, 1998.
2. Seshadri, R. et al. Comparison of the genome of the oral pathogen *Treponema denticola* with other spirochete genomes, *Proc. Natl. Acad. Sci. USA,* 101, 5646, 2004.
3. Charon, N.W. et al., Spirochete chemotaxis, motility, and the structure of the spirochetal periplasmic flagella, *Res. Microbiol.,* 143, 597, 1992.
4. Charon, N.W., and Goldstein, S.F., Genetics of motility and chemotaxis of a fascinating group of bacteria: The spirochetes, *Annu. Rev. Genet.,* 36, 47, 2002.
5. Bailey, M.J., Penn, C.W., and Cockayne, A., Evidence for the presence of lipopolysaccharide in *Treponema phagedenis* (biotype Reiterii) but not in *Treponema pallidum* (Nichols), *FEMS Microbiol. Lett.,* 27, 111, 1985.
6. Walker, E.M. et al., Demonstration of rare protein in the outer membrane of *Treponema pallidum* of subsp. *pallidum* by freeze-fracture analysis, *J. Bacteriol.,* 171, 5005, 1989.
7. Lukehart, S.A., Biologia of Treponemas. In Holmes, K.K. et al. (Editors), *Sexually Transmitted Pathogens*, 4th ed., Chapter 36, McGraw-Hill Medical, London, 2006.
8. Charon, N.W., Li, C., and Goldstein, S.F., The beguiling motility of the genus *Treponema.* In Radolf, J.D., and Lukehart, S.A. (Editors), *Pathogenic Treponema Molecular and Cellular Biology*, Chapter 6, Caister Academic Press, Norfolk, England, 2006.
9. Cox, D.L., and Radolf, J.D., Metabolism of the Treponema. In Radolf, J.D., and Lukehart, S.A. (Editors), *Pathogenic Treponema Molecular and Cellular Biology,* Chapter 4, Caister Academic Press, Norfolk, England, 2006.
10. Weinstock, G.M., et al., The genome of *Treponema pallidum*: New light on the agent of syphilis, *FEMS Microbiol. Rev.,* 22, 323, 1998.
11. World Health Organization, Department of Reproductive Health and Research. Global strategy for the prevention and control of sexually transmitted infections: 2006–2015. Breaking the chain of transmission, Geneva, 2007. http://whqlibdoc.who.int/publications/2007/9789241563475_eng.pdf. Accessed November 10, 2010.
12. World Health Organization. Global prevalence and incidence of selected curable sexually transmitted infections: overview and estimates. WHO, Geneva, WHO/HIV AIDS/2001.02.
13. Woods, CR., Syphilis in children: Congenital and acquired, *Semin. Pediatr. Infect. Dis.*, 16, 245, 2005.
14. Singh, A.E., and Romanowski, B., Syphilis: Review with emphasis on clinical, epidemiologic, and some biologic features, *Clin. Microbiol. Rev.*, 12,187, 1999.
15. Zetola, N.M., and Klausner, J.D., Syphilis and HIV infection: An update, *Clin. Infect. Dis.*, 44, 1222, 2007.
16. Schroeter, A.L. et al., Therapy for incubating syphilis. Effectiveness of gonorrhea treatment, *JAMA*, 218, 711, 1971.
17. Turner, T.B., Hardy, P.H., and Newman, B., Infectivity tests in syphilis, *Br. J. Vener. Dis.,* 45, 183, 1969.
18. Genç, M., and Ledger, W.J., Syphilis in pregnancy, *Sex. Transmit. Infect*., 76, 73, 2000.
19. Nicoll, A., and Hamers, F.F., Are trends in HIV, gonorrhoea, and syphilis worsening in western Europe? *BMJ*, 324, 1324, 2002.
20. Panchau, C. et al., Sexually transmitted diseases among adolescents in developed countries, *Fam. Plann. Perspect.,* 32, 24, 2000.
21. Yakubovsky, A. et al., Syphilis management in St. Petersburg, Russia: 1995–2001, *Sex. Transmit. Dis.,* 33, 44, 2006.
22. Chen, Z.Q. et al., Syphilis in China: Results of a national surveillance programme, *Lancet,* 369, 132, 2007.
23. Mendel, A. et al., Primary and secondary syphilis, 20 years' experience. 2. Clinical features, *Genitourin. Med.,* 65, 1, 1989.
24. Crum-Cianflone, N., Weekes, J., and Bavaro, M., Syphilitic hepatitis among HIV-infected patients, *Int. J. STD AIDS*, 20, 278, 2009.
25. Fujisaki, T., Tatewaki, M., and Fujisaki, J., A case of gastric syphilis, *Clin. Gastroenterol. Hepatol*., 6, A34, 2008.
26. Denes, E. et al., Syphilitic periostitis, *Eur. J. Intern. Med.*, 20, 78, 2009.
27. Piñón-Mosquera, R., Jiménez-Benito, J., and Olea-Cascón, J., Syphilitic bilateral panuveitis: A case report, *Arch. Soc. Esp. Oftalmol.,* 84, 101, 2009.
28. Marra, C.M., Update on neurosyphilis, *Curr. Infect. Dis. Rep.*, 11, 127, 2009.
29. Foti, C. et al., Secondary syphilis with progressive ocular involvement in an immunocompetent patient, *Eur. J. Dermatol.*, 19, 288, 2009.
30. Hunte, W., al-Ghraoui, F., and Cohen, R.J., Secondary syphilis and the nephrotic syndrome, *J. Am. Soc. Nephrol.,* 3, 1351, 1993.

31. Rolfs, R.T. et al., A randomised trial of enhanced therapy for early syphilis in patients with and without human immunodeficiency virus infection, *N. Engl. J. Med.* 337, 307, 1997.
32. Radolf, J.D., Hazlet, K.R.O., and Lukehart, S.A., Pathogenesis of syphilis. In Radolf, J.D., and Lukehart, S.A. (Editors), *Pathogenic Treponema Molecular and Cellular Biology,* pp. 193–236, Caister Academic Press, Norfolk, England, 2006.
33. Clark, E.G., and Danbolt, N., The Oslo study of the natural history of untreated syphilis; an epidemiologic investigation based on a restudy of the Boeck-Bruusgaard material; a review and appraisal, *J. Chronic Dis.*, 2, 311, 1955.
34. Lafond, R.E., and Lukehart, S.A., Biological basis for syphilis, *Clin. Microbiol. Rev*., 19, 29, 2006.
35. Morshed, M.G. et al., Neurosyphilitic gumma in a homosexual man with HIV infection confirmed by polymerase chain reaction, *Int. J. STD AIDS*, 19, 568, 2008.
36. Kampmeier, R.H., The late manifestations of syphilis: Skeletal, visceral and cardiovascular, *Med. Clin. N. Am*., 48, 667, 1964.
37. Jackman, J.D. Jr., and Radolf, J.D., Cardiovascular syphilis, *Am. J. Med*., 87, 425, 1989.
38. Golden, M.R., Marra, C.M., and Holmes, K.K., Update on syphilis: Resurgence of an old problem, *J. Am. Med. Assoc.,* 290, 1510, 2003.
39. Fleming, D.T. et al., Syphilis in Atlanta during an era of declining incidence, *Sex. Transmit. Dis*., 27, 68, 2000.
40. Reynolds, S.J. et al., High rates of syphilis among STI patients are contributing to the spread of HIV-1 in India, *Sex. Transmit. Infect*., 82, 121, 2006.
41. Zetola N.M. et al., Syphilis in the United States: An update for clinicians with an emphasis on HIV coinfection, *Mayo Clin. Proc.,* 82, 109, 2007.
42. Hutchinson, C. et al., Altered clinical presentation of early syphilis in patients with human immunodeficiency virus infection, *Ann. Intern. Med*., 121, 94, 1994.
43. Rompalo, A.M. et al., Syphilis and HIV Study Group. Clinical manifestations of early syphilis by HIV status and gender: Results of the syphilis and HIV study, *Sex. Transm. Dis.,* 28, 158, 2001.
44. Kassutto, S., and Doweiko, J.P., Syphilis in the HIV era, *Emerg. Infect. Dis.,* 10, 1471, 2004.
45. Karumudi, U., and Augenbraun, M., Syphilis and HIV: A dangerous duo, *Expert. Rev. Anti-infect. Ther*., 3, 825, 2005.
46. Fitzgerald, T.J., Miller, J.N., and Sykes, J.A., *Treponema pallidum* (Nichols strain) in tissue cultures: Cellular attachment, entry, and survival, *Infect. Immun.,* 11, 1133, 1975.
47. Cameron, C.E., Identification of a *Treponema pallidum* laminin-binding protein, *Infect. Immun.*, 71, 2525, 2003.
48. Peeling, R.W., and Hook, E.W. 3rd, The pathogenesis of syphilis: The Great Mimicker, revisited, *J. Pathol*., 208, 224, 2006.
49. Centurion-Lara, A. et al., The tprK gene is heterogeneous among *Treponema pallidum* strains and has multiple alleles, *Infect. Immun*., 68, 824, 2000.
50. Wicher, K., Horowitz, H.W., and Wicher, V., Laboratorial methods of diagnosis of syphilis for the beginning of the third millennium, *Microbes. Infect*., 1, 1035, 1999.
51. Larsen S.A. et al., Syphilis and related treponematoses. In Hausler, W.J. Jr., and Sussman, M. (Editors), *Microbiology and Microbial Infections*, Vol. 3, pp. 641–668, Oxford University Press, Inc., New York, 1998.
52. Ito, F. et al., Specific immunofluorescent staining of pathogenic treponemes with a monoclonal antibody, *J. Clin. Microbiol.*, 30, 831, 1992.
53. Fieldsteel, A.H., Cox, D.L., and Moeckli, R.A., Cultivation of virulent *Treponema pallidum* in tissue culture, *Infect. Immun.,* 32, 908, 1981.
54. Lukehart, S.A., and Marra, C.M., Isolation and laboratory maintenance of *Treponema pallidum*, *Curr. Protoc. Microbiol*., 12, 1, 2007.
55. Kingston, M. et al., Syphilis Guidelines Revision Group 2008, Clinical Effectiveness Group. UK National Guidelines on the Management of Syphilis, *Int. J. STD AIDS*, 19, 729, 2008.
56. Nelson, R.A. Jr., and Mayer, M.M., Immobilization of *Treponema pallidum* in vitro by antibody produced in syphilitic infection, *J. Exp. Med*., 89, 369, 1949.
57. Deacon, W.E., and Falcon Harris, A., A fluorescent test for treponemal antibodies, *JAMA,* 96, 477, 1975.
58. Rathlev, T., Haemagglutination tests utilizing antigens from pathogenic and apathogenic *Treponema pallidum, WHO/VDT/Res.,* 77, 65, 1965.
59. Pope, V. et al., Comparison of the serodia *Treponema pallidum* particle agglutination, Captia syphilis-G, and SpiroTek Reagin II tests with standard test techniques for diagnosis of syphilis, *J. Clin. Microbiol*., 38, 2543, 2000.
60. Castro, R.R. et al., Evaluation of the passive particle agglutination test in the serodiagnosis and follow-up of syphilis, *Am. J. Clin. Pathol*., 116, 581, 2001.
61. Castro, R. et al., Evaluation of an enzyme immunoassay technique for detection of antibodies against *Treponema pallidum*, *J. Clin. Microbiol*., 41, 250, 2003.
62. Woznicová, V. et al., Detection of *Treponema pallidum* subsp. *pallidum* from skin lesions, serum, and cerebrospinal fluid in an infant with congenital syphilis after clindamycin treatment of the mother during pregnancy, *J. Clin. Microbiol*., 45, 659, 2007.
63. Schouls, L.M. et al., Overproduction and purification of *Treponema pallidum* recombinant-DNA-derived proteins TmpA and TmpB and their potential use in serodiagnosis of syphilis, *Infect. Immun.,* 577, 2612, 1989.
64. Sambri, V. et al., Evaluation of recomWell Treponema, a novel recombinant antigen-based enzyme-linked immunosorbent assay for the diagnosis of syphilis, *Clin. Microbiol. Infect.,* 7, 200, 2001.
65. Pope, V., and Fears, M.B., Spirotek Syphilis: An enzyme immunoassay for treponemal antibodies. In Larsen, S.A., Pope, V. Jr., and Kennedy, E.J. (Editors), *A Manual of tests for Syphilis*, 9th ed., pp. 318–331, American Public Health Association, Washington, 1998.
66. Marangoni, A. et al., IgG western blot as a confirmatory test in early syphilis, *Zentralbl. Bakteriol*., 289, 125, 1999.
67. Norris, S.J., *Treponema pallidum* Polypeptide Research Group. Polypeptides of *Treponema pallidum*: Progress toward understanding their structural, functional, and immunologic roles, *Microbiol. Rev.,* 57, 750, 1993.
68. Li, J. et al., Clinical evaluation of four recombinant *Treponema pallidum* antigen-based rapid diagnostic tests for syphilis, *J. Eur. Acad. Dermatol. Venereol.,* 23, 648, 2009.
69. Burstain, J.M. et al., Sensitive detection of *Treponema pallidum* by using the polymerase chain reaction, *J. Clin. Microbiol*., 29, 62, 1991.
70. Orle, K.A. et al., Simultaneous PCR detection of *Haemophilus ducreyi, Treponema pallidum*, and herpes simplex virus types 1 and 2 from genital ulcers, *J. Clin. Microbiol*., 34, 49,1996.
71. Liu, A.Y. et al., Detection of pathogens causing genital ulcer disease by multiplex polymerase chain reaction, *Chin. Med. Sci. J.*, 20, 273, 2005.

72. Liu, H. et al., New tests for syphilis: rational design of a PCR method for detection of *Treponema pallidum* in clinical specimens using unique regions of the DNA polymerase I gene, *J. Clin. Microbiol.*, 39, 1941, 2001.
73. Castro, C. et al., Detection of *Treponema pallidum* sp *pallidum* DNA in latent syphilis, *Int. J. STD AIDS*, 18, 842, 2007.
74. Wenhai, L., Jianzhong, Z., and Cao, Y., Detection of *Treponema pallidum* in skin lesions of secondary syphilis and characterization of the inflammatory infiltrate, *Dermatology*, 208, 94, 2004.
75. Smajs, D. et al., Diagnosis of syphilis by polymerase chain reaction and molecular typing of *Treponema pallidum*, *Rev. Med. Microbiol.*, 17, 4, 2006.
76. Marfin, A.A. et al., Amplification of the DNA polymerase I gene of *Treponema pallidum* from whole blood of persons with syphilis, *Diagn. Microbiol. Infect. Dis.*, 40, 163, 2001.
77. Martin, I.E. et al., Molecular characterization of syphilis in patients in Canada: Azithromycin resistance and detection of *Treponema pallidum* DNA in whole-blood samples versus ulcerative swabs, *J. Clin. Microbiol.*, 47, 1668, 2009.
78. Sutton, M.Y. et al., Molecular subtyping of *Treponema pallidum* in Arizona County with increasing syphilis morbidity: Use of specimens from ulcers and blood, *J. Infect. Dis.*, 183, 1601, 2001.
79. Hay, P.E. et al., Detection of treponemal DNA in the CSF of patients with syphilis and HIV infection using the polymerase chart reaction, *Genitourin. Med.*, 66, 428, 1990.
80. Chung, K.Y, Lee, M.G., and Lee, J.B., Detection of *Treponema pallidum* by polymerase chain reaction in the cerebrospinal fluid of syphilis patients, *Yonsei Med. J.*, 35, 190, 1994.
81. Koek, A.G. et al., Specific and sensitive diagnosis of syphilis using a real-time PCR for *Treponema pallidum*, *Clin. Microbiol. Infect.*, 12, 1233, 2006.
82. Leslie, D.E. et al., Development of a real-time PCR assay to detect *Treponema pallidum* in clinical specimens and assessment of the assay's performance by comparison with serological testing, *J. Clin. Microbiol.*, 45, 93, 2007.
83. Gayet-Ageron, A. et al., Assessment of a real-time PCR test to diagnose syphilis from diverse biological samples, *Transm. Infect.*, 85, 264, 2009.
84. Suntoke, T.R. et al., Evaluation of multiplex real-time PCR for detection of Haemophilus ducreyi, Treponema pallidum, herpes simplex virus type 1 and 2 in the diagnosis of genital ulcer disease in the Rakai District, Uganda, *Sex. Transm. Infect.*, 85, 97, 2009.
85. Larsen, S.A. et al., Laboratory diagnosis and interpretation of tests for syphilis, *Clin. Microbiol. Rev.*, 8, 1, 1995.
86. Centers for Disease Control and Prevention. Sexually Transmitted Diseases Treatment Guidelines, 2006. *MMWR*, 55, 22, 2006.
87. Castro, R., Prieto, E.S., and Martins Pereira, F.L., Nontreponemal tests in the diagnosis of neurosyphilis: An evaluation of the Venereal Disease Research Laboratory (VDRL) and the Rapid Plasma Reagin (RPR) tests, *J. Clin. Lab. Anal.*, 22, 257, 2008.
88. Pillay, A. et al., Molecular subtyping of *Treponema pallidum* subspecies *pallidum*, *Sex. Trans. Dis.*, 8, 408, 1998.
89. Castro, R. et al., Molecular subtyping of *Treponema pallidum* subsp. *pallidum* in Lisbon, Portugal, *J. Clin. Microbiol.*, 47, 2510, 2009.

Section VI

Pan-Bacterial Detection

99 Pan-Bacterial Detection of Sepsis-Causative Pathogens

Roland P.H. Schmitz and Marc Lehmann

CONTENTS

99.1 INTRODUCTION

99.1.1 Bacteria as Sepsis-Causative Pathogens

Life-threatening bacterial (and fungal) infections and their outcomes sepsis and consecutive organ failure are frequent complications of hospitalized patients, with an increasing number of 18 million new sepsis cases each year worldwide and a mortality rate of 30%–50%.[1] Sepsis results from the host's response to bacterial and fungal infections, whereas a malfunction of the defense and repair system is responsible for the development of organ dysfunctions and, finally, multiorgan failure, as outlined by the hypothetical PIRO (predisposing factor/infection/responses/organ dysfunction) model for staging sepsis,[2] according to American College of Chest Physicians/Society of Critical Care Medicine (ACCP/SCCM) consensus criteria.[3]

In Germany, about 60,000 out of 154,000 of all septic patients die from severe sepsis and organ dysfunction, which makes it one of the most frequent causes of death in intensive care units (ICUs). The incidence of sepsis has risen due to the augmented use of invasive devices, the aging of the population, and the higher incidence of immunosuppressive conditions, such as chemotherapy for cancer and acquired immunodeficiency syndrome (AIDS).[4] About 30% of the intensive medicine budget is expended for the treatment of those patients.[5] While in 1970s through the 1990s, septical infections in Germany were attributed mainly to gram-negative bacterial species, currently gram-positives, fungi, and multiinfections are rising.[4,6,7] However, there is an obvious need for more accurate and rapid identification strategies for bacterial species.

99.1.2 Sepsis Diagnostics: Overview on Current Methods

Early detection of bacterial infection and prompt and adequate antibiosis in the first few hours of infection are the crucial steps for an effective sepsis therapy.[8–10] Epidemiological data confirm that a doubling of mortality is the consequence of inadequate therapy,[11] and an increase of mortality of more than ~7% per hour is proven in cases of delayed adequate (directed) antibiotic treatment, irrespective of the causative pathogen.[12]

While clinical manifestations of inflammations are elusive, some biochemical parameters indicate early stages of bacterial or fungal infections. The presence of three biomarkers was proven in systemic infections: C-reactive protein (CRP), procalcitonin (PCT), and D-dimer. None of them, however, functions as single independent criterion for sepsis.[13]

CRP, an acute-phase protein that plays a significant role within the complement pathway, binds to several polysaccharides present in all classes of sepsis-causative pathogens. Secretion of CRP starts within several hours of the stimulus peaking between 36 h and 50 h. After disappearance of the stimulus, CRP falls rapidly with a half-life of 19 h, but it can remain elevated, even for very long periods, if the underlying cause of the elevation persists. Only interventions affecting the inflammatory process responsible for the acute-phase reaction can change the CRP level. Changes may be very helpful in diagnosis as well as in monitoring response to therapy, as CRP levels are only determined by the rate of synthesis.[14]

PCT is a 116-amino acid prohormone that has proven to be a useful marker in sepsis and sepsis-like conditions (e.g., severe burns and mechanical trauma), and also in some infections of nonbacterial causation such as systemic fungal infection.[15] Serum levels of PCT are frequently increased in septic patients, sometimes attaining levels several thousand-fold above normal and often persisting for long periods of time. Moreover, the levels often correlate positively with the severity of the condition and mortality.[16–18] A good correlation was found between the serum procalcitonin level and the Sepsis-Related Organ Failure Assessment (SOFA) score, although no correlation was found between the latter and the CRP level.[19] It has been argued that the procalcitonin serum level may be useful for aiding the diagnosis of sepsis and discriminating between the specific phases of the disease.

Activation of the coagulation cascade is a common and early phenomenon in the development of sepsis, and this fact supports the use of anticoagulant treatments as potentially useful interventions.[20] Anticoagulation releases degradation products containing D-dimers, whereas a finding of more than 500 ng of D-dimer per milliliter is considered abnormal, and such levels are present in virtually all patients with sepsis.[21]

However, none of these potential biochemical test parameters that gain information on the status of inflammation and alter along with the development of sepsis, (especially the parameters regarding the host-response on invasive infections, pathogen cell components, or at least metabolic byproducts) permits access to a directed therapy of the disease. This exclusively demands the identification of the causative agent and putative antibiotic resistances it exhibits.

The gold standard technique, blood culture, which is the routine method in clinical microbiology laboratories to demonstrate the presence of pathogens in patients suspected of systemic infections, exhibits some drawbacks due to the patient's antibiotic/antimycotic treatment prior to sample withdrawal, a low abundance of causative agents in the blood samples, and frequently, fastidious or noncultivable organisms.[22] The yield of blood cultures alters with (i) the severity of the disease; (ii) the presence of gram-negative bacteria or polymicrobial bloodstream infections, which is seldom expected in case of sepsis; (iii) the volume of blood cultured; and (iv) the system used for culturing. Inadequate preanalytic handling may lead to false-negative results, since several microorganisms are extremely sensitive to changes in temperature and growth conditions during the exponential growth phase, with failure to be cultivated thereafter.

Various commercial cultural systems that are able to identify sepsis-causative pathogens within 4–72 h have been developed, and several advancements in blood culture have improved the detection of pathogens in the past, regarding the development of new liquid media with additional growth supplements and adsorbing agents to neutralize growth inhibitors, metabolic products, and remnant antibiotics.[23,24] Automation technology, based on analysis of the proportional release of CO_2 using fluorescent (e.g., BacT/ALERT 3D, bioMérieux, Lyon, France) or colorimetric/fluorimetric (e.g., BD BACTEC™ FX Blood Culture System/BD EpiCenter™ Operating System, Becton Dickinson) sensors enabling continuous growth monitoring of bar-code-labeled flasks remote data supply, has substantially improved yield and provision of results. However, the median time to positivity of blood cultures still amounts up to 15 h (range 2.6–127 h, whereas a sample can be reliably declared negative within up to 7 days).[25,26] The outcome depends on bacterial or fungal load, type of pathogen and its generation time, presence of growth suppressants, and the consideration of preanalytic premises (e.g., avoidance of delayed entry into blood-culture systems or inadequate timing of obtaining blood for cultivation).[27,28] The culture-based phenotypic identification of yeast species from clinical materials, for example, requires at least another day to obtain pure cultures. Positive blood cultures are, then, only the basis for further microbial diagnostics, for example, biochemical typing/species differentiation, and/or generation of antibacterial or antifungal susceptibility profiles, which is also laborious and time consuming, although the methods are partly automated.[22]

The positive predictive value of blood cultures is still high (~98%), particularly if the same pathogen is detected in repeat cultures,[29] but the overall rate of positivity of conventional blood cultures for the diagnosis of bloodstream infections attains only up to 20%,[30,31] even if recommended standard procedures are properly followed. In addition, included in this rate of positivity are false-positives mainly due to contamination with resident skin flora. The false-positive rate differs with the instruments and the media used, and ranges from 1.4% to 6.2%.[32]

Time-to-result of blood cultures is mostly too long to initiate an effective sepsis therapy. Therefore, broad-spectrum antibiotics are given simultaneously or prior to blood withdrawal without trusted microbiological findings, which enhances the broadening of multiresistant organisms and downsizes the efforts of blood cultures that are taken later.[33–35] Therapy is usually indirectly administered early in the clinical course of bloodstream infections (BSI) and prior to the reporting of positive blood-culture results. With respect to antimicrobial management, the most important information provided by the clinical microbiology laboratory appeared to be the reporting of positive blood-culture and Gram-stain results. Antimicrobial susceptibility testing data did not appear to have a comparatively important impact on antimicrobial management among patients with BSI.[36]

Many types of bacterial resistance are based on several distinct point mutations, for example, penicillin-resistance in *Pneumococcus* spp. and more than 300 extended-spectrum β-lactamases (ESBL) described so far for distinct gram-negatives. A few are single-gene encoded and therefore traceable by nucleic acid amplification techniques (NAT) per se (e.g., multiresistant *Staphylococcus aureus*, MRSA). However, the predominant number of various regulatory genes, whose role in resistance expression remains elusive, precludes simple molecular- (or proteomic-) based economic detection approaches for routine purposes. Currently, culture-based antibiotic susceptibility testing is demanded, while empirical therapy using broad-spectrum antibiotics is increasingly less predictable due to rapidly developing resistances and multiresistance types. NAT detection of the most prominent traceable resistance targets is therefore an important step for a fast confirmation or correction of the started antibiotic regimen.

99.1.3 Pathogen Detection by Nucleic Acids Amplification Techniques

NAT (e.g., polymerase chain reaction, PCR) allow more rapid target and resistance detection, within several hours, even from whole blood, compared to culture-based methods. Free bacterial and fungal DNA, as well as DNA from adherent, phagocytosed, and free intact and nonviable pathogens, are detected, whereas blood cultures contribute only in the presence of viable and metabolic active cells. On the other hand, the high sensitivity is decreased by factors like high fractions of eukaryotic bulk DNA, salts, and other blood ingredients like hemoglobin in erythrocytes and lactoferrin in leukocytes, as well as immunoglobulin G in plasma, most of which can be effectively removed by affinity chromatography steps prior to target amplification.

In recent years, numerous nucleic acid–based methods have been developed to improve the diagnosis of septic infections and the identification of pathogenic bacteria (and fungi). However, the referred NAT methods lack standardized pretreatment steps and are mostly not suited for a direct pathogen detection, especially in whole blood. The assays rely on pathogen-specific primers; multiplex assays addressing a set of different sepsis-associated pathogens; and universal broad-range assays involving conserved target sequences, such as the eubacterial 16S or 23S rDNA gene, their spacer regions, and the panfungal 8S or 18S rDNA/RNA.

Real-Time PCR (qPCR). Crucial in pathogen detection from whole blood is the low abundance of pathogen cells within the circuit. Sine qua non for a new NAT-based assay is therefore an analytical sensitivity that enters this analytical range to compete with the standard techniques in the lab. Real-time (quantitative) PCR (qPCR) devices are mostly regarded to allow for the lowest analytical sensitivity of NAT-based assays. Their analytical sensitivity is in the 5–500-CFU/mL range, depending on device and target, and is thus limited in the presence of low bacterial loads: a high percentage of adult bacteremia gained positive blood cultures from blood samples possessing only 10 CFU of sepsis-causative pathogens per ml or less,[37] a level of concentration that is below the detection limit of most classical end-point PCR assays.

In addition, qPCR requires high-priced analytical devices and its multiplex ability is usually restricted due to the limited availability of wavelength-specific fluorescence dyes (e.g., 6-carboxy-fluorescein/FAM) and/or fluorescence channels of the qPCR device used. The number of the dyes is in fact growing with the state of the art, but to achieve a still limited but increasing multiplex applicability, the fluorescence detection, for example, via dual fluorescence resonance energy transfer (FRET) probes targeting species-specific internal transcribed spacer (ITS) regions of PCR amplicons,[38,39] is combined with error-prone melting point analyses and done with a high workload.[40]

Methods are often declared to be rapid but rely in fact on preculturing of human samples. Even qPCR is often combined with preculturing, notwithstanding the designated high sensitivity. qPCR-based assays, for example, have been described for the differentiation of *Staphylococcus aureus* and coagulase-negative staphylococci (CoNS) in positive blood cultures using FRET probes,[41] and for Gram-differentiation of ICU-relevant bacteria by SYBR® Green stain after DNA spiking of body fluids.[42] Gram-specific qPCR amplification prior to sequencing was also done from spiked blood cultures, which at least reduced the time-to-result compared to the positive signaling of the BACTEC blood-culture system used for comparison.[43]

Sepsis especially causes high mortality rates in newborns, particularly in preterm, and low-birth-weight infants. Therefore, many NAT-applications, mostly real-time PCR and broad-range, primer-based PCR, are related to pathogen detection in neonatal whole blood.[31,44–46] Combination of 16S rDNA qPCR with Gram-differentiation and amplicon hybridization was also used for diagnosis and discrimination of bacterial neonatal sepsis.[47] The successful application of NAT in pediatric and neonatal sepsis has indeed attributed to higher bacterial loads of 100 CFU/mL[22] or even 300 CFU/mL blood.[48]

End-Point PCR. Standard PCR is traditionally combined with gel electrophoretic amplicon depiction (e.g., horizontal and, more often nowadays, partly microchip-based capillary gel electrophoresis). Application of the DNA microarray technique is increasingly demanded to gain elevated sensitivities and for result verification via hybridization with amplicon-specific probes, although some authors had to admit that confirmed low detection limits of up to 10^5 cells/mL in artificially spiked blood still represent a barrier for clinical application.[49] On the other hand, direct confirmation of microorganisms in blood was shown with pathogen-specific probes in a hybridization protection assay, which promises to specifically identify many pathogens up to the species level with a high sensitivity in less than an hour.[50] DNA microarrays were also applied for accelerated diagnosis of bacterial sepsis directly from blood samples.[51]

Broad-Range PCR. Beside applications relying on species- and/or genera-specific primers, universal or broad-range primers were used in a multitude of approaches combining amplicon verification via hybridization or sequencing. For example, universal amplification of 23S ribosomal DNA was combined with amplicon hybridization to an oligonucleotide array to detect the causative pathogen of bacteremia in positive blood-culture bottles.[52] The sequencing of amplicons generated with broad-range primers has been done to identify a multitude of bacterial species in whole blood,[51,53–59] but has been reported to be laborious, comparatively expensive, and rather time consuming. In general, bacterial multiinfections or the presence of more than one PCR amplicon will gain overlayed sequences and therefore nonevaluable results after sequencing. Otherwise, the addressed rDNA targets of broad-range primers are often present in multiple copies allowing for a more sensitive detection.

The use of those primers, however, should in general be discussed because the strategy might run the elevated risk of false-positive/contaminant detection. While identification of bacterial species by 16S or 23S rDNA sequencing is a rather new development, currently no consensus exists on the degree of homology required to assign a specific genus or species. Unsurprisingly, the results of studies on the diagnostic application of broad-range primers in blood are heterogeneous and difficult to interpret. On the other hand, unspecific PCR promises to facilitate the finding of microorganisms that are found less frequently, even yet unknown, or noncultivable.[22,60] Anyhow, an enhanced sensitivity of even 3 CFU *Escherichia coli*/mL whole blood could be detected at higher cycle numbers with universal primers but at the expense of specificity.[61]

However, the assays using universal primers are in part compromised by their exquisite sensitivities responsible for nucleic acid trace detection already present in associated consumables,[62–64] as reported for with β-lactamase sequences contaminated *Taq* polymerase which led to a false *Streptococcus pneumoniae* identification,[65] microbial contaminated common disinfectants and antiseptics, bacterial DNA traces in commercial blood-culture media,[66] or introduced via applying routine sample withdrawal techniques causing false-positive results.

Problematic are the clinically irrelevant results of nondisease-associated transient bacteremia, the presence of free bacterial or fungal nucleic acids due to translocation from native colonized surfaces, as well as the detection of nonviable pathogen cells. Circulation of translocated microbial DNA, or the transient presence of bacteria in blood not associated with the disease, have been proven in vivo.[67,68] The origin and clinical significance of false-positive samples are often ambiguous and might belong to as yet unknown host-pathogen interactions.[40] It should be remembered that culture-positive bacteremia was reported after tooth brushing nearly three decades ago,[69] which indicates the range of risk factors for obtaining false-positives and the difficulties of distinguishing them from true-positives. Those transient translocations of mucosal colonizing microorganisms due to minimal manipulations have been reproduced[70] and also linked to further dental measures.[71] Moreover, isolation of bacteria in asymptomatic blood donors, and even transient fungemia without apparent clinical significance has been repeatedly reported.[72,73] Pure microbial background DNA has been detected in the blood of healthy individuals,[74,75] while human bulk DNA, coisolated with microbial pathogen nucleic acids from the addressed sample material (e.g., whole blood), is additionally the cause for a minute pathogen to human DNA ratio, increased cross-reactivities with primers, and significantly reduced overall analytical assay sensitivities.[76]

In summary, the use of broad-range primers should be consciously balanced in favor of particular species detections. The absence of bacterial (and fungal) nucleic acids in diagnostic consumables has to be carefully warranted to avoid cross-reactions and/or false-positive results in general, as already considered for the use of microarrays in clinical and regulatory settings.[77]

General Fate of NAT and Current Efforts. It has been estimated from clinical/microbiological results that about two-thirds of the PCR positive results identify true causative bacteria; one-third seems to be indeterminate, presumably representing false-positives. In conclusion, despite a high positive predictive value, the negative predictive value may not be sufficient to securely exclude infection. Positive NAT-related findings therefore have to be correlated with other clinical observations and diagnostic tests by clinicians and should not be the sole reason for any therapeutical consequences. In either case of NAT (i.e., with specific or broad-range primers)/blood culture relation, the results of the molecular assays are compared with those of a poor gold standard, which is an inappropriate basis to evaluate the efficiency of the new methods.[31]

Although most of the published PCR-based approaches have been useful for the identification of bacterial species at large, they identify only one species at a time, notwithstanding that many of them rely on a multiplex-compatible probe hybridization procedure that incurs time and expense. An economically more efficient approach would be the application of protocols able of identifying broad panels of relevant species in a highly parallel fashion.[78,79] An algorithm for identifying the 15 most important bacterial pathogens in positive blood cultures by a complex qPCR/hybridization/melting curve analysis approach was presented,[80] but it showed drawbacks after efforts were made to convert the system into routine diagnostics with regard to performance and data interpretation. These handicaps may be shared with the multiplex real-time PCR-based assay SeptiFast® (Roche Diagnostics, Mannheim, Germany), which covers 25 bacterial and fungal species (approximately 90% of nosocomial-bacteremia-causative species) by FRET combined with melting point differences directly from whole blood, which promises to gain therapy-relevant information in less than 6 h (for species detections plus another 2 h for antibiotic resistances). Initial results showed detection rates between 50% and 100% for selected clinically relevant bloodstream pathogens with analytical sensitivities between 30 and 3 CFU/mL and an overall specificity of 98.8%, as compared to conventional clinical microbiology.[81] Other studies gained only marginally higher numbers of microbiologically confirmed cases of bacteremia and fungemia than detected by blood culture in their study populations using SeptiFast®.[30,79] The assay displayed further limitations with respect to high probability of false-positive results for bacteremia (above all, reported for coagulase-negative staphylococci, which could lead to an adverse antibiotic regimen), failure to detect some relevant pathogens, and overall costs.[82]

An alternative with respect to basic equipment needed and overall low assay costs seemed to be a standardized multiplex PCR-based ELISA assay (Hyplex® BloodScreen, BAG, Lich, Germany), which admittedly covers only a small panel of 10 clinically important bacteria and fungi that cause general bacteremia/fungemia at the species level, and resistance genes for identification of methicillin-resistant *Staphylococcus aureus* (MRSA) from preincubated blood cultures.[83]

Most assay protocols exclude an optimized procedure for cell lysis within the sample material, which is a crucial step in nucleic acid detection, and therefore mostly rely on in-house techniques for cell disruption.[84] Beyond any attempts for standardization, the extraction methods for pathogen DNA, that is, the combination of cell lysis and DNA isolation protocols, produce markedly differing yields of bacterial DNA and thus can significantly affect the results of NAT methods for the respective species.[85,86]

Anyway, mechanical impact has been regarded as the most efficient lysis method for fungal cells,[87,88] whereas bacterial cells are often disrupted using chaotrophic buffers or enzymes/enzyme mixtures containing lysozyme, mutanolysin, and/or (costly) lysostaphin (e.g., for the lysis of *Staphylococcus* spp.).[84] Further studies described enzymatic/thermal lysis protocols as superior to mechanical impact,[89] but it can be argued that this experience could be mainly attributed to the mechanical device used and its effective power, lysis matrix, sample to head-space volume ratio within the lysis tubes, and the protocol (device settings) applied. It has to be considered that the recently cited studies on cell lysis were done with rigid (fungal) cells devoid of human cell material, which notably affects the lysis of the target cells, but the results of Fredricks et al.,[86] which likewise support mechanical lysis, were obtained from spiked clinical samples, a technique that should basically be recommended for assay development.

However, the mechanical devices of the last generation (e.g., FastPrep®-24, MP Biomedicals, Solon, Ohio, USA), which disrupt cells for periods of several seconds to a few minutes while executing figure-8 vertical, angular motions, for the first time seem to guarantee acceptable efficiencies for standardized disruptions of designated rigid pathogen cells within clinical samples. The cells are ground between glass or ceramic-bead sand mixtures of particle sizes below 2.5 mm in diameter. Higher head-space volumes over the sample/bead-matrix suspension support the lysis efficiency, an effect that is obviously noticed in tubes of 15-mL total volume (suspension to head-space volume 2:1). Although mechanical lysis is described as an inexpensive technique, the rotating equipment, which ensures sufficient efficiency, has to be regarded as initially cost intensive. In addition, a DNA extraction method is indispensable for a PCR-based assay, and both manual and automated protocols are forthcoming,[90] the latter, often high priced.

Certainly, a recently launched diagnostic PCR-linked tool is based on a selective lysis of (human) blood cells by chaotrophic buffers and quantitative degradation of human nucleic acids via chaotroph-resistant DNase.[91] The enzyme is subsequently heat-inactivated and bacterial pathogen cells are lysed in a second step to gain their genomic DNA for NAT detection purposes. It has to be questioned whether this initially smart-appearing two-step lysis procedure is actually a drawback of the analytical approach, since sublethally, affected pathogen cells (e.g., due to prior antibiotic treatment and/or impact of the immune system) and gram-negative thin-walled cells may be disrupted within the first lysis step, which could lead to a total loss of their genomic (target) DNA.[91] Therefore, the tool may be better suited for use with rigid intact cell types.

A new multiplex PCR-based assay (VYOO®, SIRS-Lab GmbH, Jena, Germany) was launched recently and compensates the abovementioned drawbacks by combining an efficient bead-based mechanical lysis protocol with an extended multiplex PCR detection step.[91,92] The launched assay includes an unique pretreatment tool that specifically concentrates bacterial and fungal DNA by affinity chromatography and removes human background DNA. A truncated DNA-binding protein recognizes unmethylated XpYpCpGpXpY motifs within the DNA,[93] which are significantly less frequent in human than in bacterial (and yeast) DNA.[94–96] The enrichment of pathogen DNA and elimination of PCR inhibitors significantly increases the sensitivity of the chosen downstream detection method by at least one order of magnitude.[91] Casually, pre-PCR treatment methods based on the selection of DNA polymerases that are more resistant to PCR inhibitory components and the addition of amplification facilitators were tested successfully and allowed for the detection of 1 CFU of bacteria per ml of blood-culture medium.[97,98] This range of sensitivity, however, implies a precultivation step prior to PCR and does not postulate the applicability of multiplex detection.

The VYOO® standard (as compared to qPCR assays) multiplex PCR detects an optimized panel of 34 bacterial (and six fungal) species that cause life-threatening infections and make up 99% of sepsis-causative pathogens according to the results of a meta-analysis of 11 prospective clinical studies (Figure 99.1), as well as five important antibiotic resistances (e.g., *mecA, VanA, VanB, blaCTX-M15, blaSHV*). The test has been designed for the examination of the generated amplicons by gel electrophoresis or hybridization methodologies.

However, using amplicon-specific DNA length markers, the amplicon identification via electrophoresis in horizontal gels or gel capillaries is straightforward and allows for results within 8 h, an important benefit compared to assays that rely on preculturing.[99] *Enterococcus faecalis*, for example, was detected in EDTA whole blood after spiking using the new multiplex PCR tool and gel electrophoretic amplicon detection with a sensitivity of <2 CFU/mL, a level previously attained at best by qPCR-based techniques.[81]

The progress of the new standard multiplex procedure has to be attributed in part to the high sample volume of 5-mL whole blood which can be applied, regardless the benefits of using 15-mL centrifugation vials within the lysis step (see above). The importance of the blood volume cultured or processed in NAT-based methods has been discussed,[100] but the implications of reduced sample volumes have been commonly accepted for the qPCR evaluation due to the always aspired miniaturization of assay layouts. The innovative part of the new multiplex PCR assay, however, remains the protein-based pathogen-DNA affinity chromatography step, which is additionally available as an assay-independent pretreatment tool, and which may substantially enhance even home-brewed NAT-based assays.

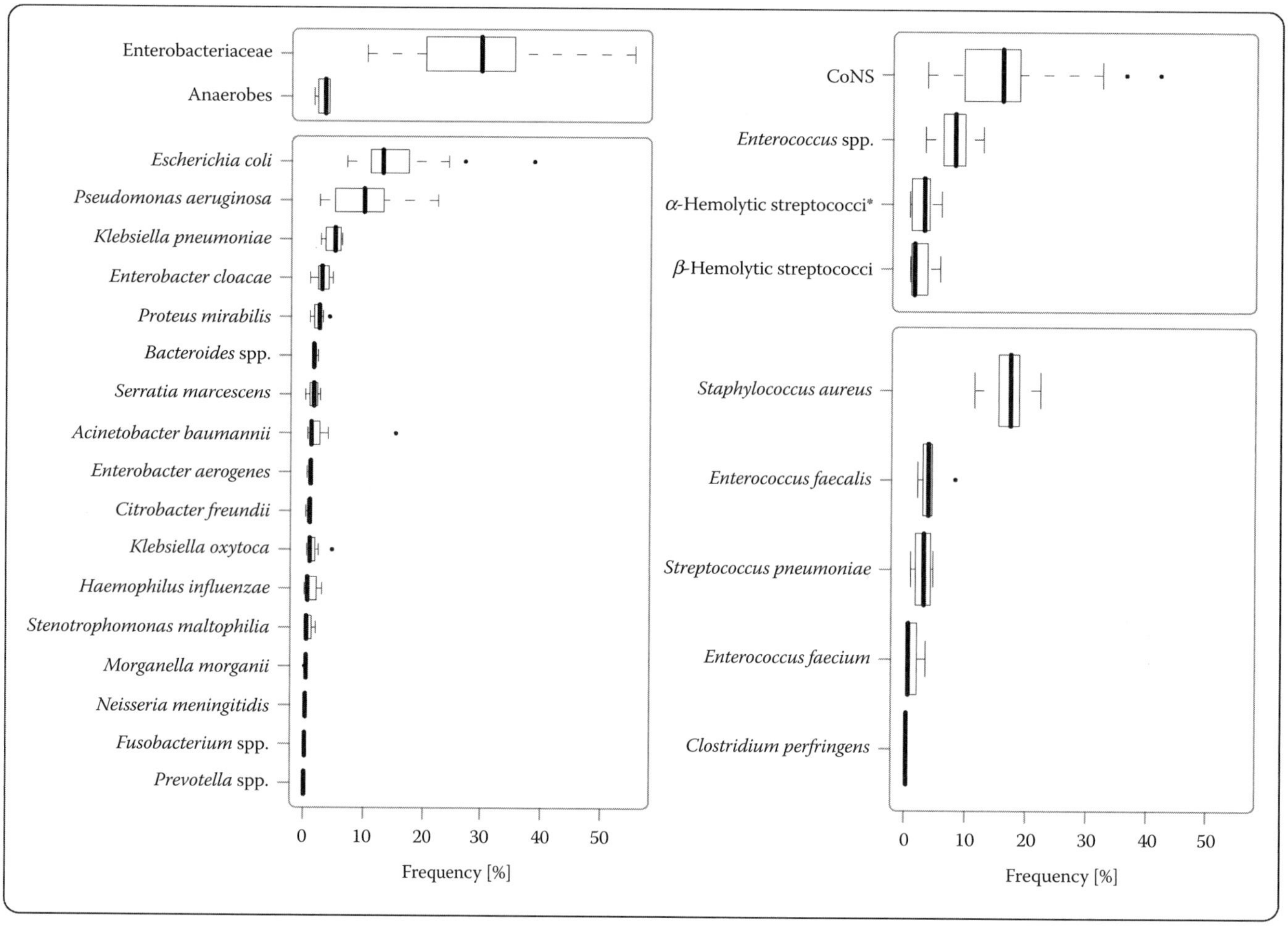

FIGURE 99.1 Frequencies of sepsis-causative pathogens as determined in 11 clinical studies. Individual species were designated only in a part of the studies, which accounts for their percental distribution as species and within genera/families and groups. The boxplot data were calculated using R (R Development Core Team, 2007). The data were taken from: Bodmann, K.-F. und Vogel, F., *Chemother. J.*, 10, 43, 2001; Cockerill, F.R. 3rd et al., *Clin. Infect. Dis.*, 24, 403, 1997; Focht, J. und Adam, D., *Klinikarzt*, 33, 35, 2004; Fridkin, S.K., and Gaynes, R.P., *Clin. Chest. Med.*, 20, 303, 1999; Geerdes, H.F. et al., *Clin. Infect. Dis.*, 15, 991, 1992; Geffers, C. et al., *Anästhesiol. Intensivmed. Notfallmed. Schmerzther.*, 39, 15, 2004; Jones, M.E. et al., *Ann. Clin. Microbiol. Antimicrobials*, 3, 1, 2004; Karlowsky, J.A. et al., *Ann. Clin. Microbiol. Antimicrobials*, 3, 1, 2004; Kübler, A. et al., *Med. Sci. Monit.*, 10, 635, 2004; Styers, D. et al., *Ann. Clin. Microbiol. Antimicrob.*, 5, 2, 2006; Vincent, J.L. et al., *Crit. Care Med.*, 34, 344, 2006; R: A language and environment for statistical computing. R Foundation for Statistical Computing (2007) Vienna, Austria, ISBN 3–900051-07–0. *without *Streptococcus pneumoniae*.

A basic overview on current diagnostic workflows for pathogen and resistance detection via NAT-based methods is depicted in Figure 99.2.

99.2 METHODS

99.2.1 Sample Preparation

Mechanical Cell Lysis (LOOXSTER®/VYOO®). The proof of blood-borne sepsis-causative microbial pathogens within the sample material demands high volumes of sample material (≤5 mL), powerful mechanical impact, and a straightforward sample processing.

1. Withdraw 5 mL of EDTA whole blood via standard techniques and process immediately. Otherwise, store the sample at 2°C–8°C until analysis.
2. Pour the homogenized liquid into a 15-mL centrifuge tube containing ~5 mL glass-bead matrix, 1:1 mixture of 0.1 mm/2.5 mm bead sizes. Ensure about 30% head-space volume.
3. Add ~1 μL of antifoam solution to minimize frothing. For mechanical impact, a FastPrep®-24 cell lysis device (MP Biomedicals, Solon, Ohio, USA), a set-up with 6.5 m/s rotation velocity: 45 s rotation—5 min intermission—45 s rotation, is recommended. The intermission is recommended for cooling down the device and sample.
4. Add 5 mL lysis buffer and 100 μL protease, vortex and incubate for 20 min at 50°C ± 2°C for proteolytic digestion of the lysate.

Total Genomic DNA Isolation. The minute quantities of pathogen DNA require a maximally efficient DNA

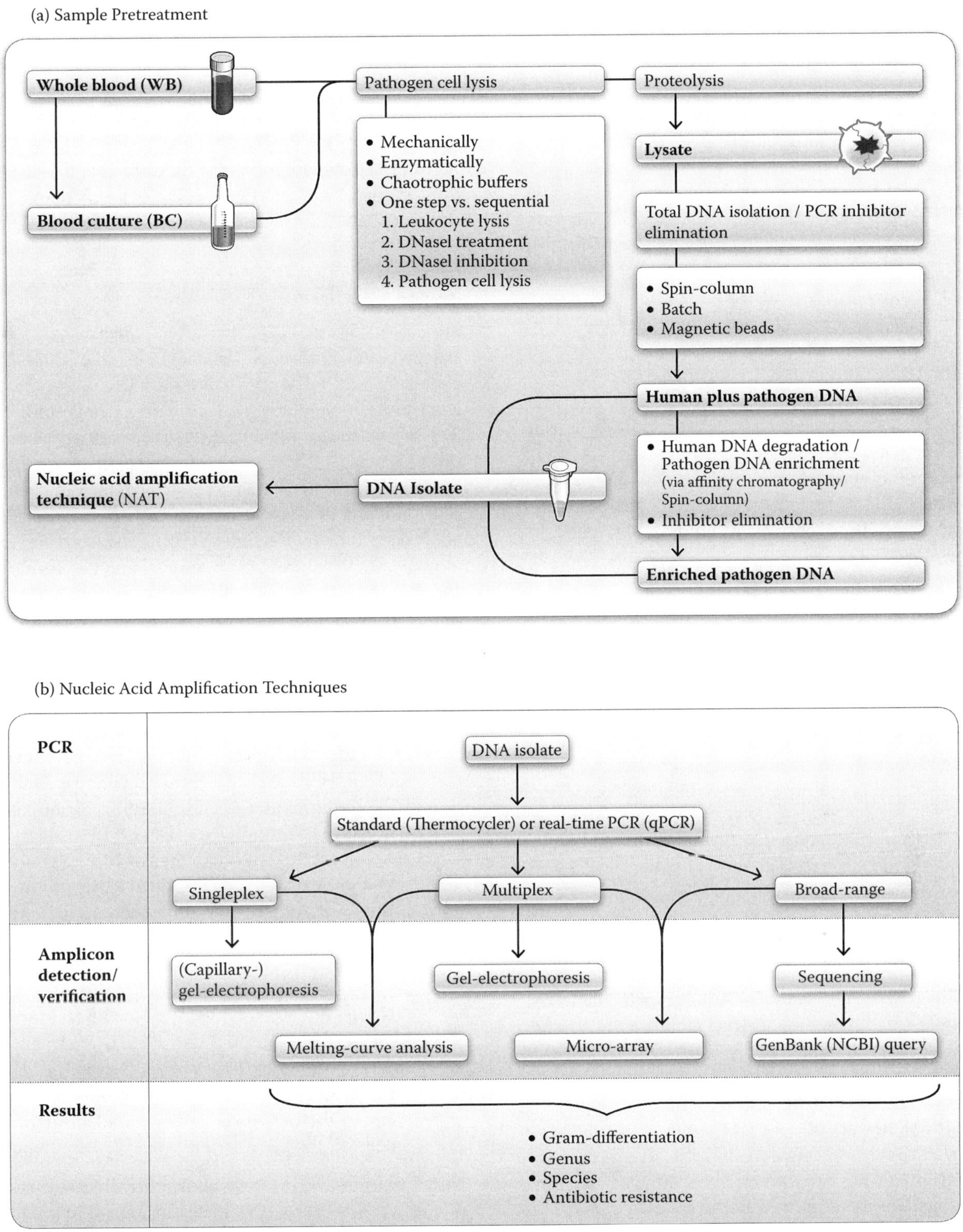

FIGURE 99.2 Workflows of sample pretreatment (a) and nucleic acid amplification techniques (NAT) (b) for the detection of sepsis-causative pathogens and antibiotic resistances in whole blood.

isolation procedure. The isolation method provided within the LOOXSTER®/VYOO® product series was proven to warrant the inevitable total-DNA yield and is compatible with the downstream pathogen DNA enrichment protocol. The lysate is loaded onto a spin-column membrane. The DNA is bound, whereas most of the added mixture ingredients easily pass through. Wash steps remove PCR inhibitors' leftovers. Subsequently, the DNA is eluted from the membrane with a high ionic strength solution. To obtain the highest purity, the eluted DNA is desalted and concentrated by precipitation. The final DNA preparation is dissolved in an appropriate low-salt buffer. The protocol was optimized for 5 mL of sample material.

1. Pour the lysate onto the membrane of a cartridge within a 50-mL centrifugation tube, retarding the glass beads. Centrifuge in a swing-out rotor at 3000 × *g* for 2 min.
2. Transfer the membrane-containing vessel into a new 50-mL tube, add 5 mL of washing buffer, and repeat the centrifugation step twice.
3. Transfer the cartridge again into a new 50-mL tube and add 2.5 mL of elution buffer. After incubation for 2–3 min at room temperature (RT) centrifuge at 3000 × *g* for 1 min.
4. Add another 2.5 mL of elution buffer and repeat centrifugation. Pour the eluate into a 15-mL tube, add 4 mL isopropanol, and centrifuge for 30 min at 5000 × *g* or 60 min at 3000 × *g* and 2°C–8°C.
5. Remove the supernatant, wash with 2 mL of cold (–20°C) 70% ethanol, and centrifuge for 5 min at 3000 × *g* and 2°C–8°C.
6. Remove the supernatant carefully, dry at 50°C ± 2°C for 15 ± 2 min, and dissolve the DNA pellet completely in 300 μL of low-salt binding buffer for 30 min at 50°C ± 2°C.

Enrichment of Bacterial and Fungal DNA. The bulk of human DNA within the total-DNA fraction minimizes or prevents trace target detection of microbial nucleic acids within the addressed range of sensitivity (<10–100 CFU/mL) applying standard PCR techniques. Therefore, human DNA has to be decreased if real-time PCR devices are not available. Even though qPCR is adopted, sensitivity will significantly be increased by diminishing bulk DNA backgrounds. However, exiguous human DNA remnants empirically enhance PCR target detection. The following LOOXSTER®/VYOO® protocol covers the enrichment of bacterial and fungal DNA from a mixture of pathogen and human DNA via protein affinity chromatography.

1. The concentration columns are to be conditioned as outlined in the instruction manuals. Pipette the DNA sample onto a prepared matrix, incubate for 30 min at RT, and centrifuge for 30 s at 1000 × *g* (RT).
2. Discard the flow-through and add 300 μL of binding buffer onto the matrix, and centrifuge subsequently again for 30 s at 1000 × *g* (RT).
3. Repeat these steps, transfer the cartridge into a new 2-mL tube and elute the enriched pathogen DNA by adding 300 μL of elution buffer (heated at 65°C ± 2°C). Pipette the matrix-DNA mixture several times carefully into the pipette tip, incubate for 5 min at RT, and centrifuge for 30 s at 1000 × *g* (RT). Repeat these steps and pool the eluate (600 μL total eluate is gained).
4. Precipitate the DNA by adding 5 μL precipitation enhancer, 60 μL 3 M sodium acetate pH 5.2, and 480 μL isopropanol. After vortexing for 10 s, centrifuge the sample for 30 min at 16,000 × *g* (RT). Wash the pellet with 1 mL ethanol (70%, –20°C), centrifuge again at 16,000 × *g* (RT), and discard the supernatants.
5. Repeat the washing/centrifugation steps, dry the pellet at RT, and dissolve it completely for 15 ± 2 min at 50°C ± 2°C in 30 μL water (DNA- and DNase-free). Determine the DNA concentration with a minimized consumption volume, for example, via the Nano-Drop® method, which uses only about 3 μL of concentrated pathogen DNA solution.

99.2.2 Detection Procedures

Principle. The multitude of different sepsis-causative bacterial and fungal species demands a parallel target detection approach. The following multiplex PCR protocol covers the amplification of species- and in part genus-specific targets of 34 bacterial and six fungal pathogens (as well as five basic antibiotic resistances) (Table 99.1). The detection of *Staphylococcus* spp. and fungal species is additionally done using pan-primers. The standard PCR is performed in two independent primer pools due to the high number of primers. However, individual in-house standard PCR or even qPCR protocols are also conceivable in combination with an efficient sample pretreatment protocol. Use of 2× QIAGEN® Multiplex PCR Master Mix (QIAGEN, Hilden, Germany), is recommended.

Procedure

1. Prepare PCR mix (25 μL) consisting of 12.5 μL 2× QIAGEN® Multiplex PCR Master Mix (which contains a HotStarTaq® DNA polymerase, multiplex PCR buffer, and a dNTP mixture), ≤1 μg pathogen enriched DNA and DNA-/DNase-free water. In the case of a DNA microarray-based amplicon detection, biotin-dUTP may be added to the mixture. The positive/inhibition control reactions additionally contain 1 μL of genomic bacterial DNA, no template controls (NTC) are devoid of DNA.
2. Denature the DNA at 95°C for 15 min; amplify with 30 cycles at 94°C for 30 s, 59°C for 1.5 min, and 72°C for 45 s; and a terminal step at 72°C for 10 min.
3. Analyze multiplex PCR amplicons on 3% TBE agarose gels stained with ethidium bromide and assign them to distinct pathogens and/or antibiotic resistances on the basis of pool-specific length markers.

Note: Alternatively, verification of PCR products may be done via DNA microarray. Hybridization of biotinylated PCR amplicons to immobilized DNA probes increases sensitivity and affords sequence result verification. The ArrayTube (AT®, Alere, Jena, Germany) platform is designed for a precise routine use of nucleic acid–based diagnostics. Unique feature of the ArrayTube® is the combination of microprobe array (biochip in the bottom) and micro reaction vial (2 mL), allowing easy and reliable array processing. For detection, colorimetric staining via a TMB-mediated color reaction, catalyzed by horseradish peroxidase, is applied. Array imaging and analysis is performed with an AT® Reader instrument.

TABLE 99.1
Species of Sepsis-Causative Pathogens (and Resistances) Specified by VYOO® Multiplex PCR, Which Represent 99% of Most Frequent Sepsis-Causative Pathogens as Determined in 11 Clinical Studies (See Figure 99.1)

Bacteria		**Fungi**
Acinetobacter baumannii	*P. melaninogenica*	*Aspergillus fumigatus*
Bacteroides fragilis	*Proteus mirabilis*	*Candida albicans*
Burkholderia cepacia	*Pseudomonas aeruginosa*	*C. glabrata*
Clostridium perfringens	*Serratia marcescens*	*C. krusei*
Enterobacter aerogenes	*Staphylococcus aureus*	*C. parapsilosis*
E. cloacae	*S. epidermidis*	*C. tropicalis*
E. faecalis	*S. haemolyticus*	**Antibiotic Resistances**
E. faecium	*S. hominis*	β-lactamase *blaCTX-M15*
Escherichia coli	*S. saprophyticus*	vancomycin *vanA*
Haemophilus influenzae	*Stenotrophomonas maltophilia*	vancomycin *vanB*
Haemophilus influenza capsular type b	*Streptococcus agalactiae*	β-lactamase *blaSHV*
Klebsiella oxytoca	*S. dysgalactiae*	methicillin *mecA*
K. pneumoniae	*S. bovis*	
Morganella morganii	*S. mutans*	
Neisseria meningitidis	*S. pneumoniae*	
Prevotella buccae	*S. pyogenes*	
P. intermedia	*S. sanguinis*	

99.3 CONCLUSION AND FUTURE PERSPECTIVES

In recent years, a broad range of PCR-based methods for the detection of sepsis-causative bacterial (and fungal) pathogens have been developed, and some of them might be suitable to support and expedite clinical findings and strengthen a directed and restricted antibiotic therapy. Standardization, however, is not yet in sight. At present, only sparse information is available on the potential cost benefits of molecular diagnosis versus conventional detection method for bloodstream infections,[82] but it has to be assumed that the high initial and running costs will be outweighed by the increased assay sensitivity and reduced time-to-result.

The amplification method itself, for example, loop-mediated isothermal amplification (LAMP),[101,102] ligase chain reaction[103,104] (e.g., Abbott LCx®, Abbott Laboratories, Illinois, USA, assay for the diagnosis of chlamydial genital tract infections), strand displacement amplification (SDA),[105] transcription-mediated amplification (TMA) (e.g., GEN-PROBE® AMPLIFIED™ products, Gen-Probe, San Diego, California, USA, for the detection of *Chlamydia trachomatis* and *Neisseria gonnorhoea*),[104] or cycling probe technology (CPT),[106] as listed and discussed for sepsis application by Schrenzel,[40] may significantly alter and improve the sensitivity of the invented verification procedures, the NAT itself will only be as suitable as the associated total-DNA release and isolation protocols are, and of course, the approach of getting rid of bulk DNA and polymerase inhibitors. Needed are combined multitarget assay systems with proven clinical utility and which offer high negative predictive values with notwithstanding high analytical sensitivities and low detection thresholds.

Further approaches in test development should focus on standardization and ongoing shortening of the clinical course, for example, via automation of test flows. Coated magnetic bead particles, for example, for covalent binding of proteins and antibodies (e.g., for affinity chromatography), capture of biotinylated biomolecules (e.g., for hybridization techniques), DNA separation, or immunological applications [e.g., enzyme-linked immunosorbent assay (ELISA)], are already used in many applications, since centrifugation is not required due to the separation of the beads, sized from 50 nm to approximately 3 μm in diameter, from aqueous phases with magnets. Time-consuming DNA-sedimentations and redilutions are circumvented. The combination of multiplex PCR with microarrays on chips, which for instance are localized at the bottom of 2-mL reaction vials (e.g., ArrayTube® System, Alere, Jena, Germany) represents a currently tedious and laborious but forward-looking development.

A new gold standard for the detection of bacterial pathogens, although not yet found, will surely be nucleic acid–based, as announced by the current diagnostic achievements. They are currently used as adjunctive diagnostic tools to traditional blood cultures, but it is just a matter of time before the new methods will play significant roles in diagnosis after being approved by the responsible regulatory agencies.

ACKNOWLEDGMENTS

The work was supported by grants from the Thuringian State Development Bank, Free State of Thuringia, Germany,

Intelligene, 2006 FE 0027 and Multileb, 2007 FE 0255, the latter promoted by the European Regional Development Fund (ERDF).

REFERENCES

1. Slade, E., Tamber, P.S., and Vincent, J.L., The Surviving Sepsis Campaign: Raising awareness to reduce mortality, *Crit. Care*, 7, 1, 2003.
2. Levy, M.M. et al., 2001 SCCM/ESICM/ACCP/ATS/SIS International Sepsis Definitions Conference, *Inten. Care Med.*, 29, 530, 2003.
3. Bone, R.C., Sibbald, W.J., and Sprung, C.L., The ACCP-SCCM consensus conference on sepsis and organ failure, *Chest*, 101, 1481, 1992.
4. Martin, G.S., and Mannino, D.M., The epidemiology of sepsis in the United States from 1979 through 2000, *N. Engl. J. Med.*, 348, 1546, 2003.
5. Reinhart, K. et al., Diagnosis and therapy of sepsis, *Clin. Res. Cardiol.*, 95, 429, 2006.
6. Bauer, M. et al., Sepsis—Aktuelle Aspekte zur Pathophysiologie, Diagnostik und Therapie, *Anaesthesist*, 55, 835, 2006.
7. Karlowsky, J.A. et al., Prevalence and antimicrobial susceptibilities of bacteria isolated from blood cultures of hospitalized patients in the United States in 2002, *Ann. Clin. Microbiol. Antimicrob.*, 3, 1, 2004.
8. Garnacho-Montero, J. et al., Impact of adequate empirical antibiotic therapy on the outcome of patients admitted to the intensive care unit with sepsis, *Crit. Care. Med.*, 31, 2742, 2003.
9. Fine, J.M. et al., Patient and hospital characteristics associated with recommended processes of care for elderly patients hospitalized with pneumonia, *Arch. Intern. Med.*, 162, 827, 2002.
10. Ibrahim, E.H. et al., The influence of inadequate antimicrobial treatment of bloodstream infections on patient outcomes in the ICU setting, *Chest*, 118, 146, 2000.
11. Vallés, J. et al., Community-acquired blood-stream infection in critically ill adult patients: Impact of shock and inappropriate antibiotic therapy on survival, *Chest*, 123, 1615, 2003.
12. Iregui, M. et al., Clinical importance of delays in the initiation of appropriate antibiotic treatment for ventilator-associated pneumonia, *Chest*, 122, 262, 2002.
13. De La Rosa, G.D. et al., Toward an operative diagnosis in sepsis: A latent class approach, *BMC Infect. Dis.*, 8, 18, 2008.
14. Povoa, P., C-reactive protein: A valuable marker of sepsis, *Intensive Care Med.*, 28, 235, 2002.
15. Becker, K.L. et al., Clinical review 167: Procalcitonin and the calcitonin gene family of peptides in inflammation, infection, and sepsis: A journey from calcitonin back to its precursors, *J. Clin. Endocrinol. Metab.*, 89, 1512, 2004.
16. Castelli, G.P., Pognani, C., Cita, M., and Paladini, R., Procalcitonin as a prognostic and diagnostic tool for septic complications after major trauma, *Crit. Care Med.*, 37, 1845, 2009.
17. Pettila, V. et al., Predictive value of procalcitonin and interleukin 6 in critically ill patients with suspected sepsis, *Intensive Care Med.*, 28, 1220, 2002.
18. Meisner, M. et al., Comparison of procalcitonin (PCT) and C-reactive protein (CRP) plasma concentrations at different SOFA scores during the course of sepsis and MODS, *Crit. Care*, 3, 45, 1999.
19. Endo, S. et al., Usefulness of procalcitonin serum level for the discrimination of severe sepsis from sepsis: A multicenter prospective study, *J. Infect. Chemother.*, 14, 244, 2008.
20. Levi, M., and Ten Cate, H., Disseminated intravascular coagulation, *N. Engl. J. Med.*, 341, 586, 1999.
21. Opal, S.M., and Esmon, C.T., Bench-to-bedside review: functional relationships between coagulation and the innate immune response and their respective roles in the pathogenesis of sepsis, *Crit. Care*, 7, 23, 2003.
22. Klouche, M., and Schröder, U., Rapid methods for diagnosis of bloodstream infections, *Clin. Chem. Lab. Med.*, 46, 888, 2008.
23. Weinstein, M.P., Current blood culture methods and systems: Clinical concepts, technology, and interpretation of results, *Clin. Infect. Dis.* 23, 40, 1996.
24. Weinstein, M.P. et al., The clinical significance of positive blood cultures in the 1990s: A prospective comprehensive evaluation of the microbiology, epidemiology, and outcome of bacteremia and fungemia in adults, *Clin. Infect. Dis.*, 24, 584, 1997.
25. Marra, A.R. et al., Time to blood culture positivity as a predictor of clinical outcome of *Staphylococcus aureus* bloodstream infection, *J. Clin. Microbiol.*, 44, 1342, 2006.
26. Peralta, G. et al., Time to positivity in blood cultures of adults with *Streptococcus pneumoniae* bacteremia, *BMC Infect. Dis.*, 6, 79, 2006.
27. Riedel, S. et al., Timing of specimen collection for blood cultures in febrile patients with bacteremia, *J. Clin. Microbiol.*, 46, 1381, 2008.
28. Akan, O.A., and Yildiz, E., Comparison of the effect of delayed entry into 2 different blood culture systems (BACTEC 9240 and BacT/ALERT 3D) on culture positivity, *Diagn. Microbiol. Infect. Dis.*, 54, 193, 2006.
29. Safdar, N., Fine, J.P., and Maki, D.G., Meta-analysis: methods for diagnosing intravascular device-related bloodstream infection, *Ann. Intern. Med.*, 142, 451, 2005.
30. Westh, H. et al., Multiplex real-time PCR and blood culture for identification of bloodstream pathogens in patients with suspected sepsis, *Clin. Microbiol. Infect.*, 15, 544, 2009.
31. Reier-Nilsen, T. et al., Comparison of broad range 16S rDNA PCR and conventional blood culture for diagnosis of sepsis in the newborn: A case control study, *BMC Pediatr.*, 9, 5, 2009.
32. Bingen, E. et al., Bacterial counts in cerebrospinal fluid of children with meningitis, *Eur. J. Clin. Microbiol. Infect. Dis.*, 9, 278, 1990.
33. Chastre, J., Evolving problems with resistant pathogens, *Clin. Microbiol. Infect.*, 14, 3, 2008.
34. Weinstein, M.P., Blood culture contamination: Persisting problems and partial progress, *Clin. Microbiol.*, 41, 2275, 2003.
35. de Kraker, M., and van de Sande-Bruinsma, N., Trends in antimicrobial resistance in Europe: Update of EARSS results, *Euro. Surveill.*, 12, E070315.3, 2007.
36. Munson, E.L. et al., Detection and treatment of bloodstream infection: Laboratory reporting and antimicrobial management, *J. Clin. Microbiol.*, 41, 495, 2003.
37. Kreger, B.E. et al., Gram-negative bacteraemia. III. Reassessment of etiology, epidemiology and ecology in 612 patients, *Am. J. Med.*, 68, 332, 1980.
38. Dunyach, C. et al., Detection and identification of *Candida* spp. in human serum by LightCycler real-time polymerase chain reaction, *Diagn. Microbiol. Infect. Dis.*, 60, 263, 2008.

39. Mancini, N. et al., Molecular diagnosis of sepsis in neutropenic patients with haematological malignancies, *J. Med. Microbiol.*, 57, 601, 2008.
40. Schrenzel, J., Clinical relevance of new diagnostic methods for bloodstream infections, *Int. J. Antimicrob. Agents*, 30, S2, 2007.
41. Sakai, H. et al., Simultaneous detection of *Staphylococcus aureus* and coagulase-negative staphylococci in positive blood cultures by real-time PCR with two fluorescence resonance energy transfer probe sets, *J. Clin. Microbiol.*, 42, 5739, 2004.
42. Klaschik, S. et al., Real-time PCR for detection and differentiation of gram-positive and gram-negative bacteria, *J. Clin. Microbiol.*, 40, 4304, 2002.
43. Gebert, S., Siegel, D., and Wellinghausen, N., Rapid detection of pathogens in blood culture bottles by real-time PCR in conjunction with the pre-analytic tool MolYsis, *J. Infect.*, 57, 307, 2008.
44. Ohlin, A. et al., Real-time PCR of the 16S-rRNA gene in the diagnosis of neonatal bacteraemia, *Acta Paediatr.*, 97, 1376, 2008.
45. Kang, S.E., Nam, Y.S., and Hong, K.W., Rapid detection of *Enterobacter sakazakii* using TaqMan real-time PCR assay, *J. Microbiol. Biotechnol.*, 17, 516, 2007.
46. Mussap, M. et al., New diagnostic tools for neonatal sepsis: the role of a real-time polymerase chain reaction for the early detection and identification of bacterial and fungal species in blood samples, *J. Chemother.*, 19 (Suppl 2), 31, 2007.
47. Wu, Y.-D. et al., Gram stain-specific-probe-based real-time PCR for diagnosis and discrimination of bacterial neonatal sepsis, *J. Clin. Microbiol.*, 46, 2613, 2008.
48. Jordan, J.A., and Durso, M.B., Real-time polymerase chain reaction for detecting bacterial DNA directly from blood of neonates being evaluated for sepsis, *J. Mol. Diagn.*, 7, 575, 2005.
49. Wiesinger-Mayr, H. et al., Identification of human pathogens isolated from blood using microarray hybridisation and signal pattern recognition, *BMC Microbiol.*, 7, 78, 2007.
50. Marlowe, E.M. et al., Application of an rRNA probe matrix for rapid identification of bacteria and fungi from routine blood cultures, *J. Clin. Microbiol.*, 41, 5127, 2003.
51. Shang, S. et al., Rapid diagnosis of bacterial sepsis with PCR amplification and microarray hybridization in 16S rRNA gene, *Pediatr. Res.*, 58, 143, 2005.
52. Anthony, R.M., Brown, T.J., and French, G.L., Rapid diagnosis of bacteraemia by universal amplification of 23S ribosomal DNA followed by hybridization to an oligonucleotide array, *J. Clin. Microbiol.*, 38, 781, 2000.
53. Millar, M.R. et al., Molecular diagnosis of vascular access device-associated infection in children being treated for cancer or leukaemia, *Clin. Microbiol. Infect.*, 14, 213, 2008.
54. Rothman, R.E., Detection of bacteremia in emergency department patients at risk for infective endocarditis using universal 16S rRNA primers in a decontaminated polymerase chain reaction assay, *J. Infect. Dis.*, 186, 1677, 2002.
55. Sleigh, J., Cursons, R., and La Pine, M., Detection of bacteraemia in critically ill patients using 16S rDNA polymerase chain reaction and DNA sequencing, *Intensive Care Med.*, 27, 1269, 2001.
56. Cursons, R.T., Jeyerajah, E., and Sleigh, J.W., The use of polymerase chain reaction to detect septicemia in critically ill patients, *Crit. Care Med.*, 27, 937, 1999.
57. Kane, T.D., Alexander, J.W., and Johannigman, J.A., The detection of microbial DNA in the blood: a sensitive method for diagnosing bacteremia and/or bacterial translocation in surgical patients, *Ann. Surg.*, 227, 1, 1998.
58. Ley, B.E. et al., Detection of bacteraemia in patients with fever and neutropenia using 16S rRNA gene amplification by polymerase chain reaction, *Eur. J. Clin. Microbiol. Infect. Dis.*, 17, 247, 1998.
59. van Burik, J.-A. et al., Panfungal PCR assay for detection of fungal infection in human blood specimens, *J. Clin. Microbiol.*, 36, 1169, 1998.
60. Fenollar, F., and Raoult, D., Molecular diagnosis of bloodstream infections caused by non-cultivable bacteria, *Int. J. Antimicrob. Agents*, 30 (Suppl 1), S7, 2007.
61. Wellinghausen, N. et al., Superiority of molecular techniques for identification of Gram-negative, oxidase-positive rods, including morphologically nontypical *Pseudomonas aeruginosa*, from patients with cystic fibrosis, *J. Clin. Microbiol.*, 43, 4070, 2005.
62. Ehricht, R. et al., Risidual DNA in thermostable DNA polymerases—a cause of irritation in diagnostic PCR and microarray assays, *Biologicals*, 35, 145, 2007.
63. Peters, R.P. et al., Detection of bacterial DNA in blood samples from febrile patients: Underestimated infection or emerging contamination? *FEMS Immunol. Med. Microbiol.*, 42, 249, 2004.
64. van der Zee, A. et al., Qiagen DNA extraction kits for sample preparation for *Legionella* PCR are not suitable for diagnostic purposes, *J. Clin. Microbiol.*, 40, 1126, 2002.
65. Koncan, R. et al., Learning from mistakes: *Taq* polymerase contaminated with β-lactamase sequences results in false emergence of *Streptococcus pneumonniae* containing TEM, *J. Antimicrob. Chemother.*, 60, 702, 2007.
66. Oie, S., and Kamiya, A., Microbial contamination of antiseptics and disinfectants, *Am. J. Infect. Control*, 24, 389, 1996.
67. Dagan, R. et al., Prospective study to determine clinical relevance of detection of pneumococcal DNA in sera of children by PCR, *J. Clin. Microbiol.*, 36, 669, 1998.
68. Isaacman, D.J. et al., Accuracy of a polymerase chain reaction-based assay for detection of pneumococcal bacteremia in children, *Pediatrics*, 101, 813, 1998.
69. Berger, S.A. et al., Bacteraemia after the use of an oral irrigation device. A controlled study in subjects with normal-appearing gingiva: Comparison with use of a toothbrush, *Ann. Intern. Med.*, 80, 510, 1974.
70. Misra, S. et al., A pilot study to assess bacteraemia associated with tooth brushing using conventional, electric or ultrasonic toothbrushes, *Eur. Arch. Paediatr. Dent.*, 8(S1), S42, 2007.
71. Tomas, I. et al., Prevalence, duration and aetiology of bacteraemia following dental extractions, *Oral Dis.*, 13, 56, 2007.
72. Davenport, P., and Land, K.J., Isolation of *Leclercia adecarboxylata* from the blood culture of an asymptomatic platelet donor, *Transfusion*, 47, 1816, 2007.
73. Rodero, L. et al., Transient fungemia caused by an amphotericin B-resistant isolate of *Candida haemulonii*, *J. Clin. Microbiol.*, 40, 2266, 2002.
74. McLaughlin, R.W. et al., Are there naturally occurring pleomorphic bacteria in the blood of healthy humans? *J. Clin. Microbiol.*, 40, 4771, 2002.
75. Nikkari, S. et al., Does blood of healthy subjects contain bacterial ribosomal DNA? *J. Clin. Microbiol.*, 39, 1956, 2001.

76. Handschur, M. et al., Preanalytic removal of human DNA eliminates false signals in general 16S rDNA PCR monitoring of bacterial pathogens in blood, *Comp. Immunol. Microbiol. Infect. Dis.*, 32, 207, 2009.
77. MAQC Consortium, The MicroArray Quality Control (MAQC) project shows inter- and intraplatform reproducibility of gene expression measurements, *Nat. Biotechnol.*, 24, 1151, 2006.
78. Palka-Santini, M. et al., Large scale multiplex PCR improves pathogen detection by DNA microarrays, *BMC Microbiol.*, 9, 1, 2009.
79. Louie, R.F. et al., Multiplex polymerase chain reaction detection enhancement of bacteraemia and fungemia, *Crit. Care Med.*, 36, 1487, 2008.
80. Wellinghausen, N. et al., Algorithm for the identification of bacterial pathogens in positive blood cultures by real-time LightCycler polymerase chain reaction (PCR) with sequence-specific probes, *Diagn. Microbiol. Infect. Dis.*, 48, 229, 2004.
81. Lehmann, L.E. et al., A multiplex real-time PCR assay for rapid detection and differentiation of 25 bacterial and fungal pathogens from whole blood samples, *Med. Microbiol. Immunol.*, 197, 313, 2008.
82. Falagas, M.E., and Panayotis, T.T., From acute renal failure to acute kidney injury: Emerging concepts, *Crit. Care Med.*, 36, 1641, 2008.
83. Wellinghausen, N. et al., Evaluation of the Hyplex BloodScreen Multiplex PCR-Enzyme-linked immunosorbent assay system for direct identification of gram-positive cocci and gram-negative bacilli from positive blood cultures, *J. Clin. Microbiol.*, 42, 3147, 2004.
84. Depardieu, F., Perichon, B., and Courvalin, P., Detection of the *van* alphabet and identification of enterococci and staphylococci at the species level by multiplex PCR, *J. Clin. Microbiol.*, 42, 5857, 2004.
85. Metwally, L. et al., Improving molecular detection of *Candida* DNA in whole blood: Comparison of seven fungal DNA extraction protocols using real-time PCR, *J. Med. Microbiol.*, 57, 296, 2008.
86. Fredricks, D.N., Smith, C., and Meier, A., Comparison of six DNA extraction methods for recovery of fungal DNA as assessed by quantitative PCR, *J. Clin. Microbiol.*, 43, 5122, 2005.
87. Wong, S.F., Mak, J.W., and Pook, P.C., New mechanical disruption method for extraction of whole cell protein from *Candida albicans*, *Southeast Asian J. Trop. Med. Public Health*, 38, 512, 2007.
88. Müller, F.M. et al., Rapid extraction of genomic DNA from medically important yeasts and filamentous fungi by high-speed cell disruption, *J. Clin. Microbiol.*, 36, 1625, 1998.
89. Lugert, R., Schettler, C., and Gross, U., Comparison of different protocols for DNA preparation and PCR for the detection of fungal pathogens in vitro, *Mycoses*, 49, 298, 2006.
90. Loeffler, J. et al., Automated extraction of genomic DNA from medically important yeast species and filamentous fungi by using the MagNA Pure LC system, *J. Clin. Microbiol.*, 40, 2240, 2002.
91. Horz, H.P. et al., Selective isolation of bacterial DNA from human clinical specimens, *J. Microbiol. Methods*, 72, 98, 2008.
92. Sachse, S. et al., Pathogen-specific DNA concentration combined with a novel multiplex PCR assay improves molecular sepsis diagnosis, *Shock*, 29, 102, 2008.
93. Sachse, S. et al., Truncated human cytidylate-phosphate-deoxyguanylate-binding protein for improved nucleic acid amplification technique-based detection of bacterial species in human samples, *J. Clin. Microbiol.*, 47, 1050, 2009.
94. Pinarbasi, E., Elliott, J., and Hornby, D.P., Activation of a yeast pseudo DNA methyltransferase by deletion of a single amino acid, *J. Mol. Biol.*, 257, 804, 1996.
95. Wilkinson, C.R. et al., The fission yeast gene pmt1+ encodes a DNA methyltransferase homologue, *Nucleic Acids Res.*, 23, 203, 1995.
96. Russell, P.J. et al., Different levels of DNA methylation in yeast and mycelial forms of *Candida albicans*, *J. Bacteriol.*, 169, 4393, 1987.
97. Al-Soud, W.A., and Rådström, P., Purification and characterization of PCR-inhibitory components in blood cells, *J. Clin. Microbiol.*, 39, 485, 2001.
98. Al-Soud, W.A., Jönsson, L.J., and Râdström, P., Identification and characterization of immunoglobulin G in blood as a major inhibitor of diagnostic PCR, *J. Clin. Microbiol.*, 38, 345, 2000.
99. Lau, A. et al., Multiplex-tandem PCR: A novel platform for the rapid detection and identification of fungal pathogens from blood culture specimens, *J. Clin. Microbiol.*, 46, 3021, 2008.
100. Arpi, M. et al., Importance of blood volume cultures in the detection of bacteraemia, *Eur. J. Clin. Microbiol. Infect. Dis.*, 8, 838, 1989.
101. Notomi, T. et al., Loop-mediated isothermal amplification of DNA, *Nucleic Acids Res.*, 28, E63, 2000.
102. Mori, Y. et al. Detection of loop-mediated isothermal amplification reaction by turbidity derived from magnesium pyrophosphate formation, *Biochem. Biophys. Res. Commun.*, 289, 150, 2001.
103. Wiedmann, M. et al., Ligase chain reaction (LCR)—overview and applications, *Genome Res.*, 3, S51, 1994.
104. Stary, A. et al., Performance of transcription-mediated amplification and ligase chain reaction assays for detection of chlamydial infection in urogenital samples obtained by invasive and noninvasive methods, *J. Clin. Microbiol.*, 36, 2666, 1998.
105. Walker, G.T. et al., Strand displacement amplification—an isothermal, in vitro DNA amplification technique, *Nucleic Acids Res.*, 20, 1691, 1992.
106. Fong, W.K. et al., Rapid solid-phase immunoassay for detection of methicillin-resistant *Staphylococcus aureus* using cycling probe technology, *J. Clin. Microbiol.*, 38, 2525, 2000.

100 Metagenomic Approaches for Bacterial Detection and Identification

Chaysavanh Manichanh

CONTENTS

100.1 INTRODUCTION

100.1.1 NEED TO STRENGTHEN TRADITIONAL METHODS

Limitations of Culture Approaches. Traditional methods for bacterial identification depend on a combination of morphological and biochemical characteristics. The analysis of these traits mostly relies on the ability to culture the microorganisms of interest. Microorganims that grow well as single cells suspended in a liquid medium and as discrete colonies on Petri plates became the model for much of modern biology. Therefore, bacterial cultures became the gold standard for experimentation and the basis of almost all recent knowledge of medical microbiology, biochemistry, and molecular biology.

Obtaining pure cultures had typically been the first step in investigating bacteria. The study of microbes will continue to be important through culture-dependent methods, but these methods do not help us describe organisms that are not easily cultured in a laboratory or allow us to characterize the biology of organisms in their natural environment. Indeed, the huge discrepancy between direct microscopic counts and numbers of culturable bacteria from human microbiota and other environmental samples is just one of several indications; what we currently know is only a tiny part of the diversity of microorganisms in nature.

Genomic Approaches. The concept of "genomics" was established by Frederick Sanger when he sequenced the first DNA-based genome of the phi X 174 (or phi X) bacteriophage in 1977.[1] With the help of bioinformatics, the development of genomics in the past three decades has permitted the molecular characterization of all the genes in a species. The precipitous decline in the cost of gene sequencing, stimulated in part by the Human Genome Project, has made it possible to generate genomic sequences for a great variety of organisms. Microbial genome sequences have since appeared at an exponentially increasing rate: the genome sequences of 887 bacteria, 65 archaea, and almost 115 eukaryotes are publicly available at the time of this writing in the Genome Online Database (Table 100.1). Comparison of the genomic sequence information has significantly enhanced our understanding of the physiology, genetics, and evolutionary development of bacteria.

Today, molecular technologies have overcome the difficulty of growing cells in culture and the inability of obtaining a complete sequence from a single cell. For instance, whole-genome amplification methods (WGA) can generate microgram quantities of genomic DNA starting from a sample of as little as a few femtograms. One of these methods is the multiple displacement amplification (MDA), which is based on the annealing of random hexamers to denatured DNA, followed by strand-displacement synthesis at constant temperature.[6]

Need to Go Beyond Traditional Microbiological and Genomic Approaches. Genomics has greatly advanced

TABLE 100.1
Published Metagenomic Human Microbiome Projects

Metagenomic Projects	Sequences (Mbp)	Reference
Fecal microbiome of human from distal gut of healthy adults	72	2
Fecal microbiome of human from obese and lean twins	>2140	3
Fecal microbiome of human healthy Japanese infants and adults	72.7	4
Fecal microbiome of human with bacterial infections	20	5

microbiology, but like laboratory bacterial cultures, traditional genomics is limited in its ability to elucidate the dynamics of microbial communities. Indeed, it is now well recognized that microbes live in hetherogeneous communities and have complex ways of interacting and communicating. Microbial communities often interact through the medium in which they grow, exchanging nutrients, biochemical products, and chemical signals without direct cell-to-cell contact. Many microbes have evolved to grow together in surface communities (biofilms). Within this group, there are those communities that live in close contact to epithelial and endothelial cells and are vital to human health.

It was at the beginning of the twentieth century that Martinus Beijerinck and Sergei Winogradsky, upon entering the era of molecular biology for a wide variety of microorganisms, initiated the field of environmental microbiology. Becking, who was inspired by the works of Beijerinck, made this remarkable observation: "Everything is everywhere, but, the environment selects."[7] He thought that the enrichment culturing technique would provide a look into the "hidden reality" of every microorganism occurring everywhere. However, although some significant advances have resulted from recent attempts to culture the as-yet-unculturable bacteria, we are still unable to mimic most microbial environments sufficiently to induce growth of bacterial strains of interest. So, nowadays, we recognize that only a very minor proportion of microorganisms can be cultured using the existing methodology.[8] As a result of the work of Winogradsky and Beijerinck, there was great enthusiasm for identifying and classifying the bacteria inhabiting our natural world. Therefore, genomics entered into play. In 1977, Carl Woese chose the sequence of 16S ribosomal RNA (rRNA) as the basis for evolutionary comparisons.[9] Woese used this technique to assert that known bacteria should be divided into two separate domains called "bacteria" and "archaea," radically changing our view of the microbial world. Using this molecular genetic tool, researchers showed that up to 90% of the organisms in the total human microbiota have not been yet cultivated.[10,11] The need to explore all microorganisms in their natural and complex environments leads, then, to a new genomics era called "metagenomics."

100.1.2 Metagenomics and High Throughput Sequencing

What Is Metagenomics? Norman Pace is the second key pioneer, after Carl Woese, of archeae research. After exploring the diversity of a microbial ecosystem using ribosomal RNA sequences, he proposed the idea of cloning DNA directly from environmental samples as early as 1985. He published the first report on the study of a marine picoplankton community by 16S rRNA gene cloning and sequencing in 1991.[12] In 1995, Healy described the metagenomic isolation of functional genes from "zoolibraries" constructed from a complex culture of environmental organisms. In 1996,[13] Ed Delong published his research in the analyses of marine samples after cloning genomic fragments of 40 kb. It is only in 1998 that the term "metagenomics" was first used by Jo Handelsman after cloning and analyzing genomic fragments from microbial soil samples.[14]

Metagenomics is described as the study of genetic material recovered directly from environmental samples bypassing the need to isolate and culture individual bacterial community members. The starting material for a metagenomic study is a mixture of DNA from a community of cells that may include bacterial, archaeal, eukaryotic, and viral species at different levels of diversity and abundance. The goal of metagenomics is to answer the questions: Who is in there and what are they doing? Metagenomics acts like a broad microscope that explores an entire environment simultaneously. To do so, it has to combine the power of genomics and bioinformatics. Metagenomics, however, is more than a just large-scale sequencing. In function-based metagenomics, millions of random DNA fragments in a library are translated into proteins by bacteria in the laboratory. Clones producing "foreign" proteins are then screened for various capabilities, such as vitamin production or antibiotic resistance.

Metagenomics can enable researchers to access the tremendous genetic diversity and metabolic activities in a complex microbial ecosystem without knowing anything about the underlying gene sequence, the structure of the desired protein, or the microbe of origin. Yet, this approach can also provide information about one single species within a community.

High-Throughput Sequencing Technologies. The advent of DNA sequencing in the 1970s, especially the Sanger method, has significantly facilitated and accelerated the whole field of molecular biology in general and studies of systems biology in particular. The Sanger method based on DNA chain terminators is very efficient, allowing the read of up to 800 bp for a single sequence. The sequencing step of several metagenomic projects has been performed using the Sanger method.[2,15] Sanger sequencing is still used daily for the study of and experiments on isolated genes of interest. However, the high demand for low-cost sequencing has driven the development of high-throughput sequencing technologies, which have succeeded in lowering the cost of DNA sequencing beyond what is possible with Sanger methods. Two such high-throughput sequencing methods that are commonly introduced in metagenomic projects are pyrosequencing, commercialized by 454 Life Sciences, and Solexa technology, commercialized by Illumina.

The pyrosequencing technology uses an in vitro cloning step, called emulsion polymerase chain reaction (PCR), to amplify individual DNA molecules. This step first isolates individual DNA molecules along with primer-coated beads in aqueous droplets within an oil phase.[16] Then, PCR amplification coats each bead with clonal copies of the DNA molecule, which remain in this way immobilized for later sequencing. The sequencing step is based on the sequencing-by-synthesis principle and relies on real-time, indirect, luminometric detection of pyrophosphate production in a cascade of enzymatic reactions. The PPi released upon incorporation of each nucleotide by DNA polymerase is converted to ATP by ATP-sulfurylase. The so-generated ATP is proportional to the amount of incorporated nucleotide. Because the added nucleotide is known, the sequence of the template can be determined by detecting on which of the four possible nucleotides serially added there is a burst of ATP production. Unincorporated nucleotides are degraded by apyrase, making the extension reaction ready for the next nucleotide addition. This routine can sequence up to 500 Mbp per run from hundreds of samples. A limitation of the method, at the time of this writing, is that the lengths of individual reads are around 300–500 nucleotides, which is twice shorter than the Sanger method.

Illumina/Solexa sequencing technology relies on the attachment of randomly fragmented genomic DNA to a planar, optically transparent surface. Attached DNA fragments are extended and bridge amplified to create an ultrahigh-density sequencing flow cell with hundreds of millions of clusters, each containing around 1000 copies of the same template. These templates are sequenced using a four-color DNA sequencing-by-synthesis technology. This method is extremely high-throughput, producing 2 Giga bp per run, and up to eight samples can be loaded in a single run. However, the current technology can generate only 25–50 bp per read, which is much shorter than pyrosequencing and even less than the Sanger method.

These two techniques eliminate the need for library construction (in other words, the community DNA can be sequenced directly, without first being cloned into a bacterial host), and also require much less starting material, which is relevant to all technologies.

100.1.3 Strategies and Methods for a Metagenomic Project

100.1.3.1 General Considerations

Metagenomics has opened avenues for studying a complex biological ecosystem allowing the characterization of single genes, pathways, organisms, and interactions inside the whole microbial community. Hence, different strategies and methods can be applied to a metagenomic project according to the final goal and, particularly, regarding how important the funding of the project is. No individual researcher is likely to have the capability and resources to achieve a comprehensive characterization of a whole microbial community. Although the cost of gene sequencing has dropped tremendously, the generation of a large-scale DNA sequence is not yet accessible to any group of research. Figure 100.1 illustrates the different molecular methods that can be applied to an environmental sample depending on the final goal of the project.

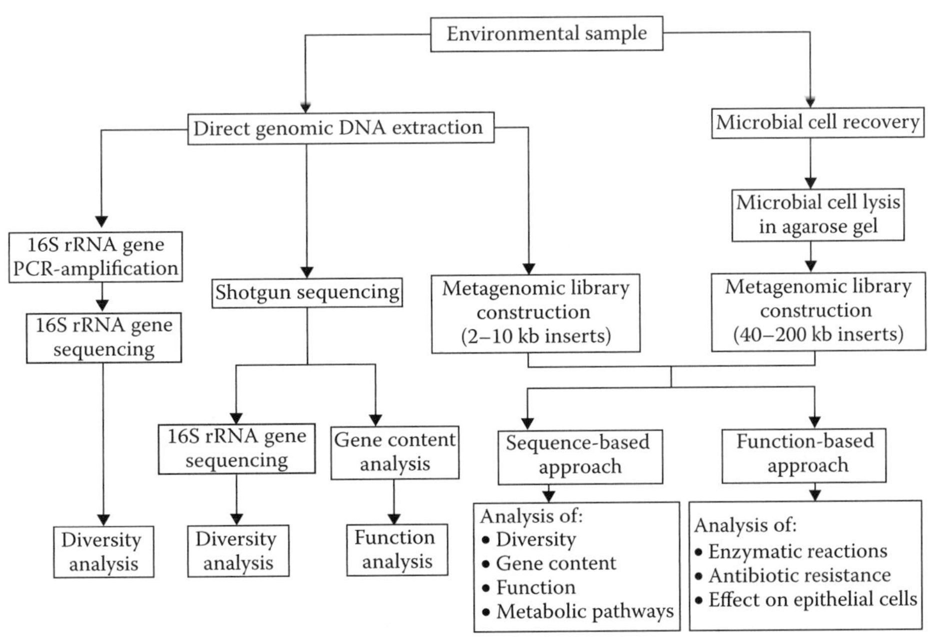

FIGURE 100.1 Molecular methods that can be applied to an environmental sample depending on the final goal of the project.

Any environment can be analyzed with metagenomic approaches as long as the nucleic acids can be extracted from the sample material. A direct diversity analysis of the microbial diversity can be performed after genomic DNA extraction using a classical technique,[17] a kit (QIAamp® DNA stool mini kit, Qiagen) or a modified adapted Qiagen kit.[18] For bacterial diversity examination, the extracted DNA can be used as template for PCR amplification of a phylogenetic marker, such as the 16S rRNA gene. Before the era of 454 and Illumina/Solexa sequencers, the PCR product was usually cloned, and a given number of clones, commonly from 100 to 1000 per sample, were then picked up for sequencing the inserts by means of a Sanger method. Other genes of interest could also be amplified from the extracted DNA as long as specific primers targeting them were known.

Following the extraction process, the full gene content of an environmental sample can also be analyzed by means of cloning a pool of short genomic fragments. The collection of clones obtained from a sample materiel is named metagenomic library. The cloning of short DNA genomic fragments (2–10 kbp) can be realized after a traditional extraction technique (as related earlier). Each DNA fragment is then inserted into a vector, such as plasmids, before amplification by a bacterial host.[15]

One way to analyze metabolic pathways is based on the cloning of higher molecular weight genomic fragments (40–200 kbp). This method requires first the recovery of bacterial cells from a material sample. Those cells are then incorporated into an agarose gel in which gentle bacterial lysis and limited DNA enzymatic digestion allows the recovery of large genomic DNA fragments.[19] The large DNA fragments released from the gel are cloned into a bacterially propagated phagemid vector system, such as cosmids or fosmids, or into bacterial artificial chromosomes (BAC).

In the sequence-based approach, the metagenomic libraries with short or large fragments can be used to directly sequence randomly the insert extremities. Also, they can be used to sequence selected genomic fragments for functions or metabolic pathway analyses. Alternatively, the collections of clones can also be deposited in membranes of macroarrays and screened for phylogenetic markers by DNA–DNA hybridization.[20]

In the functional approach, genes retrieved from the environment are heterologously expressed in a bacterial host, such as *Escherichia coli*. Sophisticated functional screens (detecting changes by enzymatic reactions, antibiotic resistance, effect on epithelial cell growth, etc.) are then employed to detect clones expressing functions of interest.[21]

Nowadays, thanks to the appearance of high-throughput sequencing technologies, the strategies to study a microbial environment have switched to another level of magnitude. These technologies permit us to read the sequence of DNA fragments without first cloning them and are well supported by the rapid development of bioinformatics tools. These strategies are leading to a new approach called "shotgun metagenomic sequencing" in which genomic DNA, obtained immediately after extraction, is directly submitted to a high-throughput sequencing using either 454 or Illumina/Solexa sequencers.

The metagenomics approach has stimulated the creation of about 160 projects from multiple microbial ecosystems, including human microbiome, as described in the GOLD (Genomes OnLine) Web site (Table 100.2). Sixty-one of these have already been completed according to the last update of the GOLD Web site in August 2009. Among the finished and published projects, eight were handling gut microbiomes of which four were human gut microbiomes (Table 100.1). Metagenomics, conducted on a massive scale, has also provided dramatic insights into the structure and metabolic potential of human microbiome. Functional screening of metagenomic libraries has led to the assignment of functions to numerous "hypothetical proteins," so far demonstrating the power of functional metagenomics.[3,4]

100.1.3.2 Sequence Surveys Based on 16S rRNA: A Starting Point of a Metagenomic Project

Why 16S rRNA Gene? The first step in the study of an environmental sample is usually the characterization of its

TABLE 100.2
List of Useful Database and Software Web Sites

Type	Database	Web Site
16S rRNA gene databases	Ribosomal Database Project–Release 10	http://rdp.cme.msu.edu/
	Silva: Comprehensive Ribosomal RNA Databases	http://www.arb-silva.de/
	greengenes.lbl.gov–Aligned 16S rRNA data and tools	http://greengenes.lbl.gov/cgi-bin/nph-NAST_align.cgi
Phylogenetics softwares		http://en.wikipedia.org/wiki/List_of_phylogenetics_software
		http://evolution.genetics.washington.edu/phylip/software.html
Others	GOLD (Genome Online Database)	http://www.genomesonline.org/gold.cgi
	IHMC (International Human Microbiome Consortium)	http://www.human-microbiome.org/
	MetaHIT	http://www.metahit.eu/
Pubmed	GenBank	http://www.pubmed.org/

composition and structure. Methods to culture for identifying and enumerating microbes in an environment had come to play a determinant role over the last decades. But, in that regard, taxonomists today consider analysis of an organism's DNA more reliable than classification based solely on phenotypes. As predominant among these methods is the phylotyping of a gene that is highly conserved in all bacteria and archaea: the 16S rRNA. This gene has become the standard for the determination of phylogenetic relationships, the assessment of diversity in the environment, and the detection and quantification of specific populations.[9,22,23] Since the first surveys of phylogenetic markers to explore the diversity of microbial ecosystems, numerous publications have been released, as illustrated by the 16,213 hits (August 2009) to the Pubmed of the NCBI (Table 100.2) with the search query of "16S gene."

The advantages provided by the 16S rRNA are several. First, while there is a homologous gene in eukaryotes (18S rRNA), it is quite distinct in its nucleotidic sequence, thereby rendering 16S rRNA a specific tool for extracting and identifying prokaryotes as separate from plant, animal, fungal, and protist DNA within a complex sample. Second, the gene is universally distributed among prokaryotes, allowing the comparison of phylogenetic relationships among all extant organisms and thus the construction of a "tree of life." Third, the gene contains conserved and variable regions that allow taxonomic identification ranging from the domain and phylum level to the species level. Fourth, the different short variable regions of this gene make it the right size for new generation sequencing, such as pyrosequencing. For all these reasons, analysis of 16S rRNA is often used as a first step in larger metagenomics projects; thus it provides relatively rapid and cost-effective assessment of bacterial diversity and abundance in samples of potential interest.

Sequencing of 16S rRNA genes directly from the environment using conserved broad-specificity PCR primers (16S surveys), has demonstrated that microbial diversity is far more extensive than we ever imagined from culture-based studies.[23]

Design of Specific and Universal Primers for 16S rRNA Gene. The identification of a whole bacterial population as well as a single bacterial species or strain requires different technical strategies based on the structure of the 16S gene. This gene has a mosaic structure made of ten highly conserved and nine hypervariable regions. Aligning the whole gene sequences from different bacterial strains allows the location of these 19 different regions for each sequence. The highly conserved regions of the gene are used to design universal primers capable of detecting all members of the environmental sample, while the highly variable regions are used to design primers that target a single bacterial species. In order to detect bacteria at the strain level, primers were often designed in the 16S-23S rRNA intergenic transcribed spacer sequence, which has greater polymorphism than adjacent 16S or 23S genes.

Bioinformatics Analyses of 16S rRNA Sequences. The study of microbial composition, diversity, and the richness of environmental samples requires powerful computational tools, particularly for high-throughput data analyses. In this last decade, specific bioinformatics tools have been extensively developed to extract biological information from the 16S sequences in environmental samples.

Since the first release of 16S rRNA sequences in the general database Genbank in the 1980s, several databases hosting this gene have emerged: Silva, Greengenes, RDP (Table 100.2).[24–26] To date, each of these databases contains several hundred thousands of the 16S rRNA sequences. They provide sequence data, aligned sequences, and phylogenetic data as well as online bioinformatics tools. Their sequences are taxonomically assigned from the species to the phylum levels, including also unknown bacterial levels. They are provided free to the international scientific community. The sample origin of these sequences varies from laboratory culture to environmental types. By means of sequence comparison tools, such as BLAST,[27] the databases can be used to identify the bacteria or population of bacteria of interest. Each of these databases mostly contains the same sequences but is complementary to the others with regard to the bioinformatic approaches and number of taxonomic divisions.

Phylogenetic trees based on 16S rRNA have met with great success in identifying the taxonomy of single microbes.[28] If the sample contains many organisms, the many different rRNA sequences will tell us "who is there" in the context of a phylogenetic tree, which describes what evolutionary relationship the bacterial strains have among them. Phylogenetic trees are constructed using computational phylogenetics methods. To construct a phylogenetic tree, the sequences have to be first aligned. For this purpose, several multiple alignment tools are available: T-Coffee,[29] M-Coffee,[30] ClustalW,[31] Muscle,[32] and so forth. The choice of one of these programs depends on the number of sequences and on the calculation power of the user computer. Then, highly variable regions of the multiple alignment were removed, and conserved regions were selected with Gblocks, using parameters optimized for rRNA alignments.[33] The use of this computerized method avoids the manual refinement of multiple alignments. Finally, from a set of aligned sequences, methods for estimating phylogenetic trees include neighbor-joining, maximum parsimony, UPGMA, Bayesian phylogenetic inference, maximum likelihood, and distance matrix methods.

In order to estimate the number of bacterial lineages in a population, sequences are assigned to operational taxonomic units (OTUs) or phylotypes based on the genetic distances between sequences. Programs like FastGroup[34] and DOTUR[35] have been developed to cluster sequences in OTUs and also to assess species richness in a sample for DOTUR. Other programs, such as LIBSHUFF,[36] UNIFRAC,[37] and TreeClimber,[38] are made available to evaluate the similarity between communities. The RDP recently developed online programs that can handle pyrosequencing datasets of several hundred thousands of sequences. Their pyrosequencing pipeline allows the taxonomical classification, clustering, or richness analyses of a huge amount of sequence data in only a few hours.

The above-cited databases also propose online bioinformatics tools to taxonomically assign, align, and cluster sequences; to compare libraries; and to estimate microbial diversity. Indeed, the dataset of interest can be compared to the available databases using sequence match programs (Seqmatch or Classifier for RDP, BLAST, Simrank, or Classify for Greengenes).

Limitation of Metagenomics Based on 16S rRNA Sequences. Despite the increasing interest in using the 16S rRNA gene to study microbial communities, this approach has several limitations. First, 16S rRNA sequences provide a phylogenetic framework into which community members can be placed, but that framework does not tell us much about the functional capabilities of the individual members or the entire community under study. Second, there is a varying number of copies of the rRNA gene between taxa (a difference of more than an order of magnitude among prokaryotes)[39] leading to a possible overestimation of microorganisms containing a high copy number and an underestimation of those harboring a low copy number. Third, the possibility that "universal" primers for PCR amplification of the 16S rRNA gene may not be as universal as they should be may also lead to a misinterpretation of the microbial richness. Indeed, considering that these primers have been designed based on the sequences present in databases, their universality is dependent on what has been already sequenced; however, this does not guarantee us the discovery of new bacterial strains.

Some of these limitations might be overcome through the use of additional, single-copy phylogenetic markers. For instance, several protein-coding regions (fusA, ileS, lepA, leuS, pyrG, recA, recG, rplB, rpoB, dnaJ) have been used as an alternative to rRNA.[40,41] These genes were selected according to two criteria: being single copy and present in most of bacteria. The analyses comparing the use of these protein-coding regions and the one of 16S rRNA has demonstrated that they could be used as an alternative to the 16S rRNA for answering questions pertaining to bacterial identification, classification, phylogenetics, and evolution.

100.1.3.3 Whole Genomes Reconstruction

Reconstruction of complete or near-complete genomes from human pathogens is possible. To date, several genomes of dominant species have been reconstructed using random sequencing: *Treponema pallidum*,[42] *Rickettsia prowazekii*,[43] *Mycobacterium leprae*,[44] and *Tropheryma whipplei*.[45] Having the complete or near-complete genome of a dominant population provides the gene inventory for the organism and allows its metabolic potential to be determined, including inferring the absence of metabolic pathways as well as their presence. However, it may not be possible, at least for now, to obtain complete genomes from more complex microbial communities, such as soil samples or even human gut microbiota because of the tremendous effort of sequencing.

Human Microbiome. The human microbiome is the collective genomes of all microorganisms living at the multiple sites of the human body (oral cavity, skin, nasal cavity, gastrointestinal tract (GI), and vagina).

Human and their commensal microorganisms have evolved together over the last two million years and have become dependent on one another.[46] Interaction between humans and their symbiotic bacteria is essential to many aspects of normal "mammalian" physiology, ranging from metabolic activity to immune homeostasis, and ecological or genetic changes that disturb this mutualism can result in disease.[47] It reflects the evolutionary selection pressures acting at both the levels of the host and the microbial cells and has a direct profound impact on human health. The commensals include eubacteria, archaebacteria, and fungi, which together form the human microbiome.[3] The goal of metagenomics of the human microbiome is to generate resources that will enable its comprehensive characterization and the analysis of its role in human health and disease.

The microorganisms that live inside and on humans are now estimated to outnumber human somatic and germ cells by a factor of 10, and the unique genes they encode are predicted to be at least 100-fold greater than the number of genes present in our own genome.[48] Sequences of 16S rRNA can be retrieved from the RDP database using queries such as "human gut," "human oral," "human vagina," or "human skin." The taxonomic classification of these sequences using the online classifier RDP tool shows differences at the phylum level in the microbial population between the different body sites, described in the Figure 100.2. The analyses down to the genus or species level still show that most of the bacteria are unclassifiable (data not shown). Despite a huge effort, until now, to decipher the human microbiome composition and function, this ecosystem remains a black box for understanding human health and disease. For this reason, the International Human Microbiome Consortium (IHMC) was launched in Heidelberg on October 2008. The objective of the IHMC is to generate and share comprehensive data resources from several body sites that will enable investigators to characterize the relationship between the composition of the human microbiome (or parts of it) and human health and disease; and to determine whether humans share core microbial-diversity profiles. This consortium is currently represented by the Canadian Institutes of Health Research (Canada), European Commission (Europe), the Institut National de la Recherche Agronomique (France), the Japan Science & Technology Agency, (JST) (Japan), the National Institutes of Health (United States of America), and the Commonwealth Scientific and Industrial Research Organisation (Australia).

Human Microbiome Associated to Human Disease. While most microorganism ecosystems colonizing our body are harmless, some of them can undergo a transition from a symbiotic to a pathogenic interaction in response to genetic or environmental changes. Such transitions could imply qualitative changes of diversity or quantitative changes in the composition of the normal population. In the last decade, most of molecular studies on human microbiome have focused on microbial-diversity exploration based on 16S rRNA gene surveys using either cloning and Sanger sequencing or, more recently, pyrosequencing techniques. The microbial

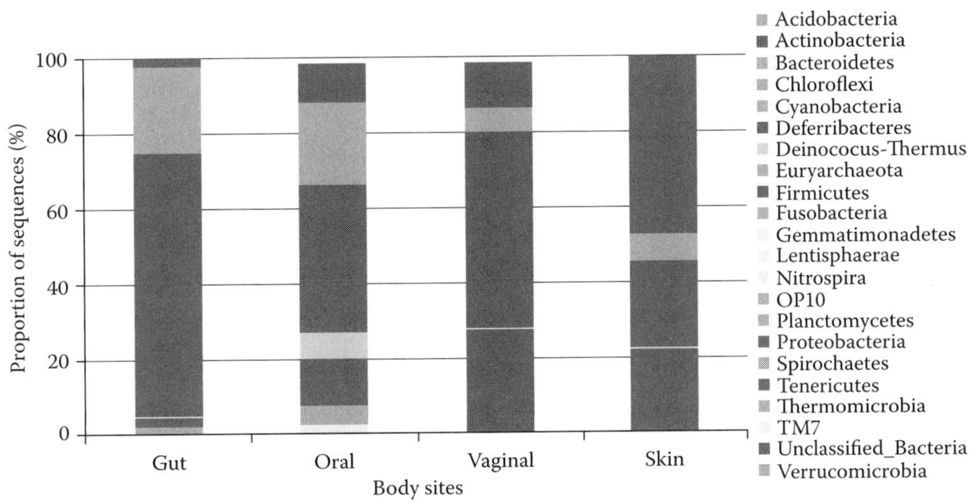

FIGURE 100.2 Taxonomic classification 16S rRNA sequences (>1200 bp) recovered from the RDP database for different sites of the human body: gut (28,667 sequences), oral (768 sequences), vaginal (485 sequences), and skin (124,808 sequences) microbiota. The sequences were first collected using the RDP Browsers tool (http://rdp.cme.msu.edu/hierarchy/hb_intro.jsp). Then, the classification was performed using the online classifier RDP tool (http://rdp.cme.msu.edu/classifier/classifier.jsp).

population from different sites of the human body (skin, oral, vaginal, and gastrointestinal tract microbiota) has been described, motivated by the need to find out what specific bacteria or group of bacteria could be associated with a given disease. Examples of small- and medium-scale projects are given below.

Many skin diseases, such as atopic dermatitis [(AD); eczema], rosacea, psoriasis, and acne have prompted the most examinations of the healthy skin microbiota to understand its role in the pathogenesis of skin disorders. A recent study comparing normal skin and skin with psoriasis showed an alteration in the composition and representation of the cutaneous bacterial biota.[19] Actinobacteria was significantly underrepresented in the psoriatic lesion samples. Representation of *Propionibacterium* species was also significantly lower in the psoriatic lesions.

Oral diseases, including dental caries, gingivitis, and periodontitis, have stimulated a comprehensive characterization of the oral commensal bacterial flora. Within the subgingival crevice, Kroes and colleagues reported five bacterial divisions: Firmicutes, Fusobacteria, Actinobacteria, Proteobacteria, and Cytophagales.[50]

In the vagina, bacterial vaginosis (BV) is marked by dramatic shifts from a poor bacterial population dominated by lactobacillus strains to a greater diversity associated with bacteria in the Clostridiales order.[51,52]

The gastrointestinal tract microbiota has been more explored than other sites for several reasons, but mainly because of the higher amount of bacteria (around 1 kg), which facilitates the extraction of the genomic bacterial DNA. The intestinal microbiota has long been recognized as playing a role in extracting nutrients from the diet, regulating host fat storage, stimulating intestinal epithelium renewal, and directing the maturation of the immune system.[53] Clearly, cross-talk between these bacteria and the human host impacts on various aspects of our physiology, such as mucin production, immune response, production of defensins by Paneth cells, angiogenesis, lipid metabolism, and so forth. Among the 70 divisions (deep evolutionary lineages) of bacteria and 13 divisions of archaea described to date, the distal gut and fecal microbiota is found to be dominated (> 99%) by four bacterial divisions, the Bacteroidetes, the Firmicutes (low GC gram-positives), the Actinobacteria (high GC gram-positive bacteria) and the Proteobacteria; and one methanogenic archaeon, *Methanobrevibacter smithii*.[54] Over the last 20 years, sequencing surveys of the 16S rRNA gene has revealed that these microbial communities are far more complex than was initially believed. Intestinal microbial diversity might be much greater than the 5000 microbial molecular species described using phenotypic features in Bergey's taxonomic manual.[55] A recent publication has identified over 5700 taxa from fecal samples[56] using pyrosequencing technology. At broad phylogenetic scales (deep evolutionary lineages), the intestinal communities of all humans appear quite similar. However, at fine scale, with respect to strain identity and relative abundance, the microbiota of an individual appears to be as personalized as a fingerprint. Today, there is no concrete data showing how different our gut microbiota is from that of our very close neighbor.

Human gut microbiome has been associated with several disorders such as obesity and inflammatory bowel disease. In both cases, the level of bacterial diversity was significantly decreased,[3,20] the bacterial composition of the two main phyla, Firmicutes and Bacteroidetes, were reshaped, and the representation of bacterial genes and metabolic pathways was altered in obese subjects.[3] In view of the intimate association of microbes and humans, we should be considered a superorganism, composed of both microbial and human parts.

Metagenomic analyses on the 16S rRNA gene have permitted the examination of structural modifications in the human microbiome involved in diseases. Thanks to the international effort through the IHMC, much larger-scale projects are also on their way to answering the questions, "who is there" at different body sites and "what are they doing" in the context of health and disease.

100.2 METHODS

100.2.1 Sample Preparation

Several protocols have been used for extracting genomic DNA from stool samples. Bacterial lysis is a critical step for a DNA-extraction protocol. Therefore, I present here a commercial protocol in which the bacterial lysis step has been reinforced. This lysis is based on two procedures. The first step is a bead beating, and the second one is an enzymatic digestion. The stool sample is first suspended in a lysis buffer provided by the QIAamp DNA Stool Mini Kit (Qiagen) and then mixed with ceramic and silica particles (Lysing Matrix E tube, QBiogen) designed to efficiently lyse all microorganisms. The mixture is then shaken in a FastPrep Bio 101 Bead shaker (QBiogen). DNA is then extracted by using the protocol for isolation of DNA for pathogen detection provided in the QIAamp DNA Stool Mini Kit from Qiagen as recommended by the manufacturer, except that a supplemental mixture of enzymes (mutamolysin and lysozyme) is added at the proteinase K step. To remove potential inhibitors, the lysed cells are incubated with an InhibitEx tablet. Then, DNA is precipitated by ethanol, applied to a column provided in the kit, followed by washes with buffers, and then eluted in buffer AE.

Sample Processing. 200 mg of stool sample are suspended in 1400 mL of ASL lysis buffer provided by the QIAamp DNA Stool Mini Kit. Transfer to a Lysing Matrix E tube. Homogenize in a FastPrep Bio 101 bead shaker at 6.5 m/s for 30 s.

DNA Extraction. Follow protocol for QIAamp® DNA stool extraction. Elute into 100 μL Qiagen buffer (incubate for 1 min before eluting).

100.2.2 Detection Procedures

(i) Protocol of Lazarevic et al.[57]

Principle. Lazarevic et al.[57] aligned 753 16S rRNA sequences from the Human Oral Microbiome Database (HOMD, October 2008) using MAFFT (-FFT-NS-2, v6.531b) for selection of primers from the conserved areas of the alignment flanking the V5 region) in 16S rRNA gene. With a 100% match, primers 784DEG (5′-GGMTTAGATACCC-3′) and 880RDEG (5′-CRTACTHCHCAGGYG-3′) sequences (covering an ~82- base segment of the V5 hypervariable region) produced 740 and 745 hits, respectively, or 732 (97.2%) of the HOMD sequences. Species coverage was within the 91%–100% range for all HOMD bacterial phyla except Chloroflexi, which is very rare in oral microbiota and has a single representative in the HOMD. The resulting 16S rRNA V5 amplicons with the primers 784DEG and 880RDEG correspond to *E. coli* positions 785–894, including primer sequence, and to positions 798–879, excluding primers. Samples from the oral cavity of three healthy individuals were amplified with this primer pair and then sequenced using the Illumina technology in a single orientation.

Procedure

1. Saliva and oropharyngeal samples are collected over a 1-week period from three adult individuals. Saliva samples are collected by expectoration into a sterile plastic 50-mL tube and kept frozen at –20°C until processing. 100 μL of each saliva sample is mixed with an equal volume of 2× lysis buffer [Tris 20 mM, EDTA 2 mM (pH 8), Tween 1%, proteinase K (Fermentas) 400 μg/mL] and incubated them for 2 h at 55°C. Proteinase K is inactivated by a 10 min incubation at 95°C and the samples are frozen at –20°C. Dry cotton swabs (Copan) are used to gently swab the posterior wall of the oropharynx, and are directly suspended in a microtube containing 200 μL of lysis buffer and processed as for the saliva samples. The saliva and oropharyngeal lysates from all three subjects are mixed in a 1:10 ratio with roughly equal contributions from the two sampling sites according to PCR yield.
2. PCR mixture (50 μL) is prepared with PrimeStar HS Premix (Takara), containing 5 μL of lysate and 0.5 μM of each forward (784DEG) and reverse (880RDEG) primer. The samples are run in two separate PCR for 15 cycles using the following parameters: 98°C for 10 s, 46°C for 15 s, and 72°C for 1 min. The two PCR are then pooled and phosphorylated with polynucleotide kinase and the Illumina paired-ends adapters are ligated with T4 DNA ligase. After PCR amplification with Phusion for 10 cycles using Illumina paired-ends PCR primers, the library is quality controlled by cloning an aliquot into a TOPO plasmid and capillary sequencing 16 clones. The library is sequenced from the forward end for 76 cycles on the Illumina Genome Analyzer system GAII using sequencing kits version 3.0.
3. Base-calling is performed with the GAPipeline 1.3.2 using standard parameters, which include purity filtering with "chastity 0.6." Sequences containing uncalled bases, or incorrect primer sequence or runs of ≥12 identical nucleotides are removed. 72-base sequence reads are trimmed to remove the 13-base forward primer sequence, yielding 59-base sequences. Taxonomy of the sequences is assigned with GAST, using a database of reference V5 rDNA sequences (RefHVR_V5) from SILVA (version 98), and taxonomy from known cultured isolates, Entrez Genome projects, the Ribosomal Database Project (RDP), Greengenes, and hand curation. GAST compares each tag to the RefHVR_V5 and aligns it to its nearest neighbors in the database and then selects the closest reference(s). The taxonomy for the tag is the lowest common ancestor for a two-thirds majority of all 16S rRNA sequences associated with the nearest V5 reference sequences. Before generating

clusters of phylotypes, all sequences that occur fewer than three times are filtered out. This reduces the number of unique sequences to a computationally manageable level, and potentially reduces the number of errors from sequencing and contamination. A multiple sequences alignment of the remaining data are created using MUSCLE with parameters -maxiters 2 and -diags, and phylotype clusters and diversity estimates are generated using MOTHUR.

Note: From 1,373,824 sequences, 135 genera or higher taxonomic ranks were identified. While the phyla Firmicutes, Proteobacteria, Actinobacteria, Fusobacteria, and TM7 are the most common and abundant, phylum Bacteroidetes is less often detected in these samples. Potential sources for this difference include classification bias in this region of the 16S rRNA gene, human sample variation, sample preparation and primer bias. Use of an Illumina sequencing approach allowed a much greater depth of coverage than previous oral microbiota studies, leading to the identification of several taxa not yet discovered in these types of samples, and the ascertation that at least 30,000 additional reads would be required to identify only one additional phylotype.

(ii) Protocol of Grice et al.[58] Grice et al.[58] determined the complexity and identity of the microbes inhabiting the skin by sequencing bacterial 16S small-subunit ribosomal RNA genes isolated from the inner elbow of five healthy human subjects. This analysis revealed 113 OTUs ("phylotypes") at the level of 97% similarity that belong to six bacterial divisions.

Procedure

1. *Sample collection.* Samples (swab, scrape, punch/shave) are collected without prior cleaning or preparation of the skin surface from nonoverlapping regions of the antecubital fossa (inner elbow) of both the left and right arm of five healthy volunteers (aged between 20 and 65 years), who are instructed not to wash for an 8-h interval prior to sampling. Swabs are obtained using a sterile cotton pledget (Johnson and Johnson) soaked in sterile 0.15 M NaCl with 0.1% Tween 20 (Fisher Scientific) and wrung of excess solution. A 4-cm^2 area of the antecubital fossa is gently rubbed using two steady strokes in a nonoverlapping manner. Superficial skin scrapings are obtained from a 4-cm^2 area with two gentle steady strokes of a sterile disposable no. 15 blade in a nonoverlapping manner. The biopsy collection site is then anesthetized with injection of 0.1 mL of 2% lidocaine with epinephrine. The punch biopsies are taken as two separate samples. First, a 4-mm-diameter punch biopsy tool is used to mark a 4-mm diameter of skin with a depth of 1 mm, and then the epidermal layer is shaved at the 1-mm depth, parallel to the skin surface using a flat razor blade. After collecting the shave biopsy specimen, a 4-mm-diameter punch biopsy is used to mark the skin at the same site of the shave biopsy to a further depth of ~3 mm. The dermal skin layer is removed using a sterile scissors and forceps. All samples are stored at −80°C until further processing.
2. *DNA extraction.* All biological specimens are first incubated in a preparation of enzymatic lysis buffer (20 mM Tris at pH 8.0, 2 mM EDTA, 1.2% Triton X-100) and lysozyme (20 mg/mL) for 30 min at 37°C. Two 5-mm sterile stainless-steel beads (Qiagen) are then added to the sample. Tubes are transferred to the TissueLyser (Qiagen) and processed for 20 s at 15 Hz. Beads are removed, and samples are incubated overnight at 56°C in Buffer AL and Proteinase K from the DNeasy DNA Extraction Kit (Qiagen). The standard protocol for the kit is followed for all subsequent steps. The purified genomic DNA is resuspended in 100 μL of buffer AE and stored at −20°C.
3. *PCR amplification of 16S rRNA genes.* 16S rRNA genes are amplified from purified genomic DNA using the primers 8F (5′-AGAGTTTGATCCTGGCTCAG-3′) and 1391R (5′-GACGGGCGGTGTGTRCA-3′). Each 50 μL reaction mixture contains 5.0 μL of 10× buffer with $MgCl_2$ (Roche Applied Science), 1 μL of dNTP mix (10 mM each; Invitrogen), 1 μL of each primer (20 μM; IDT), 3 μL of DMSO, 4 μL of bacterial genomic DNA, and 0.5 μL of FastStart High Fidelity Taq Polymerase (Roche). For each DNA sample, three replicates are performed. Negative control is included in each set of amplifications. Thermocycling consists of an initial denaturation at 95°C for 5 min; 25–30 cycles of 95°C for 30 s, 55°C for 30 s, and 72°C for 1.5 min; and a final 72°C for 10 min. Cycle number is determined on a case-by-case basis, such that amplification is still in the linear range of the reaction when stopped but sufficient PCR product for cloning is produced (usually 25–30 cycles). PCR products are then separated on an agarose gel, and bands corresponding to the ~1.3-kb product are excised with a razor blade. The PCR products are then extracted using the Qiaquick Gel Extraction kit (Qiagen), and resuspended in 30 μL of Buffer EB and stored at −20°C.
4. *Cloning and sequencing of 16S rRNA PCR products.* PCR products are cloned into the pCR2.1-TOPO vector (Invitrogen) per the manufacturer's protocol. A total of 192 of the resulting bacterial colonies per ligation are picked, plasmid DNA is purified, and plasmid inserts are sequenced bidirectionally using the M13 primers on an ABI 3730xl sequencer (Applied Biosystems Inc.). Chromatogram data quality and quantity are evaluated using *phred* Q20 counts and nonvector sequence data remaining after cross-match screening. Bidirectional reads are assembled in a pipeline developed by NHGRI's Bioinformatics and Scientific Programming Core. Sequences are extracted from chromatograms using

phred, and bidirectional pairs are assembled using *phrap*. Vector sequence detected by cross-match is trimmed off. Only assembled sequences >1250 bp are studied further. Assemblies are screened for quality, and all sequences containing >20 consecutive bases of sequence <Q20 are discarded. Sequence traces and assemblies from OTUs containing less than four sequences are further examined in *consed* to verify quality of sequence and assembly.

5. *Alignment and phylogenetic analysis of 16S rRNA sequences.* Trimmed, assembled sequences are aligned to a core set of sequences using the NAST alignment tool on the Greengenes website (http://greengenes.lbl.gov). Alignments are manually curated. All sequences are examined for chimeras using the Bellerophon tool and standard settings, also available on the Greengenes website, and 140 putative chimeras are removed from further analyses. Sequences are inserted into a reference neighbor-joining tree in ARB using the parsimony insertion tool. Hypervariable regions are masked with lanemaskPH, and the Olsen correction is used when calculating pair wise distances.
6. *OTU determination and diversity estimation.* Using the Olsen-corrected distance matrix exported from ARB, sequences are grouped into OTUs by the furthest-neighbor algorithm using DOTUR. A similarity cutoff of 97% or 99%, as indicated in the text, is used to define an OTU. The same cutoff is used for diversity indices and richness estimates that are calculated using DOTUR. Both the Shannon and Simpson (1-D) diversity indices and Chao1 richness estimates calculated by DOTUR are used to estimate microbial diversity and richness.
7. *Clustering and comparison of 16S libraries.* All calculations using the SONS package are performed using an OTU cutoff of 97% or 99% similarity. All values are reported with their SE. The UniFrac package is used to compare unique branch length between communities. The phylogenetic tree generated in ARB containing only the sequences generated, along with environmental labels, is imported into UniFrac. The UniFrac metric is calculated along with its significance value, with and without regard to relative abundance of sequences. All significance values reported are corrected for multiple comparisons by Bonferonni correction. PCA analysis is performed within the UniFrac online suite of tools. The P test, also available on the online UniFrac suite of tools, assesses trees for nonrandom distributions of lineages according to environment. All P test values reported are also corrected for multiple comparisons.
8. *Quantification of total bacteria present in samples.* A portion of the bacterial 16S gene is amplified using forward primer 63F (5′-GCAGGCCTAA CACATGCAAGTC-3′) and reverse primer 355R (5′-CTGCTGCCTCCCGTAGGAGT-3′) to yield a 292-bp PCR product. Real-time PCR is performed on an ABI 7300 system (Applied Biosystems, Inc.) using optical grade 96-well plates. Each 25 μL reaction contains 12.5 μL Platinum 2× Sybr Green Master Mix (Invitrogen), 0.25 μL $MgCl_2$ (50 mM), 0.5 μL Rox Dye, 0.25 μL of each primer (20 μM), 6.25 μL of water, and 5 μL of DNA. The function used to calculate copy number is as follows: $C_t = -3.42x + 34.06$; $R^2 = 0.99$; where C_t is the threshold cycle and x is the log copy number. Copy numbers calculated for the sample are adjusted to be relative to the surface area sampled.

Note: Of the sequences generated from 5373 16S rRNA genes from bacteria, isolated by three skin collection methods of five healthy individuals, a majority belong to the Proteobacteria division and predominantly to the genera *Pseudomonas* and *Janthinobacterium.* 113 OTUs within the divisions Proteobacteria, Actinobacteria, Firmicutes, Bacteroidetes, Cyanobacteria, *and* Acidobacteria are identified. Just 10 OTUs account for >90% of the sequences, indicating that there are a few dominant OTUs that inhabit the antecubital fossa. These results demonstrate that the skin microbiome is vastly different from the gut microbiome, which consists primarily of members of Firmicutes and Bacteroidetes divisions. Furthermore, a low level of interpersonal variation is found in skin samples, in contrast to gut samples. This study of healthy human skin microbiota serves to direct future research addressing the role of skin microbiota in health and disease, and metagenomic projects addressing the complex physiological interactions between the skin and the microbes that inhabit this environment.

(iii) Protocol of Manichanh et al.[20]

Principle. Manichanh et al.[20] used a fosmid vector to construct two libraries of genomic DNA isolated directly from fecal samples of six healthy donors and six patients with Crohn's disease (CD). By screening the two DNA libraries, each composed of 25,000 clones, for the 16S rRNA gene through DNA hybridization, the bacterial diversity of intestinal microbiota was analyzed.

Procedure

1. *Sample collection.* Fecal samples are collected from the twelve subjects (six healthy subjects and six CD patients) receiving no antibiotics for at least 3 months before sampling. Samples are processed in less than 3 h in order to preserve strictly anaerobic bacteria from oxygen. Bacterial cells are recovered by Nycodenz density gradient centrifugation and then washed twice with phosphate buffered saline (10 min at 8000 × *g*). Cell pellets are then incorporated in agarose before gentle bacterial lysis and DNA digestion by the enzyme *Sau 3AI.*
2. *Library construction.* High molecular weight bacterial DNA fragments (38–48 kb) are cloned into

fosmid, a bacterially propagated phagemid vector system suitable for cloning genomic inserts approximately 40 kb in size, using the EpiFos™ Fosmid Library Production Kit (Epicentre® Biotechnologies).

3. *Macroarray preparation.* Screening for 16S rRNA genes in the fosmid libraries involved extraction of the recombinant fosmids in order to remove the 16S rRNA gene of *Escherichia coli* used in the cloning system. Extraction is performed with a NucleoSpin 96 Flash kit (Macherey-Nagel). Between 2 and 4 ng of each extracted fosmid is spotted on nylon membranes measuring 22 × 22 cm (Amersham Biosciences) in a 5 × 5 format, using the Qbot robot (Genetix, Saint-Marcel, France); approximately 50,000 spots are deposited on each membrane.
4. *Macroarray probe preparation.* In order to screen for 16S rRNA genes using the most universal probe, whole DNA directly extracted from fecal samples is used as a template to generate a mixed probe by PCR. PCR is run with a Peltier PTC–100TM Thermal Cycler (MJ ResearchTM, Inc.) at 94°C for 10 min; 9 cycles: 92°C for 1 min; 58°C for 1 min; 72°C for 1 min; and 72°C for 10 min. Forward and reverse primers are ACM 008 F and EUB 338, respectively (Table 100.3), generating products of approximately 330 bp in length, representative of the diversity of the original ecosystem. Concentrations of PCR products are evaluated on the basis of band intensities after gel electrophoresis. PCR products (30 ng) are labeled with [α-^{33}P]dATP using the Megaprime DNA labeling System (Amersham Biosciences).
5. *Macroarray hybridization.* The spotted membranes are first incubated in prehybridization solution [6× saline sodium citrate (SSC), 1% sodium dodecyl sulfate (SDS), 10× Denhardt, 50% formamide] at 42°C for 2 h. Then, a ^{33}P-labeled probe (10^6 cpm/mL) and sonicated salmon sperm DNA (100 µg/mL, denatured for 10 min at 100°C) are added to the prehybridization solution and incubation is continued at 42°C overnight. After hybridization membranes are washed once in 2× SSC, 0.1% SDS at 42°C for 10 min, once in 2× SSC, 0.1% SDS at 60°C for 10 min, once in 0.2× SSC, 0.1% SDS at 60°C for 10 min, and once in 0.1× SSC, 0.1% SDS at 65°C for 10 min. Phosphor screens are exposed to the membranes at room temperature for 5 h and then scanned with the Molecular Dynamics Storm system and analyzed with Xdigitise software (http://www.molgen.mpg.de/˜xdigitise/).
6. *Sequencing analyses.* All sequences are determined by Genoscope (Evry, France) on an ABI 3730 DNA sequencer using four primers (ACM 008 F, ADM 330 F, SSM 1100 R, and TTM 1517 R) specific for bacteria (Table 100.3). The 16S rRNA gene is sequenced in four fragments, which are then quality controlled using the PHRED v0.020425.c program, and assembled with the PHRAP v0.990319 program. Assembled sequences covering approximately positions 200–1300 are selected for subsequent analysis. For each library, sequences are aligned using the ClustalW v1.83 program. Highly variable regions of the multiple alignment are removed, and conserved regions are selected with Gblocks, using parameters optimized for rRNA alignments. The use of this computerized method avoids the manual refinement of multiple alignments. Distance matrices are computed with DNADIST v3.6 and trees are constructed with NEIGHBOR v3.6. Unrooted trees are drawn with Treeview. An OTU is defined as a cluster of 16S rRNA sequences sharing at least 98% similarity. A single representative fecal clone is selected for each OTU or ribotype. This clone is used as a reference sequence for calculating phylogenetic distances from other aligned sequences. Chimeric sequences are checked using the RDP CHECK_CHIMERA program (RDP–II release 8.1). The stability of branches is assessed by the bootstrap method with 1000 replicates. The term "uncultured OTUs" refers to species recovered by molecular methods only. Unidentified species (or totally novel species) correspond to microorganisms that match no available sequences in public databases (NCBI and RDP–II release 9). Phylum and group level nomenclatures, as defined in *Berguey's Manual of Systematic Bacteriology* (second edition) and the RDP Naive Bayesian classifier, respectively, are used. The representative nature of our libraries is estimated using Good's coverage formula. Good's coverage is calculated as $[1 - (n/N)] \times 100$, where n is the number of single clone OTU and N is the total number of sequences for the analyzed sample.
7. Fluorescence in situ hybridization (FISH). FISH combined with flow cytometry is used to analyze the composition of fecal samples, targeting the 16S rRNA of three major groups of bacteria. The 16S rRNA specific oligonucleotide probes used in this study are listed in Table 100.3. Probes EUB 338 and NON 338 are used as positive and negative controls, respectively. Probes Bac 303, Erec 482, and Clep 866 are used to detect the *Bacteroides*, *Clostridium coccoides*, and *Clostridium leptum* groups, respectively. The specificity of probe Clep 866 is optimized by using unlabeled, mismatched oligonucleotides as competitors (Table 100.3). Fecal samples stored at –70°C are fixed in paraformaldehyde and hybridized with specific probes, and positive signals are detected by flow cytometry.

Note: Among 1,190 selected clones, 125 nonredundant ribotypes, mainly represented by the phyla Bacteroidetes and Firmicutes, were identified. Among the Firmicutes, 43 distinct ribotypes were identified in the healthy microbiota, compared with only 13 in CD ($p < 0.025$). FISH directly targeting 16S rRNA in fecal samples analyzed individually

TABLE 100.3
List of Primers and Probes for Metagenomic Analysis of Gut Microbiota

Primer or Probe	5′→3′ Sequence	Experiments
ACM 008 F	AGAGTTTGATCCTGGCTCAG	Sequencing Macroarray hybridization
ADM 330 F	ACTCCTACGGGAGGCAGC	Sequencing
SSM 1100 R	GGGCCTTGTACACACCG	Sequencing
TTM 1517 R	GYAACGAGCGCAACCC	Sequencing
NON 338[a]	ACATCCTACGGGAGGC	FISH
EUB 338[a]	GCTGCCTCCCGTAGGAGT	FISH Macroarray hybridisation
Clep 866[a]	GGTGGATWACTTATTGTG	FISH
Clep 866 competitor 1[a]	GGTGGAWACTTATTGTG	FISH
Clep 866 competitor 2[a]	GGTGGATWACTTATTGCG	FISH
Bac 303[a]	CCAATGTGGGGACCTT	FISH
Erec 482[a]	GCTTCTTAGTCARGTACCG	FISH

[a] Probes Clep 866, Erec482, and Bac 303 detect members of the *Clostridium leptum*, *Clostridium coccoides*, and *Bacteroides* groups, respectively. ACM 008 F and EUB 338 are used as primers to develop a probe for 16S rRNA library hybridization after polymerase chain reaction amplification, using DNA extracted from fecal samples and [α-^{33}P]dATP labeling.

($n = 12$) confirmed the significant reduction in the proportion of bacteria belonging to this phylum in CD patients ($p < 0.02$). The metagenomic approach allows detection of a reduced complexity of the bacterial phylum Firmicutes as a signature of the fecal microbiota in patients with CD. It also uncovers the presence of new bacterial species.

(iv) Protocol of Manichanh et al.[59]

Principle. Manichanh et al.[59] showed that the analysis of random sequence reads (RSRs) instead of 16S rRNA gene is a suitable shortcut to estimate the biodiversity of a bacterial community from metagenomic libraries. From a metagenomic library of microorganisms found in human fecal samples, 10,010 RSRs were generated. The program BLASTN search against a prokaryotic sequence database was employed to assign a taxon to each RSR. The results were compared with those obtained by screening and analyzing the clones containing 16S rRNA sequences in the whole library.

Procedure

1. *Random sequence reads.* Two human fecal metagenomic libraries (healthy and Crohn) containing 25,000 clones each were previously described.[20] Each clone contains an insert of 40 kb of prokaryotic genomic fragment. About 4500 clones have been randomly chosen to perform a one read sequencing on the two extremities of the insert (5′ and 3′). All sequences were determined by Genoscope (Evry, France) on an ABI 3730 DNA sequencer. The sequencing provided 10,010 high-quality sequences (4192 for healthy and 5818 for Crohn library) with an average size of 615 bp ranging from 43 to 880 bp. To further simplify the parsing, each RSR-5′ is then computationally attached with 10 nt to the RSR-3′ for each clone to form a RSR doublon. In this way, 1939 RSR doublons and 314 RSR singletons (for which the counterpart sequencing has failed) for the healthy library and 2750 RSR doublons and 318 RSR singletons for the Crohn library are obtained (GenBank accession nos. EU057993-EU068001).
2. *16S rRNA sequences.* To validate our computational approach, the 1200 16S sequences are obtained from a previous work.[20] These sequences ranged in size from 990 to 1330 bp.
3. *Shotgun sequences from the Sargasso Sea.* To analyze metagenomic libraries other than those from human gut, 1,982,807 sequences are downloaded from a shotgun dataset (cloning of random segments of 2–6 kb) of the Sargasso Sea. These sequences have an average length of 600 bp ranging from 100 to 1177 bp. To analyze a comparable number of RSRs than for the human gut metagenome, 5000 sequences (0.25%) are selected randomly from the downloaded dataset and used them as RSR singletons.
4. *Statistical analyses.* To compare the phylotype frequencies across different datasets and assignment procedures we computed a chi-square of homogeneity—which essentially tests whether the frequencies observed in the analyzed samples are consistent with the hypothesis of the samples being randomly drawn from the same population. *P*-values smaller than 0.05 are considered enough to reject this hypothesis and to denote therefore significant differences between the compared groups.

Note: The biodiversity observed by RSR analysis is consistent with that obtained by 16S rRNA. Furthermore, RSRs are suitable to compare the biodiversity between different metagenomic libraries. Thus, RSRs can provide a good estimate of the biodiversity of a metagenomic library and, as an alternative to 16S rRNA, and this approach is both faster and cheaper.

100.3 CONCLUSIONS

Metagenomics has revolutionized microbiology by shifting focus away from clonal isolates towards the estimated 99% of microbial species that cannot currently be cultivated. The success of metagenomics is testified by an exponentially growing sequence data in public databases, which is supported by both new high-throughput sequencing technologies and a huge effort in developing appropriate bioinformatics tools. Analysis of the 16S gene is typically used as the first metagenomic approach. Despite its multiple limitations, the 16S gene sequence stands as powerful tool for diversity, composition analyses or specific bacterial strain fishing in a complex microbial world. And, in the near future, international movement towards much larger-scale metagenomic projects will allow the characterization of the function of whole ecosystems.

REFERENCES

1. Sanger, F. et al., Nucleotide sequence of bacteriophage phi X174 DNA, *Nature*, 265, 687, 1977.
2. Gill, S.R. et al., Metagenomic analysis of the human distal gut microbiome, *Science*, 312, 1355, 2006.
3. Turnbaugh, P.J. et al., Diet-induced obesity is linked to marked but reversible alterations in the mouse distal gut microbiome, *Cell Host Microbe*, 3, 213, 2008.
4. Kurokawa, K. et al., Comparative metagenomics revealed commonly enriched gene sets in human gut microbiomes, *DNA Res.*, 14, 169, 2007.
5. Nakamura, S. et al., Metagenomic diagnosis of bacterial infections, *Emerg. Infect. Dis.*, 14, 1784, 2008.
6. Ishoey, T. et al., Genomic sequencing of single microbial cells from environmental samples, *Curr. Opin. Microbiol.*, 11, 198–204, 2008.
7. de Wit, R., and Bouvier, T., "Everything is everywhere, but, the environment selects"; what did Baas Becking and Beijerinck really say? *Environ. Microbiol.*, 8, 755, 2006.
8. Fry, J.C., Oligotrophys. In Edwards, C. (Editor), *Microbiology of Extreme Environments*, pp. 93–116, McGraw-Hill, New York, 1990.
9. Fox, G.E. et al., Classification of methanogenic bacteria by 16S ribosomal RNA characterization, *Proc. Natl. Acad. Sci. USA*, 74, 4537, 1977.
10. Langendijk, P.S. et al., Quantitative fluorescence in situ hybridization of *Bifidobacterium* spp. with genus-specific 16S rRNA-targeted probes and its application in fecal samples, *Appl. Environ. Microbiol.*, 61, 3069, 1995.
11. Zoetendal, E.G., Akkermans, A.D., and De Vos, W.M., Temperature gradient gel electrophoresis analysis of 16S rRNA from human fecal samples reveals stable and host-specific communities of active bacteria, *Appl. Environ. Microbiol.*, 64, 3854, 1998.
12. Pace, N.R., and Waugh, D.S., Design of simplified ribonuclease P RNA by phylogenetic comparison, *Methods Enzymol.*, 203, 500, 1991.
13. Stein, J.L. et al., Characterization of uncultivated prokaryotes: Isolation and analysis of a 40-kilobase-pair genome fragment from a planktonic marine archaeon, *J. Bacteriol.*, 178, 591, 1996.
14. Handelsman, J. et al., Molecular biological access to the chemistry of unknown soil microbes: A new frontier for natural products, *Chem. Biol.*, 5, R245,1998.
15. Venter, J.C. et al., Environmental genome shotgun sequencing of the Sargasso Sea, *Science*, 304, 66, 2004.
16. Ronaghi, M., Pyrosequencing sheds light on DNA sequencing, *Genome Res.*, 11, 3, 2001.
17. Godon, J.J. et al., Molecular microbial diversity of an anaerobic digestor as determined by small-subunit rDNA sequence analysis, *Appl. Environ. Microbiol.*, 63, 2802, 1997.
18. Manichanh, C. et al., The gut microbiota predispose to the pathophysiology of acute postradiotherapy diarrhea, *Am. J. Gastroenterol.*, 103, 1754, 2008.
19. Stein, J.L. et al., Characterization of uncultivated prokaryotes: isolation and analysis of a 40-kilobase-pair genome fragment from a planktonic marine archaeon, *J. Bacteriol.*, 178, 591, 1996.
20. Manichanh, C. et al., Reduced diversity of faecal microbiota in Crohn's disease revealed by a metagenomic approach, *Gut*, 55, 205, 2006.
21. Gloux, K. et al., Development of high-throughput phenotyping of metagenomic clones from the human gut microbiome for modulation of eukaryotic cell growth, *Appl. Environ. Microbiol.*, 73, 3734, 2007.
22. Gevers, D. et al., Opinion: Re-evaluating prokaryotic species, *Nat. Rev. Microbiol.*, 3, 733, 2005.
23. Tringe, S.G., and Hugenholtz, P., A renaissance for the pioneering 16S rRNA gene, *Curr. Opin. Microbiol.*, 11, 442, 2008.
24. DeSantis, T.Z. et al., Greengenes, a chimera-checked 16S rRNA gene database and workbench compatible with ARB, *Appl. Environ. Microbiol.*, 72, 5069, 2006.
25. Pruesse, E. et al., SILVA: a comprehensive online resource for quality checked and aligned ribosomal RNA sequence data compatible with ARB, *Nucleic Acids Res.*, 35, 7188–7196, 2007.
26. Cole, J.R. et al., The Ribosomal Database Project: Improved alignments and new tools for rRNA analysis, *Nucleic Acids Res.*, 37, D141, 2009.
27. Altschul, S.F. et al., Basic local alignment search tool, *J. Mol. Biol.*, 215, 403, 1990.
28. Suau, A. et al., Direct analysis of genes encoding 16S rRNA from complex communities reveals many novel molecular species within the human gut, *Appl. Environ. Microbiol.*, 65, 4799, 1999.
29. Notredame, C., Higgins, D.G., and Heringa, J., T-Coffee: A novel method for fast and accurate multiple sequence alignment, *J. Mol. Biol.*, 302, 205, 2000.
30. Wallace, I.M. et al., M-Coffee: Combining multiple sequence alignment methods with T-Coffee, *Nucleic Acids Res.*, 34, 1692, 2006.
31. Larkin, M.A. et al., Clustal W and Clustal X version 2.0, *Bioinformatics*, 23, 2947, 2007.
32. Edgar, R.C., MUSCLE: multiple sequence alignment with high accuracy and high throughput, *Nucleic Acids Res.*, 32, 1792, 2004.
33. Castresana, J., Selection of conserved blocks from multiple alignments for their use in phylogenetic analysis, *Mol. Biol. Evol.*, 17, 540, 2000.
34. Seguritan, V., and Rohwer, F., FastGroup: A program to dereplicate libraries of 16S rDNA sequences, *BMC Bioinform.*, 2, 9, 2001.

35. Schloss, P.D., and Handelsman, J., Introducing DOTUR, a computer program for defining operational taxonomic units and estimating species richness, *Appl. Environ. Microbiol.*, 71, 1501, 2005.
36. Schloss, P.D., Larget, B.R., and Handelsman, J., Integration of microbial ecology and statistics: A test to compare gene libraries, *Appl. Environ. Microbiol.*, 70, 5485, 2004.
37. Lozupone, C., and Knight, R., UniFrac: A new phylogenetic method for comparing microbial communities, *Appl. Environ. Microbiol.*, 71, 8228, 2005.
38. Schloss, P.D., and Handelsman, J., Introducing TreeClimber, a test to compare microbial community structures, *Appl. Environ. Microbiol.*, 72, 2379, 2006.
39. Acinas, S.G. et al., Divergence and redundancy of 16S rRNA sequences in genomes with multiple rrn operons, *J. Bacteriol.*, 186, 2629, 2004.
40. Santos, S.R., and Ochman, H., Identification and phylogenetic sorting of bacterial lineages with universally conserved genes and proteins, *Environ Microbiol*, 6, 754–759, 2004.
41. Alexandre, A. et al., dnaJ is a useful phylogenetic marker for alphaproteobacteria, *Int. J. Syst. Evol. Microbiol.*, 58, 2839, 2008.
42. Fraser, C.M. et al., Complete genome sequence of *Treponema pallidum*, the syphilis spirochete, *Science*, 281, 375, 1998.
43. Andersson, S.G. et al., The genome sequence of *Rickettsia prowazekii* and the origin of mitochondria, *Nature*, 396, 133, 1998.
44. Cole, S.T. et al., Massive gene decay in the leprosy bacillus, *Nature*, 409, 1007, 2001.
45. Bentley, S.D. et al., Sequencing and analysis of the genome of the Whipple's disease bacterium Tropheryma whipplei, *Lancet*, 361, 637, 2003.
46. Ley, R.E. et al., Evolution of mammals and their gut microbes, *Science*, 320, 1647, 2008.
47. Backhed, F. et al., Host-bacterial mutualism in the human intestine, *Science*, 307, 1915, 2005.
48. Turnbaugh, P.J. et al., The human microbiome project, *Nature*, 449, 804, 2007.
49. Gao, Z. et al., Molecular analysis of human forearm superficial skin bacterial biota, *Proc. Natl. Acad. Sci. USA*, 104, 2927, 2007.
50. Kroes, I., Lepp, P.W., and Relman, D.A., Bacterial diversity within the human subgingival crevice, *Proc. Natl. Acad. Sci. USA*, 96, 14547, 1999.
51. Fredricks, D.N., Fiedler, T.L., and Marrazzo, J.M., Molecular identification of bacteria associated with bacterial vaginosis, *N. Engl. J. Med.*, 353, 1899, 2005.
52. Cribby, S., Taylor, M., and Reid, G., Vaginal microbiota and the use of probiotics, *Interdiscip. Perspect. Infect. Dis.*, 256, 490, 2008.
53. Sonnenburg, J.L. et al., Glycan foraging in vivo by an intestine-adapted bacterial symbiont, *Science*, 307, 1955, 2005.
54. Eckburg, P.B. et al., Diversity of the human intestinal microbial flora, *Science*, 308, 1635, 2005.
55. Guerrero, R., Bergey's manuals and the classification of prokaryotes, *Int. Microbiol.*, 4, 103, 2001.
56. Dethlefsen, L. et al., The pervasive effects of an antibiotic on the human gut microbiota, as revealed by deep 16S rRNA sequencing, *PLoS Biol.*, 6, e280, 2008.
57. Lazarevic, V. et al., Metagenomic study of the oral microbiota by Illumina high-throughput sequencing, *J. Microbiol. Methods*, 79, 266, 2009.
58. Grice, E.A. et al., A diversity profile of the human skin microbiota, *Genome Res.*, 18, 1043, 2008.
59. Manichanh, C. et al., A comparison of random sequence reads versus 16S rDNA sequences for estimating the biodiversity of a metagenomic library, *Nucleic Acids Res.*, 36, 5180, 2008.

Index

B

F

G

J

K

L

M

N

O

P

Q

R

S

U

V

W

X

Y

Z